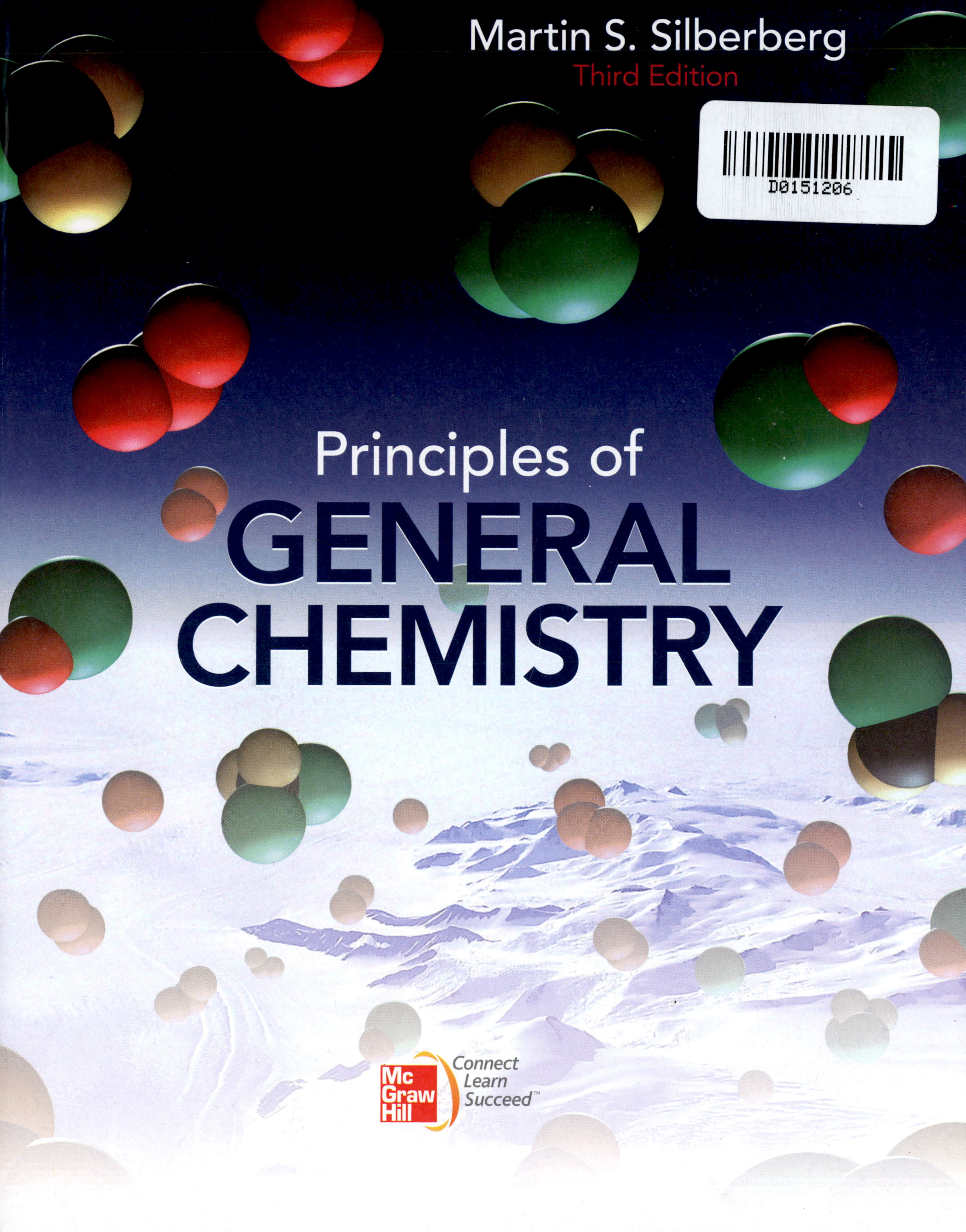
Martin S. Silberberg
Third Edition
D0151206
Principles of
GENERAL
CHEMISTRY
Mc
Graw
Hill
Connect
Learn
Succeed™

The McGraw·Hill Companies

PRINCIPLES OF GENERAL CHEMISTRY, THIRD EDITION

Published by McGraw-Hill, a business unit of The McGraw-Hill Companies, Inc., 1221 Avenue of the Americas, New York, NY 10020.

This book is printed on acid-free paper.

1 2 3 4 5 6 7 8 9 0 DOW/DOW 1 0 9 8 7 6 5 4 3 2

ISBN 978–0–07–340269–7
MHID 0–07–340269–9

Vice President, Editor-in-Chief: *Marty Lange*
Vice President, EDP: *Kimberly Meriwether David*
Senior Director of Development: *Kristine Tibbetts*
Publisher: *Ryan Blankenship*
Executive Editor: *Jeff Huettman*
Director of Digital Content Development: *David Spurgeon, Ph.D.*
Developmental Editor: *Lora Neyens*
Executive Marketing Manager: *Tamara L. Hodge*
Lead Project Manager: *Peggy J. Selle*
Senior Buyer: *Sandy Ludovissy*
Senior Media Project Manager: *Tammy Juran*
Senior Designer: *David W. Hash*
Cover Designer: *John Joran*
Cover Illustration: *Precision Graphics*
Cover Image: © *Mike Embree/National Science Foundation*
Senior Photo Research Coordinator: *Lori Hancock*
Photo Research: *Jerry Marshall/pictureresearching.com*
Compositor: *Lachina Publishing Services*
Typeface: *10/12 Times LT Std Roman*
Printer: *R.R. Donnelley*

Library of Congress Cataloging-in-Publication Data

Silberberg, Martin S. (Martin Stuart), 1945-
Principles of general chemistry / Martin S. Silberberg. — 3rd ed.
p. cm.
Includes index.
ISBN 978–0–07–340269–7 — ISBN 0–07–340269–9 (hard copy : alk. paper) 1. Chemistry—Textbooks. I. Title.
QD31.3.S55 2013
540—dc22

2011015577

www.mhhe.com

To Ruth and Daniel, with all my love
and
To the memory of my brother Bruce,
whose love, humor, and encouragement was
invaluable and will be profoundly missed.

BRIEF CONTENTS

DETAILED CONTENTS

CHAPTER 1 • Keys to the Study of Chemistry 2

CHAPTER 2 • The Components of Matter 32

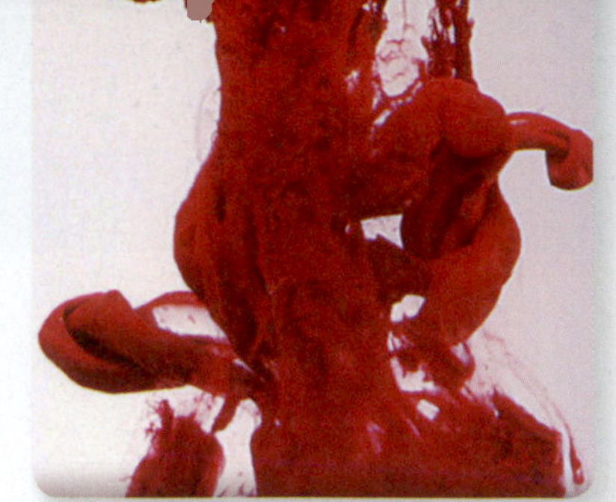

CHAPTER 3 • Stoichiometry of Formulas and Equations 71

CHAPTER 4 • Three Major Classes of Chemical Reactions 115

CHAPTER 5 • Gases and the Kinetic-Molecular Theory 148

CHAPTER 6 • Thermochemistry: Energy Flow and Chemical Change 188

CHAPTER 7 • Quantum Theory and Atomic Structure 216

CHAPTER 8 • Electron Configuration and Chemical Periodicity 245

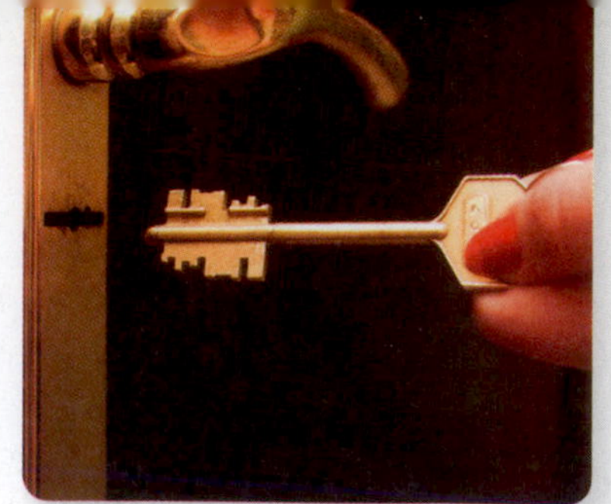

CHAPTER 9 • Models of Chemical Bonding 276

CHAPTER 10 • The Shapes of Molecules 302

CHAPTER 11 • Theories of Covalent Bonding 328

CHAPTER 12 • Intermolecular Forces: Liquids, Solids, and Phase Changes 350

CHAPTER 13 • The Properties of Solutions 391

CHAPTER 14 • Periodic Patterns in the Main-Group Elements 425

CHAPTER 15 • Organic Compounds and the Atomic Properties of Carbon 459

CHAPTER 16 • Kinetics: Rates and Mechanisms of Chemical Reactions 498

CHAPTER 17 • Equilibrium: The Extent of Chemical Reactions 542

CHAPTER 18 • Acid-Base Equilibria 579

CHAPTER 19 • Ionic Equilibria in Aqueous Systems 617

CHAPTER 20 • Thermodynamics: Entropy, Free Energy, and the Direction of Chemical Reactions 653

CHAPTER 21 • Electrochemistry: Chemical Change and Electrical Work 687

CHAPTER 22 • Transition Elements and Their Coordination Compounds 736

CHAPTER 23 • Nuclear Reactions and Their Applications 763

About the Author

Martin S. Silberberg received a B.S. in Chemistry from the City University of New York and a Ph.D. in Chemistry from the University of Oklahoma. He then accepted a research position in analytical biochemistry at the Albert Einstein College of Medicine in New York City, where he developed advanced methods to study fundamental brain mechanisms as well as neurotransmitter metabolism in Parkinson's disease. Following his years in research, Dr. Silberberg joined the faculty of Bard College at Simon's Rock, a liberal arts college known for its excellence in teaching small classes of highly motivated students. As Head of the Natural Sciences Major and Director of Premedical Studies, he taught courses in general chemistry, organic chemistry, biochemistry, and liberal arts chemistry. The close student contact afforded him insights into how students learn chemistry, where they have difficulties, and what strategies can help them succeed. Prof. Silberberg applied these insights in a broader context by establishing a text writing, editing, and consulting company. Before writing his own text, he worked as a consulting and developmental editor on chemistry, biochemistry, and physics texts for several major college publishers. He resides with his wife and son in the Pioneer Valley near Amherst, Massachusetts, where he enjoys the rich cultural and academic life of the area and relaxes by cooking, singing, and hiking.

Preface

As the new century unfolds, chemistry will play its usual, crucial role in dealing with complex environmental, medical, and industrial issues. And, as the complexities increase and more information is needed to understand them, many chemistry instructors want a more focused text to serve as the core of a powerful electronic teaching and learning package. This new, Third Edition of *Principles of General Chemistry* is the ideal choice, designed to cover key principles and skills with great readability, the most accurate molecular art available, a problem-solving approach that is universally praised, and a supporting suite of electronic products that sets a new standard in academic science.

HOW *PRINCIPLES* AND *CHEMISTRY* ARE THE SAME

Principles of General Chemistry was created from its parent text, *Chemistry: The Molecular Nature of Matter and Change,* when four expert chemistry teachers—three consulting professors and the author—joined to distill the concepts and skills at the heart of general chemistry. *Principles* covers all the material a science major needs to continue in premedical studies, engineering, or related fields. It maintains the same high standards of accuracy, clarity, and rigor as its parent and adopts the same three distinguishing hallmarks:

1. *Visualizing chemical models.* In many places in the text, concepts are explained first at the macroscopic level and then from a molecular point of view. Placed near many of these discussions, the text's celebrated graphics depict the phenomenon or change at the observable level in the lab, at the atomic level with superbly accurate molecular art, and at the symbolic level with the balanced equation.
2. *Thinking logically to solve problems.* The problem-solving approach, based on a four-step method widely approved by chemical educators, is introduced in Chapter 1 and employed *consistently* throughout the text. It encourages students to first plan a logical approach, and only then proceed to the arithmetic solution. A check step, universally recommended by instructors, fosters the habit of considering the reasonableness and magnitude of the answer. For practice and reinforcement, each worked problem has a matched follow-up problem, for which an abbreviated, multistep solution—not merely a numerical answer—appears at the end of the chapter.
3. *Applying ideas to the real world.* For today's students, who may enter one of numerous chemistry-related fields, especially important applications—such as climate change, enzyme catalysis, materials science, and others—are woven into the text discussion, and real-world scenarios are used in many worked in-chapter sample problems as well as end-of-chapter problems.

Principles and *Chemistry* also share a common topic sequence, which provides a thorough introduction to chemistry for science majors:

- Chapters 1 through 6 cover unit conversions and uncertainty, introduce atomic structure and bonding, discuss stoichiometry and reaction classes, show how gas behavior is modeled, and highlight the relation between heat and chemical change.
- Chapters 7 through 15 take an "atoms-first" approach, as they move from atomic structure and electron configuration to how atoms bond and what the resulting molecules look like and why. Intermolecular forces are covered by discussing the behavior of liquids and solids as compared with that of gases, and then leads the different behavior of solutions. These principles are then applied to the chemistry of the elements and to the compounds of carbon.
- Chapters 16 through 21 cover dynamic aspects of reaction chemistry, including kinetics, equilibrium, entropy and free energy, and electrochemistry.
- Chapters 22 and 23 cover transition elements and nuclear reactions.

HOW *PRINCIPLES* AND *CHEMISTRY* ARE DIFFERENT

Principles presents the same authoritative coverage as *Chemistry* but in 240 fewer pages. It does so by removing most of the boxed application material, thus letting instructors choose applications tailored for *their* course. Moreover, several topics that are important areas of research but not central to general chemistry were left out, including colloids, polymers, liquid crystals, and so forth. And mainstream material from the chapter on isolating the elements was blended into the chapter on electrochemistry.

Despite its much shorter length, *Principles of General Chemistry* includes *all* the pedagogy so admired in *Chemistry.* It has all the worked sample problems and about two-thirds as many end-of-chapter problems, still more than enough problems for every topic, with a high level of relevance and many real-world applications. The learning aids that students find so useful have also been retained—Concepts and Skills to Review, Section Summaries, Key Terms, Key Equations, and Brief Solutions to Follow-up Problems.

In addition, three aids not found in the parent *Chemistry* help students focus their efforts:

- *Key Principles.* At the beginning of each chapter, short bulleted paragraphs state the main concepts concisely, using many of the same phrases and terms (in *italics*) that appear in the pages to follow. A student can preview these principles before reading the chapter and then review them afterward.
- *"Think of It This Way . . ." with Analogies, Mnemonics, and Insights.* This recurring feature provides analogies for difficult concepts (e.g., the "radial probability distribution" of apples around a tree) and amazing quantities (e.g., a stadium and a marble for the relative sizes of atom and nucleus), memory shortcuts (e.g., which reaction occurs at which electrode), and useful insights (e.g., similarities between a saturated solution and a liquid-vapor system).
- *Problem-Based Learning Objectives.* The list of learning objectives at the end of each chapter includes the end-of-chapter problems that relate to each objective. Thus, a student, or instructor, can select problems that review a given topic.

WHAT'S NEW IN THE *THIRD EDITION*

To address dynamic changes in how courses are structured and how students learn—variable math and reading preparation, less time for traditional studying, electronic media as part of lectures and homework, new challenges and options in career choices—the author and publisher consulted extensively with students and faculty. Based on their input, we developed the following ways to improve the text as a whole as well as the content of individual chapters.

Global Changes to the Entire Text

Writing style and content presentation. Every line of every discussion has been revised to optimize clarity, readability, and a more direct presentation. The use of additional subheads, numbered (and titled) paragraphs, and bulleted (and titled) lists has eliminated long unbroken paragraphs. Main ideas are delineated and highlighted, making for more efficient study and lectures. As a result, the text is over 20 pages shorter than the *Second Edition.*

More worked problems. The much admired—and imitated—four-part (plan, solution, check, practice) *Sample Problems* occur in both data-based and molecular-scene format. To deepen understanding, *Follow-up Problems* have worked-out solutions at the back of each chapter, with a road map when appropriate, effectively doubling the number of worked problems. This edition has 15 more sample problems, many in the earlier chapters, where students need the most practice in order to develop confidence.

Art and figure legends. Figures have been made more realistic and modern. Figure legends have been greatly shortened, and the explanations from them have either been added to the text or included within the figures.

Page design and layout. A more open look invites the reader while maintaining the same attention to keeping text and related figures and tables near each other for easier studying.

Section summaries. This universally approved feature is even easier to use in a new bulleted format.

Chapter review. The unique *Chapter Review Guide* aids study with problem-based learning objectives, key terms, key equations, and the multistep Brief Solutions to Follow-up Problems (rather than just numerical answers).

End-of-chapter problem sets. With an enhanced design to improve readability and traditional and molecular-scene problems updated and revised, these problem sets are far more extensive than in other brief texts.

Content Changes to Individual Chapters

- Chapter 2 presents a new figure and table on molecular modeling, and it addresses the new IUPAC recommendations for atomic masses.
- Discussion of empirical formulas has been moved from Chapter 2 to Chapter 3 so that it appears just before molecular formulas.
- Chapter 3 has some sample problems from the *Second Edition* that have been divided to focus on distinct concepts, and it contains seven new sample problems.
- Chapters 3 and 4 include more extensive and consistent use of stoichiometry reaction tables in limiting-reactant problems.
- Chapter 4 presents a new molecular-scene sample problem on depicting an ionic compound in aqueous solution.
- Chapter 5 includes a new discussion on how gas laws apply to breathing.
- Chapter 5 groups stoichiometry of gaseous reactions with other rearrangements of the ideal gas law.
- Chapter 17 makes consistent use of quantitative benchmarks for determining when it is valid to assume that the amount reacting can be neglected.

ACKNOWLEDGMENTS

For the third edition of *Principles of General Chemistry,* I am once again very fortunate that Patricia Amateis of Virginia Tech prepared the *Instructors' Solutions Manual* and *Student Solutions Manual* and Libby Weberg the *Student Study Guide.*

The following individuals helped write and review goal-oriented content for LearnSmart for general chemistry: Erin Whitteck; Margaret Ruth Leslie, Kent State University; and Adam I. Keller, Columbus State Community College.

And, I greatly appreciate the efforts of all the professors who reviewed portions of the new edition or who participated in our developmental survey to assess the content needs for the text:

DeeDee A. Allen, *Wake Technical Community College*
John D. Anderson, *Midland College*
Jeanne C. Arquette, *Phoenix College*
Yiyan Bai, *Houston Community College*
Stanley A. Bajue, *Medgar Evers College, CUNY*
Jason P. Barbour, *Anne Arundel Community College*
Peter T. Bell, *Tarleton State University*
Vladimir Benin, *University of Dayton*
Paul J. Birckbichler, *Slippery Rock University*
Simon Bott, *University of Houston*
Kevin A. Boudreaux, *Angelo State University*
R. D. Braun, *University of Louisiana, Lafayette*
Stacey Buchanan, *Henry Ford Community College*
Michael E. Clay, *College of San Mateo*
Michael Columbia, *Indiana University Purdue University Fort Wayne*
Charles R. Cornett, *University of Wisconsin, Platteville*
Kevin Crawford, *The Citadel*
Mapi M. Cuevas, *Santa Fe Community College*
Kate Deline, *College of San Mateo*
Amy M. Deveau, *University of New England, Biddeford*
Jozsef Devenyi, *The University of Tennessee, Martin*
Paul A. DiMilla, *Northeastern University*
John P. DiVincenzo, *Middle Tennessee State University*
Ajit Dixit, *Wake Technical Community College*
Son Q. Do, *University of Louisiana, Lafayette*
Rosemary I. Effiong, *University of Tennessee, Martin*
Bryan Enderle, *University of California, Davis*
David K. Erwin, *Rose-Hulman Institute of Technology*
Emmanuel Ewane, *Houston Community College*
Kenneth A. French, *Blinn College*
Donna G. Friedman, *St. Louis Community College, Florissant Valley*
Herb Fynewever, *Western Michigan University*
Judy George, *Grossmont College*
Dixie J. Goss, *Hunter College City University of New York*
Ryan H. Groeneman, *Jefferson College*
Kimberly Hamilton-Wims, *Northwest Mississippi Community College*
David Hanson, *Stony Brook University*
Eric Hardegree, *Abilene Christian University*
Michael A. Hauser, *St. Louis Community College, Meramec*
Eric J. Hawrelak, *Bloomsburg University of Pennsylvania*
Monte L. Helm, *Fort Lewis College*
Sherell Hickman, *Brevard Community College*
Jeffrey Hugdahl, *Mercer University*
Michael A. Janusa, *Stephen F. Austin State University*
Richard Jarman, *College of DuPage*
Carolyn Sweeney Judd, *Houston Community College*
Bryan King, *Wytheville Community College*
Peter J. Krieger, *Palm Beach Community College*
John T. Landrum, *Florida International University, Miami*
Richard H. Langley, *Stephen F. Austin State University*
Richard Lavallee, *Santa Monica College*
Debbie Leedy, *Glendale Community College*
Alan Levine, *University of Louisiana, Lafayette*
Chunmei Li, *Stephen F. Austin State University*
Alan F. Lindmark, *Indiana University Northwest*
Donald Linn, *Indiana University Purdue University Fort Wayne*
Arthur Low, *Tarleton State University*
David Lygre, *Central Washington University*
Toni G. McCall, *Angelina College*
Debbie McClinton, *Brevard Community College*
William McHarris, *Michigan State University*
Curtis McLendon, *Saddleback College*
Lauren McMills, *Ohio University*
Jennifer E. Mihalick, *University of Wisconsin, Oshkosh*
John T. Moore, *Stephen F. Austin State University*
Brian Moulton, *Brown University*
Michael R. Mueller, *Rose-Hulman Institute of Technology*
Kathy Nabona, *Austin Community College*
Chip Nataro, *Lafayette College*
David S. Newman, *Bowling Green State University*
William J. Nixon, *St. Petersburg College*
Eileen Pérez, *Hillsborough Community College*
Richard Perkins, *University of Louisiana, Lafayette*
Eric O. Potma, *University of California, Irvine*
Nichole L. Powell, *Tuskegee University*
Parris F. Powers, *Volunteer State Community College*
Mary C. Roslonowski, *Brevard Community College*
E. Alan Sadurski, *Ohio Northern University*
G. Alan Schick, *Eastern Kentucky University*
Linda D. Schultz, *Tarleton State University*
Mary Sisak, *Slippery Rock University*
Joseph Sneddon, *McNeese State University*
Michael S. Sommer, *University of Wyoming*
Ana Maria Soto, *The College of New Jersey*
John E. Straub, *Boston University*
Richard E. Sykora, *University of South Alabama*
Robin S. Tanke, *University of Wisconsin, Stevens Point*
Maria E. Tarafa, *Miami Dade College*
Kurt Teets, *Okaloosa Walton College*
Jeffrey S. Temple, *Southeastern Louisiana University*
Lydia T. Tien, *Monroe Community College*
Thomas D. Tullius, *Boston University*
Mike Van Stipdonk, *Wichita State University*
Ramaiyer Venkatraman, *Jackson State University*
Marie Villarba, *Glendale Community College*
Kirk W. Voska, *Rogers State University*
Edward A. Walters, *University of New Mexico*
Kristine Wammer, *University of St. Thomas*
Shuhsien Wang-Batamo, *Houston Community College*
Thomas Webb, *Auburn University*
Kurt Winkelmann, *Florida Institute of Technology*
Steven G. Wood, *Brigham Young University*
Louise V. Wrensford, *Albany State University*
James A. Zimmerman, *Missouri State University*
Susan Moyer Zirpoli, *Slippery Rock University*
Tatiana M. Zuvich, *Brevard Community College*

My friends that make up the superb publishing team at McGraw-Hill Higher Education have again done an excellent job developing and producing this text. My warmest thanks for their hard work, thoughtful advice, and support go to Publisher Ryan Blankenship and Executive Editor Jeff Huettman. I lost one wonderful Senior Developmental Editor, Donna Nemmers, early in the project and found another wonderful one, Lora Neyens. Once again, Lead Project Manager Peggy Selle created a superb product, this time based on the clean, modern look of Senior Designer David Hash. Marketing Manager Tami Hodge ably presented the final text to the sales staff and academic community.

Expert freelancers made indispensable contributions as well. My superb copyeditor, Jane Hoover, continued to improve the accuracy and clarity of my writing, and proofreaders Janelle Pregler and Angie Ruden gave their consistent polish to the final manuscript. And Jerry Marshall helped me find the best photos, and Gary Hunt helped me create an exciting cover.

As always, my wife Ruth was involved every step of the way, from helping with early style decisions to checking and correcting content and layout in page proofs. And my son Daniel consulted on the choice of photos and the cover.

A Guide to Student Success: How to Get the Most Out of Your Textbook

ORGANIZING AND FOCUSING

Chapter Outline

The chapter begins with an outline that shows the sequence of topics and subtopics.

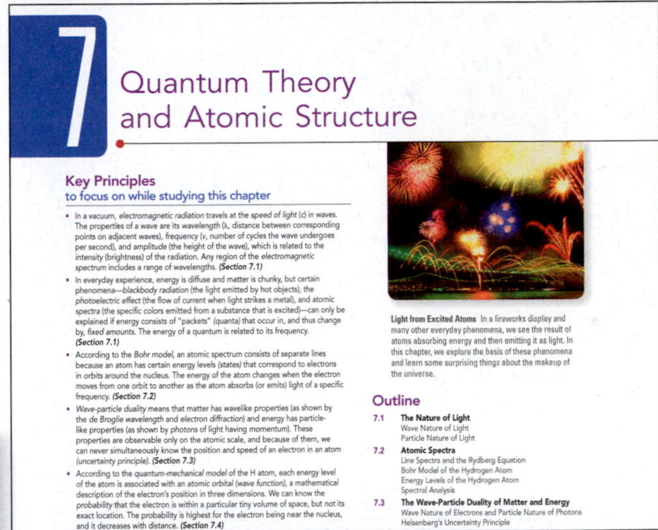

7

Quantum Theory and Atomic Structure

Key Principles
to focus on while studying this chapter

- In a vacuum, *electromagnetic radiation* travels at the *speed of light* (*c*) in waves. The properties of a wave are its *wavelength* (λ, distance between corresponding points on adjacent waves), *frequency* (ν, number of cycles the wave undergoes per second), and *amplitude* (the height of the wave), which is related to the *intensity* (brightness) of the radiation. Any region of the *electromagnetic spectrum* includes a range of wavelengths. ***(Section 7.1)***
- In everyday experience, energy is diffuse and matter is chunky, but certain phenomena—*blackbody radiation* (the light emitted by hot objects), the *photoelectric effect* (the flow of current when light strikes a metal), and *atomic spectra* (the specific colors emitted from a substance that is excited)—can only be explained if energy consists of "packets" *(quanta)* that occur in, and thus change by, *fixed amounts*. The energy of a quantum is related to its frequency. ***(Section 7.1)***
- According to the *Bohr model*, an atomic spectrum consists of separate lines because an atom has certain energy levels *(states)* that correspond to electrons in *orbits* around the nucleus. The energy of the atom changes when the electron moves from one orbit to another as the atom absorbs (or emits) light of a specific frequency. ***(Section 7.2)***
- *Wave-particle duality* means that matter has wavelike properties (as shown by the *de Broglie wavelength* and *electron diffraction*) and energy has particle-like properties (as shown by *photons* of light having *momentum*). These properties are observable only on the atomic scale, and because of them, we can never simultaneously know the position and speed of an electron in an atom *(uncertainty principle)*. ***(Section 7.3)***
- According to the *quantum-mechanical model* of the H atom, each energy level of the atom is associated with an *atomic orbital (wave function)*, a mathematical description of the electron's position in three dimensions. We can know the *probability* that the electron is within a particular tiny volume of space, but *not* its exact location. The probability is highest for the electron being near the nucleus, and it decreases with distance. ***(Section 7.4)***

Light from Excited Atoms In a fireworks display and many other everyday phenomena, we see the result of atoms absorbing energy and then emitting it as light. In this chapter, we explore the basis of these phenomena and learn some surprising things about the makeup of the universe.

Outline

Key Principles

The main principles from the chapter are given in concise, separate paragraphs so you can keep them in mind as you study. You may also want to review them when you are finished.

Concepts and Skills to Review

This unique feature helps you prepare for the upcoming chapter by referring to key material from earlier chapters that you should understand *before* you start reading the current one.

Why do substances behave as they do? That is, why is table salt (or any other ionic substance) a hard, brittle, high-melting solid that conducts a current only when molten or dissolved in water? Why is candle wax (along with most covalent substances) low melting, soft, and nonconducting, even though diamond (as well as a few other exceptions) is high melting and extremely hard? And why is copper (and most other metals) shiny, malleable, and able to conduct a current whether molten or solid? The answers lie in the *type of bonding within the substance*. In Chapter 8, we examined the properties of individual atoms and ions. But the behavior of matter really depends on how those atoms and ions bond.

9.1 • ATOMIC PROPERTIES AND CHEMICAL BONDS

Before we examine the types of chemical bonding, we should start with the most fundamental question: why do atoms bond at all? In general, *bonding lowers the*

CONCEPTS & SKILLS TO REVIEW before studying this chapter

- characteristics of ionic and covalent compounds; Coulomb's law (Section 2.7)
- polar covalent bonds and the polarity of water (Section 4.1)
- Hess's law, ΔH°_{rxn}, and ΔH°_f (Sections 6.5 and 6.6)
- atomic and ionic electron configurations (Sections 8.2 and 8.4)
- trends in atomic properties and metallic behavior (Sections 8.3 and 8.4)

Section Summaries

A bulleted list of statements conclude each section, immediately reiterating the major ideas just covered.

Summary of Section 13.3

- A solution that contains the maximum amount of dissolved solute in the presence of excess undissolved solute is saturated. A saturated solution is in equilibrium with excess solute, because solute particles are entering and leaving the solution at the same rate.
- Most solids are more soluble at higher temperatures.
- All gases have a negative ΔH_{soln} in water, so heating lowers gas solubility in water.
- Henry's law says that the solubility of a gas is directly proportional to its partial pressure above the solution.

STEP-BY-STEP PROBLEM SOLVING

Using this clear and thorough problem-solving approach, you'll learn to think through chemistry problems logically and systematically.

Sample Problems

A worked-out problem appears whenever an important new concept or skill is introduced. The step-by-step approach is shown consistently for every sample problem in the text.

- **Plan** analyzes the problem so that you can use what is known to find what is unknown. This approach develops the habit of thinking through the solution *before* performing calculations.
- In many cases, a **Road Map** specific to the problem is shown alongside the plan to lead you visually through the needed calculation steps.
- **Solution** shows the calculation steps *in the same order* as they are discussed in the plan and shown in the road map.
- **Check** fosters the habit of going over your work quickly to make sure that the answer is reasonable, chemically and mathematically—a great way to avoid careless errors.
- **Comment,** shown in many problems, provides an additional insight, and alternative approach, or a common mistake to avoid.
- **Follow-up Problem** gives you immediate practice by presenting a similar problem that requires the same approach.

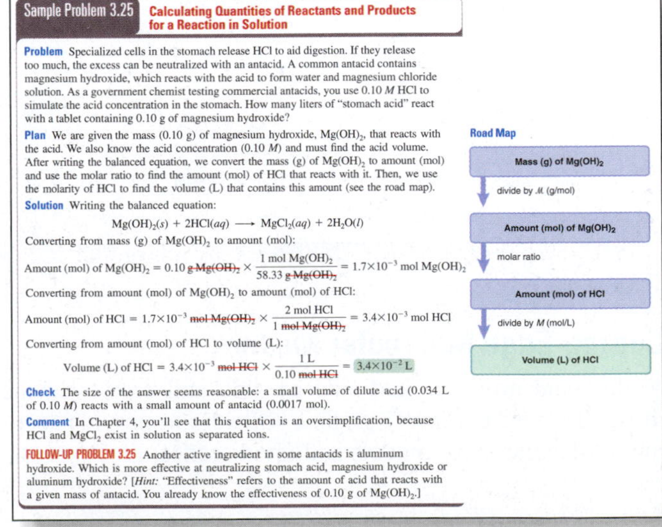

Sample Problem 3.25 **Calculating Quantities of Reactants and Products for a Reaction in Solution**

Problem Specialized cells in the stomach release HCl to aid digestion. If they release too much, the excess can be neutralized with an antacid. A common antacid contains magnesium hydroxide, which reacts with the acid to form water and magnesium chloride solution. As a government chemist testing commercial antacids, you use 0.10 *M* HCl to simulate the acid concentration in the stomach. How many liters of "stomach acid" react with a tablet containing 0.10 g of magnesium hydroxide?

Plan We are given the mass (0.10 g) of magnesium hydroxide, $Mg(OH)_2$, that reacts with the acid. We also know the acid concentration (0.10 *M*) and must find the acid volume. After writing the balanced equation, we convert the mass (g) of $Mg(OH)_2$ to amount (mol) and use the molar ratio to find the amount (mol) of HCl that reacts with it. Then, we use the molarity of HCl to find the volume (L) that contains this amount (see the road map).

Solution Writing the balanced equation:

$$Mg(OH)_2(s) + 2HCl(aq) \longrightarrow MgCl_2(aq) + 2H_2O(l)$$

Converting from mass (g) of $Mg(OH)_2$ to amount (mol):

$$\text{Amount (mol) of } Mg(OH)_2 = 0.10\ \cancel{g\ Mg(OH)_2} \times \frac{1\ mol\ Mg(OH)_2}{58.33\ \cancel{g\ Mg(OH)_2}} = 1.7\times10^{-3}\ mol\ Mg(OH)_2$$

Converting from amount (mol) of $Mg(OH)_2$ to amount (mol) of HCl:

$$\text{Amount (mol) of HCl} = 1.7\times10^{-3}\ \cancel{mol\ Mg(OH)_2} \times \frac{2\ mol\ HCl}{1\ \cancel{mol\ Mg(OH)_2}} = 3.4\times10^{-3}\ mol\ HCl$$

Converting from amount (mol) of HCl to volume (L):

$$\text{Volume (L) of HCl} = 3.4\times10^{-3}\ \cancel{mol\ HCl} \times \frac{1\ L}{0.10\ \cancel{mol\ HCl}} = 3.4\times10^{-2}\ L$$

Check The size of the answer seems reasonable: a small volume of dilute acid (0.034 L of 0.10 *M*) reacts with a small amount of antacid (0.0017 mol).

Comment In Chapter 4, you'll see that this equation is an oversimplification, because HCl and $MgCl_2$ exist in solution as separated ions.

FOLLOW-UP PROBLEM 3.25 Another active ingredient in some antacids is aluminum hydroxide. Which is more effective at neutralizing stomach acid, magnesium hydroxide or aluminum hydroxide? [*Hint:* "Effectiveness" refers to the amount of acid that reacts with a given mass of antacid. You already know the effectiveness of 0.10 g of $Mg(OH)_2$.]

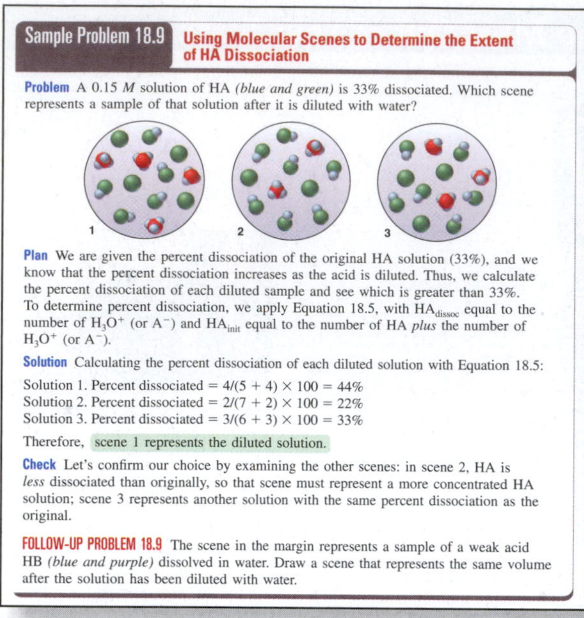

Sample Problem 18.9 **Using Molecular Scenes to Determine the Extent of HA Dissociation**

Problem A 0.15 *M* solution of HA *(blue and green)* is 33% dissociated. Which scene represents a sample of that solution after it is diluted with water?

Plan We are given the percent dissociation of the original HA solution (33%), and we know that the percent dissociation increases as the acid is diluted. Thus, we calculate the percent dissociation of each diluted sample and see which is greater than 33%. To determine percent dissociation, we apply Equation 18.5, with HA_{dissoc} equal to the number of H_3O^+ (or A^-) and HA_{init} equal to the number of HA *plus* the number of H_3O^+ (or A^-).

Solution Calculating the percent dissociation of each diluted solution with Equation 18.5:

Solution 1. Percent dissociated = $4/(5 + 4) \times 100 = 44\%$
Solution 2. Percent dissociated = $2/(7 + 2) \times 100 = 22\%$
Solution 3. Percent dissociated = $3/(6 + 3) \times 100 = 33\%$

Therefore, scene 1 represents the diluted solution.

Check Let's confirm our choice by examining the other scenes: in scene 2, HA is *less* dissociated than originally, so that scene must represent a more concentrated HA solution; scene 3 represents another solution with the same percent dissociation as the original.

FOLLOW-UP PROBLEM 18.9 The scene in the margin represents a sample of a weak acid HB *(blue and purple)* dissolved in water. Draw a scene that represents the same volume after the solution has been diluted with water.

Unique to *Principles of General Chemistry*: Molecular-Scene Sample Problems

These problems apply the same stepwise strategy to help you interpret molecular scenes and solve problems based on them.

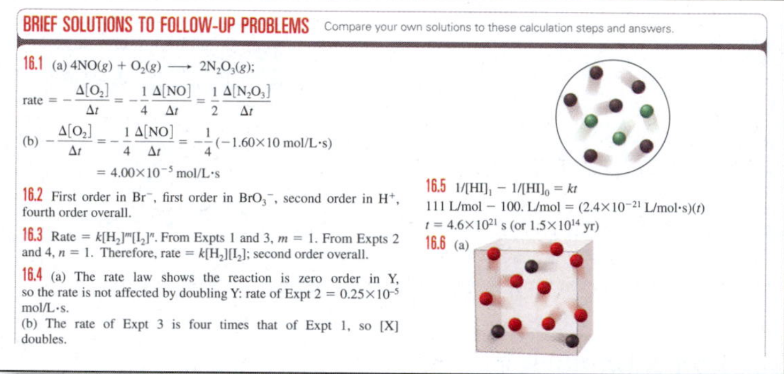

BRIEF SOLUTIONS TO FOLLOW-UP PROBLEMS Compare your own solutions to these calculation steps and answers.

16.1 (a) $4NO(g) + O_2(g) \longrightarrow 2N_2O_3(g)$;

$$\text{rate} = -\frac{\Delta[O_2]}{\Delta t} = -\frac{1}{4}\frac{\Delta[NO]}{\Delta t} = \frac{1}{2}\frac{\Delta[N_2O_3]}{\Delta t}$$

(b) $-\frac{\Delta[O_2]}{\Delta t} = -\frac{1}{4}\frac{\Delta[NO]}{\Delta t} = -\frac{1}{4}(-1.60\times10\ \text{mol/L·s})$

$= 4.00\times10^{-5}\ \text{mol/L·s}$

16.2 First order in Br^-, first order in BrO_3^-, second order in H^+, fourth order overall.

16.3 Rate = $k[H_2]^m[I_2]^n$. From Expts 1 and 3, $m = 1$. From Expts 2 and 4, $n = 1$. Therefore, rate = $k[H_2][I_2]$; second order overall.

16.4 (a) The rate law shows the reaction is zero order in Y, so the rate is not affected by doubling Y: rate of Expt 2 = 0.25×10^{-5} mol/L·s.
(b) The rate of Expt 3 is four times that of Expt 1, so [X] doubles.

16.5 $1/[HI]_1 - 1/[HI]_0 = kt$
111 L/mol − 100. L/mol = $(2.4\times10^{-21}$ L/mol·s$)(t)$
$t = 4.6\times10^{21}$ s (or 1.5×10^{14} yr)

16.6 (a)

Brief Solutions to Follow-up Problems

These provide multistep solutions at the end of the chapter, not just a one-number answer at the back of the book. This fuller treatment provides an excellent way for you to reinforce problem-solving skills.

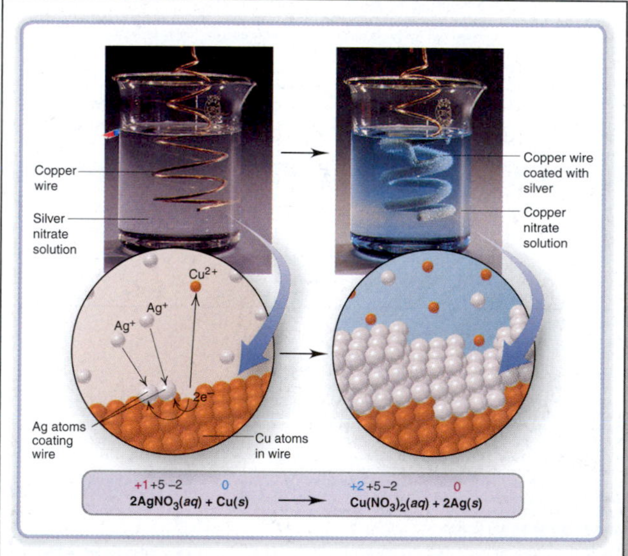

VISUALIZING CHEMISTRY

Three-Level Illustrations

A Silberberg hallmark, these illustrations provide macroscopic and molecular views of a process to help you connect these two levels of reality with each other and with the chemical equation that describes the process in symbols.

Cutting-Edge Molecular Models

Author and artist worked side by side and employed the most advanced computer-graphic software to provide accurate molecular-scale models and vivid scenes.

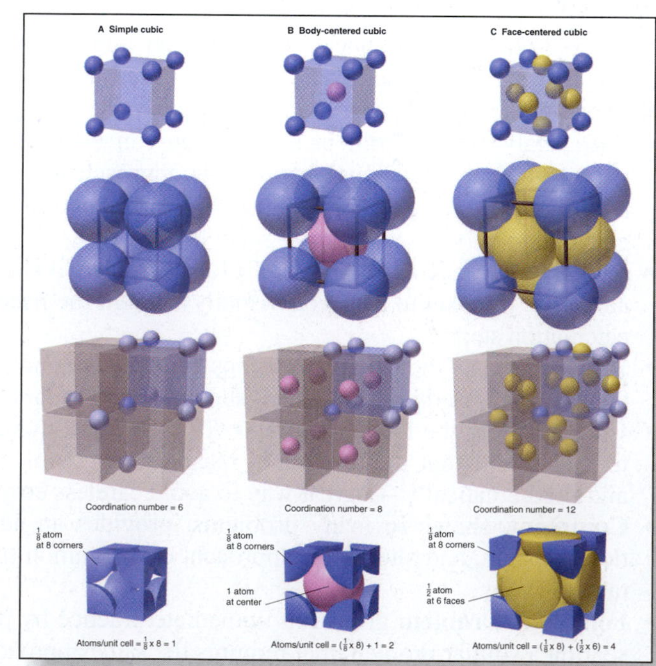

REINFORCING THE LEARNING PROCESS

Chapter Review Guide

A rich catalog of study aids ends each chapter to help you review its content:

- **Learning Objectives** are listed, with section, sample problem, and end-of-chapter problem numbers, to help you focus on key concepts and skills.
- **Key Terms** are boldfaced within the chapter and listed here by section (with page numbers); they are defined again in the Glossary.
- **Key Equations and Relationships** are highlighted and numbered within the chapter and listed here with page numbers.

CHAPTER REVIEW GUIDE The following sections provide many aids to help you study this chapter. (Numbers in parentheses refer to pages, unless noted otherwise.)

Learning Objectives These are concepts and skills to review after studying this chapter.

Related section (§), sample problem (SP), and upcoming end-of-chapter problem (EP) numbers are listed in parentheses.

1. Explain how solubility depends on the types of intermolecular forces (like-dissolves-like rule) and understand the characteristics of solutions consisting of gases, liquids, or solids (§13.1) (SP 13.1) (EPs 13.1–13.12)
2. Understand the enthalpy components of ΔH_{soln}, the dependence of ΔH_{hydr} on charge density, and why a solution process is exothermic or endothermic (§13.2) (EPs 13.13–13.15, 13.18–13.25, 13.28)
3. Comprehend the meaning of entropy and how the balance between ΔH and ΔS governs the solution process (§13.2) (EPs 13.16, 13.17, 13.26, 13.27)
4. Distinguish among saturated, unsaturated, and supersaturated solutions and explain the equilibrium nature of a saturated solution (§13.3) (EPs 13.29, 13.35)
5. Describe the effect of temperature on the solubility of solids and gases in water and the effect of pressure on the solubility of gases (Henry's law) (§13.3) (SP 13.2) (EPs 13.30–13.34, 13.36)
6. Express concentration in terms of molarity, molality, mole fraction, and parts by mass or by volume and be able to interconvert these terms (§13.4) (SPs 13.3–13.5) (EPs 13.37–13.58)
7. Describe electrolyte behavior and the four colligative properties, explain the difference between phase diagrams for a solution and a pure solvent, explain vapor-pressure lowering for non-volatile and volatile nonelectrolytes, and discuss the van't Hoff factor for colligative properties of electrolyte solutions (§13.5) (SPs 13.6–13.9) (EPs 13.59–13.83)

Key Terms These important terms appear in boldface in the chapter and are defined again in the Glossary.

Section 13.1
solute (392)
solvent (392)
miscible (392)
solubility (S) (392)
like-dissolves-like rule (393)
hydration shell (393)
ion–induced dipole force (393)
dipole–induced dipole force (393)
alloy (396)

Section 13.2
heat of solution (ΔH_{soln}) (397)
solvation (397)
hydration (398)
heat of hydration (ΔH_{hydr}) (398)
charge density (398)
entropy (S) (399)

Section 13.3
saturated solution (401)
unsaturated solution (401)
supersaturated solution (401)
Henry's law (403)

Section 13.4
molality (m) (404)
mass percent [% (w/w)] (405)
volume percent [% (v/v)] (405)
mole fraction (X) (405)

Section 13.5
colligative property (408)
electrolyte (408)
nonelectrolyte (408)
vapor pressure lowering (ΔP) (408)
Raoult's law (409)
ideal solution (409)
boiling point elevation (ΔT_b) (410)
freezing point depression (ΔT_f) (411)
semipermeable membrane (412)
osmosis (412)
osmotic pressure (Π) (413)
ionic atmosphere (415)

Key Equations and Relationships Numbered and screened concepts are listed for you to refer to or memorize.

13.1 Dividing the general heat of solution into component enthalpies (397):
$$\Delta H_{soln} = \Delta H_{solute} + \Delta H_{solvent} + \Delta H_{mix}$$

13.2 Dividing the heat of solution of an ionic compound in water into component enthalpies (398):
$$\Delta H_{soln} = \Delta H_{lattice} + \Delta H_{hydr\ of\ the\ ions}$$

13.3 Relating gas solubility to its partial pressure (Henry's law) (403):
$$S_{gas} = k_H \times P_{gas}$$

13.4 Defining concentration in terms of molarity (404):
$$\text{Molarity } (M) = \frac{\text{amount (mol) of solute}}{\text{volume (L) of solution}}$$

13.5 Defining concentration in terms of molality (404):
$$\text{Molality } (m) = \frac{\text{amount (mol) of solute}}{\text{mass (kg) of solvent}}$$

13.6 Defining concentration in terms of mass percent (405):
$$\text{Mass percent } [\%\ (w/w)] = \frac{\text{mass of solute}}{\text{mass of solution}} \times 100$$

13.7 Defining concentration in terms of volume percent (405):
$$\text{Volume percent } [\%\ (v/v)] = \frac{\text{volume of solute}}{\text{volume of solution}} \times 100$$

13.79 What is the minimum mass of glycerol ($C_3H_8O_3$) that must be dissolved in 11.0 mg of water to prevent the solution from freezing at −15°C? (Assume ideal behavior.)

180 Calculate the molality and van't Hoff factor (i) for the following aqueous solutions:
(a) 1.00 mass % NaCl, freezing point = −0.593°C
(b) 0.500 mass % CH_3COOH, freezing point = −0.159°C

13.81 Calculate the molality and van't Hoff factor (i) for the following aqueous solutions:
(a) 0.500 mass % KCl, freezing point = −0.234°C
(b) 1.00 mass % H_2SO_4, freezing point = −0.423°C

13.82 In a study designed to prepare new gasoline-resistant coatings, a polymer chemist dissolves 6.053 g of poly(vinyl alcohol) in enough water to make 100.0 mL of solution. At 25°C, the osmotic pressure of this solution is 0.272 atm. What is the molar mass of the polymer sample?

13.83 The U.S. Food and Drug Administration lists dichloromethane (CH_2Cl_2) and carbon tetrachloride (CCl_4) among the many cancer-causing chlorinated organic compounds. What are the partial pressures of these substances in the vapor above a solution of 1.60 mol of CH_2Cl_2 and 1.10 mol of CCl_4 at 23.5°C? The vapor pressures of pure CH_2Cl_2 and CCl_4 at 23.5°C are 352 torr and 118 torr, respectively. (Assume ideal behavior.)

Comprehensive Problems

13.84 The three aqueous ionic solutions represented below have total volumes of 25. mL for A, 50. mL for B, and 100. mL for C. If each sphere represents 0.010 mol of ions, calculate: (a) the total molarity of ions for each solution; (b) the highest molarity of solute; (c) the lowest molality of solute (assuming the solution densities are equal); (d) the highest osmotic pressure (assuming ideal behavior).

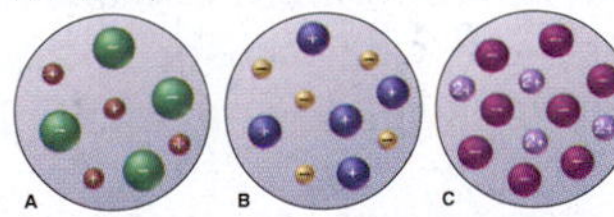

13.85 Gold occurs in seawater at an average concentration of 1.1×10^{-2} ppb. How many liters of seawater must be processed to recover 1 troy ounce of gold, assuming 81.5% efficiency (d of seawater = 1.025 g/mL; 1 troy ounce = 31.1 g)?

13.86 Use atomic properties to explain why xenon is 11 times as soluble as helium in water at 0°C on a mole basis.

13.87 Which of the following best represents a molecular-scale view of an ionic compound in aqueous solution? Explain.

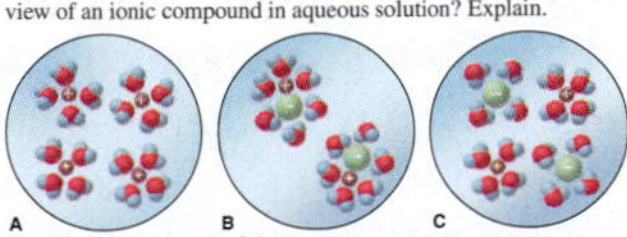

13.88 Four 0.50 m aqueous solutions are depicted. Assume the solutions behave ideally: (a) Which has the highest boiling point? (b) Which has the lowest freezing point? (c) Can you deter[mine] which one has the highest osmotic pressure? Explain.

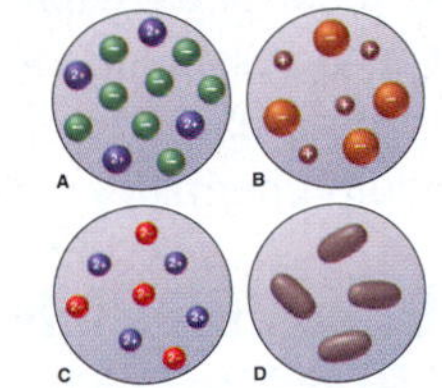

13.89 "De-icing salt" is used to melt snow and ice on streets. [The] highway department of a small town is deciding whether to [use] NaCl or $CaCl_2$, which are equally effective, to use for this purp[ose]. The town can obtain NaCl for $0.22/kg. What is the maximum [the] town should pay for $CaCl_2$ to be cost effective?

13.90 Thermal pollution from industrial wastewater causes the [tem]perature of river or lake water to increase, which can affect [fish] survival as the concentration of dissolved O_2 decreases. Use the [fol]lowing data to find the molarity of O_2 at each temperature (ass[ume] the solution density is the same as water):

Temperature (°C)	Solubility of O_2 (mg/kg H_2O)	Density of H_2O (g/mL)
0.0	14.5	0.99987
20.0	9.07	0.99823
40.0	6.44	0.99224

13.91 A chemist is studying small organic compounds for [their] potential use as an antifreeze. When 0.243 g of a compoun[d is] dissolved in 25.0 mL of water, the freezing point of the solution is −0.201°C. (a) Calculate the molar mass of the compound (d of water = 1.00 g/mL). (b) Analysis shows that the compound is 53.31 mass % C and 11.18 mass % H, the remainder being O. Calculate the empirical and molecular formulas of the compound. (c) Draw a Lewis structure for a compound with this formula that forms H bonds and another for one that does not.

13.92 Is 50% by mass of methanol dissolved in ethanol different from 50% by mass of ethanol dissolved in methanol? Explain.

13.93 Three gaseous mixtures of N_2 *(blue)*, Cl_2 *(green)*, and Ne *(purple)* are depicted below. (a) Which has the smallest mole fraction of N_2? (b) Which have the same mole fraction of Ne? (c) Rank all three in order of increasing mole fraction of Cl_2.

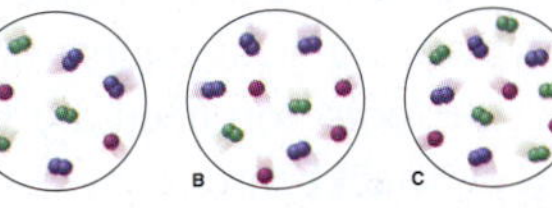

13.94 Four U tubes each have distilled water in the right arm, a solution in the left arm, and a semipermeable membrane between arms. (a) If the solute is KCl, which solution is most concentrated?

End-of-Chapter Problems

An exceptionally large number of problems ends each chapter. These are sorted by section, and many are grouped in similar pairs, with one of each pair answered in Appendix E (along with other problems having a colored number). Following these section-based problems is a large group of Comprehensive Problems, which are based on concepts and skills from any section and/or earlier chapter and are filled with applications from related sciences.

Think of It This Way

Analogies, memory shortcuts, and new insights into key ideas are provided in "Think of It This Way" features.

THINK OF IT THIS WAY
Which Half-Reaction Occurs at Which Electrode?

Here are some memory aids to help you connect the half-reaction with its electrode:

1. The words *anode* and *oxidation* start with vowels; the words *cathode* and *reduction* start with consonants.
2. Alphabetically, the *A* in anode comes before the *C* in cathode, and the *O* in oxidation comes before the *R* in reduction.
3. Look at the first syllables and use your imagination:

ANode, OXidation; REDuction, CATHode ⟹ AN OX and a RED CAT

Summary of Section 21.1

- An oxidation-reduction (redox) reaction involves the transfer of electrons from a reducing agent to an oxidizing agent.
- The half-reaction method of balancing divides the overall reaction into half-reactions that are balanced separately and then recombined.
- There are two types of electrochemical cells. In a voltaic cell, a spontaneous reaction generates electricity and does work on the surroundings. In an electrolytic cell, the

MCGRAW-HILL CONNECT®CHEMISTRY

www.mcgraw-hillconnect.com/chemistry

McGraw-Hill Connect® Chemistry is a web-based assignment and assessment platform that gives students the means to better connect with their coursework, with their instructors, and with the important concepts that they will need to know for success now and in the future. The chemical drawing tool found within Connect Chemistry is CambridgeSoft's ChemDraw®, which is widely considered the gold standard of scientific drawing programs and the cornerstone application for scientists who draw and annotate molecules, reactions, and pathways. This collaboration of Connect and ChemDraw features an easy-to-use, intuitive, and comprehensive course management and homework system with professional-grade drawing capabilities.

With Connect Chemistry, instructors can deliver assignments, quizzes, and tests online. Questions from the text are presented in an auto-gradable format and tied to the text's learning objectives. Instructors can edit existing questions and author entirely new problems. They also can track individual student performance—by question, assignment, or in relation to the class overall—with detailed grade reports; integrate grade reports easily with Learning Management Systems (LMS) such as WebCT and Blackboard; and much more.

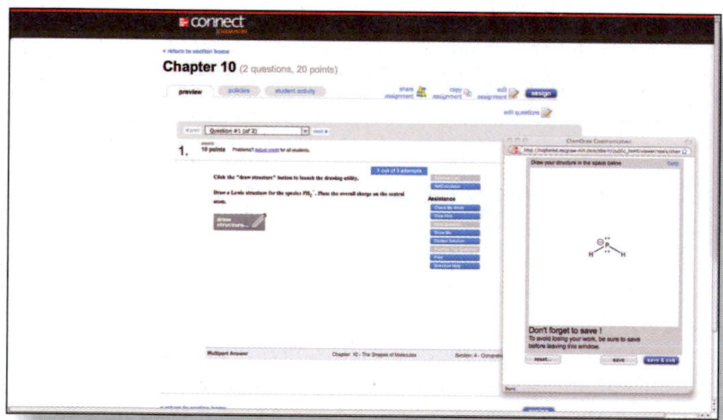

By choosing Connect Chemistry, instructors are providing their students with a powerful tool for improving academic performance and truly mastering course material. Connect Chemistry allows students to practice important skills at their own pace and on their own schedule. Importantly, students' assessment results and instructors' feedback are all saved online—so students can continually review their progress and plot their course to success.

ConnectPlus® Chemistry

Like Connect Chemistry, McGraw-Hill **ConnectPlus® Chemistry** provides students with online assignments and assessments, plus 24/7 online access to an eBook—an online edition of the text—to aid them in successfully completing their work, wherever and whenever they choose.

LearnSmart™

This adaptive diagnostic learning system, powered by Connect Chemistry and based on artificial intelligence, constantly assesses a student's knowledge of the course material. As students work within the system, McGraw-Hill LearnSmart™ develops a personal learning path adapted to what each student has actively learned and retained. This innovative study tool also has features to allow the instructor to see exactly what students have accomplished, with a built-in assessment tool for graded assignments. You can access LearnSmart for Chemistry at www.mcgrawhillconnect.com/chemistry.

McGraw-Hill Higher Education and Blackboard®

Do More

McGraw-Hill Higher Education and Blackboard have teamed up! What does this mean for you?

1. **Your life, simplified.** Now you and your students can access McGraw-Hill's Connect and Create right from within your Blackboard course—all with one single sign-on. Say goodbye to the days of logging in to multiple applications.
2. **Deep integration of content and tools.** Not only do you get single sign-on with Connect and Create, you also get deep integration of McGraw-Hill content and content engines right in Blackboard. Whether you're choosing a book for your course or building Connect assignments, all the tools you need are right where you want them—inside of Blackboard.
3. **Seamless Gradebooks.** Are you tired of keeping multiple gradebooks and manually synchronizing grades into Blackboard? We thought so. When a student completes an integrated Connect assignment, the grade for that assignment automatically (and instantly) feeds your Blackboard grade center.

4. **A solution for everyone.** Whether your institution is already using Blackboard or you just want to try Blackboard on your own, we have a solution for you. McGraw-Hill and Blackboard can now offer you easy access to industry leading technology and content, whether your campus hosts it, or we do. Be sure to ask your local McGraw-Hill representative for details.

Customizable Textbooks: Create™

Create what you've only imagined. Introducing McGraw-Hill Create—a new, self-service website that allows you to create custom course materials—print and eBooks—by drawing upon McGraw-Hill's comprehensive, cross-disciplinary content. Add your own content quickly and easily. Tap into other rights-secured third-party sources as well. Then, arrange the content in a way that makes the most sense for your course, and if you wish, personalize your book with your course name and information. Choose the best delivery format for your course: color print, black and white print, or eBook. The eBook is now viewable for the iPad! And when you are finished customizing, you will receive a free PDF review copy in just minutes! Visit McGraw-Hill Create—**www.mcgrawhillcreate.com**—today and begin building your perfect book.

Presentation Center

Within the Instructor's Presentation Center, instructors have access to editable PowerPoint lecture outlines, which appear as ready-made presentations that combine art and lecture notes for each chapter of the text. For instructors who prefer to create their lecture notes from scratch, all illustrations, photos, and tables are pre-inserted by chapter into a separate set of PowerPoint slides. An online digital library contains photos, artwork, animations, and other media types that can be used to create customized lectures, visually enhanced tests and quizzes, compelling course websites, or attractive printed support materials. All assets are copyrighted by McGraw-Hill Higher Education, but can be used by instructors for classroom purposes. The visual resources in this collection include:

- **Art** Full-color digital files of all illustrations in the book can be readily incorporated into lecture presentations, exams, or custom-made classroom materials.
- **Photos** The photo collection contains digital files of photographs from the text, which can be reproduced for multiple classroom uses.
- **Tables** Every table that appears in the text has been saved in electronic form for use in classroom presentations and/or quizzes.
- **Animations** Numerous full-color animations illustrating important processes are also provided. Harness the visual impact of concepts in motion by importing these files into classroom presentations or online course materials.

Digital Lecture Capture: Tegrity records and distributes your lecture with just a click of a button. Students can view anytime and anywhere via computer, iPod, or mobile device. *Tegrity* indexes as it records your slideshow presentations and anything shown on your computer, so students can use keywords to find exactly what they want to study.

Computerized Test Bank Prepared by Walter Orchard, Professor Emeritus of Tacoma Community College, over 2300 test questions to accompany *Principles of General Chemistry* are available utilizing Brownstone's Diploma testing software. *Diploma's* software allows you to quickly create a customized test using McGraw-Hill's supplied questions or by authoring your own. *Diploma* allows you to create your tests without an Internet connection—just download the software and question files directly to your computer.

Instructors' Solutions Manual This supplement, prepared by Patricia Amateis of Virginia Tech, contains complete, worked-out solutions for *all* the end-of-chapter problems in the text. It can be found within the Instructors Resources, on the Connect: Chemistry site.

Content Delivery Flexibility *Principles of General Chemistry,* by Martin Silberberg, is available in other formats in addition to the traditional textbook, giving instructors and students more choices for the format of their chemistry text.

Cooperative Chemistry Laboratory Manual Prepared by Melanie Cooper of Clemson University, this innovative manual features open-ended problems designed to simulate experience in a research lab. Working in groups, students investigate one problem over a period of several weeks, so they might complete three or four projects during the semester, rather than one preprogrammed experiment per class. The emphasis is on experimental design, analytic problem solving, and communication.

LEARNING SYSTEM RESOURCES FOR STUDENTS

Student Study Guide This valuable study guide, prepared by Libby Bent Weberg, is designed to help you recognize your learning style; understand how to read, classify, and create a plan for solving a problem; and practice your problem-solving skills. For each section of each chapter, the guide provides study objectives and a summary of the corresponding text. Following the summary are sample problems with detailed solutions. Each chapter has true-false questions and a self-test, with all answers provided at the end of the chapter.

Student Solutions Manual This supplement, prepared by Patricia Amateis of Virginia Tech, contains detailed solutions and explanations for all problems in the main text that have colored numbers.

Featuring ***CambridgeSoft® ChemDraw***

Connect Chemistry With Connect Chemistry, you can practice solving assigned homework problems using the Silberberg problem-solving methodology applied in the textbook. Algorithmic problems serve up multiple versions of similar problems for mastery of content, with hints and feedback for common incorrect answers to help you stay on track. Where appropriate, you engage in accurate, professional-grade chemical drawing through the use of CambridgeSoft's ChemDraw tool, which is implemented directly into homework problems. Check it out at www.mcgrawhillconnect.com/chemistry.

Mc Graw Hill LearnSmart™

LearnSmart™ This adaptive diagnostic learning system, powered by Connect: Chemistry and based on artificial intelligence, constantly assesses your knowledge of the course material. As you work within the system, LearnSmart develops a personal learning path adapted to what you have actively learned and retained. This innovative study tool also has features to allow your instructor to see exactly what you have accomplished, with a built-in assessment tool for graded assignments. You can access LearnSmart for general chemistry by going to www.mcgrawhillconnect.com/chemistry.

Animations for MP3/iPod A number of animations are available for download to your MP3/iPod through the textbook's Connect website.

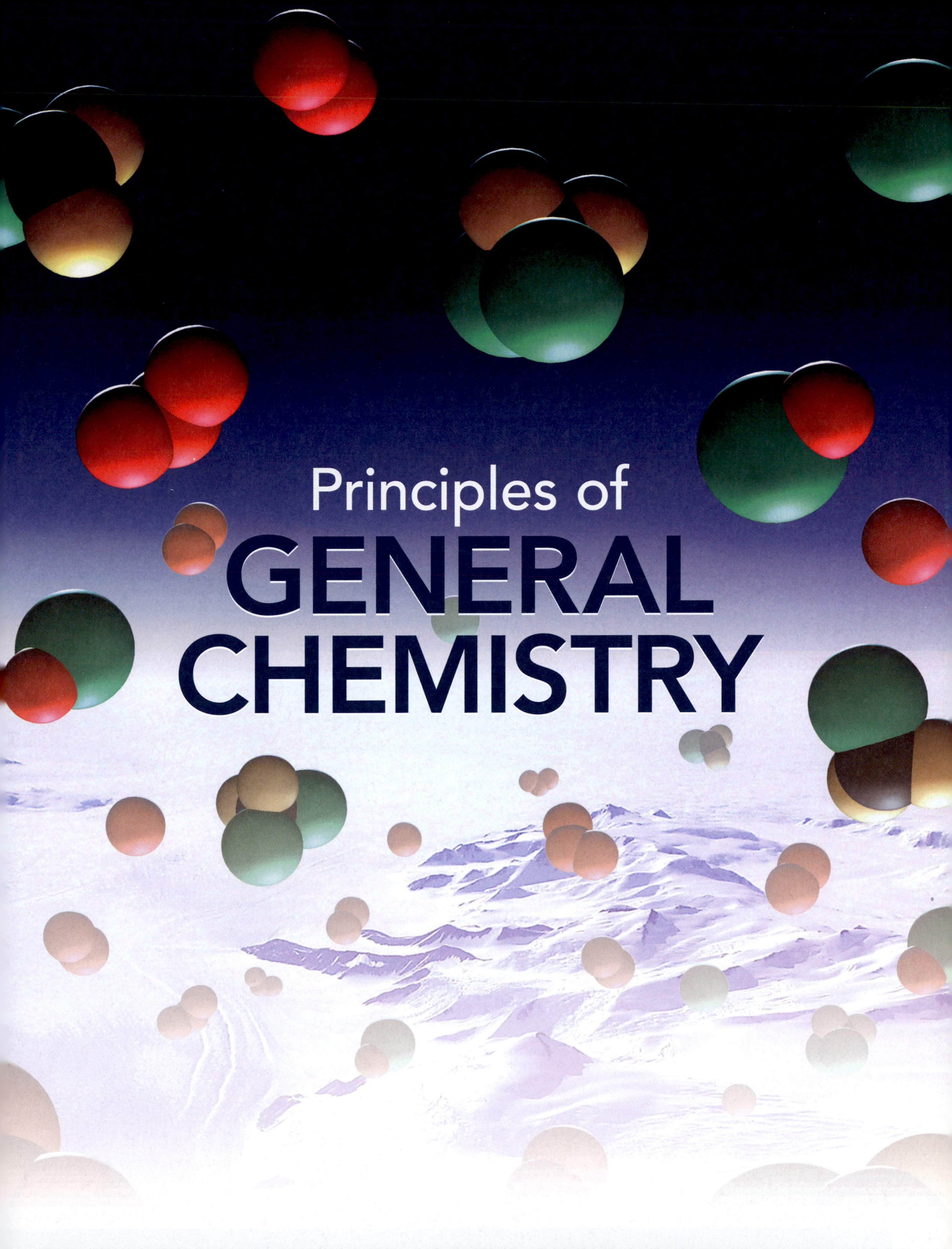
Principles of
GENERAL
CHEMISTRY

1 Keys to the Study of Chemistry

Key Principles
to focus on while studying this chapter

- *Matter* can undergo two kinds of change: *physical change* involves a change in *state*—gas, liquid, or solid—but not in ultimate makeup *(composition)*; *chemical change (reaction)* is more fundamental because it does involve a change in composition. The changes we observe result ultimately from changes too small to observe. ***(Section 1.1)***
- *Energy* occurs in different forms that are interconvertible, even as the total quantity of energy is conserved. When opposite charges are pulled apart, their *potential energy* increases; when they are released, potential energy is converted to the *kinetic energy* of the charges moving together. Matter consists of charged particles, so changes in energy accompany changes in matter. ***(Section 1.1)***
- The *scientific method* is a way of thinking that involves making *observations* and gathering *data* to develop *hypotheses* that are tested by *controlled experiments* until enough results are obtained to create a *model (theory)* that explains an aspect of nature. A sound theory can predict events but must be changed if new results conflict with it. ***(Section 1.2)***
- Any *measured quantity* is expressed by a number together with a *unit*. *Conversion factors* are ratios of equivalent quantities having different units; they are used in calculations to change the units of quantities. *Decimal prefixes* and *exponential notation* are used to express very large or very small quantities. ***(Section 1.3)***
- *The SI system* consists of seven *fundamental units*, each identifying a physical quantity such as length (meter), mass (kilogram), or temperature (kelvin). These are combined into many *derived units* used to identify quantities such as volume, density, and energy. *Extensive properties*, such as mass, depend on sample size; *intensive properties*, such as temperature, do not. ***(Section 1.4)***
- *Uncertainty* characterizes every measurement and is indicated by the number of *significant figures*. We *round* the final answer of a calculation to the same number of digits as in the least certain measurement. *Accuracy* refers to how close a measurement is to the true value; *precision* refers to how close measurements are to one another. ***(Section 1.5)***

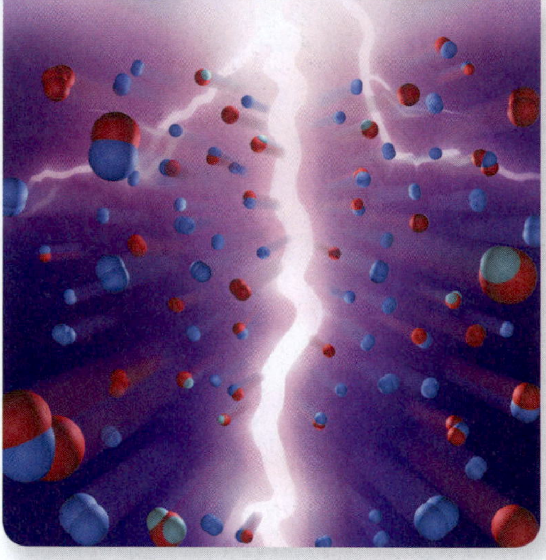

A Molecular View Within a Storm Lightning supplies the energy for many atmospheric chemical changes to occur. In fact, all the events within and around you have causes and effects at the atomic level of reality.

Outline

Maybe you're taking this course because chemistry is fundamental to understanding other natural sciences. Maybe it's required for your major. Or maybe you just want to learn more about the impact of chemistry on society or even on your everyday life. For example, did you have cereal, fruit, and coffee for breakfast today? In chemical terms, you enjoyed nutrient-enriched, spoilage-retarded carbohydrate flakes mixed in a white emulsion of fats, proteins, and monosaccharides, with a piece of fertilizer-grown, pesticide-treated fruit, and a cup of hot aqueous extract of stimulating alkaloid. Earlier, you may have been awakened by the sound created as molecules aligned in the liquid-crystal display of your clock and electrons flowed to create a noise. You might have thrown off a thermal insulator of manufactured polymer and jumped in the shower to emulsify fatty substances on your skin and hair with purified water and formulated detergents. Perhaps you next adorned yourself in an array of pleasant-smelling pigmented gels, dyed polymeric fibers, synthetic footwear, and metal-alloy jewelry. After breakfast, you probably abraded your teeth with a colloidal dispersion of artificially flavored, dental-hardening agents, grabbed your laptop (an electronic device containing ultrathin, microetched semiconductor layers powered by a series of voltaic cells), collected some books (processed cellulose and plastic, electronically printed with light- and oxygen-resistant inks), hopped in your hydrocarbon-fueled, metal-vinyl-ceramic vehicle, electrically ignited a synchronized series of controlled gaseous explosions, and took off for class!

But the true impact of chemistry extends much farther than the products we use in daily life. The most profound questions about health, climate change, even the origin of life, ultimately have chemical answers.

No matter what your reason for studying chemistry, this course will help you develop two mental skills. The first, common to all science courses, is the ability to solve problems systematically. The second is specific to chemistry, for as you comprehend its ideas, you begin to view a hidden reality filled with incredibly minute particles moving at fantastic speeds, colliding billions of times a second, and interacting in ways that determine how all the matter inside and outside of you behaves. This chapter holds the keys to enter this world.

CONCEPTS & SKILLS TO REVIEW
before studying this chapter

- exponential (scientific) notation (Appendix A)

1.1 • SOME FUNDAMENTAL DEFINITIONS

A good place to begin our exploration of chemistry is to define it and a few central concepts. **Chemistry** is *the study of matter and its properties, the changes that matter undergoes, and the energy associated with those changes.*

The Properties of Matter

Matter is the "stuff" of the universe: air, glass, planets, students—*anything that has mass and volume.* (In Section 1.4, we discuss the meanings of mass and volume in terms of how they are measured.) Chemists want to know the **composition** of matter, *the types and amounts of simpler substances that make it up.* A *substance* is a type of matter that has a defined, fixed composition.

We learn about matter by observing its **properties,** *the characteristics that give each substance its unique identity.* To identify a person, we might observe height, weight, hair and eye color, fingerprints, and even DNA pattern, until we arrive at a unique conclusion. To identify a substance, we observe two types of properties, *physical* and *chemical,* which are closely related to two types of change that matter undergoes:

- **Physical properties** are characteristics a substance shows *by itself, without changing into or interacting with another substance*. These properties include melting point, electrical conductivity, and density. A **physical change** occurs when a substance *alters its physical properties, **not** its composition.* For example, when ice melts,

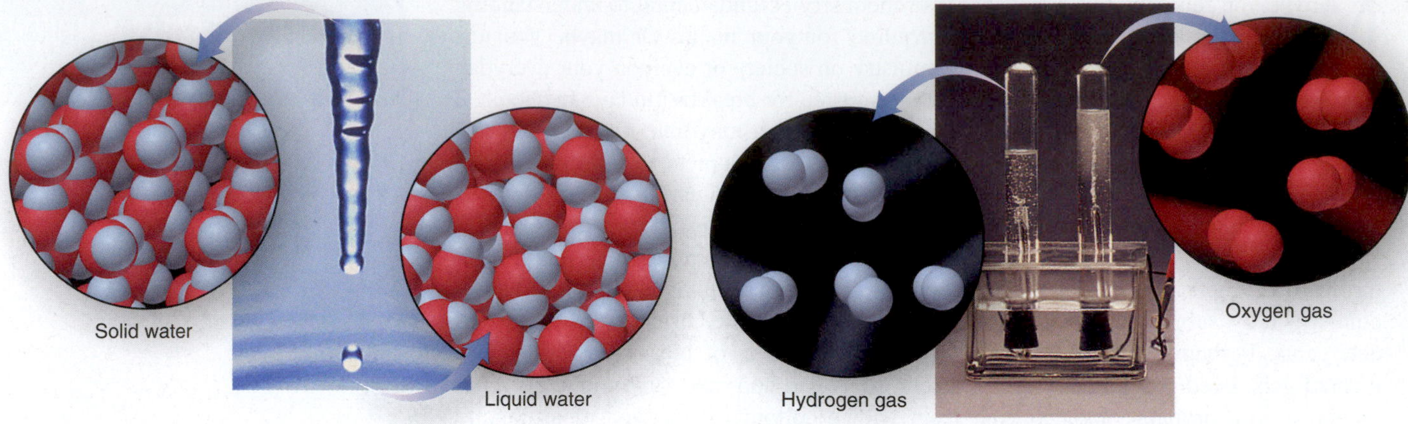

A Physical change:
*Solid form of water becomes liquid form. Particles before and after remain the same, which means composition did **not** change.*

B Chemical change:
*Electric current decomposes water into different substances (hydrogen and oxygen). Particles before and after are different, which means composition **did** change.*

Figure 1.1 The distinction between physical and chemical change.

several physical properties change, such as hardness, density, and ability to flow. But the composition of the sample does *not* change: it is still water. The photograph in Figure 1.1A shows what this change looks like in everyday life. The "blow-up" circles depict a magnified view of the particles making up the sample. In the icicle, the particles lie in a repeating pattern, whereas they are jumbled in the droplet, but *the particles are the same* in both forms of water.

Physical change (same substance before and after):

$$\text{Water (solid form)} \longrightarrow \text{water (liquid form)}$$

- **Chemical properties** are characteristics a substance shows *as it changes into or interacts with another substance (or substances)*. Chemical properties include flammability, corrosiveness, and reactivity with acids. A **chemical change,** also called a **chemical reaction,** occurs when *a substance (or substances) is converted into a different substance (or substances).* Figure 1.1B shows the chemical change (reaction) that occurs when you pass an electric current through water: the water decomposes (breaks down) into two other substances, hydrogen and oxygen, that bubble into the tubes. The composition *has* changed: the final sample is no longer water.

Chemical change (different substances before and after):

$$\text{Water} \xrightarrow{\text{electric current}} \text{hydrogen} + \text{oxygen}$$

Let's work through a sample problem that uses atomic-scale scenes to distinguish between physical and chemical change.

Sample Problem 1.1 Visualizing Change on the Atomic Scale

Problem The scenes below represent an atomic-scale view of a sample of matter, A *(center),* undergoing two different changes, to B *(left)* and to C *(right):*

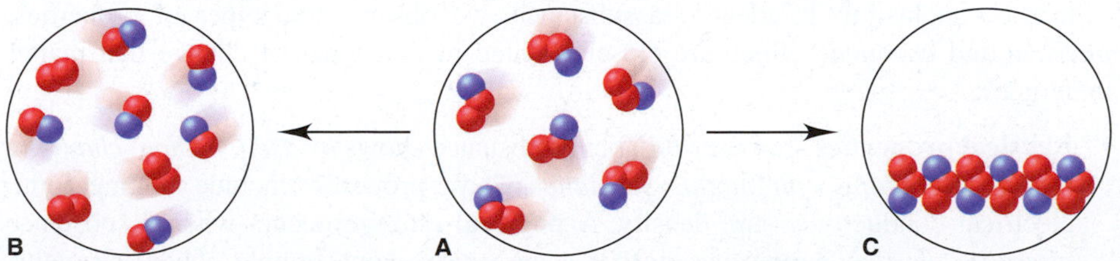

Decide whether each depiction shows a physical or a chemical change.

Plan Given depictions of two changes, we have to determine whether each represents a physical or a chemical change. The number and colors of the little spheres that make up each particle tell its "composition." Samples with particles of the *same* composition but in a different arrangement depict a *physical* change, whereas samples with particles of a *different* composition depict a *chemical* change.

Solution In A, each particle consists of one blue and two red spheres. The particles in A change into two types in B, one made of red and blue spheres and the other made of two red spheres; therefore, they have undergone a chemical change to form different particles. The particles in C are the same as those in A, but they are closer together and arranged differently; therefore, they have undergone a physical change.

FOLLOW-UP PROBLEM 1.1 Is the following change chemical or physical? (Compare your answer with the one in Brief Solutions to Follow-up Problems at the end of the chapter.)

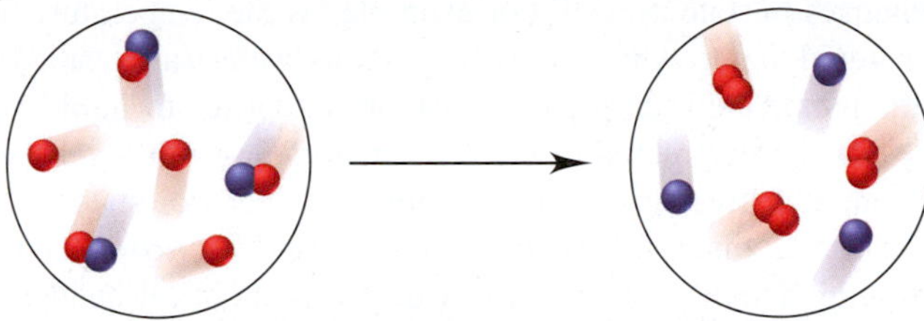

The States of Matter

Matter occurs commonly in *three physical forms* called **states:** solid, liquid, and gas. We'll define the states and see how temperature can change them.

Defining the States On the macroscopic scale, each state of matter is defined by the way the sample fills a container (Figure 1.2, *flasks at top*):

- A **solid** has a fixed shape that does not conform to the container shape. Solids are *not* defined by rigidity or hardness: solid iron is rigid and hard, but solid lead is flexible, and solid wax is soft.
- A **liquid** has a varying shape that conforms to the container shape, but only to the extent of the liquid's volume; that is, a liquid has *an upper surface.*
- A **gas** also has a varying shape that conforms to the container shape, but it fills the entire container and, thus, does *not* have a surface.

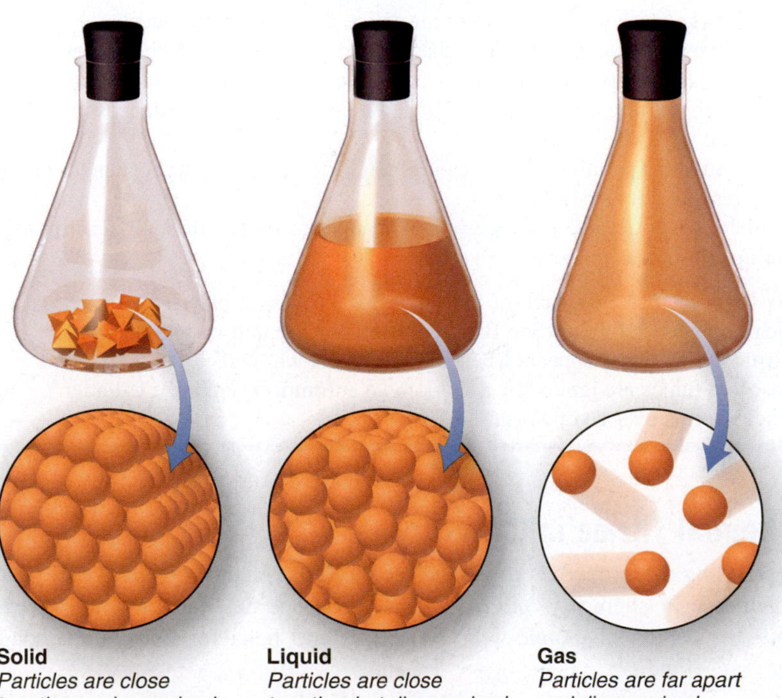
Solid
Particles are close together and organized.
Liquid
Particles are close together but disorganized.
Gas
Particles are far apart and disorganized.

Figure 1.2 The physical states of matter.

On the atomic scale, each state is defined by the relative positions of its particles (Figure 1.2, *circles at bottom*):

- In a *solid,* the particles lie next to each other in a regular, three-dimensional *array.*
- In a *liquid,* the particles also lie close together but move randomly around each other.
- In a *gas,* the particles have large distances between them and move randomly throughout the container.

Temperature and Changes of State Depending on the temperature and pressure of the surroundings, many substances can exist in each of the three physical states and undergo changes in state as well. For example, as the temperature increases, solid water melts to liquid water, which boils to gaseous water (also called *water vapor*). Similarly, as the temperature drops, water vapor condenses to liquid water, and with further cooling, the liquid freezes to ice. The majority of other substances—such as benzene, nitrogen, and iron—can undergo similar changes of state.

The main point is that *a physical change caused by heating can generally be reversed by cooling*. This is *not* generally true for a chemical change. For example, heating iron in moist air causes a chemical reaction that yields the brown, crumbly substance known as rust. Cooling does not reverse this change; rather, another chemical change (or series of them) is required.

The following sample problem provides practice in distinguishing some familiar examples of physical and chemical change.

Sample Problem 1.2 Distinguishing Between Physical and Chemical Change

Problem Decide whether each of the following processes is primarily a physical or a chemical change, and explain briefly:
(a) Frost forms as the temperature drops on a humid winter night.
(b) A cornstalk grows from a seed that is watered and fertilized.
(c) A match ignites to form ash and a mixture of gases.
(d) Perspiration evaporates when you relax after jogging.
(e) A silver fork tarnishes slowly in air.

Plan To decide whether a change is chemical or physical, we ask, "Does the substance change composition or just change form?"

Solution (a) Frost forming is a physical change: the drop in temperature changes water vapor (gaseous water) in humid air to ice crystals (solid water).
(b) A seed growing involves chemical change: the seed uses water, substances from air, fertilizer, and soil, and energy from sunlight to make complex changes in composition.
(c) The match burning is a chemical change: the combustible substances in the match head are converted into other substances.
(d) Perspiration evaporating is a physical change: the water in sweat changes its form, from liquid to gas, but not its composition.
(e) Tarnishing is a chemical change: silver changes to silver sulfide by reacting with sulfur-containing substances in the air.

FOLLOW-UP PROBLEM 1.2 Is each of the following processes primarily a physical or a chemical change? Explain. (See Brief Solutions at the end of the chapter.)
(a) Purple iodine vapor appears when solid iodine is warmed.
(b) Gasoline fumes are ignited by a spark in an automobile engine's cylinder.
(c) A scab forms over an open cut.

The Central Theme in Chemistry

Understanding the properties of a substance and the changes it undergoes leads to the central theme in chemistry: *macroscopic-scale* properties and behavior, those we can see, are the results of *atomic-scale* properties and behavior that we cannot see. The distinction between chemical and physical change is defined by composition, which we study macroscopically. But composition ultimately depends on the makeup of substances at the atomic scale. Similarly, macroscopic properties of substances in any of the three states arise from atomic-scale behavior of their particles. Picturing a chemical event on the molecular scale, even one as common as the flame of a candle, helps clarify what is taking place. What is happening when water boils or copper melts? What events occur in the invisible world of minute particles that cause a seed to grow, a neon light to glow, or a nail to rust? Throughout the text, we return to this central idea: *we study **observable** changes in matter to understand their **unobservable** causes.*

The Importance of Energy in the Study of Matter

Physical and chemical changes are accompanied by energy changes. **Energy** is often defined as *the ability to do work.* Essentially, all work involves moving something. Work is done when your arm lifts a book, when a car's engine moves the wheels, or when a falling rock moves the ground as it lands. The object doing the work (arm, engine, rock) transfers some of its energy to the object on which the work is done (book, wheels, ground).

The total energy an object possesses is the sum of its potential energy and its kinetic energy.

- **Potential energy** is *the energy due to the **position** of the object relative to other objects.*
- **Kinetic energy** is *the energy due to the **motion** of the object.*

Let's examine four systems that illustrate the relationship between these two forms of energy: a weight raised above the ground, two balls attached by a spring, two electrically charged particles, and a burning fuel and its waste products. Two concepts central to all these cases are

1. *When energy is converted from one form to the other, it is conserved, not destroyed.*
2. *Situations of lower energy are more stable, and therefore favored, over situations of higher energy (less stable).*

The four systems are depicted in Figure 1.3 on the next page:

- *A weight raised above the ground* (Figure 1.3A). The energy you exert to lift a weight against gravity increases the weight's potential energy (energy due to its position). When you drop the weight, that additional potential energy is converted to kinetic energy (energy due to motion). The situation with the weight elevated and higher in potential energy is *less stable,* so the weight will fall when released to result in a situation that is lower in potential energy and *more stable.*
- *Two balls attached by a spring* (Figure 1.3B). When you pull the balls apart, the energy you exert to stretch the relaxed spring increases the system's potential energy. This change in potential energy is converted to kinetic energy when you release the balls. The system of balls and spring is less stable (has more potential energy) when the spring is stretched than when it is relaxed.
- *Two electrically charged particles* (Figure 1.3C). Due to interactions known as *electrostatic forces, opposite charges attract each other, and like charges repel each other.* When energy is exerted to move a positive particle away from a negative one, the potential energy of the system increases, and that increase is converted to kinetic energy when the particles are pulled together by the electrostatic attraction. Similarly, when energy is used to move two positive (or two negative) particles together, their potential energy increases and changes to kinetic energy when they are pushed apart by the electrostatic repulsion. Charged particles move naturally to a more stable situation (lower energy).
- *A burning fuel and its waste products* (Figure 1.3D). Matter is composed of positively and negatively charged particles. *The chemical potential energy of a substance results from the relative positions of and the attractions and repulsions among its*

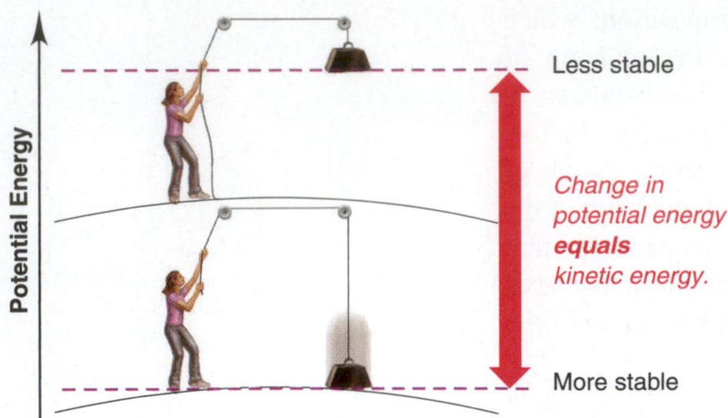

A A gravitational system. *Potential energy is gained when a weight is lifted. It is converted to kinetic energy as the weight falls.*

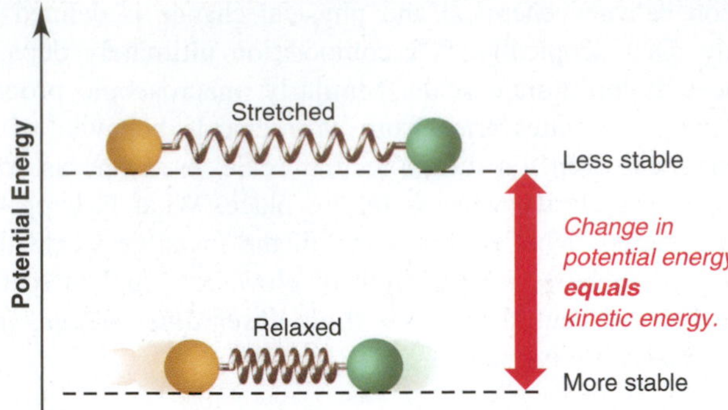

B A system of two balls attached by a spring. *Potential energy is gained when the spring is stretched. It is converted to the kinetic energy of the moving balls as the spring relaxes.*

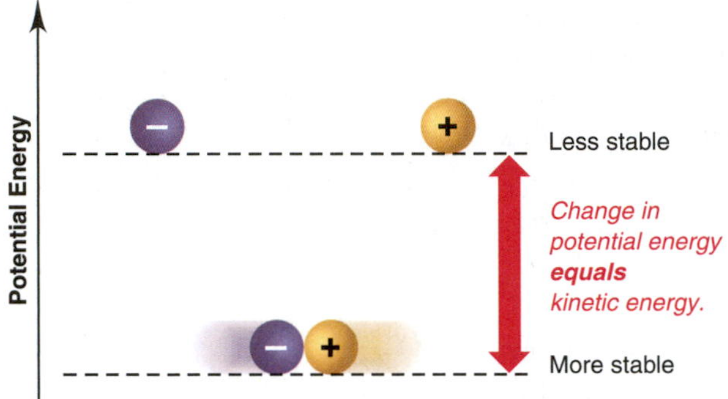

C A system of oppositely charged particles. *Potential energy is gained when the charges are separated. It is converted to kinetic energy as the attraction pulls the charges together.*

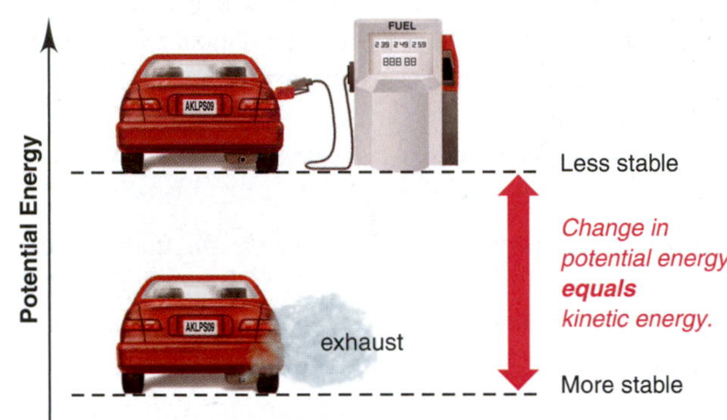

D A system of fuel and exhaust. *A fuel is higher in chemical potential energy than the exhaust. As the fuel burns, some of its potential energy is converted to the kinetic energy of the moving car.*

Figure 1.3 Potential energy is converted to kinetic energy. The dashed horizontal lines indicate the potential energy of each system before and after the change.

particles. Some substances are higher in potential energy than others. For example, gasoline and oxygen have more chemical potential energy than the exhaust gases they form. This difference is converted into kinetic energy, which moves the car, heats the interior, makes the lights shine, and so on. Similarly, the difference in potential energy between the food and air we take in and the wastes we excrete enables us to move, grow, keep warm, study chemistry, and so on.

Summary of Section 1.1

- Chemists study the composition and properties of matter and how they change.
- Each substance has a unique set of *physical* properties (attributes of the substance itself) and *chemical* properties (attributes of the substance as it interacts with or changes to other substances). Changes in matter can be *physical* (different form of the same substance) or *chemical* (different substance).
- Matter exists in three physical states—solid, liquid, and gas. The behavior of each state is due to the arrangement of the particles.
- A physical change caused by heating may be reversed by cooling. But a chemical change caused by heating can be reversed only by other chemical changes.
- Macroscopic changes result from submicroscopic changes.
- Changes in matter are accompanied by changes in energy.
- An object's potential energy is due to its position; an object's kinetic energy is due to its motion. Energy used to lift a weight, stretch a spring, or separate opposite charges increases the system's potential energy, which is converted to kinetic energy as the system returns to its original condition. Energy changes form but is conserved.
- Chemical potential energy arises from the positions and interactions of a substance's particles. When a higher energy (less stable) substance is converted into a more stable (lower energy) substance, some potential energy is converted into kinetic energy.

1.2 • THE SCIENTIFIC APPROACH: DEVELOPING A MODEL

Unlike our prehistoric ancestors, who survived through *trial and error*—gradually learning which types of stone were hard enough to shape others, which plants were edible and which poisonous—we employ the *quantitative theories* of chemistry to understand materials, make better use of them, and create new ones: specialized drugs to target diseases, advanced composites for vehicles, synthetic polymers for clothing and sports gear, liquid crystals for electronic displays, and countless others.

To understand nature, scientists use an approach called the **scientific method.** It is not a stepwise checklist, but rather a process involving creative propositions and tests aimed at objective, verifiable discoveries. There is no single procedure, and luck often plays a key role in discovery. In general terms, the scientific approach includes the following parts (Figure 1.4):

- ***Observations.*** These are the facts our ideas must explain. The most useful observations are quantitative because they can be analyzed to reveal trends. Pieces of quantitative information are **data.** When the same observation is made by many investigators in many situations with no clear exceptions, it is summarized, often in mathematical terms, as a **natural law.** The observation that mass remains constant during chemical change—made in the 18th century by the French chemist Antoine Lavoisier (1743–1794) and numerous experimenters since—is known as the law of mass conservation (Chapter 2).
- ***Hypothesis.*** Whether derived from observation or from a "spark of intuition," a hypothesis is a proposal made to explain an observation. A sound hypothesis need not be correct, but it must be *testable by experiment.* Indeed, a hypothesis is often the reason for performing an experiment: if the results do not support it, the hypothesis must be revised or discarded. Hypotheses can be altered, but experimental results cannot.
- ***Experiment.*** A set of procedural steps that tests a hypothesis, an experiment often leads to a revised hypothesis and new experiments to test it. An experiment typically contains at least two **variables,** quantities that can have more than one value. A well-designed experiment is **controlled** in that it measures the effect of one variable on another while keeping all other variables constant. Experimental results must be *reproducible* by others. Both skill and creativity play a part in experimental design.
- ***Model.*** Formulating conceptual models, or **theories,** based on *experiments* that test *hypotheses* about *observations* distinguishes scientific thinking from speculation. As hypotheses are revised according to experimental results, a model emerges to explain how the phenomenon occurs. A model is a *simplified,* not an exact, representation of some aspect of nature that we use to *predict* related phenomena. Ongoing experimentation refines the model to account for new facts.

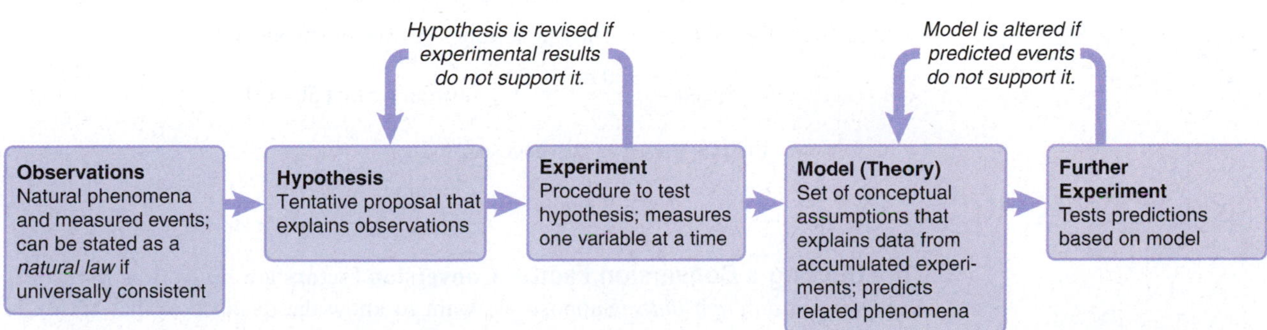

Figure 1.4 The scientific approach to understanding nature. Hypotheses and models are mental pictures that are revised to match observations and experimental results, *not* the other way around.

The following short paragraph is the first of an occasional feature that will help you learn a concept through an analogy, a unifying idea, or a memorization aid.

THINK OF IT THIS WAY
Everyday Scientific Thinking

Consider this familiar scenario. While listening to an FM station on your car's audio system, you notice the sound is garbled (observation) and assume it is caused by poor reception (hypothesis). To isolate this variable, you plug in your MP3 player and listen to a song (experiment): the sound is still garbled. If the problem is not poor reception, perhaps the speakers are at fault (new hypothesis). To isolate this variable, you listen with headphones (experiment): the sound is clear. You conclude that the speakers need to be repaired (model). The repair shop says the speakers are fine (new observation), but the car's amplifier may be at fault (new hypothesis). Repairing the amplifier corrects the garbled sound (new experiment), so the amplifier was the problem (revised model). Approaching a problem scientifically is a common practice, even if you're not aware of it.

Summary of Section 1.2

- The scientific method is a process designed to explain and predict phenomena.
- Observations lead to hypotheses about how or why a phenomenon occurs. When repeated with no exceptions, observations may be expressed as a natural law.
- Hypotheses are tested by controlled experiments and revised when necessary.
- If reproducible data support a hypothesis, a model (theory) can be developed to explain the observed phenomenon. A good model predicts related phenomena but must be refined whenever conflicting data appear.

1.3 • CHEMICAL PROBLEM SOLVING

In many ways, learning chemistry is learning how to solve chemistry problems. This section describes the problem-solving approach used throughout this book. Most problems include calculations, so let's first discuss how to handle measured quantities.

Units and Conversion Factors in Calculations

All measured quantities consist of a number *and* a unit: a person's height is "5 feet, 10 inches," not "5, 10." Ratios of quantities have ratios of units, such as miles/hour. (We discuss some important units in Section 1.4.) To minimize errors, make it a habit to *include units in all calculations.*

The arithmetic operations used with quantities are the same as those used with pure numbers; that is, units can be multiplied, divided, and canceled:

- A carpet measuring 3 feet by 4 feet (ft) has an area of

$$\text{Area} = 3 \text{ ft} \times 4 \text{ ft} = (3 \times 4)(\text{ft} \times \text{ft}) = 12 \text{ ft}^2$$

- A car traveling 350 miles (mi) in 7 hours (h) has a speed of

$$\text{Speed} = \frac{350 \text{ mi}}{7 \text{ h}} = \frac{50 \text{ mi}}{1 \text{ h}} \text{ (often written 50 mi}\cdot\text{h}^{-1}\text{)}$$

- In 3 hours, the car travels a distance of

$$\text{Distance} = 3 \cancel{\text{h}} \times \frac{50 \text{ mi}}{1 \cancel{\text{h}}} = 150 \text{ mi}$$

Constructing a Conversion Factor **Conversion factors** are *ratios used to express a quantity in different units.* Suppose we want to know the distance of that 150-mile car trip in feet. To convert miles to feet, we use *equivalent quantities,*

$$1 \text{ mi} = 5280 \text{ ft}$$

from which we can construct two conversion factors. Dividing both sides by 5280 ft gives one conversion factor (shown in blue):

$$\frac{1\ \text{mi}}{5280\ \text{ft}} = \frac{5280\ \cancel{\text{ft}}}{5280\ \cancel{\text{ft}}} = 1$$

And, dividing both sides by 1 mi gives the other conversion factor (the inverse):

$$\frac{1\ \cancel{\text{mi}}}{1\ \cancel{\text{mi}}} = \frac{5280\ \text{ft}}{1\ \text{mi}} = 1$$

Since the numerator and denominator of a conversion factor are equal, multiplying a quantity by a conversion factor is the same as multiplying by 1. Thus, *even though the number and unit change, the size of the quantity remains the same.*

To convert the distance from miles to feet, we choose the conversion factor with miles in the denominator, because it cancels miles and gives the answer in feet:

$$\text{Distance (ft)} = 150\ \cancel{\text{mi}} \times \frac{5280\ \text{ft}}{1\ \cancel{\text{mi}}} = 792{,}000\ \text{ft}$$

$$\text{mi} \quad \Rightarrow \quad \text{ft}$$

Choosing the Correct Conversion Factor It is easier to convert if you first decide whether the answer expressed in the new units should have a larger or smaller number. In the previous case, we know that a foot is *smaller* than a mile, so the distance in feet should have a *larger* number (792,000) than the distance in miles (150). The conversion factor has the larger number (5280) in the numerator, so it gave a larger number in the answer.

Most importantly, the *conversion factor you choose must cancel all units except those you want in the answer.* Therefore, set the unit you are converting *from* (beginning unit) in the *opposite position in the conversion factor* (numerator or denominator) so that it cancels and you are left with the unit you are converting *to* (final unit):

$$\cancel{\text{beginning unit}} \times \frac{\text{final unit}}{\cancel{\text{beginning unit}}} = \text{final unit} \qquad \text{as in} \qquad \cancel{\text{mi}} \times \frac{\text{ft}}{\cancel{\text{mi}}} = \text{ft}$$

Or, in cases that involve units raised to a power:

$$(\cancel{\text{beginning unit}} \times \cancel{\text{beginning unit}}) \times \frac{\text{final unit}^2}{\cancel{\text{beginning unit}^2}} = \text{final unit}^2$$

$$\text{as in} \qquad (\cancel{\text{ft}} \times \cancel{\text{ft}}) \times \frac{\text{mi}^2}{\cancel{\text{ft}^2}} = \text{mi}^2$$

Or, in cases that involve a ratio of units:

$$\frac{\cancel{\text{beginning unit}}}{\text{final unit}_1} \times \frac{\text{final unit}_2}{\cancel{\text{beginning unit}}} = \frac{\text{final unit}_2}{\text{final unit}_1} \qquad \text{as in} \qquad \frac{\cancel{\text{mi}}}{\text{h}} \times \frac{\text{ft}}{\cancel{\text{mi}}} = \frac{\text{ft}}{\text{h}}$$

Converting Between Unit Systems We use the same procedure to convert between systems of units, for example, between the English (or American) unit system and the International System (a revised metric system; Section 1.4). Suppose we know that the height of Angel Falls in Venezuela (the world's highest) is 3212 ft, and we find its height in miles as

$$\text{Height (mi)} = 3212\ \cancel{\text{ft}} \times \frac{1\ \text{mi}}{5280\ \cancel{\text{ft}}} = 0.6083\ \text{mi}$$

$$\text{ft} \quad \Rightarrow \quad \text{mi}$$

Now, we want its height in kilometers (km). The equivalent quantities are

$$1.609\ \text{km} = 1\ \text{mi}$$

Because we are converting *from* miles *to* kilometers, we use the conversion factor with miles in the denominator in order to cancel miles:

$$\text{Height (km)} = 0.6083\ \cancel{\text{mi}} \times \frac{1.609\ \text{km}}{1\ \cancel{\text{mi}}} = 0.9788\ \text{km}$$

$$\text{mi} \quad \Rightarrow \quad \text{km}$$

Notice that kilometers are *smaller* than miles, so this conversion factor gave us an answer with a *larger* number (0.9788 is larger than 0.6083).

If we want the height of Angel Falls in meters (m), we use the equivalent quantities 1 km = 1000 m to construct the conversion factor:

$$\text{Height (m)} = 0.9788\ \cancel{\text{km}} \times \frac{1000\ \text{m}}{1\ \cancel{\text{km}}} = 978.8\ \text{m}$$

$$\text{km} \Rightarrow \text{m}$$

In longer calculations, we often string together several conversion steps:

$$\text{Height (m)} = 3212\ \cancel{\text{ft}} \times \frac{1\ \cancel{\text{mi}}}{5280\ \cancel{\text{ft}}} \times \frac{1.609\ \cancel{\text{km}}}{1\ \cancel{\text{mi}}} \times \frac{1000\ \text{m}}{1\ \cancel{\text{km}}} = 978.8\ \text{m}$$

$$\text{ft} \Rightarrow \text{mi} \Rightarrow \text{km} \Rightarrow \text{m}$$

A Systematic Approach to Solving Chemistry Problems

The approach used in this book to solve problems emphasizes reasoning, not memorizing, and is based on a simple idea: plan how to solve the problem *before* you try to solve it, then check your answer, and practice with a similar follow-up problem. In general, the sample problems consist of several parts:

1. **Problem.** This part states all the information you need to solve the problem, usually framed in some interesting context.
2. **Plan.** This part helps you *think* about the solution *before* juggling numbers and pressing calculator buttons. There is often more than one way to solve a problem, and the given plan is one possibility. The plan will
   - Clarify the known and unknown: what information do you have, and what are you trying to find?
   - Suggest the steps from known to unknown: what ideas, conversions, or equations are needed?
   - Present a road map (especially in early chapters), a flow diagram of the plan. The road map has a box for each intermediate result and an arrow showing the step (conversion factor or operation) used to get to the next box.
3. **Solution.** This part shows the calculation steps in the same order as in the plan (and the road map).
4. **Check.** This part helps you check that your final answer makes sense: Are the units correct? Did the change occur in the expected direction? Is it reasonable chemically? To avoid a large math error, we also often do a rough calculation and see if we get an answer "in the same ballpark" as the actual result. Here's a typical "ballpark" calculation from everyday life. You are at a clothing store and buy three shirts at $14.97 each. With a 5% sales tax, the bill comes to $47.16. In your mind, you know that $14.97 is about $15, and 3 times $15 is $45; with the sales tax, the cost should be a bit more. So, your quick mental calculation *is* in the same ballpark as the actual cost.
5. **Comment.** This part appears occasionally to provide an application, an alternative approach, a common mistake to avoid, or an overview.
6. **Follow-up Problem.** This part presents a similar problem that requires you to apply concepts and/or methods used in solving the sample problem.

Of course, you can't learn to solve chemistry problems, any more than you can learn to swim, by reading about it, so here are a few suggestions:

- Follow along in the sample problem with pencil, paper, and calculator.
- Try the follow-up problem as soon as you finish the sample problem. A feature called Brief Solutions to Follow-up Problems appears at the end of each chapter, allowing you to compare your solution steps and answer.
- Read the sample problem and text again if you have trouble.

- Go to the Connect website for this text at www.mcgrawhillconnect.com and do the homework assignment. Hints and feedback on common incorrect answers, as well as step-by-step solutions, will help you learn to be an effective problem solver.
- The end-of-chapter problems review and extend the concepts and skills in the chapter, so work as many as you can. (Answers are given in the back of the book for problems with a colored number.)

Let's apply this systematic approach in a unit-conversion problem.

Sample Problem 1.3 Converting Units of Length

Problem To hang some paintings in your dorm, you need 325 centimeters (cm) of picture wire that sells for \$0.15/ft. How much does the wire cost?

Plan We know the length of wire in centimeters (325 cm) and the price in dollars per foot (\$0.15/ft). We can find the unknown cost of the wire by converting the length from centimeters to inches (in) and from inches to feet. The price gives us the equivalent quantities (1 ft = \$0.15) to convert feet of wire to cost in dollars. The road map starts with the known and moves through the calculation steps to the unknown.

Road Map

Solution Converting the known length from centimeters to inches: The equivalent quantities alongside the road map arrow are needed to construct the conversion factor. We choose 1 in/2.54 cm, rather than the inverse, because it gives an answer in inches:

$$\text{Length (in)} = \text{length (cm)} \times \text{conversion factor} = 325 \cancel{\text{cm}} \times \frac{1 \text{ in}}{2.54 \cancel{\text{cm}}} = 128 \text{ in}$$

Converting the length from inches to feet:

$$\text{Length (ft)} = \text{length (in)} \times \text{conversion factor} = 128 \cancel{\text{in}} \times \frac{1 \text{ ft}}{12 \cancel{\text{in}}} = 10.7 \text{ ft}$$

Converting the length in feet to cost in dollars:

$$\text{Cost (\$)} = \text{length (ft)} \times \text{conversion factor} = 10.7 \cancel{\text{ft}} \times \frac{\$0.15}{1 \cancel{\text{ft}}} = \$1.60$$

Check The units are correct for each step. The conversion factors make sense in terms of the relative unit sizes: the number of inches is *smaller* than the number of centimeters (an inch is *larger* than a centimeter), and the number of feet is *smaller* than the number of inches. The total cost seems reasonable: a little more than 10 ft of wire at \$0.15/ft should cost a little more than \$1.50.

Comment 1. We could also have strung the three steps together:

$$\text{Cost (\$)} = 325 \cancel{\text{cm}} \times \frac{1 \cancel{\text{in}}}{2.54 \cancel{\text{cm}}} \times \frac{1 \cancel{\text{ft}}}{12 \cancel{\text{in}}} \times \frac{\$0.15}{1 \cancel{\text{ft}}} = \$1.60$$

2. There are usually alternative sequences in unit-conversion problems. Here, for example, we would get the same answer if we first converted the cost of wire from \$/ft to \$/cm and kept the wire length in cm. Try it yourself.

FOLLOW-UP PROBLEM 1.3 A furniture factory needs 31.5 ft^2 of fabric to upholster one chair. Its Dutch supplier sends the fabric in bolts of exactly 200 m^2. How many chairs can be upholstered with 3 bolts of fabric (1 m = 3.281 ft)? Draw a road map to show how you plan the solution. (See Brief Solutions.)

Summary of Section 1.3

- A measured quantity consists of a number and a unit.
- A conversion factor is a ratio of equivalent quantities (and, thus, equal to 1) that is used to express a quantity in different units.

- The problem-solving approach used in this book has four parts: (1) plan the steps to the solution, which often includes a flow diagram (road map) of the steps, (2) perform the calculations according to the plan, (3) check to see if the answer makes sense, and (4) practice with a similar problem and compare your solution with the one at the end of the chapter.

1.4 • MEASUREMENT IN SCIENTIFIC STUDY

Almost everything we own is made and sold in measured amounts. The measurement systems we use have a rich history characterized by the search for *exact, invariable standards.* Measuring for purposes of trade, building, and surveying used to be based on standards that could vary: a yard was the distance from the king's nose to the tip of his outstretched arm, and an acre was the area tilled in one day by a man with a pair of oxen. Our current, far more exact system of measurement began in 1790 when a committee in France developed the original *metric system.* In 1960, another committee in France revised it to create the universally accepted **SI units** (from the French **S**ystème **I**nternational d'Unités).

General Features of SI Units

The SI system is based on seven **fundamental units,** or **base units,** each identified with a physical quantity (Table 1.1). All other units are **derived units,** combinations of the seven base units. For example, the derived unit for speed, meters per second (m/s), is the base unit for length (m) divided by the base unit for time (s). (Derived units that are a *ratio* of base units can be used as conversion factors.) For quantities much smaller or larger than the base unit, we use decimal prefixes and exponential (scientific) notation (Table 1.2). (If you need a review of exponential notation, see Appendix A.) Because the prefixes are based on powers of 10, SI units are easier to use in calculations than English units.

Table 1.1 SI Base Units

Physical Quantity (Dimension)	Unit Name	Unit Abbreviation
Mass	kilogram	kg
Length	meter	m
Time	second	s
Temperature	kelvin	K
Electric current	ampere	A
Amount of substance	mole	mol
Luminous intensity	candela	cd

Some Important SI Units in Chemistry

Here, we discuss units for length, volume, mass, density, temperature, and time; other units are presented in later chapters. Table 1.3 shows some SI quantities for length, volume, and mass, along with their English-system equivalents.

Length The SI base unit of length is the **meter (m),** which is about 2.5 times the width of this book when open. The definition is exact and invariant: 1 meter is the distance light travels in a vacuum in 1/299,792,458 of a second. A meter is a little longer than a yard (1 m = 1.094 yd); a centimeter (10^{-2} m) is about two-fifths of an inch (1 cm = 0.3937 in; 1 in = 2.54 cm). Biological cells are often measured in micrometers (1 μm = 10^{-6} m). On the atomic scale, nanometers (10^{-9} m) and picometers (10^{-12} m) are used. Many proteins have diameters of about 2 nm; atomic diameters are about 200 pm (0.2 nm). An older unit still in use is the angstrom (1 Å = 10^{-10} m = 0.1 nm = 100 pm).

Table 1.2 Common Decimal Prefixes Used with SI Units

Prefix*	Prefix Symbol	Word	Conventional Notation	Exponential Notation
tera	T	trillion	1,000,000,000,000	1×10^{12}
giga	G	billion	1,000,000,000	1×10^{9}
mega	M	million	1,000,000	1×10^{6}
kilo	k	thousand	1,000	1×10^{3}
hecto	h	hundred	100	1×10^{2}
deka	da	ten	10	1×10^{1}
—	—	one	1	1×10^{0}
deci	d	tenth	0.1	1×10^{-1}
centi	c	hundredth	0.01	1×10^{-2}
milli	m	thousandth	0.001	1×10^{-3}
micro	μ	millionth	0.000001	1×10^{-6}
nano	n	billionth	0.000000001	1×10^{-9}
pico	p	trillionth	0.000000000001	1×10^{-12}
femto	f	quadrillionth	0.000000000000001	1×10^{-15}

*The prefixes most frequently used by chemists appear in bold type.

Volume Any sample of matter has a certain **volume (*V*),** the amount of space it occupies. The SI unit of volume is the **cubic meter (m^3).** In chemistry, we often use the non-SI units **liter (L)** and **milliliter (mL)** (note the uppercase L). Medical practitioners measure body fluids in cubic decimeters (dm^3), which are equivalent to liters:

$$1\ L = 1\ dm^3 = 10^{-3}\ m^3$$

And 1 mL, or $\frac{1}{1000}$ of a liter, is equivalent to 1 cubic centimeter (cm^3):

$$1\ mL = 1\ cm^3 = 10^{-3}\ dm^3 = 10^{-3}\ L = 10^{-6}\ m^3$$

A liter is slightly larger than a quart (qt) (1 L = 1.057 qt; 1 qt = 946.4 mL); 1 fluid ounce ($\frac{1}{32}$ of a quart) equals 29.57 mL (29.57 cm^3).

Figure 1.5 shows some laboratory glassware for working with volumes. Volumetric flasks and pipets have a fixed volume indicated by a mark on the neck.

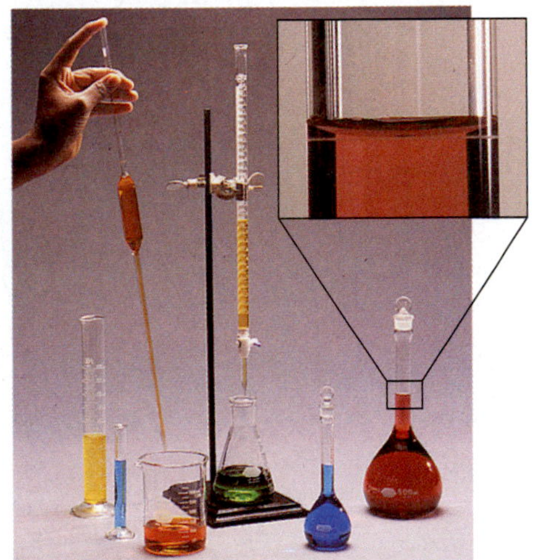

Figure 1.5 Common laboratory volumetric glassware. From left to right are two graduated cylinders, a pipet being emptied into a beaker, a buret delivering liquid to an Erlenmeyer flask, and two volumetric flasks. **Inset,** In contact with the glass neck, the liquid forms a concave meniscus (curved surface).

Table 1.3 Common SI-English Equivalent Quantities

Quantity	SI	SI Equivalents	English Equivalents	English to SI Equivalent
Length	1 kilometer (km)	1000 (10^3) meters	0.6214 mile (mi)	1 mile = 1.609 km
	1 meter (m)	100 (10^2) centimeters	1.094 yards (yd)	1 yard = 0.9144 m
		1000 millimeters (mm)	39.37 inches (in)	1 foot (ft) = 0.3048 m
	1 centimeter (cm)	0.10 (10^{-2}) meter	0.3937 inch	1 inch = 2.54 cm (exactly)
Volume	1 cubic meter (m^3)	1,000,000 (10^6) cubic centimeters	35.31 cubic feet (ft^3)	1 cubic foot = 0.02832 m^3
	1 cubic decimeter (dm^3)	1000 cubic centimeters	0.2642 gallon (gal)	1 gallon = 3.785 dm^3
			1.057 quarts (qt)	1 quart = 0.9464 dm^3
				1 quart = 946.4 cm^3
	1 cubic centimeter (cm^3)	0.001 dm^3	0.03381 fluid ounce	1 fluid ounce = 29.57 cm^3
Mass	1 kilogram (kg)	1000 grams	2.205 pounds (lb)	1 pound = 0.4536 kg
	1 gram (g)	1000 milligrams (mg)	0.03527 ounce (oz)	1 ounce = 28.35 g

Sample Problem 1.4 Converting Units of Volume

Problem The volume of an irregularly shaped solid can be determined from the volume of water it displaces. A graduated cylinder contains 19.9 mL of water. When a small piece of galena, an ore of lead, is added, it sinks and the volume increases to 24.5 mL. What is the volume of the piece of galena in cm^3 and in L?

Plan We have to find the volume of the galena from the change in volume of the cylinder contents. The volume of galena in mL is the difference before (19.9 mL) and after (24.5 mL) adding it. Since mL and cm^3 represent identical volumes, the volume in mL equals the volume in cm^3. We then use equivalent quantities (1 mL = 10^{-3} L) to convert mL to L. The road map shows these steps.

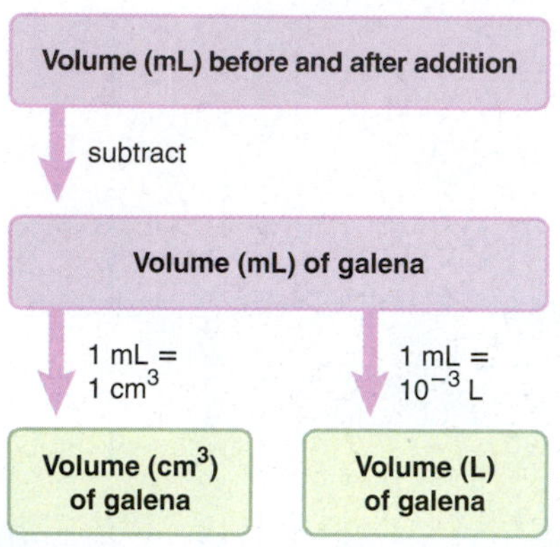

Solution Finding the volume of galena:

$$\text{Volume (mL)} = \text{volume after} - \text{volume before} = 24.5\ \text{mL} - 19.9\ \text{mL} = 4.6\ \text{mL}$$

Converting the volume from mL to cm^3:

$$\text{Volume (cm}^3) = 4.6\ \cancel{\text{mL}} \times \frac{1\ \text{cm}^3}{1\ \cancel{\text{mL}}} = 4.6\ \text{cm}^3$$

Converting the volume from mL to L:

$$\text{Volume (L)} = 4.6\ \cancel{\text{mL}} \times \frac{10^{-3}\ \text{L}}{1\ \cancel{\text{mL}}} = 4.6\times10^{-3}\ \text{L}$$

Check The units and magnitudes of the answers seem correct, and it makes sense that the volume in mL would have a number 1000 times larger than the same volume in L.

FOLLOW-UP PROBLEM 1.4 Within a cell, proteins are synthesized on particles called ribosomes. Assuming ribosomes are spherical, what is the volume (in dm^3 and μL) of a ribosome whose average diameter is 21.4 nm (V of a sphere $= \frac{4}{3}\pi r^3$)? Draw a road map to show how you plan the solution. (See Brief Solutions.)

Mass The quantity of matter an object contains is its **mass.** The SI unit of mass is the **kilogram (kg),** the only base unit whose standard is an object—a platinum-iridium cylinder kept in France—and the only one whose name has a prefix.*

The terms *mass* and *weight* have distinct meanings:

- *Mass is constant* because an object's quantity of matter cannot change.
- ***Weight** is variable* because it depends on the local gravitational field.

Because the strength of the gravitational field varies with altitude, you (and other objects) weigh slightly less on a high mountain than at sea level.

Does this mean that a sample weighed on laboratory balances in Miami (sea level) and in Denver (about 1.7 km above sea level) give different results? No, because these balances measure mass, not weight. Mechanical balances compare the object's mass with masses built into the balance, so the local gravitational field pulls on them equally. Electronic (analytical) balances generate an electric field that counteracts the local field, and the current needed to restore the pan to zero is converted to the equivalent mass and displayed.

Sample Problem 1.5 Converting Units of Mass

Problem Many international computer communications are carried by optical fibers in cables laid along the ocean floor. If one strand of optical fiber weighs 1.19×10^{-3} lb/m, what is the mass (in kg) of a cable made of six strands of optical fiber, each long enough to link New York and Paris (8.84×10^3 km)?

*The names of the other base units are used as the root words, but for units of mass we attach prefixes to the word "gram," as in "microgram" and "kilogram"; thus, we say "milligram," never "microkilogram."

Plan We have to find the mass of cable (in kg) from the given mass/length of fiber (1.19×10^{-3} lb/m), number of fibers/cable (6), and length of cable (8.84×10^{3} km). Let's first find the mass of one fiber and then the mass of cable. As shown in the road map, we convert the length of one fiber from km to m and then find its mass (in lb) by converting m to lb. Then we multiply the fiber mass by 6 to get the cable mass, and finally convert lb to kg.

Road Map

Length (km) of fiber
↓ 1 km = 10^3 m
Length (m) of fiber
↓ 1 m = 1.19×10^{-3} lb
Mass (lb) of fiber
↓ 6 fibers = 1 cable
Mass (lb) of cable
↓ 2.205 lb = 1 kg
Mass (kg) of cable

Solution Converting the fiber length from km to m:

$$\text{Length (m) of fiber} = 8.84\times10^{3}\ \cancel{\text{km}} \times \frac{10^{3}\ \text{m}}{1\ \cancel{\text{km}}} = 8.84\times10^{6}\ \text{m}$$

Converting the length of one fiber to mass (lb):

$$\text{Mass (lb) of fiber} = 8.84\times10^{6}\ \cancel{\text{m}} \times \frac{1.19\times10^{-3}\ \text{lb}}{1\ \cancel{\text{m}}} = 1.05\times10^{4}\ \text{lb}$$

Finding the mass of the cable (lb):

$$\text{Mass (lb) of cable} = \frac{1.05\times10^{4}\ \text{lb}}{1\ \cancel{\text{fiber}}} \times \frac{6\ \cancel{\text{fibers}}}{1\ \text{cable}} = 6.30\times10^{4}\ \text{lb/cable}$$

Converting the mass of the cable from lb to kg:

$$\text{Mass (kg) of cable} = \frac{6.30\times10^{4}\ \cancel{\text{lb}}}{1\ \text{cable}} \times \frac{1\ \text{kg}}{2.205\ \cancel{\text{lb}}} = 2.86\times10^{4}\ \text{kg/cable}$$

Check The units are correct. Let's think through the relative sizes of the answers to see if they make sense: The number of m should be 10^3 larger than the number of km. If 1 m of fiber weighs about 10^{-3} lb, about 10^7 m should weigh about 10^4 lb. The cable mass should be six times as much, or about 6×10^4 lb. Since 1 lb is about $\frac{1}{2}$ kg, the number of kg should be about half the number of lb.

FOLLOW-UP PROBLEM 1.5 An intravenous nutrient solution is delivered to a hospital patient at a rate of 1.5 drops per second. If a drop of solution weighs 65 mg on average, how many kilograms are delivered in 8.0 h? Draw a road map to show how you plan the solution. (See Brief Solutions.)

Density The **density (*d*)** of an object is its mass divided by its volume:

$$\text{Density} = \frac{\text{mass}}{\text{volume}} \tag{1.1}$$

We isolate each of these variables by treating density as a conversion factor:

$$\text{Mass} = \text{volume} \times \text{density} = \cancel{\text{volume}} \times \frac{\text{mass}}{\cancel{\text{volume}}}$$

or,

$$\text{Volume} = \text{mass} \times \frac{1}{\text{density}} = \cancel{\text{mass}} \times \frac{\text{volume}}{\cancel{\text{mass}}}$$

Because volume can change with temperature, so can density. But, at a given temperature and pressure, the *density of a substance is a characteristic physical property and, thus, has a specific value.*

The SI unit of density is kilograms per cubic meter (kg/m^3), but in chemistry, density has units of g/L (g/dm^3) or g/mL (g/cm^3) (Table 1.4). Note that the densities of gases are much lower than those of liquids or solids (also see Figure 1.2).

Table 1.4 Densities of Some Common Substances*

Substance	Physical State	Density (g/cm^3)
Hydrogen	gas	0.0000899
Oxygen	gas	0.00133
Grain alcohol	liquid	0.789
Water	liquid	0.998
Table salt	solid	2.16
Aluminum	solid	2.70
Lead	solid	11.3
Gold	solid	19.3

*At room temperature (20°C) and normal atmospheric pressure (1 atm).

Sample Problem 1.6 Calculating Density from Mass and Volume

Problem Lithium, a soft, gray solid with the lowest density of any metal, is a key component of advanced batteries, such as the one in your laptop. A slab of lithium weighs 1.49×10^3 mg and has sides that are 20.9 mm by 11.1 mm by 11.9 mm. Find the density of lithium in g/cm^3.

Road Map

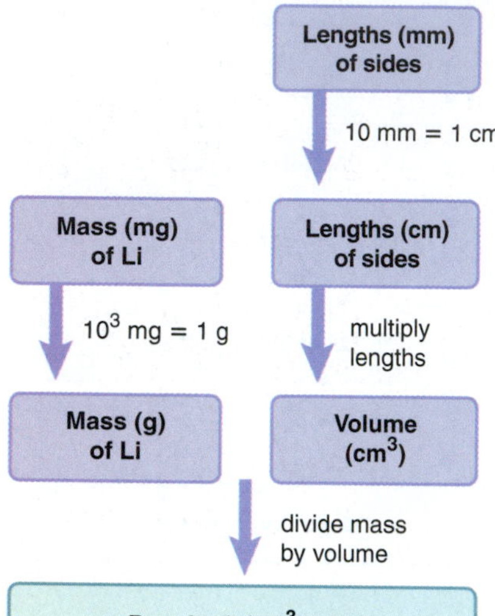

Plan To find the density in g/cm^3, we need the mass of lithium in g and the volume in cm^3. The mass is 1.49×10^3 mg, so we convert mg to g. We convert the lengths of the three sides from mm to cm, and then multiply them to find the volume in cm^3. Dividing the mass by the volume gives the density (see the road map).

Solution Converting the mass from mg to g:

$$\text{Mass (g) of lithium} = 1.49\times10^3\ \cancel{\text{mg}}\left(\frac{10^{-3}\text{ g}}{1\ \cancel{\text{mg}}}\right) = 1.49\text{ g}$$

Converting side lengths from mm to cm:

$$\text{Length (cm) of one side} = 20.9\ \cancel{\text{mm}} \times \frac{1\text{ cm}}{10\ \cancel{\text{mm}}} = 2.09\text{ cm}$$

Similarly, the other side lengths are 1.11 cm and 1.19 cm.
Multiplying the sides to get the volume:

$$\text{Volume (cm}^3) = 2.09\text{ cm} \times 1.11\text{ cm} \times 1.19\text{ cm} = 2.76\text{ cm}^3$$

Calculating the density:

$$\text{Density of lithium} = \frac{\text{mass}}{\text{volume}} = \frac{1.49\text{ g}}{2.76\text{ cm}^3} = 0.540\text{ g/cm}^3$$

Check Since 1 cm = 10 mm, the number of cm in each length should be $\frac{1}{10}$ the number of mm. The units for density are correct, and the size of the answer (~0.5 g/cm^3) seems correct since the number of g (1.49) is about half the number of cm^3 (2.76). Also, the problem states that lithium has a very low density, so this answer makes sense.

FOLLOW-UP PROBLEM 1.6 The piece of galena in Sample Problem 1.4 has a volume of 4.6 cm^3. If the density of galena is 7.5 g/cm^3, what is the mass (in kg) of that piece of galena? Draw a road map to show how you plan the solution. (See Brief Solutions.)

Temperature There is a key distinction between temperature and heat:

- **Temperature (*T*)** is *a measure* of how hot or cold one object is relative to another.
- **Heat** is *the energy* that flows from the object with the higher temperature to the object with the lower temperature. When you hold an ice cube, it feels like the "cold" flows into your hand, but actually, heat flows from your hand to the ice.

In the laboratory, we measure temperature with a **thermometer,** a narrow tube containing a fluid that expands when heated. When the thermometer is immersed in a substance hotter than itself, heat flows from the substance through the glass into the fluid, which expands and rises in the thermometer tube. If a substance is colder than the thermometer, heat flows to the substance from the fluid, which contracts and falls within the tube.

We'll consider three temperature scales: the Celsius (°C, formerly called centigrade); the Kelvin (K), which is preferred in scientific work (although the Celsius scale is still used frequently); and the Fahrenheit (°F) scales. The SI base unit of temperature is the **kelvin** (**K,** with no degree sign, °). In the United States, the Fahrenheit scale is used for weather reporting, body temperature, and so forth.

The three scales differ in the size of the unit and/or the temperature of the zero point. Figure 1.6 shows the freezing and boiling points of water in the three scales. The **Celsius scale** sets water's freezing point at 0°C and its boiling point (at normal atmospheric pressure) at 100°C. The **Kelvin (absolute) scale** uses the *same size degree* as the Celsius scale—$\frac{1}{100}$ of the difference between the freezing and boiling points of water—but it has a *different zero point;* that is, 0 K, or *absolute zero,* equals −273.15°C. Thus, in the Kelvin scale, *all temperatures are positive;* for example, water freezes at +273.15 K (0°C) and boils at +373.15 K (100°C).

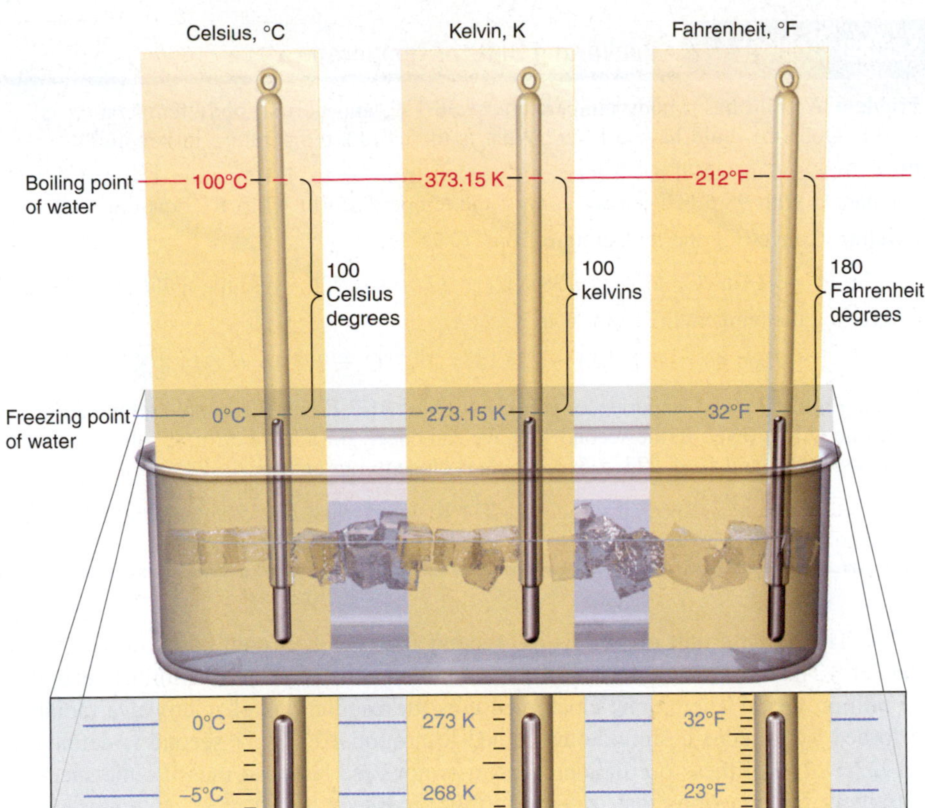

Figure 1.6 Freezing and boiling points of water in the Celsius, Kelvin (absolute), and Fahrenheit scales. At the bottom of the figure, a portion of each of the three thermometer scales is expanded to show the sizes of the units.

We convert between the Celsius and Kelvin scales by remembering the different zero points: 0°C = 273.15 K, so

$$T\text{ (in K)} = T\text{ (in °C)} + 273.15 \qquad \textbf{(1.2)}$$

And, therefore,

$$T\text{ (in °C)} = T\text{ (in K)} - 273.15 \qquad \textbf{(1.3)}$$

The Fahrenheit scale differs from the other scales in its zero point *and* in the size of its degree. Water freezes at 32°F and boils at 212°F. Therefore, 180 Fahrenheit degrees (212°F − 32°F) represents the same temperature change as 100 Celsius degrees (or 100 kelvins). Because 100 Celsius degrees equal 180 Fahrenheit degrees,

$$1 \text{ Celsius degree} = \tfrac{180}{100} \text{ Fahrenheit degrees} = \tfrac{9}{5} \text{ Fahrenheit degrees}$$

To convert a temperature from °C to °F, first change the degree size and then adjust the zero point:

$$T\text{ (in °F)} = \tfrac{9}{5}T\text{ (in °C)} + 32 \qquad \textbf{(1.4)}$$

To convert a temperature from °F to °C, do the two steps in the opposite order: adjust the zero point and then change the degree size. In other words, solve Equation 1.4 for T (in °C):

$$T\text{ (in °C)} = [T\text{ (in °F)} - 32]\tfrac{5}{9} \qquad \textbf{(1.5)}$$

Table 1.5 compares the three temperature scales.

Table 1.5 The Three Temperature Scales

Scale	Unit	Size of Degree (Relative to K)	Freezing Point of H_2O	Boiling Point of H_2O	T at Absolute Zero	Conversion
Kelvin (absolute)	kelvin (K)	—	273.15 K	373.15 K	0 K	to °C (Equation 1.2)
Celsius	Celsius degree (°C)	1	0°C	100°C	−273.15°C	to K (Equation 1.3) to °F (Equation 1.4)
Fahrenheit	Fahrenheit degree (°F)	$\frac{5}{9}$	32°F	212°F	−459.67°F	to °C (Equation 1.5)

Sample Problem 1.7 Converting Units of Temperature

Problem A child has a body temperature of 38.7°C, and normal body temperature is 98.6°F. Does the child have a fever? What is the child's temperature in kelvins?

Plan To see if the child has a fever, we convert from °C to °F (Equation 1.4) and compare it with 98.6°F. Then we convert the temperature in °C to K (Equation 1.2).

Solution Converting the temperature from °C to °F:

$$T\text{ (in °F)} = \tfrac{9}{5}T\text{ (in °C)} + 32 = \tfrac{9}{5}(38.7\text{°C}) + 32 = 101.7\text{°F; yes, the child has a fever.}$$

Converting the temperature from °C to K:

$$T\text{ (in K)} = T\text{ (in °C)} + 273.15 = 38.7\text{°C} + 273.15 = 311.8\text{ K}$$

Check From everyday experience, you know that 101.7°F is a reasonable temperature for someone with a fever. In the second step, we can check for a large error as follows: 38.7°C is almost 40°C, and 40 + 273 = 313, which is close to our answer.

FOLLOW-UP PROBLEM 1.7 Mercury melts at 234 K, lower than any other pure metal. What is its melting point in °C and °F?

Time The SI base unit of time is the **second (s),** which is now based on an atomic standard. The most recent version of the atomic clock is accurate to within 1 second in 20 million years! The atomic clock measures the oscillations of microwave radiation absorbed by gaseous cesium atoms cooled to around 10^{-6} K: 1 second is defined as 9,192,631,770 of these oscillations. Chemists now use lasers to measure the speed of extremely fast reactions that occur in a few picoseconds (10^{-12} s) or femtoseconds (10^{-15} s).

Extensive and Intensive Properties

Some variables are *dependent* on the amount of substance present; these are called **extensive properties.** On the other hand, **intensive properties** are *independent* of the amount of substance. Mass and volume, for example, are extensive properties, but density is an intensive property. Thus, a gallon of water has four times the mass of a quart of water, but it also has four times the volume, so the density, the *ratio* of mass to volume, is the same for both samples.

Another important example concerns heat, an extensive property, and temperature, an intensive property: a vat of boiling water has more heat, that is, more energy, than a cup of boiling water, but both samples have the same temperature.

Summary of Section 1.4

- The SI unit system consists of seven base units and numerous derived units.
- Exponential notation and prefixes based on powers of 10 are used to express very small and very large numbers.
- The SI base unit of length is the meter (m); on the atomic scale, the nanometer (nm) and picometer (pm) are used commonly.
- Volume (V) units are derived from length units, and the most important volume units are the cubic meter (m^3) and the liter (L).
- The *mass* of an object—the quantity of matter in it—is constant. The SI unit of mass is the kilogram (kg). The *weight* of an object varies with the gravitational field.
- Density (d) is a characteristic physical property of a substance and is the ratio of its mass to its volume.
- Temperature (T) is a measure of the relative hotness of an object. Heat is energy that flows from an object at higher T to one at lower T.
- Temperature scales differ in the size of the degree unit and/or the zero point. For scientific uses, temperature is measured in kelvins (K) or degrees Celsius (°C).
- Extensive properties, such as mass, volume, and energy, depend on the amount of a substance. Intensive properties, such as density and temperature, do not.

1.5 • UNCERTAINTY IN MEASUREMENT: SIGNIFICANT FIGURES

All measuring devices—balances, pipets, thermometers, and so forth—are made to limited specifications, and we use our imperfect senses and skills to read them. Therefore, we can *never* measure a quantity exactly; put another way, every measurement includes some **uncertainty.** The device we choose depends on how much uncertainty is acceptable. When you buy potatoes, a supermarket scale that measures in 0.1-kg increments is acceptable; it tells you that the mass is, for example, 2.0 ± 0.1 kg. The "± 0.1 kg" term expresses the uncertainty: the potatoes weigh between 1.9 and 2.1 kg. Needing more certainty than that to weigh a substance, a chemist uses a balance that measures in 0.001-kg increments and finds the substance weighs 2.036 ± 0.001 kg, that is, between 2.035 and 2.037 kg. The greater number of digits in this measurement means we know the mass of the substance with *more certainty* than we know the mass of potatoes.

We *always estimate the rightmost digit* of a measurement. The uncertainty can be expressed with the ± sign, but generally we drop the sign and *assume an uncertainty of one unit in the rightmost digit.* The digits we record, both the certain and the uncertain ones, are called **significant figures.** There are four significant figures in 2.036 kg and two in 2.0 kg. *The greater the number of significant figures, the greater is the certainty of a measurement.* Figure 1.7 shows this point for two thermometers.

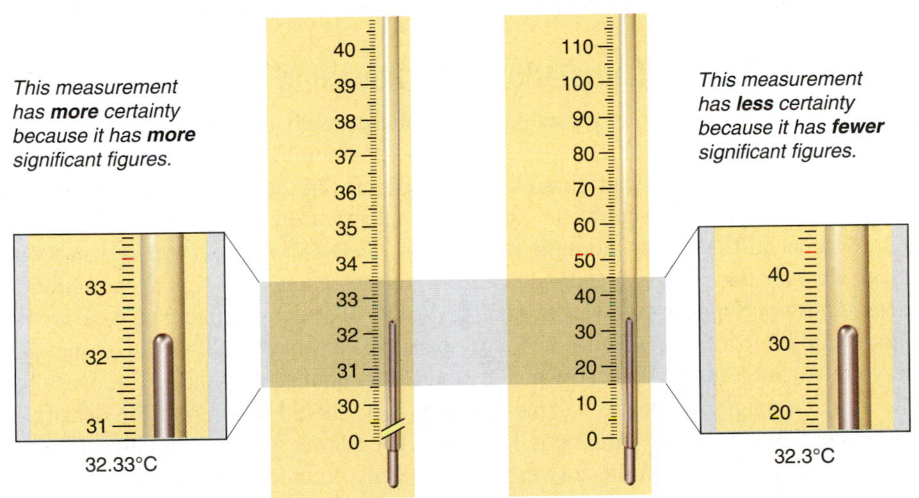

This measurement has **more** certainty because it has **more** significant figures.
32.33°C
This measurement has **less** certainty because it has **fewer** significant figures.
32.3°C

Figure 1.7 The number of significant figures in a measurement. The thermometer on the left is graduated in increments of 0.1°C; the one on the right is graduated in increments of 1°C.

Determining Which Digits Are Significant

When you take a measurement or use one in a calculation, you must know the number of digits that are significant: *all digits are significant, except zeros used only to position the decimal point.* The following procedure applies this point:

1. Make sure the measurement has a decimal point.
2. Start at the left, and move right until you reach the first nonzero digit.
3. Count that digit and every digit to its right as significant.

A complication can arise when zeros end a number:

- If there *is* a decimal point and the zeros lie either after or before it, they *are* significant: 1.1300 g has five significant figures and 6500. has four.
- If there is *no* decimal point, assume the zeros are *not* significant, unless exponential notation clarifies the quantity: 5300 L is *assumed* to have two significant figures, but 5.300×10^3 L has four, 5.30×10^3 L has three, and 5.3×10^3 L has two.
- A terminal decimal point means the zeros before it *are* significant: 500 mL has one significant figure, but 500. mL has three (as do 5.00×10^2 mL and 0.500 L).

Sample Problem 1.8 Determining the Number of Significant Figures

Problem For each of the following quantities, underline the zeros that are significant figures (sf) and determine the total number of significant figures. For (d) to (f), express each quantity in exponential notation first.

(a) 0.0030 L **(b)** 0.1044 g **(c)** 53,069 mL
(d) 0.00004715 m **(e)** 57,600. s **(f)** 0.0000007160 cm^3

Plan We determine the number of significant figures by counting digits, as just described, paying particular attention to the position of zeros in relation to the decimal point, and underline the zeros that are significant.

Solution **(a)** 0.0030 L has 2 sf
(b) 0.1044 g has 4 sf
(c) 53,069 mL has 5 sf
(d) 0.00004715 m, or 4.715×10^{-5} m, has 4 sf
(e) 57,600. s, or 5.7600×10^{4} s, has 5 sf
(f) 0.0000007160 cm^3, or 7.160×10^{-7} cm^3, has 4 sf

Check Be sure that every zero counted as significant comes *after* nonzero digit(s) in the number.

FOLLOW-UP PROBLEM 1.8 For each of the following quantities, underline the zeros that are significant figures and determine the total number of significant figures (sf). For (d) to (f), express each quantity in exponential notation first.

(a) 31.070 mg **(b)** 0.06060 g **(c)** 850.°C
(d) 200.0 mL **(e)** 0.0000039 m **(f)** 0.000401 L

Significant Figures: Calculations and Rounding Off

Measuring several quantities typically results in data with differing numbers of significant figures. In a calculation, we keep track of the number in each quantity so that we don't have more significant figures (more certainty) in the answer than in the data. If we do have too many significant figures, we must **round off** the answer.

The general rule for rounding is that *the least certain measurement sets the limit on certainty for the entire calculation and determines the number of significant figures in the final answer.* Suppose you want to find the density of a new ceramic. You measure the mass of a piece of it on a precise laboratory balance and obtain 3.8056 g; you measure the volume as 2.5 mL by displacement of water in a graduated cylinder. The mass has five significant figures, but the volume has only two. Should you report the density as 3.8056 g/2.5 mL = 1.5222 g/mL or as 1.5 g/mL? The answer with five significant figures implies more certainty than the answer with two. But you didn't measure the volume to five significant figures, so you can't possibly know the density with that much certainty. Therefore, you report 1.5 g/mL, the answer with two significant figures.

Rules for Arithmetic Operations The two rules in arithmetic calculations are

1. *For multiplication and division.* The answer contains the same number of significant figures as there are in the measurement with the *fewest significant figures*. Suppose you want to find the volume of a sheet of a new graphite composite. The length (9.2 cm) and width (6.8 cm) are obtained with a ruler, and the thickness (0.3744 cm) with a set of calipers. The calculation is

$$\text{Volume (cm}^3) = 9.2\text{ cm}\times 6.8\text{ cm}\times 0.3744\text{ cm} = 23.4225\text{ cm}^3 = 23\text{ cm}^3$$

Even though your calculator may show 23.4225 cm^3, you report 23 cm^3, the answer with two significant figures, the same as in the measurements with the lower number of significant figures. After all, if the length and width have two significant figures, you can't possibly know the volume with more certainty.

2. *For addition and subtraction.* The answer has the same number of decimal places as there are in the measurement with the *fewest decimal places*. Suppose you want the total volume after adding water to a protein solution: you have 83.5 mL of solution in a graduated cylinder and add 23.28 mL of water from a buret. The calculation is shown

in the margin. Here the calculator shows 106.78 mL, but you report the volume as 106.8 mL, because the measurement with fewer decimal places (83.5 mL) has one decimal place.

$$\begin{array}{r} 83.5\ \text{mL} \\ +\ 23.28\ \text{mL} \\ \hline 106.78\ \text{mL} \end{array}$$

Answer: Volume = 106.8 mL

Rules for Rounding Off You usually need to round off the final answer to the proper number of significant figures or decimal places. Notice that in calculating the volume of the graphite composite above, we removed the extra digits, but in calculating the total volume of the protein solution, we removed the extra digit and increased the last digit by one. The general rule for rounding is that *the least certain measurement sets the limit on the certainty of the final answer.* Here are detailed rules for rounding off:

1. If the digit removed is *more than 5,* the preceding number increases by 1: 5.379 rounds to 5.38 if you need three significant figures and to 5.4 if you need two.
2. If the digit removed is *less than 5,* the preceding number remains the same: 0.2413 rounds to 0.241 if you need three significant figures and to 0.24 if you need two.
3. If the digit removed *is 5,* the preceding number increases by 1 if it is odd and remains the same if it is even: 17.75 rounds to 17.8, but 17.65 rounds to 17.6. If the 5 is followed only by zeros, rule 3 is followed; if the 5 is followed by nonzeros, rule 1 is followed: 17.6500 rounds to 17.6, but 17.6513 rounds to 17.7.
4. *Always carry one or two additional significant figures through a multistep calculation and round off the final answer* ***only.*** Don't be concerned if you string together a calculation to check a sample or follow-up problem and find that your answer differs in the last decimal place from the one in the book. To show you the correct number of significant figures in text calculations, *we round off intermediate steps,* and that process may sometimes change the last digit.

Note that, unless you set a limit on your calculator, it gives answers with too many figures and you must round the displayed result.

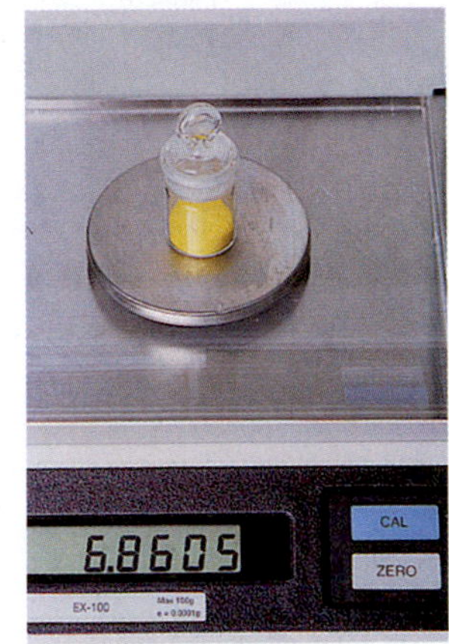

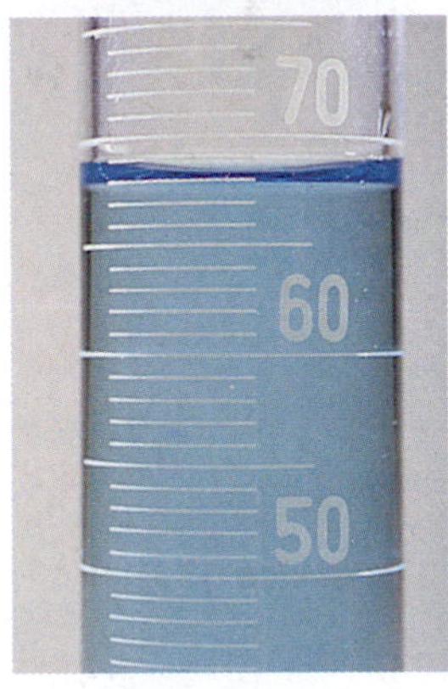

Figure 1.8 Significant figures and measuring devices. The mass measurement (6.8605 g) has more significant figures than the volume measurement (68.2 mL).

Significant Figures in the Lab The measuring device you choose determines the number of significant figures you can obtain. Suppose an experiment requires a solution made by dissolving a solid in a liquid. You weigh the solid on an analytical balance and obtain a mass with five significant figures. It would make sense to measure the liquid with a buret or a pipet, which measures volumes to more significant figures than a graduated cylinder. If you do choose the cylinder, you would have to round off more digits, and some certainty in the mass value would be wasted (Figure 1.8). With experience, you'll choose a measuring device based on the number of significant figures you need in the final answer.

Exact Numbers **Exact numbers** have no uncertainty associated with them. Some are part of a unit conversion: by definition, there are exactly 60 minutes in 1 hour, 1000 micrograms in 1 milligram, and 2.54 centimeters in 1 inch. Other exact numbers result from actually counting items: there are exactly 3 coins in my hand, 26 letters in the English alphabet, and so forth. Therefore, unlike a measured quantity, *exact numbers do not limit the number of significant figures in a calculation.*

Sample Problem 1.9 Significant Figures and Rounding

Problem Perform the following calculations and round the answers to the correct number of significant figures:

(a) $\dfrac{16.3521\ \text{cm}^2 - 1.448\ \text{cm}^2}{7.085\ \text{cm}}$ **(b)** $\dfrac{(4.80\times10^4\ \text{mg})\left(\dfrac{1\ \text{g}}{1000\ \text{mg}}\right)}{11.55\ \text{cm}^3}$

Plan We use the rules just presented in the text: **(a)** We subtract before we divide. **(b)** We note that the unit conversion involves an exact number.

Solution (a) $\dfrac{16.3521\ \text{cm}^2 - 1.448\ \text{cm}^2}{7.085\ \text{cm}} = \dfrac{14.904\ \text{cm}^{\cancel{2}}}{7.085\ \cancel{\text{cm}}} = 2.104\ \text{cm}$

(b) $\dfrac{(4.80\times10^4\ \cancel{\text{mg}})\left(\dfrac{1\ \text{g}}{1000\ \cancel{\text{mg}}}\right)}{11.55\ \text{cm}^3} = \dfrac{48.0\ \text{g}}{11.55\ \text{cm}^3} = 4.16\ \text{g/cm}^3$

Check Note that in (a) we lose a decimal place in the numerator, and in (b) we retain 3 sf in the answer because there are 3 sf in 4.80. Rounding to the nearest whole number is always a good way to check: **(a)** $(16 - 1)/7 \approx 2$; **(b)** $(5\times10^4/1\times10^3)/12 \approx 4$.

FOLLOW-UP PROBLEM 1.9 Perform the following calculation and round the answer to the correct number of significant figures: $\dfrac{25.65\ \text{mL} + 37.4\ \text{mL}}{73.55\ \text{s}\left(\dfrac{1\ \text{min}}{60\ \text{s}}\right)}$

Precision, Accuracy, and Instrument Calibration

We may use the words "precision" and "accuracy" interchangeably in everyday speech, but for scientific measurements they have distinct meanings. **Precision,** or *reproducibility,* refers to how close the measurements in a series are to each other, and **accuracy** refers to how close each measurement is to the actual value. These terms are related to two widespread types of error:

1. **Systematic error** produces values that are *either* all higher or all lower than the actual value. This type of error is part of the experimental system, often caused by a faulty device or by making the same mistake when taking a reading.
2. **Random error,** in the absence of systematic error, produces values that are higher *and* lower than the actual value. Random error *always* occurs, but its size depends on the measurer's skill and the instrument's precision.

Precise measurements have low random error, that is, small deviations from the average. *Accurate measurements have low systematic error and, generally, low random error.* In some cases, when many measurements have a high random error, the *average* may still be accurate.

Suppose each of four students measures 25.0 mL of water in a preweighed graduated cylinder and then weighs the water *plus* cylinder on a balance. If the density of water is 1.00 g/mL at the temperature of the experiment, the *actual* mass of 25.0 mL of water is 25.0 g. Each student performs the operation four times, subtracts the mass of the empty cylinder, and obtains one of four graphs (Figure 1.9). In graphs A and B, random error is small; that is, precision is high (the weighings are reproducible). In A, however, the accuracy is high as well (all the values are close to 25.0 g), whereas in B the accuracy is low (there is a systematic error). In graphs C and D, random error is large; that is, precision is low. Note, however, that in D there is also a systematic error (all the values are high), whereas in C the average of the values is close to the actual value.

Figure 1.9 Precision and accuracy in a laboratory calibration.

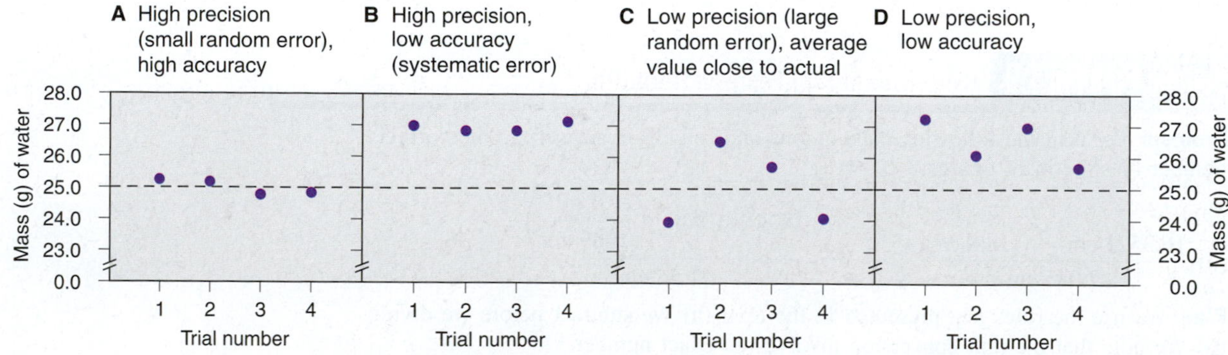

Systematic error can be taken into account through **calibration,** comparing the measuring device with a known standard. The systematic error in graph B, for example, might be caused by a poorly manufactured cylinder that reads "25.0" when it actually contains about 27 mL. If that cylinder had been calibrated, the student could have adjusted all volumes measured with it. The students also should calibrate the balance with standardized masses.

Summary of Section 1.5

- The final digit of a measurement is always estimated. Thus, all measurements have some uncertainty, which is expressed by the number of significant figures.
- The certainty of a calculated result depends on the certainty of the data, so the answer has as many significant figures as in the least certain measurement.
- Excess digits are rounded off in the final answer with a set of rules.
- The choice of laboratory device depends on the certainty needed.
- Exact numbers have as many significant figures as the calculation requires.
- Precision refers to how close values are to each other, and accuracy refers to how close values are to the actual value.
- Systematic errors give values that are either all higher or all lower than the actual value. Random errors give some values that are higher and some that are lower than the actual value.
- Precise measurements have low random error; accurate measurements have low systematic error and low random error.
- A systematic error is often caused by faulty equipment and can be compensated for by calibration.

CHAPTER REVIEW GUIDE

The following sections provide many aids to help you study this chapter. (Numbers in parentheses refer to pages, unless noted otherwise.)

Learning Objectives

These are concepts and skills to review after studying this chapter.

Related section (§), sample problem (SP), and upcoming end-of-chapter problem (EP) numbers are listed in parentheses.

1. Distinguish between physical and chemical properties and changes (§1.1) (SPs 1.1, 1.2) (EPs 1.1, 1.3–1.6)
2. Define the features of the states of matter (§1.1) (EP 1.2)
3. Understand the nature of potential energy and kinetic energy and their interconversion (§1.1) (EPs 1.7–1.8)
4. Understand the scientific approach to studying phenomena and distinguish between observation, hypothesis, experiment, and model (§1.2) (EPs 1.9–1.12)
5. Use conversion factors in calculations (§1.3) (SP 1.3) (EPs 1.13–1.15)
6. Distinguish between mass and weight, heat and temperature, and intensive and extensive properties (§1.4) (EPs 1.16, 1.17, 1.19)
7. Use numerical prefixes and common units of length, mass, volume, and temperature in unit-conversion calculations (§1.4) (SPs 1.4–1.7) (EPs 1.21–1.34, 1.36–1.40)
8. Understand scientific notation and the meaning of uncertainty; determine the number of significant figures and the number of digits after rounding (§1.5) (SPs 1.8, 1.9) (EPs 1.42–1.56)
9. Distinguish between accuracy and precision and between systematic and random error (§1.5) (EPs 1.57–1.59)

Key Terms

These important terms appear in boldface in the chapter and are defined again in the Glossary.

Section 1.1
chemistry (3)
matter (3)
composition (3)
property (3)
physical property (3)
physical change (3)
chemical property (4)
chemical change (chemical reaction) (4)
state of matter (5)
solid (5)
liquid (5)
gas (5)
energy (7)
potential energy (7)
kinetic energy (7)

Section 1.2
scientific method (9)
observation (9)
data (9)
natural law (9)
hypothesis (9)
experiment (9)
variable (9)
controlled experiment (9)
model (theory) (9)

Section 1.3
conversion factor (10)

Section 1.4
SI unit (14)
base (fundamental) unit (14)
derived unit (14)
meter (m) (14)
volume (*V*) (15)
cubic meter (m^3) (15)
liter (L) (15)
milliliter (mL) (15)
mass (16)

Key Terms continued

kilogram (kg) (16)
weight (16)
density (*d*) (17)
temperature (*T*) (18)
heat (18)
thermometer (18)
kelvin (K) (18)
Celsius scale (18)
Kelvin (absolute) scale (18)
second (s) (20)
extensive property (20)
intensive property (20)

Section 1.5
uncertainty (21)
significant figures (21)
round off (22)
exact number (23)
precision (24)
accuracy (24)
systematic error (24)
random error (24)
calibration (25)

Key Equations and Relationships

Numbered and screened concepts are listed for you to refer to or memorize.

1.1 Calculating density from mass and volume (17):

$$\text{Density} = \frac{\text{mass}}{\text{volume}}$$

1.2 Converting temperature from °C to K (19):

$$T\,(\text{in K}) = T\,(\text{in °C}) + 273.15$$

1.3 Converting temperature from K to °C (19):

$$T\,(\text{in °C}) = T\,(\text{in K}) - 273.15$$

1.4 Converting temperature from °C to °F (19):

$$T\,(\text{in °F}) = \tfrac{9}{5}T\,(\text{in °C}) + 32$$

1.5 Converting temperature from °F to °C (19):

$$T\,(\text{in °C}) = [T\,(\text{in °F}) - 32]\tfrac{5}{9}$$

BRIEF SOLUTIONS TO FOLLOW-UP PROBLEMS

Compare your own solutions to these calculation steps and answers.

1.1 Chemical. The red-and-blue and separate red particles on the left become paired red and separate blue particles on the right.

1.2 (a) Physical. Solid iodine changes to gaseous iodine.
(b) Chemical. Gasoline burns in air to form different substances.
(c) Chemical. In contact with air, substances in torn skin and blood react to form different substances.

1.3 No. of chairs

$$= 3\ \text{bolts} \times \frac{200\ \text{m}^2}{1\ \text{bolt}} \times \frac{(3.281)^2\ \text{ft}^2}{1\ \text{m}^2} \times \frac{1\ \text{chair}}{31.5\ \text{ft}^2}$$

$$= 205\ \text{chairs}$$

See Road Map 1.3.

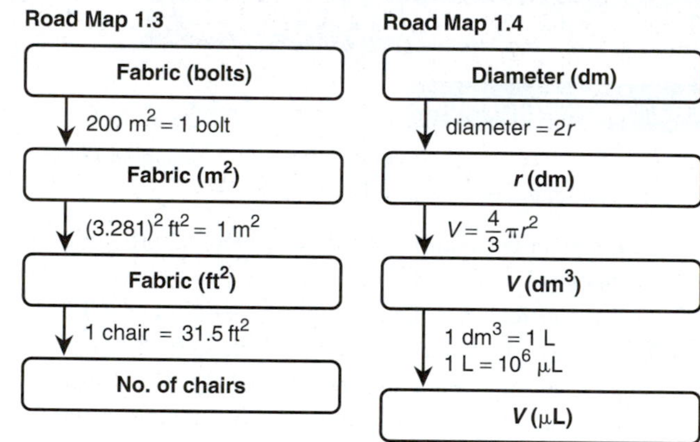

1.4 Radius of ribosome (dm) $= \dfrac{21.4\ \text{nm}}{2} \times \dfrac{1\ \text{dm}}{10^8\ \text{nm}}$

$$= 1.07\times10^{-7}\ \text{dm}$$

Volume of ribosome (dm³) $= \frac{4}{3}\pi r^3 = \frac{4}{3}(3.14)(1.07\times10^{-7}\ \text{dm})^3$

$$= 5.13\times10^{-21}\ \text{dm}^3$$

Volume of ribosome (μL) $= (5.13\times10^{-21}\ \text{dm}^3)\left(\dfrac{1\ \text{L}}{1\ \text{dm}^3}\right)\left(\dfrac{10\ \mu\text{L}}{1\ \text{L}}\right)$

$$= 5.13\times10^{-15}\ \mu\text{L}$$

See Road Map 1.4.

1.5 Mass (kg) of solution

$$= 8.0\ \text{h} \times \frac{60\ \text{min}}{1\ \text{h}} \times \frac{60\ \text{s}}{1\ \text{min}} \times \frac{1.5\ \text{drops}}{1\ \text{s}}$$

$$\times \frac{65\ \text{mg}}{1\ \text{drop}} \times \frac{1\ \text{g}}{10^3\ \text{mg}} \times \frac{1\ \text{kg}}{10^3\ \text{g}}$$

$$= 2.8\ \text{kg}$$

See Road Map 1.5.

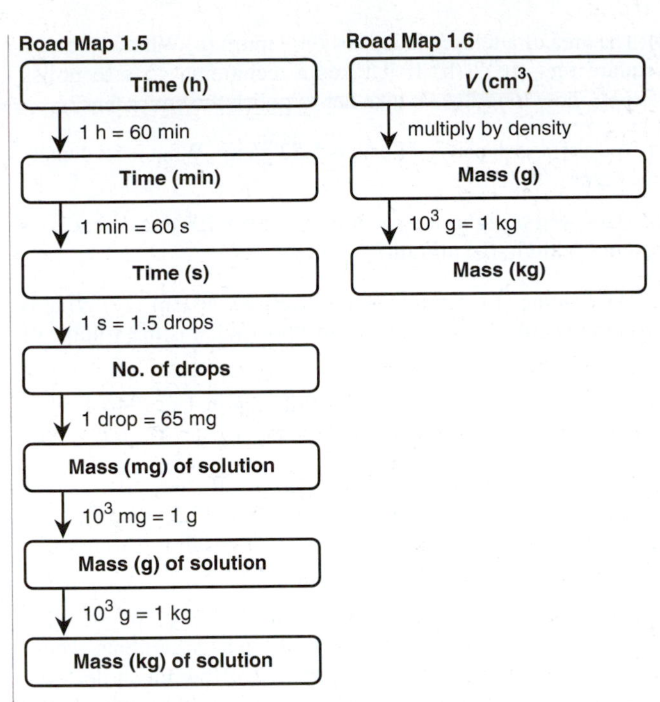

1.6 Mass (kg) of sample $= 4.6\ \cancel{cm^3} \times \dfrac{7.5\ \cancel{g}}{1\ \cancel{cm^3}} \times \dfrac{1\ kg}{10^3\ \cancel{g}}$

$= 0.034$ kg

See Road Map 1.6.

1.7 T (in °C) = 234 K − 273.15 = −39°C

T (in °F) = $\frac{9}{5}$(−39°C) + 32 = −38°F

Answer contains two significant figures (see Section 1.5).

1.8 (a) 31.070 mg, 5 sf (b) 0.06060 g, 4 sf
(c) 850.°C, 3 sf (d) 2.000×10^2 mL, 4 sf
(e) 3.9×10^{-6} m, 2 sf (f) 4.01×10^{-4} L, 3 sf

1.9 $$\frac{25.65\ \text{mL} + 37.4\ \text{mL}}{73.55\ \cancel{s}\left(\dfrac{1\ \text{min}}{60\ \cancel{s}}\right)} = 51.4\ \text{mL/min}$$

PROBLEMS

Problems with **colored** numbers are answered in Appendix E. Sections match the text and provide the numbers of relevant sample problems. Bracketed problems are grouped in pairs (indicated by a short rule) that cover the same concept. Comprehensive Problems are based on material from any section or previous chapter.

Some Fundamental Definitions

(Sample Problems 1.1 and 1.2)

1.1 Scenes A–D represent atomic-scale views of different samples of substances:

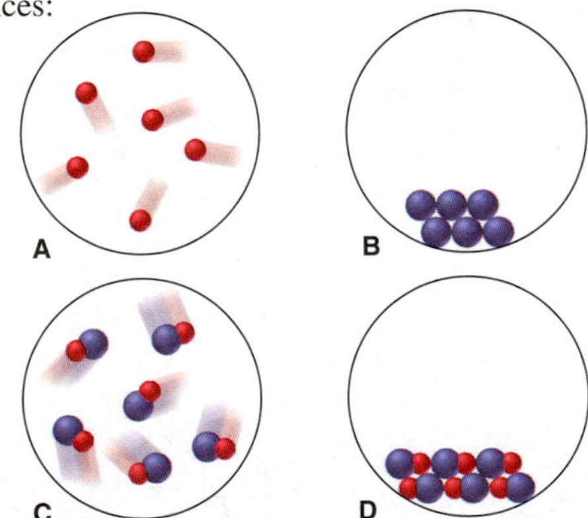

(a) Under one set of conditions, the substances in A and B mix, and the result is depicted in C. Does this represent a chemical or a physical change?
(b) Under a second set of conditions, the same substances mix, and the result is depicted in D. Does this represent a chemical or a physical change?
(c) Under a third set of conditions, the sample depicted in C changes to that in D. Does this represent a chemical or a physical change?
(d) After the change in part (c) has occurred, does the sample have different chemical properties? Physical properties?

1.2 Describe solids, liquids, and gases in terms of how they fill a container. Use your descriptions to identify the physical state (at room temperature) of the following: (a) helium in a toy balloon; (b) mercury in a thermometer; (c) soup in a bowl.

1.3 Define *physical property* and *chemical property*. Identify each type of property in the following statements:
(a) Yellow-green chlorine gas attacks silvery sodium metal to form white crystals of sodium chloride (table salt).
(b) A magnet separates a mixture of black iron shavings and white sand.

1.4 Define *physical change* and *chemical change*. State which type of change occurs in each of the following statements:
(a) Passing an electric current through molten magnesium chloride yields molten magnesium and gaseous chlorine.
(b) The iron in discarded automobiles slowly forms reddish brown, crumbly rust.

1.5 Which of the following is a chemical change? Explain your reasoning: (a) boiling canned soup; (b) toasting a slice of bread; (c) chopping a log; (d) burning a log.

1.6 Which of the following changes can be reversed by changing the temperature: (a) dew condensing on a leaf; (b) an egg turning hard when it is boiled; (c) ice cream melting; (d) a spoonful of batter cooking on a hot griddle?

1.7 For each pair, which has higher potential energy?
(a) The fuel in your car or the gaseous products in its exhaust
(b) Wood in a fire or the ashes after the wood burns

1.8 For each pair, which has higher kinetic energy?
(a) A sled resting at the top of a hill or a sled sliding down the hill
(b) Water above a dam or water falling over the dam

The Scientific Approach: Developing a Model

1.9 How are the key elements of scientific thinking used in the following scenario? While making toast, you notice it fails to pop out of the toaster. Thinking the spring mechanism is stuck, you notice that the bread is unchanged. Assuming you forgot to plug in the toaster, you check and find it is plugged in. When you take the toaster into the dining room and plug it into a different outlet, you find the toaster works. Returning to the kitchen, you turn on the switch for the overhead light and nothing happens.

1.10 Why is a quantitative observation more useful than a nonquantitative one? Which of the following is (are) quantitative? (a) The Sun rises in the east. (b) A person weighs one-sixth as much on the Moon as on Earth. (c) Ice floats on water. (d) A hand pump cannot draw water from a well more than 34 ft deep.

1.11 Describe the essential features of a well-designed experiment.

1.12 Describe the essential features of a scientific model.

Chemical Problem Solving

(Sample Problem 1.3)

1.13 When you convert feet to inches, how do you decide which part of the conversion factor should be in the numerator and which in the denominator?

1.14 Write the conversion factor(s) for
(a) in^2 to m^2 (b) km^2 to cm^2
(c) mi/h to m/s (d) lb/ft^3 to g/cm^3

1.15 Write the conversion factor(s) for
(a) cm/min to in/s (b) m^3 to in^3
(c) m/s^2 to km/h^2 (d) gal/h to L/min

Measurement in Scientific Study

(Sample Problems 1.4 to 1.7)

1.16 Describe the difference between intensive and extensive properties. Which of the following properties are intensive: (a) mass; (b) density; (c) volume; (d) melting point?

1.17 Explain the difference between mass and weight. Why is your weight on the Moon one-sixth that on Earth?

1.18 For each of the following cases, state whether the density of the object increases, decreases, or remains the same:
(a) A sample of chlorine gas is compressed.
(b) A lead weight is carried up a high mountain.
(c) A sample of water is frozen.
(d) An iron bar is cooled.
(e) A diamond is submerged in water.

1.19 Explain the difference between heat and temperature. Does 1 L of water at 65°F have more, less, or the same quantity of energy as 1 L of water at 65°C?

1.20 A one-step conversion is sufficient to convert a temperature in the Celsius scale to the Kelvin scale, but not to the Fahrenheit scale. Explain.

1.21 The average radius of a molecule of lysozyme, an enzyme in tears, is 1430 pm. What is its radius in nanometers (nm)?

1.22 The radius of a barium atom is 2.22×10^{-10} m. What is its radius in angstroms (Å)?

1.23 A small hole in the wing of a space shuttle requires a 20.7-cm^2 patch. (a) What is the patch's area in square kilometers (km^2)? (b) If the patching material costs NASA $\$3.25/in^2$, what is the cost of the patch?

1.24 The area of a telescope lens is 7903 mm^2. (a) What is the area in square feet (ft^2)? (b) If it takes a technician 45 s to polish 135 mm^2, how long does it take her to polish the entire lens?

1.25 The average density of Earth is 5.52 g/cm^3. What is its density in (a) kg/m^3; (b) lb/ft^3?

1.26 The speed of light in a vacuum is 2.998×10^8 m/s. What is its speed in (a) km/h; (b) mi/min?

1.27 The volume of a certain bacterial cell is 2.56 μm^3. (a) What is its volume in cubic millimeters (mm^3)? (b) What is the volume of 10^5 cells in liters (L)?

1.28 (a) How many cubic meters of milk are in 1 qt (946.4 mL)? (b) How many liters of milk are in 835 gal (1 gal = 4 qt)?

1.29 An empty vial weighs 55.32 g. (a) If the vial weighs 185.56 g when filled with liquid mercury (d = 13.53 g/cm^3), what is its volume? (b) How much would the vial weigh if it were filled with water (d = 0.997 g/cm^3 at 25°C)?

1.30 An empty Erlenmeyer flask weighs 241.3 g. When filled with water (d = 1.00 g/cm^3), the flask and its contents weigh 489.1 g. (a) What is the flask's volume? (b) How much does the flask weigh when filled with chloroform (d = 1.48 g/cm^3)?

1.31 A small cube of aluminum measures 15.6 mm on a side and weighs 10.25 g. What is the density of aluminum in g/cm^3?

1.32 A steel ball-bearing with a circumference of 32.5 mm weighs 4.20 g. What is the density of the steel in g/cm^3 (V of a sphere = $\frac{4}{3}\pi r^3$; circumference of a circle = $2\pi r$)?

1.33 Perform the following conversions:
(a) 68°F (a pleasant spring day) to °C and K
(b) −164°C (the boiling point of methane, the main component of natural gas) to K and °F
(c) 0 K (absolute zero, theoretically the coldest possible temperature) to °C and °F

1.34 Perform the following conversions:
(a) 106°F (the body temperature of many birds) to K and °C
(b) 3410°C (the melting point of tungsten, the highest for any metallic element) to K and °F
(c) 6.1×10^3 K (the surface temperature of the Sun) to °F and °C

1.35 A 25.0-g sample of each of three unknown metals is added to 25.0 mL of water in graduated cylinders A, B, and C, and the final volumes are depicted in the circles below. Given their densities, identify the metal in each cylinder: zinc (7.14 g/mL), iron (7.87 g/mL), or nickel (8.91 g/mL).

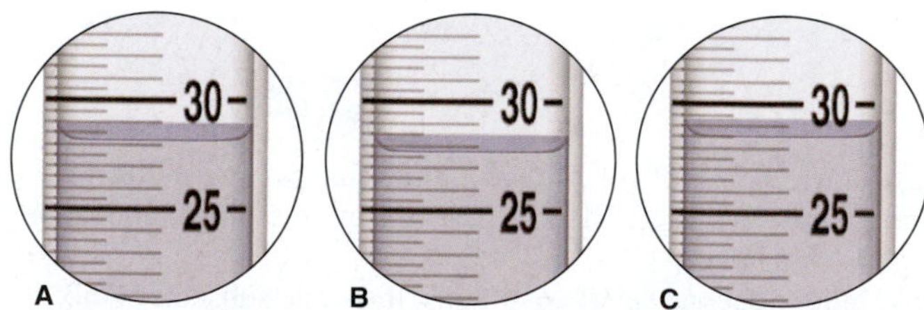

1.36 Anton van Leeuwenhoek, a 17th-century pioneer in the use of the microscope, described the microorganisms he saw as "animalcules" whose length was "25 thousandths of an inch." How long were the animalcules in meters?

1.37 The distance between two adjacent peaks on a wave is called the *wavelength*.
(a) The wavelength of a beam of ultraviolet light is 247 nanometers (nm). What is its wavelength in meters?

(b) The wavelength of a beam of red light is 6760 pm. What is its wavelength in angstroms (Å)?

1.38 In the early 20th century, thin metal foils were used to study atomic structure. (a) How many in^2 of gold foil with a thickness of 1.6×10^{-5} in could have been made from 2.0 troy oz? (b) If gold cost $20.00/troy oz at that time, how many cm^2 of gold foil could have been made from $75.00 worth of gold (1 troy oz = 31.1 g; d of gold = 19.3 g/cm^3)?

1.39 A cylindrical tube 9.5 cm high and 0.85 cm in diameter is used to collect blood samples. How many cubic decimeters (dm^3) of blood can it hold (V of a cylinder = $\pi r^2 h$)?

1.40 Copper can be drawn into thin wires. How many meters of 34-gauge wire (diameter = 6.304×10^{-3} in) can be produced from the copper in 5.01 lb of covellite, an ore of copper that is 66% copper by mass? (*Hint:* Treat the wire as a cylinder: V of cylinder = $\pi r^2 h$; d of copper = 8.95 g/cm^3.)

1.41 Each of the beakers depicted below contains two liquids that do not dissolve in each other. Three of the liquids are designated A, B, and C, and water is designated W.

(a) Which of the liquids is (are) more dense than water and which less dense?
(b) If the densities of W, C, and A are 1.0 g/mL, 0.88 g/mL, and 1.4 g/mL, respectively, which of the following densities is possible for liquid B: 0.79 g/mL, 0.86 g/mL, 0.94 g/mL, or 1.2 g/mL?

Uncertainty in Measurement: Significant Figures

(Sample Problems 1.8 and 1.9)

1.42 What is an exact number? How are exact numbers treated differently from other numbers in a calculation?

1.43 All nonzero digits are significant. State a rule that tells which zeros are significant.

1.44 Underline the significant zeros in the following numbers:
(a) 0.41 (b) 0.041 (c) 0.0410 (d) 4.0100×10^4

1.45 Underline the significant zeros in the following numbers:
(a) 5.08 (b) 508 (c) 5.080×10^3 (d) 0.05080

1.46 Carry out the following calculations, making sure that your answer has the correct number of significant figures:

(a) $\dfrac{2.795\text{ m} \times 3.10\text{ m}}{6.48\text{ m}}$

(b) $V = \frac{4}{3}\pi r^3$, where r = 17.282 mm

(c) 1.110 cm + 17.3 cm + 108.2 cm + 316 cm

1.47 Carry out the following calculations, making sure that your answer has the correct number of significant figures:

(a) $\dfrac{2.420\text{ g} + 15.6\text{ g}}{4.8\text{ g}}$ (b) $\dfrac{7.87\text{ mL}}{16.1\text{ mL} - 8.44\text{ mL}}$

(c) $V = \pi r^2 h$, where r = 6.23 cm and h = 4.630 cm

1.48 Write the following numbers in scientific notation:
(a) 131,000.0 (b) 0.00047 (c) 210,006 (d) 2160.5

1.49 Write the following numbers in scientific notation:
(a) 282.0 (b) 0.0380 (c) 4270.8 (d) 58,200.9

1.50 Write the following numbers in standard notation. Use a terminal decimal point when needed:
(a) 5.55×10^3 (b) 1.0070×10^4 (c) 8.85×10^{-7} (d) 3.004×10^{-3}

1.51 Write the following numbers in standard notation. Use a terminal decimal point when needed:
(a) 6.500×10^3 (b) 3.46×10^{-5} (c) 7.5×10^2 (d) 1.8856×10^2

1.52 Carry out each calculation, paying special attention to significant figures, rounding, and units (J = joule, the SI unit of energy; mol = mole, the SI unit for amount of substance):

(a) $\dfrac{(6.626\times10^{-34}\text{ J·s})(2.9979\times10^8\text{ m/s})}{489\times10^{-9}\text{ m}}$

(b) $\dfrac{(6.022\times10^{23}\text{ molecules/mol})(1.23\times10^2\text{ g})}{46.07\text{ g/mol}}$

(c) $(6.022\times10^{23}\text{ atoms/mol})(2.18\times10^{-18}\text{ J/atom})\left(\dfrac{1}{2^2} - \dfrac{1}{3^2}\right)$,

where the numbers 2 and 3 in the last term are exact

1.53 Carry out each calculation, paying special attention to significant figures, rounding, and units:

(a) $\dfrac{4.32\times10^7\text{ g}}{\frac{4}{3}(3.1416)(1.95\times10^2\text{ cm})^3}$ (The term $\frac{4}{3}$ is exact.)

(b) $\dfrac{(1.84\times10^2\text{ g})(44.7\text{ m/s})^2}{2}$ (The term 2 is exact.)

(c) $\dfrac{(1.07\times10^{-4}\text{ mol/L})^2(3.8\times10^{-3}\text{ mol/L})}{(8.35\times10^{-5}\text{ mol/L})(1.48\times10^{-2}\text{ mol/L})^3}$

1.54 Which statements include exact numbers?
(a) Angel Falls is 3212 ft high.
(b) There are 8 known planets in the Solar System.
(c) There are 453.59 g in 1 lb.
(d) There are 1000 mm in 1 m.

1.55 Which of the following include exact numbers?
(a) The speed of light in a vacuum is a physical constant; to six significant figures, it is 2.99792×10^8 m/s.
(b) The density of mercury at 25°C is 13.53 g/mL.
(c) There are 3600 s in 1 h.
(d) In 2010, the United States had 50 states.

1.56 How long is the metal strip shown below? Be sure to answer with the correct number of significant figures.

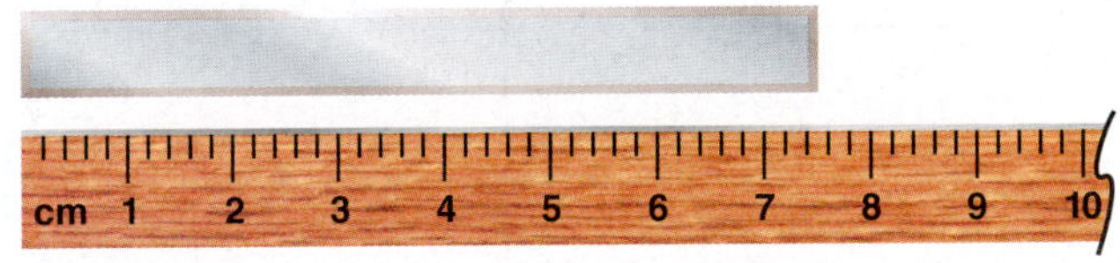

1.57 These organic solvents are used to clean compact discs:

Solvent	Density (g/mL) at 20°C
Chloroform	1.492
Diethyl ether	0.714
Ethanol	0.789
Isopropanol	0.785
Toluene	0.867

(a) If a 15.00-mL sample of CD cleaner weighs 11.775 g at 20°C, which solvent does the sample most likely contain?
(b) The chemist analyzing the cleaner calibrates her equipment and finds that the pipet is accurate to ±0.02 mL, and the balance is accurate to ±0.003 g. Is this equipment precise enough to distinguish between ethanol and isopropanol?

1.58 A laboratory instructor gives a sample of amino-acid powder to each of four students, I, II, III, and IV, and they weigh the samples. The true value is 8.72 g. Their results for three trials are

I: 8.72 g, 8.74 g, 8.70 g   II: 8.56 g, 8.77 g, 8.83 g
III: 8.50 g, 8.48 g, 8.51 g   IV: 8.41 g, 8.72 g, 8.55 g

(a) Calculate the average mass from each set of data, and tell which set is the most accurate.
(b) Precision is a measure of the average of the deviations of each piece of data from the average value. Which set of data is the most precise? Is this set also the most accurate?
(c) Which set of data is both the most accurate and the most precise?
(d) Which set of data is both the least accurate and the least precise?

1.59 The following dartboards illustrate the types of errors often seen in measurements. The bull's-eye represents the actual value, and the darts represent the data.

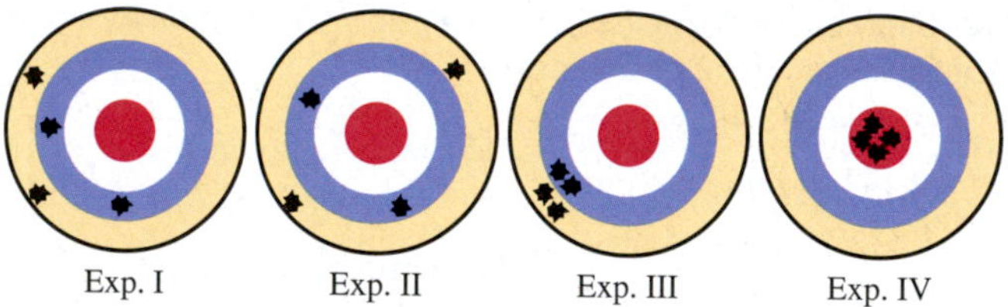

(a) Which experiments yield the same average result?
(b) Which experiment(s) display(s) high precision?
(c) Which experiment(s) display(s) high accuracy?
(d) Which experiment(s) show(s) a systematic error?

Comprehensive Problems

1.60 To make 2.000 gal of a powdered sports drink, a group of students measure out 2.000 gal of water with 500.-mL, 50.-mL, and 5-mL graduated cylinders. Show how they could get closest to 2.000 gal of water, using these cylinders the fewest times.

1.61 Two blank potential energy diagrams appear below. Beneath each diagram are objects to place in the diagram. Draw the objects on the dashed lines to indicate higher or lower potential energy and label each case as more or less stable:

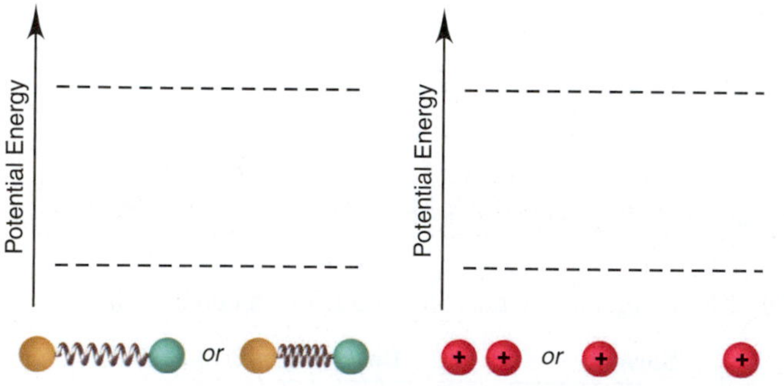

(a) Two balls attached to a relaxed *or* a compressed spring

(b) Two positive charges near *or* apart from each other

1.62 Soft drinks are about as dense as water (1.0 g/cm^3); many common metals, including iron, copper, and silver, have densities around 9.5 g/cm^3. (a) What is the mass of the liquid in a standard 12-oz bottle of diet cola? (b) What is the mass of a dime? (*Hint:* A stack of five dimes has a volume of about 1 cm^3.)

1.63 The scenes below illustrate two different mixtures. When mixture A at 273 K is heated to 473 K, mixture B results.

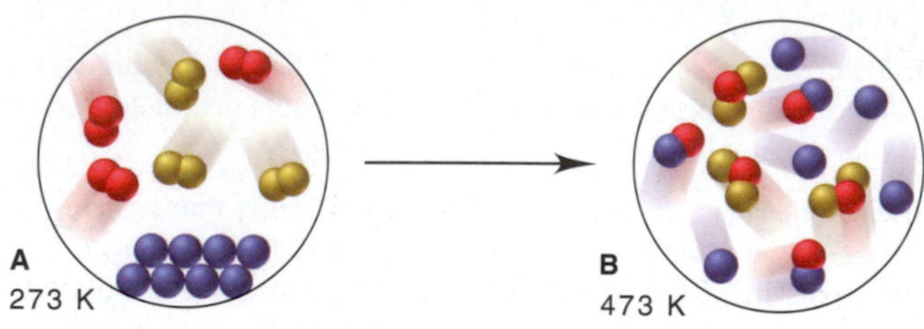

(a) How many different chemical changes occur?
(b) How many different physical changes occur?

1.64 Suppose your dorm room is 11 ft wide by 12 ft long by 8.5 ft high and has an air conditioner that exchanges air at a rate of 1200 L/min. How long would it take the air conditioner to exchange the air in your room once?

1.65 In 1933, the United States went off the international gold standard, and the price of gold increased from $20.00 to $35.00/troy oz. The twenty-dollar gold piece, known as the double eagle, weighed 33.436 g and was 90.0% gold by mass. (a) What was the value of the gold in the double eagle before and after the 1933 price change? (b) How many coins could be made from 50.0 troy oz of gold? (c) How many coins could be made from 2.00 in^3 of gold (1 troy oz = 31.1 g; d of gold = 19.3 g/cm^3)?

1.66 Bromine is used to prepare the pesticide methyl bromide and flame retardants for plastic electronic housings. It is recovered from seawater, underground brines, and the Dead Sea. The average concentrations of bromine in seawater (d = 1.024 g/mL) and the Dead Sea (d = 1.22 g/mL) are 0.065 g/L and 0.50 g/L, respectively. What is the mass ratio of bromine in the Dead Sea to that in seawater?

1.67 An Olympic-size pool is 50.0 m long and 25.0 m wide. (a) How many gallons of water (d = 1.0 g/mL) are needed to fill the pool to an average depth of 4.8 ft? (b) What is the mass (in kg) of water in the pool?

1.68 At room temperature (20°C) and pressure, the density of air is 1.189 g/L. An object will float in air if its density is less than that of air. In a buoyancy experiment with a new plastic, a chemist creates a rigid, thin-walled ball that weighs 0.12 g and has a volume of 560 cm^3.
(a) Will the ball float if it is evacuated?
(b) Will it float if filled with carbon dioxide (d = 1.830 g/L)?
(c) Will it float if filled with hydrogen (d = 0.0899 g/L)?
(d) Will it float if filled with oxygen (d = 1.330 g/L)?
(e) Will it float if filled with nitrogen (d = 1.165 g/L)?
(f) For any case in which the ball will float, how much weight must be added to make it sink?

1.69 Asbestos is a fibrous silicate mineral with remarkably high tensile strength. But it is no longer used because airborne asbestos particles can cause lung cancer. Grunerite, a type of asbestos, has a tensile strength of 3.5×10^2 kg/mm^2 (thus, a strand of grunerite with a 1-mm^2 cross-sectional area can hold up to 3.5×10^2 kg).

The tensile strengths of aluminum and Steel No. 5137 are 2.5×10^4 lb/in^2 and 5.0×10^4 lb/in^2, respectively. Calculate the cross-sectional areas (in mm^2) of wires of aluminum and of Steel No. 5137 that have the same tensile strength as a fiber of grunerite with a cross-sectional area of 1.0 μm^2.

1.70 According to the lore of ancient Greece, Archimedes discovered the displacement method of density determination while bathing and used it to find the composition of the king's crown. If a crown weighing 4 lb 13 oz displaces 186 mL of water, is the crown made of pure gold (d = 19.3 g/cm^3)?

1.71 Earth's oceans have an average depth of 3800 m, a total surface area of 3.63×10^8 km^2, and an average concentration of dissolved gold of 5.8×10^{-9} g/L. (a) How many grams of gold are in the oceans? (b) How many cubic meters of gold are in the oceans? (c) Assuming the price of gold is $370.00/troy oz, what is the value of gold in the oceans (1 troy oz = 31.1 g; d of gold = 19.3 g/cm^3)?

1.72 For the year 2007, worldwide production of aluminum was 35.1 million metric tons (t). (a) How many pounds of aluminum were produced? (b) What was its volume in cubic feet (1 t = 1000 kg; d of aluminum = 2.70 g/cm^3)?

1.73 Liquid nitrogen is obtained from liquefied air and is used industrially to prepare frozen foods. It boils at 77.36 K. (a) What is this temperature in °C? (b) What is this temperature in °F? (c) At the boiling point, the density of the liquid is 809 g/L and that of the gas is 4.566 g/L. How many liters of liquid nitrogen are produced when 895.0 L of nitrogen gas is liquefied at 77.36 K?

1.74 The speed of sound varies according to the material through which it travels. Sound travels at 5.4×10^3 cm/s through rubber and at 1.97×10^4 ft/s through granite. Calculate each of these speeds in m/s.

1.75 If a raindrop weighs 0.52 mg on average and 5.1×10^5 raindrops fall on a lawn every minute, what mass (in kg) of rain falls on the lawn in 1.5 h?

1.76 The Environmental Protection Agency (EPA) proposed a safety standard for microparticulates in air: for particles up to 2.5 μm in diameter, the maximum allowable amount is 50. μg/m^3. If your 10.0 ft $\times$ 8.25 ft $\times$ 12.5 ft dorm room just meets the EPA standard, how many of these particles are in your room? How many are in each 0.500-L breath you take? (Assume the particles are spheres of diameter 2.5 μm and made primarily of soot, a form of carbon with a density of 2.5 g/cm^3.)

1.77 Scenes A and B depict changes in matter at the atomic scale:

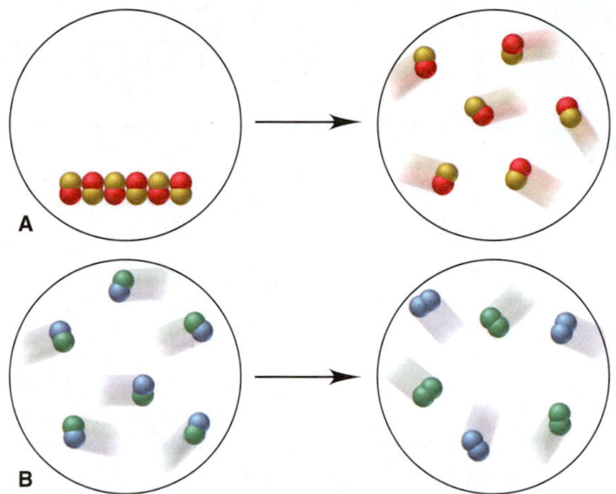

(a) Which show(s) a physical change? (b) Which show(s) a chemical change? (c) Which result(s) in different physical properties? (d) Which result(s) in different chemical properties? (e) Which result(s) in a change in state?

1.78 Earth's surface area is 5.10×10^8 km^2; its crust has a mean thickness of 35 km and a mean density of 2.8 g/cm^3. The two most abundant elements in the crust are oxygen (4.55×10^5 g/t, where t stands for "metric ton"; 1 t = 1000 kg) and silicon (2.72×10^5 g/t), and the two rarest nonradioactive elements are ruthenium and rhodium, each with an abundance of 1×10^{-4} g/t. What is the total mass of each of these elements in Earth's crust?

1.79 The three states of matter differ greatly in their viscosity, a measure of their resistance to flow. Rank the three states from highest to lowest viscosity. Explain in submicroscopic terms.

1.80 If a temperature scale were based on the freezing point (5.5°C) and boiling point (80.1°C) of benzene and the temperature difference between these points was divided into 50 units (called °X), what would be the freezing and boiling points of water in °X? (See Figure 1.6, p. 19.)

The Components of Matter

Parts Within Parts Within Parts Like this piece of granite, everyday matter consists of simpler parts made of even simpler parts. In this chapter, you'll learn about the components of matter and see how they combine.

Key Principles
to focus on while studying this chapter

- A *substance* is matter with a *fixed composition*. The two types of substances are elements and compounds: an *element* consists of a single type of *atom*, which may occur separately or as *molecules*; a *compound* consists of *molecules* (or formula units) made up of two or more different atoms combined in a fixed ratio. A *mixture* consists of two or more substances intermingled physically and, thus, has a *variable composition*. A compound's properties differ from those of its components, but a mixture's properties do not. ***(Section 2.1)***
- Three *mass laws* led to an atomic theory of matter: *mass is conserved* during a chemical change; any sample of a compound has its elements in the *same proportions by mass*; in different compounds made of the same two elements, the masses of one element that combine with a given mass of the other can be expressed as a *ratio of small integers*. ***(Section 2.2)***
- According to *Dalton's atomic theory*, atoms of a given element have a unique mass and other properties. Mass is conserved during a chemical reaction because the atoms of the reacting substances are just rearranged into different substances. ***(Section 2.3)***
- Early 20th-century experiments showed that atoms *are* divisible, consisting of a positively charged *nucleus*, which contains nearly all the atom's mass but represents a tiny fraction of its volume, and negatively charged *electrons* that surround the nucleus. ***(Section 2.4)***
- Atoms consist of three types of *subatomic particles*: positively charged *protons* and uncharged *neutrons* make up the *nucleus*, and *electrons* exist outside the nucleus. *Atoms are neutral* because the number of protons equals the number of electrons. All the atoms of an element have the same number of protons *(atomic number, Z)* and thus the same chemical behavior. *Isotopes* of an element have atoms with different masses because they have different numbers of neutrons. The *atomic mass* of an element is the weighted *average* of the masses of its naturally occurring isotopes. ***(Section 2.5)***
- In the *periodic table*, the elements are arranged by increasing atomic number into a grid of horizontal rows *(periods)* and vertical columns *(groups)*. *Metals* occupy the lower-left three-quarters of the table, and *nonmetals* are found in the upper-right corner, with *metalloids* in between. Elements in a group have similar properties. ***(Section 2.6)***
- The electrons of atoms are involved in forming compounds. In the formation of *ionic compounds*, metal atoms *transfer* electrons to nonmetal atoms, and the resulting charged particles *(ions)* attract each other into solid arrays. In the formation of *covalent compounds*, nonmetal atoms *share* electrons and usually (but not always) form individual molecules. Each compound has a unique name, formula, and mass based on its component elements. ***(Sections 2.7 and 2.8)***
- Unlike compounds, mixtures can be *separated by physical means* into their components. A *heterogeneous mixture* has a nonuniform composition with visible boundaries between the components. A *homogeneous mixture (solution)* has a uniform composition because the components (elements and/or compounds) are mixed as individual atoms, ions, or molecules. ***(Section 2.9)***

Outline

Look closely at almost any sample of matter—a rock, a piece of wood, a butterfly wing—and you'll see that it's made of smaller parts. With a microscope, you'll see still smaller parts. And, if you could zoom in a billion times closer, you'd find, on the atomic scale, the ultimate particles that make up all things.

Modern scientists are not the first to try to explain what things are made of. The philosophers of ancient Greece believed that everything was made of one or, at most, a few elemental substances (elements), whose properties gave rise to the properties of everything. But, Democritus (c. 460–370 BC), the father of atomism, took a different approach. He reasoned that if you cut a piece of, say, aluminum foil smaller and smaller, you reach a particle of aluminum too small to cut, so matter must be ultimately composed of indivisible particles with nothing but empty space between them. He called the particles *atoms* (Greek *atomos,* "uncuttable"). But, Aristotle (384–322 BC), one of the greatest philosophers of Western culture, said it was impossible for "nothing" to exist, and the concept of atoms was suppressed for 2000 years.

Finally, in the 17th century, the English scientist Robert Boyle argued that, by definition, an element is composed of "simple Bodies, not made of any other Bodies, of which all mixed Bodies are compounded, and into which they are ultimately resolved," a description remarkably close to our idea of an element, with atoms being the "simple Bodies." The next two centuries saw rapid progress in chemistry and the development of a "billiard-ball" image of the atom. Then, an early 20th-century burst of creativity led to our current model of an atom with a complex internal structure. In this chapter, we examine the properties and composition of matter on the macroscopic and atomic scales.

CONCEPTS & SKILLS TO REVIEW before studying this chapter

- physical and chemical change (Section 1.1)
- states of matter (Section 1.1)
- attraction and repulsion between charged particles (Section 1.1)
- meaning of a scientific model (Section 1.2)
- SI units and conversion factors (Section 1.3)
- significant figures in calculations (Section 1.5)

2.1 • ELEMENTS, COMPOUNDS, AND MIXTURES: AN ATOMIC OVERVIEW

Matter can be classified into three types based on its composition—elements, compounds, and mixtures. Elements and compounds are the two kinds of substances: a **substance** is matter whose composition is fixed. Mixtures are not substances because they have a variable composition.

1. *Elements.* An **element** is the simplest type of matter with unique physical and chemical properties. *It consists of only one kind of atom* and, therefore, cannot be broken down into a simpler type of matter by any physical or chemical methods. Each element has a name, such as silicon, oxygen, or copper. A sample of silicon contains only silicon atoms. The *macroscopic* properties of a piece of silicon, such as color, density, and combustibility, are different from those of a piece of copper because the *submicroscopic* properties of silicon atoms are different from those of copper atoms; that is, *each element is unique because the properties of its atoms are unique.*

In nature, most elements exist as populations of atoms, either separated or in contact with each other, depending on the physical state. Figure 2.1A depicts atoms of an element in its gaseous state. Several elements occur in molecular form: a **molecule** is an independent structure of two or more atoms bound together (Figure 2.1B). Oxygen, for example, occurs in air as *diatomic* (two-atom) molecules.

Figure 2.1 Elements, compounds, and mixtures on the atomic scale. The samples depicted here are gases, but the three types of matter also occur as liquids and solids.

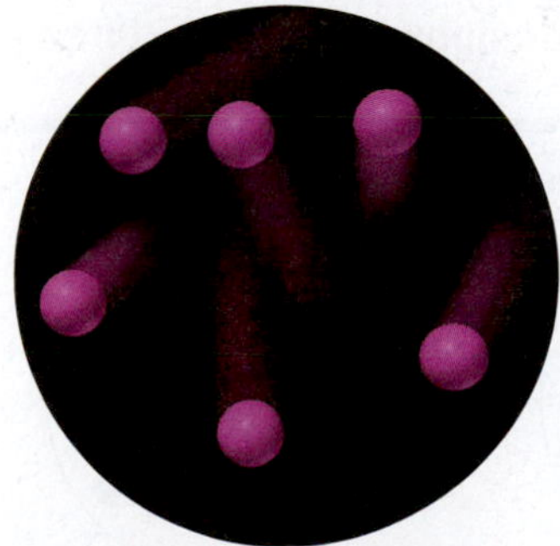

A Atoms of an element

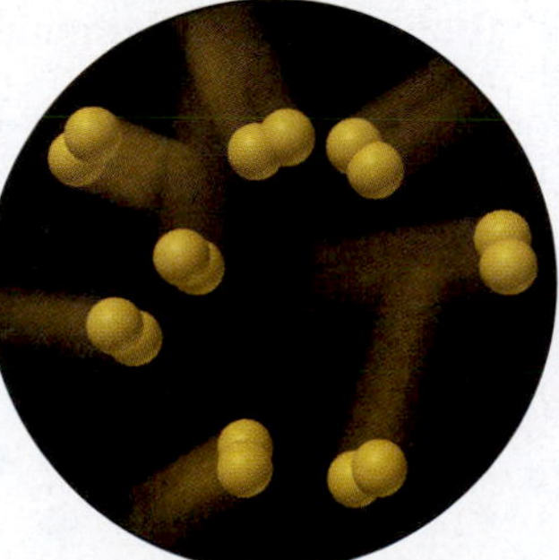

B Molecules of an element

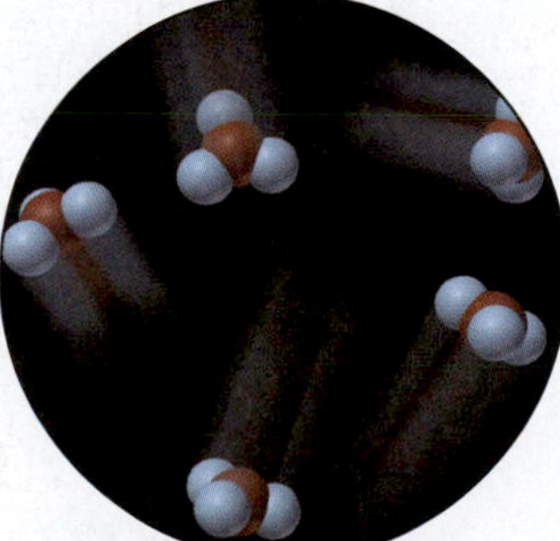

C Molecules of a compound

D Mixture of two elements and a compound

Table 2.1 Some Properties of Sodium, Chlorine, and Sodium Chloride

Property	Sodium	+	Chlorine	⟶	Sodium Chloride
Melting point	97.8°C		−101°C		801°C
Boiling point	881.4°C		−34°C		1413°C
Color	Silvery		Yellow-green		Colorless (white)
Density	0.97 g/cm^3		0.0032 g/cm^3		2.16 g/cm^3
Behavior in water	Reacts		Dissolves slightly		Dissolves freely

2. *Compounds.* A **compound** consists of *two or more different elements that are bonded chemically* (Figure 2.1C). Many compounds, such as ammonia, water, and carbon dioxide, consist of molecules. But, as we'll discuss shortly, many others, like silicon dioxide and sodium sulfate, do not. No matter what the compound, however, one defining feature is that *the elements are present in fixed parts by mass* (fixed mass ratio). This is so because *each unit of the compound consists of a fixed number of atoms of each element.* For example, a sample of ammonia is 14 parts nitrogen by mass and 3 parts hydrogen by mass *because* 1 nitrogen atom has 14 times the mass of 1 hydrogen atom; thus, each ammonia molecule consists of 1 nitrogen atom and 3 hydrogen atoms:

Ammonia gas is 14 parts N by mass and 3 parts H by mass.
1 N atom has 14 times the mass of 1 H atom.
Each ammonia molecule consists of 1 N atom and 3 H atoms.

Another defining feature of a compound is that *its properties are different from the properties of its component elements.* Table 2.1 shows a striking example: soft, silvery sodium metal and yellow-green, poisonous chlorine gas are very different from the compound they form—white, crystalline sodium chloride, or common table salt!

Unlike an element, a compound *can* be broken down into simpler substances—its component elements. For example, an electric current breaks down molten sodium chloride into metallic sodium and chlorine gas. By definition, this breakdown is a *chemical change,* not a physical one.

3. *Mixtures.* A **mixture** consists of two or more substances (elements and/or compounds) that are physically intermingled. Because a mixture is *not* a substance, in contrast to a compound, *the components of a mixture* ***can*** *vary in their parts by mass.* A mixture of the compounds sodium chloride and water, for example, can have many different parts by mass of salt to water. On the atomic scale, a mixture consists of the individual units that make up its component elements and/or compounds (Figure 2.1D). It makes sense, then, that *a mixture retains many of the properties of its components.* Saltwater, for instance, is colorless like water and tastes salty like sodium chloride.

Unlike compounds, mixtures can be separated into their components by *physical changes;* chemical changes are not needed. For example, the water in saltwater can be boiled off, a physical process that leaves behind solid sodium chloride. The following sample problem will help differentiate these types of matter.

Sample Problem 2.1 Distinguishing Elements, Compounds, and Mixtures at the Atomic Scale

Problem The scenes below represent atomic-scale views of three samples of matter:

(a)

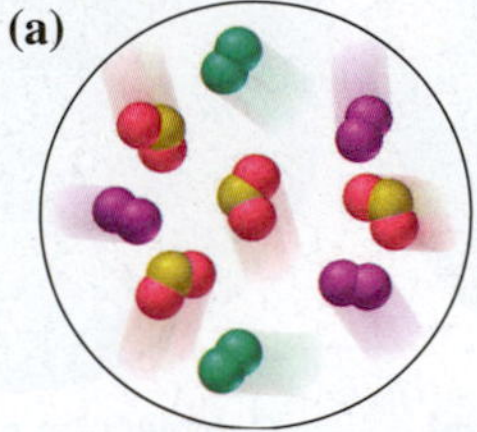

(b)

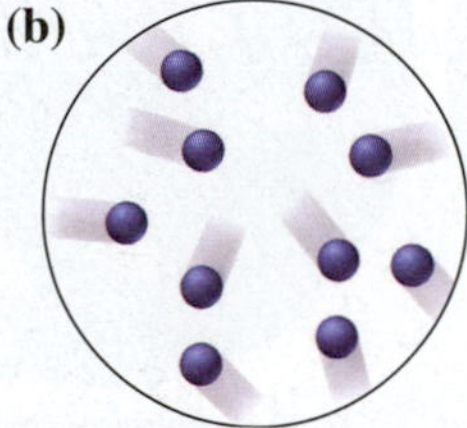

(c)

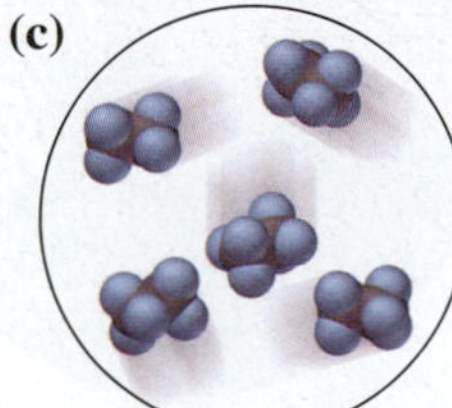

Describe each sample as an element, compound, or mixture.

Plan We have to determine the type of matter by examining the component particles. If a sample contains only one type of particle, it is either an element or a compound; if it contains more than one type, it is a mixture. Particles of an element have only one kind of atom (one color), and particles of a compound have two or more kinds of atoms.

Solution **(a)** Mixture: there are three different types of particles. Two types contain only one kind of atom, either green or purple, so they are elements, and the third type contains two red atoms for every one yellow, so it is a compound.

(b) Element: the sample consists of only blue atoms.

(c) Compound: the sample consists of molecules that each have two black and six blue atoms.

FOLLOW-UP PROBLEM 2.1 Describe the following reaction in terms of elements, compounds, and mixtures. (See Brief Solutions at the end of the chapter.)

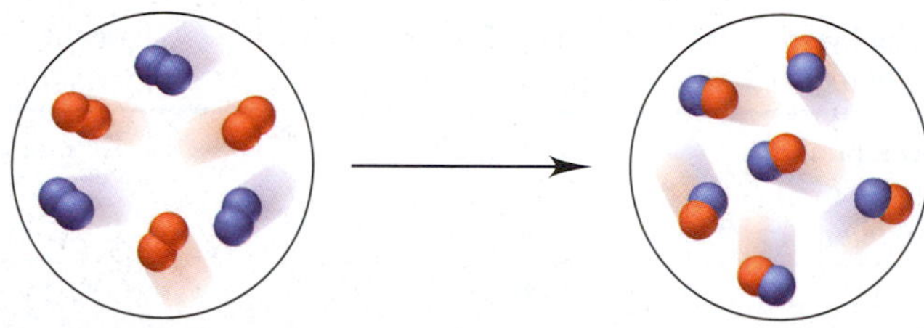

Summary of Section 2.1

- All matter exists as either elements, compounds, or mixtures.
- Every element or compound is a substance, matter with a fixed composition.
- An element consists of one type of atom and occurs as collections of individual atoms or molecules.
- A compound contains two or more elements chemically combined and exhibits different properties from its component elements. The elements occur in fixed parts by mass because each unit of the compound has a fixed number of each type of atom. Only a chemical change can break down a compound into its elements.
- A mixture consists of two or more substances mixed together, not chemically combined. The components retain their individual properties, can be present in any proportion, and can be separated by physical changes.

2.2 • THE OBSERVATIONS THAT LED TO AN ATOMIC VIEW OF MATTER

Any model of the composition of matter had to explain two widespread observations known as the *law of mass conservation* and the *law of definite (or constant) composition.* As you'll see, an atomic theory developed in the early 19th century explained these mass laws and another now known as the *law of multiple proportions.*

Mass Conservation

The most fundamental chemical observation of the 18th century was the **law of mass conservation:** *the total mass of substances does not change during a chemical reaction.* The *number* of substances may change and, by definition, their properties must, but the *total amount* of matter remains constant. The great French chemist and statesman Lavoisier first stated this law on the basis of experiments in which he reacted mercury with oxygen. He found the mass of oxygen plus the mass of mercury always equaled the mass of mercuric oxide that formed.

Even in a complex biochemical change, such as the metabolism of the sugar glucose, which involves many reactions, mass is conserved:

$$\text{180 g glucose} + \text{192 g oxygen gas} \longrightarrow \text{264 g carbon dioxide} + \text{108 g water}$$
$$\text{372 g material before} \longrightarrow \text{372 g material after}$$

Mass conservation means that, based on all chemical experience, *matter cannot be created or destroyed.* (As you'll see later, mass *does* change in nuclear reactions but *not* in chemical reactions.)

Figure 2.2 The law of definite composition. Calcium carbonate occurs in many forms (such as marble, *top*, and coral, *bottom*), but the mass percents of its elements are always the same.

Definite Composition

The sodium chloride in your salt shaker is the same substance whether it comes from a salt mine, a salt flat, or any other source. This fact is expressed in the **law of definite (or constant) composition,** which states that *no matter what its source, a particular compound is composed of the same elements in the same parts (fractions) by mass.* The **fraction by mass (mass fraction)** is the part of the compound's mass that each element contributes. It is obtained by dividing the mass of each element by the mass of the compound. The **percent by mass (mass percent, mass %)** is the fraction by mass expressed as a percentage (multiplied by 100).

Consider calcium carbonate, the major compound in seashells, marble, and coral. It is composed of three elements—calcium, carbon, and oxygen. The following results are obtained from a mass analysis of 20.0 g of calcium carbonate:

Analysis by Mass (grams/20.0 g)	Mass Fraction (parts/1.00 part)	Percent by Mass (parts/100 parts)
8.0 g calcium	0.40 calcium	40% calcium
2.4 g carbon	0.12 carbon	12% carbon
9.6 g oxygen	0.48 oxygen	48% oxygen
20.0 g	1.00 part by mass	100% by mass

The mass of each element depends on the mass of the sample—that is, more than 20.0 g of compound would contain more than 8.0 g of calcium—but *the mass fraction is fixed no matter what the size of the sample.* The sum of the mass fractions (or mass percents) equals 1.00 part (or 100%) by mass. The law of definite composition tells us that pure samples of calcium carbonate, no matter where they come from, always contain 40% calcium, 12% carbon, and 48% oxygen by mass (Figure 2.2).

Because a given element always constitutes the same mass fraction of a given compound, we can use that mass fraction to find the actual mass of the element in any sample of the compound:

$$\text{Mass of element} = \text{mass of compound} \times \frac{\text{part by mass of element}}{\text{one part by mass of compound}}$$

Or, more simply, we can skip the need to find the mass fraction first and use the results of mass analysis directly:

$$\text{Mass of element in sample} = \text{mass of compound in sample} \times \frac{\text{mass of element in compound}}{\text{mass of compound}} \qquad (2.1)$$

Sample Problem 2.2 Calculating the Mass of an Element in a Compound

Problem Pitchblende is the most important compound of uranium. Mass analysis of an 84.2-g sample shows that it contains 71.4 g of uranium, with oxygen the only other element. How many grams of uranium are in 102 kg of pitchblende?

Plan We have to find the mass of uranium in a known mass (102 kg) of pitchblende, given the mass of uranium (71.4 g) in a different mass of pitchblende (84.2 g). The mass ratio of uranium to pitchblende is the same for any sample of pitchblende. Therefore, using Equation 2.1, we multiply the mass (in kg) of the pitchblende sample by the ratio of uranium to pitchblende from the mass analysis. This gives the mass (in kg) of uranium, and we convert kilograms to grams.

Road Map

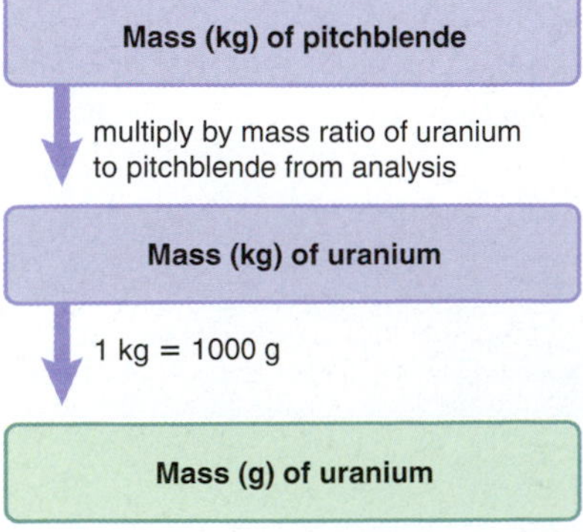

Solution Finding the mass (kg) of uranium in 102 kg of pitchblende:

$$\text{Mass (kg) of uranium} = \text{mass (kg) of pitchblende} \times \frac{\text{mass (kg) of uranium in pitchblende}}{\text{mass (kg) of pitchblende}}$$

$$= 102 \cancel{\text{kg pitchblende}} \times \frac{71.4 \text{ kg uranium}}{84.2 \cancel{\text{kg pitchblende}}} = 86.5 \text{ kg uranium}$$

Converting the mass of uranium from kg to g:

$$\text{Mass (g) of uranium} = 86.5 \cancel{\text{kg}} \text{ uranium} \times \frac{1000 \text{ g}}{1 \cancel{\text{kg}}} = 8.65{\times}10^4 \text{ g uranium}$$

Check The analysis showed that most of the mass of pitchblende is due to uranium, so the large mass of uranium makes sense. Rounding off to check the math gives:

$$\sim 100 \text{ kg pitchblende} \times \frac{70}{85} = 82 \text{ kg uranium}$$

FOLLOW-UP PROBLEM 2.2 How many metric tons (t) of oxygen are combined in a sample of pitchblende that contains 2.3 t of uranium? (*Hint:* Remember that oxygen is the only other element present.) See Brief Solutions.

Multiple Proportions

Dalton and others made an observation that applies when two elements form more than one compound, now called the **law of multiple proportions:** *if elements A and B react to form two compounds, the different masses of B that combine with a fixed mass of A can be expressed as a ratio of small whole numbers.* Consider two compounds, let's call them I and II, that carbon and oxygen form. These compounds have very different properties: the density of carbon oxide I is 1.25 g/L, whereas that of II is 1.98 g/L; I is poisonous and flammable, but II is not. Mass analysis shows that

Carbon oxide I is 57.1 mass % oxygen and 42.9 mass % carbon
Carbon oxide II is 72.7 mass % oxygen and 27.3 mass % carbon

To see the phenomenon of multiple proportions, we use the mass percents of oxygen and of carbon to find their masses in a given mass, say 100 g, of each compound. Then we divide the mass of oxygen by the mass of carbon in each compound to obtain the mass of oxygen that combines with a fixed mass of carbon:

	Carbon Oxide I	Carbon Oxide II
g oxygen/100 g compound	57.1	72.7
g carbon/100 g compound	42.9	27.3
g oxygen/g carbon	$\frac{57.1}{42.9} = 1.33$	$\frac{72.7}{27.3} = 2.66$

If we then divide the grams of oxygen per gram of carbon in II by that in I, we obtain a ratio of small whole numbers:

$$\frac{2.66 \text{ g oxygen/g carbon in II}}{1.33 \text{ g oxygen/g carbon in I}} = \frac{2}{1}$$

The law of multiple proportions tells us that in two compounds of the same elements, the mass fraction of one element relative to the other element changes in *increments based on ratios of small whole numbers.* In this case, the ratio is 2/1—for a given mass of carbon, compound II contains *2 times* as much oxygen as I, not 1.583 times, 1.716 times, or any other intermediate amount. In the next section, we'll explain the mass laws on the atomic scale.

Summary of Section 2.2

- The law of mass conservation states that the total mass remains constant during a chemical reaction.
- The law of definite composition states that any sample of a given compound has the same elements present in the same parts by mass.
- The law of multiple proportions states that, in different compounds of the same elements, the masses of one element that combine with a fixed mass of the other can be expressed as a ratio of small whole numbers.

2.3 • DALTON'S ATOMIC THEORY

With over 200 years of hindsight, it's easy to see how the mass laws could be explained by an atomic model—matter existing in indestructible units, each with a particular mass—but it was a major breakthrough in 1808 when John Dalton (1766–1844) presented his atomic theory of matter in *A New System of Chemical Philosophy.*

Postulates of the Atomic Theory

Dalton expressed his theory in a series of postulates. Like most great thinkers, he integrated the ideas of others into his own. As we go through the postulates, presented here in modern terms, we'll note which were original and which came from others.

1. All matter consists of **atoms,** tiny indivisible particles of an element that cannot be created or destroyed. (This derives from the "eternal, indestructible atoms" proposed by Democritus more than 2000 years earlier and reflects mass conservation as stated by Lavoisier.)
2. Atoms of one element *cannot* be converted into atoms of another element. In chemical reactions, the atoms of the original substances recombine to form different substances. (This rejects the belief in the magical transmutation of elements that was widely held into the 17th century.)
3. Atoms of an element are identical in mass and other properties and are different from atoms of any other element. (This contains Dalton's major new ideas: *unique mass and properties* for the atoms of a given element.)
4. Compounds result from the chemical combination of a specific ratio of atoms of different elements. (This follows directly from the law of definite composition.)

How the Theory Explains the Mass Laws

Let's see how Dalton's postulates explain the mass laws:

- *Mass conservation.* Atoms cannot be created or destroyed (postulate 1) or converted into other types of atoms (postulate 2). Therefore, a chemical reaction, in which atoms are combined differently, cannot possibly result in a mass change.
- *Definite composition.* A compound is a combination of a *specific* ratio of different atoms (postulate 4), each of which has a particular mass (postulate 3). Thus, each element in a compound constitutes a fixed fraction of the total mass.
- *Multiple proportions.* Atoms of an element have the same mass (postulate 3) and are indivisible (postulate 1). The masses of element B that combine with a fixed mass of element A give a small, whole-number ratio because different numbers of B atoms combine with each A atom in different compounds.

The *simplest* arrangement consistent with the mass data for carbon oxides I and II in our earlier example is that one atom of oxygen combines with one atom of carbon in compound I (carbon monoxide) and that two atoms of oxygen combine with one atom of carbon in compound II (carbon dioxide):

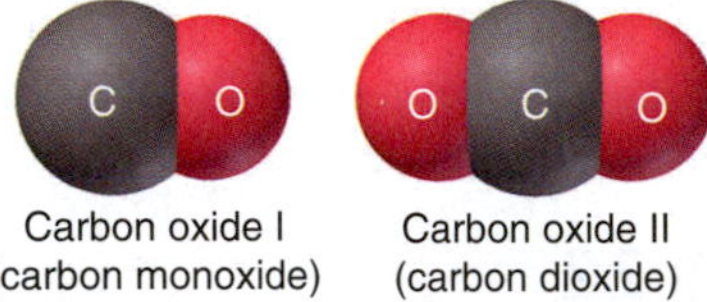

Let's work through a sample problem that reviews the mass laws.

Sample Problem 2.3 Visualizing the Mass Laws

Problem The scenes below represent an atomic-scale view of a chemical reaction:

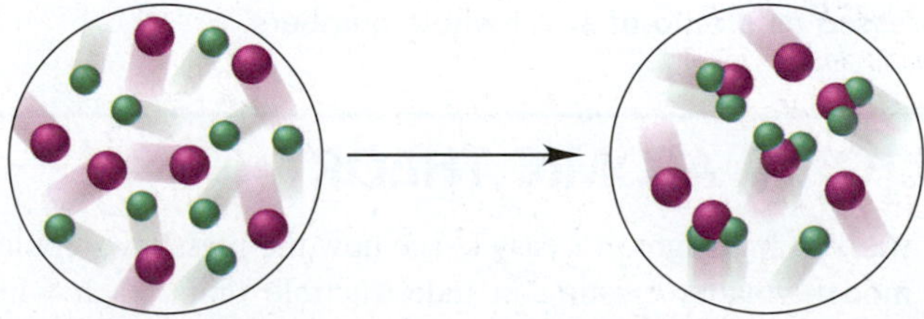

Which of the mass laws—mass conservation, definite composition, or multiple proportions—is (are) illustrated?

Plan From the depictions, we note the numbers, colors, and combinations of atoms (spheres) to see which mass laws pertain. If the numbers of each atom are the same before and after the reaction, the total mass did not change (mass conservation). If a compound forms that always has the same atom ratio, the elements are present in fixed parts by mass (definite composition). When the same elements form different compounds and the ratio of the atoms of one element that combine with one atom of the other element is a small whole number, the ratio of their masses is a small whole number as well (multiple proportions).

Solution There are seven purple and nine green atoms in each circle, so mass is conserved. The compound formed has one purple and two green atoms, so it has definite composition. Only one compound forms, so the law of multiple proportions does not pertain.

FOLLOW-UP PROBLEM 2.3 Which sample(s) best display(s) the fact that compounds of bromine (orange) and fluorine (yellow) exhibit the law of multiple proportions? Explain.

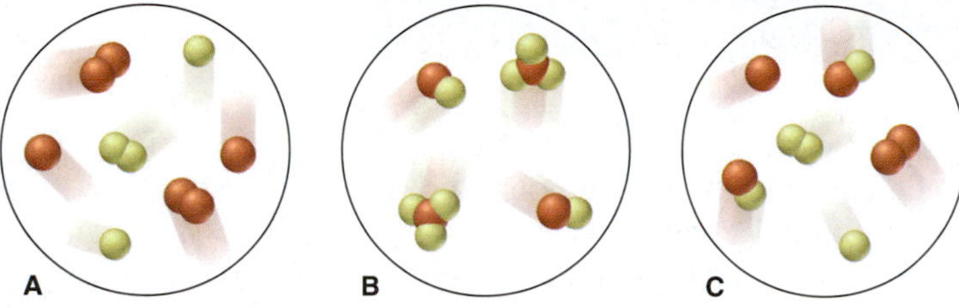

Summary of Section 2.3

- Dalton's atomic theory explained the mass laws by proposing that all matter consists of indivisible, unchangeable atoms of fixed, unique mass.
- Mass is conserved during a reaction because the atoms retain their identities but are combined differently.
- Each compound has a fixed mass fraction of each of its elements because it is composed of a fixed number of each type of atom.
- Different compounds of the same elements exhibit multiple proportions because they each consist of whole atoms.

2.4 • THE OBSERVATIONS THAT LED TO THE NUCLEAR ATOM MODEL

Dalton's model established that masses of reacting elements could be explained in terms of atoms but not why atoms bond as they do: why, for example, do two, and not three, hydrogen atoms bond with one oxygen atom in a water molecule?

Moreover, Dalton's model of the atom, which represented it as a tiny, indivisible particle, like a minute billiard ball, did not predict the existence of subatomic charged particles. These were observed in later experiments that led to the discovery of *electrons* and the atomic *nucleus*. Let's examine some of these experiments and the more complex atomic model that emerged from them.

Discovery of the Electron and Its Properties

For many years, scientists had known that matter and electric charge were related. When amber is rubbed with fur, or glass with silk, positive and negative charges form—the same charges that make your hair crackle and cling to your comb on a dry day. They also knew that an electric current could decompose certain compounds into their elements. But they did not know what a current was made of.

Cathode Rays To discover the nature of an electric current, some investigators tried passing current through nearly evacuated glass tubes fitted with metal electrodes. When the electric power source was turned on, a "ray" could be seen striking the

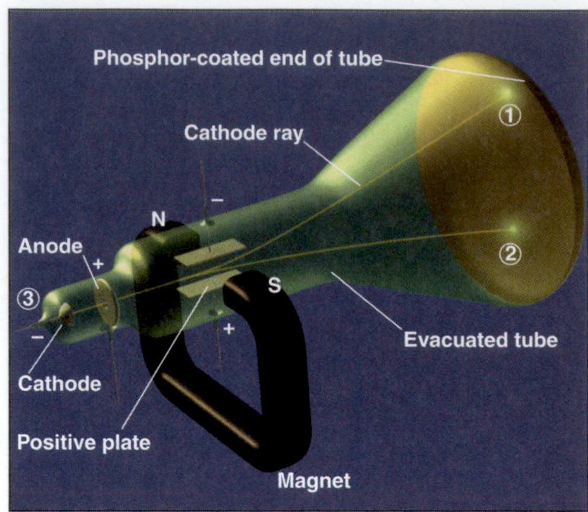

OBSERVATION	CONCLUSION
1. Ray bends in magnetic field.	Consists of charged particles
2. Ray bends toward positive plate in electric field.	Consists of negative particles
3. Ray is identical for any cathode.	Particles found in all matter

Figure 2.3 Observations that established the properties of cathode rays.

phosphor-coated end of the tube and emitting a glowing spot of light. The rays were called **cathode rays** because they originated at the negative electrode (cathode) and moved to the positive electrode (anode).

Figure 2.3 shows some properties of cathode rays based on these observations. The main conclusion was that *cathode rays consist of negatively charged particles found in all matter.* The rays appear when these particles collide with the few remaining gas molecules in the evacuated tube. Cathode ray particles were later named *electrons.*

Mass and Charge of the Electron Two classic experiments revealed the mass and charge of the electron:

1. *Mass/charge ratio.* In 1897, the British physicist J. J. Thomson (1856–1940) measured the ratio of the mass of a cathode ray particle to its charge. By comparing this value with the mass/charge ratio for the lightest charged particle in solution, Thomson estimated that the cathode ray particle weighed less than $\frac{1}{1000}$ as much as hydrogen, the lightest atom! He was shocked because this implied that, contrary to Dalton's atomic theory, *atoms contain even smaller particles.* Fellow scientists reacted with disbelief to Thomson's conclusion, thinking he was joking.

2. *Charge.* In 1909, the American physicist Robert Millikan (1868–1953) measured the *charge* of the electron. He did so by observing the movement of oil droplets in an apparatus that contained electrically charged plates and an x-ray source (Figure 2.4).

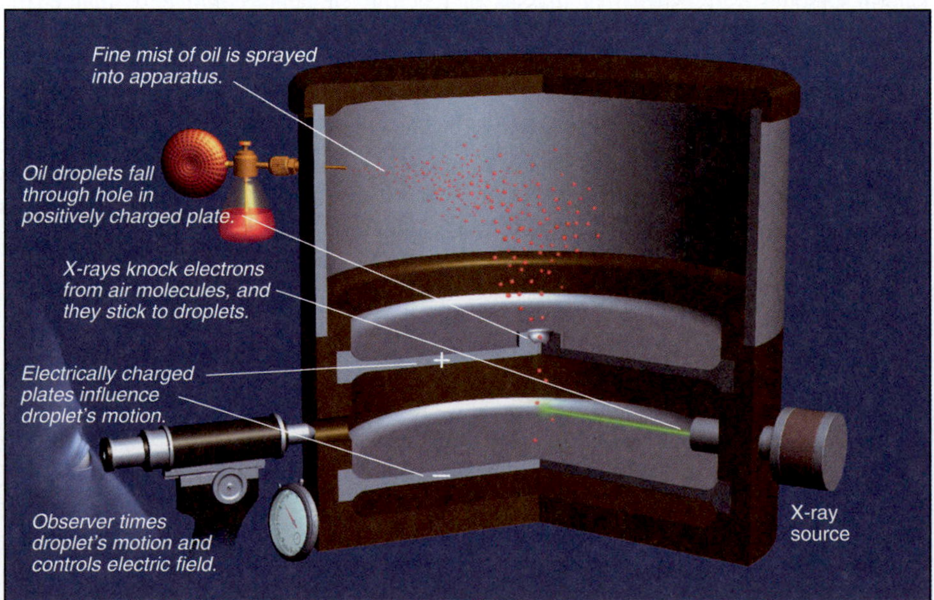

Figure 2.4 **Millikan's oil-drop experiment for measuring an electron's charge.** The total charge on an oil droplet is some whole-number multiple of the charge of the electron.

X-rays knocked electrons from gas molecules in the air within the apparatus, and the electrons stuck to an oil droplet falling through a hole in a positively charged plate. With the electric field off, Millikan measured the mass of the droplet from its rate of fall. Then, by adjusting the field's strength, he made the droplet hang suspended in the air and, thus, measured its total charge.

After many tries, Millikan found that the total charge of the various droplets was always some *whole-number multiple of a minimum charge.* If different oil droplets picked up different numbers of electrons, he reasoned that this minimum charge must be the charge of the electron itself. Remarkably, the value that he calculated over a century ago is within 1% of the modern value of the electron's charge, -1.602×10^{-19} C (C stands for *coulomb,* the SI unit of charge).

3. *Conclusion: calculating the electron's mass.* The electron's mass/charge ratio and the value for the electron's charge can be used to find the electron's mass, which is *extremely* small:

$$\text{Mass of electron} = \frac{\text{mass}}{\cancel{\text{charge}}} \times \cancel{\text{charge}} = \left(-5.686\times10^{-12}\,\frac{\text{kg}}{\cancel{\text{C}}}\right)(-1.602\times10^{-19}\,\cancel{\text{C}})$$
$$= 9.109\times10^{-31}\text{ kg} = 9.109\times10^{-28}\text{ g}$$

Discovery of the Atomic Nucleus

The presence of electrons in all matter posed some major questions about the structure of atoms. Matter is electrically neutral, so atoms must be also. But if atoms contain negatively charged electrons, what positive charges balance them? And if an electron has such a tiny mass, what accounts for an atom's much larger mass? To address these issues, Thomson proposed his "plum-pudding" model—a spherical atom composed of diffuse, positively charged matter with electrons embedded like "raisins in a plum pudding."

In 1910, New Zealand–born physicist Ernest Rutherford (1871–1937) tested this model and obtained an unexpected result (Figure 2.5):

1. *Experimental design.* Figure 2.5B shows the experimental setup, in which tiny, dense, positively charged alpha (α) particles emitted from radium are aimed at gold foil. A circular, zinc-sulfide screen registers the deflection (scattering) of the α particles by emitting light flashes when the particles strike it.

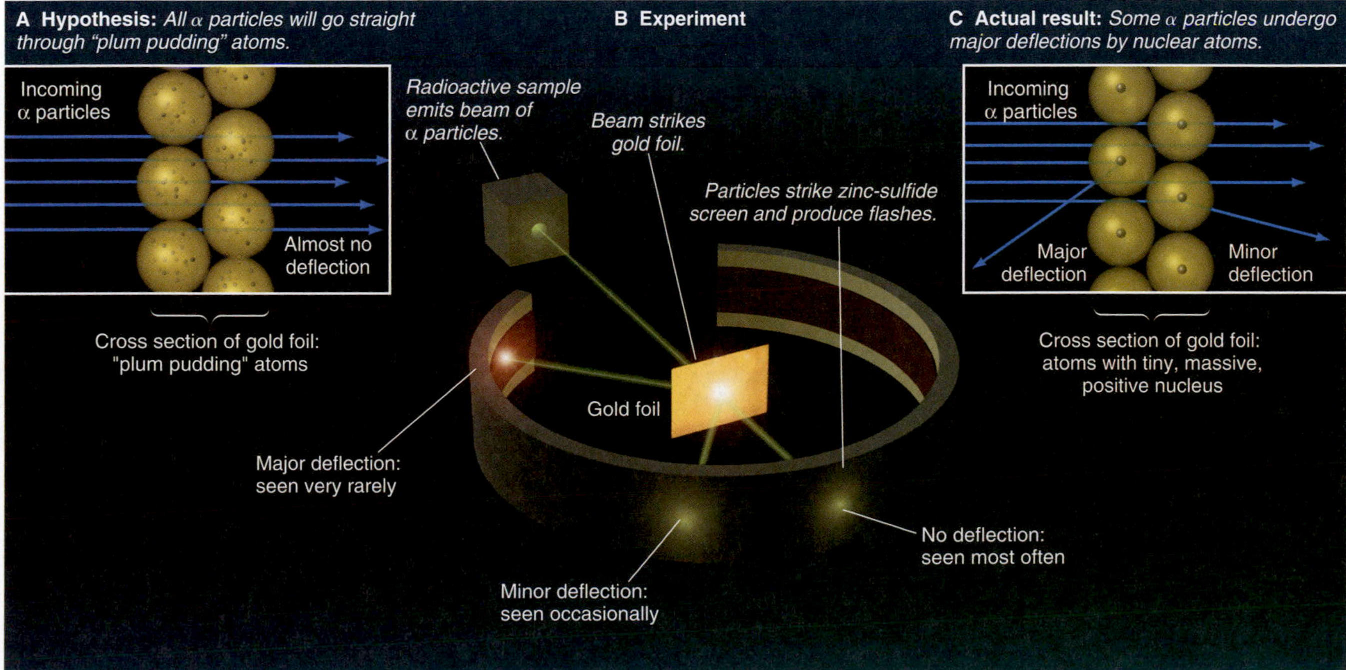

Figure 2.5 Rutherford's α-scattering experiment and discovery of the atomic nucleus.

2. *Expected results.* With Thomson's model in mind (Figure 2.5A), Rutherford expected only minor, if any, deflections of the α particles because they should act as bullets and go right through the gold atoms. After all, an electron would not deflect an α particle any more than a Ping-Pong ball would deflect a baseball.
3. *Actual results.* Initial results were consistent with this idea, but then the unexpected happened (Figure 2.5C). As Rutherford recalled: "I remember two or three days later Geiger [one of his coworkers] coming to me in great excitement and saying, 'We have been able to get some of the α particles coming backwards . . .' It was quite the most incredible event that has ever happened to me in my life. It was almost as incredible as if you fired a 15-inch shell at a piece of tissue paper and it came back and hit you." In fact, very few α particles were deflected at all, and only 1 in 20,000 had large-angle deflections of more than 90° ("coming backwards").
4. *Conclusion.* Rutherford concluded that these few α particles were being repelled by something small, dense, and positive within the gold atoms. Calculations based on the properties of α particles and the fraction of large-angle deflections showed that
   - An atom is mostly space occupied by electrons.
   - In the center is a tiny region, which Rutherford called the **nucleus,** that contains all the positive charge and essentially all the mass of the atom.

   He proposed that positive particles lay within the nucleus and called them *protons.*

Rutherford's model explained the charged nature of matter, but it could not account for all the atom's mass. After more than 20 years, in 1932, James Chadwick (1891–1974) discovered the *neutron,* an uncharged dense particle that also resides in the nucleus.

Summary of Section 2.4

- Several major discoveries at the turn of the 20th century resolved questions about Dalton's model and led to our current model of atomic structure.
- Cathode rays were shown to consist of negative particles (electrons) that exist in all matter. J. J. Thomson measured their mass/charge ratio and concluded that they are much smaller and lighter than atoms.
- Robert Millikan determined the charge of the electron, which he combined with other data to calculate its mass.
- Ernest Rutherford proposed that atoms consist of a tiny, massive, positive nucleus surrounded by electrons.

2.5 • THE ATOMIC THEORY TODAY

Dalton's model of an indivisible particle has given way to our current model of an atom with an elaborate internal architecture of subatomic particles.

Structure of the Atom

An *atom* is an electrically neutral, spherical entity composed of a positively charged central nucleus surrounded by one or more negatively charged electrons (Figure 2.6). The electrons move rapidly within the available volume, held there by the attraction of the nucleus. An atom's diameter ($\sim 1\times10^{-10}$ m) is about 20,000 times the diameter of its nucleus ($\sim 5\times10^{-15}$ m). The nucleus contributes 99.97% of the atom's mass, occupies only about 1 quadrillionth of its volume, and is incredibly dense: about 10^{14} g/mL!

THINK OF IT THIS WAY
The Tiny, Massive Nucleus

A few analogies will help you appreciate the incredible properties of the atomic nucleus. A nucleus the size of the period at the end of this sentence would weigh about 100 tons, as much as 50 cars. An atom the size of the Houston Astrodome would have a nucleus the size of a green pea that would contain virtually all the atom's mass. If a nucleus were actually the size shown in Figure 2.6 (about 1 cm across), the atom would be about 200 m across, or slightly more than twice the length of a football field!

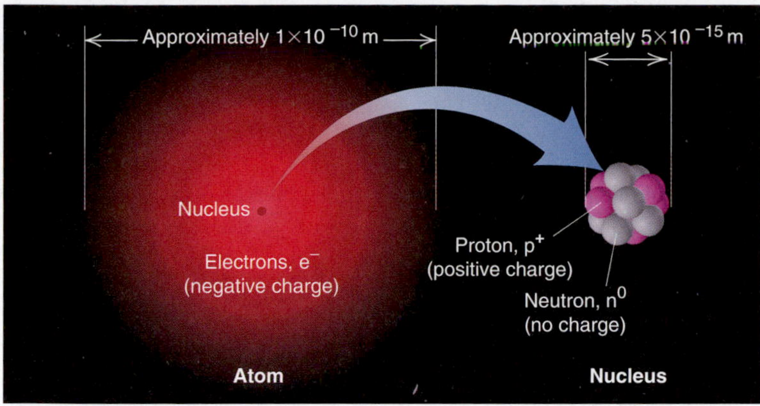

Figure 2.6 General features of the atom.

An atomic nucleus consists of protons and neutrons (the only exception is the simplest hydrogen nucleus, which is a single proton). The **proton (p^+)** has a positive charge, and the **neutron (n^0)** has no charge; thus, the positive charge of the nucleus results from its protons. The *magnitudes* of the charges possessed by a proton and by an **electron (e^-)** are equal, but the *signs* of the charges are opposite. *An atom is neutral because the number of protons in the nucleus equals the number of electrons surrounding the nucleus.* Some properties of these three subatomic particles are listed in Table 2.2.

Atomic Number, Mass Number, and Atomic Symbol

The **atomic number (Z)** of an element equals the number of protons in the nucleus of each of its atoms. *All atoms of an element have the same atomic number, and the atomic number of each element is different from that of any other element.* All carbon atoms ($Z = 6$) have 6 protons, all oxygen atoms ($Z = 8$) have 8 protons, and all uranium atoms ($Z = 92$) have 92 protons. There are currently 117 known elements, of which 90 occur in nature and 27 have been synthesized by nuclear scientists.

The **mass number (A)** is the total number of protons and neutrons in the nucleus of an atom. Each proton and each neutron contributes one unit to the mass number. Thus, a carbon atom with 6 protons and 6 neutrons in its nucleus has a mass number of 12, and a uranium atom with 92 protons and 146 neutrons in its nucleus has a mass number of 238.

The **atomic symbol** (or *element symbol*) of an element is based on its English, Latin, or Greek name, such as C for carbon, S for sulfur, and Na for sodium (Latin *natrium*). Often written with the symbol are the atomic number (Z) as a left *sub*script and the mass number (A) as a left *super*script, so element X would be ${}^{A}_{Z}X$. Since the

Table 2.2 Properties of the Three Key Subatomic Properties

	Charge		Mass		
Name (Symbol)	**Relative**	**Absolute (C)***	**Relative (amu)†**	**Absolute (g)**	**Location in Atom**
Proton (p^+)	1+	$+1.60218\times10^{-19}$	1.00727	1.67262×10^{-24}	Nucleus
Neutron (n^0)	0	0	1.00866	1.67493×10^{-24}	Nucleus
Electron (e^-)	1−	-1.60218×10^{-19}	0.00054858	9.10939×10^{-24}	Outside nucleus

*The coulomb (C) is the SI unit of charge.
†The atomic mass unit (amu) equals 1.66054×10^{-24} g; discussed later in this section.

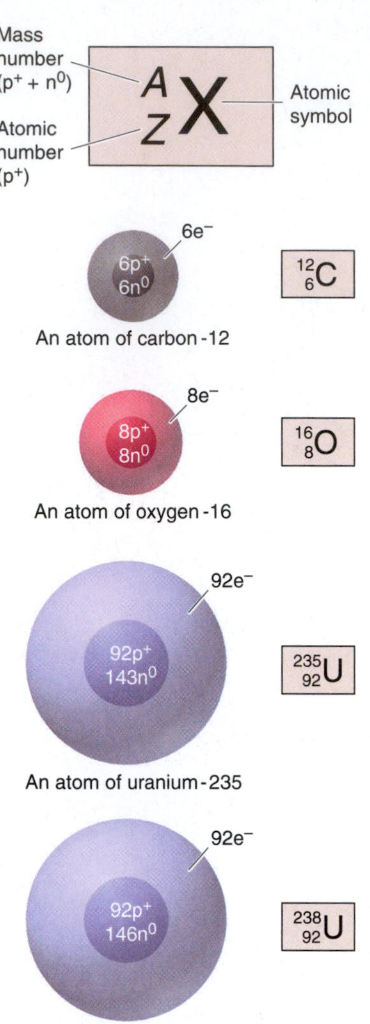

Figure 2.7 The ^A_ZX notations and spherical representations for four atoms. (The nuclei are not drawn to scale.)

mass number is the sum of protons and neutrons, the number of neutrons (N) equals the mass number minus the atomic number:

$$\text{Number of neutrons} = \text{mass number} - \text{atomic number}, \quad \text{or} \quad N = A - Z \qquad (2.2)$$

Thus, a chlorine atom, symbolized $^{35}_{17}Cl$, has $A = 35$, $Z = 17$, and $N = 35 - 17 = 18$. Because each element has its own atomic number, we also know the atomic number given the symbol. For example, instead of writing $^{12}_{6}C$ for carbon with mass number 12, we can write ^{12}C (spoken "carbon twelve"), with $Z = 6$ understood. Another way to name this atom is carbon-12.

Isotopes

All atoms of an element have the same atomic number but not the same mass number. **Isotopes** of an element are atoms that have *different numbers of neutrons* and therefore different mass numbers. For example, all carbon atoms ($Z = 6$) have 6 protons and 6 electrons, but only 98.89% of naturally occurring carbon atoms have 6 neutrons ($A = 12$). A small percentage (1.11%) have 7 neutrons ($A = 13$), and even fewer (less than 0.01%) have 8 ($A = 14$). These are carbon's three naturally occurring isotopes—^{12}C, ^{13}C, and ^{14}C. *A natural sample of carbon has these three isotopes in these relative proportions.* Five other carbon isotopes—^{9}C, ^{10}C, ^{11}C, ^{15}C, and ^{16}C—have been created in the laboratory. Figure 2.7 depicts the atomic number, mass number, and symbol for four atoms, two of which are isotopes of the element uranium.

The chemical properties of an element are primarily determined by the number of electrons, so *all isotopes of an element have nearly identical chemical behavior,* even though they have different masses.

Sample Problem 2.4 Determining the Number of Subatomic Particles in the Isotopes of an Element

Problem Silicon (Si) is a major component of semiconductor chips. It has three naturally occurring isotopes: ^{28}Si, ^{29}Si, and ^{30}Si. Determine the numbers of protons, neutrons, and electrons in each silicon isotope.

Plan The mass number (A; left superscript) of each of the three isotopes is given, which is the sum of protons and neutrons. From the elements list on this book's inside front cover, we find the atomic number (Z, number of protons), which equals the number of electrons. We obtain the number of neutrons by subtracting Z from A (Equation 2.2).

Solution From the elements list, the atomic number of silicon is 14. Therefore,

$$^{28}Si \text{ has } 14p^+, 14e^-, \text{ and } 14n^0 (28 - 14)$$
$$^{29}Si \text{ has } 14p^+, 14e^-, \text{ and } 15n^0 (29 - 14)$$
$$^{30}Si \text{ has } 14p^+, 14e^-, \text{ and } 16n^0 (30 - 14)$$

FOLLOW-UP PROBLEM 2.4 How many protons, neutrons, and electrons are in **(a)** $^{11}_{5}Q$? **(b)** $^{41}_{20}R$? **(c)** $^{131}_{53}X$? What elements do Q, R, and X represent?

Atomic Masses of the Elements; Mass Spectrometry

The mass of an atom is measured *relative* to the mass of an atomic standard. The modern standard is the carbon-12 atom, whose mass is defined as *exactly* 12 atomic mass units. Thus, the **atomic mass unit (amu)** is $\frac{1}{12}$ the mass of a carbon-12 atom. Based on this standard, the ^{1}H atom has a mass of 1.008 amu; in other words, a ^{12}C atom has almost 12 times the mass of an ^{1}H atom. We will continue to use the term *atomic mass unit* in the text, although the name of the unit has been changed to the **dalton (Da);** thus, one ^{12}C atom has a mass of 12 daltons (12 Da, or 12 amu). The atomic mass unit is a unit of relative mass, but it has an absolute mass of 1.66054×10^{-24} g.

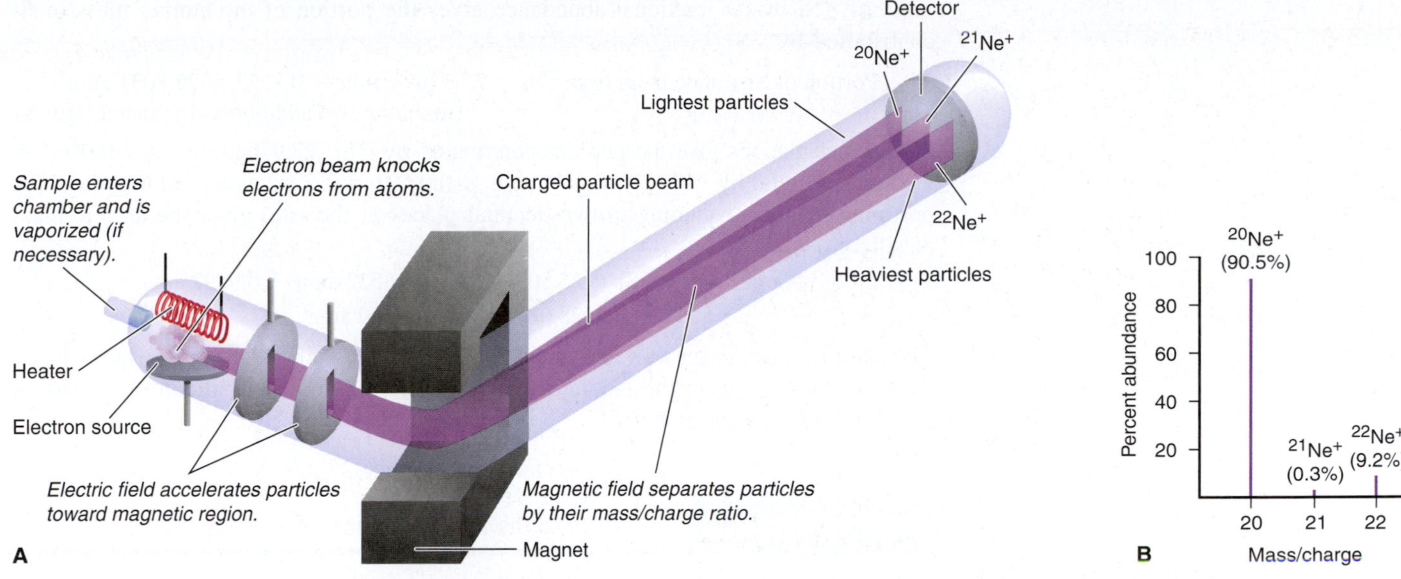

Figure 2.8 The mass spectrometer and its data. **A,** Charged particles are separated on the basis of their *m/e* values. Ne is the sample here. **B,** The percent abundance of each Ne isotope.

The isotopic makeup of an element is determined by **mass spectrometry,** a method for measuring the relative masses and abundances of atomic-scale particles and molecules very precisely. In one type of mass spectrometer, atoms of a sample of, say, elemental neon are bombarded by a high-energy electron beam (Figure 2.8A). As a result, one electron is knocked off each Ne atom, and each resulting particle has one positive charge. Thus, its mass/charge ratio (*m/e*) equals the mass of an Ne atom divided by 1+. The *m/e* values are measured to identify the masses of different isotopes of the element. The positively charged Ne particles are attracted toward a series of negatively charged plates with slits in them, and some of the particles pass through the slits into an evacuated tube exposed to a magnetic field. As the particles zoom through this region, they are deflected (their paths are bent) according to their *m/e* values: the lightest particles are deflected most and the heaviest particles least. At the end of the tube, the particles strike a detector, which records their relative positions and abundances (Figure 2.8B). Mass spectrometry is now used to measure the mass of virtually any atom or molecule. In 2002, the Nobel Prize in chemistry was awarded for the study of proteins by mass spectrometry.

Let's see how this instrument is used. With a mass spectrometer, we measure the mass ratio of ^{28}Si to ^{12}C as

$$\frac{\text{Mass of }^{28}\text{Si atom}}{\text{Mass of }^{12}\text{C standard}} = 2.331411$$

From this mass ratio, we find the **isotopic mass** of the ^{28}Si atom, the relative mass of this silicon isotope:

$$\begin{aligned}\text{Isotopic mass of }^{28}\text{Si} &= \text{measured mass ratio} \times \text{mass of }^{12}\text{C}\\ &= 2.331411 \times 12\text{ amu} = 27.97693\text{ amu}\end{aligned}$$

Along with the isotopic mass, the mass spectrometer gives the relative abundance as a percentage (or fraction) of each isotope in a sample of the element. For example, the relative abundance of ^{28}Si is 92.23% (or 0.9223).

From such data, we can obtain the **atomic mass** (also called *atomic weight*) of an element, the *average* of the masses of its naturally occurring isotopes weighted according to their abundances.* Each naturally occurring isotope of an element contributes a certain portion to the atomic mass. For instance, multiplying the isotopic

*Based on mass spectrometric analysis of isotopic abundances from various sources, the International Union of Pure and Applied Chemistry has recommended that the atomic masses of 10 elements be given as *atomic mass intervals* to indicate the range from source to source; for example, boron has its lowest atomic mass (10.806) in metamorphic rock and its highest (10.821) in surface and ground water. These changes will not affect discussions throughout the text, which use the four-digit values shown in the periodic table (Figure 2.9) and the front end sheets.

mass of ^{28}Si by its fractional abundance gives the portion of the atomic mass of Si contributed by ^{28}Si:

$$\text{Portion of Si atomic mass from } ^{28}\text{Si} = 27.97693 \text{ amu} \times 0.9223 = 25.8031 \text{ amu}$$

(retaining two additional significant figures)

Similar calculations give the portions contributed by ^{29}Si (28.976495 amu $\times$ 0.0467 = 1.3532 amu) and by ^{30}Si (29.973770 amu $\times$ 0.0310 = 0.9292 amu). Adding the three portions together (rounding to two decimal places at the end) gives the atomic mass of silicon:

$$\begin{aligned}\text{Atomic mass of Si} &= 25.8031 \text{ amu} + 1.3532 \text{ amu} + 0.9292 \text{ amu}\\ &= 28.0855 \text{ amu} = 28.09 \text{ amu}\end{aligned}$$

The atomic mass is an average value; thus, while no individual silicon atom has a mass of 28.09 amu, in the laboratory, we consider a sample of silicon to consist of atoms with this average mass.

Sample Problem 2.5 Calculating the Atomic Mass of an Element

Problem Silver (Ag; Z = 47) has 46 known isotopes, but only two occur naturally, ^{107}Ag and ^{109}Ag. Given the following data, calculate the atomic mass of Ag:

Isotope	Mass (amu)	Abundance (%)
^{107}Ag	106.90509	51.84
^{109}Ag	108.90476	48.16

Road Map

Mass (g) of each isotope

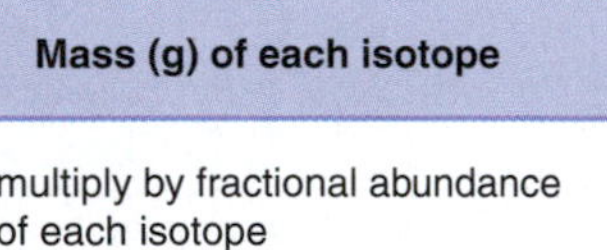

Portion of atomic mass from each isotope

add isotopic portions

Atomic mass

Plan From the mass and abundance of the two Ag isotopes, we have to find the atomic mass of Ag (weighted average of the isotopic masses). We divide each percent abundance by 100 to get the fractional abundance and then multiply that by each isotopic mass to find the portion of the atomic mass contributed by each isotope. The sum of the isotopic portions is the atomic mass.

Solution Finding the fractional abundances:

$$\text{Fractional abundance of } ^{107}\text{Ag} = 51.84/100 = 0.5184; \text{ similarly, } ^{109}\text{Ag} = 0.4816$$

Finding the portion of the atomic mass from each isotope:

$$\begin{aligned}\text{Portion of atomic mass from } ^{107}\text{Ag} &= \text{isotopic mass} \times \text{fractional abundance}\\ &= 106.90509 \text{ amu} \times 0.5184 = 55.42 \text{ amu}\\ \text{Portion of atomic mass from } ^{109}\text{Ag} &= 108.90476 \text{ amu} \times 0.4816 = 52.45 \text{ amu}\end{aligned}$$

Finding the atomic mass of silver:

$$\text{Atomic mass of Ag} = 55.42 \text{ amu} + 52.45 \text{ amu} = 107.87 \text{ amu}$$

Check The individual portions seem right: ~100 amu $\times$ 0.50 = 50 amu. The portions should be almost the same because the two isotopic abundances are almost the same. We rounded each portion to four significant figures because that is the number of significant figures in the abundance values. This is the correct atomic mass (to two decimal places); in the list of elements (inside front cover), it is rounded to 107.9 amu.

FOLLOW-UP PROBLEM 2.5 Boron (B; Z = 5) has two naturally occurring isotopes. Find the percent abundances of ^{10}B and ^{11}B given these data: atomic mass of B = 10.81 amu, isotopic mass of ^{10}B = 10.0129 amu, and isotopic mass of ^{11}B = 11.0093 amu. (*Hint:* The sum of the fractional abundances is 1. If x = abundance of ^{10}B, then $1 - x$ = abundance of ^{11}B.)

Summary of Section 2.5

- An atom has a central nucleus, which contains positively charged protons and uncharged neutrons and is surrounded by negatively charged electrons. An atom is neutral because the number of electrons equals the number of protons.

- An atom is represented by the notation ${}^{A}_{Z}X$, in which Z is the atomic number (number of protons), A the mass number (sum of protons and neutrons), and X the atomic symbol.
- An element occurs naturally as a mixture of isotopes, atoms with the same number of protons but different numbers of neutrons. Each isotope has a mass relative to the ${}^{12}C$ mass standard.
- The atomic mass of an element is the average of its isotopic masses weighted according to their natural abundances. It is determined using the mass spectrometer.

2.6 • ELEMENTS: A FIRST LOOK AT THE PERIODIC TABLE

At the end of the 18th century, Lavoisier compiled a list of the 23 elements known at that time; by 1870, 65 were known; by 1925, 88; today, there are 117 and still counting! By the mid-19th century, enormous amounts of information concerning reactions, properties, and atomic masses of the elements had been accumulated. Several researchers noted recurring, or *periodic,* patterns of behavior and proposed schemes to organize the elements according to some fundamental property.

In 1871, the Russian chemist Dmitri Mendeleev (1836–1907) published the most successful of these organizing schemes as a table of the elements listed by increasing atomic mass and arranged so that elements with similar chemical properties fell in the same column. The modern **periodic table of the elements,** based on Mendeleev's version (but arranged by *atomic number,* not mass), is one of the great classifying schemes in science and an indispensable tool to chemists—and chemistry students.

Organization of the Periodic Table One common version of the modern periodic table appears in Figure 2.9 on the next page (and inside the front cover). It is formatted as follows:

1. Each element has a box that contains its atomic number (number of protons in the nucleus), atomic symbol, and atomic mass. (A mass in parentheses is the mass number of the most stable isotope of that element.) The boxes lie, from left to right, in order of *increasing atomic number.*
2. The boxes are arranged into a grid of **periods** (horizontal rows) and **groups** (vertical columns). Each period has a number from 1 to 7. Each group has a number from 1 to 8 *and* either the letter A or B. A newer system, with group numbers from 1 to 18 but no letters, appears in parentheses under the number-letter designations. (The text uses the number-letter system and shows the newer numbering system in parentheses.)
3. The eight A groups (two on the left and six on the right) contain the *main-group elements.* The ten B groups, located between Groups 2A(2) and 3A(13), contain the *transition elements.* Two horizontal series of *inner transition elements,* the lanthanides and the actinides, fit *between* the elements in Group 3B(3) and Group 4B(4) and are placed below the main body of the table.

Classifying the Elements One of the clearest ways to classify the elements is as metals, nonmetals, and metalloids. The "staircase" line that runs from the top of Group 3A(13) to the bottom of Group 6A(16) is a dividing line:

- The **metals** (three shades of blue in Figure 2.9) lie in the large lower-left portion of the table. About three-quarters of the elements are metals, including many main-group elements and all the transition and inner transition elements. They are generally shiny solids at room temperature (mercury is the only liquid) that conduct heat and electricity well. They can be tooled into sheets (are malleable) and wires (are ductile).

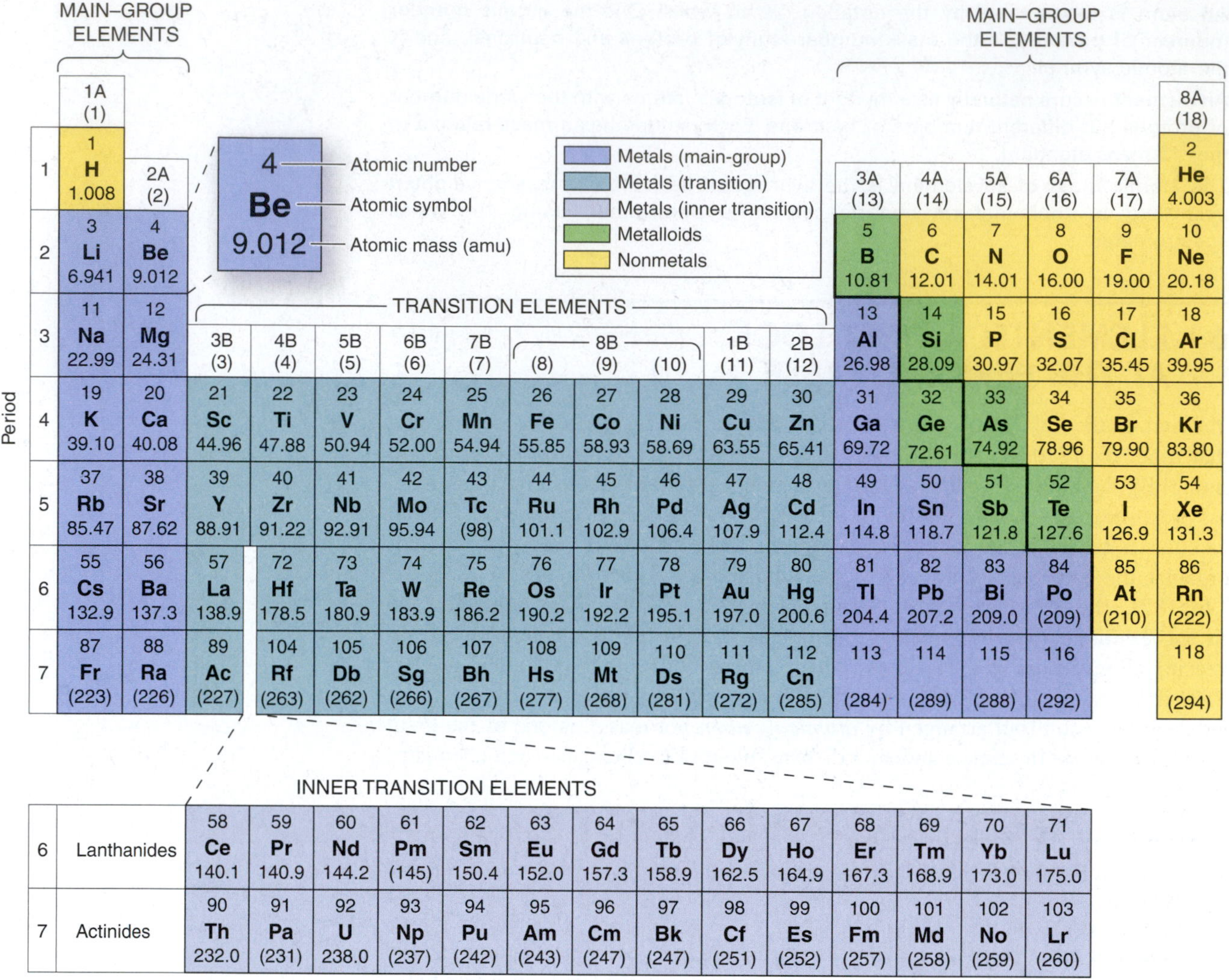

Period	1A (1)	2A (2)	3B (3)	4B (4)	5B (5)	6B (6)	7B (7)	8B (8)	8B (9)	8B (10)	1B (11)	2B (12)	3A (13)	4A (14)	5A (15)	6A (16)	7A (17)	8A (18)
1	1 H 1.008																	2 He 4.003
2	3 Li 6.941	4 Be 9.012											5 B 10.81	6 C 12.01	7 N 14.01	8 O 16.00	9 F 19.00	10 Ne 20.18
3	11 Na 22.99	12 Mg 24.31											13 Al 26.98	14 Si 28.09	15 P 30.97	16 S 32.07	17 Cl 35.45	18 Ar 39.95
4	19 K 39.10	20 Ca 40.08	21 Sc 44.96	22 Ti 47.88	23 V 50.94	24 Cr 52.00	25 Mn 54.94	26 Fe 55.85	27 Co 58.93	28 Ni 58.69	29 Cu 63.55	30 Zn 65.41	31 Ga 69.72	32 Ge 72.61	33 As 74.92	34 Se 78.96	35 Br 79.90	36 Kr 83.80
5	37 Rb 85.47	38 Sr 87.62	39 Y 88.91	40 Zr 91.22	41 Nb 92.91	42 Mo 95.94	43 Tc (98)	44 Ru 101.1	45 Rh 102.9	46 Pd 106.4	47 Ag 107.9	48 Cd 112.4	49 In 114.8	50 Sn 118.7	51 Sb 121.8	52 Te 127.6	53 I 126.9	54 Xe 131.3
6	55 Cs 132.9	56 Ba 137.3	57 La 138.9	72 Hf 178.5	73 Ta 180.9	74 W 183.9	75 Re 186.2	76 Os 190.2	77 Ir 192.2	78 Pt 195.1	79 Au 197.0	80 Hg 200.6	81 Tl 204.4	82 Pb 207.2	83 Bi 209.0	84 Po (209)	85 At (210)	86 Rn (222)
7	87 Fr (223)	88 Ra (226)	89 Ac (227)	104 Rf (263)	105 Db (262)	106 Sg (266)	107 Bh (267)	108 Hs (277)	109 Mt (268)	110 Ds (281)	111 Rg (272)	112 Cn (285)	113 (284)	114 (289)	115 (288)	116 (292)		118 (294)

Period															
6	Lanthanides	58 Ce 140.1	59 Pr 140.9	60 Nd 144.2	61 Pm (145)	62 Sm 150.4	63 Eu 152.0	64 Gd 157.3	65 Tb 158.9	66 Dy 162.5	67 Ho 164.9	68 Er 167.3	69 Tm 168.9	70 Yb 173.0	71 Lu 175.0
7	Actinides	90 Th 232.0	91 Pa (231)	92 U 238.0	93 Np (237)	94 Pu (242)	95 Am (243)	96 Cm (247)	97 Bk (247)	98 Cf (251)	99 Es (252)	100 Fm (257)	101 Md (258)	102 No (259)	103 Lr (260)

Figure 2.9 The modern periodic table. As of early 2011, elements 113–116 and 118 had not been named, and the synthesis of element 117 had not been confirmed. (See also the footnote on p. 45.)

- The **nonmetals** (yellow) lie in the small upper-right portion of the table. They are generally gases or dull, brittle solids at room temperature (bromine is the only liquid) and conduct heat and electricity poorly.
- The **metalloids** (green; also called **semimetals**), which lie along the staircase line, have properties between those of metals and nonmetals.

Two major points to keep in mind:

1. In general, elements in a group have **similar** chemical properties and elements in a period have **different** chemical properties.
2. Despite this classification of three types of elements, in reality, there is a gradation in properties from left to right and top to bottom.

It is important to learn some of the group (family) names. Group 1A(1), except for hydrogen, consists of the *alkali metals,* and Group 2A(2) consists of the *alkaline earth metals.* Both groups consist of highly reactive elements. The *halogens,*

Group 7A(17), are highly reactive nonmetals, whereas the *noble gases,* Group 8A(18), are relatively unreactive nonmetals. Other main groups [3A(13) to 6A(16)] are often named for the first element in the group; for example, Group 6A(16) is the *oxygen family.*

Summary of Section 2.6

- In the periodic table, the elements are arranged by atomic number into horizontal periods and vertical groups.
- Nonmetals appear in the upper-right portion of the table, metalloids lie along a staircase line, and metals fill the rest of the table.
- Elements within a group have similar behavior, whereas elements within a period have dissimilar behavior.

2.7 • COMPOUNDS: INTRODUCTION TO BONDING

Only a few elements occur free in nature. The noble gases—helium (He), neon (Ne), argon (Ar), krypton (Kr), xenon (Xe), and radon (Rn)—occur in air as separate atoms. In addition to occurring in compounds, oxygen (O), nitrogen (N), and sulfur (S) occur in their most common elemental form as the molecules O_2, N_2, and S_8, and carbon (C) occurs in vast, nearly pure deposits of coal. And some metals—copper (Cu), silver (Ag), gold (Au), and platinum (Pt)—are also sometimes found uncombined. But, aside from these few exceptions, *the overwhelming majority of elements occur in compounds, combined with other elements.*

Elements combine in two general ways and both involve *the electrons of the atoms of interacting elements:*

1. *Transferring electrons* from one element to another to form **ionic compounds**
2. *Sharing electrons* between atoms of different elements to form **covalent compounds**

These processes generate **chemical bonds,** the forces that hold the atoms together in a compound. This section introduces compound formation, which we'll discuss in much more detail in later chapters.

The Formation of Ionic Compounds

Ionic compounds are composed of **ions,** charged particles that form when an atom (or small group of atoms) gains or loses one or more electrons. The simplest type of ionic compound is a **binary ionic compound,** one composed of two elements. It typically forms *when a metal reacts with a nonmetal:*

- Each metal atom *loses* one or more electrons and becomes a **cation,** a positively charged ion.
- Each nonmetal atom *gains* one or more of the electrons lost by the metal atom and becomes an **anion,** a negatively charged ion.

In effect, the metal atoms *transfer electrons* to the nonmetal atoms. The resulting large numbers of cations and anions attract each other and form the ionic compound. A cation or anion derived from a single atom is called a **monatomic ion;** we'll discuss polyatomic ions, those derived from a small group of atoms, later.

The Case of Sodium Chloride *All ionic compounds are solid arrays of oppositely charged ions.* The formation of the binary ionic compound sodium chloride, common table salt, from its elements is depicted in Figure 2.10. In the electron transfer, a sodium atom *loses* one electron and forms a sodium cation, Na^+. (The charge on the ion is written as a *right superscript.*) A chlorine atom *gains* the electron and becomes a chloride anion, Cl^-. (The name change when the nonmetal atom becomes an anion is discussed in the next section.) The oppositely charged ions (Na^+ and Cl^-) attract each other, and the similarly charged ions (Na^+ and Na^+, or Cl^- and Cl^-) repel each other. The resulting solid aggregation is a regular array of alternating Na^+ and Cl^- ions that extends in all three dimensions. Even the tiniest visible grain of table salt contains an enormous number of sodium and chloride ions.

Coulomb's Law The strength of the ionic bonding depends to a great extent on the net strength of these attractions and repulsions and is described by *Coulomb's law,* which can be expressed as follows: *the energy of attraction (or repulsion) between two particles is directly proportional to the product of the charges and inversely proportional to the distance between them.*

$$\text{Energy} \propto \frac{\text{charge 1} \times \text{charge 2}}{\text{distance}}$$

In other words, as is summarized in Figure 2.11,

- Ions with higher charges attract (or repel) each other more strongly than ions with lower charges.
- Smaller ions attract (or repel) each other more strongly than larger ions, because their charges are closer together.

Predicting the Number of Electrons Lost or Gained *Ionic compounds are neutral* because they contain equal numbers of positive and negative *charges.* Thus, there are equal numbers of Na^+ and Cl^- ions in sodium chloride, because both ions are singly charged. But there are two Na^+ ions for each oxide ion, O^{2-}, in sodium oxide because two 1+ ions balance one 2− ion.

Can we predict the number of electrons a given atom will lose or gain when it forms an ion? For A-group elements, we usually find that metal atoms lose electrons and nonmetal atoms gain electrons to *form ions with the same number of electrons as in an atom of the nearest noble gas* [Group 8A(18)]. Noble gases have a stability that is related to their number (and arrangement) of electrons. Thus, a sodium

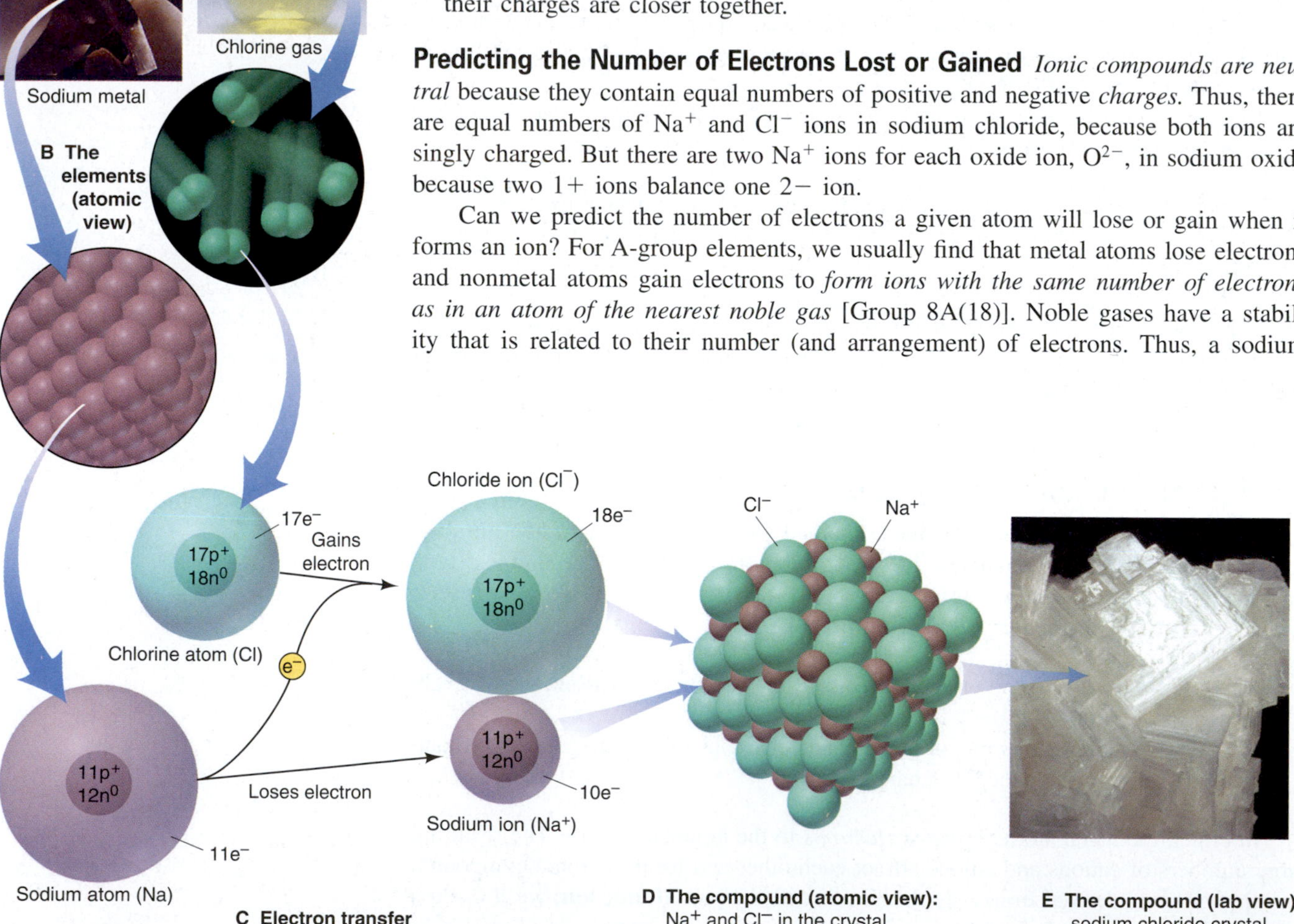

Figure 2.10 The formation of an ionic compound. **A,** The two elements as seen in the laboratory. **B,** The elements on the atomic scale. **C,** The electron transfer from Na atom to Cl atom to form Na^+ and Cl^- ions. **D,** Countless Na^+ and Cl^- ions attract each other and form a regular three-dimensional array. **E,** Crystalline NaCl occurs naturally as the mineral halite.

atom ($11e^-$) can attain the stability of a neon atom ($10e^-$), the nearest noble gas, by losing one electron. Similarly, a chlorine atom ($17e^-$) attains the stability of an argon atom ($18e^-$), its nearest noble gas, by gaining one electron. Thus, in general, when an element located near a noble gas forms a monatomic ion,

- *Metals lose electrons:* elements in Group 1A(1) lose one electron, elements in Group 2A(2) lose two, and aluminum in Group 3A(13) loses three.
- *Nonmetals gain electrons:* elements in Group 7A(17) gain one electron, oxygen and sulfur in Group 6A(16) gain two, and nitrogen in Group 5A(15) gains three.

Figure 2.11 Factors that influence the strength of ionic bonding.

Sample Problem 2.6 Predicting the Ion an Element Forms

Problem What monatomic ions do the following elements form?
(a) Iodine ($Z = 53$) **(b)** Calcium ($Z = 20$) **(c)** Aluminum ($Z = 13$)

Plan We use the given Z value to find the element in the periodic table and see where its group lies relative to the noble gases. Elements in Groups 1A, 2A, and 3A *lose* electrons to attain the same number as the nearest noble gas and become positive ions; those in Groups 5A, 6A, and 7A *gain* electrons and become negative ions.

Solution (a) I^- Iodine ($_{53}I$) is in Group 7A(17), the halogens. Like any member of this group, it gains 1 electron to attain the same number as the nearest Group 8A(18) member, in this case, $_{54}Xe$.
(b) Ca^{2+} Calcium ($_{20}Ca$) is in Group 2A(2), the alkaline earth metals. Like any Group 2A member, it loses 2 electrons to attain the same number as the nearest noble gas, $_{18}Ar$.
(c) Al^{3+} Aluminum ($_{13}Al$) is a metal in the boron family [Group 3A(13)] and thus loses 3 electrons to attain the same number as its nearest noble gas, $_{10}Ne$.

FOLLOW-UP PROBLEM 2.6 What monatomic ion is formed from **(a)** $_{16}S$; **(b)** $_{37}Rb$; **(c)** $_{56}Ba$?

The Formation of Covalent Compounds

Covalent compounds form when elements share electrons, which usually occurs between nonmetals. The simplest case of electron sharing occurs not in a compound but between two hydrogen atoms (H; $Z = 1$). Imagine two separated H atoms approaching each other (Figure 2.12). As they get closer, the nucleus of each atom attracts the electron of the other atom more strongly, but repulsions between the nuclei and between the electrons are still weak. As the separated atoms begin to interpenetrate each other, these repulsions increase. At some optimum distance between the nuclei, attractions balance repulsions, and the two atoms form a **covalent bond,** a pair of electrons mutually attracted by the two nuclei. The result is a hydrogen molecule, in which each electron no longer "belongs" to a particular H atom: the two electrons are *shared* by the two nuclei. A sample of hydrogen gas consists of these diatomic molecules (H_2)—pairs of atoms that are chemically bound and behave as an independent unit—*not* separate H atoms. Figure 2.13 shows other nonmetals that exist as molecules at room temperature.

Atoms far apart: *No interactions.*

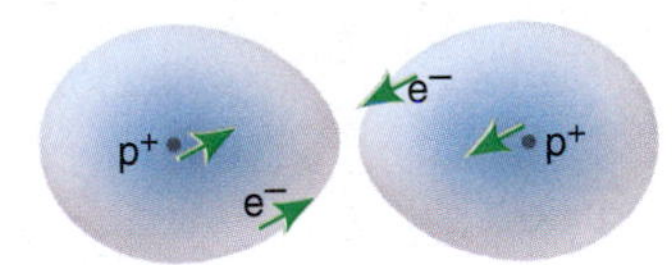

Atoms closer: *Attractions (green arrows) between nucleus of one atom and electron of the other increase. Repulsions between nuclei and between electrons are very weak.*

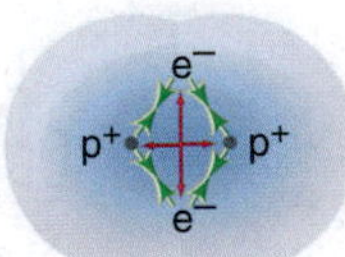

Optimum distance: *H_2 molecule forms because attractions (green arrows) balance repulsions (red arrows).*

Figure 2.12 Formation of a covalent bond between two H atoms.

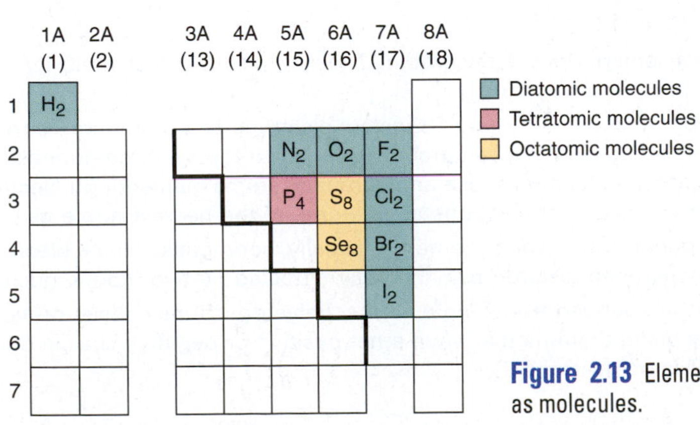

Figure 2.13 Elements that occur as molecules.

Atoms of different elements share electrons to form the molecules of a covalent compound. A sample of hydrogen fluoride, for example, consists of molecules in which one H atom forms a covalent bond with one F atom; water consists of molecules in which one O atom forms covalent bonds with two H atoms:

Hydrogen fluoride, HF

Water, H_2O

Distinguishing the Entities in Covalent and Ionic Substances There is a key distinction between the chemical entities in covalent substances and in ionic substances. *Most covalent substances consist of molecules.* A cup of water, for example, consists of individual water molecules lying near each other. In contrast, under ordinary conditions, *there are no molecules in an ionic compound.* A piece of sodium chloride, for example, is a continuous array in three dimensions of oppositely charged sodium and chloride ions, *not* a collection of individual sodium chloride "molecules."

Another key distinction between covalent and ionic substances concerns the nature of the particles attracting each other. Covalent bonding involves the mutual attraction between two (positively charged) nuclei and the two (negatively charged) electrons that reside between them. Ionic bonding involves the mutual attraction among positive and negative ions.

Polyatomic Ions: Covalent Bonds Within Ions Many ionic compounds contain **polyatomic ions,** which consist of two or more atoms bonded *covalently* and have a net positive or negative charge. For example, Figure 2.14 shows that a crystalline form of calcium carbonate *(left)* occurs on the atomic scale as an array *(center)* of polyatomic carbonate anions and monatomic calcium cations. The carbonate ion *(right)* consists of a carbon atom covalently bonded to three oxygen atoms, and two additional electrons give the ion its 2− charge. In many reactions, the polyatomic ion stays together as a unit.

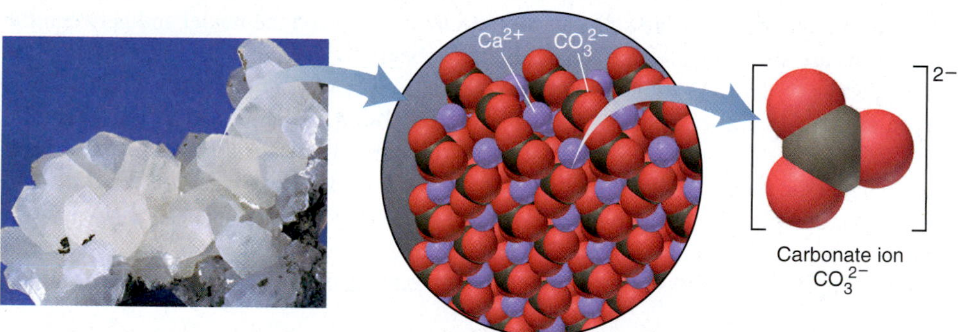

Figure 2.14 The carbonate ion in calcium carbonate.

Summary of Section 2.7

- Although a few elements occur uncombined in nature, the great majority exist in compounds.
- Ionic compounds form when a metal *transfers* electrons to a nonmetal, and the resulting positive and negative ions attract each other to form a three-dimensional array. In many cases, metal atoms lose and nonmetal atoms gain enough electrons to attain the same number of electrons as in atoms of the nearest noble gas.
- Covalent compounds form when elements, usually nonmetals, *share* electrons. Each covalent bond is an electron pair mutually attracted by two atomic nuclei.
- Monatomic ions are derived from single atoms. Polyatomic ions consist of two or more covalently bonded atoms that have a net positive or negative charge due to a deficit or excess of electrons.

2.8 • FORMULAS, NAMES, AND MASSES OF COMPOUNDS

In a **chemical formula,** element symbols and, often, numerical subscripts show the type and number of each atom in the smallest unit of the substance. In this section, you'll learn how to write the names and formulas of ionic and simple covalent compounds, how to calculate the mass of a compound from its formula, and how to visualize molecules with three-dimensional models.

Binary Ionic Compounds

Let's begin with two general rules:

- For *all* ionic compounds, *names and formulas give the positive ion (cation) first and the negative ion (anion) second.*
- For all *binary* ionic compounds, *the name of the cation is the name of the metal, and the name of the anion has the suffix -ide added to the root of the name of the nonmetal.*

For example, the anion formed from brom*ine* is named brom*ide* (brom+ide). Therefore, the compound formed from the metal calcium and the nonmetal bromine is named *calcium bromide.*

In general, if the metal of a binary ionic compound is a main-group element (A groups) it usually forms a single type of ion; if it is a transition element (B groups), it often forms more than one. We discuss each case in turn.

Compounds of Elements That Form One Ion

The periodic table presents some key points about the formulas of main-group monatomic ions:

- Monatomic ions of main-group elements have the same ionic charge; the alkali metals—Li, Na, K, Rb, Cs, and Fr—form ions with a 1+ charge; the halogens—F, Cl, Br, and I—form ions with a 1− charge; and so forth.
- For cations, ion charge equals A-group number: Na is in Group 1A and forms Na^+, Ba is in Group 2A and forms Ba^{2+}. (Exceptions in Figure 2.15 are Sn^{2+} and Pb^{2+}.)
- For anions, ion charge equals A-group number minus 8; for example, S is in Group 6A ($6 - 8 = -2$) and thus forms S^{2-}.

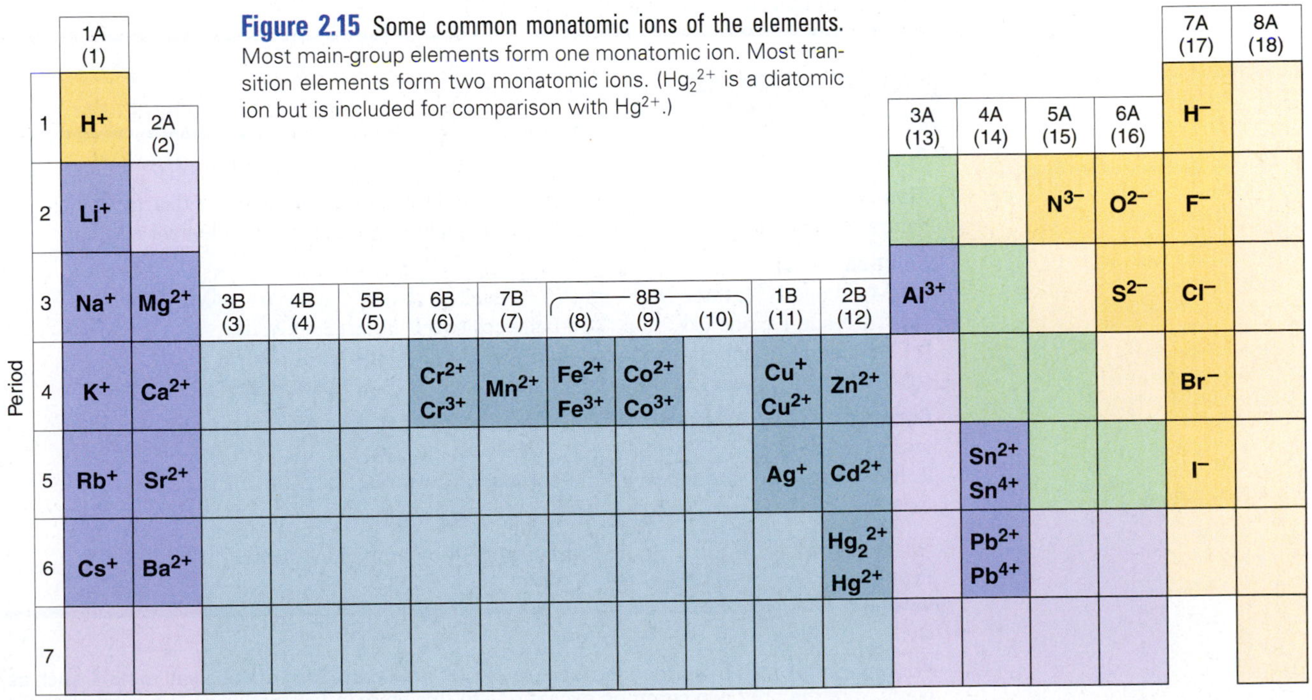

Figure 2.15 Some common monatomic ions of the elements. Most main-group elements form one monatomic ion. Most transition elements form two monatomic ions. (Hg_2^{2+} is a diatomic ion but is included for comparison with Hg^{2+}.)

Period	1A (1)	2A (2)	3B (3)	4B (4)	5B (5)	6B (6)	7B (7)	8B (8)	8B (9)	8B (10)	1B (11)	2B (12)	3A (13)	4A (14)	5A (15)	6A (16)	7A (17)	8A (18)
1	H^+																H^-	
2	Li^+														N^{3-}	O^{2-}	F^-	
3	Na^+	Mg^{2+}											Al^{3+}			S^{2-}	Cl^-	
4	K^+	Ca^{2+}				Cr^{2+} Cr^{3+}	Mn^{2+}	Fe^{2+} Fe^{3+}	Co^{2+} Co^{3+}		Cu^+ Cu^{2+}	Zn^{2+}					Br^-	
5	Rb^+	Sr^{2+}									Ag^+	Cd^{2+}		Sn^{2+} Sn^{4+}			I^-	
6	Cs^+	Ba^{2+}										Hg_2^{2+} Hg^{2+}		Pb^{2+} Pb^{4+}				
7																		

Table 2.3 Common Monatomic Ions*

Charge	Formula	Name
Cations		
1+	H^+	hydrogen
	Li^+	**lithium**
	Na^+	**sodium**
	K^+	**potassium**
	Cs^+	cesium
	Ag^+	**silver**
2+	Mg^{2+}	**magnesium**
	Ca^{2+}	**calcium**
	Sr^{2+}	strontium
	Ba^{2+}	**barium**
	Zn^{2+}	**zinc**
	Cd^{2+}	cadmium
3+	Al^{3+}	aluminum
Anions		
1−	H^-	hydride
	F^-	**fluoride**
	Cl^-	**chloride**
	Br^-	**bromide**
	I^-	**iodide**
2−	O^{2-}	**oxide**
	S^{2-}	**sulfide**
3−	N^{3-}	nitride

*Those in **boldface** are most common.

Try to memorize the A-group monatomic ions in Table 2.3 (all of these listed except Ag^+, Zn^{2+}, and Cd^{2+}) according to their positions in Figure 2.15. These ions have *the same number of electrons as an atom of the nearest noble gas.*

Because an ionic compound consists of an array of ions rather than separate molecules, its formula represents the **formula unit,** the *relative* numbers of cations and anions in the compound. The compound has zero net charge, so the positive charges of the cations balance the negative charges of the anions. For example, calcium bromide is composed of Ca^{2+} ions and Br^- ions, so two Br^- balance each Ca^{2+}. The formula is $CaBr_2$, not Ca_2Br. In this and all other formulas,

- The subscript refers to the element *preceding* it.
- The *subscript 1 is understood* from the presence of the element symbol alone (that is, we do not write Ca_1Br_2).
- The charge (without the sign) of one ion becomes the subscript of the other:

$$Ca^{2+} \quad Br^{1-} \qquad \text{gives} \qquad Ca_1Br_2 \qquad \text{or} \qquad CaBr_2$$

- Reduce the subscripts to the smallest whole numbers that retain the ratio of ions. Thus, for example, for the Ca^{2+} and O^{2-} ions in calcium oxide, we get Ca_2O_2, which we reduce to the formula CaO.*

Sample Problem 2.7 Naming Binary Ionic Compounds

Problem Name the ionic compound formed from the following pairs of elements: **(a)** magnesium and nitrogen; **(b)** iodine and cadmium; **(c)** strontium and fluorine; **(d)** sulfur and cesium.

Plan The key to naming a binary ionic compound is to recognize which element is the metal and which is the nonmetal. When in doubt, check the periodic table. We place the cation name first, add the suffix *-ide* to the nonmetal root, and place the anion name last.

Solution **(a)** Magnes*ium* is the metal; *nitr-* is the nonmetal root: magnesium nitride
(b) Cadm*ium* is the metal; *iod-* is the nonmetal root: cadmium iodide
(c) Stront*ium* is the metal; *fluor-* is the nonmetal root: strontium fluoride (Note the spelling is fl*uo*ride, not fl*ou*ride.)
(d) Ces*ium* is the metal; *sulf-* is the nonmetal root: cesium sulfide

FOLLOW-UP PROBLEM 2.7 For the following ionic compounds, give the name and periodic table group number of each element present: **(a)** zinc oxide; **(b)** silver bromide; **(c)** lithium chloride; **(d)** aluminum sulfide.

Sample Problem 2.8 Determining Formulas of Binary Ionic Compounds

Problem Write formulas for the compounds named in Sample Problem 2.7.

Plan We write the formula by finding the smallest number of each ion that gives the neutral compound. These numbers appear as *right subscripts* to the element symbol.

Solution
(a) Mg^{2+} and N^{3-}; three Mg^{2+} ions (6+) balance two N^{3-} ions (6−): Mg_3N_2
(b) Cd^{2+} and I^-; one Cd^{2+} ion (2+) balances two I^- ions (2−): CdI_2
(c) Sr^{2+} and F^-; one Sr^{2+} ion (2+) balances two F^- ions (2−): SrF_2
(d) Cs^+ and S^{2-}; two Cs^+ ions (2+) balance one S^{2-} ion (2−): Cs_2S

Comment **1.** The subscript 1 is understood and so not written; thus, in (b), we do *not* write Cd_1I_2.
2. Ion charges do *not* appear in the compound formula; thus, in (c), we do *not* write $Sr^{2+}F_2^{\,-}$.

FOLLOW-UP PROBLEM 2.8 Write the formulas of the compounds named in Follow-up Problem 2.7.

*Compounds of the mercury(I) ion, such as Hg_2Cl_2, and peroxides of the alkali metals, such as Na_2O_2, are the only two common exceptions to this step; in fact, reducing the subscripts for these compounds would give the incorrect formulas HgCl and NaO.

Table 2.4 Some Metals That Form More Than One Monatomic Ion*

Element	Ion Formula	Systematic Name	Common (Trivial) Name
Chromium	Cr^{2+}	chromium(II)	chromous
	Cr^{3+}	**chromium(III)**	chromic
Cobalt	Co^{2+}	cobalt(II)	
	Co^{3+}	cobalt(III)	
Copper	**Cu^{+}**	**copper(I)**	cuprous
	Cu^{2+}	**copper(II)**	cupric
Iron	**Fe^{2+}**	**iron(II)**	ferrous
	Fe^{3+}	**iron(III)**	ferric
Lead	**Pb^{2+}**	**lead(II)**	
	Pb^{4+}	lead(IV)	
Mercury	Hg_2^{2+}	mercury(I)	mercurous
	Hg^{2+}	**mercury(II)**	mercuric
Tin	**Sn^{2+}**	**tin(II)**	stannous
	Sn^{4+}	tin(IV)	stannic

*Listed alphabetically by metal name; the ions in **boldface** are most common.

Compounds with Metals That Form More Than One Ion As noted earlier, many metals, particularly the transition elements (B groups), can form more than one ion. Table 2.4 lists some examples; see Figure 2.15 for their placement in the periodic table. Names of compounds containing these elements include a *Roman numeral within parentheses* immediately after the metal ion's name to indicate its ionic charge. For example, iron can form Fe^{2+} and Fe^{3+} ions. The two compounds that iron forms with chlorine are $FeCl_2$, named iron(II) chloride (spoken "iron two chloride"), and $FeCl_3$, named iron(III) chloride.

We are focusing here on systematic names, but some common (trivial) names are still used. In common names for certain metal ions, the Latin root of the metal is followed by either of two suffixes (see Table 2.4):

- The suffix *-ous* for the ion with the lower charge
- The suffix *-ic* for the ion with the higher charge

Thus, iron(II) chloride is also called ferr*ous* chloride and iron(III) chloride is ferr*ic* chloride. (Memory aid: there is an ***o*** in *-ous* and *lower,* and an ***i*** in *-ic* and *higher.*)

Sample Problem 2.9 Determining Names and Formulas of Ionic Compounds of Metals That Form More Than One Ion

Problem Give the systematic names for the formulas or the formulas for the names of each compound: **(a)** tin(II) fluoride; **(b)** CrI_3; **(c)** ferric oxide; **(d)** CoS.

Solution **(a)** Tin(II) ion is Sn^{2+}; fluoride is F^-. Two F^- ions balance one Sn^{2+} ion: tin(II) fluoride is SnF_2. (The common name is stannous fluoride.)
(b) The anion is I^-, iodide, and the formula shows three I^-. Therefore, the cation must be Cr^{3+}, chromium(III) ion: CrI_3 is chromium(III) iodide. (The common name is chromic iodide.)
(c) *Ferric* is the common name for iron(III) ion, Fe^{3+}; oxide ion is O^{2-}. To balance the charges, the formula is Fe_2O_3. [The systematic name is iron(III) oxide.]
(d) The anion is sulfide, S^{2-}, which requires that the cation be Co^{2+}. The name is cobalt(II) sulfide.

FOLLOW-UP PROBLEM 2.9 Give the systematic names for the formulas or the formulas for the names of each compound: **(a)** lead(IV) oxide; **(b)** Cu_2S; **(c)** $FeBr_2$; **(d)** mercuric chloride.

Table 2.5 Common Polyatomic Ions*

Formula	Name
Cations	
NH_4^+	ammonium
H_3O^+	hydronium
Anions	
CH_3COO^- (or $C_2H_3O_2^-$)	acetate
CN^-	cyanide
OH^-	hydroxide
ClO^-	hypochlorite
ClO_2^-	chlorite
ClO_3^-	chlorate
ClO_4^-	perchlorate
NO_2^-	nitrite
NO_3^-	nitrate
MnO_4^-	permanganate
CO_3^{2-}	carbonate
HCO_3^-	hydrogen carbonate (or bicarbonate)
CrO_4^{2-}	chromate
$Cr_2O_7^{2-}$	dichromate
O_2^{2-}	peroxide
PO_4^{3-}	phosphate
HPO_4^{2-}	hydrogen phosphate
$H_2PO_4^-$	dihydrogen phosphate
SO_3^{2-}	sulfite
SO_4^{2-}	sulfate
HSO_4^-	hydrogen sulfate (or bisulfate)

*Boldface ions are most common.

No. of O atoms ↑	Prefix	Root	Suffix
	per	*root*	ate
		root	ate
		root	ite
	hypo	*root*	ite

Figure 2.16 Naming oxoanions. Prefixes and suffixes indicate the number of O atoms in the anion.

Compounds That Contain Polyatomic Ions

Many ionic compounds contain polyatomic ions. Table 2.5 shows some common polyatomic ions. Remember that *the polyatomic ion stays together as a charged unit.* The formula for potassium nitrate is KNO_3: each K^+ balances one NO_3^-. The formula for sodium carbonate is Na_2CO_3: two Na^+ balance one CO_3^{2-}. *When two or more of the same polyatomic ion are present in the formula unit, that ion appears in parentheses with the subscript written outside.* For example, calcium nitrate contains one Ca^{2+} and two NO_3^- ions and has the formula $Ca(NO_3)_2$. Parentheses and a subscript are *only* used if *more than one* of a given polyatomic ion is present; thus, sodium nitrate is $NaNO_3$, *not* $Na(NO_3)$.

Families of Oxoanions As Table 2.5 shows, most polyatomic ions are **oxoanions** (or *oxyanions*), those in which an element, usually a nonmetal, is bonded to one or more oxygen atoms. There are several families of two or four oxoanions that differ only in the number of oxygen atoms. The following simple naming conventions are used with these ions.

With two oxoanions in the family:

- The ion with *more* O atoms takes the nonmetal root and the suffix *-ate*.
- The ion with *fewer* O atoms takes the nonmetal root and the suffix *-ite*.

For example, SO_4^{2-} is the sulf*ate* ion, and SO_3^{2-} is the sulf*ite* ion; similarly, NO_3^- is nitr*ate*, and NO_2^- is nitr*ite*.

With four oxoanions in the family (a halogen bonded to O) (Figure 2.16):

- The ion with *most* O atoms has the prefix *per-*, the nonmetal root, and the suffix *-ate*.
- The ion with *one fewer* O atom has just the root and the suffix *-ate*.
- The ion with *two fewer* O atoms has just the root and the suffix *-ite*.
- The ion with *least (three fewer)* O atoms has the prefix *hypo-*, the root, and the suffix *-ite*.

For example, for the four chlorine oxoanions,

ClO_4^- is *per*chlor*ate*, ClO_3^- is chlor*ate*, ClO_2^- is chlor*ite*, ClO^- is *hypo*chlor*ite*

Hydrated Ionic Compounds Ionic compounds called **hydrates** have a specific number of water molecules in each formula unit, which is shown after a centered dot in the formula and noted in the name by a Greek numerical prefix before the word *hydrate.* Table 2.6 shows these prefixes. For example, Epsom salt has seven water molecules in each formula unit: the formula is $MgSO_4{\cdot}7H_2O$, and the name is magnesium sulfate *hepta*hydrate. Similarly, the mineral gypsum has the formula $CaSO_4{\cdot}2H_2O$ and the name calcium sulfate *di*hydrate. The water molecules, referred to as "waters of hydration," are part of the hydrate's structure. Heating can remove some or all of them, leading to a different substance. For example, when heated strongly, blue copper(II) sulfate pentahydrate ($CuSO_4{\cdot}5H_2O$) is converted to white copper(II) sulfate ($CuSO_4$).

Sample Problem 2.10 Determining Names and Formulas of Ionic Compounds Containing Polyatomic Ions

Problem Give the systematic names for the formulas or the formulas for the names of the following compounds: **(a)** $Fe(ClO_4)_2$; **(b)** sodium sulfite; **(c)** $Ba(OH)_2{\cdot}8H_2O$.

Solution **(a)** ClO_4^- is perchlorate, which has a 1− charge, so the cation must be Fe^{2+}. The name is iron(II) perchlorate. (The common name is ferrous perchlorate.)

(b) Sodium is Na^+; sulfite is SO_3^{2-}, and two Na^+ ions balance one SO_3^{2-} ion. The formula is Na_2SO_3.

(c) Ba^{2+} is barium; OH^- is hydroxide. There are eight *(octa-)* water molecules in each formula unit. The name is barium hydroxide octahydrate.

FOLLOW-UP PROBLEM 2.10 Give the systematic names for the formulas or the formulas for the names of the following compounds: **(a)** cupric nitrate trihydrate; **(b)** zinc hydroxide; **(c)** LiCN.

Sample Problem 2.11 Recognizing Incorrect Names and Formulas of Ionic Compounds

Problem Explain what is wrong with the name or formula at the end of each statement, and correct it:
(a) $Ba(C_2H_3O_2)_2$ is called barium diacetate.
(b) Sodium sulfide has the formula $(Na)_2SO_3$.
(c) Iron(II) sulfate has the formula $Fe_2(SO_4)_3$.
(d) Cesium carbonate has the formula $Cs_2(CO_3)$.

Solution **(a)** The charge of the Ba^{2+} ion *must* be balanced by *two* $C_2H_3O_2^-$ ions, so the prefix *di-* is unnecessary. For ionic compounds, we do not indicate the number of ions with numerical prefixes. The correct name is barium acetate.
(b) Two mistakes occur here. The sodium ion is monatomic, so it does *not* require parentheses. The sulfide ion is S^{2-}, *not* SO_3^{2-} (which is sulfite). The correct formula is Na_2S.
(c) The Roman numeral refers to the charge of the ion, *not* the number of ions in the formula. Fe^{2+} is the cation, so it requires one SO_4^{2-} to balance its charge. The correct formula is $FeSO_4$. [$Fe_2(SO_4)_3$ is the formula for iron(III) sulfate.]
(d) Parentheses are *not* required when only one polyatomic ion of a kind is present. The correct formula is Cs_2CO_3.

FOLLOW-UP PROBLEM 2.11 State why the formula or name at the end of each statement is incorrect, and correct it:
(a) Ammonium phosphate is $(NH_3)_4PO_4$.
(b) Aluminum hydroxide is $AlOH_3$.
(c) $Mg(HCO_3)_2$ is manganese(II) carbonate.
(d) $Cr(NO_3)_3$ is chromic(III) nitride.
(e) $Ca(NO_2)_2$ is cadmium nitrate.

Table 2.6 Numerical Prefixes* for Ionic Hydrates and Binary Covalent Compounds

Number	Prefix
1	mono-
2	di-
3	tri-
4	tetra-
5	penta-
6	hexa-
7	hepta-
8	octa-
9	nona-
10	deca-

*It is common practice to drop the final "a" in names of oxides, for example, tetroxide, *not* tetraoxide.

Acid Names from Anion Names

Acids are an important group of hydrogen-containing compounds that have been used in chemical reactions for many centuries. In the laboratory, acids are typically used in water solution. When naming them and writing their formulas, we consider acids as anions that are connected to the number of hydrogen ions (H^+) needed for charge neutrality. The two common types of acids are binary acids and oxoacids:

1. *Binary acid* solutions form when certain gaseous compounds dissolve in water. For example, when gaseous hydrogen chloride (HCl) dissolves in water, it forms hydrochloric acid, that is,

   Prefix *hydro-* + nonmetal *root* + suffix *-ic* + separate word *acid*
   hydro + chlor + ic + acid

   or *hydrochloric acid.* This naming pattern holds for many compounds in which hydrogen combines with an anion that has an *-ide* suffix.

2. *Oxoacid* names are similar to those of the oxoanions, except for two suffix changes:
   - *-ate* in the anion becomes *-ic* in the acid
   - *-ite* in the anion becomes *-ous* in the acid

   The oxoanion prefixes *hypo-* and *per-* are retained. Thus,

   BrO_4^- is *per*brom*ate*, and $HBrO_4$ is *per*brom*ic* acid.
   IO_2^- is iod*ite*, and HIO_2 is iod*ous* acid.

Sample Problem 2.12 Determining Names and Formulas of Anions and Acids

Problem Name each of the following anions and give the name and formula of the acid derived from it: **(a)** Br^-; **(b)** IO_3^-; **(c)** CN^-; **(d)** SO_4^{2-}; **(e)** NO_2^-.

Solution **(a)** The anion is bromide; the acid is hydrobromic acid, HBr.
(b) The anion is iodate; the acid is iodic acid, HIO_3.
(c) The anion is cyanide; the acid is hydrocyanic acid, HCN.
(d) The anion is sulfate; the acid is sulfuric acid, H_2SO_4. (In this case, the suffix is added to the element name *sulfur,* not to the root, *sulf-*.)
(e) The anion is nitrite; the acid is nitrous acid, HNO_2.

Comment We must add *two* H^+ ions to the sulfate ion to obtain sulfuric acid because SO_4^{2-} has a 2− charge.

FOLLOW-UP PROBLEM 2.12 Write the formula for the name or name for the formula of each acid: **(a)** chloric acid; **(b)** HF; **(c)** acetic acid; **(d)** sulfurous acid; **(e)** HBrO.

Binary Covalent Compounds

Binary covalent compounds are typically formed by the combination of two nonmetals. Some are so familiar that we use their common names, such as ammonia (NH_3), methane (CH_4), and water (H_2O), but most are named systematically:

- The element with the lower group number in the periodic table comes first in the name. The element with the higher group number comes second and is named with its root and the suffix *-ide*. (*Exception:* When the compound contains oxygen and any of the halogens chlorine, bromine, or iodine, the halogen is named first.)
- If both elements are in the same group, the one with the higher period number is named first.
- Covalent compounds use Greek numerical prefixes (see Table 2.6) to indicate the number of atoms of each element. The first element in the name has a prefix *only* when more than one atom of it is present; the second element *usually* has a prefix.

Sample Problem 2.13 Determining Names and Formulas of Binary Covalent Compounds

Problem **(a)** What is the formula of carbon disulfide?
(b) What is the name of PCl_5?
(c) Give the name and formula of the compound whose molecules each consist of two N atoms and four O atoms.

Solution **(a)** The prefix *di-* means "two." The formula is CS_2.
(b) P is the symbol for phosphorus; there are five chlorine atoms, which is indicated by the prefix *penta-*. The name is phosphorus pentachloride.
(c) Nitrogen (N) comes first in the name (lower group number). The compound is dinitrogen tetroxide, N_2O_4.

FOLLOW-UP PROBLEM 2.13 Give the name or formula for **(a)** SO_3; **(b)** SiO_2; **(c)** dinitrogen monoxide; **(d)** selenium hexafluoride.

Sample Problem 2.14 Recognizing Incorrect Names and Formulas of Binary Covalent Compounds

Problem Explain what is wrong with the name or formula at the end of each statement, and correct it: **(a)** SF_4 is monosulfur pentafluoride. **(b)** Dichlorine heptoxide is Cl_2O_6. **(c)** N_2O_3 is dinitrotrioxide.

Solution **(a)** There are two mistakes. *Mono-* is not needed if there is only one atom of the first element, and the prefix for four is *tetra-*, not *penta-*. The correct name is sulfur tetrafluoride.
(b) The prefix *hepta-* indicates seven, not six. The correct formula is Cl_2O_7.
(c) The full name of the first element is needed, and a space separates the two element names. The correct name is dinitrogen trioxide.

FOLLOW-UP PROBLEM 2.14 Explain what is wrong with the name or formula at the end of each statement, and correct it: **(a)** S_2Cl_2 is disulfurous dichloride. **(b)** Nitrogen monoxide is N_2O. **(c)** $BrCl_3$ is trichlorine bromide.

The Simplest Organic Compounds: Straight-Chain Alkanes

Organic compounds typically have complex structures that consist of chains, branches, and/or rings of carbon atoms bonded to hydrogen atoms and, often, to atoms of oxygen, nitrogen, and a few other elements. At this point, we'll lay the groundwork for naming organic compounds by focusing on the simplest ones. Rules for naming more complex ones are detailed in Chapter 15.

Hydrocarbons, the simplest type of organic compound, contain *only* carbon and hydrogen. *Alkanes* are the simplest type of hydrocarbon; many function as important fuels, such as methane, propane, butane, and the mixture that makes up gasoline. The simplest alkanes to name are the *straight-chain alkanes* because the carbon chains have no branches. Alkanes are named with a *root,* based on the number of C atoms in the chain, followed by the suffix *-ane.* Table 2.7 gives the names, molecular formulas, and space-filling models (discussed shortly) of the first 10 straight-chain alkanes. Note that the roots of the four smallest ones are new, but those for the larger ones are the same as the Greek prefixes shown in Table 2.6.

Table 2.7 The First 10 Straight-Chain Alkanes

Name (Formula)	Model
Methane (CH_4)	
Ethane (C_2H_6)	
Propane (C_3H_8)	
Butane (C_4H_{10})	
Pentane (C_5H_{12})	
Hexane (C_6H_{14})	
Heptane (C_7H_{16})	
Octane (C_8H_{18})	
Nonane (C_9H_{20})	
Decane ($C_{10}H_{22}$)	

Molecular Masses from Chemical Formulas

In Section 2.5, we calculated the atomic mass of an element. Using the periodic table and the formula of a compound, we calculate the **molecular mass** (also called *molecular weight*) of a formula unit of the compound as the sum of the atomic masses:

$$\text{Molecular mass} = \text{sum of atomic masses} \tag{2.3}$$

The molecular mass of a water molecule (using atomic masses to four significant figures from the periodic table) is

$$\begin{aligned}\text{Molecular mass of } H_2O &= (2 \times \text{atomic mass of H}) + (1 \times \text{atomic mass of O})\\ &= (2 \times 1.008 \text{ amu}) + 16.00 \text{ amu} = 18.02 \text{ amu}\end{aligned}$$

Ionic compounds don't consist of molecules, so the mass of a formula unit is termed the **formula mass** instead of *molecular mass.* To calculate the formula mass of a compound with a polyatomic ion, *the number of atoms of each element inside the parentheses is multiplied by the subscript outside the parentheses.* For barium nitrate, $Ba(NO_3)_2$,

$$\begin{aligned}&\text{Formula mass of } Ba(NO_3)_2\\ &= (1 \times \text{atomic mass of Ba}) + (2 \times \text{atomic mass of N}) + (6 \times \text{atomic mass of O})\\ &= 137.3 \text{ amu} + (2 \times 14.01 \text{ amu}) + (6 \times 16.00 \text{ amu}) = 261.3 \text{ amu}\end{aligned}$$

We can use atomic masses, not ionic masses, because electron loss equals electron gain, so electron mass is balanced. In the next two sample problems, the name or molecular depiction is used to find a compound's molecular or formula mass.

Sample Problem 2.15 Calculating the Molecular Mass of a Compound

Problem Using the periodic table, calculate the molecular (or formula) mass of **(a)** tetraphosphorus trisulfide; **(b)** ammonium nitrate.

Plan We first write the formula, then multiply the number of atoms (or ions) of each element by its atomic mass (from the periodic table), and find the sum.

Solution **(a)** The formula is P_4S_3.

$$\begin{aligned}\text{Molecular mass} &= (4 \times \text{atomic mass of P}) + (3 \times \text{atomic mass of S})\\ &= (4 \times 30.97 \text{ amu}) + (3 \times 32.07 \text{ amu}) = 220.09 \text{ amu}\end{aligned}$$

(b) The formula is NH_4NO_3. We count the total number of N atoms even though they belong to different ions:

$$\begin{aligned}&\text{Formula mass}\\ &= (2 \times \text{atomic mass of N}) + (4 \times \text{atomic mass of H}) + (3 \times \text{atomic mass of O})\\ &= (2 \times 14.01 \text{ amu}) + (4 \times 1.007 \text{ amu}) + (3 \times 16.00 \text{ amu}) = 80.05 \text{ amu}\end{aligned}$$

Check You can often find large errors by rounding atomic masses to the nearest 5 and adding: **(a)** $(4 \times 30) + (3 \times 30) = 210 \approx 220.09$. The sum has two decimal places because the atomic masses have two. **(b)** $(2 \times 15) + 4 + (3 \times 15) = 79 \approx 80.05$.

FOLLOW-UP PROBLEM 2.15 What is the molecular (or formula) mass of **(a)** hydrogen peroxide; **(b)** cesium chloride; **(c)** sulfuric acid; **(d)** potassium sulfate?

Sample Problem 2.16 Using Molecular Depictions to Determine Formula, Name, and Mass

Problem Each scene represents a binary compound. Determine its formula, name, and molecular (or formula) mass.

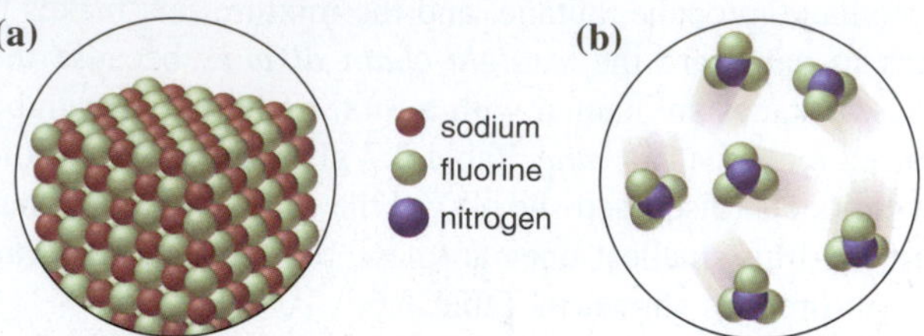

Plan Each of the compounds contains only two elements, so to find the formula, we find the simplest whole-number ratio of one atom to the other. From the formula, we determine the name and the molecular (or formula) mass.

Solution **(a)** There is one brown sphere (sodium) for each green sphere (fluorine), so the formula is NaF. A metal and nonmetal form an ionic compound, in which the metal is named first: sodium fluoride.

$$\text{Formula mass} = (1 \times \text{atomic mass of Na}) + (1 \times \text{atomic mass of F})$$
$$= 22.99 \text{ amu} + 19.00 \text{ amu} = 41.99 \text{ amu}$$

(b) There are three green spheres (fluorine) for each blue sphere (nitrogen), so the formula is NF_3. Two nonmetals form a covalent compound. Nitrogen has a lower group number, so it is named first: nitrogen trifluoride.

$$\text{Molecular mass} = (1 \times \text{atomic mass of N}) + (3 \times \text{atomic mass of F})$$
$$= 14.01 \text{ amu} + (3 \times 19.00) = 71.01 \text{ amu}$$

Check **(a)** For binary ionic compounds, we predict ionic charges from the periodic table. Na forms a 1+ ion, and F forms a 1− ion, so the charges balance with one Na^+ per F^-. Also, ionic compounds are solids, consistent with the picture. **(b)** Covalent compounds often occur as individual molecules, as in the picture. Rounding in (a) gives 25 + 20 = 45; in (b), we get 15 + (3 × 20) = 75, so there are no large errors.

FOLLOW-UP PROBLEM 2.16 Each scene represents a binary compound. Determine its name, formula, and molecular (or formula) mass.

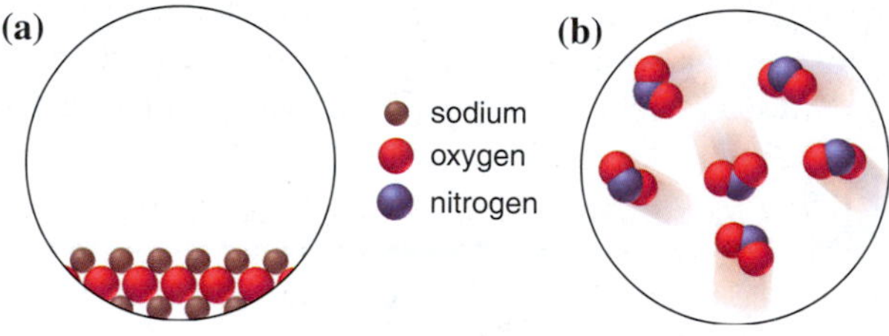

Representing Molecules with Formulas and Models

In order to represent objects too small to see, chemists employ a variety of formulas and models. Each conveys different information, as shown for water below:

- A **molecular formula** uses element symbols and, often, numerical subscripts to give the *actual* number of atoms of each element in a molecule of the compound. (Recall that, for ionic compounds, the *formula unit* gives the *relative* number of each type of ion.) The molecular formula of water is H_2O: there are two H atoms and one O atom in each molecule.

  H_2O

- A **structural formula** shows the relative placement and connections of atoms in the molecule. It uses symbols for the atoms *and* either a pair of dots *(electron-dot formula)* or a line *(bond-line formula)* to show the electron pairs in bonds between the atoms. In water, each H atom is bonded to the O atom, but not to the other H atom.

  H:O:H

  H–O–H

In models, colored balls represent atoms *(see margin)*.

- A *ball-and-stick model* shows atoms as balls and bonds as sticks, and the angles between the bonds are accurate. Note that water is a bent molecule (with a bond angle of 104.5°). This type of model exaggerates the distance between bonded atoms.
- A *space-filling model* is an accurately scaled-up image of the molecule, but bonds are not shown, and it can be difficult to see each atom in a complex molecule.

Summary of Section 2.8

- An ionic compound is named with cation first and anion second. For metals that can form more than one ion, the charge is shown with a Roman numeral.
- Oxoanions have suffixes, and sometimes prefixes, attached to the root of the element name to indicate the number of oxygen atoms.
- Names of hydrates have a numerical prefix indicating the number of associated water molecules.
- Acid names are based on anion names.
- For binary covalent compounds, the first word of the name is the element farther left or lower down in the periodic table, and prefixes show the numbers of each atom.
- The molecular (or formula) mass of a compound is the sum of the atomic masses.
- Chemical formulas give the number of atoms (molecular) or the arrangement of atoms (structural) of one unit of a compound.
- Molecular models convey information about bond angles (ball-and-stick) and relative atomic sizes and distances between atoms (space-filling).

2.9 • CLASSIFICATION OF MIXTURES

In the natural world, *matter usually occurs as mixtures.* Air, seawater, soil, and organisms are all complex mixtures of elements and compounds. There are two broad classes of mixtures:

- A **heterogeneous mixture** has one or more visible boundaries between the components. Thus, its composition is *not* uniform, but rather varies from one region to another. Many rocks are heterogeneous, having individual grains of different minerals. In some heterogeneous mixtures, such as milk and blood, the boundaries can be seen only with a microscope.
- A **homogeneous mixture** (or **solution**) has no visible boundaries because the components are individual atoms, ions, or molecules. Thus, its composition *is* uniform. A mixture of sugar dissolved in water is homogeneous, for example, because the sugar molecules and water molecules are uniformly intermingled on the molecular level. We have no way to tell visually whether a sample of matter is a substance (element or compound) or a homogeneous mixture.

Although we usually think of solutions as liquid, they can exist in all three physical states. For example, air is a gaseous solution of mostly oxygen and nitrogen molecules, and wax is a solid solution of several fatty substances. Solutions in water, called **aqueous solutions,** are especially important in the chemistry lab and comprise a major portion of the environment and of all organisms.

Recall that mixtures differ from compounds in three major ways:

1. The proportions of the components can vary.
2. The individual properties of the components are observable.
3. The components can be separated by physical means.

The difference between a mixture and a compound is well illustrated using iron and sulfur as components (Figure 2.17). Any proportion of iron metal filings and powdered sulfur forms a mixture. The iron can be separated from the sulfur with a magnet. But if we heat the container strongly, the components combine in fixed proportions by mass to form the compound iron(II) sulfide (FeS). The magnet can no longer remove the iron because it exists as Fe^{2+} ions chemically bound to S^{2-} ions.

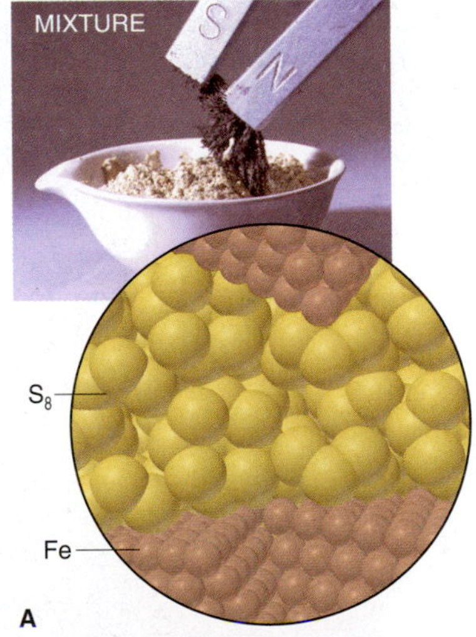

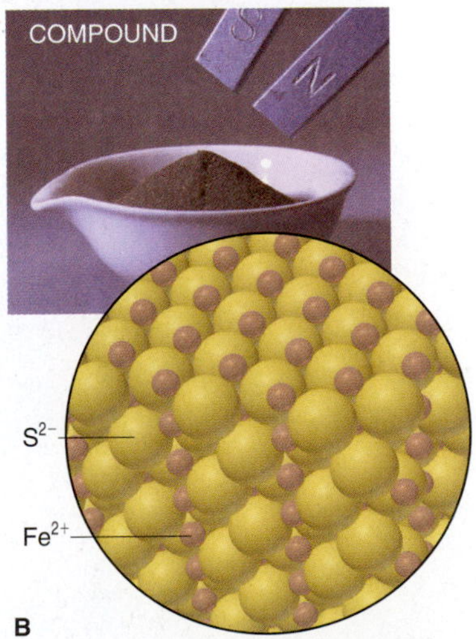

Figure 2.17 The distinction between mixtures and compounds. **A,** A *mixture* of iron and sulfur consists of the two elements. **B,** The *compound* iron(II) sulfide consists of an array of Fe^{2+} and S^{2-} ions.

An Overview of the Components of Matter

Understanding matter at the observable and atomic scales is the essence of chemistry. Figure 2.18 is a visual overview of many key terms and ideas in this chapter.

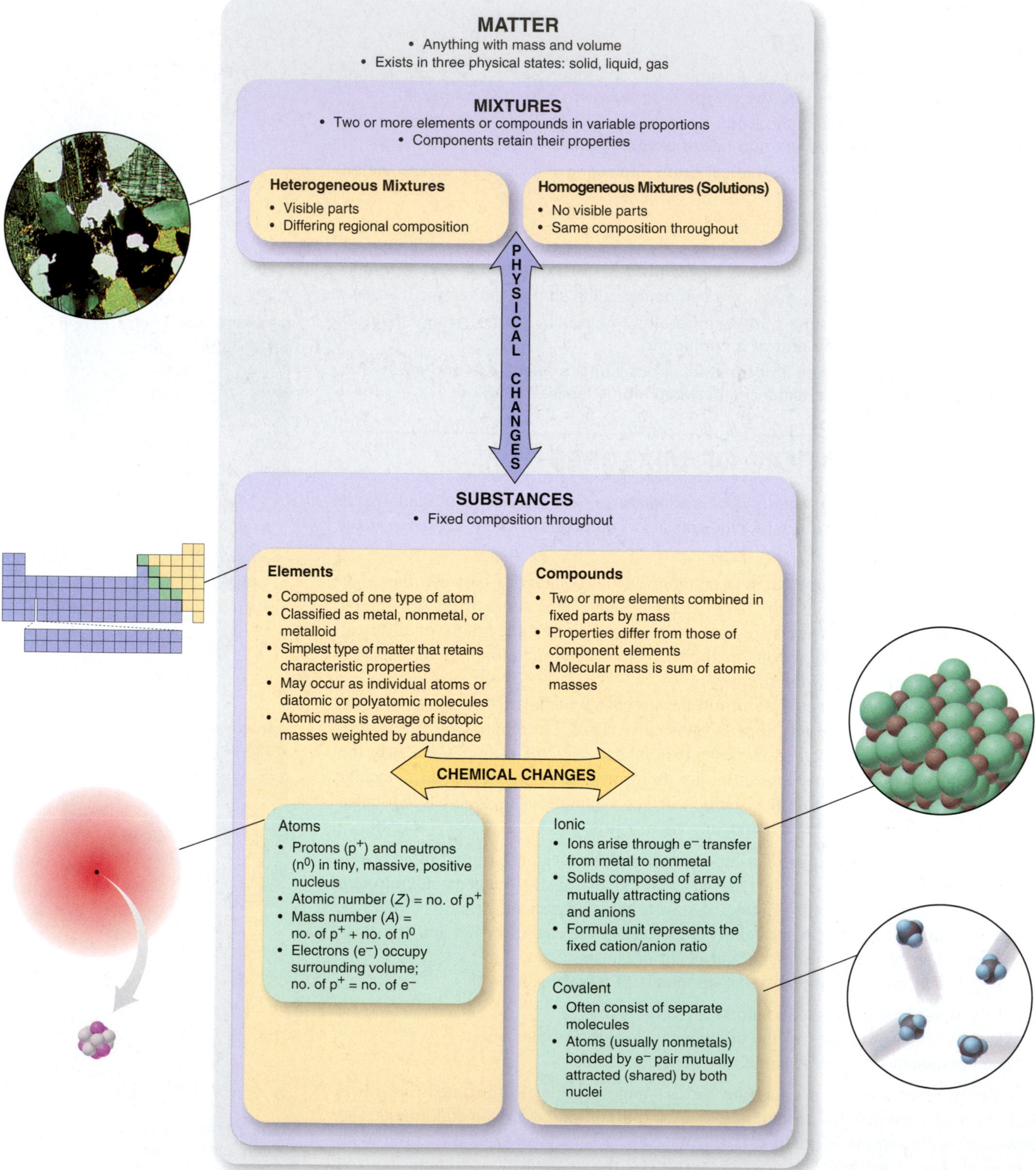

Figure 2.18 The classification of matter from a chemical point of view.

Summary of Section 2.9

- Heterogeneous mixtures have visible boundaries between the components.
- Homogeneous mixtures (solutions) have no visible boundaries because mixing occurs at the molecular level. They can occur in any physical state.
- Components of mixtures (unlike those of compounds) can have variable proportions, can be separated physically, and retain their properties.

CHAPTER REVIEW GUIDE

The following sections provide many aids to help you study this chapter. (Numbers in parentheses refer to pages, unless noted otherwise.)

Learning Objectives

These are concepts and skills to review after studying this chapter.

Related section (§), sample problem (SP), and upcoming end-of-chapter problem (EP) numbers are listed in parentheses.

1. Define the characteristics of the three types of matter—element, compound, and mixture—on the macroscopic and atomic levels (§2.1) (SP 2.1) (EPs 2.1–2.5)
2. Understand the laws of mass conservation, definite composition, and multiple proportions; use the mass ratio of element-to-compound to find the mass of an element in a compound (§2.2) (SP 2.2) (EPs 2.6–2.20, 2.80)
3. Understand Dalton's atomic theory and how it explains the mass laws (§2.3) (SP 2.3) (EP 2.21)
4. Describe the results of the key experiments by Thomson, Millikan, and Rutherford concerning atomic structure (§2.4) (EPs 2.22–2.24)
5. Explain the structure of the atom, the main features of the subatomic particles, and the significance of isotopes; use atomic notation to express the subatomic makeup of an isotope; calculate the atomic mass of an element from its isotopic composition (§2.5) (SPs 2.4, 2.5) (EPs 2.25–2.37)
6. Describe the format of the periodic table and the general location and characteristics of metals, metalloids, and nonmetals (§2.6) (EPs 2.38–2.44)
7. Explain the essential features of ionic and covalent compounds and distinguish between them; predict the monatomic ion formed from a main-group element (§2.7) (SP 2.6) (EPs 2.45–2.57)
8. Name, write the formula, and calculate the molecular (or formula) mass of ionic and binary covalent compounds (§2.8) (SPs 2.7–2.16) (EPs 2.58–2.79, 2.81)
9. Describe the types of mixtures and their properties (§2.9) (EPs 2.82–2.86)

Key Terms

These important terms appear in boldface in the chapter and are defined again in the Glossary.

Section 2.1
substance (33)
element (33)
molecule (33)
compound (34)
mixture (34)

Section 2.2
law of mass conservation (35)
law of definite (or constant) composition (36)
fraction by mass (mass fraction) (36)
percent by mass (mass percent, mass %) (36)
law of multiple proportions (37)

Section 2.3
atom (38)

Section 2.4
cathode ray (40)
nucleus (42)

Section 2.5
proton (p^+) (43)
neutron (n^0) (43)
electron (e^-) (43)
atomic number (Z) (43)
mass number (A) (43)
atomic symbol (43)
isotope (44)
atomic mass unit (amu) (44)
dalton (Da) (44)
mass spectrometry (45)
isotopic mass (45)
atomic mass (45)

Section 2.6
periodic table of the elements (47)
period (47)
group (47)
metal (47)
nonmetal (48)
metalloid (semimetal) (48)

Section 2.7
ionic compound (49)
covalent compound (49)
chemical bond (49)
ion (49)
binary ionic compound (49)
cation (49)
anion (49)
monatomic ion (49)
covalent bond (51)
polyatomic ion (52)

Section 2.8
chemical formula (53)
formula unit (54)
oxoanion (56)
hydrate (56)
binary covalent compound (58)
molecular mass (59)
formula mass (59)
molecular formula (60)
structural formula (60)

Section 2.9
heterogeneous mixture (61)
homogeneous mixture (solution) (61)
aqueous solution (61)

Key Equations and Relationships

Numbered and screened concepts are listed for you to refer to or memorize.

2.1 Finding the mass of an element in a given mass of compound (36):

Mass of element in sample

$$= \text{mass of compound in sample} \times \frac{\text{mass of element}}{\text{mass of compound}}$$

2.2 Calculating the number of neutrons in an atom (44):

$$\text{Number of neutrons} = \text{mass number} - \text{atomic number}$$

or

$$N = A - Z$$

2.3 Determining the molecular mass of a formula unit of a compound (59):

$$\text{Molecular mass} = \text{sum of atomic masses}$$

BRIEF SOLUTIONS TO FOLLOW-UP PROBLEMS

Compare your own solutions to these calculation steps and answers.

2.1 There are two types of particles reacting (left circle), one with two blue atoms and the other with two orange; the depiction shows a mixture of two elements. In the product (right circle), all the particles have one blue atom and one orange; this is a compound.

2.2 Mass (t) of pitchblende

$$= 2.3 \text{ ~~t uranium~~} \times \frac{84.2 \text{ t pitchblende}}{71.4 \text{ ~~t uranium~~}} = 2.7 \text{ t pitchblende}$$

Mass (t) of oxygen

$$= 2.7 \text{ ~~t pitchblende~~} \times \frac{(84.2 - 71.4 \text{ t oxygen})}{84.2 \text{ ~~t pitchblende~~}} = 0.41 \text{ t oxygen}$$

2.3 Sample B. Two bromine-fluorine compounds appear. In one, there are three fluorine atoms for each bromine; in the other, there is one fluorine for each bromine. Therefore, in the two compounds, the ratio of fluorines combining with one bromine is 3/1.

2.4 (a) $5p^+$, $6n^0$, $5e^-$; Q = B
(b) $20p^+$, $21n^0$, $20e^-$; R = Ca
(c) $53p^+$, $78n^0$, $53e^-$; X = I

2.5 $10.0129x + [11.0093(1 - x)] = 10.81$; $0.9964x = 0.1993$; $x = 0.2000$ and $1 - x = 0.8000$; % abundance of $^{10}B = 20.00\%$; % abundance of $^{11}B = 80.00\%$

2.6 (a) S^{2-}; (b) Rb^+; (c) Ba^{2+}

2.7 (a) Zinc [Group 2B(12)] and oxygen [Group 6A(16)]
(b) Silver [Group 1B(11)] and bromine [Group 7A(17)]
(c) Lithium [Group 1A(1)] and chlorine [Group 7A(17)]
(d) Aluminum [Group 3A(13)] and sulfur [Group 6A(16)]

2.8 (a) ZnO; (b) AgBr; (c) LiCl; (d) Al_2S_3

2.9 (a) PbO_2; (b) copper(I) sulfide (cuprous sulfide); (c) iron(II) bromide (ferrous bromide); (d) $HgCl_2$

2.10 (a) $Cu(NO_3)_2{\cdot}3H_2O$; (b) $Zn(OH)_2$; (c) lithium cyanide

2.11 (a) $(NH_4)_3PO_4$; ammonium is NH_4^+ and phosphate is PO_4^{3-}.
(b) $Al(OH)_3$; parentheses are needed around the polyatomic ion OH^-.
(c) Magnesium hydrogen carbonate; Mg^{2+} is magnesium and can have only a 2+ charge, so the Roman numeral II is not needed; HCO_3^- is hydrogen carbonate (or bicarbonate).
(d) Chromium(III) nitrate; the *-ic* ending is not used with Roman numerals; NO_3^- is nitr*ate*.
(e) Calcium nitrite; Ca^{2+} is calcium and NO_2^- is nitr*ite*.

2.12 (a) $HClO_3$; (b) hydrofluoric acid; (c) CH_3COOH (or $HC_2H_3O_2$); (d) H_2SO_3; (e) hypobromous acid

2.13 (a) Sulfur trioxide; (b) silicon dioxide; (c) N_2O; (d) SeF_6

2.14 (a) Disulfur dichloride; the *-ous* suffix is not used.
(b) NO; the name indicates one nitrogen.
(c) Bromine trichloride; Br is in a higher period in Group 7A(17), so it is named first.

2.15 (a) H_2O_2, 34.02 amu; (b) CsCl, 168.4 amu; (c) H_2SO_4, 98.09 amu; (d) K_2SO_4, 174.27 amu

2.16 (a) Na_2O. This is an ionic compound, so the name is sodium oxide.

Formula mass

$$= (2 \times \text{atomic mass of Na}) + (1 \times \text{atomic mass of O})$$
$$= (2 \times 22.99 \text{ amu}) + 16.00 \text{ amu} = 61.98 \text{ amu}$$

(b) NO_2. This is a covalent compound, and N has the lower group number, so the name is nitrogen dioxide.

Molecular mass

$$= (1 \times \text{atomic mass of N}) + (2 \times \text{atomic mass of O})$$
$$= 14.01 \text{ amu} + (2 \times 16.00 \text{ amu}) = 46.01 \text{ amu}$$

PROBLEMS

Problems with **colored** numbers are answered in Appendix E. Sections match the text and provide the numbers of relevant sample problems. Bracketed problems are grouped in pairs (indicated by a short rule) that cover the same concept. Comprehensive Problems are based on material from any section or previous chapter.

Elements, Compounds, and Mixtures: An Atomic Overview

(Sample Problem 2.1)

2.1 What is the key difference between an element and a compound?

2.2 List two differences between a compound and a mixture.

2.3 Which of the following are pure substances? Explain.
(a) Calcium chloride, used to melt ice on roads, consists of two elements, calcium and chlorine, in a fixed mass ratio.
(b) Sulfur consists of sulfur atoms combined into octatomic molecules.
(c) Baking powder, a leavening agent, contains 26% to 30% sodium hydrogen carbonate and 30% to 35% calcium dihydrogen phosphate by mass.
(d) Cytosine, a component of DNA, consists of H, C, N, and O atoms bonded in a specific arrangement.

2.4 Classify each substance in Problem 2.3 as an element, compound, or mixture, and explain your answers.

2.5 Each scene below represents a mixture. Describe each one in terms of the number(s) of elements and/or compounds present.

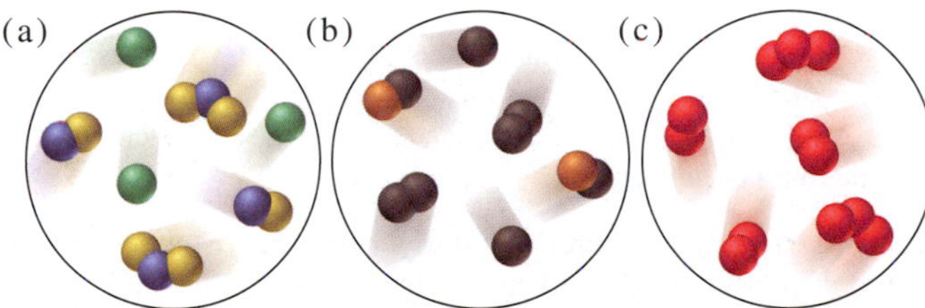

The Observations That Led to an Atomic View of Matter

(Sample Problem 2.2)

2.6 To which classes of matter—element, compound, and/or mixture—do the following apply: (a) law of mass conservation; (b) law of definite composition; (c) law of multiple proportions?

2.7 Identify the mass law that each of the following observations demonstrates, and explain your reasoning:
(a) A sample of potassium chloride from Chile contains the same percent by mass of potassium as one from Poland.
(b) A flashbulb contains magnesium and oxygen before use and magnesium oxide afterward, but its mass does not change.
(c) Arsenic and oxygen form one compound that is 65.2 mass % arsenic and another that is 75.8 mass % arsenic.

2.8 Which of the following scenes illustrate(s) the fact that compounds of chlorine (green) and oxygen (red) exhibit the law of multiple proportions? Name the compounds.

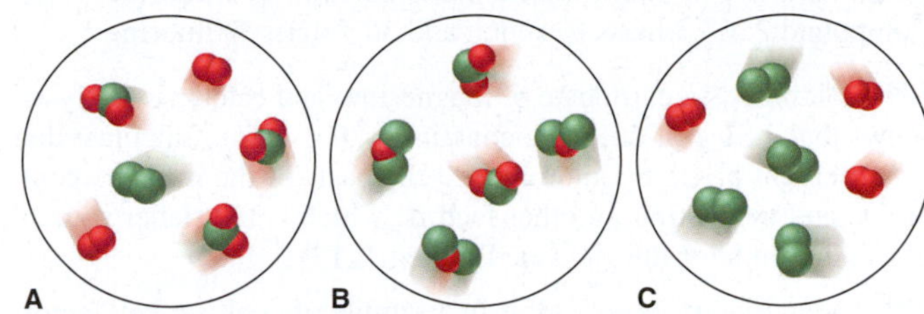

2.9 (a) Does the percent by mass of each element in a compound depend on the amount of compound? Explain.
(b) Does the mass of each element in a compound depend on the amount of compound? Explain.

2.10 Does the percent by mass of each element in a compound depend on the amount of that element used to make the compound? Explain.

2.11 State the mass law(s) demonstrated by the following experimental results, and explain your reasoning:

Experiment 1: A student heats 1.00 g of a blue compound and obtains 0.64 g of a white compound and 0.36 g of a colorless gas.

Experiment 2: A second student heats 3.25 g of the same blue compound and obtains 2.08 g of a white compound and 1.17 g of a colorless gas.

2.12 State the mass law(s) demonstrated by the following experimental results, and explain your reasoning:

Experiment 1: A student heats 1.27 g of copper and 3.50 g of iodine to produce 3.81 g of a white compound; 0.96 g of iodine remains.

Experiment 2: A second student heats 2.55 g of copper and 3.50 g of iodine to form 5.25 g of a white compound, and 0.80 g of copper remains.

2.13 Fluorite, a mineral of calcium, is a compound of the metal with fluorine. Analysis shows that a 2.76-g sample of fluorite contains 1.42 g of calcium. Calculate the (a) mass of fluorine in the sample; (b) mass fractions of calcium and fluorine in fluorite; (c) mass percents of calcium and fluorine in fluorite.

2.14 Galena, a mineral of lead, is a compound of the metal with sulfur. Analysis shows that a 2.34-g sample of galena contains 2.03 g of lead. Calculate the (a) mass of sulfur in the sample; (b) mass fractions of lead and sulfur in galena; (c) mass percents of lead and sulfur in galena.

2.15 A compound of copper and sulfur contains 88.39 g of metal and 44.61 g of nonmetal. How many grams of copper are in 5264 kg of compound? How many grams of sulfur?

2.16 A compound of iodine and cesium contains 63.94 g of metal and 61.06 g of nonmetal. How many grams of cesium are in 38.77 g of compound? How many grams of iodine?

2.17 Show, with calculations, how the following data illustrate the law of multiple proportions:
Compound 1: 47.5 mass % sulfur and 52.5 mass % chlorine
Compound 2: 31.1 mass % sulfur and 68.9 mass % chlorine

2.18 Show, with calculations, how the following data illustrate the law of multiple proportions:
Compound 1: 77.6 mass % xenon and 22.4 mass % fluorine
Compound 2: 63.3 mass % xenon and 36.7 mass % fluorine

2.19 Dolomite is a carbonate of magnesium and calcium. Analysis shows that 7.81 g of dolomite contains 1.70 g of Ca. Calculate the mass percent of Ca in dolomite. On the basis of the mass percent of Ca, and neglecting all other factors, which is the richer source of Ca, dolomite or fluorite (see Problem 2.13)?

2.20 The mass percent of sulfur in a sample of coal is a key factor in the environmental impact of the coal because the sulfur combines with oxygen when the coal is burned and the oxide can then be incorporated into acid rain. Which of the following coals would have the smallest environmental impact?

	Mass (g) of Sample	Mass (g) of Sulfur in Sample
Coal A	378	11.3
Coal B	495	19.0
Coal C	675	20.6

Dalton's Atomic Theory

(Sample Problem 2.3)

2.21 Use Dalton's theory to explain why potassium nitrate from India or Italy has the same mass percents of K, N, and O.

The Observations That Led to the Nuclear Atom Model

2.22 Thomson was able to determine the mass/charge ratio of the electron but not its mass. How did Millikan's experiment allow determination of the electron's mass?

2.23 The following charges on individual oil droplets were obtained during an experiment similar to Millikan's. Determine a charge for the electron (in C, coulombs), and explain your answer: -3.204×10^{-19} C; -4.806×10^{-19} C; -8.010×10^{-19} C; -1.442×10^{-18} C.

2.24 When Rutherford's coworkers bombarded gold foil with α particles, they obtained results that overturned the existing (Thomson) model of the atom. Explain.

The Atomic Theory Today

(Sample Problems 2.4 and 2.5)

2.25 Choose the correct answer. The difference between the mass number of an isotope and its atomic number is (a) directly related to the identity of the element; (b) the number of electrons; (c) the number of neutrons; (d) the number of isotopes.

2.26 Argon has three naturally occurring isotopes, ^{36}Ar, ^{38}Ar, and ^{40}Ar. What is the mass number of each? How many protons, neutrons, and electrons are present in each?

2.27 Chlorine has two naturally occurring isotopes, ^{35}Cl and ^{37}Cl. What is the mass number of each isotope? How many protons, neutrons, and electrons are present in each?

2.28 Do both members of the following pairs have the same number of protons? Neutrons? Electrons?
(a) $^{16}_{8}$O and $^{17}_{8}$O (b) $^{40}_{18}$Ar and $^{41}_{19}$K (c) $^{60}_{27}$Co and $^{60}_{28}$Ni
Which pair(s) consist(s) of atoms with the same *Z* value? *N* value? *A* value?

2.29 Do both members of the following pairs have the same number of protons? Neutrons? Electrons?
(a) $^{3}_{1}$H and $^{3}_{2}$He (b) $^{14}_{6}$C and $^{15}_{7}$N (c) $^{19}_{9}$F and $^{18}_{9}$F
Which pair(s) consist(s) of atoms with the same *Z* value? *N* value? *A* value?

2.30 Write the $^{A}_{Z}$X notation for each atomic depiction:

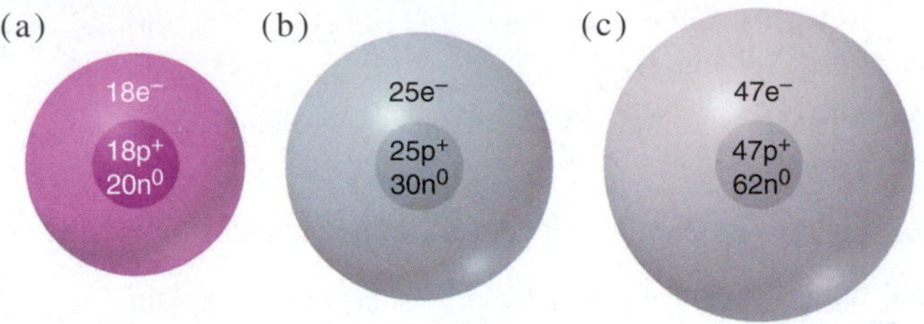

2.31 Write the $^{A}_{Z}$X notation for each atomic depiction:

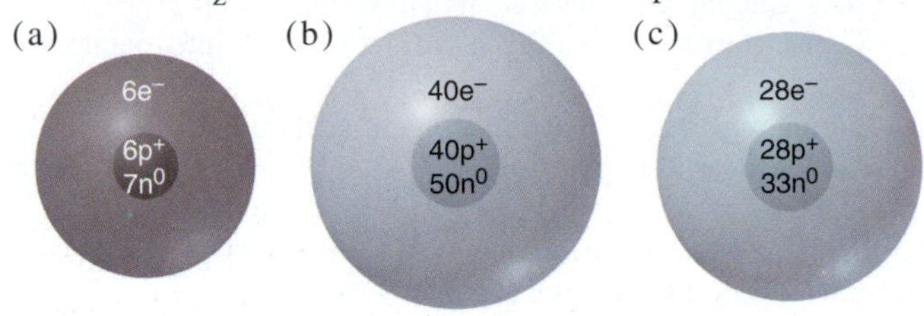

2.32 Draw atomic depictions similar to those in Problem 2.30 for (a) $^{48}_{22}$Ti; (b) $^{79}_{34}$Se; (c) $^{11}_{5}$B.

2.33 Draw atomic depictions similar to those in Problem 2.30 for (a) $^{207}_{82}$Pb; $^{9}_{4}$Be; $^{75}_{33}$As.

2.34 Gallium has two naturally occurring isotopes, ^{69}Ga (isotopic mass = 68.9256 amu, abundance = 60.11%) and ^{71}Ga (isotopic mass = 70.9247 amu, abundance = 39.89%). Calculate the atomic mass of gallium.

2.35 Magnesium has three naturally occurring isotopes, ^{24}Mg (isotopic mass = 23.9850 amu, abundance = 78.99%), ^{25}Mg (isotopic mass = 24.9858 amu, abundance = 10.00%), and ^{26}Mg (isotopic mass = 25.9826 amu, abundance = 11.01%). Calculate the atomic mass of magnesium.

2.36 Chlorine has two naturally occurring isotopes, ^{35}Cl (isotopic mass = 34.9689 amu) and ^{37}Cl (isotopic mass = 36.9659 amu). If chlorine has an atomic mass of 35.4527 amu, what is the percent abundance of each isotope?

2.37 Copper has two naturally occurring isotopes, ^{63}Cu (isotopic mass = 62.9396 amu) and ^{65}Cu (isotopic mass = 64.9278 amu). If copper has an atomic mass of 63.546 amu, what is the percent abundance of each isotope?

Elements: A First Look at the Periodic Table

2.38 Correct each of the following statements:
(a) In the modern periodic table, the elements are arranged in order of increasing atomic mass.
(b) Elements in a period have similar chemical properties.
(c) Elements can be classified as either metalloids or nonmetals.

2.39 What class of elements lies along the "staircase" line in the periodic table? How do the properties of these elements compare with those of metals and nonmetals?

2.40 What are some characteristic properties of elements to the left of the elements along the "staircase"? To the right?

2.41 Give the name, atomic symbol, and group number of the element with each Z value, and classify it as a metal, metalloid, or nonmetal:
(a) $Z = 32$ (b) $Z = 15$ (c) $Z = 2$
(d) $Z = 3$ (e) $Z = 42$

2.42 Give the name, atomic symbol, and group number of the element with each Z value, and classify it as a metal, metalloid, or nonmetal:
(a) $Z = 33$ (b) $Z = 20$ (c) $Z = 35$
(d) $Z = 19$ (e) $Z = 13$

2.43 Fill in the blanks:
(a) The symbol and atomic number of the heaviest alkaline earth metal are ______ and ______.
(b) The symbol and atomic number of the lightest metalloid in Group 4A(14) are ______ and ______.
(c) Group 1B(11) consists of the *coinage metals.* The symbol and atomic mass of the coinage metal whose atoms have the fewest electrons are ______ and ______.
(d) The symbol and atomic mass of the halogen in Period 4 are ______ and ______.

2.44 Fill in the blanks:
(a) The symbol and atomic number of the heaviest nonradioactive noble gas are ______ and ______.
(b) The symbol and group number of the Period 5 transition element whose atoms have the fewest protons are ______ and ______.
(c) The elements in Group 6A(16) are sometimes called the *chalcogens.* The symbol and atomic number of the first metallic chalcogen are ______ and ______.
(d) The symbol and number of protons of the Period 4 alkali metal atom are ______ and ______.

Compounds: Introduction to Bonding

(Sample Problem 2.6)

2.45 Describe the type and nature of the bonding that occurs between reactive metals and nonmetals.

2.46 Describe the type and nature of the bonding that often occurs between two nonmetals.

2.47 Given that the ions in LiF and in MgO are of similar size, which compound has stronger ionic bonding? Use Coulomb's law in your explanation.

2.48 Describe the formation of solid magnesium chloride ($MgCl_2$) from large numbers of magnesium and chlorine atoms.

2.49 Does potassium nitrate (KNO_3) incorporate ionic bonding, covalent bonding, or both? Explain.

2.50 What monatomic ions do potassium ($Z = 19$) and iodine ($Z = 53$) form?

2.51 What monatomic ions do barium ($Z = 56$) and selenium ($Z = 34$) form?

2.52 For each ionic depiction, give the name of the parent atom, its mass number, and its group and period numbers:

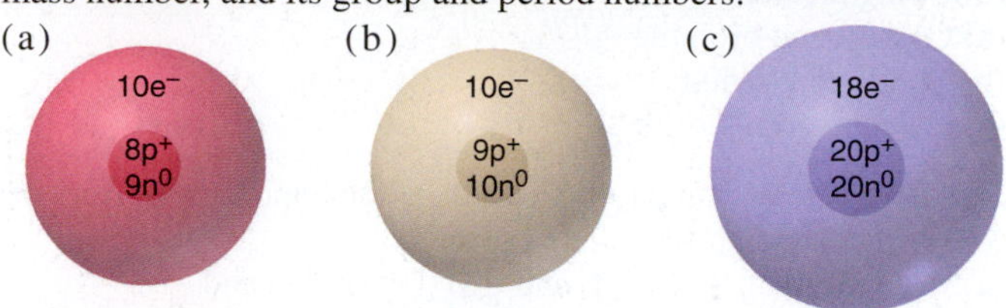

62.53 For each ionic depiction, give the name of the parent atom, its mass number, and its group and period numbers:

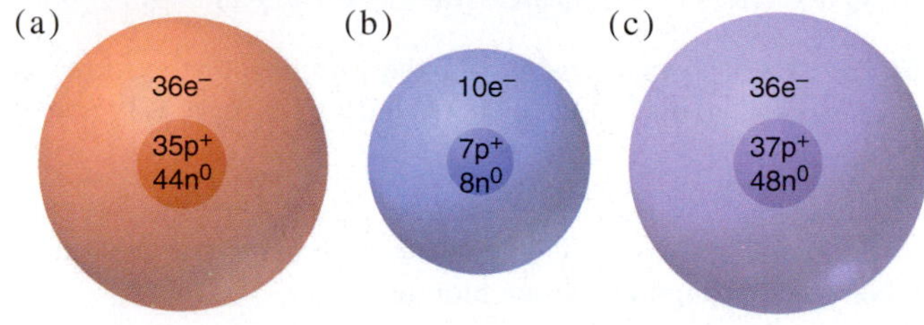

2.54 An ionic compound forms when lithium ($Z = 3$) reacts with oxygen ($Z = 8$). If a sample of the compound contains 8.4×10^{21} lithium ions, how many oxide ions does it contain?

2.55 An ionic compound forms when calcium ($Z = 20$) reacts with iodine ($Z = 53$). If a sample of the compound contains 7.4×10^{21} calcium ions, how many iodide ions does it contain?

2.56 The radii of the sodium and potassium ions are 102 pm and 138 pm, respectively. Which compound has stronger ionic attractions, sodium chloride or potassium chloride?

2.57 The radii of the lithium and magnesium ions are 76 pm and 72 pm, respectively. Which compound has stronger ionic attractions, lithium oxide or magnesium oxide?

Formulas, Names, and Masses of Compounds

(Sample Problems 2.7 to 2.16)

2.58 How is a structural formula similar to a molecular formula? How is it different?

2.59 Consider a mixture of 10 billion O_2 molecules and 10 billion H_2 molecules. In what way is this mixture similar to a sample containing 10 billion hydrogen peroxide (H_2O_2) molecules? In what way is it different?

2.60 Write a formula for each of the following compounds:
(a) Hydrazine, a rocket fuel, consists of two nitrogen atoms and four hydrogen atoms.
(b) Glucose, a sugar, consists of six carbon atoms, twelve hydrogen atoms, and six oxygen atoms.

2.61 Write a formula for each of the following compounds:
(a) Ethylene glycol, car antifreeze, consists of two carbon atoms, six hydrogen atoms, and two oxygen atoms.
(b) Peroxodisulfuric acid, a compound used to make bleaching agents, consists of two hydrogen atoms, two sulfur atoms, and eight oxygen atoms.

2.62 Give the name and formula of the compound formed from the following elements:
(a) Sodium and nitrogen
(b) Oxygen and strontium
(c) Aluminum and chlorine

2.63 Give the name and formula of the compound formed from the following elements:
(a) Cesium and bromine
(b) Sulfur and barium
(c) Calcium and fluorine

2.64 Give the name and formula of the compound formed from the following elements:
(a) $_{12}L$ and $_{9}M$ (b) $_{30}L$ and $_{16}M$ (c) $_{17}L$ and $_{38}M$

2.65 Give the name and formula of the compound formed from the following elements:
(a) $_{37}Q$ and $_{35}R$ (b) $_{8}Q$ and $_{13}R$ (c) $_{20}Q$ and $_{53}R$

2.66 Give the systematic names for the formulas or the formulas for the names: (a) tin(IV) chloride; (b) $FeBr_3$; (c) cuprous bromide; (d) Mn_2O_3.

2.67 Give the systematic names for the formulas or the formulas for the names: (a) Na_2HPO_4; (b) potassium carbonate dihydrate; (c) $NaNO_2$; (d) ammonium perchlorate.

2.68 Correct each of the following formulas:
(a) Barium oxide is BaO_2
(b) Iron(II) nitrate is $Fe(NO_3)_3$
(c) Magnesium sulfide is $MnSO_3$

2.69 Correct each of the following names:
(a) CuI is cobalt(II) iodide
(b) $Fe(HSO_4)_3$ is iron(II) sulfate
(c) $MgCr_2O_7$ is magnesium dichromium heptoxide

2.70 Give the name and formula for the acid derived from each of the following anions:
(a) hydrogen sulfate (b) IO_3^- (c) cyanide (d) HS^-

2.71 Give the name and formula for the acid derived from each of the following anions:
(a) perchlorate (b) NO_3^- (c) bromite (d) F^-

2.72 Give the name and formula of the compound whose molecules consist of two sulfur atoms and four fluorine atoms.

2.73 Give the name and formula of the compound whose molecules consist of two chlorine atoms and one oxygen atom.

2.74 Give the number of atoms of the specified element in a formula unit of each of the following compounds, and calculate the molecular (formula) mass:
(a) Oxygen in aluminum sulfate, $Al_2(SO_4)_3$
(b) Hydrogen in ammonium hydrogen phosphate, $(NH_4)_2HPO_4$
(c) Oxygen in the mineral azurite, $Cu_3(OH)_2(CO_3)_2$

2.75 Give the number of atoms of the specified element in a formula unit of each of the following compounds, and calculate the molecular (formula) mass:
(a) Hydrogen in ammonium benzoate, $C_6H_5COONH_4$
(b) Nitrogen in hydrazinium sulfate, $N_2H_6SO_4$
(c) Oxygen in the mineral leadhillite, $Pb_4SO_4(CO_3)_2(OH)_2$

2.76 Write the formula of each compound, and determine its molecular (formula) mass: (a) ammonium sulfate; (b) sodium dihydrogen phosphate; (c) potassium bicarbonate.

2.77 Write the formula of each compound, and determine its molecular (formula) mass: (a) sodium dichromate; (b) ammonium perchlorate; (c) magnesium nitrite trihydrate.

2.78 Give the formula, name, and molecular mass of the following molecules:

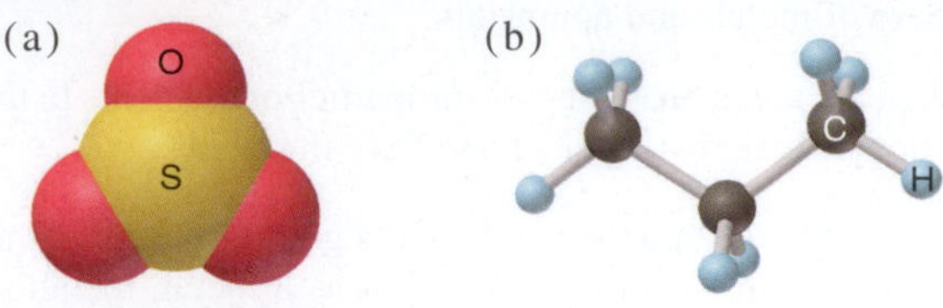

2.79 Give the formula, name, and molecular mass of the following molecules:

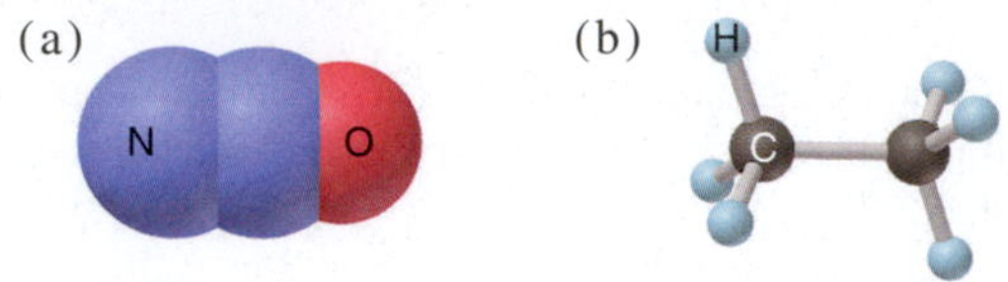

2.80 You are working in the laboratory preparing sodium chloride. Consider the following results for three preparations of the compound:

Case 1: 39.34 g Na + 60.66 g Cl_2 ⟶ 100.00 g NaCl
Case 2: 39.34 g Na + 70.00 g Cl_2 ⟶ 100.00 g NaCl + 9.34 g Cl_2
Case 3: 50.00 g Na + 50.00 g Cl_2 ⟶ 82.43 g NaCl + 17.57 g Na

Explain these results in terms of the laws of conservation of mass and definite composition.

2.81 Before the use of systematic names, many compounds had common names. Give the systematic name for each of the following:
(a) Blue vitriol, $CuSO_4{\cdot}5H_2O$
(b) Slaked lime, $Ca(OH)_2$
(c) Oil of vitriol, H_2SO_4
(d) Washing soda, Na_2CO_3
(e) Muriatic acid, HCl
(f) Epsom salt, $MgSO_4{\cdot}7H_2O$
(g) Chalk, $CaCO_3$
(h) Dry ice, CO_2
(i) Baking soda, $NaHCO_3$
(j) Lye, NaOH

Classification of Mixtures

2.82 In what main way is separating the components of a mixture different from separating the components of a compound?

2.83 What is the difference between a homogeneous and a heterogeneous mixture?

2.84 Is a solution a homogeneous or a heterogeneous mixture? Give an example of an aqueous solution.

2.85 Classify each of the following as a compound, a homogeneous mixture, or a heterogeneous mixture: (a) distilled water; (b) gasoline; (c) beach sand; (d) wine; (e) air.

2.86 Classify each of the following as a compound, a homogeneous mixture, or a heterogeneous mixture: (a) orange juice; (b) vegetable soup; (c) cement; (d) calcium sulfate; (e) tea.

Comprehensive Problems

2.87 Helium is the lightest noble gas and the second most abundant element (after hydrogen) in the universe.
(a) The radius of a helium atom is 3.1×10^{-11} m; the radius of its nucleus is 2.5×10^{-15} m. What fraction of the spherical atomic volume is occupied by the nucleus (V of a sphere $= \frac{4}{3}\pi r^3$)?

(b) The mass of a helium-4 atom is 6.64648×10^{-24} g, and each of its two electrons has a mass of 9.10939×10^{-28} g. What fraction of this atom's mass is contributed by its nucleus?

2.88 Give the molecular mass of each compound depicted below, and provide a correct name for any that are named incorrectly.

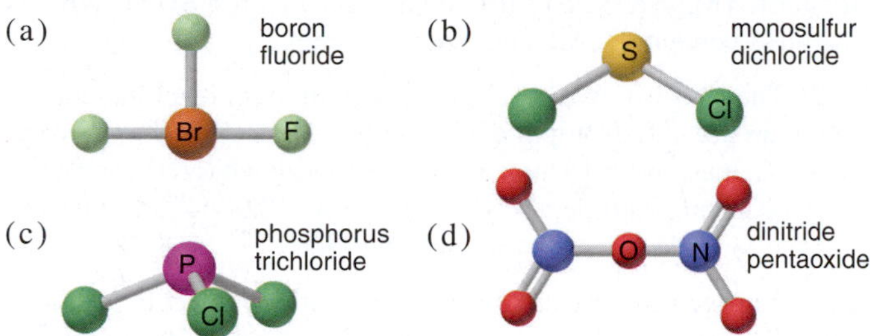

2.89 Nitrogen forms more oxides than any other element. The percents by mass of N in three different nitrogen oxides are (I) 46.69%; (II) 36.85%; (III) 25.94%. For each compound, determine (a) the simplest whole-number ratio of N to O, and (b) the number of grams of oxygen per 1.00 g of nitrogen.

2.90 Scenes A–I depict various types of matter on the atomic scale. Choose the correct scene(s) for each of the following:
(a) A mixture that fills its container
(b) A substance that cannot be broken down into simpler ones
(c) An element with a very high resistance to flow
(d) A homogeneous mixture
(e) An element that conforms to the walls of its container and displays an upper surface
(f) A gas consisting of diatomic particles
(g) A gas that can be broken down into simpler substances
(h) A substance with a 2/1 ratio of its component atoms
(i) Matter that can be separated into its component substances by physical means
(j) A heterogeneous mixture
(k) Matter that obeys the law of definite composition

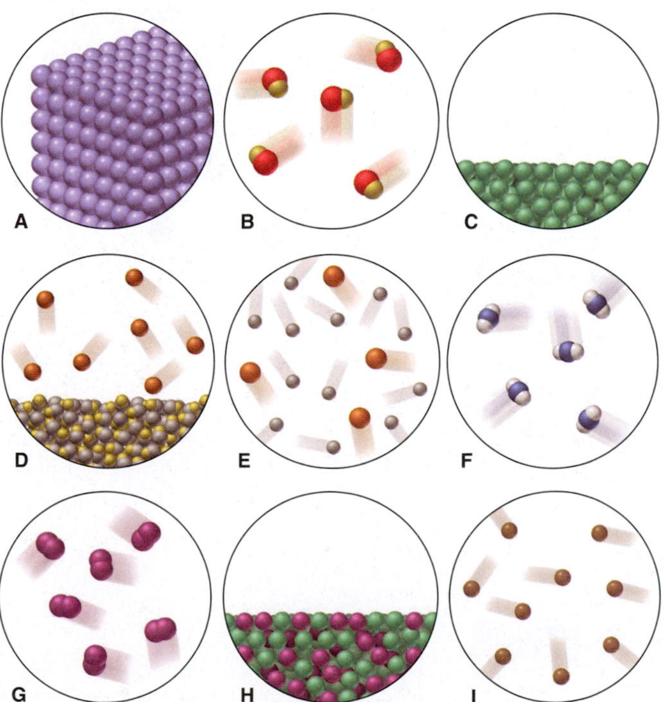

2.91 The seven most abundant ions in seawater make up more than 99% by mass of the dissolved compounds. Here are their abundances in units of mg ion/kg seawater: chloride 18,980; sodium 10,560; sulfate 2650; magnesium 1270; calcium 400; potassium 380; hydrogen carbonate 140.
(a) What is the mass % of each ion in seawater?
(b) What percent of the total mass of ions is sodium ion?
(c) How does the total mass % of alkaline earth metal ions compare with the total mass % of alkali metal ions?
(d) Which make up the larger mass fraction of dissolved components, anions or cations?

2.92 The scenes below represent a mixture of two monatomic gases undergoing a reaction when heated. Which mass law(s) is (are) illustrated by this change?

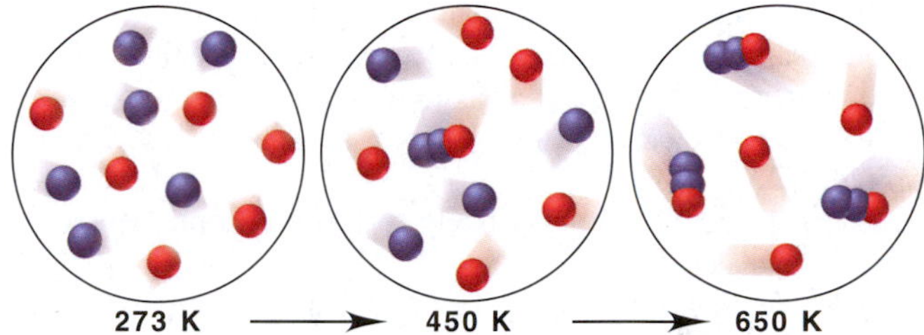

2.93 When barium (Ba) reacts with sulfur (S) to form barium sulfide (BaS), each Ba atom reacts with an S atom. If 2.50 cm^3 of Ba reacts with 1.75 cm^3 of S, are there enough Ba atoms to react with the S atoms (d of Ba = 3.51 g/cm^3; d of S = 2.07 g/cm^3)?

2.94 Succinic acid *(below)* is an important metabolite in biological energy production. Give the molecular formula, molecular mass, and the mass percent of each element in succinic acid.

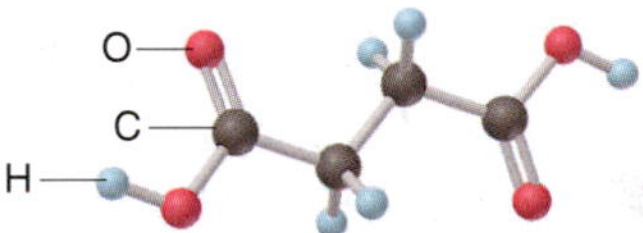

2.95 Fluoride ion is poisonous in relatively low amounts: 0.2 g of F^- per 70 kg of body weight can cause death. Nevertheless, in order to prevent tooth decay, F^- ions are added to drinking water at a concentration of 1 mg of F^- ion per L of water. How many liters of fluoridated drinking water would a 70-kg person have to consume in one day to reach this toxic level? How many kilograms of sodium fluoride would be needed to treat a 8.50×10^7-gal reservoir?

2.96 Antimony has many uses, for example, in infrared devices and as part of an alloy in lead storage batteries. The element has two naturally occurring isotopes, one with mass 120.904 amu, the other with mass 122.904 amu. (a) Write the $^{A}_{Z}X$ notation for each isotope. (b) Use the atomic mass of antimony from the periodic table to calculate the natural abundance of each isotope.

2.97 Dinitrogen monoxide (N_2O; nitrous oxide) is a greenhouse gas that enters the atmosphere principally from natural fertilizer breakdown. Some studies have shown that the isotope ratios of ^{15}N to ^{14}N and of ^{18}O to ^{16}O in N_2O depend on the source, which can thus be determined by measuring the relative abundance of molecular masses in a sample of N_2O.
(a) What different molecular masses are possible for N_2O?
(b) The percent abundance of ^{14}N is 99.6%, and that of ^{16}O is 99.8%. Which molecular mass of N_2O is least common, and which is most common?

2.98 Use the box color(s) in the periodic table below to identify the element(s) described by each of the following:

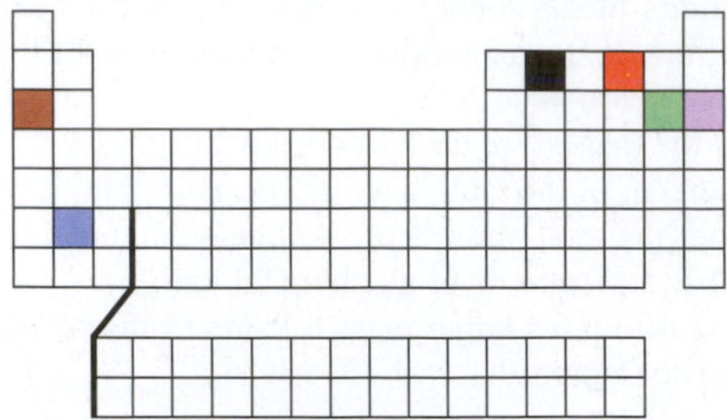

(a) Four elements that are nonmetals
(b) Two elements that are metals
(c) Three elements that are gases at room temperature
(d) Three elements that are solid at room temperature
(e) One pair of elements likely to form a covalent compound
(f) Another pair of elements likely to form a covalent compound
(g) One pair of elements likely to form an ionic compound with formula MX
(h) Another pair of elements likely to form an ionic compound with formula MX
(i) Two elements likely to form an ionic compound with formula M_2X
(j) Two elements likely to form an ionic compound with formula MX_2
(k) An element that forms no compounds
(l) A pair of elements whose compounds exhibit the law of multiple proportions

2.99 Dimercaprol ($HSCH_2CHSHCH_2OH$) is a complexing agent developed during World War I as an antidote to arsenic-based poison gas and used today to treat heavy-metal poisoning. Such an agent binds and removes the toxic element from the body.
(a) If each molecule of dimercaprol binds one arsenic (As) atom, how many atoms of As can be removed by 250. mg of dimercaprol?
(b) If one molecule binds one metal atom, calculate the mass % of each of the following metals in a metal-dimercaprol complex: mercury, thallium, chromium.

2.100 From the following ions and their radii (in pm), choose a pair that gives the strongest ionic bonding and a pair that gives the weakest: Mg^{2+}, 72; K^+, 138; Rb^+, 152; Ba^{2+}, 135; Cl^-, 181; O^{2-}, 140; I^-, 220.

2.101 A rock is 5.0% by mass fayalite (Fe_2SiO_4), 7.0% by mass forsterite (Mg_2SiO_4), and the remainder silicon dioxide. What is the mass percent of each element in the rock?

2.102 The two isotopes of potassium with significant abundance in nature are ^{39}K (isotopic mass 38.9637 amu, 93.258%) and ^{41}K (isotopic mass 40.9618 amu, 6.730%). Fluorine has only one naturally occurring isotope, ^{19}F (isotopic mass 18.9984 amu). Calculate the formula mass of potassium fluoride.

2.103 Nitrogen monoxide (NO) is a bioactive molecule in blood. Low NO concentrations cause respiratory distress and the formation of blood clots. Doctors prescribe nitroglycerin, $C_3H_5N_3O_9$, and isoamyl nitrate, $(CH_3)_2CHCH_2CH_2ONO_2$, to increase NO. If each compound releases one molecule of NO per atom of N it contains, calculate the mass percent of NO in each.

2.104 TNT (trinitrotoluene; *below*) is used as an explosive in construction. Calculate the mass of each element in 1.00 lb of TNT.

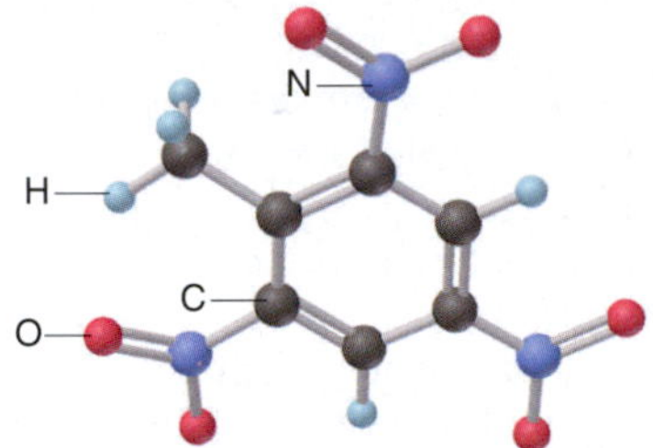

2.105 The anticancer drug Platinol (Cisplatin), $Pt(NH_3)_2Cl_2$, reacts with the cancer cell's DNA and interferes with its growth. (a) What is the mass % of platinum (Pt) in Platinol? (b) If Pt costs \$32/g, how many grams of Platinol can be made for \$1.00 million (assume that the cost of Pt determines the cost of the drug)?

2.106 Which of the following steps in an overall process involve(s) a physical change and which involve(s) a chemical change?

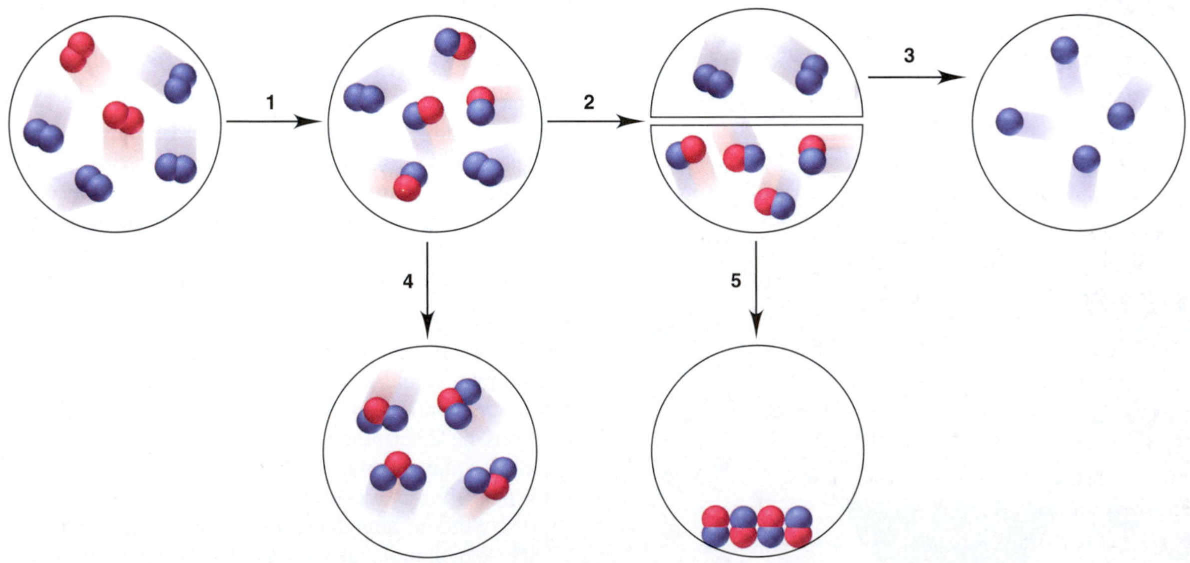

Stoichiometry of Formulas and Equations

Measuring Natural Substances Taxol, a substance with strong anticancer activity, was discovered in the bark of the Pacific yew using principles you'll learn in this chapter. Here we discuss how to determine a formula and find the amount of substance involved in a chemical change.

Outline

Key Principles
to focus on while studying this chapter

- The *mole* (mol) is the SI unit for *amount of substance* and contains *Avogadro's number* (6.022×10^{23}) of chemical entities (atoms, molecules, or ions). It has the same numerical value in grams as a single entity of the substance has in atomic mass units; for example, 1 molecule of H_2O weighs 18.02 amu and 1 mol of H_2O molecules weighs 18.02 g. Therefore, if the amount of a substance is expressed in moles, we know the number of entities in a given mass of it, which means that amount, mass, and number are interconvertible. ***(Section 3.1)***
- The subscripts in the chemical formula for a compound provide quantitative information about the amounts of each element in one mole of the compound. In an *empirical formula*, the subscripts show the *relative* numbers of moles of each element in the compound; in a *molecular formula*, they show the *actual* numbers. Isomers are different compounds with the same molecular formula. ***(Section 3.2)***
- In a *balanced equation*, chemical formulas preceded by integer (whole-number) coefficients give the same number of each kind of atom on the left *(reactants)* as on the right *(products)* but with atoms in different combinations. ***(Section 3.3)***
- Using molar ratios from the balanced equation, we calculate the amount of one substance from the amount of any other involved in the reaction. During a typical reaction, one substance (the *limiting reactant*) is used up, so it limits the amount of product that can form; the other reactant(s) is (are) in excess. The *theoretical yield*, the amount indicated by the balanced equation, is never obtained in the lab because of competing *side reactions*, incompleteness of the main reaction, and inability to collect all of the product. ***(Section 3.4)***
- For reactions in solution, we determine amounts of substances from their *concentration (molarity)* and the solution volume. To dilute a solution, we add *solvent*, which lowers the amount of *solute* dissolved in each unit volume. ***(Section 3.5)***

CONCEPTS & SKILLS TO REVIEW before studying this chapter

- atomic mass (Section 2.5)
- names and formulas of compounds (Section 2.8)
- molecular (or formula) mass (Section 2.8)
- molecular and structural formulas, and ball-and-stick and space-filling models (Section 2.8)
- mass laws in chemical reactions (Section 2.2)

Chemistry is, above all, a practical science. Imagine that you're a biochemist who has extracted a substance with medicinal activity from a natural source: what is its formula, and what quantity of metabolic products will establish a safe dosage level? Or, suppose you're a chemical engineer studying rocket-fuel thrust: what amount of propulsive gases will a fuel produce? Perhaps you're on a team of environmental chemists examining coal samples: what quantity of air pollutants will a sample produce when burned? Or, maybe you're a polymer chemist preparing a plastic with unusual properties: how much of this new material will the polymerization reaction yield? You can answer countless questions like these with a knowledge of **stoichiometry** (pronounced "stoy-key-AHM-uh-tree"; from the Greek *stoicheion,* "element or part," and *metron,* "measure"), the study of the quantitative aspects of formulas and reactions.

3.1 • THE MOLE

All the ideas in this chapter rely on understanding a key concept related to a unit called the *mole*. In daily life, we often measure things by counting or by weighing: we weigh beans or rice, but we count eggs or pencils. And we use counting units (a dozen pencils) or mass units (a kilogram of beans) to express the amount. Similarly, daily life in the laboratory involves measuring substances. We want to know the numbers of chemical entities—atoms, ions, molecules, or formula units—that react with each other, but how can we possibly count or weigh such minute objects? As you'll see, chemists have devised a unit, called the *mole,* to *count chemical entities by weighing them.*

Defining the Mole

The **mole** (abbreviated **mol**) is the SI unit for *amount of substance*. It is defined as *the amount of a substance that contains the same number of entities as the number of atoms in 12 g of carbon-12.* This number, called **Avogadro's number** (in honor of the 19th-century Italian physicist Amedeo Avogadro), is enormous:

One mole (1 mol) contains 6.022×10^{23} entities (to four significant figures) **(3.1)**

Thus,

1 mol of carbon-12	contains	6.022×10^{23} carbon-12 atoms
1 mol of H_2O	contains	6.022×10^{23} H_2O molecules
1 mol of NaCl	contains	6.022×10^{23} NaCl formula units

THINK OF IT THIS WAY
Imagine a Mole of . . .

A mole of any ordinary object is a staggering amount: a mole of periods (.) lined up side by side would equal the radius of our galaxy; a mole of marbles stacked tightly together would cover the United States 70 miles deep. However, atoms and molecules are not ordinary objects: a mole of water molecules (about 18 mL) can be swallowed in one gulp!

A counting unit, like *dozen,* tells you the number of objects but not their mass; a mass unit, like *kilogram,* tells you the mass of objects but not their number. The mole tells you both—the *number* of objects in a given *mass* of substance:

1 mol of carbon-12 contains 6.022×10^{23} carbon-12 atoms *and* has a mass of 12 g

What does it mean that the mole unit allows you to count entities by weighing the sample? Suppose you have a sample of carbon-12 and want to know the number of atoms present. You find that the sample weighs 6 g, so it is 0.5 mol of carbon-12 and, thus, contains 3.011×10^{23} atoms:

6 g of carbon-12 is 0.5 mol of carbon-12 and contains 3.011×10^{23} atoms

Knowing the amount (in moles), the mass (in grams), and the number of entities becomes very important when we mix different substances to run a reaction. The central relationship between masses on the atomic scale and on the macroscopic scale is the same for elements and compounds:

- *Elements.* The mass in *atomic mass units (amu)* of one atom of an element is the *same numerically* as the mass in *grams (g)* of 1 mole of atoms of the element. Recall from Chapter 2 that each atom of an element is considered to have the *atomic mass* given in the periodic table *(see margin)*. Thus,

1 atom of S has a mass of 32.07 amu and 1 mol (6.022×10^{23} atoms) of S has a mass of 32.07 g
1 atom of Fe has a mass of 55.85 amu and 1 mol (6.022×10^{23} atoms) of Fe has a mass of 55.85 g

  Note, also, that since atomic masses are relative, 1 Fe atom weighs 55.85/32.07 as much as 1 S atom, and 1 mol of Fe weighs 55.85/32.07 as much as 1 mol of S.

- *Compounds.* The mass in *atomic mass units (amu)* of one molecule (or formula unit) of a compound is the *same numerically* as the mass in *grams (g)* of 1 mole of the compound. Thus, for example,

1 molecule of H_2O has a mass of 18.02 amu and 1 mol (6.022×10^{23} molecules) of H_2O has a mass of 18.02 g
1 formula unit of NaCl has a mass of 58.44 amu and 1 mol (6.022×10^{23} formula units) of NaCl has a mass of 58.44 g

  Here, too, because masses are relative, 1 H_2O molecule weighs 18.02/58.44 as much as 1 NaCl formula unit, and 1 mol of H_2O weighs 18.02/58.44 as much as 1 mol of NaCl.

The two key points to remember about the importance of the mole unit are

- The *mole* lets us relate the *number* of entities to the *mass* of a sample of those entities.
- The mole maintains the *same numerical relationship* between mass on the atomic scale (atomic mass units, amu) and mass on the macroscopic scale (grams, g).

In everyday terms, a grocer *does not* know there are 1 dozen eggs from their weight or that there is 1 kilogram of beans from their count, because eggs and beans do not have fixed masses. But, by weighing out 63.55 g (1 mol) of copper, a chemist *does* know that there are 6.022×10^{23} copper atoms, because all copper atoms have an atomic mass of 63.55 amu. Figure 3.1 shows 1 mole of some familiar elements and compounds.

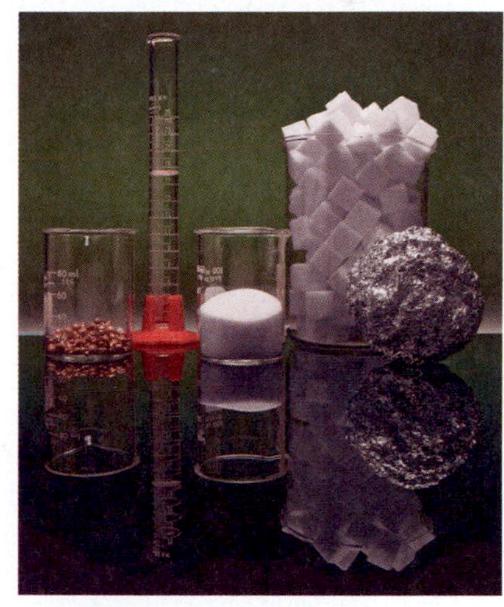

Figure 3.1 One mole (6.022×10^{23} entities) of some familiar substances. From left to right: 1 mol of copper (63.55 g), of liquid H_2O (18.02 g), of sodium chloride (table salt, 58.44 g), of sucrose (table sugar, 342.3 g), and of aluminum (26.98 g).

Determining Molar Mass

The **molar mass ($\mathcal{M}$)** of a substance is the mass per mole of its entities (atoms, molecules, or formula units) and has units of grams per mole (g/mol). The periodic table is indispensable for calculating molar mass:

1. *Elements.* To find the molar mass, look up the atomic mass and note whether the element is monatomic or molecular.

- *Monatomic elements.* The molar mass is the periodic-table value in grams per mole.* For example, the molar mass of neon is 20.18 g/mol, and the molar mass of gold is 197.0 g/mol.
- *Molecular elements.* You must know the formula to determine the molar mass (see Figure 2.13). For example, in air, oxygen exists most commonly as diatomic molecules, so the molar mass of O_2 is twice that of O:

$$\text{Molar mass } (\mathcal{M}) \text{ of } O_2 = 2 \times \mathcal{M} \text{ of O} = 2 \times 16.00 \text{ g/mol} = 32.00 \text{ g/mol}$$

  The most common form of sulfur exists as octatomic molecules, S_8:

$$\mathcal{M} \text{ of } S_8 = 8 \times \mathcal{M} \text{ of S} = 8 \times 32.07 \text{ g/mol} = 256.6 \text{ g/mol}$$

2. *Compounds. The molar mass is the sum of the molar masses of the atoms in the formula.* Thus, from the formula of sulfur dioxide, SO_2, we know that 1 mol of SO_2 molecules contains 1 mol of S atoms and 2 mol of O atoms:

$$\mathcal{M} \text{ of } SO_2 = \mathcal{M} \text{ of S} + (2 \times \mathcal{M} \text{ of O}) = 32.07 \text{ g/mol} + (2 \times 16.00 \text{ g/mol}) = 64.07 \text{ g/mol}$$

Similarly, for ionic compounds, such as potassium sulfide (K_2S), we have

$$\mathcal{M} \text{ of } K_2S = (2 \times \mathcal{M} \text{ of K}) + \mathcal{M} \text{ of S} = (2 \times 39.10 \text{ g/mol}) + 32.07 \text{ g/mol} = 110.27 \text{ g/mol}$$

*The mass value in the periodic table has no units because it is a *relative* atomic mass, given by the atomic mass (in amu) divided by 1 amu ($\frac{1}{12}$ mass of one ^{12}C atom in amu):

$$\text{Relative atomic mass} = \frac{\text{atomic mass } (\cancel{\text{amu}})}{\frac{1}{12} \text{ mass of } ^{12}\text{C } (\cancel{\text{amu}})}$$

Therefore, you use the same number for the atomic mass and for the molar mass.

Table 3.1 Information Contained in the Chemical Formula of Glucose, $C_6H_{12}O_6$ ($\mathcal{M}$ = 180.16 g/mol)

	Carbon (C)	Hydrogen (H)	Oxygen (O)
Atoms/molecule of compound	6 atoms	12 atoms	6 atoms
Moles of atoms/mole of compound	6 mol of atoms	12 mol of atoms	6 mol of atoms
Atoms/mole of compound	$6(6.022\times10^{23})$ atoms	$12(6.022\times10^{23})$ atoms	$6(6.022\times10^{23})$ atoms
Mass/molecule of compound	6(12.01 amu) = 72.06 amu	12(1.008 amu) = 12.10 amu	6(16.00 amu) = 96.00 amu
Mass/mole of compound	72.06 g	12.10 g	96.00 g

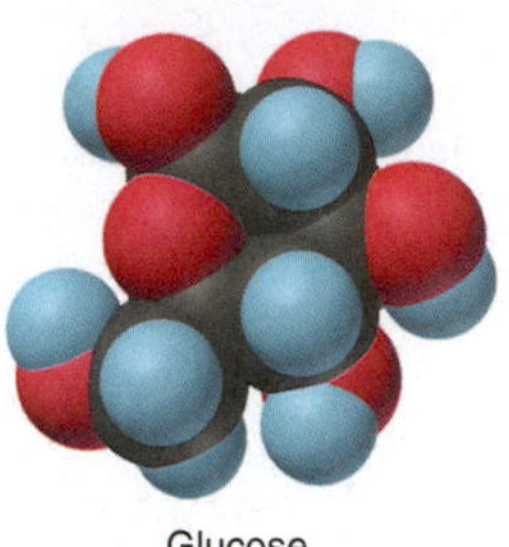
Glucose

Thus, *subscripts in a formula refer to individual atoms (or ions) as well as to moles of atoms (or ions).* Table 3.1 summarizes these ideas for glucose, $C_6H_{12}O_6$ *(see margin),* the essential sugar in energy metabolism.

Converting Between Amount, Mass, and Number of Chemical Entities

One of the most common skills in the lab—and on exams—is converting between amount (mol), mass (g), and number of entities of a substance.

1. *Converting between amount and mass.* If you know the amount of a substance, you can find its mass, and vice versa. The molar mass ($\mathcal{M}$), which expresses the equivalence between 1 mole of a substance and its mass in grams, is the conversion factor.
   - *From amount (mol) to mass (g),* multiply by the molar mass:

$$\text{Mass (g)} = \text{amount (mol)} \times \frac{\text{no. of grams}}{1 \text{ mol}} \tag{3.2}$$

   - *From mass (g) to amount (mol),* divide by the molar mass (multiply by $1/\mathcal{M}$):

$$\text{Amount (mol)} = \text{mass (g)} \times \frac{1 \text{ mol}}{\text{no. of grams}} \tag{3.3}$$

2. *Converting between amount and number.* Similarly, if you know the amount (mol), you can find the number of entities, and vice versa. Avogadro's number, which expresses the equivalence between 1 mole of a substance and the number of entities it contains, is the conversion factor.
   - *From amount (mol) to number of entities,* multiply by Avogadro's number:

$$\text{No. of entities} = \text{amount (mol)} \times \frac{6.022\times10^{23} \text{ entities}}{1 \text{ mol}} \tag{3.4}$$

   - *From number of entities to amount (mol),* divide by Avogadro's number:

$$\text{Amount (mol)} = \text{no. of entities} \times \frac{1 \text{ mol}}{6.022\times10^{23} \text{ entities}} \tag{3.5}$$

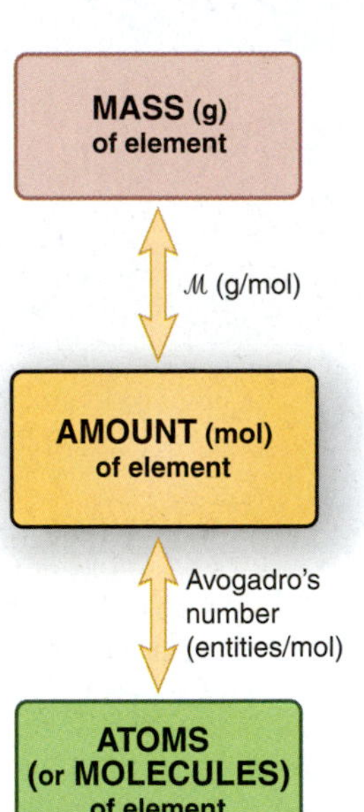

Figure 3.2 Mass-mole-number relationships for elements.

Amount-Mass-Number Conversions Involving Elements We begin with amount-mass-number relationships of elements. As Figure 3.2 shows, *convert mass (in grams) or number of entities (atoms or molecules) to amount (mol) first.* For molecular elements, Avogadro's number gives *molecules* per mole.

Let's work through a series of sample problems that show these conversions for both elements and compounds.

Sample Problem 3.1 Calculating the Mass of a Given Amount of an Element

Problem Silver (Ag) is used in jewelry and tableware but no longer in U.S. coins. How many grams of Ag are in 0.0342 mol of Ag?

Plan We know the amount of Ag (0.0342 mol) and have to find the mass (g). To convert units of *moles* of Ag to *grams* of Ag, we multiply by the *molar mass* of Ag, which we find in the periodic table (see the road map).

Road Map

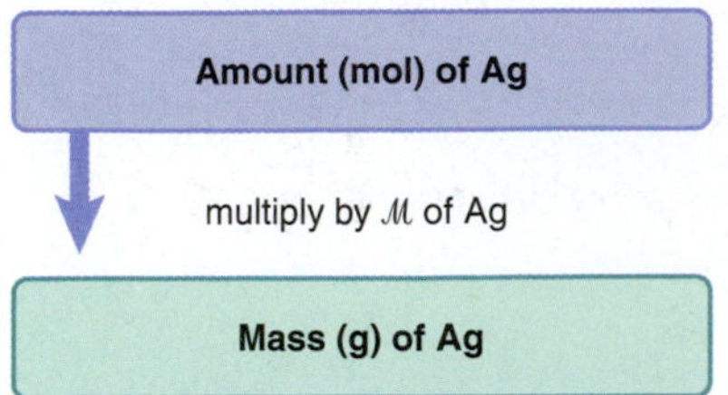

Solution Converting from amount (mol) of Ag to mass (g):

$$\text{Mass (g) of Ag} = 0.0342 \cancel{\text{mol Ag}} \times \frac{107.9 \text{ g Ag}}{1 \cancel{\text{mol Ag}}} = \boxed{3.69 \text{ g Ag}}$$

Check We rounded the mass to three significant figures because the amount (mol) has three. The units are correct. About 0.03 mol × 100 g/mol gives 3 g; the small mass makes sense because 0.0342 is a small fraction of a mole.

FOLLOW-UP PROBLEM 3.1 Graphite is the crystalline form of carbon used in "lead" pencils. How many moles of carbon are in 315 mg of graphite? Include a road map that shows how you planned the solution. (See Brief Solutions at the end of the chapter for the solution to this and all other follow-up problems.)

Sample Problem 3.2 Calculating the Number of Entities in a Given Amount of an Element

Problem Gallium (Ga) is a key element in solar panels, calculators, and other light-sensitive electronic devices. How many Ga atoms are in 2.85×10^{-3} mol of gallium?

Plan We know the amount of gallium (2.85×10^{-3} mol) and need the number of Ga atoms. We multiply amount (mol) by Avogadro's number to find number of atoms (see the road map).

Road Map

Amount (mol) of Ga

↓ multiply by 6.022×10^{23} atoms/mol

Number of Ga atoms

Solution Converting from amount (mol) of Ga to number of atoms:

$$\text{No. of Ga atoms} = 2.85\times10^{-3} \cancel{\text{mol Ga}} \times \frac{6.022\times10^{23} \text{ Ga atoms}}{1 \cancel{\text{mol Ga}}}$$

$$= \boxed{1.72\times10^{21} \text{ Ga atoms}}$$

Check The number of atoms has three significant figures because the number of moles does. When we round amount (mol) of Ga and Avogadro's number, we have $\sim(3\times10^{-3} \text{ mol})(6\times10^{23} \text{ atoms/mol}) = 18\times10^{20}$, or 1.8×10^{21} atoms, so our answer seems correct.

FOLLOW-UP PROBLEM 3.2 At rest, a person inhales 9.72×10^{21} nitrogen molecules in an average breath of air. How many moles of nitrogen atoms are inhaled? (*Hint:* In air, nitrogen occurs as a diatomic molecule.) Include a road map that shows how you planned the solution.

For the next sample problem, note that mass and number of entities relate directly to amount (mol), but *not* to each other. Therefore, *to convert between mass and number, first convert to amount.*

Sample Problem 3.3 Calculating the Number of Entities in a Given Mass of an Element

Problem Iron (Fe) is the main component of steel and, thus, the most important metal in industrial society; it is also essential in the body. How many Fe atoms are in 95.8 g of Fe?

Plan We know the mass of Fe (95.8 g) and need the number of Fe atoms. We cannot convert directly from mass to number, so we first convert to amount (mol) by dividing mass of Fe by its molar mass. Then, we multiply amount (mol) by Avogadro's number to find number of atoms (see the road map).

Road Map

Mass (g) of Fe

↓ divide by $\mathcal{M}$ of Fe

Amount (mol) of Fe

↓ multiply by 6.022×10^{23} atoms/mol

Number of Fe atoms

Solution Converting from mass (g) of Fe to amount (mol):

$$\text{Moles of Fe} = 95.8 \cancel{\text{g Fe}} \times \frac{1 \text{ mol Fe}}{55.85 \cancel{\text{g Fe}}} = 1.72 \text{ mol Fe}$$

Converting from amount (mol) of Fe to number of Fe atoms:

$$\text{No. of Fe atoms} = 1.72 \cancel{\text{mol Fe}} \times \frac{6.022\times10^{23} \text{ atoms Fe}}{1 \cancel{\text{mol Fe}}}$$

$$= 10.4\times10^{23} \text{ atoms Fe}$$

$$= \boxed{1.04\times10^{24} \text{ atoms Fe}}$$

Check Rounding the mass and the molar mass of Fe, we have ~100 g/(~60 g/mol) = 1.7 mol. Therefore, the number of atoms should be a bit less than twice Avogadro's number: $<2(6\times10^{23}) = <1.2\times10^{24}$, so the answer seems correct.

FOLLOW-UP PROBLEM 3.3 Manganese (Mn) is a transition element essential for the growth of bones. What is the mass in grams of 3.22×10^{20} Mn atoms, the number found in 1 kg of bone? Include a road map that shows how you planned the solution.

Amount-Mass-Number Conversions Involving Compounds Only one new step is needed to solve amount-mass-number problems involving compounds: we need the chemical formula to find the molar mass and the amount of each element in the compound. The relationships are shown in Figure 3.3, and Sample Problems 3.4 and 3.5 apply them.

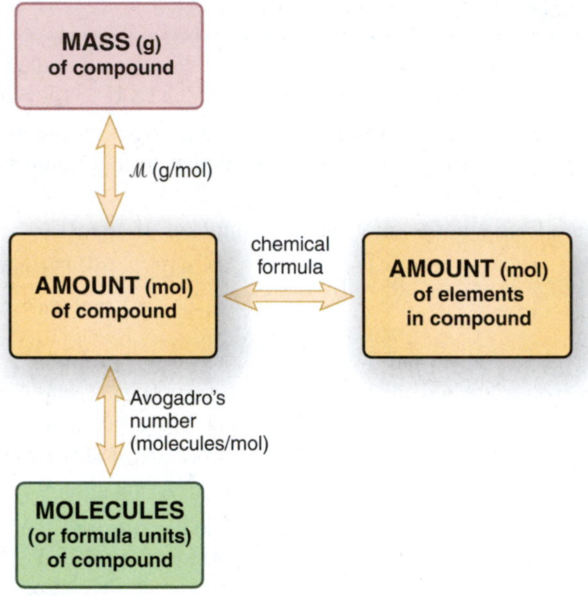

Figure 3.3 Amount-mass-number relationships for compounds. Use the chemical formula to find the amount (mol) of each element in a compound.

Sample Problem 3.4 Calculating the Number of Chemical Entities in a Given Mass of a Compound I

Problem Nitrogen dioxide is a component of urban smog that forms from gases in car exhaust. How many molecules are in 8.92 g of nitrogen dioxide?

Plan We know the mass of compound (8.92 g) and need to find the number of molecules. As you just saw in Sample Problem 3.3, to convert mass to number of entities, we have to find the amount (mol). To do so, we divide the mass by the molar mass ($\mathcal{M}$), which we calculate from the molecular formula (see Sample Problem 2.15). Once we have the amount (mol), we multiply by Avogadro's number to find the number of molecules (see the road map).

Road Map

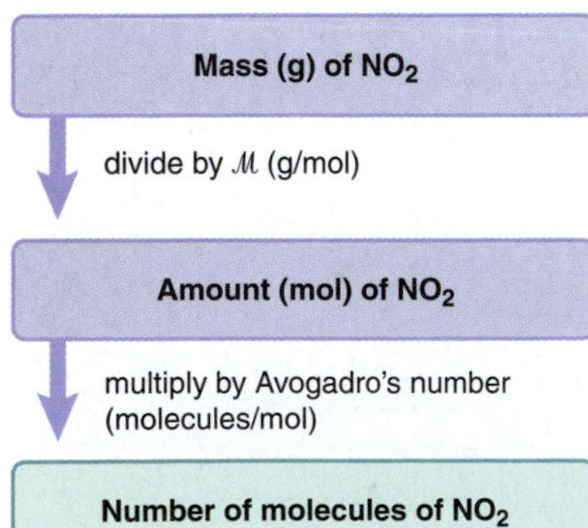

Solution The formula is NO_2. Calculating the molar mass:

$$\mathcal{M} = (1 \times \mathcal{M} \text{ of N}) + (2 \times \mathcal{M} \text{ of O}) = 14.01\text{ g/mol} + (2 \times 16.00\text{ g/mol}) = 46.01\text{ g/mol}$$

Converting from mass (g) of NO_2 to amount (mol):

$$\text{Amount (mol) of NO}_2 = 8.92\ \cancel{\text{g NO}_2} \times \frac{1\text{ mol NO}_2}{46.01\ \cancel{\text{g NO}_2}} = 0.194\text{ mol NO}_2$$

Converting from amount (mol) of NO_2 to number of molecules:

$$\text{No. of molecules} = 0.194\ \cancel{\text{mol NO}_2} \times \frac{6.022\times10^{23}\text{ NO}_2\text{ molecules}}{1\ \cancel{\text{mol NO}_2}} = 1.17\times10^{23}\text{ NO}_2\text{ molecules}$$

Check Rounding, we get $(\sim0.2\text{ mol})(6\times10^{23}) = 1.2\times10^{23}$, so the answer seems correct.

FOLLOW-UP PROBLEM 3.4 Fluoride ion is added to drinking water to prevent tooth decay. What is the mass (g) of sodium fluoride in a liter of water that contains 1.19×10^{19} formula units of the compound? Include a road map that shows how you planned the solution.

In Sample Problem 3.5, we go a step further and find the number of atoms of an element in the sample of a compound.

Sample Problem 3.5 Calculating the Number of Chemical Entities in a Given Mass of a Compound II

Problem Ammonium carbonate is a white solid that decomposes with warming. It has many uses, for example, as a component in baking powder, fire extinguishers, and smelling salts.
(a) How many formula units are in 41.6 g of ammonium carbonate?
(b) How many O atoms are in this sample?

Plan **(a)** We know the mass of compound (41.6 g) and need to find the number of formula units. As in Sample Problem 3.4, we need the formula to find the amount (mol) and then multiply by Avogadro's number to find the number of formula units. [The road map steps are for part (a).] **(b)** To find the number of O atoms, we multiply the number of formula units by the number of O atoms in one formula unit.

Road Map

Mass (g) of $(NH_4)_2CO_3$

↓ divide by $\mathcal{M}$ (g/mol)

Amount (mol) of $(NH_4)_2CO_3$

↓ multiply by 6.022×10^{23} formula units/mol

Number of formula units of $(NH_4)_2CO_3$

Solution **(a)** The formula is $(NH_4)_2CO_3$ (see Table 2.5). Calculating the molar mass:

$$\begin{aligned}\mathcal{M} &= (2 \times \mathcal{M} \text{ of N}) + (8 \times \mathcal{M} \text{ of H}) + (1 \times \mathcal{M} \text{ of C}) + (3 \times \mathcal{M} \text{ of O}) \\ &= (2 \times 14.01 \text{ g/mol N}) + (8 \times 1.008 \text{ g/mol H}) + 12.01 \text{ g/mol C} \\ &\quad + (3 \times 16.00 \text{ g/mol O}) \\ &= 96.09 \text{ g/mol } (NH_4)_2CO_3\end{aligned}$$

Converting from mass (g) to amount (mol):

$$\begin{aligned}\text{Amount (mol) of } (NH_4)_2CO_3 &= 41.6 \cancel{\text{g } (NH_4)_2CO_3} \times \frac{1 \text{ mol } (NH_4)_2CO_3}{96.09 \cancel{\text{g } (NH_4)_2CO_3}} \\ &= 0.433 \text{ mol } (NH_4)_2CO_3\end{aligned}$$

Converting from amount (mol) to formula units:

$$\begin{aligned}\text{Formula units of } (NH_4)_2CO_3 &= 0.433 \cancel{\text{mol } (NH_4)_2CO_3} \\ &\quad \times \frac{6.022 \times 10^{23} \text{ formula units } (NH_4)_2CO_3}{1 \cancel{\text{mol } (NH_4)_2CO_3}} \\ &= 2.61 \times 10^{23} \text{ formula units } (NH_4)_2CO_3\end{aligned}$$

(b) Finding the number of O atoms:

$$\begin{aligned}\text{No. of O atoms} &= 2.61 \times 10^{23} \cancel{\text{formula units } (NH_4)_2CO_3} \times \frac{3 \text{ O atoms}}{1 \cancel{\text{formula unit } (NH_4)_2CO_3}} \\ &= 7.83 \times 10^{23} \text{ O atoms}\end{aligned}$$

Check In (a), the units are correct. Since the mass is less than half the molar mass ($\sim 42/96 < 0.5$), the number of formula units should be less than half Avogadro's number ($\sim 2.6 \times 10^{23}/6.0 \times 10^{23} < 0.5$).

Comment A *common mistake* is to forget the subscript 2 outside the parentheses in $(NH_4)_2CO_3$, which would give a much lower molar mass.

FOLLOW-UP PROBLEM 3.5 Tetraphosphorus decoxide reacts with water to form phosphoric acid, a major industrial acid. In the laboratory, the oxide is a drying agent. **(a)** What is the mass (g) of 4.65×10^{22} molecules of tetraphosphorus decoxide?
(b) How many P atoms are present in this sample?

The Importance of Mass Percent

For many purposes, it is important to know how much of an element is present in a given amount of compound. In this section, we find the composition of a compound in terms of mass percent and use it to find the mass of each element in the compound.

Determining Mass Percent from a Chemical Formula Each element contributes a fraction of a compound's mass, and that fraction multiplied by 100 gives the element's mass percent. Finding the mass percent is similar on the molecular and molar scales:

- *For a molecule (or formula unit) of compound,* use the molecular (or formula) mass and chemical formula to find the mass percent of any element X in the compound:

$$\text{Mass \% of element X} = \frac{\text{atoms of X in formula} \times \text{atomic mass of X(amu)}}{\text{molecular (or formula) mass of compound (amu)}} \times 100$$

- *For a mole of compound,* use the molar mass and formula to find the mass percent of each element on a mole basis:

$$\text{Mass \% of element X} = \frac{\text{moles of X in formula} \times \text{molar mass of X (g/mol)}}{\text{mass (g) of 1 mol of compound}} \times 100 \tag{3.6}$$

As always, the individual mass percents add up to 100% (within rounding). In Sample Problem 3.6, we determine the mass percent of each element in a compound.

Sample Problem 3.6 Calculating the Mass Percent of Each Element in a Compound from the Formula

Problem In mammals, lactose (milk sugar) is metabolized to glucose ($C_6H_{12}O_6$), the key nutrient for generating chemical potential energy. What is the mass percent of each element in glucose?

Plan We know the relative amounts (mol) of the elements from the formula (6 C, 12 H, 6 O), and we have to find the mass % of each element. We multiply the amount of each by its molar mass to find its mass. Dividing each mass by the mass of 1 mol of glucose gives the mass fraction of each element, and multiplying by 100 gives each mass %. The calculation steps for any element (X) are shown in the road map.

Road Map

Amount (mol) of element X in 1 mol of glucose

↓ multiply by $\mathcal{M}$ (g/mol) of X

Mass (g) of X in 1 mol of glucose

↓ divide by mass (g) of 1 mol of compound

Mass fraction of X in glucose

↓ multiply by 100

Mass % of X in glucose

Solution Converting amount (mol) of C to mass (g):
We have 6 mol of C in 1 mol of glucose, so

$$\text{Mass (g) of C} = 6\ \cancel{\text{mol C}} \times \frac{12.01\text{ g C}}{1\ \cancel{\text{mol C}}} = 72.06\text{ g C}$$

Calculating the mass of 1 mol of glucose ($C_6H_{12}O_6$):

$$\begin{aligned}\mathcal{M} &= (6 \times \mathcal{M}\text{ of C}) + (12 \times \mathcal{M}\text{ of H}) + (6 \times \mathcal{M}\text{ of O})\\ &= (6 \times 12.01\text{ g/mol C}) + (12 \times 1.008\text{ g/mol H}) + (6 \times 16.00\text{ g/mol O})\\ &= 180.16\text{ g/mol } C_6H_{12}O_6\end{aligned}$$

Finding the mass fraction of C in glucose:

$$\text{Mass fraction of C} = \frac{\text{total mass of C}}{\text{mass of 1 mol glucose}} = \frac{72.06\text{ g C}}{180.16\text{ g glucose}} = 0.4000$$

Changing to mass %:

$$\begin{aligned}\text{Mass \% of C} &= \text{mass fraction of C} \times 100 = 0.4000 \times 100\\ &= 40.00\text{ mass \% C}\end{aligned}$$

Combining the steps for each of the other elements in glucose:

$$\text{Mass \% of H} = \frac{\text{mol H} \times \mathcal{M}\text{ of H}}{\text{mass of 1 mol glucose}} \times 100 = \frac{12\ \cancel{\text{mol H}} \times \dfrac{1.008\text{ g H}}{1\ \cancel{\text{mol H}}}}{180.16\text{ g glucose}} \times 100$$

$$= 6.714\text{ mass \% H}$$

$$\text{Mass \% of O} = \frac{\text{mol O} \times \mathcal{M}\text{ of O}}{\text{mass of 1 mol glucose}} \times 100 = \frac{6\ \cancel{\text{mol O}} \times \dfrac{16.00\text{ g O}}{1\ \cancel{\text{mol O}}}}{180.16\text{ g glucose}} \times 100$$

$$= 53.29\text{ mass \% O}$$

Check The answers make sense: even though there are equal numbers of moles of O and of C in the compound, the mass % of O is greater than the mass % of C because the molar mass of O is greater than the molar mass of C. The mass % of H is small because the molar mass of H is small. The sum of the mass percents is 100.00%.

Comment From here on, you should be able to determine the molar mass of a compound, so that calculation will no longer be shown.

FOLLOW-UP PROBLEM 3.6 Agronomists base the effectiveness of fertilizers on their nitrogen content. Ammonium nitrate is a common fertilizer. Calculate the mass percent of N in ammonium nitrate.

Determining the Mass of an Element from Its Mass Percent Sample Problem 3.6 shows that *an element always constitutes the same fraction of the mass of a given compound* (see Equation 3.6). We can use that fraction to find the mass of element in any mass of a compound:

$$\text{Mass of element} = \text{mass of compound} \times \frac{\text{mass of element in 1 mol of compound}}{\text{mass of 1 mol of compound}} \quad (3.7)$$

For example, to find the mass of oxygen in 15.5 g of nitrogen dioxide, we have

$$\text{Mass (g) of O} = 15.5 \text{ g } NO_2 \times \frac{2 \text{ mol} \times \mathcal{M} \text{ of O (g/mol)}}{\text{mass (g) of 1 mol } NO_2}$$

$$= 15.5 \text{ g } \cancel{NO_2} \times \frac{32.00 \text{ g O}}{46.01 \text{ g } \cancel{NO_2}} = 10.8 \text{ g O}$$

Sample Problem 3.7 Calculating the Mass of an Element in a Compound

Problem Use the information in Sample Problem 3.6 to determine the mass (g) of carbon in 16.55 g of glucose.

Plan To find the mass of C in the sample of glucose, we multiply the mass of the sample by the mass of 6 mol of C divided by the mass of 1 mol of glucose.

Solution Finding the mass of C in a given mass of glucose:

$$\text{Mass (g) of C} = \text{mass (g) of glucose} \times \frac{6 \text{ mol C} \times \mathcal{M} \text{ of C (g/mol)}}{\text{mass (g) of 1 mol glucose}}$$

$$= 16.55 \, \cancel{\text{g glucose}} \times \frac{72.06 \text{ g C}}{180.16 \, \cancel{\text{g glucose}}} = 6.620 \text{ g C}$$

Check Rounding shows that the answer is "in the right ballpark": 16 g times less than 0.5 parts by mass should be less than 8 g.

FOLLOW-UP PROBLEM 3.7 Use the information in Follow-up Problem 3.6 to find the mass (g) of N in 35.8 kg of ammonium nitrate.

Summary of Section 3.1

- A mole of substance is the amount that contains Avogadro's number (6.022×10^{23}) of chemical entities (atoms, molecules, or formula units).
- The mass (in grams) of a mole of the entity has the same numerical value as the mass (in amu) of the individual entity. Thus, the mole allows us to count entities by weighing them.
- Using the molar mass ($\mathcal{M}$, g/mol) of an element (or compound) and Avogadro's number as conversion factors, we can convert among amount (mol), mass (g), and number of entities.
- The mass fraction of element X in a compound is used to find the mass of X in a given amount of the compound.

3.2 • DETERMINING THE FORMULA OF AN UNKNOWN COMPOUND

In Sample Problems 3.6 and 3.7, we used a compound's formula to find the mass percent (or mass fraction) of each element in it *and* the mass of an element in any size sample of it. In this section, we do the reverse: we use the masses of elements in a compound to find the formula. Then, we look briefly at the relationship between molecular formula and molecular structure.

Let's compare three common types of formula, using hydrogen peroxide as an example:

- The **empirical formula** is derived from mass analysis. It shows the *lowest* whole numbers of moles, and thus the *relative* numbers of atoms, of each element in the compound. For example, in hydrogen peroxide, there is 1 part by mass of hydrogen for every 16 parts by mass of oxygen. Because the atomic mass of hydrogen is 1.008 amu and that of oxygen is 16.00 amu, there is one H atom for every O atom. Thus, the empirical formula is HO.

Recall from Section 2.8 that

- The **molecular formula** shows the *actual* number of atoms of each element in a molecule: the molecular formula is H_2O_2, twice the empirical formula.
- The **structural formula** shows the relative *placement and connections of atoms* in the molecule: the structural formula is H—O—O—H.

Let's see how to determine empirical and molecular formulas.

Empirical Formulas

A chemist studying an unknown compound goes through a three-step process to find the empirical formula:

1. Determine the mass (g) of each component element.
2. Convert each mass (g) to amount (mol), and write a preliminary formula.
3. Convert the amounts (mol) mathematically to whole-number (integer) subscripts. To accomplish this math conversion,
   - Divide each subscript by the smallest subscript, and
   - If necessary, multiply through by the *smallest integer* that turns all subscripts into integers.

Sample Problem 3.8 demonstrates these math steps.

Sample Problem 3.8 Determining an Empirical Formula from Amounts of Elements

Problem A sample of an unknown compound contains 0.21 mol of zinc, 0.14 mol of phosphorus, and 0.56 mol of oxygen. What is the empirical formula?

Plan We are given the amount (mol) of each element as fractions. We use these fractional amounts directly in a preliminary formula as subscripts of the element symbols. Then, we convert the fractions to whole numbers.

Road Map

Amount (mol) of each element

↓ use nos. of moles as subscripts

Preliminary formula

↓ change to integer subscripts

Empirical formula

Solution Using the fractions to write a preliminary formula, with the symbols Zn for zinc, P for phosphorus, and O for oxygen:

$$Zn_{0.21}P_{0.14}O_{0.56}$$

Converting the fractions to whole numbers:

1. Divide each subscript by the smallest one, which in this case is 0.14:

$$Zn_{\frac{0.21}{0.14}}P_{\frac{0.14}{0.14}}O_{\frac{0.56}{0.14}} \longrightarrow Zn_{1.5}P_{1.0}O_{4.0}$$

2. Multiply through by the *smallest integer* that turns all subscripts into integers. We multiply by 2 to make 1.5 (the subscript for Zn) into an integer:

$$Zn_{(1.5\times2)}P_{(1.0\times2)}O_{(4.0\times2)} \longrightarrow Zn_{3.0}P_{2.0}O_{8.0}, \text{ or } Zn_3P_2O_8$$

Check The integer subscripts must be the smallest integers with the same ratio as the original fractional numbers of moles: 3/2/8 is *the same ratio* as 0.21/0.14/0.56.

Comment A more conventional way to write this formula is $Zn_3(PO_4)_2$; this compound is zinc phosphate, formerly used widely as a dental cement.

FOLLOW-UP PROBLEM 3.8 A sample of a white solid contains 0.170 mol of boron and 0.255 mol of oxygen. What is the empirical formula?

Sample Problems 3.9–3.11 show how other types of compositional data are used to determine chemical formulas.

Sample Problem 3.9 Determining an Empirical Formula from Masses of Elements

Problem Analysis of a sample of an ionic compound yields 2.82 g of Na, 4.35 g of Cl, and 7.83 g of O. What is the empirical formula and the name of the compound?

Plan This problem is similar to Sample Problem 3.8, except that we are given element *masses* that we must convert into integer subscripts. We first divide each mass by the element's molar mass to find amount (mol). Then we construct a preliminary formula and convert the amounts (mol) to integers.

Solution Finding amount (mol) of each element:

$$\text{Amount (mol) of Na} = 2.82\ \cancel{\text{g Na}} \times \frac{1\text{ mol Na}}{22.99\ \cancel{\text{g Na}}} = 0.123\text{ mol Na}$$

$$\text{Amount (mol) of Cl} = 4.35\ \cancel{\text{g Cl}} \times \frac{1\text{ mol Cl}}{35.45\ \cancel{\text{g Cl}}} = 0.123\text{ mol Cl}$$

$$\text{Amount (mol) of O} = 7.83\ \cancel{\text{g O}} \times \frac{1\text{ mol O}}{16.00\ \cancel{\text{g O}}} = 0.489\text{ mol O}$$

Constructing a preliminary formula: $Na_{0.123}Cl_{0.123}O_{0.489}$

Converting to integer subscripts (dividing all by the smallest subscript):

$$Na_{\frac{0.123}{0.123}}Cl_{\frac{0.123}{0.123}}O_{\frac{0.489}{0.123}} \longrightarrow Na_{1.00}Cl_{1.00}O_{3.98} \approx Na_1Cl_1O_4, \quad \text{or} \quad NaClO_4$$

The empirical formula is $NaClO_4$; the name is sodium perchlorate.

Check The numbers of moles seem correct because the masses of Na and Cl are slightly more than 0.1 of their molar masses. The mass of O is greatest and its molar mass is smallest, so it should have the greatest number of moles. The ratio of subscripts, 1/1/4, is the same as the ratio of moles, 0.123/0.123/0.489 (within rounding).

FOLLOW-UP PROBLEM 3.9 An unknown metal M reacts with sulfur to form a compound with the formula M_2S_3. If 3.12 g of M reacts with 2.88 g of S, what are the names of M and M_2S_3? [*Hint:* Determine the amount (mol) of S and use the formula to find the amount (mol) of M.]

Molecular Formulas

If we know the molar mass of a compound, we can use the empirical formula to obtain the molecular formula, which uses as subscripts the *actual* numbers of moles of each element in 1 mol of compound. For some compounds, such as water (H_2O), ammonia (NH_3), and methane (CH_4), the empirical and molecular formulas are identical, but for many others, the molecular formula is a *whole-number multiple* of the empirical formula. As you saw, hydrogen peroxide has the empirical formula HO. Dividing the molar mass of hydrogen peroxide (34.02 g/mol) by the empirical formula mass of HO (17.01 g/mol) gives the whole-number multiple:

$$\text{Whole-number multiple} = \frac{\text{molar mass (g/mol)}}{\text{empirical formula mass (g/mol)}} = \frac{34.02\ \cancel{\text{g/mol}}}{17.01\ \cancel{\text{g/mol}}} = 2.000 = 2$$

Multiplying the empirical formula subscripts by 2 gives the molecular formula:

$$H_{(1\times2)}O_{(1\times2)} \text{ gives } H_2O_2$$

From Mass Percents to Formula Instead of giving masses of each element, analytical laboratories provide mass percents. We use these steps to find a formula:

1. Assume 100.0 g of compound to express each mass percent directly as mass (g).
2. Convert each mass (g) to amount (mol).
3. Derive the empirical formula.
4. Proceed as above to find the whole-number multiple and the molecular formula.

Sample Problem 3.10 Determining a Molecular Formula from Elemental Analysis and Molar Mass

Problem During excessive physical activity, lactic acid ($\mathcal{M}$ = 90.08 g/mol) forms in muscle tissue and is responsible for muscle soreness. Elemental analysis shows that this compound contains 40.0 mass % C, 6.71 mass % H, and 53.3 mass % O.
(a) Determine the empirical formula of lactic acid.
(b) Determine the molecular formula.

(a) Determining the empirical formula

Plan We know the mass % of each element and must convert each to an integer subscript. The mass of the sample of lactic acid is not given, but the mass percents are the same for any sample of it. Therefore, we assume there is 100.0 g of lactic acid and express each mass % as a number of grams. Then, we construct the empirical formula as in Sample Problem 3.9.

Solution Expressing mass % as mass (g) by assuming 100.0 g of lactic acid:

$$\text{Mass (g) of C} = \frac{40.0 \text{ parts C by mass}}{100 \text{ parts by mass}} \times 100.0 \text{ g} = 40.0 \text{ g C}$$

Similarly, we have 6.71 g of H and 53.3 g of O.

Converting from mass (g) of each element to amount (mol):

$$\text{Amount (mol) of C} = \text{mass of C} \times \frac{1}{\mathcal{M} \text{ of C}} = 40.0 \cancel{\text{g C}} \times \frac{1 \text{ mol C}}{12.01 \cancel{\text{g C}}} = 3.33 \text{ mol C}$$

Similarly, we have 6.66 mol of H and 3.33 mol of O.

Constructing the preliminary formula: $C_{3.33}H_{6.66}O_{3.33}$

Converting to integer subscripts:

$$C_{\frac{3.33}{3.33}}H_{\frac{6.66}{3.33}}O_{\frac{3.33}{3.33}} \longrightarrow C_{1.00}H_{2.00}O_{1.00} = C_1H_2O_1, \text{ the empirical formula is } \boxed{CH_2O}$$

Check The numbers of moles seem correct: the masses of C and O are each slightly more than 3 times their molar masses (e.g., for C, 40 g/(12 g/mol) > 3 mol), and the mass of H is over 6 times its molar mass of 1.

(b) Determining the molecular formula

Plan The molecular formula subscripts are whole-number multiples of the empirical formula subscripts. To find this multiple, we divide the given molar mass (90.08 g/mol) by the empirical formula mass, which we find from the sum of the elements' molar masses. Then we multiply each subscript in the empirical formula by the multiple.

Solution The empirical formula mass is 30.03 g/mol. Finding the whole-number multiple:

$$\text{Whole-number multiple} = \frac{\mathcal{M} \text{ of lactic acid}}{\mathcal{M} \text{ of empirical formula}} = \frac{90.08 \cancel{\text{g/mol}}}{30.03 \cancel{\text{g/mol}}} = 3.000 = 3$$

Determining the molecular formula:

$$C_{(1\times3)}H_{(2\times3)}O_{(1\times3)} = \boxed{C_3H_6O_3}$$

Check The calculated molecular formula has the same ratio of moles of elements (3/6/3) as the empirical formula (1/2/1) and corresponds to the given molar mass:

$$\begin{aligned}\mathcal{M} \text{ of lactic acid} &= (3 \times \mathcal{M} \text{ of C}) + (6 \times \mathcal{M} \text{ of H}) + (3 \times \mathcal{M} \text{ of O}) \\ &= (3 \times 12.01 \text{ g/mol}) + (6 \times 1.008 \text{ g/mol}) + (3 \times 16.00 \text{ g/mol}) \\ &= 90.08 \text{ g/mol}\end{aligned}$$

FOLLOW-UP PROBLEM 3.10 One of the most widespread environmental carcinogens (cancer-causing agents) is benzo[*a*]pyrene ($\mathcal{M}$ = 252.30 g/mol). It is found in coal dust, cigarette smoke, and even charcoal-grilled meat. Analysis of this hydrocarbon shows 95.21 mass % C and 4.79 mass % H. What is the molecular formula of benzo[*a*]pyrene?

Combustion Analysis of Organic Compounds Still another type of compositional data is obtained through **combustion analysis,** used to measure the amounts of carbon and hydrogen in a combustible organic compound. The unknown compound is burned in an excess of pure O_2, and the H_2O and CO_2 that form are absorbed in separate containers (Figure 3.4). By weighing the absorbers before and after combustion, we find the masses of CO_2 and H_2O and use them to find the masses of C and H in the compound; from these results, we find the empirical formula. Many organic compounds also contain oxygen, nitrogen, or a halogen. As long as the third element doesn't interfere with the absorption of H_2O and CO_2, we calculate its mass by subtracting the masses of C and H from the original mass of the compound.

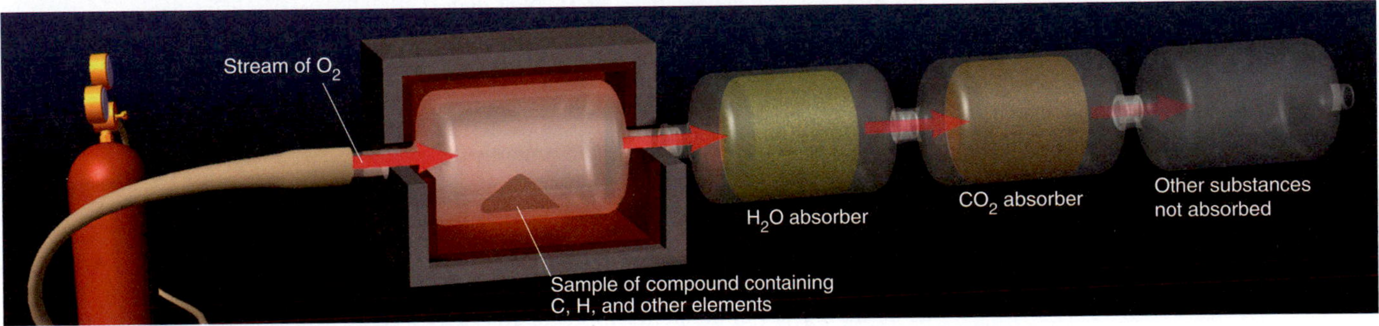

Figure 3.4 Combustion apparatus for determining formulas of organic compounds. A sample of an organic compound is burned in a stream of O_2. The resulting H_2O is absorbed by $Mg(ClO_4)_2$, and the CO_2 is absorbed by NaOH on asbestos.

Sample Problem 3.11 Determining a Molecular Formula from Combustion Analysis

Problem Vitamin C ($\mathcal{M}$ = 176.12 g/mol) is a compound of C, H, and O found in many natural sources, especially citrus fruits. When a 1.000-g sample of vitamin C is burned in a combustion apparatus, the following data are obtained:

$$\text{Mass of } CO_2 \text{ absorber after combustion} = 85.35 \text{ g}$$
$$\text{Mass of } CO_2 \text{ absorber before combustion} = 83.85 \text{ g}$$
$$\text{Mass of } H_2O \text{ absorber after combustion} = 37.96 \text{ g}$$
$$\text{Mass of } H_2O \text{ absorber before combustion} = 37.55 \text{ g}$$

What is the molecular formula of vitamin C?

Plan We find the masses of CO_2 and H_2O by subtracting the masses of the absorbers before and after the combustion. From the mass of CO_2, we use Equation 3.7 to find the mass of C. Similarly, we find the mass of H from the mass of H_2O. The mass of vitamin C (1.000 g) minus the sum of the masses of C and H gives the mass of O, the third element present. Then, we proceed as in Sample Problem 3.10: calculate amount (mol) of each element using its molar mass, construct the empirical formula, determine the whole-number multiple from the given molar mass, and construct the molecular formula.

Solution Finding the masses of combustion products:

$$\begin{aligned}\text{Mass (g) of } CO_2 &= \text{mass of } CO_2 \text{ absorber after} - \text{mass before}\\ &= 85.35 \text{ g} - 83.85 \text{ g} = 1.50 \text{ g } CO_2\end{aligned}$$

$$\begin{aligned}\text{Mass (g) of } H_2O &= \text{mass of } H_2O \text{ absorber after} - \text{mass before}\\ &= 37.96 \text{ g} - 37.55 \text{ g} = 0.41 \text{ g } H_2O\end{aligned}$$

Calculating masses (g) of C and H using Equation 3.7:

$$\text{Mass of element} = \text{mass of compound} \times \frac{\text{mass of element in 1 mol of compound}}{\text{mass of 1 mol of compound}}$$

$$\text{Mass (g) of C} = \text{mass of } CO_2 \times \frac{1 \text{ mol C} \times \mathcal{M} \text{ of C}}{\text{mass of 1 mol } CO_2} = 1.50 \cancel{\text{g } CO_2} \times \frac{12.01 \text{ g C}}{44.01 \cancel{\text{g } CO_2}}$$
$$= 0.409 \text{ g C}$$

$$\text{Mass (g) of H} = \text{mass of } H_2O \times \frac{2 \text{ mol H} \times \mathcal{M} \text{ of H}}{\text{mass of 1 mol } H_2O} = 0.41 \cancel{\text{g } H_2O} \times \frac{2.016 \text{ g H}}{18.02 \cancel{\text{g } H_2O}}$$
$$= 0.046 \text{ g H}$$

Calculating mass (g) of O:

$$\text{Mass (g) of O} = \text{mass of vitamin C sample} - (\text{mass of C} + \text{mass of H})$$
$$= 1.000 \text{ g} - (0.409 \text{ g} + 0.046 \text{ g}) = 0.545 \text{ g O}$$

Finding the amounts (mol) of elements: Dividing the mass (g) of each element by its molar mass gives 0.0341 mol of C, 0.046 mol of H, and 0.0341 mol of O.

Constructing the preliminary formula: $C_{0.0341}H_{0.046}O_{0.0341}$

Determining the empirical formula: Dividing through by the smallest subscript gives

$$C_{\frac{0.0341}{0.0341}}H_{\frac{0.046}{0.0341}}O_{\frac{0.0341}{0.0341}} = C_{1.00}H_{1.3}O_{1.00}$$

We find that 3 is the smallest integer that makes all subscripts into integers:

$$C_{(1.00\times3)}H_{(1.3\times3)}O_{(1.00\times3)} = C_{3.00}H_{3.9}O_{3.00} \approx C_3H_4O_3$$

Determining the molecular formula:

$$\text{Whole-number multiple} = \frac{\mathcal{M} \text{ of vitamin C}}{\mathcal{M} \text{ of empirical formula}} = \frac{176.12 \cancel{\text{g/mol}}}{88.06 \cancel{\text{g/mol}}} = 2.000 = 2$$

$$C_{(3\times2)}H_{(4\times2)}O_{(3\times2)} = \boxed{C_6H_8O_6}$$

Check The element masses seem correct: carbon makes up slightly more than 0.25 of the mass of CO_2 (12 g/44 g > 0.25), as do the masses in the problem (0.409 g/1.50 g > 0.25). Hydrogen makes up slightly more than 0.10 of the mass of H_2O (2 g/18 g > 0.10), as do the masses in the problem (0.046 g/0.41 g > 0.10). The molecular formula has the same ratio of subscripts (6/8/6) as the empirical formula (3/4/3) and the preliminary formula (0.0341/0.046/0.0341), and it gives the known molar mass:

$$(6 \times \mathcal{M} \text{ of C}) + (8 \times \mathcal{M} \text{ of H}) + (6 \times \mathcal{M} \text{ of O}) = \mathcal{M} \text{ of vitamin C}$$
$$(6 \times 12.01 \text{ g/mol}) + (8 \times 1.008 \text{ g/mol}) + (6 \times 16.00 \text{ g/mol}) = 176.12 \text{ g/mol}$$

Comment The subscript we calculated for H was 3.9, which we rounded to 4. But, if we had strung the calculation steps together, we would have obtained 4.0:

$$\text{Subscript of H} = 0.41 \text{ g } H_2O \times \frac{2.016 \text{ g H}}{18.02 \text{ g } H_2O} \times \frac{1 \text{ mol H}}{1.008 \text{ g H}} \times \frac{1}{0.0341 \text{ mol}} \times 3 = 4.0$$

FOLLOW-UP PROBLEM 3.11 A dry-cleaning solvent ($\mathcal{M}$ = 146.99 g/mol) that contains C, H, and Cl is suspected to be a cancer-causing agent. When a 0.250-g sample was studied by combustion analysis, 0.451 g of CO_2 and 0.0617 g of H_2O formed. Find the molecular formula.

Isomers

A molecular formula tells the *actual* number of each type of atom, providing as much information as possible from mass analysis. Yet *different compounds can have the* ***same*** *molecular formula* because the atoms can bond to each other in different arrangements to give more than one *structural formula.* **Isomers** are two or more compounds with the same molecular formula but different properties. The simplest type of isomerism, called *constitutional,* or *structural, isomerism,* occurs when the atoms link together in different arrangements. The pair of constitutional isomers shown in Table 3.2 share the molecular formula C_2H_6O but have very different properties because they are different compounds. In this case, they are even different *types* of compounds—one is an alcohol, and the other an ether.

As the number and kinds of atoms increase, the number of constitutional isomers—that is, the number of structural formulas that can be written for a given molecular formula—also increases: C_2H_6O has the two isomers that you've seen, C_3H_8O has three, and $C_4H_{10}O$ seven. We'll discuss this and other types of isomerism fully later in the text.

Table 3.2 Two Constitutional Isomers of C_2H_6O

Property	Ethanol	Dimethyl Ether
$\mathcal{M}$ (g/mol)	46.07	46.07
Boiling point	78.5°C	−25°C
Density (at 20°C)	0.789 g/mL (liquid)	0.00195 g/mL (gas)
Structural formula	CH_3CH_2OH (H—C(H)(H)—C(H)(H)—O—H)	CH_3OCH_3 (H—C(H)(H)—O—C(H)(H)—H)
Space-filling model		

Summary of Section 3.2

- From the masses of elements in a compound, their relative numbers of moles are found, which gives the empirical formula.
- If the molar mass of the compound is known, the molecular formula, the actual numbers of moles of each element, can also be determined.
- Combustion analysis provides data on the masses of C and H in an organic compound, which are used to obtain the formula.
- Atoms can bond in different arrangements (structural formulas). Two or more compounds with the same molecular formula are constitutional isomers.

3.3 • WRITING AND BALANCING CHEMICAL EQUATIONS

Thinking in terms of amounts rather than masses allows us to view reactions as large populations of interacting particles rather than as grams of material. For example, for the formation of HF from H_2 and F_2, if we weigh the substances, we find that

Macroscopic level (grams): 2.016 g of H_2 and 38.00 g of F_2 react to form 40.02 g of HF

This information tells us little except that mass is conserved. However, if we convert these masses (g) to amounts (mol), we find that

Macroscopic level (moles): 1 mol of H_2 and 1 mol of F_2 react to form 2 mol of HF

This information reveals that an enormous number of H_2 molecules react with just as many F_2 molecules to form twice as many HF molecules. Dividing by Avogadro's number gives the reaction involving a small group of molecules:

Molecular level: 1 molecule of H_2 and 1 molecule of F_2 react to form 2 molecules of HF

Thus, *the macroscopic (molar) change corresponds to the submicroscopic (molecular) change* (Figure 3.5). This information is expressed by a **chemical equation,** which shows the identities and quantities of substances in a chemical or physical change.

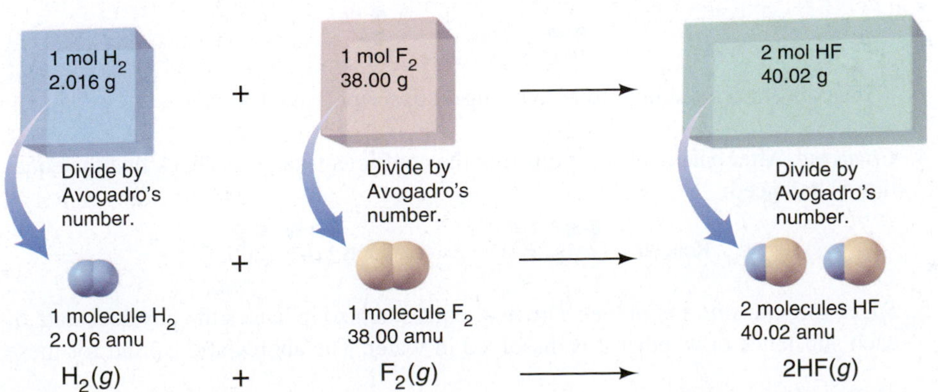

Figure 3.5 The formation of HF on the macroscopic and molecular levels.

Steps for Balancing an Equation To present a chemical change quantitatively, the equation must be *balanced: the same number of each type of atom must appear on both sides.* As an example, here is a description of a chemical change that occurs in many fireworks and in a common lecture demonstration: a magnesium strip burns in oxygen gas to yield powdery magnesium oxide. (Light and heat are also produced, but we are concerned here only with substances.) Converting this description into a balanced equation involves the following steps:

1. *Translating the statement.* We first translate the chemical statement into a "skeleton" equation: the substances present *before* the change, called **reactants,** are placed to the left of a yield arrow, which points to the substances produced *during* the change, called **products:**

$$\underbrace{\_\_Mg \quad + \quad \_\_O_2}_{\text{reactants}} \xrightarrow{\text{yield}} \underbrace{\_\_MgO}_{\text{product}}$$

magnesium and oxygen   yield   magnesium oxide

At the beginning of the balancing process, we put a blank *in front of* each formula to remind us that we have to account for its atoms.

2. *Balancing the atoms.* By shifting our attention back and forth, we *match the numbers of each type of atom on the left and the right of the yield arrow.* In each blank, we place a **balancing (stoichiometric) coefficient,** a numerical multiplier of *all the atoms* in the formula that follows it. In general, balancing is easiest when we
   - Start with the most complex substance, the one with the largest number of different types of atoms.
   - End with the least complex substance, such as an element by itself.

   In this case, MgO is the most complex, so we place a coefficient 1 in that blank:

$$\_\_Mg + \_\_O_2 \longrightarrow \underline{1}\,MgO$$

   To balance the Mg in MgO, we place a 1 in front of Mg on the left:

$$\underline{1}\,Mg + \_\_O_2 \longrightarrow \underline{1}\,MgO$$

   The O atom in MgO must be balanced by one O atom on the left. One-half an O_2 molecule provides one O atom:

$$\underline{1}\,Mg + \underline{\tfrac{1}{2}}\,O_2 \longrightarrow \underline{1}\,MgO$$

   In terms of numbers of each type of atom, the equation is balanced.

3. *Adjusting the coefficients.* There are several conventions about the final coefficients:
   - In most cases, *the smallest whole-number coefficients are preferred.* In this case, one-half of an O_2 molecule cannot exist, so we multiply the equation by 2:

$$2Mg + 1O_2 \longrightarrow 2MgO$$

   - We used the coefficient 1 to remind us to balance each substance. But, a coefficient of 1 is implied by the presence of the formula, so we don't write it:

$$2Mg + O_2 \longrightarrow 2MgO$$

   (This convention is similar to not writing a subscript 1 in a formula.)

4. *Checking.* After balancing and adjusting the coefficients, always check that the equation is balanced:

$$\text{Reactants (2Mg, 2O)} \longrightarrow \text{products (2Mg, 2O)}$$

5. *Specifying the states of matter.* The final equation also indicates the physical state of each substance or whether it is dissolved in water. The abbreviations used for these

states are shown in the margin. From the original statement, we know that a Mg "strip" is solid, O_2 is a gas, and "powdery" MgO is also solid. The balanced equation, therefore, is

$$2Mg(s) + O_2(g) \longrightarrow 2MgO(s)$$

g for gas
l for liquid
s for solid
aq for aqueous solution

As you saw in Figure 3.5, *balancing coefficients refer to both individual chemical entities and moles of entities*. Thus,

2 atoms of Mg and 1 molecule of O_2 yield 2 formula units of MgO
2 moles of Mg and 1 mole of O_2 yield 2 moles of MgO

Figure 3.6 depicts this reaction on three levels:

- *Macroscopic level (photos),* as it appears in the laboratory
- *Molecular level (blow-up circles),* as chemists imagine it (with darker colored atoms representing the stoichiometry)
- *Symbolic level,* in the form of the balanced chemical equation

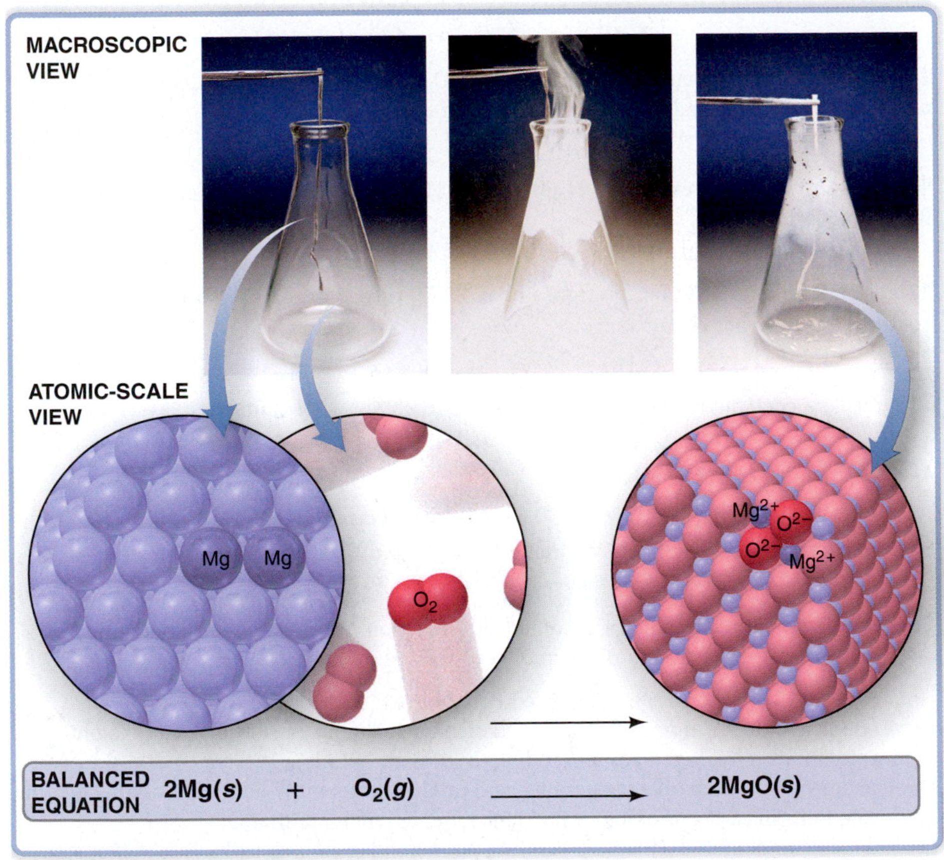

Figure 3.6 A three-level view of the reaction between magnesium and oxygen.

Keep in mind several key points about the balancing process:

- A coefficient operates on *all* the atoms in the formula that follows it:

  2MgO *means* 2 × (MgO), or 2 Mg atoms + 2 O atoms
  $2Ca(NO_3)_2$ *means* 2 × $[Ca(NO_3)_2]$, or 2 Ca atoms + 4 N atoms + 12 O atoms

- Chemical formulas *cannot* be altered. In step 2 of the example, we *cannot* balance the O atoms by changing MgO to MgO_2 because MgO_2 is a different compound.
- Other reactants or products *cannot* be added. Thus, we *cannot* balance the O atoms by changing the reactant from O_2 molecules to O atoms or by adding an O atom to the products. The description of the reaction mentions oxygen gas, which consists of O_2 molecules, *not* separate O atoms.

- A balanced equation remains balanced if you multiply all the coefficients by the same number. For example,

$$4Mg(s) + 2O_2(g) \longrightarrow 4MgO(s)$$

is also balanced because the coefficients have just been multiplied by 2. However, *by convention,* we balance an equation with the *smallest* whole-number coefficients.

Sample Problem 3.12 Balancing Chemical Equations

Problem Within the cylinders of a car's engine, the hydrocarbon octane (C_8H_{18}), one of many components of gasoline, mixes with oxygen from the air and burns to form carbon dioxide and water vapor. Write a balanced equation for this reaction.

Solution 1. *Translate* the statement into a skeleton equation (with coefficient blanks). Octane and oxygen are reactants; "oxygen from the air" implies molecular oxygen, O_2. Carbon dioxide and water vapor are products:

$$\_\_C_8H_{18} + \_\_O_2 \longrightarrow \_\_CO_2 + \_\_H_2O$$

2. *Balance the atoms.* Start with the most complex substance, C_8H_{18}, and balance O_2 last:

$$\underline{1}\,C_8H_{18} + \_\_O_2 \longrightarrow \_\_CO_2 + \_\_H_2O$$

The C atoms in C_8H_{18} end up in CO_2. Each CO_2 contains one C atom, so 8 molecules of CO_2 are needed to balance the 8 C atoms in each C_8H_{18}:

$$\underline{1}\,C_8H_{18} + \_\_O_2 \longrightarrow \underline{8}\,CO_2 + \_\_H_2O$$

The H atoms in C_8H_{18} end up in H_2O. The 18 H atoms in C_8H_{18} require the coefficient 9 in front of H_2O:

$$\underline{1}\,C_8H_{18} + \_\_O_2 \longrightarrow \underline{8}\,CO_2 + \underline{9}\,H_2O$$

There are 25 atoms of O on the right (16 in $8CO_2$ plus 9 in $9H_2O$), so we place the coefficient $\frac{25}{2}$ in front of O_2:

$$\underline{1}\,C_8H_{18} + \underline{\tfrac{25}{2}}\,O_2 \longrightarrow \underline{8}\,CO_2 + \underline{9}\,H_2O$$

3. *Adjust the coefficients.* Multiply through by 2 to obtain whole numbers:

$$2C_8H_{18} + 25O_2 \longrightarrow 16CO_2 + 18H_2O$$

4. *Check* that the equation is balanced:

$$\text{Reactants } (16\text{ C}, 36\text{ H}, 50\text{ O}) \longrightarrow \text{products } (16\text{ C}, 36\text{ H}, 50\text{ O})$$

5. *Specify* states of matter. C_8H_{18} is liquid; O_2, CO_2, and H_2O vapor are gases:

$$2C_8H_{18}(l) + 25O_2(g) \longrightarrow 16CO_2(g) + 18H_2O(g)$$

Comment This is an example of a combustion reaction. *Any* compound containing C and H that burns in an excess of air produces CO_2 and H_2O.

FOLLOW-UP PROBLEM 3.12 Write a balanced equation for each of the following:
(a) A characteristic reaction of Group 1A(1) elements: chunks of sodium react violently with water to form hydrogen gas and sodium hydroxide solution.
(b) The destruction of marble statuary by acid rain: aqueous nitric acid reacts with calcium carbonate to form carbon dioxide, water, and aqueous calcium nitrate.
(c) Halogen compounds exchanging bonding partners: phosphorus trifluoride is prepared by the reaction of phosphorus trichloride and hydrogen fluoride; hydrogen chloride is the other product. The reaction involves gases only.
(d) Explosive decomposition of dynamite: liquid nitroglycerine ($C_3H_5N_3O_9$) explodes to produce a mixture of gases—carbon dioxide, water vapor, nitrogen, and oxygen.

Visualizing a Reaction with a Molecular Scene A great way to focus on the rearrangement of atoms from reactants to products is by visualizing an equation as a molecular scene. Here's a representation of the combustion of octane we just balanced:

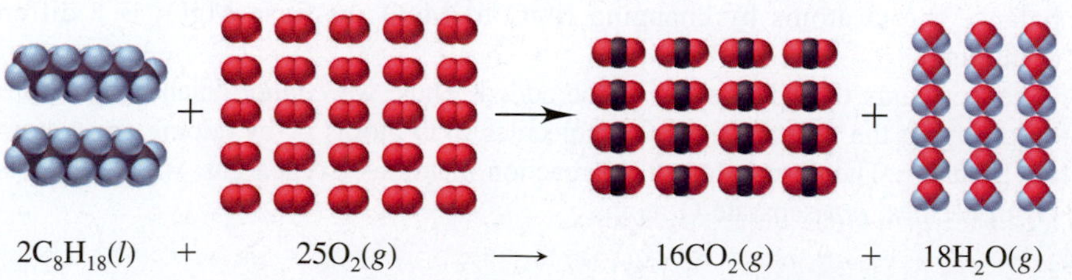

Now let's work through a sample problem to do the reverse—derive a balanced equation from a molecular scene.

Sample Problem 3.13 Balancing an Equation from a Molecular Scene

Problem The following molecular scenes depict an important reaction in nitrogen chemistry (nitrogen is blue; oxygen is red):

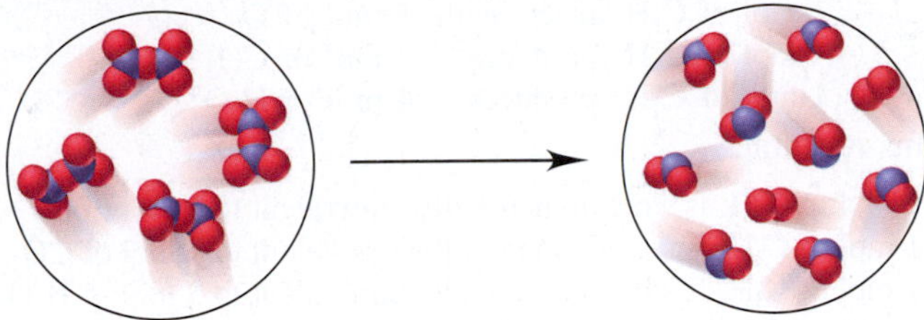

Write a balanced equation for this reaction.

Plan To write a balanced equation, we first have to determine the formulas of the molecules and obtain coefficients by counting the number of each molecule. Then, we arrange this information in the correct equation format, using the smallest whole-number coefficients and including states of matter.

Solution The reactant circle shows only one type of molecule. It has two N and five O atoms, so the formula is N_2O_5; there are four of these molecules. The product circle shows two different molecules, one with one N and two O atoms, and the other with two O atoms; there are eight NO_2 and two O_2. Thus, we have:

$$4N_2O_5 \longrightarrow 8NO_2 + 2O_2$$

Writing the balanced equation with the smallest whole-number coefficients and all substances as gases:

$$2N_2O_5(g) \longrightarrow 4NO_2(g) + O_2(g)$$

Check Reactant (4 N, 10 O) $\longrightarrow$ products (4 N, 8 + 2 = 10 O)

FOLLOW-UP PROBLEM 3.13 Write a balanced equation for the important atmospheric reaction depicted below (carbon is black; oxygen is red):

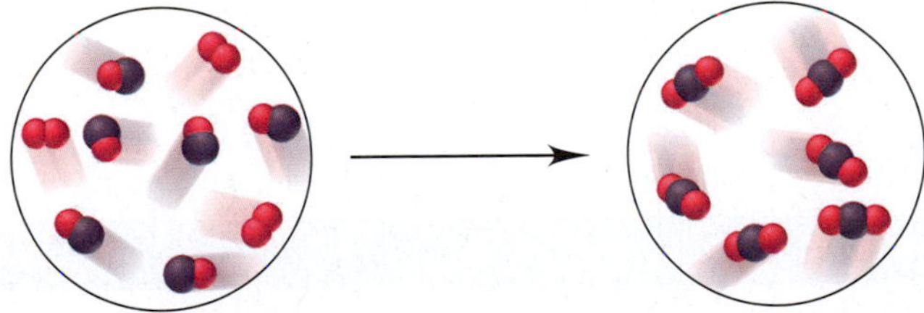

Summary of Section 3.3

- A chemical equation has reactant formulas on the left of a yield arrow and product formulas on the right.
- A balanced equation has the same number of each type of atom on both sides.
- Balancing coefficients are integer multipliers for *all* the atoms in a formula and apply to the individual entity or to moles of entities.

3.4 • CALCULATING QUANTITIES OF REACTANT AND PRODUCT

A balanced equation is essential for all calculations involving chemical change: *if you know the number of moles of one substance, the balanced equation tells you the number of moles of the others.*

Stoichiometrically Equivalent Molar Ratios from the Balanced Equation

In a balanced equation, *the amounts (mol) of substances are stoichiometrically equivalent to each other,* which means that a specific amount of one substance is formed from, produces, or reacts with a specific amount of the other. The quantitative relationships are expressed as *stoichiometrically equivalent molar ratios* that we can use as conversion factors to calculate the amounts. For example, consider the equation for the combustion of propane, a hydrocarbon fuel used in cooking and water heating:

$$C_3H_8(g) + 5O_2(g) \longrightarrow 3CO_2(g) + 4H_2O(g)$$

If we view the reaction quantitatively in terms of C_3H_8, we see that

1 mol of C_3H_8 reacts with 5 mol of O_2
1 mol of C_3H_8 produces 3 mol of CO_2
1 mol of C_3H_8 produces 4 mol of H_2O

Therefore, in this reaction,

1 mol of C_3H_8 is stoichiometrically equivalent to 5 mol of O_2
1 mol of C_3H_8 is stoichiometrically equivalent to 3 mol of CO_2
1 mol of C_3H_8 is stoichiometrically equivalent to 4 mol of H_2O

We chose to look at C_3H_8, but any two of the substances are stoichiometrically equivalent to each other. Thus,

3 mol of CO_2 is stoichiometrically equivalent to 4 mol of H_2O
5 mol of O_2 is stoichiometrically equivalent to 3 mol of CO_2

and so on. A balanced equation contains a wealth of quantitative information relating individual chemical entities, amounts (mol) of substances, and masses of substances; Table 3.3 presents the quantitative information contained in this equation.

Here's a typical problem that shows how stoichiometric equivalence is used to create conversion factors: in the combustion of propane, how many moles of O_2 are consumed when 10.0 mol of H_2O is produced? To solve this problem, we have to find the molar ratio between O_2 and H_2O. From the balanced equation, we see that for every 5 mol of O_2 consumed, 4 mol of H_2O is formed:

5 mol of O_2 is stoichiometrically equivalent to 4 mol of H_2O

As with any equivalent quantities, we can construct two conversion factors, depending on the quantity we want to find:

$$\frac{5 \text{ mol } O_2}{4 \text{ mol } H_2O} \quad \text{or} \quad \frac{4 \text{ mol } H_2O}{5 \text{ mol } O_2}$$

Table 3.3 Information Contained in a Balanced Equation

Viewed in Terms of	Reactants $C_3H_8(g)$ + $5O_2(g)$		Products $3CO_2(g)$ + $4H_2O(g)$
Molecules	1 molecule C_3H_8 + 5 molecules O_2	⟶	3 molecules CO_2 + 4 molecules H_2O
	+	⟶	+
Amount (mol)	1 mol C_3H_8 + 5 mol O_2	⟶	3 mol CO_2 + 4 mol H_2O
Mass (amu)	44.09 amu C_3H_8 + 160.00 amu O_2	⟶	132.03 amu CO_2 + 72.06 amu H_2O
Mass (g)	44.09 g C_3H_8 + 160.00 g O_2	⟶	132.03 g CO_2 + 72.06 g H_2O
Total mass (g)	204.09 g	⟶	204.09 g

Since we want to find the amount (mol) of O_2 and we know the amount (mol) of H_2O, we choose "5 mol O_2/4 mol H_2O" *(the factor at left)* to cancel "mol H_2O":

$$\text{Amount (mol) of } O_2 \text{ consumed} = 10.0 \cancel{\text{mol } H_2O} \times \frac{5 \text{ mol } O_2}{4 \cancel{\text{mol } H_2O}} = 12.5 \text{ mol } O_2$$

$$\text{mol } H_2O \xrightarrow[\text{conversion factor}]{\text{molar ratio as}} \text{mol } O_2$$

You ***cannot*** *solve this type of problem without the balanced equation.* Here is an approach for solving *any* stoichiometry problem that involves a reaction:

1. Write the balanced equation.
2. When necessary, convert the known mass (or number of entities) of one substance to amount (mol) using its molar mass (or Avogadro's number).
3. Use the molar ratio to calculate the unknown amount (mol) of the other substance.
4. When necessary, convert the amount of that other substance to the desired mass (or number of entities) using its molar mass (or Avogadro's number).

Figure 3.7 summarizes the possible relationships among quantities of substances in a reaction, and Sample Problems 3.14–3.16 apply three of them in the first chemical step of converting copper ore to copper metal.

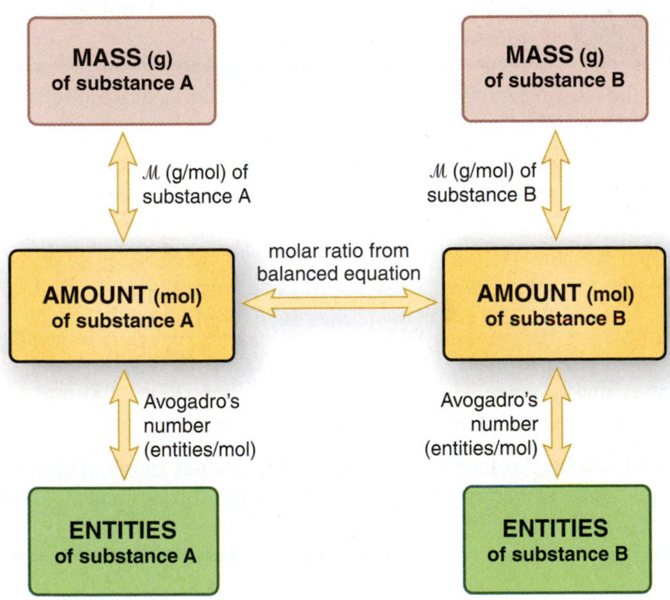

Figure 3.7 Summary of amount-mass-number relationships in a chemical equation. Start at any box (known) and move to any other (unknown) by using the conversion factor on the arrow. As always, convert to amount (mol) first.

Sample Problem 3.14 Calculating Quantities of Reactants and Products: Amount (mol) to Amount (mol)

Problem In a lifetime, the average American uses more than half a ton (> 500 kg) of copper in coins, plumbing, and wiring. Copper is obtained from sulfide ores, such as chalcocite [copper(I) sulfide] by a multistep process. After initial grinding, the ore is "roasted" (heated strongly with oxygen gas) to form powdered copper(I) oxide and gaseous sulfur dioxide. How many moles of oxygen are required to roast 10.0 mol of copper(I) sulfide?

Plan We *always* write the balanced equation first. The formulas of the reactants are Cu_2S and O_2, and the formulas of the products are Cu_2O and SO_2, so we have

$$2Cu_2S(s) + 3O_2(g) \longrightarrow 2Cu_2O(s) + 2SO_2(g)$$

We know the amount of Cu_2S (10.0 mol) and must find the amount (mol) of O_2 that is needed to roast it. The balanced equation shows that 3 mol of O_2 is needed for 2 mol of Cu_2S, so the conversion factor for finding amount (mol) of O_2 is "3 mol O_2/2 mol Cu_2S" (see the road map).

Road Map

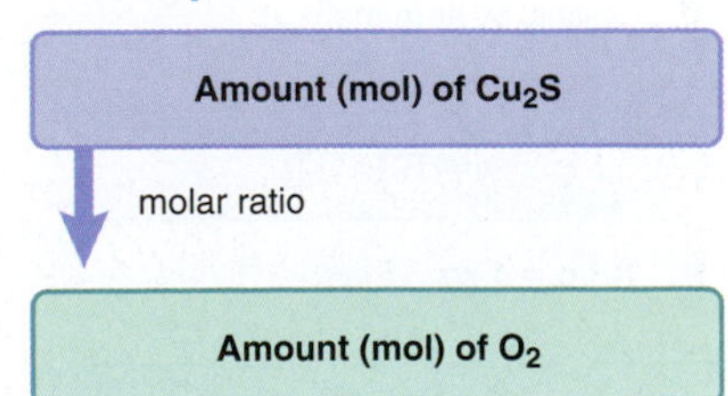

Solution Calculating the amount of O_2:

$$\text{Amount (mol) of } O_2 = 10.0 \text{ mol } Cu_2S \times \frac{3 \text{ mol } O_2}{2 \text{ mol } Cu_2S} = 15.0 \text{ mol } O_2$$

Check The units are correct, and the answer is reasonable because this molar ratio of O_2 to Cu_2S (15/10) is identical to the ratio in the balanced equation (3/2).

Comment A *common mistake* is to invert the conversion factor; that calculation would be

$$\text{Amount (mol) of } O_2 = 10.0 \text{ mol } Cu_2S \times \frac{2 \text{ mol } Cu_2S}{3 \text{ mol } O_2} = \frac{6.67 \text{ mol}^2 \ Cu_2S}{1 \text{ mol } O_2}$$

The strange units should alert you that an error was made in setting up the conversion factor. Also note that this answer, 6.67, is *less* than 10.0, whereas the equation shows that there should be *more* moles of O_2 (3 mol) than moles of Cu_2S (2 mol). Be sure to think through the calculation when setting up the conversion factor and canceling units.

FOLLOW-UP PROBLEM 3.14 Thermite is a mixture of iron(III) oxide and aluminum powders that was once used to weld railroad tracks. It undergoes a spectacular reaction to yield solid aluminum oxide and molten iron. How many moles of iron(III) oxide are needed to form 3.60×10^3 mol of iron? Include a road map that shows how you planned the solution.

Road Map 3.15

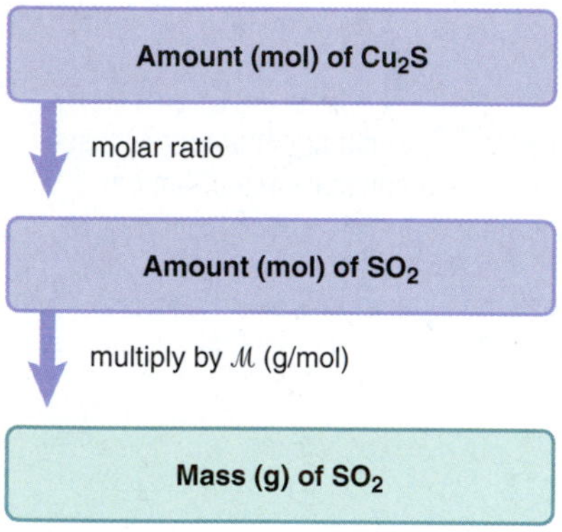

Amount (mol) of Cu_2S
↓ molar ratio
Amount (mol) of SO_2
↓ multiply by $\mathcal{M}$ (g/mol)
Mass (g) of SO_2

Sample Problem 3.15 Calculating Quantities of Reactants and Products: Amount (mol) to Mass (g)

Problem During the roasting process, how many grams of sulfur dioxide form when 10.0 mol of copper(I) sulfide reacts?

Plan Referring to the balanced equation in Sample Problem 3.14, here we are given amount of reactant (10.0 mol of Cu_2S) and need the mass (g) of product (SO_2) that forms. We find the amount (mol) of SO_2 using the molar ratio (2 mol SO_2/2 mol Cu_2S) and then multiply by its molar mass (64.07 g/mol) to find the mass (g) of SO_2 (see top road map).

Solution Combining the two conversion steps into one calculation, we have

$$\text{Mass (g) of } SO_2 = 10.0 \text{ mol } Cu_2S \times \frac{2 \text{ mol } SO_2}{2 \text{ mol } Cu_2S} \times \frac{64.07 \text{ g } SO_2}{1 \text{ mol } SO_2} = 641 \text{ g } SO_2$$

Check The answer makes sense, since the molar ratio shows that 10.0 mol of SO_2 is formed and each mole weighs about 64 g. We rounded to three significant figures.

FOLLOW-UP PROBLEM 3.15 In the thermite reaction, what amount (mol) of iron forms when 1.85×10^{25} formula units of aluminum oxide reacts? Write a road map to show how to plan the solution.

Road Map 3.16

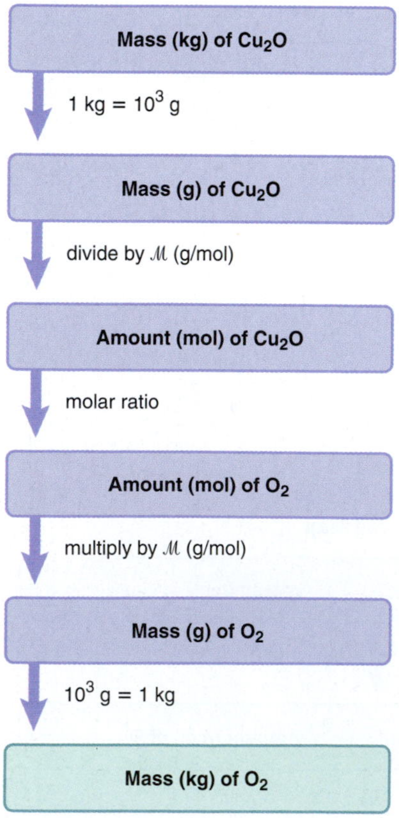

Mass (kg) of Cu_2O
↓ 1 kg = 10^3 g
Mass (g) of Cu_2O
↓ divide by $\mathcal{M}$ (g/mol)
Amount (mol) of Cu_2O
↓ molar ratio
Amount (mol) of O_2
↓ multiply by $\mathcal{M}$ (g/mol)
Mass (g) of O_2
↓ 10^3 g = 1 kg
Mass (kg) of O_2

Sample Problem 3.16 Calculating Quantities of Reactants and Products: Mass to Mass

Problem During the roasting of chalcocite, how many kilograms of oxygen are required to form 2.86 kg of copper(I) oxide?

Plan In this problem, we know the mass of the product, Cu_2O (2.86 kg), and we need the mass (kg) of O_2 that reacts to form it. Therefore, we must convert from mass of product to amount of product to amount of reactant to mass of reactant. We convert the mass of Cu_2O from kg to g and then to amount (mol). Then, we use the molar ratio (3 mol O_2/2 mol Cu_2O) to find the amount (mol) of O_2 required. Finally, we convert amount of O_2 to g and then kg (see bottom road map).

Solution Converting from kilograms of Cu_2O to moles of Cu_2O: Combining the mass unit conversion with the mass-to-amount conversion gives

$$\text{Amount (mol) of } Cu_2O = 2.86 \text{ kg } Cu_2O \times \frac{10^3 \text{ g}}{1 \text{ kg}} \times \frac{1 \text{ mol } Cu_2O}{143.10 \text{ g } Cu_2O} = 20.0 \text{ mol } Cu_2O$$

Converting from moles of Cu_2O to moles of O_2:

$$\text{Amount (mol) of } O_2 = 20.0 \text{ mol } Cu_2O \times \frac{3 \text{ mol } O_2}{2 \text{ mol } Cu_2O} = 30.0 \text{ mol } O_2$$

Converting from moles of O_2 to kilograms of O_2: Combining the amount-to-mass conversion with the mass unit conversion gives

$$\text{Mass (kg) of } O_2 = 30.0 \text{ mol } O_2 \times \frac{32.00 \text{ g } O_2}{1 \text{ mol } O_2} \times \frac{1 \text{ kg}}{10^3 \text{ g}} = 0.960 \text{ kg } O_2$$

Check The units are correct. Rounding to check the math, for example, in the final step, ~30 mol × 30 g/mol × 1 kg/10^3 g = 0.90 kg. The answer seems reasonable: even though the amount (mol) of O_2 is greater than the amount (mol) of Cu_2O, the mass of O_2 is less than the mass of Cu_2O because $\mathcal{M}$ of O_2 is less than $\mathcal{M}$ of Cu_2O.

Comment The three related sample problems (3.14–3.16) highlight the main point for solving stoichiometry problems: *convert the information given into amount (mol).* Then, use the appropriate molar ratio and any other conversion factors to complete the solution.

FOLLOW-UP PROBLEM 3.16 During the thermite reaction, how many atoms of aluminum react for every 1.00 g of aluminum oxide that forms? Include a road map that shows how you planned the solution.

Reactions That Involve a Limiting Reactant

In problems up to now, the amount of *one* reactant was given, and we assumed there was enough of the other reactant(s) to react with it completely. For example, suppose we want the amount (mol) of SO_2 that forms when 5.2 mol of Cu_2S reacts with O_2:

$$2Cu_2S(s) + 3O_2(g) \longrightarrow 2Cu_2O(s) + 2SO_2(g) \quad \text{[equation 1; see Sample Problem 3.14]}$$

We assume the 5.2 mol of Cu_2S reacts with as much O_2 as needed. Because all the Cu_2S reacts, its initial amount of 5.2 mol determines, or *limits,* the amount of SO_2 that can form, no matter how much more O_2 is present. In this situation, we call Cu_2S the **limiting reactant** (or *limiting reagent*).

Suppose, however, you know the amounts of both Cu_2S *and* O_2 and need to find out how much SO_2 forms. You first have to determine whether Cu_2S *or* O_2 is the limiting reactant—that is, which one is completely used up—because it limits how much SO_2 can form. The reactant that is *not* limiting is present *in excess,* which means the amount that doesn't react is left over. To determine which is the limiting reactant, we use the molar ratios in the balanced equation to perform a series of calculations to see *which reactant forms **less** product.*

Determining the Limiting Reactant To clarify the idea of a limiting reactant, let's consider a situation from real life. A car assembly plant has 1500 car bodies and 4000 tires. How many cars can be made with the supplies on hand? Does the plant manager need to order more car bodies or more tires? Obviously, 4 tires are required for each car body, so the "balanced equation" is

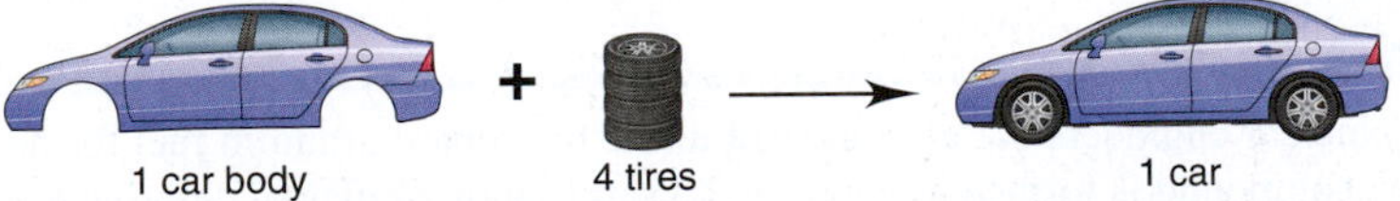

How much "product" (cars) can we make from the amount of each "reactant"?

$$1500 \text{ car bodies} \times \frac{1 \text{ car}}{1 \text{ car body}} = 1500 \text{ cars}$$

$$4000 \text{ tires} \times \frac{1 \text{ car}}{4 \text{ tires}} = 1000 \text{ cars}$$

The number of tires limits the number of cars because less "product" (fewer cars) can be produced from the available tires. There will be 1500 − 1000 = 500 car bodies in excess, and they cannot be turned into cars until more tires are delivered.

Using Reaction Tables in Limiting-Reactant Problems A good way to keep track of the quantities in a limiting-reactant problem is with a *reaction table*. The balanced equation appears at the top for the column heads. The table shows the

- *Initial* quantities of reactants and products *before* the reaction
- *Change* in the quantities of reactants and products *during* the reaction
- *Final* quantities of reactants and products remaining *after* the reaction

For example, for the car assembly "reaction," the reaction table would be

Quantity	1 car body	+	4 tires	⟶	1 car
Initial	1500		4000		0
Change	−1000		−4000		+1000
Final	500		0		1000

The body of the table shows the following important points:

- In the Initial line, "product" has not yet formed, so the entry is "0 cars."
- In the Change line, since the reactants (car bodies and tires) are used during the reaction, their quantities decrease, so the changes in their quantities have a *negative* sign. At the same time, the quantity of product (cars) increases, so the change in its quantity has a *positive* sign.
- For the Final line, we *add* the Change and Initial lines. Notice that one reactant (car bodies) is in excess, while the limiting reactant (tires) is used up.

THINK OF IT THIS WAY

Limiting Reactants in Everyday Life

In addition to the industrial setting of a car assembly plant, limiting-"reactant" situations arise in daily life all the time. A muffin recipe calls for 2 cups of flour and 1 cup of sugar, but you have 3 cups of flour and only $\frac{3}{4}$ cup of sugar. Clearly, the flour is in excess, and the sugar limits the number of muffins you can make. Or, you're making cheeseburgers for a picnic, and you have 10 buns, 12 meat patties, and 8 slices of cheese. Here, the cheese limits the number of cheeseburgers you can make. Or, there are 16 students in a cell biology lab but only 13 microscopes. The number of times limiting-reactant situations arise is almost limitless.

Solving Limiting-Reactant Problems In limiting-reactant problems, *the quantities of two (or more) reactants are given, and we first determine which is limiting.* To do this, just as we did with the cars, we use the balanced equation to solve a series of calculations to see how much product forms from the given quantity of each reactant: the limiting reactant is the one that yields the *least* quantity of product.

The following problems examine these ideas from several aspects. In Sample Problem 3.17, we solve the problem by looking at a molecular scene; in Sample Problem 3.18, we start with the amounts (mol) of two reactants, and in Sample Problem 3.19, we start with masses of two reactants.

Sample Problem 3.17 Using Molecular Depictions in a Limiting-Reactant Problem

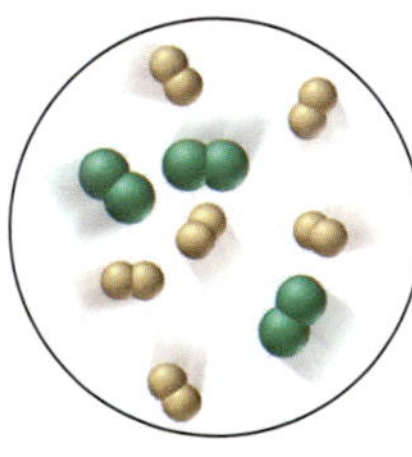

Problem Nuclear engineers use chlorine trifluoride to prepare uranium fuel for power plants. The compound is formed as a gas by the reaction of elemental chlorine and fluorine. The circle in the margin shows a representative portion of the reaction mixture before the reaction starts (chlorine is *green;* fluorine is *yellow*).
(a) Find the limiting reactant.
(b) Write a reaction table for the process.
(c) Draw a representative portion of the mixture after the reaction is complete. (*Hint:* The ClF_3 molecule has Cl bonded to three individual F atoms.)

Plan **(a)** We have to find the limiting reactant. The first step is to write the balanced equation, so we need the formulas and states of matter. From the name, chlorine trifluoride, we know the product consists of one Cl atom bonded to three F atoms, or ClF_3. Elemental chlorine and fluorine are the diatomic molecules Cl_2 and F_2, and all three substances are gases. To find the limiting reactant, we find the number of molecules of product that would form from the numbers of molecules of each reactant: whichever forms less product is the limiting reactant. **(b)** We use these numbers of molecules to write a reaction table. **(c)** We use the numbers in the Final line of the table to draw the scene.

Solution **(a)** The balanced equation is

$$Cl_2(g) + 3F_2(g) \longrightarrow 2ClF_3(g)$$

$$\text{For } Cl_2\text{: Molecules of } ClF_3 = 3 \cancel{\text{ molecules of } Cl_2} \times \frac{2 \text{ molecules of } ClF_3}{1 \cancel{\text{ molecule of } Cl_2}}$$

$$= 6 \text{ molecules of } ClF_3$$

$$\text{For } F_2\text{: Molecules of } ClF_3 = 6 \cancel{\text{ molecules of } F_2} \times \frac{2 \text{ molecules of } ClF_3}{3 \cancel{\text{ molecules of } F_2}}$$

$$= \tfrac{12}{3} \text{ molecules of } ClF_3 = 4 \text{ molecules of } ClF_3$$

Because it forms less product, F_2 is the limiting reactant.

(b) Since F_2 is the limiting reactant, all of it (6 molecules) is used in the Change line of the reaction table:

Molecules	$Cl_2(g)$	+	$3F_2(g)$	$\longrightarrow$	$2ClF_3(g)$
Initial	3		6		0
Change	−2		−6		+4
Final	1		0		4

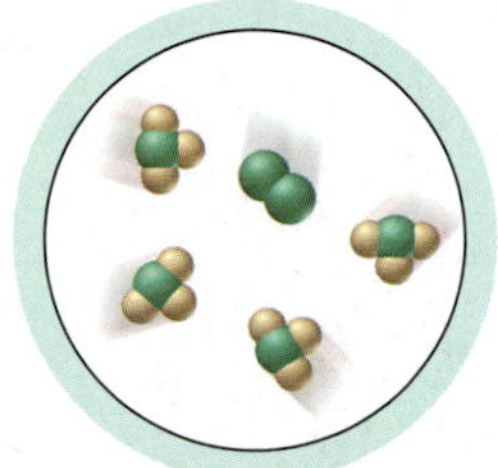

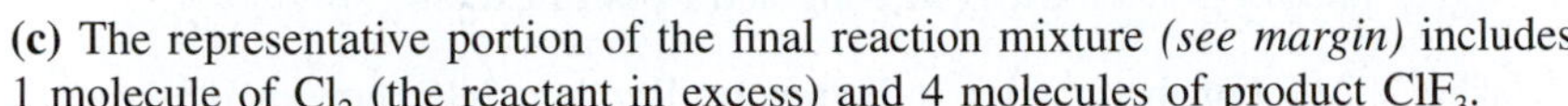

(c) The representative portion of the final reaction mixture *(see margin)* includes 1 molecule of Cl_2 (the reactant in excess) and 4 molecules of product ClF_3.

Check The equation is balanced: reactants (2Cl, 6F) $\longrightarrow$ products (2Cl, 6F). And, as shown in the circles, the numbers of each type of atom before and after the reaction are equal. Let's think through our choice of limiting reactant. From the equation, one Cl_2 needs three F_2 to form two ClF_3. Therefore, the three Cl_2 molecules in the circle depicting reactants need nine (3 × 3) F_2. But there are only six F_2, so there is not enough F_2 to react with the available Cl_2; or put the other way, there is too much Cl_2 to react with the available F_2. From either point of view, F_2 is the limiting reactant.

FOLLOW-UP PROBLEM 3.17 B_2 (B is *red*) reacts with AB as shown below:

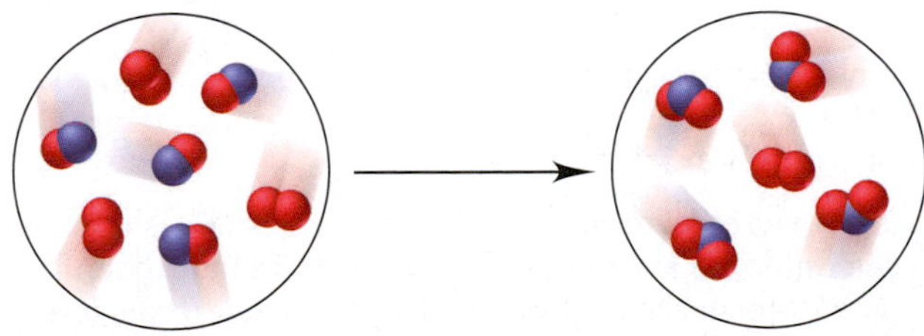

Write a balanced equation for the reaction, and determine the limiting reactant.

Sample Problem 3.18 Calculating Quantities in a Limiting-Reactant Problem: Amount to Amount

Problem In another preparation of ClF_3 (see Sample Problem 3.17), 0.750 mol of Cl_2 reacts with 3.00 mol of F_2. **(a)** Find the limiting reactant. **(b)** Write a reaction table.

Plan **(a)** We find the limiting reactant by calculating the amount (mol) of ClF_3 formed from the amount (mol) of each reactant: the reactant that forms fewer moles of ClF_3 is limiting. **(b)** We enter those values into the reaction table.

Solution **(a)** Determining the limiting reactant:

Finding amount (mol) of ClF_3 from amount (mol) of Cl_2:

$$\text{Amount (mol) of } ClF_3 = 0.750 \cancel{\text{ mol } Cl_2} \times \frac{2 \text{ mol } ClF_3}{1 \cancel{\text{ mol } Cl_2}} = 1.50 \text{ mol } ClF_3$$

Finding amount (mol) of ClF_3 from amount (mol) of F_2:

$$\text{Amount (mol) of } ClF_3 = 3.00 \cancel{\text{ mol } F_2} \times \frac{2 \text{ mol } ClF_3}{3 \cancel{\text{ mol } F_2}} = 2.00 \text{ mol } ClF_3$$

In this case, Cl_2 is limiting because it forms fewer moles of ClF_3.

(b) Writing the reaction table, with Cl_2 limiting:

Amount (mol)	$Cl_2(g)$	+	$3F_2(g)$	$\longrightarrow$	$2ClF_3(g)$
Initial	0.750		3.00		0
Change	−0.750		−2.25		+1.50
Final	0		0.75		1.50

Check Let's check that Cl_2 is the limiting reactant by assuming, for the moment, that F_2 is limiting. If that were true, all 3.00 mol of F_2 would react to form 2.00 mol of ClF_3. However, based on the balanced equation, obtaining 2.00 mol of ClF_3 would require 1.00 mol of Cl_2, and only 0.750 mol of Cl_2 is present. Thus, Cl_2 must be the limiting reactant.

Comment A major point to note from Sample Problems 3.17 and 3.18 is that the relative quantities of reactants *do not* determine which is limiting, but rather the quantity of product formed, which is based on the *molar ratio in the balanced equation*. In both problems, there is more F_2 than Cl_2. However,

- Sample Problem 3.17 has an F_2/Cl_2 ratio of 6/3, or 2/1, which is less than the required molar ratio of 3/1, so F_2 is limiting and Cl_2 is in excess.
- Sample Problem 3.18 has an F_2/Cl_2 ratio of 3.00/0.750, which is greater than the required molar ratio of 3/1, so Cl_2 is limiting and F_2 is in excess.

FOLLOW-UP PROBLEM 3.18 For the reaction in Follow-up Problem 3.17, how many moles of product form from 1.5 mol of each reactant?

Sample Problem 3.19 Calculating Quantities in a Limiting-Reactant Problem: Mass to Mass

Problem A fuel mixture used in the early days of rocketry consisted of two liquids, hydrazine (N_2H_4) and dinitrogen tetraoxide (N_2O_4), which ignite on contact to form nitrogen gas and water vapor.
(a) How many grams of nitrogen gas form when 1.00×10^2 g of N_2H_4 and 2.00×10^2 g of N_2O_4 are mixed?
(b) Write a reaction table for this process.

Road Map

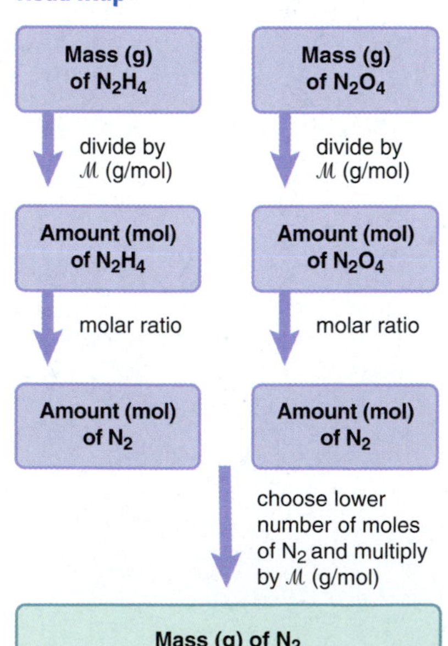

Plan The amounts of two reactants are given, which means this is a limiting-reactant problem. **(a)** To determine the mass of product formed, we must find the limiting reactant by calculating which of the given masses of reactant forms *less* nitrogen gas. As always, we first write the balanced equation. We convert the grams of each reactant to moles using its molar mass and then use the molar ratio from the balanced equation to find the number of moles of N_2 each reactant forms. Next, we convert the lower amount of N_2 to mass (see the road map). **(b)** We use the values based on the limiting reactant for the reaction table.

Solution **(a)** Writing the balanced equation:

$$2N_2H_4(l) + N_2O_4(l) \longrightarrow 3N_2(g) + 4H_2O(g)$$

Finding the amount (mol) of N_2 from the amount (mol) of each reactant

For N_2H_4: $$\text{Amount (mol) of } N_2H_4 = 1.00\times10^2\ \cancel{\text{g } N_2H_4} \times \frac{1 \text{ mol } N_2H_4}{32.05\ \cancel{\text{g } N_2H_4}} = 3.12 \text{ mol } N_2H_4$$

$$\text{Amount (mol) of } N_2 = 3.12\ \cancel{\text{mol } N_2H_4} \times \frac{3 \text{ mol } N_2}{2\ \cancel{\text{mol } N_2H_4}} = \mathbf{4.68\ mol\ N_2}$$

For N_2O_4: $$\text{Amount (mol) of } N_2O_4 = 2.00\times10^2\ \cancel{\text{g } N_2O_4} \times \frac{1 \text{ mol } N_2O_4}{92.02\ \cancel{\text{g } N_2O_4}} = 2.17 \text{ mol } N_2O_4$$

$$\text{Amount (mol) of } N_2 = 2.17\ \cancel{\text{mol } N_2O_4} \times \frac{3 \text{ mol } N_2}{1\ \cancel{\text{mol } N_2O_4}} = \mathbf{6.51\ mol\ N_2}$$

Thus, N_2H_4 is the limiting reactant because it yields less N_2.

Converting from amount (mol) of N_2 to mass (g):

$$\text{Mass (g) of } N_2 = 4.68\ \cancel{\text{mol } N_2} \times \frac{28.02 \text{ g } N_2}{1\ \cancel{\text{mol } N_2}} = \boxed{131 \text{ g } N_2}$$

(b) With N_2H_4 as the limiting reactant, the reaction table is

Amount (mol)	$2N_2H_4(g)$	+ $N_2O_4(g)$	⟶ $3N_2(g)$	+ $4H_2O(g)$
Initial	3.12	2.17	0	0
Change	−3.12	−1.56	+4.68	+6.24
Final	0	0.61	4.68	6.24

Check The number of grams of N_2O_4 is more than that of N_2H_4, but there are fewer moles of N_2O_4 because its $\mathcal{M}$ is much higher. Rounding for N_2H_4: 100 g N_2H_4 × 1 mol/32 g ≈ 3 mol; ~3 mol × $\frac{3}{2}$ ≈ 4.5 mol N_2; ~4.5 mol × 30 g/mol ≈ 135 g N_2.

Comment 1. Recall this *common mistake* in solving limiting-reactant problems: The limiting reactant is not the *reactant* present in fewer moles (or grams). Rather, it is the reactant that forms fewer moles (or grams) of *product.*

2. An *alternative approach* to finding the limiting reactant compares "How much is needed?" with "How much is given?" That is, based on the balanced equation,

- Find the amount (mol) of each reactant needed to react with the other reactant.
- Compare that *needed* amount with the *given* amount in the problem statement. There will be *more* than enough of one reactant (excess) and *less* than enough of the other (limiting).

For example, the balanced equation for this problem shows that 2 mol of N_2H_4 reacts with 1 mol of N_2O_4. The amount (mol) of N_2O_4 needed to react with the given 3.12 mol of N_2H_4 is

$$\text{Amount (mol) of } N_2O_4 \text{ needed} = 3.12 \cancel{\text{ mol } N_2H_4} \times \frac{1 \text{ mol } N_2O_4}{2 \cancel{\text{ mol } N_2H_4}} = 1.56 \text{ mol } N_2O_4$$

The amount of N_2H_4 needed to react with the given 2.17 mol of N_2O_4 is

$$\text{Amount (mol) of } N_2H_4 \text{ needed} = 2.17 \cancel{\text{ mol } N_2O_4} \times \frac{2 \text{ mol } N_2H_4}{1 \cancel{\text{ mol } N_2O_4}} = 4.34 \text{ mol } N_2H_4$$

We are given 2.17 mol of N_2O_4, which is *more* than the 1.56 mol of N_2O_4 needed, and we are given 3.12 mol of N_2H_4, which is *less* than the 4.34 mol of N_2H_4 needed. Therefore, N_2H_4 is limiting, and N_2O_4 is in excess.

FOLLOW-UP PROBLEM 3.19 How many grams of solid aluminum sulfide can be prepared by the reaction of 10.0 g of aluminum and 15.0 g of sulfur? How many grams of the nonlimiting reactant are in excess?

Theoretical, Actual, and Percent Reaction Yields

Up until now, we've assumed that 100% of the limiting reactant becomes product and that we use perfect lab technique to collect all the product. In theory, this may happen, but in reality, it doesn't, and chemists recognize three types of reaction yield:

1. *Theoretical yield.* The amount of product calculated from the molar ratio in the balanced equation is the **theoretical yield.** But, there are several reasons why the theoretical yield is *never* obtained:
   - Reactant mixtures often proceed through **side reactions** that form different products (Figure 3.8). In the rocket fuel reaction in Sample Problem 3.19, for example, the reactants might form some NO in the following side reaction:
   $$N_2H_4(l) + 2N_2O_4(l) \longrightarrow 6NO(g) + 2H_2O(g)$$
   This reaction decreases the amounts of reactants available for N_2 production.
   - Even more important, many reactions seem to stop before they are complete, so some limiting reactant is unused. (We'll see one example in Chapter 4 and many more later in the book.)
   - Isolating a product often requires several steps, and some product is lost during each one. With careful lab technique, we can minimize, but never eliminate, such losses.

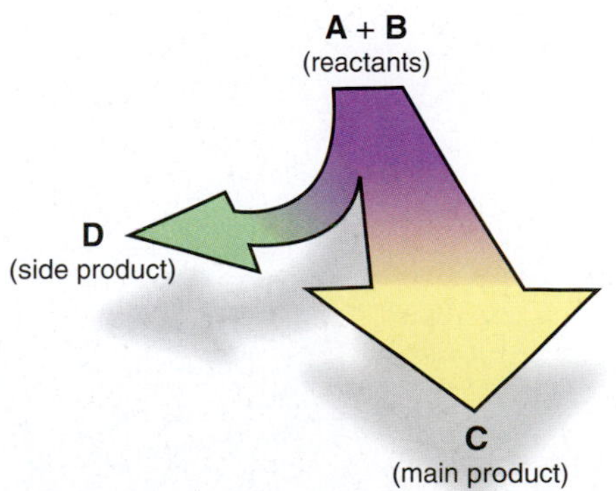

Figure 3.8 The effect of a side reaction on the yield of the main product.

2. *Actual yield.* Given these reasons for obtaining less than the theoretical yield, the amount of product actually obtained is the **actual yield.** Theoretical and actual yields are expressed in units of amount (moles) or mass (grams).

3. *Percent yield.* The **percent yield (% yield)** is the actual yield expressed as a percentage of the theoretical yield:

$$\% \text{ yield} = \frac{\text{actual yield}}{\text{theoretical yield}} \times 100 \quad (3.8)$$

By definition, the actual yield is less than the theoretical yield, so the percent yield is *always* less than 100%.

In the multistep synthesis of a complex compound, the overall yield can be surprisingly low, even if the yield of each step is high. For example, suppose a six-step synthesis has a 90.0% yield for each step. To find the overall percent yield, *express the yield of each step as a decimal, multiply them together, and then convert back to percent*. The overall yield is only slightly more than 50%:

$$\text{Overall \% yield} = (0.900 \times 0.900 \times 0.900 \times 0.900 \times 0.900 \times 0.900) \times 100 = 53.1\%$$

Sample Problem 3.20 Calculating Percent Yield

Problem Silicon carbide (SiC) is an important ceramic material made by reacting sand (silicon dioxide, SiO_2) with powdered carbon at a high temperature. Carbon monoxide is also formed. When 100.0 kg of sand is processed, 51.4 kg of SiC is recovered. What is the percent yield of SiC from this process?

Road Map

Mass (kg) of SiO_2
↓ 1. convert kg to g 2. divide by $\mathcal{M}$ (g/mol)
Amount (mol) of SiO_2
↓ molar ratio
Amount (mol) of SiC
↓ 1. multiply by $\mathcal{M}$ (g/mol) 2. convert g to kg
Mass (kg) of SiC
↓ Eq. 3.8
% Yield of SiC

Plan We are given the actual yield of SiC (51.4 kg), so we need the theoretical yield to calculate the percent yield. After writing the balanced equation, we convert the given mass of SiO_2 (100.0 kg) to amount (mol). We use the molar ratio to find the amount of SiC formed and convert it to mass (kg) to obtain the theoretical yield. Then, we use Equation 3.8 to find the percent yield (see the road map).

Solution Writing the balanced equation:

$$SiO_2(s) + 3C(s) \longrightarrow SiC(s) + 2CO(g)$$

Converting from mass (kg) of SiO_2 to amount (mol):

$$\text{Amount (mol) of } SiO_2 = 100.0 \text{ kg } SiO_2 \times \frac{1000 \text{ g}}{1 \text{ kg}} \times \frac{1 \text{ mol } SiO_2}{60.09 \text{ g } SiO_2} = 1664 \text{ mol } SiO_2$$

Converting from amount (mol) of SiO_2 to amount (mol) of SiC:
The molar ratio is 1 mol SiC/1 mol SiO_2, so

$$\text{Amount (mol) of } SiO_2 = \text{moles of SiC} = 1664 \text{ mol SiC}$$

Converting from amount (mol) of SiC to mass (kg):

$$\text{Mass (kg) of SiC} = 1664 \text{ mol SiC} \times \frac{40.10 \text{ g SiC}}{1 \text{ mol SiC}} \times \frac{1 \text{ kg}}{1000 \text{ g}} = 66.73 \text{ kg SiC}$$

Calculating the percent yield:

$$\% \text{ yield of SiC} = \frac{\text{actual yield}}{\text{theoretical yield}} \times 100 = \frac{51.4 \text{ kg SiC}}{66.73 \text{ kg SiC}} \times 100 = 77.0\%$$

Check Rounding shows that the mass of SiC seems correct: ~1500 mol × 40 g/mol × 1 kg/1000 g = 60 kg. The molar ratio of SiC/SiO_2 is 1/1, and $\mathcal{M}$ of SiC is about two-thirds ($\sim\frac{40}{60}$) of $\mathcal{M}$ of SiO_2, so 100 kg of SiO_2 should form about 66 kg of SiC.

FOLLOW-UP PROBLEM 3.20 Marble (calcium carbonate) reacts with hydrochloric acid solution to form calcium chloride solution, water, and carbon dioxide. Find the percent yield of carbon dioxide if 3.65 g is collected when 10.0 g of marble reacts.

Summary of Section 3.4

- The substances in a balanced equation are related to each other by stoichiometrically equivalent molar ratios, which are used as conversion factors to find the amount (mol) of one substance given the amount of another.

- In limiting-reactant problems, the quantities of two (or more) reactants are given, and the limiting reactant is the one that forms the lower quantity of product. Reaction tables show the initial and final quantities of all reactants and products, as well as the changes in those quantities.
- In practice, side reactions, incomplete reactions, and physical losses result in an actual yield of product that is less than the theoretical yield (the quantity based on the molar ratio from the balanced equation), giving a percent yield less than 100%. In multistep reaction sequences, the overall yield is found by multiplying the yields for each step.

3.5 • FUNDAMENTALS OF SOLUTION STOICHIOMETRY

In popular media, you may see a chemist in a lab coat, surrounded by glassware, mixing colored solutions that froth and give off billowing fumes. Most reactions in solution are not this dramatic, and good technique requires safer mixing procedures; however, it is true that aqueous solution chemistry is central to laboratory chemistry. And many environmental reactions and almost all biochemical reactions occur in solution. Liquid solutions are easier to store than gases and easier to mix than solids, and the amounts of substances in solution can be measured precisely.

We know the amounts of pure substances by converting their masses into number of moles. But for dissolved substances, we need the *concentration*—the number of moles per volume of solution—to find the volume that contains a given number of moles. In this section, we first discuss *molarity,* the most common way to express concentration (Chapter 13 covers others). Then, we see how to dilute a concentrated solution and how to use stoichiometric calculations for reactions in solution.

Expressing Concentration in Terms of Molarity

A solution consists of a smaller quantity of one substance, the **solute,** dissolved in a larger quantity of another, the **solvent.** When it dissolves, the solute's chemical entities become evenly dispersed throughout the solvent. The **concentration** of a solution is often expressed as *the quantity of solute dissolved in a given quantity of solution.*

Concentration is an *intensive* property (like density or temperature; Section 1.4) and thus is independent of the solution volume: a 50-L tank of a solution has the *same concentration* (solute quantity/solution quantity) as a 50-mL beaker of the solution. **Molarity (*M*)** expresses the concentration in units of *moles of solute per liter of solution:*

$$\text{Molarity} = \frac{\text{moles of solute}}{\text{liters of solution}} \quad \text{or} \quad M = \frac{\text{mol solute}}{\text{L soln}} \tag{3.9}$$

Sample Problem 3.21 Calculating the Molarity of a Solution

Problem Glycine has the simplest structure of the 20 amino acids that make up proteins. What is the molarity of a solution that contains 0.715 mol of glycine in 495 mL?

Plan The molarity is the number of moles of solute in each liter of solution. We divide the number of moles (0.715 mol) by the volume (495 mL) and convert the volume to liters to find the molarity (see the road map).

Road Map

Amount (mol) of glycine

↓ divide by volume (mL)

Concentration (mol/mL) of glycine

↓ 10^3 mL = 1 L

Molarity (mol/L) of glycine

Solution
$$\text{Molarity} = \frac{0.715 \text{ mol glycine}}{495 \cancel{\text{mL soln}}} \times \frac{1000 \cancel{\text{mL}}}{1 \text{ L}} = 1.44\ M \text{ glycine}$$

Check A quick look at the math shows about 0.7 mol of glycine in about 0.5 L of solution, so the concentration should be about 1.4 mol/L, or 1.4 *M*.

FOLLOW-UP PROBLEM 3.21 How many moles of KI are in 84 mL of 0.50 *M* KI? Include a road map that shows how you planned the solution.

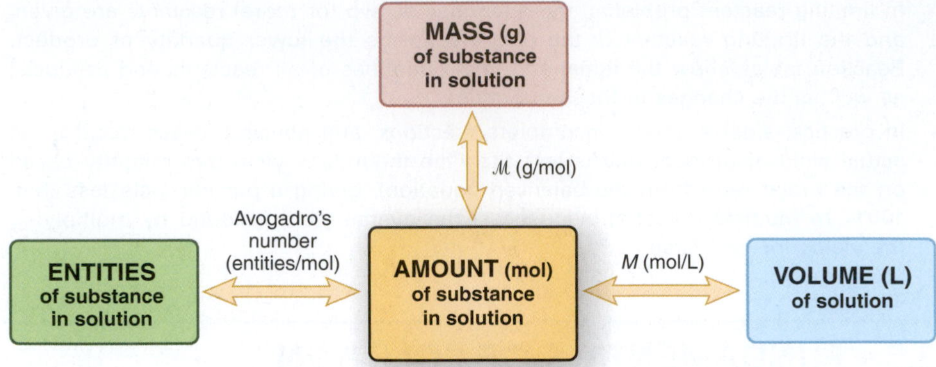

Figure 3.9 Summary of amount-mass-number relationships in solution. The amount (mol) of a substance in solution is related to the volume (L) of solution through the molarity (*M*; mol/L). As always, convert the given quantity to amount (mol) first.

Amount-Mass-Number Conversions Involving Solutions

Like many intensive properties, molarity can be used as a *conversion factor* between volume (L) of solution and amount (mol) of solute, from which we can find the mass or the number of entities of solute (Figure 3.9), as applied in Sample Problem 3.22.

Sample Problem 3.22 Calculating Mass of Solute in a Given Volume of Solution

Problem Biochemists often study reactions in solutions containing phosphate ion, commonly found in cells. How many grams of solute are in 1.75 L of 0.460 *M* sodium hydrogen phosphate?

Plan We need the mass (g) of solute, so we multiply the known solution volume (1.75 L) by the known molarity (0.460 *M*) to find the amount (mol) of solute and convert it to mass (g) using the solute's molar mass (see the road map).

Road Map

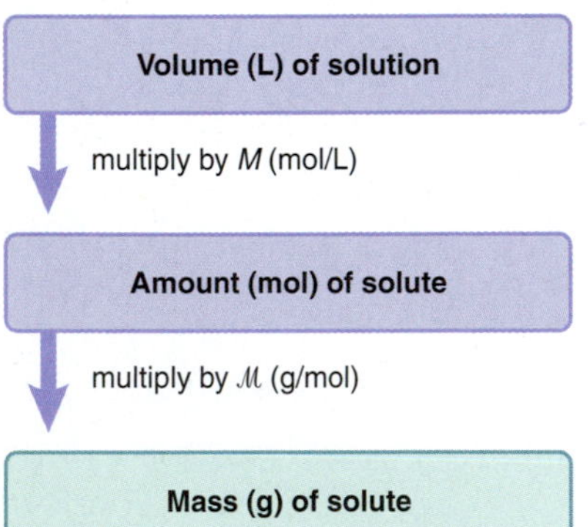

Solution Calculating amount (mol) of solute in solution:

$$\text{Amount (mol) of } Na_2HPO_4 = 1.75\ \cancel{\text{L soln}} \times \frac{0.460 \text{ mol } Na_2HPO_4}{1\ \cancel{\text{L soln}}} = 0.805 \text{ mol } Na_2HPO_4$$

Converting from amount (mol) of solute to mass (g):

$$\text{Mass (g) } Na_2HPO_4 = 0.805\ \cancel{\text{mol } Na_2HPO_4} \times \frac{141.96 \text{ g } Na_2HPO_4}{1\ \cancel{\text{mol } Na_2HPO_4}} = \boxed{114 \text{ g } Na_2HPO_4}$$

Check The answer seems to be correct: ~1.8 L of 0.5 mol/L solution contains 0.9 mol, and 150 g/mol × 0.9 mol = 135 g, which is close to 114 g of solute.

FOLLOW-UP PROBLEM 3.22 In biochemistry laboratories, solutions of sucrose (table sugar, $C_{12}H_{22}O_{11}$) are used in high-speed centrifuges to separate the parts of a biological cell. How many liters of 3.30 *M* sucrose contain 135 g of solute? Include a road map that shows how you planned the solution.

Diluting a Solution

A concentrated solution (higher molarity) is converted to a dilute solution (lower molarity) by adding solvent, which means the solution volume increases but the amount (mol) of solute stays the same. As a result, the dilute solution contains *fewer solute particles per unit volume* and, thus, has a lower concentration than the concentrated solution (Figure 3.10). If you need several different dilute solutions, prepare a concentrated solution *(stock solution)* and store it and dilute it as needed.

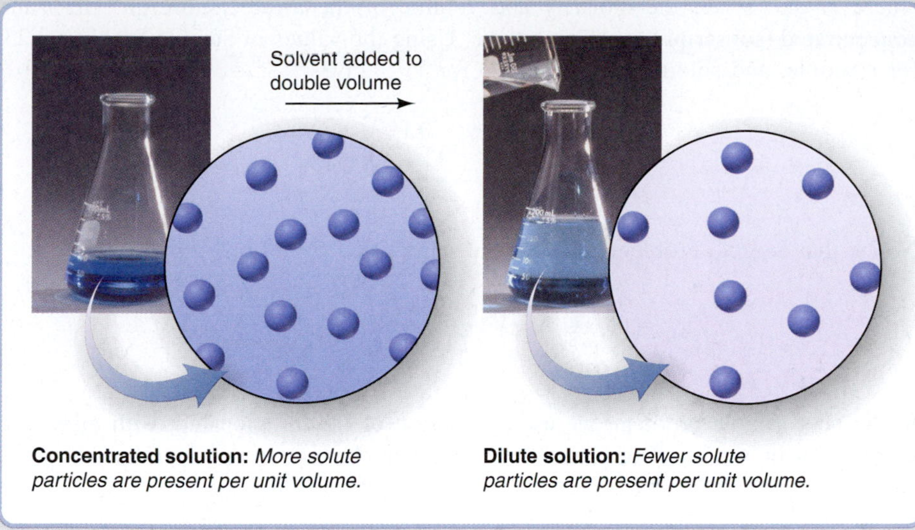
Solvent added to double volume
Concentrated solution: More solute particles are present per unit volume.
Dilute solution: Fewer solute particles are present per unit volume.

Figure 3.10 Converting a concentrated solution to a dilute solution.

Sample Problem 3.23 Preparing a Dilute Solution from a Concentrated Solution

Problem Isotonic saline is 0.15 M aqueous NaCl. It simulates the total concentration of ions in many cellular fluids, and its uses range from cleaning contact lenses to washing red blood cells. How would you prepare 0.80 L of isotonic saline from a 6.0 M stock solution?

Plan To dilute a concentrated solution, we add only solvent, so the *moles of solute are the same in both solutions.* We know the volume (0.80 L) and molarity (0.15 M) of the dilute (dil) NaCl solution needed, so we find the amount (mol) of NaCl it contains. Then, we find the volume (L) of concentrated (conc; 6.0 M) NaCl solution that contains the same amount (mol). We add solvent *up to* the final volume (see the road map).

Road Map

Volume (L) of dilute solution

↓ multiply by M (mol/L) of dilute solution

Amount (mol) of NaCl in dilute solution = Amount (mol) of NaCl in concentrated solution

↓ divide by M (mol/L) of concentrated solution

Volume (L) of concentrated solution

Solution Finding amount (mol) of solute in dilute solution:

$$\text{Amount (mol) of NaCl in dil soln} = 0.80\ \cancel{\text{L soln}} \times \frac{0.15\ \text{mol NaCl}}{1\ \cancel{\text{L soln}}} = 0.12\ \text{mol NaCl}$$

Finding amount (mol) of solute in concentrated solution:

Because we add only solvent to dilute the solution,

$$\text{Amount (mol) of NaCl in dil soln} = \text{amount (mol) of NaCl in conc soln} = 0.12\ \text{mol NaCl}$$

Finding the volume (L) of concentrated solution that contains 0.12 mol of NaCl:

$$\text{Volume (L) of conc NaCl soln} = 0.12\ \cancel{\text{mol NaCl}} \times \frac{1\ \text{L soln}}{6.0\ \cancel{\text{mol NaCl}}} = 0.020\ \text{L soln}$$

To prepare 0.80 L of dilute solution, place 0.020 L of 6.0 M NaCl in a 1.0-L graduated cylinder, add distilled water (~780 mL) to the 0.80-L mark, and stir thoroughly.

Check The answer seems reasonable because a small volume of concentrated solution is used to prepare a large volume of dilute solution. Also, the ratio of volumes (0.020 L/0.80 L) is the same as the ratio of concentrations (0.15 M/6.0 M).

FOLLOW-UP PROBLEM 3.23 A chemical engineer dilutes a stock solution of sulfuric acid by adding 25.0 m^3 of 7.50 M acid to enough water to make 500. m^3. What is the concentration of sulfuric acid in the diluted solution in g/mL?

To solve dilution problems and others involving a change in concentration, apply the following relationship:

$$M_{dil} \times V_{dil} = \text{amount (mol)} = M_{conc} \times V_{conc} \quad (3.10)$$

where M and V are the molarity and volume of the *dil*ute (subscript "dil") and *conc*entrated (subscript "conc") solutions. Using the values in Sample Problem 3.23, for example, and solving Equation 3.10 for V_{conc} gives

$$V_{conc} = \frac{M_{dil} \times V_{dil}}{M_{conc}} = \frac{0.15\ \cancel{M} \times 0.80\ \text{L}}{6.0\ \cancel{M}}$$
$$= 0.020\ \text{L}$$

Notice that Sample Problem 3.23 had the same calculation broken into two parts:

$$V_{conc} = 0.80\ \cancel{\text{L}} \times \frac{0.15\ \cancel{\text{mol NaCl}}}{1\ \cancel{\text{L}}} \times \frac{1\ \text{L}}{6.0\ \cancel{\text{mol NaCl}}}$$
$$= 0.020\ \text{L}$$

In the next sample problem, we use a variation of this relationship, with molecular scenes showing numbers of particles, to visualize changes in concentration.

Sample Problem 3.24 Visualizing Changes in Concentration

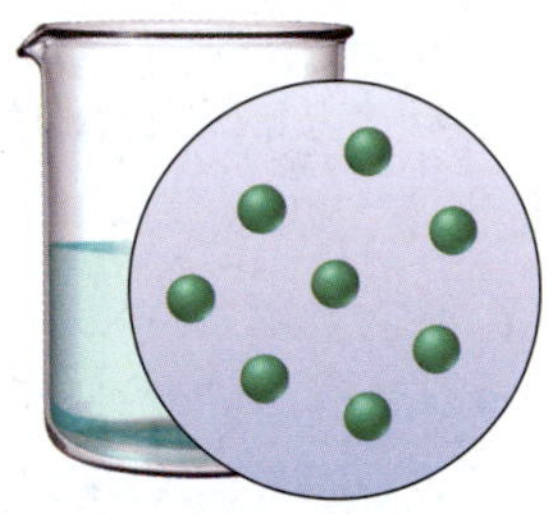

Problem The top circle at left represents a unit volume of a solution. Draw a circle representing a unit volume of the solution after each of these changes:
(a) For every 1 mL of solution, 1 mL of solvent is added.
(b) One-third of the solvent is boiled off.

Plan Given the starting solution, we have to find the number of solute particles in a unit volume after each change. The number of particles per unit volume, N, is directly related to the number of moles per unit volume, M, so we can use a relationship similar to Equation 3.10 to find the number of particles. **(a)** The volume increases, so the final solution is more dilute—fewer particles per unit volume. **(b)** Some solvent is lost, so the final solution is more concentrated—more particles per unit volume.

(a)

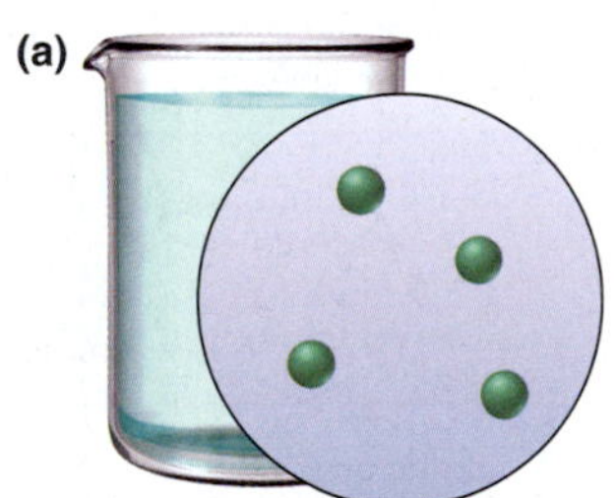

Solution **(a)** Finding the number of particles in the dilute solution, N_{dil}:

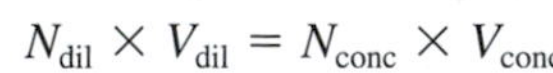

$$N_{dil} \times V_{dil} = N_{conc} \times V_{conc}$$

thus, $$N_{dil} = N_{conc} \times \frac{V_{conc}}{V_{dil}} = 8 \text{ particles} \times \frac{1\ \cancel{\text{mL}}}{2\ \cancel{\text{mL}}} = 4 \text{ particles}$$

(b)

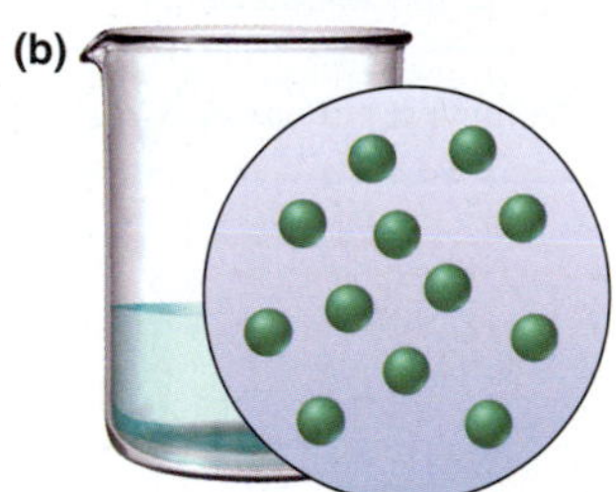

(b) Finding the number of particles in the concentrated solution, N_{conc}:

$$N_{dil} \times V_{dil} = N_{conc} \times V_{conc}$$

thus, $$N_{conc} = N_{dil} \times \frac{V_{dil}}{V_{conc}} = 8 \text{ particles} \times \frac{1\ \cancel{\text{mL}}}{\frac{2}{3}\ \cancel{\text{mL}}} = 12 \text{ particles}$$

Check In (a), the volume is doubled (from 1 mL to 2 mL), so the number of particles should be halved; $\frac{1}{2}$ of 8 is 4. In (b), the volume is $\frac{2}{3}$ of the original, so the number of particles should be $\frac{3}{2}$ of the original; $\frac{3}{2}$ of 8 is 12.

Comment In (b), we assumed that only solvent boils off. This is true with nonvolatile solutes, such as ionic compounds, but in Chapter 13, we'll encounter solutions in which both solvent *and* solute are volatile.

FOLLOW-UP PROBLEM 3.24 The circle labeled A represents a unit volume of a solution. Explain the changes that must be made to A to obtain the unit volumes in B and C.

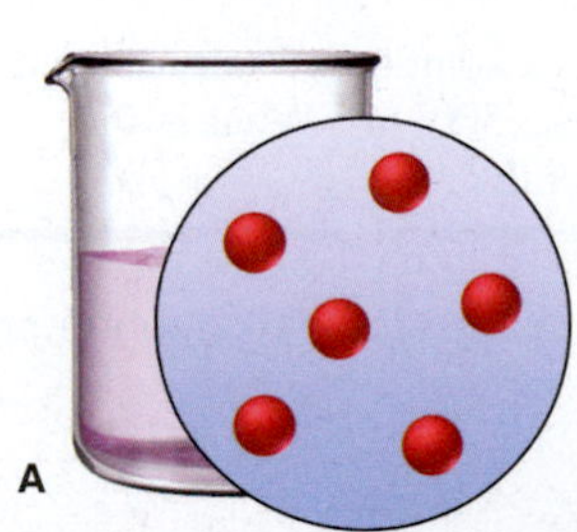

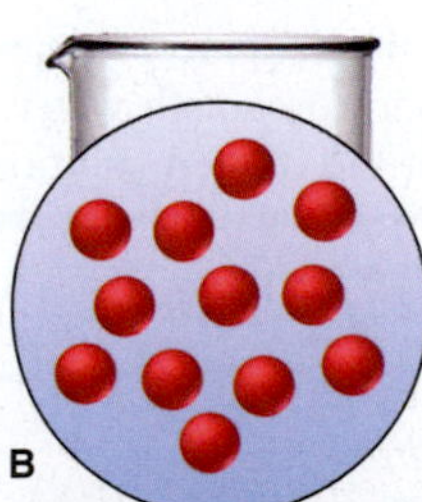

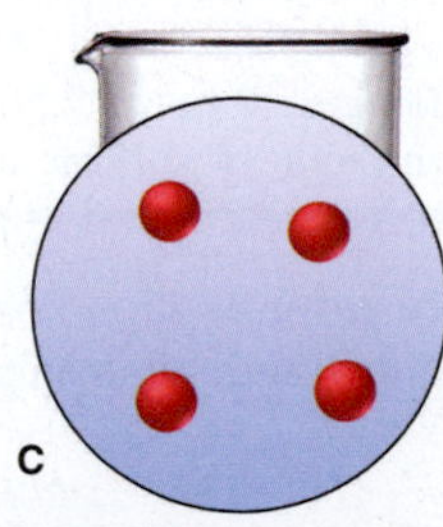

Stoichiometry of Reactions in Solution

Solving stoichiometry problems for reactions in solution requires the additional step of converting the volume of reactant or product in solution to amount (mol):

1. Balance the equation.
2. Find the amount (mol) of one substance from the volume and molarity.
3. Relate it to the stoichiometrically equivalent amount of another substance.
4. Convert to the desired units.

Sample Problem 3.25 Calculating Quantities of Reactants and Products for a Reaction in Solution

Problem Specialized cells in the stomach release HCl to aid digestion. If they release too much, the excess can be neutralized with an antacid. A common antacid contains magnesium hydroxide, which reacts with the acid to form water and magnesium chloride solution. As a government chemist testing commercial antacids, you use 0.10 *M* HCl to simulate the acid concentration in the stomach. How many liters of "stomach acid" react with a tablet containing 0.10 g of magnesium hydroxide?

Plan We are given the mass (0.10 g) of magnesium hydroxide, $Mg(OH)_2$, that reacts with the acid. We also know the acid concentration (0.10 *M*) and must find the acid volume. After writing the balanced equation, we convert the mass (g) of $Mg(OH)_2$ to amount (mol) and use the molar ratio to find the amount (mol) of HCl that reacts with it. Then, we use the molarity of HCl to find the volume (L) that contains this amount (see the road map).

Road Map

Mass (g) of $Mg(OH)_2$

↓ divide by $\mathcal{M}$ (g/mol)

Amount (mol) of $Mg(OH)_2$

↓ molar ratio

Amount (mol) of HCl

↓ divide by M (mol/L)

Volume (L) of HCl

Solution Writing the balanced equation:

$$Mg(OH)_2(s) + 2HCl(aq) \longrightarrow MgCl_2(aq) + 2H_2O(l)$$

Converting from mass (g) of $Mg(OH)_2$ to amount (mol):

$$\text{Amount (mol) of } Mg(OH)_2 = 0.10\ \cancel{\text{g } Mg(OH)_2} \times \frac{1 \text{ mol } Mg(OH)_2}{58.33\ \cancel{\text{g } Mg(OH)_2}} = 1.7\times10^{-3} \text{ mol } Mg(OH)_2$$

Converting from amount (mol) of $Mg(OH)_2$ to amount (mol) of HCl:

$$\text{Amount (mol) of HCl} = 1.7\times10^{-3}\ \cancel{\text{mol } Mg(OH)_2} \times \frac{2 \text{ mol HCl}}{1\ \cancel{\text{mol } Mg(OH)_2}} = 3.4\times10^{-3} \text{ mol HCl}$$

Converting from amount (mol) of HCl to volume (L):

$$\text{Volume (L) of HCl} = 3.4\times10^{-3}\ \cancel{\text{mol HCl}} \times \frac{1 \text{ L}}{0.10\ \cancel{\text{mol HCl}}} = \boxed{3.4\times10^{-2} \text{ L}}$$

Check The size of the answer seems reasonable: a small volume of dilute acid (0.034 L of 0.10 *M*) reacts with a small amount of antacid (0.0017 mol).

Comment In Chapter 4, you'll see that this equation is an oversimplification, because HCl and $MgCl_2$ exist in solution as separated ions.

FOLLOW-UP PROBLEM 3.25 Another active ingredient in some antacids is aluminum hydroxide. Which is more effective at neutralizing stomach acid, magnesium hydroxide or aluminum hydroxide? [*Hint:* "Effectiveness" refers to the amount of acid that reacts with a given mass of antacid. You already know the effectiveness of 0.10 g of $Mg(OH)_2$.]

Except for the additional step of finding amounts (mol) in solution, limiting-reactant problems for reactions in solution are handled just like other such problems.

Sample Problem 3.26 Solving Limiting-Reactant Problems for Reactions in Solution

Problem Mercury and its compounds have uses from fillings for teeth (as a mixture with silver, copper, and tin) to the production of chlorine. Because of their toxicity, however, soluble mercury compounds, such as mercury(II) nitrate, must be removed from industrial wastewater. One removal method reacts the wastewater with sodium sulfide solution to produce solid mercury(II) sulfide and sodium nitrate solution. In a laboratory simulation, 0.050 L of 0.010 *M* mercury(II) nitrate reacts with 0.020 L of 0.10 *M* sodium sulfide. **(a)** How many grams of mercury(II) sulfide form? **(b)** Write a reaction table for this process.

Road Map

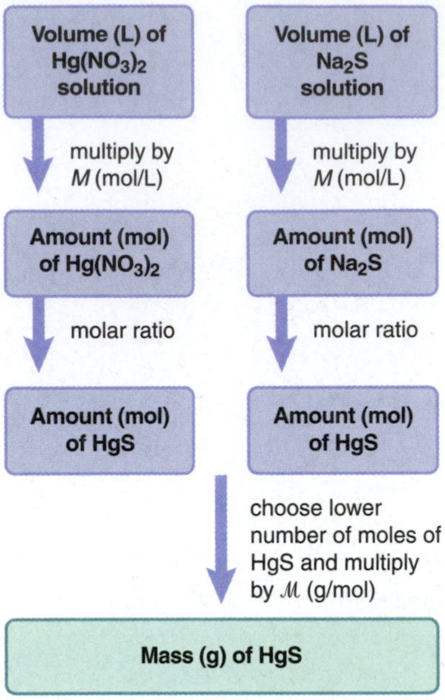

Volume (L) of $Hg(NO_3)_2$ solution → multiply by M (mol/L) → Amount (mol) of $Hg(NO_3)_2$ → molar ratio → Amount (mol) of HgS

Volume (L) of Na_2S solution → multiply by M (mol/L) → Amount (mol) of Na_2S → molar ratio → Amount (mol) of HgS

choose lower number of moles of HgS and multiply by $\mathcal{M}$ (g/mol) → Mass (g) of HgS

Plan This is a limiting-reactant problem because *the quantities of two reactants are given.* After balancing the equation, we determine the limiting reactant. From the molarity and volume of each solution, we calculate the amount (mol) of each reactant. Then, we use the molar ratio to find the amount of product (HgS) that each reactant forms. The limiting reactant forms fewer moles of HgS, which we convert to mass (g) of HgS using its molar mass (see the road map). We use the amount of HgS formed from the limiting reactant in the reaction table.

Solution **(a)** Writing the balanced equation:

$$Hg(NO_3)_2(aq) + Na_2S(aq) \longrightarrow HgS(s) + 2NaNO_3(aq)$$

Finding the amount (mol) of HgS formed from $Hg(NO_3)_2$: Combining the steps gives

$$\text{Amount (mol) of HgS} = 0.050\ \cancel{\text{L soln}} \times \frac{0.010\ \cancel{\text{mol Hg(NO}_3)_2}}{1\ \cancel{\text{L soln}}} \times \frac{1\ \text{mol HgS}}{1\ \cancel{\text{mol Hg(NO}_3)_2}}$$
$$= \mathbf{5.0{\times}10^{-4}\ mol\ HgS}$$

Finding the amount (mol) of HgS from Na_2S: Combining the steps gives

$$\text{Amount (mol) of HgS} = 0.020\ \cancel{\text{L soln}} \times \frac{0.10\ \cancel{\text{mol Na}_2\text{S}}}{1\ \cancel{\text{L soln}}} \times \frac{1\ \text{mol HgS}}{1\ \cancel{\text{mol Na}_2\text{S}}}$$
$$= \mathbf{2.0{\times}10^{-3}\ mol\ HgS}$$

$Hg(NO_3)_2$ is the limiting reactant because it forms fewer moles of HgS. Converting the amount (mol) of HgS formed from $Hg(NO_3)_2$ to mass (g):

$$\text{Mass (g) of HgS} = 5.0{\times}10^{-4}\ \cancel{\text{mol HgS}} \times \frac{232.7\ \text{g HgS}}{1\ \cancel{\text{mol HgS}}} = \boxed{0.12\ \text{g HgS}}$$

(b) With $Hg(NO_3)_2$ as the limiting reactant, the reaction table is

Amount (mol)	$Hg(NO_3)_2(aq)$	+	$Na_2S(aq)$	$\longrightarrow$	$HgS(s)$	+	$2NaNO_3(aq)$
Initial	$5.0{\times}10^{-4}$		$2.0{\times}10^{-3}$		0		0
Change	$-5.0{\times}10^{-4}$		$-5.0{\times}10^{-4}$		$+5.0{\times}10^{-4}$		$+1.0{\times}10^{-3}$
Final	0		$1.5{\times}10^{-3}$		$5.0{\times}10^{-4}$		$1.0{\times}10^{-3}$

A large excess of Na_2S remains after the reaction. Note that the amount of $NaNO_3$ formed is twice the amount of $Hg(NO_3)_2$ consumed, as the balanced equation shows.

Check As a check on our choice of the limiting reactant, let's use the alternative method in Sample Problem 3.19 (see Comment, p. 97).

Finding amount (mol) of reactants given:

$$\text{Amount (mol) of Hg(NO}_3)_2 = 0.050\ \cancel{\text{L soln}} \times \frac{0.010\ \text{mol Hg(NO}_3)_2}{1\ \cancel{\text{L soln}}} = 5.0{\times}10^{-4}\ \text{mol Hg(NO}_3)_2$$

$$\text{Amount (mol) of Na}_2\text{S} = 0.020\ \cancel{\text{L soln}} \times \frac{0.10\ \text{mol Na}_2\text{S}}{1\ \cancel{\text{L soln}}} = 2.0{\times}10^{-3}\ \text{mol Na}_2\text{S}$$

The molar ratio of the reactants is 1 $Hg(NO_3)_2$/1 Na_2S. Therefore, $Hg(NO_3)_2$ is limiting because there is less of it than we would need to react with all of the available Na_2S.

FOLLOW-UP PROBLEM 3.26 Despite the toxicity of lead, many of its compounds are still used to make pigments. **(a)** When 268 mL of 1.50 *M* lead(II) acetate reacts with 130. mL of 3.40 *M* sodium chloride, how many grams of solid lead(II) chloride can form? (Sodium acetate solution also forms.) **(b)** Using the abbreviation "Ac" for the acetate ion, write a reaction table for the process.

Summary of Section 3.5

- When reactions occur in solution, amounts of reactants and products are given in terms of concentration and volume.
- Molarity is the number of moles of solute dissolved in one liter of solution. A concentrated solution (higher molarity) is converted to a dilute solution (lower molarity) by adding solvent.
- By using molarity as a conversion factor, we can apply the principles of stoichiometry to reactions in solution.

CHAPTER REVIEW GUIDE

The following sections provide many aids to help you study this chapter. (Numbers in parentheses refer to pages, unless noted otherwise.)

Learning Objectives

These are concepts and skills to review after studying this chapter.

Related section (§), sample problem (SP), and upcoming end-of-chapter problem (EP) numbers are listed in parentheses.

1. Realize the usefulness of the mole concept, and use the relation between molecular (or formula) mass and molar mass to calculate the molar mass of any substance (§3.1) (EPs 3.1–3.5, 3.7–3.10)
2. Understand the relationships among amount of substance (in moles), mass (in grams), and number of chemical entities and convert from one to any other (§3.1) (SPs 3.1–3.5) (EPs 3.6, 3.11–3.16, 3.19)
3. Find a mass percent and use it to find the mass of element in a given mass of compound (§3.1) (SPs 3.6, 3.7) (EPs 3.17, 3.18, 3.20–3.23)
4. Determine the empirical and molecular formulas of a compound from mass analysis of its elements (§3.2) (SPs 3.8–3.11) (EPs 3.24–3.34)
5. Balance an equation given formulas or names, and use molar ratios to calculate amounts of reactants and products for reactions of pure or dissolved substances (§3.3, 3.5) (SPs 3.12–3.16, 3.25) (EPs 3.35–3.40, 3.63, 3.76, 3.77)
6. Understand why one reactant limits the yield of product, and solve limiting-reactant problems for reactions of pure or dissolved substances (§3.4, 3.5) (SPs 3.17–3.19, 3.26) (EPs 3.41–3.55, 3.62, 3.64, 3.74, 3.75)
7. Explain the reasons for lower-than-expected yields and the distinction between theoretical and actual yields, and calculate percent yield (§3.4) (SP 3.20) (EPs 3.56–3.61, 3.65)
8. Understand the meaning of concentration and the effect of dilution, and calculate molarity or mass of dissolved solute (§3.5) (SPs 3.21–3.24) (EPs 3.66–3.73, 3.78)

Key Terms

These important terms appear in boldface in the chapter and are defined again in the Glossary.

stoichiometry (72)

Section 3.1
mole (mol) (72)
Avogadro's number (72)
molar mass ($\mathcal{M}$) (73)

Section 3.2
empirical formula (80)
molecular formula (80)
structural formula (80)
combustion analysis (83)
isomer (84)

Section 3.3
chemical equation (85)
reactant (86)
product (86)
balancing (stoichiometric) coefficient (86)

Section 3.4
limiting reactant (93)
theoretical yield (97)
side reaction (97)
actual yield (97)
percent yield (% yield) (98)

Section 3.5
solute (99)
solvent (99)
concentration (99)
molarity (M) (99)

Key Equations and Relationships

Numbered and screened concepts are listed for you to refer to or memorize.

3.1 Number of entities in one mole (72):

$$1 \text{ mol contains } 6.022\times10^{23} \text{ entities (to 4 sf)}$$

3.2 Converting amount (mol) to mass (g) using $\mathcal{M}$ (74):

$$\text{Mass (g)} = \text{amount (mol)} \times \frac{\text{no. of grams}}{\text{1 mol}}$$

3.3 Converting mass (g) to amount (mol) using $1/\mathcal{M}$ (74):

$$\text{Amount (mol)} = \text{mass (g)} \times \frac{\text{1 mol}}{\text{no. of grams}}$$

3.4 Converting amount (mol) to number of entities (74):

$$\text{No. of entities} = \text{amount (mol)} \times \frac{6.022\times10^{23} \text{ entities}}{\text{1 mol}}$$

3.5 Converting number of entities to amount (mol) (74):

$$\text{Amount (mol)} = \text{no. of entities} \times \frac{\text{1 mol}}{6.022\times10^{23} \text{ entities}}$$

3.6 Calculating mass % (78):

Mass % of element X

$$= \frac{\text{moles of X in formula} \times \text{molar mass of X (g/mol)}}{\text{mass (g) of 1 mol of compound}} \times 100$$

3.7 Finding the mass of an element in any mass of compound (79):

Mass of element = mass of compound

$$\times \frac{\text{mass of element in 1 mol of compound}}{\text{mass of 1 mol of compound}}$$

3.8 Calculating percent yield (98):

$$\%\text{ yield} = \frac{\text{actual yield}}{\text{theoretical yield}} \times 100$$

3.9 Defining molarity (99):

$$\text{Molarity} = \frac{\text{moles of solute}}{\text{liters of solution}} \quad \text{or} \quad M = \frac{\text{mol solute}}{\text{L soln}}$$

3.10 Diluting a concentrated solution (101):

$$M_{\text{dil}} \times V_{\text{dil}} = \text{amount (mol)} = M_{\text{conc}} \times V_{\text{conc}}$$

BRIEF SOLUTIONS TO FOLLOW-UP PROBLEMS

Compare your own solutions to these calculation steps and answers.

3.1

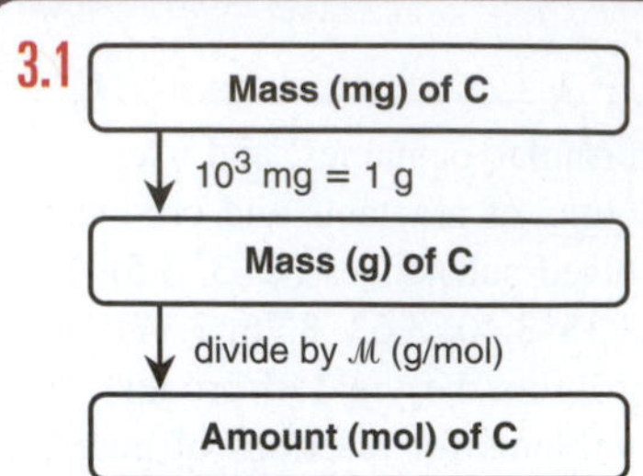

$$\text{Amount (mol) of C} = 315\ \cancel{\text{mg C}} \times \frac{1\ \cancel{\text{g}}}{10^3\ \cancel{\text{mg}}} \times \frac{1\ \text{mol C}}{12.01\ \cancel{\text{g C}}} = 2.62\times10^{-2}\ \text{mol C}$$

3.2

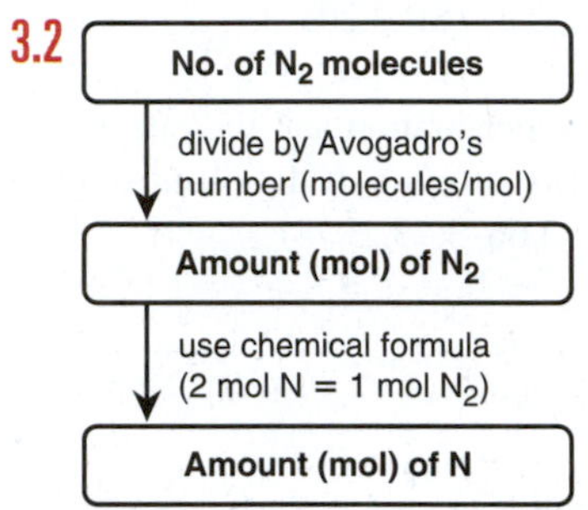

$$\text{Amount (mol) of N} = 9.72\times10^{21}\ \cancel{N_2\ \text{molecules}} \times \frac{1\ \cancel{\text{mol}\ N_2}}{6.022\times10^{23}\ \cancel{N_2\ \text{molecules}}} \times \frac{2\ \text{mol N}}{1\ \cancel{\text{mol}\ N_2}} = 3.23\times10^{-2}\ \text{mol N}$$

3.3

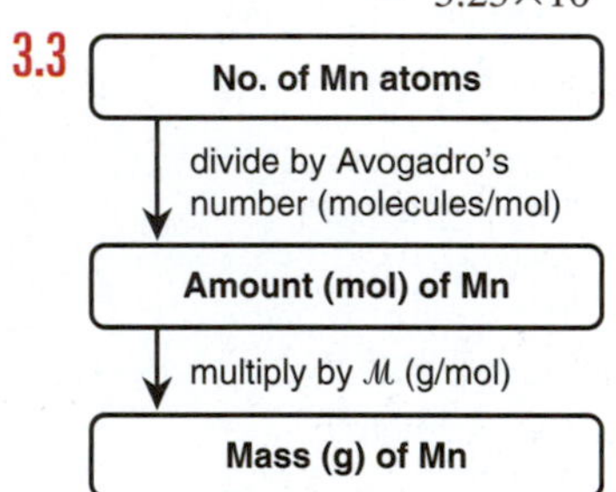

$$\text{Mass (g) of Mn} = 3.22\times10^{20}\ \cancel{\text{Mn atoms}} \times \frac{1\ \cancel{\text{mol Mn}}}{6.022\times10^{23}\ \cancel{\text{Mn atoms}}} \times \frac{54.94\ \text{g Mn}}{1\ \cancel{\text{mol Mn}}} = 2.94\times10^{-2}\ \text{g Mn}$$

3.4

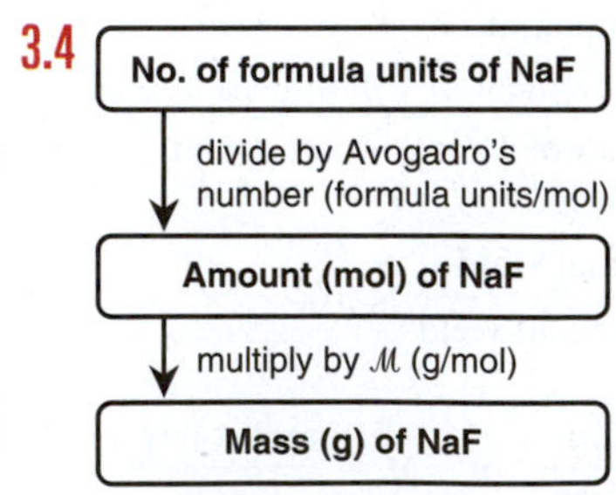

$$\mathcal{M} = (1 \times \mathcal{M}\ \text{of Na}) + (1 \times \mathcal{M}\ \text{of F}) = 22.99\ \text{g/mol} + 19.00\ \text{g/mol} = 41.99\ \text{g/mol}$$

$$\text{Mass (g) of NaF} = 1.19\times10^{19}\ \cancel{\text{NaF formula units}} \times \frac{1\ \cancel{\text{mol NaF}}}{6.022\times10^{23}\ \cancel{\text{NaF formula units}}} \times \frac{41.99\ \text{g NaF}}{1\ \cancel{\text{mol NaF}}} = 8.30\times10^{-4}\ \text{NaF}$$

3.5 (a) Mass (g) of P_4O_{10}

$$= 4.65\times10^{22}\ \cancel{\text{molecules}\ P_4O_{10}} \times \frac{1\ \cancel{\text{mol}\ P_4O_{10}}}{6.022\times10^{23}\ \cancel{\text{molecules}\ P_4O_{10}}} \times \frac{283.88\ \text{g}\ P_4O_{10}}{1\ \cancel{\text{mol}\ P_4O_{10}}} = 21.9\ \text{g}\ P_4O_{10}$$

(b) $$\text{No. of P atoms} = 4.65\times10^{22}\ \cancel{\text{molecules}\ P_4O_{10}} \times \frac{4\ \text{P atoms}}{1\ \cancel{\text{molecule}\ P_4O_{10}}} = 1.86\times10^{23}\ \text{P atoms}$$

3.6 $$\text{Mass \% of N} = \frac{2\ \cancel{\text{mol N}} \times \dfrac{14.01\ \cancel{\text{g N}}}{1\ \cancel{\text{mol N}}}}{80.05\ \cancel{\text{g}\ NH_4NO_3}} \times 100 = 35.00\ \text{mass \% N}$$

3.7 $$\text{Mass (g) of N} = 35.8\ \cancel{\text{kg}\ NH_4NO_3} \times \frac{10^3\ \cancel{\text{g}}}{1\ \cancel{\text{kg}}} \times \frac{0.3500\ \text{g N}}{1\ \cancel{\text{g}\ NH_4NO_3}} = 1.25\times10^4\ \text{g N}$$

3.8 Preliminary formula: $B_{0.170}O_{0.255}$

Divide by smaller subscript: $B_{\frac{0.170}{0.170}}O_{\frac{0.255}{0.170}} = B_{1.00}O_{1.50}$

Multiply by 2: $B_{2\times1.00}O_{2\times1.50} = B_{2.00}O_{3.00} = B_2O_3$

3.9 $$\text{Amount (mol) of S} = 2.88\ \cancel{\text{g S}} \times \frac{1\ \text{mol S}}{32.07\ \cancel{\text{g S}}} = 0.0898\ \text{mol S}$$

$$\text{Amount (mol) of M} = 0.0898\ \cancel{\text{mol S}} \times \frac{2\ \text{mol M}}{3\ \cancel{\text{mol S}}} = 0.0599\ \text{mol M}$$

$$\text{Molar mass of M} = \frac{3.12\ \text{g M}}{0.0599\ \text{mol M}} = 52.1\ \text{g/mol}$$

M is chromium, and M_2S_3 is chromium(III) sulfide.

3.10 Assuming 100.00 g of compound, we have 95.21 g of C and 4.79 g of H:

$$\text{Amount (mol) of C} = 95.21\ \cancel{\text{g C}} \times \frac{1\ \text{mol C}}{12.01\ \cancel{\text{g C}}} = 7.928\ \text{mol C}$$

Similarly, there is 4.75 mol H.

Preliminary formula: $C_{7.928}H_{4.75} \approx C_{1.67}H_{1.00}$

Empirical formula: C_5H_3

$$\text{Whole-number multiple} = \frac{252.30\ \cancel{\text{g/mol}}}{63.07\ \cancel{\text{g/mol}}} = 4$$

Molecular formula: $C_{20}H_{12}$

3.11 $$\text{Mass (g) of C} = 0.451\ \cancel{\text{g}\ CO_2} \times \frac{12.01\ \text{g C}}{44.01\ \cancel{\text{g}\ CO_2}} = 0.123\ \text{g C}$$

Similarly, there is 0.00690 g H.

Mass (g) of Cl = 0.250 g − (0.123 g + 0.00690 g) = 0.120 g Cl

Amount (mol) of elements: 0.0102 mol C; 0.00685 mol H; 0.00339 mol Cl

Empirical formula: C_3H_2Cl

Whole-number multiple = 2

Molecular formula: $C_6H_4Cl_2$

3.12 (a) $2Na(s) + 2H_2O(l) \longrightarrow H_2(g) + 2NaOH(aq)$

(b) $2HNO_3(aq) + CaCO_3(s) \longrightarrow CO_2(g) + H_2O(l) + Ca(NO_3)_2(aq)$

(c) $PCl_3(g) + 3HF(g) \longrightarrow PF_3(g) + 3HCl(g)$

(d) $4C_3H_5N_3O_9(l) \longrightarrow 12CO_2(g) + 10H_2O(g) + 6N_2(g) + O_2(g)$

3.13 From the depiction, we have

$$6CO + 3O_2 \longrightarrow 6CO_2$$

Or,

$$2CO(g) + O_2(g) \longrightarrow 2CO_2(g)$$

3.14

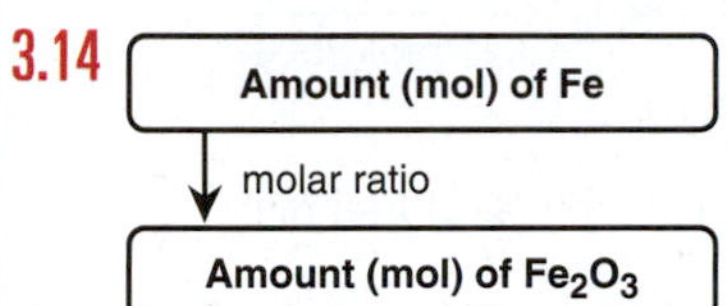

$$Fe_2O_3(s) + 2Al(s) \longrightarrow Al_2O_3(s) + 2Fe(l)$$

$$\text{Amount (mol) of } Fe_2O_3 = 3.60\times10^3 \text{ mol Fe} \times \frac{1 \text{ mol } Fe_2O_3}{2 \text{ mol Fe}} = 1.80\times10^3 \text{ mol } Fe_2O_3$$

3.15

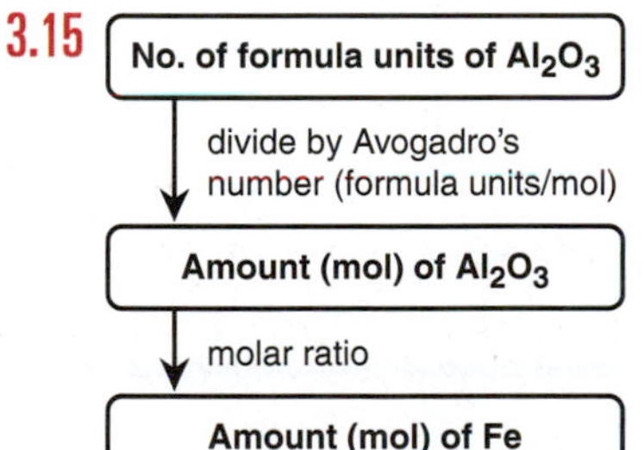

Amount (mol) of Fe

$$= 1.85\times10^{25} \text{ formula units } Al_2O_3 \times \frac{1 \text{ mol } Al_2O_3}{6.022\times10^{23} \text{ formula units } Al_2O_3} \times \frac{2 \text{ mol Fe}}{1 \text{ mol } Al_2O_3}$$

$$= 61.4 \text{ mol Fe}$$

3.16

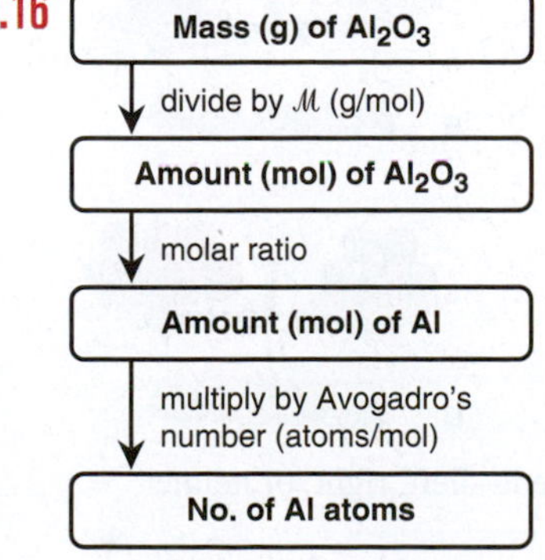

$$\text{No. of Al atoms} = 1.00 \text{ g } Al_2O_3 \times \frac{1 \text{ mol } Al_2O_3}{101.96 \text{ g } Al_2O_3} \times \frac{2 \text{ mol Al}}{1 \text{ mol } Al_2O_3} \times \frac{6.022\times10^{23} \text{ Al atoms}}{1 \text{ mol Al}}$$

$$= 1.18\times10^{22} \text{ Al atoms}$$

3.17 $4AB + 2B_2 \longrightarrow 4AB_2$, or $2AB(g) + B_2(g) \longrightarrow 2AB_2(g)$

$$\text{For AB: Molecules of } AB_2 = 4AB \times \frac{2AB_2}{2AB} = 4AB_2$$

$$\text{For } B_2\text{: Molecules of } AB_2 = 3B_2 \times \frac{2AB_2}{1B_2} = 6AB_2$$

Thus, AB_2 is the limiting reactant; one B_2 molecule is in excess.

3.18

$$\text{Amount (mol) of } AB_2 = 1.5 \text{ mol AB} \times \frac{2 \text{ mol } AB_2}{2 \text{ mol AB}} = 1.5 \text{ mol } AB_2$$

$$\text{Amount (mol) of } AB_2 = 1.5 \text{ mol } B_2 \times \frac{2 \text{ mol } AB_2}{1 \text{ mol } B_2} = 3.0 \text{ mol } AB_2$$

Therefore, 1.5 mol of AB_2 can form.

3.19 $2Al(s) + 3S(s) \longrightarrow Al_2S_3(s)$

Mass (g) of Al_2S_3 formed from 10.0 g of Al

$$= 10.0 \text{ g Al} \times \frac{1 \text{ mol Al}}{26.98 \text{ g Al}} \times \frac{1 \text{ mol } Al_2S_3}{2 \text{ mol Al}} \times \frac{150.17 \text{ g } Al_2S_3}{1 \text{ mol } Al_2S_3}$$

$$= 27.8 \text{ g } Al_2S_3$$

Similarly, mass (g) of Al_2S_3 formed from 15.0 g of S = 23.4 g Al_2S_3. Thus, S is the limiting reactant, and 23.4 g of Al_2S_3 forms.

Mass (g) of Al in excess

= total mass of Al − mass of Al used

$$= 10.0 \text{ g Al} - \left(15.0 \text{ g S} \times \frac{1 \text{ mol S}}{32.07 \text{ g S}} \times \frac{2 \text{ mol Al}}{3 \text{ mol S}} \times \frac{26.98 \text{ g Al}}{1 \text{ mol Al}}\right)$$

$$= 1.6 \text{ g Al}$$

(We would obtain the same answer if sulfur were shown more correctly as S_8.)

3.20 $CaCO_3(s) + 2HCl(aq) \longrightarrow CaCl_2(aq) + H_2O(l) + CO_2(g)$

$$\text{Theoretical yield (g) of } CO_2 = 10.0 \text{ g } CaCO_3 \times \frac{1 \text{ mol } CaCO_3}{100.09 \text{ g } CaCO_3} \times \frac{1 \text{ mol } CO_2}{1 \text{ mol } CaCO_3} \times \frac{44.01 \text{ g } CO_2}{1 \text{ mol } CO_2}$$

$$= 4.40 \text{ g } CO_2$$

$$\% \text{ yield} = \frac{3.65 \text{ g } CO_2}{4.40 \text{ g } CO_2} \times 100 = 83.0\%$$

3.21

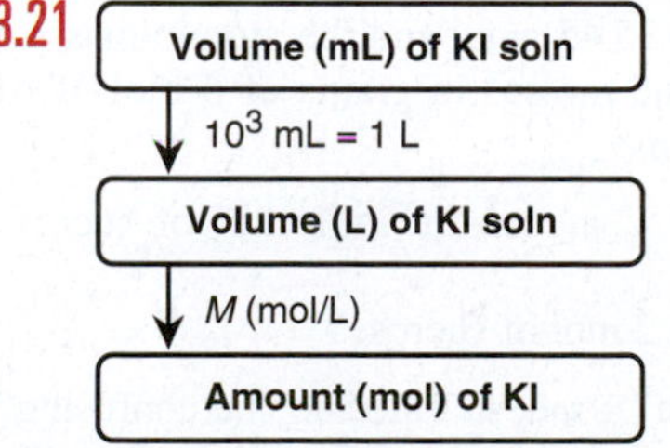

$$\text{Amount (mol) of KI} = 84 \text{ mL soln} \times \frac{1 \text{ L}}{10^3 \text{ mL}} \times \frac{0.50 \text{ mol KI}}{1 \text{ L soln}}$$

$$= 0.042 \text{ mol KI}$$

BRIEF SOLUTIONS TO FOLLOW-UP PROBLEMS (continued)

3.22

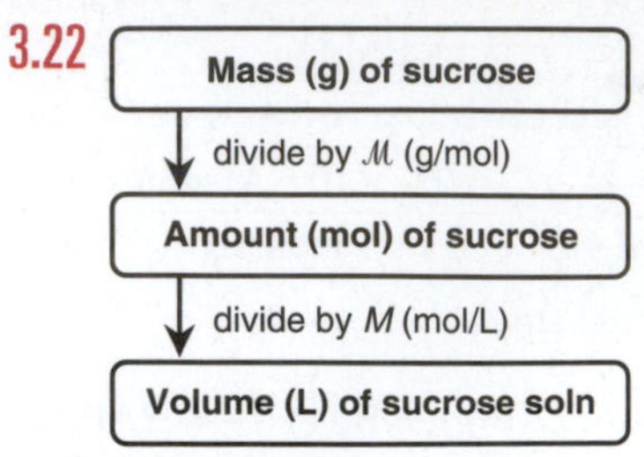

Volume (L) of sucrose soln

$$= 135\ \cancel{\text{g sucrose}} \times \frac{1\ \cancel{\text{mol sucrose}}}{342.30\ \cancel{\text{g sucrose}}} \times \frac{1\ \text{L soln}}{3.30\ \cancel{\text{mol sucrose}}}$$

$$= 0.120\ \text{L soln}$$

3.23 M_{dil} of $H_2SO_4 = \dfrac{7.50\ M \times 25.0\ \cancel{\text{m}^3}}{500.\ \cancel{\text{m}^3}} = 0.375\ M\ H_2SO_4$

Mass (g) of H_2SO_4/mL soln

$$= \frac{0.375\ \cancel{\text{mol H}_2\text{SO}_4}}{1\ \cancel{\text{L soln}}} \times \frac{1\ \cancel{\text{L}}}{10^3\ \text{mL}} \times \frac{98.09\ \text{g H}_2\text{SO}_4}{1\ \cancel{\text{mol H}_2\text{SO}_4}}$$

$$= 3.68\times10^{-2}\ \text{g/mL soln}$$

3.24 To obtain B, the total volume of sol A was reduced by half:

$$V_{\text{conc}} = V_{\text{dil}} \times \frac{N_{\text{dil}}}{N_{\text{conc}}} = 1.0\ \text{mL} \times \frac{6\ \cancel{\text{particles}}}{12\ \cancel{\text{particles}}} = 0.50\ \text{mL}$$

To obtain C, $\frac{1}{2}$ of a volume of solvent was added for every volume of A:

$$V_{\text{dil}} = V_{\text{conc}} \times \frac{N_{\text{conc}}}{N_{\text{dil}}} = 1.0\ \text{mL} \times \frac{6\ \cancel{\text{particles}}}{4\ \cancel{\text{particles}}} = 1.5\ \text{mL}$$

3.25 $Al(OH)_3(s) + 3HCl(aq) \longrightarrow AlCl_3(aq) + 3H_2O(l)$

Volume (L) of HCl neutralized

$$= 0.10\ \cancel{\text{g Al(OH)}_3} \times \frac{1\ \cancel{\text{mol Al(OH)}_3}}{78.00\ \cancel{\text{g Al(OH)}_3}}$$

$$\times \frac{3\ \cancel{\text{mol HCl}}}{1\ \cancel{\text{mol Al(OH)}_3}} \times \frac{1\ \text{L soln}}{0.10\ \cancel{\text{mol HCl}}}$$

$$= 3.8\times10^{-2}\ \text{L soln}$$

Therefore, $Al(OH)_3$ is more effective than $Mg(OH)_2$.

3.26 (a) For $Pb(C_2H_3O_2)_2$:

$$\text{Amount (mol) of Pb(C}_2\text{H}_3\text{O}_2)_2 = 0.268\ \cancel{\text{L}} \times \frac{1.50\ \text{mol Pb(C}_2\text{H}_3\text{O}_2)_2}{1\ \cancel{\text{L}}}$$

$$= 0.402\ \text{mol Pb(C}_2\text{H}_3\text{O}_2)_2$$

$$\text{Amount (mol) of PbCl}_2 = 0.402\ \cancel{\text{mol Pb(C}_2\text{H}_3\text{O}_2)_2} \times \frac{1\ \text{mol PbCl}_2}{1\ \cancel{\text{mol Pb(C}_2\text{H}_3\text{O}_2)_2}} = 0.402\ \text{mol PbCl}_2$$

$$\text{For NaCl: Amount (mol) of NaCl} = \frac{3.40\ \text{mol NaCl}}{1\ \cancel{\text{L}}} \times 0.130\ \cancel{\text{L}}$$

$$= 0.442\ \text{mol NaCl}$$

$$\text{Amount (mol) of PbCl}_2 = 0.442\ \cancel{\text{mol NaCl}} \times \frac{1\ \text{mol PbCl}_2}{2\ \cancel{\text{mol NaCl}}}$$

$$= 0.221\ \text{mol PbCl}_2$$

Thus, NaCl is the limiting reactant.

$$\text{Mass (g) of PbCl}_2 = 0.22\ \cancel{\text{mol PbCl}_2} \times \frac{278.1\ \text{g PbCl}_2}{1\ \cancel{\text{mol PbCl}_2}}$$

$$= 61.5\ \text{g PbCl}_2$$

(b) Using "Ac" for acetate ion and with NaCl limiting:

Amount (mol)	$Pb(Ac)_2$	+	2NaCl	$\longrightarrow$	$PbCl_2$	+	2NaAc
Initial	0.402		0.442		0		0
Change	−0.221		−0.442		+0.221		+0.442
Final	0.181		0		0.221		0.442

PROBLEMS

Problems with **colored** numbers are answered in Appendix E. Sections match the text and provide the numbers of relevant sample problems. Bracketed problems are grouped in pairs (indicated by a short rule) that cover the same concept. Comprehensive Problems are based on material from any section or previous chapter.

The Mole

(Sample Problems 3.1 to 3.7)

3.1 The atomic mass of Cl is 35.45 amu, and the atomic mass of Al is 26.98 amu. What are the masses in grams of 3 mol of Al atoms and of 2 mol of Cl atoms?

3.2 (a) How many moles of C atoms are in 1 mol of sucrose ($C_{12}H_{22}O_{11}$)?
(b) How many C atoms are in 2 mol of sucrose?

3.3 Why might the expression "1 mol of chlorine" be confusing? What change would remove any uncertainty? For what other elements might a similar confusion exist? Why?

3.4 How is the molecular mass of a compound the same as the molar mass, and how is it different?

3.5 What advantage is there to using a counting unit (the mole) for amount of substance rather than a mass unit?

3.6 Each of the following balances weighs the indicated numbers of atoms of two elements:

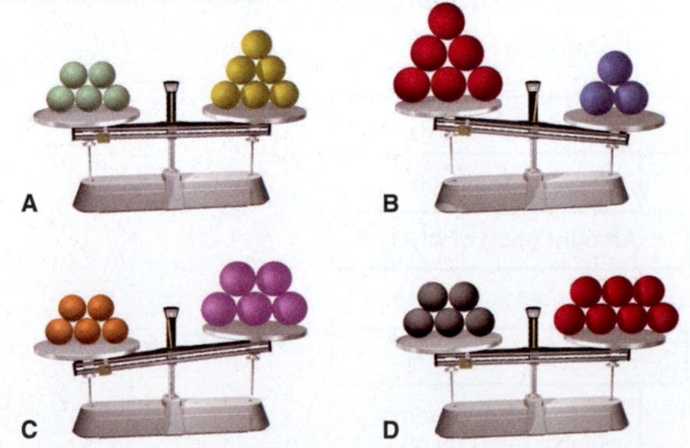

For each balance, which element—left, right, or neither,
(a) Has the higher molar mass?
(b) Has more atoms per gram?

(c) Has fewer atoms per gram?
(d) Has more atoms per mole?

3.7 Calculate the molar mass of each of the following:
(a) $Sr(OH)_2$ (b) N_2O_3 (c) $NaClO_3$ (d) Cr_2O_3

3.8 Calculate the molar mass of each of the following:
(a) $(NH_4)_3PO_4$ (b) CH_2Cl_2 (c) $CuSO_4{\cdot}5H_2O$ (d) BrF_3

3.9 Calculate the molar mass of each of the following:
(a) SnO (b) BaF_2 (c) $Al_2(SO_4)_3$ (d) $MnCl_2$

3.10 Calculate the molar mass of each of the following:
(a) N_2O_4 (b) C_4H_9OH (c) $MgSO_4{\cdot}7H_2O$ (d) $Ca(C_2H_3O_2)_2$

3.11 Calculate each of the following quantities:
(a) Mass (g) of 0.68 mol of $KMnO_4$
(b) Amount (mol) of O atoms in 8.18 g of $Ba(NO_3)_2$
(c) Number of O atoms in 7.3×10^{-3} g of $CaSO_4{\cdot}2H_2O$

3.12 Calculate each of the following quantities:
(a) Mass (kg) of 4.6×10^{21} molecules of NO_2
(b) Amount (mol) of Cl atoms in 0.0615 g of $C_2H_4Cl_2$
(c) Number of H^- ions in 5.82 g of SrH_2

3.13 Calculate each of the following quantities:
(a) Mass (g) of 6.44×10^{-2} mol of $MnSO_4$
(b) Amount (mol) of compound in 15.8 kg of $Fe(ClO_4)_3$
(c) Number of N atoms in 92.6 mg of NH_4NO_2

3.14 Calculate each of the following quantities:
(a) Total number of ions in 38.1 g of SrF_2
(b) Mass (kg) of 3.58 mol of $CuCl_2{\cdot}2H_2O$
(c) Mass (mg) of 2.88×10^{22} formula units of $Bi(NO_3)_3{\cdot}5H_2O$

3.15 Calculate each of the following quantities:
(a) Mass (g) of 8.35 mol of copper(I) carbonate
(b) Mass (g) of 4.04×10^{20} molecules of dinitrogen pentoxide
(c) Amount (mol) and number of formula units in 78.9 g of sodium perchlorate
(d) Number of sodium ions, perchlorate ions, chlorine atoms, and oxygen atoms in the mass of compound in part (c)

3.16 Calculate each of the following quantities:
(a) Mass (g) of 8.42 mol of chromium(III) sulfate decahydrate
(b) Mass (g) of 1.83×10^{24} molecules of dichlorine heptoxide
(c) Amount (mol) and number of formula units in 6.2 g of lithium sulfate
(d) Number of lithium ions, sulfate ions, sulfur atoms, and oxygen atoms in the mass of compound in part (c)

3.17 Calculate each of the following:
(a) Mass % of H in ammonium bicarbonate
(b) Mass % of O in sodium dihydrogen phosphate heptahydrate

3.18 Calculate each of the following:
(a) Mass % of I in strontium periodate
(b) Mass % of Mn in potassium permanganate

3.19 Cisplatin *(right)*, or Platinol, is used in the treatment of certain cancers. Calculate (a) the amount (mol) of compound in 285.3 g of cisplatin; (b) the number of hydrogen atoms in 0.98 mol of cisplatin.

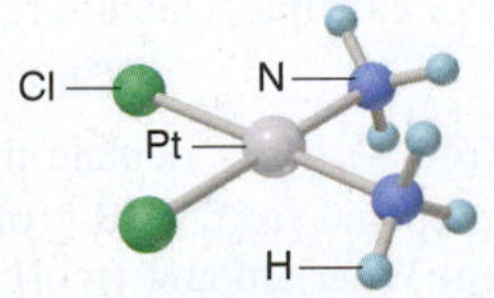

3.20 Propane is widely used in liquid form as a fuel for barbecue grills and camp stoves. For 85.5 g of propane, calculate (a) moles of compound; (b) grams of carbon.

3.21 The effectiveness of a nitrogen fertilizer is determined mainly by its mass % N. Rank the following fertilizers, most effective first: potassium nitrate; ammonium nitrate; ammonium sulfate; urea, $CO(NH_2)_2$.

3.22 The mineral galena is composed of lead(II) sulfide and has an average density of 7.46 g/cm^3. (a) How many moles of lead(II) sulfide are in 1.00 ft^3 of galena? (b) How many lead atoms are in 1.00 dm^3 of galena?

3.23 Hemoglobin, a protein in red blood cells, carries O_2 from the lungs to the body's cells. Iron (as ferrous ion, Fe^{2+}) makes up 0.33 mass % of hemoglobin. If the molar mass of hemoglobin is 6.8×10^4 g/mol, how many Fe^{2+} ions are in one molecule?

Determining the Formula of an Unknown Compound

(Sample Problems 3.8 to 3.11)

3.24 Which of the following sets of information allows you to obtain the molecular formula of a covalent compound? In each case that allows it, explain how you would proceed (draw a road map and write a Plan for a solution).
(a) Number of moles of each type of atom in a given sample of the compound
(b) Mass % of each element and the total number of atoms in a molecule of the compound
(c) Mass % of each element and the number of atoms of one element in a molecule of the compound
(d) Empirical formula and mass % of each element
(e) Structural formula

3.25 What is the empirical formula and empirical formula mass for each of the following compounds?
(a) C_2H_4 (b) $C_2H_6O_2$ (c) N_2O_5 (d) $Ba_3(PO_4)_2$ (e) Te_4I_{16}

3.26 What is the empirical formula and empirical formula mass for each of the following compounds?
(a) C_4H_8 (b) $C_3H_6O_3$ (c) P_4O_{10} (d) $Ga_2(SO_4)_3$ (e) Al_2Br_6

3.27 What is the molecular formula of each compound?
(a) Empirical formula CH_2 ($\mathcal{M}$ = 42.08 g/mol)
(b) Empirical formula NH_2 ($\mathcal{M}$ = 32.05 g/mol)
(c) Empirical formula NO_2 ($\mathcal{M}$ = 92.02 g/mol)
(d) Empirical formula CHN ($\mathcal{M}$ = 135.14 g/mol)

3.28 What is the molecular formula of each compound?
(a) Empirical formula CH ($\mathcal{M}$ = 78.11 g/mol)
(b) Empirical formula $C_3H_6O_2$ ($\mathcal{M}$ = 74.08 g/mol)
(c) Empirical formula HgCl ($\mathcal{M}$ = 472.1 g/mol)
(d) Empirical formula $C_7H_4O_2$ ($\mathcal{M}$ = 240.20 g/mol)

3.29 Find the empirical formula of the following compounds:
(a) 0.063 mol of chlorine atoms combined with 0.22 mol of oxygen atoms
(b) 2.45 g of silicon combined with 12.4 g of chlorine
(c) 27.3 mass % carbon and 72.7 mass % oxygen

3.30 Find the empirical formula of the following compounds:
(a) 0.039 mol of iron atoms combined with 0.052 mol of oxygen atoms
(b) 0.903 g of phosphorus combined with 6.99 g of bromine
(c) A hydrocarbon with 79.9 mass % carbon

3.31 A sample of 0.600 mol of a metal M reacts completely with excess fluorine to form 46.8 g of MF_2.
(a) How many moles of F are in the sample of MF_2 that forms?
(b) How many grams of M are in this sample of MF_2?
(c) What element is represented by the symbol M?

3.32 A 0.370-mol sample of a metal oxide (M_2O_3) weighs 55.4 g.
(a) How many moles of O are in the sample?
(b) How many grams of M are in the sample?
(c) What element is represented by the symbol M?

3.33 Cortisol ($\mathcal{M}$ = 362.47 g/mol) is a steroid hormone involved in protein synthesis. Medically, it has a major use in reducing inflammation from rheumatoid arthritis. Cortisol is 69.6% C, 8.34% H, and 22.1% O by mass. What is its molecular formula?

3.34 Menthol ($\mathcal{M}$ = 156.3 g/mol), the strong-smelling substance in many cough drops, is a compound of carbon, hydrogen, and oxygen. When 0.1595 g of menthol was burned in a combustion apparatus, 0.449 g of CO_2 and 0.184 g of H_2O formed. What is menthol's molecular formula?

Writing and Balancing Chemical Equations

(Sample Problems 3.12 and 3.13)

3.35 In the process of balancing the equation

$$Al + Cl_2 \longrightarrow AlCl_3$$

Student I writes: $Al + Cl_2 \longrightarrow AlCl_2$
Student II writes: $Al + Cl_2 + Cl \longrightarrow AlCl_3$
Student III writes: $2Al + 3Cl_2 \longrightarrow 2AlCl_3$
Is the approach of Student I valid? Student II? Student III? Explain.

3.36 The scenes below represent a chemical reaction between elements A (red) and B (green):

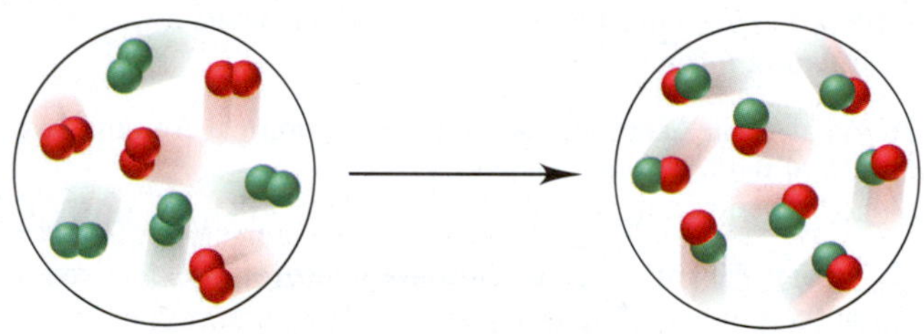

Which best represents the balanced equation for the reaction?
(a) $2A + 2B \longrightarrow A_2 + B_2$ (b) $A_2 + B_2 \longrightarrow 2AB$
(c) $B_2 + 2AB \longrightarrow 2B_2 + A_2$ (d) $4A_2 + 4B_2 \longrightarrow 8AB$

3.37 Write balanced equations for each of the following by inserting the correct coefficients in the blanks:
(a) $\_Cu(s) + \_S_8(s) \longrightarrow \_Cu_2S(s)$
(b) $\_P_4O_{10}(s) + \_H_2O(l) \longrightarrow \_H_3PO_4(l)$
(c) $\_B_2O_3(s) + \_NaOH(aq) \longrightarrow \_Na_3BO_3(aq) + \_H_2O(l)$
(d) $\_CH_3NH_2(g) + \_O_2(g) \longrightarrow \_CO_2(g) + \_H_2O(g) + \_N_2(g)$

3.38 Write balanced equations for each of the following by inserting the correct coefficients in the blanks:
(a) $\_Cu(NO_3)_2(aq) + \_KOH(aq) \longrightarrow \_Cu(OH)_2(s) + \_KNO_3(aq)$
(b) $\_BCl_3(g) + \_H_2O(l) \longrightarrow \_H_3BO_3(s) + \_HCl(g)$
(c) $\_CaSiO_3(s) + \_HF(g) \longrightarrow \_SiF_4(g) + \_CaF_2(s) + \_H_2O(l)$
(d) $\_(CN)_2(g) + \_H_2O(l) \longrightarrow \_H_2C_2O_4(aq) + \_NH_3(g)$

3.39 Convert the following into balanced equations:
(a) When gallium metal is heated in oxygen gas, it melts and forms solid gallium(III) oxide.
(b) Liquid hexane burns in oxygen gas to form carbon dioxide gas and water vapor.
(c) When solutions of calcium chloride and sodium phosphate are mixed, solid calcium phosphate forms and sodium chloride remains in solution.

3.40 Convert the following into balanced equations:
(a) When lead(II) nitrate solution is added to potassium iodide solution, solid lead(II) iodide forms and potassium nitrate solution remains.
(b) Liquid disilicon hexachloride reacts with water to form solid silicon dioxide, hydrogen chloride gas, and hydrogen gas.
(c) When nitrogen dioxide is bubbled into water, a solution of nitric acid forms and gaseous nitrogen monoxide is released.

Calculating Quantities of Reactant and Product

(Sample Problems 3.14 to 3.20)

3.41 The circle below represents a mixture of A_2 and B_2 before they react to form AB_3.

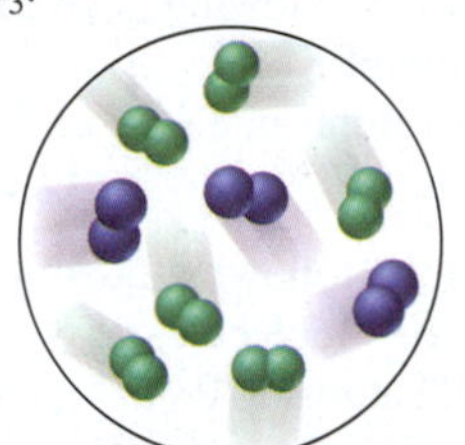

(a) What is the limiting reactant?
(b) How many molecules of product can form?

3.42 Potassium nitrate decomposes on heating, producing potassium oxide and gaseous nitrogen and oxygen:

$$4KNO_3(s) \longrightarrow 2K_2O(s) + 2N_2(g) + 5O_2(g)$$

To produce 56.6 kg of oxygen, how many (a) moles of KNO_3 and (b) grams of KNO_3 must be heated?

3.43 Chromium(III) oxide reacts with hydrogen sulfide (H_2S) gas to form chromium(III) sulfide and water:

$$Cr_2O_3(s) + 3H_2S(g) \longrightarrow Cr_2S_3(s) + 3H_2O(l)$$

To produce 421 g of Cr_2S_3, how many (a) moles of Cr_2O_3 and (b) grams of Cr_2O_3 are required?

3.44 Calculate the mass (g) of each product formed when 43.82 g of diborane (B_2H_6) reacts with excess water:

$$B_2H_6(g) + H_2O(l) \longrightarrow H_3BO_3(s) + H_2(g) \text{ [unbalanced]}$$

3.45 Calculate the mass (g) of each product formed when 174 g of silver sulfide reacts with excess hydrochloric acid:

$$Ag_2S(s) + HCl(aq) \longrightarrow AgCl(s) + H_2S(g) \text{ [unbalanced]}$$

3.46 Elemental phosphorus occurs as tetratomic molecules, P_4. What mass (g) of chlorine gas is needed to react completely with 455 g of phosphorus to form phosphorus pentachloride?

3.47 Elemental sulfur occurs as octatomic molecules, S_8. What mass (g) of fluorine gas is needed to react completely with 17.8 g of sulfur to form sulfur hexafluoride?

3.48 Many metals react with oxygen gas to form the metal oxide. For example, calcium reacts as follows:

$$2Ca(s) + O_2(g) \longrightarrow 2CaO(s)$$

You wish to calculate the mass (g) of calcium oxide that can be prepared from 4.20 g of Ca and 2.80 g of O_2.
(a) What amount (mol) of CaO can be produced from the given mass of Ca?
(b) What amount (mol) of CaO can be produced from the given mass of O_2?
(c) Which is the limiting reactant?
(d) How many grams of CaO can be produced?

3.49 Metal hydrides react with water to form hydrogen gas and the metal hydroxide. For example,

$$SrH_2(s) + 2H_2O(l) \longrightarrow Sr(OH)_2(s) + 2H_2(g)$$

You wish to calculate the mass (g) of hydrogen gas that can be prepared from 5.70 g of SrH_2 and 4.75 g of H_2O.
(a) What amount (mol) of H_2 can be produced from the given mass of SrH_2?
(b) What amount (mol) of H_2 can be produced from the given mass of H_2O?
(c) Which is the limiting reactant?
(d) How many grams of H_2 can be produced?

3.50 Calculate the maximum numbers of moles and grams of iodic acid (HIO_3) that can form when 635 g of iodine trichloride reacts with 118.5 g of water:

$$ICl_3 + H_2O \longrightarrow ICl + HIO_3 + HCl \text{ [unbalanced]}$$

How many grams of the excess reactant remains?

3.51 Calculate the maximum numbers of moles and grams of H_2S that can form when 158 g of aluminum sulfide reacts with 131 g of water:

$$Al_2S_3 + H_2O \longrightarrow Al(OH)_3 + H_2S \text{ [unbalanced]}$$

How many grams of the excess reactant remain?

3.52 When 0.100 mol of carbon is burned in a closed vessel with 8.00 g of oxygen, how many grams of carbon dioxide can form? Which reactant is in excess, and how many grams of it remain after the reaction?

3.53 A mixture of 0.0375 g of hydrogen and 0.0185 mol of oxygen in a closed container is sparked to initiate a reaction. How many grams of water can form? Which reactant is in excess, and how many grams of it remain after the reaction?

3.54 Aluminum nitrite and ammonium chloride react to form aluminum chloride, nitrogen, and water. How many grams of each substance are present after 72.5 g of aluminum nitrite and 58.6 g of ammonium chloride react completely?

3.55 Calcium nitrate and ammonium fluoride react to form calcium fluoride, dinitrogen monoxide, and water vapor. How many grams of each substance are present after 16.8 g of calcium nitrate and 17.50 g of ammonium fluoride react completely?

3.56 Two successive reactions, A $\longrightarrow$ B and B $\longrightarrow$ C, have yields of 73% and 68%, respectively. What is the overall percent yield for conversion of A to C?

3.57 Two successive reactions, D $\longrightarrow$ E and E $\longrightarrow$ F, have yields of 48% and 73%, respectively. What is the overall percent yield for conversion of D to F?

3.58 What is the percent yield of a reaction in which 45.5 g of tungsten(VI) oxide (WO_3) reacts with excess hydrogen gas to produce metallic tungsten and 9.60 mL of water (d = 1.00 g/mL)?

3.59 What is the percent yield of a reaction in which 200. g of phosphorus trichloride reacts with excess water to form 128 g of HCl and aqueous phosphorous acid (H_3PO_3)?

3.60 When 20.5 g of methane and 45.0 g of chlorine gas undergo a reaction that has a 75.0% yield, what mass (g) of chloromethane (CH_3Cl) forms? Hydrogen chloride also forms.

3.61 When 56.6 g of calcium and 30.5 g of nitrogen gas undergo a reaction that has a 93.0% yield, what mass (g) of calcium nitride forms?

3.62 Cyanogen, $(CN)_2$, has been observed in the atmosphere of Titan, Saturn's largest moon, and in the gases of interstellar nebulas. On Earth, it is used as a welding gas and a fumigant. In its reaction with fluorine gas, carbon tetrafluoride and nitrogen trifluoride gases are produced. What mass (g) of carbon tetrafluoride forms when 60.0 g of each reactant is used?

3.63 Gaseous dichlorine monoxide decomposes readily to chlorine and oxygen gases.
(a) Which scene best depicts the product mixture after the decomposition?

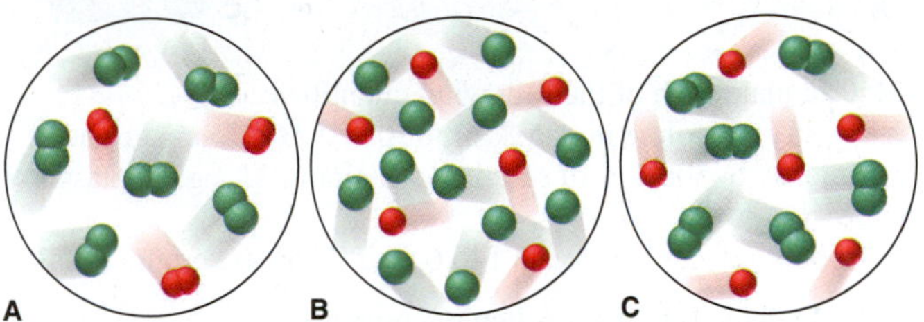

(b) Write the balanced equation for the decomposition.
(c) If each oxygen atom represents 0.050 mol, how many molecules of dichlorine monoxide were present before the decomposition?

3.64 Butane gas is compressed and used as a liquid fuel in disposable cigarette lighters and lightweight camping stoves. Suppose a lighter contains 5.50 mL of butane (d = 0.579 g/mL).
(a) How many grams of oxygen are needed to burn the butane completely?
(b) How many moles of H_2O form when all the butane burns?
(c) How many total molecules of gas form when the butane burns completely?

3.65 Sodium borohydride ($NaBH_4$) is used industrially in many organic syntheses. One way to prepare it is by reacting sodium hydride with gaseous diborane (B_2H_6). Assuming an 88.5% yield, how many grams of $NaBH_4$ can be prepared by reacting 7.98 g of sodium hydride and 8.16 g of diborane?

Fundamentals of Solution Stoichiometry

(Sample Problems 3.21 to 3.26)

3.66 Six different aqueous solutions (with solvent molecules omitted for clarity) are represented in the beakers below, and their total volumes are noted.

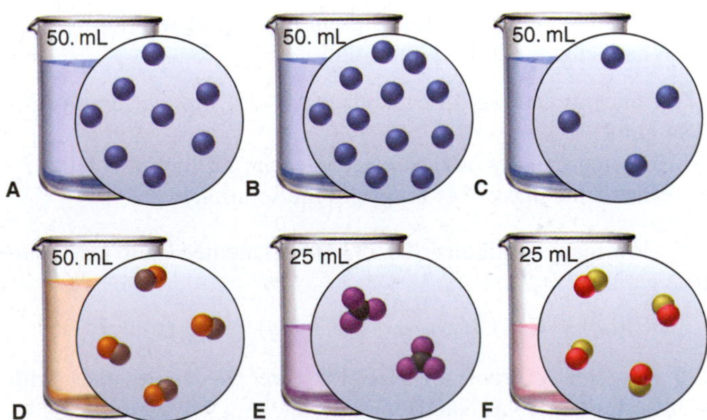

(a) Which solution has the highest molarity? (b) Which solutions have the same molarity? (c) If you mix solutions A and C, does the resulting solution have a higher, a lower, or the same molarity as solution B? (d) After 50. mL of water is added to solution D, is its molarity higher, lower, or the same as the molarity of solution F after 75 mL is added to it? (e) How much solvent must be evaporated from solution E for it to have the same molarity as solution A?

3.67 Boxes A, B, and C represent a unit volume of three solutions of the same solute. Which box, B or C, represents the solution that has (a) more solute added; (b) more solvent added; (c) higher molarity; (d) lower concentration?

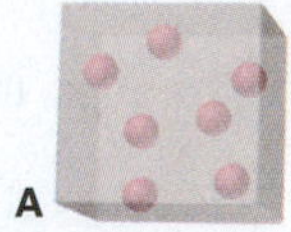

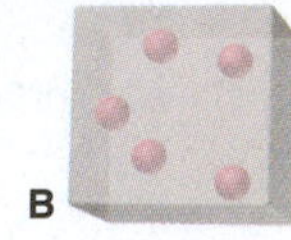

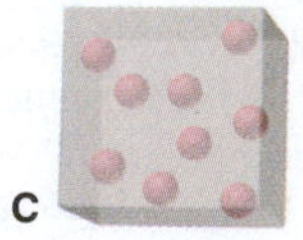

3.68 Calculate each of the following quantities:
(a) Mass (g) of solute in 185.8 mL of 0.267 *M* calcium acetate
(b) Molarity of 500. mL of solution containing 21.1 g of potassium iodide
(c) Amount (mol) of solute in 145.6 L of 0.850 *M* sodium cyanide

3.69 Calculate each of the following quantities:
(a) Volume (mL) of 2.26 *M* potassium hydroxide that contains 8.42 g of solute
(b) Number of Cu^{2+} ions in 52 L of 2.3 *M* copper(II) chloride
(c) Molarity of 275 mL of solution containing 135 mmol of glucose

3.70 Calculate each of the following quantities:
(a) Molarity of a solution prepared by diluting 37.00 mL of 0.250 *M* potassium chloride to 150.00 mL
(b) Molarity of a solution prepared by diluting 25.71 mL of 0.0706 *M* ammonium sulfate to 500.00 mL
(c) Molarity of sodium ion in a solution made by mixing 3.58 mL of 0.348 *M* sodium chloride with 500. mL of 6.81×10^{-2} *M* sodium sulfate (assume volumes are additive)

3.71 Calculate each of the following quantities:
(a) Volume (L) of 2.050 *M* copper(II) nitrate that must be diluted with water to prepare 750.0 mL of a 0.8543 *M* solution
(b) Volume (L) of 1.63 *M* calcium chloride that must be diluted with water to prepare 350. mL of a 2.86×10^{-2} *M* chloride ion solution
(c) Final volume (L) of a 0.0700 *M* solution prepared by diluting 18.0 mL of 0.155 *M* lithium carbonate with water

3.72 A sample of concentrated nitric acid has a density of 1.41 g/mL and contains 70.0% HNO_3 by mass.
(a) What mass (g) of HNO_3 is present per liter of solution?
(b) What is the molarity of the solution?

3.73 Concentrated sulfuric acid (18.3 *M*) has a density of 1.84 g/mL.
(a) How many moles of H_2SO_4 are in each milliliter of solution?
(b) What is the mass % of H_2SO_4 in the solution?

3.74 How many milliliters of 0.383 *M* HCl are needed to react with 16.2 g of $CaCO_3$?

$$2HCl(aq) + CaCO_3(s) \longrightarrow CaCl_2(aq) + CO_2(g) + H_2O(l)$$

3.75 How many grams of NaH_2PO_4 are needed to react with 43.74 mL of 0.285 *M* NaOH?

$$NaH_2PO_4(s) + 2NaOH(aq) \longrightarrow Na_3PO_4(aq) + 2H_2O(l)$$

3.76 How many grams of solid barium sulfate form when 35.0 mL of 0.160 *M* barium chloride reacts with 58.0 mL of 0.065 *M* sodium sulfate? Aqueous sodium chloride forms also.

3.77 How many moles of excess reactant are present when 350. mL of 0.210 *M* sulfuric acid reacts with 0.500 L of 0.196 *M* sodium hydroxide to form water and aqueous sodium sulfate?

3.78 Muriatic acid, an industrial grade of concentrated HCl, is used to clean masonry and cement. Its concentration is 11.7 *M*. (a) Write instructions for diluting the concentrated acid to make 3.0 gallons of 3.5 *M* acid for routine use (1 gal = 4 qt; 1 qt = 0.946 L). (b) How many milliliters of the muriatic acid solution contain 9.66 g of HCl?

Comprehensive Problems

3.79 Narceine is a narcotic in opium that crystallizes from solution as a hydrate that contains 10.8 mass % water and has a molar mass of 499.52 g/mol. Determine *x* in narceine·xH_2O.

3.80 Hydrogen-containing fuels have a "fuel value" based on their mass % H. Rank the following compounds from highest fuel value to lowest: ethane, propane, benzene, ethanol, cetyl palmitate (whale oil, $C_{32}H_{64}O_2$).

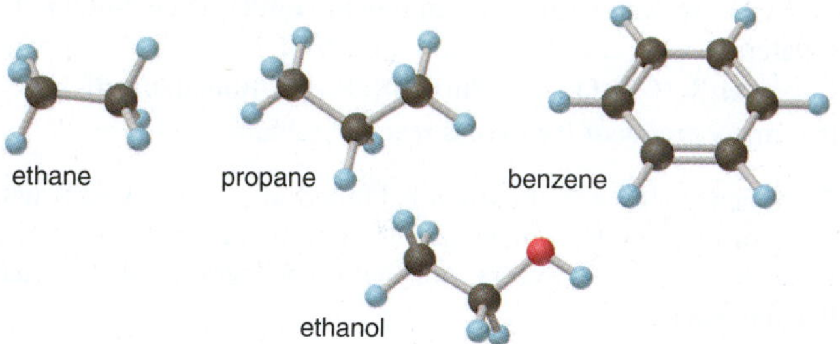

3.81 Convert the following descriptions into balanced equations:
(a) In a gaseous reaction, hydrogen sulfide burns in oxygen to form sulfur dioxide and water vapor.
(b) When crystalline potassium chlorate is heated to just above its melting point, it reacts to form two different crystalline compounds, potassium chloride and potassium perchlorate.
(c) When hydrogen gas is passed over powdered iron(III) oxide, iron metal and water vapor form.
(d) The combustion of gaseous ethane in air forms carbon dioxide and water vapor.
(e) Iron(II) chloride is converted to iron(III) fluoride by treatment with chlorine trifluoride gas. Chlorine gas is also formed.

3.82 Isobutylene is a hydrocarbon used in the manufacture of synthetic rubber. When 0.847 g of isobutylene was subjected to combustion analysis, the gain in mass of the CO_2 absorber was 2.657 g and that of the H_2O absorber was 1.089 g. What is the empirical formula of isobutylene?

3.83 One of the compounds used to increase the octane rating of gasoline is toluene *(right)*. Suppose 20.0 mL of toluene (d = 0.867 g/mL) is consumed when a sample of gasoline burns in air. (a) How many grams of oxygen are needed for complete combustion of the toluene? (b) How many total moles of gaseous products form? (c) How many molecules of water vapor form?

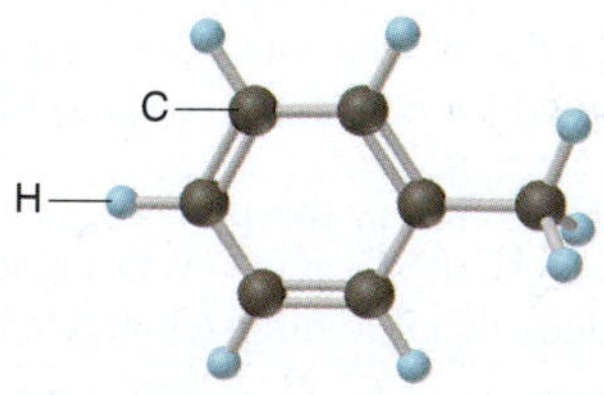

3.84 During studies of the reaction in Sample Problem 3.19,

$$2N_2H_4(l) + N_2O_4(l) \longrightarrow 3N_2(g) + 4H_2O(g)$$

a chemical engineer measured a less-than-expected yield of N_2 and discovered that the following side reaction occurs:

$$N_2H_4(l) + 2N_2O_4(l) \longrightarrow 6NO(g) + 2H_2O(g)$$

In one experiment, 10.0 g of NO formed when 100.0 g of each reactant was used. What is the highest percent yield of N_2 that can be expected?

3.85 The following circles represent a chemical reaction between AB_2 and B_2:

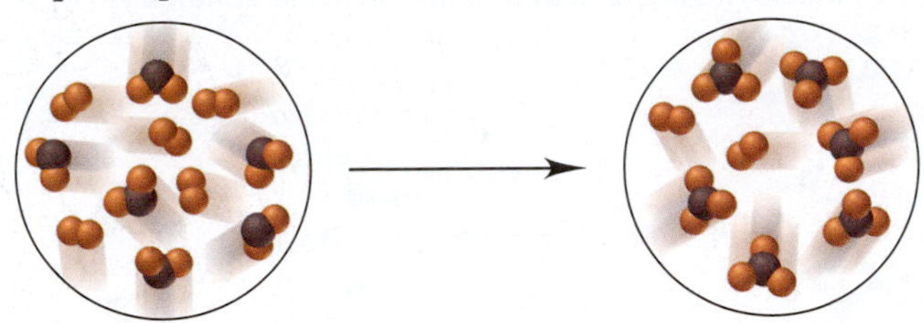

(a) Write a balanced equation for the reaction. (b) What is the limiting reactant? (c) How many moles of product can be made from 3.0 mol of B_2 and 5.0 mol of AB_2? (d) How many moles of excess reactant remain after the reaction in part (c)?

3.86 Seawater is approximately 4.0% by mass dissolved ions, 85% of which are from NaCl. (a) Find the mass % of NaCl in seawater. (b) Find the mass % of Na^+ ions and of Cl^- ions in seawater. (c) Find the molarity of NaCl in seawater at 15°C (*d* of seawater at 15°C = 1.025 g/mL).

3.87 Is each of the following statements true or false? Correct any that are false.
(a) A mole of one substance has the same number of atoms as a mole of any other substance.
(b) The theoretical yield for a reaction is based on the balanced chemical equation.
(c) A limiting-reactant problem is being stated when the available quantity of one of the reactants is given in moles.
(d) The concentration of a solution is an intensive property, but the amount of solute in a solution is an extensive property.

3.88 Box A represents one unit volume of solution A. Which box—B, C, or D—represents one unit volume after adding enough solvent to solution A to (a) triple its volume; (b) double its volume; (c) quadruple its volume?

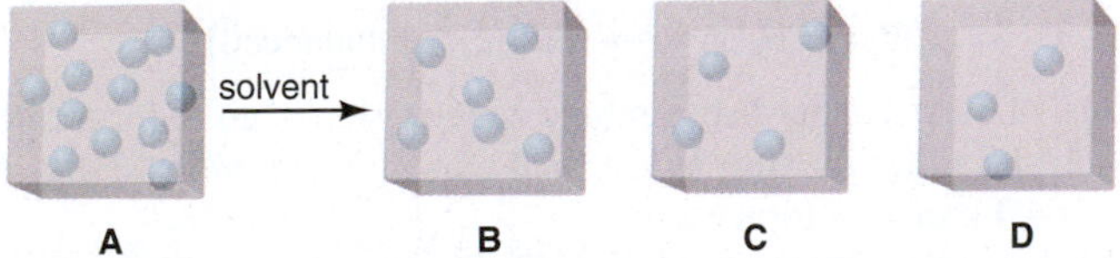

3.89 In each pair, choose the larger of the indicated quantities or state that the samples are equal:
(a) Entities: 0.4 mol of O_3 molecules or 0.4 mol of O atoms
(b) Grams: 0.4 mol of O_3 molecules or 0.4 mol of O atoms
(c) Moles: 4.0 g of N_2O_4 or 3.3 g of SO_2
(d) Grams: 0.6 mol of C_2H_4 or 0.6 mol of F_2
(e) Total ions: 2.3 mol of sodium chlorate or 2.2 mol of magnesium chloride
(f) Molecules: 1.0 g of H_2O or 1.0 g of H_2O_2
(g) Na^+ ions: 0.500 L of 0.500 *M* NaBr or 0.0146 kg of NaCl
(h) Grams: 6.02×10^{23} atoms of ^{235}U or 6.02×10^{23} atoms of ^{238}U

3.90 For the reaction between solid tetraphosphorus trisulfide and oxygen gas to form solid tetraphosphorus decoxide and sulfur dioxide gas, write a balanced equation. Show the equation (see Table 3.3) in terms of (a) molecules, (b) moles, and (c) grams.

3.91 Hydrogen gas is considered a clean fuel because it produces only water vapor when it burns. If the reaction has a 98.8% yield, what mass (g) of hydrogen forms 105 kg of water?

3.92 Assuming that the volumes are additive, what is the concentration of KBr in a solution prepared by mixing 0.200 L of 0.053 *M* KBr with 0.550 L of 0.078 *M* KBr?

3.93 Calculate each of the following quantities:
(a) Amount (mol) of 0.588 g of ammonium bromide
(b) Number of potassium ions in 88.5 g of potassium nitrate
(c) Mass (g) of 5.85 mol of glycerol ($C_3H_8O_3$)
(d) Volume (L) of 2.85 mol of chloroform ($CHCl_3$; d = 1.48 g/mL)
(e) Number of sodium ions in 2.11 mol of sodium carbonate
(f) Number of atoms in 25.0 μg of cadmium
(g) Number of atoms in 0.0015 mol of fluorine gas

3.94 Elements X (green) and Y (purple) react according to the following equation: $X_2 + 3Y_2 \longrightarrow 2XY_3$. Which molecular scene represents the product of the reaction?

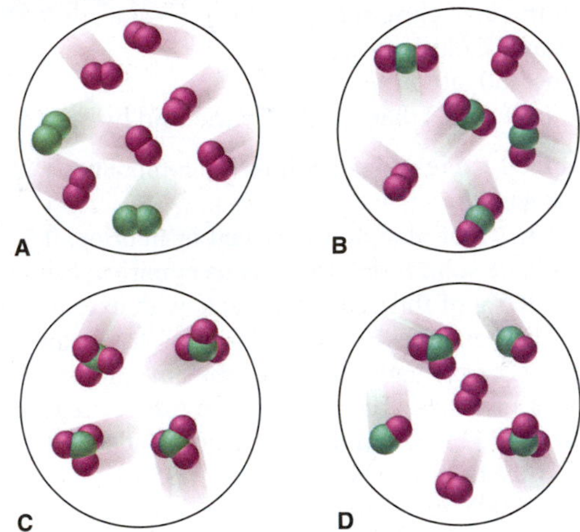

3.95 Hydrocarbon mixtures are used as fuels. (a) How many grams of $CO_2(g)$ are produced by the combustion of 200. g of a mixture that is 25.0% CH_4 and 75.0% C_3H_8 by mass? (b) A 252-g gaseous mixture of CH_4 and C_3H_8 burns in excess O_2, and 748 g of CO_2 gas is collected. What is the mass % of CH_4 in the mixture?

3.96 Nitrogen (N), phosphorus (P), and potassium (K) are the main nutrients in plant fertilizers. By industry convention, the numbers on a label refer to the mass percents of N, P_2O_5, and K_2O, in that order. Calculate the N/P/K ratio of a 30/10/10 fertilizer in terms of moles of each element, and express it as *x*/*y*/1.0.

3.97 What mass percents of ammonium sulfate, ammonium hydrogen phosphate, and potassium chloride would you use to prepare 10/10/10 plant fertilizer (see Problem 3.96)?

3.98 A 0.652-g sample of a pure strontium halide reacts with excess sulfuric acid, and the solid strontium sulfate formed is separated, dried, and found to weigh 0.755 g. What is the formula of the original halide?

3.99 When carbon-containing compounds are burned in a limited amount of air, some $CO(g)$ as well as $CO_2(g)$ is produced. A gaseous product mixture is 35.0 mass % CO and 65.0 mass % CO_2. What is the mass % of C in the mixture?

3.100 Write a balanced equation for the reaction depicted below:

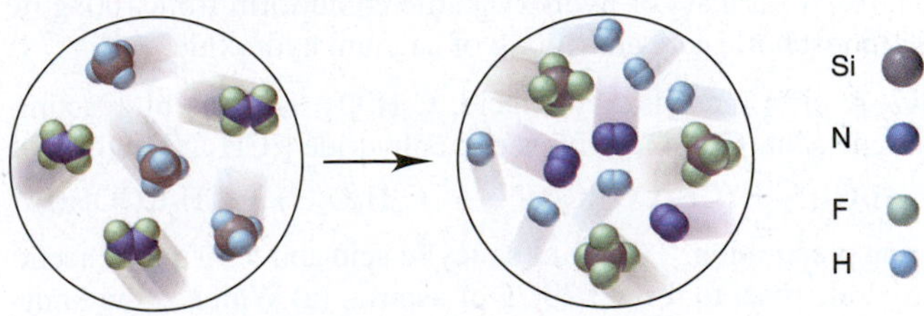

If each reactant molecule represents 1.25×10^{-2} mol and the reaction yield is 87%, how many grams of Si-containing product form?

3.101 Ferrocene, synthesized in 1951, was the first organic iron compound with Fe—C bonds. An understanding of the structure of ferrocene gave rise to new ideas about chemical bonding and led to the preparation of many useful compounds. In the combustion analysis of ferrocene, which contains only Fe, C, and H, a 0.9437-g sample produced 2.233 g of CO_2 and 0.457 g of H_2O. What is the empirical formula of ferrocene?

3.102 Citric acid *(right)* is concentrated in citrus fruits and plays a central metabolic role in nearly every animal and plant cell. (a) What are the molar mass and formula of citric acid? (b) How many moles of citric acid are in 1.50 qt of lemon juice (d = 1.09 g/mL) that is 6.82% citric acid by mass?

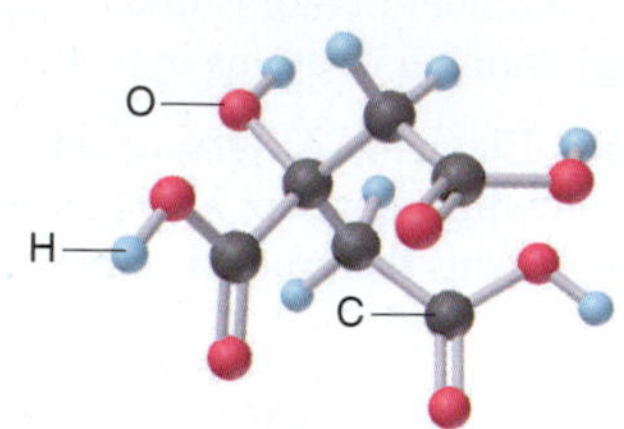

3.103 Fluorine is so reactive that it forms compounds with several of the noble gases.
(a) When 0.327 g of platinum is heated in fluorine, 0.519 g of a dark red, volatile solid forms. What is its empirical formula?
(b) When 0.265 g of this red solid reacts with excess xenon gas, 0.378 g of an orange-yellow solid forms. What is the empirical formula of this compound, the first to contain a noble gas?
(c) Fluorides of xenon can be formed by direct reaction of the elements at high pressure and temperature. Under conditions that produce only the tetra- and hexafluorides, 1.85×10^{-4} mol of xenon reacted with 5.00×10^{-4} mol of fluorine, and 9.00×10^{-6} mol of xenon was found in excess. What are the mass percents of each xenon fluoride in the product mixture?

3.104 Hemoglobin is 6.0% heme ($C_{34}H_{32}FeN_4O_4$) by mass. To remove the heme, hemoglobin is treated with acetic acid and NaCl, which forms hemin ($C_{34}H_{32}N_4O_4FeCl$). A blood sample from a crime scene contains 0.65 g of hemoglobin. (a) How many grams of heme are in the sample? (b) How many moles of heme? (c) How many grams of Fe? (d) How many grams of hemin could be formed for a forensic chemist to measure?

3.105 Manganese is a key component of extremely hard steel. The element occurs naturally in many oxides. A 542.3-g sample of a manganese oxide has an Mn/O ratio of 1.00/1.42 and consists of braunite (Mn_2O_3) and manganosite (MnO). (a) How many grams of braunite and of manganosite are in the ore? (b) What is the Mn^{3+}/Mn^{2+} ratio in the ore?

3.106 Hydroxyapatite, $Ca_5(PO_4)_3(OH)$, is the main mineral component of dental enamel, dentin, and bone, and thus has many medical uses. Coating it on metallic implants (such as titanium alloys and stainless steels) helps the body accept the implant. In the form of powder and beads, it is used to fill bone voids, which encourages natural bone to grow into the void. Hydroxyapatite is prepared by adding aqueous phosphoric acid to a dilute slurry of calcium hydroxide. (a) Write a balanced equation for this preparation. (b) What mass of hydroxyapatite could form from 100. g of 85% phosphoric acid and 100. g of calcium hydroxide?

3.107 Aspirin (acetylsalicylic acid, $C_9H_8O_4$) is made by reacting salicylic acid ($C_7H_6O_3$) with acetic anhydride [$(CH_3CO)_2O$]:

$$C_7H_6O_3(s) + (CH_3CO)_2O(l) \longrightarrow C_9H_8O_4(s) + CH_3COOH(l)$$

In one preparation, 3.077 g of salicylic acid and 5.50 mL of acetic anhydride react to form 3.281 g of aspirin. (a) Which is the limiting reactant (d of acetic anhydride = 1.080 g/mL)? (b) What is the percent yield of this reaction?

3.108 The human body excretes nitrogen in the form of urea, NH_2CONH_2. The key step in its biochemical formation is the reaction of water with arginine to produce urea and ornithine:

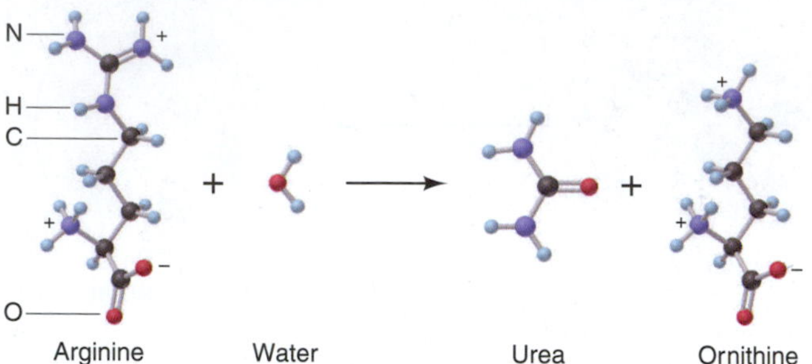

(a) What is the mass % of nitrogen in urea, in arginine, and in ornithine? (b) How many grams of nitrogen can be excreted as urea when 135.2 g of ornithine is produced?

3.109 Nitrogen monoxide reacts with elemental oxygen to form nitrogen dioxide. The scene at right represents an initial mixture of reactants. If the reaction has a 66% yield, which of the scenes below (A, B, or C) best represents the final product mixture?

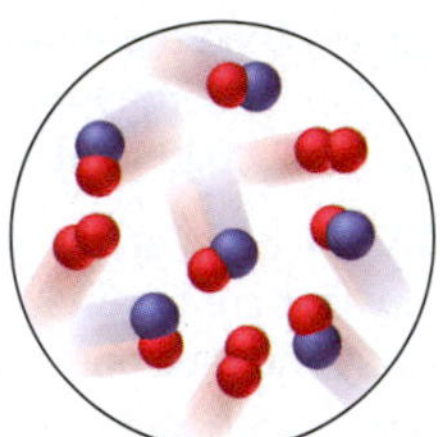

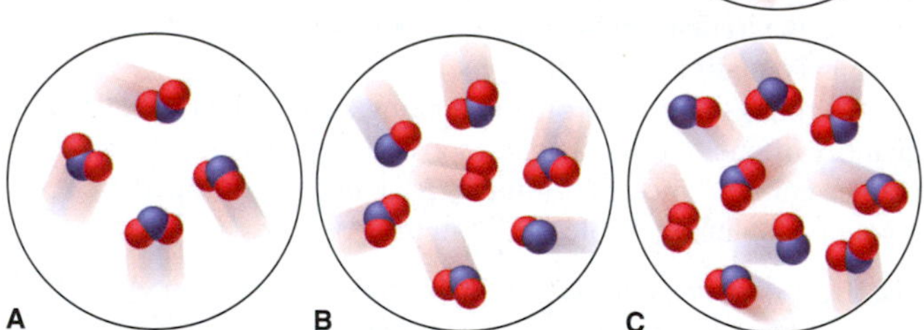

3.110 When powdered zinc is heated with sulfur, a violent reaction occurs, and zinc sulfide forms:

$$Zn(s) + S_8(s) \longrightarrow ZnS(s) \text{ [unbalanced]}$$

Some of the reactants also combine with oxygen in air to form zinc oxide and sulfur dioxide. When 83.2 g of Zn reacts with 52.4 g of S_8, 104.4 g of ZnS forms.
(a) What is the percent yield of ZnS?
(b) If all the remaining reactants combine with oxygen, how many grams of each of the two oxides form?

3.111 High-temperature superconducting oxides hold great promise in the utility, transportation, and computer industries.
(a) One superconductor is $La_{2-x}Sr_xCuO_4$. Calculate the molar masses of this oxide when $x = 0$, $x = 1$, and $x = 0.163$.
(b) Another common superconducting oxide is made by heating a mixture of barium carbonate, copper(II) oxide, and yttrium(III) oxide, followed by further heating in O_2:

$$4BaCO_3(s) + 6CuO(s) + Y_2O_3(s) \longrightarrow$$

$$2YBa_2Cu_3O_{6.5}(s) + 4CO_2(g)2YBa_2Cu_3O_{6.5}(s) + \tfrac{1}{2}O_2(g) \longrightarrow 2YBa_2Cu_3O_7(s)$$

When equal masses of the three reactants are heated, which reactant is limiting?
(c) After the product in part (b) is removed, what is the mass % of each reactant in the remaining solid mixture?

Three Major Classes of Chemical Reactions

Classifying the Countless Despite countless individual reactions, these silver chromate particles form through one of the three major classes of reactions discussed in this chapter.

Key Principles
to focus on while studying this chapter

- *Aqueous* chemical reactions are those that occur in water. Because of its *molecular shape* and uneven *distribution of electrons*, water dissolves many ionic and covalent substances. In water, many ionic compounds and a few simple, H-containing covalent compounds, such as HCl, *dissociate* into ions. ***(Section 4.1)***
- Three types of equations describe an aqueous reaction. A *molecular equation* shows all substances as intact compounds. A *total ionic equation* shows ions for all soluble substances. A *net ionic equation* is more useful because it omits *spectator ions* (those not involved in the reaction) and shows the *actual* chemical change taking place. ***(Section 4.2)***
- *Precipitation reactions* occur when soluble ionic compounds exchange ions *(metathesis)* and form an insoluble product *(precipitate)*, in which the ions attract each other so strongly that their attraction to water molecules cannot pull them apart. ***(Section 4.3)***
- An *acid* produces H^+ ions in solution, and a *base* produces OH^- ions. In an *acid-base (neutralization) reaction*, the H^+ and the OH^- ions form water. Another way to view this process is that an acid *transfers a proton* to a base. An acid-base *titration* is used to measure the amount (mol) of acid (or base). ***(Section 4.4)***
- *Oxidation* is defined as *electron loss*, and *reduction* as *electron gain*. In an *oxidation-reduction (redox) reaction*, electrons move from one reactant to the other: the *reducing agent* is oxidized (loses the electrons), and the *oxidizing agent* is reduced (gains the electrons). Chemists use *oxidation number*, the number of electrons "owned" by each atom in a substance, to follow the change. ***(Section 4.5)***
- Many common redox reactions (which are sometimes classified as *combination, decomposition, displacement, or combustion*) involve elements as reactants or products. In an *activity series*, metals are ranked by their ability to reduce H^+ or displace the ion of a different metal from an aqueous solution. ***(Section 4.6)***

Outline

CONCEPTS & SKILLS TO REVIEW before studying this chapter

- names and formulas of compounds (Section 2.8)
- nature of ionic and covalent bonding (Section 2.7)
- amount-mass-number conversions (Section 3.1)
- molarity and amount-mass-number conversions in solution (Section 3.5)
- balancing chemical equations (Section 3.3)
- calculating quantities of reactants and products (Section 3.4)

Rapid chemical changes occur among gas molecules as sunlight bathes the atmosphere or lightning rips through a stormy sky and strikes the sea. Aqueous reactions occur unceasingly in lakes, rivers, and oceans. And, in every cell of your body, thousands of reactions taking place right now enable you to function. Indeed, the amazing variety in nature is largely a consequence of the amazing variety of chemical reactions. With millions of reactions occurring in and around you, it would be impossible to describe them all. Fortunately, it isn't necessary because, when we survey even a small percentage of reactions, especially those in aqueous solution, a few major classes emerge.

4.1 • THE ROLE OF WATER AS A SOLVENT

For any reaction in solution, the solvent plays a role that depends on its chemical nature. Some solvents passively disperse the substances into individual molecules. But water is much more active, interacting strongly with the substances and even reacting with them in some cases. Nearly all environmental and biological reactions occur in water, so let's focus on how the water molecule interacts with both ionic and covalent solutes.

The Polar Nature of Water

On the atomic scale, water's great solvent power arises from the *uneven distribution of electron charge* and a *bent molecular shape,* which create a *polar molecule:*

1. *Uneven charge distribution.* Recall from Section 2.7 that the electrons in a covalent bond are shared between the atoms. In a bond between identical atoms—as in H_2, Cl_2, O_2—the sharing is equal and electron charge is distributed evenly between the two nuclei (symmetrical shading in the space-filling model of Figure 4.1A). In covalent bonds between different atoms, the sharing is uneven because one atom attracts the electron pair more strongly than the other atom does.

For example, in each O—H bond of water, the shared electrons are closer to the O atom because an O atom attracts electrons more strongly than an H atom does. (You'll see why in Chapter 9.) This uneven charge distribution creates a *polar bond,* one with partially charged "poles." In Figure 4.1B, the asymmetrical shading shows this distribution, and the δ symbol indicates a partial charge. The O end is partially negative, represented by red shading and δ−, and the H end is partially positive, represented by blue shading and δ+.

2. *Bent molecular shape.* The sequence of the H—O—H atoms in water is not linear: the water molecule is bent with a bond angle of 104.5°. In the ball-and-stick model of Figure 4.1C, the *polar arrow* points to the negative pole, and the tail, shaped like a plus sign, marks the positive pole.

3. *Molecular polarity.* The combination of polar bonds and bent shape makes water a **polar molecule:** the region near the O atom is partially negative, and the region between the H atoms is partially positive (Figure 4.1D).

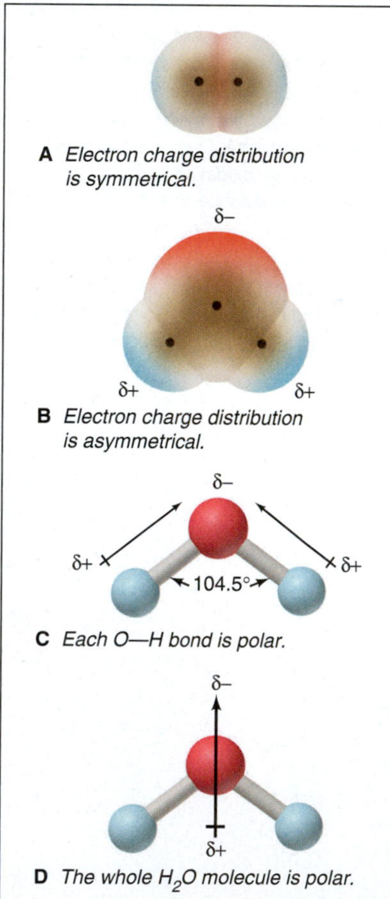

A Electron charge distribution is symmetrical.
δ−
δ+
δ+
B Electron charge distribution is asymmetrical.
δ−
δ+
δ+
104.5°
C Each O—H bond is polar.
δ−
δ+
D The whole H2O molecule is polar.

Figure 4.1 Electron distribution in molecules of H_2 and H_2O.

Ionic Compounds in Water

In this subsection, we consider two closely related aspects of aqueous solutions of ionic compounds—how they occur and how they behave. We also use a compound's formula to calculate the amount (mol) of each ion in solution.

How Ionic Compounds Dissolve: Replacement of Charge Attractions In an ionic solid, oppositely charged ions are held together by electrostatic attractions (see Figure 1.3C and Section 2.7). Water separates the ions by *replacing these attractions with others between several water molecules and each ion.* Picture a granule of a soluble ionic compound in water: the negative ends of some water molecules are

Figure 4.2 An ionic compound dissolving in water. The inset shows the polar arrow and partial charges (not shown in the rest of the scene) of each water molecule.

attracted to the cations, and the positive ends of other water molecules are attracted to the anions (Figure 4.2). Dissolution occurs because the *attractions between each type of ion and several water molecules outweigh the attractions between the ions.* Also, as the ions separate (dissociate) and become **solvated**—surrounded closely by solvent molecules—they *disperse randomly* in the solution.

For an ionic compound that doesn't dissolve in water, the attraction between ions is greater than the attraction between the ions and water. Actually, these so-called insoluble substances *do* dissolve to a very small extent, usually several orders of magnitude less than so-called soluble substances. For example, NaCl (a "soluble" compound) is over 4×10^4 times more soluble than AgCl (an "insoluble" compound):

$$\text{Solubility of NaCl in } H_2O \text{ at } 20°C = 365 \text{ g/L}$$
$$\text{Solubility of AgCl in } H_2O \text{ at } 20°C = 0.009 \text{ g/L}$$

In Chapter 13, we'll highlight that dissolving involves more than a contest between the relative energies of attraction of ions for each other or for water. And we'll emphasize that it is favored by the greater freedom of motion the ions have when they leave the solid and disperse randomly through the solution.

How Ionic Solutions Behave: Electrolytes and Electrical Conductivity When an ionic compound dissolves, the solution's *electrical conductivity,* the flow of electric current, increases dramatically. When electrodes are immersed in distilled water (Figure 4.3A, *next page*) or pushed into an ionic solid (Figure 4.3B), no current flows, as shown by the unlit bulb. But in an aqueous solution of the compound, a large current flows, as shown by the lit bulb (Figure 4.3C). Current flow implies the *movement of charged particles:* when the ionic compound dissolves, the separate solvated ions move toward the electrode of opposite charge. A substance that conducts a current when dissolved in water is an **electrolyte.** Soluble ionic compounds are *strong* electrolytes because they dissociate completely and conduct a large current.

Calculating the Number of Moles of Ions in Solution *From the formula of the soluble ionic compound, we know the number of moles of each ion in solution.* For example, the equation for dissolving KBr in water to form solvated ions is

$$KBr(s) \xrightarrow{H_2O} K^+(aq) + Br^-(aq)$$

("H_2O" above the arrow means that water is the solvent, not a reactant.) Note that 1 mol of KBr dissociates into 2 mol of ions—1 mol of K^+ and 1 mol of Br^-. Sample Problems 4.1 and 4.2 apply these ideas, first with molecular scenes and then in calculations.

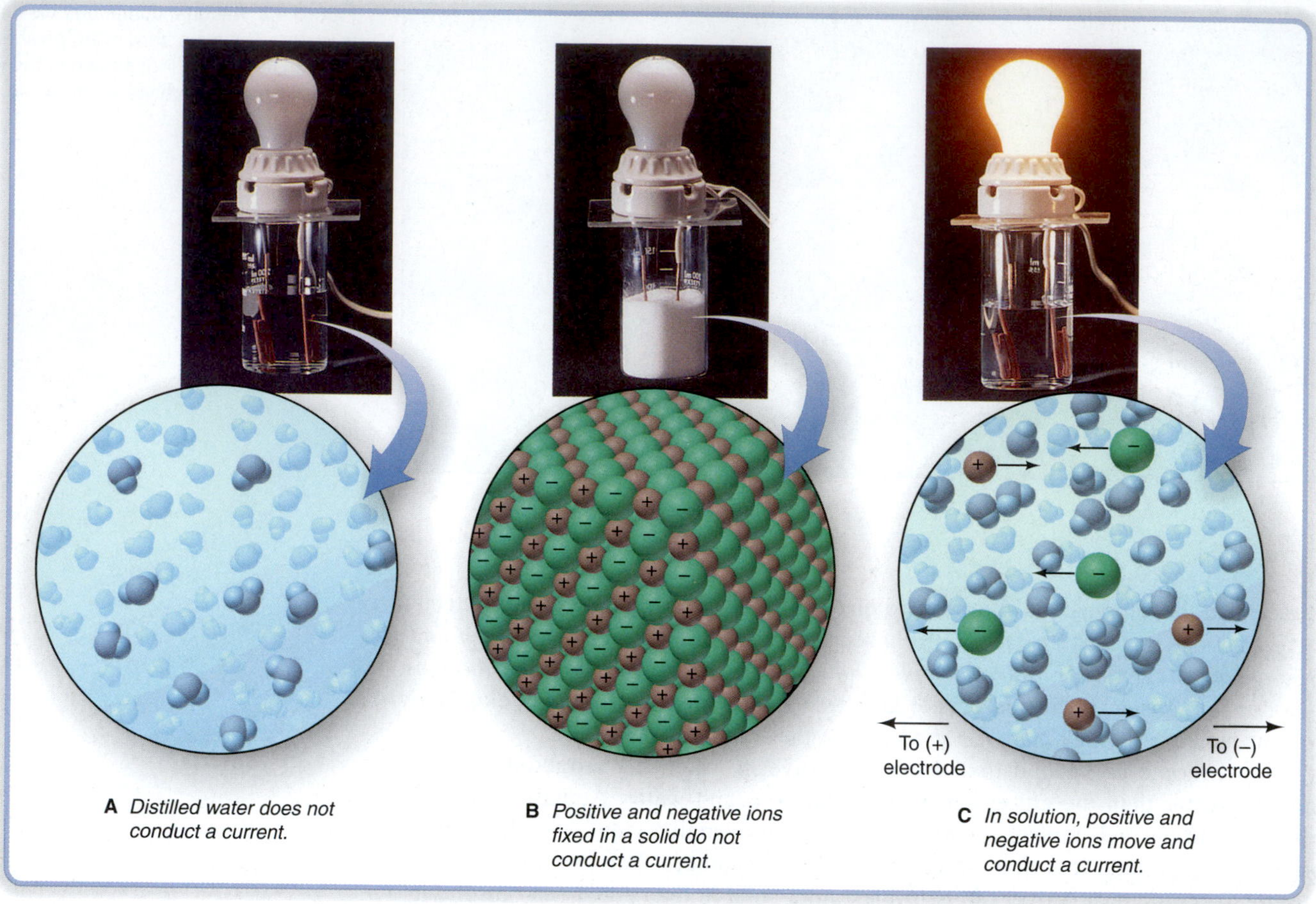

To (+) electrode

To (–) electrode

A *Distilled water does not conduct a current.*

B *Positive and negative ions fixed in a solid do not conduct a current.*

C *In solution, positive and negative ions move and conduct a current.*

Figure 4.3 The electrical conductivity of ionic solutions.

Sample Problem 4.1 Using Molecular Scenes to Depict an Ionic Compound in Aqueous Solution

Problem **(a)** Which beaker best depicts the strong electrolyte potassium sulfate in aqueous solution (water molecules are not shown)?
(b) If each particle represents 0.1 mol, what is the total number of particles in solution?

Plan **(a)** We determine the formula and write an equation for 1 mol of compound dissociating into ions. Potassium sulfate is a strong electrolyte, so it dissociates completely, but, in general, *polyatomic ions remain intact in solution.* **(b)** We count the number of separate particles, and then multiply by 0.1 mol and by Avogadro's number.

Solution **(a)** The formula is K_2SO_4, so the equation is

$$K_2SO_4(s) \xrightarrow{H_2O} 2K^+(aq) + SO_4^{2-}(aq)$$

There are two separate 1+ particles for every 2− particle, so beaker C is best.
(b) There are 9 particles, so the total amount (mol) of particles is 0.9 mol, and we have

$$\text{No. of particles} = 0.9\ \cancel{\text{mol}} \times \frac{6.022\times10^{23}\ \text{particles}}{1\ \cancel{\text{mol}}} = 5.420\times10^{23}\ \text{particles}$$

Check Rounding to check the math in (b) gives $0.9 \times 6 = 5.4$, so the answer seems correct. The number of particles is an exact number since we actually counted them; thus, the answer can have as many significant figures as in Avogadro's number.

FOLLOW-UP PROBLEM 4.1 **(a)** Which strong electrolyte is dissolved in water (water molecules not shown) in the beaker at right: LiBr, Cs_2CO_3, or $BaCl_2$? **(b)** If each particle represents 0.05 mol, what mass (g) of compound was dissolved?

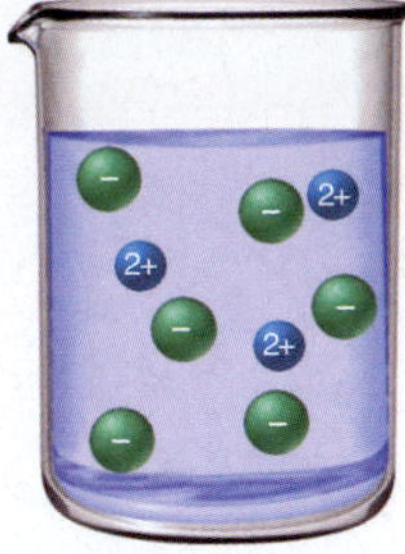

Sample Problem 4.2 Determining Amounts (mol) of Ions in Solution

Problem What amount (mol) of each ion is in each solution?
(a) 5.0 mol of ammonium sulfate dissolved in water
(b) 78.5 g of cesium bromide dissolved in water
(c) 7.42×10^{22} formula units of copper(II) nitrate dissolved in water
(d) 35 mL of 0.84 *M* zinc chloride

Plan We write an equation that shows 1 mol of compound dissociating into ions. **(a)** We multiply the number of moles of ions by 5.0. **(b)** We first convert grams to moles. **(c)** We first convert formula units to moles. **(d)** We first convert molarity and volume to moles.

Solution **(a)** $(NH_4)_2SO_4(s) \xrightarrow{H_2O} 2NH_4^+(aq) + SO_4^{2-}(aq)$

Calculating amount (mol) of NH_4^+ ions:

$$\text{Amount (mol) of } NH_4^+ = 5.0 \text{ mol } (NH_4)_2SO_4 \times \frac{2 \text{ mol } NH_4^+}{1 \text{ mol } (NH_4)_2SO_4} = 10. \text{ mol } NH_4^+$$

The formula shows 1 mol of SO_4^{2-} per mole of $(NH_4)_2SO_4$, so 5.0 mol of SO_4^{2-} is also present.

(b) $CsBr(s) \xrightarrow{H_2O} Cs^+(aq) + Br^-(aq)$

Converting from mass (g) to amount (mol):

$$\text{Amount (mol) of CsBr} = 78.5 \text{ g CsBr} \times \frac{1 \text{ mol CsBr}}{212.8 \text{ g CsBr}} = 0.369 \text{ mol CsBr}$$

Thus, 0.369 mol of Cs^+ and 0.369 mol of Br^- are present.

(c) $Cu(NO_3)_2(s) \xrightarrow{H_2O} Cu^{2+}(aq) + 2NO_3^-(aq)$

Converting from formula units to amount (mol):

$$\text{Amount (mol) of } Cu(NO_3)_2 = 7.42\times10^{22} \text{ formula units } Cu(NO_3)_2 \times \frac{1 \text{ mol } Cu(NO_3)_2}{6.022\times10^{23} \text{ formula units } Cu(NO_3)_2} = 0.123 \text{ mol } Cu(NO_3)_2$$

$$\text{Amount (mol) of } NO_3^- = 0.123 \text{ mol } Cu(NO_3)_2 \times \frac{2 \text{ mol } NO_3^-}{1 \text{ mol } Cu(NO_3)_2} = 0.246 \text{ mol } NO_3^-$$

0.123 mol of Cu^{2+} is also present.

(d) $ZnCl_2(aq) \longrightarrow Zn^{2+}(aq) + 2Cl^-(aq)$

Converting from volume (mL) and molarity (mol/L) to amount (mol):

$$\text{Amount (mol) of } ZnCl_2 = 35 \text{ mL} \times \frac{1 \text{ L}}{10^3 \text{ mL}} \times \frac{0.84 \text{ mol } ZnCl_2}{1 \text{ L}} = 2.9\times10^{-2} \text{ mol } ZnCl_2$$

$$\text{Amount (mol) of } Cl^- = 2.9\times10^{-2} \text{ mol } ZnCl_2 \times \frac{2 \text{ mol } Cl^-}{1 \text{ mol } ZnCl_2} = 5.8\times10^{-2} \text{ mol } Cl^-$$

2.9×10^{-2} mol of Zn^{2+} is also present.

Check Round off to check the math and see if the relative numbers of moles of ions are consistent with the formula. For instance, in (a), 10 mol NH_4^+/5.0 mol SO_4^{2-} = $2NH_4^+/1SO_4^{2-}$, or $(NH_4)_2SO_4$. In (d), 0.029 mol Zn^{2+}/0.058 mol Cl^- = $1Zn^{2+}/2Cl^-$, or $ZnCl_2$.

FOLLOW-UP PROBLEM 4.2 What amount (mol) of each ion is in each solution?
(a) 2 mol of potassium perchlorate dissolved in water
(b) 354 g of magnesium acetate dissolved in water
(c) 1.88×10^{24} formula units of ammonium chromate dissolved in water
(d) 1.32 L of 0.55 *M* sodium bisulfate

Covalent Compounds in Water

Water dissolves many covalent (molecular) compounds also. Table sugar (sucrose, $C_{12}H_{22}O_{11}$), beverage (grain) alcohol (ethanol, CH_3CH_2OH), and automobile antifreeze (ethylene glycol, $HOCH_2CH_2OH$) are some familiar examples. All contain their own polar bonds, which interact with the bonds of water. However, most soluble covalent substances *do not* separate into ions, but remain intact molecules. For example,

$$HOCH_2CH_2OH(l) \xrightarrow{H_2O} HOCH_2CH_2OH(aq)$$

As a result, their aqueous solutions do not conduct an electric current, and these substances are **nonelectrolytes.** (As you'll see shortly, however, a small group of *H-containing molecules* that act as acids in aqueous solution *do* dissociate into ions.) Many other covalent substances, such as benzene (C_6H_6) and octane (C_8H_{18}), do not contain polar bonds, and these substances do not dissolve appreciably in water.

Summary of Section 4.1

- Because of polar bonds and a bent shape, the water molecule is polar, and water dissolves many ionic and covalent compounds.
- When an ionic compound dissolves, the attraction between each ion and water replaces the attraction between ions. Soluble ionic compounds are electrolytes because the ions are free to move and, thus, the solution conducts electricity.
- The formula of a soluble ionic compound shows the number of moles of each ion in solution per mole of compound dissolved.
- Water dissolves many covalent substances with polar bonds. These compounds are nonelectrolytes because the molecules remain intact and, thus, the solution does not conduct electricity. Covalent compounds without polar bonds are mostly insoluble in water.

4.2 • WRITING EQUATIONS FOR AQUEOUS IONIC REACTIONS

Chemists use three types of equations to represent aqueous ionic reactions. Let's examine a reaction to see what each type shows. When solutions of silver nitrate and sodium chromate are mixed, brick-red, solid silver chromate (Ag_2CrO_4) forms. Figure 4.4 depicts the reaction at the macroscopic level *(photos),* the atomic level *(blow-up circles),* and the symbolic level with the three types of equations (reacting ions are in red type):

- The **molecular equation** *(top)* reveals the least about the species that are actually in solution because *it shows all the reactants and products as if they were intact, undissociated compounds.* Only the designation for solid, (*s*), tells that a change has occurred:

$$2AgNO_3(aq) + Na_2CrO_4(aq) \longrightarrow Ag_2CrO_4(s) + 2NaNO_3(aq)$$

- The **total ionic equation** *(middle)* is much more accurate because *it shows all the soluble ionic substances dissociated into ions.* The $Ag_2CrO_4(s)$ stands out as the only undissociated substance:

$$2Ag^+(aq) + 2NO_3^-(aq) + 2Na^+(aq) + CrO_4^{2-}(aq) \longrightarrow Ag_2CrO_4(s) + 2Na^+(aq) + 2NO_3^-(aq)$$

The charges balance: four positive and four negative for a net zero charge on the left, and two positive and two negative for a net zero charge on the right.

Notice that $Na^+(aq)$ and $NO_3^-(aq)$ appear unchanged on both sides of the equation. These are called **spectator ions** (shown with pale colors in the atomic-level scenes). They are not involved in the actual chemical change but are present only as part of the reactants; that is, we can't add an Ag^+ ion without also adding an anion, in this case, the NO_3^- ion.

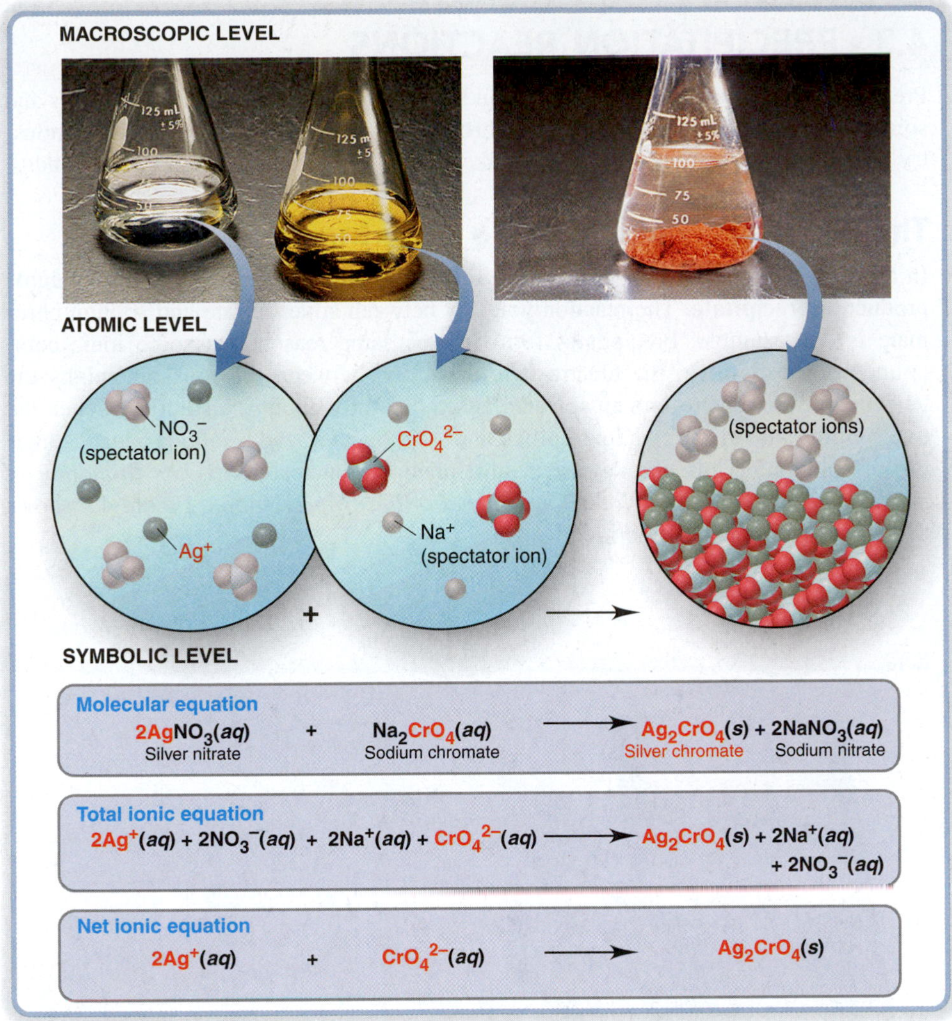

Figure 4.4 An aqueous ionic reaction and the three types of equations.

- The **net ionic equation** *(bottom)* is very useful because *it eliminates the spectator ions and shows only the actual chemical change:*

$$2Ag^+(aq) + CrO_4^{2-}(aq) \longrightarrow Ag_2CrO_4(s)$$

The formation of solid silver chromate from silver ions and chromate ions *is* the only change. To make that point clearly, suppose we mixed solutions of potassium chromate, $K_2CrO_4(aq)$, and silver acetate, $AgC_2H_3O_2(aq)$, instead of sodium chromate and silver nitrate. In that case, only the spectator ions would differ—$K^+(aq)$ and $C_2H_3O_2^-(aq)$ instead of $Na^+(aq)$ and $NO_3^-(aq)$.

Next, we'll apply these types of equations to three important classes of chemical reactions—precipitation, acid-base, and oxidation-reduction.

Summary of Section 4.2

- A molecular equation shows all substances intact and undissociated into ions.
- A total ionic equation shows all soluble ionic compounds as separate, solvated ions. Spectator ions appear unchanged on both sides of the equation.
- A net ionic equation eliminates the spectator ions and, thus, shows only the actual chemical change.

4.3 • PRECIPITATION REACTIONS

Precipitation reactions occur commonly in both nature and commerce. Coral reefs and some gems and minerals form, in part, through this process. And the chemical industry employs precipitation methods to make several important inorganic compounds.

The Key Event: Formation of a Solid from Dissolved Ions

In a **precipitation reaction,** two soluble ionic compounds react to form an insoluble product, a **precipitate.** The reaction you saw between silver nitrate and sodium chromate is one example. Precipitates form for the same reason that some ionic compounds don't dissolve: the electrostatic attraction between the ions outweighs the tendency of the ions to remain solvated and move throughout the solution. When the two solutions are mixed, the ions collide and stay together, and a solid product "comes out of solution." Thus, the key event in a precipitation reaction is *the formation of an insoluble product through the net removal of ions from solution.* Figure 4.5 shows the process for calcium fluoride.

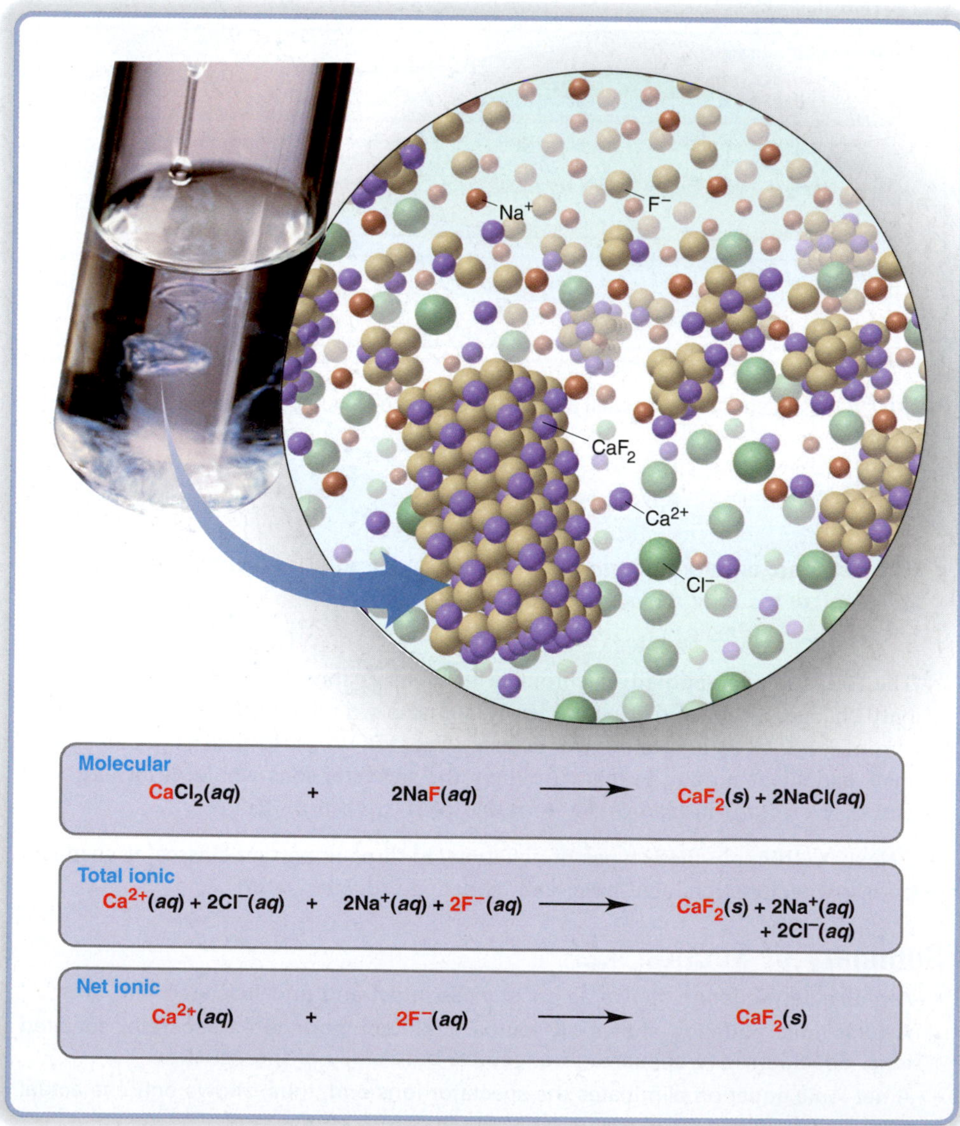

Figure 4.5 The precipitation of calcium fluoride. When aqueous solutions of NaF (from pipet) and $CaCl_2$ (in test tube) react, solid CaF_2 forms (water molecules are omitted for clarity).

Table 4.1 Solubility Rules for Ionic Compounds in Water

Soluble Ionic Compounds	Insoluble Ionic Compounds
1. All common compounds of Group 1A(1) ions (Li^+, Na^+, K^+, etc.) and ammonium ion (NH_4^+) are soluble. 2. All common nitrates (NO_3^-), acetates (CH_3COO^- or $C_2H_3O_2^-$), and most perchlorates (ClO_4^-) are soluble. 3. All common chlorides (Cl^-), bromides (Br^-), and iodides (I^-) are soluble, *except* those of Ag^+, Pb^{2+}, Cu^+, and Hg_2^{2+}. All common fluorides (F^-) are soluble, *except* those of Pb^{2+} and Group 2A(2). 4. All common sulfates (SO_4^{2-}) are soluble, except those of Ca^{2+}, Sr^{2+}, Ba^{2+}, Ag^+, and Pb^{2+}.	1. All common metal hydroxides are insoluble, *except* those of Group 1A(1) and the larger members of Group 2A(2) (beginning with Ca^{2+}). 2. All common carbonates (CO_3^{2-}) and phosphates (PO_4^{3-}) are insoluble, *except* those of Group 1A(1) and NH_4^+. 3. All common sulfides are insoluble except those of Group 1A(1), Group 2A(2), and NH_4^+.

Predicting Whether a Precipitate Will Form

To predict whether a precipitate will form when we mix two aqueous ionic solutions, we refer to the short list of solubility rules in Table 4.1. Let's see how to apply these rules. When sodium iodide and potassium nitrate are each dissolved in water, their solutions consist of solvated, dispersed ions:

$$NaI(s) \xrightarrow{H_2O} Na^+(aq) + I^-(aq)$$

$$KNO_3(s) \xrightarrow{H_2O} K^+(aq) + NO_3^-(aq)$$

Three steps help us predict if a precipitate forms:

1. *Note the ions in the reactants.* The reactant ions are

$$Na^+(aq) + I^-(aq) + K^+(aq) + NO_3^-(aq) \longrightarrow ?$$

2. *Consider all possible cation-anion combinations.* In addition to NaI and KNO_3, which we know are soluble, the other cation-anion combinations are $NaNO_3$ and KI.
3. *Decide whether any combination is insoluble.* According to Table 4.1, no reaction occurs because all the combinations—NaI, KNO_3, $NaNO_3$, and KI—are soluble: Rules 1 and 2 say all compounds of Group 1A(1) ions and all nitrates are soluble.

Now, what happens if we substitute a solution of lead(II) nitrate, $Pb(NO_3)_2$, for the KNO_3 solution? The reactant ions are Na^+, I^-, Pb^{2+}, and NO_3^-. In addition to the two soluble reactants, NaI and $Pb(NO_3)_2$, the other two possible cation-anion combinations are $NaNO_3$ and PbI_2. According to Table 4.1, $NaNO_3$ is soluble (Rule 1), but PbI_2 is *not* (Rule 3). The total ionic equation shows the reaction that occurs as Pb^{2+} and I^- ions collide and form a precipitate:

$$2Na^+(aq) + 2I^-(aq) + Pb^{2+}(aq) + 2NO_3^-(aq) \longrightarrow 2Na^+(aq) + 2NO_3^-(aq) + PbI_2(s)$$

And the net ionic equation confirms it:

$$Pb^{2+}(aq) + 2I^-(aq) \longrightarrow PbI_2(s)$$

Metathesis Reactions The molecular equation for the reaction between $Pb(NO_3)_2$ and NaI shows *the ions exchanging partners* (Figure 4.6):

$$2NaI(aq) + Pb(NO_3)_2(aq) \longrightarrow PbI_2(s) + 2NaNO_3(aq)$$

Such reactions are called *double-displacement reactions,* or **metathesis** (pronounced *meh-TA-thuh-sis*) **reactions.** The processes that form Ag_2CrO_4 (Figure 4.4) and CaF_2 (Figure 4.5) are metathesis reactions, too. Sample Problems 4.2 and 4.3 provide practice in predicting if a precipitate forms.

$2NaI(aq) + Pb(NO_3)_2(aq) \longrightarrow PbI_2(s) + 2NaNO_3(aq)$

Figure 4.6 The precipitation of PbI_2, a metathesis reaction.

Sample Problem 4.3 Predicting Whether a Precipitation Reaction Occurs; Writing Ionic Equations

Problem Does a reaction occur when each of these pairs of solutions is mixed? If so, write balanced molecular, total ionic, and net ionic equations, and identify the spectator ions.
(a) Potassium fluoride(aq) + strontium nitrate(aq) $\longrightarrow$
(b) Ammonium perchlorate(aq) + sodium bromide(aq) $\longrightarrow$

Plan We note the reactant ions, write the cation-anion combinations, and refer to Table 4.1 to see if any are insoluble. For the molecular equation, we predict the products and write them all as intact compounds. For the total ionic equation, we write the soluble compounds as separate ions. For the net ionic equation, we eliminate the spectator ions.

Solution **(a)** In addition to the reactants, the two other ion combinations are strontium fluoride and potassium nitrate. Table 4.1 shows that strontium fluoride is insoluble, so a reaction *does* occur. Writing the molecular equation:

$$2KF(aq) + Sr(NO_3)_2(aq) \longrightarrow SrF_2(s) + 2KNO_3(aq)$$

Writing the total ionic equation:

$$2K^+(aq) + 2F^-(aq) + Sr^{2+}(aq) + 2NO_3^-(aq) \longrightarrow SrF_2(s) + 2K^+(aq) + 2NO_3^-(aq)$$

Writing the net ionic equation:

$$Sr^{2+}(aq) + 2F^-(aq) \longrightarrow SrF_2(s)$$

The spectator ions are K^+ and NO_3^-.

(b) The other ion combinations are ammonium bromide and sodium perchlorate. Table 4.1 shows that all ammonium, sodium, and most perchlorate compounds are soluble, and all bromides are soluble except those of Ag^+, Pb^{2+}, Cu^+, and Hg_2^{2+}. Therefore, ***no*** **reaction occurs.** The compounds remain as solvated ions.

FOLLOW-UP PROBLEM 4.3 Predict whether a reaction occurs, and if so, write balanced total and net ionic equations:

(a) Iron(III) chloride(*aq*) + cesium phosphate(*aq*) ⟶

(b) Sodium hydroxide(*aq*) + cadmium nitrate(*aq*) ⟶

(c) Magnesium bromide(*aq*) + potassium acetate(*aq*) ⟶

(d) Silver nitrate(*aq*) + barium chloride(*aq*) ⟶

Sample Problem 4.4 Using Molecular Depictions in Precipitation Reactions

Problem The molecular views below depict reactant solutions for a precipitation reaction (with ions shown as colored spheres and water omitted for clarity):

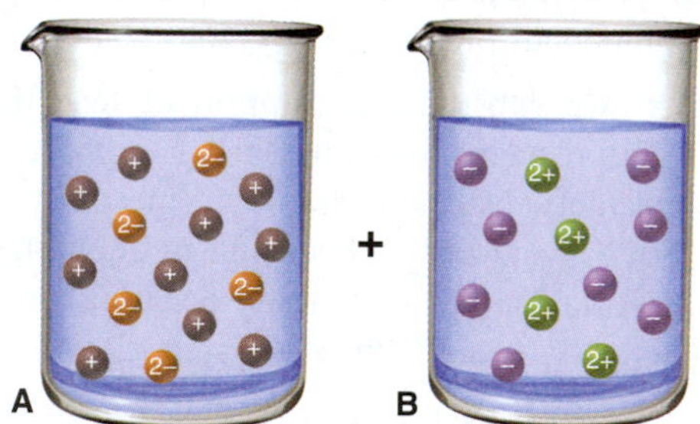

(a) Which compound is dissolved in beaker A: KCl, Na_2SO_4, $MgBr_2$, or Ag_2SO_4?

(b) Which compound is dissolved in beaker B: NH_4NO_3, $MgSO_4$, $Ba(NO_3)_2$, or CaF_2?

(c) Name the precipitate and the spectator ions when solutions A and B are mixed, and write balanced molecular, total ionic, and net ionic equations for any reaction.

(d) If each particle represents 0.010 mol of ions, what is the maximum mass (g) of precipitate that can form (assuming complete reaction)?

Plan **(a)** and **(b)** From the depictions, we note the charge and number of each kind of ion and use Table 4.1 to determine the ion combinations that are soluble. **(c)** Once we know the combinations, Table 4.1 tells which two ions form the solid, so the other two are spectator ions. **(d)** This part is a limiting-reactant problem because the amounts of two species are involved. We count the number of each kind of ion that formed the solid. We multiply the number of each reactant ion by 0.010 mol and calculate the amount (mol) of product formed from each. Whichever ion forms less is limiting, so we use the molar mass of the precipitate to find mass (g).

Solution **(a)** In solution A, there are two 1+ particles for each 2− particle. Therefore, the dissolved compound cannot be KCl or $MgBr_2$. Of the remaining two choices, Ag_2SO_4 is insoluble, so the dissolved compound must be Na_2SO_4.

(b) In solution B, there are two 1− particles for each 2+ particle. Therefore, the dissolved compound cannot be NH_4NO_3 or $MgSO_4$. Of the remaining two choices, CaF_2 is insoluble, so the dissolved compound must be $Ba(NO_3)_2$.

(c) Of the two possible ion combinations, $BaSO_4$ and $NaNO_3$, $BaSO_4$ is insoluble, so Na^+ and NO_3^- are spectator ions.

Molecular: $Ba(NO_3)_2(aq) + Na_2SO_4(aq) \longrightarrow BaSO_4(s) + 2NaNO_3(aq)$

Total ionic: $Ba^{2+}(aq) + 2NO_3^-(aq) + 2Na^+(aq) + SO_4^{2-}(aq) \longrightarrow BaSO_4(s) + 2NO_3^-(aq) + 2Na^+(aq)$

Net ionic: $Ba^{2+}(aq) + SO_4^{2-}(aq) \longrightarrow BaSO_4(s)$

(d) Finding the ion that is the limiting reactant:

For Ba^{2+}:

$$\text{Amount (mol) of } BaSO_4 = 4\ \cancel{Ba^{2+}\text{ particles}} \times \frac{0.010\ \cancel{\text{mol }Ba^{2+}\text{ ion}}}{1\ \cancel{\text{particle}}} \times \frac{1\text{ mol }BaSO_4}{1\ \cancel{\text{mol }Ba^{2+}\text{ ions}}}$$

$$= 0.040\text{ mol }BaSO_4$$

For SO_4^{2-}:

$$\text{Amount (mol) of } BaSO_4 = 5\ \cancel{SO_4^{2-}\text{ particles}} \times \frac{0.010\ \cancel{\text{mol }SO_4^{2-}\text{ ions}}}{1\ \cancel{\text{particle}}} \times \frac{1\text{ mol }BaSO_4}{1\ \cancel{\text{mol }SO_4^{2-}\text{ ions}}}$$

$$= 0.050\text{ mol }BaSO_4$$

Therefore, Ba^{2+} ion is the limiting reactant.

Calculating the mass (g) of product ($\mathcal{M}$ of $BaSO_4$ = 233.4 g/mol):

$$\text{Mass (g) of } BaSO_4 = 0.040\ \cancel{\text{mol }BaSO_4} \times \frac{233.4\text{ g }BaSO_4}{1\ \cancel{\text{mol }BaSO_4}} = 9.3\text{ g }BaSO_4$$

Check Counting the number of Ba^{2+} ions allows a more direct calculation for a check: four Ba^{2+} particles means the maximum mass of $BaSO_4$ that can form is

$$\text{Mass (g) of } BaSO_4 = 4\ \cancel{Ba^{2+}\text{ particles}} \times \frac{0.010\ \cancel{\text{mol }Ba^{2+}\text{ ions}}}{1\ \cancel{\text{particle}}} \times \frac{1\ \cancel{\text{mol }BaSO_4}}{1\ \cancel{\text{mol }Ba^{2+}\text{ ions}}} \times \frac{233.4\text{ g }BaSO_4}{1\ \cancel{\text{mol}}} = 9.3\text{ g }BaSO_4$$

FOLLOW-UP PROBLEM 4.4 Molecular views of the reactant solutions for a precipitation reaction are shown below (with ions represented as spheres and water molecules omitted):

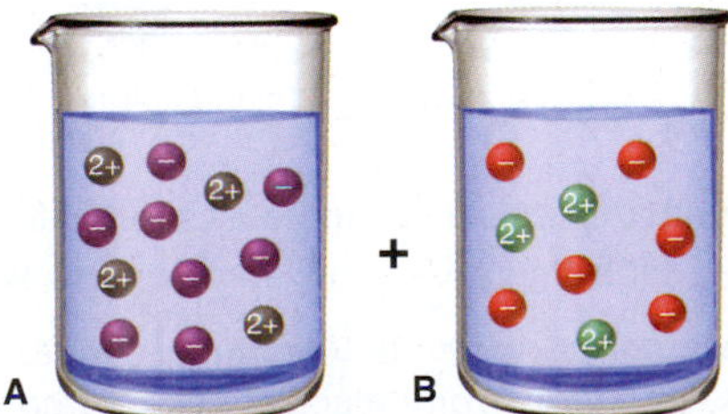

(a) Which compound is dissolved in beaker A: $Zn(NO_3)_2$, KCl, Na_2SO_4, or $PbCl_2$?
(b) Which compound is dissolved in beaker B: $(NH_4)_2SO_4$, $Cd(OH)_2$, $Ba(OH)_2$, or KNO_3?
(c) Name the precipitate and the spectator ions when solutions A and B are mixed, and write balanced molecular, total ionic, and net ionic equations for the reaction.
(d) If each particle represents 0.050 mol of ions, what is the maximum mass (g) of precipitate that can form (assuming complete reaction)?

Summary of Section 4.3

- In a precipitation reaction, an insoluble ionic compound forms when solutions of two soluble ones are mixed. The electrostatic attraction between certain pairs of solvated ions is strong enough to overcome the attraction of each ion for water.
- Based on a set of solubility rules, we can predict whether a precipitate will form by noting which of all possible cation-anion combinations is insoluble.

4.4 • ACID-BASE REACTIONS

Aqueous acid-base reactions occur in processes as diverse as the metabolic action of proteins and carbohydrates, the industrial production of fertilizer, and the revitalization of lakes damaged by acid rain.

These reactions involve water as reactant or product, in addition to its common role as solvent. Of course, an **acid-base reaction** (also called a **neutralization reaction**) occurs when an acid reacts with a base, but the definitions of these terms and the scope of this reaction class have changed over the years. For our purposes at this point, we'll use definitions that apply to substances found commonly in the lab:

- *An **acid** is a substance that produces H^+ ions when dissolved in water.*

$$HX \xrightarrow{H_2O} H^+(aq) + X^-(aq)$$

- *A **base** is a substance that produces OH^- ions when dissolved in water.*

$$MOH \xrightarrow{H_2O} M^+(aq) + OH^-(aq)$$

(Other definitions are presented in Chapter 18.)

Table 4.2 Strong and Weak Acids and Bases

Acids

Strong
Hydrochloric acid, HCl
Hydrobromic acid, HBr
Hydriodic acid, HI
Nitric acid, HNO_3
Sulfuric acid, H_2SO_4
Perchloric acid, $HClO_4$

Weak (a few of many examples)
Hydrofluoric acid, HF
Phosphoric acid, H_3PO_4
Acetic acid, CH_3COOH (or $HC_2H_3O_2$)

Bases

Strong
Group 1A(1) hydroxides:
Lithium hydroxide, LiOH
Sodium hydroxide, NaOH
Potassium hydroxide, KOH
Rubidium hydroxide, RbOH
Cesium hydroxide, CsOH

Heavy Group 2A(2) hydroxides:
Calcium hydroxide, $Ca(OH)_2$
Strontium hydroxide, $Sr(OH)_2$
Barium hydroxide, $Ba(OH)_2$

Weak (one of many examples)
Ammonia, NH_3

Acids and the Solvated Proton *Acidic solutions arise when certain covalent H-containing molecules dissociate into ions in water.* In every case, these molecules contain a *polar bond to H* in which the other atom pulls much more strongly on the electron pair. A good example is HBr. The Br end of the H—Br bond is partially negative, and the H end is partially positive. When hydrogen bromide gas dissolves in water, the poles of H_2O molecules are attracted to the oppositely charged poles of the HBr. The bond breaks, with H becoming the solvated cation $H^+(aq)$ and Br becoming the solvated anion $Br^-(aq)$:

$$HBr(g) \xrightarrow{H_2O} H^+(aq) + Br^-(aq)$$

The solvated H^+ ion is a very unusual species. The H atom is a proton surrounded by an electron, so H^+ is just a proton. With a full positive charge concentrated in such a tiny volume, H^+ attracts the negative pole of water molecules so strongly that it forms a covalent bond to one of them. We'll emphasize this interaction by writing the solvated H^+ ion as a solvated H_3O^+ ion (hydronium ion):

$$HBr(g) + H_2O(l) \longrightarrow H_3O^+(aq) + Br^-(aq)$$

Acids and Bases as Electrolytes Acids and bases are categorized in terms of their "strength," the degree to which they dissociate into ions in water (Table 4.2):

- *Strong acids and strong bases dissociate completely into ions.* Therefore, like soluble ionic compounds, they are *strong* electrolytes and conduct a large current, as shown by the brightly lit bulb (Figure 4.7A).
- *Weak acids and weak bases dissociate very little into ions.* Most of their molecules remain intact. Therefore, they are *weak* electrolytes, which means they conduct a small current (Figure 4.7B).

A Strong acid (or base) = strong electrolyte

B Weak acid (or base) = weak electrolyte

Figure 4.7 Acids and bases as electrolytes.

Because a strong acid (or strong base) dissociates completely, we can find the molarity of H^+ (or OH^-) and the amount (mol) or number of each ion in solution. (You'll see how to determine these quantities for weak acids in Chapter 18.)

Sample Problem 4.5 Determining the Number of H^+ (or OH^-) Ions in Solution

Problem Nitric acid is a major chemical in the fertilizer and explosives industries. How many $H^+(aq)$ ions are in 25.3 mL of 1.4 *M* nitric acid?

Plan We know the volume (25.3 mL) and molarity (1.4 *M*) of nitric acid, and we need the number of $H^+(aq)$. We convert from mL to L and multiply by the molarity to find the amount (mol) of acid. Table 4.2 shows that nitric acid is a strong acid, so it dissociates completely. With the formula, we write a dissociation equation, which shows the amount (mol) of H^+ ions per mole of acid. We multiply that amount by Avogadro's number to find the number of $H^+(aq)$ ions (see the road map).

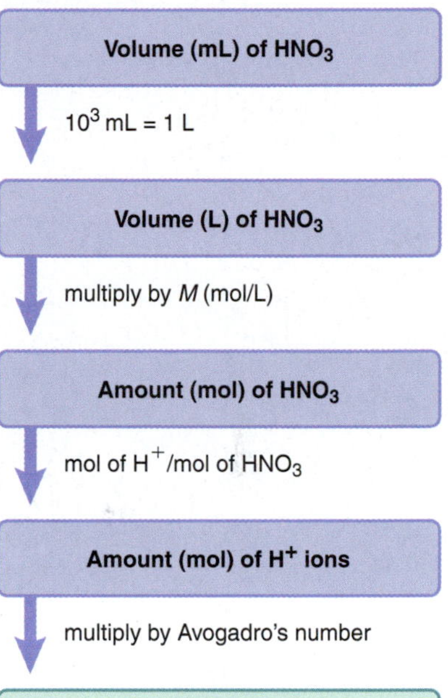

Solution Finding the amount (mol) of nitric acid:

$$\text{Amount (mol) of HNO}_3 = 25.3\ \cancel{\text{mL}} \times \frac{1\ \cancel{\text{L}}}{1000\ \cancel{\text{mL}}} \times \frac{1.4\ \text{mol}}{1\ \cancel{\text{L}}} = 0.035\ \text{mol}$$

Nitric acid is HNO_3, so we have:

$$HNO_3(l) \xrightarrow{H_2O} H^+(aq) + NO_3^-(aq)$$

Finding the number of H^+ ions:

$$\text{No. of H}^+ \text{ ions} = 0.035\ \cancel{\text{mol HNO}_3} \times \frac{1\ \cancel{\text{mol H}^+}}{1\ \cancel{\text{mol HNO}_3}} \times \frac{6.022\times10^{23}\ \text{H}^+\ \text{ions}}{1\ \cancel{\text{mol H}^+}}$$

$$= 2.1\times10^{22}\ \text{H}^+\ \text{ions}$$

Check The number of moles seems correct: 0.025 L × 1.4 mol/L = 0.035 mol, and multiplying by 6×10^{23} ions/mol gives 2×10^{22} ions.

FOLLOW-UP PROBLEM 4.5 How many $OH^-(aq)$ ions are present in 451 mL of 1.20 *M* potassium hydroxide?

Structural Features of Acids and Bases A key structural feature appears in common laboratory acids and bases:

- *Acids*. Strong acids, such as HNO_3 and H_2SO_4, and weak acids, such as HF and H_3PO_4, have one or more H atoms as part of their structure, which are either completely released (strong) or partially released (weak) as protons in water.
- *Bases*. Strong bases have either OH^- (e.g., NaOH) or O^{2-} (e.g., K_2O) as part of their structure. The oxide ion is not stable in water and reacts to form OH^- ion:

$$O^{2-}(s) + H_2O(l) \longrightarrow 2OH^-(aq) \quad \text{so} \quad K_2O(s) + H_2O(l) \longrightarrow 2K^+(aq) + 2OH^-(aq)$$

Weak bases, such as ammonia, do not contain OH^- ions, but, as you'll see in later chapters, they all have an electron pair on N. Weak bases produce OH^- ions in a reaction that occurs to a small extent in water:

$$NH_3(g) + H_2O(l) \rightleftharpoons NH_4^+(aq) + OH^-(aq)$$

Weak Acids and Bases and the Equilibrium State The unusual yield arrow in the equation above for ammonia's reaction with water indicates that the *reaction proceeds in both directions.* Indeed, as we'll discuss in later chapters, most reactions behave this way: they seem to stop before they are complete (that is, before the limiting reactant is used up) because another reaction, the reverse of the first one, is taking place just as fast. As a result, *no further change in the amounts of reactants and products occurs,* and we say the reaction has reached a state of *equilibrium.*

The reversibility of reactions explains why weak acids and bases dissociate into ions to only a small extent: the dissociation becomes balanced by a reassociation. For example, when acetic acid dissolves in water, some of the CH_3COOH molecules react with water and form H_3O^+ and CH_3COO^- ions. As more ions form, they react with each other more often to re-form acetic acid and water, and this state is indicated by the special (equilibrium) arrow:

$$CH_3COOH(aq) + H_2O(l) \rightleftharpoons H_3O^+(aq) + CH_3COO^-(aq)$$

In fact, in 0.1 *M* CH_3COOH at 25°C, only about 1.3% of the acid molecules dissociate into ions. A similarly small percentage of ammonia molecules form ions when 0.1 *M* NH_3 reacts with water. We discuss the central idea of equilibrium and its applications for chemical and physical systems in Chapters 12, 13, and 17 through 21.

The Key Event: Formation of H_2O from H^+ and OH^-

To see the key event in acid-base reactions, we'll write the three types of aqueous ionic equations (with color) and focus on the reaction between the strong acid HCl and the strong base $Ba(OH)_2$:

- The molecular equation is

$$2HCl(aq) + Ba(OH)_2(aq) \longrightarrow BaCl_2(aq) + 2H_2O(l)$$

- HCl and $Ba(OH)_2$ dissociate completely, so the total ionic equation is

$$2H^+(aq) + 2Cl^-(aq) + Ba^{2+}(aq) + 2OH^-(aq) \longrightarrow Ba^{2+}(aq) + 2Cl^-(aq) + 2H_2O(l)$$

- In the net ionic equation, we eliminate the spectator ions, $Ba^{2+}(aq)$ and $Cl^-(aq)$:

$$2H^+(aq) + 2OH^-(aq) \longrightarrow 2H_2O(l) \quad \text{or} \quad H^+(aq) + OH^-(aq) \longrightarrow H_2O(l)$$

Thus, *the key event in aqueous reactions between a strong acid and a strong base is that an* H^+ *ion from the acid and an* OH^- *ion from the base form a water molecule.* Only the spectator ions differ from one strong acid–strong base reaction to another.

Like precipitation reactions, acid-base reactions occur through *the electrostatic attraction of ions and their removal from solution as the product.* In this case, rather than an insoluble ionic solid, the product is H_2O, which consists almost entirely of undissociated molecules. Actually, water molecules dissociate *very* slightly (which, as you'll see in Chapter 18, is very important), but the formation of water in an acid-base reaction results in an enormous net removal of H^+ and OH^- ions.

The molecular and total ionic equations above show that if you evaporate the water, the spectator ions remain: the ionic compound that results from the reaction of an acid and a base is called a **salt,** which in this case is barium chloride. Thus, in an aqueous neutralization reaction, *an acid and a base form a salt solution and water:*

$$\underset{\text{acid}}{HX(aq)} + \underset{\text{base}}{MOH(aq)} \longrightarrow \underset{\text{salt}}{MX(aq)} + \underset{\text{water}}{H_2O(l)}$$

Note that *the cation of the salt comes from the base and the anion from the acid.* Also note that acid-base reactions, like precipitation reactions, are metathesis (double-displacement) reactions.

Sample Problem 4.6 Writing Ionic Equations for Acid-Base Reactions

Problem Write balanced molecular, total ionic, and net ionic equations for each of the following acid-base reactions and identify the spectator ions:
(a) Hydrochloric acid(*aq*) + potassium hydroxide(*aq*) ⟶
(b) Strontium hydroxide(*aq*) + perchloric acid(*aq*) ⟶
(c) Barium hydroxide(*aq*) + sulfuric acid(*aq*) ⟶

Plan All are strong acids and bases (see Table 4.2), so the actual reaction is between H^+ and OH^-. The products are H_2O and a salt solution of spectator ions. In (c), we note that the salt ($BaSO_4$) is insoluble (see Table 4.1), so there are no spectator ions.

Solution **(a)** Writing the molecular equation:

$$HCl(aq) + KOH(aq) \longrightarrow KCl(aq) + H_2O(l)$$

Writing the total ionic equation:

$$H^+(aq) + Cl^-(aq) + K^+(aq) + OH^-(aq) \longrightarrow K^+(aq) + Cl^-(aq) + H_2O(l)$$

Writing the net ionic equation:

$$H^+(aq) + OH^-(aq) \longrightarrow H_2O(l)$$

$K^+(aq)$ and $Cl^-(aq)$ are the spectator ions.

(b) Writing the molecular equation:

$$Sr(OH)_2(aq) + 2HClO_4(aq) \longrightarrow Sr(ClO_4)_2(aq) + 2H_2O(l)$$

Writing the total ionic equation:

$$Sr^{2+}(aq) + 2OH^-(aq) + 2H^+(aq) + 2ClO_4^-(aq) \longrightarrow Sr^{2+}(aq) + 2ClO_4^-(aq) + 2H_2O(l)$$

Writing the net ionic equation:

$$2OH^-(aq) + 2H^+(aq) \longrightarrow 2H_2O(l) \quad \text{or} \quad OH^-(aq) + H^+(aq) \longrightarrow H_2O(l)$$

$Sr^{2+}(aq)$ and $ClO_4^-(aq)$ are the spectator ions.

(c) Writing the molecular equation:

$$Ba(OH)_2(aq) + H_2SO_4(aq) \longrightarrow BaSO_4(s) + 2H_2O(l)$$

Writing the total ionic equation:

$$Ba^{2+}(aq) + 2OH^-(aq) + 2H^+(aq) + SO_4^{2-}(aq) \longrightarrow BaSO_4(s) + 2H_2O(l)$$

This is a neutralization *and* a precipitation reaction, so the net ionic equation is the same as the total ionic. There are no spectator ions.

FOLLOW-UP PROBLEM 4.6 Write balanced molecular, total ionic, and net ionic equations for the reaction between aqueous solutions of calcium hydroxide and nitric acid.

Proton Transfer in Acid-Base Reactions

When we take a closer look (with color) at the reaction between a strong acid and strong base, as well as reactions of weak acids with bases, a unifying pattern appears.

Reaction Between Strong Acid and Strong Base When HCl gas dissolves in water, the H^+ ion ends up bonded to a water molecule. Thus, hydrochloric acid actually consists of solvated H_3O^+ and Cl^- ions:

$$HCl(g) + H_2O(l) \longrightarrow H_3O^+(aq) + Cl^-(aq)$$

If we add NaOH solution, the total ionic equation shows that H_3O^+ *transfers a proton* to OH^- (leaving a water molecule written as H_2O, and forming a water molecule written as HOH):

H⁺ transfer

$$[H_3O^+(aq) + Cl^-(aq)] + [Na^+(aq) + OH^-(aq)] \longrightarrow H_2O(l) + Cl^-(aq) + Na^+(aq) + HOH(l)$$

Without the spectator ions, the net ionic equation shows more clearly the *transfer of a proton* from H_3O^+ to OH^-:

H⁺ transfer

$$H_3O^+(aq) + OH^-(aq) \longrightarrow H_2O(l) + HOH(l) \quad [\text{or } 2H_2O(l)]$$

This equation is identical to the one we saw earlier (see p. 128),

$$H^+(aq) + OH^-(aq) \longrightarrow H_2O(l)$$

with the additional H_2O molecule coming from the H_3O^+. Thus, an *acid-base reaction is a proton-transfer process.* In this case, the Cl^- and Na^+ ions remain in solution, and if the water is evaporated, they crystallize as the salt NaCl. Figure 4.8 on the next page shows this process on the atomic level. We'll discuss the proton-transfer concept thoroughly in Chapter 18.

Reactions of Weak Acids with Bases When solutions of sodium hydroxide and the weak acid acetic acid (CH_3COOH, Table 4.2) are mixed, the molecular equation is

$$CH_3COOH(aq) + NaOH(aq) \longrightarrow CH_3COONa(aq) + H_2O(l)$$

The total ionic equation is written differently for the reaction of a weak acid. Because acetic acid is weak and, thus, dissociates very little, it appears as *undissociated, intact molecules:*

$$CH_3COOH(aq) + Na^+(aq) + OH^-(aq) \longrightarrow CH_3COO^-(aq) + Na^+(aq) + H_2O(l)$$

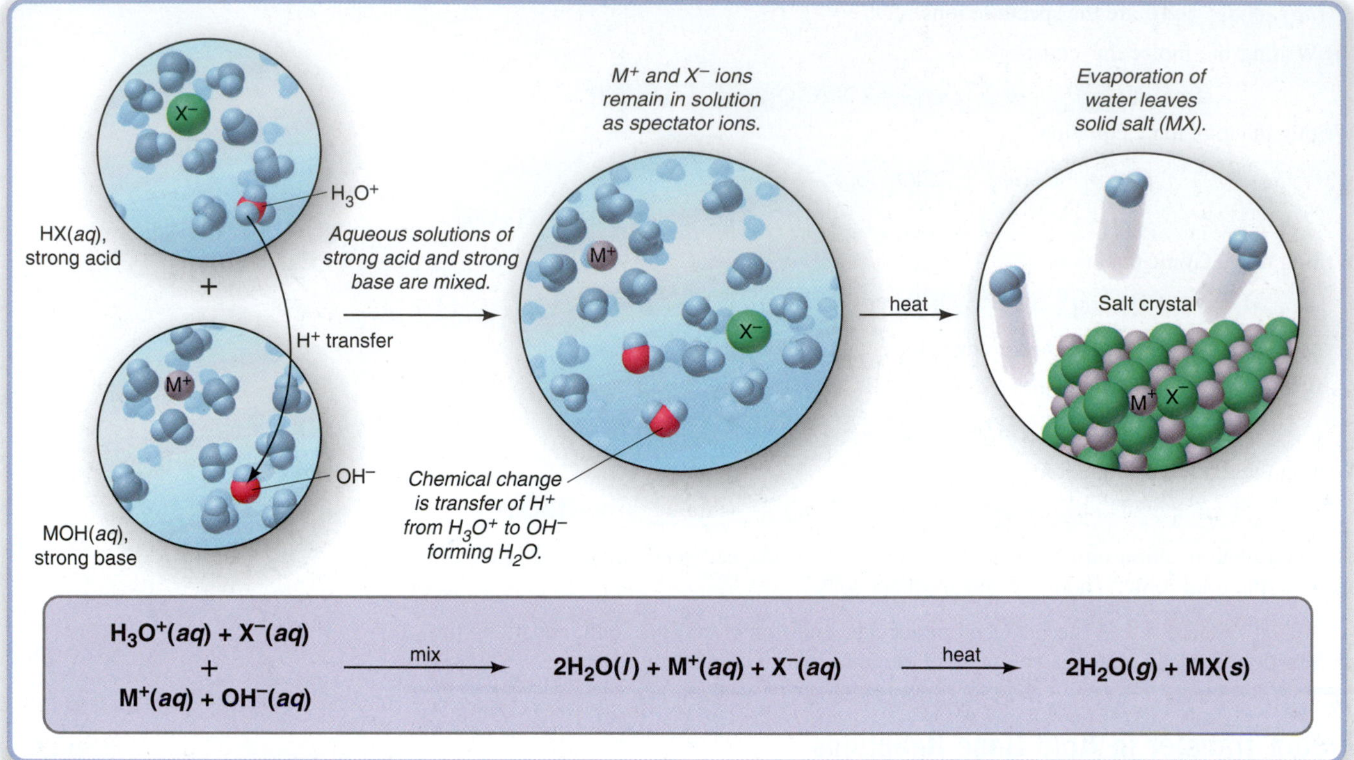

Figure 4.8 An aqueous strong acid–strong base reaction as a proton-transfer process.

The net ionic equation reveals that proton transfer occurs here as well, but directly from the weak acid rather than from H_3O^+:

$$CH_3COOH(aq) + OH^-(aq) \longrightarrow CH_3COO^-(aq) + H_2O(l)$$

(H^+ transfer)

The only spectator ion is $Na^+(aq)$.

Quantifying Acid-Base Reactions by Titration

In any **titration,** *the known concentration of one solution is used to determine the unknown concentration of another.* In a typical acid-base titration, a *standardized* solution of base, one whose concentration is *known,* is added to a solution of acid whose concentration is *unknown (or vice versa).*

Figure 4.9A shows the laboratory setup for an acid-base titration with a known volume of acid and a few drops of indicator in a flask. An *acid-base indicator* is a substance whose color is different in acid than in base; the indicator used in the figure is phenolphthalein, which is pink in base and colorless in acid. (We examine indicators in Chapters 18 and 19.) Base is added from a buret, and the OH^- ions react with the H^+ ions. As the titration nears its end (Figure 4.9B), the drop of added base creates a temporary excess of OH^-, causing some indicator molecules to change to the basic color; they return to the acidic color when the flask is swirled. There are two key stages in the titration:

- The **equivalence point** occurs when *the amount (mol) of H^+ ions in the original volume of acid has reacted with the same amount (mol) of OH^- ions from the buret:*

  Amount (mol) of H^+ (originally in flask) = amount (mol) of OH^- (added from buret)

- The **end point** occurs when a tiny excess of OH^- ions changes the indicator permanently to its basic color (Figure 4.9C).

In calculations, such as in Sample Problem 4.7, we assume that this tiny excess of OH^- ions is insignificant and that *the amount of base needed to reach the end point is the same as the amount needed to reach the equivalence point.*

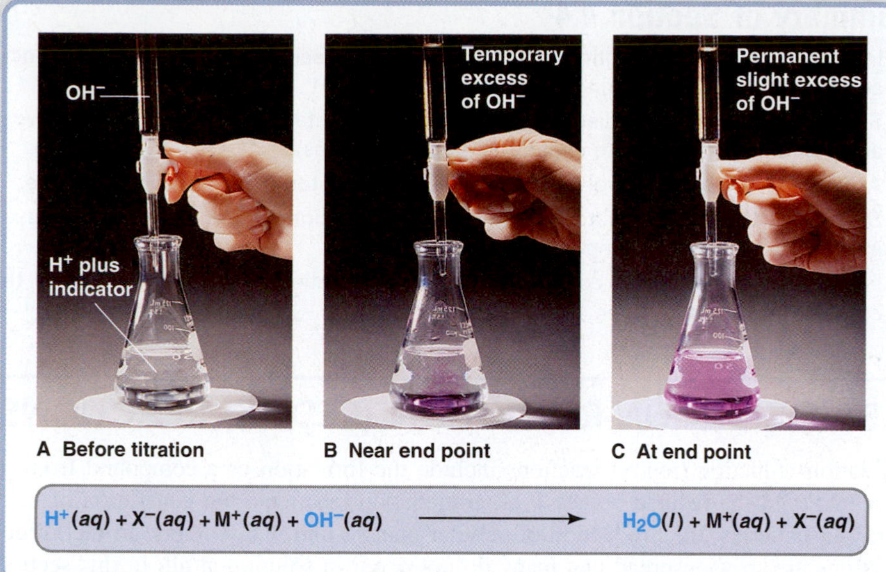

Figure 4.9 An acid-base titration.

Sample Problem 4.7 Finding the Concentration of an Acid from a Titration

Problem To standardize an HCl solution, you put 50.00 mL of it in a flask with a few drops of indicator and put 0.1524 *M* NaOH in a buret. The buret reads 0.55 mL at the start and 33.87 mL at the end point. Find the molarity of the HCl solution.

Plan We have to find the molarity of the acid from the volume of acid (50.00 mL), the initial (0.55 mL) and final (33.87 mL) volumes of base, and the molarity of the base (0.1524 *M*). First, we balance the equation. The volume of added base is the difference in buret readings, and we use the base's molarity to calculate the amount (mol) of base. Then, we use the molar ratio from the balanced equation to find the amount (mol) of acid originally present and divide by the acid's original volume to find the molarity (see the road map).

Road Map

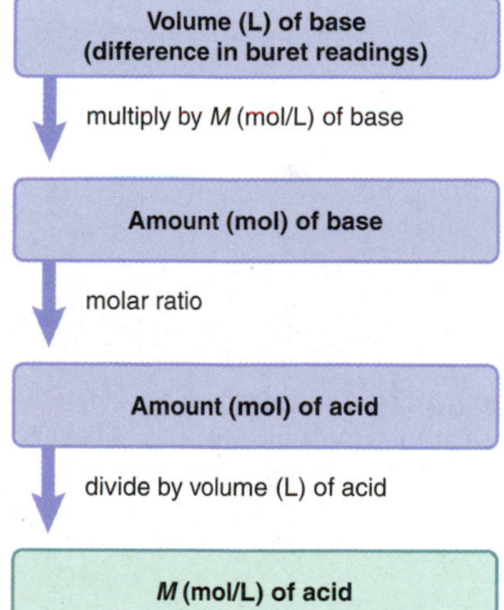

Solution Writing the balanced equation:

$$\text{NaOH}(aq) + \text{HCl}(aq) \longrightarrow \text{NaCl}(aq) + \text{H}_2\text{O}(l)$$

Finding the volume (L) of NaOH solution added:

$$\text{Volume (L) of solution} = (33.87 \cancel{\text{mL}}\ \text{soln} - 0.55 \cancel{\text{mL}}\ \text{soln}) \times \frac{1\ \text{L}}{1000 \cancel{\text{mL}}}$$
$$= 0.03332\ \text{L soln}$$

Finding the amount (mol) of NaOH added:

$$\text{Amount (mol) of NaOH} = 0.03332 \cancel{\text{L soln}} \times \frac{0.1524\ \text{mol NaOH}}{1 \cancel{\text{L soln}}}$$
$$= 5.078\times10^{-3}\ \text{mol NaOH}$$

Finding the amount (mol) of HCl originally present: Since the molar ratio is 1/1,

$$\text{Amount (mol) of HCl} = 5.078\times10^{-3} \cancel{\text{mol NaOH}} \times \frac{1\ \text{mol HCl}}{1 \cancel{\text{mol NaOH}}} = 5.078\times10^{-3}\ \text{mol HCl}$$

Calculating the molarity of HCl:

$$\text{Molarity of HCl} = \frac{5.078\times10^{-3}\ \text{mol HCl}}{50.00 \cancel{\text{mL}}} \times \frac{1000 \cancel{\text{mL}}}{1\ \text{L}} = 0.1016\ M\ \text{HCl}$$

Check The answer makes sense: a large volume of less concentrated acid neutralized a small volume of more concentrated base. With rounding, the numbers of moles of H^+ and OH^- are about equal: 50 mL × 0.1 *M* H^+ = 0.005 mol = 33 mL × 0.15 *M* OH^-.

FOLLOW-UP PROBLEM 4.7 What volume of 0.1292 *M* $Ba(OH)_2$ would neutralize 50.00 mL of the HCl solution standardized in Sample Problem 4.7?

Summary of Section 4.4

- In an acid-base (neutralization) reaction between an acid (an H^+-yielding substance) and a base (an OH^--yielding substance), H^+ and OH^- ions form H_2O.
- Strong acids and bases dissociate completely in water (strong electrolytes); weak acids and bases dissociate slightly (weak electrolytes).
- An acid-base reaction involves the transfer of a proton from an acid to a base.
- Since weak acids dissociate very little, ionic equations show a weak acid as an intact molecule transferring its proton to the base.
- In a titration, the known concentration of one solution is used to determine the concentration of the other.

4.5 • OXIDATION-REDUCTION (REDOX) REACTIONS

Oxidation-reduction (redox) reactions include the formation of a compound from its elements (and the reverse process), all combustion reactions, the generation of electricity in batteries, the production of cellular energy, and many others. In fact, redox reactions are so widespread that many do not occur in solution at all. In this section, we examine the key event in the redox process and discuss important terminology.

The Key Event: Net Movement of Electrons Between Reactants

The key chemical event in an **oxidation-reduction** (or **redox**) **reaction** is the *net movement of electrons from one reactant to another*. The movement occurs from the reactant (or atom in the reactant) with *less* attraction for electrons to the reactant (or atom) with *more* attraction for electrons.

This process occurs in the formation of both ionic and covalent compounds:

- *Ionic compounds: transfer of electrons.* In the reaction that forms MgO from its elements (see Figure 3.6), the balanced equation is

$$2Mg(s) + O_2(g) \longrightarrow 2MgO(s)$$

Figure 4.10A shows that each Mg atom loses two electrons and each O atom gains them. This loss and gain is a ***transfer*** *of electrons* away from each Mg atom to each O atom. The resulting Mg^{2+} and O^{2-} ions aggregate into an ionic solid.

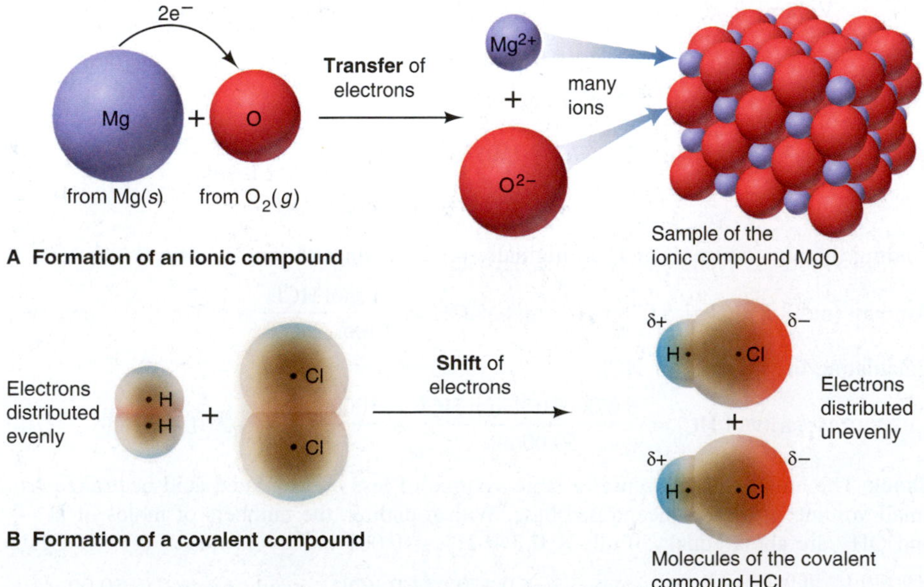

Figure 4.10 The redox process in the formation of (A) ionic and (B) covalent compounds from their elements.

- *Covalent compounds: shift of electrons.* During the formation of a covalent compound from its elements, there is more of a ***shift** of electrons* than a full transfer. Thus, *ions do not form.* Consider the formation of HCl gas:

$$H_2(g) + Cl_2(g) \longrightarrow 2HCl(g)$$

To see the electron movement, we'll compare the electron distribution in reactant and product. As Figure 4.10B shows, in H_2 and Cl_2, the electrons are shared equally between the atoms *(symmetrical shading).* Because the Cl atom attracts electrons more than the H atom, in HCl, the electrons are shared unequally *(asymmetrical shading).* Electrons shift away from H and toward Cl, so the Cl atom has more negative charge (*red* and $\delta-$) than it had in Cl_2, and the H atom has less negative charge (*blue* and $\delta+$) than it had in H_2.

Some Essential Redox Terminology

This subsection introduces certain key terms that describe central ideas of the redox process.

- **Oxidation** is the *loss* of electrons.
- **Reduction** is the *gain* of electrons.

During the formation of MgO, Mg undergoes oxidation (loss of electrons), and O_2 undergoes reduction (gain of electrons). The loss and gain are simultaneous, but we can imagine them occurring separately:

Oxidation (electron loss by Mg): $Mg \longrightarrow Mg^{2+} + 2e^-$

Reduction (electron gain by O_2): $\frac{1}{2}O_2 + 2e^- \longrightarrow O^{2-}$

(Throughout the chapter, blue type indicates oxidation, and red type reduction.)

- The **oxidizing agent** is the species doing the oxidizing (causing the electron loss).
- The **reducing agent** is the species doing the reducing (causing the electron gain).

One reactant acts on the other. During the reaction that forms MgO, *O_2 oxidizes Mg,* so O_2 is the oxidizing agent, and *Mg reduces O_2,* so Mg is the reducing agent.

Note the give and take of electrons: O_2 takes the electrons that Mg gives up, or, put the other way around, Mg gives up the electrons that O_2 takes. This means that

- *The oxidizing agent is reduced:* it takes electrons (gains them).
- *The reducing agent is oxidized:* it gives up electrons (loses them).

In the formation of HCl, *Cl_2 oxidizes H_2* (H loses some electron charge and Cl gains it), which is the same as saying that *H_2 reduces Cl_2.* The reducing agent, H_2, is oxidized, and the oxidizing agent, Cl_2, is reduced.

Using Oxidation Numbers to Monitor Electron Charge

Chemists have devised a "bookkeeping" system to monitor which atom loses electron charge and which atom gains it: each atom in a molecule (or formula unit) is assigned an **oxidation number (O.N.),** or *oxidation state,* which is the charge the atom would have *if* electrons were transferred completely, not shared.

Each element in a binary *ionic* compound has a full charge because the atom transferred its electron(s), and so the atom's oxidation number equals the ionic charge. But, each element in a *covalent* compound (or in a polyatomic ion) has a partial charge because the electrons shifted away from one atom and toward the other. For these cases, we determine oxidation number by a set of rules (Table 4.3, *next page;* you'll learn the atomic basis of the rules in Chapters 8 and 9).

An O.N. has the sign *before* the number (as in +2), whereas an ionic charge has the sign *after* the number (as in 2+). Also, unlike a unitary ionic charge, as in Na^+ or Cl^-, an O.N. of +1 or −1 retains the numeral. For example, we don't write the sodium ion as Na^{+1}, but the O.N. of the Na^+ ion is +1, not merely +.

Table 4.3 Rules for Assigning an Oxidation Number (O.N.)

General Rules

1. For an atom in its elemental form (Na, O_2, Cl_2, etc.): O.N. = 0
2. For a monatomic ion: O.N. = ion charge (with the sign *before* the numeral)
3. The sum of O.N. values for the atoms in a molecule or formula unit of a compound equals zero. The sum of O.N. values for the atoms in a polyatomic ion equals the ion's charge.

Rules for Specific Atoms or Periodic Table Groups

1. For Group 1A(1):	O.N. = +1 in all compounds
2. For Group 2A(2):	O.N. = +2 in all compounds
3. For hydrogen:	O.N. = +1 in combination with nonmetals
	O.N. = −1 in combination with metals and boron
4. For fluorine:	O.N. = −1 in all compounds
5. For oxygen:	O.N. = −1 in peroxides
	O.N. = −2 in all other compounds (except with F)
6. For Group 7A(17):	O.N. = −1 in combination with metals, nonmetals (except O), and other halogens lower in the group

Sample Problem 4.8 Determining the Oxidation Number of Each Element in a Compound (or Ion)

Problem Determine the oxidation number (O.N.) of each element in these species:
(a) Zinc chloride **(b)** Sulfur trioxide **(c)** Nitric acid **(d)** Dichromate ion

Plan We determine the formulas and consult Table 4.3, including the general rules that the O.N. values for a compound add up to zero and those for a polyatomic ion add up to the ion's charge.

Solution **(a)** $ZnCl_2$. The sum of O.N.s must equal zero. The O.N. of the Zn^{2+} ion is +2, so the O.N. of each Cl^- ion is −1, for a total of −2.
(b) SO_3. The O.N. of each oxygen is −2, for a total of −6. The O.N.s must add up to zero, so the O.N. of S is +6.
(c) HNO_3. The O.N. of H is +1, so the O.N.s of the atoms in NO_3^- must add up to −1 to equal the charge of the polyatomic ion and give zero for the compound. The O.N. of each O is −2, for a total of −6. Therefore, the O.N. of N is +5.
(d) $Cr_2O_7^{2-}$. The O.N. of each O is −2, so the total for seven O atoms is −14. Therefore, each Cr must have an O.N. of +6 in order for the sum of the O.N.s to equal the charge of the ion: $+12 + (-14) = -2$.

FOLLOW-UP PROBLEM 4.8 Determine the O.N. of each element in the following:
(a) Scandium oxide (Sc_2O_3) **(b)** Gallium chloride ($GaCl_3$)
(c) Hydrogen phosphate ion **(d)** Iodine trifluoride

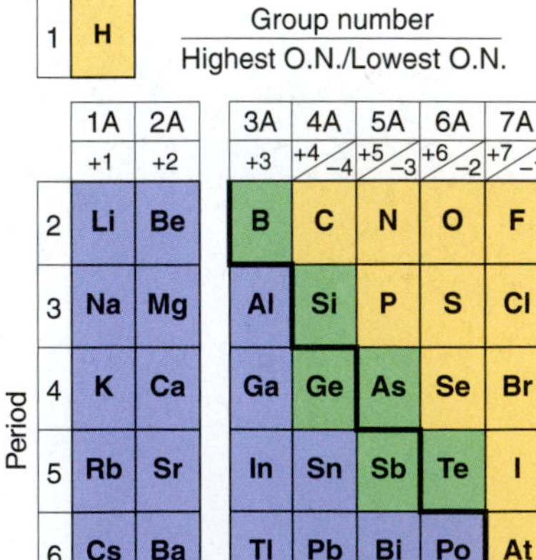

Figure 4.11 Highest and lowest oxidation numbers of reactive main-group elements.

The periodic table is a great help in learning the highest and lowest oxidation numbers of most main-group elements (Figure 4.11):

- For most main-group elements, the A-group number (1A, 2A, and so on) is the *highest* oxidation number (always positive) of any element in the group. The only exceptions are O, which never has an O.N. of +6, and F, which never has an O.N. of +7.
- For main-group nonmetals and some metalloids, the A-group number minus 8 is the *lowest* oxidation number (always negative) of any element in the group.

For example, the highest oxidation number of S (Group 6A) is +6, as in SF_6, and the lowest is (6 − 8), or −2, as in FeS and other metal sulfides.

Using O.N.s to Identify Oxidizing and Reducing Agents By assigning an oxidation number to each atom, we can see which species was oxidized and which reduced and, thus, which is the oxidizing agent and which the reducing agent:

- If an atom has a higher (more positive or less negative) O.N. in the product than it had in the reactant, the reactant that contains that atom was oxidized (lost electrons) and is the reducing agent. Thus, *oxidation is shown by an increase in O.N.*
- If an atom has a lower (more negative or less positive) O.N. in the product than it had in the reactant, the reactant that contains that atom was reduced (gained electrons) and is the oxidizing agent. Thus, *reduction is shown by a decrease in O.N.*

Sample Problem 4.9 Identifying Oxidizing and Reducing Agents

Problem Identify the oxidizing agent and the reducing agent in each of the following:
(a) $2Al(s) + 3H_2SO_4(aq) \longrightarrow Al_2(SO_4)_3(aq) + 3H_2(g)$
(b) $PbO(s) + CO(g) \longrightarrow Pb(s) + CO_2(g)$
(c) $2H_2(g) + O_2(g) \longrightarrow 2H_2O(g)$

Plan We assign an O.N. to each atom (or ion). The reducing agent contains an atom that is oxidized (O.N. *increased* from left to right in the equation). The oxidizing agent contains an atom that is reduced (O.N. *decreased*). We mark the changes with tie-lines.

Solution (a) Assigning oxidation numbers:

oxidation

$$\overset{0}{2Al}(s) + 3\overset{+1}{H_2}\overset{+6}{S}\overset{-2}{O_4}(aq) \longrightarrow \overset{+3}{Al_2}(\overset{+6}{S}\overset{-2}{O_4})_3(aq) + 3\overset{0}{H_2}(g)$$

reduction

The O.N. of Al increased from 0 to +3 (Al lost electrons), so Al was oxidized; Al is the reducing agent.

The O.N. of H decreased from +1 to 0 (H gained electrons), so H^+ was reduced; H_2SO_4 is the oxidizing agent.

(b) Assigning oxidation numbers:

oxidation

$$\overset{+2}{Pb}\overset{-2}{O}(s) + \overset{+2}{C}\overset{-2}{O}(g) \longrightarrow \overset{0}{Pb}(s) + \overset{+4}{C}\overset{-2}{O_2}(g)$$

reduction

Pb decreased its O.N. from +2 to 0, so PbO was reduced; PbO is the oxidizing agent.
C increased its O.N. from +2 to +4, so CO was oxidized; CO is the reducing agent.
When a reactant (in this case, CO) becomes a product with more O atoms (CO_2), it is oxidized; when a reactant (PbO) becomes a product with fewer O atoms (Pb), it is reduced.

(c) Assigning oxidation numbers:

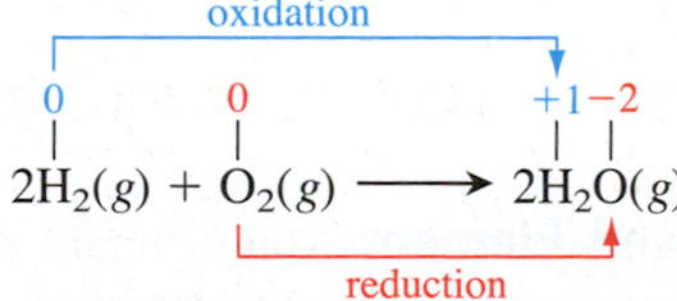

O_2 was reduced (O.N. of O decreased from 0 to −2); O_2 is the oxidizing agent.
H_2 was oxidized (O.N. of H increased from 0 to +1); H_2 is the reducing agent.
Oxygen is always the oxidizing agent in a combustion reaction.

FOLLOW-UP PROBLEM 4.9 Identify each oxidizing agent and each reducing agent:
(a) $2Fe(s) + 3Cl_2(g) \longrightarrow 2FeCl_3(s)$ **(b)** $2C_2H_6(g) + 7O_2(g) \longrightarrow 4CO_2(g) + 6H_2O(g)$
(c) $5CO(g) + I_2O_5(s) \longrightarrow I_2(s) + 5CO_2(g)$

Be sure to remember that *transferred electrons are never free because the reducing agent loses electrons and the oxidizing agent gains them* ***simultaneously.*** In other words, a complete reaction *cannot* be "an oxidation" *or* "a reduction"; it must be an oxidation-reduction. Figure 4.12 summarizes redox terminology.

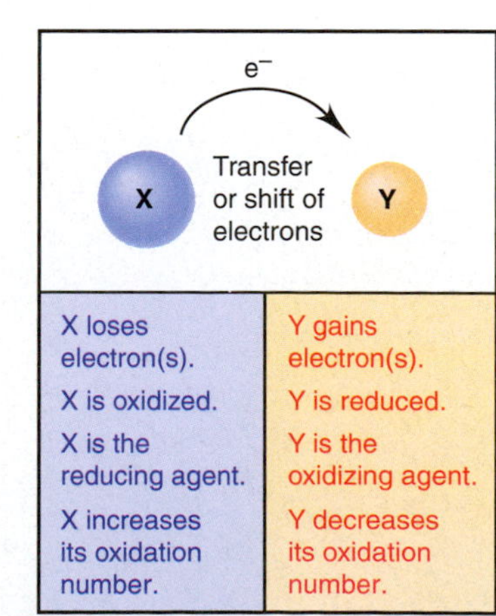

e⁻
X
Transfer or shift of electrons
Y
X loses electron(s).
X is oxidized.
X is the reducing agent.
X increases its oxidation number.
Y gains electron(s).
Y is reduced.
Y is the oxidizing agent.
Y decreases its oxidation number.

Figure 4.12 A summary of terminology for redox reactions.

Summary of Section 4.5

- When one reactant has a greater attraction for electrons than another, there is a net movement of electrons, and a redox reaction takes place. Electron gain (reduction) and electron loss (oxidation) occur simultaneously.
- Assigning oxidation numbers to all atoms in a reaction is a method for identifying a redox reaction. The species that is oxidized (contains an atom that increases in oxidation number) is the reducing agent; the species that is reduced (contains an atom that decreases in oxidation number) is the oxidizing agent.

4.6 • ELEMENTS IN REDOX REACTIONS

In many redox reactions, such as those in Sample Problem 4.9, *atoms occur as an element on one side of an equation and as part of a compound on the other.* While there are many redox reactions that do *not* involve free elements, we'll focus here on the many others that do. One way to classify these is by comparing the *numbers* of reactants and products. With that approach, we have three types—*combination, decomposition,* and *displacement;* one other type involving elements is *combustion.* In this section, we survey each type with several examples.

Combination Redox Reactions

In a combination redox reaction, two or more reactants, at least one of which is an element, form a compound:

$$X + Y \longrightarrow Z$$

Combining Two Elements Two elements may react to form binary ionic or covalent compounds. Here are some important examples:

1. *Metal and nonmetal form an ionic compound.* A metal, such as aluminum, reacts with a nonmetal, such as oxygen. The change in O.N.s shows that the metal is oxidized, so it is the reducing agent; the nonmental is reduced, so it is the oxidizing agent:

$$\overset{0}{4\text{Al}}(s) + \overset{0}{3\text{O}_2}(g) \longrightarrow 2\overset{+3}{\text{Al}_2}\overset{-2}{\text{O}_3}(s)$$

Figure 3.6 (p. 87) shows the redox reaction between magnesium metal and oxygen on the macroscopic and atomic scales.

2. *Two nonmetals form a covalent compound.* In one of thousands of examples, ammonia forms from nitrogen and hydrogen in a reaction that occurs in industry on an enormous scale:

$$\overset{0}{\text{N}_2}(g) + 3\overset{0}{\text{H}_2}(g) \longrightarrow 2\overset{-3}{\text{N}}\overset{+1}{\text{H}_3}(g)$$

Combining Compound and Element Many binary covalent compounds react with nonmetals to form larger compounds. Many nonmetal oxides react with additional O_2 to form "higher" oxides (those with more O atoms in each molecule). For example,

$$2\overset{+2}{\text{N}}\overset{-2}{\text{O}}(g) + \overset{0}{\text{O}_2}(g) \longrightarrow 2\overset{+4}{\text{N}}\overset{-2}{\text{O}_2}(g)$$

Similarly, many nonmetal halides combine with additional halogen to form "higher" halides:

$$\overset{+3}{\text{P}}\overset{-1}{\text{Cl}_3}(l) + \overset{0}{\text{Cl}_2}(g) \longrightarrow \overset{+5}{\text{P}}\overset{-1}{\text{Cl}_5}(s)$$

Decomposition Redox Reactions

In a *decomposition redox reaction,* a compound forms two or more products, at least one of which is an element:

$$Z \longrightarrow X + Y$$

In any decomposition reaction, the reactant absorbs enough energy for one or more bonds to break. The energy can take several forms, but the most important are decomposition by heat (thermal) and by electricity (electrolytic). The products are either elements or elements and smaller compounds.

Thermal Decomposition When the energy absorbed is heat, the reaction is called a *thermal decomposition.* (A Greek delta, Δ, above the yield arrow indicates strong heating is required.) Many metal oxides, chlorates, and perchlorates release oxygen when strongly heated. Heating potassium chlorate is a method for forming small amounts of oxygen in the laboratory; the same reaction occurs in some explosives and fireworks:

$$2\overset{+1}{K}\overset{+5}{Cl}\overset{-2}{O_3}(s) \xrightarrow{\Delta} 2\overset{+1}{K}\overset{-1}{Cl}(s) + 3\overset{0}{O_2}(g)$$

Notice that the lone reactant is the oxidizing *and* the reducing agent.

Electrolytic Decomposition In the process of electrolysis, a compound absorbs electrical energy and decomposes into its elements. In the early 19th century, the observation of the electrolysis of water was crucial for establishing atomic masses:

$$2\overset{+1}{H_2}\overset{-2}{O}(l) \xrightarrow{\text{electricity}} 2\overset{0}{H_2}(g) + \overset{0}{O_2}(g)$$

Many active metals, such as sodium, magnesium, and calcium, are produced industrially by electrolysis of their molten halides:

$$\overset{+2}{Mg}\overset{-1}{Cl_2}(l) \xrightarrow{\text{electricity}} \overset{0}{Mg}(l) + \overset{0}{Cl_2}(g)$$

(We examine the details of electrolysis and its role in the industrial recovery of several elements in Chapter 21.)

Displacement Redox Reactions and Activity Series

In any *displacement reaction,* the number of substances on the two sides of the equation remains the same, but atoms (or ions) exchange places. There are two types:

1. In *double*-displacement (metathesis) reactions, such as precipitation and acid-base reactions (Sections 4.3 and 4.4), atoms (or ions) of two *compounds* exchange places; these reactions are *not* redox processes:

$$AB + CD \longrightarrow AC + BD$$

2. In *single*-displacement reactions, one of the substances is an *element;* therefore, *all single-displacement reactions **are** redox processes:*

$$X + YZ \longrightarrow XZ + Y$$

In solution, single-displacement reactions occur when an atom of one element displaces the ion of another: if the displacement involves metals, the atom *reduces* the ion; if it involves nonmetals (specifically halogens), the atom *oxidizes* the ion. In two *activity series*—one for metals and one for halogens—the elements are ranked in order of their ability to displace hydrogen (for metals) and one another.

The Activity Series of the Metals Metals are ranked by their ability to displace H_2 from various sources and another metal from solution. In all displacements of H_2, the metal is the reducing agent (O.N. increases), and water or acid is the oxidizing agent (O.N. of H decreases). The activity series of the metals is based on these facts:

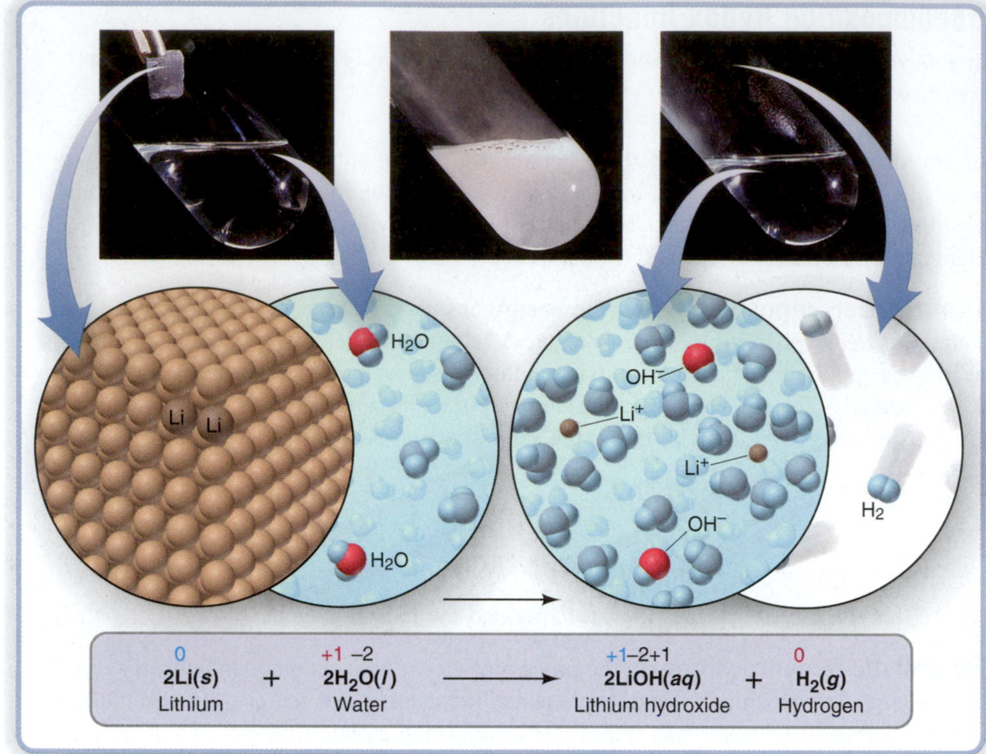

Figure 4.13 The active metal lithium displaces hydrogen from water. Only water molecules involved as products of the reaction are red and blue.

- *The most reactive metals displace H_2 from liquid water.* Group 1A(1) metals and Ca, Sr, and Ba from Group 2A(2) displace H_2 from water (Figure 4.13).
- *Slightly less reactive metals displace H_2 from steam.* Heat supplied by steam is needed for less reactive metals such as Al and Zn to displace H_2:

$$\overset{0}{2\text{Al}(s)} + 6\overset{+1\,-2}{\text{H}_2\text{O}}(g) \xrightarrow{\Delta} 2\overset{+3\,-2\,+1}{\text{Al(OH)}_3}(s) + 3\overset{0}{\text{H}_2}(g)$$

- *Still less reactive metals displace H_2 from acids.* The higher concentration of H^+ in acid solutions is needed for even less reactive metals such as Ni and Sn to displace H_2 (Figure 4.14). For nickel, the net ionic equation is

$$\overset{0}{\text{Ni}}(s) + 2\overset{+1}{\text{H}^+}(aq) \longrightarrow \overset{+2}{\text{Ni}^{2+}}(aq) + \overset{0}{\text{H}_2}(g)$$

- *The least reactive metals cannot displace H_2 from any source.* Fortunately, precious metals, such as silver, gold, and platinum, do *not* react with water or acid.
- *An atom of one metal displaces the ion of another.* Comparisons of metal reactivity show, for example, that Zn metal displaces Cu^{2+} ion from aqueous $CuSO_4$:

$$\overset{+2}{\text{Cu}^{2+}}(aq) + \overset{+6\,-2}{\text{SO}_4^{2-}}(aq) + \overset{0}{\text{Zn}}(s) \longrightarrow \overset{0}{\text{Cu}}(s) + \overset{+2}{\text{Zn}^{2+}}(aq) + \overset{+6\,-2}{\text{SO}_4^{2-}}(aq)$$

And Figure 4.15 shows that Cu metal displaces silver (Ag^+) ion from solution; therefore, zinc is more reactive than copper, which is more reactive than silver.

Figure 4.14 The displacement of H_2 from acid by nickel.

Many such reactions form the basis of the **activity series of the metals.** Note these two points from Figure 4.16:

- Elements higher on the list are stronger reducing agents than elements lower down; in other words, any metal can reduce the ions of metals below it.
- From higher to lower, a metal can displace H_2 (reduce H^+) from water, steam, acid, or not at all.

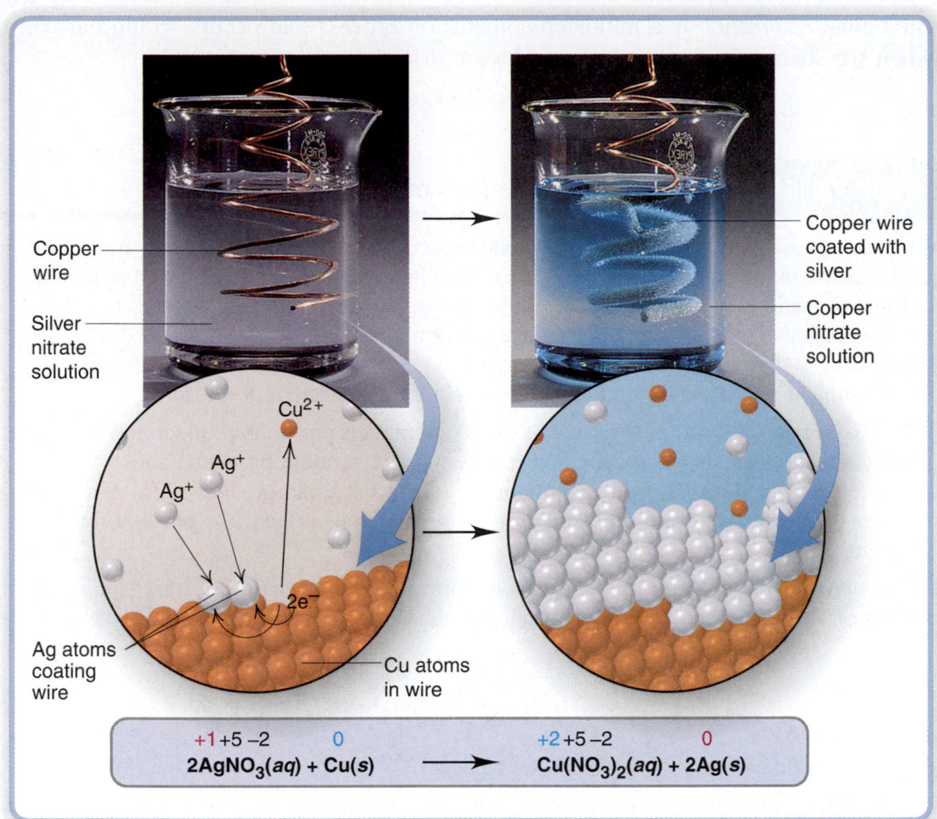

Figure 4.15 A more reactive metal (Cu) displacing the ion of a less reactive metal (Ag^+) from solution.

Consider the metals we've just discussed: Li, Al, and Ni lie above H_2, while Ag lies below it; also, Zn lies above Cu, which lies above Ag. The most reactive metals on the list are in Groups 1A(1) and 2A(2) of the periodic table, and the least reactive are Cu, Ag, and Au in Group 1B(11) and Hg in 2B(12).

The Activity Series of the Halogens Reactivity of the elements decreases down Group 7A(17), so we have

$$F_2 > Cl_2 > Br_2 > I_2$$

A halogen higher in the group is a stronger oxidizing agent than one lower down. Thus, elemental chlorine can oxidize bromide ions *(below)* or iodide ions from solution, and elemental bromine can oxidize iodide ions:

$$\overset{-1}{2Br^-(aq)} + \overset{0}{Cl_2(aq)} \longrightarrow \overset{0}{Br_2(aq)} + \overset{-1}{2Cl^-(aq)}$$

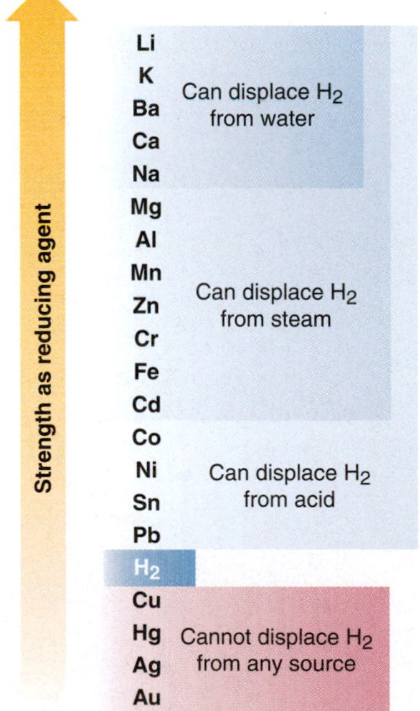

Figure 4.16 The activity series of the metals. The most active metal (strongest reducing agent) is at the top, and the least active metal (weakest reducing agent) is at the bottom.

Combustion Reactions

Combustion is the process of combining with oxygen, most commonly with the release of heat and the production of light, as in a flame. Combustion reactions are not classified by the number of reactants and products, but *all of these reactions are redox processes* because elemental oxygen is a reactant:

$$2CO(g) + O_2(g) \longrightarrow 2CO_2(g)$$

The combustion reactions that we commonly use to produce energy involve coal, petroleum, gasoline, natural gas, or wood as a reactant. These mixtures consist of substances with many C—C and C—H bonds, which break during the reaction, and each C and H atom combines with oxygen to form CO_2 and H_2O. The combustion of butane is typical:

$$2C_4H_{10}(g) + 13O_2(g) \longrightarrow 8CO_2(g) + 10H_2O(g)$$

Biological *respiration* is a multistep combustion process that occurs within our cells when we "burn" foodstuffs, such as glucose, for energy:

$$C_6H_{12}O_6(s) + 6O_2(g) \longrightarrow 6CO_2(g) + 6H_2O(g) + \text{energy}$$

Sample Problem 4.10 Identifying the Type of Redox Reaction

Problem Classify each of the following redox reactions as a combination, decomposition, or displacement reaction, write a balanced molecular equation for each, as well as total and net ionic equations for part (c), and identify the oxidizing and reducing agents:

(a) Magnesium(s) + nitrogen(g) ⟶ magnesium nitride(s)
(b) Hydrogen peroxide(l) ⟶ water + oxygen gas
(c) Aluminum(s) + lead(II) nitrate(aq) ⟶ aluminum nitrate(aq) + lead(s)

Plan To decide on reaction type, recall that combination reactions produce fewer products than reactants, decomposition reactions produce more products, and displacement reactions have the same number of reactants and products. The oxidation number (O.N.) becomes more positive for the reducing agent and less positive for the oxidizing agent.

Solution **(a)** Combination: two substances form one. This reaction occurs, along with formation of magnesium oxide, when magnesium burns in air, which is mostly N_2:

$$\overset{0}{3Mg}(s) + \overset{0}{N_2}(g) \longrightarrow \overset{+2}{Mg_3}\overset{-3}{N_2}(s)$$

Mg is the reducing agent; N_2 is the oxidizing agent.

(b) Decomposition: one substance forms two. Because hydrogen peroxide is very unstable and breaks down from heat, light, or just shaking, this reaction occurs within every bottle of this common household antiseptic:

$$2\overset{+1}{H_2}\overset{-1}{O_2}(l) \longrightarrow 2\overset{+1}{H_2}\overset{-2}{O}(l) + \overset{0}{O_2}(g)$$

H_2O_2 is both the oxidizing *and* the reducing agent. The O.N. of O in peroxides is −1. It is shown in blue *and* red because it both increases to 0 in O_2 and decreases to −2 in H_2O.

(c) Displacement: two substances form two others. As Figure 4.16 shows, Al is more active than Pb and, thus, displaces it from aqueous solution:

$$2\overset{0}{Al}(s) + 3\overset{+2}{Pb}(\overset{+5}{N}\overset{-2}{O_3})_2(aq) \longrightarrow 2\overset{+3}{Al}(\overset{+5}{N}\overset{-2}{O_3})_3(aq) + 3\overset{0}{Pb}(s)$$

Al is the reducing agent; $Pb(NO_3)_2$ is the oxidizing agent.

The total ionic equation is

$$2Al(s) + 3Pb^{2+}(aq) + 6NO_3^-(aq) \longrightarrow 2Al^{3+}(aq) + 6NO_3^-(aq) + 3Pb(s)$$

The net ionic equation is

$$2Al(s) + 3Pb^{2+}(aq) \longrightarrow 2Al^{3+}(aq) + 3Pb(s)$$

FOLLOW-UP PROBLEM 4.10 Classify each of the following redox reactions as a combination, decomposition, or displacement reaction, write a balanced molecular equation for each, as well as total and net ionic equations for parts (b) and (c), and identify the oxidizing and reducing agents:

(a) $S_8(s) + F_2(g) \longrightarrow SF_4(g)$
(b) $CsI(aq) + Cl_2(aq) \longrightarrow CsCl(aq) + I_2(aq)$
(c) $Ni(NO_3)_2(aq) + Cr(s) \longrightarrow Ni(s) + Cr(NO_3)_3(aq)$

Summary of Section 4.6

- A reaction that has an element as reactant or product is a redox reaction.
- In combination redox reactions, elements combine to form a compound, or a compound and an element combine.

- In decomposition redox reactions, compounds break down by absorption of heat or electricity into elements or into a compound and an element.
- In displacement redox reactions, one element displaces the ion of another from solution.
- Activity series rank elements in order of ability to displace each other. A more reactive metal can displace (reduce) hydrogen ion or the ion of a less reactive metal from solution. A more reactive halogen can displace (oxidize) the ion of a less reactive halogen from solution.
- Combustion releases heat through a redox reaction of a substance with O_2.

CHAPTER REVIEW GUIDE

The following sections provide many aids to help you study this chapter. (Numbers in parentheses refer to pages, unless noted otherwise.)

Learning Objectives

These are concepts and skills to review after studying this chapter.

Related section (§), sample problem (SP), and upcoming end-of-chapter problem (EP) numbers are listed in parentheses.

1. Understand how water dissolves an ionic compound compared to a covalent compound and which solution contains an electrolyte; use a compound's formula to find moles of ions in solution (§4.1) (SPs 4.1, 4.2) (EPs 4.1–4.19)
2. Understand the key events in precipitation and acid-base reactions and use ionic equations to describe them; distinguish between strong and weak acids and bases and calculate an unknown concentration from a titration (§4.2–4.4) (SPs 4.3–4.7) (EPs 4.20–4.47)
3. Understand the key event in the redox process; determine the oxidation number of any element in a compound; identify the oxidizing and reducing agents in a reaction (§4.5) (SPs 4.8, 4.9) (EPs 4.48–4.60)
4. Identify three important types of redox reactions that involve elements: combination, decomposition, displacement (§4.6) (SP 4.10) (EPs 4.61–4.75)

Key Terms

These important terms appear in boldface in the chapter and are defined again in the Glossary.

Section 4.1
polar molecule (116)
solvated (117)
electrolyte (117)
nonelectrolyte (120)

Section 4.2
molecular equation (120)
total ionic equation (120)
spectator ion (120)
net ionic equation (121)

Section 4.3
precipitation reaction (122)
precipitate (122)
metathesis reaction (123)

Section 4.4
acid-base (neutralization) reaction (126)
acid (126)
base (126)
salt (128)
titration (130)
equivalence point (130)
end point (130)

Section 4.5
oxidation-reduction (redox) reaction (132)
oxidation (133)
reduction (133)
oxidizing agent (133)
reducing agent (133)
oxidation number (O.N.) (or oxidation state) (133)

Section 4.6
activity series of the metals (138)

BRIEF SOLUTIONS TO FOLLOW-UP PROBLEMS

Compare your own solutions to these calculation steps and answers.

4.1 The compound is $BaCl_2$.

$$\text{Mass (g) of BaCl}_2 = 9\ \cancel{\text{particles}} \times \frac{0.05\ \cancel{\text{mol particles}}}{1\ \cancel{\text{particle}}} \times \frac{1\ \cancel{\text{mol BaCl}_2}}{3\ \cancel{\text{mol particles}}} \times \frac{208.2\ \text{g BaCl}_2}{1\ \cancel{\text{mol BaCl}_2}} = 31.2\ \text{g BaCl}_2$$

4.2 (a) $KClO_4(s) \xrightarrow{H_2O} K^+(aq) + ClO_4^-(aq)$;
2 mol of K^+ and 2 mol of ClO_4^-

(b) $Mg(C_2H_3O_2)_2(s) \xrightarrow{H_2O} Mg^{2+}(aq) + 2C_2H_3O_2^-(aq)$;
2.49 mol of Mg^{2+} and 4.97 mol of $C_2H_3O_2^-$

(c) $(NH_4)_2CrO_4(s) \xrightarrow{H_2O} 2NH_4^+(aq) + CrO_4^{2-}(aq)$;
6.24 mol of NH_4^+ and 3.12 mol of CrO_4^{2-}

(d) $NaHSO_4(s) \xrightarrow{H_2O} Na^+(aq) + HSO_4^-(aq)$;
0.73 mol of Na^+ and 0.73 mol of HSO_4^-

4.3 (a) $Fe^{3+}(aq) + 3Cl^-(aq) + 3Cs^+(aq) + PO_4^{3-}(aq) \longrightarrow FePO_4(s) + 3Cl^-(aq) + 3Cs^+(aq)$

$Fe^{3+}(aq) + PO_4^{3-}(aq) \longrightarrow FePO_4(s)$

(b) $2Na^+(aq) + 2OH^-(aq) + Cd^{2+}(aq) + 2NO_3^-(aq) \longrightarrow 2Na^+(aq) + 2NO_3^-(aq) + Cd(OH)_2(s)$

$2OH^-(aq) + Cd^{2+}(aq) \longrightarrow Cd(OH)_2(s)$

(c) No reaction occurs

(d) $2Ag^+(aq) + 2NO_3^-(aq) + Ba^{2+}(aq) + 2Cl^-(aq) \longrightarrow 2AgCl(s) + 2NO_3^-(aq) + Ba^{2+}(aq)$

$Ag^+(aq) + Cl^-(aq) \longrightarrow AgCl(s)$

BRIEF SOLUTIONS TO FOLLOW-UP PROBLEMS (continued)

4.4 (a) Beaker A contains a solution of $Zn(NO_3)_2$.
(b) Beaker B contains a solution of $Ba(OH)_2$.
(c) The precipitate is zinc hydroxide, and the spectator ions are Ba^{2+} and NO_3^-.

Molecular: $Zn(NO_3)_2(aq) + Ba(OH)_2(aq) \longrightarrow Zn(OH)_2(s) + Ba(NO_3)_2(aq)$

Total ionic:

$$Zn^{2+}(aq) + 2NO_3^-(aq) + Ba^{2+}(aq) + 2OH^-(aq) \longrightarrow Zn(OH)_2(s) + Ba^{2+}(aq) + 2NO_3^-(aq)$$

Net ionic: $Zn^{2+}(aq) + 2OH^-(aq) \longrightarrow Zn(OH)_2(s)$

(d) The OH^- ion is limiting.

Mass (g) of $Zn(OH)_2$

$$= 6\ \cancel{\text{OH}^-\text{ particles}} \times \frac{0.050\ \cancel{\text{mol OH}^-\text{ ions}}}{1\ \cancel{\text{OH}^-\text{ particle}}} \times \frac{1\ \cancel{\text{mol Zn(OH)}_2}}{2\ \cancel{\text{mol OH}^-\text{ ions}}} \times \frac{99.43\text{ g Zn(OH)}_2}{1\ \cancel{\text{mol Zn(OH)}_2}}$$

$$= 15\text{ g Zn(OH)}_2$$

4.5 No. of OH^- ions $= 451\ \cancel{\text{mL}} \times \frac{1\ \cancel{\text{L}}}{10^3\ \cancel{\text{mL}}} \times \frac{1.20\ \cancel{\text{mol KOH}}}{1\ \cancel{\text{L soln}}} \times \frac{1\ \cancel{\text{mol OH}^-}}{1\ \cancel{\text{mol KOH}}} \times \frac{6.022\times10^{23}\text{ OH}^-\text{ ions}}{1\ \cancel{\text{mol OH}^-}}$

$$= 3.26\times10^{23}\text{ OH}^-$$

4.6 $Ca(OH)_2(aq) + 2HNO_3(aq) \longrightarrow Ca(NO_3)_2(aq) + 2H_2O(l)$

$$Ca^{2+}(aq) + 2OH^-(aq) + 2H^+(aq) + 2NO_3^-(aq) \longrightarrow Ca^{2+}(aq) + 2NO_3^-(aq) + 2H_2O(l)$$

$H^+(aq) + OH^-(aq) \longrightarrow H_2O(l)$

4.7 $Ba(OH)_2(aq) + 2HCl(aq) \longrightarrow BaCl_2(aq) + 2H_2O(l)$

Volume (L) of soln

$$= 50.00\ \cancel{\text{mL HCl soln}} \times \frac{1\text{ L}}{10^3\ \cancel{\text{mL}}} \times \frac{0.1016\ \cancel{\text{mol HCl}}}{1\ \cancel{\text{L soln}}} \times \frac{1\ \cancel{\text{mol Ba(OH)}_2}}{2\ \cancel{\text{mol HCl}}} \times \frac{1\ \cancel{\text{L soln}}}{0.1292\ \cancel{\text{mol Ba(OH)}_2}}$$

$$= 0.01966\text{ L}$$

4.8 (a) O.N. of Sc = +3; O.N. of O = −2
(b) O.N. of Ga = +3; O.N. of Cl = −1
(c) O.N. of H = +1; O.N. of P = +5; O.N. of O = −2
(d) O.N. of I = +3; O.N. of F = −1

4.9 (a) Fe is the reducing agent; Cl_2 is the oxidizing agent.
(b) C_2H_6 is the reducing agent; O_2 is the oxidizing agent.
(c) CO is the reducing agent; I_2O_5 is the oxidizing agent.

4.10 (a) Combination: $S_8(s) + 16F_2(g) \longrightarrow 8SF_4(g)$
S_8 is the reducing agent; F_2 is the oxidizing agent.
(b) Displacement:

$2CsI(aq) + Cl_2(aq) \longrightarrow 2CsCl(aq) + I_2(aq)$

$2Cs^+(aq) + 2I^-(aq) + Cl_2(aq) \longrightarrow 2Cs^+(aq) + 2Cl^-(aq) + I_2(aq)$

$2I^-(aq) + Cl_2(aq) \longrightarrow 2Cl^-(aq) + I_2(aq)$

Cl_2 is the oxidizing agent; CsI is the reducing agent.
(c) Displacement:

$3Ni(NO_3)_2(aq) + 2Cr(s) \longrightarrow 3Ni(s) + 2Cr(NO_3)_3(aq)$

$3Ni^{2+}(aq) + 6NO_3^-(aq) + 2Cr(s) \longrightarrow 3Ni(s) + 2Cr^{3+}(aq) + 6NO_3^-(aq)$

$3Ni^{2+}(aq) + 2Cr(s) \longrightarrow 3Ni(s) + 2Cr^{3+}(aq)$

Cr is the reducing agent; $Ni(NO_3)_2$ is the oxidizing agent.

PROBLEMS

Problems with **colored** numbers are answered in Appendix E. Sections match the text and provide the numbers of relevant sample problems. Bracketed problems are grouped in pairs (indicated by a short rule) that cover the same concept. Comprehensive Problems are based on material from any section or previous chapter.

The Role of Water as a Solvent

(Sample Problems 4.1 and 4.2)

4.1 What two factors cause water to be polar?

4.2 What must be present in an aqueous solution for it to conduct an electric current? What general classes of compounds form solutions that conduct?

4.3 What occurs on the molecular level when an ionic compound dissolves in water?

4.4 Which of the following scenes best represents how the ions occur in an aqueous solution of: (a) $CaCl_2$; (b) Li_2SO_4; (c) NH_4Br?

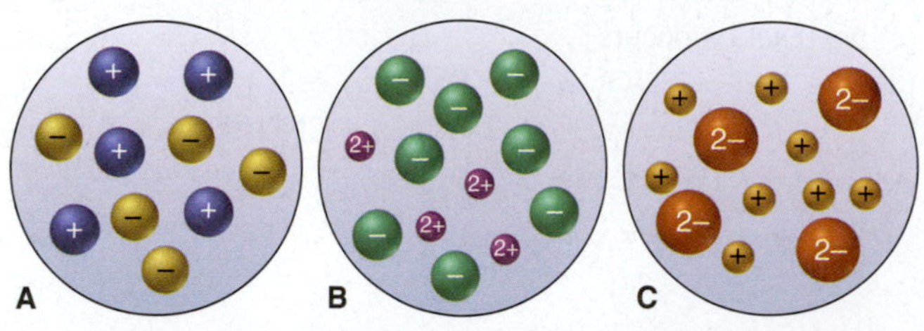

4.5 Which of the following scenes best represents a volume from a solution of magnesium nitrate?

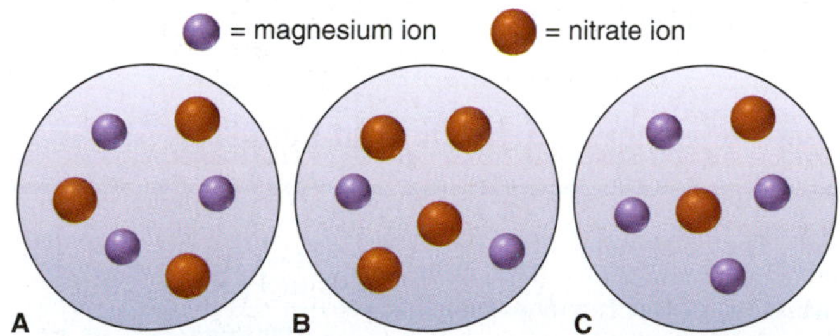

4.6 Why are some ionic compounds soluble in water and others are not?

4.7 Some covalent compounds dissociate into ions in water. What atom do these compounds have in their structures? What type of solution do they form? Name three examples of such a solution.

4.8 Is each of the following very soluble in water? Explain.
(a) Benzene, C_6H_6
(b) Sodium hydroxide
(c) Ethanol, CH_3CH_2OH
(d) Potassium acetate

4.9 Is each of the following very soluble in water? Explain.
(a) Lithium nitrate
(b) Glycine, H_2NCH_2COOH
(c) Pentane
(d) Ethylene glycol, $HOCH_2CH_2OH$

4.10 Does an aqueous solution of each of the following conduct an electric current? Explain.
(a) Cesium bromide (b) Hydrogen iodide

4.11 Does an aqueous solution of each of the following conduct an electric current? Explain.
(a) Potassium sulfate (b) Sucrose, $C_{12}H_{22}O_{11}$

4.12 How many total moles of ions are released when each of the following dissolves in water?
(a) 0.75 mol of K_3PO_4 (b) 6.88×10^{-3} g of $NiBr_2{\cdot}3H_2O$
(c) 2.23×10^{22} formula units of $FeCl_3$

4.13 How many total moles of ions are released when each of the following dissolves in water?
(a) 0.734 mol of Na_2HPO_4 (b) 3.86 g of $CuSO_4{\cdot}5H_2O$
(c) 8.66×10^{20} formula units of $NiCl_2$

4.14 How many moles and how many ions of each type are present in each of the following?
(a) 130. mL of 0.45 *M* aluminum chloride
(b) 9.80 mL of a solution containing 2.59 g lithium sulfate/L
(c) 245 mL of a solution containing 3.68×10^{22} formula units of potassium bromide per liter

4.15 How many moles and how many ions of each type are present in each of the following?
(a) 88 mL of 1.75 *M* magnesium chloride
(b) 321 mL of a solution containing 0.22 g aluminum sulfate/L
(c) 1.65 L of a solution containing 8.83×10^{21} formula units of cesium nitrate per liter

4.16 How many moles of H^+ ions are present in the following aqueous solutions?
(a) 1.40 L of 0.25 *M* perchloric acid
(b) 6.8 mL of 0.92 *M* nitric acid
(c) 2.6 L of 0.085 *M* hydrochloric acid

4.17 How many moles of H^+ ions are present in the following aqueous solutions?
(a) 1.4 mL of 0.75 *M* hydrobromic acid
(b) 2.47 mL of 1.98 *M* hydriodic acid
(c) 395 mL of 0.270 *M* nitric acid

4.18 To study a marine organism, a biologist prepares a 1.00-kg sample to simulate the ion concentrations in seawater. She mixes 26.5 g of NaCl, 2.40 g of $MgCl_2$, 3.35 g of $MgSO_4$, 1.20 g of $CaCl_2$, 1.05 g of KCl, 0.315 g of $NaHCO_3$, and 0.098 g of NaBr in distilled water. (a) If the density of the solution is 1.025 g/cm^3, what is the molarity of each ion? (b) What is the total molarity of alkali metal ions? (c) What is the total molarity of alkaline earth metal ions? (d) What is the total molarity of anions?

4.19 Water "softeners" remove metal ions such as Ca^{2+} and Fe^{3+} by replacing them with enough Na^+ ions to maintain the same number of positive charges in the solution. If 1.0×10^3 L of "hard" water is 0.015 *M* Ca^{2+} and 0.0010 *M* Fe^{3+}, how many moles of Na^+ are needed to replace these ions?

Writing Equations for Aqueous Ionic Reactions

4.20 Write two sets of equations (both molecular and total ionic) with different reactants that have the same net ionic equation as the following equation:

$$Ba(NO_3)_2(aq) + Na_2CO_3(aq) \longrightarrow BaCO_3(s) + 2NaNO_3(aq)$$

Precipitation Reactions

(Sample Problems 4.3 and 4.4)

4.21 Why do some pairs of ions precipitate and others do not?

4.22 Use Table 4.1 to determine which of the following combinations leads to a precipitation reaction. How can you identify the spectator ions in the reaction?
(a) Calcium nitrate(*aq*) + sodium chloride(*aq*) $\longrightarrow$
(b) Potassium chloride(*aq*) + lead(II) nitrate(*aq*) $\longrightarrow$

4.23 The beakers represent the aqueous reaction of $AgNO_3$ and NaCl. Silver ions are gray. What colors are used to represent NO_3^-, Na^+, and Cl^-? Write molecular, total ionic, and net ionic equations for the reaction.

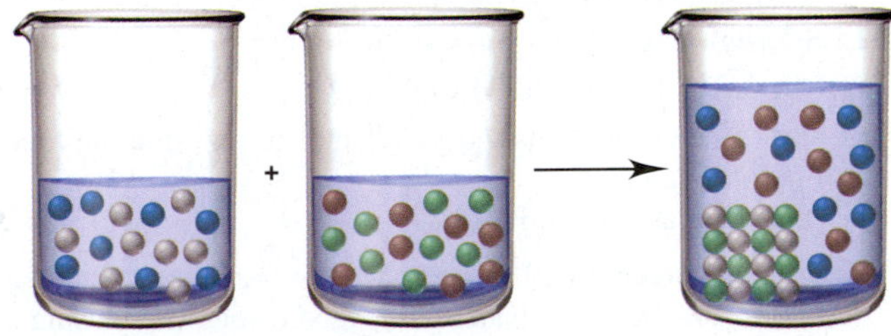

4.24 Complete the following precipitation reactions with balanced molecular, total ionic, and net ionic equations:
(a) $Hg_2(NO_3)_2(aq) + KI(aq) \longrightarrow$
(b) $FeSO_4(aq) + Sr(OH)_2(aq) \longrightarrow$

4.25 Complete the following precipitation reactions with balanced molecular, total ionic, and net ionic equations:
(a) $CaCl_2(aq) + Cs_3PO_4(aq) \longrightarrow$
(b) $Na_2S(aq) + ZnSO_4(aq) \longrightarrow$

4.26 When each of the following pairs of aqueous solutions is mixed, does a precipitation reaction occur? If so, write balanced molecular, total ionic, and net ionic equations:
(a) Sodium nitrate + copper(II) sulfate
(b) Ammonium bromide + silver nitrate

4.27 When each of the following pairs of aqueous solutions is mixed, does a precipitation reaction occur? If so, write balanced molecular, total ionic, and net ionic equations:
(a) Potassium carbonate + barium hydroxide
(b) Aluminum nitrate + sodium phosphate

4.28 If 38.5 mL of lead(II) nitrate solution reacts completely with excess sodium iodide solution to yield 0.628 g of precipitate, what is the molarity of lead(II) ion in the original solution?

4.29 If 25.0 mL of silver nitrate solution reacts with excess potassium chloride solution to yield 0.842 g of precipitate, what is the molarity of silver ion in the original solution?

4.30 With ions shown as spheres and solvent molecules omitted for clarity, the circle *(right)* illustrates the solid formed when a solution containing K^+, Mg^{2+}, Ag^+, or Pb^{2+} *(blue)* is mixed with one containing ClO_4^-, NO_3^-, or SO_4^{2-} *(yellow)*. (a) Identify the solid. (b) Write a balanced net ionic equation for the reaction. (c) If each sphere represents 5.0×10^{-4} mol of ion, what mass of product forms?

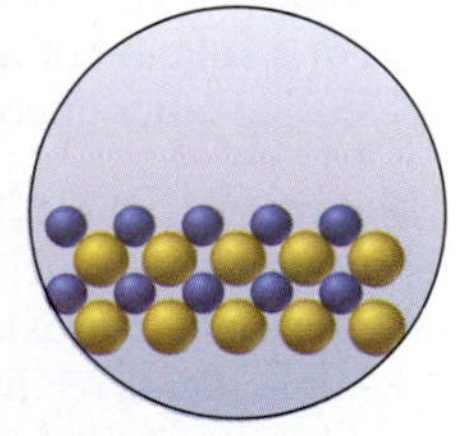

4.31 The precipitation reaction between 25.0 mL of a solution containing a cation *(purple)* and 35.0 mL of a solution containing an anion *(green)* is depicted below (with ions shown as spheres and solvent molecules omitted for clarity).

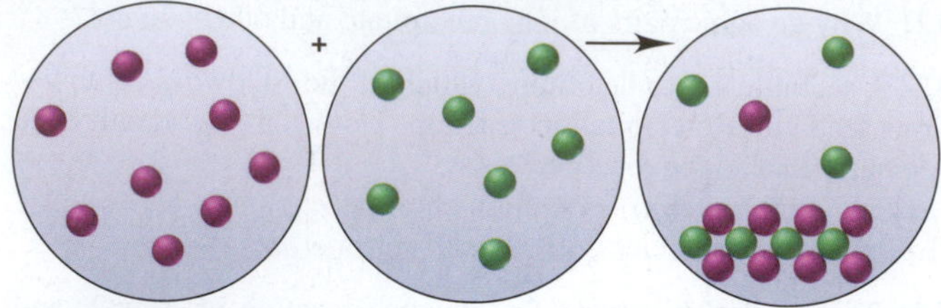

(a) Given the following choices of reactants, write balanced total ionic and net ionic equations that best represent the reaction:
(1) $KNO_3(aq) + CuCl_2(aq) \longrightarrow$
(2) $NaClO_4(aq) + CaCl_2(aq) \longrightarrow$
(3) $Li_2SO_4(aq) + AgNO_3(aq) \longrightarrow$
(4) $NH_4Br(aq) + Pb(CH_3COO)_2(aq) \longrightarrow$
(b) If each sphere represents 2.5×10^{-3} mol of ion, find the total number of ions present.
(c) What is the mass of solid formed?

4.32 The mass percent of Cl^- in a seawater sample is determined by titrating 25.00 mL of seawater with $AgNO_3$ solution, causing a precipitation reaction. An indicator is used to detect the end point, which occurs when free Ag^+ ion is present in solution after all the Cl^- has reacted. If 53.63 mL of 0.2970 *M* $AgNO_3$ is required to reach the end point, what is the mass percent of Cl^- in the seawater (*d* of seawater = 1.024 g/mL)?

4.33 Aluminum sulfate, known as *cake alum,* has a wide range of uses, from dyeing leather and cloth to purifying sewage. In aqueous solution, it reacts with base to form a white precipitate. (a) Write balanced total and net ionic equations for its reaction with aqueous NaOH. (b) What mass of precipitate forms when 185.5 mL of 0.533 *M* NaOH is added to 627 mL of a solution that contains 15.8 g of aluminum sulfate per liter?

Acid-Base Reactions

(Sample Problems 4.5 to 4.7)

4.34 Is the total ionic equation the same as the net ionic equation when $Sr(OH)_2(aq)$ and $H_2SO_4(aq)$ react? Explain.

4.35 (a) Name three common strong acids. (b) Name three common strong bases. (c) What is a characteristic behavior of a strong acid or a strong base?

4.36 (a) Name three common weak acids. (b) Name one common weak base. (c) What is the major difference between a weak acid and a strong acid or between a weak base and a strong base, and what experiment would you perform to observe it?

4.37 (a) The net ionic equation for the aqueous neutralization reaction between acetic acid and sodium hydroxide is different from that for the reaction between hydrochloric acid and sodium hydroxide. Explain by writing balanced net ionic equations. (b) For a solution of acetic acid in water, list the major species in decreasing order of concentration.

4.38 Complete the following acid-base reactions with balanced molecular, total ionic, and net ionic equations:
(a) Potassium hydroxide(*aq*) + hydrobromic acid(*aq*) $\longrightarrow$
(b) Ammonia(*aq*) + hydrochloric acid(*aq*) $\longrightarrow$

4.39 Complete the following acid-base reactions with balanced molecular, total ionic, and net ionic equations:
(a) Cesium hydroxide(*aq*) + nitric acid(*aq*) $\longrightarrow$
(b) Calcium hydroxide(*aq*) + acetic acid(*aq*) $\longrightarrow$

4.40 Limestone (calcium carbonate) is insoluble in water but dissolves when a hydrochloric acid solution is added. Write balanced total ionic and net ionic equations, showing hydrochloric acid as it actually exists in water and the reaction as a proton-transfer process.

4.41 Zinc hydroxide is insoluble in water but dissolves when a nitric acid solution is added. Why? Write balanced total ionic and net ionic equations, showing nitric acid as it actually exists in water and the reaction as a proton-transfer process.

4.42 If 25.98 mL of 0.1180 *M* KOH solution reacts with 52.50 mL of CH_3COOH solution, what is the molarity of the acid solution?

4.43 If 26.25 mL of 0.1850 *M* NaOH solution reacts with 25.00 mL of H_2SO_4, what is the molarity of the acid solution?

4.44 An auto mechanic spills 88 mL of 2.6 *M* H_2SO_4 solution from an auto battery. How many milliliters of 1.6 *M* $NaHCO_3$ must be poured on the spill to react completely with the sulfuric acid?

4.45 One of the first steps in the enrichment of uranium for use in nuclear power plants involves a displacement reaction between UO_2 and aqueous HF:

$$UO_2(s) + HF(aq) \longrightarrow UF_4(s) + H_2O(l) \text{ [unbalanced]}$$

How many liters of 2.40 *M* HF will react with 2.15 kg of UO_2?

4.46 An unknown amount of acid can often be determined by adding an excess of base and then "back-titrating" the excess. A 0.3471-g sample of a mixture of oxalic acid, which has two ionizable protons, and benzoic acid, which has one, is treated with 100.0 mL of 0.1000 *M* NaOH. The excess NaOH is titrated with 20.00 mL of 0.2000 *M* HCl. Find the mass % of benzoic acid.

4.47 A mixture of bases can sometimes be the active ingredient in antacid tablets. If 0.4826 g of a mixture of $Al(OH)_3$ and $Mg(OH)_2$ is neutralized with 17.30 mL of 1.000 *M* HNO_3, what is the mass % of $Al(OH)_3$ in the mixture?

Oxidation-Reduction (Redox) Reactions

(Sample Problems 4.8 to 4.9)

4.48 Why must every redox reaction involve an oxidizing agent and a reducing agent?

4.49 In which of the following equations does sulfuric acid act as an oxidizing agent? In which does it act as an acid? Explain.
(a) $4H^+(aq) + SO_4^{2-}(aq) + 2NaI(s) \longrightarrow 2Na^+(aq) + I_2(s) + SO_2(g) + 2H_2O(l)$
(b) $BaF_2(s) + 2H^+(aq) + SO_4^{2-}(aq) \longrightarrow 2HF(aq) + BaSO_4(s)$

4.50 Give the oxidation number of nitrogen in the following:
(a) NH_2OH (b) N_2F_4 (c) NH_4^+ (d) HNO_2

4.51 Give the oxidation number of sulfur in the following:
(a) $SOCl_2$ (b) H_2S_2 (c) H_2SO_3 (d) Na_2S

4.52 Give the oxidation number of arsenic in the following:
(a) AsH_3 (b) $H_2AsO_4^-$ (c) $AsCl_3$

4.53 Give the oxidation number of phosphorus in the following:
(a) $H_2P_2O_7^{2-}$ (b) PH_4^+ (c) PCl_5

4.54 Give the oxidation number of manganese in the following:
(a) MnO_4^{2-} (b) Mn_2O_3 (c) $KMnO_4$

4.55 Give the oxidation number of chromium in the following:
(a) CrO_3 (b) $Cr_2O_7^{2-}$ (c) $Cr_2(SO_4)_3$

4.56 Identify the oxidizing and reducing agents in the following:
(a) $5H_2C_2O_4(aq) + 2MnO_4^-(aq) + 6H^+(aq) \longrightarrow 2Mn^{2+}(aq) + 10CO_2(g) + 8H_2O(l)$

(b) $3Cu(s) + 8H^+(aq) + 2NO_3^-(aq) \longrightarrow$
$3Cu^{2+}(aq) + 2NO(g) + 4H_2O(l)$

4.57 Identify the oxidizing and reducing agents in the following:
(a) $Sn(s) + 2H^+(aq) \longrightarrow Sn^{2+}(aq) + H_2(g)$
(b) $2H^+(aq) + H_2O_2(aq) + 2Fe^{2+}(aq) \longrightarrow$
$2Fe^{3+}(aq) + 2H_2O(l)$

4.58 Identify the oxidizing and reducing agents in the following:
(a) $8H^+(aq) + 6Cl^-(aq) + Sn(s) + 4NO_3^-(aq) \longrightarrow$
$SnCl_6^{2-}(aq) + 4NO_2(g) + 4H_2O(l)$
(b) $2MnO_4^-(aq) + 10Cl^-(aq) + 16H^+(aq) \longrightarrow$
$5Cl_2(g) + 2Mn^{2+}(aq) + 8H_2O(l)$

4.59 Identify the oxidizing and reducing agents in the following:
(a) $8H^+(aq) + Cr_2O_7^{2-}(aq) + 3SO_3^{2-}(aq) \longrightarrow$
$2Cr^{3+}(aq) + 3SO_4^{2-}(aq) + 4H_2O(l)$
(b) $NO_3^-(aq) + 4Zn(s) + 7OH^-(aq) + 6H_2O(l) \longrightarrow$
$4Zn(OH)_4^{2-}(aq) + NH_3(aq)$

4.60 A person's blood alcohol (C_2H_5OH) level can be determined by titrating a sample of blood plasma with a potassium dichromate solution. The balanced equation is

$$16H^+(aq) + 2Cr_2O_7^{2-}(aq) + C_2H_5OH(aq) \longrightarrow 4Cr^{3+}(aq) + 2CO_2(g) + 11H_2O(l)$$

If 35.46 mL of 0.05961 M $Cr_2O_7^{2-}$ is required to titrate 28.00 g of plasma, what is the mass percent of alcohol in the blood?

Elements in Redox Reactions

(Sample Problem 4.10)

4.61 Which type of redox reaction leads to each of the following?
(a) An increase in the number of substances
(b) A decrease in the number of substances
(c) No change in the number of substances

4.62 Why do decomposition redox reactions typically have compounds as reactants, whereas combination redox and displacement redox reactions have one or more elements?

4.63 Which of the types of reactions discussed in Section 4.6 commonly produce more than one compound?

4.64 Balance each of the following redox reactions and classify it as a combination, decomposition, or displacement reaction:
(a) $Sb(s) + Cl_2(g) \longrightarrow SbCl_3(s)$
(b) $AsH_3(g) \longrightarrow As(s) + H_2(g)$
(c) $Zn(s) + Fe(NO_3)_2(aq) \longrightarrow Zn(NO_3)_2(aq) + Fe(s)$

4.65 Balance each of the following redox reactions and classify it as a combination, decomposition, or displacement reaction:
(a) $Mg(s) + H_2O(g) \longrightarrow Mg(OH)_2(s) + H_2(g)$
(b) $Cr(NO_3)_3(aq) + Al(s) \longrightarrow Al(NO_3)_3(aq) + Cr(s)$
(c) $PF_3(g) + F_2(g) \longrightarrow PF_5(g)$

4.66 Predict the product(s) and write a balanced equation for each of the following redox reactions:
(a) $N_2(g) + H_2(g) \longrightarrow$
(b) $NaClO_3(s) \xrightarrow{\Delta}$
(c) $Ba(s) + H_2O(l) \longrightarrow$

4.67 Predict the product(s) and write a balanced equation for each of the following redox reactions:
(a) $Fe(s) + HClO_4(aq) \longrightarrow$
(b) $S_8(s) + O_2(g) \longrightarrow$
(c) $BaCl_2(l) \xrightarrow{\text{electricity}}$

4.68 Predict the product(s) and write a balanced equation for each of the following redox reactions:
(a) Cesium + iodine $\longrightarrow$
(b) Aluminum + aqueous manganese(II) sulfate $\longrightarrow$
(c) Sulfur dioxide + oxygen $\longrightarrow$
(d) Butane and oxygen $\longrightarrow$
(e) Write a balanced net ionic equation for (b).

4.69 Predict the product(s) and write a balanced equation for each of the following redox reactions:
(a) Pentane (C_5H_{12}) + oxygen $\longrightarrow$
(b) Phosphorus trichloride + chlorine $\longrightarrow$
(c) Zinc + hydrobromic acid $\longrightarrow$
(d) Aqueous potassium iodide + bromine $\longrightarrow$
(e) Write a balanced net ionic equation for (d).

4.70 How many grams of O_2 can be prepared from the thermal decomposition of 4.27 kg of HgO? Name and calculate the mass (in kg) of the other product.

4.71 How many grams of chlorine gas can be produced from the electrolytic decomposition of 874 g of calcium chloride? Name and calculate the mass (in g) of the other product.

4.72 In a combination reaction, 1.62 g of lithium is mixed with 6.50 g of oxygen.
(a) Which reactant is present in excess?
(b) How many moles of product are formed?
(c) After reaction, how many grams of each reactant and product are present?

4.73 In a combination reaction, 2.22 g of magnesium is heated with 3.75 g of nitrogen.
(a) Which reactant is present in excess?
(b) How many moles of product are formed?
(c) After reaction, how many grams of each reactant and product are present?

4.74 A mixture of $CaCO_3$ and CaO weighing 0.693 g was heated to produce gaseous CO_2. After heating, the remaining solid weighed 0.508 g. Assuming all the $CaCO_3$ broke down to CaO and CO_2, calculate the mass percent of $CaCO_3$ in the original mixture.

4.75 Before arc welding was developed, a displacement reaction involving aluminum and iron(III) oxide was commonly used to produce molten iron (the thermite process). This reaction was used, for example, to connect sections of iron rails for train tracks. Calculate the mass of molten iron produced when 1.50 kg of aluminum reacts with 25.0 mol of iron(III) oxide.

Comprehensive Problems

4.76 Nutritional biochemists have known for decades that acidic foods cooked in cast-iron cookware can supply significant amounts of dietary iron (ferrous ion). (a) Write a balanced net ionic equation, with oxidation numbers, that supports this fact. (b) Measurements show an increase from 3.3 mg of iron to 49 mg of iron per $\frac{1}{2}$-cup (125-g) serving during the slow preparation of tomato sauce in a cast-iron pot. How many ferrous ions are present in a 26-oz (737-g) jar of the tomato sauce?

4.77 The brewing industry uses yeast to convert glucose to ethanol. The baking industry uses the carbon dioxide produced in the same reaction to make bread rise:

$$C_6H_{12}O_6(s) \xrightarrow{\text{yeast}} 2C_2H_5OH(l) + 2CO_2(g)$$

How many grams of ethanol can be produced from 100. g of glucose? What volume of CO_2 is produced? (Assume 1 mol of gas occupies 22.4 L at the conditions used.)

4.78 A chemical engineer determines the mass percent of iron in an ore sample by converting the Fe to Fe^{2+} in acid and then reacting the Fe^{2+} with MnO_4^-. A 1.1081-g sample was dissolved in acid, and it reacted completely with 39.32 mL of 0.03190 *M* $KMnO_4$. The balanced equation is

$$8H^+(aq) + 5Fe^{2+}(aq) + MnO_4^-(aq) \longrightarrow 5Fe^{3+}(aq) + Mn^{2+}(aq) + 4H_2O(l)$$

Calculate the mass percent of iron in the ore.

4.79 You are given solutions of HCl and NaOH and must determine their concentrations. You use 27.5 mL of NaOH to titrate 100. mL of HCl and 18.4 mL of NaOH to titrate 50.0 mL of 0.0782 *M* H_2SO_4. Find the unknown concentrations.

4.80 The flask *(right)* represents the products of the titration of 25 mL of sulfuric acid with 25 mL of sodium hydroxide.
(a) Write balanced molecular, total ionic, and net ionic equations for the reaction.
(b) If each orange sphere represents 0.010 mol of sulfate ion, how many moles of acid and of base reacted?
(c) What are the molarities of the acid and the base?

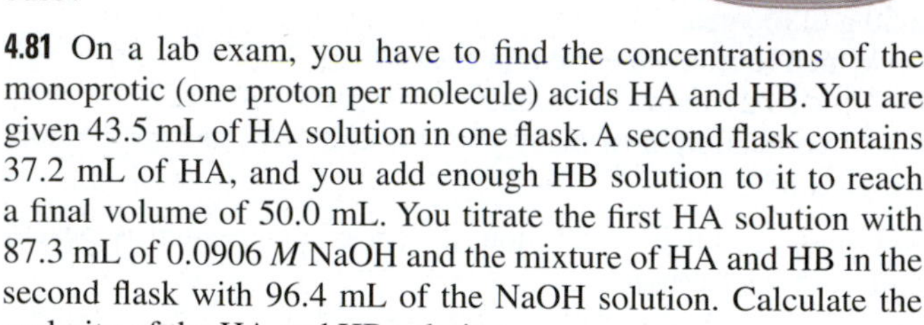

4.81 On a lab exam, you have to find the concentrations of the monoprotic (one proton per molecule) acids HA and HB. You are given 43.5 mL of HA solution in one flask. A second flask contains 37.2 mL of HA, and you add enough HB solution to it to reach a final volume of 50.0 mL. You titrate the first HA solution with 87.3 mL of 0.0906 *M* NaOH and the mixture of HA and HB in the second flask with 96.4 mL of the NaOH solution. Calculate the molarity of the HA and HB solutions.

4.82 Nitric acid, a major industrial and laboratory acid, is produced commercially by the multistep Ostwald process, which begins with the oxidation of ammonia:

Step 1. $4NH_3(g) + 5O_2(g) \longrightarrow 4NO(g) + 6H_2O(l)$
Step 2. $2NO(g) + O_2(g) \longrightarrow 2NO_2(g)$
Step 3. $3NO_2(g) + H_2O(l) \longrightarrow 2HNO_3(l) + NO(g)$

(a) What are the oxidizing and reducing agents in each step?
(b) Assuming 100% yield in each step, what mass (in kg) of ammonia must be used to produce 3.0×10^4 kg of HNO_3?

4.83 For the following aqueous reactions, complete and balance the molecular equation and write a net ionic equation:
(a) Manganese(II) sulfide + hydrobromic acid
(b) Potassium carbonate + strontium nitrate
(c) Potassium nitrite + hydrochloric acid
(d) Calcium hydroxide + nitric acid
(e) Barium acetate + iron(II) sulfate
(f) Barium hydroxide + hydrocyanic acid
(g) Copper(II) nitrate + hydrosulfuric acid
(h) Magnesium hydroxide + chloric acid

4.84 Various data can be used to find the composition of an alloy (a metallic mixture). Show that calculating the mass % of Mg in a magnesium-aluminum alloy (d = 2.40 g/cm^3) using each of the following pieces of data gives the same answer (within rounding): (a) a sample of the alloy has a mass of 0.263 g (d of Mg = 1.74 g/cm^3; d of Al = 2.70 g/cm^3); (b) an identical sample reacting with excess aqueous HCl forms 1.38×10^{-2} mol of H_2; (c) an identical sample reacting with excess O_2 forms 0.483 g of oxide.

4.85 Sodium peroxide (Na_2O_2) is often used in self-contained breathing devices, such as those used in fire emergencies, because it reacts with exhaled CO_2 to form Na_2CO_3 and O_2. How many liters of respired air can react with 80.0 g of Na_2O_2 if each liter of respired air contains 0.0720 g of CO_2?

4.86 Magnesium is used in airplane bodies and other lightweight alloys. The metal is obtained from seawater in a process that includes precipitation, neutralization, evaporation, and electrolysis. How many kilograms of magnesium can be obtained from 1.00 km^3 of seawater if the initial Mg^{2+} concentration is 0.13% by mass (d of seawater = 1.04 g/mL)?

4.87 Physicians who specialize in sports medicine routinely treat athletes and dancers. Ethyl chloride, a local anesthetic commonly used for simple injuries, is the product of the combination of ethylene with hydrogen chloride:

$$C_2H_4(g) + HCl(g) \longrightarrow C_2H_5Cl(g)$$

If 0.100 kg of C_2H_4 and 0.100 kg of HCl react: (a) How many molecules of gas (reactants plus products) are present when the reaction is complete? (b) How many moles of gas are present when half the product forms?

4.88 Carbon dioxide is removed from the atmosphere of space capsules by reaction with a solid metal hydroxide. The products are water and the metal carbonate.
(a) Calculate the mass of CO_2 that can be removed by reaction with 3.50 kg of lithium hydroxide.
(b) How many grams of CO_2 can be removed by 1.00 g of each of the following: lithium hydroxide, magnesium hydroxide, and aluminum hydroxide?

4.89 Calcium dihydrogen phosphate, $Ca(H_2PO_4)_2$, and sodium hydrogen carbonate, $NaHCO_3$, are ingredients of baking powder that react to produce CO_2, which causes dough or batter to rise:

$$Ca(H_2PO_4)_2(s) + NaHCO_3(s) \longrightarrow CO_2(g) + H_2O(g) + CaHPO_4(s) + Na_2HPO_4(s)$$
[unbalanced]

If the baking powder contains 31% $NaHCO_3$ and 35% $Ca(H_2PO_4)_2$ by mass:
(a) How many moles of CO_2 are produced from 1.00 g of baking powder?
(b) If 1 mol of CO_2 occupies 37.0 L at 350°F (a typical baking temperature), what volume of CO_2 is produced from 1.00 g of baking powder?

4.90 In a titration of HNO_3, you add a few drops of phenolphthalein indicator to 50.00 mL of acid in a flask. You quickly add 20.00 mL of 0.0502 *M* NaOH but overshoot the end point, and the solution turns deep pink. Instead of starting over, you add 30.00 mL of the acid, and the solution turns colorless. Then, it takes 3.22 mL of the NaOH to reach the end point.
(a) What is the concentration of the HNO_3 solution?
(b) How many moles of NaOH were in excess after the first addition?

4.91 The active compound in Pepto-Bismol contains C, H, O, and Bi.
(a) When 0.22105 g of it was burned in excess O_2, 0.1422 g of bismuth(III) oxide, 0.1880 g of carbon dioxide, and 0.02750 g of water were formed. What is the empirical formula of this compound?
(b) Given a molar mass of 1086 g/mol, determine the molecular formula.

(c) Complete and balance the acid-base reaction between bismuth(III) hydroxide and salicylic acid ($HC_7H_5O_3$), which is used to form this compound.
(d) A dose of Pepto-Bismol contains 0.600 mg of active ingredient. If the yield of the reaction in part (c) is 88.0%, what mass (in mg) of bismuth(III) hydroxide is required to prepare one dose?

4.92 Two aqueous solutions contain the ions indicated below.

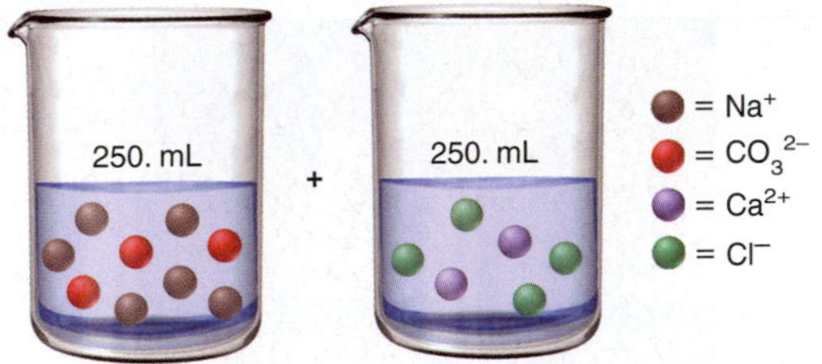

(a) Write balanced molecular, total ionic, and net ionic equations for the reaction that occurs when the solutions are mixed.
(b) If each sphere represents 0.050 mol of ion, what mass (in g) of precipitate forms, assuming 100% yield?
(c) What is the concentration of each ion in solution after reaction?

4.93 In 1997 and 2009, at United Nations conferences on climate change, many nations agreed to expand their research efforts to develop renewable sources of carbon-based fuels. For more than a quarter century, Brazil has been engaged in a program to replace gasoline with ethanol derived from the root crop manioc (cassava).
(a) Write separate balanced equations for the complete combustion of ethanol (C_2H_5OH) and of gasoline (represented by the formula C_8H_{18}).
(b) What mass of oxygen is required to burn completely 1.00 L of a mixture that is 90.0% gasoline ($d = 0.742$ g/mL) and 10.0% ethanol ($d = 0.789$ g/mL) by volume?
(c) If 1.00 mol of O_2 occupies 22.4 L, what volume of O_2 is needed to burn 1.00 L of the mixture?
(d) Air is 20.9% O_2 by volume. What volume of air is needed to burn 1.00 L of the mixture?

4.94 In a car engine, gasoline (represented by C_8H_{18}) does not burn completely, and some CO, a toxic pollutant, forms along with CO_2 and H_2O. If 5.0% of the gasoline forms CO:
(a) What is the ratio of CO_2 to CO molecules in the exhaust?
(b) What is the mass ratio of CO_2 to CO?
(c) What percentage of the gasoline must form CO for the mass ratio of CO_2 to CO to be exactly 1/1?

4.95 The amount of ascorbic acid (vitamin C; $C_6H_8O_6$) in tablets is determined by reaction with bromine and then titration of the hydrobromic acid with standard base:

$$C_6H_8O_6(aq) + Br_2(aq) \longrightarrow C_6H_6O_6(aq) + 2HBr(aq)$$
$$HBr(aq) + NaOH(aq) \longrightarrow NaBr(aq) + H_2O(l)$$

A certain tablet is advertised as containing 500 mg of vitamin C. One tablet was dissolved in water and reacted with Br_2. The solution was then titrated with 43.20 mL of 0.1350 *M* NaOH. Did the tablet contain the advertised quantity of vitamin C?

4.96 In the process of *pickling,* rust is removed from newly produced steel by washing the steel in hydrochloric acid:
(1) $6HCl(aq) + Fe_2O_3(s) \longrightarrow 2FeCl_3(aq) + 3H_2O(l)$
During the process, some iron is lost as well:
(2) $2HCl(aq) + Fe(s) \longrightarrow FeCl_2(aq) + H_2(g)$
(a) Which reaction, if either, is a redox process? (b) If reaction 2 did not occur and all the HCl were used, how many grams of Fe_2O_3 could be removed and $FeCl_3$ produced in a 2.50×10^3-L bath of 3.00 *M* HCl? (c) If reaction 1 did not occur and all the HCl were used, how many grams of Fe could be lost and $FeCl_2$ produced in a 2.50×10^3-L bath of 3.00 *M* HCl? (d) If 0.280 g of Fe is lost per gram of Fe_2O_3 removed, what is the mass ratio of $FeCl_2$ to $FeCl_3$?

4.97 At liftoff, a space shuttle uses a solid mixture of ammonium perchlorate and aluminum powder to obtain great thrust from the volume change of solid to gas. In the presence of a catalyst, the mixture forms solid aluminum oxide and aluminum trichloride and gaseous water and nitrogen monoxide.
(a) Write a balanced equation for the reaction, and identify the reducing and oxidizing agents.
(b) How many total moles of gas (water vapor and nitrogen monoxide) are produced when 50.0 kg of ammonium perchlorate reacts with a stoichiometric amount of Al?
(c) What is the change in volume from this reaction? (*d* of $NH_4ClO_4 = 1.95$ g/cc, Al = 2.70 g/cc, $Al_2O_3 = 3.97$ g/cc, and $AlCl_3 = 2.44$ g/cc; assume 1 mol of gas occupies 22.4 L.)

Gases and the Kinetic-Molecular Theory

Key Principles
to focus on while studying this chapter

- The physical properties of gases differ significantly from those of liquids and solids because gas particles are much farther apart. ***(Section 5.1)***
- *Pressure* is a force acting on an area; the atmosphere's gases exert a pressure on Earth's surface that is measured with a *barometer.* ***(Section 5.2)***
- Four gas variables—*volume* (V), *pressure* (P), *temperature* (T), and *amount* (n)—are *interdependent.* For a hypothetical *ideal gas,* volume changes *linearly* with a change in any one of the other variables, as long as the remaining two are held constant. These behaviors are described by gas laws *(Boyle's, Charles's, and Avogadro's),* which are combined into the *ideal gas law* ($PV = nRT$). Most simple gases behave ideally at ordinary pressures and temperatures. ***(Section 5.3)***
- Rearrangements of the ideal gas law are used to calculate the *density* and *molar mass* of a gas and the *partial pressure* of each gas in a gas mixture *(Dalton's law).* We use gas variables (P, V, and T) in stoichiometry problems to find the amounts (n) of gaseous reactants or products in a reaction. ***(Section 5.4)***
- To explain the behavior of gases, the *kinetic-molecular theory* postulates that an ideal gas consists of *points of mass moving in straight lines between elastic collisions* (no loss of energy). A key result of the theory is that, at a given temperature, the particles of a gas have a range of speeds, but all gases have the same average kinetic energy. Thus, temperature is a measure of *molecular motion,* as given by the average kinetic energy of gas particles. ***(Section 5.5)***
- The theory also predicts that heavier gas particles move more slowly on average than lighter ones, and, thus, two gases *effuse* (move through a tiny hole into a vacuum) or *diffuse* (move through one another) at rates inversely proportional to the square roots of their molar masses *(Graham's law).* ***(Section 5.5)***
- At extremely low temperature and high pressure, *real gas behavior deviates* from ideal behavior because the actual volume of the gas particles and the attractions (and repulsions) they experience during collisions become important factors. To account for real gas behavior, the ideal gas law is revised to the more accurate *van der Waals equation.* ***(Section 5.6)***

The Power of Expanding Gas At 180°C, the bit of water inside a kernel of popcorn vaporizes, and the gas pressure reaches more than nine times that of the atmosphere. The hull ruptures, and the corn's starch and proteins form an expanded foamy mass.

Outline

People have been studying the behavior of gases and the other states of matter throughout history; in fact, three of the four "elements" proposed by the ancient Greeks were air (gas), water (liquid), and earth (solid). Yet, despite millennia of observations, many questions remain. In this chapter and its companion, Chapter 12, we examine the physical states and their interrelations. Here, we highlight the gaseous state, the one we understand best. However, we'll put aside the *chemical* behavior unique to each specific gas and focus instead *on the* ***physical*** *behavior common to all gases.* For instance, although the particular gases differ, the same physical behaviors underlie the operation of a car and the baking of bread, the thrust of a rocket engine and the explosion of a kernel of popcorn, the process of breathing and the creation of thunder.

CONCEPTS & SKILLS TO REVIEW
before studying this chapter

- physical states of matter (Section 1.1)
- SI unit conversions (Section 1.4)
- amount-mass-number conversions (Section 3.1)

5.1 • AN OVERVIEW OF THE PHYSICAL STATES OF MATTER

Most substances can exist as a solid, a liquid, or a gas under appropriate conditions of pressure and temperature. In Chapter 1, we used the relative position and motion of the particles of a substance to distinguish how each state fills a container (see Figure 1.2). Recall that:

- A gas adopts the container shape and fills it, because its particles are far apart and move randomly.
- A liquid adopts the container shape to the extent of its volume and forms an upper surface because its particles are close together but free to move around each other.
- A solid has a fixed shape regardless of the container shape, because its particles are close together and held rigidly in place.

Figure 5.1 focuses on the three states of bromine.

Several other aspects of their behaviors distinguish gases from liquids and solids:

1. *Gas volume changes significantly with pressure.* When a sample of gas is confined to a container of variable volume, such as a cylinder with a piston, *increasing* the force on the piston *decreases* the gas volume. Removing the force allows the volume to

Gas: *Particles are far apart, move freely, and fill the available space.*

Liquid: *Particles are close together but move around one another.*

Solid: *Particles are close together in a regular array and do not move around one another.*

Figure 5.1 The three states of matter. Many substances, such as bromine (Br_2), can exist under appropriate conditions of pressure and temperature as a gas, liquid, or solid.

increase again. Gases under pressure can do work: compressed air in a jackhammer breaks rock and cement, and in tires it lifts the weight of a car. In contrast, the volume of a liquid or a solid does not change significantly under pressure.

2. *Gas volume changes significantly with temperature.* When a sample of gas at constant pressure is heated, it expands; when it is cooled, it shrinks. This volume change is 50 to 100 times greater for gases than for liquids or solids. The expansion that occurs when gases are rapidly heated can lift a rocket or pop corn.
3. *Gases flow very freely.* Gases flow much more freely than liquids and solids. This behavior allows gases to be transported more easily through pipes, but it also means they leak more rapidly out of small holes and cracks.
4. *Gases have relatively low densities.* Gas density is usually measured in units of grams per liter (g/L), whereas liquid and solid densities are in grams per milliliter (g/mL), about 1000 times as dense. For example, at 20°C and normal atmospheric pressure, the density of $O_2(g)$ is 1.3 g/**L**, whereas the density of $H_2O(l)$ is 1.0 g/**mL** and the density of NaCl(*s*) is 2.2 g/**mL.** When a gas cools, its density *increases* because its volume *decreases:* on cooling from 20°C to 0°C, the density of $O_2(g)$ increases from 1.3 to 1.4 g/L.
5. *Gases form a solution in any proportions.* Air is a solution of 18 gases. Two liquids, however, may or may not form a solution: water and ethanol do, but water and gasoline do not. Two solids generally do not form a solution unless they are melted, mixed, and then allowed to solidify (as when the alloy bronze is made from copper and tin).

These macroscopic properties arise because the particles in a gas are much farther apart than those in a liquid or a solid.

Summary of Section 5.1

- The volume of a gas can be altered significantly by changing the applied force or the temperature. Corresponding changes for liquids and solids are *much* smaller.
- Gases flow more freely and have much lower densities than liquids and solids.
- Gases mix in any proportions to form solutions; liquids and solids generally do not.
- Differences in the physical states are due to the greater average distance between particles in a gas than in a liquid or a solid.

5.2 • GAS PRESSURE AND ITS MEASUREMENT

You can blow up a balloon or pump up a tire because *a gas exerts pressure on the walls of its container.* **Pressure (*P*)** is defined as the force exerted per unit of surface area:

$$\text{Pressure} = \frac{\text{force}}{\text{area}}$$

Earth's gravity attracts the atmospheric gases, and they exert a force uniformly on *all* surfaces. The force, or *weight,* of these gases creates a pressure of about 14.7 pounds per square inch (lb/in^2; psi) of surface. Thus, a pressure of 14.7 lb/in^2 exists on the outside of your room (or your body), and it equals the pressure on the inside. What would happen if the pressures were *not* equal? Consider the empty can attached to a vacuum pump in Figure 5.2. With the pump off *(left)* the can maintains its shape because the pressure on the outside is equal to the pressure on the inside. With the pump on *(right),* nearly all of the air inside is removed, decreasing the internal pressure greatly, and the pressure of the atmosphere easily crushes the can. The vacuum-filtration flasks and tubing that you may have used in the lab have thick walls that withstand the enormous difference in pressure that arises as the flask is evacuated.

Measuring Atmospheric Pressure

The **barometer** is used to measure atmospheric pressure. The device is still essentially the same as it was when invented in 1643 by the Italian physicist Evangelista Torricelli: a tube about 1 m long, closed at one end, filled with mercury (Hg), and inverted into a dish containing more mercury. When the tube is inverted, some of the mercury

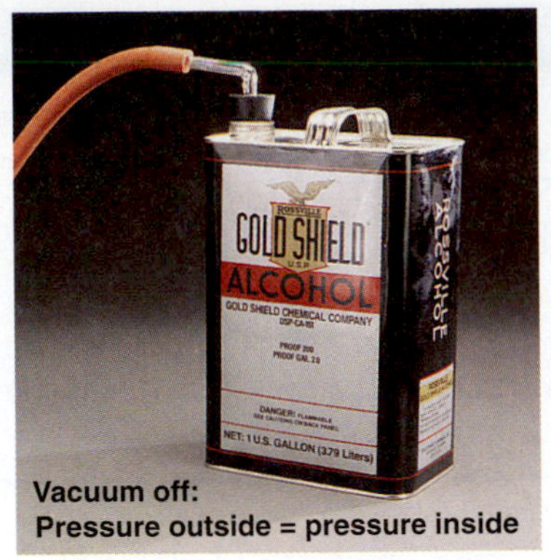

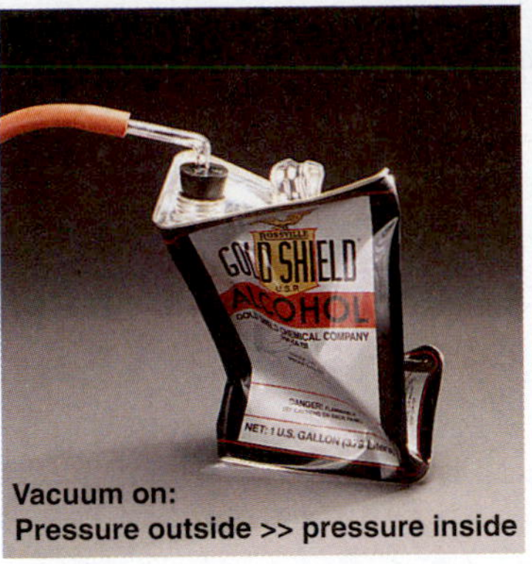

Figure 5.2 Effect of atmospheric pressure when it is exerted equally on inner and outer surfaces *(left)* and when the internal pressure is greatly decreased *(right)*.

flows out into the dish, and a vacuum forms above the mercury remaining in the tube (Figure 5.3). At sea level, under ordinary atmospheric conditions, the mercury stops flowing out when the surface of the mercury in the tube is about 760 mm above the surface of the mercury in the dish. At that height, the column of mercury exerts the same pressure (weight/area) on the mercury surface in the dish as the atmosphere does: $P_{Hg} = P_{atm}$. Likewise, if you evacuate a closed tube and invert it into a dish of mercury, the atmosphere pushes the mercury up to a height of about 760 mm.

Notice that we did not specify the diameter of the barometer tube. If the mercury in a 1-cm diameter tube rises to a height of 760 mm, the mercury in a 2-cm diameter tube will rise to that height also. The *weight* of mercury is greater in the wider tube, but so is the area; thus, the *pressure,* the *ratio* of weight to area, is the same.

Because the pressure of the mercury column is directly proportional to its height, a unit commonly used for pressure is millimeters of mercury (mmHg). We discuss other units of pressure shortly. *At sea level and 0°C, normal atmospheric pressure is 760 mmHg;* at the top of Mt. Everest (elevation 29,028 ft, or 8848 m), the atmospheric pressure is only about 270 mmHg. Thus, *pressure decreases with altitude:* the column of air above the sea is taller, so it weighs more than the column of air above Mt. Everest.

Laboratory barometers contain mercury because its high density allows these instruments to be a convenient size. If a barometer contained water instead, it would have to be more than 34 ft high, because the pressure of the atmosphere equals the pressure of a column of water about 10,300 mm (almost 34 ft) high. For a given pressure, the ratio of heights (h) of the liquid columns is inversely related to the ratio of the densities (d) of the liquids:

$$\frac{h_{H_2O}}{h_{Hg}} = \frac{d_{Hg}}{d_{H_2O}}$$

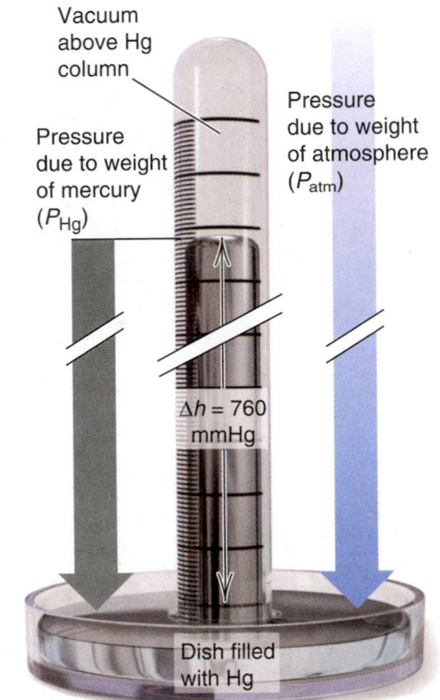

Figure 5.3 A mercury barometer. The pressure of the atmosphere, P_{atm}, balances the pressure of the mercury column, P_{Hg}.

Units of Pressure

Pressure results from a force exerted on an area. The SI unit of force is the newton (N): 1 N = 1 kg·m/s^2 (about the weight of an apple). The SI unit of pressure is the **pascal (Pa),** which equals a force of one newton exerted on an area of one square meter:

$$1 \text{ Pa} = 1 \text{ N/m}^2$$

A much larger unit is the **standard atmosphere (atm),** the average atmospheric pressure measured at sea level and 0°C. It is defined in terms of the pascal:

$$1 \text{ atm} = 101.325 \text{ kilopascals (kPa)} = 1.01325\times10^5 \text{ Pa}$$

Another common unit is the **millimeter of mercury (mmHg),** mentioned earlier; in honor of Torricelli, this unit has been renamed the **torr:**

$$1 \text{ torr} = 1 \text{ mmHg} = \frac{1}{760} \text{ atm} = \frac{101.325}{760} \text{ kPa} = 133.322 \text{ Pa}$$

The *bar* is coming into more common use in chemistry:

$$1 \text{ bar} = 1\times10^2 \text{ kPa} = 1\times10^5 \text{ Pa}$$

Table 5.1 Common Units of Pressure

Unit	Normal Atmospheric Pressure at Sea Level and 0°C
pascal (Pa); kilopascal (kPa)	1.01325×10^5 Pa; 101.325 kPa
atmosphere (atm)	1 atm*
millimeters of mercury (mmHg)	760 mmHg*
torr	760 torr*
pounds per square inch (lb/in^2 or psi)	14.7 lb/in^2
bar	1.01325 bar

*This is an exact quantity; in calculations, we use as many significant figures as necessary.

Despite a gradual change to SI units, many chemists still express pressure in torrs and atmospheres, so those units are used in this book, with reference to pascals and bars. Table 5.1 lists some important pressure units with the corresponding values for normal atmospheric pressure.

Sample Problem 5.1 Converting Units of Pressure

Problem A geochemist heats a limestone ($CaCO_3$) sample and collects the CO_2 released in an evacuated flask. After the system comes to room temperature, Δh = 291.4 mmHg. Calculate the CO_2 pressure in torrs, atmospheres, and kilopascals.

Plan The CO_2 pressure is given in units of mmHg, so we construct conversion factors from Table 5.1 to find the pressure in the other units.

Solution Converting from mmHg to torr:

$$P_{CO_2}\text{ (torr)} = 291.4\ \cancel{\text{mmHg}} \times \frac{1\text{ torr}}{1\ \cancel{\text{mmHg}}} = 291.4\text{ torr}$$

Converting from torr to atm:

$$P_{CO_2}\text{ (atm)} = 291.4\ \cancel{\text{torr}} \times \frac{1\text{ atm}}{760\ \cancel{\text{torr}}} = 0.3834\text{ atm}$$

Converting from atm to kPa:

$$P_{CO_2}\text{ (kPa)} = 0.3834\ \cancel{\text{atm}} \times \frac{101.325\text{ kPa}}{1\ \cancel{\text{atm}}} = 38.85\text{ kPa}$$

Check There are 760 torr in 1 atm, so ~300 torr should be <0.5 atm. There are ~100 kPa in 1 atm, so <0.5 atm should be <50 kPa.

Comment **1.** In the conversion from torr to atm, we retained four significant figures because the conversion factor (1 atm = 760 torr) is an *exact* number and has as many significant figures as required (see the footnote on Table 5.1).
2. From here on, except in complex situations, *unit canceling will no longer be shown in equations.*

FOLLOW-UP PROBLEM 5.1 The CO_2 released from another mineral sample is collected in an evacuated flask, and the measured P_{CO_2} is 579.6 torr. What is this pressure in torrs, pascals, and lb/in^2?

Summary of Section 5.2

- Gases exert pressure (force/area) on all surfaces they contact.
- A barometer measures atmospheric pressure based on the height of a mercury column that the atmosphere can support (760 mmHg at sea level and 0°C).
- Pressure units include the atmosphere (atm), torr (identical to mmHg), and pascal (Pa, the SI unit).

5.3 • THE GAS LAWS AND THEIR EXPERIMENTAL FOUNDATIONS

The physical behavior of a sample of gas can be described completely by four variables: pressure (P), volume (V), temperature (T), and amount (number of moles, n). These variables are interdependent, which means that *any one of them can be determined by measuring the other three.* Three key relationships exist among the four gas variables—Boyle's, Charles's, and Avogadro's laws. Each of these *gas laws expresses the effect of one variable on another, with the remaining two variables held constant.* Because volume is so easy to measure, the laws are expressed as the effect on gas volume of a change in the pressure, temperature, or amount of the gas.

The individual gas laws are special cases of a unifying relationship called the *ideal gas law,* which quantitatively describes the behavior of an **ideal gas,** one that exhibits linear relationships among volume, pressure, temperature, and amount. Although *no ideal gas actually exists,* most simple gases behave nearly ideally at ordinary temperatures and pressures. We discuss the ideal gas law after the three individual laws.

The Relationship Between Volume and Pressure: Boyle's Law

Following Torricelli's invention of the barometer, the great 17th-century English chemist Robert Boyle studied the effect of pressure on the volume of a sample of gas.

1. *The experiment.* Figure 5.4 illustrates the setup Boyle might have used in his experiments (parts A and B), the data he might have collected (part C), and graphs of the data (parts D and E). Boyle sealed the shorter leg of a J-shaped glass tube and poured mercury into the longer open leg, thereby trapping some air (the gas in the experiment) in the shorter leg. He calculated the gas volume (V_{gas}) from the height of the trapped air and the diameter of the tube. The total pressure, P_{total}, applied to the trapped gas is the pressure of the atmosphere, P_{atm} (760 mm, measured with a barometer), plus the difference in the heights of the mercury columns (Δh) in the two legs of the J tube, 20 mm (Figure 5.4A); thus, P_{total} is 780 torr. By adding mercury, Boyle increased P_{total}, and the gas volume decreased. In Figure 5.4B, more mercury has been added to the longer leg of the tube, increasing Δh to 800 mm, so P_{total} doubles to 1560 torr; note that V_{gas} is halved from 20 mL to 10 mL. In this way, by keeping the temperature and amount of gas constant, Boyle was able to measure the effect of the applied pressure on gas volume.

Note the following results in Figure 5.4:

- The product of corresponding P and V values is a constant (part C, rightmost column).
- V is *inversely* proportional to P (part D).
- V is *directly* proportional to $1/P$ (part E), and a plot of V versus $1/P$ is linear. This *linear relationship between two gas variables* is a hallmark of ideal gas behavior.

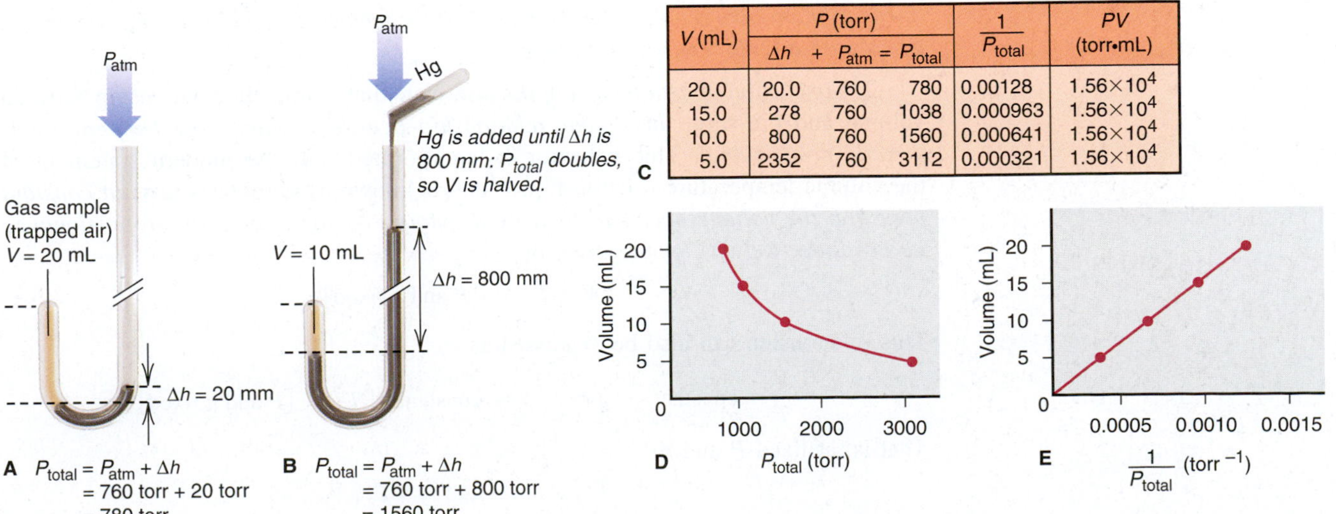

V (mL)	P (torr) Δh	+ P_{atm}	= P_{total}	$\frac{1}{P_{total}}$	PV (torr•mL)
20.0	20.0	760	780	0.00128	1.56×10^4
15.0	278	760	1038	0.000963	1.56×10^4
10.0	800	760	1560	0.000641	1.56×10^4
5.0	2352	760	3112	0.000321	1.56×10^4

Figure 5.4 Boyle's law, the relationship between the volume and pressure of a gas.

2. *Conclusion and statement of the law.* The generalization of Boyle's observations is known as **Boyle's law:** *at constant temperature, the volume occupied by a fixed amount of gas is* ***inversely*** *proportional to the applied (external) pressure,* or

$$V \propto \frac{1}{P} \qquad [T \text{ and } n \text{ fixed}] \tag{5.1}$$

This relationship can also be expressed as

$$PV = \text{constant} \qquad \text{or} \qquad V = \frac{\text{constant}}{P} \qquad [T \text{ and } n \text{ fixed}]$$

That is, at fixed T and n,

$$P\uparrow, V\downarrow \qquad \text{and} \qquad P\downarrow, V\uparrow$$

The constant is the same for most simple gases under ordinary conditions. Thus, tripling the external pressure reduces the volume of a gas to a third of its initial value; halving the pressure doubles the volume; and so forth.

The wording of Boyle's law focuses on *external* pressure. But notice that, as mercury is added, the mercury level rises until the pressure of the trapped gas *on* the mercury increases enough to stop its rise. At that point, the pressure exerted *on* the gas equals the pressure exerted *by* the gas (P_{gas}). Thus, in general, if V_{gas} increases, P_{gas} decreases, and vice versa.

The Relationship Between Volume and Temperature: Charles's Law

Boyle's work showed that the pressure-volume relationship holds only at constant temperature, but why should that be so? It would take more than a century, until the work of French scientists J. A. C. Charles and J. L. Gay-Lussac around 1800, for the relationship between gas volume and temperature to be understood.

1. *The experiment.* Let's examine this relationship by measuring the volume at different temperatures of a fixed amount of a gas under constant pressure. A straight tube, closed at one end, traps a fixed amount of gas (air) under a small mercury plug. The tube is immersed in a water bath that is warmed with a heater or cooled with ice. After each change of temperature, we measure the length of the gas column, which is proportional to its volume. The total pressure exerted on the gas is constant because the mercury plug and the atmospheric pressure do not change (Figure 5.5, parts A and B).

Figure 5.5C shows some typical data. Consider the red line, which shows how the volume of 0.04 mol of gas at 1 atm pressure changes with temperature. Extrapolating that line to lower temperatures *(dashed portion)* shows that, in theory, the gas occupies zero volume at −273.15°C (the intercept on the temperature axis). Plots for a different amount of gas *(green)* or a different gas pressure *(blue)* have different slopes, but they all converge at this temperature, called *absolute zero* (0 K, or −273.15°C). And *this linear relation between gas volume and absolute temperature holds for most common gases over a wide temperature range.*

2. *Conclusion and statement of the law.* Note that, unlike the relationship between volume and pressure, *this linear relationship between volume and temperature is directly proportional.* This behavior is incorporated into the modern statement of the volume-temperature relationship, which is known as **Charles's law:** *at constant pressure, the volume occupied by a fixed amount of gas is* ***directly*** *proportional to its absolute (Kelvin) temperature,* or

$$V \propto T \qquad [P \text{ and } n \text{ fixed}] \tag{5.2}$$

This relationship can also be expressed as

$$\frac{V}{T} = \text{constant} \qquad \text{or} \qquad V = \text{constant} \times T \qquad [P \text{ and } n \text{ fixed}]$$

That is, at fixed P and n,

$$T\uparrow, V\uparrow \qquad \text{and} \qquad T\downarrow, V\downarrow$$

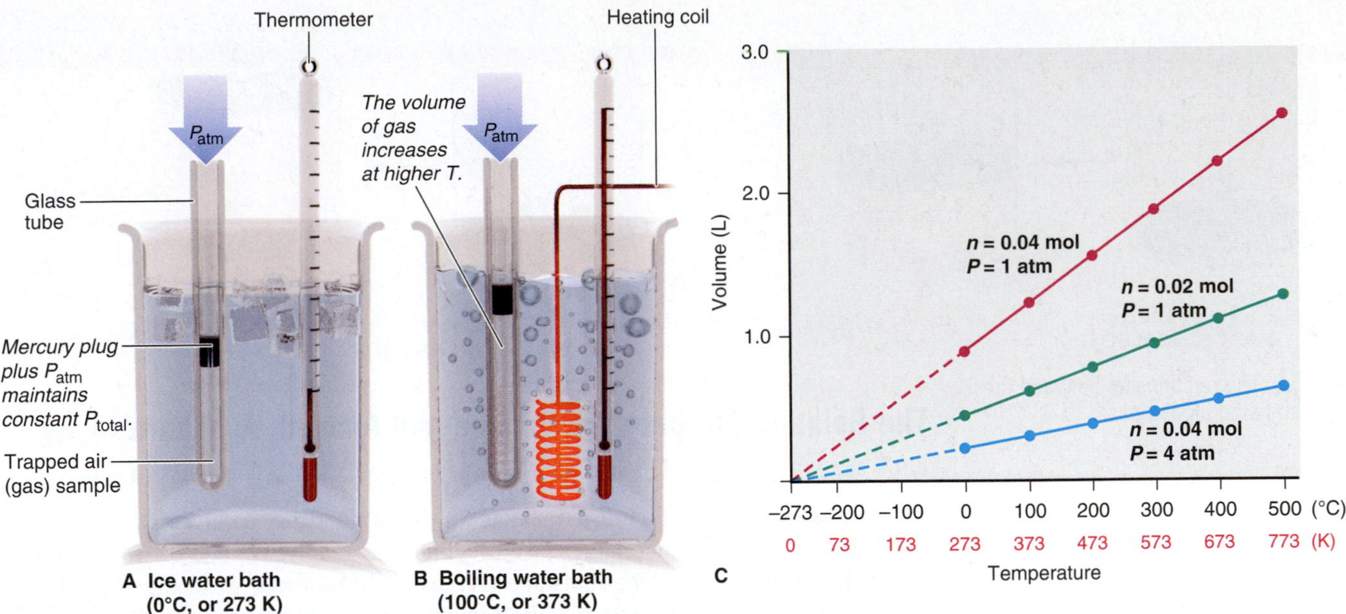

Figure 5.5 Charles's law, the relationship between the volume and temperature of a gas.

If T increases, V increases, and vice versa. Once again, for any given P and n, the constant is the same for most simple gases under ordinary conditions.

The dependence of gas volume on the *absolute* temperature means that you *must use the Kelvin scale in gas law calculations.* For instance, if the temperature changes from 200 K to 400 K, the volume of gas doubles. But, if the temperature changes from 200°C to 400°C, the volume increases by a factor of 1.42; that is,

$$\left(\frac{400°\text{C} + 273.15}{200°\text{C} + 273.15}\right) = \frac{673}{473} = 1.42$$

Other Relationships Based on Boyle's and Charles's Laws Two other important relationships arise from Boyle's and Charles's laws:

1. *The pressure-temperature relationship.* Charles's law is expressed as the effect of temperature on gas *volume* at constant pressure. But volume and pressure are interdependent, so a similar relationship can be expressed for the effect of temperature on pressure (sometimes referred to as *Amontons's law*). Measure the pressure in your car or bike tires before and after a long ride, and you'll find that it increases. Heating due to friction between the tires and the road increases the air temperature inside the tires, but since a tire's volume can't increase very much, the air pressure does. Thus, *at constant volume, the pressure exerted by a fixed amount of gas is directly proportional to the absolute temperature:*

$$P \propto T \qquad [V \text{ and } n \text{ fixed}] \tag{5.3}$$

or

$$\frac{P}{T} = \text{constant} \qquad \text{or} \qquad P = \text{constant} \times T$$

That is, at fixed V and n,

$$T\uparrow, P\uparrow \qquad \text{and} \qquad T\downarrow, P\downarrow$$

2. *The combined gas law.* Combining Boyle's and Charles's laws gives the *combined gas law,* which applies to cases when changes in *two* of the three variables (V, P, T) affect the third:

$$V \propto \frac{T}{P} \qquad \text{or} \qquad V = \text{constant} \times \frac{T}{P} \qquad \text{or} \qquad \frac{PV}{T} = \text{constant}$$

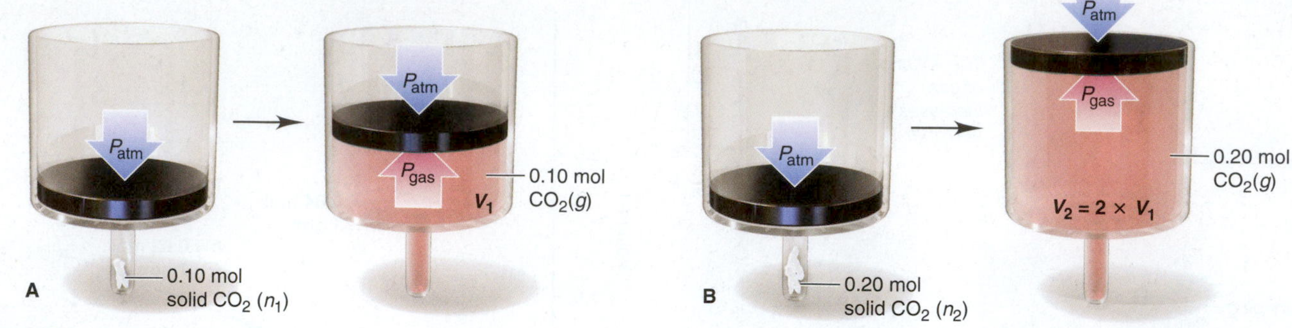

Figure 5.6 The relationship between the volume and amount of a gas.

The Relationship Between Volume and Amount: Avogadro's Law

Let's see why both Boyle's and Charles's laws specify a fixed amount of gas.

1. *The experiment.* Figure 5.6 shows an experiment that involves two small test tubes, each fitted to a much larger piston-cylinder assembly. We add 0.10 mol (4.4 g) of dry ice (solid CO_2) to the first tube (A) and 0.20 mol (8.8 g) to the second tube (B). As the solid CO_2 warms to room temperature, it changes to gaseous CO_2, and the volume increases until $P_{gas} = P_{atm}$. At constant temperature, when all the solid has changed to gas, cylinder B has twice the volume of cylinder A.

2. *Conclusion and statement of the law.* Thus, *at fixed temperature and pressure, the volume occupied by a gas is directly proportional to the amount (mol) of gas:*

$$V \propto n \qquad [P \text{ and } T \text{ fixed}] \tag{5.4}$$

That is, as n increases, V increases, and vice versa. This relationship is also expressed as

$$\frac{V}{n} = \text{constant} \qquad \text{or} \qquad V = \text{constant} \times n$$

That is, at fixed P and T,

$$n\uparrow,\ V\uparrow \qquad \text{and} \qquad n\downarrow,\ V\downarrow$$

The constant is the same for all simple gases at ordinary temperature and pressure. This relationship is another way of expressing **Avogadro's law,** which states that *at fixed temperature and pressure, equal volumes of* ***any*** *ideal gas contain equal numbers of particles (or moles).*

Familiar Applications of the Gas Laws The gas laws apply to countless familiar phenomena: gasoline burning in a car engine, dough rising and baking—even the act of breathing (Figure 5.7). When you inhale, the downward movement of your diaphragm and the expansion of your rib cage increase your lung volume, which decreases the air pressure inside, so air rushes in (Boyle's law). The greater amount of air stretches the lung tissue, which expands the volume further (Avogadro's law), and the air expands slightly as it warms to body temperature (Charles's law). When you exhale, these steps occur in reverse.

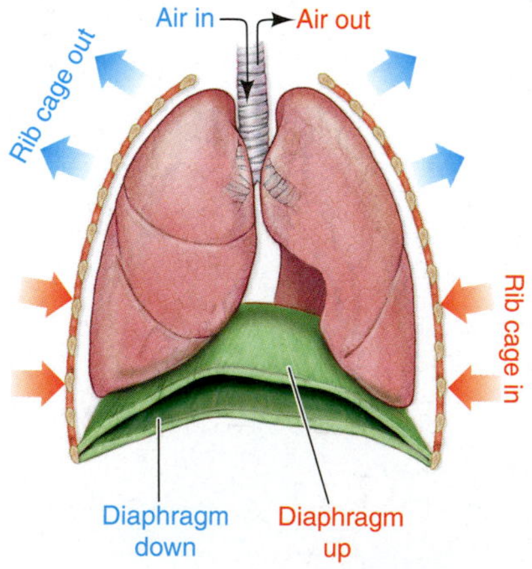

Air in
Air out
Rib cage out
Rib cage in
Diaphragm down
Diaphragm up

Figure 5.7 The process of breathing applies the gas laws.

Gas Behavior at Standard Conditions

To better understand the factors that influence gas behavior, chemists have assigned a baseline set of *standard conditions* called **standard temperature and pressure (STP):**

$$\text{STP:} \quad 0°\text{C } (273.15 \text{ K}) \text{ and } 1 \text{ atm } (760 \text{ torr}) \tag{5.5}$$

Under these conditions, the volume of 1 mol of an ideal gas is called the **standard molar volume:**

$$\text{Standard molar volume} = 22.4141 \text{ L or } 22.4 \text{ L } [\text{to 3 sf}] \tag{5.6}$$

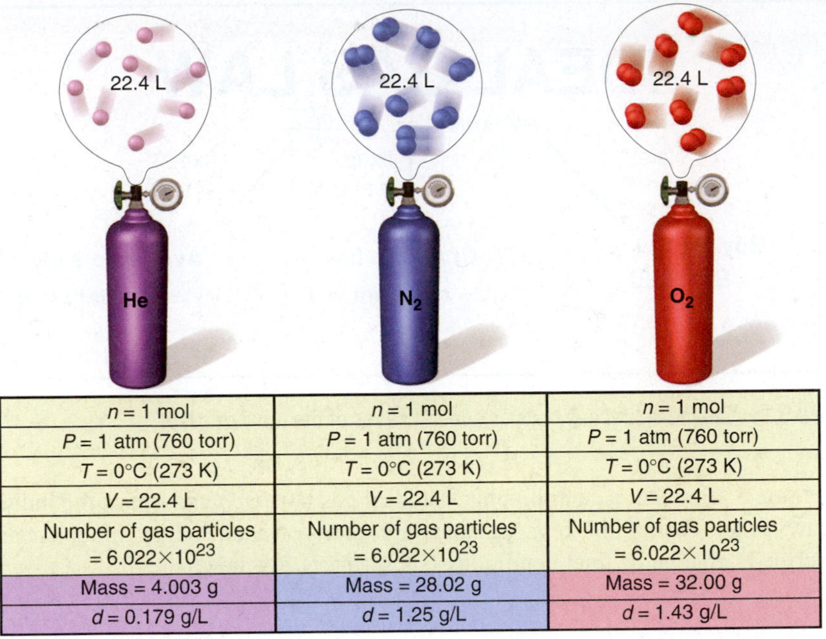

He	N_2	O_2
$n = 1$ mol	$n = 1$ mol	$n = 1$ mol
$P = 1$ atm (760 torr)	$P = 1$ atm (760 torr)	$P = 1$ atm (760 torr)
$T = 0$°C (273 K)	$T = 0$°C (273 K)	$T = 0$°C (273 K)
$V = 22.4$ L	$V = 22.4$ L	$V = 22.4$ L
Number of gas particles $= 6.022\times10^{23}$	Number of gas particles $= 6.022\times10^{23}$	Number of gas particles $= 6.022\times10^{23}$
Mass = 4.003 g	Mass = 28.02 g	Mass = 32.00 g
$d = 0.179$ g/L	$d = 1.25$ g/L	$d = 1.43$ g/L

Figure 5.8 **Standard molar volume.** One mole of an ideal gas occupies 22.4 L at STP (0°C and 1 atm).

Figure 5.8 compares the properties of 1 mol of three simple gases—helium, nitrogen, and oxygen—at STP. Note that they differ only in terms of mass and, thus, density. Figure 5.9 compares the volumes of some familiar objects with the standard molar volume of an ideal gas.

Figure 5.9 The volumes of 1 mol (22.4 L) of an ideal gas and of some familiar objects: 1 gal of milk (3.79 L), a basketball (7.50 L), and 2.00 L of a carbonated drink.

The Ideal Gas Law

Each of the three gas laws shows how one of the three other gas variables affects gas volume:

- Boyle's law focuses on pressure ($V \propto 1/P$).
- Charles's law focuses on temperature ($V \propto T$).
- Avogadro's law focuses on amount (mol) of gas ($V \propto n$).

By combining these individual effects, we obtain the **ideal gas law** (or *ideal gas equation*):

$$V \propto \frac{nT}{P} \quad \text{or} \quad PV \propto nT \quad \text{or} \quad \frac{PV}{nT} = R$$

where R is a proportionality constant known as the **universal gas constant.** Rearranging the equation on the right above gives the most common form of the ideal gas law:

$$PV = nRT \tag{5.7}$$

We obtain a value of R by measuring the volume, temperature, and pressure of a given amount of gas and substituting the values into the ideal gas law. For example, using standard conditions for the gas variables and 1 mol of gas, we have

$$R = \frac{PV}{nT} = \frac{1 \text{ atm} \times 22.4141 \text{ L}}{1 \text{ mol} \times 273.15 \text{ K}} = 0.082058 \frac{\text{atm}\cdot\text{L}}{\text{mol}\cdot\text{K}} = 0.0821 \frac{\text{atm}\cdot\text{L}}{\text{mol}\cdot\text{K}} \text{ [to 3 sf]} \tag{5.8}$$

This numerical value of R corresponds to P, V, and T expressed *in these units; R has a different numerical value when different units are used.* For example, on p. 174, R has the value 8.314 J/mol·K (J stands for joule, the SI unit of energy).

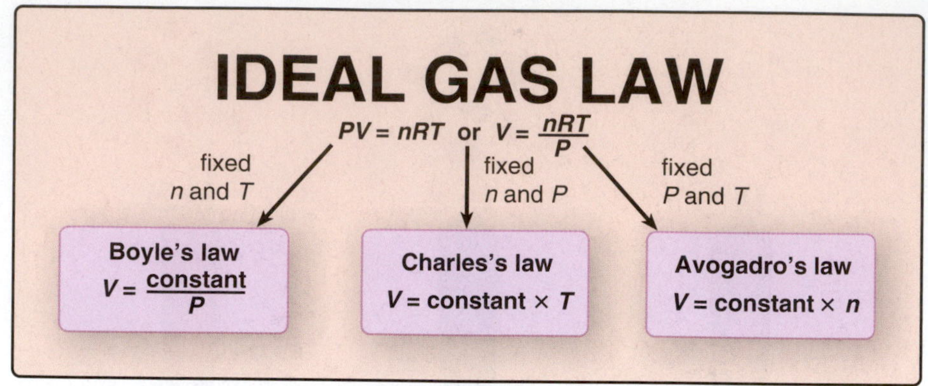

Figure 5.10 The individual gas laws as special cases of the ideal gas law.

Figure 5.10 makes a central point: the ideal gas law *becomes* one of the individual gas laws when two of the four variables are kept constant. When initial conditions (subscript 1) change to final conditions (subscript 2), we have

$$P_1V_1 = n_1RT_1 \quad \text{and} \quad P_2V_2 = n_2RT_2$$

Thus

$$\frac{P_1V_1}{n_1T_1} = R \quad \text{and} \quad \frac{P_2V_2}{n_2T_2} = R$$

so

$$\frac{P_1V_1}{n_1T_1} = \frac{P_2V_2}{n_2T_2}$$

Notice that if, for example, the two variables P and T remain constant, then $P_1 = P_2$ and $T_1 = T_2$, and we obtain an expression for Avogadro's law:

$$\frac{\cancel{P_1}V_1}{n_1\cancel{T_1}} = \frac{\cancel{P_2}V_2}{n_2\cancel{T_2}} \quad \text{or} \quad \frac{V_1}{n_1} = \frac{V_2}{n_2}$$

As you'll see next, you can use a similar approach to solve gas law problems. Thus, by keeping track of the initial and final values of the gas variables, you avoid the need to memorize the three individual gas laws.

Solving Gas Law Problems

Gas law problems are stated in many ways but are usually one of two types:

1. *A change in one of the four variables causes a change in another, while the two other variables remain constant.* In this type, the ideal gas law reduces to one of the individual gas laws, and you solve for the new value of the affected variable. Units must be consistent and T must always be in kelvins, but R is not involved. Sample Problems 5.2 to 5.4 and 5.6 are of this type. [A variation on this type involves the combined gas law (p. 156) when simultaneous changes in two of the variables cause a change in a third.]

2. *One variable is unknown, but the other three are known and no change occurs.* In this type, exemplified by Sample Problem 5.5, you apply the ideal gas law directly to find the unknown, and the units must conform to those in R.

Solving these problems requires a systematic approach:

- Summarize the changing gas variables—knowns and unknown—and those held constant.
- Convert units, if necessary.
- Rearrange the ideal gas law to obtain the needed relationship of variables, and solve for the unknown.

Sample Problem 5.2 Applying the Volume-Pressure Relationship

Problem Boyle's apprentice finds that the air trapped in a J tube occupies 24.8 cm^3 at 1.12 atm. By adding mercury to the tube, he increases the pressure on the trapped air to 2.64 atm. Assuming constant temperature, what is the new volume of air (in L)?

Plan We must find the final volume (V_2) in liters, given the initial volume (V_1), initial pressure (P_1), and final pressure (P_2). The temperature and amount of gas are fixed. We convert the units of V_1 from cm^3 to mL and then to L, rearrange the ideal gas law to the appropriate form, and solve for V_2. (Note that the road map has two parts.)

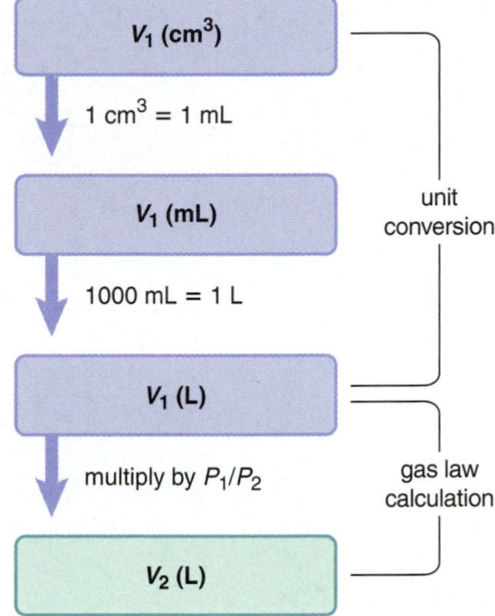

Solution Summarizing the gas variables:

$P_1 = 1.12$ atm $\qquad P_2 = 2.64$ atm

$V_1 = 24.8$ cm^3 (convert to L) $\qquad V_2 =$ unknown $\qquad$ T and n remain constant

Converting V_1 from cm^3 to L:

$$V_1 = 24.8\ \text{cm}^3 \times \frac{1\ \text{mL}}{1\ \text{cm}^3} \times \frac{1\ \text{L}}{1000\ \text{mL}} = 0.0248\ \text{L}$$

Rearranging the ideal gas law and solving for V_2: At fixed n and T, we have

$$\frac{P_1V_1}{\cancel{n_1T_1}} = \frac{P_2V_2}{\cancel{n_2T_2}} \quad \text{or} \quad P_1V_1 = P_2V_2$$

$$V_2 = V_1 \times \frac{P_1}{P_2} = 0.0248\ \text{L} \times \frac{1.12\ \text{atm}}{2.64\ \text{atm}} = 0.0105\ \text{L}$$

Check The relative values of P and V can help us check the math: P more than doubled, so V_2 should be less than $\frac{1}{2}V_1$ ($0.0105/0.0248 < \frac{1}{2}$).

Comment Predicting the direction of the change provides another check on the problem setup: Since P increases, V will decrease; thus, V_2 should be less than V_1. To make $V_2 < V_1$, we must multiply V_1 by a number *less than* 1. This means the ratio of pressures must be *less than* 1, so the larger pressure (P_2) must be in the denominator, or P_1/P_2.

FOLLOW-UP PROBLEM 5.2 A sample of argon gas occupies 105 mL at 0.871 atm. If the temperature remains constant, what is the volume (in L) at 26.3 kPa?

Sample Problem 5.3 Applying the Pressure-Temperature Relationship

Problem A steel tank used for fuel delivery is fitted with a safety valve that opens if the internal pressure exceeds 1.00×10^3 torr. It is filled with methane at 23°C and 0.991 atm and placed in boiling water 100.°C. Will the safety valve open?

Plan The question "Will the safety valve open?" translates to "Is P_2 greater than 1.00×10^3 torr at T_2?" Thus, P_2 is the unknown, and T_1, T_2, and P_1 are given, with V (steel tank) and n fixed. We convert both T values to kelvins and P_1 to torrs in order to compare P_2 with the safety-limit pressure. We rearrange the ideal gas law and solve for P_2.

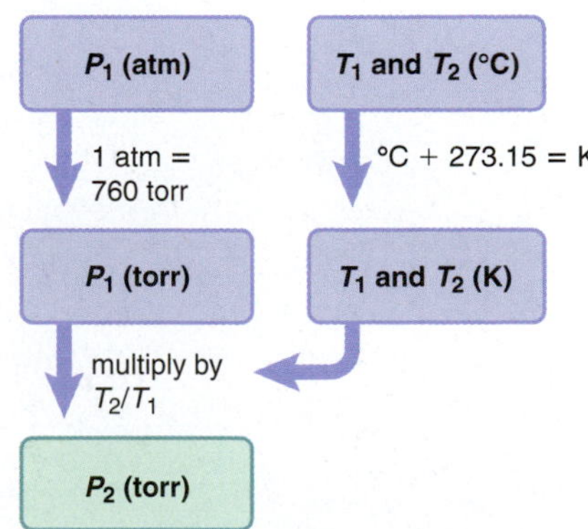

Solution Summarizing the gas variables:

$P_1 = 0.991$ atm (convert to torr) $\qquad P_2 =$ unknown

$T_1 = 23$°C (convert to K) $\qquad T_2 = 100.$°C (convert to K)

V and n remain constant

Converting T from °C to K:

$$T_1\ (\text{K}) = 23^\circ\text{C} + 273.15 = 296\ \text{K} \qquad T_2\ (\text{K}) = 100.^\circ\text{C} + 273.15 = 373\ \text{K}$$

Converting P from atm to torr:

$$P_1\ (\text{torr}) = 0.991\ \text{atm} \times \frac{760\ \text{torr}}{1\ \text{atm}} = 753\ \text{torr}$$

Rearranging the ideal gas law and solving for P_2: At fixed n and V, we have

$$\frac{P_1\cancel{V_1}}{\cancel{n_1}T_1} = \frac{P_2\cancel{V_2}}{\cancel{n_2}T_2} \quad \text{or} \quad \frac{P_1}{T_1} = \frac{P_2}{T_2}$$

$$P_2 = P_1 \times \frac{T_2}{T_1} = 753 \text{ torr} \times \frac{373 \text{ K}}{296 \text{ K}} = 949 \text{ torr}$$

P_2 is less than 1.00×10^3 torr, so the valve will *not* open.

Check Let's predict the change to check the math: Because $T_2 > T_1$, we expect $P_2 > P_1$. Thus, the temperature ratio should be >1 (T_2 in the numerator). The T ratio is about 1.25 (373/296), so the P ratio should also be about 1.25 (950/750 ≈ 1.25).

FOLLOW-UP PROBLEM 5.3 An engineer pumps air at 0°C into a newly designed piston-cylinder assembly. The volume measures 6.83 cm³. At what temperature (in K) will the volume be 9.75 cm³?

Sample Problem 5.4 Applying the Volume-Amount Relationship

Problem A scale model of a blimp rises when it is filled with helium to a volume of 55.0 dm³. When 1.10 mol of He is added to the blimp, the volume is 26.2 dm³. How many more grams of He must be added to make it rise? Assume constant T and P.

Plan We are given the initial amount of helium (n_1), the initial volume of the blimp (V_1), and the volume needed for it to rise (V_2), and we need the additional mass of helium to make it rise. So we first need to find n_2. We rearrange the ideal gas law to the appropriate form, solve for n_2, subtract n_1 to find the additional amount ($n_{\text{add'l}}$), and then convert moles to grams.

Road Map

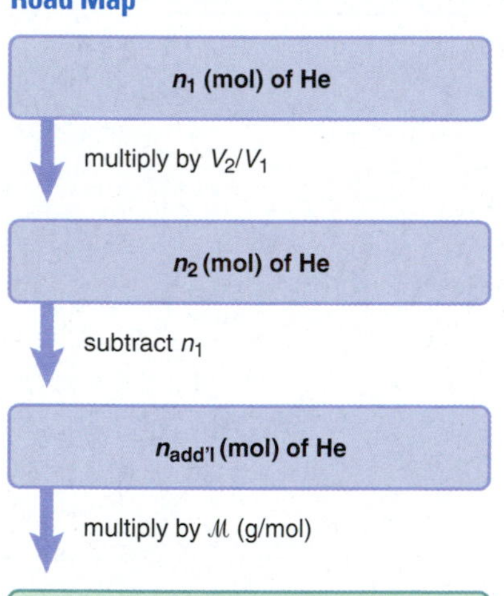

Solution Summarizing the gas variables:

$n_1 = 1.10$ mol $\quad n_2 =$ unknown (find, and then subtract n_1)
$V_1 = 26.2$ dm³ $\quad V_2 = 55.0$ dm³
P and T remain constant

Rearranging the ideal gas law and solving for n_2: At fixed P and T, we have

$$\frac{\cancel{P_1}V_1}{n_1\cancel{T_1}} = \frac{\cancel{P_2}V_2}{n_2\cancel{T_2}} \quad \text{or} \quad \frac{V_1}{n_1} = \frac{V_2}{n_2}$$

$$n_2 = n_1 \times \frac{V_2}{V_1} = 1.10 \text{ mol He} \times \frac{55.0 \text{ dm}^3}{26.2 \text{ dm}^3} = 2.31 \text{ mol He}$$

Finding the additional amount of He:

$$n_{\text{add'l}} = n_2 - n_1 = 2.31 \text{ mol He} - 1.10 \text{ mol He} = 1.21 \text{ mol He}$$

Converting amount (mol) of He to mass (g):

$$\text{Mass (g) of He} = 1.21 \text{ mol He} \times \frac{4.003 \text{ g He}}{1 \text{ mol He}} = 4.84 \text{ g He}$$

Check We predict that $n_2 > n_1$ because $V_2 > V_1$: since V_2 is about twice V_1 (55/26 ≈ 2), n_2 should be about twice n_1 (2.3/1.1 ≈ 2). Since $n_2 > n_1$, we were right to multiply n_1 by a number >1 (that is, V_2/V_1). About 1.2 mol × 4 g/mol ≈ 4.8 g.

Comment **1.** A different sequence of steps will give the same answer: first find the additional volume ($V_{\text{add'l}} = V_2 - V_1$), and then solve directly for $n_{\text{add'l}}$. Try it yourself. **2.** You saw that Charles's law ($V \propto T$ at fixed P and n) becomes a similar relationship between P and T at fixed V and n. The follow-up problem demonstrates that Avogadro's law ($V \propto n$ at fixed P and T) becomes a similar relationship at fixed V and T.

FOLLOW-UP PROBLEM 5.4 A rigid plastic container holds 35.0 g of ethylene gas (C_2H_4) at a pressure of 793 torr. What is the pressure if 5.0 g of ethylene is removed at constant temperature?

Sample Problem 5.5 Solving for an Unknown Gas Variable at Fixed Conditions

Problem A steel tank has a volume of 438 L and is filled with 0.885 kg of O_2. Calculate the pressure of O_2 at 21°C.

Plan We are given V, T, and the mass of O_2, and we must find P. Since conditions are not changing, we apply the ideal gas law without rearranging it. We use the given V in liters, convert T to kelvins and mass (kg) of O_2 to amount (mol), and solve for P.

Solution Summarizing the gas variables:

$V = 438$ L $\qquad T = 21°C$ (convert to K)

$n = 0.885$ kg O_2 (convert to mol) $\qquad P =$ unknown

Converting T from °C to K:

$$T\text{ (K)} = 21°\text{C} + 273.15 = 294\text{ K}$$

Converting from mass (g) of O_2 to amount (mol):

$$n = \text{mol of } O_2 = 0.885\text{ kg } O_2 \times \frac{1000\text{ g}}{1\text{ kg}} \times \frac{1\text{ mol } O_2}{32.00\text{ g } O_2} = 27.7\text{ mol } O_2$$

Solving for P (note the unit canceling here):

$$P = \frac{nRT}{V} = \frac{27.7\ \cancel{\text{mol}} \times 0.0821\ \dfrac{\text{atm}\cdot\cancel{\text{L}}}{\cancel{\text{mol}}\cdot\cancel{\text{K}}} \times 294\ \cancel{\text{K}}}{438\ \cancel{\text{L}}}$$

$$= \boxed{1.53\text{ atm}}$$

Check The amount of O_2 seems correct: ~900 g/(30 g/mol) = 30 mol. To check the approximate size of the final calculation, round off the values, including that for R:

$$P = \frac{30\text{ mol } O_2 \times 0.1\ \dfrac{\text{atm}\cdot\text{L}}{\text{mol}\cdot\text{K}} \times 300\text{ K}}{450\text{ L}} = 2\text{ atm}$$

which is reasonably close to 1.53 atm.

FOLLOW-UP PROBLEM 5.5 The steel tank in the sample problem develops a slow leak that is discovered and sealed. A pressure gauge fitted on the tank shows the new pressure is 1.37 atm. How many grams of O_2 remain?

Finally, in a picture problem, we apply the gas laws to determine the balanced equation for a gaseous reaction.

Sample Problem 5.6 Using Gas Laws to Determine a Balanced Equation

Problem The piston-cylinder below is depicted before and after a gaseous reaction that is carried out in it at constant pressure: the temperature is 150 K before and 300 K after the reaction. (Assume the cylinder is insulated.)

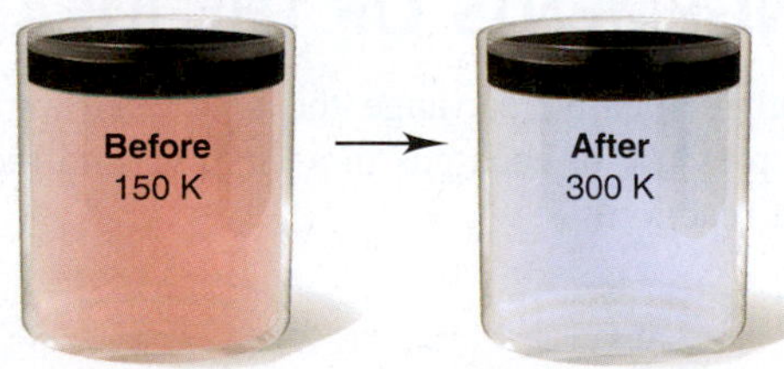

Which of the following balanced equations describes the reaction?

(1) $A_2(g) + B_2(g) \longrightarrow 2AB(g)$

(2) $2AB(g) + B_2(g) \longrightarrow 2AB_2(g)$

(3) $A(g) + B_2(g) \longrightarrow AB_2(g)$

(4) $2AB_2(g) \longrightarrow A_2(g) + 2B_2(g)$

Plan We are shown a depiction of the volume and temperature of a gas mixture before and after a reaction and must deduce the balanced equation. The problem says that P is constant, and the picture shows that, when T doubles, V stays the same. If n were also constant, Charles's law tells us that V should double when T doubles. But, since V does not change, n cannot be constant. From Avogadro's law, the only way to maintain V constant, with P constant and T doubling, is for n to be halved. So we examine the four balanced equations and count the number of moles on each side to see in which equation n is halved.

Solution In equation (1), n does not change, so doubling T would double V.
In equation (2), n decreases from 3 mol to 2 mol, so doubling T would increase V by one-third.
In equation (3), n decreases from 2 mol to 1 mol. Doubling T would exactly balance the decrease from halving n, so V would stay the same.
In equation (4), n increases, so doubling T would more than double V.
Therefore, equation (3) is correct:

$$A(g) + B_2(g) \longrightarrow AB_2(g)$$

FOLLOW-UP PROBLEM 5.6 The piston-cylinder below shows the volumes of a gaseous reaction mixture before and after a reaction that takes place at constant pressure and an initial temperature of −73°C.

If the *unbalanced* equation is $CD(g) \longrightarrow C_2(g) + D_2(g)$, what is the final temperature (in °C)?

Summary of Section 5.3

- Four interdependent variables define the physical behavior of an ideal gas: volume (V), pressure (P), temperature (T), and amount (number of moles, n).
- Most simple gases display nearly ideal behavior at ordinary temperatures and pressures.
- Boyle's, Charles's, and Avogadro's laws refer to the linear relationships between the volume of a gas and the pressure, temperature, and amount of gas, respectively.
- At STP (0°C and 1 atm), 1 mol of an ideal gas occupies 22.4 L.
- The ideal gas law incorporates the individual gas laws into one equation: $PV = nRT$, where R is the universal gas constant.

5.4 • REARRANGEMENTS OF THE IDEAL GAS LAW

In this section, we mathematically rearrange the ideal gas law to find gas density, molar mass, the partial pressure of each gas in a mixture, and the amount of gaseous reactant or product in a reaction.

The Density of a Gas

One mole of any gas behaving ideally occupies the same volume at a given temperature and pressure, so differences in gas density ($d = m/V$) depend on differences in molar mass (see Figure 5.8). For example, at STP, 1 mol of O_2 occupies the same volume as 1 mol of N_2; however, O_2 is denser because each O_2 molecule has a greater mass (32.00 amu) than each N_2 molecule (28.02 amu). Thus, d of O_2 is $\frac{32.00}{28.02} \times d$ of N_2.

We can rearrange the ideal gas law to calculate the density of a gas from its molar mass. Recall that the number of moles (n) is the mass (m) divided by the molar mass ($\mathcal{M}$), $n = m/\mathcal{M}$. Substituting for n in the ideal gas law gives

$$PV = \frac{m}{\mathcal{M}}RT$$

Rearranging to isolate m/V gives

$$\frac{m}{V} = d = \frac{\mathcal{M} \times P}{RT} \qquad (5.9)$$

Two important ideas are expressed by Equation 5.9:

- *The density of a gas is directly proportional to its molar mass.* The volume of a given amount of a heavier gas equals the volume of the same amount of a lighter gas (Avogadro's law), so the density of the heavier gas is higher (as you just saw for O_2 and N_2).
- *The density of a gas is inversely proportional to the temperature.* As the volume of a gas increases with temperature (Charles's law), the same mass occupies more space, so the density of the gas is lower.

We use Equation 5.9 to find the density of a gas at any temperature and pressure near standard conditions.

Sample Problem 5.7 Calculating Gas Density

Problem A chemical engineer uses waste CO_2 from a manufacturing process, instead of chlorofluorocarbons, as a "blowing agent" in the production of polystyrene. Find the density (in g/L) of CO_2 and the number of molecules per liter **(a)** at STP (0°C and 1 atm) and **(b)** at room conditions (20.°C and 1.00 atm).

Plan We must find the density (d) and the number of molecules of CO_2, given two sets of P and T data. We find $\mathcal{M}$, convert T to kelvins, and calculate d with Equation 5.9. Then we convert the mass per liter to molecules per liter with Avogadro's number.

Solution **(a)** Density and molecules per liter of CO_2 at STP. Summary of gas properties:

$$T = 0°C + 273.15 = 273\ K \qquad P = 1\ atm \qquad \mathcal{M}\text{ of } CO_2 = 44.01\ g/mol$$

Calculating density (note the unit canceling here):

$$d = \frac{\mathcal{M} \times P}{RT} = \frac{44.01\ g/\cancel{mol} \times 1.00\ \cancel{atm}}{0.0821\ \frac{\cancel{atm}\cdot L}{\cancel{mol}\cdot \cancel{K}} \times 273\ \cancel{K}} = \boxed{1.96\ g/L}$$

Converting from mass/L to molecules/L:

$$\text{Molecules } CO_2/L = \frac{1.96\ g\ CO_2}{1\ L} \times \frac{1\ mol\ CO_2}{44.01\ g\ CO_2} \times \frac{6.022\times10^{23}\ \text{molecules } CO_2}{1\ mol\ CO_2}$$

$$= \boxed{2.68\times10^{22}\ \text{molecules } CO_2/L}$$

(b) Density and molecules of CO_2 per liter at room conditions. Summary of gas properties:

$$T = 20.°C + 273.15 = 293\ K \qquad P = 1.00\ atm \qquad \mathcal{M}\text{ of } CO_2 = 44.01\ g/mol$$

Calculating density:

$$d = \frac{\mathcal{M} \times P}{RT} = \frac{44.01\ g/mol \times 1.00\ atm}{0.0821\ \frac{atm\cdot L}{mol\cdot K} \times 293\ K} = \boxed{1.83\ g/L}$$

Converting from mass/L to molecules/L:

$$\text{Molecules } CO_2/L = \frac{1.83\ g\ CO_2}{1\ L} \times \frac{1\ mol\ CO_2}{44.01\ g\ CO_2} \times \frac{6.022\times10^{23}\ \text{molecules } CO_2}{1\ mol\ CO_2}$$

$$= \boxed{2.50\times10^{22}\ \text{molecules } CO_2/L}$$

Check Round off to check the density values; for example, in (a), at STP:

$$\frac{50 \text{ g/mol} \times 1 \text{ atm}}{0.1 \dfrac{\text{atm}\cdot\text{L}}{\text{mol}\cdot\text{K}} \times 250 \text{ K}} = 2 \text{ g/L} \approx 1.96 \text{ g/L}$$

At the higher temperature in (b), the density should decrease, which can happen only if there are fewer molecules per liter, so the answer is reasonable.

Comment 1. An *alternative approach* for finding the density of most simple gases, but *at STP only,* is to divide the molar mass by the standard molar volume, 22.4 L:

$$d = \frac{\mathcal{M}}{V} = \frac{44.01 \text{ g/mol}}{22.4 \text{ L/mol}} = 1.96 \text{ g/L}$$

Then, once you know the density at one temperature (0°C), you can find it at any other temperature with the following relationship: $d_1/d_2 = T_2/T_1$.
2. Note that we have different numbers of significant figures for the pressure values. In (a), "1 atm" is part of the definition of STP, so it is an exact number. In (b), "1.00 atm" is specified to allow three significant figures in the answer.

FOLLOW-UP PROBLEM 5.7 Compare the density of CO_2 at 0°C and 380. torr with its density at STP.

The Molar Mass of a Gas

Through another rearrangement of the ideal gas law, we can determine the molar mass of an unknown gas or a volatile liquid (one that is easily vaporized):

$$n = \frac{m}{\mathcal{M}} = \frac{PV}{RT} \quad \text{so} \quad \mathcal{M} = \frac{mRT}{PV} \tag{5.10}$$

Notice that this equation is just a rearrangement of Equation 5.9.

Sample Problem 5.8 Finding the Molar Mass of a Volatile Liquid

Problem An organic chemist isolates a colorless liquid from a petroleum sample. She places the liquid in a preweighed flask and puts the flask in boiling water, which vaporizes the liquid and fills the flask with gas. She closes the flask and reweighs it. She obtains the following data:

Volume (V) of flask = 213 mL T = 100.0°C P = 754 torr
Mass of flask + gas = 78.416 g Mass of flask = 77.834 g

Calculate the molar mass of the liquid.

Plan We are given V, T, P, and mass data and must find the molar mass ($\mathcal{M}$) of the liquid. We convert V to liters, T to kelvins, and P to atmospheres, find the mass of gas by subtracting the mass of the flask from the mass of the flask plus gas, and use Equation 5.10 to calculate $\mathcal{M}$.

Solution Summarizing and converting the gas variables:

$$V(\text{L}) = 213 \text{ mL} \times \frac{1 \text{ L}}{1000 \text{ mL}} = 0.213 \text{ L} \qquad T(\text{K}) = 100.0°\text{C} + 273.15 = 373.2 \text{ K}$$

$$P(\text{atm}) = 754 \text{ torr} \times \frac{1 \text{ atm}}{760 \text{ torr}} = 0.992 \text{ atm} \qquad m = 78.416 \text{ g} - 77.834 \text{ g} = 0.582 \text{ g}$$

Calculating $\mathcal{M}$:

$$\mathcal{M} = \frac{mRT}{PV} = \frac{0.582 \text{ g} \times 0.0821 \dfrac{\text{atm}\cdot\text{L}}{\text{mol}\cdot\text{K}} \times 373.2 \text{ K}}{0.992 \text{ atm} \times 0.213 \text{ L}} = 84.4 \text{ g/mol}$$

Check Rounding to check the arithmetic, we have

$$\frac{0.6 \text{ g} \times 0.08 \dfrac{\text{atm}\cdot\text{L}}{\text{mol}\cdot\text{K}} \times 375 \text{ K}}{1 \text{ atm} \times 0.2 \text{ L}} = 90 \text{ g/mol} \qquad \text{(which is close to 84.4 g/mol)}$$

FOLLOW-UP PROBLEM 5.8 An empty 149-mL flask weighs 68.322 g before a sample of volatile liquid is added. The flask is then placed in a hot (95.0°C) water bath; the barometric pressure is 740. torr. The liquid vaporizes and the gas fills the flask. After cooling, flask and condensed liquid together weigh 68.697 g. What is the molar mass of the liquid?

The Partial Pressure of Each Gas in a Mixture of Gases

The gas behaviors we've discussed so far were observed in experiments with air, which is a mixture of gases; thus, the ideal gas law holds for virtually any simple gas at ordinary conditions, whether pure or a mixture, because

- Gases mix homogeneously (form a solution) in any proportions.
- Each gas in a mixture behaves as if it were the only gas present (assuming no chemical interactions).

Dalton's Law of Partial Pressures The second point above was discovered by John Dalton. During a nearly lifelong study of humidity, he observed that when water vapor is added to dry air, the total air pressure increases by the pressure of the water vapor:

$$P_{\text{humid air}} = P_{\text{dry air}} + P_{\text{added water vapor}}$$

He concluded that each gas in the mixture exerts a **partial pressure** equal to the pressure it would exert *by itself.* Stated as **Dalton's law of partial pressures,** his discovery was that *in a mixture of unreacting gases, the total pressure is the sum of the partial pressures of the individual gases:*

$$P_{\text{total}} = P_1 + P_2 + P_3 + \cdots \qquad \textbf{(5.11)}$$

As an example, suppose we have a tank of fixed volume that contains nitrogen gas at a certain pressure, and we introduce a sample of hydrogen gas into the tank. Each gas behaves independently, so we can write an ideal gas law expression for each:

$$P_{N_2} = \frac{n_{N_2}RT}{V} \quad \text{and} \quad P_{H_2} = \frac{n_{H_2}RT}{V}$$

Because each gas occupies the same total volume and is at the same temperature, the pressure of each gas depends only on its amount, n. Thus, the total pressure is

$$P_{\text{total}} = P_{N_2} + P_{H_2} = \frac{n_{N_2}RT}{V} + \frac{n_{H_2}RT}{V} = \frac{(n_{N_2} + n_{H_2})RT}{V} = \frac{n_{\text{total}}RT}{V}$$

where $n_{\text{total}} = n_{N_2} + n_{H_2}$.

Each component in a mixture contributes a fraction of the total number of moles in the mixture; this portion is the **mole fraction (X)** of that component. Multiplying X by 100 gives the mole percent. The sum of the mole fractions of all components must be 1, and the sum of the mole percents must be 100%. For N_2 in our mixture, the mole fraction is

$$X_{N_2} = \frac{n_{N_2}}{n_{\text{total}}} = \frac{n_{N_2}}{n_{N_2} + n_{H_2}}$$

If the total pressure is due to the total number of moles, the partial pressure of gas A is the total pressure multiplied by the mole fraction of A, X_A:

$$P_A = X_A \times P_{\text{total}} \qquad \textbf{(5.12)}$$

Equation 5.12 is a very useful result. To see that it is valid for the mixture of N_2 and H_2, we recall that $X_{N_2} + X_{H_2} = 1$; then we obtain

$$P_{\text{total}} = P_{N_2} + P_{H_2} = (X_{N_2} \times P_{\text{total}}) + (X_{H_2} \times P_{\text{total}}) = (X_{N_2} + X_{H_2})P_{\text{total}} = 1 \times P_{\text{total}}$$

Road Map

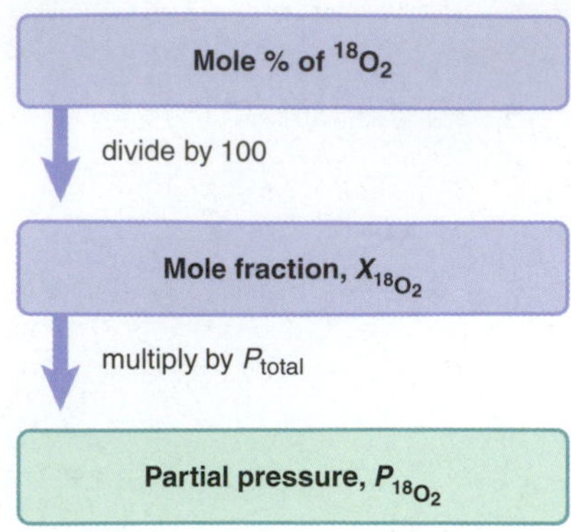
Mole % of $^{18}O_2$
divide by 100
Mole fraction, $X_{^{18}O_2}$
multiply by P_{total}
Partial pressure, $P_{^{18}O_2}$

Sample Problem 5.9 Applying Dalton's Law of Partial Pressures

Problem To study O_2 uptake by muscle at high altitude, a physiologist prepares an atmosphere consisting of 79 mole % N_2, 17 mole % $^{16}O_2$, and 4.0 mole % $^{18}O_2$. (The isotope ^{18}O will be measured to determine O_2 uptake.) The total pressure is 0.75 atm to simulate high altitude. Find the mole fraction and partial pressure of $^{18}O_2$ in the mixture.

Plan We must find $X_{^{18}O_2}$ and $P_{^{18}O_2}$ from P_{total} (0.75 atm) and the mole % of $^{18}O_2$ (4.0). Dividing the mole % by 100 gives the mole fraction, $X_{^{18}O_2}$. Then, using Equation 5.12, we multiply $X_{^{18}O_2}$ by P_{total} to find $P_{^{18}O_2}$.

Solution Calculating the mole fraction of $^{18}O_2$:

$$X_{^{18}O_2} = \frac{4.0 \text{ mol \% } ^{18}O_2}{100} = \boxed{0.040}$$

Solving for the partial pressure of $^{18}O_2$:

$$P_{^{18}O_2} = X_{^{18}O_2} \times P_{total} = 0.040 \times 0.75 \text{ atm} = \boxed{0.030 \text{ atm}}$$

Check $X_{^{18}O_2}$ is small because the mole % is small, so $P_{^{18}O_2}$ should be small also.

Comment At high altitudes, specialized brain cells that are sensitive to O_2 and CO_2 levels in the blood trigger an increase in rate and depth of breathing for several days, until a person becomes acclimated.

FOLLOW-UP PROBLEM 5.9 To prevent the presence of air, noble gases are placed over highly reactive chemicals to act as inert "blanketing" gases. A chemical engineer puts a mixture of noble gases consisting of 5.50 g of He, 15.0 g of Ne, and 35.0 g of Kr in a piston-cylinder assembly at STP. Calculate the partial pressure of each gas.

Table 5.2 Vapor Pressure of Water (P_{H_2O}) at Different T

T (°C)	P_{H_2O} (torr)	T (°C)	P_{H_2O} (torr)
0	4.6	40	55.3
5	6.5	45	71.9
10	9.2	50	92.5
12	10.5	55	118.0
14	12.0	60	149.4
16	13.6	65	187.5
18	15.5	70	233.7
20	17.5	75	289.1
22	19.8	80	355.1
24	22.4	85	433.6
26	25.2	90	525.8
28	28.3	95	633.9
30	31.8	100	760.0
35	42.2		

Collecting a Gas over Water Whenever a gas is in contact with water, some of the water vaporizes into the gas. The water vapor that mixes with the gas contributes the *vapor pressure,* a portion of the total pressure that depends only on the water temperature (Table 5.2). A common use of the law of partial pressures is to determine the yield of a water-insoluble gas formed in a reaction: the gaseous product bubbles through water, some water vaporizes into the bubbles, and the mixture of product gas and water vapor is collected into an inverted container (Figure 5.11).

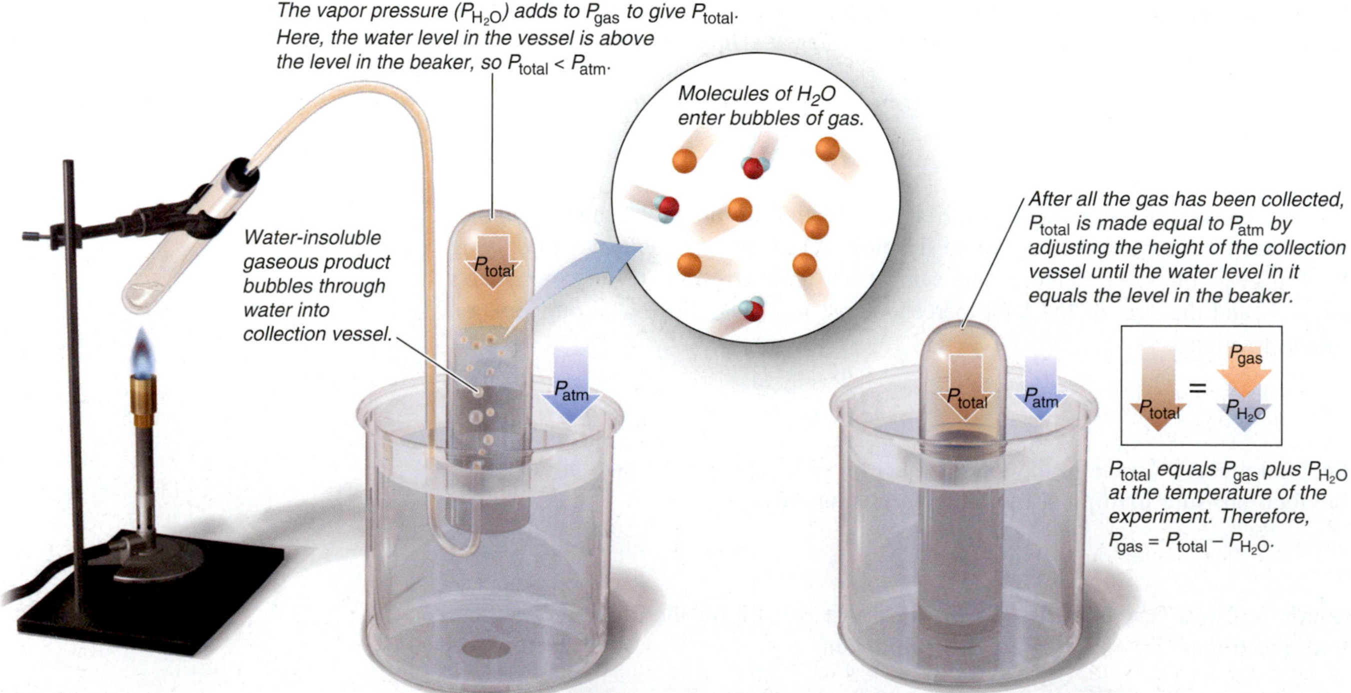
The vapor pressure (P_{H_2O}) adds to P_{gas} to give P_{total}. Here, the water level in the vessel is above the level in the beaker, so $P_{total} < P_{atm}$.
Molecules of H_2O enter bubbles of gas.
Water-insoluble gaseous product bubbles through water into collection vessel.
P_{total}
P_{atm}
After all the gas has been collected, P_{total} is made equal to P_{atm} by adjusting the height of the collection vessel until the water level in it equals the level in the beaker.
P_{total}
P_{atm}
P_{total} = P_{gas} P_{H_2O}
P_{total} equals P_{gas} plus P_{H_2O} at the temperature of the experiment. Therefore, $P_{gas} = P_{total} - P_{H_2O}$.

Figure 5.11 Collecting a water-insoluble gaseous product and determining its pressure.

To determine the yield, we look up the vapor pressure (P_{H_2O}) at the temperature of the experiment in Table 5.2 and subtract it from the total gas pressure (P_{total}, corrected for barometric pressure) to get the partial pressure of the gaseous product (P_{gas}). With V and T known, we can calculate the amount of product.

Sample Problem 5.10 Calculating the Amount of Gas Collected over Water

Problem Acetylene (C_2H_2), an important fuel in welding, is produced in the laboratory when calcium carbide (CaC_2) reacts with water:

$$CaC_2(s) + 2H_2O(l) \longrightarrow C_2H_2(g) + Ca(OH)_2(aq)$$

For a sample of acetylene collected over water, total gas pressure (adjusted to barometric pressure) is 738 torr and the volume is 523 mL. At the temperature of the gas (23°C), the vapor pressure of water is 21 torr. How many grams of acetylene are collected?

Plan In order to find the mass of C_2H_2, we first need to find the number of moles of C_2H_2, $n_{C_2H_2}$, which we can obtain from the ideal gas law by calculating $P_{C_2H_2}$. The barometer reading gives us P_{total}, which is the sum of $P_{C_2H_2}$ and P_{H_2O}, and we are given P_{H_2O}, so we subtract to find $P_{C_2H_2}$. We are also given V and T, so we convert to consistent units, and find $n_{C_2H_2}$ from the ideal gas law. Then we convert moles to grams using the molar mass from the formula, as shown in the road map.

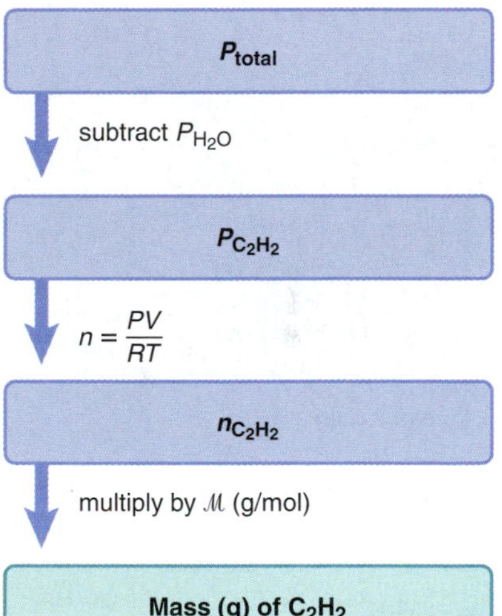

Solution Summarizing and converting the gas variables:

$$P_{C_2H_2}\ (\text{torr}) = P_{total} - P_{H_2O} = 738\ \text{torr} - 21\ \text{torr} = 717\ \text{torr}$$

$$P_{C_2H_2}\ (\text{atm}) = 717\ \text{torr} \times \frac{1\ \text{atm}}{760\ \text{torr}} = 0.943\ \text{atm}$$

$$V\ (\text{L}) = 523\ \text{mL} \times \frac{1\ \text{L}}{1000\ \text{mL}} = 0.523\ \text{L}$$

$$T\ (\text{K}) = 23°\text{C} + 273.15 = 296\ \text{K}$$

$$n_{C_2H_2} = \text{unknown}$$

Solving for $n_{C_2H_2}$:

$$n_{C_2H_2} = \frac{PV}{RT} = \frac{0.943\ \text{atm} \times 0.523\ \text{L}}{0.0821\ \frac{\text{atm·L}}{\text{mol·K}} \times 296\ \text{K}} = 0.0203\ \text{mol}$$

Converting $n_{C_2H_2}$ to mass (g):

$$\text{Mass (g) of } C_2H_2 = 0.0203\ \text{mol } C_2H_2 \times \frac{26.04\ \text{g } C_2H_2}{1\ \text{mol } C_2H_2} = \boxed{0.529\ \text{g } C_2H_2}$$

Check Rounding to one significant figure and doing a quick arithmetic check for n gives

$$n \approx \frac{1\ \text{atm} \times 0.5\ \text{L}}{0.08\ \frac{\text{atm·L}}{\text{mol·K}} \times 300\ \text{K}} = 0.02\ \text{mol} \approx 0.0203\ \text{mol}$$

Comment The C_2^{2-} ion (called the *carbide,* or *acetylide, ion*) is $^-C{\equiv}C^-$, which acts as a base in water, removing an H^+ ion from two H_2O molecules to form acetylene, $H{-}C{\equiv}C{-}H$.

FOLLOW-UP PROBLEM 5.10 A small piece of zinc reacts with dilute HCl to form H_2, which is collected over water at 16°C into a large flask. The total pressure is adjusted to barometric pressure (752 torr), and the volume is 1495 mL. Use Table 5.2 to help calculate the partial pressure and mass of H_2.

The Ideal Gas Law and Reaction Stoichiometry

As you saw in Chapters 3 and 4, and in the preceding discussion of collecting a gas over water, many reactions involve gases as reactants or products. From the balanced equation for such a reaction, you can calculate the amounts (mol) of reactants and products and convert these quantities into masses or numbers of molecules. Figure 5.12 *(next page)* shows how you use the ideal gas law to convert between gas variables (P, T, and V) and amounts (mol) of gaseous reactants and products. In effect, you combine a gas law problem with a stoichiometry problem, as you'll see in Sample Problems 5.11 and 5.12.

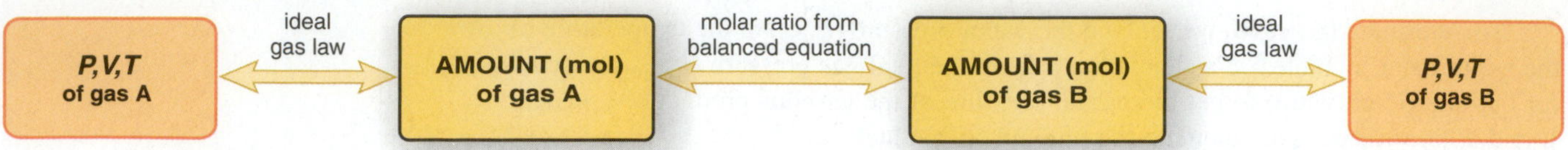

Figure 5.12 The relationships among the amount (mol, n) of gaseous reactant (or product) and the gas pressure (P), volume (V), and temperature (T).

Sample Problem 5.11 Using Gas Variables to Find Amounts of Reactants or Products I

Problem Copper reacts with any oxygen present as an impurity in the ethylene used to make polyethylene. The copper is regenerated when hot H_2 reduces the copper(II) oxide, forming the pure metal and H_2O. What volume of H_2 at 765 torr and 225°C is needed to reduce 35.5 g of copper(II) oxide?

Road Map

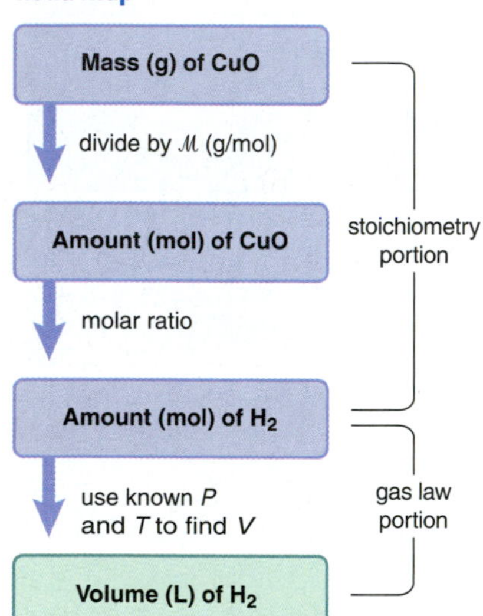

Plan This is a stoichiometry *and* gas law problem. To find V_{H_2}, we first need n_{H_2}. We write and balance the equation. Next, we convert the given mass (35.5 g) of copper(II) oxide, CuO, to amount (mol) and use the molar ratio to find amount (mol) of H_2 needed (stoichiometry portion). Then, we use the ideal gas law to convert moles of H_2 to liters (gas law portion). A road map is shown, but you are familiar with all the steps.

Solution Writing the balanced equation:

$$CuO(s) + H_2(g) \longrightarrow Cu(s) + H_2O(g)$$

Calculating n_{H_2}:

$$n_{H_2} = 35.5 \text{ g CuO} \times \frac{1 \text{ mol CuO}}{79.55 \text{ g CuO}} \times \frac{1 \text{ mol } H_2}{1 \text{ mol CuO}} = 0.446 \text{ mol } H_2$$

Summarizing and converting other gas variables:

$$V = \text{unknown} \qquad P\,(\text{atm}) = 765 \text{ torr} \times \frac{1 \text{ atm}}{760 \text{ torr}} = 1.01 \text{ atm}$$

$$T\,(\text{K}) = 225°\text{C} + 273.15 = 498 \text{ K}$$

Solving for V_{H_2}:

$$V = \frac{nRT}{P} = \frac{0.446 \text{ mol} \times 0.0821 \dfrac{\text{atm·L}}{\text{mol·K}} \times 498 \text{ K}}{1.01 \text{ atm}} = \boxed{18.1 \text{ L}}$$

Check One way to check the answer is to compare it with the molar volume of an ideal gas at STP (22.4 L at 273.15 K and 1 atm). One mole of H_2 at STP occupies about 22 L, so less than 0.5 mol occupies less than 11 L. T is less than twice 273 K, so V should be less than twice 11 L.

Comment The main point here is that the stoichiometry provides one gas variable (n), two more are given, and the ideal gas law is used to find the fourth.

FOLLOW-UP PROBLEM 5.11 Sulfuric acid reacts with sodium chloride to form aqueous sodium sulfate and hydrogen chloride gas. How many milliliters of gas form at STP when 0.117 kg of sodium chloride reacts with excess sulfuric acid?

Sample Problem 5.12 Using Gas Variables to Find Amounts of Reactants or Products II

Chlorine gas reacting with potassium.

Problem The alkali metals [Group 1A(1)] react with the halogens [Group 7A(17)] to form ionic metal halides. What mass of potassium chloride forms when 5.25 L of chlorine gas at 0.950 atm and 293 K reacts with 17.0 g of potassium *(see photo)*?

Plan The amounts of two reactants are given, so this is a limiting-reactant problem. The only difference between this and previous limiting-reactant problems (see Sample

Problem 3.19, p. 96) is that here we use the ideal gas law to find the amount (n) of gaseous reactant from the known V, P, and T. We first write the balanced equation and then use it to find the limiting reactant and the amount and mass of product.

Solution Writing the balanced equation:

$$2K(s) + Cl_2(g) \longrightarrow 2KCl(s)$$

Summarizing the gas variables:

$$P = 0.950 \text{ atm} \qquad V = 5.25 \text{ L}$$
$$T = 293 \text{ K} \qquad n = \text{unknown}$$

Solving for n_{Cl_2}:

$$n_{Cl_2} = \frac{PV}{RT} = \frac{0.950 \text{ atm} \times 5.25 \text{ L}}{0.0821 \frac{\text{atm}\cdot\text{L}}{\text{mol}\cdot\text{K}} \times 293 \text{ K}} = 0.207 \text{ mol}$$

Converting from mass (g) of potassium (K) to amount (mol):

$$\text{Amount (mol) of K} = 17.0 \text{ g K} \times \frac{1 \text{ mol K}}{39.10 \text{ g K}} = 0.435 \text{ mol K}$$

Determining the limiting reactant: If Cl_2 is limiting,

$$\text{Amount (mol) of KCl} = 0.207 \text{ mol } Cl_2 \times \frac{2 \text{ mol KCl}}{1 \text{ mol } Cl_2} = 0.414 \text{ mol KCl}$$

If K is limiting,

$$\text{Amount (mol) of KCl} = 0.435 \text{ mol K} \times \frac{2 \text{ mol KCl}}{2 \text{ mol K}} = 0.435 \text{ mol KCl}$$

Cl_2 is the limiting reactant because it forms less KCl.
Converting from amount (mol) of KCl to mass (g):

$$\text{Mass (g) of KCl} = 0.414 \text{ mol KCl} \times \frac{74.55 \text{ g KCl}}{1 \text{ mol KCl}} = 30.9 \text{ g KCl}$$

Check The gas law calculation seems correct. At STP, 22 L of Cl_2 gas contains about 1 mol, so a 5-L volume will contain a bit less than 0.25 mol of Cl_2. Moreover, since P (in numerator) is slightly lower than STP, and T (in denominator) is slightly higher than STP, these values should lower the calculated n further below the ideal value. The mass of KCl seems correct: less than 0.5 mol of KCl gives $<0.5 \text{ mol} \times \mathcal{M}$ (~75 g/mol), and $30.9 \text{ g} < 0.5 \text{ mol} \times 75 \text{ g/mol}$.

FOLLOW-UP PROBLEM 5.12 Ammonia and hydrogen chloride gases react to form solid ammonium chloride. A 10.0-L reaction flask contains ammonia at 0.452 atm and 22°C, and 155 mL of hydrogen chloride gas at 7.50 atm and 271 K is introduced. After the reaction occurs and the temperature returns to 22°C, what is the pressure inside the flask? (Neglect the volume of the solid product.)

Summary of Section 5.4

- Gas density is inversely related to temperature: higher T causes lower d, and vice versa. At the same P and T, gases with larger $\mathcal{M}$ have higher d.
- In a mixture of gases, each component contributes its partial pressure to the total pressure (Dalton's law of partial pressures). The mole fraction of each component is the ratio of its partial pressure to the total pressure.
- When a gaseous reaction product is collected by bubbling it through water, the total pressure is the sum of the gas pressure and the vapor pressure of water at the given temperature.
- By converting the variables P, V, and T for a gaseous reactant (or product) to amount (n, mol), we can solve stoichiometry problems for gaseous reactions.

5.5 • THE KINETIC-MOLECULAR THEORY: A MODEL FOR GAS BEHAVIOR

So far we have discussed behaviors of various gas samples under different conditions: decreasing cylinder volume, increasing tank pressure, and so forth. This section presents the central model that explains macroscopic gas behavior at the level of individul particles: the **kinetic-molecular theory.** The theory draws quantitative conclusions based on a few postulates (assumptions), but our discussion will be largely qualitative.

How the Kinetic-Molecular Theory Explains the Gas Laws

Let's address some questions the theory must answer, state the postulates, and then draw conclusions that explain the gas laws and related phenomena.

Questions Concerning Gas Behavior Observing gas behavior at the macroscopic level, we must derive a molecular model that explains it:

1. *Origin of pressure.* Pressure is a measure of the force a gas exerts on a surface. How do individual gas particles create this force?
2. *Boyle's law ($V \propto 1/P$).* A change in gas pressure in one direction causes a change in gas volume in the other. What happens to the particles when external pressure compresses the gas volume? And why aren't liquids and solids compressible?
3. *Dalton's law ($P_{total} = P_1 + P_2 + P_3 + \cdots$).* The pressure of a gas mixture is the sum of the pressures of the individual gases. Why does each gas contribute to the total pressure in proportion to its number of particles?
4. *Charles's law ($V \propto T$).* A change in temperature causes a corresponding change in volume. What effect does higher temperature have on gas particles that increases gas volume? This question raises a more fundamental one: what does temperature measure on the molecular scale?
5. *Avogadro's law ($V \propto n$).* Gas volume depends on the number of moles present, not on the chemical nature of the gas. But shouldn't 1 mol of heavier particles exert more pressure, and thus take up more space, than 1 mol of lighter ones?

Postulates of the Kinetic-Molecular Theory The theory is based on three postulates:

Postulate 1. *Particle volume.* A gas consists of a large collection of individual particles with empty space between them. The volume of each particle is so small compared with the volume of the whole sample that it is assumed to be zero; each particle is essentially a point of mass.

Postulate 2. *Particle motion.* The particles are in constant, random, straight-line motion, except when they collide with the container walls or with each other.

Postulate 3. *Particle collisions.* The collisions are *elastic,* which means that, like minute billiard balls, the colliding particles exchange energy but do not lose any energy through friction. Thus, *their total kinetic energy (E_k) is constant.* Between collisions, the particles do not influence each other by attractive or repulsive forces.

Gas behavior that conforms to these postulates is called *ideal.* (As you'll see in Section 5.6, most real gases behave almost ideally at ordinary temperatures and pressures.)

Imagine what a sample of gas in a container looks like. Countless minute particles move in every direction, smashing into the container walls and each other. Any given particle changes its speed often—at one moment standing still from a head-on collision and the next moment zooming away from a smash on the side.

In the sample as a whole, *each particle has a molecular speed (u); most are moving near the most probable speed, but some are much faster and others much slower.* Figure 5.13 depicts this distribution of molecular speeds for N_2 gas at three temperatures.

Note that the curves flatten and spread at higher temperatures and that the *most probable speed (the peak of each curve) increases as the temperature increases.* This

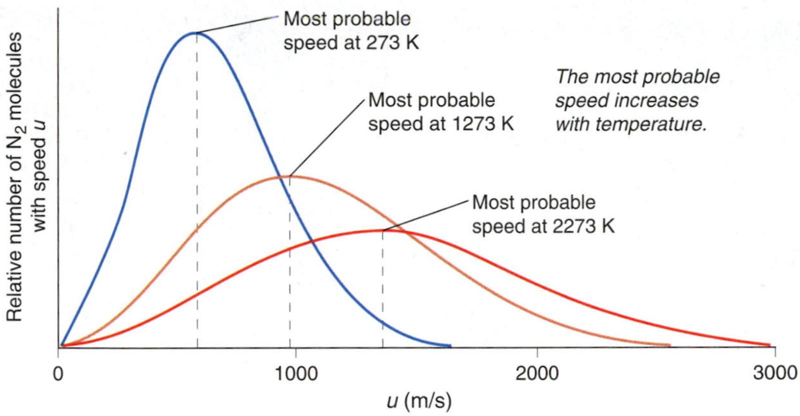
Most probable speed at 273 K
Most probable speed at 1273 K
The most probable speed increases with temperature.
Most probable speed at 2273 K
Relative number of N2 molecules with speed u
0
1000
2000
3000
u (m/s)

Figure 5.13 Distribution of molecular speeds for N_2 at three temperatures.

increase occurs because the average kinetic energy of the molecules, which is related to the most probable speed, is proportional to the absolute temperature: $\overline{E_k} \propto T$, or $\overline{E_k} = cT$, where $\overline{E_k}$ is the average kinetic energy of the molecules (an overbar indicates the average value of a quantity) and c is a constant that is the same for any gas. (We'll return to this equation shortly.) Thus, a major conclusion based on the distribution of speeds, which arises directly from postulate 3, is that *at a given temperature, all gases have the same average kinetic energy.*

A Molecular View of the Gas Laws Let's keep visualizing gas particles in a container to see how the theory explains the macroscopic behavior of gases and answers the questions we posed above:

1. *Origin of pressure* (Figure 5.14). From postulates 1 and 2, each gas particle (point of mass) colliding with the container walls (and bottom of piston) exerts a force. Countless collisions over the inner surface of the container result in a pressure. The greater the number of particles, the more frequently they collide with the container, and so the greater the pressure.

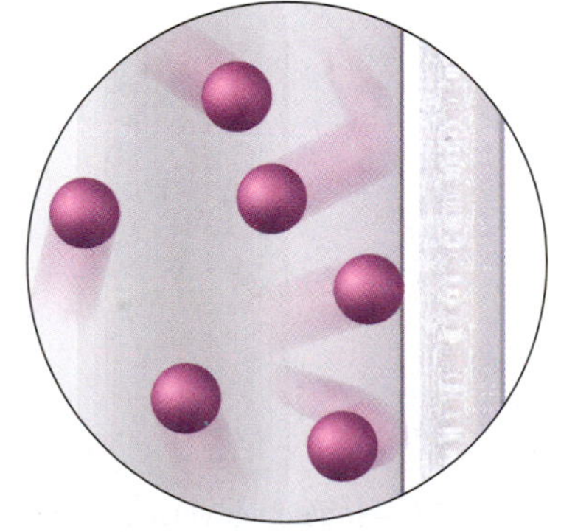

Figure 5.14 Pressure arises from countless collisions between gas particles and walls.

2. *Boyle's law* ($V \propto 1/P$) (Figure 5.15). The particles in a gas are points of mass with empty space between them (postulate 1). Before any change in pressure, the pressure exerted *by* the gas (P_{gas}) equals the pressure exerted *on* the gas (P_{ext}), and there is some average distance (d_1) between the particles and the container walls. As P_{ext} increases at constant temperature, the average distance (d_2) between the particles and the walls decreases (that is, $d_2 < d_1$), and so the sample volume decreases. Collisions of the particles with the walls become more frequent over the shorter average distance, which causes P_{gas} to increase until it again equals P_{ext}. The fact that liquids and solids cannot be compressed implies there is little, if any, free space between their particles.

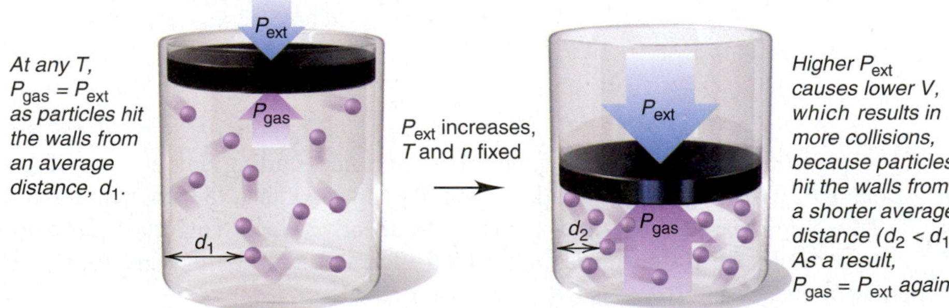
At any T, Pgas = Pext as particles hit the walls from an average distance, d1.
Pext
Pgas
d1
Pext increases, T and n fixed
Pext
Pgas
d2
Higher Pext causes lower V, which results in more collisions, because particles hit the walls from a shorter average distance (d2 < d1). As a result, Pgas = Pext again.

Figure 5.15 A molecular view of Boyle's law.

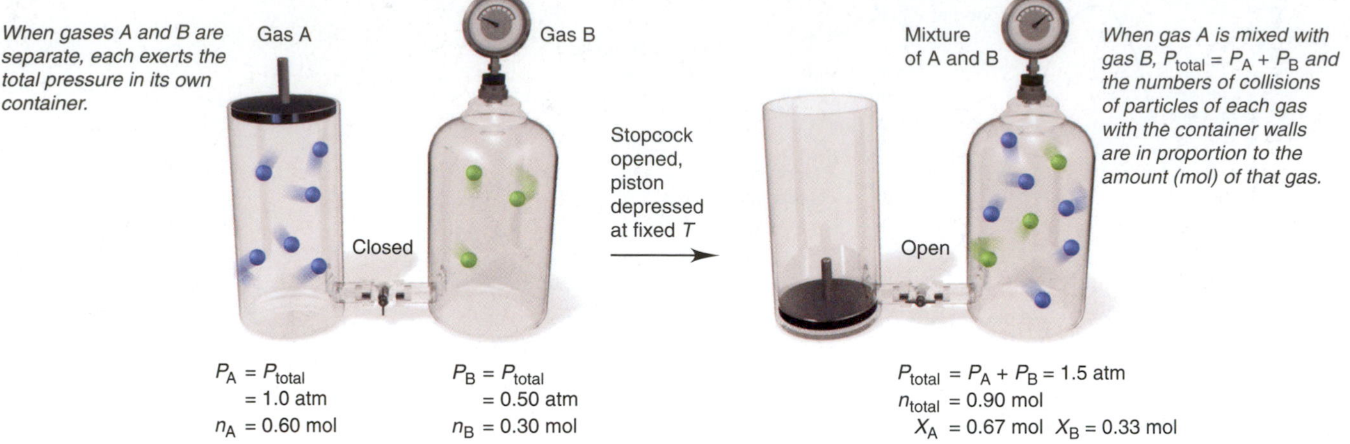
When gases A and B are separate, each exerts the total pressure in its own container.
Gas A
Gas B
Closed
Stopcock opened, piston depressed at fixed T
Mixture of A and B
Open
When gas A is mixed with gas B, $P_{total} = P_A + P_B$ and the numbers of collisions of particles of each gas with the container walls are in proportion to the amount (mol) of that gas.
$P_A = P_{total}$ = 1.0 atm
n_A = 0.60 mol
$P_B = P_{total}$ = 0.50 atm
n_B = 0.30 mol
$P_{total} = P_A + P_B$ = 1.5 atm
n_{total} = 0.90 mol
X_A = 0.67 mol X_B = 0.33 mol

Figure 5.16 A molecular view of Dalton's law.

3. *Dalton's law of partial pressures* ($P_{total} = P_A + P_B$) (Figure 5.16). Adding a given amount (mol) of gas A to a given amount of gas B causes an increase in the total number of particles, in proportion to the particles of A added. This increase causes a corresponding increase in the total number of collisions with the walls per second (postulate 2), which causes a corresponding increase in the total pressure of the gas mixture (P_{total}). Each gas exerts a fraction of P_{total} in proportion to its fraction of the total number of particles (or equivalently, its fraction of the total number of moles, that is, the mole fraction).

4. *Charles's law* ($V \propto T$) (Figure 5.17). At some starting temperature, T_1, the external (atmospheric) pressure (P_{atm}) equals the pressure of the gas (P_{gas}). When the gas is heated and the temperature increases to T_2, the most probable molecular speed and the average kinetic energy increase (postulate 3). Thus, the particles hit the walls more frequently *and* more energetically. This change temporarily increases P_{gas}. As a result, the piston moves up, which increases the volume and lowers the collision frequency until P_{atm} and P_{gas} are again equal.

Figure 5.17 A molecular view of Charles's law.

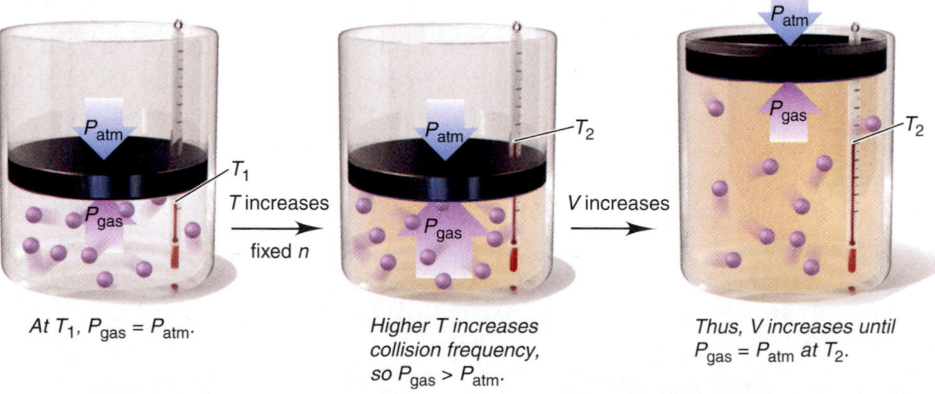
P_{atm}
P_{gas}
T_1
T increases
fixed n
T_2
V increases
At T_1, $P_{gas} = P_{atm}$.
Higher T increases collision frequency, so $P_{gas} > P_{atm}$.
Thus, V increases until $P_{gas} = P_{atm}$ at T_2.

5. *Avogadro's law* ($V \propto n$) (Figure 5.18). At some starting amount, n_1, of gas, P_{atm} equals P_{gas}. When more gas is added from the attached tank, the amount increases to n_2. Thus, more particles hit the walls more frequently, which temporarily increases P_{gas}. As a result, the piston moves up, which increases the volume and lowers the collision frequency until P_{atm} and P_{gas} are again equal.

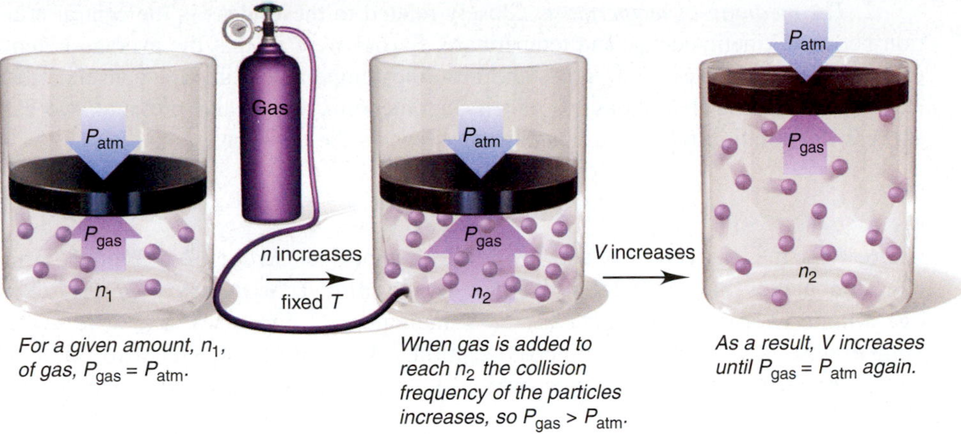

Figure 5.18 A molecular view of Avogadro's law.

The Central Importance of Kinetic Energy Recall from Chapter 1 that the kinetic energy of an object is the energy associated with its motion. It is key to explaining some implications of Avogadro's law and, most importantly, the meaning of temperature.

1. *Implications of Avogadro's law.* As we just saw, Avogadro's law says that, at any given T and P, the volume of a gas depends only on the number of moles—that is, number of particles—in the sample. The law doesn't mention the chemical nature of the gas, so equal numbers of particles of any two gases, say O_2 and H_2, should occupy the same volume. But, why don't the heavier O_2 molecules exert more pressure on the container walls, and thus take up more volume, than the lighter H_2 molecules? To answer this, we'll show one way to express kinetic energy mathematically:

$$E_k = \frac{1}{2}\text{mass} \times \text{speed}^2$$

This equation says that, for a given E_k, an object's mass and speed are inversely related, which means that *a heavier object moving slower can have the same kinetic energy as a lighter object moving faster.* Figure 5.19 shows that, for several gases, *the most probable speed (top of each curve) increases as the molar mass (number in parentheses) decreases.*

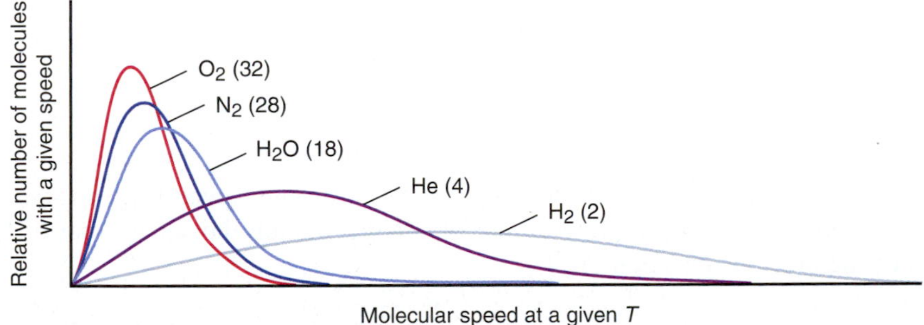

Figure 5.19 The relationship between molar mass and molecular speed. At a given temperature, gases with lower molar masses (numbers in parentheses) have higher most probable speeds (peak of each curve).

As we saw earlier, postulate 3 of the kinetic-molecular theory directly implies that, at a given T, all gases have the same average kinetic energy. From Figure 5.19, we see that O_2 molecules move more slowly, on average, than H_2 molecules. With their higher most probable speed, H_2 molecules collide with the walls of a container more often than O_2 molecules do, but their lower mass means that each collision has less force. Therefore, at a given T, equimolar samples of H_2 and O_2 (or any other gas) exert the same pressure and, thus, occupy the same volume because, on average, *their molecules hit the walls with the same kinetic energy.*

2. *The meaning of temperature.* Closely related to these ideas is the central relation between kinetic energy and temperature. Earlier we said that the average kinetic energy of the particles ($\overline{E_k}$) equals the absolute temperature times a constant; that is, $\overline{E_k} = cT$. Using definitions of velocity, momentum, force, and pressure, we can express this relationship by an equation that reveals the constant c:

$$\overline{E_k} = \frac{3}{2}\left(\frac{R}{N_A}\right)T$$

where R is the gas constant and N_A is the symbol for Avogadro's number. This equation makes the essential point that *temperature is a* ***measure*** *of the average kinetic energy of the particles:* as T increases, $\overline{E_k}$ increases, and vice versa. Temperature is an intensive property (Section 1.4), so it is not related to the *total* energy of motion of the particles, which depends on the size of the sample, but to the *average* energy.

Thus, for example, in the macroscopic world, we heat a beaker of water over a flame and see the mercury rise inside a thermometer we put in the beaker. We see this because, in the molecular world, kinetic energy transfers from higher energy gas particles in the flame, in turn, to lower energy particles in the beaker glass, the water molecules, the particles in the thermometer glass, and the atoms of mercury.

Root-Mean-Square Speed Finally, let's derive an expression for the speed of a gas particle that has the average kinetic energy of the particles in a sample. From the general expression for kinetic energy of an object,

$$E_k = \tfrac{1}{2}\text{mass} \times \text{speed}^2$$

the average kinetic energy of each particle in a large population is

$$\overline{E_k} = \tfrac{1}{2}m\overline{u^2}$$

where m is the particle's mass (atomic or molecular) and $\overline{u^2}$ is the average of the squares of the molecular speeds. Setting these two expressions for average kinetic energy equal to each other gives

$$\tfrac{1}{2}m\overline{u^2} = \frac{3}{2}\left(\frac{R}{N_A}\right)T$$

Multiplying through by Avogadro's number, N_A, gives the average kinetic energy for a mole of gas particles:

$$\tfrac{1}{2}N_A\, m\overline{u^2} = \tfrac{3}{2}RT$$

Avogadro's number times the molecular mass, $N_A \times m$, is the molar mass, $\mathcal{M}$, and solving for $\overline{u^2}$, we have

$$\overline{u^2} = \frac{2}{\mathcal{M}} \times \frac{3}{2}RT = \frac{3RT}{\mathcal{M}}$$

The square root of $\overline{u^2}$ is the root-mean-square speed, or **rms speed (u_{rms}):** *a particle moving at this speed has the average kinetic energy.* Taking the square root of both sides of the previous equation gives

$$u_{rms} = \sqrt{\frac{3RT}{\mathcal{M}}} \quad \textbf{(5.13)}$$

where R is the gas constant, T is the absolute temperature, and $\mathcal{M}$ is the molar mass. (Because we want u in m/s and R includes the joule, which has units of $\text{kg}\cdot\text{m}^2/\text{s}^2$, we use the value 8.314 J/mol·K for R and express $\mathcal{M}$ in kg/mol.)

Thus, as an example, the root-mean-square speed of an O_2 molecule ($\mathcal{M} = 3.200\times10^{-2}$ kg/mol) at room temperature (20°C, or 293 K) in the air you're breathing right now is

$$u_{rms} = \sqrt{\frac{3RT}{\mathcal{M}}} = \sqrt{\frac{3(8.314\ \text{J/mol}\cdot\text{K})(293\ \text{K})}{3.200\times10^{-2}\ \text{kg/mol}}} = \sqrt{\frac{3(8.314\ \cancel{\text{kg}}\cdot\text{m}^2/\text{s}^2/\cancel{\text{mol}}\cdot\cancel{\text{K}})(293\ \cancel{\text{K}})}{3.200\times10^{-2}\ \cancel{\text{kg/mol}}}}$$

$$= 478\ \text{m/s (about 1070 mi/hr)}$$

Effusion and Diffusion

The movement of a gas into a vacuum and the movement of gases through one another are phenomena with some vital applications.

The Process of Effusion One of the early triumphs of the kinetic-molecular theory was an explanation of **effusion,** the process by which a gas escapes through a tiny hole in its container into an evacuated space. In 1846, Thomas Graham studied the effusion rate of a gas, the number of molecules escaping per unit time, and found that it was inversely proportional to the square root of the gas density. But, density is directly related to molar mass, so **Graham's law of effusion** is stated as follows: *the rate of effusion of a gas is inversely proportional to the square root of its molar mass,* or

$$\text{Rate of effusion} \propto \frac{1}{\sqrt{\mathcal{M}}}$$

Argon (Ar) is lighter than krypton (Kr), so it effuses faster, assuming equal pressures of the two gases (Figure 5.20). Thus, the ratio of the rates is

$$\frac{\text{Rate}_{\text{Ar}}}{\text{Rate}_{\text{Kr}}} = \frac{\sqrt{\mathcal{M}_{\text{Kr}}}}{\sqrt{\mathcal{M}_{\text{Ar}}}} \quad \text{or, in general,} \quad \frac{\text{Rate}_{\text{A}}}{\text{Rate}_{\text{B}}} = \frac{\sqrt{\mathcal{M}_{\text{B}}}}{\sqrt{\mathcal{M}_{\text{A}}}} = \sqrt{\frac{\mathcal{M}_{\text{B}}}{\mathcal{M}_{\text{A}}}} \qquad (5.14)$$

The kinetic-molecular theory explains that, at a given temperature and pressure, *the gas with the lower molar mass effuses faster because the rms speed of its molecules is higher; therefore, more molecules collide with the hole and escape per unit time.*

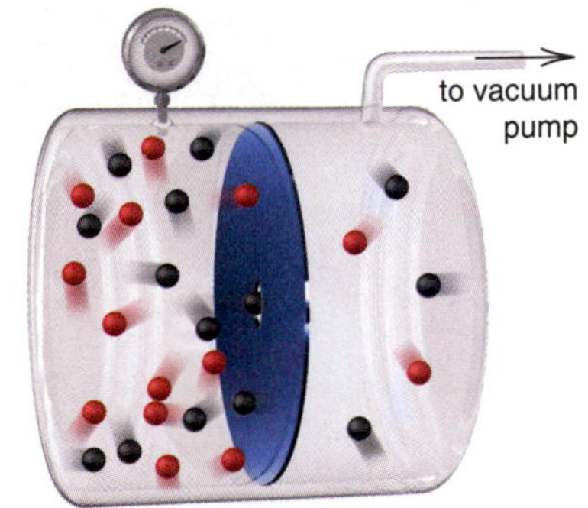

Figure 5.20 Effusion. Lighter *(black)* particles effuse faster than heavier *(red)* particles.

Sample Problem 5.13 Applying Graham's Law of Effusion

Problem A mixture of helium (He) and methane (CH_4) is placed in an effusion apparatus. Calculate the ratio of their effusion rates.

Plan The effusion rate is inversely proportional to $\sqrt{\mathcal{M}}$, so we find the molar mass of each substance from the formula and take its square root. The inverse of the ratio of the square roots is the ratio of the effusion rates.

Solution $\mathcal{M}$ of CH_4 = 16.04 g/mol $\quad$ $\mathcal{M}$ of He = 4.003 g/mol

Calculating the ratio of the effusion rates:

$$\frac{\text{Rate}_{\text{He}}}{\text{Rate}_{\text{CH}_4}} = \sqrt{\frac{\mathcal{M}_{\text{CH}_4}}{\mathcal{M}_{\text{He}}}} = \sqrt{\frac{16.04\ \text{g/mol}}{4.003\ \text{g/mol}}} = \sqrt{4.007} = 2.002$$

Check A ratio >1 makes sense because the lighter He should effuse faster than the heavier CH_4. Because the molar mass of CH_4 is about four times the molar mass of He, He should effuse about twice as fast as CH_4 ($\sqrt{4}$).

FOLLOW-UP PROBLEM 5.13 If it takes 1.25 min for 0.010 mol of He to effuse, how long will it take for the same amount of ethane (C_2H_6) to effuse?

Applications of Effusion The process of effusion has two important uses.

1. *Determination of molar mass.* We can also use Graham's law to *determine the molar mass of an unknown gas.* By comparing the effusion rate of gas X with that of a known gas, such as He, we can solve for the molar mass of X:

$$\frac{\text{Rate}_{\text{X}}}{\text{Rate}_{\text{He}}} = \sqrt{\frac{\mathcal{M}_{\text{He}}}{\mathcal{M}_{\text{X}}}}$$

Squaring both sides and solving for the molar mass of X gives

$$\mathcal{M}_{\text{X}} = \mathcal{M}_{\text{He}} \times \left(\frac{\text{rate}_{\text{He}}}{\text{rate}_{\text{X}}}\right)^2$$

2. *Preparation of nuclear fuel.* By far the most important application of Graham's law is in the preparation of fuel for nuclear energy reactors. The process of *isotope enrichment* increases the proportion of fissionable, but rarer, ^{235}U (only 0.7% by mass of naturally occurring uranium) to the nonfissionable, more abundant ^{238}U (99.3% by mass). Because the two isotopes have identical chemical properties, they are extremely difficult to separate chemically. But, one way to separate them takes advantage of a difference in a physical property—the effusion rate of their gaseous compounds. Uranium ore is treated with fluorine to yield a gaseous mixture of $^{238}UF_6$ and $^{235}UF_6$ that is pumped through a series of chambers separated by porous barriers. Molecules of $^{235}UF_6$ are slightly lighter ($\mathcal{M} = 349.03$) than molecules of $^{238}UF_6$ ($\mathcal{M} = 352.04$), so they move slightly faster and effuse through each barrier 1.0043 times faster. Many passes must be made, each one increasing the fraction of $^{235}UF_6$, until the mixture obtained is 3–5% by mass $^{235}UF_6$. This process was developed during the latter years of World War II and produced enough ^{235}U for two of the world's first atomic bombs.

The Process of Diffusion Closely related to effusion is the process of gaseous **diffusion,** the movement of one gas through another. Diffusion rates are also described generally by Graham's law:

$$\text{Rate of diffusion} \propto \frac{1}{\sqrt{\mathcal{M}}}$$

For two gases at equal pressures, such as NH_3 and HCl, moving through another gas or a mixture of gases, such as air, we find

$$\frac{\text{Rate}_{NH_3}}{\text{Rate}_{HCl}} = \sqrt{\frac{\mathcal{M}_{HCl}}{\mathcal{M}_{NH_3}}}$$

The reason for this dependence on molar mass is the same as for effusion rates: *lighter molecules have higher average speeds than heavier molecules, so they move farther in a given time.*

THINK OF IT THIS WAY
A Bizarre (and Dangerous) Molecular Highway

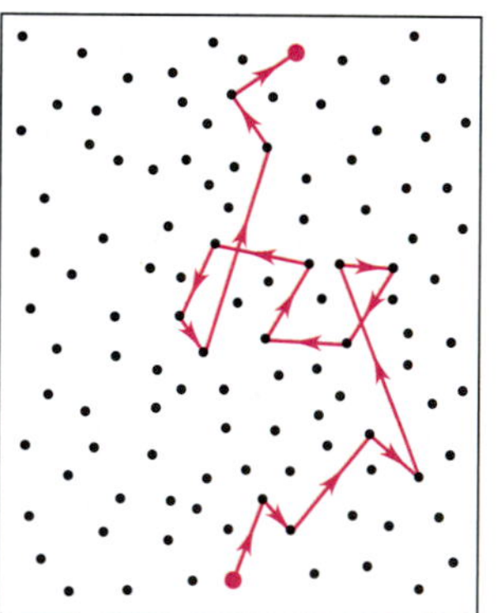

Why don't you smell perfume until a few seconds after the bottle has been opened? One reason is that a molecule of the scent doesn't travel very far before it collides with another molecule in the air. Look at the tortuous path of the red "odor" molecule, and imagine how much faster you could walk through an empty room than through a room full of moving people. Behavior in the molecular world is almost inconceivable in everyday life: to match the number of collisions per second of an N_2 molecule in room air, a bumper car in an enormous amusement-park ride would travel at 2.8 billion mi/s (4.5 billion km/s, much faster than the speed of light!) and would smash into another car every 700 yd (640 m).

Diffusion also occurs when a gas enters a liquid (and to a small extent in a solid). However, the average distances between molecules are so much shorter in a liquid that collisions are much more frequent; thus, diffusion of a gas through a liquid is *much* slower than through a gas. Nevertheless, this type of diffusion is vital in organisms, for example, in the movement of O_2 from lungs to blood.

Summary of Section 5.5

- The kinetic-molecular theory postulates that gas particles have no volume, move in straight-line paths between elastic (energy-conserving) collisions, and have average kinetic energies proportional to the absolute temperature of the gas.
- This theory explains the gas laws in terms of changes in distances between particles and the container walls, changes in molecular speed, and the energy of collisions.
- Temperature is a measure of the average kinetic energy of the particles.
- Molecular motion is characterized by a temperature-dependent, most probable speed (within a range of speeds).
- Effusion and diffusion rates are *inversely* proportional to the square root of the molar mass (Graham's law) because they are *directly* proportional to molecular speed.

5.6 • REAL GASES: DEVIATIONS FROM IDEAL BEHAVIOR

A fundamental principle of science is that simpler models are more useful than complex ones, as long as they explain the data. With only a few postulates, the kinetic-molecular theory explains the behavior of most gases under ordinary conditions. But two of the postulates are useful approximations that do not reflect reality:

1. *Gas particles are* ***not*** *points of mass* but have volumes determined by the sizes of their atoms and the lengths and directions of their bonds.
2. *Attractive and repulsive forces* ***do*** *exist among gas particles* because atoms contain charged subatomic particles and many bonds are polar. (As you'll see in Chapter 12, such forces lead to changes of physical state.)

These two features cause deviations from ideal behavior under *extreme conditions of low temperature and high pressure.* These deviations mean that we must alter the simple model and the ideal gas law to predict the behavior of real gases.

Effects of Extreme Conditions on Gas Behavior

At ordinary conditions—relatively high temperatures and low pressures—most real gases exhibit nearly ideal behavior. Yet, even at STP (0°C and 1 atm), gases deviate *slightly* from ideal behavior. Table 5.3 shows that the standard molar volumes of several gases, when measured to five significant figures, do not equal the ideal value. Note that the deviations increase as the boiling point rises.

The phenomena that cause slight deviations under standard conditions exert more influence as the temperature decreases and pressure increases. Figure 5.21 shows a plot of PV/RT versus external pressure (P_{ext}) for 1 mol of several real gases and an ideal gas. The values on the horizontal axis are the external pressures at which the PV/RT ratios were calculated. The PV/RT values range from normal (at P_{ext} = 1 atm, PV/RT = 1) to very high (at $P_{ext} \approx$ 1000 atm, $PV/RT \approx$ 1.6 to 2.3). For the *ideal* gas, PV/RT is 1 at any P_{ext}.

The PV/RT curve for methane (CH_4) is typical of most gases: it decreases *below* the ideal value at moderately high P_{ext} and then rises *above* the ideal value as P_{ext} increases to very high values. This shape arises from two overlapping effects:

- At moderately high P_{ext}, PV/RT values are lower than ideal values (less than 1) because of *interparticle attractions.*
- At very high P_{ext}, PV/RT values are greater than ideal values (more than 1) because of *particle volume.*

Table 5.3 Molar Volume of Some Common Gases at STP (0°C and 1 atm)

Gas	Molar Volume (L/mol)	Boiling Point (°C)
He	22.435	−268.9
H_2	22.432	−252.8
Ne	22.422	−246.1
Ideal gas	**22.414**	—
Ar	22.397	−185.9
N_2	22.396	−195.8
O_2	22.390	−183.0
CO	22.388	−191.5
Cl_2	22.184	−34.0
NH_3	22.079	−33.4

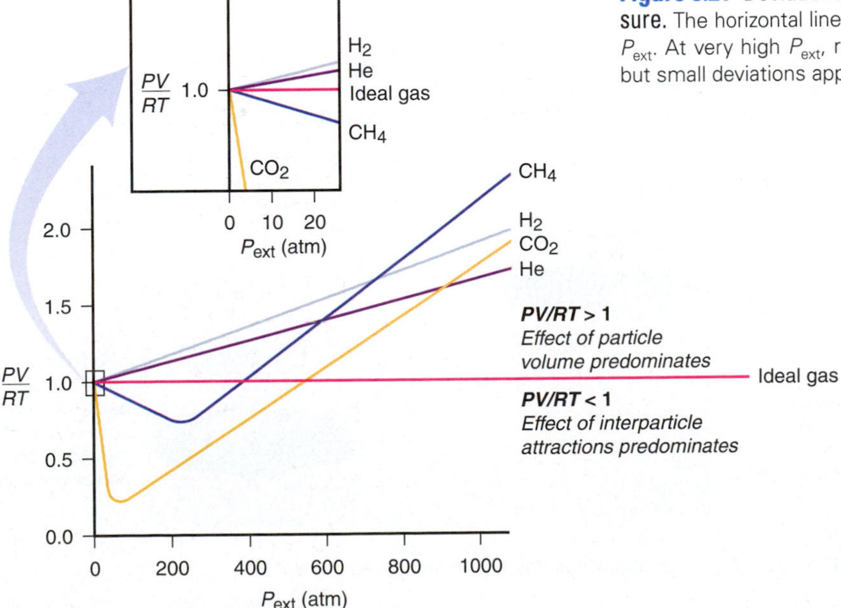

Figure 5.21 Deviations from ideal behavior with increasing external pressure. The horizontal line shows that, for 1 mol of ideal gas, PV/RT = 1 at all P_{ext}. At very high P_{ext}, real gases deviate significantly from ideal behavior, but small deviations appear even at ordinary pressures (expanded portion).

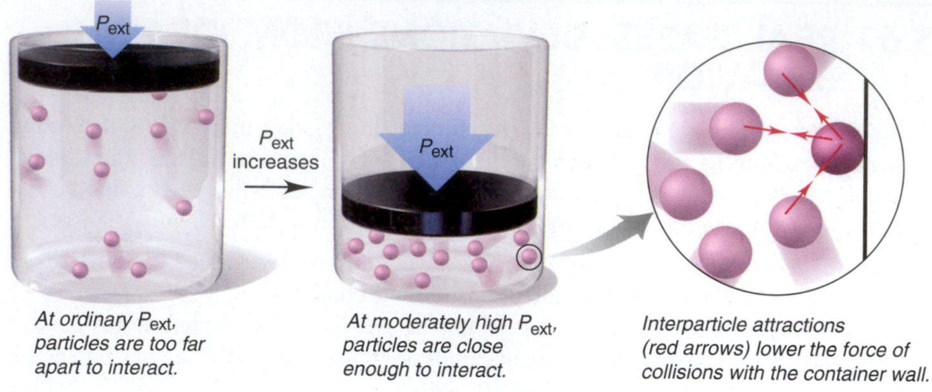

Figure 5.22 The effect of interparticle attractions on measured gas pressure.

Let's examine these effects on the molecular level:

1. *Effect of interparticle attractions.* Interparticle attractions occur between separate atoms or molecules and are caused by imbalances in electron distributions. They are important only over *very* short distances and are *much* weaker than the covalent bonding forces that hold a molecule together. At normal P_{ext}, the spaces between the particles are so large that attractions are negligible and the gas behaves nearly ideally. As P_{ext} rises, the volume of the sample decreases and the particles get closer together, so interparticle attractions have a greater effect. As a particle approaches the container wall under these higher pressures, nearby particles attract it, which lessens the force of its impact (Figure 5.22). *Repeated throughout the sample, this effect results in decreased gas pressure and, thus, a smaller numerator in PV/RT.* Similarly, lowering the temperature slows the particles, so they attract each other for a longer time.

2. *Effect of particle volume.* At normal P_{ext}, the space between particles (free volume) is enormous compared with the volume of the particles *themselves* (particle volume); thus, the free volume is essentially equal to V, the container volume in PV/RT. At *moderately* high P_{ext} and as free volume decreases, the particle volume makes up an increasing proportion of the container volume. At *extremely* high pressures, the particle volume makes the free volume significantly *less* than the container volume (Figure 5.23). But, we continue to use the container volume for V in PV/RT. Thus, the numerator *and* the ratio become artificially high. This particle volume effect increases as P_{ext} increases, eventually outweighing the effect of interparticle attractions and causing PV/RT to rise above the ideal value.

In Figure 5.21, note that the H_2 and He curves do not show the typical dip at moderate pressures. These gases consist of particles with such weak interparticle attractions that the particle volume effect predominates at all pressures.

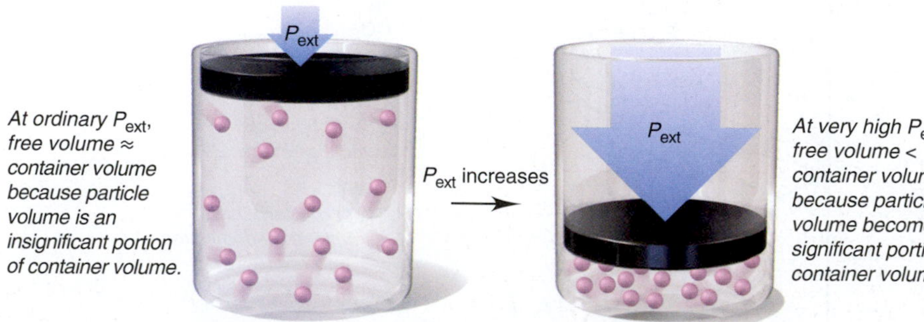

Figure 5.23 The effect of particle volume on measured gas volume.

The van der Waals Equation: Adjusting the Ideal Gas Law

To describe real gas behavior more accurately, we need to adjust the ideal gas equation in two ways:

1. Adjust P *up* by adding a factor that accounts for interparticle attractions.
2. Adjust V *down* by subtracting a factor that accounts for particle volume.

In 1873, Johannes van der Waals revised the ideal gas equation to account for the behavior of real gases. The **van der Waals equation** for n moles of a real gas is

$$\underbrace{\left(P + \frac{n^2a}{V^2}\right)}_{\text{adjusts } P \text{ up}}\underbrace{(V - nb)}_{\text{adjusts } V \text{ down}} = nRT \qquad (5.15)$$

where P is the measured pressure, V is the known container volume, n and T have their usual meanings, and a and b are **van der Waals constants,** experimentally determined and specific for a given gas (Table 5.4). The constant a depends on the number and distribution of electrons, which relate to the complexity of a particle and the strength of its interparticle attractions. The constant b relates to particle volume. For instance, CO_2 is both more complex and larger than H_2, and the values of their constants reflect this.

Here is a typical application of the van der Waals equation. A 1.98-L vessel contains 215 g (4.89 mol) of dry ice. After standing at 26°C (299 K), the $CO_2(s)$ changes to $CO_2(g)$. The pressure is measured (P_{real}) and then calculated by the ideal gas law (P_{IGL}) and, using the appropriate values of a and b, by the van der Waals equation (P_{VDW}). The results are revealing:

$$P_{real} = 44.8 \text{ atm} \qquad P_{IGL} = 60.6 \text{ atm} \qquad P_{VDW} = 45.9 \text{ atm}$$

Comparing the real value with each calculated value shows that P_{IGL} is 35.3% greater than P_{real}, but P_{VDW} is only 2.5% greater than P_{real}. At these conditions, CO_2 deviates so much from ideal behavior that the ideal gas law is not very useful.

Table 5.4 Van der Waals Constants for Some Common Gases

Gas	$a\left(\frac{atm \cdot L^2}{mol^2}\right)$	$b\left(\frac{L}{mol}\right)$
He	0.034	0.0237
Ne	0.211	0.0171
Ar	1.35	0.0322
Kr	2.32	0.0398
Xe	4.19	0.0511
H_2	0.244	0.0266
N_2	1.39	0.0391
O_2	1.36	0.0318
Cl_2	6.49	0.0562
CH_4	2.25	0.0428
CO	1.45	0.0395
CO_2	3.59	0.0427
NH_3	4.17	0.0371
H_2O	5.46	0.0305

Summary of Section 5.6

- At very high P or low T, all gases deviate significantly from ideal behavior.
- As external pressure increases, most real gases exhibit first a lower and then a higher PV/RT; for 1 mol of an ideal gas, this ratio remains constant at 1.
- The deviations from ideal behavior are due to (1) attractions between particles, which lower the pressure (and decrease PV/RT), and (2) the volume of the particles themselves, which takes up an increasingly larger fraction of the container volume (and increases PV/RT).
- The van der Waals equation includes constants specific for a given gas to correct for deviations from ideal behavior.

CHAPTER REVIEW GUIDE

The following sections provide many aids to help you study this chapter. (Numbers in parentheses refer to pages, unless noted otherwise.)

Learning Objectives

These are concepts and skills to review after studying this chapter.

Related section (§), sample problem (SP), and end-of-chapter problem (EP) numbers are listed in parentheses.

1. Explain how gases differ from liquids and solids (§5.1) (EPs 5.1, 5.2)
2. Understand how a barometer works and interconvert units of pressure (§5.2) (SP 5.1) (EPs 5.3–5.9)
3. Describe Boyle's, Charles's, and Avogadro's laws, understand how they relate to the ideal gas law, and apply them in calculations (§5.3) (SPs 5.2–5.6) (EPs 5.10–5.24)
4. Apply the ideal gas law to determine the density of a gas at different temperatures, the molar mass of a gas, and the partial pressure (or mole fraction) of each gas in a mixture (Dalton's law) (§5.4) (SPs 5.7–5.10) (EPs 5.25–5.37, 5.44–5.47)
5. Use stoichiometry and the gas laws to calculate amounts of reactants and products (§5.4) (SPs 5.11, 5.12) (EPs 5.38–5.43, 5.48–5.51)
6. Understand the kinetic-molecular theory and how it explains the gas laws, average molecular speed and kinetic energy, and the processes of effusion and diffusion (§5.5) (SP 5.13) (EPs 5.52–5.63)
7. Explain why interparticle attractions and particle volume cause real gases to deviate from ideal behavior and how the van der Waals equation corrects for the deviations (§5.6) (EPs 5.64–5.67)

Key Terms

These important terms appear in boldface in the chapter and are defined again in the Glossary.

Section 5.2

pressure (P) (150)
barometer (150)
pascal (Pa) (151)
standard atmosphere (atm) (151)
millimeter of mercury (mmHg) (151)
torr (151)

Section 5.3

ideal gas (153)
Boyle's law (154)
Charles's law (154)
Avogadro's law (156)
standard temperature and pressure (STP) (156)
standard molar volume (156)
ideal gas law (157)
universal gas constant (R) (157)

Section 5.4

partial pressure (165)
Dalton's law of partial pressures (165)
mole fraction (X) (165)

Section 5.5

kinetic-molecular theory (170)
rms speed (u_{rms}) (174)
effusion (175)
Graham's law of effusion (175)
diffusion (176)

Section 5.6

van der Waals equation (179)
van der Waals constants (179)

Key Equations and Relationships

Numbered and screened concepts are listed for you to refer to or memorize.

5.1 Expressing the volume-pressure relationship (Boyle's law) (154):

$$V \propto \frac{1}{P} \quad \text{or} \quad PV = \text{constant} \quad [T \text{ and } n \text{ fixed}]$$

5.2 Expressing the volume-temperature relationship (Charles's law) (154):

$$V \propto T \quad \text{or} \quad \frac{V}{T} = \text{constant} \quad [P \text{ and } n \text{ fixed}]$$

5.3 Expressing the pressure-temperature relationship (Amontons's law) (155):

$$P \propto T \quad \text{or} \quad \frac{P}{T} = \text{constant} \quad [V \text{ and } n \text{ fixed}]$$

5.4 Expressing the volume-amount relationship (Avogadro's law) (156):

$$V \propto n \quad \text{or} \quad \frac{V}{n} = \text{constant} \quad [P \text{ and } T \text{ fixed}]$$

5.5 Defining standard temperature and pressure (156):

STP: 0°C (273.15 K) and 1 atm (760 torr)

5.6 Stating the volume of 1 mol of an ideal gas at STP (156):

Standard molar volume = 22.4141 L = 22.4 L [3 sf]

5.7 Relating volume to pressure, temperature, and amount (ideal gas law) (157):

$$PV = nRT \quad \text{and} \quad \frac{P_1V_1}{n_1T_1} = \frac{P_2V_2}{n_2T_2}$$

5.8 Calculating the value of R (157):

$$R = \frac{PV}{nT} = \frac{1 \text{ atm} \times 22.4141 \text{ L}}{1 \text{ mol} \times 273.15 \text{ K}}$$

$$= 0.082058 \frac{\text{atm·L}}{\text{mol·K}} = 0.0821 \frac{\text{atm·L}}{\text{mol·K}} \quad [3 \text{ sf}]$$

5.9 Rearranging the ideal gas law to find gas density (163):

$$PV = \frac{m}{\mathcal{M}} RT$$

so

$$\frac{m}{V} = d = \frac{\mathcal{M} \times P}{RT}$$

5.10 Rearranging the ideal gas law to find molar mass (164):

$$n = \frac{m}{\mathcal{M}} = \frac{PV}{RT} \quad \text{so} \quad \mathcal{M} = \frac{mRT}{PV}$$

5.11 Relating the total pressure of a gas mixture to the partial pressures of the components (Dalton's law of partial pressures) (165):

$$P_{total} = P_1 + P_2 + P_3 + \cdots$$

5.12 Relating partial pressure to mole fraction (165):

$$P_A = X_A \times P_{total}$$

5.13 Defining rms speed as a function of molar mass and temperature (174):

$$u_{rms} = \sqrt{\frac{3RT}{\mathcal{M}}}$$

5.14 Applying Graham's law of effusion (175):

$$\frac{\text{Rate}_A}{\text{Rate}_B} = \frac{\sqrt{\mathcal{M}_B}}{\sqrt{\mathcal{M}_A}} = \sqrt{\frac{\mathcal{M}_B}{\mathcal{M}_A}}$$

5.15 Applying the van der Waals equation to find the pressure or volume of a gas under extreme conditions (179):

$$\left(P + \frac{n^2a}{V^2}\right)(V - nb) = nRT$$

BRIEF SOLUTIONS TO FOLLOW-UP PROBLEMS

Compare your own solutions to these calculation steps and answers.

5.1 $P_{CO_2}\text{ (torr)} = (753.6\text{ mmHg} - 174.0\text{ mmHg}) \times \frac{1\text{ torr}}{1\text{ mmHg}}$

$= 579.6\text{ torr}$

$P_{CO_2}\text{ (Pa)} = 579.6\text{ torr} \times \frac{1\text{ atm}}{760\text{ torr}} \times \frac{1.01325\times10^5\text{ Pa}}{1\text{ atm}}$

$= 7.727\times10^4\text{ Pa}$

$P_{CO_2}\text{ (lb/in}^2) = 579.6\text{ torr} \times \frac{1\text{ atm}}{760\text{ torr}} \times \frac{14.7\text{ lb/in}^2}{1\text{ atm}}$

$= 11.2\text{ lb/in}^2$

5.2 $P_2\text{ (atm)} = 26.3\text{ kPa} \times \frac{1\text{ atm}}{101.325\text{ kPa}} = 0.260\text{ atm}$

$V_2\text{ (L)} = 105\text{ mL} \times \frac{1\text{ L}}{1000\text{ mL}} \times \frac{0.871\text{ atm}}{0.260\text{ atm}} = 0.352\text{ L}$

5.3 $T_2\text{ (K)} = 273\text{ K} \times \frac{9.75\text{ cm}^3}{6.83\text{ cm}^3} = 390.\text{ K}$

5.4 $P_2\text{ (torr)} = 793\text{ torr} \times \frac{35.0\text{ g} - 5.0\text{ g}}{35.0\text{ g}} = 680.\text{ torr}$

(There is no need to convert mass to moles because the ratio of masses equals the ratio of moles.)

5.5 $n = \frac{PV}{RT} = \frac{1.37\text{ atm} \times 438\text{ L}}{0.0821\frac{\text{atm·L}}{\text{mol·K}} \times 294\text{ K}} = 24.9\text{ mol }O_2$

$\text{Mass (g) of }O_2 = 24.9\text{ mol }O_2 \times \frac{32.00\text{ g }O_2}{1\text{ mol }O_2} = 7.97\times10^2\text{ g }O_2$

5.6 The balanced equation is $2CD(g) \longrightarrow C_2(g) + D_2(g)$, so n does not change. Therefore, given constant P, the temperature, T, must double: $T_1 = -73°C + 273.15 = 200\text{ K}$; so $T_2 = 400\text{ K}$, or $400\text{ K} - 273.15 = 127°C$.

5.7 $d\text{ (at 0°C and 380 torr)} = \frac{44.01\text{ g/mol} \times \frac{380\text{ torr}}{760\text{ torr/atm}}}{0.0821\frac{\text{atm·L}}{\text{mol·K}} \times 273\text{ K}}$

$= 0.982\text{ g/L}$

The density is lower at the smaller P because V is larger. In this case, d is lowered by one-half because P is one-half as much.

5.8

$\mathcal{M} = \frac{(68.697\text{ g} - 68.322\text{ g}) \times 0.0821\frac{\text{atm·L}}{\text{mol·K}} \times (273.15 + 95.0)\text{ K}}{\frac{740.\text{ torr}}{760\text{ torr/atm}} \times \frac{149\text{ mL}}{1000\text{ mL/L}}}$

$= 78.1\text{ g/mol}$

5.9 $n_{total} = \left(5.50\text{ g He} \times \frac{1\text{ mol He}}{4.003\text{ g He}}\right)$

$+ \left(15.0\text{ g Ne} \times \frac{1\text{ mol Ne}}{20.18\text{ g Ne}}\right)$

$+ \left(35.0\text{ g Kr} \times \frac{1\text{ mol Kr}}{83.80\text{ g Kr}}\right)$

$= 2.53\text{ mol}$

$P_{He} = \left(\frac{5.50\text{ g He} \times \frac{1\text{ mol He}}{4.003\text{ g He}}}{2.53\text{ mol}}\right) \times 1\text{ atm} = 0.543\text{ atm}$

$P_{Ne} = 0.294\text{ atm} \qquad P_{Kr} = 0.165\text{ atm}$

5.10 $P_{H_2} = 752\text{ torr} - 13.6\text{ torr} = 738\text{ torr}$

$\text{Mass (g) of }H_2 = \left(\frac{\frac{738\text{ torr}}{760\text{ torr/atm}} \times 1.495\text{ L}}{0.0821\frac{\text{atm·L}}{\text{mol·K}} \times 289\text{ K}}\right) \times \frac{2.016\text{ g }H_2}{1\text{ mol }H_2}$

$= 0.123\text{ g }H_2$

5.11 $H_2SO_4(aq) + 2NaCl(s) \longrightarrow Na_2SO_4(aq) + 2HCl(g)$

$n_{HCl} = 0.117\text{ kg NaCl} \times \frac{10^3\text{ g}}{1\text{ kg}} \times \frac{1\text{ mol NaCl}}{58.44\text{ g NaCl}} \times \frac{2\text{ mol HCl}}{2\text{ mol NaCl}}$

$= 2.00\text{ mol HCl}$

At STP, $V\text{ (mL)} = 2.00\text{ mol} \times \frac{22.4\text{ L}}{1\text{ mol}} \times \frac{10^3\text{ mL}}{1\text{ L}}$

$= 4.48\times10^4\text{ mL}$

5.12 $NH_3(g) + HCl(g) \longrightarrow NH_4Cl(s)$

$n_{NH_3} = 0.187$ mol and $n_{HCl} = 0.0522$ mol; thus, HCl is the limiting reactant.

n_{NH_3} after reaction

$= 0.187\text{ mol }NH_3 - \left(0.0522\text{ mol HCl} \times \frac{1\text{ mol }NH_3}{1\text{ mol HCl}}\right)$

$= 0.135\text{ mol }NH_3$

$P = \frac{0.135\text{ mol} \times 0.0821\frac{\text{atm·L}}{\text{mol·K}} \times 295\text{ K}}{10.0\text{ L}} = 0.327\text{ atm}$

5.13 $\frac{\text{Rate of He}}{\text{Rate of }C_2H_6} = \sqrt{\frac{30.07\text{ g/mol}}{4.003\text{ g/mol}}} = 2.741$

Time for C_2H_6 to effuse $= 1.25\text{ min} \times 2.741 = 3.43\text{ min}$

PROBLEMS

Problems with **colored** numbers are answered in Appendix E. Sections match the text and provide the numbers of relevant sample problems. Bracketed problems are grouped in pairs (indicated by a short rule) that cover the same concept. Comprehensive Problems are based on material from any section or previous chapter.

An Overview of the Physical States of Matter

5.1 How does a sample of gas differ in its behavior from a sample of liquid in each of the following situations?
(a) The sample is transferred from one container to a larger one.
(b) The sample is heated in an expandable container, but no change of state occurs.
(c) The sample is placed in a cylinder with a piston, and an external force is applied.

5.2 Are the particles in a gas farther apart or closer together than the particles in a liquid? Use your answer to explain each of the following general observations:
(a) Gases are more compressible than liquids.
(b) Gases flow much more freely than liquids.
(c) After thorough stirring, all gas mixtures are solutions.
(d) The density of a substance in the gas state is lower than in the liquid state.

Gas Pressure and Its Measurement

(Sample Problem 5.1)

5.3 How does a barometer work? Is the column of mercury in a barometer shorter when it is on a mountaintop or at sea level? Explain.

5.4 On a cool, rainy day, the barometric pressure is 730 mmHg. Calculate the barometric pressure in centimeters of water (cmH_2O) (d of Hg = 13.5 g/mL; d of H_2O = 1.00 g/mL).

5.5 A long glass tube, sealed at one end, has an inner diameter of 10.0 mm. The tube is filled with water and inverted into a pail of water. If the atmospheric pressure is 755 mmHg, how high (in mmH_2O) is the column of water in the tube (d of Hg = 13.5 g/mL; d of H_2O = 1.00 g/mL)?

5.6 Convert the following:
(a) 0.745 atm to mmHg
(b) 992 torr to bar
(c) 365 kPa to atm
(d) 804 mmHg to kPa

5.7 Convert the following:
(a) 76.8 cmHg to atm
(b) 27.5 atm to kPa
(c) 6.50 atm to bar
(d) 0.937 kPa to torr

5.8 Convert each of the pressures described below to atm:
(a) At the peak of Mt. Everest, atmospheric pressure is only 2.75×10^2 mmHg.
(b) A cyclist fills her bike tires to 86 psi.
(c) The surface of Venus has an atmospheric pressure of 9.15×10^6 Pa.
(d) At 100 ft below sea level, a scuba diver experiences a pressure of 2.54×10^4 torr.

5.9 The gravitational force exerted by Earth on an object is given by $F = mg$, where F is the force in newtons, m is the mass in kilograms, and g is the acceleration due to gravity (9.81 m/s^2).
(a) Use the definition of the pascal to calculate the mass (in kg) of the atmosphere above 1 m^2 of ocean.
(b) Osmium ($Z = 76$) is a transition metal in Group 8B(8) and has the highest density of any element (22.6 g/mL). If an osmium column is 1 m^2 in area, how high must it be for its pressure to equal atmospheric pressure? [Use the answer from part (a) in your calculation.]

The Gas Laws and Their Experimental Foundations

(Sample Problems 5.2 to 5.6)

5.10 A student states Boyle's law as follows: "The volume of a gas is inversely proportional to its pressure." How is this statement incomplete? Give a correct statement of Boyle's law.

5.11 In the following relationships, which quantities are variables and which are fixed: (a) Charles's law; (b) Avogadro's law; (c) Amontons's law?

5.12 Boyle's law relates gas volume to pressure, and Avogadro's law relates gas volume to amount (mol). State a relationship between gas pressure and amount (mol).

5.13 Each of the following processes caused the gas volume to double, as shown. For each process, tell how the remaining gas variable changed or state that it remained fixed:
(a) T doubles at fixed P.
(b) T and n are fixed.
(c) At fixed T, the reaction is $CD_2(g) \longrightarrow C(g) + D_2(g)$.
(d) At fixed P, the reaction is $A_2(g) + B_2(g) \longrightarrow 2AB(g)$.

5.14 What is the effect of the following on the volume of 1 mol of an ideal gas?
(a) The pressure is tripled (at constant T).
(b) The absolute temperature is increased by a factor of 3.0 (at constant P).
(c) Three more moles of the gas are added (at constant P and T).

5.15 What is the effect of the following on the volume of 1 mol of an ideal gas?
(a) The pressure is reduced by a factor of 4 (at constant T).
(b) The pressure changes from 760 torr to 202 kPa, and the temperature changes from 37°C to 155 K.
(c) The temperature changes from 305 K to 32°C, and the pressure changes from 2 atm to 101 kPa.

5.16 A sample of sulfur hexafluoride gas occupies 9.10 L at 198°C. Assuming that the pressure remains constant, what temperature (in °C) is needed to reduce the volume to 2.50 L?

5.17 A 93-L sample of dry air cools from 145°C to −22°C while the pressure is maintained at 2.85 atm. What is the final volume?

5.18 A sample of Freon-12 (CF_2Cl_2) occupies 25.5 L at 298 K and 153.3 kPa. Find its volume at STP.

5.19 A sample of carbon monoxide occupies 3.65 L at 298 K and 745 torr. Find its volume at −14°C and 367 torr.

5.20 A sample of chlorine gas is confined in a 5.0-L container at 328 torr and 37°C. How many moles of gas are in the sample?

5.21 If 1.47×10^{-3} mol of argon occupies a 75.0-mL container at 26°C, what is the pressure (in torr)?

5.22 You have 357 mL of chlorine trifluoride gas at 699 mmHg and 45°C. What is the mass (in g) of the sample?

5.23 A 75.0-g sample of dinitrogen monoxide is confined in a 3.1-L vessel. What is the pressure (in atm) at 115°C?

5.24 In preparation for a demonstration, your professor brings a 1.5-L bottle of sulfur dioxide into the lecture hall before class to allow the gas to reach room temperature. If the pressure gauge reads 85 psi and the temperature in the hall is 23°C, how many moles of sulfur dioxide are in the bottle? (*Hint:* The gauge reads zero when 14.7 psi of gas remains.)

Rearrangements of the Ideal Gas Law

(Sample Problems 5.7 to 5.12)

5.25 Why is moist air less dense than dry air?

5.26 To collect a beaker of H_2 gas by displacing the air already in the beaker, would you hold the beaker upright or inverted? Why? How would you hold the beaker to collect CO_2?

5.27 Why can we use a gas mixture, such as air, to study the general behavior of an ideal gas under ordinary conditions?

5.28 How does the partial pressure of gas A in a mixture compare to its mole fraction in the mixture? Explain.

5.29 The scene at right represents a portion of a mixture of four gases A *(purple)*, B *(black)*, C *(green)*, and D_2 *(orange)*.
(a) Which has the highest partial pressure?
(b) Which has the lowest partial pressure?
(c) If the total pressure is 0.75 atm, what is the partial pressure of D_2?

5.30 What is the density of Xe gas at STP?

5.31 Find the density of Freon-11 ($CFCl_3$) at 120°C and 1.5 atm.

5.32 How many moles of gaseous arsine (AsH_3) occupy 0.0400 L at STP? What is the density of gaseous arsine?

5.33 The density of a noble gas is 2.71 g/L at 3.00 atm and 0°C. Identify the gas.

5.34 Calculate the molar mass of a gas at 388 torr and 45°C if 206 ng occupies 0.206 μL.

5.35 When an evacuated 63.8-mL glass bulb is filled with a gas at 22°C and 747 mmHg, the bulb gains 0.103 g in mass. Is the gas N_2, Ne, or Ar?

5.36 After 0.600 L of Ar at 1.20 atm and 227°C is mixed with 0.200 L of O_2 at 501 torr and 127°C in a 400-mL flask at 27°C, what is the pressure in the flask?

5.37 A 355-mL container holds 0.146 g of Ne and an unknown amount of Ar at 35°C and a total pressure of 626 mmHg. Calculate the number of moles of Ar present.

5.38 How many grams of phosphorus react with 35.5 L of O_2 at STP to form tetraphosphorus decaoxide?

$$P_4(s) + 5O_2(g) \longrightarrow P_4O_{10}(s)$$

5.39 How many grams of potassium chlorate decompose to potassium chloride and 638 mL of O_2 at 128°C and 752 torr?

$$2KClO_3(s) \longrightarrow 2KCl(s) + 3O_2(g)$$

5.40 How many grams of phosphine (PH_3) can form when 37.5 g of phosphorus and 83.0 L of hydrogen gas react at STP?

$$P_4(s) + H_2(g) \longrightarrow PH_3(g) \quad \text{[unbalanced]}$$

5.41 When 35.6 L of ammonia and 40.5 L of oxygen gas at STP burn, nitrogen monoxide and water form. After the products return to STP, how many grams of nitrogen monoxide are present?

$$NH_3(g) + O_2(g) \longrightarrow NO(g) + H_2O(l) \quad \text{[unbalanced]}$$

5.42 Aluminum reacts with excess hydrochloric acid to form aqueous aluminum chloride and 35.8 mL of hydrogen gas over water at 27°C and 751 mmHg. How many grams of aluminum reacted?

5.43 How many liters of hydrogen gas are collected over water at 18°C and 725 mmHg when 0.84 g of lithium reacts with water? Aqueous lithium hydroxide also forms.

5.44 The air in a hot-air balloon at 744 torr is heated from 17°C to 60.0°C. Assuming that the amount (mol) of air and the pressure remain constant, what is the density of the air at each temperature? (The average molar mass of air is 28.8 g/mol.)

5.45 A sample of a liquid hydrocarbon known to consist of molecules with five carbon atoms is vaporized in a 0.204-L flask by immersion in a water bath at 101°C. The barometric pressure is 767 torr, and the remaining gas weighs 0.482 g. What is the molecular formula of the hydrocarbon?

5.46 A sample of air contains 78.08% nitrogen, 20.94% oxygen, 0.05% carbon dioxide, and 0.93% argon, by volume. How many molecules of each gas are present in 1.00 L of the sample at 25°C and 1.00 atm?

5.47 An environmental chemist sampling industrial exhaust gases from a coal-burning plant collects a CO_2-SO_2-H_2O mixture in a 21-L steel tank until the pressure reaches 850. torr at 45°C.
(a) How many moles of gas are collected?
(b) If the SO_2 concentration in the mixture is 7.95×10^3 parts per million by volume (ppmv), what is its partial pressure? [*Hint:* ppmv = (volume of component/volume of mixture) $\times 10^6$.]

5.48 "Strike anywhere" matches contain the compound tetraphosphorus trisulfide, which burns to form tetraphosphorus decoxide and sulfur dioxide gas. How many milliliters of sulfur dioxide, measured at 725 torr and 32°C, can be produced from burning 0.800 g of tetraphosphorus trisulfide?

5.49 Xenon hexafluoride was one of the first noble gas compounds synthesized. The solid reacts rapidly with the silicon dioxide in glass or quartz containers to form liquid $XeOF_4$ and gaseous silicon tetrafluoride. What is the pressure in a 1.00-L container at 25°C after 2.00 g of xenon hexafluoride reacts? (Assume that silicon tetrafluoride is the only gas present and that it occupies the entire volume.)

5.50 In the four piston-cylinder assemblies below, the reactant in the left cylinder is about to undergo a reaction at constant T and P:

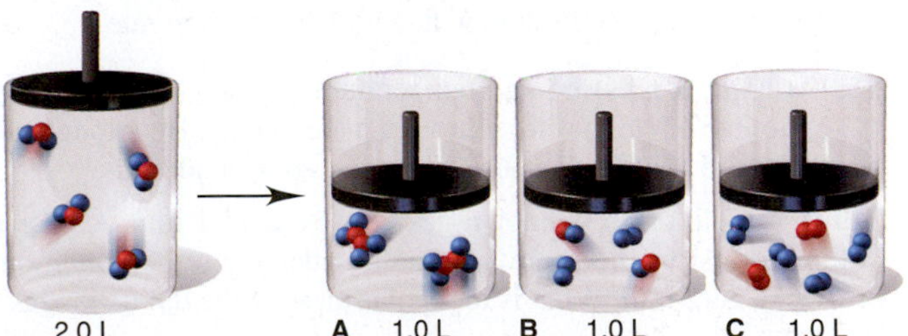

Which of the other three depictions best represents the products of the reaction?

5.51 Roasting galena [lead(II) sulfide] is a step in the industrial isolation of lead. How many liters of sulfur dioxide, measured at STP, are produced by the reaction of 3.75 kg of galena with 228 L of oxygen gas at 220°C and 2.0 atm? Lead(II) oxide also forms.

The Kinetic-Molecular Theory: A Model for Gas Behavior

(Sample Problem 5.13)

5.52 Use the kinetic-molecular theory to explain the change in gas pressure that results from warming a sample of gas.

5.53 How does the kinetic-molecular theory explain why 1 mol of krypton and 1 mol of helium have the same volume at STP?

5.54 Three 5-L flasks, fixed with pressure gauges and small valves, each contain 4 g of gas at 273 K. Flask A contains H_2, flask B contains He, and flask C contains CH_4. Rank the flask contents in terms of (a) pressure, (b) average molecular kinetic energy, (c) diffusion rate after the valve is opened, (d) total kinetic energy of the molecules, (e) density, and (f) collision frequency.

5.55 What is the ratio of effusion rates for the lightest gas, H_2, and the heaviest known gas, UF_6?

5.56 What is the ratio of effusion rates for O_2 and Kr?

5.57 The graph below shows the distribution of molecular speeds for argon and helium at the same temperature.

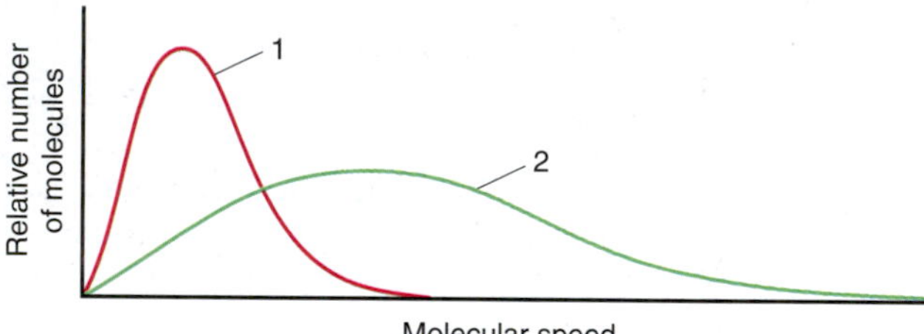

(a) Does curve 1 or 2 better represent the behavior of argon?
(b) Which curve represents the gas that effuses more slowly?
(c) Which curve more closely represents the behavior of fluorine gas? Explain.

5.58 The graph below shows the distribution of molecular speeds for a gas at two different temperatures.

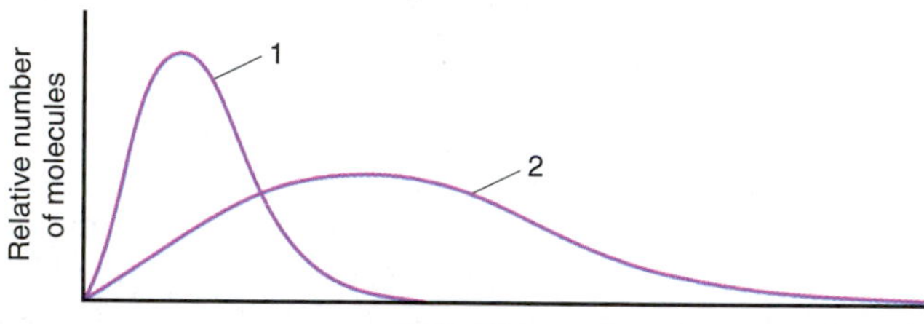

(a) Does curve 1 or 2 better represent the behavior of the gas at the lower temperature?
(b) Which curve represents the gas when it has a higher $\overline{E}_k$?
(c) Which curve is consistent with a higher diffusion rate?

5.59 At a given pressure and temperature, it takes 4.85 min for a 1.5-L sample of He to effuse through a membrane. How long does it take for 1.5 L of F_2 to effuse under the same conditions?

5.60 A sample of an unknown gas effuses in 11.1 min. An equal volume of H_2 in the same apparatus under the same conditions effuses in 2.42 min. What is the molar mass of the unknown gas?

5.61 White phosphorus melts and then vaporizes at high temperature. The gas effuses at a rate that is 0.404 times that of neon in the same apparatus under the same conditions. How many atoms are in a molecule of gaseous white phosphorus?

5.62 Helium (He) is the lightest noble gas component of air, and xenon (Xe) is the heaviest. [For this problem, use $R = 8.314$ J/(mol·K) and $\mathcal{M}$ in kg/mol.] (a) Find the rms speed of He in winter (0.°C) and in summer (30.°C). (b) Compare the rms speed of He with that of Xe at 30.°C. (c) Find the average kinetic energy per mole of He and of Xe at 30.°C. (d) Find the average kinetic energy per molecule of He at 30.°C.

5.63 A mixture of gaseous disulfur difluoride, dinitrogen tetrafluoride, and sulfur tetrafluoride is placed in an effusion apparatus. (a) Rank the gases in order of increasing effusion rate. (b) Find the ratio of effusion rates of disulfur difluoride and dinitrogen tetrafluoride. (c) If gas X is added, and it effuses at 0.935 times the rate of sulfur tetrafluoride, find the molar mass of X.

Real Gases: Deviations from Ideal Behavior

5.64 Do interparticle attractions cause negative or positive deviations from the *PV/RT* ratio of an ideal gas? Use Table 5.3 to rank Kr, CO_2, and N_2 in order of increasing magnitude of these deviations.

5.65 Does particle volume cause negative or positive deviations from the *PV/RT* ratio of an ideal gas? Use Table 5.3 to rank Cl_2, H_2, and O_2 in order of increasing magnitude of these deviations.

5.66 Does N_2 behave more ideally at 1 atm or at 500 atm? Explain.

5.67 Does SF_6 (boiling point = 16°C at 1 atm) behave more ideally at 150°C or at 20°C? Explain.

Comprehensive Problems

5.68 Hemoglobin is the protein that transports O_2 through the blood from the lungs to the rest of the body. In doing so, each molecule of hemoglobin combines with four molecules of O_2. If 1.00 g of hemoglobin combines with 1.53 mL of O_2 at 37°C and 743 torr, what is the molar mass of hemoglobin?

5.69 A baker uses sodium hydrogen carbonate (baking soda) as the leavening agent in a banana-nut quickbread. The baking soda decomposes in either of two possible reactions:

(1) $2NaHCO_3(s) \longrightarrow Na_2CO_3(s) + H_2O(l) + CO_2(g)$

(2) $NaHCO_3(s) + H^+(aq) \longrightarrow H_2O(l) + CO_2(g) + Na^+(aq)$

Calculate the volume (in mL) of CO_2 that forms at 200.°C and 0.975 atm per gram of $NaHCO_3$ by each of the reaction processes.

5.70 Chlorine is produced from sodium chloride by the electrochemical chlor-alkali process. During the process, the chlorine is collected in a container that is isolated from the other products to prevent unwanted (and explosive) reactions. If a 15.50-L container holds 0.5950 kg of Cl_2 gas at 225°C, calculate:

(a) P_{IGL} (b) $P_{VDW}\left(\text{use } R = 0.08206 \dfrac{\text{atm·L}}{\text{mol·K}}\right)$

5.71 In a certain experiment, magnesium boride (Mg_3B_2) reacted with acid to form a mixture of four boron hydrides (B_xH_y), three as liquids (labeled I, II, and III) and one as a gas (IV).
(a) When a 0.1000-g sample of each liquid was transferred to an evacuated 750.0-mL container and volatilized at 70.00°C,

sample I had a pressure of 0.05951 atm; sample II, 0.07045 atm; and sample III, 0.05767 atm. What is the molar mass of each liquid? (b) Boron is 85.63% by mass in sample I, 81.10% in II, and 82.98% in III. What is the molecular formula of each sample? (c) Sample IV was found to be 78.14% boron. Its rate of effusion was compared to that of sulfur dioxide; under identical conditions, 350.0 mL of sample IV effused in 12.00 min and 250.0 mL of sulfur dioxide effused in 13.04 min. What is the molecular formula of sample IV?

5.72 Three equal volumes of gas mixtures, all at the same T, are depicted below (with gas A *red*, gas B *green*, and gas C *blue*):

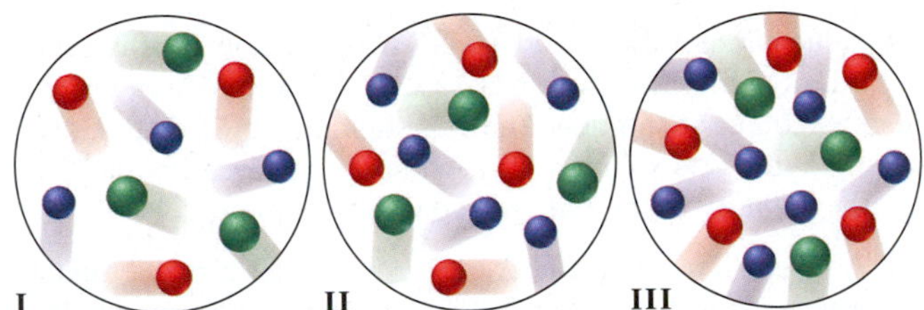

(a) Which sample, if any, has the highest partial pressure of A?
(b) Which sample, if any, has the lowest partial pressure of B?
(c) In which sample, if any, do the gas particles have the highest average kinetic energy?

5.73 When air is inhaled, it enters the alveoli of the lungs, and varying amounts of the component gases exchange with dissolved gases in the blood. The resulting alveolar gas mixture is quite different from the atmospheric mixture. The following table presents selected data on the composition and partial pressure of four gases in the atmosphere and in the alveoli:

	Atmosphere (sea level)		Alveoli	
Gas	**Mole %**	**Partial Pressure (torr)**	**Mole %**	**Partial Pressure (torr)**
N_2	78.6	—	—	569
O_2	20.9	—	—	104
CO_2	00.04	—	—	40
H_2O	00.46	—	—	47

If the total pressure of each gas mixture is 1.00 atm, calculate:
(a) The partial pressure (in torr) of each gas in the atmosphere
(b) The mole % of each gas in the alveoli
(c) The number of O_2 molecules in 0.50 L of alveolar air (volume of an average breath of a person at rest) at 37°C

5.74 Radon (Rn) is the heaviest, and only radioactive, member of Group 8A(18) (noble gases). It is a product of the disintegration of heavier radioactive nuclei found in minute concentrations in many common rocks used for construction. In recent years, concern has arisen about the incidence of lung cancer due to inhaled residential radon. If 1.0×10^{15} atoms of radium (Ra) produce an average of 1.373×10^{4} atoms of Rn per second, how many liters of Rn, measured at STP, are produced per day by 1.0 g of Ra?

5.75 At 1450. mmHg and 286 K, a skin diver exhales a 208-mL bubble of air that is 77% N_2, 17% O_2, and 6.0% CO_2 by volume.
(a) How many milliliters would the volume of the bubble be if it were exhaled at the surface at 1 atm and 298 K?
(b) How many moles of N_2 are in the bubble?

5.76 Nitrogen dioxide is used industrially to produce nitric acid, but it contributes to acid rain and photochemical smog. What volume (in L) of nitrogen dioxide is formed at 735 torr and 28.2°C by reacting 4.95 cm^3 of copper (d = 8.95 g/cm^3) with 230.0 mL of nitric acid (d = 1.42 g/cm^3, 68.0% HNO_3 by mass)?

$$Cu(s) + 4HNO_3(aq) \longrightarrow Cu(NO_3)_2(aq) + 2NO_2(g) + 2H_2O(l)$$

5.77 In a collision of sufficient force, automobile air bags respond by electrically triggering the explosive decomposition of sodium azide (NaN_3) to its elements. A 50.0-g sample of sodium azide was decomposed, and the nitrogen gas generated was collected over water at 26°C. The total pressure was 745.5 mmHg. How many liters of dry N_2 were generated?

5.78 An anesthetic gas contains 64.81% carbon, 13.60% hydrogen, and 21.59% oxygen, by mass. If 2.00 L of the gas at 25°C and 0.420 atm weighs 2.57 g, what is the molecular formula of the anesthetic?

5.79 Aluminum chloride is easily vaporized above 180°C. The gas escapes through a pinhole 0.122 times as fast as helium at the same conditions of temperature and pressure in the same apparatus. What is the molecular formula of aluminum chloride gas?

5.80 An atmospheric chemist studying the pollutant SO_2 places a mixture of SO_2 and O_2 in a 2.00-L container at 800. K and 1.90 atm. When the reaction occurs, gaseous SO_3 forms, and the pressure falls to 1.65 atm. How many moles of SO_3 form?

5.81 The thermal decomposition of ethylene occurs during the compound's transit in pipelines and during the formation of polyethylene. The decomposition reaction is

$$CH_2{=}CH_2(g) \longrightarrow CH_4(g) + C(\text{graphite})$$

If the decomposition begins at 10°C and 50.0 atm with a gas density of 0.215 g/mL and the temperature increases by 950 K,
(a) What is the final pressure of the confined gas (ignore the volume of graphite and use the van der Waals equation)?
(b) How does the PV/RT value of CH_4 compare to that in Figure 5.21? Explain.

5.82 Liquid nitrogen trichloride is heated in a 2.50-L closed reaction vessel until it decomposes completely to gaseous elements. The resulting mixture exerts a pressure of 754 mmHg at 95°C.
(a) What is the partial pressure of each gas in the container?
(b) What is the mass of the original sample?

5.83 Analysis of a newly discovered gaseous silicon-fluorine compound shows that it contains 33.01 mass % silicon. At 27°C, 2.60 g of the compound exerts a pressure of 1.50 atm in a 0.250-L vessel. What is the molecular formula of the compound?

5.84 Azodicarbonamide, $NH_2CON{=}NCONH_2$, is a blowing (foaming) agent for sponge rubber and expanded plastics. Its decomposition at 195–202°C is given by

$$NH_2CON{=}NCONH_2(s) \longrightarrow NH_3(g) + CO(g) + N_2(g) + HCNO(g)$$

$$NH_3(g) + HCNO(g) \longrightarrow \text{nonvolatile polymers}(s)$$

Calculate the volume (in mL) of gas, corrected to STP, in the final mixture from decomposition of 1.00 g of azodicarbonamide.

5.85 A gaseous organic compound containing only carbon, hydrogen, and nitrogen is burned in oxygen gas, and the volume of each reactant and product is measured under the same conditions of temperature and pressure. Reaction of four volumes of the compound produces four volumes of CO_2, two volumes of N_2, and ten volumes of water vapor. (a) How many volumes of O_2 were required? (b) What is the empirical formula of the compound?

5.86 Containers A, B, and C are attached by closed stopcocks of negligible volume.

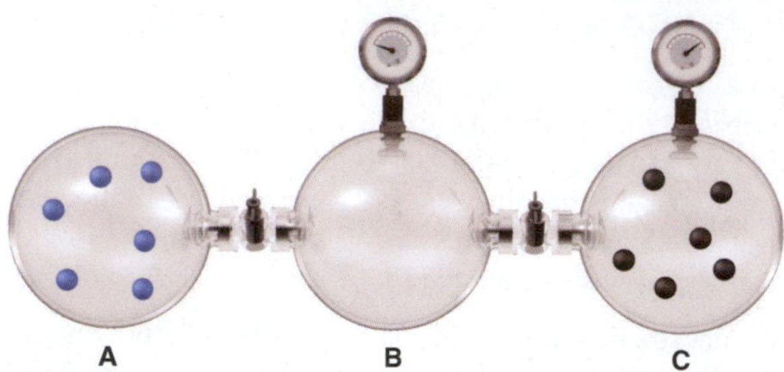

If each particle shown in the picture represents 10^6 particles,
(a) How many blue particles and black particles are in B after the stopcocks are opened and the system reaches equilibrium?
(b) How many blue particles and black particles are in A after the stopcocks are opened and the system reaches equilibrium?
(c) If the pressure in C, P_C, is 750 torr before the stopcocks are opened, what is P_C afterward? (d) What is P_B afterward?

5.87 Combustible vapor-air mixtures are flammable over a limited range of concentrations. The minimum volume % of vapor that gives a combustible mixture is called the *lower flammable limit* (LFL). Generally, the LFL is about half the stoichiometric mixture, the concentration required for complete combustion of the vapor in air. (a) If oxygen is 20.9 vol % of air, estimate the LFL for *n*-hexane, C_6H_{14}. (b) What volume (in mL) of *n*-hexane ($d = 0.660$ g/cm^3) is required to produce a flammable mixture of hexane in 1.000 m^3 of air at STP?

5.88 By what factor would a scuba diver's lungs expand if she ascended rapidly to the surface from a depth of 125 ft without inhaling or exhaling? If an expansion factor greater than 1.5 causes lung rupture, how far could she safely ascend from 125 ft without breathing? Assume constant temperature (d of seawater = 1.04 g/mL; d of Hg = 13.5 g/mL).

5.89 At a height of 300 km above Earth's surface, an astronaut finds that the atmospheric pressure is about 10^{-8} mmHg and the temperature is 500 K. How many molecules of gas are there per milliliter at this altitude?

5.90 (a) What is the rms speed of O_2 at STP? (b) If the average distance between O_2 molecules at STP is 6.33×10^{-8} m, how many collisions are there per second? [Use $R = 8.314$ J/(mol·K) and $\mathcal{M}$ in kg/mol.]

5.91 Standard conditions are based on relevant environmental conditions. If normal average surface temperature and pressure on Venus are 730. K and 90 atm, respectively, what is the standard molar volume of an ideal gas on Venus?

5.92 Each day, the Hawaiian volcano Kilauea emits an average of 1.5×10^3 m^3 of gas, when corrected to 298 K and 1.00 atm. The mixture contains gases that contribute to global warming and acid rain, and some are toxic. An atmospheric chemist analyzes a sample and finds the following mole fractions: 0.4896 CO_2, 0.0146 CO, 0.3710 H_2O, 0.1185 SO_2, 0.0003 S_2, 0.0047 H_2, 0.0008 HCl, and 0.0003 H_2S. How many metric tons (t) of each gas are emitted per year (1 t = 1000 kg)?

5.93 To study a key fuel-cell reaction, a chemical engineer has 20.0-L tanks of H_2 and of O_2 and wants to use up both tanks to form 28.0 mol of water at 23.8°C. (a) Use the ideal gas law to find the pressure needed in each tank. (b) Use the van der Waals equation to find the pressure needed in each tank. (c) Compare the results from the two equations.

5.94 How many liters of gaseous hydrogen bromide at 29°C and 0.965 atm will a chemist need if she wishes to prepare 3.50 L of 1.20 *M* hydrobromic acid?

5.95 A mixture of CO_2 and Kr weighs 35.0 g and exerts a pressure of 0.708 atm in its container. Since Kr is expensive, you wish to recover it from the mixture. After the CO_2 is completely removed by absorption with NaOH(*s*), the pressure in the container is 0.250 atm. How many grams of CO_2 were originally present? How many grams of Kr can you recover?

5.96 Aqueous sulfurous acid (H_2SO_3) was made by dissolving 0.200 L of sulfur dioxide gas at 19°C and 745 mmHg in water to yield 500.0 mL of solution. The acid solution required 10.0 mL of sodium hydroxide solution to reach the titration end point. What was the molarity of the sodium hydroxide solution?

5.97 A person inhales air richer in O_2 and exhales air richer in CO_2 and water vapor. During each hour of sleep, a person exhales a total of about 300 L of this CO_2-enriched and H_2O-enriched air. (a) If the partial pressures of CO_2 and H_2O in exhaled air are each 30.0 torr at 37.0°C, calculate the mass (g) of CO_2 and of H_2O exhaled in 1 h of sleep. (b) How many grams of body mass does the person lose in an 8-h sleep if all the CO_2 and H_2O exhaled come from the metabolism of glucose?

$$C_6H_{12}O_6(s) + 6O_2(g) \longrightarrow 6CO_2(g) + 6H_2O(g)$$

5.98 Given these relationships for average kinetic energy,

$$\overline{E_k} = \tfrac{1}{2}m\overline{u^2} \qquad \text{and} \qquad \overline{E_k} = \tfrac{3}{2}\left(\frac{R}{N_A}\right)T$$

where m is molecular mass, u is rms speed, R is the gas constant [in J/(mol·K)], N_A is Avogadro's number, and T is absolute temperature: (a) derive Equation 5.13; (b) derive Equation 5.14.

5.99 Many water treatment plants use chlorine gas to kill microorganisms before the water is released for residential use. A plant engineer has to maintain the chlorine pressure in a tank below the 85.0-atm rating and, to be safe, decides to fill the tank to 80.0% of this maximum pressure.
(a) How many moles of Cl_2 gas can be kept in an 850.-L tank at 298 K if she uses the ideal gas law in the calculation?
(b) What is the tank pressure if she uses the van der Waals equation for this amount of gas? (c) Did the engineer fill the tank to the desired pressure?

5.100 In A, the picture shows a cylinder with 0.1 mol of a gas that behaves ideally. Choose the cylinder (B, C, or D) that correctly represents the volume of the gas after each of the following changes. If none of the cylinders is correct, specify "none."
(a) P is doubled at fixed n and T.
(b) T is reduced from 400 K to 200 K at fixed n and P.
(c) T is increased from 100°C to 200°C at fixed n and P.
(d) 0.1 mol of gas is added at fixed P and T.
(e) 0.1 mol of gas is added and P is doubled at fixed T.

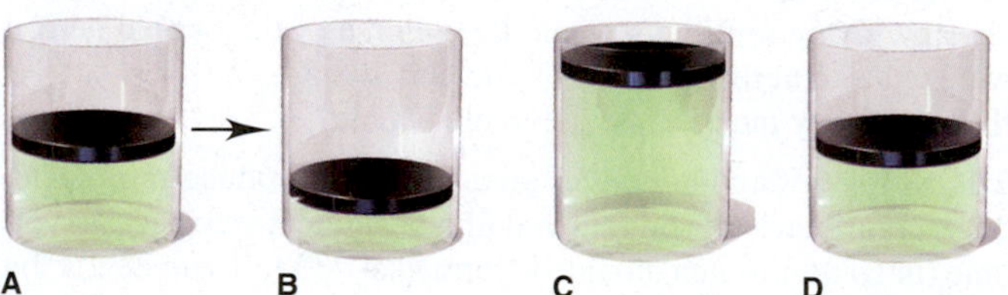

5.101 A 6.0-L flask contains a mixture of methane (CH_4), argon, and helium at 45°C and 1.75 atm. If the mole fractions of helium and argon are 0.25 and 0.35, respectively, how many molecules of methane are present?

5.102 A large portion of metabolic energy arises from the biological combustion of glucose:

$$C_6H_{12}O_6(s) + 6O_2(g) \longrightarrow 6CO_2(g) + 6H_2O(g)$$

(a) If this reaction is carried out in an expandable container at 37°C and 780. torr, what volume of CO_2 is produced from 20.0 g of glucose and excess O_2? (b) If the reaction is carried out at the same conditions with the stoichiometric amount of O_2, what is the partial pressure of each gas when the reaction is 50% complete (10.0 g of glucose remains)?

5.103 According to government standards, the *8-h threshold limit value* is 5000 ppmv for CO_2 and 0.1 ppmv for Br_2 (1 ppmv is 1 part by volume in 10^6 parts by volume). Exposure to either gas for 8 h above these limits is unsafe. At STP, which of the following would be unsafe for 8 h of exposure?
(a) Air with a partial pressure of 0.2 torr of Br_2
(b) Air with a partial pressure of 0.2 torr of CO_2
(c) 1000 L of air containing 0.0004 g of Br_2 gas
(d) 1000 L of air containing 2.8×10^{22} molecules of CO_2

5.104 One way to prevent emission of the pollutant NO from industrial plants is by a catalyzed reaction with NH_3:

$$4NH_3(g) + 4NO(g) + O_2(g) \xrightarrow{\text{catalyst}} 4N_2(g) + 6H_2O(g)$$

(a) If the NO has a partial pressure of 4.5×10^{-5} atm in the flue gas, how many liters of NH_3 are needed per liter of flue gas at 1.00 atm? (b) If the reaction takes place at 1.00 atm and 365°C, how many grams of NH_3 are needed per kL of flue gas?

5.105 An equimolar mixture of Ne and Xe is accidentally placed in a container that has a tiny leak. After a short while, a very small proportion of the mixture has escaped. What is the mole fraction of Ne in the effusing gas?

5.106 One way to utilize naturally occurring uranium (0.72% ^{235}U and 99.27% ^{238}U) as a nuclear fuel is to enrich it (increase its ^{235}U content) by allowing gaseous UF_6 to effuse through a porous membrane (see p. 176). From the relative rates of effusion of $^{235}UF_6$ and $^{238}UF_6$, find the number of steps needed to produce uranium that is 3.0 mole % ^{235}U, the enriched fuel used in many nuclear reactors.

5.107 In preparation for a combustion demonstration, a professor fills a balloon with equal molar amounts of H_2 and O_2, but the demonstration has to be postponed until the next day. During the night, both gases leak through pores in the balloon. If 35% of the H_2 leaks, what is the O_2/H_2 ratio in the balloon the next day?

5.108 Phosphorus trichloride is important in the manufacture of insecticides, fuel additives, and flame retardants. Phosphorus has only one naturally occurring isotope, ^{31}P, whereas chlorine has two, ^{35}Cl (75%) and ^{37}Cl (25%). (a) What different molecular masses (in amu) can be found for PCl_3? (b) Which is the most abundant? (c) What is the ratio of the effusion rates of the heaviest and the lightest PCl_3 molecules?

Thermochemistry: Energy Flow and Chemical Change

Key Principles
to focus on while studying this chapter

- Chemical or physical change is *always* accompanied by a change in the energy content of the matter. ***(Introduction)***
- To study a *change in energy* (ΔE), scientists conceptually divide the universe into the *system* (the part being studied) and the *surroundings* (everything else). All energy changes occur as *heat* (q) and/or *work* (w), transferred either from the surroundings to the system or from the system to the surroundings ($\Delta E = q + w$). Thus, the total energy of the universe is constant *(law of conservation of energy, or first law of thermodynamics)*. ***(Section 6.1)***
- Because E is a *state function*—a property that depends *only* on the current state of the system—ΔE depends only on the difference between the initial and final values of E. Therefore, the magnitude of ΔE is the same no matter how a given change in energy occurs. ***(Section 6.1)***
- *Enthalpy* (H) is also a state function and is related to E. The *change in enthalpy* (ΔH) equals the heat transferred at constant pressure, q_P. Most laboratory, environmental, and biological changes occur at constant P, so ΔH is more relevant than ΔE and easier to measure. ***(Section 6.2)***
- The *change in the enthalpy of a reaction* (ΔH_{rxn}) is negative (<0) if the reaction releases heat *(exothermic)* and positive (>0) if it absorbs heat *(endothermic)*. ***(Section 6.2)***
- The more heat a substance absorbs, the higher its temperature becomes, but each material has its own *capacity* for absorbing heat. Knowing this capacity and measuring the change in temperature in a *calorimeter*, we can find ΔH_{rxn}. ***(Section 6.3)***
- The quantity of heat released or absorbed in a reaction is related *stoichiometrically* to the amounts (mol) of reactants and products. ***(Section 6.4)***
- Because H is a state function, we can find ΔH of any reaction by imagining the reaction occurring as the sum of other reactions whose enthalpies of reaction we know or can measure *(Hess's law)*. ***(Section 6.5)***
- Chemists have specified a set of conditions, called *standard states*, to determine and compare enthalpies (and other thermodynamic variables) of different reactions. Each substance has a *standard enthalpy of formation* (ΔH°_f), the enthalpy of reaction when 1 mol of the substance is formed from its elements under these conditions. The ΔH°_f values for each substance in a reaction are used to calculate the *standard enthalpy of reaction* (ΔH°_{rxn}). ***(Section 6.6)***

Heat Released and Absorbed As you digest food, heat is released, and some of it is absorbed by ice you hold in your hand. In this chapter, we explore energy flow during chemical or physical change.

Outline

All matter contains energy, so *whenever matter undergoes a change, the quantity of energy that the matter contains also changes.* If a burning candle melts a piece of ice, both chemical and physical changes occur. In the chemical change, the higher energy reactants (wax and O_2) form lower energy products (CO_2 and H_2O), and the difference in energy is *released* as heat and light. Then, during the physical change, some of that heat is *absorbed* when lower energy ice becomes higher energy water. In a thunderstorm, the changes in energy are reversed: lower energy N_2 and O_2 *absorb* energy from lightning to form higher energy NO, and higher energy water vapor *releases* energy as it condenses to lower energy liquid water that falls as rain.

This interplay of matter and energy has an enormous impact on society. On an everyday level, many familiar materials release, absorb, or alter the flow of energy. Fuels such as oil and wood release energy to warm our homes or power our vehicles. Fertilizers help crops convert solar energy into the energy in food. Metal wires speed the flow of electrical energy, and polymer fibers in winter clothing limit the flow of thermal energy away from our bodies.

Thermodynamics is the study of energy and its transformations, and two chapters in this text address this central topic. Our focus here is on **thermochemistry,** the branch of thermodynamics that deals with heat in chemical and physical change.

CONCEPTS & SKILLS TO REVIEW
before studying this chapter

- energy and its interconversion (Section 1.1)
- distinction between heat and temperature (Section 1.4)
- nature of chemical bonding (Section 2.7)
- calculations of reaction stoichiometry (Section 3.4)
- properties of the gaseous state (Section 5.1)
- relation between kinetic energy and temperature (Section 5.5)

6.1 • FORMS OF ENERGY AND THEIR INTERCONVERSION

In Chapter 1, we saw that all energy is either potential or kinetic, and that these forms are interconvertible. An object has potential energy by virtue of its position and kinetic energy by virtue of its motion. Let's re-examine these ideas by considering a weight raised above the ground. As your muscles or a motor raise the weight, its potential energy increases; this energy is converted to kinetic energy as the weight falls (see Figure 1.3). When it hits the ground, some of that kinetic energy appears as *work* done when the weight moves the soil and pebbles slightly, and some appears as *heat* when it warms them slightly. Thus, in this situation, *potential energy is converted to kinetic energy, which appears as work and heat.*

Several other forms of energy—solar, electrical, nuclear, and chemical—are examples of potential and kinetic energy on the atomic scale. No matter what the form of energy or the situation, *when energy is transferred from one object to another, it appears as work and/or heat.* In this section, we examine this idea in terms of the release or absorption of energy during a chemical or physical change.

Defining the System and Its Surroundings

In order to observe and measure a change in energy, we must first define the **system**—the part of the universe we are focusing on. The moment we define the system, everything else is defined as the **surroundings.**

For example, for a flask containing a solution, if we define the system as the contents of the flask, then the flask itself, other nearby equipment, and perhaps the rest of the laboratory are the surroundings. In principle, the rest of the universe is the surroundings, but in practice, we consider only the parts that are relevant to the system: it's not likely that a thunderstorm in central Asia or a methane blizzard on Neptune will affect the contents of the flask, but the temperature and pressure of the lab might.

THINK OF IT THIS WAY
Wherever You Look, There Is a System

If we define a weight falling to the ground as the system, then the soil and pebbles that are moved and warmed when it lands are the surroundings. An astronomer may define a galaxy as the system and nearby galaxies as the surroundings. An ecologist in Africa can define a zebra herd as the system and the animals, plants, and water supplies that the herd has contact with as the surroundings. A microbiologist may define a bacterial cell as the system and the extracellular solution as the surroundings. Thus, in general, the nature of the experiment and the focus of the experimenter define system and surroundings.

Energy Transfer to and from a System

Each particle in a system has potential energy and kinetic energy, and the sum of all these energies is the **internal energy, *E*,** of the system (some texts use the symbol *U*). When the reactants in a chemical system change to products, the system's internal energy has changed. This change, ΔE, is the difference between the internal energy *after* the change (E_{final}) and *before* the change ($E_{initial}$):

$$\Delta E = E_{final} - E_{initial} = E_{products} - E_{reactants} \quad (6.1)$$

where Δ (Greek *delta*) means "change (or difference) in" and refers to the *final state* ***minus*** *the initial state.* Thus, ΔE is the final quantity of energy of the system *minus* the initial quantity.

Because the universe consists of only system and surroundings, *a change in the energy of the system must be accompanied by an* ***equal*** *and* ***opposite*** *change in the energy of the surroundings.* In an *energy diagram,* the final and initial states are horizontal lines along a vertical energy axis, with ΔE the difference in the heights of the lines. A system can change its internal energy in one of two ways:

- By releasing some energy in a transfer *to* the surroundings (Figure 6.1A):

$$E_{final} < E_{initial} \quad \text{so} \quad \Delta E < 0$$

- By absorbing some energy in a transfer *from* the surroundings (Figure 6.1B):

$$E_{final} > E_{initial} \quad \text{so} \quad \Delta E > 0$$

Thus, ΔE is a *transfer* of energy from system to surroundings, or vice versa.

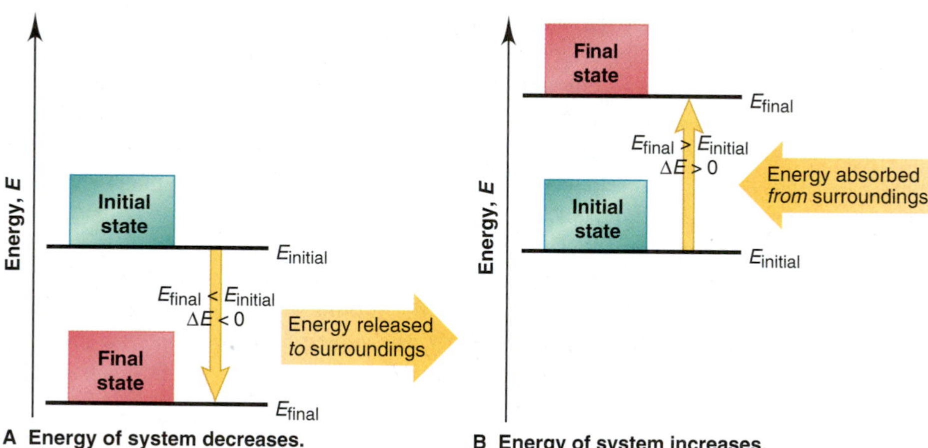

Figure 6.1 Energy diagrams for the transfer of internal energy (E) between a system and its surroundings. **A,** When the system releases energy, ΔE ($E_{final} - E_{initial}$) is negative. **B,** When the system absorbs energy, ΔE ($E_{final} - E_{initial}$) is positive. (The vertical yellow arrow always has its tail at the initial state.)

Heat and Work: Two Forms of Energy Transfer

Energy transferred from system to surroundings or vice versa appears in two forms:

1. *Heat.* **Heat** or *thermal energy* (symbolized by ***q***) is the energy transferred as a result of a difference in temperature between the system and the surroundings. For example, energy in the form of heat is transferred from hot soup (system) to the bowl, air, and table (surroundings) because they are at a lower temperature.

2. *Work.* All other forms of energy transfer involve some type of **work (*w*),** the energy transferred when an object is moved by a force. When you (system) kick a football, energy is transferred as work because the force of the kick moves the ball and air around it (surroundings). When you pump up a ball, energy is transferred as work because the added air (system) exerts a force on the inner wall of the ball (surroundings) and moves it outward.

The total change in a system's internal energy is the sum of the energy transferred as heat and/or work:

$$\Delta E = q + w \quad (6.2)$$

The values of q and w (and, therefore, of ΔE) can have either a positive or negative sign. *We define the sign of the energy change from the* ***system's*** *perspective:*

- Energy transferred *into* the system is *positive* because the *system ends up with more* energy.
- Energy transferred *out from* the system is *negative,* because the *system ends up with less* energy.

Innumerable combinations of heat and/or work can change a system's internal energy. In the rest of this subsection, we'll examine the four simplest cases—two that involve only heat and two that involve only work.

Energy Transferred as Heat Only For a system that transfers energy only as heat (q) and does no work ($w = 0$), we have, from Equation 6.2, $\Delta E = q + 0 = q$. There are two ways this transfer can happen:

1. *Heat flowing* ***out*** *from a system.* Suppose hot water is the system, and the beaker holding it and the rest of the lab are the surroundings. The water transfers energy as heat outward until the temperature of the water and the surroundings are equal. Since heat flows *out* from the system, the final energy of the system is less than the initial energy. Heat is released, so *q is negative,* and therefore *ΔE is negative* (Figure 6.2A).

2. *Heat flowing* ***into*** *a system.* If the system consists of ice water, the surroundings transfer energy as heat *into* the system, once again until the ice melts and the temperature of the water and the surroundings become equal. In this case, heat flows *in,* so the final energy of the system is higher than its initial energy. Heat is absorbed, so *q is positive,* and therefore *ΔE is positive* (Figure 6.2B).

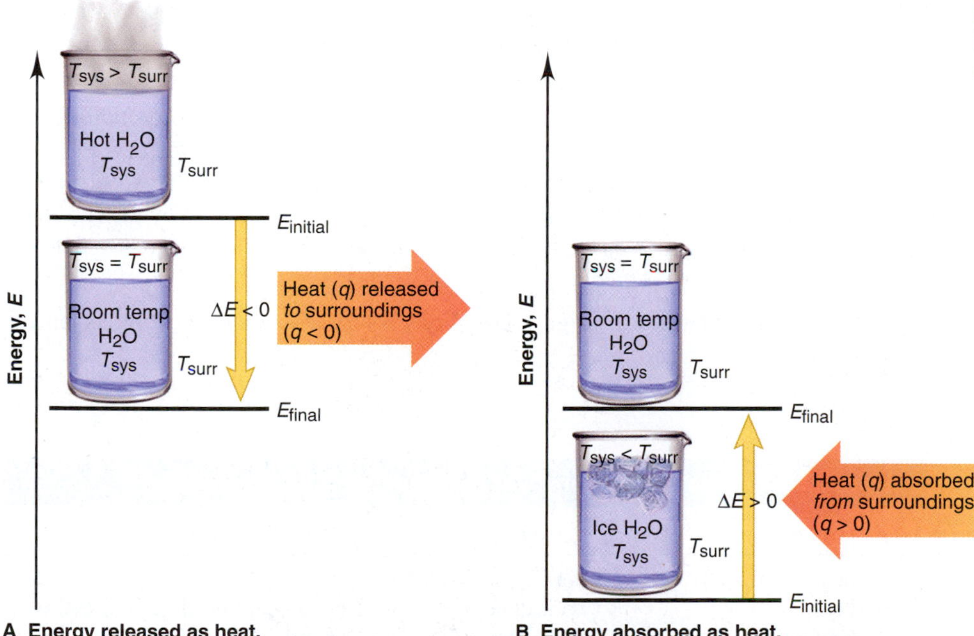

Figure 6.2 The two cases where energy is transferred as heat only. **A,** The system releases heat. **B,** The system absorbs heat.

THINK OF IT THIS WAY
Thermodynamics in the Kitchen

A new view of two familiar kitchen appliances can clarify the sign of q. The air in a refrigerator (surroundings) has a lower temperature than a newly added piece of food (system), so the food releases energy as heat to the refrigerator air, $q < 0$. The air in a hot oven (surroundings) has a higher temperature than a newly added piece of food (system), so the food absorbs energy as heat from the oven air, $q > 0$.

Energy Transferred as Work Only For a system that transfers energy only as work, $q = 0$; therefore, $\Delta E = 0 + w = w$. There are two ways this transfer can happen:

1. *Work done* ***by*** *a system* (Figure 6.3A, *next page*). Consider the reaction between zinc and hydrochloric acid as it takes place in a nearly evacuated, insulated container attached to a piston-cylinder assembly. We define the system as the reaction mixture, and the surroundings as the container, piston-cylinder, outside air, and so forth. In the initial state, the internal energy is the energy of the reactants (metallic zinc and

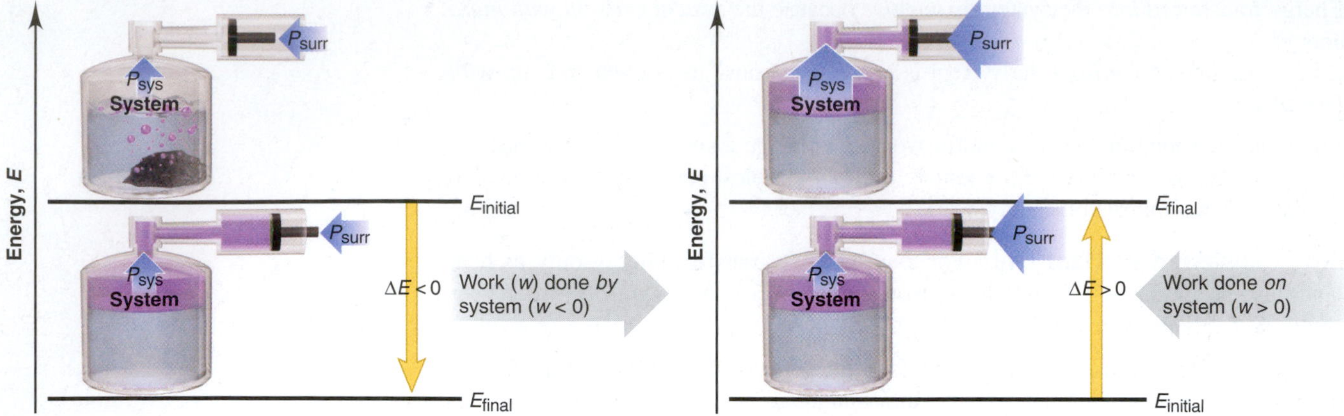

Figure 6.3 The two cases where energy is transferred as work only. **A,** The system does work *on* the surroundings. **B,** The system has work done on it *by* the surroundings.

aqueous H^+ and Cl^- ions), and in the final state, it is the energy of the products (H_2 gas and aqueous Zn^{2+} and Cl^- ions). As the gas forms, it pushes back the piston. Thus, energy is transferred as work done *by* the system *on* the surroundings. Since the system releases energy as work, *w is negative;* the final energy of the system is less than the initial energy, so *ΔE is negative*. (The work done here is not very useful because it just pushes back the piston and outside air. But, if the reaction mixture is gasoline and oxygen and the surroundings are an automobile engine, much of the internal energy is transferred as the work done to move the car.)

2. *Work done **on** a system* (Figure 6.3B). Suppose that, after the reaction of zinc and hydrochloric acid is over, we make another change (which makes the new $E_{initial}$ the same as the previous E_{final}): we increase the pressure of the surroundings (wider P_{surr} arrow) so that the piston moves in. Energy is transferred as work done *by* the surroundings *on* the system, so *w is positive.* The final energy of the system is greater than the initial energy, so *ΔE is positive.*

Table 6.1 summarizes the sign conventions for *q* and *w* and their effect on the sign of Δ*E*.

Table 6.1 The Sign Conventions* for *q*, *w*, and Δ*E*

q	+	*w*	=	Δ*E*
+ (heat *absorbed*)		+ (work done *on*)		+ (energy *absorbed*)
+ (heat *absorbed*)		− (work done *by*)		Depends on the ***sizes*** of *q* and *w*
− (heat *released*)		+ (work done *on*)		Depends on the ***sizes*** of *q* and *w*
− (heat *released*)		− (work done *by*)		− (energy *released*)

*From the perspective of the system.

The Law of Energy Conservation

As you've seen, when a system absorbs energy, the surroundings release it, and when a system releases energy, the surroundings absorb it. Energy transferred between system and surroundings can be in the form of heat and/or various types of work—mechanical, electrical, radiant, chemical, and so forth.

Indeed, energy is often converted from one form to another during transfers. For example, when gasoline burns in a car engine, the reaction releases energy that is transferred as heat and work. The heat warms the car parts, passenger compartment, and surrounding air. The work is done when mechanical energy turns the car's wheels and belts. That energy is converted into the electrical energy of the sound system, the radiant energy of the headlights, the chemical energy of the battery, and so forth. The sum of all these forms equals the change in energy between reactants and products as the gasoline is burned.

Complex biological processes follow the same general pattern. During photosynthesis, green plants use solar energy to convert the lower energy bonds in CO_2 and H_2O into the higher energy bonds in starch and O_2; when you digest starch, this energy is converted into the muscular (mechanical) energy needed to run a marathon.

Thus, energy changes form but does not simply appear or disappear—energy cannot be created or destroyed. Put another way, *energy is conserved: the total energy of the system plus the surroundings remains constant.* The **law of conservation of energy,** also known as the **first law of thermodynamics,** restates this basic observation: *the total energy of the universe is constant.* It is expressed mathematically as

$$\Delta E_{\text{universe}} = \Delta E_{\text{system}} + \Delta E_{\text{surroundings}} = 0 \tag{6.3}$$

This law applies, as far as we know, to all systems, from a burning match to continental drift, from the pumping of your heart to the formation of the Solar System.

Units of Energy

The SI unit of energy is the **joule (J),** a derived unit composed of three base units:

$$1\text{ J} = 1\text{ kg}\cdot\text{m}^2/\text{s}^2$$

Both heat and work are expressed in joules. Let's see why the joule is the unit for work. The work (w) done on a mass is the force (F) times the distance (d) that the mass moves: $w = F \times d$. A *force* changes the velocity of a mass over time; that is, a force *accelerates* a mass. Velocity has units of meters per second (m/s), so acceleration (a) has units of meters per second per second (m/s^2). Force, therefore, has units of mass (m, in kilograms) times acceleration:

$$F = m \times a \quad \text{has units of} \quad \text{kg}\cdot\text{m/s}^2$$

Therefore, $$w = F \times d \quad \text{has units of} \quad (\text{kg}\cdot\text{m/s}^2) \times \text{m} = \text{kg}\cdot\text{m}^2/\text{s}^2 = \text{J}$$

The **calorie (cal)** is an older unit defined originally as the quantity of energy needed to raise the temperature of 1 g of water by 1°C (from 14.5°C to 15.5°C). The calorie is now defined in terms of the joule:

$$1\text{ cal} \equiv 4.184\text{ J} \qquad \text{or} \qquad 1\text{ J} = \frac{1}{4.184}\text{ cal} = 0.2390\text{ cal}$$

Since the quantities of energy involved in chemical reactions are usually quite large, chemists use the kilojoule (kJ), or, in earlier sources, the kilocalorie (kcal):

$$1\text{ kJ} = 1000\text{ J} = 0.2390\text{ kcal} = 239.0\text{ cal}$$

The nutritional Calorie (note the capital C), the unit that shows the energy available from food, is actually a kilocalorie.

The *British thermal unit (Btu),* a unit that you may have seen used for the energy output of appliances, is the quantity of energy required to raise the temperature of 1 lb of water by 1°F; *1 Btu is equivalent to 1055 J.* In general, the SI unit (J or kJ) is used in this text.

Sample Problem 6.1 Determining the Change in Internal Energy of a System

Problem When gasoline burns in a car engine, the heat released causes the products CO_2 and H_2O to expand, which pushes the pistons outward. Excess heat is removed by the car's radiator. If the expanding gases do 451 J of work on the pistons and the system releases 325 J to the surroundings as heat, calculate the change in energy (ΔE) in J, kJ, and kcal.

Plan We must define system and surroundings to choose signs for q and w, and then we calculate ΔE with Equation 6.2. The system is the reactants and products, and the surroundings are the pistons, the radiator, and the rest of the car. Heat is released by the system, so q is negative. Work is done by the system to push the pistons outward, so w is also negative. We obtain the answer in J and then convert it to kJ and kcal.

Solution Calculating ΔE (from Equation 6.2) in J:

$$q = -325 \text{ J}$$
$$w = -451 \text{ J}$$
$$\Delta E = q + w = -325 \text{ J} + (-451 \text{ J}) = -776 \text{ J}$$

Converting from J to kJ:

$$\Delta E = -776 \text{ J} \times \frac{1 \text{ kJ}}{1000 \text{ J}} = -0.776 \text{ kJ}$$

Converting from kJ to kcal:

$$\Delta E = -0.776 \text{ kJ} \times \frac{1 \text{ kcal}}{4.184 \text{ kJ}} = -0.185 \text{ kcal}$$

Check The answer is reasonable: combustion of gasoline releases energy from the system, so $E_{\text{final}} < E_{\text{initial}}$ and ΔE should be negative. Rounding shows that, since 4 kJ $\approx$ 1 kcal, nearly 0.8 kJ should be nearly 0.2 kcal.

FOLLOW-UP PROBLEM 6.1 In a reaction, gaseous reactants form a liquid product. The heat absorbed by the surroundings is 26.0 kcal, and the work done on the system is 15.0 Btu. Calculate ΔE (in kJ).

State Functions and the Path Independence of the Energy Change

The internal energy (E) of a system is called a **state function,** a property dependent only on the *current* state of the system (its composition, volume, pressure, and temperature), *not* on the path the system takes to reach that state.

THINK OF IT THIS WAY
Your Personal Financial State Function

The *balance* in your checkbook is a state function of your personal financial system. You can open a new checking account with a birthday gift of \$50, or you can open one with a deposit of a \$100 paycheck and then write two \$25 checks. The two paths to the balance are different, but the balance (current state) is the same.

Thus, as in the financial analogy, the energy change of a system can occur by countless combinations of heat (q) and/or work (w). However, because E is a state function, the overall ΔE is the same no matter what the specific combination may be. That is, *ΔE does **not** depend on how the change takes place, but only on the **difference** between the final and initial states.*

As an example, let's define a system in its initial state as 1 mol of octane (a component of gasoline) together with enough O_2 to burn it completely. In its final state, the system is the CO_2 and H_2O that form:

$$C_8H_{18}(l) + \tfrac{25}{2}O_2(g) \longrightarrow 8CO_2(g) + 9H_2O(g)$$

initial state (E_{initial}) final state (E_{final})

Energy is transferred *out* from the system as heat and/or work, so ΔE is negative. Figure 6.4 shows just two of the many ways the change can occur. If we burn the octane in an open container *(left)*, ΔE is transferred almost completely as heat (with

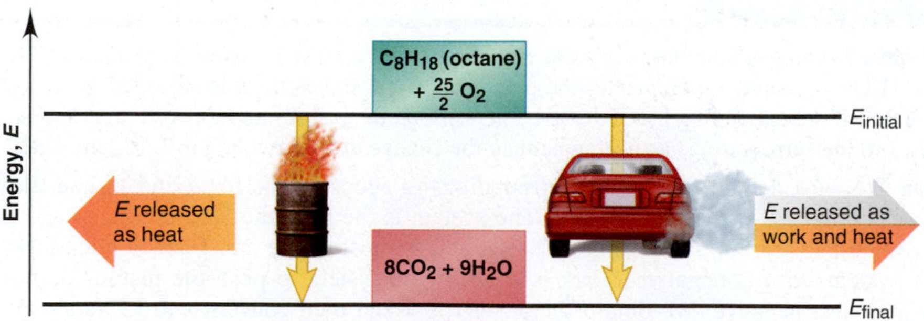

Figure 6.4 Two different paths for the energy change of a system. Even though q and w for the two paths are different, ΔE is the same.

a small quantity of work done to push back the atmosphere). If we burn it in a car engine *(right)*, a much larger portion ($\sim$30%) of ΔE is transferred as work that moves the car, with the rest released as heat that warms the car, exhaust gases, and surrounding air. If we burn the octane in a lawn mower or a motorcycle, ΔE appears as other combinations of work and heat. Thus, q and w are *not* state functions because their values *do* depend on the path the system takes, but ΔE (the *sum* of q and w) *does not.*

Pressure (P), volume (V), and temperature (T) are some other state functions. Path independence means that *changes in state functions—ΔE, ΔP, ΔV, and ΔT—depend only on the initial and final states.*

Summary of Section 6.1

- Internal energy (E) is transferred as heat (q) when system and surroundings are at different temperatures or as work (w) when an object is moved by a force.
- Heat absorbed by a system ($q > 0$) or work done on a system ($w > 0$) increases the system's E; heat released by a system ($q < 0$) or work done by a system ($w < 0$) decreases its E. The change in the internal energy is the sum of the heat and work: $\Delta E = q + w$. Heat and work are measured in joules (J).
- Energy is always conserved: it can change from one form to another and move into or out of the system, but the total quantity of energy in the universe (system *plus* surroundings) is constant.
- The internal energy of a system is a state function and, thus, is independent of how the system attained that energy; therefore, the same overall ΔE can occur through any combination of q and w.

6.2 • ENTHALPY: CHEMICAL CHANGE AT CONSTANT PRESSURE

Most physical and chemical changes occur at nearly constant atmospheric pressure—a reaction in an open flask, the freezing of a lake, a biochemical process in an organism. In this section, we discuss *enthalpy,* a thermodynamic variable that relates directly to energy changes at constant pressure.

The Meaning of Enthalpy

To determine ΔE, we must measure both heat and work. The two most important types of chemical work are electrical work, done by moving charged particles (Chapter 21), and **pressure-volume work (*PV* work),** the mechanical work done when the volume of the system changes in the presence of an external pressure (P). The quantity of PV work equals P times the change in volume (ΔV, or $V_{\text{final}} - V_{\text{initial}}$). In an open flask (or in a cylinder with a weightless, frictionless piston) (Figure 6.5), a system of an expanding gas does PV work *on* the surroundings, so it has a negative sign:

$$w = -P\Delta V \quad (6.4)$$

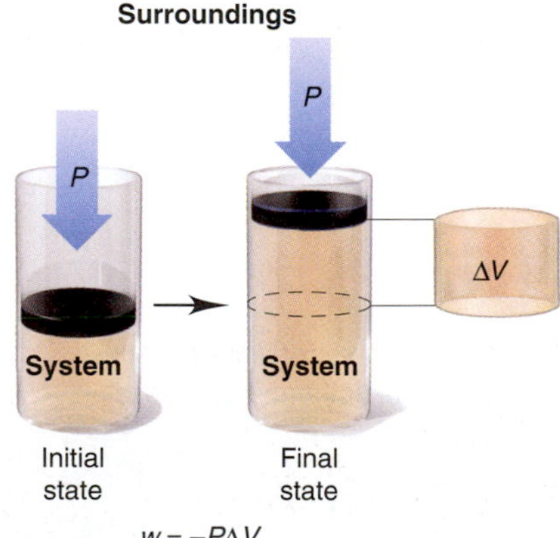

Figure 6.5 Pressure-volume work. An expanding gas pushing back the atmosphere does PV work ($w = -P\Delta V$).

For reactions *at constant pressure,* a thermodynamic variable called **enthalpy (*H*)** eliminates the need to measure PV work. The enthalpy of a system is defined as the internal energy *plus* the product of the pressure and volume:

$$H = E + PV$$

The **change in enthalpy (Δ*H*)** is the change in internal energy *plus* the product of the pressure, which is constant, and the change in volume (ΔV):

$$\Delta H = \Delta E + P\Delta V \quad (6.5)$$

Combining Equations 6.2 ($\Delta E = q + w$) and 6.4 ($w = -P\Delta V$) gives

$$\Delta E = q + w = q + (-P\Delta V) = q - P\Delta V$$

At constant pressure, we denote q as q_P, giving $\Delta E = q_P - P\Delta V$, which we can use to solve for q_P:

$$q_P = \Delta E + P\Delta V$$

Notice that the right sides of this equation and of Equation 6.5 are the same:

$$q_P = \Delta E + P\Delta V = \Delta H \qquad \textbf{(6.6)}$$

Thus, *the change in enthalpy equals the heat absorbed or released at* ***constant*** *pressure*. For most changes at constant pressure, ΔH is more relevant than ΔE *and* easier to obtain: *to find* ΔH, *measure* q_P. We discuss the laboratory method in Section 6.3.

Exothermic and Endothermic Processes

Because H is a combination of the three state functions E, P, and V, it is also a state function. Therefore, ΔH equals H_{final} ***minus*** $H_{initial}$. For a reaction, H_{final} is $H_{products}$ and $H_{initial}$ is $H_{reactants}$, so the enthalpy change of a reaction is

$$\Delta H = H_{final} - H_{initial} = H_{products} - H_{reactants}$$

Since $H_{products}$ can be either more or less than $H_{reactants}$, the *sign* of ΔH indicates whether heat is absorbed or released during the reaction. We determine the sign of ΔH *by imagining the heat as "reactant" or "product."* The two possibilities are

1. *Exothermic process: heat as product.* An **exothermic** ("heat out") **process** *releases* heat and results in a *decrease* in the enthalpy of the system:

$$\text{Exothermic:} \quad H_{products} < H_{reactants} \quad \text{so} \quad \Delta H < 0$$

For example, when methane burns in air, heat flows *out of* the system into the surroundings, so we show it as a product:

$$CH_4(g) + 2O_2(g) \longrightarrow CO_2(g) + 2H_2O(g) + \textit{heat}$$

The reactants (1 mol of CH_4 and 2 mol of O_2) release heat during the reaction, so they originally had more enthalpy than the products (1 mol of CO_2 and 2 mol of H_2O):

$$H_{reactants} > H_{products} \quad \text{so} \quad \Delta H\ (H_{products} - H_{reactants}) < 0$$

This exothermic change is shown in the **enthalpy diagram** in Figure 6.6A.

2. *Endothermic process: heat as reactant.* An **endothermic** ("heat in") **process** *absorbs* heat and results in an *increase* in the enthalpy of the system:

$$\text{Endothermic:} \quad H_{products} > H_{reactants} \quad \text{so} \quad \Delta H > 0$$

When ice melts, for instance, heat flows *into* the ice from the surroundings, so we show the heat as a reactant:

$$\textit{heat} + H_2O(s) \longrightarrow H_2O(l)$$

Because heat is absorbed, the enthalpy of the product (water) must be higher than that of the reactant (ice):

$$H_{water} > H_{ice} \quad \text{so} \quad \Delta H\ (H_{water} - H_{ice}) > 0$$

This endothermic change is shown in Figure 6.6B. (In general, the value of an enthalpy change assumes reactants and products are at the same temperature.)

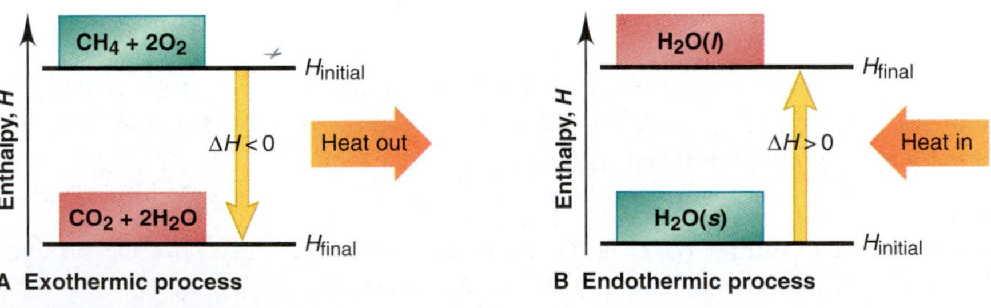

Figure 6.6 Enthalpy diagrams for exothermic and endothermic processes. **A,** The combustion of methane is exothermic: $\Delta H < 0$. **B,** The melting of ice is endothermic: $\Delta H > 0$.

Sample Problem 6.2 Drawing Enthalpy Diagrams and Determining the Sign of ΔH

Problem In each of the following cases, determine the sign of ΔH, state whether the reaction is exothermic or endothermic, and draw an enthalpy diagram:

(a) $H_2(g) + \frac{1}{2}O_2(g) \longrightarrow H_2O(l) + 285.8\ \text{kJ}$

(b) $40.7\ \text{kJ} + H_2O(l) \longrightarrow H_2O(g)$

Plan From each equation, we note whether heat is a "product" (exothermic; $\Delta H < 0$) or a "reactant" (endothermic; $\Delta H > 0$). For exothermic reactions, reactants are above products on the enthalpy diagram; for endothermic reactions, reactants are below products. The ΔH arrow *always* points from reactants to products.

Solution **(a)** Heat is a product (on the right), so $\Delta H < 0$ and the reaction is exothermic. The enthalpy diagram appears in the margin *(top)*.
(b) Heat is a reactant (on the left), so $\Delta H > 0$ and the reaction is endothermic. The enthalpy diagram appears in the margin *(bottom)*.

Check Substances that are on the same side of the equation as the heat have less enthalpy than substances on the other side, so make sure those substances are placed on the lower line of the diagram.

FOLLOW-UP PROBLEM 6.2 When nitroglycerine decomposes, the reaction creates a violent explosion and releases 5.72×10^3 kJ of heat per mole:

$$C_3H_5(NO_3)_3(l) \longrightarrow 3CO_2(g) + \tfrac{5}{2}H_2O(g) + \tfrac{1}{4}O_2(g) + \tfrac{3}{2}N_2(g)$$

Is the decomposition of nitroglycerine exothermic or endothermic? Draw an enthalpy diagram for the process.

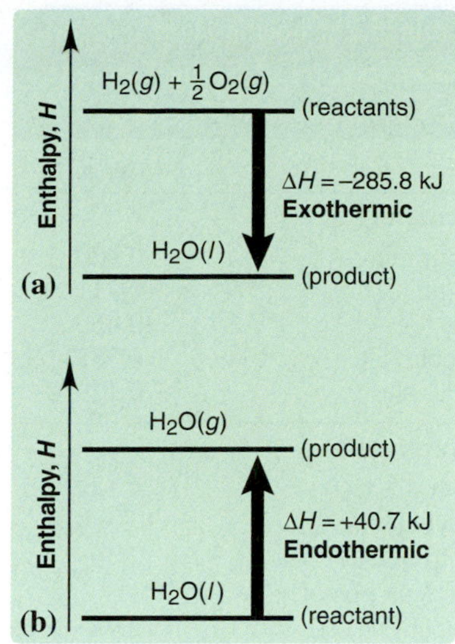

Summary of Section 6.2

- Enthalpy (H) is a state function, so any change in enthalpy (ΔH) is independent of how the change occurred. At constant P, the value of ΔH equals $\Delta E + PV$ work, which occurs when the volume of the system changes in the presence of an external pressure.
- ΔH equals q_P, the heat released or absorbed during a chemical or physical change that takes place at constant pressure.
- A change that releases heat is exothermic ($\Delta H < 0$); a change that absorbs heat is endothermic ($\Delta H > 0$).

6.3 • CALORIMETRY: MEASURING THE HEAT OF A CHEMICAL OR PHYSICAL CHANGE

Data about energy content and usage are everywhere, from the caloric value of a slice of bread to the gas mileage of a new car. In this section, we consider the basic ideas that allow these values to be determined.

Specific Heat Capacity

To find the energy change during a process, we measure the quantity of heat released or absorbed by relating it to the change in temperature. You know from everyday experience that the more you heat an object, the higher its temperature, and the more you cool it, the lower its temperature; in other words, the quantity of heat (q) absorbed or released by an object is proportional to its temperature change:

$$q \propto \Delta T \quad \text{or} \quad q = \text{constant} \times \Delta T \quad \text{or} \quad \frac{q}{\Delta T} = \text{constant}$$

Every object has its own **heat capacity,** the quantity of heat required to change its temperature by 1 K. Heat capacity is the proportionality constant in the preceding equation:

$$\text{Heat capacity} = \frac{q}{\Delta T} \quad [\text{in units of J/K}]$$

A related property is **specific heat capacity (c),** the quantity of heat required to change the temperature of 1 *gram* of the object by 1 K:

$$\text{Specific heat capacity } (c) = \frac{q}{\text{mass} \times \Delta T} \quad [\text{in units of J/g·K}]$$

If we know c of the object being heated (or cooled), we can measure the mass and the temperature change and calculate the heat absorbed (or released):

$$q = c \times \text{mass} \times \Delta T \tag{6.7}$$

Table 6.2 Specific Heat Capacities (c) of Some Elements, Compounds, and Materials

	c (J/g·K)*
Elements	
Aluminum, Al	0.900
Graphite, C	0.711
Iron, Fe	0.450
Copper, Cu	0.387
Gold, Au	0.129
Compounds	
Water, $H_2O(l)$	4.184
Ethyl alcohol, $C_2H_5OH(l)$	2.46
Ethylene glycol, $(CH_2OH)_2(l)$	2.42
Carbon tetrachloride, $CCl_4(l)$	0.862
Solid Materials	
Wood	1.76
Cement	0.88
Glass	0.84
Granite	0.79
Steel	0.45

*At 298 K (25°C).

Equation 6.7 says that when the object gets hotter, that is, when ΔT ($T_{\text{final}} - T_{\text{initial}}$) is positive, $q > 0$ (the object absorbs heat). And when the object gets cooler, that is, when ΔT is negative, $q < 0$ (the object releases heat). Table 6.2 lists the specific heat capacities of some representative substances and materials. Notice that metals have relatively low values of c and water has a very high value: for instance, it takes over 30 times as much energy to increase the temperature of a gram of water by 1 K as it does a gram of gold!

Closely related to the specific heat capacity (but reserved for *substances*) is the **molar heat capacity** (C; note the capital letter), the quantity of heat required to change the temperature of 1 *mole* of a substance by 1 K:

$$\text{Molar heat capacity } (C) = \frac{q}{\text{amount (mol)} \times \Delta T} \qquad [\text{in units of J/mol·K}]$$

To find C of liquid H_2O, we multiply c of liquid H_2O (4.184 J/g·K) by the molar mass of H_2O (18.02 g/mol):

$$C \text{ of } H_2O(l) = 4.184 \frac{\text{J}}{\text{g·K}} \times \frac{18.02 \text{ g}}{1 \text{ mol}} = 75.40 \frac{\text{J}}{\text{mol·K}}$$

Sample Problem 6.3 Finding the Quantity of Heat from a Temperature Change

Problem A layer of copper welded to the bottom of a skillet weighs 125 g. How much heat is needed to raise the temperature of the copper layer from 25°C to 300.°C? The specific heat capacity (c) of Cu is given in Table 6.2.

Plan We know the mass (125 g) and c (0.387 J/g·K) of Cu and can find ΔT in °C, which equals ΔT in K. We then use Equation 6.7 to calculate the heat.

Solution Calculating ΔT and then q:

$$\Delta T = T_{\text{final}} - T_{\text{initial}} = 300.°\text{C} - 25°\text{C} = 275°\text{C} = 275 \text{ K}$$

$$q = c \times \text{mass (g)} \times \Delta T = 0.387 \text{ J/g·K} \times 125 \text{ g} \times 275 \text{ K} = 1.33\times10^4 \text{ J}$$

Check Heat is absorbed by the copper layer (system), so q is positive. Rounding shows the arithmetic is reasonable: $q \approx 0.4 \text{ J/g·K} \times 100 \text{ g} \times 300 \text{ K} = 1.2\times10^4 \text{ J}$.

FOLLOW-UP PROBLEM 6.3 Find the heat released (in kJ) when 5.50 L of ethylene glycol (d = 1.11 g/mL; see Table 6.2) in a car radiator cools from 37.0°C to 25.0°C.

The Two Common Types of Calorimetry

A **calorimeter** is used to measure the heat released (or absorbed) by a physical or chemical process. This apparatus is the "surroundings" that changes temperature when heat is transferred to or from the system. Let's look at two types of calorimeter—one designed to measure the heat at constant pressure and the other at constant volume.

Constant-Pressure Calorimetry For processes that take place at constant pressure, the heat transferred (q_P) can be measured in a simple *coffee-cup calorimeter* (Figure 6.7). This device is used to find the specific heat capacity of a solid that does not react with or dissolve in water. The solid (system) is weighed, heated to some known temperature, and added to a known mass and temperature of water (surroundings) in the calorimeter. After being stirred, the final water temperature, which is also the final temperature of the solid, is measured. Assuming no heat escapes the calorimeter, the heat released by the system ($-q_{\text{sys}}$, or $-q_{\text{solid}}$) is equal in magnitude but opposite in sign to the heat absorbed by the surroundings ($+q_{\text{surr}}$, or $+q_{H_2O}$):

$$-q_{\text{solid}} = q_{H_2O}$$

Substituting from Equation 6.7 on each side of this equation gives

$$-(c_{\text{solid}} \times \text{mass}_{\text{solid}} \times \Delta T_{\text{solid}}) = c_{H_2O} \times \text{mass}_{H_2O} \times \Delta T_{H_2O}$$

All the quantities are known or measured except c_{solid}:

$$c_{\text{solid}} = -\frac{c_{H_2O} \times \text{mass}_{H_2O} \times \Delta T_{H_2O}}{\text{mass}_{\text{solid}} \times \Delta T_{\text{solid}}}$$

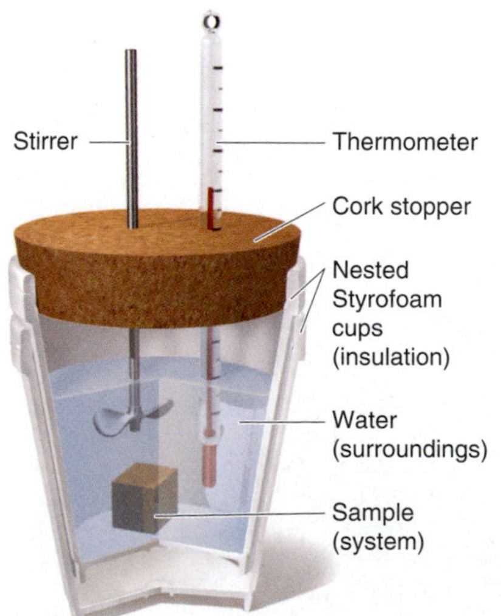

Figure 6.7 Coffee-cup calorimeter. This device measures the heat transferred at constant pressure (q_P).

Sample Problem 6.4 Determining the Specific Heat Capacity of a Solid

Problem You heat 22.05 g of a solid in a test tube to 100.00°C and add it to 50.00 g of water in a coffee-cup calorimeter. The water temperature changes from 25.10°C to 28.49°C. Find the specific heat capacity of the solid.

Plan We are given the masses of the solid (22.05 g) and of H_2O (50.00 g), and we can find the temperature changes of the water and of the solid by subtracting the given values, always using $T_{final} - T_{initial}$. Using Equation 6.7, we set the heat released by the solid ($-q_{solid}$) equal to the heat absorbed by the water (q_{water}). The specific heat of water is known, and we solve for c_{solid}.

Solution Finding ΔT_{solid} and ΔT_{water}:

$$\Delta T_{water} = T_{final} - T_{initial} = (28.49°C - 25.10°C) = 3.39°C = 3.39\ K$$
$$\Delta T_{solid} = T_{final} - T_{initial} = (28.49°C - 100.00°C) = -71.51°C = -71.51\ K$$

Solving for c_{solid}:

$$c_{solid} = -\frac{c_{H_2O} \times mass_{H_2O} \times \Delta T_{H_2O}}{mass_{solid} \times \Delta T_{solid}} = -\frac{4.184\ J/g{\cdot}K \times 50.00\ g \times 3.39\ K}{22.05\ g \times (-71.51\ K)} = 0.450\ J/g{\cdot}K$$

Check Rounding gives -4 J/g·K × 50 g × 3 K/[20 g × (−70°C)] = 0.4 J/g·K, so the answer seems correct.

FOLLOW-UP PROBLEM 6.4 A 12.18-g sample of a shiny, orange-brown metal is heated to 65.00°C in a controlled water bath. The metal is then added to 25.00 g of water in a coffee-cup calorimeter, and the water temperature changes from 25.55°C to 27.25°C. What is the unknown metal (see Table 6.2)?

In the next sample problem, the calorimeter is used to study the change in heat during an aqueous acid-base reaction. Recall that, if a reaction takes place at constant pressure, the heat of the reaction (q_{rxn}) is equal to its enthalpy change (ΔH).

Sample Problem 6.5 Determining the Enthalpy Change of an Aqueous Reaction

Problem You place 50.0 mL of 0.500 *M* NaOH in a coffee-cup calorimeter at 25.00°C and add 25.0 mL of 0.500 *M* HCl, also at 25.00°C. After stirring, the final temperature is 27.21°C. [Assume that the total volume is the sum of the individual volumes and that the final solution has the same density (1.00 g/mL) and specific heat capacity (Table 6.2) as water.]
(a) Calculate q_{soln} (in J).
(b) Calculate the change in enthalpy, ΔH, of the reaction (in kJ/mol of H_2O formed).

(a) Calculate q_{soln}

Plan The solution mixture is the surroundings, and as the reaction takes place, heat flows into it. To find q_{soln}, we use Equation 6.7, so we need the mass of solution, the change in temperature, and the specific heat capacity. We know the solutions' volumes (25.0 mL and 50.0 mL), so we find their masses with the given density (1.00 g/mL). Then, to find q_{soln}, we multiply the total mass by the given c (4.184 J/g·K) and the change in T, which we find from $T_{final} - T_{initial}$.

Solution Finding $mass_{soln}$ and ΔT_{soln}:

$$\text{Total mass (g) of solution} = (25.0\ mL + 50.0\ mL) \times 1.00\ g/mL = 75.0\ g$$
$$\Delta T = 27.21°C - 25.00°C = 2.21°C = 2.21\ K$$

Finding q_{soln}:

$$q_{soln} = c_{soln} \times mass_{soln} \times \Delta T_{soln} = (4.184\ J/g{\cdot}K)(75.0\ g)(2.21\ K) = 693\ J$$

Check Rounding to check q_{soln} gives 4 J/g·K × 75 g × 2 K = 600 J, which is close to the answer.

(b) Calculate the change in enthalpy (ΔH)

Plan To find ΔH, we write the balanced equation for the acid-base reaction and use the volumes and the concentrations (0.500 *M*) to find amount (mol) of each reactant (H^+ and OH^-). Since the amounts of two reactants are given, we determine which is limiting, that is, which gives less product (H_2O). The heat of the surroundings is q_{soln}, and it is the negative of the heat of the reaction (q_{rxn}), which equals ΔH. And dividing q_{rxn} by the amount (mol) of water formed gives ΔH in kJ/mol of water formed.

Solution Writing the balanced equation:

$$HCl(aq) + NaOH(aq) \longrightarrow H_2O(l) + NaCl(aq)$$

Finding amounts (mol) of reactants:

$$\text{Amount (mol) of HCl} = 0.500 \text{ mol HCl/L} \times 0.0250 \text{ L} = 0.0125 \text{ mol HCl}$$

$$\text{Amount (mol) of NaOH} = 0.500 \text{ mol NaOH/L} \times 0.0500 \text{ L} = 0.0250 \text{ mol NaOH}$$

Finding the amount (mol) of product: All the coefficients in the equation are 1, which means that the amount (mol) of reactant yields that amount of product. Therefore, HCl is limiting because it yields less product: 0.0125 mol of H_2O.

Finding ΔH: Heat absorbed by the solution was released by the reaction; that is,

$$q_{soln} = -q_{rxn} = 693 \text{ J} \quad \text{so} \quad q_{rxn} = -693 \text{ J}$$

$$\Delta H\text{(kJ/mol)} = \frac{q_{rxn}}{\text{mol } H_2O} \times \frac{1 \text{ kJ}}{1000 \text{ J}} = \frac{-693 \text{ J}}{0.0125 \text{ mol } H_2O} \times \frac{1 \text{ kJ}}{1000 \text{ J}} = -55.4 \text{ kJ/mol } H_2O$$

Check We check for the limiting reactant: The volume of H^+ is half the volume of OH^-, but they have the same concentration and a 1/1 stoichiometric ratio. Therefore, the amount (mol) of H^+ determines the amount of product. Rounding and taking the negative of q_{soln} to find ΔH gives -600 J/0.012 mol $= -5\times10^4$ J/mol, or -50 kJ/mol, so the answer seems correct.

FOLLOW-UP PROBLEM 6.5 After 50.0 mL of 0.500 *M* $Ba(OH)_2$ and the same volume and concentration of HCl react in a coffee-cup calorimeter, you find q_{soln} to be 1.386 kJ. Calculate ΔH of the reaction in kJ/mol of H_2O formed.

Constant-Volume Calorimetry Constant-volume calorimetry is often carried out in a *bomb calorimeter,* a device commonly used to measure the heat of combustion reactions, such as for fuels and foods. In the coffee-cup calorimeter, we assume all the heat is absorbed by the water, but in reality, some must be absorbed by the stirrer, thermometer, and so forth. With the much more precise bomb calorimeter, the *heat capacity of the entire calorimeter* is known (or can be determined).

Figure 6.8 depicts the preweighed combustible sample in a metal-walled chamber (the bomb), which is filled with oxygen gas and immersed in an insulated water bath fitted with motorized stirrer and thermometer. A heating coil connected to an electrical source ignites the sample, and the heat released raises the temperature of the bomb,

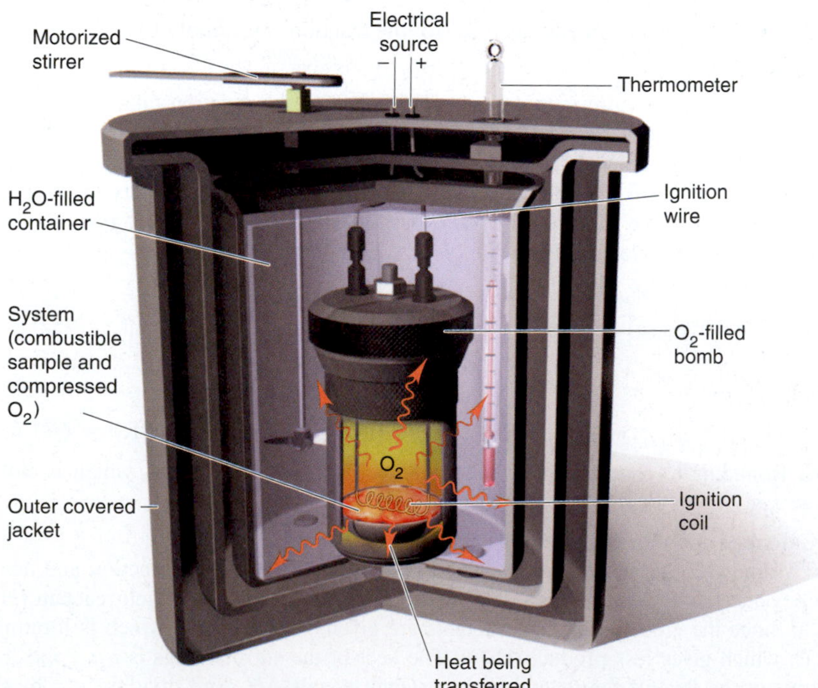

Figure 6.8 A bomb calorimeter. This device measures the heat released at constant volume (q_V). It is often used to study combustion reactions.

water, and other calorimeter parts. Because we know the mass of the sample and the heat capacity of the entire calorimeter, we can use the measured ΔT to calculate the heat released.

Sample Problem 6.6 Calculating the Heat of a Combustion Reaction

Problem A manufacturer claims that its new dietetic dessert has "fewer than 10 Calories per serving." To test the claim, a chemist at the Department of Consumer Affairs places one serving in a bomb calorimeter and burns it in O_2. The initial temperature is 21.862°C and the temperature rises to 26.799°C. If the heat capacity of the calorimeter is 8.151 kJ/K, is the manufacturer's claim correct?

Plan When the dessert (system) burns, the heat released is absorbed by the calorimeter:

$$-q_{\text{system}} = q_{\text{calorimeter}}$$

The claim is correct if the heat of the system is less than 10 Calories. To find the heat, we multiply the given heat capacity of the calorimeter (8.151 kJ/K) by ΔT and then convert to Calories.

Solution Finding ΔT:

$$\Delta T = T_{\text{final}} - T_{\text{initial}} = 26.799°\text{C} - 21.862°\text{C} = 4.937°\text{C} = 4.937\ \text{K}$$

Calculating the heat absorbed by the calorimeter:

$$q_{\text{calorimeter}} = \text{heat capacity} \times \Delta T = 8.151\ \text{kJ/K} \times 4.937\ \text{K} = 40.24\ \text{kJ}$$

Recall that 1 Calorie = 1 kcal = 4.184 kJ. Therefore, 10 Calories = 41.84 kJ, so the claim is correct.

Check A quick math check shows that the answer is reasonable: 8 kJ/K × 5 K = 40 kJ.

Comment The volume of the metal-walled bomb is fixed, so $\Delta V = 0$, and $P\Delta V = 0$. Thus, the energy change measured is the heat at constant volume (q_V), which equals ΔE, not ΔH:

$$\Delta E = q + w = q_V + 0 = q_V$$

However, in most cases, ΔH is usually very close to ΔE. For example, ΔH is only 0.5% larger than ΔE for the combustion of H_2 and only 0.2% smaller for the combustion of octane.

FOLLOW-UP PROBLEM 6.6 A chemist burns 0.8650 g of graphite (a form of carbon) in a new bomb calorimeter, and CO_2 forms. If 393.5 kJ of heat is released per mole of graphite and ΔT is 2.613 K, what is the heat capacity of the bomb calorimeter?

Summary of Section 6.3

- We calculate ΔH of a process by measuring the energy transferred as heat at constant pressure (q_P). To do this, we determine ΔT and multiply it by the mass of the substance and by its specific heat capacity (c), which is the quantity of energy needed to raise the temperature of 1 g of the substance by 1 K.
- Calorimeters measure the heat released (or absorbed) during a process either at constant pressure (coffee cup; $q_P = \Delta H$) or at constant volume (bomb; $q_V = \Delta E$).

6.4 • STOICHIOMETRY OF THERMOCHEMICAL EQUATIONS

A **thermochemical equation** is a balanced equation that includes the enthalpy change of the reaction (ΔH). Keep in mind that a given ΔH refers only to the *amounts (mol) of substances **and** their states of matter in that equation.* The enthalpy change of any process has two aspects:

- *Sign.* The sign of ΔH depends on whether the reaction is exothermic (−) or endothermic (+). A forward reaction has the *opposite* sign of the reverse reaction.
  Decomposition of 2 mol of water to its elements (endothermic):

$$2H_2O(l) \longrightarrow 2H_2(g) + O_2(g) \qquad \Delta H = 572\ \text{kJ}$$

Formation of 2 mol of water from its elements (exothermic):

$$2H_2(g) + O_2(g) \longrightarrow 2H_2O(l) \quad \Delta H = -572 \text{ kJ}$$

- *Magnitude.* The magnitude of ΔH is *proportional to the amount of substance.* Formation of 1 mol of water from its elements (half the preceding amount):

$$H_2(g) + \tfrac{1}{2}O_2(g) \longrightarrow H_2O(l) \quad \Delta H = -286 \text{ kJ}$$

Two key points to understand about thermochemical equations are as follows:

1. *Balancing coefficients.* When necessary, we use fractional coefficients to balance an equation, because we are specifying the magnitude of ΔH for a *particular amount* (often 1 mol) *of substance,* say gaseous SO_2:

$$\tfrac{1}{8}S_8(s) + O_2(g) \longrightarrow SO_2(g) \quad \Delta H = -296.8 \text{ kJ}$$

2. *Thermochemical equivalence.* For a *particular reaction,* a certain amount of substance is thermochemically equivalent to a certain quantity of energy. Some examples from the previous two reactions are

286 kJ is thermochemically equivalent to $\frac{1}{2}$ mol of $O_2(g)$
286 kJ is thermochemically equivalent to 1 mol of $H_2O(l)$
296.8 kJ is thermochemically equivalent to $\frac{1}{8}$ mol of $S_8(s)$

Just as we use stoichiometrically equivalent molar ratios to find amounts of substances, we use thermochemically equivalent quantities to find the ΔH of a reaction for a given amount of substance. Also, just as we use molar mass (in g/mol) to convert an amount (mol) of a substance to mass (g), we use the ΔH (in kJ/mol) to convert an amount of a substance to an equivalent quantity of heat (in kJ). Figure 6.9 shows this new relationship, and Sample Problem 6.7 applies it.

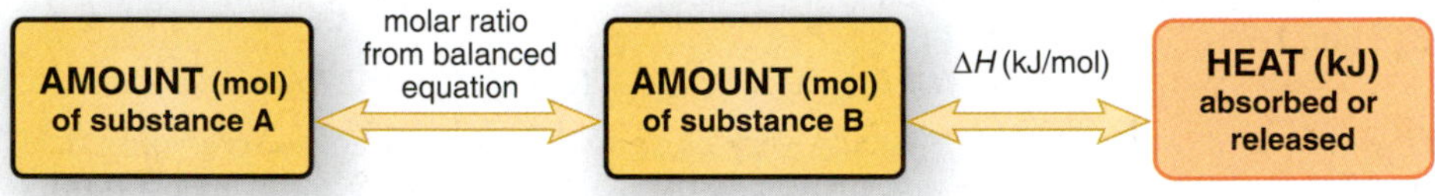

Figure 6.9 The relationship between amount (mol) of substance and the energy (kJ) transferred as heat during a reaction.

Sample Problem 6.7 Using the Enthalpy Change of a Reaction (ΔH) to Find Amounts of Substance

Problem The major source of aluminum in the world is bauxite (mostly aluminum oxide). Its thermal decomposition can be written as

$$Al_2O_3(s) \xrightarrow{\Delta} 2Al(s) + \tfrac{3}{2}O_2(g) \quad \Delta H = 1676 \text{ kJ}$$

If aluminum is produced this way (see Comment), how many grams of aluminum can form when 1.000×10^3 kJ of heat is transferred?

Road Map

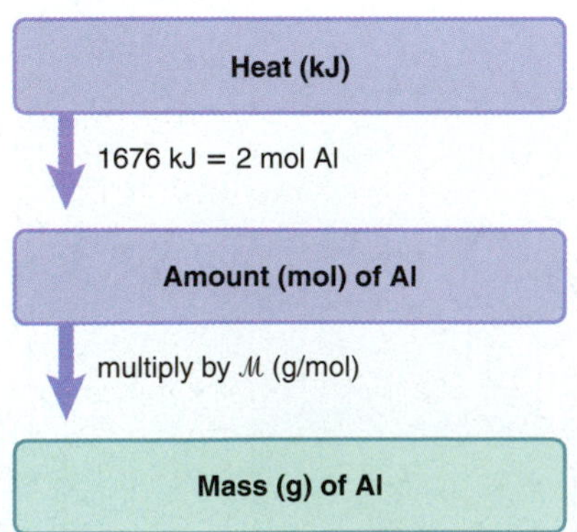

Plan From the balanced equation and the enthalpy change, we see that 1676 kJ of heat is thermochemically equivalent to 2 mol of Al. We convert the given number of kJ to amount (mol) formed and then to mass (g).

Solution Combining steps to convert from heat transferred to mass of Al:

$$\text{Mass (g) of Al} = (1.000\times10^3 \text{ kJ}) \times \frac{2 \text{ mol Al}}{1676 \text{ kJ}} \times \frac{26.98 \text{ g Al}}{1 \text{ mol Al}} = 32.20 \text{ g Al}$$

Check The mass of aluminum seems correct: ~1700 kJ forms about 2 mol of Al (54 g), so 1000 kJ should form a bit more than half that amount (27 g).

Comment In practice, aluminum is not obtained by heating bauxite but by supplying electrical energy (as you'll see in Chapter 22). Because H is a state function, however, the total energy required for the change, ΔH, is the same no matter how it occurs.

FOLLOW-UP PROBLEM 6.7 Hydrogenation reactions, in which H_2 and an "unsaturated" organic compound combine, are used in the food, fuel, and polymer industries. In the simplest case, ethene (C_2H_4) and H_2 form ethane (C_2H_6). If 137 kJ is given off per mole of C_2H_4 reacting, how much heat is released when 15.0 kg of C_2H_6 forms?

Summary of Section 6.4

- A thermochemical equation shows a balanced reaction *and* its ΔH value. The sign of ΔH for a forward reaction is opposite that for the reverse reaction. The magnitude of ΔH is specific for the given equation.
- The amount of a substance and the quantity of heat specified by the balanced equation are thermochemically equivalent and act as conversion factors to find the quantity of heat transferred when any amount of the substance reacts.

6.5 • HESS'S LAW: FINDING ΔH OF ANY REACTION

In some cases, a reaction is difficult, even impossible, to carry out individually: it may be part of a complex biochemical process, or take place under extreme conditions, or require a change in conditions to occur. Even if we can't run a reaction in the lab, we can still find its enthalpy change. In fact, the state-function property of enthalpy (H) allows us to find ΔH of *any* reaction for which we can write an equation.

This application is based on **Hess's law:** *the enthalpy change of an overall process is the sum of the enthalpy changes of its individual steps:*

$$\Delta H_{\text{overall}} = \Delta H_1 + \Delta H_2 + \cdots + \Delta H_n \qquad \textbf{(6.8)}$$

This law follows from the fact that ΔH for a process depends only on the difference between the final and initial states. We apply Hess's law by

- Imagining that an overall reaction occurs through a series of individual reaction steps, whether or not it actually does. Adding the steps must give the overall reaction.
- Choosing individual reaction steps that each has a known ΔH.
- Adding the known ΔH values for the steps to get the unknown ΔH of the overall reaction. We can also find an unknown ΔH of one of the steps by subtraction, if we know the ΔH values for the overall reaction and all the other steps.

Let's apply Hess's law to the oxidation of sulfur to sulfur trioxide, the key change in the industrial production of sulfuric acid and in the formation of acid rain. (To introduce this approach, we'll use S as the formula for sulfur, rather than the more correct S_8.) When we burn S in an excess of O_2, sulfur dioxide (SO_2) forms, *not* sulfur trioxide (SO_3) (equation 1). After a change in conditions, we add more O_2 and oxidize SO_2 to SO_3 (equation 2). Thus, we cannot put S and O_2 in a calorimeter and find ΔH for the overall reaction of S to SO_3 (equation 3). But, we *can* find it with Hess's law. The three equations are

$$\begin{array}{llll} \text{Equation 1:} & S(s) + O_2(g) \longrightarrow SO_2(g) & \Delta H_1 = -296.8\text{ kJ} \\ \text{Equation 2:} & 2SO_2(g) + O_2(g) \longrightarrow 2SO_3(g) & \Delta H_2 = -198.4\text{ kJ} \\ \text{Equation 3:} & S(s) + \tfrac{3}{2}O_2(g) \longrightarrow SO_3(g) & \Delta H_3 = ? \end{array}$$

If we can manipulate equation 1 and/or equation 2, *along with their ΔH value(s),* so the equations add up to equation 3, their ΔH values will add up to the unknown ΔH_3.

First, we identify our "target" equation, the one whose ΔH we want to find, and note the amount (mol) of each reactant and product; our "target" is equation 3. Then, we manipulate equations 1 and/or 2 to make them add up to equation 3:

- Equations 3 and 1 have the same amount of S, so we don't change equation 1.
- Equation 3 has half as much SO_3 as equation 2, so we multiply equation 2, ***and*** $\boldsymbol{\Delta H_2}$, by $\frac{1}{2}$; that is, we always treat the equation and its ΔH value the same.
- With the targeted amounts of reactants and products present, we add equation 1 to the halved equation 2 and cancel terms that appear on both sides:

$$\begin{array}{lll} \text{Equation 1:} & S(s) + O_2(g) \longrightarrow SO_2(g) & \Delta H_1 = -296.8\text{ kJ} \\ \tfrac{1}{2}(\text{Equation 2}): & SO_2(g) + \tfrac{1}{2}O_2(g) \longrightarrow SO_3(g) & \tfrac{1}{2}(\Delta H_2) = -99.2\text{ kJ} \\ \hline \text{Equation 3:} & S(s) + O_2(g) + \cancel{SO_2(g)} + \tfrac{1}{2}O_2(g) \longrightarrow \cancel{SO_2(g)} + SO_3(g) & \\ \text{or,} & S(s) + \tfrac{3}{2}O_2(g) \longrightarrow SO_3(g) & \Delta H_3 = -396.0\text{ kJ} \end{array}$$

Because ΔH depends only on the difference between H_{final} and $H_{initial}$, Hess's law tells us that the difference between the enthalpies of the reactants (1 mol of S and $\frac{3}{2}$ mol of O_2) and the product (1 mol of SO_3) is the same, whether S is oxidized directly to SO_3 (impossible) or through the intermediate formation of SO_2 (actual).

To summarize, calculating an unknown ΔH involves three steps:

1. Identify the target equation, the step whose ΔH is unknown, and note the amount (mol) of each reactant and product.
2. Manipulate each equation with known ΔH values so that the target amount (mol) of each substance is on the correct side of the equation. Remember to:
   - Change the sign of ΔH when you reverse an equation.
   - Multiply amount (mol) and ΔH by the same factor.
3. Add the manipulated equations and their resulting ΔH values to get the target equation and its ΔH. All substances except those in the target equation must cancel.

Sample Problem 6.8 Using Hess's Law to Calculate an Unknown ΔH

Problem Two pollutants that form in auto exhaust are CO and NO. An environmental chemist must convert these pollutants to less harmful gases through the following:

$$CO(g) + NO(g) \longrightarrow CO_2(g) + \tfrac{1}{2}N_2(g) \qquad \Delta H = ?$$

Given the following information, calculate the unknown ΔH:

$$\text{Equation A:} \quad CO(g) + \tfrac{1}{2}O_2(g) \longrightarrow CO_2(g) \qquad \Delta H = -283.0 \text{ kJ}$$

$$\text{Equation B:} \quad N_2(g) + O_2(g) \longrightarrow 2NO(g) \qquad \Delta H = 180.6 \text{ kJ}$$

Plan We note the amount (mol) of each substance and on which side each appears in the target equation. We manipulate equations A and/or B *and* their ΔH values as needed and add them together to obtain the target equation and the unknown ΔH.

Solution Noting substances in the target equation: For reactants, there are 1 mol of CO and 1 mol of NO; for products, there are 1 mol of CO_2 and $\frac{1}{2}$ mol of N_2.
Manipulating the given equations:

- Equation A has the same amounts of CO and CO_2 on the same sides of the arrow as in the target, so we leave it as written.
- Equation B has twice as much N_2 and NO as the target, and they are on the opposite sides in the target. Thus, we reverse equation B, change the sign of its ΔH, and multiply both by $\frac{1}{2}$:

$$\tfrac{1}{2}[2NO(g) \longrightarrow N_2(g) + O_2(g)] \qquad \Delta H = -\tfrac{1}{2}(\Delta H) = -\tfrac{1}{2}(180.6 \text{ kJ})$$

or

$$NO(g) \longrightarrow \tfrac{1}{2}N_2(g) + \tfrac{1}{2}O_2(g) \qquad \Delta H = -90.3 \text{ kJ}$$

Adding the manipulated equations to obtain the target equation:

Equation A:	$CO(g) + \cancel{\tfrac{1}{2}O_2(g)} \longrightarrow CO_2(g)$	$\Delta H = -283.0$ kJ
$\frac{1}{2}$(Equation B reversed)	$NO(g) \longrightarrow \tfrac{1}{2}N_2(g) + \cancel{\tfrac{1}{2}O_2(g)}$	$\Delta H = -90.3$ kJ
Target equation:	$CO(g) + NO(g) \longrightarrow CO_2(g) + \tfrac{1}{2}N_2(g)$	$\Delta H = -373.3$ kJ

Check Obtaining the desired target equation is a sufficient check. Be sure to remember to change the *sign* of ΔH for any equation you reverse.

FOLLOW-UP PROBLEM 6.8 Nitrogen oxides undergo many reactions in the environment and in industry. Given the following information, calculate ΔH for the overall equation, $2NO_2(g) + \frac{1}{2}O_2(g) \longrightarrow N_2O_5(s)$:

$$N_2O_5(s) \longrightarrow 2NO(g) + \tfrac{3}{2}O_2(g) \qquad \Delta H = 223.7 \text{ kJ}$$

$$NO(g) + \tfrac{1}{2}O_2(g) \longrightarrow NO_2(g) \qquad \Delta H = -57.1 \text{ kJ}$$

Summary of Section 6.5

- Because H is a state function, we can use Hess's law to determine ΔH of any reaction by assuming that it is the sum of other reactions.
- After manipulating the equations of those reactions and their ΔH values to match the substances in the target equation, we add the manipulated ΔH values to find the unknown ΔH.

6.6 • STANDARD ENTHALPIES OF REACTION (ΔH°_{rxn})

In this section, we see how Hess's law allows us to determine the ΔH values of an enormous number of reactions. Thermodynamic variables, such as ΔH, vary somewhat with conditions. Therefore, in order to study and compare reactions, chemists have established a set of specific conditions called **standard states:**

- For a *gas,* the standard state is 1 atm* and ideal behavior.
- For a substance in *aqueous solution,* the standard state is 1 *M* concentration.
- For a *pure substance* (element or compound), the standard state is usually the most stable form of the substance at 1 atm and the temperature of interest. In this text (and in most thermodynamic tables), that temperature is usually 25°C (298 K).

The standard-state symbol (shown as a degree sign) indicates that the variable has been measured with *all the substances in their standard states.* For example, when the enthalpy change of a reaction is measured at the standard state, it is the **standard enthalpy of reaction, ΔH°_{rxn}** (also called *standard heat of reaction*).

Formation Equations and Their Standard Enthalpy Changes

In a **formation equation,** 1 mol of a compound forms from its elements. The **standard enthalpy of formation** (**ΔH°_f;** or *standard heat of formation*) is the enthalpy change for the formation equation when all the substances are in their standard states. For instance, the formation equation for methane (CH_4) is

$$C(\text{graphite}) + 2H_2(g) \longrightarrow CH_4(g) \quad \Delta H^\circ_f = -74.9 \text{ kJ}$$

Fractional coefficients are often used with reactants to obtain 1 mol of the product:

$$Na(s) + \tfrac{1}{2}Cl_2(g) \longrightarrow NaCl(s) \qquad \Delta H^\circ_f = -411.1 \text{ kJ}$$

$$2C(\text{graphite}) + 3H_2(g) + \tfrac{1}{2}O_2(g) \longrightarrow C_2H_5OH(l) \quad \Delta H^\circ_f = -277.6 \text{ kJ}$$

Standard enthalpies of formation have been tabulated for many substances. Table 6.3 shows several, and a much more extensive table appears in Appendix B. The values in Table 6.3 make two points:

1. *For an element in its standard state,* $\Delta H^\circ_f = 0$.
   - The standard state for sodium is the solid ($\Delta H^\circ_f = 0$); it takes 107.8 kJ of heat to form 1 mol of gaseous Na ($\Delta H^\circ_f = 107.8$ kJ/mol).
   - The standard state for molecular elements, such as the halogens, is the molecular form, not separate atoms: for Cl_2, $\Delta H^\circ_f = 0$, but for Cl, $\Delta H^\circ_f = 121.0$ kJ/mol.
   - Some elements exist in different forms (called *allotropes;* Chapter 14), but only one is the standard state. The standard state of carbon is graphite ($\Delta H^\circ_f = 0$), not diamond ($\Delta H^\circ_f = 1.9$ kJ/mol), the standard state of oxygen is O_2 ($\Delta H^\circ_f = 0$), not ozone (O_3; $\Delta H^\circ_f = 143$ kJ/mol), and the standard state of sulfur is S_8 in its rhombic crystal form ($\Delta H^\circ_f = 0$), not in its monoclinic form ($\Delta H^\circ_f = 0.3$ kJ/mol).
2. *Most compounds have a negative ΔH°_f.* That is, most compounds have exothermic formation reactions: *under standard conditions, heat is released when most compounds form from their elements.*

Table 6.3 Selected Standard Enthalpies of Formation at 25°C (298 K)

Formula	ΔH°_f (kJ/mol)
Calcium	
$Ca(s)$	0
$CaO(s)$	−635.1
$CaCO_3(s)$	−1206.9
Carbon	
C(graphite)	0
C(diamond)	1.9
$CO(g)$	−110.5
$CO_2(g)$	−393.5
$CH_4(g)$	−74.9
$CH_3OH(l)$	−238.6
$HCN(g)$	135
$CS_2(l)$	87.9
Chlorine	
$Cl(g)$	121.0
$Cl_2(g)$	0
$HCl(g)$	−92.3
Hydrogen	
$H(g)$	218.0
$H_2(g)$	0
Nitrogen	
$N_2(g)$	0
$NH_3(g)$	−45.9
$NO(g)$	90.3
Oxygen	
$O_2(g)$	0
$O_3(g)$	143
$H_2O(g)$	−241.8
$H_2O(l)$	−285.8
Silver	
$Ag(s)$	0
$AgCl(s)$	−127.0
Sodium	
$Na(s)$	0
$Na(g)$	107.8
$NaCl(s)$	−411.1
Sulfur	
S_8(rhombic)	0
S_8(monoclinic)	0.3
$SO_2(g)$	−296.8
$SO_3(g)$	−396.0

Sample Problem 6.9 Writing Formation Equations

Problem Write a balanced formation equation for each of the following and include the value of ΔH°_f:
(a) $AgCl(s)$ **(b)** $CaCO_3(s)$ **(c)** $HCN(g)$

Plan We write the elements as the reactants and 1 mol of the compound as the product, being sure all substances are in their standard states. Then, we balance the equations and find the ΔH°_f values in Table 6.3 or Appendix B.

*The definition of the standard state for gases has been changed to 1 bar, a slightly lower pressure than the 1 atm standard on which the data in this book are based (1 atm = 101.3 kPa = 1.013 bar). Except for very precise experimental work, this makes very little difference in the standard enthalpy values.

Solution **(a)** $Ag(s) + \frac{1}{2}Cl_2(g) \longrightarrow AgCl(s) \quad \Delta H_f^\circ = -127.0\text{ kJ}$

(b) $Ca(s) + C(\text{graphite}) + \frac{3}{2}O_2(g) \longrightarrow CaCO_3(s) \quad \Delta H_f^\circ = -1206.9\text{ kJ}$

(c) $\frac{1}{2}H_2(g) + C(\text{graphite}) + \frac{1}{2}N_2(g) \longrightarrow HCN(g) \quad \Delta H_f^\circ = 135\text{ kJ}$

FOLLOW-UP PROBLEM 6.9 Write a balanced formation equation for each of the following and include its ΔH_f°: **(a)** $CH_3OH(l)$, **(b)** $CaO(s)$, **(c)** $CS_2(l)$.

Determining ΔH°_{rxn} from ΔH°_f Values for Reactants and Products

We can use ΔH°_f values to determine any ΔH°_{rxn}. Applying Hess's law, we imagine the reaction occurring in two steps (Figure 6.10):

Step 1. Each reactant decomposes to its elements. This is the *reverse* of the formation reaction for the *reactant,* so the standard enthalpy change is $-\Delta H^\circ_f$.

Step 2. Each product forms from its elements. This step *is* the formation reaction for the *product,* so the standard enthalpy change is ΔH°_f.

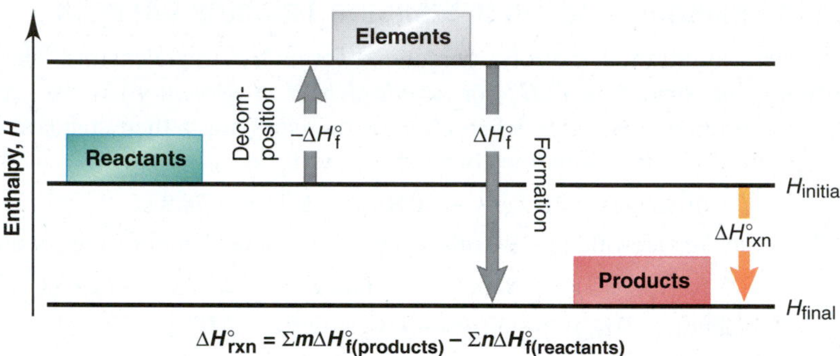

Figure 6.10 The two-step process for determining ΔH°_{rxn} from ΔH°_f values.

According to Hess's law, we add the enthalpy changes for these steps to obtain the overall enthalpy change for the reaction (ΔH°_{rxn}). Suppose we want ΔH°_{rxn} for

$$TiCl_4(l) + 2H_2O(g) \longrightarrow TiO_2(s) + 4HCl(g)$$

We write this equation as though it were the sum of four individual equations, one for each compound. The first two show step 1, the decomposition of the reactants to their elements (*reverse* of their formation); the second two show step 2, the formation of the products from their elements:

$$\begin{array}{rcll} TiCl_4(l) & \longrightarrow & Ti(s) + 2Cl_2(g) & -\Delta H^\circ_f[TiCl_4(l)] \\ 2H_2O(g) & \longrightarrow & 2H_2(g) + O_2(g) & -2\Delta H^\circ_f[H_2O(g)] \\ Ti(s) + O_2(g) & \longrightarrow & TiO_2(s) & \Delta H^\circ_f[TiO_2(s)] \\ 2H_2(g) + 2Cl_2(g) & \longrightarrow & 4HCl(g) & 4\Delta H^\circ_f[HCl(g)] \end{array}$$

$$TiCl_4(l) + 2H_2O(g) + \cancel{Ti(s)} + \cancel{O_2(g)} + \cancel{2H_2(g)} + \cancel{2Cl_2(g)} \longrightarrow \cancel{Ti(s)} + \cancel{2Cl_2(g)} + \cancel{2H_2(g)} + \cancel{O_2(g)} + TiO_2(s) + 4HCl(g)$$

or $$TiCl_4(l) + 2H_2O(g) \longrightarrow TiO_2(s) + 4HCl(g)$$

It's important to realize that when titanium(IV) chloride and water react, the reactants don't *actually* decompose to their elements, which then recombine to form the products. But the great usefulness of Hess's law and the state-function concept is that ΔH°_{rxn} is the difference between two state functions, $H^\circ_{products}$ minus $H^\circ_{reactants}$, so it doesn't matter how the reaction *actually* occurs. We add the individual enthalpy changes to find ΔH°_{rxn}:

$$\begin{aligned} \Delta H^\circ_{rxn} &= \Delta H^\circ_f[TiO_2(s)] + 4\Delta H^\circ_f[HCl(g)] + \{-\Delta H^\circ_f[TiCl_4(l)]\} + \{-2\Delta H^\circ_f[H_2O(g)]\} \\ &= \underbrace{\{\Delta H^\circ_f[TiO_2(s)] + 4\Delta H^\circ_f[HCl(g)]\}}_{\text{Products}} - \underbrace{\{\Delta H^\circ_f[TiCl_4(l)] + 2\Delta H^\circ_f[H_2O(g)]\}}_{\text{Reactants}} \end{aligned}$$

The arithmetic in this case gives $\Delta H^\circ_{rxn} = -25.39$ kJ, but more importantly, when we generalize the pattern, we see that *the standard enthalpy of reaction is the sum of the standard enthalpies of formation of the* **products** *minus the sum of the standard enthalpies of formation of the* **reactants** (see Figure 6.10):

$$\Delta H^\circ_{rxn} = \Sigma m\Delta H^\circ_{f\,(products)} - \Sigma n\Delta H^\circ_{f\,(reactants)} \quad (6.9)$$

where Σ means "sum of," and m and n are the amounts (mol) of the products and reactants given by the coefficients in the balanced equation.

Sample Problem 6.10 Calculating ΔH°_{rxn} from ΔH°_f Values

Problem Nitric acid is used to make many products, including fertilizers, dyes, and explosives. The first step in its production is the oxidation of ammonia:

$$4NH_3(g) + 5O_2(g) \longrightarrow 4NO(g) + 6H_2O(g)$$

Calculate ΔH°_{rxn} from ΔH°_f values.

Plan We use values from Table 6.3 (or Appendix B) and apply Equation 6.9.

Solution Calculating ΔH°_{rxn}:

$$\begin{aligned}\Delta H^\circ_{rxn} &= \Sigma m\Delta H^\circ_{f\,(products)} - \Sigma n\Delta H^\circ_{f\,(reactants)}\\ &= \{4\Delta H^\circ_f[NO(g)] + 6\Delta H^\circ_f[H_2O(g)]\} - \{4\Delta H^\circ_f[NH_3(g)] + 5\Delta H^\circ_f[O_2(g)]\}\\ &= (4\text{ mol})(90.3\text{ kJ/mol}) + (6\text{ mol})(-241.8\text{ kJ/mol})\\ &\quad - [(4\text{ mol})(-45.9\text{ kJ/mol}) + (5\text{ mol})(0\text{ kJ/mol})]\\ &= 361\text{ kJ} - 1451\text{ kJ} + 184\text{ kJ} - 0\text{ kJ} = -906\text{ kJ}\end{aligned}$$

Check We write formation equations, with ΔH°_f values for the amounts of each compound, in the correct direction (forward for products and reverse for reactants) and find the sum:

$$\begin{array}{llr} 4NH_3(g) \longrightarrow \cancel{2N_2(g)} + \cancel{6H_2(g)} & -4\Delta H^\circ_f = -4(-45.9\text{ kJ}) = & 184\text{ kJ}\\ \cancel{2N_2(g)} + 2O_2(g) \longrightarrow 4NO(g) & 4\Delta H^\circ_f = 4(90.3\text{ kJ}) = & 361\text{ kJ}\\ \cancel{6H_2(g)} + 3O_2(g) \longrightarrow 6H_2O(g) & 6\Delta H^\circ_f = 6(-241.8\text{ kJ}) = & -1451\text{ kJ}\\ \hline 4NH_3(g) + 5O_2(g) \longrightarrow 4NO(g) + 6H_2O(g) & \Delta H^\circ_{rxn} = & -906\text{ kJ}\end{array}$$

Comment In this problem, we know the individual ΔH°_f values and find the sum, ΔH°_{rxn}. In the follow-up problem, we know the sum and want to find one of the ΔH°_f values.

FOLLOW-UP PROBLEM 6.10 Use the following to find ΔH°_f of methanol [$CH_3OH(l)$]:

$$CH_3OH(l) + \tfrac{3}{2}O_2(g) \longrightarrow CO_2(g) + 2H_2O(g) \quad \Delta H^\circ_{rxn} = -638.5\text{ kJ}$$

$$\Delta H^\circ_f \text{ of } CO_2(g) = -393.5\text{ kJ/mol}$$

$$\Delta H^\circ_f \text{ of } H_2O(g) = -241.8\text{ kJ/mol}$$

Fossil Fuels and Climate Change

Out of necessity, the nations of the world are finally beginning to radically rethink the issue of energy use. No scientific challenge today is greater than reversing the climatic effects of our increasing dependence on the combustion of **fossil fuels**—coal, petroleum, and natural gas. Because these fuels form so much more slowly than we consume them, they are *nonrenewable*. In contrast, wood and other fuels derived from plant and animal matter are *renewable*.

All carbon-based fuels release CO_2 when burned, and in the past few decades it has become increasingly clear that our use of these fuels is changing Earth's climate. The ability of CO_2 to absorb heat plays a vital temperature-regulating role in the atmosphere. Much of the sunlight that shines on Earth is absorbed by the land and oceans and converted to heat. Like the glass of a greenhouse, atmospheric CO_2 does not absorb visible light from the Sun, but it traps some of the heat (infrared light) radiating back from Earth's surface and, thus, helps warm the atmosphere. This process is called the natural *greenhouse effect* (Figure 6.11, *left; next page*).

Over several billion years, due largely to the spread of plant life, which uses CO_2 in photosynthesis, the amount of CO_2 originally present in Earth's atmosphere decreased to 0.028% by volume. However, today (mid-2011), as a result of the human use of fossil fuels for the past 200 years, and especially the past 65, this amount has

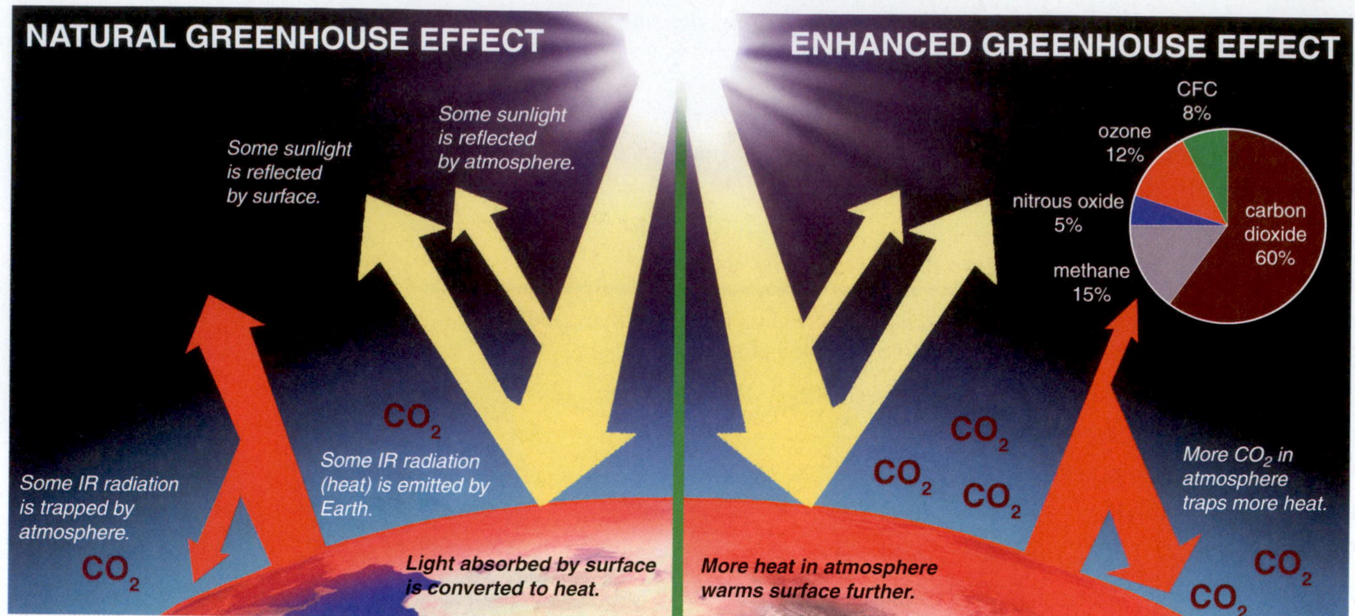

Figure 6.11 The trapping of heat by the atmosphere. Of the total sunlight reaching Earth, some is reflected and some is absorbed and converted to infrared (IR) radiation (heat). Some heat emitted by the surface is trapped by atmospheric CO_2, creating a *natural* greenhouse effect *(left)* that has been essential to life. But, largely as a result of human activity in the past 150 years, and especially the past several decades, the buildup of CO_2 and several other greenhouse gases *(pie chart)* has created an *enhanced* greenhouse effect *(right)*.

increased to slightly over 0.039%. Thus, although the same amount of solar energy passes through the atmosphere, more is trapped as heat, which has created an *enhanced* greenhouse effect that is changing the climate through *global warming* (Figure 6.11, *right*). Based on current trends in fossil fuel use, various estimates show that the CO_2 concentration will increase to between 0.049% and 0.126% by 2100.

Computer-based models that simulate the climate's behavior are used to predict how much the temperature will rise. Despite several complicating factors, the best models predicted a net warming of the atmosphere, and for the past several years, scientists have documented the predicted effects. The average temperature has increased by 0.6 ± 0.2°C since the late 19th century, of which 0.2–0.3°C has occurred over just the past 25 years. Globally, the decade from 2001 to 2010 was the warmest ever recorded. Snow cover and glacier extent in the Northern Hemisphere and floating ice in the Arctic Ocean have decreased dramatically. Globally, sea level has risen an average of 17 cm (6.7 in) over the past century, and flooding and other extreme weather events have increased worldwide.

Today, the models predict a future temperature rise more than 50% higher than the 1.0–3.5°C rise predicted only 10 years ago. Such increases would significantly alter rainfall patterns and crop yields throughout the world and could increase sea level as much as 1 meter, thereby flooding low-lying regions, such as the Netherlands, half of Florida, much of southern Asia, and many Pacific island nations. To make matters worse, as we burn fossil fuels that *release* CO_2, we cut down the forests that *absorb* it.

In addition to developing alternative energy sources to reduce fossil-fuel consumption, some researchers are studying *carbon capture and sequestration (CCS),* accomplished by large-scale tree planting and by liquefying CO_2 released from coal-fired power plants and burying it underground or injecting it deep into the oceans. But, CCS has not been effective, and most other scientists are pressing for more efficient usage and conserving current supplies.

In 1997, the United Nations Conference on Climate Change in Kyoto, Japan, created an international treaty that set legally binding limits on release of greenhouse gases. It was ratified by 189 countries, but not by the largest emitter of CO_2, the United States. The 2005 conference in Montreal, Canada, presented overwhelming scientific evidence that confirmed the human impact on climate change, but the 2009 conference in Copenhagen, Denmark, which was attended by representatives from 192 nations, failed to produce a legally binding agreement on emission targets. In late 2010, however, the conference in Cancun, Mexico, resulted in some financial commitments to help developing countries rely on alternative energy sources.

Summary of Section 6.6

- Standard states are a set of specific conditions used for determining thermodynamic variables for all substances.
- When 1 mol of a compound forms from its elements with all substances in their standard states, the enthalpy change is the standard enthalpy of formation, ΔH_f°.
- Hess's law allows us to picture a reaction as the decomposition of the reactants to their elements followed by the formation of the products from their elements.
- We use tabulated ΔH_f° values to find ΔH_{rxn}° or use known ΔH_{rxn}° and ΔH_f° values to find an unknown ΔH_f°.
- Because of major concerns about climate change, nations are beginning to develop alternative means of producing energy.

CHAPTER REVIEW GUIDE

The following sections provide many aids to help you study this chapter. (Numbers in parentheses refer to pages, unless noted otherwise.)

Learning Objectives

These are concepts and skills to review after studying this chapter.

Related section (§), sample problem (SP), and end-of-chapter problem (EP) numbers are listed in parentheses.

1. Interconvert energy units; understand that ΔE of a system appears as the total heat and/or work transferred to or from its surroundings; understand the meaning of a state function (§6.1) (SP 6.1) (EPs 6.1–6.9)
2. Understand the meaning of H, why we measure ΔH, and the distinction between exothermic and endothermic reactions; draw enthalpy diagrams for chemical and physical changes (§6.2) (SP 6.2) (EPs 6.10–6.19)
3. Understand the relation between specific heat capacity and heat transferred in both constant-pressure (coffee-cup) and constant-volume (bomb) calorimeters (§6.3) (SPs 6.3–6.6) (EPs 6.20–6.32)
4. Understand the relation between heat of reaction and amount of substance (§6.4) (SP 6.7) (EPs 6.33–6.42)
5. Explain the importance of Hess's law and use it to find an unknown ΔH (§6.5) (SP 6.8) (EPs 6.43–6.48)
6. View a reaction as the decomposition of reactants followed by the formation of products; understand formation equations and how to use ΔH_f° values to find ΔH_{rxn}° (§6.6) (SPs 6.9, 6.10) (EPs 6.49–6.58)

Key Terms

These important terms appear in boldface in the chapter and are defined again in the Glossary.

thermodynamics (189)
thermochemistry (189)

Section 6.1
system (189)
surroundings (189)
internal energy (E) (190)
heat (q) (190)
work (w) (190)
law of conservation of energy (first law of thermodynamics) (193)
joule (J) (193)
calorie (cal) (193)
state function (194)

Section 6.2
pressure-volume work (PV work) (195)
enthalpy (H) (195)
change in enthalpy (ΔH) (195)
exothermic process (196)
enthalpy diagram (196)
endothermic process (196)

Section 6.3
heat capacity (197)
specific heat capacity (c) (197)
molar heat capacity (C) (198)
calorimeter (198)

Section 6.4
thermochemical equation (201)

Section 6.5
Hess's law (203)

Section 6.6
standard state (205)
standard enthalpy of reaction (ΔH_{rxn}°) (205)
formation equation (205)
standard enthalpy of formation (ΔH_f°) (205)
fossil fuel (207)

Key Equations and Relationships

Numbered and screened concepts are listed for you to refer to or memorize.

6.1 Defining the change in internal energy (190):

$$\Delta E = E_{final} - E_{initial} = E_{products} - E_{reactants}$$

6.2 Expressing the change in internal energy in terms of heat and work (190):

$$\Delta E = q + w$$

6.3 Stating the first law of thermodynamics (law of conservation of energy) (193):

$$\Delta E_{universe} = \Delta E_{system} + \Delta E_{surroundings} = 0$$

6.4 Determining the work due to a change in volume at constant pressure (*PV* work) (195):

$$w = -P\Delta V$$

6.5 Relating the enthalpy change to the internal energy change at constant pressure (195):

$$\Delta H = \Delta E + P\Delta V$$

6.6 Identifying the enthalpy change with the heat absorbed or released at constant pressure (196):

$$q_P = \Delta E + P\Delta V = \Delta H$$

6.7 Calculating the heat absorbed or released when a substance undergoes a temperature change or a reaction occurs (197):

$$q = c \times \text{mass} \times \Delta T$$

6.8 Calculating the overall enthalpy change of a reaction (Hess's law) (203):

$$\Delta H_{overall} = \Delta H_1 + \Delta H_2 + \cdots + \Delta H_n$$

6.9 Calculating the standard enthalpy of reaction (207):

$$\Delta H^\circ_{rxn} = \Sigma m\Delta H^\circ_{f\,(products)} - \Sigma n\Delta H^\circ_{f\,(reactants)}$$

BRIEF SOLUTIONS TO FOLLOW-UP PROBLEMS

Compare your own solutions to these calculation steps and answers.

6.1 $\Delta E = q + w$

$$= \left(-26.0\text{ kcal} \times \frac{4.184\text{ kJ}}{1\text{ kcal}}\right) + \left(15.0\text{ Btu} \times \frac{1.055\text{ kJ}}{1\text{ Btu}}\right)$$

$$= -93\text{ kJ}$$

6.2 The reaction is exothermic.

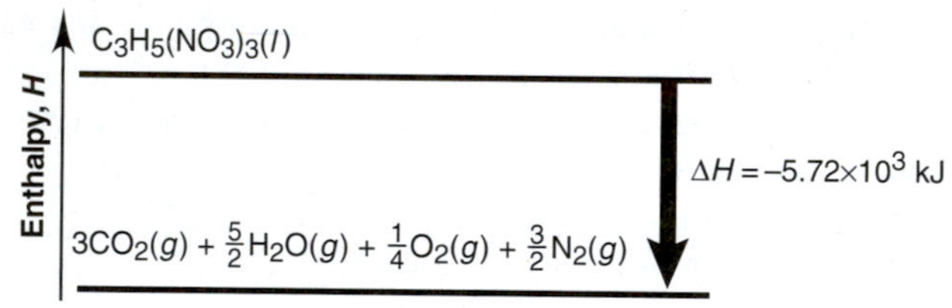

6.3 $\Delta T = 25.0°\text{C} - 37.0°\text{C} = -12.0°\text{C} = -12.0\text{ K}$

$$\text{Mass (g)} = 1.11\text{ g/mL} \times \frac{1000\text{ mL}}{1\text{ L}} \times 5.50\text{ L} = 6.10\times10^3\text{ g}$$

$$q = c \times \text{mass} \times \Delta T$$

$$= (2.42\text{ J/g·K})\left(\frac{1\text{ kJ}}{1000\text{ J}}\right)(6.10\times10^3\text{ g})(-12.0\text{ K}) = -177\text{ kJ}$$

6.4 $c_{solid} = -\dfrac{4.184\text{ J/g·K} \times 25.00\text{ g} \times 1.70\text{ K}}{12.18\text{ g} \times (-37.75\text{ K})} = 0.387\text{ J/g·K}$

From Table 6.2, the metal is copper.

6.5 $2HCl(aq) + Ba(OH)_2(aq) \longrightarrow 2H_2O(l) + BaCl_2(aq)$

Amount (mol) of HCl $= 0.500\text{ mol HCl/L} \times 0.0500\text{ L}$
$= 0.0250\text{ mol HCl}$

Similarly, we have 0.0250 mol $BaCl_2$.

Finding the limiting reactant from the balanced equation:

$$\text{Amount (mol) of } H_2O = 0.0250\text{ mol HCl} \times \frac{2\text{ mol } H_2O}{2\text{ mol HCl}} = 0.0250\text{ mol } H_2O$$

$$\text{Amount (mol) of } H_2O = 0.0250\text{ mol } Ba(OH)_2 \times \frac{2\text{ mol } H_2O}{1\text{ mol } Ba(OH)_2} = 0.0500\text{ mol } H_2O$$

Thus, HCl is limiting.

$$\Delta H\text{ (kJ/mol } H_2O) = \frac{q_{rxn}}{\text{mol } H_2O} = \frac{-1.386\text{ kJ}}{0.0250\text{ mol } H_2O} = -55.4\text{ kJ/mol } H_2O$$

6.6

$$-q_{sample} = q_{calorimeter}$$

$$-(0.8650\text{ g C})\left(\frac{1\text{ mol C}}{12.01\text{ g C}}\right)(-393.5\text{ kJ/mol C}) = (2.613\text{ K})x$$

x = heat capacity of calorimeter = 10.85 kJ/K

6.7 $C_2H_4(g) + H_2(g) \longrightarrow C_2H_6(g) + 137\text{ kJ}$

$$\text{Heat (kJ)} = 15.0\text{ kg} \times \frac{1000\text{ g}}{1\text{ kg}} \times \frac{1\text{ mol } C_2H_6}{30.07\text{ g } C_2H_6} \times \frac{137\text{ kJ}}{1\text{ mol}} = 6.83\times10^4\text{ kJ}$$

6.8

$2NO(g) + \frac{3}{2}O_2(g) \longrightarrow N_2O_5(s)$ $\quad\Delta H = -223.7\text{ kJ}$

$2NO_2(g) \longrightarrow 2NO(g) + O_2(g)$ $\quad\Delta H = 114.2\text{ kJ}$

$\cancel{2NO(g)} + \frac{1}{2}\cancel{\frac{3}{2}}O_2(g) + 2NO_2(g) \longrightarrow N_2O_5(s) + \cancel{2NO(g)} + \cancel{O_2(g)}$

$2NO_2(g) + \frac{1}{2}O_2(g) \longrightarrow N_2O_5(s)$ $\quad\Delta H = -109.5\text{ kJ}$

6.9 (a) $C(\text{graphite}) + 2H_2(g) + \frac{1}{2}O_2(g) \longrightarrow CH_3OH(l)$ $\quad\Delta H^\circ_f = -238.6\text{ kJ}$

(b) $Ca(s) + \frac{1}{2}O_2(g) \longrightarrow CaO(s)$ $\quad\Delta H^\circ_f = -635.1\text{ kJ}$

(c) $C(\text{graphite}) + \frac{1}{4}S_8(\text{rhombic}) \longrightarrow CS_2(l)$ $\quad\Delta H^\circ_f = 87.9\text{ kJ}$

6.10 ΔH°_f of $CH_3OH(l)$

$$= -\Delta H^\circ_{rxn} + 2\Delta H^\circ_f[H_2O(g)] + \Delta H^\circ_f[CO_2(g)]$$

$$= 638.5\text{ kJ} + (2\text{ mol})(-241.8\text{ kJ/mol}) + (1\text{ mol})(-393.5\text{ kJ/mol})$$

$$= -238.6\text{ kJ}$$

PROBLEMS

Problems with **colored** numbers are answered in Appendix E. Sections match the text and provide the numbers of relevant sample problems. Bracketed problems are grouped in pairs (indicated by a short rule) that cover the same concept. Comprehensive Problems are based on material from any section or previous chapter.

Forms of Energy and Their Interconversion

(Sample Problem 6.1)

6.1 If you feel warm after exercising, have you increased the internal energy of your body? Explain.

6.2 An *adiabatic* process is one that involves no heat transfer. What is the relationship between work and the change in internal energy in an adiabatic process?

6.3 Name a common device used to accomplish each change:
(a) Electrical energy to thermal energy
(b) Electrical energy to sound energy
(c) Electrical energy to light energy
(d) Mechanical energy to electrical energy
(e) Chemical energy to electrical energy

6.4 You lift your textbook and drop it onto a desk. Describe the energy transformations (from one form to another) that occur, moving backward in time from a moment after impact.

6.5 A system receives 425 J of heat from and delivers 425 J of work to its surroundings. What is the change in internal energy of the system (in J)?

6.6 A system releases 255 cal of heat to the surroundings and delivers 428 cal of work. What is the change in internal energy of the system (in cal)?

6.7 Complete combustion of 2.0 metric tons of coal to gaseous carbon dioxide releases 6.6×10^{10} J of heat. Convert this energy to (a) kilojoules; (b) kilocalories; (c) British thermal units.

6.8 Thermal decomposition of 5.0 metric tons of limestone to lime and carbon dioxide absorbs 9.0×10^{6} kJ of heat. Convert this energy to (a) joules; (b) calories; (c) British thermal units.

6.9 The nutritional calorie (Calorie) is equivalent to 1 kcal. One pound of body fat is equivalent to about 4.1×10^{3} Calories. Express this quantity of energy in joules and kilojoules.

Enthalpy: Chemical Change at Constant Pressure

(Sample Problem 6.2)

6.10 Classify the following processes as exothermic or endothermic: (a) freezing of water; (b) boiling of water; (c) digestion of food; (d) a person running; (e) a person growing; (f) wood being chopped; (g) heating with a furnace.

6.11 Why can we measure only *changes* in enthalpy, not absolute enthalpy values?

6.12 Draw an enthalpy diagram for a general exothermic reaction; label the axis, reactants, products, and ΔH with its sign.

6.13 Draw an enthalpy diagram for a general endothermic reaction; label the axis, reactants, products, and ΔH with its sign.

6.14 Write a balanced equation and draw an approximate enthalpy diagram for: (a) combustion of 1 mol of ethane; (b) freezing of liquid water.

6.15 Write a balanced equation and draw an approximate enthalpy diagram for (a) formation of 1 mol of sodium chloride from its elements (heat is released); (b) vaporization of liquid benzene.

6.16 Write a balanced equation and draw an approximate enthalpy diagram for (a) combustion of 1 mol of liquid methanol (CH_3OH); (b) formation of 1 mol of NO_2 from its elements (heat is absorbed).

6.17 Write a balanced equation and draw an approximate enthalpy diagram for (a) sublimation of dry ice [conversion of $CO_2(s)$ directly to $CO_2(g)$]; (b) reaction of 1 mol of SO_2 with O_2.

6.18 The circles represent a phase change at constant temperature:

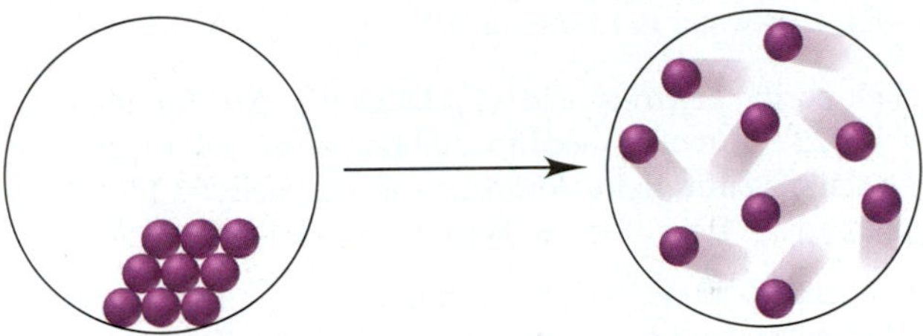

Is the value of each of the following positive (+), negative (−), or zero: (a) q_{sys}; (b) ΔE_{sys}; (c) ΔE_{univ}?

6.19 The scenes below represent a physical change taking place in a piston-cylinder assembly:

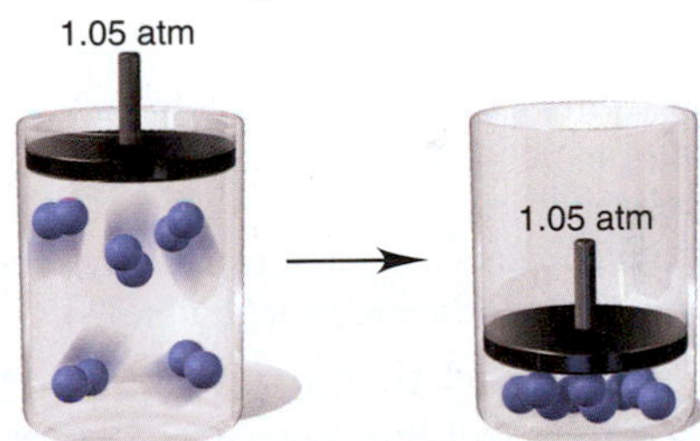

(a) Is w_{sys} +, −, or 0? (b) Is ΔH_{sys} +, −, or 0? (c) Can you determine whether ΔE_{surr} is +, −, or 0? Explain.

Calorimetry: Measuring the Heat of a Chemical or Physical Change

(Sample Problems 6.3 to 6.6)

6.20 What data do you need to determine the specific heat capacity of a substance?

6.21 Is the specific heat capacity of a substance an intensive or extensive property? Explain.

6.22 Find q when 22.0 g of water is heated from 25.0°C to 100.°C.

6.23 Calculate q when 0.10 g of ice is cooled from 10.°C to −75°C (c_{ice} = 2.087 J/g·K).

6.24 A 295-g aluminum engine part at an initial temperature of 13.00°C absorbs 75.0 kJ of heat. What is the final temperature of the part (c of Al = 0.900 J/g·K)?

6.25 A 27.7-g sample of the radiator coolant ethylene glycol releases 688 J of heat. What was the initial temperature of the sample if the final temperature is 32.5°C (*c* of ethylene glycol = 2.42 J/g·K)?

6.26 Two iron bolts of equal mass—one at 100.°C, the other at 55°C—are placed in an insulated container. Assuming the heat capacity of the container is negligible, what is the final temperature inside the container (*c* of iron = 0.450 J/g·K)?

6.27 One piece of copper jewelry at 105°C has twice the mass of another piece at 45°C. Both are placed in a calorimeter of negligible heat capacity. What is the final temperature inside the calorimeter (*c* of copper = 0.387 J/g·K)?

6.28 When 155 mL of water at 26°C is mixed with 75 mL of water at 85°C, what is the final temperature? (Assume that no heat is released to the surroundings; *d* of water is 1.00 g/mL.)

6.29 An unknown volume of water at 18.2°C is added to 24.4 mL of water at 35.0°C. If the final temperature is 23.5°C, what was the unknown volume? (Assume that no heat is released to the surroundings; *d* of water is 1.00 g/mL.)

6.30 High-purity benzoic acid (C_6H_5COOH; ΔH for combustion = −3227 kJ/mol) is used to calibrate bomb calorimeters. A 1.221-g sample burns in a calorimeter (heat capacity = 1365 J/°C) that contains 1.200 kg of water. What is the temperature change?

6.31 Two aircraft rivets, one iron and the other copper, are placed in a calorimeter that has an initial temperature of 20.°C. The data for the rivets are as follows:

	Iron	Copper
Mass (g)	30.0	20.0
Initial T (°C)	0.0	100.0
c (J/g·K)	0.450	0.387

(a) Will heat flow from Fe to Cu or from Cu to Fe?
(b) What other information is needed to correct any measurements in an actual experiment?
(c) What is the maximum final temperature of the system (assuming the heat capacity of the calorimeter is negligible)?

6.32 When 25.0 mL of 0.500 *M* H_2SO_4 is added to 25.0 mL of 1.00 *M* KOH in a coffee-cup calorimeter at 23.50°C, the temperature rises to 30.17°C. Calculate ΔH of this reaction. (Assume the total volume is the sum of the volumes and the density and specific heat capacity of the solution are the same as for water.)

Stoichiometry of Thermochemical Equations

(Sample Problem 6.7)

6.33 Would you expect $O_2(g) \longrightarrow 2O(g)$ to have a positive or a negative ΔH? Explain.

6.34 Is ΔH positive or negative when 1 mol of water vapor condenses to liquid water? Why? How does this value compare with ΔH for the vaporization of 2 mol of liquid water to water vapor?

6.35 Consider the following balanced thermochemical equation for a reaction sometimes used for H_2S production:

$$\tfrac{1}{8}S_8(s) + H_2(g) \longrightarrow H_2S(g) \quad \Delta H = -20.2 \text{ kJ}$$

(a) Is this an exothermic or endothermic reaction?
(b) What is ΔH for the reverse reaction?
(c) What is ΔH when 2.6 mol of S_8 reacts?
(d) What is ΔH when 25.0 g of S_8 reacts?

6.36 Consider the following balanced thermochemical equation for the decomposition of the mineral magnesite:

$$MgCO_3(s) \longrightarrow MgO(s) + CO_2(g) \quad \Delta H = 117.3 \text{ kJ}$$

(a) Is heat absorbed or released in the reaction?
(b) What is ΔH for the reverse reaction?
(c) What is ΔH when 5.35 mol of CO_2 reacts with excess MgO?
(d) What is ΔH when 35.5 g of CO_2 reacts with excess MgO?

6.37 When 1 mol of NO(*g*) forms from its elements, 90.29 kJ of heat is absorbed. (a) Write a balanced thermochemical equation. (b) What is ΔH when 3.50 g of NO decomposes to its elements?

6.38 When 1 mol of KBr(*s*) decomposes to its elements, 394 kJ of heat is absorbed. (a) Write a balanced thermochemical equation. (b) What is ΔH when 10.0 kg of KBr forms from its elements?

6.39 Liquid hydrogen peroxide, an oxidizing agent in many rocket fuel mixtures, releases oxygen gas on decomposition:

$$2H_2O_2(l) \longrightarrow 2H_2O(l) + O_2(g) \quad \Delta H = -196.1 \text{ kJ}$$

How much heat is released when 652 kg of H_2O_2 decomposes?

6.40 Compounds of boron and hydrogen are remarkable for their unusual bonding (described in Section 14.5) and also for their reactivity. With the more reactive halogens, for example, diborane (B_2H_6) forms trihalides even at low temperatures:

$$B_2H_6(g) + 6Cl_2(g) \longrightarrow 2BCl_3(g) + 6HCl(g) \quad \Delta H = -755.4 \text{ kJ}$$

What is ΔH per kilogram of diborane that reacts?

6.41 Most ethylene (C_2H_4), the starting material for producing polyethylene, comes from petroleum processing. It also occurs naturally as a fruit-ripening hormone and as a component of natural gas. (a) The heat transferred during combustion of C_2H_4 is −1411 kJ/mol. Write a balanced thermochemical equation. (b) How many grams of C_2H_4 must burn to give 70.0 kJ of heat?

6.42 Sucrose ($C_{12}H_{22}O_{11}$, table sugar) is oxidized in the body by O_2 via a complex set of reactions that produces $CO_2(g)$ and $H_2O(g)$ and releases 5.64×10^3 kJ/mol of sucrose. (a) Write a balanced thermochemical equation for the overall process. (b) How much heat is released per gram of sucrose oxidized?

Hess's Law: Finding ΔH of Any Reaction

(Sample Problem 6.8)

6.43 Express Hess's law in your own words.

6.44 Calculate ΔH for

$$Ca(s) + \tfrac{1}{2}O_2(g) + CO_2(g) \longrightarrow CaCO_3(s)$$

given the following reactions:

$$Ca(s) + \tfrac{1}{2}O_2(g) \longrightarrow CaO(s) \quad \Delta H = -635.1 \text{ kJ}$$
$$CaCO_3(s) \longrightarrow CaO(s) + CO_2(g) \quad \Delta H = 178.3 \text{ kJ}$$

6.45 Calculate ΔH for

$$2NOCl(g) \longrightarrow N_2(g) + O_2(g) + Cl_2(g)$$

given the following reactions:

$$\tfrac{1}{2}N_2(g) + \tfrac{1}{2}O_2(g) \longrightarrow NO(g) \quad \Delta H = 90.3 \text{ kJ}$$
$$NO(g) + \tfrac{1}{2}Cl_2(g) \longrightarrow NOCl(g) \quad \Delta H = -38.6 \text{ kJ}$$

6.46 Write the balanced overall equation (equation 3) for the following process, calculate $\Delta H_{overall}$, and match the number of each equation with the letter of the appropriate arrow in Figure P6.46 (*facing page*):

(1) $N_2(g) + O_2(g) \longrightarrow 2NO(g)$ $\quad \Delta H = 180.6$ kJ
(2) $2NO(g) + O_2(g) \longrightarrow 2NO_2(g)$ $\quad \Delta H = -114.2$ kJ
(3) $\quad \Delta H_{overall} = ?$

6.47 Write the balanced overall equation (equation 3) for the following process, calculate $\Delta H_{overall}$, and match the number of each equation with the letter of the appropriate arrow in Figure P6.47:

$$(1)\quad P_4(s) + 6Cl_2(g) \longrightarrow 4PCl_3(g) \qquad \Delta H = -1148\text{ kJ}$$
$$(2)\quad 4PCl_3(g) + 4Cl_2(g) \longrightarrow 4PCl_5(g) \qquad \Delta H = -460\text{ kJ}$$
$$(3)\qquad\qquad \Delta H_{overall} = ?$$

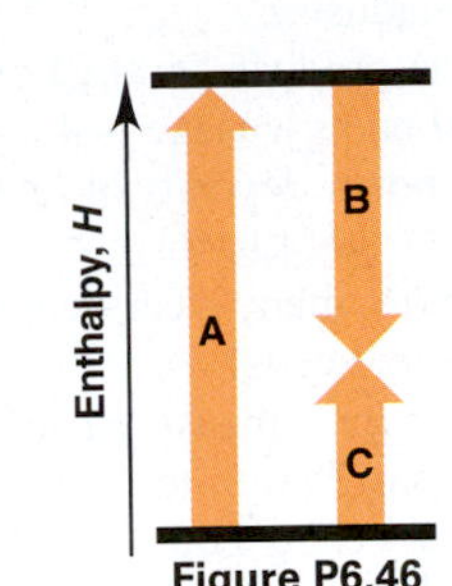

Figure P6.46

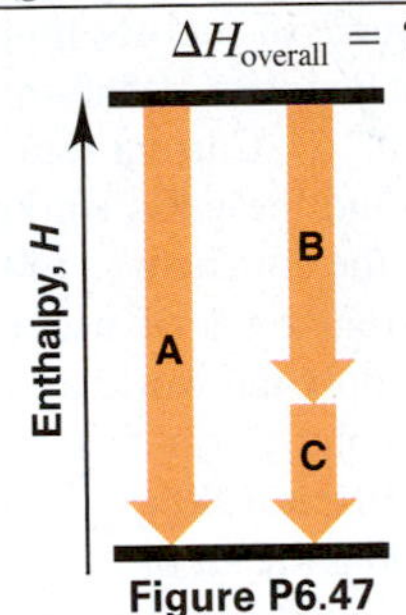

Figure P6.47

6.48 Diamond and graphite are two crystalline forms of carbon. At 1 atm and 25°C, diamond changes to graphite so slowly that the enthalpy change of the process must be obtained indirectly. Using equations from the numbered list below, determine ΔH for

$$\text{C(diamond)} \longrightarrow \text{C(graphite)}$$

$$(1)\ \text{C(diamond)} + O_2(g) \longrightarrow CO_2(g) \quad \Delta H = -395.4\text{ kJ}$$
$$(2)\ 2CO_2(g) \longrightarrow 2CO(g) + O_2(g) \quad \Delta H = 566.0\text{ kJ}$$
$$(3)\ \text{C(graphite)} + O_2(g) \longrightarrow CO_2(g) \quad \Delta H = -393.5\text{ kJ}$$
$$(4)\ 2CO(g) \longrightarrow \text{C(graphite)} + CO_2(g) \quad \Delta H = -172.5\text{ kJ}$$

Standard Enthalpies of Reaction (ΔH°_{rxn})

(Sample Problems 6.9 and 6.10)

6.49 What is the difference between the standard enthalpy of formation and the standard enthalpy of reaction?

6.50 Make any changes needed in each of the following equations to make the enthalpy change equal to ΔH°_f for the compound:
(a) $Cl(g) + Na(s) \longrightarrow NaCl(s)$
(b) $H_2O(g) \longrightarrow 2H(g) + \frac{1}{2}O_2(g)$
(c) $\frac{1}{2}N_2(g) + \frac{3}{2}H_2(g) \longrightarrow NH_3(g)$

6.51 Use Table 6.3 or Appendix B to write a balanced formation equation at standard conditions for each of the following compounds: (a) $CaCl_2$; (b) $NaHCO_3$; (c) CCl_4; (d) HNO_3.

6.52 Use Table 6.3 or Appendix B to write a balanced formation equation at standard conditions for each of the following compounds: (a) HI; (b) SiF_4; (c) O_3; (d) $Ca_3(PO_4)_2$.

6.53 Calculate ΔH°_{rxn} for each of the following:
(a) $2H_2S(g) + 3O_2(g) \longrightarrow 2SO_2(g) + 2H_2O(g)$
(b) $CH_4(g) + Cl_2(g) \longrightarrow CCl_4(l) + HCl(g)$ [unbalanced]

6.54 Calculate ΔH°_{rxn} for each of the following:
(a) $SiO_2(s) + 4HF(g) \longrightarrow SiF_4(g) + 2H_2O(l)$
(b) $C_2H_6(g) + O_2(g) \longrightarrow CO_2(g) + H_2O(g)$ [unbalanced]

6.55 Copper(I) oxide can be oxidized to copper(II) oxide:

$$Cu_2O(s) + \frac{1}{2}O_2(g) \longrightarrow 2CuO(s) \quad \Delta H^\circ_{rxn} = -146.0\text{ kJ}$$

Given ΔH°_f of $Cu_2O(s) = -168.6$ kJ/mol, find ΔH°_f of CuO(s).

6.56 Acetylene burns in air by the following equation:

$$C_2H_2(g) + \frac{5}{2}O_2(g) \longrightarrow 2CO_2(g) + H_2O(g) \qquad \Delta H^\circ_{rxn} = -1255.8\text{ kJ}$$

Given ΔH°_f of $CO_2(g) = -393.5$ kJ/mol and ΔH°_f of $H_2O(g) = -241.8$ kJ/mol, find ΔH°_f of $C_2H_2(g)$.

6.57 Nitroglycerine, $C_3H_5(NO_3)_3(l)$, a powerful explosive used in mining, detonates to produce a hot gaseous mixture of nitrogen, water, carbon dioxide, and oxygen.
(a) Write a balanced equation for this reaction using the smallest whole-number coefficients.
(b) If $\Delta H^\circ_{rxn} = -2.29\times10^4$ kJ for the equation as written in part (a), calculate ΔH°_f of nitroglycerine.

6.58 The common lead-acid car battery produces a large burst of current, even at low temperatures, and is rechargeable. The reaction that occurs while recharging a "dead" battery is

$$2PbSO_4(s) + 2H_2O(l) \longrightarrow Pb(s) + PbO_2(s) + 2H_2SO_4(l)$$

(a) Use ΔH°_f values from Appendix B to calculate ΔH°_{rxn}.
(b) Use the following equations to check your answer to part (a):

$$(1)\ Pb(s) + PbO_2(s) + 2SO_3(g) \longrightarrow 2PbSO_4(s) \qquad \Delta H^\circ_{rxn} = -768\text{ kJ}$$
$$(2)\ SO_3(g) + H_2O(l) \longrightarrow H_2SO_4(l) \qquad \Delta H^\circ_{rxn} = -132\text{ kJ}$$

Comprehensive Problems

6.59 Stearic acid ($C_{18}H_{36}O_2$) is a fatty acid, a molecule with a long hydrocarbon chain and an organic acid group (COOH) at the end. It is used to make cosmetics, ointments, soaps, and candles and is found in animal tissue as part of many saturated fats. In fact, when you eat meat, you are ingesting some fats containing stearic acid.
(a) Write a balanced equation for the combustion of stearic acid to gaseous products. (b) Calculate ΔH°_{rxn} for this combustion (ΔH°_f of $C_{18}H_{36}O_2 = -948$ kJ/mol). (c) Calculate the heat (q) released in kJ and kcal when 1.00 g of stearic acid is burned completely. (d) A candy bar contains 11.0 g of fat and 100. Cal from fat; is this consistent with your answer for part (c)?

6.60 A balloonist begins a trip in a helium-filled balloon in early morning when the temperature is 15°C. By mid-afternoon, the temperature is 30.°C. Assuming the pressure remains at 1.00 atm, for each mole of helium, calculate the following:
(a) The initial and final volumes
(b) The change in internal energy, ΔE (*Hint:* Helium behaves like an ideal gas, so $E = \frac{3}{2}nRT$. Be sure the units of R are consistent with those of E.)
(c) The work (w) done by the helium (in J)
(d) The heat (q) transferred (in J)
(e) ΔH for the process (in J)
(f) Explain the relationship between the answers to parts (d) and (e).

6.61 In winemaking, the sugars in grapes undergo *fermentation* by yeast to yield CH_3CH_2OH and CO_2. During cellular *respiration* (combustion), sugar and ethanol yield water vapor and CO_2.
(a) Using $C_6H_{12}O_6$ for sugar, calculate ΔH°_{rxn} of fermentation and of respiration.
(b) Write a combustion reaction for ethanol. Which has a higher ΔH°_{rxn} for combustion per mole of C, sugar or ethanol?

6.62 The following scenes represent a gaseous reaction between compounds of nitrogen (blue) and oxygen (red) at 298 K:

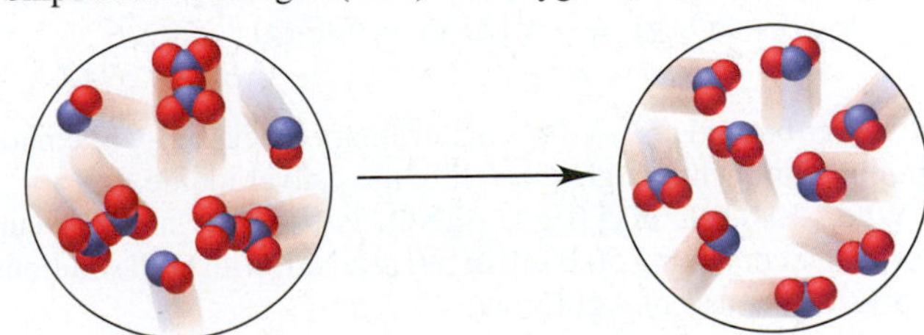

(a) Write a balanced equation and use Appendix B to calculate ΔH°_{rxn}.
(b) If each molecule of product represents 1.50×10^{-2} mol, what quantity of heat (in J) is released or absorbed?

6.63 Iron metal is produced in a blast furnace through a complex series of reactions that involve reduction of iron(III) oxide with carbon monoxide.
(a) Write a balanced overall equation for the process, including the other product.
(b) Use the equations below to calculate ΔH°_{rxn} for the overall equation:
(1) $3Fe_2O_3(s) + CO(g) \longrightarrow 2Fe_3O_4(s) + CO_2(g)$ $\Delta H^\circ = -48.5$ kJ
(2) $Fe(s) + CO_2(g) \longrightarrow FeO(s) + CO(g)$ $\Delta H^\circ = -11.0$ kJ
(3) $Fe_3O_4(s) + CO(g) \longrightarrow 3FeO(s) + CO_2(g)$ $\Delta H^\circ = 22$ kJ

6.64 Pure liquid octane (C_8H_{18}; $d = 0.702$ g/mL) is used as the fuel in a test of a new automobile drive train.
(a) How much energy (in kJ) is released by complete combustion of the octane in a 20.4-gal fuel tank to gases ($\Delta H^\circ_{rxn} = -5.45\times10^3$ kJ/mol)?
(b) The energy delivered to the wheels at 65 mph is 5.5×10^4 kJ/h. Assuming all the energy is transferred to the wheels, what is the cruising range (in km) of the car on a full tank?
(c) If the actual cruising range is 455 miles, explain your answer to part (b).

6.65 Four 50.-g samples of different liquids are placed in separate beakers at $T_{initial} = 25.00°C$. Each liquid is heated until 450. J of heat has been absorbed; T_{final} is shown on each beaker below. Rank the liquids in order of increasing specific heat capacity.

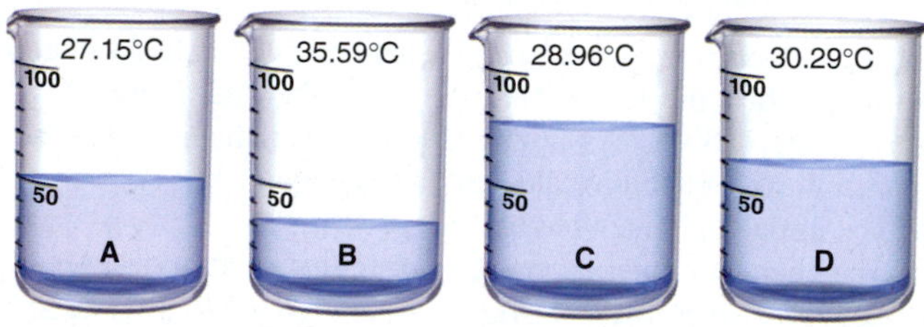

6.66 When simple sugars, called *monosaccharides,* link together, they form a variety of complex sugars and, ultimately, *polysaccharides,* such as starch, glycogen, and cellulose. Glucose and fructose have the same molecular formula, $C_6H_{12}O_6$, but different arrangements of atoms. A molecule of glucose links with one of fructose to form a molecule of sucrose (table sugar) and a molecule of liquid water. The ΔH°_f values for glucose, fructose, and sucrose are -1273 kJ/mol, -1266 kJ/mol, and -2226 kJ/mol, respectively. Write a balanced equation for this reaction and calculate ΔH°_{rxn}.

6.67 Reaction of gaseous ClF with F_2 yields liquid ClF_3, an important fluorinating agent. Use the following thermochemical equations to calculate ΔH°_{rxn} for this reaction:
(1) $2ClF(g) + O_2(g) \longrightarrow Cl_2O(g) + OF_2(g)$ $\Delta H^\circ_{rxn} = 167.5$ kJ
(2) $2F_2(g) + O_2(g) \longrightarrow 2OF_2(g)$ $\Delta H^\circ_{rxn} = -43.5$ kJ
(3) $2ClF_3(l) + 2O_2(g) \longrightarrow Cl_2O(g) + 3OF_2(g)$ $\Delta H^\circ_{rxn} = 394.1$ kJ

6.68 Silver bromide is used to coat ordinary black-and-white photographic film, while high-speed film uses silver iodide.
(a) When 50.0 mL of 5.0 g/L $AgNO_3$ is added to a coffee-cup calorimeter containing 50.0 mL of 5.0 g/L NaI, with both solutions at 25°C, what mass of AgI forms?
(b) Use Appendix B to find ΔH°_{rxn}.
(c) What is ΔT_{soln} (assuming the volumes are additive and the solution has the density and specific heat capacity of water)?

6.69 When organic matter decomposes under oxygen-free (anaerobic) conditions, methane is one of the products. Thus, enormous deposits of natural gas, which is almost entirely methane, serve as a major source of fuel for home and industry.
(a) Known deposits of natural gas can produce 5600 EJ of energy (1 EJ = 10^{18} J). Current total global energy usage is 4.0×10^2 EJ per year. Find the mass (in kg) of known deposits of natural gas (ΔH°_{rxn} for the combustion of $CH_4 = -802$ kJ/mol).
(b) At current rates of usage, for how many years could these deposits supply the world's total energy needs?
(c) What volume (in ft^3) of natural gas, measured at STP, is required to heat 1.00 qt of water from 25.0°C to 100.0°C (d of $H_2O = 1.00$ g/mL; d of CH_4 at STP = 0.72 g/L)?
(d) The fission of 1 mol of uranium (about 4×10^{-4} ft^3) in a nuclear reactor produces 2×10^{13} J. What volume (in ft^3) of natural gas would produce the same amount of energy?

6.70 The heat of atomization (ΔH°_{atom}) is the heat needed to form separated gaseous atoms from a substance in its standard state. The equation for the atomization of graphite is

$$C(graphite) \longrightarrow C(g)$$

Use Hess's law to calculate ΔH°_{atom} of graphite from these data:
(1) ΔH°_f of $CH_4 = -74.9$ kJ/mol
(2) ΔH°_{atom} of $CH_4 = 1660$ kJ/mol
(3) ΔH°_{atom} of $H_2 = 432$ kJ/mol

6.71 A reaction takes place in a steel vessel within a chamber filled with argon gas. Shown below are molecular views of the argon adjacent to the surface of the reaction vessel before and after the reaction. Was the reaction exothermic or endothermic? Explain.

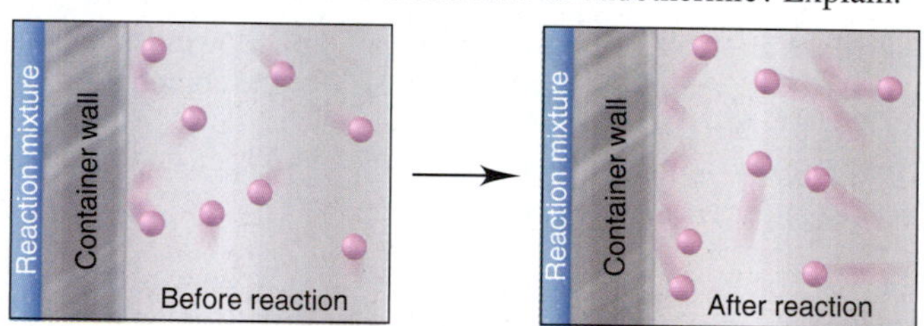

6.72 Benzene (C_6H_6) and acetylene (C_2H_2) have the same empirical formula, CH. Which releases more energy per mole of CH (ΔH°_f of gaseous $C_6H_6 = 82.9$ kJ/mol)?

6.73 An aqueous waste stream with a maximum concentration of 0.50 M H_2SO_4 ($d = 1.030$ g/mL at 25°C) is neutralized by controlled addition of 40% NaOH ($d = 1.430$ g/L) before it goes to the process sewer and then to the chemical plant waste treatment facility. A safety review finds that the waste stream could meet a small stream of an immiscible organic compound, which could form a flammable vapor in air at 40.°C. The maximum temperature reached by the NaOH solution and the waste stream is 31°C. Could the temperature increase due to the heat transferred by the neutralization cause the organic vapor to explode? Assume the specific heat capacity of each solution is 4.184 J/g·K.

6.74 Kerosene, a common space-heater fuel, is a mixture of hydrocarbons whose "average" formula is $C_{12}H_{26}$.
(a) Write a balanced equation, using the simplest whole-number coefficients, for the complete combustion of kerosene to gases.
(b) If $\Delta H^\circ_{rxn} = -1.50\times10^4$ kJ for the combustion equation as written in part (a), determine ΔH°_f of kerosene.
(c) Calculate the heat released by combustion of 0.50 gal of kerosene (d of kerosene = 0.749 g/mL).

(d) How many gallons of kerosene must be burned for a kerosene furnace to produce 1250. Btu (1 Btu = 1.055 kJ)?

6.75 Coal gasification is a multistep process to convert coal into cleaner-burning gaseous fuels. In one step, a certain coal sample reacts with superheated steam:

$$\text{C(coal)} + \text{H}_2\text{O}(g) \longrightarrow \text{CO}(g) + \text{H}_2(g) \qquad \Delta H^\circ_{\text{rxn}} = 129.7\text{ kJ}$$

(a) Combine this reaction with the following two to write an overall reaction for the production of methane:

$$\text{CO}(g) + \text{H}_2\text{O}(g) \longrightarrow \text{CO}_2(g) + \text{H}_2(g) \qquad \Delta H^\circ_{\text{rxn}} = -41\text{ kJ}$$

$$\text{CO}(g) + 3\text{H}_2(g) \longrightarrow \text{CH}_4(g) + \text{H}_2\text{O}(g) \qquad \Delta H^\circ_{\text{rxn}} = -206\text{ kJ}$$

(b) Calculate $\Delta H^\circ_{\text{rxn}}$ for this overall change.
(c) Using the value in (b) and calculating the $\Delta H^\circ_{\text{rxn}}$ for combustion of methane, find the total heat for gasifying 1.00 kg of coal and burning the methane formed (assume water forms as a gas and $\mathcal{M}$ of coal = 12.00 g/mol).

6.76 Phosphorus pentachloride is used in the industrial preparation of organic phosphorus compounds. Equation 1 shows its preparation from PCl_3 and Cl_2:

$$(1)\ \text{PCl}_3(l) + \text{Cl}_2(g) \longrightarrow \text{PCl}_5(s)$$

Use equations 2 and 3 to calculate ΔH_{rxn} for equation 1:

$$(2)\ \text{P}_4(s) + 6\text{Cl}_2(g) \longrightarrow 4\text{PCl}_3(l) \qquad \Delta H = -1280\text{ kJ}$$

$$(3)\ \text{P}_4(s) + 10\text{Cl}_2(g) \longrightarrow 4\text{PCl}_5(s) \qquad \Delta H = -1774\text{ kJ}$$

6.77 A typical candy bar weighs about 2 oz (1.00 oz = 28.4 g).
(a) Assuming that a candy bar is 100% sugar and that 1.0 g of sugar is equivalent to about 4.0 Calories of energy, calculate the energy (in kJ) contained in a typical candy bar.
(b) Assuming that your mass is 58 kg and you convert chemical potential energy to work with 100% efficiency, how high would you have to climb to work off the energy in a candy bar? (Potential energy = mass × g × height, where g = 9.8 m/s^2.)
(c) Why is your actual conversion of potential energy to work less than 100% efficient?

6.78 Silicon tetrachloride is produced annually on the multikiloton scale and used in making transistor-grade silicon. It can be produced directly from the elements (reaction 1) or, more cheaply, by heating sand and graphite with chlorine gas (reaction 2). If water is present in reaction 2, some tetrachloride may be lost in an unwanted side reaction (reaction 3):

$$(1)\ \text{Si}(s) + 2\text{Cl}_2(g) \longrightarrow \text{SiCl}_4(g)$$

$$(2)\ \text{SiO}_2(s) + 2\text{C(graphite)} + 2\text{Cl}_2(g) \longrightarrow \text{SiCl}_4(g) + 2\text{CO}(g)$$

$$(3)\ \text{SiCl}_4(g) + 2\text{H}_2\text{O}(g) \longrightarrow \text{SiO}_2(s) + 4\text{HCl}(g) \qquad \Delta H^\circ_{\text{rxn}} = -139.5\text{ kJ}$$

(a) Use reaction 3 to calculate the standard enthalpies of reaction of reactions 1 and 2. (b) What is the standard enthalpy of reaction for a fourth reaction that is the sum of reactions 2 and 3?

6.79 Use the following information to find ΔH°_f of gaseous HCl:

$$\text{N}_2(g) + 3\text{H}_2(g) \longrightarrow 2\text{NH}_3(g) \qquad \Delta H^\circ_{\text{rxn}} = -91.8\text{ kJ}$$

$$\text{N}_2(g) + 4\text{H}_2(g) + \text{Cl}_2(g) \longrightarrow 2\text{NH}_4\text{Cl}(s) \qquad \Delta H^\circ_{\text{rxn}} = -628.8\text{ kJ}$$

$$\text{NH}_3(g) + \text{HCl}(g) \longrightarrow \text{NH}_4\text{Cl}(s) \qquad \Delta H^\circ_{\text{rxn}} = -176.2\text{ kJ}$$

6.80 You want to determine ΔH° for the reaction

$$\text{Zn}(s) + 2\text{HCl}(aq) \longrightarrow \text{ZnCl}_2(aq) + \text{H}_2(g)$$

(a) To do so, you first determine the heat capacity of a calorimeter using the following reaction, whose ΔH is known:

$$\text{NaOH}(aq) + \text{HCl}(aq) \longrightarrow \text{NaCl}(aq) + \text{H}_2\text{O}(l) \qquad \Delta H^\circ = -57.32\text{ kJ}$$

Calculate the heat capacity of the calorimeter from these data:
Amounts used: 50.0 mL of 2.00 M HCl and
50.0 mL of 2.00 M NaOH
Initial T of both solutions: 16.9°C
Maximum T recorded during reaction: 30.4°C
Density of resulting NaCl solution: 1.04 g/mL
c of 1.00 M NaCl(aq) = 3.93 J/g·K

(b) Use the result from part (a) and the following data to determine $\Delta H^\circ_{\text{rxn}}$ for the reaction between zinc and HCl(aq):
Amounts used: 100.0 mL of 1.00 M HCl and 1.3078 g of Zn
Initial T of HCl solution and Zn: 16.8°C
Maximum T recorded during reaction: 24.1°C
Density of 1.0 M HCl solution = 1.015 g/mL
c of resulting $ZnCl_2(aq)$ = 3.95 J/g·K

(c) Given the values below, what is the error in your experiment?

$$\Delta H^\circ_f \text{ of HCl}(aq) = -1.652\times10^2\text{ kJ/mol}$$

$$\Delta H^\circ_f \text{ of ZnCl}_2(aq) = -4.822\times10^2\text{ kJ/mol}$$

6.81 One mole of nitrogen gas confined within a cylinder by a piston is heated from 0°C to 819°C at 1.00 atm.
(a) Calculate the work done by the expanding gas in joules (1 J = 9.87×10^{-3} atm·L). Assume all the energy is used to do work.
(b) What would be the temperature change if the gas were heated using the same amount of energy in a container of fixed volume? (Assume the specific heat capacity of N_2 is 1.00 J/g·K.)

6.82 The chemistry of nitrogen oxides is very versatile. Given the following reactions and their standard enthalpy changes,

$$(1)\ \text{NO}(g) + \text{NO}_2(g) \longrightarrow \text{N}_2\text{O}_3(g) \qquad \Delta H^\circ_{\text{rxn}} = -39.8\text{ kJ}$$

$$(2)\ \text{NO}(g) + \text{NO}_2(g) + \text{O}_2(g) \longrightarrow \text{N}_2\text{O}_5(g) \qquad \Delta H^\circ_{\text{rxn}} = -112.5\text{ kJ}$$

$$(3)\ 2\text{NO}_2(g) \longrightarrow \text{N}_2\text{O}_4(g) \qquad \Delta H^\circ_{\text{rxn}} = -57.2\text{ kJ}$$

$$(4)\ 2\text{NO}(g) + \text{O}_2(g) \longrightarrow 2\text{NO}_2(g) \qquad \Delta H^\circ_{\text{rxn}} = -114.2\text{ kJ}$$

$$(5)\ \text{N}_2\text{O}_5(s) \longrightarrow \text{N}_2\text{O}_5(g) \qquad \Delta H^\circ_{\text{rxn}} = 54.1\text{ kJ}$$

calculate the standard enthalpy of reaction for

$$\text{N}_2\text{O}_3(g) + \text{N}_2\text{O}_5(s) \longrightarrow 2\text{N}_2\text{O}_4(g)$$

6.83 Liquid methanol (CH_3OH) can be used as an alternative fuel in pickup and SUV engines. An industrial method for preparing it involves the catalytic hydrogenation of carbon monoxide:

$$\text{CO}(g) + 2\text{H}_2(g) \xrightarrow{\text{catalyst}} \text{CH}_3\text{OH}(l)$$

How much heat (in kJ) is released when 15.0 L of CO at 85°C and 112 kPa reacts with 18.5 L of H_2 at 75°C and 744 torr?

6.84 (a) How much heat is released when 25.0 g of methane burns in excess O_2 to form gaseous CO_2 and H_2O?
(b) Calculate the temperature of the product mixture if the methane and air are both at an initial temperature of 0.0°C. Assume a stoichiometric ratio of methane to oxygen from the air, with air being 21% O_2 by volume (c of CO_2 = 57.2 J/mol·K; c of $H_2O(g)$ = 36.0 J/mol·K; c of N_2 = 30.5 J/mol·K).

7 Quantum Theory and Atomic Structure

Light from Excited Atoms In a fireworks display and many other everyday phenomena, we see the result of atoms absorbing energy and then emitting it as light. In this chapter, we explore the basis of these phenomena and learn some surprising things about the makeup of the universe.

Key Principles
to focus on while studying this chapter

- In a vacuum, *electromagnetic radiation* travels at the *speed of light* (c) in waves. The properties of a wave are its *wavelength* (λ, distance between corresponding points on adjacent waves), *frequency* (ν, number of cycles the wave undergoes per second), and *amplitude* (the height of the wave), which is related to the *intensity* (brightness) of the radiation. Any region of the *electromagnetic spectrum* includes a range of wavelengths. ***(Section 7.1)***
- In everyday experience, energy is diffuse and matter is chunky, but certain phenomena—*blackbody radiation* (the light emitted by hot objects), the *photoelectric effect* (the flow of current when light strikes a metal), and *atomic spectra* (the specific colors emitted from a substance that is excited)—can only be explained if energy consists of "packets" *(quanta)* that occur in, and thus change by, *fixed amounts*. The energy of a quantum is related to its frequency. ***(Section 7.1)***
- According to the *Bohr model*, an atomic spectrum consists of separate lines because an atom has certain energy levels *(states)* that correspond to electrons in *orbits* around the nucleus. The energy of the atom changes when the electron moves from one orbit to another as the atom absorbs (or emits) light of a specific frequency. ***(Section 7.2)***
- *Wave-particle duality* means that matter has wavelike properties (as shown by the *de Broglie wavelength* and *electron diffraction*) and energy has particle-like properties (as shown by *photons* of light having *momentum*). These properties are observable only on the atomic scale, and because of them, we can never simultaneously know the position and speed of an electron in an atom *(uncertainty principle)*. ***(Section 7.3)***
- According to the *quantum-mechanical model* of the H atom, each energy level of the atom is associated with an *atomic orbital (wave function)*, a mathematical description of the electron's position in three dimensions. We can know the *probability* that the electron is within a particular tiny volume of space, but *not* its exact location. The probability is highest for the electron being near the nucleus, and it decreases with distance. ***(Section 7.4)***
- *Quantum numbers* denote each atomic orbital's energy *(n, principal)*, shape *(l, angular momentum)*, and spatial orientation *(m_l, magnetic)*. An *energy level* consists of *sublevels*, which consist of *orbitals*. There is a *hierarchy* of quantum numbers: n limits l, which limits m_l. ***(Section 7.4)***
- In the H atom, there is only one type of electrostatic interaction: the attraction between nucleus and electron. Thus, for the H atom *only*, the energy levels depend solely on the principal quantum number (n). ***(Section 7.4)***

Outline

Over a few remarkable decades—from around 1890 to 1930—a revolution took place in how we view the makeup of the universe. Earlier, Dalton's atomic theory had established the idea of individual units of matter, and then Rutherford's model substituted nuclear atoms for "billiard balls" or "plum puddings." However, almost as soon as Rutherford proposed his model, a major problem arose. If a nucleus and an electron are to remain apart, the energy of the electron's motion (kinetic energy) must balance the energy of attraction (potential energy). But the laws of classical physics said that a negative particle moving in a curved path around a positive one *must* emit radiation and thus lose energy. If these laws applied to atoms, why didn't the electron lose energy and spiral into the nucleus? The behavior of subatomic matter seemed to violate real-world experience and accepted principles.

The breakthroughs that occurred in the early 20th century forced a complete rethinking of the classical picture of matter and energy. In the macroscopic world, matter occurs in chunks you can hold and weigh, and the amount of matter in a sample changes piece by piece. In contrast, energy is "massless," and its quantity can change continuously. Matter moves in specific paths, whereas light and other types of energy travel in diffuse waves. As you'll see in this chapter, however, as soon as scientists probed the subatomic world, these distinctions between particulate matter and wavelike energy began to fade, revealing a much more amazing reality.

CONCEPTS & SKILLS TO REVIEW before studying this chapter

- discovery of the electron and atomic nucleus (Section 2.4)
- major features of atomic structure (Section 2.5)
- changes in the energy state of a system (Section 6.1)

7.1 • THE NATURE OF LIGHT

Visible light, x-rays, and microwaves are some of the types of **electromagnetic radiation** (also called *electromagnetic energy* or *radiant energy*). All electromagnetic radiation consists of energy propagated by electric and magnetic fields that increase and decrease in intensity as they move through space. This *classical wave model* explains why rainbows form, how magnifying glasses work, and many other familiar observations. But, it cannot explain observations on the very *un*familiar atomic scale because, in that realm, energy behaves as if it consists of particles!

The Wave Nature of Light

The wave properties of electromagnetic radiation are described by three variables and one constant (Figure 7.1):

1. *Frequency* (ν, Greek *nu*). The **frequency** of a wave is the number of cycles it undergoes per second, expressed by the unit 1/second [s^{-1}; also called a *hertz* (Hz)].
2. *Wavelength* (λ, Greek *lambda*). The **wavelength** is the distance between any point on a wave and the corresponding point on the next crest (or trough) of the wave, that is, the distance the wave travels during one cycle. Wavelength has units of meters or, for very short wavelengths, nanometers (nm, 10^{-9} m), picometers (pm, 10^{-12} m), or the non-SI unit angstroms (Å, 10^{-10} m).
3. *Speed*. The speed of a wave is the distance it moves per unit time (meters per second), the product of its frequency (cycles per second) and wavelength (meters per cycle):

   Units for speed of wave: $\frac{\cancel{\text{cycles}}}{\text{s}} \times \frac{\text{m}}{\cancel{\text{cycle}}} = \frac{\text{m}}{\text{s}}$

   In a vacuum, electromagnetic radiation moves at 2.99792458×10^8 m/s (3.00×10^8 m/s to three significant figures), a *physical constant* called the **speed of light (*c*)**:

$$c = \nu \times \lambda \quad (7.1)$$

   Since the product of ν and λ is a constant, they have a reciprocal relationship—*radiation with a high frequency has a short wavelength, and vice versa:*

   $\nu\uparrow\lambda\downarrow$ and $\nu\downarrow\lambda\uparrow$

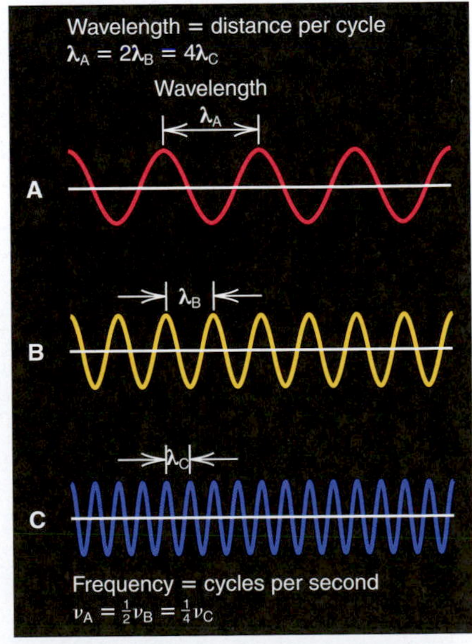

Figure 7.1 The reciprocal relationship of frequency and wavelength.

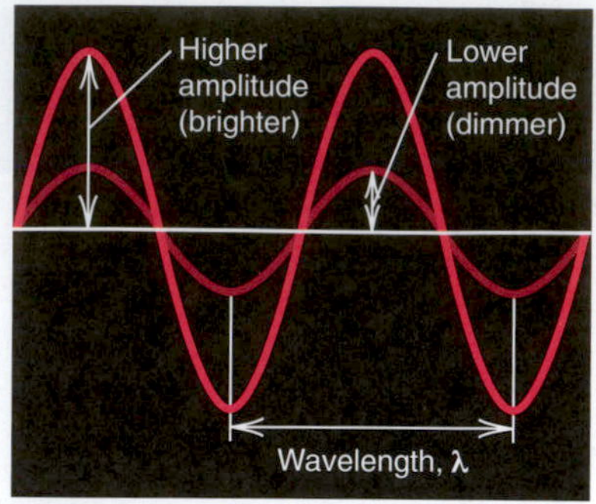

Figure 7.2 Differing amplitude (brightness, or intensity) of a wave.

4. *Amplitude.* The **amplitude** of a wave is the *height* of the crest (or *depth* of the trough). For an electromagnetic wave, the amplitude is related to the *intensity* of the radiation, or its brightness in the case of visible light. Light of a particular color has a specific frequency (and thus, wavelength) but, as Figure 7.2 shows, it can be dimmer (lower amplitude, less intense) or brighter (higher amplitude, more intense).

The Electromagnetic Spectrum Visible light represents a small region of the **electromagnetic spectrum** (Figure 7.3). *All waves in the spectrum travel at the same speed through a vacuum but differ in frequency and, therefore, wavelength.*

The spectrum is a *continuum of radiant energy,* so each region meets the next. For instance, the **infrared (IR)** region meets the *microwave* region on one end and the *visible* region on the other. We perceive different wavelengths (or frequencies) of visible light as colors, from red ($\lambda \approx 750$ nm) to violet ($\lambda \approx 400$ nm). Light of a single wavelength is called *monochromatic* (Greek, "one color"), whereas light of many wavelengths is *polychromatic.* White light is polychromatic. The region adjacent to visible light on the short-wavelength end consists of **ultraviolet (UV)** radiation (also called *ultraviolet light*). Still shorter wavelengths (higher frequencies) make up the x-ray and gamma (γ) ray regions. Some types of electromagnetic radiation are utilized by familiar devices; for example, long-wavelength, low-frequency radiation is used by microwave ovens, radios, and cell phones.

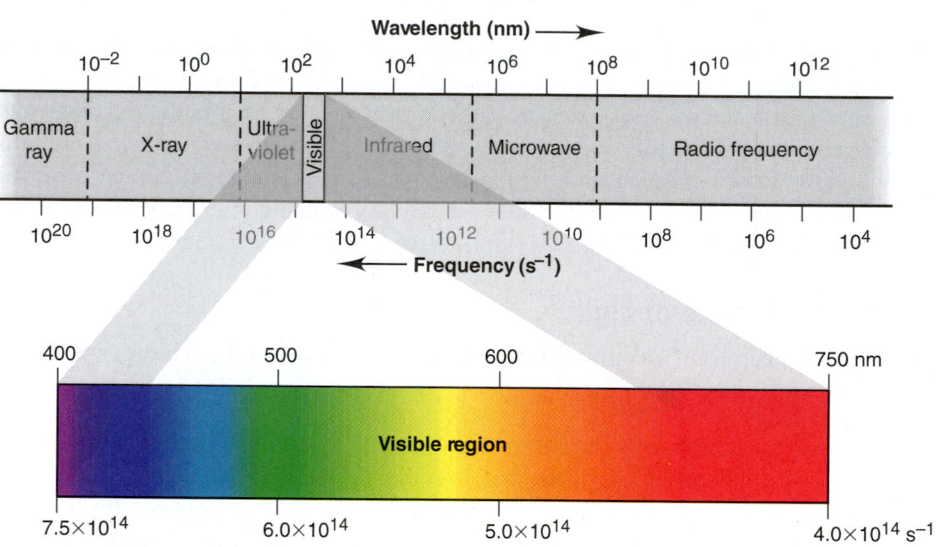

Figure 7.3 Regions of the electromagnetic spectrum. The visible region is expanded (and the scale made linear) to show the component colors.

Sample Problem 7.1 Interconverting Wavelength and Frequency

Problem A dental hygienist uses x-rays ($\lambda = 1.00$ Å) to take a series of dental radiographs while the patient listens to a radio station ($\lambda = 325$ cm) and looks out the window at the blue sky ($\lambda = 473$ nm). What is the frequency (in s^{-1}) of the electromagnetic radiation from each source? (Assume that the radiation travels at the speed of light, 3.00×10^8 m/s.)

Plan We are given the wavelengths, so we use Equation 7.1 to find the frequencies. However, we must first convert the wavelengths to meters because c has units of m/s.

Road Map

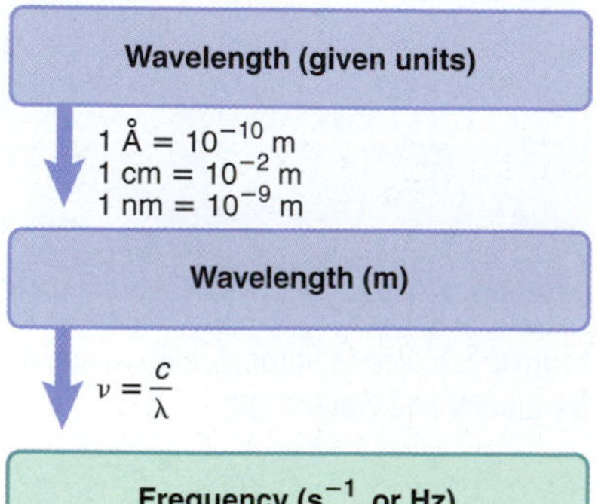

Solution For the x-rays: Converting from angstroms to meters,

$$\lambda = 1.00\ \text{Å} \times \frac{10^{-10}\ \text{m}}{1\ \text{Å}} = 1.00\times10^{-10}\ \text{m}$$

Calculating the frequency:

$$\nu = \frac{c}{\lambda} = \frac{3.00\times10^{8}\ \text{m/s}}{1.00\times10^{-10}\ \text{m}} = 3.00\times10^{18}\ \text{s}^{-1}$$

For the radio signal: Combining steps to calculate the frequency,

$$\nu = \frac{c}{\lambda} = \frac{3.00\times10^{8}\ \text{m/s}}{325\ \text{cm} \times \dfrac{10^{-2}\ \text{m}}{1\ \text{cm}}} = 9.23\times10^{7}\ \text{s}^{-1}$$

For the blue sky: Combining steps to calculate the frequency,

$$\nu = \frac{c}{\lambda} = \frac{3.00\times10^{8}\ \text{m/s}}{473\ \text{nm} \times \dfrac{10^{-9}\ \text{m}}{1\ \text{nm}}} = 6.34\times10^{14}\ \text{s}^{-1}$$

Check The orders of magnitude are correct for the regions of the electromagnetic spectrum (see Figure 7.3): x-rays (10^{19} to $10^{16}\ \text{s}^{-1}$), radio waves (10^{9} to $10^{4}\ \text{s}^{-1}$), and visible light (7.5×10^{14} to $4.0\times10^{14}\ \text{s}^{-1}$).

Comment The radio station here is broadcasting at $92.3\times10^{6}\ \text{s}^{-1}$, or 92.3 million Hz (92.3 MHz), about midway in the FM range.

FOLLOW-UP PROBLEM 7.1 Some diamonds appear yellow because they contain nitrogen compounds that absorb purple light of frequency 7.23×10^{14} Hz. Calculate the wavelength (in nm and Å) of the absorbed light.

The Classical Distinction Between Energy and Matter In our everyday world, matter and energy behave very differently. Let's examine some distinctions between the behavior of waves of energy and particles of matter.

1. *Refraction and dispersion.* Light of a given wavelength travels at different speeds through various transparent media—vacuum, air, water, quartz, and so forth. Therefore, when a light wave passes from one medium into another, the speed of the wave changes. Figure 7.4A shows the phenomenon known as **refraction.** If the wave strikes the boundary between media, say, between air and water, at an angle other than 90°, the change in speed causes a change in direction, and the wave continues at a different angle. The angle of refraction depends on the two media and the wavelength of the light. In the related process of *dispersion,* white light separates (disperses) into its component colors when it passes through a prism (or other refracting object) because

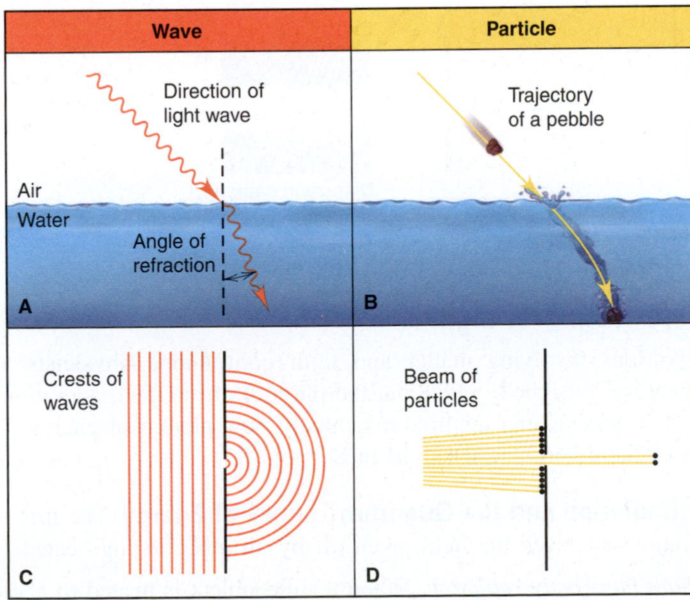

Figure 7.4 Different behaviors of waves and particles. **A,** Refraction: The speed of a light wave passing between media changes immediately, which bends its path. **B,** The speed of a particle continues changing gradually. **C,** Diffraction: A wave bends around both edges of a small opening, forming a semicircular wave. **D,** Particles either enter a small opening or not.

each incoming wave is refracted at a slightly different angle. In fact, rainbows appear when sunlight is dispersed through water droplets.

In contrast to a wave of light, a particle of matter, like a pebble, does not undergo refraction. If you throw a pebble through the air into a pond, it continues to slow down gradually along a curved path after entering the water (Figure 7.4B).

2. *Diffraction and interference*. When a wave strikes the edge of an object, it bends around it in a phenomenon called **diffraction.** If the wave passes through a slit about as wide as its wavelength, it bends around both edges of the slit and forms a semicircular wave on the other side of the opening (Figure 7.4C).

In contrast, when you throw a collection of particles, like a handful of sand, at a small opening, some particles hit the edge, while others go through the opening and continue in a narrower group (Figure 7.4D).

When waves of light pass through two adjacent slits, the nearby emerging circular waves interact through the process of *interference.* If the crests of the waves coincide *(in phase),* they interfere *constructively*—the amplitudes add together to form a brighter region. If crests coincide with troughs *(out of phase),* they interfere *destructively*—the amplitudes cancel to form a darker region. The result is a *diffraction pattern* (Figure 7.5).

In contrast, particles passing through adjacent openings continue in straight paths, some colliding and moving at different angles.

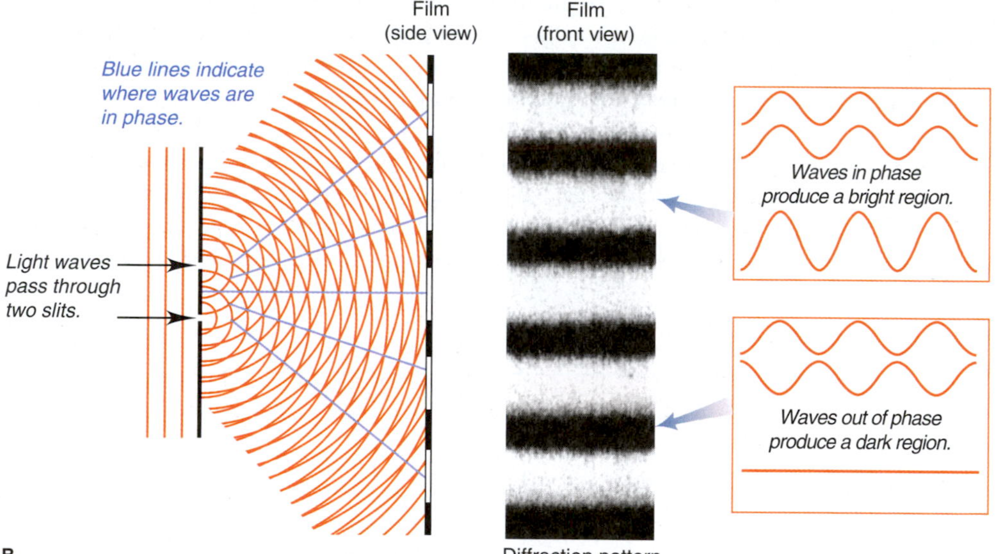

Figure 7.5 Formation of a diffraction pattern. **A,** Constructive interference and destructive interference occur as water waves pass through two adjacent slits. **B,** Light waves passing through two slits also emerge as circular waves and create a diffraction pattern.

The Particle Nature of Light

Three observations involving matter and light confounded physicists at the turn of the 20th century: blackbody radiation, the photoelectric effect, and atomic spectra. Explaining these phenomena required a radically new picture of energy. We discuss the first two of them here and the third in Section 7.2.

Blackbody Radiation and the Quantum Theory of Energy The first of the puzzling observations involved the light given off by an object being heated.

- *Observation: blackbody radiation.* When a solid object is heated to about 1000 K, it begins to emit visible light, as you can see in the red glow of smoldering coal. At about 1500 K, the light is brighter and more orange, like that from an electric heating coil. At temperatures greater than 2000 K, the light is still brighter and whiter, like that emitted by the filament of a lightbulb. These changes in intensity and wavelength of emitted light as an object is heated are characteristic of *blackbody radiation,* light given off by a hot *blackbody.** All attempts to account for these changes by applying classical electromagnetic theory failed.

- *Explanation: the quantum theory.* In 1900, the German physicist Max Planck (1858–1947) made a radical assumption that eventually led to an entirely new view of energy. He proposed that a hot, glowing object could emit (or absorb) only *certain* quantities of energy:

  $$E = nh\nu$$

  where E is the energy of the radiation, ν is its frequency, n is a positive integer (1, 2, 3, and so on) called a **quantum number,** and h is **Planck's constant.** With energy in joules (J) and frequency in s^{-1}, h has units of J·s:

  $$h = 6.62606876\times10^{-34}\ \text{J·s} = 6.626\times10^{-34}\ \text{J·s} \quad (4\ \text{sf})$$

  A hot object's radiation must be emitted by its atoms. If each atom can *emit* only certain quantities of energy, it follows that each atom *has* only certain quantities of energy. Thus, the energy of an atom is *quantized:* it occurs in fixed quantities, rather than being continuous. Each change in an atom's energy occurs when the atom absorbs or emits one or more "packets," or definite amounts, of energy. Each energy packet is called a **quantum** ("fixed quantity"; plural, *quanta*). A quantum of energy is equal to $h\nu$. Thus, *an atom changes its energy state by emitting (or absorbing) one or more quanta,* and the energy of the emitted (or absorbed) radiation is equal to the *difference in the atom's energy states:*

  $$\Delta E_{\text{atom}} = E_{\text{emitted (or absorbed) radiation}} = \Delta nh\nu$$

  Because the atom can change its energy only by integer multiples of $h\nu$, the smallest change occurs when an atom in a given energy state changes to an adjacent state, that is, when $\Delta n = 1$:

  $$\Delta E = h\nu \qquad \textbf{(7.2)}$$

The Photoelectric Effect and the Photon Theory of Light Despite the idea of quantization, physicists still pictured energy as traveling in waves. But, the wave model could not explain the second confusing observation, the flow of current when light strikes a metal.

- *Observation: the **photoelectric effect.*** When monochromatic light of sufficient frequency shines on a metal plate, a current flows (Figure 7.6). It was thought that the current arises because light transfers energy that frees electrons from the metal surface. However, the effect had two confusing features, the presence of a threshold frequency and the absence of a time lag:
  1. *Presence of a threshold frequency.* For current to flow, the light shining on the metal must have a minimum, or threshold, *frequency,* and different metals have different minimum frequencies. But the wave theory associates the energy of light with its *amplitude* (intensity), not its frequency (color). Thus, the theory predicts that an electron would break free when it absorbed enough energy from light of *any* color.
  2. *Absence of a time lag.* Current flows the moment light of the minimum frequency shines on the metal, regardless of the light's intensity. But the wave theory predicts that with dim light there would be a time lag before the current flows, because the electrons would have to absorb enough energy to break free.

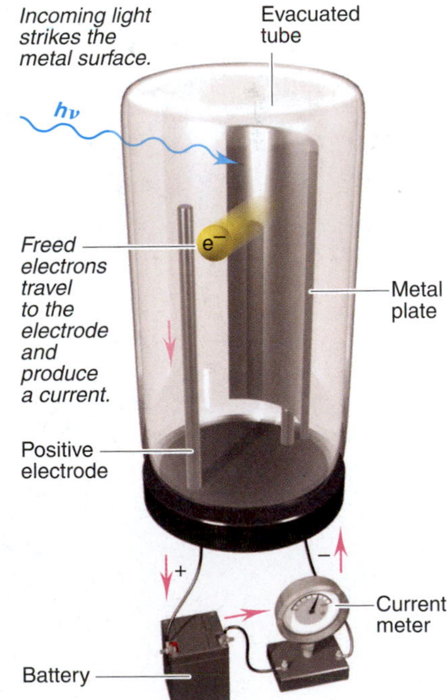

Figure 7.6 The photoelectric effect.

- *Explanation: the photon theory.* Building on Planck's ideas, Albert Einstein proposed in 1905 that light itself is particulate, quantized into tiny "bundles" of energy, later called **photons.** Each atom changes its energy, ΔE_{atom}, when it absorbs or emits one photon, one "particle" of light, whose energy is related to its *frequency,* not its amplitude:

  $$E_{\text{photon}} = h\nu = \Delta E_{\text{atom}}$$

*A blackbody is an idealized object that absorbs all the radiation incident on it. A hollow cube with a small hole in one wall approximates a blackbody.

Let's see how the photon theory explains the two features of the photoelectric effect:

1. *Why there is a threshold frequency.* A beam of light consists of an enormous number of photons. The intensity (brightness) is related to the *number* of photons, but *not* to the energy of each. Therefore, a photon of a certain *minimum* energy must be absorbed to free an electron from the surface (see Figure 7.6). Since energy depends on frequency ($h\nu$), the theory predicts a threshold frequency.
2. *Why there is no time lag.* An electron breaks free when it absorbs a photon of *enough* energy; it cannot break free by "saving up" energy from several photons, each having less than the minimum energy. The current is weak in dim light because fewer photons of enough energy can free fewer electrons per unit time, but some current flows *as soon as* light of sufficient energy (frequency) strikes the metal.

THINK OF IT THIS WAY
Ping-Pong Photons

This analogy helps explain why light of *insufficient* energy *cannot* free an electron from the metal surface. If one Ping-Pong ball doesn't have enough energy to knock a book off a shelf, neither does a series of Ping-Pong balls, because the book can't save up the energy from the individual impacts. But one baseball traveling at the same speed *does* have enough energy to move the book. Whereas the energy of a baseball is related to its mass and velocity, the energy of a photon is related to its frequency.

Sample Problem 7.2 Calculating the Energy of Radiation from Its Wavelength

Problem A student uses a microwave oven to heat a meal. The wavelength of the radiation is 1.20 cm. What is the energy of one photon of this microwave radiation?

Plan We know λ in centimeters (1.20 cm) so we convert to meters, find the frequency with Equation 7.1, and then find the energy of one photon with Equation 7.2.

Solution Combining steps to find the energy of a photon:

$$E = h\nu = \frac{hc}{\lambda} = \frac{(6.626\times10^{-34}\ \text{J·s})(3.00\times10^{8}\ \text{m/s})}{(1.20\ \text{cm})\left(\dfrac{10^{-2}\ \text{m}}{1\ \text{cm}}\right)} = 1.66\times10^{-23}\ \text{J}$$

Check Checking the order of magnitude gives

$$\frac{10^{-33}\ \text{J·s}\times10^{8}\ \text{m/s}}{10^{-2}\ \text{m}} = 10^{-23}\ \text{J}$$

FOLLOW-UP PROBLEM 7.2 Calculate the energy of one photon of **(a)** ultraviolet light ($\lambda = 1\times10^{-8}$ m); **(b)** visible light ($\lambda = 5\times10^{-7}$ m); and **(c)** infrared light ($\lambda = 1\times10^{-4}$ m). What do the answers indicate about the relationship between the wavelength and energy of light?

Summary of Section 7.1

- Electromagnetic radiation travels in waves characterized by a given wavelength (λ) and frequency (ν).
- Electromagnetic waves travel through a vacuum at the speed of light, c (3.00×10^8 m/s), which equals $\nu \times \lambda$. Therefore, wavelength and frequency have a reciprocal relationship.
- The intensity (brightness) of light is related to the amplitude of its waves.
- The electromagnetic spectrum ranges from very long radio waves to very short gamma rays and includes the visible region between wavelengths 750 nm (red) and 400 nm (violet).
- Refraction (change in a wave's speed when entering a different medium) and diffraction (bend of a wave around an edge of an object) indicate that energy is wavelike, with properties distinct from those of particles of matter.
- Blackbody radiation and the photoelectric effect, however, are consistent with energy occurring in discrete packets, like particles.

- Light exists as photons (quanta) whose energy is proportional to their frequency.
- According to quantum theory, an atom has only certain quantities of energy ($E = nh\nu$), and it can change its energy only by absorbing or emitting a photon whose energy equals the change in the atom's energy.

7.2 • ATOMIC SPECTRA

The third confusing observation about matter and energy involved the light emitted when an element is vaporized and then excited electrically, as occurs in a neon sign. In this section, we discuss the nature of that light, why it created a problem for the existing atomic model, and how a new model solved the problem.

Line Spectra and the Rydberg Equation

When light from electrically excited gaseous atoms passes through a slit and is refracted by a prism, it does not create a *continuous spectrum,* or rainbow, as sunlight does. Instead, it creates a **line spectrum,** a series of fine lines at specific frequencies separated by black spaces. Figure 7.7A shows the apparatus and the line spectrum of atomic hydrogen. Figure 7.7B shows that each spectrum is *characteristic* of the element producing it.

Spectroscopists studying atomic hydrogen identified several series of spectral lines in different regions of the electromagnetic spectrum. Figure 7.8 on the next page shows three of them. Using data, not theory, the Swedish physicist Johannes Rydberg (1854–1919) developed a relationship, called the *Rydberg equation,* that predicted the position and wavelength of any line in a given series:

$$\frac{1}{\lambda} = R\left(\frac{1}{n_1^2} - \frac{1}{n_2^2}\right) \qquad (7.3)$$

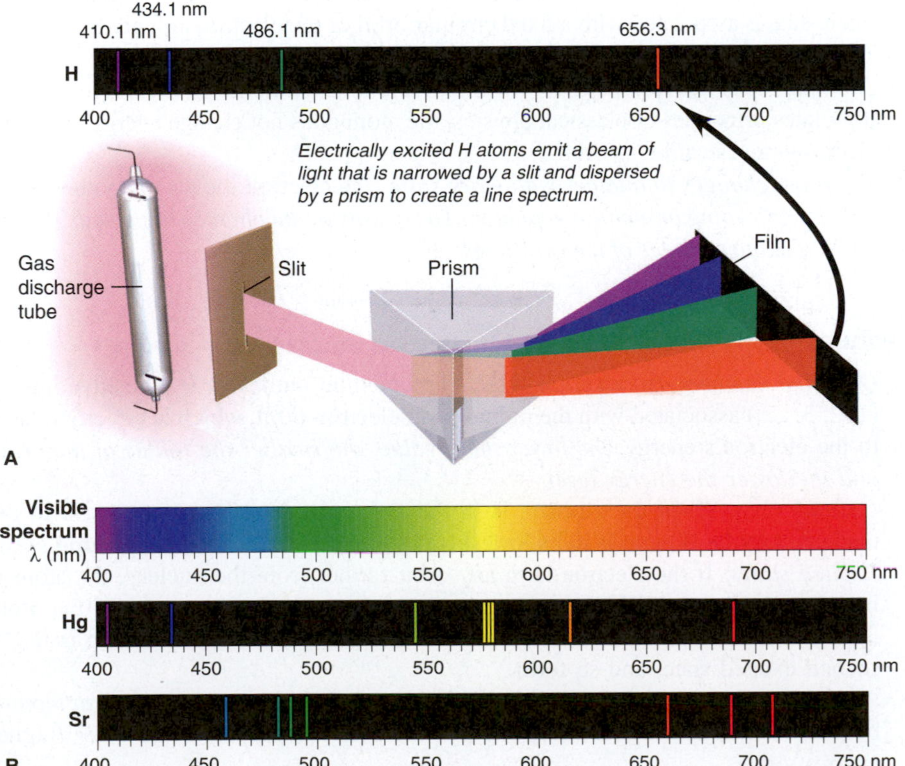

Figure 7.7 The line spectra of several elements. **A,** The line spectrum of atomic hydrogen. **B,** Unlike the continuous spectrum of white light, emission spectra of elements, such as mercury and strontium, appear as characteristic series of colored lines.

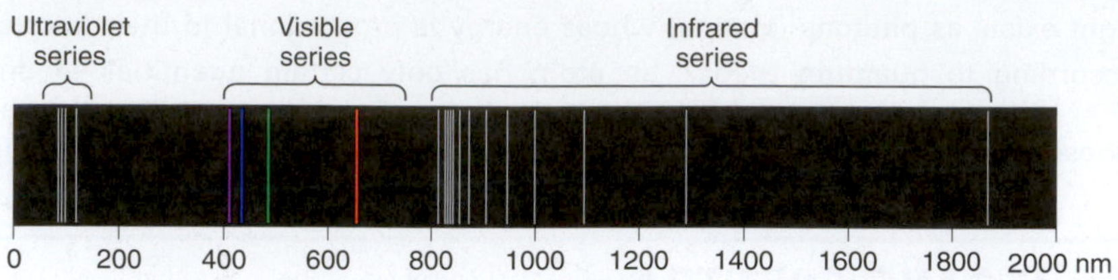

Figure 7.8 Three series of spectral lines of atomic hydrogen.

where λ is the wavelength of the line, n_1 and n_2 are positive integers with $n_2 > n_1$, and R is the Rydberg constant ($1.096776 \times 10^7\ m^{-1}$). For the visible series, $n_1 = 2$:

$$\frac{1}{\lambda} = R\left(\frac{1}{2^2} - \frac{1}{n_2^2}\right), \quad \text{with } n_2 = 3, 4, 5, \ldots$$

(Problems 7.19 and 7.20 are two of several at the end of the chapter that apply the Rydberg equation.)

The occurrence of line spectra did not correlate with classical theory. If an electron spiraled closer to the nucleus, the frequency of the radiation it emitted should be related to the time of revolution. That time changes smoothly on a spiral path, so the frequency of the radiation should change smoothly and, thus, create a continuous spectrum. Rutherford's model of a nuclear atom did not predict line spectra!

The Bohr Model of the Hydrogen Atom

Two years after the nuclear model was proposed, Niels Bohr (1885–1962), a young Danish physicist working in Rutherford's laboratory, suggested a model for the H atom that *did* predict the existence of line spectra.

Postulates of the Model In his model, Bohr used Planck's and Einstein's ideas about quantized energy and proposed three postulates:

1. *The H atom has only certain energy levels,* which Bohr called **stationary states.** Each state is associated with a fixed circular orbit of the electron around the nucleus. The higher the energy level, the farther the orbit is from the nucleus.
2. *The atom does **not** radiate energy while in one of its stationary states.* Even though it violates principles of classical physics, the atom does not change energy while the electron moves *within* an orbit.
3. *The atom changes to another stationary state* (the electron moves to another orbit) *only by absorbing or emitting a photon. The energy of the photon ($h\nu$) equals the difference in the energies of the two states:*

$$E_{\text{photon}} = \Delta E_{\text{atom}} = E_{\text{final}} - E_{\text{initial}} = h\nu$$

Features of the Model The model has several key features:

- *Quantum numbers and electron orbit.* The quantum number n is a positive integer (1, 2, 3, . . .) associated with the radius of an electron orbit, which is directly related to the electron's energy: *the lower the n value, the smaller the radius of the orbit, and the lower the energy level.*
- *Ground state.* When the electron is in the first orbit ($n = 1$), it is closest to the nucleus, and the H atom is in its lowest (first) energy level, called the **ground state.**
- *Excited states.* If the electron is in any orbit farther from the nucleus, the atom is in an **excited state.** When the electron is in the second orbit ($n = 2$), the atom is in the first excited state; when it's in the third orbit ($n = 3$), the atom is in the second excited state, and so forth.
- *Absorption.* If an H atom *absorbs* a photon whose energy equals the *difference* between lower and higher energy levels, the electron moves to the outer (higher energy) orbit.
- *Emission.* If an H atom in a higher energy level (electron in farther orbit) returns to a lower energy level (electron in closer orbit), the atom *emits* a photon whose energy equals the difference between the two levels. Figure 7.9 shows an analogy that illustrates absorption and emission.

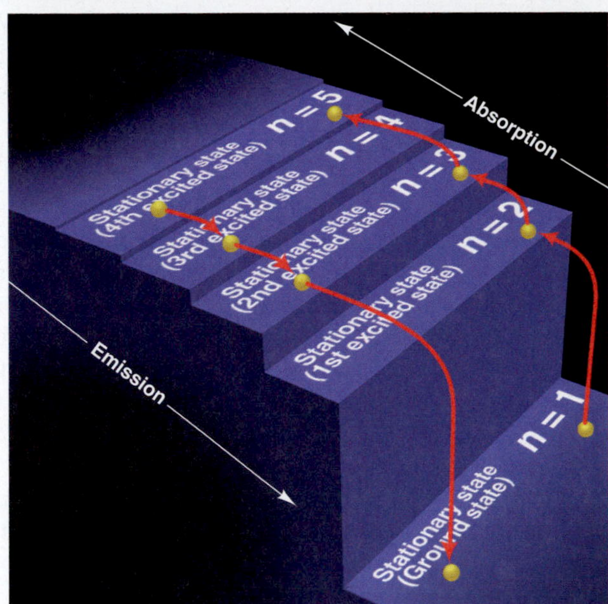

Figure 7.9 A quantum "staircase" as an analogy for atomic energy levels. Note that the electron can move up or down one or more steps at a time but cannot lie *between* steps.

How the Model Explains Line Spectra A spectral line results when a photon of specific energy (and thus frequency) is *emitted.* The emission occurs when the electron moves to an orbit closer to the nucleus as the atom's energy changes from a higher state to a lower one. Therefore, *an atomic spectrum is not continuous because the atom's energy is not continuous, but rather has only certain states.*

Figure 7.10A shows how Bohr's model accounts for three series of spectral lines of hydrogen. When a sample of gaseous H atoms is excited, different atoms absorb different quantities of energy. Each atom has one electron, but there are so many

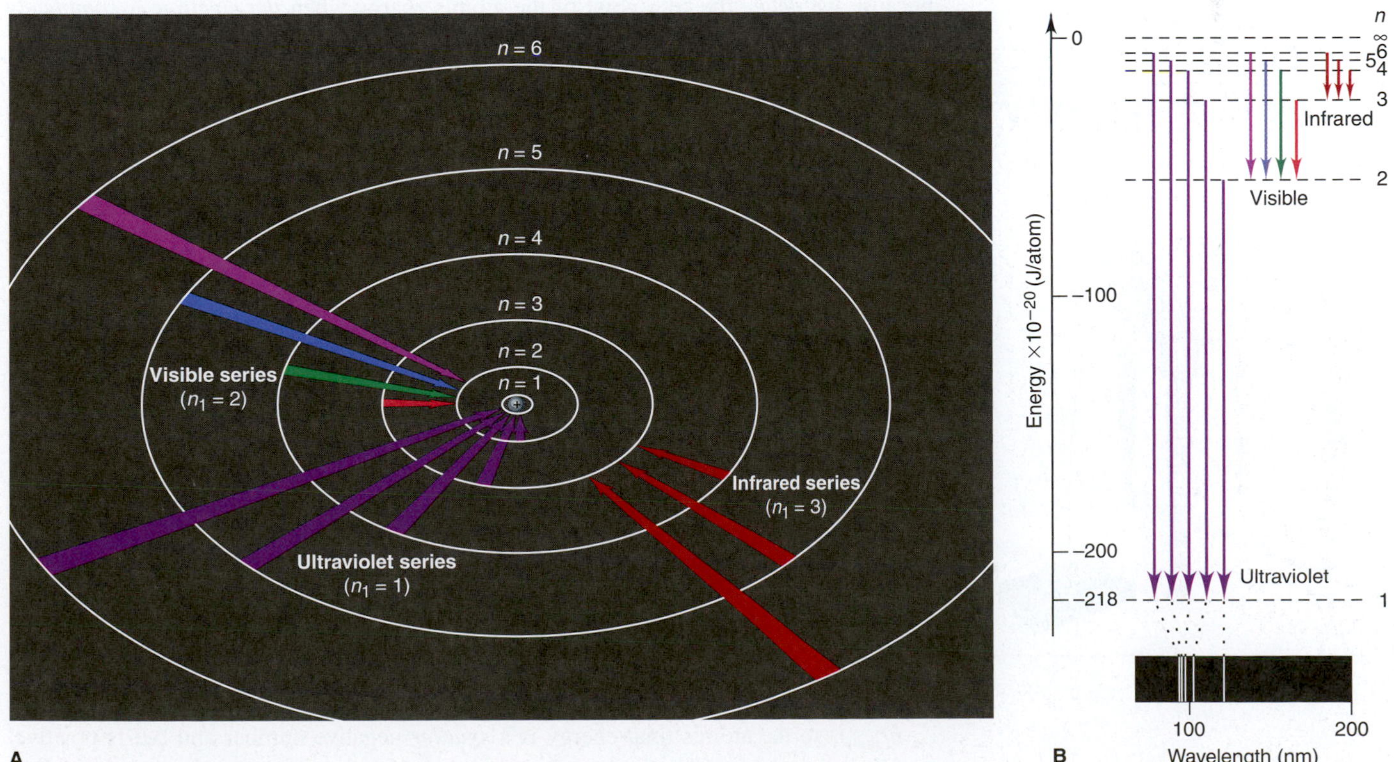

Figure 7.10 The Bohr explanation of three series of spectral lines emitted by the H atom. A, In a given series, the outer electron drops to the same inner orbit (the same value of n_1 in the Rydberg equation). **B,** An energy diagram shows how the ultraviolet series arises.

atoms in the sample that all the energy levels (orbits) have electrons. When electrons drop from outer orbits to the $n = 3$ orbit (second excited state), the emitted photons create the *infrared* series of lines. The *visible* series arises when electrons drop to the $n = 2$ orbit (first excited state), and the *ultraviolet* series arises when electrons drop to the $n = 1$ orbit (ground state). Figure 7.10B shows how the specific lines in the ultraviolet series appear.

Limitations of the Model Despite its great success in predicting the spectral lines of H, the Bohr model failed with every other atom. The reason is that it is a *one-electron model:* it works beautifully for the H atom and for one-electron ions, such as He^+ (Z = 2), Li^{2+} (Z = 3), and Be^{3+} (Z = 4). But it fails completely for species with more than one electron because electron-electron repulsions and additional nucleus-electron attractions create much more complex interactions. Even more fundamentally, as we'll see in Section 7.4, *electrons do not move in fixed, defined orbits.* As a picture of the atom, the Bohr model is incorrect, but we still retain the terms "ground state" and "excited state" and the model's central idea: *the energy of an atom occurs in discrete levels, and an atom changes energy by absorbing or emitting a photon of specific energy.*

The Energy Levels of the Hydrogen Atom

Bohr's work leads to an equation for calculating the energy levels of an atom:

$$E = -2.18\times10^{-18}\text{ J}\left(\frac{Z^2}{n^2}\right)$$

where Z is the charge of the nucleus. For the H atom, $Z = 1$, so we have

$$E = -2.18\times10^{-18}\text{ J}\left(\frac{1^2}{n^2}\right) = -2.18\times10^{-18}\text{ J}\left(\frac{1}{n^2}\right)$$

Therefore, the energy of the ground state ($n = 1$) of the H atom is

$$E = -2.18\times10^{-18}\text{ J}\left(\frac{1}{1^2}\right) = -2.18\times10^{-18}\text{ J}$$

The negative sign for the energy (also used for the axis values in Figure 7.10B) appears because we *define* the zero point of the atom's energy when *the electron is completely removed from the nucleus.* Thus, $E = 0$ when $n = \infty$, so $E < 0$ for any smaller n.

THINK OF IT THIS WAY

A Book on a Desk and the H Atom's Energy

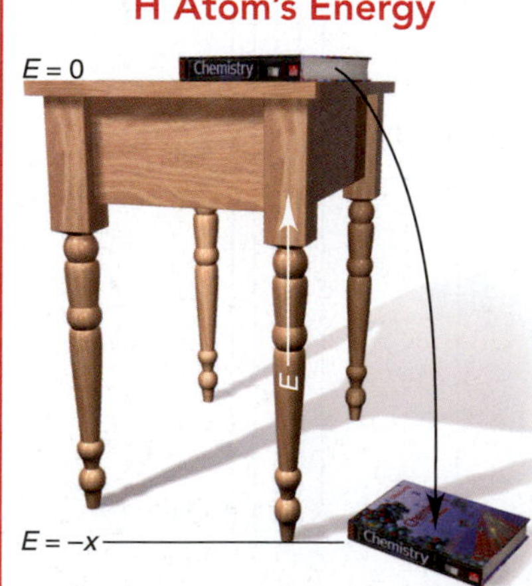

If you *define* the potential energy of a book-desk system as zero when the book rests on the desk, the system has negative energy when the book lies on the floor. Similarly, the nucleus-electron system of the H atom is defined as having zero energy when the electron is completely separated from the nucleus. Thus, this system's energy is negative when the electron is close enough to the nucleus to be attracted by it.

Applying Bohr's Equation for the Energy Levels of an Atom We can use the equation for the energy levels in several ways:

1. *Finding the difference in energy between two levels.* By subtracting the initial energy level of the atom from its final energy level, we find the change in energy when the electron moves between the two levels:

$$\Delta E = E_{\text{final}} - E_{\text{initial}} = -2.18\times10^{-18}\text{ J}\left(\frac{1}{n^2_{\text{final}}} - \frac{1}{n^2_{\text{initial}}}\right) \quad \textbf{(7.4)}$$

Note that, since n is in the denominator,

- *When the atom emits energy,* the electron moves closer to the nucleus ($n_{\text{final}} < n_{\text{initial}}$), so the atom's final energy is a *larger* negative number and ΔE is negative.
- *When the atom absorbs energy,* the electron moves away from the nucleus ($n_{\text{final}} > n_{\text{initial}}$), so the atom's final energy is a *smaller* negative number and ΔE is positive. (Analogously, in Chapter 6, you saw that, when the system releases heat, ΔH is negative, and when it absorbs heat, ΔH is positive.)

2. *Finding the energy needed to ionize the H atom.* We can also find the energy needed to remove the electron completely, that is, find ΔE for the following change:

$$H(g) \longrightarrow H^+(g) + e^-$$

We substitute $n_{final} = \infty$ and $n_{initial} = 1$ into Equation 7.4 and obtain

$$\Delta E = E_{final} - E_{initial} = -2.18\times10^{-18}\ J\left(\frac{1}{\infty^2} - \frac{1}{1^2}\right)$$
$$= -2.18\times10^{-18}\ J\,(0 - 1) = 2.18\times10^{-18}\ J$$

Energy must be *absorbed* to remove the electron from the nucleus, so ΔE is positive.

The *ionization energy* of hydrogen is the energy required to form 1 mol of gaseous H^+ ions from 1 mol of gaseous H atoms. Thus, for 1 mol of H atoms,

$$\Delta E = \left(2.18\times10^{-18}\ \frac{J}{atom}\right)\left(6.022\times10^{23}\ \frac{atoms}{mol}\right)\left(\frac{1\ kJ}{10^3\ J}\right) = 1.31\times10^3\ kJ/mol$$

Ionization energy is a key atomic property, and we'll return to it in Chapter 8.

3. *Finding the wavelength of a spectral line.* Once we know ΔE from Equation 7.4, we find the wavelengths of the spectral lines of the H atom by combining the relation between frequency and wavelength (Equation 7.1) with Planck's expression for the change in energy of an atom (Equation 7.2) and solving for λ:

$$\Delta E = h\nu = \frac{hc}{\lambda} \qquad \text{or} \qquad \lambda = \frac{hc}{\Delta E}$$

In Sample Problem 7.3, we find the energy change when an H atom absorbs a photon.

Sample Problem 7.3 Determining ΔE and λ of an Electron Transition

Problem A hydrogen atom absorbs a photon of UV light (see Figure 7.10), and its electron enters the $n = 4$ energy level. Calculate **(a)** the change in energy of the atom and **(b)** the wavelength (in nm) of the photon.

Plan **(a)** The H atom absorbs energy, so $E_{final} > E_{initial}$. We are given $n_{final} = 4$, and Figure 7.10 shows that $n_{initial} = 1$ because a UV photon is absorbed. We apply Equation 7.4 to find ΔE. **(b)** Once we know ΔE, we find the frequency with Equation 7.2 and the wavelength (in m) with Equation 7.1. Then we convert from meters to nanometers.

Solution **(a)** Substituting the known values into Equation 7.4:

$$\Delta E = -2.18\times10^{-18}\ J\left(\frac{1}{n^2_{final}} - \frac{1}{n^2_{initial}}\right) = -2.18\times10^{-18}\ J\left(\frac{1}{4^2} - \frac{1}{1^2}\right)$$
$$= -2.18\times10^{-18}\ J\left(\frac{1}{16} - \frac{1}{1}\right) = 2.04\times10^{-18}\ J$$

(b) Using Equations 7.2 and 7.1 to solve for λ:

$$\Delta E = h\nu = \frac{hc}{\lambda}$$

therefore,

$$\lambda = \frac{hc}{\Delta E} = \frac{(6.626\times10^{-34}\ J\cdot s)(3.00\times10^8\ m/s)}{2.04\times10^{-18}\ J} = 9.74\times10^{-8}\ m$$

Converting m to nm:

$$\lambda = 9.74\times10^{-8}\ m \times \frac{1\ nm}{10^{-9}\ m} = 97.4\ nm$$

Check **(a)** The energy change is positive, which is consistent with absorption. **(b)** The wavelength is within the UV region (about 10–380 nm).

Comment In the follow-up problem, note that if ΔE is negative (the atom loses energy), we use its absolute value, $|\Delta E|$, because λ must have a positive value.

FOLLOW-UP PROBLEM 7.3 A hydrogen atom with its electron in the $n = 6$ energy level emits a photon of IR light. Calculate **(a)** the change in energy of the atom and **(b)** the wavelength (in Å) of the photon.

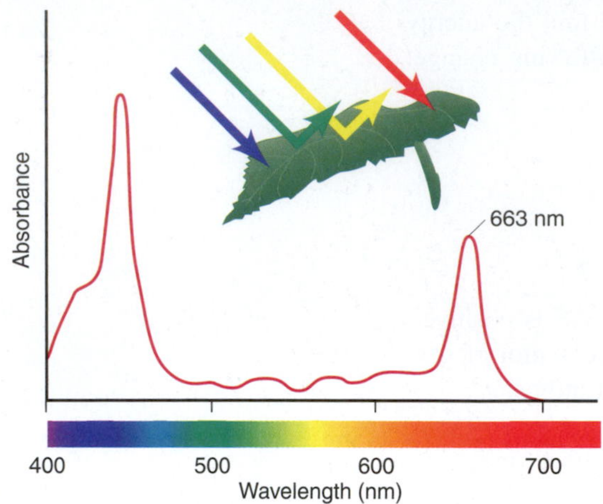

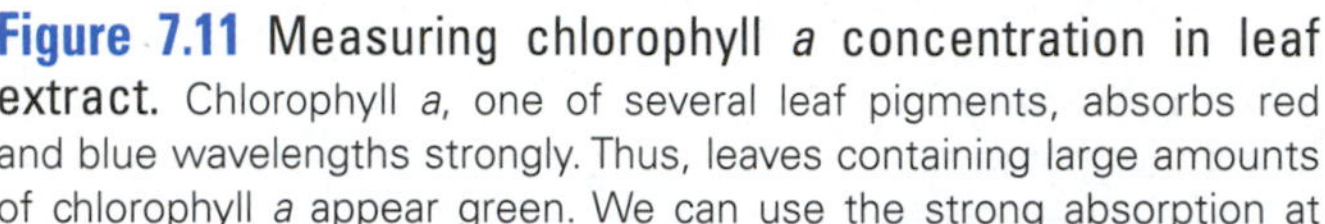

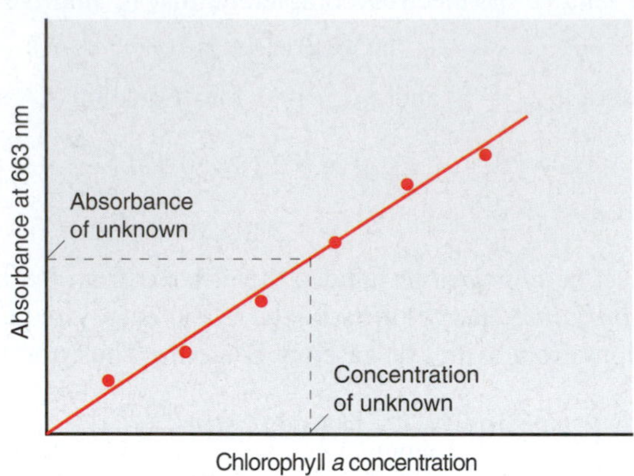

Figure 7.11 Measuring chlorophyll *a* concentration in leaf extract. Chlorophyll *a*, one of several leaf pigments, absorbs red and blue wavelengths strongly. Thus, leaves containing large amounts of chlorophyll *a* appear green. We can use the strong absorption at 663 nm in the spectrum **(A)** to quantify the amount of chlorophyll *a* present in a plant extract by comparing that absorbance to a series of known standards **(B).**

Spectral Analysis in the Laboratory

Analysis of the spectrum of the H atom led to the Bohr model, the first step toward our current model of the atom. From its use by 19th-century chemists as a means of identifying elements and compounds, spectrometry has developed into a major tool of modern chemistry. The terms *spectroscopy, spectrophotometry,* and **spectrometry** refer to a large group of instrumental techniques that obtain spectra that correspond to a substance's atomic or molecular energy levels. (Elements produce lines, but complex molecules produce spectral peaks.) The two types of spectra most often obtained are emission and absorption spectra:

- An **emission spectrum** is produced when atoms in an excited state emit photons characteristic of the element as they return to lower energy states. The characteristic colors of fireworks and sodium-vapor streetlights are due to one or a few prominent lines in the emission spectra of the atoms present.
- An **absorption spectrum** is produced when atoms absorb photons of certain wavelengths and become excited from lower to higher energy states. Therefore, the absorption spectrum of an element appears as dark lines against a bright background.

A spectrometer can also be used to measure the concentration of a substance in a solution because *the absorbance, the amount of light of a given wavelength absorbed by a substance, is proportional to the number of molecules.* Suppose, for example, you want to determine the concentration of chlorophyll in an ether solution of leaf extract. You select a strongly absorbed wavelength in a peak of the chlorophyll spectrum (such as 663 nm in Figure 7.11A), measure the absorbance of the leaf-extract solution, and compare it with the absorbances of a series of ether solutions with known chlorophyll concentrations (Figure 7.11B).

Summary of Section 7.2

- Unlike sunlight, light emitted by electrically excited atoms of an element appears as separate spectral lines.
- Spectroscopists use an empirical formula (the Rydberg equation) to determine the wavelength of a spectral line. Atomic hydrogen displays several series of lines.
- To explain the existence of line spectra, Bohr proposed that an electron moves in fixed orbits. It moves from one orbit to another when the atom absorbs or emits a photon whose energy equals the difference in energy levels (orbits).
- Bohr's model predicts only the spectrum of the H atom and other one-electron species. Despite this, Bohr was correct that an atom's energy is quantized.
- Spectrometry is an instrumental technique that obtains emission and absorption spectra used to identify substances and measure their concentrations.

7.3 • THE WAVE-PARTICLE DUALITY OF MATTER AND ENERGY

The early proponents of quantum theory demonstrated that *energy is particle-like*. Physicists who further developed the theory turned this proposition upside down and showed that *matter is wavelike*. The sharp divisions we perceive in everyday life between matter and energy have been completely blurred. Strange as this idea may seem, it is the key to our modern atomic model.

The Wave Nature of Electrons and the Particle Nature of Photons

Bohr's model was a perfect case of fitting theory to data: he *assumed* that an atom has only certain energy levels in order to *explain* line spectra. However, Bohr had no theoretical basis for the assumption. Several breakthroughs in the early 1920s provided that basis and blurred the distinction between matter (chunky and massive) and energy (diffuse and massless).

The Wave Nature of Electrons Attempting to explain why an atom has fixed energy levels, a French physics student, Louis de Broglie, considered other systems that display only certain allowed motions, such as the vibrations of a plucked guitar string. Figure 7.12 shows that, because the end of the string is fixed, only certain vibrational frequencies (and wavelengths) can occur. De Broglie proposed that *if energy is particle-like, perhaps matter is wavelike.* He reasoned that *if electrons have wavelike motion* in orbits of fixed radii, they would have only certain allowable frequencies and energies.

Combining Einstein's famous equation for mass-energy equivalence ($E = mc^2$) with the equation for the energy of a photon ($E = h\nu = hc/\lambda$), de Broglie derived an equation for the wavelength of any particle of mass m—whether planet, baseball, or electron—moving at speed u:

$$\lambda = \frac{h}{mu} \tag{7.5}$$

According to this equation for the **de Broglie wavelength,** *matter behaves as though it moves in a wave.* An object's wavelength is *inversely* proportional to its mass, so heavy objects such as planets and baseballs have wavelengths *many* orders of magnitude smaller than the object itself (Table 7.1, *next page*).

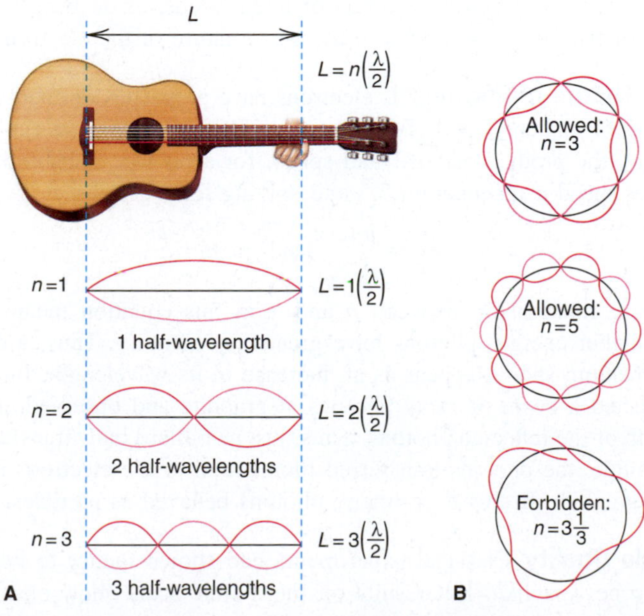

Figure 7.12 Wave motion in restricted systems. **A,** One half-wavelength ($\lambda/2$) is the "quantum" of the guitar string's vibration. With string length L fixed by a finger on the fret, allowed vibrations occur when L is a whole-number multiple (n) of $\lambda/2$. **B,** In a circular electron orbit, only whole numbers of wavelengths are allowed (n = 3 and n = 5 are shown). A wave with a fractional number of wavelengths (such as $n = 3\frac{1}{3}$) is "forbidden" because it will die out through overlap of crests and troughs.

Table 7.1 The de Broglie Wavelengths of Several Objects

Substance	Mass (g)	Speed (m/s)	λ (m)
Slow electron	9×10^{-28}	1.0	7×10^{-4}
Fast electron	9×10^{-28}	5.9×10^{6}	1×10^{-10}
Alpha particle	6.6×10^{-24}	1.5×10^{7}	7×10^{-15}
1-gram mass	1.0	0.01	7×10^{-29}
Baseball	142	40.0	1×10^{-34}
Earth	6.0×10^{27}	3.0×10^{4}	4×10^{-63}

Sample Problem 7.4 Calculating the de Broglie Wavelength of an Electron

Problem Find the de Broglie wavelength of an electron with a speed of 1.00×10^{6} m/s (electron mass $= 9.11\times10^{-31}$ kg; $h = 6.626\times10^{-34}$ kg·m²/s).

Plan We know the speed (1.00×10^{6} m/s) and mass (9.11×10^{-31} kg) of the electron, so we substitute these into Equation 7.5 to find λ.

Solution

$$\lambda = \frac{h}{mu} = \frac{6.626\times10^{-34}\ \text{kg}\cdot\text{m}^2/\text{s}}{(9.11\times10^{-31}\ \text{kg})(1.00\times10^{6}\ \text{m/s})} = 7.27\times10^{-10}\ \text{m}$$

Check The order of magnitude and units seem correct:

$$\lambda \approx \frac{10^{-33}\ \text{kg}\cdot\text{m}^2/\text{s}}{(10^{-30}\ \text{kg})(10^{6}\ \text{m/s})} = 10^{-9}\ \text{m}$$

FOLLOW-UP PROBLEM 7.4 What is the speed of an electron that has a de Broglie wavelength of 100. nm?

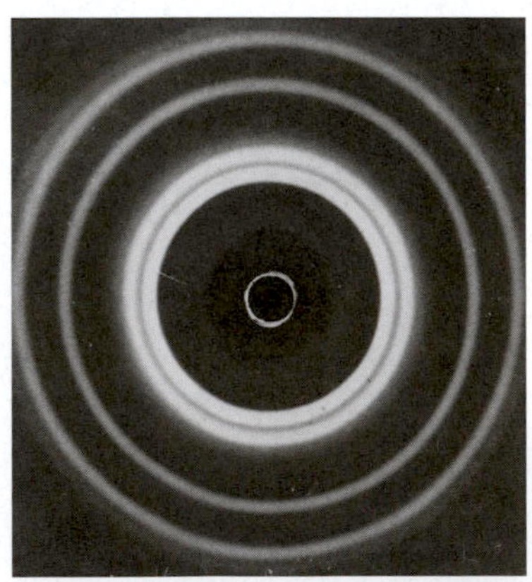

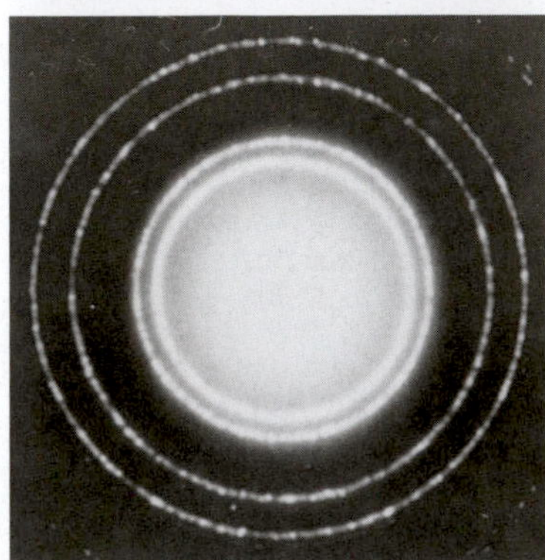

Figure 7.13 Diffraction patterns of aluminum with x-rays *(top)* and electrons *(bottom)*.

If electrons travel in waves, they should exhibit diffraction and interference. A fast-moving electron has a wavelength of about 10^{-10} m, so a beam of such electrons should be diffracted by the spaces between atoms in a crystal—about 10^{-10} m. In 1927, C. Davisson and L. Germer guided a beam of x-rays and then a beam of electrons at a nickel crystal and obtained two diffraction patterns; Figure 7.13 shows the patterns for aluminum. Thus, electrons—particles with mass and charge—create diffraction patterns, just as electromagnetic waves do. (Indeed, the electron microscope has had a revolutionary impact on modern biology due to its ability to magnify objects up to 200,000 times, which depends on the wavelike behavior of electrons.) Even though electrons do not have orbits of fixed radius, as de Broglie thought, the energy levels of the atom *are* related to the wave nature of the electron.

The Particle Nature of Photons If electrons have properties of energy, do photons have properties of matter? The de Broglie equation suggests that we can calculate the momentum (p), the product of mass and speed, for a photon. Substituting the speed of light (c) for speed u in Equation 7.5 and solving for p gives

$$\lambda = \frac{h}{mc} = \frac{h}{p} \qquad \text{and} \qquad p = \frac{h}{\lambda}$$

The inverse relationship between p and λ in this equation means that shorter wavelength (higher energy) photons have greater momentum. Thus, a decrease in a photon's momentum should appear as an increase in its wavelength. In 1923, Arthur Compton directed a beam of x-ray photons at graphite and observed an increase in the wavelength of the reflected photons. Thus, just as billiard balls transfer momentum when they collide, the photons transferred momentum to the electrons in the carbon atoms of the graphite. In this experiment, photons behaved as particles.

Wave-Particle Duality Classical experiments had shown matter to be particle-like and energy to be wavelike. But results on the atomic scale show electrons moving

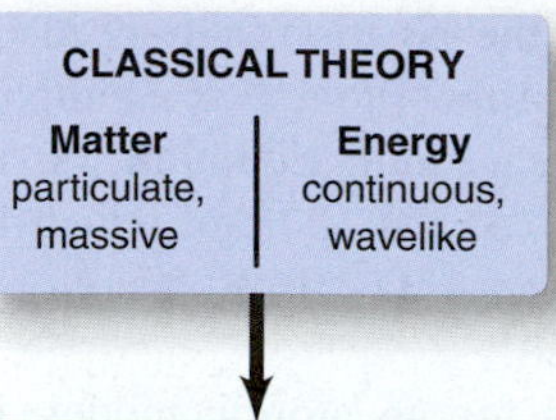

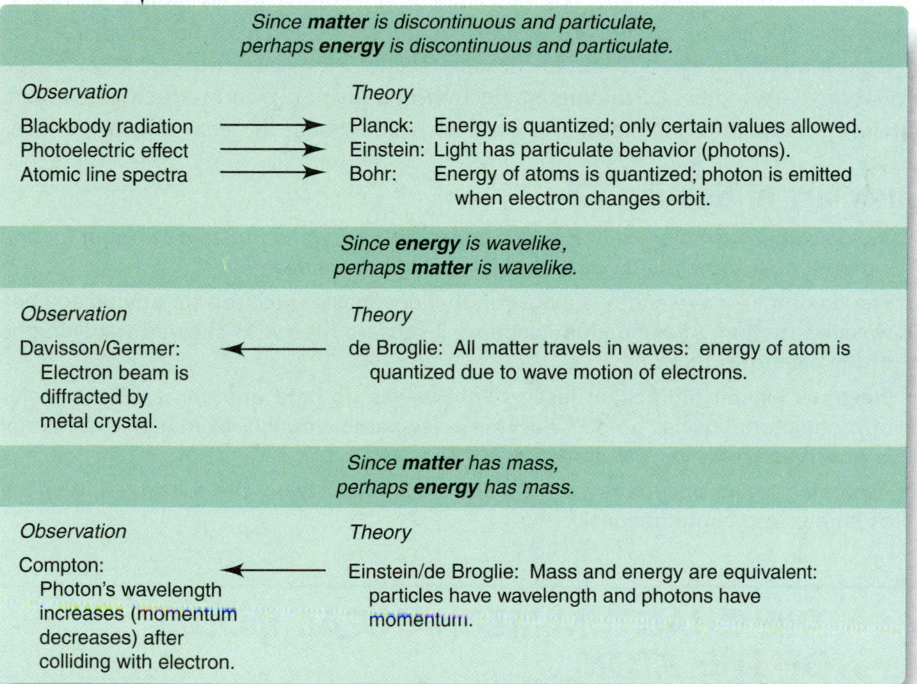

Figure 7.14 Major observations and theories leading from classical theory to quantum theory.

in waves and photons having momentum. Thus, every property of matter was also a property of energy. The truth is that *both* matter and energy show *both* behaviors: each possesses both "faces." In some experiments, we observe one face; in other experiments, we observe the other face. Our everyday distinction between matter and energy is meaningful in the macroscopic world, *not* in the atomic world. The distinction is in our minds and the limited definitions we have created, not inherent in nature. This dual character of matter and energy is known as the **wave-particle duality.** Figure 7.14 summarizes the theories and observations that led to this new understanding.

Heisenberg's Uncertainty Principle

In classical physics, a moving particle has a definite location at any instant, whereas a wave is spread out in space. If an electron has the properties of *both* a particle and a wave, can we determine its position in the atom? In 1927, the German physicist Werner Heisenberg postulated the **uncertainty principle,** which states that it is impossible to know simultaneously the position *and* momentum (mass times speed) of a particle. For a particle with constant mass m, the principle is expressed mathematically as

$$\Delta x \cdot m\Delta u \geq \frac{h}{4\pi} \tag{7.6}$$

where Δx is the uncertainty in position, Δu is the uncertainty in speed, and h is Planck's constant. The more accurately we know the position of the particle (smaller Δx), the less accurately we know its speed (larger Δu), and vice versa.

For a macroscopic object like a baseball, Δx and Δu are insignificant because the mass is enormous compared with $h/4\pi$. Thus, by knowing the position and speed of a pitched baseball, we can use the laws of motion to predict its trajectory and whether it will be a ball or a strike. However, using the position and speed of an electron to predict its trajectory is a very different proposition. For example, if we take an electron's speed as 6×10^6 m/s $\pm$ 1%, then Δu in Equation 7.6 is 6×10^4 m/s, and the uncertainty in the electron's position (Δx) is 10^{-9} m, which is about 10 times greater than the diameter of the entire atom (10^{-10} m). Therefore, we have no idea where in the atom the electron is located!

The uncertainty principle has profound implications for an atomic model. It means that *we cannot assign fixed paths for electrons,* such as the circular orbits of Bohr's model. As you'll see in the next section, the most we can know is the *probability*—the odds—of finding an electron in a given region of space; but we are not *sure* it is there any more than a gambler is sure about the next roll of the dice.

Summary of Section 7.3

- As a result of work based on Planck's quantum theory and Einstein's photon theory, we no longer view matter and energy as distinct entities.
- The de Broglie wavelength is based on the idea that an electron (or any object) has wavelike motion. Allowed atomic energy levels are related to allowed wavelengths of the electron's motion.
- Electrons exhibit diffraction, just as light waves do, and photons exhibit transfer of momentum, just as objects do. This wave-particle duality of matter and energy is observable only on the atomic scale.
- According to the uncertainty principle, we can never know the position and speed of an electron simultaneously.

7.4 • THE QUANTUM-MECHANICAL MODEL OF THE ATOM

Acceptance of the dual nature of matter and energy and of the uncertainty principle culminated in the field of **quantum mechanics,** which examines the wave nature of objects on the atomic scale. In 1926, Erwin Schrödinger derived an equation that is the basis for the *quantum-mechanical model* of the H atom. The model describes an atom with specific quantities of energy that result from allowed frequencies of its electron's wavelike motion. The electron's position can only be known within a certain probability. Key features of the model are described in the following subsections.

The Atomic Orbital and the Probable Location of the Electron

Two central aspects of the quantum-mechanical model concern the atomic orbital and the electron's probable location.

The Schrödinger Equation and the Atomic Orbital The electron's matter-wave occupies the space near the nucleus and is continuously influenced by it. The **Schrödinger equation** is quite complex but can be represented in simpler form as

$$\mathcal{H}\psi = E\psi$$

where E is the energy of the atom. The symbol ψ (Greek *psi,* pronounced "sigh") is called a **wave function,** or **atomic orbital,** a mathematical description of the electron's matter-wave in three dimensions. The symbol $\mathcal{H}$, called the Hamiltonian operator, represents a set of mathematical operations that, when carried out with a particular ψ, yields one of the allowed energy states of the atom.* Thus, *each solution of the equation gives an energy state associated with a given atomic orbital.*

*The complete form of the Schrödinger equation in terms of the three linear axes is

$$\left[-\frac{h^2}{8\pi^2 m_e}\left(\frac{d^2}{dx^2}+\frac{d^2}{dy^2}+\frac{d^2}{dz^2}\right)+V(x,y,z)\right]\psi(x,y,z) = E\psi(x,y,z)$$

where ψ is the wave function; m_e is the electron's mass; E is the total quantized energy of the atomic system; and V is the potential energy at point (x, y, z).

An important point to keep in mind throughout this discussion is that an "orbital" in the quantum-mechanical model *bears no resemblance* to an "orbit" in the Bohr model: an *orbit* is an electron's actual path around the nucleus, whereas an *orbital* is a mathematical function that describes the electron's matter-wave but has no physical meaning.

The Probable Location of the Electron While we cannot know *exactly* where the electron is at any moment, we can know where it *probably* is, that is, where it spends most of its time. We get this information by squaring the wave function. Thus, even though ψ has no physical meaning, ψ^2 does and is called the *probability density,* a measure of the probability of finding the electron in some tiny volume of the atom. We depict the electron's probable location in several ways, which we'll look at first for the H atom's *ground state:*

1. *Probability of the electron being in some tiny volume of the atom.* For each energy level, we can create an *electron probability density diagram,* or more simply, an **electron density diagram.** The value of ψ^2 for a given volume is shown with dots: the greater the density of dots, the higher the probability of finding the electron in that volume. Note, that for the ground state of the H atom, *the electron probability density decreases with distance from the nucleus* along a line, *r* (Figure 7.15A).

These diagrams are also called **electron cloud depictions** because, if we *could* take a time-exposure photograph of the electron in wavelike motion around the nucleus, it would appear as a "cloud" of positions. The electron cloud is an *imaginary* picture of the electron changing its position rapidly over time; it does *not* mean that an electron is a diffuse cloud of charge.

Figure 7.15B shows a plot of ψ^2 vs. *r*. Due to the thickness of the printed line, the curve appears to touch the axis; however, in the blow-up circle, we see that *the probability of the electron being far from the nucleus is very small, but not zero.*

2. *Total probability density at some distance from the nucleus.* To find *radial probability distribution,* that is, the *total* probability of finding the electron at some distance *r* from the nucleus, we first mentally divide the volume around the nucleus into thin, concentric, spherical layers, like the layers of an onion (shown in cross section in Figure 7.15C). Then, we find the *sum of ψ^2 values* in each layer to see which is most likely to contain the electron.

The falloff in probability density with distance has an important effect. Near the nucleus, *the volume of each layer increases faster than its density of dots decreases.* The result of these opposing effects is that the *total* probability peaks in a layer *near,*

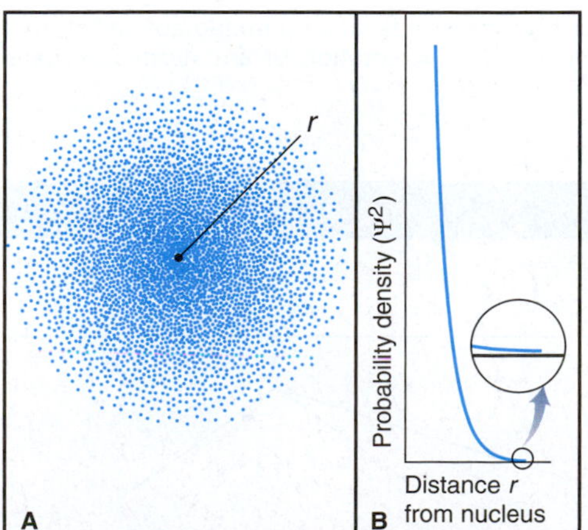

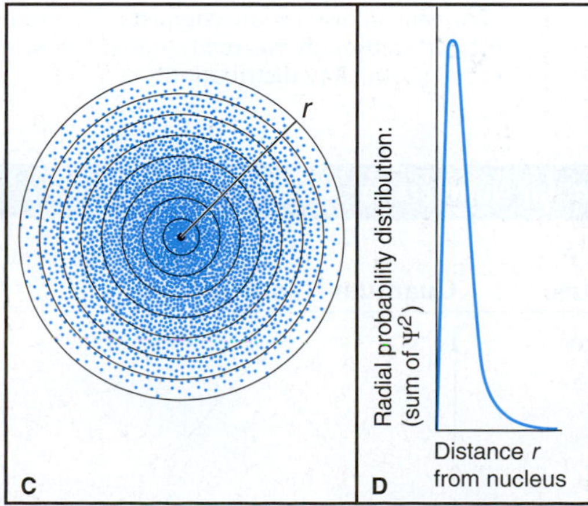

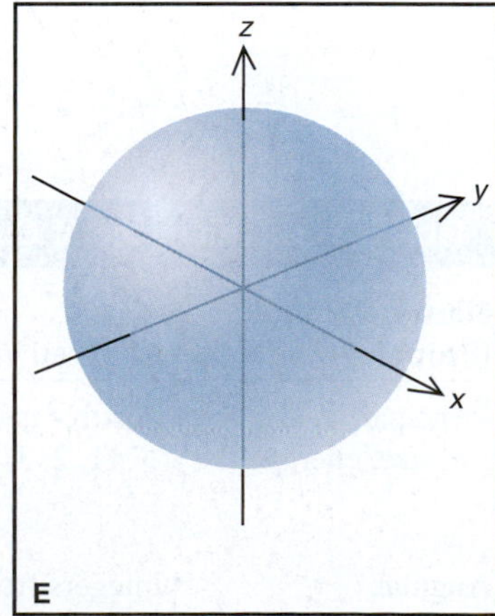

Figure 7.15 Electron probability density in the ground-state H atom. **A,** In the electron density diagram, the density of dots represents the probability of the electron within a tiny volume and decreases with distance, *r,* from the nucleus. **B,** The probability density (ψ^2) decreases with *r* but does not reach zero *(blow-up circle).* **C,** Counting dots within each layer gives the total probability of the electron being in that layer. **D,** A radial probability distribution plot shows that total electron density peaks *near,* but not *at,* the nucleus. **E,** A 90% probability contour for the ground state of the H atom.

but not *at,* the nucleus. For example, the total probability in the second layer is higher than in the first, but this result disappears with greater distance. Figure 7.15D shows this result as a **radial probability distribution plot.**

THINK OF IT THIS WAY

A Radial Probability Distribution of Apples

An analogy might clarify why the curve in the radial probability distribution plot peaks and then falls off. Picture fallen apples around the base of an apple tree: the density of apples is greatest near the trunk and decreases with distance. Divide the ground under the tree into foot-wide concentric rings and collect the apples within each ring. Apple density is greatest in the first ring, but the area of the second ring is larger, and so it contains a greater *total* number of apples. Farther out near the edge of the tree, rings have more area but lower apple "density," so the total number of apples decreases. A plot of "number of apples in each ring" vs. "distance from trunk" shows a peak at some distance close to the trunk.

3. *Probability contour and the size of the atom.* How far away from the nucleus can we find the electron? This is the same as asking "How big is the H atom?" Recall from Figure 7.15B that the probability of finding the electron far from the nucleus is not zero. Therefore, we *cannot* assign a definite volume to an atom. However, we can visualize an atom with a 90% **probability contour:** the electron is somewhere within that volume 90% of the time (Figure 7.15E).

As you'll see later in this section, each atomic orbital has a distinctive radial probability distribution and 90% probability contour.

Quantum Numbers of an Atomic Orbital

An atomic orbital is specified by three quantum numbers (Table 7.2) that are part of the solution of the Schrödinger equation and indicate the size, shape, and orientation in space of the orbital.*

1. The **principal quantum number (*n*)** *is a positive integer* (1, 2, 3, and so forth). It indicates the relative *size* of the orbital and therefore the relative *distance from the nucleus* of the peak in the radial probability distribution plot. The principal quantum number specifies the *energy level* of the H atom: *the higher the n value, the higher the energy level.* When the electron occupies an orbital with $n = 1$, the H atom is in its ground state and has its lowest energy. When the electron occupies an orbital with $n = 2$ (first excited state), the atom has more energy.

*For ease in discussion, chemists often refer to the size, shape, and orientation of an "atomic orbital," although we really mean the size, shape, and orientation of an "atomic orbital's radial probability distribution."

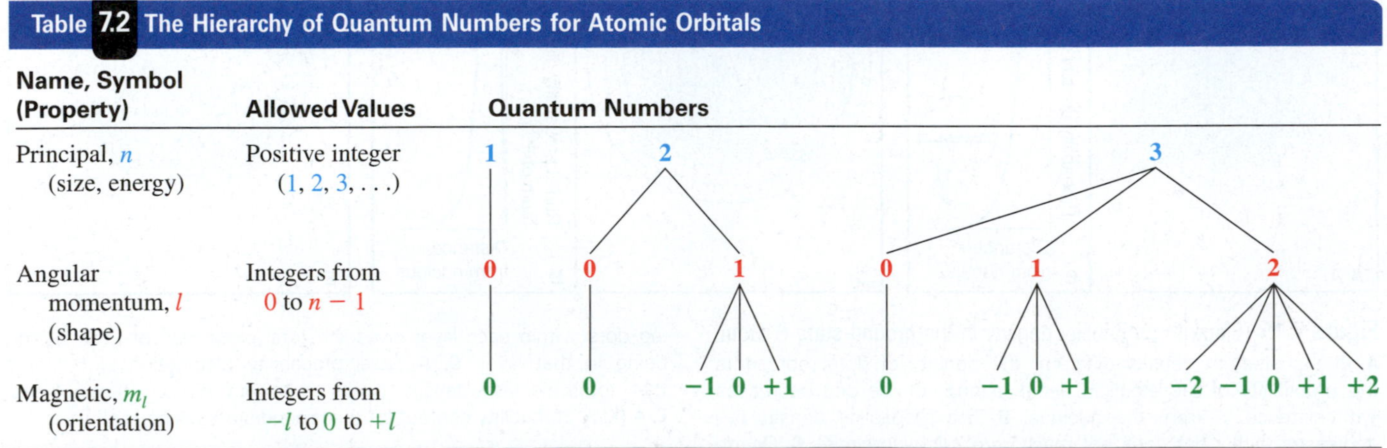

Table 7.2 The Hierarchy of Quantum Numbers for Atomic Orbitals

Name, Symbol (Property)	Allowed Values	Quantum Numbers					
Principal, n (size, energy)	Positive integer (1, 2, 3, . . .)	1	2		3		
Angular momentum, l (shape)	Integers from 0 to $n - 1$	0	0	1	0	1	2
Magnetic, m_l (orientation)	Integers from $-l$ to 0 to $+l$	0	0	−1 0 +1	0	−1 0 +1	−2 −1 0 +1 +2

2. The **angular momentum quantum number (l)** *is an integer from 0 to $n - 1$*. It is related to the *shape* of the orbital. Note that the principal quantum number sets a limit on the angular momentum quantum number: n limits l. For an orbital with $n = 1$, l can have only one value, 0. For orbitals with $n = 2$, l can have two values, 0 or 1. For orbitals with $n = 3$, l can have three values, 0, 1, or 2; and so forth. Thus, the number of possible l values equals the value of n.

3. The **magnetic quantum number (m_l)** *is an integer from $-l$ through 0 to $+l$*. It prescribes the three-dimensional *orientation* of the orbital in the space around the nucleus. The angular momentum quantum number sets a limit on the magnetic quantum number: l limits m_l. An orbital with $l = 0$ can have only $m_l = 0$. However, an orbital with $l = 1$ can have one of three m_l values, -1, 0, or $+1$; that is, there are three possible orbitals with $l = 1$, each with its own orientation. Note that the number of m_l values *equals* $2l + 1$, which is the number of orbitals for a given l. The total number of m_l values, that is, the total number of orbitals, for a given n value is n^2.

Sample Problem 7.5 Determining Quantum Numbers for an Energy Level

Problem What values of the angular momentum (l) and magnetic (m_l) quantum numbers are allowed for a principal quantum number (n) of 3? How many orbitals are allowed?

Plan We determine allowable quantum numbers with the rules from the text: l values are integers from 0 to $n - 1$, and m_l values are integers from $-l$ to 0 to $+l$. One m_l value is assigned to each orbital, so the number of m_l values gives the number of orbitals.

Solution Determining l values: for $n = 3$, $l = 0, 1, 2$.

Determining m_l for each l value:

For $l = 0$, $m_l = 0$

For $l = 1$, $m_l = -1, 0, +1$

For $l = 2$, $m_l = -2, -1, 0, +1, +2$

There are nine m_l values, so there are nine orbitals with $n = 3$.

Check Table 7.2 shows that we are correct. As we saw, the total number of orbitals for a given n value is n^2, and for $n = 3$, $n^2 = 9$.

FOLLOW-UP PROBLEM 7.5 What are the possible l and m_l values for $n = 4$?

Quantum Numbers and Energy Levels

The energy states and orbitals of the atom are described with specific terms and are associated with one or more quantum numbers:

1. *Level.* The atom's energy **levels,** or *shells,* are given by the n value: the smaller the n value, the lower the energy level and the greater the probability that the electron is closer to the nucleus.

2. *Sublevel.* The atom's levels are divided into **sublevels,** or *subshells,* that are given by the l value. Each designates the orbital shape with a letter:

$l = 0$ is an s sublevel.
$l = 1$ is a p sublevel.
$l = 2$ is a d sublevel.
$l = 3$ is an f sublevel.

(The letters derive from names of spectroscopic lines: *s*harp, *p*rincipal, *d*iffuse, and *f*undamental.) Sublevels with l values greater than 3 are designated by consecutive letters after f: g sublevel, h sublevel, and so on. A sublevel is named with its n value and letter designation; for example, the sublevel (subshell) with $n = 2$ and $l = 0$ is called 2s. (We discuss orbital shapes in the next subsection and in Chapter 8.)

3. *Orbital.* Each combination of n, l, and m_l specifies the size (energy), shape, and spatial orientation of one of the atom's orbitals. We know the quantum numbers of the orbitals in a sublevel from the sublevel name and the quantum-number hierarchy. For example, any orbital in the 2*s* sublevel has $n = 2$ and $l = 0$, and given that l value, it can have only $m_l = 0$; thus, the 2*s* sublevel has only one orbital. Any orbital in the 3*p* sublevel has $n = 3$ and $l = 1$, and given that l value, one orbital has $m_l = -1$, another has $m_l = 0$, and a third has $m_l = +1$; thus, the 3*p* sublevel has three orbitals.

Sample Problem 7.6 Determining Sublevel Names and Orbital Quantum Numbers

Problem Give the name, magnetic quantum numbers, and number of orbitals for each sublevel with the given n and l quantum numbers:
(a) $n = 3, l = 2$ **(b)** $n = 2, l = 0$ **(c)** $n = 5, l = 1$ **(d)** $n = 4, l = 3$

Plan We name the sublevel (subshell) with the n value and the letter designation of the l value. From the l value, we find the number of possible m_l values, which equals the number of orbitals in that sublevel.

Solution

n	l	Sublevel Name	Possible m_l Values	No. of Orbitals
(a) 3	2	3*d*	−2, −1, 0, +1, +2	5
(b) 2	0	2*s*	0	1
(c) 5	1	5*p*	−1, 0, +1	3
(d) 4	3	4*f*	−3, −2, −1, 0, +1, +2, +3	7

Check Check the number of orbitals in each sublevel using

$$\text{No. of orbitals} = \text{no. of } m_l \text{ values} = 2l + 1$$

FOLLOW-UP PROBLEM 7.6 What are the n, l, and possible m_l values for the 2*p* and 5*f* sublevels?

Sample Problem 7.7 Identifying Incorrect Quantum Numbers

Problem What is wrong with each of the following quantum number designations and/or sublevel names?

n	l	m_l	Name
(a) 1	1	0	1*p*
(b) 4	3	+1	4*d*
(c) 3	1	−2	3*p*

Solution **(a)** A sublevel with $n = 1$ can have only $l = 0$, not $l = 1$. The only possible sublevel name is 1*s*.
(b) A sublevel with $l = 3$ is an *f* sublevel, not a *d* sublevel. The name should be 4*f*.
(c) A sublevel with $l = 1$ can have only −1, 0, or +1 for m_l, not −2.

Check Check that l is always less than n and that m_l is always $\geq -l$ and $\leq +l$.

FOLLOW-UP PROBLEM 7.7 Supply the missing quantum numbers and sublevel names.

n	l	m_l	Name
(a) ?	?	0	4*p*
(b) 2	1	0	?
(c) 3	2	−2	?
(d) ?	?	?	2*s*

Shapes of Atomic Orbitals

Each sublevel of the H atom consists of a set of orbitals with characteristic shapes. As you'll see in Chapter 8, orbitals for the other atoms have similar shapes.

The *s* Orbital An orbital with $l = 0$ has a *spherical* shape with the nucleus at its center and is called an ***s* orbital.** Because a sphere has only one orientation, an *s* orbital has only one m_l value: for any *s* orbital, $m_l = 0$.

1. *The 1s orbital* holds the electron in the H atom's ground state. *The electron probability density is highest at the nucleus.* Figure 7.16A shows this graphically *(top)*, and an electron density *relief map (inset)* depicts the graph's curve in three dimensions. Note the quarter-section of a three-dimensional electron cloud depiction *(middle)* has the darkest shading at the nucleus. For reasons discussed earlier (see Figure 7.16D), the radial probability distribution plot *(bottom)* has its peak slightly out from the nucleus. Both plots fall off smoothly with distance.

2. *The 2s orbital* (Figure 7.16B) has two regions of higher electron density. The radial probability distribution (Figure 7.16B, *bottom*) of the more distant region is *higher* than that of the closer one because the sum of ψ^2 for it is taken over a much larger volume. Between the two regions is a spherical **node,** where the probability of finding the electron drops to zero ($\psi^2 = 0$ at the node, analogous to zero amplitude

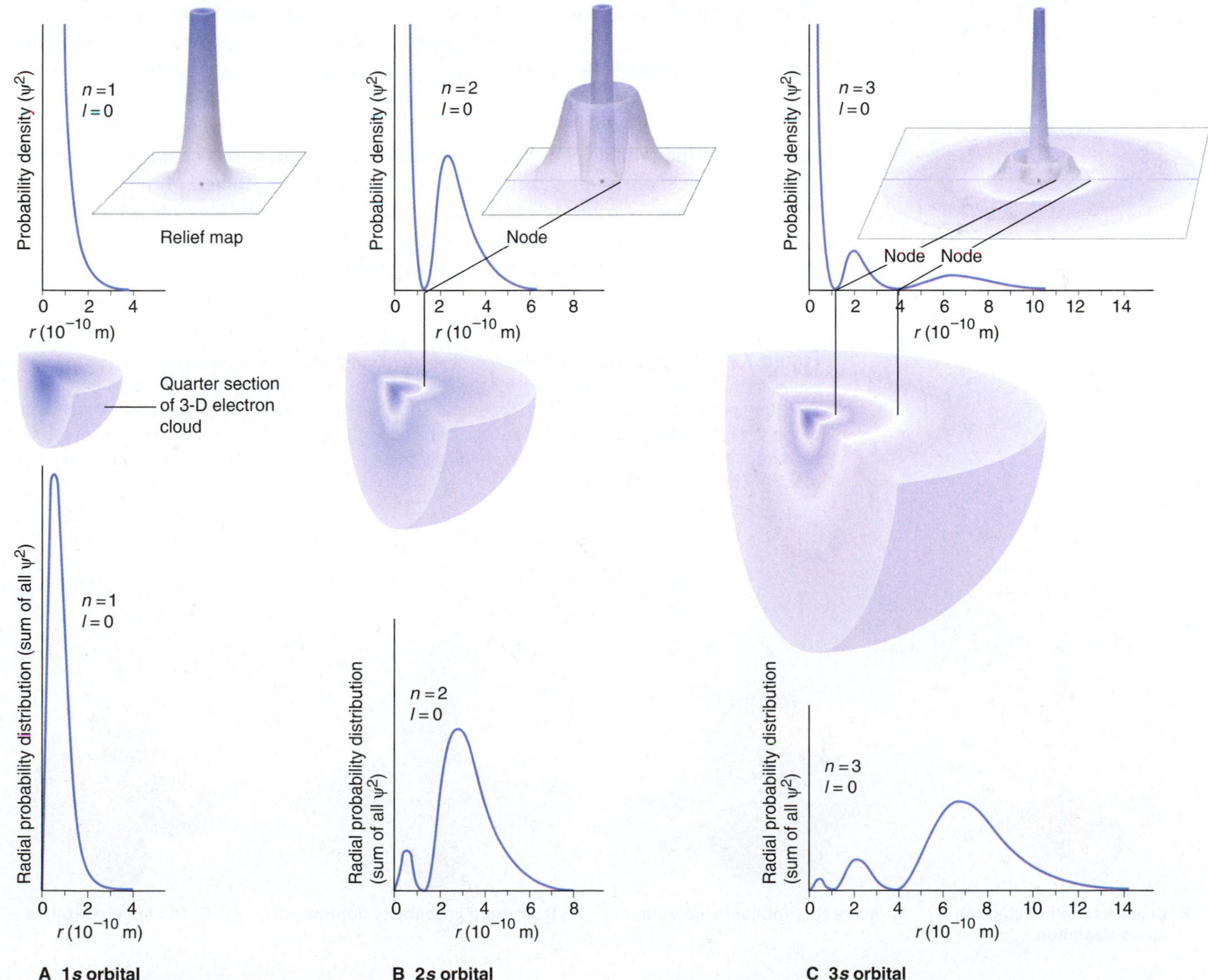

Figure 7.16 The 1*s*, 2*s*, and 3*s* orbitals. For each of the *s* orbitals, a plot of probability density vs. distance (*top*, with the relief map, *inset*, showing the plot in three dimensions) lies above a quarter section of an electron cloud depiction of the 90% probability contour *(middle)*, which lies above a radial probability distribution plot *(bottom)*. **A,** The 1*s* orbital. **B,** The 2*s* orbital. **C,** The 3*s* orbital.

of a wave exactly between the peak and trough). Because the 2*s* orbital is larger than the 1*s*, an electron in the 2*s* spends more time farther from the nucleus (in the larger of the two regions) than it does when it occupies the 1*s*.

3. *The 3s orbital* (Figure 7.16C) has three regions of high electron density and two nodes. Here again, the highest radial probability is at the greatest distance from the nucleus. This pattern of more nodes and higher probability with distance from the nucleus continues with the 4*s*, 5*s*, and so forth.

The *p* Orbital An orbital with $l = 1$ is called a ***p* orbital** and has two regions (lobes) of high probability, one on *either side* of the nucleus (Figure 7.17). The *nucleus lies at the nodal plane* of this dumbbell-shaped orbital. Since the maximum value of l is $n - 1$, only levels with $n = 2$ or higher have a *p* orbital: the lowest energy *p* orbital (the one closest to the nucleus) is the 2*p*. *One p* orbital consists of *two* lobes, and the electron spends *equal* time in both. Similar to the pattern for *s* orbitals, a 3*p* orbital is larger than a 2*p*, a 4*p* is larger than a 3*p*, and so forth.

Unlike *s* orbitals, *p* orbitals *have* different spatial orientations. The three possible m_l values of -1, 0, and $+1$ refer to three *mutually perpendicular* orientations; that is, while identical in size, shape, and energy, the three *p* orbitals differ in orientation. We associate *p* orbitals with the *x*, *y*, and *z* axes: the p_x orbital lies along the *x* axis, the p_y along the *y* axis, and the p_z along the *z* axis. (There is no relationship between a particular axis and a given m_l value.)

The *d* Orbital An orbital with $l = 2$ is called a ***d* orbital.** There are five possible m_l values for $l = 2$: -2, -1, 0, $+1$, and $+2$. Thus, a *d* orbital has any one of five orientations (Figure 7.18). Four of the five *d* orbitals have four lobes (a cloverleaf shape) with two mutually perpendicular nodal planes between them and the nucleus at the junction of the lobes (Figure 7.18C). (The orientation of the nodal planes always lies between the orbital lobes.) Three of these orbitals lie in the *xy*, *xz*, and *yz* planes, with their lobes *between* the axes, and are called the d_{xy}, d_{xz}, and d_{yz} orbitals. A fourth, the $d_{x^2-y^2}$ orbital, also lies in the *xy* plane, but its lobes are *along* the axes. The fifth *d* orbital, the d_{z^2}, has two major lobes *along* the *z* axis, and a donut-shaped region girdles the center. An electron in a *d* orbital spends equal time in all of its lobes.

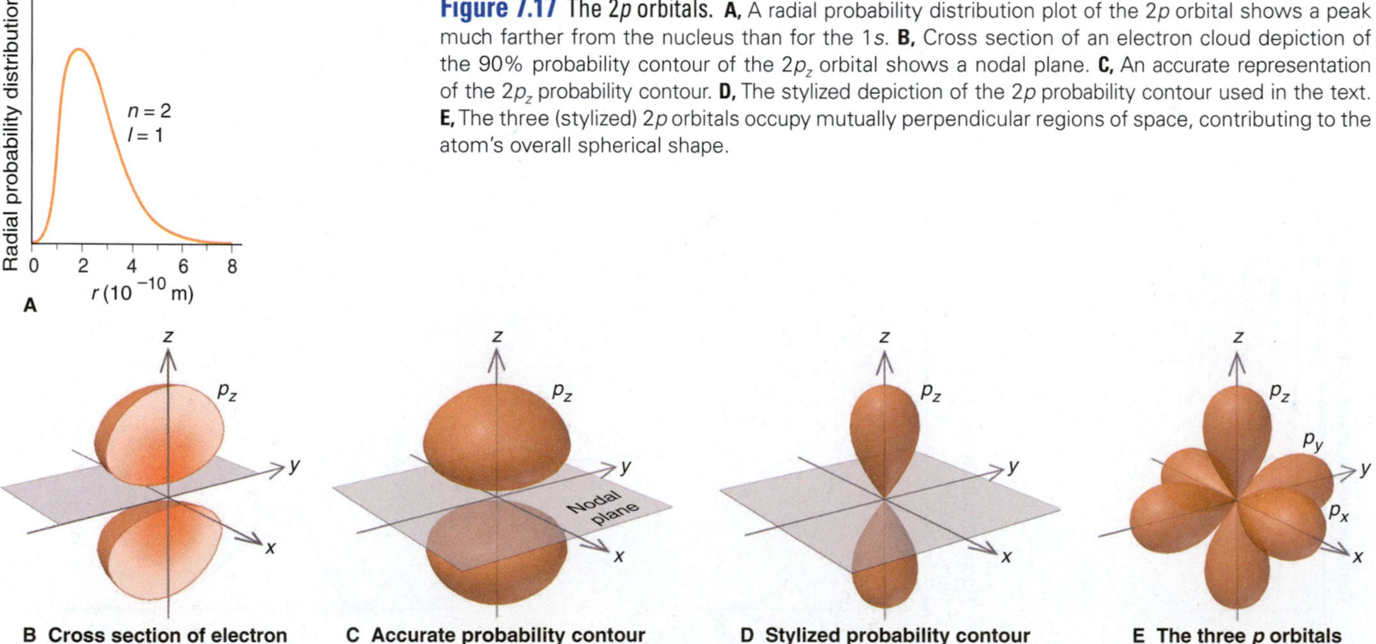

Figure 7.17 The 2*p* orbitals. **A,** A radial probability distribution plot of the 2*p* orbital shows a peak much farther from the nucleus than for the 1*s*. **B,** Cross section of an electron cloud depiction of the 90% probability contour of the $2p_z$ orbital shows a nodal plane. **C,** An accurate representation of the $2p_z$ probability contour. **D,** The stylized depiction of the 2*p* probability contour used in the text. **E,** The three (stylized) 2*p* orbitals occupy mutually perpendicular regions of space, contributing to the atom's overall spherical shape.

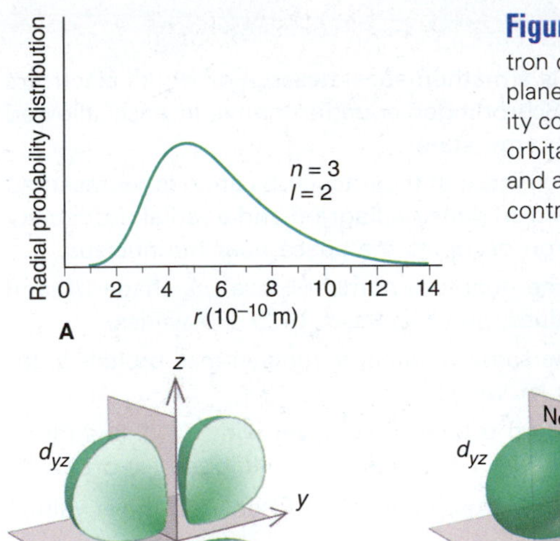

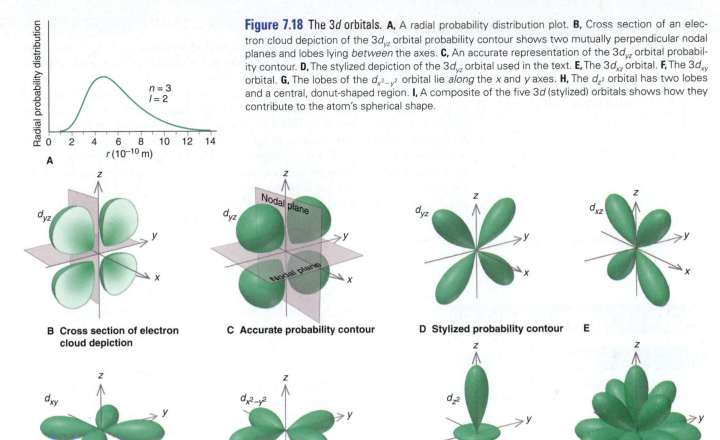

Figure 7.18 **The 3*d* orbitals.** **A,** A radial probability distribution plot. **B,** Cross section of an electron cloud depiction of the $3d_{yz}$ orbital probability contour shows two mutually perpendicular nodal planes and lobes lying *between* the axes. **C,** An accurate representation of the $3d_{yz}$ orbital probability contour. **D,** The stylized depiction of the $3d_{yz}$ orbital used in the text. **E,** The $3d_{xz}$ orbital. **F,** The $3d_{xy}$ orbital. **G,** The lobes of the $d_{x^2-y^2}$ orbital lie *along* the *x* and *y* axes. **H,** The d_{z^2} orbital has two lobes and a central, donut-shaped region. **I,** A composite of the five 3*d* (stylized) orbitals shows how they contribute to the atom's spherical shape.

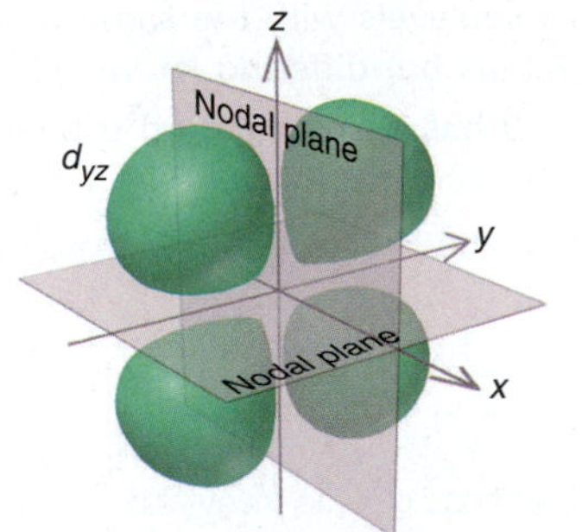

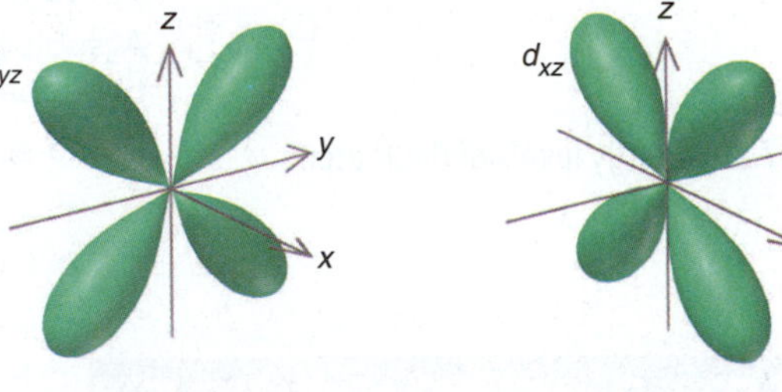

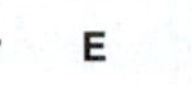

B Cross section of electron cloud depiction

C Accurate probability contour

D Stylized probability contour

E

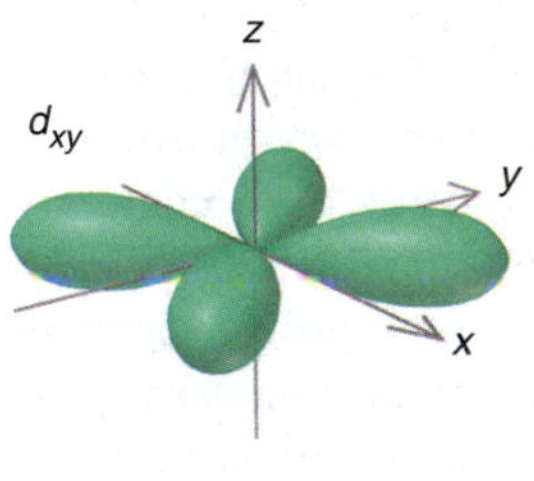

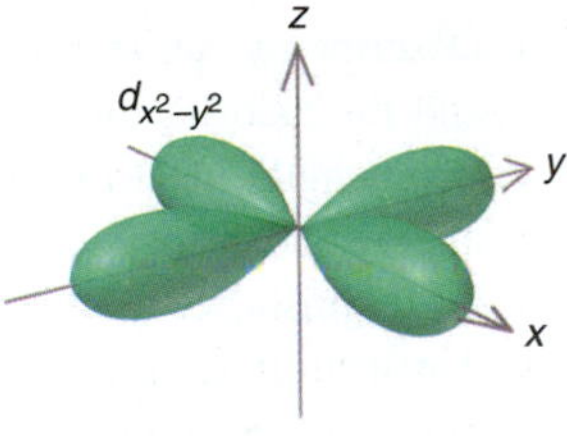

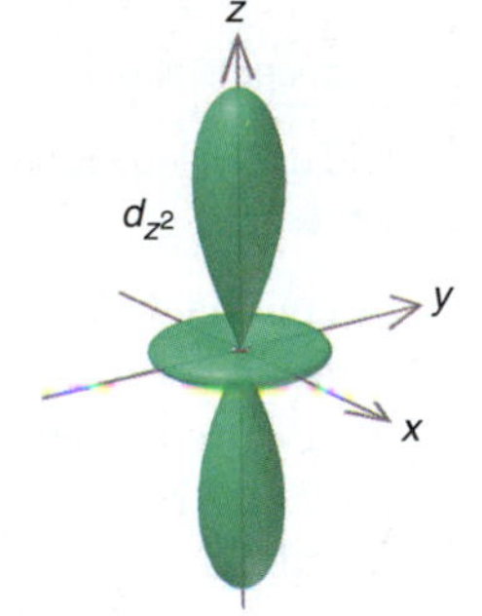

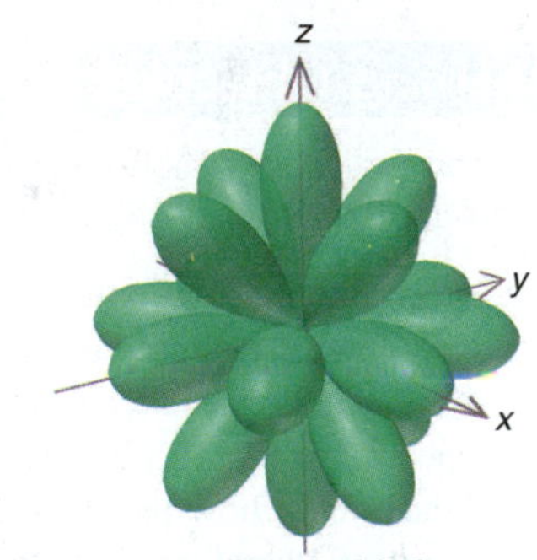

F

G

H

I The five *d* orbitals

In keeping with the quantum-number hierarchy, a *d* orbital ($l = 2$) must have a principal quantum number of $n = 3$ or higher, so 3*d* is the lowest energy *d* sublevel. Orbitals in the 4*d* sublevel are larger (extend farther from the nucleus) than the 3*d*, and the 5*d* are larger still.

Orbitals with Higher *l* Values Orbitals with $l = 3$ are *f* orbitals and have a principal quantum number of at least $n = 4$. Figure 7.19 shows one of the seven *f* orbitals ($2l + 1 = 7$); each *f* orbital has a complex, multilobed shape with several nodal planes. Orbitals with $l = 4$ are *g* orbitals, but they play no known role in chemical bonding.

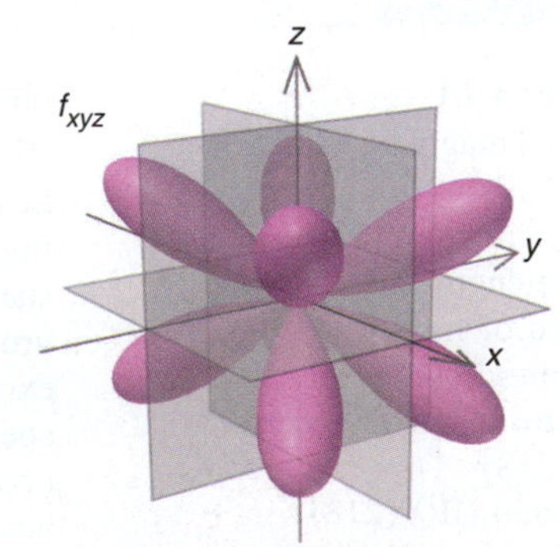

Figure 7.19 The $4f_{xyz}$ orbital, one of the seven 4*f* orbitals.

The Special Case of Energy Levels in the H Atom

With regard to energy levels and sublevels, the H atom is a special case. When an H atom gains energy, its electron occupies an orbital of higher *n* value, which is (on average) farther from the nucleus. But, because it has just one electron, *hydrogen is the only atom whose energy state depends completely on the principal quantum number, n.* As you'll see in Chapter 8, because of additional nucleus-electron attractions and electron-electron repulsions, the energy states of all other atoms depend on the *n and l* values of the occupied orbitals. Thus, *for the H atom only,* all four $n = 2$ orbitals (one 2*s* and three 2*p*) have the same energy, all nine $n = 3$ orbitals (one 3*s*, three 3*p*, and five 3*d*) have the same energy (Figure 7.20, *next page*), and so forth.

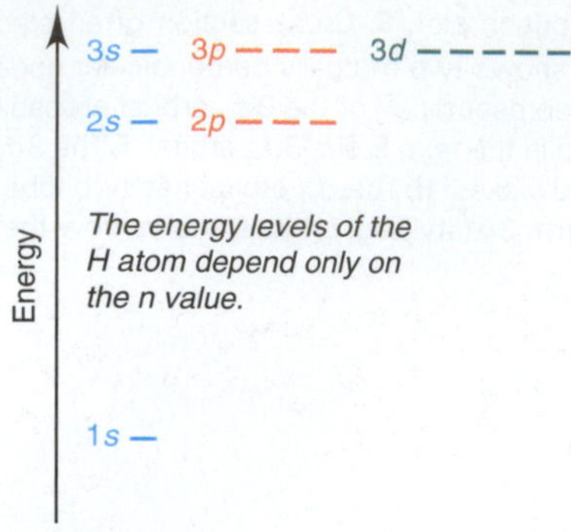

Figure 7.20 Energy levels of the H atom.

Summary of Section 7.4

- The atomic orbital (ψ, wave function) is a mathematical description of the electron's wavelike behavior in an atom. The Schrödinger equation converts each allowed wave function to one of the atom's energy states.
- The probability density of finding the electron at a particular location is represented by ψ^2. For a given energy level, an electron density diagram and a radial probability distribution plot show how the electron occupies the space near the nucleus.
- An atomic orbital is described by three quantum numbers: size (n), shape (l), and orientation (m_l): n limits l to $n - 1$ values, and l limits m_l to $2l + 1$ values.
- An energy level has sublevels with the same n value; a sublevel has orbitals with the same n and l values but differing m_l values.
- A sublevel with $l = 0$ has a spherical (s) orbital; a sublevel with $l = 1$ has three, two-lobed (p) orbitals; and a sublevel with $l = 2$ has five multilobed (d) orbitals.
- In the special case of the H atom, the energy levels depend only on the n value.

CHAPTER REVIEW GUIDE

The following sections provide many aids to help you study this chapter. (Numbers in parentheses refer to pages, unless noted otherwise.)

Learning Objectives

These are concepts and skills to review after studying this chapter.

Related section (§), sample problem (SP), and upcoming end-of-chapter problem (EP) numbers are listed in parentheses.

1. Describe the relationships among frequency, wavelength, and energy of light, and know the meaning of amplitude; have a general understanding of the electromagnetic spectrum (§7.1) (SPs 7.1, 7.2) (EPs 7.1, 7.2, 7.5, 7.7–7.14)
2. Understand how particles and waves differ and how the work of Planck (quantization of energy) and Einstein (photon theory) changed thinking about it (§7.1) (EPs 7.3, 7.4, 7.6)
3. Explain the Bohr model and the importance of discrete atomic energy levels (§7.2) (SP 7.3) (EPs 7.15–7.28)
4. Describe the wave-particle duality of matter and energy and the theories and experiments that revealed it (particle wavelength, electron diffraction, photon momentum, uncertainty principle) (§7.3) (SP 7.4) (EPs 7.29–7.36)
5. Distinguish between ψ (wave function) and ψ^2 (probability density); understand the meaning of electron density diagrams and radial probability distribution plots; describe the hierarchy of quantum numbers, the hierarchy of levels, sublevels, and orbitals, and the shapes and nodes of *s*, *p*, and *d* orbitals; and determine quantum numbers and sublevel designations (§7.4) (SPs 7.5–7.7) (EPs 7.37–7.49)

Key Terms

These important terms appear in boldface in the chapter and are defined again in the Glossary.

Section 7.1
electromagnetic radiation (217)
frequency (ν) (217)
wavelength (λ) (217)
speed of light (*c*) (217)
amplitude (218)
electromagnetic spectrum (218)
infrared (IR) (218)
ultraviolet (UV) (218)
refraction (219)
diffraction (220)
quantum number (221)
Planck's constant (*h*) (221)
quantum (221)
photoelectric effect (221)
photon (221)

Section 7.2
line spectrum (223)
stationary state (224)
ground state (224)
excited state (224)
spectrometry (228)
emission spectrum (228)
absorption spectrum (228)

Section 7.3
de Broglie wavelength (229)
wave-particle duality (231)
uncertainty principle (231)

Section 7.4
quantum mechanics (232)
Schrödinger equation (232)
atomic orbital (wave function) (232)
electron density diagram (233)
electron cloud depiction (233)
radial probability distribution plot (234)
probability contour (234)
principal quantum number (n) (234)
angular momentum quantum number (l) (235)
magnetic quantum number (m_l) (235)
level (shell) (235)
sublevel (subshell) (235)
s orbital (237)
node (237)
p orbital (238)
d orbital (238)

Key Equations and Relationships

Numbered and screened concepts are listed for you to refer to or memorize.

7.1 Relating the speed of light to its frequency and wavelength (217):

$$c = \nu \times \lambda$$

7.2 Determining the smallest change in an atom's energy (221):

$$\Delta E = h\nu$$

7.3 Calculating the wavelength of any line in the H atom spectrum (Rydberg equation) (223):

$$\frac{1}{\lambda} = R\left(\frac{1}{n_1^2} - \frac{1}{n_2^2}\right)$$

where n_1 and n_2 are positive integers and $n_2 > n_1$

7.4 Finding the difference between two energy levels in the H atom (226):

$$\Delta E = E_{\text{final}} - E_{\text{initial}} = -2.18\times10^{-18}\text{ J}\left(\frac{1}{n_{\text{final}}^2} - \frac{1}{n_{\text{initial}}^2}\right)$$

7.5 Calculating the wavelength of any moving particle (de Broglie wavelength) (229):

$$\lambda = \frac{h}{mu}$$

7.6 Finding the uncertainty in position or speed of a particle (Heisenberg's uncertainty principle) (231):

$$\Delta x \cdot m\Delta u \geq \frac{h}{4\pi}$$

BRIEF SOLUTIONS TO FOLLOW-UP PROBLEMS

Compare your own solutions to these calculation steps and answers.

7.1 $\lambda\text{ (nm)} = \dfrac{3.00\times10^8\text{ m/s}}{7.23\times10^{14}\text{ s}^{-1}} \times \dfrac{10^9\text{ nm}}{1\text{ m}} = 415\text{ nm}$

$\lambda\text{ (Å)} = 415\text{ nm} \times \dfrac{10\text{ Å}}{1\text{ nm}} = 4150\text{ Å}$

7.2 (a) UV: $E = hc/\lambda$

$$= \frac{(6.626\times10^{-34}\text{ J·s})(3.00\times10^8\text{ m/s})}{1\times10^{-8}\text{ m}}$$

$$= 2\times10^{-17}\text{ J}$$

(b) Visible: $E = 4\times10^{-19}$ J; (c) IR: $E = 2\times10^{-21}$ J

As λ increases, E decreases.

7.3 (a) With $n_{\text{final}} = 3$ for an IR photon:

$$\Delta E = -2.18\times10^{-18}\text{ J}\left(\frac{1}{n_{\text{final}}^2} - \frac{1}{n_{\text{initial}}^2}\right)$$

$$= -2.18\times10^{-18}\text{ J}\left(\frac{1}{3^2} - \frac{1}{6^2}\right)$$

$$= -2.18\times10^{-18}\text{ J}\left(\frac{1}{9} - \frac{1}{36}\right) = -1.82\times10^{-19}\text{ J}$$

(b) $\lambda = \dfrac{hc}{|\Delta E|} = \dfrac{(6.626\times10^{-34}\text{ J·s})(3.00\times10^8\text{ m/s})}{1.82\times10^{-19}\text{ J}} \times \dfrac{1\text{ Å}}{10^{-10}\text{ m}}$

$= 1.09\times10^4$ Å

7.4 $u = \dfrac{h}{m\lambda} = \dfrac{6.626\times10^{-34}\text{ kg·m}^2\text{/s}}{(9.11\times10^{-31}\text{ kg})\left(100\text{ nm} \times \dfrac{1\text{ m}}{10^9\text{ nm}}\right)}$

$= 7.27\times10^3$ m/s

7.5 $n = 4$, so $l = 0, 1, 2, 3$. In addition to the nine m_l values in Sample Problem 7.5 there are those for $l = 3$:

$$m_l = -3, -2, -1, 0, +1, +2, +3$$

7.6 For 2*p*: $n = 2$, $l = 1$, $m_l = -1, 0, +1$

For 5*f*: $n = 5$, $l = 3$, $m_l = -3, -2, -1, 0, +1, +2, +3$

7.7 (a) $n = 4$, $l = 1$; (b) name is 2*p*; (c) name is 3*d*; (d) $n = 2$, $l = 0$, $m_l = 0$

PROBLEMS

Problems with **colored** numbers are answered in Appendix E. Sections match the text and provide the numbers of relevant sample problems. Bracketed problems are grouped in pairs (indicated by a short rule) that cover the same concept. Comprehensive Problems are based on material from any section or previous chapter.

The Nature of Light

(Sample Problems 7.1 and 7.2)

7.1 In what ways are microwave and ultraviolet radiation the same? In what ways are they different?

7.2 Consider the following types of electromagnetic radiation:

(1) Microwave (2) Ultraviolet (3) Radio waves
(4) Infrared (5) X-ray (6) Visible

(a) Arrange them in order of increasing wavelength.
(b) Arrange them in order of increasing frequency.
(c) Arrange them in order of increasing energy.

7.3 In the 17th century, Newton proposed that light was a stream of particles. The wave-particle debate continued for over 250 years until Planck and Einstein presented their ideas. Give two pieces of evidence for the wave model and two for the particle model.

7.4 What new idea about energy did Planck use to explain blackbody radiation?

7.5 Portions of electromagnetic waves A, B, and C are represented below (not drawn to scale):

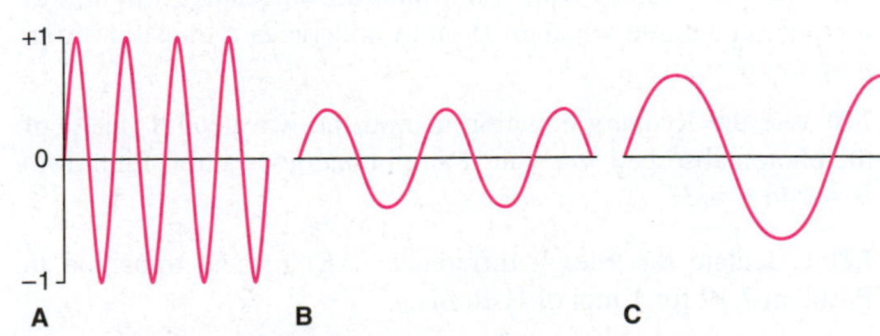

Rank them in order of (a) increasing frequency; (b) increasing energy; (c) increasing amplitude. (d) If wave B just barely fails to cause a current when shining on a metal, is wave A or C more likely to do so? (e) If wave B represents visible radiation, is wave A or C more likely to be IR radiation?

7.6 What new idea about light did Einstein use to explain the photoelectric effect? Why does the photoelectric effect exhibit a threshold frequency but not a time lag?

7.7 An AM station broadcasts rock music at "950 on your radio dial." Units for AM frequencies are given in kilohertz (kHz). Find the wavelength of the station's radio waves in meters (m), nanometers (nm), and angstroms (Å).

7.8 An FM station broadcasts music at 93.5 MHz (megahertz, or 10^6 Hz). Find the wavelength (in m, nm, and Å) of these waves.

7.9 A radio wave has a frequency of 3.8×10^{10} Hz. What is the energy (in J) of one photon of this radiation?

7.10 An x-ray has a wavelength of 1.3 Å. Calculate the energy (in J) of one photon of this radiation.

7.11 Rank these photons in terms of increasing energy: (a) blue ($\lambda = 453$ nm); (b) red ($\lambda = 660$ nm); (c) yellow ($\lambda = 595$ nm).

7.12 Rank these photons in terms of decreasing energy: (a) IR ($\nu = 6.5\times10^{13}\ s^{-1}$); (b) microwave ($\nu = 9.8\times10^{11}\ s^{-1}$); (c) UV ($\nu = 8.0\times10^{15}\ s^{-1}$).

7.13 Cobalt-60 is a radioactive isotope used to treat cancers. A gamma ray emitted by this isotope has an energy of 1.33 MeV (million electron volts; 1 eV $= 1.602\times10^{-19}$ J). What is the frequency (in Hz) and the wavelength (in m) of this gamma ray?

7.14 (a) Ozone formation in the upper atmosphere starts when oxygen molecules absorb UV radiation of wavelengths ≤ 242 nm. Find the frequency and energy of the least energetic of these photons. (b) Ozone absorbs radiation of wavelengths 2200–2900 Å, thus protecting organisms from this radiation. Find the frequency and energy of the most energetic of these photons.

Atomic Spectra

(Sample Problem 7.3)

7.15 How is n_1 in the Rydberg equation (Equation 7.3) related to the quantum number n in the Bohr model?

7.16 Distinguish between an absorption spectrum and an emission spectrum. With which did Bohr work?

7.17 Which of these electron transitions correspond to absorption of energy and which to emission?
(a) $n = 2$ to $n = 4$ (b) $n = 3$ to $n = 1$
(c) $n = 5$ to $n = 2$ (d) $n = 3$ to $n = 4$

7.18 Why couldn't the Bohr model predict spectra for atoms other than hydrogen?

7.19 Use the Rydberg equation to find the wavelength (in nm) of the photon emitted when an H atom undergoes a transition from $n = 5$ to $n = 2$.

7.20 Use the Rydberg equation to find the wavelength (in Å) of the photon absorbed when an H atom undergoes a transition from $n = 1$ to $n = 3$.

7.21 Calculate the energy difference (ΔE) for the transition in Problem 7.19 for 1 mol of H atoms.

7.22 Calculate the energy difference (ΔE) for the transition in Problem 7.20 for 1 mol of H atoms.

7.23 Arrange the following H atom electron transitions in order of *increasing* frequency of the photon absorbed or emitted:
(a) $n = 2$ to $n = 4$ (b) $n = 2$ to $n = 1$
(c) $n = 2$ to $n = 5$ (d) $n = 4$ to $n = 3$

7.24 Arrange the following H atom electron transitions in order of *decreasing* wavelength of the photon absorbed or emitted:
(a) $n = 2$ to $n = \infty$ (b) $n = 4$ to $n = 20$
(c) $n = 3$ to $n = 10$ (d) $n = 2$ to $n = 1$

7.25 The electron in a ground-state H atom absorbs a photon of wavelength 97.20 nm. To what energy level does it move?

7.26 An electron in the $n = 5$ level of an H atom emits a photon of wavelength 1281 nm. To what energy level does it move?

7.27 In addition to continuous radiation, fluorescent lamps emit some visible lines from mercury. A prominent line has a wavelength of 436 nm. What is the energy (in J) of one photon of it?

7.28 A Bohr-model representation of the H atom is shown below with six electron transitions depicted by arrows:

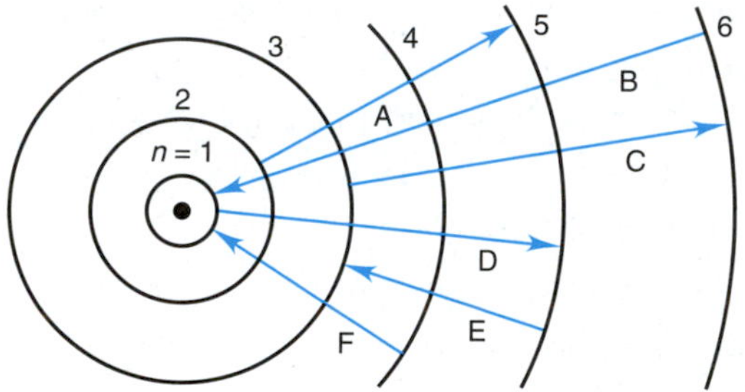

(a) Which transitions are absorptions and which are emissions?
(b) Rank the emissions in terms of increasing energy.
(c) Rank the absorptions in terms of increasing wavelength of light absorbed.

The Wave-Particle Duality of Matter and Energy

(Sample Problem 7.4)

7.29 If particles have wavelike motion, why don't we observe that motion in the macroscopic world?

7.30 Why can't we overcome the uncertainty predicted by Heisenberg's principle by building more precise instruments to reduce the error in measurements below the $h/4\pi$ limit?

7.31 A 232-lb fullback runs 40 yd at 19.8 ± 0.1 mi/h.
(a) What is his de Broglie wavelength (in meters)?
(b) What is the uncertainty in his position?

7.32 An alpha particle (mass $= 6.6\times10^{-24}$ g) emitted by a radium isotope travels at $3.4\times10^7 \pm 0.1\times10^7$ mi/h.
(a) What is its de Broglie wavelength (in meters)?
(b) What is the uncertainty in its position?

7.33 How fast must a 56.5-g tennis ball travel to have a de Broglie wavelength equal to that of a photon of green light (5400 Å)?

7.34 How fast must a 142-g baseball travel to have a de Broglie wavelength equal to that of an x-ray photon with $\lambda = 100.$ pm?

7.35 A sodium flame has a characteristic yellow color due to emissions of wavelength 589 nm. What is the mass equivalence of one photon of this wavelength ($1\ J = 1\ kg{\cdot}m^2/s^2$)?

7.36 A lithium flame has a characteristic red color due to emissions of wavelength 671 nm. What is the mass equivalence of 1 mol of photons of this wavelength (1 J = 1 kg·m^2/s^2)?

The Quantum-Mechanical Model of the Atom

(Sample Problems 7.5 to 7.7)

7.37 What physical meaning is attributed to ψ^2?

7.38 What does "electron density in a tiny volume of space" mean?

7.39 What feature of an orbital is related to each of the following?
(a) Principal quantum number (n)
(b) Angular momentum quantum number (l)
(c) Magnetic quantum number (m_l)

7.40 How many orbitals in an atom can have each of the following designations: (a) 1*s;* (b) 4*d;* (c) 3*p;* (d) $n = 3$?

7.41 How many orbitals in an atom can have each of the following designations: (a) 5*f;* (b) 4*p;* (c) 5*d;* (d) $n = 2$?

7.42 Give all possible m_l values for orbitals that have each of the following: (a) $l = 2$; (b) $n = 1$; (c) $n = 4$, $l = 3$.

7.43 Give all possible m_l values for orbitals that have each of the following: (a) $l = 3$; (b) $n = 2$; (c) $n = 6$, $l = 1$.

7.44 For each of the following, give the sublevel designation, the allowable m_l values, and the number of orbitals:
(a) $n = 4$, $l = 2$ (b) $n = 5$, $l = 1$ (c) $n = 6$, $l = 3$

7.45 For each of the following, give the sublevel designation, the allowable m_l values, and the number of orbitals:
(a) $n = 2$, $l = 0$ (b) $n = 3$, $l = 2$ (c) $n = 5$, $l = 1$

7.46 For each of the following sublevels, give the n and l values and the number of orbitals: (a) 5*s;* (b) 3*p;* (c) 4*f.*

7.47 For each of the following sublevels, give the n and l values and the number of orbitals: (a) 6*g;* (b) 4*s;* (c) 3*d.*

7.48 Are the following combinations allowed? If not, show two ways to correct them:
(a) $n = 2$; $l = 0$; $m_l = -1$ (b) $n = 4$; $l = 3$; $m_l = -1$
(c) $n = 3$; $l = 1$; $m_l = 0$ (d) $n = 5$; $l = 2$; $m_l = +3$

7.49 Are the following combinations allowed? If not, show two ways to correct them:
(a) $n = 1$; $l = 0$; $m_l = 0$ (b) $n = 2$; $l = 2$; $m_l = +1$
(c) $n = 7$; $l = 1$; $m_l = +2$ (d) $n = 3$; $l = 1$; $m_l = -2$

Comprehensive Problems

7.50 The photoelectric effect is illustrated in a plot of the kinetic energies of electrons ejected from the surface of potassium metal or silver metal at different frequencies of incident light.

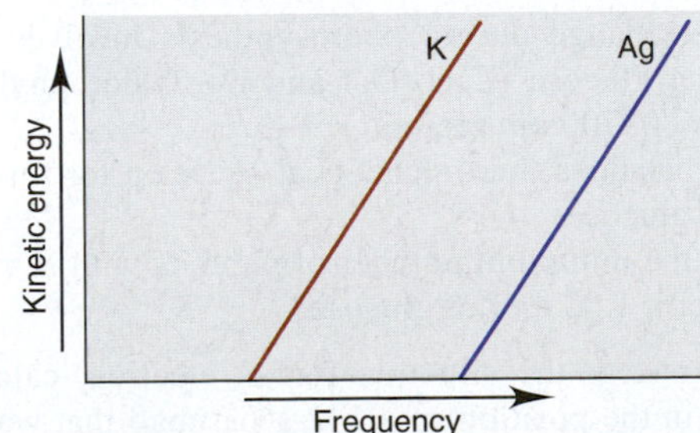

(a) Why don't the lines begin at the origin? (b) Why don't the lines begin at the same point? (c) From which metal will light of shorter wavelength eject an electron? (d) Why are the slopes equal?

7.51 A minimum of 2.0×10^{-17} J of energy is needed to trigger a series of impulses in the optic nerve that eventually reach the brain. (a) How many photons of red light (700. nm) are needed? (b) How many photons of blue light (475 nm)?

7.52 One reason carbon monoxide (CO) is toxic is that it binds to the blood protein hemoglobin more strongly than oxygen does. The bond between hemoglobin and CO absorbs radiation of 1953 cm^{-1}. (The units are the reciprocal of the wavelength in centimeters.) Calculate the wavelength (in nm and Å) and the frequency (in Hz) of the absorbed radiation.

7.53 A metal ion M^{n+} has a single electron. The highest energy line in its emission spectrum has a frequency of 2.961×10^{16} Hz. Identify the ion.

7.54 TV and radio stations transmit in specific frequency bands of the radio region of the electromagnetic spectrum.
(a) TV channels 2 to 13 (VHF) broadcast signals between the frequencies of 59.5 and 215.8 MHz, whereas FM radio stations broadcast signals with wavelengths between 2.78 and 3.41 m. Do these bands of signals overlap?
(b) AM radio signals have frequencies between 550 and 1600 kHz. Which has a broader transmission band, AM or FM?

7.55 In his explanation of the threshold frequency in the photoelectric effect, Einstein reasoned that the absorbed photon must have a minimum energy to dislodge an electron from the metal surface. This energy is called the *work function* (ϕ) of that metal. What is the longest wavelength of radiation (in nm) that could cause the photoelectric effect in each of these metals: (a) calcium, $\phi = 4.60\times10^{-19}$ J; (b) titanium, $\phi = 6.94\times10^{-19}$ J; (c) sodium, $\phi = 4.41\times10^{-19}$ J?

7.56 You have three metal samples—A, B, and C—that are tantalum (Ta), barium (Ba), and tungsten (W), but you don't know which is which. Metal A emits electrons in response to visible light; metals B and C require UV light. (a) Identify metal A, and find the longest wavelength that removes an electron. (b) What range of wavelengths would distinguish B and C? [The work functions are Ta (6.81×10^{-19} J), Ba (4.30×10^{-19} J), and W (7.16×10^{-19} J); work function is explained in Problem 7.55.]

7.57 A laser (*l*ight *a*mplification by *s*timulated *e*mission of *r*adiation) provides nearly monochromatic high-intensity light. Lasers are used in eye surgery, CD/DVD players, basic research, and many other areas. Some dye lasers can be "tuned" to emit a desired wavelength. Fill in the blanks in the following table of the properties of some common lasers:

Type	λ (nm)	ν (s^{-1})	*E* (J)	Color
He-Ne	632.8	?	?	?
Ar	?	6.148×10^{14}	?	?
Ar-Kr	?	?	3.499×10^{-19}	?
Dye	663.7	?	?	?

7.58 As space exploration increases, means of communication with humans and probes on other planets are being developed.
(a) How much time (in s) does it take for a radio wave of frequency 8.93×10^7 s^{-1} to reach Mars, which is 8.1×10^7 km from Earth?
(b) If it takes this radiation 1.2 s to reach the Moon, how far (in m) is the Moon from Earth?

7.59 A ground-state H atom absorbs a photon of wavelength 94.91 nm, and its electron attains a higher energy level. The atom then emits two photons: one of wavelength 1281 nm to reach an

intermediate energy level, and a second to return to the ground state.
(a) What higher level did the electron reach?
(b) What intermediate level did the electron reach?
(c) What was the wavelength of the second photon emitted?

7.60 Use the relative size of the 3*s* orbital below to answer the following questions about orbitals A–D.

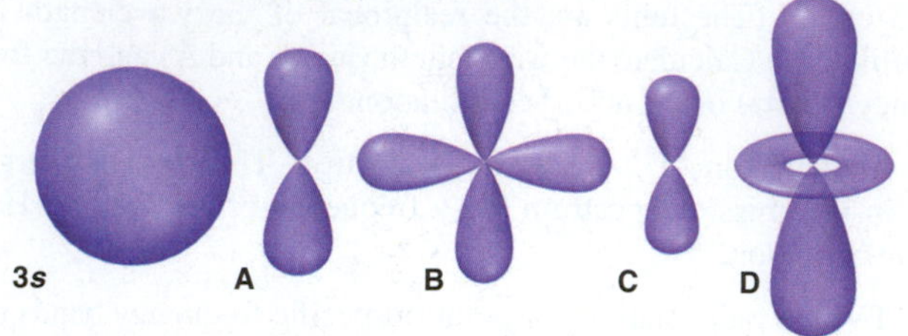

(a) Which orbital has the highest value of *n*? (b) Which orbital(s) have a value of $l = 1$? $l = 2$? (c) How many other orbitals with the same value of *n* have the same shape as orbital B? Orbital C? (d) Which orbital has the highest energy? Lowest energy?

7.61 Why do the spaces between spectral lines within a series decrease as wavelength becomes shorter?

7.62 Enormous numbers of microwave photons are needed to warm macroscopic samples of matter. A portion of soup containing 252 g of water is heated in a microwave oven from 20.°C to 98°C, with radiation of wavelength 1.55×10^{-2} m. How many photons are absorbed by the water in the soup?

7.63 The quantum-mechanical treatment of the hydrogen atom gives this expression for the wave function, ψ, of the 1*s* orbital:

$$\psi = \frac{1}{\sqrt{\pi}}\left(\frac{1}{a_0}\right)^{3/2} e^{-r/a_0}$$

where *r* is the distance from the nucleus and a_0 is 52.92 pm. The probability of finding the electron in a tiny volume at distance *r* from the nucleus is proportional to ψ^2. The total probability of finding the electron at all points at distance *r* from the nucleus is proportional to $4\pi r^2\psi^2$. Calculate the values (to three significant figures) of ψ, ψ^2, and $4\pi r^2\psi^2$ to fill in the following table and sketch a plot of each set of values versus *r*. Compare the latter two plots with those in Figure 7.16A (p. 237):

r (pm)	ψ (pm$^{-3/2}$)	ψ^2 (pm^{-3})	$4\pi r^2\psi^2$ (pm^{-1})
0			
50			
100			
200			

7.64 Photoelectron spectroscopy applies the principle of the photoelectric effect to study orbital energies of atoms and molecules. High-energy radiation (usually UV or x-ray) is absorbed by a sample and an electron is ejected. The orbital energy can be calculated from the known energy of the radiation and the measured energy of the electron lost. The following energy differences were determined for several electron transitions:

$$\Delta E_{2\longrightarrow1} = 4.098\times10^{-17}\text{ J} \qquad \Delta E_{3\longrightarrow1} = 4.854\times10^{-17}\text{ J}$$
$$\Delta E_{5\longrightarrow1} = 5.242\times10^{-17}\text{ J} \qquad \Delta E_{4\longrightarrow2} = 1.024\times10^{-17}\text{ J}$$

Calculate ΔE and λ of a photon emitted in the following transitions: (a) level 3 $\longrightarrow$ 2; (b) level 4 $\longrightarrow$ 1; (c) level 5 $\longrightarrow$ 4.

7.65 For any microscope, the smallest object observable is one-half the wavelength of the radiation used. For example, the smallest object observable with light of 400 nm is 2×10^{-7} m. (a) What is the smallest object observable with an electron microscope using electrons moving at 5.5×10^4 m/s? (b) At 3.0×10^7 m/s?

7.66 In fireworks, the heat of the reaction of an oxidizing agent, such as $KClO_4$, with an organic compound excites certain salts, which emit specific colors. Strontium salts have an intense emission at 641 nm, and barium salts have one at 493 nm. (a) What colors do these emissions produce? (b) What is the energy (in kJ) of these emissions for 5.00 g each of the chloride salts of Sr and Ba? (Assume that all the heat produced is converted to emitted light.)

7.67 Atomic hydrogen produces several series of spectral lines. Each series fits the Rydberg equation with its own particular n_1 value. Calculate the value of n_1 (by trial and error if necessary) that would produce a series of lines in which:
(a) The *highest* energy line has a wavelength of 3282 nm.
(b) The *lowest* energy line has a wavelength of 7460 nm.

7.68 Fish-liver oil is a good source of vitamin A, whose concentration is measured spectrometrically at a wavelength of 329 nm.
(a) Suggest a reason for using this wavelength.
(b) In what region of the spectrum does this wavelength lie?
(c) When 0.1232 g of fish-liver oil is dissolved in 500. mL of solvent, the absorbance is 0.724 units. When 1.67×10^{-3} g of vitamin A is dissolved in 250. mL of solvent, the absorbance is 1.018 units. Calculate the vitamin A concentration in the fish-liver oil.

7.69 Many calculators use photocells as their energy source. Find the maximum wavelength needed to remove an electron from silver ($\phi = 7.59\times10^{-19}$ J). Is silver a good choice for a photocell that uses visible light? [The concept of the work function (ϕ) is explained in Problem 7.55.]

7.70 In a game of "Clue," Ms. White is killed in the conservatory. A spectrometer in each room records who is present to help find the murderer. For example, if someone wearing yellow is in a room, light at 580 nm is reflected. The suspects are Col. Mustard, Prof. Plum, Mr. Green, Ms. Peacock (blue), and Ms. Scarlet. At the time of the murder, the spectrometer in the dining room shows a reflection at 520 nm, those in the lounge and study record lower frequencies, and the one in the library records the shortest possible wavelength. Who killed Ms. White? Explain.

7.71 Technetium (Tc; $Z = 43$) is a synthetic element used as a radioactive tracer in medical studies. A Tc atom emits a beta particle (electron) with a kinetic energy (E_k) of 4.71×10^{-15} J. What is the de Broglie wavelength of this electron ($E_k = \frac{1}{2}mv^2$)?

7.72 Electric power is measured in watts (1 W = 1 J/s). About 95% of the power output of an incandescent bulb is converted to heat and 5% to light. If 10% of that light shines on your chemistry text, how many photons per second shine on the book from a 75-W bulb? (Assume the photons have a wavelength of 550 nm.)

7.73 The net change during photosynthesis involves CO_2 and H_2O forming glucose ($C_6H_{12}O_6$) and O_2. Chlorophyll absorbs light in the 600–700 nm region.
(a) Write a balanced thermochemical equation for formation of 1.00 mol of glucose.
(b) What is the minimum number of photons with $\lambda = 680.$ nm needed to form 1.00 mol of glucose?

7.74 In contrast to the situation for an electron, calculate the uncertainty in the position of a 142-g baseball that was pitched in 2010 by a Cincinnati Reds reliever and traveled at 100.0 mi/h $\pm$ 1.00%.

Electron Configuration and Chemical Periodicity

Expanding Shells of Electrons A nautilus's shell, with its recurring and increasing chambers, is an analogy for the increasing numbers of electrons in the energy levels of an atom. Recurring patterns of electrons within atoms correlate with recurring chemical behavior of the elements.

Outline

Key Principles
to focus on while studying this chapter

- Arranging the elements by atomic number reveals a periodic recurrence of similar properties (*periodic law*). **(Introduction)**
- The *electron configuration* of an atom, the distribution of electrons into its energy levels and sublevels, ultimately determines the behavior of the element. **(Section 8.1)**
- There are three new characteristics required to explain the behavior of atoms with more than one electron (all elements except hydrogen): a fourth quantum number (m_s) specifies *electron spin;* an orbital holds no more than two electrons (*exclusion principle*); and interactions between electrons and between nucleus and electrons (*shielding* and *penetration*) cause energy levels to split into sublevels of different energy. **(Section 8.1)**
- The periodic table is "*built up*" by adding one electron (and one proton and one or more neutrons) to each preceding atom. This method results in vertical groups of elements having *identical outer electron configurations* and, thus, similar behavior. The few seemingly *anomalous configurations* are explained by factors that affect the order of sublevel energies. **(Section 8.2)**
- Three atomic properties—*atomic size, ionization energy* (energy involved in removing an electron from an atom), and *electron affinity* (energy involved in adding an electron to an atom)—exhibit recurring trends throughout the periodic table. **(Section 8.3)**
- These atomic properties have a profound effect on many macroscopic properties, including *metallic behavior, acid-base behavior of oxides, ionic behavior,* and *magnetic behavior* of the elements and their compounds. **(Section 8.4)**

CONCEPTS & SKILLS TO REVIEW before studying this chapter

- format of the periodic table (Section 2.6)
- characteristics of metals and nonmetals (Section 2.6)
- attractions, repulsions, and Coulomb's law (Section 2.7)
- characteristics of acids and bases (Section 4.4)
- rules for assigning quantum numbers (Section 7.4)

We often see recurring patterns in nature—day and night, the seasons, solar and lunar eclipses, the rhythm of a heartbeat. Elements exhibit recurring patterns in properties because their atoms do.

The outpouring of scientific creativity by early 20th-century physicists that led to the new quantum-mechanical model of the atom was preceded by countless hours of laboratory work by 19th-century chemists who were exploring the nature of electrolytes, the kinetic-molecular theory, and chemical thermodynamics. Condensed from all these efforts, an enormous body of facts about the elements became organized into the periodic table:

- *The original periodic table.* In 1870, the Russian chemist Dmitri Mendeleev arranged the 65 elements known at the time into a table and summarized their behavior in the **periodic law:** when arranged by *atomic mass,* the elements exhibit a periodic recurrence of similar properties. Mendeleev left blank spaces in his table and was even able to *predict* the properties of several elements, for example, germanium, that were not discovered until later.
- *The modern periodic table.* Today's table *(inside front cover)* includes 52 elements not known in 1870 and, most importantly, arranges the elements by *atomic number* (number of protons) *not* atomic mass.

The great test for the new atomic model, which relates directly to the order of elements in the table, was to answer a central question: *why* do the elements behave as they do? Or, rephrasing to fit the main topic of this chapter, how does the **electron configuration** of an element—*the distribution of electrons within the levels and sublevels of its atoms*—relate to its chemical and physical properties?

8.1 • CHARACTERISTICS OF MANY-ELECTRON ATOMS

The Schrödinger equation (introduced in Chapter 7) does not give *exact* solutions for the energy levels of *many-electron atoms,* those with more than one electron—that is, all atoms except hydrogen. However, unlike the Bohr model, it gives excellent *approximate* solutions. Three additional features become important in many-electron atoms: (1) a fourth quantum number, (2) the number of electrons that can occupy an orbital, and (3) a splitting of energy levels into sublevels.

The Electron-Spin Quantum Number

The three quantum numbers n, l, and m_l describe the size (energy), shape, and orientation, respectively, of an atomic orbital. An additional quantum number describes a property called *spin,* which is a property of the electron and not the orbital.

When a beam of atoms that have one or more lone electrons passes through a nonuniform magnetic field (created by magnet faces with different shapes), it splits into two beams; Figure 8.1 shows this for a beam of H atoms. Each electron behaves like a

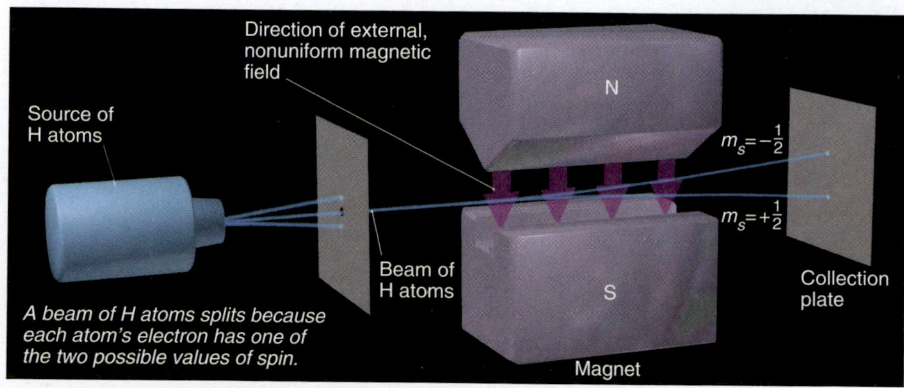

Figure 8.1 The effect of electron spin.

spinning charge and generates a tiny magnetic field, which can have one of two values of *spin*. The two electron fields have opposing directions, so half of the electrons are *attracted* by the large external magnetic field while the other half are *repelled* by it.

Corresponding to the two directions of the electron's field, the **spin quantum number (m_s)** has two possible values, $+\frac{1}{2}$ or $-\frac{1}{2}$. Thus, *each electron in an atom is described completely by a set of* ***four*** *quantum numbers: the first three describe its orbital, and the fourth describes its spin.* The quantum numbers are summarized in Table 8.1.

Now we can write a set of four quantum numbers for any electron in the ground state of any atom. For example, the set of quantum numbers for the lone electron in hydrogen (H; $Z = 1$) is $n = 1$, $l = 0$, $m_l = 0$, and $m_s = +\frac{1}{2}$. (The spin quantum number could just as well have been $-\frac{1}{2}$, but by convention, we assign $+\frac{1}{2}$ to the first electron in an orbital.)

Table 8.1 Summary of Quantum Numbers of Electrons in Atoms

Name	Symbol	Permitted Values	Property
Principal	n	Positive integers (1, 2, 3, . . .)	Orbital energy (size)
Angular momentum	l	Integers from 0 to $n - 1$	Orbital shape (the l values 0, 1, 2, and 3 correspond to *s*, *p*, *d*, and *f* orbitals, respectively)
Magnetic	m_l	Integers from $-l$ to 0 to $+l$	Orbital orientation
Spin	m_s	$+\frac{1}{2}$ or $-\frac{1}{2}$	Direction of e^- spin

The Exclusion Principle and Orbital Occupancy

The element after hydrogen is helium (He; $Z = 2$), the first with atoms having more than one electron. The first electron in the He ground state has the same set of quantum numbers as the electron in the H atom, but the second He electron does not. Based on observations of excited states, the Austrian physicist Wolfgang Pauli formulated the **exclusion principle:** *no two electrons in the same atom can have the same four quantum numbers.* Therefore, the second He electron occupies the same orbital as the first but has an opposite spin: $n = 1$, $l = 0$, $m_l = 0$, and $m_s = -\frac{1}{2}$.

THINK OF IT THIS WAY
Baseball Quantum Numbers

The unique set of quantum numbers that describes an electron is analogous to the unique location of a box seat at a baseball game. The stadium (atom) is divided into section (n, level), box (l, sublevel), row (m_l, orbital), and seat (m_s, spin). Only one person (electron) can have this particular set of stadium "quantum numbers."

The major consequence of the exclusion principle involves orbital occupancy: *an atomic orbital can hold a maximum of two electrons and they must have opposing spins.* The 1*s* orbital in He is *filled,* and the electrons have *paired spins.* Thus, a beam of He atoms is not split in an experiment like that in Figure 8.1.

Electrostatic Effects and Energy-Level Splitting

Electrostatic effects—attraction of opposite charges and repulsion of like charges—play a major role in determining the energy states of many-electron atoms. Unlike the H atom, in which there is only the attraction between nucleus and electron and the energy state is determined *only* by the n value, the energy states of many-electron atoms are also affected by electron-electron repulsions. You'll see shortly how these additional interactions give rise to *the splitting of energy levels into sublevels of differing energies: the energy of an orbital in a many-electron atom depends mostly on its n value (size) and to a lesser extent on its l value (shape).*

Our first encounter with energy-level splitting occurs with lithium (Li; $Z = 3$). The first two electrons of Li fill its 1*s* orbital, so the third Li electron must go into the $n = 2$ level. But, this level has 2*s and* 2*p* sublevels: which does the third electron enter? For reasons we discuss below, the 2*s* is lower in energy than the 2*p*, so the ground state of Li has its third electron in the 2*s*.

The energy differences among sublevels arise from three factors—*nuclear attraction, electron repulsions,* and *orbital shape* (i.e., radial probability distribution). Their interplay leads to two phenomena—*shielding* and *penetration*—that occur in all atoms *except* hydrogen. [In the following discussion, keep in mind that more energy is needed to remove an electron from a more stable (lower energy) sublevel than from a less stable (higher energy) sublevel.]

The Effect of Nuclear Charge (*Z*) on Sublevel Energy Higher charges interact more strongly than lower charges (Coulomb's law, Section 2.7). Therefore, *a higher nuclear charge increases nucleus-electron attractions and, thus, lowers sublevel energy (stabilizes the atom).* We see this effect by comparing the 1*s* sublevel energies of three species with one electron—H atom ($Z = 1$), He^+ ion ($Z = 2$), and Li^{2+} ion ($Z = 3$): the 1*s* sublevel in H is the least stable (highest energy), and the 1*s* sublevel in Li^{2+} is the most stable.

Shielding: The Effect of Electron Repulsions on Sublevel Energy In many-electron atoms, each electron "feels" not only the attraction to the nucleus but also repulsions from other electrons. Repulsions counteract the nuclear attraction somewhat, making each electron easier to remove by, in effect, helping to push it away. We speak of each electron "shielding" the other electrons to some extent from the nuclear charge. **Shielding** (also called *screening*) reduces the full nuclear charge to an **effective nuclear charge (Z_{eff})**, the nuclear charge an electron *actually experiences, and this lower nuclear charge makes the electron easier to remove.*

1. *Shielding by other electrons in a given energy level.* Electrons in the *same* energy level shield each other somewhat. Compare the He atom and He^+ ion: both have a 2+ nuclear charge, but He has two electrons in the 1*s* sublevel and He^+ has only one. It takes less than half as much energy to remove an electron from He (2372 kJ/mol) than from He^+ (5250 kJ/mol) because the second electron in He repels the first, in effect causing a lower Z_{eff}.

2. *Shielding by electrons in inner energy levels.* Because inner electrons spend nearly all their time *between* the outer electrons and the nucleus, they shield outer electrons *much more effectively* than do electrons in the same level. *Shielding by inner electrons greatly lowers Z_{eff} for outer electrons.*

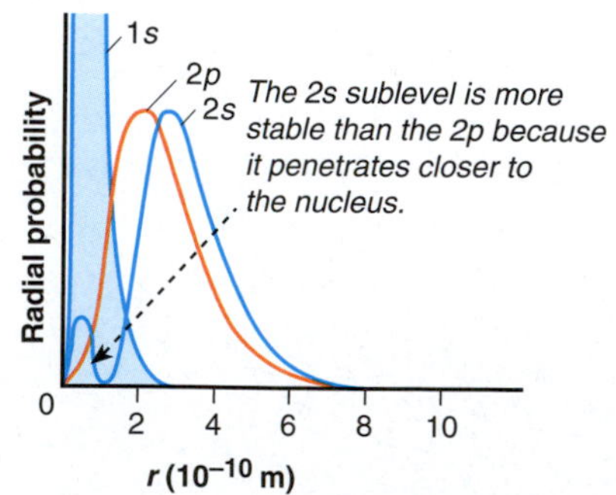

Figure 8.2 Penetration and sublevel energy.

Penetration: The Effect of Orbital Shape on Sublevel Energy To see why the third Li electron occupies the 2*s* sublevel rather than the 2*p*, we have to consider orbital shapes, that is, radial probability distributions (Figure 8.2). A 2*p* orbital *(orange curve)* is slightly closer to the nucleus, on average, than the major portion of the 2*s* orbital *(blue curve)*. But a small portion of the 2*s* radial probability distribution peaks within the 1*s* region. Thus, an electron in the 2*s* orbital spends part of its time "penetrating" very close to the nucleus. **Penetration** has two effects:

- It *increases the nuclear attraction* for a 2*s* electron over that for a 2*p* electron.
- It *decreases the shielding* of a 2*s* electron by the 1*s* electrons.

Thus, since it takes more energy to remove a 2*s* electron (520 kJ/mol) than a 2*p* (341 kJ/mol), the 2*s* sublevel is lower in energy than the 2*p*.

Splitting of Levels into Sublevels In general, *penetration and the resulting effects on shielding cause an energy level to split into sublevels of differing energy.* The lower the *l* value of a sublevel, the more its electrons penetrate, and so the greater their attraction to the nucleus. Therefore, *for a given n value, a lower l value indicates a more stable (lower energy) sublevel:*

$$\text{Order of sublevel energies: } s < p < d < f \tag{8.1}$$

Thus, the $2s$ ($l = 0$) is lower in energy than the $2p$ ($l = 1$), the $3p$ ($l = 1$) is lower than the $3d$ ($l = 2$), and so forth.

Figure 8.3 shows the general energy order of levels (n value) and how they are split into sublevels (l values) of differing energies. (Compare this with the H atom energy levels in Figure 7.20.) Next, we'll use this energy order to construct a periodic table of ground-state atoms.

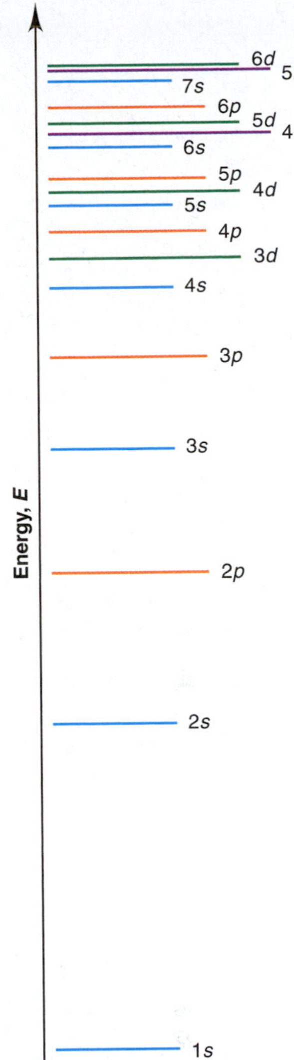

Figure 8.3 Order for filling energy sublevels with electrons. In general, energies of sublevels increase with the principal quantum number n ($1 < 2 < 3$, etc.) and the angular momentum quantum number l ($s < p < d < f$). As n increases, some sublevels overlap; for example, the $4s$ sublevel is lower in energy than the $3d$. (Line color indicates sublevel type.)

Summary of Section 8.1

- Identifying electrons in many-electron atoms requires four quantum numbers: three (n, l, m_l) describe the orbital, and a fourth (m_s) describes electron spin.
- The exclusion principle requires each electron to have a unique set of four quantum numbers; therefore, an orbital can hold no more than two electrons, and their spins must be paired (opposite).
- Electrostatic interactions determine sublevel energies as follows:
  1. Greater nuclear charge lowers sublevel energy, making electrons harder to remove.
  2. Electron-electron repulsions raise sublevel energy, making electrons easier to remove. Repulsions shield electrons from the full nuclear charge, reducing it to an effective nuclear charge, Z_{eff}. Inner electrons shield outer electrons very effectively.
  3. Penetration makes an electron harder to remove because nuclear attraction increases and shielding decreases. As a result, an energy level is split into sublevels with the energy order $s < p < d < f$.

8.2 • THE QUANTUM-MECHANICAL MODEL AND THE PERIODIC TABLE

Quantum mechanics provides the theoretical foundation for the experimentally based periodic table. In this section, we fill the table by determining the ground-state electron configuration of each element—*the lowest-energy distribution of electrons in the sublevels of its atoms*. Note especially the *recurring pattern in electron configurations, which is the basis for recurring patterns in chemical behavior.*

A useful way to determine electron configurations is based on the **aufbau principle** (German *aufbauen,* "to build up"). We start at the beginning of the periodic table and add one proton to the nucleus and one electron to the *lowest energy sublevel available.* (Of course, one or more neutrons are also added to the nucleus.)

There are two common ways to indicate the distribution of electrons:

- *The electron configuration.* This shorthand notation consists of the principal energy level (n value), the letter designation of the sublevel (l value), and the number of electrons (#) in the sublevel, written as a superscript: $\boldsymbol{nl^{\#}}$.
- *The orbital diagram.* An **orbital diagram** consists of a box (or circle, or just a line) for each orbital in a given energy level, grouped by sublevel (with nl designation shown beneath), with an arrow representing an electron *and* its spin: ↑ is $+\frac{1}{2}$ and ↓ is $-\frac{1}{2}$. (Throughout the text, orbital occupancy is also indicated by color intensity: no color is empty, pale color is half-filled, and full color is filled.)

Building Up Period 1

Let's begin by applying the aufbau principle to Period 1, whose ground-state elements have only the $n = 1$ level and, thus, only the $1s$ sublevel, which consists of only the $1s$ orbital. We'll also assign a set of four quantum numbers to each element's *last added* electron.

1. *Hydrogen.* For the electron in H, as you've seen, the set of quantum numbers is H ($Z = 1$): $n = 1, l = 0, m_l = 0, m_s = +\frac{1}{2}$. The electron configuration (spoken "one-ess-one") and orbital diagram are

H (Z = 1) $1s^1$ [↑]
$1s$

2. *Helium.* Recall that the first electron in He has the same quantum numbers as the electron in H, but the second He electron has opposing spin (exclusion principle) He ($Z = 2$): $n = 1, l = 0, m_l = 0, m_s = -\frac{1}{2}$. The electron configuration (spoken "one-ess-two," *not* "one-ess-squared") and orbital diagram are

He (Z = 2) $1s^2$ [↑↓]
$1s$

Building Up Period 2

The exclusion principle says an orbital can hold no more than two electrons. Therefore, with He, the $1s$ orbital, the $1s$ sublevel, the $n = 1$ level, and Period 1 are filled. Filling the $n = 2$ level builds up Period 2 and begins with the $2s$ sublevel, which is the next lowest in energy (see Figure 8.2) and consists of only the $2s$ orbital. When the $2s$ sublevel is filled, we proceed to fill the $2p$.

1. *Lithium.* The first two electrons in Li fill the $1s$ sublevel, and the last added Li electron has quantum numbers $n = 2, l = 0, m_l = 0, m_s = +\frac{1}{2}$. The electron configuration and orbital diagram are

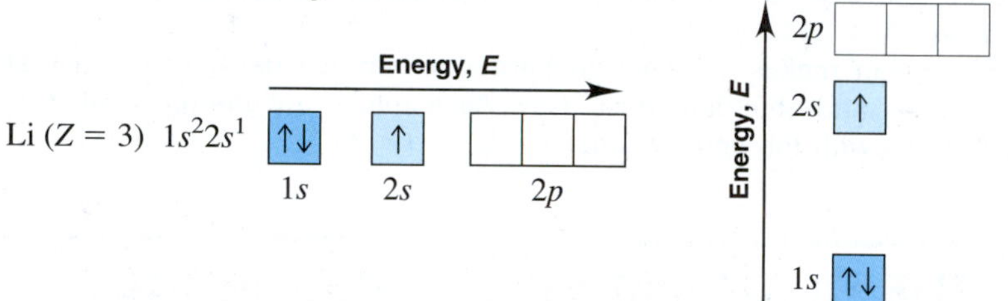

(Note that a complete orbital diagram shows all the orbitals for the given n value, whether or not they are occupied.) To save space on a page, orbital diagrams are written horizontally, with *the sublevel energy increasing left to right.* The orbital diagram to the right above shows the sublevels vertically.

2. *Beryllium.* The $2s$ orbital is only half-filled in Li, and the fourth electron of beryllium fills it with the electron's spin paired: $n = 2, l = 0, m_l = 0, m_s = -\frac{1}{2}$.

Be (Z = 4) $1s^22s^2$ [↑↓] [↑↓] [| |]
$1s$ $2s$ $2p$

3. *Boron.* The next lowest energy sublevel is the $2p$. A p sublevel has $l = 1$, so the m_l (orientation) value can be -1, 0, or $+1$. The three orbitals in the $2p$ sublevel have *equal energy* (same n and l values), which means that the fifth electron of boron can go into *any one of the 2p orbitals.* For convenience, let's label the boxes from left to right: -1, 0, $+1$. By convention, we start on the left and place the fifth electron in the $m_l = -1$ orbital: $n = 2, l = 1, m_l = -1, m_s = +\frac{1}{2}$.

-1 0 $+1$
B (Z = 5) $1s^22s^22p^1$ [↑↓] [↑↓] [↑| |]
$1s$ $2s$ $2p$

4. *Carbon.* To minimize electron-electron repulsions, the sixth electron of carbon enters one of the *unoccupied* $2p$ orbitals; by convention, we place it in the $m_l = 0$ orbital. Experiment shows that the spin of this electron is *parallel* to (the same as) the spin of the other $2p$ electron. This result exemplifies **Hund's rule:** *when orbitals of equal energy are available, the electron configuration of lowest energy has the maximum number*

of unpaired electrons with parallel spins. Thus, the sixth electron has $n = 2$, $l = 1$, $m_l = 0$, and $m_s = +\frac{1}{2}$.

C (Z = 6) $1s^22s^22p^2$ [↑↓] [↑↓] [↑ ↑] (1s, 2s, 2p)

5. *Nitrogen*. Based on Hund's rule, nitrogen's seventh electron enters the last empty $2p$ orbital, with its spin parallel to the other two: $n = 2$, $l = 1$, $m_l = +1$, $m_s = +\frac{1}{2}$.

N (Z = 7) $1s^22s^22p^3$ [↑↓] [↑↓] [↑ ↑ ↑] (1s, 2s, 2p)

6. *Oxygen*. The eighth electron in oxygen must enter one of the three half-filled $2p$ orbitals and "pair up" with (oppose the spin of) the electron present. We place the electron in a half-filled $2p$ orbital: $n = 2$, $l = 1$, $m_l = -1$, $m_s = -\frac{1}{2}$.

O (Z = 8) $1s^22s^22p^4$ [↑↓] [↑↓] [↑↓ ↑ ↑] (1s, 2s, 2p)

7. *Fluorine*. Fluorine's ninth electron enters the next of the two remaining half-filled $2p$ orbitals: $n = 2$, $l = 1$, $m_l = 0$, $m_s = -\frac{1}{2}$.

F (Z = 9) $1s^22s^22p^5$ [↑↓] [↑↓] [↑↓ ↑↓ ↑] (1s, 2s, 2p)

8. *Neon*. Only one unfilled $2p$ orbital remains, so the tenth electron of neon occupies it: $n = 2$, $l = 1$, $m_l = +1$, $m_s = -\frac{1}{2}$. With neon, the $n = 2$ level is filled.

Ne (Z = 10) $1s^22s^22p^6$ [↑↓] [↑↓] [↑↓ ↑↓ ↑↓] (1s, 2s, 2p)

Sample Problem 8.1 Determining Quantum Numbers from Orbital Diagrams

Problem Use the orbital diagram shown above to write the set of quantum numbers for the third electron and the set for the eighth electron of the F atom.

Plan Referring to the orbital diagram, we identify the electron of interest and note its level (n), sublevel (l), orbital (m_l), and spin (m_s).

Solution The third electron is in the $2s$ orbital. The upward arrow indicates a spin of $+\frac{1}{2}$:

$$n = 2, l = 0, m_l = 0, m_s = +\tfrac{1}{2}$$

The eighth electron is in the first $2p$ orbital, which is designated $m_l = -1$, and has a downward arrow:

$$n = 2, l = 1, m_l = -1, m_s = -\tfrac{1}{2}$$

FOLLOW-UP PROBLEM 8.1 Use the periodic table to identify the element with the electron configuration $1s^22s^22p^4$. Write its orbital diagram and the set of quantum numbers for its sixth electron.

Building Up Period 3

The Period 3 elements, Na through Ar, lie directly under the Period 2 elements, Li through Ne. That is, even though the $n = 3$ level splits into $3s$, $3p$, and $3d$ sublevels, Period 3 fills only $3s$ and $3p$; as you'll see shortly, the $3d$ is filled in Period 4. Table 8.2 introduces two ways to present electron distributions more concisely:

- *Partial orbital diagrams* show only the sublevels being filled, here the $3s$ and $3p$.
- *Condensed electron configurations (rightmost column)* have the element symbol of the previous noble gas in brackets, to stand for its configuration, followed by the electron configuration of filled inner sublevels and the energy level being filled. For example, the condensed electron configuration of sulfur is [Ne] $3s^23p^4$, where [Ne] stands for $1s^22s^22p^6$.

Table 8.2 Partial Orbital Diagrams and Electron Configurations* for the Elements in Period 3

Atomic Number	Element	Partial Orbital Diagram (3s and 3p Sublevels Only) 3s	3p	Full Electron Configuration†	Condensed Electron Configuration
11	Na	↑	[ ] [ ] [ ]	$[1s^22s^22p^6]\ 3s^1$	$[Ne]\ 3s^1$
12	Mg	↑↓	[ ] [ ] [ ]	$[1s^22s^22p^6]\ 3s^2$	$[Ne]\ 3s^2$
13	Al	↑↓	↑ [ ] [ ]	$[1s^22s^22p^6]\ 3s^23p^1$	$[Ne]\ 3s^23p^1$
14	Si	↑↓	↑ ↑ [ ]	$[1s^22s^22p^6]\ 3s^23p^2$	$[Ne]\ 3s^23p^2$
15	P	↑↓	↑ ↑ ↑	$[1s^22s^22p^6]\ 3s^23p^3$	$[Ne]\ 3s^23p^3$
16	S	↑↓	↑↓ ↑ ↑	$[1s^22s^22p^6]\ 3s^23p^4$	$[Ne]\ 3s^23p^4$
17	Cl	↑↓	↑↓ ↑↓ ↑	$[1s^22s^22p^6]\ 3s^23p^5$	$[Ne]\ 3s^23p^5$
18	Ar	↑↓	↑↓ ↑↓ ↑↓	$[1s^22s^22p^6]\ 3s^23p^6$	$[Ne]\ 3s^23p^6$

*Colored type indicates the sublevel to which the last electron is added.
†The full configuration is not usually written with square brackets; they are included here to show how the [Ne] designation arises.

In Na (the second alkali metal) and Mg (the second alkaline earth metal), electrons are added to the 3*s* sublevel, which contains only the 3*s* orbital; this is directly comparable to the filling of the 2*s* sublevel in Li and Be in Period 2. Then, in the same way as the 2*p* orbitals of B, C, and N in Period 2 are half-filled, the last electrons added to Al, Si, and P in Period 3 half-fill successive 3*p* orbitals with spins parallel (Hund's rule). The last electrons added to S, Cl, and Ar then successively pair up to fill those 3*p* orbitals, which fills the 3*p* sublevel. Thus, Ag, the next noble gas after He and Ne, ends Period 3. (As you'll see shortly, the 3*d* orbitals are filled in Period 4.)

Similar Electron Configurations Within Groups

One of the central points in chemistry is that *similar outer electron configurations correlate with similar chemical behavior.* Figure 8.4 shows the condensed electron configurations of the first 18 elements. Note the similarity within each group. Here are examples from three of the groups:

- In Group 1A(1), Li and Na have the outer electron configuration ns^1 (where n is the quantum number of the highest energy level), as do the other alkali metals (K, Rb, Cs, and Fr). All are highly reactive metals whose atoms lose the outer electron when they form ionic compounds with nonmetals, and all react vigorously with water to displace H_2.
- In Group 7A(17), F and Cl have the outer electron configuration ns^2np^5, as do the other halogens (Br, I, and At). All are reactive nonmetals that occur as diatomic molecules (X_2), and all form ionic compounds with metals (KX, MgX_2), covalent compounds with hydrogen (HX) that yield acidic solutions in water, and covalent compounds with carbon (CX_4).
- In Group 8A(18), He has the electron configuration ns^2, and all the other elements in the group have the outer configuration ns^2np^6. Consistent with their *filled* energy levels, all members of this group are very unreactive monatomic gases.

To summarize the major connection between quantum mechanics and chemical periodicity: *sublevels are filled in order of increasing energy, which leads to outer electron configurations that recur periodically, which leads to chemical properties that recur periodically.*

Period	1A (1)	2A (2)	3A (13)	4A (14)	5A (15)	6A (16)	7A (17)	8A (18)
1	1 **H** $1s^1$							2 **He** $1s^2$
2	3 **Li** [He] $2s^1$	4 **Be** [He] $2s^2$	5 **B** [He] $2s^22p^1$	6 **C** [He] $2s^22p^2$	7 **N** [He] $2s^22p^3$	8 **O** [He] $2s^22p^4$	9 **F** [He] $2s^22p^5$	10 **Ne** [He] $2s^22p^6$
3	11 **Na** [Ne] $3s^1$	12 **Mg** [Ne] $3s^2$	13 **Al** [Ne] $3s^23p^1$	14 **Si** [Ne] $3s^23p^2$	15 **P** [Ne] $3s^23p^3$	16 **S** [Ne] $3s^23p^4$	17 **Cl** [Ne] $3s^23p^5$	18 **Ar** [Ne] $3s^23p^6$

Figure 8.4 Condensed electron configurations in the first three periods. Elements in a group have similar outer electron configurations *(color)*.

Building Up Period 4: The First Transition Series

Period 4 contains the first series of **transition elements,** those in which *d* orbitals are being filled. Let's examine three factors that affect the filling pattern in a period with a transition series. (We'll return to the first in Section 8.4.)

1. *Effects of shielding and penetration on sublevel energy.* The 3*d* sublevel is filled in Period 4, but *the 4s sublevel is filled first.* This switch in filling order is due to shielding and penetration effects. Based on the 3*d* radial probability distribution (see Figure 7.18), a 3*d* electron spends most of the time outside the filled inner $n = 1$ and $n = 2$ levels, so it is shielded very effectively from the nuclear charge. However, the outermost 4*s* electron penetrates close to the nucleus part of the time, so it is subject to a greater attraction. As a result, the 4*s* orbital is slightly *lower* in energy than the 3*d* and fills first. In any period, *the ns sublevel fills before the* $(n - 1)d$ *sublevel.* Other variations in the filling pattern occur at higher values of *n* because sublevel energies become very close together (see Figure 8.3).

2. *Filling the 4s and 3d sublevels.* Table 8.3 *(next page)* shows the partial orbital diagrams and full and condensed electron configurations for the 18 elements in Period 4. The first two elements, K and Ca, are the next alkali and alkaline earth metals, respectively; the last electron of K half-fills and that of Ca fills the 4*s* sublevel. The last electron of scandium (Sc; $Z = 21$), the first transition element, occupies any one of the five 3*d* orbitals because they are equal in energy; Sc has the electron configuration [Ar] $4s^23d^1$. Filling of the 3*d* orbitals proceeds one electron at a time, as with the *p* orbitals, except in two cases, chromium (Cr; $Z = 24$) and copper (Cu; $Z = 29$), discussed next.

3. *Stability of half-filled and filled sublevels.* Vanadium (V; $Z = 23$) has three half-filled *d* orbitals ([Ar] $4s^23d^3$). However, the last electron of the next element, Cr, does not enter a fourth empty *d* orbital to give [Ar] $4s^23d^4$; instead, Cr has one electron in the 4*s* sublevel and five in the 3*d* sublevel, making both sublevels half filled: [Ar] $4s^13d^5$ *(see margin).* In the next element, manganese (Mn; $Z = 25$), the 4*s* sublevel is filled again ([Ar] $4s^23d^5$).

Because it follows nickel (Ni; [Ar] $4s^23d^8$), copper would be expected to have the configuration [Ar] $4s^23d^9$. Instead, the 4*s* sublevel of Cu is *half-filled* with 1 electron, and the 3*d* sublevel is *filled* with 10 *(see margin).* From these two exceptions, Cr and Cu, we conclude that *half-filled and filled sublevels are unexpectedly stable (low in energy);* we see this pattern with many other elements.

With zinc (Zn; $Z = 30$), the 4*s* sublevel is filled ([Ar] $4s^23d^{10}$), and the first transition series ends. The 4*p* sublevel is filled by the next six elements, and Period 4 ends with the noble gas krypton (Kr; $Z = 36$).

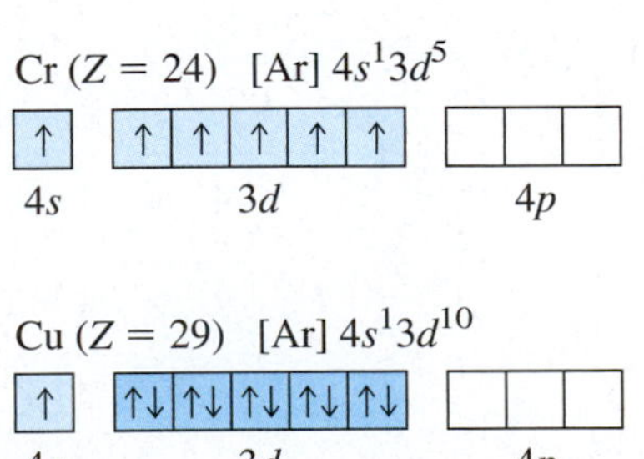

Table 8.3 Partial Orbital Diagrams and Electron Configurations* for the Elements in Period 4

Atomic Number	Element	Partial Orbital Diagram (4s, 3d, and 4p Sublevels Only) 4s	3d	4p	Full Electron Configuration	Condensed Electron Configuration
19	K	[↑]	[ ][ ][ ][ ][ ]	[ ][ ][ ]	$1s^22s^22p^63s^23p^64s^1$	[Ar] $4s^1$
20	Ca	[↑↓]	[ ][ ][ ][ ][ ]	[ ][ ][ ]	$1s^22s^22p^63s^23p^64s^2$	[Ar] $4s^2$
21	Sc	[↑↓]	[↑][ ][ ][ ][ ]	[ ][ ][ ]	$1s^22s^22p^63s^23p^64s^23d^1$	[Ar] $4s^23d^1$
22	Ti	[↑↓]	[↑][↑][ ][ ][ ]	[ ][ ][ ]	$1s^22s^22p^63s^23p^64s^23d^2$	[Ar] $4s^23d^2$
23	V	[↑↓]	[↑][↑][↑][ ][ ]	[ ][ ][ ]	$1s^22s^22p^63s^23p^64s^23d^3$	[Ar] $4s^23d^3$
24	Cr	[↑]	[↑][↑][↑][↑][↑]	[ ][ ][ ]	$1s^22s^22p^63s^23p^64s^13d^5$	[Ar] $4s^13d^5$
25	Mn	[↑↓]	[↑][↑][↑][↑][↑]	[ ][ ][ ]	$1s^22s^22p^63s^23p^64s^23d^5$	[Ar] $4s^23d^5$
26	Fe	[↑↓]	[↑↓][↑][↑][↑][↑]	[ ][ ][ ]	$1s^22s^22p^63s^23p^64s^23d^6$	[Ar] $4s^23d^6$
27	Co	[↑↓]	[↑↓][↑↓][↑][↑][↑]	[ ][ ][ ]	$1s^22s^22p^63s^23p^64s^23d^7$	[Ar] $4s^23d^7$
28	Ni	[↑↓]	[↑↓][↑↓][↑↓][↑][↑]	[ ][ ][ ]	$1s^22s^22p^63s^23p^64s^23d^8$	[Ar] $4s^23d^8$
29	Cu	[↑]	[↑↓][↑↓][↑↓][↑↓][↑↓]	[ ][ ][ ]	$1s^22s^22p^63s^23p^64s^13d^{10}$	[Ar] $4s^13d^{10}$
30	Zn	[↑↓]	[↑↓][↑↓][↑↓][↑↓][↑↓]	[ ][ ][ ]	$1s^22s^22p^63s^23p^64s^23d^{10}$	[Ar] $4s^23d^{10}$
31	Ga	[↑↓]	[↑↓][↑↓][↑↓][↑↓][↑↓]	[↑][ ][ ]	$1s^22s^22p^63s^23p^64s^23d^{10}4p^1$	[Ar] $4s^23d^{10}4p^1$
32	Ge	[↑↓]	[↑↓][↑↓][↑↓][↑↓][↑↓]	[↑][↑][ ]	$1s^22s^22p^63s^23p^64s^23d^{10}4p^2$	[Ar] $4s^23d^{10}4p^2$
33	As	[↑↓]	[↑↓][↑↓][↑↓][↑↓][↑↓]	[↑][↑][↑]	$1s^22s^22p^63s^23p^64s^23d^{10}4p^3$	[Ar] $4s^23d^{10}4p^3$
34	Se	[↑↓]	[↑↓][↑↓][↑↓][↑↓][↑↓]	[↑↓][↑][↑]	$1s^22s^22p^63s^23p^64s^23d^{10}4p^4$	[Ar] $4s^23d^{10}4p^4$
35	Br	[↑↓]	[↑↓][↑↓][↑↓][↑↓][↑↓]	[↑↓][↑↓][↑]	$1s^22s^22p^63s^23p^64s^23d^{10}4p^5$	[Ar] $4s^23d^{10}4p^5$
36	Kr	[↑↓]	[↑↓][↑↓][↑↓][↑↓][↑↓]	[↑↓][↑↓][↑↓]	$1s^22s^22p^63s^23p^64s^23d^{10}4p^6$	[Ar] $4s^23d^{10}4p^6$

*Colored type indicates sublevel(s) whose occupancy changes when the last electron is added.

General Principles of Electron Configurations

Figure 8.5 shows the partial (highest energy sublevels being filled) ground-state electron configurations of the known elements. Let's highlight some key relationships among them.

Similar Outer Electron Configurations Within a Group Among the main-group elements (A groups)—the *s*-block and *p*-block elements—outer electron configurations within a group are identical. Some variations in the transition elements (B groups, *d* block) and inner transition elements (*f* block) occur, as we'll see.

Orbital Filling Order When the elements are "built up" by filling levels and sublevels in order of increasing energy, we obtain the sequence in the periodic table. Reading the table from left to right, like words on a page, gives the energy order of levels and sublevels (Figure 8.6); the following is a memory aid for sublevel filling order when a periodic table is not available.

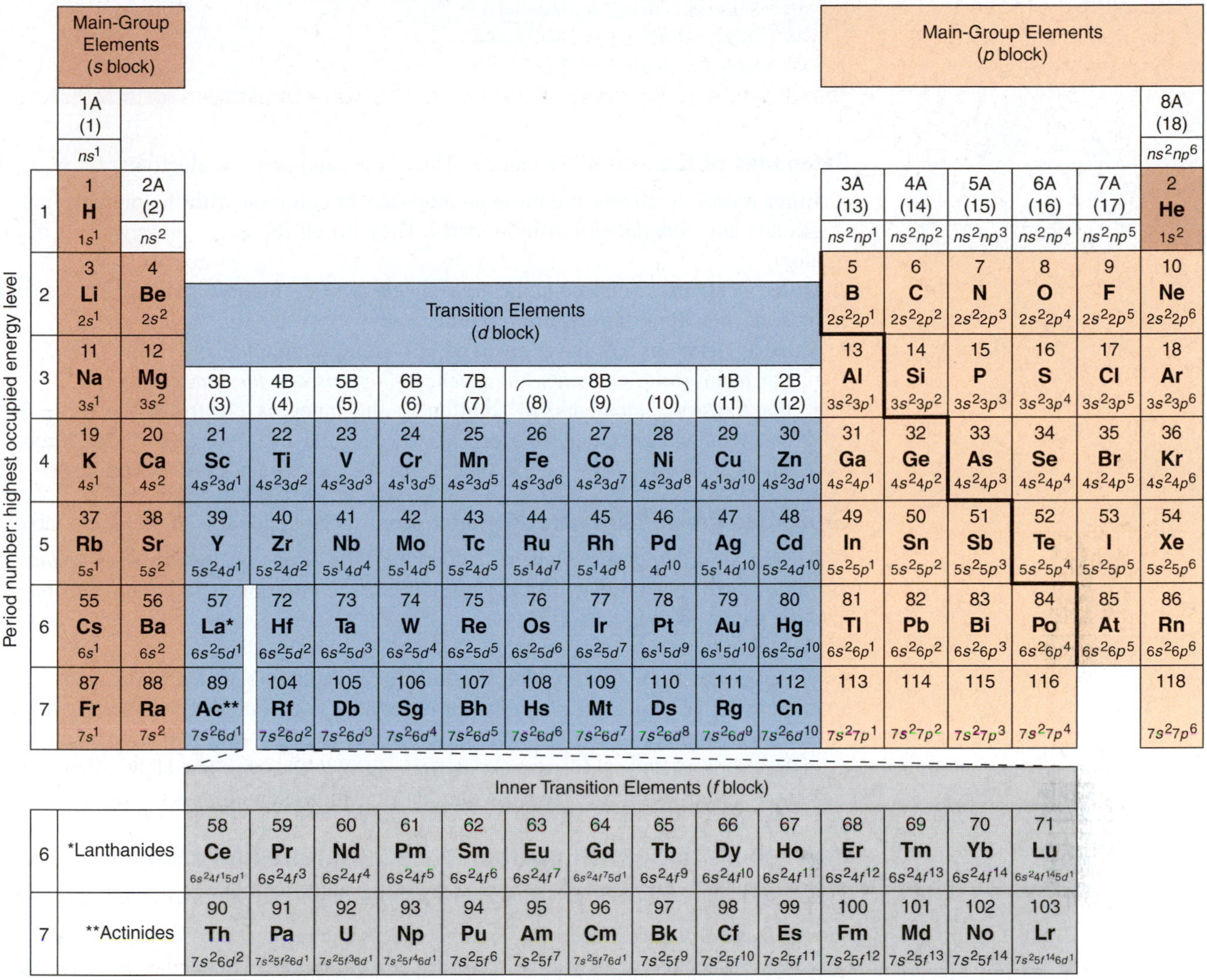

Main-Group Elements (s block) | Transition Elements (d block) | Main-Group Elements (p block)

Period number: highest occupied energy level

Period	1A (1) ns^1	2A (2) ns^2	3B (3)	4B (4)	5B (5)	6B (6)	7B (7)	8B (8)	8B (9)	8B (10)	1B (11)	2B (12)	3A (13) ns^2np^1	4A (14) ns^2np^2	5A (15) ns^2np^3	6A (16) ns^2np^4	7A (17) ns^2np^5	8A (18) ns^2np^6
1	1 **H** $1s^1$																	2 **He** $1s^2$
2	3 **Li** $2s^1$	4 **Be** $2s^2$											5 **B** $2s^22p^1$	6 **C** $2s^22p^2$	7 **N** $2s^22p^3$	8 **O** $2s^22p^4$	9 **F** $2s^22p^5$	10 **Ne** $2s^22p^6$
3	11 **Na** $3s^1$	12 **Mg** $3s^2$											13 **Al** $3s^23p^1$	14 **Si** $3s^23p^2$	15 **P** $3s^23p^3$	16 **S** $3s^23p^4$	17 **Cl** $3s^23p^5$	18 **Ar** $3s^23p^6$
4	19 **K** $4s^1$	20 **Ca** $4s^2$	21 **Sc** $4s^23d^1$	22 **Ti** $4s^23d^2$	23 **V** $4s^23d^3$	24 **Cr** $4s^13d^5$	25 **Mn** $4s^23d^5$	26 **Fe** $4s^23d^6$	27 **Co** $4s^23d^7$	28 **Ni** $4s^23d^8$	29 **Cu** $4s^13d^{10}$	30 **Zn** $4s^23d^{10}$	31 **Ga** $4s^24p^1$	32 **Ge** $4s^24p^2$	33 **As** $4s^24p^3$	34 **Se** $4s^24p^4$	35 **Br** $4s^24p^5$	36 **Kr** $4s^24p^6$
5	37 **Rb** $5s^1$	38 **Sr** $5s^2$	39 **Y** $5s^24d^1$	40 **Zr** $5s^24d^2$	41 **Nb** $5s^14d^4$	42 **Mo** $5s^14d^5$	43 **Tc** $5s^24d^5$	44 **Ru** $5s^14d^7$	45 **Rh** $5s^14d^8$	46 **Pd** $4d^{10}$	47 **Ag** $5s^14d^{10}$	48 **Cd** $5s^24d^{10}$	49 **In** $5s^25p^1$	50 **Sn** $5s^25p^2$	51 **Sb** $5s^25p^3$	52 **Te** $5s^25p^4$	53 **I** $5s^25p^5$	54 **Xe** $5s^25p^6$
6	55 **Cs** $6s^1$	56 **Ba** $6s^2$	57 **La*** $6s^25d^1$	72 **Hf** $6s^25d^2$	73 **Ta** $6s^25d^3$	74 **W** $6s^25d^4$	75 **Re** $6s^25d^5$	76 **Os** $6s^25d^6$	77 **Ir** $6s^25d^7$	78 **Pt** $6s^15d^9$	79 **Au** $6s^15d^{10}$	80 **Hg** $6s^25d^{10}$	81 **Tl** $6s^26p^1$	82 **Pb** $6s^26p^2$	83 **Bi** $6s^26p^3$	84 **Po** $6s^26p^4$	85 **At** $6s^26p^5$	86 **Rn** $6s^26p^6$
7	87 **Fr** $7s^1$	88 **Ra** $7s^2$	89 **Ac**** $7s^26d^1$	104 **Rf** $7s^26d^2$	105 **Db** $7s^26d^3$	106 **Sg** $7s^26d^4$	107 **Bh** $7s^26d^5$	108 **Hs** $7s^26d^6$	109 **Mt** $7s^26d^7$	110 **Ds** $7s^26d^8$	111 **Rg** $7s^26d^9$	112 **Cn** $7s^26d^{10}$	113 $7s^27p^1$	114 $7s^27p^2$	115 $7s^27p^3$	116 $7s^27p^4$		118 $7s^27p^6$

Inner Transition Elements (f block)

Period	Series														
6	*Lanthanides	58 **Ce** $6s^24f^15d^1$	59 **Pr** $6s^24f^3$	60 **Nd** $6s^24f^4$	61 **Pm** $6s^24f^5$	62 **Sm** $6s^24f^6$	63 **Eu** $6s^24f^7$	64 **Gd** $6s^24f^75d^1$	65 **Tb** $6s^24f^9$	66 **Dy** $6s^24f^{10}$	67 **Ho** $6s^24f^{11}$	68 **Er** $6s^24f^{12}$	69 **Tm** $6s^24f^{13}$	70 **Yb** $6s^24f^{14}$	71 **Lu** $6s^24f^{14}5d^1$
7	**Actinides	90 **Th** $7s^26d^2$	91 **Pa** $7s^25f^26d^1$	92 **U** $7s^25f^36d^1$	93 **Np** $7s^25f^46d^1$	94 **Pu** $7s^25f^6$	95 **Am** $7s^25f^7$	96 **Cm** $7s^25f^76d^1$	97 **Bk** $7s^25f^9$	98 **Cf** $7s^25f^{10}$	99 **Es** $7s^25f^{11}$	100 **Fm** $7s^25f^{12}$	101 **Md** $7s^25f^{13}$	102 **No** $7s^25f^{14}$	103 **Lr** $7s^25f^{14}6d^1$

Figure 8.5 A periodic table of partial ground-state electron configurations.

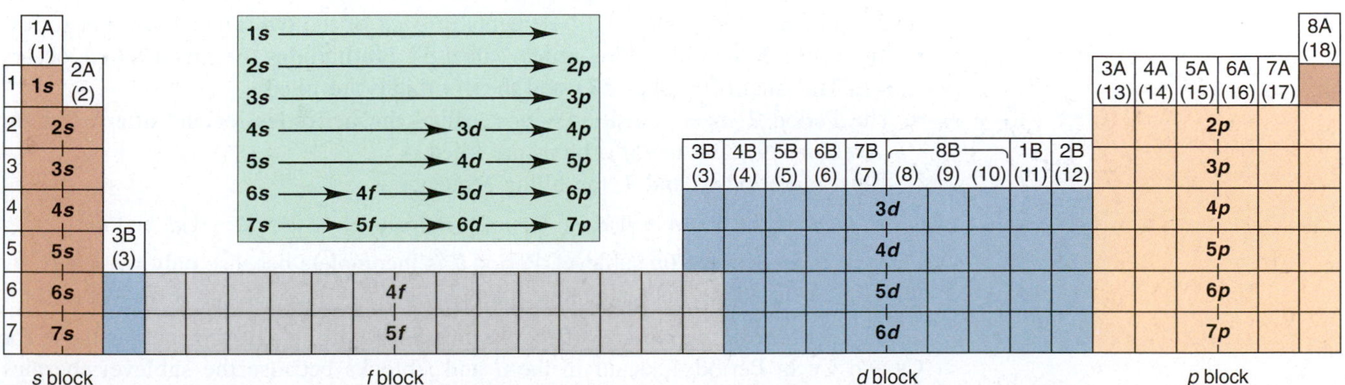

Figure 8.6 **Orbital filling and the periodic table.** This form of the periodic table shows the sublevel blocks. *Pale green inset:* A summary of the sublevel filling order.

THINK OF IT THIS WAY
The Filling Order of the Periodic Table

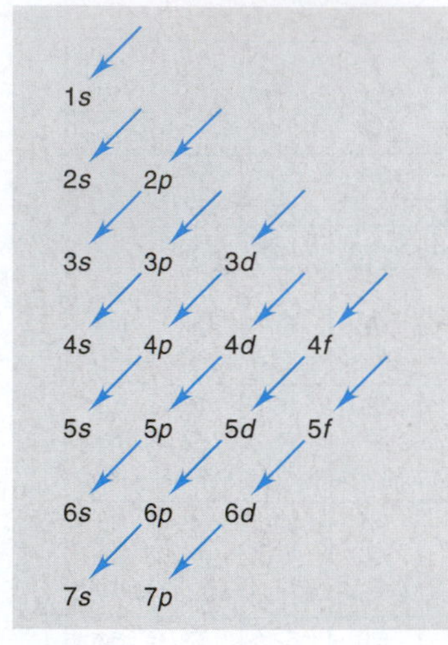

If you list the sublevels as shown at right and follow the direction of the arrows, starting from 1*s*, you obtain the sublevels in order of increasing energy. Note that

- the *n* value is constant horizontally,
- the *l* value is constant vertically, and
- the sum $n + l$ is constant diagonally.

Categories of Electrons We can describe three categories of electrons:

1. **Inner (core) electrons** are those an atom has in common with the previous noble gas and any *completed* transition series. They fill all the *lower energy levels* of an atom.
2. **Outer electrons** are those in the *highest energy level* (highest *n* value). They spend most of their time farthest from the nucleus.
3. **Valence electrons** are those involved in forming compounds:
   - For main-group elements, the *valence electrons* ***are*** *the outer electrons.*
   - For transition elements, in addition to the outer *ns* electrons, the $(n - 1)d$ electrons are also valence electrons, though the metals Fe ($Z = 26$) through Zn ($Z = 30$) may use only a few, if any, of their *d* electrons in bonding.

Group and Period Numbers Key information is embedded in the periodic table:

- Among the main-group elements (A groups), *the A number equals the number of outer electrons* (those with the highest *n*); thus, chlorine (Cl; Group **7**A) has 7 outer electrons, and so forth.
- *The period number is the n value of the highest energy level.*
- For an energy level, the *n* value squared (n^2) is the number of *orbitals,* and $2n^2$ is the maximum number of *electrons* (or elements). For example, consider the $n = 3$ level. The number of orbitals is $n^2 = 9$: one 3*s,* three 3*p,* and five 3*d*. The number of electrons is $2n^2 = 18$: two 3*s* and six 3*p* electrons for the eight elements of Period 3, and ten 3*d* electrons for the ten transition elements of Period 4.

Intervening Series: Transition and Inner Transition Elements

The *d* block and *f* block occur between the main-group *s* and *p* blocks (see Figure 8.6).

1. *Transition series.* Periods 4, 5, 6, and 7 incorporate the 3*d,* 4*d,* 5*d,* and 6*d* sublevels, respectively. As you've seen, the general pattern is that the $(n - 1)d$ sublevel is filled between the *ns* and *np* sublevels. Thus, in Period 5, the filling order is 5*s,* then 4*d,* and then 5*p*.

2. *Inner transition series.* Period 6 holds the first of two series of **inner transition elements,** those in which *f* orbitals are being filled (Figure 8.6). The *f* orbitals have $l = 3$, so the possible m_l values are $-3, -2, -1, 0, +1, +2$, and $+3$; that is, there are seven *f* orbitals, for a total of 14 elements in *each* of the two inner transition series:

- The Period 6 inner transition series, called the **lanthanides** (or **rare earths**), occurs after lanthanum (La; $Z = 57$), and the 4*f* orbitals are filled.
- The Period 7 inner transition series, called the **actinides,** occurs after actinium (Ac; $Z = 89$), and the 5*f* orbitals are filled.

Thus, in Periods 6 and 7, the filling sequence is

ns, first of the $(n - 1)d$, all $(n - 2)f$, remainder of the $(n - 1)d$, and *np*

Period 6 ends with the 6*p* sublevel. Period 7 is incomplete because only five elements with 7*p* electrons are known at this time; element 117 has not yet been synthesized.

3. *Irregular filling patterns.* Irregularities in the filling pattern, such as those for Cr and Cu in Period 4, occur in the *d* and *f* blocks because the sublevel energies in these larger atoms differ very little. Even though occasional deviations occur in the *d* block, the sum of *ns* electrons and $(n - 1)d$ electrons always equals the new group number (in parentheses). For instance, despite variations in Group 6B(**6**)—Cr, Mo, W, and Sg—the sum of *ns* and $(n - 1)d$ electrons is 6; for Group 8B(**10**)—Ni, Pd, Pt, and Ds—the sum is 10. (We discuss the electron configurations of transition elements further in Chapter 22.)

Sample Problem 8.2 Determining Electron Configurations

Problem Using the periodic table (not Table 8.3 or Figure 8.5) and assuming a regular filling pattern, give the full and condensed electron configurations, partial orbital diagrams showing valence electrons only, and number of inner electrons for the following elements:
(a) Potassium (K; $Z = 19$) **(b)** Technetium (Tc; $Z = 43$) **(c)** Lead (Pb; $Z = 82$)

Plan The atomic number tells us the number of electrons, and the periodic table shows the order for filling sublevels. In the partial orbital diagrams, we include all electrons added after the previous noble gas *except* those in *filled* inner sublevels. The number of inner electrons is the sum of those in the previous noble gas and in filled *d* and *f* sublevels.

Solution **(a)** For K ($Z = 19$), the full electron configuration is $1s^22s^22p^63s^23p^64s^1$.
The condensed configuration is $[Ar]\ 4s^1$.
The partial orbital diagram showing valence electrons is

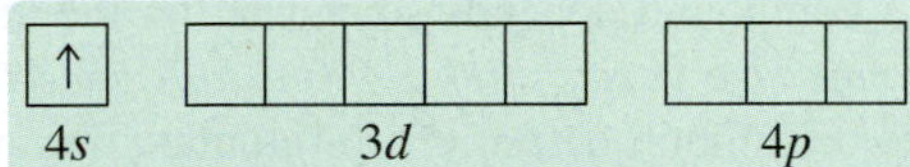

K is a main-group element in Group 1A(1) and Period 4, so there are 18 inner electrons.
(b) For Tc ($Z = 43$), assuming the expected pattern, the full electron configuration is $1s^22s^22p^63s^23p^64s^23d^{10}4p^65s^24d^5$.
The condensed electron configuration is $[Kr]\ 5s^24d^5$.
The partial orbital diagram showing valence electrons is

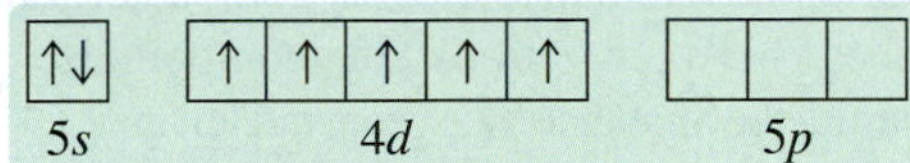

Tc is a transition element in Group 7B(7) and Period 5, so there are 36 inner electrons.
(c) For Pb ($Z = 82$), the full electron configuration is $1s^22s^22p^63s^23p^64s^23d^{10}4p^65s^24d^{10}5p^66s^24f^{14}5d^{10}6p^2$.
The condensed electron configuration is $[Xe]\ 6s^24f^{14}5d^{10}6p^2$.
The partial orbital diagram showing valence electrons (no filled inner sublevels) is

↑↓	↑ ↑
6s	6p

Pb is a main-group element in Group 4A(14) and Period 6, so there are 54 (in Xe) + 14 (in the 4*f* sublevel) + 10 (in the 5*d* sublevel) = 78 inner electrons.

Check Be sure that the sum of the superscripts (numbers of electrons) in the full electron configuration equals the atomic number and that the number of *valence* electrons in the condensed configuration equals the number of electrons in the partial orbital diagram.

FOLLOW-UP PROBLEM 8.2 Without referring to Table 8.3 or Figure 8.5, give full and condensed electron configurations, partial orbital diagrams showing valence electrons only, and the number of inner electrons for the following elements:
(a) Ni ($Z = 28$) **(b)** Sr ($Z = 38$) **(c)** Po ($Z = 84$)

Summary of Section 8.2

- By the aufbau principle, one electron is added to an atom of each successive element in accord with the exclusion principle (no two electrons can have the same set of quantum numbers) and Hund's rule (orbitals of equal energy become half-filled, with electron spins parallel, before any pairing of spins occurs).
- The elements of a group have similar outer electron configurations and similar chemical behavior.
- For the main-group elements, valence electrons (those involved in reactions) are in the outer (highest energy) level only. For transition elements, $(n - 1)d$ electrons are also considered valence electrons.
- Because of shielding of *d* electrons by electrons in inner sublevels and penetration by the *ns* electron, the $(n - 1)d$ sublevel fills after the *ns* and before the *np* sublevels.
- In Periods 6 and 7, $(n - 2)f$ orbitals fill between the first and second $(n - 1)d$ orbitals.

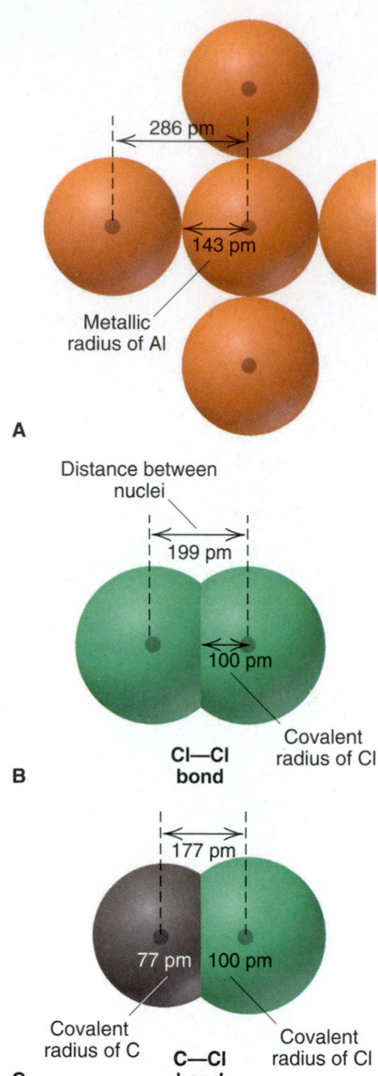

Figure 8.7 Defining atomic size. **A,** The metallic radius of aluminum. **B,** The covalent radius of chlorine. **C,** Known covalent radii and distances between nuclei can be used to find unknown radii.

8.3 • TRENDS IN THREE ATOMIC PROPERTIES

In this section, we focus on three atomic properties that reflect the central importance of electron configuration and effective nuclear charge: atomic size, ionization energy, and electron affinity. Most notably, these properties are *periodic,* which means they generally exhibit consistent changes, or *trends,* within a group or period.

Trends in Atomic Size

Recall from Chapter 7 that we often represent atoms with spherical contours in which the electrons spend 90% of their time. We *define* **atomic size** (the extent of the contour) in terms of how closely one atom lies next to another. But, in practice, as we discuss in Chapter 12, we measure the distance between atomic nuclei in a sample of an element and divide that distance in half. Because atoms do not have hard surfaces, the size of an atom in a compound depends somewhat on the atoms near it. In other words, *an element's atomic size varies slightly from substance to substance.*

Figure 8.7 shows two common definitions of atomic size:

1. **Metallic radius.** Used mostly for *metals,* it is one-half the shortest distance between nuclei of adjacent, individual atoms in a crystal of the element (Figure 8.7A).
2. **Covalent radius.** Used for elements occurring as molecules, mostly *nonmetals,* it is one-half the shortest distance between nuclei of bonded atoms (Figure 8.7B).

Radii measured for some elements are used to determine the radii of other elements from distances between atoms in compounds. For instance, in a carbon-chlorine compound, the distance between nuclei in a C—Cl bond is 177 pm. Using the known covalent radius of Cl (100 pm), we find the covalent radius of C (177 pm − 100 pm = 77 pm) (Figure 8.7C).

Main-Group Elements Figure 8.8 shows the atomic radii of the main-group elements and most of the transition elements. Among the main-group elements, atomic size varies within both groups and periods as a result of two opposing influences:

1. *Changes in n.* As the principal quantum number (n) increases, the probability that outer electrons spend most of their time farther from the nucleus increases as well; thus, atomic size increases.
2. *Changes in* Z_{eff}. As the effective nuclear charge (Z_{eff}) increases, outer electrons are pulled closer to the nucleus; thus, atomic size decreases.

The net effect of these influences depends on how effectively the inner electrons shield the increasing nuclear charge:

1. *Down a group, n dominates.* As we move down a main group, each member has *one more level of inner electrons that shield the outer electrons very effectively.* Even though additional protons do moderately increase Z_{eff} for the outer electrons, the atoms get larger as a result of the increasing n value:

*Atomic radius generally **increases** down a group.*

2. *Across a period,* Z_{eff} *dominates.* Across a period from left to right, electrons are added to the *same* outer level, so the shielding by inner electrons does not change. Despite greater electron repulsions, outer electrons shield each other only slightly, so Z_{eff} rises significantly, and the outer electrons are pulled closer to the nucleus:

*Atomic radius generally **decreases** across a period.*

Transition Elements As Figure 8.8 shows, size trends are *not* as consistent for the transition elements:

1. *Down a transition group,* n increases, but shielding by an additional level of inner electrons results in only a small size increase from Period 4 to 5 and none from 5 to 6.

2. *Across a transition series,* atomic size shrinks through the first two or three elements because of the increasing nuclear charge. But, from then on, *size remains relatively constant* because shielding by the inner d electrons counteracts the increase in Z_{eff}. Thus, for example, in Period 4, the third transition element, vanadium (V;

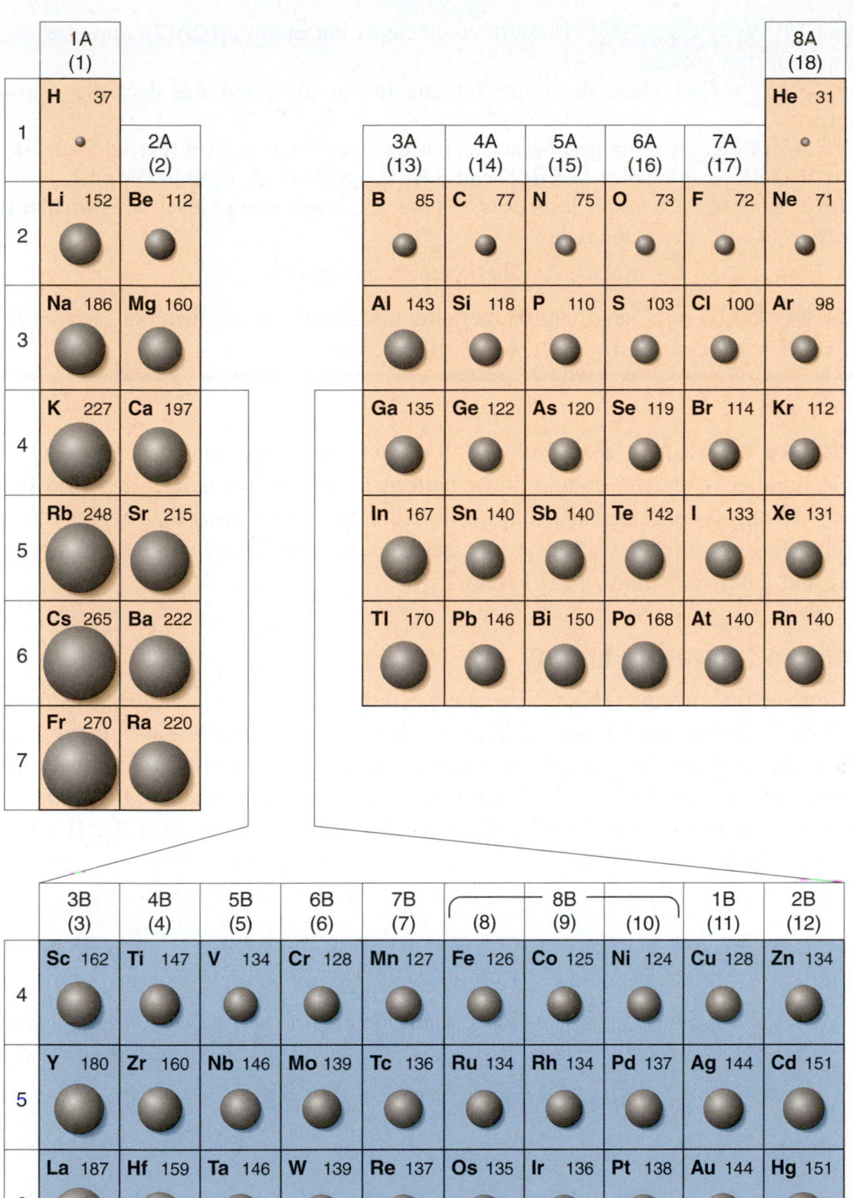

Figure 8.8 Atomic radii of the main-group and transition elements. Atomic radii (in picometers) are shown for the main-group elements *(tan)* and the transition elements *(blue)*. (Values for the noble gases are calculated.)

$Z = 23$), has the same radius as the last, zinc (Zn; $Z = 30$). This pattern also appears in Periods 5 and 6 in the transition and both inner transition series.

3. *A transition series affects atomic size in neighboring main groups.* Shielding by *d* electrons causes a *major size decrease from Group 2A(2) to Group 3A(13)* in Periods 4 through 6. Because the *np* sublevel has more penetration than the $(n - 1)d$, the first *np* electron [added in Group 3A(13)] "feels" a much greater Z_{eff}, due to all the protons added in the intervening transition elements. The greatest decrease occurs in Period 4: calcium (Ca; $Z = 20$) in Group 2A(2) is nearly 50% larger than gallium (Ga; $Z = 31$) in 3A(13). In fact, *d*-orbital shielding causes gallium to be slightly *smaller* than aluminum (Al; $Z = 13$), the element above it!

Sample Problem 8.3 Ranking Elements by Atomic Size

Problem Using only the periodic table (not Figure 8.8), rank each set of main-group elements in order of *decreasing* atomic size:
(a) Ca, Mg, Sr **(b)** K, Ga, Ca **(c)** Br, Rb, Kr **(d)** Sr, Ca, Rb

Plan To rank the elements by atomic size, we find them in the periodic table. They are main-group elements, so size increases down a group and decreases across a period.

Solution **(a)** Sr > Ca > Mg. These three elements are in Group 2A(2), and size decreases up the group.
(b) K > Ca > Ga. These three elements are in Period 4, and size decreases across a period.
(c) Rb > Br > Kr. Rb is largest because it has one more energy level (Period 5) and is farthest to the left. Kr is smaller than Br because Kr is farther to the right in Period 4.
(d) Rb > Sr > Ca. Ca is smallest because it has one fewer energy level. Sr is smaller than Rb because it is farther to the right.

Check From Figure 8.8, we see that the rankings are correct.

FOLLOW-UP PROBLEM 8.3 Using only the periodic table, rank the elements in each set in order of *increasing* size: **(a)** Se, Br, Cl; **(b)** I, Xe, Ba.

Periodicity of Atomic Size Figure 8.9 shows the variation in atomic size with atomic number. Note the up-and-down pattern as size drops across a period to the noble gas *(purple)* and then leaps up to the alkali metal *(brown)* that begins the next period. Also note the deviations from the smooth size decrease in each transition *(blue)* and inner transition *(green)* series.

Trends in Ionization Energy

The **ionization energy (IE)** is the energy required for the *complete removal* of 1 mol of electrons from 1 mol of gaseous atoms or ions. Pulling an electron away from a nucleus *requires* energy to overcome their electrostatic attraction. Because energy flows *into* the system, the ionization energy is always positive (like ΔH of an endothermic reaction). (Chapter 7 presented the ionization energy of the H atom as the energy difference between $n = 1$ and $n = \infty$, where the electron is completely removed.) The ionization energy is a key factor in an element's reactivity:

- *Atoms with a low IE tend to form cations during reactions.*
- *Atoms with a high IE (except the noble gases) tend to form anions.*

Many-electron atoms can lose more than one electron. The *first* ionization energy (IE_1) removes an outermost electron (highest energy sublevel) from a gaseous atom:

$$\text{Atom}(g) \longrightarrow \text{ion}^+(g) + e^- \qquad \Delta E = IE_1 > 0 \tag{8.2}$$

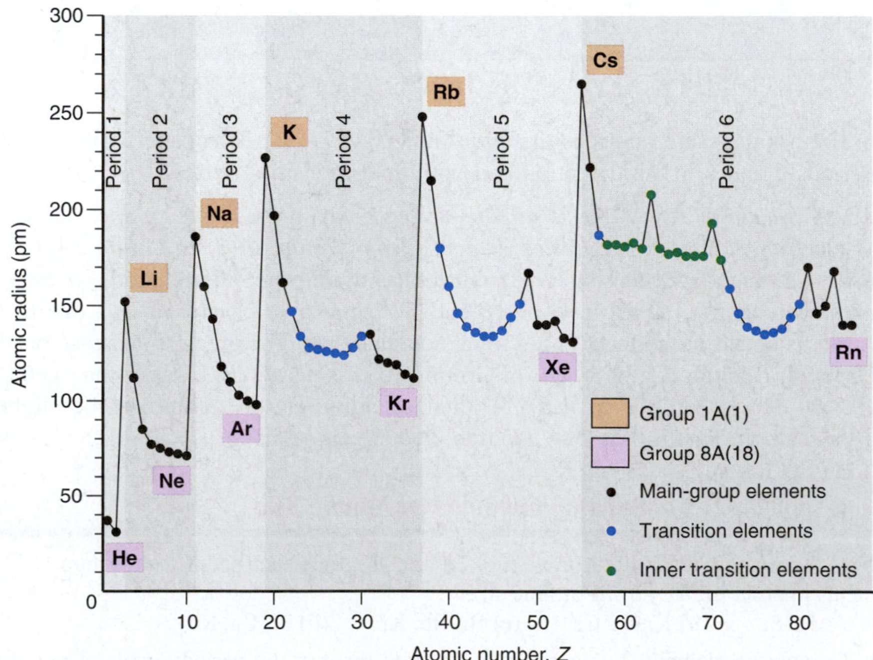

Figure 8.9 Periodicity of atomic radius.

The *second* ionization energy (IE_2) removes a second electron. Since this electron is pulled away from a positive ion, IE_2 is always larger than IE_1:

$$\text{Ion}^+(g) \longrightarrow \text{ion}^{2+}(g) + e^- \qquad \Delta E = IE_2 \text{ (always} > IE_1)$$

Periodicity of First Ionization Energy Figure 8.10 shows the variation in first ionization energy with atomic number. This up-and-down pattern—IE_1 rising across a period to the noble gas *(purple)* and then dropping down to the next alkali metal *(brown)*—is the *inverse* of the variation in atomic size (Figure 8.9): *as size decreases, it takes more energy to remove an electron because the nucleus is closer, so IE_1 increases.*

Let's examine the group and period trends and their exceptions:

1. *Down a group.* As we move *down* a main group, the *n* value increases, so atomic size does as well. As the distance from nucleus to outer electron increases, their attraction lessens, so the electron is easier to remove (Figure 8.11, *next page*):

Ionization energy generally ***decreases*** *down a group.*

The only significant exception occurs in Group 3A(13): IE_1 decreases from boron (B) to aluminum (Al), but not for the rest of the group. Filling the transition series in Periods 4, 5, and 6 causes a much higher Z_{eff} and an unusually small change in size, so outer electrons in the larger Group 3A members are held tighter.

2. *Across a period.* As we move left to right across a period, Z_{eff} increases and atomic size decreases. The attraction between nucleus and outer electron increases, so the electron is harder to remove:

Ionization energy generally ***increases*** *across a period.*

There are two exceptions to the otherwise smooth increase in IE_1 across periods:

- In Periods 2 and 3, there are dips at the Group 3A(13) elements, B and Al. These elements have the first *np* electrons, which are removed more easily because the resulting ion has a filled (stable) *ns* sublevel.
- In Periods 2 and 3, once again, there are dips at the Group 6A(16) elements, O and S. These elements have a fourth *np* electron, the first to pair up with another *np* electron, and electron-electron repulsions raise the orbital energy. The fourth *np* electron is easier to remove because doing so relieves the repulsions and leaves a half-filled (stable) *np* sublevel.

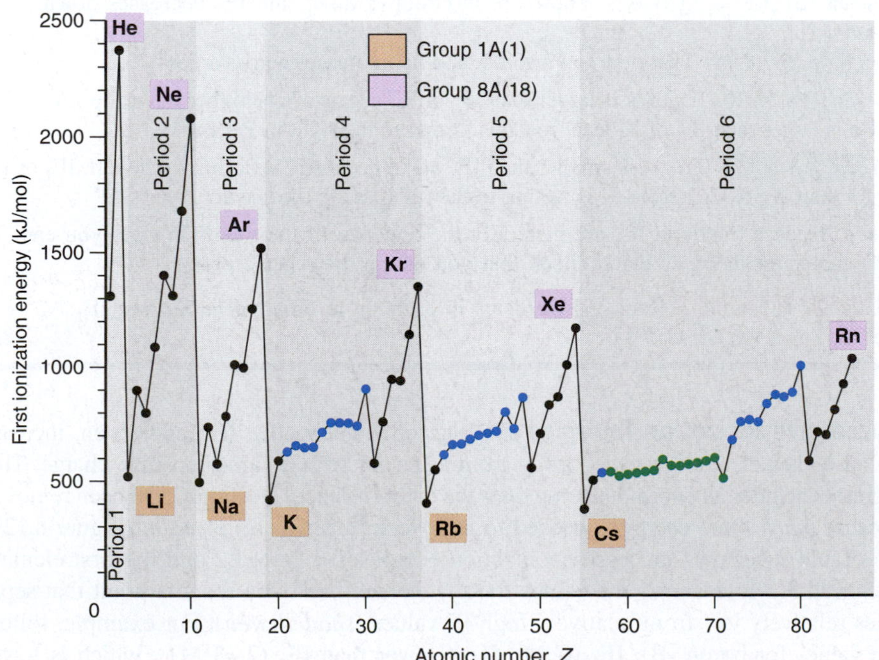

Figure 8.10 Periodicity of first ionization energy (IE_1). This trend is the *inverse* of the trend in atomic size (see Figure 8.9).

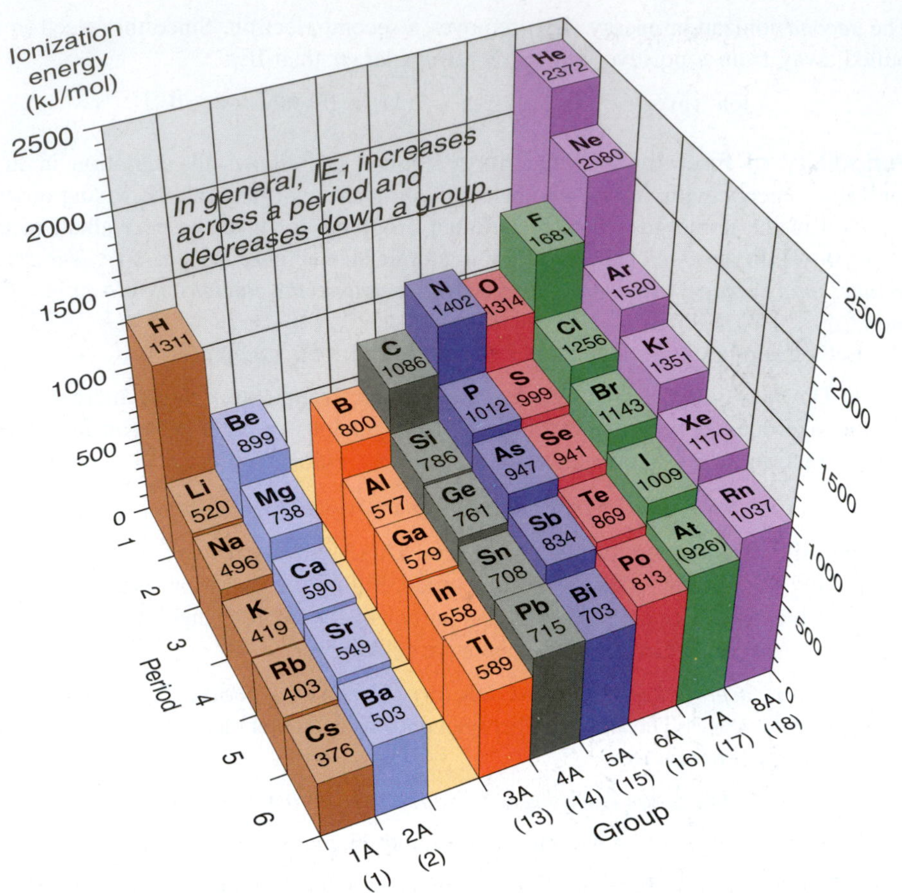

Figure 8.11 First ionization energies of the main-group elements.

Sample Problem 8.4 Ranking Elements by First Ionization Energy

Problem Using the periodic table only, rank the elements in each set in order of *decreasing* IE_1: **(a)** Kr, He, Ar; **(b)** Sb, Te, Sn; **(c)** K, Ca, Rb; **(d)** I, Xe, Cs.

Plan We find the elements in the periodic table and then apply the general trends of decreasing IE_1 down a group and increasing IE_1 across a period.

Solution **(a)** He > Ar > Kr. These are in Group 8A(18), and IE_1 decreases down a group.

(b) Te > Sb > Sn. These are in Period 5, and IE_1 increases across a period.

(c) Ca > K > Rb. IE_1 of K is larger than IE_1 of Rb because K is higher in Group 1A(1). IE_1 of Ca is larger than IE_1 of K because Ca is farther to the right in Period 4.

(d) Xe > I > Cs. IE_1 of I is smaller than IE_1 of Xe because I is farther to the left. IE_1 of I is larger than IE_1 of Cs because I is farther to the right and in the previous period.

Check Because trends in IE_1 are generally the opposite of the trends in size, you can rank the elements by size and check that you obtain the reverse order.

FOLLOW-UP PROBLEM 8.4 Rank the elements in each set in order of *increasing* IE_1: **(a)** Sb, Sn, I; **(b)** Sr, Ca, Ba.

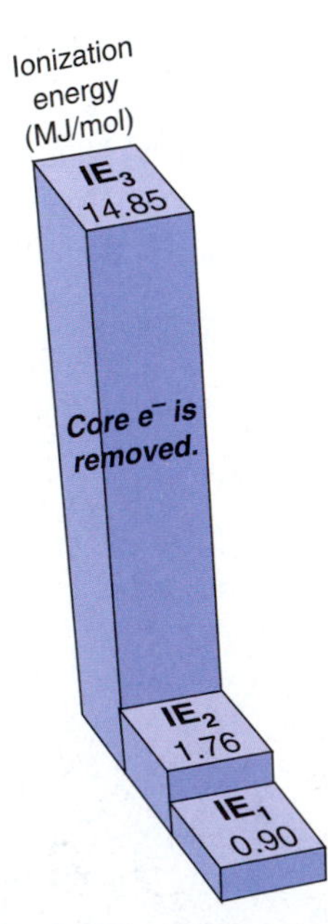

Figure 8.12 The first three ionization energies of beryllium. Beryllium has two valence electrons, so IE_3 is much larger than IE_2.

Successive Ionization Energies For a given element, IE_1, IE_2, and so on, increase because each electron is pulled away from a species with a higher positive charge. This increase includes an enormous jump *after* the outer (valence) electrons have been removed because *much* more energy is needed to remove an inner (core) electron (Figure 8.12).

Table 8.4 shows successive ionization energies for Period 2 and the first element in Period 3. Move across the values for any element, and you reach a point that separates relatively low from relatively high IE values *(shaded area)*. For example, follow the values for boron (B): IE_1 (0.80 MJ) is lower than IE_2 (2.43 MJ), which is lower

Table 8.4 Successive Ionization Energies of the Elements Lithium Through Sodium

Z	Element	Number of Valence Electrons	Ionization Energy (MJ/mol)* IE_1	IE_2	IE_3	IE_4	IE_5	IE_6	IE_7	IE_8	IE_9	IE_{10}
3	Li	1	0.52	7.30	11.81							
4	Be	2	0.90	1.76	14.85	21.01			CORE ELECTRONS			
5	B	3	0.80	2.43	3.66	25.02	32.82					
6	C	4	1.09	2.35	4.62	6.22	37.83	47.28				
7	N	5	1.40	2.86	4.58	7.48	9.44	53.27	64.36			
8	O	6	1.31	3.39	5.30	7.47	10.98	13.33	71.33	84.08		
9	F	7	1.68	3.37	6.05	8.41	11.02	15.16	17.87	92.04	106.43	
10	Ne	8	2.08	3.95	6.12	9.37	12.18	15.24	20.00	23.07	115.38	131.43
11	Na	1	0.50	4.56	6.91	9.54	13.35	16.61	20.11	25.49	28.93	141.37

*MJ/mol, or megajoules per mole = 10^3 kJ/mol.

than IE_3 (3.66 MJ), which is *much* lower than IE_4 (25.02 MJ). From this jump, we know that boron has three electrons in the highest energy level ($1s^22s^22p^1$). Because they are so difficult to remove, *core electrons are not involved in reactions.*

Sample Problem 8.5 Identifying an Element from Its Ionization Energies

Problem Name the Period 3 element with the following ionization energies (kJ/mol), and write its full electron configuration:

IE_1	IE_2	IE_3	IE_4	IE_5	IE_6
1012	1903	2910	4956	6278	22,230

Plan We look for a large jump in the IE values, which occurs after all valence electrons have been removed. Then we refer to the periodic table to find the Period 3 element with this number of valence electrons and write its electron configuration.

Solution The large jump occurs after IE_5, indicating that the element has five valence electrons and, thus, is in Group 5A(15). This Period 3 element is phosphorus (P; Z = 15). Its electron configuration is $1s^22s^22p^63s^23p^3$.

FOLLOW-UP PROBLEM 8.5 Element Q is in Period 3 and has the following ionization energies (in kJ/mol):

IE_1	IE_2	IE_3	IE_4	IE_5	IE_6
577	1816	2744	11,576	14,829	18,375

Name element Q, and write its full electron configuration.

Trends in Electron Affinity

The **electron affinity (EA)** is the energy change (kJ/mol) accompanying the *addition* of 1 mol of electrons to 1 mol of gaseous atoms or ions. The *first electron affinity* (EA_1) refers to the formation of 1 mol of monovalent (1−) gaseous anions:

$$\text{Atom}(g) + e^- \longrightarrow \text{ion}^-(g) \qquad \Delta E = EA_1$$

As with ionization energy, there is a first electron affinity, a second, and so forth. The first electron is *attracted* by the atom's nucleus, so in most cases, *EA_1 is negative* (energy is released), analogous to the negative ΔH for an exothermic reaction.* But the second electron affinity (EA_2) is always positive because energy must be *absorbed* to overcome electrostatic repulsions and add another electron to a negative ion.

*Some tables of EA_1 list them as positive values because energy would be *absorbed* to remove an electron from the anion.

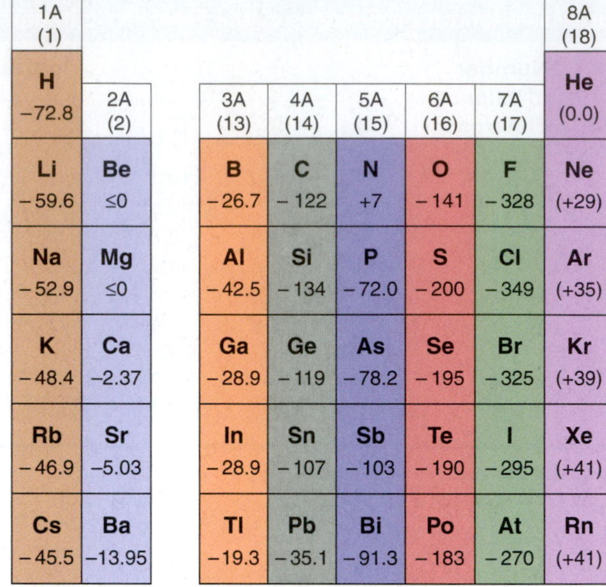

1A (1)	2A (2)	3A (13)	4A (14)	5A (15)	6A (16)	7A (17)	8A (18)
H −72.8							He (0.0)
Li −59.6	Be ≤0	B −26.7	C −122	N +7	O −141	F −328	Ne (+29)
Na −52.9	Mg ≤0	Al −42.5	Si −134	P −72.0	S −200	Cl −349	Ar (+35)
K −48.4	Ca −2.37	Ga −28.9	Ge −119	As −78.2	Se −195	Br −325	Kr (+39)
Rb −46.9	Sr −5.03	In −28.9	Sn −107	Sb −103	Te −190	I −295	Xe (+41)
Cs −45.5	Ba −13.95	Tl −19.3	Pb −35.1	Bi −91.3	Po −183	At −270	Rn (+41)

Figure 8.13 **Electron affinities of the main-group elements (in kJ/mol).** Values for Group 8A(18) are estimates, which is indicated by parentheses.

Factors other than Z_{eff} and atomic size affect electron affinities, so trends are not regular, as are those for size and IE_1. The many exceptions arise from changes in sublevel energy and electron-electron repulsion:

- *Down a group.* We might expect a smooth decrease (smaller negative number) down a group because size increases, and the nucleus is farther away from an electron being added. But only Group 1A(1) exhibits this behavior (Figure 8.13).
- *Across a period.* We might expect a regular increase (larger negative number) across a period because size decreases, and higher Z_{eff} should attract the electron being added more strongly. There is an overall left-to-right increase, but it is not at all regular.

Despite irregularities, relative values of IE and EA show three general patterns:

1. *Reactive nonmetals.* Members of Group 6A(16) and especially Group 7A(17) (the halogens) have high IEs and highly negative (exothermic) EAs: these elements lose electrons with difficulty but attract them strongly. Therefore, *in their ionic compounds, they form negative ions.*
2. *Reactive metals.* Members of Groups 1A(1) and 2A(2) have low IEs and slightly negative (exothermic) EAs: they lose electrons easily but attract them weakly, if at all. Therefore, *in their ionic compounds, they form positive ions.*
3. *Noble gases.* Members of Group 8A(18) have very high IEs and slightly positive (endothermic) EAs: *they tend **not** to lose or gain electrons.* In fact, only the larger members of the group (Kr, Xe, and Rn) form compounds at all.

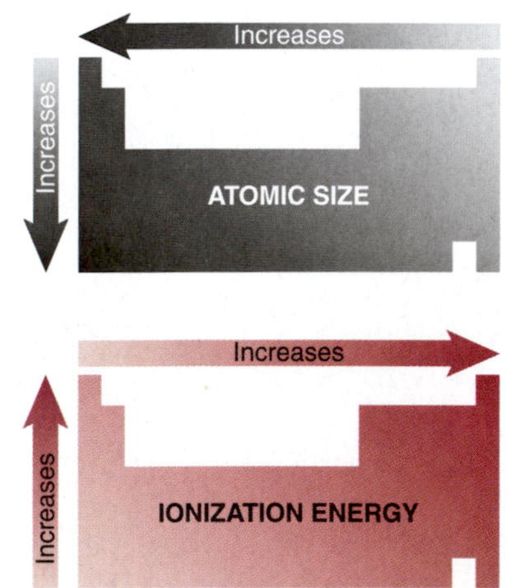

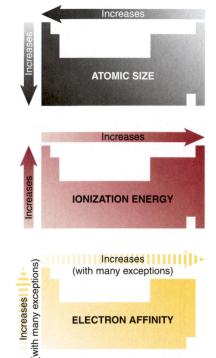

Figure 8.14 **Trends in three atomic properties.** For electron affinity, the dashed arrows indicate that there are numerous exceptions to expected trends.

Summary of Section 8.3

- Trends in three atomic properties are summarized in Figure 8.14.
- Atomic size (half the distance between nuclei of adjacent atoms) increases down a main group and decreases across a period. In a transition series, size remains relatively constant.
- First ionization energy (the energy required to remove the outermost electron from a mole of gaseous atoms) is inversely related to atomic size: IE_1 decreases down a main group and increases across a period.
- Successive ionization energies of an element show a very large increase after all valence electrons have been removed, because the first inner (core) electron is in an orbital of much lower energy and so is held very tightly.
- Electron affinity (the energy involved in adding an electron to a mole of gaseous atoms) shows many variations from expected trends.
- Based on the relative sizes of IEs and EAs, Group 1A(1) and 2A(2) elements tend to form cations and Group 6A(16) and 7A(17) elements tend to form anions in ionic compounds. Group 8A(18) elements are very unreactive.

8.4 • ATOMIC PROPERTIES AND CHEMICAL REACTIVITY

All physical and chemical behaviors of the elements and their compounds are based on electron configuration and effective nuclear charge. In this section, you'll see how atomic properties determine metallic behavior and the properties of ions.

Trends in Metallic Behavior

The three general classes of elements have distinguishing properties:

- *Metals,* found in the left and lower three-quarters of the periodic table, are typically shiny solids, have moderate to high melting points, are good conductors of heat and electricity, can be machined into wires and sheets, and lose electrons to nonmetals.
- *Nonmetals,* found in the upper right quarter of the table, are typically not shiny, have relatively low melting points, are poor conductors, are mostly crumbly solids or gases, and tend to gain electrons from metals.
- *Metalloids,* found between the other two classes, have intermediate properties.

Thus, *metallic behavior decreases left-to-right across a period and increases down a group in the periodic table* (Figure 8.15).

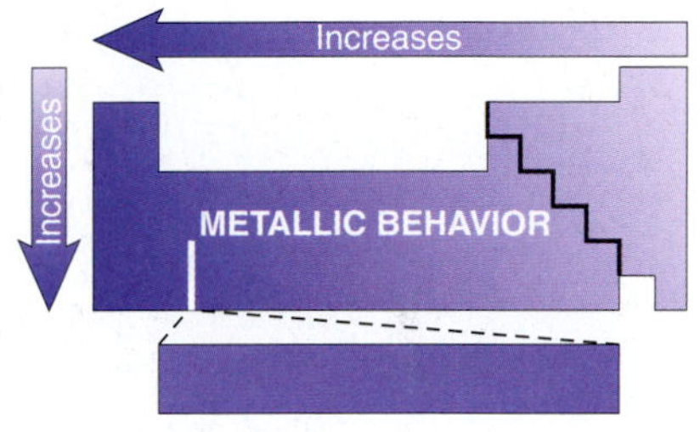

Figure 8.15 Trends in metallic behavior. (Hydrogen appears next to helium.)

Remember, though, that some elements don't fit these categories: as graphite, nonmetallic carbon is a good electrical conductor; the nonmetal iodine is shiny; metallic gallium melts in your hand; mercury is a liquid; and iron is brittle. Despite such exceptions, in this discussion, we'll make several generalizations about metallic behavior and its application to the acid-base behavior of oxides.

Relative Tendency to Lose or Gain Electrons Metals tend to lose electrons to nonmetals during reactions:

1. *Down a main group.* The increase in metallic behavior down a group is consistent with an increase in size and a decrease in IE and is most obvious in groups with more than one class of element, such as Group 5A(15): *elements at the top can form anions, and those at the bottom can form cations.* Nitrogen (N) is a gaseous nonmetal, and phosphorus (P) is a soft nonmetal; both occur occasionally as 3− anions in their compounds. Arsenic (As) and antimony (Sb) are metalloids, with Sb the more metallic, and neither forms ions readily. Bismuth (Bi) is a typical metal, forming a 3+ cation in its mostly ionic compounds. Groups 3A(13), 4A(14), and 6A(16) show a similar trend. But even in Group 2A(2), which contains only metals, the tendency to form cations increases down the group: beryllium (Be) forms covalent compounds with nonmetals, whereas all compounds of barium (Ba) are ionic.

2. *Across a period.* The decrease in metallic behavior across a period is consistent with a decrease in size, an increase in IE, and a more favorable (more negative) EA. Consider Period 3: *elements at the left tend to form cations, and those at the right tend to form anions.* Sodium and magnesium are metals that occur as Na^+ and Mg^{2+} in seawater, minerals, and organisms. Aluminum is a metallic element and occurs as Al^{3+} in some compounds, but it bonds covalently in most others. Silicon (Si) is a shiny metalloid that does not occur as a monatomic ion. Phosphorus is a white, waxy nonmetal that occurs rarely as P^{3-}, whereas crumbly, yellow sulfur forms S^{2-} in many compounds, and gaseous, yellow-green chlorine occurs in nature almost always as Cl^-.

Acid-Base Behavior of Oxides Metals are also distinguished from nonmetals by the acid-base behavior of their oxides in water:

- Most main-group metals *transfer* electrons to oxygen, so their *oxides are ionic. In water, these oxides act as bases,* producing OH^- ions from O^{2-} and reacting with acids.
- Nonmetals *share* electrons with oxygen, so *nonmetal oxides are covalent. In water, these oxides act as acids,* producing H^+ ions and reacting with bases.

Some metals and many metalloids form oxides that are **amphoteric:** they can act as acids *or* bases in water.

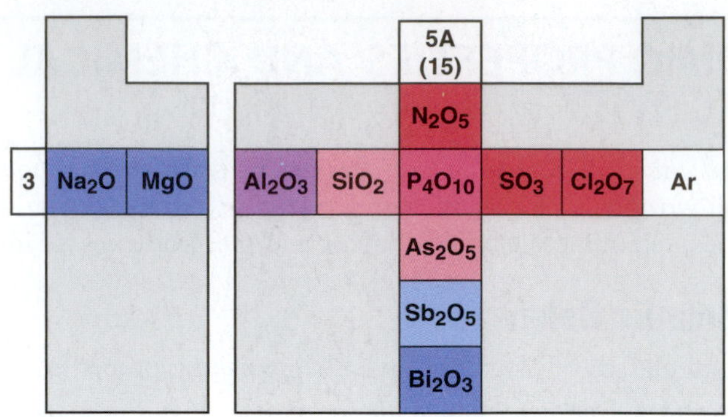

Figure 8.16 Acid-base behavior of some element oxides.

In Figure 8.16, the acid-base behavior of some common oxides of elements in Group 5A(15) and Period 3 is shown with a gradient from blue (basic) to red (acidic):

1. *As elements become more metallic down a group (larger size and smaller IE), their oxides become more basic.* In Group 5A(15), dinitrogen pentoxide, N_2O_5, forms the strong acid HNO_3:

$$N_2O_5(s) + H_2O(l) \longrightarrow 2HNO_3(aq)$$

Tetraphosphorus decoxide, P_4O_{10}, forms the weaker acid H_3PO_4:

$$P_4O_{10}(s) + 6H_2O(l) \longrightarrow 4H_3PO_4(aq)$$

The oxide of the metalloid arsenic is weakly acidic, whereas that of the metalloid antimony is weakly basic. Bismuth, the most metallic of the group, forms a basic oxide that is insoluble in water but reacts with acid to yield a salt and water:

$$Bi_2O_3(s) + 6HNO_3(aq) \longrightarrow 2Bi(NO_3)_3(aq) + 3H_2O(l)$$

2. *As the elements become less metallic across a period (smaller size and higher IE), their oxides become more acidic.* In Period 3, Na_2O and MgO are strongly basic, and amphoteric aluminum oxide (Al_2O_3) reacts with acid or with base:

$$Al_2O_3(s) + 6HCl(aq) \longrightarrow 2AlCl_3(aq) + 3H_2O(l)$$

$$Al_2O_3(s) + 2NaOH(aq) + 3H_2O(l) \longrightarrow 2NaAl(OH)_4(aq)$$

Silicon dioxide is weakly acidic, forming a salt and water with base:

$$SiO_2(s) + 2NaOH(aq) \longrightarrow Na_2SiO_3(aq) + H_2O(l)$$

The common oxides of phosphorus, sulfur, and chlorine form acids of increasing strength: H_3PO_4, H_2SO_4, and $HClO_4$.

Properties of Monatomic Ions

So far we have focused on the reactants—the atoms—in the process of electron loss and gain. Now we focus on the products—the ions—considering their electron configurations, magnetic properties, and sizes.

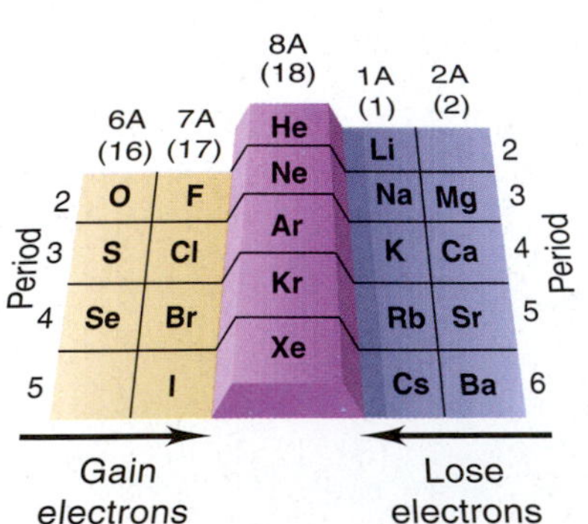

Figure 8.17 Main-group elements whose ions have noble gas electron configurations.

Electron Configurations of Main-Group Ions Why does an ion have a particular charge: Na^+ not Na^{2+}, or F^- not F^{2-}? Why do some metals form two ions, such as Sn^{2+} and Sn^{4+}? The answer relates to the location of the element in the periodic table and the energy associated with losing or gaining electrons:

1. *Ions with a noble gas configuration.* Atoms of the noble gases have very low reactivity because their highest energy level is *filled* (ns^2np^6). Thus, *when elements at either end of a period form ions, they attain a filled outer level—a noble gas configuration.* These elements lie on either side of Group 8A(18), and their ions are **isoelectronic** (Greek *iso,* "same") with the nearest noble gas (Figure 8.17).

- Elements in Groups 1A(1) and 2A(2) *lose* electrons and become isoelectronic with the *previous* noble gas. The Na^+ ion, for example, is isoelectronic with neon (Ne):

$$\text{Na } (1s^22s^22p^63s^1) \longrightarrow e^- + Na^+ \text{ ([He] } 2s^22p^6) \text{ [isoelectronic with Ne ([He] } 2s^22p^6)]$$

- Elements in Groups 6A(16) and 7A(17) *gain* electrons and become isoelectronic with the *next* noble gas. The Br^- ion, for example, is isoelectronic with krypton (Kr):

$$\text{Br ([Ar] } 4s^23d^{10}4p^5) + e^- \longrightarrow Br^- \text{ ([Ar] } 4s^23d^{10}4p^6)$$

$$\text{[isoelectronic with Kr ([Ar] } 4s^23d^{10}4p^6)]$$

The energy needed to remove electrons from metals or add them to nonmetals determines the charges of the resulting ions:

- *Cations*. Removing another electron from Na^+ or from Mg^{2+} means removing a core electron, which requires too much energy: thus, $NaCl_2$ and MgF_3 do *not* exist.
- *Anions*. Similarly, adding another electron to F^- or to O^{2-} means putting it into the next higher energy level ($n = 3$). With 10 electrons ($1s^22s^22p^6$) acting as inner electrons, the nuclear charge would be shielded very effectively, and adding an outer electron would require too much energy: thus, we never see Na_2F or Mg_3O_2.

2. *Ions without a noble gas configuration*. Except for aluminum, the metals of Groups 3A(13) to 5A(15) do not form ions with noble gas configurations. Instead, they form cations with two different stable configurations:

- *Pseudo–noble gas configuration.* If the metal atom empties its highest energy level, it attains the stability of empty *ns* and *np* sublevels and filled inner $(n - 1)d$ sublevel. This $(n - 1)d^{10}$ configuration is called a **pseudo–noble gas configuration.** For example, tin (Sn; $Z = 50$) loses four electrons to form the tin(IV) ion (Sn^{4+}), which has empty 5*s* and 5*p* sublevels and a filled inner 4*d* sublevel:

$$\text{Sn ([Kr] } 5s^24d^{10}5p^2) \longrightarrow Sn^{4+} \text{ ([Kr] } 4d^{10}) + 4e^-$$

- *Inert pair configuration*. Alternatively, the metal atom loses just its *np* electrons and attains a stable configuration with filled *ns* and $(n - 1)d$ sublevels. The retained ns^2 electrons are sometimes called an *inert pair*. For example, in the more common tin(II) ion (Sn^{2+}), the atom loses the two 5*p* electrons and has filled 5*s* and 4*d* sublevels:

$$\text{Sn ([Kr] } 5s^24d^{10}5p^2) \longrightarrow Sn^{2+} \text{ ([Kr] } 5s^24d^{10}) + 2e^-$$

Thallium, lead, and bismuth, the largest and, thus, most metallic members of Groups 3A(13) to 5A(15), form ions that retain the ns^2 pair: Tl^+, Pb^{2+}, and Bi^{3+}.

Once again, energy considerations explain these configurations. It would be energetically impossible for metals in Groups 3A(13) to 5A(15) to achieve noble gas configurations: tin, for example, would have to lose 14 electrons—ten 4*d* in addition to the two 5*p* and two 5*s*—to be isoelectronic with krypton (Kr; $Z = 36$), the previous noble gas.

Sample Problem 8.6 Writing Electron Configurations of Main-Group Ions

Problem Using condensed electron configurations, write equations representing the formation of the ion(s) of the following elements:

(a) Iodine ($Z = 53$) **(b)** Potassium ($Z = 19$) **(c)** Indium ($Z = 49$)

Plan We identify the element's position in the periodic table and recall that

- Ions of elements in Groups 1A(1), 2A(2), 6A(16), and 7A(17) are isoelectronic with the nearest noble gas.
- Metals in Groups 3A(13) to 5A(15) lose the *ns* and *np* electrons or just the *np*.

Solution (a) Iodine is in Group 7A(17), so it gains one electron, and I^- is isoelectronic with xenon:

$$\text{I ([Kr] } 5s^24d^{10}5p^5) + e^- \longrightarrow I^- \text{ ([Kr] } 5s^24d^{10}5p^6) \qquad \text{(same as Xe)}$$

(b) Potassium is in Group 1A(1), so it loses one electron; K^+ is isoelectronic with argon:

$$\text{K ([Ar] } 4s^1) \longrightarrow K^+ \text{ ([Ar])} + e^-$$

(c) Indium is in Group 3A(13), so it loses either three electrons to form In^{3+} (with a pseudo–noble gas configuration) or one to form In^{+} (with an inert pair):

$$\text{In ([Kr] } 5s^2 4d^{10} 5p^1) \longrightarrow In^{3+} \text{ ([Kr] } 4d^{10}) + 3e^-$$

$$\text{In ([Kr] } 5s^2 4d^{10} 5p^1) \longrightarrow In^{+} \text{ ([Kr] } 5s^2 4d^{10}) + e^-$$

Check Be sure that the number of electrons in the ion's electron configuration, plus the number gained or lost to form the ion, equals Z.

FOLLOW-UP PROBLEM 8.6 Using condensed electron configurations, write equations representing the formation of the ion(s) of the following elements:
(a) Ba (Z = 56) **(b)** O (Z = 8) **(c)** Pb (Z = 82)

Electron Configurations of Transition Metal Ions In contrast to many main-group ions, *transition metal ions rarely attain a noble gas configuration.* Aside from the Period 4 elements scandium, which forms Sc^{3+}, and titanium, which occasionally forms Ti^{4+}, *a transition element typically forms more than one cation by losing all of its ns and some of its (n − 1)d electrons.*

The reason, once again, is that energy costs are too high. Let's consider again the filling of Period 4. At the beginning of Period 4 (the same point holds in other periods), penetration makes the 4*s* sublevel *more stable* than the 3*d*. Therefore, the first and second electrons added enter the 4*s*, which is the outer sublevel. But, the 3*d* is an *inner* sublevel, so as it begins to fill, its electrons are not well shielded from the increasing nuclear charge.

A *crossover in sublevel energy* results: the 3*d* becomes *more stable* than the 4*s* in the transition series (Figure 8.18). This crossover has a major effect on the formation of Period 4 transition metal ions: because the 3*d* electrons are held tightly and shield those in the outer sublevel, *the 4s electrons of a transition metal are lost **before** the 3d electrons.* Thus, 4*s* electrons are added before 3*d* electrons to form the *atom* and are lost before them to form the *ion,* the so-called "first-in, first-out" rule.

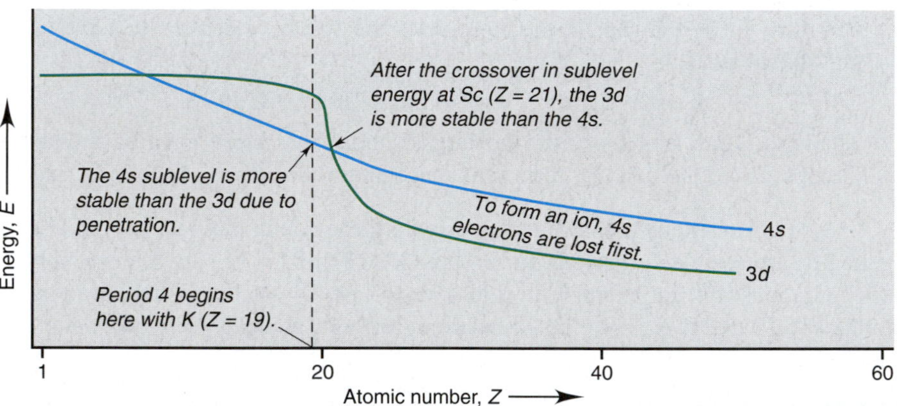

Figure 8.18 The crossover of sublevel energies in Period 4.

Ion Formation: A Summary of Electron Loss or Gain The various ways that cations form have one point in common—*outer electrons are removed first.* Here is a summary of the rules for formation of any main-group or transition metal ion:

- Main-group *s*-block metals lose all electrons with the highest *n* value.
- Main-group *p*-block metals lose *np* electrons before *ns* electrons.
- Transition (*d*-block) metals lose *ns* electrons before (*n* − 1)*d* electrons.
- Nonmetals gain electrons in the *p* orbitals of highest *n* value.

Magnetic Properties of Transition Metal Ions We learn a great deal about an element's electron configuration from atomic spectra, and magnetic studies provide additional evidence.

Recall that electron spin generates a tiny magnetic field, which causes a beam of H atoms to split in an external magnetic field (see Figure 8.1). Only a beam of a species (atoms, ions, or molecules) with one or more *unpaired* electrons will split. A beam of silver atoms (Ag; $Z = 47$) was used in the original 1921 experiment:

Ag ($Z = 47$) [Kr] $5s^1 4d^{10}$ [↑] [↑↓|↑↓|↑↓|↑↓|↑↓] [| |]
5s 4d 5p

Note the unpaired 5*s* electron. A beam of cadmium atoms (Cd; $Z = 48$) is not split because their 5*s* electrons are *paired* ([Kr] $5s^2 4d^{10}$).

A species with unpaired electrons exhibits **paramagnetism:** it is attracted by an external field. A species with all of its electrons paired exhibits **diamagnetism:** it is not attracted (and is slightly repelled) by the field (Figure 8.19). Many transition metals and their compounds are paramagnetic because their atoms and ions have unpaired electrons.

Let's see how magnetic studies might provide evidence for a proposed electron configuration. Spectral analysis of titanium metal yields the electron configuration [Ar] $4s^2 3d^2$, and experiment shows that the metal is paramagnetic, which indicates the presence of unpaired electrons. Spectral analysis shows that the Ti^{2+} ion is [Ar] $3d^2$, indicating loss of the 4*s* electrons. In support of the spectra, magnetic studies show that Ti^{2+} compounds are paramagnetic. If Ti had lost its 3*d* electrons to form Ti^{2+}, its compounds would be diamagnetic:

$$\text{Ti ([Ar] } 4s^2 3d^2) \longrightarrow \text{Ti}^{2+} \text{ ([Ar] } 3d^2) + 2e^-$$

The partial orbital diagrams are

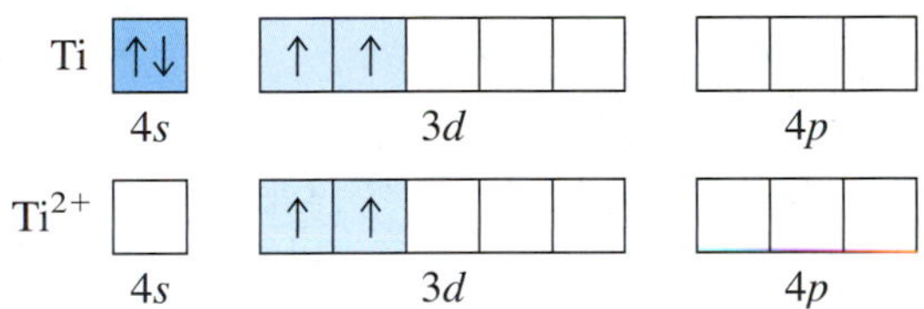

Sample Problem 8.7 Writing Electron Configurations and Predicting Magnetic Behavior of Transition Metal Ions

Problem Use condensed electron configurations to write an equation for the formation of each transition metal ion, and predict whether it is paramagnetic:
(a) Mn^{2+} ($Z = 25$) **(b)** Cr^{3+} ($Z = 24$) **(c)** Hg^{2+} ($Z = 80$)

Plan We first write the condensed electron configuration of the atom, recalling the irregularity for Cr. Then we remove electrons, beginning with *ns* electrons, to attain the ion charge. If unpaired electrons are present, the ion is paramagnetic.

Solution (a) Mn ([Ar] $4s^2 3d^5$) $\longrightarrow$ Mn^{2+} ([Ar] $3d^5$) + $2e^-$
There are five unpaired e^-, so Mn^{2+} is paramagnetic.

(b) Cr ([Ar] $4s^1 3d^5$) $\longrightarrow$ Cr^{3+} ([Ar] $3d^3$) + $3e^-$
There are three unpaired e^-, so Cr^{3+} is paramagnetic.

(c) Hg ([Xe] $6s^2 4f^{14} 5d^{10}$) $\longrightarrow$ Hg^{2+} ([Xe] $4f^{14} 5d^{10}$) + $2e^-$
The 4*f* and 5*d* sublevels are filled, so there are no unpaired e^-: Hg^{2+} is *not* paramagnetic.

Check We removed the *ns* electrons first, and the sum of the lost electrons and those in the electron configuration of the ion equals *Z*.

FOLLOW-UP PROBLEM 8.7 Write the condensed electron configuration of each transition metal ion, and predict whether it is paramagnetic:
(a) V^{3+} ($Z = 23$) **(b)** Ni^{2+} ($Z = 28$) **(c)** La^{3+} ($Z = 57$)

Ionic Size vs. Atomic Size The **ionic radius** is a measure of the size of an ion and is obtained from the distance between the nuclei of adjacent ions in a crystalline ionic compound (Figure 8.20). From the relation between effective nuclear charge (Z_{eff}) and atomic size, we can predict the size of an ion relative to its parent atom:

- *Cations are smaller than parent atoms.* When a cation forms, electrons are *removed from* the outer level. The resulting decrease in shielding due to fewer electron repulsions allows the nucleus to pull the remaining electrons closer.

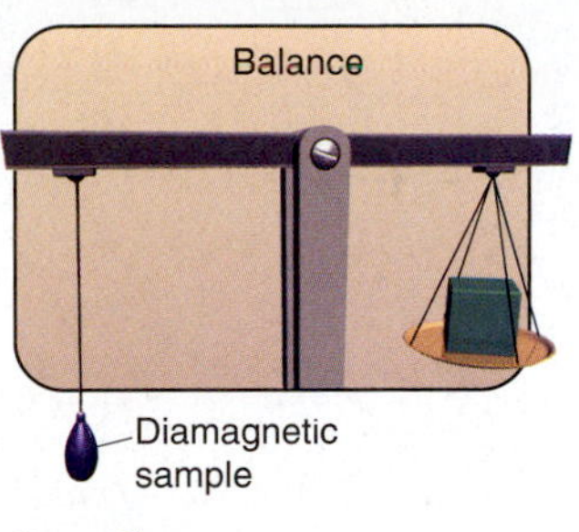

N S
A Electromagnet

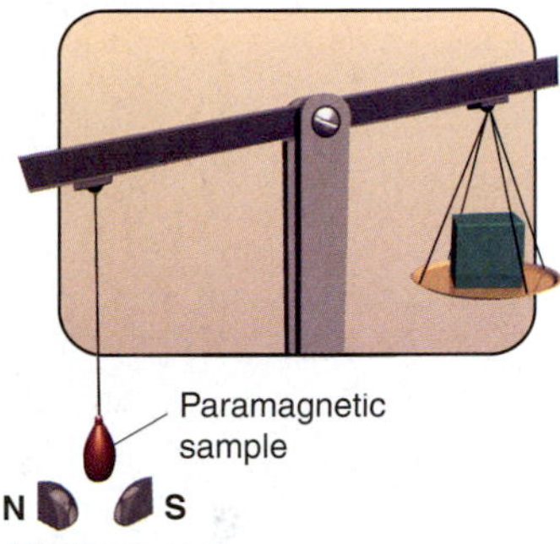

B Electromagnet

Figure 8.19 Measuring the magnetic behavior of a sample. The substance is weighed with the external magnetic field "off." **A,** If the substance is diamagnetic (has all *paired* electrons), its apparent mass is unaffected (or slightly reduced) with the field "on." **B,** If the substance is paramagnetic (has *unpaired* electrons), its apparent mass increases.

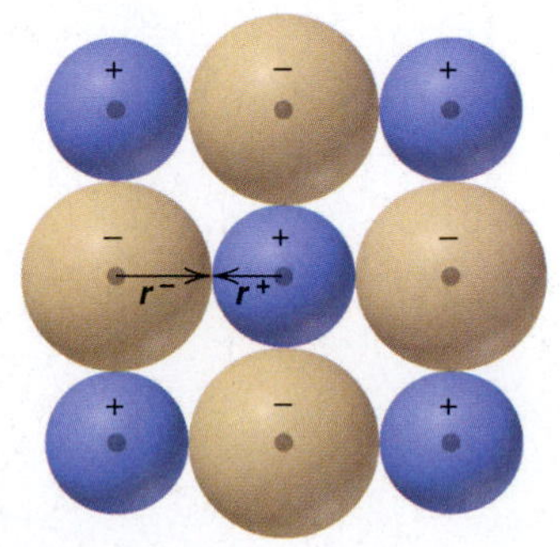

Figure 8.20 Ionic radius. Cation radius (r^+) and anion radius (r^-) together make up the distance between nuclei.

- *Anions are larger than parent atoms.* When an anion forms, electrons are *added to* the outer level. The increase in shielding due to greater electron repulsions means the electrons occupy more space.

Figure 8.21 shows the radii of some main-group ions and their parent atoms:

1. *Down a group, ionic size increases* because *n* increases.
2. *Across a period,* for instance, Period 3, the pattern is complex:
   - *Among cations,* the increase in Z_{eff} from left to right makes Na^+ larger than Mg^{2+}, which is larger than Al^{3+}.
   - *From last cation to first anion,* a great jump in size occurs: we are *adding* electrons rather than removing them, so repulsions increase sharply. For instance, P^{3-} has eight more electrons than Al^{3+}.
   - *Among anions,* the increase in Z_{eff} from left to right makes P^{3-} larger than S^{2-}, which is larger than Cl^-.
   - *Within an isoelectronic series,* these factors have striking results. Within the dashed outline in Figure 8.21, the ions are isoelectronic with neon. Period 2 anions are much larger than Period 3 cations because the same number of electrons are attracted by an increasing nuclear charge. The pattern is

$$3- > 2- > 1- > 1+ > 2+ > 3+$$

3. *Cation size decreases with charge.* When a metal forms more than one cation, *the greater the ionic charge, the smaller the ionic radius.* Of the two ions of iron, for example, Fe^{3+} has one fewer electron, so shielding is reduced somewhat, and the same nucleus is attracting fewer electrons. As a result, Z_{eff} increases, so Fe^{3+} (65 pm) is smaller than Fe^{2+} (78 pm).

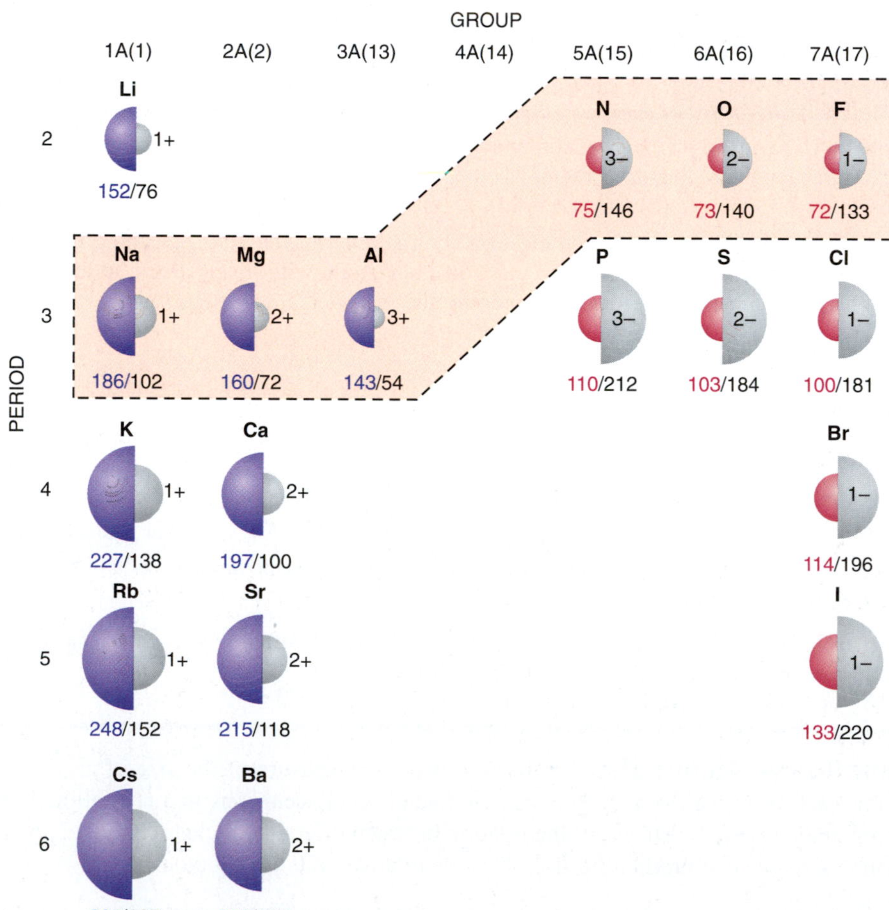

Figure 8.21 Ionic vs. atomic radii. Atomic radii *(color)* and ionic radii *(gray)* are given in picometers. Metal atoms *(blue)* form *smaller* positive ions, and nonmetal atoms *(red)* form *larger* negative ions. Ions in the dashed outline are *isoelectronic* with neon.

Sample Problem 8.8 Ranking Ions by Size

Problem Rank each set of ions in order of *decreasing* size, and explain your ranking:
(a) Ca^{2+}, Sr^{2+}, Mg^{2+} **(b)** K^+, S^{2-}, Cl^- **(c)** Au^+, Au^{3+}

Plan We find the position of each element in the periodic table and apply the ideas presented in the text.

Solution **(a)** Mg^{2+}, Ca^{2+}, and Sr^{2+} are all from Group 2A(2), so their sizes decrease up the group: $Sr^{2+} > Ca^{2+} > Mg^{2+}$.
(b) The ions K^+, S^{2-}, and Cl^- are isoelectronic. S^{2-} has a lower Z_{eff} than Cl^-, so it is larger. K^+ is a cation and has the highest Z_{eff}, so it is smallest: $S^{2-} > Cl^- > K^+$.
(c) Au^+ has a lower charge than Au^{3+}, so it is larger: $Au^+ > Au^{3+}$.

FOLLOW-UP PROBLEM 8.8 Rank the ions in each set in order of *increasing* size:
(a) Cl^-, Br^-, F^- **(b)** Na^+, Mg^{2+}, F^- **(c)** Cr^{2+}, Cr^{3+}

Summary of Section 8.4

- Metallic behavior correlates with large atomic size and low ionization energy. Thus, metallic behavior increases down a group and decreases across a period.
- Within the main groups, metal oxides are basic and nonmetal oxides acidic. Thus, oxides become more acidic across a period and more basic down a group.
- Many main-group elements form ions that are isoelectronic with the nearest noble gas. Removing (or adding) more electrons than needed to attain the noble gas configuration requires a prohibitive amount of energy.
- Metals in Groups 3A(13) to 5A(15) lose either their *np* electrons or both their *ns* and *np* electrons.
- Transition metals lose *ns* electrons before $(n - 1)d$ electrons and commonly form more than one ion.
- Many transition metals and their compounds are paramagnetic because their atoms (or ions) have unpaired electrons.
- Cations are smaller and anions larger than their parent atoms. Ionic radius increases down a group. Across a period, ionic radii generally decrease, but a large increase occurs from the last cation to the first anion.

CHAPTER REVIEW GUIDE

The following sections provide many aids to help you study this chapter. (Numbers in parentheses refer to pages, unless noted otherwise.)

Learning Objectives

These are concepts and skills to review after studying this chapter.

Related section (§), sample problem (SP), and upcoming end-of-chapter problem (EP) numbers are listed in parentheses.

1. Understand the periodic law and the arrangement of elements by atomic number (Introduction) (EPs 8.1–8.3)
2. Describe the importance of the spin quantum number (m_s) and the exclusion principle for populating an orbital; understand how shielding and penetration lead to the splitting of energy levels into sublevels (§8.1) (EPs 8.4–8.13)
3. Understand orbital filling order, how outer configuration correlates with chemical behavior, and the distinction among inner, outer, and valence electrons; write the set of quantum numbers for any electron in an atom as well as full and condensed electron configurations and orbital diagrams for the atoms of any element (§8.2) (SPs 8.1, 8.2) (EPs 8.14–8.32)
4. Describe atomic size, ionization energy, and electron affinity and their periodic trends; explain patterns in successive ionization energies and identify which electrons are involved in ion formation (to yield a noble gas or pseudo–noble gas electron configuration) (§8.3) (SPs 8.3–8.5) (EPs 8.33–8.47)
5. Describe the general properties of metals and nonmetals and understand how trends in metallic behavior relate to ion formation, oxide acidity, and magnetic behavior; understand the relation between atomic and ionic size and write ion electron configurations (§8.4) (SPs 8.6–8.8) (EPs 8.48–8.64)

Key Terms

These important terms appear in boldface in the chapter and are defined again in the Glossary.

Introduction
periodic law (246)
electron configuration (246)

Section 8.1
spin quantum number (m_s) (247)
exclusion principle (247)
shielding (248)
effective nuclear charge (Z_{eff}) (248)
penetration (248)

Section 8.2
aufbau principle (249)
orbital diagram (249)
Hund's rule (250)
transition elements (253)
inner (core) electrons (256)
outer electrons (256)
valence electrons (256)
inner transition elements (256)
lanthanides (rare earths) (256)
actinides (256)

Section 8.3
atomic size (258)
metallic radius (258)
covalent radius (258)
ionization energy (IE) (260)
electron affinity (EA) (263)

Section 8.4
amphoteric (265)
isoelectronic (266)
pseudo–noble gas configuration (267)
paramagnetism (269)
diamagnetism (269)
ionic radius (269)

Key Equations and Relationships

Numbered and screened concepts are listed for you to refer to or memorize.

8.1 Defining the energy order of sublevels in terms of the angular momentum quantum number (l value) (249):

Order of sublevel energies: $s < p < d < f$

8.2 Meaning of the first ionization energy (260):

$$\text{Atom}(g) \longrightarrow \text{ion}^+(g) + e^- \qquad \Delta E = IE_1 > 0$$

BRIEF SOLUTIONS TO FOLLOW-UP PROBLEMS

Compare your own solutions to these calculation steps and answers.

8.1 The element has eight electrons, so $Z = 8$: oxygen.

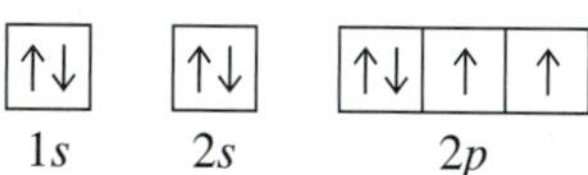

Sixth electron: $n = 2, l = 1, m_l = 0, +\frac{1}{2}$

8.2 (a) For Ni, $1s^22s^22p^63s^23p^64s^23d^8$; [Ar] $4s^23d^8$

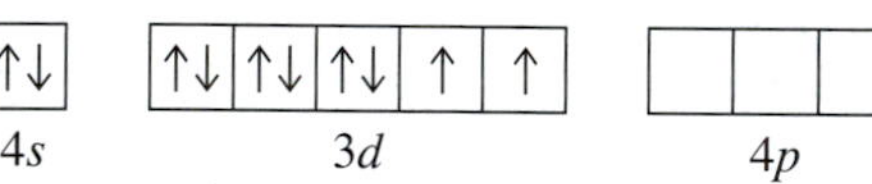

Ni has 18 inner electrons.

(b) For Sr, $1s^22s^22p^63s^23p^64s^23d^{10}4p^65s^2$; [Kr] $5s^2$

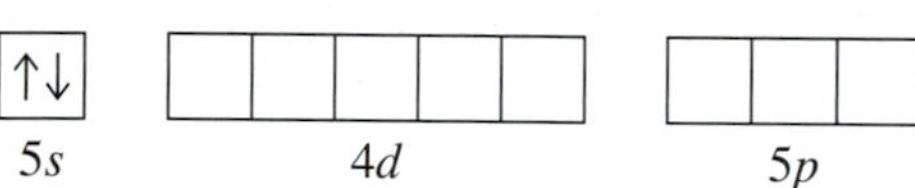

Sr has 36 inner electrons.

(c) For Po, $1s^22s^22p^63s^23p^64s^23d^{10}4p^65s^24d^{10}5p^66s^24f^{14}5d^{10}6p^4$; [Xe] $6s^24f^{14}5d^{10}6p^4$

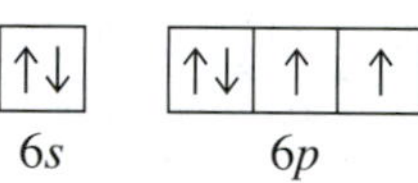

Po has 78 inner electrons.

8.3 (a) Cl < Br < Se; (b) Xe < I < Ba

8.4 (a) Sn < Sb < I; (b) Ba < Sr < Ca

8.5 Q is aluminum: $1s^22s^22p^63s^23p^1$

8.6 (a) Ba ([Xe] $6s^2$) $\longrightarrow$ Ba^{2+} ([Xe]) + $2e^-$

(b) O ([He] $2s^22p^4$) + $2e^-$ $\longrightarrow$ O^{2-} ([He] $2s^22p^6$) (same as Ne)

(c) Pb ([Xe] $6s^24f^{14}5d^{10}6p^2$) $\longrightarrow$ Pb^{2+} ([Xe] $6s^24f^{14}5d^{10}$) + $2e^-$

Pb ([Xe] $6s^24f^{14}5d^{10}6p^2$) $\longrightarrow$ Pb^{4+} ([Xe] $4f^{14}5d^{10}$) + $4e^-$

8.7 (a) V^{3+}: [Ar] $3d^2$; paramagnetic

(b) Ni^{2+}: [Ar] $3d^8$; paramagnetic

(c) La^{3+}: [Xe]; not paramagnetic (diamagnetic)

8.8 (a) $F^- < Cl^- < Br^-$

(b) $Mg^{2+} < Na^+ < F^-$

(c) $Cr^{3+} < Cr^{2+}$

PROBLEMS

Problems with **colored** numbers are answered in Appendix E. Sections match the text and provide the numbers of relevant sample problems. Bracketed problems are grouped in pairs (indicated by a short rule) that cover the same concept. Comprehensive Problems are based on material from any section or previous chapter.

Introduction

8.1 What would be your reaction to a claim that a new element had been discovered and it fit between tin (Sn) and antimony (Sb) in the periodic table?

8.2 In an early 20th century study of atomic x-ray spectra, British physicist Henry Moseley discovered a relationship that replaced atomic mass as the criterion for ordering the elements. By what criterion are the elements now ordered in the periodic table? Give an example of a sequence of element order that was confirmed by Moseley's findings.

8.3 Before Mendeleev published his periodic table, German chemist Johann Döbereiner grouped elements with similar properties into "triads," in which the unknown properties of one member

could be predicted by averaging known values of the properties of the others. To test this idea, predict the values of the following quantities:
(a) The atomic mass of K from the atomic masses of Na and Rb
(b) The melting point of Br_2 from the melting points of Cl_2 (−101.0°C) and I_2 (113.6°C) (actual value = −7.2°C)

Characteristics of Many-Electron Atoms

8.4 Summarize the rules for the allowable values of the four quantum numbers of an electron in an atom.

8.5 Which of the quantum numbers relate(s) to the electron only? Which relate(s) to the orbital?

8.6 State the exclusion principle. What does it imply about the number and spin of electrons in an atomic orbital?

8.7 What is the key distinction between sublevel energies in one-electron species, such as the H atom, and those in many-electron species, such as the C atom? What factors lead to this distinction? Would you expect the pattern of sublevel energies in Be^{3+} to be more like that in H or that in C? Explain.

8.8 Define *shielding* and *effective nuclear charge*. What is the connection between the two?

8.9 What is penetration? How is it related to shielding? Use the penetration effect to explain the difference in relative orbital energies of a 3*p* and a 3*d* electron in the same atom.

8.10 How many electrons in an atom can have each of the following quantum number or sublevel designations?
(a) $n = 2, l = 1$ (b) 3*d* (c) 4*s*

8.11 How many electrons in an atom can have each of the following quantum number or sublevel designations?
(a) $n = 2, l = 1, m_l = 0$ (b) 5*p* (c) $n = 4, l = 3$

8.12 How many electrons in an atom can have each of the following quantum number or sublevel designations?
(a) 4*p* (b) $n = 3, l = 1, m_l = +1$ (c) $n = 5, l = 3$

8.13 How many electrons in an atom can have each of the following quantum number or sublevel designations?
(a) 2*s* (b) $n = 3, l = 2$ (c) 6*d*

The Quantum-Mechanical Model and the Periodic Table

(Sample Problems 8.1 and 8.2)

8.14 State the periodic law, and explain its relation to electron configuration. (Use Na and K in your explanation.)

8.15 State Hund's rule in your own words, and show its application in the orbital diagram of the nitrogen atom.

8.16 How does the aufbau principle, in connection with the periodic law, lead to the format of the periodic table?

8.17 For main-group elements, are outer electron configurations similar or different within a group? Within a period? Explain.

8.18 Write a full set of quantum numbers for the following:
(a) The outermost electron in an Rb atom
(b) The electron gained when an S^- ion becomes an S^{2-} ion
(c) The electron lost when an Ag atom ionizes
(d) The electron gained when an F^- ion forms from an F atom

8.19 Write a full set of quantum numbers for the following:
(a) The outermost electron in an Li atom
(b) The electron gained when a Br atom becomes a Br^- ion
(c) The electron lost when a Cs atom ionizes
(d) The highest energy electron in the ground-state B atom

8.20 Write the full ground-state electron configuration for each:
(a) Rb (b) Ge (c) Ar

8.21 Write the full ground-state electron configuration for each:
(a) Br (b) Mg (c) Se

8.22 Draw a partial (valence-level) orbital diagram, and write the condensed ground-state electron configuration for each:
(a) Ti (b) Cl (c) V

8.23 Draw a partial (valence-level) orbital diagram, and write the condensed ground-state electron configuration for each:
(a) Ba (b) Co (c) Ag

8.24 Draw the partial (valence-level) orbital diagram, and write the symbol, group number, and period number of the element:
(a) [He] $2s^22p^4$ (b) [Ne] $3s^23p^3$

8.25 Draw the partial (valence-level) orbital diagram, and write the symbol, group number, and period number of the element:
(a) [Kr] $5s^24d^{10}$ (b) [Ar] $4s^23d^8$

8.26 From each partial (valence-level) orbital diagram, write the condensed electron configuration and group number:

(a) 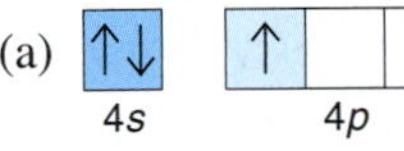

(b) 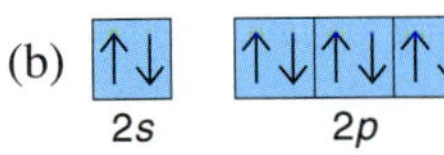

8.27 From each partial (valence-level) orbital diagram, write the condensed electron configuration and group number:

(a) 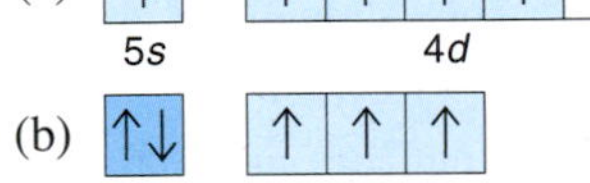

(b) ↑↓ | ↑ ↑ ↑
2s 2p

8.28 How many inner, outer, and valence electrons are present in an atom of each of the following elements?
(a) O (b) Sn (c) Ca (d) Fe (e) Se

8.29 How many inner, outer, and valence electrons are present in an atom of each of the following elements?
(a) Br (b) Cs (c) Cr (d) Sr (e) F

8.30 Identify each element below, and give the symbols of the other elements in its group:
(a) [He] $2s^22p^1$ (b) [Ne] $3s^23p^4$ (c) [Xe] $6s^25d^1$

8.31 Identify each element below, and give the symbols of the other elements in its group:
(a) [Ar] $4s^23d^{10}4p^4$ (b) [Xe] $6s^24f^{14}5d^2$ (c) [Ar] $4s^23d^5$

8.32 One reason spectroscopists study excited states is to gain information about the energies of orbitals that are unoccupied in an atom's ground state. Each of the following electron configurations represents an atom in an excited state. Identify the element, and write its condensed ground-state configuration:
(a) $1s^22s^22p^63s^13p^1$ (b) $1s^22s^22p^63s^23p^44s^1$
(c) $1s^22s^22p^63s^23p^64s^23d^44p^1$ (d) $1s^22s^22p^53s^1$

Trends in Three Atomic Properties

(Sample Problems 8.3 to 8.5)

8.33 Explain the relationship between the trends in atomic size and in ionization energy within the main groups.

8.34 Given the following partial (valence-level) electron configurations, (a) identify each element, (b) rank the four elements in order of increasing atomic size, and (c) rank them in order of increasing ionization energy:

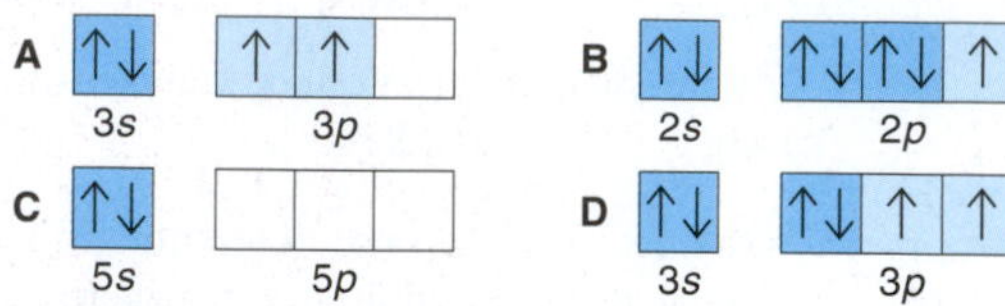

8.35 In what region of the periodic table will you find elements with relatively high IEs? With relatively low IEs?

8.36 (a) Why do successive IEs of a given element always increase? (b) When the difference between successive IEs of a given element is exceptionally large (for example, between IE_1 and IE_2 of K), what do we learn about its electron configuration? (c) The bars represent the relative magnitudes of the first five ionization energies of an atom:

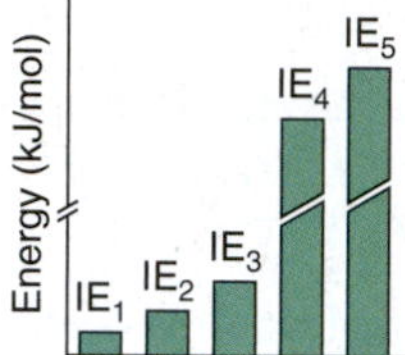

Identify the element and write its complete electron configuration, assuming it comes from (a) Period 2; (b) Period 3; (c) Period 4.

8.37 In a plot of IE_1 for the Period 3 elements (see Figure 8.10, p. 261), why do the values for elements in Groups 3A(13) and 6A(16) drop slightly below the generally increasing trend?

8.38 Which group in the periodic table has elements with high (endothermic) IE_1 and very negative (exothermic) first electron affinities (EA_1)? Give the charge on the ions these atoms form.

8.39 How does *d*-electron shielding influence atomic size among the Period 4 transition elements?

8.40 Arrange each set in order of *increasing* atomic size:
(a) Rb, K, Cs (b) C, O, Be (c) Cl, K, S (d) Mg, K, Ca

8.41 Arrange each set in order of *decreasing* atomic size:
(a) Ge, Pb, Sn (b) Sn, Te, Sr (c) F, Ne, Na (d) Be, Mg, Na

8.42 Arrange each set of atoms in order of *increasing* IE_1:
(a) Sr, Ca, Ba (b) N, B, Ne (c) Br, Rb, Se (d) As, Sb, Sn

8.43 Arrange each set of atoms in order of *decreasing* IE_1:
(a) Na, Li, K (b) Be, F, C (c) Cl, Ar, Na (d) Cl, Br, Se

8.44 Write the full electron configuration of the Period 2 element with the following successive IEs (in kJ/mol):

$IE_1 = 801$ $IE_2 = 2427$ $IE_3 = 3659$
$IE_4 = 25{,}022$ $IE_5 = 32{,}822$

8.45 Write the full electron configuration of the Period 3 element with the following successive IEs (in kJ/mol):

$IE_1 = 738$ $IE_2 = 1450$ $IE_3 = 7732$
$IE_4 = 10{,}539$ $IE_5 = 13{,}628$

8.46 Which element in each of the following sets would you expect to have the *highest* IE_2?
(a) Na, Mg, Al (b) Na, K, Fe (c) Sc, Be, Mg

8.47 Which element in each of the following sets would you expect to have the *lowest* IE_3?
(a) Na, Mg, Al (b) K, Ca, Sc (c) Li, Al, B

Atomic Properties and Chemical Reactivity

(Sample Problems 8.6 to 8.8)

8.48 List three ways in which metals and nonmetals differ.

8.49 Summarize the trend in metallic character as a function of position in the periodic table. Is it the same as the trend in atomic size? The trend in ionization energy?

8.50 Summarize the acid-base behavior of the main-group metal and nonmetal oxides in water. How does oxide acidity in water change down a group and across a period?

8.51 What is a pseudo–noble gas configuration? Give an example of one ion from Group 3A(13) that has it.

8.52 The charges of a set of isoelectronic ions vary from 3+ to 3−. Place the ions in order of increasing size.

8.53 Which element would you expect to be *more* metallic?
(a) Ca or Rb (b) Mg or Ra (c) Br or I

8.54 Which element would you expect to be *more* metallic?
(a) S or Cl (b) In or Al (c) As or Br

8.55 Write the charge and full ground-state electron configuration of the monatomic ion most likely to be formed by each:
(a) Cl (b) Na (c) Ca

8.56 Write the charge and full ground-state electron configuration of the monatomic ion most likely to be formed by each:
(a) Rb (b) N (c) Br

8.57 How many unpaired electrons are present in a ground-state atom from each of the following groups?
(a) 2A(2) (b) 5A(15) (c) 8A(18) (d) 3A(13)

8.58 How many unpaired electrons are present in a ground-state atom from each of the following groups?
(a) 4A(14) (b) 7A(17) (c) 1A(1) (d) 6A(16)

8.59 Write the condensed ground-state electron configurations of these transition metal ions, and state which are paramagnetic:
(a) V^{3+} (b) Cd^{2+} (c) Co^{3+} (d) Ag^{+}

8.60 Write the condensed ground-state electron configurations of these transition metal ions, and state which are paramagnetic:
(a) Mo^{3+} (b) Au^{+} (c) Mn^{2+} (d) Hf^{2+}

8.61 Palladium (Pd; $Z = 46$) is diamagnetic. Draw partial orbital diagrams to show which of the following electron configurations is consistent with this fact:
(a) [Kr] $5s^24d^8$ (b) [Kr] $4d^{10}$ (c) [Kr] $5s^14d^9$

8.62 Niobium (Nb; $Z = 41$) has an anomalous ground-state electron configuration for a Group 5B(5) element: [Kr] $5s^14d^4$. What is the expected electron configuration for elements in this group? Draw partial orbital diagrams to show how paramagnetic measurements could support niobium's actual configuration.

8.63 Rank the ions in each set in order of *increasing* size, and explain your ranking:
(a) Li^+, K^+, Na^+ (b) Se^{2-}, Rb^+, Br^- (c) O^{2-}, F^-, N^{3-}

8.64 Rank the ions in each set in order of *decreasing* size, and explain your ranking:
(a) Se^{2-}, S^{2-}, O^{2-} (b) Te^{2-}, Cs^+, I^- (c) Sr^{2+}, Ba^{2+}, Cs^+

Comprehensive Problems

8.65 Name the element described in each of the following:
(a) Smallest atomic radius in Group 6A(16)
(b) Largest atomic radius in Period 6
(c) Smallest metal in Period 3
(d) Highest IE_1 in Group 4A(14)
(e) Lowest IE_1 in Period 5
(f) Most metallic in Group 5A(15)
(g) Group 3A(13) element that forms the most basic oxide
(h) Period 4 element with highest energy level filled
(i) Condensed ground-state electron configuration of [Ne] $3s^2 3p^2$
(j) Condensed ground-state electron configuration of [Kr] $5s^2 4d^6$
(k) Forms 2+ ion with electron configuration [Ar] $3d^3$
(l) Period 5 element that forms 3+ ion with pseudo–noble gas configuration
(m) Period 4 transition element that forms 3+ diamagnetic ion
(n) Period 4 transition element that forms 2+ ion with a half-filled *d* sublevel
(o) Heaviest lanthanide
(p) Period 3 element whose 2− ion is isoelectronic with Ar
(q) Alkaline earth metal whose cation is isoelectronic with Kr
(r) Group 5A(15) metalloid with the most acidic oxide

8.66 When a nonmetal oxide reacts with water, it forms an oxoacid with the same oxidation number as the nonmetal. Give the name and formula of the oxide used to prepare each of these oxoacids: (a) hypochlorous acid; (b) chlorous acid; (c) chloric acid; (d) perchloric acid; (e) sulfuric acid; (f) sulfurous acid; (g) nitric acid; (h) nitrous acid; (i) carbonic acid; (j) phosphoric acid.

8.67 The energy difference between the 5*d* and 6*s* sublevels in gold accounts for its color. Assuming this energy difference is about 2.7 eV (electron volt; 1 eV $= 1.602 \times 10^{-19}$ J), explain why gold has a warm yellow color.

8.68 Write the formula and name of the compound formed from the following ionic interactions: (a) The 2+ ion and the 1− ion are both isoelectronic with the atoms of a chemically unreactive Period 4 element. (b) The 2+ ion and the 2− ion are both isoelectronic with the Period 3 noble gas. (c) The 2+ ion is the smallest with a filled *d* sublevel; the anion forms from the smallest halogen. (d) The ions form from the largest and smallest ionizable atoms in Period 2.

8.69 The hot glowing gases around the Sun, the *corona*, can reach millions of degrees Celsius, high enough to remove many electrons from gaseous atoms. Iron ions with charges as high as 14+ have been observed in the corona. Which ions from Fe^+ to Fe^{14+} are paramagnetic? Which would be most strongly attracted to a magnetic field?

8.70 Rubidium and bromine atoms are depicted at right. (a) What monatomic ions do they form? (b) What electronic feature characterizes this pair of ions, and which noble gas are they related to? (c) Which pair best represents the relative ionic sizes?

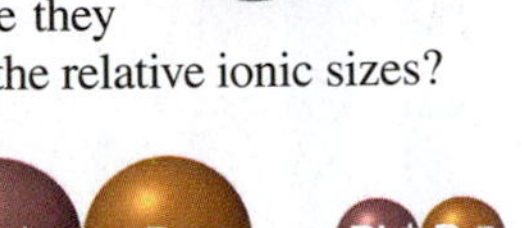

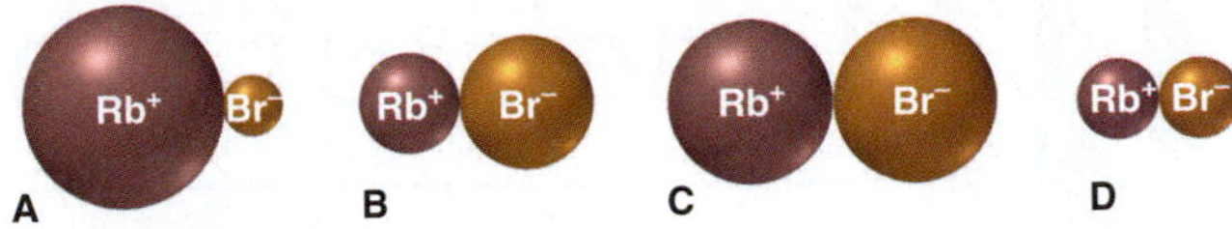

8.71 Partial (valence-level) electron configurations for four different ions are shown below:

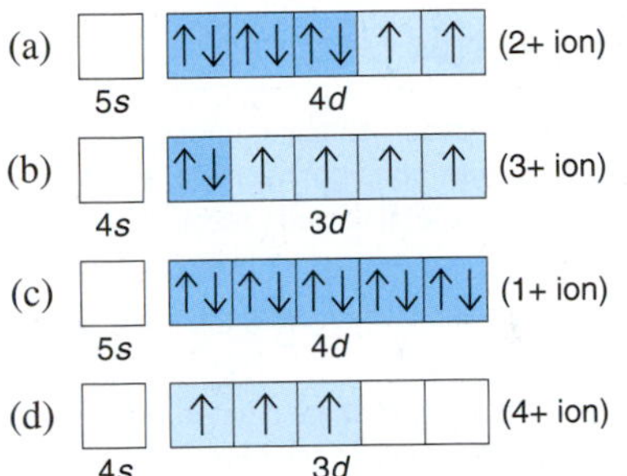

Identify the elements from which the ions are derived, and write the formula of the oxide each ion forms.

8.72 The bars in the graph at right represent the relative magnitudes of the first five ionization energies of an atom. Identify the element and write its complete electron configuration, assuming it comes from (a) Period 2; (b) Period 3; (c) Period 4.

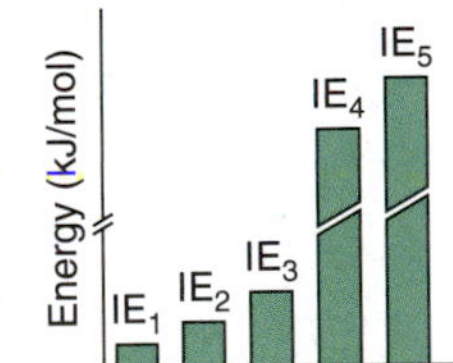

8.73 Data from the planet Zog for some main-group elements are shown below (Zoggian units are linearly related to Earth units but are not shown). Radio signals from Zog reveal that balloonium is a monatomic gas with two positive nuclear charges. Use the data to deduce the names that Earthlings give to these elements:

Name	Atomic Radius	IE_1	EA_1
Balloonium	10	339	0
Inertium	24	297	+4.1
Allotropium	34	143	−28.6
Brinium	63	70.9	−7.6
Canium	47	101	−15.3
Fertilium	25	200	0
Liquidium	38	163	−46.4
Utilium	48	82.4	−6.1
Crimsonium	72	78.4	−2.9

Models of Chemical Bonding

Key Principles
to focus on while studying this chapter

- Two classes of elements, metals and nonmetals, combine through three types of bonding: metal and nonmetal through *ionic bonding*, nonmetal and nonmetal through *covalent bonding*, and metal and metal through *metallic bonding*. ***(Section 9.1)***
- *Lewis symbols* predict bonding behavior and show how atoms obey the octet rule. ***(Section 9.1)***
- Ionic bonding is the attraction among the ions that are created when metal atoms *transfer* electrons to nonmetal atoms. Ionic compounds form because of the large magnitude of the *lattice energy*. That is, even though energy is required to form the ions, much more energy is released when the ions attract each other to form a solid. ***(Section 9.2)***
- The strong attractions among their ions make ionic compounds hard, high-melting solids that conduct a current only when melted or dissolved. ***(Section 9.2)***
- A covalent bond is the attraction between the nuclei of two nonmetal atoms and the electron pair they share. Each covalent bond has a specific *energy* and *length* that depend on the bonded atoms and an *order* that depends on the number of electron pairs shared. Chemists study the types of covalent bonds in a compound spectroscopically. ***(Section 9.3)***
- Most covalent compounds consist of separate molecules, so they have low melting and boiling points. These physical changes disrupt the weak attractions *between* the molecules while leaving the strong covalent bonds *within* the molecules intact. Some substances have covalent bonds throughout, and they are very hard and high melting. ***(Section 9.3)***
- During a reaction, energy is *absorbed* to break certain bonds in the reactant molecules and is *released* to form the bonds in the product molecules; the enthalpy of reaction is the *difference* between the energy absorbed and the energy released. ***(Section 9.4)***
- Each atom in a covalent bond attracts the shared electron pair according to its *electronegativity (EN)*. A covalent bond is *polar* if the two atoms have considerably different EN values; therefore, the *ionic character* of a bond—from highly ionic to nonpolar covalent—varies with the difference in EN values of the bonded atoms. ***(Section 9.5)***

The Hard Cutting the Strong Only after some diamond dust forms a paste on the oil-coated saw blade can a diamond be cut in a reasonable time. Diamonds are so hard because of the number and strength of their covalent bonds.

Outline

Why do substances behave as they do? That is, why is table salt (or any other ionic substance) a hard, brittle, high-melting solid that conducts a current only when molten or dissolved in water? Why is candle wax (along with most covalent substances) low melting, soft, and nonconducting, even though diamond (as well as a few other exceptions) is high melting and extremely hard? And why is copper (and most other metals) shiny, malleable, and able to conduct a current whether molten or solid? The answers lie in the *type of bonding within the substance*. In Chapter 8, we examined the properties of individual atoms and ions. But the behavior of matter really depends on how those atoms and ions bond.

CONCEPTS & SKILLS TO REVIEW before studying this chapter

- characteristics of ionic and covalent compounds; Coulomb's law (Section 2.7)
- polar covalent bonds and the polarity of water (Section 4.1)
- Hess's law, ΔH°_{rxn}, and ΔH°_f (Sections 6.5 and 6.6)
- atomic and ionic electron configurations (Sections 8.2 and 8.4)
- trends in atomic properties and metallic behavior (Sections 8.3 and 8.4)

9.1 • ATOMIC PROPERTIES AND CHEMICAL BONDS

Before we examine the types of chemical bonding, we should start with the most fundamental question: why do atoms bond at all? In general, *bonding lowers the potential energy between positive and negative particles* (see Figure 1.3), whether they are oppositely charged ions or nuclei and electron pairs. Just as the strength of attractions and repulsions among nucleus and electrons determines the properties of an atom, the type and strength of chemical bonds determine the properties of a substance.

Types of Bonding: Three Ways Metals and Nonmetals Combine

In general, there is a gradation from atoms of more metallic elements to atoms of more nonmetallic elements across a period *and* up a group (Figure 9.1). Three types of bonding result from the three ways these two types of atoms can combine:

1. *Metal with nonmetal: electron transfer and ionic bonding* (Figure 9.2A, *next page*). We observe **ionic bonding** between atoms with large differences in their tendencies to lose or gain electrons. Such differences occur between reactive metals [Groups 1A(1) and 2A(2)] and nonmetals [Group 7A(17) and the top of Group 6A(16)]. A metal atom (low IE) loses its one or two valence electrons, and a nonmetal atom (highly negative EA) gains the electron(s). *Electron transfer* from metal to nonmetal occurs, and each atom forms an ion with a noble gas electron configuration. The electrostatic attractions between these positive and negative ions draw them into a three-dimensional array to form an ionic solid. Note that the chemical formula of an ionic compound is the *empirical formula* because it gives the cation-to-anion ratio.

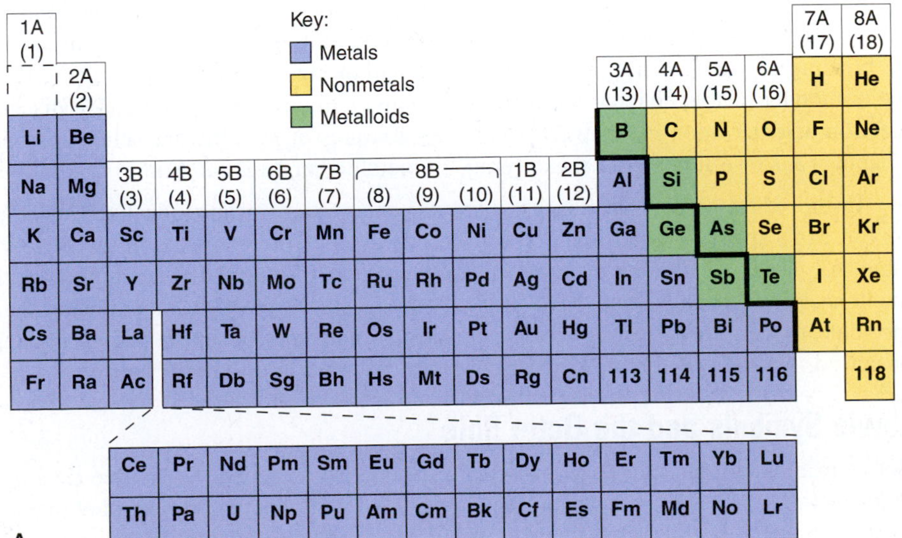

Figure 9.1 A comparison of metals and nonmetals. **A,** Location within the periodic table. **B,** Relative magnitudes of some atomic properties across a period.

PROPERTY	METAL ATOM	NONMETAL ATOM
Atomic size	Larger	Smaller
Z_{eff}	Lower	Higher
IE	Lower	Higher
EA	Less negative	More negative

B

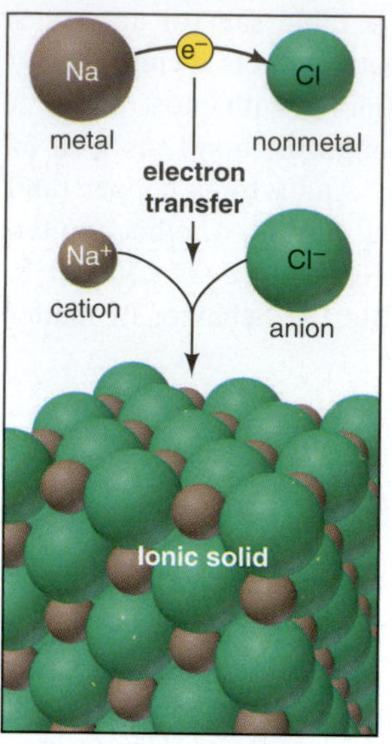

A Ionic bonding

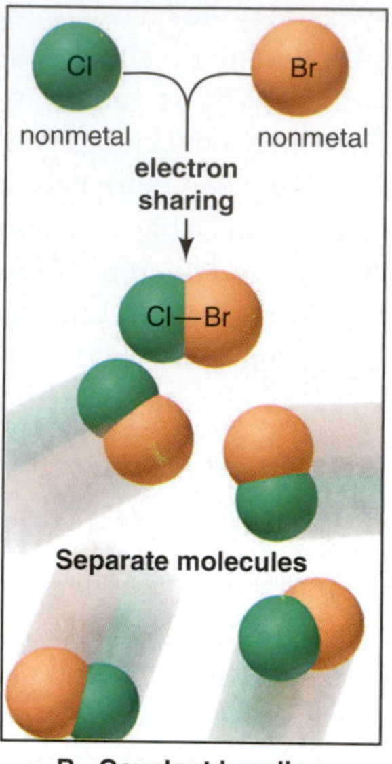

B Covalent bonding

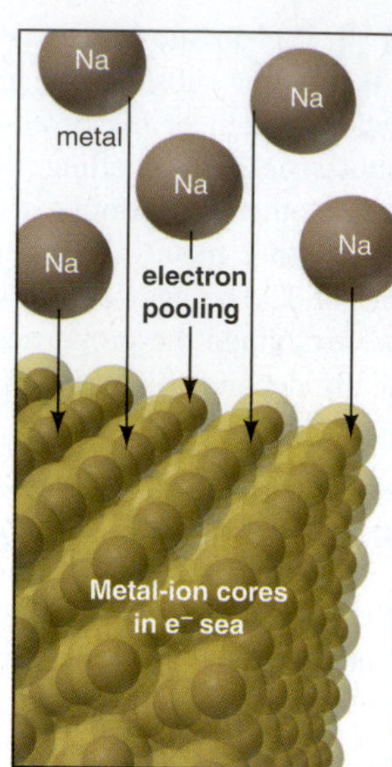

C Metallic bonding

Figure 9.2 Three models of chemical bonding.

2. *Nonmetal with nonmetal: electron sharing and covalent bonding* (Figure 9.2B). When two atoms differ little, or not at all, in their tendencies to lose or gain electrons, we observe *electron sharing* and **covalent bonding,** which occurs most commonly between nonmetal atoms. Each atom holds onto its own electrons tightly (high IE) and attracts other electrons (highly negative EA). The nucleus of each atom attracts the valence electrons of the other, which draws the atoms together. The shared electron pair is typically *localized* between the two atoms, linking them in a covalent bond of a particular length and strength. In most cases, separate molecules result when atoms bond covalently. Note that the chemical formula of a covalent compound is the *molecular formula* because it gives the actual numbers of atoms in each molecule.

3. *Metal with metal: electron pooling and metallic bonding* (Figure 9.2C). Metal atoms are relatively large, and their few outer electrons are well shielded by filled inner levels (core electrons). Thus, they lose outer electrons easily (low IE) and do not gain them readily (slightly negative or positive EA). These properties lead metal atoms to share their valence electrons, but not by covalent bonding. In the simplest model of **metallic bonding,** the enormous number of atoms in a sample of a metal *pool* their valence electrons into a "sea" of electrons that "flows" between and around each metal-ion core (nucleus plus inner electrons), thereby attracting them and holding them together. Unlike the localized electrons in covalent bonding, electrons in metallic bonding are *delocalized,* moving freely throughout the entire piece of metal. (For the remainder of this chapter, we'll focus on ionic and covalent bonding. We discuss electron delocalization in Chapter 11 and the structures of metallic and other solids in Chapter 12, so we'll postpone the coverage of metallic bonding until then.)

In the world of real substances, there are exceptions to these idealized models, so you can't always predict bond type from positions of the elements in the periodic table. As just one example, when the metal beryllium [Group 2A(2)] combines with the nonmetal chlorine [Group 7A(17)], the bonding fits the covalent model better than the ionic model. Thus, just as we see gradations in atomic behavior within a group or period, we see a gradation in bonding from one type to another (Figure 9.3).

Lewis Symbols and the Octet Rule

Before examining each model, let's discuss a method for depicting the valence electrons of interacting atoms that predicts how they bond. In a **Lewis electron-dot symbol** (named for the American chemist G. N. Lewis), the element symbol represents the nucleus *and* inner electrons, and dots around the symbol represent the valence electrons (Figure 9.4). Note that the pattern of dots is the same for elements within a group.

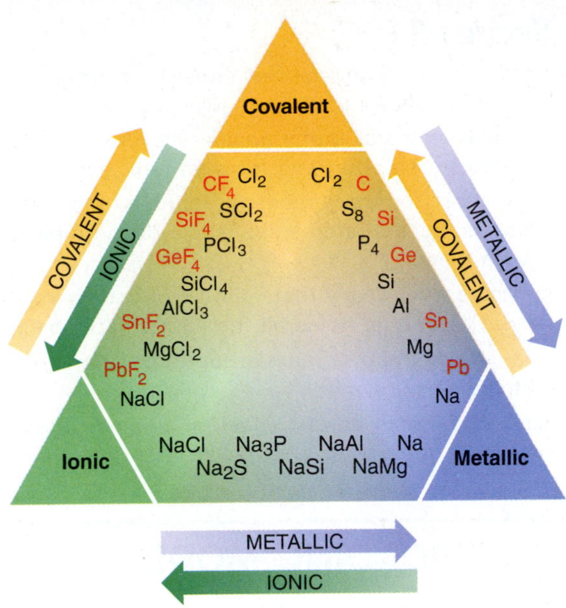

Figure 9.3 Gradations in bond type among Period 3 *(black type)* and Group 4A *(red type)* elements.

We use these steps to write the Lewis symbol for any main-group element:

1. Note its A-group number (1A to 8A), which tells the number of valence electrons.
2. Place one dot at a time on each of the four sides (top, right, bottom, left) of the element symbol.
3. Keep adding dots, pairing them, until all are used up.

The specific placement of dots is not important; that is, in addition to the one shown in Figure 9.4, the Lewis symbol for nitrogen can *also* be written as

·N̈: or ·Ṅ̤· or :N̈·

The Lewis symbol provides information about an element's bonding behavior:

- For a *metal,* the *total* number of dots is the number of electrons an atom loses to form a cation.
- For a *nonmetal,* the number of *unpaired* dots equals either the number of electrons an atom *gains* to form an anion or the number it *shares* to form covalent bonds.

The Lewis symbol for carbon illustrates the last point. Rather than one pair of dots and two unpaired dots, as its electron configuration seems to call for ([He] $2s^22p^2$), carbon has four unpaired dots because it forms four bonds. Larger nonmetals can form as many bonds as the number of dots in their Lewis symbol (Chapter 10).

In his pioneering studies, Lewis generalized much of bonding behavior into the **octet rule:** *when atoms bond, they lose, gain, or share electrons to attain a filled outer level of eight electrons* (*or two,* for H and Li). The octet rule holds for nearly all of the compounds of Period 2 elements and a large number of others as well.

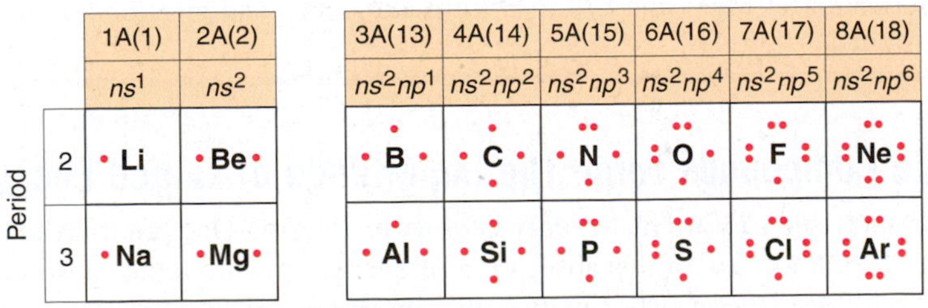

Period	1A(1)	2A(2)	3A(13)	4A(14)	5A(15)	6A(16)	7A(17)	8A(18)
	ns^1	ns^2	ns^2np^1	ns^2np^2	ns^2np^3	ns^2np^4	ns^2np^5	ns^2np^6
2	·Li	·Be·	·Ḃ·	·Ċ̣·	·N̈̇·	:Ö̤·	:F̤̈:	:N̤̈e:
3	·Na	·Mg·	·Ȧl·	·Ṩi·	·P̈̇·	:S̤̈·	:C̤̈l:	:Ä̤r:

Figure 9.4 Lewis electron-dot symbols for elements in Periods 2 and 3.

Summary of Section 9.1

- Nearly all naturally occurring substances consist of atoms or ions bonded to others. Chemical bonding allows atoms to lower their energy.
- Ionic bonding occurs when metal atoms transfer electrons to nonmetal atoms, and the resulting ions attract each other and form an ionic solid.
- Covalent bonding is most common between nonmetal atoms and usually results in individual molecules. Bonded atoms share one or more pairs of electrons that are localized between them.
- Metallic bonding occurs when many metal atoms pool their valence electrons into a delocalized electron "sea" that holds all the atoms in the sample together.
- The Lewis electron-dot symbol of a main-group atom shows valence electrons as dots surrounding the element symbol.
- The octet rule says that atoms bond by losing, gaining, or sharing electrons to attain a filled outer level of eight (or two) electrons.

9.2 • THE IONIC BONDING MODEL

The central idea of the ionic bonding model is the *transfer of electrons from metal atoms to nonmetal atoms to form ions that attract each other into a solid compound.* In most cases, for the main groups, the ion that forms has a filled outer level of either two or eight electrons, the number in the nearest noble gas (octet rule).

The transfer of an electron from a lithium atom to a fluorine atom is depicted in three ways in Figure 9.5. In each, Li loses its single outer electron and is left with a filled $n = 1$ level (two e^-), while F gains a single electron to fill its $n = 2$ level (eight e^-). In this case, each atom is one electron away from the configuration of its nearest noble gas, so the number of electrons lost by each Li equals the number gained by each F. Therefore, equal numbers of Li^+ and F^- ions form, as the formula LiF indicates. That is, in ionic bonding, *the total number of electrons lost by the metal atom(s) equals the total number of electrons gained by the nonmetal atom(s).*

Sample Problem 9.1 Depicting Ion Formation

Problem Use partial orbital diagrams and Lewis symbols to depict the formation of Na^+ and O^{2-} ions from the atoms, and give the formula of the compound formed.

Plan First we draw the orbital diagrams and Lewis symbols for Na and O atoms. To attain filled outer levels, Na loses one electron and O gains two. To make the number of electrons lost equal the number gained, two Na atoms are needed for each O atom.

Solution

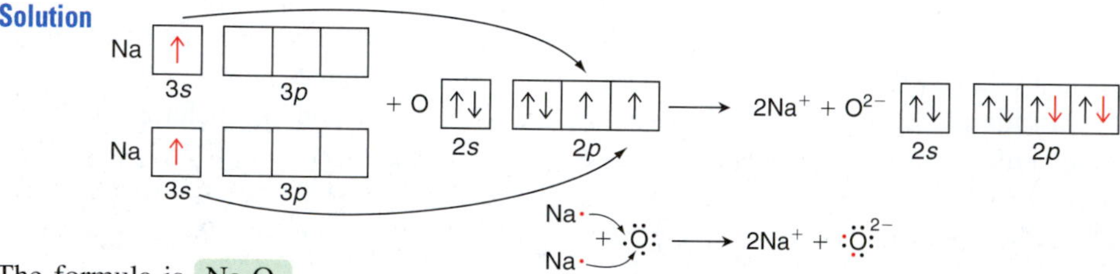

The formula is Na_2O.

FOLLOW-UP PROBLEM 9.1 Use condensed electron configurations and Lewis symbols to depict the formation of Mg^{2+} and Cl^- ions from the atoms, and give the formula of the compound formed.

Why Ionic Compounds Form: The Importance of Lattice Energy

You may be surprised to learn that energy is *absorbed* during electron transfer. So why does it occur? And, in view of this absorption of energy, why do ionic substances exist at all? As you'll see, the answer involves the enormous quantity of energy *released* as the ions that form coalesce into a solid.

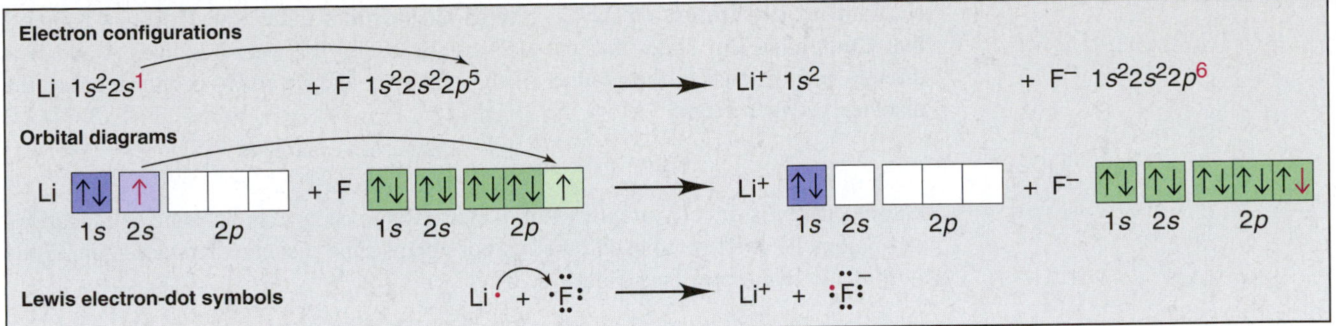

Figure 9.5 Three ways to depict electron transfer in the formation of Li^+ and F^-. The electron being transferred is shown in red.

1. *The electron-transfer process.* Consider the electron-transfer process for the formation of lithium fluoride, which involves a gaseous Li atom losing an electron, and a gaseous F atom gaining it:

- The first ionization energy (IE_1) of Li is the energy absorbed when 1 mol of gaseous Li atoms loses 1 mol of valence electrons:

$$Li(g) \longrightarrow Li^+(g) + e^- \qquad IE_1 = 520 \text{ kJ}$$

- The first electron affinity (EA_1) of F is the energy released when 1 mol of gaseous F atoms gains 1 mol of electrons:

$$F(g) + e^- \longrightarrow F^-(g) \qquad EA_1 = -328 \text{ kJ}$$

- Taking the sum shows that electron transfer *by itself* requires energy:

$$Li(g) + F(g) \longrightarrow Li^+(g) + F^-(g) \qquad IE_1 + EA_1 = 192 \text{ kJ}$$

2. *Other steps that absorb energy.* The total energy needed prior to ion formation adds to the sum of IE_1 and EA_1: metallic lithium must be made into gaseous atoms (161 kJ/mol), and fluorine molecules must be broken into separate atoms (79.5 kJ/mol).

3. *Steps that release energy.* Despite these endothermic steps, the standard enthalpy of formation (ΔH°_f) of solid LiF is −617 kJ/mol; that is, 617 kJ is *released* when 1 mol of LiF(*s*) forms from its elements. Formation of LiF is typical of reactions between active metals and nonmetals: ionic solids form readily (Figure 9.6).

A

B

Figure 9.6 The exothermic formation of sodium bromide. **A,** Sodium (in beaker under mineral oil) and bromine. **B,** The reaction is rapid and vigorous.

If the overall reaction releases energy, there must be some step that is exothermic enough to outweigh the endothermic steps. This step involves the *strong attraction between pairs of oppositely charged ions.* When 1 mol of $Li^+(g)$ and 1 mol of $F^-(g)$ form 1 mol of gaseous LiF molecules, a large quantity of heat is released:

$$Li^+(g) + F^-(g) \longrightarrow LiF(g) \qquad \Delta H^\circ = -755 \text{ kJ}$$

But, as you know, under ordinary conditions, LiF does not exist as gaseous molecules: *even more energy is released when the separate gaseous ions coalesce into a crystalline solid* because each ion attracts *several* oppositely charged ions:

$$Li^+(g) + F^-(g) \longrightarrow LiF(s) \qquad \Delta H^\circ = -1050 \text{ kJ}$$

The negative of this enthalpy change is 1050 kJ, the lattice energy of LiF. The **lattice energy** ($\Delta H^\circ_{lattice}$) is the enthalpy change that occurs when 1 mol of ionic solid separates into gaseous ions. It indicates the strength of ionic interactions, which influence melting point, hardness, solubility, and other properties.

A key point to keep in mind is that *ionic solids exist only because the lattice energy exceeds the energy required for the electron transfer.* In other words, the energy *required* for elements to lose or gain electrons is *supplied* by the attraction between the ions they form: energy is expended to form the ions, but it is more than regained when they attract each other and form a solid.

Periodic Trends in Lattice Energy

The lattice energy results from electrostatic interactions among ions, so its magnitude depends on ionic size, ionic charge, and ionic arrangement in the solid. Therefore, we expect to see periodic trends in lattice energy.

Explaining the Trends in $\Delta H^\circ_{\text{lattice}}$ with Coulomb's Law Recall from Chapter 2 that **Coulomb's law** states that the electrostatic energy between particles A and B is directly proportional to the product of their charges and inversely proportional to the distance between them:

$$\text{Electrostatic energy} \propto \frac{\text{charge A} \times \text{charge B}}{\text{distance}}$$

Lattice energy is directly proportional to electrostatic energy. In an ionic solid, cations and anions lie as close to each other as possible, so the distance between them is the sum of the ionic radii (see Figure 8.20):

$$\text{Electrostatic energy} \propto \frac{\text{cation charge} \times \text{anion charge}}{\text{cation radius} + \text{anion radius}} \propto \Delta H^\circ_{\text{lattice}} \qquad \textbf{(9.1)}$$

This relationship helps us explain the effects of ionic size and charge on trends in lattice energy:

1. *Effect of ionic size.* As we move down a group, ionic radii increase, so the electrostatic energy between cations and anions decreases; thus, lattice energies should decrease as well. Figure 9.7 shows that, for the alkali-metal halides, lattice energy decreases down the group whether we hold the cation constant (LiF to LiI) or the anion constant (LiF to RbF).

2. *Effect of ionic charge.* Across a period, ionic charge changes. For example, lithium fluoride and magnesium oxide have cations and anions of about equal radii (Li^+ = 76 pm and Mg^{2+} = 72 pm; F^- = 133 pm and O^{2-} = 140 pm). The major difference is between singly charged Li^+ and F^- ions and doubly charged Mg^{2+} and O^{2-} ions. The difference in the lattice energies of the two compounds is striking:

$$\Delta H^\circ_{\text{lattice}} \text{ of LiF} = 1050 \text{ kJ/mol} \qquad \text{and} \qquad \Delta H^\circ_{\text{lattice}} \text{ of MgO} = 3923 \text{ kJ/mol}$$

This nearly fourfold increase in $\Delta H^\circ_{\text{lattice}}$ reflects the fourfold increase in the product of the charges (1 × 1 vs. 2 × 2) in the numerator in Equation 9.1. The very large lattice energy of MgO more than compensates for the energy required to form the Mg^{2+} and O^{2-} ions. In fact, the lattice energy is the reason that compounds with 2+ cations and 2− anions even exist.

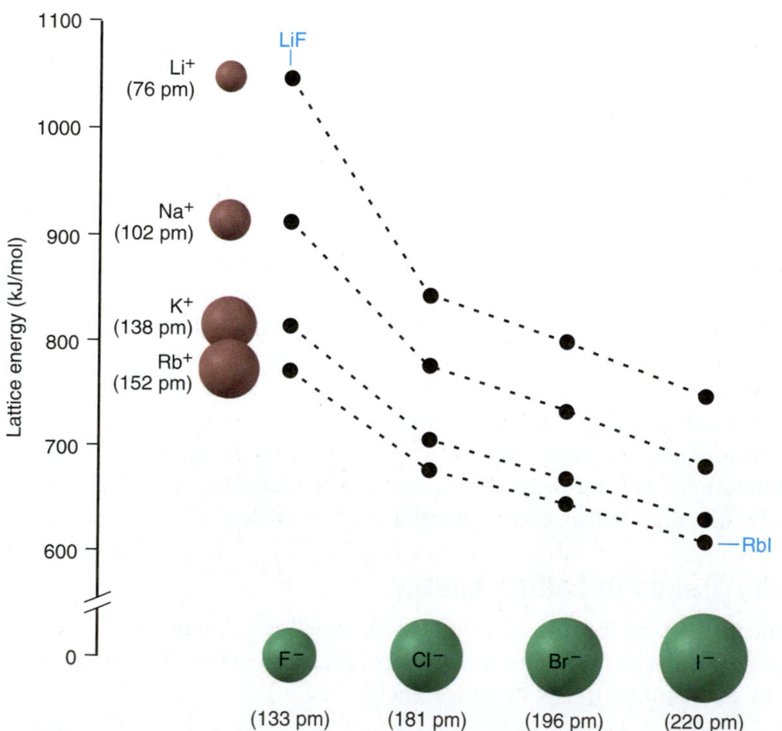

Figure 9.7 Trends in lattice energy. The lattice energies are shown for compounds formed from a given Group 1A(1) cation *(left side)* and one of the Group 7A(17) anions *(bottom)*. LiF (smallest ions) has the highest lattice energy, and RbI (largest ions) has the lowest.

A

B

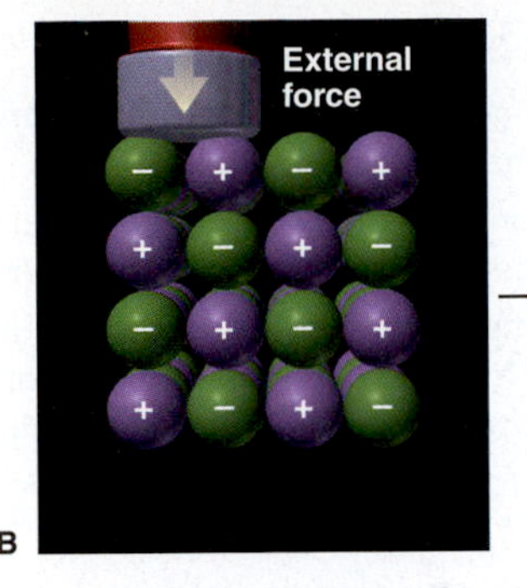

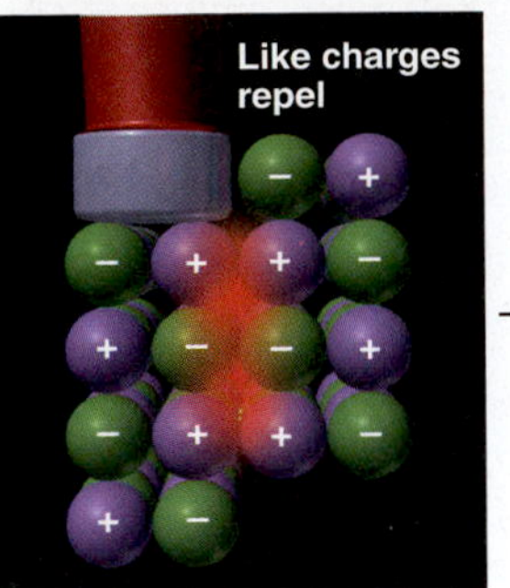

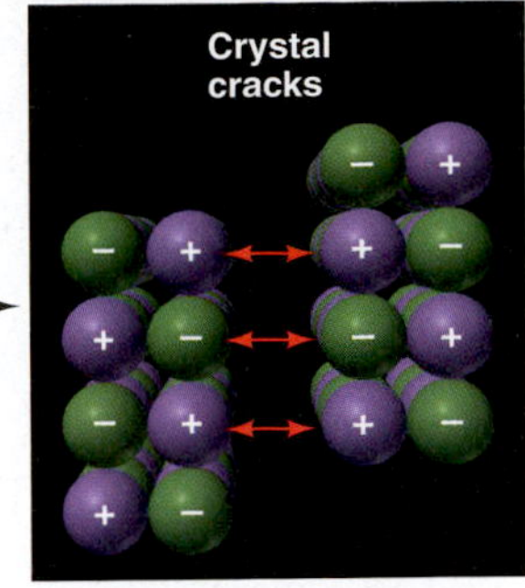

Figure 9.8 Why ionic compounds crack. **A,** Ionic compounds crack when struck with enough force. **B,** When a force moves like charges near each other, repulsions cause a crack.

How the Model Explains the Properties of Ionic Compounds

The central role of any model is to explain the facts. With atomic-level views, we can see how the ionic bonding model accounts for the properties of ionic solids:

1. *Physical behavior.* As a typical ionic compound, a piece of rock salt (NaCl) is *hard* (does not dent), *rigid* (does not bend), and *brittle* (cracks without deforming). These properties arise from the strong attractive forces that hold the ions *in specific positions*. Moving them out of position requires overcoming these forces, so rock salt does not dent or bend. If enough force *is* applied, ions of like charge are brought next to each other, and repulsions between them crack the sample suddenly (Figure 9.8).

2. *Electrical conductivity.* Ionic compounds typically *do not* conduct electricity in the solid state but *do* conduct when melted or dissolved. According to the model, the solid consists of fixed ions, but when it melts or dissolves, the ions can move and carry a current (Figure 9.9).

3. *Thermal conductivity.* Large amounts of energy are needed to free the ions from their fixed positions and separate them. Thus, we expect ionic compounds to have high melting points and much higher boiling points (Table 9.1). In fact, the interionic attraction is so strong that the vapor consists of **ion pairs,** gaseous ionic molecules, rather than individual ions. In their normal state, as you know, ionic compounds are solid arrays of ions, and *no separate molecules exist.*

Table 9.1 Melting and Boiling Points of Some Ionic Compounds

Compound	mp (°C)	bp (°C)
CsBr	636	1300
NaI	661	1304
$MgCl_2$	714	1412
KBr	734	1435
$CaCl_2$	782	>1600
NaCl	801	1413
LiF	845	1676
KF	858	1505
MgO	2852	3600

Solid ionic compound

Molten ionic compound

Ionic compound dissolved in water

Figure 9.9 Electrical conductance and ion mobility.

Summary of Section 9.2

- In ionic bonding, a metal transfers electrons to a nonmetal, and the resulting ions attract each other to form a solid.
- Main-group elements often attain a filled outer level (either eight electrons or two) by forming ions with the electron configuration of the nearest noble gas.
- Ion formation by itself *absorbs* energy, but more than enough energy is *released* when the ions form a solid. The lattice energy, the energy required to separate the solid into gaseous ions, is the reason ionic solids exist.
- Lattice energies increase with higher ionic charge and decrease with larger ionic radius.
- According to the ionic bonding model, the strong electrostatic attractions that keep ions in position explain why ionic solids are hard, conduct a current only when melted or dissolved, and have high melting and boiling points.
- Ion pairs form when an ionic compound vaporizes.

9.3 • THE COVALENT BONDING MODEL

Look through the *Handbook of Chemistry and Physics,* and you'll find that the number of covalent compounds dwarfs the number of ionic compounds. Molecules held together by covalent bonds range from tiny, diatomic hydrogen to biological and synthetic macromolecules with thousands of atoms. And covalent bonds occur in all polyatomic ions, too. Without doubt, *sharing electrons is the main way that atoms interact.*

The Formation of a Covalent Bond

Why does hydrogen gas consist of H_2 molecules and not separate H atoms? Figure 9.10 plots the potential energy of a system of two isolated H atoms versus the distance between their nuclei (see also Figure 2.12). Let's start at the right end of the curve and move along it as the atoms get closer:

- *At point 1,* the atoms are far apart, and each acts as though the other were not present.

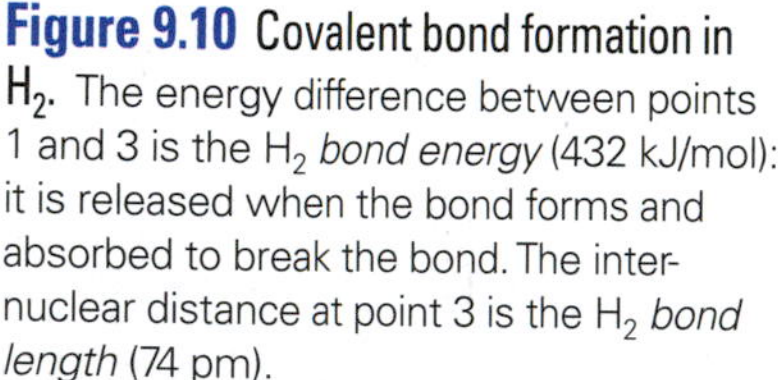

Figure 9.10 Covalent bond formation in H_2. The energy difference between points 1 and 3 is the H_2 *bond energy* (432 kJ/mol): it is released when the bond forms and absorbed to break the bond. The internuclear distance at point 3 is the H_2 *bond length* (74 pm).

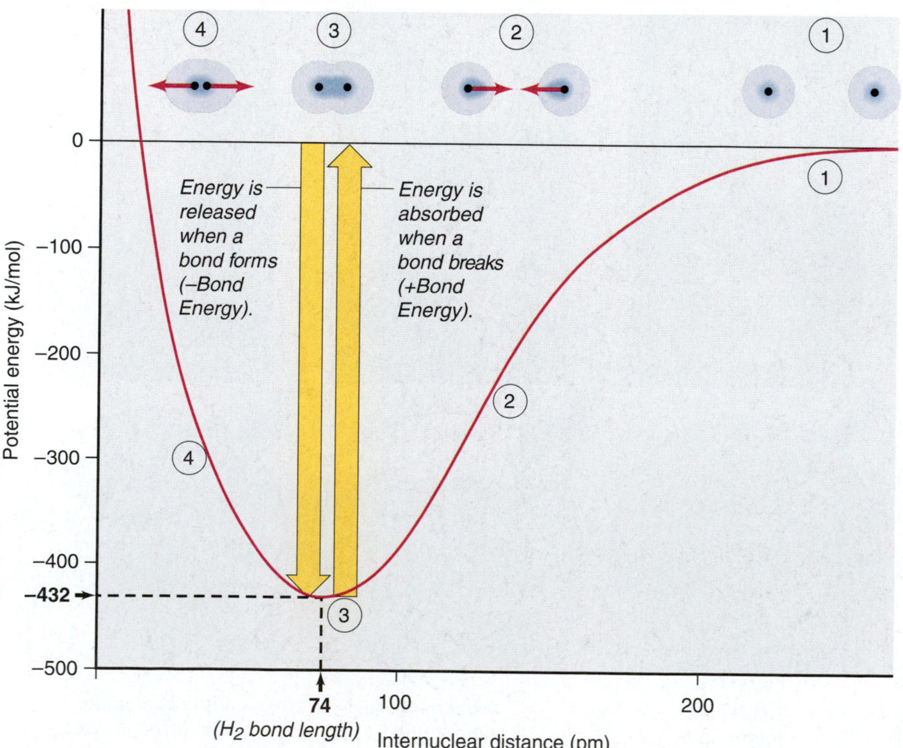

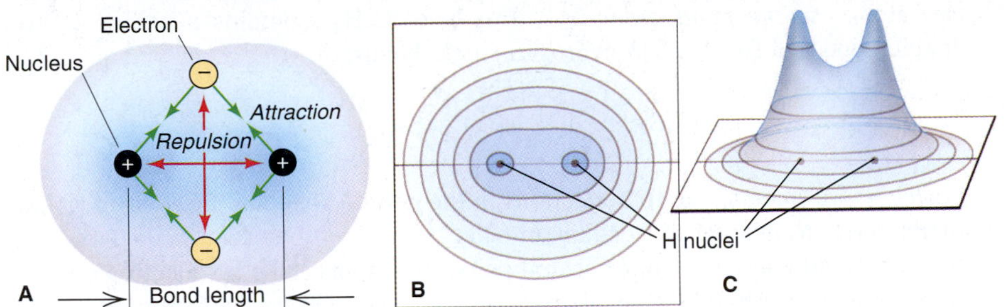

Figure 9.11 Distribution of electron density in H_2. **A,** At some distance (bond length), attractions balance repulsions. Electron density *(blue shading)* is high around and between the nuclei. **B,** Electron density doubles with each concentric curve. **C,** The highest regions of electron density are shown as peaks.

- *At point 2,* the distance between the atoms has decreased enough for each nucleus to start attracting the other atom's electron, which lowers the potential energy. As the atoms get closer, these attractions increase, but so do repulsions between the nuclei and between the electrons.
- *At point 3* (bottom of the energy "well"), the maximum attraction is achieved in the face of the increasing repulsion, and the system has its minimum energy.
- *At point 4,* if it were reached, the atoms would be too close, and the rise in potential energy from increasing repulsions would push them apart toward point 3 again.

Thus, a **covalent bond** arises from the balance between the nuclei attracting the electrons and electrons and nuclei repelling each other. (We'll return to Figure 9.10 shortly.)

Formation of a covalent bond always results in greater electron density ***between the nuclei.*** Figure 9.11 depicts this fact with a cross section of a space-filling model (A), an *electron density contour map* (B), and an *electron density relief map* (C).

Bonding Pairs and Lone Pairs

To achieve a full outer (valence) level of electrons, *each atom in a covalent bond "counts" the shared electrons as belonging entirely to itself.* Thus, the two shared electrons in H_2 simultaneously fill the outer level of *both* H atoms, as clarified by the blue circles added below (which are *not* part of the Lewis structures). The **shared pair,** or **bonding pair,** is represented by a pair of dots or a line:

bonding pair H:H or H—H

An outer-level electron pair that is *not* involved in bonding is called a **lone pair,** or **unshared pair.** The bonding pair in HF fills the outer level of the H atom *and,* together with three lone pairs, fills the outer level of the F atom as well:

bonding pair H:F: lone pairs or H—F:

In F_2 the bonding pair and three lone pairs fill the outer level of *each* F atom:

:F:F: or :F—F:

(This text generally shows bonding pairs as lines and lone pairs as dots.)

Properties of a Covalent Bond: Order, Energy, and Length

A covalent bond has three important properties that are closely related to one another and to the compound's reactivity—bond order, bond energy, and bond length.

1. *Bond order.* The **bond order** is the number of electron pairs being shared by a given pair of atoms:

- A **single bond,** as shown above in H_2, HF, or F_2, is the most common bond and consists of one bonding pair of electrons. Thus, a *single bond has a bond order of 1.*
- Many molecules (and ions) contain *multiple bonds,* in which more than one pair is shared between two atoms. Multiple bonds usually involve C, O, and/or N atoms. A **double bond** consists of two bonding electron pairs, four electrons shared between

two atoms, so *the bond order is 2.* Ethylene (C_2H_4) contains a carbon-carbon double bond and four carbon-hydrogen single bonds:

H:C::C:H (with H on each carbon) or H₂C=CH₂

Each carbon "counts" the four electrons in the double bond and the four in its two single bonds to hydrogens to attain an octet.

- A **triple bond** consists of three shared pairs: two atoms share six electrons, so *the bond order is 3.* The N_2 molecule has a triple bond, and each N atom also has a lone pair. Six shared and two unshared electrons give *each* N atom an octet:

:N⋮⋮N: or :N≡N:

2. *Bond energy.* The strength of a covalent bond depends on the magnitude of the attraction between the nuclei and shared electrons. The **bond energy (BE)** (also called *bond enthalpy* or *bond strength*) is the energy needed to overcome this attraction and is defined as the standard enthalpy change for breaking the bond in 1 mol of gaseous molecules. Bond breakage is an *endothermic* process, so *bond energy is always positive:*

$$\text{A—B}(g) \longrightarrow \text{A}(g) + \text{B}(g) \qquad \Delta H^\circ_{\text{bond breaking}} = \text{BE}_{\text{A—B}} \text{ (always} > 0)$$

The bond energy is the difference in energy between separated and bonded atoms (the potential energy difference between points 1 and 3, the energy "well" in Figure 9.10). *The same quantity of energy absorbed to break the bond is released when the bond forms.* Bond formation is an *exothermic* process, so *the sign of its enthalpy change is always negative:*

$$\text{A}(g) + \text{B}(g) \longrightarrow \text{A—B}(g) \qquad \Delta H^\circ_{\text{bond forming}} = -\text{BE}_{\text{A—B}} \text{ (always} < 0)$$

Table 9.2 lists the energies of some common bonds. By definition,

- *Stronger bonds are lower in energy (have a deeper energy well).*
- *Weaker bonds are higher in energy (have a shallower energy well).*

The energy of a given bond varies slightly from molecule to molecule and even within the same molecule, so each value is an *average* bond energy.

3. *Bond length.* A covalent bond has a **bond length,** the distance between the nuclei of two bonded atoms. In Figure 9.10, bond length is the distance between the nuclei at the point of minimum energy (bottom of the well), and Table 9.2 shows the lengths of some covalent bonds. Like bond energies, these values are *average* bond lengths for a bond in different substances. Bond length is related to the sum of the radii of the bonded atoms. In fact, most atomic radii are calculated from measured bond lengths (see Figure 8.9C). Bond lengths for a series of similar bonds, as in the halogens, increase with atomic size (Figure 9.12).

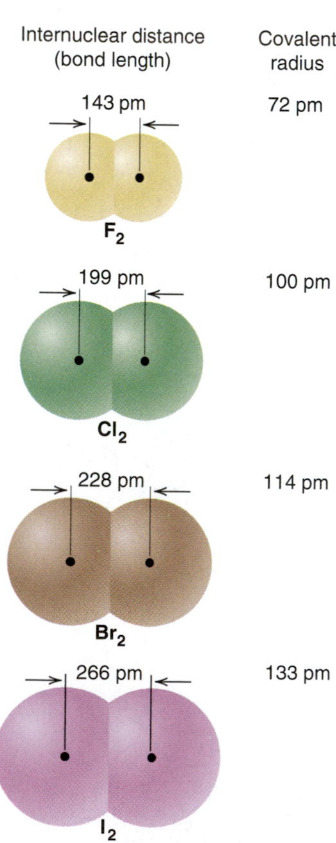

Figure 9.12 Bond length and covalent radius.

The order, energy, and length of a covalent bond are interrelated. Two nuclei are more strongly attracted to two shared pairs than to one, so double-bonded atoms are drawn closer together *and* are more difficult to pull apart than single-bonded atoms: *for a given pair of atoms, a higher bond order results in a shorter bond length and a higher bond energy.* Thus, as Table 9.3 shows, for a given pair of atoms, *a shorter bond is a stronger bond.*

In some cases, we can see a relation among atomic size, bond length, and bond energy by varying one of the atoms in a single bond while holding the other constant:

- *Variation within a group.* The trend in carbon-halogen single bond lengths, C—I > C—Br > C—Cl, parallels the trend in atomic size, I > Br > Cl, and is opposite to the trend in bond energy, C—Cl > C—Br > C—I.
- *Variation within a period.* Looking again at single bonds involving carbon, the trend in bond lengths, C—N > C—O > C—F, is opposite to the trend in bond energy, C—F > C—O > C—N.

Table 9.2 Average Bond Energies (kJ/mol) and Bond Lengths (pm)

Bond	Energy	Length	Bond	Energy	Length	Bond	Energy	Length	Bond	Energy	Length
Single Bonds											
H—H	432	74	N—H	391	101	Si—H	323	148	S—H	347	134
H—F	565	92	N—N	160	146	Si—Si	226	234	S—S	266	204
H—Cl	427	127	N—P	209	177	Si—O	368	161	S—F	327	158
H—Br	363	141	N—O	201	144	Si—S	226	210	S—Cl	271	201
H—I	295	161	N—F	272	139	Si—F	565	156	S—Br	218	225
			N—Cl	200	191	Si—Cl	381	204	S—I	~170	234
C—H	413	109	N—Br	243	214	Si—Br	310	216			
C—C	347	154	N—I	159	222	Si—I	234	240	F—F	159	143
C—Si	301	186							F—Cl	193	166
C—N	305	147	O—H	467	96	P—H	320	142	F—Br	212	178
C—O	358	143	O—P	351	160	P—Si	213	227	F—I	263	187
C—P	264	187	O—O	204	148	P—P	200	221	Cl—Cl	243	199
C—S	259	181	O—S	265	151	P—F	490	156	Cl—Br	215	214
C—F	453	133	O—F	190	142	P—Cl	331	204	Cl—I	208	243
C—Cl	339	177	O—Cl	203	164	P—Br	272	222	Br—Br	193	228
C—Br	276	194	O—Br	234	172	P—I	184	246	Br—I	175	248
C—I	216	213	O—I	234	194				I—I	151	266
Multiple Bonds											
C═C	614	134	N═N	418	122	C≡C	839	121	N≡N	945	110
C═N	615	127	N═O	607	120	C≡N	891	115	N≡O	631	106
C═O	745 (799 in CO_2)	123	O_2	498	121	C≡O	1070	113			

Table 9.3 The Relation of Bond Order, Bond Length, and Bond Energy

Bond	Bond Order	Average Bond Length (pm)	Average Bond Energy (kJ/mol)
C—O	1	143	358
C═O	2	123	745
C≡O	3	113	1070
C—C	1	154	347
C═C	2	134	614
C≡C	3	121	839
N—N	1	146	160
N═N	2	122	418
N≡N	3	110	945

Sample Problem 9.2 Comparing Bond Length and Bond Strength

Problem Without referring to Table 9.2, rank the bonds in each set in order of *decreasing* bond length and *decreasing* bond strength:
(a) S—F, S—Br, S—Cl **(b)** C=O, C—O, C≡O

Plan **(a)** S is singly bonded to three different halogen atoms, so the bond order is the same. Bond length increases and bond strength decreases as the halogen's atomic radius increases. **(b)** The same two atoms are bonded, but the bond orders differ. In this case, bond strength increases and bond length decreases as bond order increases.

Solution **(a)** Atomic size increases down a group, so F < Cl < Br.

Bond length: S—Br > S—Cl > S—F

Bond strength: S—F > S—Cl > S—Br

(b) By ranking the bond orders, C≡O > C=O > C—O, we obtain

Bond length: C—O > C=O > C≡O

Bond strength: C≡O > C=O > C—O

Check From Table 9.2, we see that the rankings are correct.

Comment Remember that for bonds involving pairs of different atoms, as in part (a), *the relationship between length and strength holds* only *for single bonds* and not in every case, so apply it carefully.

FOLLOW-UP PROBLEM 9.2 Rank the bonds in each set in order of *increasing* bond length and *increasing* bond strength: **(a)** Si—F, Si—C, Si—O; **(b)** N=N, N—N, N≡N.

How the Model Explains the Properties of Covalent Substances

The covalent bonding model proposes that electron sharing between pairs of atoms leads to *strong, localized bonds.* Most, but not all, covalent substances consist of individual molecules. These *molecular* covalent substances have very different *physical* properties than *network* covalent solids because different types of forces give rise to them.

1. *Physical properties of molecular covalent substances.* At first glance, the model seems inconsistent with physical properties of covalent substances. Most are gases (such as methane and ammonia), liquids (such as benzene and water), or low-melting solids (such as sulfur and paraffin wax). If covalent bonds are so strong (~200 to 500 kJ/mol), why do covalent substances melt and boil at such low temperatures?

To answer this, we'll focus on two different forces: (1) *strong bonding forces* hold the atoms together within the molecule, and (2) *weak intermolecular forces* act between separate molecules in the sample. It is the weak forces *between* molecules that account for the physical properties of *molecular* covalent substances. For example, look what happens when pentane (C_5H_{12}) boils (Figure 9.13): weak forces *between* pentane molecules are overcome, not the strong C—C and C—H bonds *within* each pentane molecule.

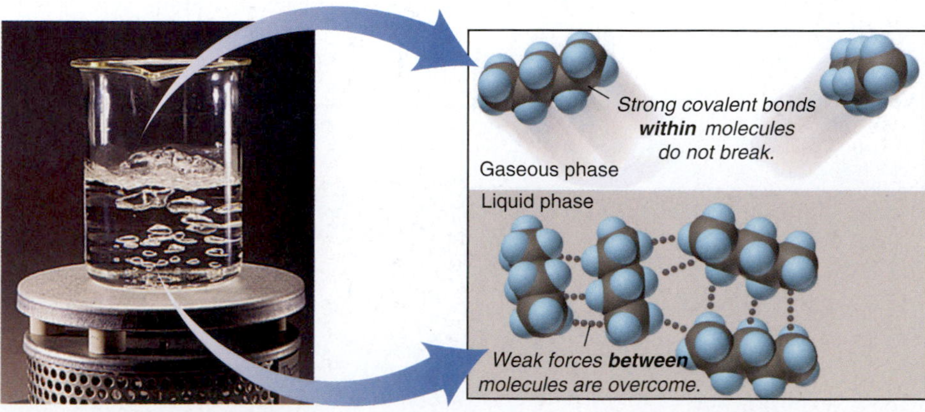
Strong covalent bonds **within** molecules do not break.
Gaseous phase
Liquid phase
Weak forces **between** molecules are overcome.

Figure 9.13 Strong forces within molecules and weak forces between them.

2. *Physical properties of network covalent solids.* Some covalent substances do not consist of separate molecules. Rather, these *network covalent solids* are held together by covalent bonds *between atoms throughout the sample,* and their properties *do* reflect the strength of covalent bonds. Two examples are quartz and diamond (Figure 9.14). Quartz (SiO_2; *top*) has silicon-oxygen covalent bonds in three dimensions; no separate SiO_2 molecules exist. It is very hard and melts at 1550°C. Diamond (*bottom;* see also the chapter-opening photo) has covalent bonds connecting each carbon atom to four others. It is the hardest natural substance known and melts at around 3550°C. Thus, covalent bonds *are* strong, but most covalent substances consist of separate molecules with weak forces between them. (We discuss intermolecular forces in detail in Chapter 12.)

3. *Electrical conductivity.* An electric current is carried by either mobile electrons or mobile ions. Most covalent substances are poor electrical conductors, whether melted or dissolved, because their electrons are localized as either shared or unshared pairs, so they are not free to move, and no ions are present.

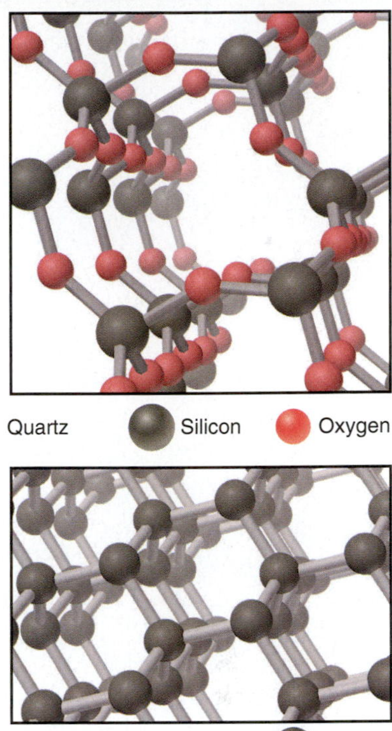

Figure 9.14 Covalent bonds of network covalent solids: quartz and diamond.

Using IR Spectroscopy to Study Covalent Compounds

Chemists often study the types of covalent bonds in a molecule using a technique called **infrared (IR) spectroscopy.** The bonds in all molecules, whether in a gas, a liquid, or a solid, undergo continual vibrations. We can think of any covalent bond between two atoms, say, the C—C bond in ethane ($H_3C—CH_3$), as a spring that is continually stretching, twisting, and bending. Each motion occurs at a particular frequency, which depends on the "stiffness" of the spring (the bond energy), the type of motion, and the masses of the atoms. The frequencies of these vibrational motions correspond to the wavelengths of photons that lie within the IR region of the electromagnetic spectrum. Thus, the energies of these motions are quantized. And, just as an atom can absorb a photon of a particular energy and attain a different electron energy level (Chapter 7), a molecule can absorb an IR photon of a particular energy and attain a different vibrational energy level.

Each kind of bond (C—C, C═C, C—O, etc.) absorbs a characteristic range of IR wavelengths and quantity of radiation, which depends on the molecule's overall structure. The absorptions by all the bonds in a given molecule create a unique pattern that appears as downward pointing peaks of varying depth and sharpness. Thus, *each compound has a characteristic IR spectrum* that can be used to identify it, much like a fingerprint is used to identify a person. As an example, consider the compounds 2-butanol and diethyl ether. These compounds have the same molecular formula ($C_4H_{10}O$) but different structural formulas and, therefore, are constitutional (structural) isomers. Figure 9.15 shows that they have very different IR spectra.

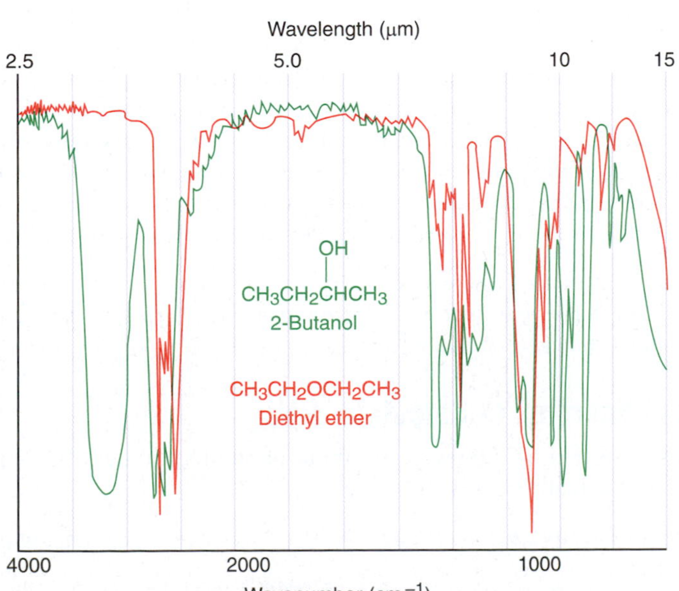

Figure 9.15 The infrared spectra of 2-butanol *(green)* and diethyl ether *(red).*

Summary of Section 9.3

- A shared, localized pair of valence electrons holds the nuclei of two atoms together in a covalent bond, filling each atom's outer level.
- Bond order is the number of shared pairs between two atoms. Bond energy is the energy absorbed to separate the atoms; the same quantity of energy is released when the bond forms. Bond length is the distance between their nuclei.
- For a given pair of atoms, bond order is directly related to bond energy and inversely related to bond length.
- Molecular covalent substances are soft and low melting because of the weak forces *between* the molecules. Network covalent solids are hard and high melting because covalent bonds join all the atoms in the sample.
- Most covalent substances have low electrical conductivity because their electrons are localized and ions are absent.
- Atoms in a covalent bond vibrate, and the energies of these vibrations are studied with IR spectroscopy.

9.4 • BOND ENERGY AND CHEMICAL CHANGE

The relative strengths of the bonds in reactants and products determine whether heat is released or absorbed in a chemical reaction. In Chapter 20, you'll see that the change in bond energy is one of two factors determining whether the reaction occurs at all. In this section, we discuss the origin of the enthalpy of reaction (ΔH°_{rxn}), use bond energies to calculate it, and look at the energy available from fuels and foods.

Changes in Bond Energy: Where Does ΔH°_{rxn} Come From?

In Chapter 6, we discussed the heat involved in a chemical change but never asked a central question: where does the enthalpy of reaction (ΔH°_{rxn}) come from? For example, when 1 mol of H_2 and 1 mol of F_2 react to form 2 mol of HF at 1 atm and 298 K,

$$H_2(g) + F_2(g) \longrightarrow 2HF(g) + 546\ \text{kJ}$$

where does the 546 kJ come from? We find the answer by looking closely at the energies of the molecules involved. A system's total internal energy is composed of its kinetic energy and potential energy. Let's see how these change during the formation of HF:

- *Kinetic energy.* The most important contributions to the kinetic energy are the molecules' movements in space, rotations, and vibrations. However, since kinetic energy is proportional to temperature, which is constant at 298 K, it doesn't change during the reaction.
- *Potential energy.* The most important contributions to the potential energy are phase changes and changes in the attraction between vibrating atoms, between nucleus and electrons (and between electrons) in each atom, between protons and neutrons in each nucleus, and between nuclei and the shared electron pair in each bond. Of these, there are no phase changes, vibrational forces vary only slightly as the bonded atoms change, and forces within the atoms and nuclei don't change at all. The only significant change in potential energy comes from changes in the attraction between the nuclei and the shared electron pair—the bond energy.

Thus, our answer to "where does ΔH°_{rxn} come from?" is that it doesn't really "come from" anywhere: *the heat released or absorbed during a chemical change is due to differences between reactant bond energies and product bond energies.*

Using Bond Energies to Calculate ΔH°_{rxn}

Hess's law (see Section 6.5) allows us to think of any reaction as a two-step process, whether or not it actually occurs that way:

1. *A quantity of heat is absorbed ($\Delta H^\circ > 0$) to break the reactant bonds and form separate atoms.*

2. *A different quantity of heat is then released ($\Delta H^\circ < 0$) when the atoms form product bonds.*

The sum (symbolized by Σ) of these enthalpy changes is the enthalpy of reaction, ΔH°_{rxn}:

$$\Delta H^\circ_{rxn} = \Sigma\Delta H^\circ_{reactant\ bonds\ broken} + \Sigma\Delta H^\circ_{product\ bonds\ formed} \quad (9.2)$$

- In an exothermic reaction, the magnitude of $\Delta H^\circ_{product\ bonds\ formed}$ is *greater* than that of $\Delta H^\circ_{reactant\ bonds\ broken}$, so the sum, ΔH°_{rxn}, is *negative* (heat is released).
- In an endothermic reaction, the opposite situation is true, the magnitude of $\Delta H^\circ_{product\ bonds\ formed}$ is *smaller* than that of $\Delta H^\circ_{reactant\ bonds\ broken}$, so ΔH°_{rxn} is *positive* (heat is absorbed).

An equivalent form of Equation 9.2 uses bond energies:

$$\Delta H^\circ_{rxn} = \Sigma BE_{reactant\ bonds\ broken} - \Sigma BE_{product\ bonds\ formed}$$

(We need the minus sign because all bond energies are positive.)

Typically, only certain bonds break and form during a reaction, but with Hess's law, the following is a simpler method of calculating ΔH°_{rxn}:

1. Break **all** the reactant bonds to obtain individual atoms.
2. Use the atoms to form **all** the product bonds.
3. Add the bond energies, with appropriate signs, to obtain the enthalpy of reaction.

(This method assumes reactants and products do not change physical state; additional heat is involved when phase changes occur. We address this topic in Chapter 12.)

Let's use the method to calculate ΔH°_{rxn} for two reactions:

1. *Formation of HF.* When 1 mol of H—H bonds and 1 mol of F—F bonds absorb energy and break, the 2 mol of H atoms and 2 mol of F atoms form 2 mol of H—F bonds, which releases energy (Figure 9.16). We find the bond energy values in Table 9.2 and use a positive sign for bonds broken and a negative sign for bonds formed:

Bonds broken:

$$1 \times \text{H—H} = (1\ \text{mol})(432\ \text{kJ/mol}) = 432\ \text{kJ}$$
$$1 \times \text{F—F} = (1\ \text{mol})(159\ \text{kJ/mol}) = 159\ \text{kJ}$$
$$\Sigma\Delta H^\circ_{reactant\ bonds\ broken} = 591\ \text{kJ}$$

Bonds formed:

$$2 \times \text{H—F} = (2\ \text{mol})(-565\ \text{kJ/mol}) = \Sigma\Delta H^\circ_{product\ bonds\ formed} = -1130\ \text{kJ}$$

Applying Equation 9.2 gives

$$\Delta H^\circ_{rxn} = \Sigma\Delta H^\circ_{reactant\ bonds\ broken} + \Sigma\Delta H^\circ_{product\ bonds\ formed}$$
$$= 591\ \text{kJ} + (-1130\ \text{kJ}) = -539\ \text{kJ}$$

The small discrepancy between this bond energy value (-539 kJ) and the value from tabulated ΔH° values (-546 kJ) is due to variations in experimental method.

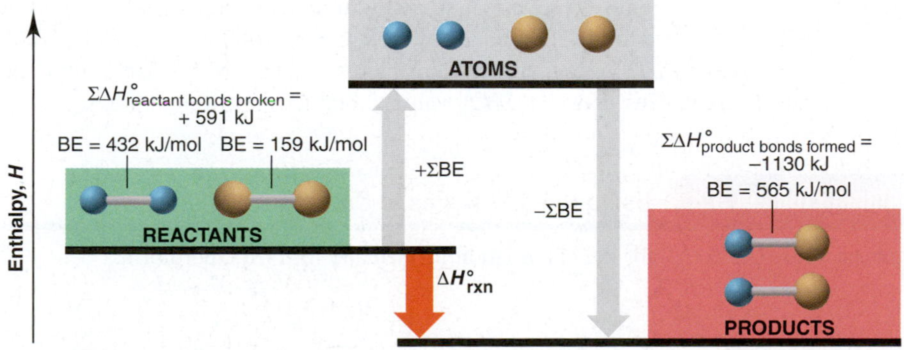

Figure 9.16 Using bond energies to calculate ΔH°_{rxn} for HF formation.

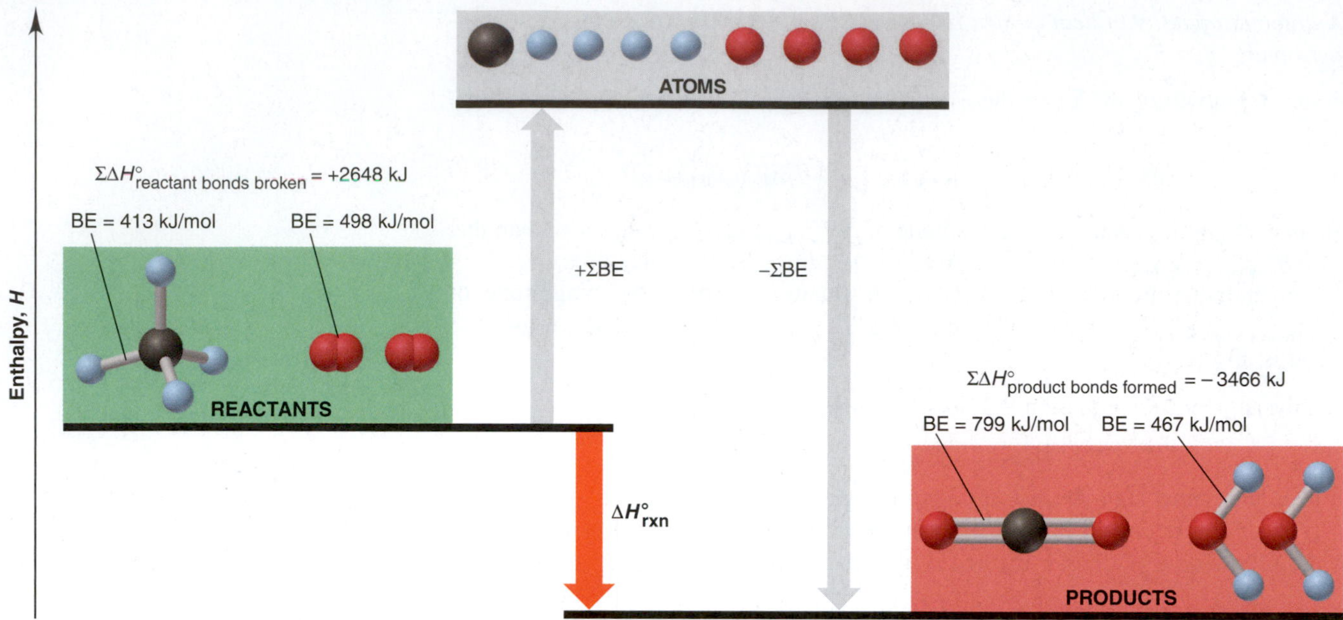

Figure 9.17 Using bond energies to calculate $\Delta H^\circ_{\text{rxn}}$ for the combustion of methane.

2. *Combustion of CH_4.* In this more complicated reaction, all the bonds in CH_4 and O_2 break, and the atoms form all the bonds in CO_2 and H_2O (Figure 9.17). Once again, we use Table 9.2 and appropriate signs for bonds broken and bonds formed:

Bonds broken:

$$\begin{aligned} 4 \times \text{C—H} &= (4\ \text{mol})(413\ \text{kJ/mol}) = 1652\ \text{kJ} \\ 2 \times \text{O}_2 &= (2\ \text{mol})(498\ \text{kJ/mol}) = 996\ \text{kJ} \\ \hline \Sigma\Delta H^\circ_{\text{reactant bonds broken}} &= 2648\ \text{kJ} \end{aligned}$$

Bonds formed:

$$\begin{aligned} 2 \times \text{C=O} &= (2\ \text{mol})(-799\ \text{kJ/mol}) = -1598\ \text{kJ} \\ 4 \times \text{O—H} &= (4\ \text{mol})(-467\ \text{kJ/mol}) = -1868\ \text{kJ} \\ \hline \Sigma\Delta H^\circ_{\text{product bonds formed}} &= -3466\ \text{kJ} \end{aligned}$$

Applying Equation 9.2 gives

$$\begin{aligned} \Delta H^\circ_{\text{rxn}} &= \Sigma\Delta H^\circ_{\text{reactant bonds broken}} + \Sigma\Delta H^\circ_{\text{product bonds formed}} \\ &= 2648\ \text{kJ} + (-3466\ \text{kJ}) = -818\ \text{kJ} \end{aligned}$$

In addition to variations in experimental method, there is a more basic reason for the discrepancy between the $\Delta H^\circ_{\text{rxn}}$ obtained from bond energies (−818 kJ) and the value obtained by calorimetry (−802 kJ; Section 6.3). A bond energy is an *average* value for a given bond over many compounds. The value *in a particular substance* is usually close, but not equal, to the average. For example, the C—H bond energy of 413 kJ/mol is the average for C—H bonds in many molecules. In fact, 415 kJ is actually required to break 1 mol of C—H bonds in methane, or 1660 kJ for 4 mol of these bonds, which gives a $\Delta H^\circ_{\text{rxn}}$ closer to the calorimetric value. Thus, it isn't surprising to find small discrepancies between $\Delta H^\circ_{\text{rxn}}$ values obtained in different ways.

Sample Problem 9.3 Using Bond Energies to Calculate $\Delta H^\circ_{\text{rxn}}$

Problem Calculate $\Delta H^\circ_{\text{rxn}}$ for the chlorination of methane to form chloroform:

$$\text{H—}\underset{\text{H}}{\overset{\text{H}}{\text{C}}}\text{—H} + 3\ \text{Cl—Cl} \longrightarrow \text{Cl—}\underset{\text{Cl}}{\overset{\text{H}}{\text{C}}}\text{—Cl} + 3\ \text{H—Cl}$$

Plan *All* the reactant bonds break, and *all* the product bonds form. We find the bond energies in Table 9.2 and substitute the two sums, with correct signs, into Equation 9.2.

Solution Finding the standard enthalpy changes for bonds broken and for bonds formed: For bonds broken, the bond energy values are

$$\begin{aligned} 4 \times \text{C—H} &= (4\ \text{mol})(413\ \text{kJ/mol}) = 1652\ \text{kJ} \\ 3 \times \text{Cl—Cl} &= (3\ \text{mol})(243\ \text{kJ/mol}) = 729\ \text{kJ} \\ \hline \Sigma\Delta H^\circ_{\text{bonds broken}} &= 2381\ \text{kJ} \end{aligned}$$

For bonds formed, the values are

$$\begin{aligned} 3 \times \text{C—Cl} &= (3\ \text{mol})(-339\ \text{kJ/mol}) = -1017\ \text{kJ} \\ 1 \times \text{C—H} &= (1\ \text{mol})(-413\ \text{kJ/mol}) = -413\ \text{kJ} \\ 3 \times \text{H—Cl} &= (3\ \text{mol})(-427\ \text{kJ/mol}) = -1281\ \text{kJ} \\ \hline \Sigma\Delta H^\circ_{\text{bonds formed}} &= -2711\ \text{kJ} \end{aligned}$$

Calculating $\Delta H^\circ_{\text{rxn}}$:

$$\Delta H^\circ_{\text{rxn}} = \Sigma\Delta H^\circ_{\text{bonds broken}} + \Sigma\Delta H^\circ_{\text{bonds formed}} = 2381\ \text{kJ} + (-2711\ \text{kJ}) = -330\ \text{kJ}$$

Check The signs of the enthalpy changes are correct: $\Sigma\Delta H^\circ_{\text{bonds broken}} > 0$, and $\Sigma\Delta H^\circ_{\text{bonds formed}} < 0$. More energy is released than absorbed, so $\Delta H^\circ_{\text{rxn}}$ is negative:

$$\sim 2400\ \text{kJ} + [\sim(-2700\ \text{kJ})] = -300\ \text{kJ}$$

FOLLOW-UP PROBLEM 9.3 One of the most important industrial reactions is the formation of ammonia from its elements:

$$\text{N}{\equiv}\text{N} + 3\,\text{H—H} \longrightarrow 2\,\text{H—}\underset{\text{H}}{\underset{|}{\text{N}}}\text{—H}$$

Use bond energies to calculate $\Delta H^\circ_{\text{rxn}}$.

Summary of Section 9.4

- The only component of internal energy that changes significantly during a reaction is the bond energies of reactants and products, and this change appears as the enthalpy of reaction, $\Delta H^\circ_{\text{rxn}}$.
- A reaction involves breaking reactant bonds and forming product bonds. Applying Hess's law, we use tabulated bond energies to calculate $\Delta H^\circ_{\text{rxn}}$.

9.5 • BETWEEN THE EXTREMES: ELECTRONEGATIVITY AND BOND POLARITY

Scientific models are idealized descriptions of reality. The ionic and covalent bonding models portray compounds as formed by *either* complete electron transfer *or* complete electron sharing. But, in real substances, most atoms are joined by *polar covalent bonds*—partly ionic and partly covalent (Figure 9.18). In this section, we explore the "in-between" nature of these bonds and its importance in the properties of substances.

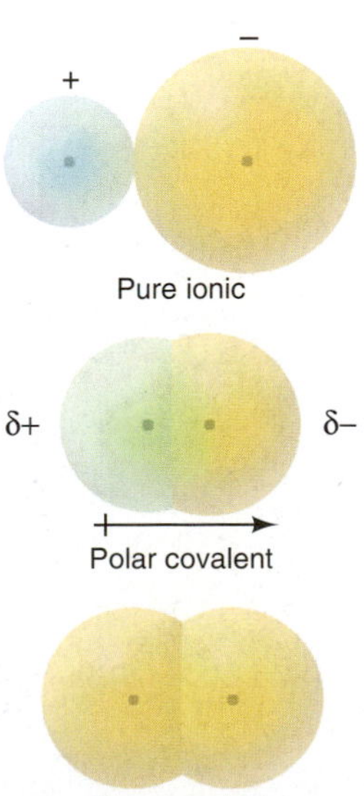

Figure 9.18 Bonding between the models. Pure ionic bonding *(top)* and pure covalent bonding *(bottom)* are far less common than polar covalent bonding *(middle)*.

Electronegativity

Electronegativity (EN) is the relative ability of a bonded atom to attract shared electrons.* We might expect the H—F bond energy to be the average of an H—H bond (432 kJ/mol) and an F—F bond (159 kJ/mol), or 296 kJ/mol. But, the actual HF bond energy is 565 kJ/mol, or 269 kJ/mol *higher!* To explain this difference, the American chemist Linus Pauling reasoned that it is due to an *electrostatic (charge) contribution* to the H—F bond energy. If F attracts the shared electron pair more strongly than H, that is, if F is more *electronegative* than H, the electrons will spend more time

*Electronegativity refers to a *bonded* atom attracting a shared pair; electron affinity refers to a gaseous atom gaining an electron to form an anion. Elements with a high EN also have a highly negative EA.

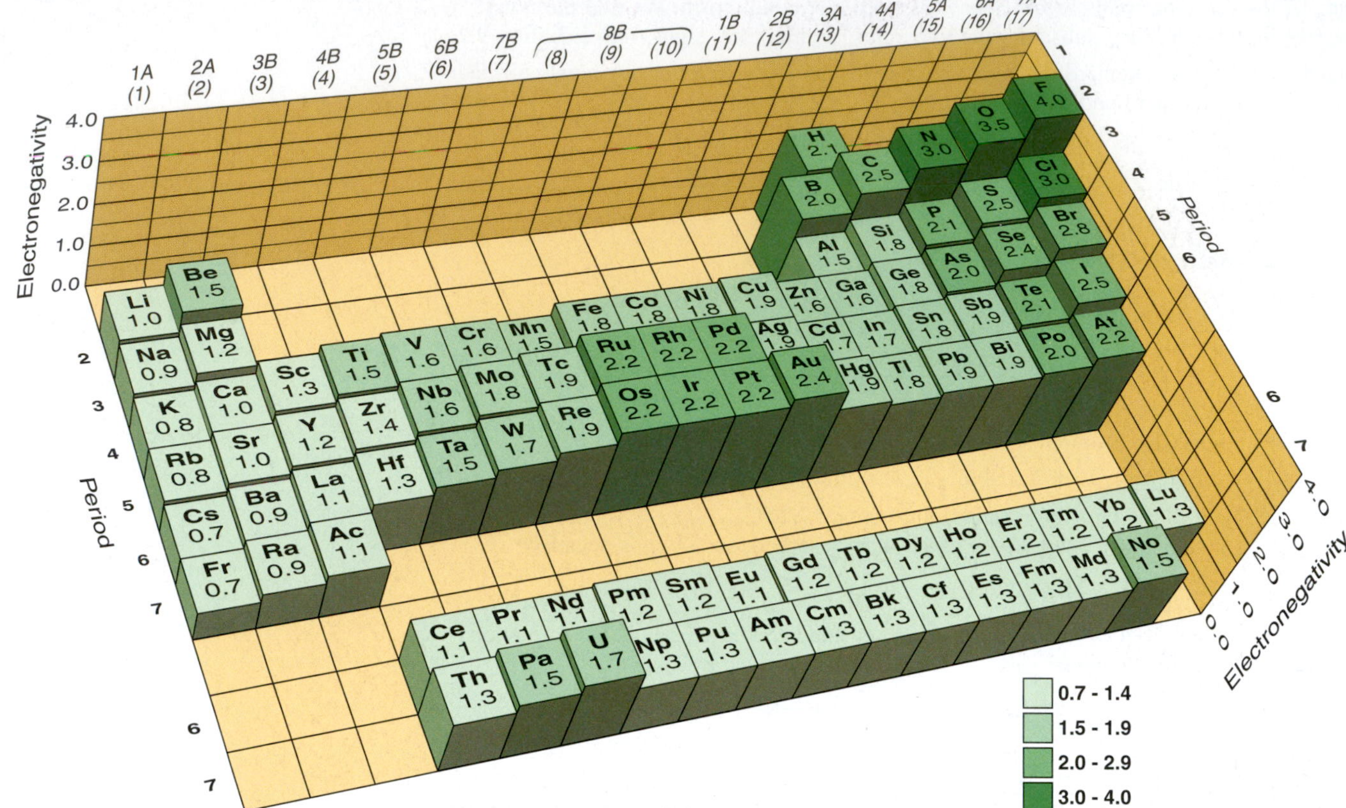

Figure 9.19 The Pauling electronegativity (EN) scale. The height of each post is proportional to the EN, which is shown on top. The key has several EN cutoffs. In the main groups, EN *increases* across and *decreases* down. The transition and inner transition elements show little change in EN. Here hydrogen is placed near elements with similar EN values.

closer to F. This unequal sharing makes the F end of the bond partially negative and the H end partially positive. The electrostatic attraction between these partial charges *increases* the energy required to break the bond. From studies with many other compounds, Pauling derived a scale of *relative EN values* based on fluorine having the highest EN value, 4.0 (Figure 9.19).

Trends in Electronegativity Because the nucleus of a smaller atom is closer to the shared pair than that of a larger atom, it attracts the bonding electrons more strongly. So, in general, electronegativity is inversely related to atomic size. Thus, for the main-group elements, *electronegativity generally increases up a group and across a period.*

Electronegativity and Oxidation Number An important use of electronegativity is in determining an atom's oxidation number (O.N.; see Section 4.5):

1. The more electronegative atom in a bond is assigned *all* the *shared* electrons; the less electronegative atom is assigned *none.*
2. Each atom in a bond is assigned *all* of its *unshared* electrons.
3. The oxidation number is given by

$$\text{O.N.} = \text{no. of valence } e^- - (\text{no. of shared } e^- + \text{no. of unshared } e^-)$$

In HCl, for example, Cl is more electronegative than H. Cl has 7 valence electrons and is assigned 8 (2 shared + 6 unshared), so its O.N. is $7 - 8 = -1$. The H atom has 1 valence electron and is assigned none, so its O.N. is $1 - 0 = +1$.

Bond Polarity and Partial Ionic Character

Whenever atoms of different electronegativities form a bond, such as H (2.1) and F (4.0) in HF, the bonding pair is shared *unequally*. This unequal distribution of electron density results in a **polar covalent bond.** It is depicted by a polar arrow ($\mapsto$) pointing toward the partially negative pole or by $\delta+$ and $\delta-$ symbols (see Figure 4.1):

$$\overset{\longmapsto}{\text{H—}\ddot{\text{F}}\text{:}} \quad \text{or} \quad \overset{\delta+}{\text{H}}\text{—}\overset{\delta-}{\ddot{\text{F}}}\text{:}$$

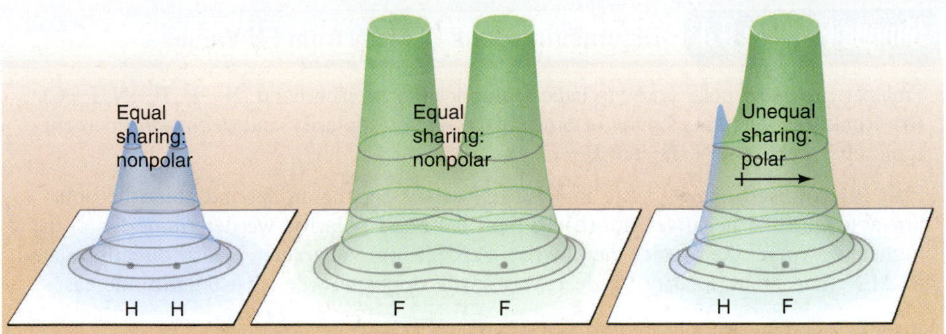

Figure 9.20 Electron density distributions in H_2, F_2, and HF. In HF, the electron density shifts from H to F. (The electron density peaks for F have been cut off in the relief maps to limit the figure height.)

In the H—H and F—F bonds, where the atoms are identical, the bonding pair is shared *equally,* and a **nonpolar covalent bond** results. In Figure 9.20, relief maps show the distribution of electron density in H_2, F_2, and HF.

The Importance of Electronegativity Difference (ΔEN) The **electronegativity difference (ΔEN),** the difference between the EN values of the bonded atoms, is directly related to a bond's polarity. It ranges from 0.0 in a diatomic element, such as H_2, O_2, or Cl_2, all the way up to 3.3, the difference between the most electronegative atom, F (4.0), and the most electropositive, Cs (0.7), in the ionic compound CsF.

Another parameter closely related to ΔEN is the **partial ionic character** of a bond: *a greater ΔEN results in larger partial charges and higher partial ionic character.* Consider three Cl-containing gaseous molecules: ΔEN for LiCl(*g*) is $3.0 - 1.0 = 2.0$; for HCl(*g*), it is $3.0 - 2.1 = 0.9$; and for Cl_2(*g*), it is $3.0 - 3.0 = 0.0$. Thus, the bond in LiCl has more ionic character than the one in HCl, which has more than the one in Cl_2.

Here are two approaches that quantify ionic character. Both use arbitrary cutoffs, which is not really consistent with the actual gradation in bonding:

1. *ΔEN range*. This approach divides bonds into mostly ionic, polar covalent, mostly covalent, and nonpolar covalent based on a range of ΔEN values (Figure 9.21).

2. *Percent ionic character*. This approach is based on the behavior of a diatomic molecule in an electric field. A plot of *percent ionic character* vs. ΔEN for several gaseous molecules shows that, as expected, *percent ionic character generally increases with ΔEN* (Figure 9.22A). A value of 50% divides ionic from covalent bonds. Note that a substance like Cl_2(*g*) has 0% ionic character, but none has 100% ionic character: *electron sharing occurs to some extent in every bond,* even the bond in an alkali halide (Figure 9.22B).

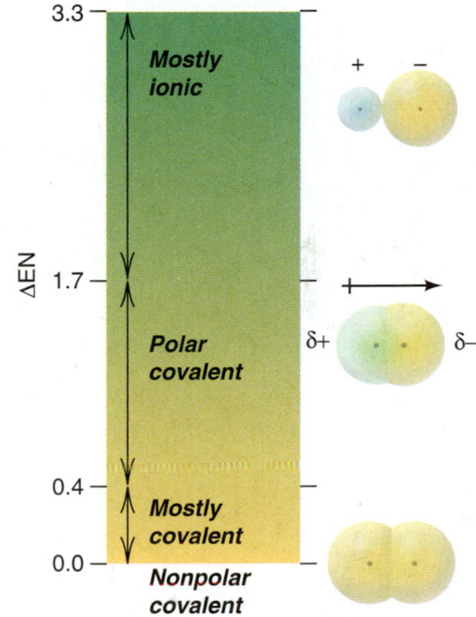

Figure 9.21 ΔEN ranges for classifying the partial ionic character of bonds.

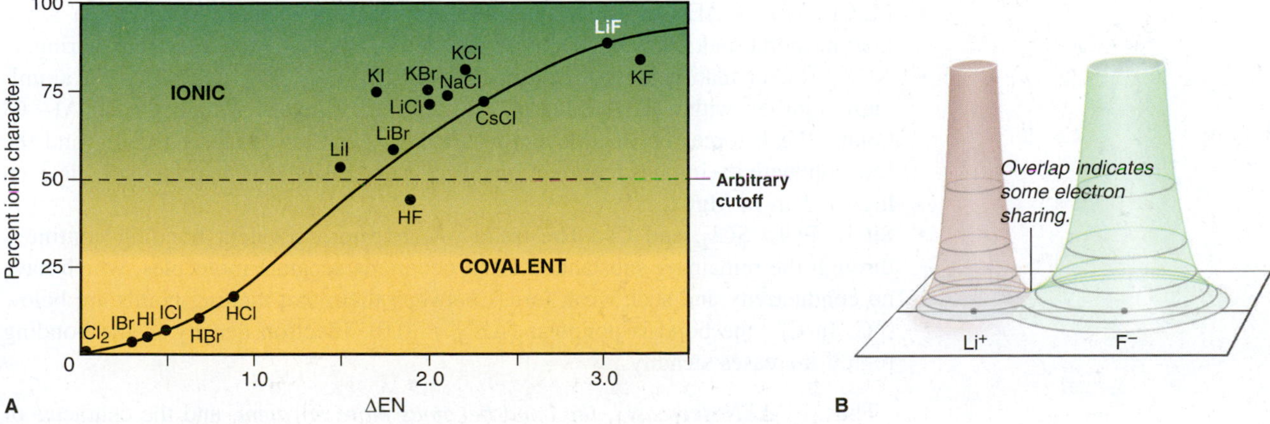

Figure 9.22 Percent ionic character as a function of ΔEN. **A,** ΔEN correlates with ionic character. **B,** Even for highly ionic LiF (ΔEN = 3.0), the relief map shows some electron sharing between the ions.

Sample Problem 9.4 Determining Bond Polarity from EN Values

Problem **(a)** Use a polar arrow to indicate the polarity of each bond: N—H, F—N, I—Cl. **(b)** Rank the following bonds in order of *increasing* polarity and *decreasing* percent ionic character: H—N, H—O, H—C.

Plan **(a)** We use Figure 9.19 to find the EN values for the atoms and point the polar arrow toward the negative end. **(b)** To rank the bond polarity, we determine ΔEN: the higher the value, the greater the polarity. Percent ionic character is also directly related to ΔEN (and bond polarity); it decreases in the opposite order that polarity increases.

Solution **(a)** The EN of N = 3.0 and the EN of H = 2.1, so $\overset{\longleftarrow\!+}{\text{N—H}}$

The EN of F = 4.0 and the EN of N = 3.0, so $\overset{\longleftarrow\!+}{\text{F—N}}$

The EN of I = 2.5 and the EN of Cl = 3.0, so $\overset{+\!\longrightarrow}{\text{I—Cl}}$

(b) The ΔEN values are 0.9 for H—N, 1.4 for H—O, and 0.4 for H—C.
The order of *increasing* bond polarity is H—C < H—N < H—O.
The order of *decreasing* percent ionic character is H—O > H—N > H—C.

Check In (b), we can check the order of bond polarity using periodic trends. Each bond involves H and a Period 2 atom. Since size decreases and EN increases across a period, the polarity is greatest for the bond to O (farthest to the right in Period 2).

Comment In Chapter 10, you'll see that the bond polarity contributes to the overall polarity of the molecule, which is a major factor determining behavior.

FOLLOW-UP PROBLEM 9.4 Arrange each set of bonds in order of increasing polarity, and indicate bond polarity with δ+ and δ− symbols:
(a) Cl—F, Br—Cl, Cl—Cl **(b)** Si—Cl, P—Cl, S—Cl, Si—Si

The Gradation in Bonding Across a Period

A metal and a nonmetal—elements from the left and right sides of the periodic table—have a relatively large ΔEN and typically form an ionic compound. Two nonmetals—both from the right side of the table—have a small ΔEN and form a covalent compound. When we combine chlorine with each of the other Period 3 elements, starting with sodium, we observe a steady decrease in ΔEN and a gradation in bond type from ionic through polar covalent to nonpolar covalent.

Figure 9.23 shows samples of common Period 3 chlorides—NaCl, $MgCl_2$, $AlCl_3$, $SiCl_4$, PCl_3, and SCl_2, as well as Cl_2—along with the change in ΔEN and two physical properties:

- NaCl. Sodium chloride is a white (colorless) crystalline solid with a ΔEN of 2.1, a high melting point, and high electrical conductivity when molten—ionic by any criterion. Nevertheless, just as for LiF (Figure 9.22B), a small but significant amount of electron sharing appears in the NaCl electron density relief map.
- $MgCl_2$. With a ΔEN of 1.8, magnesium chloride is still ionic, but it has a lower melting point and lower conductivity, as well as slightly more electron sharing.
- $AlCl_3$. Rather than being a three-dimensional lattice of Al^{3+} and Cl^- ions, aluminum chloride, with a ΔEN value of 1.5, consists of layers of highly polar Al—Cl bonds. Weak forces between layers result in a much lower melting point, and the low conductivity implies few free ions. Electron density between the nuclei is even higher than in $MgCl_2$.
- $SiCl_4$, PCl_3, SCl_2, and Cl_2. The trend toward more covalent bonding continues through the remaining substances. Each occurs as separate molecules, which have no conductivity and such weak forces *between* them that melting points are below 0°C. In Cl_2, the bond is nonpolar (ΔEN = 0.0). Electron density in the bonding region increases steadily.

Thus, *as ΔEN decreases, the bond becomes more covalent,* and the character of the substance changes from ionic solid to covalent gas.

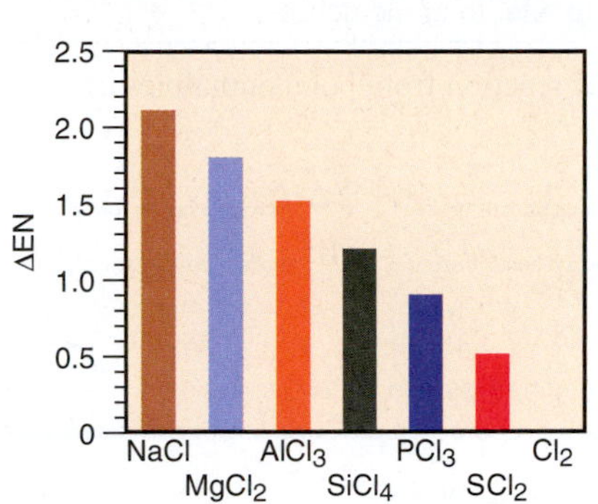

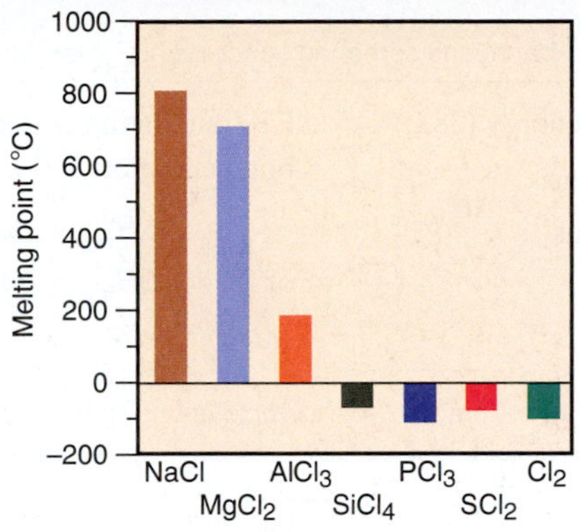

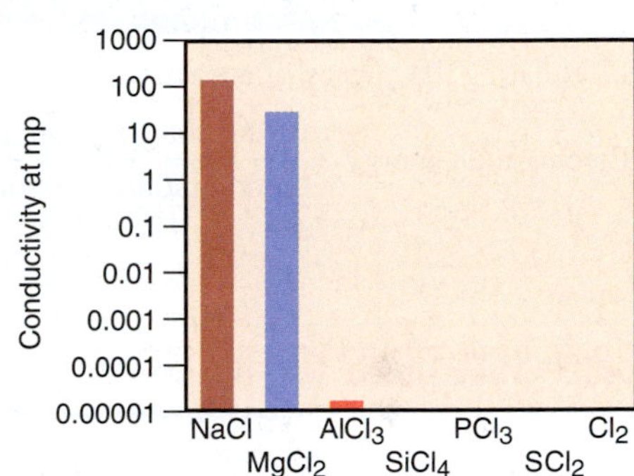

Figure 9.23 Properties of the Period 3 chlorides. As ΔEN decreases, melting point and electrical conductivity decrease because the bond type changes from ionic to polar covalent to nonpolar covalent.

Summary of Section 9.5

- Electronegativity is the ability of a bonded atom to attract shared electrons, which generates opposite partial charges at the ends of the bond and contributes to the bond energy.
- Electronegativity increases across a period and decreases down a group, the reverse of the trends in atomic size.
- The larger the ΔEN for two bonded atoms, the more polar the bond is and the greater its ionic character.
- For Period 3 chlorides, there is a gradation in bond type from ionic to polar covalent to nonpolar covalent.

CHAPTER REVIEW GUIDE

The following sections provide many aids to help you study this chapter. (Numbers in parentheses refer to pages, unless noted otherwise.)

Learning Objectives

These are concepts and skills to review after studying this chapter.

Related section (§), sample problem (SP), and upcoming end-of-chapter problem (EP) numbers are listed in parentheses.

1. Explain how differences in atomic properties lead to the three types of chemical bonding (§9.1) (EPs 9.1–9.7)
2. Depict main-group atoms with Lewis electron-dot symbols (§9.1) (EPs 9.8–9.11)
3. Understand the key features of ionic bonding, the significance of the lattice energy, and how the model explains the properties of ionic compounds (§9.2) (EPs 9.12–9.15, 9.20–9.22)
4. Depict the formation of binary ionic compounds with electron configurations, partial orbital diagrams, and Lewis electron-dot symbols (§9.2) (SP 9.1) (EPs 9.16–9.19)
5. Describe the formation of a covalent bond, the interrelationship among bond length, strength, and order, and how the model explains the properties of covalent compounds (§9.3) (SP 9.2) (EPs 9.23–9.31)
6. Understand how changes in bond energy account for ΔH°_{rxn} and be able to divide a reaction into bond-breaking and bond-forming steps (§9.4) (SP 9.3) (EPs 9.32–9.39)
7. Describe the trends in electronegativity, and understand how the polarity of a bond and the partial ionic character of a compound relate to ΔEN of the bonded atoms (§9.5) (SP 9.4) (EPs 9.40–9.55)

Key Terms

These important terms appear in boldface in the chapter and are defined again in the Glossary.

Section 9.1
ionic bonding (277)
covalent bonding (278)
metallic bonding (278)
Lewis electron-dot symbol (278)
octet rule (279)

Section 9.2
lattice energy ($\Delta H^\circ_{lattice}$) (281)
Coulomb's law (282)
ion pair (283)

Section 9.3
covalent bond (285)
bonding (shared) pair (285)
lone (unshared) pair (285)
bond order (285)
single bond (285)
double bond (285)
triple bond (286)
bond energy (BE) (286)
bond length (286)
infrared (IR) spectroscopy (289)

Section 9.5
electronegativity (EN) (293)
polar covalent bond (294)
nonpolar covalent bond (295)
electronegativity difference (ΔEN) (295)
partial ionic character (295)

Key Equations and Relationships

Numbered and screened concepts are listed for you to refer to or memorize.

9.1 Relating the energy of attraction to the lattice energy (282):

$$\text{Electrostatic energy} \propto \frac{\text{cation charge} \times \text{anion charge}}{\text{cation radius} + \text{anion radius}} \propto \Delta H^\circ_{lattice}$$

9.2 Calculating enthalpy of reaction from bond enthalpies or bond energies (291):

$$\Delta H^\circ_{rxn} = \Sigma \Delta H^\circ_{\text{reactant bonds broken}} + \Sigma \Delta H^\circ_{\text{product bonds formed}}$$

or

$$\Delta H^\circ_{rxn} = \Sigma BE_{\text{reactant bonds broken}} - \Sigma BE_{\text{product bonds formed}}$$

BRIEF SOLUTIONS TO FOLLOW-UP PROBLEMS

Compare your own solutions to these calculation steps and answers.

9.1 Mg ([Ne] $3s^2$) + 2Cl ([Ne] $3s^2 3p^5$) $\longrightarrow$

Mg^{2+} ([Ne]) + 2Cl$^-$ ([Ne] $3s^2 3p^6$)

$\cdot$Mg$\cdot$ + 2 $\cdot\ddot{\text{Cl}}$: $\longrightarrow$ Mg^{2+} + 2 :$\ddot{\text{Cl}}$:$^-$

Formula: $MgCl_2$

9.2 (a) Bond length: Si—F < Si—O < Si—C
Bond strength: Si—C < Si—O < Si—F
(b) Bond length: N≡N < N=N < N—N
Bond strength: N—N < N=N < N≡N

9.3 N≡N + 3 H—H $\longrightarrow$ 2 H—N(—H)—H

$\Sigma \Delta H^\circ_{\text{bonds broken}}$ = 1 N≡N + 3 H—H
= 945 kJ + 1296 kJ = 2241 kJ
$\Sigma \Delta H^\circ_{\text{bonds formed}}$ = 6 N—H = −2346 kJ
ΔH°_{rxn} = −105 kJ

9.4 (a) Cl—Cl < $\overset{\delta+}{\text{Br}}$—$\overset{\delta-}{\text{Cl}}$ < $\overset{\delta+}{\text{Cl}}$—$\overset{\delta-}{\text{F}}$
(b) Si—Si < $\overset{\delta+}{\text{S}}$—$\overset{\delta-}{\text{Cl}}$ < $\overset{\delta+}{\text{P}}$—$\overset{\delta-}{\text{Cl}}$ < $\overset{\delta+}{\text{Si}}$—$\overset{\delta-}{\text{Cl}}$

PROBLEMS

Problems with **colored** numbers are answered in Appendix E. Sections match the text and provide the numbers of relevant sample problems. Bracketed problems are grouped in pairs (indicated by a short rule) that cover the same concept. Comprehensive Problems are based on material from any section or previous chapter.

Atomic Properties and Chemical Bonds

9.1 In general terms, how does each of the following atomic properties influence the metallic character of the main-group elements in a period?
(a) Ionization energy
(b) Atomic radius
(c) Number of outer electrons
(d) Effective nuclear charge

9.2 Three solids are represented below. What is the predominant type of intramolecular bonding in each?

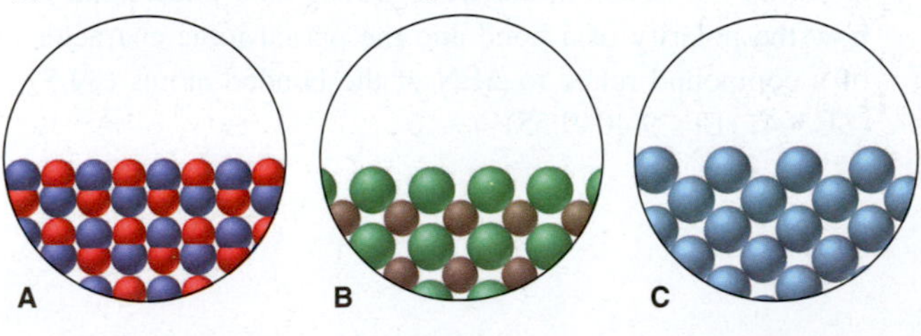

9.3 What is the relationship between the tendency of a main-group element to form a monatomic ion and its position in the periodic table? In what part of the table are the main-group elements that typically form cations? Anions?

9.4 Which member of each pair is *more* metallic?
(a) Na or Cs (b) Mg or Rb (c) As or N

9.5 Which member of each pair is *less* metallic?
(a) I or O (b) Be or Ba (c) Se or Ge

9.6 State the type of bonding—ionic, covalent, or metallic—you would expect in each: (a) CsF(s); (b) N_2(g); (c) Na(s).

9.7 State the type of bonding—ionic, covalent, or metallic—you would expect in each: (a) ICl_3(g); (b) N_2O(g); (c) LiCl(s).

9.8 Draw a Lewis electron-dot symbol for (a) Rb; (b) Si; (c) I.

9.9 Draw a Lewis electron-dot symbol for (a) Ba; (b) Kr; (c) Br.

9.10 Give the group number and general electron configuration of an element with each electron-dot symbol: (a) $\cdot\ddot{\text{X}}$: (b) $\dot{\text{X}}\cdot$

9.11 Give the group number and general electron configuration of an element with each electron-dot symbol: (a) $\cdot\dot{\text{X}}$: (b) $\cdot\dot{\text{X}}\cdot$

The Ionic Bonding Model

(Sample Problem 9.1)

9.12 If energy is required to form monatomic ions from metals and nonmetals, why do ionic compounds exist?

9.13 (a) In general, how does the lattice energy of an ionic compound depend on the charges and sizes of the ions? (b) Ion arrangements of three general salts are represented below. Rank them in order of increasing lattice energy.

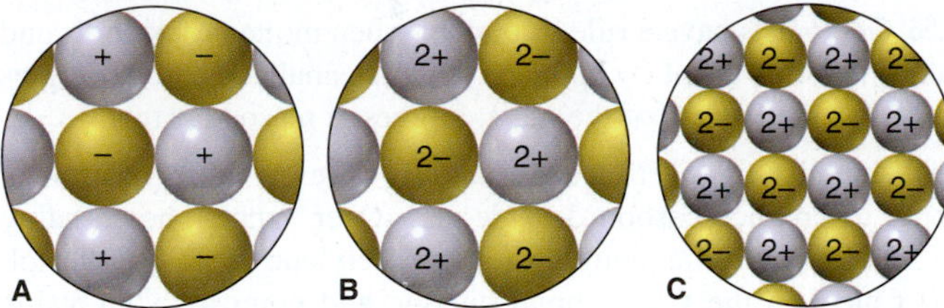

9.14 When gaseous Na^+ and Cl^- ions form gaseous NaCl ion pairs, 548 kJ/mol of energy is released. Why, then, does NaCl occur as a solid under ordinary conditions?

9.15 To form S^{2-} ions from gaseous sulfur atoms requires 214 kJ/mol, but these ions exist in solids such as K_2S. Explain.

9.16 Use condensed electron configurations and Lewis electron-dot symbols to depict the ions formed from each of the following atoms, and predict the formula of their compound:
(a) Ba and Cl (b) Sr and O (c) Al and F (d) Rb and O

9.17 Use condensed electron configurations and Lewis electron-dot symbols to depict the ions formed from each of the following atoms, and predict the formula of their compound:
(a) Cs and S (b) O and Ga (c) N and Mg (d) Br and Li

9.18 For each ionic compound formula, identify the main group to which X belongs: (a) X_2O_3; (b) XCO_3; (c) Na_2X.

9.19 For each ionic compound formula, identify the main group to which X belongs: (a) CaX_2; (b) Al_2X_3; (c) XPO_4.

9.20 For each pair, choose the compound with the lower lattice energy, and explain your choice: (a) CaS or BaS; (b) NaF or MgO.

9.21 For each pair, choose the compound with the lower lattice energy, and explain your choice: (a) NaF or NaCl; (b) K_2O or K_2S.

9.22 Aluminum oxide (Al_2O_3) is a widely used industrial abrasive (emery, corundum), for which the specific application depends on the hardness of the crystal. What does this hardness imply about the magnitude of the lattice energy? Would you have predicted from the chemical formula that Al_2O_3 is hard? Explain.

The Covalent Bonding Model

(Sample Problem 9.2)

9.23 Describe the interactions that occur between individual chlorine atoms as they approach each other and form Cl_2. What combination of forces gives rise to the energy holding the atoms together and to the final internuclear distance?

9.24 Define bond energy using the H—Cl bond as an example. When this bond breaks, is energy absorbed or released? Is the accompanying ΔH value positive or negative? How do the magnitude and sign of this ΔH value relate to the value that accompanies H—Cl bond formation?

9.25 For single bonds between similar types of atoms, how does the strength of the bond relate to the sizes of the atoms? Explain.

9.26 How does the energy of the bond between a given pair of atoms relate to the bond order? Why?

9.27 When liquid benzene (C_6H_6) boils, does the gas consist of molecules, ions, or separate atoms? Explain.

9.28 Using the periodic table only, arrange the members of each of the following sets in order of increasing bond *strength:*
(a) Br—Br, Cl—Cl, I—I (b) S—H, S—Br, S—Cl
(c) C═N, C—N, C≡N

9.29 Using the periodic table only, arrange the members of each of the following sets in order of increasing bond *length:*
(a) H—F, H—I, H—Cl (b) C—S, C═O, C—O
(c) N—H, N—S, N—O

9.30 Formic acid (HCOOH; structural formula shown below) is secreted by certain species of ants when they bite.

H—C(═O)—O—H

Rank the relative strengths of (a) the C—O and C═O bonds and (b) the H—C and H—O bonds. Explain these rankings.

9.31 In IR spectra, the stretching of a C═C bond appears at a shorter wavelength than that of a C—C bond. Would you expect the wavelength for the stretching of a C≡C bond to be shorter or longer than that for a C═C bond? Explain.

Bond Energy and Chemical Change

(Sample Problem 9.3)

9.32 Write a solution plan (without actual numbers, but including the bond energies you would use and how you would combine them algebraically) for calculating the total enthalpy change of the following reaction:

$$H_2(g) + O_2(g) \longrightarrow H_2O_2(g) \qquad (\text{H—O—O—H})$$

9.33 The text points out that, for similar types of substances, one with weaker bonds is usually more reactive than one with stronger bonds. Why is this generally true?

9.34 Why is there a discrepancy between an enthalpy of reaction obtained from calorimetry and one obtained from bond energies?

9.35 Which of the following gases would you expect to have the greater enthalpy of reaction per mole for combustion? Why?

methane (H—CH₂—H, i.e., CH_4) or formaldehyde (H—C(═O)—H)

9.36 Use bond energies to calculate the enthalpy of reaction:

$H_2C{=}CH_2$ + Cl—Cl ⟶ H—CH(Cl)—CH(Cl)—H

9.37 Use bond energies to calculate the enthalpy of reaction:

O═C═O + 2 H—N(H)—H ⟶ H—N(H)—C(═O)—N(H)—H + H—O—H

9.38 An important industrial route to extremely pure acetic acid is the reaction of methanol with carbon monoxide:

H—C(H)(H)—O—H + C≡O ⟶ H—C(H)(H)—C(═O)—O—H

Use bond energies to calculate the enthalpy of reaction.

9.39 Sports trainers treat sprains and soreness with ethyl bromide. It is manufactured by reacting ethylene with hydrogen bromide:

$$H_2C{=}CH_2 + H{-}Br \longrightarrow H{-}\underset{H}{\overset{H}{C}}{-}\underset{H}{\overset{H}{C}}{-}Br$$

Use bond energies to find the enthalpy of reaction.

Between the Extremes: Electronegativity and Bond Polarity

(Sample Problem 9.4)

9.40 Describe the vertical and horizontal trends in electronegativity (EN) among the main-group elements. According to Pauling's scale, what are the two most electronegative elements? The two least electronegative elements?

9.41 What is the general relationship between IE_1 and EN for the elements? Why?

9.42 Is the H—O bond in water nonpolar covalent, polar covalent, or ionic? Define each term, and explain your choice.

9.43 How does electronegativity differ from electron affinity?

9.44 How is the partial ionic character of a bond in a diatomic molecule related to ΔEN for the bonded atoms? Why?

9.45 Using the periodic table only, arrange the elements in each set in order of *increasing* EN: (a) S, O, Si; (b) Mg, P, As.

9.46 Using the periodic table only, arrange the elements in each set in order of *increasing* EN: (a) I, Br, N; (b) Ca, H, F.

9.47 Use Figure 9.19 (p. 294) to indicate the polarity of each bond with a *polar arrow:* (a) N—B; (b) N—O; (c) C—S; (d) S—O; (e) N—H; (f) Cl—O.

9.48 Use Figure 9.19 (p. 294) to indicate the polarity of each bond with *partial charges:* (a) Br—Cl; (b) F—Cl; (c) H—O; (d) Se—H; (e) As—H; (f) S—N.

9.49 Which is the more polar bond in each of the following pairs from Problem 9.47: (a) or (b); (c) or (d); (e) or (f)?

9.50 Which is the more polar bond in each of the following pairs from Problem 9.48: (a) or (b); (c) or (d); (e) or (f)?

9.51 Are the bonds in each of the following substances ionic, nonpolar covalent, or polar covalent? Arrange the substances with polar covalent bonds in order of increasing bond polarity:
(a) S_8 (b) RbCl (c) PF_3 (d) SCl_2 (e) F_2 (f) SF_2

9.52 Are the bonds in each of the following substances ionic, nonpolar covalent, or polar covalent? Arrange the substances with polar covalent bonds in order of increasing bond polarity:
(a) KCl (b) P_4 (c) BF_3 (d) SO_2 (e) Br_2 (f) NO_2

9.53 Rank the members of each set of compounds in order of *increasing* ionic character of their bonds. Use *polar arrows* to indicate the bond polarity of each:
(a) HBr, HCl, HI (b) H_2O, CH_4, HF (c) SCl_2, PCl_3, $SiCl_4$

9.54 Rank the members of each set of compounds in order of *decreasing* ionic character of their bonds. Use *partial charges* to indicate the bond polarity of each:
(a) PCl_3, PBr_3, PF_3 (b) BF_3, NF_3, CF_4 (c) SeF_4, TeF_4, BrF_3

9.55 The energy of the C—C bond is 347 kJ/mol, and that of the Cl—Cl bond is 243 kJ/mol. Which of the following values might you expect for the C—Cl bond energy? Explain.
(a) 590 kJ/mol (sum of the values given)
(b) 104 kJ/mol (difference of the values given)
(c) 295 kJ/mol (average of the values given)
(d) 339 kJ/mol (greater than the average of the values given)

Comprehensive Problems

9.56 Geologists have a rule of thumb: when molten rock cools and solidifies, crystals of compounds with the smallest lattice energies appear at the bottom of the mass. Suggest a reason for this.

9.57 Acetylene gas (ethyne; HC≡CH) burns in an oxyacetylene torch to produce carbon dioxide and water vapor. The enthalpy of reaction for the combustion of acetylene is 1259 kJ/mol. (a) Calculate the C≡C bond energy, and compare your value with that in Table 9.2 (p. 287). (b) When 500.0 g of acetylene burns, how many kilojoules of heat are given off? (c) How many grams of CO_2 form? (d) How many liters of O_2 at 298 K and 18.0 atm are consumed?

9.58 Even though so much energy is required to form a metal cation with a 2+ charge, the alkaline earth metals form halides with general formula MX_2, rather than MX.
(a) Use the following data to calculate the ΔH°_f of MgCl:

$$Mg(s) \longrightarrow Mg(g) \qquad \Delta H^\circ = 148\ \text{kJ}$$
$$Cl_2(g) \longrightarrow 2Cl(g) \qquad \Delta H^\circ = 243\ \text{kJ}$$
$$Mg(g) \longrightarrow Mg^+(g) + e^- \qquad \Delta H^\circ = 738\ \text{kJ}$$
$$Cl(g) + e^- \longrightarrow Cl^-(g) \qquad \Delta H^\circ = -349\ \text{kJ}$$
$$\Delta H^\circ_{\text{lattice}}\ \text{of MgCl} = 783.5\ \text{kJ/mol}$$

(b) Is MgCl favored energetically relative to Mg and Cl_2? Explain.
(c) Use Hess's law to calculate ΔH° for the conversion of MgCl to $MgCl_2$ and Mg (ΔH°_f of $MgCl_2 = -641.6$ kJ/mol).
(d) Is MgCl favored energetically relative to $MgCl_2$? Explain.

9.59 By using photons of specific wavelengths, chemists can dissociate gaseous HI to produce H atoms with certain speeds. When HI dissociates, the H atoms move away rapidly, whereas the heavier I atoms move more slowly. (a) What is the longest wavelength (in nm) that can dissociate a molecule of HI? (b) If a photon of 254 nm is used, what is the excess energy (in J) over that needed for dissociation? (c) If this excess energy is carried away by the H atom as kinetic energy, what is its speed (in m/s)?

9.60 Carbon dioxide is a linear molecule. Its vibrational motions include symmetrical stretching, bending, and asymmetrical stretching, and the frequencies are $4.02\times10^{13}\ s^{-1}$, $2.00\times10^{13}\ s^{-1}$, and $7.05\times10^{13}\ s^{-1}$, respectively. (a) In what region of the electromagnetic spectrum are these frequencies? (b) Calculate the energy (in J) of each vibration. Which occurs most readily (takes the least energy)?

9.61 In developing the concept of electronegativity, Pauling used the term *excess bond energy* for the difference between the actual bond energy of X—Y and the average bond energies of X—X and Y—Y (see text discussion for the case of HF). Based on the values in Figure 9.19 (p. 294), which of the following substances contains bonds with no excess bond energy?
(a) PH_3 (b) CS_2 (c) BrCl (d) BH_3 (e) Se_8

9.62 Without stratospheric ozone (O_3), harmful solar radiation would cause gene alterations. Ozone forms when the bond in O_2

breaks and each O atom reacts with another O_2 molecule. It is destroyed by reaction with Cl atoms formed when the C—Cl bond in synthetic chemicals breaks. Find the wavelengths of light that can break the C—Cl bond and the bond in O_2.

9.63 "Inert" xenon actually forms many compounds, especially with highly electronegative fluorine. The ΔH°_f values for xenon difluoride, tetrafluoride, and hexafluoride are −105, −284, and −402 kJ/mol, respectively. Find the average bond energy of the Xe—F bonds in each fluoride.

9.64 The HF bond length is 92 pm, 16% shorter than the sum of the covalent radii of H (37 pm) and F (72 pm). Suggest a reason for this difference. Similar data show that the difference becomes smaller down the group, from HF to HI. Explain.

9.65 There are two main types of covalent bond breakage. In homolytic breakage (as in Table 9.2, p. 287), each atom in the bond gets one of the shared electrons. In some cases, the electronegativity of adjacent atoms affects the bond energy. In heterolytic breakage, one atom gets both electrons and the other gets none; thus, a cation and an anion form. (a) Why is the C—C bond in $H_3C—CF_3$ (423 kJ/mol) stronger than that in $H_3C—CH_3$ (376 kJ/mol)? (b) Use bond energy and any other data to calculate the enthalpy of reaction for the heterolytic cleavage of O_2.

9.66 Find the longest wavelengths of light that can cleave the bonds in elemental nitrogen, oxygen, and fluorine.

9.67 The average C—H bond energy in CH_4 is 415 kJ/mol. Use Table 9.2 (p. 287) and the following to calculate the average C—H bond energy in ethane (C_2H_6; C—C bond), in ethene (C_2H_4; C═C bond), and in ethyne (C_2H_2; C≡C bond):

$$C_2H_6(g) + H_2(g) \longrightarrow 2CH_4(g) \qquad \Delta H^\circ_{rxn} = -65.07 \text{ kJ/mol}$$
$$C_2H_4(g) + 2H_2(g) \longrightarrow 2CH_4(g) \qquad \Delta H^\circ_{rxn} = -202.21 \text{ kJ/mol}$$
$$C_2H_2(g) + 3H_2(g) \longrightarrow 2CH_4(g) \qquad \Delta H^\circ_{rxn} = -376.74 \text{ kJ/mol}$$

9.68 Carbon-carbon bonds form the "backbone" of nearly every organic and biological molecule. The average bond energy of the C—C bond is 347 kJ/mol. Calculate the frequency and wavelength of the least energetic photon that can break an average C—C bond. In what region of the electromagnetic spectrum is this radiation?

9.69 In a future hydrogen-fuel economy, the cheapest source of H_2 will certainly be water. It takes 467 kJ to produce 1 mol of H atoms from water. What is the frequency, wavelength, and minimum energy of a photon that can free an H atom from water?

9.70 Dimethyl ether (CH_3OCH_3) and ethanol (CH_3CH_2OH) are constitutional isomers (see Table 3.2, p. 85). (a) Use Table 9.2 (p. 287) to calculate ΔH°_{rxn} for the formation of each compound as a gas from methane and oxygen; water vapor also forms. (b) State which reaction is more exothermic. (c) Calculate ΔH°_{rxn} for the conversion of ethanol to dimethyl ether.

10 The Shapes of Molecules

Shape Matters Nearly every biochemical reaction occurs because the shape of one molecule in some way matches the shape of another—only one key fits the lock. In this chapter, you'll learn how the placement of bonded atoms gives rise to a molecule's shape.

Key Principles
to focus on while studying this chapter

- A *Lewis structure* shows the relative positions of the atoms in a molecule (or polyatomic ion), as well as the placement of all the shared and unshared electron pairs. It is generated from the molecular formula through a series of steps that often apply the octet rule. ***(Section 10.1)***
- In many molecules or ions, one electron pair in a double bond spreads over an adjacent single bond, thereby stabilizing the system by *delocalizing* its charge. In such cases, more than one Lewis structure, each called a *resonance form,* can be drawn, and the species exists as a *resonance hybrid,* a mixture of the resonance forms. ***(Section 10.1)***
- By assigning to each atom a *formal charge* based on the electrons belonging to the atom and those shared by it, we can select the most important of the various resonance forms, the one that is most like the hybrid. ***(Section 10.1)***
- Many compounds do not obey the octet rule, by having either fewer than or more than eight electrons around the central atom. ***(Section 10.1)***
- According to VSEPR theory, each group of valence electrons, whether a bonding pair or a lone pair, around a central atom repels the others. These repulsions give rise to five geometric arrangements—*linear, trigonal planar, tetrahedral, trigonal bipyramidal,* and *octahedral.* Various molecular shapes, with characteristic bond angles, arise from these arrangements. ***(Section 10.2)***
- A molecule may be polar or nonpolar, depending on its shape and the polarities of its bonds. ***(Section 10.3)***

Outline

In the everyday world, the shapes of interacting objects often reflect their function—cup and saucer, hand and glove, or the complementary shapes of lock and key. The same is true in the molecular world. But, with all the symbols, lines, and dots you've been seeing, it's easy to forget that every molecule has a characteristic minute architecture. Each atom, bonding pair, and lone pair has its own position relative to the others, with angles and distances between them that are determined by the attractive and repulsive forces governing all matter. Molecular shape plays a crucial role in the interactions of reactants, the behavior of synthetic materials, and especially in the life-sustaining processes in cells. In this chapter, we see how to depict molecules, first as two-dimensional drawings and then as three-dimensional objects.

CONCEPTS & SKILLS TO REVIEW
before studying this chapter

- electron configurations of main-group elements (Section 8.2)
- electron-dot symbols (Section 9.1)
- octet rule (Section 9.1)
- bond order, bond length, and bond energy (Sections 9.3 and 9.4)
- polar covalent bonds and bond polarity (Section 9.5)

10.1 • DEPICTING MOLECULES AND IONS WITH LEWIS STRUCTURES

The first step toward visualizing a molecule is to convert its molecular formula to its **Lewis structure** (or **Lewis formula***), which shows electron-dot symbols for the atoms, the bonding pairs as lines, and the lone pairs that fill each atom's outer level (valence shell) as pairs of dots. In many cases, the octet rule (Section 9.1) guides us in allotting electrons to the atoms in a Lewis structure; in many other cases, as you'll see later in this section, we set the rule aside.

Applying the Octet Rule to Write Lewis Structures

To draw a Lewis structure, we decide on the relative placement of the atoms in the molecule or polyatomic ion and then distribute the total number of valence electrons as bonding and lone pairs. We begin with species that "obey" the octet rule, in which each atom fills its outer level with eight electrons (or two for hydrogen).

Molecules with Single Bonds Figure 10.1 lays out the steps for drawing Lewis structures for species with only single bonds. Let's use nitrogen trifluoride, NF_3, to introduce the steps:

Step 1. Place the atoms relative to each other. For compounds with the general molecular formula AB_n, place the atom with the *lower group number* in the center because it needs more electrons to attain an octet; usually, this is also the atom with the *lower electronegativity*. In NF_3, the N (Group 5A; EN = 3.0) has five electrons and so needs three, whereas each F (Group 7A; EN = 4.0) has seven and needs only one; thus, N goes in the center with the three F atoms around it:

F
F N F

*A Lewis *structure* does ***not*** indicate the three-dimensional shape, so it may be more correct to call it a Lewis *formula*, but we follow convention and use the term *structure*.

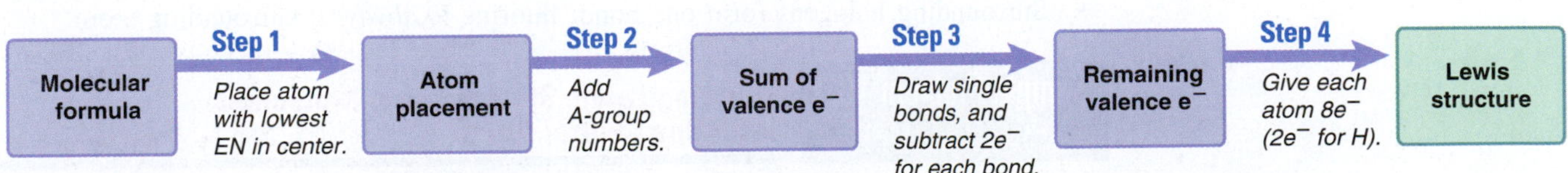

Figure 10.1 The steps in converting a molecular formula into a Lewis structure.

If the atoms have the same group number, as in SO_3 or ClF_3, place the atom with the *higher period number* (also lower EN) in the center. H can form only one bond, so it is *never* a central atom.

Step 2. Determine the total number of valence electrons.

- For molecules, add up the valence electrons of the atoms. (Recall that the number of valence electrons equals the A-group number.) In NF_3, N has five valence electrons, and each F has seven:

$$[1 \times N(5e^-)] + [3 \times F(7e^-)] = 5e^- + 21e^- = 26 \text{ valence } e^-$$

- For polyatomic ions, *add* one e^- for each negative charge, or *subtract* one e^- for each positive charge.

Step 3. Draw a single bond from each surrounding atom to the central atom, and subtract $2e^-$ from the total for each bond to find the number of e^- remaining:

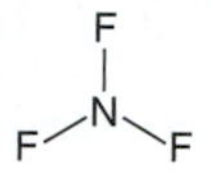

F
|
F—N—F

$$3 \text{ N—F bonds} \times 2e^- = 6e^- \quad \text{so} \quad 26e^- - 6e^- = 20e^- \text{ remaining}$$

Step 4. Distribute the remaining electrons in pairs so that each atom ends up with $8e^-$ (or $2e^-$ for H). First, place lone pairs on the *surrounding (more electronegative) atoms* to give each an octet. If any electrons remain, place them around the central atom—each F gets 3 pairs ($3 \times 6e^- = 18e^-$) and the N gets 1 ($2e^-$), for a total of $20e^-$. Then check that each atom has $8e^-$:

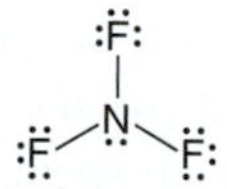

:F̤̈:
|
:F̤̈—N̈—F̤̈:

This is the Lewis structure for NF_3. It is a neutral species, so the total number of electrons (bonds plus lone pairs) equals the sum of the valence electrons:

$$6e^- \text{ in three bonds} + 20e^- \text{ in ten lone pairs} = 26 \text{ valence } e^-$$

If we were writing a Lewis structure for a polyatomic ion, we would also take the charge into account, as described in step 2.

Since Lewis structures do not indicate shape, an equally correct depiction of NF_3 is

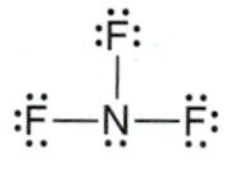

:F̤̈:
|
:F̤̈—N̤—F̤̈:

or any other that retains the *same connections among the atoms*—a central N atom connected by single bonds to each of three surrounding F atoms.

Using these four steps, you can draw a Lewis structure for any singly bonded species with a central C, N, or O atom, as well as for some species with central atoms from higher periods. In nearly all their compounds,

- Hydrogen atoms form one bond.
- Carbon atoms form four bonds.
- Nitrogen atoms form three bonds.
- Oxygen atoms form two bonds.
- Surrounding halogens form one bond; fluorine is *always* a surrounding atom.

Sample Problem 10.1 Drawing Lewis Structures for Molecules with One Central Atom

Problem Draw a Lewis structure for CCl_2F_2, one of the compounds responsible for the depletion of stratospheric ozone.

Solution *Step 1.* Place the atoms relative to each other. In CCl_2F_2, carbon has the lowest group number and EN, so it is the central atom *(see margin)*. The halogen atoms surround it, but their specific positions are not important.

Step 1:

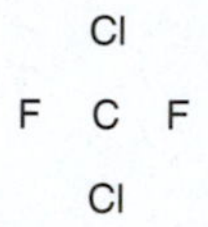

Cl
F C F
Cl

Step 2. Determine the total number of valence electrons (from A-group numbers): C is in Group 4A; F and Cl are in Group 7A. Therefore, we have

$$[1 \times C(4e^-)] + [2 \times F(7e^-)] + [2 \times Cl(7e^-)] = 32 \text{ valence } e^-$$

Step 3. Draw single bonds to the central atom *(see margin)* and subtract $2e^-$ for each bond:

$$4 \text{ bonds} \times 2e^- = 8e^- \quad \text{so} \quad 32e^- - 8e^- = 24e^- \text{ remaining}$$

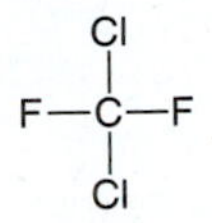

Step 4. Distribute the remaining electrons in pairs, beginning with the surrounding atoms, so that each atom has an octet. Each surrounding halogen gets 3 pairs *(see margin)*.

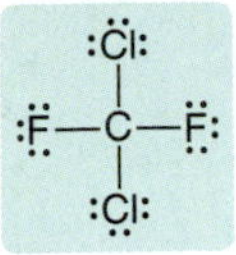

Check Always check that each atom has an octet. Bonding electrons belong to each atom in the bond. The total number in bonds ($8e^-$) and lone pairs ($24e^-$) equals 32 valence e^-. As expected, C has four bonds and the surrounding halogens have one each.

Step 3: F—C—F with Cl above and Cl below C

Step 4: :F̤̈—C—F̤̈: with :C̤̈l: above and :C̤̈l: below C

FOLLOW-UP PROBLEM 10.1 Draw a Lewis structure for **(a)** H_2S; **(b)** OF_2; **(c)** $SOCl_2$.

In molecules with two or more central atoms bonded to each other, it is usually clear which atoms are surrounding.

Sample Problem 10.2 Drawing Lewis Structures for Molecules with More Than One Central Atom

Problem Draw the Lewis structure for methanol (molecular formula CH_4O), an important industrial alcohol that can be used as a gasoline alternative in cars.

Solution *Step 1.* Place the atoms relative to each other. The H atoms can have only one bond, so C and O must be central and adjacent to each other. C has four bonds and O has two, so we arrange the H atoms accordingly *(see margin)*.

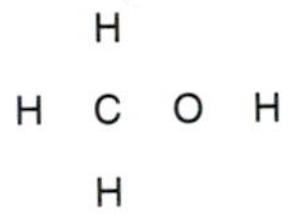

Step 2. Find the sum of valence electrons (C is in Group 4A, O is in Group 6A):

$$[1 \times C(4e^-)] + [1 \times O(6e^-)] + [4 \times H(1e^-)] = 14e^-$$

Step 3. Add single bonds *(see margin)* and subtract $2e^-$ for each bond:

$$5 \text{ bonds} \times 2e^- = 10e^- \quad \text{so} \quad 14e^- - 10e^- = 4e^- \text{ remaining}$$

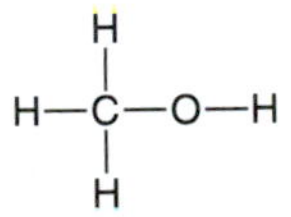

Step 4. Add the remaining electrons in pairs to fill each valence level. C already has an octet, and each H shares $2e^-$ with the C; so the remaining $4e^-$ form two lone pairs on O to give the Lewis structure for methanol *(see margin)*.

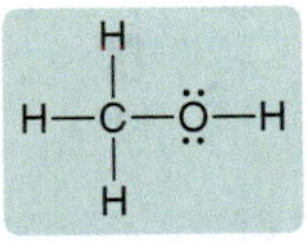

Check Each H atom has $2e^-$, and C and O each have $8e^-$. The total number of valence electrons is $14e^-$, which equals $10e^-$ in bonds plus $4e^-$ in two lone pairs. Each H has one bond, C has four, and O has two.

Step 1: H C O H, with H above and H below C

Step 3: H—C—O—H, with H above and H below C

Step 4: H—C—Ö̤—H, with H above and H below C

FOLLOW-UP PROBLEM 10.2 Draw a Lewis structure for **(a)** hydroxylamine (NH_3O); **(b)** dimethyl ether (C_2H_6O; no O—H bonds).

Molecules with Multiple Bonds In most cases, if there are not enough electrons for the central atom(s) to attain an octet, a multiple bond is present, and we add the following step to the procedure for drawing a Lewis structure:

Step 5. Cases involving multiple bonds. If a central atom does not end up with an octet, change a lone pair on a surrounding atom into another bonding pair to the central atom, thus forming a multiple bond.

Sample Problem 10.3 Drawing Lewis Structures for Molecules with Multiple Bonds

Problem Draw Lewis structures for the following:
(a) Ethylene (C_2H_4), the most important reactant in the manufacture of polymers
(b) Nitrogen (N_2), the most abundant atmospheric gas

Plan We show the structure resulting from steps 1 to 4: placing the atoms, counting the total valence electrons, making single bonds, and distributing the remaining valence electrons in pairs to attain octets. Then we continue with step 5, if needed.

Solution (a) For C_2H_4. After steps 1 to 4, we have

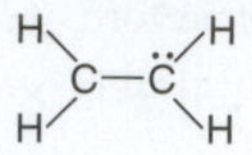

Step 5. Change a lone pair to a bonding pair. The right C has an octet, but the left C has only $6e^-$, so we change the lone pair to another bonding pair between the two C atoms:

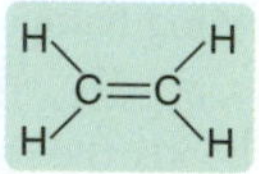

(b) For N_2. After steps 1 to 4, we have :N̈—N̈:
Step 5. Neither N has an octet, so we change a lone pair to a bonding pair, :N̈=N:
In this case, moving one lone pair to make a double bond still does not give the right N an octet, so we move a lone pair from the left N to make a triple bond, :N≡N:

Check (a) Each C has four bonds and counts the $4e^-$ in the double bond as part of its own octet. The valence electron total is $12e^-$, all in six bonds. (b) Each N counts the $6e^-$ in the triple bond as part of its own octet. The valence electron total is $10e^-$, which equals the electrons in three bonds and two lone pairs.

FOLLOW-UP PROBLEM 10.3 Draw Lewis structures for (a) CO (the only common molecule in which C has three bonds); (b) HCN; (c) CO_2.

Resonance: Delocalized Electron-Pair Bonding

We often find that, for a molecule or polyatomic ion with *a double bond next to a single bond,* we can convert the molecular formula to more than one Lewis structure. Which, if any, is correct?

The Need for Resonance Structures To understand this issue, consider ozone (O_3), an air pollutant at ground level but an absorber of harmful ultraviolet (UV) radiation in the stratosphere. Two Lewis structures (with lettered O atoms for clarity) are

I II

In structure I, oxygen B has a double bond to oxygen A and a single bond to oxygen C. In structure II, the single and double bonds are reversed. You can rotate I to get II, so these are *not* different types of ozone molecules, but different Lewis structures for the *same* molecule.

In fact, *neither* Lewis structure depicts O_3 accurately, because the two oxygen-oxygen bonds in O_3 are actually identical in length and energy. The bonds in O_3 have properties between an O—O bond and an O=O bond, something like a "one-and-a-half" bond. The molecule is shown more correctly with two Lewis structures, called **resonance structures** (or **resonance forms**), and a two-headed resonance arrow (⟷) between them. Resonance structures *have the same relative placement of atoms but different locations of bonding and lone electron pairs.* You can convert one resonance form to another by moving a lone pair to a bonding position, and vice versa:

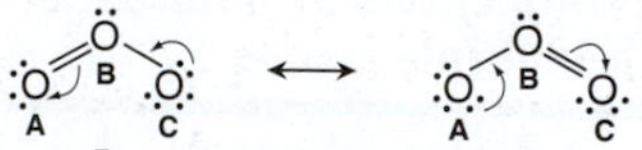

I ⟷ II

Resonance structures are not real bonding depictions: O_3 does *not* change back and forth quickly from structure I to structure II. The actual molecule is a **resonance hybrid,** an average of the resonance forms.

THINK OF IT THIS WAY

A Purple Mule, Not a Blue Horse and a Red Donkey

Consider these analogies for a resonance hybrid. A mule is a genetic mix, a hybrid, of a horse and a donkey; it is not a horse one instant and a donkey the next. Similarly, the color purple is a mix of two other colors, red and blue, not red one instant and blue the next. In the same sense, a resonance hybrid is one molecular species, not one resonance form this instant and another resonance form the next. The problem is that we cannot depict the hybrid accurately with a single Lewis structure.

Blue horse ⟷ Red donkey

≡

Purple mule

Electron Delocalization Our need for more than one Lewis structure to depict O_3 is due to **electron-pair delocalization.** In a single, double, or triple bond, each electron pair is *localized* between the bonded atoms. In a resonance hybrid, two of the electron pairs (one bonding and one lone pair) are *delocalized:* their density is "spread" over a few adjacent atoms.

In O_3, the result is two identical bonds, each consisting of a single bond (the localized pair) and a *partial bond* (the contribution from one of the delocalized pairs). We draw the resonance hybrid with a curved dashed line to show the delocalized pairs:

resonance hybrid

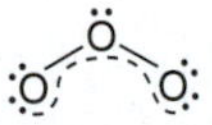

Resonance is very common. For example, benzene (C_6H_6, *shown below*) has two important resonance forms in which alternating single and double bonds have different positions. The actual molecule is an average of the two forms with six C—C bonds and three electron pairs delocalized over all six C atoms. The delocalized pairs are often shown as a dashed circle (or sometimes simply a circle):

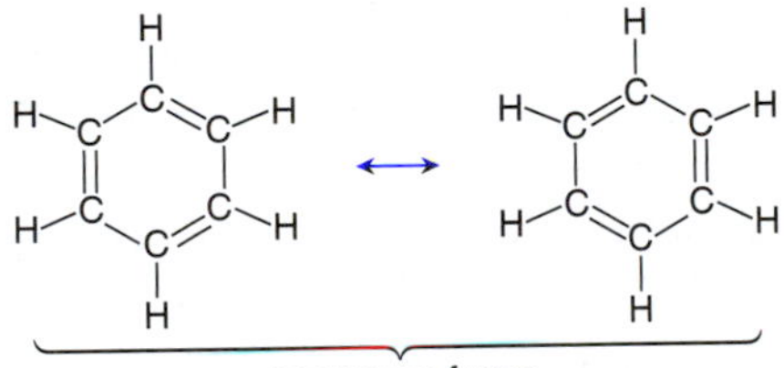

resonance forms

or

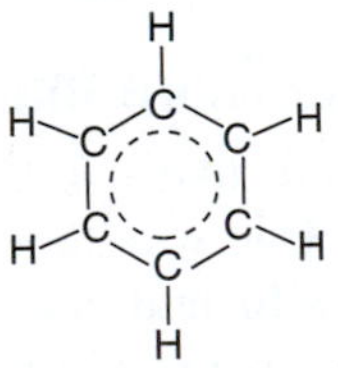

resonance hybrid

Fractional Bond Orders Partial bonding, as in resonance hybrids, often leads to fractional bond orders. For the oxygen-oxygen bonds in O_3, we have

$$\text{Bond order} = \frac{3 \text{ electron pairs}}{2 \text{ bonded-atom pairs}} = 1\tfrac{1}{2}$$

The carbon-carbon bond order in benzene is 9 electron pairs/6 bonded-atom pairs, or $1\frac{1}{2}$ also. For the carbonate ion, CO_3^{2-}, three resonance structures can be drawn. Each has 4 electron pairs shared among 3 bonded-atom pairs, so the bond order is $\frac{4}{3}$, or $1\frac{1}{3}$. One of the three resonance structures for CO_3^{2-} is

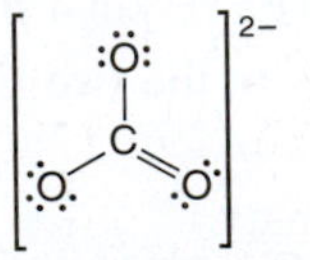

Note here (and in Sample Problem 10.4) that the Lewis structure of a polyatomic ion is drawn in *square brackets with the ion charge outside the brackets.*

Sample Problem 10.4 **Drawing Resonance Structures**

Problem Draw resonance structures for the nitrate ion, NO_3^-, and find the bond order.

Plan We draw a Lewis structure, remembering to add $1e^-$ to the total number of valence electrons because of the 1− ionic charge. Then we move lone and bonding pairs to draw other resonance forms and connect them with the resonance arrow. The bond order is the number of shared electron pairs divided by the number of atom pairs.

Solution After steps 1 to 4, we have

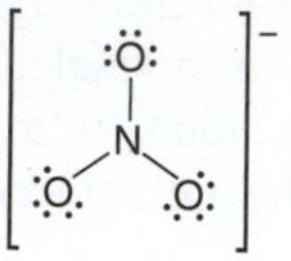

Step 5. Since N has only $6e^-$, we change a lone pair on one of the O atoms to a bonding pair to form a double bond, which gives each atom an octet. All the O atoms are equivalent, however, so we can move a lone pair from any one of the three and obtain three resonance structures:

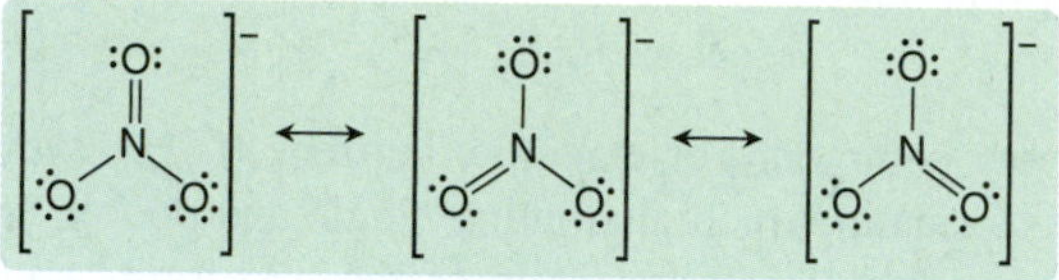

The bond order is

$$\frac{4 \text{ shared electron pairs}}{3 \text{ bonded-atom pairs}} = 1\tfrac{1}{3}$$

Check Each structure has the same relative placement of atoms, an octet around each atom, and $24e^-$ (the sum of the valence electron total and $1e^-$ from the ionic charge, distributed in four bonds and eight lone pairs).

Comment These three resonance forms contribute equally to the resonance hybrid because all of the surrounding atoms are identical. This is not always the case, as you'll see next.

FOLLOW-UP PROBLEM 10.4 One of the three resonance structures for CO_3^{2-} was shown just before Sample Problem 10.4. Draw the other two.

Formal Charge: Selecting the More Important Resonance Structure

If one resonance form "looks" more like the resonance hybrid than the others, it "weights" the average in its favor. One way to select the more important resonance form is by determining each atom's **formal charge,** the charge it would have *if the bonding electrons were shared equally*. Let's examine this concept and then see how formal charge compares with oxidation number.

Determining Formal Charge An atom's formal charge is its total number of valence electrons minus *all* of its unshared valence electrons and *half* of its shared valence electrons. Thus,

$$\text{Formal charge of atom} = \text{no. of valence } e^- - (\text{no. of unshared valence } e^- + \tfrac{1}{2} \text{ no. of shared valence } e^-) \quad \textbf{(10.1)}$$

For example, in O_3, the formal charge of oxygen A in resonance form I is

$$6 \text{ valence } e^- - (4 \text{ unshared } e^- + \tfrac{1}{2} \text{ of } 4 \text{ shared } e^-) = 6 - 4 - 2 = 0$$

The formal charges of all the atoms in the two O_3 resonance forms are

Form I		Form II
$O_A[6 - 4 - \frac{1}{2}(4)] = 0$	(0) A, (+1) B, (−1) C ⟷ (−1) A, (+1) B, (0) C	$O_A[6 - 6 - \frac{1}{2}(2)] = -1$
$O_B[6 - 2 - \frac{1}{2}(6)] = +1$		$O_B[6 - 2 - \frac{1}{2}(6)] = +1$
$O_C[6 - 6 - \frac{1}{2}(2)] = -1$	I II	$O_C[6 - 4 - \frac{1}{2}(4)] = 0$

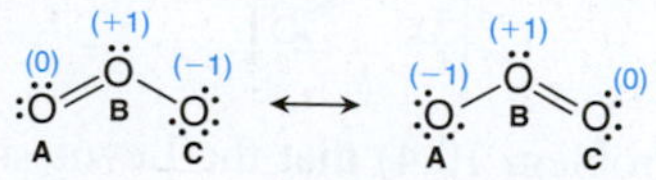

Forms I and II have the same formal charges but on different O atoms, so they contribute equally to the resonance hybrid. *Formal charges must sum to the actual charge on the species:* zero for a molecule or the ionic charge for an ion.

Note that, in I, instead of oxygen's usual two bonds, O_B has three bonds and O_C has one. Only when an atom has a zero formal charge does it have its usual number of bonds; the same holds for C in CO_3^{2-}, N in NO_3^-, and so forth.

Choosing the More Important Resonance Form Three criteria help us choose the more important resonance structure:

- Smaller formal charges (positive *or* negative) are preferable to larger ones.
- The *same* nonzero formal charges on adjacent atoms are not preferred.
- A more negative formal charge should reside on a more electronegative atom.

As in the case of O_3, the resonance forms for CO_3^{2-}, NO_3^-, and benzene all have identical atoms surrounding the central atom(s) and, thus, have identical formal charges and are equally important contributors to the resonance hybrid. But, let's apply these criteria to the cyanate ion, NCO^-, which has two *different* atoms around the central one. Three resonance forms with formal charges are

Formal charges: (−2) (0) (+1) | (−1) (0) (0) | (0) (0) (−1)

Resonance forms: $[:\ddot{\underset{..}{N}}-C\equiv O:]^- \longleftrightarrow [\ddot{\underset{..}{N}}=C=\ddot{\underset{..}{O}}]^- \longleftrightarrow [:N\equiv C-\ddot{\underset{..}{O}}:]^-$

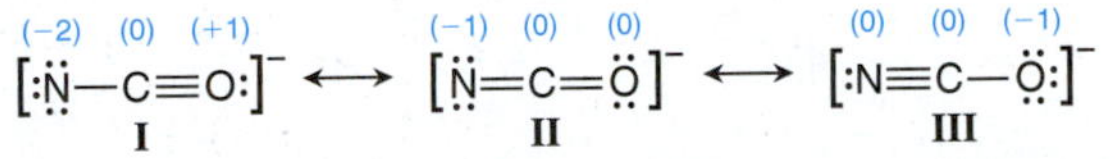

I II III

Form I is *not* an important contributor to the hybrid because it has a larger formal charge on N and a positive formal charge on the more electronegative O. Forms II and III have the same magnitude of charges, but III has a −1 charge on O, the more electronegative atom. Therefore, II and III are more important than I, and III is more important than II.

Formal Charge Versus Oxidation Number Formal charge (used to examine resonance structures) is *not* the same as oxidation number (used to monitor redox reactions):

- For a *formal charge,* bonding electrons are *shared equally* by the atoms (as if the bonding were *nonpolar covalent*), so each atom has half of them:

$$\text{Formal charge} = \text{valence } e^- - (\text{lone pair } e^- + \tfrac{1}{2}\text{ bonding } e^-)$$

- For an *oxidation number,* bonding electrons are *transferred completely* to the more electronegative atom (as if the bonding were *pure ionic*):

$$\text{Oxidation number} = \text{valence } e^- - (\text{lone pair } e^- + \text{bonding } e^-)$$

For the three cyanate ion resonance structures,

Formal charges: (−2) (0) (+1) | (−1) (0) (0) | (0) (0) (−1)

$[:\ddot{\underset{..}{N}}-C\equiv O:]^- \longleftrightarrow [\ddot{\underset{..}{N}}=C=\ddot{\underset{..}{O}}]^- \longleftrightarrow [:N\equiv C-\ddot{\underset{..}{O}}:]^-$

Oxidation numbers: −3 +4 −2 | −3 +4 −2 | −3 +4 −2

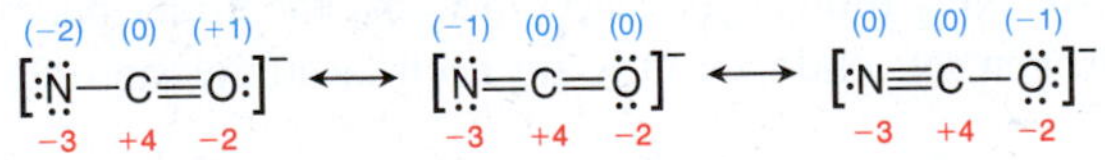

Notice that the oxidation numbers *do not* change from one resonance form to another (because the electronegativities *do not* change), but the formal charges *do* change (because the numbers of bonding and lone pairs *do* change).

Lewis Structures for Exceptions to the Octet Rule

The octet rule applies to most molecules (and ions) with Period 2 central atoms, but not every one, and not to many with central atoms from Period 3 and higher. Three important exceptions occur in molecules with (1) electron-deficient atoms, (2) odd-electron atoms, and (3) atoms with expanded valence shells. In this discussion, you'll also see that formal charge has limitations for selecting the best resonance form.

Molecules with Electron-Deficient Atoms Gaseous molecules containing either beryllium or boron as the central atom are often **electron deficient:** they have *fewer* than eight electrons around the central atom. The Lewis structures, with formal charges, of gaseous beryllium chloride* and boron trifluoride are

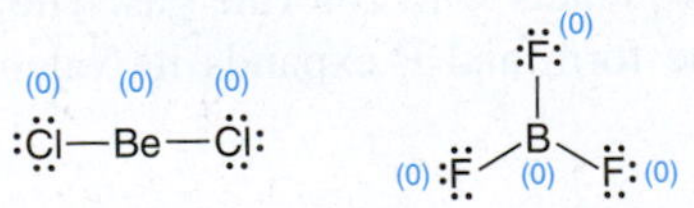

(0) (0) (0)

$:\ddot{\underset{..}{Cl}}-Be-\ddot{\underset{..}{Cl}}:$

BF_3: B (0) bonded to three $:\ddot{\underset{..}{F}}:$ atoms, each (0)

*Even though beryllium is in Group 2A(2), most Be compounds have considerable covalent bonding. For example, molten $BeCl_2$ does not conduct electricity, indicating a lack of ions.

There are only four electrons around Be and six around B. Surrounding halogen atoms don't form multiple bonds to the central atoms to give them an octet, because the halogens are much more electronegative. Formal charges make the following structures unlikely:

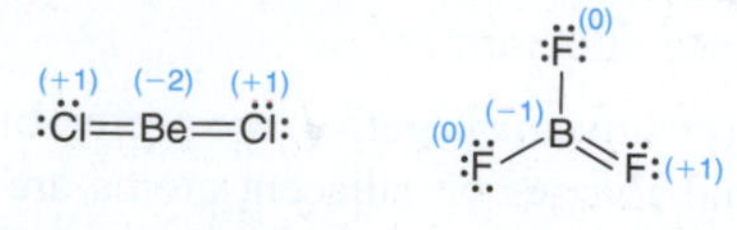

(+1) (−2) (+1)
:C̈l=Be=C̈l:

(0) :F: (−1) B (0) :F: =F: (+1)

(Some data for BF_3 show a shorter than expected B—F bond. Shorter bonds indicate double-bond character, so the structure with the B=F bond may be a minor contributor to a resonance hybrid.) Electron-deficient atoms often attain an octet by forming additional bonds in reactions. When BF_3 reacts with ammonia, for instance, a compound forms in which boron attains an octet:*

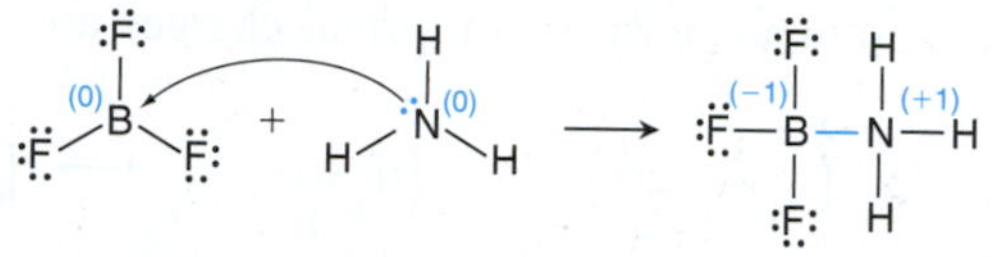

$$BF_3 + NH_3 \longrightarrow F_3B{-}NH_3$$

Molecules with Odd-Electron Atoms A few molecules contain a central atom with an odd number of valence electrons, so they cannot have all their electrons in pairs. Most have a central atom from an odd-numbered group, such as N [Group 5A(15)] or Cl [Group 7A(17)]. These are called **free radicals,** species that contain a lone (unpaired) electron, which makes them paramagnetic (Section 8.4) and extremely reactive (such reactivity can be especially harmful to biomolecules).

Consider the free radical nitrogen dioxide, NO_2, a major contributor to urban smog that is formed when the NO in auto exhaust is oxidized. NO_2 has several resonance forms. Two differ in terms of which O atom is doubly bonded, as in the case of ozone. Two others have the lone electron residing on the N or on an O, so the resonance hybrid has the lone electron delocalized over these two atoms:

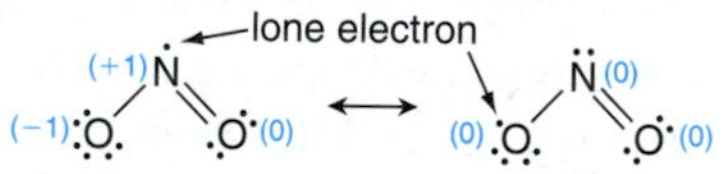

lone electron
(+1) N, (−1) O, O (0) ⟷ N (0), (0) O, O (0)

But the form with the lone electron on N *(left)* may be more important because of the way NO_2 reacts. Free radicals react with each other to pair up their lone electrons. When two NO_2 molecules react, the lone electrons pair up to form the N—N bond in dinitrogen tetroxide (N_2O_4) and each N attains an octet:

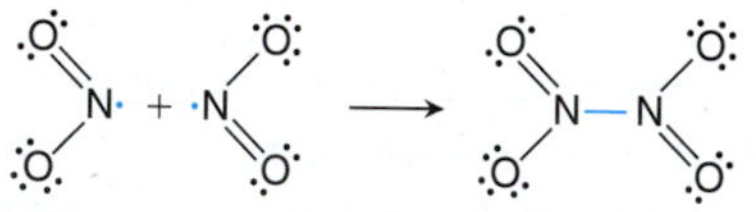

$$NO_2 + NO_2 \longrightarrow O_2N{-}NO_2$$

Apparently, in this case, the lone electron spends most of its time on N, so formal charge is not very useful for picking the most important resonance form; we'll see other cases below.

Expanded Valence Shells Many molecules (and ions) have more than eight valence electrons around the central atom. *An atom expands its valence shell to form more bonds, which releases energy.* The central atom must be large and have empty orbitals that can hold the additional pairs. Therefore, **expanded valence shells** (levels) occur only with *nonmetals from Period 3 or higher because they have d orbitals available.* Such a central atom may be bonded to more than four atoms or to four or fewer.

1. *Central atom bonded to more than four atoms.* Phosphorus pentachloride, PCl_5, is a fuming yellow-white solid used to manufacture lacquers and films. It forms when phosphorus trichloride, PCl_3, reacts with chlorine gas. The P in PCl_3 has an octet, but two more bonds to chlorine form and P expands its valence shell to 10 electrons in

*Reactions in which one species "donates" an electron pair to another to form a covalent bond are Lewis acid-base reactions, which we discuss fully in Chapter 18.

PCl_5. Note that when PCl_5 forms, *one* Cl—Cl bond breaks *(left side of the equation)*, and *two* P—Cl bonds form *(right side)*, for a net increase of one bond:

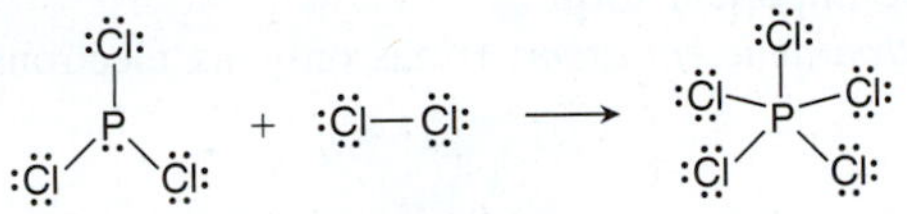

Another example of a central atom with an expanded valence shell is sulfur hexafluoride, SF_6, a remarkably dense and inert gas used as an insulator in electrical equipment. The central sulfur is surrounded by six single bonds, one to each fluorine, for a total of 12 electrons:

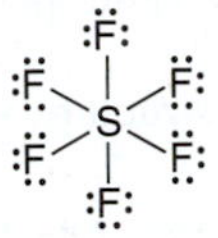

2. *Central atom bonded to four or fewer atoms.* By applying the concept of formal charge, we can draw Lewis structures in which a central atom has an expanded valence shell and is bonded to *four or fewer* atoms. Consider sulfuric acid, the industrial chemical produced in the greatest quantity. Two resonance forms of H_2SO_4, with formal charges, are

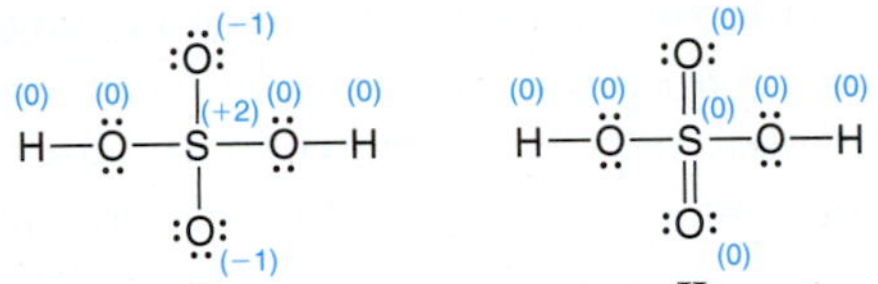

Form I obeys the octet rule, but it has several nonzero formal charges. In form II, sulfur has 12 electrons (6 bonds) around it, but all zero formal charges. Thus, based on the formal charge rules alone, II contributes more than I to the resonance hybrid. More important than whether rules are followed, form II is consistent with observation. In gaseous H_2SO_4, the two sulfur-oxygen bonds *with* an H atom attached to the O are 157 pm long, whereas the two sulfur-oxygen bonds *without* an H atom are 142 pm long. This shorter bond indicates double-bond character, and other measurements indicate greater electron density in the bonds without the attached H.

3. *Limitations of formal charges.* It's important to realize that determining formal charges is a useful, but not perfect, tool for assessing the importance of contributions to a resonance hybrid. You've already seen that it does not predict an important resonance form of NO_2. In fact, theoretical calculations indicate that, for many species with central atoms from Period 3 or higher, such as H_2SO_4, forms with expanded valence shells and zero formal charges (form II above) may actually be *less* important than forms with higher formal charges that follow the octet rule (form I). But we will continue to apply the formal charge rules because it is usually the simplest approach consistent with experimental data.

Sample Problem 10.5 Drawing Lewis Structures for Octet-Rule Exceptions

Problem Draw a Lewis structure and identify the octet-rule exception for **(a)** H_3PO_4 (draw two resonance forms and select the more important); **(b)** $BFCl_2$.

Plan We draw each Lewis structure and examine it for exceptions to the octet rule. **(a)** The central atom is in Period 3, so it can have more than an octet. **(b)** The central atom is B, which can have fewer than an octet of electrons.

Solution (a) H_3PO_4 has two resonance forms. The structures, with formal charges, are

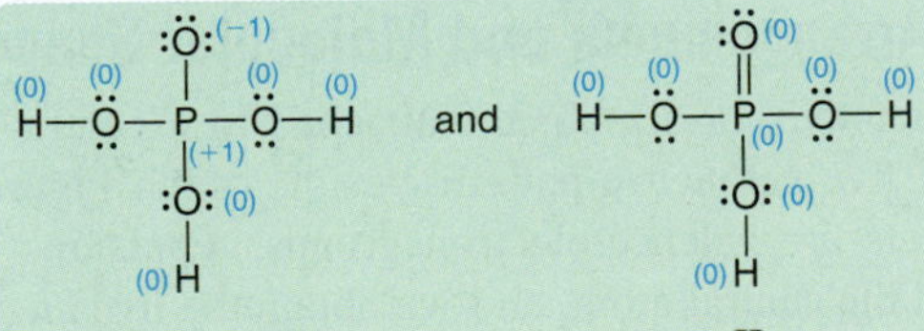

Structure I obeys the octet rule but has nonzero formal charges. Structure II has an expanded valence shell with zero formal charges. According to formal charge rules, structure II is the more important form.

(b) $BFCl_2$ has an *electron-deficient atom;* B has only six electrons surrounding it:

Comment In (a), structure II is consistent with bond-length measurements, which show one shorter (152 pm) phosphorus-oxygen bond and three longer (157 pm) ones. Nevertheless, as for H_2SO_4, calculations show that structure I may be more important.

FOLLOW-UP PROBLEM 10.5 Draw a Lewis structure with minimal formal charges for **(a)** $POCl_3$; **(b)** ClO_2; **(c)** XeF_4.

Summary of Section 10.1

- A stepwise process converts a molecular formula into a Lewis structure, a two-dimensional representation of a molecule (or ion) that shows the placement of atoms and distribution of valence electrons among bonding and lone pairs.
- When two or more Lewis structures can be drawn for the same relative placement of atoms, the actual molecule (or ion) is a hybrid of those resonance forms.
- Formal charges can be useful for choosing the more important contributor to the hybrid, but experimental data always determine the choice.
- Molecules with an electron-deficient atom (central Be or B) or an odd-electron atom (free radicals) have less than an octet around the central atom but often attain an octet in reactions.
- In a molecule (or ion) with a central atom from Period 3 or higher, that atom can have more than eight valence electrons because it is larger and has empty *d* orbitals for expanding its valence shell.

10.2 • VALENCE-SHELL ELECTRON-PAIR REPULSION (VSEPR) THEORY AND MOLECULAR SHAPE

Virtually every biochemical process hinges to a great extent on the shapes of interacting molecules. Every medicine you take, odor you smell, or flavor you taste depends on part or all of one molecule fitting together with another. Biologists have found that complex behaviors in many organisms, such as mating, defense, navigation, and feeding, often depend on one molecule's shape matching that of another. In this section, we discuss a model for predicting the shape of a molecule.

To obtain the molecular shape, chemists start with the Lewis structure and apply ***v*alence-*s*hell *e*lectron-*p*air *r*epulsion (VSEPR) theory.** Its basic principle is that, *to minimize repulsions, each group of valence electrons around a central atom is located as far as possible from the others.* A "group" of electrons is any number that occupies a localized region around an atom: single bond, double bond, triple bond, lone pair, or even lone electron. (The two electron pairs in a double bond or the three pairs in a triple bond occupy separate orbitals, so they remain near each other and act as one electron group, as you'll see in Chapter 11.) The **molecular shape** is the three-dimensional arrangement of nuclei joined by the bonding groups.

Electron-Group Arrangements and Molecular Shapes

When two, three, four, five, or six objects attached to a central point maximize the space between them, five geometric patterns result, which Figure 10.2A shows with balloons. If the objects are valence-electron groups, repulsions maximize the space each occupies around the central atom, and we obtain the five *electron-group arrangements* seen in the great majority of molecules and polyatomic ions.

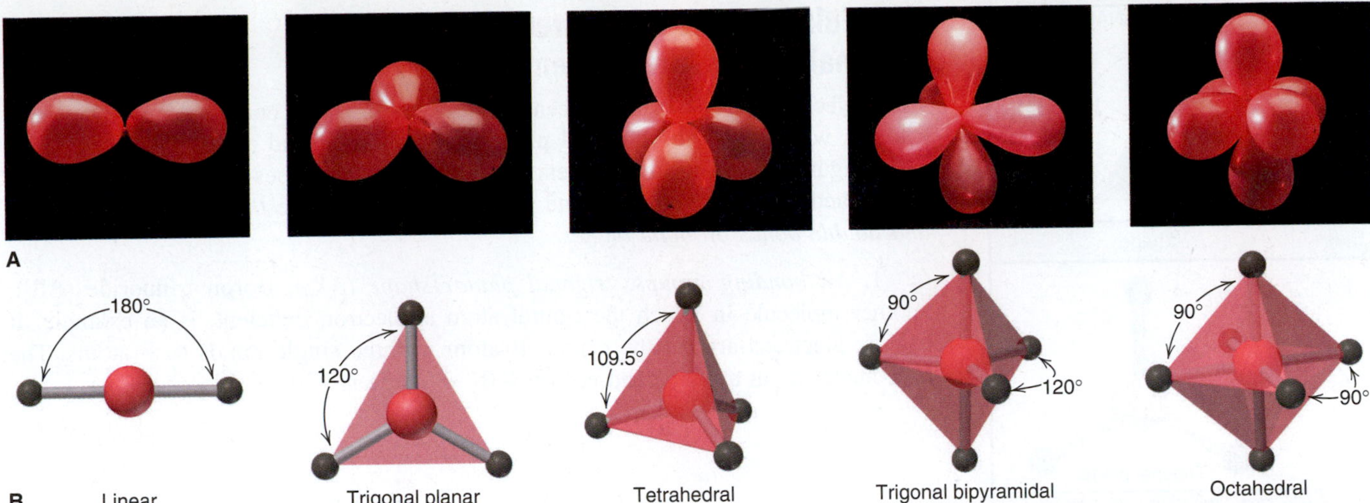

Figure 10.2 Electron-group repulsions and molecular shapes. **A,** Five geometric orientations arise when each balloon occupies as much space as possible. **B,** Mutually repelling bonding groups *(gray sticks)* attach a surrounding atom *(dark gray)* to the central atom *(red)*. The name is the electron-group arrangement.

Classifying Molecular Shapes *The electron-group arrangement* is defined by the bonding *and* nonbonding electron groups, but the *molecular shape* is defined by the relative positions of the nuclei, which are connected by the bonding groups only. Figure 10.2B shows the molecular shapes that occur when *all* the surrounding electron groups are *bonding* groups. When some are *nonbonding* groups, no nucleus is attached so different molecular shapes occur. Thus, *the same electron-group arrangement can give rise to different molecular shapes:* some with all bonding groups (as in Figure 10.2B) and others with bonding and nonbonding groups. To classify molecular shapes, we assign each a specific AX_mE_n designation, where m and n are integers, A is the central atom, X is a surrounding atom, and E is a nonbonding valence-electron group (usually a lone pair).

The Importance of Bond Angle The **bond angle** is the angle formed by the nuclei of two surrounding atoms with the nucleus of the central atom at the vertex. The angles shown for the shapes in Figure 10.2B are *ideal* bond angles determined by basic geometry alone. We observe them when all the bonding groups are the same and connected to the same type of atom. When this is not the case, the *real* bond angles are not equal to the ideal angles:

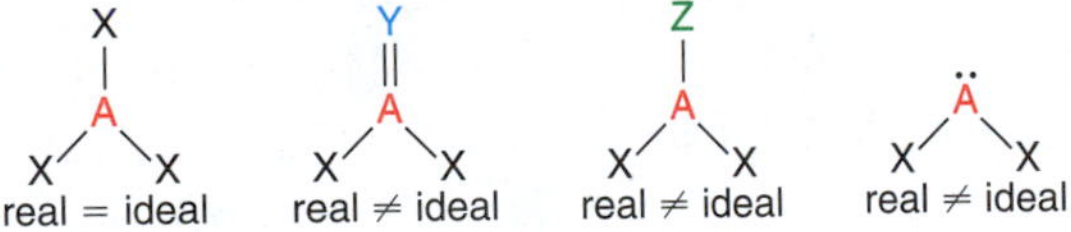

The Molecular Shape with Two Electron Groups (Linear Arrangement)

Two electron groups attached to a central atom point in opposite directions. This **linear arrangement** of electron groups results in a molecule with a **linear shape** and a bond angle of 180°. Figure 10.3 shows the general form *(top)* and shape *(middle)* with VSEPR shape class (AX_2), and gives the formulas of some linear molecules.

Gaseous beryllium chloride ($BeCl_2$) is a linear molecule. Recall that the central Be atom is electron deficient, with two electron pairs around it:

180°
:C̤̈l—Be—C̤̈l:

In carbon dioxide, the central C atom forms two double bonds with the O atoms:

180°
Ö=C=Ö

Each double bond acts as one electron group and is 180° away from the other. The lone pairs on the O atoms of CO_2 or on the Cl atoms of $BeCl_2$ are *not* involved in the molecular shape: only electron groups around the *central* atom affect shape.

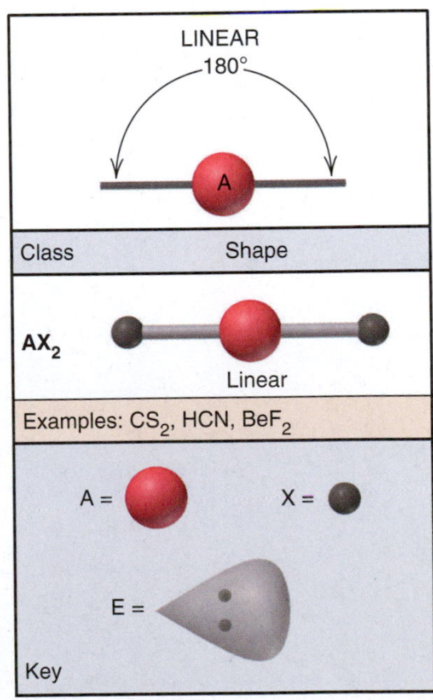

Figure 10.3 The single molecular shape of the linear electron-group arrangement. The key *(bottom)* for A, X, and E also refers to Figures 10.4, 10.5, 10.7, and 10.8.

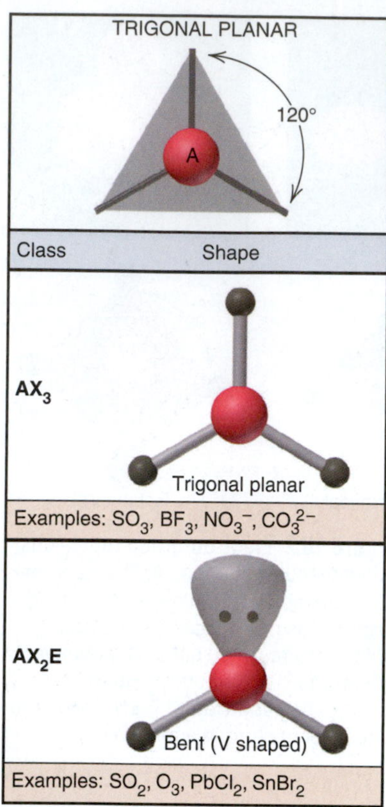

Figure 10.4 The two molecular shapes of the trigonal planar electron-group arrangement.

Molecular Shapes with Three Electron Groups (Trigonal Planar Arrangement)

Three electron groups around a central atom point to the corners of an equilateral triangle, which gives the **trigonal planar arrangement** and an ideal bond angle of 120° (Figure 10.4). This arrangement has two molecular shapes—one with all bonding groups and the other with one lone pair. It allows us to see *the effects of lone pairs and double bonds on bond angles.*

1. *All bonding groups: trigonal planar shape (AX_3).* Boron trifluoride (BF_3), another molecule in which the central atom is electron deficient, is an example. It has six electrons around the central B atom in three single bonds to F atoms. The four nuclei lie in a plane, and each F—B—F angle is 120°:

The nitrate ion (NO_3^-) is one of several polyatomic ions with the trigonal planar shape. One of three resonance forms of the nitrate ion (Sample Problem 10.4) is

The resonance hybrid has three identical bonds of bond order $1\frac{1}{3}$, so the ideal bond angle is observed.

2. *One lone pair: bent or V shape (AX_2E).* Gaseous tin(II) chloride is a molecule with a **bent shape,** or **V shape.** The three electron groups are in a trigonal plane with the lone pair at one of the triangle's corners. A lone pair often has a major effect on bond angle. Because it is held by only one nucleus, a lone pair is less confined than a bonding pair and so exerts stronger repulsions. In general, *a lone pair repels bonding pairs more than bonding pairs repel each other, so it decreases the angle between bonding pairs.* Note the 95° bond angle in $SnCl_2$, which is considerably less than the ideal 120°:

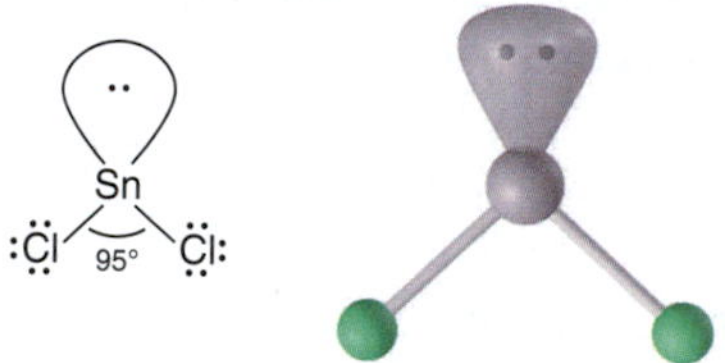

Effect of Double Bonds on Bond Angle When the surrounding atoms and electron groups are not identical, the bond angles may also be affected. Consider formaldehyde (CH_2O), whose uses include the manufacture of countertops and the production of methanol. Its trigonal planar shape includes two types of surrounding atoms (O and H) and two types of electron groups (single and double bonds):

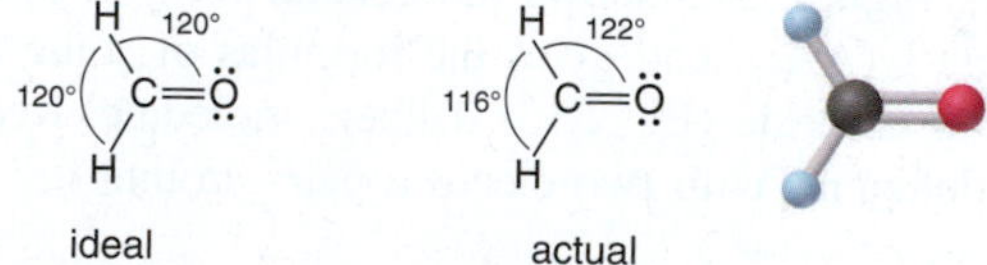

The actual H—C—H bond angle is less than the ideal 120° because *the greater electron density of a double bond repels electrons in single bonds more than the single bonds repel each other.*

Molecular Shapes with Four Electron Groups (Tetrahedral Arrangement)

Shapes based on two or three electron groups lie in a plane, but four electron groups require three dimensions to maximize separation. Consider methane, whose Lewis structure *(below, left)* shows four bonds pointing to the corners of a square, which suggests 90° bond angles. But, *Lewis structures do **not** depict shape*. In three dimensions, the four electron groups lie at the corners of a *tetrahedron,* a polyhedron with four faces made of equilateral triangles, giving bond angles of 109.5° (Figure 10.5).

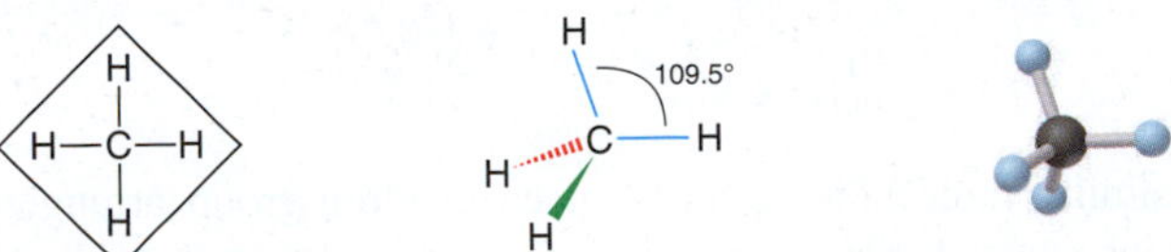

A *wedge-bond perspective drawing* for methane *(above, middle)* indicates depth by using solid and dashed wedges for bonds out of the plane of the page. The normal bond lines *(blue)* are in the plane of the page; the solid wedge *(green)* is the bond from the C atom in the plane of the page to the H above that plane; and the dashed wedge *(red)* is the bond from the C to the H below the plane of the page. The ball-and-stick model *(above, right)* shows the tetrahedral shape more clearly.

*All molecules or ions with four electron groups around a central atom adopt the **tetrahedral arrangement**.* There are three shapes with this arrangement:

1. *All bonding groups: tetrahedral shape (AX_4).* Methane has a tetrahedral shape, a very common geometry in organic molecules. In Sample Problem 10.1, we drew the Lewis structure for CCl_2F_2, without regard to relative placement of the four halogen atoms around the carbon atom. Because Lewis structures are flat, it may seem like you can write two different ones for CCl_2F_2, but they represent the same molecule, as a twist of the wrist reveals (Figure 10.6).

2. *One lone pair: trigonal pyramidal shape (AX_3E).* Ammonia (NH_3) is an example of a molecule with a **trigonal pyramidal shape,** a tetrahedron with one vertex "missing." Stronger repulsions by the lone pair make the H—N—H bond angle slightly less than the ideal 109.5°. The lone pair forces the N—H pairs closer to each other, and the bond angle is 107.3°.

Picturing shapes is a great way to visualize a reaction. For instance, when ammonia reacts with an acid, the lone pair on N forms a bond to the H^+ and yields the ammonium ion (NH_4^+), one of many tetrahedral polyatomic ions. As the lone pair becomes a bonding pair, the H—N—H angle expands from 107.3° to 109.5°:

$$NH_3\ (107.3°) + H^+ \longrightarrow [NH_4]^+\ (109.5°)$$

3. *Two lone pairs: bent or V shape (AX_2E_2).* Water is the most important V-shaped molecule with the tetrahedral arrangement. [Note that, in the trigonal planar arrangement, the V shape has two bonding groups and *one* lone pair (AX_2E), and its ideal

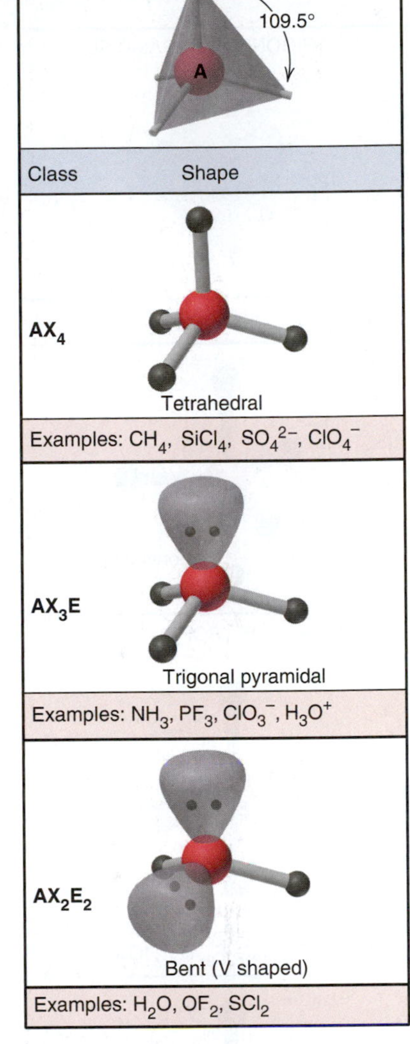

Figure 10.5 The three molecular shapes of the tetrahedral electron-group arrangement.

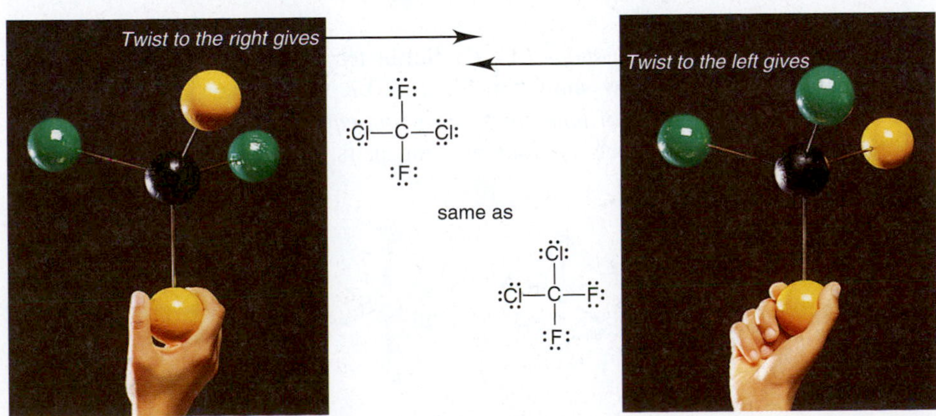

Figure 10.6 Lewis structures do not indicate molecular shape. In this model, Cl is green and F is yellow.

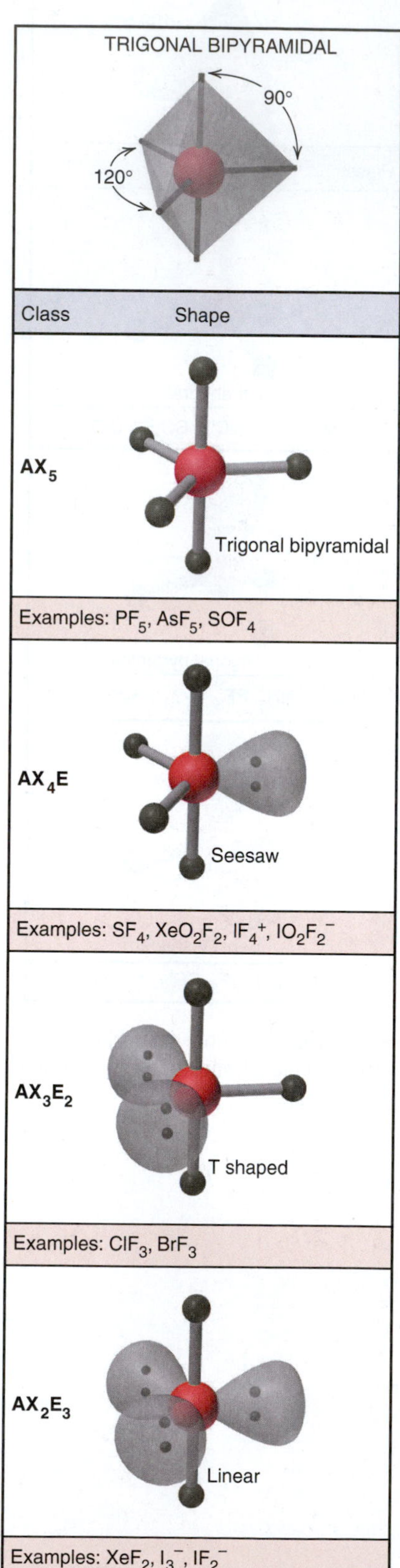

Figure 10.7 The four molecular shapes of the trigonal bipyramidal electron-group arrangement.

bond angle is 120°, not 109.5°.] Repulsions from two lone pairs are greater than from one, and the H—O—H bond angle is 104.5°, less than the H—N—H angle in NH_3:

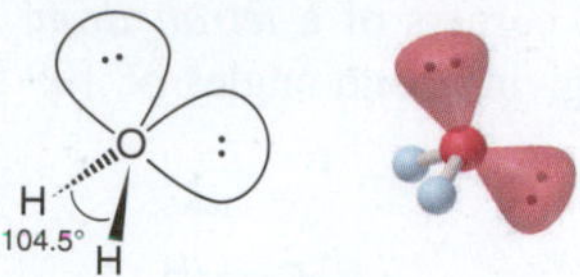

Thus, for similar molecules within a given electron-group arrangement, electron-electron repulsions cause deviations from ideal bond angles in the following order:

Lone pair–lone pair > lone pair–bonding pair > bonding pair–bonding pair **(10.2)**

Molecular Shapes with Five Electron Groups (Trigonal Bipyramidal Arrangement)

All molecules with five or six electron groups have a central atom from Period 3 or higher because only those atoms have *d* orbitals available to expand the valence shell.

Relative Positions of Electron Groups Five mutually repelling electron groups form the **trigonal bipyramidal arrangement,** in which two trigonal pyramids share a common base (Figure 10.7). This is the only case in which *there are two different positions for electron groups and two ideal bond angles.* Three **equatorial groups** lie in a trigonal plane that includes the central atom, and two **axial groups** lie above and below this plane. Therefore, a 120° bond angle separates equatorial groups, and a 90° angle separates axial from equatorial groups. Two factors come into play:

- The greater the bond angle, the weaker the repulsions, so *equatorial-equatorial (120°) repulsions are weaker than axial-equatorial (90°) repulsions.*
- The stronger repulsions from lone pairs means that, when possible, *lone pairs occupy equatorial positions.*

Shapes for the Trigonal Bipyramidal Arrangement The tendency for lone pairs to occupy equatorial positions, and thus minimize stronger axial-equatorial repulsions, governs three of the four shapes for this arrangement.

1. *All bonding groups: trigonal bipyramidal shape (AX_5).* Phosphorus pentachloride (PCl_5) has a trigonal bipyramidal shape. With five identical surrounding atoms, the bond angles are ideal:

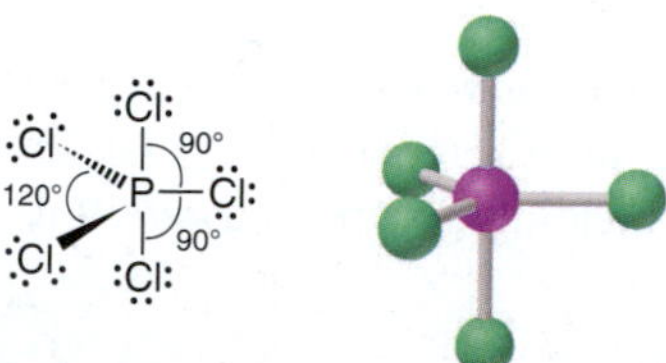

2. *One lone pair: seesaw shape (AX_4E).* Sulfur tetrafluoride (SF_4), a strong fluorinating agent, has the **seesaw shape;** in Figure 10.7, the "seesaw" is tipped on an end. This is the first example of *lone pairs occupying equatorial positions* to minimize repulsions. The lone pair repels all four bonding pairs, reducing the bond angles to 101.5° and 86.8°:

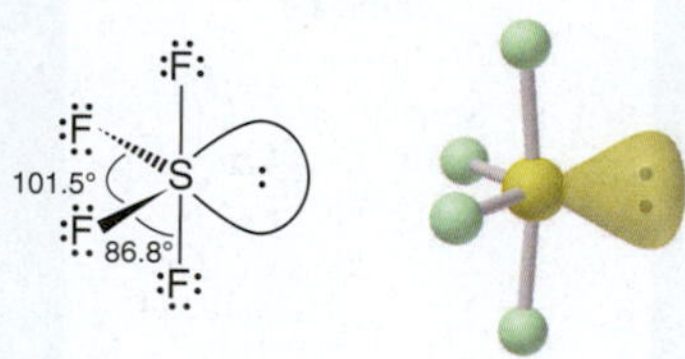

3. *Two lone pairs: T shape (AX_3E_2).* Bromine trifluoride (BrF_3), one of many compounds with fluorine bonded to a larger halogen, has a **T shape.** Since both lone pairs occupy equatorial positions, we see a greater decrease in the axial-equatorial bond angle, down to 86.2°:

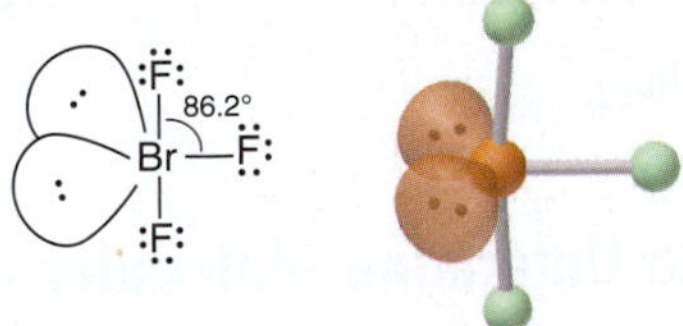

4. *Three lone pairs: linear shape (AX_2E_3).* The triiodide ion (I_3^-), which forms when I_2 dissolves in aqueous I^- solution, is linear. With three equatorial lone pairs and two axial bonding pairs, the three nuclei form a straight line and a 180° X—A—X bond angle:

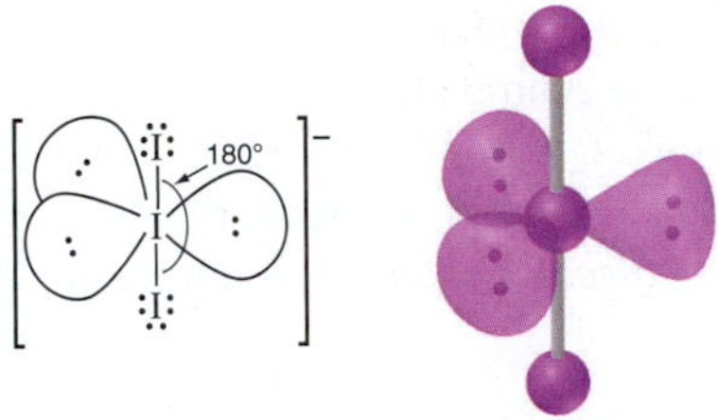

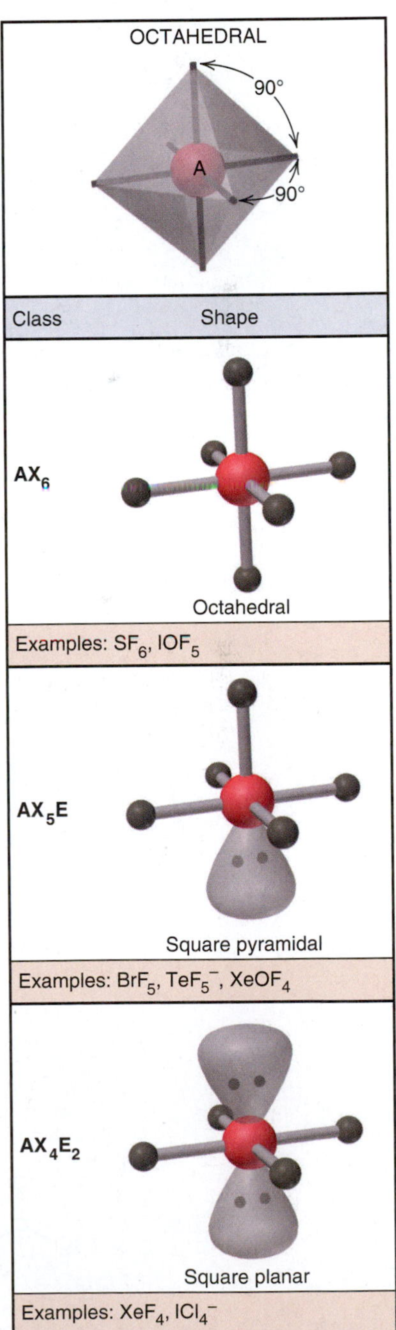

Figure 10.8 The three molecular shapes of the octahedral electron-group arrangement.

Molecular Shapes with Six Electron Groups (Octahedral Arrangement)

Six electron groups form the **octahedral arrangement.** An *octahedron* is a polyhedron with eight equilateral triangles for faces and six identical vertices (Figure 10.8). Each of the six groups points to a corner, which gives a 90° ideal bond angle.

1. *All bonding groups: octahedral shape (AX_6).* When seesaw-shaped SF_4 (discussed above) reacts with more F_2, the central S atom expands its valence shell further to form octahedral sulfur hexafluoride (SF_6):

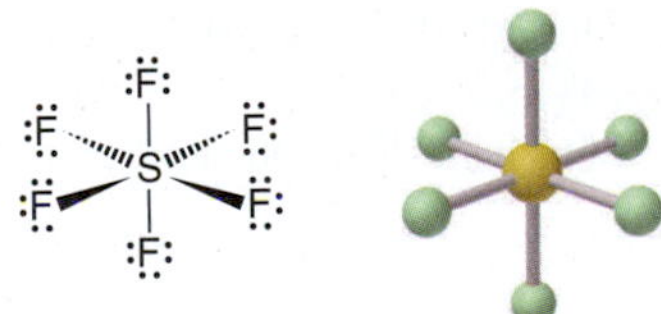

2. *One lone pair: square pyramidal shape (AX_5E).* Iodine pentafluoride (IF_5) has a **square pyramidal shape.** Note that it makes no difference where one lone pair resides because all the bond angles are 90°. The lone pair reduces the bond angles to 81.9°:

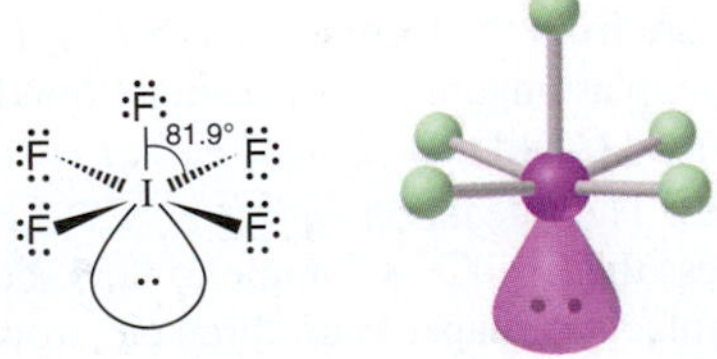

3. *Two lone pairs: square planar shape (AX_4E_2).* Xenon tetrafluoride (XeF_4) has a **square planar shape.** To avoid stronger 90° lone pair–lone pair repulsions, two lone pairs lie *opposite* each other:

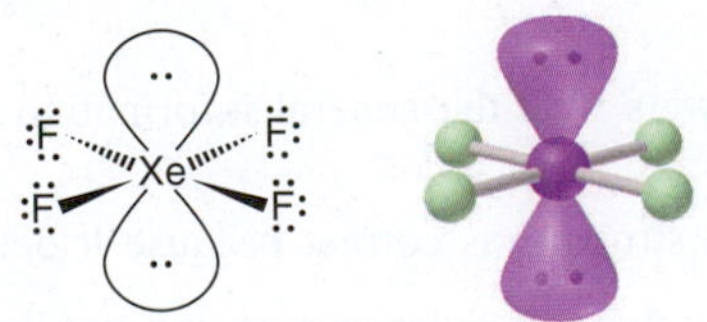

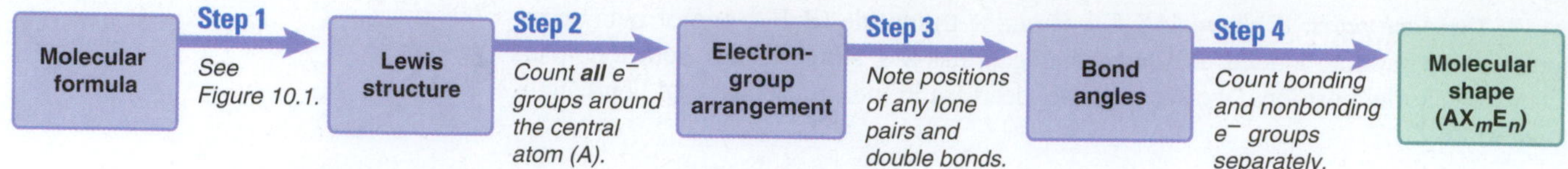

Figure 10.9 The four steps in converting a molecular formula to a molecular shape.

Using VSEPR Theory to Determine Molecular Shape

Let's apply a stepwise method for using VSEPR theory to determine a molecular shape from a molecular formula (Figure 10.9):

Step 1. Write the Lewis structure from the molecular formula (see Figure 10.1) to show the relative placement of atoms and the number of electron groups.

Step 2. Assign an electron-group arrangement by counting *all* electron groups (bonding plus nonbonding) around the central atom.

Step 3. Predict the ideal bond angle from the electron-group arrangement and *the effect of any deviation* caused by lone pairs or double bonds.

Step 4. Draw and name the molecular shape by counting bonding groups and nonbonding groups separately.

The next two sample problems apply these steps.

Sample Problem 10.6 Examining Shapes with Two, Three, or Four Electron Groups

Problem Draw the molecular shapes and predict the bond angles (relative to the ideal angles) of **(a)** PF_3 and **(b)** $COCl_2$.

Solution (a) For PF_3:

Step 1. Write the Lewis structure from the formula *(see below, left).*

Step 2. Assign the electron-group arrangement: Three bonding groups and one lone pair give four electron groups around P and the *tetrahedral arrangement.*

Step 3. Predict the bond angle: The ideal bond angle is 109.5°. There is one lone pair, so the actual bond angle will be less than 109.5°.

Step 4. Draw and name the molecular shape: With one lone pair, PF_3 has a trigonal pyramidal shape (AX_3E):

:F—P—F: (with :F: below P) ⟹ (4 e⁻ groups) Tetrahedral arrangement ⟹ (1 lone pair) <109.5° ⟹ (3 bonding groups) [trigonal pyramidal P with 3 F, 96.3°] AX_3E

(b) For $COCl_2$:

Step 1. Write the Lewis structure from the formula *(see below, left).*

Step 2. Assign the electron-group arrangement: Two single bonds and one double bond give three electron groups around C and the *trigonal planar arrangement.*

Step 3. Predict the bond angles: The ideal bond angle is 120°, but the double bond between C and O will compress the Cl—C—Cl angle to less than 120°.

Step 4. Draw and name the molecular shape: With three electron groups and no lone pairs, $COCl_2$ has a trigonal planar shape (AX_3):

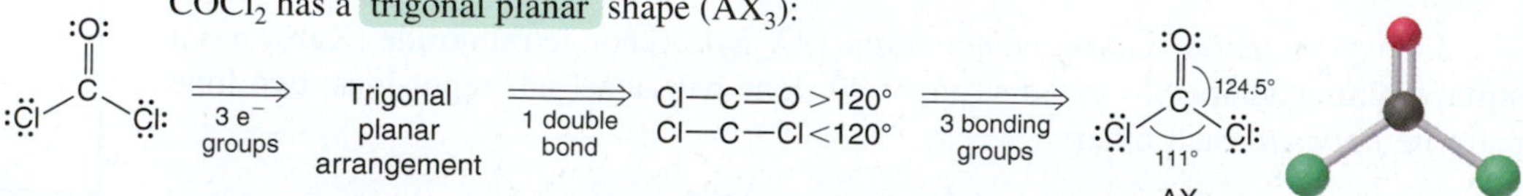

Check We compare the answers with the general information in Figures 10.5 and 10.4, respectively.

Comment Be sure the Lewis structure is correct because it determines the other steps.

FOLLOW-UP PROBLEM 10.6 Draw the molecular shapes and predict the bond angles (relative to the ideal angles) of **(a)** CS_2; **(b)** $PbCl_2$; **(c)** CBr_4; **(d)** SF_2.

Sample Problem 10.7 Examining Shapes with Five or Six Electron Groups

Problem Draw the molecular shapes and predict the bond angles (relative to the ideal angles) of **(a)** SbF_5 and **(b)** BrF_5.

Plan We proceed as in Sample Problem 10.6, being sure to minimize the number of axial-equatorial repulsions.

Solution **(a)** For SbF_5:
Step 1. Lewis structure *(see below, left).*
Step 2. Electron-group arrangement: With five electron groups, this is the *trigonal bipyramidal* arrangement.
Step 3. Bond angles: All the groups and surrounding atoms are identical, so the bond angles are ideal: 120° between equatorial groups and 90° between axial and equatorial groups.
Step 4. Molecular shape: Five electron groups and no lone pairs give the trigonal bipyramidal shape (AX_5):

5 e⁻ groups ⟹ Trigonal bipyramidal arrangement ⟹ no lone pairs or double bonds ⟹ Ideal bond angles ⟹ 5 bonding groups ⟹ AX_5 (90°, 120°)

(b) For BrF_5:
Step 1. Lewis structure *(see below, left).*
Step 2. Electron-group arrangement: Six electron groups give the *octahedral* arrangement.
Step 3. Bond angles: The lone pair will make all bond angles less than the ideal 90°.
Step 4. Molecular shape: With one lone pair, BrF_5 has the square pyramidal shape (AX_5E):

6 e⁻ groups ⟹ Octahedral arrangement ⟹ 1 lone pair ⟹ <90° ⟹ 5 bonding groups ⟹ AX_5E (84.8°)

Check We compare our answers with Figures 10.7 and 10.8.

Comment We will also see the linear, tetrahedral, square planar, and octahedral shapes in an important group of substances, called *coordination compounds,* in Chapter 23.

FOLLOW-UP PROBLEM 10.7 Draw the molecular shapes and predict the bond angles (relative to the ideal angles) of **(a)** ICl_2^-; **(b)** ClF_3; **(c)** SOF_4.

Molecular Shapes with More Than One Central Atom

The shapes of molecules with more than one central atom are composites of the shapes around each of the central atoms. Here are two examples:

1. Ethane (CH_3CH_3; molecular formula C_2H_6) is a component of natural gas. With four bonding groups and no lone pairs around the two central carbons, ethane is shaped like two overlapping tetrahedra (Figure 10.10A).

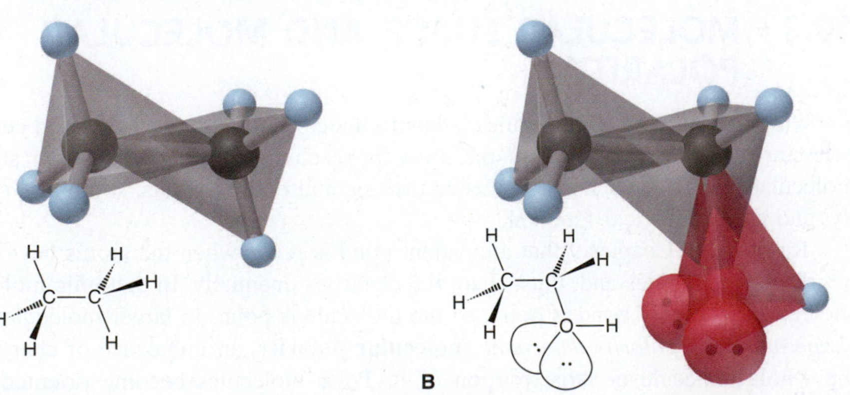

Figure 10.10 The tetrahedral shapes around the central atoms and the overall shapes of ethane **(A)** and ethanol **(B)**.

2. Ethanol (CH_3CH_2OH; molecular formula C_2H_6O), the intoxicating substance in beer, wine, and whiskey, has three central atoms (Figure 10.10B). The CH_3— group is tetrahedrally shaped, and the —CH_2— group has four bonding groups around its central C atom, so it is also tetrahedrally shaped. The O atom has two bonding groups and two lone pairs around it, so the —OH group has a V shape (AX_2E_2).

Sample Problem 10.8 Predicting Molecular Shapes with More Than One Central Atom

Problem Determine the shape around each central atom in acetone, $(CH_3)_2CO$.

Plan There are three central C atoms, two of which are in CH_3— groups. We determine the shape around one central atom at a time.

Solution *Step 1.* Lewis structure *(see below, left).*
Step 2. Electron-group arrangement: Each CH_3— group has four electron groups around its central C, so its electron-group arrangement is *tetrahedral*. The third C atom has three electron groups around it, so it has the *trigonal planar arrangement.*
Step 3. Bond angles: The H—C—H angle in CH_3— should be near the ideal 109.5°. The C═O double bond will compress the C—C—C angle to less than the ideal 120°.
Step 4. Shapes around central atoms: With four electron groups and no lone pairs, the shape around C in each CH_3— is tetrahedral (AX_4). With three electron groups and no lone pairs, the shape around the middle C is trigonal planar (AX_3):

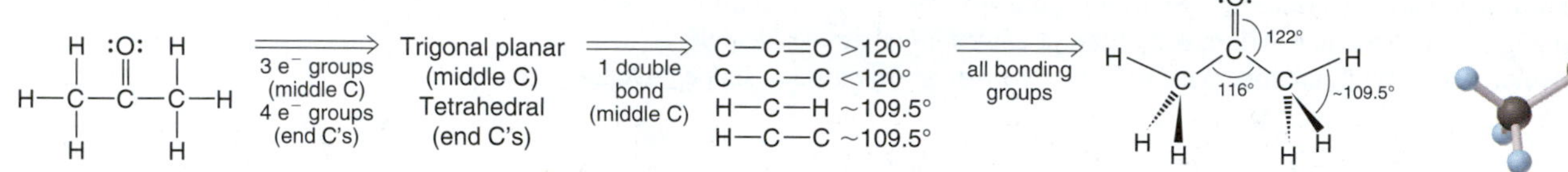

FOLLOW-UP PROBLEM 10.8 Determine the shape around each central atom and predict any deviations from ideal bond angles in the following: **(a)** H_2SO_4; **(b)** propyne (C_3H_4; there is one C≡C bond); **(c)** S_2F_2.

Summary of Section 10.2

- VSEPR theory proposes that each electron group (single bond, multiple bond, lone pair, or lone electron) around a central atom remains as far from the others as possible.
- Five electron-group arrangements are possible when two, three, four, five, or six electron groups surround a central atom. Each arrangement is associated with one or more molecular shapes, depending on the numbers of bonding and lone pairs.
- Ideal bond angles are based on the regular geometric arrangements. Deviations from them occur when surrounding atoms and/or electron groups are not identical.
- Lone pairs and double bonds exert stronger repulsions than single bonds.
- Shapes of larger molecules are composites of the shapes around each central atom.

10.3 • MOLECULAR SHAPE AND MOLECULAR POLARITY

Knowing the shape of its molecules is key to understanding the physical and chemical behavior of a substance. One of the most far-reaching effects of molecular shape is molecular polarity, which can influence melting and boiling points, solubility, reactivity, and even biological function.

Recall from Chapter 9 that a covalent bond is *polar* when the atoms have different electronegativities and, thus, share the electrons unequally. In diatomic molecules, such as HF, the only bond is polar, so the molecule is polar. In larger molecules, *both shape and bond polarity determine* **molecular polarity,** an imbalance of charge over the whole molecule or large portion of it. Polar molecules become oriented in an electric field, with their partially charged ends pointing toward the oppositely charged plates (Figure 10.11). **Dipole moment (μ)** is a measure of molecular polarity, given in the unit *debye* (D) derived from SI units of charge (coulomb, C) and length (m): 1 D = 3.34×10^{-30} C·m. [The unit is named for the Dutch-American scientist Peter Debye (1884–1966) who won the Nobel Prize in 1936 for contributions to the fields of molecular structure and solution behavior.]

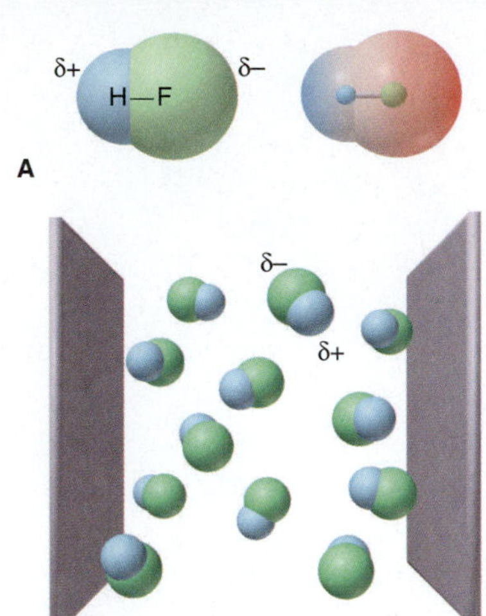

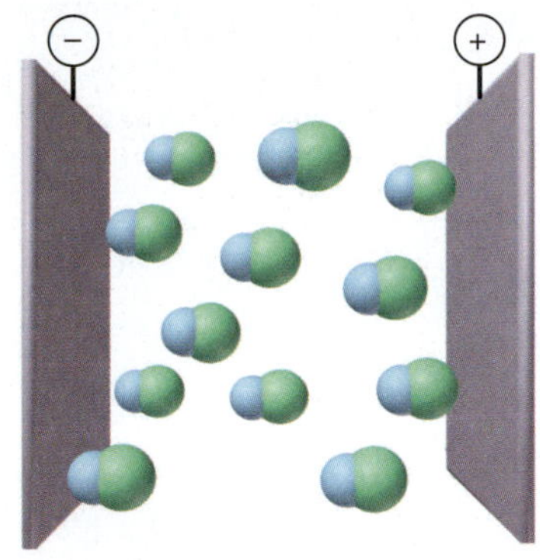

Figure 10.11 The orientation of polar molecules in an electric field. **A,** Space-filling *(left)* and electron density models (*right*, with red negative and blue positive) of the polar HF molecule. **B,** With the external electric field off, HF molecules are oriented randomly. **C,** With the field on, the molecules, on average, become oriented.

Bond Polarity, Bond Angle, and Dipole Moment

The presence of polar bonds does not *always* result in a polar molecule; we must also consider shape and the atoms surrounding the central atom. Here are three cases:

1. *CO_2: polar bonds, nonpolar molecule.* In carbon dioxide, the electronegativity difference between C (EN = 2.5) and O (EN = 3.5) makes each C═O bond polar. But CO_2 is linear, so the bonds point 180° from each other. The two bond polarities are counterbalanced, and the molecule has *no net dipole moment* (μ = 0 D). The electron density model shows regions of high negative charge *(red)* distributed equally on either side of the central region of high positive charge *(blue)*:

2. *H_2O: polar bonds, polar molecule.* Water also has two polar bonds, but it *is* polar (μ = 1.85 D). In each O—H bond, electron density is pulled from H (EN = 2.1) toward O (EN = 3.5). Bond polarities are *not* counterbalanced because the molecule is V shaped (see also Figure 4.1). The bond polarities are partially reinforced, making the O end partially negative and the other end (the region between the H atoms) partially positive:

(The molecular polarity of water has some amazing effects, from determining the composition of the oceans to supporting life itself, as you'll see in Chapter 12.)

3. *Same shapes, different polarities.* When different molecules have the same shape, the identities of the surrounding atoms affect polarity. Carbon tetrachloride (CCl_4) and chloroform ($CHCl_3$) are tetrahedral molecules with very different polarities. In CCl_4, all the surrounding atoms are Cl atoms. Each C—Cl bond is polar (ΔEN = 0.5), but the molecule is nonpolar (μ = 0 D) because the bond polarities counterbalance each other. In $CHCl_3$, an H replaces one Cl, disrupting the balance and giving chloroform a significant dipole moment (μ = 1.01 D):

Sample Problem 10.9 Predicting the Polarity of Molecules

Problem For each compound, use the molecular shape and EN values and trends (Figure 9.18) to predict the direction of bond and molecular polarity, if present: **(a)** ammonia, NH_3; **(b)** boron trifluoride, BF_3; **(c)** carbonyl sulfide, COS (atom sequence SCO).

Plan We draw and name the molecular shape and point a polar arrow toward the atom with higher EN in each bond. If the bond polarities balance one another, the molecule is nonpolar; if they reinforce each other, we show the direction of the molecular polarity.

Solution **(a)** For NH_3: The molecular shape is trigonal pyramidal. N (EN = 3.0) is more electronegative than H (EN = 2.1), so the bond polarities point toward N and partially reinforce each other; thus the molecular polarity points toward N:

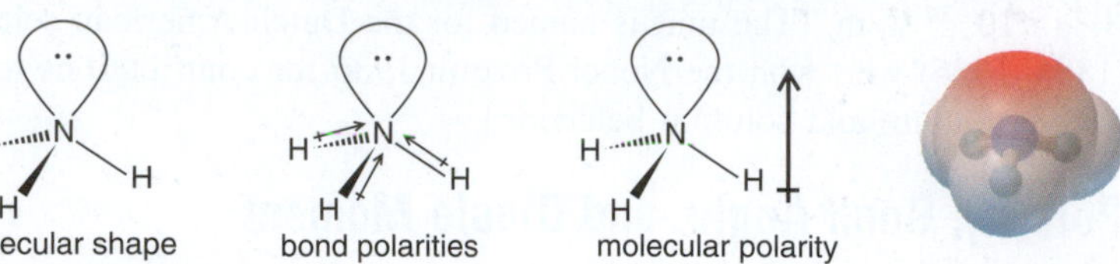

Therefore, ammonia is polar.

(b) For BF_3: The molecular shape is trigonal planar. F (EN = 4.0) is farther to the right in Period 2 than B (EN = 2.0), so it is more electronegative; thus, each bond polarity points toward F. However, the bond angle is 120°, so the three bond polarities balance each other, and BF_3 has no molecular polarity:

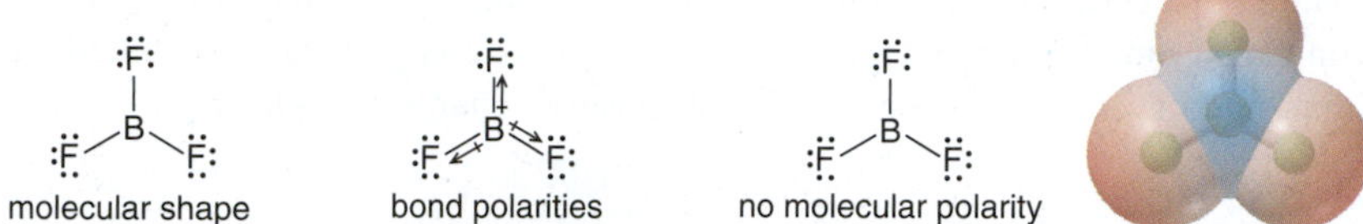

Therefore, boron trifluoride is nonpolar.

(c) For COS: The molecular shape is linear. With C and S having the same EN, the C=S bond is nonpolar, but the C=O bond is quite polar ($\Delta EN = 1.0$), so there is a net molecular polarity toward the O:

S=C=O molecular shape   S=C=O bond polarity   S=C=O molecular polarity

Therefore, carbonyl sulfide is polar.

Check The electron density models confirm our conclusions. Note that, in (b), the negative *(red)* regions surround the central B atom *(blue)* symmetrically.

FOLLOW-UP PROBLEM 10.9 Show the bond polarities and molecular polarity, if any, for each compound: **(a)** dichloromethane (CH_2Cl_2); **(b)** iodine oxide pentafluoride (IOF_5); **(c)** nitrogen tribromide (NBr_3).

Summary of Section 10.3

- Bond polarity and molecular shape determine molecular polarity, which is measured as a dipole moment.
- A molecule with polar bonds is not necessarily a polar molecule. When bond polarities counterbalance each other, the molecule is nonpolar; when they reinforce each other, the molecule is polar.

CHAPTER REVIEW GUIDE

The following sections provide many aids to help you study this chapter. (Numbers in parentheses refer to pages, unless noted otherwise.)

Learning Objectives

These are concepts and skills to review after studying this chapter.

Related section (§), sample problem (SP), and upcoming end-of-chapter problem (EP) numbers are listed in parentheses.

1. Use the octet rule to write a Lewis structure from a molecular formula (§10.1) (SPs 10.1–10.3) (EPs 10.1, 10.5–10.8)
2. Understand how electron delocalization explains bond properties, and write resonance structures (§10.1) (SP 10.4) (EPs 10.2, 10.9–10.12)
3. Describe the three types of exceptions to the octet rule, draw Lewis structures for such molecules, and use formal charges to select the most important resonance structure (§10.1) (SP 10.5) (EPs 10.3, 10.4, 10.13–10.24)
4. Describe the five electron-group arrangements and associated molecular shapes, predict molecular shapes from Lewis structures, and explain deviations from ideal bond angles (§10.2) (SPs 10.6–10.8) (EPs 10.25–10.50)
5. Understand how a molecule's polarity arises, and use molecular shape and EN values to predict the direction of a dipole (§10.3) (SP 10.9) (EPs 10.51–10.56)

Key Terms

These important terms appear in boldface in the chapter and are defined again in the Glossary.

Section 10.1
Lewis structure (Lewis formula) (303)
resonance structure (resonance form) (306)
resonance hybrid (306)
electron-pair delocalization (307)
formal charge (308)
electron deficient (309)
free radical (310)
expanded valence shell (310)

Section 10.2
valence-shell electron-pair repulsion (VSEPR) theory (312)
molecular shape (312)
bond angle (313)
linear arrangement (313)
linear shape (313)
trigonal planar arrangement (314)
bent shape (V shape) (314)
tetrahedral arrangement (315)
trigonal pyramidal shape (315)
trigonal bipyramidal arrangement (316)
equatorial group (316)
axial group (316)
seesaw shape (316)
T shape (317)
octahedral arrangement (317)
square pyramidal shape (317)
square planar shape (317)

Section 10.3
molecular polarity (320)
dipole moment (μ) (321)

Key Equations and Relationships

Numbered and screened concepts are listed for you to refer to or memorize.

10.1 Calculating the formal charge on an atom (308):

Formal charge of atom
$= \text{no. of valence } e^-$
$- (\text{no. of unshared valence } e^- + \frac{1}{2} \text{ no. of shared valence } e^-)$

10.2 Ranking the effect of electron-pair repulsions on bond angle (316):
Lone pair–lone pair > lone pair–bonding pair > bonding pair–bonding pair

BRIEF SOLUTIONS TO FOLLOW-UP PROBLEMS

Compare your own solutions to these calculation steps and answers.

10.1 (a) H—S̈: with H (b) :Ö: with :F̤: and :F̤: (c) :Ö: on S̈ with :C̤l: and :C̤l:

10.2 (a) H—N̈—Ö—H with H (b) H—C—Ö—C—H with H on each C

10.3 (a) :C≡O: (b) H—C≡N: (c) :Ö=C=Ö:

10.4 $[CO_3]^{2-} \longleftrightarrow [CO_3]^{2-}$ resonance structures

10.5 (a) :Ö: =P with three :C̤l: (b) O, Cl, O (c) Xe with four :F̤:

10.6 (a) S̈=C=S̈ Linear, 180° (b) Pb with two :C̤l: V shaped, <120°

(c) C with four :B̤r: Tetrahedral, 109.5° (d) S with two :F̤: V shaped, <109.5°

10.7
(a) $[ICl_2]^-$ Linear, 180°
(b) ClF_3 T shaped, <90°
(c) SOF_4 Trigonal bipyramidal
F_{eq}—S—F_{eq} angle <120°
F_{ax}—S—F_{eq} angle <90°

10.8 (a) S is tetrahedral; double bonds compress O—S—O angle to <109.5°. Shape around each O in an —OH is V shape; lone pairs compress H—O—S angle to <109.5°.

(b) H—C≡C—H with CH_3 group. Shape around C in CH_3— is tetrahedral, with angles ~109.5°; other C atoms are linear, 180°.

(c) F—S=S—F. V shape around each S; F—S—S angle <109.5°.

10.9

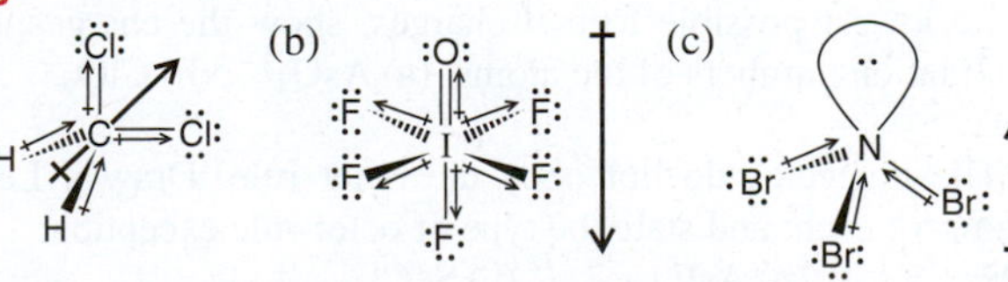

PROBLEMS

Problems with **colored** numbers are answered in Appendix E. Sections match the text and provide the numbers of relevant sample problems. Bracketed problems are grouped in pairs (indicated by a short rule) that cover the same concept. Comprehensive Problems are based on material from any section or previous chapter.

Depicting Molecules and Ions with Lewis Structures

(Sample Problems 10.1 to 10.5)

10.1 Which of these atoms *cannot* serve as a central atom in a Lewis structure: (a) O; (b) He; (c) F; (d) H; (e) P? Explain.

10.2 When is a resonance hybrid needed to adequately depict the bonding in a molecule? Using NO_2 as an example, explain how a resonance hybrid is consistent with the actual bond length, bond strength, and bond order.

10.3 In which of these bonding patterns does X obey the octet rule?

(a) (b) (c) (d) (e) (f) (g) (h)

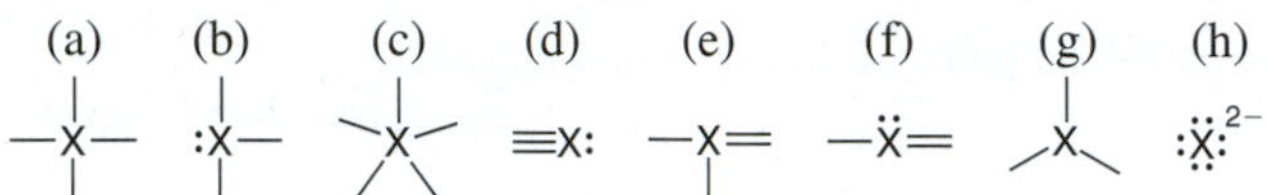

10.4 What is required for an atom to expand its valence shell? Which of the following atoms can expand its valence shell: F, S, H, Al, Se, Cl?

10.5 Draw a Lewis structure for (a) SiF_4; (b) $SeCl_2$; (c) COF_2 (C is central).

10.6 Draw a Lewis structure for (a) PH_4^+; (b) C_2F_4; (c) SbH_3.

10.7 Draw a Lewis structure for (a) PF_3; (b) H_2CO_3 (both H atoms are attached to O atoms); (c) CS_2.

10.8 Draw a Lewis structure for (a) CH_4S; (b) S_2Cl_2; (c) $CHCl_3$.

10.9 Draw Lewis structures of all the important resonance forms of (a) NO_2^+; (b) NO_2F (N is central).

10.10 Draw Lewis structures of all the important resonance forms of (a) HNO_3 ($HONO_2$); (b) $HAsO_4^{2-}$ ($HOAsO_3^{2-}$).

10.11 Draw Lewis structures of all the important resonance forms of (a) N_3^-; (b) NO_2^-.

10.12 Draw Lewis structures of all the important resonance forms of (a) HCO_2^- (H is attached to C); (b) $HBrO_4$ ($HOBrO_3$).

10.13 Draw the Lewis structure with lowest formal charges, and determine the charge of each atom in (a) IF_5; (b) AlH_4^-.

10.14 Draw the Lewis structure with lowest formal charges, and determine the charge of each atom in (a) OCS; (b) NO.

10.15 Draw a Lewis structure for a resonance form of each ion with the lowest possible formal charges, show the charges, and give oxidation numbers of the atoms: (a) BrO_3^-; (b) SO_3^{2-}.

10.16 Draw a Lewis structure for a resonance form of each ion with the lowest possible formal charges, show the charges, and give oxidation numbers of the atoms: (a) AsO_4^{3-}; (b) ClO_2^-.

10.17 These species do not obey the octet rule. Draw a Lewis structure for each, and state the type of octet-rule exception:
(a) BH_3 (b) AsF_4^- (c) $SeCl_4$

10.18 These species do not obey the octet rule. Draw a Lewis structure for each, and state the type of octet-rule exception:
(a) PF_6^- (b) ClO_3 (c) H_3PO_3 (one P—H bond)

10.19 These species do not obey the octet rule. Draw a Lewis structure for each, and state the type of octet-rule exception:
(a) BrF_3 (b) ICl_2^- (c) BeF_2

10.20 These species do not obey the octet rule. Draw a Lewis structure for each, and state the type of octet-rule exception:
(a) O_3^- (b) XeF_2 (c) SbF_4^-

10.21 Molten beryllium chloride reacts with chloride ion from molten NaCl to form the $BeCl_4^{2-}$ ion, in which the Be atom attains an octet. Show the net ionic reaction with Lewis structures.

10.22 Despite many attempts, the perbromate ion (BrO_4^-) was not prepared in the laboratory until about 1970. (In fact, articles were published explaining theoretically why it could never be prepared!) Draw a Lewis structure for BrO_4^- in which all atoms have lowest formal charges.

10.23 Cryolite (Na_3AlF_6) is an indispensable component in the electrochemical production of aluminum. Draw a Lewis structure for the AlF_6^{3-} ion.

10.24 Phosgene is a colorless, highly toxic gas that was employed against troops in World War I and it is used today as a key reactant in organic syntheses. From the following resonance structures, select the one with the lowest formal charges:

A B C

Valence-Shell Electron-Pair Repulsion (VSEPR) Theory and Molecular Shape

(Sample Problems 10.6 to 10.8)

10.25 If you know the formula of a molecule or ion, what is the first step in predicting its shape?

10.26 In what situation is the name of the molecular shape the same as the name of the electron-group arrangement?

10.27 Which of the following numbers of electron groups can give rise to a bent (V-shaped) molecule: two, three, four, five, six? Draw an example for each case, showing the shape classification (AX_mE_n) and the ideal bond angle.

10.28 Name all the molecular shapes that have a tetrahedral electron-group arrangement.

10.29 Consider the following molecular shapes:

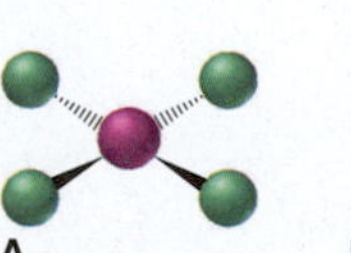
A

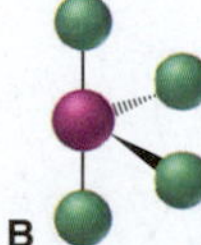
B

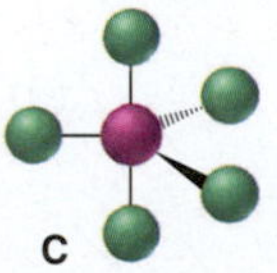
C

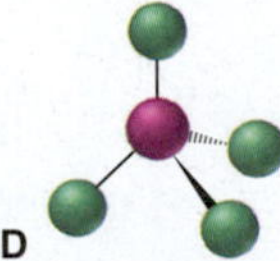
D

(a) Which has the most electron pairs (both shared and unshared) around the central atom?
(b) Which has the most unshared pairs around the central atom?
(c) Do any have only shared pairs around the central atom?

10.30 Why aren't lone pairs considered along with surrounding bonding groups when describing the molecular shape?

10.31 Use wedge-bond perspective drawings (if necessary) to sketch the atom positions in a general molecule of formula (not shape class) AX_n that has each of the following shapes:
(a) V shaped (b) trigonal planar (c) trigonal bipyramidal
(d) T shaped (e) trigonal pyramidal (f) square pyramidal

10.32 What would you expect to be the electron-group arrangement around atom A in each of the following cases? For each arrangement, give the ideal bond angle and the direction of any expected deviation:

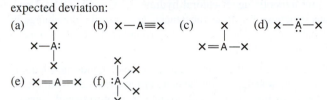

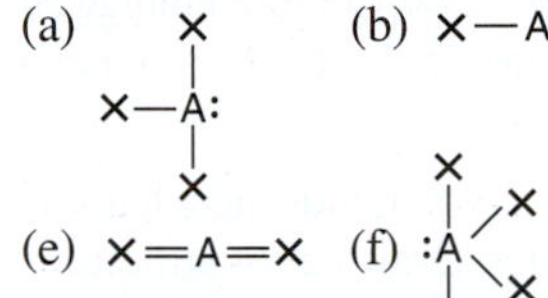

(c) X=A—X with X above A (d) X—Ä—X (f) :A with four X

10.33 Determine the electron-group arrangement, molecular shape, and ideal bond angle(s) for each of the following:
(a) O_3 (b) H_3O^+ (c) NF_3

10.34 Determine the electron-group arrangement, molecular shape, and ideal bond angle(s) for each of the following:
(a) SO_4^{2-} (b) NO_2^- (c) PH_3

10.35 Determine the electron-group arrangement, molecular shape, and ideal bond angle(s) for each of the following:
(a) CO_3^{2-} (b) SO_2 (c) CF_4

10.36 Determine the electron-group arrangement, molecular shape, and ideal bond angle(s) for each of the following:
(a) SO_3 (b) N_2O (N is central) (c) CH_2Cl_2

10.37 Name the shape and give the AX_mE_n classification and ideal bond angle(s) for each of the following general molecules:

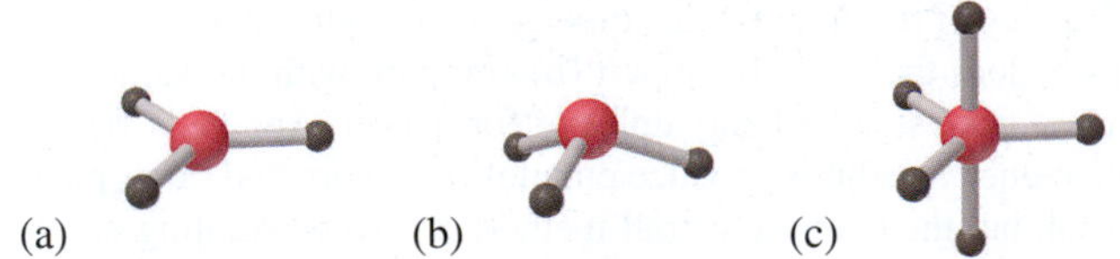

10.38 Name the shape and give the AX_mE_n classification and ideal bond angle(s) for each of the following general molecules:

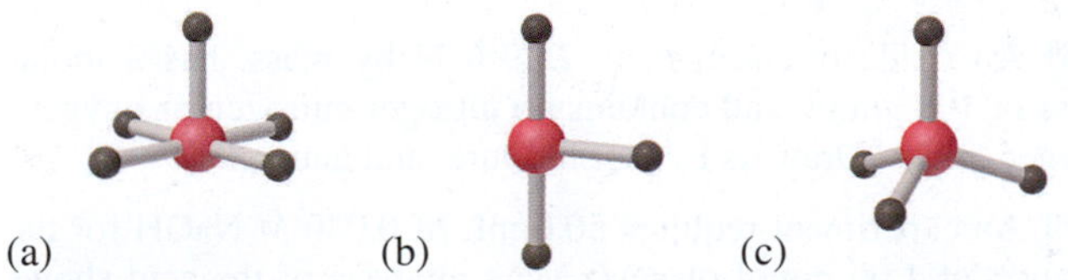

10.39 Determine the shape, ideal bond angle(s), and direction of any deviation from the ideal angle(s) for each of the following:
(a) ClO_2^- (b) PF_5 (c) SeF_4 (d) KrF_2

10.40 Determine the shape, ideal bond angle(s), and direction of any deviation from the ideal angle(s) for each of the following:
(a) ClO_3^- (b) IF_4^- (c) $SeOF_2$ (d) TeF_5^-

10.41 Determine the shape around each central atom in each molecule, and explain any deviation from ideal bond angles:
(a) CH_3OH (b) N_2O_4 (O_2NNO_2)

10.42 Determine the shape around each central atom in each molecule, and explain any deviation from ideal bond angles:
(a) H_3PO_4 (no H—P bond) (b) CH_3—O—CH_2CH_3

10.43 Determine the shape around each central atom in each molecule, and explain any deviation from ideal bond angles:
(a) CH_3COOH (b) H_2O_2

10.44 Determine the shape around each central atom in each molecule, and explain any deviation from ideal bond angles:
(a) H_2SO_3 (no H—S bond) (b) N_2O_3 ($ONNO_2$)

10.45 Arrange the following AF_n species in order of *increasing* F—A—F bond angles: BF_3, BeF_2, CF_4, NF_3, OF_2.

10.46 Arrange the following ACl_n species in order of *decreasing* Cl—A—Cl bond angles: SCl_2, OCl_2, PCl_3, $SiCl_4$, $SiCl_6^{2-}$.

10.47 State an ideal value for each of the bond angles in each molecule, and note where you expect deviations:

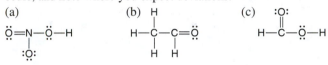

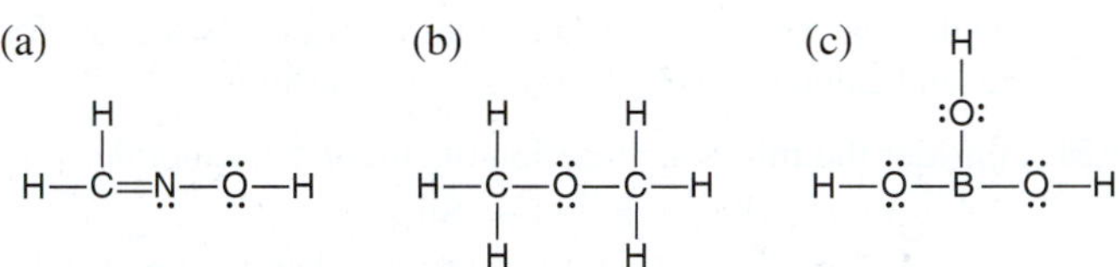

10.48 State an ideal value for each of the bond angles in each molecule, and note where you expect deviations:
(a) Ö=N—Ö—H (with :O: on N) (b) H—C—C=Ö (with H atoms on C) (c) H—C—Ö—H (with :O: double-bonded to C)

10.49 Because both tin and carbon are members of Group 4A(14), they form structurally similar compounds. However, tin exhibits a greater variety of structures because it forms several ionic species. Predict the shapes and ideal bond angles, including any deviations, for the following:
(a) $Sn(CH_3)_2$ (b) $SnCl_3^-$ (c) $Sn(CH_3)_4$
(d) SnF_5^- (e) SnF_6^{2-}

10.50 In the gas phase, phosphorus pentachloride exists as separate molecules. In the solid phase, however, the compound is composed of alternating PCl_4^+ and PCl_6^- ions. (a) What change(s) in molecular shape occur(s) as PCl_5 solidifies? (b) How does the Cl—P—Cl angle change?

Molecular Shape and Molecular Polarity

(Sample Problem 10.9)

10.51 How can a molecule with polar covalent bonds not be polar? Give an example.

10.52 Consider the molecules SCl_2, F_2, CS_2, CF_4, and BrCl.
(a) Which has bonds that are the most polar?
(b) Which molecules have a dipole moment?

10.53 Consider the molecules BF_3, PF_3, BrF_3, SF_4, and SF_6.
(a) Which has bonds that are the most polar?
(b) Which molecules have a dipole moment?

10.54 Which molecule in each pair has the greater dipole moment? Give the reason for your choice.
(a) SO_2 or SO_3 (b) ICl or IF
(c) SiF_4 or SF_4 (d) H_2O or H_2S

10.55 Which molecule in each pair has the greater dipole moment? Give the reason for your choice.
(a) ClO_2 or SO_2 (b) HBr or HCl
(c) $BeCl_2$ or SCl_2 (d) AsF_3 or AsF_5

10.56 There are three different dichloroethylenes (molecular formula $C_2H_2Cl_2$), which we can designate X, Y, and Z. Compound X has no dipole moment, but compound Z does. Compounds X and Z each combine with hydrogen to give the same product:

$$C_2H_2Cl_2\,(\text{X or Z}) + H_2 \longrightarrow ClCH_2{-}CH_2Cl$$

What are the structures of X, Y, and Z? Would you expect compound Y to have a dipole moment?

Comprehensive Problems

10.57 In addition to ammonia, nitrogen forms three other hydrides: hydrazine (N_2H_4), diazene (N_2H_2), and tetrazene (N_4H_4).
(a) Use Lewis structures to compare the strength, length, and order of the nitrogen-nitrogen bonds in hydrazine, diazene, and N_2.
(b) Tetrazene (atom sequence H_2NNNNH_2) decomposes above 0°C to hydrazine and nitrogen gas. Draw a Lewis structure for tetrazene, and calculate ΔH°_{rxn} for this decomposition.

10.58 Consider the following reaction of silicon tetrafluoride:

$$SiF_4 + F^- \longrightarrow SiF_5^-$$

(a) Which depiction below best illustrates the change in molecular shape around Si? (b) Give the name and AX_mE_n designation of each shape in the depiction chosen in part (a).

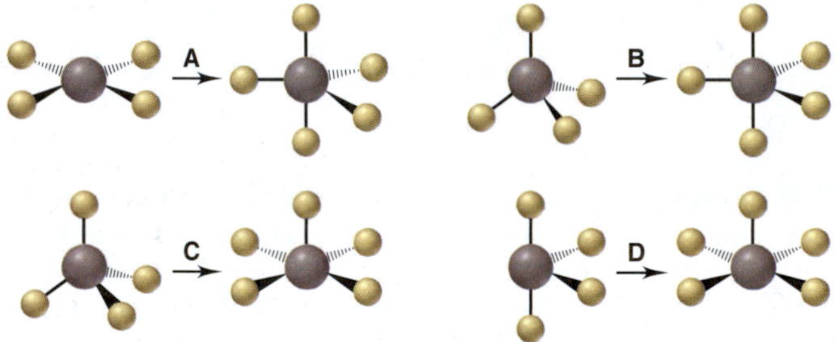

10.59 Nitrosyl fluoride (NOF) has an atom sequence in which all atoms have formal charges of zero. Write the Lewis structure consistent with this fact.

10.60 Both aluminum and iodine form chlorides, Al_2Cl_6 and I_2Cl_6, with "bridging" Cl atoms. The Lewis structures are

:Cl̤ Cl̤ Cl̤: Al Al :Cl̤ Cl̤ Cl̤:   :Cl̤ Cl̤ Cl̤: I I :Cl̤ Cl̤ Cl̤:

(a) What is the formal charge on each atom? (b) Which of these molecules has a planar shape? Explain.

10.61 The VSEPR model was developed before any xenon compounds had been prepared. Thus, these compounds provided an excellent test of the model's predictive power. What would you predict for the shapes of XeF_2, XeF_4, and XeF_6?

10.62 When SO_3 gains two electrons, SO_3^{2-} forms. (a) Which depiction shown below best illustrates the change in molecular shape around S? (b) Does molecular polarity change during this reaction?

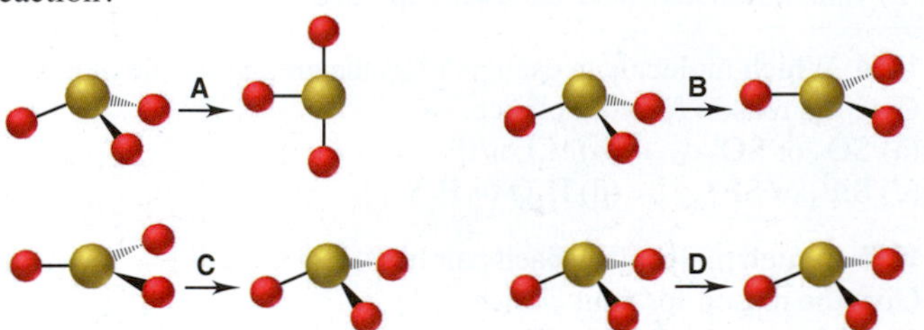

10.63 The actual bond angle in NO_2 is 134.3°, and in NO_2^- it is 115.4°, although the ideal bond angle is 120° in both. Explain.

10.64 "Inert" xenon actually forms several compounds, especially with the highly electronegative elements oxygen and fluorine. The simple fluorides XeF_2, XeF_4, and XeF_6 are all formed by direct reaction of the elements. As you might expect from the size of the xenon atom, the Xe—F bond is not a strong one. Calculate the Xe—F bond energy in XeF_6, given that the enthalpy of formation is −402 kJ/mol.

10.65 Chloral, $Cl_3C{-}CH{=}O$, reacts with water to form the sedative and hypnotic agent chloral hydrate, $Cl_3C{-}CH(OH)_2$. Draw Lewis structures for these substances, and describe the change in molecular shape, if any, that occurs around each of the carbon atoms during the reaction.

10.66 Like several other bonds, carbon-oxygen bonds have lengths and strengths that depend on the bond order. Draw Lewis structures for the following species, and arrange them in order of increasing carbon-oxygen bond length and then by increasing carbon-oxygen bond strength: (a) CO; (b) CO_3^{2-}; (c) H_2CO; (d) CH_4O; (e) HCO_3^- (H attached to O).

10.67 The four bonds of carbon tetrachloride (CCl_4) are polar, but the molecule is nonpolar because the bond polarity is canceled by the symmetric tetrahedral shape. When other atoms substitute for some of the Cl atoms, the symmetry is broken and the molecule becomes polar. Use Figure 9.19 (p. 294) to rank the following molecules from the least polar to the most polar: CH_2Br_2, CF_2Cl_2, CH_2F_2, CH_2Cl_2, CBr_4, CF_2Br_2.

10.68 Ethanol (CH_3CH_2OH) is being used as a gasoline additive or alternative in many parts of the world.
(a) Use bond energies to find ΔH°_{rxn} for the combustion of gaseous ethanol. (Assume H_2O forms as a gas.)
(b) In its standard state at 25°C, ethanol is a liquid. Its vaporization requires 40.5 kJ/mol. Correct the value from part (a) to find the enthalpy of reaction for the combustion of liquid ethanol.
(c) How does the value from part (b) compare with the value you calculate from standard enthalpies of formation (Appendix B)?
(d) "Greener" methods produce ethanol from corn and other plant material, but the main industrial method involves hydrating ethylene from petroleum. Use Lewis structures and bond energies to calculate ΔH°_{rxn} for the formation of gaseous ethanol from ethylene gas with water vapor.

10.69 An oxide of nitrogen is 25.9% N by mass, has a molar mass of 108 g/mol, and contains no nitrogen-nitrogen or oxygen-oxygen bonds. Draw its Lewis structure, and name it.

10.70 An experiment requires 50.0 mL of 0.040 *M* NaOH for the titration of 1.00 mmol of acid. Mass analysis of the acid shows 2.24% hydrogen, 26.7% carbon, and 71.1% oxygen. Draw the Lewis structure of the acid.

10.71 A major short-lived, neutral species in flames is OH.
(a) What is unusual about the electronic structure of OH?
(b) Use the standard enthalpy of formation of OH(*g*) (39.0 kJ/mol) and bond energies to calculate the O—H bond energy in OH(*g*).
(c) From the average value for the O—H bond energy in Table 9.2 (p. 287) and your value for the O—H bond energy in OH(*g*), find the energy needed to break the first O—H bond in water.

10.72 Pure HN_3 (atom sequence HNNN) is explosive. In aqueous solution, it is a weak acid that yields the azide ion, N_3^-. Draw resonance structures to explain why the nitrogen-nitrogen bond lengths are equal in N_3^- but unequal in HN_3.

10.73 Except for nitrogen, the elements of Group 5A(15) all form pentafluorides, and most form pentachlorides. The chlorine atoms of PCl_5 can be replaced with fluorine atoms one at a time to give, successively, PCl_4F, PCl_3F_2, . . . , PF_5. (a) Given the sizes of F and Cl, would you expect the first two F substitutions to be at axial or equatorial positions? Explain. (b) Which of the five fluorine-containing molecules have no dipole moment?

10.74 Dinitrogen monoxide (N_2O) supports combustion in a manner similar to oxygen, with the nitrogen atoms forming N_2. Draw three resonance structures for N_2O (one N is central), and use formal charges to decide the relative importance of each. What correlation can you suggest between the most important structure and the observation that N_2O supports combustion?

10.75 The Murchison meteorite that landed in Australia in 1969 contained 92 different amino acids, including 21 found in Earth organisms. A skeleton structure (single bonds only) of one of these extraterrestrial amino acids is shown below.

```
H3N—CH—C—O
     |  |
    CH2 O
     |
    CH3
```

Draw a Lewis structure, and identify any atoms having a nonzero formal charge.

10.76 A student isolates a product with the molecular shape shown at right (F is orange). (a) If the species is a neutral compound, can the black sphere represent selenium (Se)? (b) If the species is an anion, can the black sphere represent N? (c) If the black sphere represents Br, what is the charge of the species?

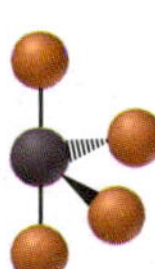

10.77 When gaseous sulfur trioxide is dissolved in concentrated sulfuric acid, disulfuric acid forms:

$$SO_3(g) + H_2SO_4(l) \longrightarrow H_2S_2O_7(l)$$

Use bond energies (Table 9.2, p. 287) to determine ΔH°_{rxn}. (The S atoms in $H_2S_2O_7$ are bonded through an O atom. Assume Lewis structures with zero formal charges; BE of S=O is 552 kJ/mol.)

10.78 In addition to propyne (see Follow-up Problem 10.8), there are two other constitutional isomers of formula C_3H_4. Draw a Lewis structure for each, determine the shape around each carbon, and predict any deviations from ideal bond angles.

10.79 A molecule of formula AY_3 is found experimentally to be polar. Which molecular shapes are possible and which are impossible for AY_3?

10.80 In contrast to the cyanate ion (NCO^-), which is stable and found in many compounds, the fulminate ion (CNO^-), with its different atom sequence, is unstable and forms compounds with heavy metal ions, such as Ag^+ and Hg^{2+}, that are explosive. Like the cyanate ion, the fulminate ion has three resonance structures. Which is the most important contributor to the resonance hybrid? Suggest a reason for the instability of fulminate.

10.81 Consider the following molecular shapes:

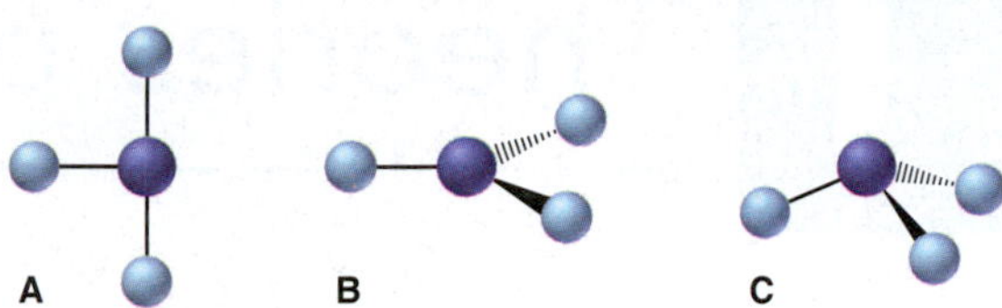

(a) Match each shape with one of the following species: XeF_3^+, $SbBr_3$, $GaCl_3$.
(b) Which, if any, is polar?
(c) Which has the most valence electrons around the central atom?

10.82 Hydrogen cyanide can be catalytically reduced with hydrogen to form methylamine. Use Lewis structures and bond energies to determine ΔH°_{rxn} for

$$HCN(g) + 2H_2(g) \longrightarrow CH_3NH_2(g)$$

10.83 Ethylene, C_2H_4, and tetrafluoroethylene, C_2F_4, are used to make the polymers polyethylene and polytetrafluoroethylene (Teflon), respectively.
(a) Draw the Lewis structures for C_2H_4 and C_2F_4, and give the ideal H—C—H and F—C—F bond angles.
(b) The actual H—C—H and F—C—F bond angles are 117.4° and 112.4°, respectively. Explain these deviations.

10.84 Lewis structures of mescaline, a hallucinogenic compound in peyote cactus, and dopamine, a neurotransmitter in the mammalian brain, appear below. Suggest a reason for mescaline's ability to disrupt nerve impulses.

mescaline

dopamine

10.85 Phosphorus pentachloride, a key industrial compound with annual world production of about 2×10^7 kg, is used to make other compounds. It reacts with sulfur dioxide to produce phosphorus oxychloride ($POCl_3$) and thionyl chloride ($SOCl_2$). Draw a Lewis structure, and name the molecular shape of each product.

Theories of Covalent Bonding

What Is a Covalent Bond? How do molecular shapes, like this model of sulfur hexafluoride, emerge from the orbitals of interacting atoms? And how do bond strength and other properties arise? In this chapter, we examine two theories that rationalize the covalent bond.

Key Principles
to focus on while studying this chapter

- According to *valence bond (VB) theory,* a covalent bond forms when two electrons with *opposing spins* are localized in the region where orbitals on the bonding atoms *overlap*. Bond strength is related to the *extent* and *direction* of overlap. ***(Section 11.1)***
- To account for the shapes of molecules, VB theory proposes that the orbitals of an isolated atom mix together and become *hybrid orbitals,* which have the same orientation in space as the electron-group arrangements of VSEPR theory. Each hybrid orbital has a shape that optimizes overlap and, thus, maximizes bond strength. ***(Section 11.1)***
- The *mode* of orbital overlap determines the type of covalent bond: a *sigma* (σ) *bond* results from *end-to-end* overlap, and a *pi* (π) *bond* results from *side-to-side* overlap. A single bond is a σ bond, a double bond consists of a σ bond and a π bond, and a triple bond consists of a σ bond and two π bonds. Atoms connected by a σ bond can rotate with respect to each other, but a π bond restricts rotation. ***(Section 11.2)***
- According to *molecular orbital (MO) theory, atomic orbitals (AOs)* combine to form *molecular orbitals (MOs),* which spread over the whole molecule. Adding AOs together gives a *bonding MO;* subtracting them gives an *antibonding MO.* Each type of MO has its own shape and energy. A molecule is stabilized when electrons occupy a bonding MO. ***(Section 11.3)***
- Like AOs, MOs have specific energy levels and become occupied one electron at a time by a total of two electrons. MO theory proposes that bond strength and length, as well as magnetic properties, depend on the number of electrons in a molecule's orbitals. ***(Section 11.3)***

Outline

All scientific models have limitations because they are simplifications of reality. The VSEPR theory accounts for molecular shapes, but it doesn't explain how they arise from interactions of atomic orbitals. After all, the orbitals described in Chapter 7 aren't oriented toward the corners of, say, an octahedron, like those for the sulfur in SF_6.

In this chapter, we discuss two theories of bonding in molecules that complement each other. Valence bond (VB) theory rationalizes observed molecular shapes through interactions of atomic orbitals; molecular orbital (MO) theory explains molecular energy levels and related properties.

CONCEPTS & SKILLS TO REVIEW
before studying this chapter

- atomic orbital shapes (Section 7.4)
- the exclusion principle (Section 8.1)
- Hund's rule (Section 8.2)
- Lewis structures (Section 10.1)
- resonance in covalent bonding (Section 10.1)
- molecular shapes (Section 10.2)
- molecular polarity (Section 10.3)

11.1 • VALENCE BOND (VB) THEORY AND ORBITAL HYBRIDIZATION

What *is* a covalent bond, and what characteristic gives it strength? And how can we explain *molecular* shapes based on the interactions of *atomic* orbitals? The most useful approach for answering these questions is **valence bond (VB) theory.**

The Central Themes of VB Theory

The basic principle of VB theory is that *a covalent bond forms when orbitals of two atoms overlap and a pair of electrons occupy the overlap region.* In the terminology of quantum mechanics (Chapter 7), overlap of the two orbitals means their wave functions are *in phase,* so the amplitude between the nuclei increases (see Figure 7.5). The central themes of VB theory derive from this principle:

1. *Opposing spins of the electron pair.* As the exclusion principle (Section 8.1) requires, the space formed by the overlapping orbitals *has a maximum capacity for two electrons that have opposite (paired) spins.* In the simplest case, a molecule of H_2 forms when the 1*s* orbitals of two H atoms overlap, and the electrons, with their spins paired, spend more time in the overlap region (up and down arrows in Figure 11.1A).

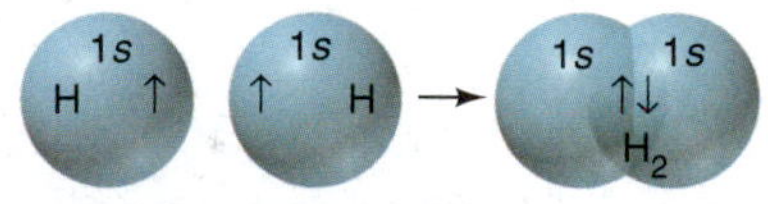

A Hydrogen, H_2

2. *Maximum overlap of bonding orbitals.* Bond strength depends on the attraction between nuclei and shared electrons, so *the greater the orbital overlap, the stronger the bond.* Extent of overlap depends on orbital shape and direction. An *s* orbital is spherical, so its orientation is the same in any direction, but *p* and *d* orbitals have specified directions. Thus, a *p* or *d* orbital involved in a bond is oriented to *maximize overlap.* In HF, for example, the 1*s* orbital of H overlaps a half-filled 2*p* orbital of F *along its long axis* (Figure 11.1B). In F_2, the two half-filled 2*p* orbitals interact end to end, that is, *along the long axes* of the orbitals (Figure 11.1C).

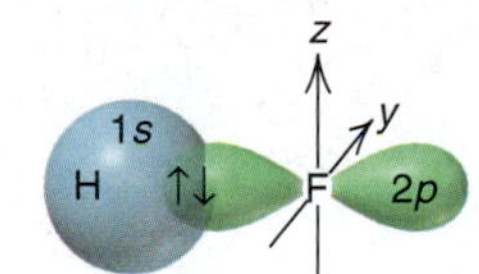

B Hydrogen fluoride, HF

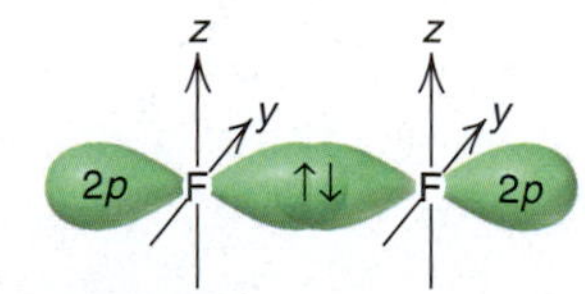

C Fluorine, F_2

Figure 11.1 Orbital overlap and spin pairing in three diatomic molecules. (In **B** and **C**, the $2p_x$ orbital is shown bonding; the other two 2*p* orbitals of F are omitted for clarity.)

3. *Hybridization of atomic orbitals.* To account for the bonding in diatomic molecules like HF or F_2, we picture direct overlap of *s* and/or *p* orbitals of isolated atoms. But how can we account for the shape of a molecule like methane from the shapes and orientations of C and H atomic orbitals? A C atom ([He] $2s^22p^2$) has two valence electrons in the spherical 2*s* orbital and one each in two of the three mutually perpendicular 2*p* orbitals. If the half-filled *p* orbitals overlap the 1*s* orbitals of two H atoms, *two* C—H bonds would form with a 90° H—C—H bond angle. But methane has the formula CH_4, not CH_2, and its bond angles are 109.5°.

To explain such facts, Linus Pauling proposed that, during bonding, *the valence atomic orbitals in the isolated atoms become* ***different*** *when they are in the molecule.* Quantum-mechanical calculations show that if we mathematically "mix" certain combinations of orbitals, we form new ones whose spatial orientations *do* match the observed molecular shapes. The process of orbital mixing is called **hybridization,** and the new atomic orbitals are called **hybrid orbitals.**

4. *Features of hybrid orbitals.* Here are some central points about hybrid orbitals that form during bonding:

- The *number* of hybrid orbitals formed *equals* the number of atomic orbitals mixed.
- The *type* of hybrid orbitals formed *varies* with the types of atomic orbitals mixed.
- The *shape* and *orientation* of a hybrid orbital *maximizes* overlap with the orbital of the other atom in the bond.

You can think of hybridization as a process in which atomic orbitals mix, hybrid orbitals form, they overlap other orbitals, and electrons enter the overlap region with opposing spins, thus forming stable bonds. In truth, hybridization is a concept that helps us explain the molecular shapes we observe.

Types of Hybrid Orbitals

We postulate the type of hybrid orbitals in a molecule *after* we observe its shape. Note that the orientations of the five types of hybrid orbitals we discuss next correspond to the five electron-group arrangements in VSEPR theory.

sp Hybridization When two electron groups surround the central atom, we observe a linear shape, which means the bonding orbitals must have a linear orientation.

1. *Orbitals mixed and orbitals formed.* VB theory explains the linear orientation by proposing that two *nonequivalent* orbitals of a central atom, one *s* and one *p*, mix and form two *equivalent* **sp hybrid orbitals** that are oriented 180° apart (Figure 11.2A). The shape of these hybrid orbitals, with one large and one small lobe, differs markedly from the shapes of the atomic orbitals that were mixed. The orbital orientations increase electron density *in the bonding direction* and minimize repulsions between electrons that occupy them. Thus, *both shape and orientation maximize overlap with the orbital of the other atom in the bond.*

2. *Overlap of orbitals from central and surrounding atoms:* $BeCl_2$. In beryllium chloride, the Be atom is *sp* hybridized. Figure 11.2B depicts the hybridization of Be in a vertical orbital box diagram, and part C shows the diagram with shaded contours. Bond formation with Cl is shown in part D. Two empty unhybridized 2*p* orbitals of Be lie perpendicular to each other and to the *sp* hybrids. The hybrid orbitals overlap the half-filled 3*p* orbital in each of two Cl atoms. (The 3*p* and *sp* hybrid orbitals that are partially colored on the left become fully colored on the right, after each orbital is filled with two electrons.)

sp^2 Hybridization We use this type of hybridization to rationalize the two shapes possible for the trigonal planar electron-group arrangement.

1. *Orbitals mixed and orbitals formed.* Mixing one *s* and two *p* orbitals gives three **sp^2 hybrid orbitals** that point to the corners of an equilateral triangle, their axes 120° apart. (In hybrid orbital notations, unlike in electron configurations, superscripts refer to the number of *atomic orbitals* of a given type, *not* the number of *electrons* in the orbital: thus, one *s* and two *p* orbitals give s^1p^2, or sp^2.) The third 2*p* orbital remains unhybridized.

2. *Overlap of orbitals from central and surrounding atoms:* BF_3. The central B atom in BF_3 is sp^2 hybridized, with the three sp^2 orbitals in a trigonal plane and the third 2*p* orbital unhybridized and perpendicular to this plane (Figure 11.3, p. 332). Each half-filled sp^2 orbital overlaps the half-filled 2*p* orbital of an F atom, and the six valence electrons—three from B and one from each of the three F atoms—form three bonding pairs.

3. *Placement of lone pairs.* To account for other molecular shapes within a given electron-group arrangement, one or more hybrid orbitals contains a lone pair. In ozone (O_3), for example, the central O is sp^2 hybridized and a lone pair fills one of the three sp^2 orbitals.

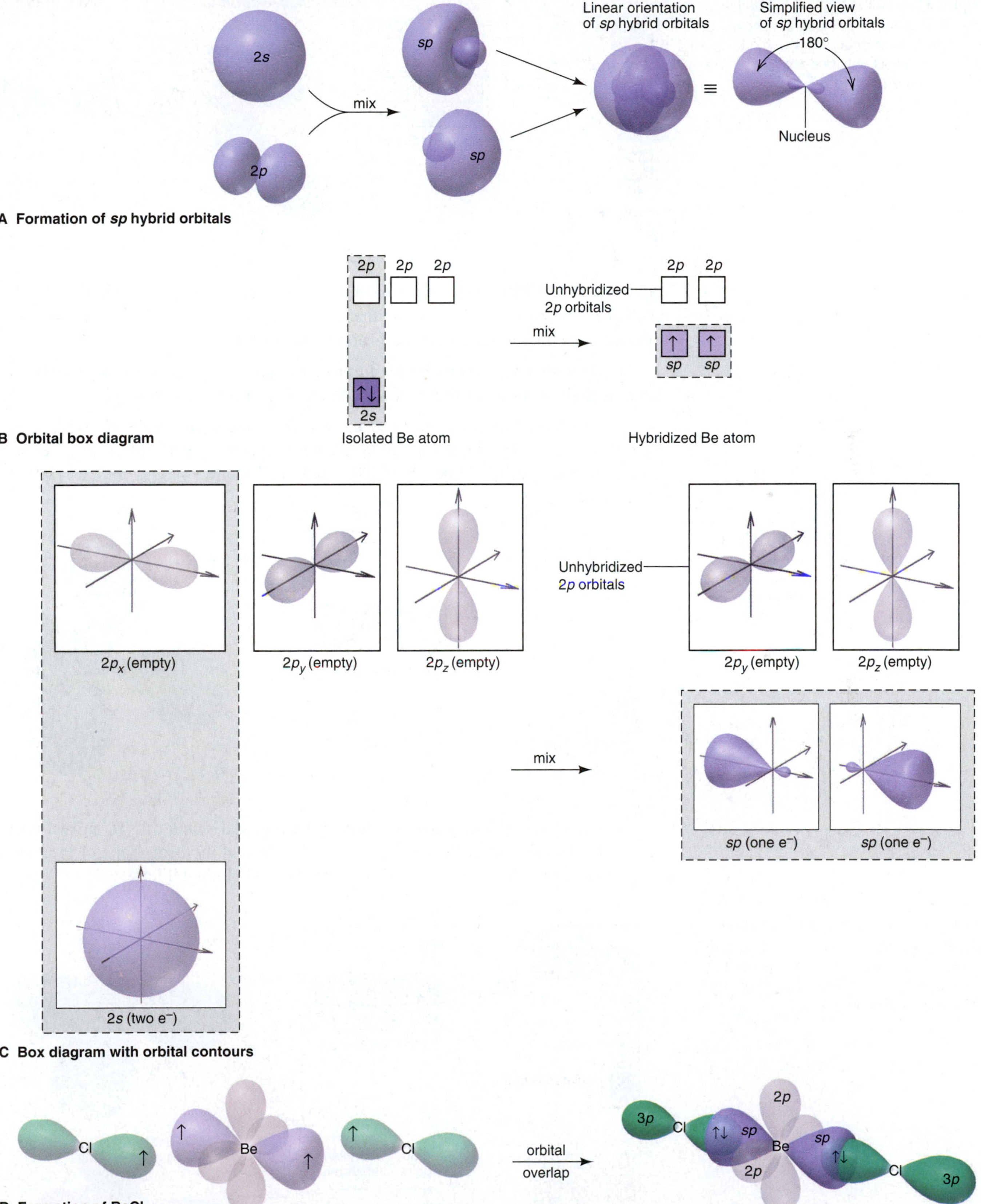

Figure 11.2 Formation and orientation of *sp* hybrid orbitals and the bonding in $BeCl_2$. **A,** One 2*s* and one 2*p* atomic orbital mix to form two *sp* hybrid orbitals. (The simplified hybrid orbitals at far right are used elsewhere, often without the small lobe.) **B,** The orbital box diagram for hybridization of Be, drawn vertically. **C,** The orbital box diagram with orbital contours. **D,** Overlap of Be and Cl orbitals to form $BeCl_2$. (Only the 3*p* orbital of Cl involved in bonding is shown.)

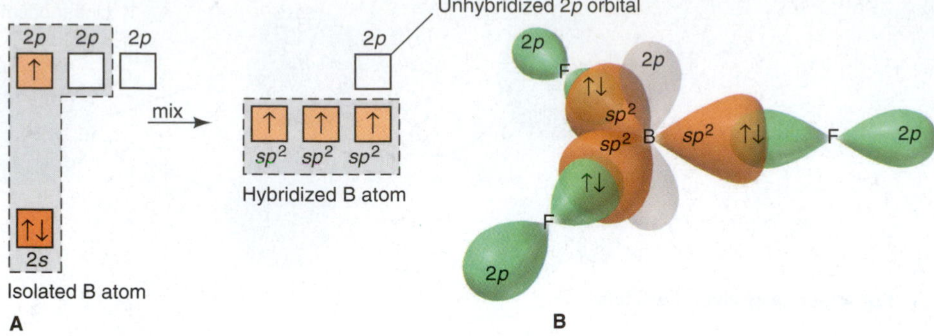

Figure 11.3 The sp^2 hybrid orbitals in BF_3. **A,** The orbital box diagram shows the formation of three sp^2 hybrid orbitals. One $2p$ orbital is unhybridized and empty. **B,** Contour depiction of BF_3.

sp^3 Hybridization Let's return to the question posed earlier about the shape of methane. The type of hybridization that accounts for the shape of CH_4 applies to any species with a tetrahedral electron-group arrangement.

1. *Orbitals mixed and orbitals formed*. Mixing one s and three p orbitals gives four $\boldsymbol{sp^3}$ **hybrid orbitals** that point to the corners of a tetrahedron.

2. *Overlap of orbitals from central and surrounding atoms: CH_4*. The C atom in methane is sp^3 hybridized. Its four valence electrons half-fill the four sp^3 hybrids, which overlap the half-filled $1s$ orbitals of four H atoms to form four C—H bonds (Figure 11.4).

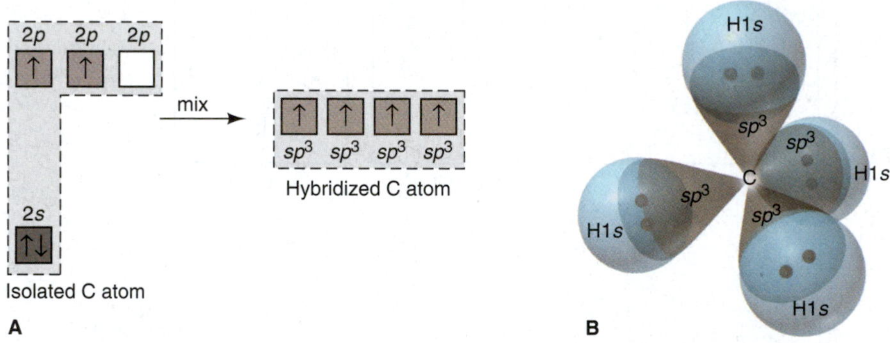

Figure 11.4 The sp^3 hybrid orbitals in CH_4. **A,** The orbital box diagram shows formation of four sp^3 hybrids. **B,** Contour depiction of CH_4, with electron pairs shown as dots.

3. *Placement of lone pairs*. The trigonal pyramidal shape of NH_3 arises when a lone pair fills any one of the four sp^3 orbitals of N, and the bent shape of H_2O arises when lone pairs fill any two of the sp^3 orbitals of O (Figure 11.5).

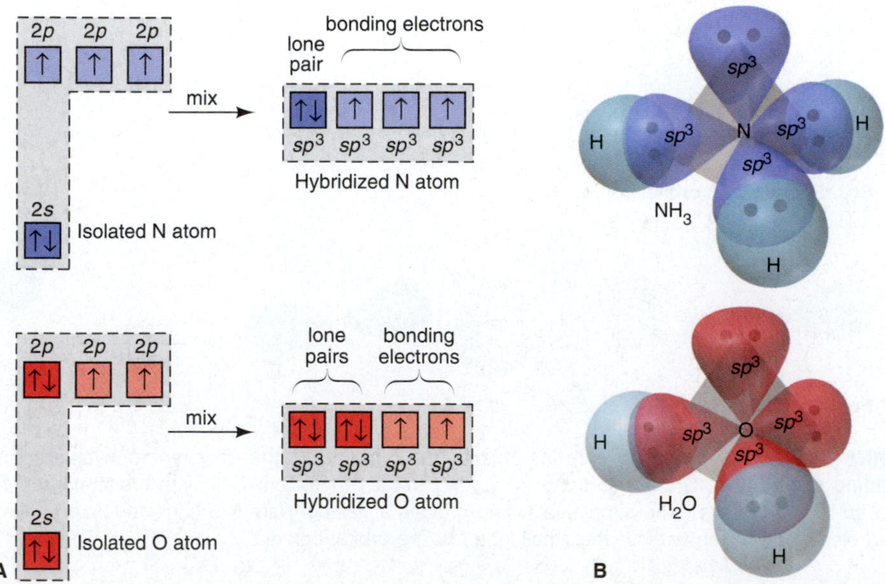

Figure 11.5 The sp^3 hybrid orbitals in NH_3 and H_2O. **A,** The orbital box diagrams show sp^3 hybridization, with lone pairs filling one (NH_3) or two (H_2O) hybrid orbitals. **B,** Contour depictions.

sp^3d Hybridization Molecules with shapes due to the trigonal bipyramidal electron-group arrangement have central atoms from Period 3 or higher because VB theory proposes that *d* orbitals, as well as *s* and *p* orbitals, are mixed to form hybrid orbitals.

1. *Orbitals mixed and orbitals formed.* Mixing one 3*s*, the three 3*p*, and one of the five 3*d* orbitals gives five **sp^3d hybrid orbitals,** which point to the corners of a trigonal bipyramid.

2. *Overlap of orbitals from central and surrounding atoms: PCl_5.* The P atom in PCl_5 is sp^3d hybridized. Each hybrid orbital overlaps a 3*p* orbital of a Cl atom, and ten valence electrons—five from P and one from each of the five Cl atoms—form five P—Cl bonds (Figure 11.6).

3. *Placement of lone pairs.* Seesaw, T-shaped, and linear molecules have lone pairs in, respectively, one, two, or three of the central atom's sp^3d orbitals.

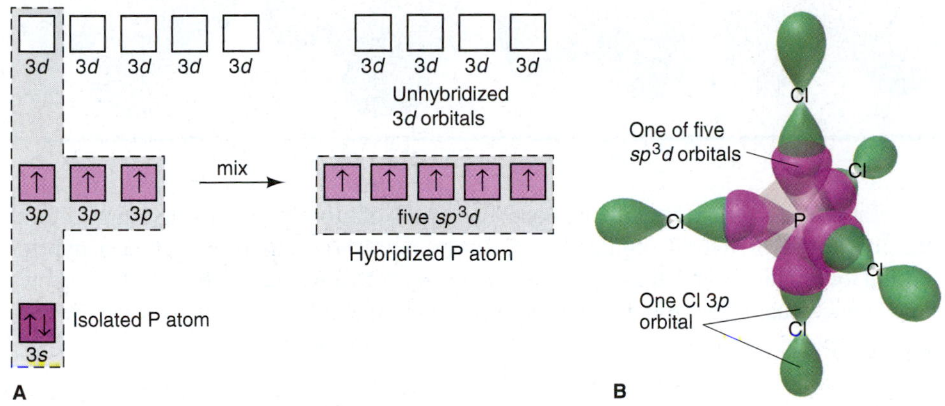

Figure 11.6 The sp^3d hybrid orbitals in PCl_5. **A,** The orbital box diagram shows the formation of five half-filled sp^3d orbitals. Four 3*d* orbitals are unhybridized and empty. **B,** Contour depiction of PCl_5. (For clarity, unhybridized 3*d* orbitals of P, the other two 3*p* orbitals of Cl, and the five bonding P—Cl pairs are not shown.)

sp^3d^2 Hybridization The VB model proposes that molecules with shapes having the octahedral electron-group arrangement also use *d* orbitals to form hybrids.

1. *Orbitals mixed and orbitals formed.* Mixing one 3*s*, the three 3*p*, and two 3*d* orbitals gives six **sp^3d^2 hybrid orbitals,** which point to the corners of an octahedron.

2. *Overlap of orbitals from central and surrounding atoms: SF_6.* The S atom in SF_6 is sp^3d^2 hybridized. Each half-filled hybrid orbital overlaps a half-filled 2*p* orbital of an F atom, and 12 valence electrons—six from S and one from each of the six F atoms—form six S—F bonds (Figure 11.7).

3. *Placement of lone pairs.* Square pyramidal and square planar molecules have lone pairs in one and two of the central atom's sp^3d^2 orbitals, respectively.

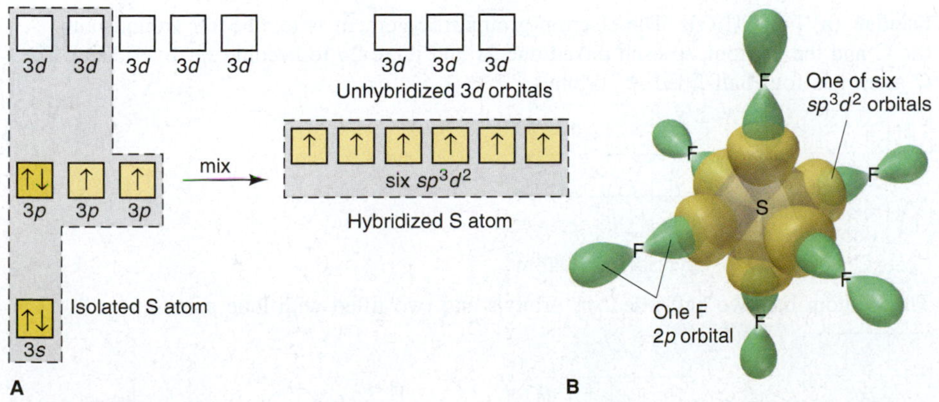

Figure 11.7 The sp^3d^2 hybrid orbitals in SF_6. **A,** The orbital box diagram shows formation of six half-filled sp^3d^2 orbitals; three 3*d* orbitals remain unhybridized and empty. **B,** Contour depiction of SF_6.

Table 11.1 Composition and Orientation of Hybrid Orbitals

	Linear	Trigonal Planar	Tetrahedral	Trigonal Bipyramidal	Octahedral
Atomic orbitals mixed	one *s* one *p*	one *s* two *p*	one *s* three *p*	one *s* three *p* one *d*	one *s* three *p* two *d*
Hybrid orbitals formed	two *p*	three sp^2	four sp^3	five sp^3d	six sp^3d^2
Unhybridized orbitals remaining	two *p*	one *p*	none	four *d*	three *d*
Orientation					

Table 11.1 summarizes the numbers and types of atomic orbitals that mix to form the five types of hybrid orbitals. Once again, note the similarities between hybrid-orbital orientations and the shapes predicted by VSEPR theory. Figure 11.8 shows three conceptual steps from a molecular formula to the hybrid orbitals in the molecule, and Sample Problem 11.1 focuses on the last step of that process.

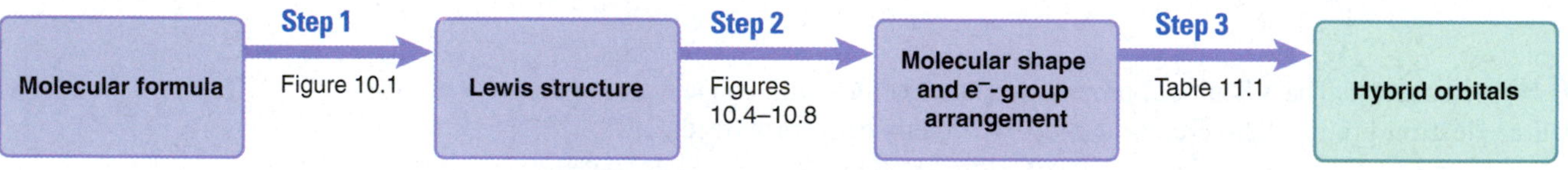

Figure 11.8 From molecular formula to hybrid orbitals. (See Figures 10.1 and 10.4–10.8.)

Sample Problem 11.1 Proposing the Hybrid Orbitals in a Molecule

Problem Use partial orbital diagrams to describe how mixing the atomic orbitals of the central atom(s) leads to the hybrid orbitals in each of the following:
(a) Methanol, CH_3OH **(b)** Sulfur tetrafluoride, SF_4

Plan Prior to the Solution steps below, we use the molecular formula to draw the Lewis structure and determine the electron-group arrangement of each central atom. Then, starting from there, we use Table 11.1 to propose the type of hybrid orbitals. We write the partial orbital diagram for each central atom before and after the orbitals are hybridized.

Solution **(a)** For CH_3OH. The electron-group arrangement is tetrahedral around both the C and the O atom, so each mixed one 2*s* and three 2*p* to become sp^3 hybridized. The C atom has four half-filled sp^3 orbitals:

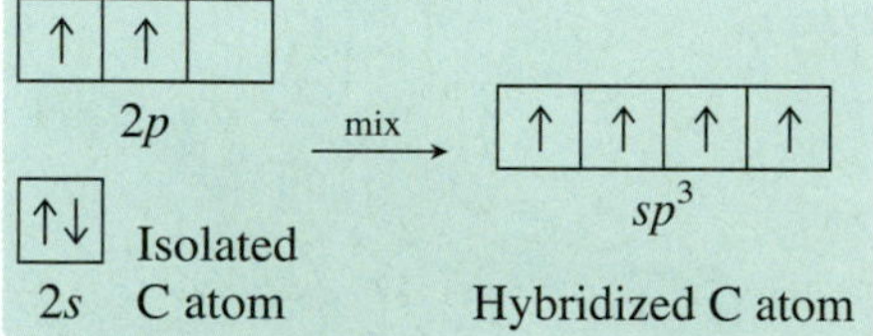

The O atom has two half-filled sp^3 orbitals and two filled with lone pairs:

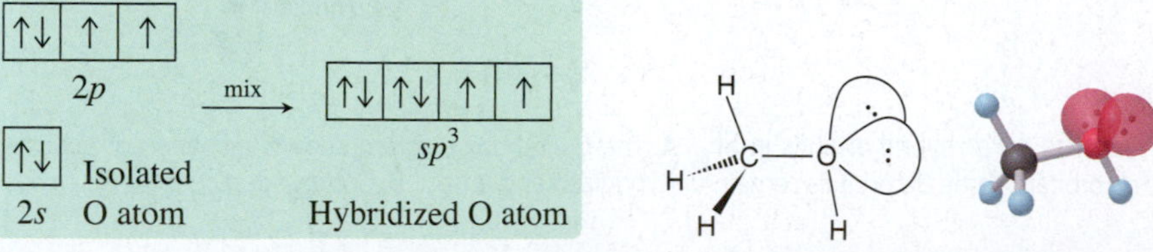

(b) For SF_4. The electron-group arrangement is trigonal bipyramidal, so the central S atom is sp^3d hybridized, which means one 3*s*, three 3*p*, and one 3*d* orbital are mixed. One hybrid orbital is filled with a lone pair, and four are half-filled. Four unhybridized 3*d* orbitals remain empty:

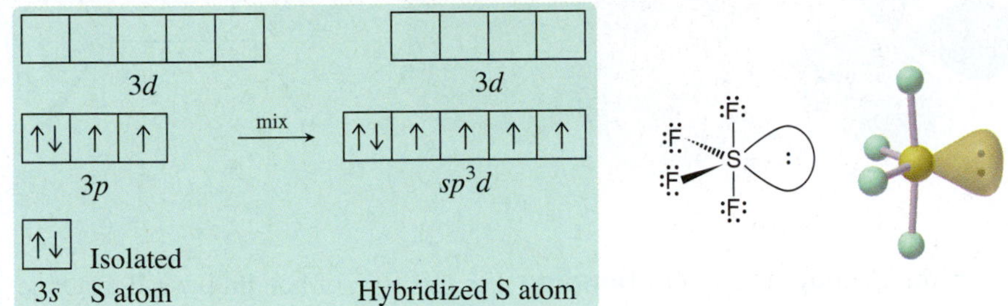

FOLLOW-UP PROBLEM 11.1 Use partial orbital diagrams to propose how the atomic orbitals of the central atom mix to form hybrid orbitals in each of the following:
(a) Beryllium fluoride, BeF_2
(b) Silicon tetrachloride, $SiCl_4$
(c) Xenon tetrafluoride, XeF_4

Summary of Section 11.1

- VB theory explains that a covalent bond forms when two atomic orbitals overlap and two electrons with paired (opposite) spins occupy the overlap region.
- To explain molecular shape, the theory proposes that, during bonding, atomic orbitals mix to form hybrid orbitals with a different shape and direction. This process gives rise to greater overlap and, thus, stronger bonds.
- Based on the observed molecular shape (and the related electron-group arrangement), we postulate the type of hybrid orbital that accounts for the shape.

11.2 • MODES OF ORBITAL OVERLAP AND THE TYPES OF COVALENT BONDS

Orbitals can overlap by two *modes*—end to end or side to side—which gives rise to two types of covalent bonds—*sigma* and *pi*. We'll use VB theory to describe the two types here and, as you'll see, they can be described by molecular orbital theory as well.

Orbital Overlap in Single and Multiple Bonds

Ethane (C_2H_6), ethylene (C_2H_4), and acetylene (C_2H_2) have different shapes. Ethane is tetrahedral at both carbons with bond angles near the ideal 109.5°. Ethylene is trigonal planar at both carbons with bond angles near the ideal 120°. Acetylene is linear with bond angles of 180°:

~109.5° ethane ~120° ethylene 180° H—C≡C—H acetylene

In these molecules, two modes of orbital overlap result in two types of bonds.

End-to-End Overlap and Sigma (σ) Bonding Both C atoms of ethane are sp^3 hybridized (Figure 11.9, *next page*). The C—C bond arises from overlap of the end of one sp^3 orbital with the end of the other. *End-to-end* overlap forms a **sigma (σ) bond,** which has its *highest electron density along the bond axis* and is shaped like an ellipse rotated about its long axis (like a football). *All single bonds are σ bonds,* including the six C—H bonds in ethane.

Figure 11.9 The σ bonds in ethane (C_2H_6). **A,** Depiction using atomic contours. **B,** An electron density model shows very slightly positive *(blue)* and negative *(red)* regions. **C,** Wedge-bond perspective drawing.

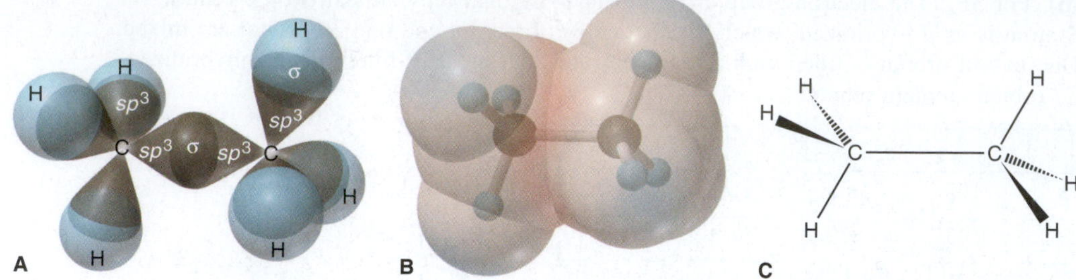

Side-to-Side Overlap and Pi (π) Bonding For this type of bonding, we'll examine ethylene and acetylene:

1. *In ethylene,* each C atom is sp^2 hybridized. The four valence electrons of C half-fill the three sp^2 orbitals *and* the unhybridized 2*p* orbital, which lies perpendicular to the sp^2 plane (Figure 11.10). Two sp^2 orbitals of each C form C—H σ bonds. The third sp^2 orbital forms a σ bond with the other C. With the σ-bonded C atoms near each other, their half-filled 2*p* orbitals overlap *side to side,* which forms a **pi (π) bond.** It has *two regions (lobes) of electron density,* one above and one below the σ-bond axis. *The two electrons in one π bond occupy both lobes. Any double bond consists of one σ bond and one π bond,* which increases electron density between the nuclei (Figure 11.10D). The two electron pairs act as one electron group because each pair occupies a different orbital, which reduces repulsions.

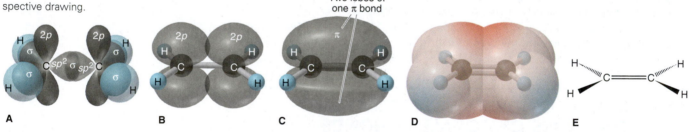

Figure 11.10 The σ and π bonds in ethylene (C_2H_4). **A,** The C—C σ bond and the four C—H σ bonds are shown with the unhybridized 2*p* orbitals. **B,** An accurate depiction of the 2*p* orbitals shows the side-to-side overlap; σ bonds are shown in ball-and-stick form. **C,** Two overlapping regions comprise *one* π bond, which is occupied by two electrons. **D,** With four electrons (one σ bond and one π bond) between the C atoms, electron density *(red)* is higher there. **E,** Wedge-bond perspective drawing.

2. *In acetylene,* each C atom is *sp* hybridized, and its four valence electrons half-fill the two *sp* hybrids *and* the two unhybridized 2*p* orbitals (Figure 11.11). Each C forms a C—H σ bond with one *sp* orbital and a C—C σ bond with the other. Side-to-side overlap of one pair of 2*p* orbitals gives one π bond, with electron density above and below the σ bond. Side-to-side overlap of the other pair of 2*p* orbitals gives another π bond, 90° away from the first, with electron density in front and back of the σ bond. The result is a *cylindrically symmetrical* H—C≡C—H molecule. Note the greater electron density between the C atoms created by the six bonding electrons. *Any triple bond consists of one σ and two π bonds.*

Figure 11.11 The σ bonds and π bonds in acetylene (C_2H_2). **A,** A contour depiction shows the C—C σ bond, the two C—H σ bonds, and two unhybridized 2*p* orbitals (one pair is in *red* for clarity). **B,** The 2*p* orbitals (accurate form) overlap side to side; σ bonds are shown by a ball-and-stick model. **C,** Overlapping regions of two π bonds, one black and the other red. **D,** The molecule has cylindrical symmetry. Six electrons (one σ bond and two π bonds) create even higher electron density *(red)* between the C atoms. **E,** Bond-line drawing.

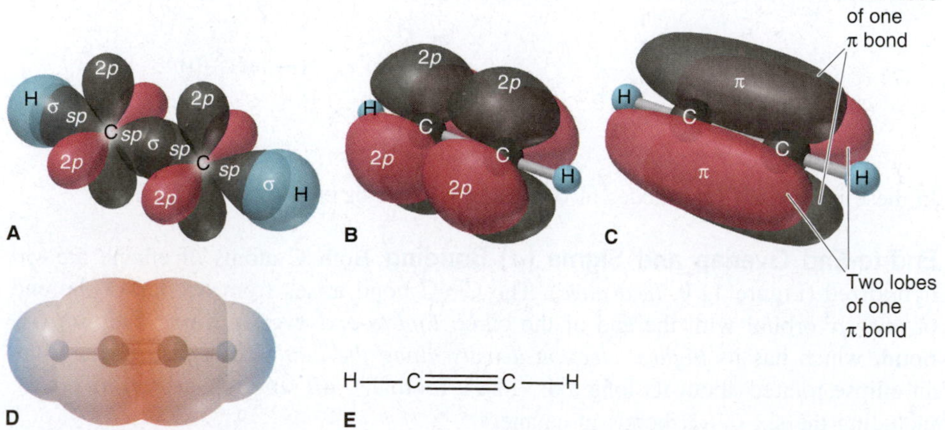

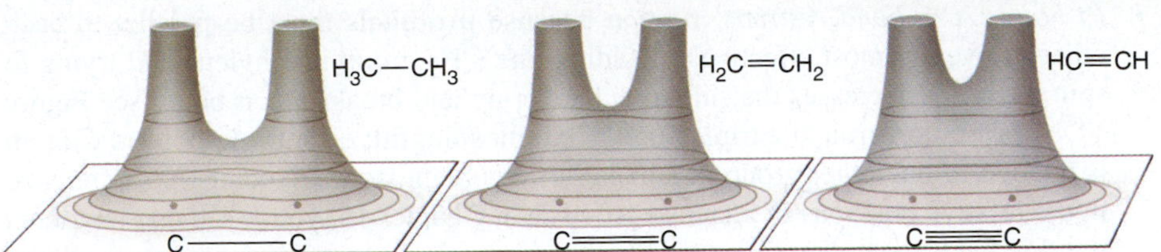

Figure 11.12 Electron density and bond order in ethane, ethylene, and acetylene.

Mode of Overlap, Bond Strength, and Bond Order Because orbitals overlap less side to side than end to end, a π bond is weaker than a σ bond; thus, for carbon-carbon bonds, a double bond is less than twice as strong as a single bond (Table 9.2). Figure 11.12 shows electron density relief maps of the three types of carbon-carbon bonds; note the increasing electron density between the nuclei from single to double to triple bond.

Lone-pair repulsions, bond polarities, and other factors affect overlap between other pairs of atoms. Nevertheless, as a rough approximation, in terms of bond order (BO), a double bond (BO = 2) is about twice as strong as a single bond (BO = 1), and a triple bond (BO = 3) is about three times as strong.

Sample Problem 11.2 Describing the Types of Bonds in Molecules

Problem Describe the types of bonds and orbitals in acetone, $(CH_3)_2CO$.

Plan We use the shape around each central atom to propose the hybrid orbitals used and use unhybridized orbitals to form the C=O bond.

Solution The shapes are tetrahedral around each C of the two CH_3 (methyl) groups and trigonal planar around the middle C (see Sample Problem 10.8). Thus, the middle C has three sp^2 orbitals and one unhybridized p orbital. Each of the two methyl C atoms has four sp^3 orbitals. Three of these form σ bonds with the 1s orbitals of H atoms; the fourth forms a σ bond with an sp^2 orbital of the middle C. Thus, two of the three sp^2 orbitals of the middle C form σ bonds to the other two C atoms.

The O atom is also sp^2 hybridized and has an unhybridized p orbital that can form a π bond. Two of the O atom's sp^2 orbitals hold lone pairs, and the third forms a σ bond with the third sp^2 orbital of the middle C atom. The unhybridized, half-filled 2p orbitals of C and O form a π bond. The σ and π bonds constitute the C=O bond:

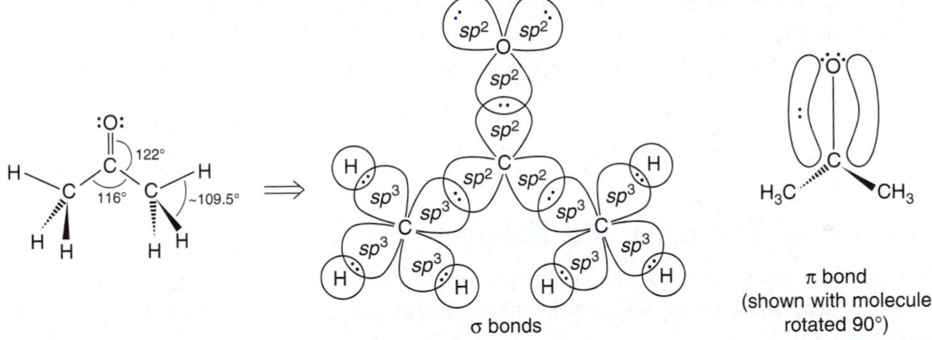

Comment Why is the O atom in acetone shown hybridized? After all, it could use two perpendicular p orbitals for the σ and π bonds with C and leave the other p and the s orbital to hold the two lone pairs. However, having each lone pair in an sp^2 orbital oriented away from the C=O bond lowers electron-electron repulsions.

FOLLOW-UP PROBLEM 11.2 Describe the types of bonds and orbitals in **(a)** hydrogen cyanide, HCN, and **(b)** carbon dioxide, CO_2.

Orbital Overlap and Molecular Rotation

The type of overlap—end-to-end or side-to-side—affects rotation around the bond:

- *Sigma bond.* A σ *bond allows free rotation* because the extent of overlap is not affected. If you could hold one CH_3 group of ethane, the other CH_3 could spin without affecting the overlap of the C—C σ bond (see Figure 11.9).

- *Pi bond.* A π *bond restricts rotation* because p orbitals must be parallel to each other to overlap most effectively. Holding one CH_2 group in ethylene and trying to spin the other decreases the side-to-side overlap and breaks the π bond (see Figure 11.10). Rotation around a triple bond is not meaningful: each triple-bonded C atom is bonded to one other group in a linear arrangement, so there can be no difference in the relative positions of attached groups. (In Chapter 15, you'll see that restricted rotation leads to another type of isomerism.)

Summary of Section 11.2

- End-to-end overlap of atomic orbitals forms a σ bond, which allows free rotation of the bonded parts of the molecule.
- Side-to-side overlap forms a π bond, which restricts rotation.
- A multiple bond consists of a σ bond and either one π bond (double bond) or two (triple bond). Multiple bonds have greater electron density between the nuclei than single bonds do and, thus, higher bond energies.

11.3 • MOLECULAR ORBITAL (MO) THEORY AND ELECTRON DELOCALIZATION

Scientists choose the model that best answers a question: VSEPR for one about molecular shape and VB theory for one about orbital overlap. But neither adequately explains magnetic and spectral properties, and both understate the importance of electron delocalization. To deal with phenomena like these, which involve molecular energy levels, chemists choose **molecular orbital (MO) theory.** The MO model is a quantum-mechanical treatment for molecules similar to the one for atoms (Chapter 8): just as an atom has atomic orbitals (AOs) of given energies and shapes that are occupied by the atom's electrons, a molecule has **molecular orbitals (MOs)** of given energies and shapes that are occupied by the molecule's electrons.

There is a key distinction between VB and MO theories in how they picture a molecule:

- VB theory pictures a molecule as a group of atoms bonded through *localized* overlapping of valence-shell atomic and/or hybrid orbitals occupied by electrons.
- MO theory pictures a molecule as a collection of nuclei with orbitals *delocalized* over the whole molecule and occupied by electrons.

Despite its usefulness, MO theory has a drawback: MOs are more difficult to visualize than the shapes of VSEPR theory or the hybrid orbitals of VB theory.

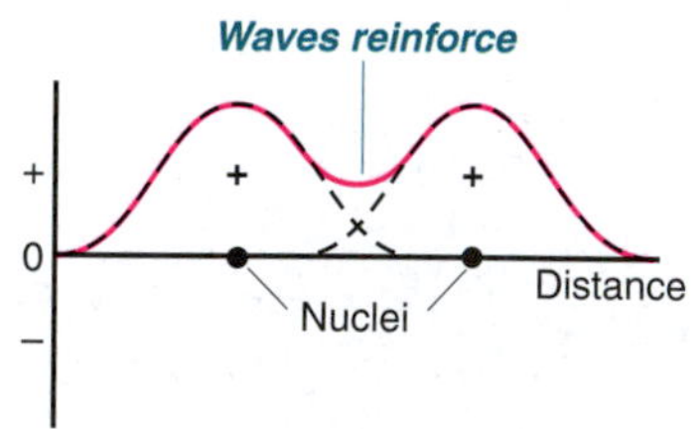

A Amplitudes of wave functions added

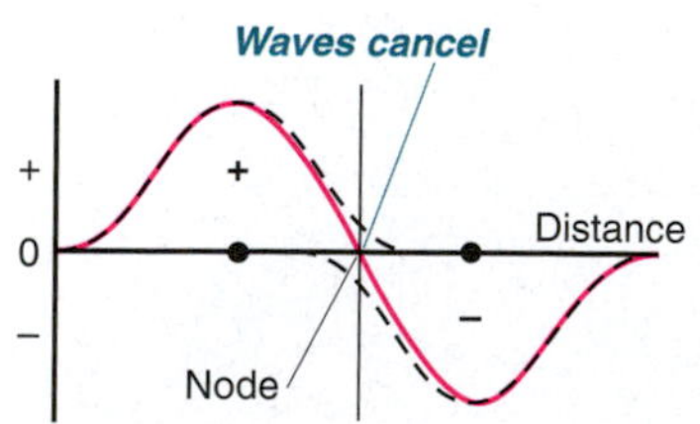

B Amplitudes of wave functions subtracted

Figure 11.13 An analogy between light waves and atomic wave functions.

The Central Themes of MO Theory

Several key ideas of MO theory appear in its description of H_2 and other simple species: how MOs form, what their energies and shapes are, and how they fill with electrons.

Formation of Molecular Orbitals Just as we need approximations to solve the Schrödinger equation for any atom with more than one electron, we need an approximation to determine the MOs of even the simplest molecule, H_2. The approximation is to mathematically *combine* (add or subtract) AOs (atomic wave functions) of nearby atoms to form MOs (molecular wave functions). Thus, when two H nuclei lie near each other, their AOs overlap and combine in two ways:

- *Adding the wave functions together.* This combination forms a **bonding MO,** which has *a region of high electron density between the nuclei.* Additive overlap is analogous to light waves reinforcing each other, which makes the amplitude higher and the light brighter. For electron waves, the overlap *increases* the probability that the electrons are between the nuclei (Figure 11.13A).

- *Subtracting the wave functions from each other.* This combination forms an **antibonding MO,** which has *a node, a region of zero electron density, between the nuclei* (Figure 11.13B). Subtractive overlap is analogous to light waves canceling each other, causing the light to disappear. With electron waves, the probability that the electrons lie between the nuclei *decreases* to zero.

The two possible combinations for hydrogen atoms H_A and H_B are

AO of H_A + AO of H_B = bonding MO of H_2 (more e^- density between nuclei)

AO of H_A − AO of H_B = antibonding MO of H_2 (less e^- density between nuclei)

Notice that *the number of AOs combined always equals the number of MOs formed:* two H atomic orbitals combine to form two H_2 molecular orbitals.

Shape and Energy of H_2 Molecular Orbitals Bonding and antibonding MOs have different shapes and energies. Figure 11.14 shows these orbitals for H_2:

- *Bonding MO.* A bonding MO is *lower in energy* than the AOs that form it. Because it is spread mostly *between* the nuclei, nuclear repulsions decrease while nucleus-electron attractions increase. Moreover, two electrons in this MO can delocalize their charges over a larger volume than in nearby, separate AOs, which lowers electron repulsions. Because of these electrostatic effects, when electrons occupy this orbital, the H_2 molecule is *more stable* than the separate H atoms.
- *Antibonding MO.* An antibonding MO is *higher in energy* than the AOs that form it. With most of the electron density *outside* the internuclear region and a *node between the nuclei,* nuclear repulsions increase. Therefore, when electrons occupy this orbital, the H_2 molecule is *less stable* than the separate H atoms.

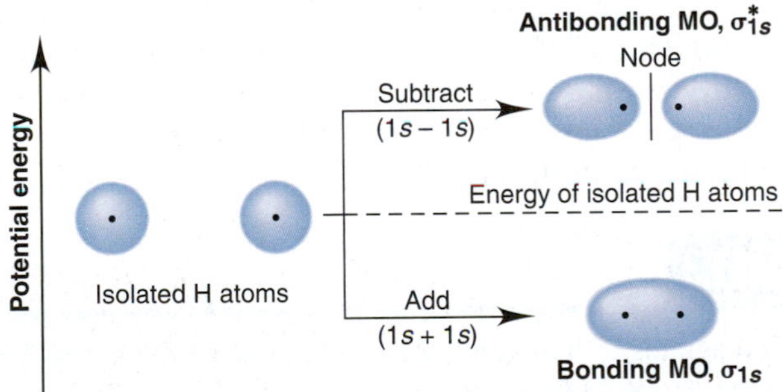

Figure 11.14 Contours and energies of H_2 bonding and antibonding MOs.

Both the bonding and antibonding MOs of H_2 are **sigma (σ) MOs** because they are cylindrically symmetrical about an imaginary line between the nuclei. The bonding MO is denoted by σ_{1s}; that is, a σ MO derived from 1*s* AOs. Antibonding orbitals are denoted with a superscript star: the antibonding MO derived from 1*s* AOs is σ^*_{1s} (spoken "sigma, one ess, star").

For AOs to interact enough to form MOs, they must be similar in *energy* and *orientation.* The 1*s* orbitals of two H atoms have identical energy and orientation, so they interact strongly.

Electrons in Molecular Orbitals Several aspects of MO theory—filling of MOs, energy-level diagrams, electron configuration, and bond order—relate to earlier ideas:

1. *Filling MOs with electrons.* Electrons enter MOs just as they do AOs:

- MOs are filled in order of increasing energy (aufbau principle).
- An MO can hold a maximum of two electrons with opposite spins (exclusion principle).
- Orbitals of equal energy are half-filled, with spins parallel, before any of them are filled (Hund's rule).

2. *MO energy-level diagrams.* A **molecular orbital (MO) diagram** shows the relative energy and number of electrons for each MO, as well as for the AOs from which they formed. In the MO diagram for H_2 (Figure 11.15), two electrons, one from the AO of each H, fill the H_2 bonding MO, while the antibonding MO remains empty.

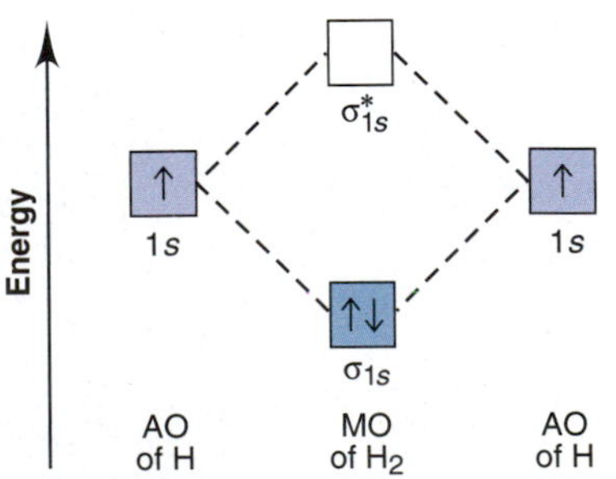

Figure 11.15 MO diagram for H_2. The vertical placement of the boxes indicates relative energies. Orbital occupancy is shown with arrows and color (dark = full, pale = half-filled, none = empty).

3. *Electron configuration.* Just as we write an electron configuration for an atom, we can write one for a molecule. The symbol of each occupied MO is written in parentheses, with the number of electrons in it as a superscript outside: the electron configuration of H_2 is $(\sigma_{1s})^2$.

4. *Bond order.* In a Lewis structure, bond order is the number of electron pairs per atom-to-atom linkage. The **MO bond order** is the number of electrons in bonding MOs minus the number in antibonding MOs, multiplied by $\frac{1}{2}$:

$$\text{Bond order} = \tfrac{1}{2}[(\text{no. of e}^- \text{ in bonding MO}) - (\text{no. of e}^- \text{ in antibonding MO})] \quad \textbf{(11.1)}$$

Here are three key points about MO bond order:

- Bond order > 0: the molecule is more stable than the separate atoms, so it *will* form. For H_2, the bond order is $\frac{1}{2}(2 - 0) = 1$.
- Bond order $= 0$: the molecule is as stable as the separate atoms, so it will *not* form (occurs when equal numbers of electrons occupy bonding and antibonding MOs).
- In general, the *higher* the bond order, the *stronger* the bond is.

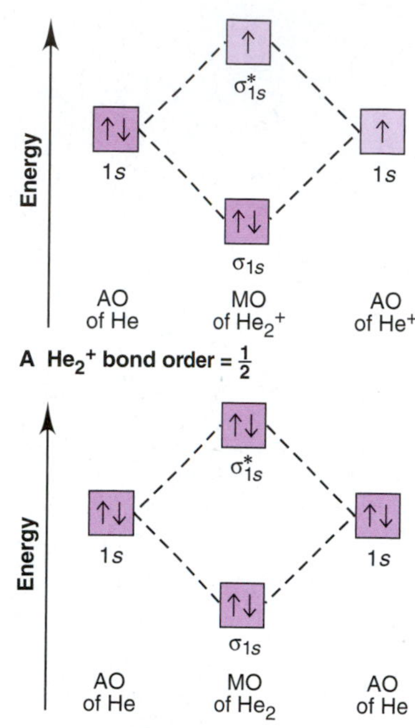

Figure 11.16 MO diagrams for He_2^+ and He_2.

Do He_2^+ and He_2 Exist? One of the early triumphs of MO theory was its ability to *predict* the existence of He_2^+, the helium molecule-ion, which consists of two He nuclei and three electrons. Let's use MO diagrams to see why He_2^+ exists but He_2 doesn't:

- In He_2^+, the 1*s* AOs form MOs (Figure 11.16A). The three electrons are distributed as a pair in the σ_{1s} MO and a lone electron in the σ^*_{1s} MO. The bond order is $\frac{1}{2}(2 - 1) = \frac{1}{2}$. Thus, He_2^+ has a relatively weak bond, but it should exist. Indeed, this species has been observed frequently when He atoms collide with He^+ ions. Its electron configuration is $(\sigma_{1s})^2(\sigma^*_{1s})^1$.
- In He_2, with four electrons in the σ_{1s} and σ^*_{1s} MOs, both the bonding and antibonding orbitals are filled (Figure 11.16B). Stabilization from the electron pair in the bonding MO is canceled by destabilization from the electron pair in the antibonding MO. With a zero bond order $[\frac{1}{2}(2 - 2) = 0]$, we predict, and experiment has so far confirmed, that a covalent He_2 molecule does not exist.

Sample Problem 11.3 Predicting Stability of Species Using MO Diagrams

Problem Use MO diagrams to find bond orders and predict whether H_2^+ and H_2^- exist. If either exists, write its electron configuration.

Plan Since the 1*s* AOs form the MOs, the MO diagrams are similar to the one for H_2. We find the number of electrons in each species and distribute them one at a time to the MOs in order of increasing energy. We obtain the bond order with Equation 11.1 and write the electron configuration as described in the text.

Solution For H_2^+. H_2 has two e^-, so H_2^+ has only one, which enters the bonding MO *(see diagram, below left)*. The bond order is $\frac{1}{2}(1 - 0) = \frac{1}{2}$, so we predict that H_2^+ exists. The electron configuration is $(\sigma_{1s})^1$.

For H_2^-. H_2 has two e^-, so H_2^- has three. We place two paired e^- in the bonding MO and one unpaired e^- in the antibonding MO *(see diagram, below right)*. The bond order is $\frac{1}{2}(2 - 1) = \frac{1}{2}$, so we predict that H_2^- exists. The electron configuration is $(\sigma_{1s})^2(\sigma^*_{1s})^1$.

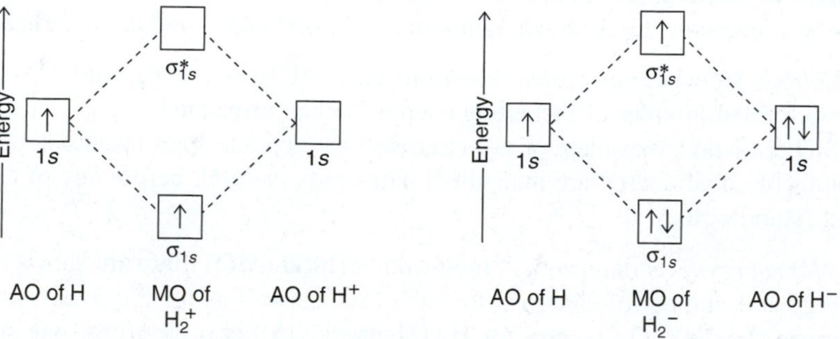

Check The number of electrons in the MOs equals the number of electrons in the AOs.

Comment Both species have been detected spectroscopically, H_2^+ in the material around stars and H_2^- in the laboratory.

FOLLOW-UP PROBLEM 11.3 Use an MO diagram to find the bond order and predict whether two H^- ions could form H_2^{2-}. If H_2^{2-} does exist, write its electron configuration.

Homonuclear Diatomic Molecules of Period 2 Elements

Homonuclear diatomic molecules are composed of two identical atoms. In addition to H_2 from Period 1, you're familiar with N_2, O_2, and F_2 from Period 2 as the elemental forms under standard conditions. Others in Period 2—Li_2, Be_2, B_2, C_2, and Ne_2—are observed, if at all, at high temperatures. We'll divide them into molecules from the *s* block, Groups 1A(1) and 2A(2), and the *p* block, Groups 3A(13) through 8A(18).

Bonding in the *s*-Block Homonuclear Diatomic Molecules Both Li and Be occur as metals under normal conditions, but MO theory can examine their stability as diatomic gases, dilithium (Li_2) and diberyllium (Be_2).

These atoms have electrons in inner (1*s*) and outer (2*s*) AOs, but we ignore the inner ones because, in general, *only outer (valence) AOs interact enough to form MOs.* Like 1*s* AOs, these 2*s* AOs form σ MOs, cylindrically symmetrical around the internuclear axis.

- In Li_2, the two valence electrons fill the bonding (σ_{2s}) MO, with opposing spins, leaving the antibonding (σ^*_{2s}) MO empty (Figure 11.17A). The bond order is $\frac{1}{2}(2 - 0) = 1$. In fact, Li_2 *has* been observed; the electron configuration is $(\sigma_{2s})^2$.
- In Be_2, the four valence electrons fill the σ_{2s} and σ^*_{2s} MOs (Figure 11.17B), giving an orbital occupancy similar to that in He_2. The bond order is $\frac{1}{2}(2 - 2) = 0$, and the ground state of Be_2 has never been observed.

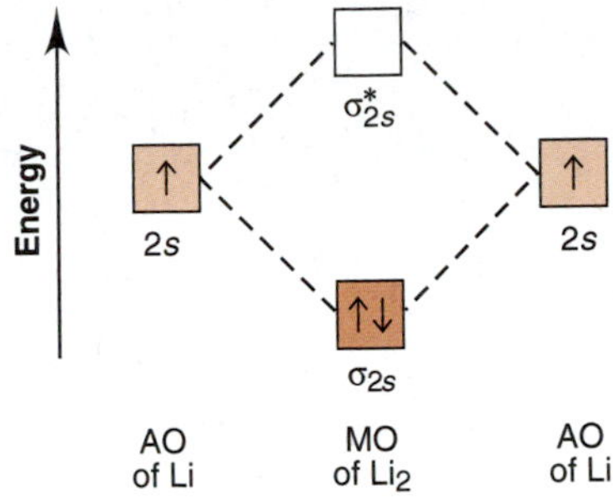

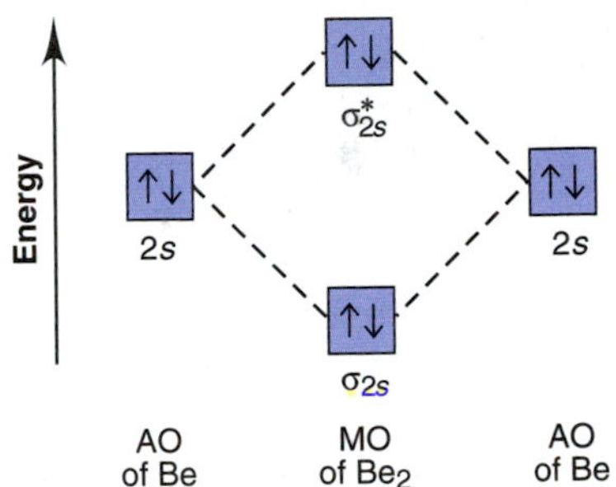

Figure 11.17 **Bonding in *s*-block homonuclear diatomic molecules.** Only outer (valence) AOs interact enough to form MOs.

Shape and Energy of MOs from Atomic *p*-Orbital Combinations In boron, atomic 2*p* orbitals are occupied. Recall that *p* orbitals can overlap by two modes, which correspond to two ways their wave functions combine (Figure 11.18):

- End-to-end combination gives a pair of σ MOs, the σ_{2p} and σ^*_{2p}.
- Side-to-side combination gives a pair of **pi (π) MOs,** π_{2p} and π^*_{2p}.

Despite the different shapes, MOs derived from *p* orbitals are like those from *s* orbitals. Bonding MOs have most of the electron density *between* the nuclei, and antibonding MOs have most *outside* the internuclear region, with a node between the nuclei.

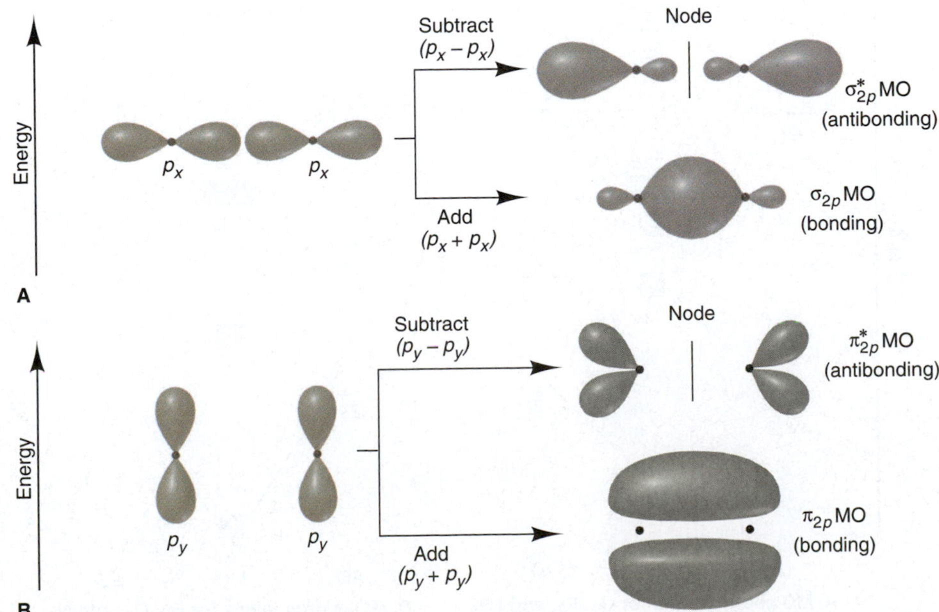

Figure 11.18 **Shapes and energies of σ and π MOs from combinations of 2*p* atomic orbitals.** **A,** The *p* orbitals lying along the internuclear axis (designated p_x) overlap end to end and form σ_{2p} and σ^*_{2p} MOs. **B,** The *p* orbitals perpendicular to the internuclear axis overlap side to side and form two π MOs. (The p_y interactions, shown here, are the same as for the p_z orbitals, for a total of four π MOs.)

The order of energy levels for MOs, whether bonding or antibonding, is based on the order of AO energy levels *and* on the mode of the *p*-orbital overlap:

- MOs formed from 2*s* orbitals are *lower in energy* than MOs formed from 2*p* orbitals because 2*s* AOs are lower in energy than 2*p* AOs.
- Bonding MOs are *lower in energy* than antibonding MOs: σ_{2p} is lower in energy than σ^*_{2p} and π_{2p} is lower than π^*_{2p}.
- Atomic *p* orbitals overlap more extensively end to end than side to side. Thus, the σ_{2p} MO is usually lower in energy than the π_{2p} MO. We also find that the destabilizing effect of the σ^*_{2p} MO is greater than that of the π^*_{2p} MO.

Thus, the energy order for MOs derived from 2*p* orbitals is typically

$$\sigma_{2p} < \pi_{2p} < \pi^*_{2p} < \sigma^*_{2p}$$

Each atom has three mutually perpendicular 2*p* orbitals. When the six *p* orbitals in two atoms combine, the two that interact end to end form one σ and one σ^* MO, and the two pairs of orbitals that interact side to side form two π MOs and two π^* MOs. Placing these orientations within the energy order gives the *expected* MO diagram for the *p*-block Period 2 homonuclear diatomic molecules (Figure 11.19A).

Recall that only AOs of similar energy interact enough to form MOs. This fact leads to two energy orders:

1. ***Without** s and p orbital mixing: O_2, F_2, and Ne_2.* The order in Figure 11.19A assumes that *s* and *p* AOs are so different in energy that they do not interact; we say the orbitals do not *mix*. Lying at the right of Period 2, O, F, and Ne are relatively small. Thus, as electrons start to pair up in the half-filled 2*p* orbitals, strong repulsions raise the energy of the 2*p* orbitals high enough above the 2*s* orbitals to prevent orbital mixing.

2. ***With** s and p orbital mixing: B_2, C_2, and N_2.* B, C, and N atoms are relatively large, with 2*p* AOs only half-filled, so repulsions are weaker. As a result, orbital energies are close enough for some mixing to occur between the 2*s* of one atom and the end-on 2*p* of the other. The effect is to *lower* the energy of the σ_{2s} and σ^*_{2s} MOs and *raise* the energy of the σ_{2p} and σ^*_{2p} MOs; the π MOs are not affected. The MO diagram for B_2, C_2, and N_2 reflects this mixing (Figure 11.19B). The only qualitative difference from the MO diagram for O_2, F_2, and Ne_2 is the *reverse in energy order* of the σ_{2p} and π_{2p} MOs.

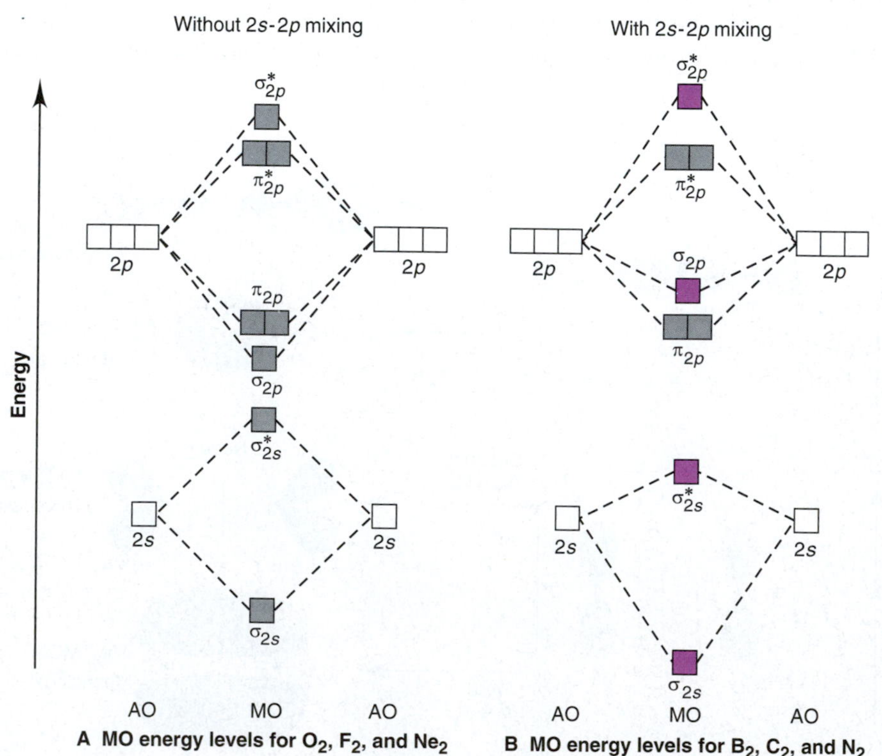

Figure 11.19 Relative MO energy levels for Period 2 homonuclear diatomic molecules. (For clarity, the MOs affected by mixing are shown in *purple*.)

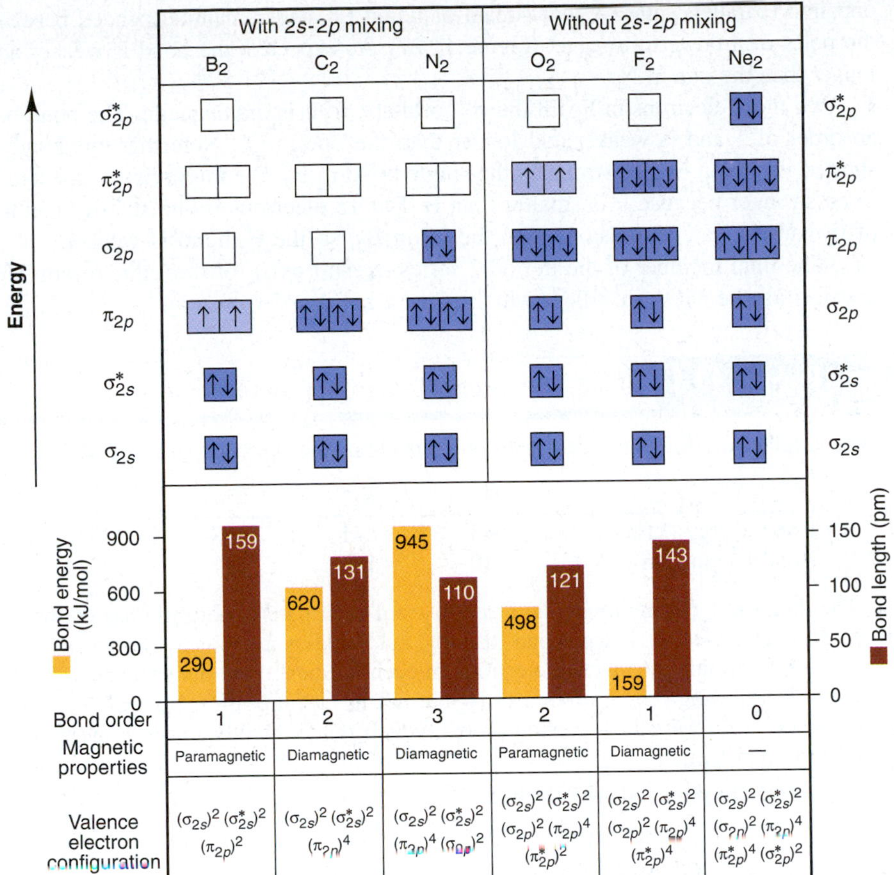

Figure 11.20 MO occupancy and some properties of B_2 through Ne_2. Energy order and occupancy of MOs are above bar graphs of bond energy and length, bond order, magnetic properties, and outer (valence) electron configuration.

Bonding in the *p*-Block Homonuclear Diatomic Molecules Figure 11.20 shows the MOs, electron occupancy, and some properties of B_2 through Ne_2. Note that

1. *Higher* bond order correlates with *greater* bond energy and *shorter* bond length.

2. Orbital occupancy correlates with magnetic properties. Recall from Chapter 8 that if a substance has unpaired electrons, it is *paramagnetic,* and if all electrons are paired, the substance is *diamagnetic;* the same distinction applies to molecules. Let's examine the MO occupancy and some properties of these molecules:

- *B_2*. The B_2 molecule has six outer electrons: four fill the σ_{2s} and σ^*_{2s} MOs. The remaining two electrons occupy the two π_{2p} MOs, one in each orbital, in keeping with Hund's rule. With four electrons in bonding MOs and two in antibonding MOs, the bond order of B_2 is $\frac{1}{2}(4 - 2) = 1$. As expected from the two lone electrons, B_2 is paramagnetic.
- *C_2*. Two additional electrons in C_2 fill the two π_{2p} MOs. With two more bonding electrons than B_2, the bond order of C_2 is 2 and the bond is stronger and shorter. But with all the electrons paired, C_2 is diamagnetic.
- *N_2*. Two more electrons in N_2 fill the σ_{2p} MO, so the molecule is also diamagnetic. The bond order of 3 is consistent with the triple bond in the Lewis structure and with a stronger, shorter bond.
- *O_2*. With this molecule, we see the power of MO theory over VB theory and others based on electrons in localized orbitals. It is impossible to write one Lewis structure consistent with the fact that O_2 is double bonded and paramagnetic. We can write one with a double bond and paired electrons, and another with a single bond and two unpaired electrons:

  Ö=Ö or :Ö—Ö:

  MO theory resolves this paradox beautifully: with eight electrons in bonding MOs and four in antibonding MOs, the bond order is $\frac{1}{2}(8 - 4) = 2$, and it is paramagnetic because *one* electron occupies each of *two* π^*_{2p} MOs and these two electrons have

Figure 11.21 The paramagnetic properties of O_2.

unpaired (parallel) spins. A thin stream of liquid O_2 will remain suspended between the poles of a powerful magnet (Figure 11.21). As expected, the bond is weaker and longer than the one in N_2.

- *F_2*. Two more electrons in F_2 fill the π^*_{2p} orbitals, so it is diamagnetic. The bond has an order of 1 and is weaker and longer than the one in O_2. Note that this bond is shorter, yet about half as strong as the single bond in B_2. We might have expected it to be stronger because F is smaller than B. But 18 electrons in the smaller volume of F_2 cause greater repulsions than the 10 in B_2, so the F_2 bond is weaker.
- *Ne_2*. The final member of the Period 2 series doesn't exist for the same reason He_2 doesn't: all the MOs are filled, which gives a bond order of zero.

Sample Problem 11.4 Using MO Theory to Explain Bond Properties

Problem Explain the following data with diagrams showing the occupancy of MOs:

	N_2	N_2^+	O_2	O_2^+
Bond energy (kJ/mol)	945	841	498	623
Bond length (pm)	110	112	121	112

Plan The data show that removing an electron from each parent molecule has opposite effects: N_2^+ has a weaker, longer bond than N_2, but O_2^+ has a stronger, shorter bond than O_2. We determine the valence electrons in each species, draw the sequence of MO energy levels (showing orbital mixing in N_2 but not in O_2), and fill them with electrons. To explain the data, we calculate bond orders, which relate directly to bond energy and inversely to bond length.

Solution Determining the valence electrons:

N has 5 valence e^-, so N_2 has 10 and N_2^+ has 9

O has 6 valence e^-, so O_2 has 12 and O_2^+ has 11

Drawing and filling the MO diagrams:

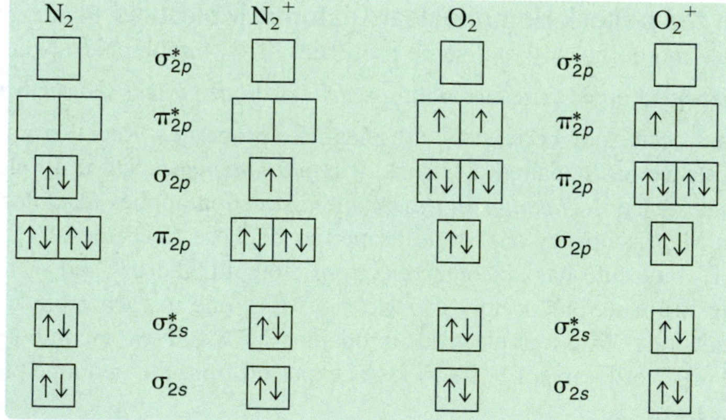

Calculating bond orders:

$\frac{1}{2}(8-2)=3$ $\quad\frac{1}{2}(7-2)=2.5$ $\quad\frac{1}{2}(8-4)=2$ $\quad\frac{1}{2}(8-3)=2.5$

Explaining the data:

1. When N_2 becomes N_2^+, a *bonding* electron is removed, so the bond order decreases. Thus, N_2^+ has a weaker, longer bond than N_2.
2. When O_2 becomes O_2^+, an *antibonding* electron is removed, so the bond order increases. Thus, O_2^+ has a stronger, shorter bond than O_2.

Check The answers make sense in terms of bond order, bond energy, and bond length. Check that the total number of bonding and antibonding electrons equals the number of valence electrons calculated.

FOLLOW-UP PROBLEM 11.4 Determine the bond orders for the following species: F_2^{2-}, F_2^-, F_2, F_2^+, F_2^{2+}. List the species in order of increasing bond energy and in order of increasing bond length.

Summary of Section 11.3

- Molecular orbital (MO) theory treats a molecule as a collection of nuclei with MOs delocalized over the entire structure.
- Atomic orbitals of comparable energy can be added or subtracted to obtain bonding or antibonding MOs, respectively.
- Bonding MOs, whether σ or π, have most of the electron density between the nuclei and are lower in energy than the AOs; most of the electron density in antibonding MOs does not lie between the nuclei, so these MOs are higher in energy.
- MOs are filled in order of their energy with paired electrons having opposite spins.
- MO diagrams show energy levels and orbital occupancy. Diagrams for the Period 2 homonuclear diatomic molecules explain bond energy, bond length, and magnetic behavior.

CHAPTER REVIEW GUIDE

The following sections provide many aids to help you study this chapter. (Numbers in parentheses refer to pages, unless noted otherwise.)

Learning Objectives

These are concepts and skills to review after studying this chapter.

Related section (§), sample problem (SP), and upcoming end-of-chapter problem (EP) numbers are listed in parentheses.

1. Describe how isolated atomic orbitals mix to form hybrid atomic orbitals, and use the electron-group arrangement from VSEPR theory to postulate the hybrid orbitals of a central atom (§11.1) (SP 11.1) (EPs 11.1–11.19)
2. Describe the modes of overlap that give sigma (σ) or pi (π) bonds, and explain the makeup of single and multiple bonds (§11.2) (SP 11.2) (EPs 11.20–11.24)
3. Understand how MOs arise from AOs; describe the shapes of MOs, and draw MO diagrams, with electron configurations and bond orders; and explain properties of homonuclear diatomic species from Periods 1 and 2 (§11.3) (SPs 11.3, 11.4) (EPs 11.25–11.38)

Key Terms

These important terms appear in boldface in the chapter and are defined again in the Glossary.

Section 11.1
valence bond (VB) theory (329)
hybridization (329)
hybrid orbital (329)
sp hybrid orbital (330)
sp^2 hybrid orbital (330)
sp^3 hybrid orbital (332)
sp^3d hybrid orbital (333)
sp^3d^2 hybrid orbital (333)

Section 11.2
sigma (σ) bond (336)
pi (π) bond (336)

Section 11.3
molecular orbital (MO) theory (338)
molecular orbital (MO) (338)
bonding MO (338)
antibonding MO (339)
sigma (σ) MO (339)
molecular orbital (MO) diagram (340)
MO bond order (340)
homonuclear diatomic molecule (341)
pi (π) MO (341)

Key Equations and Relationships

Numbered and screened concepts are listed for you to refer to or memorize.

11.1 Calculating the MO bond order (340):

$$\text{Bond order} = \tfrac{1}{2}[(\text{no. of } e^- \text{ in bonding MO}) - (\text{no. of } e^- \text{ in antibonding MO})]$$

BRIEF SOLUTIONS TO FOLLOW-UP PROBLEMS

Compare your own solutions to these calculation steps and answers.

11.1 (a) Shape is linear, so Be is *sp* hybridized.

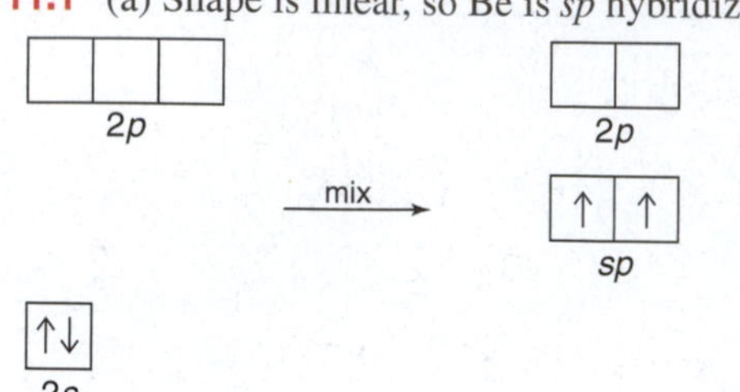

(b) Shape is tetrahedral, so Si is sp^3 hybridized.

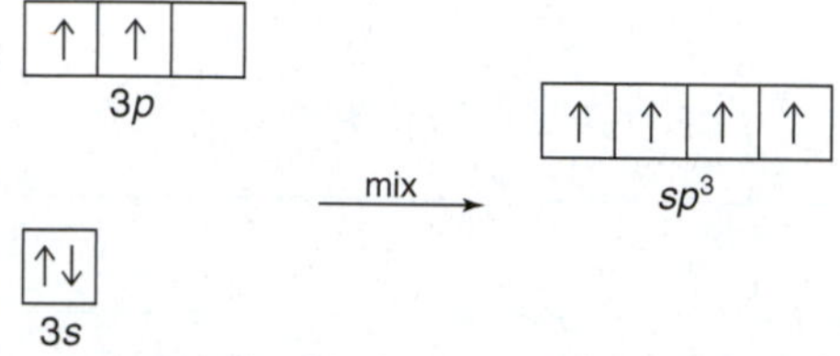

(c) Shape is square planar, so Xe is sp^3d^2 hybridized.

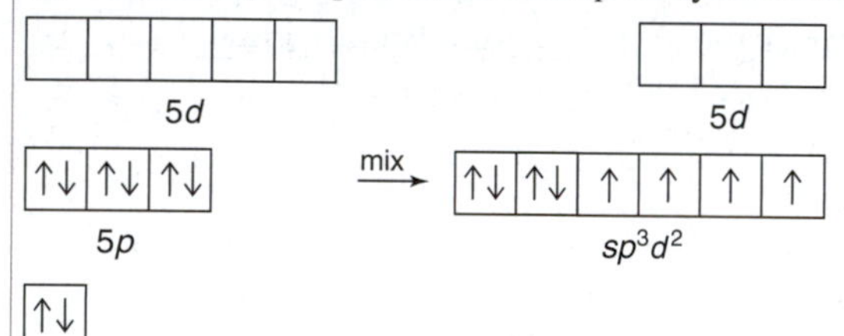

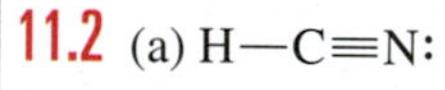

Isolated Xe atom Hybridized Xe atom

11.2 (a) H—C≡N:

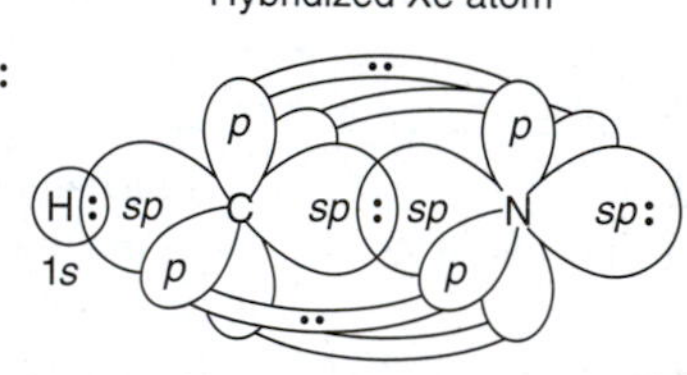

HCN is linear, so C is *sp* hybridized. N is also *sp* hybridized. One *sp* of C overlaps the 1*s* of H to form a σ bond. The other *sp* of C overlaps one *sp* of N to form a σ bond. The other *sp* of N holds a lone pair. Two unhybridized *p* orbitals of N and two of C overlap to form two π bonds.

(b) Ö=C=Ö

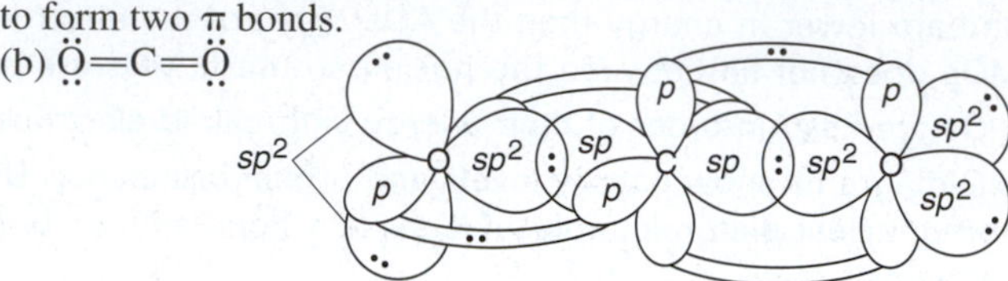

CO_2 is linear, so C is *sp* hybridized. Both O atoms are sp^2 hybridized. Each *sp* of C overlaps one sp^2 of an O to form two σ bonds. Each of the two unhybridized *p* orbitals of C forms a π bond with the unhybridized *p* of one of the two O atoms. Two sp^2 of each O hold lone pairs.

11.3

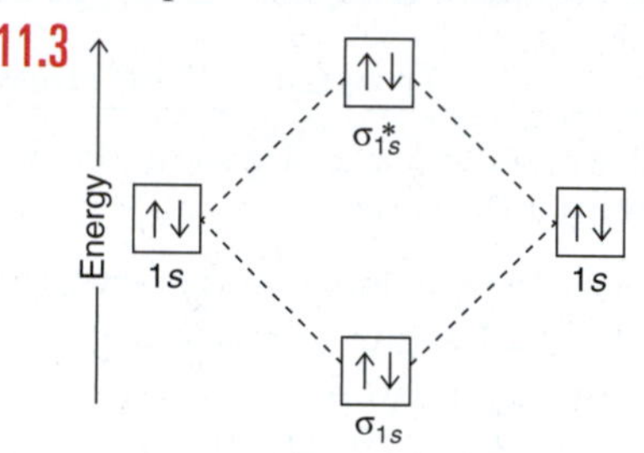

Does not exist: bond order $= \frac{1}{2}(2 - 2) = 0$.

11.4 Bond orders: $F_2^{2-} = 0$; $F_2^- = \frac{1}{2}$; $F_2 = 1$; $F_2^+ = 1\frac{1}{2}$; $F_2^{2+} = 2$

Bond energy: $F_2^{2-} < F_2^- < F_2 < F_2^+ < F_2^{2+}$

Bond length: $F_2^{2+} < F_2^+ < F_2 < F_2^-$; F_2^{2-} does not exist

PROBLEMS

Problems with **colored** numbers are answered in Appendix E. Sections match the text and provide the numbers of relevant sample problems. Bracketed problems are grouped in pairs (indicated by a short rule) that cover the same concept. Comprehensive Problems are based on material from any section or previous chapter.

Valence Bond (VB) Theory and Orbital Hybridization

(Sample Problem 11.1)

11.1 What type of central-atom orbital hybridization corresponds to each electron-group arrangement: (a) trigonal planar; (b) octahedral; (c) linear; (d) tetrahedral; (e) trigonal bipyramidal?

11.2 What is the orbital hybridization of a central atom that has one lone pair and bonds to: (a) two other atoms; (b) three other atoms; (c) four other atoms; (d) five other atoms?

11.3 How do carbon and silicon differ with regard to the *types* of orbitals available for hybridization? Explain.

11.4 How many hybrid orbitals form when four atomic orbitals of a central atom mix? Explain.

11.5 Give the number and type of hybrid orbital that forms when each set of atomic orbitals mixes:
(a) two *d*, one *s*, and three *p* (b) three *p* and one *s*

11.6 Give the number and type of hybrid orbital that forms when each set of atomic orbitals mixes:
(a) one *p* and one *s* (b) three *p*, one *d*, and one *s*

11.7 What is the hybridization of nitrogen in each of the following: (a) NO; (b) NO_2; (c) NO_2^-?

11.8 What is the hybridization of carbon in each of the following: (a) CO_3^{2-}; (b) $C_2O_4^{2-}$; (c) NCO^-?

11.9 What is the hybridization of chlorine in each of the following: (a) ClO_2; (b) ClO_3^-; (c) ClO_4^-?

11.10 What is the hybridization of bromine in each of the following: (a) BrF_3; (b) BrO_2^-; (c) BrF_5?

11.11 Which types of atomic orbitals of the central atom mix to form hybrid orbitals in (a) $SiClH_3$; (b) CS_2; (c) SCl_3F; (d) NF_3?

11.12 Which types of atomic orbitals of the central atom mix to form hybrid orbitals in (a) Cl_2O; (b) $BrCl_3$; (c) PF_5; (d) SO_3^{2-}?

11.13 Phosphine (PH_3) reacts with borane (BH_3) as follows:

$$PH_3 + BH_3 \longrightarrow H_3P—BH_3$$

(a) Which of the illustrations below depicts the change, if any, in the orbital hybridization of P during this reaction?
(b) Which depicts the change, if any, in the orbital hybridization of B?

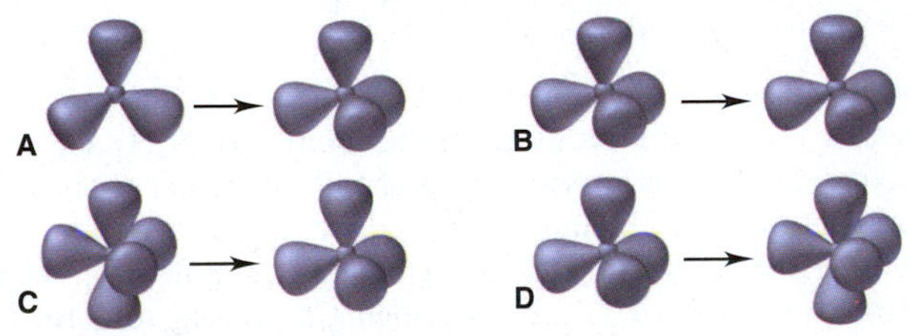

11.14 The illustrations below depict differences in orbital hybridization of some tellurium (Te) fluorides.
(a) Which depicts the difference, if any, between TeF_6 *(left)* and TeF_5^- *(right)?*
(b) Which depicts the difference, if any, between TeF_4 *(left)* and TeF_6 *(right)*?

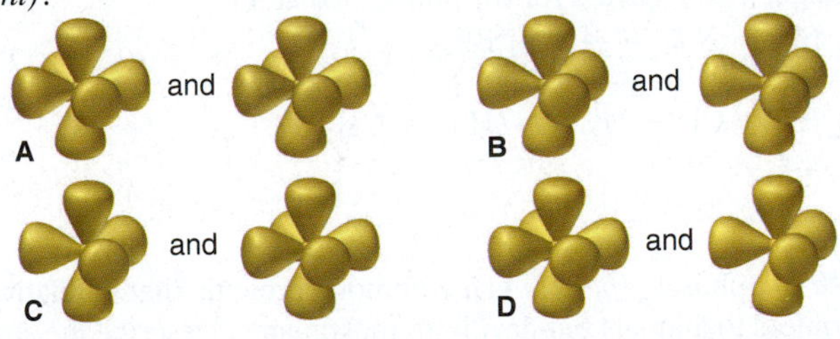

11.15 Use partial orbital diagrams to show how the atomic orbitals of the central atom lead to hybrid orbitals in (a) $GeCl_4$; (b) BCl_3; (c) CH_3^+.

11.16 Use partial orbital diagrams to show how the atomic orbitals of the central atom lead to hybrid orbitals in (a) BF_4^-; (b) PO_4^{3-}; (c) SO_3.

11.17 Use partial orbital diagrams to show how the atomic orbitals of the central atom lead to hybrid orbitals in (a) $SeCl_2$; (b) H_3O^+; (c) IF_4^-.

11.18 Use partial orbital diagrams to show how the atomic orbitals of the central atom lead to hybrid orbitals in (a) $AsCl_3$; (b) $SnCl_2$; (c) PF_6^-.

11.19 Methyl isocyanate, $CH_3—\ddot{N}=C=\ddot{O}$:, is an intermediate in the manufacture of many pesticides. In 1984 a leak from a manufacturing plant resulted in the death of more than 2000 people in Bhopal, India. What are the hybridizations of the N atom and the two C atoms in methyl isocyanate? Sketch the molecular shape.

Modes of Orbital Overlap and the Types of Covalent Bonds

(Sample Problem 11.2)

11.20 Are these statements true or false? Correct any false ones.
(a) Two σ bonds comprise a double bond.
(b) A triple bond consists of one π bond and two σ bonds.
(c) Bonds formed from atomic *s* orbitals are always σ bonds.
(d) A π bond restricts rotation about the σ-bond axis.
(e) A π bond consists of two pairs of electrons.
(f) End-to-end overlap results in a bond with electron density above and below the bond axis.

11.21 Describe the hybrid orbitals used by the central atom and the type(s) of bonds formed in (a) NO_3^-; (b) CS_2; (c) CH_2O.

11.22 Describe the hybrid orbitals used by the central atom and the type(s) of bonds formed in (a) O_3; (b) I_3^-; (c) $COCl_2$ (C is central).

11.23 Describe the hybrid orbitals used by the central atom(s) and the type(s) of bonds formed in (a) FNO; (b) C_2F_4; (c) $(CN)_2$.

11.24 Describe the hybrid orbitals used by the central atom(s) and the type(s) of bonds formed in (a) BrF_3; (b) $CH_3C\equiv CH$; (c) SO_2.

Molecular Orbital (MO) Theory and Electron Delocalization

(Sample Problems 11.3 and 11.4)

11.25 Two *p* orbitals from one atom and two *p* orbitals from another atom are combined to form molecular orbitals for the joined atoms. How many MOs will result from this combination? Explain.

11.26 Certain atomic orbitals on two atoms were combined to form the following MOs. Name the atomic orbitals used and the MOs formed, and explain which MO has higher energy:

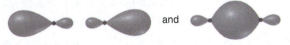

11.27 How do the bonding and antibonding MOs formed from a given pair of AOs compare to each other with respect to (a) energy; (b) presence of nodes; (c) internuclear electron density?

11.28 Antibonding MOs always have at least one node. Can a bonding MO have a node? If so, draw an example.

11.29 How many electrons does it take to fill each of the following?
(a) A σ bonding MO
(b) A π antibonding MO
(c) The MOs formed from combination of the 1*s* orbitals of two atoms?

11.30 How many electrons does it take to fill each of the following?
(a) The MOs formed from combination of the 2*p* orbitals of two atoms
(b) A σ^*_{2p} MO
(c) The MOs formed from combination of the 2*s* orbitals of two atoms?

11.31 The molecular orbitals depicted below are derived from $2p$ atomic orbitals in F_2^+.
(a) Give the orbital designations.
(b) Which is occupied by at least one electron in F_2^+?
(c) Which is occupied by only one electron in F_2^+?

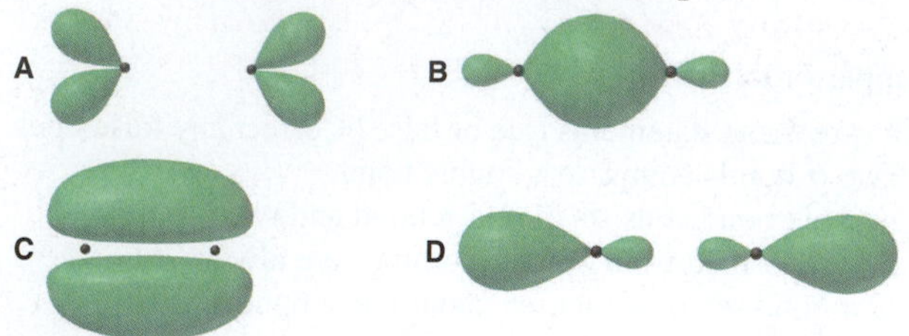

11.32 The molecular orbitals depicted below are derived from $n = 2$ atomic orbitals.
(a) Give the orbital designations.
(b) Which is highest in energy?
(c) Lowest in energy?
(d) Rank the MOs in order of increasing energy for B_2.

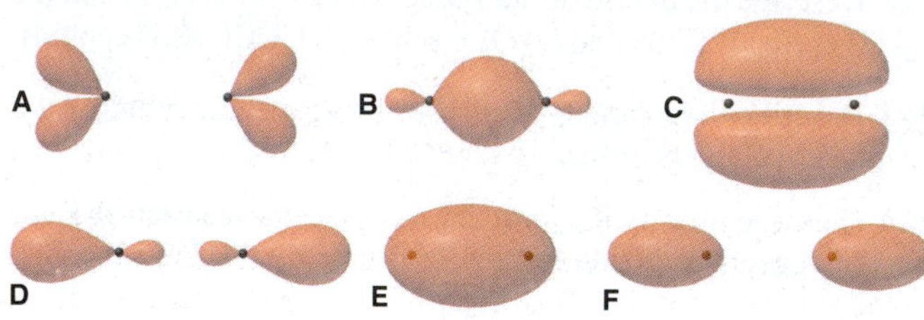

11.33 Show the shapes of bonding and antibonding MOs formed by combination of (a) an *s* orbital and a *p* orbital; (b) two *p* orbitals (end to end).

11.34 Show the shapes of bonding and antibonding MOs formed by combination of (a) two *s* orbitals; (b) two *p* orbitals (side to side).

11.35 Use MO diagrams and the bond orders you obtain from them to answer: (a) Is Be_2^+ stable? (b) Is Be_2^+ diamagnetic? (c) What is the outer (valence) electron configuration of Be_2^+?

11.36 Use MO diagrams and the bond orders you obtain from them to answer: (a) Is O_2^- stable? (b) Is O_2^- paramagnetic? (c) What is the outer (valence) electron configuration of O_2^-?

11.37 Use MO diagrams to place C_2^-, C_2, and C_2^+ in order of (a) increasing bond energy; (b) increasing bond length.

11.38 Use MO diagrams to place B_2^+, B_2, and B_2^- in order of (a) decreasing bond energy; (b) decreasing bond length.

Comprehensive Problems

11.39 Predict the shape, state the hybridization of the central atom, and give the ideal bond angle(s) and any expected deviations for:
(a) BrO_3^- (b) $AsCl_4^-$ (c) SeO_4^{2-} (d) BiF_5^{2-}
(e) SbF_4^+ (f) AlF_6^{3-} (g) IF_4^+

11.40 Butadiene *(right)* is a colorless gas used to make synthetic rubber and many other compounds. How many σ bonds and π bonds does the molecule have?

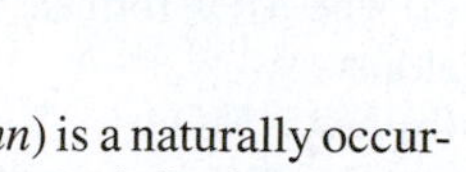

11.41 Epinephrine (or adrenaline; *next column*) is a naturally occurring hormone that is also manufactured commercially for use as a heart stimulant, a nasal decongestant, and a glaucoma treatment.

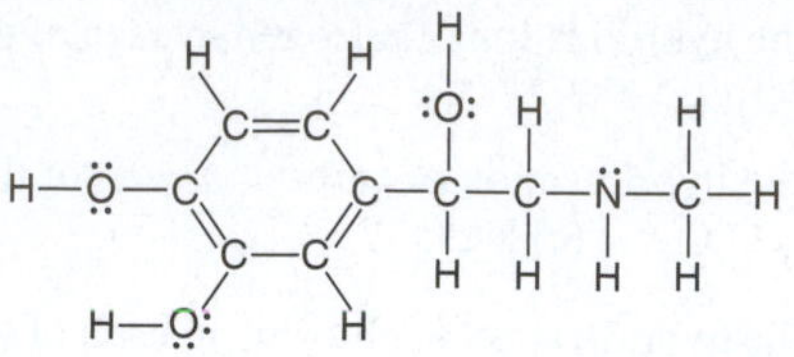

(a) What is the hybridization of each C, O, and N atom? (b) How many σ bonds does the molecule have? (c) How many π electrons are delocalized in the ring?

11.42 Use partial orbital diagrams to show how the atomic orbitals of the central atom lead to the hybrid orbitals in:
(a) IF_2^- (b) ICl_3 (c) $XeOF_4$ (d) BHF_2

11.43 Isoniazid *(below)* is an antibacterial agent that is effective against many common strains of tuberculosis. (a) How many σ bonds are in the molecule? (b) What is the hybridization of each C and N atom?

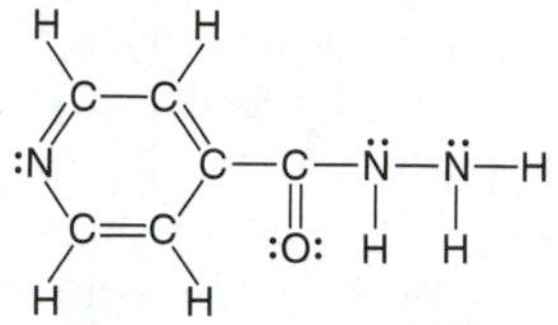

11.44 Hydrazine, N_2H_4, and carbon disulfide, CS_2, form a cyclic molecule *(below)*. (a) Draw Lewis structures for N_2H_4 and CS_2. (b) How do electron-group arrangement, molecular shape, and hybridization of N change when N_2H_4 reacts to form the product? (c) How do electron-group arrangement, molecular shape, and hybridization of C change when CS_2 reacts to form the product?

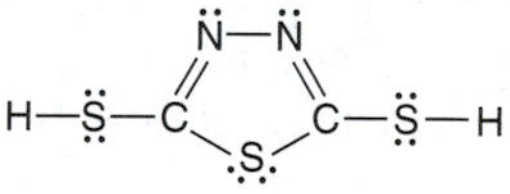

11.45 In each of the following equations, what hybridization change, if any, occurs for the underlined atom?
(a) $\underline{B}F_3 + NaF \longrightarrow Na^+BF_4^-$
(b) $\underline{P}Cl_3 + Cl_2 \longrightarrow PCl_5$
(c) $H\underline{C}{\equiv}CH + H_2 \longrightarrow H_2C{=}CH_2$
(d) $\underline{Si}F_4 + 2F^- \longrightarrow SiF_6^{2-}$
(e) $\underline{S}O_2 + \frac{1}{2}O_2 \longrightarrow SO_3$

11.46 Glyphosate *(below)* is a common herbicide that is relatively harmless to animals but deadly to most plants. Describe the shape around and the hybridization of the P, N, and three numbered C atoms.

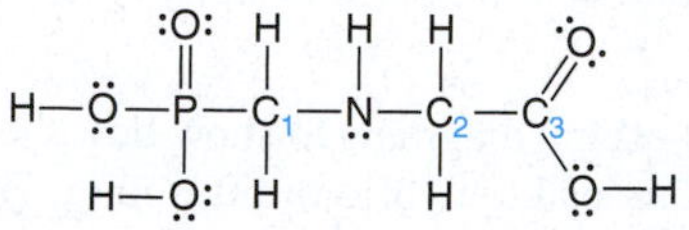

11.47 The sulfate ion can be represented with four S—O bonds or with two S—O and two S=O bonds.
(a) Which representation is better from the standpoint of formal charges?
(b) What is the shape of the sulfate ion, and what hybrid orbitals of S are postulated for the σ bonding?
(c) In view of the answer to part (b), what orbitals of S must be used for the π bonds? What orbitals of O?
(d) Draw a diagram to show how one atomic orbital from S and one from O overlap to form a π bond.

11.48 Tryptophan is one of the amino acids found in proteins:

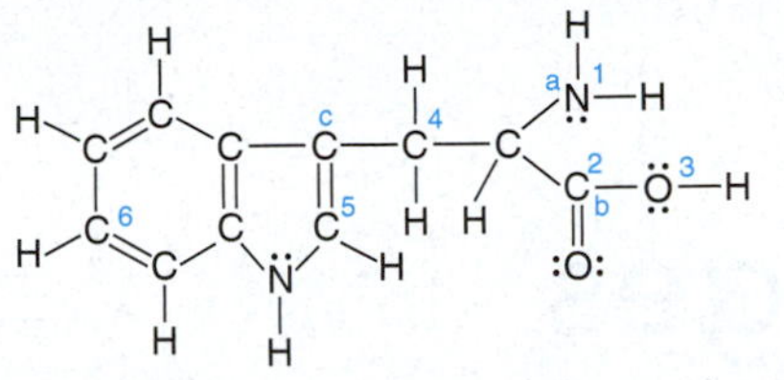

(a) What is the hybridization of each of the numbered C, N, and O atoms? (b) How many σ bonds are present in tryptophan? (c) Predict the bond angles at points a, b, and c.

11.49 Some species with two oxygen atoms only are the oxygen molecule, O_2; the peroxide ion, O_2^{2-}; the superoxide ion, O_2^-; and the dioxygenyl ion, O_2^+. Draw an MO diagram for each, rank them in order of increasing bond length, and find the number of unpaired electrons in each.

11.50 There is concern in health-related government agencies that the American diet contains too much meat, and numerous recommendations have been made urging people to consume more fruit and vegetables. One of the richest sources of vegetable protein is soy, available in many forms. Among these is soybean curd, or tofu, which is a staple of many Asian diets. Chemists have isolated an anticancer agent called *genistein* from tofu, which may explain the much lower incidence of cancer among people in the Far East. A valid Lewis structure for genistein is

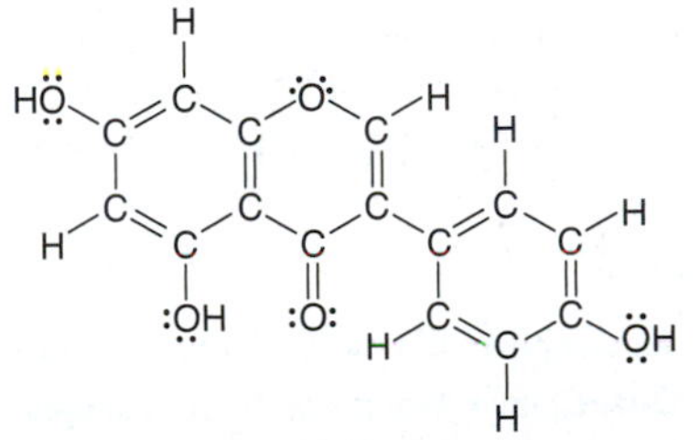

(a) Is the hybridization of each C in the right-hand ring the same? Explain. (b) Is the hybridization of the O atom in the center ring the same as that of the O atoms in OH groups? Explain. (c) How many carbon-oxygen σ bonds are there? How many carbon-oxygen π bonds? (d) Do all the lone pairs on oxygens occupy the same type of hybrid orbital? Explain.

11.51 An organic chemist synthesizes the molecule below:

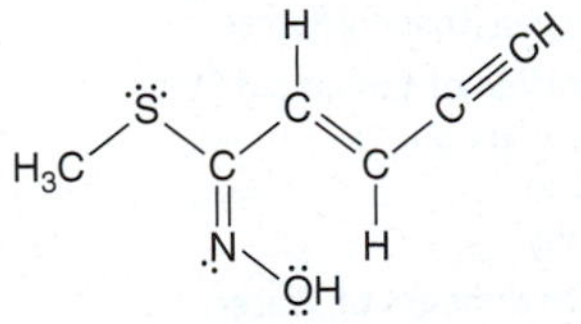

(a) Which of the orientations of hybrid orbitals shown below are present in the molecule? (b) Are there any present that are not shown below? If so, what are they? (c) How many of each type of hybrid orbital are present?

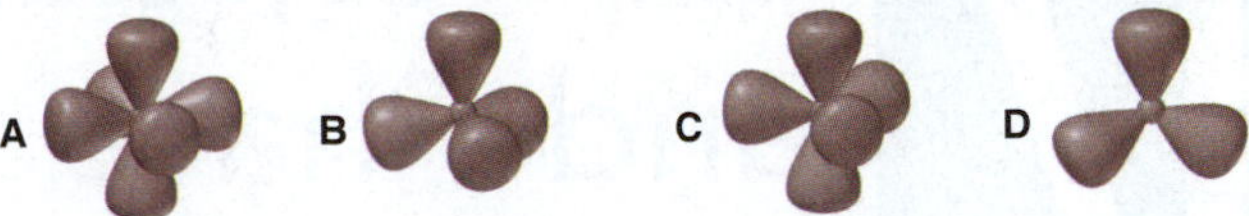

11.52 The hydrocarbon allene, $H_2C{=}C{=}CH_2$, is obtained indirectly from petroleum and used as a precursor for several types of plastics. What is the hybridization of each C atom in allene? Draw a bonding picture for allene with lines for σ bonds, and show the arrangement of the π bonds. Be sure to represent the geometry of the molecule in three dimensions.

11.53 Sulfur forms oxides, oxoanions, and halides. What is the hybridization of the central S in SO_2, SO_3, SO_3^{2-}, SCl_4, SCl_6, and S_2Cl_2 (atom sequence Cl—S—S—Cl)?

11.54 Silicon tetrafluoride reacts with F^- to produce the hexafluorosilicate ion, SiF_6^{2-}; GeF_4 behaves similarly, but CF_4 does not. (a) Draw Lewis structures for SiF_4, GeF_6^{2-}, and CF_4. (b) What is the hybridization of the central atom in each species? (c) Why doesn't CF_4 react with F^- to form CF_6^{2-}?

11.55 The compound 2,6-dimethylpyrazine *(below)* gives chocolate its odor and is used in flavorings. (a) Which atomic orbitals mix to form the hybrid orbitals of N? (b) In what type of hybrid orbital do the lone pairs of N reside? (c) Is C in CH_3 hybridized the same as any C in the ring? Explain.

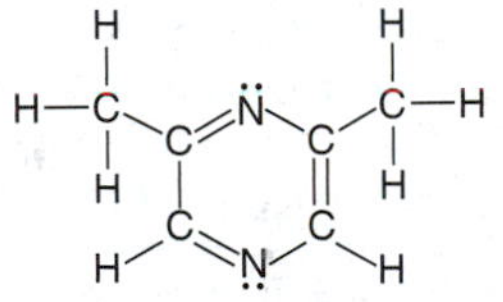

11.56 Acetylsalicylic acid (aspirin), the most widely used medicine in the world, has the Lewis structure shown below. (a) What is the hybridization of each C and each O atom? (b) How many localized π bonds are present? (c) How many C atoms have a trigonal planar shape around them? A tetrahedral shape?

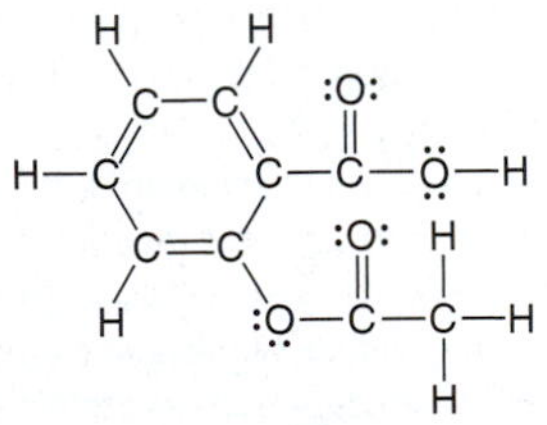

12 Intermolecular Forces: Liquids, Solids, and Phase Changes

Key Principles
to focus on while studying this chapter

- The relative magnitudes of the energy of motion *(kinetic)* of the particles and the energy of attraction *(potential)* among them determine whether a substance is a gas, a liquid, or a solid under a given set of conditions: in a gas, kinetic energy is much greater than potential energy; in a solid, the relative magnitudes are reversed. ***(Section 12.1)***
- The same factors determine changes of state *(vaporization–condensation, melting–freezing, sublimation–deposition)*, and each type of change is associated with an enthalpy change that is positive in one direction and negative in the other. It takes more energy to convert a liquid to a gas than a solid to a liquid. ***(Section 12.1)***
- A *heating-cooling curve* shows the changes of state that occur when heat is added to or removed from a substance at a constant rate. Within a state *(phase)*, a change in temperature accompanies a change in heat (*kinetic energy*, that is, in molecular motion). A change in state *(phase change)* occurs at constant temperature and arises from a change in average interparticle distance, which relates to attractions *(potential energy)*. ***(Section 12.2)***
- In a closed system, a phase change is *reversible*, and the system reaches a state of *dynamic equilibrium*. As a liquid vaporizes in a closed container at a given temperature, the rates of vaporization and condensation become equal, so the pressure of the gas *(vapor pressure)* becomes constant. The vapor pressure increases with temperature and decreases with stronger intermolecular forces. The *Clausius-Clapeyron equation* relates vapor pressure to temperature. A *phase diagram* shows the range of pressure and temperature at which each phase is stable and at which phase changes occur. ***(Section 12.2)***
- Bonding forces are stronger than nonbonding forces. A charged region of one molecule attracts an oppositely charged region of another, and the strength of these *intermolecular forces* determines many physical properties. A charged region of one molecule can *induce* a region in another molecule to become charged, depending on the *polarizability* of the electron clouds. *Hydrogen bonding* requires an H atom bonded to N, O, or F; *dispersion forces* exist between all molecules. ***(Section 12.3)***
- Combinations of intermolecular forces determine the properties of liquids—*surface tension, capillarity,* and *viscosity.* ***(Section 12.4)***
- The physical properties of water—great solvent power, high specific heat capacity, high heat of vaporization, high surface tension, high capillarity, and lower density of the solid—emerge from its atomic and molecular properties and play vital roles in biology and the environment. ***(Section 12.5)***
- Crystalline solids consist of particles tightly packed into a regular array called a *lattice*. The simplest repeating portion of the lattice is the *unit cell*. Many substances crystallize in one of three *cubic* unit cells. These differ in the arrangement of the particles and, therefore, in the number of particles per unit cell and how efficiently they are packed. ***(Section 12.6)***
- The properties of five types of crystalline solids—*atomic, molecular, ionic, metallic,* and *network covalent*—depend on the type(s) of particles in the crystal and the resultant interparticle forces. ***(Section 12.6)***
- The *electron-sea model* explains that the properties of metals are due to highly *delocalized* valence electrons.
- *Band theory* explains many properties of metals *(conductors)*, metalloids *(semiconductors)*, and nonmetals *(insulators)* in terms of bands of overlapping molecular orbitals and the size of the *energy gap* between those filled with valence electrons and those that are unfilled.

Three Forms of One. Mist over a stream in winter completes this scene of the three phases of water. In this chapter we discuss the forces *between* molecules that give rise to the phases and phase changes.

Outline

All the matter in and around you occurs in one or more of the three physical states—gas, liquid, or solid. Under different conditions, many substances can occur in any of the states. The three states were introduced in Chapter 1, and their properties were compared when we examined gases in Chapter 5. Now, we examine liquids and solids, which are called *condensed* states because, unlike in gases, their *particles are very close together.*

CONCEPTS & SKILLS TO REVIEW
before studying this chapter

- properties of gases, liquids, and solids (Section 5.1)
- kinetic-molecular theory of gases (Section 5.5)
- kinetic and potential energy (Section 6.1)
- enthalpy change, heat capacity, and Hess's law (Sections 6.2, 6.3, and 6.5)
- diffraction of light (Section 7.1)
- Coulomb's law (Section 9.2)
- chemical bonding models (Chapter 9)
- molecular polarity (Section 10.3)
- molecular orbital treatment of diatomic molecules (Section 11.3)

12.1 • AN OVERVIEW OF PHYSICAL STATES AND PHASE CHANGES

Each physical state is called a **phase,** a physically distinct, homogeneous part of a system. The water in a closed container constitutes one phase; the water vapor above the liquid is a second phase; add some ice, and there are three.

In this section, you'll see that interactions between the potential energy and the kinetic energy of the particles give rise to the properties of each phase:

- The *potential energy*, in the form of **intermolecular forces** (or, more generally, *interparticle forces*), tends to draw the molecules together. According to Coulomb's law, the electrostatic potential energy depends on the charges of the particles and the distances between them (Section 9.2).
- The *kinetic energy* associated with the random motion of the molecules tends to disperse them. It is related to their average speed and is proportional to the absolute temperature (Section 5.5).

These interactions also explain **phase changes,** changes in physical state from one phase to another—liquid to solid, solid to gas, and so forth.

A Kinetic-Molecular View of the Three States Imagine yourself among the particles in any of the three states of water. Look closely and you'll discover two types of electrostatic forces at work:

1. *Intra*molecular (bonding) forces exist *within* each molecule. The *chemical* behavior of the three states is identical because each consists of the same bent, polar H—O—H molecules held together by identical covalent bonding forces.
2. *Inter*molecular (nonbonding) forces exist *between* the molecules. The *physical* behavior of the states is different because the strengths of these forces differ from state to state.

Whether a substance occurs as a gas, liquid, or solid depends on the interplay of the potential and kinetic energy:

- *In a gas,* the potential energy (energy of attraction) is small relative to the kinetic energy (energy of motion); thus, on average, the particles are far apart. This large distance has several macroscopic consequences: a gas fills its container, is highly compressible, and flows easily through another gas (Table 12.1).
- *In a liquid,* attractions are stronger because the particles are touching, but they have enough kinetic energy to move randomly around each other. Thus, a liquid conforms to the shape of its container but has a surface; it resists an applied force and thus compresses very slightly; and it flows, but *much* more slowly than gases.

Table 12.1 A Macroscopic Comparison of Gases, Liquids, and Solids

State	Shape and Volume	Compressibility	Ability to Flow
Gas	Conforms to shape and volume of container	High	High
Liquid	Conforms to shape of container; volume limited by surface	Very low	Moderate
Solid	Maintains its own shape and volume	Almost none	Almost none

- *In a solid,* the attractions dominate the motion so much that the particles are fixed in position relative to one another, just jiggling in place. Thus, a solid has its own shape, compresses even less than liquids, and does not flow significantly.

THINK OF IT THIS WAY
Environmental Flow

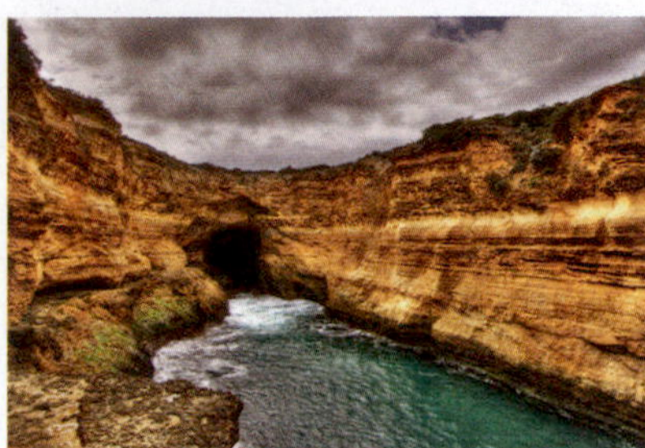

The environment provides a perfect demonstration of these differences in ability to flow. Atmospheric gases mix so well that the lowest 80 km of air has a uniform composition. Much less mixing in the oceans allows the composition at various depths to support different species. And rocks intermingle so little that adjacent strata remain separated for millions of years.

Types of Phase Changes and Their Enthalpies When we consider phase changes, understanding the effect of temperature is critical:

- As *temperature increases,* the average kinetic energy does too, so the faster moving particles overcome attractions more easily.
- As *temperature decreases,* particles slow, so attractions can pull them together.

Each phase change has a name and an associated enthalpy change:

1. *Gas to liquid, and vice versa.* As the temperature drops, the molecules in the gas phase come together and form a liquid in the process of **condensation;** the opposite process, changing from a liquid to a gas, is **vaporization.**

2. *Liquid to solid, and vice versa.* As the temperature drops further, the particles move slower and become fixed in position in the process of **freezing;** the opposite change is called **melting,** or **fusion.** In common speech, *freezing* implies low temperature because we think of water. But, molten metals, for example, freeze (solidify) at much higher temperatures and have medical, industrial, and artistic applications, such as gold dental crowns, steel auto bodies, and bronze statues.

3. *Gas to solid, and vice versa.* All three states of water are familiar because they are stable under ordinary conditions. Carbon dioxide, on the other hand, is familiar as a gas and a solid (dry ice), but liquid CO_2 occurs only at pressures of 5.1 atm or greater. At ordinary conditions, solid CO_2 changes directly to a gas, a process called **sublimation.** Freeze-dried foods are prepared by sublimation. The opposite process, changing from a gas directly into a solid, is called **deposition**—ice crystals form on a cold window from the deposition of water vapor.

The accompanying enthalpy changes are either exothermic or endothermic:

- *Exothermic changes.* As the molecules of a gas attract each other into a liquid, and then become fixed in a solid, the system of particles *loses* energy, which is released as heat. Thus, *condensing, freezing, and depositing are exothermic changes.*
- *Endothermic changes.* Heat must be absorbed by the system to overcome the attractive forces that keep the particles fixed in place in a solid or near each other in a liquid. Thus, *melting, vaporizing, and subliming are endothermic changes.*

Sweating has a cooling effect because heat from your body vaporizes the water. To achieve this cooling, cats lick themselves and dogs pant.

For a pure substance, each phase change is accompanied by a standard enthalpy change, given in units of *kilojoules per mole* (measured at 1 atm and the temperature of the change). For vaporization, it is the **heat** (or *enthalpy*) **of vaporization (ΔH°_{vap}),** and for fusion (melting), it is the **heat** (or *enthalpy*) **of fusion (ΔH°_{fus}).** In the case of water, we have

$$H_2O(l) \longrightarrow H_2O(g) \qquad \Delta H = \Delta H^\circ_{vap} = 40.7 \text{ kJ/mol (at } 100^\circ\text{C)}$$
$$H_2O(s) \longrightarrow H_2O(l) \qquad \Delta H = \Delta H^\circ_{fus} = 6.02 \text{ kJ/mol (at } 0^\circ\text{C)}$$

The reverse processes, condensing and freezing, have enthalpy changes of the *same magnitude but opposite sign:*

$$H_2O(g) \longrightarrow H_2O(l) \qquad \Delta H = -\Delta H^\circ_{vap} = -40.7 \text{ kJ/mol}$$
$$H_2O(l) \longrightarrow H_2O(s) \qquad \Delta H = -\Delta H^\circ_{fus} = -6.02 \text{ kJ/mol}$$

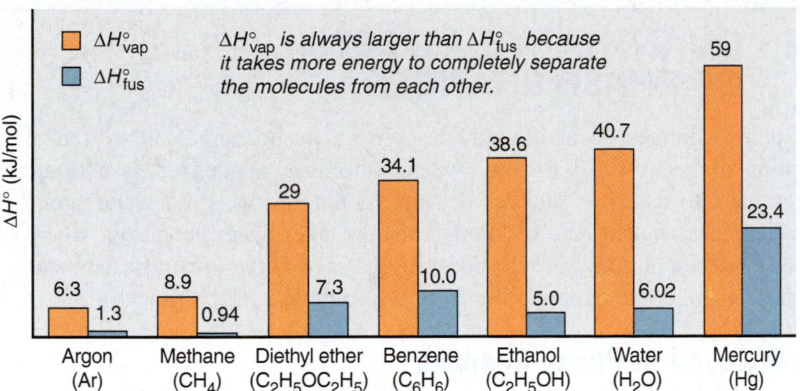

Figure 12.1 Heats of vaporization and fusion for several common substances.

Water behaves typically in that it takes much less energy to melt the solid than to vaporize the liquid: $\Delta H^\circ_{fus} < \Delta H^\circ_{vap}$; that is, it takes less energy to reduce the intermolecular forces enough for the molecules to move out of their fixed positions (melt a solid) than to separate them completely (vaporize a liquid) (Figure 12.1).

The **heat** (or *enthalpy*) **of sublimation** ($\boldsymbol{\Delta H^\circ_{subl}}$) is the enthalpy change when 1 mol of a substance sublimes, and the negative of this value is the change when 1 mol of the substance deposits. Since sublimation can be thought of as a combination of melting and vaporizing, Hess's law (Section 6.5) says that the heat of sublimation equals the sum of the heats of fusion and vaporization:

$$\begin{array}{lll} \text{Solid} \longrightarrow \text{liquid} & & \Delta H^\circ_{fus} \\ \text{Liquid} \longrightarrow \text{gas} & & \Delta H^\circ_{vap} \\ \hline \text{Solid} \longrightarrow \text{gas} & & \Delta H^\circ_{subl} \end{array}$$

Figure 12.2 summarizes the phase changes and their enthalpy changes.

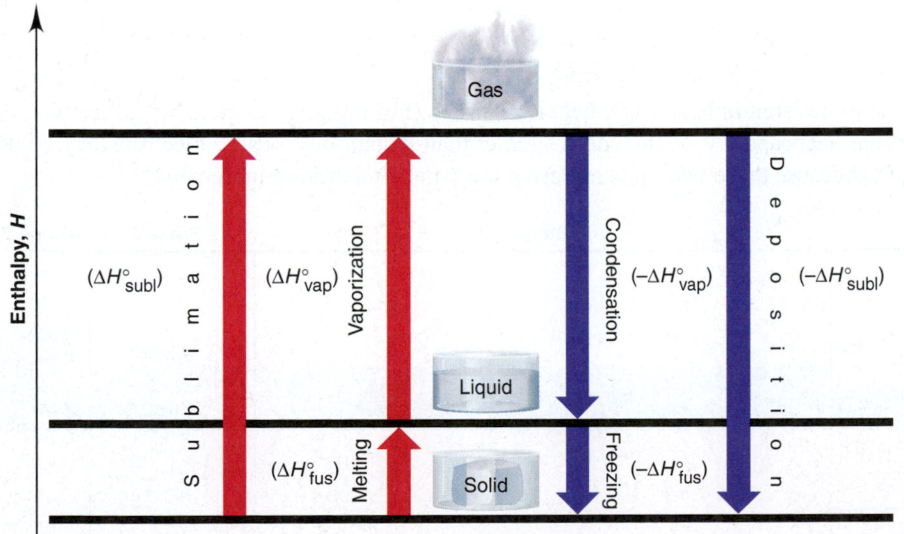

Figure 12.2 Phase changes and their enthalpy changes. Fusion (or melting), vaporization, and sublimation are endothermic changes (positive ΔH°), whereas freezing, condensation, and deposition are exothermic changes (negative ΔH°).

Summary of Section 12.1

- Because of the relative magnitudes of intermolecular forces (potential energy) and average speed (kinetic energy), the particles in a gas are far apart and moving randomly, those in a liquid are in contact and moving relative to each other, and those in a solid are in contact and in fixed positions. These molecular-level differences account for macroscopic differences in shape, compressibility, and ability to flow.
- When a solid becomes a liquid (melting, or fusion), a liquid becomes a gas (vaporization), or a solid becomes a gas (sublimation), energy is absorbed to overcome intermolecular forces and increase the average distance between particles. As particles come closer together in the reverse changes (freezing, condensation, and deposition), energy is released. Each phase change is associated with a given enthalpy change under specified conditions.

12.2 • QUANTITATIVE ASPECTS OF PHASE CHANGES

Many phase changes occur around you every day, accompanied by the release or absorption of heat. When it rains, water vapor has condensed to a liquid, which changes back to a gas as puddles dry up. In the spring, solid water melts, and in winter, it freezes again. And the same changes take place, but faster, whenever you make a pot of tea or a tray of ice cubes. In this section, we quantify the heat involved in a phase change and examine the equilibrium nature of the process.

Heat Involved in Phase Changes

We apply a kinetic-molecular approach to phase changes with a **heating-cooling curve,** which shows the changes in temperature of a sample when heat is absorbed or released at a constant rate. Let's examine what happens when 2.50 mol of gaseous water in a closed container undergoes a change from 130°C to −40°C at a constant pressure of 1 atm. We divide this process into five heat-releasing (exothermic) stages (Figure 12.3):

Stage 1. Gaseous water cools. Water molecules zoom chaotically at a range of speeds, smashing into each other and the container walls. At the starting temperature, the most probable speed of the molecules, and thus their average kinetic energy (E_k), is high enough to overcome the potential energy (E_p) of attractions. As the temperature falls, the average E_k decreases and attractions become more important. The change is

$$H_2O(g)\,[130°C] \longrightarrow H_2O(g)\,[100°C]$$

The heat (q) is the product of the amount (number of moles, n) of water, the molar heat capacity of *gaseous* water, $C_{water(g)}$, and the temperature change during this step, ΔT ($T_{final} - T_{initial}$):

$$\begin{aligned} q &= n \times C_{water(g)} \times \Delta T = (2.50\text{ mol})(33.1\text{ J/mol}\cdot°C)(100°C - 130°C) \\ &= -2482\text{ J} = -2.48\text{ kJ} \end{aligned}$$

The minus sign indicates that heat is released. (For purposes of canceling, the units for molar heat capacity, C, include °C, rather than K, but this doesn't affect the magnitude of C because these two units represent the same temperature increment.)

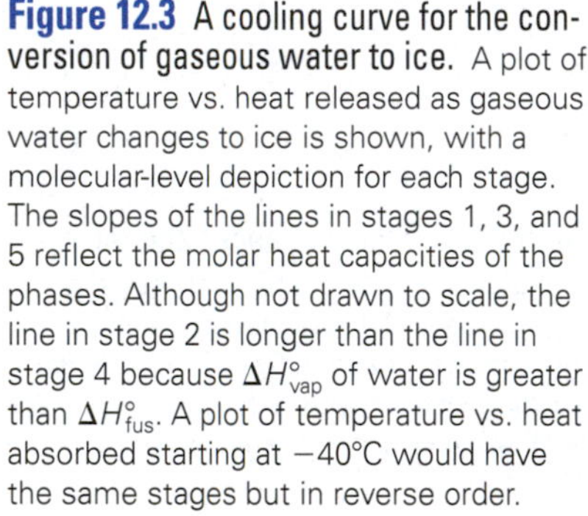

Figure 12.3 A cooling curve for the conversion of gaseous water to ice. A plot of temperature vs. heat released as gaseous water changes to ice is shown, with a molecular-level depiction for each stage. The slopes of the lines in stages 1, 3, and 5 reflect the molar heat capacities of the phases. Although not drawn to scale, the line in stage 2 is longer than the line in stage 4 because ΔH°_{vap} of water is greater than ΔH°_{fus}. A plot of temperature vs. heat absorbed starting at −40°C would have the same stages but in reverse order.

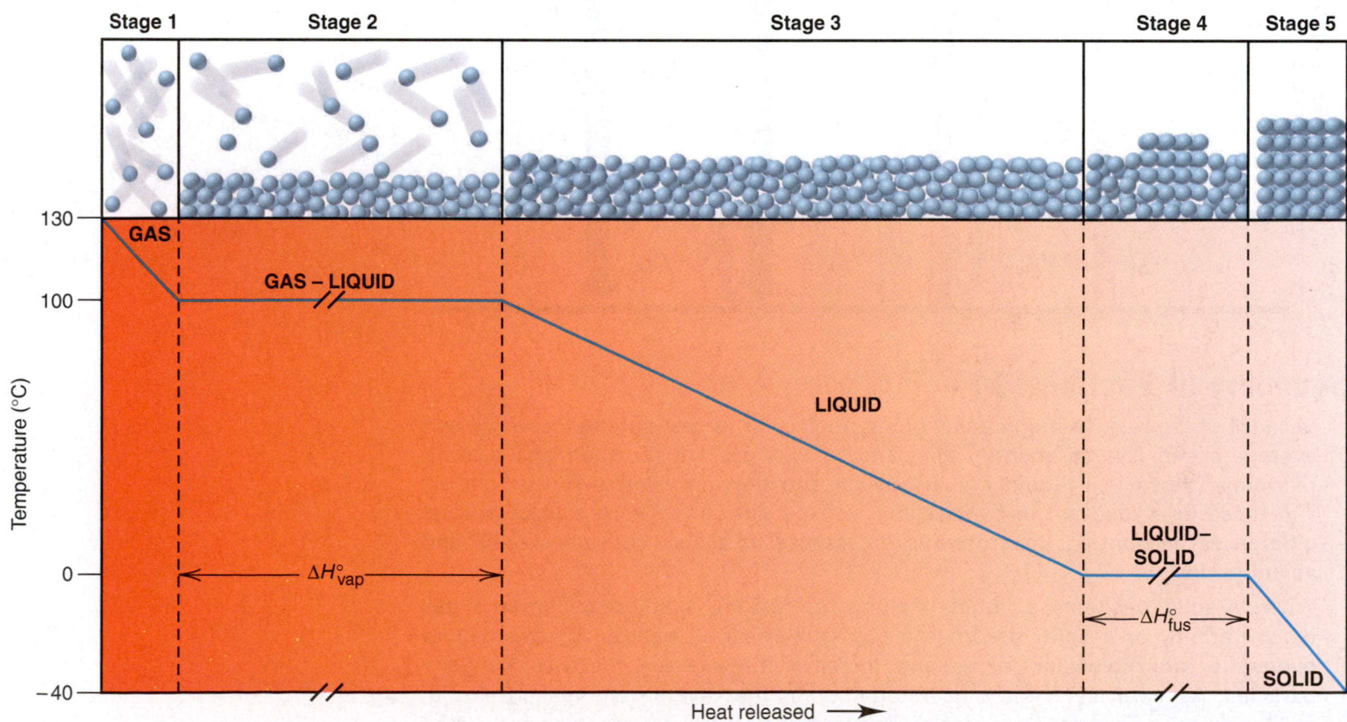

Stage 2. Gaseous water condenses. At the condensation point, intermolecular attractions cause the slowest of the molecules to aggregate into microdroplets and then a bulk liquid. Note that, during the phase change,

- The temperature of the sample, and thus its average E_k, is constant. At the same temperature, molecules move farther between collisions in a gas than in a liquid, but their *average* speed is the same.
- Releasing heat from the sample decreases the average E_p as the molecules approach each other.

Thus, at 100°C, gaseous and liquid water have the same average E_k, but the liquid has lower average E_p. The change is

$$H_2O(g)\ [100°C] \longrightarrow H_2O(l)\ [100°C]$$

The heat is the amount (n) times the negative of the heat of vaporization ($-\Delta H°_{vap}$):

$$q = n(-\Delta H°_{vap}) = (2.50\ \text{mol})(-40.7\ \text{kJ/mol}) = -102\ \text{kJ}$$

This stage contributes the *greatest portion of the total heat released* because of the large decrease in E_p as the molecules become so much closer in the liquid than they were in the gas.

Stage 3. Liquid water cools. The molecules in the liquid state continue to lose heat, which appears as a decrease in temperature, that is, as a decrease in the most probable molecular speed and, thus, the average E_k. The temperature decreases as long as the sample remains liquid. The change is

$$H_2O(l)\ [100°C] \longrightarrow H_2O(l)\ [0°C]$$

The heat depends on amount (n), the molar heat capacity of *liquid* water, and ΔT:

$$q = n \times C_{water(l)} \times \Delta T = (2.50\ \text{mol})(75.4\ \text{J/mol·°C})(0°C - 100°C)$$
$$= -18{,}850\ \text{J} = -18.8\ \text{kJ}$$

Stage 4. Liquid water freezes. At 0°C, the sample loses E_p as increasing intermolecular attractions cause the molecules to align themselves into the crystalline structure of ice. Molecular motion continues only as random jiggling about fixed positions. As we saw during condensation, temperature and average E_k are constant during freezing. The change is

$$H_2O(l)\ [0°C] \longrightarrow H_2O(s)\ [0°C]$$

The heat is equal to n times the negative of the heat of fusion ($-\Delta H°_{fus}$):

$$q = n(-\Delta H°_{fus}) = (2.50\ \text{mol})(-6.02\ \text{kJ/mol}) = -15.0\ \text{kJ}$$

Stage 5. Solid water cools. With motion restricted to jiggling in place, further cooling merely reduces the average speed of this jiggling. The change is

$$H_2O(s)\ [0°C] \longrightarrow H_2O(s)\ [-40°C]$$

The heat depends on n, the molar heat capacity of *solid* water, and ΔT:

$$q = n \times C_{water(s)} \times \Delta T = (2.50\ \text{mol})(37.6\ \text{J/mol·°C})(-40°C - 0°C)$$
$$= -3760\ \text{J} = -3.76\ \text{kJ}$$

According to Hess's law, the total heat released is the sum of the heats released for the individual stages. The sum of q for stages 1 to 5 is -142 kJ. Two key points stand out for this or any similar process, whether exothermic or endothermic:

- *Within a phase,* heat flow is accompanied by *a change in temperature,* which is associated with a change in average E_k as *the most probable speed of the molecules changes.* The heat released or absorbed depends on the amount of substance, the molar heat capacity *for that phase,* and the change in temperature.
- *During a phase change,* heat flow occurs at a *constant temperature,* which is associated with a change in average E_p as *the average distance between molecules changes.* Both phases are present and (as you'll see below) are in equilibrium during the change. The heat released or absorbed depends on the amount of substance and the enthalpy change for that phase change.

Let's work a molecular-scene sample problem to make sure these ideas are clear.

Sample Problem 12.1 Finding the Heat of a Phase Change Depicted by Molecular Scenes

Problem The scenes below represent a phase change of water. Select data from the previous discussion to find the heat (in kJ) released or absorbed when 24.3 g of H_2O undergoes this change.

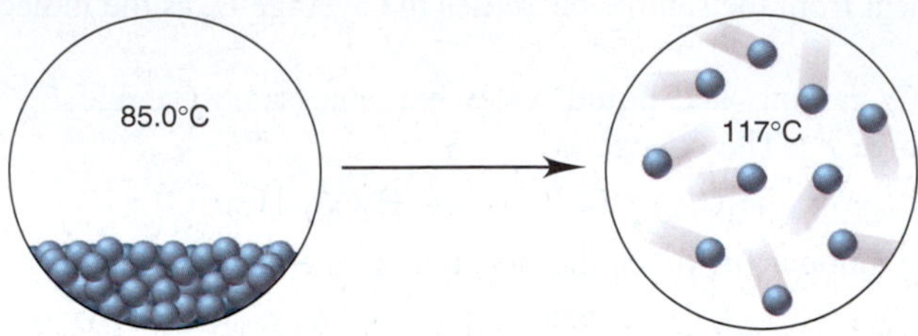

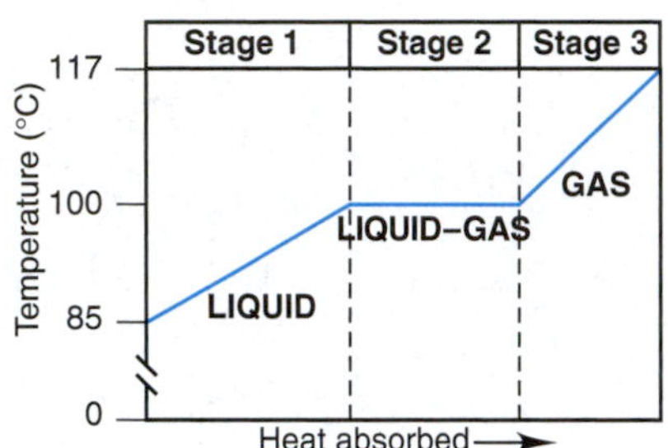

Plan From the molecular scenes, data in the text, and the given mass (24.3 g) of water, we have to find the heat that accompanies this change. The scenes show a disorderly, condensed phase at 85.0°C changing to separate molecules at 117°C. Thus, the phase change they depict is vaporization, an endothermic process. The text data are given per mole, so we first convert the mass (g) of water to amount (mol). There are three stages: (1) heating the liquid from 85.0°C to 100.°C, (2) converting liquid water at 100.°C to gaseous water at 100.°C, and (3) heating the gas from 100.°C to 117°C *(see margin)*. We add the values of q for these stages to obtain the total heat.

Solution Converting from mass (g) of H_2O to amount (mol):

$$\text{Amount (mol) of } H_2O = 24.3 \text{ g } H_2O \times \frac{1 \text{ mol}}{18.02 \text{ g } H_2O} = 1.35 \text{ mol}$$

Finding the heat accompanying stage 1, $H_2O(l)\ [85.0°C] \longrightarrow H_2O(l)\ [100.°C]$:

$$q = n \times C_{\text{water}(l)} \times \Delta T = (1.35 \text{ mol})(75.4 \text{ J/mol·°C})(100.°C - 85.0°C)$$
$$= 1527 \text{ J} = 1.53 \text{ kJ}$$

Finding the heat accompanying stage 2, $H_2O(l)\ [100.°C] \longrightarrow H_2O(g)\ [100.°C]$:

$$q = n(\Delta H°_{\text{vap}}) = (1.35 \text{ mol})(40.7 \text{ kJ/mol}) = 54.9 \text{ kJ}$$

Finding the heat accompanying stage 3, $H_2O(g)\ [100.°C] \longrightarrow H_2O(g)\ [117°C]$:

$$q = n \times C_{\text{water}(g)} \times \Delta T = (1.35 \text{ mol})(33.1 \text{ J/mol·°C})(117°C - 100.°C)$$
$$= 759.6 \text{ J} = 0.760 \text{ kJ}$$

Adding the three heats together to find the total heat of the process:

$$\text{Total heat (kJ)} = 1.53 \text{ kJ} + 54.9 \text{ kJ} + 0.760 \text{ kJ} = 57.2 \text{ kJ}$$

Check The heat should have a positive value because it is absorbed. Be sure to round to check each value of q; for example, in stage 1, 1.33 mol × 75 J/mol·°C × 15°C = 1500 J. Note that the phase change itself (stage 2) requires the most energy and, thus, dominates the final answer. The $\Delta H°_{\text{vap}}$ units include kJ, whereas the molar heat capacity units include J, which is a thousandth as large.

FOLLOW-UP PROBLEM 12.1 The scenes below represent a phase change of water. Select data from the text discussion to find the heat (in kJ) released or absorbed when 2.25 mol of H_2O undergoes this change.

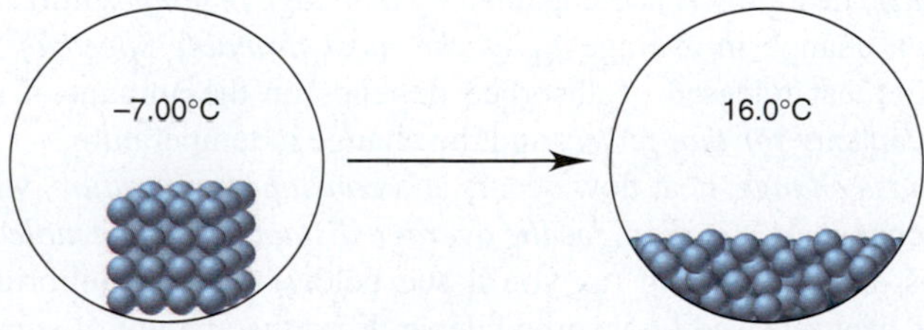

The Equilibrium Nature of Phase Changes

In everyday experience, phase changes take place in *open* containers—the outdoors, a pot on a stove, the freezer compartment of a refrigerator—so they are not reversible. But, in a *closed* container, *phase changes* **are** *reversible and reach equilibrium,* just as chemical changes do. In this discussion, we examine the three phase equilibria.

Liquid-Gas Equilibria Vaporization and condensation are familiar events. Let's see how these processes differ in open and closed systems of a liquid in a flask:

1. *Open system: nonequilibrium process.* Picture an *open* flask containing a pure liquid at constant temperature. Within their range of speeds, some molecules at the surface have a high enough E_k to overcome attractions and vaporize. Nearby molecules fill the gap, and with heat supplied by the constant-temperature surroundings, the process continues until the entire liquid phase is gone.

2. *Closed system: equilibrium process.* Now picture a *closed* flask at constant temperature and assume a vacuum exists above a liquid (Figure 12.4A). Two processes take place: Some molecules at the surface have a high enough E_k to *vaporize.* After a short time, molecules in the vapor collide with the surface, and the slower ones are attracted strongly enough to *condense.*

At first, these two processes occur at different rates. The number of molecules in a given surface area is constant, so the number of molecules leaving the surface per unit time—the rate of vaporization—is also constant, and the pressure increases. With time, the number of molecules colliding with and entering the surface—the rate of condensation—increases as the vapor becomes more populated, so the increase in pressure slows. Eventually, the rate of condensation equals the rate of vaporization; from this time onward, *the pressure is constant* (Figure 12.4B).

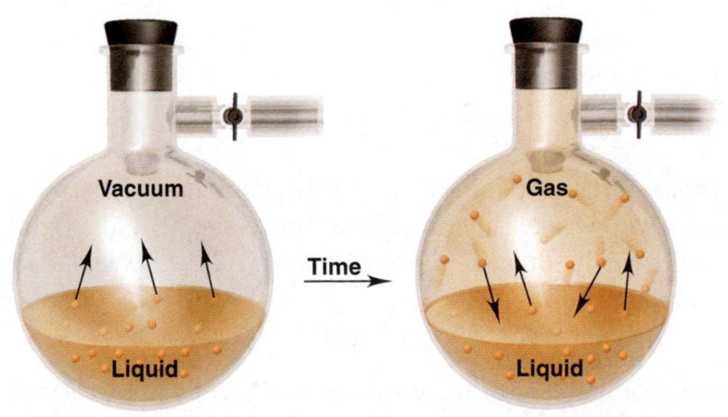

A Molecules in the liquid vaporize.

B Molecules vaporize and condense at the same rate.

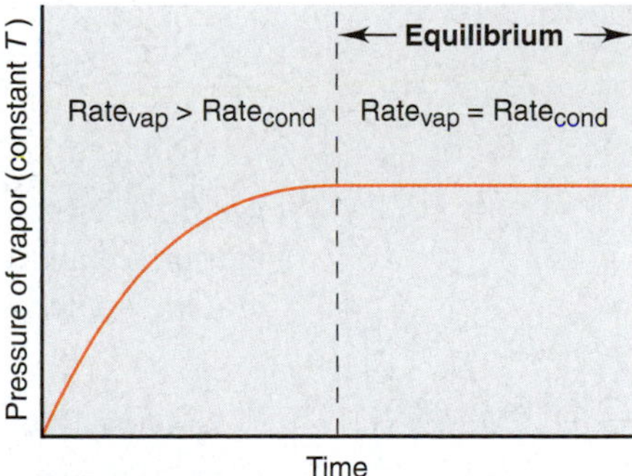

C Plot of pressure vs. time.

Figure 12.4 Liquid-gas equilibrium. **A,** Molecules leave the surface at a constant rate and the pressure rises. **B,** At equilibrium, the same number of molecules leave and enter the liquid in a given time. **C,** Pressure increases until, at equilibrium, it is constant.

Macroscopically, the situation at this point seems static, but at the molecular level, molecules are entering and leaving the liquid at equal rates. The system has reached a state of **dynamic equilibrium:**

$$\text{liquid} \rightleftharpoons \text{gas}$$

The pressure exerted by the vapor at equilibrium is called the *equilibrium vapor pressure,* or just the **vapor pressure,** of the liquid at that temperature. Figure 12.4C depicts the entire process graphically. (In later chapters, beginning with Chapter 17, you'll see that reactions also reach a state of equilibrium, in which reactants are changing into products and products into reactants at the same rate. Thus, the yield of product becomes constant, and no further changes in concentration occur.)

The Effects of Temperature and Intermolecular Forces on Vapor Pressure

The vapor pressure is affected by two factors—a change in temperature and a change in the gas itself, that is, in the type and/or strength of intermolecular forces:

1. *Effect of temperature*. Temperature has a major effect on vapor pressure because it changes the fraction of molecules moving fast enough to escape the liquid and, by the same token, the fraction moving slowly enough to be recaptured (Figure 12.5; see also Figure 5.5). At the higher temperature, T_2, more molecules have enough energy to leave the surface. Thus, in general, *the higher the temperature is, the higher the vapor pressure:*

$$\text{higher } T \Longrightarrow \text{higher } P$$

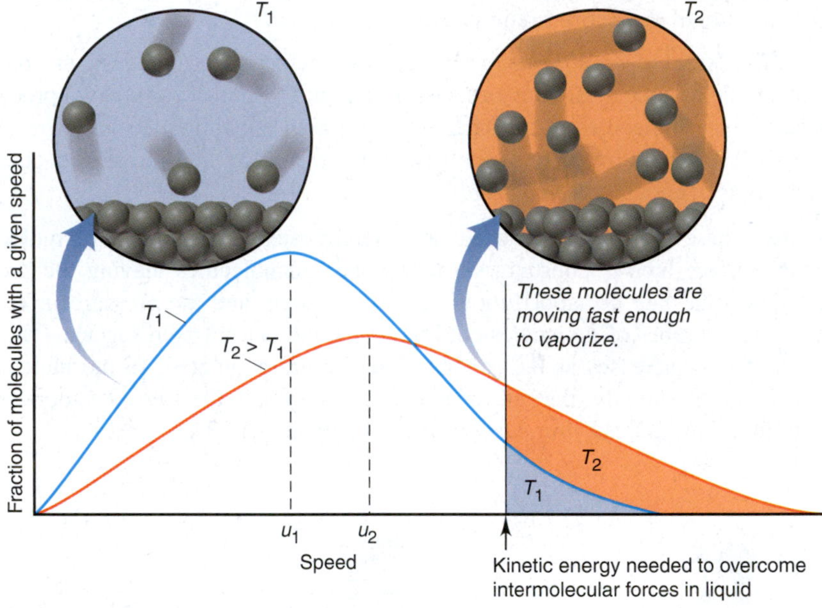

Figure 12.5 The effect of temperature on the distribution of molecular speeds.

2. *Effect of intermolecular forces*. At a given T, all substances have the same average E_k. Therefore, molecules with weaker intermolecular forces are held less tightly at the surface and vaporize more easily. In general, *the weaker the intermolecular forces are, the higher the vapor pressure:*

$$\text{weaker forces} \Longrightarrow \text{higher } P$$

Figure 12.6 shows the vapor pressure of three liquids as a function of temperature:

- The effect of temperature is seen in the steeper rise as the temperature increases.
- The effect of intermolecular forces is seen in the values of the vapor pressure, the short, horizontal dashed lines intersecting the vertical (pressure) axis at a given temperature *(vertical dashed line at 20°C):* the intermolecular forces in diethyl ether (highest vapor pressure) are weaker than those in ethanol, which are weaker than those in water (lowest vapor pressure).

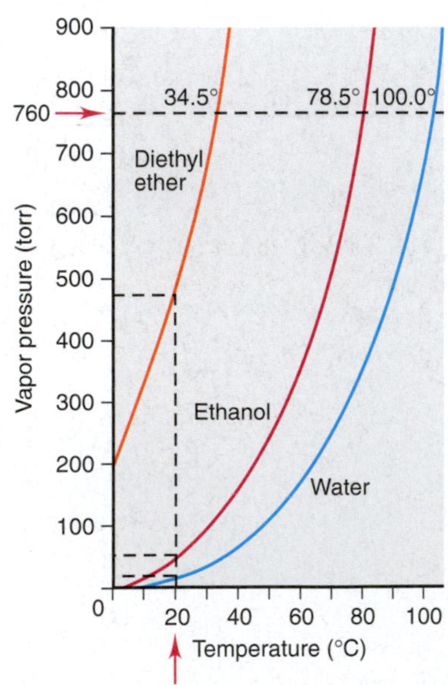

Figure 12.6 Vapor pressure as a function of temperature and intermolecular forces.

Quantifying the Effect of Temperature The nonlinear relationship between P and T is converted to a linear one with the **Clausius-Clapeyron equation:**

$$\ln P = \frac{-\Delta H_{\text{vap}}}{R}\left(\frac{1}{T}\right) + C$$
$$y = \quad m \quad x \; + b$$

where ln P is the natural logarithm of the vapor pressure, ΔH_{vap} is the heat of vaporization, R is the universal gas constant (8.314 J/mol·K), T is the absolute temperature, and C is a constant (not related to heat capacity). The equation is often used to find the heat of vaporization. The equation for a straight line is shown under it in blue, with $y = \ln P$, $x = 1/T$, m (the slope) $= -\Delta H_{\text{vap}}/R$, and b (the y-axis intercept) $= C$.

A plot of ln P vs. $1/T$ gives a straight line, as shown for diethyl ether and water in Figure 12.7.

A two-point version of the Clausius-Clapeyron equation allows us to calculate ΔH_{vap} if the vapor pressures at two temperatures are known:

$$\ln \frac{P_2}{P_1} = \frac{-\Delta H_{vap}}{R}\left(\frac{1}{T_2} - \frac{1}{T_1}\right) \qquad (12.1)$$

If ΔH_{vap} and P_1 at T_1 are known, we can also calculate the vapor pressure (P_2) at any other temperature (T_2) or the temperature at any other pressure.

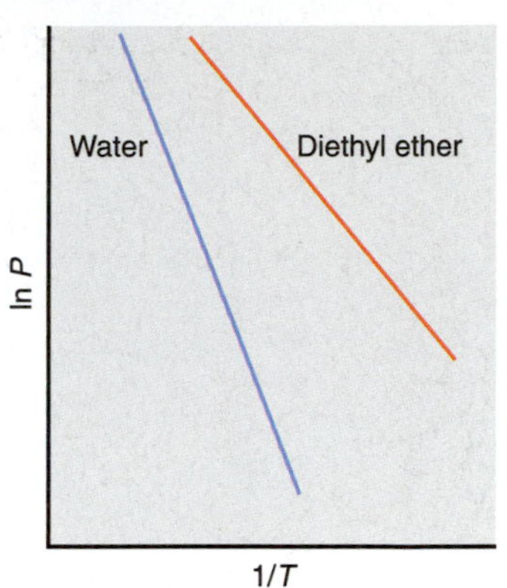

Figure 12.7 Linear plots of the relationship between vapor pressure and temperature. The slope is steeper for water because its ΔH_{vap} is greater.

Sample Problem 12.2 Applying the Clausius-Clapeyron Equation

Problem The vapor pressure of ethanol is 115 torr at 34.9°C. If ΔH_{vap} of ethanol is 38.6 kJ/mol, calculate the temperature (in °C) when the vapor pressure is 760 torr.

Plan We are given ΔH_{vap}, P_1, P_2, and T_1 and substitute them into Equation 12.1 to solve for T_2. The value of R here is 8.314 J/mol·K, so we must convert T_1 to K to obtain T_2, and then convert T_2 back to °C.

Solution Substituting the values into Equation 12.1 and solving for T_2:

$$\ln \frac{P_2}{P_1} = \frac{-\Delta H_{vap}}{R}\left(\frac{1}{T_2} - \frac{1}{T_1}\right)$$

$$T_1 = 34.9°\text{C} + 273.15 = 308.0\ \text{K}$$

$$\ln \frac{760\ \text{torr}}{115\ \text{torr}} = \left(-\frac{38.6\times10^3\ \text{J/mol}}{8.314\ \text{J/mol·K}}\right)\left(\frac{1}{T_2} - \frac{1}{308.0\ \text{K}}\right)$$

$$1.888 = (-4.64\times10^3)\left[\frac{1}{T_2} - (3.247\times10^{-3})\right]$$

$$T_2 = 352\ \text{K}$$

Converting T_2 from K to °C:

$$T_2 = 352\ \text{K} - 273.15 = 79°\text{C}$$

Check Round off to check the math. The change is in the right direction: higher P should occur at higher T. As we discuss next, a substance has a vapor pressure of 760 torr at its *normal boiling point*. Checking the *CRC Handbook of Chemistry and Physics* shows that the boiling point of ethanol is 78.5°C, very close to our answer.

FOLLOW-UP PROBLEM 12.2 At 34.1°C, the vapor pressure of water is 40.1 torr. What is the vapor pressure at 85.5°C? The ΔH_{vap} of water is 40.7 kJ/mol.

Vapor Pressure and Boiling Point Let's discuss what is happening when a liquid boils and then see the effect of pressure on boiling point.

1. *How a liquid boils*. In an *open* container, the weight of the atmosphere bears down on a liquid surface. As the temperature rises, molecules move more quickly throughout the liquid. At some temperature, the average E_k of the molecules in the liquid is great enough for them to form bubbles of vapor *in the interior,* and the liquid boils. At any lower temperature, the bubbles collapse as soon as they start to form because the external pressure is greater than the vapor pressure inside the bubbles. Thus, the **boiling point** *is the temperature at which the vapor pressure equals the external pressure,* which is usually that of the atmosphere. As in condensation and freezing, once boiling begins, the temperature of the liquid remains constant until all of the liquid is gone.

2. *Effect of pressure on boiling point.* The boiling point of a liquid varies with elevation. At high elevations, a lower atmospheric pressure is exerted on the liquid surface, so molecules in the interior need less kinetic energy to form bubbles. Thus, *the boiling point depends on the applied pressure.*

In mountainous regions, food takes *more* time to cook because the boiling point is lower and the boiling liquid is not as hot; for instance, in Boulder, Colorado (elevation 5430 ft, or 1655 m), water boils at 94°C. On the other hand, in a pressure cooker, food takes *less* time to cook because the boiling point is higher at the higher pressure. The *normal boiling point* is observed at standard atmospheric pressure (760 torr, or 101.3 kPa; *long, horizontal dashed line* in Figure 12.6).

Solid-Liquid Equilibria The particles in a crystal are continually jiggling about their fixed positions. As the temperature rises, the particles jiggle more vigorously, until some have enough kinetic energy to break free of their positions. At this point, melting begins. As more molecules enter the liquid (molten) phase, some collide with the solid and become fixed in position again. Because the phases remain in contact, a dynamic equilibrium is established when the melting rate equals the freezing rate. The temperature at which this occurs is called the **melting point.** The temperature remains fixed at the melting point until all the solid melts.

Because liquids and solids are nearly incompressible, pressure has little effect on the rates of melting and freezing: a plot of pressure vs. temperature for a solid-liquid phase change is typically a *nearly* vertical straight line.

Solid-Gas Equilibria Sublimation is not very familiar because solids have *much* lower vapor pressures than liquids. A substance sublimes rather than melts because the intermolecular attractions are not great enough to keep the molecules near each other when they leave the solid state. Some solids *do* have high enough vapor pressures to sublime at ordinary conditions, including dry ice (carbon dioxide), solid iodine, and moth repellants, all nonpolar molecules with weak intermolecular forces.

The plot of pressure vs. temperature for a solid-gas phase change reflects the large effect of temperature on vapor pressure; thus, it resembles the liquid-gas curve in rising steeply with higher temperatures.

Phase Diagrams: Effect of Pressure and Temperature on Physical State

The **phase diagram** of a substance combines the liquid-gas, solid-liquid, and solid-gas curves and gives the conditions of temperature and pressure at which each phase is stable and where phase changes occur.

The Phase Diagram for CO_2 and Most Substances The diagram for CO_2, which is typical of most substances, has four general features (Figure 12.8A):

1. *Regions of the diagram.* Each region presents the conditions of pressure and temperature for which the phase is stable. If another phase is placed under those conditions, it will change to the stable phase. In general, the solid is stable at low temperature and high pressure, the gas at high temperature and low pressure, and the liquid at intermediate conditions.

2. *Lines between regions.* The lines are the phase-transition curves discussed earlier. Any point along a line shows the pressure and temperature at which the phases are in equilibrium. The solid-liquid line has a slightly *positive* slope (slants to the *right* with increasing pressure) because, for most substances, the solid is more dense than the liquid: an increase in pressure converts the liquid to the solid. (Water is *the* major exception.)

3. *The triple point.* The three phase-transition curves meet at the **triple point,** at which all three phases are in equilibrium. As strange as it sounds, at the triple point in Figure 12.8A, CO_2 is subliming and depositing, melting and freezing, and vaporizing and condensing simultaneously! Substances with several solid and/or liquid forms can have more than one triple point.

The CO_2 phase diagram shows why dry ice (solid CO_2) doesn't melt under ordinary conditions. The triple-point pressure is 5.1 atm, so liquid CO_2 doesn't occur at 1 atm because it is not stable. The horizontal dashed line at 1.0 atm crosses the solid-gas line, so when solid CO_2 is heated, it sublimes at −78°C rather than melts. If our normal atmospheric pressure were 5.2 atm, liquid CO_2 *would* occur.

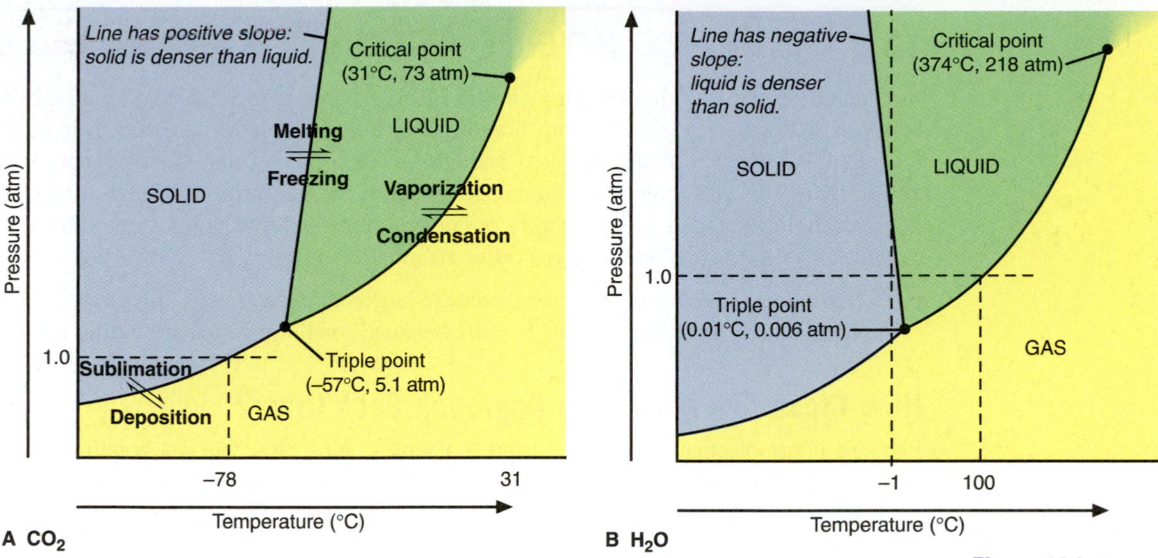

Figure 12.8 Phase diagrams for CO_2 and H_2O. (The slope of each solid-liquid line is exaggerated, and the axes are not linear.)

4. *The critical point.* Heat a liquid in a closed container and its density decreases. At the same time, more of the liquid vaporizes, so the density of the vapor increases. At the **critical point,** the two densities become equal and the phase boundary disappears. The temperature at the critical point is the *critical temperature* (T_c), and the pressure is the *critical pressure* (P_c). The average E_k is so high at this point that the vapor cannot be condensed at any pressure. The two most common gases in air have critical temperatures far below room temperature: O_2 cannot be condensed above -119°C, and N_2 above -147°C.

The Solid-Liquid Line for Water The phase diagram for water has the same four features but differs from others in one major respect that reveals a key property. Unlike almost any other substance, the solid form is *less dense* than the liquid; that is, *water expands upon freezing.* Thus, the solid-liquid line has a *negative* slope (slants to the *left* with increasing pressure): an increase in pressure converts the solid to the liquid, and the higher the pressure, the lower the temperature at which water freezes (Figure 12.8B). The vertical dashed line at -1°C crosses the solid-liquid line, which means that ice melts with only an increase in pressure.

The triple point of water occurs at low pressure (0.006 atm). Therefore, when solid water is heated at 1.0 atm *(horizontal dashed line),* the solid-liquid line is crossed at 0°C, the normal melting point. Thus, ice melts rather than sublimes. The horizontal dashed line then crosses the liquid-gas curve at 100°C, the normal boiling point.

Summary of Section 12.2

- A heating-cooling curve depicts the change in temperature when a substance absorbs or releases heat at a constant rate. Within a phase, temperature (and average E_k) changes. During a phase change, temperature (and average E_k) is constant, but E_p changes. The total enthalpy change for the system is found using Hess's law.
- In a closed container, the liquid and gas phases of a substance reach equilibrium. The vapor pressure, the pressure of the gas at equilibrium, is related directly to temperature and inversely to the strength of the intermolecular forces.
- The Clausius-Clapeyron equation relates the vapor pressure to the temperature and is often used to find ΔH_{vap}.
- A liquid in an open container boils when its vapor pressure equals the external pressure.
- Solid-liquid equilibrium occurs at the melting point. Some solids sublime because they have very weak intermolecular forces.
- The phase diagram of a substance shows the phase that is stable at any P and T, the conditions at which phase changes occur, and the conditions at the critical point and the triple point. Water differs from most substances in that its solid phase is less dense than its liquid phase, so its solid-liquid line has a negative slope.

12.3 • TYPES OF INTERMOLECULAR FORCES

In Chapter 9, we saw that bonding (*intra*molecular) forces are due to the attraction between cations and anions (ionic bonding) or between nuclei and electron pairs (covalent bonding). But the physical behavior of the phases and their changes are due primarily to *inter*molecular (nonbonding) forces, which arise from the attraction *between* molecules with partial charges or between ions and molecules. Coulomb's law explains why the two types of forces differ so much in magnitude:

- *Bonding forces are relatively strong* because larger charges are closer together.
- *Intermolecular forces are relatively weak* because smaller charges are farther apart.

How Close Can Molecules Approach Each Other?

Because intermolecular forces arise from molecules being close but not bonding, we'd like to know how close molecules can get to each other. To see the minimum distance *between* molecules, consider solid Cl_2. When we measure the distances between two Cl nuclei, we obtain two different values (Figure 12.9A):

- *Bond length and covalent radius*. The shorter distance, called the *bond length,* is between *two bonded Cl atoms in the **same** molecule*. One-half this distance is the *covalent radius.*
- *Van der Waals distance and radius*. The longer distance is between *two nonbonded Cl atoms in **adjacent** molecules*. It is called the *van der Waals (VDW) distance*. At this distance, intermolecular attractions balance electron-cloud repulsions; thus, the VDW distance is as close as one Cl_2 molecule can approach another. The **van der Waals radius** is one-half the closest distance between nuclei of identical *nonbonded* atoms. The VDW radius of an atom is *always larger than its covalent radius.* Like covalent radii, VDW radii decrease across a period and increase down a group (Figure 12.9B).

As we discuss intermolecular forces (also called *van der Waals forces*), consult Table 12.2, which compares them with bonding forces.

Figure 12.9 Covalent and van der Waals radii and their periodic trends. **A,** The van der Waals (VDW) radius is one-half the distance between adjacent *nonbonded* atoms ($\frac{1}{2}$ × VDW distance). **B,** Like covalent radii *(blue quarter-circles and numbers),* VDW radii *(red quarter-circles and numbers)* increase down a group and decrease across a period.

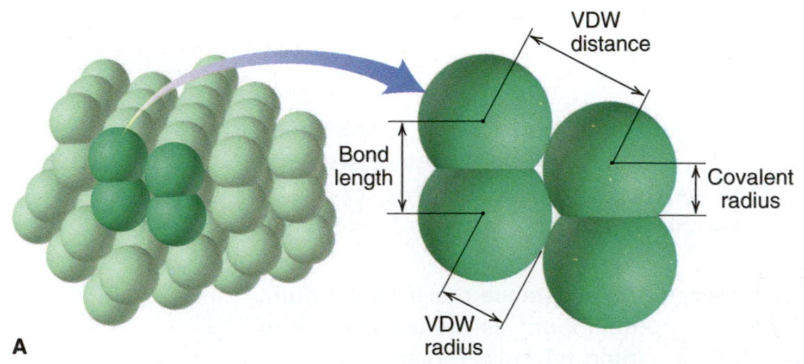

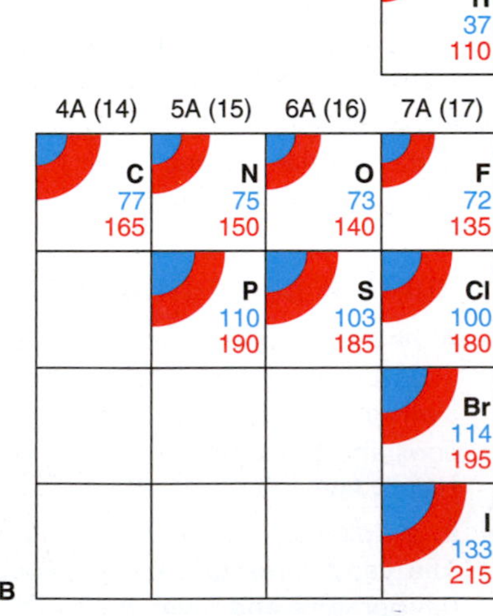

Ion-Dipole Forces

When an ion and a nearby polar molecule (dipole) attract each other, an **ion-dipole force** results. The most important example takes place when an ionic compound dissolves in water. The ions become separated because the attractions between the ions and the oppositely charged poles of the H_2O molecules are stronger than the attractions between the ions themselves. Ion-dipole forces in solutions and their associated energy are discussed fully in Chapter 13.

Table 12.2 Comparison of Bonding and Nonbonding (Intermolecular) Forces

Force	Model	Basis of Attraction	Energy (kJ/mol)	Example
Bonding				
Ionic		Cation–anion	400–4000	NaCl
Covalent		Nuclei–shared e^- pair	150–1100	H—H
Metallic		Cations–delocalized electrons	75–1000	Fe
Nonbonding (Intermolecular)				
Ion-dipole		Ion charge–dipole charge	40–600	$Na^{+}\cdots O$(H)(H)
H bond	δ− δ+ δ− —A—H·······:B—	Polar bond to H–dipole charge (high EN of N, O, F)	10–40	:Ö(H)—H···:Ö(H)—H
Dipole-dipole		Dipole charges	5–25	I—Cl···I—Cl
Ion–induced dipole		Ion charge–polarizable e^- cloud	3–15	$Fe^{2+}\cdots O_2$
Dipole–induced dipole		Dipole charge–polarizable e^- cloud	2–10	H—Cl···Cl—Cl
Dispersion (London)		Polarizable e^- cloud	0.05–40	F—F···F—F

Dipole-Dipole Forces

In Figure 10.11, you saw that an external electric field orients gaseous polar molecules. The polar molecules in liquids and solids lie near each other, and their partial charges act as tiny electric fields that give rise to **dipole-dipole forces:** the positive pole of one molecule attracts the negative pole of another (Figure 12.10). The orientation is more orderly in a solid than in a liquid because the average kinetic energy is lower.

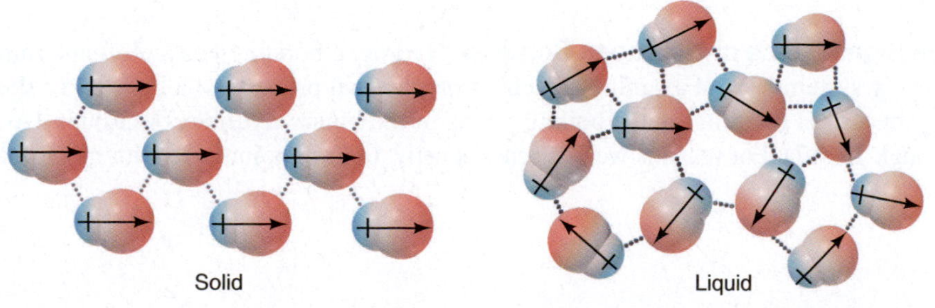

Figure 12.10 Polar molecules and dipole-dipole forces. (Spaces between the molecules are exaggerated.)

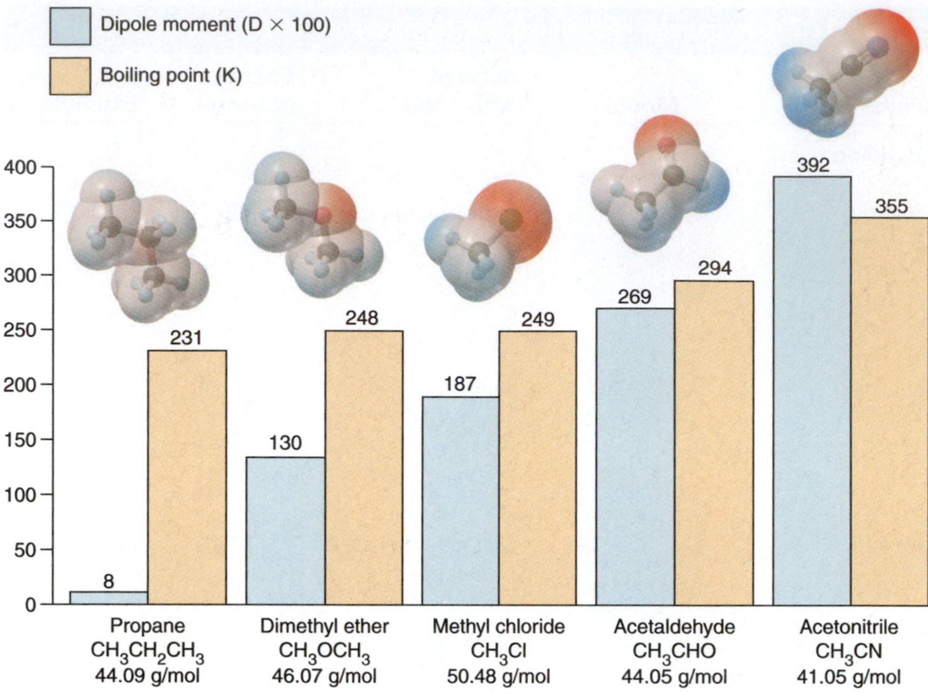

Figure 12.11 Dipole moment and boiling point. (Relative strength of the dipole moments is indicated by color intensities in the electron-density models.)

These forces depend on the magnitude of the molecular dipole moment. For compounds of similar molar mass, the greater the dipole moment, the greater the dipole-dipole forces, so the more energy it takes to separate the molecules; thus the boiling point is higher. Methyl chloride, for instance, has a smaller dipole moment than acetaldehyde and boils at a lower temperature (Figure 12.11).

The Hydrogen Bond

A special type of dipole-dipole force arises between molecules that have *an H atom bonded to a small, highly electronegative atom with lone electron pairs, specifically N, O, or F*. The basis of this force is that the H—N, H—O, and H—F bonds are very polar. When the partially positive H of one molecule is attracted to the partially negative lone pair on the N, O, or F of another molecule, a **hydrogen bond (H bond)** forms. Thus, the atom sequence of an H bond (dotted line) is —B:····H—A—, where A *and* B are N, O, or F. Some examples are

—F:····H—F: —N:····H—N: —F:····H—O— —O:····H—N— —N:····H—O:

The first two are found in samples of pure HF and NH_3.

The small sizes of N, O, and F are essential to H bonding for two reasons:

1. The atoms are so electronegative that their covalently bonded H is highly positive.
2. The lone pair on the N, O, or F of the other molecule can come close to the H.

The Significance of Hydrogen Bonding Hydrogen bonding has a profound impact in many systems. We'll examine one effect on physical properties. Figure 12.12 shows the effect of H bonding on the boiling points of the binary hydrides of Groups 4A(14) through 7A(17). For reasons we'll discuss shortly, boiling points rise with molar mass,

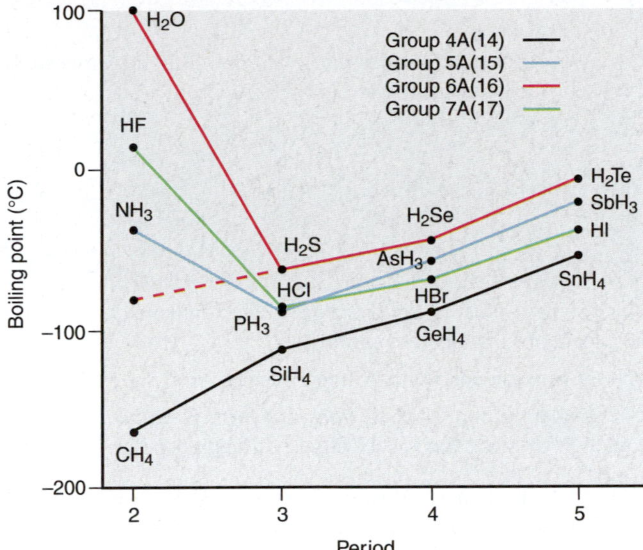

Figure 12.12 **Hydrogen bonding and boiling point.** NH_3, H_2O, and HF have exceptionally high boiling points because they form H bonds.

as the Group 4A(14) hydrides show. However, the first member in each of the other groups—NH_3, H_2O, and HF—deviates enormously from this expected trend. Within samples of these substances, the molecules form strong H bonds, so it takes more energy for the molecules to separate and enter the gas phase. For example, on the basis of molar mass alone, we would expect water to boil about 200°C lower than it actually does *(red dashed line).*

In Section 12.5, we'll discuss the effects that H bonds in water have in nature. In fact, the significance of hydrogen bonding in biological systems cannot be emphasized too strongly. It is a key feature in the structure and function of proteins and nucleic acids and is responsible for the action of many *enzymes,* the proteins that speed metabolic reactions, as well as for the functioning of genes.

Sample Problem 12.3 Drawing Hydrogen Bonds Between Molecules of a Substance

Problem Which of the following substances exhibits H bonding? For any that do, show the H bonds between two of its molecules.

(a) C_2H_6 **(b)** CH_3OH **(c)** $CH_3\overset{\displaystyle O}{\overset{\|}{C}}{-}NH_2$

Plan If the molecule does *not* contain N, O, or F, it cannot form H bonds. If it contains any of these atoms covalently bonded to H, we draw two molecules in the —B:····H—A— pattern.

Solution **(a)** For C_2H_6. No N, O, or F, so no H bonds can form.
(b) For CH_3OH. The H covalently bonded to the O in one molecule forms an H bond to the lone pair on the O of an adjacent molecule:

```
      H
      |
   H—C—H          H
      |           |
   H—Ö:····H—Ö—C—H
                  |
                  H
```

(c) For $CH_3\overset{\overset{O}{\|}}{C}{-}NH_2$. Two of these molecules can form one H bond between an H bonded to N and the O, or they can form two such H bonds:

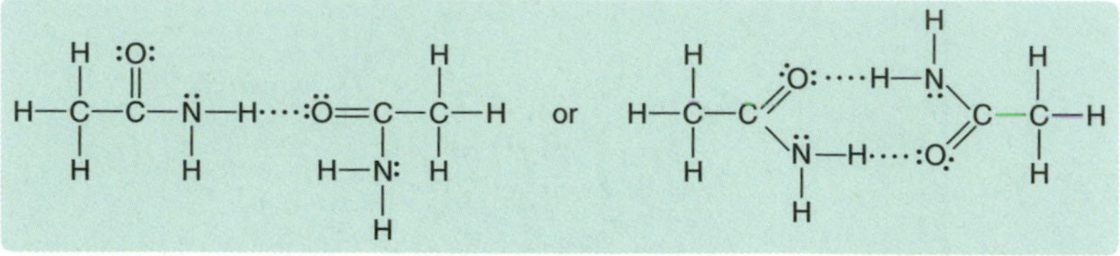

A third possibility (not shown) is an H bond between an H attached to N in one molecule and the lone pair of N in another molecule.

Check The —B:····H—A— sequence (with A and B either N, O, or F) is present.

Comment Note that H covalently bonded to C *does not form H bonds* because carbon is not electronegative enough to make the C—H bond sufficiently polar.

FOLLOW-UP PROBLEM 12.3 Which of these substances exhibits H bonding? Draw the H bond(s) between two molecules of the substance where appropriate.

(a) $CH_3\overset{\overset{O}{\|}}{C}{-}OH$ **(b)** CH_3CH_2OH **(c)** $CH_3\overset{\overset{O}{\|}}{C}CH_3$

Polarizability and Induced Dipole Forces

Even though electrons are attracted to nuclei and localized in bonding and lone pairs, we often picture them as "clouds" of negative charge because they are in constant motion. A nearby electric field can *induce* a distortion in the cloud, pulling electron density toward a positive pole of a field or pushing it away from a negative one:

- *For a nonpolar molecule,* the distortion induces a temporary dipole moment.
- *For a polar molecule,* it enhances the dipole moment already present.

In addition to charged plates connected to a battery, the source of the electric field can be the charge of an ion or the partial charges of a polar molecule.

How easily the electron cloud of an atom (or ion) can be distorted is called its **polarizability.** Smaller particles are less polarizable than larger ones because their electrons are closer to the nucleus and therefore held more tightly. Thus, we observe several trends:

- Polarizability *increases down a group* because atomic size increases and larger electron clouds are easier to distort.
- Polarizability *decreases across a period* because increasing Z_{eff} makes the atoms smaller and holds the electrons more tightly.
- Cations are *less* polarizable than their parent atoms because they are smaller; anions are *more* polarizable because they are larger.

Ion–induced dipole and dipole–induced dipole forces are the two types of charge-induced dipole forces; they are most important in solution, so we'll focus on them in Chapter 13. Nevertheless, *polarizability affects all intermolecular forces.*

Dispersion (London) Forces

So far, we've discussed forces that depend on the existing charge of an ion or a polar molecule. But what forces cause nonpolar substances like octane, chlorine, and argon to condense and solidify? As you'll see, polarizability plays the central role in the most universal intermolecular force.

The intermolecular force responsible for the condensed states of nonpolar substances is the **dispersion force** (or **London force,** named for Fritz London, the physicist who explained its quantum-mechanical basis around 1930). Dispersion forces *are present between all atoms, ions, and molecules* because they are caused by the *motion of electrons in atoms.* Let's examine their key aspects:

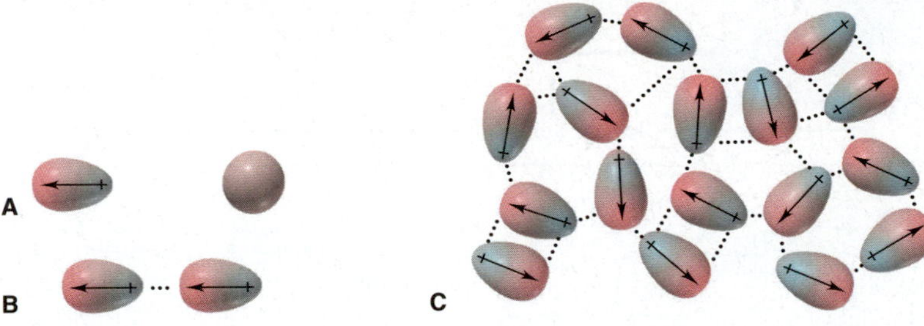

Figure 12.13 Dispersion forces among nonpolar particles. **A,** When atoms are far apart, an instantaneous dipole in one atom *(left)* doesn't influence another. **B,** But, when atoms are close together, the instantaneous dipole induces a dipole in the other. **C,** The process occurs through the sample.

1. *Source*. Picture one atom in a sample of, say, argon gas. Over time, its 18 electrons are distributed uniformly, so the atom is nonpolar. But at any instant, there may be more electrons on one side of the nucleus than the other, which gives the atom an *instantaneous dipole*. When a pair of argon atoms is far apart, they don't influence each other, but when close together, *the instantaneous dipole in one atom induces a dipole in its neighbor,* and they attract each other. This process spreads to other atoms and throughout the sample. At low temperatures, these attractions keep the atoms together (Figure 12.13). Thus, dispersion forces are *instantaneous dipole–induced dipole forces*.

2. *Prevalence*. While they are the *only* force existing between nonpolar particles, *dispersion forces contribute to the energy of attraction in all substances* because they exist between *all* particles. In fact, except for the forces between small, highly polar molecules or between molecules forming H bonds, the *dispersion force is the dominant intermolecular force.* Calculations show, for example, that 85% of the attraction between HCl molecules is due to dispersion forces and only 15% to dipole-dipole forces. Even for water, 75% of the intermolecular attraction comes from H bonds and 25% from dispersion forces.

3. *Relative strength*. The relative strength of dispersion forces depends on the polarizability of the particles, so they are weak for small particles, like H_2 and He, but stronger for larger particles, like I_2 and Xe. *Polarizability depends on the number of electrons, which correlates closely with molar mass* because heavier particles are either larger atoms or molecules with more atoms and, thus, more electrons. For this reason, as molar mass increases down the Group 4A(14) hydrides (see Figure 12.12) or down the halogens or the noble gases, dispersion forces increase and so do boiling points (Figure 12.14).

4. *Effect of molecular shape*. For a pair of nonpolar substances with the same molar mass, stronger attractions occur for a molecular shape that has more area over which the electrons can be distorted. For example, the two five-carbon alkanes, *n*-pentane and neopentane (2,2-dimethylpropane) are structural isomers—same molecular formula (C_5H_{12}) but different properties. *n*-Pentane is more cylindrical and neopentane more spherical (Figure 12.15). Thus, two *n*-pentane molecules make more contact than do two neopentane molecules, so dispersion forces act at more points, and *n*-pentane has a higher boiling point.

Figure 12.16 *(next page)* shows how to determine the intermolecular forces in a sample.

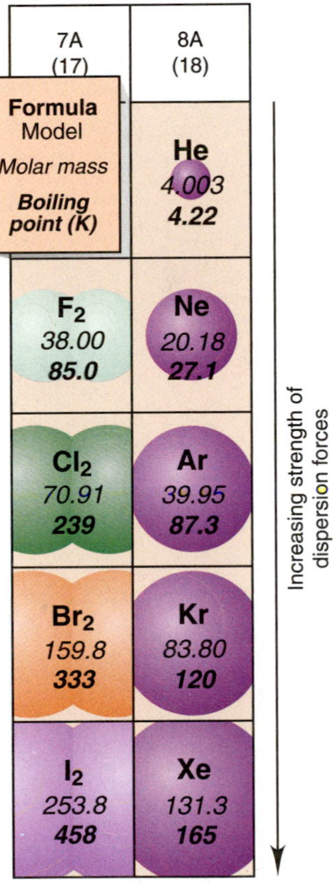

7A (17)
8A (18)
Formula
Model
Molar mass
Boiling point (K)
He
4.003
4.22
F2
38.00
85.0
Ne
20.18
27.1
Cl2
70.91
239
Ar
39.95
87.3
Br2
159.8
333
Kr
83.80
120
I2
253.8
458
Xe
131.3
165
Increasing strength of dispersion forces

Figure 12.14 Molar mass and trends in boiling point.

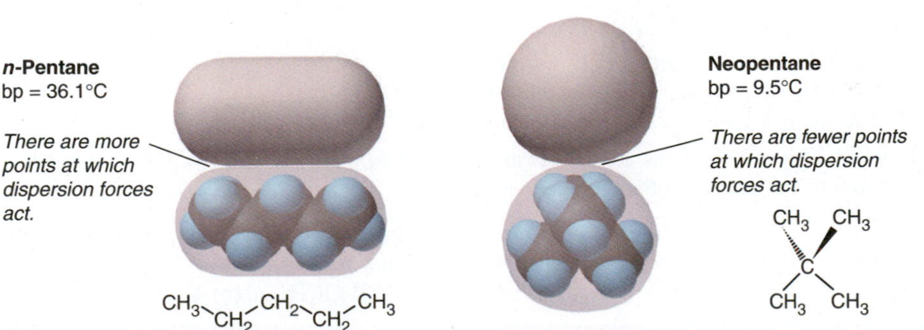

Figure 12.15 Molecular shape, intermolecular contact, and boiling point.

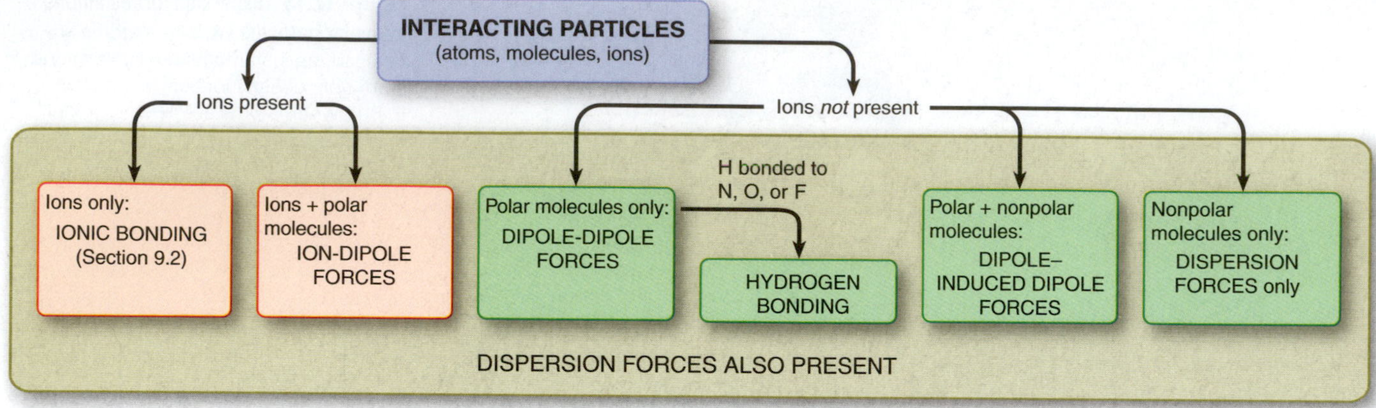

Figure 12.16 Determining the intermolecular forces in a sample.

Sample Problem 12.4 Predicting the Types of Intermolecular Forces

Problem For each substance, identify the key bonding and/or intermolecular force(s), and predict which one of the pair has the higher boiling point:

(a) $MgCl_2$ or PCl_3
(b) CH_3NH_2 or CH_3F
(c) CH_3OH or CH_3CH_2OH
(d) Hexane ($CH_3CH_2CH_2CH_2CH_2CH_3$) or 2,2-dimethylbutane $\left(CH_3C(CH_3)_2CH_2CH_3\right)$

Plan We examine the formulas and structures for key differences between members of the pair: Are ions present? Are molecules polar or nonpolar? Is N, O, or F bonded to H? Do the molecules have different masses or shapes?

To rank boiling points, we consult Figure 12.16 and Table 12.2. Remember that

- Bonding forces are stronger than intermolecular forces.
- Hydrogen bonding is a strong type of dipole-dipole force.
- Dispersion forces are decisive when the difference is molar mass or molecular shape.

Solution **(a)** $MgCl_2$ consists of Mg^{2+} and Cl^- ions held together by ionic bonding forces; PCl_3 consists of polar molecules, so intermolecular dipole-dipole forces are present. The forces in $MgCl_2$ are stronger, so it should have a higher boiling point.

(b) CH_3NH_2 and CH_3F both consist of polar molecules of about the same molar mass. CH_3NH_2 has N—H bonds, so it can form H bonds *(see margin)*. CH_3F contains a C—F bond but no H—F bond, so dipole-dipole forces occur but not H bonds. Therefore, CH_3NH_2 should have the higher boiling point.

(c) CH_3OH and CH_3CH_2OH molecules both contain an O—H bond, so they can form H bonds *(see margin)*. CH_3CH_2OH has an additional —CH_2— group and thus a higher molar mass, which correlates with stronger dispersion forces; therefore, it should have a higher boiling point.

(d) Hexane and 2,2-dimethylbutane are nonpolar molecules of the same molar mass but different molecular shapes *(see margin)*. Cylindrical hexane molecules make more intermolecular contact than more compact 2,2-dimethylbutane molecules do, so hexane should have stronger dispersion forces and a higher boiling point.

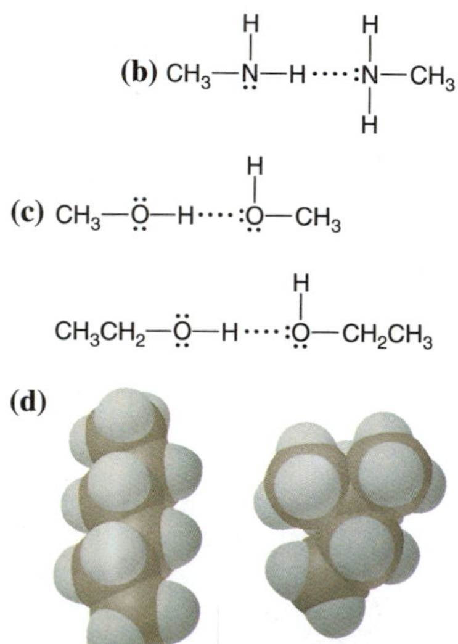

Check The actual boiling points show that our predictions are correct:

(a) $MgCl_2$ (1412°C) and PCl_3 (76°C)
(b) CH_3NH_2 (−6.3°C) and CH_3F (−78.4°C)
(c) CH_3OH (64.7°C) and CH_3CH_2OH (78.5°C)
(d) Hexane (69°C) and 2,2-dimethylbutane (49.7°C)

Comment Dispersion forces are *always* present, but in parts (a) and (b), they are much less significant than the other forces that occur.

FOLLOW-UP PROBLEM 12.4 In each pair, identify the intermolecular forces present for each substance, and predict which substance has the higher boiling point:
(a) CH_3Br or CH_3F **(b)** $CH_3CH_2CH_2OH$ or $CH_3CH_2OCH_3$ **(c)** C_2H_6 or C_3H_8

Summary of Section 12.3

- The van der Waals radius determines the shortest distance over which intermolecular forces operate; it is always larger than the covalent radius.
- Intermolecular forces are much weaker than bonding (intramolecular) forces.
- Ion-dipole forces occur between ions and polar molecules.
- Dipole-dipole forces occur between oppositely charged poles on polar molecules.
- Hydrogen bonding, a special type of dipole-dipole force, occurs when H bonded to N, O, or F is attracted to the lone pair of N, O, or F in another molecule.
- Electron clouds can be distorted (polarized) in an electric field.
- Ion– and dipole–induced dipole forces arise between a charge and the dipole it induces in another molecule.
- Dispersion (London) forces are instantaneous dipole–induced dipole forces that occur among all particles and increase with number of electrons (molar mass). Molecular shape determines the extent of contact between molecules and can be a factor in the strength of dispersion forces.

12.4 • PROPERTIES OF THE LIQUID STATE

Of the three states, only liquids combine the ability to flow with the effects of strong intermolecular forces. We understand this state least at the molecular level. Because of the *random* arrangement of the particles in a gas, any region of the sample is virtually identical to any other. And, different regions of a crystalline solid are identical because of the *orderly* arrangement of the particles (as you'll see in Section 12.6). Liquids, however, have regions that are orderly one moment and random the next. Despite this complexity, many macroscopic properties, such as surface tension, capillarity, and viscosity are well understood.

Surface Tension

Intermolecular forces have different effects on a molecule at the surface compared with one in the interior (Figure 12.17):

- An interior molecule is attracted by others on all sides.
- A surface molecule is attracted only by others below and to the sides, so it experiences a *net attraction downward.*

Therefore, to increase attractions and become more stable, a surface molecule tends to move into the interior. For this reason, *a liquid surface has the fewest molecules and, thus, the smallest area possible.* In effect, the surface behaves like a "taut skin" covering the interior.

The only way to increase the surface area is for molecules to move up by breaking attractions in the interior, which requires energy. The **surface tension** is the energy required to increase the surface area and has units of J/m^2 (Table 12.3). In general, *the stronger the forces between particles, the more energy it takes to increase the surface area, so the greater the surface tension.* Water has a high surface tension because its molecules form multiple H bonds. *Surfactants* (*surf*ace-*act*ive agents), such as soaps, petroleum recovery agents, and fat emulsifiers, decrease the surface tension of water by congregating at the surface and disrupting the H bonds.

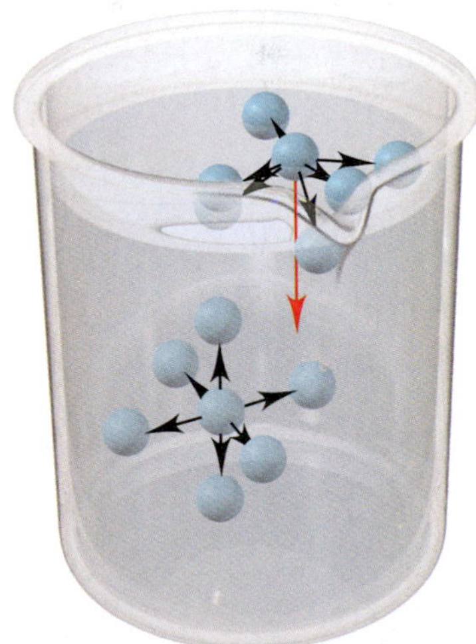

Figure 12.17 The molecular basis of surface tension.

Table 12.3 Surface Tension and Forces Between Particles

Substance	Formula	Surface Tension (J/m^2) at 20°C	Major Force(s)
Diethyl ether	$CH_3CH_2OCH_2CH_3$	1.7×10^{-2}	Dipole-dipole; dispersion
Ethanol	CH_3CH_2OH	2.3×10^{-2}	H bonding
Butanol	$CH_3CH_2CH_2CH_2OH$	2.5×10^{-2}	H bonding; dispersion
Water	H_2O	7.3×10^{-2}	H bonding
Mercury	Hg	48×10^{-2}	Metallic bonding

Capillarity

The rising of a liquid against the pull of gravity through a narrow space, such as a thin tube, is called *capillary action,* or **capillarity.** Capillarity results from a competition between the intermolecular forces within the liquid (cohesive forces) and those between the liquid and the tube walls (adhesive forces). Let's look at the difference between the capillarities of water and mercury in glass:

1. *Water in glass.* When you place a narrow glass tube in water, why does the liquid rise up the tube and form a concave meniscus? Glass is mostly silicon dioxide (SiO_2), so water molecules form adhesive H bonding forces with the O atoms of the glass. As a result, a thin film of water creeps up the wall. At the same time, cohesive H bonding forces between water molecules, which give rise to surface tension, make the surface taut. These adhesive and cohesive forces combine to raise the water level and produce the concave meniscus (Figure 12.18A). The liquid rises until gravity pulling down is balanced by adhesive forces pulling up.

2. *Mercury in glass.* When you place a glass tube in a dish of mercury, why does the liquid drop below the level in the dish and form a convex meniscus? The cohesive forces among the mercury atoms are metallic bonds, so they are *much* stronger than the mostly dispersion adhesive forces between mercury and glass. As a result, the liquid pulls away from the walls. At the same time, the surface atoms are being pulled toward the interior by mercury's high surface tension, so the level drops. These combined forces produce the convex meniscus seen in a laboratory barometer (Figure 12.18B).

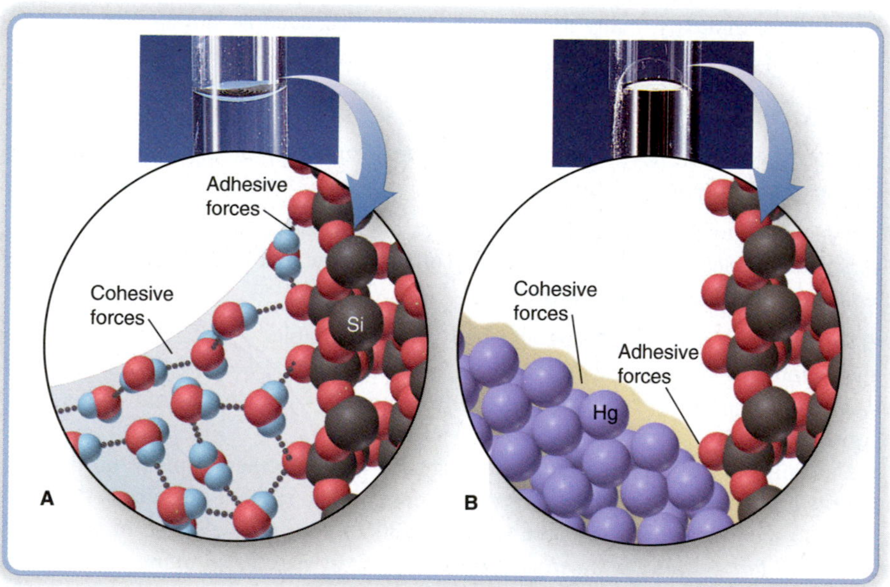

Figure 12.18 Capillary action and the shape of the water or mercury meniscus in glass. **A,** Water displays a concave meniscus. **B,** Mercury displays a convex meniscus.

Table 12.4 Viscosity of Water at Several Temperatures

Temperature (°C)	Viscosity (N·s/m²)*
20	1.00×10^{-3}
40	0.65×10^{-3}
60	0.47×10^{-3}
80	0.35×10^{-3}

*The units of viscosity are newton-seconds per square meter.

Viscosity

Viscosity is the resistance of a fluid to flow, and it results from intermolecular attractions that impede the movement of molecules around and past each other. Both gases and liquids flow, but liquid viscosities are *much* higher because the much shorter distances between their particles result in many more points for intermolecular forces to resist the flow of nearby molecules. Let's examine two factors—temperature and molecular shape—that influence viscosity:

- *Effect of temperature. Viscosity decreases with heating* (Table 12.4). Faster moving molecules overcome intermolecular forces more easily, so the resistance to flow decreases. Next time you heat cooking oil, watch the oil flow more easily and spread out in the pan as it warms.

- *Effect of molecular shape*. Small, spherical molecules make little contact and pour easily, like buckshot from a bowl. Long molecules make more contact and become entangled and pour slowly, like cooked spaghetti from a bowl. Thus, given the same types of intermolecular forces, liquids consisting of longer molecules have higher viscosities.

A striking example of a change in viscosity occurs during the making of syrup. Even at room temperature, a concentrated aqueous sugar solution has a higher viscosity than water because of H bonding among the many hydroxyl (—OH) groups on the ring-shaped sugar molecules. When the solution is slowly heated to boiling, the sugar molecules react with each other and link covalently, gradually forming long chains. Hydrogen bonds and dispersion forces occur at many points along the chains, and the resulting syrup is a viscous liquid that pours slowly and clings to a spoon. When a viscous syrup is cooled, it may become stiff enough to be picked up and stretched—into taffy candy.

Summary of Section 12.4

- Surface tension is a measure of the energy required to increase a liquid's surface area. Greater intermolecular forces within a liquid create higher surface tension.
- Capillarity, the rising of a liquid through a narrow space, occurs when the forces between a liquid and a surface (adhesive) are greater than those within the liquid (cohesive).
- Viscosity, the resistance to flow, depends on molecular shape and decreases with temperature. Stronger intermolecular forces create higher viscosity.

12.5 • THE UNIQUENESS OF WATER

Water is absolutely amazing stuff, with some of the most unusual properties of any substance, but it is so familiar we take it for granted. Like any substance, its properties arise inevitably from those of its atoms. Each O and H atom attains a filled outer level by sharing electrons in single bonds. With two bonding pairs and two lone pairs around O and a large electronegativity difference in each O—H bond, the H_2O molecule is bent and highly polar. This arrangement is crucial because it allows each molecule to engage in four H bonds with its neighbors (Figure 12.19). From these basic atomic and molecular facts emerges unique and remarkable macroscopic behavior.

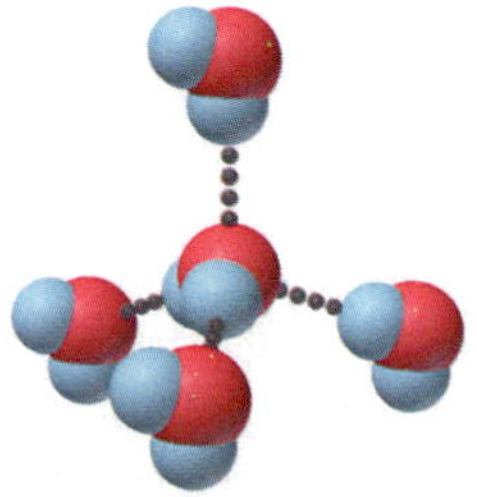

Figure 12.19 H-bonding ability of water. One H_2O molecule can form four H bonds to other molecules, resulting in a tetrahedral arrangement.

Solvent Properties of Water

The *great solvent power* of water results from its polarity and H-bonding ability:

- It dissolves ionic compounds through ion-dipole forces that separate the ions from the solid and keep them in solution (see Figure 4.2).
- It dissolves polar nonionic substances, such as ethanol (CH_3CH_2OH) and glucose ($C_6H_{12}O_6$), by H bonding.
- It dissolves nonpolar atmospheric gases to a limited extent through dipole–induced dipole and dispersion forces.

Water is the environmental and biological solvent, forming the complex solutions we know as oceans, lakes, and cellular fluid. From a chemical point of view, all organisms, from bacteria to humans, are highly organized systems of membranes enclosing and compartmentalizing complex aqueous solutions.

Thermal Properties of Water

When a substance is heated, some of the added energy increases average molecular speed, some increases vibration and rotation, and some overcomes intermolecular forces. The values for two of water's thermal properties have global impacts.

1. *Specific heat capacity.* Because water has so many strong H bonds, its *specific heat capacity* is higher than any common liquid. With oceans covering 70% of Earth's surface, daytime energy from the Sun causes relatively small changes in temperature, allowing life to survive. On the waterless, airless Moon, temperatures range from 100°C to −150°C during a complete lunar day. Even in Earth's deserts, day-night temperature differences of 40°C are common.

2. *Heat of vaporization.* Numerous strong H bonds give water a very *high heat of vaporization.* Two examples show why this is crucial. The average adult has 40 kg of body water and generates about 10,000 kJ of heat a day from metabolism. If this heat were used to increase the average E_k of water molecules in the body, the rise in body temperaure of tens of degrees would mean immediate death. Instead, the heat is converted to E_p as it breaks H bonds and evaporates sweat, resulting in a stable body temperature and minimal loss of body fluid. On a planetary scale, the Sun's energy vaporizes ocean water in warm latitudes, and the potential energy is released as heat to warm cooler regions when the vapor condenses to rain. This global-scale cycling of water powers many weather patterns.

Surface Properties of Water

Hydrogen bonding is also responsible for water's *high surface tension* and *high capillarity.* Except for some molten metals and salts, water has the highest surface tension of any liquid. It keeps plant debris resting on a pond surface, providing shelter and nutrients for fish and insects. High capillarity means water rises through the tiny spaces between soil particles, so plant roots can absorb deep groundwater during dry periods.

The Unusual Density of Solid Water

In the solid state, the tetrahedral arrangement of H-bonded water molecules (see Figure 12.19) leads to the hexagonal, *open structure* of ice (Figure 12.20A), and the symmetrical beauty of snowflakes (Figure 12.20B) reflects this hexagonal organization. The large spaces within ice make *the solid less dense than the liquid* and explain the negative slope of the solid-liquid line in the phase diagram for water (see Figure 12.8B). As pressure is applied, some H bonds break, so the ordered crystal structure is disrupted, and the ice liquefies. When ice melts at 0°C, the loosened molecules pack much more closely, filling spaces in the collapsing solid structure. As a result, liquid water is most dense (1.000 g/mL) at around 4°C (3.98°C). With more heating, the density decreases through normal thermal expansion.

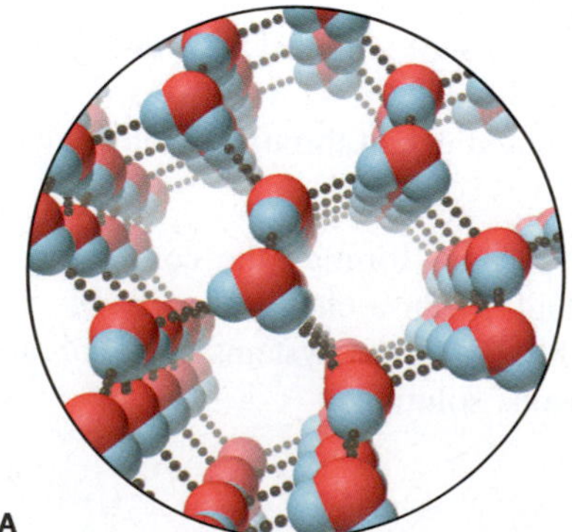

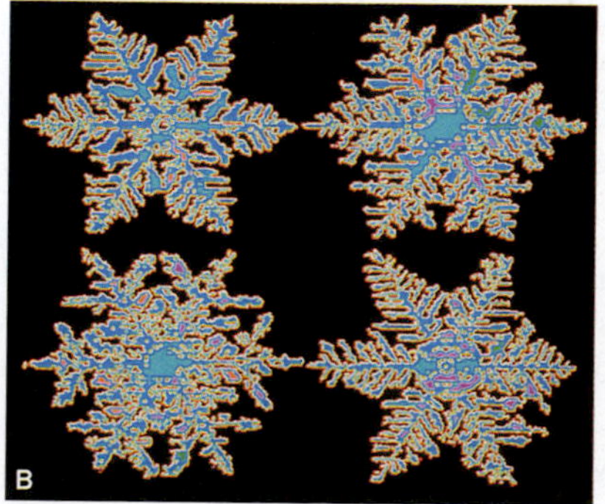

Figure 12.20 The hexagonal structure of ice. **A,** The open, hexagonal molecular structure of ice. **B,** The beauty of six-pointed snowflakes reflects this hexagonal structure.

This change in density is vital for freshwater life. When the surface of a lake freezes in winter, the ice floats. If the solid were denser than the liquid, as is true for nearly every other substance, the surface water would freeze and sink until the entire lake was solid. Aquatic life would not survive from year to year. As lake water becomes colder in the fall and early winter, it becomes more dense *before* it freezes. Similarly, in spring, less dense ice thaws to form more dense water *before* the water expands. During both of these seasonal density changes, the top layer of water reaches the high-density point first and sinks. The next layer of water rises because it is slightly less dense, reaches 4°C, and likewise sinks. This sinking and rising distribute nutrients and dissolved oxygen.

Summary of Section 12.5

- The atomic properties of H and O result in water's bent molecular shape, polarity, and H-bonding ability.
- These properties give water the ability to dissolve many ionic and polar compounds.
- Water's high specific heat capacity and heat of vaporization give Earth and its organisms a narrow temperature range.
- Water's exceptionally high surface tension and capillarity are essential to plants and animals.
- Because water expands on freezing, lake life survives in winter and nutrients mix from seasonal density changes.

12.6 • THE SOLID STATE: STRUCTURE, PROPERTIES, AND BONDING

Stroll through a museum's mineral collection, and you'll be struck by the variety and beauty of crystalline solids. In this section, we discuss the structural features of these and other solids and the intermolecular forces that create them. We also consider the main bonding model that explains many properties of solids.

Structural Features of Solids

We can divide solids into two broad categories:

- **Crystalline solids** have well defined shapes because their particles—atoms, molecules, or ions—occur in an orderly arrangement (Figure 12.21).
- **Amorphous solids** have poorly defined shapes because their particles lack an orderly arrangement throughout the sample. Examples are rubber and glass.

Figure 12.21 The beauty of crystalline solids. **A,** Wulfenite. **B,** Barite *(left)* on calcite *(right)*. **C,** Beryl (emerald). **D,** Quartz (amethyst).

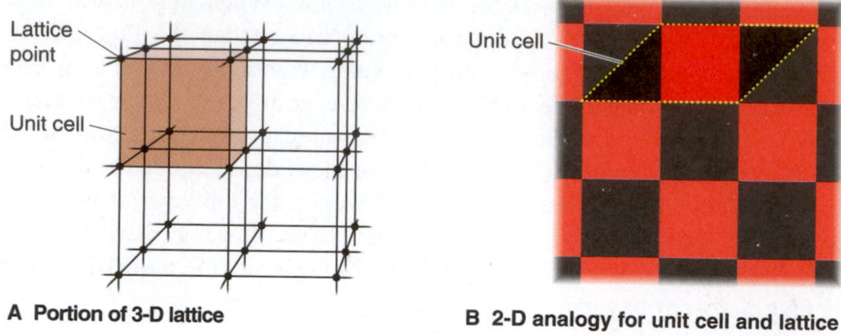

Figure 12.22 The crystal lattice and the unit cell. **A,** A small portion of a lattice is shown as points connected by lines, with a unit cell *(colored)*. **B,** A checkerboard analogy for a lattice.

The Crystal Lattice and the Unit Cell The particles in a crystal are packed tightly in an orderly, three-dimensional array. As the simplest case, consider the particles as *identical* spherical atoms, and imagine a point at the center of each. The collection of points forms a regular pattern called the crystal **lattice.** The lattice consists of *all points with identical surroundings;* that is, there would be no way to tell if you moved from one lattice point to another.

Figure 12.22A shows a portion of a lattice and the **unit cell,** the *smallest* portion that gives the crystal if it is repeated in all directions. A two-dimensional analogy for a unit cell and the resulting crystal lattice appears in a checkerboard (Figure 12.22B), a section of tiled floor, a strip of wallpaper, or any other pattern that is constructed from a repeating unit.

There are 7 crystal systems and 14 types of unit cells that occur in nature, but we will be concerned primarily with the *cubic system*. The solid states of a majority of metallic elements, some covalent compounds, and many ionic compounds occur as cubic lattices. A key parameter of any lattice is the **coordination number,** the number of *nearest* neighbors of a particle. There are three types of cubic unit cells:

1. In the **simple cubic unit cell** (Figure 12.23A), the centers of eight identical particles define the corners of a cube (shown in the expanded view, *top row*). The particles touch along the cube edges (see the space-filling view, *second row*), but they do not touch diagonally along the cube faces or through its center. An expanded portion of the crystal *(third row)* shows that the coordination number of each particle is 6: four in its own layer, one in the layer above, and one in the layer below.
2. In the **body-centered cubic unit cell** (Figure 12.23B), identical particles lie at each corner *and* in the center of the cube. Those at the corners do not touch each other, but they all touch the one in the center. Each particle is surrounded by eight nearest neighbors, four above and four below, so the coordination number is 8.
3. In the **face-centered cubic unit cell** (Figure 12.23C), identical particles lie at each corner *and* in the center of each face but not in the center of the cube. Particles at the corners touch those in the faces but not each other. The coordination number is 12.

How many particles make up a unit cell? For particles of the same size, *the higher the coordination number is, the greater the number of particles in a given volume.* Since one unit cell touches another, with no gaps, a particle at a corner or face is *shared* by adjacent cells. In the cubic unit cells, the particle at each corner is part of eight adjacent cells (Figure 12.23, *third row*), so one-eighth of each particle belongs to each cell *(bottom row)*. There are eight corners in a cube, so

- A simple cubic unit cell contains $8 \times \frac{1}{8}$ particle $= 1$ particle.
- A body-centered cubic unit cell contains $8 \times \frac{1}{8}$ particle $= 1$ particle plus 1 particle in the center, for a total of 2 particles.
- A face-centered cubic unit cell contains $8 \times \frac{1}{8}$ particle $= 1$ particle plus one-half particle in each of the six faces, or $6 \times \frac{1}{2}$ particle $= 3$ particles, for a total of 4 particles.

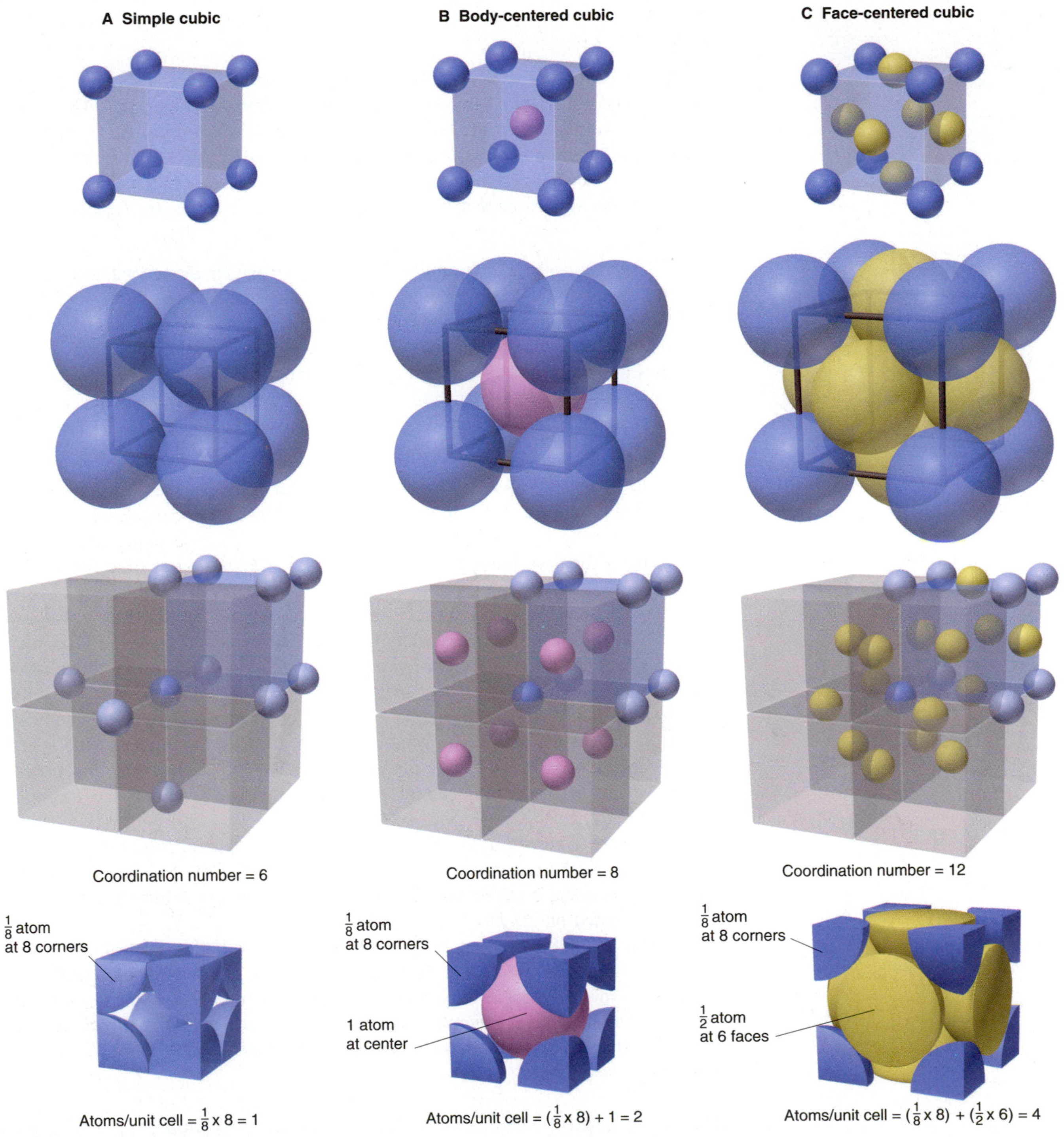

Figure 12.23 The three cubic unit cells. **A,** Simple cubic unit cell. **B,** Body-centered cubic unit cell. **C,** Face-centered cubic unit cell. *Top row:* Cubic arrangements of atoms in expanded view. *Second row:* Space-filling view of these cubic arrangements. All atoms are identical but, for clarity, corner atoms are blue, body-centered atoms pink, and face-centered atoms yellow. *Third row:* A unit cell *(shaded blue)* in an expanded portion of the crystal. The number of nearest neighbors around one particle *(dark blue in center)* is the coordination number. *Bottom row:* The total numbers of atoms in the actual unit cell. The simple cubic has one atom, the body-centered has two, and the face-centered has four.

The efficient packing of fruit.

Packing Efficiency and the Creation of Unit Cells Unit cells result from the ways atoms pack together, which are similar to the ways that macroscopic spheres—marbles, golf balls, fruit—are packed *(see photo)*. Let's pack *identical* spheres to create the three cubic unit cells and the hexagonal unit cell and determine the **packing efficiency,** the percentage of the total volume occupied by the spheres themselves:

1. *The simple cubic unit cell.* When we arrange the first layer of spheres in vertical and horizontal rows, large diamond-shaped spaces are formed (Figure 12.24A, *cutaway portion*). If we place the next layer of spheres *directly above* the first, we obtain an arrangement based on the *simple* cubic unit cell (Figure 12.24B). The spheres occupy only 52% of the unit-cell volume, so 48% is empty space between them. This is a very inefficient way to pack spheres, so neither fruit nor atoms are typically packed this way.

2. *The body-centered cubic unit cell.* Rather than placing the second layer (colored *green* for clarity) directly above the first, we use space more efficiently by placing it on the diamond-shaped spaces in the first layer (Figure 12.24C). Then we pack the third layer onto the diamond-shaped spaces in the second, which makes the first and third layers line up vertically. This arrangement is based on the *body-centered* cubic unit cell, and its packing efficiency is much higher at 68%. Several metallic elements, including chromium, iron, and all the Group 1A(1) elements, have a crystal structure based on this unit cell.

3. *The hexagonal and face-centered cubic unit cells.* Spheres are packed most efficiently in these cells. First, in the bottom layer (labeled *a, orange*), we shift every other row laterally so that the large diamond-shaped spaces become smaller triangular spaces. Then we place the second layer (labeled *b, green*) over these spaces (Figure 12.24D).

In layer *b,* notice that some spaces are orange because they lie above *spheres* in layer *a,* whereas other spaces are white because they lie above *spaces* in layer *a*. We can place the third layer in either of two ways, which gives rise to two different unit cells:

- *Hexagonal unit cell.* If we place the third layer of spheres *(orange)* over the orange spaces (look down and left to Figure 12.24E), they lie directly over the spheres in layer *a*. Every other layer is placed identically (an *abab. . .* layering pattern), and we obtain **hexagonal closest packing,** which is based on the *hexagonal* unit cell.
- *Face-centered unit cell.* If we place the third layer of spheres *(blue)* over the white spaces in layer *b* (look down and right to Figure 12.24F), the placement is different from layers *a* and *b* (an *abcabc. . .* pattern) and we obtain **cubic closest packing,** which is based on the *face-centered cubic* unit cell.

The packing efficiency of both hexagonal and cubic closest packing is 74%, and the coordination number of both is 12. Most metallic elements crystallize in one or the other of these arrangements. Magnesium, titanium, and zinc are some that adopt the hexagonal structure; nickel, copper, and lead adopt the cubic structure, as do many ionic compounds and other substances, such as CO_2, CH_4, and most noble gases.

Observing Crystal Structures: X-Ray Diffraction Analysis Our understanding of crystalline solids is based on the ability to "see" their crystal structures. One of the most powerful tools for doing this is **x-ray diffraction analysis.** In Chapter 7, we discussed wave diffraction and saw how interference patterns of bright and dark regions appear when light passes through closely spaced slits, especially those that are spaced at the distance of the light's wavelength (see Figure 7.5). Because x-ray wavelengths are about the same size as the spaces between layers of particles in many solids, the layers diffract x-rays.

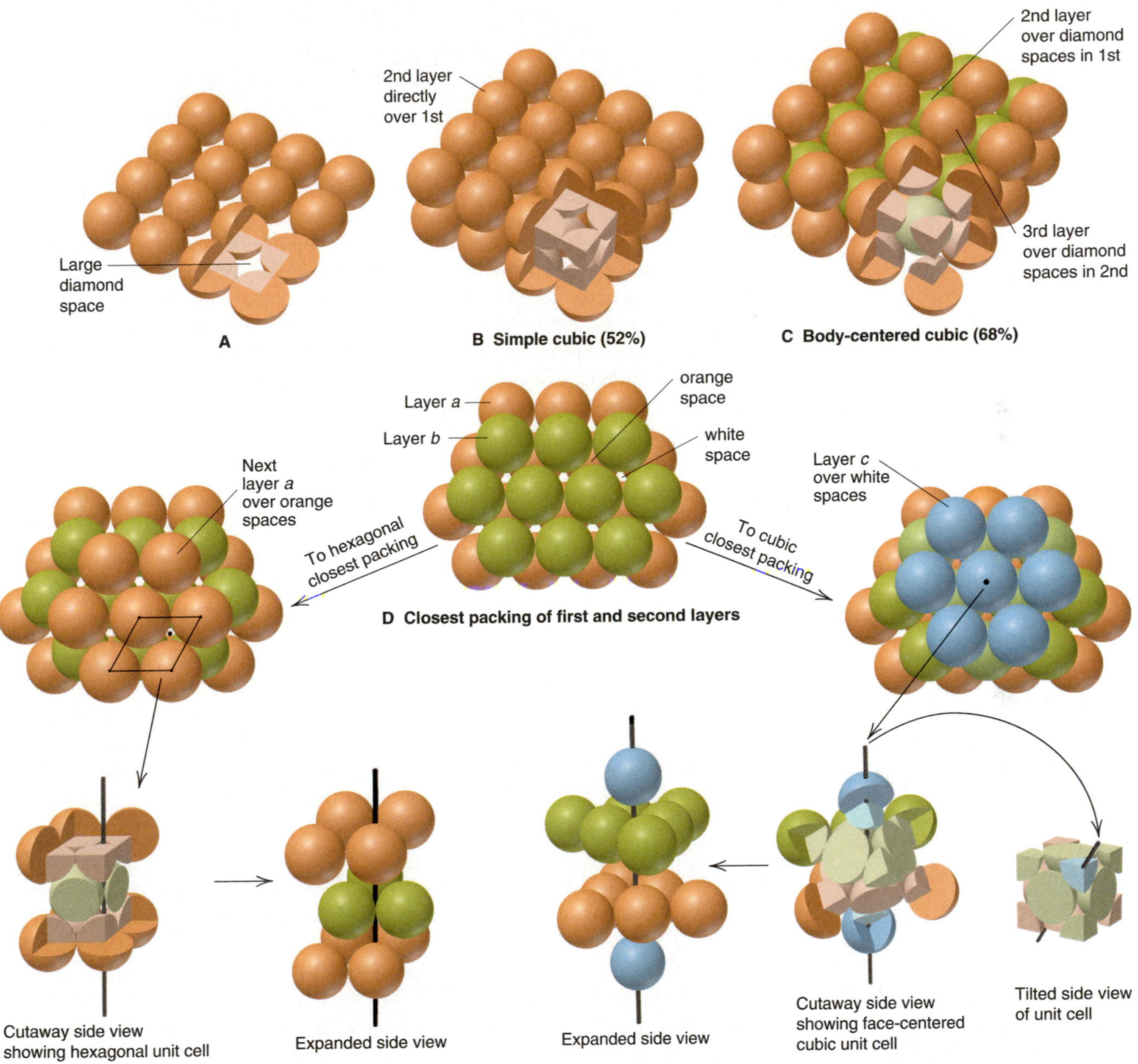

Figure 12.24 **Packing spheres to obtain three cubic and hexagonal unit cells.** **A,** In the first layer, each sphere lies next to another horizontally and vertically; note the large diamond-shaped spaces *(see cutaway)*. **B,** If the spheres in the next layer lie directly over those in the first, the packing is based on the simple cubic unit cell *(pale orange cube, lower right corner)*. **C,** If the spheres in the next layer lie in the diamond-shaped spaces of the first layer, the packing is based on the *body-centered cubic* unit cell *(lower right corner)*. **D,** The closest possible packing of the first layer *(layer a, orange)* is obtained by shifting every other row in part A to obtain smaller triangular spaces. The spheres of the second layer *(layer b, green)* are placed above these spaces; note the orange and white spaces that result. **E,** When the third layer *(next layer a, orange)* is placed over the orange spaces, we obtain an *abab. . .* pattern and the hexagonal unit cell. **F,** When the third layer *(layer c, blue)* covers the white spaces, we get an *abcabc. . .* pattern and the face-centered cubic unit cell.

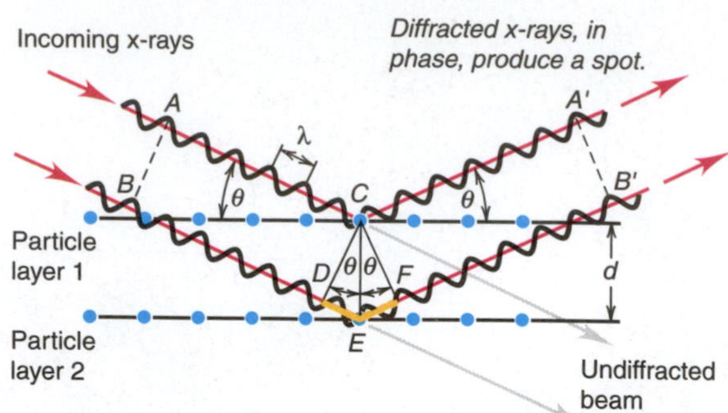

Figure 12.25 Diffraction of x-rays by crystal planes.

Let's see how this technique is used to measure a key parameter in a crystal structure: the distance (d) between layers of atoms. Figure 12.25 depicts a side view of two layers in a simplified lattice. Two waves impinge on the crystal at an angle θ and are diffracted at the same angle by adjacent layers. When the first wave strikes the top layer and the second strikes the next layer, the waves are *in phase* (peaks aligned with peaks and troughs with troughs). If they are still in phase after being diffracted, a spot appears on a nearby photographic plate. Note that this will occur only if the additional distance traveled by the second wave ($DE + EF$ in the figure) is a whole number of wavelengths, $n\lambda$, where n is an integer (1, 2, 3, and so on). From trigonometry, we find that

$$n\lambda = 2d \sin \theta$$

where θ is the known angle of incoming light, λ is its known wavelength, and d is the unknown distance between the layers in the crystal. This relationship is the *Bragg equation,* named for W. H. Bragg and his son W. L. Bragg, who shared the Nobel Prize in physics in 1915 for their work on crystal structure analysis.

Rotating the crystal changes the angle of incoming radiation and produces a different set of spots. Modern x-ray diffraction equipment automatically rotates the crystal and measures thousands of diffractions, and a computer calculates the distances and angles within the lattice.

In addition to the distance between layers, another piece of data obtained from x-ray crystallography is the edge length of the unit cell. Figure 12.26 shows how the edge length and basic geometry are used to find the atomic (or ionic) radii in the three cubic unit cells. Sample Problem 12.5 demonstrates this approach.

Sample Problem 12.5 Determining Atomic Radius from the Unit Cell

Problem Copper adopts cubic closest packing, and the edge length of the unit cell is 361.5 pm. What is the atomic radius of copper?

Plan Cubic closest packing gives a face-centered cubic unit cell, and we know the edge length. With Figure 12.26 and $A = 361.5$ pm, we solve for r *(see margin).*

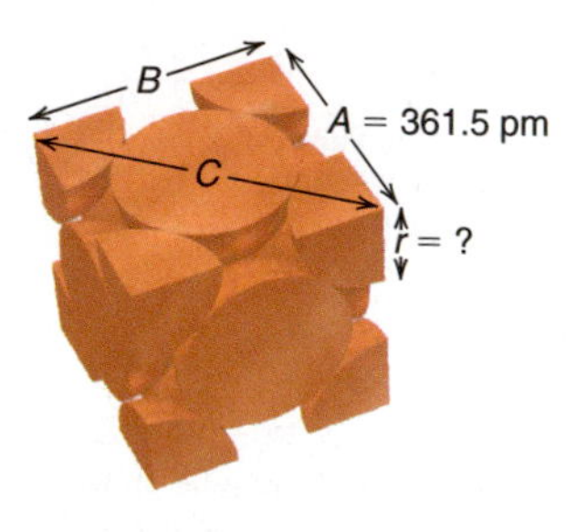

Solution Using the Pythagorean theorem to find C, the diagonal of the cell's face:

$$C = \sqrt{A^2 + B^2}$$

The unit cell is a cube, so $A = B$. Therefore,

$$C = \sqrt{2A^2} = \sqrt{2(361.5\ \text{pm})^2} = 511.2\ \text{pm}$$

Finding r: $C = 4r$. Therefore, $r = 511.2\ \text{pm}/4 =$ 127.8 pm.

Check Rounding and quickly checking the math gives

$$C = \sqrt{2(4\times10^2\ \text{pm})^2} = \sqrt{2(16\times10^4\ \text{pm}^2)}$$

or ~500–600 pm; thus, $r \approx$ 125–150 pm. The actual value for copper is 128 pm (see Figure 8.10).

FOLLOW-UP PROBLEM 12.5 Iron crystallizes in a body-centered cubic structure. If the atomic radius of Fe is 126 pm, find the edge length (in nm) of the unit cell.

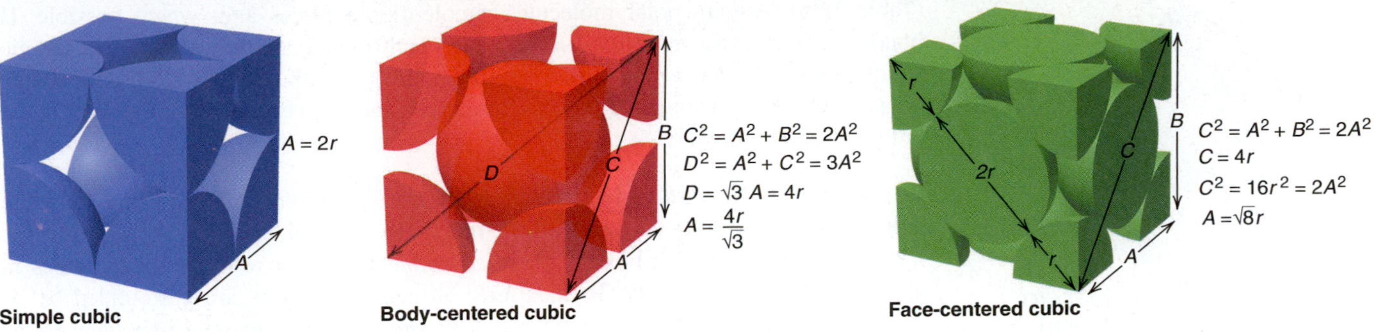

Figure 12.26 Edge length and atomic (ionic) radius in the three cubic unit cells.

Types and Properties of Crystalline Solids

The five most important types of solids are defined by the type(s) of particle(s) in the crystal (Table 12.5). We'll highlight interparticle forces and physical properties.

Atomic Solids Individual atoms held together only by *dispersion forces* form an **atomic solid,** and the noble gases [Group 8A(18)] are the only substances that form such solids. The very weak forces among the atoms mean melting and boiling points and heats of vaporization and fusion are all very low, rising smoothly with increasing molar mass. Argon crystallizes in a cubic closest packed structure (Figure 12.27), as do the other noble gases.

Figure 12.27 Cubic closest packing of frozen argon (face-centered cubic unit cell).

Molecular Solids In the many thousands of **molecular solids,** individual molecules occupy the lattice points. Various combinations of dipole-dipole, dispersion, and H-bonding forces account for a wide range of physical properties. Dispersion forces in nonpolar substances lead to melting points that generally increase with molar mass

Table 12.5 Characteristics of the Major Types of Crystalline Solids

Type	Particle(s)	Interparticle Forces	Physical Properties	Examples [mp, °C]
Atomic	Atoms	Dispersion	Soft, very low mp, poor thermal and electrical conductors	Group 8A(18) (Ne [−249] to Rn [−71])
Molecular	Molecules	Dispersion, dipole-dipole, H bonds	Fairly soft, low to moderate mp, poor thermal and electrical conductors	*Nonpolar** O_2 [−219], C_4H_{10} [−138] Cl_2 [−101], C_6H_{14} [−95], P_4 [44.1] *Polar* SO_2 [−73], $CHCl_3$ [−64], HNO_3 [−42], H_2O [0.0], CH_3COOH [17]
Ionic	Positive and negative ions	Ion-ion attraction	Hard and brittle, high mp, good thermal and electrical conductors when molten	NaCl [801] CaF_2 [1423] MgO [2852]
Metallic	Atoms	Metallic bond	Soft to hard, low to very high mp, excellent thermal and electrical conductors, malleable and ductile	Na [97.8] Zn [420] Fe [1535]
Network covalent	Atoms	Covalent bond	Very hard, very high mp, usually poor thermal and electrical conductors	SiO_2 (quartz) [1610] C (diamond) [~4000]

*Nonpolar molecular solids are arranged in order of increasing molar mass. Note the correlation with increasing melting point (mp).

Figure 12.28 Cubic closest packing (face-centered unit cell) of frozen CH_4.

(Table 12.5). Among polar molecules, dipole-dipole forces and, where possible, H bonding occur. Most molecular solids have much higher melting points than atomic solids (noble gases) but much lower melting points than other types of solids. Methane crystallizes with a face-centered cubic unit cell, and the center of each carbon is the lattice point (Figure 12.28).

Ionic Solids: The Example of Sodium Chloride In crystalline **ionic solids,** the unit cell contains particles with whole, rather than partial, charges. As a result, the interparticle forces (ionic bonds) are *much* stronger than the van der Waals forces in atomic or molecular solids. To maximize attractions, cations are surrounded by as many anions as possible, and vice versa, with *the smaller of the two ions lying in the spaces (holes) formed by the packing of the larger.* Because the unit cell is the smallest portion of the crystal that maintains the overall spatial arrangement, it is also the smallest portion that maintains the overall chemical composition. In other words, *the unit cell has the same cation/anion ratio as the empirical formula.*

Ionic compounds adopt several different crystal structures, but many use cubic closest packing. As an example, let's consider a structure that has a 1/1 ratio of ions. The *sodium chloride structure* is found in many compounds, including most of the alkali metal [Group 1A(1)] halides and hydrides, the alkaline earth metal [Group 2A(2)] oxides and sulfides, several transition-metal oxides and sulfides, and most of the silver halides. To visualize this structure, first imagine Cl^- anions and Na^+ cations organized separately in face-centered cubic (cubic closest packed) arrays. The crystal structure arises when these two arrays penetrate each other such that the smaller Na^+ ions end up in the holes between the larger Cl^- ions, as shown in Figure 12.29A. Thus, each Na^+ is surrounded by six Cl^-, and vice versa (coordination number = 6). Figure 12.29B is a space-filling depiction of the unit cell showing a face-centered cube of Cl^- ions with Na^+ ions between them. Note the four Cl^- $[(8 \times \frac{1}{8}) + (6 \times \frac{1}{2}) = 4\ Cl^-]$ and four Na^+ $[(12 \times \frac{1}{4}) + 1$ in the center $= 4\ Na^+]$, giving a 1/1 ion ratio.

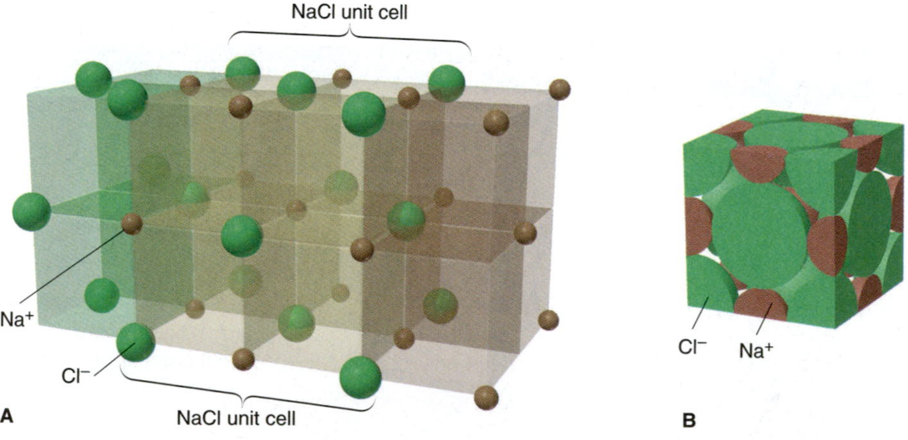

Figure 12.29 The sodium chloride structure. **A,** Expanded view. **B,** Space-filling model of the NaCl unit cell.

The properties of ionic solids are a direct consequence of the *fixed ion positions* and *very strong interionic forces,* which create a high lattice energy. Thus, ionic solids typically have high melting points and low electrical conductivities. When a large amount of heat is supplied and the ions gain enough kinetic energy to break free of their positions, the solid melts and the mobile ions conduct a current. Ionic compounds are hard because only a strong external force can change the relative positions of many trillions of interacting ions. If enough force *is* applied to move them, ions of like charge are brought near each other, and their repulsions crack the crystal (see Figure 9.8).

Metallic Solids Most metallic elements crystallize in one of the two closest packed structures (Figure 12.30). In contrast to the weak dispersion forces in atomic solids, powerful metallic bonding forces hold atoms together in **metallic solids.** The properties of metals—high electrical and thermal conductivity, luster, and malleability—result from their delocalized electrons (Section 9.1). Melting points and hardnesses of metallic solids are also related to packing efficiency and number of valence electrons. We discuss bonding models that explain these metallic properties in the next two subsections.

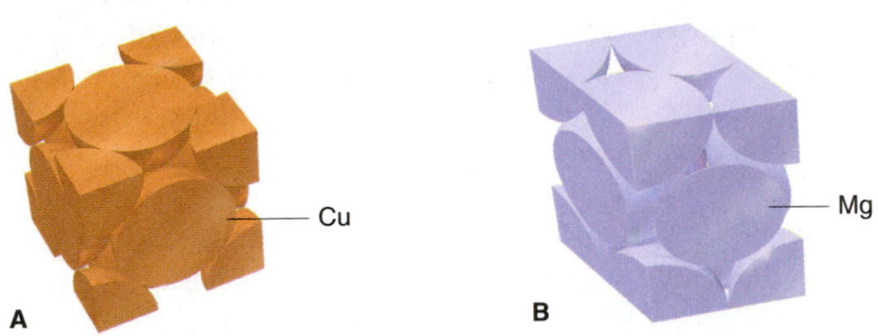

Figure 12.30 Crystal structures of metals. **A,** Copper adopts cubic closest packing. **B,** Magnesium adopts hexagonal closest packing.

Network Covalent Solids Strong covalent bonds link the atoms together in a **network covalent solid;** thus, separate particles are not present. These substances adopt a variety of crystal structures depending on the details of their bonding.

As a consequence of strong bonding, all network covalent solids have extremely high melting and boiling points, but their conductivity and hardness vary. Two examples with the same composition but strikingly different properties are the two common crystalline forms of elemental carbon, graphite and diamond (Table 12.6):

- *Graphite* occurs as stacked flat sheets of hexagonal carbon rings with a strong σ-bond framework and delocalized π bonds, reminiscent of benzene (Section 10.1); the arrangement looks like chicken wire or honeycomb. Whereas the π-bonding electrons of benzene are delocalized over one ring, those of graphite are delocalized over the entire sheet. Thus, graphite conducts electricity well—it is a common electrode material—but only in the plane of the sheets. The sheets interact via dispersion forces, and impurities, such as O_2, between the sheets allow them to slide past each other, which explains why graphite is soft and used as a lubricant.
- *Diamond* adopts a face-centered cubic unit cell, with each C tetrahedrally bonded to four others in an endless array. Throughout the crystal, strong single bonds make diamond the hardest natural substance known. Like most network covalent solids, diamond does not conduct electricity because the bonding electrons are localized.

Table 12.6 Comparison of the Properties of Diamond and Graphite

Property	Graphite	Diamond
Density (g/cm^3)	2.27	3.51
Hardness	<1 (very soft)	10 (hardest)
Melting point (K)	4100	4100
Color	Shiny black	Colorless transparent
Electrical conductivity	High (along sheet)	None
ΔH°_{rxn} for combustion (kJ/mol)	−393.5	−395.4
ΔH°_f (kJ/mol)	0 (standard state)	1.90

The most important network covalent solids are the *silicates,* which consist of extended arrays of covalently bonded silicon and oxygen atoms. Quartz (SiO_2) is a common example. We'll discuss silicates, which form the structure of clays, rocks, and many minerals, when we consider the chemistry of silicon in Chapter 14.

Bonding in Solids I: The Electron-Sea Model of Metallic Bonding

The simplest model that accounts for the properties of metals is the **electron-sea model.** It proposes that all the metal atoms in a sample pool their valence electrons to form an electron "sea" that is delocalized throughout the piece. The metal ions (nuclei plus core electrons) are submerged within this electron sea in an orderly array (see Figure 9.2C). They are not held in place as rigidly as the ions in an ionic solid, and no two metal atoms are bonded through a localized pair of electrons as in a covalent bond. Rather, *the valence electrons are shared among all the atoms in the sample,* and the piece of metal is held together by the mutual attraction of the metal cations for the mobile, highly delocalized valence electrons. The *regularity,* but not rigidity, of the metal-ion array and the *mobility* of the valence electrons account for three major physical properties of metals:

1. *Phase changes.* Metals have moderate to high melting points because the attractions between the cations and the delocalized electrons are not broken during melting, but boiling points are very high because each cation and its electron(s) must break away from the others. Gallium provides a striking example: it melts in your hand (mp 29.8°C) but doesn't boil until 2403°C. The alkaline earth metals [Group 2A(2)] have higher melting points than the alkali metals [Group 1A(1)] because of greater attraction between their 2+ cations and twice the number of valence electrons.

2. *Mechanical properties.* When struck by a hammer, metals usually bend or dent rather than crack or shatter. Instead of repelling each other, the metal cations slide past each other through the electron sea and end up in new positions (Figure 12.31). Compare this behavior with that of an ionic solid (see Figure 9.8). As a result, many metals can be flattened into sheets (malleable) and pulled into wires (ductile). Gold is in a class by itself: 1 g of gold (a cube 0.37 cm on a side) can be hammered into a 1.0-m^2 sheet that is only 230 atoms (50 nm) thick or drawn into a wire 165 m long and 20 μm thick!

3. *Conductivity.* Metals are good electrical conductors because the mobile electrons carry current. Metals conduct heat well because the mobile electrons disperse heat more quickly than do the localized electron pairs or fixed ions in other materials.

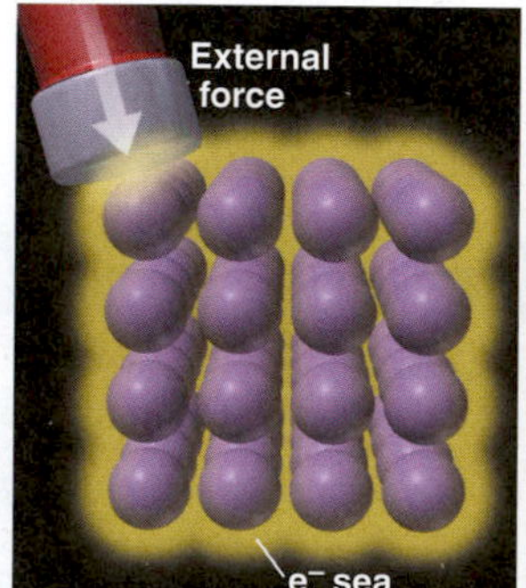

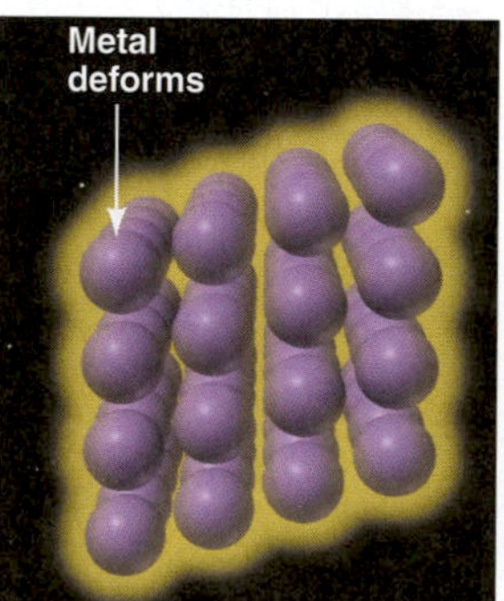

Figure 12.31 Why metals dent and bend rather than crack.

Bonding in Solids II: Band Theory

Molecular orbital (MO) theory offers another explanation of bonding in solids. It is a more quantitative, and therefore more useful, model called **band theory.** We'll focus on bonding in metals and the conductivity of metals, metalloids, and nonmetals.

Formation of Valence and Conduction Bands Recall from Section 11.3 that when two atoms form a diatomic molecule, their atomic orbitals (AOs) combine to

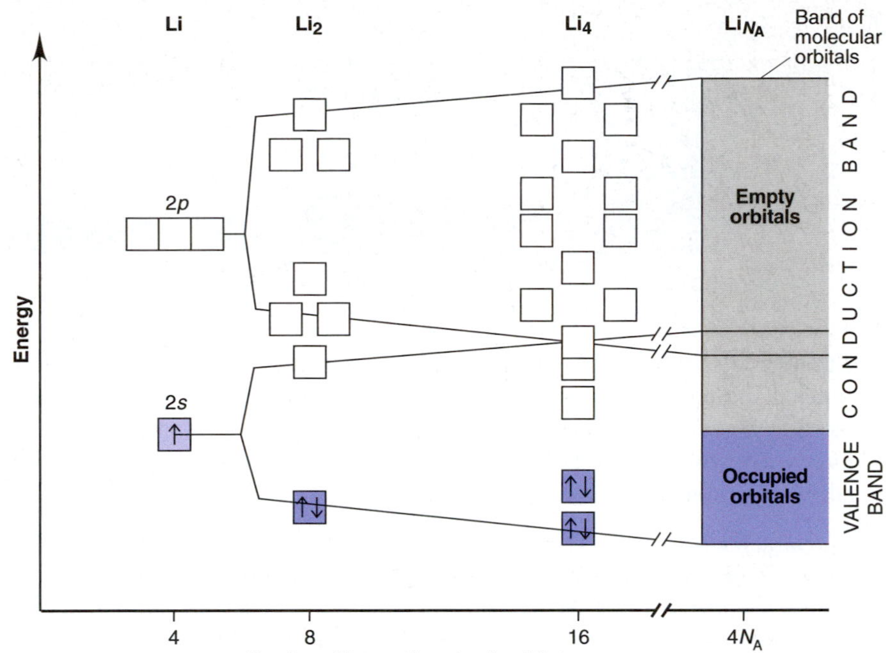

Figure 12.32 The band of molecular orbitals in lithium metal.

form an equal number of molecular orbitals (MOs). Figure 12.32 shows the formation of MOs in lithium. To form Li_2, each Li atom has four valence orbitals (one $2s$ and three $2p$) that combine to form eight MOs, four bonding and four antibonding. The order of MOs shows $2s$-$2p$ mixing (Figure 11.19B). Two more Li atoms form Li_4, a slightly larger aggregate, with 16 delocalized MOs. As more Li atoms join the cluster, more MOs are created, and their energy levels lie closer and closer together. Extending this process to 7 g (1 mol) of lithium results in 6×10^{23} Li atoms (Li_{N_A}) combining to form an extremely large number (4 × Avogadro's number) of delocalized MOs. *The energies of the MOs are so close that they form a continuum, or band, of MOs.*

The lower energy MOs are occupied by the $2s^1$ valence electrons and make up the **valence band.** The empty MOs that are higher in energy make up the **conduction band.** In Li metal, the valence band is derived from the $2s$ AOs, and the conduction band is derived from $2s$ and mostly $2p$ AOs. In Li_2, two valence electrons fill the lowest energy bonding MO and leave the antibonding MO empty. In Li metal, 1 mol of valence electrons fills the valence band and leaves the conduction band empty.

How Band Theory Explains Metallic Properties The key to understanding the properties of metals is that *the valence and conduction bands are contiguous,* that is, the highest level of one touches the lowest of the other. This means that, given an infinitesimal quantity of energy, electrons jump from the filled valence band to the unfilled conduction band: *the electrons are completely delocalized and thus free to move throughout the piece of metal.*

- *Electrical conductivity.* Metals conduct so well because an applied field easily excites the highest energy valence electrons into empty conduction orbitals, allowing them to move through the sample.
- *Luster.* With so many closely spaced levels available, electrons absorb and release photons of many frequencies as they move between the valence and conduction bands.
- *Malleability.* Under an applied force, layers of positive metal ions move past each other, always protected from mutual repulsions by the delocalized electrons (see Figure 12.31).
- *Thermal conductivity.* When a metal wire is heated, the highest energy electrons are excited and their extra energy is transferred as kinetic energy along the wire's length.

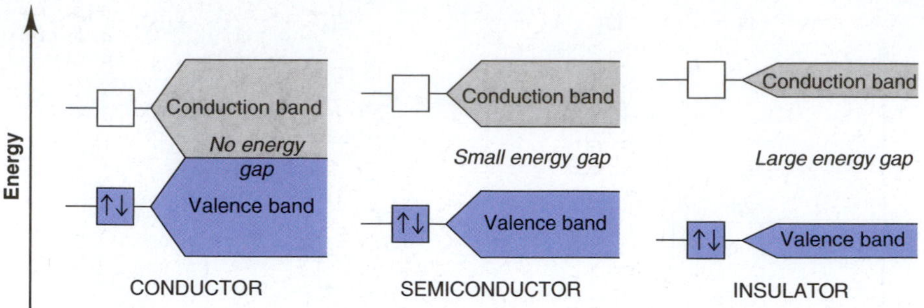

Figure 12.33 Electrical conductivity in a conductor, semiconductor, and insulator.

Conductivity of Solids and the Size of the Energy Gap Like metal atoms, large numbers of nonmetal and metalloid atoms can form bands of MOs. Band theory explains differences in electrical conductivity and the effect of temperature among these three classes of substances in terms of the presence of an energy gap between their valence and conduction bands (Figure 12.33):

1. *Conductors (metals).* The valence and conduction bands of a **conductor** have *no energy gap* between them, so electrons flow when a tiny electrical potential difference is applied. When the temperature is raised, greater random motion of the atoms hinders electron movement: conductivity *decreases* when a metal is heated.
2. *Semiconductors (metalloids).* In a **semiconductor,** a *small energy gap* exists between the valence and conduction bands. Thermally excited electrons can cross the gap, allowing a small current to flow: in contrast to a conductor, conductivity *increases* when a semiconductor is heated.
3. *Insulators (nonmetals).* In an **insulator,** a *large energy gap* exists between the bands: no current is observed even when the substance is heated.

Another type of electrical conductivity, called **superconductivity,** has been the focus of intensive research for the past few decades. When metals conduct at ordinary temperatures, moving electrons collide with vibrating atoms; the reduction in their flow appears as resistive heating and represents a loss of energy. For many years, to conduct with no energy loss—to superconduct—required minimizing this vibrational movement by cooling with liquid helium (bp = 4 K; price ≈ \$11/L). Then, in 1986, certain ionic oxides that superconduct near the boiling point of liquid nitrogen (bp = 77 K; price = \$0.20/L) were prepared.

Like metal conductors, oxide superconductors, such as $YBa_2Cu_3O_7$, have no gap between bands. In 1989, oxides with Bi and Tl instead of Y and Ba were synthesized and found to superconduct at 125 K; and in 1993, an oxide with Hg, Ba, and Ca, in addition to Cu and O, superconducted at 133 K. Such materials could transmit electricity with no loss of energy, allowing power plants to be located far from cities. They could be part of ultrasmall microchips for ultrafast computers, electromagnets to levitate superfast trains, and inexpensive medical diagnostic equipment with superb image clarity. However, the oxides are brittle and not easy to machine, and when warmed, the superconductivity may disappear and not return on cooling. These and related problems will involve chemists, physicists, and engineers for many years.

Summary of Section 12.6

- Particles in crystalline solids lie at points that form a structure of repeating unit cells.
- The three types of unit cells of the cubic system are simple, body-centered, and face-centered. The highest packing efficiency occurs with cubic (face-centered) and hexagonal closest packing.
- Bond angles and distances in a crystal are determined with x-ray diffraction analysis. These data are used to determine atomic radii.
- Atomic [Group 8A(18)] solids adopt cubic closest packing, with atoms held together by weak dispersion forces.
- Molecular solids have molecules at the lattice points and often adopt cubic closest packing. Combinations of intermolecular forces (dispersion, dipole-dipole, and H bonding) result in physical properties that vary greatly.

- Ionic solids crystallize with one ion filling holes in the cubic closest packed array of the other. High melting points, hardness, and low conductivity arise from strong ionic attractions.
- Most metals have a closest packed structure. Their physical properties result from the high packing efficiency and the presence of delocalized electrons.
- Atoms of network covalent solids are covalently bonded throughout the sample, so these substances have very high melting and boiling points.
- The electron-sea model of metallic bonding pictures metal atom cores aligned in a "sea" of delocalized electrons. The model accounts for phase changes, mechanical behavior, and electrical and thermal conductivity of metals.
- Band theory proposes that atomic orbitals of many atoms combine to form a continuum, or band, of molecular orbitals. Metals are electrical conductors because electrons move freely from the filled (valence) band to the empty (conduction) band. Insulators have a large energy gap between the two bands, and semiconductors have a small gap, which can be bridged by heating. Superconductors conduct with no loss of energy.

CHAPTER REVIEW GUIDE

The following sections provide many aids to help you study this chapter. (Numbers in parentheses refer to pages, unless noted otherwise.)

Learning Objectives

These are concepts and skills to review after studying this chapter.

Related section (§), sample problem (SP), and upcoming end-of-chapter problem (EP) numbers are listed in parentheses.

1. Explain how kinetic energy and potential energy determine the properties of the three states and phase changes, what occurs when heat is added or removed from a substance, and how to calculate the enthalpy change (§12.1, §12.2) (SP 12.1) (EPs 12.1–12.9, 12.15, 12.16)
2. Understand that phase changes are equilibrium processes and how vapor pressure and boiling point are related (§12.2) (SP 12.2) (EPs 12.10, 12.11, 12.13, 12.17, 12.18, 12.21)
3. Use a phase diagram to show the phases and phase changes of a substance at different conditions of pressure and temperature (§12.2) (EPs 12.12, 12.14, 12.19, 12.20, 12.22)
4. Distinguish between bonding and molecular forces, predict the relative strengths of intermolecular forces acting in a substance, and understand the impact of H bonding on physical properties (§12.3) (SPs 12.3, 12.4) (EPs 12.23–12.45)
5. Define surface tension, capillarity, and viscosity, and describe how intermolecular forces influence their magnitudes (§12.4) (EPs 12.46–12.52)
6. Understand how the macroscopic properties of water arise from its molecular properties (§12.5) (EPs 12.53–12.58)
7. Describe the three types of cubic unit cells and explain how to find the number of particles in each and how packing of spheres gives rise to each; calculate the atomic radius of an element from its crystal structure; distinguish the types of crystalline solids; explain how the electron-sea model and band theory account for the properties of metals and how the size of the energy gap explains the conductivity of substances (§12.6) (SP 12.5) (EPs 12.59–12.79)

Key Terms

These important terms appear in boldface in the chapter and are defined again in the Glossary.

Section 12.1
phase (351)
intermolecular forces (351)
phase change (351)
condensation (352)
vaporization (352)
freezing (352)
melting (fusion) (352)
sublimation (352)
deposition (352)
heat of vaporization (ΔH°_{vap}) (352)
heat of fusion (ΔH°_{fus}) (352)
heat of sublimation (ΔH°_{subl}) (353)

Section 12.2
heating-cooling curve (354)
dynamic equilibrium (357)
vapor pressure (357)
Clausius-Clapeyron equation (358)
boiling point (359)
melting point (360)
phase diagram (360)
triple point (360)
critical point (361)

Section 12.3
van der Waals radius (362)
ion-dipole force (362)
dipole-dipole force (363)
hydrogen bond (H bond) (364)
polarizability (366)
dispersion (London) force (366)

Section 12.4
surface tension (369)
capillarity (370)
viscosity (370)

Section 12.6
crystalline solid (373)
amorphous solid (373)
lattice (374)
unit cell (374)
coordination number (374)
simple cubic unit cell (374)
body-centered cubic unit cell (374)
face-centered cubic unit cell (374)
packing efficiency (376)
hexagonal closest packing (376)
cubic closest packing (376)
x-ray diffraction analysis (376)
atomic solid (379)
molecular solid (379)
ionic solid (380)
metallic solid (381)
network covalent solid (381)
electron-sea model (382)
band theory (382)
valence band (383)
conduction band (383)
conductor (384)
semiconductor (384)
insulator (384)
superconductivity (384)

Key Equations and Relationships

Numbered and screened concepts are listed for you to refer to or memorize.

12.1 Using the vapor pressure at one temperature to find the vapor pressure at another temperature (two-point form of the Clausius-Clapeyron equation) (359):

$$\ln \frac{P_2}{P_1} = \frac{-\Delta H_{vap}}{R}\left(\frac{1}{T_2} - \frac{1}{T_1}\right)$$

BRIEF SOLUTIONS TO FOLLOW-UP PROBLEMS

Compare your own solutions to these calculation steps and answers.

12.1 The scenes represent solid water at $-7.00°C$ melting to liquid water at 16.0°C, so there are three stages.

Stage 1, $H_2O(s)$ [$-7.00°C$] $\longrightarrow$ $H_2O(s)$ [0.00°C]:

$q = n \times C_{water(s)} \times \Delta T = (2.25 \text{ mol})(37.6 \text{ J/mol·°C})(7.00°C) = 592 \text{ J} = 0.592 \text{ kJ}$

Stage 2, $H_2O(s)$ [0.00°C] $\longrightarrow$ $H_2O(l)$ [0.00°C]:

$q = n(\Delta H^\circ_{fus}) = (2.25 \text{ mol})(6.02 \text{ kJ/mol}) = 13.5 \text{ kJ}$

Stage 3, $H_2O(l)$ [0.00°C] $\longrightarrow$ $H_2O(l)$ [16.0°C]:

$q = n \times C_{water(l)} \times \Delta T = (2.25 \text{ mol})(75.4 \text{ J/mol·°C})(16.0°C) = 2714 \text{ J} = 2.71 \text{ kJ}$

Total heat (kJ) = 0.592 kJ + 13.5 kJ + 2.71 kJ = 16.8 kJ

12.2
$$\ln \frac{P_2}{P_1} = \left(\frac{-40.7\times10^3 \text{ J/mol}}{8.314 \text{ J/mol·K}}\right) \times \left(\frac{1}{273.15 + 85.5 \text{ K}} - \frac{1}{273.15 + 34.1 \text{ K}}\right)$$
$$= (-4.90\times10^3 \text{ K})(-4.6\times10^{-4} \text{ K}^{-1}) = 2.28$$

$\frac{P_2}{P_1} = 9.8$; thus, $P_2 = 40.1 \text{ torr} \times 9.8 = 3.9\times10^2 \text{ torr}$

12.3 (a) [structure: two acetic acid molecules H-bonded as a dimer, O:····H–O] (b) [structure: two ethanol molecules, O–H of one H-bonded to :Ö of the other]

(c) No H bonding

12.4 (a) Dipole-dipole, dispersion; CH_3Br
(b) H bonds, dipole-dipole, dispersion; $CH_3CH_2CH_2OH$
(c) Dispersion; C_3H_8

12.5 From Figure 12.26 *(middle)*,

$$A = \frac{4r}{\sqrt{3}} = \frac{4(126 \text{ pm})}{\sqrt{3}} = 291 \text{ pm}$$

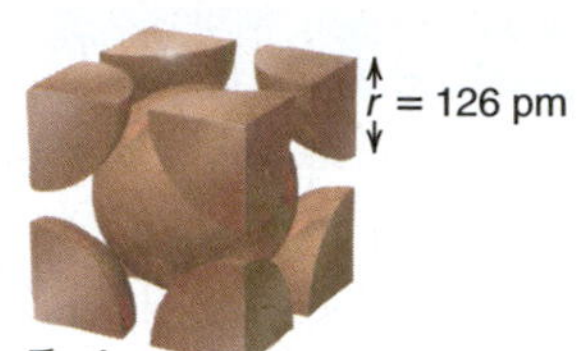

PROBLEMS

Problems with **colored** numbers are answered in Appendix E. Sections match the text and provide the numbers of relevant sample problems. Bracketed problems are grouped in pairs (indicated by a short rule) that cover the same concept. Comprehensive Problems are based on material from any section or previous chapter.

An Overview of Physical States and Phase Changes

12.1 How does the energy of attraction between particles compare with their energy of motion in a gas and in a solid? As part of your answer, identify two macroscopic properties that differ between a gas and a solid.

12.2 (a) Why are gases more easily compressed than liquids? (b) Why do liquids have a greater ability to flow than solids?

12.3 What type of forces, intramolecular or intermolecular:
(a) Prevent ice cubes from adopting the shape of their container?
(b) Are overcome when ice melts?
(c) Are overcome when liquid water is vaporized?
(d) Are overcome when gaseous water is converted to hydrogen gas and oxygen gas?

12.4 (a) Why is the heat of fusion (ΔH_{fus}) of a substance smaller than its heat of vaporization (ΔH_{vap})?
(b) Why is the heat of sublimation (ΔH_{subl}) of a substance greater than its ΔH_{vap}?
(c) At a given temperature and pressure, how does the magnitude of the heat of vaporization of a substance compare with that of its heat of condensation?

12.5 Name the phase change in each of these events: (a) Dew appears on a lawn in the morning. (b) Icicles change into liquid water. (c) Wet clothes dry on a summer day.

12.6 Name the phase change in each of these events: (a) A diamond film forms on a surface from gaseous carbon atoms in a vacuum. (b) Mothballs in a bureau drawer disappear over time. (c) Molten iron from a blast furnace is cast into ingots ("pigs").

12.7 Liquid propane, a widely used fuel, is produced by compressing gaseous propane. During the process, approximately 15 kJ of energy is released for each mole of gas liquefied. Where does this energy come from?

12.8 Many heat-sensitive and oxygen-sensitive solids, such as camphor, are purified by warming under vacuum. The solid vaporizes directly, and the vapor crystallizes on a cool surface. What phase changes are involved in this method?

Quantitative Aspects of Phase Changes

(Sample Problems 12.1 and 12.2)

12.9 Describe the changes (if any) in potential energy and in kinetic energy among the molecules when gaseous PCl_3 condenses to a liquid at a fixed temperature.

12.10 When benzene is at its melting point, two processes occur simultaneously and balance each other. Describe these processes on the macroscopic and molecular levels.

12.11 Liquid hexane (bp = 69°C) is placed in a closed container at room temperature. At first, the pressure of the vapor phase increases, but after a short time, it stops changing. Why?

12.12 Match each numbered point in the phase diagram for compound Q with the correct molecular scene:

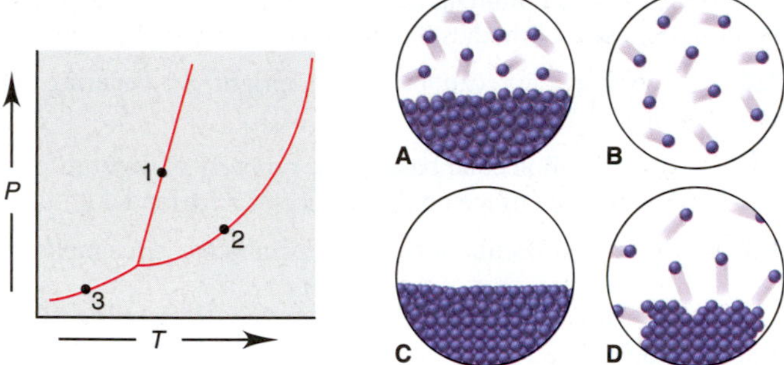

12.13 At 1.1 atm, will water boil at 100.°C? Explain.

12.14 The phase diagram for substance A has a solid-liquid line with a positive slope, and that for substance B has a solid-liquid line with a negative slope. What macroscopic property can distinguish A from B?

12.15 From the data below, calculate the total heat (in J) needed to convert 22.00 g of ice at −6.00°C to liquid water at 0.500°C:

mp at 1 atm:	0.0°C	ΔH°_{fus}:	6.02 kJ/mol
c_{liquid}:	4.21 J/g·°C	c_{solid}:	2.09 J/g·°C

12.16 From the data below, calculate the total heat (in J) needed to convert 0.333 mol of gaseous ethanol at 300°C and 1 atm to liquid ethanol at 25.0°C and 1 atm:

bp at 1 atm:	78.5°C	ΔH°_{vap}:	38.6 kJ/mol
c_{gas}:	1.43 J/g·°C	c_{liquid}:	2.45 J/g·°C

12.17 A liquid has a ΔH°_{vap} of 35.5 kJ/mol and a boiling point of 122°C at 1.00 atm. What is its vapor pressure at 113°C?

12.18 What is the ΔH°_{vap} of a liquid that has a vapor pressure of 621 torr at 85.2°C and a boiling point of 95.6°C at 1 atm?

12.19 Use these data to draw a qualitative phase diagram for ethylene (C_2H_4). Is $C_2H_4(s)$ more or less dense than $C_2H_4(l)$?

bp at 1 atm:	−103.7°C
mp at 1 atm:	−169.16°C
Critical point:	9.9°C and 50.5 atm
Triple point:	−169.17°C and 1.20×10^{-3} atm

12.20 Use these data to draw a qualitative phase diagram for H_2. Does H_2 sublime at 0.05 atm? Explain.

mp at 1 atm:	13.96 K
bp at 1 atm:	20.39 K
Triple point:	13.95 K and 0.07 atm
Critical point:	33.2 K and 13.0 atm
Vapor pressure of solid at 10 K:	0.001 atm

12.21 Butane is a common fuel used in cigarette lighters and camping stoves. Normally supplied in metal containers under pressure, the fuel exists as a mixture of liquid and gas, so high temperatures may cause the container to explode. At 25.0°C, the vapor pressure of butane is 2.3 atm. What is the pressure in the container at 135°C (ΔH°_{vap} = 24.3 kJ/mol)?

12.22 Use Figure 12.8A, p. 361, to answer the following:
(a) Carbon dioxide is sold in steel cylinders under pressures of approximately 20 atm. Is there liquid CO_2 in the cylinder at room temperature (~20°C)? At 40°C? At −40°C? At −120°C?
(b) Carbon dioxide is also sold as solid chunks, called *dry ice,* in insulated containers. If the chunks are warmed by leaving them in an open container at room temperature, will they melt?
(c) If a container is nearly filled with dry ice and then sealed and warmed to room temperature, will the dry ice melt?
(d) If dry ice is compressed at a temperature below its triple point, will it melt?

Types of Intermolecular Forces

(Sample Problems 12.3 and 12.4)

12.23 Why are covalent bonds typically much stronger than intermolecular forces?

12.24 Even though molecules are neutral, the dipole-dipole force is an important interparticle force that exists among them. Explain.

12.25 (a) Name the type of force depicted in each scene below.
(b) Rank the forces in order of increasing strength.

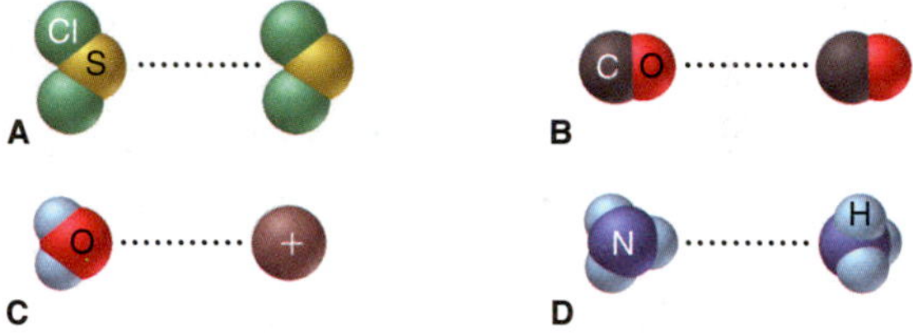

12.26 Oxygen and selenium are members of Group 6A(16). Water forms H bonds, but H_2Se does not. Explain.

12.27 Polar molecules exhibit dipole-dipole forces. Do they also exhibit dispersion forces? Explain.

12.28 Distinguish between *polarizability* and *polarity*. How does each influence intermolecular forces?

12.29 How can one nonpolar molecule induce a dipole in a nearby nonpolar molecule?

12.30 What is the strongest interparticle force in each substance?
(a) CH_3OH (b) CCl_4 (c) Cl_2

12.31 What is the strongest interparticle force in each substance?
(a) H_3PO_4 (b) SO_2 (c) $MgCl_2$

12.32 What is the strongest interparticle force in each substance?
(a) CH_3Cl (b) CH_3CH_3 (c) NH_3

12.33 What is the strongest interparticle force in each substance?
(a) Kr (b) BrF (c) H_2SO_4

12.34 Which member of each pair of compounds forms intermolecular H bonds? Draw the H-bonded structures in each case:
(a) $CH_3CH(OH)CH_3$ or CH_3SCH_3 (b) HF or HBr

12.35 Which member of each pair of compounds forms intermolecular H bonds? Draw the H-bonded structures in each case:
(a) $(CH_3)_2NH$ or $(CH_3)_3N$ (b) $HOCH_2CH_2OH$ or FCH_2CH_2F

12.36 Which has the greater polarizability? Explain.
(a) Br^- or I^- (b) $CH_2{=}CH_2$ or $CH_3{-}CH_3$ (c) H_2O or H_2Se

12.37 Which has the greater polarizability? Explain.
(a) Ca^{2+} or Ca (b) CH_3CH_3 or $CH_3CH_2CH_3$ (c) CCl_4 or CF_4

12.38 Which member in each pair of liquids has the *higher* vapor pressure at a given temperature? Explain.
(a) C_2H_6 or C_4H_{10} (b) CH_3CH_2OH or CH_3CH_2F
(c) NH_3 or PH_3

12.39 Which member in each pair of liquids has the *lower* vapor pressure at a given temperature? Explain.
(a) $HOCH_2CH_2OH$ or $CH_3CH_2CH_2OH$
(b) CH_3COOH or $(CH_3)_2C{=}O$ (c) HF or HCl

12.40 Which substance has the *lower* boiling point? Explain.
(a) LiCl or HCl (b) NH_3 or PH_3 (c) Xe or I_2

12.41 Which substance has the *higher* boiling point? Explain.
(a) CH_3CH_2OH or $CH_3CH_2CH_3$ (b) NO or N_2
(c) H_2S or H_2Te

12.42 Which substance has the *lower* boiling point? Explain.
(a) $CH_3CH_2CH_2CH_3$ or $CH_2{-}CH_2$ / $CH_2{-}CH_2$ (cyclobutane ring)
(b) NaBr or PBr_3
(c) H_2O or HBr

12.43 Which substance has the *higher* boiling point? Explain.
(a) CH_3OH or CH_3CH_3
(b) FNO or ClNO
(c) F(H)C=C(F)(H) [F and H on first carbon; F and H on second carbon, trans] or H(F)C=C(F)(H) [cis]

12.44 Dispersion forces are the only intermolecular forces present in motor oil, yet it has a high boiling point. Explain.

12.45 Why does the antifreeze ingredient ethylene glycol ($HOCH_2CH_2OH$; $\mathcal{M} = 62.07$ g/mol) have a boiling point of 197.6°C, whereas propanol ($CH_3CH_2CH_2OH$; $\mathcal{M} = 60.09$ g/mol), a compound with a similar molar mass, has a boiling point of only 97.4°C?

Properties of the Liquid State

12.46 Before the phenomenon of surface tension was understood, physicists described the surface of water as being covered with a "skin." What causes this skinlike phenomenon?

12.47 Small, equal-sized drops of oil, water, and mercury lie on a waxed floor. How does each liquid behave? Explain.

12.48 Does the *strength* of the intermolecular forces in a liquid change as the liquid is heated? Explain. Why does liquid viscosity decrease with rising temperature?

12.49 Rank the following in order of *increasing* surface tension at a given temperature, and explain your ranking:
(a) $CH_3CH_2CH_2OH$ (b) $HOCH_2CH(OH)CH_2OH$
(c) $HOCH_2CH_2OH$

12.50 Rank the following in order of *decreasing* surface tension at a given temperature, and explain your ranking:
(a) CH_3OH (b) CH_3CH_3 (c) $H_2C{=}O$

12.51 Use Figure 12.1, p. 353 to answer the following: (a) Does it take more heat to melt 12.0 g of CH_4 or 12.0 g of Hg? (b) Does it take more heat to vaporize 12.0 g of CH_4 or 12.0 g of Hg? (c) What is the principal intermolecular force in each sample?

12.52 Pentanol ($C_5H_{11}OH$; $\mathcal{M} = 88.15$ g/mol) has nearly the same molar mass as hexane (C_6H_{14}; $\mathcal{M} = 86.17$ g/mol) but is more than 12 times as viscous at 20°C. Explain.

The Uniqueness of Water

12.53 For what types of substances is water a good solvent? For what types is it a poor solvent? Explain.

12.54 A water molecule can engage in as many as four H bonds. Explain.

12.55 Warm-blooded animals have a narrow range of body temperature because their bodies have a high water content. Explain.

12.56 A drooping plant can be made upright by watering the ground around it. Explain.

12.57 Describe the molecular basis of the property of water responsible for the presence of ice on the surface of a frozen lake.

12.58 Describe in molecular terms what occurs when ice melts.

The Solid State: Structure, Properties, and Bonding

(Sample Problem 12.5)

12.59 What is the difference between an amorphous solid and a crystalline solid on the macroscopic and molecular levels? Give an example of each.

12.60 How are a solid's unit cell and crystal structure related?

12.61 For structures consisting of identical atoms, how many atoms are contained in the simple, body-centered, and face-centered cubic unit cells? Explain how you obtained the values.

12.62 List four physical characteristics of a solid metal.

12.63 Briefly account for the following relative values:
(a) The melting point of sodium is 89°C, whereas that of potassium is 63°C.
(b) The melting points of Li and Be are 180°C and 1287°C, respectively.
(c) Lithium boils more than 1100°C higher than it melts.

12.64 Magnesium metal is easily deformed by an applied force, whereas magnesium fluoride is shattered. Why do these two solids behave so differently?

12.65 What is the energy gap in band theory? Compare its size in superconductors, conductors, semiconductors, and insulators.

12.66 What type of crystal lattice does each metal form? (The number of atoms per unit cell is given in parentheses.)
(a) Ni (4) (b) Cr (2) (c) Ca (4)

12.67 What is the number of atoms per unit cell for each metal?
(a) Polonium, Po (b) Manganese, Mn (c) Silver, Ag

12.68 When cadmium oxide (A) reacts to form cadmium selenide (B), a change in unit cell occurs, as depicted below:

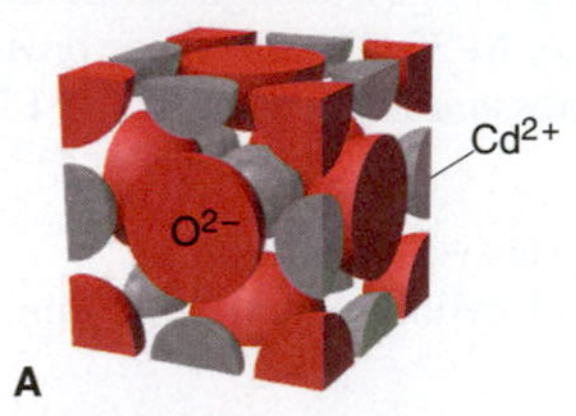

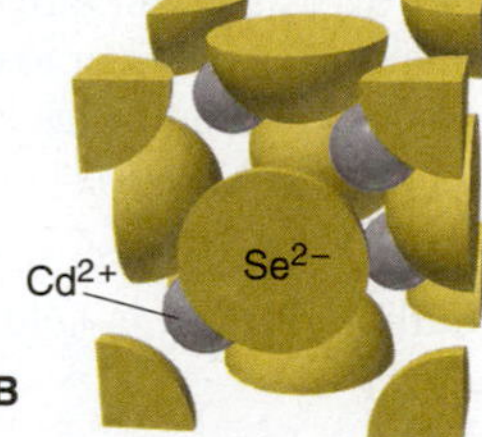

(a) What is the change in unit cell?
(b) Does the coordination number of cadmium change? Explain.

12.69 As molten iron cools to 1674 K, it adopts one type of cubic unit cell (A); then, as the temperature drops below 1181 K, it changes to another (B), as depicted below:

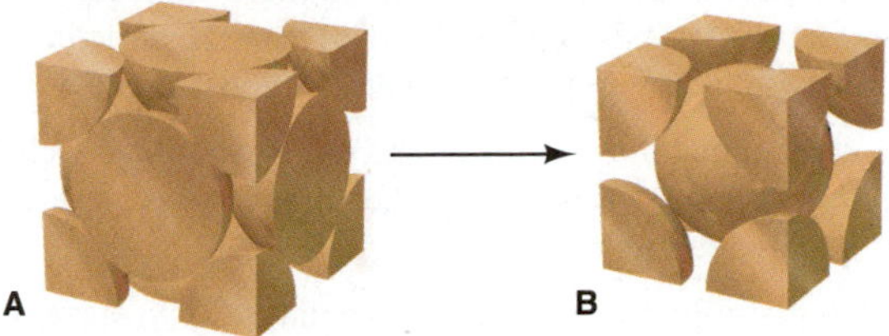

(a) What is the change in unit cell?
(b) Which crystal structure has the greater packing efficiency?

12.70 Of the five major types of crystalline solid, which does each of the following form, and why: (a) Ni; (b) F_2; (c) CH_3OH; (d) Sn; (e) Si; (f) Xe?

12.71 Of the five major types of crystalline solid, which does each of the following form, and why: (a) SiC; (b) Na_2SO_4; (c) SF_6; (d) cholesterol ($C_{27}H_{45}OH$); (e) KCl; (f) BN?

12.72 Zinc oxide adopts the zinc blende crystal structure (Figure P12.72). How many Zn^{2+} ions are in the ZnO unit cell?

12.73 Calcium sulfide adopts the sodium chloride crystal structure (Figure P12.73). How many S^{2-} ions are in the CaS unit cell?

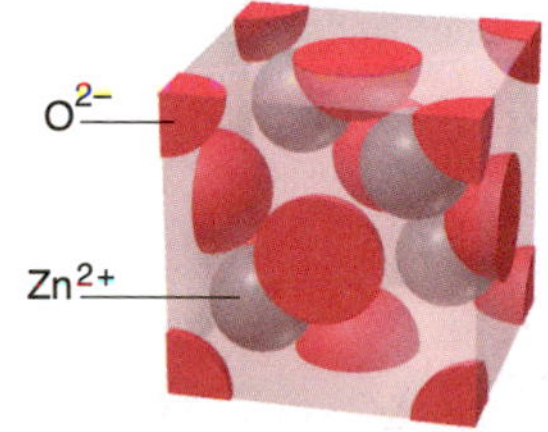

Figure P12.72

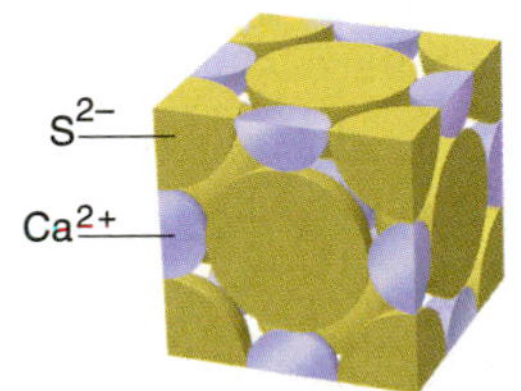

Figure P12.73

12.74 Zinc selenide (ZnSe) crystallizes in the zinc blende structure (see Figure P12.72) and has a density of 5.42 g/cm^3.
(a) How many Zn and Se ions are in each unit cell?
(b) What is the mass of a unit cell?
(c) What is the volume of a unit cell?
(d) What is the edge length of a unit cell?

12.75 An element crystallizes in a face-centered cubic lattice and has a density of 1.45 g/cm^3. The edge of its unit cell is 4.52×10^{-8} cm.
(a) How many atoms are in each unit cell?
(b) What is the volume of a unit cell?
(c) What is the mass of a unit cell?
(d) Calculate an approximate atomic mass for the element.

12.76 Classify each of the following as a conductor, insulator, or semiconductor: (a) phosphorus; (b) mercury; (c) germanium.

12.77 Predict the effect (if any) of an increase in temperature on the electrical conductivity of (a) antimony; (b) tellurium; (c) bismuth.

12.78 Use condensed electron configurations to predict the relative hardnesses and melting points of rubidium ($Z = 37$), vanadium ($Z = 23$), and cadmium ($Z = 48$).

12.79 One of the most important enzymes in the world—nitrogenase, the plant protein that catalyzes nitrogen fixation—contains active clusters of iron, sulfur, and molybdenum atoms. Crystalline molybdenum (Mo) has a body-centered cubic unit cell (d of Mo = 10.28 g/cm^3). (a) Determine the edge length of the unit cell. (b) Calculate the atomic radius of Mo.

12.80 Barium is the largest nonradioactive alkaline earth metal. It has a body-centered cubic unit cell and a density of 3.62 g/cm^3. What is the atomic radius of barium? (*Hint:* The reciprocal of a metal's density multiplied by its molar mass gives cm^3/mol of the metal, which includes the space between atoms; volume of a sphere: $V = \frac{4}{3}\pi r^3$.)

Comprehensive Problems

12.81 Bismuth is used to calibrate instruments employed in high-pressure studies because it has several well-characterized crystalline phases. Its phase diagram *(right)* shows the liquid phase and five solid phases that are stable above 1 katm (1000 atm) and up to 300°C. (a) Which solid phases are stable at 25°C? (b) Which phase is stable at 50 katm and 175°C? (c) As the pressure is reduced from 100 to 1 katm at 200°C, what phase transitions does bismuth undergo? (d) What phases are present at each of the triple points?

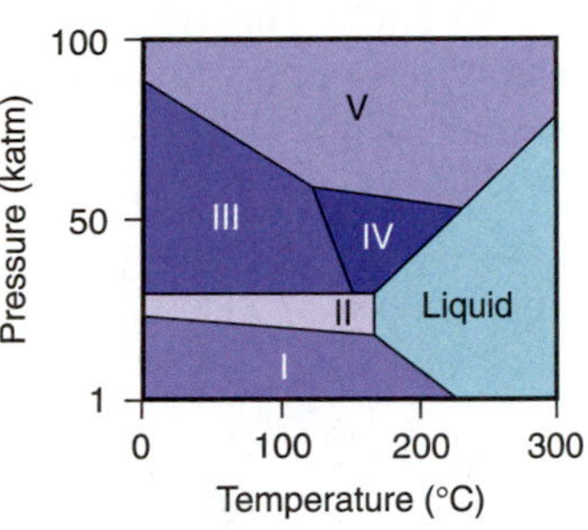

12.82 Mercury (Hg) vapor is toxic and readily absorbed from the lungs. At 20.°C, mercury (ΔH_{vap} = 59.1 kJ/mol) has a vapor pressure of 1.20×10^{-3} torr, which is high enough to be hazardous. To reduce the danger to workers in processing plants, Hg is cooled to lower its vapor pressure. At what temperature would the vapor pressure of Hg be at the safer level of 5.0×10^{-5} torr?

12.83 Consider the phase diagram shown for substance X.
(a) What phase(s) is (are) present at point A? E? F? H? B? C?
(b) Which point corresponds to the critical point? Which point corresponds to the triple point?
(c) What curve corresponds to conditions at which the solid and gas are in equilibrium?
(d) Describe what happens when you start at point A and increase the temperature at constant pressure.
(e) Describe what happens when you start at point H and decrease the pressure at constant temperature.
(f) Is liquid X more or less dense than solid X?

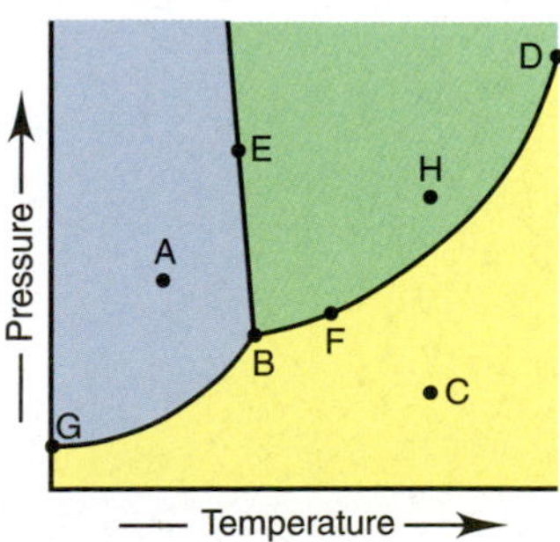

12.84 Some oxide superconductors adopt a crystal structure similar to that of *perovskite* ($CaTiO_3$). The unit cell is cubic with a Ti^{4+} ion in each corner, a Ca^{2+} ion in the body center, and O^{2-} ions at the midpoint of each edge. (a) Is this unit cell simple, body-centered, or face-centered? (b) If the unit cell edge length is 3.84 Å, what is the density of perovskite (in g/cm^3)?

12.85 Iron crystallizes in a body-centered cubic structure. The volume of one Fe atom is 8.38×10^{-24} cm^3, and the density of Fe is 7.874 g/cm^3. Find an approximate value for Avogadro's number.

12.86 The only alkali metal halides that do not adopt the NaCl structure are CsCl, CsBr, and CsI, formed from the largest alkali metal cation and the three largest halide ions. These crystallize in the *cesium chloride structure* (shown here for CsCl). This structure has been used as an example of how dispersion forces can dominate in the presence of ionic forces. Use the ideas of coordination number and polarizability to explain why the CsCl structure exists.

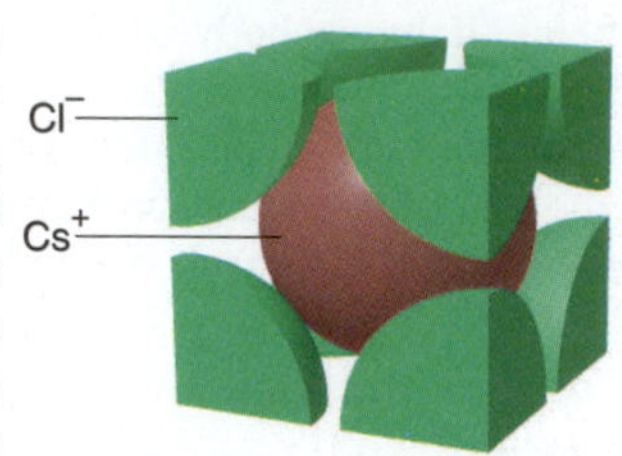

12.87 Like most transition metals, tantalum (Ta) exhibits several oxidation states. Give the formula of each tantalum compound whose unit cell is depicted below:

(a)

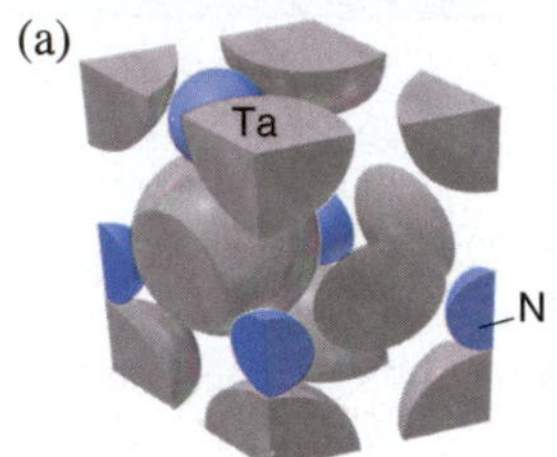

(b)

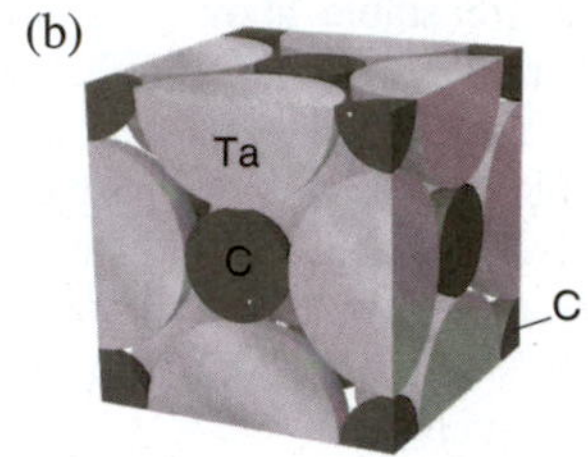

12.88 KF has the same type of crystal structure as NaCl. The unit cell of KF has an edge length of 5.39 Å. Find the density of KF.

12.89 Furfural, which is prepared from corncobs, is an important solvent in synthetic rubber manufacturing, and it is reduced to furfuryl alcohol, which is used to make polymer resins. Furfural can also be oxidized to 2-furoic acid.

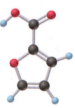

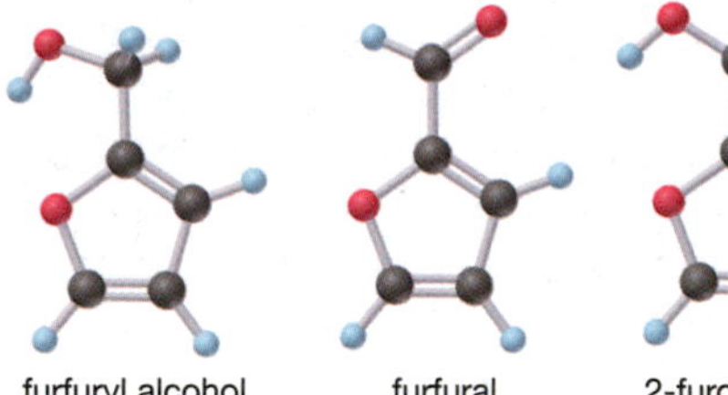

(a) Which of these compounds can form H bonds? Draw structures in each case.

(b) The molecules of some substances can form an "internal" H bond, that is, an H bond *within* a molecule. This takes the form of a polygon with atoms as corners and bonds as sides and an H bond as one of the sides. Which of these molecules is (are) likely to form a stable internal H bond? Draw the structure. (*Hint:* Structures with 5 or 6 atoms as corners are most stable.)

12.90 A cubic unit cell contains atoms of element A at each corner and atoms of element Z on each face. What is the empirical formula of the compound?

12.91 Is it possible for a salt of formula AB_3 to have a face-centered cubic unit cell of anions with cations in all eight of the available holes? Explain.

12.92 A 4.7-L sealed bottle containing 0.33 g of liquid ethanol, C_2H_6O, is placed in a refrigerator and reaches equilibrium with its vapor at −11°C. (a) What mass of ethanol is present in the vapor? (b) When the container is removed and warmed to room temperature, 20.°C, will all the ethanol vaporize? (c) How much liquid ethanol would be present at 0.0°C? The vapor pressure of ethanol is 10. torr at −2.3°C and 40. torr at 19°C.

12.93 In a body-centered cubic unit cell, the central atom lies on an internal diagonal of the cell and touches the corner atoms.

(a) Find the length of the diagonal in terms of r, the atomic radius. (b) If the edge length of the cube is a, what is the length of a *face* diagonal? (c) Derive an expression for a in terms of r. (d) How many atoms are in this unit cell? (e) What fraction of the unit cell volume is filled with spheres?

12.94 The ball-and-stick models below represent three compounds with the same molecular formula, $C_4H_8O_2$:

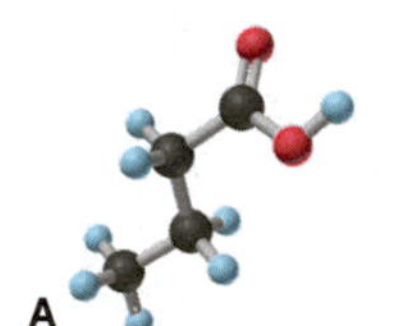

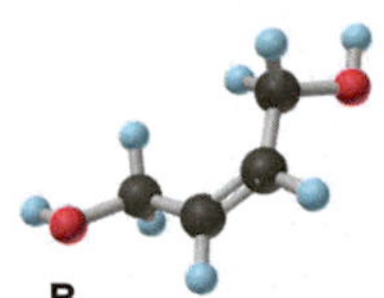

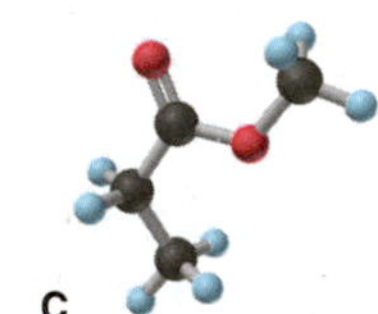

(a) Which compound(s) can form intermolecular H bonds?
(b) Which has the highest viscosity?

The Properties of Solutions

13

Your Morning Solution Even the simple act of adding sugar to a cup of tea or coffee involves several aspects of solid-liquid solution formation, including the effect of temperature on solubility.

Outline

Key Principles
to focus on while studying this chapter

- A *solution* is a *homogeneous* mixture in which a *solute* dissolves in a *solvent,* and the separate particles occur as individual atoms, ions, or molecules. *Solubility* refers to the amount of solute that dissolves in a fixed amount of solvent at a given temperature. Intermolecular forces occur between solute and solvent particles, and the *like-dissolves-like rule* refers to the fact that solutions form when solute and solvent have similar types and strengths of intermolecular forces. Solutions occur with substances in any combination of physical states. ***(Section 13.1)***
- Dissolving involves enthalpy changes: heat is absorbed to separate solute particles as well as solvent particles, and it is released when they mix. The relative magnitudes of these individual quantities of heat determine whether the overall solution formation is exothermic (negative *heat of solution,* ΔH_{soln}) or endothermic (positive ΔH_{soln}). ***(Section 13.2)***
- Dissolving also involves changes in the freedom of motion of the particles and, thus, in the dispersal of their kinetic energy; the number of ways in which a system can disperse its energy is related to a quantity called *entropy.* ***(Section 13.2)***
- A *saturated solution* contains the maximum amount of solute that can dissolve at a given temperature. Any excess undissolved solute is in *equilibrium* with dissolved solute. Because of the equilibrium nature of solubility, most solids are *more* soluble in water at higher temperatures, all gases are *less* soluble in water at higher temperatures, and the solubility of a gas is directly proportional to its pressure *(Henry's law).* ***(Section 13.3)***
- The concentration of a solution is expressed through different concentration terms, including *molarity, molality, mass percent, volume percent,* and *mole fraction.* These terms are *interconvertible.* ***(Section 13.4)***
- The physical properties of a solution differ from those of the solvent. *Vapor pressure lowering, boiling point elevation, freezing point depression,* and *osmotic pressure* are known as *colligative properties* because they depend on the number, not the chemical nature, of the solute particles. ***(Section 13.5)***
- The vapor pressure above an *ideal solution* of a *nonvolatile nonelectrolyte* is lowered by an amount proportional to the mole fraction of the solute *(Raoult's law).* For a *volatile nonelectrolyte,* the vapor has a higher proportion of the more volatile solute than the solution does. ***(Section 13.5)***
- In a solution of a strong *electrolyte,* interactions among ions cause deviations from ideal behavior and, thus, from the expected magnitudes of the colligative properties. ***(Section 13.5)***

CONCEPTS & SKILLS TO REVIEW before studying this chapter

- calculations involving mass percent (Section 3.1) and molarity (Section 3.5)
- electrolytes; water as solvent (Sections 4.1 and 12.5)
- mole fraction and Dalton's law (Section 5.4)
- intermolecular forces and polarizability (Section 12.3)
- vapor pressure of liquids (Section 12.2)

Virtually all the gases, liquids, and solids in the real world are *mixtures*—two or more substances mixed together physically, not combined chemically. Synthetic mixtures, such as glass and soap, usually contain only a dozen or so components, but natural mixtures, such as seawater and soil, often contain over 50. Living mixtures, such as trees and students, are the most complex—even a simple bacterial cell contains well over 5000 different compounds (Table 13.1).

Table 13.1 Approximate Composition of a Bacterium

Substance	Mass % of Cell	Number of Types	Number of Molecules
Water	~70	1	5×10^{10}
Ions	1	20	?
Sugars*	3	200	3×10^{8}
Amino acids*	0.4	100	5×10^{7}
Lipids*	2	50	3×10^{7}
Nucleotides*	0.4	200	1×10^{7}
Other small molecules	0.2	~200	?
Macromolecules (proteins, nucleic acids, polysaccharides)	23	~5000	6×10^{6}

*Includes precursors and metabolites.

Recall from Chapter 2 that a mixture has two defining characteristics: *its composition can be variable,* and *it retains some properties of its components.* In this chapter, we focus on solutions, the most common type of mixture. *A solution is a homogeneous mixture,* one with no boundaries separating its components; in other words, a solution exists as one phase. A *heterogeneous mixture* has two or more phases. The pebbles in concrete or the bubbles in champagne are obvious indications that these are heterogeneous mixtures. But smoke and milk are heterogeneous mixtures with very small component particles and thus no visibly distinct phases. The essential distinction is that in a solution *all the particles are individual atoms, ions, or molecules.*

In this discussion, we consider the types of solutions, why they form, the different concentration units that describe them, and how their properties differ from those of pure substances.

13.1 • TYPES OF SOLUTIONS: INTERMOLECULAR FORCES AND SOLUBILITY

A **solute** dissolves in a **solvent** to form a solution. In general, *the solvent is the most abundant component,* but in some cases, the substances are **miscible**—soluble in each other in any proportion—so the terms "solute" and "solvent" lose their meaning. *The physical state of the solvent usually determines the physical state of the solution.* Solutions can be gaseous, liquid, or solid, but we focus mostly on liquid solutions because they are by far the most important.

The **solubility (*S*)** of a solute is the maximum amount that dissolves in a fixed quantity of a given solvent at a given temperature, where an excess of the solute is present. Different solutes have different solubilities:

- Sodium chloride (NaCl), S = 39.12 g/100. mL water at 100.°C.
- Silver chloride (AgCl), S = 0.0021 g/100. mL water at 100.°C.

Solubility is a *quantitative* term, but *dilute* and *concentrated* are qualitative, referring to the *relative* amounts of dissolved solute; the NaCl solution above is concentrated, and the AgCl solution is dilute.

A given solute may dissolve in one solvent and not another. One explanation lies in the relative strengths of the intermolecular forces within both solute and solvent and between them. The useful rule-of-thumb **"like dissolves like"** says that *substances with similar types of intermolecular forces dissolve in each other.* Thus, by knowing the forces, we can often predict whether a solute will dissolve in a solvent.

Intermolecular Forces in Solution

All the intermolecular forces we discussed for pure substances also occur in solutions. Figure 13.1 shows some examples, and others are presented below (see also Section 12.3):

1. *Ion-dipole forces* are the principal force involved when an ionic compound dissolves in water. Two events occur simultaneously:
   - *Forces compete*. When a salt is added to water, an ion and the oppositely charged pole of a water molecule attract each other. These attractions compete with and overcome attractions between the ions, and the crystal structure breaks down.
   - *Hydration shells form*. As an ion separates, water molecules cluster around it in **hydration shells.** The number of water molecules in the innermost shell depends on the ion's size: four fit tetrahedrally around small ions like Li^+, while the larger Na^+ and F^- have six water molecules surrounding them octahedrally (Figure 13.2). In the innermost shell, normal H bonding is disrupted to form the ion-dipole forces. But these water molecules are H bonded to others in the next shell, and those are H bonded to others still farther away.
2. *Hydrogen bonding* is the principal force in solutions of polar, O- and N-containing organic and biological compounds, such as alcohols, amines, and amino acids.
3. *Dipole-dipole forces,* in the absence of H bonding, allow polar molecules like propanal (CH_3CH_2CHO) to dissolve in polar solvents like dichloromethane (CH_2Cl_2).
4. **Ion–induced dipole forces,** one type of *charge-induced dipole force,* rely on polarizability. They arise when an ion's charge distorts the electron cloud of a nearby nonpolar molecule. This type of force initiates the binding of the Fe^{2+} ion in hemoglobin to an O_2 molecule entering a red blood cell.
5. **Dipole–induced dipole forces,** also based on polarizability, arise when a polar molecule distorts the electron cloud of a nonpolar molecule. They are weaker than ion–induced dipole forces because the charge of each pole is less than an ion's (Coulomb's law). The solubility in water of atmospheric O_2, N_2, and noble gases, while limited, is due in part to these forces. Paint thinners and grease solvents also use them.
6. *Dispersion forces* contribute to the solubility of all solutes in all solvents, but they are the *principal* intermolecular force in solutions of nonpolar substances, such as those in petroleum or gasoline.

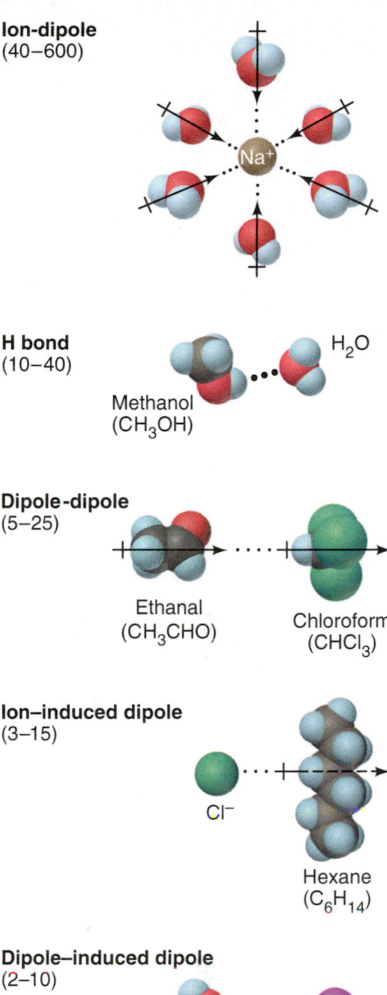

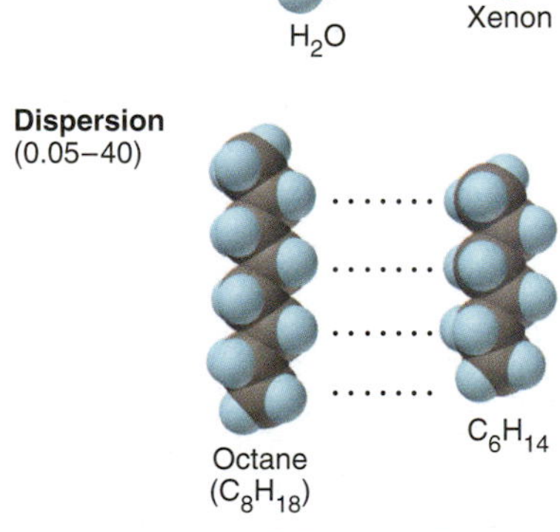

Figure 13.1 Types of intermolecular forces in solutions. Forces are listed in order of decreasing strength (values are in kJ/mol), with an example of each.

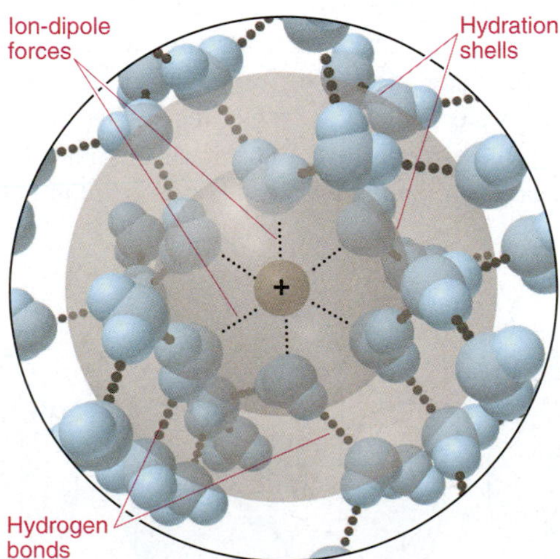

Figure 13.2 Hydration shells around an Na^+ ion. Ion-dipole forces orient water molecules around an ion. In the innermost shell here, six water molecules surround the cation octahedrally.

Liquid Solutions and the Role of Molecular Polarity

From cytoplasm to tree sap, gasoline to cleaning fluid, iced tea to urine, liquid solutions are very familiar. Water is the most prominent solvent, but there are many other liquid solvents, with polarities from very polar to nonpolar.

Applying the Like-Dissolves-Like Rule The like-dissolves-like rule means that when the forces *within* the solute are *similar* to those *within* the solvent, the forces *replace* each other and a solution forms. Thus, for example,

- *Salts are soluble in water* because the strong ion-dipole attractions between ion and water are similar to the strong attractions between the ions and the strong H bonds between water molecules, so they *can* replace each other.
- *Salts are insoluble in hexane* (C_6H_{14}) because the weak ion–induced dipole forces between ion and nonpolar hexane *cannot* replace attractions between the ions.
- *Oil is insoluble in water* because the weak dipole–induced dipole forces between oil and water molecules cannot replace the strong H bonds within water molecules and the extensive dispersion forces within the oil.
- *Oil is soluble in hexane* because dispersion forces in one readily replace dispersion forces in the other.

Dual Polarity and Effects on Solubility To examine these ideas further, let's compare the solubilities of a series of alcohols in water and hexane, solvents with very different intermolecular forces. Alcohols are organic compounds that have a dual polarity, a polar hydroxyl (—OH) group bonded to a nonpolar hydrocarbon group:

- The —OH portion interacts through strong H bonds with water and through weak dipole–induced dipole forces with hexane.
- The hydrocarbon portion interacts through many dispersion forces with hexane and through few, and weak, dipole–induced dipole forces with water.

The general formula for an alcohol is $CH_3(CH_2)_nOH$, and we'll look at straight-chain examples with one to six carbons ($n = 0$ to 5):

1. *Solubility in water is high for smaller alcohols*. From the models in Table 13.2, we see that the —OH group is a relatively large portion of the alcohols with one to three carbons ($n = 0$ to 2). These molecules interact with each other through H bonding, just as water molecules do. When they mix with water, H bonding in solute and in solvent is replaced by H bonding *between* solute and solvent (Figure 13.3). As a result, these smaller alcohols are miscible with water.

2. *Solubility in water is low for larger alcohols*. Solubility decreases dramatically for alcohols larger than three carbons ($n > 2$); in fact, those with chains longer than six carbons are insoluble in water. For larger alcohols to dissolve, the nonpolar chains have to move among the water molecules, replacing weak attractions with water for strong H bonds between water molecules. While the —OH portion of such an alcohol forms H bonds to water, these cannot make up for all the other H bonds between water molecules that have to break to make room for the hydrocarbon portion.

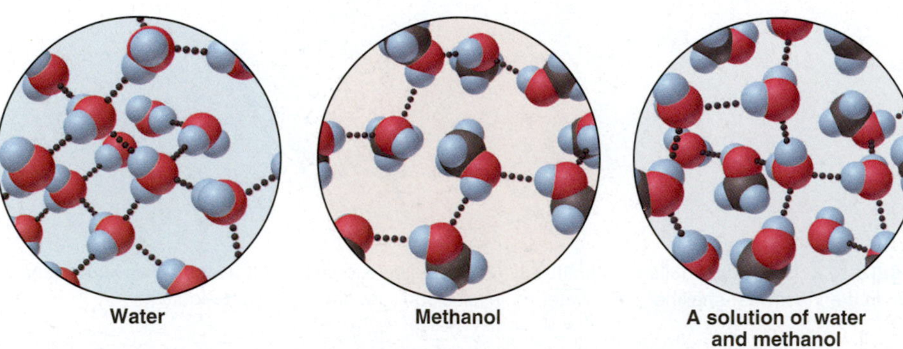

Figure 13.3 Like dissolves like: solubility of methanol in water. H bonds in water and in methanol are similar, so they replace one another when the two substances form a solution.

Table 13.2 Solubility* of a Series of Alcohols in Water and in Hexane

Alcohol	Model	Solubility in Water	Solubility in Hexane
CH_3OH (methanol)		∞	1.2
CH_3CH_2OH (ethanol)		∞	∞
$CH_3(CH_2)_2OH$ (1-propanol)		∞	∞
$CH_3(CH_2)_3OH$ (1-butanol)		1.1	∞
$CH_3(CH_2)_4OH$ (1-pentanol)		0.30	∞
$CH_3(CH_2)_5OH$ (1-hexanol)		0.058	∞

*Expressed in mol alcohol/1000 g solvent at 20°C.

Table 13.2 shows that the opposite trend occurs with hexane:

1. *Solubility in hexane is low for the smallest alcohol.* In addition to dispersion forces, weak dipole–induced dipole forces exist between the —OH of methanol (CH_3OH) and hexane. These cannot replace the strong H bonding between CH_3OH molecules, so its solubility is low.

2. *Solubility in hexane is high for larger alcohols.* In any larger alcohol ($n > 0$), dispersion forces between the hydrocarbon portion and hexane *can* replace dispersion forces between hexane molecules. With only weak forces within the solvent to be replaced, even ethanol, with a two-carbon chain, has enough dispersion forces between it and hexane to be miscible.

Many organic molecules have polar and nonpolar portions, which determine their solubility. For example, carboxylic acids and amines behave like alcohols: methanoic acid (HCOOH, formic acid) and methanamine (CH_3NH_2) are miscible with water and slightly soluble in hexane, whereas hexanoic acid [$CH_3(CH_2)_4COOH$] and 1-hexanamine [$CH_3(CH_2)_5NH_2$] are slightly soluble in water and very soluble in hexane.

Sample Problem 13.1 Predicting Relative Solubilities

Problem Predict which solvent will dissolve more of the given solute:
(a) Sodium chloride in methanol (CH_3OH) or in 1-propanol ($CH_3CH_2CH_2OH$)
(b) Ethylene glycol ($HOCH_2CH_2OH$) in hexane ($CH_3CH_2CH_2CH_2CH_2CH_3$) or in water
(c) Diethyl ether ($CH_3CH_2OCH_2CH_3$) in water or in ethanol (CH_3CH_2OH)

Plan We examine the formulas of solute and solvent to determine the forces in and between solute and solvent. A solute is more soluble in a solvent whose intermolecular forces are similar to, and therefore can replace, its own.

Solution (a) Methanol. NaCl is ionic, so it dissolves through ion-dipole forces. Both methanol and 1-propanol have a polar —OH group, but the hydrocarbon portion of each alcohol interacts only weakly with the ions and 1-propanol has a longer hydrocarbon portion than methanol.
(b) Water. Ethylene glycol molecules have two —OH groups, so they interact with each other through H bonding. H bonds formed with H_2O can replace these H bonds between solute molecules better than dipole–induced dipole forces with hexane can.
(c) Ethanol. Diethyl ether molecules interact through dipole-dipole and dispersion forces. They can form H bonds to H_2O or to ethanol. But ethanol can also interact with the ether effectively through dispersion forces because it has a hydrocarbon chain.

FOLLOW-UP PROBLEM 13.1 Which solute is more soluble in the given solvent: (a) 1-butanol ($CH_3CH_2CH_2CH_2OH$) or 1,4-butanediol ($HOCH_2CH_2CH_2CH_2OH$) in water; (b) chloroform ($CHCl_3$) or carbon tetrachloride (CCl_4) in water?

Table 13.3 Correlation Between Boiling Point and Solubility in Water

Gas	Solubility (*M*)*	bp (K)
He	4.2×10^{-4}	4.2
Ne	6.6×10^{-4}	27.1
N_2	10.4×10^{-4}	77.4
CO	15.6×10^{-4}	81.6
O_2	21.8×10^{-4}	90.2
NO	32.7×10^{-4}	121.4

*At 273 K and 1 atm.

Gas-Liquid Solutions A substance with very weak intermolecular attractions has a low boiling point and is a gas under ordinary conditions. Likewise, it is not very soluble in water because solute-solvent forces are weak. Thus, for nonpolar or slightly polar gases, boiling point generally correlates with solubility in water (Table 13.3).

The small amount of a nonpolar gas that *does* dissolve may be vital. At 25°C and 1 atm, the solubility of O_2 is only 3.2 mL/100. mL of water, but aquatic animal life requires it. At times, the solubility of a nonpolar gas may *seem* high because it is also *reacting* with solvent. Oxygen seems more soluble in blood than in water because it bonds to hemoglobin in red blood cells. Carbon dioxide, which is essential for aquatic plants and coral-reef growth, seems very soluble in water (~81 mL of CO_2/100. mL of H_2O at 25°C and 1 atm) because it is dissolving *and* reacting:

$$CO_2(g) + H_2O(l) \rightleftharpoons H^+(aq) + HCO_3^-(aq)$$

Gas Solutions and Solid Solutions

Gas solutions and solid solutions also have vital importance and numerous applications.

Gas-Gas Solutions *All gases are miscible with each other.* Air is the classic example of a gaseous solution, consisting of about 18 gases in widely differing proportions. Anesthetic gas proportions are finely adjusted to the needs of the patient and the length of the surgical procedure.

Gas-Solid Solutions *When a gas dissolves in a solid, it occupies the spaces between the closely packed particles.* Hydrogen gas can be purified by passing an impure sample through palladium. Only H_2 molecules are small enough to fit between the Pd atoms, where they form Pd—H bonds. The H atoms move from one Pd atom to another and emerge from the metal as H_2 molecules.

Solid-Solid Solutions Solids diffuse so little that their mixtures are usually heterogeneous. Some solid-solid solutions can be formed by melting the solids and then mixing them and allowing them to freeze. Many **alloys,** mixtures of substances that have a metallic character, are solid-solid solutions (although several have microscopic heterogeneous regions). Brass, a familiar example of an alloy, is a mixture of zinc and copper. Waxes are amorphous solid-solid solutions that may contain small regions of crystalline regularity. A natural *wax* is a solid of biological origin that is insoluble in water but dissolves in nonpolar solvents.

Summary of Section 13.1

- A solution is a homogeneous mixture of a solute dissolved in a solvent through the action of intermolecular forces.
- Ion-dipole, ion–induced dipole, and dipole–induced dipole forces occur in solutions, in addition to all the intermolecular forces that also occur in pure substances.

- If similar intermolecular forces occur in solute and solvent, they replace each other when the substances mix and a solution is likely to form ("like dissolves like").
- When ionic compounds dissolve in water, the ions become surrounded by hydration shells of H-bonded water molecules.
- Solubility of organic molecules in various solvents depends on the relative sizes of their polar and nonpolar portions.
- The solubility of nonpolar gases in water is low because of weak intermolecular forces. Gases are miscible with one another, and they dissolve in solids by fitting into spaces between the closely packed particles.
- Solid-solid solutions include alloys (some of which are formed by mixing molten components) and waxes.

13.2 • WHY SUBSTANCES DISSOLVE: UNDERSTANDING THE SOLUTION PROCESS

The qualitative *macroscopic* rule "like dissolves like" is based on *molecular* interactions between solute and solvent. To see *why* like dissolves like, we'll break down the solution process conceptually into steps and examine them quantitatively.

Heat of Solution: Solution Cycles

Before a solution forms, solute particles are attracting each other, as are solvent particles. For one to dissolve in the other, three steps must take place, each accompanied by an enthalpy change:

Step 1. Solute particles separate from each other. This step involves overcoming intermolecular attractions, so it is *endothermic:*

$$\text{Solute (aggregated)} + heat \longrightarrow \text{solute (separated)} \qquad \Delta H_{\text{solute}} > 0$$

Step 2. Solvent particles separate from each other. This step also involves overcoming attractions, so it is *endothermic,* too:

$$\text{Solvent (aggregated)} + heat \longrightarrow \text{solvent (separated)} \qquad \Delta H_{\text{solvent}} > 0$$

Step 3. Solute and solvent particles mix and form a solution. The different particles attract each other and come together, so this step is *exothermic:*

$$\text{Solute (separated)} + \text{solvent (separated)} \longrightarrow \text{solution} + heat \qquad \Delta H_{\text{mix}} < 0$$

The overall process is called a *thermochemical solution cycle,* and in yet another application of Hess's law, we combine the three individual enthalpy changes to find the **heat** (or *enthalpy*) **of solution (ΔH_{soln}),** the total enthalpy change that occurs when solute and solvent form a solution:

$$\Delta H_{\text{soln}} = \Delta H_{\text{solute}} + \Delta H_{\text{solvent}} + \Delta H_{\text{mix}} \qquad \textbf{(13.1)}$$

Overall solution formation is exothermic or endothermic, and ΔH_{soln} is either positive or negative, depending on the relative sizes of the individual ΔH values:

- *Exothermic process:* $\Delta H_{\text{soln}} < 0$. If the sum of the endothermic terms ($\Delta H_{\text{solute}} + \Delta H_{\text{solvent}}$) is *smaller* than the exothermic term (ΔH_{mix}), the process is exothermic and ΔH_{soln} is negative (Figure 13.4A, *next page*).
- *Endothermic process:* $\Delta H_{\text{soln}} > 0$. If the sum of the endothermic terms is *larger* than the exothermic term, the process is endothermic and ΔH_{soln} is positive (Figure 13.4B, *next page*). If ΔH_{soln} is highly positive, the solute may not dissolve significantly in that solvent.

Heat of Hydration: Ionic Solids in Water

The $\Delta H_{\text{solvent}}$ and ΔH_{mix} components of the solution cycle are difficult to measure individually. Combined, they equal the enthalpy change for **solvation,** the process of surrounding a solute particle with solvent particles. Solvation in water is called

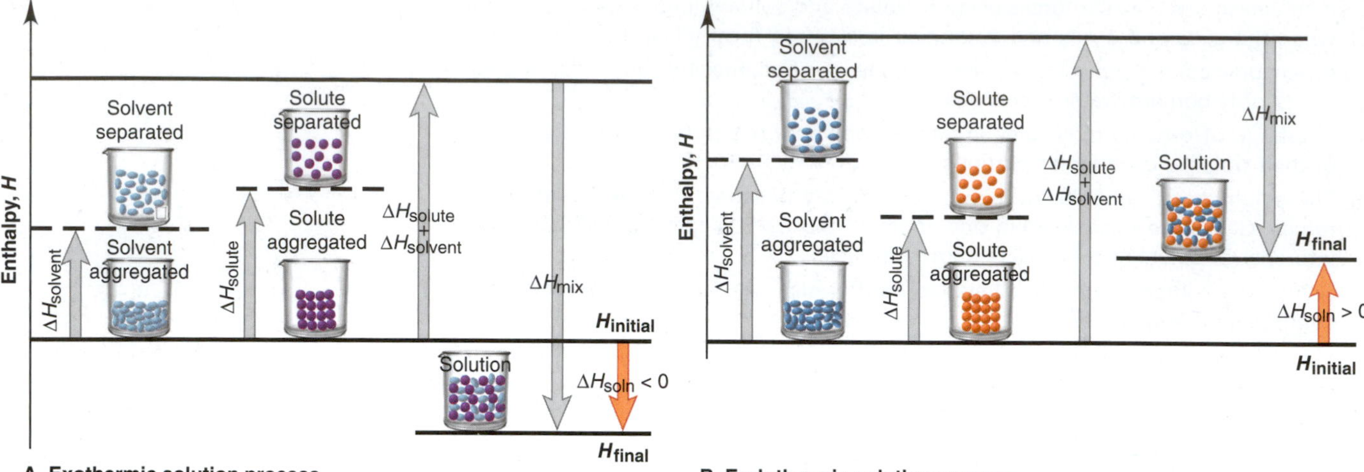

Figure 13.4 Enthalpy components of the heat of solution. **A,** ΔH_{mix} is larger than the sum of ΔH_{solute} and $\Delta H_{solvent}$, so ΔH_{soln} is negative. **B,** ΔH_{mix} is smaller than the sum of ΔH_{solute} and $\Delta H_{solvent}$, so ΔH_{soln} is positive.

hydration. Thus, enthalpy changes for separating the water molecules ($\Delta H_{solvent}$) and mixing the separated solute with them (ΔH_{mix}) are combined into the **heat** (or *enthalpy*) **of hydration (ΔH_{hydr}).** In water, Equation 13.1 becomes

$$\Delta H_{soln} = \Delta H_{solute} + \Delta H_{hydr}$$

The heat of hydration is a key factor in dissolving an ionic solid. Breaking H bonds in water is more than compensated for by forming the stronger ion-dipole forces, so hydration of an ion is *always* exothermic. The ΔH_{hydr} of an ion is defined as the enthalpy change for the hydration of 1 mol of separated (gaseous) ions:

$$M^+(g)\ [\text{or } X^-(g)] \xrightarrow{H_2O} M^+(aq)\ [\text{or } X^-(aq)] \qquad \Delta H_{\text{hydr of the ion}}\ (\text{always} <0)$$

Importance of Charge Density Heats of hydration exhibit trends based on the ion's **charge density,** the ratio of its charge to its volume. In general, the higher the charge density, the more negative ΔH_{hydr} is. Coulomb's law explains why: the higher the charge of an ion and the smaller its radius, the closer it gets to the oppositely charged pole of an H_2O molecule (see Figure 2.11), and the stronger the attraction. Thus,

- A 2+ ion attracts H_2O molecules more strongly than a 1+ ion of similar size.
- A small 1+ ion attracts H_2O molecules more strongly than a large 1+ ion.

Periodic trends in ΔH_{hydr} values are based on trends in charge density:

- *Down a group,* the charge stays the same and the size increases; thus, the charge densities decrease, as do the ΔH_{hydr} values.
- *Across a period,* say, from Group 1A(1) to Group 2A(2), the 2A ion has a smaller radius *and* a higher charge, so its charge density and ΔH_{hydr} are greater.

Components of Aqueous Heats of Solution To separate an ionic solute, MX, into gaseous ions requires a lot of energy (ΔH_{solute}); recall from Chapter 9 that this is the lattice energy, and it is highly positive:

$$MX(s) \longrightarrow M^+(g) + X^-(g) \qquad \Delta H_{solute}\ (\text{always} > 0) = \Delta H_{lattice}$$

Thus, for ionic compounds in water, the heat of solution is the lattice energy (always positive) plus the combined heats of hydration of the ions (always negative):

$$\Delta H_{soln} = \Delta H_{lattice} + \Delta H_{\text{hydr of the ions}} \qquad (13.2)$$

Once again, the sizes of the individual terms determine the sign of ΔH_{soln}.

Figure 13.5 shows enthalpy diagrams for three ionic solutes dissolving in water:

- *NaCl.* Sodium chloride has a small positive ΔH_{soln} (3.9 kJ/mol) because its lattice energy is only slightly greater than the combined ionic heats of hydration: if you dissolve NaCl in water in a flask, you don't feel any temperature change.

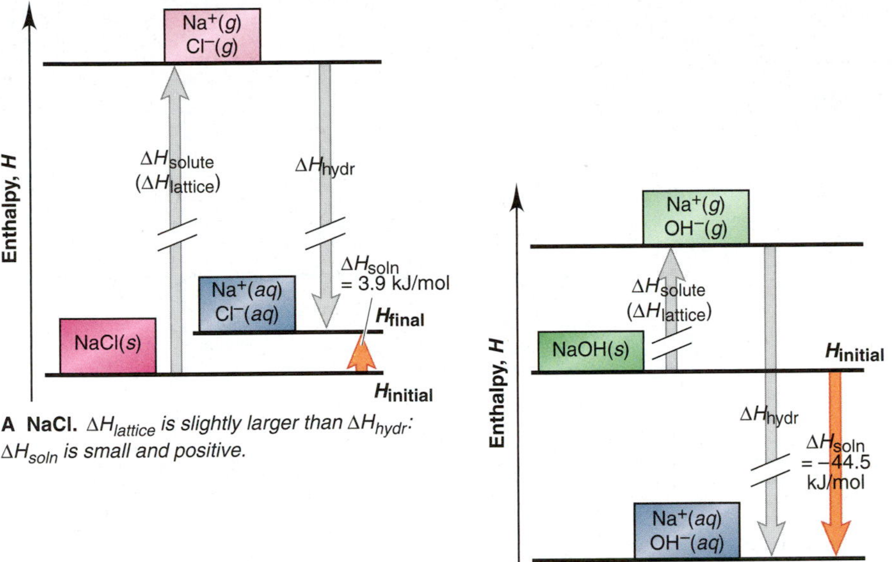

A NaCl. *$\Delta H_{lattice}$ is slightly larger than ΔH_{hydr}: ΔH_{soln} is small and positive.*

B NaOH. *ΔH_{hydr} dominates: ΔH_{soln} is large and negative.*

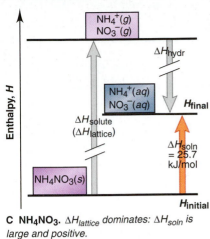

C NH_4NO_3. *$\Delta H_{lattice}$ dominates: ΔH_{soln} is large and positive.*

Figure 13.5 Enthalpy diagrams for three ionic compounds dissolving in water.

- *NaOH.* Sodium hydroxide has a large negative ΔH_{soln} (−44.5 kJ/mol) because its lattice energy is much *smaller* than the combined ionic heats of hydration: if you dissolve NaOH in water, the flask feels hot.
- *NH_4NO_3.* Ammonium nitrate has a large positive ΔH_{soln} (25.7 kJ/mol) because its lattice energy is much *larger* than the combined ionic heats of hydration: if you dissolve NH_4NO_3 in water, the flask feels cold.

The Solution Process and the Change in Entropy

The heat of solution (ΔH_{soln}) is one of two factors that determine whether a solute dissolves. The other factor is the natural tendency of a system of particles to spread out, so the system's kinetic energy becomes more dispersed or distributed. A thermodynamic variable called **entropy (*S*)** is directly related to the number of ways a system can disperse its energy, which is closely related to the freedom of motion of the particles.

Let's see what it means for a system to "distribute" its energy. We'll first compare the three physical states and then compare solute and solvent with solution.

Entropy and the Three Physical States The states of matter differ significantly in their entropy.

- In a solid, the particles are fixed in their positions with little freedom of motion. In a liquid, they can move around each other and so have greater freedom of motion. And in a gas, the particles have little restriction and much more freedom of motion.
- The more freedom of motion the particles have, the more ways they can distribute their kinetic energy; thus, a liquid has higher entropy than a solid, and a gas has higher entropy than a liquid:

$$S_{gas} > S_{liquid} > S_{solid}$$

- Thus, there is a *change* in entropy (ΔS) associated with a phase change, and it can be positive or negative. For example, the change in entropy when a liquid vaporizes ($\Delta S_{vap} = S_{gas} - S_{liquid}$) is positive ($\Delta S_{vap} > 0$), the change in entropy when a liquid freezes (fusion) ($\Delta S_{fus} = S_{solid} - S_{liquid}$) is negative ($\Delta S_{fus} < 0$), and so forth.

Entropy and the Formation of Solutions The formation of solutions also involves a change in entropy. *A solution usually has higher entropy than the pure solute and pure solvent* because the number of ways to distribute the energy is related to the number of interactions between different molecules. There are far more interactions possible when solute and solvent are mixed than when they are pure; thus,

$$S_{soln} > (S_{solute} + S_{solvent}) \qquad \text{or} \qquad \Delta S_{soln} > 0$$

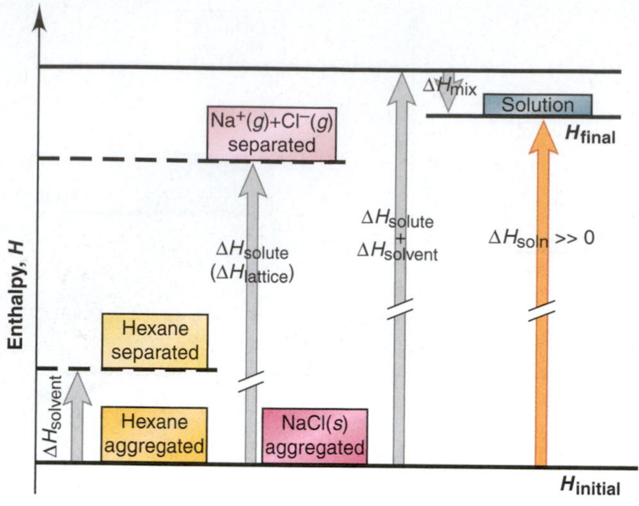

A NaCl. *ΔH_{mix} is* ***much*** *smaller than ΔH_{solute}; ΔH_{soln} is so much larger than the entropy increase due to mixing that NaCl does* ***not*** *dissolve.*

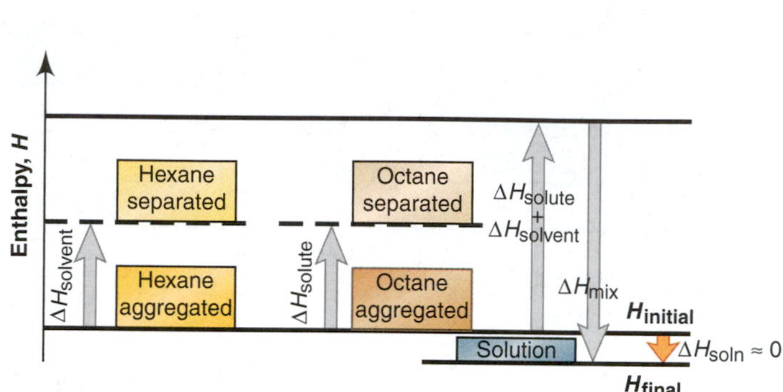

B Octane. *ΔH_{soln} is very small, but the entropy increase due to mixing is large, so octane dissolves.*

Figure 13.6 Enthalpy diagrams for dissolving **(A)** NaCl and **(B)** octane in hexane.

Enthalpy vs. Entropy Changes in Solution Formation Solution formation involves two factors: *systems change naturally toward a state of lower enthalpy and higher entropy,* so the relative sizes of ΔH_{soln} and ΔS_{soln} determine whether a solution forms. Let's consider three solute-solvent pairs to see which factor dominates:

1. *NaCl in hexane.* Given their very different intermolecular forces, we predict that sodium chloride does *not* dissolve in hexane (C_6H_{14}) (Figure 13.6A). Separating the solvent is easy because the dispersion forces are weak ($\Delta H_{solvent} \geq 0$), but separating the solute requires supplying the very large $\Delta H_{lattice}$ ($\Delta H_{solute} >> 0$). Mixing releases little heat because ion–induced dipole forces between Na^+ (or Cl^-) and hexane are weak ($\Delta H_{mix} \leq 0$). Because the sum of the endothermic terms is *much* larger than the exothermic term, $\Delta H_{soln} >> 0$. *A solution does not form because the entropy increase from mixing solute and solvent would be much smaller than the enthalpy increase required to separate the solute: $\Delta S_{mix} << \Delta H_{solute}$.*

2. *Octane in hexane.* We predict that octane (C_8H_{18}) is soluble in hexane because both are held together by dispersion forces of similar strength; in fact, they are miscible. Both ΔH_{solute} and $\Delta H_{solvent}$ are small because the forces are weak. The similar forces mean that ΔH_{mix} is small also. Thus, a lot of heat is not released; in fact, ΔH_{soln} is around zero (Figure 13.6B). So why does a solution form so readily? *With no enthalpy change driving the process, octane dissolves in hexane because the entropy increases greatly when the pure substances mix: $\Delta S_{mix} >> \Delta H_{soln}$.*

3. *NH_4NO_3 in water.* A large enough increase in entropy can sometimes cause a solution to form even when the enthalpy increase is large ($\Delta H_{soln} >> 0$). As we saw in Figure 13.5C, when ammonium nitrate (NH_4NO_3) dissolves in water, the process is highly endothermic; that is, $\Delta H_{lattice} >> \Delta H_{\text{hydr of the ions}}$. Nevertheless, *the increase in entropy that occurs when the crystal breaks down and the ions mix with water molecules is greater than the increase in enthalpy: $\Delta S_{soln} > \Delta H_{soln}$.*

In Chapter 20, we'll look in more depth at the relation between enthalpy and entropy to understand physical and chemical systems.

Summary of Section 13.2

- In a thermochemical solution cycle, the heat of solution is the sum of the endothermic separations of solute and of solvent and the exothermic mixing of their particles.
- In water, the combination of solvent separation and mixing with solute particles is hydration. For ions, heats of hydration depend on the ion's charge density but are always negative because ion-dipole forces are strong. Charge density exhibits periodic trends.

- Systems naturally increase their entropy (distribute their energy in more ways). A gas has higher entropy than a liquid, which has higher entropy than a solid, and a solution has higher entropy than the pure solute and solvent.
- Relative sizes of the enthalpy and entropy changes determine solution formation. A substance with a positive ΔH_{soln} dissolves *only* if ΔS_{soln} is larger than ΔH_{soln}.

13.3 • SOLUBILITY AS AN EQUILIBRIUM PROCESS

When an excess amount of solid is added to a solvent, particles leave the crystal, are surrounded by solvent, and move away. Some dissolved solute particles collide with undissolved solute and recrystallize, but, as long as the rate of dissolving is greater than the rate of recrystallizing, the concentration rises. At a given temperature, when solid is dissolving at the same rate as dissolved particles are recrystallizing, the concentration remains constant and *undissolved solute is in equilibrium with dissolved solute:*

$$\text{Solute (undissolved)} \rightleftharpoons \text{solute (dissolved)}$$

Figure 13.7 shows an ionic solid in equilibrium with dissolved cations and anions.

Three terms express the extent of this solution process:

- A **saturated solution** is at equilibrium and contains the maximum amount of dissolved solute at a given temperature in the presence of undissolved solute. Therefore, if you filter off the solution and add more solute, it doesn't dissolve.
- An **unsaturated solution** contains *less* than the equilibrium concentration of dissolved solute; add more solute, and more dissolves until the solution is saturated.
- A **supersaturated solution** contains *more* than the equilibrium concentration and is unstable relative to the saturated solution; add a "seed" crystal of solute or tap the container, and the excess solute crystallizes immediately, leaving a saturated solution (Figure 13.8).

Figure 13.7 Equilibrium in a saturated solution. At some temperature, the number of solute particles dissolving *(white arrows)* per unit time equals the number recrystallizing *(black arrows).*

Figure 13.8 Sodium acetate crystallizing from a supersaturated solution. When a seed crystal of sodium acetate is added to a supersaturated solution of it **(A)**, solute begins to crystallize **(B)** and continues until the remaining solution is saturated **(C)**.

THINK OF IT THIS WAY
A Saturated Solution Is Like a Liquid and Its Vapor

Many similarities exist between a saturated solution and a pure liquid and its vapor in a closed flask (see Section 12.2). For the liquid-vapor system, rates of vaporizing and condensing are equal; for the solution, rates of dissolving and recrystallizing are equal. In the liquid-vapor system, particles leave the liquid to enter the space above it, and their concentration (pressure) increases until, at equilibrium, the space is "saturated" with vapor at a given temperature. In the solution, particles leave the solute to enter the solvent, and their concentration increases until, at equilibrium, the solvent is saturated with solute at a given temperature.

Effect of Temperature on Solubility

You know that more sugar dissolves in hot tea than in iced tea; in fact, temperature affects the solubility of most substances. Let's examine the effects of temperature on the solubility of solids and gases.

Temperature and the Solubility of Solids in Water Like sugar, *most solids are more soluble at higher temperatures* (Figure 13.9, *next page*). Cerium sulfate is the

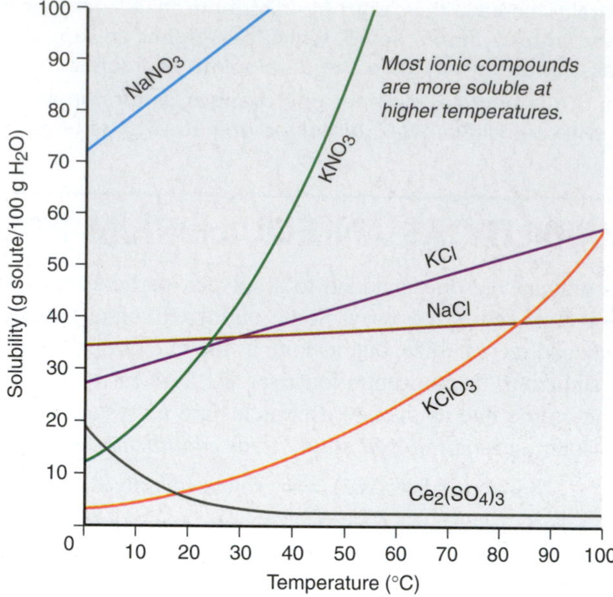

Figure 13.9 Relation between solubility and temperature for several ionic compounds.

only exception included in the figure, but several other salts, mostly sulfates, behave similarly. Some salts have higher solubility up to a certain temperature and then lower solubility at still higher temperatures.

Unfortunately, the effect of temperature on solubility is a complex phenomenon, and the sign of ΔH_{soln} does not reflect that complexity. Tabulated ΔH_{soln} values give the enthalpy change for making a solution at the standard state of 1 *M;* but in order to understand the effect of temperature, we need to know the sign of the enthalpy change very close to the point of saturation, which may differ from the sign of the tabulated value. For example, tables give a negative ΔH_{soln} for NaOH and a positive one for NH_4NO_3, yet both compounds are more soluble at higher temperatures. The point is that, although the effect of temperature reflects the equilibrium nature of solubility, no single measure can predict the effect for a given solute.

Temperature and the Solubility of Gases in Water The effect of temperature on *gas* solubility is much more predictable. When a solid dissolves in a liquid, the solute particles must separate, so $\Delta H_{solute} > 0$. In contrast, gas particles are already separated, so $\Delta H_{solute} \approx 0$. Because hydration is exothermic ($\Delta H_{hydr} < 0$), the sum of these two terms is negative. Thus, $\Delta H_{soln} < 0$ for all gases in water:

$$\text{Solute}(g) + \text{water}(l) \rightleftharpoons \text{saturated solution}(aq) + \text{heat}$$

Thus, *the solubility of any gas in water* ***decreases*** *with rising temperature (addition of heat).* Gases have weak intermolecular forces with water. When the temperature rises, the average kinetic energy increases, allowing the gas particles to easily overcome these forces and re-enter the gas phase.

One effect of this behavior is *thermal pollution.* Many electric power plants withdraw large amounts of water from a nearby river or lake for cooling, and the warmed water is returned to the source. The metabolic rates of fish and other aquatic animals increase in this warmer water, increasing their need for O_2. But the concentration of dissolved O_2 is lower in warm water, so the animals become "oxygen deprived." Also, the less dense warm water floats and prevents O_2 from reaching the cooler water below. Thus, even creatures at deeper levels become oxygen deprived. Farther from the plant, the water temperature and O_2 solubility return to normal. To mitigate the problem, cooling towers lower the temperature of the water before it exits the plant; nuclear power plants use a similar approach (Section 23.7).

Effect of Pressure on Solubility

Pressure has little effect on the solubility of liquids and solids because they are almost incompressible. But it has a *major* effect on the solubility of gases. Consider

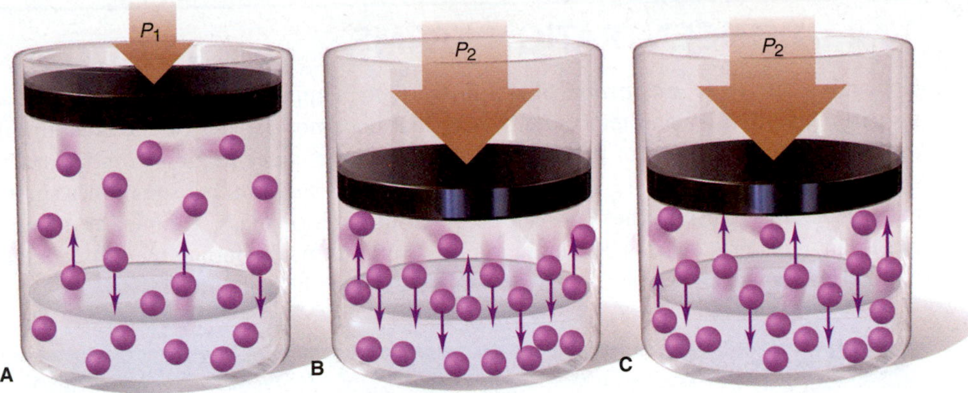

Figure 13.10 The effect of pressure on gas solubility.

a piston-cylinder assembly with a gas above a saturated aqueous solution of the gas (Figure 13.10A). At equilibrium, at a given pressure, the same number of gas molecules enter and leave the solution per unit time:

$$\text{Gas} + \text{solvent} \rightleftharpoons \text{saturated solution}$$

Push down on the piston, and you disturb the equilibrium: gas volume decreases, so gas pressure (and concentration) increases, and gas particles collide with the liquid surface more often. Thus, more particles enter than leave the solution per unit time (Figure 13.10B). More gas dissolves to reduce this disturbance (a shift to the right in the preceding equation) until the system re-establishes equilibrium (Figure 13.10C).

Henry's law expresses the quantitative relationship between gas pressure and solubility: *the solubility of a gas (S_{gas}) is directly proportional to the partial pressure of the gas (P_{gas}) above the solution:*

$$S_{gas} = k_H \times P_{gas} \qquad (13.3)$$

where k_H is the *Henry's law constant* and is specific for a given gas-solvent combination at a given temperature. With S_{gas} in mol/L and P_{gas} in atm, the units of k_H are mol/L·atm (that is, mol·L^{-1}·atm^{-1}).

Sample Problem 13.2 Using Henry's Law to Calculate Gas Solubility

Problem The partial pressure of carbon dioxide gas inside a bottle of cola is 4 atm at 25°C. What is the solubility of CO_2? The Henry's law constant for CO_2 in water is 3.3×10^{-2} mol/L·atm at 25°C.

Plan We know P_{CO_2} (4 atm) and the value of k_H (3.3×10^{-2} mol/L·atm), so we substitute them into Equation 13.3 to find S_{CO_2}.

Solution $S_{CO_2} = k_H \times P_{CO_2} = (3.3\times10^{-2}\ \text{mol/L·atm})(4\ \text{atm}) = 0.1\ \text{mol/L}$

Check The units are correct. We rounded to one significant figure to match the number in the pressure. A 0.5-L bottle of cola has about 2 g (0.05 mol) of dissolved CO_2.

FOLLOW-UP PROBLEM 13.2 If air contains 78% N_2 by volume, what is the solubility of N_2 in water at 25°C and 1 atm (k_H for N_2 in H_2O at 25°C = 7×10^{-4} mol/L·atm)?

Summary of Section 13.3

- A solution that contains the maximum amount of dissolved solute in the presence of excess undissolved solute is saturated. A saturated solution is in equilibrium with excess solute, because solute particles are entering and leaving the solution at the same rate.
- Most solids are more soluble at higher temperatures.
- All gases have a negative ΔH_{soln} in water, so heating lowers gas solubility in water.
- Henry's law says that the solubility of a gas is directly proportional to its partial pressure above the solution.

13.4 • CONCENTRATION TERMS

Concentration is the *proportion* of a substance in a mixture, so it is an *intensive* property (like density and temperature), one that does not depend on the quantity of mixture: 1.0 L or 1.0 mL of 0.1 *M* NaCl have the same concentration. Concentration is a ratio of quantities (Table 13.4), most often solute to *solution,* but sometimes solute to *solvent.* Both parts of the ratio can be given in units of mass, volume, or amount (mol), and chemists express concentration by several terms, including molarity, molality, and various expressions of "parts of solute per part by solution."

Table 13.4 Concentration Definitions

Concentration Term	Ratio
Molarity (M)	$\dfrac{\text{amount (mol) of solute}}{\text{volume (L) of solution}}$
Molality (m)	$\dfrac{\text{amount (mol) of solute}}{\text{mass (kg) of solvent}}$
Parts by mass	$\dfrac{\text{mass of solute}}{\text{mass of solution}}$
Parts by volume	$\dfrac{\text{volume of solute}}{\text{volume of solution}}$
Mole fraction (X)	$\dfrac{\text{amount (mol) of solute}}{\text{amount (mol) of solute} + \text{amount (mol) of solvent}}$

Molarity and Molality

Two very common concentration terms are molarity and molality:

1. Molarity (M) is the *number of moles of solute dissolved in 1 L of solution:*

$$\text{Molarity } (M) = \frac{\text{amount (mol) of solute}}{\text{volume (L) of solution}} \qquad \textbf{(13.4)}$$

In Chapter 3, we used molarity to convert liters of solution into amount of dissolved solute. Molarity has two drawbacks that affect its use in precise work:

- *Effect of temperature*. A liquid expands when heated, so a unit volume of hot solution contains less solute than one of cold solution; thus, the molarity is different.
- *Effect of mixing*. Because of solute-solvent interactions that are difficult to predict, *volumes may not be additive:* adding 500. mL of one solution to 500. mL of another may not give 1000. mL of final solution.

2. **Molality (m)** does not contain volume in its ratio; it is *the number of moles of solute dissolved in 1000 g (1 kg) of solvent:*

$$\text{Molality } (m) = \frac{\text{amount (mol) of solute}}{\text{mass (kg) of solvent}} \qquad \textbf{(13.5)}$$

Note that molality includes the quantity of *solvent,* not solution. Molality has two advantages over molarity for precise work:

- *Effect of temperature*. Molal solutions are based on *masses* of components, not *volume*. And since mass does not change with temperature, neither does molality.
- *Effect of mixing*. Unlike volumes, masses *are* additive: adding 500. g of one solution to 500. g of another *does* give 1000. g of final solution.

For these reasons, molality is the preferred term when temperature, and hence density, may change, as in a study of physical properties. Note that, in the case of water, 1 L has a mass of 1 kg, so *molality and molarity are nearly the same for dilute aqueous solutions.*

Sample Problem 13.3 Calculating Molality

Problem What is the molality of a solution prepared by dissolving 32.0 g of $CaCl_2$ in 271 g of water?

Plan To use Equation 13.5, we convert mass of $CaCl_2$ (32.0 g) to amount (mol) with the molar mass (g/mol) and then divide by the mass of water (271 g), being sure to convert from grams to kilograms (see the road map).

Road Map

Mass (g) of $CaCl_2$
↓ divide by $\mathcal{M}$ (g/mol)
Amount (mol) of $CaCl_2$
↓ divide by kg of water
Molality (m) of $CaCl_2$ solution

Solution Converting from grams of solute to moles:

$$\text{Amount (mol) of CaCl}_2 = 32.0\text{ g CaCl}_2 \times \frac{1\text{ mol of CaCl}_2}{110.98\text{ g CaCl}_2} = 0.288\text{ mol CaCl}_2$$

Finding molality:

$$\text{Molality} = \frac{\text{mol solute}}{\text{kg solvent}} = \frac{0.288\text{ mol CaCl}_2}{271\text{ g} \times \dfrac{1\text{ kg}}{10^3\text{ g}}} = 1.06\ m\text{ CaCl}_2$$

Check The answer seems reasonable: the amount (mol) of $CaCl_2$ and mass (kg) of H_2O are about the same, so their ratio is about 1.

FOLLOW-UP PROBLEM 13.3 How many grams of glucose ($C_6H_{12}O_6$) must be dissolved in 563 g of ethanol (C_2H_5OH) to prepare a 2.40×10^{-2} m solution?

Parts of Solute by Parts of Solution

Several concentration terms relate the number of solute (or solvent) parts to a number of *solution* parts. Both can be expressed in units of mass, volume, or amount (mol).

Parts by Mass The most common of these terms is **mass percent, % (w/w),** which you encountered in Chapter 3. The word *percent* means "per hundred," so mass percent means mass of solute dissolved in 100. parts by mass of solution:

$$\text{Mass percent} = \frac{\text{mass of solute}}{\text{mass of solute} + \text{mass of solvent}} \times 100 = \frac{\text{mass of solute}}{\text{mass of solution}} \times 100 \quad (13.6)$$

Mass percent values appear on jars of solid chemicals to indicate impurities.

Two very similar terms are parts per million (ppm) by mass and parts per billion (ppb) by mass, or grams of solute per million or per billion grams of solution: in Equation 13.6, you multiply by 10^6 or by 10^9, respectively, instead of by 100.

Parts by Volume The most common parts-by-volume term is **volume percent, % (v/v),** the volume of solute in 100. volumes of solution:

$$\text{Volume percent} = \frac{\text{volume of solute}}{\text{volume of solution}} \times 100 \quad (13.7)$$

For example, rubbing alcohol is an aqueous solution of isopropanol (a three-carbon alcohol) that contains 70 volumes of alcohol in 100. volumes of solution, or 70% (v/v).

Parts by volume is often used to express tiny concentrations of liquids or gases. Minor atmospheric components occur in parts per million by volume (ppmv). For example, about 0.05 ppmv of the toxic gas carbon monoxide (CO) is in clean air, 1000 times as much (50 ppmv of CO) in air over urban traffic, and 10,000 times as much (500 ppmv of CO) in cigarette smoke.

A concentration term often used in health-related facilities for aqueous solutions is % (w/v), a ratio of solute *weight* (actually mass) to solution *volume.* Thus, a 1.5% (w/v) NaCl solution contains 1.5 g of NaCl per 100. mL of *solution.*

Mole Fraction The **mole fraction (X)** of a solute is the ratio of number of moles of solute to the total number of moles (solute plus solvent), that is, parts by mole:

$$\text{Mole fraction } (X) = \frac{\text{amount (mol) of solute}}{\text{amount (mol) of solute} + \text{amount (mol) of solvent}} \quad (13.8)$$

Put another way, the mole fraction gives the proportion of solute (or solvent) particles in solution. The *mole percent* is the mole fraction expressed as a percentage:

$$\text{Mole percent (mol \%)} = \text{mole fraction} \times 100$$

Sample Problem 13.4 Expressing Concentrations in Parts by Mass, Parts by Volume, and Mole Fraction

Problem **(a)** Find the concentration of calcium ion (in ppm) in a 3.50-g pill that contains 40.5 mg of Ca^{2+}.
(b) The label on a 0.750-L bottle of Italian chianti indicates "11.5% alcohol by volume." How many liters of alcohol does the bottle of wine contain?
(c) A sample of rubbing alcohol contains 142 g of isopropyl alcohol (C_3H_7OH) and 58.0 g of water. What are the mole fractions of alcohol and water?

Plan **(a)** We know the mass of Ca^{2+} (40.5 mg) and the mass of the pill (3.50 g). We convert the mass of Ca^{2+} from mg to g, find the mass ratio of Ca^{2+} to pill, and multiply by 10^6 to obtain ppm. **(b)** We know the volume % (11.5%, or 11.5 parts by volume of alcohol to 100. parts of chianti) and the total volume (0.750 L), so we use Equation 13.7 to find liters of alcohol. **(c)** We know the mass and formula of each component, so we convert masses to amounts (mol) and apply Equation 13.8 to find the mole fractions.

Solution **(a)** Finding parts per million by mass of Ca^{2+}. Combining the steps, we have

$$\text{ppm Ca}^{2+} = \frac{\text{mass of Ca}^{2+}}{\text{mass of pill}} \times 10^6 = \frac{40.5 \text{ mg Ca}^{2+} \times \dfrac{1 \text{ g}}{10^3 \text{ mg}}}{3.50 \text{ g}} \times 10^6$$
$$= \boxed{1.16{\times}10^4 \text{ ppm Ca}^{2+}}$$

(b) Finding volume (L) of alcohol:

$$\text{Volume (L) of alcohol} = 0.750 \text{ L chianti} \times \frac{11.5 \text{ L alcohol}}{100. \text{ L chianti}}$$
$$= \boxed{0.0862 \text{ L}}$$

(c) Finding mole fractions. Converting from mass (g) to amount (mol):

$$\text{Amount (mol) of C}_3\text{H}_7\text{OH} = 142 \text{ g C}_3\text{H}_7\text{OH} \times \frac{1 \text{ mol C}_3\text{H}_7\text{OH}}{60.09 \text{ g C}_3\text{H}_7\text{OH}} = 2.36 \text{ mol C}_3\text{H}_7\text{OH}$$

$$\text{Amount (mol) of H}_2\text{O} = 58.0 \text{ g H}_2\text{O} \times \frac{1 \text{ mol H}_2\text{O}}{18.02 \text{ g H}_2\text{O}} = 3.22 \text{ mol H}_2\text{O}$$

Calculating mole fractions:

$$X_{C_3H_7OH} = \frac{\text{moles of C}_3\text{H}_7\text{OH}}{\text{total moles}} = \frac{2.36 \text{ mol}}{2.36 \text{ mol} + 3.22 \text{ mol}} = \boxed{0.423}$$

$$X_{H_2O} = \frac{\text{moles of H}_2\text{O}}{\text{total moles}} = \frac{3.22 \text{ mol}}{2.36 \text{ mol} + 3.22 \text{ mol}} = \boxed{0.577}$$

Check **(a)** The mass ratio is about 0.04 g/4 g $= 10^{-2}$, and $10^{-2}{\times}10^6 = 10^4$ ppm, so it seems correct. **(b)** The volume % is a bit more than 10%, so the volume of alcohol should be a bit more than 75 mL (0.075 L). **(c)** Always check that the *mole fractions add up to 1:* $0.423 + 0.577 = 1.000$.

FOLLOW-UP PROBLEM 13.4 An alcohol solution contains 35.0 g of 1-propanol (C_3H_7OH) and 150. g of ethanol (C_2H_5OH). Calculate the mass percent and the mole fraction of each alcohol.

Interconverting Concentration Terms

All the terms we just discussed represent different ways of expressing concentration, so they are interconvertible. Keep these points in mind:

- To convert a term based on amount to one based on mass, you need the molar mass. These conversions are similar to the mass-mole conversions you've done earlier.

- To convert a term based on mass to one based on volume, you need the solution *density.* Given the mass of solution, the density (mass/volume) gives the volume, or vice versa.
- Molality includes quantity of *solvent;* the other terms include quantity of *solution.*

Sample Problem 13.5 Interconverting Concentration Terms

Problem Hydrogen peroxide is a powerful oxidizing agent; it is used in concentrated solution in rocket fuel, but in dilute solution in hair bleach. An aqueous solution of H_2O_2 is 30.0% by mass and has a density of 1.11 g/mL. Calculate the **(a)** molality, **(b)** mole fraction of H_2O_2, and **(c)** molarity.

Plan We know the mass % (30.0) and the density (1.11 g/mL). **(a)** For molality, we need the amount (mol) of solute and the mass (kg) of *solvent.* If we assume 100.0 g of solution, the mass % equals the grams of H_2O_2, which we subtract to obtain the grams of solvent. To find molality, we convert grams of H_2O_2 to moles and divide by mass of solvent (converting g to kg). **(b)** To find the mole fraction, we use the moles of H_2O_2 [from part (a)] and convert the grams of H_2O to moles. Then we divide the moles of H_2O_2 by the total moles. **(c)** To find molarity, we assume 100.0 g of solution and use the solution density to find the volume. Then we divide the moles of H_2O_2 [from part (a)] by *solution* volume (in L).

Solution **(a)** From mass % to molality. Finding mass of solvent (assuming 100.0 g of solution):

$$\text{Mass (g) of } H_2O = 100.0 \text{ g solution} - 30.0 \text{ g } H_2O_2 = 70.0 \text{ g } H_2O$$

Converting from grams of H_2O_2 to moles:

$$\text{Amount (mol) of } H_2O_2 = 30.0 \text{ g } H_2O_2 \times \frac{1 \text{ mol } H_2O_2}{34.02 \text{ g } H_2O_2} = 0.882 \text{ mol } H_2O_2$$

Calculating molality:

$$\text{Molality of } H_2O_2 = \frac{0.882 \text{ mol } H_2O_2}{70.0 \text{ g} \times \dfrac{1 \text{ kg}}{10^3 \text{ g}}} = 12.6\ m\ H_2O_2$$

(b) From mass % to mole fraction:

$$\text{Amount (mol) of } H_2O_2 = 0.882 \text{ mol } H_2O_2 \text{ [from part (a)]}$$

$$\text{Amount (mol) of } H_2O = 70.0 \text{ g } H_2O \times \frac{1 \text{ mol } H_2O}{18.02 \text{ g } H_2O} = 3.88 \text{ mol } H_2O$$

$$X_{H_2O_2} = \frac{0.882 \text{ mol}}{0.882 \text{ mol} + 3.88 \text{ mol}} = 0.185$$

(c) From mass % and density to molarity. Converting from solution mass to volume:

$$\text{Volume (mL) of solution} = 100.0 \text{ g} \times \frac{1 \text{ mL}}{1.11 \text{ g}} = 90.1 \text{ mL}$$

Calculating molarity:

$$\text{Molarity} = \frac{\text{mol } H_2O_2}{\text{L soln}} = \frac{0.882 \text{ mol } H_2O_2}{90.1 \text{ mL} \times \dfrac{1 \text{ L soln}}{10^3 \text{ mL}}} = 9.79\ M\ H_2O_2$$

Check Rounding shows the answers to be reasonable: **(a)** The ratio of ~0.9 mol/0.07 kg is greater than 10. **(b)** ~0.9 mol H_2O_2/(1 mol + 4 mol) ≈ 0.2. **(c)** The ratio of moles to liters (0.9/0.09) is around 10.

FOLLOW-UP PROBLEM 13.5 Concentrated hydrochloric acid is 11.8 *M* HCl and has a density of 1.190 g/mL. Calculate the mass %, molality, and mole fraction of HCl.

Summary of Section 13.4

- The concentration of a solution is independent of the quantity of solution and can be expressed as molarity (mol solute/L solution), molality (mol solute/kg solvent), parts by mass (mass solute/mass solution), parts by volume (volume solute/volume solution), or mole fraction [mol solute/(mol solute + mol solvent)] (Table 13.4).

- Molality is based on mass, so it is independent of temperature; the mole fraction gives the proportion of dissolved particles.
- If, in addition to the quantities of solute and solution, the solution density is known, the various ways of expressing concentration are interconvertible.

13.5 • COLLIGATIVE PROPERTIES OF SOLUTIONS

The presence of solute gives a solution different physical properties than the pure solvent. But, in the case of four important properties, it is the *number* of solute particles, *not* their chemical identity, that makes the difference. These **colligative properties** (*colligative* means "collective") are vapor pressure lowering, boiling point elevation, freezing point depression, and osmotic pressure. Most of the effects are small, but they have many applications, including some vital to organisms.

We predict the magnitude of a colligative property from the solute formula, which shows the number of particles in solution and is closely related to our classification of solutes by their ability to conduct an electric current (Chapter 4):

1. *Electrolytes.* An aqueous solution of an **electrolyte** conducts because the solute separates into ions as it dissolves.
   - *Strong electrolytes*—soluble salts, strong acids, and strong bases—dissociate completely, so their solutions conduct well.
   - *Weak electrolytes*—weak acids and weak bases—dissociate very little, so their solutions conduct poorly.
2. *Nonelectrolytes.* Compounds such as sugar and alcohol do not dissociate into ions at all. They are **nonelectrolytes** because their solutions do not conduct a current.

Thus we can predict that

- *For nonelectrolytes,* 1 mol of compound yields 1 mol of particles in solution. For example, 0.35 M glucose contains 0.35 mol of solute particles per liter.
- *For strong electrolytes,* 1 mol of compound yields the amount (mol) of ions in the formula unit: 0.4 M Na_2SO_4 has 0.8 mol of Na^+ ions and 0.4 mol of SO_4^{2-} ions, or 1.2 mol of particles, per liter (see Sample Problem 4.1).
- *For weak electrolytes,* the calculation is complicated because the solution reaches equilibrium; we examine these systems in Chapters 18 and 19.

In this section, we discuss colligative properties of three types of solute—nonvolatile nonelectrolytes, volatile nonelectrolytes, and strong electrolytes.

Nonvolatile Nonelectrolyte Solutions

We start with solutions of *nonvolatile nonelectrolytes* because they provide the clearest examples of the colligative properties. These solutions contain solutes that are not ionic and so do not dissociate, and they have negligible vapor pressure at the boiling point of the solvent; sucrose (table sugar) dissolved in water is an example.

Vapor Pressure Lowering *The vapor pressure of a nonvolatile nonelectrolyte solution is always lower than the vapor pressure of the pure solvent.* The difference in vapor pressures is the **vapor pressure lowering (ΔP).**

1. *Why the vapor pressure of a solution is lower.* The fundamental reason for this lowering involves entropy, specifically the relative entropies of vaporization of solvent versus solution. A liquid vaporizes because a gas has higher entropy. In a closed container, vaporization continues until the numbers of particles leaving and entering the liquid phase per unit time are equal, that is, when the system reaches equilibrium. But, as we said, the entropy of a solution is higher than that of pure solvent, so fewer

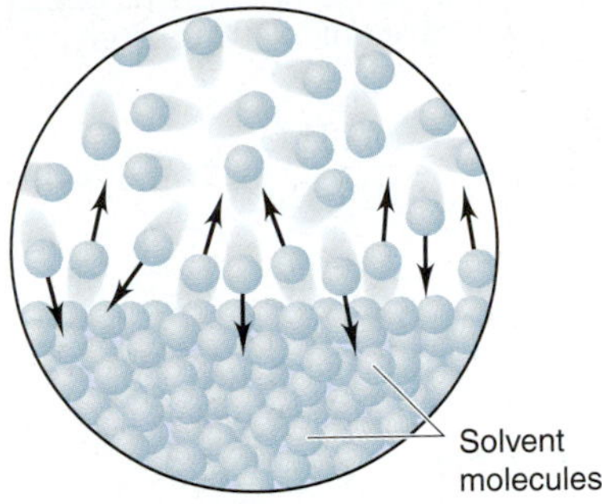
Equilibrium is reached with a given number of particles in the vapor.
Solvent molecules
A Pure solvent

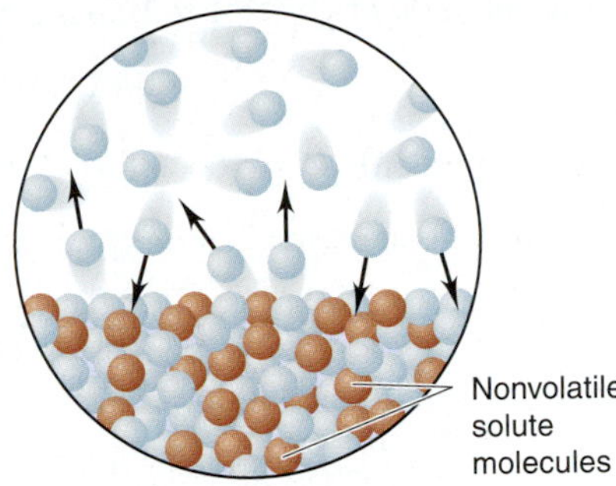
Equilibrium is reached with fewer particles in the vapor.
Nonvolatile solute molecules
B Solvent and dissolved solute

Figure 13.11 Effect of solute on the vapor pressure of solution. **A,** Equilibrium between a liquid and its vapor occurs when equal numbers of molecules vaporize and condense in a given time. **B,** Solute molecules increase the entropy of the solution, so equilibrium occurs at a lower vapor pressure.

solvent particles need to vaporize to reach the same entropy. With fewer particles in the gas phase, the vapor above a solution has lower pressure (Figure 13.11).

2. *Quantifying vapor pressure lowering.* **Raoult's law** says that the vapor pressure of solvent above a solution ($P_{solvent}$) equals the mole fraction of solvent ($X_{solvent}$) times the vapor pressure of the pure solvent ($P^\circ_{solvent}$):

$$P_{solvent} = X_{solvent} \times P^\circ_{solvent} \qquad (13.9)$$

Since $X_{solvent}$ is less than 1 in a solution, $P_{solvent}$ is less than $P^\circ_{solvent}$. An **ideal solution** is one that follows Raoult's law at any concentration. However, just as real gases deviate from ideality, so do real solutions. In practice, Raoult's law works reasonably well for *dilute* solutions and becomes exact at infinite dilution.

Let's see how the *amount* of dissolved solute affects ΔP. The solution consists of solvent and solute, so the sum of their mole fractions equals 1:

$$X_{solvent} + X_{solute} = 1; \quad \text{thus,} \quad X_{solvent} = 1 - X_{solute}$$

From Raoult's law, we have

$$P_{solvent} = X_{solvent} \times P^\circ_{solvent} = (1 - X_{solute}) \times P^\circ_{solvent}$$

Multiplying through on the right side gives

$$P_{solvent} = P^\circ_{solvent} - (X_{solute} \times P^\circ_{solvent})$$

Rearranging and introducing ΔP gives

$$P^\circ_{solvent} - P_{solvent} = \Delta P = X_{solute} \times P^\circ_{solvent} \qquad (13.10)$$

Thus, ΔP equals the mole fraction of solute times the vapor pressure of the pure solvent—a relationship applied in the next sample problem.

Sample Problem 13.6 Using Raoult's Law to Find ΔP

Problem Find the vapor pressure lowering, ΔP, when 10.0 mL of glycerol ($C_3H_8O_3$) is added to 500. mL of water at 50.°C. At this temperature, the vapor pressure of pure water is 92.5 torr and its density is 0.988 g/mL. The density of glycerol is 1.26 g/mL.

Plan To calculate ΔP, we use Equation 13.10. We are given the vapor pressure of pure water ($P^\circ_{H_2O} = 92.5$ torr), so we just need the mole fraction of glycerol, $X_{glycerol}$. We convert the given volume of glycerol (10.0 mL) to mass using the given density (1.26 g/L), find the molar mass from the formula, and convert mass (g) to amount (mol). The same procedure gives amount of H_2O. From these amounts, we find $X_{glycerol}$ and ΔP.

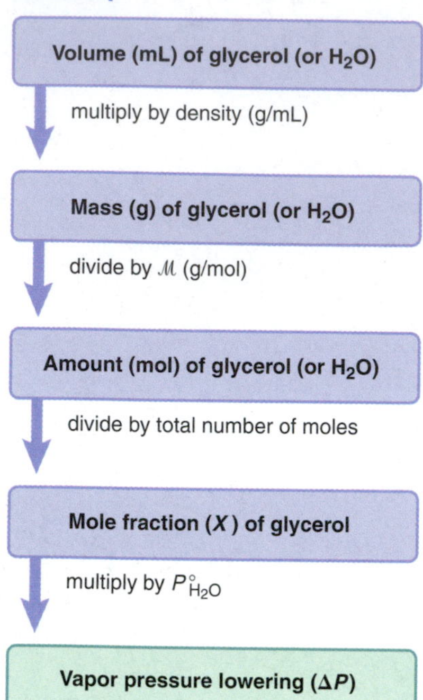

Solution Calculating the amounts (mol) of glycerol and of water:

$$\text{Amount (mol) of glycerol} = 10.0\ \text{mL glycerol} \times \frac{1.26\ \text{g glycerol}}{1\ \text{mL glycerol}} \times \frac{1\ \text{mol glycerol}}{92.09\ \text{g glycerol}} = 0.137\ \text{mol glycerol}$$

$$\text{Amount (mol) of } H_2O = 500.\ \text{mL } H_2O \times \frac{0.988\ \text{g } H_2O}{1\ \text{mL } H_2O} \times \frac{1\ \text{mol } H_2O}{18.02\ \text{g } H_2O} = 27.4\ \text{mol } H_2O$$

Calculating the mole fraction of glycerol:

$$X_{\text{glycerol}} = \frac{0.137\ \text{mol}}{0.137\ \text{mol} + 27.4\ \text{mol}} = 0.00498$$

Finding the vapor pressure lowering:

$$\Delta P = X_{\text{glycerol}} \times P^\circ_{H_2O} = 0.00498 \times 92.5\ \text{torr} = 0.461\ \text{torr}$$

Check The amount of each component seems correct: for glycerol, ~10 mL × 1.25 g/mL ÷ 100 g/mol = 0.125 mol; for H_2O, ~500 mL × 1 g/mL ÷ 20 g/mol = 25 mol. The small ΔP is reasonable because the mole fraction of solute is small.

Comment **1.** The calculation assumes that glycerol is nonvolatile. At 1 atm, glycerol boils at 290.0°C, so the vapor pressure of glycerol at 50°C is negligible. **2.** We have assumed that Raoult's law applies because the solution is dilute.

FOLLOW-UP PROBLEM 13.6 Calculate the vapor pressure lowering of a solution of 2.00 g of aspirin ($\mathcal{M}$ = 180.15 g/mol) in 50.0 g of methanol (CH_3OH) at 21.2°C. Pure methanol has a vapor pressure of 101 torr at this temperature.

Boiling Point Elevation *A solution boils at a higher temperature than the pure solvent.* This colligative property results from the previous vapor pressure lowering:

1. *Why a solution boils at a higher T.* Recall that the boiling point, T_b, of a liquid is the temperature at which its vapor pressure equals the external pressure, P_{ext}. But, the vapor pressure of a solution is always lower than that of the pure solvent. Therefore, it is lower than P_{ext} at T_b of the solvent, so the solution doesn't boil. Thus, the **boiling point elevation (ΔT_b)** results because a higher temperature is needed to raise the solution's vapor pressure to equal P_{ext}.

We superimpose a phase diagram for the solution on one for the solvent to see ΔT_b (Figure 13.12): the gas-liquid line for the solution lies *below* the line for the solvent at any T and to the right of it at any P, and the line crosses 1 atm (P_{ext}, or P_{atm}) at a higher T.

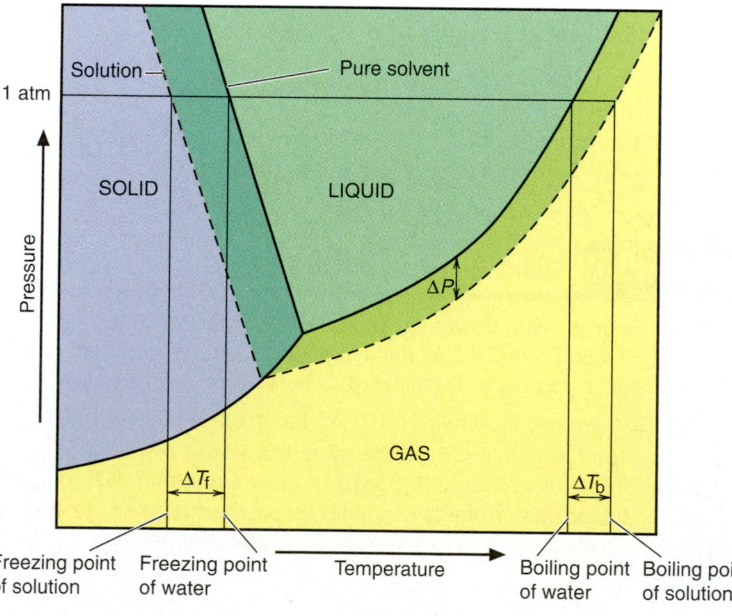

Figure 13.12 Boiling and freezing points of solvent and solution. Phase diagrams of an aqueous solution *(dashed lines)* and of pure water *(solid lines)* show that, by lowering the vapor pressure (ΔP), a solute elevates the boiling point (ΔT_b) and depresses the freezing point (ΔT_f). (The slope of the solid-liquid line is exaggerated.)

2. *Quantifying boiling point elevation.* Like vapor pressure lowering, boiling point elevation is proportional to the concentration of solute:

$$\Delta T_b \propto m \quad \text{or} \quad \Delta T_b = K_b m \qquad \textbf{(13.11)}$$

where m is the solution molality and K_b is the *molal boiling point elevation constant.* Since $\Delta T_b > 0$, we subtract the lower solvent T_b from the higher solution T_b:

$$\Delta T_b = T_{b(\text{solution})} - T_{b(\text{solvent})}$$

Molality is used because it relates to mole fraction and, thus, to particles of solute, and it is not affected by temperature.

The constant K_b has units of degrees Celsius per molal unit (°C/m) and is specific for a given solvent (Table 13.5). The K_b for water is 0.512°C/m, so ΔT_b for aqueous solutions is quite small. For example, if you dissolve 1.00 mol of glucose (180. g; 1.00 mol of particles) or 0.500 mol of NaCl (29.2 g; also 1.00 mol of particles) in 1.00 kg of water at 1 atm, the solutions will boil at 100.512°C instead of 100.000°C.

Table 13.5 Molal Boiling Point Elevation and Freezing Point Depression Constants of Several Solvents

Solvent	Boiling Point (°C)*	K_b (°C/m)	Melting Point (°C)	K_f (°C/m)
Acetic acid	117.9	3.07	16.6	3.90
Benzene	80.1	2.53	5.5	4.90
Carbon disulfide	46.2	2.34	−111.5	3.83
Carbon tetrachloride	76.5	5.03	−23	30.
Chloroform	61.7	3.63	−63.5	4.70
Diethyl ether	34.5	2.02	−116.2	1.79
Ethanol	78.5	1.22	−117.3	1.99
Water	100.0	0.512	0.0	1.86

*At 1 atm.

Freezing Point Depression *A solution freezes at a lower temperature than the pure solvent,* and this colligative property also results from vapor pressure lowering:

1. *Why a solution freezes at a lower T.* Only solvent vaporizes from solution, so solute molecules are left behind. Similarly, *only solvent freezes,* again leaving solute molecules behind. The freezing point of a solution is the temperature at which its vapor pressure equals that of the pure solvent, that is, when solid solvent and liquid solution are in equilibrium. The **freezing point depression** (ΔT_f) occurs because the vapor pressure of the solution is always lower than that of the solvent, so the solution freezes at a lower temperature; that is, only at a lower temperature will solvent particles leave and enter the solid at the same rate. In Figure 13.12, the solid-liquid line for the solution is to the left of the pure solvent line at 1 atm and at every other pressure.

2. *Quantifying freezing point depression.* Like ΔT_b, the freezing point depression is proportional to the molal concentration of solute:

$$\Delta T_f \propto m \quad \text{or} \quad \Delta T_f = K_f m \qquad \textbf{(13.12)}$$

where K_f is the *molal freezing point depression constant,* which also has units of °C/m (see Table 13.5). Also, like ΔT_b, $\Delta T_f > 0$, but in this case, we subtract the lower solution T_f from the higher solvent T_f:

$$\Delta T_f = T_{f(\text{solvent})} - T_{f(\text{solution})}$$

Here, too, the effect in aqueous solution is small because K_f for water is just 1.86°C/m. Thus, at 1 atm, 1 m glucose, 0.5 m NaCl, and 0.33 m K_2SO_4, all solutions with 1 mol of particles per kilogram of water, freeze at −1.86°C instead of at 0.00°C.

You have encountered practical applications of freezing point depression if you have added antifreeze—a solution of ethylene glycol in water—to your car's radiator or have seen airplace de-icers used before takeoff. Also, roads are "salted" with NaCl and $CaCl_2$ in winter to lower the freezing point of water, causing road ice to melt.

Sample Problem 13.7 Determining Boiling and Freezing Points of a Solution

Problem You add 1.00 kg of ethylene glycol ($C_2H_6O_2$) antifreeze to 4450 g of water in your car's radiator. What are the boiling and freezing points of the solution?

Plan To find the boiling and freezing points, we need ΔT_b and ΔT_f. We first find the molality by converting mass of solute (1.00 kg) to amount (mol) and dividing by mass of solvent (4450 g, converted to kg). Then we calculate ΔT_b and ΔT_f from Equations 13.11 and 13.12 (using constants from Table 13.5). We add ΔT_b to the solvent boiling point and subtract ΔT_f from its freezing point. The road map shows the steps.

Road Map

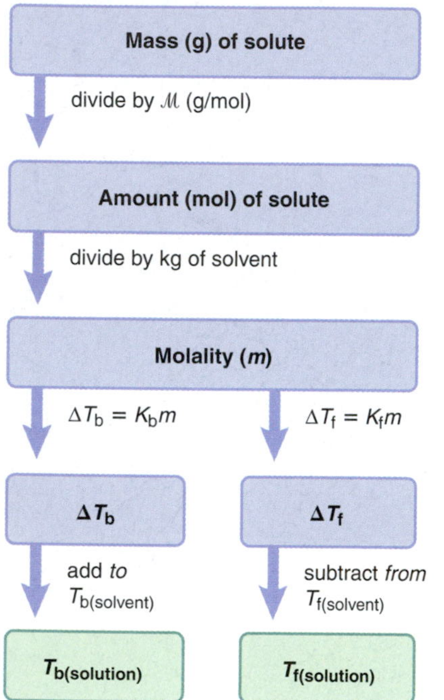

Solution Calculating the molality:

$$\text{Amount (mol) of } C_2H_6O_2 = 1.00 \text{ kg } C_2H_6O_2 \times \frac{10^3 \text{ g}}{1 \text{ kg}} \times \frac{1 \text{ mol } C_2H_6O_2}{62.07 \text{ g } C_2H_6O_2}$$
$$= 16.1 \text{ mol } C_2H_6O_2$$

$$\text{Molality} = \frac{\text{mol solute}}{\text{kg solvent}} = \frac{16.1 \text{ mol } C_2H_6O_2}{4450 \text{ g } H_2O \times \dfrac{1 \text{ kg}}{10^3 \text{ g}}} = 3.62\ m\ C_2H_6O_2$$

Finding the boiling point elevation and $T_{b(solution)}$, with $K_b = 0.512°C/m$:

$$\Delta T_b = \frac{0.512°C}{m} \times 3.62\ m = 1.85°C$$

$$T_{b(solution)} = T_{b(solvent)} + \Delta T_b = 100.00°C + 1.85°C = 101.85°C$$

Finding the freezing point depression and $T_{f(solution)}$, with $K_f = 1.86°C/m$:

$$\Delta T_f = \frac{1.86°C}{m} \times 3.62\ m = 6.73°C$$

$$T_{f(solution)} = T_{f(solvent)} - \Delta T_f = 0.00°C - 6.73°C = -6.73°C$$

Check The changes in boiling and freezing points should be in the same proportion as the constants used. That is, $\Delta T_b/\Delta T_f$ should equal K_b/K_f: $1.85/6.73 = 0.275 = 0.512/1.86$.

Comment These answers are approximate because the concentration far exceeds that of a *dilute* solution, for which Raoult's law is most accurate.

FOLLOW-UP PROBLEM 13.6 What is the lowest molality of ethylene glycol solution that will protect your car's coolant from freezing at 0.00°F? (Assume the solution is ideal.)

Osmotic Pressure This colligative property occurs when solutions of different concentrations are separated by a **semipermeable membrane,** one that allows solvent, but *not* solute, to pass through. The phenomenon is called **osmosis:** a net flow of solvent into the more concentrated solution causes a pressure difference known as *osmotic pressure*. Many organisms regulate internal concentrations by osmosis.

1. *Why osmotic pressure arises*. Consider the simplest case with an apparatus in which a semipermeable membrane lies at the curve of a U tube and separates an aqueous sugar solution from pure water. Water molecules pass in *either* direction, but the larger sugar molecules do not. Because sugar molecules are on the solution side of the membrane, fewer water molecules touch that side, so fewer leave the solution than enter it in a given time (Figure 13.13A). The *net flow of water into the solution* increases its volume and thus decreases its concentration.

As the height of the solution rises and the height of the water falls, a pressure difference forms that resists more water entering and pushes some water back through the membrane. When water is being pushed out of the solution at the same rate it is entering, the system is at equilibrium (Figure 13.13B). The pressure difference at

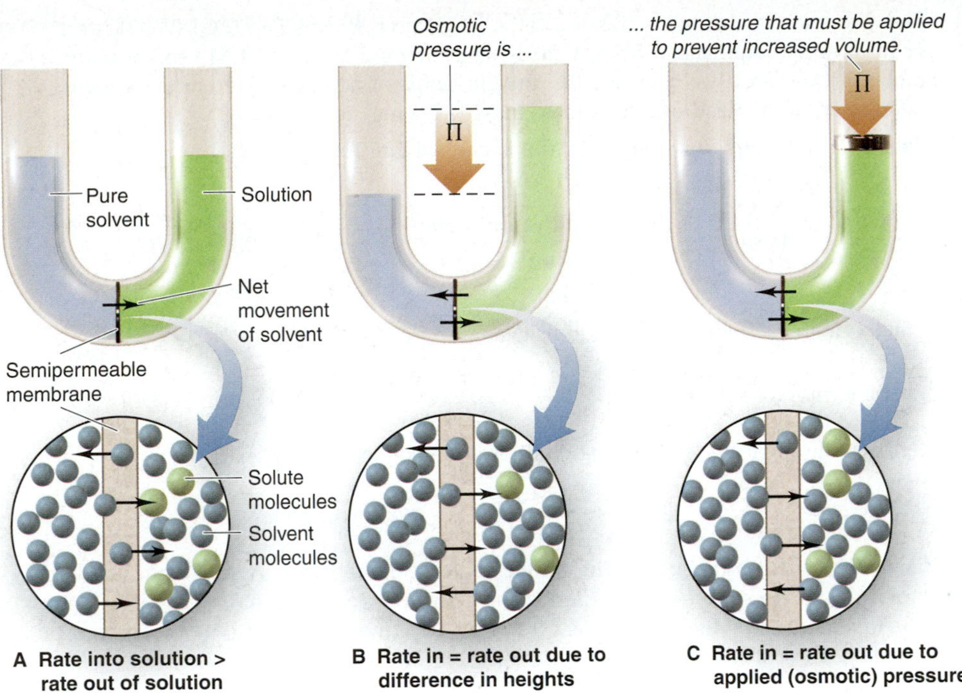

Figure 13.13 The development of osmotic pressure. **A,** In a given time, more solvent enters the solution through the membrane than leaves. **B,** At equilibrium, solvent flow is equalized. **C,** The *osmotic pressure* (Π) *prevents* the volume change.

this point is the **osmotic pressure (Π),** which is the same as the pressure that must be applied to *prevent* net movement of water from solvent to solution (or from lower to higher concentration, Figure 13.13C).

2. *Quantifying osmotic pressure.* The osmotic pressure is proportional to the number of solute particles in a given solution *volume,* that is, to the molarity (*M*):

$$\Pi \propto \frac{n_{\text{solute}}}{V_{\text{soln}}} \qquad \text{or} \qquad \Pi \propto M$$

The proportionality constant is *R* times the absolute temperature *T*. Thus,

$$\Pi = \frac{n_{\text{solute}}}{V_{\text{soln}}} RT = MRT \tag{13.13}$$

The similarity of Equation 13.13 to the ideal gas law ($P = nRT/V$) is not surprising, because both relate the pressure of a system to its concentration and temperature.

Using Colligative Properties to Find Solute Molar Mass

Each colligative property is proportional to solute concentration. Thus, by measuring the property—lower freezing point, higher boiling point, and so forth—we determine the amount (mol) of solute particles and, given the mass of solute, the molar mass.

In principle, any of the colligative properties can be used but, of the four, osmotic pressure creates the largest changes and, thus, the most precise measurements. Polymer chemists and biochemists estimate molar masses as great as 10^5 g/mol by measuring osmotic pressure. Because only a tiny fraction of a mole of a macromolecular solute dissolves, the change in the other colligative properties would be too small.

Sample Problem 13.8 Determining Molar Mass from Osmotic Pressure

Problem Biochemists have discovered more than 400 mutant varieties of hemoglobin, the blood protein that carries O_2. A physician dissolves 21.5 mg of one variety in water to make 1.50 mL of solution at 5.0°C. She measures an osmotic pressure of 3.61 torr. What is the molar mass of the protein?

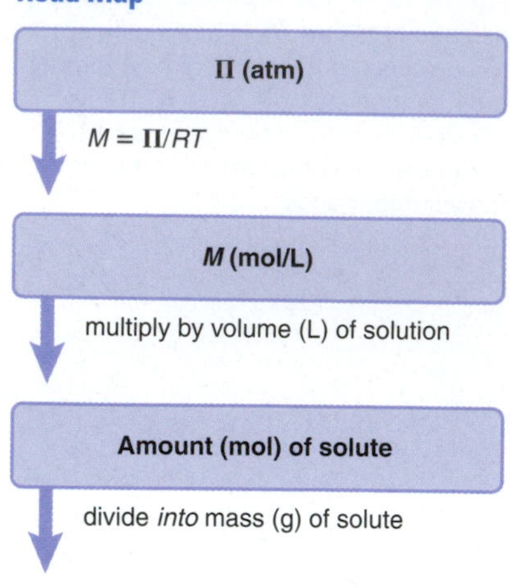

Plan We know the osmotic pressure ($\Pi = 3.61$ torr), R, and T (5.0°C). We convert Π from torr to atm, and T from °C to K, and then use Equation 13.13 to solve for molarity (M). Then we calculate the amount (mol) of hemoglobin from the known volume (1.50 mL) and use the known mass (21.5 mg) to find $\mathcal{M}$.

Solution Combining unit conversion steps and solving for molarity:

$$M = \frac{\Pi}{RT} = \frac{\dfrac{3.61\ \text{torr}}{760\ \text{torr/1 atm}}}{\left(0.0821\ \dfrac{\text{atm·L}}{\text{mol·K}}\right)(273.15\ \text{K} + 5.0)} = 2.08\times10^{-4}\ M$$

Finding amount (mol) of solute (after changing mL to L):

$$\text{Amount (mol) of solute} = M \times V = \frac{2.08\times10^{-4}\ \text{mol}}{1\ \text{L soln}} \times 0.00150\ L\ \text{soln} = 3.12\times10^{-7}\ \text{mol}$$

Calculating molar mass of hemoglobin (after changing mg to g):

$$\mathcal{M} = \frac{0.0215\ \text{g}}{3.12\times10^{-7}\ \text{mol}} = 6.89\times10^{4}\ \text{g/mol}$$

Check The small osmotic pressure implies a very low molarity. Hemoglobin is a large molecule, so we expect a small number of moles [($\sim2\times10^{-4}$ mol/L)(1.5×10^{-3} L) = 3×10^{-7} mol] and a high $\mathcal{M}$ ($\sim21\times10^{-3}$ g/3×10^{-7} mol = 7×10^{4} g/mol). The most common form of mammalian hemoglobin has a molar mass of 64,500 g/mol.

FOLLOW-UP PROBLEM 13.8 At 37°C, 0.30 M sucrose has about the same osmotic pressure as blood. What is the osmotic pressure of blood?

Volatile Nonelectrolyte Solutions

What is the effect on vapor pressure when the solute *is* volatile, that is, when the vapor consists of solute *and* solvent molecules? From Raoult's law (Equation 13.9),

$$P_{\text{solvent}} = X_{\text{solvent}} \times P^{\circ}_{\text{solvent}} \quad \text{and} \quad P_{\text{solute}} = X_{\text{solute}} \times P^{\circ}_{\text{solute}}$$

where X_{solvent} and X_{solute} are the mole fractions in the *liquid* phase. According to Dalton's law (Section 5.4), the total vapor pressure is the sum of the partial vapor pressures:

$$P_{\text{total}} = P_{\text{solvent}} + P_{\text{solute}} = (X_{\text{solvent}} \times P^{\circ}_{\text{solvent}}) + (X_{\text{solute}} \times P^{\circ}_{\text{solute}})$$

Thus, just as a nonvolatile solute lowers the vapor pressure of the solvent by making the solvent's mole fraction less than 1, *the presence of each volatile component lowers the vapor pressure of the other* by making each mole fraction less than 1.

Let's examine this idea with a solution of benzene (C_6H_6) and toluene (C_7H_8), which are miscible, in which the mole fractions in the liquid are equal: $X_{\text{ben}} = X_{\text{tol}} = 0.500$. At 25°C, the vapor pressures of the pure substances are 95.1 torr for benzene (P°_{ben}) and 28.4 torr for toluene (P°_{tol}). Note that benzene is *more volatile* than toluene. We find the partial pressures from Raoult's law:

$$P_{\text{ben}} = X_{\text{ben}} \times P^{\circ}_{\text{ben}} = 0.500 \times 95.1\ \text{torr} = 47.6\ \text{torr}$$
$$P_{\text{tol}} = X_{\text{tol}} \times P^{\circ}_{\text{tol}} = 0.500 \times 28.4\ \text{torr} = 14.2\ \text{torr}$$

Thus, the presence of benzene in the liquid lowers the vapor pressure of toluene, and vice versa.

Now let's calculate the mole fraction of each substance *in the vapor* by applying Dalton's law. Recall from Section 5.4 that $X_A = P_A/P_{\text{total}}$. Therefore, for benzene and toluene in the vapor,

$$X_{\text{ben}} = \frac{P_{\text{ben}}}{P_{\text{total}}} = \frac{47.6\ \text{torr}}{47.6\ \text{torr} + 14.2\ \text{torr}} = 0.770$$

$$X_{\text{tol}} = \frac{P_{\text{tol}}}{P_{\text{total}}} = \frac{14.2\ \text{torr}}{47.6\ \text{torr} + 14.2\ \text{torr}} = 0.230$$

The key point is that *the vapor has a higher mole fraction of the* ***more*** *volatile component.* Through a single vaporization-condensation step, a 50/50 *liquid* ratio of benzene-to-toluene created a 77/23 *vapor* ratio. Condense this vapor into a separate container, and the new *liquid* would have this 77/23 composition, and the new *vapor* above it would be enriched still further in the more volatile benzene.

In *fractional distillation,* numerous vaporization-condensation steps are carried out on a solution of two or more volatile components to continually enrich the vapor until it consists solely of the most volatile component. Fractional distillation is used in the industrial process of petroleum refining to separate hundreds of individual compounds in crude oil into a small number of "fractions" based on boiling point range.

Strong Electrolyte Solutions

For colligative properties of strong electrolyte solutions, the solute formula tells us the number of particles. For instance, the boiling point elevation (ΔT_b) of 0.050 *m* NaCl should be 2 × ΔT_b of 0.050 *m* glucose ($C_6H_{12}O_6$), because NaCl dissociates into two particles per formula unit. Thus, we use a multiplying factor called the *van't Hoff factor (i),* named after the Dutch chemist Jacobus van't Hoff (1852–1911):

$$i = \frac{\text{measured value for electrolyte solution}}{\text{expected value for nonelectrolyte solution}}$$

To calculate colligative properties for strong electrolyte solutions, we include *i:*

For vapor pressure lowering: $\Delta P = i(X_{\text{solute}} \times P^{\circ}_{\text{solvent}})$

For boiling point elevation: $\Delta T_b = i(K_b m)$

For freezing point depression: $\Delta T_f = i(K_f m)$

For osmotic pressure: $\Pi = i(MRT)$

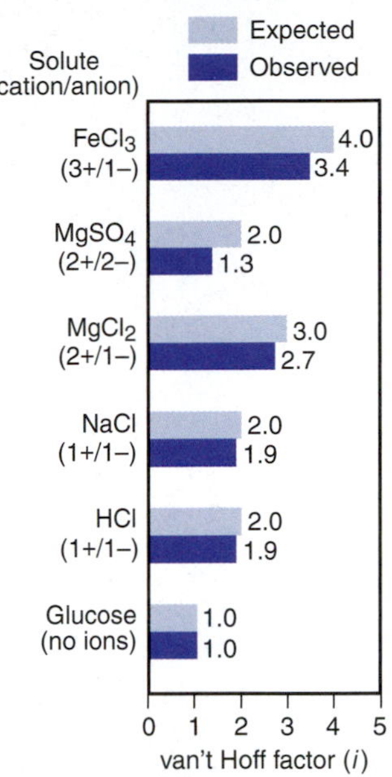

Figure 13.14 Nonideal behavior of strong electrolyte solutions. Van't Hoff factors (*i*) for 0.050 *m* solutions show the largest deviation for salts with multiply charged ions. Glucose (a nonelectrolyte) behaves as expected.

Nonideal Solutions and Ionic Atmospheres *If* strong electrolyte solutions behaved ideally, the factor *i* would be the amount (mol) of particles in solution divided by the amount (mol) of dissolved solute; that is, *i* would be 2 for KBr, 3 for $Mg(NO_3)_2$, and so forth. However, *most strong electrolyte solutions are **not** ideal,* and the measured value of *i* is typically *lower* than the value expected from the formula. For example, for the boiling point elevation of 0.050 *m* NaCl, the expected value is 2.0, but, from experiment, we have

$$i = \frac{\Delta T_b \text{ of } 0.050\ m \text{ NaCl}}{\Delta T_b \text{ of } 0.050\ m \text{ glucose}} = \frac{0.049°\text{C}}{0.026°\text{C}} = 1.9$$

It seems as though the ions are not behaving independently, even though other evidence indicates that soluble salts dissociate completely. One clue to understanding the results is that multiply charged ions cause a larger deviation (Figure 13.14).

To explain this nonideal behavior, we picture positive ions clustered, on average, near negative ions, and vice versa, to form an **ionic atmosphere** of net opposite charge (Figure 13.15). In effect, each type of ion acts "tied up," so its actual concentration seems *lower*. The *effective* concentration is the *stoichiometric* concentration, which is based on the formula, multiplied by *i*. The greater the charge, the stronger the attractions, which explains the larger deviation for compounds that dissociate into multiply charged ions. Just as gases display nearly ideal behavior at low pressures because the particles are far apart, solutions display nearly ideal behavior at low concentrations.

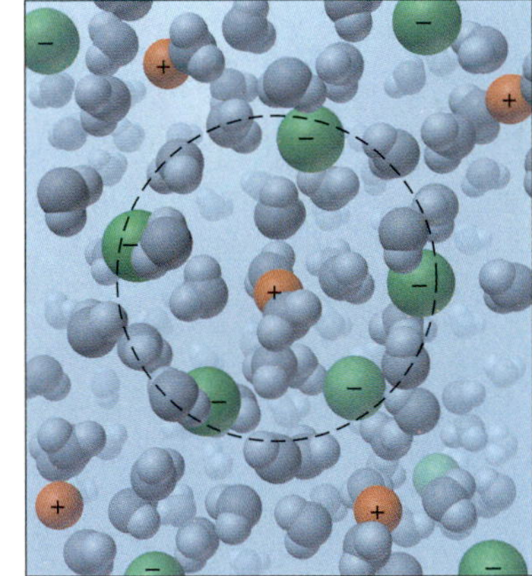

Figure 13.15 An ionic atmosphere model for nonideal behavior of electrolyte solutions.

Sample Problem 13.9 Depicting Strong Electrolyte Solutions

Problem A 0.952-g sample of magnesium chloride dissolves in 100. g of water in a flask.

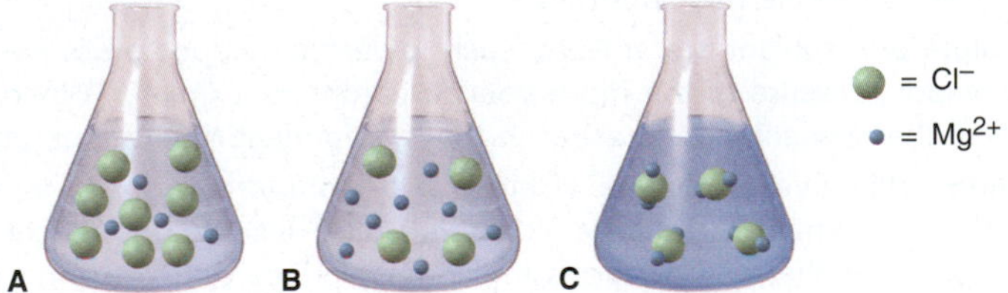

(a) Which scene depicts the solution best?
(b) What is the amount (mol) represented by each green sphere?
(c) Assuming the solution is ideal, what is its freezing point (at 1 atm)?

Plan **(a)** We find the numbers of cations and anions per formula unit from the name and compare it with the three scenes. **(b)** We convert the given mass to amount (mol), use the answer from part (a) to find moles of chloride ions *(green spheres)*, and divide by the number of green spheres to get moles/sphere. **(c)** We find the molality (m) from amount (mol) of solute divided by the given mass of water (changed to kg). We multiply K_f for water (Table 13.5) by m to get ΔT_f, and then subtract that from 0.000°C to get the solution freezing point.

Solution **(a)** The formula is $MgCl_2$; only scene A has 1 Mg^{2+} for every 2 Cl^-.

(b)
$$\text{Amount (mol) of } MgCl_2 = \frac{0.952 \text{ g } MgCl_2}{95.21 \text{ g/mol } MgCl_2} = 0.0100 \text{ mol } MgCl_2$$

Therefore,
$$\text{Amount (mol) of } Cl^- = 0.0100 \text{ mol } MgCl_2 \times \frac{2\ Cl^-}{1\ MgCl_2} = 0.0200 \text{ mol } Cl^-$$

$$\text{Moles/sphere} = \frac{0.0200 \text{ mol } Cl^-}{8 \text{ spheres}} = 2.50 \times 10^{-3} \text{ mol/sphere}$$

(c)
$$\text{Molality } (m) = \frac{\text{mol of solute}}{\text{kg of solvent}} = \frac{0.0100 \text{ mol } MgCl_2}{100.\text{ g} \times \dfrac{1 \text{ kg}}{1000 \text{ g}}} = 0.100\ m\ MgCl_2$$

Assuming an ideal solution, $i = 3$ for $MgCl_2$ (3 ions per formula unit), so we have

$$\Delta T_f = i(K_f m) = 3(1.86°C/m \times 0.100\ m) = 0.558°C$$

and

$$T_f = 0.000°C - 0.558°C = -0.558°C$$

Check Let's quickly check the ΔT_f in part (c): We have 0.01 mol dissolved in 0.1 kg, or 0.1 m. Then, rounding K_f, we have about 3(2°C/m × 0.1 m) = 0.6°C.

FOLLOW-UP PROBLEM 13.9 The $MgCl_2$ solution in the sample problem has a density of 1.006 g/mL at 20.0°C.
(a) What is the osmotic pressure of the solution?
(b) A U tube, with a semipermeable membrane, is filled with the solution in the left arm and a glucose solution of equal molarity in the right. After time, which scene depicts the U tube best?

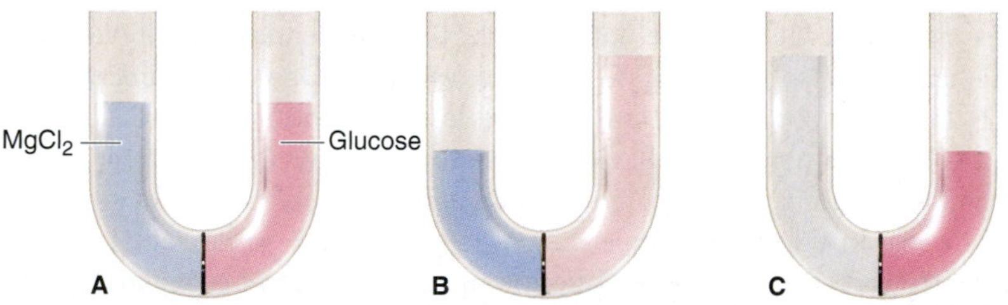

Summary of Section 13.5

- Colligative properties arise from the number, not the type, of solute particles.
- Compared to pure solvent, a solution has lower vapor pressure (Raoult's law), elevated boiling point, and depressed freezing point, and it gives rise to osmotic pressure.
- Colligative properties are used to determine solute molar mass; osmotic pressure gives the most precise measurements.
- When solute *and* solvent are volatile, each lowers the vapor pressure of the other, with the vapor pressure of the more volatile component greater. When the vapor is condensed, the new solution is richer in that component than the original solution.
- Calculating colligative properties of electrolyte solutions requires a factor (i) that adjusts for the number of ions per formula unit. These solutions exhibit nonideal behavior because charge attractions effectively reduce the concentration of ions.

CHAPTER REVIEW GUIDE

The following sections provide many aids to help you study this chapter. (Numbers in parentheses refer to pages, unless noted otherwise.)

Learning Objectives

These are concepts and skills to review after studying this chapter.

Related section (§), sample problem (SP), and upcoming end-of-chapter problem (EP) numbers are listed in parentheses.

1. Explain how solubility depends on the types of intermolecular forces (like-dissolves-like rule) and understand the characteristics of solutions consisting of gases, liquids, or solids (§13.1) (SP 13.1) (EPs 13.1–13.12)
2. Understand the enthalpy components of ΔH_{soln}, the dependence of ΔH_{hydr} on charge density, and why a solution process is exothermic or endothermic (§13.2) (EPs 13.13–13.15, 13.18–13.25, 13.28)
3. Comprehend the meaning of entropy and how the balance between ΔH and ΔS governs the solution process (§13.2) (EPs 13.16, 13.17, 13.26, 13.27)
4. Distinguish among saturated, unsaturated, and supersaturated solutions and explain the equilibrium nature of a saturated solution (§13.3) (EPs 13.29, 13.35)
5. Describe the effect of temperature on the solubility of solids and gases in water and the effect of pressure on the solubility of gases (Henry's law) (§13.3) (SP 13.2) (EPs 13.30–13.34, 13.36)
6. Express concentration in terms of molarity, molality, mole fraction, and parts by mass or by volume and be able to interconvert these terms (§13.4) (SPs 13.3–13.5) (EPs 13.37–13.58)
7. Describe electrolyte behavior and the four colligative properties, explain the difference between phase diagrams for a solution and a pure solvent, explain vapor-pressure lowering for non-volatile and volatile nonelectrolytes, and discuss the van't Hoff factor for colligative properties of electrolyte solutions (§13.5) (SPs 13.6–13.9) (EPs 13.59–13.83)

Key Terms

These important terms appear in boldface in the chapter and are defined again in the Glossary.

Section 13.1
solute (392)
solvent (392)
miscible (392)
solubility (S) (392)
like-dissolves-like rule (393)
hydration shell (393)
ion–induced dipole force (393)
dipole–induced dipole force (393)
alloy (396)

Section 13.2
heat of solution (ΔH_{soln}) (397)
solvation (397)
hydration (398)
heat of hydration (ΔH_{hydr}) (398)
charge density (398)
entropy (S) (399)

Section 13.3
saturated solution (401)
unsaturated solution (401)
supersaturated solution (401)
Henry's law (403)

Section 13.4
molality (m) (404)
mass percent [% (w/w)] (405)
volume percent [% (v/v)] (405)
mole fraction (X) (405)

Section 13.5
colligative property (408)
electrolyte (408)
nonelectrolyte (408)
vapor pressure lowering (ΔP) (408)
Raoult's law (409)
ideal solution (409)
boiling point elevation (ΔT_b) (410)
freezing point depression (ΔT_f) (411)
semipermeable membrane (412)
osmosis (412)
osmotic pressure (Π) (413)
ionic atmosphere (415)

Key Equations and Relationships

Numbered and screened concepts are listed for you to refer to or memorize.

13.1 Dividing the general heat of solution into component enthalpies (397):

$$\Delta H_{soln} = \Delta H_{solute} + \Delta H_{solvent} + \Delta H_{mix}$$

13.2 Dividing the heat of solution of an ionic compound in water into component enthalpies (398):

$$\Delta H_{soln} = \Delta H_{lattice} + \Delta H_{hydr\ of\ the\ ions}$$

13.3 Relating gas solubility to its partial pressure (Henry's law) (403):

$$S_{gas} = k_H \times P_{gas}$$

13.4 Defining concentration in terms of molarity (404):

$$\text{Molarity } (M) = \frac{\text{amount (mol) of solute}}{\text{volume (L) of solution}}$$

13.5 Defining concentration in terms of molality (404):

$$\text{Molality } (m) = \frac{\text{amount (mol) of solute}}{\text{mass (kg) of solvent}}$$

13.6 Defining concentration in terms of mass percent (405):

$$\text{Mass percent } [\% \text{ (w/w)}] = \frac{\text{mass of solute}}{\text{mass of solution}} \times 100$$

13.7 Defining concentration in terms of volume percent (405):

$$\text{Volume percent } [\% \text{ (v/v)}] = \frac{\text{volume of solute}}{\text{volume of solution}} \times 100$$

Key Equations and Relationships continued

13.8 Defining concentration in terms of mole fraction (405):

Mole fraction (X)

$$= \frac{\text{amount (mol) of solute}}{\text{amount (mol) of solute} + \text{amount (mol) of solvent}}$$

13.9 Expressing the relationship between the vapor pressure of solvent above a solution and its mole fraction in the solution (Raoult's law) (409):

$$P_{\text{solvent}} = X_{\text{solvent}} \times P^{\circ}_{\text{solvent}}$$

13.10 Calculating the vapor pressure lowering due to solute (409):

$$\Delta P = X_{\text{solute}} \times P^{\circ}_{\text{solvent}}$$

13.11 Calculating the boiling point elevation of a solution (411):

$$\Delta T_b = K_b m$$

13.12 Calculating the freezing point depression of a solution (411):

$$\Delta T_f = K_f m$$

13.13 Calculating the osmotic pressure of a solution (413):

$$\Pi = \frac{n_{\text{solute}}}{V_{\text{soln}}} RT = MRT$$

BRIEF SOLUTIONS TO FOLLOW-UP PROBLEMS

Compare your own solutions to these calculation steps and answers.

13.1 (a) 1,4-Butanediol is more soluble in water because it can form more H bonds.
(b) Chloroform is more soluble in water because of dipole-dipole forces.

13.2 $S_{N_2} = (7\times10^{-4}\ \text{mol/L}\cdot\text{atm})(0.78\ \text{atm})$
$= 5\times10^{-4}\ \text{mol/L}$

13.3 Mass (g) of glucose

$$= 563\ \text{g ethanol} \times \frac{1\ \text{kg}}{10^3\ \text{g}} \times \frac{2.40\times10^{-2}\ \text{mol glucose}}{1\ \text{kg ethanol}} \times \frac{180.16\ \text{g glucose}}{1\ \text{mol glucose}}$$

$$= 2.43\ \text{g glucose}$$

13.4 Mass % $C_3H_7OH = \dfrac{35.0\ \text{g}}{35.0\ \text{g} + 150.\ \text{g}} \times 100$

$= 18.9$ mass %

Mass % $C_2H_5OH = 100.0 - 18.9$

$= 81.1$ mass %

$$X_{C_3H_7OH} = \frac{35.0\ \text{g}\ C_3H_7OH \times \dfrac{1\ \text{mol}\ C_3H_7OH}{60.09\ \text{g}\ C_3H_7OH}}{\left(35.0\ \text{g}\ C_3H_7OH \times \dfrac{1\ \text{mol}\ C_3H_7OH}{60.09\ \text{g}\ C_3H_7OH}\right) + \left(150.\ \text{g}\ C_2H_5OH \times \dfrac{1\ \text{mol}\ C_2H_5OH}{46.07\ \text{g}\ C_2H_5OH}\right)} = 0.152$$

$X_{C_2H_5OH} = 1.000 - 0.152 = 0.848$

13.5 Mass % HCl $= \dfrac{\text{mass of HCl}}{\text{mass of soln}} \times 100$

$$= \frac{\dfrac{11.8\ \text{mol HCl}}{1\ \text{L soln}} \times \dfrac{36.46\ \text{g HCl}}{1\ \text{mol HCl}}}{\dfrac{1.190\ \text{g}}{1\ \text{mL soln}} \times \dfrac{10^3\ \text{mL}}{1\ \text{L}}} \times 100$$

$= 36.2$ mass % HCl

$$\text{Mass (kg) of soln} = 1\ \text{L soln} \times \frac{1.190\times10^{-3}\ \text{kg soln}}{1\times10^{-3}\ \text{L soln}}$$

$= 1.190$ kg soln

$$\text{Mass (kg) of HCl} = 11.8\ \text{mol HCl} \times \frac{36.46\ \text{g HCl}}{1\ \text{mol HCl}} \times \frac{1\ \text{kg}}{10^3\ \text{g}}$$

$= 0.430$ kg HCl

$$\text{Molality of HCl} = \frac{\text{mol HCl}}{\text{kg water}} = \frac{\text{mol HCl}}{\text{kg soln} - \text{kg HCl}}$$

$$= \frac{11.8\ \text{mol HCl}}{0.760\ \text{kg}\ H_2O} = 15.5\ m\ \text{HCl}$$

$$X_{HCl} = \frac{\text{mol HCl}}{\text{mol HCl} + \text{mol}\ H_2O}$$

$$= \frac{11.8\ \text{mol}}{11.8\ \text{mol} + \left(760\ \text{g}\ H_2O \times \dfrac{1\ \text{mol}}{18.02\ \text{g}\ H_2O}\right)} = 0.219$$

13.6 $\Delta P = X_{\text{aspirin}} \times P^{\circ}_{\text{methanol}}$

$$= \frac{\dfrac{2.00\ \text{g}}{180.15\ \text{g/mol}}}{\dfrac{2.00\ \text{g}}{180.15\ \text{g/mol}} + \dfrac{50.0\ \text{g}}{32.04\ \text{g/mol}}} \times 101\ \text{torr}$$

$= 0.713$ torr

13.7 Molality of $C_2H_6O_2 = \dfrac{(32°\text{F} - 0.00°\text{F})\left(\dfrac{5°\text{C}}{9°\text{F}}\right)}{1.86°\text{C}/m} = 9.56\ m$

13.8 $\Pi = MRT = (0.30\ \text{mol/L})\left(0.0821\ \dfrac{\text{atm}\cdot\text{L}}{\text{mol}\cdot\text{K}}\right)(37°\text{C} + 273.15)$

$= 7.6$ atm

13.9 (a) Mass of 0.100 m solution
$= 1$ kg water $+ 0.100$ mol $MgCl_2$
$= 1000$ g $+ 9.52$ g $= 1009.52$ g

$$\text{Volume of solution} = 1009.52 \text{ g} \times \frac{1 \text{ mL}}{1.006 \text{ g}} = 1003 \text{ mL}$$

$$\text{Molarity} = \frac{9.52 \text{ g MgCl}_2}{1003 \text{ mL soln}} \times \frac{1 \text{ mol}}{95.21 \text{ g MgCl}_2} \times \frac{10^3 \text{ mL}}{1 \text{ L}}$$
$= 9.97 \times 10^{-2}$ M

Osmotic pressure (Π)
$= i(MRT)$

$$= 3(9.97\times10^{-2} \text{ mol/L})\left(0.0821 \frac{\text{atm}\cdot\text{L}}{\text{mol}\cdot\text{K}}\right)(293 \text{ K})$$
$= 7.19$ atm

(b) Scene C

PROBLEMS

Problems with **colored** numbers are answered in Appendix E. Sections match the text and provide the numbers of relevant sample problems. Bracketed problems are grouped in pairs (indicated by a short rule) that cover the same concept. Comprehensive Problems are based on material from any section or previous chapter.

Types of Solutions: Intermolecular Forces and Solubility

(Sample Problem 13.1)

13.1 Describe how properties of seawater illustrate the two characteristics that define mixtures.

13.2 What types of intermolecular forces give rise to hydration shells in an aqueous solution of sodium chloride?

13.3 Acetic acid (CH_3COOH) is miscible with water. Would you expect carboxylic acids with general formula $CH_3(CH_2)_nCOOH$ to become more or less water soluble as n increases? Explain.

13.4 Which gives the more concentrated solution, (a) KNO_3 in H_2O or (b) KNO_3 in carbon tetrachloride (CCl_4)? Explain.

13.5 Which gives the more concentrated solution, stearic acid [$CH_3(CH_2)_{16}COOH$] in (a) H_2O or (b) CCl_4? Explain.

13.6 What is the strongest type of intermolecular force between solute and solvent in each solution?
(a) $CsCl(s)$ in $H_2O(l)$ (b) $CH_3C(=O)CH_3(l)$ in $H_2O(l)$
(c) $CH_3OH(l)$ in $CCl_4(l)$

13.7 What is the strongest type of intermolecular force between solute and solvent in each solution?
(a) $Cu(s)$ in $Ag(s)$ (b) $CH_3Cl(g)$ in $CH_3OCH_3(g)$
(c) $CH_3CH_3(g)$ in $CH_3CH_2CH_2NH_2(l)$

13.8 What is the strongest type of intermolecular force between solute and solvent in each solution?
(a) $CH_3OCH_3(g)$ in $H_2O(l)$ (b) $Ne(g)$ in $H_2O(l)$
(c) $N_2(g)$ in $C_4H_{10}(g)$

13.9 What is the strongest type of intermolecular force between solute and solvent in each solution?
(a) $C_6H_{14}(l)$ in $C_8H_{18}(l)$ (b) $H_2C{=}O(g)$ in $CH_3OH(l)$
(c) $Br_2(l)$ in $CCl_4(l)$

13.10 Which member of each pair is more soluble in diethyl ether? Why?
(a) $NaCl(s)$ or $HCl(g)$ (b) $H_2O(l)$ or $CH_3CH(=O)(l)$
(c) $MgBr_2(s)$ or $CH_3CH_2MgBr(s)$

13.11 Which member of each pair is more soluble in water? Why?
(a) $CH_3CH_2OCH_2CH_3(l)$ or $CH_3CH_2OCH_3(g)$
(b) $CH_2Cl_2(l)$ or $CCl_4(l)$
(c) cyclohexane or tetrahydropyran

13.12 Gluconic acid is a derivative of glucose used in cleaners and in the dairy and brewing industries. Caproic acid is a carboxylic acid used in the flavoring industry. Although both are six-carbon acids *(see structures below)*, gluconic acid is soluble in water and nearly insoluble in hexane, whereas caproic acid has the opposite solubility behavior. Explain.

$CH_2(OH)—CH(OH)—CH(OH)—CH(OH)—CH(OH)—COOH$
gluconic acid

$CH_3—CH_2—CH_2—CH_2—CH_2—COOH$
caproic acid

Why Substances Dissolve: Understanding the Solution Process

13.13 What is the relationship between solvation and hydration?

13.14 (a) What is the charge density of an ion, and what two properties of an ion affect it?
(b) Arrange the following in order of increasing charge density:

(c) How do the two properties in part (a) affect the ionic heat of hydration, ΔH_{hydr}?

13.15 For ΔH_{soln} to be very small, what quantities must be nearly equal in magnitude? Will their signs be the same or opposite?

13.16 Water is added to a flask containing solid NH_4Cl. As the salt dissolves, the solution becomes colder.
(a) Is the dissolving of NH_4Cl exothermic or endothermic?
(b) Is the magnitude of $\Delta H_{lattice}$ of NH_4Cl larger or smaller than the combined ΔH_{hydr} of the ions? Explain.
(c) Given the answer to (a), why does NH_4Cl dissolve in water?

13.17 An ionic compound has a highly negative ΔH_{soln} in water. Would you expect it to be very soluble or nearly insoluble in water? Explain in terms of enthalpy and entropy changes.

13.18 Sketch an enthalpy diagram for the process of dissolving $KCl(s)$ in H_2O (endothermic).

13.19 Sketch an enthalpy diagram for the process of dissolving $NaI(s)$ in H_2O (exothermic).

13.20 Which ion in each pair has greater charge density? Explain.
(a) Na^+ or Cs^+ (b) Sr^{2+} or Rb^+ (c) Na^+ or Cl^-
(d) O^{2-} or F^- (e) OH^- or SH^- (f) Mg^{2+} or Ba^{2+}
(g) Mg^{2+} or Na^+ (h) NO_3^- or CO_3^{2-}

13.21 Which ion has the lower ratio of charge to volume? Explain.
(a) Br^- or I^- (b) Sc^{3+} or Ca^{2+} (c) Br^- or K^+
(d) S^{2-} or Cl^- (e) Sc^{3+} or Al^{3+} (f) SO_4^{2-} or ClO_4^-
(g) Fe^{3+} or Fe^{2+} (h) Ca^{2+} or K^+

13.22 Which has the *larger* ΔH_{hydr} in each pair of Problem 13.20?

13.23 Which has the *smaller* ΔH_{hydr} in each pair of Problem 13.21?

13.24 (a) Use the following data to calculate the combined heat of hydration for the ions in potassium bromate ($KBrO_3$):

$\Delta H_{lattice} = 745$ kJ/mol $\quad \Delta H_{soln} = 41.1$ kJ/mol

(b) Which ion contributes more to the answer to part (a)? Why?

13.25 (a) Use the following data to calculate the combined heat of hydration for the ions in sodium acetate ($NaC_2H_3O_2$):

$\Delta H_{lattice} = 763$ kJ/mol $\quad \Delta H_{soln} = 17.3$ kJ/mol

(b) Which ion contributes more to the answer to part (a)? Why?

13.26 State whether the entropy of the system increases or decreases in each of the following processes:
(a) Gasoline burns in a car engine.
(b) Gold is extracted and purified from its ore.
(c) Ethanol (CH_3CH_2OH) dissolves in 1-propanol ($CH_3CH_2CH_2OH$).

13.27 State whether the entropy of the system increases or decreases in each of the following processes:
(a) Pure gases are mixed to prepare an anesthetic.
(b) Electronic-grade silicon is prepared from sand.
(c) Dry ice (solid CO_2) sublimes.

13.28 Besides its use in making black-and-white film, silver nitrate ($AgNO_3$) is used similarly in forensic science. The NaCl left behind in the sweat of a fingerprint is treated with $AgNO_3$ solution to form AgCl. This precipitate is developed to show the black-and-white fingerprint pattern. Given $\Delta H_{lattice}$ of $AgNO_3$ = 822 kJ/mol and $\Delta H_{hydr} = -799$ kJ/mol, calculate its ΔH_{soln}.

Solubility as an Equilibrium Process

(Sample Problem 13.2)

13.29 You are given a bottle of solid X and three aqueous solutions of X—one saturated, one unsaturated, and one supersaturated. How would you determine which solution is which?

13.30 Why does the solubility of any gas in water decrease with rising temperature?

13.31 For a saturated aqueous solution of each of the following at 20°C and 1 atm, will the solubility increase, decrease, or stay the same when the indicated change occurs?
(a) $O_2(g)$, increase P (b) $N_2(g)$, increase V

13.32 For a saturated aqueous solution of each of the following at 20°C and 1 atm, will the solubility increase, decrease, or stay the same when the indicated change occurs?
(a) $He(g)$, decrease T (b) $RbI(s)$, increase P

13.33 The Henry's law constant (k_H) for O_2 in water at 20°C is 1.28×10^{-3} mol/L·atm. (a) How many grams of O_2 will dissolve in 2.50 L of H_2O that is in contact with pure O_2 at 1.00 atm? (b) How many grams of O_2 will dissolve in 2.50 L of H_2O that is in contact with air, where the partial pressure of O_2 is 0.209 atm?

13.34 Argon makes up 0.93% by volume of air. Calculate its solubility (mol/L) in water at 20°C and 1.0 atm. The Henry's law constant for Ar under these conditions is 1.5×10^{-3} mol/L·atm.

13.35 Caffeine is about 10 times as soluble in hot water as in cold water. A chemist puts a hot-water extract of caffeine into an ice bath, and some caffeine crystallizes. Is the remaining solution saturated, unsaturated, or supersaturated?

13.36 The partial pressure of CO_2 gas above the liquid in a bottle of champagne at 20°C is 5.5 atm. What is the solubility of CO_2 in champagne? Assume Henry's law constant is the same for champagne as for water: at 20°C, $k_H = 3.7\times10^{-2}$ mol/L·atm.

Concentration Terms

(Sample Problems 13.3 to 13.5)

13.37 Explain the difference between molarity and molality. Under what circumstances would molality be a more accurate measure of the concentration of a prepared solution than molarity? Why?

13.38 A solute has a solubility in water of 21 g/kg solvent. Is this value the same as 21 g/kg solution? Explain.

13.39 You want to convert among molarity, molality, and mole fraction of a solution. You know the masses of solute and solvent and the volume of solution. Is this enough information to carry out all the conversions? Explain.

13.40 When a solution is heated, which ways of expressing concentration change in value? Which remain unchanged? Explain.

13.41 Calculate the molarity of each aqueous solution:
(a) 32.3 g of table sugar ($C_{12}H_{22}O_{11}$) in 100. mL of solution
(b) 5.80 g of $LiNO_3$ in 505 mL of solution

13.42 Calculate the molarity of each aqueous solution:
(a) 0.82 g of ethanol (C_2H_5OH) in 10.5 mL of solution
(b) 1.27 g of gaseous NH_3 in 33.5 mL of solution

13.43 Calculate the molarity of each aqueous solution:
(a) 78.0 mL of 0.240 *M* NaOH diluted to 0.250 L with water
(b) 38.5 mL of 1.2 *M* HNO_3 diluted to 0.130 L with water

13.44 Calculate the molarity of each aqueous solution:
(a) 25.5 mL of 6.25 *M* HCl diluted to 0.500 L with water
(b) 8.25 mL of 2.00×10^{-2} *M* KI diluted to 12.0 mL with water

13.45 How would you prepare the following aqueous solutions?
(a) 365 mL of 8.55×10^{-2} *M* KH_2PO_4 from solid KH_2PO_4
(b) 465 mL of 0.335 *M* NaOH from 1.25 *M* NaOH

13.46 How would you prepare the following aqueous solutions?
(a) 2.5 L of 0.65 *M* NaCl from solid NaCl
(b) 15.5 L of 0.3 *M* urea [$(NH_2)_2C{=}O$] from 2.1 *M* urea

13.47 Calculate the molality of the following:
(a) A solution containing 85.4 g of glycine (NH_2CH_2COOH) dissolved in 1.270 kg of H_2O
(b) A solution containing 8.59 g of glycerol ($C_3H_8O_3$) in 77.0 g of ethanol (C_2H_5OH)

13.48 Calculate the molality of the following:
(a) A solution containing 174 g of HCl in 757 g of H_2O
(b) A solution containing 16.5 g of naphthalene ($C_{10}H_8$) in 53.3 g of benzene (C_6H_6)

13.49 What is the molality of a solution consisting of 44.0 mL of benzene (C_6H_6; d = 0.877 g/mL) in 167 mL of hexane (C_6H_{14}; d = 0.660 g/mL)?

13.50 What is the molality of a solution consisting of 2.66 mL of carbon tetrachloride (CCl_4; d = 1.59 g/mL) in 76.5 mL of methylene chloride (CH_2Cl_2; d = 1.33 g/mL)?

13.51 How would you prepare the following aqueous solutions?
(a) 3.10×10^2 g of 0.125 m ethylene glycol ($C_2H_6O_2$) from ethylene glycol and water
(b) 1.20 kg of 2.20 mass % HNO_3 from 52.0 mass % HNO_3

13.52 How would you prepare the following aqueous solutions?
(a) 1.50 kg of 0.0355 m ethanol (C_2H_5OH) from ethanol and water
(b) 445 g of 13.0 mass % HCl from 34.1 mass % HCl

13.53 A solution contains 0.35 mol of isopropanol (C_3H_7OH) dissolved in 0.85 mol of water. (a) What is the mole fraction of isopropanol? (b) The mass percent? (c) The molality?

13.54 A solution contains 0.100 mol of NaCl dissolved in 8.60 mol of water. (a) What is the mole fraction of NaCl? (b) The mass percent? (c) The molality?

13.55 Calculate the molality, molarity, and mole fraction of NH_3 in an 8.00 mass % aqueous solution (d = 0.9651 g/mL).

13.56 Calculate the molality, molarity, and mole fraction of $FeCl_3$ in a 28.8 mass % aqueous solution (d = 1.280 g/mL).

13.57 Wastewater from a cement factory contains 0.25 g of Ca^{2+} ion and 0.056 g of Mg^{2+} ion per 100.0 L of solution. The solution density is 1.001 g/mL. Calculate the Ca^{2+} and Mg^{2+} concentrations in ppm (by mass).

13.58 An automobile antifreeze mixture is made by mixing equal volumes of ethylene glycol (d = 1.114 g/mL; $\mathcal{M}$ = 62.07 g/mol) and water (d = 1.00 g/mL) at 20°C. The density of the mixture is 1.070 g/mL. Express the concentration of ethylene glycol as:
(a) Volume percent (b) Mass percent (c) Molarity
(d) Molality (e) Mole fraction

Colligative Properties of Solutions

(Sample Problems 13.6 to 13.9)

13.59 Express Raoult's law in words. Is Raoult's law valid for a solution of a volatile solute? Explain.

13.60 What are the most important differences between the phase diagram of a pure solvent and the phase diagram of a solution of that solvent?

13.61 Is the boiling point of 0.01 m KF(aq) higher or lower than that of 0.01 m glucose(aq)? Explain.

13.62 Which aqueous solution has a freezing point closer to its predicted value, 0.01 m NaBr or 0.01 m $MgCl_2$? Explain.

13.63 The freezing point depression constants of the solvents cyclohexane and naphthalene are 20.1°C/m and 6.94°C/m, respectively. Which solvent will give a more accurate result if you are using freezing point depression to determine the molar mass of a substance that is soluble in either one? Why?

13.64 Classify the following substances as strong electrolytes, weak electrolytes, or nonelectrolytes:
(a) Hydrogen chloride (HCl) (b) Potassium nitrate (KNO_3)
(c) Glucose ($C_6H_{12}O_6$) (d) Ammonia (NH_3)

13.65 Classify the following substances as strong electrolytes, weak electrolytes, or nonelectrolytes:
(a) Sodium permanganate ($NaMnO_4$)
(b) Acetic acid (CH_3COOH)
(c) Methanol (CH_3OH)
(d) Calcium acetate [$Ca(C_2H_3O_2)_2$]

13.66 How many moles of solute particles are present in 1 L of each of the following aqueous solutions?
(a) 0.3 M KBr (b) 0.065 M HNO_3
(c) 10^{-4} M $KHSO_4$ (d) 0.06 M ethanol (C_2H_5OH)

13.67 How many moles of solute particles are present in 1 mL of each of the following aqueous solutions?
(a) 0.02 M $CuSO_4$ (b) 0.004 M $Ba(OH)_2$
(c) 0.08 M pyridine (C_5H_5N) (d) 0.05 M $(NH_4)_2CO_3$

13.68 Which solution has the lower freezing point?
(a) 11.0 g of CH_3OH in 100. g of H_2O *or*
22.0 g of CH_3CH_2OH in 200. g of H_2O
(b) 20.0 g of H_2O in 1.00 kg of CH_3OH *or*
20.0 g of CH_3CH_2OH in 1.00 kg of CH_3OH

13.69 Which solution has the higher boiling point?
(a) 38.0 g of $C_3H_8O_3$ in 250. g of ethanol *or*
38.0 g of $C_2H_6O_2$ in 250. g of ethanol
(b) 15 g of $C_2H_6O_2$ in 0.50 kg of H_2O *or*
15 g of NaCl in 0.50 kg of H_2O

13.70 Rank the following aqueous solutions in order of increasing (a) osmotic pressure; (b) boiling point; (c) freezing point; (d) vapor pressure at 50°C:
(I) 0.100 m $NaNO_3$
(II) 0.100 m glucose
(III) 0.100 m $CaCl_2$

13.71 Rank the following aqueous solutions in order of decreasing (a) osmotic pressure; (b) boiling point; (c) freezing point; (d) vapor pressure at 298 K:
(I) 0.04 m urea [$(NH_2)_2C{=}O$]
(II) 0.01 m $AgNO_3$
(III) 0.03 m $CuSO_4$

13.72 Calculate the vapor pressure of a solution of 34.0 g of glycerol ($C_3H_8O_3$) in 500.0 g of water at 25°C. The vapor pressure of water at 25°C is 23.76 torr. (Assume ideal behavior.)

13.73 Calculate the vapor pressure of a solution of 0.39 mol of cholesterol in 5.4 mol of toluene at 32°C. Pure toluene has a vapor pressure of 41 torr at 32°C. (Assume ideal behavior.)

13.74 What is the freezing point of 0.251 m urea in water?

13.75 What is the boiling point of 0.200 m lactose in water?

13.76 The boiling point of ethanol (C_2H_5OH) is 78.5°C. What is the boiling point of a solution of 6.4 g of vanillin ($\mathcal{M}$ = 152.14 g/mol) in 50.0 g of ethanol (K_b of ethanol = 1.22°C/m)?

13.77 The freezing point of benzene is 5.5°C. What is the freezing point of a solution of 5.00 g of naphthalene ($C_{10}H_8$) in 444 g of benzene (K_f of benzene = 4.90°C/m)?

13.78 What is the minimum mass of ethylene glycol ($C_2H_6O_2$) that must be dissolved in 14.5 kg of water to prevent the solution from freezing at −12.0°F? (Assume ideal behavior.)

13.79 What is the minimum mass of glycerol ($C_3H_8O_3$) that must be dissolved in 11.0 mg of water to prevent the solution from freezing at −15°C? (Assume ideal behavior.)

180 Calculate the molality and van't Hoff factor (i) for the following aqueous solutions:
(a) 1.00 mass % NaCl, freezing point = −0.593°C
(b) 0.500 mass % CH_3COOH, freezing point = −0.159°C

13.81 Calculate the molality and van't Hoff factor (i) for the following aqueous solutions:
(a) 0.500 mass % KCl, freezing point = −0.234°C
(b) 1.00 mass % H_2SO_4, freezing point = −0.423°C

13.82 In a study designed to prepare new gasoline-resistant coatings, a polymer chemist dissolves 6.053 g of poly(vinyl alcohol) in enough water to make 100.0 mL of solution. At 25°C, the osmotic pressure of this solution is 0.272 atm. What is the molar mass of the polymer sample?

13.83 The U.S. Food and Drug Administration lists dichloromethane (CH_2Cl_2) and carbon tetrachloride (CCl_4) among the many cancer-causing chlorinated organic compounds. What are the partial pressures of these substances in the vapor above a solution of 1.60 mol of CH_2Cl_2 and 1.10 mol of CCl_4 at 23.5°C? The vapor pressures of pure CH_2Cl_2 and CCl_4 at 23.5°C are 352 torr and 118 torr, respectively. (Assume ideal behavior.)

Comprehensive Problems

13.84 The three aqueous ionic solutions represented below have total volumes of 25. mL for A, 50. mL for B, and 100. mL for C. If each sphere represents 0.010 mol of ions, calculate: (a) the total molarity of ions for each solution; (b) the highest molarity of solute; (c) the lowest molality of solute (assuming the solution densities are equal); (d) the highest osmotic pressure (assuming ideal behavior).

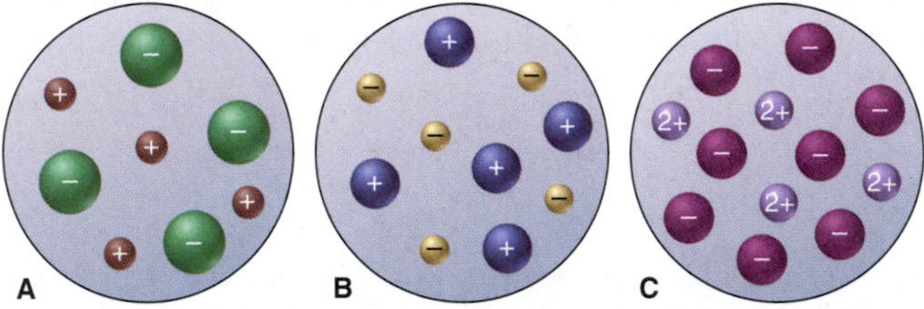

13.85 Gold occurs in seawater at an average concentration of 1.1×10^{-2} ppb. How many liters of seawater must be processed to recover 1 troy ounce of gold, assuming 81.5% efficiency (d of seawater = 1.025 g/mL; 1 troy ounce = 31.1 g)?

13.86 Use atomic properties to explain why xenon is 11 times as soluble as helium in water at 0°C on a mole basis.

13.87 Which of the following best represents a molecular-scale view of an ionic compound in aqueous solution? Explain.

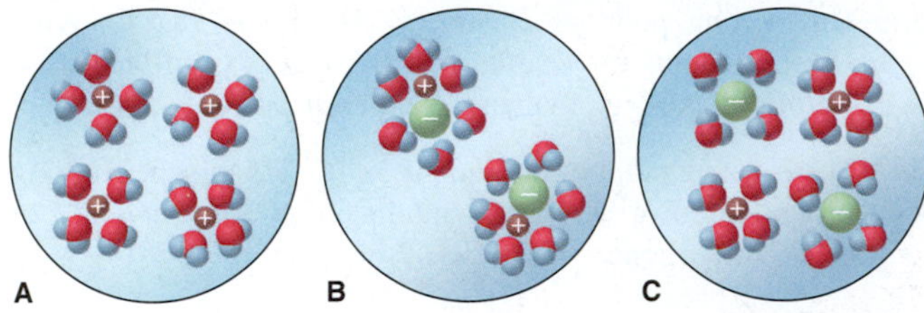

13.88 Four 0.50 m aqueous solutions are depicted. Assume the solutions behave ideally: (a) Which has the highest boiling point? (b) Which has the lowest freezing point? (c) Can you determine which one has the highest osmotic pressure? Explain.

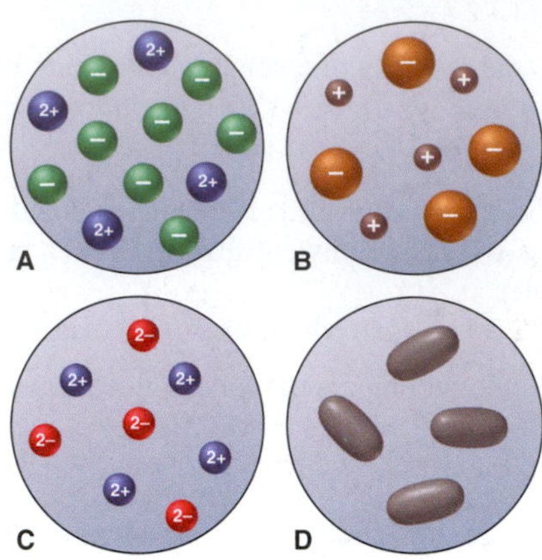

13.89 "De-icing salt" is used to melt snow and ice on streets. The highway department of a small town is deciding whether to buy NaCl or $CaCl_2$, which are equally effective, to use for this purpose. The town can obtain NaCl for \$0.22/kg. What is the maximum the town should pay for $CaCl_2$ to be cost effective?

13.90 Thermal pollution from industrial wastewater causes the temperature of river or lake water to increase, which can affect fish survival as the concentration of dissolved O_2 decreases. Use the following data to find the molarity of O_2 at each temperature (assume the solution density is the same as water):

Temperature (°C)	Solubility of O_2 (mg/kg H_2O)	Density of H_2O (g/mL)
0.0	14.5	0.99987
20.0	9.07	0.99823
40.0	6.44	0.99224

13.91 A chemist is studying small organic compounds for their potential use as an antifreeze. When 0.243 g of a compound is dissolved in 25.0 mL of water, the freezing point of the solution is −0.201°C. (a) Calculate the molar mass of the compound (d of water = 1.00 g/mL). (b) Analysis shows that the compound is 53.31 mass % C and 11.18 mass % H, the remainder being O. Calculate the empirical and molecular formulas of the compound. (c) Draw a Lewis structure for a compound with this formula that forms H bonds and another for one that does not.

13.92 Is 50% by mass of methanol dissolved in ethanol different from 50% by mass of ethanol dissolved in methanol? Explain.

13.93 Three gaseous mixtures of N_2 *(blue)*, Cl_2 *(green)*, and Ne *(purple)* are depicted below. (a) Which has the smallest mole fraction of N_2? (b) Which have the same mole fraction of Ne? (c) Rank all three in order of increasing mole fraction of Cl_2.

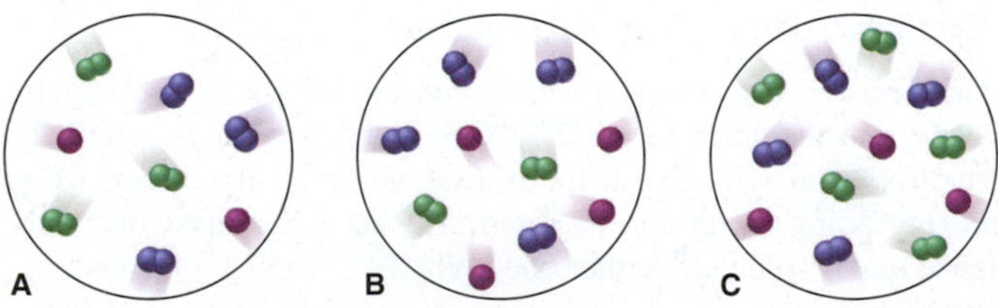

13.94 Four U tubes each have distilled water in the right arm, a solution in the left arm, and a semipermeable membrane between arms. (a) If the solute is KCl, which solution is most concentrated?

(b) If each solute is different but all the solutions have the same molarity, which contains the smallest number of dissolved ions?

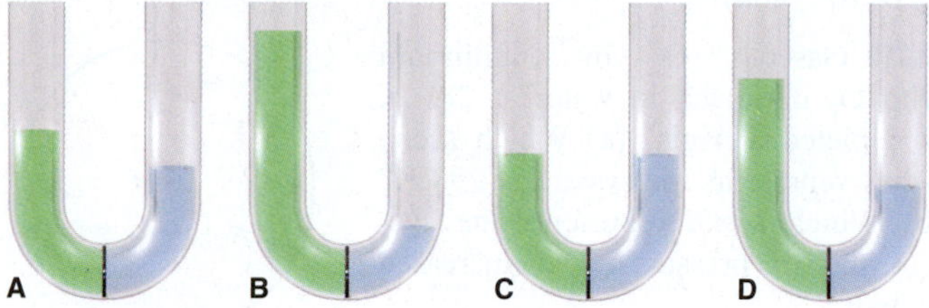

13.95 β-Pinene ($C_{10}H_{16}$) and α-terpineol ($C_{10}H_{18}O$) are used in cosmetics to provide a "fresh pine" scent. At 367 K, the pure substances have vapor pressures of 100.3 torr and 9.8 torr, respectively. What is the composition of the vapor (in terms of mole fractions) above a solution containing equal masses of these compounds at 367 K? (Assume ideal behavior.)

13.96 A solution of 1.50 g of solute dissolved in 25.0 mL of H_2O at 25°C has a boiling point of 100.45°C. (a) What is the molar mass of the solute if it is a nonvolatile nonelectrolyte and the solution behaves ideally (d of H_2O at 25°C = 0.997 g/mL)? (b) Conductivity measurements show the solute to be ionic of general formula AB_2 or A_2B. What is the molar mass if the solution behaves ideally? (c) Analysis indicates an empirical formula of CaN_2O_6. Explain the difference between the actual formula mass and that calculated from the boiling point elevation. (d) Find the van't Hoff factor (i) for this solution.

13.97 A pharmaceutical preparation made with ethanol (C_2H_5OH) is contaminated with methanol (CH_3OH). A sample of vapor above the liquid mixture contains a 97/1 mass ratio of C_2H_5OH to CH_3OH. What is the mass ratio of these alcohols in the liquid? At the temperature of the liquid, the vapor pressures of C_2H_5OH and CH_3OH are 60.5 torr and 126.0 torr, respectively.

13.98 Water treatment plants commonly use chlorination to destroy bacteria. A byproduct is chloroform ($CHCl_3$), a suspected carcinogen, produced when HOCl, formed by reaction of Cl_2 and water, reacts with dissolved organic matter. The United States, Canada, and the World Health Organization have set a limit of 100. ppb of $CHCl_3$ in drinking water. Convert this concentration into molarity, molality, mole fraction, and mass percent.

13.99 A biochemical engineer isolates a bacterial gene fragment and dissolves a 10.0-mg sample in enough water to make 30.0 mL of solution. The osmotic pressure of the solution is 0.340 torr at 25°C. (a) What is the molar mass of the gene fragment? (b) If the solution density is 0.997 g/mL, how large is the freezing point depression for this solution (K_f of water = 1.86°C/m)?

13.100 Glyphosate is the active ingredient in a common weed and grass killer. It is sold as an 18.0% by mass solution with a density of 8.94 lb/gal. (a) How many grams of Glyphosate are in a 16.0 fl oz container (1 gal = 128 fl oz)? (b) To treat a patio area of 300. ft^2, it is recommended that 3.00 fl oz be diluted with water to 1.00 gal. What is the mass percent of Glyphosate in the diluted solution (1 gal = 3.785 L)?

13.101 Two beakers are placed in a closed container (*left*). One beaker contains water, the other a concentrated aqueous sugar solution. With time, the solution volume increases and the water volume decreases (*right*). Explain on the molecular level.

13.102 Although other solvents are available, dichloromethane (CH_2Cl_2) is still often used to "decaffeinate" drinks because the solubility of caffeine in CH_2Cl_2 is 8.35 times that in water.
(a) A 100.0-mL sample of cola containing 10.0 mg of caffeine is extracted with 60.0 mL of CH_2Cl_2. What mass of caffeine remains in the aqueous phase?
(b) A second identical cola sample is extracted with two successive 30.0-mL portions of CH_2Cl_2. What mass of caffeine remains in the aqueous phase after each extraction?
(c) Which approach extracts more caffeine?

13.103 Tartaric acid occurs in crystalline residues found in wine vats. It is used in baking powders and as an additive in foods. It contains 32.3% by mass carbon and 3.97% by mass hydrogen; the balance is oxygen. When 0.981 g of tartaric acid is dissolved in 11.23 g of water, the solution freezes at −1.26°C. Find the empirical and molecular formulas of tartaric acid.

13.104 A florist prepares a solution of nitrogen-phosphorus fertilizer by dissolving 5.66 g of NH_4NO_3 and 4.42 g of $(NH_4)_3PO_4$ in enough water to make 20.0 L of solution. What are the molarities of NH_4^+ and of PO_4^{3-} in the solution?

13.105 Urea is a white crystalline solid used as a fertilizer, in the pharmaceutical industry, and in the manufacture of certain polymer resins. Analysis of urea reveals that, by mass, it is 20.1% carbon, 6.7% hydrogen, 46.5% nitrogen, and the balance oxygen.
(a) Find the empirical formula of urea.
(b) A 5.0 g/L solution of urea in water has an osmotic pressure of 2.04 atm, measured at 25°C. What is the molar mass and molecular formula of urea?

13.106 The total concentration of dissolved particles in blood is 0.30 M. An intravenous (IV) solution must be isotonic with blood, which means it must have the same concentration. (a) To relieve dehydration, a patient is given 100. mL/h of IV glucose ($C_6H_{12}O_6$) for 2.5 h. What mass (g) of glucose did she receive? (b) If isotonic saline (NaCl) is used, what is the molarity of the solution? (c) If the patient is given 150. mL/h of IV saline for 1.5 h, how many grams of NaCl did she receive?

13.107 Deviations from Raoult's law lead to the formation of *azeotropes,* constant boiling mixtures that cannot be separated by distillation, making industrial separations difficult. For components A and B, there is a positive deviation if the A-B attraction is less than A-A and B-B attractions (A and B reject each other), and a negative deviation if the A-B attraction is greater than A-A and B-B attractions. If the A-B attraction is nearly equal to the A-A and B-B attractions, the solution obeys Raoult's law. Explain

whether the behavior of each pair will be nearly ideal, have a positive deviation, or have a negative deviation:
(a) Benzene (C_6H_6) and methanol
(b) Water and ethyl acetate
(c) Hexane and heptane
(d) Methanol and water
(e) Water and hydrochloric acid

13.108 In ice-cream making, the ingredients are kept below 0.0°C in an ice-salt bath. (a) Assuming that NaCl dissolves completely and forms an ideal solution, what mass of it is needed to lower the melting point of 5.5 kg of ice to −5.0°C? (b) Given the same assumptions as in part (a), what mass of $CaCl_2$ is needed?

13.109 Soft drinks are canned under 4 atm of CO_2 and release CO_2 when opened.
(a) How many moles of CO_2 are dissolved in a 355-mL can of soda before it is opened?
(b) After it has gone flat?
(c) What volume (in L) would the released CO_2 occupy at 1.00 atm and 25°C (k_H for CO_2 at 25°C is 3.3×10^{-2} mol/L·atm; P_{CO_2} in air is 3×10^{-4} atm)?

13.110 Gaseous O_2 in equilibrium with O_2 dissolved in water at 283 K is depicted at right. (a) Which scene below represents the system at 298 K? (b) Which scene represents the system when the pressure of O_2 is increased by half?

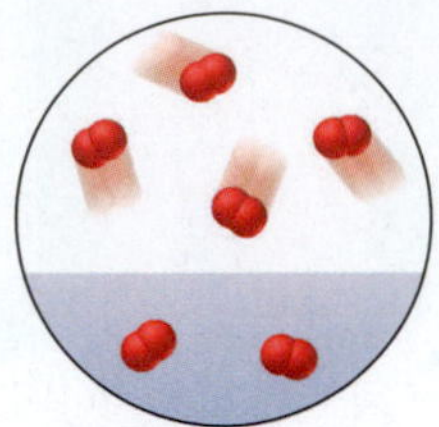

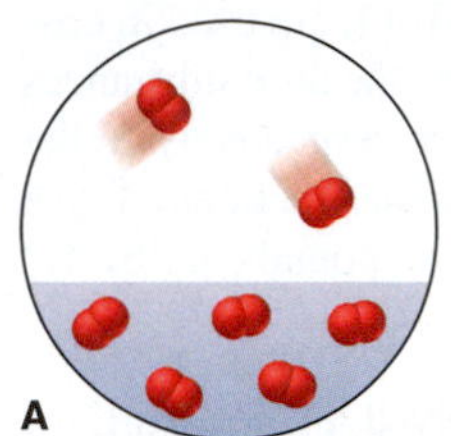

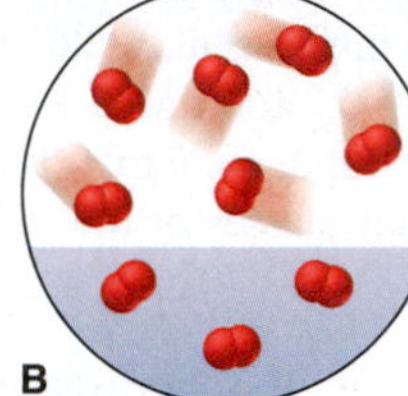

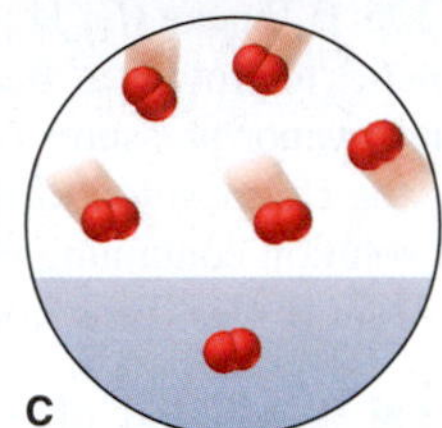

Periodic Patterns in the Main-Group Elements

Patterns in Nature Recurring patterns are everywhere—in the timing of an eclipse, the rows of seeds in a sunflower, and the wavelengths of sound made by a piano's strings. In this chapter, we explore how the periodic properties of the main-group elements emerge from patterns of electrons in atoms.

Key Principles
to focus on while studying this chapter

- Hydrogen does not fit into any particular family (group) because its tiny size and simple structure give it unique properties.
- Within a family of elements, *similar behavior* results from a *similar outer electron configuration.*
- Because the Period 2 elements have a small atomic size and only four outer-level orbitals, they exhibit some behavior that is *anomalous* within their groups.
- In Period 4 and higher, Group 3A(13) and 4A(14) elements deviate from expected trends because their nuclei attract outer *s* and *p* electrons very strongly due to poor shielding by their inner *d* and *f* electrons.
- Because atoms get larger down a group, *metallic behavior* (such as ability to form cations and basicity of oxides) increases, and this trend becomes especially apparent in Groups 3A(13) to 6A(16).
- In Groups 3A(13) to 6A(16), nearly every element exhibits *more than one oxidation state*, and the lower state becomes more common going down the group.
- Many elements occur in different forms *(allotropes)*, each with its own properties.
- Group 1A(1) and 7A(17) elements are very reactive because each is one electron away from having a filled outer level; Group 8A(18) elements have a filled outer level and thus are very unreactive.

Outline

CONCEPTS & SKILLS TO REVIEW before studying this chapter

- redox behavior and oxidation states (Section 4.5)
- electron configurations (Section 8.2)
- trends in atomic size, ionization energy, metallic behavior, and electronegativity (Sections 8.3, 8.4, and 9.5)
- trends in element properties and type of bonding (Sections 8.4 and 9.5)
- models of ionic, covalent, and metallic bonding (Sections 9.2, 9.3, and 12.6)
- resonance and formal charge (Section 10.1)
- molecular shape and polarity (Sections 10.2 and 10.3)
- orbital hybridization and modes of orbital overlap (Sections 11.1 and 11.2)
- phase changes and phase diagrams, intermolecular forces, and crystalline solids (Sections 12.2, 12.3, and 12.6)

In your study of chemistry so far, you've learned how to name compounds, balance equations, and calculate reaction yields. You've seen how heat is related to chemical and physical change, how electron configuration influences atomic properties, how elements bond to form compounds, and how the arrangement of bonding and lone pairs accounts for molecular shapes. You've learned modern theories of bonding and, most recently, seen how atomic-scale properties give rise to the macroscopic properties of gases, liquids, solids, and solutions.

The purpose of this knowledge, of course, is to make sense of the magnificent diversity of chemical and physical behavior around you. The periodic table, which organizes much of this diversity, was derived from chemical facts observed in countless hours of 18th- and 19th-century research. One of the greatest achievements in science is 20th-century quantum theory, which provides a theoretical basis for the periodic table's arrangement. In this chapter, we apply general ideas of bonding and structure from earlier chapters to the main-group elements to see how their position in the table allows us to explain their behavior.

14.1 • HYDROGEN, THE SIMPLEST ATOM

A hydrogen atom consists of a nucleus with a single positive charge, surrounded by a single electron. Despite this simple structure, or perhaps because of it, hydrogen may be the most important element of all. In the Sun, hydrogen (H) nuclei combine to form helium (He) nuclei in a process that provides nearly all Earth's energy. (We discuss this process in Chapter 23.) About 90% of all the matter in the universe consists of H atoms, making it the most abundant element by far. On Earth, only tiny amounts of the free, diatomic element occur naturally, but hydrogen is abundant in combination with oxygen in water. Because of its simple structure and low molar mass, nonpolar gaseous H_2 is colorless and odorless, and its extremely weak dispersion forces result in very low melting (−259°C) and boiling points (−253°C).

Where Hydrogen Fits in the Periodic Table

Hydrogen has no perfectly suitable position in the periodic table. Depending on the property, it may fit better in Group 1A(1), 4A(14), or 7A(17):

- *Like the Group 1A(1) elements,* hydrogen has an outer electron configuration of ns^1, a single valence electron, and a common +1 oxidation state. However, unlike the alkali metals, hydrogen *shares* its electron with nonmetals rather than transferring it to them. Moreover, hydrogen has a much higher ionization energy and electronegativity than any of the alkali metals.
- *Like the Group 4A(14) elements,* hydrogen has a half-filled valence level (but with only one electron) and similar ionization energy, electron affinity, electronegativity, and bond energies.
- *Like the Group 7A(17) elements,* hydrogen occurs as diatomic molecules and fills its outer (valence) level either by electron sharing or by gaining one electron from a metal to form an anion (hydride, H^-) with a 1− charge. However, while the monatomic halide ions (X^-) are common and stable, the H^- ion is rare and reactive.

Hydrogen's unique behavior is attributable to its tiny size. Hydrogen has a high ionization energy (IE) because its electron is very close to the nucleus, with no inner electrons to shield it from the positive charge. It has a low electronegativity (EN) for a nonmetal because it has only one proton to attract bonding electrons. In this chapter, hydrogen appears in either Group 1A(1) or 7A(17) depending on the property being considered.

Highlights of Hydrogen Chemistry

We discussed the enormous impact of hydrogen bonding on physical properties in Chapters 12 and 13. In terms of its chemical properties as an element, hydrogen is very reactive, combining with nearly every other element to form ionic or covalent hydrides.

Ionic (Saltlike) Hydrides With very reactive metals, such as those in Group 1A(1) and the larger members of Group 2A(2) (Ca, Sr, and Ba), hydrogen forms *saltlike hydrides*—white, crystalline solids composed of the metal cation and the hydride ion:

$$2Li(s) + H_2(g) \longrightarrow 2LiH(s)$$
$$Ca(s) + H_2(g) \longrightarrow CaH_2(s)$$

In water, H^- is a strong base that pulls H^+ from surrounding H_2O molecules to form H_2 and OH^-:

$$NaH(s) + H_2O(l) \longrightarrow Na^+(aq) + OH^-(aq) + H_2(g)$$

The hydride ion is also a powerful reducing agent; for example, it reduces Ti(IV) to the free metal:

$$TiCl_4(l) + 4LiH(s) \longrightarrow Ti(s) + 4LiCl(s) + 2H_2(g)$$

Covalent (Molecular) Hydrides Hydrogen reacts with nonmetals to form many *covalent hydrides.* In most of them, hydrogen has an oxidation number of +1 because the other nonmetal has a higher electronegativity.

Conditions for preparing the covalent hydrides depend on the reactivity of the other nonmetal. For example, with stable, triple-bonded N_2, hydrogen reacts at high temperatures (~400°C) and pressures (~250 atm), and the reaction needs a catalyst to proceed at any practical speed:

$$N_2(g) + 3H_2(g) \xrightarrow{\text{catalyst}} 2NH_3(g) \qquad \Delta H^\circ_{\text{rxn}} = -91.8 \text{ kJ}$$

Industrial facilities throughout the world use this reaction to produce millions of tons of ammonia each year for fertilizers, explosives, and synthetic fibers. On the other hand, hydrogen combines rapidly with reactive, single-bonded F_2, even at extremely low temperatures (−196°C):

$$F_2(g) + H_2(g) \longrightarrow 2HF(g) \qquad \Delta H^\circ_{\text{rxn}} = -546 \text{ kJ}$$

14.2 • GROUP 1A(1): THE ALKALI METALS

The first group of elements in the periodic table is named for the alkaline (basic) nature of their oxides and for the basic solutions the elements form in water. Group 1A(1) provides the best example of regular trends with no significant exceptions. All the elements in the group—lithium (Li), sodium (Na), potassium (K), rubidium (Rb), cesium (Cs), and rare, radioactive francium (Fr)—are very reactive metals. The Family Portrait of Group 1A(1) (*next page*) is the first in a series that provides an overview of each of the main groups, summarizing key atomic, physical, and chemical properties. (E represents any member of the family.)

Why the Alkali Metals Have Unusual Physical Properties

The alkali metals are softer and have lower melting and boiling points and lower densities than nearly any other metals. This unusual physical behavior can be traced to their atomic size, the largest in their respective periods, and to the ns^1 valence electron configuration. Because there is only one valence electron and it is relatively far from the nucleus, metallic bonding is weak, which results in a soft consistency (K can be squeezed like clay) and low melting point. The low densities of the alkali metals (Li floats on light oil) result from their having the lowest molar masses and largest atomic radii (and, thus, volumes) in their periods.

Why the Alkali Metals Are So Reactive

The alkali metals are extremely reactive elements. They are *powerful reducing agents,* always occurring in nature as 1+ cations rather than as free metals. (As we discuss in Section 21.7, highly endothermic reduction processes are required to prepare the free metals industrially from their molten salts.)

Group 1A(1): The Alkali Metals

KEY ATOMIC PROPERTIES, PHYSICAL PROPERTIES, AND REACTIONS

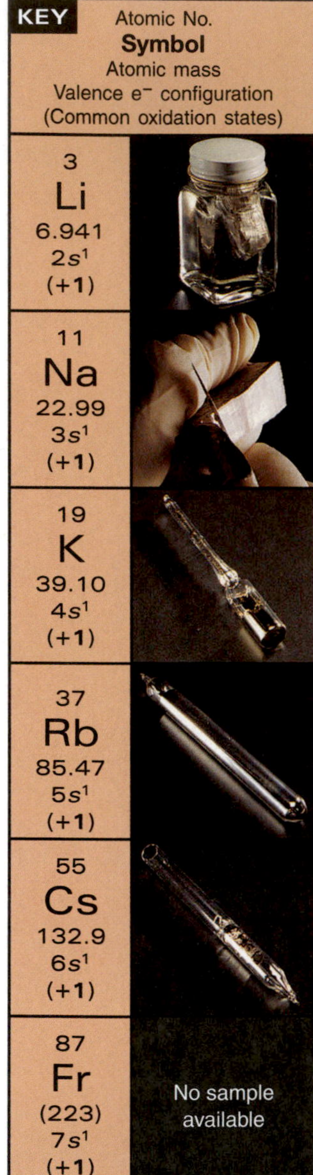

KEY
Atomic No.
Symbol
Atomic mass
Valence e⁻ configuration
(Common oxidation states)
3
Li
6.941
2s1
(+1)
11
Na
22.99
3s1
(+1)
19
K
39.10
4s1
(+1)
37
Rb
85.47
5s1
(+1)
55
Cs
132.9
6s1
(+1)
87
Fr
(223)
7s1
(+1)
No sample available

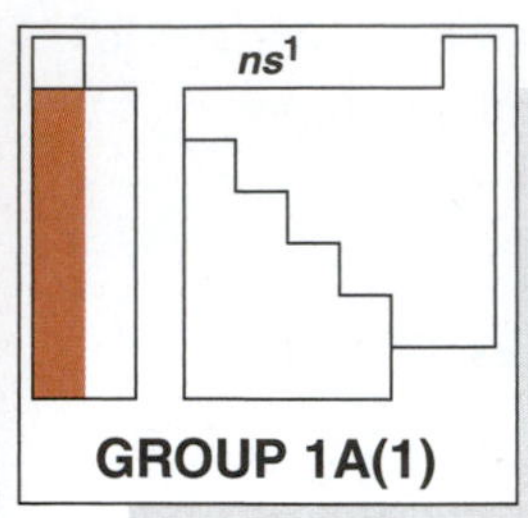

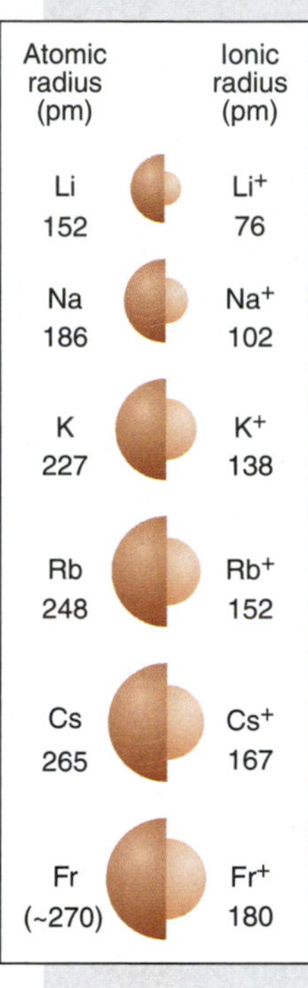

Atomic Properties

Group electron configuration is ns^1. All members have the +1 oxidation state and form an E^+ ion. Atoms have the largest size and lowest IE and EN in their periods. Down the group, atomic and ionic size increase, while IE and EN decrease.

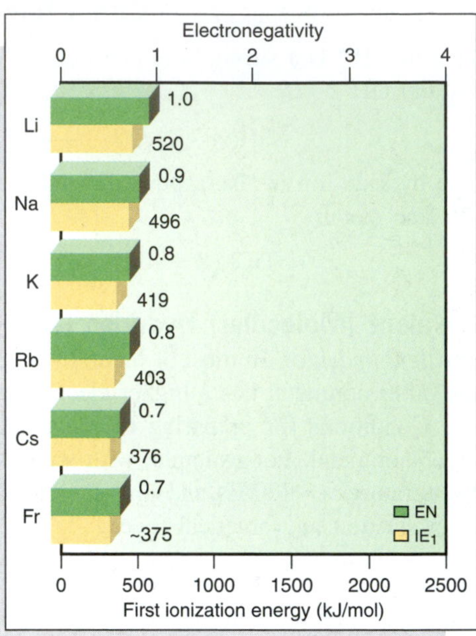

Physical Properties

Metallic bonding is relatively weak because there is only one valence electron. Therefore, these metals are soft with relatively low melting and boiling points. These values decrease down the group because larger atom cores attract delocalized electrons less strongly. Large atomic size and low atomic mass result in low density; thus, density generally increases down the group because mass increases more than size.

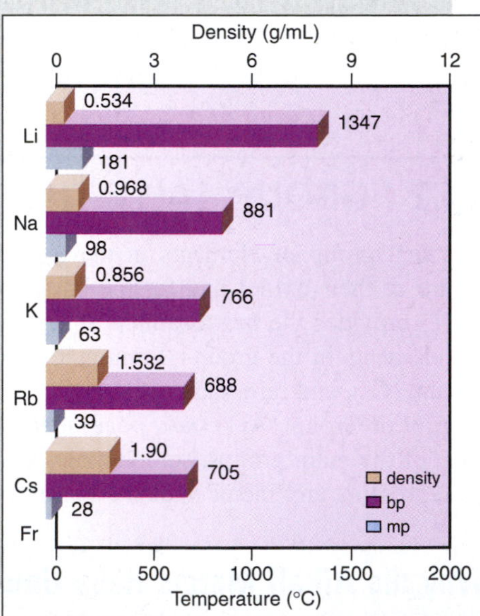

Reactions

1. The alkali metals reduce H in H_2O from the +1 to the 0 oxidation state:

$$2E(s) + 2H_2O(l) \longrightarrow 2E^+(aq) + 2OH^-(aq) + H_2(g)$$

The reaction becomes more vigorous down the group.

2. The alkali metals reduce oxygen, but the product depends on the metal. Li forms the oxide, Li_2O; Na forms the peroxide (O.N. of O = −1), Na_2O_2; K, Rb, and Cs form the superoxide (O.N. of O = $-\frac{1}{2}$), EO_2:

$$4Li(s) + O_2(g) \longrightarrow 2Li_2O(s)$$
$$2Na(s) + O_2(g) \longrightarrow Na_2O_2(s)$$
$$K(s) + O_2(g) \longrightarrow KO_2(s)$$

3. The alkali metals reduce hydrogen to form ionic (saltlike) hydrides:

$$2E(s) + H_2(g) \longrightarrow 2EH(s)$$

NaH is an industrial base and reducing agent that is used to prepare other reducing agents, such as $NaBH_4$.

4. The alkali metals reduce halogens to form ionic halides:

$$2E(s) + X_2 \longrightarrow 2EX(s) \qquad (X = F, Cl, Br, I)$$

The ns^1 configuration, which is the basis for their physical properties, is also the reason alkali metals form salts so readily. Their low ionization energies give rise to small cations, which can lie close to anions, resulting in high lattice energies. Some examples of this reactivity occur with halogens, water, oxygen, and hydrogen.

- The alkali metals (E)* reduce halogens to form ionic solids in highly exothermic reactions:

$$2E(s) + X_2 \longrightarrow 2EX(s) \quad (X = F, Cl, Br, I)$$

- They reduce the hydrogen in water, reacting vigorously (Rb and Cs explosively) to form H_2 and a metal hydroxide solution:

$$2E(s) + 2H_2O(l) \longrightarrow 2E^+(aq) + 2OH^-(aq) + H_2(g)$$

- They reduce O_2, but the product depends on the metal. Li forms the oxide, Li_2O; Na the peroxide (O.N. of O = −1), Na_2O_2; and K, Rb, and Cs the superoxides (O.N. of O = $-\frac{1}{2}$), EO_2:

$$4Li(s) + O_2(g) \longrightarrow 2Li_2O(s)$$
$$2Na(s) + O_2(g) \longrightarrow Na_2O_2(s)$$
$$K(s) + O_2(g) \longrightarrow KO_2(s)$$

  Thus, in air, the metals tarnish rapidly, so Na and K are usually kept under mineral oil (an unreactive liquid) in the laboratory, and Rb and Cs are handled with gloves under an inert argon atmosphere.

- They reduce molecular hydrogen to form ionic (saltlike) hydrides:

$$2E(s) + H_2(g) \longrightarrow 2EH(s)$$

For a given anion, the trend in lattice energy is the inverse of the trend in cation size and, thus, follows Coulomb's law: *as the cation becomes larger, the lattice energy becomes smaller.* Figure 14.1 shows this steady decrease in lattice energy within the Group 1A(1) and 2A(2) chlorides. Despite these strong ionic attractions in the solid, *nearly all Group 1A salts are water soluble* because the ions attract water molecules to create a highly exothermic heat of hydration (ΔH_{hydr}).

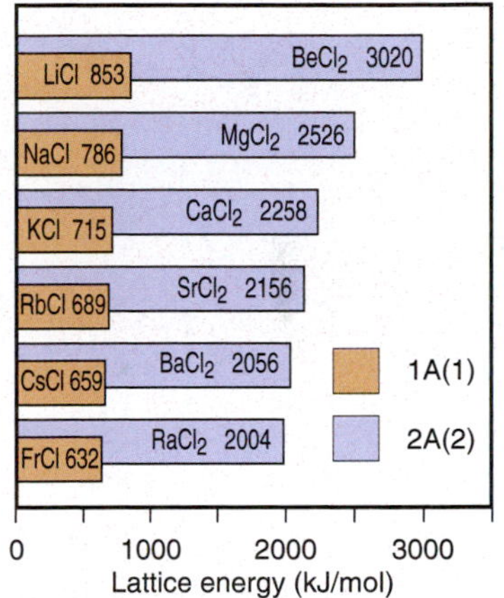

Figure 14.1 Lattice energies of the Group 1A(1) and 2A(2) chlorides.

The Anomalous Behavior of Period 2 Members

A consistent feature within the main groups is that, as a result of their *small atomic size* and *small number of outer-level orbitals,* all the Period 2 members display some anomalous (unrepresentative) behavior within their groups.

- *Lithium, Group 1A(1):* Li is the only member of its group that forms a simple oxide and nitride, Li_2O and Li_3N, with O_2 and N_2 in air, and only Li forms molecular compounds with organic halides:

$$2Li(s) + CH_3CH_2Cl(g) \longrightarrow CH_3CH_2Li(s) + LiCl(s)$$

  Moreover, because of the high charge density of Li^+, many lithium salts have significant covalent character. Thus, halides of Li are more soluble in polar organic solvents than halides of Na and K.

- *Beryllium, Group 2A(2):* Be displays even more anomalous behavior than Li. Because of the extremely high charge density of Be^{2+}, the discrete ion does not exist, and all Be compounds exhibit covalent bonding.
- *Boron, Group 3A(13):* B is the only member of its group to form a complex family of compounds with metals and covalent compounds with hydrogen (boranes).
- *Carbon, Group 4A(14):* C shows extremely unusual behavior: it bonds to itself (and a small number of other elements) so extensively and diversely that it gives rise to countless organic and biological compounds.
- *Nitrogen, Group 5A(15):* Triple-bonded, gaseous N_2 is dramatically different from its reactive, solid family members.
- *Oxygen, Group 6A(16):* The only gas in its group, O_2 is much more reactive than sulfur and the other members.

*Throughout this chapter, E represents any element in a given group.

- *Fluorine, Group 7A(17):* F_2 is so electronegative that it reacts violently with water and is the only halogen that forms a weak hydrohalic acid, HF.
- *Helium, Group 8A(18):* He has the lowest melting and boiling points and the smallest heats of phase change of any noble gas, indeed of any element.

14.3 • GROUP 2A(2): THE ALKALINE EARTH METALS

The Group 2A(2) elements are called *alkaline earth metals* because their oxides give basic (alkaline) solutions and melt at such high temperatures that they remained as solids ("earths") in the alchemists' fires. The group is a fascinating collection of elements: rare beryllium (Be), common and essential magnesium (Mg) and calcium (Ca), less familiar strontium (Sr) and barium (Ba), and radioactive radium (Ra). The Group 2A(2) Family Portrait presents an overview of these elements.

How the Alkaline Earth and Alkali Metals Compare Physically

In general terms, whatever differences occur between the elements in Groups 1A(1) and 2A(2) are those of degree, not kind, and are due to the change in outer electron configuration: ns^2 vs. ns^1. Two valence electrons and a nucleus with one additional positive charge allow much stronger metallic bonding. Consequently, melting and boiling points are much higher for 2A metals than for the corresponding 1A metals. Compared to transition metals such as iron and chromium, the alkaline earths are soft and lightweight, but their stronger metallic bonding and smaller atomic sizes make them harder and denser than the alkali metals.

Magnesium is a particularly versatile member of Group 2A(2). Because it forms a tough oxide layer that prevents further reaction in air, it is alloyed with aluminum for camera bodies and luggage and with the lanthanides for auto engine blocks and missile parts.

How the Alkaline Earth and Alkali Metals Compare Chemically

The alkaline earth metals display a wider range of chemical behavior than the alkali metals, largely because of the unrepresentative covalent bonding of beryllium. The second valence electron lies in the same sublevel as the first, so it is poorly shielded and Z_{eff} is greater. Therefore, Group 2A(2) elements have smaller atomic radii and higher ionization energies than Group 1A(1) elements. Yet, despite the higher second IEs required to form the 2+ cations, *all the alkaline earths (except Be) form ionic compounds* because the resulting high lattice energies more than compensate for the large total IEs.

Like the alkali metals, the alkaline earth metals are *strong reducing agents:*

- Each reduces O_2 in air to form the oxide (Ba also forms the peroxide, BaO_2).
- Except for Be and Mg, which form adherent oxide coatings, the alkaline earths reduce H_2O at room temperature to form H_2.
- Except for Be, the alkaline earths reduce the halogens, N_2, and H_2 to form ionic compounds.

The Group 2A oxides are strongly basic (except for amphoteric BeO) and react with acidic oxides to form salts, such as sulfites and carbonates; for example,

$$SrO(s) + CO_2(g) \longrightarrow SrCO_3(s)$$

The natural carbonates limestone and marble are major structural materials and the commercial sources for most 2A compounds.

One of the main differences between Groups 1A and 2A is the lower solubility of 2A salts. With such high lattice energies, most 2A fluorides, carbonates, phosphates, and sulfates are insoluble, unlike the corresponding 1A compounds.

Group 2A(2): The Alkaline Earth Metals — Family Portrait

KEY ATOMIC PROPERTIES, PHYSICAL PROPERTIES, AND REACTIONS

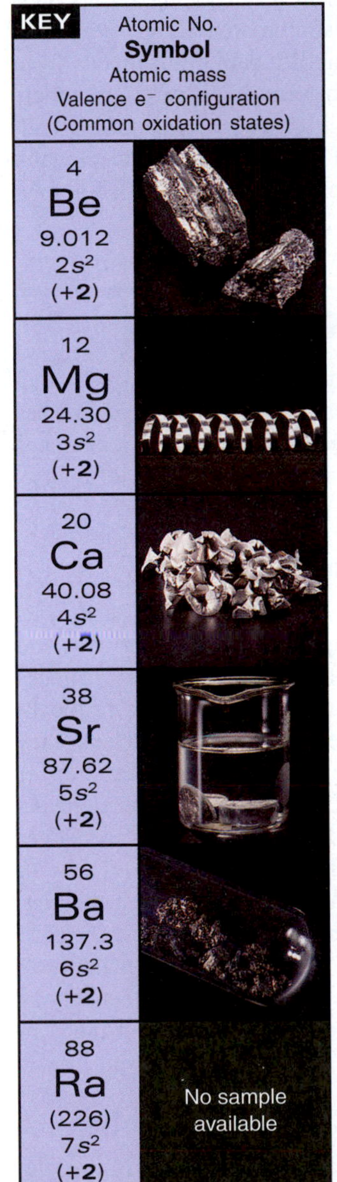

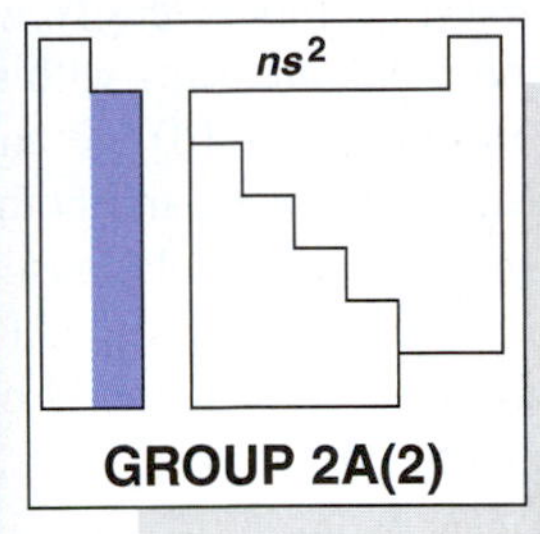

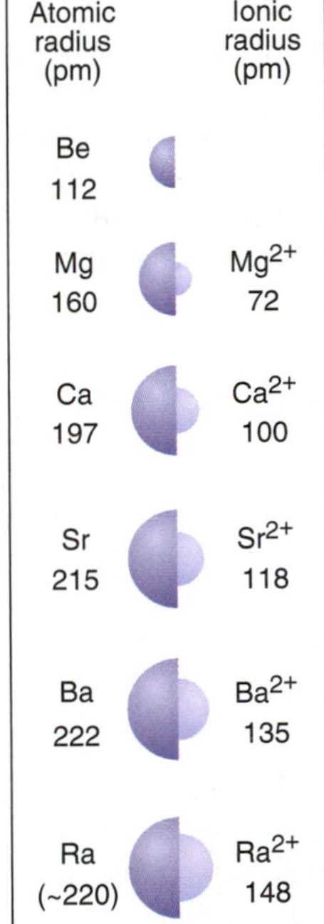

Atomic Properties

Group electron configuration is ns^2 (filled ns sublevel). All members have the +2 oxidation state and, except for Be, form compounds with an E^{2+} ion. Atomic and ionic sizes increase down the group but are smaller than for the corresponding 1A(1) elements. IE and EN decrease down the group but are higher than for the corresponding 1A(1) elements.

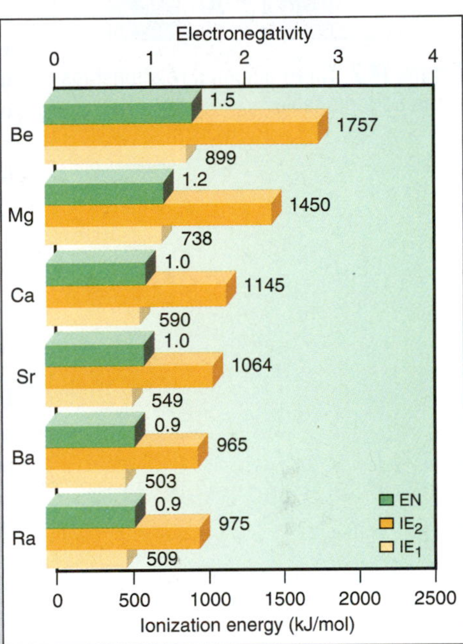

Physical Properties

Metallic bonding involves two valence electrons. These metals are still relatively soft but are much harder than the 1A(1) metals. Melting and boiling points generally decrease, and densities generally increase down the group. These values are much higher than for 1A(1) elements, and the trend is not as regular.

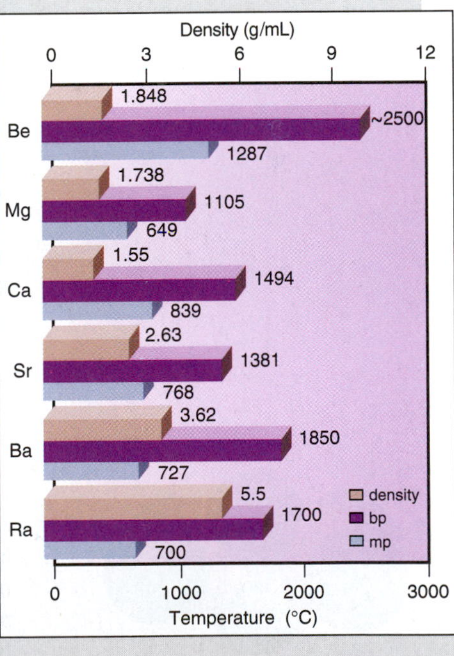

Reactions

1. The metals reduce O_2 to form the oxides:

$$2E(s) + O_2(g) \longrightarrow 2EO(s)$$

Ba also forms the peroxide, BaO_2.

2. The larger metals reduce water to form hydrogen gas:

$$E(s) + 2H_2O(l) \longrightarrow E^{2+}(aq) + 2OH^-(aq) + H_2(g) \quad (E = Ca, Sr, Ba)$$

Be and Mg form an oxide coating that allows only slight reaction.

3. The metals reduce halogens to form ionic halides:

$$E(s) + X_2 \longrightarrow EX_2(s) \qquad [X = F \text{ (not with Be), Cl, Br, I}]$$

4. Most of the elements reduce hydrogen to form ionic hydrides:

$$E(s) + H_2(g) \longrightarrow EH_2(s) \qquad (E = \text{all except Be})$$

5. The elements reduce nitrogen to form ionic nitrides:

$$3E(s) + N_2(g) \longrightarrow E_3N_2(s)$$

6. Except for amphoteric BeO, the element oxides are basic:

$$EO(s) + H_2O(l) \longrightarrow E^{2+}(aq) + 2OH^-(aq)$$

$Ca(OH)_2$ is a component of cement and mortar.

7. All carbonates undergo thermal decomposition to the oxide:

$$ECO_3(s) \xrightarrow{\Delta} EO(s) + CO_2(g)$$

This reaction is used to produce CaO (lime) in huge amounts from naturally occurring limestone.

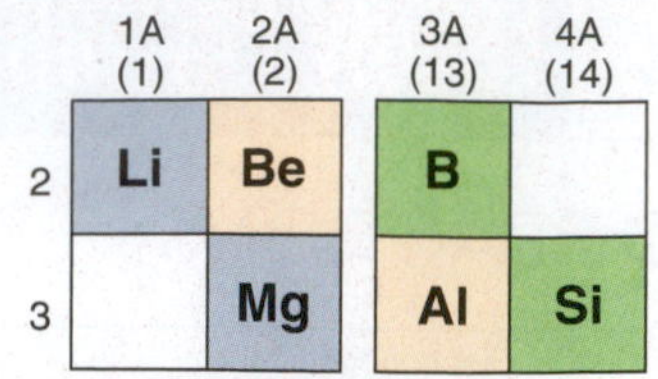

Figure 14.2 Three diagonal relationships in the periodic table.

Diagonal Relationships

Diagonal relationships are *similarities between a Period 2 element and one diagonally down and to the right in Period 3.* Three such relationships are of interest to us here (Figure 14.2). The first occurs between Li and Mg. Both form nitrides with N_2, hydroxides and carbonates that decompose easily with heat, organic compounds with a polar covalent metal-carbon bond, and salts with similar solubilities. Beryllium in Group 2A(2) and aluminum in Group 3A(13) are another pair. Both metals form oxide coatings, so they don't react with water, and both form amphoteric, extremely hard, high-melting oxides. The third diagonal relationship occurs between the metalloids boron in Group 3A(13) and silicon in Group 4A(14). Both behave electrically as semiconductors and both form weakly acidic, solid oxoacids and flammable, low-melting, strongly reducing, covalent hydrides.

14.4 • GROUP 3A(13): THE BORON FAMILY

Boron (B) heads the third family of main-group elements, but its properties are not representative, as the Group 3A(13) Family Portrait shows. Metallic aluminum (Al) has properties that are more typical of the group, but its great abundance and importance contrast with the rareness of gallium (Ga), indium (In), thallium (Tl), and, of course, the recently synthesized element 113.

How the Transition Elements Influence Properties

Group 3A(13) is the first in the *p* block of the periodic table. In Periods 2 and 3, its members lie just one element away from those in Group 2A(2), but in Period 4 and higher, a large gap separates the two groups, with 10 transition elements (*d* block) each in Periods 4, 5, and 6 and an additional 14 inner transition elements (*f* block) in Period 6. Because *d* and *f* electrons spend very little time near the nucleus, they shield the outer (*s* and *p*) electrons in Ga, In, and Tl very little from the stronger nuclear attraction (greater Z_{eff}) (Sections 7.4 and 8.4). As a result, these elements deviate from the usual trends down a group, having smaller atomic radii and larger ionization energies and electronegativities than expected.

With respect to physical properties, boron is a black, hard, very high-melting, network covalent metalloid, but the other 3A members are shiny, relatively soft, low-melting metals. Aluminum's low density and three valence electrons make it an exceptional conductor: for a given mass, it conducts a current twice as effectively as copper. Gallium has the largest liquid temperature range of any element: it melts in your hand but does not boil until 2403°C.

Features That First Appear in This Group's Chemical Properties

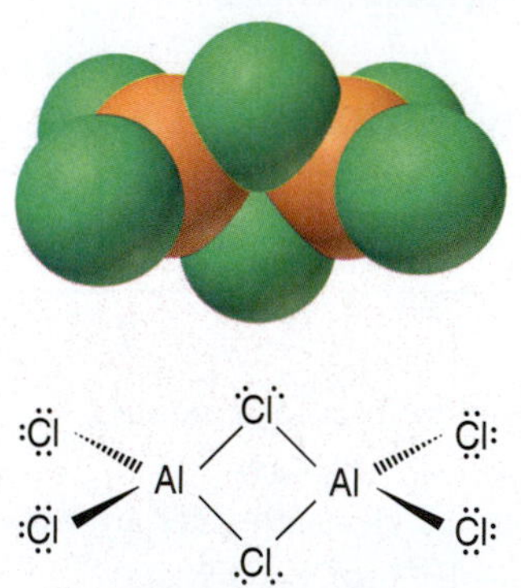

Figure 14.3 The dimeric structure of gaseous aluminum chloride.

Looking down Group 3A(13), we see a wide range of chemical behavior. Boron, the anomalous member from Period 2, is the first metalloid we've encountered so far and the only one in the group. It is much less reactive at room temperature than the other members and forms covalent bonds exclusively. Although aluminum acts like a metal physically, its halides exist in the gas phase as covalent *dimers*—molecules formed by joining two identical smaller molecules (Figure 14.3)—and its oxide is amphoteric rather than basic. Most of the other 3A compounds are ionic, but they have more covalent character than similar 2A compounds because the 3A cations can polarize nearby electron clouds more effectively.

Three features are common to the elements in Groups 3A(13) to 6A(16):

1. *Presence of multiple oxidation states.* Many of the larger elements in these groups also have an important oxidation state *two lower than the A-group number.* The lower state occurs when the atoms lose their *np* electrons only, not their two *ns* electrons. This fact is often called the *inert-pair effect* (Section 8.4).

2. *Increasing prominence of the lower oxidation state.* When a group exhibits more than one oxidation state, *the lower state becomes more prominent going down the group.* In Group 3A(13), for instance, all members exhibit the +3 state, but the +1

Group 3A(13): The Boron Family

Family Portrait

KEY ATOMIC PROPERTIES, PHYSICAL PROPERTIES, AND REACTIONS

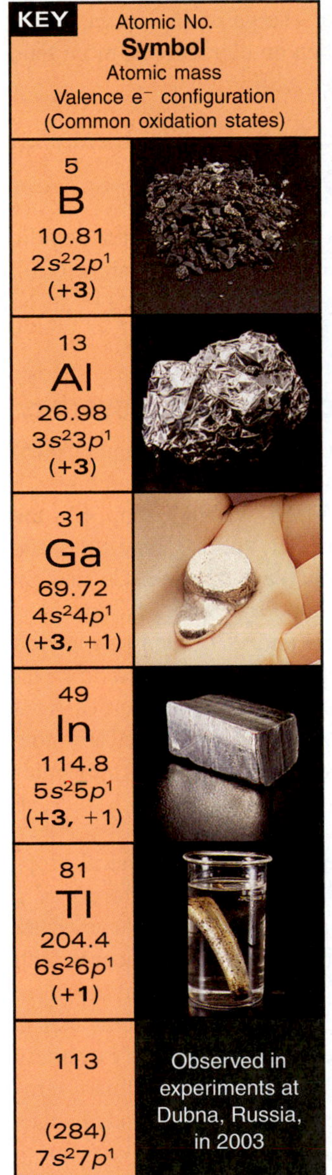

KEY
Atomic No.
Symbol
Atomic mass
Valence e⁻ configuration
(Common oxidation states)
5
B
10.81
2s22p1
(+3)
13
Al
26.98
3s23p1
(+3)
31
Ga
69.72
4s24p1
(+3, +1)
49
In
114.8
5s25p1
(+3, +1)
81
Tl
204.4
6s26p1
(+1)
113
(284)
7s27p1
Observed in experiments at Dubna, Russia, in 2003

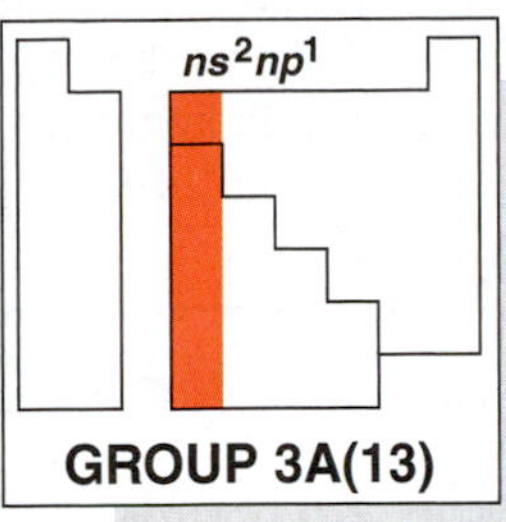

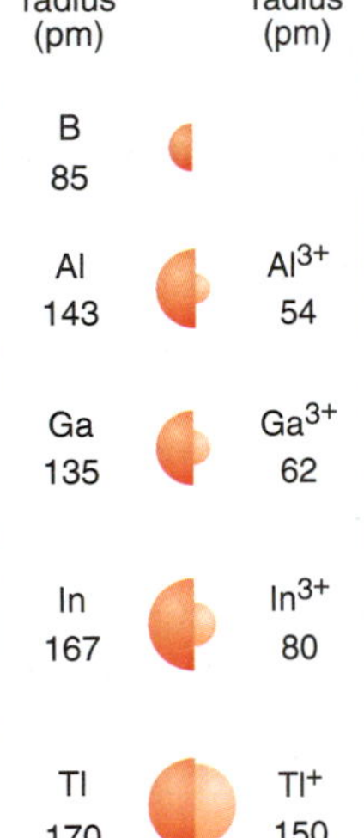

Atomic Properties

Group electron configuration is ns^2np^1. All except Tl commonly display the +3 oxidation state. The +1 state becomes more common down the group. Atomic size is smaller and EN is higher than for 2A(2) elements; IE is lower, however, because it is easier to remove an electron from the higher energy *p* sublevel. Atomic size, IE, and EN do not change as expected down the group because there are intervening transition and inner transition elements.

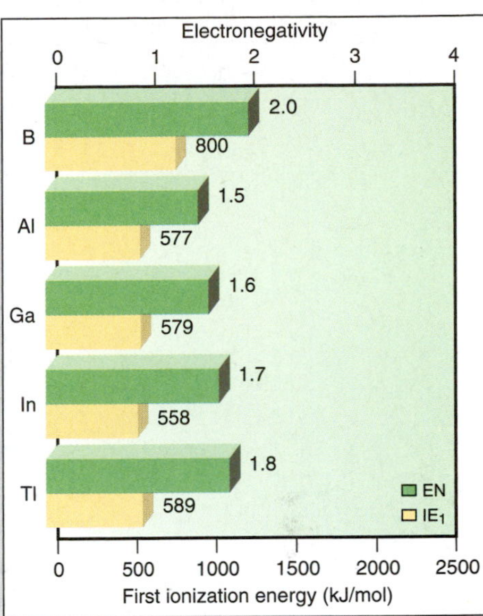

Physical Properties

Bonding changes from network covalent in B to metallic in the rest of the group. Thus, B has a much higher melting point than the others, but there is no overall trend. Boiling points decrease down the group. Densities increase down the group.

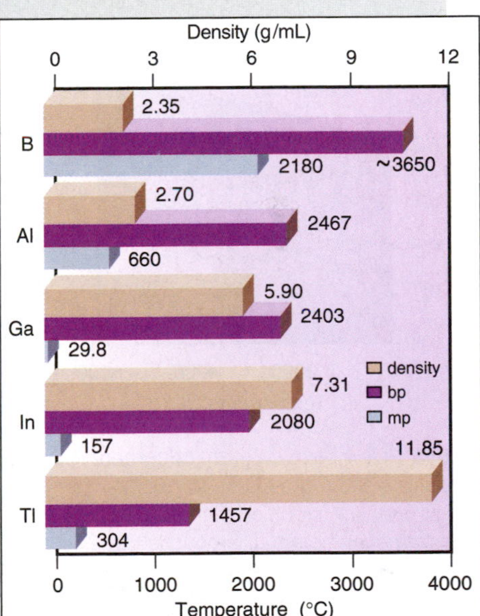

Reactions

1. The elements react sluggishly, if at all, with water:

$$2Ga(s) + 6H_2O(hot) \longrightarrow 2Ga^{3+}(aq) + 6OH^-(aq) + 3H_2(g)$$

$$2Tl(s) + 2H_2O(steam) \longrightarrow 2Tl^+(aq) + 2OH^-(aq) + H_2(g)$$

Al becomes covered with a layer of Al_2O_3 that prevents further reaction.

2. When strongly heated in pure O_2, all members form oxides:

$$4E(s) + 3O_2(g) \xrightarrow{\Delta} 2E_2O_3(s) \qquad (E = B, Al, Ga, In)$$

$$4Tl(s) + O_2(g) \xrightarrow{\Delta} 2Tl_2O(s)$$

Oxide acidity decreases down the group: B_2O_3 (weakly acidic) > Al_2O_3 > Ga_2O_3 > In_2O_3 > Tl_2O (strongly basic), and the +1 oxide is more basic than the +3 oxide.

3. All members reduce halogens (X_2):

$$2E(s) + 3X_2 \longrightarrow 2EX_3 \qquad (E = B, Al, Ga, In)$$

$$2Tl(s) + X_2 \longrightarrow 2TlX(s)$$

The BX_3 compounds are volatile and covalent. Trihalides of Al, Ga, and In are (mostly) ionic solids.

state first appears with some compounds of gallium and becomes the only important state of thallium.

3. *Relative basicity of oxides.* In general, *oxides of the element in the lower oxidation state are more basic than oxides of the element in the higher oxidation state.* For example, in Group 3A(13), In_2O is more basic than In_2O_3. The lower charge of In^+ does not polarize the O^{2-} ion as much as the higher charge of In^{3+} does, so the O^{2-} ion is more available to act as a base. In general, when an element has more than one oxidation state, *it acts more like a metal in its lower state.*

14.5 • GROUP 4A(14): THE CARBON FAMILY

The whole range of elemental behavior occurs within Group 4A(14): nonmetallic carbon (C) leads off, followed by the metalloids silicon (Si) and germanium (Ge), with metallic tin (Sn) and lead (Pb) next, and recently synthesized element 114 at the bottom of the group [Group 4A(14) Family Portrait].

How the Type of Bonding in an Element Affects Physical Properties

The elements of Group 4A(14) and their neighbors in Groups 3A(13) and 5A(15) illustrate how physical properties, such as melting point and heat of fusion (ΔH_{fus}), depend on the type of bonding in an element (Table 14.1). Within Group 4A, the large decrease in melting point between the network covalent solids C and Si is due to longer, weaker bonds in the Si structure; the large decrease between Ge and Sn is due to the change from covalent network to metallic bonding. Similarly, considering horizontal trends, the large increases in melting point and ΔH_{fus} across a period between Al and Si and between Ga and Ge reflect the change from metallic to covalent network bonding. Note the abrupt rises in the values for these properties from metallic Al, Ga, and Sn to the network covalent metalloids Si, Ge, and Sb, and note the abrupt drops from the covalent networks of C and Si to the individual molecules of N and P in Group 5A.

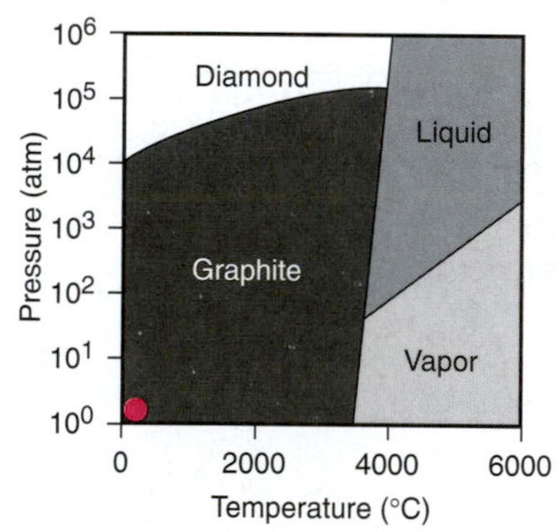

Figure 14.4 Phase diagram of carbon. Graphite is the more stable form of carbon at ordinary conditions *(small red circle at extreme lower left).* Diamond is more stable at very high pressure.

Allotropism: Different Forms of an Element Striking variations in physical properties often appear among **allotropes,** different crystalline or molecular forms of a substance. One allotrope is usually more stable than another at a particular pressure and temperature. An important example of allotropism in Group 4A(14) occurs with carbon. It is difficult to imagine two substances made of the same atom that are more different than graphite and diamond. Graphite is a black electrical conductor that is soft and "greasy," whereas diamond is a colorless electrical insulator that is extremely hard. Graphite is the standard state of carbon, the more stable form at ordinary temperatures and pressure (Figure 14.4). Fortunately for jewelry owners, diamond changes to graphite at a negligible rate under normal conditions.

Table 14.1 Bond Type and the Melting Process in Groups 3A(13) to 5A(15)

Period	Group 3A(13) Element	Bond Type	Melting Point (°C)	ΔH_{fus} (kJ/mol)	Group 4A(14) Element	Bond Type	Melting Point (°C)	ΔH_{fus} (kJ/mol)	Group 5A(15) Element	Bond Type	Melting Point (°C)	ΔH_{fus} (kJ/mol)
2	B	Covalent network	2180	23.6	C	Covalent network	4100	Very high	N	Covalent molecule	−210	0.7
3	Al	Metallic	660	10.5	Si	Covalent network	1420	50.6	P	Covalent molecule	44.1	2.5
4	Ga	Metallic	30	5.6	Ge	Covalent network	945	36.8	As	Covalent network	816	27.7
5	In	Metallic	157	3.3	Sn	Metallic	232	7.1	Sb	Covalent network	631	20.0
6	Tl	Metallic	304	4.3	Pb	Metallic	327	4.8	Bi	Metallic	271	10.5

Key: Metallic; Covalent network; Covalent molecule; Metal; Metalloid; Nonmetal

Group 4A(14): The Carbon Family

Family Portrait

KEY ATOMIC PROPERTIES, PHYSICAL PROPERTIES, AND REACTIONS

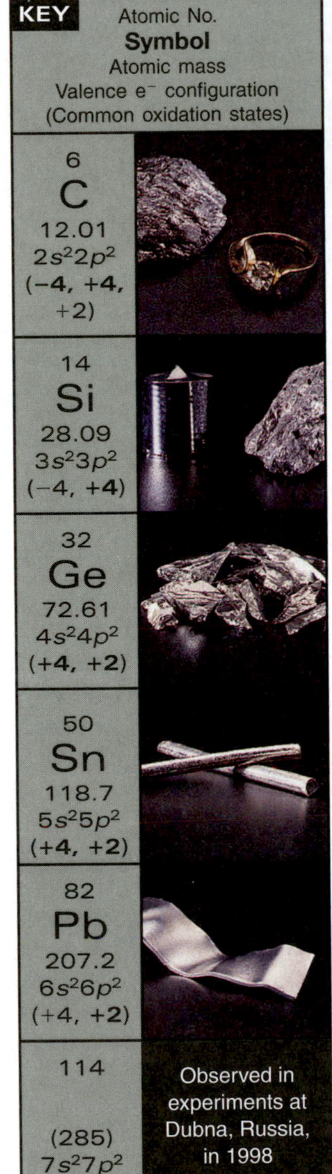

KEY
Atomic No.
Symbol
Atomic mass
Valence e⁻ configuration
(Common oxidation states)
6
C
12.01
$2s^22p^2$
(−4, +4, +2)
14
Si
28.09
$3s^23p^2$
(−4, +4)
32
Ge
72.61
$4s^24p^2$
(+4, +2)
50
Sn
118.7
$5s^25p^2$
(+4, +2)
82
Pb
207.2
$6s^26p^2$
(+4, +2)
114
(285)
$7s^27p^2$
Observed in experiments at Dubna, Russia, in 1998

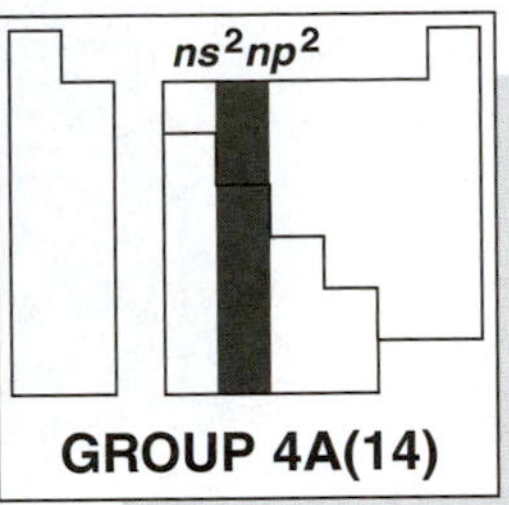

Atomic radius (pm)	Ionic radius (pm)
C 77	
Si 118	
Ge 122	
Sn 140	Sn^{2+} 118
Pb 146	Pb^{2+} 119

Atomic Properties

Group electron configuration is ns^2np^2. Down the group, the number of oxidation states decreases, and the lower (+2) state becomes more common. Down the group, size increases. Because transition and inner transition elements intervene, IE and EN do not decrease smoothly.

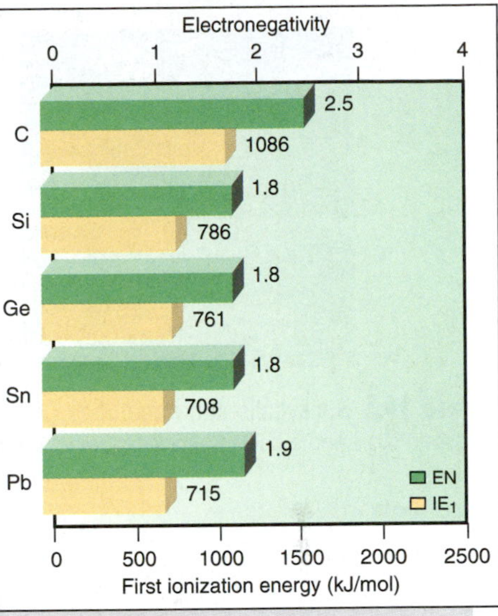

Physical Properties

Trends in properties, such as decreasing hardness and melting point, are due to changes in types of bonding within the solid: covalent network in C, Si, and Ge; metallic in Sn and Pb. Down the group, density increases because of several factors, including differences in crystal packing.

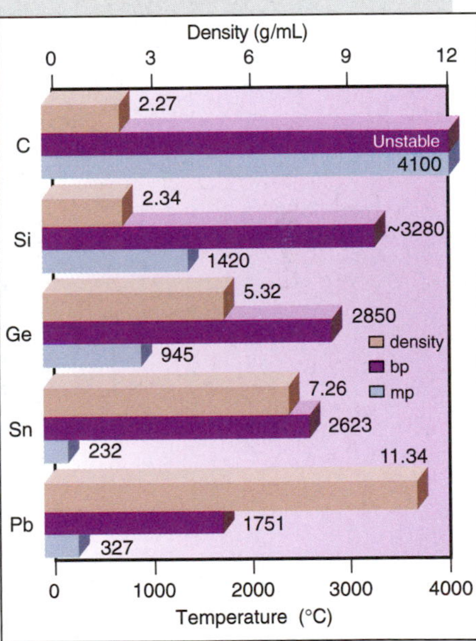

Reactions

1. The elements are oxidized by halogens:

$$E(s) + 2X_2 \longrightarrow EX_4 \quad (E = C, Si, Ge)$$

The +2 halides are more stable for tin and lead, SnX_2 and PbX_2.

2. The elements are oxidized by O_2:

$$E(s) + O_2(g) \longrightarrow EO_2 \quad (E = C, Si, Ge, Sn)$$

Pb forms the +2 oxide, PbO. Oxides become more basic down the group. The reaction of CO_2 and H_2O provides the weak acidity of natural unpolluted waters:

$$CO_2(g) + H_2O(l) \rightleftharpoons [H_2CO_3(aq)] \rightleftharpoons H^+(aq) + HCO_3^-(aq)$$

3. Hydrocarbons react with O_2 to form CO_2 and H_2O. The reaction for methane is adapted to yield heat or electricity:

$$CH_4(g) + 2O_2(g) \longrightarrow CO_2(g) + 2H_2O(g)$$

4. Silica is reduced to form elemental silicon:

$$SiO_2(s) + 2C(s) \longrightarrow Si(s) + 2CO(g)$$

This crude silicon is made ultrapure through zone refining for use in the manufacture of computer chips.

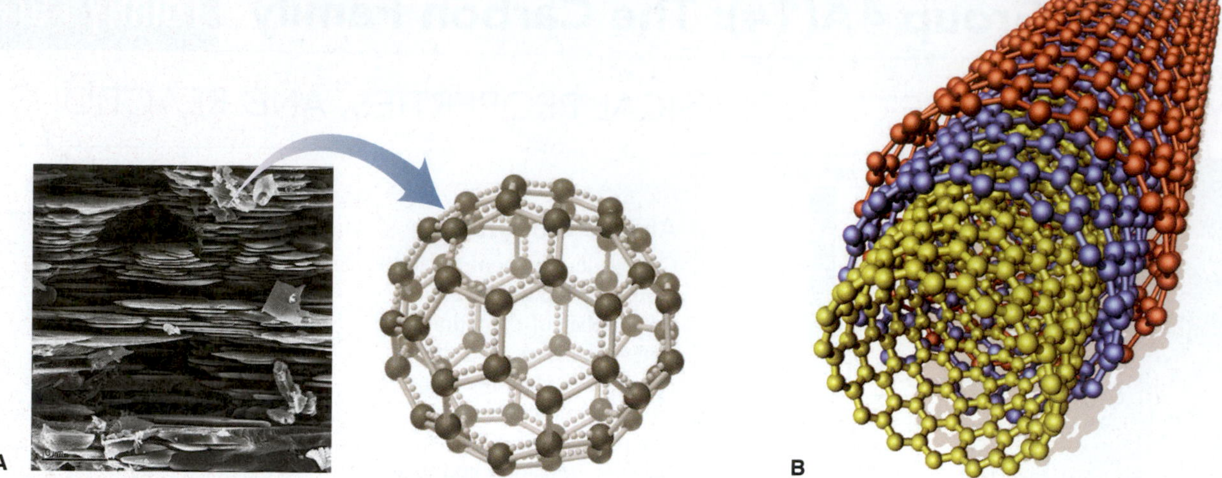

Figure 14.5 **Buckyballs and nanotubes.** **A,** Crystals of buckminsterfullerene (C_{60}) and a ball-and-stick model of the molecule. **B,** Nanotubes are single or, as shown in this colorized model, concentric graphite-like tubes (shown without the fullerene ends).

In the mid-1980s, a newly discovered allotrope of carbon began generating great interest. Mass spectrometric analysis of soot had shown evidence for a soccer ball–shaped molecule of formula C_{60} dubbed *buckminsterfullerene* (informally called a "buckyball") (Figure 14.5A).

Since 1990, when multigram quantities of C_{60} (Figure 14.5A, *photo*) and related fullerenes were prepared, metal atoms have been incorporated into the structure and many different atoms and groups (fluorine, hydroxyl groups, sugars, etc.) have been attached, resulting in compounds with a range of useful properties. In 1991, extremely thin (~1 nm in diameter) graphite-like tubes (Figure 14.5B) with fullerene ends were prepared. Along their length, these *nanotubes* are much stronger than steel (on a mass basis), and they conduct electricity. With potential applications in nanoscale electronics, energy storage, catalysis, polymers, and medicine, fullerene and nanotube chemistry is an active area in materials research. The 2010 Nobel Prize in physics was awarded for research into a new form of carbon called *graphene.* It has remarkable conductivity and strength, yet exists as extended graphite-like sheets only one atom thick.

How Bonding Changes in the Carbon Family's Compounds

The Group 4A(14) elements display a wide range of chemical behavior, from the covalent compounds of carbon to the ionic compounds of lead. Carbon's intermediate EN of 2.5 ensures that it virtually always forms covalent bonds, but the larger members of the group form bonds with increasing ionic character. With nonmetals, Si and Ge form strong polar covalent bonds, such as the Si—O bond, which is one of the strongest involving any Period 3 element (BE = 368 kJ/mol) and is responsible for the physical and chemical stability of Earth's solid surface.

The pattern of elements having more than one oxidation state observed in Group 3A(13) also appears here. Thus, compounds with Si in the +4 state are much more stable than those with Si in the +2 state, whereas compounds with Pb in the +2 state are more stable than those with Pb in the +4 state. These elements behave more like metals in the lower oxidation state. For example, $SnCl_2$ and $PbCl_2$ are white, relatively high-melting, water-soluble crystals—typical properties of a salt—whereas $SnCl_4$ is a volatile, benzene-soluble liquid, and $PbCl_4$ is a thermally unstable oil. Similarly, SnO and PbO are more basic than SnO_2 and PbO_2.

Highlights of Carbon Chemistry

Like the other Period 2 elements, carbon is an anomaly in its group, and its bonding ability makes it an anomaly in the entire periodic table.

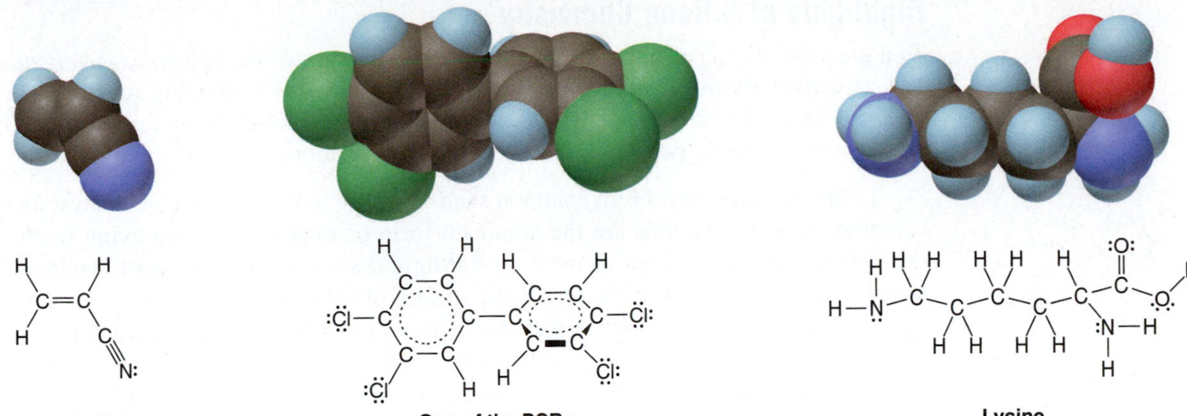

Figure 14.6 Three organic compounds. Acrylonitrile, a precursor of acrylic fibers. A typical PCB (one of 209 different polychlorinated biphenyls). Lysine, one of about 20 amino acids that occur in proteins.

Organic Compounds Two major properties of carbon give rise to the enormous field of organic chemistry.

- Carbon has the ability to bond to itself—a process known as *catenation.* As a result of its small size and its capacity for four bonds, carbon can form chains, branches, and rings that lead to myriad structures. Add a lot of H, some O and N, a bit of S, P, halogens, and a few metals, and you have the whole organic world! Figure 14.6 shows three of the several million organic compounds known. Monocarbon halides (or halomethanes) are tetrahedral molecules *(see margin).* The short, strong bonds in chlorofluorocarbons (CFCs, or Freons) make their long-term persistence in the upper atmosphere a major environmental problem (Chapter 16).
- Carbon has the ability to form multiple bonds. Multiple bonds are common in carbon structures because the C—C bond is short enough for side-to-side overlap of two half-filled $2p$ orbitals to form π bonds. (In Chapter 15, we discuss in detail how the atomic properties of carbon give rise to the diverse structures and reactivities of organic compounds.)

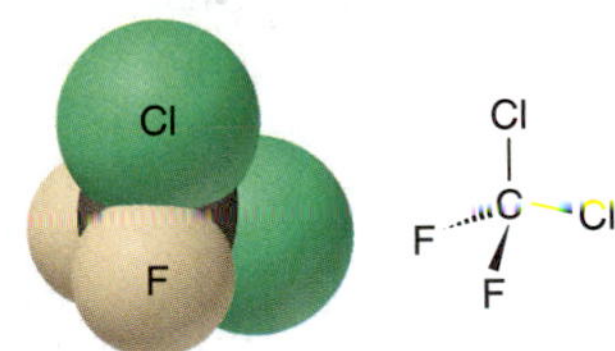

Freon-12, CCl_2F_2

Because the other 4A members are larger, E—E bonds become longer and weaker, so catenation and multiple bonding are much less important down the group.

Inorganic Compounds In contrast to its organic compounds, carbon's inorganic compounds are simple. We'll look at just the two most important types.

1. *Carbonates.* Metal carbonates are the main mineral form. Marble, limestone, chalk, coral, and several other types are found in enormous deposits throughout the world. Carbonates are used in several common antacids because they react with the HCl in stomach acid:

$$CaCO_3(s) + 2HCl(aq) \longrightarrow CaCl_2(aq) + CO_2(g) + H_2O(l)$$

Identical net ionic reactions with sulfuric and nitric acids protect lakes bounded by limestone deposits from the harmful effects of acid rain.

2. *Oxides.* Carbon forms two common gaseous oxides, CO_2 and CO.

- Carbon dioxide is essential to all life: it is the primary source of carbon in plants through photosynthesis—and animals eat the plants. Its aqueous solution is the cause of mild acidity in natural waters. However, its atmospheric buildup from deforestation and excessive use of fossil fuels is severely affecting the global climate.
- Carbon monoxide forms when carbon or its organic compounds burn in an inadequate supply of O_2:

$$2C(s) + O_2(g) \longrightarrow 2CO(g)$$

Carbon monoxide is a key component of fuel mixtures and is widely used in the production of methanol, formaldehyde, and other major industrial compounds. When inhaled in cigarette smoke or polluted air, it enters the blood and binds strongly to the Fe(II) in hemoglobin, preventing the normal binding of O_2. The cyanide ion (CN^-), which is *isoelectronic* with CO,

$$[:C{\equiv}N:]^- \quad \text{same electronic structure as} \quad :C{\equiv}O:$$

is toxic because it binds to a key iron-containing protein in energy production.

Highlights of Silicon Chemistry

To a great extent, the chemistry of silicon is the chemistry of the *silicon-oxygen bond.* Just as carbon forms unending C—C chains, the —Si—O— grouping repeats itself endlessly in a wide variety of **silicates,** the most important minerals on the planet, and in **silicones,** synthetic polymers that have many applications:

1. *Silicate minerals.* From common sand and clay to semiprecious amethyst and carnelian, silicate minerals are the dominant form of matter in the nonliving world. Oxygen, the most abundant element on Earth, and silicon, the next most abundant, account for four of every five atoms on the surface of the planet!

The silicate building unit is the *orthosilicate grouping,* $—SiO_4—$, a tetrahedral arrangement of four oxygens around a central silicon. Several well-known minerals, such as zircon and beryl (the natural source of beryllium), contain SiO_4^{4-} ions or small groups of them linked together (Figure 14.7). In extended structures, one of the O atoms links the next Si—O group to form chains, a second one forms crosslinks to neighboring chains to form sheets, and the third forms more crosslinks to create three-dimensional frameworks. Chains of silicate groups compose the asbestos minerals, sheets give rise to talc and mica, and frameworks occur in feldspar and quartz.

2. *Silicone polymers.* Unlike the naturally occurring silicates, silicone polymers are manufactured substances, consisting of alternating Si and O atoms with two organic groups also bonded to each Si atom in a very long Si—O chain, as in *poly(dimethyl siloxane):*

$$\begin{array}{ccccccccccc} & CH_3 & & CH_3 & & CH_3 & & CH_3 & & CH_3 & \\ & | & & | & & | & & | & & | & \\ \cdots O— & Si & —O— & Si & —O— & Si & —O— & Si & —O— & Si & —O\cdots \\ & | & & | & & | & & | & & | & \\ & CH_3 & & CH_3 & & CH_3 & & CH_3 & & CH_3 & \end{array}$$

Silicones have properties of both plastics and minerals. The organic groups give the chains flexibility and produce the weak intermolecular forces between chains that are characteristic of a plastic, while the O—Si—O backbone confers the thermal stability and nonflammability of a mineral. Structures similar to those of the silicates can be created by adding various reactants to form silicone chains, sheets, and frameworks. Chains are oily liquids used as lubricants and as components of car polish and makeup. Sheets are components of gaskets, space suits, and contact lenses. Frameworks find uses as laminates on circuit boards, in nonstick cookware, and in artificial skin and bone.

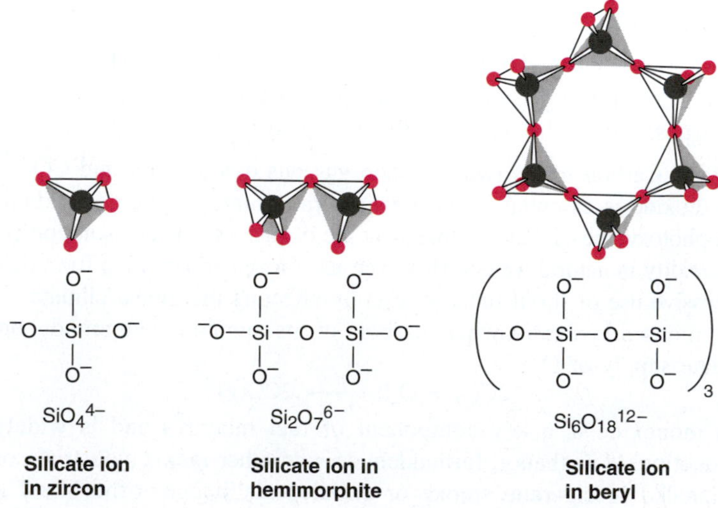

Figure 14.7 Structures of the silicate anions in some minerals.

14.6 • GROUP 5A(15): THE NITROGEN FAMILY

The first two elements of Group 5A(15), gaseous nonmetallic nitrogen (N) and solid nonmetallic phosphorus (P), play major roles in both nature and industry. Below these nonmetals are two metalloids, arsenic (As) and antimony (Sb), followed by the metal bismuth (Bi), and the recently synthesized element 115. The Group 5A(15) Family Portrait on the next page provides an overview.

The Wide Range of Physical Behavior

Group 5A(15) displays the widest range of physical behavior we've seen so far because there are large changes in bonding and intermolecular forces:

- *Nitrogen* occurs as a gas consisting of N_2 molecules with such weak intermolecular forces that the element *boils* more than 200°C below room temperature.
- *Phosphorus* exists most commonly as tetrahedral P_4 molecules in the solid phase. Because P is heavier and more polarizable than N, it has stronger dispersion forces and melts about 25°C above room temperature. Phosphorus exhibits allotropism. The white form consists of individual tetrahedral molecules (Figure 14.8A), making it low melting and soluble in nonpolar solvents; with a small 60° bond angle and, thus, weak P—P bonds, it is highly reactive (Figure 14.8B). In the red form, the P_4 units exist in chains, which make this allotrope much less reactive, high melting, and insoluble (Figure 14.8C).
- *Arsenic* consists of extended, puckered sheets in which each As atom is covalently bonded to three others and has nonbonding interactions with three nearest neighbors in adjacent sheets, giving As the highest melting point in the group.
- *Antimony* has a similar covalent network, also resulting in a high melting point.
- *Bismuth* has metallic bonding and thus a lower melting point than As and Sb.

Patterns in Chemical Behavior

The same general pattern of chemical behavior that we discussed for Group 4A(14) appears again in this group, reflected in the change from nonmetallic N to metallic Bi.

1. *Ion formation.* Nearly all Group 5A(15) compounds have *covalent bonds* because a 5A element must *gain* three electrons to form an ion with a noble gas electron configuration. Enormous lattice energy results when 3− anions attract cations, but this occurs for N only with active metals, such as Li_3N and Mg_3N_2 (and perhaps with P in Na_3P).

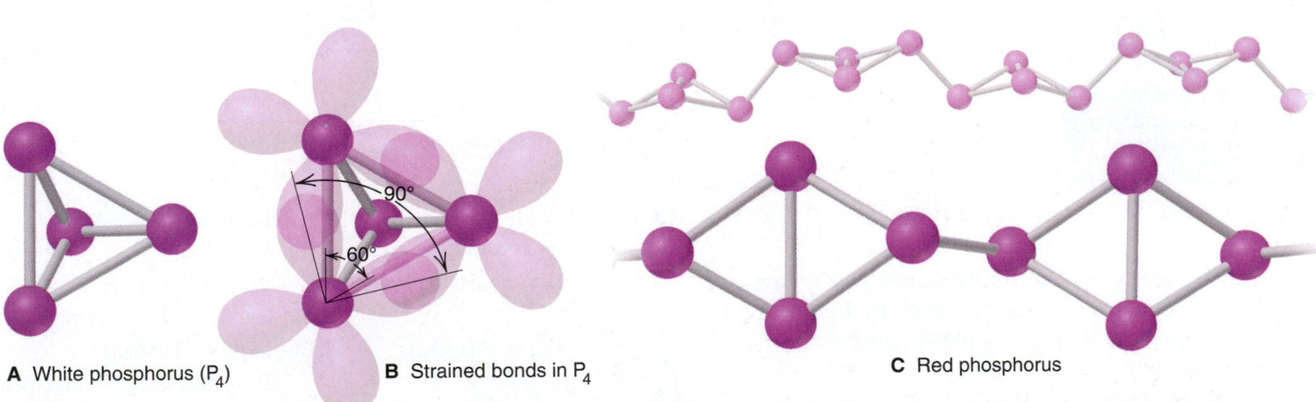

Figure 14.8 Two allotropes of phosphorus. **A,** White phosphorus exists as individual P_4 molecules. **B,** The reactivity of P_4 is due in part to the bond strain that arises from the 60° bond angle. **C,** In red phosphorus, one of the P—P bonds of the white form has broken and links the P_4 units together. Lone pairs (not shown) reside in *s* orbitals in both allotropes.

Family Portrait

Group 5A(15): The Nitrogen Family

KEY ATOMIC PROPERTIES, PHYSICAL PROPERTIES, AND REACTIONS

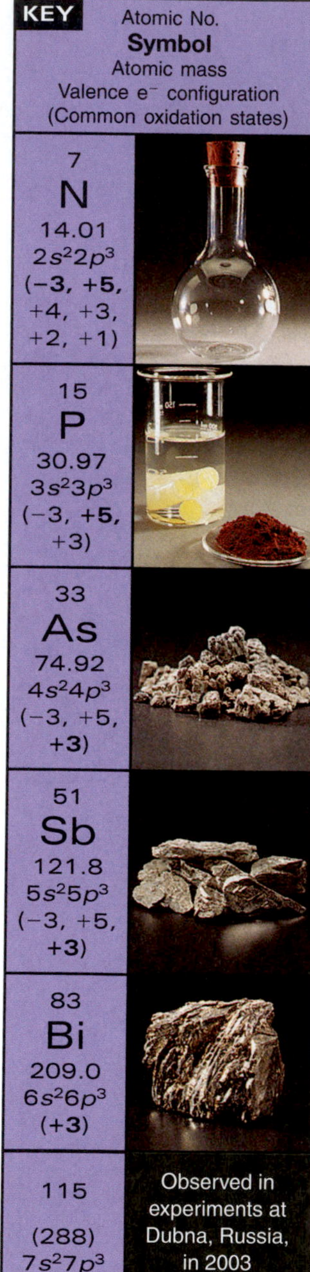

KEY
Atomic No.
Symbol
Atomic mass
Valence e⁻ configuration
(Common oxidation states)
7
N
14.01
$2s^22p^3$
(−3, +5, +4, +3, +2, +1)
15
P
30.97
$3s^23p^3$
(−3, +5, +3)
33
As
74.92
$4s^24p^3$
(−3, +5, +3)
51
Sb
121.8
$5s^25p^3$
(−3, +5, +3)
83
Bi
209.0
$6s^26p^3$
(+3)
115
(288)
$7s^27p^3$
Observed in experiments at Dubna, Russia, in 2003

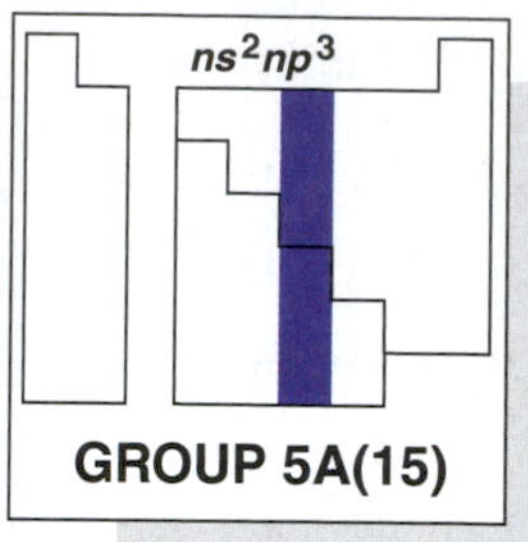

Atomic radius (pm)	Ionic radius (pm)
N 75	N^{3-} 146
P 110	P^{3-} 212
As 120	
Sb 140	
Bi 150	Bi^{3+} 103

Atomic Properties

Group electron configuration is ns^2np^3. The np sublevel is half-filled, with each p orbital containing one electron (parallel spin). The number of oxidation states decreases down the group, and the lower (+3) state becomes more common. Atomic properties follow generally expected trends. The large (~50%) increase in size from N to P correlates with the much lower IE and EN of P.

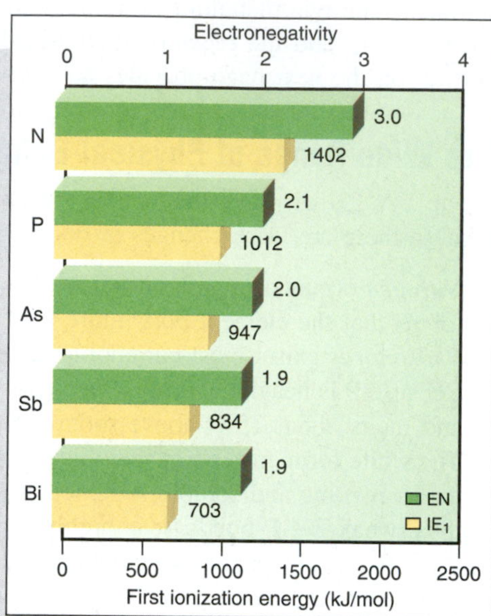

Physical Properties

Physical properties reflect the change from individual molecules (N, P) to network covalent solid (As, Sb) to metal (Bi). Thus, melting points increase and then decrease. Large atomic size and low atomic mass result in low density. Because mass increases more than size down the group, the density of the elements as solids increases. The dramatic increase in density from P to As is due to the intervening transition elements.

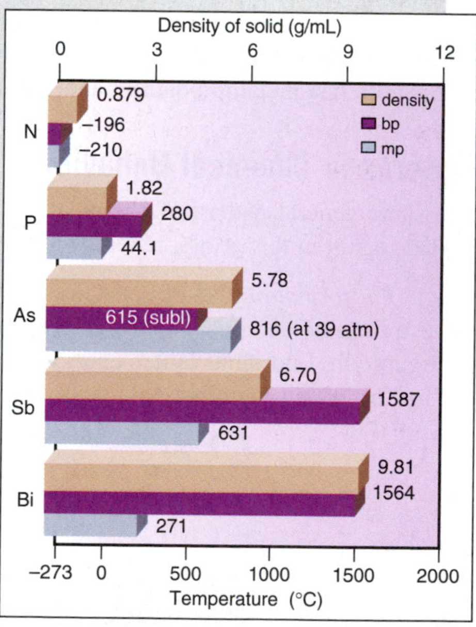

Reactions

1. Nitrogen is "fixed" industrially in the Haber process:

$$N_2(g) + 3H_2(g) \rightleftharpoons 2NH_3(g)$$

Further reactions convert NH_3 to NO, NO_2, and HNO_3 *(see text)*. Hydrides of some other group members are formed from reaction in water (or with H_3O^+) of a metal phosphide, arsenide, and so forth:

$$Ca_3P_2(s) + 6H_2O(l) \longrightarrow 2PH_3(g) + 3Ca(OH)_2(aq)$$

2. Halides are formed by direct combination of the elements:

$$2E(s) + 3X_2 \longrightarrow 2EX_3 \qquad \text{(E = all except N)}$$

$$EX_3 + X_2 \longrightarrow EX_5 \qquad \text{(E = all except N and Bi with X = F and Cl, but no } BiCl_5\text{; E = P for X = Br)}$$

3. Oxoacids are formed from the halides in a reaction with water that is common to many nonmetal halides:

$$EX_3 + 3H_2O(l) \longrightarrow H_3EO_3(aq) + 3HX(aq) \qquad \text{(E = all except N)}$$

$$EX_5 + 4H_2O(l) \longrightarrow H_3EO_4(aq) + 5HX(aq) \qquad \text{(E = all except N and Bi)}$$

Note that the oxidation number of E does *not* change.

2. *Oxidation state and oxide acidity.* As in Groups 3A and 4A, as we move down the group there are fewer oxidation states and the lower state becomes more prominent: N exhibits every state possible for a 5A element, from +5, as in HNO_3, to −3, as in NH_3; only the +5 and +3 states are common for P, As, and Sb; and +3 is the only common state of Bi. The oxides change from acidic to amphoteric to basic, and the lower oxide (for example, Sb_2O_3) is more basic than the higher oxide (Sb_2O_5) because the lower oxide's E-to-O bond is more ionic.

3. *Hydrides.* All the elements form gaseous hydrides of formula EH_3. Ammonia is made industrially at high pressure and moderately high temperature:

$$N_2(g) + 3H_2(g) \rightleftharpoons 2NH_3(g)$$

The other hydrides are very poisonous and form by reaction in water of a metal phosphide, arsenide, and so forth, which acts as a strong base; for example,

$$Ca_3As_2(s) + 6H_2O(l) \longrightarrow 2AsH_3(g) + 3Ca(OH)_2(aq)$$

The Group 5A(15) hydrides show several interesting bonding and structural patterns:

- Despite its much lower molar mass, NH_3 melts and boils at higher temperatures than the other 5A hydrides as a result of H *bonding.*
- Bond angles decrease from 107.3° for NH_3 to around 90° for the other hydrides, which suggests that the larger atoms use unhybridized *p* orbitals.
- E—H bond lengths increase down the group, so bond strength and thermal stability decrease: AsH_3 decomposes at 250°C, SbH_3 at 20°C, and BiH_3 at −45°C.

4. *Halides.* Through direct combination of the elements, the Group 5A(15) members form all possible trihalides (EX_3) and pentafluorides (EF_5), but few other pentahalides. As with the hydrides, stability of the halides decreases as the E—X bond becomes longer with larger halogens.

In an aqueous reaction pattern *typical of many nonmetal halides,* each 5A halide reacts with water to yield the hydrogen halide and the oxoacid, in which E has the *same* O.N. as it had in the original halide. For example, PX_5 (O.N. of P = +5) produces phosphoric acid (O.N. of P = +5) and HX:

$$PCl_5(s) + 4H_2O(l) \longrightarrow H_3PO_4(l) + 5HCl(g)$$

Highlights of Nitrogen Chemistry

The most striking highlight of nitrogen chemistry is the *inertness* of N_2. Even though the atmosphere consists of nearly four-fifths N_2 and one-fifth O_2, the searing temperature of a lightning bolt is needed to form significant amounts of nitrogen oxides. Indeed, N_2 reacts at high temperatures with H_2, Li, Group 2A(2) members, B, Al, C, Si, Ge, O_2, and many transition elements. Here we focus on the nitrogen oxides and the oxoacids and their salts.

Nitrogen Oxides Nitrogen is remarkable for having six stable oxides, each with a *positive* enthalpy of formation because of the great strength of the N≡N bond (Table 14.2, *next page*). Unlike the hydrides and halides of nitrogen, the oxides are planar. Nitrogen displays all its positive oxidation states in these compounds, and in N_2O and N_2O_3, the two N atoms have different states. Of special interest are NO and NO_2.

1. *Nitrogen monoxide* (NO; also called *nitrogen oxide* or *nitric oxide*) is a molecule with an odd electron that has biochemical functions ranging from neurotransmission to control of blood flow. The commercial preparation of NO through the oxidation of ammonia occurs as a first step in the production of nitric acid:

$$4NH_3(g) + 5O_2(g) \longrightarrow 4NO(g) + 6H_2O(g)$$

Nitrogen monoxide is also produced whenever air is heated to high temperatures, as in a car engine or a lightning storm:

$$N_2(g) + O_2(g) \xrightarrow{\text{high } T} 2NO(g)$$

Heating converts NO to two other oxides:

$$3NO(g) \xrightarrow{\Delta} N_2O(g) + NO_2(g)$$

Table 14.2 Structures and Properties of the Nitrogen Oxides

Formula	Name	Space-filling Model	Lewis Structure	Oxidation State of N	ΔH°_f (kJ/mol) at 298 K	Comment
N_2O	Dinitrogen monoxide (dinitrogen oxide; nitrous oxide)		:N≡N—Ö:	+1 (0, +2)	82.0	Colorless gas; used as dental anesthetic ("laughing gas") and aerosol propellant
NO	Nitrogen monoxide (nitrogen oxide; nitric oxide)		·N=Ö:	+2	90.3	Colorless, paramagnetic gas; biochemical messenger; air pollutant
N_2O_3	Dinitrogen trioxide			+3 (+2, +4)	83.7	Reddish brown gas (reversibly dissociates to NO and NO_2)
NO_2	Nitrogen dioxide			+4	33.2	Orange-brown, paramagnetic gas formed during HNO_3 manufacture; poisonous air pollutant
N_2O_4	Dinitrogen tetroxide			+4	9.16	Colorless to yellow liquid (reversibly dissociates to NO_2)
N_2O_5	Dinitrogen pentoxide			+5	11.3	Colorless, volatile solid consisting of NO_2^+ and NO_3^-; gas consists of N_2O_5 molecules

This type of redox reaction is called a **disproportionation.** It occurs when a substance *acts as both the oxidizing and reducing agents in a reaction*. In the process, an atom in the reactant occurs in both lower and higher states in the products: the oxidation state of N in NO (+2) becomes +1 in N_2O and +4 in NO_2.

2. *Nitrogen dioxide* (NO_2), a brown poisonous gas, forms to a small extent when NO reacts with additional oxygen:

$$2NO(g) + O_2(g) \rightleftharpoons 2NO_2(g)$$

Like NO, NO_2 has an odd electron that is more localized on the N atom. Thus, NO_2 dimerizes reversibly to *dinitrogen tetroxide:*

$$O_2N\cdot(g) + \cdot NO_2(g) \rightleftharpoons O_2N{-}NO_2(g) \quad (\text{or } N_2O_4)$$

Thunderstorms form NO and NO_2 and carry them down to the soil, where they act as natural fertilizers. In urban settings, however, their formation leads to *photochemical smog* in a series of reactions also involving sunlight, ozone (O_3), unburned gasoline, and various other components.

Nitrogen Oxoacids and Oxoanions There are two common oxoacids of nitrogen (Figure 14.9):

1. *Nitric acid* (HNO_3) is produced in the *Ostwald process*. The first two steps are the oxidations of NH_3 to NO and of NO to NO_2. The final step is a disproportionation, as the oxidation numbers show:

$$3\overset{+4}{N}O_2(g) + H_2O(l) \longrightarrow 2H\overset{+5}{N}O_3(aq) + \overset{+2}{N}O(g)$$

The NO is recycled to make more NO_2.

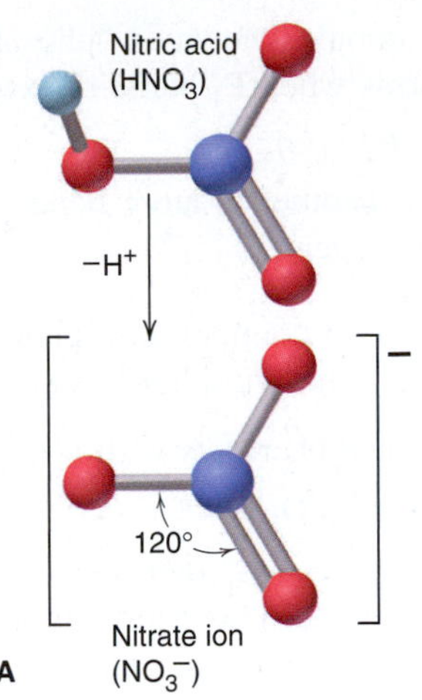

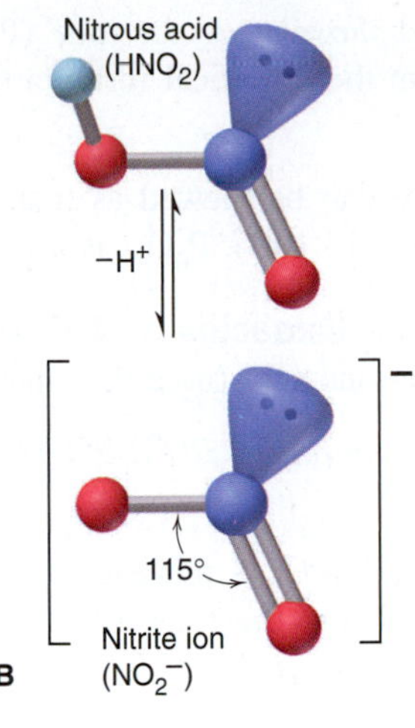

Figure 14.9 The structures of nitric and nitrous acids and their oxoanions. **A,** Nitric acid loses a proton (H^+) to form the trigonal planar nitrate ion (one of three resonance forms is shown). **B,** Nitrous acid, a much weaker acid, forms the planar nitrite ion. Note the effect of nitrogen's lone pair in reducing the ideal 120° bond angle to 115° (one of two resonance forms is shown).

In nitric acid, as in all oxoacids, *the acidic H is attached to one of the O atoms* (Figure 14.9A). In the laboratory, nitric acid is used as a strong oxidizing acid. The products of its reactions with metals vary with the metal's reactivity and the acid's concentration. In the following examples, notice from the net ionic equations that *N is reduced, so the NO_3^- ion is the oxidizing agent.* (Nitrate ion that is not reduced is a spectator ion and does not appear in the net ionic equations):

- With an active metal, such as Al, and dilute acid, N is reduced from the +5 state all the way to the −3 state in the ammonium ion, NH_4^+:

$$8Al(s) + 30HNO_3(aq;\ 1\ M) \longrightarrow 8Al(NO_3)_3(aq) + 3NH_4NO_3(aq) + 9H_2O(l)$$
$$8Al(s) + 30H^+(aq) + 3NO_3^-(aq) \longrightarrow 8Al^{3+}(aq) + 3NH_4^+(aq) + 9H_2O(l)$$

- With a less reactive metal, such as Cu, and more concentrated acid, N is reduced to the +2 state in NO:

$$3Cu(s) + 8HNO_3(aq;\ 3\text{ to }6\ M) \longrightarrow 3Cu(NO_3)_2(aq) + 4H_2O(l) + 2NO(g)$$
$$3Cu(s) + 8H^+(aq) + 2NO_3^-(aq) \longrightarrow 3Cu^{2+}(aq) + 4H_2O(l) + 2NO(g)$$

- With still more concentrated acid, N is reduced only to the +4 state in NO_2:

$$Cu(s) + 4HNO_3(aq;\ 12\ M) \longrightarrow Cu(NO_3)_2(aq) + 2H_2O(l) + 2NO_2(g)$$
$$Cu(s) + 4H^+(aq) + 2NO_3^-(aq) \longrightarrow Cu^{2+}(aq) + 2H_2O(l) + 2NO_2(g)$$

Nitrates form when HNO_3 reacts with metals or with their hydroxides, oxides, or carbonates. *All nitrates are soluble in water.*

2. *Nitrous acid* (HNO_2; Figure 14.9B), a much weaker acid than HNO_3, forms when metal nitrites are treated with a strong acid:

$$NaNO_2(aq) + HCl(aq) \longrightarrow HNO_2(aq) + NaCl(aq)$$

These two acids reveal a *general pattern in relative acid strength among oxoacids:* the more O atoms bonded to the central nonmetal, the stronger the acid. We'll discuss the pattern quantitatively in Chapter 18.

Highlights of Phosphorus Chemistry

Phosphorus forms two important oxides and two common oxoacids, as well as polyphosphates.

Phosphorus Oxides Phosphorus forms two important oxides.

1. *Tetraphosphorus hexoxide* (P_4O_6) forms when white P_4 reacts with limited oxygen:

$$P_4(s) + 3O_2(g) \longrightarrow P_4O_6(s)$$

P_4O_6 has the same tetrahedral orientation of the P atoms in P_4, with an O atom between each pair of P atoms (Figure 14.10A).

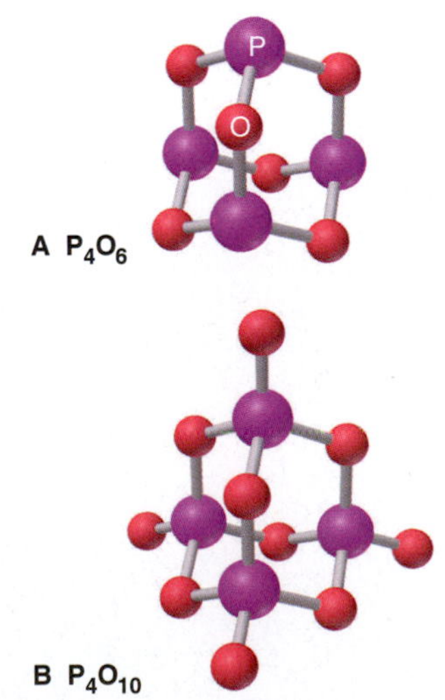

Figure 14.10 Important oxides of phosphorus.

2. *Tetraphosphorus decoxide* (P_4O_{10}) is commonly known as "phosphorus pentoxide" from the empirical formula (P_2O_5). It forms when P_4 burns in excess O_2:

$$P_4(s) + 5O_2(g) \longrightarrow P_4O_{10}(s)$$

Its structure can be viewed as that of P_4O_6 with another O atom bonded to each P atom (Figure 14.10B). P_4O_{10} is a powerful drying agent.

Phosphorus Oxoacids and Oxoanions The two common phosphorus oxoacids are phosphorous acid (note the change in spelling) and phosphoric acid.

1. *Phosphorous acid* (H_3PO_3) is formed when P_4O_6 reacts with water:

$$P_4O_6(s) + 6H_2O(l) \longrightarrow 4H_3PO_3(l)$$

The formula H_3PO_3 is misleading because the acid has only two acidic H atoms; the third is bonded to the central P and does not dissociate. Phosphorous acid is a weak acid in water but reacts completely in two steps with excess strong base:

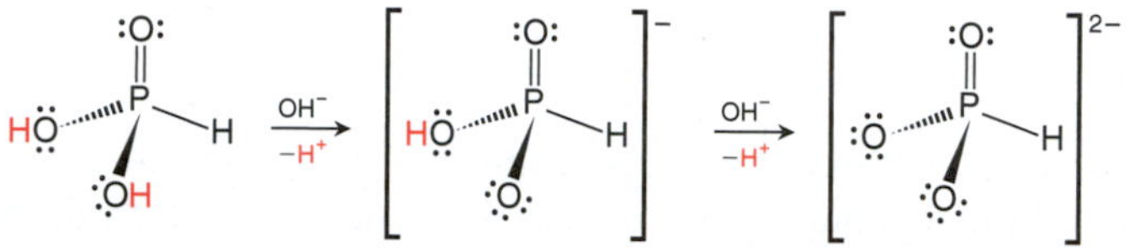

Salts of phosphorous acid contain the phosphite ion, HPO_3^{2-}.

2. *Phosphoric acid* (H_3PO_4), one of the "top-10" most important compounds in chemical manufacturing, is formed in an exothermic reaction of P_4O_{10} with water:

$$P_4O_{10}(s) + 6H_2O(l) \longrightarrow 4H_3PO_4(l)$$

The presence of many H bonds makes pure H_3PO_4 syrupy, more than 75 times as viscous as water. H_3PO_4 is a weak triprotic acid; in water, it loses one proton:

$$H_3PO_4(l) + H_2O(l) \rightleftharpoons H_2PO_4^-(aq) + H_3O^+(aq)$$

In excess strong base, however, it dissociates completely to give the three phosphate oxoanions:

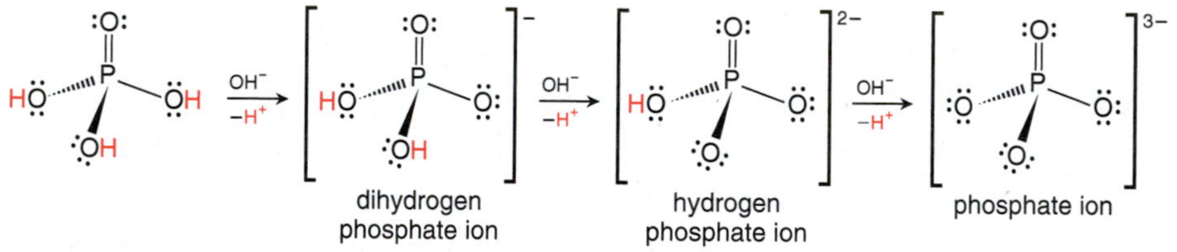

dihydrogen phosphate ion | hydrogen phosphate ion | phosphate ion

Phosphoric acid has a central role in fertilizer production and is also used in soft drinks to add tartness. The various phosphate salts also have numerous essential applications from paint stripper (Na_3PO_4) to rubber stabilizer (K_3PO_4) to fertilizer [$Ca(H_2PO_4)_2$ and $(NH_4)_2HPO_4$].

Polyphosphates Polyphosphates are formed by heating hydrogen phosphates, which lose water as they form P—O—P linkages. This type of reaction, in which an H_2O molecule is lost for every pair of —OH groups that join, is called a **dehydration-condensation;** it occurs in the formation of polyoxoanion chains and other very large molecules, both synthetic and natural, made of repeating units.

14.7 • GROUP 6A(16): THE OXYGEN FAMILY

Oxygen (O) and sulfur (S) are among the most important elements in industry, the environment, and living things. Selenium (Se), tellurium (Te), radioactive polonium (Po), and recently synthesized element 116 lie beneath them in Group 6A(16), as shown in the Family Portrait.

Group 6A(16): The Oxygen Family

Family Portrait

KEY ATOMIC PROPERTIES, PHYSICAL PROPERTIES, AND REACTIONS

KEY
Atomic No.
Symbol
Atomic mass
Valence e⁻ configuration
(Common oxidation states)
8
O
16.00
2s²2p⁴
(−1, −2)
16
S
32.07
3s²3p⁴
(−2, +6, +4, +2)
34
Se
78.96
4s²4p⁴
(−2, +6, +4, +2)
52
Te
127.6
5s²5p⁴
(−2, +6, +4, +2)
84
Po
(209)
6s²6p⁴
(+4, +2)
116
(292)
7s²7p⁴
Observed in experiments at Dubna, Russia, in 2004

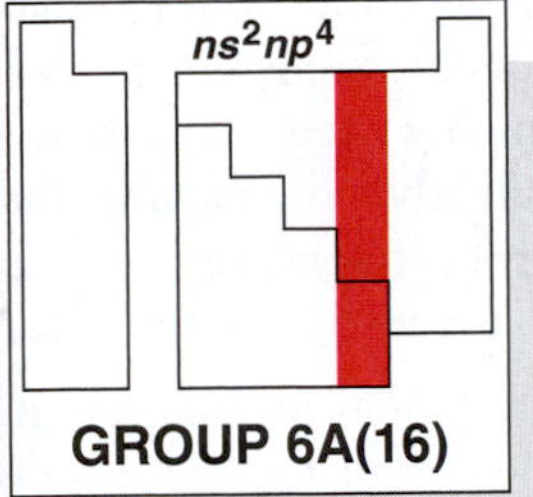

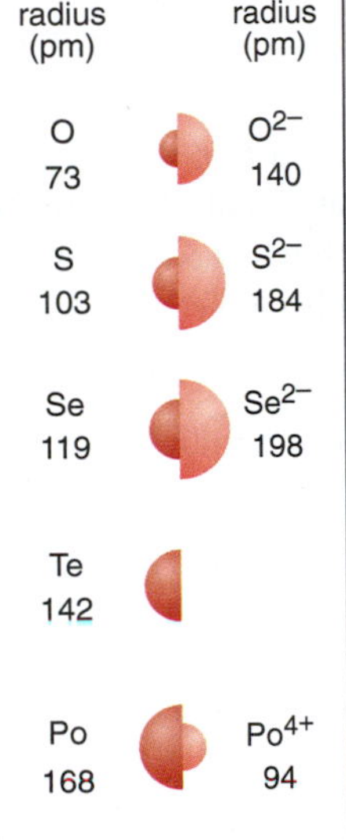

Atomic Properties

Group electron configuration is ns^2np^4. As in Groups 3A(13) and 5A(15), a lower (+4) oxidation state becomes more common down the group. Down the group, atomic and ionic sizes increase, and IE and EN decrease.

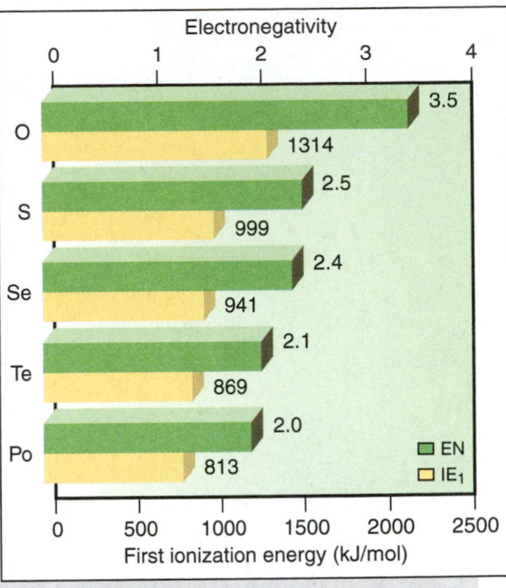

Physical Properties

Melting points increase through Te, which has covalent bonding, and then decrease for Po, which has metallic bonding. Densities of the elements as solids increase steadily.

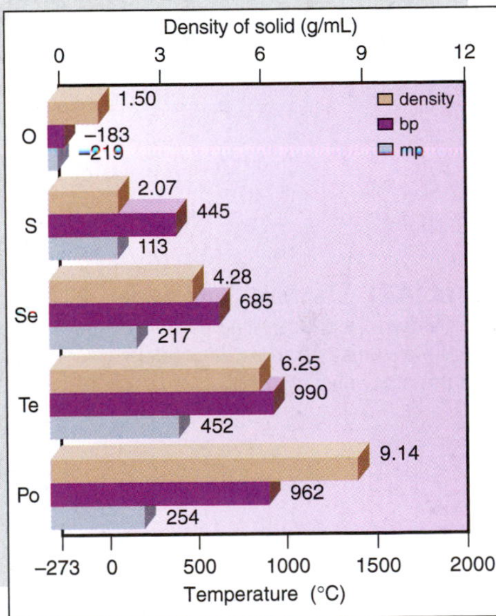

Reactions

1. Halides are formed by direct combination:

$$E(s) + X_2(g) \longrightarrow \text{various halides}$$

(E = S, Se, Te; X = F, Cl)

2. The other elements in the group are oxidized by O_2:

$$E(s) + O_2(g) \longrightarrow EO_2 \quad (E = S, Se, Te, Po)$$

SO_2 is oxidized further, and the product is used in the final step of H_2SO_4 manufacture *(see text)*:

$$2SO_2(g) + O_2(g) \longrightarrow 2SO_3(g)$$

3. The thiosulfate ion is formed when an alkali metal sulfite reacts with sulfur, as in the preparation of "hypo," photographers' developing solution:

$$S_8(s) + 8Na_2SO_3(s) \longrightarrow 8Na_2S_2O_3(aq)$$

How the Oxygen and Nitrogen Families Compare Physically

Group 6A(16) resembles Group 5A(15) in many respects, so let's look at some common themes. The pattern of physical properties we saw in Group 5A appears again in this group:

- *Oxygen,* like nitrogen, occurs as a low-boiling diatomic gas.
- *Sulfur,* like phosphorus, occurs as a polyatomic molecular solid.
- *Selenium,* like arsenic, commonly occurs as a gray metalloid.
- *Tellurium,* like antimony, is slightly more metallic than the preceding group member but still displays network covalent bonding.
- *Polonium,* like bismuth, has a metallic crystal structure.

As in Group 5A, electrical conductivities increase steadily down Group 6A as bonding changes from nonmetal molecules (insulators) to metalloid networks (semiconductors) to a metallic solid (conductor).

Allotropism Multiple forms of an element are much more common in Group 6A(16) than in Group 5A(15).

1. *Oxygen.* Oxygen has two allotropes: life-giving dioxygen (O_2) and poisonous triatomic ozone (O_3). Oxygen gas is colorless, odorless, paramagnetic, and thermally stable. In contrast, ozone gas is bluish, has a pungent odor, is diamagnetic, and decomposes in heat and especially in ultraviolet (UV) light:

$$2O_3(g) \xrightarrow{UV} 3O_2(g)$$

This ability to absorb high-energy photons makes stratospheric ozone vital to life; we'll discuss its depletion in Chapter 16.

2. *Sulfur.* The S atom's ability to bond to itself over a wide range of bond lengths and angles makes sulfur the allotrope "champion" of the periodic table, with more than 10 forms. The most stable is orthorhombic α-S_8, a crown-shaped ring of eight atoms, also called *cyclo-*S_8 (Figure 14.11); all other S allotropes eventually revert to this one.

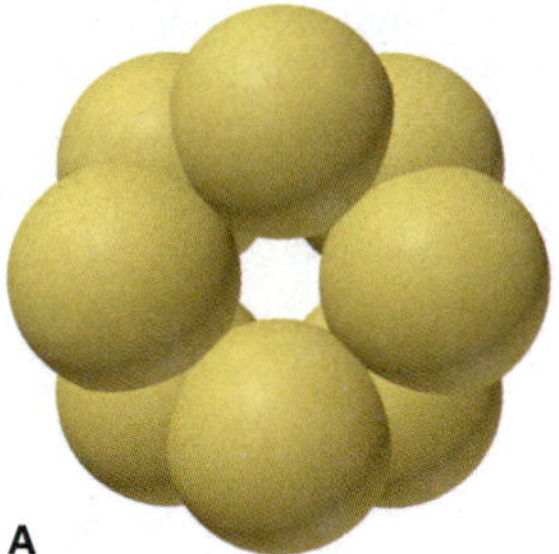

Figure 14.11 The cyclo-S_8 molecule. **A,** Top view of a space-filling model. **B,** Side view of a ball-and-stick model; note the crownlike shape.

3. *Selenium.* Selenium also has several allotropes, some consisting of crown-shaped Se_8 molecules. Gray Se is composed of layers of helical chains. Its ability to conduct a current when exposed to visible light gave birth to the photocopying industry.

How the Oxygen and Nitrogen Families Compare Chemically

Changes in chemical behavior in Group 6A(16) are also similar to those in Group 5A(15). Even though O and S occur as anions much more often than do N and P, like N and P, they bond covalently with almost every other nonmetal. Covalent bonds appear in the compounds of Se and Te (as in those of As and Sb), whereas Po behaves like a metal (as does Bi) in some of its saltlike compounds. In contrast to nitrogen, oxygen has few common oxidation states, but the earlier pattern returns with the other Group 6A members: among the common positive (+6 and +4) states, the +4 state is seen more often in Te and Po.

Oxygen's high EN (3.5) and great oxidizing strength are second only to those of fluorine. But the other members of Group 6A are much less electronegative, form anions much less often, and form hydrides that exhibit no H bonding.

1. *Types and properties of hydrides.* All the Group 6A(16) elements except O form foul-smelling, poisonous, gaseous hydrides (H_2E) when the metal sulfide, selenide, and so forth is treated with acid. For example,

$$FeSe(s) + 2HCl(aq) \longrightarrow H_2Se(g) + FeCl_2(aq)$$

In their bonding and thermal stability, 6A hydrides are similar to 5A hydrides:

- Only water can form H bonds, so it melts and boils at much higher temperatures than the other hydrides (see Figure 12.12).
- Bond angles drop from the nearly tetrahedral value for H_2O (104.5°) to around 90° for the larger 6A hydrides, suggesting that the central atom uses *un*hybridized *p* orbitals.
- E—H bond length increases and bond energy decreases down the group. One consequence is that the 6A hydrides are acids in water, as we discuss in Chapter 18.

2. *Types and properties of halides.* Except for O, the Group 6A(16) elements form a wide range of halides, whose structure and reactivity patterns depend on the *sizes of the central atom and the surrounding halogens (X):* With increasing size of E and X, E—X bond length increases, electron repulsions between lone pairs and X atoms weaken, and a greater number of stable halides form. Thus, S, Se, and Te form hexafluorides; Se, Te, and Po form tetrachlorides and tetrabromides; and Te and Po form tetraiodides.

Highlights of Oxygen Chemistry

Oxygen is the most abundant element on Earth's surface, occurring both as the free element and in innumerable oxides, silicates, carbonates, and phosphates, as well as in water. Virtually all free O_2 has a biological origin, having been formed for billions of years by photosynthetic algae and multicellular plants in an overall equation that looks simple but represents many steps:

$$nH_2O(l) + nCO_2(g) \xrightarrow{\text{light}} nO_2(g) + (CH_2O)_n \text{ (carbohydrates)}$$

The reverse process occurs during combustion and respiration.

Every element (except He, Ne, and Ar) forms at least one oxide, many by direct combination. For this reason, a useful way to classify element oxides is by their acid-base properties. The oxides of Group 6A(16) exhibit expected trends in acidity down the group, with SO_3 the most acidic and PoO_2 the most basic.

Highlights of Sulfur Chemistry

Like phosphorus, sulfur forms two common oxides and two oxoacids, one of which is essential to a wide variety of industries.

Sulfur Oxides Sulfur forms two important oxides.

1. *Sulfur dioxide* (SO_2) is a colorless, choking gas that forms when S, H_2S, or a metal sulfide burns in air:

$$2H_2S(g) + 3O_2(g) \longrightarrow 2H_2O(g) + 2SO_2(g)$$
$$4FeS_2(s) + 11O_2(g) \longrightarrow 2Fe_2O_3(s) + 8SO_2(g)$$

2. *Sulfur trioxide* (SO_3) is produced when sulfur dioxide reacts in oxygen. A catalyst (Chapter 16) must be used to speed up this very slow reaction. En route to the production of sulfuric acid, a vanadium(V) oxide catalyst is used:

$$SO_2(g) + \tfrac{1}{2}O_2(g) \xrightarrow{V_2O_5/K_2O \text{ catalyst}} SO_3(g)$$

These two sulfur oxides also form when sulfur impurities in coal burn and then oxidize further. In contact with rain, they form H_2SO_3 and H_2SO_4 and contribute to a major pollution problem that we discuss in Chapter 19.

Sulfur Oxoacids Sulfur forms two important oxoacids.

1. *Sulfurous acid* (H_2SO_3), formed when sulfur dioxide dissolves in water, exists in equilibrium with hydrated SO_2 rather than as stable H_2SO_3 molecules:

$$SO_2(aq) + H_2O(l) \rightleftharpoons [H_2SO_3(aq)] \rightleftharpoons H^+(aq) + HSO_3^-(aq)$$

Sulfurous acid is weak and has two acidic protons, forming the hydrogen sulfite (bisulfite, HSO_3^-) and sulfite (SO_3^{2-}) ions with strong base. Because the S in SO_3^{2-} is in the +4 state and is easily oxidized to the +6 state, sulfites are good reducing agents and are used to preserve foods and wine by eliminating undesirable products of air oxidation.

2. *Sulfuric acid* (H_2SO_4) is produced when SO_2 is oxidized catalytically to SO_3, which is then absorbed into concentrated H_2SO_4 and treated with additional H_2O:

$$SO_3 \text{ (in concentrated } H_2SO_4) + H_2O(l) \longrightarrow H_2SO_4(l)$$

With more than 40 million tons produced each year in the United States alone, H_2SO_4 ranks first among all industrial chemicals; it is vital to fertilizer production, metal, pigment, and textile processing, and soap and detergent manufacturing.

Due to extensive H-bonding, concentrated sulfuric acid is viscous, a colorless liquid that is 98% H_2SO_4 by mass. Like other strong acids, H_2SO_4 dissociates completely in water, forming the hydrogen sulfate (or bisulfate) ion, a much weaker acid:

$$H_2SO_4 \xrightarrow{-H^+} [HSO_4]^- \underset{}{\overset{-H^+}{\rightleftharpoons}} [SO_4]^{2-}$$

hydrogen sulfate ion — sulfate ion

14.8 • GROUP 7A(17): THE HALOGENS

As we move across the table, the last elements of great reactivity are found in Group 7A(17). The halogens begin with fluorine (F), the strongest electron "grabber" of all. Chlorine (Cl, by far the most important industrially), bromine (Br), and iodine (I) also form compounds with most elements, and even rare, radioactive astatine (At) is thought to be reactive [Group 7A(17) Family Portrait].

How the Halogens and the Alkali Metals Contrast Physically

Like the alkali metals at the other end of the periodic table, the halogens display regular trends in their physical properties. But the 7A trends are opposite to the 1A trends because of the differences in bonding. The alkali metals consist of atoms held together by metallic bonding, which *decreases* in strength as the atoms become larger down the group. The halogens, on the other hand, exist as diatomic molecules that interact through dispersion forces, which *increase* in strength as the atoms become larger and more easily polarized. Thus, at room temperature, F_2 is a very pale yellow gas, Cl_2 a yellow-green gas, Br_2 a brown-orange liquid, and I_2 a purple-black solid.

Why the Halogens Are So Reactive

The Group 7A(17) elements react with most metals and nonmetals to form many ionic and covalent compounds: metal and nonmetal halides, halogen oxides, and oxoacids. The main reason for halogen reactivity is the same as for alkali metal reactivity—an electron configuration one electron away from that of a noble gas. Whereas a 1A metal atom must *lose* one electron to attain a filled outer level, *a 7A nonmetal atom must gain one electron to fill its outer level.* It fills this outer level in one of two ways:

1. Gaining an electron from a metal atom, thus forming a negative ion as the metal forms a positive one.
2. Sharing an electron pair with a nonmetal atom, thus forming a covalent bond.

Electronegativity and Bond Properties The halogens' reactivity reflects the decrease in electronegativity down the group, but the exceptional reactivity of elemental F_2 is also related to the weakness of the F—F bond. This bond is short, but F is so small that lone pairs on one atom repel those on the other, weakening the bond

Group 7A(17): The Halogens

Family Portrait

KEY ATOMIC PROPERTIES, PHYSICAL PROPERTIES, AND REACTIONS

KEY
Atomic No.
Symbol
Atomic mass
Valence e⁻ configuration
(Common oxidation states)
9
F
19.00
2s²2p⁵
(−1)
Photograph not available
17
Cl
35.45
3s²3p⁵
(−1, +7, +5, +3, +1)
35
Br
79.90
4s²4p⁵
(−1, +7, +5, +3, +1)
53
I
126.9
5s²5p⁵
(−1, +7, +5, +3, +1)
85
At
(210)
6s²6p⁵
(−1)
Extremely rare, no sample available

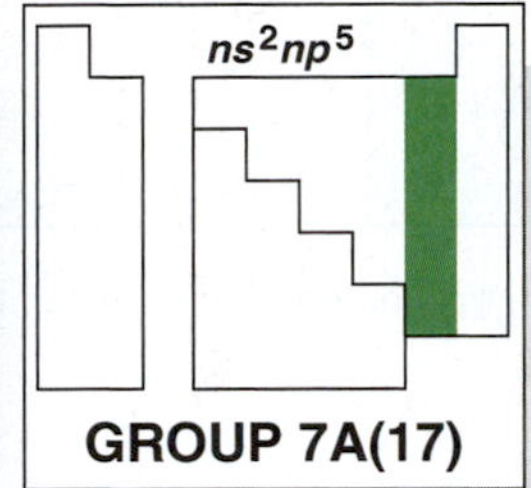

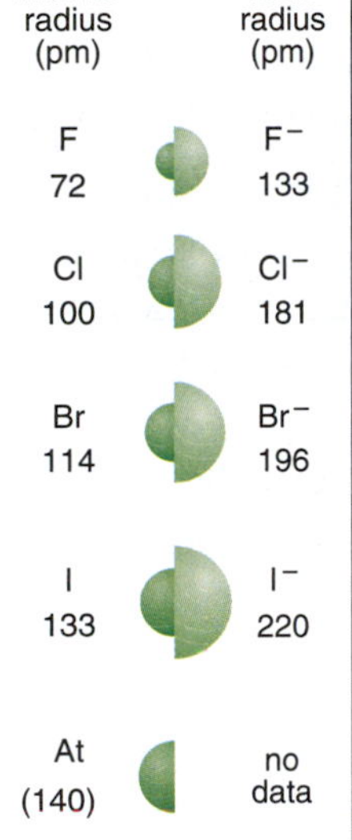

Atomic Properties

Group electron configuration is ns^2np^5; elements lack one electron to complete their outer level. The −1 oxidation state is the most common for all members. Except for F, the halogens exhibit all odd-numbered states (+7 through −1). Down the group, atomic and ionic sizes increase steadily, as IE and EN decrease.

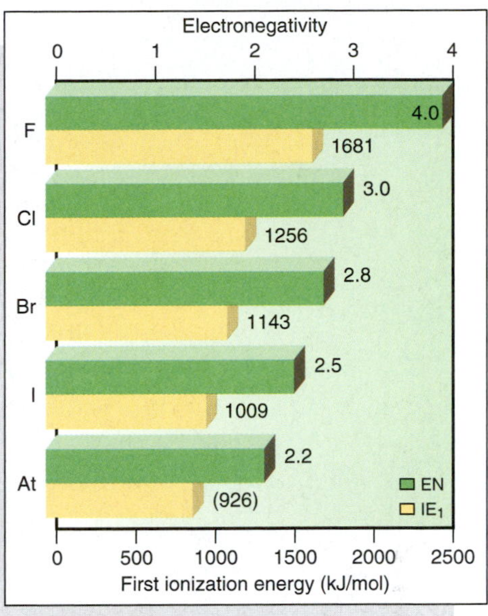

Physical Properties

Down the group, melting and boiling points increase smoothly as a result of stronger dispersion forces between larger molecules. The densities of the elements as liquids (at the given T) increase steadily with molar mass.

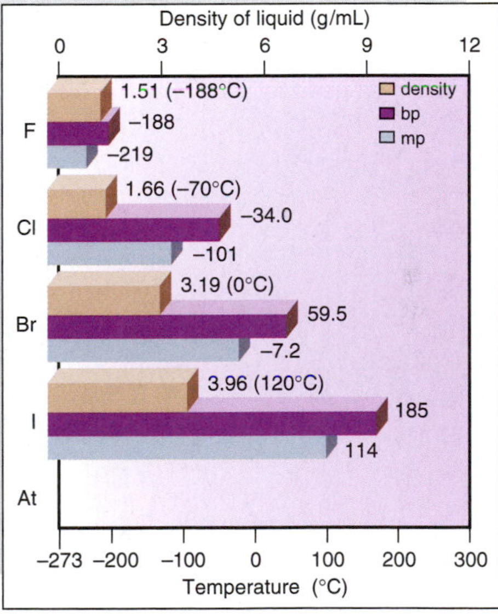

Reactions

1. The halogens (X_2) oxidize many metals and nonmetals. The reaction with hydrogen, although not used commercially for HX production (except for high-purity HCl), is characteristic of these strong oxidizing agents:

$$X_2 + H_2(g) \longrightarrow 2HX(g)$$

2. The halogens undergo disproportionation in water:

$$X_2 + H_2O(l) \rightleftharpoons HX(aq) + HXO(aq) \qquad (X = Cl, Br, I)$$

In aqueous base, the reaction goes to completion to form hypohalites and, at higher temperatures, halates, for example:

$$3Cl_2(g) + 6OH^-(aq) \xrightarrow{\Delta} ClO_3^-(aq) + 5Cl^-(aq) + 3H_2O(l)$$

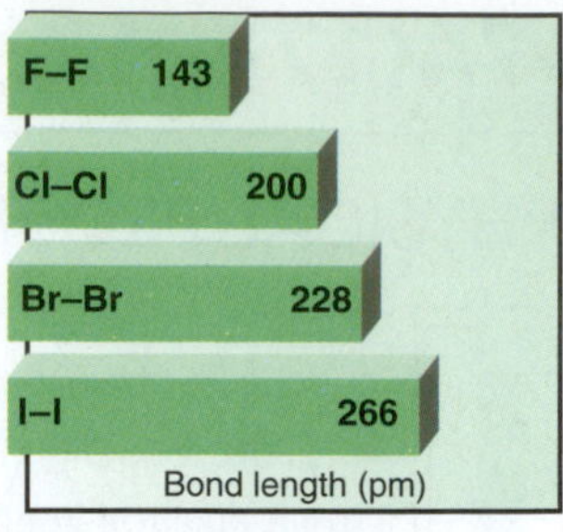

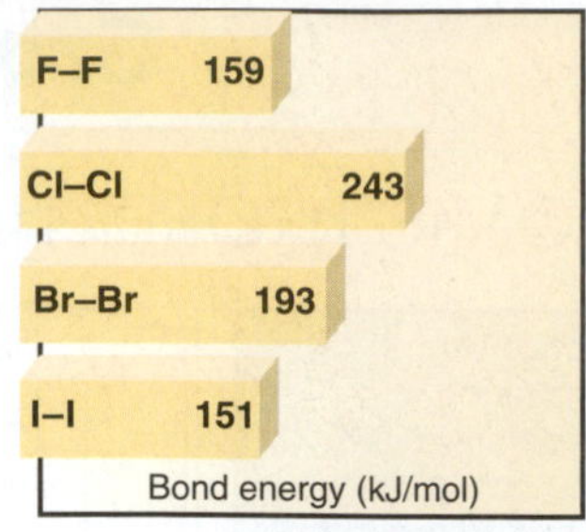

Figure 14.12 Bond lengths and bond energies of the halogens.

(Figure 14.12). As a result, F_2 reacts with every element except He, Ne, and Ar—in many cases, explosively.

Redox Behavior The halogens act as *oxidizing agents* in the majority of their reactions, and halogens higher in the group can oxidize halide ions lower down:

$$F_2(g) + 2X^-(aq) \longrightarrow 2F^-(aq) + X_2(aq) \qquad (X = Cl, Br, I)$$

Thus, the oxidizing ability of X_2 increases *up* the group: the higher the EN, the more strongly each X atom pulls electrons away. Similarly, the reducing ability of X^- increases *down* the group: the larger the ion, the more easily it gives up its electron (Figure 14.13).

Highlights of Halogen Chemistry

In this section, we examine the compounds the halogens form with hydrogen and with each other, as well as their oxides, oxoanions, and oxoacids.

The Hydrogen Halides The halogens form gaseous hydrogen halides (HX) through direct combination with H_2 or through the action of a concentrated acid on the metal halide. Commercially, HCl forms as a byproduct during the chlorination of hydrocarbons to form useful materials, such as poly(vinyl chloride), or PVC, a polymer used commonly for plumbing pipes.

In water, gaseous HX molecules form a *hydrohalic acid*. Only HF, with its relatively short, strong bond, forms a weak acid:

$$HF(g) + H_2O(l) \rightleftharpoons H_3O^+(aq) + F^-(aq)$$

The other hydrohalic acids dissociate completely to form the stoichiometric amount of H_3O^+ ions:

$$HBr(g) + H_2O(l) \longrightarrow H_3O^+(aq) + Br^-(aq)$$

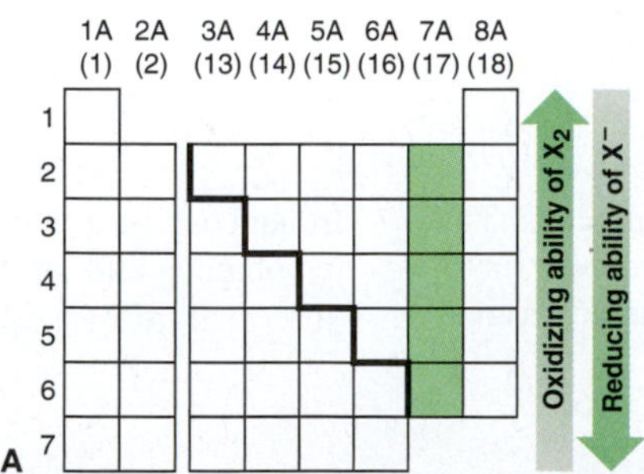

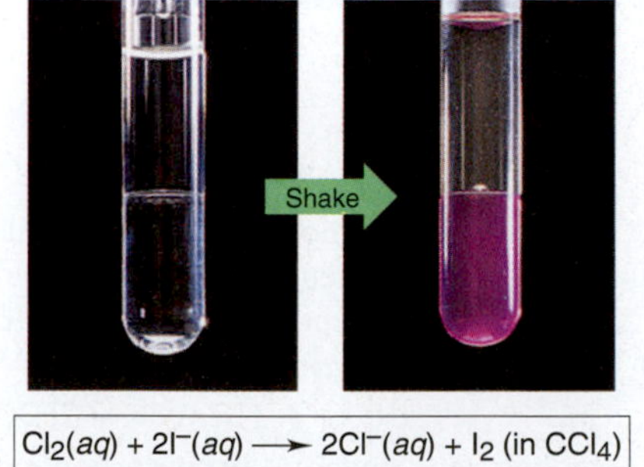

Figure 14.13 **The relative oxidizing ability of the halogens.** **A,** Halogen redox behavior is based on atomic properties such as electron affinity, ionic charge density, and electronegativity. A halogen (X_2) higher in the group can oxidize a halide ion (X^-) lower down. **B,** As an example, when aqueous Cl_2 is added to a solution of I^- *(top layer)*, it oxidizes the I^- to I_2, which dissolves in the CCl_4 solvent *(bottom layer)* to give a purple solution.

(We saw reactions similar to these in Chapter 4. They involve *transfer* of a proton from acid to H_2O and can be considered a type of acid-base reaction. We discuss such reactions further in Chapter 18.)

Interhalogen Compounds: The "Halogen Halides" Halogens react exothermically with one another to form many *interhalogen compounds.* The simplest are diatomic molecules, such as ClF or BrCl. Every binary combination of the four common halogens is known. The more electronegative halogen is in the −1 oxidation state, and the less electronegative is in the +1 state. Interhalogens of general formula XY_n (n = 3, 5, 7) form through a variety of reactions, including direct reaction of the elements. In every case, the central atom has the lower *electronegativity* and a positive oxidation state.

Halogen Oxides, Oxoacids, and Oxoanions The Group 7A(17) elements form many oxides that are *powerful oxidizing agents and acids in water.* Dichlorine monoxide (Cl_2O), chlorine dioxide (ClO_2), and dichlorine heptoxide (Cl_2O_7) are important examples.

The halogen oxoacids and oxoanions are produced by reactions of the halogens and their oxides with water. Most of the oxoacids are stable only in solution. Table 14.3 shows ball-and-stick models of the acids in which each atom has its lowest formal charge; note the formulas, which emphasize that H is bonded to O. The hypohalites (XO^-), halites (XO_2^-), and halates (XO_3^-) are oxidizing agents formed by aqueous disproportionation reactions [see the Group 7A(17) Family Portrait, reaction 2]. You may have heated solid alkali chlorates in the laboratory to form small amounts of O_2:

$$2MClO_3(s) \xrightarrow{\Delta} 2MCl(s) + 3O_2(g)$$

Potassium chlorate is the oxidizer in "safety" matches.

Several perhalates (XO_4^-) are also strong oxidizing agents. Ammonium perchlorate, prepared from sodium perchlorate, was the oxidizing agent for the aluminum powder in the solid-fuel booster rockets of the space shuttles; each launch used more than 700 tons of NH_4ClO_4:

$$10Al(s) + 6NH_4ClO_4(s) \longrightarrow 4Al_2O_3(s) + 12H_2O(g) + 3N_2(g) + 2AlCl_3(g)$$

The relative strengths of the halogen oxoacids depend on the eletronegativity and the oxidation state of the halogen. We consider these factors quantitatively in Chapter 18.

Table 14.3 The Known Halogen Oxoacids*

Central Atom	Hypohalous Acid (HOX)	Halous Acid (HOXO)	Halic Acid ($HOXO_2$)	Perhalic Acid ($HOXO_3$)
Fluorine	HOF	—	—	—
Chlorine	HOCl	HOClO	$HOClO_2$	$HOClO_3$
Bromine	HOBr	(HOBrO)?	$HOBrO_2$	$HOBrO_3$
Iodine	HOI	—	$HOIO_2$	$HOIO_3$, $(HO)_5IO$
Oxoanion	Hypohalite	Halite	Halate	Perhalate

*Lone pairs are shown only on the halogen atom, and each atom has its lowest formal charge.

14.9 • GROUP 8A(18): THE NOBLE GASES

The Group 8A(18) elements are helium (He), the second most abundant element in the universe; neon (Ne); argon (Ar); krypton (Kr); xenon (Xe); and radioactive radon (Rn). Only the last three form compounds [Group 8A(18) Family Portrait]. The noble gases make up about 1% by volume of the atmosphere, primarily due to the abundance of Ar.

Physical Properties

Lying at the far right side of the periodic table, the Group 8A(18) elements consist of individual atoms with filled outer levels and the smallest radii in their periods: even Li, the smallest alkali metal (152 pm), is bigger than Rn, the largest noble gas (140 pm). These elements come as close to behaving as ideal gases as any substance. Only at very low temperatures do they condense and solidify. In fact, He requires an increase in pressure to solidify: to 25 atm at −272.2°C. Weak dispersion forces hold these elements in condensed states, with melting and boiling points that increase, as expected, with molar mass.

Why the Noble Gases Can Form Compounds

After their discovery in the late 19th century, the Group 8A(18) elements had been considered, and even formerly named, the "inert" gases. Atomic theory and, more important, all experiments had supported this idea. Then, in 1962, all this changed when the first noble gas compound was prepared.

The discovery of noble gas reactivity is a classic example of clear thinking in the face of an unexpected event. At the time, a young inorganic chemist named Neil Bartlett was studying platinum fluorides. When he accidentally exposed PtF_6 to air, its deep-red color lightened slightly, and analysis showed that the PtF_6 had oxidized O_2 to form the ionic compound $[O_2]^+[PtF_6]^-$. Knowing that the ionization energy of the oxygen molecule ($O_2 \longrightarrow O_2^+ + e^-$; IE = 1175 kJ/mol) is very close to IE_1 of xenon (1170 kJ/mol), Bartlett reasoned that PtF_6 might also be able to oxidize xenon. He prepared $XePtF_6$, an orange-yellow solid and, within a few months, XeF_2 and XeF_4. In addition to its +2 and +4 oxidation states, Xe has the +6 state in several compounds, such as XeF_6, and the +8 state in the unstable oxide, XeO_4. A few compounds of Kr and Rn have also been made.

CHAPTER REVIEW GUIDE

The following sections provide many aids to help you study this chapter. (Numbers in parentheses refer to pages, unless noted otherwise.)

Learning Objectives

These are concepts and skills to review after studying this chapter.

Related section (§), sample problem (SP), and upcoming end-of-chapter problem (EP) numbers are listed in parentheses.

1. Compare hydrogen with alkali metals and halogens, and distinguish saltlike from covalent hydrides (§14.1) (EPs 14.1–14.5)
2. Discuss key features of Group 1A(1), and understand how the ns^1 configuration explains physical and chemical properties (§14.2) (EPs 14.7–14.16)
3. Understand the anomalous behaviors of the Period 2 elements within their groups (§14.2) (EPs 14.6, 14.9, 14.10, 14.22)
4. Discuss key features of Group 2A(2), and understand how the ns^2 configuration explains differences between Groups 1A(1) and 2A(2) (§14.3) (EPs 14.17, 14.19–14.21)
5. Describe the three main diagonal relationships among main-group elements (§14.3) (EPs 14.18, 14.45)

Group 8A(18): The Noble Gases

KEY ATOMIC AND PHYSICAL PROPERTIES

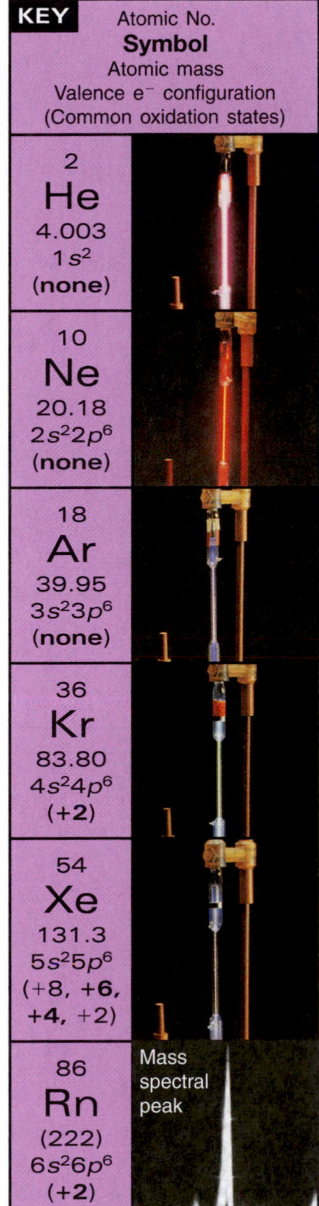

KEY
Atomic No.
Symbol
Atomic mass
Valence e⁻ configuration
(Common oxidation states)
2
He
4.003
$1s^2$
(none)
10
Ne
20.18
$2s^22p^6$
(none)
18
Ar
39.95
$3s^23p^6$
(none)
36
Kr
83.80
$4s^24p^6$
(+2)
54
Xe
131.3
$5s^25p^6$
(+8, +6, +4, +2)
86
Rn
(222)
$6s^26p^6$
(+2)
Mass spectral peak

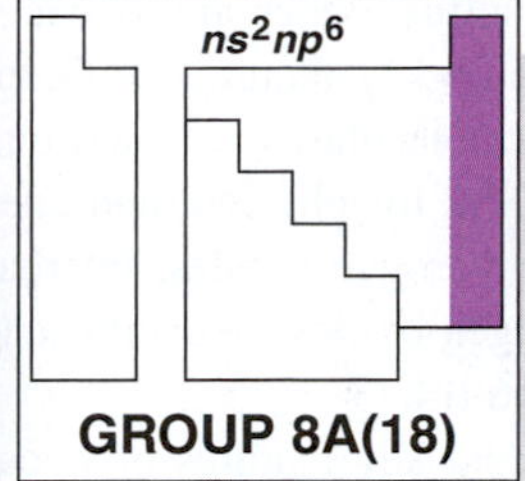

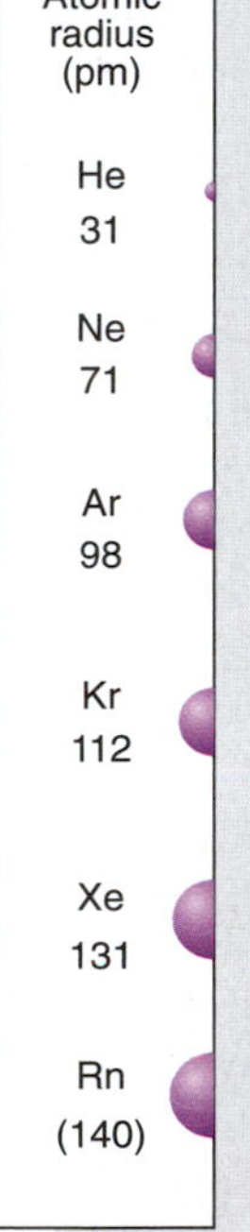

Atomic Properties

Group electron configuration is $1s^2$ for He and ns^2np^6 for the others. The valence shell is filled. Only Kr, Xe, and Rn are known to form compounds. The more reactive Xe exhibits all even oxidation states (+2 to +8). This group contains the smallest atoms with the highest IEs in their periods. Down the group, atomic size increases and IE decreases steadily. (EN values are given only for Kr and Xe.)

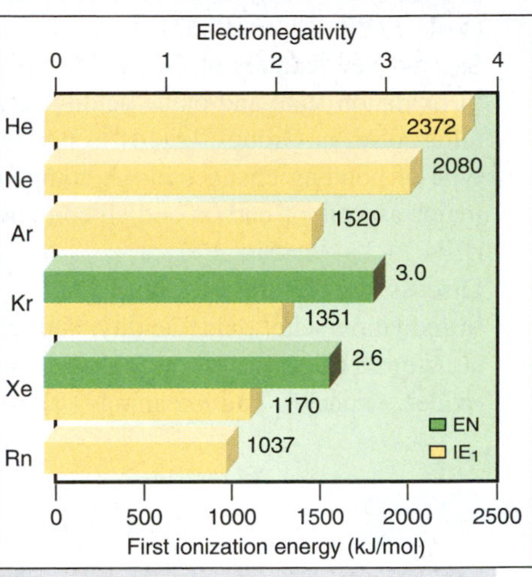

Physical Properties

Melting and boiling points of these gaseous elements are extremely low but increase down the group because of stronger dispersion forces. Note the extremely small liquid ranges. Densities (at STP) increase steadily, as expected.

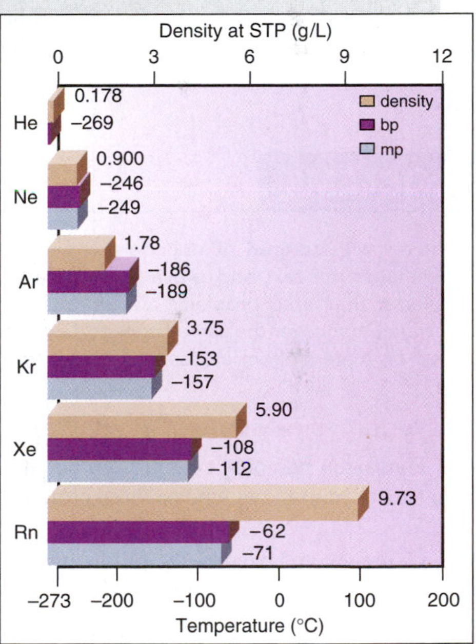

Learning Objectives continued

6. Discuss key features of Group 3A(13), especially patterns in oxidation state and oxide acidity; understand how the presence of *d* and *f* electrons affects group properties (§14.4) (EPs 14.23–14.32)
7. Discuss key features of Group 4A(14), especially patterns in oxidation state and oxide acidity; give examples of allotropism in Groups 4A(14) to 6A(16); describe how the type of bonding changes in 4A compounds; and describe major aspects of carbon and silicon chemistry (§14.5) (EPs 14.33–14.45)
8. Discuss key features of Group 5A(15), especially patterns in oxidation state, oxide acidity, and hydride and halide structures; and describe the nitrogen and phosphorus oxides, oxoacids, and oxoanions (§14.6) (EPs 14.46–14.56)
9. Discuss key features of Group 6A(16); compare the patterns in oxidation state, oxide acidity, and hydride and halide structures with those of Group 5A(15); and describe the sulfur oxides and oxoacids (§14.7) (EPs 14.57–14.65)
10. Discuss key features of Group 7A(17), understand how intermolecular forces and outer electron configuration account for physical and chemical properties, and describe the hydrogen halides, interhalogen compounds, and halogen oxides, oxoacids, and oxoanions (§14.8) (EPs 14.66–14.73)
11. Discuss key features of Group 8A(18), understand how intermolecular forces and a filled outer shell account for physical and chemical properties, and give examples of noble gas compounds (§14.9) (EPs 14.74–14.76)

Key Terms

These important terms appear in boldface in the chapter and are defined again in the Glossary.

Section 14.3
diagonal relationship (432)

Section 14.5
allotrope (434)
silicate (438)
silicone (438)

Section 14.6
disproportionation reaction (442)
dehydration-condensation reaction (444)

PROBLEMS

Problems with **colored** numbers are answered in Appendix E. Sections match the text and provide the numbers of relevant sample problems. Bracketed problems are grouped in pairs (indicated by a short rule) that cover the same concept. Comprehensive Problems are based on material from any section or previous chapter.

Hydrogen, the Simplest Atom

14.1 Hydrogen has only one proton, but its IE_1 is much greater than that of lithium, which has three protons. Explain.

14.2 Complete and balance the following equations:
(a) An active metal reacting with acid,

$$Al(s) + HCl(aq) \longrightarrow$$

(b) A saltlike (alkali metal) hydride reacting with water,

$$LiH(s) + H_2O(l) \longrightarrow$$

14.3 Complete and balance the following equations:
(a) A saltlike (alkaline earth metal) hydride reacting with water,

$$CaH_2(s) + H_2O(l) \longrightarrow$$

(b) Reduction of a metal halide by hydrogen to form a metal,

$$PdCl_2(aq) + H_2(g) \longrightarrow$$

14.4 Compounds such as $NaBH_4$, $Al(BH_4)_3$, and $LiAlH_4$ are complex hydrides used as reducing agents in many syntheses.
(a) Give the oxidation state of each element in these compounds.
(b) Write a Lewis structure for the polyatomic anion in $NaBH_4$, and predict its shape.

14.5 Unlike the F^- ion, which has an ionic radius close to 133 pm in all alkali metal fluorides, the ionic radius of H^- varies from 137 pm in LiH to 152 pm in CsH. Suggest an explanation for the large variability in the size of H^- but not F^-.

Group 1A(1): The Alkali Metals

14.6 Lithium salts are often much less soluble in water than the corresponding salts of other alkali metals. For example, at 18°C, the concentration of a saturated LiF solution is 1.0×10^{-2} *M*, whereas that of a saturated KF solution is 1.6 *M*. How would you explain this behavior?

14.7 The alkali metals play virtually the same general chemical role in all their reactions. (a) What is this role? (b) How is it based on atomic properties? (c) Using sodium, write two balanced equations that illustrate this role.

14.8 How do atomic properties account for the low densities of the Group 1A(1) elements?

14.9 How does the maximum oxidation number vary across a period in the main groups? Is the pattern in Period 2 different?

14.10 What correlation, if any, exists for the Period 2 elements between group number and the number of covalent bonds the element typically forms? How is the correlation different for elements in Periods 3 to 6?

14.11 Each of the following properties shows a regular trend in Group 1A(1). Predict whether each increases or decreases *down* the group: (a) density; (b) ionic size; (c) E—E bond energy; (d) IE_1; (e) magnitude of ΔH_{hydr} of E^+ ion.

14.12 Each of the following properties shows a regular trend in Group 1A(1). Predict whether each increases or decreases *up* the group: (a) melting point; (b) E—E bond length; (c) hardness; (d) molar volume; (e) lattice energy of EBr.

14.13 Write a balanced equation for the formation from its elements of sodium peroxide, an industrial bleach.

14.14 Write a balanced equation for the formation of rubidium bromide through a reaction of a strong acid and a strong base.

14.15 Although the alkali metal halides can be prepared directly from the elements, the far less expensive industrial route is treatment of the carbonate or hydroxide with aqueous hydrohalic acid (HX) followed by recrystallization. Balance the reaction between potassium carbonate and aqueous hydriodic acid.

14.16 Lithium forms several useful organolithium compounds. Calculate the mass percent of Li in the following:
(a) Lithium stearate ($C_{17}H_{35}COOLi$), a water-resistant grease used in cars because it does not harden at cold temperatures
(b) Butyllithium (LiC_4H_9), a reagent in organic syntheses

Group 2A(2): The Alkaline Earth Metals

14.17 How do Groups 1A(1) and 2A(2) compare with respect to reaction of the metals with water?

14.18 Alkaline earth metals are involved in two key diagonal relationships in the periodic table.
(a) Give the two pairs of elements in these diagonal relationships.
(b) For each pair, cite two similarities that demonstrate the relationship.
(c) Using one pair, suggest why the members are so similar in behavior. (*Hint:* Consider ion size and/or charge.)

14.19 The melting points of alkaline earth metals are many times higher than those of the alkali metals. Explain this difference on the basis of atomic properties. Name three other physical properties for which Group 2A(2) metals have higher values than the corresponding 1A(1) metals.

14.20 Write a balanced equation for each reaction:
(a) "Slaking" of lime (treatment with water)
(b) Combustion of calcium in air

14.21 Write a balanced equation for each reaction:
(a) Thermal decomposition of witherite (barium carbonate)
(b) Neutralization of stomach acid (HCl) by milk of magnesia (magnesium hydroxide)

14.22 In some reactions, Be behaves like a typical alkaline earth metal; in others, it does not. Complete and balance the following:
(a) $BeO(s) + H_2O(l) \longrightarrow$
(b) $BeCl_2(l) + Cl^-(l;$ from molten NaCl$) \longrightarrow$
In which reaction does Be behave like the other Group 2A(2) members?

Group 3A(13): The Boron Family

14.23 How do the transition metals in Period 4 affect the pattern of ionization energies in Group 3A(13)? How does this pattern compare with that in Group 3B(3)?

14.24 How do the acidities of aqueous solutions of Tl_2O and Tl_2O_3 compare with each other? Explain.

14.25 Despite the expected decrease in atomic size, there is an unexpected drop in IE_1 values between Groups 2A(2) and 3A(13) in Periods 2 through 4. Explain this pattern in terms of electron configurations and orbital energies.

14.26 Rank the following oxides in order of increasing aqueous *acidity:* Ga_2O_3, Al_2O_3, In_2O_3.

14.27 Rank the following hydroxides in order of increasing aqueous *basicity:* $Al(OH)_3$, $B(OH)_3$, $In(OH)_3$.

14.28 Thallium forms the compound TlI_3. What is the apparent oxidation state of Tl in this compound? Given that the anion is I_3^-, what is the actual oxidation state of Tl? Draw the shape of the anion, giving its VSEPR class and bond angles. Propose a reason why the compound does not exist as $(Tl^{3+})(I^-)_3$.

14.29 Very stable dihalides of the Group 3A(13) metals are known. What is the apparent oxidation state of Ga in $GaCl_2$? Given that $GaCl_2$ consists of a Ga^+ cation and a $GaCl_4^-$ anion, what are the actual oxidation states of Ga? Draw the shape of the anion, giving its VSEPR class and bond angles.

14.30 Give the name and symbol or formula of a Group 3A(13) element or compound that fits each description or use:
(a) Largest temperature range for liquid state of an element
(b) Metal protected from oxidation by adherent oxide coat

14.31 Indium (In) reacts with HCl to form a diamagnetic solid with the formula $InCl_2$.
(a) Write condensed electron configurations for In, In^+, In^{2+}, and In^{3+}.
(b) Which of these species is (are) diamagnetic and which paramagnetic?
(c) What is the apparent oxidation state of In in $InCl_2$?
(d) Given your answers to parts (b) and (c), explain how $InCl_2$ can be diamagnetic.

14.32 Boric acid, $B(OH)_3$ (or H_3BO_3), does not lose a proton in water, but rather bonds to the O atom of an H_2O molecule, which then releases an H^+ ion to form the $B(OH)_4^-$ ion. Use VESPR theory to draw structures, with ideal bond angles, for boric acid and the anion it forms when it reacts with water.

Group 4A(14): The Carbon Family

14.33 How do the physical properties of a network covalent solid and a molecular covalent solid differ? Why?

14.34 How does the basicity of SnO_2 in water compare with that of CO_2? Explain.

14.35 Nearly every compound of silicon has the element in the +4 oxidation state. In contrast, most compounds of lead have the element in the +2 state.
(a) What general observation do these facts illustrate?
(b) Explain in terms of atomic and molecular properties.
(c) Give an analogous example from Group 3A(13).

14.36 The sum of IE_1 through IE_4 for Group 4A(14) elements shows a decrease from C to Si, a slight increase from Si to Ge, a decrease from Ge to Sn, and an increase from Sn to Pb.
(a) What is the expected trend for IEs down a group?
(b) Suggest a reason for the deviations in Group 4A(14).
(c) Which group might show even greater deviations?

14.37 Give explanations for the large drops in melting point from C to Si and from Ge to Sn.

14.38 What is an allotrope? Name a Group 4A(14) element that exhibits allotropism, and identify two of its allotropes.

14.39 Even though EN values vary relatively little down Group 4A(14), the elements change from nonmetal to metal. Explain.

14.40 How do atomic properties account for the enormous number of carbon compounds? Why don't other Group 4A(14) elements behave similarly?

14.41 Draw a Lewis structure for each species:
(a) The cyclic silicate ion $Si_4O_{12}^{8-}$
(b) A cyclic hydrocarbon with formula C_4H_8

14.42 Draw a Lewis structure for each species:
(a) The cyclic silicate ion $Si_6O_{18}^{12-}$
(b) A cyclic hydrocarbon with formula C_6H_{12}

14.43 Zeolite A, $Na_{12}[(AlO_2)_{12}(SiO_2)_{12}]\cdot 27H_2O$, is used to soften water by replacing Ca^{2+} and Mg^{2+} with Na^+. Hard water from a certain source is 4.5×10^{-3} *M* Ca^{2+} and 9.2×10^{-4} *M* Mg^{2+}, and a pipe delivers 25,000 L of this hard water per day. What mass (in kg) of zeolite A is needed to soften a week's supply of the water? (Assume zeolite A loses its capacity to exchange ions when 85 mol % of its Na^+ has been lost.)

14.44 Give the name and symbol or formula of a Group 4A(14) element or compound that fits each description or use:
(a) Hardest known natural substance
(b) Medicinal antacid
(c) Atmospheric gas implicated in climate change
(d) Gas that binds to Fe(II) in blood
(e) Element used in the manufacture of computer chips

14.45 One similarity between B and Si is the explosive combustion of their hydrides in air. Write balanced equations for the combustion of B_2H_6 and of Si_4H_{10}.

Group 5A(15): The Nitrogen Family

14.46 Which Group 5A(15) elements form trihalides? Pentahalides? Explain.

14.47 As you move down Group 5A(15), the melting points of the elements increase and then decrease. Explain.

14.48 (a) What is the range of oxidation states shown by the elements of Group 5A(15) as you move down the group?
(b) How does this range illustrate the general rule for the range of oxidation states in the groups on the right side of the periodic table?

14.49 (a) How does the type of bonding in element oxides correlate with the electronegativity of the elements?
(b) How does the acid-base behavior of element oxides correlate with the electronegativity of the elements?

14.50 (a) How does the metallic character of an element correlate with the *acidity* of its oxide?
(b) What trends, if any, exist in oxide *basicity* across a period and down a group?

14.51 Rank the following oxides in order of increasing acidity in water: Sb_2O_3, Bi_2O_3, P_4O_{10}, Sb_2O_5.

14.52 Complete and balance the following:
(a) $As(s) + \text{excess } O_2(g) \longrightarrow$
(b) $Bi(s) + \text{excess } F_2(g) \longrightarrow$
(c) $Ca_3As_2(s) + H_2O(l) \longrightarrow$

14.53 Complete and balance the following:
(a) $\text{Excess } Sb(s) + Br_2(l) \longrightarrow$
(b) $HNO_3(aq) + MgCO_3(s) \longrightarrow$
(c) $K_2HPO_4(s) \xrightarrow{\Delta}$

14.54 The pentafluorides of the larger members of Group 5A(15) have been prepared, but N can have only eight electrons. A claim has been made that, at low temperatures, a compound with the empirical formula NF_5 forms. Draw a possible Lewis structure for this compound. (*Hint:* NF_5 is ionic.)

14.55 Give the name and symbol or formula of a Group 5A(15) element or compound that fits each description or use:
(a) Hydride that exhibits hydrogen bonding
(b) Oxide used as a drying agent
(c) Odd-electron molecule (two examples)
(d) Compound used as an additive in soft drinks

14.56 Nitrous oxide (N_2O), the "laughing gas" used as an anesthetic by dentists, is made by thermal decomposition of solid NH_4NO_3. Write a balanced equation for this reaction. What are the oxidation states of N in NH_4NO_3 and in N_2O?

Group 6A(16): The Oxygen Family

14.57 Rank the following in order of increasing electrical conductivity, and explain your ranking: Po, S, Se.

14.58 The oxygen and nitrogen families have some obvious similarities and differences.
(a) State two general physical similarities between Group 5A(15) and 6A(16) elements.
(b) State two general chemical similarities between Group 5A(15) and 6A(16) elements.
(c) State two chemical similarities between P and S.
(d) State two physical similarities between N and O.
(e) State two chemical differences between N and O.

14.59 A molecular property of the Group 6A(16) hydrides changes abruptly down the group. This change has been explained in terms of a change in orbital hybridization.
(a) Between what periods does the change occur?
(b) What is the change in the molecular property?
(c) What is the change in hybridization?
(d) What other group displays a similar change?

14.60 Complete and balance the following:
(a) $NaHSO_4(aq) + NaOH(aq) \longrightarrow$
(b) $S_8(s) + \text{excess } F_2(g) \longrightarrow$
(c) $FeS(s) + HCl(aq) \longrightarrow$
(d) $Te(s) + I_2(s) \longrightarrow$

14.61 Complete and balance the following:
(a) $H_2S(g) + O_2(g) \longrightarrow$
(b) $SO_3(g) + H_2O(l) \longrightarrow$
(c) $SF_4(g) + H_2O(l) \longrightarrow$
(d) $Al_2Se_3(s) + H_2O(l) \longrightarrow$

14.62 Is each oxide basic, acidic, or amphoteric in water: (a) SeO_2; (b) N_2O_3; (c) K_2O; (d) BeO; (e) BaO?

14.63 Is each oxide basic, acidic, or amphoteric in water: (a) MgO; (b) N_2O_5; (c) CaO; (d) CO_2; (e) TeO_2?

14.64 Give the name and symbol or formula of a Group 6A(16) element or compound that fits each description or use:
(a) Unstable allotrope of oxygen
(b) Oxide having sulfur with the same O.N. as in sulfuric acid
(c) Air pollutant produced by burning sulfur-containing coal
(d) Compound used in solution in the photographic process

14.65 Disulfur decafluoride is intermediate in reactivity between SF_4 and SF_6. It disproportionates at 150°C to these monosulfur fluorides. Write a balanced equation for this reaction, and give the oxidation state of S in each compound.

Group 7A(17): The Halogens

14.66 Iodine monochloride and elemental bromine have nearly the same molar mass and liquid density but very different boiling points.
(a) What molecular property is primarily responsible for this difference in boiling point? What atomic property gives rise to it? Explain.
(b) Which substance has a higher boiling point? Why?

14.67 (a) Give the physical state and color of the common halogens at STP. (b) Explain the change in physical state down Group 7A(17) in terms of molecular properties.

14.68 (a) What are the common oxidation states of the halogens?
(b) Give an explanation based on electron configuration for the range and values of the oxidation states of chlorine.
(c) Why is fluorine an exception to the pattern of oxidation states found for the other group members?

14.69 Select the stronger bond in each pair:
(a) Cl—Cl or Br—Br (b) Br—Br or I—I
(c) F—F or Cl—Cl. Why doesn't the F—F bond strength follow the group trend?

14.70 A halogen (X_2) disproportionates in base in several steps to X^- and XO_3^-. Write the overall equation for the disproportionation of Br_2 to Br^- and BrO_3^-.

14.71 Complete and balance the following equations. If no reaction occurs, write NR:
(a) $I_2(s) + H_2O(l) \longrightarrow$
(b) $Br_2(l) + I^-(aq) \longrightarrow$
(c) $CaF_2(s) + H_2SO_4(l) \longrightarrow$

14.72 Complete and balance the following equations. If no reaction occurs, write NR:
(a) $Cl_2(g) + I^-(aq) \longrightarrow$
(b) $Br_2(l) + Cl^-(aq) \longrightarrow$
(c) $ClF(g) + F_2(g) \longrightarrow$

14.73 An industrial chemist treats solid NaCl with concentrated H_2SO_4 and obtains gaseous HCl and $NaHSO_4$. When she substitutes solid NaI for NaCl, gaseous H_2S, solid I_2, and S_8 are obtained but no HI. (a) What type of reaction did the H_2SO_4 undergo with NaI? (b) Why does NaI, but not NaCl, cause this type of reaction? (c) To produce HI(*g*) by the reaction of NaI with an acid, how does the acid have to differ from sulfuric acid?

Group 8A(18): The Noble Gases

14.74 Which of the noble gases is the most abundant in the universe? In Earth's atmosphere?

14.75 Why do the noble gases have such low boiling points?

14.76 Explain why Xe and, to a limited extent, Kr and Rn form compounds, whereas He, Ne, and Ar do not.

Comprehensive Problems

14.77 Xenon tetrafluoride reacts with antimony pentafluoride to form the ionic complex $[XeF_3]^+[SbF_6]^-$. (a) Which depiction shows the molecular shapes of the reactants and product? (b) How, if at all, does the hybridization of xenon change in the reaction?

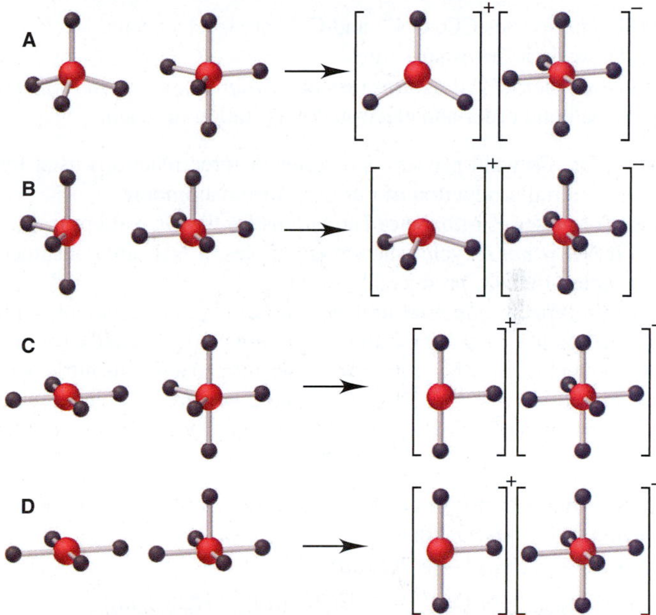

14.78 The interhalogen compound IF undergoes the reaction depicted below (I is purple and F is green):

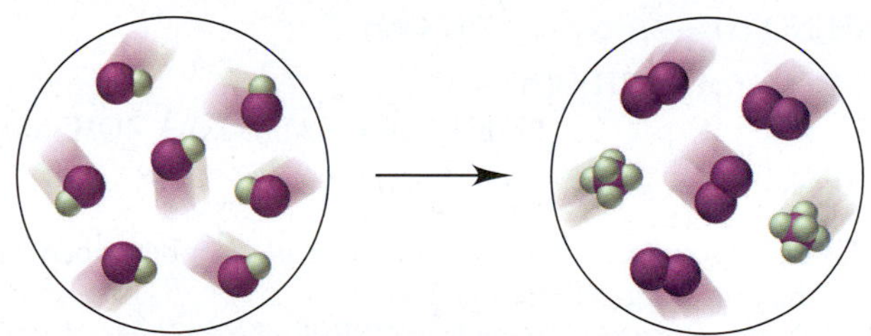

(a) Write the balanced equation. (b) Name the interhalogen product. (c) What type of reaction is shown? (d) If each molecule of IF represents 2.50×10^{-3} mol, what mass of each product forms?

14.79 The main reason alkali metal dihalides (MX_2) do *not* form is the high IE_2 of the metal. (a) Why is IE_2 so high for alkali metals? (b) The IE_2 for Cs is 2255 kJ/mol, low enough for CsF_2 to form exothermically ($\Delta H_f^\circ = -125$ kJ/mol). This compound cannot be synthesized, however, because CsF forms with a much greater release of heat ($\Delta H_f^\circ = -530$ kJ/mol). Thus, the breakdown of CsF_2 to CsF happens readily. Write the equation for this breakdown, and calculate the enthalpy of reaction per mole of CsF.

14.80 Semiconductors made from elements in Groups 3A(13) and 5A(15) are typically prepared by direct reaction of the elements at high temperature. An engineer treats 32.5 g of molten gallium with 20.4 L of white phosphorus vapor at 515 K and 195 kPa. If purification losses are 7.2% by mass, how many grams of gallium phosphide will be prepared?

14.81 Two substances with empirical formula HNO are hyponitrous acid ($\mathcal{M}$ = 62.04 g/mol) and nitroxyl ($\mathcal{M}$ = 31.02 g/mol).
(a) What is the molecular formula of each species?
(b) For each species, draw the Lewis structure having the lowest formal charges. (*Hint:* Hyponitrous acid has an N=N bond.)
(c) Predict the shape around the N atoms of each species.
(d) When hyponitrous acid loses two protons, it forms the hyponitrite ion. Draw *cis* and *trans* forms of this ion.

14.82 The species CO, CN^-, and C_2^{2-} are isoelectronic.
(a) Draw their Lewis structures.
(b) Draw their MO diagrams (assume 2*s*-2*p* mixing, as in N_2), and give the bond order and electron configuration for each.

14.83 The Ostwald process is a series of three reactions used for the industrial production of nitric acid from ammonia.
(a) Write a series of balanced equations for the Ostwald process.
(b) If NO is *not* recycled, how many moles of NH_3 are consumed per mole of HNO_3 produced?
(c) In a typical industrial unit, the process is very efficient, with a 96% yield for the first step. Assuming 100% yields for the subsequent steps, what volume of concentrated aqueous nitric acid (60.% by mass; *d* = 1.37 g/mL) can be prepared for each cubic meter of a gas mixture that is 90.% air and 10.% NH_3 by volume at the industrial conditions of 5.0 atm and 850.°C?

14.84 What is a disproportionation reaction, and which of the following fit the description?
(a) $I_2(s) + KI(aq) \longrightarrow KI_3(aq)$
(b) $2ClO_2(g) + H_2O(l) \longrightarrow HClO_3(aq) + HClO_2(aq)$
(c) $Cl_2(g) + 2NaOH(aq) \longrightarrow NaCl(aq) + NaClO(aq) + H_2O(l)$
(d) $NH_4NO_2(s) \longrightarrow N_2(g) + 2H_2O(g)$
(e) $3MnO_4^{2-}(aq) + 2H_2O(l) \longrightarrow 2MnO_4^-(aq) + MnO_2(s) + 4OH^-(aq)$
(f) $3AuCl(s) \longrightarrow AuCl_3(s) + 2Au(s)$

14.85 Which group(s) of the periodic table is (are) described by each of the following general statements?
(a) The elements form compounds of VSEPR class AX_3E.
(b) The free elements are strong oxidizing agents and form monatomic ions and oxoanions.
(c) The atoms form compounds by combining with two other atoms that donate one electron each.
(d) The free elements are strong reducing agents, show only one nonzero oxidation state, and form mainly ionic compounds.
(e) The elements can form stable compounds with only three bonds, but as a central atom, they can accept a pair of electrons from a fourth atom without expanding their valence shell.
(f) Only larger members of the group are chemically active.

14.86 Bromine monofluoride (BrF) disproportionates to bromine gas and bromine tri- and pentafluorides. Use the following to find ΔH°_{rxn} for the decomposition of BrF to its elements:

$$3BrF(g) \longrightarrow Br_2(g) + BrF_3(l) \quad \Delta H_{rxn} = -125.3 \text{ kJ}$$
$$5BrF(g) \longrightarrow 2Br_2(g) + BrF_5(l) \quad \Delta H_{rxn} = -166.1 \text{ kJ}$$
$$BrF_3(l) + F_2(g) \longrightarrow BrF_5(l) \quad \Delta H_{rxn} = -158.0 \text{ kJ}$$

14.87 In addition to Al_2Cl_6, aluminum forms other species with halide ions joining two aluminum atoms. One such species is the ion $Al_2Cl_7^-$. The ion is symmetrical, with a 180° Al—Cl—Al bond angle.
(a) What orbitals does Al use to bond with the Cl atoms?
(b) What is the shape around each Al?
(c) What is the hybridization of the central Cl?
(d) What do the shape and hybridization suggest about the presence of lone pairs of electrons on the central Cl?

14.88 Element E forms an oxide of general structure A and a chloride of general structure B, shown at right. For the anion EF_5^-, what is (a) the molecular shape; (b) the hybridization of E; (c) the O.N. of E?

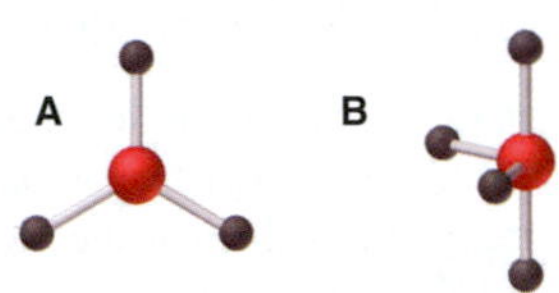

14.89 The bond angles in the nitrite ion, nitrogen dioxide, and the nitronium ion (NO_2^+) are 115°, 134°, and 180°, respectively. Explain these values using Lewis structures and VSEPR theory.

14.90 The triatomic molecular ion H_3^+ was first detected and characterized by J. J. Thomson using mass spectrometry. Use the bond energy of H_2 (432 kJ/mol) and the proton affinity of H_2 ($H_2 + H^+ \longrightarrow H_3^+$; $\Delta H = -337$ kJ/mol) to calculate the enthalpy of reaction for $H + H + H^+ \longrightarrow H_3^+$.

14.91 Copper(II) hydrogen arsenite ($CuHAsO_3$) is a green pigment once used in wallpaper. In damp conditions, mold metabolizes this compound to trimethylarsine [$(CH_3)_3As$], a highly toxic gas. (a) Calculate the mass percent of As in each compound. (b) How much $CuHAsO_3$ must react to reach a toxic level in a room that measures 12.35 m × 7.52 m × 2.98 m (arsenic is toxic at 0.50 mg/m³)?

Organic Compounds and the Atomic Properties of Carbon

Preparing an Organic Snack Except for the metal knife, everything in this scene—bread and wrapper, peanut butter, jelly, plate, and snacker—are composed of organic compounds. In this chapter, you'll see that all the properties of these compounds emerge from those of the carbon atom.

Outline

Key Principles
to focus on while studying this chapter

- Carbon's unusual ability to form single and multiple bonds to other carbons and to a few other nonmetals gives its compounds *structural complexity* and *chemical diversity*. The diversity arises from the presence of *functional groups*, specific combinations of bonded atoms that react in characteristic ways. ***(Section 15.1)***
- *Hydrocarbons* (compounds containing only C and H) are classified as *alkanes* (all single bonds), *alkenes* (at least one C=C bond), *alkynes* (at least one C≡C bond), and *aromatic* hydrocarbons (at least one planar ring with delocalized π electrons). The C=C and C≡C bonds are functional groups. ***(Section 15.2)***
- *Constitutional (structural) isomers* have different arrangements of atoms. *Stereoisomers* have the same atom arrangement but different spatial orientations. The two types of stereoisomers are *optical isomers*, mirror images that cannot be superimposed, and *geometric (cis-trans) isomers*, which have different orientations of groups around a C=C bond. ***(Section 15.2)***
- Three common types of organic reactions are *addition* (two atoms or groups are added and a C=C bond is converted to a C—C bond), *elimination* (two atoms or groups are removed and a C—C bond is converted to a C=C bond), and *substitution* (one atom or group replaces another). ***(Section 15.3)***
- *Functional groups* undergo characteristic reactions: groups with only single bonds (*alcohol, haloalkane,* and *amine*) undergo substitution or elimination; groups with double bonds (*alkene, aldehyde,* and *ketone*) and those with triple bonds (*alkyne* and *nitrile*) undergo addition; and groups with both single and double bonds (*carboxylic acid, ester,* and *amide*) undergo substitution. ***(Section 15.4)***
- *Polymers* are made by covalently linking many small repeat units (*monomers*). Particular monomers are used to give *synthetic polymers* desired properties. *Addition polymers* form through a *free-radical chain reaction* involving monomers with a C=C group. *Condensation polymers* consist of monomers with two functional groups that link together by *dehydration-condensation reactions*. ***(Section 15.5)***
- *Polysaccharides* (monomers, *sugars*), *proteins* (monomers, *amino acids*), and *nucleic acids* (monomers, *nucleotides*) are *natural polymers*. *DNA* occurs as a *double helix*, with bases in each strand H-bonded to specific bases in the other. The *base sequence* of an organism's DNA determines the *amino-acid sequences* of its proteins, which determine the proteins' structure and function. ***(Section 15.6)***

CONCEPTS & SKILLS TO REVIEW
before studying this chapter

- naming straight-chain alkanes (Section 2.8)
- constitutional isomerism (Section 3.2)
- ΔEN and bond polarity (Section 9.5)
- resonance structures (Section 10.1)
- VSEPR theory (Section 10.2)
- orbital hybridization (Section 11.1)
- σ and π bonding (Section 11.2)
- types of intermolecular forces (Section 12.3)
- properties of the Period 2 elements (Section 14.2)
- properties of the Group 4A(14) elements (Section 14.5)

Through delicately controlled mechanisms, a living cell oxidizes food for energy, maintains the concentrations of thousands of aqueous components, interacts continuously with its environment, synthesizes both simple and complex molecules, and even reproduces itself! For all our technological prowess, no human-made system even approaches the cell in its complexity and sheer elegance of function.

This amazing chemical machine consumes, creates, and consists largely of *organic compounds.* In addition, except for a few inorganic salts and ever-present water, nearly everything you put into or on your body—food, medicine, cosmetics, and clothing—consists of organic compounds. Organic fuels warm our homes, cook our meals, and power our vehicles. Major industries are devoted to producing organic compounds, including plastics, pharmaceuticals, and insecticides.

What *is* an organic compound? The dictionary definition is "a compound of carbon," but that definition is too general because it includes carbonates, cyanides, carbides, cyanates, and other carbon-containing ionic compounds that most chemists classify as inorganic. Here is a more specific definition: all **organic compounds** contain carbon, nearly always bonded to other carbons and hydrogen, and often to other elements.

In the early 19th century, organic compounds were usually obtained from living things, so they were thought to possess a spiritual "vital force" that made them impossible to synthesize and fundamentally different from inorganic compounds. Today, we know that *the same chemical principles govern organic and inorganic systems* because the behavior of a compound—no matter how marvelous—arises from the properties of its elements.

15.1 • THE SPECIAL NATURE OF CARBON AND THE CHARACTERISTICS OF ORGANIC MOLECULES

Although there is nothing mystical about organic molecules, their indispensable role in biology and industry leads us to ask if carbon has some extraordinary attributes that give it a special chemical "personality." Of course, each element has its own specific properties, but the atomic properties of carbon do give it bonding capabilities beyond those of any other element. This exceptional behavior leads to the two obvious characteristics of organic molecules—structural complexity and chemical diversity.

The Structural Complexity of Organic Molecules

Most organic molecules have more complex structures than most inorganic molecules. A quick review of carbon's atomic properties and bonding behavior shows why.

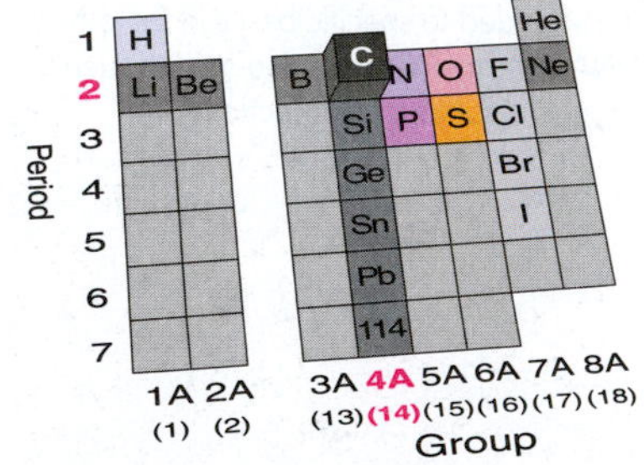

Figure 15.1 The position of carbon in the periodic table. Other elements common in organic compounds are H, N, O, P, S, and the halogens.

1. *Electron configuration, electronegativity, and covalent bonding.* Carbon's ground-state electron configuration of [He] $2s^22p^2$—four electrons more than He and four fewer than Ne—means that the formation of a carbon ion (C^{4+} or C^{4-}) takes much too much energy under ordinary conditions. Lying at the center of Period 2, carbon has an electronegativity (EN = 2.5) that is midway between that of the most metallic element (Li, EN = 1.0) and the most nonmetallic element (F, EN = 4.0) of Period 2 (Figure 15.1). Therefore, *carbon shares electrons to attain a filled outer (valence) level,* bonding covalently in all its elemental forms and compounds.

2. *Bond properties, catenation, and molecular shape.* The *number* and *strength* of carbon's bonds lead to the property of **catenation,** the ability to bond to itself, which allows it to form a multitude of chemically and thermally stable chain, ring, and branched compounds.

- Through the process of orbital hybridization (Section 11.1), carbon forms four bonds in virtually all its compounds, and they point in as many as four different directions.
- The small size of carbon allows close approach to another atom and thus greater orbital overlap, so carbon forms relatively short, strong bonds.
- The C—C bond is short enough to allow side-to-side overlap of half-filled, unhybridized *p* orbitals and the formation of multiple bonds, which restrict rotation of attached groups.

3. *Molecular stability.* Although silicon and several other elements also catenate, none can form chains as stable as those of carbon. Atomic and bonding properties confer three crucial differences between C and Si chains that explain why C chains are so stable and, therefore, so common:

- *Atomic size and bond strength.* As atomic size increases down Group 4A(14), bonds between identical atoms become longer and weaker. Thus, a C—C bond (347 kJ/mol) is much stronger than an Si—Si bond (226 kJ/mol).
- *Relative enthalpies of reaction.* A C—C bond (347 kJ/mol) and a C—O bond (358 kJ/mol) have nearly the same energy, so relatively little heat is released when a C chain reacts and one bond replaces the other. In contrast, an Si—O bond (368 kJ/mol) is much stronger than an Si—Si bond (226 kJ/mol), so a large quantity of heat is released when an Si chain reacts.
- *Orbitals available for reaction.* Unlike C, Si has low-energy *d* orbitals that can be attacked (occupied) by the lone pairs of incoming reactants. For example, ethane (CH_3—CH_3) is stable in water and does not react in air unless sparked, whereas disilane (SiH_3—SiH_3) breaks down in water and ignites spontaneously in air.

The Chemical Diversity of Organic Molecules

In addition to their complex geometries, organic compounds are noted for their sheer number and diverse chemical behavior. Several million organic compounds are known, and thousands more are discovered or synthesized each year. This incredible diversity is also founded on atomic and bonding behavior and is due to three interrelated factors, discussed below.

1. *Bonding to heteroatoms.* Many organic compounds contain **heteroatoms,** atoms other than C or H. The most common heteroatoms are N and O, but S, P, and the halogens often occur, and organic compounds with other elements are known as well. Figure 15.2 shows that 23 different molecular structures are possible from various arrangements of just four C atoms singly bonded to each other, one O atom (either singly or doubly bonded), and the necessary number of H atoms.

2. *Electron density and reactivity.* Most reactions start—that is, a new bond begins to form—*when a region of high electron density on one molecule meets a region of low electron density on another.* These regions may be due to the presence of a multiple bond or to the properties of carbon-heteroatom bonds. Consider the reactivities of four bonds commonly found in organic molecules:

- *The C—C bond.* When C is singly bonded to another C, as occurs in portions of nearly every organic molecule, the EN values are equal and the bond is nonpolar. Therefore, in general, *C—C bonds are unreactive.*
- *The C—H bond.* This bond, which also occurs in nearly every organic molecule, is short (109 pm) and very nearly nonpolar, with EN values of H (2.1) and C (2.5). Thus, *C—H bonds are largely unreactive.*
- *The C—O bond.* This bond, which occurs in many types of organic molecules, is highly polar ($\Delta EN = 1.0$), with the O end electron rich and the C end electron poor. As a result of this imbalance, *the C—O bond is reactive* (easy to break), and, given appropriate conditions, a reaction will occur there.
- *Bonds to other heteroatoms.* Even when a carbon-heteroatom bond has a small ΔEN, such as that for C—Br ($\Delta EN = 0.3$), or none at all, as for C—S ($\Delta EN = 0$), the heteroatoms are generally large, so their bonds to carbon are long, weak, and thus *reactive.*

3. *Importance of functional groups.* One of the most important ideas in organic chemistry is that of the **functional group,** a specific combination of bonded atoms that reacts in a *characteristic* way, no matter what molecule it occurs in. In nearly every case, *the reaction of an organic compound takes place at the functional group.* Functional groups vary from carbon-carbon multiple bonds to several combinations of carbon-heteroatom bonds, and each has its own pattern of reactivity. A particular bond may be a functional group itself or *part* of one or more functional groups. For

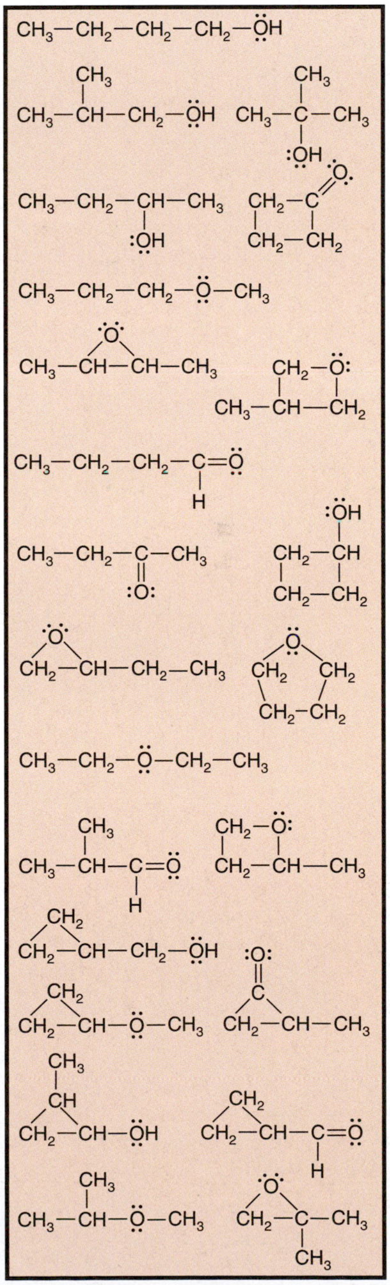

Figure 15.2 Heteroatoms and different bonding arrangements lead to great chemical diversity.

example, the C—O bond occurs in four functional groups. We discuss the reactivity of three of these groups in this chapter:

$-\overset{|}{\underset{|}{C}}-\ddot{\underset{..}{O}}-H$	$-\overset{:O:}{\overset{\|}{C}}-\ddot{\underset{..}{O}}-H$	$-\overset{:O:}{\overset{\|}{C}}-\ddot{\underset{..}{O}}-\overset{|}{\underset{|}{C}}-$
alcohol group	carboxylic acid group	ester group

Summary of Section 15.1

- Carbon's small size, intermediate electronegativity, four valence electrons, and ability to form multiple bonds result in the structural complexity of organic compounds.
- These factors lead to carbon's ability to catenate, which creates chains, branches, and rings of C atoms. Small size and absence of *d* orbitals in the valence level lead to strong, chemically resistant bonds that point in as many as four directions from each C.
- Carbon's ability to bond to many other elements, including O and N, creating polar bonds and resulting in greater reactivity, leads to the chemical diversity of organic compounds.
- Most organic compounds contain functional groups, specific combinations of bonded atoms that react in characteristic ways.

15.2 • THE STRUCTURES AND CLASSES OF HYDROCARBONS

A fanciful, anatomical analogy can be made between an organic molecule and an animal. The carbon-carbon bonds form the skeleton: the longest continual chain is the backbone, and any branches are the limbs. Covering the skeleton is a skin of hydrogen atoms, with functional groups protruding at specific locations, like chemical hands ready to grab an incoming reactant. In this section, we "dissect" one group of compounds down to their skeletons and see how to name and draw them.

Hydrocarbons, the simplest type of organic compound, are a large group of substances containing only H and C atoms. Some common fuels, such as natural gas and gasoline, are hydrocarbon mixtures. Some hydrocarbons, such as ethylene, acetylene, and benzene, are important *feedstocks,* precursor reactants used to make other compounds.

Carbon Skeletons and Hydrogen Skins

Let's begin by examining the possible bonding arrangements of C atoms only (leaving off the H atoms for now) in simple skeletons without multiple bonds or rings. To distinguish different skeletons, focus on the *arrangement* of C atoms (that is, the successive linkages of one to another) and keep in mind that *groups joined by a single (sigma) bond are relatively free to rotate* (Section 11.2).

Structures with one, two, or three carbons can be arranged in only one way. Whether you draw three C atoms in a line or with a bend, the arrangement is the same. Four C atoms, however, have two possible arrangements—a four-C chain or a three-C chain with a one-C branch at the central C:

C—C—C—C same as C—C—C(—C) C—C(—C)—C same as C—C(—C)—C

As the total number of C atoms increases, the number of different arrangements increases as well. Five C atoms have 3 possible arrangements (Figure 15.3A); 6 C

A C—C—C—C—C

B C=C—C—C—C

C—C=C—C—C

C

Figure 15.3 Some five-carbon skeletons.

atoms can be arranged in 5 ways, 7 C atoms in 9 ways, 10 C atoms in 75 ways, and 20 C atoms in more than 300,000 ways! If we include multiple bonds and rings, the number of arrangements increases further. For example, including one C=C bond in the five-C skeletons creates 5 more arrangements (Figure 15.3B), and including one ring creates 5 more (Figure 15.3C).

When determining the number of different skeletons for a given number of C atoms, remember that

- Each C atom can form a *maximum* of four single bonds, or two single and one double bond, or one single and one triple bond.
- The *arrangement* of C atoms determines the skeleton, so a straight chain and a bent chain represent the same skeleton.
- Groups joined by a single bond can *rotate,* so a branch pointing down is the same as one pointing up. (Recall that a double bond restricts rotation.)

If we put a hydrogen "skin" on a carbon skeleton, we obtain a hydrocarbon. Figure 15.4 shows that a skeleton has the correct number of H atoms when each C has four bonds. Sample Problem 15.1 provides practice drawing hydrocarbons.

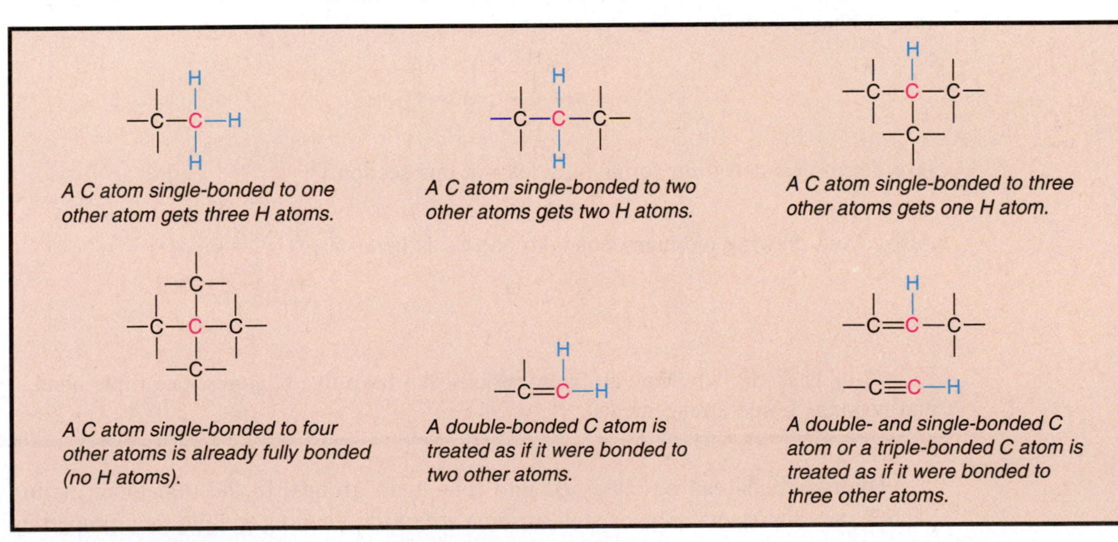

A C atom single-bonded to one other atom gets three H atoms.
A C atom single-bonded to two other atoms gets two H atoms.
A C atom single-bonded to three other atoms gets one H atom.
A C atom single-bonded to four other atoms is already fully bonded (no H atoms).
A double-bonded C atom is treated as if it were bonded to two other atoms.
A double- and single-bonded C atom or a triple-bonded C atom is treated as if it were bonded to three other atoms.

Figure 15.4 Adding the H-atom skin to the C-atom skeleton.

Sample Problem 15.1 Drawing Hydrocarbons

Problem Draw structures that have different atom arrangements for hydrocarbons with:

(a) Six C atoms, no multiple bonds, and no rings

(b) Four C atoms, one double bond, and no rings

Plan In each case, we draw the longest carbon chain first and then work down to smaller chains with branches at different points along them. The process typically involves trial and error. Then, we add H atoms to give each C a total of four bonds.

Solution **(a)** Compounds with six C atoms:

```
            H  H  H  H  H  H
            |  |  |  |  |  |
6-C chain: H—C—C—C—C—C—C—H
            |  |  |  |  |  |
            H  H  H  H  H  H
```

```
                                                   H
                                                   |
                                                 H—C—H
             H  H  H  H  H                H  H    |  H  H
             |  |  |  |  |                |  |    |  |  |
5-C chains: H—C—C—C—C—C—H              H—C—C—C—C—C—H
             |  |  |     |                |  |  |  |  |
             H  H  H     H                H  H  H  H  H
                 H—C—H
                   |
                   H
```

```
               H                      H
               |                      |
             H—C—H                  H—C—H
             H |  H  H             H  H  |  H
             | |  |  |             |  |  |  |
4-C chains: H—C—C—C—C—H          H—C—C—C—C—H
             | |  |  |             |  |  |  |
             H |  H  H             H  |  H  H
             H—C—H                  H—C—H
               |                      |
               H                      H
```

(b) Compounds with four C atoms and one double bond:

```
             H  H  H  H          H  H  H  H                   H     H
             |  |  |  |          |  |  |  |                   |     |
4-C chains: H—C—C=C—C—H        H—C—C—C=C—H     3-C chain: H—C—C=C—H
             |        |          |  |                         |  |
             H        H          H  H                         H  |
                                                               H—C—H
                                                                 |
                                                                 H
```

Check Be sure that each skeleton has the correct number of C atoms and multiple bonds and that no arrangements are repeated or omitted; remember a double bond counts as two bonds.

Comment Avoid some *common mistakes:*

```
                                                          C
                                                          |
In (a): C—C—C—C—C  is the same skeleton as  C—C—C—C—C
          |
          C

              C                                  C  C
              |                                  |  |
          C—C—C—C  is the same skeleton as  C—C—C—C
            |
            C
```

In (b): C—C—C=C is the same skeleton as C=C—C—C

The double bond restricts rotation, so, in addition to the *cis* form shown leftmost in part (b), another possibility is the *trans* form:

```
    H  H     H
    |  |     |
  H—C—C=C—C—H
    |     |  |
    H     H  H
```

(We discuss *cis* and *trans* forms fully later in this section.)

```
                                                     5 bonds
                                               H  H  H
                                               |  |  |
Also, avoid drawing too many bonds to one C, as here: H—C—C=C—H
                                               |  |
                                               H  |
                                                H—C—H
                                                  |
                                                  H
```

FOLLOW-UP PROBLEM 15.1 Draw all hydrocarbons that have five C atoms, one triple bond, and no rings (three arrangements).

Hydrocarbons can be classified into four main groups. In the remainder of this section, we discuss naming, as well as some structural features and physical properties of each group. Later, we'll discuss the chemical behavior of hydrocarbons.

Alkanes: Hydrocarbons with Only Single Bonds

A hydrocarbon that contains only single bonds is an **alkane** (general formula $\mathbf{C_nH_{2n+2}}$, where n is a positive integer). For example, if $n = 5$, the formula is $C_5H_{[(2\times5)+2]}$, or C_5H_{12}. The alkanes comprise a **homologous series,** one in which each member differs from the next by a —CH_2— (methylene) group. In an alkane, each C is sp^3 hybridized. Because each C is bonded to the *maximum number of other atoms* (C or H), alkanes are referred to as **saturated hydrocarbons.**

Naming Alkanes You learned the names of the first 10 straight-chain alkanes in Section 2.8. Here we discuss general rules for naming any alkane and, by extension, other organic compounds as well. *Each chain, branch, or ring has a name based on the **number** of C atoms.* The name of a compound has three portions:

PREFIX + ROOT + SUFFIX

- *Root:* The root tells the number of C atoms in the longest *continuous* chain in the molecule. The roots for the ten smallest alkanes are shown in Table 15.1. As you can see, there are special roots for compounds with chains of one to four C atoms; roots of longer chains are based on Greek numbers.
- *Suffix:* The suffix tells the *type of organic compound* that is being named; that is, it identifies the key functional group the molecule possesses. The suffix is placed *after* the root.
- *Prefix:* Each prefix identifies a *group attached to the main chain* and the number of the carbon to which it is attached. Prefixes identifying hydrocarbon branches are the same as root names (Table 15.1) but have *-yl* as their ending. Each prefix is placed *before* the root.

For example, in the name 2-methylbutane, *2-methyl-* is the prefix (a one-carbon branch is attached to C-2 of the main chain), *-but-* is the root (the main chain has four C atoms), and *-ane* is the suffix (the compound is an alkane).

To obtain the systematic name of a compound,

1. Name the longest chain (root).
2. Add the compound type (suffix).
3. Name any branches (prefix).

Table 15.2 presents the rules for naming any organic compound and applies them to an alkane component of gasoline. Other organic compounds are named with a variety of other prefixes and suffixes (see Table 15.5). In addition to these *systematic* names, some *common* names are still in use.

Table 15.1 Numerical Roots for Carbon Chains and Branches

Roots	Number of C Atoms
meth-	1
eth-	2
prop-	3
but-	4
pent-	5
hex-	6
hept-	7
oct-	8
non-	9
dec-	10

Depicting Alkanes with Formulas and Models Chemists have several ways to depict organic compounds. Expanded, condensed, and carbon-skeleton formulas are easy to draw; ball-and-stick and space-filling models show the actual shapes.

Table 15.2 Rules for Naming an Organic Compound

1. Naming the longest chain (root)
   (a) Find the longest *continuous* chain of C atoms.
   (b) Select the root that corresponds to the number of C atoms in this chain.

CH_3 (on C-2)
CH_3—CH—CH—CH_2—CH_2—CH_3
CH_2—CH_3 (on C-3)
6 carbons ⟹ hex-

2. Naming the compound type (suffix)
   (a) For alkanes, add the suffix *-ane* to the chain root. (Other suffixes appear in Table 15.5 with their functional group and compound type.)
   (b) If the chain forms a ring, the name is preceded by *cyclo-*.

hex- + -ane ⟹ hexane

3. Naming the branches (prefixes) (If the compound has no branches, the name consists of the root and suffix.)
   (a) Each branch name consists of a subroot (number of C atoms) and the ending *-yl* to signify that it is not part of the main chain.
   (b) Branch names precede the chain name. When two or more branches are present, their names appear in *alphabetical* order.
   (c) To specify where the branch occurs along the chain, number the main-chain C atoms consecutively, starting at the end *closer* to a branch, to achieve the *lowest* numbers for the branches. Precede each branch name with the number of the main-chain C to which that branch is attached.

CH_3 methyl
CH_3—CH—CH—CH_2—CH_2—CH_3
CH_2—CH_3 ethyl
ethylmethylhexane

CH_3
1 2 3 4 5 6
CH_3—CH—CH—CH_2—CH_2—CH_3
CH_2—CH_3
3-ethyl-2-methylhexane

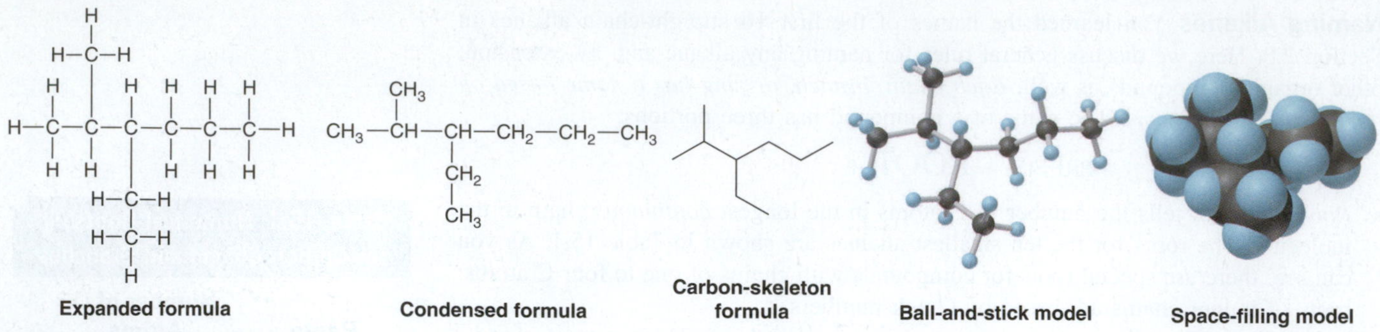

Expanded formula

$CH_3-CH(CH_3)-CH(CH_2CH_3)-CH_2-CH_2-CH_3$

Condensed formula

Carbon-skeleton formula

Ball-and-stick model

Space-filling model

Figure 15.5 Ways of depicting the alkane 3-ethyl-2-methylhexane.

The *expanded formula* is a Lewis structure, so it shows each atom and bond. One type of *condensed formula* groups each C atom with its H atoms. *Carbon-skeleton formulas* show only carbon-carbon bonds and appear as zig-zag lines, often with branches. *Each end or bend of a zig-zag line or branch represents a C atom attached to the number of H atoms that gives it a total of four bonds:*

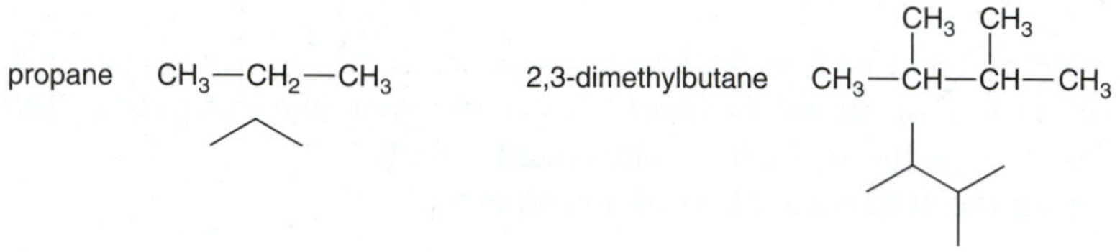

propane $CH_3-CH_2-CH_3$

2,3-dimethylbutane $CH_3-CH(CH_3)-CH(CH_3)-CH_3$

Figure 15.5 shows these formulas (and models) of the compound named in Table 15.2.

Cyclic Hydrocarbons A **cyclic hydrocarbon** contains one or more rings in its structure. When a straight-chain alkane (C_nH_{2n+2}) forms a ring, two H atoms are lost as the C—C bond forms to join the two ends of the chain. Thus, *cycloalkanes* have the general formula C_nH_{2n}. Cyclic hydrocarbons are often drawn with carbon-skeleton formulas (Figure 15.6, *top row*). Except for three-carbon rings, *cycloalkanes are nonplanar,* as the models show. This structural feature arises from the tetrahedral shape around each C atom and the need to minimize electron repulsions between adjacent H atoms. As a result, orbital overlap of adjacent C atoms is maximized. The most stable form of cyclohexane is called the *chair conformation* (Figure 15.6D).

Figure 15.6 Depicting cycloalkanes.

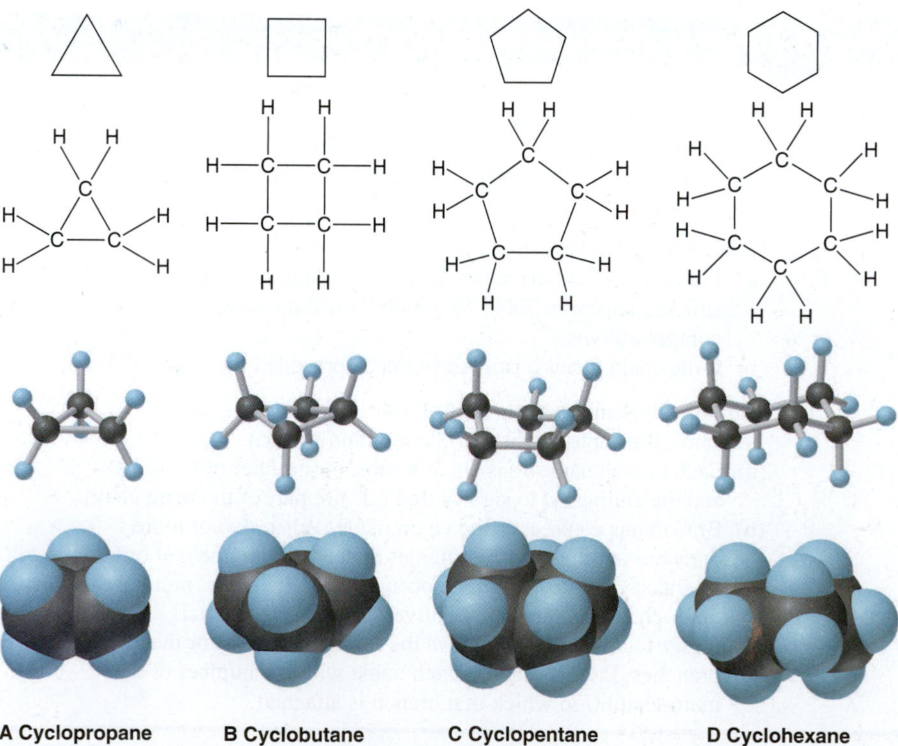

Constitutional Isomerism and the Physical Properties of Alkanes

Recall from Section 3.2 that two or more compounds with the same molecular formula but different properties are called **isomers.** Those with *different arrangements of bonded atoms* are **constitutional** (or **structural**) **isomers;** alkanes with the same number of C atoms but different skeletons are examples. The smallest alkane to exhibit constitutional isomerism has four C atoms: two different compounds have the formula C_4H_{10} (Table 15.3). The unbranched one is butane (common name, *n*-butane; *n*- stands for "normal," or having a straight chain), and the other is 2-methylpropane (common name, *iso*butane). Similarly, three compounds have the formula C_5H_{12}. The unbranched isomer is pentane (common name, *n*-pentane); the one with a methyl group at C-2 of a four-C chain is 2-methylbutane (common name, *iso*pentane). The third isomer has two methyl branches on C-2 of a three-C chain, so its name is 2,2-dimethylpropane (common name, *neo*pentane).

Dispersion Forces and Boiling Points Because alkanes are nearly nonpolar, we expect their physical properties to be determined by dispersion forces, and the boiling points in Table 15.3 bear this out. The four-C alkanes boil lower than the five-C compounds. Moreover, within each group of isomers, the more spherical member (isobutane or neopentane) boils lower than the more elongated one (*n*-butane or *n*-pentane). As you saw in Chapter 12, this trend occurs because a spherical shape leads to less intermolecular contact, and thus lower total dispersion forces, than does an elongated shape.

A particularly clear example of the effect of dispersion forces on physical properties occurs among the unbranched alkanes (*n*-alkanes). Among these compounds, boiling points increase steadily with chain length: the longer the chain,

Table 15.3 The Constitutional Isomers of C_4H_{10} and C_5H_{12}

Systematic Name (Common Name)	Expanded Formula	Condensed and Skeleton Formulas	Space-Filling Model	Density (g/mL)	Boiling Point (°C)
Butane (*n*-butane)	H H H H H—C—C—C—C—H H H H H	$CH_3—CH_2—CH_2—CH_3$		0.579	−0.5
2-Methylpropane (isobutane)	H H H H—C—C—C—H H H H—C—H H	$CH_3—CH—CH_3$ CH_3		0.549	−11.6
Pentane (*n*-pentane)	H H H H H H—C—C—C—C—C—H H H H H H	$CH_3—CH_2—CH_2—CH_2—CH_3$		0.626	36.1
2-Methylbutane (isopentane)	H H H H H—C—C—C—C—H H H H H—C—H H	$CH_3—CH—CH_2—CH_3$ CH_3		0.620	27.8
2,2-Dimethylpropane (neopentane)	H H—C—H H H H—C—C—C—H H H H—C—H H	CH_3 $CH_3—C—CH_3$ CH_3		0.614	9.5

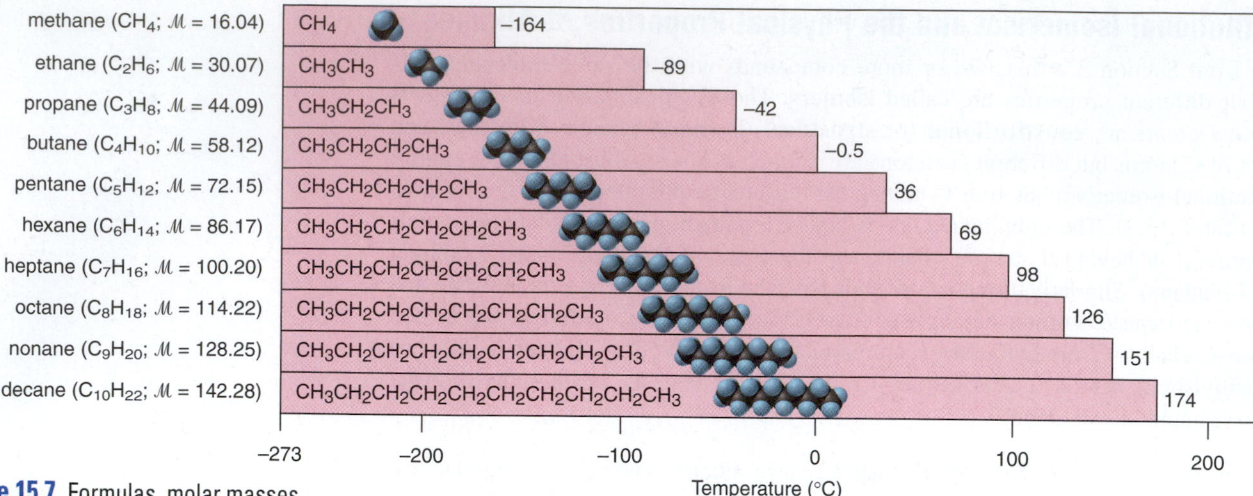

Figure 15.7 Formulas, molar masses (in g/mol), structures, and boiling points (in °C at 1 atm pressure) of the first 10 unbranched alkanes.

the greater the intermolecular contact, the stronger the dispersion forces, and the higher the boiling point (Figure 15.7). Pentane (five C atoms) is the smallest *n*-alkane that exists as a liquid at room temperature. The solubility of alkanes, and of all hydrocarbons, is easy to predict from the like-dissolves-like rule (Section 13.1). Alkanes are miscible in each other and in other nonpolar solvents, such as benzene, but are nearly insoluble in water. The solubility of pentane in water, for example, is only 0.36 g/L at room temperature.

Chiral Molecules and Optical Isomerism

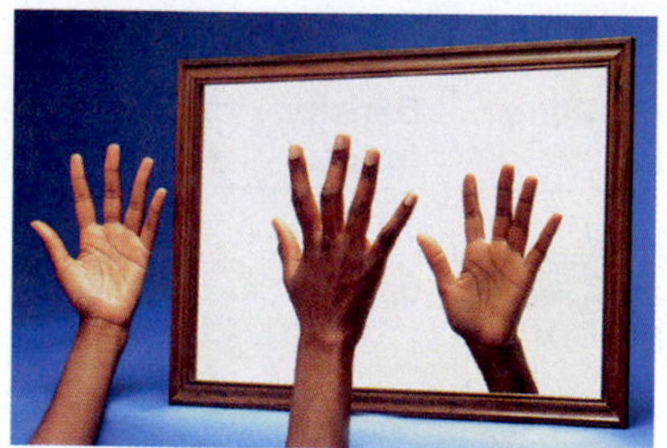

Figure 15.8 An analogy for optical isomers.

Another type of isomerism exhibited by some alkanes and many other organic (as well as some inorganic) compounds is called *stereoisomerism.* **Stereoisomers** are molecules with the same arrangement of atoms *but different orientations of groups in space. Optical isomerism* is one type of stereoisomerism: *when two objects are mirror images of each other and cannot be superimposed, they are* **optical isomers,** also called *enantiomers.* To use a familiar example, your right hand is an optical isomer of your left. Look at your right hand in a mirror: the *image* is identical to your left hand (Figure 15.8). No matter how you twist your arms around, however, your hands cannot lie on top of each other with your palms facing in the same direction and be superimposed. They are not superimposable because each is *asymmetric:* there is no plane of symmetry that divides your hand into two identical parts.

Asymmetry and Chirality An asymmetric molecule is called **chiral** (Greek *cheir,* "hand"). Typically, an organic molecule *is chiral if it contains a carbon atom that is bonded to four **different** groups.* This C atom is called a *chiral center,* or an asymmetric carbon. In 3-methylhexane, for example, C-3 is a chiral center, because it is bonded to four different groups: H—, CH_3—, CH_3—CH_2—, and CH_3—CH_2—CH_2— (Figure 15.9A). Like your two hands, the two forms are mirror images and cannot be superimposed on each other: when two of the groups are superimposed, the other two are opposite each other. Thus, the two forms are optical isomers. The central C atom in the amino acid alanine is also a chiral center (Figure 15.9B).

Figure 15.9 Two chiral molecules.

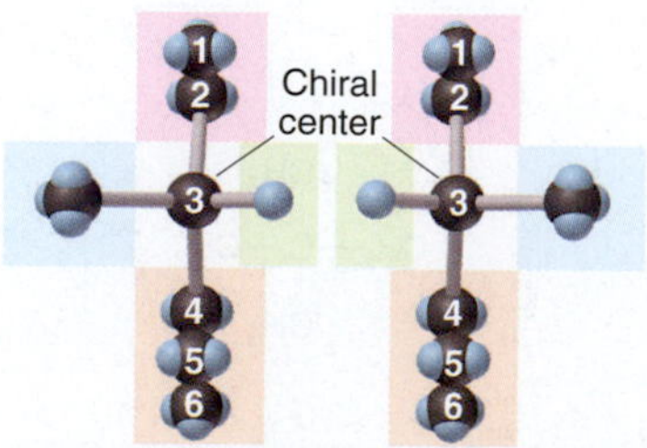

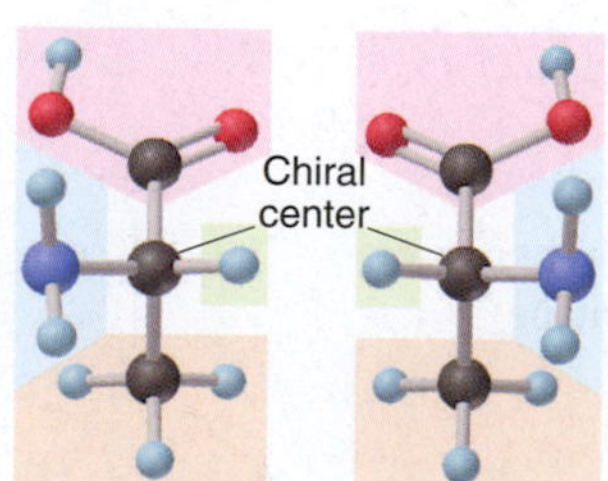

Properties of Optical Isomers Unlike constitutional isomers, which have different physical properties such as boiling point, optical isomers are identical in all but three respects:

1. In their *physical* properties, *optical isomers differ only in the direction that each isomer rotates the plane of polarized light.* An instrument called a *polarimeter* is used to measure the angle that the plane is rotated. A beam of light consists of waves that oscillate in all planes. A polarizing filter blocks all waves except those in one plane, so the light emerging through the filter is *plane-polarized.* An optical isomer is **optically active** because it rotates the plane of this polarized light. The *dextrorotatory* isomer (designated *d* or +) rotates the plane of light clockwise; the *levorotatory* isomer (designated *l* or −) is the mirror image of the *d* isomer and rotates the plane counterclockwise. An equimolar mixture of the two isomers does not rotate the plane of light because the opposing rotations cancel each other.

2. In their *chemical* properties, *optical isomers differ only in a chiral (asymmetric) chemical environment,* one that distinguishes "right-handed" from "left-handed" molecules. As an analogy, your right hand fits well in your right glove but not in your left glove.

3. In their *biological* properties, which arise directly from their chemical properties, *optical isomers differ in usability by an organism.* Nearly all carbohydrates and amino acids are optically active, but only one of the isomers is biologically usable. For example, you metabolize *d*-glucose for energy but excrete *l*-glucose unused. Similarly, *l*-alanine is incorporated naturally into your proteins, but *d*-alanine is not. An organism can utilize only one of a pair of optical isomers because of its enzymes (Section 16.7), proteins that speed virtually every reaction in a cell. Many drugs are chiral molecules: one optical isomer has a certain biological activity, and the other has either a different type of activity or none at all.

Alkenes: Hydrocarbons with Double Bonds

A hydrocarbon that contains at least one C=C bond is called an **alkene.** With two H atoms removed to make the double bond, alkenes have the general formula $\mathbf{C_nH_{2n}}$. The double-bonded C atoms are sp^2 hybridized. Because their carbon atoms are bonded to fewer than the maximum of four atoms each, alkenes are considered **unsaturated hydrocarbons.**

Alkene names differ from those of alkanes in two respects:

1. The main chain (root) *must* contain both C atoms of the double bond, even if it is not the longest chain. The chain is numbered from the end *closer* to the C=C bond, and the position of the bond is indicated by the number of the *first* C atom in it.
2. The suffix for alkenes is *-ene.*

For example, there are three four-C alkenes (C_4H_8), two unbranched and one branched (see Sample Problem 15.1b). The branched isomer is 2-methylpropene; the unbranched isomer with the C=C bond between C-1 and C-2 is 1-butene; the unbranched isomer with the C=C bond between C-2 and C-3 is 2-butene. As you'll see next, there are two isomers of 2-butene, but they are of a different sort.

The C=C Bond and Geometric (*cis-trans*) Isomerism There are two major structural differences between alkanes and alkenes.

- Alkanes have a *tetrahedral* geometry (bond angles of ~109.5°) around each C atom, whereas the double-bonded C atoms in alkenes are *trigonal planar* (~120°).
- The C—C bond *allows* rotation of bonded groups, so the atoms in an alkane continually change their relative positions; in contrast, the π bond of the alkene C=C bond *restricts* rotation, which fixes the relative positions of the atoms bonded to it.

This rotational restriction leads to another type of stereoisomerism. **Geometric isomers** (also called ***cis-trans*** **isomers**) have different orientations of groups around

Table 15.4 The Geometric Isomers of 2-Butene

Systematic Name	Condensed and Skeleton Formulas	Space-Filling Model	Density (g/mL)	Boiling Point (°C)
cis-2-Butene	$CH_3(H)C{=}C(H)CH_3$ (cis)		0.621	3.7
trans-2-Butene	$CH_3(H)C{=}C(H)CH_3$ (trans)		0.604	0.9

a double bond (or similar structural feature). Table 15.4 shows the two geometric isomers of 2-butene (also see Comment, Sample Problem 15.1), *cis*-2-butene and *trans*-2-butene. In general, the *cis* isomer has the *larger portions of the main chain* (in this case, two CH_3 groups) *on the same side* of the double bond, and the *trans* isomer has them on *opposite sides.* For a molecule to have geometric isomers, *each C atom in the C=C bond must be bonded to two different groups,* in this case, CH_3 and H.

Like structural isomers, geometric isomers have different physical properties. Note in Table 15.4 that the two 2-butenes differ in molecular shape *and* physical properties. The *cis* isomer has a bend in the chain that the *trans* isomer lacks.

Geometric Isomers and the Chemistry of Vision The first step in the sequence of events that allows us to see relies on the different shapes of a pair of geometric isomers. *Retinal,* a 20-C compound consisting of a 15-C chain with five 1-C branches and 5 C=C bonds, is part of the molecule responsible for receiving light energy. There are two biologically occurring isomers of retinal, which have very different shapes: the all-*trans* isomer is elongated, and the 11-*cis* isomer is sharply bent around the double bond between C-11 and C-12.

Certain cells of the retina are densely packed with *rhodopsin,* a large molecule consisting of a protein covalently bonded to 11-*cis* retinal. The initial chemical event in vision occurs when this molecule absorbs a photon of visible light. The energy range of photons of visible light (165–293 kJ/mol) spans the energy needed to break a C=C π bond (about 250 kJ/mol). Within a few millionths of a second after the absorption of a photon, the 11-*cis* π bond of retinal breaks, the intact σ bond between C-11 and C-12 rotates, and the π bond re-forms to produce all-*trans* retinal.

The rapid and significant change in shape of retinal causes the attached protein to change its shape as well, triggering a flow of ions into the retina's cells and initiating electrical impulses, which the optic nerve conducts to the brain. Because of the speed and efficiency with which light causes such a large structural change in retinal, natural selection has chosen it to be the photon absorber in organisms as different as purple bacteria, mollusks, insects, and vertebrates.

Alkynes: Hydrocarbons with Triple Bonds

Hydrocarbons that contain at least one C≡C bond are called **alkynes.** Their general formula is $\mathbf{C_nH_{2n-2}}$ because they have two H atoms fewer than alkenes with the same number of carbons. Because a carbon involved in a C≡C bond can bond to only one other atom, the geometry around each C atom is linear (180°): each C is *sp* hybridized. Alkynes are named in the same way as alkenes, except that the suffix is *-yne*. Because of their localized π electrons, C=C and C≡C bonds are electron rich and act as functional groups. Thus, alkenes and alkynes are much more reactive than alkanes (we discuss some reactions of alkenes and alkynes in Section 15.4).

Sample Problem 15.2 Naming Alkanes, Alkenes, and Alkynes

Problem Give the systematic name for each of the following, indicate the chiral center in part (d), and draw two geometric isomers for part (e):

(a) 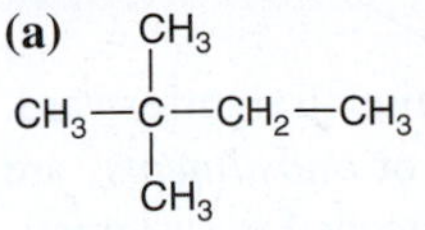
(b) 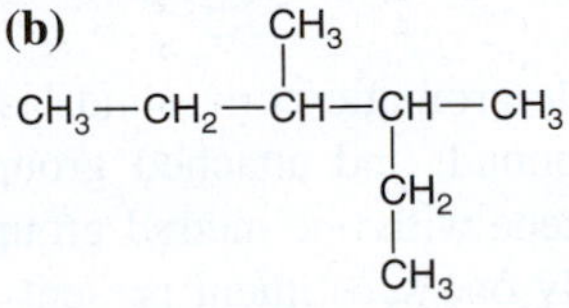
(c) 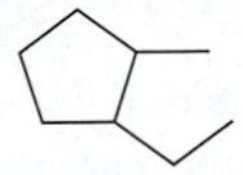
(d) 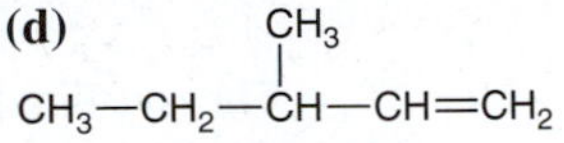
(e) 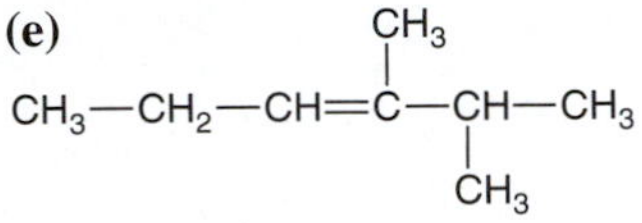

Plan For (a) to (c), we refer to Table 15.2, p. 465. We first name the longest chain *(root-)* and add the suffix *-ane* because there are only single bonds. Then we find the *lowest* branch numbers by counting C atoms from the end *closer* to a branch. Finally, we name each branch *(root- + -yl)* and put the names alphabetically before the root name. For (d) and (e), the longest chain that *includes* the multiple bond is numbered from the end closer to it. For (d), the chiral center is the C atom bonded to four different groups. In (e), the *cis* isomer has larger groups on the same side of the double bond, and the *trans* isomer has them on opposite sides.

Solution

(a) 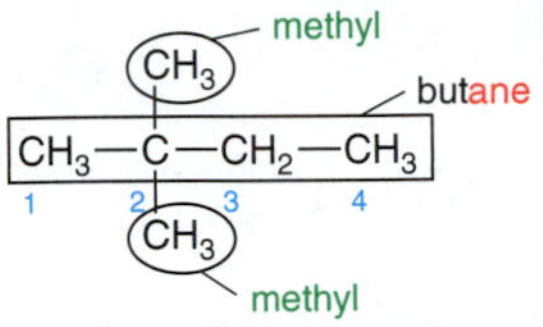

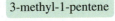

2,2-dimethylbutane

When a type of branch appears more than once, we group the branch numbers and indicate the number of branches with a prefix, as in 2,2-*di*methyl.

(b) 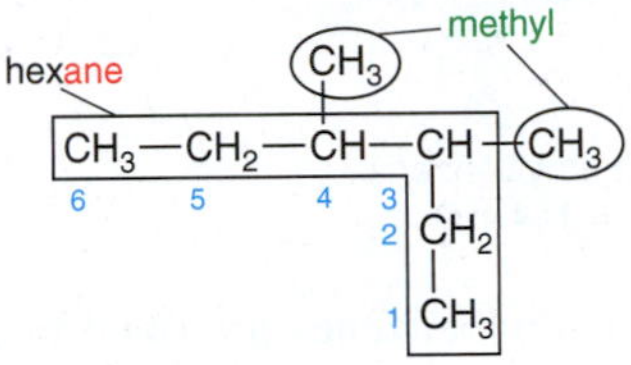

3,4-dimethylhexane

In this case, we can number the chain from either end because the branches are the same and are attached to the two central C atoms.

(c) 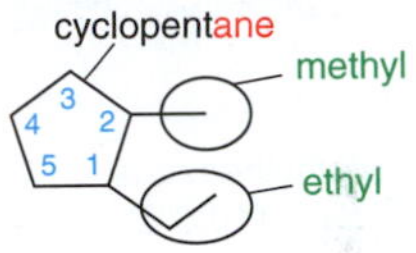

1-ethyl-2-methylcyclopentane

We number the ring C atoms so that a branch is attached to C-1.

(d) 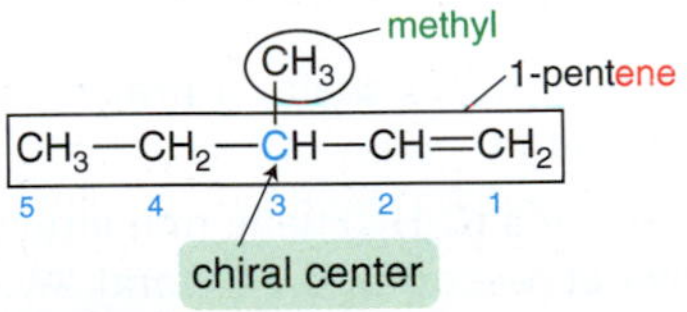

3-methyl-1-pentene

(e) 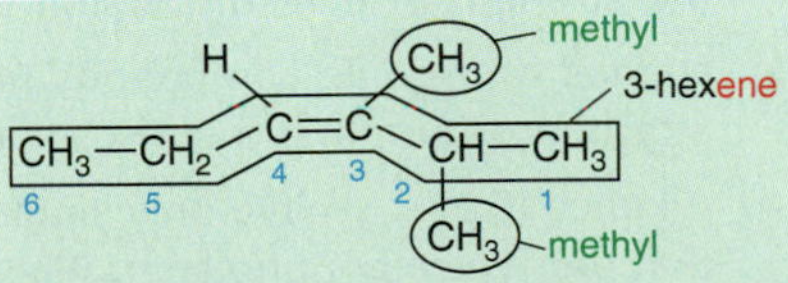

cis-2,3-dimethyl-3-hexene

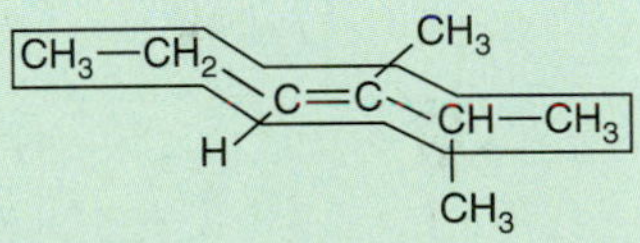

trans-2,3-dimethyl-3-hexene

Check A good check (and excellent practice) is to reverse the process by drawing structures for the names to see if you come up with the structures given in the problem.

Comment 1. In (b), C-3 and C-4 are chiral centers, as are C-1 and C-2 in (c). However, in (b) the molecule is not chiral: it has a plane of symmetry between C-3 and C-4, so each half of the molecule rotates light in opposite directions. **2.** Avoid these common mistakes: In (b), 2-ethyl-3-methylpentane is wrong: the longest chain is *hexane.* In (c), 1-methyl-2-ethylcyclopentane is wrong: the branch names should appear *alphabetically.*

FOLLOW-UP PROBLEM 15.2 Draw condensed formulas for the following compounds: **(a)** 3-ethyl-3-methyloctane; **(b)** 1-ethyl-3-propylcyclohexane (also draw a carbon-skeleton formula for this compound); **(c)** 3,3-diethyl-1-hexyne; **(d)** *trans*-3-methyl-3-heptene.

Aromatic Hydrocarbons: Cyclic Molecules with Delocalized π Electrons

Unlike the cycloalkanes, **aromatic hydrocarbons** are planar molecules, usually with one or more rings of six C atoms, and are often drawn with alternating single and double bonds. As you learned for benzene (Section 10.1), however, all the ring bonds are identical, with values of length and strength *between* those of a C—C and a C═C bond. To indicate this, benzene is also shown as a resonance hybrid, with a circle (or dashed circle) representing the delocalized character of the π electrons:

The systematic naming of simple aromatic compounds is quite straightforward. Usually, benzene is the parent compound, and attached groups, or *substituents,* are named as prefixes. For example, benzene with one methyl group attached is systematically named *methylbenzene.* With only one substituent present, we do not number the ring C atoms; when two or more groups are attached, however, we number in such a way that one of the groups is attached to ring C-1. Thus, methylbenzene and the three structural isomers with two methyl groups attached are

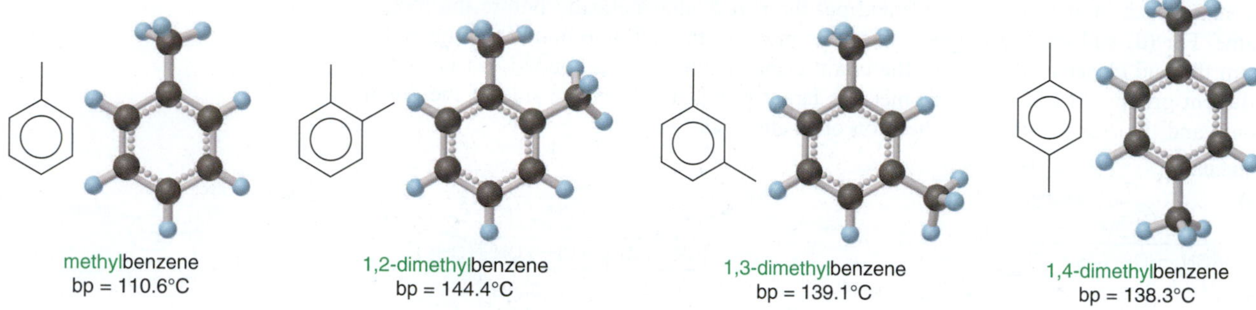

methylbenzene bp = 110.6°C | 1,2-dimethylbenzene bp = 144.4°C | 1,3-dimethylbenzene bp = 139.1°C | 1,4-dimethylbenzene bp = 138.3°C

The dimethylbenzenes are important solvents and feedstocks for polyester fibers and dyes. Benzene and many other aromatic hydrocarbons have been shown to have carcinogenic (cancer-causing) activity.

Summary of Section 15.2

- Hydrocarbons contain only C and H atoms, so their physical properties depend on the strength of their dispersion forces.
- Names of organic compounds have a root for the longest chain, a prefix for any attached group, and a suffix for the type of compound.
- Alkanes (C_nH_{2n+2}) have only single bonds. Cycloalkanes (C_nH_{2n}) have ring structures that are typically nonplanar. Alkenes (C_nH_{2n}) have at least one C═C bond. Alkynes (C_nH_{2n-2}) have at least one C≡C bond. Aromatic hydrocarbons have at least one planar ring with delocalized π electrons.
- Constitutional (structural) isomers have different atom arrangements.
- Stereoisomers (optical and geometric) have the same arrangement of atoms, but their atoms are oriented differently in space. Optical isomers cannot be superimposed on each other because they are asymmetric, with four different groups bonded to the C that is the chiral center. They have identical physical and chemical properties except in their rotation of plane-polarized light, their reactivity in a chiral environment, and their biological activity in cells. Geometric *(cis-trans)* isomers have groups oriented differently around a C═C bond, which restricts rotation.
- Light converts a *cis* isomer of retinal to the all-*trans* form, initiating the process of vision.

15.3 • SOME IMPORTANT CLASSES OF ORGANIC REACTIONS

Organic reactions are classified according to the chemical process involved. Three important classes are *addition, elimination,* and *substitution* reactions.

To depict reactions more generally, it's common practice to use an uppercase R with a single bond, R—, to signify a nonreacting organic group attached to one of the atoms shown; you can usually picture R— as an **alkyl group,** a saturated hydrocarbon group with one bond available to link to another atom. Thus, R—CH_2—Br has an alkyl group attached to a CH_2 group bonded to a Br atom; R—CH═CH_2 is an alkene with an alkyl group attached to one of the carbons in the double bond; and so forth. (Often, when more than one R group is present, we write R, R′, R″, and so forth, to indicate that these groups may be different.)

The three classes of organic reactions we discuss here can be identified by comparing the *number of bonds to C* in reactants and products.

1. An **addition reaction** occurs when an unsaturated reactant becomes a saturated product:

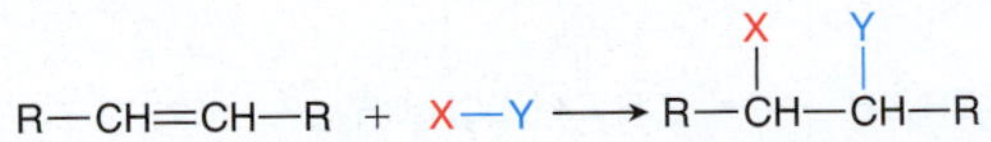

$$R{-}CH{=}CH{-}R + X{-}Y \longrightarrow R{-}\underset{}{\overset{X}{\overset{|}{C}}}H{-}\overset{Y}{\overset{|}{C}}H{-}R$$

Note the C atoms are bonded to *more* atoms in the product than in the reactant.

The C=C and C≡C bonds and the C=O bond commonly undergo addition reactions. In each case, the π bond breaks, leaving the σ bond intact. In the product, the two C atoms (or C and O) form two additional σ bonds. A good example is the addition reaction in which H and Cl from HCl add to the double bond in ethylene:

$$CH_2{=}CH_2 \quad + \quad H{-}Cl \quad \longrightarrow \quad H{-}CH_2{-}CH_2{-}Cl$$

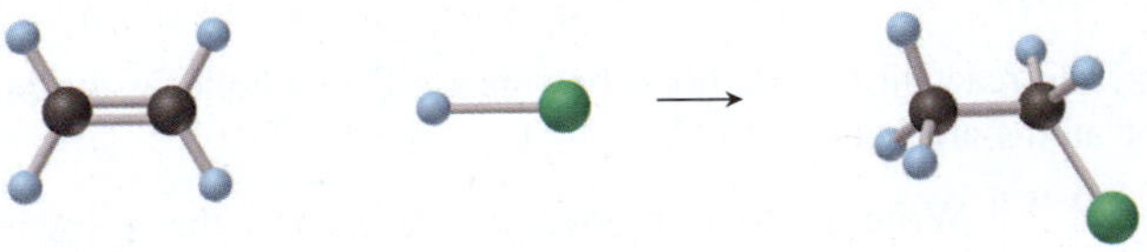

2. An **elimination reaction** is the opposite of an addition reaction. It occurs when a saturated reactant becomes an unsaturated product:

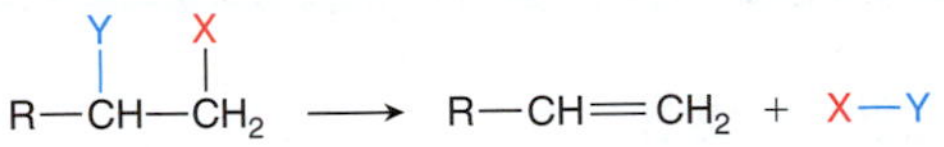

$$R{-}\overset{Y}{\overset{|}{C}}H{-}\overset{X}{\overset{|}{C}}H_2 \longrightarrow R{-}CH{=}CH_2 + X{-}Y$$

Note that the C atoms are bonded to *fewer* atoms in the product than in the reactant. A pair of halogen atoms, an H atom and a halogen atom, or an H atom and an —OH group are typically eliminated, but C atoms are not:

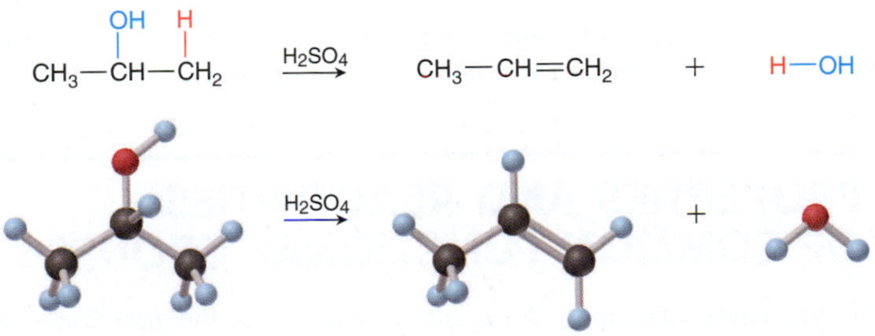

$$CH_3{-}\overset{OH}{\overset{|}{C}}H{-}\overset{H}{\overset{|}{C}}H_2 \xrightarrow{H_2SO_4} CH_3{-}CH{=}CH_2 \quad + \quad H{-}OH$$

$$\xrightarrow{H_2SO_4} \quad +$$

3. A **substitution reaction** occurs when an atom (or group) from an added reactant substitutes for one in the organic reactant:

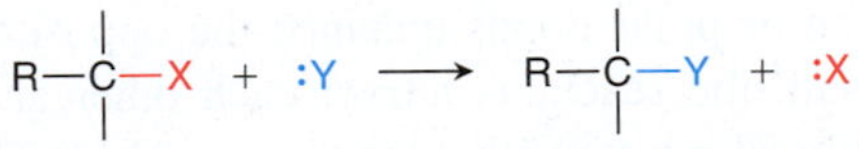

$$R{-}\underset{|}{\overset{|}{C}}{-}X + :Y \longrightarrow R{-}\underset{|}{\overset{|}{C}}{-}Y + :X$$

Note that the C atom is bonded to the *same number* of atoms in the product as in the reactant. The C atom may be saturated or unsaturated, and X and Y can be many different atoms, but generally *not* C. The main flavor ingredient in banana oil, for instance, forms through a substitution reaction; note that the O substitutes for the Cl:

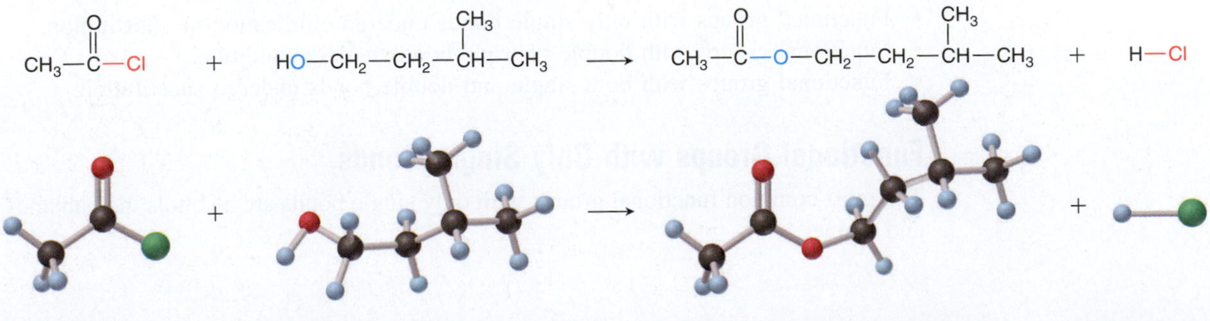

$$CH_3{-}\overset{O}{\overset{\|}{C}}{-}Cl \quad + \quad HO{-}CH_2{-}CH_2{-}\overset{CH_3}{\overset{|}{C}}H{-}CH_3 \longrightarrow CH_3{-}\overset{O}{\overset{\|}{C}}{-}O{-}CH_2{-}CH_2{-}\overset{CH_3}{\overset{|}{C}}H{-}CH_3 \quad + \quad H{-}Cl$$

$$+ \qquad \longrightarrow \qquad +$$

Sample Problem 15.3 Recognizing the Type of Organic Reaction

Problem State whether each reaction is an addition, elimination, or substitution:

(a) $CH_3—CH_2—CH_2—Br \longrightarrow CH_3—CH{=}CH_2 + HBr$

(b) cyclopentene $+ H_2 \longrightarrow$ cyclopentane

(c) $CH_3C({=}O)—Br + CH_3CH_2OH \longrightarrow CH_3C({=}O)—OCH_2CH_3 + HBr$

Plan We determine the type of reaction by looking for any change in the number of atoms bonded to C:

- More atoms bonded to C indicates an *addition.*
- Fewer atoms bonded to C indicates an *elimination.*
- Same number of atoms bonded to C indicates a *substitution.*

Solution **(a)** Elimination: two bonds in the reactant, C—H and C—Br, are absent in the product, so fewer atoms are bonded to C.

(b) Addition: two more C—H bonds have formed in the product, so more atoms are bonded to C.

(c) Substitution: the reactant C—Br bond becomes a C—O bond in the product, so the same number of atoms are bonded to C.

FOLLOW-UP PROBLEM 15.3 Write a balanced equation for each of the following:

(a) An addition reaction between 2-butene and Cl_2

(b) A substitution reaction between $CH_3—CH_2—CH_2—Br$ and OH^-

(c) The elimination of H_2O from $(CH_3)_3C—OH$

Summary of Section 15.3

- In an addition reaction, a π bond breaks, and the two C atoms (or one C and one O) are bonded to more atoms in the product.
- In an elimination reaction, a π bond forms, with the two C atoms bonded to fewer atoms in the product.
- In a substitution reaction, one atom bonded to C is replaced by another, but the total number of atoms bonded to the C does not change.

15.4 • PROPERTIES AND REACTIVITIES OF COMMON FUNCTIONAL GROUPS

The central organizing principle of organic chemistry is the *functional group. The distribution of electron density in a functional group is the key to the reactivity of the compound.* The electron density can be high, as in the C=C and C≡C bonds, or it can be low at one end of a bond and high at the other, as in the C—Cl and C—O bonds. Such electron-rich or polar bonds enhance the oppositely charged pole in the other reactant. As a result, the reactants attract each other and begin a sequence of bond-forming and bond-breaking steps that lead to a product. Thus, *the intermolecular forces that affect physical properties and solubility also affect reactivity.* Table 15.5 lists some of the important functional groups in organic compounds.

When we classify functional groups by bond order, they follow certain patterns of reactivity:

- Functional groups with only single bonds undergo elimination or substitution.
- Functional groups with double or triple bonds undergo addition.
- Functional groups with both single and double bonds undergo substitution.

Functional Groups with Only Single Bonds

The most common functional groups with only single bonds are alcohols, haloalkanes, and amines.

Table 15.5 Important Functional Groups in Organic Compounds

Functional Group	Compound Type	Prefix or Suffix of Name	Example: Lewis Structure	Example: Ball-and-Stick Model	Example: Systematic Name (Common Name)
$>C=C<$	alkene	-ene	$H_2C=CH_2$		ethene (ethylene)
$—C≡C—$	alkyne	-yne	$H—C≡C—H$		ethyne (acetylene)
$—C—\ddot{O}—H$	alcohol	-ol	$H_3C—\ddot{O}—H$		methanol (methyl alcohol)
$—C—\ddot{X}:$ (X = halogen)	haloalkane	halo-	$H_3C—\ddot{Cl}:$		chloromethane (methyl chloride)
$—C—\ddot{N}—$	amine	-amine	$H_3C—CH_2—\ddot{N}H_2$		ethanamine (ethylamine)
$—C(=\ddot{O})—H$	aldehyde	-al	$H_3C—C(=\ddot{O})—H$		ethanal (acetaldehyde)
$—C—C(=\ddot{O})—C—$	ketone	-one	$H_3C—C(=\ddot{O})—CH_3$		2-propanone (acetone)
$—C(=\ddot{O})—\ddot{O}—H$	carboxylic acid	-oic acid	$H_3C—C(=\ddot{O})—\ddot{O}—H$		ethanoic acid (acetic acid)
$—C(=\ddot{O})—\ddot{O}—C—$	ester	-oate	$H_3C—C(=\ddot{O})—\ddot{O}—CH_3$		methyl ethanoate (methyl acetate)
$—C(=\ddot{O})—\ddot{N}—$	amide	-amide	$H_3C—C(=\ddot{O})—\ddot{N}H_2$		ethanamide (acetamide)
$—C≡N:$	nitrile	-nitrile	$H_3C—C≡N:$		ethanenitrile (acetonitrile, methyl cyanide)

Methanol (methyl alcohol)
Byproduct in coal gasification; de-icing agent; gasoline substitute; precursor of organic compounds

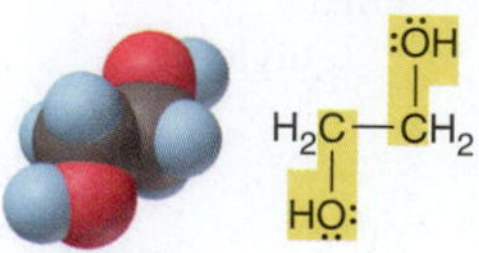

1,2-Ethanediol (ethylene glycol)
Main component of auto antifreeze

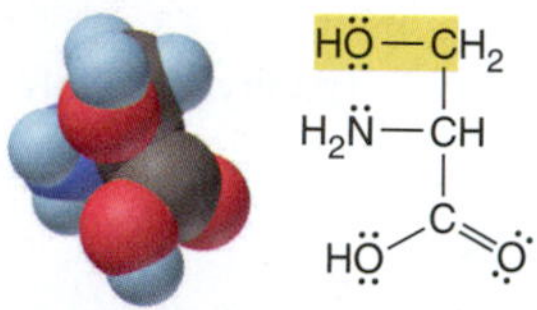

Serine
Amino acid found in most proteins

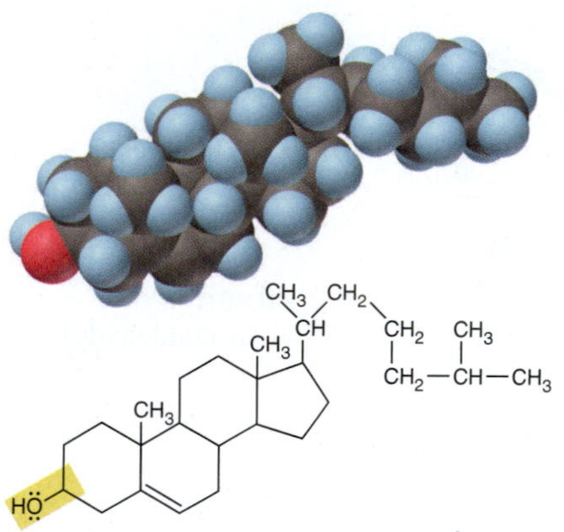

Cholesterol
Major sterol in animals; essential for cell membranes; precursor of steroid hormones

Figure 15.10 Some molecules with the alcohol functional group.

Alcohols The **alcohol** functional group consists of a carbon bonded to an —OH group, $-\overset{|}{\underset{|}{C}}-\ddot{O}-H$, and the general formula of an alcohol is R—OH. Alcohols are named by dropping the final *-e* from the parent hydrocarbon name and adding the suffix *-ol*. Thus, the two-carbon alcohol is ethanol (ethan- + -ol). The common name is the hydrocarbon *root-* + *-yl,* followed by "alcohol"; thus, the common name of ethanol is ethyl alcohol. (This substance, obtained from fermented grain, has been consumed by people as an intoxicant in beverages since ancient times; today, it is recognized as the most abused drug in the world.) Alcohols are common organic reactants, and the functional group occurs in many biomolecules, including carbohydrates, sterols, and some amino acids. Figure 15.10 shows the names, structures, and uses of some important compounds that contain the alcohol group.

The physical properties of the smaller alcohols are similar to those of water. They have high melting and boiling points as a result of hydrogen bonding, and they dissolve polar molecules and some salts.

Alcohols undergo elimination and substitution reactions.

- Elimination of H and OH, called *dehydration,* requires acid and forms an alkene:

$$\text{cyclohexanol} \xrightarrow{H^+} \text{cyclohexene} + H_2O$$

Elimination of two H atoms is an *oxidation* which typically requires an inorganic oxidizing agent, such as potassium dichromate ($K_2Cr_2O_7$) in aqueous H_2SO_4. The product has a C═O group:

$$CH_3-CH_2-\overset{OH}{\overset{|}{C}H}-CH_3 \xrightarrow[H_2SO_4]{K_2Cr_2O_7} CH_3-CH_2-\overset{O}{\overset{\|}{C}}-CH_3$$

2-butanol 2-butanone

Alcohols with an OH group at the end of the chain ($R-CH_2-OH$) can be oxidized further to acids. (This is *not* an elimination reaction.) Wine turns sour, for example, when the ethanol in contact with air is oxidized to acetic acid (vinegar):

$$CH_3-\overset{OH}{\overset{|}{C}H_2} \xrightarrow[-H_2O]{\frac{1}{2}O_2} CH_3-\overset{O}{\overset{\|}{C}}H \xrightarrow{\frac{1}{2}O_2} CH_3-\overset{O}{\overset{\|}{C}}-OH$$

- Substitution yields products with other single-bonded functional groups. Reactions of hydrohalic acids with many alcohols give haloalkanes:

$$R_2CH-OH + HBr \longrightarrow R_2CH-Br + HOH$$

As you'll see shortly, *the C atom undergoing the change in a substitution is bonded to a more electronegative element,* which makes it partially positive and, thus, a target for a negatively charged or electron-rich group of an incoming reactant.

Haloalkanes A *halogen* atom (X) bonded to a carbon gives the **haloalkane** functional group, $-\overset{|}{\underset{|}{C}}-\ddot{X}:$, and compounds with the general formula R—X. Haloalkanes (common name, **alkyl halides**) are named by identifying the halogen with a prefix on the hydrocarbon name and numbering the C atom to which the halogen is attached, as in bromomethane, 2-chloropropane, or 1,3-diiodohexane.

Like alcohols, haloalkanes undergo substitution and elimination reactions.

- Just as many alcohols undergo substitution to form alkyl halides when treated with halide ions in acid, many halides undergo substitution to form alcohols in base. For example, OH^- attacks the positive C end of the C—X bond and displaces X^-:

$$\underset{\text{1-bromobutane}}{CH_3—CH_2—CH_2—CH_2—Br} + OH^- \longrightarrow \underset{\text{1-butanol}}{CH_3—CH_2—CH_2—CH_2—OH} + Br^-$$

  Substitutions by groups such as —CN, —SH, —OR, and $—NH_2$ allow chemists to convert alkyl halides to a host of other families of compounds.
- Just as addition of HX *to* an alkene produces haloalkanes, elimination of HX *from* a haloalkane by reaction with a strong base, such as potassium ethoxide, produces an alkene:

$$\underset{\text{2-chloro-2-methylpropane}}{CH_3—\overset{\overset{CH_3}{|}}{\underset{\underset{Cl}{|}}{C}}—CH_3} + \underset{\text{potassium ethoxide}}{CH_3—CH_2—OK} \longrightarrow \underset{\text{2-methylpropene}}{CH_3—\overset{\overset{CH_3}{|}}{C}=CH_2} + KCl + CH_3—CH_2—OH$$

Haloalkanes have many important uses, but many are carcinogenic in mammals, have severe neurological effects in humans, and, to make matters worse, are very stable and accumulate in the environment.

Amines The **amine** functional group is $—\overset{|}{\underset{|}{C}}—\overset{\cdot\cdot}{\underset{|}{N}}—$. Chemists classify amines as derivatives of ammonia, with R groups in place of one or more of the H atoms in NH_3. *Primary* (1°) amines are RNH_2, *secondary* (2°) amines are R_2NH, and *tertiary* (3°) amines are R_3N. Like ammonia, amines have trigonal pyramidal shapes, and the lone pair of electrons results in a partially negative N atom (Figure 15.11).

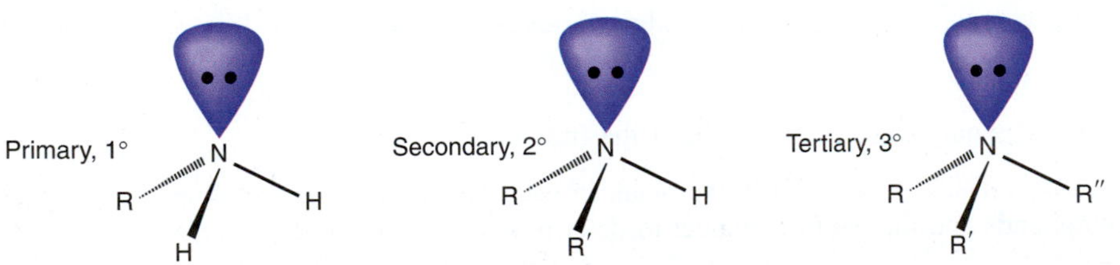

Figure 15.11 General structures of amines.

Systematic names drop the final *-e* of the alkane and add the suffix *-amine,* as in ethanamine. However, there is still wide usage of common names, in which the suffix *-amine* follows the name of the alkyl group; thus, methylamine has one methyl group attached to N, diethylamine has two ethyl groups attached, and so forth. Figure 15.12 shows that the amine functional group occurs in many biomolecules.

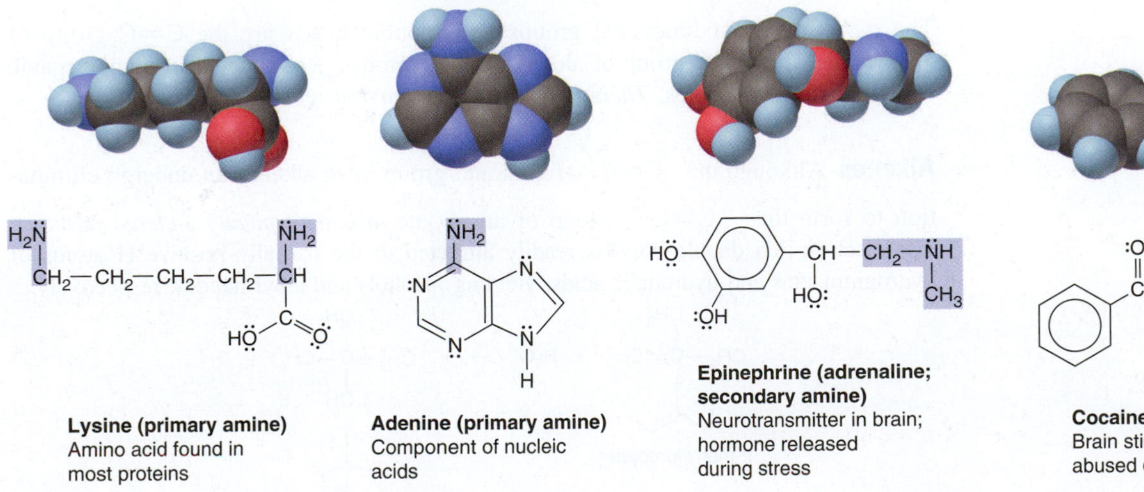

Figure 15.12 Some biomolecules with the amine functional group.

Amines undergo substitution reactions in which the lone pair of N attacks the partially positive C in an alkyl halide to displace X^- and form a larger amine:

$$2\,CH_3{-}CH_2{-}\ddot{N}H_2 + CH_3{-}CH_2{-}Cl \longrightarrow CH_3{-}CH_2{-}\underset{\displaystyle CH_3{-}CH_2}{\underset{|}{\ddot{N}H}} + CH_3{-}CH_2{-}\overset{+}{N}H_3Cl^-$$

ethylamine — chloroethane — diethylamine — ethylammonium chloride

(Two molecules of amine are needed: one attacks the chloroethane, and the other binds the released H^+ to prevent it from remaining on the diethylamine product.)

Sample Problem 15.4 Predicting the Reactions of Alcohols, Alkyl Halides, and Amines

Problem Determine the reaction type and predict the product(s) for each reaction:

(a) $CH_3{-}CH_2{-}CH_2{-}I + NaOH \longrightarrow$

(b) $CH_3{-}CH_2{-}Br + 2\,CH_3{-}CH_2{-}CH_2{-}NH_2 \longrightarrow$

(c) $CH_3{-}\underset{\displaystyle OH}{\underset{|}{C}H}{-}CH_3 \xrightarrow[H_2SO_4]{Cr_2O_7^{2-}}$

Plan We first determine the functional group(s) of the organic reactant(s) and then examine any inorganic reactant(s) to decide on the reaction type, keeping in mind that, in general, these functional groups undergo substitution or elimination. **(a)** The reactant is an alkyl halide, so the OH^- of the inorganic base NaOH substitutes for I^-. **(b)** The reactants are an amine and an alkyl halide, so the N of the amine substitutes for the Br. **(c)** The organic reactant is an alcohol, the inorganic reactants comprise a strong oxidizing agent, and the alcohol group undergoes elimination to form a C=O.

Solution **(a)** Substitution: The products are $CH_3{-}CH_2{-}CH_2{-}OH + NaI$

(b) Substitution: The products are $CH_3{-}CH_2{-}CH_2{-}\underset{\displaystyle CH_2{-}CH_3}{\underset{|}{NH}} + CH_3{-}CH_2{-}CH_2{-}\overset{+}{N}H_3Br^-$

(c) Elimination (oxidation): The product is $CH_3{-}\underset{\displaystyle O}{\underset{\|}{C}}{-}CH_3$

Check The only changes should be at the functional group.

FOLLOW-UP PROBLEM 15.4 Fill in the blank in each reaction. (*Hint:* Examine any inorganic compounds and the organic product to determine the organic reactant.)

(a) ____ $+ CH_3{-}ONa \longrightarrow CH_3{-}CH{=}\overset{\displaystyle CH_3}{\overset{|}{C}}{-}CH_3 + NaCl + CH_3{-}OH$

(b) ____ $\xrightarrow[H_2SO_4]{Cr_2O_7^{\,2-}} CH_3{-}CH_2{-}\overset{\displaystyle O}{\overset{\|}{C}}{-}OH$

Functional Groups with Double Bonds

The most important functional groups with double bonds are the C=C group of alkenes and the C=O group of aldehydes and ketones. Both appear in many organic and biological molecules. *Their most common reaction type is addition.*

Alkenes Although the $\rangle C{=}C\langle$ functional group in an alkene can undergo elimination to form the $-C{\equiv}C-$ group of an alkyne, *alkenes typically undergo addition.* The electron-rich double bond is readily attracted to the partially positive H atoms of hydronium ions and hydrohalic acids, yielding alcohols and alkyl halides, respectively:

$$CH_3{-}\overset{\displaystyle CH_3}{\overset{|}{C}}{=}CH_2 + H_3O^+ \longrightarrow CH_3{-}\overset{\displaystyle CH_3}{\overset{|}{\underset{\displaystyle OH}{\underset{|}{C}}}}{-}CH_3 + H^+$$

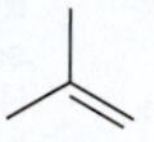

2-methylpropene 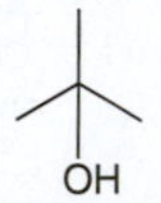 — 2-methyl-2-propanol

$$CH_3{-}CH{=}CH_2 + HCl \longrightarrow CH_3{-}CHCl{-}CH_3$$

propene → 2-chloropropane

Aldehydes and Ketones The C=O bond, or **carbonyl group,** is one of the most chemically versatile.

- In the **aldehyde** functional group, the carbonyl C is bonded to H (and often to another C), so it occurs *at the end of a chain,* R—C(H)=Ö. Aldehyde names drop the final *-e* from the alkane name and add *-al;* thus, the three-C aldehyde is propanal.
- In the **ketone** functional group, the carbonyl C is bonded to two other C atoms, —C—C(=:O:)—C—, so it occurs *within the chain.* Ketones, R—C(=O)—R′, are named by numbering the carbonyl C, dropping the final *-e* from the alkane name, and adding *-one.* For example, the unbranched, five-C ketone with the carbonyl C as C-2 in the chain is named 2-pentanone. Figure 15.13 shows some common carbonyl compounds.

Figure 15.13 Some common aldehydes and ketones.

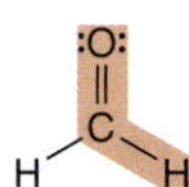

Methanal (formaldehyde)
Used to make resins in plywood, dishware, countertops; biological preservative

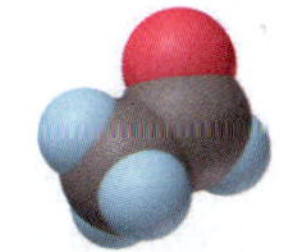
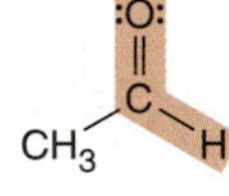

Ethanal (acetaldehyde)
Narcotic product of ethanol metabolism; used to make perfumes, flavors, plastics, other chemicals

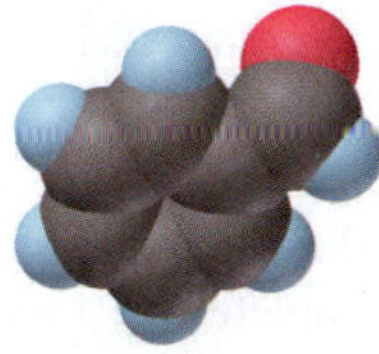
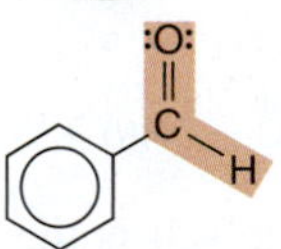

Benzaldehyde
Artificial almond flavoring

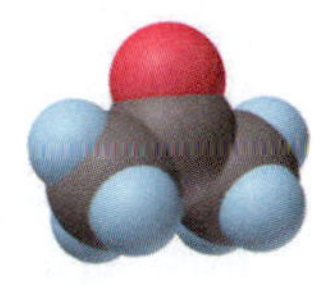
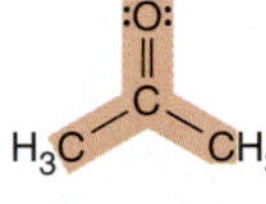

2-Propanone (acetone)
Solvent for fat, rubber, plastic, varnish, lacquer; chemical feedstock

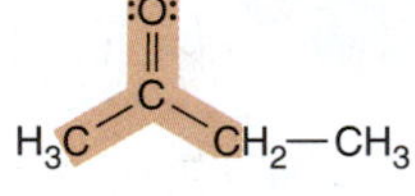

2-Butanone (methyl ethyl ketone)
Important solvent

Aldehydes and ketones are formed by the oxidation of alcohols:

$$CH_3{-}CH_2{-}OH \xrightarrow{\text{oxidation}} CH_3{-}C(=O){-}H$$

ethanol → ethanal (common name, acetaldehyde)

3-pentanol →(oxidation) 3-pentanone (common name, diethylketone)

Conversely, as a result of their unsaturation, carbonyl compounds can undergo *addition* and be reduced to alcohols:

cyclobutanone →(reduction) cyclobutanol

Like the C=C bond, the C=O bond is *electron rich;* unlike the C=C bond, it is *highly polar* ($\Delta EN = 1.0$). Figure 15.14 shows an orbital contour model (Figure 15.14A) and depicts this polarity with a charged resonance form (Figure 15.14B).

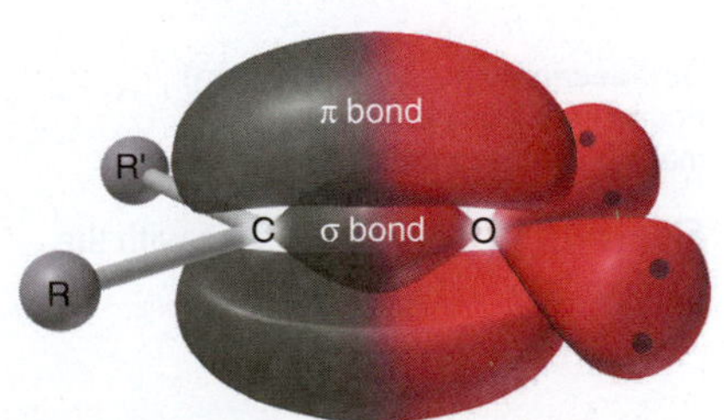

A Orbital contour model

R′R C=Ö ⟷ R′R C^{+}—Ö:$^{-}$

B Charged resonance form

Figure 15.14 The polar carbonyl group.

Functional Groups with Both Single and Double Bonds

A family of three functional groups contains C double bonded to O (a carbonyl group) *and* single bonded to O or N. The parent of the family is the **carboxylic acid** group, $-\overset{\overset{:\text{O}:}{\|}}{\text{C}}-\ddot{\underset{..}{\text{O}}}\text{H}$, also called the *carboxyl group* and written —COOH. *The most important reaction type of this family is substitution from one member to another.* Substitution for the —OH by the —OR of an alcohol gives the **ester** group, $-\overset{\overset{:\text{O}:}{\|}}{\text{C}}-\ddot{\underset{..}{\text{O}}}-\text{R}$; substitution by the $-\underset{|}{\ddot{\text{N}}}-$ of an amine gives the **amide** group, $-\overset{\overset{:\text{O}:}{\|}}{\text{C}}-\underset{..}{\overset{|}{\text{N}}}-$.

Carboxylic Acids Carboxylic acids, $\text{R}-\overset{\overset{\text{O}}{\|}}{\text{C}}-\text{OH}$, are named by dropping the *-e* from the alkane name and adding *-oic acid;* however, many common names are used. For example, the systematic name of the four-C acid is butanoic acid (the carboxyl C is counted when choosing the root); its common name is butyric acid. Figure 15.15 shows some important carboxylic acids. The carboxyl C already has three bonds, so it forms only one other. In formic acid (methanoic acid), the carboxyl C is bonded to an H, but in all other carboxylic acids it is bonded to a chain or ring.

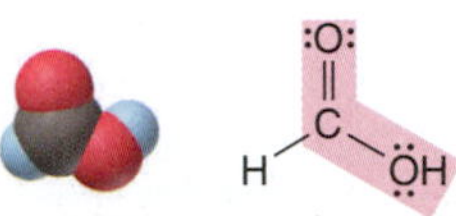

Methanoic acid (formic acid)
An irritating component of ant and bee stings

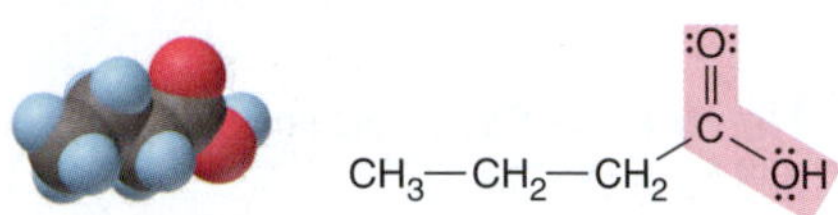

Butanoic acid (butyric acid)
Odor of rancid butter; suspected component of monkey sex attractant

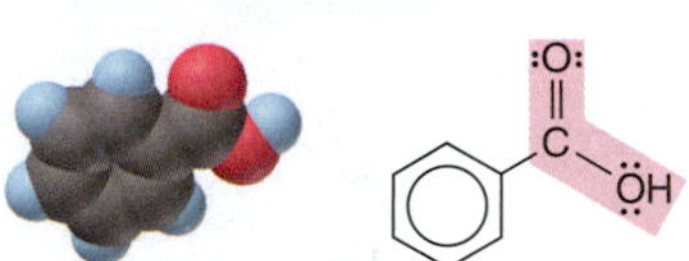

Benzoic acid
Calorimetric standard; used in preserving food, dyeing fabric, curing tobacco

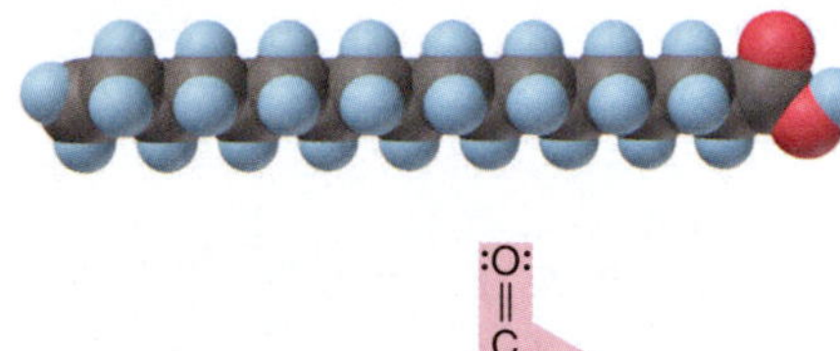

Octadecanoic acid (stearic acid)
Found in animal fats; used in making candles and soaps

Figure 15.15 Some molecules with the carboxylic acid functional group.

Carboxylic acids are weak acids in water:

$$\underset{\substack{\text{ethanoic acid}\\ \text{(acetic acid)}}}{\text{CH}_3-\overset{\overset{\text{O}}{\|}}{\text{C}}-\text{OH}(l)} + \text{H}_2\text{O}(l) \rightleftharpoons \text{CH}_3-\overset{\overset{\text{O}}{\|}}{\text{C}}-\text{O}^-(aq) + \text{H}_3\text{O}^+(aq)$$

At equilibrium in a solution of typical concentration, more than 99% of the acid molecules are undissociated at any given moment. In strong base, however, a carboxylic acid reacts completely to form a salt and water:

$$\text{CH}_3-\overset{\overset{\text{O}}{\|}}{\text{C}}-\text{OH}(l) + \text{NaOH}(aq) \longrightarrow \text{CH}_3-\overset{\overset{\text{O}}{\|}}{\text{C}}-\text{O}^-(aq) + \text{Na}^+(aq) + \text{H}_2\text{O}(l)$$

The anion of the salt is the *carboxylate ion,* named by dropping *-oic acid* and adding *-oate;* the sodium salt of butanoic acid, for instance, is sodium butanoate.

Carboxylic acids with long hydrocarbon chains (usually with an even number of C atoms) are **fatty acids,** an essential group of compounds found in all cells. Animal fatty acids have saturated chains (see stearic acid, Figure 15.15, *bottom*), whereas many from plants are unsaturated. Fatty acid salts, usually having a cation from Group 1A(1) or 2A(2), are soaps. When clothes with greasy spots are immersed in soapy water, the nonpolar "tails" of the soap molecules interact with the grease, while the ionic "heads" interact with the water. Agitation of the water rinses the grease away.

Substitution of carboxylic acids occurs through a two-step sequence: *addition plus elimination equals substitution.* Addition to the trigonal planar C atom gives an unstable tetrahedral intermediate, which immediately undergoes elimination to revert to a trigonal planar product (in this case, X is OH):

$$\text{R}-\overset{\overset{\text{O}}{\|}}{\text{C}}-\text{X} + \text{Z}-\text{Y} \underset{}{\overset{\text{addition}}{\rightleftharpoons}} \left[\text{R}-\underset{\underset{\text{Y}}{|}}{\overset{\overset{\text{O}-\text{Z}}{|}}{\text{C}}}-\text{X}\right] \overset{\text{elimination}}{\rightleftharpoons} \text{R}-\overset{\overset{\text{O}}{\|}}{\text{C}}-\text{Y} + \text{Z}-\text{X}$$

Strong heating of a carboxylic acid forms an **acid anhydride** through a type of substitution called a *dehydration-condensation reaction* (Section 14.6), in which two molecules condense into one with loss of water:

$$\text{R}-\overset{\overset{\text{O}}{\|}}{\text{C}}\boxed{-\text{OH} + \text{H}}\text{O}-\overset{\overset{\text{O}}{\|}}{\text{C}}-\text{R} \xrightarrow{\Delta} \text{R}-\overset{\overset{\text{O}}{\|}}{\text{C}}-\text{O}-\overset{\overset{\text{O}}{\|}}{\text{C}}-\text{R} + \text{HOH}$$

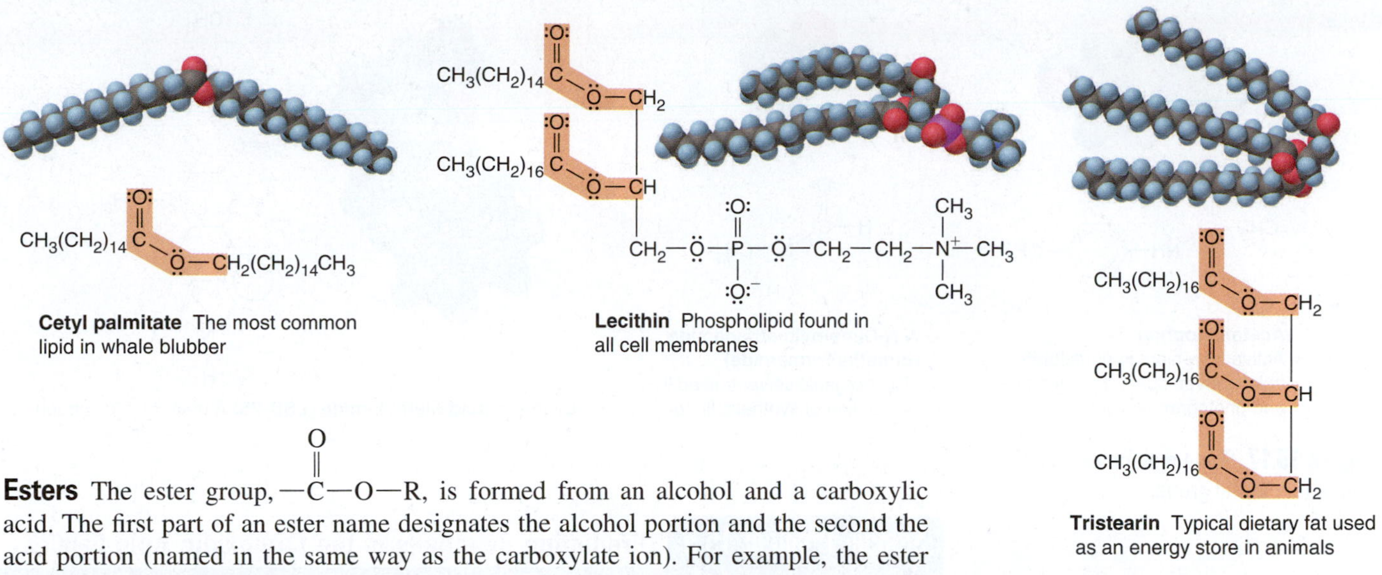

$$CH_3(CH_2)_{14}C(=\ddot{O}:)-\ddot{O}-CH_2(CH_2)_{14}CH_3$$

Cetyl palmitate The most common lipid in whale blubber

$$CH_3(CH_2)_{14}C(=\ddot{O}:)-\ddot{O}-CH_2$$
$$CH_3(CH_2)_{16}C(=\ddot{O}:)-\ddot{O}-CH$$
$$CH_2-\ddot{O}-P(=\ddot{O}:)(-\ddot{O}:^-)-\ddot{O}-CH_2-CH_2-N^+(CH_3)_2-CH_3$$

Lecithin Phospholipid found in all cell membranes

$$CH_3(CH_2)_{16}C(=\ddot{O}:)-\ddot{O}-CH_2$$
$$CH_3(CH_2)_{16}C(=\ddot{O}:)-\ddot{O}-CH$$
$$CH_3(CH_2)_{16}C(=\ddot{O}:)-\ddot{O}-CH_2$$

Tristearin Typical dietary fat used as an energy store in animals

Figure 15.16 Some lipid molecules with the ester functional group.

Esters The ester group, $-\overset{\overset{\large O}{\|}}{C}-O-R$, is formed from an alcohol and a carboxylic acid. The first part of an ester name designates the alcohol portion and the second the acid portion (named in the same way as the carboxylate ion). For example, the ester formed between ethanol and ethanoic acid is ethyl ethanoate (common name, ethyl acetate), a solvent for nail polish and model glue.

The ester group occurs commonly in **lipids,** a large group of fatty biological substances. Most dietary fats are *triglycerides,* esters that are composed of three fatty acids linked to the alcohol 1,2,3-trihydroxypropane (common name, glycerol) and that function as energy stores. Some important lipids are shown in Figure 15.16; lecithin is one of several phospholipids that make up the lipid bilayer in all cell membranes.

Esters, like acid anhydrides, form through a dehydration-condensation reaction; in this case, it is called an *esterification*:

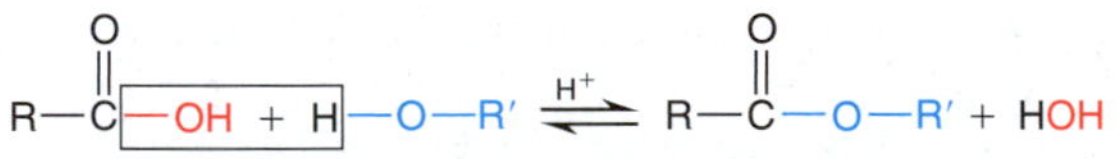

$$R-\overset{\overset{\large O}{\|}}{C}-OH + H-O-R' \overset{H^+}{\rightleftharpoons} R-\overset{\overset{\large O}{\|}}{C}-O-R' + HOH$$

Note that the esterification reaction is reversible. The opposite of dehydration-condensation is **hydrolysis,** in which the O atom of water is attracted to the partially positive C atom of the ester, cleaving (lysing) the ester molecule into two parts. One part bonds to water's —OH, and the other part to water's other H.

Amides The product of a substitution between an amine (or NH_3) and an ester is an amide with the functional group $-\overset{\overset{\large O}{\|}}{C}-\overset{|}{N}-$. The partially negative N of the amine is attracted to the partially positive C of the ester, an alcohol (ROH) is lost, and an amide forms:

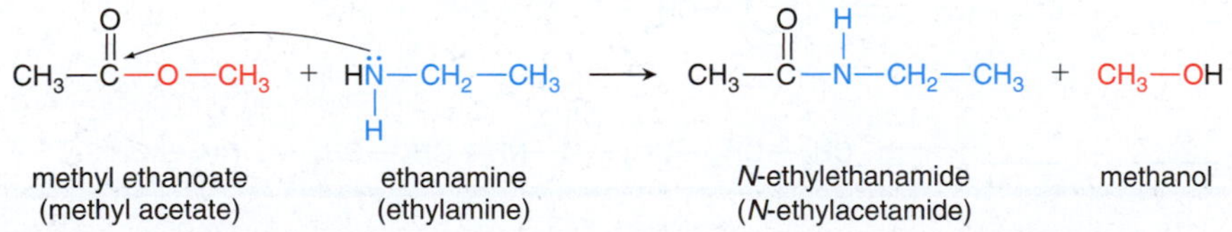

$$CH_3-\overset{\overset{\large O}{\|}}{C}-O-CH_3 + H\ddot{N}(H)-CH_2-CH_3 \longrightarrow CH_3-\overset{\overset{\large O}{\|}}{C}-N(H)-CH_2-CH_3 + CH_3-OH$$

methyl ethanoate (methyl acetate)   ethanamine (ethylamine)   *N*-ethylethanamide (*N*-ethylacetamide)   methanol

Amides are named by denoting the amine portion with *N*- and replacing *-oic acid* from the parent carboxylic acid with *-amide.* In the amide produced in the preceding reaction, the ethyl group comes from the amine, and the acid portion comes from ethanoic acid (acetic acid). Some amides are shown in Figure 15.17 on the next page. The most important example of the amide group is the *peptide bond* (discussed in Section 15.6), which links amino acids in a protein.

An amide is hydrolyzed in hot water (or base) to a carboxylic acid (or carboxylate ion) and an amine. Thus, even though amides are not normally formed in the following way, they can be viewed as the result of a reversible dehydration-condensation:

$$R-\overset{\overset{\large O}{\|}}{C}-OH + H-N(H)-R' \rightleftharpoons R-\overset{\overset{\large O}{\|}}{C}-N(H)-R' + HOH$$

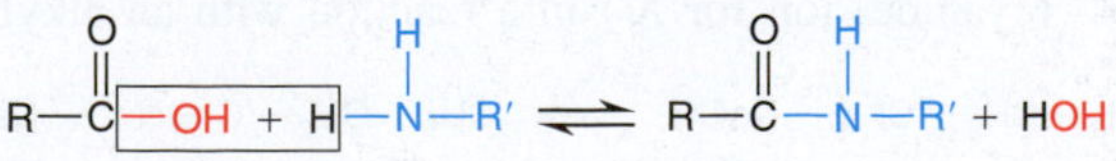

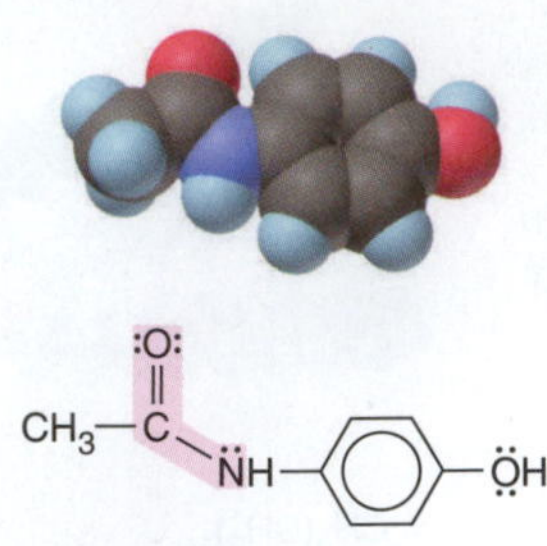

Acetaminophen
Active ingredient in nonaspirin pain relievers; used to make dyes and photographic chemicals

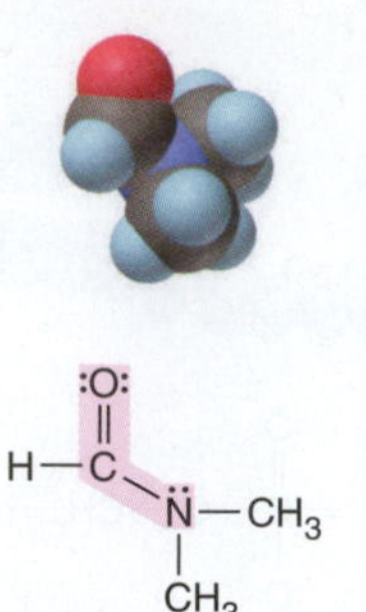

***N,N*-Dimethylmethanamide (dimethylformamide)**
Major organic solvent; used in production of synthetic fibers

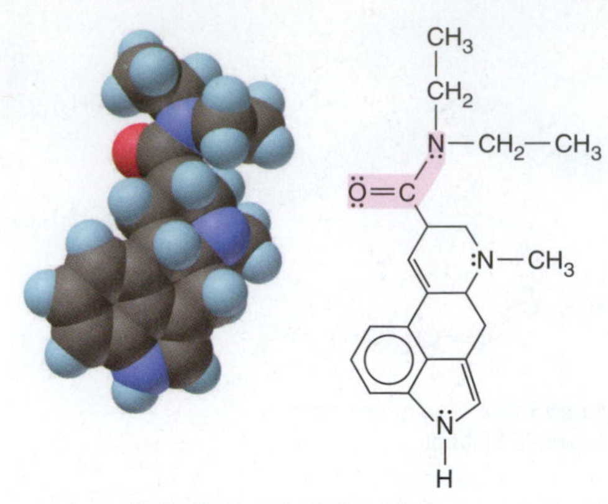

Lysergic acid diethylamide (LSD-25) A potent hallucinogen

Figure 15.17 Some molecules with the amide functional group.

Sample Problem 15.5 Predicting Reactions of the Carboxylic Acid Family

Problem Predict the product(s) of the following reactions:

$$\textbf{(a)}\ CH_3-CH_2-CH_2-\overset{\overset{O}{\|}}{C}-OH + CH_3-\overset{\overset{OH}{|}}{C}H-CH_3 \xrightleftharpoons{H^+}$$

$$\textbf{(b)}\ CH_3-\overset{\overset{CH_3}{|}}{C}H-CH_2-CH_2-\overset{\overset{O}{\|}}{C}-NH-CH_2-CH_3 \xrightarrow[H_2O]{NaOH}$$

Plan We discussed substitution reactions (including addition-elimination and dehydration-condensation) and hydrolysis. **(a)** A carboxylic acid and an alcohol react, so the reaction must be a substitution to form an ester and water. **(b)** An amide reacts with OH^-, so it is hydrolyzed to an amine and a sodium carboxylate.

Solution (a) Formation of an ester:

$$CH_3-CH_2-CH_2-\overset{\overset{O}{\|}}{C}-O-\overset{\overset{CH_3}{|}}{C}H-CH_3 + H_2O$$

(b) Basic hydrolysis of an amide:

$$CH_3-\overset{\overset{CH_3}{|}}{C}H-CH_2-CH_2-\overset{\overset{O}{\|}}{C}-O^- + Na^+ + H_2N-CH_2-CH_3$$

Check Note that in part (b), the carboxylate ion forms, rather than the acid, because the aqueous NaOH present reacts with any carboxylic acid as it forms.

FOLLOW-UP PROBLEM 15.5 Fill in the blanks in the following reactions:

$$\textbf{(a)}\ \_\_\_\_ + CH_3-OH \xrightleftharpoons{H^+} C_6H_5-CH_2-\overset{\overset{O}{\|}}{C}-O-CH_3 + H_2O$$

$$\textbf{(b)}\ \_\_\_\_ + \_\_\_\_ \longrightarrow CH_3-CH_2-CH_2-\overset{\overset{O}{\|}}{C}-NH-CH_2-CH_3 + CH_3-OH$$

Functional Groups with Triple Bonds

Alkynes and nitriles are the only two important functional groups with triple bonds.

Alkynes Alkynes, with an electron-rich $-C{\equiv}C-$ group, undergo addition (by H_2O, H_2, HX, X_2, and so forth) to form double-bonded or saturated compounds:

$$\underset{\text{propyne}}{CH_3-C{\equiv}CH} \xrightarrow{H_2} \underset{\text{propene}}{CH_3-CH{=}CH_2} \xrightarrow{H_2} \underset{\text{propane}}{CH_3-CH_2-CH_3}$$

Nitriles Nitriles ($R-C{\equiv}N$) contain the **nitrile** group ($-C{\equiv}N\!:$) and are made by substituting a CN^- (cyanide) ion for X^- in a reaction with an alkyl halide:

$$CH_3-CH_2-Cl + NaCN \longrightarrow CH_3-CH_2-C{\equiv}N + NaCl$$

This reaction is useful because it *increases the chain by one C atom.* Nitriles are versatile because once they are formed, they can be reduced to amines or hydrolyzed to carboxylic acids:

$$CH_3-CH_2-CH_2-NH_2 \xleftarrow{\text{reduction}} CH_3-CH_2-C\equiv N \xrightarrow[\text{hydrolysis}]{H_3O^+,\ H_2O} CH_3-CH_2-\overset{\overset{O}{\|}}{C}-OH + NH_4^+$$

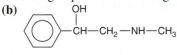

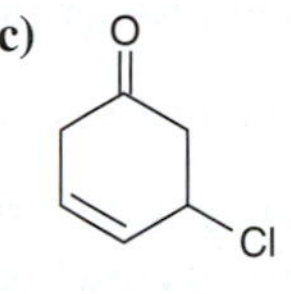

Sample Problem 15.6 Recognizing Functional Groups

Problem Circle and name the functional groups in the following molecules:

(a) (b) (c)

Plan We use Table 15.5 (p. 475) to identify the various functional groups.

Solution

(a)

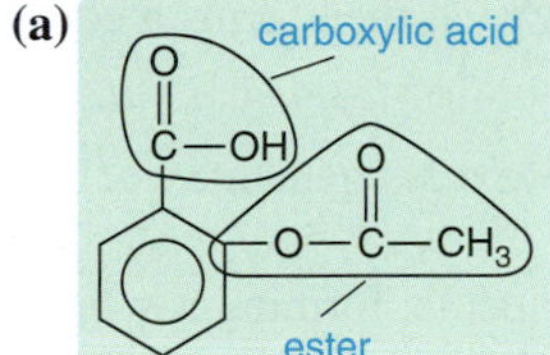

(b)

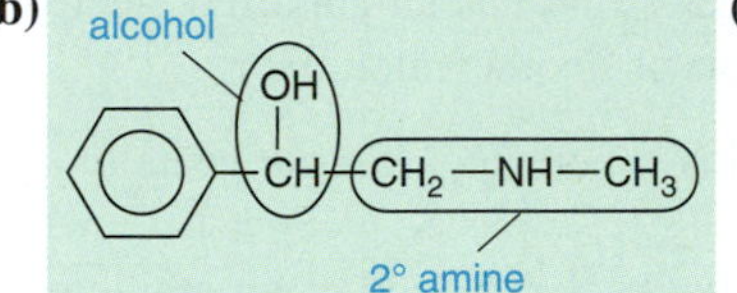

(c)

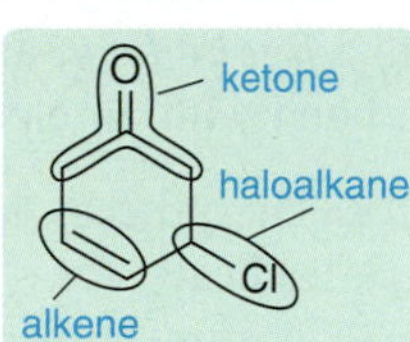

FOLLOW-UP PROBLEM 15.6 Circle and name the functional groups:

(a)

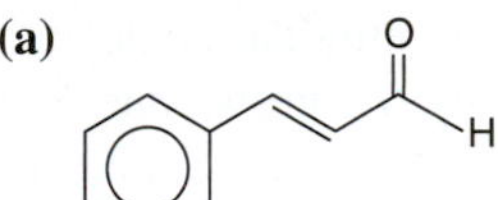

(b) $H_2N-\overset{\overset{O}{\|}}{C}-CH_2-\underset{\underset{Br}{|}}{CH}-CH_3$

Summary of Section 15.4

- Organic reactions are initiated when regions of high and low electron density in reactant molecules attract each other.
- Functional groups containing only single bonds—alcohols, amines, and alkyl halides—take part in substitution and elimination reactions.
- Functional groups with double or triple bonds—alkenes, aldehydes, ketones, alkynes, and nitriles—generally take part in addition reactions.
- Functional groups with both double and single bonds—carboxylic acids, esters, and amides—generally take part in substitution reactions.
- Many reactions change one functional group to another, but some, such as those involving the cyanide ion, change the carbon skeleton.

15.5 • THE MONOMER-POLYMER THEME I: SYNTHETIC MACROMOLECULES

In its simplest form, a **polymer** (Greek, "many parts") is an extremely large molecule, or **macromolecule,** consisting of a covalently linked chain of smaller molecules, called **monomers** (Greek, "one part"). The monomer is the *repeat unit* of the polymer, and a typical polymer may have from hundreds to hundreds of thousands of repeat units. There are many types of monomers, and their chemical structures allow for the complete repertoire of intermolecular forces. *Synthetic* polymers are created by chemical reactions in the laboratory; *natural* polymers (or *biopolymers*) are created

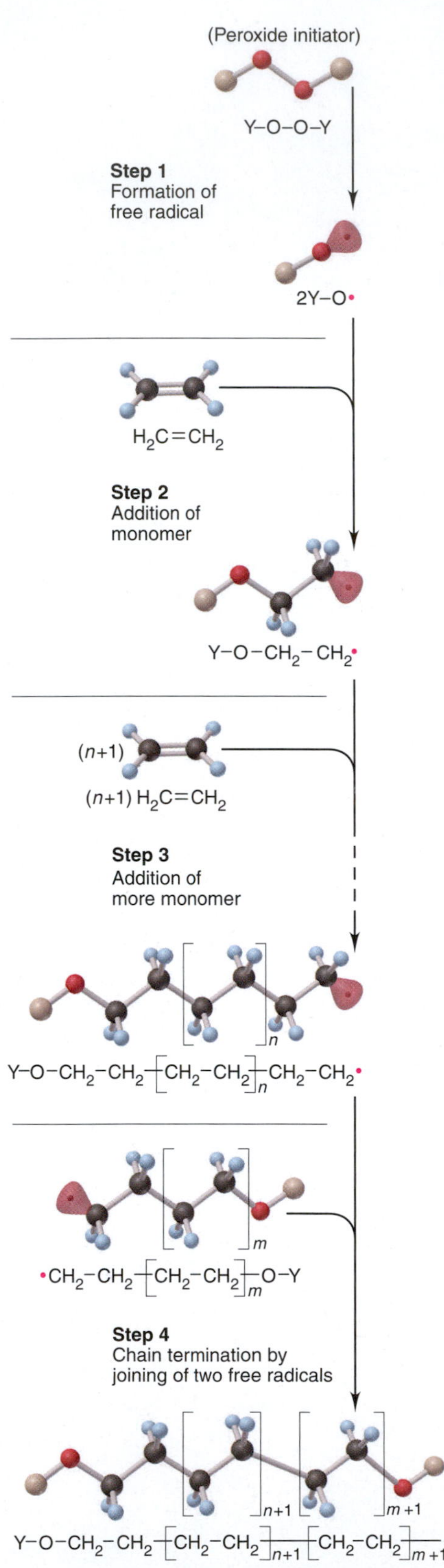

Figure 15.18 Steps in the free-radical polymerization of ethylene.

by chemical reactions within organisms, and we'll discuss them in the next section. In this section, we see how synthetic polymers are named and discuss the two types of reactions that link monomers covalently into a chain.

To name a polymer, we just add the prefix *poly-* to the monomer name, as in *polyethylene* or *polystyrene*. When the monomer has a two-word name, parentheses are used, as in *poly(vinyl chloride).*

The two major types of reaction processes that form synthetic polymers lend their names to the resulting polymer classes—addition and condensation.

Addition Polymers

Addition polymers form when monomers undergo an addition reaction with one another. These substances are also called *chain-reaction* (or *chain-growth*) *polymers* because as each monomer adds to the chain, it forms a new reactive site to continue the process. The monomers of most addition polymers have the $\gt C{=}C\lt$ grouping. As you can see from Table 15.6, the essential chemical differences between an acrylic sweater, a plastic grocery bag, and a bowling ball are due to the different groups that are attached to the double-bonded C atoms of the monomer.

The *free-radical polymerization* of ethene (ethylene, $CH_2{=}CH_2$) to polyethylene is a simple example of the addition process (Figure 15.18). The monomer reacts to form a *free radical,* a species that has an unpaired electron, which then forms a covalent bond with an electron of another monomer:

Step 1. The process begins when an *initiator,* usually a peroxide, generates a free radical.

Step 2. The free radical attacks the π bond of an ethylene molecule, forming a σ bond with one of the *p* electrons and leaving the other unpaired, creating a new free radical.

Step 3. This new free radical then attacks the π bond of another ethylene molecule, joining it to the chain end, and the backbone of the polymer grows one unit longer.

Step 4. This process stops when two free radicals form a covalent bond or when a very stable free radical is formed by addition of an *inhibitor* molecule.

The most important polymerization reactions are *stereoselective* and create polymers whose repeat units have groups spatially oriented in particular ways. Through the use of these reactions, polyethylene chains with molar masses of 10^4 to 10^5 g/mol are made by varying conditions and reagents.

Similar methods are used to make polypropylenes, $\left[CH_2{-}\underset{\displaystyle CH_3}{\underset{|}{CH}}\right]_n$, that have all the CH_3 groups of the repeat units oriented either on one side of the chain or on alternating sides. The different orientations lead to different packing efficiencies of the chains and, thus, different degrees of crystallinity, which lead to differences in such physical properties as density, rigidity, and elasticity.

Condensation Polymers

The monomers of **condensation polymers** must have *two functional groups;* we can designate such a monomer as A—**R**—B (where A and B may or may not be the same, and **R** is the rest of the molecule). Most commonly, the monomers link when an A group on one undergoes a *dehydration-condensation reaction* with a B group on another:

$$\tfrac{1}{2}nH{-}A{-}\mathbf{R}{-}B{-}OH + \tfrac{1}{2}nH{-}A{-}\mathbf{R}{-}B{-}OH \xrightarrow{-(n-1)HOH} H{+}A{-}\mathbf{R}{-}B{+}_nOH$$

Many condensation polymers are *copolymers,* those consisting of two or more different repeat units. Two major types are polyamides and polyesters.

Table 15.6 Some Major Addition Polymers

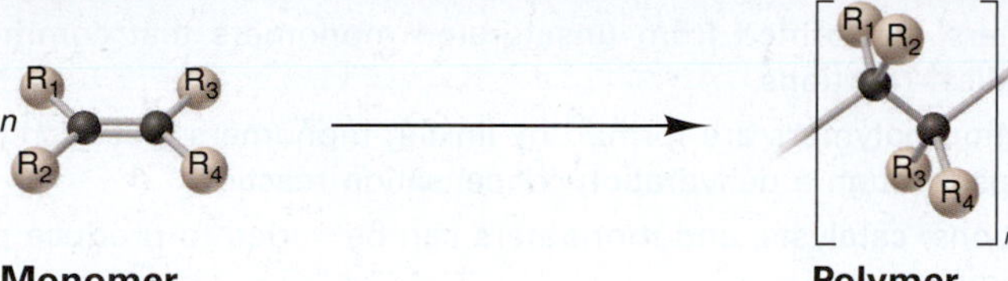

Monomer	Polymer	Applications
$H_2C=CH_2$	polyethylene	Plastic bags; bottles; toys
$F_2C=CF_2$	polytetrafluoroethylene	Cooking utensils (e.g., Teflon)
$H_2C=CH-CH_3$	polypropylene	Carpeting (indoor-outdoor); bottles
$H_2C=CH-Cl$	poly(vinyl chloride)	Plastic wrap; garden hose; indoor plumbing
$H_2C=CH-C_6H_5$	polystyrene	Insulation; furniture; packing materials
$H_2C=CH-CN$	polyacrylonitrile	Yarns (e.g., Orlon, Acrilan), fabrics, and wigs
$H_2C=CH-O-C(=O)-CH_3$	poly(vinyl acetate)	Adhesives; paints; textile coatings; computer disks
$H_2C=CCl_2$	poly(vinylidene chloride)	Food wrap (e.g., Saran)
$H_2C=C(CH_3)-C(=O)-O-CH_3$	poly(methyl methacrylate)	Glass substitute (e.g., Lucite, Plexiglas); bowling balls; paint

1. *Polyamides.* Condensation of carboxylic acid and amine monomers forms *polyamides (nylons).* One of the most common is *nylon-66,* manufactured by mixing equimolar amounts of a six-C diamine (1,6-diaminohexane) and a six-C diacid (1,6-hexanedioic acid). The basic amine reacts with the acid to form a "nylon salt." Heating drives off water and forms the amide bonds:

$$n\text{HO}-\overset{\text{O}}{\overset{\|}{\text{C}}}-(\text{CH}_2)_4-\overset{\text{O}}{\overset{\|}{\text{C}}}-\text{OH} + n\text{H}_2\text{N}-(\text{CH}_2)_6-\text{NH}_2 \xrightarrow[-(2n-1)\text{H}_2\text{O}]{\Delta} \text{HO}\left[\overset{\text{O}}{\overset{\|}{\text{C}}}-(\text{CH}_2)_4-\overset{\text{O}}{\overset{\|}{\text{C}}}-\text{NH}-(\text{CH}_2)_6-\text{NH}\right]_n\text{H}$$

Covalent bonds within the chains give nylons great strength, and H bonds between chains give them great flexibility. About half of all nylons are made to reinforce automobile tires; the others are used for rugs, clothing, fishing line, and so forth.

2. *Polyesters.* Condensation of carboxylic acid and alcohol monomers forms *polyesters.* Dacron, a popular polyester fiber, is woven from polymer strands formed when equimolar amounts of 1,4-benzenedicarboxylic acid and 1,2-ethanediol react. Blending these polyester fibers with various amounts of cotton gives fabrics that are durable, easily dyed, and crease resistant. Extremely thin Mylar films, used for recording tape and food packaging, are also made from this polymer.

Summary of Section 15.5

- Polymers are extremely large molecules that are made of repeat units called monomers.
- Addition polymers are formed from unsaturated monomers that commonly link through free-radical reactions.
- Most condensation polymers are formed by linking monomers that each have two functional groups through a dehydration-condensation reaction.
- Reaction conditions, catalysts, and monomers can be varied to produce polymers with different properties.

15.6 • THE MONOMER-POLYMER THEME II: BIOLOGICAL MACROMOLECULES

The monomer-polymer theme was being played out in nature eons before humans employed it to such great advantage. Biological macromolecules are condensation polymers created by nature's reaction chemistry and improved through evolution. These remarkable molecules are the best demonstration of the versatility of carbon and its handful of atomic partners.

Natural polymers, such as polysaccharides, proteins, and nucleic acids, are the "stuff of life." Some have structures that make wood strong, fingernails hard, and wool flexible. Others speed up the myriad reactions that occur in every cell or defend the body against infection. Still others possess the genetic information organisms need to forge other biomolecules. Remarkable as these giant molecules are, the functional groups of their monomers and the reactions that link them are identical to those of other, smaller organic molecules, and the same intermolecular forces that dissolve smaller molecules stabilize these giant molecules in the aqueous medium of the cell.

Figure 15.19 **A,** The structure of glucose in aqueous solution. **B,** The formation of a disaccharide.

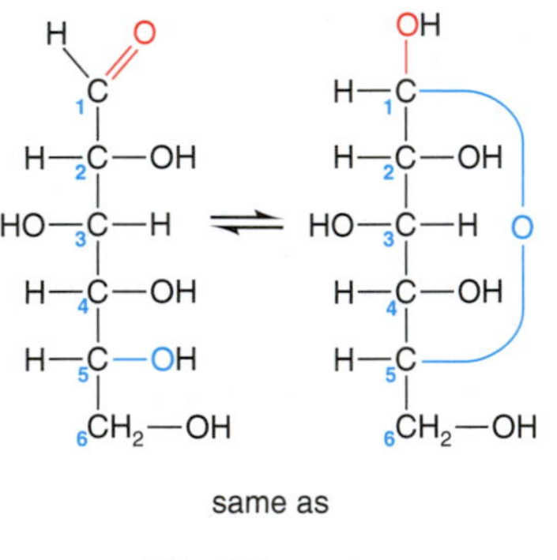

Sugars and Polysaccharides

In essence, the same chemical change occurs when you burn a piece of wood or eat a piece of bread. Wood and bread are mixtures of *carbohydrates,* substances that provide energy through oxidation.

Monomer Structure and Linkage Glucose and other simple sugars, from the three-C *trioses* to the seven-C *heptoses,* are called **monosaccharides** and consist of carbon chains with attached hydroxyl and carbonyl groups. In addition to their roles as individual molecules engaged in energy metabolism, they serve as the monomer units of **polysaccharides.** Most natural polysaccharides are formed from five- and six-C units. In aqueous solution, *an alcohol group and the aldehyde (or ketone) group of the same monosaccharide react with each other to form a cyclic molecule with either a five- or six-membered ring* (Figure 15.19A).

When two monosaccharides undergo a dehydration-condensation reaction, a **disaccharide** forms. For example, sucrose (table sugar) is a disaccharide of glucose and fructose (Figure 15.19B).

Types of Polysaccharides A polysaccharide consists of *many* monosaccharide units linked together. The three major natural polysaccharides consist entirely of glucose units, but they differ with respect to how these units are linked.

- *Cellulose* is the most abundant organic chemical on Earth. More than 50% of the carbon in plants occurs in the cellulose of stems and leaves; wood is largely

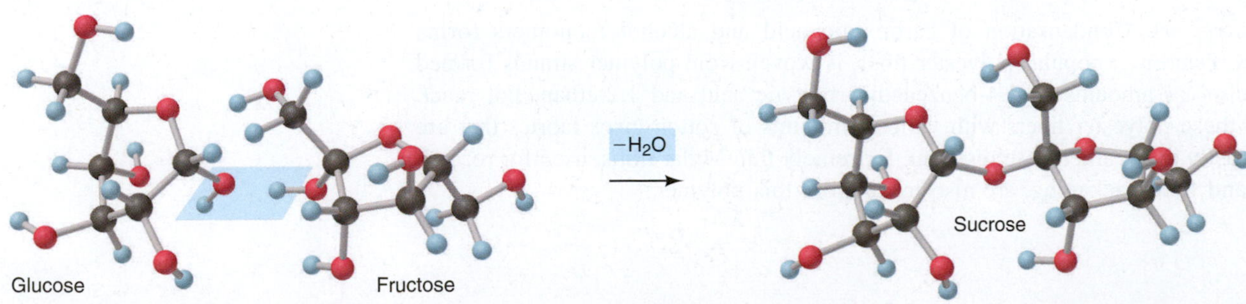

cellulose, and cotton is more than 90% cellulose. Cellulose consists of long chains of glucose. The great strength of wood is due largely to the countless H bonds between cellulose chains.

- *Starch* is a mixture of polysaccharides of glucose and serves as the energy storage molecule in plants. It occurs as a helical molecule of several thousand glucose units mixed with a highly branched, bushlike molecule of up to a million glucose units.
- *Glycogen* functions as the energy storage molecule in animals. It occurs in liver and muscle cells as large, insoluble granules consisting of glycogen molecules made from 1000 to more than 500,000 glucose units.

The carbons involved in the bonds between glucose units in these polysaccharides differ in their chirality. Humans lack the enzyme to break the particular link in cellulose, so we cannot digest it (unfortunately!), but we can break the links in starch and glycogen.

Amino Acids and Proteins

As you saw in Section 15.5, synthetic polyamides (such as nylon-66) are formed from two monomers, one with a carboxyl group at each end and the other with an amine group at each end. **Proteins,** the polyamides of nature, are unbranched polymers formed from monomers called **amino acids**, *each of which has both a carboxyl group and an amine group.*

Monomer Structure and Linkage An amino acid has both its carboxyl group and its amine group attached to the α*-carbon,* the second C atom in the chain. Proteins are made up of about 20 different types of amino acids, each with its own particular R group (gray-screened group in Figure 15.20).

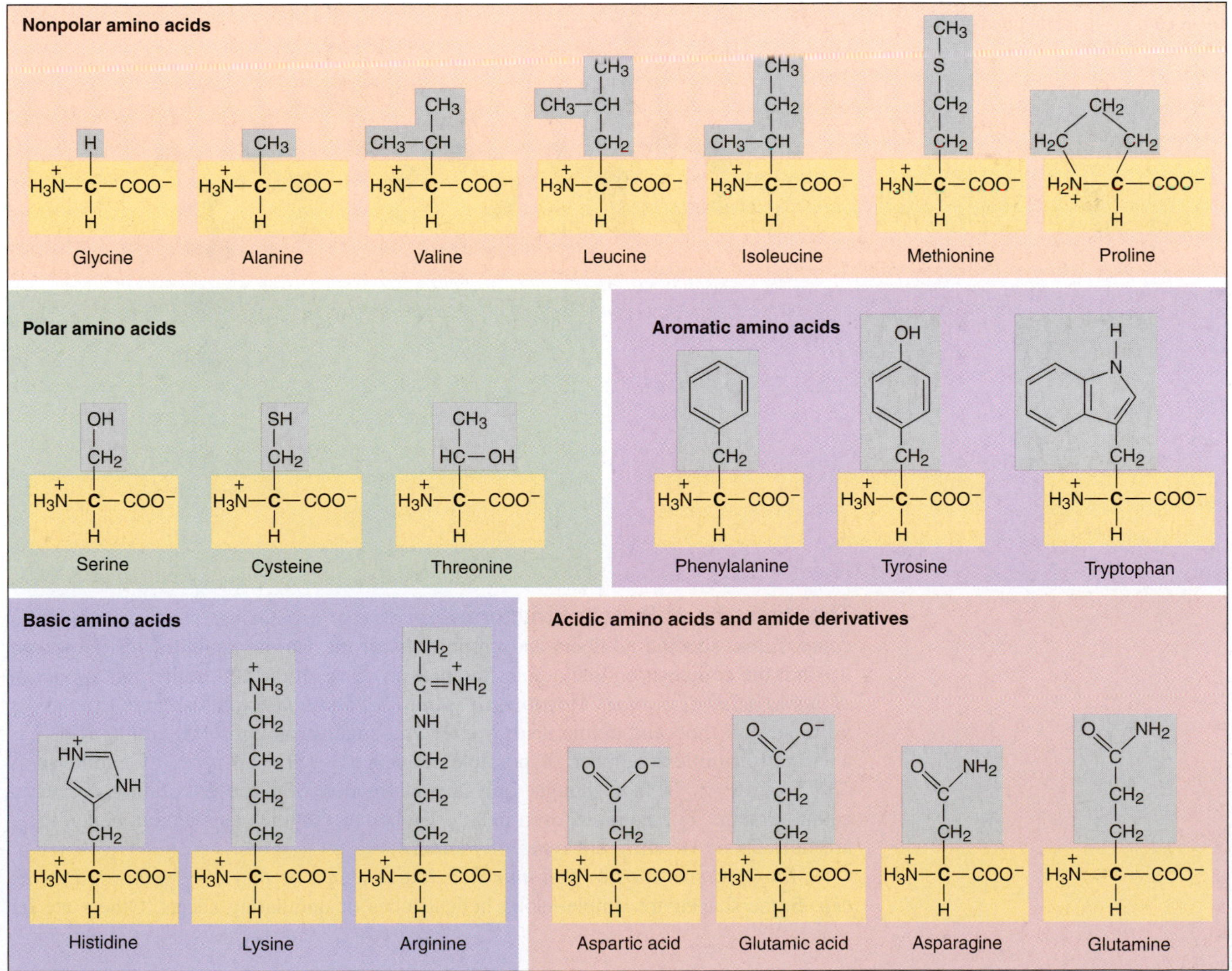

Figure 15.20 The common amino acids. The R groups are screened gray, and the α-carbons *(boldface)*, with carboxyl and amino groups, are screened yellow. Here the amino acids are shown with the charges they have under physiological conditions.

In the aqueous cell fluid, the NH_2 and COOH groups of amino acids are charged because the carboxyl groups each transfer an H^+ ion to H_2O to form H_3O^+. These H_3O^+ ions transfer the H^+ to the amine groups. The overall process can be viewed as an intramolecular acid-base reaction:

$$H_2N-\underset{H}{\overset{R}{C}}-COOH \quad \text{gives} \quad H_3\overset{+}{N}-\underset{H}{\overset{R}{C}}-COO^-$$

($-H_2O$; H_3O^+; $+H_2O$; α-carbon; R = one of 20 different groups)

An H atom is the third group bonded to the α-carbon, and the fourth is the R group (also called the *side chain*).

Each amino acid is linked to the next through a *peptide (amide) bond* formed by a dehydration-condensation reaction in which the carboxyl group of one monomer reacts with the amine group of the next. Therefore, the polypeptide chain—the backbone of the protein—has a repeating structure that consists of an *α-carbon bonded to a peptide group bonded to the next α-carbon bonded to the next peptide group,* and so forth (Figure 15.21). The various R groups (side chains; *gray screens in the figure*) dangle from the α-carbons on alternate sides of the main chain.

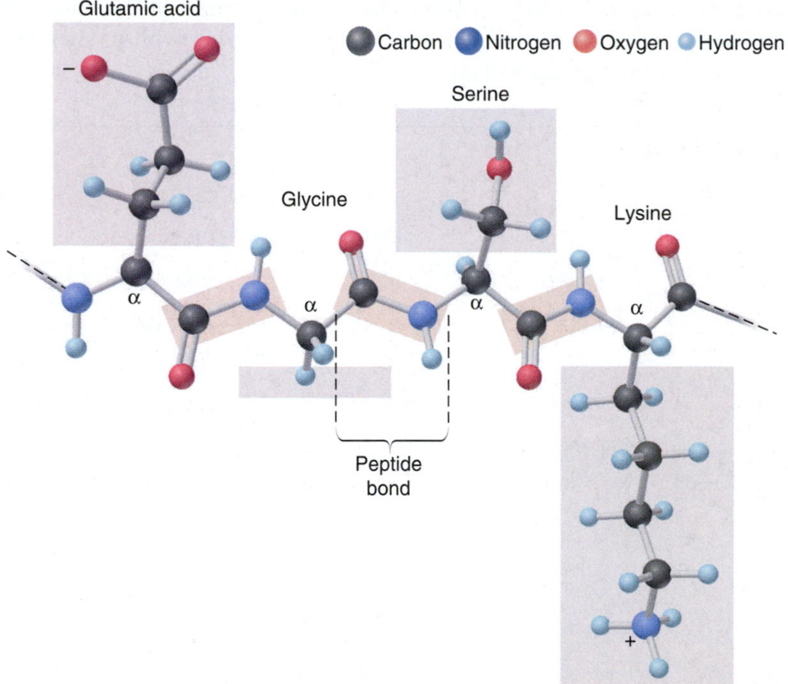

Figure 15.21 A portion of a polypeptide chain. Three peptide bonds *(orange screens)* join four amino acids in this portion of a polypeptide chain. Note the repeating pattern of the chain: peptide bond—α-carbon—peptide bond—α-carbon—and so on.

The Hierarchy of Protein Structure *Each type of protein has its own amino acid composition,* specific numbers and proportions of the various amino acids. However, it is not the composition that defines the protein's role in the cell; rather, *the sequence of amino acids determines the protein's shape and function.* Proteins range from about 50 to several thousand amino acids, yet even a small protein of 100 amino acids has a virtually limitless number of possible sequences of the 20 types of amino acids ($20^{100} \approx 10^{130}$). In fact, though, only a tiny fraction of these possibilities occur in actual proteins. For example, even in an organism as complex as a human being, there are only about 10^5 different types of protein.

Many proteins start to fold into their native shape as they are synthesized in the cell. Some shapes are simple—long helical tubes or undulating sheets. Others are far

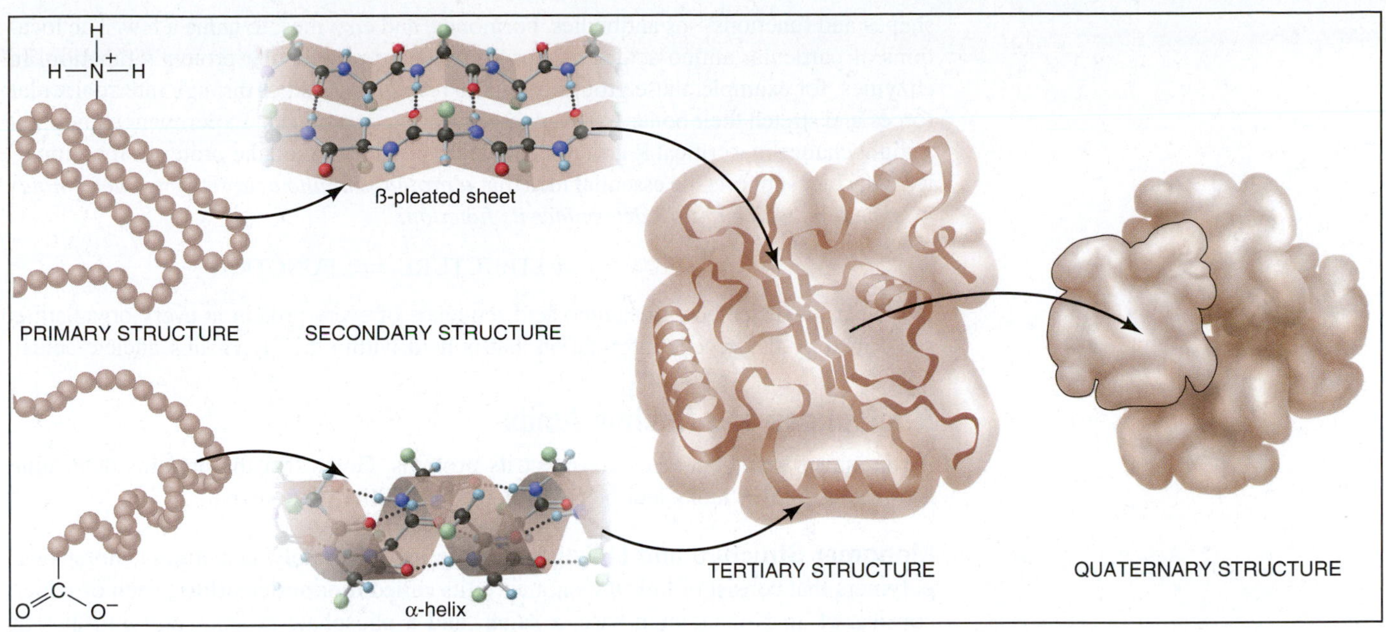

Figure 15.22 The structural hierarchy of proteins.

more complex—baskets, Y shapes, spheroid blobs, and countless other globular forms. Biochemists define a hierarchy for the overall structure of a protein (Figure 15.22):

1. *Primary (1°) structure,* the most basic level, refers to the sequence of covalently bonded amino acids in the polypeptide chain.
2. *Secondary (2°) structure* refers to sections of the chain that, as a result of H bonding between nearby peptide groupings, adopt shapes called α-helices and β-pleated sheets.
3. *Tertiary (3°) structure* refers to the three-dimensional folding of the whole polypeptide chain, which results from many forces. The —SH ends of two cysteine side chains form a covalent *disulfide* bridge (—S—S—) that brings together distant parts of the chain. Polar and ionic side chains interact with surrounding water through ion-dipole forces and H bonds. And nonpolar side chains interact through dispersion forces within the nonaqueous protein interior. Thus, *soluble proteins have polar-ionic exteriors and nonpolar interiors.*
4. *Quaternary (4°) structure,* the highest level, occurs in proteins made up of several polypeptide chains (subunits) and refers to the way the chains assemble into the overall multi-subunit protein. Hemoglobin, for example, consists of four subunits arranged as shown in Figure 15.22 *(right).*

Note that *only the 1° structure involves covalent bonds; the 2°, 3°, and 4° structures rely primarily on intermolecular forces.*

The Relation Between Structure and Function Two broad classes of proteins differ in the complexity of their amino acid compositions and sequences and, therefore, in their structure and function:

1. *Fibrous proteins* are key components of biological materials that require strength and flexibility. They have simple amino acid compositions, repetitive structures, and extended shapes. Consider collagen, the most common animal protein, which makes up as much as 40% of human body weight. More than 30% of its amino acids are glycine, and another 20% are proline. It exists as long "cables" consisting of three intertwined chains, with the peptide C═O groups in one chain H bonding with the peptide N—H groups in another. As the main component of tendons, skin, and blood vessels, collagen has a high tensile strength; in fact, a 1-mm thick strand can support a 10-kg weight!

2. *Globular proteins* have complex compositions; they often contain all 20 common amino acids in varying proportions. They are typically compact, with a wide variety of

shapes and functions—as antibodies, hormones, and enzymes, to name a few. The locations of particular amino-acid R groups are crucial to a globular protein's function. In enzymes, for example, these groups bring the reactants together through intermolecular forces and stretch their bonds to speed their reaction to products. Experiment shows that a slight change in a critical R group dramatically reduces a globular protein's functionality. This fact supports the essential idea that *a protein's amino acid sequence determines its structure, which in turn determines its function:*

SEQUENCE ⟹ STRUCTURE ⟹ FUNCTION

Next, we'll see how the amino acid sequence of every protein in every organism is prescribed by the genetic information that is held within the organism's nucleic acids.

Nucleotides and Nucleic Acids

An organism's nucleic acids construct its proteins. Given that the proteins determine how the organism looks and behaves, no job could be more essential.

Monomer Structure and Linkage **Nucleic acids** are polynucleotides, unbranched polymers that consist of linked monomer units called **mononucleotides,** each of which consists of an N-containing base, a sugar, and a phosphate group. The two types of nucleic acid, ***ribo**nucleic **a**cid* (RNA) and ***d**eoxyribo**n**ucleic **a**cid* (DNA), differ in the sugar portions of their mononucleotides. RNA contains *ribose*, and DNA contains *deoxyribose*, in which —H substitutes for —OH on the second C of ribose.

The cellular precursors that form a nucleic acid are *nucleoside triphosphates* (Figure 15.23A). Dehydration-condensation reactions between them create a chain

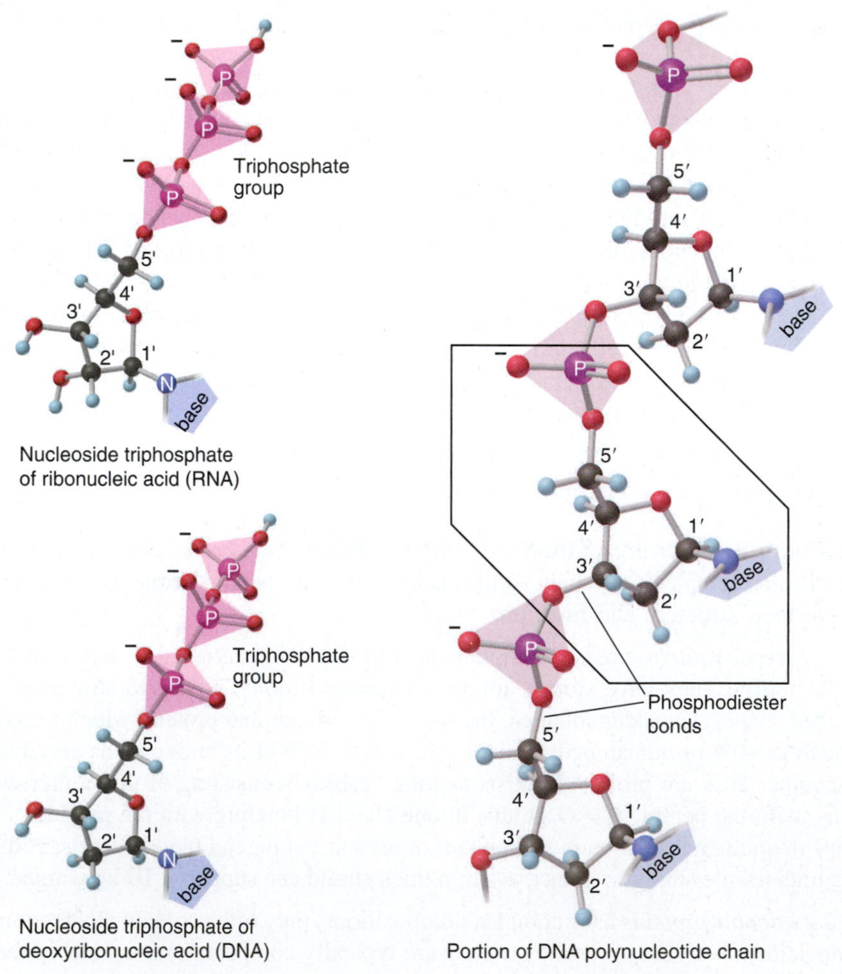

Figure 15.23 Nucleic acid precursors and their linkage.

with the repeating pattern —*sugar—phosphate—sugar—phosphate,* and so on (Figure 15.23B). Attached to each sugar is one of four N-containing bases—thymine (T), cytosine (C), guanine (G), and adenine (A). In RNA, uracil (U) substitutes for thymine. The bases dangle off the chain, much like the R groups dangle off the polypeptide chain of a protein.

DNA Structure and Base Pairing In the cell nucleus, the many millions of nucleotides in DNA occur as two chains wrapped around each other in a **double helix** (Figure 15.24). Intermolecular forces play a central role in stabilizing this structure. On the exterior, negatively charged sugar-phosphate chains form ion-dipole and H bonds with the aqueous surroundings. In the interior, the flat, N-containing bases stack above each other, which allows extensive interaction through dispersion forces.

Most important, *each base in one chain "pairs" with a base in the other through H bonding.* The essential feature of these **base pairs,** which is crucial to the structure and function of DNA, is that *each base is always paired with the same partner:* A with T, and G with C. Thus, *the base sequence on one chain is the complement of the sequence on the other.* For example, the sequence A—C—T on one chain is *always* paired with T—G—A on the other.

Each DNA molecule is folded into a tangled mass that forms one of the cell's *chromosomes.* The DNA molecule is amazingly long and thin: if the largest human chromosome were stretched out, it would be 4 cm (more than 1.5 in) long; in the cell nucleus, however, it is wound into a structure only 5 nm in diameter—8 million times shorter!

From DNA to Protein In the **genetic code,** each base in a DNA chain acts as a "letter," each three-base sequence as a "word," and *each word codes for a specific amino acid.* For example, the sequence C—A—C codes for the amino acid histidine, A—A—G codes for lysine, and so on. In a complex process that occurs largely

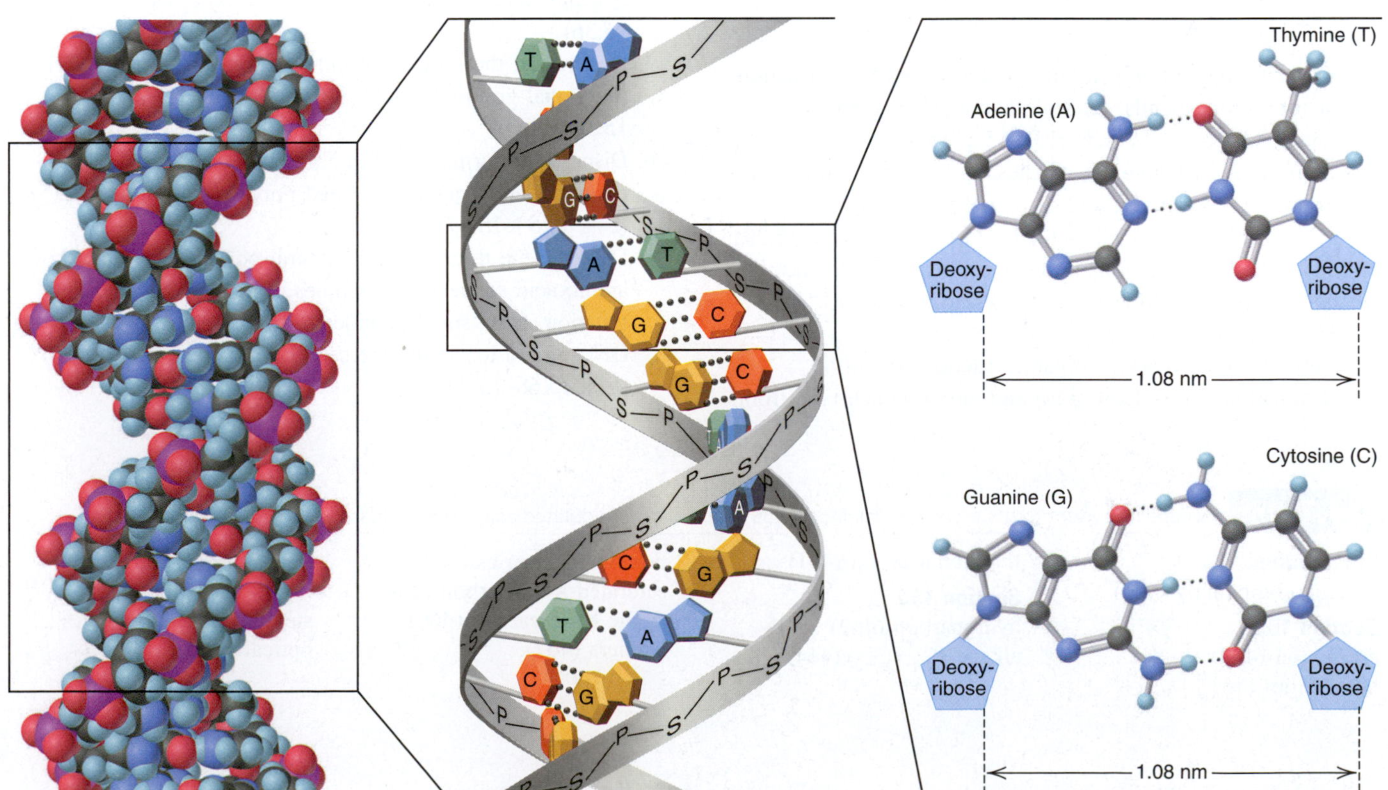

Figure 15.24 The double helix of DNA and a section showing base pairs.

through *H bonding between base pairs,* the DNA message of three-base words is transcribed into an RNA message of three-base words, which is then translated into the sequence of linked amino acids that make up a protein:

DNA BASE SEQUENCE ⇒ RNA BASE SEQUENCE ⇒ PROTEIN AMINO ACID SEQUENCE

The biopolymers provide striking evidence that the same atomic properties that give rise to covalent bonds, molecular shape, and intermolecular forces provide the means for all life forms to flourish.

Summary of Section 15.6

- The three types of natural polymers—polysaccharides, proteins, and nucleic acids—are formed by dehydration-condensation reactions.
- Polysaccharides are formed from cyclic monosaccharides, such as glucose. Cellulose, starch, and glycogen have structural or energy-storage roles.
- Proteins are polyamides formed from as many as 20 different types of amino acids. Fibrous proteins have extended shapes and play structural roles. Globular proteins have compact shapes and function as antibodies, hormones, and enzymes. The amino acid sequence of a protein determines its shape and function.
- Nucleic acids (DNA and RNA) are polynucleotides consisting of four different mononucleotides. The base sequence of the DNA chain determines the sequence of amino acids in an organism's proteins. Hydrogen bonding between specific base pairs results in DNA's double helical structure and is the key to the process of protein synthesis.

CHAPTER REVIEW GUIDE

The following sections provide many aids to help you study this chapter. (Numbers in parentheses refer to pages, unless noted otherwise.)

Learning Objectives

These are concepts and skills to review after studying this chapter.

Related section (§), sample problem (SP), and upcoming end-of-chapter problem (EP) numbers are listed in parentheses.

1. Explain why carbon's atomic properties lead to formation of four strong bonds, multiple bonds, chains, and functional groups (§15.1) (EPs 15.1–15.4)
2. Name and draw alkanes, alkenes, and alkynes with expanded, condensed, and carbon-skeleton formulas (§15.2) (SPs 15.1, 15.2a–c) (EPs 15.5, 15.9–15.18, 15.29)
3. Distinguish among constitutional, optical, and geometric isomers (§15.2) (SP 15.2d, e) (EPs 15.6–15.8, 15.19–15.28, 15.30, 15.31)
4. Describe three types of organic reactions (addition, elimination, and substitution) and identify each type from reactants and products (§15.3) (SP 15.3) (EPs 15.32–15.36)
5. Understand the properties and reaction types of the various functional groups (§15.4) (SPs 15.4–15.6) (EPs 15.37–15.58)
6. Discuss the formation of addition and condensation polymers and draw abbreviated polymer formulas (§15.5) (EPs 15.59–15.67)
7. Describe the three types of natural polymers, explain how amino-acid sequence determines protein shape, and thus function, draw small peptides, and use the sequence of one DNA strand to predict the sequence of the other (§15.6) (EPs 15.68–15.79)

Key Terms

These important terms appear in boldface in the chapter and are defined again in the Glossary.

Introduction
organic compounds (460)

Section 15.1
catenation (460)
heteroatom (461)
functional group (461)

Section 15.2
hydrocarbon (462)
alkane (C_nH_{2n+2}) (464)
homologous series (464)
saturated hydrocarbon (464)
cyclic hydrocarbon (466)
isomers (467)
constitutional (structural) isomers (467)
stereoisomers (468)
optical isomers (468)

chiral molecule (468)
optically active (469)
alkene (C_nH_{2n}) (469)
unsaturated hydrocarbon (469)
geometric (*cis-trans*) isomers (469)
alkyne (C_nH_{2n-2}) (470)
aromatic hydrocarbon (471)

Section 15.3
alkyl group (472)
addition reaction (473)
elimination reaction (473)
substitution reaction (473)

Section 15.4
alcohol (476)
haloalkane (alkyl halide) (476)
amine (477)
carbonyl group (479)
aldehyde (479)
ketone (479)
carboxylic acid (480)
ester (480)
amide (480)
fatty acid (480)
acid anhydride (480)
lipid (481)
hydrolysis (481)
nitrile (482)

Section 15.5
polymer (483)
macromolecule (483)
monomer (483)
addition polymer (484)
condensation polymer (484)

Section 15.6
monosaccharide (486)
polysaccharide (486)
disaccharide (486)
protein (487)
amino acid (487)
nucleic acid (490)
mononucleotide (490)
double helix (491)
base pair (491)
genetic code (491)

BRIEF SOLUTIONS TO FOLLOW-UP PROBLEMS

Compare your own solutions to these calculation steps and answers.

15.1 $H-C(H)(H)-C(H)(H)-C(H)(H)-C\equiv C-H$ $\quad$ $H-C(H)(H)-C(H)(H)-C\equiv C-C(H)(H)-H$

$H-C(H)(H)-C(H)(CH_3)-C\equiv C-H$ (with the branch drawn as $H-C(H)(H)-$ on the second carbon)

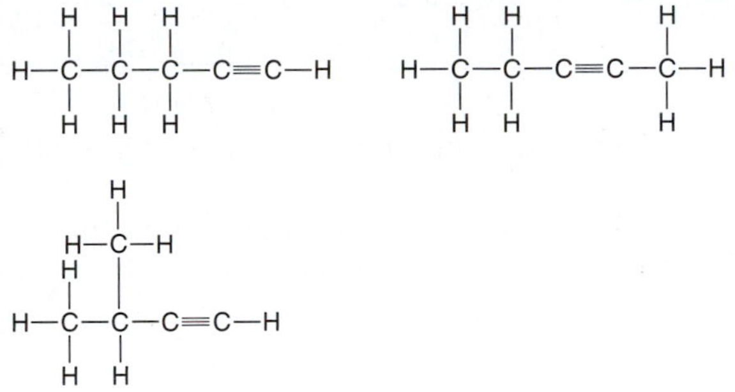

15.2 (a) $CH_3-CH_2-CH_2-CH_2-CH_2-C(CH_2-CH_3)(CH_3)-CH_2-CH_3$

(b) Cyclohexane ring bearing $CH_2-CH_2-CH_3$ and, on the 1,3-position, CH_2-CH_3 and (skeletal form of the same structure)

(c) $HC\equiv C-C(CH_2-CH_3)(CH_2-CH_3)-CH_2-CH_2-CH_3$

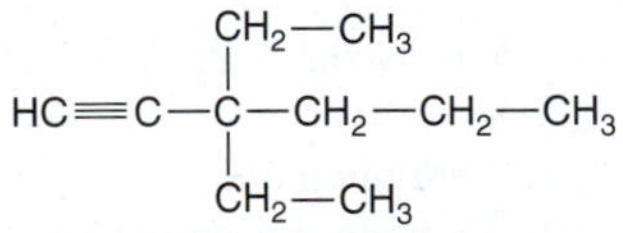

(d) $(CH_3-CH_2)(CH_3)C=C(H)(CH_2-CH_2-CH_3)$

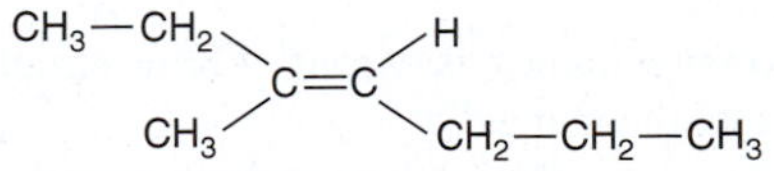

15.3 (a) $CH_3-CH=CH-CH_3 + Cl_2 \longrightarrow CH_3-CH(Cl)-CH(Cl)-CH_3$

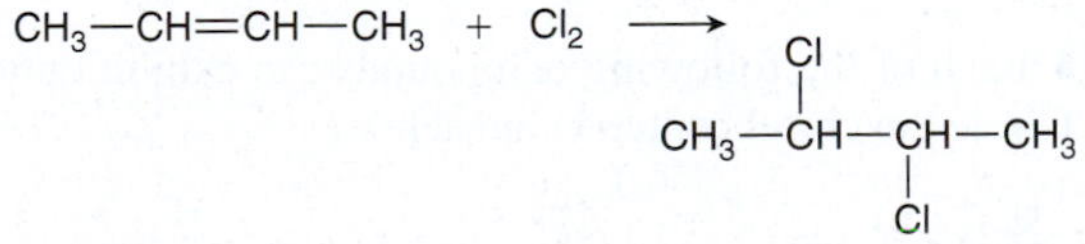

(b) $CH_3-CH_2-CH_2-Br + OH^- \longrightarrow CH_3-CH_2-CH_2-OH + Br^-$

(c) $CH_3-C(CH_3)(OH)-CH_3 \xrightarrow{-H_2O} CH_3-C(CH_3)=CH_2$

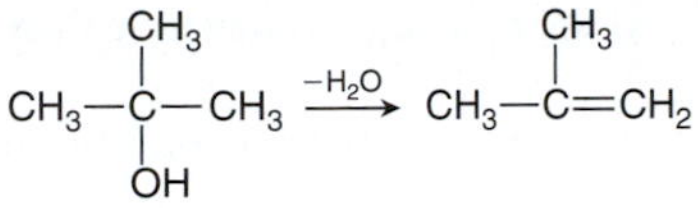

15.4 (a) $CH_3-CH(Cl)-CH(CH_3)-CH_3$ or $CH_3-CH_2-C(CH_3)(Cl)-CH_3$

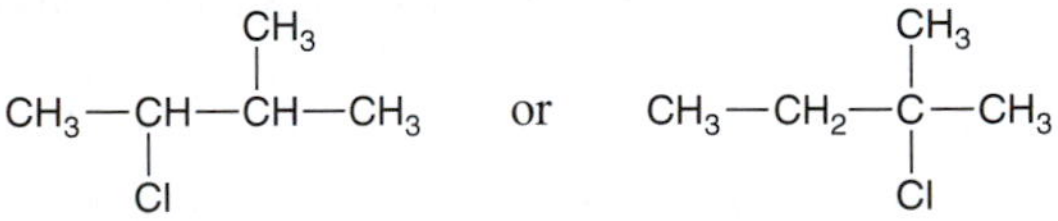

(b) $CH_3-CH_2-CH_2-OH$

15.5 (a) $C_6H_5-CH_2-C(=O)-OH$ (benzene ring)

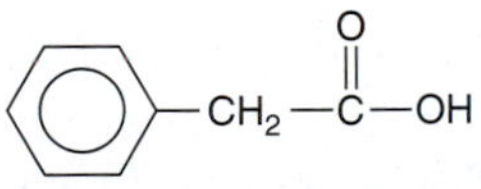

(b) $CH_3-CH_2-CH_2-C(=O)-O-CH_3 + CH_3-CH_2-NH_2$

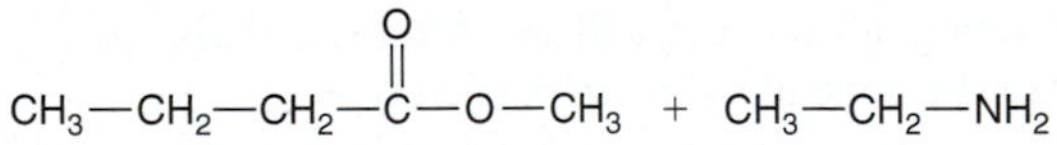

15.6 (a) $C_6H_5-CH=CH-C(=O)H$ (skeletal structure; circled groups labelled: alkene, aldehyde)

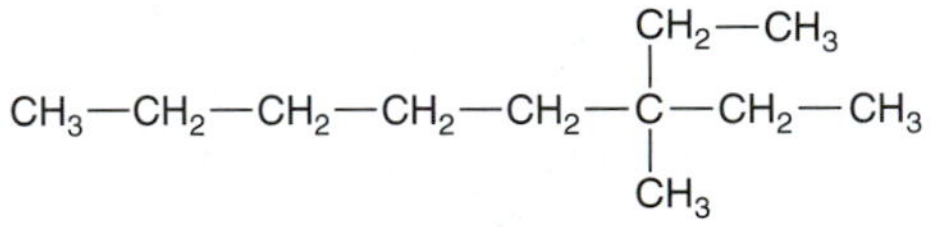

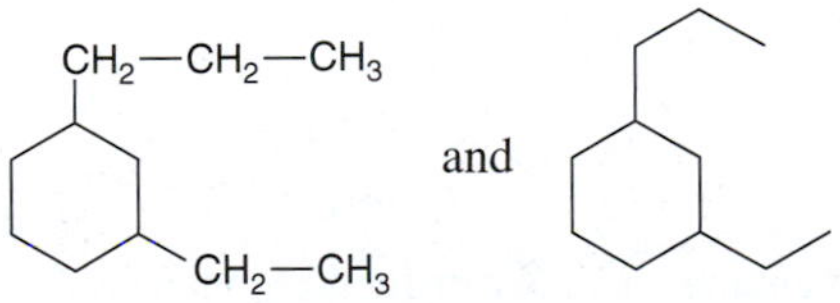

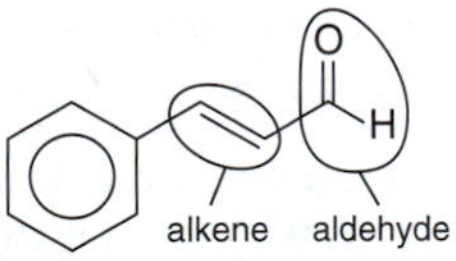

(b) $H_2N-C(=O)-CH_2-CH(Br)-CH_3$ (circled groups labelled: amide, haloalkane)

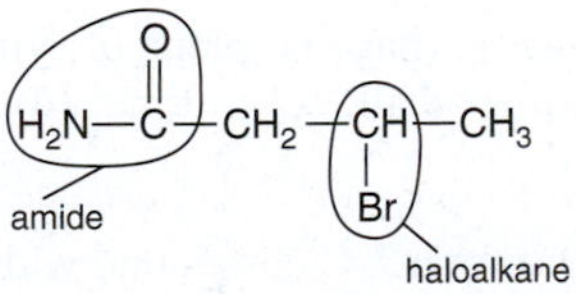

PROBLEMS

Problems with **colored** numbers are answered in Appendix E. Sections match the text and provide the numbers of relevant sample problems. Bracketed problems are grouped in pairs (indicated by a short rule) that cover the same concept. Comprehensive Problems are based on material from any section or previous chapter.

The Special Nature of Carbon and the Characteristics of Organic Molecules

15.1 Explain each statement in terms of atomic properties:
(a) Carbon engages in covalent rather than ionic bonding.
(b) Carbon has four bonds in all its organic compounds.
(c) Carbon forms neither stable cations, like many metals, nor stable anions, like many nonmetals.
(d) Carbon bonds to itself more extensively than does any other element.
(e) Carbon forms stable multiple bonds.

15.2 Carbon bonds to many elements other than itself.
(a) Name six elements that commonly bond to carbon in organic compounds.
(b) Which of these elements are heteroatoms?
(c) Which of these elements are more electronegative than carbon? Less electronegative?
(d) How does bonding of carbon to heteroatoms increase the number of organic compounds?

15.3 Silicon lies just below carbon in Group 4A(14) and also forms four covalent bonds. Why aren't there as many silicon compounds as carbon compounds?

15.4 Which of these bonds to carbon would you expect to be relatively reactive: C—H, C—C, C—I, C=O, C—Li? Explain.

The Structures and Classes of Hydrocarbons

(Sample Problems 15.1 and 15.2)

15.5 (a) What structural feature is associated with each type of hydrocarbon: alkane, cycloalkane, alkene, and alkyne?
(b) Give the general formula for each type.
(c) Which hydrocarbons are considered saturated?

15.6 Define each type of isomer: (a) constitutional; (b) geometric; (c) optical. Which types of isomers are stereoisomers?

15.7 Among alkenes, alkynes, and aromatic hydrocarbons, only alkenes exhibit *cis-trans* isomerism. Why don't the others?

15.8 Which objects are asymmetric (have no plane of symmetry): (a) a circular clock face; (b) a football; (c) a dime; (d) a brick; (e) a hammer; (f) a spring?

15.9 Draw all possible skeletons for a 7-C compound with
(a) A 6-C chain and 1 double bond
(b) A 5-C chain and 1 double bond
(c) A 5-C ring and no double bonds

15.10 Draw all possible skeletons for a 6-C compound with
(a) A 5-C chain and 2 double bonds
(b) A 5-C chain and 1 triple bond
(c) A 4-C ring and no double bonds

15.11 Add the correct number of hydrogens to each of the skeletons in Problem 15.9.

15.12 Add the correct number of hydrogens to each of the skeletons in Problem 15.10.

15.13 Draw correct structures, by making a single change, for any that are incorrect:

(a) 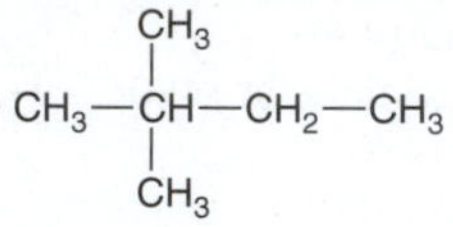 $CH_3—CH(CH_3)(CH_3)—CH_2—CH_3$ (one CH_3 above and one below the CH)
(b) $CH_3{=}CH—CH_2—CH_3$
(c) 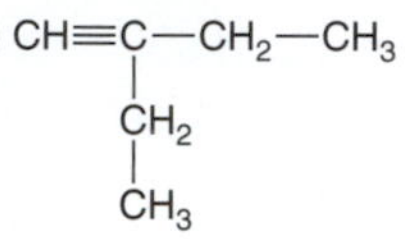 $CH{\equiv}C(CH_2CH_3)—CH_2—CH_3$ (with $CH_2—CH_3$ below the C)
(d)  CH_3—(benzene ring)—CH_3

15.14 Draw correct structures, by making a single change, for any that are incorrect:

(a) $CH_3—CH{=}CH—CH_2—CH_3$
(b) 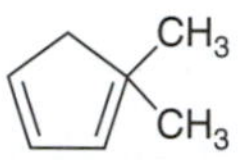 cyclopentadiene ring bearing two CH_3 groups on the same carbon
(c) $CH_3—C{\equiv}CH—CH_2—CH_3$
(d) 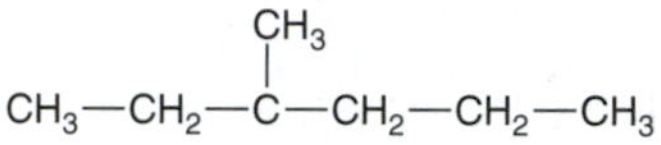 $CH_3—CH_2—C(CH_3)—CH_2—CH_2—CH_3$ (CH_3 above the C)

15.15 Draw the structure or give the name of each compound:
(a) 2,3-dimethyloctane (b) 1-ethyl-3-methylcyclohexane
(c) 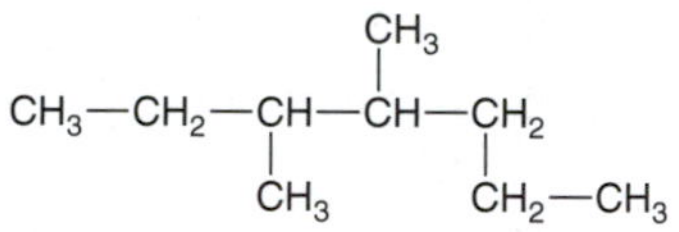 $CH_3—CH_2—CH(CH_3)—CH(CH_3)—CH_2$, with $CH_2—CH_3$ below the final CH_2 (CH_3 below the first CH, CH_3 above the second CH)
(d) 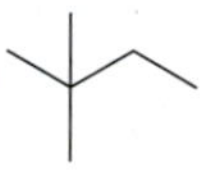 [skeletal structure]

15.16 Draw the structure or give the name of each compound:
(a) 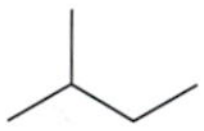 [skeletal structure]
(b) 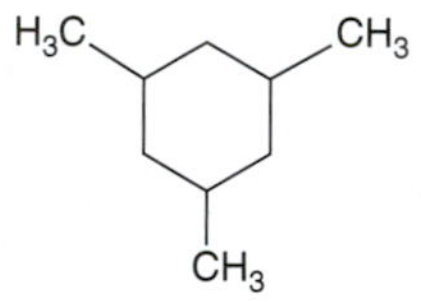 cyclohexane ring with H_3C, CH_3, and CH_3 substituents
(c) 1,2-diethylcyclopentane (d) 2,4,5-trimethylnonane

15.17 Each of the following names is wrong. Draw structures based on them, and correct the names:
(a) 4-methylhexane (b) 2-ethylpentane
(c) 2-methylcyclohexane (d) 3,3-methyl-4-ethyloctane

15.18 Each of the following names is wrong. Draw structures based on them, and correct the names:
(a) 3,3-dimethylbutane (b) 1,1,1-trimethylheptane
(c) 1,4-diethylcyclopentane (d) 1-propylcyclohexane

15.19 Each of the following compounds can exhibit optical activity. Circle the chiral center(s) in each:

(a) 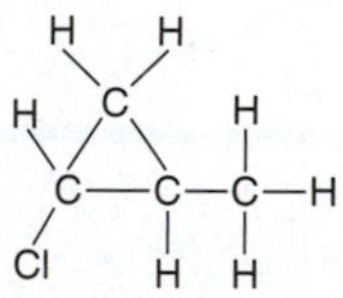 three-membered ring: CH_2 (H, H) bonded to C(H)(Cl) and to C(H), which bears CH_3 (C with H, H, H)
(b) 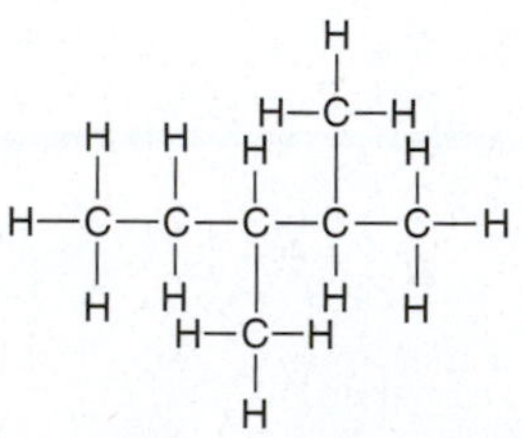 $H—C—C—C—C—C—H$ chain with H atoms and two CH_3 groups (H—C—H above and H—C—H below the central carbons)

15.20 Each of the following compounds can exhibit optical activity. Circle the chiral center(s) in each:

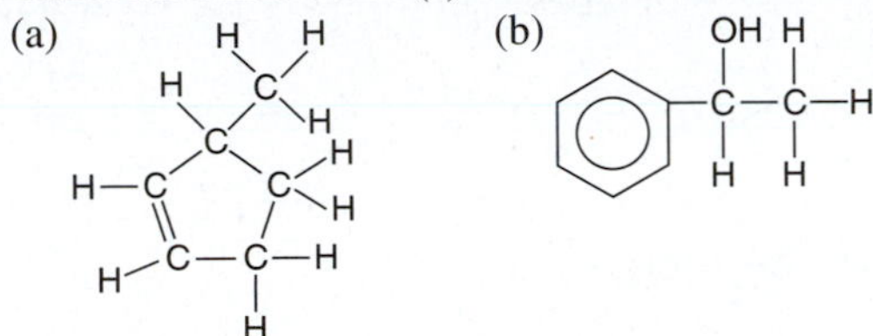

15.21 Draw structures from the following names, and determine which compounds are optically active:
(a) 3-bromohexane (b) 3-chloro-3-methylpentane
(c) 1,2-dibromo-2-methylbutane

15.22 Draw structures from the following names, and determine which compounds are optically active:
(a) 1,3-dichloropentane (b) 3-chloro-2,2,5-trimethylhexane
(c) 1-bromo-1-chlorobutane

15.23 Which of the following structures exhibit geometric isomerism? Draw and name the two isomers in each case:

(a) $CH_3—CH_2—CH{=}CH—CH_3$ (b)

(c) $CH_3—\overset{CH_3}{\overset{|}{C}}{=}CH—\overset{CH_3}{\overset{|}{C}H}—CH_2—CH_3$

15.24 Which of the following structures exhibit geometric isomerism? Draw and name the two isomers in each case:

(a) $CH_3—\underset{CH_3}{\underset{|}{\overset{CH_3}{\overset{|}{C}}}}—CH{=}CH—CH_3$ (b)

(c) $Cl—CH_2—CH{=}\overset{CH_3}{\overset{|}{C}}—CH_2—CH_2—CH_2—CH_3$

15.25 Which compounds exhibit geometric isomerism? Draw and name the two isomers in each case:
(a) propene (b) 3-hexene
(c) 1,1-dichloroethene (d) 1,2-dichloroethene

15.26 Which compounds exhibit geometric isomerism? Draw and name the two isomers in each case:
(a) 1-pentene (b) 2-pentene
(c) 1-chloropropene (d) 2-chloropropene

15.27 Draw and name all the constitutional isomers of dichlorobenzene.

15.28 Draw and name all the constitutional isomers of trimethylbenzene.

15.29 Butylated hydroxytoluene (BHT) is a common preservative added to cereals and other dry foods. Its systematic name is 1-hydroxy-2,6-di-*tert*-butyl-4-methylbenzene (where "*tert*-butyl" is 1,1-dimethylethyl). Draw the structure of BHT.

15.30 There are two compounds with the name 2-methyl-3-hexene, but only one with the name 2-methyl-2-hexene. Explain with structures.

15.31 Any tetrahedral atom with four different groups attached can be a chiral center. Which of these species is optically active?
(a) CHClBrF (b) $NBrCl_2H^+$
(c) $PFClBrI^+$ (d) SeFClBrH

Some Important Classes of Organic Reactions

(Sample Problem 15.3)

15.32 Determine the type of each of the following reactions:

(a) $CH_3—CH_2—\overset{Br}{\overset{|}{C}H}—CH_3 \xrightarrow[\Delta]{NaOH} CH_3—CH{=}CH—CH_3 + NaBr + H_2O$

(b) $CH_3—CH{=}CH—CH_2—CH_3 + H_2 \xrightarrow{Pt} CH_3—CH_2—CH_2—CH_2—CH_3$

15.33 Determine the type of each of the following reactions:

(a) $CH_3—\overset{O}{\overset{\|}{C}H} + HCN \longrightarrow CH_3—\overset{OH}{\overset{|}{C}H}—CN$

(b) $CH_3—\overset{O}{\overset{\|}{C}}—O—CH_3 + CH_3—NH_2 \xrightarrow{H^+} CH_3—\overset{O}{\overset{\|}{C}}—NH—CH_3 + CH_3—OH$

15.34 Write equations for the following: (a) an addition reaction between H_2O and 3-hexene (H^+ is required and shown above the yield arrow); (b) an elimination reaction between 2-bromopropane and hot potassium ethoxide, $CH_3—CH_2—OK$ (KBr and ethanol are also products); (c) a light-induced substitution reaction between Cl_2 and ethane to form 1,1-dichloroethane.

15.35 Write equations for the following: (a) a substitution reaction between 2-bromopropane and KI; (b) an addition reaction between cyclohexene and Cl_2; (c) an addition reaction between 2-propanone and H_2 (Ni metal is required and shown above the yield arrow).

15.36 Phenylethylamine is a natural substance that is structurally similar to amphetamine. It is found in sources as diverse as almond oil and human urine, where it occurs at elevated concentrations as a result of stress and certain forms of schizophrenia. One method of synthesizing the compound for pharmacological and psychiatric studies involves two steps:

$$C_6H_5—CH_2—Cl + NaCN \longrightarrow C_6H_5—CH_2—C{\equiv}N \xrightarrow[Pt]{H_2} C_6H_5—CH_2—CH_2—NH_2$$

phenylethylamine

Classify each step as an addition, elimination, or substitution.

Properties and Reactivities of Common Functional Groups

(Sample Problems 15.4 to 15.6)

15.37 Compounds with nearly identical molar masses often have very different physical properties. Choose the compound with the higher value for each of the following properties, and explain your choice.
(a) Solubility in water: chloroethane or methylethylamine
(b) Melting point: diethyl ether ($C_2H_5—O—C_2H_5$) or 1-butanol
(c) Boiling point: trimethylamine or propylamine

15.38 Fill in each blank with a general formula for the type of compound formed: $R—CH{=}CH_2 \underset{H^+}{\overset{H_2O,\ H^+}{\rightleftharpoons}}$ ______

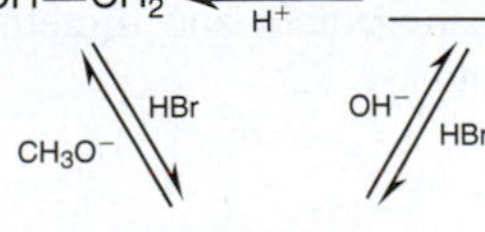

15.39 Why does the C=O group react differently from the C=C group? Show an example of the difference.

15.40 Many substitution reactions are initiated by electrostatic attraction between reactants. Show where this attraction arises in the formation of an amide from an amine and an ester.

15.41 What reaction type is common to the formation of esters and acid anhydrides? What is the other product?

15.42 Both alcohols and carboxylic acids undergo substitution, but the processes are very different. Explain.

15.43 Name the type of organic compound from each description of the functional group: (a) polar group that has only single bonds and does not include O or N; (b) group that is polar and has a triple bond; (c) group that has single and double bonds and is acidic in water; (d) group that has a double bond and must be at the end of a C chain.

15.44 Name the type of organic compound from each description of the functional group: (a) N-containing group with single and double bonds; (b) group that is not polar and has a double bond; (c) polar group that has a double bond and cannot be at the end of a C chain; (d) group that has only single bonds and is basic in water.

15.45 Circle and name the functional group(s) in each compound:

(a) $CH_3—CH=CH—CH_2—OH$

(b) $Cl—CH_2—C_6H_4—\overset{O}{\overset{\|}{C}}—OH$

(c) [cyclohexene ring]$—\overset{O}{\overset{\|}{C}}—N(CH_3)_2$

(d) $N\equiv C—CH_2—\underset{O}{\underset{\|}{C}}—CH_3$

(e) [cyclobutyl]$—\overset{O}{\overset{\|}{C}}—O—CH_2—CH_3$

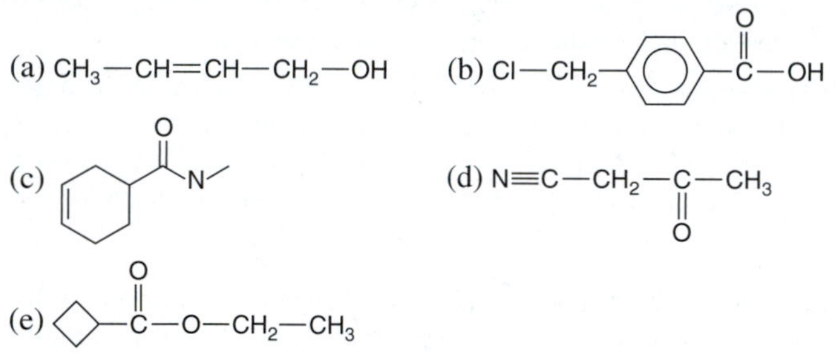

15.46 Circle and name the functional group(s) in each compound:

(a) $HO—\overset{O}{\overset{\|}{C}}—CH_2—\overset{O}{\overset{\|}{C}}—H$

(b) $I—CH_2—CH_2—C\equiv CH$

(c) $CH_2=CH—CH_2—\overset{O}{\overset{\|}{C}}—O—CH_3$

(d) $CH_3—NH—\overset{O}{\overset{\|}{C}}—\overset{O}{\overset{\|}{C}}—O—CH_3$

(e) $CH_3—\underset{}{\overset{Br}{\overset{|}{C}}}H—CH=CH—CH_2—NH—CH_3$

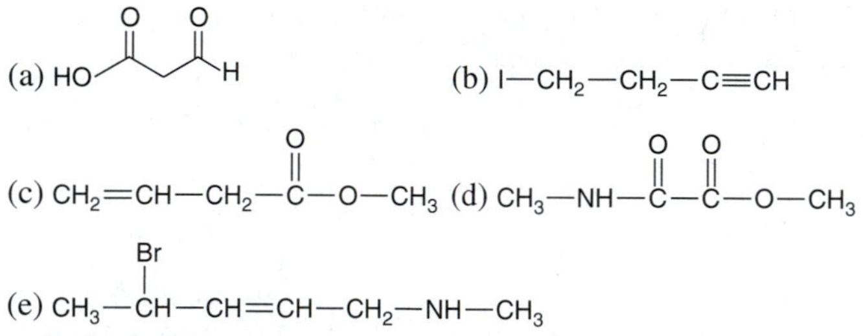

15.47 Draw all alcohols with the formula $C_5H_{12}O$.

15.48 Draw all aldehydes and ketones with the formula $C_5H_{10}O$.

15.49 Draw all amines with the formula $C_4H_{11}N$.

15.50 Draw all carboxylic acids with the formula $C_5H_{10}O_2$.

15.51 Draw the organic product formed when the following compounds undergo a substitution reaction: (a) acetic acid and methylamine; (b) butanoic acid and 2-propanol; (c) formic acid and 2-methyl-1-propanol.

15.52 Draw the organic product formed when the following compounds undergo a substitution reaction: (a) acetic acid and 1-hexanol; (b) propanoic acid and dimethylamine; (c) ethanoic acid and diethylamine.

15.53 Draw condensed formulas for the carboxylic acid and alcohol that form the following esters:

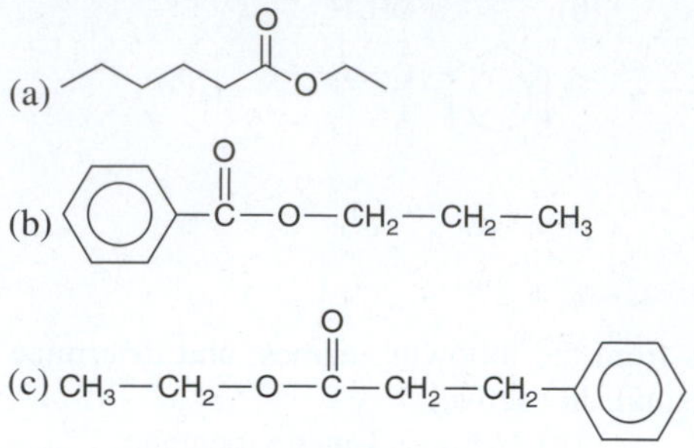

(a) [skeletal: $CH_3CH_2CH_2CH_2—\overset{O}{\overset{\|}{C}}—O—CH_2CH_3$]

(b) $C_6H_5—\overset{O}{\overset{\|}{C}}—O—CH_2—CH_2—CH_3$

(c) $CH_3—CH_2—O—\overset{O}{\overset{\|}{C}}—CH_2—CH_2—C_6H_5$

15.54 Draw condensed formulas for the carboxylic acid and amine that form the following amides:

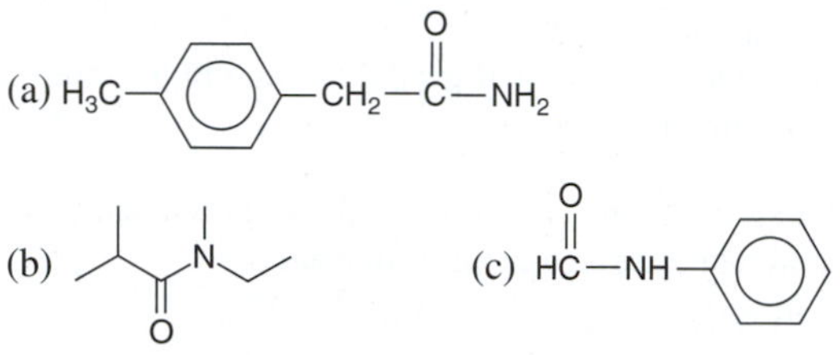

(a) $H_3C—C_6H_4—CH_2—\overset{O}{\overset{\|}{C}}—NH_2$

(b) [skeletal: $(CH_3)_2CH—\underset{O}{\underset{\|}{C}}—N(CH_3)—CH_2CH_3$]

(c) $H\overset{O}{\overset{\|}{C}}—NH—C_6H_5$

15.55 Fill in the expected organic substances:

(a) $CH_3—CH_2—Br \xrightarrow{OH^-}$ ____ $\xrightarrow[H^+]{CH_3—CH_2—\overset{O}{\overset{\|}{C}}—OH}$ ____

(b) $CH_3—CH_2—\overset{Br}{\overset{|}{C}}H—CH_3 \xrightarrow{CN^-}$ ____ $\xrightarrow{H_3O^+,\ H_2O}$ ____

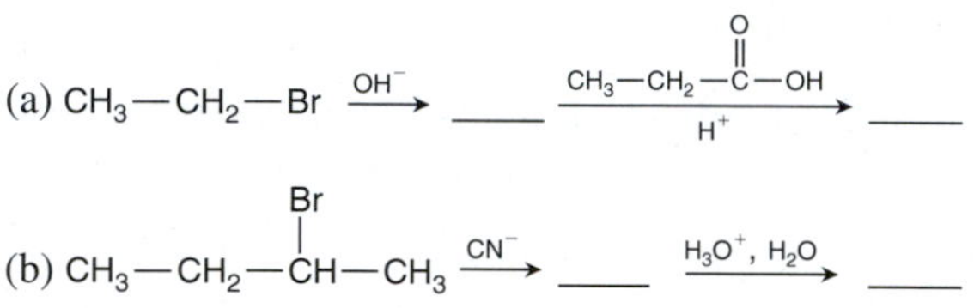

15.56 Fill in the expected organic substances:

(a) $CH_3—CH_2—CH=CH_2 \xrightarrow{H^+,\ H_2O}$ ____ $\xrightarrow{Cr_2O_7^{2-},\ H^+}$ ____

(b) $CH_3—CH_2—\overset{O}{\overset{\|}{C}}—CH_3 \xrightarrow{CH_3—CH_2—Li} \xrightarrow{H_2O}$ ____

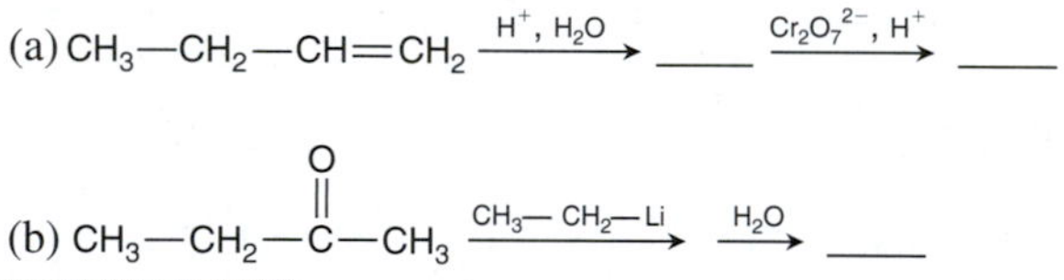

15.57 (a) Draw the four isomers of $C_5H_{12}O$ that can be oxidized to an aldehyde. (b) Draw the three isomers of $C_5H_{12}O$ that can be oxidized to a ketone. (c) Draw the isomers of $C_5H_{12}O$ that cannot be easily oxidized to an aldehyde or ketone. (d) Name any isomer that is an alcohol.

15.58 Ethyl formate ($H\overset{O}{\overset{\|}{C}}—O—CH_2—CH_3$) is added to foods to give them the flavor of rum. How would you synthesize ethyl formate from ethanol, methanol, and any inorganic reagents?

The Monomer-Polymer Theme I: Synthetic Macromolecules

15.59 Name the reaction processes that lead to the two types of synthetic polymers.

15.60 Which functional group occurs in the monomers of addition polymers? How are these polymers different from one another?

15.61 Which intermolecular force is primarily responsible for the different types of polyethylene? Explain.

15.62 Which of the two types of synthetic polymer is more similar chemically to biopolymers? Explain.

15.63 Which functional groups react to form nylons? Polyesters?

15.64 Draw an abbreviated formula for the following polymers, with brackets around the repeat unit:
(a) Poly(vinyl chloride) (PVC) from (b) Polypropylene from

H(H)C=C(H)(Cl)   H(H)C=C(H)(CH_3)

15.65 Draw an abbreviated formula for the following polymers, with brackets around the repeat unit:
(a) Teflon from (b) Polystyrene from

F(F)C=C(F)(F)   H(H)C=C(H)(C$_6$H$_5$)

15.66 Write a balanced equation for the reaction between 1,4-benzenedicarboxylic acid and 1,2-dihydroxyethane to form the polyester Dacron. Draw an abbreviated structure for the polymer, with brackets around the repeat unit.

15.67 Write a balanced equation for the reaction of dihydroxydimethylsilane *(right)* to form the condensation polymer sold as Silly Putty.

HO—Si(CH_3)$_2$—OH

The Monomer-Polymer Theme II: Biological Macromolecules

15.68 Which type of polymer is formed from each of the following monomers: (a) amino acids; (b) alkenes; (c) simple sugars; (d) mononucleotides?

15.69 What is the key structural difference between fibrous and globular proteins? How is it related, in general, to the proteins' amino acid composition?

15.70 Protein shape, function, and amino acid sequence are interrelated. Which determines which?

15.71 What is base pairing? How does it pertain to DNA structure?

15.72 Draw the R group of (a) alanine; (b) histidine; (c) methionine.

15.73 Draw the R group of (a) glycine; (b) isoleucine; (c) tyrosine.

15.74 Draw the formula of each of the following tripeptides:
(a) Aspartic acid-histidine-tryptophan
(b) Glycine-cysteine-tyrosine with the charges that exist in cell fluid

15.75 Draw the formula of each of the following tripeptides:
(a) Lysine-phenylalanine-threonine
(b) Alanine-leucine-valine with the charges that exist in cell fluid

15.76 Write the sequence of the complementary DNA strand that pairs with each of the following DNA base sequences:
(a) TTAGCC (b) AGACAT

15.77 Write the sequence of the complementary DNA strand that pairs with each of the following DNA base sequences:
(a) GGTTAC (b) CCCGAA

15.78 Protein shapes are maintained by a variety of forces that arise from interactions between the amino-acid R groups. Name the amino acid that possesses each R group and the force that could arise in each of the following interactions:
(a) —CH_2—SH with HS—CH_2—
(b) —$(CH_2)_4$—NH_3^+ with $^-$O—C(=O)—CH_2—
(c) —CH_2—C(=O)—NH_2 with HO—CH_2—
(d) —CH(CH_3)—CH_3 with C_6H_5—CH_2—

15.79 Amino acids have an average molar mass of 100 g/mol. How many bases on a single strand of DNA are needed to code for a protein with a molar mass of 5×10^5 g/mol?

Comprehensive Problems

15.80 A synthesis of 2-butanol was performed by treating 2-bromobutane with hot sodium hydroxide solution. The yield was 60%, indicating that a significant portion of the reactant was converted into a second product. Predict what this other product might be.

15.81 Pyrethrins, such as jasmolin II *(below),* are a group of natural compounds synthesized by flowers of the genus *Chrysanthemum* (known as pyrethrum flowers) to act as insecticides.
(a) Circle and name the functional groups in jasmolin II.
(b) What is the hybridization of the numbered carbons?
(c) Which, if any, of the numbered carbons are chiral centers?

[Structure of jasmolin II with carbons numbered 1–7: CH_2=CH—CH (cyclopropane with C(CH_3)$_2$)—CH—C(=O)—O—HC (five-membered ring with C(CH_3), C—C(=O)—O—CH_3, H_2C, C=O)]

15.82 Compound A is branched and optically active and contains C, H, and O. (a) A 0.500-g sample burns in excess O_2 to yield 1.25 g of CO_2 and 0.613 g of H_2O. Determine the empirical formula. (b) When 0.225 g of compound A vaporizes at 755 torr and 97°C, the vapor occupies 78.0 mL. Determine the molecular formula. (c) Careful oxidation of the compound yields a ketone. Name and draw compound A, and circle the chiral center.

15.83 The genetic code consists of a series of three-base words that each code for a given amino acid.
(a) Using the selections from the genetic code shown below, determine the amino acid sequence coded by the following segment of RNA:

UCCACAGCCUAUAUGGCAAACUUGAAG

AUG = methionine	CCU = proline	CAU = histidine
UGG = tryptophan	AAG = lysine	UAU = tyrosine
GCC = alanine	UUG = leucine	CGG = arginine
UGU = cysteine	AAC = asparagine	ACA = threonine
UCC = serine	GCA = alanine	UCA = serine

(b) What is the complementary DNA sequence from which this RNA sequence was made? (*Hint:* Uracil, U, in RNA pairs with adenine, A, in DNA.)

15.84 Sodium propanoate (CH_3—CH_2—C(=O)—ONa) is a common preservative found in breads, cheeses, and pies. How would you synthesize sodium propanoate from 1-propanol and any inorganic reactants?

15.85 Supply the missing organic and/or inorganic substances:

(a) CH_3—CH(Cl)—CH_3 $\xrightarrow{?}$ CH_3—CH=CH_2 $\xrightarrow{?}$ CH_3—CH(Br)—CH_2(Br)

(b) CH_3—CH_2—CH_2—OH $\xrightarrow{?}$ CH_3—CH_2—C(=O)—OH + ? $\xrightarrow{?}$ CH_3—CH_2—C(=O)—O—CH_2—C_6H_5

16 Kinetics: Rates and Mechanisms of Chemical Reactions

Warming Up to Face a New Day The rates of metabolic processes of cold-blooded animals, like this desert tortoise, increase as temperatures rise toward midday. In this chapter, you'll see how temperature and several other factors affect the speed of a reaction.

Key Principles
to focus on while studying this chapter

- The *rate of a reaction* is the *change* in the concentration of reactant (or product) per unit of time. Reaction rates vary over a wide range, but each reaction has a specific rate under a given set of conditions. The rate depends on *concentration* and *physical state* because reactants must collide to react. It depends even more on *temperature* because the collisions must occur with enough kinetic energy. ***(Section 16.1)***
- The rate changes as the reaction proceeds: fastest at the beginning, when reactant concentration is highest, and slowest at the end. *Average rate* is the concentration change over a period of time, and *instantaneous rate* is the change at any instant. Kinetic studies typically measure the *initial rate*, the rate at the moment the reactants are mixed, so product is absent. ***(Section 16.2)***
- The rate of a reaction is given by a *rate law* (or *rate equation*). This expression includes a temperature-dependent *rate constant* and one or more concentration terms raised to an exponent, called a *reaction order*, that defines how the concentration of that reactant affects the rate. The rate law must be determined by experiment, *not* from the balanced equation, and several methods exist for measuring initial rates. ***(Section 16.3)***
- An *integrated rate law* includes concentration *and* time as variables. In addition to another way to find the reaction order, it is used to find the *half-life*, the time required for half of a reactant to be used up. The half-life of a first-order reaction does *not* depend on reactant concentration. ***(Section 16.4)***
- *Collision theory* proposes that reactant molecules must collide with a minimum energy, the *energy of activation* (E_a), in order to react. The *Arrhenius equation* shows that rate increases with temperature and decreases with E_a by affecting the rate constant. ***(Section 16.5)***
- Higher temperature increases the frequency of collisions and, more importantly, the *fraction* of collisions with energy greater than E_a. For a collision to be *effective*, the atoms in the colliding molecules must be oriented correctly for a bond to form between them. ***(Section 16.5)***
- *Transition state theory* explains that the E_a is the energy needed to form a high-energy species that exists only momentarily and includes partially broken reactant bonds and partially formed product bonds. Every step in a reaction has such a *transition state (activated complex)*. ***(Section 16.5)***
- Chemists explain the rate law for an overall reaction by proposing a *reaction mechanism* that consists of several *elementary steps*, each with its own rate law. To be a valid mechanism, the sum of the elementary steps must give the balanced equation, the steps must be physically reasonable, and the mechanism must correlate with the rate law. The rate law of the slowest step (the *rate-determining step*) must give the overall rate law. ***(Section 16.6)***
- A *catalyst* is a component of a reaction mixture that speeds the reaction (in both directions) but is not consumed. It functions by *lowering the* E_a of the rate-determining step of an alternative mechanism for the same overall reaction. Catalysts can function in the same *(homogeneous)* or a different *(heterogeneous)* phase from the reactants and products. They are essential components of many industrial, and nearly all biological, reactions. ***(Section 16.7)***

Outline

Until now, we've focused on quantitative factors affecting reactions only in terms of the amounts of reactants and products. Yet, while a balanced equation is essential for calculating yields, it tells us nothing about three dynamic aspects of chemical change:

- How fast is the reaction proceeding?
- How far will the reaction proceed toward completion?
- Does the reaction proceed by releasing energy or by absorbing it?

This chapter addresses the first of these questions and focuses on the field of *kinetics*. We'll address the other two questions in upcoming chapters. Answering all three is crucial to understanding modern technology, as well as reactions in the environment and those in living things.

CONCEPTS & SKILLS TO REVIEW before studying this chapter

- influence of temperature on molecular speed (Section 5.5)

16.1 • FOCUSING ON REACTION RATE

By definition, in a chemical reaction, reactants change into products. **Chemical kinetics,** the study of how fast that change occurs, focuses on the **reaction rate,** the change in the concentrations of reactants (or products) as a function of time. Different reactions have different rates: *in a faster reaction (higher rate), the reactant concentration decreases quickly, whereas in a slower reaction (lower rate), it decreases slowly* (Figure 16.1).

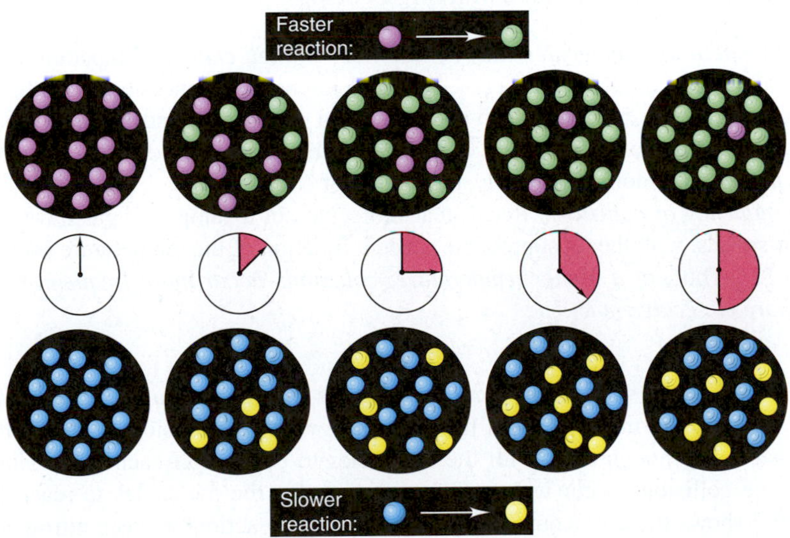

Figure 16.1 A faster reaction *(top)* and a slower reaction *(bottom).* As time elapses, reactant *decreases* and product *increases*.

Under any given set of conditions, a rate is determined by the nature of the reactants. At room temperature, for example, hydrogen reacts explosively with fluorine but extremely slowly with nitrogen:

$$H_2(g) + F_2(g) \longrightarrow 2HF(g) \quad \text{[very fast]}$$
$$3H_2(g) + N_2(g) \longrightarrow 2NH_3(g) \quad \text{[very slow]}$$

Furthermore, any given reaction has a different rate under different conditions.

Chemical processes occur over a wide range of rates. Some—like a neutralization, a precipitation, or an explosion—may take a second or less. Processes that include many reactions—like the ripening of fruit—take days to months. Human aging continues for decades, and the formation of coal from dead plants takes hundreds of millions of years. Knowing the reaction rate can be essential: how quickly a medicine acts can make the difference between life and death, and how long an industrial product takes to form can make the difference between profit and loss.

We can control four factors that affect rate: the concentrations of reactants, their physical state, the temperature of the reaction, and the use of a catalyst. We consider the first three here and the fourth in Section 16.7.

1. *Concentration: molecules must collide to react.* A major factor influencing reaction rate is reactant concentration, because a reaction can occur only if the reactant molecules collide. The more molecules present in the container, the more frequently they collide, and so the more often they react. Thus, *reaction rate is proportional to the concentration of reactants:*

$$\text{Rate} \propto \text{collision frequency} \propto \text{concentration}$$

2. *Physical state: molecules must mix to collide.* Collision frequency also depends on physical state, which determines how easily the reactants mix. When the reactants are in the same phase, as in an aqueous solution, random thermal motion brings them into contact, but gentle stirring mixes them further. When the reactants are in different phases, contact occurs only at the interface between the phases, so vigorous stirring or even grinding may be needed. Thus, *the more finely divided a solid or liquid reactant, the greater its surface area, the more contact it makes with the other reactant, and the faster the reaction occurs.* For example, a thick steel nail heated in oxygen glows feebly, but the same mass of steel wool under the same conditions bursts into flame. For the same reason, you start a campfire with twigs, not logs.

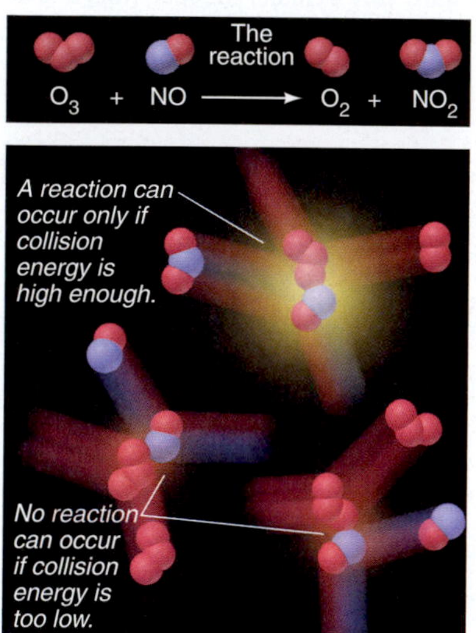
The reaction
O_3 + NO ⟶ O_2 + NO_2
A reaction can occur only if collision energy is high enough.
No reaction can occur if collision energy is too low.

Figure 16.2 Sufficient collision energy is required for a reaction to occur.

3. *Temperature: molecules must collide with enough energy.* Temperature usually has a major effect on the rate of a reaction. Two kitchen appliances employ this effect: a refrigerator slows down chemical processes that spoil food, and an oven speeds up other chemical processes that cook it. Temperature affects reaction rate by increasing the *frequency* and, more importantly, the *energy* of collisions:

- *Frequency of collisions.* Recall that molecules in a sample of gas have a range of speeds, with the most probable speed a function of the temperature (see Figure 5.13). Thus, *at a higher temperature, collisions occur more frequently, and so more molecules react:*

$$\text{Rate} \propto \text{collision } \textbf{frequency} \propto \text{temperature}$$

- *Energy of collisions.* Even more important is that *temperature affects the kinetic energy* of the molecules and thus the *energy* of the collisions. Most collisions have only enough energy for the molecules to bounce off each other. However, some collisions occur with sufficient energy for the molecules to react. Figure 16.2 shows these outcomes of collisions for the reaction between nitrogen monoxide (NO) and ozone (O_3). *At a higher temperature, more sufficiently energetic collisions occur, and so more molecules react:*

$$\text{Rate} \propto \text{collision } \textbf{energy} \propto \text{temperature}$$

Summary of Section 16.1

- Chemical kinetics focuses on reaction rate, the change in concentration over time.
- Under a given set of conditions, each reaction has its own rate.
- Concentration affects rate by influencing the frequency of collisions between reactant molecules.
- Physical state affects rate by determining how well reactants can mix.
- Temperature affects rate by influencing the frequency and, more importantly, the energy of the collisions between reactant molecules.

16.2 • EXPRESSING THE REACTION RATE

A *rate* is a change in some variable per unit of time. The most familiar examples relate to speed, the change in position of an object divided by the change in time. For instance, if we measure a runner's initial position, x_1, at time t_1, and final position, x_2, at time t_2, the average speed is

$$\text{Rate of motion (speed)} = \frac{\text{change in position}}{\text{change in time}} = \frac{x_2 - x_1}{t_2 - t_1} = \frac{\Delta x}{\Delta t}$$

For the rate of a *reaction,* we measure the changes in concentrations of reactants or products per unit time: *reactant concentrations decrease while product concentrations increase.* For the general reaction A $\longrightarrow$ B, we measure the initial reactant concentration (A_1) at t_1, allow the reaction to proceed, and then quickly measure the final reactant concentration (A_2) at t_2. The change in concentration divided by the change in time gives the rate (actually, the *average* rate over the specified time period, as you'll see shortly):

$$\text{Rate} = -\frac{\text{change in concentration of A}}{\text{change in time}} = -\frac{\text{conc } A_2 - \text{conc } A_1}{t_2 - t_1} = -\frac{\Delta(\text{conc A})}{\Delta t}$$

The negative sign is important because, by convention, reaction rate is a *positive* number. But, since conc A_2 must be *lower* than conc A_1, the *change in concentration (final − initial) of reactant A is negative.* We use the negative sign to convert the negative change in reactant concentration to a positive value for the rate. Suppose the concentration of A changes from 1.2 mol/L (conc A_1) to 0.75 mol/L (conc A_2) over a 125-s period. The rate is

$$\text{Rate} = -\frac{0.75 \text{ mol/L} - 1.2 \text{ mol/L}}{125 \text{ s} - 0 \text{ s}} = 3.6\times10^{-3} \text{ mol/L·s}$$

Square brackets, [ ], *indicate a concentration in moles per liter.* For example, [A] is the concentration of A in mol/L, and the rate expressed in terms of A is

$$\text{Rate} = -\frac{\Delta[\text{A}]}{\Delta t} \qquad \textbf{(16.1)}$$

The units for the rate are moles per liter per second (mol L^{-1} s^{-1}, or mol/L·s), or any time unit convenient for the reaction (minutes, years, and so on).

If instead we measure the *product* concentrations to determine the rate, we find that conc B_2 is always *higher* than conc B_1. Thus, the *change* in product concentration, Δ[B], is *positive,* and the reaction rate for A $\longrightarrow$ B expressed in terms of B is

$$\text{Rate} = +\frac{\Delta[\text{B}]}{\Delta t}$$

(The plus sign is usually understood and not shown.)

Average, Instantaneous, and Initial Reaction Rates

In most cases, *the rate varies as a reaction proceeds.* Consider the reversible gas-phase reaction between ethylene and ozone, one of many reactions that may be involved in the formation of smog:

$$C_2H_4(g) + O_3(g) \rightleftharpoons C_2H_4O(g) + O_2(g)$$

The equation shows that for every molecule of C_2H_4 that reacts, a molecule of O_3 reacts; thus, $[O_3]$ and $[C_2H_4]$ decrease at the same rate:

$$\text{Rate} = -\frac{\Delta[C_2H_4]}{\Delta t} = -\frac{\Delta[O_3]}{\Delta t}$$

Table 16.1 Concentration of O_3 at Various Times in Its Reaction with C_2H_4 at 303 K

Time (s)	Concentration of O_3 (mol/L)
0.0	3.20×10^{-5}
10.0	2.42×10^{-5}
20.0	1.95×10^{-5}
30.0	1.63×10^{-5}
40.0	1.40×10^{-5}
50.0	1.23×10^{-5}
60.0	1.10×10^{-5}

When we start with a known $[O_3]$ in a closed vessel at 30°C (303 K) and measure $[O_3]$ at 10.0-s intervals during the first minute after adding C_2H_4, we obtain the data in Table 16.1 and the plot of $[O_3]$ vs. t in Figure 16.3 *(red curve)*. Two key points are

- The data points in Figure 16.3 result in a curved line, which means that the rate is changing (a straight line would mean that the rate is constant).
- The rate *decreases* during the course of the reaction because we are plotting *reactant* concentration versus time: as O_3 molecules react, fewer are present to collide with C_2H_4 molecules, and the rate, the change in $[O_3]$ over time, therefore decreases.

Three types of reaction rates are shown in the figure:

1. *Average rate.* Over a given period of time, the **average rate** is the slope of the line joining two points along the curve. The rate over the entire 60.0 s is the total change in concentration divided by the total change in time (Figure 16.3, *line* ***a***):

$$\text{Rate} = -\frac{\Delta[O_3]}{\Delta t} = -\frac{(1.10\times10^{-5}\text{ mol/L}) - (3.20\times10^{-5}\text{ mol/L})}{60.0\text{ s} - 0.0\text{ s}} = 3.50\times10^{-7}\text{ mol/L·s}$$

This quantity is the average rate over the entire period: during the first 60.0 s of the reaction, $[O_3]$ decreases an *average* of 3.50×10^{-7} mol/L each second.

But, when you drive a car for a few miles, your speed over shorter distances may be lower or higher than your average speed. Similarly, the decrease in $[O_3]$ over the whole time period may not be the same over any shorter time period. For example, for the first 10.0 s, between 0.0 s and 10.0 s, the average rate (Figure 16.3, *line* ***b***) is

$$\text{Rate} = -\frac{\Delta[O_3]}{\Delta t} = -\frac{(2.42\times10^{-5}\text{ mol/L}) - (3.20\times10^{-5}\text{ mol/L})}{10.0\text{ s} - 0.0\text{ s}} = 7.80\times10^{-7}\text{ mol/L·s}$$

And, for the last 10.0 s, between 50.0 s and 60.0 s, the average rate (Figure 16.3, *line* ***c***) is

$$\text{Rate} = -\frac{\Delta[O_3]}{\Delta t} = -\frac{(1.10\times10^{-5}\text{ mol/L}) - (1.23\times10^{-5}\text{ mol/L})}{60.0\text{ s} - 50.0\text{ s}} = 1.30\times10^{-7}\text{ mol/L·s}$$

The earlier rate is six times faster than the later rate.

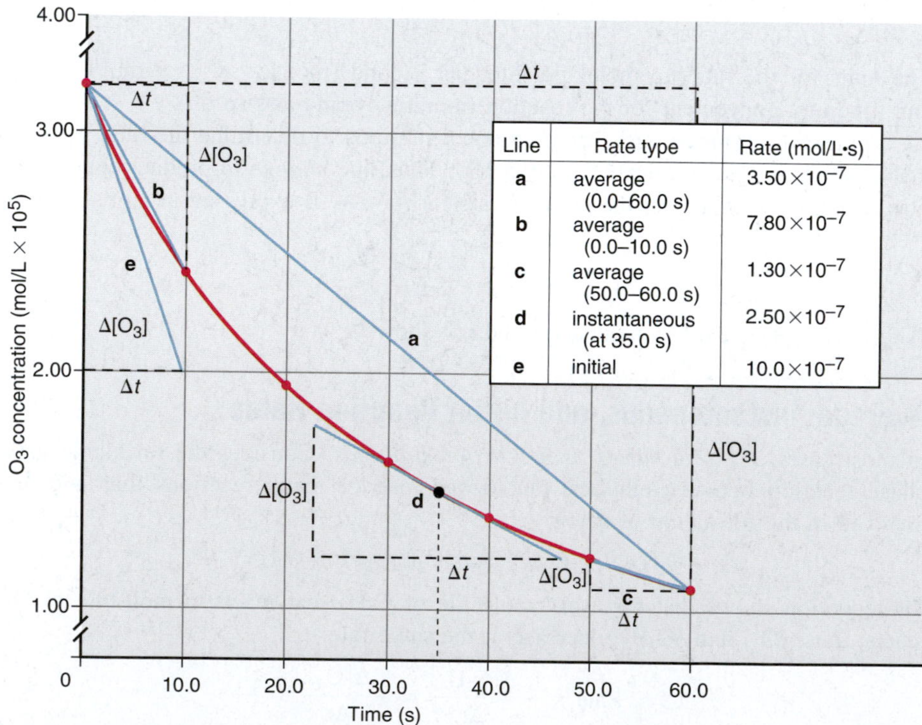

Figure 16.3 Three types of reaction rates for the reaction of O_3 and C_2H_4.

2. *Instantaneous rate.* The shorter the time period we choose, the closer we come to the **instantaneous rate,** the rate at a particular instant during the reaction. *The slope of a line tangent to the curve at any point gives the instantaneous rate at that time.* For example, the rate at 35.0 s is 2.50×10^{-7} mol/L·s, the slope of the line tangent to the curve through the point at $t = 35.0$ s (Figure 16.3, *line **d***). In general, we use the term *reaction rate* to mean *instantaneous* reaction rate.

3. *Initial rate.* The instantaneous rate at the moment the reactants are mixed (that is, at $t = 0$) is the **initial rate.** We use this rate to avoid a complication: as a reaction proceeds in the *forward* direction (reactants $\longrightarrow$ products), product increases, causing the *reverse* reaction (reactants $\longleftarrow$ products), to occur more quickly. To find the overall (net) rate, we would have to calculate the difference between the forward and reverse rates. But, for the initial rate, $t = 0$, so product concentrations are negligible, and so is the reverse rate. We find the initial rate from the slope of the line tangent to the curve at $t = 0$ s (Figure 16.3, *line **e***). We typically use initial rates to find other kinetic parameters.

Expressing Rate in Terms of Reactant and Product Concentrations

So far, for the reaction between C_2H_4 and O_3, we've expressed the rate in terms of $[O_3]$, which is *decreasing.* The rate is expressed the same way in terms of $[C_2H_4]$, but it is exactly the opposite in terms of the product concentrations because they are *increasing*. From the balanced equation, we see that one molecule each of C_2H_4O and of O_2 appear for every molecule of C_2H_4 and of O_3 that disappear. We can express the rate in terms of any of the four substances involved:

$$\text{Rate} = -\frac{\Delta[C_2H_4]}{\Delta t} = -\frac{\Delta[O_3]}{\Delta t} = \frac{\Delta[C_2H_4O]}{\Delta t} = \frac{\Delta[O_2]}{\Delta t}$$

(Note the negative values for the rate in terms of reactants.) Figure 16.4 plots the changes in concentrations of one reactant (C_2H_4) and one product (O_2) simultaneously. The curves have the same shape but are inverted relative to each other, because, *for this reaction,* product appears at the same rate that reactant disappears.

For many other reactions, though, reactants disappear and products appear at different rates. Consider the reaction between hydrogen and iodine to form hydrogen iodide:

$$H_2(g) + I_2(g) \longrightarrow 2HI(g)$$

For every molecule of H_2 that disappears, one molecule of I_2 disappears and *two* molecules of HI appear. In other words, the rate of $[H_2]$ decrease is the same as the rate of $[I_2]$ decrease, but both are only half the rate of [HI] increase. If we refer the changes in $[I_2]$ and [HI] to the change in $[H_2]$, we have

$$\text{Rate} = -\frac{\Delta[H_2]}{\Delta t} = -\frac{\Delta[I_2]}{\Delta t} = \frac{1}{2}\frac{\Delta[HI]}{\Delta t}$$

If, instead, we refer the changes in $[H_2]$ and $[I_2]$ to the change in [HI], we obtain

$$\text{Rate} = \frac{\Delta[HI]}{\Delta t} = -2\frac{\Delta[H_2]}{\Delta t} = -2\frac{\Delta[I_2]}{\Delta t}$$

Note that this expression gives a rate that is double the previous one. Thus, *the expression for the rate of a reaction and its numerical value depend on which substance serves as the reference.*

We can summarize these results for any reaction,

$$aA + bB \longrightarrow cC + dD$$

where a, b, c, and d are coefficients of the balanced equation, as follows:

$$\text{Rate} = -\frac{1}{a}\frac{\Delta[A]}{\Delta t} = -\frac{1}{b}\frac{\Delta[B]}{\Delta t} = \frac{1}{c}\frac{\Delta[C]}{\Delta t} = \frac{1}{d}\frac{\Delta[D]}{\Delta t} \quad (16.2)$$

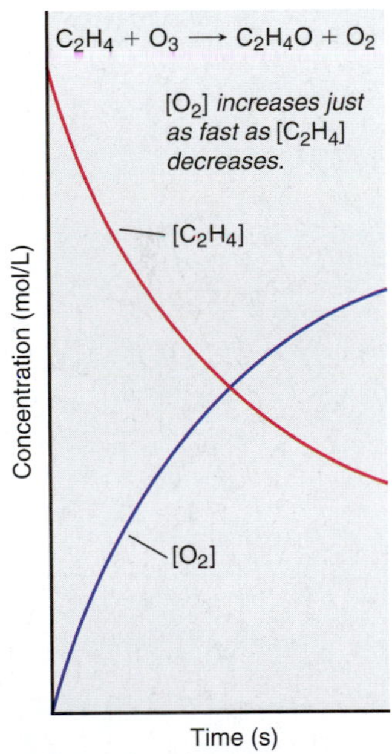

Figure 16.4 Plots of $[C_2H_4]$ and $[O_2]$ vs. time.

Sample Problem 16.1 Expressing Rate in Terms of Changes in Concentration with Time

Problem Hydrogen gas has a nonpolluting combustion product (water vapor). It is used as a fuel aboard the space shuttle and in earthbound cars with prototype engines:

$$2H_2(g) + O_2(g) \longrightarrow 2H_2O(g)$$

(a) Express the rate in terms of changes in $[H_2]$, $[O_2]$, and $[H_2O]$ with time.
(b) When $[O_2]$ is decreasing at 0.23 mol/L·s, at what rate is $[H_2O]$ increasing?

Plan **(a)** Of the three substances in the equation, let's choose O_2 as the reference because its coefficient is 1. For every molecule of O_2 that disappears, two molecules of H_2 disappear. Thus, the rate of $[O_2]$ decrease is one-half the rate of $[H_2]$ decrease. By similar reasoning, the rate of $[O_2]$ decrease is one-half the rate of $[H_2O]$ increase. **(b)** Because $[O_2]$ is decreasing, the change in its concentration must be negative. We substitute the given rate as a negative value (−0.23 mol/L·s) into the expression and solve for $\Delta[H_2O]/\Delta t$.

Solution **(a)** Expressing the rate in terms of each component:

$$\text{Rate} = -\frac{\Delta[O_2]}{\Delta t} = -\frac{1}{2}\frac{\Delta[H_2]}{\Delta t} = \frac{1}{2}\frac{\Delta[H_2O]}{\Delta t}$$

(b) Calculating the rate of change of $[H_2O]$:

$$\frac{1}{2}\frac{\Delta[H_2O]}{\Delta t} = -\frac{\Delta[O_2]}{\Delta t} = -(-0.23\ \text{mol/L·s})$$

$$\frac{\Delta[H_2O]}{\Delta t} = 2(0.23\ \text{mol/L·s}) = 0.46\ \text{mol/L·s}$$

Check **(a)** A good check is to use the rate expression to obtain the balanced equation: $[H_2]$ changes twice as fast as $[O_2]$, so two H_2 molecules react for each O_2. $[H_2O]$ changes twice as fast as $[O_2]$, so two H_2O molecules form from each O_2. Thus, we get $2H_2 + O_2 \longrightarrow 2H_2O$. The values of $[H_2]$ and $[O_2]$ decrease, so they have minus signs; $[H_2O]$ increases, so it has a plus sign. Another check is to use Equation 16.2, with A = H_2, $a = 2$; B = O_2, $b = 1$; and C = H_2O, $c = 2$:

$$\text{Rate} = -\frac{1}{a}\frac{\Delta[A]}{\Delta t} = -\frac{1}{b}\frac{\Delta[B]}{\Delta t} = \frac{1}{c}\frac{\Delta[C]}{\Delta t}$$

or

$$\text{Rate} = -\frac{1}{2}\frac{\Delta[H_2]}{\Delta t} = -\frac{\Delta[O_2]}{\Delta t} = \frac{1}{2}\frac{\Delta[H_2O]}{\Delta t}$$

(b) Given the rate expression, it makes sense that the numerical value of the rate of $[H_2O]$ increase is twice that of $[O_2]$ decrease.

Comment Thinking through this type of problem at the molecular level is the best approach, but use Equation 16.2 to confirm your answer.

FOLLOW-UP PROBLEM 16.1 **(a)** Balance the following equation and express the rate in terms of the change in concentration with time for each substance:

$$NO(g) + O_2(g) \longrightarrow N_2O_3(g)$$

(b) How fast is $[O_2]$ decreasing when $[NO]$ is decreasing at a rate of 1.60×10^{-4} mol/L·s?

Summary of Section 16.2

- The average reaction rate is the change in reactant (or product) concentration over a change in time, Δt. The rate slows as the reaction proceeds because reactants are used up.
- The instantaneous rate at time t is the slope of the tangent to a curve that plots concentration vs. time.
- The initial rate, the instantaneous rate at $t = 0$, occurs when reactants have just been mixed and before any product accumulates.
- The expression for a reaction rate and its numerical value depend on which reaction component is being referenced.

16.3 • THE RATE LAW AND ITS COMPONENTS

The centerpiece of any kinetic study of a reaction is the **rate law** (or **rate equation**), which expresses the rate as a function of concentrations and temperature. The rate law is based on experiment, so any hypothesis about how the reaction occurs on the molecular level must conform to it.

In this discussion, we generally consider reactions for which the products do not appear in the rate law, so the rate depends only on *reactant* concentrations and temperature. For a general reaction occurring at a fixed temperature,

$$aA + bB + \cdots \longrightarrow cC + dD + \cdots$$

the rate law is

$$\text{Rate} = k[A]^m[B]^n \cdots \quad \textbf{(16.3)}$$

The term k is a proportionality constant, called the **rate constant,** that is specific for a given reaction at a given temperature and does *not* change as the reaction proceeds. (As we'll see in Section 16.5, *k does* change with temperature.) The exponents m and n, called the **reaction orders,** define how the rate is affected by reactant concentration; we'll see how to determine them shortly. Two key points to remember are

- *The balancing coefficients a and b in the reaction equation are* ***not*** *necessarily related in any way to the reaction orders m and n.*
- *The components of the rate law—rate, reaction orders, and rate constant—**must** be found by experiment.*

In the remainder of this section, we'll find the components of the rate law by measuring concentrations to determine the *initial rate,* using initial rates to determine the *reaction orders,* and using these values to calculate the *rate constant.* With the rate law for a reaction, we can predict the rate for any initial concentrations.

Some Laboratory Methods for Determining the Initial Rate

We determine an initial rate from a plot of concentration vs. time, so we need a quick, accurate method for measuring concentration. Let's briefly discuss three common approaches.

- *Change in color.* For reactions that involve a colored substance, *spectroscopic methods* can be used to measure concentration. For example, in the oxidation of nitrogen monoxide, only the product, nitrogen dioxide, is colored:

  $$2NO(g) + O_2(g) \longrightarrow 2NO_2(g;\ \text{brown})$$

  As time proceeds, the brown color of the reaction mixture deepens.
- *Change in pressure.* Reactions that involve a change in number of moles of gas can be monitored by measuring the change in pressure. Consider the reaction between zinc and acetic acid:

  $$Zn(s) + 2CH_3COOH(aq) \longrightarrow Zn^{2+}(aq) + 2CH_3COO^-(aq) + H_2(g)$$

  The rate is directly proportional to the increase in H_2 gas pressure.
- *Change in conductivity.* In the reaction between an organic halide (2-bromo-2-methylpropane) and water,

  $$(CH_3)_3C\text{—}Br(l) + H_2O(l) \longrightarrow (CH_3)_3C\text{—}OH(l) + H^+(aq) + Br^-(aq)$$

  the HBr that forms is a strong acid and dissociates completely into ions; thus, the conductivity of the reaction mixture increases as time proceeds.

Determining Reaction Orders

With the initial rate in hand, we can determine reaction orders. Let's first discuss what reaction orders are and then see how to determine them by controlling reactant concentrations.

Meaning and Terminology A reaction has an *individual* order "with respect to" or "in" each reactant, and an *overall* order, the sum of the individual orders.

Consider first the simplest case, a reaction with only one reactant, A:

$$A \longrightarrow \text{products}$$

- *First order.* The reaction is *first order* overall if the rate is directly proportional to [A]. That is, if the rate doubles when [A] doubles, the rate depends on [A] raised to the first power, $[A]^1$ (the ^{1} is omitted). Thus, the reaction is *first order* in (or with respect to) A:

$$\text{Rate} = k[A]^1 = k[A]$$

- *Second order.* The reaction is *second order* overall if the rate is directly proportional to the square of [A]. That is, if the rate quadruples when [A] doubles, the rate depends on [A] squared, $[A]^2$:

$$\text{Rate} = k[A]^2$$

- *Zero order.* The reaction is *zero order* overall if the rate is *not* dependent on [A] at all (a situation that is common in metal-catalyzed and biochemical processes, as you'll see later). If the rate does not change when [A] changes, we express this fact mathematically by saying that the rate depends on [A] raised to the zero power, $[A]^0$:

$$\text{Rate} = k[A]^0 = k(1) = k$$

Let's look at some examples of observed rate laws and note the reaction orders. For the reaction between nitrogen monoxide and hydrogen gas,

$$2NO(g) + 2H_2(g) \longrightarrow N_2(g) + 2H_2O(g)$$

the rate law is

$$\text{Rate} = k[NO]^2[H_2]$$

This reaction is second order in NO. And, even though H_2 has a coefficient of 2 in the balanced equation, the reaction is first order in H_2. It is third order overall ($2 + 1 = 3$).

For the reaction between nitrogen monoxide and ozone,

$$NO(g) + O_3(g) \longrightarrow NO_2(g) + O_2(g)$$

the rate law is

$$\text{Rate} = k[NO][O_3]$$

This reaction is first order with respect to NO and first order with respect to O_3, so it is second order overall ($1 + 1 = 2$).

Finally, for the hydrolysis of 2-bromo-2-methylpropane,

$$(CH_3)_3C\text{—}Br(l) + H_2O(l) \longrightarrow (CH_3)_3C\text{—}OH(l) + H^+(aq) + Br^-(aq)$$

the rate law is

$$\text{Rate} = k[(CH_3)_3CBr]$$

This reaction is first order in 2-bromo-2-methylpropane and zero order with respect to H_2O, despite its coefficient of 1 in the balanced equation. If we want to note that water is a reactant, we can write

$$\text{Rate} = k[(CH_3)_3CBr][H_2O]^0$$

Overall, this is a first-order reaction ($1 + 0 = 1$). These examples reiterate an important point: *reaction orders* ***cannot*** *be deduced from the balanced equation but* ***must*** *be determined from experimental data.*

Reaction orders are usually positive integers or zero, but they can also be fractional or negative. For the reaction

$$CHCl_3(g) + Cl_2(g) \longrightarrow CCl_4(g) + HCl(g)$$

a fractional order appears in the rate law:

$$\text{Rate} = k[CHCl_3][Cl_2]^{1/2}$$

This reaction order means that the rate depends on the square root of $[Cl_2]$. For example, if $[Cl_2]$ increases by a factor of 4, the rate increases by a factor of 2, the square root of the change in $[Cl_2]$. The overall order of this reaction is $\frac{3}{2}$.

A negative reaction order means the rate *decreases* when the concentration of that component increases. Negative orders are often seen when the rate law includes products. For example, for the atmospheric reaction

$$2O_3(g) \rightleftharpoons 3O_2(g)$$

the rate law is

$$\text{Rate} = k[O_3]^2[O_2]^{-1} = k\frac{[O_3]^2}{[O_2]}$$

If $[O_2]$ doubles, the reaction proceeds half as fast. This reaction is second order in O_3 and negative first order in O_2, so it is first order overall $[2 + (-1) = 1]$.

Sample Problem 16.2 Determining Reaction Orders from Rate Laws

Problem For each of the following reactions, use the given rate law to determine the reaction order with respect to each reactant and the overall order:
(a) $2NO(g) + O_2(g) \longrightarrow 2NO_2(g)$; rate $= k[NO]^2[O_2]$
(b) $CH_3CHO(g) \longrightarrow CH_4(g) + CO(g)$; rate $= k[CH_3CHO]^{3/2}$
(c) $H_2O_2(aq) + 3I^-(aq) + 2H^+(aq) \longrightarrow I_3^-(aq) + 2H_2O(l)$; rate $= k[H_2O_2][I^-]$

Plan We inspect the exponents in the rate law, *not* the coefficients of the balanced equation, to find the individual orders, and then take their sum to find the overall reaction order.

Solution **(a)** The exponent of [NO] is 2, so the reaction is second order with respect to NO, first order with respect to O_2, and third order overall.

(b) The reaction is $\frac{3}{2}$ order in CH_3CHO and $\frac{3}{2}$ order overall.

(c) The reaction is first order in H_2O_2, first order in I^-, and second order overall.

The reactant H^+ does not appear in the rate law, so the reaction is zero order in H^+.

Check Be sure that each reactant has an order and that the sum of the individual orders gives the overall order.

FOLLOW-UP PROBLEM 16.2 Experiment shows that the reaction

$$5Br^-(aq) + BrO_3^-(aq) + 6H^+(aq) \longrightarrow 3Br_2(l) + 3H_2O(l)$$

obeys this rate law: rate $= k[Br^-][BrO_3^-][H^+]^2$. What is the reaction order in each reactant and the overall reaction order?

Determining Reaction Orders by Changing the Reactant Concentrations

Now let's see how reaction orders are found *before* the rate law is known. Before looking at a real reaction, we'll go through the process for substances A and B in this reaction:

$$A + 2B \longrightarrow C + D$$

The rate law, expressed in general terms, is

$$\text{Rate} = k[A]^m[B]^n$$

To find the values of *m* and *n*, *we run a series of experiments in which one reactant concentration changes while the other is kept constant, and we measure the effect on the initial rate in each case.* Table 16.2 on the next page shows the results.

Table 16.2 Initial Rates for the Reaction Between A and B

Experiment	Initial Rate (mol/L·s)	Initial [A] (mol/L)	Initial [B] (mol/L)
1	1.75×10^{-3}	2.50×10^{-2}	3.00×10^{-2}
2	3.50×10^{-3}	5.00×10^{-2}	3.00×10^{-2}
3	3.50×10^{-3}	2.50×10^{-2}	6.00×10^{-2}
4	7.00×10^{-3}	5.00×10^{-2}	6.00×10^{-2}

1. *Finding m, the order with respect to A.* By comparing experiments 1 and 2, in which [A] doubles and [B] is constant, we can obtain m. First, we take the ratio of the general rate laws for these two experiments:

$$\frac{\text{Rate 2}}{\text{Rate 1}} = \frac{k[\text{A}]_2^m[\text{B}]_2^n}{k[\text{A}]_1^m[\text{B}]_1^n}$$

where $[\text{A}]_2$ is the concentration of A in experiment 2, $[\text{B}]_1$ is the concentration of B in experiment 1, and so forth. Because k is a constant and [B] does not change between these two experiments, those quantities cancel:

$$\frac{\text{Rate 2}}{\text{Rate 1}} = \frac{\cancel{k}[\text{A}]_2^m\cancel{[\text{B}]_2^n}}{\cancel{k}[\text{A}]_1^m\cancel{[\text{B}]_1^n}} = \frac{[\text{A}]_2^m}{[\text{A}]_1^m} = \left(\frac{[\text{A}]_2}{[\text{A}]_1}\right)^m$$

Substituting the values from Table 16.2, we have

$$\frac{3.50\times10^{-3}\ \text{mol/L·s}}{1.75\times10^{-3}\ \text{mol/L·s}} = \left(\frac{5.00\times10^{-2}\ \text{mol/L}}{2.50\times10^{-2}\ \text{mol/L}}\right)^m$$

Dividing, we obtain

$$2.00 = (2.00)^m \quad \text{so} \quad m = 1$$

Thus, the reaction is first order in A, because when [A] doubles, the rate doubles.

2. *Finding n, the order with respect to B.* To find n, we compare experiments 3 and 1 in which [A] is held constant and [B] doubles:

$$\frac{\text{Rate 3}}{\text{Rate 1}} = \frac{k[\text{A}]_3^m[\text{B}]_3^n}{k[\text{A}]_1^m[\text{B}]_1^n}$$

As before, k is a constant, and in this pair of experiments, [A] does not change, so those quantities cancel, and we have

$$\frac{\text{Rate 3}}{\text{Rate 1}} = \frac{\cancel{k}\cancel{[\text{A}]_3^m}[\text{B}]_3^n}{\cancel{k}\cancel{[\text{A}]_1^m}[\text{B}]_1^n} = \frac{[\text{B}]_3^n}{[\text{B}]_1^n} = \left(\frac{[\text{B}]_3}{[\text{B}]_1}\right)^n$$

The actual values give

$$\frac{3.50\times10^{-3}\ \text{mol/L·s}}{1.75\times10^{-3}\ \text{mol/L·s}} = \left(\frac{6.00\times10^{-2}\ \text{mol/L}}{3.00\times10^{-2}\ \text{mol/L}}\right)^n$$

Dividing, we obtain

$$2.00 = (2.00)^n \quad \text{so} \quad n = 1$$

Thus, the reaction is also first order in B because when [B] doubles, the rate doubles. We can check this conclusion from experiment 4: when *both* [A] and [B] double, the rate should quadruple, and it does. Thus, the rate law, with m and n equal to 1, is

$$\text{Rate} = k[\text{A}][\text{B}]$$

Note, especially, that while the order with respect to B is 1, the coefficient of B in the balanced equation is 2. This demonstrates again that *reaction orders must be determined from experiment.*

Next, let's go through this process for a real reaction, the one between oxygen and nitrogen monoxide, a key step in the formation of acid rain and in the industrial production of nitric acid:

$$O_2(g) + 2NO(g) \longrightarrow 2NO_2(g)$$

Table 16.3 Initial Rates for the Reaction Between O_2 and NO

Experiment	Initial Rate (mol/L·s)	Initial Reactant Concentrations (mol/L) $[O_2]$	$[NO]$
1	3.21×10^{-3}	1.10×10^{-2}	1.30×10^{-2}
2	6.40×10^{-3}	2.20×10^{-2}	1.30×10^{-2}
3	12.8×10^{-3}	1.10×10^{-2}	2.60×10^{-2}
4	9.60×10^{-3}	3.30×10^{-2}	1.30×10^{-2}
5	28.8×10^{-3}	1.10×10^{-2}	3.90×10^{-2}

The general rate law is

$$\text{Rate} = k[O_2]^m[NO]^n$$

Table 16.3 shows experiments that change one reactant concentration while keeping the other constant. If we compare experiments 1 and 2, we see the effect of doubling $[O_2]$ on the rate. First, we take the ratio of their rate laws:

$$\frac{\text{Rate 2}}{\text{Rate 1}} = \frac{k[O_2]_2^m[NO]_2^n}{k[O_2]_1^m[NO]_1^n}$$

As before, the constant quantities—k and [NO]—cancel:

$$\frac{\text{Rate 2}}{\text{Rate 1}} = \frac{\cancel{k}[O_2]_2^m\cancel{[NO]_2^n}}{\cancel{k}[O_2]_1^m\cancel{[NO]_1^n}} = \frac{[O_2]_2^m}{[O_2]_1^m} = \left(\frac{[O_2]_2}{[O_2]_1}\right)^m$$

Substituting the values from Table 16.3, we obtain

$$\frac{6.40\times10^{-3}\ \text{mol/L·s}}{3.21\times10^{-3}\ \text{mol/L·s}} = \left(\frac{2.20\times10^{-2}\ \text{mol/L}}{1.10\times10^{-2}\ \text{mol/L}}\right)^m$$

Dividing, we obtain

$$1.99 = (2.00)^m$$

Rounding to one significant figure gives

$$2 = 2^m \qquad \text{so} \qquad m = 1$$

Sometimes, the exponent is not as easy to find by inspection as it is here. In those cases, we solve for m starting with an equation of the form $a = b^m$, which gives

$$\log a = m \log b \qquad \text{or} \qquad m = \frac{\log a}{\log b} = \frac{\log 1.99}{\log 2.00} = 0.993$$

which rounds to 1. Thus, the reaction is first order in O_2: when $[O_2]$ doubles, the rate doubles.

To find the order with respect to NO, we compare experiments 3 and 1, in which $[O_2]$ is held constant and [NO] is doubled:

$$\frac{\text{Rate 3}}{\text{Rate 1}} = \frac{k[O_2]_3^m[NO]_3^n}{k[O_2]_1^m[NO]_1^n}$$

Canceling the constant k and the unchanging $[O_2]$, we have

$$\frac{\text{Rate 3}}{\text{Rate 1}} = \left(\frac{[NO]_3}{[NO]_1}\right)^n$$

The actual values give

$$\frac{12.8\times10^{-3}\ \text{mol/L·s}}{3.21\times10^{-3}\ \text{mol/L·s}} = \left(\frac{2.60\times10^{-2}\ \text{mol/L}}{1.30\times10^{-2}\ \text{mol/L}}\right)^n$$

Dividing, we obtain

$$3.99 = (2.00)^n$$

Solving for n:

$$n = \frac{\log 3.99}{\log 2.00} = 2.00 \quad \text{(or 2)}$$

The reaction is second order in NO: when [NO] doubles, the rate quadruples. Thus, the complete rate law is

$$\text{Rate} = k[O_2][NO]^2$$

In this case, the reaction orders happen to be the same as the equation coefficients; nevertheless, they must *always* be determined by experiment.

The next two sample problems offer practice with this approach; the first is based on data and the second on molecular scenes.

Sample Problem 16.3 Determining Reaction Orders from Rate Data

Problem Many gaseous reactions occur in car engines and exhaust systems. One of these is

$$NO_2(g) + CO(g) \longrightarrow NO(g) + CO_2(g) \qquad \text{rate} = k[NO_2]^m[CO]^n$$

Use the following data to determine the individual and overall reaction orders:

Experiment	Initial Rate (mol/L·s)	Initial $[NO_2]$ (mol/L)	Initial [CO] (mol/L)
1	0.0050	0.10	0.10
2	0.080	0.40	0.10
3	0.0050	0.10	0.20

Plan We need to solve the general rate law for m and for n and then add those orders to get the overall order. To solve for each exponent, we proceed as in the text, taking the ratio of the rate laws for two experiments in which only the reactant in question changes.

Solution Calculating m in $[NO_2]^m$: We take the ratio of the rate laws for experiments 1 and 2, in which $[NO_2]$ varies but [CO] is constant:

$$\frac{\text{Rate 2}}{\text{Rate 1}} = \frac{k[NO_2]_2^m[CO]_2^n}{k[NO_2]_1^m[CO]_1^n} = \left(\frac{[NO_2]_2}{[NO_2]_1}\right)^m \quad \text{or} \quad \frac{0.080 \text{ mol/L·s}}{0.0050 \text{ mol/L·s}} = \left(\frac{0.40 \text{ mol/L}}{0.10 \text{ mol/L}}\right)^m$$

This gives $16 = (4.0)^m$, so we have $m = \log 16/\log 4.0 = 2.0$. The reaction is second order in NO_2.

Calculating n in $[CO]^n$: We take the ratio of the rate laws for experiments 1 and 3, in which [CO] varies but $[NO_2]$ is constant:

$$\frac{\text{Rate 3}}{\text{Rate 1}} = \frac{k[NO_2]_3^2[CO]_3^n}{k[NO_2]_1^2[CO]_1^n} = \left(\frac{[CO]_3}{[CO]_1}\right)^n \quad \text{or} \quad \frac{0.0050 \text{ mol/L·s}}{0.0050 \text{ mol/L·s}} = \left(\frac{0.20 \text{ mol/L}}{0.10 \text{ mol/L}}\right)^n$$

We have $1.0 = (2.0)^n$, so $n = 0$. The rate does not change when [CO] varies, so the reaction is zero order in CO.

Therefore, the rate law is

$$\text{Rate} = k[NO_2]^2[CO]^0 = k[NO_2]^2(1) = k[NO_2]^2$$

The reaction is second order overall.

Check A good check is to reason through the orders. If $m = 1$, quadrupling $[NO_2]$ would quadruple the rate; but the rate *more* than quadruples, so $m > 1$. If $m = 2$, quadrupling $[NO_2]$ would increase the rate by a factor of 16 (4^2). The ratio of rates is $0.080/0.005 = 16$, so $m = 2$. In contrast, increasing [CO] has no effect on the rate, which can happen only if $[CO]^n = 1$, so $n = 0$.

FOLLOW-UP PROBLEM 16.3 Find the rate law and the individual and overall reaction orders for the reaction $H_2 + I_2 \longrightarrow 2HI$ using the following data at 450°C:

Experiment	Initial Rate (mol/L·s)	Initial $[H_2]$ (mol/L)	Initial $[I_2]$ (mol/L)
1	1.9×10^{-23}	0.0113	0.0011
2	1.1×10^{-22}	0.0220	0.0033
3	9.3×10^{-23}	0.0550	0.0011
4	1.9×10^{-22}	0.0220	0.0056

Sample Problem 16.4 Determining Reaction Orders from Molecular Scenes

Problem At a particular temperature and volume, two gases, A *(red)* and B *(blue)*, react. The following molecular scenes represent starting mixtures for four experiments:

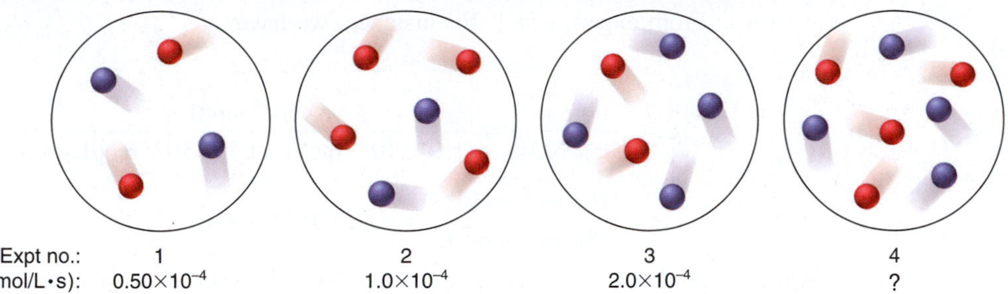

(a) What is the reaction order with respect to A? With respect to B? The overall order?
(b) Write the rate law for the reaction.
(c) Predict the initial rate of Expt 4.

Plan **(a)** As before, we find the individual reaction orders by seeing how a change in each reactant changes the rate. In this case, however, instead of using concentration data, we count numbers of particles. The sum of the individual orders is the overall order. **(b)** To write the rate law, we use the orders from part (a) as exponents in the general rate law. **(c)** Using the results from Expts 1 through 3 and the rate law from part (b), we find the unknown initial rate of Expt 4.

Solution **(a)** Finding the individual and overall orders. For reactant A *(red):* Expts 1 and 2 show that when the number of particles of A doubles (from 2 to 4), with the number of particles of B constant (at 2), the rate doubles (from 0.5×10^{-4} mol/L·s to 1.0×10^{-4} mol/L·s). Thus, the order with respect to A is 1. For reactant B *(blue):* Expts 1 and 3 show that when the number of particles of B doubles (from 2 to 4), with the number of particles of A constant (at 2), the rate quadruples (from 0.5×10^{-4} mol/L·s to 2.0×10^{-4} mol/L·s). Thus, the order with respect to B is 2. The overall order is $1 + 2 = 3$.

(b) Writing the rate law: The general rate law is rate $= k[A]^m[B]^n$, so we have

$$\text{Rate} = k[A][B]^2$$

(c) Finding the initial rate of Expt 4: There are several possibilities, but let's compare Expts 3 and 4, in which the number of particles of A doubles (from 2 to 4) and the number of particles of B doesn't change. Since the rate law shows that the reaction is first order in A, the initial rate in Expt 4 should be double the initial rate in Expt 3, or 4.0×10^{-4} mol/L·s.

Check A good check is to compare other pairs of experiments. **(a)** Comparing Expts 2 and 3 shows that the number of B doubles, which causes the rate to quadruple, and the number of A decreases by half, which causes the rate to halve; so the overall rate change should double (from 1.0×10^{-4} mol/L·s to 2.0×10^{-4} mol/L·s), which it does. **(c)** Comparing Expts 2 and 4, in which the number of A is constant and the number of B doubles, the rate should quadruple, which means the initial rate of Expt 4 would be 4.0×10^{-4} mol/L·s, as we found.

FOLLOW-UP PROBLEM 16.4 The scenes below show three experiments at a given temperature and volume involving reactants X *(black)* and Y *(green):*

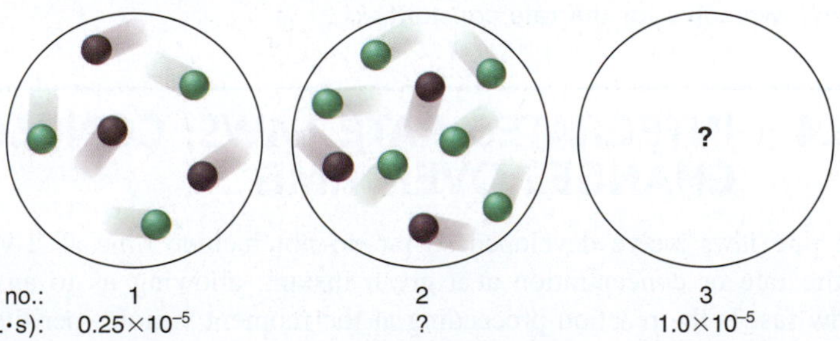

If the rate law for the reaction is rate $= k[X]^2$: **(a)** What is the initial rate of Expt 2? **(b)** Draw a scene for Expt 3 that involves a single change to the scene for Expt 1.

Table 16.4 Units of the Rate Constant *k* for Several Overall Reaction Orders

Overall Reaction Order	Units of *k* (*t* in seconds)
0	mol/L·s (or mol L^{-1} s^{-1})
1	1/s (or s^{-1})
2	L/mol·s (or L mol^{-1} s^{-1})
3	L^2/mol^2·s (or L^2 mol^{-2} s^{-1})

General formula:

$$\text{Units of } k = \frac{\left(\dfrac{\text{L}}{\text{mol}}\right)^{\text{order}-1}}{\text{unit of } t}$$

Determining the Rate Constant

Let's find the rate constant for the reaction of O_2 and NO. With the rate, reactant concentrations, and reaction orders known, the sole remaining unknown in the rate law is the rate constant, *k*. We can use data from any of the experiments in Table 16.3 to solve for *k*. From experiment 1, for instance, we have

$$\text{Rate} = k[O_2]_1[NO]_1^2$$

$$k = \frac{\text{rate 1}}{[O_2]_1[NO]_1^2} = \frac{3.21\times10^{-3}\ \text{mol/L·s}}{(1.10\times10^{-2}\ \text{mol/L})(1.30\times10^{-2}\ \text{mol/L})^2}$$

$$= \frac{3.21\times10^{-3}\ \text{mol/L·s}}{1.86\times10^{-6}\ \text{mol}^3/\text{L}^3} = 1.73\times10^3\ \text{L}^2/\text{mol}^2\text{·s}$$

Always check that the values of *k* for a series of experiments are constant within experimental error. To three significant figures, the average value of *k* for the five experiments in Table 16.3 is 1.72×10^3 L^2/mol^2·s.

With concentrations in mol/L and the reaction rate in units of mol/L·time, the units for *k* depend on the order of the reaction and, of course, the time unit. For this reaction, the units for *k*, L^2/mol^2·s, are required to give a rate with units of mol/L·s:

$$\frac{\text{mol}}{\text{L·s}} = \frac{\text{L}^2}{\text{mol}^2\text{·s}} \times \frac{\text{mol}}{\text{L}} \times \left(\frac{\text{mol}}{\text{L}}\right)^2$$

The rate constant will *always* have these units for an overall third-order reaction with the time unit in seconds. Table 16.4 shows the units of *k* for common integer overall orders, but you can always determine the units mathematically.

Figure 16.5 summarizes the steps for studying the kinetics of a reaction.

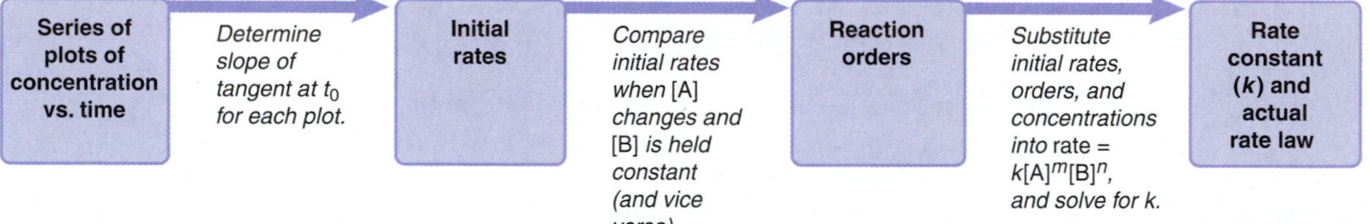

Figure 16.5 Information sequence to determine the kinetic parameters of a reaction.

Summary of Section 16.3

- An experimentally determined rate law shows how the rate of a reaction depends on concentration. Considering only initial rates (that is, no products), the expression for a general rate law is rate = $k[A]^m[B]^n \cdots$. This reaction is *m*th order with respect to A and *n*th order with respect to B; the overall reaction order is $m + n$.
- With an accurate method for obtaining initial rates, reaction orders are determined experimentally by varying the concentration of one reactant at a time to see its effect on the rate.
- By substituting the known rate, concentrations, and reaction orders into the rate law, we solve for the rate constant, *k*.

16.4 • INTEGRATED RATE LAWS: CONCENTRATION CHANGES OVER TIME

The rate laws we've developed so far do not include time as a variable. They tell us the rate or concentration at a given instant, allowing us to answer the question "How fast is the reaction proceeding at the moment *y* moles per liter of A are mixed with *z* moles per liter of B?" By employing different forms of the rate laws, called **integrated rate laws,** we can include time as a variable and answer other questions, such as "How long will it take to use up *x* moles per liter of A?" and "What is [A] after *y* minutes of reaction?"

Integrated Rate Laws for First-, Second-, and Zero-Order Reactions

As we've seen, for a *general first-order reaction,* A $\longrightarrow$ B, the rate is the negative of the change in [A] divided by the change in time:

$$\text{Rate} = -\frac{\Delta[\text{A}]}{\Delta t}$$

It can also be expressed in terms of the rate law:

$$\text{Rate} = k[\text{A}]$$

Setting these two expressions equal to each other gives

$$-\frac{\Delta[\text{A}]}{\Delta t} = k[\text{A}] \qquad \text{or} \qquad -\frac{\Delta[\text{A}]}{[\text{A}]} = k\Delta t$$

Through calculus, integrating over time gives the integrated rate law for a first-order reaction:

$$\ln\frac{[\text{A}]_0}{[\text{A}]_t} = kt \qquad (\text{first-order reaction; rate} = k[\text{A}]) \qquad \textbf{(16.4)}$$

where ln is the natural logarithm, $[\text{A}]_0$ is the concentration of A at $t = 0$, and $[\text{A}]_t$ is the concentration of A at any time t.

For a *general second-order reaction* involving two reactants, A and B, the expression including time is complex. In the simpler case, the rate law contains only reactant A. Setting the two rate expressions equal to each other gives

$$\text{Rate} = -\frac{\Delta[\text{A}]}{\Delta t} = k[\text{A}]^2 \qquad \text{or} \qquad -\frac{\Delta[\text{A}]}{[\text{A}]^2} = k\Delta t$$

Integrating over time gives the integrated rate law for a second-order reaction involving one reactant:

$$\frac{1}{[\text{A}]_t} - \frac{1}{[\text{A}]_0} = kt \qquad (\text{second-order reaction; rate} = k[\text{A}]^2) \qquad \textbf{(16.5)}$$

For a *general zero-order reaction*, setting the two rate expressions equal to each other gives

$$\text{Rate} = -\frac{\Delta[\text{A}]}{\Delta t} = k[\text{A}]^0 = k \qquad \text{or} \qquad -\Delta[\text{A}] = k\Delta t$$

Integrating over time gives the integrated rate law for a zero-order reaction:

$$[\text{A}]_t - [\text{A}]_0 = -kt \qquad (\text{zero-order reaction; rate} = k[\text{A}]^0 = k) \qquad \textbf{(16.6)}$$

Sample Problem 16.5 shows one way integrated rate laws are applied.

Sample Problem 16.5 Determining the Reactant Concentration after a Given Time

Problem At 1000°C, cyclobutane (C_4H_8) decomposes to two molecules of ethylene (C_2H_4), in a first-order reaction with the very high rate constant of 87 s^{-1}.
(a) The initial C_4H_8 concentration is 2.00 *M*. What is the concentration after 0.010 s?
(b) What fraction of C_4H_8 has decomposed in this time?

Plan (a) We must find the concentration of cyclobutane at time t, $[C_4H_8]_t$. The problem tells us this is a first-order reaction, so we use the integrated first-order rate law:

$$\ln\frac{[C_4H_8]_0}{[C_4H_8]_t} = kt$$

We know k (87 s^{-1}), t (0.010 s), and $[C_4H_8]_0$ (2.00 *M*), so we can solve for $[C_4H_8]_t$.
(b) The fraction decomposed is the concentration that has decomposed divided by the initial concentration:

$$\text{Fraction decomposed} = \frac{[C_4H_8]_0 - [C_4H_8]_t}{[C_4H_8]_0}$$

Solution (a) Substituting the data into the integrated rate law:

$$\ln \frac{2.00 \text{ mol/L}}{[C_4H_8]_t} = (87 \text{ s}^{-1})(0.010 \text{ s}) = 0.87$$

Taking the antilog of both sides:

$$\frac{2.00 \text{ mol/L}}{[C_4H_8]_t} = e^{0.87} = 2.4$$

Solving for $[C_4H_8]_t$:

$$[C_4H_8]_t = \frac{2.00 \text{ mol/L}}{2.4} = 0.83 \text{ mol/L}$$

(b) Finding the fraction that has decomposed after 0.010 s:

$$\frac{[C_4H_8]_0 - [C_4H_8]_t}{[C_4H_8]_0} = \frac{2.00 \text{ mol/L} - 0.83 \text{ mol/L}}{2.00 \text{ mol/L}} = 0.58$$

Check The concentration remaining after 0.010 s (0.83 mol/L) is less than the starting concentration (2.00 mol/L), which makes sense. Raising e to an exponent slightly less than 1 should give a number (2.4) slightly less than the value of e (2.718). Moreover, the final result makes sense: a high rate constant indicates a fast reaction, so it's not surprising that so much decomposes in such a short time.

Comment Integrated rate laws are also used to solve for the time it takes to reach a certain reactant concentration, as in the follow-up problem.

FOLLOW-UP PROBLEM 16.5 At 25°C, hydrogen iodide breaks down very slowly to hydrogen and iodine: rate $= k[HI]^2$. The rate constant at 25°C is 2.4×10^{-21} L/mol·s. If 0.0100 mol of HI(g) is placed in a 1.0-L container, how long will it take for the concentration of HI to reach 0.00900 mol/L (10.0% reacted)?

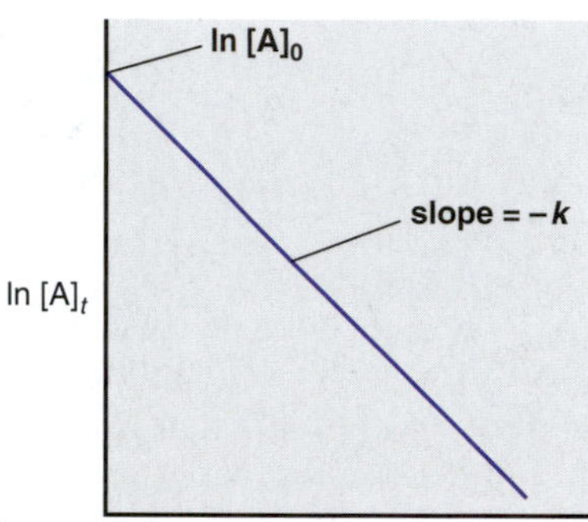

A First-order reaction

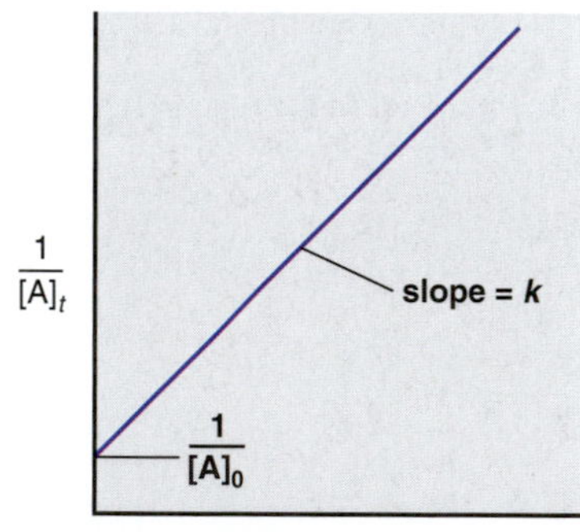

B Second-order reaction

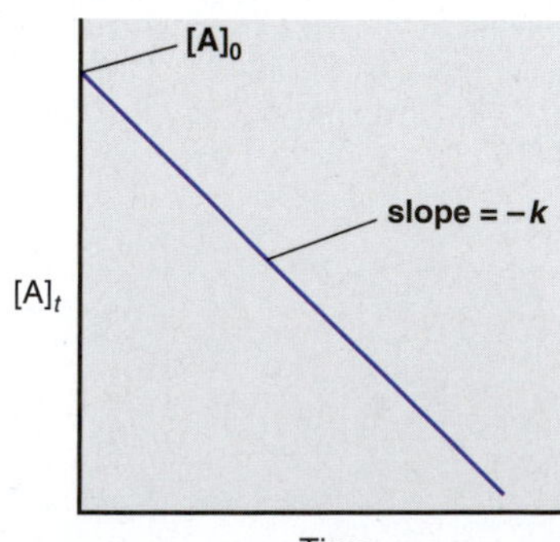

C Zero-order reaction

Figure 16.6 Graphical method for finding the reaction order from the integrated rate law.

Determining Reaction Orders from an Integrated Rate Law

In Sample Problem 16.3, we found the reaction orders using rate data. If rate data are not available, we can rearrange the integrated rate law into an equation for a straight line, $y = mx + b$, where m is the slope and b is the y-axis intercept, and then use a graphical method to find the order:

- For a *first-order reaction,* we have

$$\ln \frac{[A]_0}{[A]_t} = kt$$

From Appendix A, we know that $\ln \frac{a}{b} = \ln a - \ln b$, so we have

$$\ln [A]_0 - \ln [A]_t = kt$$

Rearranging gives

$$\ln [A]_t = -kt + \ln [A]_0$$
$$y = mx + b$$

Therefore, a plot of $\ln [A]_t$ vs. t gives a straight line with slope $= -k$ and y intercept $= \ln [A]_0$ (Figure 16.6A).

- For a *second-order reaction* with one reactant, we have

$$\frac{1}{[A]_t} - \frac{1}{[A]_0} = kt$$

Rearranging gives

$$\frac{1}{[A]_t} = kt + \frac{1}{[A]_0}$$
$$y = mx + b$$

In this case, a plot of $1/[A]_t$ vs. t gives a straight line with slope $= k$ and y intercept $= 1/[A]_0$ (Figure 16.6B).

- For a *zero-order reaction,* we have

$$[A]_t - [A]_0 = -kt$$

Rearranging gives

$$[A]_t = -kt + [A]_0$$
$$y = mx + b$$

Here a plot of $[A]_t$ vs. t gives a straight line with slope $= -k$ and y intercept $= [A]_0$ (Figure 16.6C).

Therefore, some trial-and-error graphical plotting is required to find the reaction order from the concentration and time data:

- If you obtain a straight line when you plot ln [reactant] vs. t, the reaction is *first order* with respect to that reactant.
- If you obtain a straight line when you plot 1/[reactant] vs. t, the reaction is *second order* with respect to that reactant.
- If you obtain a straight line when you plot [reactant] vs. t, the reaction is *zero order* with respect to that reactant.

Figure 16.7 shows how to use this approach to determine the order for the decomposition of N_2O_5. Because the plot of ln $[N_2O_5]$ vs. t *is* linear (part B), while the plots of $[N_2O_5]$ vs. t (part A) and of $1/[N_2O_5]$ vs. t (part C) are *not*, the decomposition of N_2O_5 must be first order in N_2O_5.

Time (min)	$[N_2O_5]$	ln $[N_2O_5]$	$1/[N_2O_5]$
0	0.0165	−4.104	60.6
10	0.0124	−4.390	80.6
20	0.0093	−4.68	1.1×10^2
30	0.0071	−4.95	1.4×10^2
40	0.0053	−5.24	1.9×10^2
50	0.0039	−5.55	2.6×10^2
60	0.0029	−5.84	3.4×10^2

Figure 16.7 Graphical determination of the reaction order for the decomposition of N_2O_5. The time and concentration data in the table are used to obtain the three plots, **A, B,** and **C.**

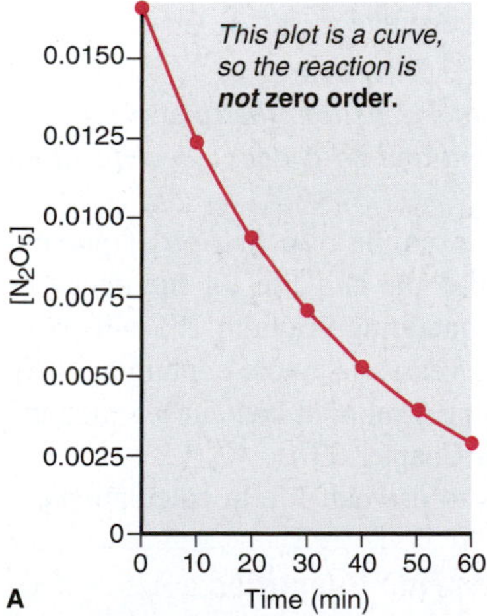

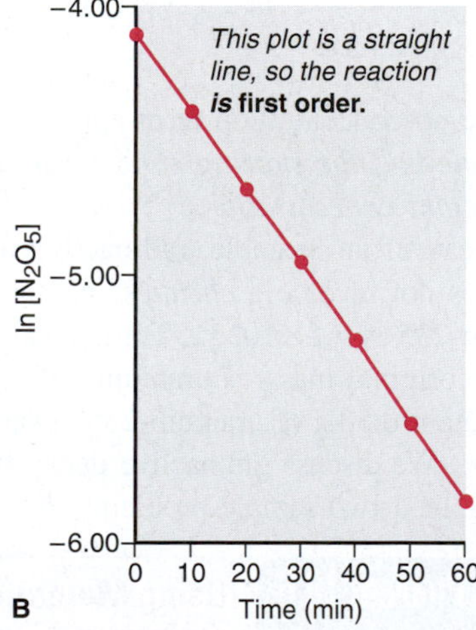

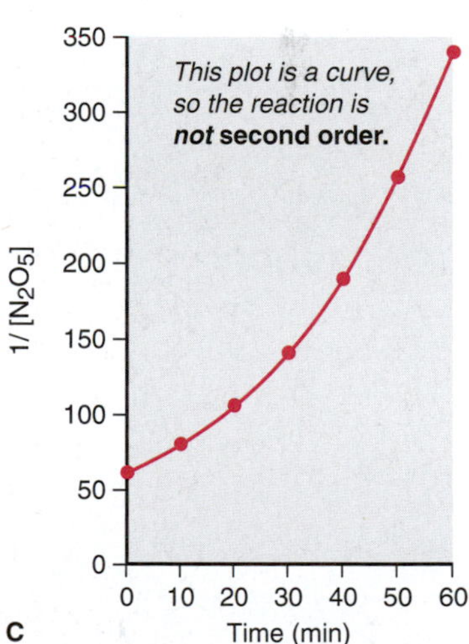

Reaction Half-Life

The **half-life ($t_{1/2}$)** is the time it takes a given reactant concentration to reach *half its initial value.* A half-life has time units appropriate for a given reaction and is characteristic of that reaction at a given temperature. For example, the half-life for the first-order decomposition of N_2O_5 at 45°C is 24.0 min. Therefore, if we start with, say, 0.0600 mol/L of N_2O_5 at 45°C, 0.0300 mol/L will have reacted after 24 min (one half-life), and 0.0300 mol/L will remain; after 48 min (two half-lives), 0.0150 mol/L will remain; after 72 min (three half-lives), 0.0075 mol/L will remain, and so forth (Figure 16.8, *next page*). The mathematical expression for the half-life depends on the overall order of the reaction.

Figure 16.8 A plot of $[N_2O_5]$ vs. time for three reaction half-lives.

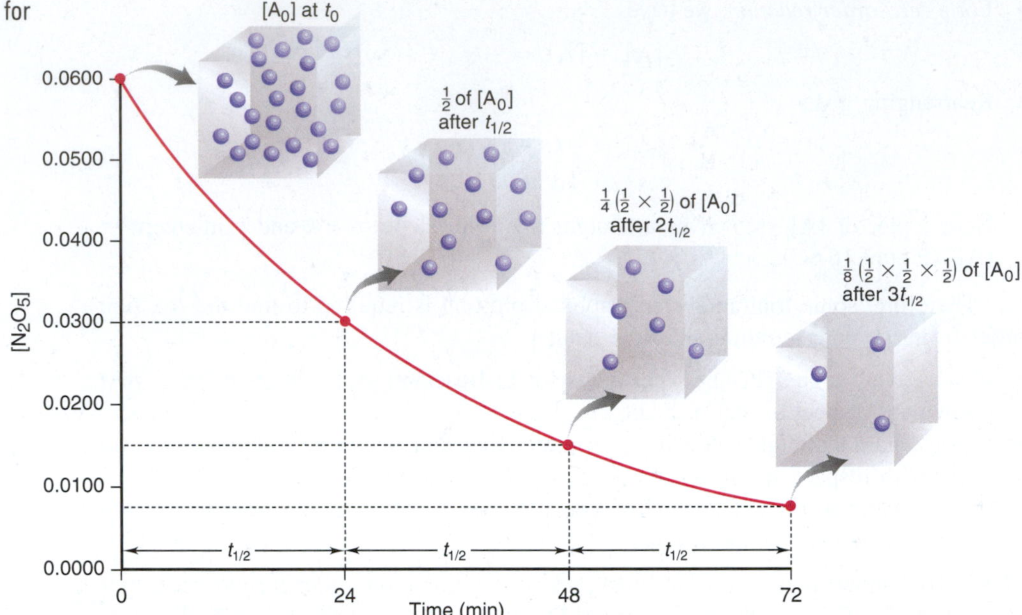

First-Order Reactions We can derive an expression for the half-life of a first-order reaction from the integrated rate law, which is

$$\ln \frac{[A]_0}{[A]_t} = kt$$

By definition, after one half-life, $t = t_{1/2}$, and $[A]_t = \frac{1}{2}[A]_0$. Substituting and canceling $[A]_0$ gives

$$\ln \frac{[A]_0}{\frac{1}{2}[A]_0} = kt_{1/2} \qquad \text{or} \qquad \ln 2 = kt_{1/2}$$

Then, solving for $t_{1/2}$, we have

$$t_{1/2} = \frac{\ln 2}{k} = \frac{0.693}{k} \qquad \text{(first-order process; rate} = k[A]) \qquad \mathbf{(16.7)}$$

Because no concentration term appears, *for a first-order reaction, the time it takes to reach one-half the starting concentration is a constant and so it does not depend on the reactant concentration.*

Decay of an unstable, radioactive nucleus is an example of a first-order process that does not involve a *chemical* change. For example, the half-life for the decay of uranium-235 is 7.1×10^8 yr. Thus, a sample of ore containing uranium-235 will have half the original mass of uranium-235 after 7.1×10^8 years: a sample containing 1 kg will contain 0.5 kg of uranium-235, a sample containing 1 mg will contain 0.5 mg, and so forth. (We discuss radioactive decay thoroughly in Chapter 23.)

The next two sample problems show some ways to use half-life in calculations.

Sample Problem 16.6 Using Molecular Scenes to Find Quantities at Various Times

Problem Substance A *(green)* decomposes to two other substances, B *(blue)* and C *(yellow)*, in a first-order gaseous reaction. The molecular scenes below show a portion of the reaction mixture at two different times:

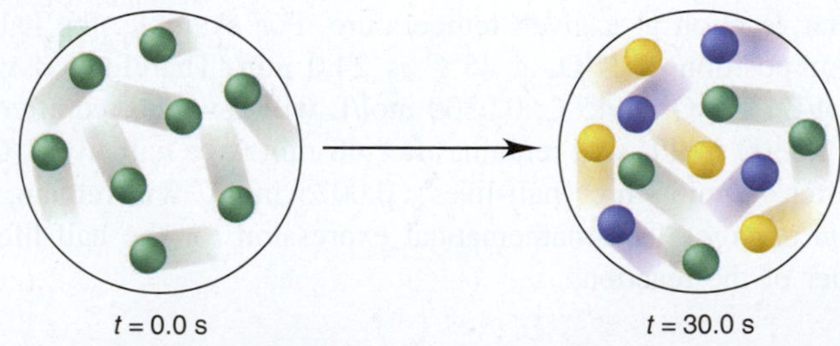

(a) Draw a similar molecular scene of the reaction mixture at $t = 60.0$ s.
(b) Find the rate constant of the reaction.
(c) If the total pressure (P_{total}) of the mixture is 5.00 atm at 90.0 s, what is the partial pressure of substance B (P_B)?

Plan We are shown molecular scenes of a reaction at two times with various numbers of reactant and product particles and have to predict quantities at two later times. **(a)** We count the number of A particles and see that A has decreased by half after 30.0 s; thus, the half-life is 30.0 s. The time $t = 60.0$ s represents two half-lives, so the number of A will decrease by half again, and each A forms one B and one C. **(b)** We substitute the value of the half-life in Equation 16.7 to find k. **(c)** First, we find the numbers of particles at $t = 90.0$ s, which represents three half-lives. To find P_B, we multiply the mole fraction of B, X_B, by P_{total} (5.00 atm) (Chapter 5). To find X_B, we know that the number of particles is equivalent to the number of moles, so we divide the number of B particles by the total number of particles.

Solution **(a)** The number of A particles decreased from 8 to 4 in one half-life (30.0 s), so after two half-lives (60.0 s), the number will be 2 (1/2 of 4) A particles. Each A decomposes to 1 B and 1 C, so 6 (8 − 2) particles of A form 6 particles of B and 6 of C *(see margin)*.

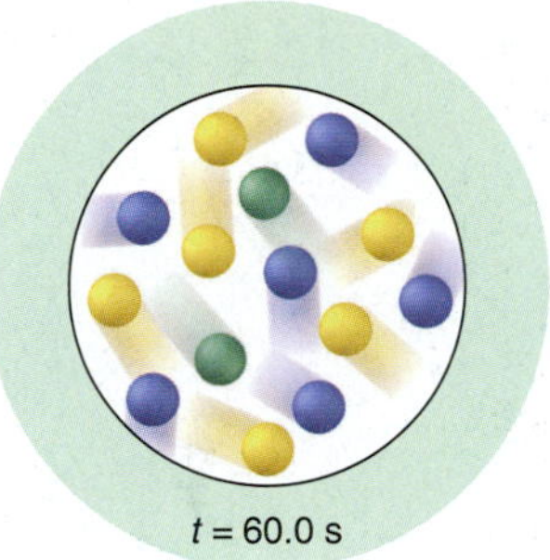
$t = 60.0$ s

(b) Finding the rate constant, k:

$$t_{1/2} = \frac{0.693}{k} \quad \text{so} \quad k = \frac{0.693}{t_{1/2}} = \frac{0.693}{30.0\text{ s}} = 2.31\times10^{-2}\text{ s}^{-1}$$

(c) Finding the number of particles after 90.0 s: After a third half-life, there will be 1 A, 7 B, and 7 C particles.
Finding the mole fraction of B, X_B:

$$X_B = \frac{7}{1 + 7 + 7} = \frac{7}{15} = 0.467$$

Finding the partial pressure of B, P_B:

$$P_B = X_B \times P_{total} = 0.467 \times 5.00\text{ atm} = 2.33\text{ atm}$$

Check For (b), rounding gives 0.7/30, which is a bit over 0.02, so the answer seems correct. For (c), X_B is almost 0.5, so P_B is a bit less than half of 5 atm, or <2.5 atm.

FOLLOW-UP PROBLEM 16.6 Substance X *(black)* changes to substance Y *(red)* in a first-order gaseous reaction. The scenes below represent the reaction mixture in a cubic container at two different times:

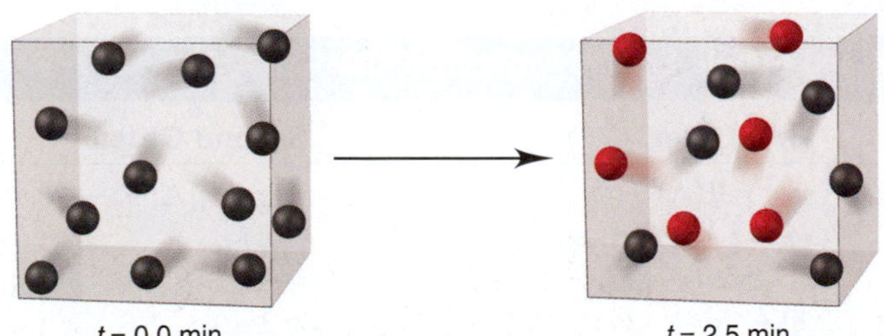

(a) Draw a scene that represents the mixture at 5.0 min.
(b) If each sphere represents 0.20 mol of particles and the volume of the cubic container is 0.50 L, what is the molarity of X at 10.0 min?

Sample Problem 16.7 Determining the Half-Life of a First-Order Reaction

Problem Cyclopropane is the smallest cyclic hydrocarbon. Because its 60° bond angles reduce orbital overlap, its bonds are weak. As a result, it is thermally unstable and rearranges to propene at 1000°C via the following first-order reaction:

$$\text{cyclo-}(CH_2)_3(g) \xrightarrow{\Delta} CH_3{-}CH{=}CH_2(g)$$

The rate constant is 9.2 s^{-1}. **(a)** What is the half-life of the reaction? **(b)** How long does it take for the concentration of cyclopropane to reach one-quarter of the initial value?

Plan **(a)** The cyclopropane rearrangement is first order, so to find $t_{1/2}$ we use Equation 16.7 and substitute for k ($9.2\ s^{-1}$). **(b)** Each half-life decreases the concentration to one-half of its initial value, so two half-lives decrease it to one-quarter.

Solution **(a)** Solving for $t_{1/2}$:

$$t_{1/2} = \frac{\ln 2}{k} = \frac{0.693}{9.2\ \text{s}^{-1}} = 0.075\ \text{s}$$

It takes 0.075 s for half the cyclopropane to form propene at this temperature.
(b) Finding the time to reach one-quarter of the initial concentration:

$$\text{Time} = 2(t_{1/2}) = 2(0.075\ \text{s}) = 0.15\ \text{s}$$

Check For part (a), rounding gives $0.7/9\ s^{-1} = 0.08$ s, so the answer seems correct.

FOLLOW-UP PROBLEM 16.7 Iodine-123 is used to study thyroid gland function. This radioactive isotope breaks down in a first-order process with a half-life of 13.1 h. What is the rate constant for the process?

Second-Order Reactions In contrast to the half-life of a first-order reaction, the half-life of a second-order reaction *does* depend on reactant concentration:

$$t_{1/2} = \frac{1}{k[\text{A}]_0} \qquad \text{(second-order process; rate} = k[\text{A}]^2)$$

Note that here *the half-life is **inversely** proportional to the initial reactant concentration.* This relationship means that a second-order reaction with a high initial reactant concentration has a *shorter* half-life, and one with a low initial reactant concentration has a *longer* half-life.

Zero-Order Reactions In contrast to the half-life of a second-order reaction, *the half-life of a zero-order reaction is **directly** proportional to the initial reactant concentration:*

$$t_{1/2} = \frac{[\text{A}]_0}{2k} \qquad \text{(zero-order process; rate} = k)$$

Thus, if a zero-order reaction begins with a high reactant concentration, it has a longer half-life than if it begins with a low reactant concentration.

Table 16.5 summarizes the features of zero-, first-, and second-order reactions.

Table 16.5 An Overview of Zero-Order, First-Order, and Simple Second-Order Reactions

	Zero Order	First Order	Second Order
Rate law	rate $= k$	rate $= k[\text{A}]$	rate $= k[\text{A}]^2$
Units for k	mol/L·s	1/s	L/mol·s
Half-life	$\frac{[\text{A}]_0}{2k}$	$\frac{\ln 2}{k}$	$\frac{1}{k[\text{A}]_0}$
Integrated rate law in straight-line form	$[\text{A}]_t = -kt + [\text{A}]_0$	$\ln [\text{A}]_t = -kt + \ln [\text{A}]_0$	$1/[\text{A}]_t = kt + 1/[\text{A}]_0$
Plot for straight line	$[\text{A}]_t$ vs. t	$\ln [\text{A}]_t$ vs. t	$1/[\text{A}]_t$ vs. t
Slope, y intercept	$-k$, $[\text{A}]_0$	$-k$, $\ln [\text{A}]_0$	k, $1/[\text{A}]_0$

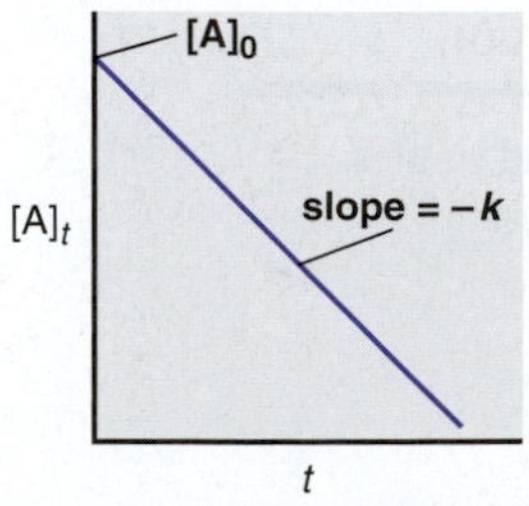

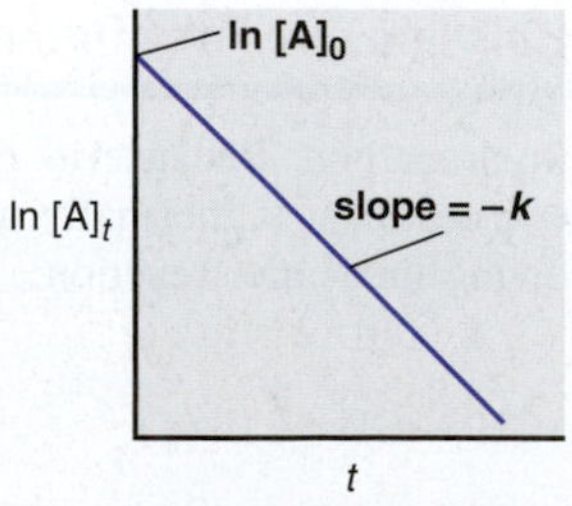

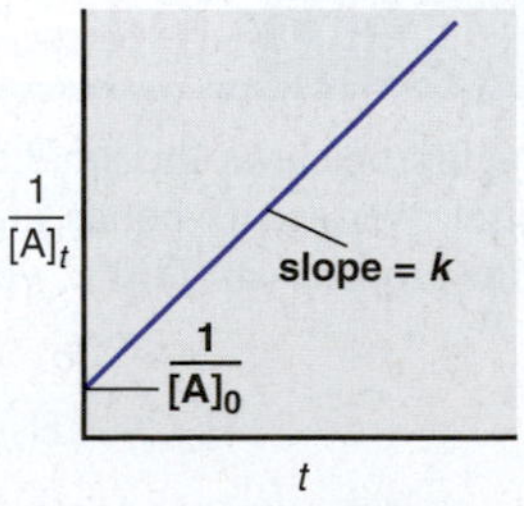

Summary of Section 16.4

- Integrated rate laws are used to find either the time needed to reach a certain concentration of reactant or the concentration present after a given time.
- Rearrangements of the integrated rate laws that give equations in the form of a straight line allow us to determine reaction orders and rate constants graphically.
- The half-life is the time needed for the reactant concentration to reach half its initial value; for first-order reactions, the half-life is constant—that is, it is independent of concentration.

16.5 • THEORIES OF CHEMICAL KINETICS

As was pointed out at the beginning of the chapter, concentration and temperature have major effects on reaction rate. Chemists employ two models—*collision theory* and *transition state theory*—to explain these effects.

Collision Theory: Basis of the Rate Law

The basic tenet of **collision theory** is that particles—atoms, molecules, or ions—must collide to react. But number of collisions can't be the only factor determining rate, or all reactions would be over in an instant. For example, at 1 atm and 20°C, the N_2 and O_2 molecules in 1 mL of air experience about 10^{27} collisions per second. If all that was needed for a reaction to occur was an N_2 molecule colliding with an O_2 molecule, our atmosphere would consist almost entirely of NO; in fact, only traces are present. Thus, this theory also relies on the concepts of collision energy and molecular structure to explain the effects of concentration and temperature on rate.

Why Concentrations Are Multiplied in the Rate Law Particles must collide to react, and the laws of probability tell us that the number of collisions depends on the *product* of the numbers of reactant particles, not their sum. Suppose we have only two particles of A and two of B confined in a reaction vessel. Figure 16.9 shows that four A-B collisions are possible. When we add another particle of A, six A-B collisions (3 × 2) are possible, not just five (3 + 2). Similarly, when we add another particle of B, nine A-B collisions (3 × 3) are possible, not just six (3 + 3). Thus, collision theory explains why we *multiply* the concentrations in the rate law to obtain the observed rate.

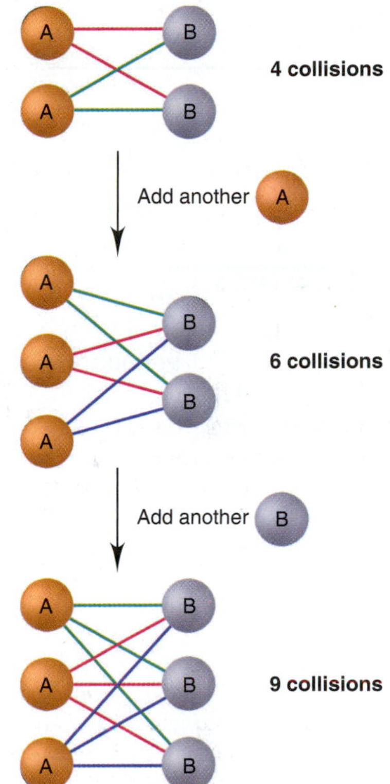

Figure 16.9 The number of possible collisions is the product, not the sum, of reactant concentrations.

The Effect of Temperature on the Rate Constant and the Rate Temperature typically has a dramatic effect on reaction rate: for many reactions near room temperature, an increase of 10 K (10°C) doubles or triples the rate. Figure 16.10A shows kinetic data for an organic reaction—hydrolysis, or reaction with water, of the ester ethyl acetate. To understand the effect of temperature, we measure concentrations and times for the reaction run at different temperatures. Solving each rate expression for k and plotting the results (Figure 16.10B), we find that *k increases exponentially as T increases.*

These results are consistent with findings obtained in 1889 by the Swedish chemist Svante Arrhenius. In its modern form, the **Arrhenius equation** is

$$k = Ae^{-E_a/RT} \tag{16.8}$$

Expt	[Ester]	[H_2O]	T (K)	Rate (mol/L·s)	k (L/mol·s)
1	0.100	0.200	288	1.04×10^{-3}	0.0521
2	0.100	0.200	298	2.02×10^{-3}	0.101
3	0.100	0.200	308	3.68×10^{-3}	0.184
4	0.100	0.200	318	6.64×10^{-3}	0.332

A

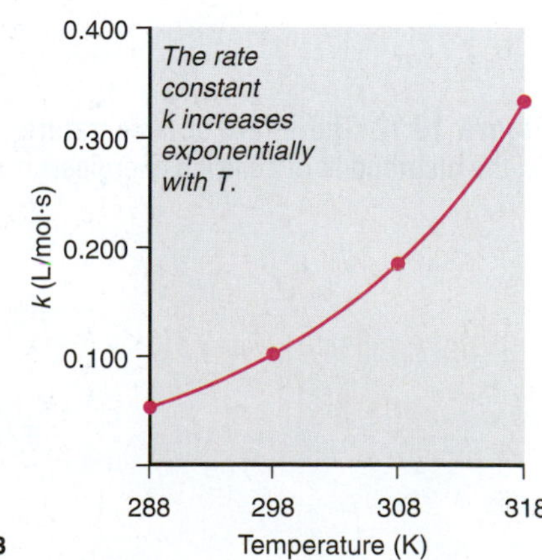

Figure 16.10 Increase of the rate constant with temperature for the hydrolysis of an ester.

where k is the rate constant, e is the base of natural logarithms, T is the absolute temperature, and R is the universal gas constant. (We'll focus on the term E_a in the next subsection and the term A a bit later.) Note the relationship between k and T, especially that T is in the denominator of a negative exponent. Thus, *as T increases, the value of the negative exponent becomes smaller, which means k becomes larger, so the rate increases:*

$$\text{Higher } T \Longrightarrow \text{larger } k \Longrightarrow \text{increased rate}$$

The Central Importance of Activation Energy The effect of temperature on k is closely related to the **activation energy (E_a)** of a reaction, an energy *threshold* that the colliding molecules must exceed in order to react. As an analogy, in order to succeed at the high jump, an athlete must exert at least enough energy to get over the bar. Similarly, reactant molecules must collide with a certain minimum energy to reach an *activated state,* from which reactant bonds can change to product bonds (Figure 16.11).

As you can see from Figure 16.11, a reversible reaction has two activation energies. The activation energy for the forward reaction, $E_{a(fwd)}$, is the energy difference between the activated state and the reactants; the activation energy for the reverse reaction, $E_{a(rev)}$, is the energy difference between the activated state and the products. The reaction represented in the diagram is exothermic ($\Delta H_{rxn} < 0$) in the forward direction. Thus, the products are at a lower energy than the reactants, and $E_{a(fwd)}$ is less than $E_{a(rev)}$. This difference equals the enthalpy of reaction, ΔH_{rxn}:

$$\Delta H_{rxn} = E_{a(fwd)} - E_{a(rev)} \tag{16.9}$$

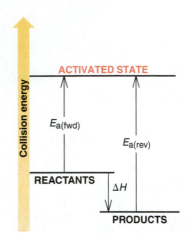

Figure 16.11 Energy-level diagram for a reaction. Like a high jumper with enough energy to go over the bar, molecules must collide with enough energy, E_a, to reach an activated state.

The Effect of Temperature on Collision Energy A rise in temperature has two effects on moving particles: it causes a higher collision *frequency* and a higher collision *energy.* Let's see how each affects rate.

- *Collision frequency*. If particles move faster, they collide more often. Calculations show that a 10 K rise in temperature from, say, 288 K to 298 K, increases the average molecular speed by 2%, which would lead to, at most, a 4% increase in rate (because the expression for the particles' average kinetic energy, $\frac{1}{2}m\overline{u}^2$, includes their average speed *squared*). Thus, higher collision frequency cannot possibly account for the doubling or tripling of rates observed with a 10 K rise. Indeed, the effect of temperature on collision *frequency* is only a minor factor.
- *Collision energy.* On the other hand, the effect of temperature on collision *energy* is a major factor. At a given temperature, the fraction f of collisions with energy equal to or greater than E_a is

$$f = e^{-E_a/RT}$$

where e is the base of natural logarithms, T is the absolute temperature, and R is the universal gas constant. The right side of this expression appears in the Arrhenius equation (Equation 16.8), which shows that a rise in T causes a larger k. We now see why—because *a rise in temperature enlarges the fraction of collisions with enough energy to exceed E_a* (Figure 16.12).

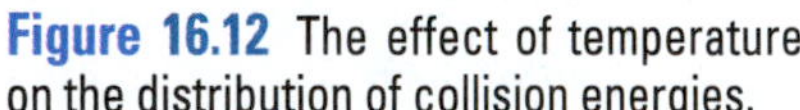

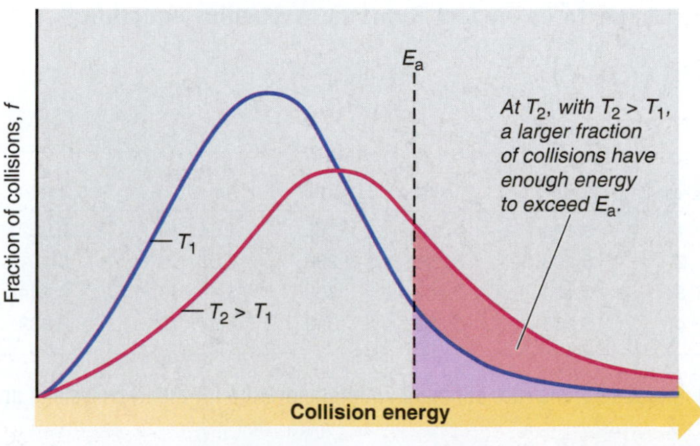

Figure 16.12 The effect of temperature on the distribution of collision energies.

Table 16.6 shows that the magnitudes of *both* E_a *and* T *affect the size of this fraction* for the hydrolysis of the ester we discussed in Figure 16.10. In the top portion, temperature is held constant, and the fraction of sufficiently energetic collisions shrinks several orders of magnitude with each 25-kJ/mol increase in E_a. (To extend the high jump analogy, as the height of the bar is raised, a smaller fraction of the athletes have enough energy to jump over it.) In the bottom portion, E_a is held constant, and the fraction nearly doubles for each 10 K (10°C) rise in temperature.

Therefore, *the smaller the activation energy (or the higher the temperature), the larger the fraction of sufficiently energetic collisions, the larger the value of k, and the higher the reaction rate:*

Smaller E_a (or higher T) $\Longrightarrow$ larger f $\Longrightarrow$ larger k $\Longrightarrow$ higher rate

Table 16.6 The Effect of E_a and T on the Fraction (f) of Collisions with Sufficient Energy to Allow Reaction

E_a (kJ/mol)	f (at T = 298 K)
50	1.70×10^{-9}
75	7.03×10^{-14}
100	2.90×10^{-18}
T	**f (at E_a = 50 kJ/mol)**
25°C (298 K)	1.70×10^{-9}
35°C (308 K)	3.29×10^{-9}
45°C (318 K)	6.12×10^{-9}

Calculating the Activation Energy We can calculate E_a from the Arrhenius equation by taking the natural logarithm of both sides of the equation, which recasts it into the form of an equation for a straight line:

$$k = Ae^{-E_a/RT}$$

$$\ln k = \ln A - \frac{E_a}{R}\left(\frac{1}{T}\right)$$

$$y = b + mx$$

A plot of ln k vs. $1/T$ gives a straight line whose slope is $-E_a/R$ and whose y intercept is ln A (Figure 16.13). We know the constant R, so we can determine E_a graphically from a series of k values at different temperatures.

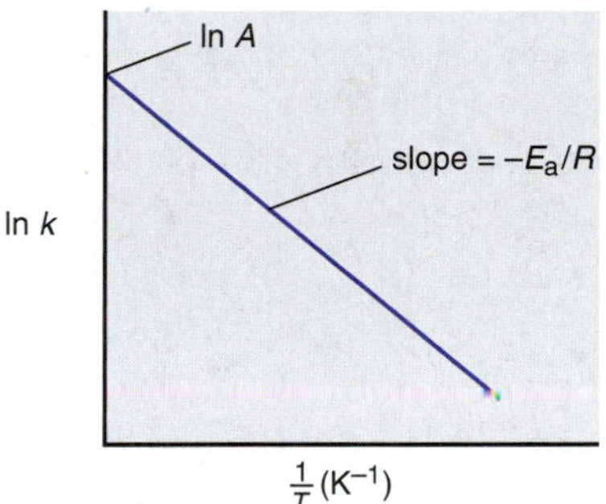

Figure 16.13 Graphical determination of the activation energy.

We can find E_a in another way if we know the rate constants at two temperatures, T_2 and T_1:

$$\ln k_2 = \ln A - \frac{E_a}{R}\left(\frac{1}{T_2}\right) \qquad \ln k_1 = \ln A - \frac{E_a}{R}\left(\frac{1}{T_1}\right)$$

When we subtract ln k_1 from ln k_2, the term "ln A" drops out and the other terms can be rearranged to give

$$\ln\frac{k_2}{k_1} = -\frac{E_a}{R}\left(\frac{1}{T_2} - \frac{1}{T_1}\right) \qquad (16.10)$$

Then, we can solve for E_a, as in the next sample problem.

Sample Problem 16.8 Determining the Energy of Activation

Problem The decomposition of hydrogen iodide, $2HI(g) \longrightarrow H_2(g) + I_2(g)$, has rate constants of 9.51×10^{-9} L/mol·s at 500. K and 1.10×10^{-5} L/mol·s at 600. K. Find E_a.

Plan We are given the rate constants, k_1 and k_2, at two temperatures, T_1 and T_2, so we substitute into Equation 16.10 and solve for E_a.

Solution Rearranging Equation 16.10 to solve for E_a:

$$\ln\frac{k_2}{k_1} = -\frac{E_a}{R}\left(\frac{1}{T_2} - \frac{1}{T_1}\right)$$

$$E_a = -R\left(\ln\frac{k_2}{k_1}\right)\left(\frac{1}{T_2} - \frac{1}{T_1}\right)^{-1}$$

$$= -(8.314\ \text{J/mol·K})\left(\ln\frac{1.10\times10^{-5}\ \text{L/mol·s}}{9.51\times10^{-9}\ \text{L/mol·s}}\right)\left(\frac{1}{600.\ \text{K}} - \frac{1}{500.\ \text{K}}\right)^{-1}$$

$$= 1.76\times10^{5}\ \text{J/mol} = 1.76\times10^{2}\ \text{kJ/mol}$$

Comment Be sure to retain the same number of significant figures in $1/T$ as you have in T, or a significant error could be introduced. Round to the correct number of significant figures only at the final answer. On most pocket calculators, the expression $(1/T_2 - 1/T_1)$ is entered as follows: $(T_2)(1/x) - (T_1)(1/x) =$

FOLLOW-UP PROBLEM 16.8 The reaction $2NOCl(g) \longrightarrow 2NO(g) + Cl_2(g)$ has an E_a of 1.00×10^2 kJ/mol and a rate constant of 0.286 L/mol·s at 500. K. What is the rate constant at 490. K?

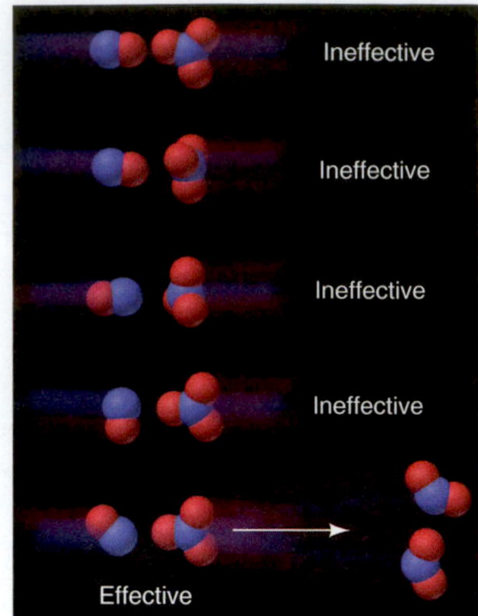

Figure 16.14 The importance of molecular orientation to an effective collision. In the one effective orientation *(bottom)*, contact occurs between the atoms that become bonded in the product.

The Effect of Molecular Structure on Rate At ordinary temperatures, the enormous number of collisions per second between reactant particles is reduced by six or more orders of magnitude by counting only those with enough energy to react. And, even this tiny fraction of all collisions is typically much larger than the number of **effective collisions,** those that actually lead to product because *the atoms that become bonded in the product make contact.* Thus, to be effective, a collision must have enough energy *and* the appropriate *molecular orientation.*

In the Arrhenius equation, molecular orientation is contained in the term A:

$$k = Ae^{-E_a/RT}$$

This term is the **frequency factor,** the product of the collision frequency Z and an *orientation probability factor, p,* which is specific for each reaction:

$$A = pZ$$

The factor p is related to the structural complexity of the reactants. You can think of p as the ratio of effectively oriented collisions to all possible collisions. Figure 16.14 shows a few of the collision orientations for the following reaction:

$$NO(g) + NO_3(g) \longrightarrow 2NO_2(g)$$

Of the five orientations shown, only one has the effective orientation, with the N of NO making contact with an O of NO_3. Actually, the p value for this reaction is 0.006: only 6 collisions in every 1000 (1 in 167) have the correct orientation.

The more complex the molecular structure, the smaller the p value is. Individual atoms are spherical, so reactions between them have p values near 1: as long as reacting atoms collide with enough energy, the product forms. At the other extreme are biochemical reactions, in which two small molecules (or portions of larger molecules) react only when they collide with enough energy on a specific tiny region of a giant protein. The p value for these reactions is often much less than 10^{-6}, fewer than one in a million. The fact that countless such biochemical reactions are occurring in you right now attests to the astounding number of collisions per second!

Transition State Theory: What the Activation Energy Is Used For

Collision theory explains the importance of effective collisions, and **transition state theory** focuses on the high-energy species that exists at the moment of an effective collision when reactants are becoming products.

Visualizing the Transition State As two molecules approach each other, repulsions between their electron clouds continually increase, so they slow down as some of their kinetic energy is converted to potential energy. If they collide, but the energy of the collision is *less* than the activation energy, the molecules bounce off each other.

However, in a tiny fraction of collisions in which the molecules are moving fast enough, *their kinetic energies push them together with enough force to overcome the repulsions and surpass the activation energy.* And, in an even tinier fraction of these sufficiently energetic collisions, the molecules are oriented effectively. In those cases, nuclei in one molecule attract electrons in the other, atomic orbitals overlap, electron densities shift, and some bonds lengthen and weaken while others shorten and strengthen. At some point during this smooth transformation, *a species with partial bonds exists* that is neither reactant nor product. This very unstable species, called the **transition state** (or **activated complex**) exists only at the instant of highest potential energy. Thus, *the activation energy of a reaction is used to reach the transition state.*

Consider the reaction between methyl bromide and hydroxide ion:

$$BrCH_3 + OH^- \longrightarrow Br^- + CH_3OH$$

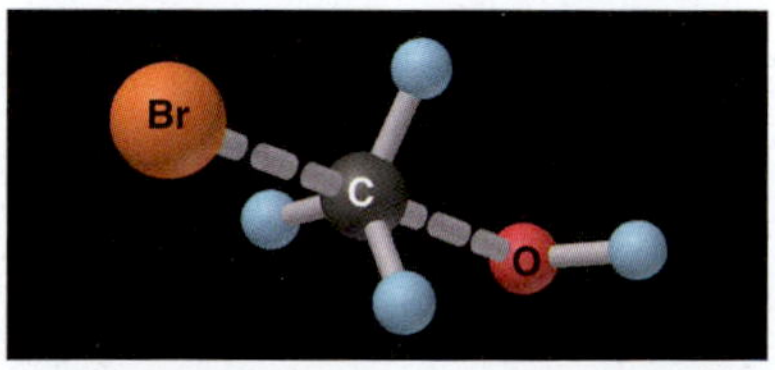

Figure 16.15 The transition state of the reaction between $BrCH_3$ and OH^-. Note the partial *(dashed)* C—O and Br—C bonds and the trigonal bipyramidal shape.

The electronegative bromine makes the carbon in $BrCH_3$ partially positive, and the carbon attracts the negatively charged oxygen in OH^-. As a C—O bond begins to form, the Br—C bond begins to weaken. In the transition state (Figure 16.15), C is surrounded by five atoms (trigonal bipyramidal; Section 10.2), which never occurs in its stable compounds. This high-energy species has three normal C—H bonds and two partial bonds, one from C to O and the other from Br to C.

Reaching the transition state does not guarantee that a reaction will proceed to products because *a transition state can change in either direction.* In this case, if the C—O bond continues to strengthen, products form; but, if the Br—C bond becomes stronger again, the transition state reverts to reactants.

Depicting the Change with a Reaction Energy Diagram A useful way to depict these events is with a **reaction energy diagram,** which shows the potential energy of the system during the reaction as a smooth curve. Figure 16.16 shows the reaction energy diagram for the reaction of $BrCH_3$ and OH^-, and also includes electron density relief maps, structural formulas, and molecular-scale views at various points during the change.

The horizontal axis, labeled "Reaction progress," indicates that reactants change to products from left to right. This reaction is exothermic, so reactants are higher in energy than products, which means $E_{a(fwd)}$ is less than $E_{a(rev)}$.

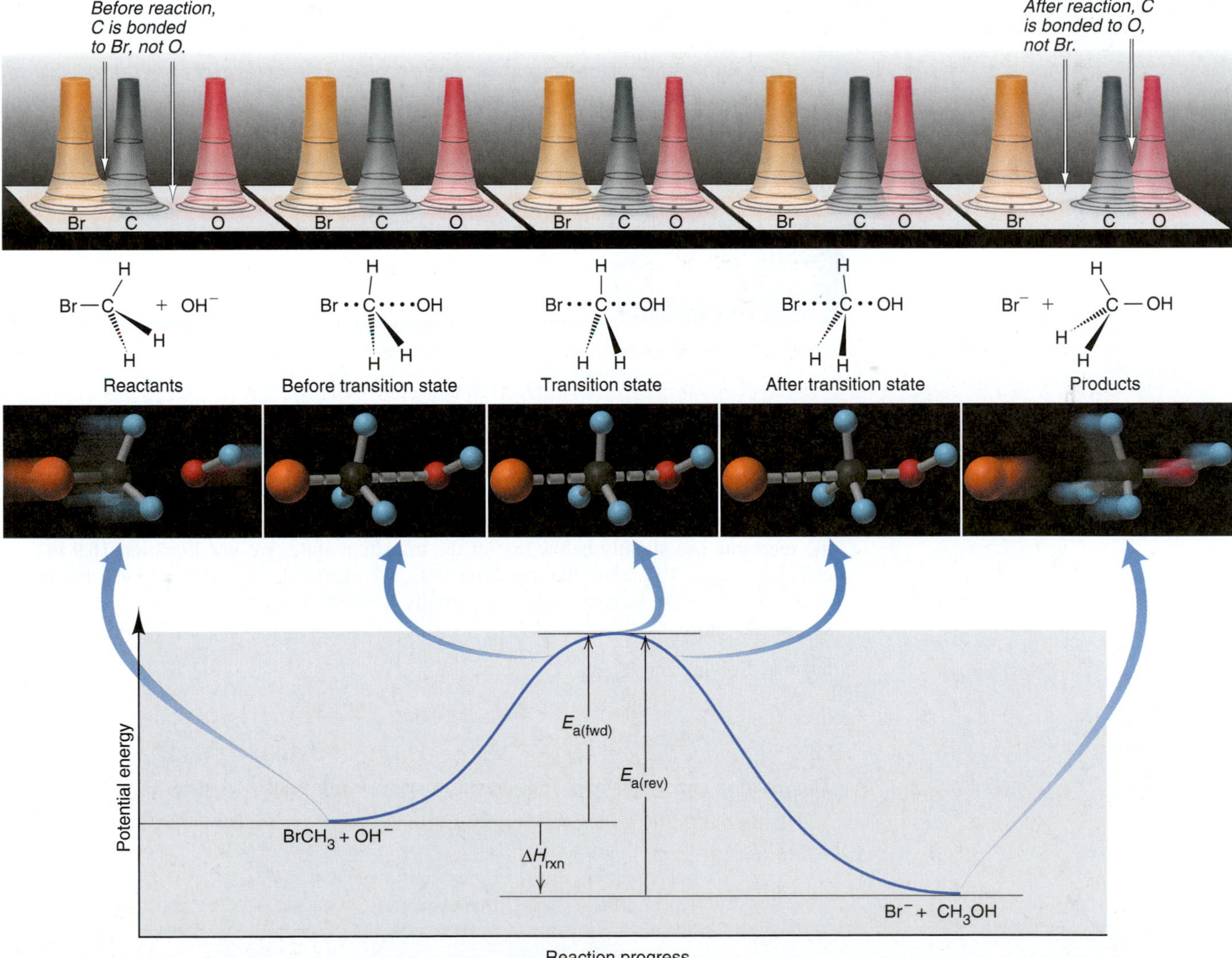

Figure 16.16 Depicting the reaction between $BrCH_3$ and OH^-. A plot of potential energy vs. reaction progress shows the relative energy levels of reactants, products, and transition state joined by a curved line, as well as the activation energies of the forward and reverse steps and the heat of reaction. The electron density relief maps, structural formulas, and molecular-scale views depict the change at five points. Note the gradual bond forming and bond breaking as the system goes through the transition state.

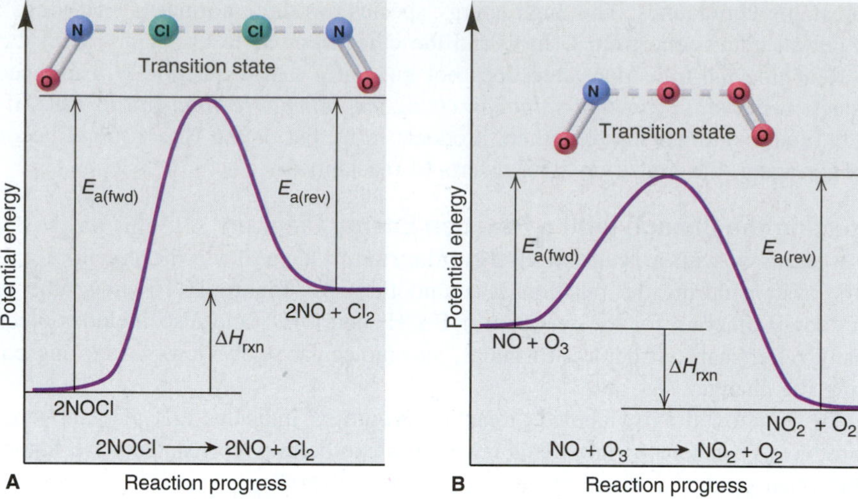

Figure 16.17 Reaction energy diagrams and possible transition states for two reactions. **A,** Endothermic reaction. **B,** Exothermic reaction.

Transition state theory proposes that *every reaction (or every step in an overall reaction) goes through its own transition state.* Figure 16.17 presents reaction energy diagrams for two gas-phase reactions. In each case, the structure of the transition state is predicted from the orientations of the reactant atoms that must become bonded in the product.

Sample Problem 16.9 Drawing Reaction Energy Diagrams and Transition States

Problem A key reaction in the upper atmosphere is

$$O_3(g) + O(g) \longrightarrow 2O_2(g)$$

The $E_{a(fwd)}$ is 19 kJ, and the ΔH_{rxn} for the reaction as written is −392 kJ. Draw a reaction energy diagram, predict a structure for the transition state, and calculate $E_{a(rev)}$.

Plan The reaction is highly exothermic ($\Delta H_{rxn} = -392$ kJ), so the products are much lower in energy than the reactants. The small $E_{a(fwd)}$ (19 kJ) means the energy of the reactants lies slightly below that of the transition state. We use Equation 16.9 to calculate $E_{a(rev)}$. To predict the transition state, we sketch the species and note that one of the bonds in O_3 weakens, and this partially bonded O begins forming a bond to the separate O atom.

Solution Solving for $E_{a(rev)}$:

$$\Delta H_{rxn} = E_{a(fwd)} - E_{a(rev)}$$

So, $$E_{a(rev)} = E_{a(fwd)} - \Delta H_{rxn} = 19 \text{ kJ} - (-392 \text{ kJ}) = 411 \text{ kJ}$$

The reaction energy diagram (not drawn to scale), with transition state, is

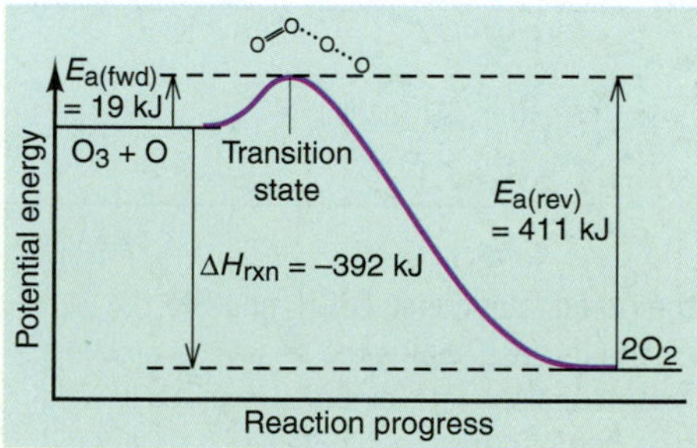

Check Rounding to find $E_{a(rev)}$ gives ~20 + 390 = 410.

FOLLOW-UP PROBLEM 16.9 The following reaction energy diagram depicts another key atmospheric reaction. Label the axes, identify $E_{a(fwd)}$, $E_{a(rev)}$, and ΔH_{rxn}, draw and label the transition state, and calculate $E_{a(rev)}$ for the reaction.

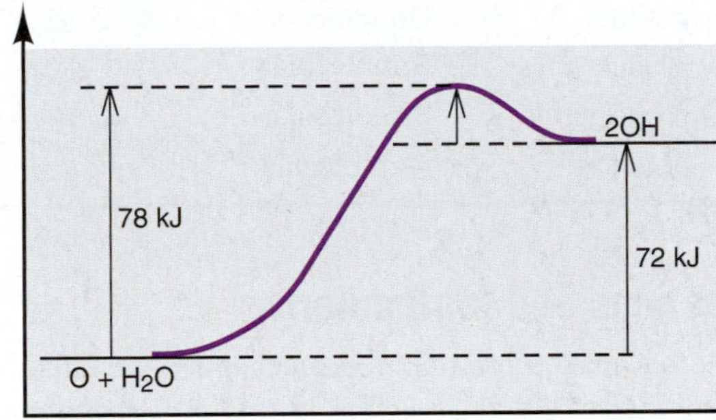

Summary of Section 16.5

- According to collision theory, reactant particles must collide to react, and the number of possible collisions is found by multiplying, not adding, the numbers of different reactant particles.
- As the Arrhenius equation shows, a rise in temperature increases the rate because it increases the rate constant.
- The activation energy, E_a, is the minimum energy needed for colliding particles to react.
- The relative E_a values for the forward and reverse reactions depend on whether the overall reaction is exothermic or endothermic.
- At higher temperatures, more collisions have enough energy to exceed E_a.
- E_a can be determined graphically from k values obtained at different T values.
- Molecules must collide with an effective orientation for them to react, so structural complexity decreases rate.
- Transition state theory focuses on the change of kinetic energy to potential energy as reactant particles collide and form an unstable transition state.
- Given a sufficiently energetic collision and an effective molecular orientation, the reactant species form the transition state, which either continues toward product(s) or reverts to reactant(s).
- A reaction energy diagram depicts the changing potential energy throughout a reaction's progress from reactants through transition states to products.

16.6 • REACTION MECHANISMS: THE STEPS FROM REACTANT TO PRODUCT

You can't understand how a car works by examining the body and wheels, or even the engine as a whole. You need to look inside the engine to see how its parts fit together and function. Similarly, by examining the overall balanced equation, we can't know how a reaction works. We must look "inside" to see how reactants change into products.

Most reactions occur through a **reaction mechanism,** a sequence of single reaction steps that sum to the overall equation. For example, a possible mechanism for the overall reaction

$$2A + B \longrightarrow E + F$$

might involve these three simpler steps:

(1) $A + B \longrightarrow C$
(2) $C + A \longrightarrow D$
(3) $\quad D \longrightarrow E + F$

Adding the steps and canceling common substances gives the overall equation:

$$A + B + \cancel{C} + A + \cancel{D} \longrightarrow \cancel{C} + \cancel{D} + E + F \qquad \text{or} \qquad 2A + B \longrightarrow E + F$$

A mechanism is a hypothesis about how a reaction occurs; chemists *propose* a mechanism and then *test* to see that it fits with the observed rate law.

Table 16.7 Rate Laws for General Elementary Steps

Elementary Step	Molecularity	Rate Law
$A \longrightarrow$ product	Unimolecular	Rate $= k[A]$
$2A \longrightarrow$ product	Bimolecular	Rate $= k[A]^2$
$A + B \longrightarrow$ product	Bimolecular	Rate $= k[A][B]$
$2A + B \longrightarrow$ product	Termolecular	Rate $= k[A]^2[B]$

Elementary Reactions and Molecularity

The individual steps that make up a reaction mechanism are called **elementary reactions** (or **elementary steps**). Each describes a *single molecular event*—one particle decomposing, two particles combining, and so forth. An elementary step is characterized by its **molecularity,** the number of *reactant* particles in the step. Consider the mechanism for the breakdown of ozone in the stratosphere. The overall equation is

$$2O_3(g) \longrightarrow 3O_2(g)$$

A two-step mechanism has been proposed:

(1) $O_3(g) \longrightarrow O_2(g) + O(g)$
(2) $O_3(g) + O(g) \longrightarrow 2O_2(g)$

The first step is a **unimolecular reaction,** one that involves the decomposition or rearrangement of a single particle (O_3). The second step is a **bimolecular reaction,** one in which two particles (O_3 and O) react. Some *termolecular* elementary steps occur, but they are extremely rare because the probability of three particles colliding simultaneously with enough energy and an effective orientation is very small. Higher molecularities are not known. Therefore, in general, *we propose unimolecular and/or bimolecular reactions as steps in a chemically reasonable mechanism.*

The rate law for an elementary reaction, unlike that for an overall reaction, *can* be deduced from the reaction stoichiometry. An elementary reaction occurs in one step, so its rate must be proportional to the product of the reactant concentrations. Therefore, *only* for an elementary step, *we use the equation coefficients as the reaction orders in the rate law; that is, reaction order equals molecularity* (Table 16.7).

Sample Problem 16.10 Determining Molecularity and Rate Laws for Elementary Steps

Problem The following elementary steps are proposed for a reaction mechanism:
(1) $NO_2Cl(g) \longrightarrow NO_2(g) + Cl(g)$
(2) $NO_2Cl(g) + Cl(g) \longrightarrow NO_2(g) + Cl_2(g)$
(a) Write the overall balanced equation.
(b) Determine the molecularity of each step.
(c) Write the rate law for each step.

Plan We find the overall equation from the sum of the elementary steps. The molecularity of each step equals the total number of *reactant* particles. We write the rate law for each step using the molecularities as reaction orders.

Solution **(a)** Writing the overall balanced equation:

$$NO_2Cl(g) \longrightarrow NO_2(g) + Cl(g)$$
$$NO_2Cl(g) + Cl(g) \longrightarrow NO_2(g) + Cl_2(g)$$
$$NO_2Cl(g) + NO_2Cl(g) + \cancel{Cl(g)} \longrightarrow NO_2(g) + \cancel{Cl(g)} + NO_2(g) + Cl_2(g)$$
$$2NO_2Cl(g) \longrightarrow 2NO_2(g) + Cl_2(g)$$

(b) Finding the molecularity of each step: The first step has one reactant, NO_2Cl, so it is unimolecular. The second step has two reactants, NO_2Cl and Cl, so it is bimolecular.
(c) Writing rate laws for the elementary steps:
(1) $\text{Rate}_1 = k_1[NO_2Cl]$
(2) $\text{Rate}_2 = k_2[NO_2Cl][Cl]$

Check In part (a), be sure the equation is balanced; in part (c), be sure the substances in brackets are the *reactants* of each elementary step.

FOLLOW-UP PROBLEM 16.10 These elementary steps are proposed for a mechanism:

(1) $2NO(g) \longrightarrow N_2O_2(g)$
(2) $2H_2(g) \longrightarrow 4H(g)$
(3) $N_2O_2(g) + H(g) \longrightarrow N_2O(g) + HO(g)$
(4) $HO(g) + H(g) \longrightarrow H_2O(g)$
(5) $H(g) + N_2O(g) \longrightarrow HO(g) + N_2(g)$

(a) Write the balanced equation for the overall reaction.
(b) Determine the molecularity of each step.
(c) Write the rate law for each step.

The Rate-Determining Step of a Reaction Mechanism

All the elementary steps in a mechanism have their own rates. However, one step is usually *much* slower than the others. This step, called the **rate-determining step** (or **rate-limiting step**), limits how fast the overall reaction proceeds. Therefore, *the rate law of the rate-determining step becomes the rate law for the overall reaction.*

THINK OF IT THIS WAY
The Rate-Determining Step for Traffic Flow

As an analogy for the rate-determining step, imagine driving home on a wide avenue that leads to a bridge with a tollbooth at the end. Traffic is flowing smoothly until the road narrows as it approaches the bridge. Traffic then slows so much that it takes longer to get over the bridge than the rest of the trip combined. The bottleneck over the bridge determines how long the overall trip home takes.

Consider the reaction between nitrogen dioxide and carbon monoxide:

$$NO_2(g) + CO(g) \longrightarrow NO(g) + CO_2(g)$$

If this reaction were an elementary step—that is, if the mechanism consisted of only one step—we could immediately write the overall rate law as

$$\text{Rate} = k[NO_2][CO]$$

But, as you saw in Sample Problem 16.3, experimental data show the rate law is

$$\text{Rate} = k[NO_2]^2$$

Thus, the overall reaction cannot be elementary.

A proposed two-step mechanism is

(1) $NO_2(g) + NO_2(g) \longrightarrow NO_3(g) + NO(g)$ [slow; rate determining]
(2) $NO_3(g) + CO(g) \longrightarrow NO_2(g) + CO_2(g)$ [fast]

In this mechanism, NO_3 functions as a **reaction intermediate,** a substance formed and used up during the reaction. Even though it does not appear in the overall balanced equation, a reaction intermediate is essential for the reaction to occur. Intermediates are less stable than the reactants and products, but unlike *much* less stable transition states, they have normal bonds and can sometimes be isolated.

Rate laws for the two elementary steps listed above are

(1) $\text{Rate}_1 = k_1[NO_2][NO_2] = k_1[NO_2]^2$
(2) $\text{Rate}_2 = k_2[NO_3][CO]$

Three key points to notice about this mechanism are

- If $k_1 = k$, the rate law for the rate-determining step (step 1) becomes identical to the observed rate law.
- Because the first step is slow, $[NO_3]$ is low. As soon as any NO_3 forms, it is consumed by the fast second step, so the reaction takes as long as the first step does.
- CO does not appear in the rate law (reaction order = 0) because it takes part in the mechanism *after* the rate-determining step.

Correlating the Mechanism with the Rate Law

Coming up with a reasonable reaction mechanism is a classic demonstration of the scientific method. Using observations and data from rate experiments, we hypothesize the individual steps and then test our hypothesis with further evidence. If the evidence supports our mechanism, we can accept it; if not, we must propose a different one. We can never *prove* that a mechanism represents the *actual* chemical change, only that it is consistent with the data.

A valid mechanism must meet three criteria:

1. *The elementary steps must add up to the overall balanced equation.*
2. *The elementary steps must be reasonable.* They should generally involve one reactant particle (unimolecular) or two (bimolecular).
3. *The mechanism must correlate with the rate law,* not the other way around.

Mechanisms with a Slow Initial Step The reaction between NO_2 and CO that we considered earlier has a mechanism with a slow initial step; that is, the first step is rate-determining. The reaction between nitrogen dioxide and fluorine is another example:

$$2NO_2(g) + F_2(g) \longrightarrow 2NO_2F(g)$$

The experimental rate law is first order in NO_2 and in F_2,

$$\text{Rate} = k[NO_2][F_2]$$

and the accepted mechanism is

(1) $NO_2(g) + F_2(g) \longrightarrow NO_2F(g) + F(g)$ [slow; rate determining]

(2) $NO_2(g) + F(g) \longrightarrow NO_2F(g)$ [fast]

Note that the free fluorine atom is a reaction intermediate.

Let's see how this mechanism meets the three criteria.

1. The elementary reactions sum to the balanced equation:

$$NO_2(g) + NO_2(g) + F_2(g) + \cancel{F(g)} \longrightarrow NO_2F(g) + NO_2F(g) + \cancel{F(g)}$$

or

$$2NO_2(g) + F_2(g) \longrightarrow 2NO_2F(g)$$

2. Both steps are bimolecular and, thus, reasonable.
3. To determine whether the mechanism is consistent with the observed rate law, we first write the rate laws for the elementary steps:

   (1) $\text{Rate}_1 = k_1[NO_2][F_2]$

   (2) $\text{Rate}_2 = k_2[NO_2][F]$

   Step 1 is the rate-determining step, and with $k_1 = k$, it is the same as the overall rate law, so the third criterion is met.

Note that the second molecule of NO_2 is involved *after* the rate-determining step, so it does not appear in the overall rate law. Thus, as in the mechanism for the reaction of NO_2 and CO, *the overall rate law includes all the reactants involved in the rate-determining step.*

Figure 16.18 is a reaction energy diagram for the reaction of NO_2 and F_2. Note the following:

- *Each step in the mechanism has its own transition state.* (Note that only one molecule of NO_2 reacts in step 1, and only the first transition state is depicted.)
- The F atom (the reaction intermediate) is a reactive, unstable species (as you know from halogen chemistry), so it is higher in energy than the reactants or product.
- The first step is slower (rate limiting), so its activation energy is *larger* than that of the second step.
- The overall reaction is exothermic, so the product is lower in energy than the reactants.

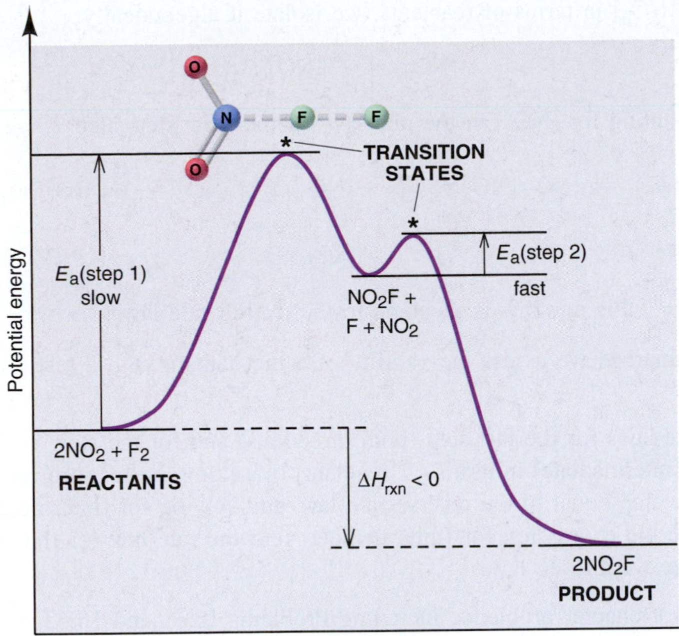

Figure 16.18 Reaction energy diagram for the two-step reaction of NO_2 and F_2. The proposed transition state is shown for step 1. Reactants for the second step are the F atom (the reaction intermediate) and the second molecule of NO_2.

Mechanisms with a Fast Initial Step If the rate-limiting step in a mechanism is *not* the initial step, the product of the fast initial step builds up and starts reverting to reactant. With time, this *fast, reversible step reaches equilibrium,* as product changes to reactant as fast as it forms. As you'll see, this situation allows us to fit the mechanism to the overall rate law.

Consider once again the oxidation of nitrogen monoxide:

$$2NO(g) + O_2(g) \longrightarrow 2NO_2(g)$$

The observed rate law is

$$\text{Rate} = k[NO]^2[O_2]$$

and a proposed mechanism is

(1) $NO(g) + O_2(g) \rightleftharpoons NO_3(g)$ [fast, reversible]
(2) $NO_3(g) + NO(g) \longrightarrow 2NO_2(g)$ [slow; rate determining]

Let's go through the three criteria to see if this mechanism is valid. With cancellation of the reaction intermediate, NO_3, the sum of the steps gives the overall equation, so the first criterion is met:

$$NO(g) + O_2(g) + \cancel{NO_3(g)} + NO(g) \longrightarrow \cancel{NO_3(g)} + 2NO_2(g)$$

or

$$2NO(g) + O_2(g) \longrightarrow 2NO_2(g)$$

Both steps are bimolecular, so the second criterion is met.

To see whether the third criterion—that the mechanism is consistent with the observed rate law—is met, we first write rate laws for the elementary steps:

(1) $\text{Rate}_{1(\text{fwd})} = k_1[NO][O_2]$
$\text{Rate}_{1(\text{rev})} = k_{-1}[NO_3]$

where k_{-1} is the rate constant and NO_3 is the reactant for the reverse reaction.

(2) $\text{Rate}_2 = k_2[NO_3][NO]$

Next, we show that the rate law for the rate-determining step (step 2) gives the overall rate law. As written, it does not, because it contains the intermediate NO_3, and *an overall rate law includes only reactants (and products).* We eliminate $[NO_3]$ from the rate law for step 2 by expressing it in terms of reactants, as follows. Step 1 reaches equilibrium when the forward and reverse rates are equal:

$$\text{Rate}_{1(\text{fwd})} = \text{Rate}_{1(\text{rev})} \quad \text{or} \quad k_1[NO][O_2] = k_{-1}[NO_3]$$

To express $[NO_3]$ in terms of reactants, we isolate it algebraically:

$$[NO_3] = \frac{k_1}{k_{-1}}[NO][O_2]$$

Then, substituting for $[NO_3]$ in the rate law for the slow step, step 2, we obtain

$$\text{Rate}_2 = k_2[NO_3][NO] = k_2\underbrace{\left(\frac{k_1}{k_{-1}}[NO][O_2]\right)}_{[NO_3]}[NO] = \frac{k_2k_1}{k_{-1}}[NO]^2[O_2]$$

With $k = \dfrac{k_2k_1}{k_{-1}}$, this rate law is identical to the overall rate law.

To summarize, we assess the validity of a mechanism with a fast initial step as follows:

1. Write rate laws for the fast step (both directions) and for the slow step.
2. Express [intermediate] in terms of [reactant] by setting the forward rate law of the reversible step equal to the reverse rate law, and solving for [intermediate].
3. Substitute the expression for [intermediate] into the rate law for the slow step to obtain the overall rate law.

Several end-of-chapter problems, including Problems 16.61 and 16.62, provide additional examples of this approach.

To restate a key point, *for any mechanism, only reactants involved up to and including the slow (rate-determining) step appear in the overall rate law.*

Summary of Section 16.6

- The mechanisms of most common reactions consist of two or more elementary steps, each of which shows a single molecular event.
- The molecularity of an elementary step equals the number of reactant particles and is the same as the total reaction order of the step. Only unimolecular and bimolecular steps are chemically reasonable.
- The rate-determining (slowest) step in a mechanism determines how fast the overall reaction occurs, and its rate law is equivalent to the overall rate law.
- Reaction intermediates are species that form in one step and react in a later one.
- For a mechanism to be valid, (1) the elementary steps must add up to the overall balanced equation, (2) the steps must be reasonable, and (3) the mechanism must correlate with the rate law.
- If a mechanism begins with a slow step, only those reactants involved in the slow step appear in the overall rate law.
- If a mechanism begins with a fast step, the product of the fast step accumulates as an intermediate, and the step reaches equilibrium. To show that the mechanism is valid, we express [intermediate] in terms of [reactant].
- Only reactants involved in steps up to and including the slow (rate-determining) step appear in the overall rate law.

16.7 • CATALYSIS: SPEEDING UP A REACTION

Increasing the rate of a reaction has countless applications in both engineering and biology. Raising the temperature can speed up a reaction but energy for industrial processes is costly and many substances, especially organic and biological ones, are heat sensitive. More commonly, by far, a reaction is accelerated by a **catalyst,** a substance that increases the rate *without* being consumed in the reaction. Because of this fundamental property of catalysts, only a small, nonstoichiometric amount is required to speed the reaction. Despite this, catalysts are employed in so many processes that several million tons are produced annually in the United States alone. Nature is the master designer and user of catalysts: every organism relies on protein catalysts, known as *enzymes,* to speed up life-sustaining reactions; even the simplest bacterium employs thousands of them.

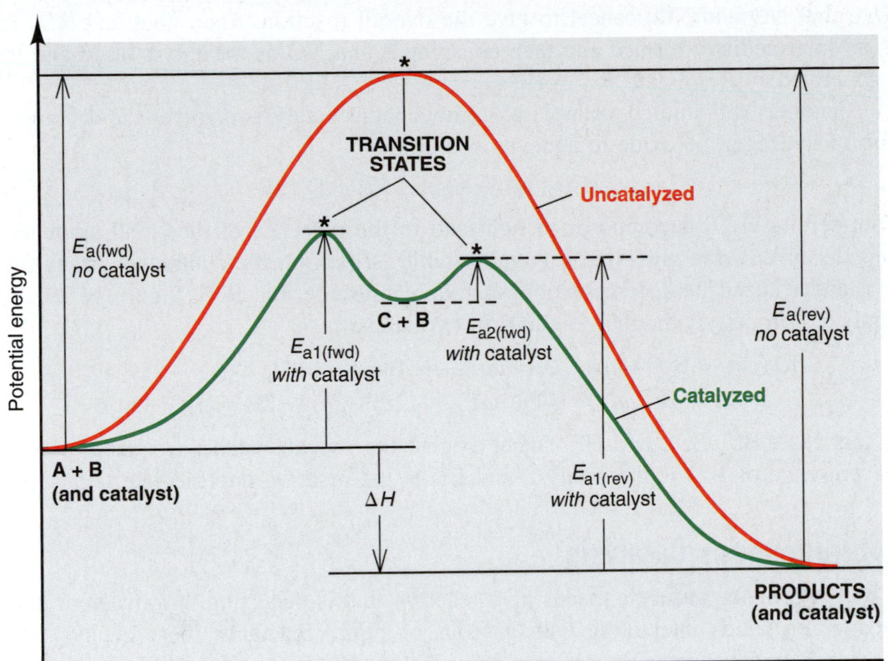

Figure 16.19 **Reaction energy diagram for a catalyzed *(green)* and an uncatalyzed *(red)* process.** A catalyst speeds a reaction by providing a different, lower energy pathway *(green)*. (Only the first step of the catalyzed reverse reaction is labeled.)

The Basis of Catalytic Action

Each catalyst has a specific way of acting, but in general, *a catalyst causes a lower activation energy, which makes the rate constant larger and, thus, the reaction rate higher:*

$$\text{Catalyst} \Longrightarrow \text{lower } E_a \Longrightarrow \text{larger } k \Longrightarrow \text{higher rate}$$

Consider a general *uncatalyzed* reaction that proceeds by a one-step mechanism involving a bimolecular collision between the reactants A and B (Figure 16.19). The activation energy is relatively large, so the rate is relatively low:

$$A + B \longrightarrow \text{product} \qquad [\text{larger } E_a \Longrightarrow \text{lower rate}]$$

In the *catalyzed* reaction, reactant A interacts with the catalyst in one step to form the intermediate C, and then C reacts with B in a second step to form product and regenerate the catalyst:

$$(1)\ A + \text{catalyst} \longrightarrow C \qquad [\text{smaller } E_a \Longrightarrow \text{higher rate}]$$
$$(2)\ C + B \longrightarrow \text{product} + \text{catalyst} \qquad [\text{smaller } E_a \Longrightarrow \text{higher rate}]$$

Three points to note in Figure 16.19 are

- A catalyst speeds up the forward *and* reverse reactions. Thus, a reaction has the *same yield* with or without a catalyst, but the product forms *faster.*
- A catalyst causes a *lower total activation energy* by providing a *different mechanism* for the reaction. The total of the activation energies for both steps of the catalyzed pathway $[E_{a1(fwd)} + E_{a2(fwd)}]$ is less than the forward activation energy of the uncatalyzed pathway.
- The catalyst is *not* consumed, but rather used and then regenerated.

Homogeneous Catalysis

Chemists classify catalysts based on whether or not they act in the same phase as the reactants and products. A **homogeneous catalyst** exists in solution with the reaction mixture, so it must be a gas, liquid, or soluble solid.

A thoroughly studied example of homogeneous catalysis in the gas phase was formerly used in sulfuric acid manufacture. The key step, the oxidation of sulfur dioxide to sulfur trioxide, occurs so slowly that it is not economical:

$$SO_2(g) + \tfrac{1}{2}O_2(g) \longrightarrow SO_3(g)$$

In the presence of nitrogen monoxide, however, the reaction speeds up dramatically:

$$NO(g) + \tfrac{1}{2}O_2(g) \longrightarrow NO_2(g)$$
$$NO_2(g) + SO_2(g) \longrightarrow NO(g) + SO_3(g)$$

A

B

Figure 16.20 The catalyzed decomposition of H_2O_2. **A,** A small amount of NaBr is added to a solution of H_2O_2. **B,** Oxygen gas forms quickly as $Br^-(aq)$ catalyzes the H_2O_2 decomposition; the reaction intermediate, Br_2, turns the solution orange.

Note that NO and NO_2 cancel to give the overall reaction. Also, note that NO_2 acts as an intermediate (formed and then consumed) and NO as a catalyst (used and then regenerated).

Another well-studied example of homogeneous catalysis involves the decomposition of hydrogen peroxide in aqueous solution:

$$2H_2O_2(aq) \longrightarrow 2H_2O(l) + O_2(g)$$

Commercial H_2O_2 decomposes in light and in the presence of the small amounts of ions dissolved from glass, but it is quite stable in dark plastic containers. Many other substances speed its decomposition, including bromide ion, Br^- (Figure 16.20). The catalyzed process is thought to occur in two steps:

$$2Br^-(aq) + H_2O_2(aq) + 2H^+(aq) \longrightarrow Br_2(aq) + 2H_2O(l)$$
$$Br_2(aq) + H_2O_2(aq) \longrightarrow 2Br^-(aq) + 2H^+(aq) + O_2(g)$$

In this case, Br_2, Br^-, and H^+ cancel to give the overall balanced equation, Br^- (in the presence of H^+) is the catalyst, and Br_2 is the reaction intermediate.

Heterogeneous Catalysis

A **heterogeneous catalyst** speeds up a reaction that is occurring in a different phase. Most often solids interacting with gaseous or liquid reactants, these catalysts have enormous surface areas, sometimes as much as 500 m^2/g. Very early in the reaction, the rate depends on reactant concentration, but almost immediately, the reaction becomes zero order: since the rate-determining step occurs on the catalyst's surface, once the reactant covers it, adding more reactant cannot increase the rate further.

A very important organic example of heterogeneous catalysis is **hydrogenation,** the addition of H_2 to C═C bonds to form C—C bonds. The petroleum, plastics, and food industries employ this process on an enormous scale. The conversion of vegetable oil into margarine is one example. The simplest hydrogenation converts ethylene (ethene) to ethane:

$$H_2C{=}CH_2(g) + H_2(g) \longrightarrow H_3C{-}CH_3(g)$$

In the absence of a catalyst, the reaction is very slow. But, at high H_2 pressure (high $[H_2]$) and in the presence of finely divided Ni, Pd, or Pt metal, it is rapid even at ordinary temperatures. The Group 8B(10) metals catalyze by *chemically adsorbing the reactants onto their surface* (Figure 16.21). The adsorbed H_2 splits into two H atoms that become weakly bonded to the catalyst's surface (catM):

$$H{-}H(g) + 2\text{catM}(s) \longrightarrow 2\text{catM}{-}H \text{ (H atoms bound to metal surface)}$$

Then, C_2H_4 adsorbs and reacts with the H atoms, one at a time, to form C_2H_6. Thus, the catalyst acts by lowering the activation energy of the slow step as part of a different mechanism.

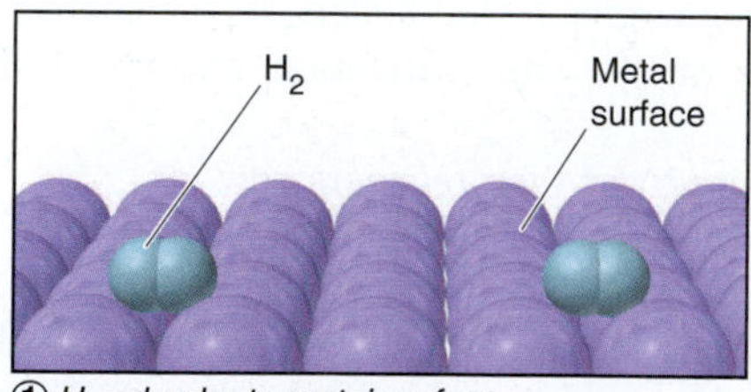

① H_2 adsorbs to metal surface.

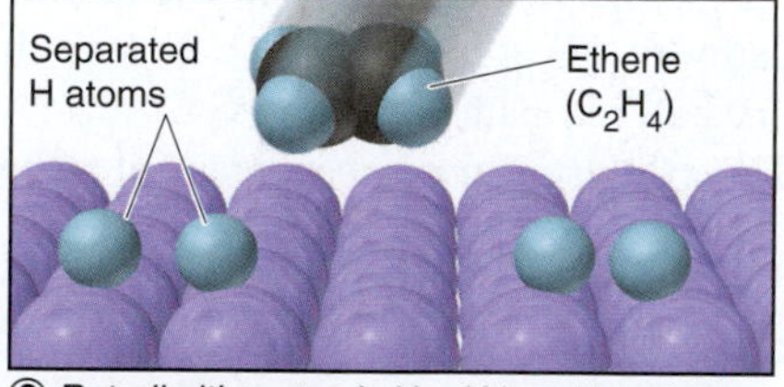

② Rate-limiting step is H—H bond breakage.

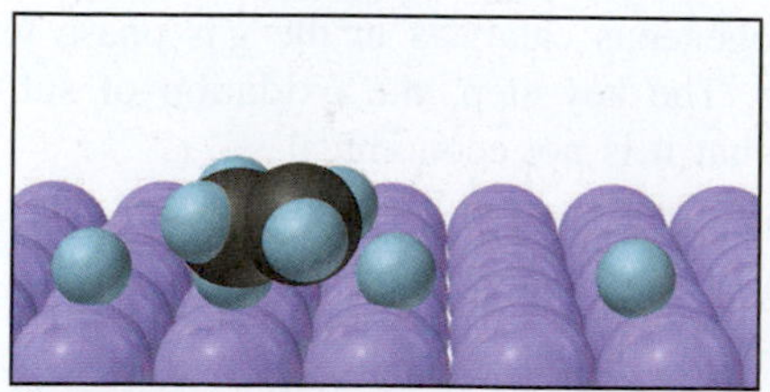
③ After C_2H_4 adsorbs, one C—H forms.

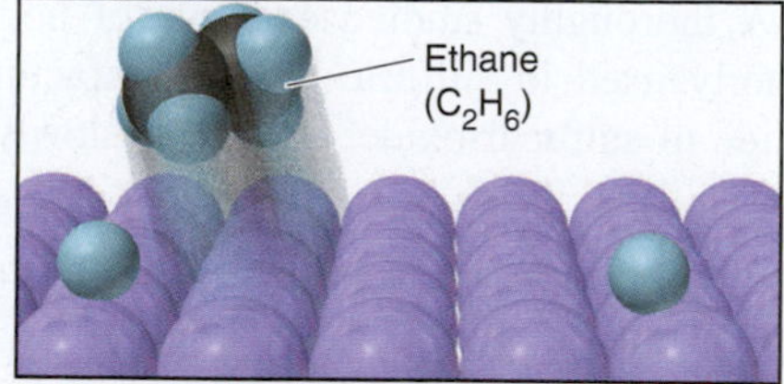

④ Another C—H bond forms; C_2H_6 leaves surface.

Figure 16.21 The metal-catalyzed hydrogenation of ethene.

Catalysis in Nature

Many catalytic processes occur in nature. We discuss two examples here: the first occurs within all organisms, and the second takes place in the upper atmosphere.

Cellular Catalysis: The Function of Enzymes Within every living cell, thousands of individual reactions occur in dilute solution at ordinary temperatures and pressures. The rates of these reactions respond smoothly to various factors, including tiny concentration changes, signals from other cells, and environmental stresses. Virtually every cell reaction is catalyzed by its own specific **enzyme,** a protein whose complex three-dimensional shape (Section 15.6)—and thus its function—has been perfected through natural selection. Several features account for enzyme activity:

1. *Structure.* Every enzyme has an **active site,** a small region whose shape results from those of the side chains (R groups) of the amino acids that make it up. When reactant molecules, called *substrates,* bind to the active site, usually through *intermolecular forces,* the chemical change begins. With molar masses ranging from 15,000 to 1,000,000 g/mol, most enzymes are enormous relative to their substrates, and they are often embedded within membranes. Thus, like a heterogeneous catalyst, an enzyme provides a surface on which a substrate is immobilized temporarily, waiting for another reactant to land nearby. Like a homogeneous catalyst, the active-site R groups interact directly with the substrates in multistep sequences.

2. *Efficiency and specificity.* Enzymes are incredibly *efficient* catalysts. Consider the hydrolysis of urea, a key component in amino-acid metabolism:

$$(NH_2)_2C{=}O(aq) + 2H_2O(l) + H^+(aq) \longrightarrow 2NH_4^+(aq) + HCO_3^-(aq)$$

In water at room temperature, the rate constant for the uncatalyzed reaction is 3×10^{-10} s^{-1}. Under the same conditions in the presence of the enzyme urease (pronounced "*yur*-ee-ase"), the rate constant increases 10^{14}-fold, to 3×10^4 s^{-1}! Enzymes are also extremely *specific:* urease catalyzes *only* this hydrolysis reaction, and no other enzyme does so.

3. *Models.* There are two main models of enzyme action. In the *lock-and-key model,* when the "key" (substrate) fits the "lock" (active site), the chemical change begins. However, experiments show that, in many cases, the enzyme changes shape when the substrate lands at its active site. Thus, rather than a rigidly shaped lock in which a particular key fits, the *induced-fit model* pictures a "hand" (substrate) entering a "glove" (active site), causing it to attain its functional shape.

4. *Mechanisms.* Enzymes act through a variety of catalytic mechanisms. In some cases, the active-site R groups bring the reacting atoms of the substrates closer together. In other cases, the R groups stretch the substrate bond that is to be broken. Some R groups provide an H^+ ion that increases the speed of a rate-determining step; others remove an H^+ ion at a critical step. Regardless of their specific mode of action, *all enzymes function by binding to the reaction's transition state and thus stabilizing it.* In this way, the enzyme lowers the activation energy, increasing the reaction rate.

Atmospheric Catalysis: Industrial Chemicals and Depletion of the Ozone Layer Both homogeneous and heterogeneous catalysis play key roles in the depletion of ozone from the stratosphere. At Earth's surface, ozone is an air pollutant, contributing to smog and other problems. In the stratosphere, however, a natural layer of ozone absorbs UV radiation from the Sun. If this radiation were to reach the surface, it could break bonds in DNA, promote skin cancer, and damage simple life forms at the base of the food chain.

Stratospheric ozone concentrations are maintained by a sequence of reactions:

$$O_2 \xrightarrow{UV} 2O$$

$$O + O_2 \longrightarrow O_3 \qquad \text{[ozone formation]}$$

$$O + O_3 \longrightarrow 2O_2 \qquad \text{[ozone breakdown]}$$

In 1995, Paul J. Crutzen, Mario J. Molina, and F. Sherwood Rowland received the Nobel Prize in chemistry for showing that chlorofluorocarbons (CFCs), used as aerosol propellants and air-conditioning coolants, were disrupting this sequence by catalyzing the breakdown reaction. CFCs are stable in the lower atmosphere, but as they slowly rise to the stratosphere, UV radiation cleaves them:

$$CF_2Cl_2 \xrightarrow{UV} CF_2Cl\cdot + Cl\cdot$$

(The dots are unpaired electrons resulting from bond cleavage.) Like many species with unpaired electrons (free radicals), atomic Cl is very reactive. It reacts with ozone to produce chlorine monoxide (ClO·), which then regenerates Cl atoms:

$$O_3 + Cl\cdot \longrightarrow ClO\cdot + O_2$$
$$ClO\cdot + O \longrightarrow \cdot Cl + O_2$$

The sum of these steps is the ozone breakdown reaction:

$$O_3 + \cancel{\cdot Cl} + \cancel{\cdot ClO} + O \longrightarrow \cancel{\cdot ClO} + O_2 + \cancel{\cdot Cl} + O_2$$

or

$$O_3 + O \longrightarrow 2O_2$$

Thus, the Cl atom is a homogeneous catalyst: it exists in the same phase as the reactants, speeds the reaction by allowing a different mechanism, and is regenerated. During its stratospheric half-life of about 2 years, each Cl atom speeds the breakdown of about 100,000 ozone molecules.

High levels of ClO over Antarctica have given rise to an *ozone hole,* an area of the stratosphere showing a severe reduction of ozone. The hole enlarges by heterogeneous catalysis, as stratospheric clouds and dust from volcanic activity provide a surface that speeds formation of Cl atoms by other mechanisms. Despite international agreements that phased out production of CFCs in 2010 and similar compounds by 2040, full recovery of the ozone layer is likely to take the rest of this century! The good news is that halogen levels in the lower atmosphere have begun to fall significantly.

Summary of Section 16.7

- A catalyst increases the rate of a reaction without being consumed. It accomplishes this by providing another mechanism with a lower activation energy.
- Homogeneous catalysts function in the same phase as the reactants. Heterogeneous catalysts act in a different phase from the reactants.
- The hydrogenation of carbon-carbon double bonds takes place on a solid metal catalyst, which speeds up the rate-determining step, the breakage of the H_2 bond.
- Enzymes are protein catalysts of high efficiency and specificity that act by stabilizing the transition state and, thus, lowering the activation energy.
- Chlorine atoms from CFC molecules speed the breakdown of stratospheric ozone.

CHAPTER REVIEW GUIDE

The following sections provide many aids to help you study this chapter. (Numbers in parentheses refer to pages, unless noted otherwise.)

Learning Objectives

These are concepts and skills to review after studying this chapter.

Related section (§), sample problem (SP), and upcoming end-of-chapter problem (EP) numbers are listed in parentheses.

1. Explain why reaction rate depends on concentration, physical state, and temperature (§16.1) (EPs 16.1–16.6)
2. Understand how reaction rate is expressed in terms of changing reactant and product concentrations over time, and distinguish among average, instantaneous, and initial rates (§16.2) (SP 16.1) (EPs 16.7–16.19)
3. Describe the information needed to determine the rate law, and explain how to calculate reaction orders and rate constant (§16.3) (SPs 16.2–16.4) (EPs 16.20–16.28)
4. Understand how to use integrated rate laws to find concentration at a given time (or vice versa) and reaction order, and explain the meaning of half-life (§16.4) (SPs 16.5–16.7) (EPs 16.29–16.34)
5. Explain the effect of temperature on the rate constant (Arrhenius equation) and the importance of activation energy (§16.5) (SP 16.8) (EPs 16.37–16.42, 16.45–16.48, 16.51)
6. Understand collision theory (why concentrations are multiplied, how temperature affects the fraction of collisions exceeding E_a, and how rate depends on the number of effective collisions) and transition state theory (how E_a is used to form the transition state and how a reaction energy diagram depicts the progress of a reaction) (§16.5) (SP 16.9) (EPs 16.35, 16.36, 16.43, 16.44, 16.49, 16.50, 16.52)
7. Understand elementary steps and molecularity, and be able to construct a valid reaction mechanism with either a slow or a fast initial step (§16.6) (SP 16.10) (EPs 16.53–16.64)
8. Explain how a catalyst speeds a reaction by lowering E_a, and distinguish between homogeneous and heterogeneous catalysis. Understand how enzymes function and how stratospheric ozone depletion is catalyzed. (§16.7) (EPs 16.65, 16.66)

Key Terms

These important terms appear in boldface in the chapter and are defined again in the Glossary.

Section 16.1
chemical kinetics (499)
reaction rate (499)

Section 16.2
average rate (502)
instantaneous rate (503)
initial rate (503)

Section 16.3
rate law (rate equation) (505)
rate constant (505)
reaction orders (505)

Section 16.4
integrated rate law (512)
half-life ($t_{1/2}$) (515)

Section 16.5
collision theory (519)
Arrhenius equation (519)
activation energy (E_a) (520)
effective collision (522)
frequency factor (522)
transition state theory (522)
transition state (activated complex) (522)
reaction energy diagram (523)

Section 16.6
reaction mechanism (525)
elementary reaction (elementary step) (526)
molecularity (526)
unimolecular reaction (526)
bimolecular reaction (526)
rate-determining (rate-limiting) step (527)
reaction intermediate (527)

Section 16.7
catalyst (530)
homogeneous catalyst (531)
heterogeneous catalyst (532)
hydrogenation (532)
enzyme (533)
active site (533)

Key Equations and Relationships

Numbered and screened concepts are listed for you to refer to or memorize.

16.1 Expressing reaction rate in terms of reactant A (501):

$$\text{Rate} = -\frac{\Delta[\text{A}]}{\Delta t}$$

16.2 Expressing the rate of a general reaction (503):

$$\text{Rate} = -\frac{1}{a}\frac{\Delta[\text{A}]}{\Delta t} = -\frac{1}{b}\frac{\Delta[\text{B}]}{\Delta t} = \frac{1}{c}\frac{\Delta[\text{C}]}{\Delta t} = \frac{1}{d}\frac{\Delta[\text{D}]}{\Delta t}$$

16.3 Writing a general rate law (in which products do not appear) (505):

$$\text{Rate} = k[\text{A}]^m[\text{B}]^n \cdots$$

16.4 Calculating the time to reach a given [A] in a first-order reaction (rate = k[A]) (513):

$$\ln\frac{[\text{A}]_0}{[\text{A}]_t} = kt$$

16.5 Calculating the time to reach a given [A] in a simple second-order reaction (rate = $k[\text{A}]^2$) (513):

$$\frac{1}{[\text{A}]_t} - \frac{1}{[\text{A}]_0} = kt$$

16.6 Calculating the time to reach a given [A] in a zero-order reaction (rate = k) (513):

$$[\text{A}]_t - [\text{A}]_0 = -kt$$

16.7 Finding the half-life of a first-order process (516):

$$t_{1/2} = \frac{\ln 2}{k} = \frac{0.693}{k}$$

16.8 Relating the rate constant to the temperature (Arrhenius equation) (519):

$$k = Ae^{-E_a/RT}$$

16.9 Relating the heat of reaction to the forward and reverse activation energies (520):

$$\Delta H_{\text{rxn}} = E_{\text{a(fwd)}} - E_{\text{a(rev)}}$$

16.10 Calculating the activation energy (rearranged form of Arrhenius equation) (521):

$$\ln\frac{k_2}{k_1} = -\frac{E_a}{R}\left(\frac{1}{T_2} - \frac{1}{T_1}\right)$$

BRIEF SOLUTIONS TO FOLLOW-UP PROBLEMS

Compare your own solutions to these calculation steps and answers.

16.1 (a) $4\text{NO}(g) + \text{O}_2(g) \longrightarrow 2\text{N}_2\text{O}_3(g)$;

$$\text{rate} = -\frac{\Delta[\text{O}_2]}{\Delta t} = -\frac{1}{4}\frac{\Delta[\text{NO}]}{\Delta t} = \frac{1}{2}\frac{\Delta[\text{N}_2\text{O}_3]}{\Delta t}$$

(b) $$-\frac{\Delta[\text{O}_2]}{\Delta t} = -\frac{1}{4}\frac{\Delta[\text{NO}]}{\Delta t} = -\frac{1}{4}(-1.60\times10\ \text{mol/L·s})$$

$$= 4.00\times10^{-5}\ \text{mol/L·s}$$

16.2 First order in Br^-, first order in BrO_3^-, second order in H^+, fourth order overall.

16.3 Rate = $k[\text{H}_2]^m[\text{I}_2]^n$. From Expts 1 and 3, $m = 1$. From Expts 2 and 4, $n = 1$. Therefore, rate = $k[\text{H}_2][\text{I}_2]$; second order overall.

16.4 (a) The rate law shows the reaction is zero order in Y, so the rate is not affected by doubling Y: rate of Expt 2 = 0.25×10^{-5} mol/L·s.
(b) The rate of Expt 3 is four times that of Expt 1, so [X] doubles.

16.5 $1/[\text{HI}]_1 - 1/[\text{HI}]_0 = kt$
111 L/mol − 100. L/mol = $(2.4\times10^{-21}$ L/mol·s$)(t)$
$t = 4.6\times10^{21}$ s (or 1.5×10^{14} yr)

16.6 (a)

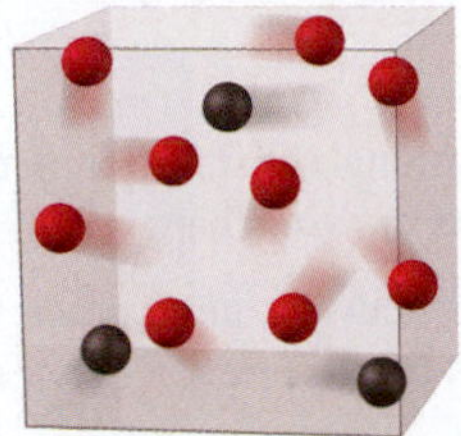

BRIEF SOLUTIONS TO FOLLOW-UP PROBLEMS (CONTINUED)

(b) At 10.0 min (4 half-lives), there are 0.75 particles of X.

$$\text{Amount (mol)} = 0.75 \text{ particles} \times \frac{0.20 \text{ mol X}}{1 \text{ particle}} = 0.15 \text{ mol X}$$

$$M = \frac{0.15 \text{ mol X}}{0.50 \text{ L}} = 0.30\ M$$

16.7 $t_{1/2} = (\ln 2)/k$; $k = 0.693/13.1 \text{ h} = 5.29\times10^{-2} \text{ h}^{-1}$

16.8 $$\ln \frac{0.286 \text{ L/mol·s}}{k_1} = -\frac{1.00\times10^5 \text{ J/mol}}{8.314 \text{ J/mol·K}} \times \left(\frac{1}{500.\text{ K}} - \frac{1}{490.\text{ K}}\right) = 0.491$$

$k_1 = 0.175$ L/mol·s

16.9

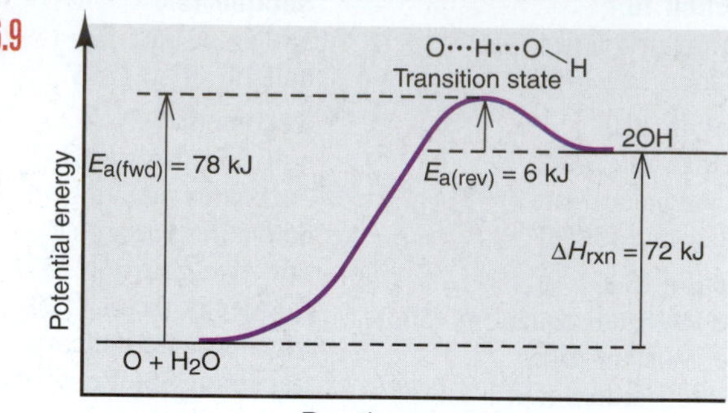

16.10 (a) Balanced equation (after doubling step 4):
$2NO(g) + 2H_2(g) \longrightarrow N_2(g) + 2H_2O(g)$
(b) Step 2 is unimolecular. All others are bimolecular.
(c) $\text{Rate}_1 = k_1[NO]^2$; $\text{rate}_2 = k_2[H_2]$; $\text{rate}_3 = k_3[N_2O_2][H]$; $\text{rate}_4 = k_4[HO][H]$; $\text{rate}_5 = k_5[H][N_2O]$.

PROBLEMS

Problems with **colored** numbers are answered in Appendix E. Sections match the text and provide the numbers of relevant sample problems. Bracketed problems are grouped in pairs (indicated by a short rule) that cover the same concept. Comprehensive Problems are based on material from any section or previous chapter.

Focusing on Reaction Rate

16.1 What variable of a chemical reaction is measured over time to obtain the reaction rate?

16.2 How does an increase in pressure affect the rate of a gas-phase reaction? Explain.

16.3 A reaction is carried out with water as the solvent. How does the addition of more water to the reaction vessel affect the rate of the reaction? Explain.

16.4 A gas reacts with a solid that is present in large chunks. Then the reaction is run again with the solid pulverized. How does the increase in the surface area of the solid affect the rate of its reaction with the gas? Explain.

16.5 How does an increase in temperature affect the rate of a reaction? Explain the two factors involved.

16.6 In a kinetics experiment, a chemist places crystals of iodine in a closed reaction vessel, introduces a given quantity of H_2 gas, and obtains data to calculate the rate of HI formation. In a second experiment, she uses the same amounts of iodine and hydrogen, but first warms the flask to 130°C, a temperature above the sublimation point of iodine. In which of these experiments does the reaction proceed at a higher rate? Explain.

Expressing the Reaction Rate

(Sample Problem 16.1)

16.7 Define *reaction rate*. Assuming constant temperature and a closed reaction vessel, why does the rate change with time?

16.8 (a) What is the difference between an average rate and an instantaneous rate? (b) What is the difference between an initial rate and an instantaneous rate?

16.9 Give two reasons to measure initial rates in a kinetics study.

16.10 For the reaction $A(g) \longrightarrow B(g)$, sketch two curves on the same set of axes that show
(a) The formation of product as a function of time
(b) The consumption of reactant as a function of time

16.11 For the reaction $C(g) \longrightarrow D(g)$, [C] vs. time is plotted:

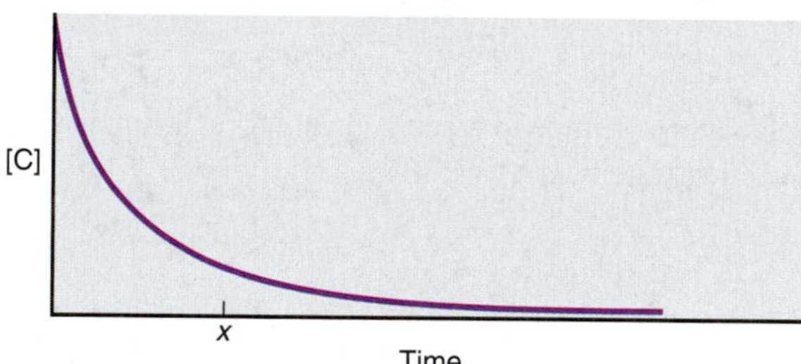

How do you determine each of the following?
(a) The average rate over the entire experiment
(b) The reaction rate at time x
(c) The initial reaction rate
(d) Would the values in parts (a), (b), and (c) be different if you plotted [D] vs. time? Explain.

16.12 The compound AX_2 decomposes according to the equation $2AX_2(g) \longrightarrow 2AX(g) + X_2(g)$. In one experiment, $[AX_2]$ was measured at various times and these data were obtained:

Time (s)	$[AX_2]$ (mol/L)
0.0	0.0500
2.0	0.0448
6.0	0.0300
8.0	0.0249
10.0	0.0209
20.0	0.0088

(a) Find the average rate over the entire experiment.
(b) Is the initial rate higher or lower than the rate in part (a)? Use graphical methods to estimate the initial rate.

16.13 (a) Use the data from Problem 16.12 to calculate the average rate from 8.0 to 20.0 s.
(b) Is the rate at exactly 5.0 s higher or lower than the rate in part (a)? Use graphical methods to estimate the rate at 5.0 s.

16.14 Express the rate of reaction in terms of the change in concentration of each of the reactants and products:

$$A(g) + 2B(g) \longrightarrow C(g)$$

When [B] is decreasing at 0.5 mol/L·s, how fast is [A] decreasing?

16.15 Express the rate of reaction in terms of the change in concentration of each of the reactants and products:

$$2D(g) + 3E(g) + F(g) \longrightarrow 2G(g) + H(g)$$

When [D] is decreasing at 0.1 mol/L·s, how fast is [H] increasing?

16.16 Reaction rate is expressed in terms of changes in concentration of reactants and products. Write a balanced equation for

$$\text{Rate} = -\frac{1}{2}\frac{\Delta[N_2O_5]}{\Delta t} = \frac{1}{4}\frac{\Delta[NO_2]}{\Delta t} = \frac{\Delta[O_2]}{\Delta t}$$

16.17 Reaction rate is expressed in terms of changes in concentration of reactants and products. Write a balanced equation for

$$\text{Rate} = -\frac{\Delta[CH_4]}{\Delta t} = -\frac{1}{2}\frac{\Delta[O_2]}{\Delta t} = \frac{1}{2}\frac{\Delta[H_2O]}{\Delta t} = \frac{\Delta[CO_2]}{\Delta t}$$

16.18 The decomposition of NOBr is studied by measuring pressure because the number of moles of gas changes; it cannot be studied colorimetrically because both NOBr and Br_2 are reddish brown:

$$2NOBr(g) \longrightarrow 2NO(g) + Br_2(g)$$

Use the data below to answer the following:
(a) Determine the average rate over the entire experiment.
(b) Determine the average rate between 2.00 and 4.00 s.
(c) Use graphical methods to estimate the initial reaction rate.
(d) Use graphical methods to estimate the rate at 7.00 s.
(e) At what time does the instantaneous rate equal the average rate over the entire experiment?

Time (s)	[NOBr] (mol/L)
0.00	0.0100
2.00	0.0071
4.00	0.0055
6.00	0.0045
8.00	0.0038
10.00	0.0033

16.19 Although the depletion of stratospheric ozone threatens life on Earth today, its accumulation was one of the crucial processes that allowed life to develop in prehistoric times:

$$3O_2(g) \longrightarrow 2O_3(g)$$

(a) Express the reaction rate in terms of $[O_2]$ and $[O_3]$.
(b) At a given instant, the reaction rate in terms of $[O_2]$ is 2.17×10^{-5} mol/L·s. What is it in terms of $[O_3]$?

The Rate Law and Its Components

(Sample Problems 16.2 to 16.4)

16.20 The rate law for the general reaction

$$aA + bB + \cdots \longrightarrow cC + dD + \cdots$$

is rate $= k[A]^m[B]^n \cdots$. (a) Explain the meaning of k. (b) Explain the meanings of m and n. Does $m = a$ and $n = b$? Explain. (c) If the reaction is first order in A and second order in B, and time is measured in minutes (min), what are the units for k?

16.21 By what factor does the rate change in each of the following cases (assuming constant temperature)?
(a) A reaction is first order in reactant A, and [A] is doubled.
(b) A reaction is second order in reactant B, and [B] is halved.
(c) A reaction is second order in reactant C, and [C] is tripled.

16.22 Give the individual reaction orders for all substances and the overall reaction order from the following rate law:

$$\text{Rate} = k[BrO_3^-][Br^-][H^+]^2$$

16.23 Give the individual reaction orders for all substances and the overall reaction order from the following rate law:

$$\text{Rate} = k\frac{[O_3]^2}{[O_2]}$$

16.24 By what factor does the rate in Problem 16.22 change if each of the following changes occurs: (a) $[BrO_3^-]$ is doubled; (b) $[Br^-]$ is halved; (c) $[H^+]$ is quadrupled?

16.25 By what factor does the rate in Problem 16.23 change if each of the following changes occurs: (a) $[O_3]$ is doubled; (b) $[O_2]$ is doubled; (c) $[O_2]$ is halved?

16.26 For the reaction

$$4A(g) + 3B(g) \longrightarrow 2C(g)$$

the following data were obtained at constant temperature:

Experiment	Initial Rate (mol/L·min)	Initial [A] (mol/L)	Initial [B] (mol/L)
1	5.00	0.100	0.100
2	45.0	0.300	0.100
3	10.0	0.100	0.200
4	90.0	0.300	0.200

(a) What is the order with respect to each reactant? (b) Write the rate law. (c) Calculate k (using the data from Expt 1).

16.27 For the reaction

$$A(g) + B(g) + C(g) \longrightarrow D(g)$$

the following data were obtained at constant temperature:

Expt	Initial Rate (mol/L·s)	Initial [A] (mol/L)	Initial [B] (mol/L)	Initial [C] (mol/L)
1	6.25×10^{-3}	0.0500	0.0500	0.0100
2	1.25×10^{-2}	0.1000	0.0500	0.0100
3	5.00×10^{-2}	0.1000	0.1000	0.0100
4	6.25×10^{-3}	0.0500	0.0500	0.0200

(a) What is the order with respect to each reactant? (b) Write the rate law. (c) Calculate k (using the data from Expt 1).

16.28 Phosgene is a toxic gas prepared by the reaction of carbon monoxide with chlorine:

$$CO(g) + Cl_2(g) \longrightarrow COCl_2(g)$$

These data were obtained in a kinetics study of its formation:

Experiment	Initial Rate (mol/L·s)	Initial [CO] (mol/L)	Initial [Cl_2] (mol/L)
1	1.29×10^{-29}	1.00	0.100
2	1.33×10^{-30}	0.100	0.100
3	1.30×10^{-29}	0.100	1.00
4	1.32×10^{-31}	0.100	0.0100

(a) Write the rate law for the formation of phosgene.
(b) Calculate the average value of the rate constant.

Integrated Rate Laws: Concentration Changes over Time

(Sample Problems 16.5 to 16.7)

16.29 How are integrated rate laws used to determine reaction order? What is the order with respect to reactant if a plot of
(a) The natural logarithm of [reactant] vs. time is linear?
(b) The inverse of [reactant] vs. time is linear?
(c) [Reactant] vs. time is linear?

16.30 Define the *half-life* of a reaction. Explain on the molecular level why the half-life of a first-order reaction is constant.

16.31 For the simple decomposition reaction

$$AB(g) \longrightarrow A(g) + B(g)$$

rate = $k[AB]^2$ and k = 0.2 L/mol·s. How long will it take for [AB] to reach one-third of its initial concentration of 1.50 M?

16.32 For the reaction in Problem 16.31, what is [AB] after 10.0 s?

16.33 In a first-order decomposition reaction, 50.0% of a compound decomposes in 10.5 min. (a) What is the rate constant of the reaction? (b) How long does it take for 75.0% of the compound to decompose?

16.34 A decomposition reaction has a rate constant of 0.0012 yr^{-1}. (a) What is the half-life of the reaction? (b) How long does it take for [reactant] to reach 12.5% of its original value?

Theories of Chemical Kinetics

(Sample Problems 16.8 and 16.9)

16.35 What is the central idea of collision theory? How does this model explain the effect of concentration on reaction rate?

16.36 Is collision frequency the only factor affecting rate? Explain.

16.37 Arrhenius proposed that each reaction has an energy threshold that must be reached for the particles to react. The kinetic theory of gases proposes that the average kinetic energy of the particles is proportional to the absolute temperature. How do these concepts relate to the effect of temperature on rate?

16.38 Use the exponential term in the Arrhenius equation to explain how temperature affects reaction rate.

16.39 How is the activation energy determined from the Arrhenius equation?

16.40 (a) Graph the relationship between k (vertical axis) and T (horizontal axis). (b) Graph the relationship between ln k (vertical axis) and $1/T$ (horizontal axis). How is the activation energy determined from this graph?

16.41 (a) For a reaction with a given E_a, how does an increase in T affect the rate? (b) For a reaction at a given T, how does a decrease in E_a affect the rate?

16.42 Assuming the activation energies are equal, which of the following reactions will proceed at a higher rate at 50°C? Explain.

$$NH_3(g) + HCl(g) \longrightarrow NH_4Cl(s)$$
$$N(CH_3)_3(g) + HCl(g) \longrightarrow (CH_3)_3NHCl(s)$$

16.43 For the reaction $A(g) + B(g) \longrightarrow AB(g)$, how many unique collisions between A and B are possible if there are four particles of A and three particles of B present in the vessel?

16.44 For the reaction $A(g) + B(g) \longrightarrow AB(g)$, how many unique collisions between A and B are possible if 1.01 mol of A(g) and 2.12 mol of B(g) are present in the vessel?

16.45 At 25°C, what is the fraction of collisions with energy equal to or greater than an activation energy of 100. kJ/mol?

16.46 If the temperature in Problem 16.45 is increased to 50.°C, by what factor does the fraction of collisions with energy equal to or greater than the activation energy change?

16.47 The rate constant of a reaction is 4.7×10^{-3} s^{-1} at 25°C, and the activation energy is 33.6 kJ/mol. What is k at 75°C?

16.48 The rate constant of a reaction is 4.50×10^{-5} L/mol·s at 195°C and 3.20×10^{-3} L/mol·s at 258°C. What is the activation energy of the reaction?

16.49 For the reaction $ABC + D \rightleftharpoons AB + CD$, ΔH°_{rxn} = −55 kJ/mol and $E_{a(fwd)}$ = 215 kJ/mol. Assuming a one-step reaction, (a) draw a reaction energy diagram; (b) calculate $E_{a(rev)}$; and (c) sketch a possible transition state if ABC is V-shaped.

16.50 For the reaction $A_2 + B_2 \longrightarrow 2AB$, $E_{a(fwd)}$ = 125 kJ/mol and $E_{a(rev)}$ = 85 kJ/mol. Assuming the reaction occurs in one step, (a) draw a reaction energy diagram; (b) calculate ΔH°_{rxn}; and (c) sketch a possible transition state.

16.51 Understanding the high-temperature formation and breakdown of the nitrogen oxides is essential for controlling the pollutants generated from power plants and cars. The first-order breakdown of dinitrogen monoxide to its elements has rate constants of 0.76/s at 727°C and 0.87/s at 757°C. What is the activation energy of this reaction?

16.52 Aqua regia, a mixture of HCl and HNO_3, has been used since alchemical times to dissolve many metals, including gold. Its orange color is due to the presence of nitrosyl chloride. Consider this one-step reaction for the formation of this compound:

$$NO(g) + Cl_2(g) \longrightarrow NOCl(g) + Cl(g) \qquad \Delta H^\circ = 83\ kJ$$

(a) Draw a reaction energy diagram, given $E_{a(fwd)}$ is 86 kJ/mol.
(b) Calculate $E_{a(rev)}$.
(c) Sketch a possible transition state for the reaction. (*Note:* The atom sequence of nitrosyl chloride is Cl—N—O.)

Reaction Mechanisms: The Steps from Reactant to Product

(Sample Problem 16.10)

16.53 Is the rate of an overall reaction lower, higher, or equal to the average rate of the individual steps? Explain.

16.54 Explain why the coefficients of an elementary step equal the reaction orders of its rate law but those of an overall reaction do not.

16.55 Is it possible for more than one mechanism to be consistent with the rate law of a given reaction? Explain.

16.56 What is the difference between a reaction intermediate and a transition state?

16.57 Why is a bimolecular step more reasonable physically than a termolecular step?

16.58 If a slow step precedes a fast step in a two-step mechanism, do the substances in the fast step appear in the rate law? Explain.

16.59 A proposed mechanism for the reaction of carbon dioxide with hydroxide ion in aqueous solution is
(1) $CO_2(aq) + OH^-(aq) \longrightarrow HCO_3^-(aq)$ [slow]
(2) $HCO_3^-(aq) + OH^-(aq) \longrightarrow CO_3^{2-}(aq) + H_2O(l)$ [fast]

(a) What is the overall equation?
(b) Identify the reaction intermediate(s), if any.
(c) What are the molecularity and the rate law for each step?
(d) Is the mechanism consistent with the actual rate law: rate = $k[CO_2][OH^-]$?

16.60 A proposed mechanism for the gas-phase reaction between chlorine and nitrogen dioxide is
(1) $Cl_2(g) + NO_2(g) \longrightarrow Cl(g) + NO_2Cl(g)$ [slow]
(2) $Cl(g) + NO_2(g) \longrightarrow NO_2Cl(g)$ [fast]
(a) What is the overall equation?
(b) Identify the reaction intermediate(s), if any.
(c) What are the molecularity and the rate law for each step?
(d) Is the mechanism consistent with the actual rate law: rate = $k[Cl_2][NO_2]$?

16.61 The proposed mechanism for a reaction is
(1) $A(g) + B(g) \rightleftharpoons X(g)$ [fast]
(2) $X(g) + C(g) \longrightarrow Y(g)$ [slow]
(3) $Y(g) \longrightarrow D(g)$ [fast]
(a) What is the overall equation?
(b) Identify the intermediate(s), if any.
(c) What are the molecularity and the rate law for each step?
(d) Is the mechanism consistent with the actual rate law: rate = $k[A][B][C]$?
(e) Is the following one-step mechanism equally valid: $A(g) + B(g) + C(g) \longrightarrow D(g)$?

16.62 Consider the following mechanism:
(1) $ClO^-(aq) + H_2O(l) \rightleftharpoons HClO(aq) + OH^-(aq)$ [fast]
(2) $I^-(aq) + HClO(aq) \longrightarrow HIO(aq) + Cl^-(aq)$ [slow]
(3) $OH^-(aq) + HIO(aq) \longrightarrow H_2O(l) + IO^-(aq)$ [fast]
(a) What is the overall equation?
(b) Identify the intermediate(s), if any.
(c) What are the molecularity and the rate law for each step?
(d) Is the mechanism consistent with the actual rate law: rate = $k[ClO^-][I^-]$?

16.63 In a study of nitrosyl halides, a chemist proposes the following mechanism for the synthesis of nitrosyl bromide:

$$NO(g) + Br_2(g) \rightleftharpoons NOBr_2(g) \text{ [fast]}$$
$$NOBr_2(g) + NO(g) \longrightarrow 2NOBr(g) \text{ [slow]}$$

If the rate law is rate = $k[NO]^2[Br_2]$, is the proposed mechanism valid? If so, show that it satisfies the three criteria for validity.

16.64 The rate law for $2NO(g) + O_2(g) \longrightarrow 2NO_2(g)$ is rate = $k[NO]^2[O_2]$. In addition to the mechanism in the text (p. 529), the following three have been proposed:

I $2NO(g) + O_2(g) \longrightarrow 2NO_2(g)$

II $2NO(g) \rightleftharpoons N_2O_2(g)$ [fast]
$N_2O_2(g) + O_2(g) \longrightarrow 2NO_2(g)$ [slow]

III $2NO(g) \rightleftharpoons N_2(g) + O_2(g)$ [fast]
$N_2(g) + 2O_2(g) \longrightarrow 2NO_2(g)$ [slow]

(a) Which of these mechanisms is consistent with the rate law?
(b) Which of these mechanisms is most reasonable? Why?

Catalysis: Speeding Up a Reaction

16.65 Consider the reaction $N_2O(g) \xrightarrow{Au} N_2(g) + \frac{1}{2}O_2(g)$.
(a) Does the gold catalyst (Au, above the yield arrow) act as a homogeneous or a heterogeneous catalyst?
(b) On the same set of axes, sketch the reaction energy diagrams for the catalyzed and the uncatalyzed reaction.

16.66 Does a catalyst increase reaction rate by the same means as a rise in temperature does? Explain.

Comprehensive Problems

16.67 Consider the following reaction energy diagram:

(a) How many elementary steps are in the reaction mechanism?
(b) Which step is rate limiting?
(c) Is the overall reaction exothermic or endothermic?

16.68 The catalytic destruction of ozone occurs via a two-step mechanism, where X can be any of several species:
(1) $X + O_3 \longrightarrow XO + O_2$ [slow]
(2) $XO + O \longrightarrow X + O_2$ [fast]
(a) Write the overall reaction.
(b) Write the rate law for each step.
(c) X acts as ______, and XO acts as ______.
(d) High-flying aircraft release NO, which catalyzes this process, into the stratosphere. When O_3 and NO concentrations are 5×10^{12} molecules/cm^3 and 1.0×10^9 molecules/cm^3, respectively, what is the rate of O_3 depletion (k for the rate-determining step is 6×10^{-15} cm^3/molecule·s)?

16.69 A slightly bruised apple will rot extensively in about 4 days at room temperature (20°C). If it is kept in the refrigerator at 0°C, the same extent of rotting takes about 16 days. What is the activation energy for the rotting reaction?

16.70 Benzoyl peroxide, the substance most widely used against acne, has a half-life of 9.8×10^3 days when refrigerated. How long will it take to lose 5% of its potency (95% remaining)?

16.71 The rate law for the reaction

$$NO_2(g) + CO(g) \longrightarrow NO(g) + CO_2(g)$$

is rate = $k[NO_2]^2$; one possible mechanism is shown on p. 527.
(a) Draw a reaction energy diagram for that mechanism, given that $\Delta H^\circ_{overall} = -226$ kJ/mol.
(b) Consider the following alternative mechanism:
(1) $2NO_2(g) \longrightarrow N_2(g) + 2O_2(g)$ [slow]
(2) $2CO(g) + O_2(g) \longrightarrow 2CO_2(g)$ [fast]
(3) $N_2(g) + O_2(g) \longrightarrow 2NO(g)$ [fast]
Is the alternative mechanism consistent with the rate law? Is one mechanism more reasonable physically? Explain.

16.72 In acidic solution, the breakdown of sucrose into glucose and fructose has this rate law: rate = $k[H^+][\text{sucrose}]$. The initial rate of sucrose breakdown is measured in a solution that is 0.01 M H^+, 1.0 M sucrose, 0.1 M fructose, and 0.1 M glucose. How does the rate change if
(a) [Sucrose] is changed to 2.5 M?
(b) [Sucrose], [fructose], and [glucose] are all changed to 0.5 M?
(c) $[H^+]$ is changed to 0.0001 M?
(d) [Sucrose] and $[H^+]$ are both changed to 0.1 M?

16.73 The following molecular scenes represent starting mixtures I and II for the reaction of A *(black)* with B *(orange)*:

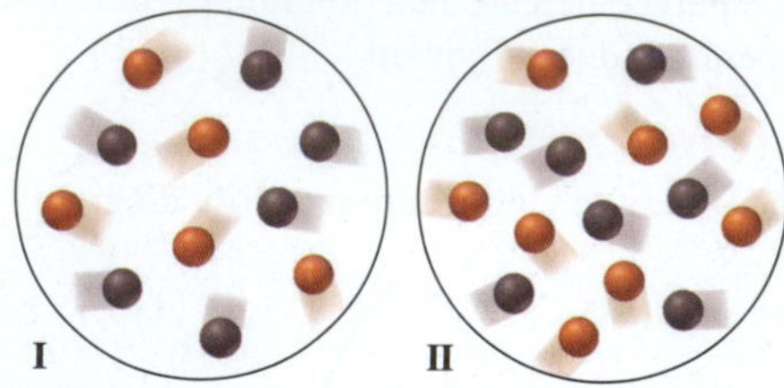

Each sphere represents 0.010 mol, and the volume is 0.50 L. If the reaction is first order in A and first order in B and the initial rate for I is 8.3×10^{-4} mol/L·min, what is the initial rate for II?

16.74 Biacetyl, the flavoring that makes margarine taste "just like butter," is extremely stable at room temperature, but at 200°C it undergoes a first-order breakdown with a half-life of 9.0 min. An industrial flavor-enhancing process requires that a biacetyl-flavored food be heated briefly at 200°C. How long can the food be heated and retain 85% of its buttery flavor?

16.75 A biochemist studying breakdown of the insecticide DDT finds that it decomposes by a first-order reaction with a half-life of 12 yr. How long does it take DDT in a soil sample to decompose from 275 ppbm to 10. ppbm (parts per billion by mass)?

16.76 The acid-catalyzed hydrolysis of sucrose occurs by the following overall reaction, whose kinetic data are given below:

$$\underset{\text{sucrose}}{C_{12}H_{22}O_{11}(s)} + H_2O(l) \longrightarrow \underset{\text{glucose}}{C_6H_{12}O_6(aq)} + \underset{\text{fructose}}{C_6H_{12}O_6(aq)}$$

[Sucrose] (mol/L)	Time (h)
0.501	0.00
0.451	0.50
0.404	1.00
0.363	1.50
0.267	3.00

(a) Determine the rate constant and the half-life of the reaction.
(b) How long does it take to hydrolyze 75% of the sucrose?
(c) Other studies have shown that this reaction is actually second order overall but appears to follow first-order kinetics. (Such a reaction is called a *pseudo–first-order reaction.*) Suggest a reason for this apparent first-order behavior.

16.77 Proteins in the body undergo continual breakdown and synthesis. Insulin is a polypeptide hormone that stimulates fat and muscle to take up glucose. Once released from the pancreas, it has a first-order half-life in the blood of 8.0 min. To maintain an adequate blood concentration of insulin, it must be replenished in a time interval equal to $1/k$. How long is this interval?

16.78 At body temperature (37°C), the rate constant of an enzyme-catalyzed decomposition is 2.3×10^{14} times that of the uncatalyzed reaction. If the frequency factor, A, is the same for both processes, by how much does the enzyme lower the E_a?

16.79 Is each of these statements true? If not, explain why.
(a) At a given T, all molecules have the same kinetic energy.
(b) Halving the P of a gaseous reaction doubles the rate.
(c) A higher activation energy gives a lower reaction rate.
(d) A temperature rise of 10°C doubles the rate of any reaction.
(e) If reactant molecules collide with greater energy than the activation energy, they change into product molecules.
(f) The activation energy of a reaction depends on temperature.
(g) The rate of a reaction increases as the reaction proceeds.
(h) Activation energy depends on collision frequency.
(i) A catalyst increases the rate by increasing collision frequency.
(j) Exothermic reactions are faster than endothermic reactions.
(k) Temperature has no effect on the frequency factor (A).
(l) The activation energy of a reaction is lowered by a catalyst.
(m) For most reactions, ΔH_{rxn} is lowered by a catalyst.
(n) The orientation probability factor (p) is near 1 for reactions between single atoms.
(o) The initial rate of a reaction is its maximum rate.
(p) A bimolecular reaction is generally twice as fast as a unimolecular reaction.
(q) The molecularity of an elementary reaction is proportional to the molecular complexity of the reactant(s).

16.80 Even when a mechanism is consistent with the rate law, later work may show it to be incorrect. For example, the reaction between hydrogen and iodine has this rate law: rate = $k[H_2][I_2]$. The long-accepted mechanism had a single bimolecular step; that is, the overall reaction was thought to be elementary:

$$H_2(g) + I_2(g) \longrightarrow 2HI(g)$$

In the 1960s, however, spectroscopic evidence showed the presence of free I atoms during the reaction. Kineticists have since proposed a three-step mechanism:

(1) $I_2(g) \rightleftharpoons 2I(g)$ [fast]
(2) $H_2(g) + I(g) \rightleftharpoons H_2I(g)$ [fast]
(3) $H_2I(g) + I(g) \longrightarrow 2HI(g)$ [slow]

Show that this mechanism is consistent with the rate law.

16.81 Many drugs decompose in blood by a first-order process.
(a) Two tablets of aspirin supply 0.60 g of the active compound. After 30 min, this compound reaches a maximum concentration of 2 mg/100 mL of blood. If the half-life for its breakdown is 90 min, what is its concentration (in mg/100 mL) 2.5 h after it reaches its maximum concentration?
(b) For the decomposition of an antibiotic in a person with a normal temperature (98.6°F), $k = 3.1\times10^{-5}\ s^{-1}$; for a person with a fever (temperature of 101.9°F), $k = 3.9\times10^{-5}\ s^{-1}$. If the person with the fever must take another pill when $\frac{2}{3}$ of the first pill has decomposed, how many hours should she wait to take a second pill? A third pill? (Assume the pill is effective immediately.)
(c) Calculate E_a for decomposition of the antibiotic in part (b).

16.82 In the lowest region of the atmosphere, ozone is one of the components of photochemical smog. It is generated in air when nitrogen dioxide, formed by the oxidation of nitrogen monoxide from car exhaust, reacts by the following mechanism:

(1) $NO_2(g) \xrightarrow[h\nu]{k_1} NO(g) + O(g)$
(2) $O(g) + O_2(g) \xrightarrow{k_2} O_3(g)$

Assuming the rate of formation of atomic oxygen in step 1 equals the rate of its consumption in step 2, use the data below to calculate (a) the concentration of atomic oxygen [O]; (b) the rate of ozone formation.

$$k_1 = 6.0\times10^{-3}\ s^{-1} \qquad [NO_2] = 4.0\times10^{-9}\ M$$
$$k_2 = 1.0\times10^{6}\ L/mol{\cdot}s \qquad [O_2] = 1.0\times10^{-2}\ M$$

16.83 In Houston (near sea level), water boils at 100.0°C. In Cripple Creek, Colorado (elevation about 9500 ft), it boils at 90.0°C. If it takes 4.8 min to cook an egg in Cripple Creek and 4.5 min in Houston, what is E_a for this process?

16.84 Chlorine is commonly used to disinfect drinking water, and inactivation of pathogens by chlorine follows first-order kinetics. The following data show *E. coli* inactivation:

Contact Time (min)	Percent (%) Inactivation
0.00	0.0
0.50	68.3
1.00	90.0
1.50	96.8
2.00	99.0
2.50	99.7
3.00	99.9

(a) Determine the first-order inactivation constant, k. [*Hint:* % inactivation $= 100 \times (1 - [A]_t/[A]_0)$.]
(b) How much contact time is required for 95% inactivation?

16.85 The reaction and rate law for the gas-phase decomposition of dinitrogen pentaoxide are

$$2N_2O_5(g) \longrightarrow 4NO_2(g) + O_2(g) \qquad \text{rate} = k[N_2O_5]$$

Which of the following can be considered valid mechanisms for the reaction?

I One-step collision

	Step	
II	$2N_2O_5(g) \longrightarrow 2NO_3(g) + 2NO_2(g)$	[slow]
	$2NO_3(g) \longrightarrow 2NO_2(g) + 2O(g)$	[fast]
	$2O(g) \longrightarrow O_2(g)$	[fast]
III	$N_2O_5(g) \rightleftharpoons NO_3(g) + NO_2(g)$	[fast]
	$NO_2(g) + N_2O_5(g) \longrightarrow 3NO_2(g) + O(g)$	[slow]
	$NO_3(g) + O(g) \longrightarrow NO_2(g) + O_2(g)$	[fast]
IV	$2N_2O_5(g) \rightleftharpoons 2NO_2(g) + N_2O_3(g) + 3O(g)$	[fast]
	$N_2O_3(g) + O(g) \longrightarrow 2NO_2(g)$	[slow]
	$2O(g) \longrightarrow O_2(g)$	[fast]
V	$2N_2O_5(g) \longrightarrow N_4O_{10}(g)$	[slow]
	$N_4O_{10}(g) \longrightarrow 4NO_2(g) + O_2(g)$	[fast]

16.86 Like any catalyst, palladium, platinum, and nickel catalyze both directions of a reaction: addition of hydrogen to (hydrogenation) and its elimination from (dehydrogenation) carbon-carbon double bonds.
(a) Which variable determines whether an alkene will be hydrogenated or dehydrogenated?
(b) Which reaction requires a higher temperature?
(c) How can all-*trans* fats arise during hydrogenation of fats that contain some *cis*- double bonds?

16.87 The molecular scenes below represent the first-order reaction in which cyclopropane (*red*) is converted to propene (*green*). Determine (a) the half-life and (b) the rate constant.

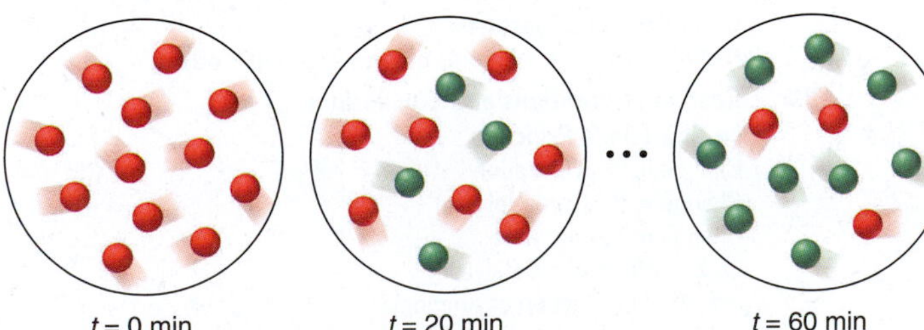

16.88 Consider the following organic reaction, in which one halogen replaces another in an alkyl halide:

$$BrCH_2CH_3 + KI \longrightarrow CH_3CH_2I + KBr$$

In acetone, this particular reaction goes to completion because KI is soluble in acetone but KBr is not. In the mechanism, I^- approaches the carbon *opposite* to the Br (see Figure 16.16, with I^- instead of OH^-). After Br^- has been replaced by I^- and precipitates as KBr, other I^- ions react with the ethyl iodide by the same mechanism.
(a) If we designate the C bonded to the halogen as C-1, what is the shape around C-1 and the hybridization of C-1 in ethyl iodide?
(b) In the transition state, one of the two lobes of the unhybridized $2p$ orbital of C-1 overlaps a p orbital of I, while the other lobe overlaps a p orbital of Br. What is the shape around C-1 and the hybridization of C-1 in the transition state?
(c) The deuterated reactant, $BrCHDCH_3$ (where D is deuterium, ^2H), has two optical isomers because C-1 is chiral. If the reaction is run with one of the isomers, the ethyl iodide is *not* optically active. Explain.

16.89 The scenes depict four initial reaction mixtures for the reaction of A (*blue*) and B (*yellow*), with and without a solid present (*gray cubes*). The initial rate, $-\Delta[A]/\Delta t$ (in mol/L·s), is shown, with each sphere representing 0.010 mol and the container volume at 0.50 L.

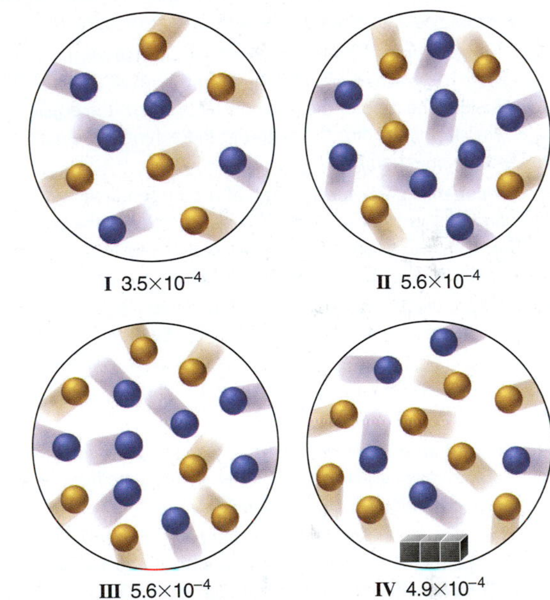

(a) What is the rate law in the absence of a catalyst?
(b) What is the overall reaction order?
(c) Find the rate constant.
(d) Do the gray cubes have a catalytic effect? Explain.

16.90 Figure 16.21 (p. 532) shows key steps in the metal-catalyzed (M) hydrogenation of ethylene:

$$C_2H_4(g) + H_2(g) \xrightarrow{M} C_2H_6(g)$$

Use the following symbols to write a mechanism that gives the overall equation:

H_2(ads)	adsorbed hydrogen molecules
M—H	hydrogen atoms bonded to metal atoms
C_2H_4(ads)	adsorbed ethylene molecules
C_2H_5(ads)	adsorbed ethyl radicals

16.91 A (*green*), B (*blue*), and C (*red*) are structural isomers. The molecular filmstrip depicts them undergoing a chemical change as time proceeds.
(a) Write a mechanism for the reaction.
(b) What role does C play?

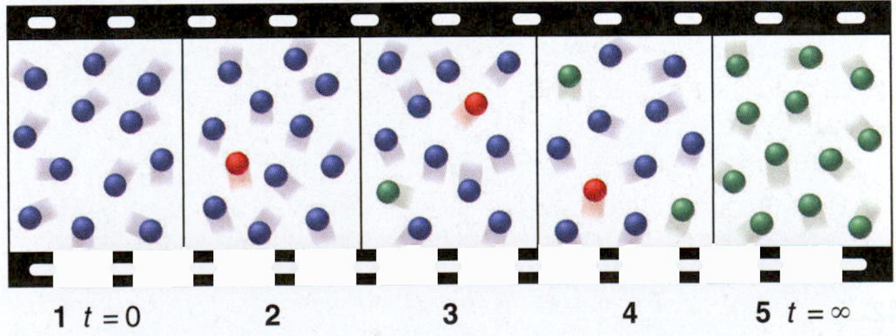

17 Equilibrium: The Extent of Chemical Reactions

Key Principles
to focus on while studying this chapter

- The principles of equilibrium and kinetics apply to different aspects of a chemical change: the extent *(yield)* of a reaction is not related to its *rate*. ***(Introduction)***
- All reactions are *reversible*. When the forward and reverse reaction rates are *equal*, the system has reached *equilibrium*. After this point, there is no further *observable* change. The *ratio* of the rate constants equals the *equilibrium constant, K*. The size of K is directly related to the *extent* of the reaction at a given temperature. ***(Section 17.1)***
- The *reaction quotient, Q*, is a specific ratio of product and reactant concentration terms. The various ways to write Q are all based *directly* on the balanced equation. The value of Q *changes continually* until the system reaches equilibrium, at which point $Q = K$. ***(Section 17.2)***
- The ideal gas law is used to quantitatively relate an equilibrium constant based on concentrations, K_c, to one based on pressures, K_p. ***(Section 17.3)***
- At any point in a reaction, we can predict its direction by comparing Q and K: if $Q < K$, the reaction will form more product; if $Q > K$, the reaction will form more reactant; if $Q = K$, the reaction is at equilibrium. ***(Section 17.4)***
- If the initial concentration of a reactant, $[A]_{init}$, is much larger than the *change* in its concentration to reach equilibrium, x, we make the *simplifying assumption* that x can be neglected in calculations. ***(Section 17.5)***
- If a system at equilibrium is *disturbed* by a change in conditions (concentration, pressure, or temperature), it will temporarily *not* be at equilibrium, but will then undergo a *net reaction* to reach equilibrium again *(Le Châtelier's principle)*. A change in concentration, pressure, or the presence of a catalyst does not affect K, but a change in temperature does. ***(Section 17.6)***

Back and Forth Like these balls passing between the hands of a juggler, the opposing steps of a reversible reaction occur at equal rates. In this chapter, we focus mostly on equilibrium in gaseous systems.

Outline

CONCEPTS & SKILLS TO REVIEW before studying this chapter

- equilibrium vapor pressure (Section 12.2)
- equilibrium nature of a saturated solution (Section 13.3)
- dependence of rate on concentration (Sections 16.2 and 16.5)
- rate laws for elementary reactions (Section 16.6)
- function of a catalyst (Section 16.7)

Just as reactions vary greatly in their speed, they also vary greatly in their extent. Indeed, kinetics and equilibrium apply to different aspects of a reaction:

- Kinetics applies to the *speed* (or rate) of a reaction, the concentration of reactant that disappears (or of product that appears) per unit time.
- Equilibrium applies to the *extent* (or yield) of a reaction, the concentrations of reactant and product present after unlimited time, or when no further change occurs.

As you'll see in this chapter, at equilibrium, no further *net* change occurs because the forward and reverse reactions reach a *balance*.

In fact, a fast reaction may go almost completely, just partially, or only slightly toward products before this balance is reached. Consider acid dissociation in water. In 1 *M* HCl, virtually all the hydrogen chloride molecules are dissociated into ions. In contrast, in 1 *M* CH_3COOH, fewer than 1% of the acetic acid molecules are dissociated at any given time. Yet both reactions take less than a second to reach completion. Similarly, some slow reactions yield a large amount of product, whereas others yield very little. After a few years, a steel water-storage tank will rust, and it will do so completely given enough time. But, no matter how long you wait, the water inside the tank will not decompose to hydrogen and oxygen.

Knowing the extent of a given reaction is crucial. How much product—medicine, polymer, fuel—can you obtain from a particular reaction mixture? How can you adjust conditions to obtain more? If a reaction is slow but has a good yield, will a catalyst speed it up enough to make it useful?

In this chapter, we consider equilibrium principles in systems of gases and pure liquids and solids; we'll discuss various solution equilibria in the next two chapters.

17.1 • THE EQUILIBRIUM STATE AND THE EQUILIBRIUM CONSTANT

Experimental results from countless reactions have shown that, *given sufficient time, the concentrations of reactants and products no longer change.* This apparent cessation of chemical change occurs because *all reactions are reversible and reach a state of equilibrium.* Let's examine a chemical system at the macroscopic and molecular levels to see how equilibrium arises and then consider some quantitative aspects of the process:

1. *A macroscopic view of equilibrium.* The system we'll consider is the reversible gaseous reaction between colorless dinitrogen tetroxide and brown nitrogen dioxide:

$$N_2O_4(g;\ \text{colorless}) \rightleftharpoons 2NO_2(g;\ \text{brown})$$

As soon as we introduce some liquid N_2O_4 (bp = 21°C) into a sealed container kept at 200°C, it vaporizes, and the gas begins to turn pale brown. As time passes, the brown darkens, until, after less than 30 seconds, the color stops changing. The first three photos in Figure 17.1 *(next page)* show the color change, and the last photo shows no further change.

2. *A molecular view of equilibrium.* On the molecular level, as shown in the blow-up circles of Figure 17.1, a dynamic scene unfolds. The N_2O_4 molecules fly wildly throughout the container, a few splitting into two NO_2 molecules. As time passes, more N_2O_4 molecules decompose and the concentration of NO_2 rises. As the number of N_2O_4 molecules decreases, N_2O_4 decomposition slows down. At the same time, increasing numbers of NO_2 molecules collide and combine, and re-formation of N_2O_4 speeds up. Eventually, the system reaches equilibrium: N_2O_4 molecules are decomposing into NO_2 molecules just as fast as NO_2 molecules are combining into N_2O_4.

Thus, at equilibrium, *reactant and product concentrations are constant because a change in one direction is balanced by a change in the other as the forward and reverse rates become equal:*

$$\text{At equilibrium: rate}_{\text{fwd}} = \text{rate}_{\text{rev}} \qquad \textbf{(17.1)}$$

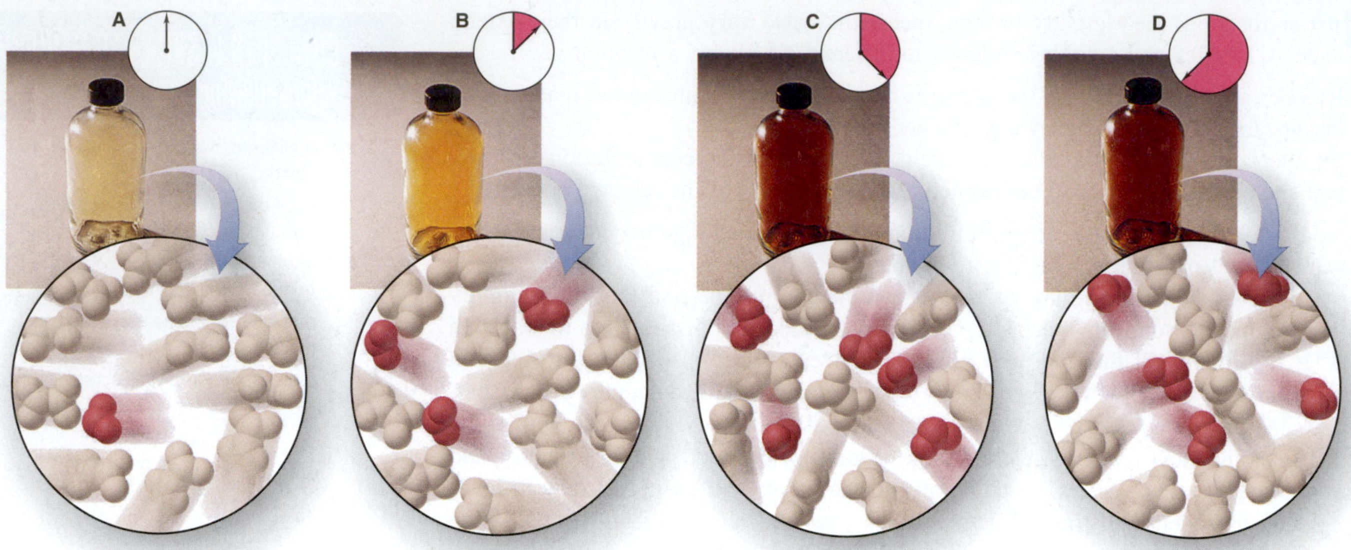

Figure 17.1 Reaching equilibrium on the macroscopic and molecular levels. **A,** The reaction mixture consists mostly of colorless N_2O_4. **B,** As N_2O_4 decomposes to NO_2, the mixture becomes pale brown. **C,** At equilibrium, the color and the concentrations of NO_2 and N_2O_4 no longer change. **D,** The reaction continues in both directions at equal rates, so the concentrations (and color) remain constant.

3. *A quantitiative view of equilibrium: a constant ratio of constants.* Let's see how reactant and product concentrations affect this process. At a particular temperature, when the system reaches equilibrium, we have

$$\text{rate}_{\text{fwd}} = \text{rate}_{\text{rev}}$$

In this reaction system, both forward and reverse reactions are elementary steps (Section 16.6), so we can write their rate laws directly from the balanced equation:

$$k_{\text{fwd}}[N_2O_4]_{\text{eq}} = k_{\text{rev}}[NO_2]^2_{\text{eq}}$$

where k_{fwd} and k_{rev} are the forward and reverse rate constants, respectively, and the subscript "eq" refers to concentrations at equilibrium. By rearranging, we set the ratio of the rate constants equal to the ratio of the concentration terms:

$$\frac{k_{\text{fwd}}}{k_{\text{rev}}} = \frac{[NO_2]^2_{\text{eq}}}{[N_2O_4]_{\text{eq}}}$$

The ratio of constants creates a new constant called the **equilibrium constant (*K*):**

$$K = \frac{k_{\text{fwd}}}{k_{\text{rev}}} = \frac{[NO_2]^2_{\text{eq}}}{[N_2O_4]_{\text{eq}}} \qquad \textbf{(17.2)}$$

The equilibrium constant K is a number equal to a particular ratio of equilibrium concentrations of product(s) to reactant(s) at a particular temperature.

4. *K as a measure of reaction extent.* The *magnitude* of *K* is an indication of *how far a reaction proceeds toward product at a given temperature*. Different reactions, even at the same temperature, have a wide range of concentrations at equilibrium—from almost all reactant to almost all product—so they have a wide range of equilibrium constants. Here are three examples of different magnitudes of *K*:

- *Small K* (Figure 17.2A). If a reaction yields little product before reaching equilibrium, it has a small *K*; if *K* is very small, we may say there is "no reaction." For example, there is "no reaction" between nitrogen and oxygen at 1000 K:*

$$N_2(g) + O_2(g) \rightleftharpoons 2NO(g) \qquad K = 1\times10^{-30}$$

- *Large K* (Figure 17.2B). Conversely, if a reaction reaches equilibrium with little reactant remaining, it has a large *K;* if *K* is very large, we say the reaction "goes to completion." The oxidation of carbon monoxide "goes to completion" at 1000 K:

$$2CO(g) + O_2(g) \rightleftharpoons 2CO_2(g) \qquad K = 2.2\times10^{22}$$

*To distinguish them in print, the equilibrium constant is represented by a capital italic *K*, whereas the temperature unit, the kelvin, is a capital roman K. Also, since the kelvin is a unit, it always follows a number.

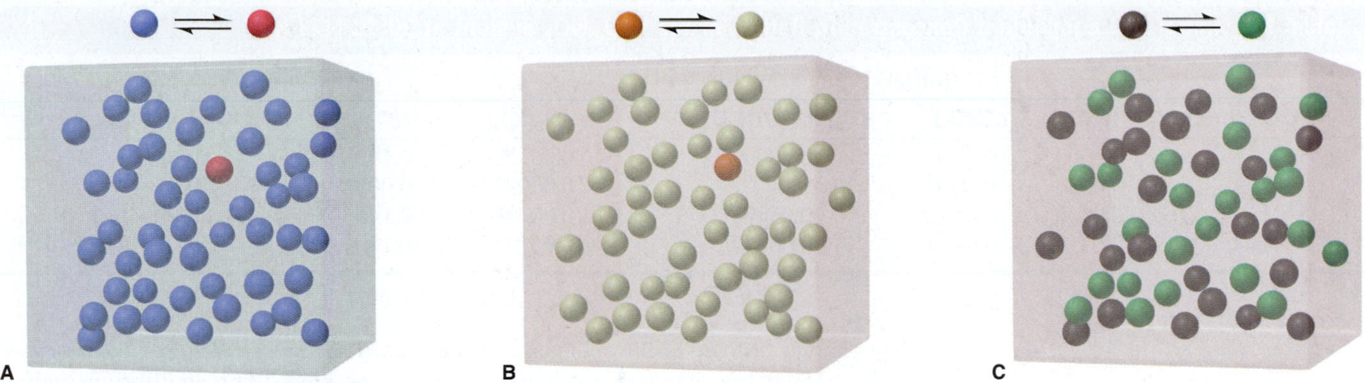

Figure 17.2 The range of equilibrium constants. **A,** For this reaction, $K = 1/49 = 0.020$. **B,** For this reaction, $K = 49/1 = 49$. **C,** For this reaction, $K = 25/25 = 1.0$.

- *Intermediate K* (Figure 17.2C). When significant amounts of both reactant and product are present at equilibrium, K has an intermediate value, as when bromine monochloride breaks down to its elements at 1000 K:

$$2BrCl(g) \rightleftharpoons Br_2(g) + Cl_2(g) \qquad K = 5$$

Summary of Section 17.1

- Kinetics and equilibrium are distinct aspects of a reaction system, and the rate and extent of a reaction are not necessarily related.
- When the forward and reverse reactions occur at the same rate, concentrations no longer change and the system has reached equilibrium.
- The equilibrium constant (K) is a number equal to a particular ratio of product to reactant concentrations at a certain temperature: K is small if a high concentration of reactant(s) is present at equilibrium, and it is large if a high concentration of product(s) is present at equilibrium.

17.2 • THE REACTION QUOTIENT AND THE EQUILIBRIUM CONSTANT

We have introduced the equilibrium constant in terms of a ratio of rate constants, but the original research on chemical equilibrium was developed many years before the principles of kinetics. In 1864, two Norwegian chemists, Cato Guldberg and Peter Waage, observed that *at a given temperature, a chemical system reaches a state in which a particular ratio of product to reactant concentrations has a constant value.* This is a statement of the **law of chemical equilibrium,** or the **law of mass action.**

Changing Value of the Reaction Quotient

The particular ratio of concentration terms that we write for a given reaction is called the **reaction quotient** (**Q,** also known as the *mass-action expression*). For the reversible breakdown of N_2O_4 to NO_2, the reaction quotient, which is based directly on the balanced equation, is

$$N_2O_4(g) \rightleftharpoons 2NO_2(g) \qquad Q = \frac{[NO_2]^2}{[N_2O_4]}$$

As the reaction proceeds toward equilibrium, the concentrations of reactants and products change continually, and so does their ratio, the value of Q: at a given temperature, at the beginning of the reaction, the concentrations have initial values, and Q has an initial value; a moment later, the concentrations have slightly different values, and so does Q; after another moment, the concentrations and the value

Table 17.1 Initial and Equilibrium Concentration Ratios for the N_2O_4-NO_2 System at 200°C (473 K)

	Initial			Equilibrium		
Experiment	$[N_2O_4]$	$[NO_2]$	Q, $[NO_2]^2/[N_2O_4]$	$[N_2O_4]_{eq}$	$[NO_2]_{eq}$	K, $[NO_2]^2_{eq}/[N_2O_4]_{eq}$
1	0.1000	0.0000	0.0000	0.00357	0.193	10.4
2	0.0000	0.1000	∞	0.000924	0.0982	10.4
3	0.0500	0.0500	0.0500	0.00204	0.146	10.4
4	0.0750	0.0250	0.00833	0.00275	0.170	10.5

of Q change further. These changes continue, *until the system reaches equilibrium.* At that point, reactant and product concentrations have their equilibrium values and no longer change. Thus, the value of Q no longer changes and equals K at that temperature:

$$\text{At equilibrium: } Q = K \tag{17.3}$$

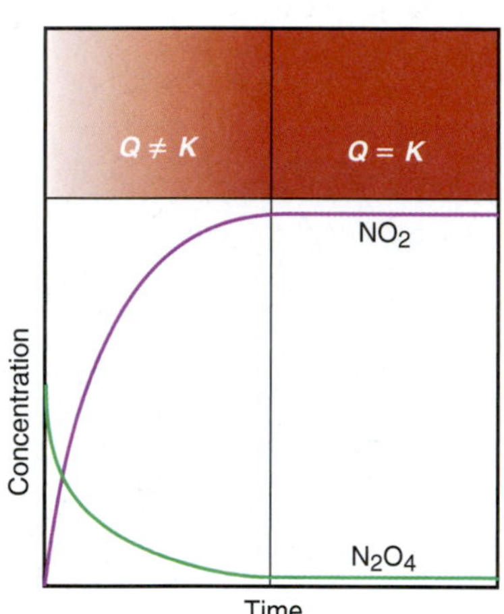

Figure 17.3 The change in Q during the N_2O_4-NO_2 reaction. Based on data from Experiment 1 in Table 17.1.

In formulating the law of mass action, Guldberg and Waage found that, *for a particular system and temperature, the same equilibrium state is attained regardless of starting concentrations.* For example, for the N_2O_4-NO_2 system at 200°C, the data from four experiments appear in Table 17.1. Two essential points stand out:

- The ratio of *initial* concentrations *(fourth column)* varies widely but always gives the same ratio of *equilibrium* concentrations *(rightmost column).*
- The *individual* equilibrium concentrations are different in each case, but the *ratio* of these equilibrium concentrations is constant.

Thus, monitoring Q tells whether the system has reached equilibrium or, if it hasn't, how far away it is and in which direction it is changing.

The curves in Figure 17.3 show experiment 1 in Table 17.1. Note that $[N_2O_4]$ and $[NO_2]$ change smoothly during the course of the reaction (as indicated by the changing brown color at the top), and so does the value of Q. Once the system reaches equilibrium (constant brown color), the concentrations no longer change and Q equals K. In other words, for any given chemical system, *K is a special value of Q that occurs when the reactant and product concentrations have their equilibrium values.*

Writing the Reaction Quotient

In Chapter 16, you saw that the rate law for an overall reaction *cannot* be written from the balanced equation, but must be determined from rate data. In contrast, the reaction quotient *must* be written directly from the balanced equation.

Constructing the Reaction Quotient The most common form of the reaction quotient shows reactant and product terms as molar concentrations, which are designated by square brackets, [ ]. Thus, from now on, we designate the *reaction quotient based on concentrations* as Q_c. (We also designate the *equilibrium constant based on concentrations* as K_c.) For the general equation

$$a\text{A} + b\text{B} \rightleftharpoons c\text{C} + d\text{D}$$

where a, b, c, and d are the stoichiometric coefficients, the reaction quotient is

$$Q_c = \frac{[\text{C}]^c[\text{D}]^d}{[\text{A}]^a[\text{B}]^b} \tag{17.4}$$

Thus, *Q is a ratio of product concentration terms multiplied together and divided by reactant concentration terms multiplied together, with each term raised to the power of its balancing coefficient.*

Two steps are needed to write a reaction quotient for any chemical system:

1. *Start with the balanced equation.* For example, for the formation of ammonia from its elements, the balanced equation (with colored coefficients for reference) is

$$N_2(g) + 3H_2(g) \rightleftharpoons 2NH_3(g)$$

2. *Arrange the terms and exponents.* Place the product terms in the numerator and the reactant terms in the denominator, multiplied by each other, and raise each term to the power of its balancing coefficient (colored as in the balanced equation):

$$Q_c = \frac{[NH_3]^2}{[N_2][H_2]^3}$$

Sample Problem 17.1 Writing the Reaction Quotient from the Balanced Equation

Problem Write the reaction quotient, Q_c, for each of the following reactions:
(a) The decomposition of dinitrogen pentoxide, $N_2O_5(g) \rightleftharpoons NO_2(g) + O_2(g)$
(b) The combustion of propane, $C_3H_8(g) + O_2(g) \rightleftharpoons CO_2(g) + H_2O(g)$

Plan We balance the equations and then construct the reaction quotient (Equation 17.4).

Solution **(a)** $2N_2O_5(g) \rightleftharpoons 4NO_2(g) + O_2(g)$

$$Q_c = \frac{[NO_2]^4[O_2]}{[N_2O_5]^2}$$

(b) $C_3H_8(g) + 5O_2(g) \rightleftharpoons 3CO_2(g) + 4H_2O(g)$

$$Q_c = \frac{[CO_2]^3[H_2O]^4}{[C_3H_8][O_2]^5}$$

Check Be sure that the exponents in Q_c are the same as the balancing coefficients. A good check is to reverse the process and see if you obtain the balanced equation: change the numerator to products, the denominator to reactants, and the exponents to coefficients.

FOLLOW-UP PROBLEM 17.1 Write the reaction quotient, Q_c, for each of the following reactions:
(a) The first step in nitric acid production, $NH_3(g) + O_2(g) \rightleftharpoons NO(g) + H_2O(g)$
(b) The disproportionation of nitrogen monoxide, $NO(g) \rightleftharpoons N_2O(g) + NO_2(g)$

Why *Q* and *K* Are Unitless In this text (and most others), *the values of Q and K are unitless numbers.* The reason is that each term in Q represents the *ratio* of the quantity of the substance (molar concentration or pressure) to its thermodynamic standard-state quantity. Recall from Section 6.6 that the standard states are 1 *M* for a substance in solution, 1 atm for a gas, and the pure substance for a liquid or solid. Thus, a concentration of 1.20 *M* becomes

$$\frac{1.20\ M\ \text{(measured quantity)}}{1\ M\ \text{(standard-state quantity)}} = 1.20$$

Similarly, a pressure of 0.53 atm becomes $\frac{0.53\ \text{atm}}{1\ \text{atm}} = 0.53$. Since the quantity terms are unitless, their ratio, which gives the value of Q (or K), is also unitless.

Writing *Q* for an Overall Reaction We follow the same procedure for writing a reaction quotient whether an equation represents an individual reaction step or an overall multistep reaction. *If an overall reaction is the* **sum** *of two or more reactions, the overall reaction quotient (or equilibrium constant) is the* **product** *of the reaction quotients (or equilibrium constants) for the steps:*

$$Q_{overall} = Q_1 \times Q_2 \times Q_3 \times \cdots$$

and

$$K_{overall} = K_1 \times K_2 \times K_3 \times \cdots \qquad (17.5)$$

Sample Problem 17.2 demonstrates this point.

Sample Problem 17.2 Writing the Reaction Quotient and Finding *K* for an Overall Reaction

Problem Nitrogen dioxide is a toxic pollutant that contributes to photochemical smog. One way it forms is through the following sequence:

(1) $N_2(g) + O_2(g) \rightleftharpoons 2NO(g)$ $\quad K_{c1} = 4.3\times10^{-25}$

(2) $2NO(g) + O_2(g) \rightleftharpoons 2NO_2(g)$ $\quad K_{c2} = 6.4\times10^{9}$

(a) Show that the overall Q_c for this reaction sequence is the same as the product of the Q_c's for the individual reactions.
(b) Given that both reactions occur at the same temperature, find K_c for the overall reaction.

Plan **(a)** We first write the overall reaction by adding the individual reactions and then write the overall Q_c. Next, we write the Q_c for each step. We *add* the steps and *multiply* their Q_c's, canceling common terms, to obtain the overall Q_c. **(b)** We know the individual K_c's (4.3×10^{-25} and 6.4×10^{9}), so we multiply them to find $K_{c(overall)}$.

Solution **(a)** Writing the overall reaction and its reaction quotient:

$$\begin{array}{rl} (1) & N_2(g) + O_2(g) \rightleftharpoons \cancel{2NO(g)} \\ (2) & \cancel{2NO(g)} + O_2(g) \rightleftharpoons 2NO_2(g) \\ \hline \text{Overall:} & N_2(g) + 2O_2(g) \rightleftharpoons 2NO_2(g) \end{array}$$

$$Q_{c(overall)} = \frac{[NO_2]^2}{[N_2][O_2]^2}$$

Writing the reaction quotients for the individual steps:

For step 1, $$Q_{c1} = \frac{[NO]^2}{[N_2][O_2]}$$

For step 2, $$Q_{c2} = \frac{[NO_2]^2}{[NO]^2[O_2]}$$

Multiplying the individual reaction quotients and canceling:

$$Q_{c1} \times Q_{c2} = \frac{\cancel{[NO]^2}}{[N_2][O_2]} \times \frac{[NO_2]^2}{\cancel{[NO]^2}[O_2]} = \frac{[NO_2]^2}{[N_2][O_2]^2} = Q_{c(overall)}$$

(b) Calculating the overall K_c:

$$K_{c(overall)} = K_{c1} \times K_{c2} = (4.3\times10^{-25})(6.4\times10^{9}) = 2.8\times10^{-15}$$

Check Round off and check the calculation in part (b):

$$K_c \approx (4\times10^{-25})(6\times10^{9}) = 24\times10^{-16} = 2.4\times10^{-15}$$

FOLLOW-UP PROBLEM 17.2 The following sequence of steps has been proposed for the overall reaction between H_2 and Br_2 to form HBr:

(1) $Br_2(g) \rightleftharpoons 2Br(g)$

(2) $Br(g) + H_2(g) \rightleftharpoons HBr(g) + H(g)$

(3) $H(g) + Br(g) \rightleftharpoons HBr(g)$

Write the overall balanced equation and show that the overall Q_c is the product of the Q_c's for the individual steps.

Writing *Q* for a Forward and a Reverse Reaction A given reaction quotient depends on the *direction* in which the balanced equation is written. Consider, for example, the oxidation of sulfur dioxide to sulfur trioxide. This reaction is a key step in acid rain formation and sulfuric acid production. The balanced equation is

$$2SO_2(g) + O_2(g) \rightleftharpoons 2SO_3(g)$$

The reaction quotient for this equation *as written* is

$$Q_{c(fwd)} = \frac{[SO_3]^2}{[SO_2]^2[O_2]}$$

If we had written the reverse reaction, the decomposition of sulfur trioxide,

$$2SO_3(g) \rightleftharpoons 2SO_2(g) + O_2(g)$$

the reaction quotient would be the *reciprocal* of $Q_{c(fwd)}$:

$$Q_{c(rev)} = \frac{[SO_2]^2[O_2]}{[SO_3]^2} = \frac{1}{Q_{c(fwd)}}$$

Thus, *a reaction quotient (or equilibrium constant) for a forward reaction is the* ***reciprocal*** *of the reaction quotient (or equilibrium constant) for the reverse reaction:*

$$Q_{c(fwd)} = \frac{1}{Q_{c(rev)}} \quad \text{and} \quad K_{c(fwd)} = \frac{1}{K_{c(rev)}} \tag{17.6}$$

The K_c values for the forward and reverse reactions shown above at 1000 K are

$$K_{c(fwd)} = 261 \quad \text{and} \quad K_{c(rev)} = \frac{1}{K_{c(fwd)}} = \frac{1}{261} = 3.83\times10^{-3}$$

These values make sense: if the forward reaction goes far to the right (high K_c), the reverse reaction does not (low K_c).

Writing Q for a Reaction with Coefficients Multiplied by a Common Factor

Multiplying all the coefficients of the equation by some factor also changes Q. For example, multiplying all the coefficients in the equation for the formation of SO_3 by $\frac{1}{2}$ gives

$$SO_2(g) + \tfrac{1}{2}O_2(g) \rightleftharpoons SO_3(g)$$

For this equation, the reaction quotient is

$$Q'_{c(fwd)} = \frac{[SO_3]}{[SO_2][O_2]^{1/2}}$$

Notice that Q_c for the halved equation equals Q_c for the original equation raised to the $\frac{1}{2}$ power:

$$Q'_{c(fwd)} = Q^{1/2}_{c(fwd)} = \left(\frac{[SO_3]^2}{[SO_2]^2[O_2]}\right)^{1/2} = \frac{[SO_3]}{[SO_2][O_2]^{1/2}}$$

Once again, the same property holds for the equilibrium constants. Relating the halved reaction to the original, we have

$$K'_{c(fwd)} = K^{1/2}_{c(fwd)} = (261)^{1/2} = 16.2$$

In general, *if all the coefficients of the balanced equation are multiplied by some factor, that factor becomes the exponent for relating the reaction quotients and the equilibrium constants.* For a multiplying factor n, which we can write as

$$n(aA + bB \rightleftharpoons cC + dD)$$

the reaction quotient and equilibrium constant are

$$Q' = Q^n = \left(\frac{[C]^c[D]^d}{[A]^a[B]^b}\right)^n \quad \text{and} \quad K' = K^n \tag{17.7}$$

Writing Q for a Reaction Involving Pure Liquids or Solids Until now, we've looked at reactions involving only gases. These are *homogeneous* equilibria, systems in which all the components are in the same phase. When the components are in different phases, the system reaches *heterogeneous* equilibrium.

Consider the decomposition of limestone to lime and carbon dioxide:

$$CaCO_3(s) \rightleftharpoons CaO(s) + CO_2(g)$$

Based on the rules for writing the reaction quotient, we have

$$Q_c = \frac{[CaO][CO_2]}{[CaCO_3]}$$

However, like its density, at a given temperature, a pure solid always has the same "concentration," that is, the same number of moles per liter of solid. Therefore, the concentration of a pure solid is constant, as is the concentration of a pure liquid.

Table 17.2 Ways of Expressing Q and Calculating K

Form of Chemical Equations	Form of Q	Value of K
Reference reaction: $A \rightleftharpoons B$	$Q_{(ref)} = \frac{[B]}{[A]}$	$K_{(ref)} = \frac{[B]_{eq}}{[A]_{eq}}$
Reference reaction: $B \rightleftharpoons A$	$Q = \frac{1}{Q_{(ref)}} = \frac{[A]}{[B]}$	$K = \frac{1}{K_{(ref)}}$
Reaction as sum of two steps: (1) $A \rightleftharpoons C$	$Q_1 = \frac{[C]}{[A]}; Q_2 = \frac{[B]}{[C]}$	
(2) $C \rightleftharpoons B$	$Q_{overall} = Q_1 \times Q_2 = Q_{(ref)}$ $= \frac{\cancel{[C]}}{[A]} \times \frac{[B]}{\cancel{[C]}} = \frac{[B]}{[A]}$	$K_{overall} = K_1 \times K_2$ $= K_{(ref)}$
Coefficients multiplied by n	$Q = Q^n_{(ref)}$	$K = K^n_{(ref)}$
Reaction with pure solid or liquid component, such as A(*s*)	$Q = Q_{(ref)}[A] = [B]$	$K = K_{(ref)}[A] = [B]$

Because we are concerned with concentrations that *change* as they approach equilibrium, *we eliminate the terms for pure liquids and solids from the reaction quotient.* To do this, we multiply both sides of the above equation by $[CaCO_3]$ and divide both sides by [CaO] and get a new reaction quotient, Q'_c:

$$Q'_c = Q_c\frac{[CaCO_3]}{[CaO]} = [CO_2]$$

Thus, only the CO_2 concentration changes, so no matter what the amounts of CaO and $CaCO_3$, the reaction quotient equals the CO_2 concentration.

Table 17.2 summarizes the ways of writing Q and calculating K.

Summary of Section 17.2

- The reaction quotient based on concentration, Q_c, is a particular ratio of product to reactant concentrations. The value of Q_c changes as the reaction proceeds. When the system reaches equilibrium at a particular temperature, $Q_c = K_c$.
- If a reaction is the sum of two or more steps, the overall Q (or K) is the product of the individual Q's (or K's).
- Q is based on the balanced equation exactly as written, so it changes if the equation is reversed or multiplied by some factor, and K changes accordingly.
- The concentrations of pure liquids or solids do not appear in Q because they are constant.
- There are three criteria that define a system at equilibrium:
  1. Reactant and product concentrations are constant over time.
  2. The opposing reaction rates are equal: $rate_{fwd} = rate_{rev}$.
  3. The reaction quotient equals the equilibrium constant: $Q = K$.

17.3 • EXPRESSING EQUILIBRIA WITH PRESSURE TERMS: RELATION BETWEEN K_c AND K_p

It is easier to measure gas pressure than concentration, so when a reaction involves gases, we often express the reaction quotient in terms of partial pressures instead of concentrations. As long as the gases behave nearly ideally during the experiment, the ideal gas law (Section 5.3) allows us to relate pressure (P) to concentration (n/V):

$$PV = nRT, \quad \text{so} \quad P = \frac{n}{V}RT \quad \text{or} \quad \frac{P}{RT} = \frac{n}{V}$$

Thus, at constant T, *pressure is directly proportional to molar concentration.* For example, in the reaction between gaseous NO and O_2,

$$2NO(g) + O_2(g) \rightleftharpoons 2NO_2(g)$$

the *reaction quotient based on pressures,* Q_p, is

$$Q_p = \frac{P^2_{NO_2}}{P^2_{NO} \times P_{O_2}}$$

The equilibrium constant obtained when all components are present at their equilibrium partial pressures is designated K_p, the *equilibrium constant based on pressures*. All the relationships in Table 17.2 hold if we substitute partial pressures for concentrations.

In many cases, K_p has a value different from K_c. But if you know one, the *change in amount (mol) of gas,* Δn_{gas}, from the balanced equation allows you to calculate the other. Let's see how this conversion works for the oxidation of NO:

$$2NO(g) + O_2(g) \rightleftharpoons 2NO_2(g)$$

As the balanced equation shows,

$$3 \text{ mol } (2 \text{ mol} + 1 \text{ mol}) \text{ gaseous reactants} \rightleftharpoons 2 \text{ mol gaseous products}$$

With Δ meaning final *minus* initial (products *minus* reactants), we have

$$\Delta n_{gas} = \text{moles of gaseous product} - \text{moles of gaseous reactant} = 2 - 3 = -1$$

Keep this value of Δn_{gas} in mind because it appears in the conversion that follows.

The reaction quotient based on concentrations is

$$Q_c = \frac{[NO_2]^2}{[NO]^2[O_2]}$$

Rearranging the ideal gas law to $n/V = P/RT$, we write the terms in square brackets as n/V and convert them to partial pressures, P; then we collect the RT terms and cancel:

$$Q_c = \frac{\dfrac{n^2_{NO_2}}{V^2}}{\dfrac{n^2_{NO}}{V^2} \times \dfrac{n_{O_2}}{V}} = \frac{\dfrac{P^2_{NO_2}}{(RT)^2}}{\dfrac{P^2_{NO}}{(RT)^2} \times \dfrac{P_{O_2}}{RT}} = \frac{P^2_{NO_2}}{P^2_{NO} \times P_{O_2}} \times \frac{\dfrac{1}{\cancel{(RT)^2}}}{\dfrac{1}{\cancel{(RT)^2}} \times \dfrac{1}{RT}} = \frac{P^2_{NO_2}}{P^2_{NO} \times P_{O_2}} \times RT$$

Note that the far right side of the previous expression is Q_p multiplied by RT. Thus,

$$Q_c = Q_p(RT)$$

Also, at equilibrium, $K_c = K_p(RT)$; thus,

$$K_p = \frac{K_c}{RT} \quad \text{or} \quad K_c(RT)^{-1}$$

Note especially that *the exponent of the RT term equals the change in the amount (mol) of gas (Δn_{gas}) from the balanced equation,* −1. Thus, in general, we have

$$K_p = K_c(RT)^{\Delta n_{gas}} \qquad \textbf{(17.8)}$$

Based on Equation 17.8, *if the amount (mol) of gas does not change in the reaction,* $\Delta n_{gas} = 0$, so the RT term drops out (i.e., equals 1), and $K_p = K_c$. (In calculations, be sure the units for R are consistent with the units for pressure.)

Sample Problem 17.3 Converting Between K_c and K_p

Problem A chemical engineer injects limestone ($CaCO_3$) into the hot flue gas of a coal-burning power plant to form lime (CaO), which scrubs SO_2 from the gas and forms gypsum ($CaSO_4 \cdot 2H_2O$). Find K_c for the following reaction:

$$CaCO_3(s) \rightleftharpoons CaO(s) + CO_2(g) \qquad K_p = 2.1\times10^{-4} \text{ (at 1000. K)}$$

Plan We know K_p (2.1×10^{-4}), so to convert between K_p and K_c, we must first determine Δn_{gas} from the balanced equation. Then we rearrange Equation 17.8. (If we assume that pressure is measured in atmospheres, $R = 0.0821$ atm·L/mol·K.)

Solution Determining Δn_{gas}: There is 1 mol of gaseous product and no gaseous reactant, so $\Delta n_{gas} = 1 - 0 = 1$.

Rearranging Equation 17.8 and calculating K_c:

$$K_p = K_c(RT)^1$$

so

$$K_c = K_p(RT)^{-1} = (2.1\times10^{-4})(0.0821 \times 1000.)^{-1} = 2.6\times10^{-6}$$

Check Rounding gives

$$(2\times10^{-4})(0.1\times10^{3})^{-1} = (2\times10^{-4})(10^{-2}) = 2\times10^{-6}$$

FOLLOW-UP PROBLEM 17.3 Calculate K_p for the following reaction:

$$PCl_3(g) + Cl_2(g) \rightleftharpoons PCl_5(g) \qquad K_c = 1.67 \text{ (at 500. K)}$$

Summary of Section 17.3

- For a gaseous reaction, the reaction quotient and the equilibrium constant can be expressed in terms of partial pressures (Q_p and K_p).
- If you know K_p, you can find K_c, and vice versa: $K_p = K_c(RT)^{\Delta n_{gas}}$.

17.4 • COMPARING *Q* AND *K* TO PREDICT REACTION DIRECTION

Suppose you have a mixture of reactants and products and you know K at the temperature of the reaction. By comparing the value of Q with K, you can predict whether the reaction has reached equilibrium or, if not, in which direction it will proceed. With product terms in the numerator of Q and reactant terms in the denominator, *more product makes Q larger, and more reactant makes Q smaller.*

There are three possibilities for the relative sizes of Q and K (Figure 17.4):

- $Q < K$. If Q is smaller than K, the denominator (reactants) is large relative to the numerator (products). For Q to equal K, the reactants must decrease and the products increase. The reaction will proceed to the right, toward products:

$$\text{If } Q < K, \text{ reactants} \longrightarrow \text{products}$$

- $Q > K$. If Q is larger than K, the numerator (products) will decrease and the denominator (reactants) will increase. The reaction will proceed to the left, toward reactants:

$$\text{If } Q > K, \text{ reactants} \longleftarrow \text{products}$$

- $Q = K$. This situation occurs when the reactant and product terms equal their equilibrium values. No further net change takes place:

$$\text{If } Q = K, \text{ reactants} \rightleftharpoons \text{products}$$

Sample Problem 17.4 relies on molecular scenes to predict reaction direction, and Sample Problem 17.5 relies on concentration data.

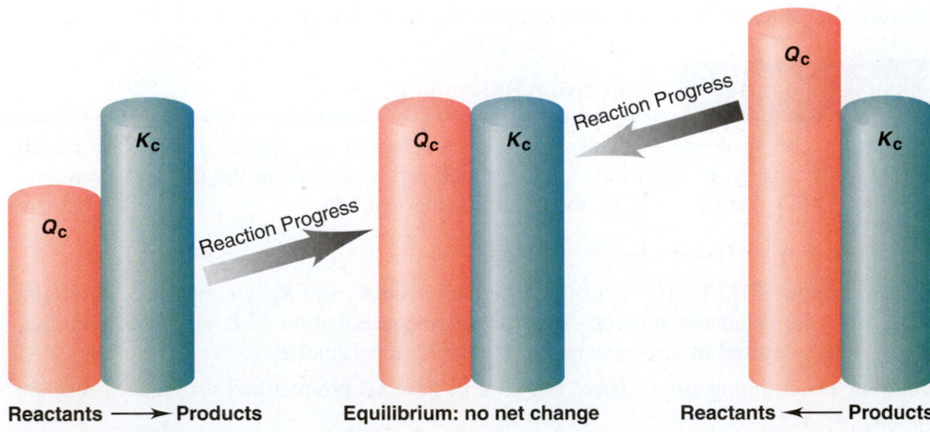

Figure 17.4 Reaction direction and the relative sizes of *Q* and *K*. When Q_c is smaller or larger than K_c, the reaction continues until $Q_c = K_c$. Note that K_c remains the same throughout.

Sample Problem 17.4 Using Molecular Scenes to Predict Reaction Direction

Problem For the reaction $A(g) \rightleftharpoons B(g)$, the equilibrium mixture at 175°C is [A] = 2.8×10^{-4} *M* and [B] = 1.2×10^{-4} *M*. The molecular scenes below represent mixtures at various times during runs 1–4 of this reaction (A is *red*; B is *blue*). Will the reaction proceed to the right or left or not at all for each mixture to reach equilibrium?

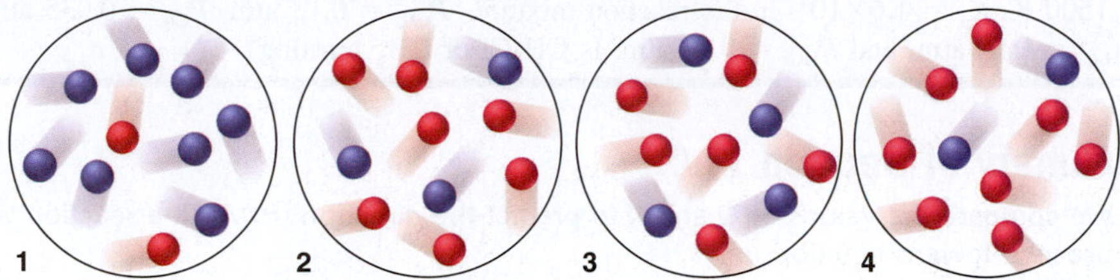

Plan We must compare Q_c with K_c to predict the reaction direction, so we first use the given equilibrium concentrations to find K_c. Then we count spheres and calculate Q_c for each mixture. If $Q_c < K_c$, the reaction proceeds to the right (reactants to products); if $Q_c > K_c$, it proceeds to the left (products to reactants); and if $Q_c = K_c$, there is no further net change.

Solution Writing the reaction quotient and using the data to find K_c:

$$Q_c = \frac{[B]}{[A]} = \frac{1.2\times10^{-4}}{2.8\times10^{-4}} = 0.43 = K_c$$

Counting red (A) and blue (B) spheres to calculate Q_c for each mixture:

1. $Q_c = 8/2 = 4.0$ 2. $Q_c = 3/7 = 0.43$ 3. $Q_c = 4/6 = 0.67$ 4. $Q_c = 2/8 = 0.25$

Comparing Q_c with K_c to predict reaction direction:

1. $Q_c > K_c$: left 2. $Q_c = K_c$: no net change 3. $Q_c > K_c$: left 4. $Q_c < K_c$: right

Check Making an error in the calculation for K_c would lead to incorrect conclusions throughout, so check that step: the exponents are the same, and 1.2/2.8 is a bit less than 0.5, as is the calculated K_c. You can check the final answers by inspection; for example, for the number of B (8) in mixture 1 to equal the number at equilibrium (3), more B must change to A, so the reaction must proceed to the left.

FOLLOW-UP PROBLEM 17.4 At 338 K, the reaction $X(g) \rightleftharpoons Y(g)$ has a K_c of 1.4. The scenes represent different mixtures at 338 K, with X orange and Y green. In which direction will the reaction proceed (if at all) for each mixture to reach equilibrium?

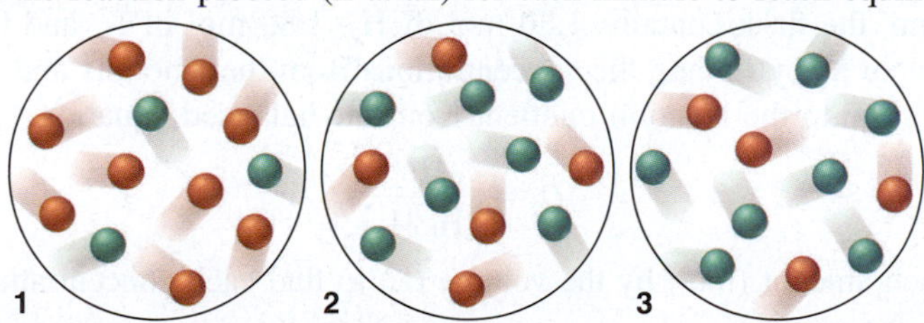

Sample Problem 17.5 Using Concentrations to Predict Reaction Direction

Problem For the reaction $N_2O_4(g) \rightleftharpoons 2NO_2(g)$, $K_c = 0.21$ at 100°C. At a point during the reaction, $[N_2O_4] = 0.12$ *M* and $[NO_2] = 0.55$ *M*. Is the reaction at equilibrium? If not, in which direction will it proceed?

Plan We write Q_c, find its value by substituting the given concentrations, and compare its value with the given K_c.

Solution Writing the reaction quotient and solving for Q_c:

$$Q_c = \frac{[NO_2]^2}{[N_2O_4]} = \frac{0.55^2}{0.12} = 2.5$$

With $Q_c > K_c$, the reaction is not at equilibrium and will proceed to the left until $Q_c = K_c$.

Check With $[NO_2] > [N_2O_4]$, we expect to obtain a value for Q_c that is greater than 0.21. If $Q_c > K_c$, the numerator will decrease and the denominator will increase until $Q_c = K_c$; that is, this reaction will proceed toward reactants.

FOLLOW-UP PROBLEM 17.5 Chloromethane forms by the reaction

$$CH_4(g) + Cl_2(g) \rightleftharpoons CH_3Cl(g) + HCl(g)$$

At 1500 K, $K_p = 1.6\times10^4$. In the reaction mixture, $P_{CH_4} = 0.13$ atm, $P_{Cl_2} = 0.035$ atm, $P_{CH_3Cl} = 0.24$ atm, and $P_{HCl} = 0.47$ atm. Is CH_3Cl or CH_4 forming?

Summary of Section 17.4

- We compare the values of Q and K to predict the direction in which a reaction will proceed toward equilibrium.
  - If $Q_c < K_c$, more product forms.
  - If $Q_c > K_c$, more reactant forms.
  - If $Q_c = K_c$, there is no net change.

17.5 • HOW TO SOLVE EQUILIBRIUM PROBLEMS

Many kinds of equilibrium problems arise in the real world—and on chemistry exams—but we can group most of them into two types:

1. We know the equilibrium quantities and solve for K.
2. We know K and the initial quantities and solve for the equilibrium quantities.

Using Quantities to Find the Equilibrium Constant

There are two common variations on this type of problem: one involves substituting quantities to solve for K, and the other requires first finding some of the quantities.

Substituting Equilibrium Quantities into Q to Find K In this type of problem, we use given equilibrium quantities to calculate K. Suppose, for example, that equal amounts of gaseous hydrogen and iodine are injected into a 1.50-L flask at a fixed temperature. In time, the following equilibrium is attained:

$$H_2(g) + I_2(g) \rightleftharpoons 2HI(g)$$

At equilibrium, the flask contains 1.80 mol of H_2, 1.80 mol of I_2, and 0.520 mol of HI. We calculate K_c by finding the concentrations from the amounts and flask volume and substituting into the reaction quotient from the balanced equation:

$$Q_c = \frac{[HI]^2}{[H_2][I_2]}$$

We divide each amount (mol) by the volume (L) to find each concentration (mol/L):

$$[H_2] = \frac{1.80\text{ mol}}{1.50\text{ L}} = 1.20\ M$$

Similarly, $[I_2] = 1.20\ M$, and $[HI] = 0.347\ M$. Substituting these values into the expression for Q_c gives K_c:

$$K_c = \frac{(0.347)^2}{(1.20)(1.20)} = 8.36\times10^{-2}$$

Using a Reaction Table to Find Equilibrium Quantities and K When some quantities are not given, we determine them from the reaction stoichiometry and then find K. In the following example, note the use of the *reaction table*.

In a study of carbon oxidation, an evacuated vessel containing a small amount of powdered graphite is heated to 1080 K. Gaseous CO_2 is added to a pressure of 0.458 atm, and CO forms. At equilibrium, the total pressure is 0.757 atm. Calculate K_p.

As always, we start by writing the balanced equation and the reaction quotient:

$$CO_2(g) + C(\text{graphite}) \rightleftharpoons 2CO(g)$$

The data are given in atmospheres and we must find K_p, so we write the expression for Q_p; note that it does *not* include a term for the solid, C(graphite):

$$Q_p = \frac{P_{CO}^2}{P_{CO_2}}$$

We are given the initial P_{CO_2} and the P_{total} at equilibrium. To find K_p, we must find the equilibrium pressures of CO_2 and CO and then substitute them into Q_p.

Let's think through what happened in the vessel. An unknown portion of the CO_2 reacted with graphite to form an unknown amount of CO. We already know the *relative* amounts of CO_2 and CO from the balanced equation: for each mole of CO_2 that reacts, 2 mol of CO forms, which means that when x atm of CO_2 reacts, $2x$ atm of CO forms:

$$x \text{ atm } CO_2 \longrightarrow 2x \text{ atm CO}$$

The pressure of CO_2 at equilibrium, $P_{CO_2(eq)}$, is the initial pressure, $P_{CO_2(init)}$, *minus* x, the CO_2 that reacts (the change in P_{CO_2} due to the reaction):

$$P_{CO_2(init)} - x = P_{CO_2(eq)} = 0.458 - x$$

Similarly, the pressure of CO at equilibrium, $P_{CO(eq)}$, is the initial pressure, $P_{CO(init)}$, *plus* $2x$, the CO that forms (the change in P_{CO} due to the reaction). Because $P_{CO(init)}$ is zero, we have

$$P_{CO(init)} + 2x = 0 + 2x = 2x = P_{CO(eq)}$$

We can summarize this information in a reaction table, similar to the tables introduced in Chapter 3, but for the "Final" quantities, we use those at equilibrium. The table shows the balanced equation and

- the *initial* quantities (concentrations or pressures) of reactants and products
- the *changes* in these quantities during the reaction
- the *equilibrium* quantities

Pressure (atm)	**$CO_2(g)$**	**+**	**C(graphite)**	**$\rightleftharpoons$**	**$2CO(g)$**
Initial	0.458		—		0
Change	$-x$		—		$+2x$
Equilibrium	$0.458 - x$		—		$2x$

We treat each column like a list of numbers to add together: *the initial quantity plus the change in that quantity gives the equilibrium quantity.* Note that we *only* include substances whose concentrations change; thus, the column for C(graphite) is blank.

To solve for K_p, we substitute equilibrium values into Q_p, so we first have to find x. To do this, we use the other piece of data given, $P_{total(eq)}$. According to Dalton's law of partial pressures (Section 5.4) and using the equilibrium quantities from the reaction table,

$$P_{total(eq)} = 0.757 \text{ atm} = P_{CO_2(eq)} + P_{CO(eq)} = (0.458 \text{ atm} - x) + 2x$$

Thus, $\quad 0.757 \text{ atm} = 0.458 \text{ atm} + x \quad$ and $\quad x = 0.299 \text{ atm}$

With x known, we determine the equilibrium partial pressures:

$$P_{CO_2(eq)} = 0.458 \text{ atm} - x = 0.458 \text{ atm} - 0.299 \text{ atm} = 0.159 \text{ atm}$$

$$P_{CO(eq)} = 2x = 2(0.299 \text{ atm}) = 0.598 \text{ atm}$$

Then we substitute them into Q_p to find K_p:

$$Q_p = \frac{P_{CO(eq)}^2}{P_{CO_2(eq)}} = \frac{0.598^2}{0.159} = 2.25 = K_p$$

(From now on, the subscripts "init" and "eq" appear only when it is not clear whether a quantity is an initial or equilibrium value.)

Sample Problem 17.6 Calculating K_c from Concentration Data

Problem In order to study hydrogen halide decomposition, a researcher fills an evacuated 2.00-L flask with 0.200 mol of HI gas and allows the reaction to proceed at 453°C:

$$2HI(g) \rightleftharpoons H_2(g) + I_2(g)$$

At equilibrium, [HI] = 0.078 *M*. Calculate K_c.

Plan To calculate K_c, we need the equilibrium concentrations. We can find the initial [HI] from the amount (0.200 mol) and the flask volume (2.00 L), and we are given [HI] at equilibrium (0.078 *M*). From the balanced equation, when $2x$ mol of HI reacts, x mol of H_2 and x mol of I_2 form. We set up a reaction table, use the known [HI] at equilibrium to solve for x (the change in $[H_2]$ or $[I_2]$), and substitute the concentrations into Q_c.

Solution Calculating initial [HI]:

$$[HI] = \frac{0.200 \text{ mol}}{2.00 \text{ L}} = 0.100\ M$$

Setting up the reaction table, with $x = [H_2]$ or $[I_2]$ that forms and $2x$ = [HI] that reacts:

Concentration (*M*)	2HI(*g*)	$\rightleftharpoons$	$H_2(g)$	+	$I_2(g)$
Initial	0.100		0		0
Change	$-2x$		$+x$		$+x$
Equilibrium	$0.100 - 2x$		x		x

Solving for x, using the known [HI] at equilibrium:

$$[HI] = 0.100\ M - 2x = 0.078\ M \qquad \text{so} \qquad x = 0.011\ M$$

Therefore, the equilibrium concentrations are

$$[H_2] = [I_2] = 0.011\ M \qquad \text{and we are given} \qquad [HI] = 0.078\ M$$

Substituting into the reaction quotient:

$$Q_c = \frac{[H_2][I_2]}{[HI]^2}$$

Thus,

$$K_c = \frac{(0.011)(0.011)}{0.078^2} = 0.020$$

Check Rounding gives $\sim 0.01^2/0.08^2 = 0.02$. Because the initial [HI] of 0.100 *M* fell slightly at equilibrium to 0.078 *M*, relatively little product formed; so we expect $K_c < 1$.

FOLLOW-UP PROBLEM 17.6 The atmospheric oxidation of nitrogen monoxide, $2NO(g) + O_2(g) \rightleftharpoons 2NO_2(g)$, was studied at 184°C with initial pressures of 1.000 atm of NO and 1.000 atm of O_2. At equilibrium, $P_{O_2} = 0.506$ atm. Calculate K_p.

Using the Equilibrium Constant to Find Quantities

The type of problem that involves finding equilibrium quantities also has several variations. Sample Problem 17.7 is one variation, in which we know *K* and some of the equilibrium concentrations and must find another equilibrium concentration.

Sample Problem 17.7 Determining Equilibrium Concentrations from K_c

Problem In a study of the conversion of methane to other fuels, a chemical engineer mixes gaseous CH_4 and H_2O in a 0.32-L flask at 1200 K. At equilibrium, the flask contains 0.26 mol of CO, 0.091 mol of H_2, 0.041 mol of CH_4, and some H_2O. What is $[H_2O]$ at equilibrium? $K_c = 0.26$ for this process at 1200 K.

Plan First, we write the balanced equation and the reaction quotient. We calculate the equilibrium concentrations from the given numbers of moles and the flask volume (0.32 L). Substituting these into Q_c and setting it equal to the given K_c (0.26), we solve for the unknown equilibrium concentration, $[H_2O]$.

Solution Writing the balanced equation and reaction quotient:

$$CH_4(g) + H_2O(g) \rightleftharpoons CO(g) + 3H_2(g) \qquad Q_c = \frac{[CO][H_2]^3}{[CH_4][H_2O]}$$

Determining the equilibrium concentrations:

$$[CH_4] = \frac{0.041 \text{ mol}}{0.32 \text{ L}} = 0.13\ M$$

Similarly, $[CO] = 0.81\ M$ and $[H_2] = 0.28\ M$.
Calculating $[H_2O]$ at equilibrium: Since $Q_c = K_c$, rearranging gives

$$[H_2O] = \frac{[CO][H_2]^3}{[CH_4]K_c} = \frac{(0.81)(0.28)^3}{(0.13)(0.26)} = 0.53\ M$$

Check Always check by substituting the concentrations into Q_c to confirm that the result is equal to K_c:

$$Q_c = \frac{[CO][H_2]^3}{[CH_4][H_2O]} = \frac{(0.81)(0.28)^3}{(0.13)(0.53)} = 0.26 = K_c$$

FOLLOW-UP PROBLEM 17.7 Nitrogen monoxide decomposes by the following equation: $2NO(g) \rightleftharpoons N_2(g) + O_2(g)$; $K_c = 2.3\times10^{30}$ at 298 K. In the atmosphere, $P_{O_2} = 0.209$ atm and $P_{N_2} = 0.781$ atm. What is the equilibrium partial pressure of NO in the air we breathe? (*Hint*: You need K_p to find the partial pressure.)

In a somewhat more involved variation, we know K and *initial* quantities and must find *equilibrium* quantities, for which we use a reaction table. Sample Problem 17.8 focuses on this approach.

Sample Problem 17.8 Determining Equilibrium Concentrations from Initial Concentrations and K_c

Problem Fuel engineers use the extent of the change from CO and H_2O to CO_2 and H_2 to regulate the proportions of synthetic fuel mixtures. If 0.250 mol of CO and 0.250 mol of H_2O gases are placed in a 125-mL flask at 900 K, what is the composition of the equilibrium mixture? At this temperature, K_c is 1.56.

Plan We have to find the "composition" of the equilibrium mixture, in other words, the equilibrium concentrations. As always, we write the balanced equation and use it to write the reaction quotient. We find the initial [CO] and $[H_2O]$ from the amounts (0.250 mol of each) and volume (0.125 L), use the balanced equation to define x and set up a reaction table, substitute into Q_c, and solve for x, from which we calculate the concentrations.

Solution Writing the balanced equation and reaction quotient:

$$CO(g) + H_2O(g) \rightleftharpoons CO_2(g) + H_2(g) \qquad Q_c = \frac{[CO_2][H_2]}{[CO][H_2O]}$$

Calculating initial reactant concentrations:

$$[CO] = [H_2O] = \frac{0.250 \text{ mol}}{0.125 \text{ L}} = 2.00\ M$$

Setting up the reaction table, with x = [CO] and $[H_2O]$ that react:

Concentration (M)	$CO(g)$	+	$H_2O(g)$	$\rightleftharpoons$	$CO_2(g)$	+	$H_2(g)$
Initial	2.00		2.00		0		0
Change	$-x$		$-x$		$+x$		$+x$
Equilibrium	$2.00 - x$		$2.00 - x$		x		x

Substituting into the reaction quotient and solving for x:

$$Q_c = \frac{[CO_2][H_2]}{[CO][H_2O]} = \frac{(x)(x)}{(2.00 - x)(2.00 - x)} = \frac{x^2}{(2.00 - x)^2}$$

At equilibrium, we have

$$Q_c = K_c = 1.56 = \frac{x^2}{(2.00 - x)^2}$$

We can apply the following math shortcut in this case *but not in general:* Because the right side of the equation is a perfect square, we take the square root of both sides:

$$\sqrt{1.56} = \frac{x}{2.00 - x} = \pm 1.25$$

A positive number (1.56) has a positive *and* a negative root, but in this case, only the positive root has any chemical meaning, so we ignore the negative root:*

$$1.25 = \frac{x}{2.00 - x} \quad \text{or} \quad 2.50 - 1.25x = x$$

So $\quad 2.50 = 2.25x \quad$ therefore $\quad x = 1.11\ M$

Calculating equilibrium concentrations:

$$[CO] = [H_2O] = 2.00\ M - x = 2.00\ M - 1.11\ M = 0.89\ M$$

$$[CO_2] = [H_2] = x = 1.11\ M$$

Check Given the intermediate size of K_c (1.56), it makes sense that the changes in concentration are moderate. It's a good idea to check that the sign of x in the reaction table is correct—only reactants were initially present, so x has a negative sign for reactants and a positive sign for products. Also check that the equilibrium concentrations give the known K_c: $\frac{(1.11)(1.11)}{(0.89)(0.89)} = 1.56$.

FOLLOW-UP PROBLEM 17.8 The decomposition of HI at low temperature was studied by injecting 2.50 mol of HI into a 10.32-L vessel at 25°C. What is $[H_2]$ at equilibrium for the reaction $2HI(g) \rightleftharpoons H_2(g) + I_2(g)$; $K_c = 1.26\times10^{-3}$?

Using the Quadratic Formula to Find the Unknown The shortcut we used to simplify the math in Sample Problem 17.8 is a special case that occurred because we started with equal concentrations of the reactants, but typically, we start with different concentrations of reactants. Suppose, for example, we start with 2.00 M CO and 1.00 M H_2O. Now, the reaction table becomes

Concentration (M)	$CO(g)$	+	$H_2O(g)$	$\rightleftharpoons$	$CO_2(g)$	+	$H_2(g)$
Initial	2.00		1.00		0		0
Change	$-x$		$-x$		$+x$		$+x$
Equilibrium	$2.00 - x$		$1.00 - x$		x		x

Substituting these values into Q_c, we obtain

$$Q_c = \frac{[CO_2][H_2]}{[CO][H_2O]} = \frac{(x)(x)}{(2.00 - x)(1.00 - x)} = \frac{x^2}{x^2 - 3.00x + 2.00}$$

At equilibrium, we have

$$1.56 = \frac{x^2}{x^2 - 3.00x + 2.00}$$

To solve for x in this case, we rearrange the previous expression into the form of a *quadratic equation:*

$$a\,x^2 + \quad b\,x + \quad c = 0$$

$$0.56x^2 - 4.68x + 3.12 = 0$$

where $a = 0.56$, $b = -4.68$, and $c = 3.12$. Then we find x with the quadratic formula (Appendix A):

$$x = \frac{-b \pm \sqrt{b^2 - 4ac}}{2a}$$

*The negative root gives $-1.25 = \frac{x}{2.00 - x}$, or $-2.50 + 1.25x = x$.

So $\quad -2.50 = 0.25x$ and $x = 10.\ M$

This value has no chemical meaning because we started with 2.00 M as the concentration of each reactant, so it is impossible for x to be 10. M. Moreover, the square root of an equilibrium constant is another equilibrium constant, which cannot have a negative value.

The ± sign means that we obtain two possible values for x:

$$x = \frac{4.68 \pm \sqrt{(-4.68)^2 - 4(0.56)(3.12)}}{2(0.56)}$$

$$x = 7.6\ M \qquad \text{and} \qquad x = 0.73\ M$$

But only one of the values makes sense chemically. The larger value gives negative concentrations at equilibrium (for example, for CO, $2.00\ M - 7.6\ M = -5.6\ M$), which have no meaning. Therefore, $x = 0.73\ M$, and we have

$$[CO] = 2.00\ M - x = 2.00\ M - 0.73\ M = 1.27\ M$$
$$[H_2O] = 1.00\ M - x = 0.27\ M$$
$$[CO_2] = [H_2] = x = 0.73\ M$$

Checking to see if these values give the known K_c, we have

$$K_c = \frac{(0.73)(0.73)}{(1.27)(0.27)} = 1.6 \text{ (within rounding of 1.56)}$$

A Simplifying Assumption for Finding the Unknown In many cases, we can use chemical "common sense" to make an assumption that simplifies the math by avoiding the need to use the quadratic formula to find x. In general, *if a reaction has a relatively small K and a relatively large initial reactant concentration, the concentration change (x) can often be neglected.* This assumption does not mean that $x = 0$, because then there would be no reaction. It means that if a reaction starts with a high $[\text{reactant}]_{\text{init}}$ and proceeds very little to reach equilibrium (small K), the reactant concentration at equilibrium, $[\text{reactant}]_{\text{eq}}$, will be nearly the same as $[\text{reactant}]_{\text{init}}$.

Here's an everyday analogy for this assumption. On a bathroom scale, you weigh 158 lb. Take off your wristwatch, and you still weigh 158 lb. Within the accuracy of the scale, the weight of the watch is so small compared with your body weight that it can be neglected:

Initial body weight − weight of watch = final body weight ≈ initial body weight

Similarly, let's say the initial concentration of A is 0.500 M and, because of a small K_c, the concentration of A that reacts is 0.002 M. We can then assume that

$$0.500\ M - 0.002\ M = 0.498\ M \approx 0.500\ M$$

that is,

$$[A]_{\text{init}} - [A]_{\text{reacting}} = [A]_{\text{eq}} \approx [A]_{\text{init}} \qquad \textbf{(17.9)}$$

To justify the assumption that x is negligible, we make sure the error introduced is not significant. One common criterion for "significant" is the 5% rule: *if the assumption results in a change that is less than 5% of the initial concentration, the error is not significant, and the assumption is justified.* Let's see how making this assumption simplifies the math and if it is justified in the cases of two different $[\text{reactant}]_{\text{init}}$ values.

Sample Problem 17.9 Making a Simplifying Assumption to Calculate Equilibrium Concentrations

Problem Phosgene is a potent chemical warfare agent that is now outlawed by international agreement. It decomposes by the reaction

$$COCl_2(g) \rightleftharpoons CO(g) + Cl_2(g) \qquad K_c = 8.3\times10^{-4} \text{ (at 360°C)}$$

Calculate [CO], $[Cl_2]$, and $[COCl_2]$ when each of the following amounts of phosgene decomposes and reaches equilibrium in a 10.0-L flask:

(a) 5.00 mol of $COCl_2$ **(b)** 0.100 mol of $COCl_2$

Plan We know from the balanced equation that when x mol of $COCl_2$ decomposes, x mol of CO and x mol of Cl_2 form. We use the volume (10.0 L) to convert amount (5.00 mol or 0.100 mol) to molar concentration, define x and set up the reaction table, and substitute the values into Q_c. Before using the quadratic formula, we assume that x is negligibly small. After solving for x, we check the assumption and find the equilibrium concentrations. If the assumption is not justified, we use the quadratic formula to find x.

Solution **(a)** For 5.00 mol of $COCl_2$. Writing the reaction quotient:

$$Q_c = \frac{[CO][Cl_2]}{[COCl_2]}$$

Calculating the initial reactant concentration, $[COCl_2]_{init}$:

$$[COCl_2]_{init} = \frac{5.00\ mol}{10.0\ L} = 0.500\ M$$

Setting up the reaction table, with x equal to $[COCl_2]_{reacting}$:

Concentration (M)	$COCl_2(g)$	$\rightleftharpoons$	$CO(g)$	+	$Cl_2(g)$
Initial	0.500		0		0
Change	$-x$		$+x$		$+x$
Equilibrium	$0.500 - x$		x		x

If we use the equilibrium values in Q_c with the given K_c, we obtain

$$Q_c = \frac{[CO][Cl_2]}{[COCl_2]} = \frac{x^2}{0.500 - x} = K_c = 8.3\times10^{-4}$$

Because K_c is small, the reaction does not proceed very far to the right, so let's assume that x ($[COCl_2]_{reacting}$) can be neglected. In other words, we assume that the equilibrium concentration is nearly the same as the initial concentration, 0.500 M:

$$[COCl_2]_{init} - [COCl_2]_{reacting} \approx [COCl_2]_{eq}$$

$$0.500\ M - x \approx 0.500\ M$$

Using this assumption, we substitute and solve for x:

$$K_c = 8.3\times10^{-4} \approx \frac{x^2}{0.500}$$

$$x^2 \approx (8.3\times10^{-4})(0.500) \qquad \text{so} \qquad x \approx 2.0\times10^{-2}$$

Checking the assumption by seeing if the error is <5%:

$$\frac{[\text{Change}]}{[\text{Initial}]} \times 100 = \frac{2.0\times10^{-2}}{0.500} \times 100 = 4\% \ (<5\%, \text{so the assumption is justified})$$

Solving for the equilibrium concentrations:

$$[CO] = [Cl_2] = x = 2.0\times10^{-2}\ M$$

$$[COCl_2] = 0.500\ M - x = 0.480\ M$$

(b) For 0.100 mol of $COCl_2$. The calculation in this case is the same as the calculation in part (a), except that $[COCl_2]_{init}$ = 0.100 mol/10.0 L = 0.0100 M. Thus, at equilibrium,

$$Q_c = \frac{[CO][Cl_2]}{[COCl_2]} = \frac{x^2}{0.0100 - x} = K_c = 8.3\times10^{-4}$$

Making the assumption that $0.0100\ M - x \approx 0.0100\ M$ and solving for x:

$$K_c = 8.3\times10^{-4} \approx \frac{x^2}{0.0100} \qquad x \approx 2.9\times10^{-3}$$

Checking the assumption:

$$\frac{2.9\times10^{-3}}{0.0100} \times 100 = 29\% \qquad (>5\%, \text{so the assumption is } \textit{not} \text{ justified})$$

We must solve the quadratic equation:

$$x^2 + (8.3\times10^{-4})x - (8.3\times10^{-6}) = 0$$

The only meaningful value of x is 2.5×10^{-3}.
Solving for the equilibrium concentrations:

$$[CO] = [Cl_2] = 2.5\times10^{-3}\ M$$

$$[COCl_2] = 1.00\times10^{-2}\ M - x = 7.5\times10^{-3}\ M$$

Check Once again, use the calculated values to be sure you obtain the given K_c.

Comment The main point is that the simplifying assumption was justified at the high $[COCl_2]_{init}$ but *not* at the low $[COCl_2]_{init}$.

FOLLOW-UP PROBLEM 17.9 In a study of the effect of temperature on halogen decomposition, 0.50 mol of I_2 was heated in a 2.5-L vessel, and the following reaction occurred: $I_2(g) \rightleftharpoons 2I(g)$.
(a) Calculate $[I_2]$ and [I] at equilibrium at 600 K; $K_c = 2.94\times10^{-10}$.
(b) Calculate $[I_2]$ and [I] at equilibrium at 2000 K; $K_c = 0.209$.

Predicting When the Assumption Will Be Justified To summarize, we assume that x ($[A]_{reacting}$) can be neglected if K_c is relatively small and/or $[A]_{init}$ is relatively large. The same holds for K_p and $P_{A(init)}$. But *how* small or large must these variables be? Here's a benchmark for deciding when to make the assumption:

- If $\frac{[A]_{init}}{K_c} > 400$, the assumption is justified: neglecting x introduces an error $<5\%$.
- If $\frac{[A]_{init}}{K_c} < 400$, the assumption is *not* justified; neglecting x introduces an error $>5\%$, so we solve a quadratic equation to find x.

For example, using the values from Sample Problem 17.9, we have

Part (a): For $[A]_{init} = 0.500\ M$, $\frac{0.500}{8.3\times10^{-4}} = 6.0\times10^2$, which is greater than 400.

Part (b): For $[A]_{init} = 0.0100\ M$, $\frac{0.0100}{8.3\times10^{-4}} = 12$, which is less than 400.

We will make a similar assumption in many problems in Chapters 18 and 19.

Problems Involving Mixtures of Reactants and Products

In the problems so far, the reaction *had* to go toward products because it started with only reactants. Therefore, in the reaction tables, we knew that the unknown change in reactant concentration had a negative sign ($-x$) and the change in product concentration had a positive sign ($+x$). If, however, we start with a *mixture* of reactants and products, the reaction direction is not obvious. In those cases, we apply the idea from Section 17.4 and first *compare the values of Q and K to find the direction the reaction is proceeding to reach equilibrium.* (To focus on this idea, Sample Problem 17.10 uses concentrations that avoid the need for the quadratic formula.)

Sample Problem 17.10 Predicting Reaction Direction and Calculating Equilibrium Concentrations

Problem The research and development unit of a chemical company is studying the reaction of CH_4 and H_2S, two components of natural gas:

$$CH_4(g) + 2H_2S(g) \rightleftharpoons CS_2(g) + 4H_2(g)$$

In one experiment, 1.00 mol of CH_4, 1.00 mol of CS_2, 2.00 mol of H_2S, and 2.00 mol of H_2 are mixed in a 250.-mL vessel at 960°C. At this temperature, $K_c = 0.036$.
(a) In which direction will the reaction proceed to reach equilibrium?
(b) If $[CH_4] = 5.56\ M$ at equilibrium, what are the equilibrium concentrations of the other substances?

Plan **(a)** To find the direction, we convert the given initial amounts and volume (0.250 L) to concentrations, calculate Q_c, and compare it with K_c. **(b)** Based on this information, we determine the sign of each concentration change for the reaction table and then use the known $[CH_4]$ at equilibrium (5.56 M) to determine x and the other equilibrium concentrations.

Solution **(a)** Calculating the initial concentrations:

$$[CH_4] = \frac{1.00\ \text{mol}}{0.250\ \text{L}} = 4.00\ M$$

Similarly, $[H_2S] = 8.00\ M$, $[CS_2] = 4.00\ M$, and $[H_2] = 8.00\ M$.

Calculating the value of Q_c:

$$Q_c = \frac{[CS_2][H_2]^4}{[CH_4][H_2S]^2} = \frac{(4.00)(8.00)^4}{(4.00)(8.00)^2} = 64.0$$

Comparing Q_c and K_c: $Q_c > K_c$ ($64.0 > 0.036$), so the reaction proceeds to the left. Therefore, concentrations of reactants increase and those of products decrease.

(b) Setting up a reaction table, with $x = [CS_2]$ that reacts, which equals $[CH_4]$ that forms:

Concentration (M)	$CH_4(g)$	+	$2H_2S(g)$	$\rightleftharpoons$	$CS_2(g)$	+	$4H_2(g)$
Initial	4.00		8.00		4.00		8.00
Change	$+x$		$+2x$		$-x$		$-4x$
Equilibrium	$4.00 + x$		$8.00 + 2x$		$4.00 - x$		$8.00 - 4x$

Solving for x: At equilibrium,

$$[CH_4] = 5.56\ M = 4.00\ M + x \qquad \text{so} \qquad x = 1.56\ M$$

Thus,

$$[H_2S] = 8.00\ M + 2x = 8.00\ M + 2(1.56\ M) = 11.12\ M$$

$$[CS_2] = 4.00\ M - x = 2.44\ M$$

$$[H_2] = 8.00\ M - 4x = 1.76\ M$$

Check The comparison of Q_c and K_c showed the reaction proceeding to the left. The given data from part (b) confirm this because $[CH_4]$ increases from 4.00 M to 5.56 M during the reaction. Check that the concentrations give the known K_c:

$$\frac{(2.44)(1.76)^4}{(5.56)(11.12)^2} = 0.0341, \text{ which is close to } 0.036$$

FOLLOW-UP PROBLEM 17.10 An inorganic chemist studying the reactions of phosphorus halides mixes 0.1050 mol of PCl_5 with 0.0450 mol of Cl_2 and 0.0450 mol of PCl_3 in a 0.5000-L flask at 250°C: $PCl_5(g) \rightleftharpoons PCl_3(g) + Cl_2(g)$; $K_c = 4.2\times10^{-2}$.

(a) In which direction will the reaction proceed?

(b) If $[PCl_5] = 0.2065\ M$ at equilibrium, what are the equilibrium concentrations of the other components?

Figure 17.5 makes three groups of the steps for solving equilibrium problems when you know K and some initial quantities and must find the equilibrium quantities.

SOLVING EQUILIBRIUM PROBLEMS

PRELIMINARY SETTING UP

1. Write the balanced equation.
2. Write the reaction quotient, Q.
3. Convert all amounts into the correct units (M or atm).

WORKING ON THE REACTION TABLE

4. When reaction direction is not known, compare Q with K.
5. Construct a reaction table.

✓Check the sign of x, the change in the concentration (or pressure).

SOLVING FOR x AND EQUILIBRIUM QUANTITIES

6. Substitute the quantities into Q.
7. To simplify the math, assume that x is negligible: ($[A]_{init} - x = [A]_{eq} \approx [A]_{init}$)
8. Solve for x.

✓Check that assumption is justified (<5% error). If not, solve quadratic equation for x.

9. Find the equilibrium quantities.

✓Check to see that calculated values give the known K.

Figure 17.5 Steps in solving equilibrium problems.

Summary of Section 17.5

- In equilibrium problems, we typically use quantities (concentrations or pressures) of reactants and products to find K, or we use K to find quantities.
- Reaction tables summarize the initial quantities, their changes during the reaction, and the equilibrium quantities.
- To simplify calculations, we assume that if K is small and the initial quantity of reactant is large, the unknown change in reactant (x) can be neglected. If this assumption is not justified (that is, if the error is greater than 5%), we use the quadratic formula to find x.
- For reactions that start with a mixture of reactants and products, we first determine reaction direction by comparing Q and K to decide on the sign of x.

17.6 • REACTION CONDITIONS AND EQUILIBRIUM: LE CHÂTELIER'S PRINCIPLE

Change conditions so that a system is no longer at equilibrium, and it has the remarkable ability to adjust itself and reattain equilibrium. This phenomenon is described by **Le Châtelier's principle:** when a chemical system at equilibrium is disturbed, it reattains equilibrium by undergoing a net reaction that reduces the effect of the disturbance.

Two phrases in this statement need further explanation:

1. How is a system "disturbed"? At equilibrium, Q equals K. The system is disturbed when a change in conditions forces it temporarily out of equilibrium ($Q \neq K$). Three common disturbances are a change in concentration, a change in pressure (caused by a change in volume), or a change in temperature. We'll discuss each below.

2. What does a "net reaction" mean? This phrase refers to a shift in the *equilibrium position* to the right or left. The *equilibrium position* is defined by the specific equilibrium concentrations (or pressures). A shift to the right is a net reaction from reactant to product until equilibrium is reattained; a shift to the left is a net reaction from product to reactant. Thus, *concentrations (or pressures) change in a way that reduces the effect of the change in conditions, and the system attains a new equilibrium position.*

For the remainder of this section, we'll examine a system at equilibrium to see how it responds to changes in concentration, pressure (volume), or temperature; then, we'll see what happens when we add a catalyst. Although Le Châtelier's principle holds for any system at equilibrium, our example will be the gaseous reaction between phosphorus trichloride and chlorine to produce phosphorus pentachloride:

$$PCl_3(g) + Cl_2(g) \rightleftharpoons PCl_5(g)$$

The Effect of a Change in Concentration

When a system at equilibrium is disturbed by a change in concentration of one of the components, it reacts in the direction that reduces the change:

- If the concentration of A increases, the system reacts to consume some of it.
- If the concentration of B decreases, the system reacts to produce some of it.

Only components that appear in Q can have an effect, so changes in the amounts of pure liquids and solids cannot.

A Qualitative View of a Concentration Change At 523 K, the PCl_3-Cl_2-PCl_5 system reaches equilibrium when

$$Q_c = \frac{[PCl_5]}{[PCl_3][Cl_2]} = 24.0 = K_c$$

Starting with Q_c equal to K_c, let's think through some changes in concentration:

1. *Adding a reactant.* What happens if we disturb the system by adding some Cl_2 gas? To reduce the disturbance, the system will consume some of the added Cl_2 by shifting toward product. With regard to the reaction quotient, when the $[Cl_2]$ term increases, the value of Q_c decreases; thus, $Q_c \neq K_c$. As some of the added Cl_2 reacts with some of the PCl_3 to form more PCl_5, the denominator becomes smaller again and the numerator larger, until eventually $Q_c = K_c$ again. Notice the changes in the new equilibrium concentrations: $[Cl_2]$ and $[PCl_5]$ are higher than in the original equilibrium position, and $[PCl_3]$ is lower. Nevertheless, the ratio of values gives the same K_c. Thus, *the equilibrium position shifts to the right when a component on the left is added:*

$$PCl_3 + Cl_2(\text{added}) \longrightarrow PCl_5$$

2. *Removing a reactant.* What happens if we disturb the system by removing some PCl_3? To reduce this disturbance, the system will replace the PCl_3 by consuming some PCl_5 and proceeding toward reactants. With regard to Q_c, when the $[PCl_3]$ term decreases, Q_c increases, so $Q_c \neq K_c$. As some PCl_5 decomposes to PCl_3 and Cl_2, the numerator decreases and the denominator increases until $Q_c = K_c$ again. Once again, the new and old equilibrium concentrations are different, but the value of K_c is not. Thus, *the equilibrium position shifts to the left when a component on the left is removed:*

$$PCl_3(\text{removed}) + Cl_2 \longleftarrow PCl_5$$

3. *Adding or removing a product.* The same points we just made for a reactant hold for a product. If we add PCl_5, the equilibrium position shifts to the left; if we remove some PCl_5, the equilibrium position shifts to the right.

*ANY OF THESE CHANGES CAUSES A SHIFT TO THE **RIGHT**.*

increase increase decrease

$PCl_3 + Cl_2 \rightleftharpoons PCl_5$

decrease decrease increase

*ANY OF THESE CHANGES CAUSES A SHIFT TO THE **LEFT**.*

Figure 17.6 The effect of a change in concentration on a system at equilibrium.

In other words, no matter how the disturbance in concentration comes about, *the system reacts to consume some of the added substance or produce some of the removed substance* to make $Q_c = K_c$ again (Figure 17.6):

- The equilibrium position shifts to the *right* if a reactant is added or a product is removed: [reactant] increases or [product] decreases.
- The equilibrium position shifts to the *left* if a reactant is removed or a product is added: [reactant] decreases or [product] increases.

A Quantitative View of a Concentration Change As you saw, the system "reduces the effect of the disturbance." But the effect is not completely eliminated, as we can see from a quantitative comparison of original and new equilibrium positions.

Consider what happens when we add Cl_2 to a system whose original equilibrium position was established with $[PCl_3] = 0.200\ M$, $[Cl_2] = 0.125\ M$, and $[PCl_5] = 0.600\ M$. That is,

$$Q_c = \frac{[PCl_5]}{[PCl_3][Cl_2]} = \frac{0.600}{(0.200)(0.125)} = 24.0 = K_c$$

Now we add enough Cl_2 to increase its concentration by 0.075 M to a new $[Cl_2]_{init}$ of 0.200 M. The reaction proceeds, and the system comes to a new equilibrium position. From Le Châtelier's principle, we predict that adding more reactant will shift the equilibrium position to the right. Experimental measurement shows that the new $[PCl_5]_{eq}$ is 0.637 M.

Table 17.3 shows a reaction table of the entire process: the original equilibrium position, the disturbance, the (new) initial concentrations, the direction of x (the change needed to reattain equilibrium), and the new equilibrium position. Figure 17.7 depicts the process.

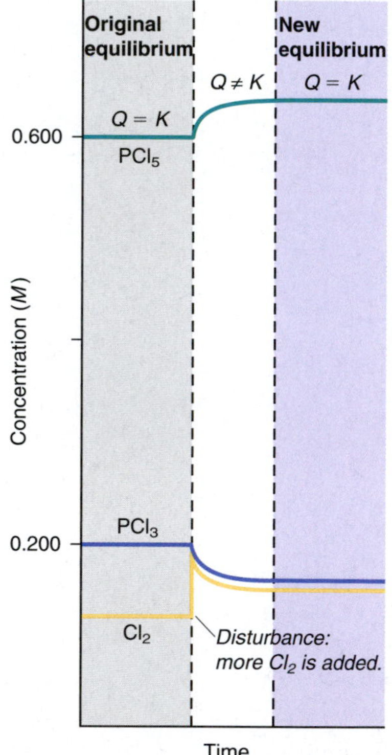

Figure 17.7 The effect of added Cl_2 on the PCl_3-Cl_2-PCl_5 system. The original equilibrium concentrations are shown at left *(gray region)*. When Cl_2 *(yellow curve)* is added, its concentration increases instantly *(vertical part of yellow curve)* and then falls gradually as it reacts with PCl_3 to form more PCl_5. Soon, equilibrium is re-established at new concentrations *(blue region)* but with the same K.

Let's determine the new equilibrium concentrations. From Table 17.3,

$$[PCl_5] = 0.600\ M + x = 0.637\ M \quad \text{so} \quad x = 0.037\ M$$

Thus, $$[PCl_3] = [Cl_2] = 0.200\ M - x = 0.200\ M - 0.037\ M = 0.163\ M$$

Therefore, at equilibrium,

$$K_{c(\text{original})} = \frac{0.600}{(0.200)(0.125)} = 24.0$$

$$K_{c(\text{new})} = \frac{0.637}{(0.163)(0.163)} = 24.0$$

There are several key points to notice about the new equilibrium concentrations:

- As we predicted, $[PCl_5]$ (0.637 M) is higher than the original concentration (0.600 M).
- $[Cl_2]$ (0.163 M) is higher than the *original* equilibrium concentration (0.125 M), but lower than the new initial concentration (0.200 M); thus, the disturbance is *reduced but not eliminated.*
- $[PCl_3]$ (0.163 M), the other reactant, is lower than the original equilibrium concentration (0.200 M) because some reacted with the added Cl_2.
- Most importantly, although the position of equilibrium shifted to the right, *at a given temperature, K_c does **not** change with a change in concentration.*

Table 17.3 The Effect of Added Cl_2 on the PCl_3-Cl_2-PCl_5 System

Concentration (M)	$PCl_3(g)$	+	$Cl_2(g)$	$\rightleftharpoons$	$PCl_5(g)$
Original equilibrium	0.200		0.125		0.600
Disturbance			+0.075		
New initial	0.200		0.200		0.600
Change	$-x$		$-x$		$+x$
New equilibrium	$0.200 - x$		$0.200 - x$		$0.600 + x$ (0.637)*

*Experimentally determined value.

Sample Problem 17.11 Predicting the Effect of a Change in Concentration on the Equilibrium Position

Problem To improve air quality and obtain a useful product, chemists often remove sulfur from coal and natural gas by treating the contaminant hydrogen sulfide with O_2:

$$2H_2S(g) + O_2(g) \rightleftharpoons 2S(s) + 2H_2O(g)$$

What happens to
(a) $[H_2O]$ if O_2 is added?
(b) $[H_2S]$ if O_2 is added?
(c) $[O_2]$ if H_2S is removed?
(d) $[H_2S]$ if sulfur is added?

Plan We write the reaction quotient to see how Q_c is affected by each disturbance, relative to K_c. This effect tells us the direction in which the reaction proceeds for the system to reattain equilibrium and how each concentration changes.

Solution Writing the reaction quotient: $Q_c = \dfrac{[H_2O]^2}{[H_2S]^2[O_2]}$

(a) When O_2 is added, the denominator of Q_c increases, so $Q_c < K_c$. The reaction proceeds to the right until $Q_c = K_c$ again, so $[H_2O]$ increases.

(b) As in part (a), when O_2 is added, $Q_c < K_c$. Some H_2S reacts with the added O_2 as the reaction proceeds to the right, so $[H_2S]$ decreases.

(c) When H_2S is removed, the denominator of Q_c decreases, so $Q_c > K_c$. As the reaction proceeds to the left to re-form H_2S, more O_2 forms as well, so $[O_2]$ increases.

(d) The *concentration* of solid S does not change (even though it must be present), so it does not appear in the reaction quotient. Adding more S has no effect, so $[H_2S]$ is unchanged (but see Comment 2 below).

Check Apply Le Châtelier's principle to see that the reaction proceeds in the direction that lowers the increased concentration or raises the decreased concentration.

Comment 1. As you know, sulfur exists most commonly as S_8. How would this change in formula affect the answers? The balanced equation and Q_c would be

$$8H_2S(g) + 4O_2(g) \rightleftharpoons S_8(s) + 8H_2O(g) \qquad Q_c = \frac{[H_2O]^8}{[H_2S]^8[O_2]^4}$$

The value of K_c is different for this equation, but the changes described in the problem have the same effects. Thus, shifts in equilibrium position predicted by Le Châtelier's principle are not affected by a change in the balancing coefficients.
2. In (d), you saw that adding a solid has no effect on the concentrations of other components: because *the **concentration** of the solid cannot change,* it does not appear in Q. But *the **amount** of solid can change.* Adding H_2S shifts the reaction to the right, so more S forms.

FOLLOW-UP PROBLEM 17.11 In a study of glass etching, a chemist examines the reaction between sand (SiO_2) and hydrogen fluoride at 150°C:

$$SiO_2(s) + 4HF(g) \rightleftharpoons SiF_4(g) + 2H_2O(g)$$

Predict the effect on $[SiF_4]$ when **(a)** $H_2O(g)$ is removed; **(b)** some liquid water is added; **(c)** HF is removed; **(d)** some sand is removed.

The Effect of a Change in Pressure (Volume)

Changes in pressure can have a large effect on equilibrium systems containing gaseous components. (A change in pressure has a negligible effect on liquids and solids because they are nearly incompressible.) Pressure changes can occur in three ways:

- *Changing the concentration of a gaseous component.* We just considered the effect of changing the concentration of a component, and that reasoning applies here.
- *Adding an inert gas (one that does not take part in the reaction).* As long as the volume of the system is constant, adding an inert gas has no effect on the equilibrium position because *all concentrations, and thus partial pressures, remain the same.* Moreover, the inert gas does not appear in Q, so it cannot have an effect.

- *Changing the volume of the reaction vessel.* This change can cause a large shift in equilibrium position, but only for reactions in which the number of moles of gas, n_{gas}, changes.

Let's consider the two possible situations for the third way: changing the volume of the reaction vessel.

1. *Reactions in which n_{gas} changes.* Suppose the PCl_3-Cl_2-PCl_5 system is in a piston-cylinder assembly. We press down on the piston to halve the volume, so the gas pressure doubles. To reduce this disturbance, the system responds by *reducing the number of gas molecules.* And the only way to do that is through a net reaction toward the side with *fewer moles of gas,* in this case, toward product:

$$PCl_3(g) + Cl_2(g) \longrightarrow PCl_5(g)$$
$$\text{2 mol gas} \longrightarrow \text{1 mol gas}$$

Recall that $Q_c = \frac{[PCl_5]}{[PCl_3][Cl_2]}$. When the volume is halved, the concentrations double, but the denominator of Q_c is the product of two concentrations, so it quadruples while the numerator only doubles. Thus, Q_c becomes less than K_c. As a result, the system forms more PCl_5 and a new equilibrium position is reached.

Thus, for a system that consists of gases at equilibrium, in which the amount (mol) of gas, n_{gas}, changes during the reaction (Figure 17.8):

- If the volume becomes smaller (pressure is higher), the reaction shifts so that the total number of gas molecules decreases.
- If the volume becomes larger (pressure is lower), the reaction shifts so that the total number of gas molecules increases.

2. *Reactions in which n_{gas} does **not** change.* For the formation of hydrogen iodide from its elements, we have the same amount (mol) of gas on both sides:

$$H_2(g) + I_2(g) \rightleftharpoons 2HI(g)$$
$$\text{2 mol gas} \longrightarrow \text{2 mol gas}$$

Therefore, Q_c has the same number of terms in the numerator and denominator:

$$Q_c = \frac{[HI]^2}{[H_2][I_2]} = \frac{[HI][HI]}{[H_2][I_2]}$$

Because a change in volume has the same effect on the numerator and denominator, *there is **no** effect on the equilibrium position.*

In terms of the equilibrium constant, *a change in volume is, in effect, a change in concentration:* a decrease in volume raises the concentration, and vice versa. Therefore, like other changes in concentration, *a change in pressure due to a change in volume does **not** alter K_c.*

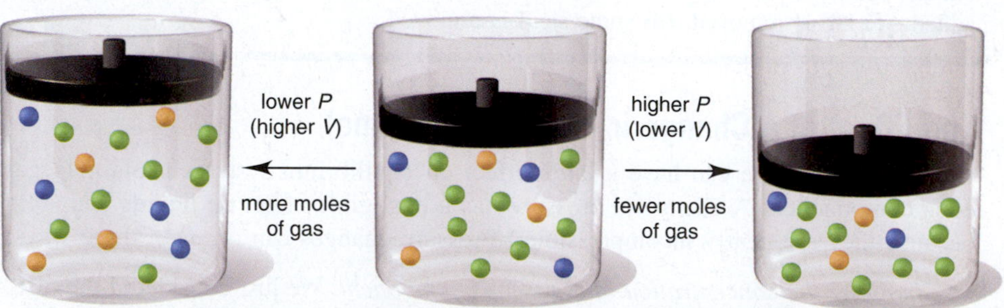

Figure 17.8 The effect of a change in pressure (volume) on a system at equilibrium. The system of gases *(center)* is at equilibrium. For the reaction

● + ● ⇌ ●

two moles of gas form one. An increase in pressure *(right)* decreases the volume, so the equilibrium shifts to form *fewer* molecules. A decrease in pressure *(left)* increases the volume, so the equilibrium shifts to form *more* molecules.

Sample Problem 17.12 **Predicting the Effect of a Change in Volume (Pressure) on the Equilibrium Position**

Problem How would you change the volume of each of the following reactions to *increase* the yield of the product(s)?
(a) $CaCO_3(s) \rightleftharpoons CaO(s) + CO_2(g)$
(b) $S(s) + 3F_6(g) \rightleftharpoons SF_6(g)$
(c) $Cl_2(g) + I_2(g) \rightleftharpoons 2ICl(g)$

Plan Whenever gases are present, a change in volume causes a change in concentration. For reactions in which the number of moles of gas changes, if the volume decreases (pressure increases), the equilibrium position shifts to lower the pressure by reducing the number of moles of gas. A volume increase (pressure decrease) has the opposite effect.

Solution **(a)** The only gas is the product CO_2. To make the system produce more molecules of gas, that is, more CO_2, we increase the volume (decrease the pressure).
(b) With 3 mol of gas on the left and only 1 mol on the right, we decrease the volume (increase the pressure) to form fewer molecules of gas and, thus, more SF_6.
(c) The number of moles of gas is the same on both sides of the equation, so a change in volume (pressure) will have no effect on the yield of ICl.

Check Let's predict the relative values of Q_c and K_c.
(a) $Q_c = [CO_2]$, so increasing the volume will make $Q_c < K_c$, and the system will yield more CO_2.
(b) $Q_c = [SF_6]/[F_2]^3$. Lowering the volume increases $[F_2]$ and $[SF_6]$ proportionately, but Q_c decreases because of the exponent 3 in the denominator. To make $Q_c = K_c$ again, $[SF_6]$ must increase.
(c) $Q_c = [ICl]^2/[Cl_2][I_2]$. A change in volume (pressure) affects the numerator (2 mol) and denominator (2 mol) equally, so it will have no effect.

FOLLOW-UP PROBLEM 17.12 Would you increase or decrease the pressure (via a volume change) of each of the following reaction mixtures to *decrease* the yield of products?
(a) $2SO_2(g) + O_2(g) \rightleftharpoons 2SO_3(g)$
(b) $4NH_3(g) + 5O_2(g) \rightleftharpoons 4NO(g) + 6H_2O(g)$
(c) $CaC_2O_4(s) \rightleftharpoons CaCO_3(s) + CO(g)$

The Effect of a Change in Temperature

Of the three types of disturbances that may occur—a change in concentration, pressure, or temperature—*only temperature changes alter K.* To see why, let's focus on the sign of ΔH°_{rxn}:

$$PCl_3(g) + Cl_2(g) \rightleftharpoons PCl_5(g) \qquad \Delta H^\circ_{rxn} = -111 \text{ kJ}$$

The forward reaction is exothermic (releases heat; $\Delta H^\circ_{rxn} < 0$), so the reverse reaction is endothermic (absorbs heat; $\Delta H^\circ_{rxn} > 0$):

$$PCl_3(g) + Cl_2(g) \longrightarrow PCl_5(g) + \textbf{heat} \text{ (exothermic)}$$
$$PCl_3(g) + Cl_2(g) \longleftarrow PCl_5(g) + \textbf{heat} \text{ (endothermic)}$$

If we consider *heat as a component of the equilibrium system,* a rise in temperature occurs when heat is "added" to the system and a drop in temperature occurs when heat is "removed" from the system. As with a change in any other component, the system shifts to reduce the effect of the change. Therefore, *a temperature increase (adding heat) favors the endothermic (heat-absorbing) direction, and a temperature decrease (removing heat) favors the exothermic (heat-releasing) direction.*

If we start with the system at equilibrium, Q_c equals K_c. Increase the temperature, and the system absorbs the added heat by decomposing some PCl_5 to PCl_3 and Cl_2. The denominator of Q_c becomes larger and the numerator smaller, so the system reaches a new equilibrium position at a smaller ratio of concentration terms, that is, a lower K_c. Similarly, if the temperature drops, the system releases more heat by forming more PCl_5 from some PCl_3 and Cl_2. The numerator of Q_c becomes larger, the denominator smaller, and the new equilibrium position has a higher K_c. Thus,

- *A temperature rise will increase K_c for a system with a positive ΔH°_{rxn}.*
- *A temperature rise will decrease K_c for a system with a negative ΔH°_{rxn}.*

Let's review these ideas with a sample problem.

Sample Problem 17.13 Predicting the Effect of a Change in Temperature on the Equilibrium Position

Problem How does an *increase* in temperature affect the equilibrium concentration of the underlined substance and K for each of the following reactions?

(a) $CaO(s) + H_2O(l) \rightleftharpoons \underline{Ca(OH)_2(aq)}$ $\quad \Delta H° = -82$ kJ

(b) $CaCO_3(s) \rightleftharpoons CaO(s) + \underline{CO_2(g)}$ $\quad \Delta H° = 178$ kJ

(c) $\underline{SO_2(g)} \rightleftharpoons S(s) + O_2(g)$ $\quad \Delta H° = 297$ kJ

Plan We write each equation to show heat as a reactant or product. The temperature increases when we add heat, so the system shifts to absorb the heat; that is, the endothermic reaction occurs. Thus, K will increase if the forward reaction is endothermic and decrease if it is exothermic.

Solution (a) $CaO(s) + H_2O(l) \rightleftharpoons Ca(OH)_2(aq) +$ ***heat***

Adding heat shifts the system to the left: $[Ca(OH)_2]$ and K will decrease.

(b) $CaCO_3(s) +$ ***heat*** $\rightleftharpoons CaO(s) + CO_2(g)$

Adding heat shifts the system to the right: $[CO_2]$ and K will increase.

(c) $SO_2(g) +$ ***heat*** $\rightleftharpoons S(s) + O_2(g)$

Adding heat shifts the system to the right: $[SO_2]$ will decrease and K will increase.

Check Check your answers by reasoning through a *decrease* in temperature: heat is removed and the exothermic direction is favored. All the answers should be opposite.

Comment Note that in (a) these ideas hold for solutions as well.

FOLLOW-UP PROBLEM 17.13 How does a *decrease* in temperature affect the partial pressure of the underlined substance and the value of K for each of the following reactions?

(a) $C(graphite) + \underline{2H_2(g)} \rightleftharpoons CH_4(g)$ $\quad \Delta H° = -75$ kJ

(b) $\underline{N_2(g)} + O_2(g) \rightleftharpoons 2NO(g)$ $\quad \Delta H° = 181$ kJ

(c) $P_4(s) + 10Cl_2(g) \rightleftharpoons \underline{4PCl_5(g)}$ $\quad \Delta H° = -1528$ kJ

The Lack of Effect of a Catalyst

Let's briefly consider what effect, if any, adding a catalyst would have on the system. Recall from Chapter 16 that a catalyst speeds up a reaction by lowering the activation energy, thereby increasing the forward *and* reverse rates to the same extent. Thus, *a catalyst shortens the time it takes to reach equilibrium but has* ***no*** *effect on the equilibrium position.* That is, if we add a catalyst to a mixture of PCl_3 and Cl_2 at 523 K, the system attains the *same* equilibrium concentrations of PCl_3, Cl_2, and PCl_5 *more quickly* than it does without the catalyst.

THINK OF IT THIS WAY
Catalyzed Perpetual Motion?

Catalyst
PCl_3
Cl_2
PCl_5

An imaginary engine shows why a catalyst must speed a reaction in *both* directions. It consists of a piston attached to a flywheel, whose rocker arm holds the catalyst and moves it in and out of the reaction in a cylinder. Suppose the catalyst *could* increase the rate of PCl_5 breakdown but not PCl_5 formation. When the catalyst is in the cylinder, PCl_5 breaks down to PCl_3 and Cl_2 faster than it forms from them. Thus, 1 mol of gas forms 2 mol, which raises the pressure and the piston moves out. With the catalyst out of the cylinder, PCl_3 and Cl_2 re-form PCl_5, which lowers gas pressure and the piston moves in. If (and this is the big *if*) this catalyst could change the rate in only one direction, its presence would change K and the process would supply power with no external input of energy!

Table 17.4 summarizes the effects of changing conditions. Many changes alter the equilibrium *position*, but only temperature changes alter the equilibrium *constant*. Sample Problem 17.14 shows how to visualize equilibrium at the molecular level.

Even though catalysts cannot change the reaction yield, they often play key roles in optimizing it. The industrial production of ammonia, described in the next subsection, provides an example of a catalyzed improvement of yield.

Table 17.4 Effects of Various Disturbances on a System at Equilibrium

Disturbance	Effect on Equilibrium Position	Effect on Value of *K*
Concentration		
Increase [reactant]	Toward formation of product	None
Decrease [reactant]	Toward formation of reactant	None
Increase [product]	Toward formation of reactant	None
Decrease [product]	Toward formation of product	None
Pressure		
Increase P (decrease V)	Toward formation of fewer moles of gas	None
Decrease P (increase V)	Toward formation of more moles of gas	None
Increase P (add inert gas, no change in V)	None; concentrations unchanged	None
Temperature		
Increase T	Toward absorption of heat	Increases if $\Delta H^\circ_{rxn} > 0$ Decreases if $\Delta H^\circ_{rxn} < 0$
Decrease T	Toward release of heat	Increases if $\Delta H^\circ_{rxn} < 0$ Decreases if $\Delta H^\circ_{rxn} > 0$
Catalyst added	None; forward and reverse rates increase equally; equilibrium attained sooner	None

Sample Problem 17.14 Determining Equilibrium Parameters from Molecular Scenes

Problem For the reaction,

$$X(g) + Y_2(g) \rightleftharpoons XY(g) + Y(g) \qquad \Delta H > 0$$

the following molecular scenes depict different reaction mixtures (X is *green,* Y is *purple*):

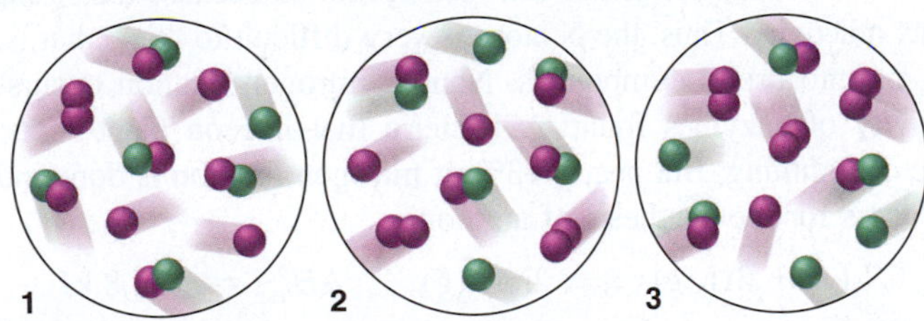

(a) If $K = 2$ at the temperature of the reaction, which scene represents the mixture at equilibrium?
(b) Will the reaction mixtures in the other two scenes proceed toward reactants or toward products to reach equilibrium?
(c) For the mixture at equilibrium, how will a rise in temperature affect $[Y_2]$?

Plan (a) We are given the balanced equation and K and must choose the scene that represents the mixture at equilibrium. We write Q, and for each scene, count particles and find the value of Q. Whichever scene gives a Q equal to K (2) represents the mixture at equilibrium. **(b)** For each of the other two reaction mixtures, we compare the value of Q with 2. If $Q > K$, the numerator (product side) is too high, so the reaction proceeds toward reactants; if $Q < K$, the reaction proceeds toward products. **(c)** We are given that $\Delta H > 0$, so we must see whether a rise in T increases or decreases $[Y_2]$, one of the reactants.

Solution (a) For the reaction, we have $Q = \dfrac{[XY][Y]}{[X][Y_2]}$. Thus,

$$\text{scene 1: } Q = \frac{5 \times 3}{1 \times 1} = 15 \qquad \text{scene 2: } Q = \frac{4 \times 2}{2 \times 2} = 2 \qquad \text{scene 3: } Q = \frac{3 \times 1}{3 \times 3} = \frac{1}{3}$$

For scene 2, $Q = K$, so scene 2 represents the mixture at equilibrium.

(b) For scene 1, Q (15) > K (2), so the reaction proceeds toward reactants.
For scene 3, Q ($\frac{1}{3}$) < K (2), so the reaction proceeds toward products.
(c) The reaction is endothermic, so heat acts as a reactant:

$$X(g) + Y_2(g) + heat \rightleftharpoons XY(g) + Y(g)$$

Therefore, adding heat to the left shifts the reaction to the right, so $[Y_2]$ decreases.

Check **(a)** Remember that quantities in the numerator (or denominator) of Q are multiplied, not added. For example, the denominator for scene 1 is $1 \times 1 = 1$, not $1 + 1 = 2$.
(c) A good check is to imagine that $\Delta H < 0$ and see if you get the opposite result:

$$X(g) + Y_2(g) \rightleftharpoons XY(g) + Y(g) + heat$$

If $\Delta H < 0$, adding heat would shift the reaction to the left and increase $[Y_2]$.

FOLLOW-UP PROBLEM 17.14 For the reaction $C_2(g) + D_2(g) \rightleftharpoons 2CD(g)$; $\Delta H < 0$, these molecular scenes depict different reaction mixtures (C is *red,* D is *blue*):

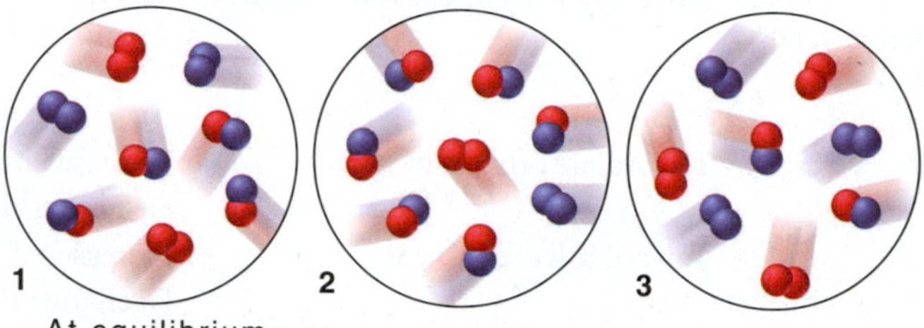

(a) Calculate the value of K_p. **(b)** In which direction will the reaction proceed for the mixtures *not* at equilibrium? **(c)** For the mixture at equilibrium, what effect will a rise in T have on the total amount (mol) of gas (increase, decrease, no effect)? Explain.

The Industrial Production of Ammonia

Le Châtelier's principle has countless applications in natural systems and in the chemical industry. As a case study, we'll look at the synthesis of ammonia, which, on a mole basis, is produced industrially in greater amount than any other compound.

Nitrogen occurs in many compounds, and four of every five molecules in the atmosphere are N_2. Yet the supply of *usable* nitrogen is limited because the strong triple bond in N_2 lowers its reactivity. Thus, the N atom is very difficult to "fix," that is, to combine with other atoms into useful compounds. Natural nitrogen fixation occurs through the fine-tuned activity of enzymes found in bacteria that live on plant roots, or through the brute force of lightning. But nearly 13% of nitrogen fixation is done industrially via the **Haber process** for the synthesis of ammonia:

$$N_2(g) + 3H_2(g) \rightleftharpoons 2NH_3(g) \qquad \Delta H^\circ_{rxn} = -91.8 \text{ kJ}$$

Developed by the German chemist Fritz Haber in 1913 and first used in a plant making 12,000 tons of ammonia a year, the process now yields over 110 million tons a year. Over 80% of this is used as fertilizer, with most of the remainder used to make explosives and nylons and other polymers, and smaller amounts going into production of refrigerants, rubber stabilizers, household cleaners, and pharmaceuticals.

Optimizing Reaction Conditions: Yield vs. Rate The Haber process applies equilibrium *and* kinetics principles to produce ammonia economically. From the balanced equation, we see three ways to maximize NH_3 yield:

1. *Decrease [NH_3].* Removing NH_3 as it forms will make the system shift toward producing more to reattain equilibrium.
2. *Decrease volume (increase pressure).* Since 4 mol of gas reacts to form 2 mol of gas, decreasing the volume will shift the system toward making fewer moles of gas.
3. *Decrease temperature.* Because the formation of NH_3 is exothermic, decreasing the temperature (removing heat) will shift the equilibrium position toward product, thereby increasing K_c (Table 17.5).

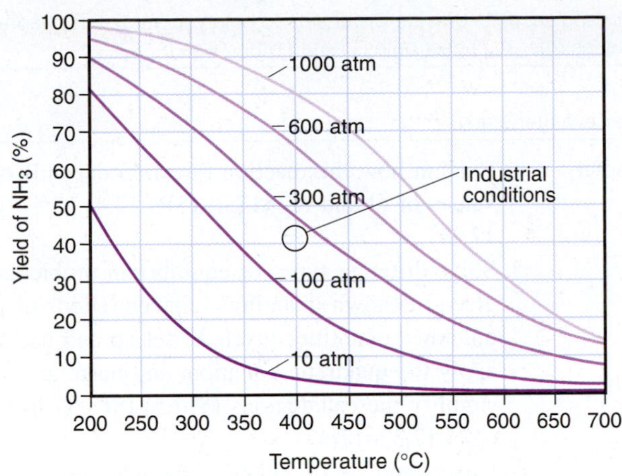

Figure 17.9 **Percent yield of ammonia vs. temperature at five different pressures.** At very high *P* and low *T (top left)*, the yield is high, but the rate is low. Industrial conditions *(circle)* are between 200 and 300 atm at about 400°C.

Table 17.5 Effect of *T* on K_c for NH_3 Synthesis

T(K)	K_c
200.	7.17×10^{15}
300.	2.69×10^{8}
400.	3.94×10^{4}
500.	1.72×10^{2}
600.	4.53×10^{0}
700.	2.96×10^{-1}
800.	3.96×10^{-2}

Therefore, the conditions for maximizing the yield of product are continual removal of NH_3, high pressure, and low temperature. Figure 17.9 shows the percent yield of NH_3 at various combinations of pressure and temperature. Note the almost complete conversion (98.3%) to product at 1000 atm and 473 K (200.°C).

Although the *yield* is favored at this relatively low temperature, the *rate* of formation is too slow to be economical. In practice, a compromise optimizes yield *and* rate. High pressure and continuous removal are used to increase yield, but the temperature is raised to a moderate level and *a catalyst is used to increase the rate.* Achieving the same rate without a catalyst would require much higher temperatures, which would reduce the yield.

The Actual Industrial Conditions Modern plants operate at about 200 to 300 atm and around 673 K (400.°C). The stoichiometric ratio of reactant gases (N_2/H_2 = 1/3 by volume) is injected into the heated, pressurized reaction chamber. The gases flow over catalyst beds that consist of 5-mm to 10-mm chunks of iron crystals embedded in a fused mixture of MgO, Al_2O_3, and SiO_2. The emerging equilibrium mixture contains about 35% NH_3 and is cooled until the NH_3 (b.p. −33.4°C) condenses, which is removed and stored. Because N_2 and H_2 have much lower boiling points, they are recycled by pumping them back into the reaction chamber.

Summary of Section 17.6

- If a system at equilibrium is disturbed, it undergoes a net reaction that reduces the disturbance and returns the system to equilibrium.
- Changes in concentration cause a net reaction to consume the added component or to produce the removed component.
- For a reaction that involves a change in amount (mol) of gas, an increase in pressure (decrease in volume) causes a net reaction toward fewer moles of gas, and a decrease in pressure causes the opposite change.
- Although the equilibrium position changes as a result of a concentration or volume change, *K* does not.
- A temperature change affects *K*: higher *T* increases *K* for an endothermic reaction (positive ΔH°_{rxn}) and decreases *K* for an exothermic reaction (negative ΔH°_{rxn}).
- A catalyst causes the system to reach equilibrium more quickly by speeding forward and reverse reactions equally, but it does not affect the equilibrium position.
- Ammonia is produced in a process favored by high pressure, low temperature, and continual removal of product. To make the process economical, intermediate temperature and pressure and a catalyst are used.

CHAPTER REVIEW GUIDE

The following sections provide many aids to help you study this chapter. (Numbers in parentheses refer to pages, unless noted otherwise.)

Learning Objectives

These are concepts and skills to review after studying this chapter.

Related section (§), sample problem (SP), and upcoming end-of-chapter problem (EP) numbers are listed in parentheses.

1. Distinguish between the rate and the extent of a reaction; understand that the equilibrium constant (K) is a number whose magnitude is related to the extent of the reaction (§17.1) (EPs 17.1–17.5)
2. Understand that the reaction quotient (Q) changes until the system reaches equilibrium, when it equals K; write Q for any balanced equation, and calculate K given concentrations (§17.2) (SPs 17.1, 17.2) (EPs 17.6–17.19)
3. Use the ideal gas law and Δn_{gas} to convert between K_c and K_p (§17.3) (SP 17.3) (EPs 17.20–17.26)
4. Explain how the reaction direction depends on the relative values of Q and K (§17.4) (SPs 17.4, 17.5) (EPs 17.27–17.31)
5. Solve different types of equilibrium problems; calculate K given unknown quantities (concentrations or pressures), or unknown quantities given K; set up and use a reaction table, apply the quadratic equation, and make an assumption to simplify the calculations (§17.5) (SPs 17.6–17.10) (EPs 17.32–17.47)
6. Understand Le Châtelier's principle, and predict the effects of concentration, pressure (volume), temperature, and a catalyst on equilibrium position and on K (§17.6) (SPs 17.11–17.14) (EPs 17.48–17.64)

Key Terms

These important terms appear in boldface in the chapter and are defined again in the Glossary.

Section 17.1
equilibrium constant (K) (544)

Section 17.2
law of chemical equilibrium (law of mass action) (545)
reaction quotient (Q) (545)

Section 17.6
Le Châtelier's principle (562)
Haber process (570)

Key Equations and Relationships

Numbered and screened concepts are listed for you to refer to or memorize.

17.1 Defining equilibrium in terms of reaction rates (543):

$$\text{At equilibrium: rate}_{fwd} = \text{rate}_{rev}$$

17.2 Defining the equilibrium constant for the reaction $A \rightleftharpoons 2B$ (544):

$$K = \frac{k_{fwd}}{k_{rev}} = \frac{[B]^2_{eq}}{[A]_{eq}}$$

17.3 Defining the equilibrium constant in terms of the reaction quotient (546):

$$\text{At equilibrium: } Q = K$$

17.4 Expressing Q_c for the reaction $aA + bB \rightleftharpoons cC + dD$ (546):

$$Q_c = \frac{[C]^c[D]^d}{[A]^a[B]^b}$$

17.5 Finding the overall K for a reaction sequence (547):

$$K_{overall} = K_1 \times K_2 \times K_3 \times \cdots$$

17.6 Finding K of a reaction from K of the reverse reaction (549):

$$K_{fwd} = \frac{1}{K_{rev}}$$

17.7 Finding K of a reaction multiplied by a factor n (549):

$$Q' = Q^n = \left(\frac{[C]^c[D]^d}{[A]^a[B]^b}\right)^n \qquad K' = K^n$$

17.8 Relating K based on pressures to K based on concentrations (551):

$$K_p = K_c(RT)^{\Delta n_{gas}}$$

17.9 Assuming that ignoring the concentration that reacts introduces no significant error (559):

$$[A]_{init} - [A]_{reacting} = [A]_{eq} \approx [A]_{init}$$

BRIEF SOLUTIONS TO FOLLOW-UP PROBLEMS

Compare your own solutions to these calculation steps and answers.

17.1 (a) $Q_c = \dfrac{[NO]^4[H_2O]^6}{[NH_3]^4[O_2]^5}$ (b) $Q_c = \dfrac{[N_2O][NO_2]}{[NO]^3}$

17.2 $H_2(g) + Br_2(g) \rightleftharpoons 2HBr(g)$;

$$Q_{c(overall)} = \frac{[HBr]^2}{[H_2][Br_2]}$$

$$Q_{c(overall)} = Q_{c1} \times Q_{c2} \times Q_{c3}$$

$$= \frac{\cancel{[Br]^2}}{[Br_2]} \times \frac{[HBr]\cancel{[H]}}{\cancel{[Br]}[H_2]} \times \frac{[HBr]}{\cancel{[H]}\cancel{[Br]}} = \frac{[HBr]^2}{[H_2][Br_2]}$$

17.3 $K_p = K_c(RT)^{-1} = 1.67\left(0.0821 \dfrac{\text{atm·L}}{\text{mol·K}} \times 500.\text{ K}\right)^{-1}$

$= 4.07\times10^{-2}$

17.4 $K_c = \dfrac{[Y]}{[X]} = 1.4$

1. $Q = 0.33$, right 2. $Q = 1.4$, no net change 3. $Q = 2.0$, left

17.5 $Q_p = \dfrac{(P_{CH_3Cl})(P_{HCl})}{(P_{CH_4})(P_{Cl_2})} = \dfrac{(0.24)(0.47)}{(0.13)(0.035)} = 25$;

$Q_p < K_p$, so CH_3Cl is forming.

17.6 From the reaction table for $2NO + O_2 \rightleftharpoons 2NO_2$,
$P_{O_2} = 1.000 \text{ atm} - x = 0.506 \text{ atm} \qquad x = 0.494 \text{ atm}$
Also, $P_{NO} = 0.012$ atm and $P_{NO_2} = 0.988$ atm, so

$$K_p = \frac{0.988^2}{0.012^2(0.506)} = 1.3\times10^4$$

17.7 Since $\Delta n_{gas} = 0$, $K_p = K_c = 2.3\times10^{30} = \dfrac{(0.781)(0.209)}{P^2_{NO}}$

Thus, $\qquad P_{NO} = 2.7\times10^{-16}$ atm

17.8 From the reaction table, $[H_2] = [I_2] = x$; $[HI] = 0.242 - 2x$.

Thus, $\qquad K_c = 1.26\times10^{-3} = \dfrac{x^2}{(0.242 - 2x)^2}$

Taking the square root of both sides, ignoring the negative root, and solving gives $x = [H_2] = 8.02\times10^{-3}\ M$.

17.9 (a) Based on the reaction table, and assuming that $0.20\ M - x \approx 0.20\ M$,

$$K_c = 2.94\times10^{-10} \approx \frac{4x^2}{0.20} \qquad x \approx 3.8\times10^{-6}$$

Error $= 1.9\times10^{-3}\%$, so assumption is justified; therefore, at equilibrium, $[I_2] = 0.20\ M$ and $[I] = 7.6\times10^{-6}\ M$.
(b) Based on the same reaction table and assumption, $x \approx 0.10$; error is 50%, so assumption is *not* justified. Solve equation:

$$4x^2 + 0.209x - 0.042 = 0 \qquad x = 0.080\ M$$

Therefore, at equilibrium, $[I_2] = 0.12\ M$ and $[I] = 0.16\ M$.

17.10 (a) $Q_c = \dfrac{(0.0900)(0.0900)}{0.2100} = 3.86\times10^{-2}$
$Q_c < K_c$, so reaction proceeds to the right.
(b) From the reaction table,
$[PCl_5] = 0.2100\ M - x = 0.2065\ M \qquad$ so $\qquad x = 0.0035\ M$
Thus, $[Cl_2] = [PCl_3] = 0.0900\ M + x = 0.0935\ M$.

17.11 (a) $[SiF_4]$ increases; (b) decreases; (c) decreases; (d) no effect.

17.12 (a) Decrease *P*; (b) increase *P*; (c) increase *P*.

17.13 (a) P_{H_2} will decrease; *K* will increase; (b) P_{N_2} will increase; *K* will decrease; (c) P_{PCl_5} will increase; *K* will increase.

17.14 (a) Since $P = \dfrac{n}{V}RT$ and, in this case, *V, R,* and *T* cancel,

$$K_p = \frac{n^2_{CD}}{n_{C_2}\times n_{D_2}} = \frac{16}{(2)(2)} = 4$$

(b) Scene 2, to the left; scene 3, to the right. (c) There are 2 mol of gas on each side of the balanced equation, so increasing *T* has no effect on total moles of gas.

PROBLEMS

Problems with **colored** numbers are answered in Appendix E. Sections match the text and provide the numbers of relevant sample problems. Bracketed problems are grouped in pairs (indicated by a short rule) that cover the same concept. Comprehensive Problems are based on material from any section or previous chapter.

The Equilibrium State and the Equilibrium Constant

17.1 A change in reaction conditions increases the rate of a certain forward reaction more than that of the reverse reaction. What is the effect on the equilibrium constant and on the concentrations of reactants and products at equilibrium?

17.2 When a chemical company employs a new reaction to manufacture a product, the chemists consider its rate (kinetics) and yield (equilibrium). How do each of these affect the usefulness of a manufacturing process?

17.3 If there is no change in concentrations, why is the equilibrium state considered dynamic?

17.4 Is *K* very large or very small for a reaction that goes essentially to completion? Explain.

17.5 White phosphorus, P_4, is produced by the reduction of phosphate rock, $Ca_3(PO_4)_2$. If exposed to oxygen, the waxy, white solid smokes, bursts into flames, and releases a large quantity of heat:

$$P_4(g) + 5O_2(g) \rightleftharpoons P_4O_{10}(s) + \textit{heat}$$

Does this reaction have a large or small equilibrium constant? Explain.

The Reaction Quotient and the Equilibrium Constant

(Sample Problems 17.1 and 17.2)

17.6 For a given reaction at a given temperature, the value of *K* is constant. Is the value of *Q* also constant? Explain.

17.7 In a study of the thermal decomposition of lithium peroxide,

$$2Li_2O_2(s) \rightleftharpoons 2Li_2O(s) + O_2(g)$$

a chemist finds that, as long as some Li_2O_2 is present at the end of the experiment, the amount of O_2 obtained in a given container at a given *T* is the same. Explain.

17.8 In a study of the formation of HI from its elements,

$$H_2(g) + I_2(g) \rightleftharpoons 2HI(g)$$

equal amounts of H_2 and I_2 were placed in a container, which was then sealed and heated.
(a) On one set of axes, sketch concentration vs. time curves for H_2 and HI, and explain how *Q* changes as a function of time.
(b) Is the value of *Q* different if $[I_2]$ is plotted instead of $[H_2]$?

17.9 Explain the difference between a heterogeneous and a homogeneous equilibrium. Give an example of each.

17.10 Does *Q* for the formation of 1 mol of NO from its elements differ from *Q* for the decomposition of 1 mol of NO to its elements? Explain and give the relationship between the two *Q*'s.

17.11 Does Q for the formation of 1 mol of NH_3 from H_2 and N_2 differ from Q for the formation of NH_3 from H_2 and 1 mol of N_2? Explain and give the relationship between the two Q's.

17.12 Balance each reaction and write its reaction quotient, Q_c:
(a) $NO(g) + O_2(g) \rightleftharpoons N_2O_3(g)$
(b) $SF_6(g) + SO_3(g) \rightleftharpoons SO_2F_2(g)$
(c) $SClF_5(g) + H_2(g) \rightleftharpoons S_2F_{10}(g) + HCl(g)$

17.13 Balance each reaction and write its reaction quotient, Q_c:
(a) $C_2H_6(g) + O_2(g) \rightleftharpoons CO_2(g) + H_2O(g)$
(b) $CH_4(g) + F_2(g) \rightleftharpoons CF_4(g) + HF(g)$
(c) $SO_3(g) \rightleftharpoons SO_2(g) + O_2(g)$

17.14 At a particular temperature, $K_c = 1.6\times10^{-2}$ for

$$2H_2S(g) \rightleftharpoons 2H_2(g) + S_2(g)$$

Calculate K_c for each of the following reactions:
(a) $\frac{1}{2}S_2(g) + H_2(g) \rightleftharpoons H_2S(g)$
(b) $5H_2S(g) \rightleftharpoons 5H_2(g) + \frac{5}{2}S_2(g)$

17.15 At a particular temperature, $K_c = 6.5\times10^2$ for

$$2NO(g) + 2H_2(g) \rightleftharpoons N_2(g) + 2H_2O(g)$$

Calculate K_c for each of the following reactions:
(a) $NO(g) + H_2(g) \rightleftharpoons \frac{1}{2}N_2(g) + H_2O(g)$
(b) $2N_2(g) + 4H_2O(g) \rightleftharpoons 4NO(g) + 4H_2(g)$

17.16 Balance each of the following examples of heterogeneous equilibria and write each reaction quotient, Q_c:
(a) $Na_2O_2(s) + CO_2(g) \rightleftharpoons Na_2CO_3(s) + O_2(g)$
(b) $H_2O(l) \rightleftharpoons H_2O(g)$
(c) $NH_4Cl(s) \rightleftharpoons NH_3(g) + HCl(g)$

17.17 Balance each of the following examples of heterogeneous equilibria and write each reaction quotient, Q_c:
(a) $H_2O(l) + SO_3(g) \rightleftharpoons H_2SO_4(aq)$
(b) $KNO_3(s) \rightleftharpoons KNO_2(s) + O_2(g)$
(c) $S_8(s) + F_2(g) \rightleftharpoons SF_6(g)$

17.18 Write Q_c for each of the following:
(a) Hydrogen chloride gas reacts with oxygen gas to produce chlorine gas and water vapor.
(b) Solid diarsenic trioxide reacts with fluorine gas to produce liquid arsenic pentafluoride and oxygen gas.
(c) Gaseous sulfur tetrafluoride reacts with liquid water to produce gaseous sulfur dioxide and hydrogen fluoride gas.
(d) Solid molybdenum(VI) oxide reacts with gaseous xenon difluoride to form liquid molybdenum(VI) fluoride, xenon gas, and oxygen gas.

17.19 The interhalogen compound ClF_3 is prepared in a two-step fluorination of chlorine gas:

$$Cl_2(g) + F_2(g) \rightleftharpoons ClF(g)$$
$$ClF(g) + F_2(g) \rightleftharpoons ClF_3(g)$$

(a) Balance each step and write the overall equation.
(b) Show that the overall Q_c equals the product of the Q_c's for the individual steps.

Expressing Equilibria with Pressure Terms: Relation Between K_c and K_p
(Sample Problem 17.3)

17.20 Guldberg and Waage proposed the definition of the equilibrium constant as a certain ratio of *concentrations*. What relationship allows us to use a particular ratio of *partial pressures* (for a gaseous reaction) to express an equilibrium constant? Explain.

17.21 When are K_c and K_p equal, and when are they not?

17.22 A certain reaction at equilibrium has more moles of gaseous products than of gaseous reactants.
(a) Is K_c larger or smaller than K_p?
(b) Write a statement about the relative sizes of K_c and K_p for any gaseous equilibrium.

17.23 Determine Δn_{gas} for each of the following reactions:
(a) $2KClO_3(s) \rightleftharpoons 2KCl(s) + 3O_2(g)$
(b) $2PbO(s) + O_2(g) \rightleftharpoons 2PbO_2(s)$
(c) $I_2(s) + 3XeF_2(s) \rightleftharpoons 2IF_3(s) + 3Xe(g)$

17.24 Determine Δn_{gas} for each of the following reactions:
(a) $MgCO_3(s) \rightleftharpoons MgO(s) + CO_2(g)$
(b) $2H_2(g) + O_2(g) \rightleftharpoons 2H_2O(l)$
(c) $HNO_3(l) + ClF(g) \rightleftharpoons ClONO_2(g) + HF(g)$

17.25 Calculate K_c for each of the following equilibria:
(a) $CO(g) + Cl_2(g) \rightleftharpoons COCl_2(g)$; $K_p = 3.9\times10^{-2}$ at 1000. K
(b) $S_2(g) + C(s) \rightleftharpoons CS_2(g)$; $K_p = 28.5$ at 500. K

17.26 Calculate K_c for each of the following equilibria:
(a) $H_2(g) + I_2(g) \rightleftharpoons 2HI(g)$; $K_p = 49$ at 730. K
(b) $2SO_2(g) + O_2(g) \rightleftharpoons 2SO_3(g)$; $K_p = 2.5\times10^{10}$ at 500. K

Comparing Q and K to Predict Reaction Direction
(Sample Problems 17.4 and 17.5)

17.27 When the numerical value of Q is less than K, in which direction does the reaction proceed to reach equilibrium? Explain.

17.28 The following molecular scenes depict the aqueous reaction $2D \rightleftharpoons E$, with D *red* and E *blue*. Each sphere represents 0.0100 mol, but the volume is 1.00 L in scene A, whereas in scenes B and C, it is 0.500 L.

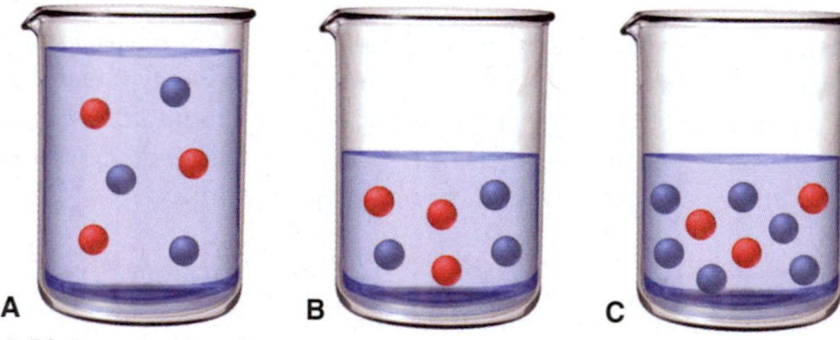

(a) If the reaction in scene A is at equilibrium, calculate K_c.
(b) Are the reactions in scenes B and C at equilibrium? Which, if either, is not, and in which direction will it proceed?

17.29 At 425°C, $K_p = 4.18\times10^{-9}$ for the reaction

$$2HBr(g) \rightleftharpoons H_2(g) + Br_2(g)$$

In one experiment, 0.20 atm of HBr(g), 0.010 atm of $H_2(g)$, and 0.010 atm of $Br_2(g)$ are introduced into a container. Is the reaction at equilibrium? If not, in which direction will it proceed?

17.30 At 100°C, $K_p = 60.6$ for the reaction

$$2NOBr(g) \rightleftharpoons 2NO(g) + Br_2(g)$$

In a given experiment, 0.10 atm of each component is placed in a container. Is the system at equilibrium? If not, in which direction will the reaction proceed?

17.31 The water-gas shift reaction plays a central role in the chemical methods for obtaining cleaner fuels from coal:

$$CO(g) + H_2O(g) \rightleftharpoons CO_2(g) + H_2(g)$$

At a given temperature, $K_p = 2.7$. If 0.13 mol of CO, 0.56 mol of H_2O, 0.62 mol of CO_2, and 0.43 mol of H_2 are put in a 2.0-L flask, in which direction does the reaction proceed?

How to Solve Equilibrium Problems

(Sample Problems 17.6 to 17.10)

17.32 For a problem involving the catalyzed reaction of methane and steam, the following reaction table was prepared:

Pressure (atm)	$CH_4(g)$ +	$2H_2O(g)$ ⇌	$CO_2(g)$ +	$4H_2(g)$
Initial	0.30	0.40	0	0
Change	$-x$	$-2x$	$+x$	$+4x$
Equilibrium	$0.30 - x$	$0.40 - 2x$	x	$4x$

Explain the entries in the "Change" and "Equilibrium" rows.

17.33 (a) What is the basis of the approximation that avoids using the quadratic formula to find an equilibrium concentration? (b) When should this approximation *not* be made?

17.34 In an experiment to study the formation of HI(*g*),

$$H_2(g) + I_2(g) \rightleftharpoons 2HI(g)$$

$H_2(g)$ and $I_2(g)$ were placed in a sealed container at a certain temperature. At equilibrium, $[H_2] = 6.50\times10^{-5}$ *M*, $[I_2] = 1.06\times10^{-3}$ *M*, and $[HI] = 1.87\times10^{-3}$ *M*. Calculate K_c for the reaction at this temperature.

17.35 Gaseous ammonia was introduced into a sealed container and heated to a certain temperature:

$$2NH_3(g) \rightleftharpoons N_2(g) + 3H_2(g)$$

At equilibrium, $[NH_3] = 0.0225$ *M*, $[N_2] = 0.114$ *M*, and $[H_2] = 0.342$ *M*. Calculate K_c for the reaction at this temperature.

17.36 Gaseous PCl_5 decomposes according to the reaction

$$PCl_5(g) \rightleftharpoons PCl_3(g) + Cl_2(g)$$

In one experiment, 0.15 mol of $PCl_5(g)$ was introduced into a 2.0-L container. Construct the reaction table for this process.

17.37 Hydrogen fluoride, HF, can be made from the reaction

$$H_2(g) + F_2(g) \rightleftharpoons 2HF(g)$$

In one experiment, 0.10 mol of $H_2(g)$ and 0.050 mol of $F_2(g)$ are added to a 0.50-L flask. Write a reaction table for this process.

17.38 For the following reaction, $K_p = 6.5\times10^4$ at 308 K:

$$2NO(g) + Cl_2(g) \rightleftharpoons 2NOCl(g)$$

At equilibrium, $P_{NO} = 0.35$ atm and $P_{Cl_2} = 0.10$ atm. What is the equilibrium partial pressure of NOCl(*g*)?

17.39 For the following reaction, $K_p = 0.262$ at 1000°C:

$$C(s) + 2H_2(g) \rightleftharpoons CH_4(g)$$

At equilibrium, P_{H_2} is 1.22 atm. What is the equilibrium partial pressure of $CH_4(g)$?

17.40 Ammonium hydrogen sulfide decomposes according to the following reaction, for which $K_p = 0.11$ at 250°C:

$$NH_4HS(s) \rightleftharpoons H_2S(g) + NH_3(g)$$

If 55.0 g of $NH_4HS(s)$ is placed in a sealed 5.0-L container, what is the partial pressure of $NH_3(g)$ at equilibrium?

17.41 Hydrogen sulfide decomposes according to the following reaction, for which $K_c = 9.30\times10^{-8}$ at 700°C:

$$2H_2S(g) \rightleftharpoons 2H_2(g) + S_2(g)$$

If 0.45 mol of H_2S is placed in a 3.0-L container, what is the equilibrium concentration of $H_2(g)$ at 700°C?

17.42 Even at high *T*, the formation of NO is not favored:

$$N_2(g) + O_2(g) \rightleftharpoons 2NO(g) \quad K_c = 4.10\times10^{-4} \text{ at } 2000°C$$

What is [NO] when a mixture of 0.20 mol of $N_2(g)$ and 0.15 mol of $O_2(g)$ reach equilibrium in a 1.0-L container at 2000°C?

17.43 Nitrogen dioxide decomposes according to the reaction

$$2NO_2(g) \rightleftharpoons 2NO(g) + O_2(g)$$

where $K_p = 4.48\times10^{-13}$ at a certain temperature. If 0.75 atm of NO_2 is added to a container and allowed to come to equilibrium, what are the equilibrium partial pressures of NO(*g*) and $O_2(g)$?

17.44 In an analysis of interhalogen reactivity, 0.500 mol of ICl was placed in a 5.00-L flask, where it decomposed at a high *T*: $2ICl(g) \rightleftharpoons I_2(g) + Cl_2(g)$. Calculate the equilibrium concentrations of I_2, Cl_2, and ICl ($K_c = 0.110$ at this temperature).

17.45 A toxicologist studying mustard gas, $S(CH_2CH_2Cl)_2$, a blistering agent, prepares a mixture of 0.675 *M* SCl_2 and 0.973 *M* C_2H_4 and allows it to react at room temperature (20.0°C):

$$SCl_2(g) + 2C_2H_4(g) \rightleftharpoons S(CH_2CH_2Cl)_2(g)$$

At equilibrium, $[S(CH_2CH_2Cl)_2] = 0.350$ *M*. Calculate K_p.

17.46 The first step in HNO_3 production is the catalyzed oxidation of NH_3. Without a catalyst, a different reaction predominates:

$$4NH_3(g) + 3O_2(g) \rightleftharpoons 2N_2(g) + 6H_2O(g)$$

When 0.0150 mol of $NH_3(g)$ and 0.0150 mol of $O_2(g)$ are placed in a 1.00-L container at a certain temperature, the N_2 concentration at equilibrium is 1.96×10^{-3} *M*. Calculate K_c.

17.47 A key step in the extraction of iron from its ore is

$$FeO(s) + CO(g) \rightleftharpoons Fe(s) + CO_2(g) \quad K_p = 0.403 \text{ at } 1000°C$$

This step occurs in the 700°C to 1200°C zone within a blast furnace. What are the equilibrium partial pressures of CO(*g*) and $CO_2(g)$ when 1.00 atm of CO(*g*) and excess FeO(*s*) react in a sealed container at 1000°C?

Reaction Conditions and Equilibrium: Le Châtelier's Principle

(Sample Problems 17.11 to 17.14)

17.48 What is the difference between the equilibrium position and the equilibrium constant of a reaction? Which changes as a result of a change in reactant concentration?

17.49 Scenes A, B, and C below depict the following reaction at three temperatures:

$$NH_4Cl(s) \rightleftharpoons NH_3(g) + HCl(g) \quad \Delta H^\circ_{rxn} = 176 \text{ kJ}$$

A

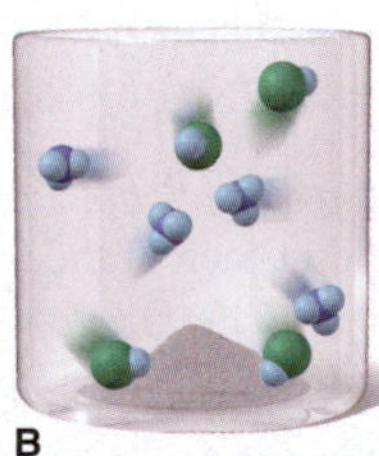
B

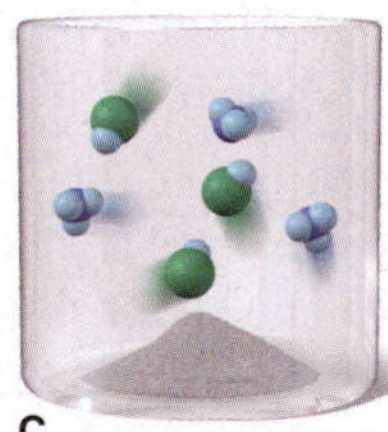
C

(a) Which best represents the reaction mixture at the highest temperature? Explain. (b) Which best represents the reaction mixture at the lowest temperature? Explain.

17.50 What is implied by the word "constant" in the term *equilibrium constant?* Give two reaction parameters that can be changed without changing the value of an equilibrium constant.

17.51 Le Châtelier's principle is related ultimately to the rates of the forward and reverse steps in a reaction. Explain (a) why an increase in reactant concentration shifts the equilibrium position to the right but does not change *K;* (b) why a decrease in *V* shifts the equilibrium position toward fewer moles of gas but does not change K.

17.52 (a) Explain why a rise in T shifts the equilibrium position of an exothermic reaction to the left and also changes K. (b) If the temperature rises from T_1 to T_2 in an endothermic reaction, is K_2 larger or smaller than K_1? Explain.

17.53 An equilibrium mixture of two solids and a gas, in the reaction $XY(s) \rightleftharpoons X(g) + Y(s)$, is depicted at right (X is *green* and Y is *black*). Does scene A, B, or C best represent the system at equilibrium after two formula units of $Y(s)$ is added? Explain.

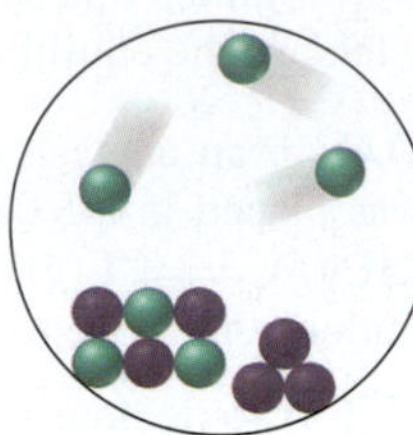

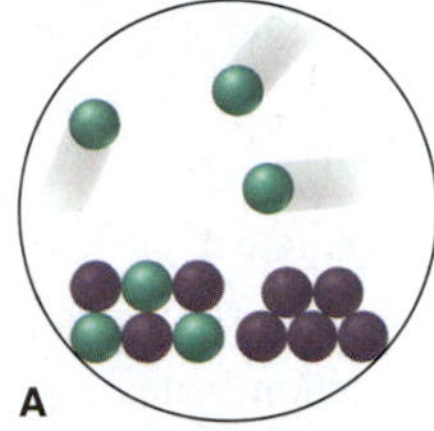

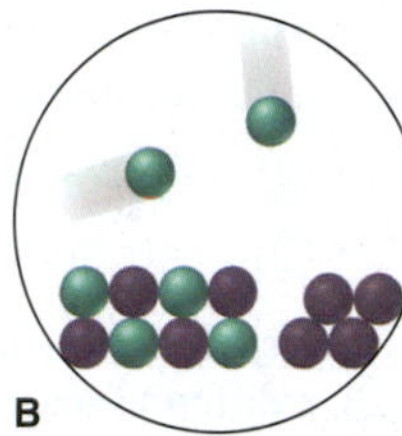

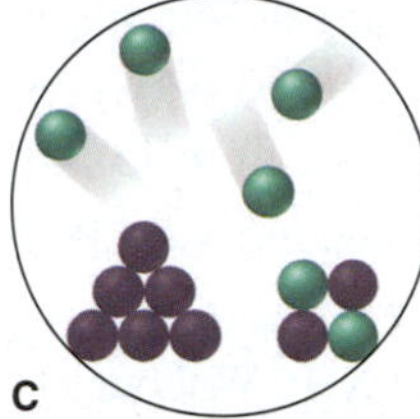

17.54 Consider this equilibrium system:

$$CO(g) + Fe_3O_4(s) \rightleftharpoons CO_2(g) + 3FeO(s)$$

How does the equilibrium position shift as a result of each of the following disturbances? (a) CO is added. (b) CO_2 is removed by adding solid NaOH. (c) Additional $Fe_3O_4(s)$ is added to the system. (d) Dry ice is added at constant temperature.

17.55 Sodium bicarbonate undergoes thermal decomposition according to the reaction

$$2NaHCO_3(s) \rightleftharpoons Na_2CO_3(s) + CO_2(g) + H_2O(g)$$

How does the equilibrium position shift as a result of each of the following disturbances? (a) 0.20 atm of argon gas is added. (b) $NaHCO_3(s)$ is added. (c) $Mg(ClO_4)_2(s)$ is added as a drying agent to remove H_2O. (d) Dry ice is added at constant T.

17.56 Predict the effect of *increasing* the container volume on the amounts of each reactant and product in the following reactions:
(a) $F_2(g) \rightleftharpoons 2F(g)$
(b) $2CH_4(g) \rightleftharpoons C_2H_2(g) + 3H_2(g)$

17.57 Predict the effect of *decreasing* the container volume on the amounts of each reactant and product in the following reactions:
(a) $C_3H_8(g) + 5O_2(g) \rightleftharpoons 3CO_2(g) + 4H_2O(l)$
(b) $4NH_3(g) + 3O_2(g) \rightleftharpoons 2N_2(g) + 6H_2O(g)$

17.58 How would you adjust the *volume* of the container in order to maximize product yield in each of the following reactions?
(a) $Fe_3O_4(s) + 4H_2(g) \rightleftharpoons 3Fe(s) + 4H_2O(g)$
(b) $2C(s) + O_2(g) \rightleftharpoons 2CO(g)$

17.59 How would you adjust the *volume* of the container in order to maximize product yield in each of the following reactions?
(a) $Na_2O_2(s) \rightleftharpoons 2Na(l) + O_2(g)$
(b) $C_2H_2(g) + 2H_2(g) \rightleftharpoons C_2H_6(g)$

17.60 Predict the effect of *increasing* the temperature on the amounts of products in the following reactions:
(a) $CO(g) + 2H_2(g) \rightleftharpoons CH_3OH(g) \quad \Delta H^\circ_{rxn} = -90.7$ kJ
(b) $C(s) + H_2O(g) \rightleftharpoons CO(g) + H_2(g) \quad \Delta H^\circ_{rxn} = 131$ kJ
(c) $2NO_2(g) \rightleftharpoons 2NO(g) + O_2(g)$ (endothermic)
(d) $2C(s) + O_2(g) \rightleftharpoons 2CO(g)$ (exothermic)

17.61 Predict the effect of *decreasing* the temperature on the amounts of reactants in the following reactions:
(a) $C_2H_2(g) + H_2O(g) \rightleftharpoons CH_3CHO(g) \quad \Delta H^\circ_{rxn} = -151$ kJ
(b) $CH_3CH_2OH(l) + O_2(g) \rightleftharpoons CH_3CO_2H(l) + H_2O(g) \quad \Delta H^\circ_{rxn} = -451$ kJ
(c) $2C_2H_4(g) + O_2(g) \rightleftharpoons 2CH_3CHO(g)$ (exothermic)
(d) $N_2O_4(g) \rightleftharpoons 2NO_2(g)$ (endothermic)

17.62 The minerals hematite (Fe_2O_3) and magnetite (Fe_3O_4) exist in equilibrium with atmospheric oxygen:

$$4Fe_3O_4(s) + O_2(g) \rightleftharpoons 6Fe_2O_3(s) \quad K_p = 2.5\times10^{87} \text{ at } 298 \text{ K}$$

(a) Determine P_{O_2} at equilibrium. (b) Given that P_{O_2} in air is 0.21 atm, in which direction will the reaction proceed to reach equilibrium? (c) Calculate K_c at 298 K.

17.63 The oxidation of SO_2 is the key step in H_2SO_4 production:

$$SO_2(g) + \tfrac{1}{2}O_2(g) \rightleftharpoons SO_3(g) \quad \Delta H^\circ_{rxn} = -99.2 \text{ kJ}$$

(a) What qualitative combination of T and P maximizes SO_3 yield?
(b) How does addition of O_2 affect Q? K?
(c) Why is catalysis used for this reaction?

17.64 You are a member of a research team of chemists discussing plans for a plant to produce ammonia:

$$N_2(g) + 3H_2(g) \rightleftharpoons 2NH_3(g)$$

(a) The plant will operate at close to 700 K, at which K_p is 1.00×10^{-4}, and employs the stoichiometric 1/3 ratio of N_2/H_2. At equilibrium, the partial pressure of NH_3 is 50. atm. Calculate the partial pressures of each reactant and P_{total}.
(b) One member of the team suggests the following: since the partial pressure of H_2 is cubed in the reaction quotient, the plant could produce the same amount of NH_3 if the reactants were in a 1/6 ratio of N_2/H_2 and could do so at a lower pressure, which would cut operating costs. Calculate the partial pressure of each reactant and P_{total} under these conditions, assuming an unchanged partial pressure of 50. atm for NH_3. Is the suggestion valid?

Comprehensive Problems

17.65 One of the most important industrial sources of ethanol is reaction of steam with ethylene derived from crude oil:

$$C_2H_4(g) + H_2O(g) \rightleftharpoons C_2H_5OH(g)$$
$$\Delta H^\circ_{rxn} = -47.8 \text{ kJ} \quad K_c = 9\times10^3 \text{ at } 600. \text{ K}$$

(a) At equilibrium, $P_{C_2H_5OH} = 200.$ atm and $P_{H_2O} = 400.$ atm. Calculate $P_{C_2H_4}$.
(b) Is the highest yield of ethanol obtained at high or low pressures? High or low temperatures?
(c) In ammonia manufacture, the yield is increased by condensing the NH_3 to a liquid and removing it from the vessel. Would condensing the C_2H_5OH have the same effect for ethanol production? Explain.

17.66 The "filmstrip" represents five molecular scenes of a gaseous mixture as it reaches equilibrium over time:

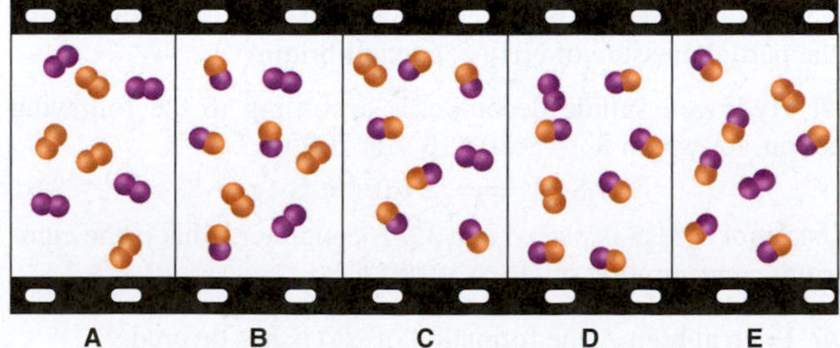

X is *purple* and Y is *orange*: $X_2(g) + Y_2(g) \rightleftharpoons 2XY(g)$.
(a) Write the reaction quotient, Q, for this reaction.
(b) If each particle represents 0.1 mol, find Q for each scene.
(c) If $K > 1$, is time progressing to the right or to the left? Explain.
(d) Calculate K at this temperature.
(e) If $\Delta H^\circ_{rxn} < 0$, which scene, if any, best represents the mixture at a higher temperature? Explain.
(f) Which scene, if any, best represents the mixture at a higher pressure (lower volume)? Explain.

17.67 An industrial chemist introduces 2.0 atm of H_2 and 2.0 atm of CO_2 into a 1.00-L container at 25.0°C and then raises the temperature to 700.°C, at which $K_c = 0.534$:

$$H_2(g) + CO_2(g) \rightleftharpoons H_2O(g) + CO(g)$$

How many grams of H_2 are present at equilibrium?

17.68 As an EPA scientist studying catalytic converters and urban smog, you want to find K_c for the following reaction:

$$2NO_2(g) \rightleftharpoons N_2(g) + 2O_2(g) \qquad K_c = ?$$

Use the following data to find the unknown K_c:

$$\tfrac{1}{2}N_2(g) + \tfrac{1}{2}O_2(g) \rightleftharpoons NO(g) \qquad K_c = 4.8\times10^{-10}$$

$$2NO_2(g) \rightleftharpoons 2NO(g) + O_2(g) \qquad K_c = 1.1\times10^{-5}$$

17.69 An engineer examining the oxidation of SO_2 in the manufacture of sulfuric acid determines that $K_c = 1.7\times10^8$ at 600. K:

$$2SO_2(g) + O_2(g) \rightleftharpoons 2SO_3(g)$$

(a) At equilibrium, $P_{SO_3} = 300.$ atm and $P_{O_2} = 100.$ atm. Calculate P_{SO_2}. (b) The engineer places a mixture of 0.0040 mol of $SO_2(g)$ and 0.0028 mol of $O_2(g)$ in a 1.0-L container and raises the temperature to 1000 K. At equilibrium, 0.0020 mol of $SO_3(g)$ is present. Calculate K_c and P_{SO_2} for this reaction at 1000. K.

17.70 When 0.100 mol of $CaCO_3(s)$ and 0.100 mol of $CaO(s)$ are placed in an evacuated sealed 10.0-L container and heated to 385 K, $P_{CO_2} = 0.220$ atm after equilibrium is established:

$$CaCO_3(s) \rightleftharpoons CaO(s) + CO_2(g)$$

An additional 0.300 atm of $CO_2(g)$ is pumped in. What is the total mass (in g) of $CaCO_3$ after equilibrium is re-established?

17.71 Use each of the following reaction quotients to write the balanced equation:

(a) $Q = \dfrac{[CO_2]^2[H_2O]^2}{[C_2H_4][O_2]^3}$ (b) $Q = \dfrac{[NH_3]^4[O_2]^7}{[NO_2]^4[H_2O]^6}$

17.72 In combustion studies of H_2 as an alternative fuel, you find evidence that the hydroxyl radical (HO) is formed in flames by the reaction $H(g) + \tfrac{1}{2}O_2(g) \rightleftharpoons HO(g)$. Use the following data to calculate K_c for the reaction:

$$\tfrac{1}{2}H_2(g) + O_2(g) \rightleftharpoons HO(g) \qquad K_c = 0.58$$

$$\tfrac{1}{2}H_2(g) \rightleftharpoons H(g) \qquad K_c = 1.6\times10^{-3}$$

17.73 An equilibrium mixture of car exhaust gases consisting of 10.0 volumes of CO_2, 1.00 volume of unreacted O_2, and 50.0 volumes of unreacted N_2 leaves the engine at 4.0 atm and 800. K.
(a) Given this equilibrium, what is the partial pressure of CO?

$$2CO_2(g) \rightleftharpoons 2CO(g) + O_2(g) \qquad K_p = 1.4\times10^{-28} \text{ at } 800. \text{ K}$$

(b) Assuming the mixture has enough time to reach equilibrium, what is the concentration in picograms per liter (pg/L) of CO in the exhaust gas? (The actual concentration of CO in car exhaust is much higher because the gases do *not* reach equilibrium in the short transit time through the engine and exhaust system.)

17.74 Consider the following reaction:

$$3Fe(s) + 4H_2O(g) \rightleftharpoons Fe_3O_4(s) + 4H_2(g)$$

(a) Fe_3O_4 is a compound of iron in which Fe occurs in two oxidation states. What are the oxidation states of Fe in Fe_3O_4?
(b) At 900°C, K_c for the reaction is 5.1. If 0.050 mol of $H_2O(g)$ and 0.100 mol of $Fe(s)$ are placed in a 1.0-L container at 900°C, how many grams of Fe_3O_4 are present at equilibrium?

17.75 For the reaction $M_2 + N_2 \rightleftharpoons 2MN$, scene A represents the mixture at equilibrium, with M *black* and N *orange*. If each molecule represents 0.10 mol and the volume is 1.0 L, how many moles of each substance will be present in scene B when that mixture reaches equilibrium?

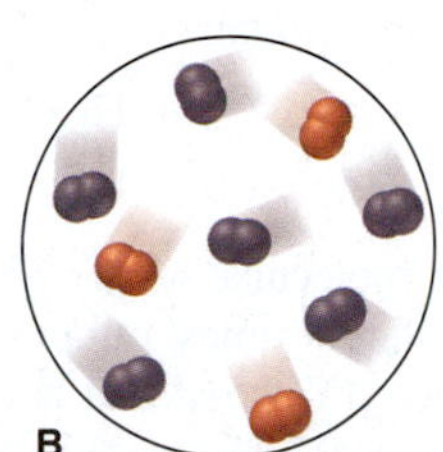

17.76 A study of the water-gas shift reaction (see Problem 17.31) was made in which equilibrium was reached with $[CO] = [H_2O] = [H_2] = 0.10\ M$ and $[CO_2] = 0.40\ M$. After 0.60 mol of H_2 is added to the 2.0-L container and equilibrium is re-established, what are the new concentrations of all the components?

Note: The synthesis of ammonia is a major process throughout the industrialized world. Problems 17.77 to 17.80 refer to various aspects of this all-important reaction:

$$N_2(g) + 3H_2(g) \rightleftharpoons 2NH_3(g) \qquad \Delta H^\circ_{rxn} = -91.8 \text{ kJ}$$

17.77 When ammonia is made industrially, the mixture of N_2, H_2, and NH_3 that emerges from the reaction chamber is far from equilibrium. Why does the plant supervisor use reaction conditions that produce less than the maximum yield of ammonia?

17.78 The following reaction can be used to make H_2 for the synthesis of ammonia from the greenhouse gases carbon dioxide and methane:

$$CH_4(g) + CO_2(g) \rightleftharpoons 2CO(g) + 2H_2(g)$$

(a) What is the percent yield of H_2 when an equimolar mixture of CH_4 and CO_2 with a total pressure of 20.0 atm reaches equilibrium at 1200. K, at which $K_p = 3.548\times10^6$?
(b) What is the percent yield of H_2 for this system at 1300. K, at which $K_p = 2.626\times10^7$?

17.79 Using CH_4 and steam as a source of H_2 for NH_3 synthesis requires high temperatures. Rather than burning CH_4 separately to heat the mixture, it is more efficient to inject some O_2 into the reaction mixture. All of the H_2 is thus released for the synthesis, and the heat of reaction for the combustion of CH_4 helps maintain the required temperature. Imagine the reaction occurring in two steps:

$$2CH_4(g) + O_2(g) \rightleftharpoons 2CO(g) + 4H_2(g) \qquad K_p = 9.34\times10^{28} \text{ at } 1000. \text{ K}$$

$$CO(g) + H_2O(g) \rightleftharpoons CO_2(g) + H_2(g) \qquad K_p = 1.374 \text{ at } 1000. \text{ K}$$

(a) Write the overall equation for the reaction of methane, steam, and oxygen to form carbon dioxide and hydrogen.
(b) What is K_p for the overall reaction?
(c) What is K_c for the overall reaction?
(d) A mixture of 2.0 mol of CH_4, 1.0 mol of O_2, and 2.0 mol of steam with a total pressure of 30. atm reacts at 1000. K at constant

volume. Assuming that the reaction is complete and the ideal gas law is a valid approximation, what is the final pressure?

17.80 One mechanism for the synthesis of ammonia proposes that N_2 and H_2 molecules catalytically dissociate into atoms:

$$N_2(g) \rightleftharpoons 2N(g) \qquad \log K_p = -43.10$$
$$H_2(g) \rightleftharpoons 2H(g) \qquad \log K_p = -17.30$$

(a) Find the partial pressure of N in N_2 at 1000. K and 200. atm.
(b) Find the partial pressure of H in H_2 at 1000. K and 600. atm.
(c) How many N atoms and H atoms are present per liter?
(d) Based on these answers, which of the following is a more reasonable step to continue the mechanism after the catalytic dissociation? Explain.

$$N(g) + H(g) \longrightarrow NH(g)$$
$$N_2(g) + H(g) \longrightarrow NH(g) + N(g)$$

17.81 The molecular scenes below depict the reaction $Y \rightleftharpoons 2Z$ at four different times, out of sequence, as it reaches equilibrium. Each sphere (Y is *red* and Z is *green*) represents 0.025 mol, and the volume is 0.40 L. (a) Which scene(s) represent(s) equilibrium? (b) List the scenes in the correct sequence. (c) Calculate K_c.

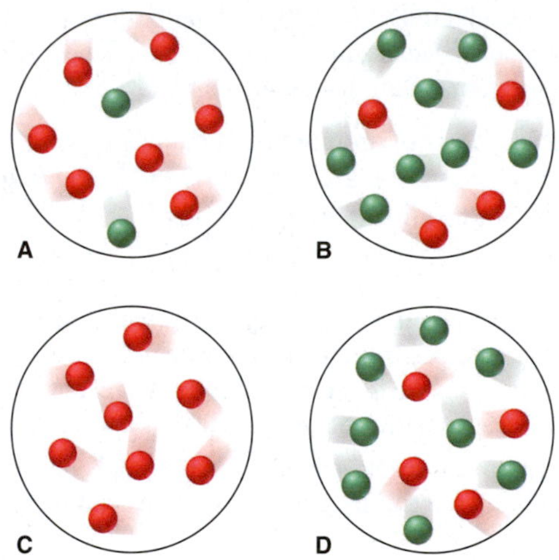

17.82 For the reaction

$$H_2S(g) \rightleftharpoons 2H_2(g) + S_2(g) \qquad K_c = 9.0\times10^{-8} \text{ at } 700^\circ C$$

the initial concentrations of the three gases are 0.300 M H_2S, 0.300 M H_2, and 0.150 M S_2. Determine the equilibrium concentrations of the gases.

17.83 The two most abundant atmospheric gases react to a tiny extent at 298 K in the presence of a catalyst:

$$N_2(g) + O_2(g) \rightleftharpoons 2NO(g) \qquad K_p = 4.35\times10^{-31}$$

(a) What are the equilibrium pressures of the three gases when the atmospheric partial pressures of O_2 (0.210 atm) and of N_2 (0.780 atm) are put into an evacuated 1.00-L flask at 298 K with the catalyst? (b) What is P_{total} in the container? (c) Find K_c at 298 K.

17.84 Isopentyl alcohol reacts with pure acetic acid to form isopentyl acetate, the essence of banana oil:

$$C_5H_{11}OH + CH_3COOH \rightleftharpoons CH_3COOC_5H_{11} + H_2O$$

A student adds a drying agent to remove H_2O and thus increase the yield of banana oil. Is this approach reasonable? Explain.

17.85 Isomers Q (*blue*) and R (*yellow*) interconvert. They are depicted in an equilibrium mixture in scene A. Scene B represents the mixture after addition of more Q. How many molecules of each isomer are present when the mixture in scene B attains equilibrium again?

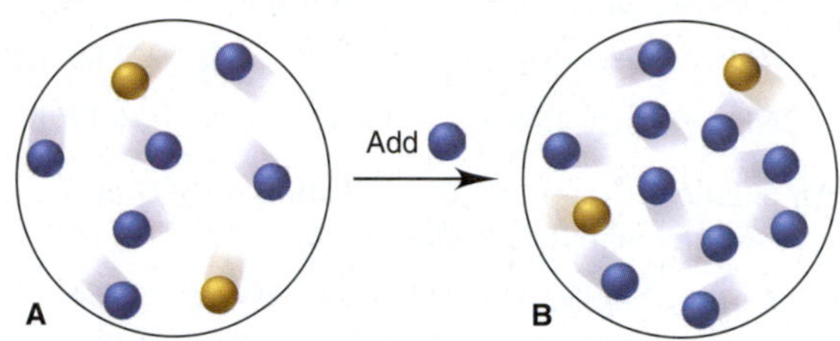

17.86 Glauber's salt, $Na_2SO_4\cdot10H_2O$, was used by J. R. Glauber in the 17th century as a medicinal agent. At 25°C, $K_p = 4.08\times10^{-25}$ for the loss of waters of hydration from Glauber's salt:

$$Na_2SO_4\cdot10H_2O(s) \rightleftharpoons Na_2SO_4(s) + 10H_2O(g)$$

(a) What is the vapor pressure of water at 25°C in a closed container holding a sample of $Na_2SO_4\cdot10H_2O(s)$?
(b) How do the following changes affect the ratio (higher, lower, same) of hydrated form to anhydrous form for the system above?
(1) Add more $Na_2SO_4(s)$ (2) Reduce the container volume
(3) Add more water vapor (4) Add N_2 gas

17.87 In a study of synthetic fuels, 0.100 mol of CO and 0.100 mol of water vapor are added to a 20.00-L container at 900.°C, and they react to form CO_2 and H_2. At equilibrium, [CO] is 2.24×10^{-3} M. (a) Calculate K_c at this temperature. (b) Calculate P_{total} in the flask at equilibrium. (c) How many moles of CO must be added to double this pressure? (d) After P_{total} is doubled and the system reattains equilibrium, what is $[CO]_{eq}$?

Acid-Base Equilibria

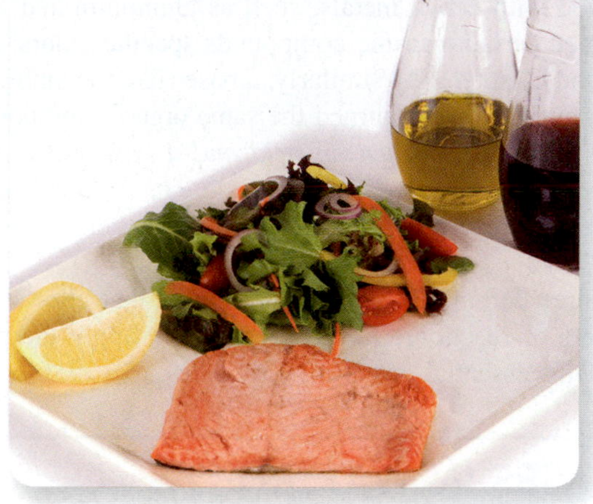

Culinary Equilibrium Weak acids in lemon and vinegar react with weak bases in fish through appetizing neutralizations. In this chapter, we examine acid-base chemistry with evolving definitions and models.

Outline

Key Principles
to focus on while studying this chapter

- In aqueous systems, the proton always exists as a *hydronium ion*, H_3O^+. In the *Arrhenius acid-base definition,* an acid is an *H-containing compound* that yields H_3O^+ in water, a base is an *OH-containing compound* that yields OH^- in water, and an acid-base *(neutralization)* reaction occurs when H_3O^+ and OH^- form H_2O. ***(Section 18.1)***
- The dissociation of a weak acid, HA, in water is reversible and is associated with an equilibrium constant called the *acid-dissociation constant*, K_a. The stronger the acid, the higher its K_a: weak acids typically have K_a values that are several orders of magnitude less than 1, falling mostly between 10^{-2} and 10^{-10}. ***(Section 18.1)***
- Water molecules dissociate *(autoionize)* reversibly to a very small extent in a process associated with the *ion-product constant for water*, K_w. Acidity or basicity is determined by the relative magnitudes of $[H_3O^+]$ and $[OH^-]$: in a neutral solution (or pure water), $[H_3O^+] = [OH^-]$, in an acidic solution, $[H_3O^+] > [OH^-]$, and in a basic solution, $[H_3O^+] < [OH^-]$. ***(Section 18.2)***
- The *pH*, the negative log of $[H_3O^+]$, is a measure of a solution's acidity: $pH < 7$ means the solution is acidic, and $pH > 7$ means it is basic. Because K_w is a constant (at a given temperature), the values for $[H_3O^+]$, $[OH^-]$, pH, and pOH are *interconvertible*. ***(Section 18.2)***
- In the *Brønsted-Lowry acid-base definition*, a base is any species that *accepts a proton;* therefore, there are many more Brønsted-Lowry bases than Arrhenius bases. When base B accepts a proton from acid HA, the species BH^+ and A^- form. HA and A^- are a *conjugate acid-base pair,* as are BH^+ and B. Thus, an acid-base reaction is a *proton-transfer process* between two conjugate acid-base pairs, with the stronger acid and base forming the weaker base and acid. ***(Section 18.3)***
- The *proportion* of molecules of a weak acid HA that dissociates increases as the initial [HA] decreases. ***(Section 18.4)***
- *Polyprotic acids* have more than one ionizable proton; in solution, essentially all the H_3O^+ comes from the first dissociation. ***(Section 18.4)***
- Ammonia and amines and the anions of weak acids behave as weak bases in a process associated with a *base-dissociation constant*, K_b. The reaction of HA with H_2O added to the reaction of A^- with H_2O gives the reaction for the autoionization of water; thus, $K_a \times K_b = K_w$. ***(Section 18.5)***
- The electronegativity of the central atom and either the strength of the bond to H (nonmetal hydrides) or the number of O atoms (oxoacids) determine acid strength. For hydrated metal ions, *charge density* is the key factor determining their acidity. ***(Section 18.6)***
- For a salt solution, the ion that reacts with water to a greater extent (higher K) determines the solution's acidity or basicity. ***(Section 18.7)***
- In the *Lewis acid-base definition*, an acid is any species that *accepts a lone pair* to form a new bond and yield an *adduct;* therefore, there are many more Lewis acids than Arrhenius or Brønsted-Lowry acids. Lewis acids include metal ions and molecules with either electron-deficient atoms or polar multiple bonds. ***(Section 18.8)***

CONCEPTS & SKILLS TO REVIEW before studying this chapter

- role of water as solvent (Section 4.1)
- writing ionic equations (Section 4.2)
- acids, bases, and acid-base reactions (Section 4.4)
- proton transfer in acid-base reactions (Section 4.4)
- properties of an equilibrium constant (Section 17.2)
- solving equilibrium problems (Section 17.5)

Acids and bases have been used as laboratory chemicals for centuries, as well as in the home. Common household acids include acetic acid (CH_3COOH, vinegar), citric acid ($H_3C_6H_5O_7$, in citrus fruits), and phosphoric acid (H_3PO_4, a flavoring in carbonated beverages). Sodium hydroxide (NaOH, drain cleaner) and ammonia (NH_3, glass cleaner), are household bases. And acid–base chemistry occurs throughout the environment and organisms.

You may have noticed that some acids (e.g., acetic and citric) have a sour taste. In fact, sourness was a defining property in the 17th century: an acid was any substance that had a sour taste; reacted with active metals, such as aluminum and zinc, to produce hydrogen gas; and turned certain organic compounds specific colors. (We discuss *indicators* in this chapter and Chapter 19.) Similarly, a base (like the amines in fish) was any substance that had a bitter taste and turned the same organic compounds different colors. Moreover, it was known that *when an acid and a base react, each cancels the properties of the other in a process called neutralization.* Although these early definitions described distinctive properties, they gave way to others based on molecular behavior.

In this chapter, we develop three definitions of acids and bases that allow us to understand ever-increasing numbers of reactions. In the process, we apply the principles of chemical equilibrium to this essential group of substances.

18.1 • ACIDS AND BASES IN WATER

Most laboratory work with acids and bases involves water, as do most environmental, biological, and industrial applications. Recall from our discussion in Chapter 4 that *water is **the** product in all reactions between strong acids and strong bases,* which the net ionic equation for a general reaction shows:

$$HX(aq) + MOH(aq) \longrightarrow MX(aq) + H_2O(l) \qquad \text{[molecular]}$$

and

$$H^+(aq) + OH^-(aq) \longrightarrow H_2O(l) \qquad \text{[net ionic]}$$

where M is a metal ion and X a nonmetal ion. Furthermore, whenever an acid dissociates in water, solvent molecules participate in the reaction:

$$HA(g \text{ or } l) + H_2O(l) \longrightarrow A^-(aq) + H_3O^+(aq)$$

Water molecules surround the proton to form H-bonded species with the general formula $H(H_2O)_n{}^+$. Because the proton is so small, its charge density is very high, so it attracts water especially strongly. The proton bonds covalently to one of the lone electron pairs of a water molecule's O atom to form a **hydronium ion, H_3O^+,** which forms H bonds to several other water molecules to give species such as $H_7O_3{}^+$, or $H(H_2O)_3{}^+$ (Figure 18.1). To emphasize the active role of water and the proton-water interaction, the hydrated proton is usually shown in the text as $H_3O^+(aq)$, although, for simplicity, it is sometimes shown as $H^+(aq)$.

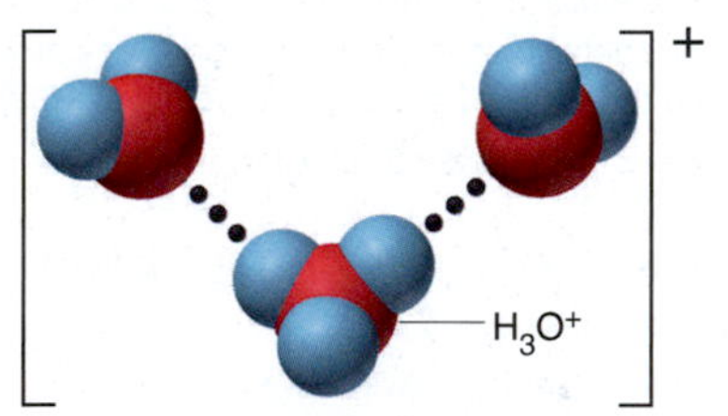

Figure 18.1 The hydrated proton.

Release of H+ or OH− and the Arrhenius Acid-Base Definition

The earliest definition that highlighted the molecular nature of acids and bases is the **Arrhenius acid-base definition,** which classifies these substances in terms of their formulas and behavior *in water:*

- An *acid* is a substance with H in its formula that dissociates in water to yield H_3O^+.
- A *base* is a substance with OH in its formula that dissociates in water to yield OH^-.

Some typical Arrhenius acids are HCl, HNO_3, and HCN, and some typical bases are NaOH, KOH, and $Ba(OH)_2$. While Arrhenius bases contain discrete OH^- ions in their structures, Arrhenius acids *never* contain discrete H^+ ions. Instead, they contain *covalently bonded H atoms that ionize when the molecule dissolves in water.*

When an acid and a base react, they undergo **neutralization.** The meaning of this term has changed, as we'll see, but in the Arrhenius sense, neutralization occurs when *the H^+ from the acid and the OH^- from the base form H_2O.* A key point about neutralization that Arrhenius was able to explain is that no matter which strong acid and strong base react, and no matter which salt results, ΔH°_{rxn} is −55.9 kJ per mole of water formed. Arrhenius suggested that the enthalpy change is always the same because the reaction is always the same—a hydrogen ion and a hydroxide ion form water:

$$H^+(aq) + OH^-(aq) \longrightarrow H_2O(l) \qquad \Delta H^\circ_{rxn} = -55.9 \text{ kJ}$$

The dissolved salt that forms along with the water exists as hydrated spectator ions and does not affect ΔH°_{rxn}.

Despite its importance at the time, limitations in the Arrhenius definition soon became apparent. Arrhenius and many others realized that even though some substances do *not* have discrete OH^- ions, they still behave as bases. For example, NH_3 and K_2CO_3 also yield OH^- in water. As you'll see shortly, broader acid-base definitions are required to include these species.

Variation in Acid Strength: The Acid-Dissociation Constant (K_a)

Acids (and bases) are classified by their *strength,* the amount of H_3O^+ (or OH^-) produced per mole of substance dissolved, in other words, by the extent of their dissociation into ions (see Table 4.2). Because acids and bases are electrolytes, their strength correlates with electrolyte strength: *strong electrolytes dissociate completely, and weak electrolytes dissociate slightly.*

1. *Strong acids dissociate* ***completely*** *into ions in water* (Figure 18.2A):

$$HA(g \text{ or } l) + H_2O(l) \longrightarrow H_3O^+(aq) + A^-(aq)$$

In a dilute solution of a strong acid, *HA molecules are no longer present:* $[H_3O^+] \approx [HA]_{init}$. In other words, $[HA]_{eq} \approx 0$, so the value of K_c is extremely large:

$$Q_c = \frac{[H_3O^+][A^-]}{[HA][H_2O]} \qquad \text{(at equilibrium, } Q_c = K_c >> 1\text{)}$$

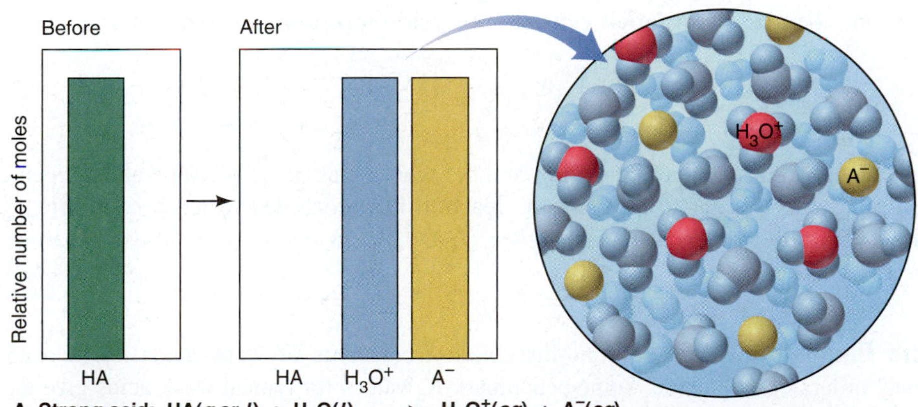

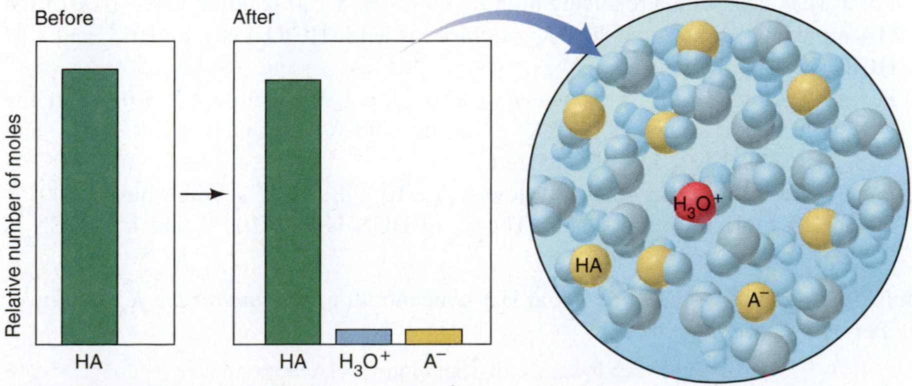

Figure 18.2 The extent of dissociation for strong acids and weak acids. The bar graphs show the relative numbers of moles of species before *(left)* and after *(right)* acid dissociation. **A,** A strong acid dissociates completely; virtually no HA molecules are present. **B,** A weak acid dissociates very little, remaining mostly as intact HA molecules.

Because the reaction is essentially complete, we usually don't express it as an equilibrium process. In dilute aqueous nitric acid, for example, there are virtually no undissociated nitric acid molecules:

$$HNO_3(l) + H_2O(l) \longrightarrow H_3O^+(aq) + NO_3^-(aq)$$

2. *Weak acids dissociate* ***slightly*** *into ions in water* (Figure 18.2B):

$$HA(aq) + H_2O(l) \rightleftharpoons H_3O^+(aq) + A^-(aq)$$

In a dilute solution of a weak acid, *the great majority of HA molecules are undissociated.* Thus, $[H_3O^+] \ll [HA]_{init}$ and $[HA]_{eq} \approx [HA]_{init}$, so K_c is very small. Hydrocyanic acid is an example of a weak acid:

$$HCN(aq) + H_2O(l) \rightleftharpoons H_3O^+(aq) + CN^-(aq)$$

$$Q_c = \frac{[H_3O^+][CN^-]}{[HCN][H_2O]} \qquad \text{(at equilibrium, } Q_c = K_c \ll 1\text{)}$$

(As in Chapter 17, brackets with no subscript mean molar concentration *at equilibrium;* that is, [X] means $[X]_{eq}$. In this chapter, we are dealing with systems *at equilibrium,* so instead of writing Q and stating that Q equals K at equilibrium, we'll express K directly as a collection of equilibrium concentration terms.)

The Meaning of K_a We write a *specific* equilibrium constant for acid dissociation that only includes species whose concentrations change to a significant extent. The equilibrium expression for the dissociation of a general *weak acid,* HA, in water is

$$K_c = \frac{[H_3O^+][A^-]}{[HA][H_2O]}$$

The concentration of water ($[H_2O] = \frac{1000\text{ g}}{18.02\text{ g/mol}} = 55.5\ M$) is typically a few orders of magnitude larger than [HA]. Therefore, $[H_2O]$ is essentially constant when HA dissociates. As you saw for pure liquids and solids (Section 17.2), we simplify the equilibrium expression by multiplying K_c by $[H_2O]$ to define a new equilibrium constant, the **acid-dissociation constant** (or **acid-ionization constant**), K_a**:**

$$K_c[H_2O] = K_a = \frac{[H_3O^+][A^-]}{[HA]} \qquad \textbf{(18.1)}$$

Like any equilibrium constant, K_a is a number whose magnitude is temperature dependent and tells how far to the right the reaction has proceeded to reach equilibrium. Thus, *the stronger the acid, the higher $[H_3O^+]$ is at equilibrium, and the larger the K_a*:

Stronger acid $\Longrightarrow$ higher $[H_3O^+]$ $\Longrightarrow$ larger K_a

The Range of K_a Values Acid-dissociation constants of weak acids range over many orders of magnitude. Some benchmark K_a values for typical weak acids give an idea of the fraction of HA molecules that dissociate into ions:

- For a weak acid with a relatively high K_a ($\sim 10^{-2}$), a 1 M solution has ~10% of the HA molecules dissociated. The K_a of chlorous acid ($HClO_2$) is 1.1×10^{-2}, and 1 M $HClO_2$ is 10.% dissociated.
- For a weak acid with a moderate K_a ($\sim 10^{-5}$), a 1 M solution has ~0.3% of the HA molecules dissociated. The K_a of acetic acid (CH_3COOH) is 1.8×10^{-5}, and 1 M CH_3COOH is 0.42% dissociated.
- For a weak acid with a relatively low K_a ($\sim 10^{-10}$), a 1 M solution has ~0.001% of the HA molecules dissociated. The K_a of HCN is 6.2×10^{-10}, and 1 M HCN is 0.0025% dissociated.

Thus, for solutions of the same initial HA concentration, *the smaller the K_a, the lower the percent dissociation of HA:*

Weaker acid $\Longrightarrow$ lower % dissociation of HA $\Longrightarrow$ smaller K_a

A list of K_a values for some weak acids appears in Appendix C.

Classifying the Relative Strengths of Acids and Bases

Referring to a list of K_a values is the surest way to quantify strengths of weak acids, but you can classify acids and bases qualitatively as strong or weak from their formulas:

- *Strong acids.* Two types of strong acids, with examples *you should memorize,* are
  1. The hydrohalic acids HCl, HBr, and HI
  2. Oxoacids in which the number of O atoms exceeds the number of ionizable protons by two or more, such as HNO_3, H_2SO_4, and $HClO_4$; for example, in the case of H_2SO_4, 4 O's − 2 H's = 2
- *Weak acids.* There are many *more* weak acids than strong ones. Four types are
  1. The hydrohalic acid HF
  2. Acids in which H is not bonded to O or to a halogen, such as HCN and H_2S
  3. Oxoacids in which the number of O atoms equals or exceeds the number of ionizable protons by one, such as HClO, HNO_2, and H_3PO_4
  4. Carboxylic acids (general formula RCOOH, with the ionizable proton shown in red), such as CH_3COOH and C_6H_5COOH
- *Strong bases.* Water-soluble compounds containing O^{2-} or OH^- ions are strong bases. The cations are usually those of the most active metals:
  1. M_2O or MOH, where M = Group 1A(1) metal (Li, Na, K, Rb, Cs)
  2. MO or $M(OH)_2$, where M = Group 2A(2) metal (Ca, Sr, Ba)

  [MgO and $Mg(OH)_2$ are only slightly soluble in water, but the portion that dissolves dissociates completely.]
- *Weak bases.* Many compounds with an electron-rich nitrogen atom are weak bases (none is an Arrhenius base). The common structural feature is an N atom with a lone electron pair (shown in blue):
  1. Ammonia ($\ddot{N}H_3$)
  2. Amines (general formula $R\ddot{N}H_2$, $R_2\ddot{N}H$, or $R_3\ddot{N}$), such as $CH_3CH_2\ddot{N}H_2$, $(CH_3)_2\ddot{N}H$, and $(C_3H_7)_3\ddot{N}$

Sample Problem 18.1 Classifying Acid and Base Strength from the Chemical Formula

Problem Classify each of the following compounds as a strong acid, weak acid, strong base, or weak base:

(a) KOH **(b)** $(CH_3)_2CHCOOH$
(c) H_2SeO_4 **(d)** $(CH_3)_2CHNH_2$

Plan We examine the formula and classify each acid or base, using the text descriptions. Particular points to note for acids are the numbers of O atoms relative to ionizable H atoms and the presence of the —COOH group. For bases, note the nature of the cation or the presence of an N atom that has a lone pair.

Solution (a) Strong base: KOH is one of the Group 1A(1) hydroxides.
(b) Weak acid: $(CH_3)_2CHCOOH$ is a carboxylic acid, as indicated by the —COOH group.
(c) Strong acid: H_2SeO_4 is an oxoacid in which the number of O atoms exceeds the number of ionizable protons by two.
(d) Weak base: $(CH_3)_2CHNH_2$ has a lone pair on the N and is an amine.

FOLLOW-UP PROBLEM 18.1 Which member of each pair is the stronger acid or base?
(a) HClO or $HClO_3$ **(b)** HCl or CH_3COOH **(c)** NaOH or CH_3NH_2

Summary of Section 18.1

- In aqueous solution, water binds the proton released from an acid to form a hydrated species represented by $H_3O^+(aq)$.
- By the Arrhenius definition, acids contain H and yield H_3O^+ in water, bases contain OH and yield OH^- in water, and an acid-base reaction (neutralization) is the reaction of H^+ and OH^- to form H_2O.

- Acid strength depends on $[H_3O^+]$ relative to [HA] in aqueous solution. Strong acids dissociate completely and weak acids slightly.
- The extent of dissociation is expressed by the acid-dissociation constant, K_a. Most weak acids have K_a values ranging from about 10^{-2} to 10^{-10}.
- Many acids and bases can be classified as strong or weak based on their formulas.

18.2 • AUTOIONIZATION OF WATER AND THE pH SCALE

Before we discuss the next major acid-base definition, let's examine a crucial property of water that enables us to quantify $[H_3O^+]$: *water dissociates very slightly into ions* in an equilibrium process known as **autoionization** (or self-ionization):

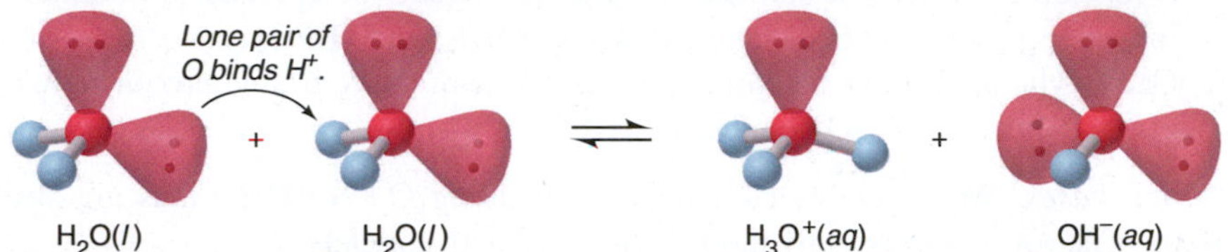

The Equilibrium Nature of Autoionization: The Ion-Product Constant for Water (K_w)

Like any equilibrium process, the autoionization of water is described quantitatively by an equilibrium constant:

$$K_c = \frac{[H_3O^+][OH^-]}{[H_2O]^2}$$

Because the concentration of H_2O (55.5 *M*) remains essentially constant, we multiply K_c by $[H_2O]^2$ to obtain a new equilibrium constant, the **ion-product constant for water, K_w:**

$$K_c[H_2O]^2 = K_w = [H_3O^+][OH^-] = 1.0\times10^{-14} \quad \text{(at 25°C)} \qquad \textbf{(18.2)}$$

Notice that *one H_3O^+ ion and one OH^- ion form for each H_2O molecule that dissociates.* Therefore, in pure water, we find that

$$[H_3O^+] = [OH^-] = \sqrt{1.0\times10^{-14}} = 1.0\times10^{-7}\,M \quad \text{(at 25°C)}$$

Since pure water has a concentration of about 55.5 *M,* these equilibrium concentrations are attained when only 1 in 555 million water molecules dissociates reversibly into ions!

Autoionization of water affects aqueous acid-base chemistry in two major ways:

1. *A change in $[H_3O^+]$ causes an inverse change in $[OH^-]$,* and vice versa:

Higher $[H_3O^+] \Longrightarrow$ lower $[OH^-]$ and Higher $[OH^-] \Longrightarrow$ lower $[H_3O^+]$

Recall from Le Châtelier's principle (Section 17.6) that a change in concentration shifts the equilibrium position but does *not* change the equilibrium constant. Therefore, if some acid is added, $[H_3O^+]$ increases and $[OH^-]$ decreases as the ions react to form water; similarly, if some base is added, $[OH^-]$ increases and $[H_3O^+]$ decreases. In both cases, as long as the temperature is constant, the value of K_w is constant.

2. *Both ions are present in all aqueous systems.* Thus, all acidic solutions contain a low $[OH^-]$, and all basic solutions contain a low $[H_3O^+]$. The equilibrium nature of autoionization allows us to define "acidic" and "basic" solutions in terms of relative magnitudes of $[H_3O^+]$ and $[OH^-]$:

In an *acidic* solution, $[H_3O^+] > [OH^-]$
In a *neutral* solution, $[H_3O^+] = [OH^-]$
In a *basic* solution, $[H_3O^+] < [OH^-]$

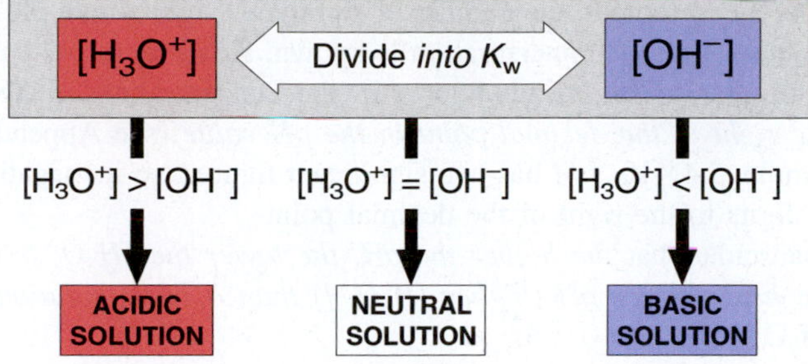

Figure 18.3 The relationship between $[H_3O^+]$ and $[OH^-]$ and the relative acidity of solutions.

Figure 18.3 summarizes these relationships. Moreover, if you know the value of K_w at a particular temperature and the concentration of one of the two ions, you can find the concentration of the other:

$$[H_3O^+] = \frac{K_w}{[OH^-]} \quad \text{or} \quad [OH^-] = \frac{K_w}{[H_3O^+]}$$

Sample Problem 18.2 Calculating $[H_3O^+]$ or $[OH^-]$ in Aqueous Solution

Problem A research chemist adds a measured amount of HCl gas to pure water at 25°C and obtains a solution with $[H_3O^+] = 3.0\times10^{-4}\ M$. Calculate $[OH^-]$. Is the solution neutral, acidic, or basic?

Plan We use the known value of K_w at 25°C (1.0×10^{-14}) and the given $[H_3O^+]$ ($3.0\times10^{-4}\ M$) to solve for $[OH^-]$. Then we compare $[H_3O^+]$ with $[OH^-]$ to determine whether the solution is acidic, basic, or neutral (see Figure 18.3).

Solution Calculating $[OH^-]$:

$$[OH^-] = \frac{K_w}{[H_3O^+]} = \frac{1.0\times10^{-14}}{3.0\times10^{-4}} = 3.3\times10^{-11}\ M$$

Because $[H_3O^+] > [OH^-]$, the solution is acidic.

Check It makes sense that adding an acid to water results in an acidic solution. Also, since $[H_3O^+]$ is greater than $10^{-7}\ M$, $[OH^-]$ must be less than $10^{-7}\ M$ to give a constant K_w.

FOLLOW-UP PROBLEM 18.2 Calculate $[H_3O^+]$ in a solution that has $[OH^-] = 6.7\times10^{-2}\ M$ at 25°C. Is the solution neutral, acidic, or basic?

Expressing the Hydronium Ion Concentration: The pH Scale

In aqueous solutions, $[H_3O^+]$ can vary from about 10 M to $10^{-15}\ M$. To handle numbers with negative exponents more conveniently in calculations, we convert them to positive numbers using a numerical system called a *p-scale,* the negative of the common (base-10) logarithm of the number. Applying this numerical system to $[H_3O^+]$ gives **pH,** the negative logarithm of $[H^+]$, or $[H_3O^+]$:

$$\text{pH} = -\log\,[H_3O^+] \tag{18.3}$$

What is the pH of $10^{-12}\ M\ H_3O^+$ solution?

$$\text{pH} = -\log\,[H_3O^+] = -\log 10^{-12} = (-1)(-12) = 12$$

Similarly, a $10^{-3}\ M\ H_3O^+$ solution has a pH of 3, and a $5.4\times10^{-4}\ M\ H_3O^+$ solution has a pH of 3.27:

$$\text{pH} = -\log\,[H_3O^+] = (-1)(\log 5.4 + \log 10^{-4}) = 3.27$$

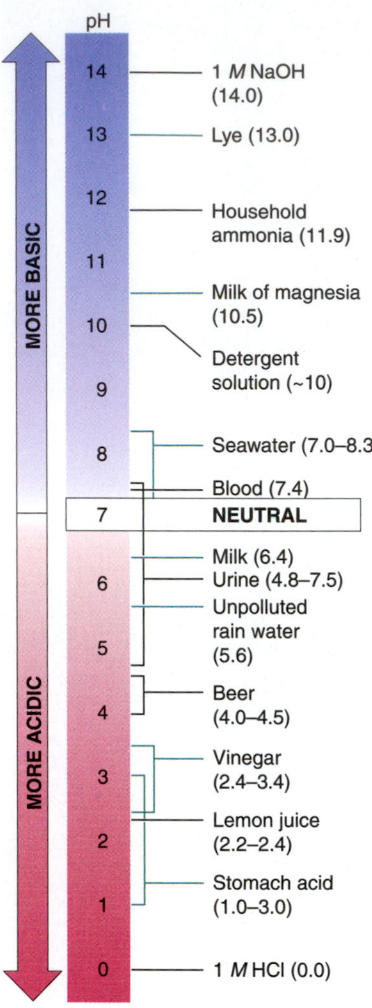

Figure 18.4 The pH values of some familiar aqueous solutions.

As with any measurement, the number of significant figures in a pH value reflects the precision with which the concentration is known. However, a pH value is a logarithm, so the number of significant figures in the concentration equals the number of digits *to the right of the decimal point in the pH value* (see Appendix A). In the preceding example, 5.4×10^{-4} *M* has two significant figures, so its negative logarithm, 3.27, has two digits to the right of the decimal point.

Note in particular that *the higher the pH, the lower the [H_3O^+]*. Therefore, *an acidic solution has a lower pH (higher [H_3O^+]) than a basic solution.* At 25°C in pure water, [H_3O^+] is 1.0×10^{-7} *M*, so

$$\text{pH of an acidic solution} < 7.00$$
$$\text{pH of a neutral solution} = 7.00$$
$$\text{pH of a basic solution} > 7.00$$

Figure 18.4 shows that the pH values of some familiar aqueous solutions fall within a range of 0 to 14.

Because the pH scale is logarithmic, a solution of pH 1.0 has an [H_3O^+] that is 10 times higher than that of a pH 2.0 solution, 100 times higher than that of a pH 3.0 solution, and so forth. To find the [H_3O^+] from the pH, you perform the opposite arithmetic process; that is, you find the negative antilog of pH:

$$[H_3O^+] = 10^{-pH}$$

A p-scale is used to express other quantities as well:

- Hydroxide ion concentration can be expressed as pOH:
$$pOH = -\log [OH^-]$$
Acidic solutions have a higher pOH (lower [OH^-]) than basic solutions.
- Equilibrium constants can be expressed as p*K*:
$$pK = -\log K$$
A low pK corresponds to a high K. A reaction that reaches equilibrium with mostly products present (proceeds far to the right) has a low p*K* (high *K*), whereas one that has mostly reactants present at equilibrium has a high p*K* (low *K*). Table 18.1 shows this relationship for aqueous equilibria of some weak acids.

Table 18.1 The Relationship Between K_a and pK_a

Acid Name (Formula)	K_a at 25°C	pK_a
Hydrogen sulfate ion (HSO_4^-)	1.0×10^{-2}	1.99
Nitrous acid (HNO_2)	7.1×10^{-4}	3.15
Acetic acid (CH_3COOH)	1.8×10^{-5}	4.75
Hypobromous acid (HBrO)	2.3×10^{-9}	8.64
Phenol (C_6H_5OH)	1.0×10^{-10}	10.00

The Relations Among pH, pOH, and pK_w Taking the negative log of both sides of the K_w expression gives a very useful relationship among pK_w, pH, and pOH:

$$K_w = [H_3O^+][OH^-] = 1.0\times10^{-14} \quad \text{(at 25°C)}$$
$$-\log K_w = (-\log [H_3O^+]) + (-\log [OH^-]) = -\log (1.0\times10^{-14})$$

$$pK_w = pH + pOH = 14.00 \quad \text{(at 25°C)} \qquad \textbf{(18.4)}$$

Note these important points:

1. The sum of pH and pOH is pK_w for any aqueous solution at any temperature, and pK_w equals 14.00 at 25°C.
2. Because K_w is constant, pH, pOH, [H_3O^+], and [OH^-] are interrelated (Figure 18.5).

	$[H_3O^+]$	pH	$[OH^-]$	pOH
MORE BASIC ↑	1.0×10^{-15}	15.00	1.0×10^{1}	−1.00
	1.0×10^{-14}	14.00	1.0×10^{0}	0.00
	1.0×10^{-13}	13.00	1.0×10^{-1}	1.00
BASIC	1.0×10^{-12}	12.00	1.0×10^{-2}	2.00
	1.0×10^{-11}	11.00	1.0×10^{-3}	3.00
	1.0×10^{-10}	10.00	1.0×10^{-4}	4.00
	1.0×10^{-9}	9.00	1.0×10^{-5}	5.00
	1.0×10^{-8}	8.00	1.0×10^{-6}	6.00
NEUTRAL	1.0×10^{-7}	7.00	1.0×10^{-7}	7.00
	1.0×10^{-6}	6.00	1.0×10^{-8}	8.00
	1.0×10^{-5}	5.00	1.0×10^{-9}	9.00
	1.0×10^{-4}	4.00	1.0×10^{-10}	10.00
ACIDIC	1.0×10^{-3}	3.00	1.0×10^{-11}	11.00
	1.0×10^{-2}	2.00	1.0×10^{-12}	12.00
	1.0×10^{-1}	1.00	1.0×10^{-13}	13.00
	1.0×10^{0}	0.00	1.0×10^{-14}	14.00
MORE ACIDIC ↓	1.0×10^{1}	−1.00	1.0×10^{-15}	15.00

Figure 18.5 The relations among $[H_3O^+]$, pH, $[OH^-]$, and pOH. At 25°C, pK_w is 14.00.

Sample Problem 18.3 Calculating $[H_3O^+]$, pH, $[OH^-]$, and pOH

Problem In an art restoration project, a conservator prepares copper-plate etching solutions by diluting concentrated HNO_3 to 2.0 *M*, 0.30 *M*, and 0.0063 *M* HNO_3. Calculate $[H_3O^+]$, pH, $[OH^-]$, and pOH of the three solutions at 25°C.

Plan We know from its formula that HNO_3 is a strong acid, so it dissociates completely; thus, $[H_3O^+] = [HNO_3]_{init}$. We use the given concentrations and the value of K_w at 25°C (1.0×10^{-14}) to find $[H_3O^+]$ and $[OH^-]$ and then use them to calculate pH and pOH.

Solution Calculating the values for 2.0 *M* HNO_3:

$$[H_3O^+] = 2.0\ M$$

$$pH = -\log [H_3O^+] = -\log 2.0 = -0.30$$

$$[OH^-] = \frac{K_w}{[H_3O^+]} = \frac{1.0\times10^{-14}}{2.0} = 5.0\times10^{-15}\ M$$

$$pOH = -\log (5.0\times10^{-15}) = 14.30$$

Calculating the values for 0.30 *M* HNO_3:

$$[H_3O^+] = 0.30\ M$$

$$pH = -\log [H_3O^+] = -\log 0.30 = 0.52$$

$$[OH^-] = \frac{K_w}{[H_3O^+]} = \frac{1.0\times10^{-14}}{0.30} = 3.3\times10^{-14}\ M$$

$$pOH = -\log (3.3\times10^{-14}) = 13.48$$

Calculating the values for 0.0063 *M* HNO_3:

$$[H_3O^+] = 6.3\times10^{-3}\ M$$

$$pH = -\log [H_3O^+] = -\log (6.3\times10^{-3}) = 2.20$$

$$[OH^-] = \frac{K_w}{[H_3O^+]} = \frac{1.0\times10^{-14}}{6.3\times10^{-3}} = 1.6\times10^{-12}\ M$$

$$pOH = -\log (1.6\times10^{-12}) = 11.80$$

Check As the solution becomes more dilute, $[H_3O^+]$ decreases, so pH increases, as we expect. An $[H_3O^+] > 1.0\ M$, as in 2.0 *M* HNO_3, gives a positive log, so it results in a negative pH. The arithmetic seems correct because pH + pOH = 14.00 in each case.

Comment On most calculators, finding the pH requires several keystrokes. For example, to find the pH of 6.3×10^{-3} *M* HNO_3 solution, you enter: 6.3, EXP, 3, +/−, log, +/−.

FOLLOW-UP PROBLEM 18.3 A solution of NaOH has a pH of 9.52. What is its pOH, $[H_3O^+]$, and $[OH^-]$ at 25°C?

A

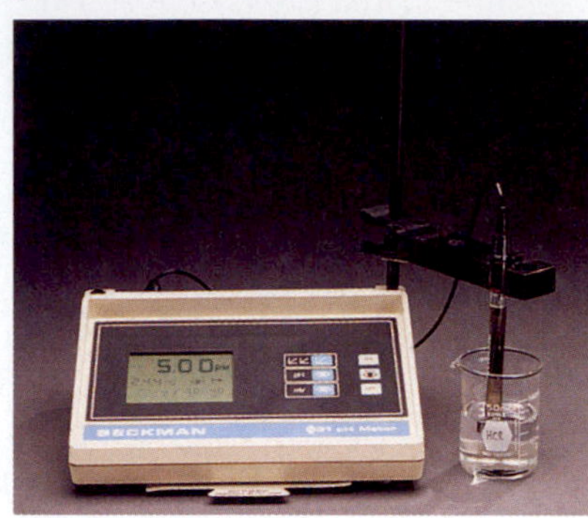

B

Figure 18.6 Methods for measuring the pH of an aqueous solution. **A,** pH paper. **B,** pH meter.

Measuring pH In the laboratory, pH values are usually obtained in two ways:

1. **Acid-base indicators** are organic molecules whose colors depend on the acidity of the solution in which they are dissolved. A pH can be estimated quickly with *pH paper,* a paper strip impregnated with one or a mixture of indicators. A drop of solution is placed on the strip, and the color is compared with the colors on a chart (Figure 18.6A).

2. A *pH meter* measures $[H_3O^+]$ by means of two electrodes immersed in the test solution. One electrode supplies a reference system; the other consists of a very thin glass membrane that separates a known internal $[H_3O^+]$ from the unknown external $[H_3O^+]$. The difference in $[H_3O^+]$ creates a voltage difference across the membrane, which is displayed in pH units (Figure 18.6B). We examine this device in Chapter 21.

Summary of Section 18.2

- Pure water autoionizes to a small extent in a process whose equilibrium constant is the ion-product constant for water, K_w (1.0×10^{-14} at 25°C).
- $[H_3O^+]$ and $[OH^-]$ are inversely related: in acidic solution, $[H_3O^+]$ is greater than $[OH^-]$; the reverse is true in basic solution; and the two are equal in neutral solution.
- To express small values of $[H_3O^+]$, we use the pH scale ($pH = -\log[H_3O^+]$). Similarly, $pOH = -\log[OH^-]$, and $pK = -\log K$.
- A high pH represents a low $[H_3O^+]$. In acidic solutions, $pH < 7.00$; in basic solutions, $pH > 7.00$; and in neutral solutions, $pH = 7.00$. The sum of pH and pOH equals pK_w (14.00 at 25°C).
- A pH is typically measured with either an acid-base indicator or a pH meter.

18.3 • PROTON TRANSFER AND THE BRØNSTED-LOWRY ACID-BASE DEFINITION

Earlier we noted a key limitation of the Arrhenius definition: many substances that yield OH^- ions in water do not contain OH in their formulas. Examples include ammonia, the amines, and many salts of weak acids, such as NaF. Another limitation is that water had to be the solvent for acid-base reactions. In the early 20th century, J. N. Brønsted and T. M. Lowry suggested definitions that remove these limitations. (We introduced some of these ideas in Section 4.4.)

According to the **Brønsted-Lowry acid-base definition,**

- *An acid is a* ***proton donor,*** *any species that donates an* H^+ *ion.* An acid must contain H in its formula; HNO_3 and $H_2PO_4^-$ are two of many examples. All Arrhenius acids are Brønsted-Lowry acids.
- *A base is a* ***proton acceptor,*** *any species that accepts an* H^+ *ion.* A base must contain a lone pair of electrons to bind H^+; a few examples are NH_3, CO_3^{2-}, and F^-, as well as OH^- itself. Brønsted-Lowry bases are not Arrhenius bases, but all Arrhenius bases contain the Brønsted-Lowry base OH^-.

From this perspective, an acid-base reaction occurs when *one species donates a proton and another species simultaneously accepts it: an acid-base reaction is thus a proton-transfer process.* Acid-base reactions can occur between gases, in nonaqueous solutions, and in heterogeneous mixtures, as well as in aqueous solutions.

According to this definition, an acid-base reaction occurs even when an acid (or a base) just dissolves in water, because water acts as the proton acceptor (or donor):

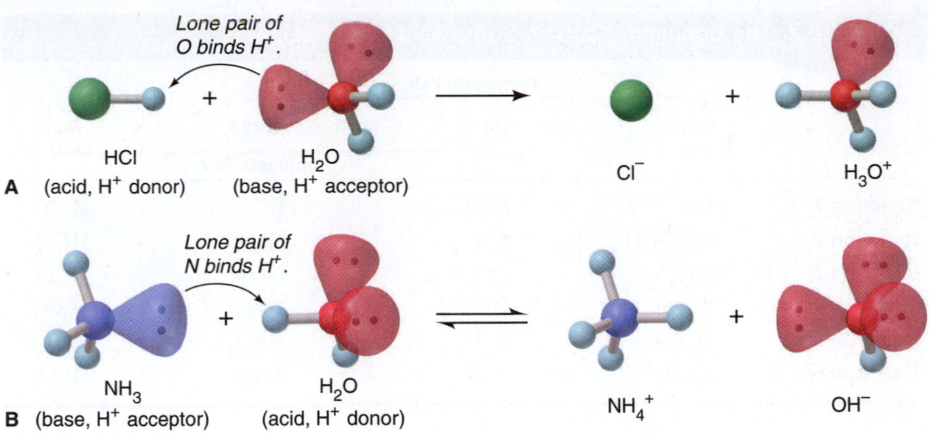

Figure 18.7 Dissolving of an acid or base in water as a Brønsted-Lowry acid-base reaction. A, The acid HCl dissolving in the base water. **B,** The base NH_3 dissolving in the acid water.

1. *Acid donates a proton to water* (Figure 18.7A). When HCl dissolves in water, an H^+ ion (a proton) is transferred from HCl to H_2O, where it becomes attached to a lone pair of electrons on the O atom, forming H_3O^+. Thus, HCl (the acid) has *donated* the H^+, and H_2O (the base) has *accepted* it:

$$HCl(g) + H_2\ddot{\underset{..}{O}}(l) \longrightarrow Cl^-(aq) + H_3\ddot{O}^+(aq)$$

2. *Base accepts a proton from water* (Figure 18.7B). When ammonia dissolves in water, an H^+ from H_2O is transferred to the lone pair of N, forming NH_4^+, and the H_2O becomes an OH^- ion:

$$\ddot{N}H_3(aq) + H_2O(l) \rightleftharpoons NH_4^+(aq) + OH^-(aq)$$

In this case, H_2O (the acid) has *donated* the H^+, and NH_3 (the base) has *accepted* it.

Note that H_2O is *amphiprotic:* it acts as a base (accepts an H^+) in one case and as an acid (donates an H^+) in the other. Many other species are amphiprotic as well.

Conjugate Acid-Base Pairs

The Brønsted-Lowry definition provides a new way to look at these reactions because it focuses on the reactants *and* the products as acids and bases. For example, let's examine the reaction between hydrogen sulfide and ammonia:

$$H_2S + NH_3 \rightleftharpoons HS^- + NH_4^+$$

In the forward reaction, H_2S acts as an acid by donating an H^+ to NH_3, which acts as a base by accepting it. In the reverse reaction, the ammonium ion, NH_4^+, acts as an acid by donating an H^+ to the hydrogen sulfide ion, HS^-, which acts as a base. Notice that the acid, H_2S, becomes a base, HS^-, and the base, NH_3, becomes an acid, NH_4^+.

In Brønsted-Lowry terminology, H_2S and HS^- are a **conjugate acid-base pair:** HS^- is the conjugate base of the acid H_2S. Similarly, NH_3 and NH_4^+ are a conjugate acid-base pair: NH_4^+ is the conjugate acid of the base NH_3. *Every acid has a conjugate base, and every base has a conjugate acid.* For any conjugate acid-base pair,

- The conjugate base has one *fewer* H and one *more* minus charge than the acid.
- The conjugate acid has one *more* H and one *fewer* minus charge than the base.

A Brønsted-Lowry acid-base reaction occurs when *an acid and a base react to form their conjugate base and conjugate acid, respectively:*

$$\text{acid}_1 + \text{base}_2 \rightleftharpoons \text{base}_1 + \text{acid}_2$$

Table 18.2 shows some Brønsted-Lowry acid-base reactions. Note these points:

- Each reaction has an acid and a base as reactants *and* as products, comprising two conjugate acid-base pairs.
- Acids and bases can be neutral, cationic, or anionic.
- The same species can be an acid or a base (amphiprotic), depending on the other species reacting. Water behaves this way in reactions 1 and 4, and HPO_4^{2-} does so in reactions 4 and 6.

Table 18.2 The Conjugate Pairs in Some Acid-Base Reactions

	Acid	+	Base	⇌	Base	+	Acid
	(Conjugate Pair: Acid / Base on right)		(Conjugate Pair: Base / Acid on right)				
Reaction 1	HF	+	H_2O		F^-	+	H_3O^+
Reaction 2	$HCOOH$	+	CN^-		$HCOO^-$	+	HCN
Reaction 3	NH_4^+	+	CO_3^{2-}		NH_3	+	HCO_3^-
Reaction 4	$H_2PO_4^-$	+	OH^-		HPO_4^{2-}	+	H_2O
Reaction 5	H_2SO_4	+	$N_2H_5^+$		HSO_4^-	+	$N_2H_6^{2+}$
Reaction 6	HPO_4^{2-}	+	SO_3^{2-}		PO_4^{3-}	+	HSO_3^-

Sample Problem 18.4 Identifying Conjugate Acid-Base Pairs

Problem The following reactions are important environmental processes. Identify the conjugate acid-base pairs.

(a) $H_2PO_4^-(aq) + CO_3^{2-}(aq) \rightleftharpoons HCO_3^-(aq) + HPO_4^{2-}(aq)$

(b) $H_2O(l) + SO_3^{2-}(aq) \rightleftharpoons OH^-(aq) + HSO_3^-(aq)$

Plan To find the conjugate pairs, we find the species that donated an H^+ (acid) and the species that accepted it (base). The acid (or base) on the left becomes its conjugate base (or conjugate acid) on the right. Remember, the conjugate acid has one more H and one fewer minus charge than its conjugate base.

Solution (a) $H_2PO_4^-$ has one more H^+ than HPO_4^{2-}; CO_3^{2-} has one fewer H^+ than HCO_3^-. Therefore, $H_2PO_4^-$ and HCO_3^- are the acids, and HPO_4^{2-} and CO_3^{2-} are the bases. The conjugate acid-base pairs are $H_2PO_4^-/HPO_4^{2-}$ and HCO_3^-/CO_3^{2-}.

(b) H_2O has one more H^+ than OH^-; SO_3^{2-} has one fewer H^+ than HSO_3^-. The acids are H_2O and HSO_3^-; the bases are OH^- and SO_3^{2-}. The conjugate acid-base pairs are H_2O/OH^- and HSO_3^-/SO_3^{2-}.

FOLLOW-UP PROBLEM 18.4 Identify the conjugate acid-base pairs:

(a) $CH_3COOH(aq) + H_2O(l) \rightleftharpoons CH_3COO^-(aq) + H_3O^+(aq)$

(b) $H_2O(l) + F^-(aq) \rightleftharpoons OH^-(aq) + HF(aq)$

Relative Acid-Base Strength and the Net Direction of Reaction

The *net* direction of an acid-base reaction depends on relative acid and base strengths: *A reaction proceeds to the greater extent in the direction in which a stronger acid and stronger base form a weaker acid and weaker base.*

Competition for the Proton The net direction of the reaction of H_2S and NH_3 is to the right ($K_c > 1$) because H_2S is a stronger acid than NH_4^+, the other acid present, and NH_3 is a stronger base than HS^-, the other base:

$$\underset{\text{stronger acid}}{H_2S} + \underset{\text{stronger base}}{NH_3} \rightleftharpoons \underset{\text{weaker base}}{HS^-} + \underset{\text{weaker acid}}{NH_4^+}$$

stronger acid + stronger base ⟶ weaker base + weaker acid

You might think of the process as *a competition for the proton between the two bases,* NH_3 and HS^-, in which NH_3 wins.

In effect, the extent of acid (HA) dissociation in water can be viewed as a competition for the proton between the two bases, A^- and H_2O. Strong and weak acids give different results:

1. *Strong acids.* When the strong acid HNO_3 dissolves, it completely transfers an H^+ to the base, H_2O, forming the conjugate base NO_3^- and the conjugate acid H_3O^+:

$$HNO_3 + H_2O \rightleftharpoons NO_3^- + H_3O^+$$

stronger acid + stronger base ⟶ weaker base + weaker acid

(Even though an equilibrium arrow is used here, the net direction is so far to the right that $K_c >> 1$ and the reaction is essentially complete.) HNO_3 is a stronger acid than

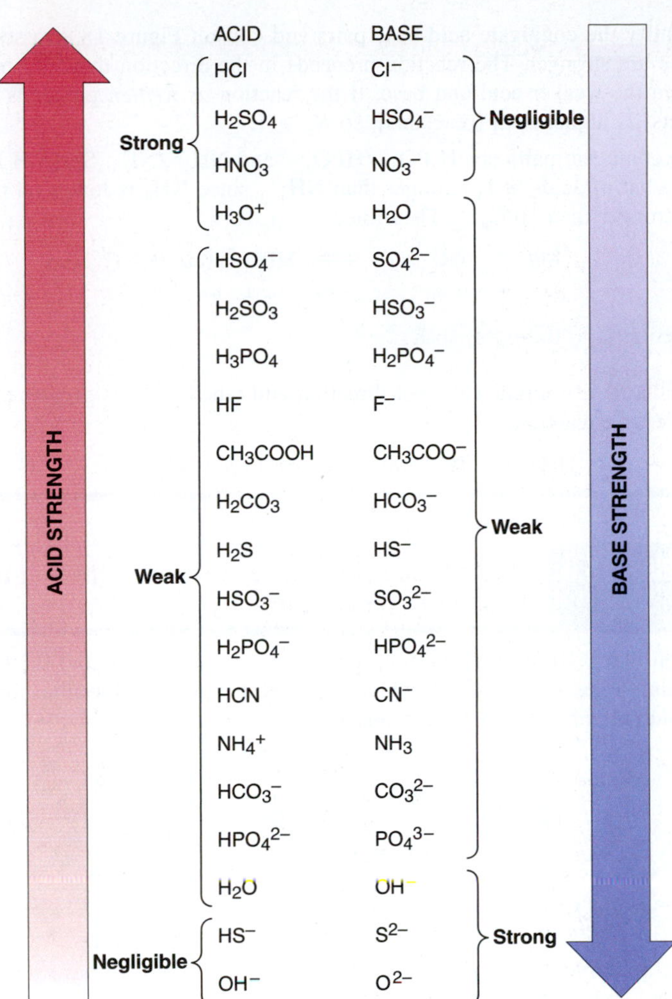

Figure 18.8 Strengths of conjugate acid-base pairs. The stronger the acid is, the weaker its conjugate base. The strongest acid is at top left and the strongest base at bottom right. When an acid reacts with a base farther down the list, the reaction proceeds to the right ($K_c > 1$).

H_3O^+, and H_2O is a stronger base than NO_3^-. Thus, with a strong HA, H_2O wins the competition for the proton because A^- is a *much* weaker base. In fact, *the only acidic species that remains in strong-acid solutions is* H_3O^+.

2. *Weak acids.* On the other hand, with weak acids such as HF, the A^- (F^-) wins the competition because it is a stronger base than H_2O ($K_c < 1$):

$$\underset{\text{weaker acid}}{HF} + \underset{\text{weaker base}}{H_2O} \rightleftharpoons \underset{\text{stronger base}}{F^-} + \underset{\text{stronger acid}}{H_3O^+}$$

weaker acid + weaker base ⟵ stronger base + stronger acid

Ranking Conjugate Pairs Based on evidence from many such reactions, we can rank conjugate pairs in terms of the ability of the acid to transfer its proton (Figure 18.8). Note that *a weaker acid has a stronger conjugate base:* the acid can't give up its proton very readily because its conjugate base is holding it too strongly.

We use Figure 18.8 to predict the net direction of a reaction between any two pairs, that is, whether the equilibrium position lies to the right ($K_c > 1$) or to the left ($K_c < 1$). *A reaction proceeds to the right if an acid reacts with a base lower on the list* because this combination produces a weaker conjugate base and a weaker conjugate acid. Here are two sample problems that demonstrate this key idea.

Sample Problem 18.5 Predicting the Net Direction of an Acid-Base Reaction

Problem Predict the net direction and whether K_c is greater or less than 1 for the following reaction (assume equal initial concentrations of all species):

$$H_2PO_4^-(aq) + NH_3(aq) \rightleftharpoons NH_4^+(aq) + HPO_4^{2-}(aq)$$

Plan We identify the conjugate acid-base pairs and consult Figure 18.8 to see which acid and base are stronger. The reaction proceeds in the direction that the stronger acid and base form the weaker acid and base. If the reaction *as written* proceeds to the right, then [products] is higher than [reactants], so $K_c > 1$.

Solution The conjugate pairs are $H_2PO_4^-/HPO_4^{2-}$ and NH_4^+/NH_3. Since $H_2PO_4^-$ is higher on the list of acids, it is stronger than NH_4^+; since NH_3 is lower on the list of bases, it is stronger than HPO_4^{2-}. Therefore,

$$\underset{\text{stronger acid}}{H_2PO_4^-(aq)} + \underset{\text{stronger base}}{NH_3(aq)} \rightleftharpoons \underset{\text{weaker base}}{HPO_4^{2-}(aq)} + \underset{\text{weaker acid}}{NH_4^+(aq)}$$

The net direction is to the right, so $K_c > 1$.

FOLLOW-UP PROBLEM 18.5 Predict the net direction and whether K_c is greater or less than 1 for the following reaction:

$$H_2O(l) + HS^-(aq) \rightleftharpoons OH^-(aq) + H_2S(aq)$$

Sample Problem 18.6 Using Molecular Scenes to Predict the Net Direction of an Acid-Base Reaction

Problem Given that 0.10 *M* HX *(blue and green)* has a pH of 2.88, and 0.10 *M* HY *(blue and orange)* has a pH of 3.52, which scene best represents the final mixture after equimolar solutions of HX and Y^- are mixed?

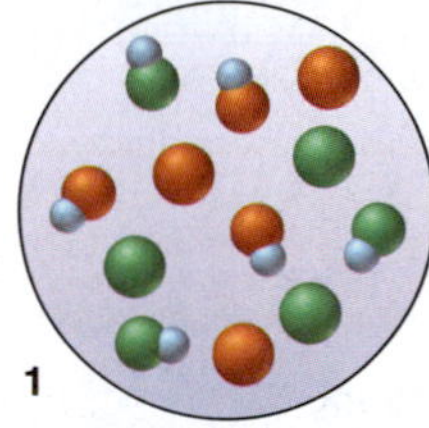
1

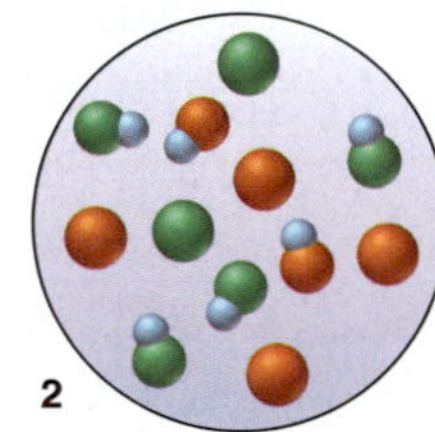
2

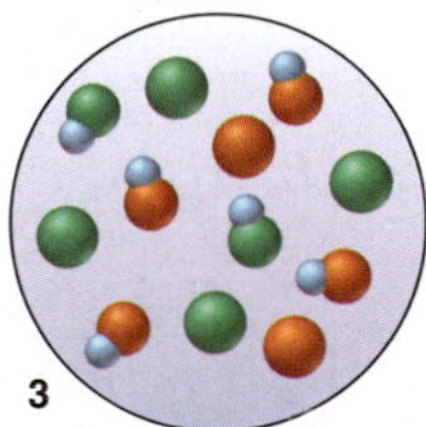
3

Plan A stronger acid and base yield a weaker acid and base, so we have to determine the relative acid strengths of HX and HY in order to choose the correct molecular scene. The concentrations of the acid solutions are equal, so we can pick the stronger acid directly from the pH values of the two acid solutions. Because the stronger acid reacts to a greater extent, fewer molecules of it will be in the scene than molecules of the weaker acid.

Solution The HX solution has a lower pH (2.88) than the HY solution (3.52), so we know right away that HX is the stronger acid and Y^- is the stronger base. Therefore, the reaction of HX and Y^- has a $K_c > 1$, which means the equilibrium mixture will have more HY than HX. Scene 1 has equal numbers of HX and HY, which would occur if the acids were of equal strength, and scene 2 shows fewer HY than HX, which would occur if HY were stronger. Therefore, only scene 3 is consistent with the relative acid strengths.

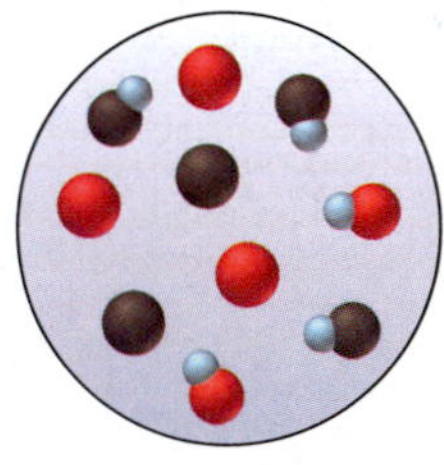

FOLLOW-UP PROBLEM 18.6 The scene in the margin represents the equilibrium mixture after 0.10 *M* solutions of HA *(blue and red)* and B^- *(black)* are combined. Does the reaction have a K_c greater or less than 1? Which acid is stronger, HA or HB?

Summary of Section 18.3

- The Brønsted-Lowry acid-base definition does not require that bases contain OH in their formula or that acid-base reactions occur in aqueous solution.
- An acid is a species that donates a proton and a base is one that accepts it, so an acid-base reaction is a proton-transfer process.
- When an acid donates a proton, it becomes the conjugate base; when a base accepts a proton, it becomes the conjugate acid. In an acid-base reaction, acids and bases form their conjugates. A stronger acid has a weaker conjugate base, and vice versa.
- An acid-base reaction proceeds to the greater extent ($K > 1$) in the direction in which a stronger acid and base form a weaker base and acid.

18.4 • SOLVING PROBLEMS INVOLVING WEAK-ACID EQUILIBRIA

Just as you saw in Chapter 17 for equilibrium problems in general, there are two general types of equilibrium problems involving weak acids and their conjugate bases:

1. Given equilibrium concentrations, find K_a.
2. Given K_a and some concentrations, find other equilibrium concentrations.

For all of these problems, we'll apply the same problem-solving approach, notation system, and assumptions:

- *The problem-solving approach.* Start with what is given in the problem and move toward what you want to find. Make a habit of applying the following steps:
  1. Write the balanced equation and K_a expression; these tell you what to find.
  2. Define x as the unknown change in concentration that occurs during the reaction. Frequently, $x = [HA]_{dissoc}$, the concentration of HA that dissociates, which, through the use of certain assumptions, also equals $[H_3O^+]$ and $[A^-]$ at equilibrium.
  3. Construct a reaction table (for most problems) that incorporates x.
  4. Make assumptions (usually that x is very small relative to the initial concentration; see below) that simplify the calculations.
  5. Substitute the values into the K_a expression, and solve for x.
  6. Check that the assumptions are justified with the 5% test first used in Sample Problem 17.9. If they are not justified, use the quadratic formula to find x.

- *The notation system.* As always, molar concentration is indicated with brackets. A subscript refers to where the species comes from or when it occurs in the reaction process. For example, $[H_3O^+]_{from\ HA}$ is the molar concentration of H_3O^+ that comes from the dissociation of HA; $[HA]_{init}$ is the initial molar concentration of HA, that is, before dissociation; $[HA]_{dissoc}$ is the molar concentration of HA that dissociates; and so forth. A bracketed formula with *no* subscript represents the molar concentration of the species *at equilibrium.*

- *The assumptions.* We make two assumptions to simplify the arithmetic:
  1. The $[H_3O^+]$ from the autoionization of water is so much smaller than the $[H_3O^+]$ from the dissociation of HA that we can neglect it:
  $$[H_3O^+] = [H_3O^+]_{from\ HA} + [H_3O^+]_{from\ H_2O} \approx [H_3O^+]_{from\ HA}$$
  Note that each molecule of HA that dissociates forms one H_3O^+ and one A^-, so $[H_3O^+] = [A^-]$.
  2. A weak acid has a small K_a. Therefore, it dissociates to such a small extent that we can neglect the change in its concentration to find its equilibrium concentration:
  $$[HA] = [HA]_{init} - [HA]_{dissoc} \approx [HA]_{init}$$

Finding K_a Given Concentrations

This type of problem involves finding K_a of a weak acid from the concentration of one of the species in solution, usually $[H_3O^+]$ from a given pH:

$$HA(aq) + H_2O(l) \rightleftharpoons H_3O^+(aq) + A^-(aq)$$

$$K_a = \frac{[H_3O^+][A^-]}{[HA]}$$

You prepare an aqueous solution of HA and measure its pH. Thus, you know $[HA]_{init}$, can calculate $[H_3O^+]$ from the pH, and then determine $[A^-]$ and $[HA]$ at equilibrium. You substitute these values into the K_a expression and solve for K_a. Let's go through the approach in Sample Problem 18.7.

Sample Problem 18.7 Finding K_a of a Weak Acid from the Solution pH

Problem Phenylacetic acid ($C_6H_5CH_2COOH$, simplified here to HPAc) builds up in the blood of persons with phenylketonuria, an inherited disorder that, if untreated, causes mental retardation and death. A study of the acid shows that the pH of 0.12 *M* HPAc is 2.62. What is the K_a of phenylacetic acid?

Plan We are given $[HPAc]_{init}$ (0.12 *M*) and the pH (2.62) and must find K_a. We first write the equation for HPAc dissociation and the expression for K_a to see which values we need to find. We assume $[H_3O^+]_{from\ H_2O}$ is negligible, so we can use the given pH to find $[H_3O^+]$, which equals $[PAc^-]$ and $[HPAc]_{dissoc}$. To find [HPAc], we assume that, because it is a weak acid, very little dissociates, so $[HPAc]_{init} - [HPAc]_{dissoc} = [HPAc] \approx [HPAc]_{init}$. We make these assumptions, substitute the equilibrium values, solve for K_a, and then check the assumptions with the 5% rule (Sample Problem 17.8). Note that we *don't* need a reaction table here because the pH will give us $[HPAc]_{dissoc}$.

Solution Writing the dissociation equation and K_a expression:

$$HPAc(aq) + H_2O(l) \rightleftharpoons H_3O^+(aq) + PAc^-(aq) \qquad K_a = \frac{[H_3O^+][PAc^-]}{[HPAc]}$$

Calculating $[H_3O^+]$:

$$[H_3O^+] = 10^{-pH} = 10^{-2.62} = 2.4\times10^{-3}\ M$$

Making the assumptions:

1. The calculated $[H_3O^+]$ (2.4×10^{-3} *M*) >> $[H_3O^+]_{from\ H_2O}$ (1.0×10^{-7} *M*), so we assume that $[H_3O^+] \approx [H_3O^+]_{from\ HPAc} = x$ (the change in [HPAc]).
2. HPAc is a weak acid, so we assume that $[HPAc] = 0.12\ M - x \approx 0.12\ M$.

Solving for the equilibrium concentrations:

$$x \approx [H_3O^+] = [PAc^-] = 2.4\times10^{-3}\ M$$

$$[HPAc] = 0.12\ M - x = 0.12\ M - (2.4\times10^{-3}\ M) \approx 0.12\ M \quad \text{(to 2 sf)}$$

Substituting these values into K_a:

$$K_a = \frac{[H_3O^+][PAc^-]}{[HPAc]} \approx \frac{(2.4\times10^{-3})(2.4\times10^{-3})}{0.12} = 4.8\times10^{-5}$$

Checking the assumptions by finding the percent error in concentration:

1. For $[H_3O^+]_{from\ H_2O}$: $\dfrac{1\times10^{-7}\ M}{2.4\times10^{-3}\ M} \times 100 = 4\times10^{-3}\%$ (<5%; assumption is justified).

2. For $[HPAc]_{dissoc}$: $\dfrac{2.4\times10^{-3}\ M}{0.12\ M} \times 100 = 2.0\%$ (<5%; assumption is justified).

Check The $[H_3O^+]$ makes sense: pH 2.62 should give $[H_3O^+]$ between 10^{-2} and 10^{-3} *M*. The K_a calculation also seems in the correct range: $(10^{-3})^2/10^{-1} = 10^{-5}$, and this value seems reasonable for a weak acid.

FOLLOW-UP PROBLEM 18.7 The conjugate acid of ammonia is the weak acid NH_4^+. If a 0.2 *M* NH_4Cl solution has a pH of 5.0, what is the K_a of NH_4^+?

Finding Concentrations Given K_a

The second type of equilibrium problem involving weak acids gives some concentration data and K_a and asks for the equilibrium concentration of some component. Such problems are very similar to those we solved in Chapter 17 in which a substance with a given initial concentration reacted to an unknown extent (see Sample Problems 17.8 to 17.10). We *will* use a reaction table in these problems to find the values, and since we now know that $[H_3O^+]_{from\ H_2O}$ is so small relative to $[H_3O^+]_{from\ HA}$, we will neglect it and enter the initial $[H_3O^+]$ in all reaction tables as zero.

Sample Problem 18.8 Determining Concentration from K_a and Initial [HA]

Problem Propanoic acid (CH_3CH_2COOH, which we simplify as HPr) is a carboxylic acid whose salts are used to retard mold growth in foods. What is the $[H_3O^+]$ of 0.10 *M* HPr ($K_a = 1.3\times10^{-5}$)?

Plan We know the initial concentration (0.10 M) and K_a (1.3×10^{-5}) of HPr, and we need to find $[H_3O^+]$. First, we write the balanced equation and the expression for K_a. We know $[HPr]_{init}$ but not [HPr] (that is, the concentration at equilibrium). If we let $x = [HPr]_{dissoc}$, x is also $[H_3O^+]_{from\ HPr}$ and $[Pr^-]$ because each HPr dissociates into one H_3O^+ and one Pr^-. With this information, we set up a reaction table. We assume that, because HPr has a small K_a, it dissociates very little. After solving for x, we check the assumption.

Solution Writing the balanced equation and expression for K_a:

$$HPr(aq) + H_2O(l) \rightleftharpoons H_3O^+(aq) + Pr^-(aq) \qquad K_a = \frac{[H_3O^+][Pr^-]}{[HPr]} = 1.3\times10^{-5}$$

Setting up a reaction table, with $x = [HPr]_{dissoc} = [H_3O^+]_{from\ HPr} = [Pr^-] = [H_3O^+]$:

Concentration (M)	$HPr(aq)$	+	$H_2O(l)$	$\rightleftharpoons$	$H_3O^+(aq)$	+	$Pr^-(aq)$
Initial	0.10		—		0		0
Change	$-x$		—		$+x$		$+x$
Equilibrium	$0.10 - x$		—		x		x

Making the assumption: K_a is small, so x is small compared with $[HPr]_{init}$; therefore, $0.10\ M - x \approx 0.10\ M$. Substituting into the K_a expression and solving for x:

$$K_a = \frac{[H_3O^+][Pr^-]}{[HPr]} = 1.3\times10^{-5} \approx \frac{(x)(x)}{0.10}$$

$$x \approx \sqrt{(0.10)(1.3\times10^{-5})} = 1.1\times10^{-3}\ M = [H_3O^+]$$

Checking the assumption for $[HPr]_{dissoc}$:

$$\frac{1.1\times10^{-3}\ M}{0.10\ M} \times 100 = 1.1\% \quad (<5\%;\ \text{assumption is justified})$$

Check The $[H_3O^+]$ seems reasonable for a dilute solution of a weak acid with a moderate K_a. By reversing the calculation, we can check the math: $(1.1\times10^{-3})^2/0.10 = 1.2\times10^{-5}$, which is within rounding of the given K_a.

Comment In Chapter 17, we used a benchmark, in addition to the 5% rule, to see if the assumption is justified (see the discussion following Sample Problem 17.9):

- If $\frac{[HA]_{init}}{K_a} > 400$, the assumption is justified: neglecting x introduces an error $<5\%$.
- If $\frac{[HA]_{init}}{K_a} < 400$, the assumption is *not* justified; neglecting x introduces an error $>5\%$, so we solve a quadratic equation to find x.

In this sample problem, we have $\frac{0.10}{1.3\times10^{-5}} = 7.7\times10^3$, which is greater than 400. In the follow-up problem, the ratio is less than 400.

FOLLOW-UP PROBLEM 18.8 Cyanic acid (HOCN) is an extremely acrid, unstable substance. What is the $[H_3O^+]$ of 0.10 M HOCN ($K_a = 3.5\times10^{-4}$)?

The Effect of Concentration on the Extent of Acid Dissociation

If we repeat the calculation in Sample Problem 18.8, but start with a lower [HPr], we observe a very interesting fact about the extent of dissociation of a weak acid. Suppose the initial concentration of HPr is one-tenth as much, 0.010 M rather than 0.10 M. After filling in the reaction table and making the same assumptions, we find that

$$x = [H_3O^+] = [HPr]_{dissoc} = 3.6\times10^{-4}\ M$$

Now, let's compare the percentages of HPr molecules dissociated at the two different initial acid concentrations, using the relationship

$$\text{Percent HA dissociated} = \frac{[HA]_{dissoc}}{[HA]_{init}} \times 100 \qquad \textbf{(18.5)}$$

Case 1: $[\text{HPr}]_{\text{init}} = 0.10\ M$

$$\text{Percent HPr dissociated} = \frac{1.1\times10^{-3}\ M}{1.0\times10^{-1}\ M}\times100 = 1.1\%$$

Case 2: $[\text{HPr}]_{\text{init}} = 0.010\ M$

$$\text{Percent HPr dissociated} = \frac{3.6\times10^{-4}\ M}{1.0\times10^{-2}\ M}\times100 = 3.6\%$$

As the initial acid concentration decreases, the percent dissociation of the acid increases. Don't confuse the *concentration* of HA dissociated with the *percent* HA dissociated. The concentration, $[\text{HA}]_{\text{dissoc}}$, is lower in the diluted HA solution because the actual *number* of dissociated HA molecules is smaller. It is the *fraction* (or the *percent*) of dissociated HA molecules that increases with dilution.

THINK OF IT THIS WAY

Are Gaseous and Weak-Acid Equilibria Alike?

Weak acids dissociating to a greater extent as they are diluted is analogous to the shift in a gaseous reaction when the container volume increases (Section 17.6). In the gaseous reaction, an increase in volume as the piston is withdrawn shifts the equilibrium position to favor more moles of gas. In the case of HA dissociation, the increase in volume as solvent is added shifts the equilibrium position to favor more moles of ions.

Sample Problem 18.9 uses molecular scenes to highlight this idea. (Note that, in order to depict the scenes practically, the acid has a much higher percent dissociation than any real weak acid does.)

Sample Problem 18.9 Using Molecular Scenes to Determine the Extent of HA Dissociation

Problem A 0.15 M solution of HA *(blue and green)* is 33% dissociated. Which scene represents a sample of that solution after it is diluted with water?

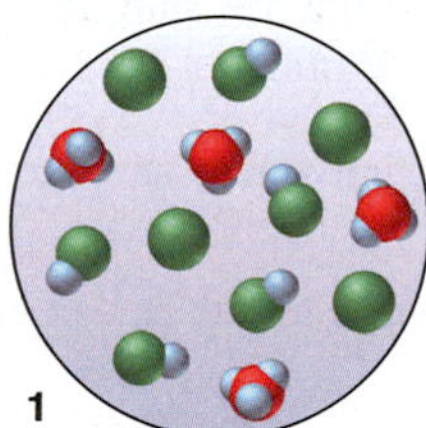

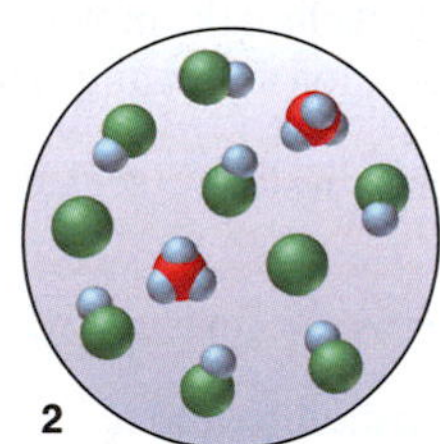

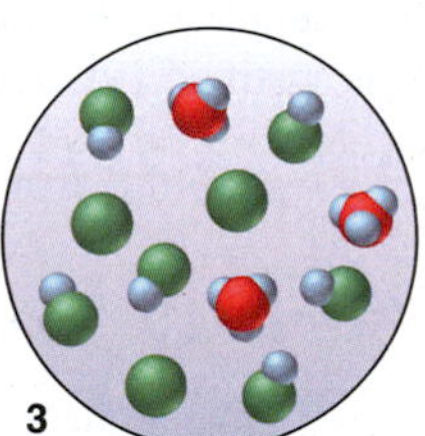

Plan We are given the percent dissociation of the original HA solution (33%), and we know that the percent dissociation increases as the acid is diluted. Thus, we calculate the percent dissociation of each diluted sample and see which is greater than 33%. To determine percent dissociation, we apply Equation 18.5, with $\text{HA}_{\text{dissoc}}$ equal to the number of H_3O^+ (or A^-) and HA_{init} equal to the number of HA *plus* the number of H_3O^+ (or A^-).

Solution Calculating the percent dissociation of each diluted solution with Equation 18.5:

Solution 1. Percent dissociated = $4/(5 + 4) \times 100 = 44\%$
Solution 2. Percent dissociated = $2/(7 + 2) \times 100 = 22\%$
Solution 3. Percent dissociated = $3/(6 + 3) \times 100 = 33\%$

Therefore, scene 1 represents the diluted solution.

Check Let's confirm our choice by examining the other scenes: in scene 2, HA is *less* dissociated than originally, so that scene must represent a more concentrated HA solution; scene 3 represents another solution with the same percent dissociation as the original.

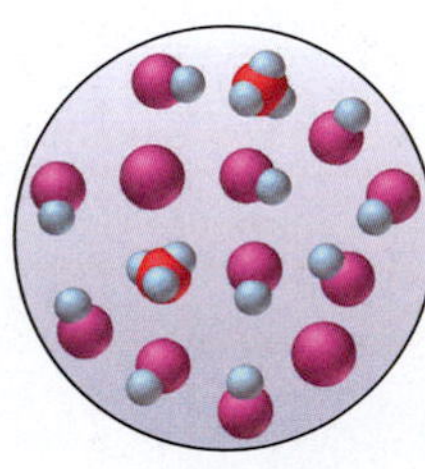

FOLLOW-UP PROBLEM 18.9 The scene in the margin represents a sample of a weak acid HB *(blue and purple)* dissolved in water. Draw a scene that represents the same volume after the solution has been diluted with water.

The Behavior of Polyprotic Acids

Acids with more than one ionizable proton are **polyprotic acids.** In solution, each dissociation step has a different K_a. For example, phosphoric acid is a triprotic acid (three ionizable protons), so it has three K_a values:

$$H_3PO_4(aq) + H_2O(l) \rightleftharpoons H_2PO_4^-(aq) + H_3O^+(aq)$$

$$K_{a1} = \frac{[H_2PO_4^-][H_3O^+]}{[H_3PO_4]} = 7.2\times10^{-3}$$

$$H_2PO_4^-(aq) + H_2O(l) \rightleftharpoons HPO_4^{2-}(aq) + H_3O^+(aq)$$

$$K_{a2} = \frac{[HPO_4^{2-}][H_3O^+]}{[H_2PO_4^-]} = 6.3\times10^{-8}$$

$$HPO_4^{2-}(aq) + H_2O(l) \rightleftharpoons PO_4^{3-}(aq) + H_3O^+(aq)$$

$$K_{a3} = \frac{[PO_4^{3-}][H_3O^+]}{[HPO_4^{2-}]} = 4.2\times10^{-13}$$

The relative K_a values show that H_3PO_4 is a much stronger acid than $H_2PO_4^-$, which is much stronger than HPO_4^{2-}.

Note that the general pattern seen for H_3PO_4 occurs for all polyprotic acids (see the list in Appendix C):

$$K_{a1} >> K_{a2} >> K_{a3}$$

This trend occurs because it is more difficult for an H^+ ion to leave a singly charged anion (such as $H_2PO_4^-$) than to leave a neutral molecule (such as H_3PO_4), and more difficult still for it to leave a doubly charged anion (such as HPO_4^{2-}). Successive K_a values typically differ by several orders of magnitude. This fact simplifies calculations because *we usually neglect the* H_3O^+ *coming from the subsequent dissociations.*

Summary of Section 18.4

- Two common types of weak-acid equilibrium problems involve finding K_a from a given concentration and finding a concentration from a given K_a.
- We summarize the given information in a reaction table and simplify the arithmetic by assuming (1) that $[H_3O^+]_{\text{from } H_2O}$ is much smaller than $[H_3O^+]_{\text{from HA}}$ and can be neglected and (2) that weak acids dissociate so little that $[HA]_{init} \approx [HA]$ at equilibrium.
- The *fraction* of weak acid molecules that dissociates is greater in a more dilute solution, even though the total $[H_3O^+]$ is lower.
- Polyprotic acids have more than one ionizable proton, but we assume that the first dissociation provides virtually all the H_3O^+.

18.5 • WEAK BASES AND THEIR RELATION TO WEAK ACIDS

The Brønsted-Lowry concept expands the definition of a base to encompass a host of species that the Arrhenius definition excludes: *to accept a proton, a base needs only a lone electron pair.*

Let's examine the equilibrium system of a weak base (B) as it dissolves in water: B accepts a proton from H_2O, which acts as an acid, leaving behind an OH^- ion:

$$B(aq) + H_2O(l) \rightleftharpoons BH^+(aq) + OH^-(aq)$$

This general reaction for a base in water is described by the equilibrium expression

$$K_c = \frac{[BH^+][OH^-]}{[B][H_2O]}$$

Based on our earlier reasoning, we incorporate $[H_2O]$ in the value of K_c and obtain the **base-dissociation constant** (or **base-ionization constant**), K_b:

$$K_b = \frac{[BH^+][OH^-]}{[B]} \quad (18.6)$$

Despite the name "base-dissociation constant," *no base dissociates in the process.*

As in the relation between pK_a and K_a, we know that a lower pK_b indicates a higher K_b, that is, a stronger base. In aqueous solution, the two large classes of weak bases are (1) the molecules ammonia and the amines and (2) the anions of weak acids.

Molecules as Weak Bases: Ammonia and the Amines

Ammonia is the simplest N-containing compound that acts as a weak base in water:

$$NH_3(aq) + H_2O(l) \rightleftharpoons NH_4^+(aq) + OH^-(aq) \qquad K_b = 1.76\times10^{-5} \text{ (at 25°C)}$$

Despite the label on the bottle in the lab that reads "ammonium hydroxide," an aqueous solution of ammonia consists largely of water and *unprotonated* NH_3 molecules, as its small K_b indicates. In a 1.0 *M* NH_3 solution, for example, $[OH^-] = [NH_4^+] = 4.2\times10^{-3}$ *M*, so about 99.6% of the NH_3 is not protonated. A list of K_b values for some molecular bases appears in Appendix C.

If one or more of the H atoms in ammonia is replaced by an organic group (designated as R), an *amine* results: RNH_2, R_2NH, or R_3N (Section 15.4; see Figure 15.11). The key structural feature of these organic compounds, as in all Brønsted-Lowry bases, is *a lone pair of electrons that can bind the proton donated by the acid.* Figure 18.9 depicts this process for methylamine, the simplest amine.

To find the pH of a solution of a molecular weak base, we use an approach very similar to that for a weak acid: write the equilibrium expression, set up a reaction table to find $[B]_{reacting}$, make the usual assumptions, and then solve for $[OH^-]$. The only additional step is to convert $[OH^-]$ to $[H_3O^+]$ in order to calculate pH.

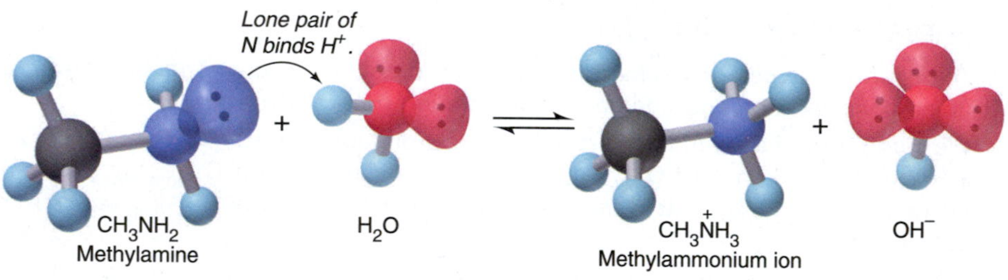

Figure 18.9 Abstraction of a proton from water by the base methylamine.

Sample Problem 18.10 Determining pH from K_b and Initial [B]

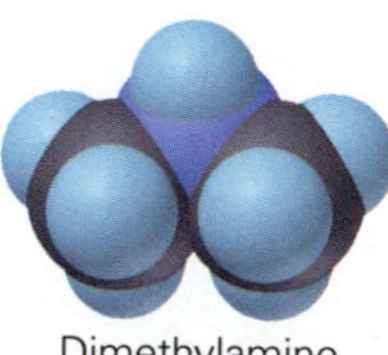
Dimethylamine

Problem Dimethylamine, $(CH_3)_2NH$ *(see margin)*, a key intermediate in detergent manufacture, has a K_b of 5.9×10^{-4}. What is the pH of 1.5 *M* $(CH_3)_2NH$?

Plan We know the initial concentration (1.5 *M*) and K_b (5.9×10^{-4}) of $(CH_3)_2NH$ and have to find the pH. The amine reacts with water to form OH^-, so we have to find $[OH^-]$ and then calculate $[H_3O^+]$ and pH. We first write the balanced equation and K_b expression. Because $K_b >> K_w$, the $[OH^-]$ from the autoionization of water is negligible, so we disregard it and assume that all the $[OH^-]$ comes from the base reacting with water. Because K_b is small, we assume that the amount of amine reacting, $[(CH_3)_2NH]_{reacting}$, can be neglected. We set up a reaction table, make the assumption, and solve for *x*. Then we check the assumption and convert $[OH^-]$ to $[H_3O^+]$ using K_w; finally, we calculate pH.

Solution Writing the balanced equation and K_b expression:

$$(CH_3)_2NH(aq) + H_2O(l) \rightleftharpoons (CH_3)_2NH_2^+(aq) + OH^-(aq)$$

$$K_b = \frac{[(CH_3)_2NH_2^+][OH^-]}{[(CH_3)_2NH]}$$

Setting up the reaction table, with $x = [(CH_3)_2NH]_{reacting} = [(CH_3)_2NH_2^+] = [OH^-]$:

Concentration (*M*)	$(CH_3)_2NH(aq)$	+	$H_2O(l)$	$\rightleftharpoons$	$(CH_3)_2NH_2^+(aq)$	+	$OH^-(aq)$
Initial	1.5		—		0		0
Change	$-x$		—		$+x$		$+x$
Equilibrium	$1.5 - x$		—		x		x

Making the assumption: K_b is small, so $[(CH_3)_2NH]_{init} - [(CH_3)_2NH]_{reacting} = [(CH_3)_2NH] \approx [(CH_3)_2NH]_{init}$; thus, $1.5\ M - x \approx 1.5\ M$.
Substituting into the K_b expression and solving for x:

$$K_b = \frac{[(CH_3)_2NH_2^+][OH^-]}{[(CH_3)_2NH]} = 5.9\times10^{-4} \approx \frac{x^2}{1.5}$$

$$x = [OH^-] \approx 3.0\times10^{-2}\ M$$

Checking the assumption:

$$\frac{3.0\times10^{-2}\ M}{1.5\ M} \times 100 = 2.0\% \quad (<5\%;\ \text{assumption is justified})$$

Note that the Comment in Sample Problem 18.8 also applies to problems involving weak bases:

$$\frac{[B]_{init}}{K_b} = \frac{1.5}{5.9\times10^{-4}} = 2.5\times10^3 > 400$$

Calculating pH:

$$[H_3O^+] = \frac{K_w}{[OH^-]} = \frac{1.0\times10^{-14}}{3.0\times10^{-2}} = 3.3\times10^{-13}\ M$$

$$pH = -\log(3.3\times10^{-13}) = 12.48$$

Check The value of x seems reasonable: $\sqrt{(\sim6\times10^{-4})(1.5)} = \sqrt{9\times10^{-4}} = 3\times10^{-2}$. Because $(CH_3)_2NH$ is a weak base, the pH should be several pH units above 7.

FOLLOW-UP PROBLEM 18.10 Pyridine (C_5H_5N, *see margin*) serves as a solvent *and* base in many organic syntheses. It has a pK_b of 8.77. What is the pH of 0.10 *M* pyridine?

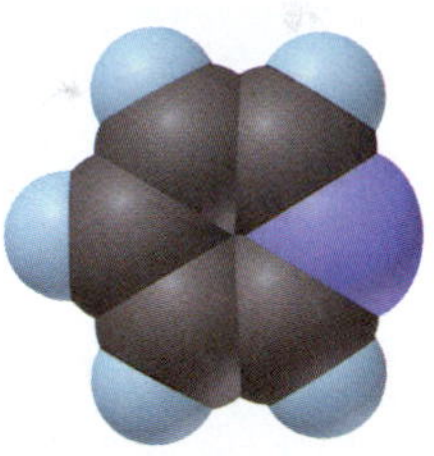

Pyridine

Anions of Weak Acids as Weak Bases

The other large group of Brønsted-Lowry bases consists of anions of weak acids:

$$A^-(aq) + H_2O(l) \rightleftharpoons HA(aq) + OH^-(aq) \qquad K_b = \frac{[HA][OH^-]}{[A^-]}$$

For example, F^-, the anion of the weak acid HF, is a weak base:

$$F^-(aq) + H_2O(l) \rightleftharpoons HF(aq) + OH^-(aq) \qquad K_b = \frac{[HF][OH^-]}{[F^-]}$$

Why is a solution of HA acidic and a solution of A^- basic? We'll find the answer from relative concentrations of species in 1 *M* HF and in 1 *M* NaF:

1. *The acidity of HA(aq).* HF is a weak acid, so the equilibrium position of the acid dissolving in water lies far to the left:

$$HF(aq) + H_2O(l) \rightleftharpoons H_3O^+(aq) + F^-(aq)$$

Water also contributes H_3O^+ and OH^-, but their concentrations are extremely small:

$$2H_2O(l) \rightleftharpoons H_3O^+(aq) + OH^-(aq)$$

Of all the species present—HF, H_2O, H_3O^+, F^-, and OH^-—the two that can influence the acidity of the solution are H_3O^+, predominantly from HF, and OH^- from water. Thus, the HF solution is acidic because $[H_3O^+]_{from\ HF} >> [OH^-]_{from\ H_2O}$.

2. *The basicity of $A^-(aq)$.* Now, consider the species present in 1 *M* NaF. The salt dissociates completely to yield 1 *M* Na^+ and 1 *M* F^-. The Na^+ behaves as a spectator ion, while some F^- reacts as a weak base to produce small amounts of HF and OH^-:

$$F^-(aq) + H_2O(l) \rightleftharpoons HF(aq) + OH^-(aq)$$

As before, water dissociation contributes minute amounts of H_3O^+ and OH^-. Thus, in addition to the Na^+ ion, the species present are the same as in the HF solution: HF, H_2O, H_3O^+, F^-, and OH^-. The two species that affect the acidity are OH^-, predominantly from F^- reacting with water, and H_3O^+ from water. In this case, $[OH^-]_{\text{from } F^-} >> [H_3O^+]_{\text{from } H_2O}$, so the solution is basic.

To summarize, *the relative concentrations of HA and A^- determine the acidity or basicity of a solution:*

- In an HA solution, $[HA] >> [A^-]$ and $[H_3O^+]_{\text{from HA}} >> [OH^-]_{\text{from } H_2O}$, so the solution is acidic.
- In an A^- solution, $[A^-] >> [HA]$ and $[OH^-]_{\text{from } A^-} >> [H_3O^+]_{\text{from } H_2O}$, so the solution is basic.

The Relation Between K_a and K_b of a Conjugate Acid-Base Pair

A key relationship exists between the K_a of HA and the K_b of A^-, which we can see by writing the two reactions as a reaction sequence and adding them:

$$\cancel{HA} + H_2O \rightleftharpoons H_3O^+ + \cancel{A^-}$$
$$\cancel{A^-} + H_2O \rightleftharpoons \cancel{HA} + OH^-$$
$$2H_2O \rightleftharpoons H_3O^+ + OH^-$$

The sum of the two dissociation reactions is the autoionization of water. Recall from Chapter 17 that, for a reaction that is the *sum* of two or more reactions, the overall equilibrium constant is the *product* of the individual equilibrium constants. Therefore, writing the equilibrium expressions for each reaction gives

$$\frac{[H_3O^+]\cancel{[A^-]}}{\cancel{[HA]}} \times \frac{\cancel{[HA]}[OH^-]}{\cancel{[A^-]}} = [H_3O^+][OH^-]$$

or

$$K_a \times K_b = K_w \qquad (18.7)$$

This relationship allows us to find K_a of the acid in a conjugate pair given K_b of the base, and vice versa. Reference tables typically have K_a and K_b values for *molecular species only.* The K_b for F^- and the K_a for $CH_3NH_3^+$, for example, do not appear in standard tables, but you can calculate them by looking up the value for the molecular conjugate species and relating it to K_w. To find the K_b value for F^-, for instance, we look up the K_a value for HF and apply Equation 18.7:

$$K_a \text{ of HF} = 6.8\times10^{-4} \quad \text{(from Appendix C)}$$

So, we have

$$K_a \text{ of HF} \times K_b \text{ of } F^- = K_w$$

or

$$K_b \text{ of } F^- = \frac{K_w}{K_a \text{ of HF}} = \frac{1.0\times10^{-14}}{6.8\times10^{-4}} = 1.5\times10^{-11}$$

Sample Problem 18.11 Determining the pH of a Solution of A^-

Problem Sodium acetate (CH_3COONa, or NaAc for this problem) is used in textile dyeing. What is the pH of 0.25 *M* NaAc at 25°C? K_a of acetic acid (HAc) is 1.8×10^{-5}.

Plan We know the initial concentration of Ac^- (0.25 *M*) and the K_a of HAc (1.8×10^{-5}), and we have to find the pH of the Ac^- solution, which acts as a base in water. We write the base dissociation equation and K_b expression. If we can find $[OH^-]$, we can use K_w to find $[H_3O^+]$ and convert it to pH. To solve for $[OH^-]$, we need the K_b of Ac^-, which we obtain from the K_a of HAc by applying Equation 18.7. We set up a reaction table to find $[OH^-]$ and make the usual assumption that K_b is small, so $[Ac^-]_{init} \approx [Ac^-]$.

Solution Writing the base dissociation equation and K_b expression:

$$Ac^-(aq) + H_2O(l) \rightleftharpoons HAc(aq) + OH^-(aq) \qquad K_b = \frac{[HAc][OH^-]}{[Ac^-]}$$

Setting up the reaction table, with $x = [Ac^-]_{reacting} = [HAc] = [OH^-]$:

Concentration (M)	$Ac^-(aq)$	+	$H_2O(l)$	$\rightleftharpoons$	$HAc(aq)$	+	$OH^-(aq)$
Initial	0.25		—		0		0
Change	$-x$		—		$+x$		$+x$
Equilibrium	$0.25 - x$		—		x		x

Solving for K_b of Ac^-:

$$K_b = \frac{K_w}{K_a} = \frac{1.0\times10^{-14}}{1.8\times10^{-5}} = 5.6\times10^{-10}$$

Making the assumption: Because K_b is small, $0.25\ M - x \approx 0.25\ M$.
Substituting into the expression for K_b and solving for x:

$$K_b = \frac{[HAc][OH^-]}{[Ac^-]} = 5.6\times10^{-10} \approx \frac{x^2}{0.25} \qquad x = [OH^-] \approx 1.2\times10^{-5}\ M$$

Checking the assumption:

$$\frac{1.2\times10^{-5}\ M}{0.25\ M} \times 100 = 4.8\times10^{-3}\% \quad (<5\%;\ \text{assumption is justified})$$

Also note that $$\frac{0.25}{5.6\times10^{-10}} = 4.5\times10^{8} > 400$$

Solving for pH:

$$[H_3O^+] = \frac{K_w}{[OH^-]} = \frac{1.0\times10^{-14}}{1.2\times10^{-5}} = 8.3\times10^{-10}\ M$$

$$\text{pH} = -\log(8.3\times10^{-10}) = 9.08$$

Check The K_b calculation seems reasonable: $\sim10\times10^{-15}/2\times10^{-5} = 5\times10^{-10}$. Because Ac^- is a weak base, $[OH^-] > [H_3O^+]$; thus, pH > 7, which makes sense.

FOLLOW-UP PROBLEM 18.11 Sodium hypochlorite (NaClO) is the active ingredient in household laundry bleach. What is the pH of 0.20 M NaClO?

Summary of Section 18.5

- The extent to which a weak base accepts a proton from water to form OH^- is expressed by a base-dissociation constant, K_b.
- Brønsted-Lowry bases include NH_3 and amines and the anions of weak acids. All produce basic solutions by accepting H^+ from water, which yields OH^-, thus making $[H_3O^+] < [OH^-]$.
- A solution of HA is acidic because $[HA] >> [A^-]$, so $[H_3O^+] > [OH^-]$. A solution of A^- is basic because $[A^-] >> [HA]$, so $[OH^-] > [H_3O^+]$.
- By multiplying the expressions for K_a of HA and K_b of A^-, we obtain K_w. This relationship allows us to calculate either K_a of BH^+ or K_b of A^-.

18.6 • MOLECULAR PROPERTIES AND ACID STRENGTH

The strength of an acid depends on its ability to donate a proton, which depends in turn on the strength of the bond to the acidic proton. In this section, we apply trends in atomic and bond properties to determine the trends in acid strength of nonmetal hydrides and oxoacids and then discuss the acidity of hydrated metal ions.

Acid Strength of Nonmetal Hydrides

Two factors determine how easily a proton is released from a nonmetal hydride:

- The electronegativity of the central nonmetal (E)
- The strength of the E—H bond

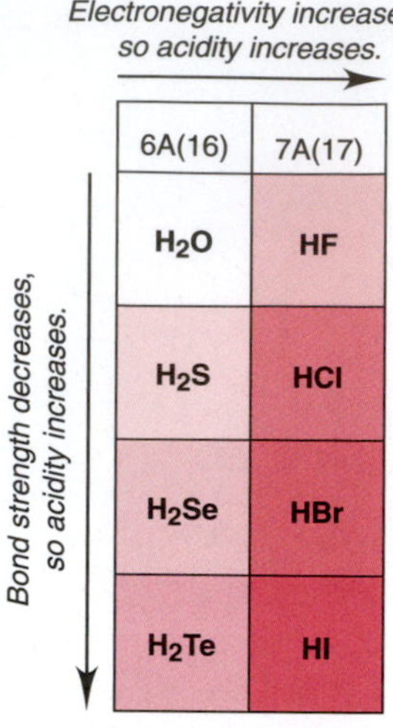

Figure 18.10 The effect of atomic and molecular properties on nonmetal hydride acidity.

Figure 18.10 displays two periodic trends among the nonmetal hydrides:

1. *Across a period, acid strength increases.* The electronegativity of the nonmetal E determines the trend. From left to right, as E becomes more electronegative, it withdraws electron density from H, and the E—H bond becomes more polar. As a result, H^+ is pulled away more easily by an O atom of a water molecule. In aqueous solution, the hydrides of Groups 3A(13) to 5A(15) do not behave as acids, but an increase in acid strength is seen from Group 6A(16) to 7A(17).

2. *Down a group, acid strength increases.* The E—H bond strength determines the trend. As E becomes larger, the E—H bond becomes longer and weaker, so H^+ comes off more easily. For example, the hydrohalic acid strength increases down the group:

$$HF \ll HCl < HBr < HI$$

Acid Strength of Oxoacids

All oxoacids have the acidic H atom bonded to an O atom, so bond length is not involved. Two other factors determine the acid strength of oxoacids (Section 14.7):

- The electronegativity of the central nonmetal (E)
- The number of O atoms around E (O.N. of E)

Figure 18.11 summarizes these trends:

1. *For oxoacids with the* ***same*** *number of O atoms, acid strength increases with the electronegativity of E.* Consider the hypohalous acids (HOE, where E is a halogen atom). The more electronegative E is, the more polar the O—H bond becomes and the more easily H^+ is lost (Figure 18.11A). Electronegativity decreases down a group, as does acid strength:

K_a of HOCl $= 2.9\times10^{-8}$ K_a of HOBr $= 2.3\times10^{-9}$ K_a of HOI $= 2.3\times10^{-11}$

Similarly, in Group 6A(16), H_2SO_4 is stronger than H_2SeO_4; in Group 5A(15), H_3PO_4 is stronger than H_3AsO_4, and so forth.

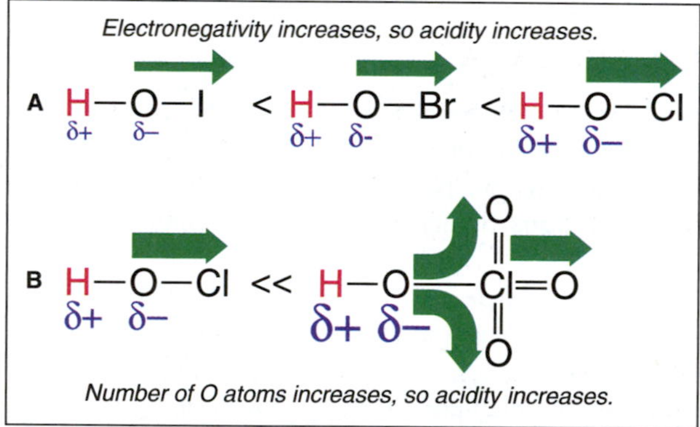

Figure 18.11 The relative strengths of oxoacids. **A,** Cl withdraws electron density (thickness of green arrow) from the O—H bond most effectively, making that bond most polar (relative size of δ symbols). **B,** Additional O atoms pull more electron density from the O—H bond.

2. *For oxoacids with* ***different*** *numbers of O atoms, acid strength increases with number of O atoms (or O.N. of central nonmetal).* The electronegative O atoms pull electron density away from E, which makes the O—H bond more polar. The more O atoms present, the greater the shift in electron density, and the more easily the H^+ ion comes off (Figure 18.11B). Therefore, the chlorine oxoacids ($HOClO_n$, with n from 0 to 3) increase in strength with the number of O atoms (and O.N. of Cl):

K_a of $HO\overset{+1}{Cl} = 2.9\times10^{-8}$ K_a of $HO\overset{+3}{Cl}O = 1.12\times10^{-2}$ K_a of $HO\overset{+5}{Cl}O_2 \approx 1$ K_a of $HO\overset{+7}{Cl}O_3 > 10^7$

Similarly, HNO_3 is stronger than HNO_2, H_2SO_4 is stronger than H_2SO_3, and so forth.

Acidity of Hydrated Metal Ions

The aqueous solutions of certain metal ions are acidic because the *hydrated* metal ion transfers an H^+ ion to water. Consider a general metal nitrate, $M(NO_3)_n$, as it dissolves in water. The ions separate and the metal ion becomes bonded to some number of H_2O

molecules. This equation shows the hydration of the cation (M^{n+}) with H_2O molecules and (*aq*); hydration of the anion (NO_3^-) is indicated by just (*aq*):

$$M(NO_3)_n(s) + xH_2O(l) \longrightarrow M(H_2O)_x^{n+}(aq) + nNO_3^-(aq)$$

If the metal ion, M^{n+}, is *small and highly charged,* its high charge density withdraws sufficient electron density from the O—H bonds of the bound water molecules that an H^+ is released. Thus, the hydrated cation, $M(H_2O)_x^{n+}$, is a typical Brønsted-Lowry acid. The bound H_2O that releases the H^+ becomes a bound OH^- ion:

$$M(H_2O)_x^{n+}(aq) + H_2O(l) \rightleftharpoons M(H_2O)_{x-1}OH^{(n-1)+}(aq) + H_3O^+(aq)$$

Salts of most M^{2+} and M^{3+} ions yield acidic aqueous solutions. The K_a values for some acidic hydrated metal ions appear in Appendix C.

Consider the small, highly charged Al^{3+} ion. When an aluminum salt, such as $Al(NO_3)_3$, dissolves in water, the following steps occur:

$$Al(NO_3)_3(s) + 6H_2O(l) \longrightarrow Al(H_2O)_6^{3+}(aq) + 3NO_3^-(aq)$$

[dissolution and hydration]

$$Al(H_2O)_6^{3+}(aq) + H_2O(l) \rightleftharpoons Al(H_2O)_5OH^{2+}(aq) + H_3O^+(aq)$$

[dissociation of weak acid]

Note the formulas of the hydrated metal ions in the last step. When H^+ is released, the number of bound H_2O molecules decreases by 1 (from 6 to 5) and the number of bound OH^- ions increases by 1 (from 0 to 1), which reduces the ion's positive charge by 1 (from 3 to 2) (Figure 18.12). This pattern of changes in the formula of the hydrated metal ion before and after it loses a proton appears for any highly charged metal ion in water.

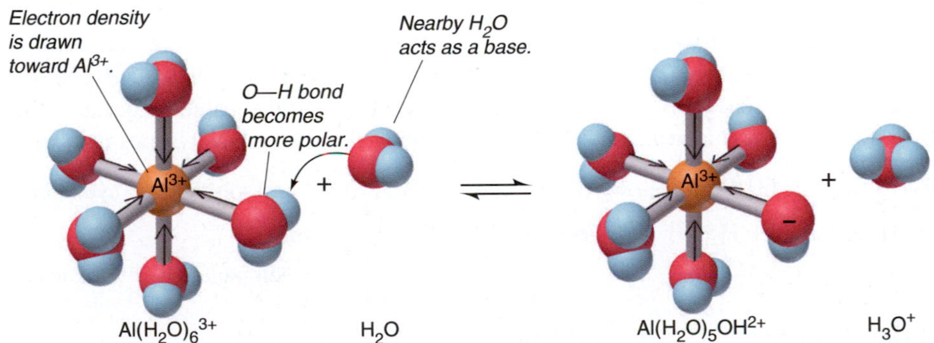

Figure 18.12 The acidic behavior of the hydrated Al^{3+} ion. The hydrated Al^{3+} ion is small and multiply charged and pulls electron density from the O—H bonds, allowing an H^+ ion to be transferred to a nearby water molecule.

Summary of Section 18.6

- For nonmetal hydrides, acid strength increases across a period, with the electronegativity of the nonmetal (E), and down a group, with the length of the E—H bond.
- For oxoacids with the same number of O atoms, acid strength increases with electronegativity of E; for oxoacids with the same E, acid strength increases with number of O atoms (or O.N. of E).
- Small, highly charged metal ions are acidic in water because they withdraw electron density from the O—H bonds of bound H_2O molecules, releasing an H^+ ion to the solution.

18.7 • ACID-BASE PROPERTIES OF SALT SOLUTIONS

In many cases, when a salt dissolves, one or both of its ions may react with water and affect the pH of the solution. You've seen that cations of weak bases (such as NH_4^+) are acidic, anions of weak acids (such as CN^-) are basic, and small, highly charged metal cations (such as Al^{3+}) are acidic. In addition, certain ions (such as $H_2PO_4^-$ and HCO_3^-) can act as an acid or a base. In this section, we classify the acid-base behavior of the various types of salt solutions.

Salts That Yield Neutral Solutions

A salt consisting of the anion of a strong acid and the cation of a strong base yields a neutral solution because the ions do not react with water. When a strong acid such as HNO_3 dissolves, the reaction goes essentially to completion because *the anion of a strong acid is a much weaker base than water.* The anion is hydrated, but *it does not react with water:*

$$HNO_3(l) + H_2O(l) \longrightarrow NO_3^-(aq) + H_3O^+(aq) \quad \text{[dissolution and hydration]}$$

Similarly, a strong base, such as NaOH, dissolves completely. The cation, in this case Na^+, is hydrated, but *it is not small and charged enough to react with water:*

$$NaOH(s) \xrightarrow{H_2O} Na^+(aq) + OH^-(aq) \quad \text{[dissolution and hydration]}$$

The anions of strong acids are the halide ions (except F^-) and ions of strong oxoacids, such as NO_3^- and ClO_4^-. The cations of strong bases are Group 1A(1) and Ca^{2+}, Sr^{2+}, and Ba^{2+} Group 2A(2). *Salts containing only these ions yield neutral solutions.*

Salts That Yield Acidic Solutions

Two types of salts yield acidic solutions:

1. *A salt consisting of the cation of a weak base and the anion of a strong acid yields an acidic solution because the cation acts as a weak acid,* and the anion does not react. For example, NH_4Cl yields an acidic solution because NH_4^+, the cation of the weak base NH_3, is a weak acid; Cl^-, the anion of the strong HCl, does not react:

$$NH_4Cl(s) \xrightarrow{H_2O} NH_4^+(aq) + Cl^-(aq) \quad \text{[dissolution and hydration]}$$
$$NH_4^+(aq) + H_2O(l) \rightleftharpoons NH_3(aq) + H_3O^+(aq) \quad \text{[dissociation of weak acid]}$$

2. *A salt consisting of a small, highly charged metal cation and the anion of a strong acid yields an acidic solution because the cation acts as a weak acid,* and the anion does not react. For example, $Fe(NO_3)_3$ yields an acidic solution because the hydrated Fe^{3+} ion is a weak acid; NO_3^-, the anion of the strong HNO_3, does not react:

$$Fe(NO_3)_3(s) + 6H_2O(l) \xrightarrow{H_2O} Fe(H_2O)_6^{3+}(aq) + 3NO_3^-(aq) \quad \text{[dissolution and hydration]}$$
$$Fe(H_2O)_6^{3+}(aq) + H_2O(l) \rightleftharpoons Fe(H_2O)_5OH^{2+}(aq) + H_3O^+(aq)$$

[dissociation of weak acid]

Salts That Yield Basic Solutions

A salt consisting of the anion of a weak acid and the cation of a strong base yields a basic solution because the anion acts as a weak base, and the cation does not react. Sodium acetate, for example, yields a basic solution because the CH_3COO^- ion, the anion of the weak acid CH_3COOH, acts as a weak base; Na^+, the cation of the strong base NaOH, does not react:

$$CH_3COONa(s) \xrightarrow{H_2O} Na^+(aq) + CH_3COO^-(aq) \quad \text{[dissolution and hydration]}$$
$$CH_3COO^-(aq) + H_2O(l) \rightleftharpoons CH_3COOH(aq) + OH^-(aq) \quad \text{[reaction of weak base]}$$

Sample Problem 18.12 Predicting Relative Acidity of Salt Solutions from Reactions of the Ions with Water

Problem Predict whether aqueous solutions of the following salts are acidic, basic, or neutral, and write an equation for the reaction of any ion with water:
(a) Potassium perchlorate, $KClO_4$
(b) Sodium benzoate, C_6H_5COONa
(c) Chromium(III) nitrate, $Cr(NO_3)_3$

Plan The formula shows the cation and anion. Depending on an ion's ability to react with water, the solution will be neutral (strong-acid anion and strong-base cation), acidic (weak-base cation or highly charged metal cation with strong-acid anion), or basic (weak-acid anion and strong-base cation).

Solution (a) Neutral. The ions are K^+ and ClO_4^-. The K^+ is from the strong base KOH, and the ClO_4^- is from the strong acid $HClO_4$. Neither ion reacts with water.
(b) Basic. The ions are Na^+ and $C_6H_5COO^-$. The Na^+ is the cation of the strong base NaOH, so it does not react with water. The benzoate ion, $C_6H_5COO^-$, is the anion of the weak acid benzoic acid, so it reacts with water to produce OH^- ion:

$$C_6H_5COO^-(aq) + H_2O(l) \rightleftharpoons C_6H_5COOH(aq) + OH^-(aq)$$

(c) Acidic. The ions are Cr^{3+} and NO_3^-. The NO_3^- is the anion of the strong acid HNO_3, so it does not react with water. The Cr^{3+} ion is small and highly charged, so the hydrated ion, $Cr(H_2O)_6^{3+}$, reacts with water to produce H_3O^+:

$$Cr(H_2O)_6^{3+}(aq) + H_2O(l) \rightleftharpoons Cr(H_2O)_5OH^{2+}(aq) + H_3O^+(aq)$$

FOLLOW-UP PROBLEM 18.12 Write equations to predict whether solutions of the following salts are acidic, basic, or neutral: **(a)** $KClO_2$; **(b)** $CH_3NH_3NO_3$; **(c)** CsI.

Salts of Weakly Acidic Cations and Weakly Basic Anions

If a salt consists of a cation that is a weak acid *and* an anion that is a weak base, the overall acidity of the solution depends on the relative acid strength (K_a) and base strength (K_b) of the separated ions. Consider a solution of ammonium cyanide, NH_4CN, and the reactions that occur between the separated ions and water. Ammonium ion is the conjugate acid of a weak base, so it is a weak acid:

$$NH_4^+(aq) + H_2O(l) \rightleftharpoons NH_3(aq) + H_3O^+(aq)$$

Cyanide ion is the anion of a weak acid, so it is a weak base:

$$CN^-(aq) + H_2O(l) \rightleftharpoons HCN(aq) + OH^-(aq)$$

The reaction that goes farther to the right determines the pH of the solution, so we compare the K_a of NH_4^+ with the K_b of CN^-. Only molecular compounds are listed in K_a and K_b tables, so we calculate these values for the ions with Equation 18.7:

$$K_a \text{ of } NH_4^+ = \frac{K_w}{K_b \text{ of } NH_3} = \frac{1.0\times10^{-14}}{1.76\times10^{-5}} = 5.7\times10^{-10}$$

$$K_b \text{ of } CN^- = \frac{K_w}{K_a \text{ of HCN}} = \frac{1.0\times10^{-14}}{6.2\times10^{-10}} = 1.6\times10^{-5}$$

Because K_b of $CN^- > K_a$ of NH_4^+, the NH_4CN solution is basic.

Salts of Amphiprotic Anions

The only salts left to consider are those in which the the cation comes from a strong base and the anion comes from a polyprotic acid with one or more ionizable protons still attached. These anions are amphiprotic—they can act as an acid and release a proton *to* water or as a base and abstract a proton *from* water. As in the previous case, to determine the overall acidity of their solutions, we compare the magnitudes of K_a and K_b, but here we compare both the K_a and K_b of the same species, the anion.

For example, Na_2HPO_4 consists of Na^+, the cation of a strong base, which does not react with water, and HPO_4^{2-}, the second anion of the weak polyprotic acid H_3PO_4. In water, the salt undergoes three steps:

1. $Na_2HPO_4(s) \xrightarrow{H_2O} 2Na^+(aq) + HPO_4^{2-}(aq)$ [dissolution and hydration]
2. $HPO_4^{2-}(aq) + H_2O(l) \rightleftharpoons PO_4^{3-}(aq) + H_3O^+(aq)$ [acting as a weak acid]
3. $HPO_4^{2-}(aq) + H_2O(l) \rightleftharpoons H_2PO_4^-(aq) + OH^-(aq)$ [acting as a weak base]

We must decide whether step 2 or step 3 goes farther to the right. From Appendix C, we find that the K_a of HPO_4^{2-} is 4.2×10^{-13}. To calculate the K_b of HPO_4^{2-}, we look up the K_a of H_2PO_4 and use Equation 18.7:

$$K_b \text{ of } HPO_4^{2-} = \frac{K_w}{K_a \text{ of } H_2PO_4^-} \quad \text{or} \quad \frac{1.0\times10^{-14}}{6.3\times10^{-8}} = 1.6\times10^{-7}$$

Because $K_b > K_a$, HPO_4^{2-} acts as a base, so a solution of Na_2HPO_4 is basic.

Table 18.3 on the next page displays the acid-base behavior of the various types of salts in water.

Table 18.3 The Acid-Base Behavior of Salts in Water

Relative Acidity: Examples	pH	Nature of Ions	Ion That Reacts with Water: Examples
Neutral: NaCl, KBr, $Ba(NO_3)_2$	7.0	Cation of strong base Anion of strong acid	None
Acidic: NH_4Cl, NH_4NO_3, CH_3NH_3Br	<7.0	Cation of weak base Anion of strong acid	Cation: $NH_4^+ + H_2O \rightleftharpoons NH_3 + H_3O^+$
Acidic: $Al(NO_3)_3$, $CrBr_3$, $FeCl_3$	<7.0	Small, highly charged cation Anion of strong acid	Cation: $Al(H_2O)_6^{3+} + H_2O \rightleftharpoons Al(H_2O)_5OH^{2+} + H_3O^+$
Acidic/Basic: NH_4ClO_2, NH_4CN, $Pb(CH_3COO)_2$	<7.0 if $K_{a(cation)} > K_{b(anion)}$ >7.0 if $K_{b(anion)} > K_{a(cation)}$	Cation of weak base (or small, highly charged cation) Anion of weak acid	Cation *and* anion: $CH_3NH_3^+ + H_2O \rightleftharpoons CH_3NH_2 + H_3O^+$ $F^- + H_2O \rightleftharpoons HF + OH^-$
Acidic/Basic: NaH_2PO_4, $KHCO_3$, $NaHSO_3$	<7.0 if $K_{a(anion)} > K_{b(anion)}$ >7.0 if $K_{b(anion)} > K_{a(anion)}$	Cation of strong base Anion of polyprotic acid	Anion: $HSO_3^- + H_2O \rightleftharpoons SO_3^{2-} + H_3O^+$ $HSO_3^- + H_2O \rightleftharpoons H_2SO_3 + OH^-$

Sample Problem 18.13 Predicting the Relative Acidity of a Salt Solution from K_a and K_b of the Ions

Problem Determine whether an aqueous solution of zinc formate, $Zn(HCOO)_2$, at 25°C is acidic, basic, or neutral.

Plan The formula consists of the small, highly charged, and therefore weakly acidic, Zn^{2+} cation and the weakly basic $HCOO^-$ anion of the weak acid HCOOH. To determine the relative acidity of the solution, we write equations that show the reactions of the ions with water, and then find K_a of Zn^{2+} (from Appendix C) and calculate K_b of $HCOO^-$ (from K_a of HCOOH in Appendix C) to see which ion reacts with water to a greater extent.

Solution Writing the reactions with water:

$$Zn(H_2O)_6^{2+}(aq) + H_2O(l) \rightleftharpoons Zn(H_2O)_5OH^+(aq) + H_3O^+(aq)$$
$$HCOO^-(aq) + H_2O(l) \rightleftharpoons HCOOH(aq) + OH^-(aq)$$

Obtaining K_a and K_b of the ions: From Appendix C, the K_a of $Zn(H_2O)_6^{2+}(aq)$ is 1×10^{-9}. We obtain K_a of HCOOH and solve for K_b of $HCOO^-$:

$$K_b \text{ of } HCOO^- = \frac{K_w}{K_a \text{ of HCOOH}} = \frac{1.0\times10^{-14}}{1.8\times10^{-4}} = 5.6\times10^{-11}$$

K_a of $Zn(H_2O)_6^{2+} > K_b$ of $HCOO^-$, so the solution is acidic.

FOLLOW-UP PROBLEM 18.13 Determine whether solutions of the following salts are acidic, basic, or neutral at 25°C: **(a)** $Cu(CH_3COO)_2$; **(b)** NH_4F; **(c)** $KHSO_3$.

Summary of Section 18.7

- Salts that yield a neutral solution consist of ions that do not react with water.
- Salts that yield an acidic solution contain an unreactive anion and a cation that releases a proton to water.
- Salts that yield a basic solution contain an unreactive cation and an anion that accepts a proton from water.

- If both cation and anion react with water, the ion that reacts to the greater extent (higher K) determines the acidity or basicity of the salt solution.
- If the anion is amphiprotic (from a polyprotic acid), the strength of the anion as an acid (K_a) or as a base (K_b) determines the acidity of the salt.

18.8 • ELECTRON-PAIR DONATION AND THE LEWIS ACID-BASE DEFINITION

The final acid-base concept we consider was developed by Gilbert N. Lewis, whose contribution to understanding valence electron pairs in bonding we discussed in Chapter 9. Whereas the Brønsted-Lowry concept focuses on the proton in defining a species as an acid or a base, the Lewis concept highlights the role of the *electron pair.* The **Lewis acid-base definition** holds that

- A *base* is any species that *donates* an electron pair to form a bond.
- An *acid* is any species that *accepts* an electron pair to form a bond.

The Lewis definition, like the Brønsted-Lowry definition, requires that a base have an electron pair to donate, so it does not expand the classes of bases. However, *this definition greatly expands the classes of acids.* Many species, such as CO_2 and Cu^{2+}, that do not contain H in their formula (and thus cannot be Brønsted-Lowry acids) are Lewis acids because they accept an electron pair in reactions. Thus, the proton itself is a Lewis acid because it accepts the electron pair donated by a base:

$$\text{B:} + \text{H}^+ \rightleftharpoons \text{B—H}^+$$

Thus, *all Brønsted-Lowry acids donate H^+, a Lewis acid.*

The product of a Lewis acid-base reaction is an **adduct,** *a single species that contains a **new** covalent bond:*

$$\text{A} + \text{:B} \rightleftharpoons \text{A—B (adduct)}$$

Thus, the Lewis concept radically broadens the idea of an acid-base reaction:

- According to Arrhenius, it is the formation of H_2O from H^+ and OH^-.
- According to Brønsted and Lowry, it is H^+ transfer from a stronger acid to a stronger base to form a weaker base and weaker acid.
- According to Lewis, it is *the donation and acceptance of an electron pair to form a covalent bond in an adduct.*

By definition, then,

- *A Lewis base must have a lone pair of electrons to donate.*
- *A Lewis acid must have a vacant orbital* (or the ability to rearrange its bonds to form one) to accept a lone pair and form a new bond.

In this section, we discuss molecules and positive ions that act as Lewis acids.

Molecules as Lewis Acids

Many molecules act as Lewis acids. In every case, the atom accepting the electron pair has low electron density due to either an electron deficiency or a polar multiple bond.

Lewis Acids with Electron-Deficient Atoms The most important of the *electron-deficient* Lewis acids are compounds of the Group 3A(13) elements boron and aluminum. Recall from Chapters 10 and 14 that these compounds have fewer than eight electrons around the central atom, so they react to complete that atom's octet. For example, BF_3 accepts an electron pair from ammonia to form a covalent bond:

$$BF_3 + :NH_3 \rightleftharpoons F_3B\text{—}NH_3$$

acid base adduct

Aluminum chloride dissolves freely in relatively nonpolar diethyl ether when the ether's O atom donates an electron pair to Al to form a covalent bond:

$$CH_3{-}CH_2{-}\ddot{O}(CH_2{-}CH_3) + AlCl_3 \rightleftharpoons CH_3{-}CH_2{-}O(CH_2{-}CH_3){-}AlCl_3$$

base acid adduct

Lewis Acids with Polar Multiple Bonds Molecules with a polar double bond also function as Lewis acids. An electron pair on the Lewis base approaches the partially positive end of the double bond to form the new bond in the adduct, as the π bond breaks. For example, consider the reaction of SO_2 in water. The electronegative O atoms in SO_2 make the central S partially positive. The O atom of water donates a lone pair to the S, thus forming an S—O bond and breaking one of the π bonds. Then, a proton is transferred from water to the O that was part of the π bond. The resulting adduct is sulfurous acid:

$$O{=}S^{\delta+}{=}O^{\delta-} + H{-}\ddot{O}{-}H \rightleftharpoons H{-}O{-}S(O{-}H){=}O$$

acid base adduct

The analogous formation of a carbonate from a metal oxide and carbon dioxide occurs in a nonaqueous system. The O^{2-} ion (shown coming from CaO) donates an electron pair to the partially positive C in CO_2, a π bond breaks, and the CO_3^{2-} ion (shown as part of $CaCO_3$) forms as the adduct:

$$Ca^{2+}\ :\ddot{O}:^{2-} + \ddot{O}{=}C^{\delta+}{=}\ddot{O}^{\delta-} \rightleftharpoons Ca^{2+}\left[\begin{array}{c}O\\ \| \\ O{-}C{-}O\end{array}\right]^{2-}$$

base acid adduct

Metal Ions as Lewis Acids

In the Lewis sense, hydration of a metal ion is itself an acid-base reaction. As electron pairs on the O atoms of H_2O molecules form covalent bonds, the hydrated cation is the adduct; thus, *a metal ion acts as a Lewis acid when it dissolves in water:*

$$M^{2+} + 6H_2O(l) \rightleftharpoons M(H_2O)_6^{2+}(aq)$$

acid base adduct

Ammonia is a stronger Lewis base than water, so adding aqueous NH_3 displaces H_2O from a hydrated cation:

$$\underset{\text{hydrated adduct}}{Ni(H_2O)_6^{2+}(aq)} + \underset{\text{base}}{6NH_3(aq)} \rightleftharpoons \underset{\text{ammoniated adduct}}{Ni(NH_3)_6^{2+}(aq)} + 6H_2O(l)$$

We discuss the equilibrium nature of these acid-base reactions in greater detail in Chapter 19, and we investigate the structures of these ions in Chapter 22.

Many biomolecules with central metal ions are Lewis adducts. Most often, O and N atoms of organic groups donate their lone pairs as the Lewis bases. Chlorophyll is a Lewis adduct of a central Mg^{2+} and four N atoms in an organic ring. Vitamin B_{12} has a similar structure with a central Co^{3+}, and so does heme with a central Fe^{2+}. Several other metal ions, such as Zn^{2+}, Mo^{2+}, and Cu^{2+}, are bound at the active sites of enzymes and function as Lewis acids in the catalytic action.

Sample Problem 18.14 Identifying Lewis Acids and Bases

Problem Identify the Lewis acids and Lewis bases in the following reactions:
(a) $H^+ + OH^- \rightleftharpoons H_2O$
(b) $Cl^- + BCl_3 \rightleftharpoons BCl_4^-$
(c) $K^+ + 6H_2O \rightleftharpoons K(H_2O)_6^+$

Plan We examine the formulas to see which species accepts the electron pair (Lewis acid) and which donates it (Lewis base) in forming the adduct.

Solution (a) The H^+ ion accepts an electron pair from the OH^- ion in forming a bond. H^+ is the acid and OH^- is the base.
(b) The Cl^- ion has four lone pairs and uses one to form a new bond to the central B. BCl_3 is the acid and Cl^- is the base.
(c) The K^+ ion does not have any valence electrons to provide, so the bond is formed when electron pairs from O atoms of water enter empty orbitals of K^+. Thus, K^+ is the acid and H_2O is the base.

Check The Lewis acids (H^+, BCl_3, and K^+) each have an unfilled valence shell that can accept an electron pair from the Lewis bases (OH^-, Cl^-, and H_2O).

FOLLOW-UP PROBLEM 18.14 Identify the Lewis acids and Lewis bases:
(a) $OH^- + Al(OH)_3 \rightleftharpoons Al(OH)_4^-$ (b) $SO_3 + H_2O \rightleftharpoons H_2SO_4$
(c) $Co^{3+} + 6NH_3 \rightleftharpoons Co(NH_3)_6^{3+}$

Summary of Section 18.8

- The Lewis acid-base definition focuses on the donation or acceptance of an electron pair to form a new covalent bond in an adduct, the product of an acid-base reaction. Lewis bases donate the electron pair, and Lewis acids accept it.
- Thus, many species that do not contain H act as Lewis acids; examples are molecules with electron-deficient atoms and those with polar double bonds.
- Metal ions act as Lewis acids when they dissolve in water, which acts as a Lewis base, to form the adduct, a hydrated cation.
- Many metal ions function as Lewis acids in biomolecules.

CHAPTER REVIEW GUIDE

The following sections provide many aids to help you study this chapter. (Numbers in parentheses refer to pages, unless noted otherwise.)

Learning Objectives

These are concepts and skills to review after studying this chapter.

Related section (§), sample problem (SP), and upcoming end-of-chapter problem (EP) numbers are listed in parentheses.

1. Understand the nature of the hydrated proton, the Arrhenius definition of an acid and a base, and why all strong acid–strong base reactions have the same ΔH°_{rxn}; describe how acid strength is expressed by K_a; classify strong and weak acids and bases from their formulas (§18.1) (SP 18.1) (EPs 18.1–18.12)
2. Describe the autoionization of water and explain the meaning of K_w and pH; understand why $[H_3O^+]$ is inversely related to $[OH^-]$ and how their relative magnitudes define the acidity of a solution; interconvert pH, pOH, $[H_3O^+]$, and $[OH^-]$ (§18.2) (SPs 18.2, 18.3) (EPs 18.13–18.24)
3. Understand the Brønsted-Lowry definitions of an acid and a base; discuss how water can act as a base or as an acid and how an acid-base reaction is a proton-transfer process involving two conjugate acid-base pairs, with the stronger acid and base forming the weaker base and acid (§18.3) (SPs 18.4–18.6) (EPs 18.25–18.40)
4. Calculate K_a from the pH of an HA solution or pH from the K_a and initial [HA]; explain why the percent dissociation of HA increases as [HA] decreases and know how to find percent dissociation; understand why the first dissociation of a polyprotic acid has a much larger K_a than the second (§18.4) (SPs 18.7–18.9) (EPs 18.41–18.55)
5. Understand the meaning of K_b, and why ammonia and amines and weak-acid anions are bases; discuss how relative [HA] and $[A^-]$ determine the acidity or basicity of a solution, and show the relationship between K_a and K_b of a conjugate acid-base pair and K_w; calculate pH from K_b and $[B]_{init}$, and find K_b of A^- from K_a of HA (§18.5) (SPs 18.10, 18.11) (EPs 18.56–18.68)
6. Understand how electronegativity and bond strength affect the acid strength of nonmetal hydrides or how number of O atoms affects the strength of oxoacids, and explain why certain metal ions form acidic solutions (§18.6) (EPs 18.69–18.76)
7. Understand how combinations of cations and anions lead to acidic, basic, or neutral solutions (§18.7) (SPs 18.12, 18.13) (EPs 18.77–18.83)
8. Describe how a Lewis acid-base reaction involves formation of a new covalent bond, and identify Lewis acids and bases (§18.8) (SP 18.14) (EPs 18.84–18.94)

Key Terms

These important terms appear in boldface in the chapter and are defined again in the Glossary.

Section 18.1
hydronium ion, H_3O^+ (580)
Arrhenius acid-base definition (580)
neutralization (580)
acid-dissociation (acid-ionization) constant (K_a) (582)

Section 18.2
autoionization (584)
ion-product constant for water (K_w) (584)
pH (585)
acid-base indicator (588)

Section 18.3
Brønsted-Lowry acid-base definition (588)
proton donor (588)
proton acceptor (588)
conjugate acid-base pair (589)

Section 18.4
polyprotic acid (597)

Section 18.5
base-dissociation (base-ionization) constant (K_b) (598)

Section 18.8
Lewis acid-base definition (607)
adduct (607)

Key Equations and Relationships

Numbered and screened concepts are listed for you to refer to or memorize.

18.1 Defining the acid-dissociation constant (582):

$$K_a = \frac{[H_3O^+][A^-]}{[HA]}$$

18.2 Defining the ion-product constant for water (584):

$$K_w = [H_3O^+][OH^-] = 1.0\times10^{-14} \quad \text{(at 25°C)}$$

18.3 Defining pH (585):

$$pH = -\log [H_3O^+]$$

18.4 Relating pK_w to pH and pOH (586):

$$pK_w = pH + pOH = 14.00 \quad \text{(at 25°C)}$$

18.5 Finding the percent dissociation of HA (595):

$$\text{Percent HA dissociated} = \frac{[HA]_{dissoc}}{[HA]_{init}} \times 100$$

18.6 Defining the base-dissociation constant (598):

$$K_b = \frac{[BH^+][OH^-]}{[B]}$$

18.7 Expressing the relationship among K_a, K_b, and K_w (600):

$$K_a \times K_b = K_w$$

BRIEF SOLUTIONS TO FOLLOW-UP PROBLEMS

Compare your own solutions to these calculation steps and answers.

18.1 (a) $HClO_3$; (b) HCl; (c) NaOH

18.2 $[H_3O^+] = \frac{1.0\times10^{-14}}{6.7\times10^{-2}} = 1.5\times10^{-13}$ *M*; basic

18.3 pOH = 14.00 − 9.52 = 4.48

$[H_3O^+] = 10^{-9.52} = 3.0\times10^{-10}$ *M*

$[OH^-] = \frac{1.0\times10^{-14}}{3.0\times10^{-10}} = 3.3\times10^{-5}$ *M*

18.4 (a) CH_3COOH/CH_3COO^- and H_3O^+/H_2O
(b) H_2O/OH^- and HF/F^-

18.5 H_2S is higher on the list of acids, and OH^- is lower on the list of bases, so

$$H_2O(l) + HS^-(aq) \rightleftharpoons OH^-(aq) + H_2S(aq)$$

weaker acid + weaker base ⟵ stronger base + stronger acid

The net direction is to the left, so $K_c < 1$.

18.6 There are more HB molecules than HA, so $K_c > 1$ and HA is the stronger acid.

18.7 $NH_4^+(aq) + H_2O(l) \rightleftharpoons NH_3(aq) + H_3O^+(aq)$

$[H_3O^+] = 10^{-pH} = 10^{-5.0} = 1\times10^{-5}\ M = [NH_3]$

And, $[NH_4^+] = 0.2\ M - (1\times10^{-5}\ M) = 0.2\ M$

$$K_a = \frac{[NH_3][H_3O^+]}{[NH_4^+]} \approx \frac{(1\times10^{-5})^2}{0.2} = 5\times10^{-10}$$

18.8 $K_a = \frac{[H_3O^+][OCN^-]}{[HOCN]} = \frac{(x)(x)}{0.10 - x} = 3.5\times10^{-4}$

Since $\frac{[HOCN]_{init}}{K_a} = \frac{0.10}{3.5\times10^{-4}} = 286 < 400$, you must solve a quadratic equation: $x^2 + (3.5\times10^{-4})x - (3.5\times10^{-5}) = 0$

$x = [H_3O^+] = 5.7\times10^{-3}$ *M*

18.9 There is no single correct scene: any scene in which the total number of HB + H_3O^+ (or HB + B^-) is less than in the original solution, yet the number of HB dissociated is greater, would be correct, for example:

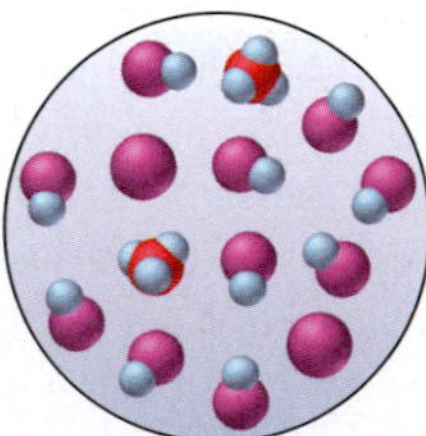

18.10 $K_b = \frac{[C_5H_5NH^+][OH^-]}{[C_5H_5N]} = 10^{-8.77} = 1.7\times10^{-9}$

Assuming $0.10\ M - x \approx 0.10\ M$, $K_b = 1.7\times10^{-9} \approx \frac{x^2}{0.10}$;

$x = [OH^-] \approx 1.3\times10^{-5}\ M$; $[H_3O^+] = 7.7\times10^{-10}\ M$; pH = 9.11

18.11 $K_b \text{ of } ClO^- = \frac{K_w}{K_a \text{ of } HClO} = \frac{1.0\times10^{-14}}{2.9\times10^{-8}} = 3.4\times10^{-7}$

Assuming $0.20\ M - x \approx 0.20\ M$,

$$K_b = 3.4\times10^{27} = \frac{[HClO][OH^-]}{[ClO^-]} \approx \frac{x^2}{0.20};$$

$x = [OH^-] \approx 2.6\times10^{-4}\ M$; $[H_3O^+] = 3.8\times10^{-11}\ M$; pH = 10.42

18.12 (a) Basic:

$$ClO_2^-(aq) + H_2O(l) \rightleftharpoons HClO(aq) + OH^-(aq)$$

K^+ is from the strong base KOH.
(b) Acidic:

$$CH_3NH_3^+(aq) + H_2O(l) \rightleftharpoons CH_3NH_2(aq) + H_3O^+(aq)$$

NO_3^- is from the strong acid HNO_3.
(c) Neutral: Cs^+ is from the strong base CsOH; I^- is from the strong acid HI.

18.13 (a) K_a of $Cu(H_2O)_6^{2+} = 3\times10^{-8}$

$$K_b \text{ of } CH_3COO^- = \frac{K_w}{K_a \text{ of } CH_3COOH} = 5.6\times10^{-10}$$

Since $K_a > K_b$, $Cu(CH_3COO)_2(aq)$ is acidic.

(b) $$K_a \text{ of } NH_4^+ = \frac{K_w}{K_b \text{ of } NH_3} = 5.7\times10^{-10}$$

$$K_b \text{ of } F^- = \frac{K_w}{K_a \text{ of HF}} = 1.5\times10^{-11}$$

Because $K_a > K_b$, $NH_4F(aq)$ is acidic.
(c) From Appendix C, K_a of $HSO_3^- = 6.5\times10^{-8}$

$$K_b \text{ of } HSO_3^- = \frac{K_w}{K_a \text{ of } H_2SO_3} = 7.1\times10^{-13}$$

Because $K_a > K_b$, $KHSO_3(aq)$ is acidic.

18.14 (a) OH^- is the Lewis base; $Al(OH)_3$ is the Lewis acid.
(b) H_2O is the Lewis base; SO_3 is the Lewis acid.
(c) NH_3 is the Lewis base; Co^{3+} is the Lewis acid.

PROBLEMS

Problems with **colored** numbers are answered in Appendix E. Sections match the text and provide the numbers of relevant sample problems. Bracketed problems are grouped in pairs (indicated by a short rule) that cover the same concept. Comprehensive Problems are based on material from any section or previous chapter.

Note: Unless stated otherwise, all problems refer to aqueous solutions at 298 K (25°C).

Acids and Bases in Water

(Sample Problem 18.1)

18.1 What is the role of water in the Arrhenius acid-base definition?

18.2 What do Arrhenius acids have in common? What do Arrhenius bases have in common? Explain neutralization in terms of the Arrhenius acid-base definition. What data led Arrhenius to propose this idea of neutralization?

18.3 Why is the Arrhenius acid-base definition too limited? Give an example for which the Arrhenius definition does not apply.

18.4 What do "strong" and "weak" mean for acids and bases? K_a values of weak acids vary over more than 10 orders of magnitude. What do the acids have in common that makes them "weak"?

18.5 Which of the following are Arrhenius acids?
(a) H_2O (b) $Ca(OH)_2$ (c) H_3PO_3 (d) HI

18.6 Which of the following are Arrhenius bases?
(a) CH_3COOH (b) HOH (c) CH_3OH (d) H_2NNH_2

18.7 Write the K_a expression for each of the following in water:
(a) HNO_2 (b) CH_3COOH (c) $HBrO_2$

18.8 Write the K_a expression for each of the following in water:
(a) $H_2PO_4^-$ (b) H_3PO_2 (c) HSO_4^-

18.9 Use Appendix C to rank the following in order of *increasing* acid strength: HIO_3, HI, CH_3COOH, HF.

18.10 Use Appendix C to rank the following in order of *decreasing* acid strength: HClO, HCl, HCN, HNO_2.

18.11 Classify each as a strong or weak acid or base:
(a) H_3AsO_4 (b) $Sr(OH)_2$ (c) HIO (d) $HClO_4$

18.12 Classify each as a strong or weak acid or base:
(a) CH_3NH_2 (b) K_2O (c) HI (d) HCOOH

Autoionization of Water and the pH Scale

(Sample Problems 18.2 and 18.3)

18.13 What is an autoionization reaction? Write equations for the autoionization reactions of H_2O and of H_2SO_4.

18.14 (a) What is the change in pH when $[OH^-]$ increases by a factor of 10? (b) What is the change in $[H_3O^+]$ when the pH decreases by 3 units?

18.15 Which solution has the higher pH? Explain.
(a) A 0.1 *M* solution of an acid with $K_a = 1\times10^{-4}$ or one with $K_a = 4\times10^{-5}$
(b) A 0.1 *M* solution of an acid with $pK_a = 3.0$ or one with $pK_a = 3.5$
(c) A 0.1 *M* solution or a 0.01 *M* solution of a weak acid
(d) A 0.1 *M* solution of a weak acid or a 0.1 *M* solution of a strong acid
(e) A 0.1 *M* solution of an acid or a 0.01 *M* solution of a base
(f) A solution of pOH 6.0 or one of pOH 8.0

18.16 (a) What is the pH of 0.0111 *M* NaOH? Is the solution neutral, acidic, or basic? (b) What is the pOH of 1.35×10^{-3} *M* HCl? Is the solution neutral, acidic, or basic?

18.17 (a) What is the pH of 0.0333 *M* HNO_3? Is the solution neutral, acidic, or basic? (b) What is the pOH of 0.0347 *M* KOH? Is the solution neutral, acidic, or basic?

18.18 (a) What are $[H_3O^+]$, $[OH^-]$, and pOH in a solution with a pH of 9.85? (b) What are $[H_3O^+]$, $[OH^-]$, and pH in a solution with a pOH of 9.43?

18.19 (a) What are $[H_3O^+]$, $[OH^-]$, and pOH in a solution with a pH of 4.77? (b) What are $[H_3O^+]$, $[OH^-]$, and pH in a solution with a pOH of 5.65?

18.20 How many moles of H_3O^+ or OH^- must you add to 5.6 L of HA solution to adjust its pH from 4.52 to 5.25? Assume a negligible volume change.

18.21 How many moles of H_3O^+ or OH^- must you add to 87.5 mL of HA solution to adjust its pH from 8.92 to 6.33? Assume a negligible volume change.

18.22 The two molecular scenes below depict the relative concentrations of H_3O^+ *(purple)* in solutions of the same volume (with anions and solvent molecules omitted for clarity). If the pH in scene A is 4.8, what is the pH in scene B?

18.23 Although pure water is an *extremely* weak electrolyte, parents commonly warn their children of the danger of swimming in a pool or lake during a lightning storm. Explain.

18.24 Like any equilibrium constant, K_w changes with temperature. (a) Given that autoionization is endothermic, how does K_w change with rising T? Explain with a reaction that includes heat as reactant or product. (b) In many medical applications, the value of K_w at 37°C (body temperature) may be more appropriate than the value at 25°C, 1.0×10^{-14}. The pH of pure water at 37°C is 6.80. Calculate K_w, pOH, and $[OH^-]$ at this temperature.

Proton Transfer and the Brønsted-Lowry Acid-Base Definition

(Sample Problems 18.4 to 18.6)

18.25 How are the Arrhenius and Brønsted-Lowry acid-base definitions different? How are they similar? Name two Brønsted-Lowry bases that are not Arrhenius bases. Can you do the same for acids? Explain.

18.26 What is a conjugate acid-base pair? What is the relationship between the two members of the pair?

18.27 (a) A Brønsted-Lowry acid-base reaction proceeds in the net direction in which a stronger acid and stronger base form a weaker acid and weaker base. Explain.
(b) The molecular scene at right depicts an aqueous solution of two conjugate acid-base pairs: HA/A^- and HB/B^-. The base in the first pair is represented by red spheres and the base in the second pair by green spheres; solvent molecules are omitted for clarity. Which is the stronger acid? Stronger base? Explain.

18.28 What is an amphiprotic species? Name one and write balanced equations that show why it is amphiprotic.

18.29 Give the formula of the conjugate base:
(a) HCl (b) H_2CO_3 (c) H_2O

18.30 Give the formula of the conjugate base:
(a) HPO_4^{2-} (b) NH_4^+ (c) HS^-

18.31 Give the formula of the conjugate acid:
(a) NH_3 (b) NH_2^- (c) nicotine, $C_{10}H_{14}N_2$

18.32 Give the formula of the conjugate acid:
(a) O^{2-} (b) SO_4^{2-} (c) H_2O

18.33 In each equation, label the acids, bases, and conjugate pairs:
(a) $NH_3 + H_3PO_4 \rightleftharpoons NH_4^+ + H_2PO_4^-$
(b) $CH_3O^- + NH_3 \rightleftharpoons CH_3OH + NH_2^-$
(c) $HPO_4^{2-} + HSO_4^- \rightleftharpoons H_2PO_4^- + SO_4^{2-}$

18.34 In each equation, label the acids, bases, and conjugate pairs:
(a) $NH_4^+ + CN^- \rightleftharpoons NH_3 + HCN$
(b) $H_2O + HS^- \rightleftharpoons OH^- + H_2S$
(c) $HSO_3^- + CH_3NH_2 \rightleftharpoons SO_3^{2-} + CH_3NH_3^+$

18.35 Write balanced net ionic equations for the following reactions, and label the conjugate acid-base pairs:
(a) $NaOH(aq) + NaH_2PO_4(aq) \rightleftharpoons H_2O(l) + Na_2HPO_4(aq)$
(b) $KHSO_4(aq) + K_2CO_3(aq) \rightleftharpoons K_2SO_4(aq) + KHCO_3(aq)$

18.36 Write balanced net ionic equations for the following reactions, and label the conjugate acid-base pairs:
(a) $HNO_3(aq) + Li_2CO_3(aq) \rightleftharpoons LiNO_3(aq) + LiHCO_3(aq)$
(b) $2NH_4Cl(aq) + Ba(OH)_2(aq) \rightleftharpoons$
$2H_2O(l) + BaCl_2(aq) + 2NH_3(aq)$

18.37 The following aqueous species constitute two conjugate acid-base pairs. Use them to write one acid-base reaction with $K_c > 1$ and another with $K_c < 1$: HS^-, Cl^-, HCl, H_2S.

18.38 The following aqueous species constitute two conjugate acid-base pairs. Use them to write one acid-base reaction with $K_c > 1$ and another with $K_c < 1$: NO_3^-, F^-, HF, HNO_3.

18.39 Use Figure 18.8 (p. 591) to determine whether $K_c > 1$ for
(a) $HCl + NH_3 \rightleftharpoons NH_4^+ + Cl^-$
(b) $H_2SO_3 + NH_3 \rightleftharpoons HSO_3^- + NH_4^+$

18.40 Use Figure 18.8 (p. 591) to determine whether $K_c < 1$ for
(a) $H_2PO_4^- + F^- \rightleftharpoons HPO_4^{2-} + HF$
(b) $CH_3COO^- + HSO_4^- \rightleftharpoons CH_3COOH + SO_4^{2-}$

Solving Problems Involving Weak-Acid Equilibria

(Sample Problems 18.7 to 18.19)

18.41 In each of the following cases, is the concentration of acid before and after dissociation nearly the same or very different? Explain your reasoning: (a) a concentrated solution of a strong acid; (b) a concentrated solution of a weak acid; (c) a dilute solution of a weak acid; (d) a dilute solution of a strong acid.

18.42 In which of the following solutions will $[H_3O^+]$ be approximately equal to $[CH_3COO^-]$: (a) 0.1 M CH_3COOH; (b) 1×10^{-7} M CH_3COOH; (c) a solution containing both 0.1 M CH_3COOH and 0.1 M CH_3COONa? Explain.

18.43 Why do successive K_a's decrease for all polyprotic acids?

18.44 A 0.15 M solution of butanoic acid, $CH_3CH_2CH_2COOH$, contains 1.51×10^{-3} M H_3O^+. What is the K_a of butanoic acid?

18.45 A 0.035 M solution of a weak acid (HA) has a pH of 4.88. What is the K_a of the acid?

18.46 Nitrous acid, HNO_2, has a K_a of 7.1×10^{-4}. What are $[H_3O^+]$, $[NO_2^-]$, and $[OH^-]$ in 0.60 M HNO_2?

18.47 Hydrofluoric acid, HF, has a K_a of 6.8×10^{-4}. What are $[H_3O^+]$, $[F^-]$, and $[OH^-]$ in 0.75 *M* HF?

18.48 Chloroacetic acid, $ClCH_2COOH$, has a pK_a of 2.87. What are $[H_3O^+]$, pH, $[ClCH_2COO^-]$, and $[ClCH_2COOH]$ in 1.25 *M* $ClCH_2COOH$?

18.49 Hypochlorous acid, HClO, has a pK_a of 7.54. What are $[H_3O^+]$, pH, $[ClO^-]$, and [HClO] in 0.115 *M* HClO?

18.50 In a 0.20 *M* solution, a weak acid is 3.0% dissociated.
(a) Calculate the $[H_3O^+]$, pH, $[OH^-]$, and pOH of the solution.
(b) Calculate K_a of the acid.

18.51 In a 0.735 *M* solution, a weak acid is 12.5% dissociated.
(a) Calculate the $[H_3O^+]$, pH, $[OH^-]$, and pOH of the solution.
(b) Calculate K_a of the acid.

18.52 The weak acid HZ has a K_a of 2.55×10^{-4}.
(a) Calculate the pH of 0.075 *M* HZ.
(b) Calculate the pOH of 0.045 *M* HZ.

18.53 The weak acid HQ has a pK_a of 4.89.
(a) Calculate the $[H_3O^+]$ of 3.5×10^{-2} *M* HQ.
(b) Calculate the $[OH^-]$ of 0.65 *M* HQ.

18.54 Acetylsalicylic acid (aspirin), $HC_9H_7O_4$, is the most widely used pain reliever and fever reducer. Find the pH of 0.018 *M* aqueous aspirin at body temperature (K_a at 37°C $= 3.6\times10^{-4}$).

18.55 Formic acid, HCOOH, the simplest carboxylic acid, is used in the textile and rubber industries and is secreted as a defense by many species of ants (family *Formicidae*). Calculate the percent dissociation of 0.75 *M* HCOOH.

Weak Bases and Their Relation to Weak Acids

(Sample Problems 18.10 and 18.11)

18.56 What is the key structural feature of all Brønsted-Lowry bases? How does this feature function in an acid-base reaction?

18.57 Why are most anions basic in H_2O? Give formulas of four anions that are not basic.

18.58 Except for the Na^+ spectator ion, aqueous solutions of CH_3COOH and CH_3COONa contain the same species. (a) What are the species (other than H_2O)? (b) Why is 0.1 *M* CH_3COOH acidic and 0.1 *M* CH_3COONa basic?

18.59 Write balanced equations and K_b expressions for these Brønsted-Lowry bases in water:
(a) Pyridine, C_5H_5N (b) CO_3^{2-}

18.60 Write balanced equations and K_b expressions for these Brønsted-Lowry bases in water:
(a) Benzoate ion, $C_6H_5COO^-$ (b) $(CH_3)_3N$

18.61 What is the pH of 0.070 *M* dimethylamine?

18.62 What is the pH of 0.12 *M* diethylamine?

18.63 (a) What is the pK_b of ClO_2^-?
(b) What is the pK_a of the dimethylammonium ion, $(CH_3)_2NH_2^+$?

18.64 (a) What is the pK_b of NO_2^-?
(b) What is the pK_a of the hydrazinium ion, $H_2N{-}NH_3^+$ (K_b of hydrazine $= 8.5\times10^{-7}$)?

18.65 (a) What is the pH of 0.150 *M* KCN?
(b) What is the pH of 0.40 *M* triethylammonium chloride, $(CH_3CH_2)_3NHCl$?

18.66 (a) What is the pH of 0.100 *M* sodium phenolate, C_6H_5ONa, the sodium salt of phenol?
(b) What is the pH of 0.15 *M* methylammonium bromide, CH_3NH_3Br (K_b of $CH_3NH_2 = 4.4\times10^{-4}$)?

18.67 Sodium hypochlorite solution, sold as chlorine bleach, is potentially dangerous because of the basicity of ClO^-, the active bleaching ingredient. What is $[OH^-]$ in an aqueous solution that is 6.5% NaClO by mass? What is the pH of the solution? (Assume *d* of solution = 1.0 g/mL.)

18.68 Codeine ($C_{18}H_{21}NO_3$) is a narcotic pain reliever that forms a salt with HCl. What is the pH of 0.050 *M* codeine hydrochloride (pK_b of codeine = 5.80)?

Molecular Properties and Acid Strength

18.69 Across a period, how does the electronegativity of a nonmetal affect the acidity of its hydride?

18.70 How does the atomic size of a nonmetal affect the acidity of its hydride?

18.71 A strong acid has a weak bond to its acidic proton, whereas a weak acid has a strong bond to its acidic proton. Explain.

18.72 Perchloric acid, $HClO_4$, is the strongest of the halogen oxoacids, and hypoiodous acid, HIO, is the weakest. What two factors govern this difference in acid strength?

18.73 Choose the *stronger* acid in each of the following pairs:
(a) H_2Se or H_3As (b) $B(OH)_3$ or $Al(OH)_3$ (c) $HBrO_2$ or HBrO

18.74 Choose the *weaker* acid in each of the following pairs:
(a) HI or HBr (b) H_3AsO_4 or H_2SeO_4 (c) HNO_3 or HNO_2

18.75 Use Appendix C to choose the solution with the *lower* pH:
(a) 0.5 *M* $CuBr_2$ or 0.5 *M* $AlBr_3$
(b) 0.3 *M* $ZnCl_2$ or 0.3 *M* $SnCl_2$

18.76 Use Appendix C to choose the solution with the *lower* pH:
(a) 0.1 *M* $FeCl_3$ or 0.1 *M* $AlCl_3$
(b) 0.1 *M* $BeCl_2$ or 0.1 *M* $CaCl_2$

Acid-Base Properties of Salt Solutions

(Sample Problems 18.12 and 18.13)

18.77 What determines whether an aqueous solution of a salt will be acidic, basic, or neutral? Give an example of each type of salt.

18.78 Why is aqueous NaF basic but aqueous NaCl neutral?

18.79 The NH_4^+ ion forms acidic solutions, and the CH_3COO^- ion forms basic solutions. However, a solution of ammonium acetate is almost neutral. Do all of the ammonium salts of weak acids form neutral solutions? Explain your answer.

18.80 Explain with equations and calculations, when necessary, whether an aqueous solution of each of these salts is acidic, basic, or neutral: (a) KBr; (b) NH_4I; (c) KCN.

18.81 Explain with equations and calculations, when necessary, whether an aqueous solution of each of these salts is acidic, basic, or neutral: (a) $Cr(NO_3)_3$; (b) NaHS; (c) $Zn(CH_3COO)_2$.

18.82 Rank the following salts in order of *increasing* pH of their 0.1 *M* aqueous solutions:
(a) KNO_3, K_2SO_3, K_2S, $Fe(NO_3)_2$
(b) NH_4NO_3, $NaHSO_4$, $NaHCO_3$, Na_2CO_3

18.83 Rank the following salts in order of *decreasing* pH of their 0.1 *M* aqueous solutions:
(a) $FeCl_2$, $FeCl_3$, $MgCl_2$, $KClO_2$
(b) NH_4Br, $NaBrO_2$, $NaBr$, $NaClO_2$

Electron-Pair Donation and the Lewis Acid-Base Definition

(Sample Problem 18.14)

18.84 What feature must a molecule or ion have for it to act as a Lewis base? A Lewis acid? Explain the roles of these features.

18.85 How do Lewis acids differ from Brønsted-Lowry acids? How are they similar? Do Lewis bases differ from Brønsted-Lowry bases? Explain.

18.86 (a) Is a weak Brønsted-Lowry base necessarily a weak Lewis base? Explain with an example.
(b) Identify the Lewis bases in the following reaction:

$$Cu(H_2O)_4^{2+}(aq) + 4CN^-(aq) \rightleftharpoons Cu(CN)_4^{2-}(aq) + 4H_2O(l)$$

(c) Given that $K_c > 1$ for the reaction in part (b), which Lewis base is stronger?

18.87 In which of the three acid-base concepts can water be a product of an acid-base reaction? In which is it the only product?

18.88 (a) Give an example of a *substance* that is a base in two of the three acid-base definitions, but not in the third.
(b) Give an example of a *substance* that is an acid in one of the three acid-base definitions, but not in the other two.

18.89 Which are Lewis acids and which are Lewis bases?
(a) Cu^{2+} (b) Cl^- (c) $SnCl_2$ (d) OF_2

18.90 Which are Lewis acids and which are Lewis bases?
(a) Na^+ (b) NH_3 (c) CN^- (d) BF_3

18.91 Identify the Lewis acid and Lewis base in each equation:
(a) $Na^+ + 6H_2O \rightleftharpoons Na(H_2O)_6^+$
(b) $CO_2 + H_2O \rightleftharpoons H_2CO_3$
(c) $F^- + BF_3 \rightleftharpoons BF_4^-$

18.92 Identify the Lewis acid and Lewis base in each equation:
(a) $Fe^{3+} + 2H_2O \rightleftharpoons FeOH^{2+} + H_3O^+$
(b) $H_2O + H^- \rightleftharpoons OH^- + H_2$
(c) $4CO + Ni \rightleftharpoons Ni(CO)_4$

18.93 Classify the following as Arrhenius, Brønsted-Lowry, or Lewis acid-base reactions. A reaction may fit all, two, one, or none of the categories:
(a) $Ag^+ + 2NH_3 \rightleftharpoons Ag(NH_3)_2^+$
(b) $H_2SO_4 + NH_3 \rightleftharpoons HSO_4^- + NH_4^+$
(c) $2HCl \rightleftharpoons H_2 + Cl_2$
(d) $AlCl_3 + Cl^- \rightleftharpoons AlCl_4^-$

18.94 Classify the following as Arrhenius, Brønsted-Lowry, or Lewis acid-base reactions. A reaction may fit all, two, one, or none of the categories:
(a) $Cu^{2+} + 4Cl^- \rightleftharpoons CuCl_4^{2-}$
(b) $Al(OH)_3 + 3HNO_3 \rightleftharpoons Al^{3+} + 3H_2O + 3NO_3^-$
(c) $N_2 + 3H_2 \rightleftharpoons 2NH_3$
(d) $CN^- + H_2O \rightleftharpoons HCN + OH^-$

Comprehensive Problems

18.95 In humans, blood pH is maintained within a narrow range: *acidosis* occurs if the blood pH is below 7.35, and *alkalosis* occurs if the pH is above 7.45. Given that the pK_w of blood is 13.63 at 37°C (body temperature), what is the normal range of $[H_3O^+]$ and of $[OH^-]$ in blood?

18.96 When carbon dioxide dissolves in water, it undergoes a multistep equilibrium process, with $K_{overall} = 4.5\times10^{-7}$, which is simplified to the following:

$$CO_2(g) + H_2O(l) \rightleftharpoons H_2CO_3(aq)$$
$$H_2CO_3(aq) + H_2O(l) \rightleftharpoons HCO_3^-(aq) + H_3O^+(aq)$$

(a) Classify each step as a Lewis or a Brønsted-Lowry acid-base reaction.
(b) What is the pH of nonpolluted rainwater in equilibrium with clean air (P_{CO_2} in clean air $= 3.2\times10^{-4}$ atm; Henry's law constant for CO_2 at 25°C is 0.033 mol/L·atm)?
(c) What is $[CO_3^{2-}]$ in rainwater (K_a of $HCO_3^- = 4.7\times10^{-11}$)?
(d) If the partial pressure of CO_2 in clean air doubles in the next few decades, what will the pH of rainwater become?

18.97 Many molecules with central atoms from Period 3 or higher take part in Lewis acid-base reactions in which the central atom expands its valence shell. $SnCl_4$ reacts with $(CH_3)_3N$ as follows:

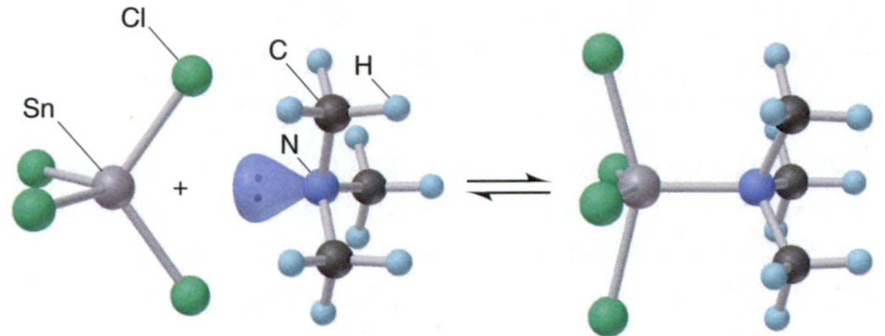

(a) Identify the Lewis acid and the Lewis base in the reaction.
(b) Give the *nl* designation of the sublevel of the central atom in the acid before it accepts the lone pair.

18.98 Use Appendix C to calculate $[H_2C_2O_4]$, $[HC_2O_4^-]$, $[C_2O_4^{2-}]$, $[H_3O^+]$, pH, $[OH^-]$, and pOH in a 0.200 *M* solution of the diprotic acid oxalic acid. (*Hint:* Assume all the $[H_3O^+]$ comes from the first dissociation.)

18.99 A chemist makes four successive 10-fold dilutions of 1.0×10^{-5} *M* HCl. Calculate the pH of the original solution and of each diluted solution (through 1.0×10^{-9} *M* HCl).

18.100 The beakers shown contain 0.300 L of aqueous solutions of a moderately weak acid HY. Each particle represents 0.010 mol; solvent molecules are omitted for clarity. (a) The reaction in beaker A is at equilibrium. Calculate *Q* for B, C, and D to determine which, if any, is (are) also at equilibrium. (b) For any not at equilibrium, in which direction does the reaction proceed? (c) Does dilution affect the extent of dissociation of a weak acid? Explain.

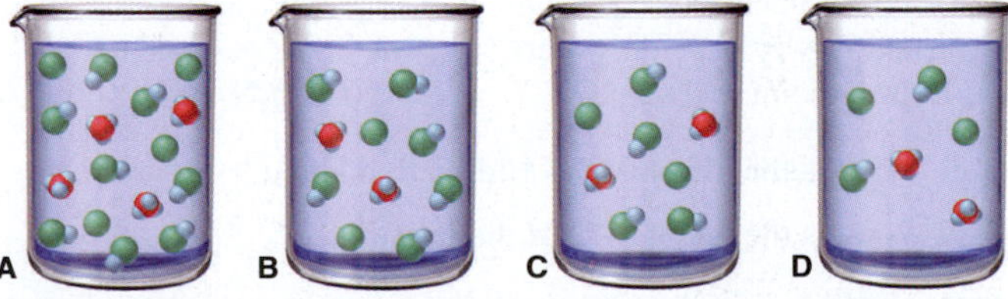

18.101 Hydrogen peroxide, H_2O_2 ($pK_a = 11.75$), is commonly used as a bleaching agent and an antiseptic. The product sold in stores is 3% H_2O_2 by mass and contains 0.001% phosphoric acid by mass to stabilize the solution. Which contributes more H_3O^+ to this commercial solution, the H_2O_2 or the H_3PO_4?

18.102 Esters, RCOOR′, are formed by the reaction of carboxylic acids, RCOOH, and alcohols, R′OH, where R and R′ are hydrocarbon groups. Many esters are responsible for the odors of fruit and, thus, have important uses in the food and cosmetics industries. The first two steps in the mechanism of ester formation are

$$(1)\ \mathrm{R{-}\overset{\overset{\displaystyle :\ddot{O}:}{\|}}{C}{-}\ddot{O}H} + \mathrm{H^+} \rightleftharpoons \mathrm{R{-}\underset{+}{\overset{\overset{\displaystyle :\ddot{O}H}{|}}{C}}{-}\ddot{O}H}$$

$$(2)\ \mathrm{R{-}\underset{+}{\overset{\overset{\displaystyle :\ddot{O}H}{|}}{C}}{-}\ddot{O}H} + \mathrm{R'{-}\ddot{O}H} \rightleftharpoons \mathrm{R{-}\overset{\overset{\displaystyle :\ddot{O}H}{|}}{\underset{\underset{\displaystyle :\overset{+}{O}{-}R'\ (\text{with H on O})}{|}}{C}}{-}\ddot{O}H}$$

Identify the Lewis acids and Lewis bases in these two steps.

18.103 Thiamine hydrochloride ($C_{12}H_{18}ON_4SCl_2$) is a water-soluble form of thiamine (vitamin B_1; $K_a = 3.37\times10^{-7}$). How many grams of the hydrochloride must be dissolved in 10.00 mL of water to give a pH of 3.50?

18.104 When an Fe^{3+} salt is dissolved in water, the solution becomes acidic due to formation of $Fe(H_2O)_5OH^{2+}$ and H_3O^+. The overall process involves both Lewis and Brønsted-Lowry acid-base reactions. Write the equations for the process.

18.105 A 1.000 *m* solution of chloroacetic acid ($ClCH_2COOH$) freezes at -1.93°C. Find the K_a of chloroacetic acid. (Assume the molarities equal the molalities.)

18.106 At 50°C and 1 atm, $K_w = 5.19\times10^{-14}$. Calculate the concentrations in parts (a)–(c) under these conditions:
(a) $[H_3O^+]$ in pure water
(b) $[H_3O^+]$ in 0.010 *M* NaOH
(c) $[OH^-]$ in 0.0010 *M* $HClO_4$
(d) Calculate $[H_3O^+]$ in 0.0100 *M* KOH at 100°C and 1000 atm pressure ($K_w = 1.10\times10^{-12}$).
(e) Calculate the pH of pure water at 100°C and 1000 atm.

18.107 Calcium propionate [$Ca(CH_3CH_2COO)_2$; calcium propanoate] is a mold inhibitor used in food, tobacco, and pharmaceuticals. (a) Use balanced equations to show whether aqueous calcium propionate is acidic, basic, or neutral. (b) Use Appendix C to find the resulting pH when 8.75 g of $Ca(CH_3CH_2COO)_2$ dissolves in enough water to give 0.500 L of solution.

18.108 A site in Pennsylvania receives a total annual deposition of 2.688 g/m^2 of sulfate from fertilizer and acid rain. The ratio by mass of ammonium sulfate to ammonium bisulfate to sulfuric acid in the deposited material is 3.0/5.5/1.0. (a) How much acid, expressed as kg of sulfuric acid, is deposited over an area of 10. km^2? (b) How many pounds of $CaCO_3$ are needed to neutralize this acid? (c) If 10. km^2 is the area of an unpolluted lake 3 m deep and there is no loss of acid, what pH would the lake's water have at the end of the year? (Assume constant volume and negligible runoff from the surrounding land.)

18.109 (a) If $K_w = 1.139\times10^{-15}$ at 0°C and 5.474×10^{-14} at 50°C, find $[H_3O^+]$ and pH of water at 0°C and 50°C.
(b) The autoionization constant for heavy water (deuterium oxide, D_2O) is 3.64×10^{-16} at 0°C and 7.89×10^{-15} at 50°C. Find $[D_3O^+]$ and pD of heavy water at 0°C and 50°C.
(c) Suggest a reason for these differences.

18.110 Carbon dioxide is less soluble in dilute HCl than in dilute NaOH. Explain.

18.111 HX ($\mathcal{M}$ = 150. g/mol) and HY ($\mathcal{M}$ = 50.0 g/mol) are weak acids. A solution of 12.0 g/L of HX has the same pH as one containing 6.00 g/L of HY. Which is the stronger acid? Why?

18.112 The beakers below depict the aqueous dissociations of weak acids HA *(blue and green)* and HB *(blue and yellow)*; solvent molecules are omitted for clarity. If the HA solution is 0.50 L, and the HB solution is 0.25 L, and each particle represents 0.010 mol, find the K_a of each acid. Which acid, if either, is stronger?

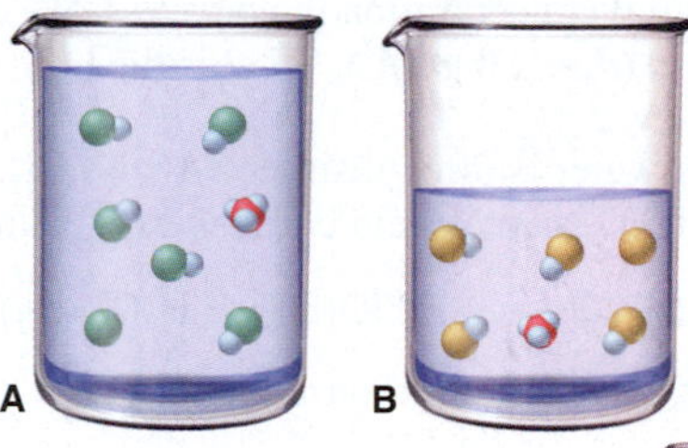

18.113 The molecular scene depicts the relative concentrations of H_3O^+ *(purple)* and OH^- *(green)* in an aqueous solution at 25°C. (Other ions and solvent molecules are omitted for clarity.) (a) Calculate the pH. (b) How many H_3O^+ ions would you have to draw for every OH^- ion to depict a solution of pH 4?

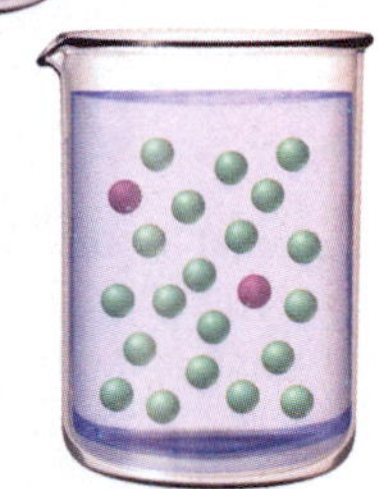

18.114 Polymers are not very soluble in water, but their solubility increases if their side chains have charged groups.
(a) Casein, a milk protein, contains many $—COO^-$ groups on its side chains. How does the solubility of casein vary with pH?
(b) Histones are proteins essential to the function of DNA. They are weakly basic due to the presence of side chains with $—NH_2$ and $=NH$ groups. How does the solubility of a histone vary with pH?

18.115 Hemoglobin (Hb) transports oxygen in the blood:

$$HbH^+(aq) + O_2(aq) + H_2O(l) \longrightarrow HbO_2(aq) + H_3O^+(aq)$$

In blood, $[H_3O^+]$ is held nearly constant at 4×10^{-8} *M*.
(a) How does the equilibrium position change in the lungs?
(b) How does it change in O_2-deficient cells?
(c) Excessive vomiting may lead to metabolic *alkalosis,* in which $[H_3O^+]$ in blood *decreases*. How does this condition affect the ability of Hb to transport O_2?
(d) Diabetes mellitus may lead to metabolic *acidosis,* in which $[H_3O^+]$ in blood *increases*. How does this condition affect the ability of Hb to transport O_2?

18.116 Nitrogen is discharged from wastewater treatment facilities into rivers and streams, usually as NH_3 and NH_4^+:

$$NH_3(aq) + H_2O(l) \rightleftharpoons NH_4^+(aq) + OH^-(aq) \qquad K_b = 1.76\times10^{-5}$$

One strategy for removing it is to raise the pH and "strip" the NH_3 from solution by bubbling air through the water. (a) At pH 7.00, what fraction of the total nitrogen in solution is NH_3, defined as $[NH_3]/([NH_3] + [NH_4^+])$? (b) What is the fraction at pH 10.00? (c) Explain the basis of ammonia stripping.

18.117 The antimalarial properties of quinine ($C_{20}H_{24}N_2O_2$) saved thousands of lives during construction of the Panama Canal. This substance is a classic example of a valuable drug sourced from tropical forests. Both N atoms are basic, but the N *(blue)* of the 3° amine group is far more basic ($pK_b = 5.1$) than the N within the aromatic ring system ($pK_b = 9.7$). (a) A saturated solution of quinine in water is only 1.6×10^{-3} *M*. What is the pH of this solution?

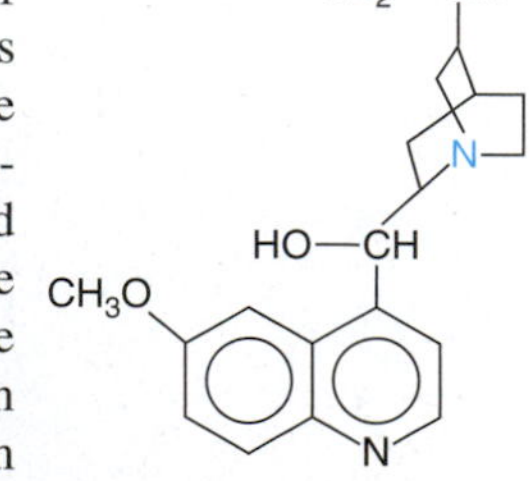

(b) Show that the aromatic N contributes negligibly to the pH of the solution.
(c) Because of its low solubility, quinine is given as the salt quinine hydrochloride ($C_{20}H_{24}N_2O_2 \cdot HCl$), which is 120 times more soluble than quinine. What is the pH of 0.33 *M* quinine hydrochloride?
(d) An antimalarial concentration in water is 1.5% quinine hydrochloride by mass (d = 1.0 g/mL). What is the pH?

18.118 Drinking water is often disinfected with Cl_2, which hydrolyzes to form HClO, a weak acid but powerful disinfectant:

$$Cl_2(aq) + 2H_2O(l) \longrightarrow HClO(aq) + H_3O^+(aq) + Cl^-(aq)$$

The fraction of HClO in solution is defined as

$$\frac{[HClO]}{[HClO] + [ClO^-]}$$

(a) What is the fraction of HClO at pH 7.00 (K_a of HClO = 2.9×10^{-8})?
(b) What is the fraction at pH 10.00?

18.119 The following scenes represent three weak acids HA (where A = X, Y, or Z) dissolved in water (H_2O is not shown):

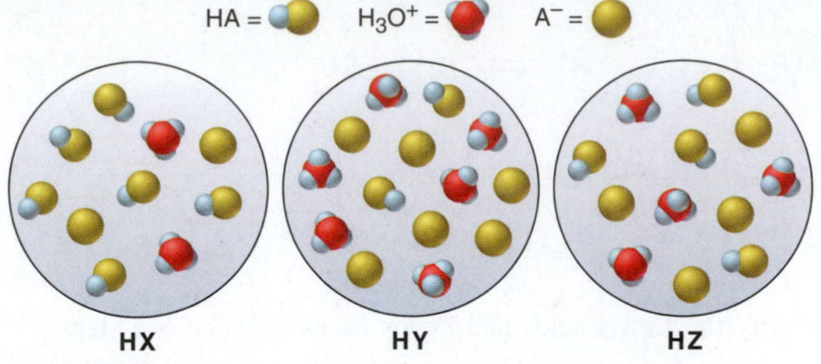

(a) Rank the acids in order of increasing K_a.
(b) Rank the acids in order of increasing pK_a.
(c) Rank the conjugate bases in order of increasing pK_b.
(d) What is the percent dissociation of HX?
(e) If equimolar amounts of the sodium salts of the acids (NaX, NaY, and NaZ) were dissolved in water, which solution would have the highest pOH? The lowest pH?

Ionic Equilibria in Aqueous Systems

Global Carbonate Equilibria The coral reefs that ring many tropical areas of ocean form through complex interactions among gaseous CO_2, solid $CaCO_3$, and dissolved Ca^{2+} and HCO_3^- ions. In this chapter, we introduce three types of aqueous ionic equilibria with major applications in industry and the environment.

Outline

Key Principles
to focus on while studying this chapter

- In an aqueous solution at equilibrium, an ionic substance dissociates less if one of the ions in the substance is already present. As a result of this *common-ion effect,* an acid (HA) dissociates more (and thus the solution has a lower pH) in water than in a solution containing A^-. ***(Section 19.1)***
- An *acid-base buffer* is a solution containing high concentrations of a conjugate acid-base pair. It resists changes in pH because of the common-ion effect: the conjugate base (or acid) component reacts with added H_3O^+ from a strong acid (or OH^- from a strong base) to keep pH relatively constant. A concentrated buffer has more *capacity* to resist a pH change than a dilute buffer. Buffers have a *range* of about 2 pH units, which corresponds to a value from 10 to 0.1 for the ratio $[A^-]/[HA]$. ***(Section 19.1)***
- The *equivalence point* of an acid-base titration occurs when the amount (mol) of acid equals the amount (mol) of base. The pH at the equivalence point depends on the acid-base properties of the cation and the anion present: in a strong acid–strong base titration, the equivalence point is at pH 7; in a weak acid–strong base titration, $pH > 7$; and in a weak base–strong acid titration, $pH < 7$. In the latter two titrations, a *buffer region* occurs before the equivalence point is reached. ***(Section 19.2)***
- The dissolution in water of a *slightly soluble ionic compound* reaches an equilibrium characterized by a *solubility-product constant,* K_{sp}, that is much *lower* than 1. Addition of a common ion lowers the compound's solubility still further. Lowering the pH (adding H_3O^+) increases the solubility if the anion of the ionic compound comes from a weak acid. ***(Section 19.3)***
- By comparing values of Q_{sp} and K_{sp}, we can determine if a precipitate will form (the solubility will be exceeded) when ionic solutions are mixed. ***(Section 19.3)***
- Acid rain occurs when nitrogen and/or sulfur oxides react with moisture in the air. Its destructive effects are lessened through the buffering action that occurs if lakes are bounded by limestone-rich soils. ***(Section 19.3)***
- A *complex ion* consists of a central metal ion bonded to molecules or anions called *ligands.* Complex ions form in a stepwise process characterized by a *formation constant,* K_f, that is much *greater* than 1. Adding an anion that can act as a ligand increases the solubility of a slightly soluble ionic compound because it forms a complex ion with the ionic compound's cation. ***(Section 19.4)***

CONCEPTS & SKILLS TO REVIEW before studying this chapter

- solubility rules for ionic compounds (Section 4.3)
- equilibrium nature of a saturated solution (Section 13.3)
- effect of concentration on equilibrium position (Section 17.6)
- conjugate acid-base pairs (Section 18.3)
- calculations for weak-acid and weak-base equilibria (Sections 18.4 and 18.5)
- acid-base properties of salt solutions (Section 18.7)
- Lewis acids and bases (Section 18.8)

Having just examined the principles of equilibrium applied to acids and bases, we now turn to their role in aqueous ionic systems. The unique formations in limestone caves and the vast expanses of oceanic coral reefs arise from subtle shifts in carbonate solubility equilibria. Similar interactions prevent acidification of some lakes. Organisms survive by maintaining cellular pH within narrow limits through complex carbonate and phosphate equilibria. In soils, equilibria involving clays control the availability of ionic nutrients for plants. In industrial settings, equilibrium principles govern the softening of water and the purification of products by precipitation of unwanted ions. And they even pertain to how the weak acids in wine and vinegar influence the delicate taste of a fine sauce. In this chapter, we explore three aqueous ionic equilibrium systems: acid-base buffers, slightly soluble salts, and complex ions.

19.1 • EQUILIBRIA OF ACID-BASE BUFFERS

Why do some lakes become acidic when showered by acid rain, while others remain unaffected? How does blood maintain a constant pH in contact with countless cellular acid-base reactions? How can a chemist sustain a nearly constant [H_3O^+] in reactions that consume or produce H_3O^+ or OH^-? The answer in each case depends on the action of a buffer.

What a Buffer Is and How It Works: The Common-Ion Effect

In everyday language, a buffer is something that lessens the impact of an external force. An **acid-base buffer** is a solution that *lessens the impact on pH from the addition of acid or base*. Add a small amount of H_3O^+ or OH^- to an unbuffered solution, and the pH changes by several units, thus, [H_3O^+] changes by *several orders of magnitude* (Figure 19.1).

The same addition of strong acid or strong base to a buffered solution causes only a minor change in pH (Figure 19.2). To withstand these additions, a buffer must

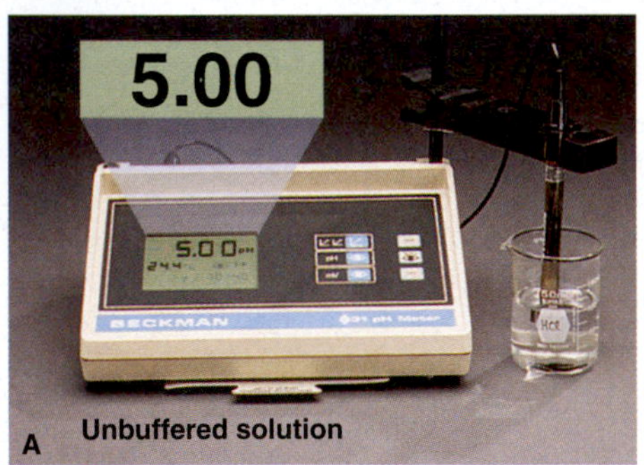

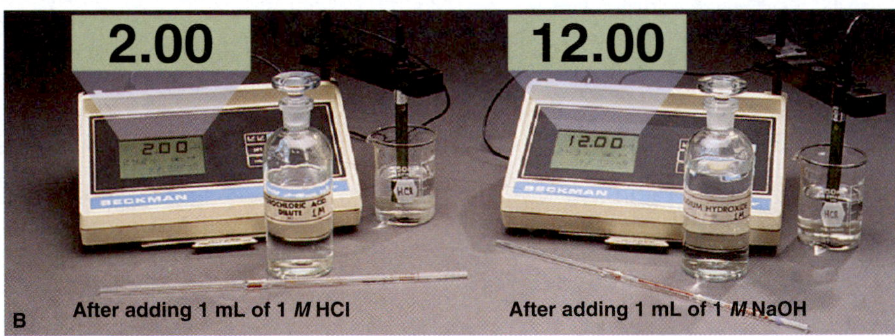

Figure 19.1 The effect of adding acid or base to an unbuffered solution. **A,** A 100-mL sample of dilute HCl is adjusted to pH 5.00. **B,** Adding 1 mL of strong acid *(left)* or of strong base *(right)* changes the pH by several units.

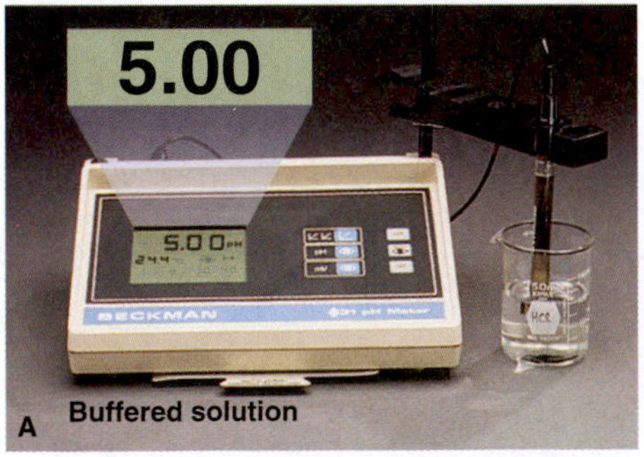

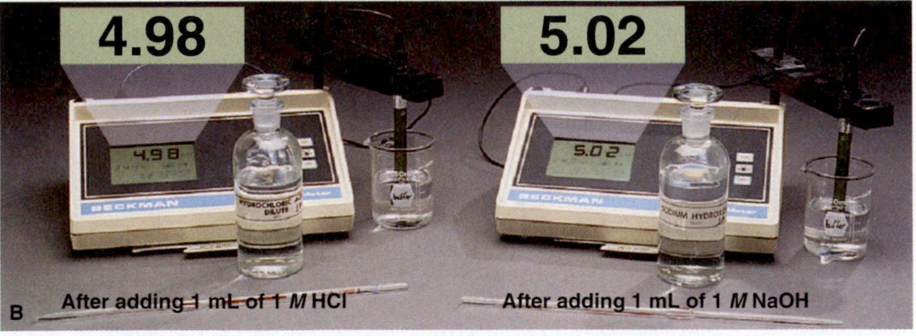

Figure 19.2 The effect of adding acid or base to a buffered solution. **A,** A 100-mL sample of an acetate buffer (1 *M* CH_3COOH mixed with 1 *M* CH_3COONa) is adjusted to pH 5.00. **B,** Adding 1 mL of strong acid *(left)* or of strong base *(right)* changes the pH negligibly.

have an acidic component that reacts with the added OH^- *and* a basic component that reacts with the added H_3O^+. However, these components can't be any acid and base because they would neutralize each other. Most often, *the components of a buffer are a conjugate acid-base pair* (weak acid and conjugate base or weak base and conjugate acid). The buffer in Figure 19.2, for example, is a mixture of acetic acid (CH_3COOH) and acetate ion (CH_3COO^-).

Presence of a Common Ion Buffers work through the **common-ion effect.** When you dissolve acetic acid in water, the acid dissociates slightly:

$$CH_3COOH(aq) + H_2O(l) \rightleftharpoons H_3O^+(aq) + CH_3COO^-(aq)$$

What happens if you now introduce acetate ion by adding the soluble salt sodium acetate? From Le Châtelier's principle (Section 17.6), we know that adding CH_3COO^- ion will shift the equilibrium position to the left; thus, $[H_3O^+]$ decreases, in effect lowering the extent of acid dissociation:

$$CH_3COOH(aq) + H_2O(l) \overset{\longleftarrow}{\rightleftharpoons} H_3O^+(aq) + CH_3COO^-(aq; \text{added})$$

We get the same result when we add acetic acid to a sodium acetate solution instead of water. The acetate ion already present suppresses the acid from dissociating as much as it does in water, thus keeping the $[H_3O^+]$ lower (and pH higher). In either case, the effect is less acid dissociation. Acetate ion is called *the common ion* because it is "common" to both the acetic acid and sodium acetate solutions. *The common-ion effect occurs when a given ion is added to an equilibrium mixture that already contains that ion, and the position of equilibrium shifts away from forming it.*

Table 19.1 shows that the percent dissociation (and the $[H_3O^+]$) of an acetic acid solution decreases as the concentration of acetate ion (supplied by dissolving sodium acetate) increases. Note that the *common ion, CH_3COO^- (or A^-), suppresses the dissociation of CH_3COOH (HA),* which makes the solution less acidic (higher pH).

Table 19.1 The Effect of Added Acetate Ion on the Dissociation of Acetic Acid

$[CH_3COOH]_{init}$	$[CH_3COO^-]_{added}$	% Dissociation*	H_3O^+	pH
0.10	0.00	1.3	1.3×10^{-3}	2.89
0.10	0.050	0.036	3.6×10^{-5}	4.44
0.10	0.10	0.018	1.8×10^{-5}	4.74
0.10	0.15	0.012	1.2×10^{-5}	4.92

$$\text{*\% Dissociation} = \frac{[CH_3COOH]_{dissoc}}{[CH_3COOH]_{init}} \times 100$$

Relative Concentrations of Buffer Components A buffer works because *large amounts of the acidic (HA) and basic (A^-) components consume small amounts of added OH^- or H_3O^+, respectively.* Consider what happens to $[H_3O^+]$ in a solution with high $[CH_3COOH]$ and high $[CH_3COO^-]$ when we add small amounts of strong acid or base. The equilibrium expression for HA dissociation is

$$K_a = \frac{[CH_3COO^-][H_3O^+]}{[CH_3COOH]}$$

Solving for $[H_3O^+]$ gives

$$[H_3O^+] = K_a \times \frac{[CH_3COOH]}{[CH_3COO^-]}$$

Since K_a is constant, *the $[H_3O^+]$ of the solution depends on the buffer-component concentration ratio,* $\dfrac{[CH_3COOH]}{[CH_3COO^-]}$:

- If the ratio $[HA]/[A^-]$ goes up, $[H_3O^+]$ goes up.
- If the ratio $[HA]/[A^-]$ goes down, $[H_3O^+]$ goes down.

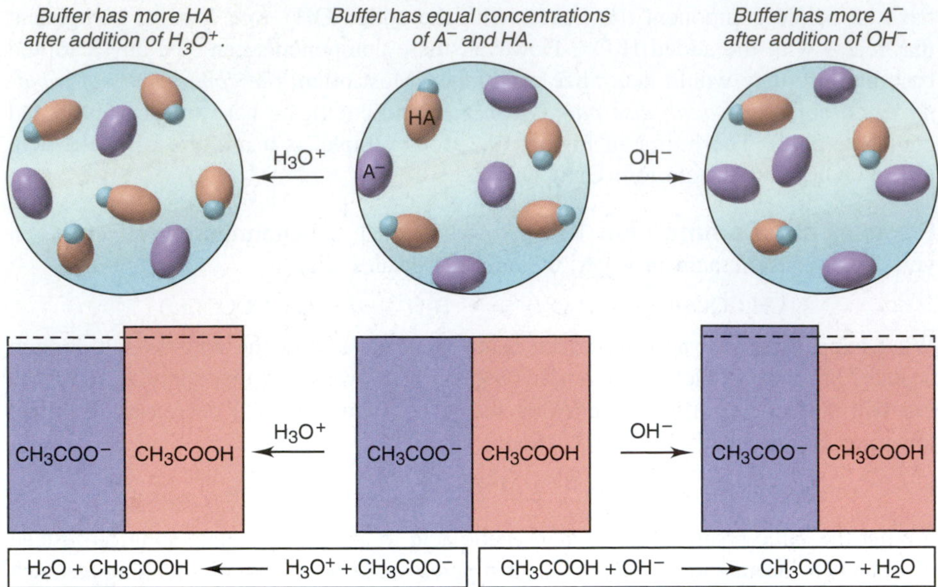

Figure 19.3 How a buffer works. The relative concentrations of the buffer components, acetic acid (CH_3COOH, *red*) and acetate ion (CH_3COO^-, *blue*) are indicated by the heights of the bars.

Let's track this ratio as we add strong acid or strong base to a buffer in which $[HA] = [A^-]$ (Figure 19.3, *middle*):

1. *Strong acid.* When we add a small amount of strong acid, the H_3O^+ ions react with an *equal (stoichiometric) amount* of CH_3COO^- from the buffer to form more CH_3COOH:

$$H_3O^+(aq;\text{ added}) + CH_3COO^-(aq;\text{ from buffer}) \longrightarrow CH_3COOH(aq) + H_2O(l)$$

As a result, $[CH_3COO^-]$ goes down by that small amount and $[CH_3COOH]$ goes up by that amount, which increases the buffer-component concentration ratio (Figure 19.3, *to the left*). The $[H_3O^+]$ increases *very* slightly.

2. *Strong base*. The addition of a small amount of strong base produces the opposite result. The OH^- ions react with an *equal (stoichiometric) amount* of CH_3COOH from the buffer to form that much more CH_3COO^- (Figure 19.3, *to the right*):

$$CH_3COOH(aq;\text{ from buffer}) + OH^-(aq;\text{ added}) \longrightarrow CH_3COO^-(aq) + H_2O(l)$$

This time, the buffer-component concentration ratio decreases, which decreases $[H_3O^+]$ *very* slightly.

Thus, the buffer components consume *nearly all* the added H_3O^+ or OH^-. To reiterate, as long as the amount of added H_3O^+ or OH^- is small compared with the amounts of the buffer components, *the conversion of one component into the other produces a small change in the buffer-component concentration ratio and, consequently, a small change in $[H_3O^+]$ and in pH.* Sample Problem 19.1 demonstrates how small these pH changes typically are. Note that the latter two parts of the problem combine a stoichiometry portion, like the problems in Chapter 3, and a weak-acid dissociation portion, like those in Chapter 18.

Sample Problem 19.1 Calculating the Effect of Added H_3O^+ or OH^- on Buffer pH

Problem Calculate the pH:
(a) Of a buffer solution consisting of 0.50 *M* CH_3COOH and 0.50 *M* CH_3COONa
(b) After adding 0.020 mol of solid NaOH to 1.0 L of the buffer solution in part (a)
(c) After adding 0.020 mol of HCl to 1.0 L of the buffer solution in part (a) K_a of $CH_3COOH = 1.8\times10^{-5}$. (Assume the additions cause negligible volume changes.)

Plan For each part, we know, or can find, $[CH_3COOH]_{init}$ and $[CH_3COO^-]_{init}$. We know the K_a of CH_3COOH (1.8×10^{-5}) and need to find $[H_3O^+]$ at equilibrium and convert it to pH. **(a)** We use the given concentrations of buffer components (each 0.50 M) as the initial values. As in earlier problems, we assume that x, the $[CH_3COOH]$ that dissociates, which equals $[H_3O^+]$, is so small relative to $[CH_3COOH]_{init}$ that it can be neglected. We set up a reaction table, solve for x, and check the assumption.
(b) and **(c)** We assume that the added OH^- or H_3O^+ reacts completely with the buffer components to yield new $[CH_3COOH]_{init}$ and $[CH_3COO^-]_{init}$, and then the acid dissociates to an unknown extent. We set up two reaction tables. The first summarizes the stoichiometry of adding strong base (0.020 mol) or acid (0.020 mol). The second summarizes the dissociation of the new $[CH_3COOH]_{init}$, and we proceed as in part (a) to find the new $[H_3O^+]$.

Solution **(a)** The original pH: $[H_3O^+]$ in the original buffer. Setting up a reaction table with $x = [CH_3COOH]_{dissoc} = [H_3O^+]$ (as in Chapter 18, we assume that $[H_3O^+]$ from H_2O is negligible and disregard it):

Concentration (M)	$CH_3COOH(aq)$ +	$H_2O(l)$ ⇌	$CH_3COO^-(aq)$ +	$H_3O^+(aq)$
Initial	0.50	—	0.50	0
Change	$-x$	—	$+x$	$+x$
Equilibrium	$0.50 - x$	—	$0.50 + x$	x

Making the assumption and finding the equilibrium $[CH_3COOH]$ and $[CH_3COO^-]$: With K_a small, x is small, so we assume

$$[CH_3COOH] = 0.50\ M - x \approx 0.50\ M \quad \text{and} \quad [CH_3COO^-] = 0.50\ M + x \approx 0.50\ M$$

Solving for x ($[H_3O^+]$ at equilibrium):

$$x = [H_3O^+] = K_a \times \frac{[CH_3COOH]}{[CH_3COO^-]} \approx (1.8\times10^{-5}) \times \frac{0.50}{0.50} = 1.8\times10^{-5}\ M$$

Checking the assumption:

$$\frac{1.8\times10^{-5}\ M}{0.50\ M} \times 100 = 3.6\times10^{-3}\% < 5\%$$

The assumption is justified, and we will use the same assumption in parts (b) and (c). Also, according to the other criterion (see Sample Problems 17.9 and 18.8),

$$\frac{[HA]_{init}}{K_a} = \frac{0.50}{1.8\times10^{-5}} = 2.8\times10^4 > 400$$

Calculating pH:

$$pH = -\log[H_3O^+] = -\log(1.8 \times 10^{-5}) = 4.74$$

(b) The pH after adding base (0.020 mol of NaOH to 1.0 L of buffer). Finding $[OH^-]_{added}$:

$$[OH^-]_{added} = \frac{0.020\ \text{mol}\ OH^-}{1.0\ \text{L soln}} = 0.020\ M\ OH^-$$

Setting up a reaction table for the *stoichiometry* of adding OH^- to CH_3COOH:

Concentration (M)	$CH_3COOH(aq)$ +	$OH^-(aq)$ ⟶	$CH_3COO^-(aq)$ +	$H_2O(l)$
Initial	0.50	0.020	0.50	—
Change	-0.020	-0.020	$+0.020$	—
Final	0.48	0	0.52	—

Setting up a reaction table for the *acid dissociation,* using these new initial concentrations. As in part (a), $x = [CH_3COOH]_{dissoc} = [H_3O^+]$:

Concentration (M)	$CH_3COOH(aq)$ +	$H_2O(l)$ ⇌	$CH_3COO^-(aq)$ +	$H_3O^+(aq)$
Initial	0.48	—	0.52	0
Change	$-x$	—	$+x$	$+x$
Equilibrium	$0.48 - x$	—	$0.52 + x$	x

Making the assumption that x is small, and solving for x:

$$[CH_3COOH] = 0.48\ M - x \approx 0.48\ M \quad \text{and} \quad [CH_3COO^-] = 0.52\ M + x \approx 0.52\ M$$

$$x = [H_3O^+] = K_a \times \frac{[CH_3COOH]}{[CH_3COO^-]} \approx (1.8\times10^{-5}) \times \frac{0.48}{0.52} = 1.7\times10^{-5}\ M$$

Calculating the pH:

$$pH = -\log[H_3O^+] = -\log(1.7\times10^{-5}) = 4.77$$

The addition of strong base increased the concentration of the basic buffer component at the expense of the acidic buffer component. Note especially that the pH *increased only slightly,* from 4.74 to 4.77.

(c) The pH after adding acid (0.020 mol of HCl to 1.0 L of buffer). Finding $[H_3O^+]_{added}$:

$$[H_3O^+]_{added} = \frac{0.020 \text{ mol } H_3O^+}{1.0 \text{ L soln}} = 0.020\ M\ H_3O^+$$

Now we proceed as in part (b), by first setting up a reaction table for the *stoichiometry* of adding H_3O^+ to CH_3COO^-:

Concentration (*M*)	$CH_3COO^-(aq)$	+ $H_3O^+(aq)$	⟶ $CH_3COOH(aq)$	+ $H_2O(l)$
Initial	0.50	0.020	0.50	—
Change	−0.020	−0.020	+0.020	—
Final	0.48	0	0.52	—

The reaction table for the *acid dissociation,* with $x = [CH_3COOH]_{dissoc} = [H_3O^+]$ is

Concentration (*M*)	$CH_3COOH(aq)$	+ $H_2O(l)$	⇌ $CH_3COO^-(aq)$	+ $H_3O^+(aq)$
Initial	0.52	—	0.48	0
Change	$-x$	—	$+x$	$+x$
Equilibrium	$0.52 - x$	—	$0.48 + x$	x

Making the assumption that x is small, and solving for x:

$$[CH_3COOH] = 0.52\ M - x \approx 0.52\ M \quad \text{and} \quad [CH_3COO^-] = 0.48\ M + x \approx 0.48\ M$$

$$x = [H_3O^+] = K_a \times \frac{[CH_3COOH]}{[CH_3COO^-]} \approx (1.8\times10^{-5}) \times \frac{0.52}{0.48} = 2.0\times10^{-5}\ M$$

Calculating the pH:

$$pH = -\log[H_3O^+] = -\log(2.0\times10^{-5}) = 4.70$$

The addition of strong acid increased the concentration of the acidic buffer component at the expense of the basic buffer component and *lowered* the pH only slightly, from 4.74 to 4.70.

Check The changes in $[CH_3COOH]$ and $[CH_3COO^-]$ occur in opposite directions in parts (b) and (c), which makes sense. The additions were of equal amounts, so the pH increase in (b) should equal the pH decrease in (c), within rounding.

Comment In part (a), we justified our assumption that x can be neglected. Therefore, in parts (b) and (c), we could have used the "Final" values from the last line of the stoichiometry reaction tables directly for the ratio of buffer components; that would have allowed us to dispense with a reaction table for the acid dissociation. In subsequent problems in this chapter, we will follow this more straightforward approach.

FOLLOW-UP PROBLEM 19.1 Calculate the pH of a buffer consisting of 0.50 M HF and 0.45 M F^- **(a)** before and **(b)** after addition of 0.40 g of NaOH to 1.0 L of the buffer (K_a of HF = 6.8×10^{-4}).

The Henderson-Hasselbalch Equation

For any weak acid, HA, the dissociation equation and K_a expression are

$$HA + H_2O \rightleftharpoons H_3O^+ + A^-$$

$$K_a = \frac{[H_3O^+][A^-]}{[HA]}$$

The key variable that determines $[H_3O^+]$ is the concentration *ratio* of acid species to base species, so, as before, rearranging to isolate $[H_3O^+]$ gives

$$[H_3O^+] = K_a \times \frac{[HA]}{[A^-]}$$

Taking the negative common (base 10) logarithm of both sides gives

$$-\log [H_3O^+] = -\log K_a - \log \left(\frac{[HA]}{[A^-]}\right)$$

from which definitions give

$$pH \quad = \quad pK_a \quad - \log \frac{[HA]}{[A^-]}$$

Then, because of the nature of logarithms, when we invert the buffer-component concentration ratio, the sign of the logarithm changes, to give $pH = pK_a + \log \left(\frac{[A^-]}{[HA]}\right)$.

Generalizing the previous equation for any conjugate acid-base pair gives the **Henderson-Hasselbalch equation:**

$$pH = pK_a + \log \left(\frac{[base]}{[acid]}\right) \tag{19.1}$$

This relationship is very useful for two reasons. First, it allows us to solve directly for pH instead of having to calculate $[H_3O^+]$ first. For instance, by applying the Henderson-Hasselbalch equation in part (b) of Sample Problem 19.1, we could have found the pH of the buffer after the addition of NaOH as follows:

$$pH = pK_a + \log \left(\frac{[CH_3COO^-]}{[CH_3COOH]}\right) = 4.74 + \log \left(\frac{0.52}{0.48}\right) = 4.77$$

Second, as you'll see shortly, it allows us to prepare a buffer of a desired pH just by mixing the appropriate amounts of A^- and HA.

Buffer Capacity and Buffer Range

Let's consider two key aspects of a buffer—its capacity and the closely related range.

Buffer Capacity **Buffer capacity** is a measure of the "strength" of the buffer, its ability to maintain the pH following addition of strong acid or base. Capacity depends ultimately on component concentrations, both the absolute and relative concentrations:

1. In terms of *absolute* concentrations, *the more concentrated the buffer components, the greater the capacity.* Thus, for a given amount of added H_3O^+ or OH^-, the pH of a higher capacity buffer changes less than the pH of a lower capacity buffer (Figure 19.4). Note that *buffer pH is independent of buffer capacity.* A buffer made of equal volumes of 1.0 *M* CH_3COOH and 1.0 *M* CH_3COO^- has the same pH (4.74) as a buffer made of equal volumes of 0.10 *M* CH_3COOH and 0.10 *M* CH_3COO^-, but the more concentrated buffer has a greater capacity.

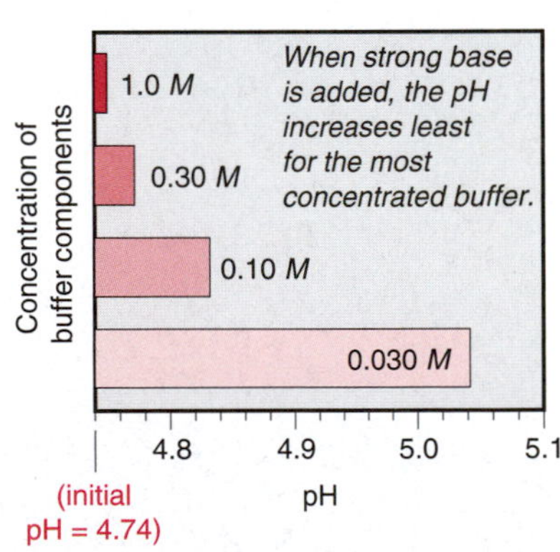

Figure 19.4 The relation between buffer capacity and pH change. The bars indicate the final pH values, after strong base was added, for four CH_3COOH/CH_3COO^- buffers with the same initial pH (4.74) and different component concentrations.

2. In terms of *relative* concentrations, *the closer the component concentrations are to each other, the greater the capacity.* As a buffer functions, the concentration of one component increases relative to the other. Because the concentration ratio determines the pH, the less the ratio changes, the less the pH changes. Let's compare the percent change in component concentration ratio for a buffer at two different initial ratios:

- Add 0.010 mol of OH^- to 1.00 L of buffer with initial concentrations $[HA] = [A^-] = 1.000\ M$: $[A^-]$ becomes 1.010 *M* and [HA] becomes 0.990 *M*,

$$\frac{[A^-]_{init}}{[HA]_{init}} = \frac{1.000\ M}{1.000\ M} = 1.000 \qquad \frac{[A^-]_{final}}{[HA]_{final}} = \frac{1.010\ M}{0.990\ M} = 1.02$$

$$\text{Percent change} = \frac{1.02 - 1.000}{1.000} \times 100 = 2\%$$

- Add 0.010 mol of OH^- to 1.00 L of buffer with initial concentrations [HA] = 0.250 *M* and $[A^-]$ = 1.750 *M*: $[A^-]$ becomes 1.760 *M* and [HA] becomes 0.240 *M*,

$$\frac{[A^-]_{init}}{[HA]_{init}} = \frac{1.750\,M}{0.250\,M} = 7.00 \qquad \frac{[A^-]_{final}}{[HA]_{final}} = \frac{1.760\,M}{0.240\,M} = 7.33$$

$$\text{Percent change} = \frac{7.33 - 7.00}{7.00} \times 100 = 4.7\%$$

Note that the change in the concentration ratio is more than twice as large when the initial concentrations of the components are very different. Thus, *a buffer has the highest capacity when the component concentrations are equal,* that is, when $[A^-]/[HA] = 1$:

$$pH = pK_a + \log\left(\frac{[A^-]}{[HA]}\right) = pK_a + \log 1 = pK_a + 0 = pK_a$$

Therefore, *a buffer whose pH is equal to or near the pK_a of its acid component has the highest capacity* for a given concentration.

Buffer Range **Buffer range** is the pH range over which the buffer is effective and is also related to the *relative* buffer-component concentrations. The further the concentration ratio is from 1, the less effective the buffer (the lower the buffer capacity). In practice, if the $[A^-]/[HA]$ ratio is greater than 10 or less than 0.1—that is, if one component concentration is more than 10 times the other—buffering action is poor. Since log 10 = +1 and log 0.1 = −1, *buffers have a usable range within ±1 pH unit of the pK_a of the acid component:*

$$pH = pK_a + \log\left(\frac{10}{1}\right) = pK_a + 1 \qquad \text{and} \qquad pH = pK_a + \log\left(\frac{1}{10}\right) = pK_a - 1$$

Sample Problem 19.2 Using Molecular Scenes to Examine Buffers

Problem The molecular scenes below represent equal volumes of four HA/A^- buffers. (HA is blue and green, A^- is green, and other ions and water are not shown.)

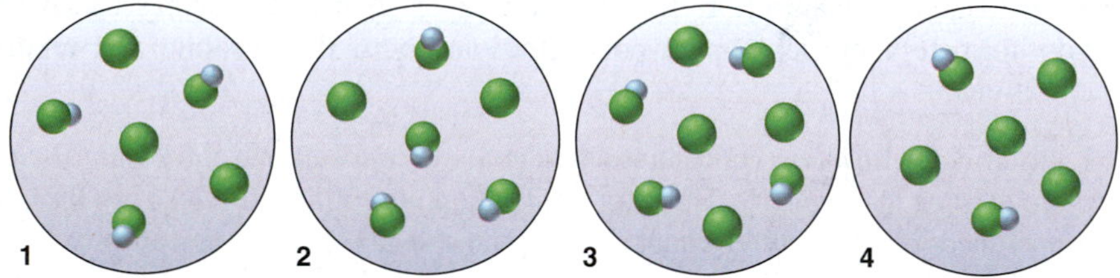

(a) Which buffer has the highest pH?
(b) Which buffer has the greatest capacity?
(c) Should we add a small amount of concentrated strong acid or strong base to convert sample 1 to sample 2 (assuming no volume change)?

Plan The molecular scenes show varying numbers of weak acid molecules (HA) and the conjugate base (A^-). Because the volumes are equal, the scenes represent molarities as well as numbers. **(a)** As the pH rises, more HA loses its H^+ and becomes A^-, so the $[A^-]/[HA]$ ratio will increase. We examine the scenes to see which has the highest ratio. **(b)** Buffer capacity depends on buffer-component concentration *and* ratio. We examine the scenes to see which has a high concentration and a ratio close to 1.
(c) Adding strong acid converts some A^- to HA, and adding strong base does the opposite. Comparing the $[A^-]/[HA]$ ratios in samples 1 and 2 tells us which to add.

Solution **(a)** The $[A^-]/[HA]$ ratios are as follows: For sample 1, $[A^-]/[HA] = 3/3 = 1$. Similarly, sample 2 = 0.5; sample 3 = 1; sample 4 = 2. Sample 4 has the highest pH because it has the highest $[A^-]/[HA]$ ratio.
(b) Samples 1 and 3 have a $[A^-]/[HA]$ ratio of 1, but sample 3 has the greater capacity because it has a higher concentration.
(c) Sample 2 has a lower $[A^-]/[HA]$ ratio than sample 1, so we would add strong acid to sample 1 to convert some A^- to HA.

FOLLOW-UP PROBLEM 19.2 The molecular scene *(see margin)* shows a sample of an HB/B^- buffer. (HB is blue and yellow, B^- is yellow; other ions and water are not shown.)
(a) Would you add a small amount of concentrated strong acid or strong base to increase the buffer capacity?
(b) Assuming no volume change, draw a scene that represents the buffer with the highest possible capacity after the addition in part (a).

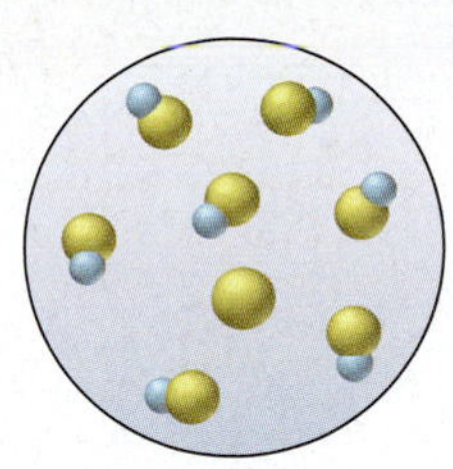

Preparing a Buffer

Even though chemical supply-houses offer buffers in a variety of pH values and concentrations, you may have to prepare a specific one, for example, in an environmental or biomedical application. Several steps are required to prepare a buffer:

1. *Decide on the conjugate acid-base pair.* The choice is determined mostly by the desired pH. Remember that a buffer is most effective when the buffer-component concentration ratio is close to 1; in that case, the pH is close to the pK_a of the acid. Convert pK_a to K_a, choose the acid from a list, such as that in Appendix C, and use the sodium salt as the conjugate base.
2. *Find the ratio $[A^-]/[HA]$ that gives the desired pH, using the Henderson-Hasselbalch equation.* Note that, because HA is a weak acid, and thus dissociates very little, the equilibrium concentrations are approximately equal to the initial concentrations; that is,

$$pH = pK_a + \log\left(\frac{[A^-]}{[HA]}\right) \approx pK_a + \log\left(\frac{[A^-]_{init}}{[HA]_{init}}\right)$$

   Therefore, you can use the ratio directly in the next step.
3. *Choose the buffer concentration and calculate the amounts to mix.* Remember that the higher the concentration, the greater the buffer capacity. For most laboratory applications, concentrations from 0.05 *M* to 0.5 *M* are suitable. From a given amount (usually in the form of concentration and volume) of one component, find the amount of the other component using the buffer-component concentration ratio.
4. *Mix the amounts together and adjust the buffer pH to the desired value.* Add small amounts of strong acid or strong base, while monitoring the solution with a pH meter.

Sample Problem 19.3 goes through steps 2 and 3.

Sample Problem 19.3 Preparing a Buffer

Problem An environmental chemist needs a carbonate buffer of pH 10.00 to study the effects of acid rain on limestone-rich soils. How many grams of Na_2CO_3 must she add to 1.5 L of 0.20 *M* $NaHCO_3$ to make the buffer? K_a of HCO_3^- is 4.7×10^{-11}.

Plan The conjugate pair is HCO_3^- (acid) and CO_3^{2-} (base), and we know the buffer volume (1.5 L) and the concentration (0.20 *M*) of HCO_3^-, so we need to find the buffer-component concentration ratio that gives pH 10.00 and the mass of Na_2CO_3 to dissolve. We convert pH to $[H_3O^+]$ and use the K_a expression to solve for $[CO_3^{2-}]$. Multiplying by the volume of solution gives the amount (mol) of CO_3^{2-} needed, which we convert to mass (g) of Na_2CO_3.

Solution Calculating $[H_3O^+]$: $[H_3O^+] = 10^{-pH} = 10^{-10.00} = 1.0\times10^{-10}\ M$

Solving for $[CO_3^{2-}]$ in the concentration ratio:

$$HCO_3^-(aq) + H_2O(l) \rightleftharpoons H_3O^+(aq) + CO_3^{2-}(aq) \qquad K_a = \frac{[H_3O^+][CO_3^{2-}]}{[HCO_3^-]}$$

So

$$[CO_3^{2-}] = \frac{K_a[HCO_3^-]}{[H_3O^+]} = \frac{(4.7\times10^{-11})(0.20)}{1.0\times10^{-10}} = 0.094\ M$$

Calculating the amount (mol) of CO_3^{2-} needed for the given volume:

$$\text{Amount (mol) of } CO_3^{2-} = 1.5\ \text{L soln} \times \frac{0.094\ \text{mol}\ CO_3^{2-}}{1\ \text{L soln}} = 0.14\ \text{mol}\ CO_3^{2-}$$

Calculating the mass (g) of Na_2CO_3 needed:

$$\text{Mass (g) of } Na_2CO_3 = 0.14 \text{ mol } Na_2CO_3 \times \frac{105.99 \text{ g } Na_2CO_3}{1 \text{ mol } Na_2CO_3} = 15 \text{ g } Na_2CO_3$$

The chemist should dissolve 15 g of Na_2CO_3 into about 1.3 L of 0.20 *M* $NaHCO_3$ and add more 0.20 *M* $NaHCO_3$ to make 1.5 L. Using a pH meter, she can then adjust the pH to 10.00 by dropwise addition of concentrated strong acid or base.

Check For a useful buffer range, the concentration of the acidic component, $[HCO_3^-]$, must be within a factor of 10 of the concentration of the basic component, $[CO_3^{2-}]$. And we have (1.5 L)(0.20 *M* HCO_3^-), or 0.30 mol of HCO_3^-, and 0.14 mol of CO_3^{2-}; $0.30/0.14 = 2.1$. Make sure the relative amounts of components are reasonable: we want a pH below the pK_a of HCO_3^- (10.33), so we want more of the acidic than the basic species.

FOLLOW-UP PROBLEM 19.3 How would you prepare a benzoic acid/benzoate buffer with pH = 4.25, starting with 5.0 L of 0.050 *M* sodium benzoate (C_6H_5COONa) solution and adding the acidic component? K_a of benzoic acid (C_6H_5COOH) is 6.3×10^{-5}.

Another way to prepare a buffer is to form one of the components by *partial neutralization* of the other. For example, you can prepare an $HCOOH/HCOO^-$ buffer by mixing aqueous solutions of HCOOH and NaOH. As OH^- reacts with HCOOH, neutralization of some of the HCOOH produces the $HCOO^-$ needed:

$$\text{HCOOH (HA total)} + OH^-\text{(amt added)} \longrightarrow$$
$$\text{HCOOH (HA total} - OH^- \text{ amt added)} + HCOO^-(OH^- \text{ amt added}) + H_2O$$

This method is based on the same chemical process that occurs when a weak acid is titrated with a strong base, as you'll see in Section 19.2.

Summary of Section 19.1

- The pH of a buffered solution changes much less than the pH of an unbuffered solution when H_3O^+ or OH^- is added.
- A buffer consists of a weak acid and its conjugate base (or a weak base and its conjugate acid). To be effective, the amounts of the components must be *much* greater than the amount of H_3O^+ or OH^- added.
- The buffer-component concentration ratio determines the pH; the ratio and the pH are related by the Henderson-Hasselbalch equation.
- When H_3O^+ or OH^- is added to a buffer, one component reacts to form the other; thus, $[H_3O^+]$ (and pH) changes only slightly.
- A concentrated (higher capacity) buffer undergoes smaller changes in pH than a dilute buffer. When the buffer pH equals the pK_a of the acid component, the buffer has its highest capacity.
- A buffer has an effective pH range of $pK_a \pm 1$ pH unit.
- To prepare a buffer, choose the conjugate acid-base pair, calculate the ratio of components, determine the buffer concentration, and adjust the final solution to the desired pH.

19.2 • ACID-BASE TITRATION CURVES

In Chapter 4, we used titrations to quantify acid-base reactions. In this section, we focus on the **acid-base titration curve,** a plot of pH vs. volume of titrant added. We discuss curves for strong acid–strong base, weak acid–strong base, and weak base–strong acid titrations. Running a titration is an exercise for the lab, but understanding the roles of acid-base indicators and of salt solutions (Section 18.7) and buffers applies key principles of acid-base equilibria.

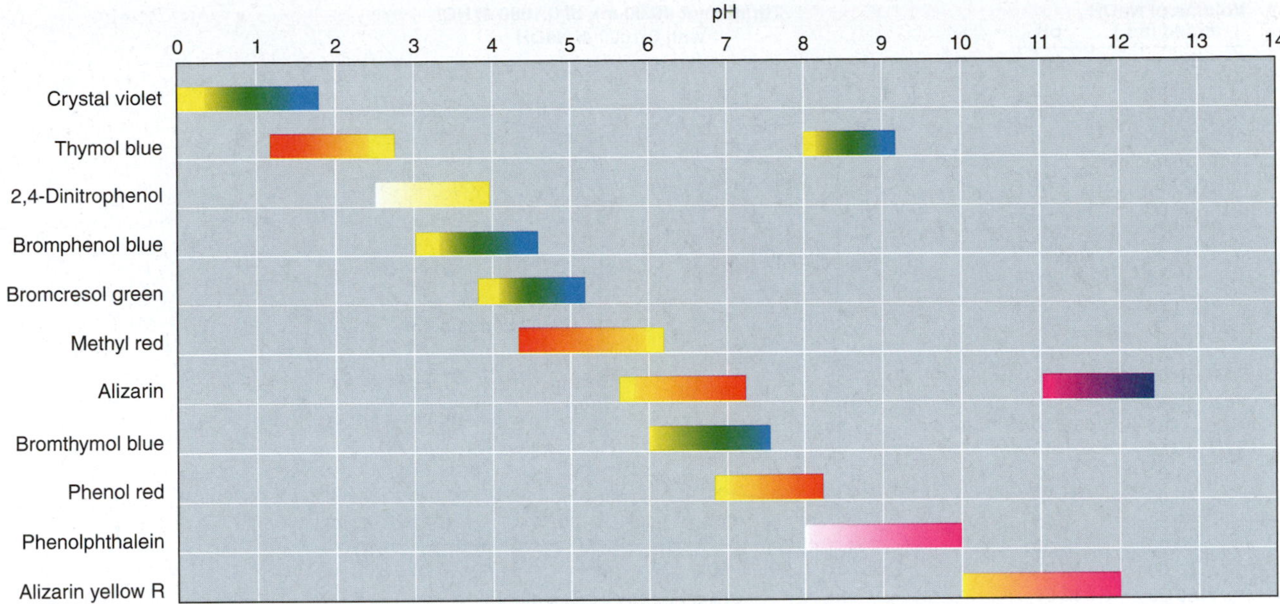

Figure 19.5 Colors and approximate pH ranges of some common acid-base indicators.

Monitoring pH with Acid-Base Indicators

An *acid-base indicator* is a weak organic acid (denoted here as HIn) whose color differs from the color of its conjugate base (In^-), with the color change occurring over a specific, narrow pH range. Typically, one or both forms are intensely colored, so only a tiny amount of indicator is needed, far too little to affect the pH during the titration.

Figure 19.5 shows the color change(s) and pH range(s) of some acid-base indicators. To select an indicator, you must know the approximate pH of the titration end point, which means you know the ionic species present. Because the indicator is a weak acid, the $[HIn]/[In^-]$ ratio is governed by the $[H_3O^+]$ of the solution:

$$HIn(aq) + H_2O(l) \rightleftharpoons H_3O^+(aq) + In^-(aq) \qquad K_a \text{ of HIn} = \frac{[H_3O^+][In^-]}{[HIn]}$$

Therefore, $$\frac{[HIn]}{[In^-]} = \frac{[H_3O^+]}{K_a}$$

An indicator's color reflects the concentration ratio:

- We see the HIn color if the $[HIn]/[In^-]$ ratio is 10/1 or greater.
- We see the In^- color if the $[HIn]/[In^-]$ ratio is 1/10 or less.
- Between these extremes, the two colors merge into an intermediate hue.

Thus, an indicator has a *color range* equal to a 10^2-fold range in the [HIn]/[In] ratio: an *indicator changes color over a range of about 2 pH units.* For example, bromthymol blue has a pH range of about 6.0 to 7.6. As Figure 19.6 shows, it is yellow below that range *(left)*, blue above it *(right)*, and greenish in between *(center)*.

Figure 19.6 The color change of the indicator bromthymol blue.

Strong Acid–Strong Base Titration Curves

Figure 19.7 on the next page shows a typical curve for the titration of a strong acid with a strong base, the data used to construct it, and molecular views of the key species in solution at various points during the titration.

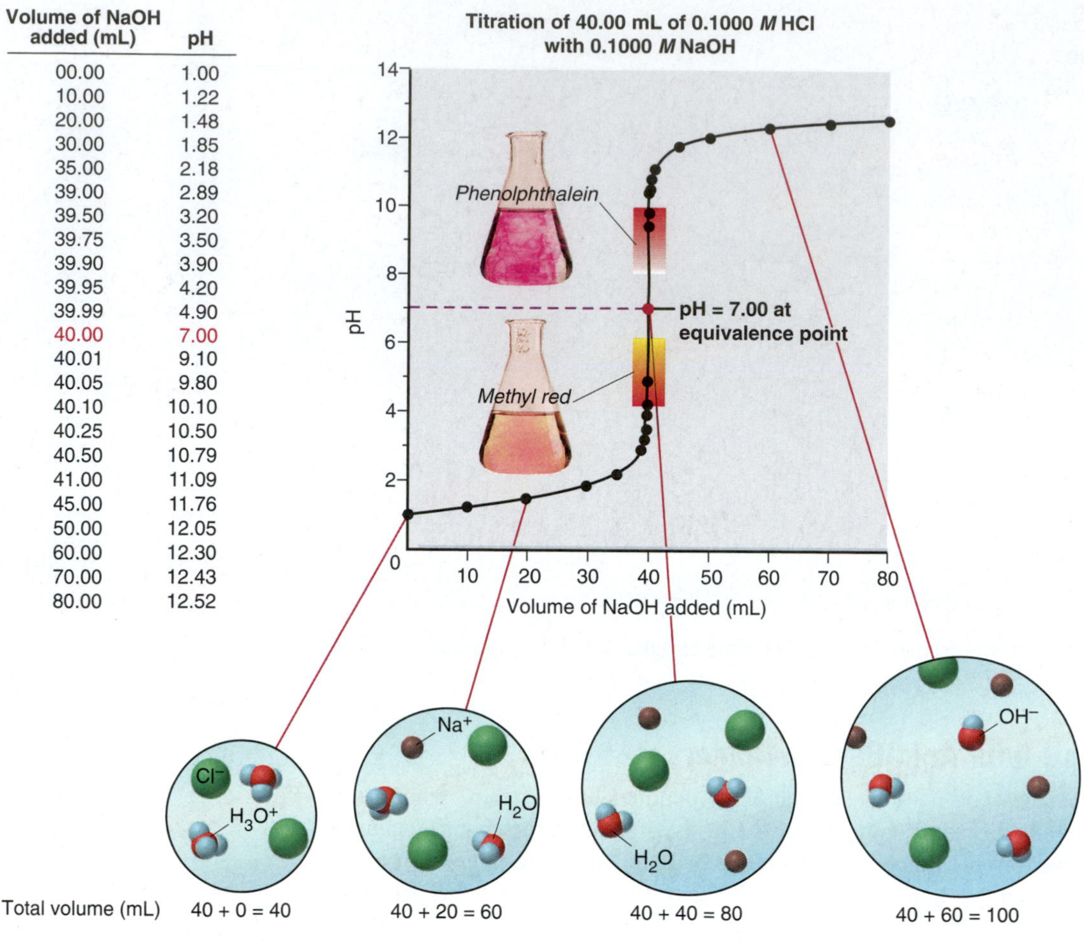

Volume of NaOH added (mL)	pH
00.00	1.00
10.00	1.22
20.00	1.48
30.00	1.85
35.00	2.18
39.00	2.89
39.50	3.20
39.75	3.50
39.90	3.90
39.95	4.20
39.99	4.90
40.00	7.00
40.01	9.10
40.05	9.80
40.10	10.10
40.25	10.50
40.50	10.79
41.00	11.09
45.00	11.76
50.00	12.05
60.00	12.30
70.00	12.43
80.00	12.52

Figure 19.7 **Curve for a strong acid–strong base titration.** Data *(at left)* for the titration of 40.00 mL of 0.1000 *M* HCl with 0.1000 *M* NaOH. In a strong acid–strong base titration, pH = 7.00 at the equivalence point. Each molecular view shows the *relative* numbers of species, other than solvent (any H_2O shown is the product of the reaction), at one point during the titration. The circles increase in size in proportion to the solution's total volume.

Features of the Curve There are three distinct regions of the titration curve, which correspond to three major changes in slope:

1. *The pH starts out low,* reflecting the high $[H_3O^+]$ of the strong acid, and increases gradually as acid is neutralized by the added base. (Cl^- is from HCl and Na^+ from NaOH.)
2. *The pH rises 6 to 8 units very rapidly.* This steep increase begins when the amount (mol) of OH^- added nearly equals the amount (mol) of H_3O^+ originally present in the acid. One or two more drops of base neutralize the remaining tiny excess of acid and introduce a tiny excess of base.
3. *The pH increases slowly* beyond the steep rise as more base (shown as OH^-) is added.

Let's distinguish between the *equivalence point* and the *end point,* both of which occur on the steep portion of the curve:

- The **equivalence point** occurs when the *number of moles of added OH^- equals the number of moles of H_3O^+* originally present. *The solution consists of the anion of the strong acid and the cation of the strong base.* Recall from Chapter 18 that *these ions do not react with water, so the solution is neutral: pH = 7.00.*
- The **end point** occurs when the indicator, which we added before the titration, changes color. *We choose an indicator with a color change close to the pH of the equivalence point.* Figure 19.5 includes two indicators suitable for a strong acid–strong base titration. Methyl red changes from red at pH 4.2 to yellow at pH 6.3, and phenolphthalein changes from colorless at pH 8.3 to pink at pH 10.0. Neither change occurs *at* the equivalence point (pH 7.00), but both occur on the vertical portion of the curve. Therefore, when methyl red turns yellow or phenolphthalein pink, we are within a drop or two of the equivalence point. In practice, then, the *visible* end point signals the *invisible* equivalence point.

Calculating the pH During the Titration By knowing the chemical species present during the titration, we can calculate the pH at various points along the way:

1. *Initial solution of strong HA.* In Figure 19.7, 40.00 mL of 0.1000 M HCl is titrated with 0.1000 M NaOH. Because a strong acid is completely dissociated, [HCl] = $[H_3O^+]$ = 0.1000 M. Therefore, the initial pH is*

$$\text{pH} = -\log [H_3O^+] = -\log (0.1000) = 1.00$$

2. *Before the equivalence point.* As we start adding base, some acid is neutralized and the volume of solution increases. To find the pH at various points up to the equivalence point, we find the *initial* amount (mol) of H_3O^+ and subtract the amount *reacted, which equals the amount (mol) of OH^- added,* to find the amount (mol) of H_3O^+ remaining. Then, we use the *total* volume to calculate $[H_3O^+]$ and convert to pH. For example, after adding 20.00 mL of 0.1000 M NaOH:

- *Find the amount (mol) of H_3O^+ remaining.*

$$\begin{array}{lr} \text{Initial amount (mol) of } H_3O^+ = 0.04000 \text{ L} \times 0.1000\, M = & 0.004000 \text{ mol } H_3O^+ \\ -\text{Amount (mol) of } OH^- \text{ added} = 0.02000 \text{ L} \times 0.1000\, M = & 0.002000 \text{ mol } OH^- \\ \hline \text{Amount (mol) of } H_3O^+ \text{ remaining} = & 0.002000 \text{ mol } H_3O^+ \end{array}$$

- *Calculate $[H_3O^+]$.* We divide by the *total volume* because one solution dilutes the ions in the other:

$$[H_3O^+] = \frac{\text{amount (mol) of } H_3O^+ \text{ remaining}}{\text{original volume of acid} + \text{volume of added base}}$$

$$= \frac{0.002000 \text{ mol } H_3O^+}{0.04000 \text{ L} + 0.02000 \text{ L}} = 0.03333\, M \qquad \text{pH} = 1.48$$

Given the amount of OH^- added, we are halfway to the equivalence point; but we are still on the initial slow rise of the curve, so the pH is still very low. Similar calculations give values up to the equivalence point.

3. *At the equivalence point.* After 40.00 mL of 0.1000 M NaOH (0.004000 mol of OH^-) has been added to the initial 0.004000 mol of H_3O^+, the equivalence point is reached. The solution contains Na^+ and Cl^-, neither of which reacts with water. Because of the autoionization of water, however,

$$[H_3O^+] = 1.0\times10^{-7}\, M \qquad \text{pH} = 7.00$$

4. *After the equivalence point.* From the equivalence point on, the pH calculation is based on the amount (mol) of *excess* OH^- present. For example, after adding 50.00 mL of NaOH, we have

$$\begin{array}{lr} \text{Total amount (mol) of } OH^- \text{ added} = 0.05000 \text{ L} \times 0.1000\, M = & 0.005000 \text{ mol } OH^- \\ -\text{Amount (mol) of } H_3O^+ \text{ consumed} = 0.04000 \text{ L} \times 0.1000\, M = & 0.004000 \text{ mol } H_3O^+ \\ \hline \text{Amount (mol) of excess } OH^- = & 0.001000 \text{ mol } OH^- \end{array}$$

$$[OH^-] = \frac{0.001000 \text{ mol } OH^-}{0.04000 \text{ L} + 0.05000 \text{ L}} = 0.01111\, M \qquad \text{pOH} = 1.95$$

$$\text{pH} = \text{p}K_w - \text{pOH} = 14.00 - 1.95 = 12.05$$

Weak Acid–Strong Base Titration Curves

Figure 19.8 on the next page shows a curve for the titration of a weak acid with strong base: 40.00 mL of 0.1000 M propanoic acid ($K_a = 1.3\times10^{-5}$) titrated with 0.1000 M NaOH. (We abbreviate the acid, CH_3CH_2COOH, as HPr and the conjugate base, $CH_3CH_2COO^-$, as Pr^-.)

Features of the Curve The dotted curve in Figure 19.8 corresponds to the bottom half of the strong acid–strong base curve (Figure 19.7). There are four key points to note for the weak acid curve, and the first three differ from the strong acid curve.

*In acid-base titrations, volumes and concentrations are usually known to four significant figures, but pH is reported to no more than two digits to the right of the decimal point.

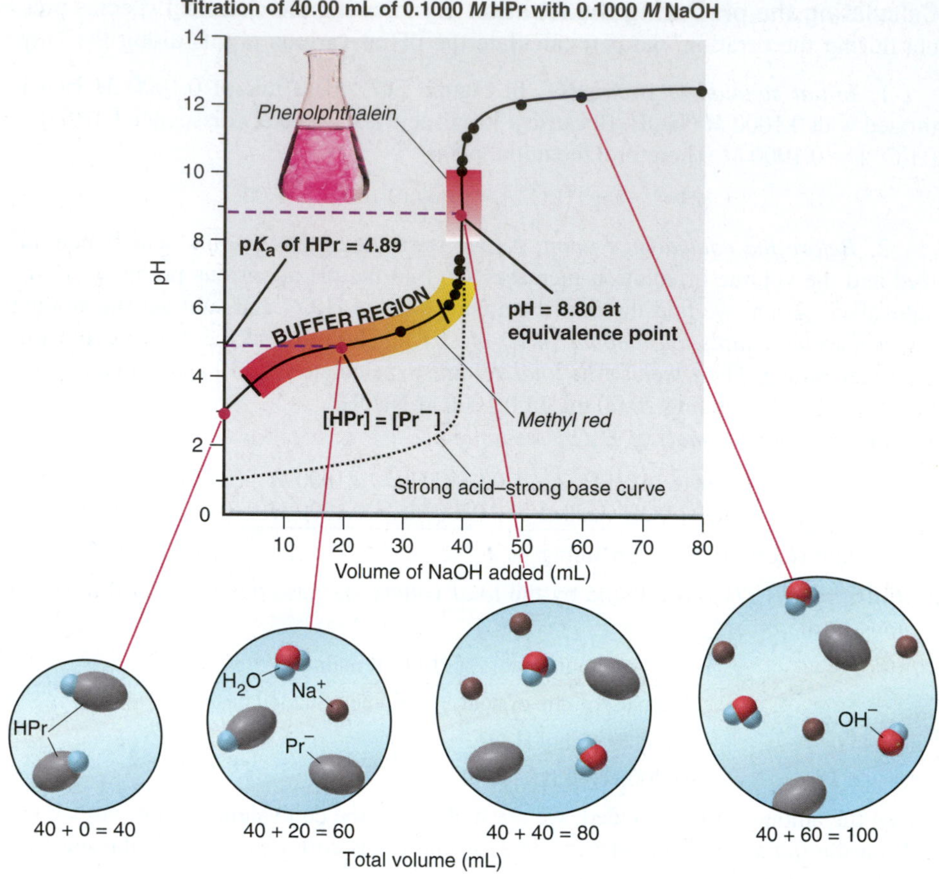

Figure 19.8 Curve for a weak acid–strong base titration.

1. *The initial pH is higher.* Because the weak acid (HPr) dissociates slightly, much less H_3O^+ is present (none shown in molecular view) than with the strong acid.
2. *The curve rises gradually in the so-called buffer region before the steep rise to the equivalence point.* As HPr reacts with strong base, more Pr^- forms, which creates an HPr/Pr^- buffer. At the midpoint of the buffer region, half the initial HPr has reacted (that is, half of the OH^- needed to reach the equivalence point has been added), so $[HPr] = [Pr^-]$, or $[Pr^-]/[HPr] = 1$. Therefore, *the pH equals the pK_a:*

   $$\text{Middle of buffer region: } pH = pK_a + \log\left(\frac{[Pr^-]}{[HPr]}\right) = pK_a + \log 1 = pK_a + 0 = pK_a$$

   The pH observed at this point is used to estimate the pK_a of an unknown acid.
3. *The pH at the equivalence point is above 7.00.* The solution contains the strong-base cation Na^+, which does not react with water, and the weak-acid anion Pr^-, which acts as a weak base to accept a proton from H_2O and yield OH^-.
4. *The pH increases slowly* beyond the equivalence point as excess OH^- is added.

The choice of indicator here is more limited than for a strong acid–strong base titration because the steep rise occurs over a smaller pH range. Phenolphthalein will work because it changes color within this range. But methyl red will not because its color change requires about 30 mL of titrant, rather than just a drop or two.

Calculating the pH During the Titration During the weak acid–strong base titration we must take into account the partial dissociation of the weak acid, the presence of the buffer, and the reaction of the conjugate base with water. The four key regions of the titration curve each require a separate calculation to find $[H_3O^+]$:

1. *Initial HA solution.* Before base is added, a weak acid dissociates, so we find $[H_3O^+]$ as in Section 18.4: we set up a reaction table with $x = [HPr]_{dissoc}$, assume $[H_3O^+]$ (and $[Pr^-]$) $= [HPr]_{dissoc} << [HPr]_{init}$, and solve for x:

$$K_a = \frac{[H_3O^+][Pr^-]}{[HPr]} \approx \frac{x^2}{[HPr]_{init}} \qquad \text{therefore,} \qquad x = [H_3O^+] \approx \sqrt{K_a \times [HPr]_{init}}$$

2. *Solution of HA and added base.* As we add NaOH, HPr forms Pr^-. Thus, for much of the titration up to the equivalence point, we have an HPr/Pr^- buffer. Therefore, we find $[H_3O^+]$ from

$$[H_3O^+] = K_a \times \frac{[HPr]}{[Pr^-]}$$

(We can also find the pH directly with the Henderson-Hasselbalch equation.) Because *volumes cancel in the concentration ratio,* that is, $[HPr]/[Pr^-]$ = amount (mol) of HPr/amount (mol) of Pr^-, we don't need volumes or have to calculate concentrations.

3. *Equivalent amounts of HA and added base.* At the equivalence point, all of the HPr has reacted, so the solution contains Pr^-, which reacts with water to form OH^-:

$$Pr^-(aq) + H_2O(l) \rightleftharpoons HPr(aq) + OH^-(aq)$$

This explains why, in a weak acid–strong base titration, the pH > 7.00 at the equivalence point. To calculate $[H_3O^+]$ (see Section 18.5), we first find K_b of Pr^- from K_a of HPr, set up a reaction table (assume $[Pr^-] >> [Pr^-]_{reacting}$), and solve for $[OH^-]$. Since we use a single concentration, $[Pr^-]$, to solve for $[OH^-]$, we *do* need the total volume. Then, we convert $[OH^-]$ to $[H_3O^+]$. These two steps are

$$(1)\ [OH^-] \approx \sqrt{K_b \times [Pr^-]}, \quad \text{where} \quad K_b = \frac{K_w}{K_a} \quad \text{and} \quad [Pr^-] = \frac{\text{amount (mol) of HPr}_{init}}{\text{total volume}}$$

$$(2)\ [H_3O^+] = \frac{K_w}{[OH^-]}$$

Combining them into one step gives

$$[H_3O^+] \approx \frac{K_w}{\sqrt{K_b \times [Pr^-]}}$$

4. *Solution of excess base.* Beyond the equivalence point, as in the strong acid–strong base titration, we are adding excess OH^-:

$$[H_3O^+] = \frac{K_w}{[OH^-]} \quad \text{where} \quad [OH^-] = \frac{\text{amount (mol) of excess OH}^-}{\text{total volume}}$$

Sample Problem 19.4 shows the overall approach.

Sample Problem 19.4 Finding the pH During a Weak Acid–Strong Base Titration

Problem Calculate the pH during the titration of 40.00 mL of 0.1000 *M* propanoic acid (HPr; $K_a = 1.3 \times 10^{-5}$) after each of the following additions of 0.1000 *M* NaOH:
(a) 0.00 mL **(b)** 30.00 mL **(c)** 40.00 mL **(d)** 50.00 mL

Plan **(a)** 0.00 mL: No base has been added yet, so this is a weak-acid dissociation. We calculate the pH as we did in Section 18.4. **(b)** 30.00 mL: We find the amount (mol) of Pr^- and of HPr, and substitute into the K_a expression to solve for $[H_3O^+]$, and convert to pH. **(c)** 40.00 mL: The amount (mol) of NaOH added equals the initial amount (mol) of HPr, so a solution of Na^+ and Pr^- exists. We calculate the pH as we did in Section 18.5, except that we need *total* volume to find $[Pr^-]$. **(d)** 50.00 mL: We calculate the amount (mol) of excess OH^- in the total volume, convert to $[H_3O^+]$ and then to pH.

Solution **(a)** 0.00 mL of 0.1000 *M* NaOH added. Following the approach used in Sample Problem 18.8 and just described in the text, we obtain

$$[H_3O^+] \approx \sqrt{K_a \times [HPr]_{init}} = \sqrt{(1.3\times10^{-5})(0.1000)} = 1.1\times10^{-3}\,M \quad \text{so} \quad pH = 2.96$$

(b) 30.00 mL of 0.1000 *M* NaOH added. Calculating the ratio of moles of HPr to Pr^-:

$$\text{Initial amount (mol) of HPr} = 0.04000\ \text{L} \times 0.1000\ M = 0.004000\ \text{mol HPr}$$

$$\text{Amount (mol) of NaOH added} = 0.03000\ \text{L} \times 0.1000\ M = 0.003000\ \text{mol OH}^-$$

For every mole of NaOH, 1 mol of Pr^- forms, so we have this reaction table:

Amount (mol)	$HPr(aq)$	+	$OH^-(aq)$	⟶	$Pr^+(aq)$	+	$H_2O(l)$
Initial	0.004000		0.003000		0		—
Change	−0.003000		−0.003000		+0.003000		—
Final	0.001000		0		0.003000		—

The last line of the table gives the new initial amounts of HPr and Pr^- that react to attain a new equilibrium. With x very small, we assume that the $[HPr]/[Pr^-]$ ratio at equilibrium is essentially equal to the ratio of these new initial amounts (see Comment in Sample Problem 19.1). Thus,

$$\frac{[HPr]}{[Pr^-]} = \frac{0.001000 \text{ mol}}{0.003000 \text{ mol}} = 0.3333$$

Solving for $[H_3O^+]$: $$[H_3O^+] = K_a \times \frac{[HPr]}{[Pr^-]} = (1.3\times10^{-5})(0.3333) = 4.3\times10^{-6}\ M$$

$$\text{pH} = 5.37$$

(c) 40.00 mL of 0.1000 M NaOH added. Calculating $[Pr^-]$ after all HPr has reacted:

$$[Pr^-] = \frac{0.004000 \text{ mol}}{0.04000 \text{ L} + 0.04000 \text{ L}} = 0.05000\ M$$

Calculating K_b using the given K_a (see Sample Problem 18.2 for similar steps):

$$K_b = \frac{K_w}{K_a} = \frac{1.0\times10^{-14}}{1.3\times10^{-5}} = 7.7\times10^{-10}$$

Solving for $[H_3O^+]$ as described in the text for region 3 of the curve:

$$[H_3O^+] \approx \frac{K_w}{\sqrt{K_b \times [Pr^-]}} = \frac{1.0\times10^{-14}}{\sqrt{(7.7\times10^{-10})(0.05000)}} = 1.6\times10^{-9}\ M$$

$$\text{pH} = 8.80$$

(d) 50.00 mL of 0.1000 M NaOH added.

$$\text{Amount (mol) of excess OH}^- = (0.1000\ M)(0.05000 \text{ L} - 0.04000 \text{ L}) = 0.001000 \text{ mol}$$

$$[OH^-] = \frac{\text{amount (mol) of excess OH}^-}{\text{total volume}} = \frac{0.001000 \text{ mol}}{0.09000 \text{ L}} = 0.01111\ M$$

$$[H_3O^+] = \frac{K_w}{[OH^-]} = \frac{1.0\times10^{-14}}{0.01111} = 9.0\times10^{-13}\ M$$

$$\text{pH} = 12.05$$

Check As expected, the pH increases through the four regions of the titration. Be sure to round off and check the arithmetic along the way.

FOLLOW-UP PROBLEM 19.4 A chemist titrates 20.00 mL of 0.2000 M HBrO ($K_a = 2.3\times10^{-9}$) with 0.1000 M NaOH. What is the pH:
(a) Before any base is added?
(b) When $[HBrO] = [BrO^-]$?
(c) At the equivalence point?
(d) When the amount (mol) of OH^- added is twice the amount of HBrO present initially?
(e) Sketch the titration curve; label the pK_a and equivalence point.

Weak Base–Strong Acid Titration Curves

The opposite of a weak acid–strong base titration is the titration of a weak base (NH_3) with a strong acid (HCl). This titration curve, shown in Figure 19.9, has the *same shape as the weak acid–strong base curve, but it is inverted.*

Thus, we have the same features as the weak acid–strong base curve, but *the pH decreases* throughout the process:

1. *The initial weak-base solution has a pH above 7.00* (see Sample Problem 18.10).
2. *The pH decreases gradually in the buffer region,* where significant amounts of NH_3 and its conjugate acid, NH_4^+, are present. At the midpoint of this region, *the pH equals the pK_a of NH_4^+.* (Cl^- is from HCl.)
3. *The curve drops steeply to the equivalence point.* All the NH_3 has reacted with added HCl, and the solution contains only NH_4^+ and Cl^-. Note that *the pH at the equivalence point is below 7.00* because Cl^- does not react with water and NH_4^+ is acidic:
$$NH_4^+(aq) + H_2O(l) \rightleftharpoons NH_3(aq) + H_3O^+(aq)$$
4. *The pH decreases slowly* beyond the equivalence point as excess H_3O^+ is added.

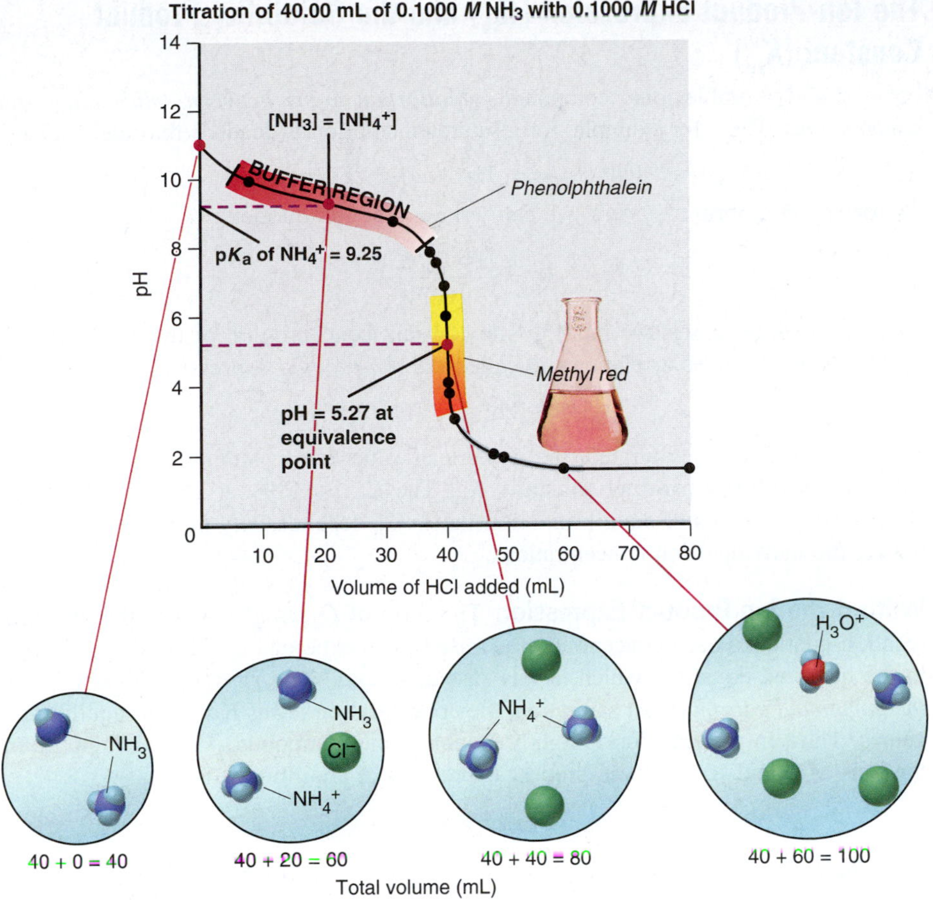

Figure 19.9 Curve for a weak base–strong acid titration.

For this titration, phenolphthalein changes color too slowly, but methyl red's change occurs on the steep portion of the curve and straddles the equivalence point, so it is the perfect choice.

Summary of Section 19.2

- An acid-base (pH) indicator is a weak acid that has a differently colored conjugate base form and changes color over about 2 pH units.
- In a strong acid–strong base titration, the pH starts out low, rises slowly, then shoots up near the equivalence point (pH = 7).
- In a weak acid–strong base titration, the pH starts out higher, rises slowly in the buffer region (pH = pK_a at the midpoint), and then rises quickly near the equivalence point (pH > 7).
- A weak base–strong acid titration curve has a shape that is the inverse of the weak acid–strong base case, with the pH decreasing to the equivalence point (pH < 7).

19.3 • EQUILIBRIA OF SLIGHTLY SOLUBLE IONIC COMPOUNDS

In this section, we explore an equilibrium system that involves the solubility of ionic compounds. Recall from Chapter 13 that most solutes, even those called "soluble," have a limited solubility in a particular solvent. In a saturated solution at a particular temperature, equilibrium exists between dissolved and undissolved solute. Slightly soluble ionic compounds, which we've been calling "insoluble," reach equilibrium with very little solute dissolved. In this introductory treatment, we will *assume* that, as with a soluble ionic compound, the small amount of a slightly soluble ionic compound that does dissolve dissociates completely into ions.

The Ion-Product Expression (Q_{sp}) and the Solubility-Product Constant (K_{sp})

For a slightly soluble ionic compound, *equilibrium exists between solid solute and aqueous ions*. Thus, for example, for a saturated solution of lead(II) fluoride, we have

$$PbF_2(s) \rightleftharpoons Pb^{2+}(aq) + 2F^-(aq)$$

As for any equilibrium system, we can write a reaction quotient:

$$Q_c = \frac{[Pb^{2+}][F^-]^2}{[PbF_2]}$$

And, as in earlier cases, we multiply the constant concentration of the solid, $[PbF_2]$, by Q_c. Doing so gives a relationship called the *ion-product expression,* Q_{sp}:

$$Q_{sp} = Q_c[PbF_2] = [Pb^{2+}][F^-]^2$$

When the solution is saturated, the numerical value of Q_{sp} attains a constant value called the **solubility-product constant, K_{sp}.** The K_{sp} for PbF_2 at 25°C, for example, is 3.6×10^{-8}. Like other equilibrium constants, K_{sp} depends *only* on the temperature, not on the individual ion concentrations.

Writing the Ion-Product Expression The form of Q_{sp} is identical to that of other reaction quotients: each concentration is raised to an exponent equal to the coefficient in the balanced equation, which in this case also *equals the subscript of each ion in the compound's formula.* At saturation, the concentration terms have their equilibrium values. Thus, in general, for a slightly soluble ionic compound, M_pX_q, composed of the ions M^{n+} and X^{z-}, the ion-product expression at equilibrium is

$$Q_{sp} = [M^{n+}]^p[X^{z-}]^q = K_{sp} \qquad (19.2)$$

From here on, we'll equate the ion-product expression directly with K_{sp}.

The Special Case of Metal Sulfides The aqueous equilibria of metal sulfides have a special form of the ion-product expression. The sulfide ion, S^{2-}, is so basic that it reacts completely with water to form the hydrogen sulfide ion (HS^-) and OH^-:

$$S^{2-}(aq) + H_2O(l) \longrightarrow HS^-(aq) + OH^-(aq)$$

Although S^{2-} is not stable in water, we can think of the dissolution process as the sum of two steps, with S^{2-} formed in the first and consumed immediately in the second. Thus, for instance, for manganese(II) sulfide, we have

$$MnS(s) \rightleftharpoons Mn^{2+}(aq) + \cancel{S^{2-}(aq)}$$
$$\cancel{S^{2-}(aq)} + H_2O(l) \longrightarrow HS^-(aq) + OH^-(aq)$$
$$MnS(s) + H_2O(l) \rightleftharpoons Mn^{2+}(aq) + HS^-(aq) + OH^-(aq)$$

Note that we have constant concentrations of water and the solid, so incorporating them into Q_c gives the ion-product expression at equilibrium:

$$Q_c[MnS][H_2O] = K_{sp} = [Mn^{2+}][HS^-][OH^-]$$

Sample Problem 19.5 Writing Ion-Product Expressions

Problem Write the ion-product expression at equilibrium for each compound:

(a) Magnesium carbonate **(b)** Iron(II) hydroxide
(c) Calcium phosphate **(d)** Silver sulfide

Plan We write an equation for a saturated solution and then write the ion-product expression at equilibrium, K_{sp} (Equation 19.2), noting the sulfide in part (d).

Solution (a) Magnesium carbonate:

$$MgCO_3(s) \rightleftharpoons Mg^{2+}(aq) + CO_3^{2-}(aq) \qquad K_{sp} = [Mg^{2+}][CO_3^{2-}]$$

(b) Iron(II) hydroxide:

$$Fe(OH)_2(s) \rightleftharpoons Fe^{2+}(aq) + 2OH^-(aq) \qquad K_{sp} = [Fe^{2+}][OH^-]^2$$

(c) Calcium phosphate:

$$Ca_3(PO_4)_2(s) \rightleftharpoons 3Ca^{2+}(aq) + 2PO_4^{3-}(aq) \qquad K_{sp} = [Ca^{2+}]^3[PO_4^{3-}]^2$$

(d) Silver sulfide:

$$Ag_2S(s) \rightleftharpoons 2Ag^+(aq) + \cancel{S^{2-}(aq)}$$
$$\cancel{S^{2-}(aq)} + H_2O(l) \longrightarrow HS^-(aq) + OH^-(aq)$$
$$Ag_2S(s) + H_2O(l) \rightleftharpoons 2Ag^+(aq) + HS^-(aq) + OH^-(aq) \qquad K_{sp} = [Ag^+]^2[HS^-][OH^-]$$

Check Except for part (d), you can check by reversing the process to see if you obtain the formula of the compound from K_{sp}.

Comment In part (d), we include H_2O as a reactant to obtain a balanced equation.

FOLLOW-UP PROBLEM 19.5 Write the ion-product expression at equilibrium for:
(a) Calcium sulfate **(b)** Chromium(III) carbonate
(c) Magnesium hydroxide **(d)** Arsenic(III) sulfide

Calculations Involving the Solubility-Product Constant

In Chapters 17 and 18, we described two types of equilibrium problems. In one type, we use concentrations (or other quantities) to find K, and in the other, we use K to find concentrations. Here we encounter the same two types.

The Problem with Assuming Complete Dissociation Before we focus on calculations, let's address a complication that affects accuracy, resulting in approximate answers. We've been assuming that the small amount of slightly soluble ionic compounds that does dissolve dissociates completely into separate ions, but this is an oversimplification. Many slightly soluble salts have polar covalent metal-nonmetal bonds (Section 9.5), and partially dissociated or even undissociated species occur in solution. Here are two of many examples:

- With slightly soluble lead(II) chloride in water, the solution contains not only the separate $Pb^{2+}(aq)$ and $Cl^-(aq)$ ions we expect from complete dissociation, but also undissociated $PbCl_2(aq)$ molecules and $PbCl^+(aq)$ ions.
- In an aqueous solution of $CaSO_4$, undissociated ion pairs, $Ca^{2+}SO_4^{2-}(aq)$, are present.

These partly dissociated and undissociated species increase the *actual* solubility of a slightly soluble salt above the value obtained by assuming complete dissociation. More advanced courses discuss these factors, but this text simply mentions them in the Comments of several sample problems. Thus, it is best to view the specific answers to the calculations as first approximations.

The Meaning of K_{sp} and How to Determine It from Solubility Values The K_{sp} value indicates *how far the dissolution proceeds at equilibrium (saturation).* Table 19.2 presents a few K_{sp} values; Appendix C lists many more. Note that all the values are low, but they range over many orders of magnitude. In Sample Problem 19.6, we find K_{sp} from the solubility of a compound.

Table 19.2 Solubility-Product Constants (K_{sp}) of Selected Ionic Compounds at 25°C

Name, Formula	K_{sp}
Aluminum hydroxide, $Al(OH)_3$	3×10^{-34}
Cobalt(II) carbonate, $CoCO_3$	1.0×10^{-10}
Iron(II) hydroxide, $Fe(OH)_2$	4.1×10^{-15}
Lead(II) fluoride, PbF_2	3.6×10^{-8}
Lead(II) sulfate, $PbSO_4$	1.6×10^{-8}
Mercury(I) iodide, Hg_2I_2	4.7×10^{-29}
Silver sulfide, Ag_2S	8×10^{-48}
Zinc iodate, $Zn(IO_3)_2$	3.9×10^{-6}

Sample Problem 19.6 Determining K_{sp} from Solubility

Problem (a) Lead(II) sulfate ($PbSO_4$) is a key component in lead-acid car batteries. Its solubility in water at 25°C is 4.25×10^{-3} g/100 mL solution. What is the K_{sp} of $PbSO_4$?
(b) When lead(II) fluoride (PbF_2) is shaken with pure water at 25°C, the solubility is found to be 0.64 g/L. Calculate the K_{sp} of PbF_2.

Plan We are given the solubilities in various units and must find K_{sp}. As always, we write a dissolution equation and the ion-product expression for each compound. This tells us the number of moles of each ion. We use the molar mass to convert the solubility of the compound from the given mass units to *molar solubility* (molarity), then use it to find the molarity of each ion, and substitute into the ion-product expression to calculate K_{sp}.

Solution (a) For $PbSO_4$. Writing the equation and ion-product (K_{sp}) expression:

$$PbSO_4(s) \rightleftharpoons Pb^{2+}(aq) + SO_4^{2-}(aq) \qquad K_{sp} = [Pb^{2+}][SO_4^{2-}]$$

Converting solubility to molar solubility:

$$\text{Molar solubility of } PbSO_4 = \frac{0.00425 \text{ g } PbSO_4}{100 \text{ mL soln}} \times \frac{1000 \text{ mL}}{1 \text{ L}} \times \frac{1 \text{ mol } PbSO_4}{303.3 \text{ g } PbSO_4}$$
$$= 1.40\times10^{-4}\ M\ PbSO_4$$

Determining molarities of the ions: Because 1 mol of Pb^{2+} and 1 mol of SO_4^{2-} form when 1 mol of $PbSO_4$ dissolves, $[Pb^{2+}] = [SO_4^{2-}] = 1.40\times10^{-4}\ M$.

Substituting these values into the ion-product expression to calculate K_{sp}:

$$K_{sp} = [Pb^{2+}][SO_4^{2-}] = (1.40\times10^{-4})^2 = 1.96\times10^{-8}$$

(b) For PbF_2. Writing the equation and K_{sp} expression:

$$PbF_2(s) \rightleftharpoons Pb^{2+}(aq) + 2F^-(aq) \qquad K_{sp} = [Pb^{2+}][F^-]^2$$

Converting solubility to molar solubility:

$$\text{Molar solubility of } PbF_2 = \frac{0.64 \text{ g } PbF_2}{1 \text{ L soln}} \times \frac{1 \text{ mol } PbF_2}{245.2 \text{ g } PbF_2} = 2.6\times10^{-3}\ M\ PbF_2$$

Determining molarities of the ions: Since 1 mol of Pb^{2+} and 2 mol of F^- form when 1 mol of PbF_2 dissolves, we have

$$[Pb^{2+}] = 2.6\times10^{-3}\ M \qquad \text{and} \qquad [F^-] = 2(2.6\times10^{-3}\ M) = 5.2\times10^{-3}\ M$$

Substituting these values into the ion-product expression to calculate K_{sp}:

$$K_{sp} = [Pb^{2+}][F^-]^2 = (2.6\times10^{-3})(5.2\times10^{-3})^2 = 7.0\times10^{-8}$$

Check The low solubilities are consistent with K_{sp} values being small. **(a)** The molar solubility seems about right: $\sim \dfrac{4\times10^{-2}\text{ g/L}}{3\times10^2\text{ g/mol}} \approx 1.3\times10^{-4}\ M$. Squaring this number gives 1.7×10^{-8}, close to the calculated K_{sp}. **(b)** Let's check the math in the final step as follows: $\sim(3\times10^{-3})(5\times10^{-3})^2 = 7.5\times10^{-8}$, close to the calculated K_{sp}.

Comment **1.** In part (b), the formula PbF_2 means that $[F^-]$ is twice $[Pb^{2+}]$. Substituting into the ion-product expression, we square this value of $[F^-]$.

2. The tabulated K_{sp} values for these compounds (Table 19.2) are lower than our calculated values. For PbF_2, for instance, the tabulated value is 3.6×10^{-8}, but we calculated 7.0×10^{-8} from solubility data. The discrepancy arises because we assumed that PbF_2 in solution dissociates completely to Pb^{2+} and F^-. Here is an example of the complication pointed out earlier. Actually, about a third of the PbF_2 dissolves as $PbF^+(aq)$ and a small amount dissolves as undissociated $PbF_2(aq)$. The solubility given in the problem statement (0.64 g/L) is determined experimentally and includes these other species, which we did *not* include in our calculation.

Fluorite.

FOLLOW-UP PROBLEM 19.6 When fluorite (CaF_2; *see photo*) is pulverized and shaken in water at 18°C, 10.0 mL of solution contains 1.5×10^{-4} g of solute. Find the K_{sp} of CaF_2 at 18°C.

Determining Solubility from K_{sp} The reverse of Sample Problem 19.6 involves finding the solubility of a compound based on its formula and K_{sp} value. We'll use an approach similar to the one we used for weak acids in Sample Problem 18.8: we define the unknown amount dissolved—the molar solubility—as *S*, include ion concentrations in terms of this unknown in a reaction table, and solve for *S*.

Sample Problem 19.7 Determining Solubility from K_{sp}

Problem Calcium hydroxide (slaked lime) is a major component of mortar, plaster, and cement, and solutions of $Ca(OH)_2$ are used in industry as a strong, inexpensive base. Calculate the molar solubility of $Ca(OH)_2$ in water if K_{sp} is 6.5×10^{-6}.

Plan We write the dissolution equation and the ion-product expression. We know K_{sp} (6.5×10^{-6}), so to find molar solubility (*S*), we set up a reaction table that expresses $[Ca^{2+}]$ and $[OH^-]$ in terms of *S*, substitute into the ion-product expression, and solve for *S*.

Solution Writing the equation and ion-product expression:

$$Ca(OH)_2(s) \rightleftharpoons Ca^{2+}(aq) + 2OH^-(aq) \qquad K_{sp} = [Ca^{2+}][OH^-]^2 = 6.5\times10^{-6}$$

Setting up a reaction table, with S = molar solubility:

Concentration (M)	$Ca(OH)_2(s)$	$\rightleftharpoons$	$Ca^{2+}(aq)$	+	$2OH^-(aq)$
Initial	—		0		0
Change	—		$+S$		$+2S$
Equilibrium	—		S		$2S$

Substituting into the ion-product expression and solving for S:

$$K_{sp} = [Ca^{2+}][OH^-]^2 = (S)(2S)^2 = (S)(4S^2) = 4S^3 = 6.5\times10^{-6}$$

$$S = \sqrt[3]{\frac{6.5\times10^{-6}}{4}} = 1.2\times10^{-2}\ M$$

Check We expect a low solubility from a slightly soluble salt. If we reverse the calculation, we should obtain the given K_{sp}: $4(1.2\times10^{-2})^3 = 6.9\times10^{-6}$, close to 6.5×10^{-6}.

Comment **1.** Note that we did not double and *then* square $[OH^-]$. $2S$ *is* the $[OH^-]$, so we just squared it, as the ion-product expression specified.
2. Once again, we assumed that the solid dissociates completely. Actually, the solubility is increased to about $2.0\times10^{-2}\ M$ by the presence of $CaOH^+(aq)$ formed in the reaction $Ca(OH)_2(s) \rightleftharpoons CaOH^+(aq) + OH^-(aq)$. Our calculated answer is only approximate because we did not take this other species into account.

FOLLOW-UP PROBLEM 19.7 Milk of magnesia, a suspension of $Mg(OH)_2$ in water, relieves indigestion by neutralizing stomach acid. What is the molar solubility of $Mg(OH)_2$ ($K_{sp} = 6.3\times10^{-10}$) in water?

Using K_{sp} Values to Compare Solubilities As long as we compare compounds with the *same total number of ions* in their formulas, K_{sp} values indicate *relative* solubility: *the higher the K_{sp}, the greater the solubility* (Table 19.3). Note that for compounds that form three ions, the relationship holds whether the cation/anion ratio is 1/2 or 2/1, because the mathematical expression containing S is the same ($4S^3$) in the calculation (see Sample Problem 19.7).

Table 19.3 Relationship Between K_{sp} and Solubility at 25°C

No. of Ions	Formula	Cation/Anion	K_{sp}	Solubility (M)
2	$MgCO_3$	1/1	3.5×10^{-8}	1.9×10^{-4}
2	$PbSO_4$	1/1	1.6×10^{-8}	1.3×10^{-4}
2	$BaCrO_4$	1/1	2.1×10^{-10}	1.4×10^{-5}
3	$Ca(OH)_2$	1/2	6.5×10^{-6}	1.2×10^{-2}
3	BaF_2	1/2	1.5×10^{-6}	7.2×10^{-3}
3	CaF_2	1/2	3.2×10^{-11}	2.0×10^{-4}
3	Ag_2CrO_4	2/1	2.6×10^{-12}	8.7×10^{-5}

Effect of a Common Ion on Solubility

From Le Châtelier's principle (Section 17.6), we know that *adding a common ion decreases the solubility of a slightly soluble ionic compound.* Consider a saturated solution of lead(II) chromate:

$$PbCrO_4(s) \rightleftharpoons Pb^{2+}(aq) + CrO_4^{2-}(aq) \qquad K_{sp} = [Pb^{2+}][CrO_4^{2-}] = 2.3\times10^{-13}$$

At a given temperature, K_{sp} depends on the product of the ion concentrations. If the concentration of either ion goes up, the other goes down to maintain K_{sp}. Suppose we add Na_2CrO_4, a soluble salt, to the saturated $PbCrO_4$ solution. The concentration of

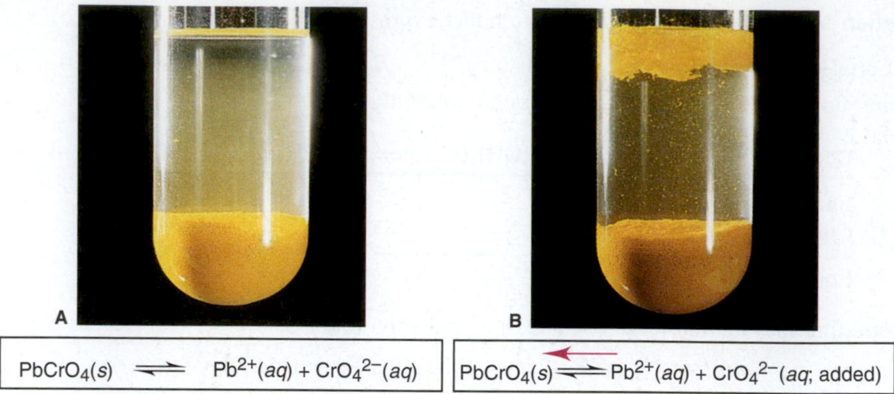

Figure 19.10 The effect of a common ion on solubility. **A,** Lead(II) chromate, a slightly soluble salt, forms a saturated solution. **B,** When Na_2CrO_4 solution is added, the amount of $PbCrO_4(s)$ increases.

the common ion, CrO_4^{2-}, increases, and some of it combines with Pb^{2+} ion to form more solid $PbCrO_4$ (Figure 19.10). That is, the equilibrium position shifts to the left:

$$PbCrO_4(s) \rightleftharpoons Pb^{2+}(aq) + CrO_4^{2-}(aq; \text{added})$$

As a result of the addition, $[Pb^{2+}]$ is lower. And, since $[Pb^{2+}]$ defines the solubility of $PbCrO_4$, in effect, the solubility of $PbCrO_4$ has decreased. Note that we would get the same result if Na_2CrO_4 solution were the solvent; that is, $PbCrO_4$ is more soluble in water than in aqueous Na_2CrO_4.

Sample Problem 19.8 Calculating the Effect of a Common Ion on Solubility

Problem In Sample Problem 19.7, we calculated the solubility of $Ca(OH)_2$ in water. What is its solubility in 0.10 M $Ca(NO_3)_2$? K_{sp} of $Ca(OH)_2$ is 6.5×10^{-6}.

Plan Addition of Ca^{2+}, the common ion, should lower the solubility. We write the equation and ion-product expression and set up a reaction table, with $[Ca^{2+}]_{init}$ reflecting the 0.10 M $Ca(NO_3)_2$ and S equal to $[Ca^{2+}]_{\text{from } Ca(OH)_2}$. To simplify the math, we assume that, because K_{sp} is low, S is so small relative to $[Ca^{2+}]_{init}$ that it can be neglected. Then we solve for S and check the assumption.

Solution Writing the equation and ion-product expression:

$$Ca(OH)_2(s) \rightleftharpoons Ca^{2+}(aq) + 2OH^-(aq) \qquad K_{sp} = [Ca^{2+}][OH^-]^2 = 6.5\times10^{-6}$$

Setting up the reaction table, with $S = [Ca^{2+}]_{\text{from } Ca(OH)_2}$:

Concentration (M)	$Ca(OH)_2(s)$	$\rightleftharpoons$	$Ca^{2+}(aq)$	+	$2OH^-(aq)$
Initial	—		0.10		0
Change	—		$+S$		$+2S$
Equilibrium	—		$0.10 + S$		$2S$

Making the assumption: K_{sp} is small, so $S << 0.10\ M$; thus, $0.10\ M + S \approx 0.10\ M$. Substituting into the ion-product expression and solving for S:

$$K_{sp} = [Ca^{2+}][OH^-]^2 = 6.5\times10^{-6} \approx (0.10)(2S)^2$$

Therefore, $$4S^2 \approx \frac{6.5\times10^{-6}}{0.10} \quad \text{so} \quad S \approx \sqrt{\frac{6.5\times10^{-5}}{4}} = 4.0\times10^{-3}\ M$$

Checking the assumption: $$\frac{4.0\times10^{-3}\ M}{0.10\ M} \times 100 = 4.0\% < 5\%$$

Check In Sample Problem 19.7, the solubility of $Ca(OH)_2$ was 0.012 M; here it is 0.0040 M, one-third as much. As expected, the solubility *decreased* in the presence of added Ca^{2+}, the common ion.

FOLLOW-UP PROBLEM 19.8 To improve the x-ray image used to diagnose intestinal disorders, a patient drinks an aqueous suspension of $BaSO_4$ before the procedure because Ba^{2+} is opaque to x-rays. However, Ba^{2+} is also toxic; thus, $[Ba^{2+}]$ is lowered by adding dilute Na_2SO_4. What is the solubility of $BaSO_4$ ($K_{sp} = 1.1\times10^{-10}$) in **(a)** water and **(b)** 0.10 M Na_2SO_4?

Effect of pH on Solubility

If a slightly soluble ionic compound contains the anion of a weak acid, *addition of H_3O^+ (from a strong acid) increases its solubility.* Once again, Le Châtelier's principle explains why. Consider a saturated solution of calcium carbonate:

$$CaCO_3(s) \rightleftharpoons Ca^{2+}(aq) + CO_3^{2-}(aq)$$

Adding strong acid introduces H_3O^+, which reacts with the anion of a weak acid, CO_3^{2-}, to form the anion of another weak acid, HCO_3^-:

$$CO_3^{2-}(aq) + H_3O^+(aq) \longrightarrow HCO_3^-(aq) + H_2O(l)$$

If enough H_3O^+ is added, carbonic acid forms, which decomposes to H_2O and CO_2, and the gas escapes the container:

$$HCO_3^-(aq) + H_3O^+(aq) \longrightarrow [H_2CO_3(aq)] + H_2O(l) \longrightarrow CO_2(g) + 2H_2O(l)$$

The net effect of adding H_3O^+ is a shift in the equilibrium position to the right, and more $CaCO_3$ dissolves:

$$CaCO_3(s) \rightleftharpoons Ca^{2+} + CO_3^{2-} \xrightarrow{H_3O^+} HCO_3^- \xrightarrow{H_3O^+} [H_2CO_3] \longrightarrow CO_2(g) + H_2O + Ca^{2+}$$

This overall equation is the basis of a qualitative field test for carbonate minerals because the CO_2 bubbles vigorously (Figure 19.11). Far more importantly, it indicates the potential destructive effect of *ocean acidification* on vast expanses of coral reefs (composed of calcium carbonate) as increased atmospheric CO_2 dissolves in seawater and lowers its pH:

$$CO_2(g) + 2H_2O(l) \rightleftharpoons HCO_3^-(aq) + H_3O^+(aq)$$

In contrast, adding H_3O^+ to a saturated solution of a slightly soluble ionic compound with a strong-acid anion, such as AgCl, has no effect on its solubility:

$$AgCl(s) \rightleftharpoons Ag^+(aq) + Cl^-(aq)$$

The Cl^- ion can coexist with high $[H_3O^+]$, so the equilibrium position is not affected.

Figure 19.11 **Test for the presence of a carbonate.** When a carbonate mineral is treated with HCl, bubbles of CO_2 form.

Sample Problem 19.9 Predicting the Effect on Solubility of Adding Strong Acid

Problem Write balanced equations to explain whether addition of H_3O^+ from a strong acid affects the solubility of each ionic compound:
(a) Lead(II) bromide **(b)** Copper(II) hydroxide **(c)** Iron(II) sulfide

Plan We write the balanced dissolution equation and note the anion:
- Weak-acid anions react with H_3O^+ and increase solubility when strong acid is added.
- Strong-acid anions do not react with H_3O^+, so addition of strong acid has no effect.

Solution **(a)** $PbBr_2(s) \rightleftharpoons Pb^{2+}(aq) + 2Br^-(aq)$
No effect. Br^- is the anion of HBr, a strong acid, so it does not react with H_3O^+.

(b) $Cu(OH)_2(s) \rightleftharpoons Cu^2(aq) + 2OH^-(aq)$ Increases solubility. OH^- is the anion of H_2O, a very weak acid, so it reacts with the added H_3O^+:

$$OH^-(aq) + H_3O^+(aq) \longrightarrow 2H_2O(l)$$

(c) $FeS(s) + H_2O(l) \rightleftharpoons Fe^{2+}(aq) + HS^-(aq) + OH^-(aq)$ Increases solubility. The S^{2-} ion reacts completely with water to form HS^- and OH^-. The added H_3O^+ reacts with both of these weak-acid anions:

$$HS^-(aq) + H_3O^+(aq) \longrightarrow H_2S(aq) + H_2O(l)$$
$$OH^-(aq) + H_3O^+(aq) \longrightarrow 2H_2O(l)$$

FOLLOW-UP PROBLEM 19.9 Write balanced equations to show how addition of $HNO_3(aq)$ affects the solubility of: **(a)** calcium fluoride; **(b)** zinc sulfide; **(c)** silver iodide.

Predicting the Formation of a Precipitate: Q_{sp} vs. K_{sp}

As in Chapter 17, here we compare the values of Q_{sp} and K_{sp} to see if a reaction has reached equilibrium and, if not, in which net direction it will move until it does.

Using solutions of *soluble* salts that contain the ions of *slightly* soluble salts, we can calculate the ion concentrations and predict the result when we mix the solutions:

- If $Q_{sp} = K_{sp}$, the solution is saturated and no change will occur.
- If $Q_{sp} > K_{sp}$, a precipitate will form until the remaining solution is saturated.
- If $Q_{sp} < K_{sp}$, no precipitate will form because the solution is unsaturated.

Sample Problem 19.10 Predicting Whether a Precipitate Will Form

Problem A common laboratory method for preparing a precipitate is to mix solutions containing the component ions. Does a precipitate form when 0.100 L of 0.30 *M* $Ca(NO_3)_2$ is mixed with 0.200 L of 0.060 *M* NaF?

Plan First, we decide which slightly soluble salt could form, look up its K_{sp} value in Appendix C, and write a dissolution equation and ion-product expression. To see whether mixing these solutions forms the precipitate, we find the initial ion concentrations by calculating the amount (mol) of each ion from its concentration and volume, and dividing by the *total* volume because each solution dilutes the other. Finally, we substitute these concentrations to calculate Q_{sp}, and compare Q_{sp} with K_{sp}.

Precipitation of CaF_2.

Solution The ions present are Ca^{2+}, Na^+, F^-, and NO_3^-. All sodium and all nitrate salts are soluble (Table 4.1), so the only possibility is CaF_2 ($K_{sp} = 3.2\times10^{-11}$). Writing the equation and ion-product expression:

$$CaF_2(s) \rightleftharpoons Ca^{2+}(aq) + 2F^-(aq) \qquad Q_{sp} = [Ca^{2+}][F^-]^2$$

Calculating the ion concentrations:

$$\text{Amount (mol) of } Ca^{2+} = 0.30\ M\ Ca^{2+} \times 0.100\ L = 0.030\ \text{mol } Ca^{2+}$$

$$[Ca^{2+}]_{init} = \frac{0.030\ \text{mol } Ca^{2+}}{0.100\ L + 0.200\ L} = 0.10\ M\ Ca^{2+}$$

$$\text{Amount (mol) of } F^- = 0.060\ M\ F^- \times 0.200\ L = 0.012\ \text{mol } F^-$$

$$[F^-]_{init} = \frac{0.012\ \text{mol } F^-}{0.100\ L + 0.200\ L} = 0.040\ M\ F^-$$

Substituting into the ion-product expression and comparing Q_{sp} with K_{sp}:

$$Q_{sp} = [Ca^{2+}]_{init}[F^-]^2_{init} = (0.10)(0.040)^2 = 1.6\times10^{-4}$$

Because $Q_{sp} > K_{sp}$, CaF_2 will precipitate until $Q_{sp} = 3.2\times10^{-11}$.

Check Make sure you round off and quickly check the math. For example, $Q_{sp} = (1\times10^{-1})(4\times10^{-2})^2 = 1.6\times10^{-4}$. With K_{sp} so low, CaF_2 must have a low solubility, and given the sizable concentrations being mixed, we would expect CaF_2 to precipitate.

FOLLOW-UP PROBLEM 19.10 As a result of mineral erosion and biological activity, phosphate ion is common in natural waters, where it often precipitates as insoluble salts such as $Ca_3(PO_4)_2$. If $[Ca^{2+}]_{init} = [PO_4^{3-}]_{init} = 1.0\times10^{-9}$ *M* in a given river, will $Ca_3(PO_4)_2$ precipitate? K_{sp} of $Ca_3(PO_4)_2$ is 1.2×10^{-29}.

Sample Problem 19.11 Using Molecular Scenes to Predict Whether a Precipitate Will Form

Problem The four scenes below represent solutions of silver *(gray)* and carbonate *(black and red)* ions above solid silver carbonate. (The solid, other ions, and water are not shown.)

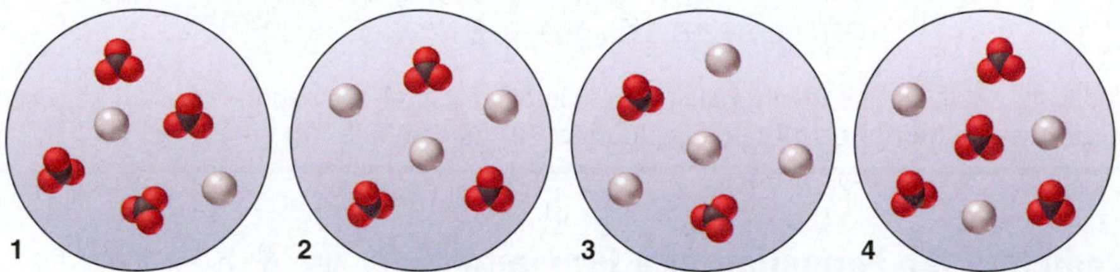

(a) Assuming the ions come *only* from the solid, which scene best represents the solution in equilibrium with the solid?
(b) In which, if any, other scene(s) would additional solid silver carbonate form?
(c) Explain how, if at all, addition of a small volume of concentrated strong acid affects the $[Ag^+]$ in scene 4 and the mass of solid present.

Plan **(a)** The solution of silver and carbonate ions in equilibrium with the solid (Ag_2CO_3) should have the same relative numbers of cations and anions as in the formula. We examine the scenes to see which has a ratio of 2 Ag^+ to 1 CO_3^{2-}. **(b)** A solid forms if the value of Q_{sp} exceeds the value of K_{sp}. We write the dissolution equation and Q_{sp} expression. Then we count ions to calculate Q_{sp} in each scene and see which, if any, exceeds the value for the solution from part (a). **(c)** The CO_3^{2-} ion reacts with added H_3O^+, so adding strong acid will shift the equilibrium to the right. We write the equations and determine how a shift to the right affects $[Ag^+]$ and the mass of solid Ag_2CO_3.

Solution **(a)** Scene 3 is the only one with an Ag^+/CO_3^{2-} ratio of 2/1, as in the solid's formula.
(b) Calculating the ion products:

$$Ag_2CO_3(s) \rightleftharpoons 2Ag^+(aq) + CO_3^{2-}(aq) \qquad Q_{sp} = [Ag^+]^2[CO_3^{2-}]$$

Scene 1: $Q_{sp} = (2)^2(4) = 16$ Scene 2: $Q_{sp} = (3)^2(3) = 27$
Scene 3: $Q_{sp} = (4)^2(2) = 32$ Scene 4: $Q_{sp} = (3)^2(4) = 36$

Therefore, from scene 3, $K_{sp} = 32$; the Q_{sp} value for scene 4 is the only other one that equals or exceeds 32, so a precipitate of Ag_2CO_3 will form there.
(c) Writing the equations:

$$(1) \qquad Ag_2CO_3(s) \rightleftharpoons 2Ag^+(aq) + CO_3^{2-}(aq)$$

$$(2) \quad CO_3^{2-}(aq) + 2H_3O^+(aq) \longrightarrow [H_2CO_3(aq)] + 2H_2O(l) \longrightarrow 3H_2O(l) + CO_2(g)$$

The CO_2 leaves as a gas, so adding H_3O^+ shifts the equilibrium position of reaction 2 to the right. This change lowers the $[CO_3^{2-}]$ in reaction 1, thereby causing more CO_3^{2-} to form. As a result, more solid dissolves, which means that the $[Ag^+]$ increases and the mass of Ag_2CO_3 decreases.

Check **(a)** In scene 1, the formula has two CO_3^{2-} per formula unit, not two Ag^+.
(b) Even though scene 4 has fewer Ag^+ ions than scene 3, its Q_{sp} value is higher and exceeds the K_{sp}.

FOLLOW-UP PROBLEM 19.11 The following scenes represent solutions of nickel(II) *(black)* and hydroxide *(red and blue)* ions that are above solid nickel(II) hydroxide. (For clarity, the solid, other ions, and water are not shown.)

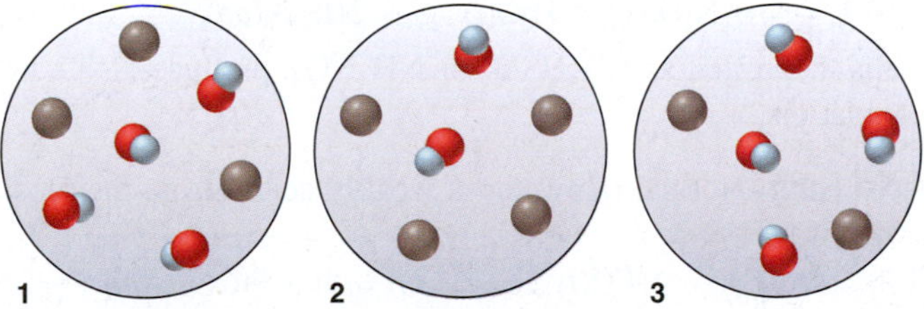

(a) Assuming the ions come only from the solid, which scene best depicts the solution at equilibrium with the solid?
(b) In which, if any, other scene(s) would additional solid form?
(c) Would addition of a small amount of concentrated strong acid or strong base affect the mass of solid present in any scene? Explain.

Ionic Equilibria and the Acid-Rain Problem

Acid rain, the deposition of acids in wet form as rain, snow, or fog or in dry form as solid particles, is a global environmental problem that involves several ionic equilibria. It has been observed in the countries of North America, the Amazon basin, Europe, Russia, many parts of Asia, and even at the North and South Poles. Let's see how it arises and how to prevent some of its harmful effects.

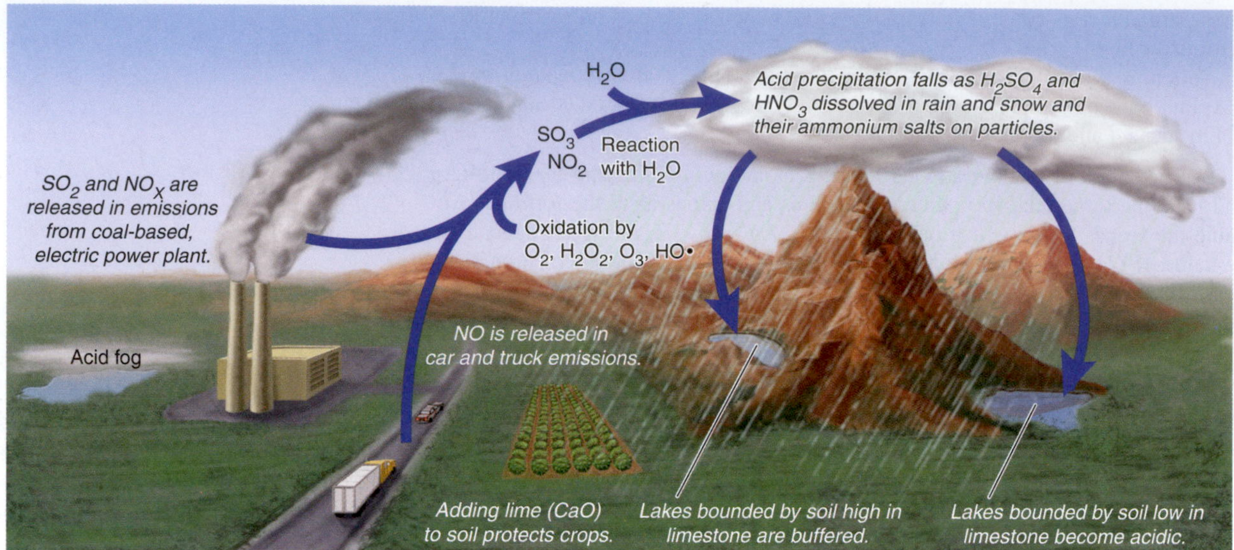

Figure 19.12 Formation of acidic precipitation. A complex interplay of human activities, atmospheric chemistry, and environmental distribution leads to acidic precipitation and its harmful effects.

Origins of Acid Rain The strong acids H_2SO_4 and HNO_3 cause the greatest concern, so let's see how they form (Figure 19.12):

1. *Sulfuric acid.* Sulfur dioxide (SO_2), formed mostly by the burning of high-sulfur coal, forms sulfurous acid (H_2SO_3) in contact with water. The atmospheric pollutants hydrogen peroxide (H_2O_2) and ozone (O_3) dissolve in the water in clouds and oxidize the sulfurous acid to sulfuric acid:

$$H_2O_2(aq) + H_2SO_3(aq) \longrightarrow H_2SO_4(aq) + H_2O(l)$$

Alternatively, SO_2 is oxidized in the atmosphere to sulfur trioxide (SO_3), which forms H_2SO_4 with water.

2. *Nitric acid.* Nitrogen oxides (NO_x) form when N_2 and O_2 react in car and truck engines and electric power plants. During the day, NO forms NO_2 and HNO_3 in a process that creates smog. At night, NO_x is converted to N_2O_5, which reacts with water to form HNO_3:

$$N_2O_5(g) + H_2O(l) \longrightarrow 2HNO_3(aq)$$

Ammonium salts, deposited as NH_4HSO_4 or NH_4NO_3, produce HNO_3 in soil through biochemical oxidation.

The pH of Acid Rain Normal rainwater is weakly acidic from the reaction of atmospheric CO_2 with water:

$$CO_2(g) + 2H_2O(l) \rightleftharpoons H_3O^+(aq) + HCO_3^-(aq)$$

Yet, as long ago as 1984, rainfall in the United States had already reached an average pH of 4.2. Rain in West Virginia has had a pH of 1.8, lower than the pH of lemon juice. And rain in industrial parts of Sweden once had a pH of 2.7, about the same as vinegar.

Effects of Acid Rain These 10- to 10,000-fold excesses of $[H_3O^+]$ are very destructive to fish (many species die at a pH below 5) and to forests. The aluminosilicates that make up most soils are nearly insoluble. But contact with H_3O^+ dissolves some of the bound Al^{3+}, which is extremely toxic to fish, and many ions that act as nutrients for plants and animals, such as Ca^{2+} and Mg^{2+}, are dissolved and carried away.

Marble and limestone (both primarily $CaCO_3$) in buildings and monuments react with sulfuric acid to form gypsum ($CaSO_4 \cdot 2H_2O$), which flakes off. Ironically, the same process that destroys these structures rescues lakes that are bounded by limestone-rich soil. Limestone dissolves sufficiently in lake water to form an HCO_3^-/CO_3^{2-} buffer capable of absorbing the incoming H_3O^+ and maintaining a mildly basic pH:

$$CO_3^{2-}(aq) + H_3O^+(aq) \rightleftharpoons HCO_3^-(aq) + H_2O(l)$$

Lakes and rivers in areas with limestone-poor bedrock can be deacidified by *liming* (treating with limestone). This approach is expensive (Sweden spent tens of millions of dollars for this purpose in the 1990s) and only a stopgap because the lakes become acidic again within several years.

Preventing Acid Rain Effective prevention of acid rain addresses the sources of the pollutants:

1. *Sulfur pollutants*. The principal means of minimizing SO_2 release is by "scrubbing" power-plant emissions with limestone. Another method reduces SO_2 with methane, coal, or H_2S, and the mixture is converted catalytically to sulfur, which is sold as a byproduct:

$$16H_2S(g) + 8SO_2(g) \xrightarrow{\text{catalyst}} 3S_8(s) + 16H_2O(l)$$

Also, coal can be converted into gaseous and liquid low-sulfur fuels.

2. *Nitrogen pollutants*. Through the use of a catalytic converter in an auto exhaust system, NO_x species are reduced to N_2. In power plant emissions, NO_x is decreased by treating hot stack gases with ammonia in the presence of a heterogeneous catalyst:

$$4NO(g) + 4NH_3(g) + O_2(g) \xrightarrow{\text{catalyst}} 4N_2(g) + 6H_2O(g)$$

Reducing power-plant emissions in North America and Europe has increased the pH of rainfall and some surface waters, and lime (CaO) is routinely used to protect croplands.

Summary of Section 19.3

- Only as a first approximation does the dissolved portion of a slightly soluble salt dissociate completely into ions.
- In a saturated solution, dissolved ions and the undissolved solid salt are in equilibrium. The product of the ion concentrations, each raised to the power of its subscript in the formula, has a constant value ($Q_{sp} = K_{sp}$).
- The value of K_{sp} can be obtained from the solubility, and vice versa.
- Adding a common ion lowers a compound's solubility.
- Adding H_3O^+ (lowering pH) increases a compound's solubility if the anion of the compound is also that of a weak acid.
- An ionic solid forms if $Q_{sp} > K_{sp}$ when solutions that each contain one of the ions are mixed.
- Acid rain is a global problem that involves several ionic equilibria. Lakes bounded by limestone-rich soils form buffer systems that prevent acidification.

19.4 • EQUILIBRIA INVOLVING COMPLEX IONS

A third kind of aqueous ionic equilibrium involves a type of ion mentioned briefly in Section 18.8. A *simple ion,* such as Na^+ or CH_3COO^-, consists of one or a few bonded atoms, with an excess or deficit of electrons. A **complex ion** consists of a central metal ion covalently bonded to two or more anions or molecules, called **ligands.** Hydroxide, chloride, and cyanide ions are examples of ionic ligands; water, carbon monoxide, and ammonia are some molecular ligands. In the complex ion $Cr(NH_3)_6^{3+}$, for example, the central Cr^{3+} is surrounded by six NH_3 ligands, giving an overall 3+ charge (Figure 19.13). Hydrated metal ions are complex ions with water as ligands (Section 18.6). In Chapter 22, we discuss the transition metals and the structures and properties of the numerous complex ions they form. Here, we focus on equilibria of hydrated ions with ligands other than water.

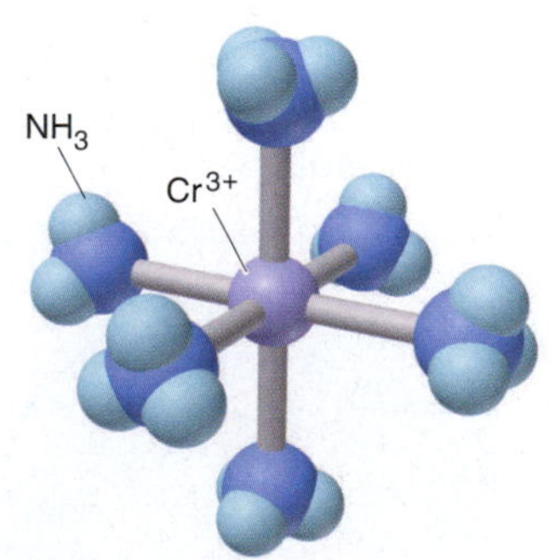

Figure 19.13 $Cr(NH_3)_6^{3+}$, a typical complex ion.

Formation of Complex Ions

When a salt like $M(NO_3)_2$ dissolves in water, a complex ion forms, with water as ligands around the metal ion. In many cases, when we treat this hydrated cation with a solution of another ligand, the bound water molecules are replaced. For example, a hydrated M^{2+} ion, $M(H_2O)_4^{2+}$, forms the ammoniated ion, $M(NH_3)_4^{2+}$, in aqueous NH_3:

$$M(H_2O)_4^{2+}(aq) + 4NH_3(aq) \rightleftharpoons M(NH_3)_4^{2+}(aq) + 4H_2O(l)$$

At equilibrium, this system is expressed by a ratio of concentration terms whose form follows that of any other equilibrium expression:

$$K_c = \frac{[M(NH_3)_4^{2+}][H_2O]^4}{[M(H_2O)_4^{2+}][NH_3]^4}$$

Once again, because the concentration of water is essentially constant in aqueous reactions, we incorporate it into K_c and obtain the expression for a new equilibrium constant called the **formation constant, K_f:**

$$K_f = \frac{K_c}{[H_2O]^4} = \frac{[M(NH_3)_4^{2+}]}{[M(H_2O)_4^{2+}][NH_3]^4}$$

Or, more generally, for the formation of a complex ion between the hydrated metal ion $M(H_2O)_x^{n+}$ and molecular ligand L, we have

$$M(H_2O)_x^{n+}(aq) + xL(aq) \rightleftharpoons ML_x^{n+}(aq) + xH_2O(l)$$

And,

$$K_f = \frac{[ML_x^{n+}]}{[M(H_2O)_x^{n+}][L]^x} \qquad \textbf{19.3}$$

At the molecular level, depicted in Figure 19.14, the actual process is stepwise, with ammonia molecules replacing water molecules one at a time. This process yields a series of intermediate species, each with its own formation constant:

$$M(H_2O)_4^{2+}(aq) + NH_3(aq) \rightleftharpoons M(H_2O)_3(NH_3)^{2+}(aq) + H_2O(l)$$

$$K_{f1} = \frac{[M(H_2O)_3(NH_3)^{2+}]}{[M(H_2O)_4^{2+}][NH_3]}$$

$$M(H_2O)_3(NH_3)^{2+}(aq) + NH_3(aq) \rightleftharpoons M(H_2O)_2(NH_3)_2^{2+}(aq) + H_2O(l)$$

$$K_{f2} = \frac{[M(H_2O)_2(NH_3)_2^{2+}]}{[M(H_2O)_3(NH_3)^{2+}][NH_3]}$$

$$M(H_2O)_2(NH_3)_2^{2+}(aq) + NH_3(aq) \rightleftharpoons M(H_2O)(NH_3)_3^{2+}(aq) + H_2O(l)$$

$$K_{f3} = \frac{[M(H_2O)(NH_3)_3^{2+}]}{[M(H_2O)_2(NH_3)_2^{2+}][NH_3]}$$

$$M(H_2O)(NH_3)_3^{2+}(aq) + NH_3(aq) \rightleftharpoons M(NH_3)_4^{2+}(aq) + H_2O(l)$$

$$K_{f4} = \frac{[M(NH_3)_4^{2+}]}{[M(H_2O)(NH_3)_3^{2+}][NH_3]}$$

The *sum* of the equations gives the overall equation (shown in Figure 19.14), so the *product* of the individual formation constants gives the overall formation constant:

$$K_f = K_{f1} \times K_{f2} \times K_{f3} \times K_{f4}$$

Figure 19.14 The stepwise exchange of NH_3 for H_2O in $M(H_2O)_4^{2+}$. The molecular views show the first exchange and the fully ammoniated ion.

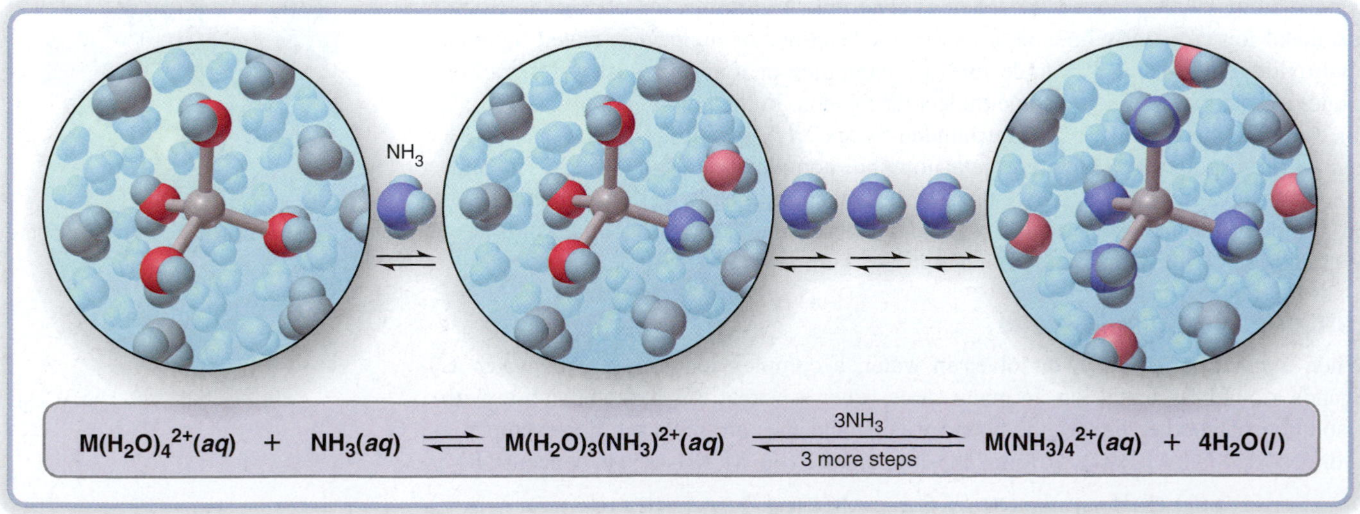

Recall that *all complex ions are Lewis adducts* (Section 18.8). The metal ion acts as a Lewis acid (accepts an electron pair), and the ligand acts as a Lewis base (donates an electron pair). In the formation of $M(NH_3)_4^{2+}$, the K_f for each step is much larger than 1 because ammonia is a stronger Lewis base than water. Therefore, if we add excess ammonia to the $M(H_2O)_4^{2+}$ solution, nearly all the M^{2+} ions exist as $M(NH_3)_4^{2+}(aq)$.

Appendix C shows K_f values of some complex ions. Notice that they are all 10^6 or greater, which means that the formation reactions proceed far to the right.

Complex Ions and the Solubility of Precipitates

In Section 19.3, you saw that H_3O^+ increases the solubility of a slightly soluble ionic compound if its anion is that of a weak acid. Similarly, *a ligand increases the solubility of a slightly soluble ionic compound if it forms a complex ion with the cation.* For example, iron(II) sulfide is very slightly soluble:

$$FeS(s) + H_2O(l) \rightleftharpoons Fe^{2+}(aq) + HS^-(aq) + OH^-(aq) \qquad K_{sp} = 8\times10^{-16}$$

When we add some 1.0 *M* NaCN, the CN^- ions act as ligands and react with the small amount of $Fe^{2+}(aq)$ to form the complex ion $Fe(CN)_6^{4-}$:

$$Fe^{2+}(aq) + 6CN^-(aq) \rightleftharpoons Fe(CN)_6^{4-}(aq) \qquad K_f = 3\times10^{35}$$

To see the effect of complex-ion formation on the solubility of FeS, we add the equations and, therefore, multiply their equilibrium constants:

$$FeS(s) + 6CN^-(aq) + H_2O(l) \rightleftharpoons Fe(CN)_6^{4-}(aq) + HS^-(aq) + OH^-(aq)$$

$$K_{overall} = K_{sp} \times K_f = (8\times10^{-16})(3\times10^{35}) = 2\times10^{20}$$

The overall dissociation of FeS into ions increased enormously in the presence of the ligand.

Sample Problem 19.12 Calculating the Effect of Complex-Ion Formation on Solubility

Problem In black-and-white film developing, excess AgBr is removed from a film negative with "hypo," an aqueous solution of sodium thiosulfate ($Na_2S_2O_3$), which forms the complex ion $Ag(S_2O_3)_2^{3-}$. Calculate the solubility of AgBr in **(a)** H_2O; **(b)** 1.0 *M* hypo. K_f of $Ag(S_2O_3)_2^{3-}$ is 4.7×10^{13} and K_{sp} of AgBr is 5.0×10^{-13}.

Plan **(a)** After writing the equation and the ion-product expression, we use the given K_{sp} to solve for *S*, the molar solubility of AgBr. **(b)** In hypo, Ag^+ forms a complex ion with $S_2O_3^{2-}$, which shifts the equilibrium and dissolves more AgBr. We write the complex-ion equation and add it to the equation for dissolving AgBr to obtain the overall equation for dissolving AgBr in hypo. We multiply K_{sp} by K_f to find $K_{overall}$. To find the solubility of AgBr in hypo, we set up a reaction table, with $S = [Ag(S_2O_3)_2^{3-}]$, substitute into the expression for $K_{overall}$, and solve for *S*.

Solution **(a)** Solubility in water. Writing the equation for the saturated solution and the ion-product expression:

$$AgBr(s) \rightleftharpoons Ag^+(aq) + Br^-(aq) \qquad K_{sp} = [Ag^+][Br^-]$$

Solving for solubility (*S*) directly from the equation: We know that

$$S = [AgBr]_{dissolved} = [Ag^+] = [Br^-]$$

Thus, $K_{sp} = [Ag^+][Br^-] = S^2 = 5.0\times10^{-13}$

so $S = 7.1\times10^{-7}\ M$

(b) Solubility in 1.0 *M* hypo. Writing the overall equation:

$$AgBr(s) \rightleftharpoons \cancel{Ag^+(aq)} + Br^-(aq)$$

$$\cancel{Ag^+(aq)} + 2S_2O_3^{2-}(aq) \rightleftharpoons Ag(S_2O_3)_2^{3-}(aq)$$

$$AgBr(s) + 2S_2O_3^{2-}(aq) \rightleftharpoons Ag(S_2O_3)_2^{3-}(aq) + Br^-(aq)$$

Calculating $K_{overall}$:

$$K_{overall} = \frac{[Ag(S_2O_3)_2^{3-}][Br^-]}{[S_2O_3^{2-}]^2} = K_{sp} \times K_f = (5.0\times10^{-13})(4.7\times10^{13}) = 24$$

Setting up a reaction table, with $S = [AgBr]_{dissolved} = [Ag(S_2O_3)_2^{3-}]$:

Concentration (M)	$AgBr(s)$	+ $2S_2O_3^{2-}(aq)$	⇌ $Ag(S_2O_3)_2^{3-}(aq)$	+ $Br^-(aq)$
Initial	—	1.0	0	0
Change	—	$-2S$	$+S$	$+S$
Equilibrium	—	$1.0 - 2S$	S	S

Substituting the values into the expression for $K_{overall}$ and solving for S:

$$K_{overall} = \frac{[Ag(S_2O_3)_2^{3-}][Br^-]}{[S_2O_3^{2-}]^2} = \frac{S^2}{(1.0\,M - 2S)^2} = 24$$

Taking the square root of both sides gives

$$\frac{S}{1.0\,M - 2S} = \sqrt{24} = 4.9 \quad \text{so} \quad S = 4.9\,M - 9.8S \quad \text{and} \quad 10.8S = 4.9\,M$$

$$[Ag(S_2O_3)_2^{3-}] = S = 0.45\,M$$

Check **(a)** From the number of ions in the formula of AgBr, we know that $S = \sqrt{K_{sp}}$, so the order of magnitude seems right: $\sim\sqrt{10^{-14}} \approx 10^{-7}$. **(b)** The $K_{overall}$ seems correct: the exponents cancel, and $5 \times 5 = 25$. Most importantly, the answer makes sense because the photographic process requires the remaining AgBr to be washed off the film and the large $K_{overall}$ confirms that. We can check S by rounding and working backward to find $K_{overall}$: from the reaction table, we find that

$$[(S_2O_3)^{2-}] = 1.0\,M - 2S = 1.0\,M - 2(0.45\,M) = 1.0\,M - 0.90\,M = 0.1\,M$$

so $K_{overall} \approx (0.45)^2/(0.1)^2 = 20$, within rounding of the calculated value.

FOLLOW-UP PROBLEM 19.12 How does the solubility of AgBr in 1.0 M NH_3 compare with its solubility in 1.0 M hypo? K_f of $Ag(NH_3)_2^+$ is 1.7×10^7.

Summary of Section 19.4

- A complex ion consists of a central metal ion covalently bonded to two or more negatively charged or neutral ligands. Its formation is described by a formation constant, K_f.
- A hydrated metal ion is a complex ion with water molecules as ligands. Other ligands (stronger Lewis bases) can displace the water in a stepwise process. In most cases, the K_f value of each step is large, so the fully substituted complex ion forms almost completely with excess ligand.
- Adding a solution containing a ligand increases the solubility of an ionic precipitate if the cation forms a complex ion with the ligand.

CHAPTER REVIEW GUIDE

The following sections provide many aids to help you study this chapter. (Numbers in parentheses refer to pages, unless noted otherwise.)

Learning Objectives

These are concepts and skills to review after studying this chapter.

Related section (§), sample problem (SP), and upcoming end-of-chapter problem (EP) numbers are listed in parentheses.

1. Explain how a common ion suppresses a reaction that forms it; describe buffer capacity and buffer range, and understand why the concentrations of buffer components must be high relative to the amount of added H_3O^+ or OH^-; derive and understand the usefulness of the Henderson–Hasselbalch equation; explain how to prepare a buffer (§19.1) (SPs 19.1–19.3) (EPs 19.1–19.25)
2. Understand how an acid-base indicator works, how the equivalence point and end point in an acid-base titration differ, and how strong acid–strong base, weak acid–strong base, and strong acid–weak base titration curves differ; explain the significance of the pH at the midpoint of the buffer region; choose an appropriate indicator, and calculate the pH at any point in a titration (§19.2) (SP 19.4) (EPs 19.26–19.44)
3. Describe the equilibrium of a slightly soluble ionic compound in water, and explain the meaning of K_{sp}; understand how a common ion and pH affect solubility and how to predict precipitate formation from the values of Q_{sp} and K_{sp}; understand the sources and remediation of acid rain (§19.3) (SPs 19.5–19.11) (EPs 19.45–19.64)
4. Describe the stepwise formation of a complex ion, and explain the meaning of K_f; calculate the effect of complex-ion formation on solubility (§19.4) (SP 19.12) (EPs 19.65–19.73)

Key Terms

These important terms appear in boldface in the chapter and are defined again in the Glossary.

Section 19.1
acid-base buffer (618)
common-ion effect (619)
Henderson-Hasselbalch equation (623)
buffer capacity (623)
buffer range (624)

Section 19.2
acid-base titration curve (626)
equivalence point (628)
end point (628)

Section 19.3
solubility-product constant (K_{sp}) (634)

Section 19.4
complex ion (643)
ligand (643)
formation constant (K_f) (644)

Key Equations and Relationships

Numbered and screened concepts are listed for you to refer to or memorize.

19.1 Finding the pH from known concentrations of a conjugate acid-base pair (Henderson-Hasselbalch equation) (623):

$$\text{pH} = \text{p}K_a + \log\left(\frac{[\text{base}]}{[\text{acid}]}\right)$$

19.2 Defining the ion-product expression at equilibrium for a saturated solution of a slightly soluble compound, M_pX_q, composed of M^{n+} and X^{z-} ions (634):

$$Q_{sp} = [M^{n+}]^p[X^{z-}]^q = K_{sp}$$

19.3 Defining the equilibrium constant for formation of the complex ion $ML_x^{\ n+}$ from $M(H_2O)_x^{\ n+}$ and a molecular ligand, L (644):

$$K_f = \frac{[ML_x^{\ n+}]}{[M(H_2O)_x^{\ n+}][L]^x}$$

BRIEF SOLUTIONS TO FOLLOW-UP PROBLEMS

Compare your own solutions to these calculation steps and answers.

19.1 (a) Before addition:
Assuming x is small enough to be neglected,

$[HF] = 0.50\ M$ and $[F^-] = 0.45\ M$

$$[H_3O^+] = K_a \times \frac{[HF]}{[F^-]} \approx (6.8\times10^{-4})\left(\frac{0.50}{0.45}\right) = 7.6\times10^{-4}\ M$$

pH = 3.12

(b) After addition of 0.40 g of NaOH (0.010 mol of NaOH) to 1.0 L of buffer,

$[HF] = 0.49\ M$ and $[F^-] = 0.46\ M$

$$[H_3O^+] \approx (6.8\times10^{-4})\left(\frac{0.49}{0.46}\right) = 7.2\times10^{-4}\ M;\ \text{pH} = 3.14$$

19.2 (a) Strong base would convert HB to B^-, thereby making the ratio closer to 1.
(b)

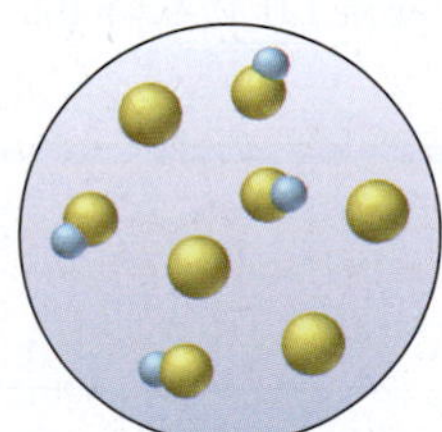

19.3 $[H_3O^+] = 10^{-pH} = 10^{-4.25} = 5.6\times10^{-5}$

$$[C_6H_5COOH] = \frac{[H_3O^+][C_6H_5COO^-]}{K_a} = \frac{(5.6\times10^{-5})(0.050)}{6.3\times10^{-5}} = 0.044\ M$$

Mass (g) of C_6H_5COOH

$$= 5.0\ \text{L soln} \times \frac{0.044\ \text{mol}\ C_6H_5COOH}{1\ \text{L soln}} \times \frac{122.12\ \text{g}\ C_6H_5COOH}{1\ \text{mol}\ C_6H_5COOH} = 27\ \text{g}\ C_6H_5COOH$$

Dissolve 27 g of C_6H_5COOH in 4.9 L of 0.050 M C_6H_5COONa and add solution to make 5.0 L. Adjust pH to 4.25 with strong acid or base.

19.4 (a) $[H_3O^+] \approx \sqrt{(2.3\times10^{-9})(0.2000)} = 2.1\times10^{-5}\ M$

pH = 4.68

(b) $$[H_3O^+] = K_a \times \frac{[HBrO]}{[BrO^-]} = (2.3\times10^{-9})(1) = 2.3\times10^{-9}\ M$$

pH = 8.64

(c)

$$[BrO^-] = \frac{\text{amount (mol) of BrO}^-}{\text{total volume (L)}} = \frac{0.004000\ \text{mol}}{0.06000\ \text{L}} = 0.06667\ M$$

$$K_b \text{ of BrO}^- = \frac{K_w}{K_a \text{ of HBrO}} = 4.3\times10^{-6}$$

$$[H_3O^+] = \frac{K_w}{\sqrt{K_b \times [BrO^-]}} \approx \frac{1.0\times10^{-14}}{\sqrt{(4.3\times10^{-6})(0.06667)}} = 1.9\times10^{-11}\ M$$

pH = 10.72

BRIEF SOLUTIONS TO FOLLOW-UP PROBLEMS (continued)

(d) Amount (mol) of OH^- added = 0.008000 mol
Volume (L) of OH^- soln = 0.08000 L

$$[OH^-] = \frac{\text{amount (mol) of } OH^- \text{ unreacted}}{\text{total volume (L)}}$$

$$= \frac{0.008000 \text{ mol} - 0.004000 \text{ mol}}{(0.02000 + 0.08000) \text{ L}} = 0.04000\ M$$

$$[H_3O^+] = \frac{K_w}{[OH^-]} = 2.5\times10^{-13}$$

pH = 12.60

(e)

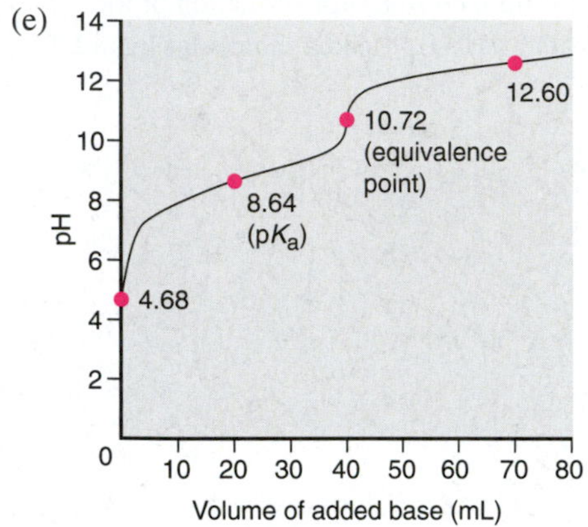

19.5 (a) $K_{sp} = [Ca^{2+}][SO_4^{2-}]$
(b) $K_{sp} = [Cr^{3+}]^2[CO_3^{2-}]^3$
(c) $K_{sp} = [Mg^{2+}][OH^-]^2$
(d) $K_{sp} = [As^{3+}]^2[HS^-]^3[OH^-]^3$

19.6 $$[CaF_2] = \frac{1.5\times10^{-4} \text{ g } CaF_2}{10.0 \text{ mL soln}} \times \frac{1000 \text{ mL}}{1 \text{ L}} \times \frac{1 \text{ mol } CaF_2}{78.08 \text{ g } CaF_2} = 1.9\times10^{-4}\ M$$

$CaF_2(s) \rightleftharpoons Ca^{2+}(aq) + 2F^-(aq)$
$[Ca^{2+}] = 1.9\times10^{-4}\ M$ and $[F^-] = 3.8\times10^{-4}\ M$
$K_{sp} = [Ca^{2+}][F^-]^2 = (1.9\times10^{-4})(3.8\times10^{-4})^2 = 2.7\times10^{-11}$

19.7 From the reaction table, $[Mg^{2+}] = S$ and $[OH^-] = 2S$
$K_{sp} = [Mg^{2+}][OH^-]^2 = 4S^3 = 6.3\times10^{-10}$; $S = 5.4\times10^{-4}\ M$

19.8 (a) In water: $K_{sp} = [Ba^{2+}][SO_4^{2-}] = S^2 = 1.1\times10^{-10}$
$S = 1.0\times10^{-5}$
(b) In 0.10 M Na_2SO_4: $[SO_4^{2-}] = 0.10\ M$
$K_{sp} = 1.1\times10^{-10} \approx S \times 0.10$; $S = 1.1\times10^{-9}\ M$
S decreases in the presence of the common ion SO_4^{2-}.

19.9 (a) Increases the solubility.
$CaF_2(s) \rightleftharpoons Ca^{2+}(aq) + 2F^-(aq)$
$F^-(aq) + H_3O^+(aq) \longrightarrow HF(aq) + H_2O(l)$
(b) Increases the solubility.
$ZnS(s) + H_2O(l) \rightleftharpoons Zn^{2+}(aq) + HS^-(aq) + OH^-(aq)$
$HS^-(aq) + H_3O^+(aq) \longrightarrow H_2S(aq) + H_2O(l)$
$OH^-(aq) + H_3O^+(aq) \longrightarrow 2H_2O(l)$
(c) No effect. $I^-(aq)$ is the conjugate base of the strong acid HI.

19.10 $Ca_3(PO_4)_2(s) \rightleftharpoons 3Ca^{2+}(aq) + 2PO_4^{3-}(aq)$
$Q_{sp} = [Ca^{2+}]^3[PO_4^{3-}]^2 = (1.0\times10^{-9})^5 = 1.0\times10^{-45}$
$Q_{sp} < K_{sp}$, so $Ca_3(PO_4)_2$ will not precipitate.

19.11 (a) Scene 3 has the same relative numbers of ions as in the formula.
(b) Based on $Ni(OH)_2(s) \rightleftharpoons Ni^{2+}(aq) + 2OH^-(aq)$ and $Q_{sp} = [Ni^{2+}][OH^-]^2$, the ion products are $(3)(4)^2 = 48$ in scene 1; $(4)(2)^2 = 16$ in scene 2; and $(2)(4)^2 = 32 = K_{sp}$ in scene 3. Q_{sp} of scene 1 exceeds K_{sp} of scene 3.
(c) Addition of strong acid will decrease mass of $Ni(OH)_2(s)$ by reacting with OH^-, thereby causing more solid to dissolve; addition of strong base will increase mass of $Ni(OH)_2(s)$ due to common-ion effect.

19.12 $AgBr(s) + 2NH_3(aq) \rightleftharpoons Ag(NH_3)_2^+(aq) + Br^-(aq)$
$K_{overall} = K_{sp}$ of AgBr $\times K_f$ of $Ag(NH_3)_2^+$
$= 8.5\times10^{-6}$

From the reaction table, $$\frac{S}{1.0 - 2S} = \sqrt{8.5\times10^{-6}} = 2.9\times10^{-3}$$

$S = [Ag(NH_3)_2^+] = 2.9\times10^{-3}\ M$
Solubility of AgBr is greater in 1.0 M hypo than in 1.0 M NH_3.

PROBLEMS

Problems with **colored** numbers are answered in Appendix E. Sections match the text and provide the numbers of relevant sample problems. Bracketed problems are grouped in pairs (indicated by a short rule) that cover the same concept. Comprehensive Problems are based on material from any section or previous chapter.

Note: Unless stated otherwise, all of the problems for this chapter refer to aqueous solutions at 298 K (25°C).

Equilibria of Acid-Base Buffers

(Sample Problems 19.1 to 19.3)

19.1 What is the purpose of an acid-base buffer?

19.2 How do the acid and base components of a buffer function? Why are they often a conjugate acid-base pair of a weak acid?

19.3 What is the common-ion effect? How is it related to Le Châtelier's principle? Explain with equations that include HF and NaF.

19.4 The scenes below depict solutions of the same HA/A^- buffer (with other ions and water molecules omitted for clarity). (a) Which solution has the greatest buffer capacity? (b) Explain how the pH ranges of the buffers compare. (c) Which solution can react with the largest amount of added strong acid?

19.5 When a small amount of H_3O^+ is added to a buffer, does the pH remain constant? Explain.

19.6 What is the difference between buffers with high and low capacities? Will adding 0.01 mol of HCl produce a greater pH change in a buffer with a high or a low capacity? Explain.

19.7 Which of these factors influence buffer capacity? How?
(a) Conjugate acid-base pair
(b) pH of the buffer
(c) Concentration of buffer components
(d) Buffer range
(e) pK_a of the acid component

19.8 What is the relationship between the buffer range and the buffer-component concentration ratio?

19.9 A chemist needs a pH 3.5 buffer. Should she use NaOH with formic acid ($K_a = 1.8\times10^{-4}$) or with acetic acid ($K_a = 1.8\times10^{-5}$)? Why? What is the disadvantage of choosing the other acid? What is the role of the NaOH?

19.10 What are the $[H_3O^+]$ and the pH of a propanoic acid–propanoate buffer that consists of 0.35 *M* CH_3CH_2COONa and 0.15 *M* CH_3CH_2COOH (K_a of propanoic acid $= 1.3\times10^{-5}$)?

19.11 What are the $[H_3O^+]$ and the pH of a benzoic acid–benzoate buffer that consists of 0.33 *M* C_6H_5COOH and 0.28 *M* C_6H_5COONa (K_a of benzoic acid $= 6.3\times10^{-5}$)?

19.12 Find the pH of a buffer that consists of 1.3 *M* sodium phenolate (C_6H_5ONa) and 1.2 *M* phenol (C_6H_5OH) (pK_a of phenol = 10.00).

19.13 Find the pH of a buffer that consists of 0.12 *M* boric acid (H_3BO_3) and 0.82 *M* sodium borate (NaH_2BO_3) (pK_a of boric acid = 9.24).

19.14 Find the pH of a buffer that consists of 0.25 *M* NH_3 and 0.15 *M* NH_4Cl (pK_b of NH_3 = 4.75).

19.15 Find the pH of a buffer that consists of 0.50 *M* methylamine (CH_3NH_2) and 0.60 *M* CH_3NH_3Cl (pK_b of CH_3NH_2 = 3.35).

19.16 What is the component concentration ratio, $[Pr^-]/[HPr]$, of a buffer that has a pH of 5.44 (K_a of HPr $= 1.3\times10^{-5}$)?

19.17 What is the component concentration ratio, $[NO_2^-]/[HNO_2]$, of a buffer that has a pH of 2.95 (K_a of $HNO_2 = 7.1\times10^{-4}$)?

19.18 A buffer containing 0.2000 *M* of acid, HA, and 0.1500 *M* of its conjugate base, A^-, has a pH of 3.35. What is the pH after 0.0015 mol of NaOH is added to 0.5000 L of this solution?

19.19 A buffer that contains 0.40 *M* base, B, and 0.25 *M* of its conjugate acid, BH^+, has a pH of 8.88. What is the pH after 0.0020 mol of HCl is added to 0.25 L of this solution?

19.20 A buffer is prepared by mixing 204 mL of 0.452 *M* HCl and 0.500 L of 0.400 *M* sodium acetate. (See Appendix C.) (a) What is the pH? (b) How many grams of KOH must be added to 0.500 L of the buffer to change the pH by 0.15 units?

19.21 A buffer is prepared by mixing 50.0 mL of 0.050 *M* sodium bicarbonate and 10.7 mL of 0.10 *M* NaOH. (See Appendix C.) (a) What is the pH? (b) How many grams of HCl must be added to 25.0 mL of the buffer to change the pH by 0.07 units?

19.22 Choose specific acid-base conjugate pairs to make the following buffers: (a) pH ≈ 4.5; (b) pH ≈ 7.0. (See Appendix C.)

19.23 Choose specific acid-base conjugate pairs to make the following buffers: (a) $[H_3O^+] \approx 1\times10^{-9}$ *M*; (b) $[OH^-] \approx 3\times10^{-5}$ *M*. (See Appendix C.)

19.24 An industrial chemist studying bleaching and sterilizing prepares several hypochlorite buffers. Find the pH of (a) 0.100 *M* HClO and 0.100 *M* NaClO; (b) 0.100 *M* HClO and 0.150 *M* NaClO; (c) 0.150 *M* HClO and 0.100 *M* NaClO; (d) 1.0 L of the solution in part (a) after 0.0050 mol of NaOH has been added.

19.25 Oxoanions of phosphorus are buffer components in blood. For a KH_2PO_4/Na_2HPO_4 solution with pH = 7.40 (pH of normal arterial blood), what is the buffer-component concentration ratio?

Acid-Base Titration Curves

(Sample Problem 19.4)

19.26 How can you estimate the pH range of an indicator's color change? Why do some indicators have two separate pH ranges?

19.27 Why does the color change of an indicator take place over a range of about 2 pH units?

19.28 Why doesn't the addition of an acid-base indicator affect the pH of the test solution?

19.29 What is the difference between the end point of a titration and the equivalence point? Is the equivalence point always reached first? Explain.

19.30 Some automatic titrators measure the slope of a titration curve to determine the equivalence point. What happens to the slope that enables the instrument to recognize this point?

19.31 The scenes below depict the relative concentrations of H_3PO_4, $H_2PO_4^-$, and HPO_4^{2-} during a titration with aqueous NaOH, but they are out of order. (Phosphate groups are purple, hydrogens are blue, and Na^+ ions and water molecules are not shown.) (a) List the scenes in the correct order. (b) What is the pH in the correctly ordered second scene (see Appendix C)? (c) If it requires 10.00 mL of the NaOH solution to reach this scene, how much more is needed to reach the last scene?

= H_3PO_4 = $H_2PO_4^-$ = HPO_4^{2-}

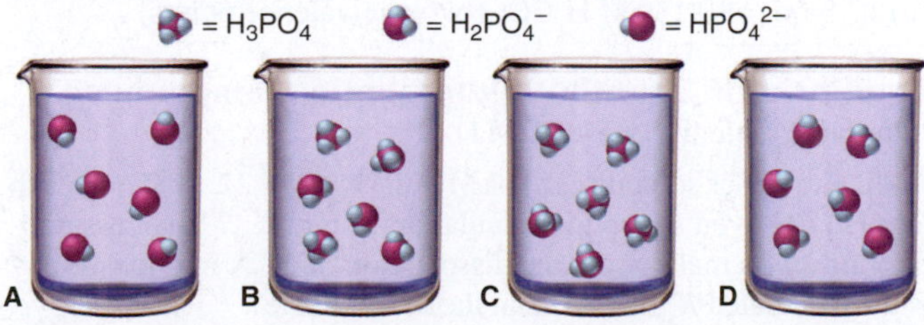

19.32 Explain how *strong acid*–strong base, *weak acid*–strong base, and *weak base*–strong acid titrations using the same concentrations differ in terms of (a) the initial pH and (b) the pH at the equivalence point. (The component in italics is in the flask.)

19.33 What species are in the buffer region of a weak acid–strong base titration? How are they different from the species at the equivalence point? How are they different from the species in the buffer region of a weak base–strong acid titration?

19.34 Why is the center of the buffer region of a weak acid–strong base titration significant?

19.35 The indicator cresol red has $K_a = 3.5\times10^{-9}$. Over what approximate pH range does it change color?

19.36 The indicator ethyl red has $K_a = 3.8\times10^{-6}$. Over what approximate pH range does it change color?

19.37 Use Figure 19.5 to find an indicator for these titrations:
(a) 0.10 M HCl with 0.10 M NaOH
(b) 0.10 M HCOOH (Appendix C) with 0.10 M NaOH

19.38 Use Figure 19.5 to find an indicator for these titrations:
(a) 0.10 M CH_3NH_2 (Appendix C) with 0.10 M HCl
(b) 0.50 M HI with 0.10 M KOH

19.39 Calculate the pH during the titration of 40.00 mL of 0.1000 M HCl with 0.1000 M NaOH solution after the following additions of base:
(a) 0 mL (b) 25.00 mL (c) 39.00 mL (d) 39.90 mL
(e) 40.00 mL (f) 40.10 mL (g) 50.00 mL

19.40 Calculate the pH during the titration of 30.00 mL of 0.1000 M KOH with 0.1000 M HBr solution after the following additions of acid:
(a) 0 mL (b) 15.00 mL (c) 29.00 mL (d) 29.90 mL
(e) 30.00 mL (f) 30.10 mL (g) 40.00 mL

19.41 Find the pH during the titration of 20.00 mL of 0.1000 M butanoic acid, $CH_3CH_2CH_2COOH$ ($K_a = 1.54\times10^{-5}$), with 0.1000 M NaOH solution after the following additions of titrant:
(a) 0 mL (b) 10.00 mL (c) 15.00 mL (d) 19.00 mL
(e) 19.95 mL (f) 20.00 mL (g) 20.05 mL (h) 25.00 mL

19.42 Find the pH during the titration of 20.00 mL of 0.1000 M triethylamine, $(CH_3CH_2)_3N$ ($K_b = 5.2\times10^{-4}$), with 0.1000 M HCl solution after the following additions of titrant:
(a) 0 mL (b) 10.00 mL (c) 15.00 mL (d) 19.00 mL
(e) 19.95 mL (f) 20.00 mL (g) 20.05 mL (h) 25.00 mL

19.43 Find the pH of the equivalence point(s) and the volume (mL) of 0.0372 M NaOH needed to reach the point(s) in titrations of
(a) 42.2 mL of 0.0520 M CH_3COOH
(b) 28.9 mL of 0.0850 M H_2SO_3 (two equivalence points)

19.44 Find the pH of the equivalence point(s) and the volume (mL) of 0.0588 M KOH needed to reach the point(s) in titrations of
(a) 23.4 mL of 0.0390 M HNO_2
(b) 17.3 mL of 0.130 M H_2CO_3 (two equivalence points)

Equilibria of Slightly Soluble Ionic Compounds

(Sample Problems 19.5 to 19.11)

19.45 The molar solubility (S) of M_2X is 5×10^{-5} M. Find S of each ion. How do you set up the calculation to find K_{sp}? What assumption must you make about the dissociation of M_2X into ions? Why is the calculated K_{sp} higher than the actual value?

19.46 Why does pH affect the solubility of BaF_2 but not of $BaCl_2$?

19.47 In a gaseous equilibrium, the reverse reaction occurs when $Q_c > K_c$. What occurs in aqueous solution when $Q_{sp} > K_{sp}$?

19.48 Write the ion-product expressions for (a) silver carbonate; (b) barium fluoride; (c) copper(II) sulfide.

19.49 Write the ion-product expressions for (a) iron(III) hydroxide; (b) barium phosphate; (c) tin(II) sulfide.

19.50 The solubility of silver carbonate is 0.032 M at 20°C. Calculate its K_{sp}.

19.51 The solubility of zinc oxalate is 7.9×10^{-3} M at 18°C. Calculate its K_{sp}.

19.52 The solubility of silver dichromate at 15°C is 8.3×10^{-3} g/100 mL solution. Calculate its K_{sp}.

19.53 The solubility of calcium sulfate at 30°C is 0.209 g/100 mL solution. Calculate its K_{sp}.

19.54 Find the molar solubility of $SrCO_3$ ($K_{sp} = 5.4\times10^{-10}$) in (a) pure water and (b) 0.13 M $Sr(NO_3)_2$.

19.55 Find the molar solubility of $BaCrO_4$ ($K_{sp} = 2.1\times10^{-10}$) in (a) pure water and (b) 1.5×10^{-3} M Na_2CrO_4.

19.56 Calculate the molar solubility of $Ca(IO_3)_2$ in (a) 0.060 M $Ca(NO_3)_2$ and (b) 0.060 M $NaIO_3$. (See Appendix C.)

19.57 Calculate the molar solubility of Ag_2SO_4 in (a) 0.22 M $AgNO_3$ and (b) 0.22 M Na_2SO_4. (See Appendix C.)

19.58 Which compound in each pair is more soluble in water?
(a) Magnesium hydroxide or nickel(II) hydroxide
(b) Lead(II) sulfide or copper(II) sulfide
(c) Silver sulfate or magnesium fluoride

19.59 Which compound in each pair is more soluble in water?
(a) Strontium sulfate or barium chromate
(b) Calcium carbonate or copper(II) carbonate
(c) Barium iodate or silver chromate

19.60 Write equations to show whether the solubility of either of the following is affected by pH: (a) AgCl; (b) $SrCO_3$.

19.61 Write equations to show whether the solubility of either of the following is affected by pH: (a) CuBr; (b) $Ca_3(PO_4)_2$.

19.62 Does any solid $Cu(OH)_2$ form when 0.075 g of KOH is dissolved in 1.0 L of 1.0×10^{-3} M $Cu(NO_3)_2$?

19.63 Does any solid $PbCl_2$ form when 3.5 mg of NaCl is dissolved in 0.250 L of 0.12 M $Pb(NO_3)_2$?

19.64 When blood is donated, sodium oxalate solution is used to precipitate Ca^{2+}, which triggers clotting. A 104-mL sample of blood contains 9.7×10^{-5} g Ca^{2+}/mL. A technologist treats the sample with 100.0 mL of 0.1550 M $Na_2C_2O_4$. Calculate $[Ca^{2+}]$ after the treatment. (See Appendix C for K_{sp} of $CaC_2O_4\cdot H_2O$.)

Equilibria Involving Complex Ions

(Sample Problem 19.12)

19.65 How can a metal cation be at the center of a complex anion?

19.66 Write equations to show the stepwise reaction of $Cd(H_2O)_4^{2+}$ in an aqueous solution of KI to form CdI_4^{-2}. Show that $K_{f(overall)} = K_{f1} \times K_{f2} \times K_{f3} \times K_{f4}$.

19.67 Consider the dissolution of PbS in water:

$$PbS(s) + H_2O(l) \rightleftharpoons Pb^{2+}(aq) + HS^-(aq) + OH^-(aq)$$

Adding aqueous NaOH causes more PbS to dissolve. Does this violate Le Châtelier's principle? Explain.

19.68 Write a balanced equation for the reaction of $Hg(H_2O)_4^{2+}$ in aqueous KCN.

19.69 Write a balanced equation for the reaction of $Zn(H_2O)_4^{2+}$ in aqueous NaCN.

19.70 Write a balanced equation for the reaction of $Ag(H_2O)_2^+$ in aqueous $Na_2S_2O_3$.

19.71 Write a balanced equation for the reaction of $Al(H_2O)_6^{3+}$ in aqueous KF.

19.72 Find the solubility of $Cr(OH)_3$ in a buffer of pH 13.0 [K_{sp} of $Cr(OH)_3 = 6.3\times10^{-31}$; K_f of $Cr(OH)_4^- = 8.0\times10^{29}$].

19.73 Find the solubility of AgI in 2.5 M NH_3 [K_{sp} of AgI $= 8.3\times10^{-17}$; K_f of $Ag(NH_3)_2^+ = 1.7\times10^7$].

Comprehensive Problems

19.74 A microbiologist is preparing a medium on which to culture *E. coli* bacteria. She buffers the medium at pH 7.00 to minimize the effect of acid-producing fermentation. What volumes of equimolar aqueous solutions of K_2HPO_4 and KH_2PO_4 must she combine to make 100. mL of the pH 7.00 buffer?

19.75 Tris(hydroxymethyl)aminomethane [$(HOCH_2)_3CNH_2$], known as TRIS, is a weak base used in biochemical experiments to make buffer solutions in the pH range of 7 to 9. A certain TRIS buffer has a pH of 8.10 at 25°C and a pH of 7.80 at 37°C. Why does the pH change with temperature?

19.76 Water flowing through pipes of carbon steel must be kept at pH 5 or greater to limit corrosion. If an 8.0×10^3 lb/hr water stream contains 10 ppm sulfuric acid and 0.015% acetic acid, how many pounds per hour of sodium acetate trihydrate must be added to maintain that pH?

19.77 Gout is caused by an error in metabolism that leads to a buildup of uric acid in body fluids, which is deposited as slightly soluble sodium urate ($C_5H_3N_4O_3Na$) in the joints. If the extracellular [Na^+] is 0.15 M and the solubility of sodium urate is 0.085 g/100. mL, what is the minimum urate ion concentration (abbreviated [Ur^-]) that will cause a deposit of sodium urate?

19.78 Cadmium ion in solution is analyzed by being precipitated as the sulfide, a yellow compound used as a pigment in everything from artists' oil paints to glass and rubber. Calculate the molar solubility of cadmium sulfide at 25°C.

19.79 The solubility of KCl is 3.7 M at 20°C. Two beakers contain 100. mL of saturated KCl solution: 100. mL of 6.0 M HCl is added to the first beaker and 100. mL of 12 M HCl to the second. (a) Find the ion-product constant of KCl at 20°C. (b) What mass, if any, of KCl will precipitate from each beaker?

19.80 Manganese(II) sulfide is one of the compounds found in the nodules on the ocean floor that may eventually be a primary source of many transition metals. The solubility of MnS is 4.7×10^{-4} g/100 mL solution. Estimate the K_{sp} of MnS.

19.81 The normal pH of blood is 7.40 ± 0.05 and is controlled in part by a H_2CO_3/HCO_3^- buffer system.
(a) Assuming that the K_a value for carbonic acid at 25°C applies to blood, what is the $[H_2CO_3]/[HCO_3^-]$ ratio in normal blood?
(b) In a condition called *acidosis,* the blood is too acidic. What is the $[H_2CO_3]/[HCO_3^-]$ ratio in a patient whose blood pH is 7.20?

19.82 The figure depicts a saturated solution of $MCl_2(s)$ in the presence of dilute aqueous NaCl; each sphere represents 1.0×10^{-6} mol of ion, and the volume is 250.0 mL (solid MCl_2 is shown as green chunks; Na^+ ions and water molecules are not shown). (a) Calculate the K_{sp} of MCl_2. (b) If $M(NO_3)_2(s)$ is added, is there an increase, decrease, or no change in the number of Cl^- particles? In K_{sp}? In mass of $MCl_2(s)$?

19.83 Tooth enamel consists of hydroxyapatite, $Ca_5(PO_4)_3OH$ ($K_{sp} = 6.8\times10^{-37}$). Fluoride ion added to drinking water reacts with $Ca_5(PO_4)_3OH$ to form the more tooth decay–resistant fluorapatite, $Ca_5(PO_4)_3F$ ($K_{sp} = 1.0\times10^{-60}$). Fluoridated water has dramatically decreased cavities among children. Calculate the solubility of $Ca_5(PO_4)_3OH$ and of $Ca_5(PO_4)_3F$ in water.

19.84 The acid-base indicator ethyl orange turns from red to yellow over the pH range 3.4 to 4.8. Estimate K_a for ethyl orange.

19.85 Instrumental acid-base titrations use a pH meter to monitor the changes in pH and volume. The equivalence point is found from the volume at which the curve has the steepest slope.
(a) Use the data in Figure 19.7 (p. 628) to calculate the slope ($\Delta pH/\Delta V$) for all pairs of adjacent points and to calculate the average volume (V_{avg}) for each interval.
(b) Plot $\Delta pH/\Delta V$ vs. V_{avg} to find the steepest slope, and thus the volume at the equivalence point. (For example, the first pair of points gives $\Delta pH = 0.22$, $\Delta V = 10.00$ mL; hence, $\Delta pH/\Delta V = 0.022$ mL^{-1}, and $V_{avg} = 5.00$ mL.)

19.86 What is the pH of a solution of 6.5×10^{-9} mol of $Ca(OH)_2$ in 10.0 L of water [K_{sp} of $Ca(OH)_2 = 6.5\times10^{-6}$]?

19.87 A student wants to dissolve the maximum amount of CaF_2 ($K_{sp} = 3.2\times10^{-11}$) to make 1 L of aqueous solution.
(a) Into which of the following solvents should she dissolve the salt?

(I) Pure water (II) 0.01 M HF
(III) 0.01 M NaOH (IV) 0.01 M HCl
(V) 0.01 M $Ca(OH)_2$

(b) Which would dissolve the least amount of salt?

19.88 The Henderson-Hasselbalch equation gives a relationship for obtaining the pH of a buffer solution consisting of HA and A^-. Derive an analogous relationship for obtaining the pOH of a buffer solution consisting of B and BH^+.

19.89 Calcium ion present in water supplies is easily precipitated as calcite ($CaCO_3$):

$$Ca^{2+}(aq) + CO_3^{2-}(aq) \rightleftharpoons CaCO_3(s)$$

Because the K_{sp} decreases with temperature, heating hard water forms a calcite "scale," which clogs pipes and water heaters. Find the solubility of calcite in water (a) at 10°C ($K_{sp} = 4.4\times10^{-9}$) and (b) at 30°C ($K_{sp} = 3.1\times10^{-9}$).

19.90 The well water in an area is "hard" because it is in equilibrium with $CaCO_3$ in the surrounding rocks. What is the concentration of Ca^{2+} in the well water (assuming the water's pH is such that the CO_3^{2-} ion is not hydrolyzed)? (See Appendix C for K_{sp} of $CaCO_3$.)

19.91 An environmental technician collects a sample of rainwater. A light on her portable pH meter indicates low battery power, so she uses indicator solutions to estimate the pH. A piece of litmus paper turns red, indicating acidity, so she divides the sample into thirds and obtains the following results: thymol blue turns yellow; bromphenol blue turns green; and methyl red turns red. Estimate the pH of the rainwater.

19.92 Calculate the molar solubility of $Hg_2C_2O_4$ (K_{sp} = 1.75×10^{-13}) in 0.13 M $Hg_2(NO_3)_2$.

19.93 Quantitative analysis of Cl^- ion is often performed by a titration with silver nitrate, using sodium chromate as an indicator. As standardized $AgNO_3$ is added, both white AgCl and red Ag_2CrO_4 precipitate, but so long as some Cl^- remains, the Ag_2CrO_4 redissolves as the mixture is stirred. When the red color persists during stirring, the equivalence point has been reached.
(a) Calculate the equilibrium constant for the reaction

$$2AgCl(s) + CrO_4^{2-}(aq) \rightleftharpoons Ag_2CrO_4(s) + 2Cl^-(aq)$$

(b) Explain why the silver chromate redissolves.
(c) If 25.00 cm^3 of 0.1000 M NaCl is mixed with 25.00 cm^3 of 0.1000 M $AgNO_3$, what is the concentration of Ag^+ remaining in solution? Is this sufficient to precipitate any silver chromate?

19.94 Some kidney stones form by the precipitation of calcium oxalate monohydrate ($CaC_2O_4\cdot H_2O$, $K_{sp} = 2.3\times10^{-9}$). The pH of urine varies from 5.5 to 7.0, and the average $[Ca^{2+}]$ in urine is 2.6×10^{-3} M.
(a) If the [oxalic acid] in urine is 3.0×10^{-13} M, will kidney stones form at pH = 5.5?
(b) At pH = 7.0?
(c) Vegetarians have a urine pH above 7. Are they more or less likely to form kidney stones?

19.95 A 35.00-mL solution of 0.2500 M HF is titrated with a standardized 0.1532 M solution of NaOH at 25°C.
(a) What is the pH of the HF solution before titrant is added?
(b) How many milliliters of titrant are required to reach the equivalence point?
(c) What is the pH at 0.50 mL before the equivalence point?
(d) What is the pH at the equivalence point?
(e) What is the pH at 0.50 mL after the equivalence point?

19.96 A lake that has a surface area of 10.0 acres (1 acre = 4.840×10^3 yd^2) receives 1.00 in. of rain of pH 4.20. (Assume the acidity of the rain is due to a strong, monoprotic acid.)
(a) How many moles of H_3O^+ are in the rain falling on the lake?
(b) If the lake is unbuffered (pH = 7.00) and its average depth is 10.0 ft before the rain, find the pH after the rain has been mixed with lake water. (Ignore runoff from the surrounding land.)
(c) If the lake contains hydrogen carbonate ions (HCO_3^-), what mass of HCO_3^- would neutralize the acid in the rain?

19.97 Sodium chloride is purified for use as table salt by adding HCl to a saturated solution of NaCl (317 g/L). Will pure NaCl precipitate when 28.5 mL of 8.65 M HCl is added to 0.100 L of saturated solution? Show calculations to explain your answer.

19.98 Because of the toxicity of mercury compounds, mercury(I) chloride is used in antibacterial salves. The mercury(I) ion (Hg_2^{2+}) consists of two bound Hg^+ ions.
(a) What is the empirical formula of mercury(I) chloride?
(b) Calculate $[Hg_2^{2+}]$ in a saturated solution of mercury(I) chloride ($K_{sp} = 1.5\times10^{-18}$).
(c) A seawater sample contains 0.20 lb of NaCl per gallon. Find $[Hg_2^{2+}]$ if the seawater is saturated with mercury(I) chloride.
(d) How many grams of mercury(I) chloride are needed to saturate 4900 km^3 of pure water (the volume of Lake Michigan)?
(e) How many grams of mercury(I) chloride are needed to saturate 4900 km^3 of seawater?

19.99 Scenes A to D represent tiny portions of 0.10 M aqueous solutions of a weak acid HA (*red and blue*; $K_a = 4.5\times10^{-5}$), its conjugate base A^- (*red*), or a mixture of the two (only these species are shown):

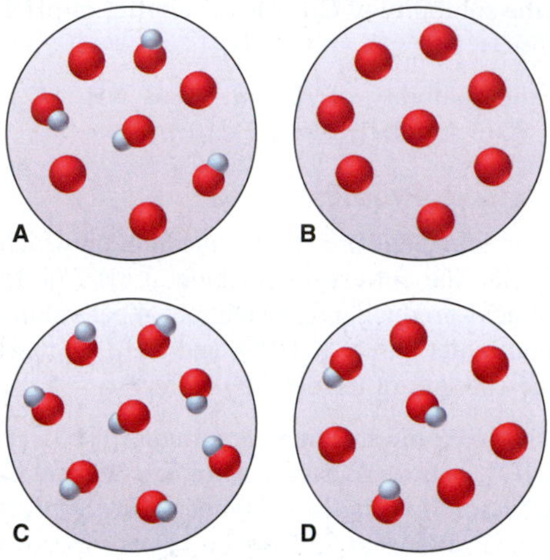

(a) Which scene(s) show(s) a buffer?
(b) What is the pH of each solution?
(c) Arrange the scenes in sequence, assuming that they represent stages in a weak acid–strong base titration.
(d) Which scene represents the titration at its equivalence point?

19.100 Scenes A to C represent aqueous solutions of the slightly soluble salt MZ (only the ions of this salt are shown):

$$MZ(s) \rightleftharpoons M^{2+}(aq) + Z^{2-}(aq)$$

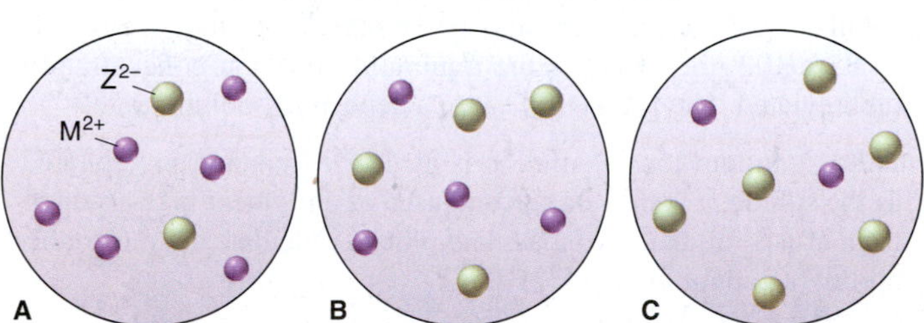

(a) Which scene represents the solution just after solid MZ is stirred thoroughly in distilled water?
(b) If each sphere represents 2.5×10^{-6} M of ions, what is the K_{sp} of MZ?
(c) Which scene represents the solution after $Na_2Z(aq)$ has been added?
(d) If Z^{2-} is CO_3^{2-}, which scene represents the solution after the pH has been lowered?

Thermodynamics: Entropy, Free Energy, and the Direction of Chemical Reactions

Down Is Easier Than Up Certain events seem to happen naturally—water flows downhill, cars rust, air escapes from a punctured tire, lit fuel keeps burning, and so forth. In this chapter, you'll learn the factors that determine why some changes occur—and why others don't.

Outline

Key Principles
to focus on while studying this chapter

- A process such as a rock falling or a fuel burning is said to be *spontaneous* because, once started, it continues by itself. Neither the first law of thermodynamics (law of conservation of energy) nor the sign of ΔH can predict which processes are spontaneous, but any *reaction proceeds spontaneously toward equilibrium. **(Section 20.1)***
- The total kinetic energy of a system consists of all the motions of its particles—rotations, vibrations, and translations—each of which is *quantized.* A *microstate* of the system is any specific combination of these quantized energy states. The *entropy* (S) of a system is directly related to the *number of microstates* (W) into which the system *disperses its energy,* which is closely associated with the *freedom of motion* of the particles. ***(Section 20.1)***
- A substance has more entropy in its gaseous than its liquid state and more in its liquid than its solid state. ***(Section 20.1)***
- The *second law of thermodynamics* states that a spontaneous process occurs in the direction that increases the entropy of the universe (system plus surroundings). In other words, a change occurs spontaneously if the energy of the universe becomes more dispersed. ***(Section 20.1)***
- The *third law of thermodynamics*—the entropy of a perfect crystal is zero at 0 K—allows us to calculate *absolute* entropies. The *standard molar entropy* (S°) of a substance is influenced by temperature, physical state, dissolution, and atomic size or molecular complexity. ***(Section 20.1)***
- Gases have such high entropy that if a reaction has a net increase in amount (mol) of gas, the standard entropy change of the reaction is positive ($\Delta S^\circ_{rxn} > 0$); if a reaction has a net decrease in amount (mol) of gas, $\Delta S^\circ_{rxn} < 0$. ***(Section 20.2)***
- For a spontaneous process, $\Delta S_{univ} > 0$, so the only way the entropy of the system can decrease is if the entropy of the surroundings increases even more. For a process at equilibrium, $\Delta S_{univ} = 0$. ***(Section 20.2)***
- The *free energy change* (ΔG) of a process is a measure of its spontaneity. Because $\Delta G = \Delta H - T\Delta S$, temperature influences reaction spontaneity by affecting the size of $T\Delta S$. The free energy of a system decreases in a spontaneous process; that is, if $\Delta S_{univ} > 0$, $\Delta G_{sys} < 0$. ***(Section 20.3)***
- The free energy change is the portion of the total energy change available to do *work.* In any real process, however, some of the free energy change is always *lost as heat.* In a *coupling of reactions,* a spontaneous step with a larger negative ΔG drives a nonspontaneous step with a smaller positive ΔG. ***(Section 20.3)***
- Reaction spontaneity ($\Delta G < 0$) depends on two factors: the free energy change at standard conditions, ΔG°, and the size of the reaction quotient, Q. No matter what the starting conditions, any process is spontaneous until $Q = K$ ($\Delta G = 0$). ***(Section 20.4)***

CONCEPTS & SKILLS TO REVIEW before studying this chapter

- internal energy, heat, and work (Section 6.1)
- state functions (Section 6.1) and standard states (Section 6.6)
- enthalpy, ΔH, and Hess's law (Sections 6.2 and 6.5)
- entropy and solution formation (Section 13.2)
- comparing Q and K to find reaction direction (Section 17.4)

In the last few chapters, we've examined some fundamental questions about change, whether chemical or physical: How fast does it occur? How far does it go toward completion? And, especially, how are rate and extent affected by concentration and temperature?

But *why* does a change occur in the first place? Some changes seem to happen by themselves in one direction but not in the other. A candle burns in air and forms carbon dioxide and water vapor, but those products will never react to form candle wax and oxygen. A new steel shovel left outside slowly rusts, but put a rusty one outside and it won't become shiny. A cube of sugar dissolves in a cup of coffee after a few seconds of stirring, but stir for another millennium and the dissolved sugar won't form the cube again. Chemists speak of these types of changes as being *spontaneous*. Most release energy, but some absorb it. In this chapter, we discuss the nature of such spontaneous changes. The principles of thermodynamics that we cover here apply, as far as we know, to every system in the universe!

20.1 • THE SECOND LAW OF THERMODYNAMICS: PREDICTING SPONTANEOUS CHANGE

A **spontaneous change** of a system is one that occurs under specified conditions without a continuous input of energy from outside the system. The freezing of water, for example, is spontaneous at 1 atm and $-5°C$. A spontaneous process, such as burning or falling, may need a little "push" to get started—a spark to ignite gasoline vapors in your car's engine, a shove to knock a book off your desk—but once the process begins, it supplies the energy to continue. In contrast, a *nonspontaneous* change occurs only if the surroundings *continuously* supply the system with energy. Under given conditions, *if a change is spontaneous in one direction, it is* ***not*** *spontaneous in the other.*

The term *spontaneous* does not mean *instantaneous* nor does it tell anything about how long a process takes to occur; it means that, given enough time, the process will happen by itself. Many processes are spontaneous but slow—ripening, rusting, and aging.

A chemical reaction proceeding toward equilibrium is a spontaneous change. Recall that we can predict the net direction of the reaction—its spontaneous direction—by comparing the reaction quotient (Q) with the equilibrium constant (K). But *why* is there a drive to attain equilibrium? And what determines the value of the equilibrium constant? And, most importantly, can we predict the direction of a spontaneous change in cases that are not as obvious as burning gasoline or falling books?

The First Law of Thermodynamics Does Not Predict Spontaneous Change

Let's see whether energy changes can clarify the criterion for spontaneity. Recall from Chapter 6 that the first law of thermodynamics (law of conservation of energy) states that the internal energy (E) of a system, the sum of the kinetic and potential energies of its particles, changes when heat (q) and/or work (w) are absorbed or released:

$$\Delta E = q + w$$

Whatever is not part of the system (sys) is part of the surroundings (surr); thus, *the change in energy and, therefore, heat and/or work absorbed by the system is released by the surroundings,* and vice versa:

$$\Delta E_{sys} = -\Delta E_{surr} \quad \text{or} \quad (q + w)_{sys} = -(q + w)_{surr}$$

Since the system plus the surroundings is the universe (univ), it follows that *the total energy of the universe is constant, so the change in energy of the universe is zero:**

$$\Delta E_{sys} + \Delta E_{surr} = 0 = \Delta E_{univ}$$

*Any modern statement of conservation of energy must take into account mass-energy equivalence and the processes in stars, which convert enormous amounts of matter into energy. Thus, the total *mass-energy* of the universe is constant.

The first law accounts for the energy of a process, but can it predict the direction? When gasoline burns in a car engine, the first law states that the potential energy difference between the bonds in the fuel mixture and those in the exhaust gases is converted to the kinetic energy of the moving car and its parts plus the heat released to the environment. But, why doesn't the heat released in the engine convert exhaust fumes back into gasoline and oxygen? When an ice cube melts in your hand, the first law tells that energy from your hand is converted to kinetic energy as the solid changes to a liquid. But, why doesn't the pool of water in your cupped fingers transfer the heat back to your hand and refreeze? Neither of these events violates the first law—if you measured the work and heat in each case, you would find that energy is conserved—but these reverse changes never happen. That is, *the first law by itself does not predict the **direction** of a spontaneous change.*

The Sign of ΔH Does Not Predict Spontaneous Change

Could the sign of the enthalpy change (ΔH), the heat gained or lost at constant pressure (q_P), be the criterion for spontaneity? If so, we would expect exothermic processes ($\Delta H < 0$) to be spontaneous and endothermic processes ($\Delta H > 0$) to be nonspontaneous. Let's examine some examples to see if this is true.

1. *Spontaneous processes with* $\Delta H < 0$. All freezing and condensing processes are exothermic and spontaneous at certain conditions:

$$H_2O(l) \longrightarrow H_2O(s) \qquad \Delta H^\circ_{rxn} = -\Delta H^\circ_{fus} = -6.02\ \text{kJ}\ (1\ \text{atm};\ T = 0^\circ\text{C})$$

All combustion reactions are spontaneous and exothermic:

$$CH_4(g) + 2O_2(g) \longrightarrow CO_2(g) + 2H_2O(g) \qquad \Delta H^\circ_{rxn} = -802\ \text{kJ}$$

Oxidation of iron and other metals occurs spontaneously and exothermically:

$$2Fe(s) + \tfrac{3}{2}O_2(g) \longrightarrow Fe_2O_3(s) \qquad \Delta H^\circ_{rxn} = -826\ \text{kJ}$$

Ionic compounds form spontaneously and exothermically from their elements:

$$Na(s) + \tfrac{1}{2}Cl_2(g) \longrightarrow NaCl(s) \qquad \Delta H^\circ_{rxn} = -411\ \text{kJ}$$

2. *Spontaneous processes with* $\Delta H > 0$. However, in many cases, an exothermic process occurs spontaneously under one set of conditions, whereas the opposite, endothermic, process occurs spontaneously under another set. All melting and vaporizing processes are endothermic *and* spontaneous at certain conditions:

$$H_2O(s) \longrightarrow H_2O(l) \qquad \Delta H^\circ_{rxn} = \Delta H^\circ_{fus} = +6.02\ \text{kJ}\ (1\ \text{atm};\ T = 0^\circ\text{C})$$

At ordinary pressure water vaporizes spontaneously:

$$H_2O(l) \longrightarrow H_2O(g) \qquad \Delta H^\circ_{rxn} = \Delta H^\circ_{vap} = +44.0\ \text{kJ}\ (1\ \text{atm};\ T = 100^\circ\text{C})$$

Most soluble salts dissolve endothermically *and* spontaneously:

$$NaCl(s) \xrightarrow{H_2O} Na^+(aq) + Cl^-(aq) \qquad \Delta H^\circ_{soln} = +3.9\ \text{kJ}$$

$$NH_4NO_3(s) \xrightarrow{H_2O} NH_4^+(aq) + NO_3^-(aq) \qquad \Delta H^\circ_{soln} = +25.7\ \text{kJ}$$

Even some endothermic reactions are spontaneous:

$$N_2O_5(s) \longrightarrow 2NO_2(g) + \tfrac{1}{2}O_2(g) \qquad \Delta H^\circ_{rxn} = +109.5\ \text{kJ}$$

$$Ba(OH)_2{\cdot}8H_2O(s) + 2NH_4NO_3(s) \longrightarrow$$
$$Ba^{2+}(aq) + 2NO_3^-(aq) + 2NH_3(aq) + 10H_2O(l) \qquad \Delta H^\circ_{rxn} = +62.3\ \text{kJ}$$

Given just these few examples, we see that, as for the first law, *the sign of* ΔH *by itself does not predict the **direction** of a spontaneous change.*

Freedom of Particle Motion and Dispersal of Particle Energy

When we look closely at the previous examples of spontaneous *endothermic* processes, they have one major feature in common: in every case, the chemical entities—atoms, molecules, or ions—have more freedom of motion *after* the change. Put another way, after the change, the particles have a wider range of energy of motion (kinetic energy); we say that the energy has become more dispersed, distributed, or spread out:

- The phase changes convert a solid, in which motion is restricted, to a liquid, in which particles have more freedom to move around each other, and then to a gas, in which the particles have much greater freedom of motion. Thus, the energy of motion is more dispersed.
- Dissolving a salt changes a crystalline solid and a pure liquid into separate ions and solvent molecules moving and interacting, so their freedom of motion is greater and their energy of motion more dispersed.
- In the chemical reactions, *fewer* moles of crystalline solids produce *more* moles of gases and/or solvated ions, so once again, the freedom of motion of the particles increases and their energy of motion is more dispersed:

	less freedom of particle motion		more freedom of particle motion
	localized energy of motion	$\longrightarrow$	dispersed energy of motion
Phase change:	solid $\longrightarrow$ liquid	$\longrightarrow$	gas
Dissolving of salt:	crystalline solid + liquid	$\longrightarrow$	ions in solution
Chemical change:	crystalline solids	$\longrightarrow$	gases + ions in solution

In thermodynamic terms, a change in the freedom of motion of particles in a system, that is, in the dispersal of their energy of motion, is a key factor affecting the ***direction*** *of a spontaneous process.*

Entropy and the Number of Microstates

Earlier, we discussed the quantized *electronic* energy levels of an atom (Chapter 7) and of a molecule (Chapter 11). In addition, the *kinetic* energy levels of a molecule's motions—vibrational, rotational, and translational—are quantized. And now we'll see that the energy state of a whole system of particles is quantized, too.

Energy Dispersal and the Meaning of Entropy Let's see why freedom of motion and dispersal of energy relate to spontaneous change:

- *Quantization of energy.* Picture a system of, say, 1 mol of N_2 gas and focus on one molecule. At any instant, it is moving through space (translating) at some speed and rotating at some frequency, and its atoms are vibrating at some frequency. In the next instant, the molecule collides with another or with the container, and these motional (kinetic) energy states change to different values. The complete quantum state of the molecule at any instant consists of its electronic states and these translational, rotational, and vibrational states. In this discussion, we focus on the latter three.
- *Number of microstates.* The energy of the other molecules is similarly quantized. Each quantized state of the system of molecules is called a *microstate,* and at any instant, the total energy of the system is dispersed throughout one microstate. In the next instant, it is dispersed throughout a different microstate. The number of microstates possible for a system of 1 mol of molecules is staggering, on the order of $10^{10^{23}}$.
- *Dispersal of energy.* At a given set of conditions, each microstate has the *same* total energy as any other. Therefore, each microstate is equally possible for the system, and the laws of probability say that, over time, all microstates are equally likely. The number of microstates for a system is the number of ways it can disperse (distribute or spread) its kinetic energy among the various motions of all its particles.

In 1877, the Austrian mathematician and physicist Ludwig Boltzmann related the number of microstates (W) to the **entropy (S)** of a system:

$$S = k \ln W \quad \textbf{(20.1)}$$

where k, the *Boltzmann constant,* is the universal gas constant (R) divided by Avogadro's number (N_A), or R/N_A, and equals 1.38×10^{-23} J/K. The term W is the number of microstates, so it has no units; therefore, S has units of joules/kelvin (J/K). Thus,

- A system with fewer microstates (smaller W) has *lower entropy (lower S).*
- A system with more microstates (larger W) has *higher entropy (higher S).*

For our earlier examples of endothermic processes,

	lower entropy (fewer microstates) $\longrightarrow$ higher entropy (more microstates)
Phase change:	solid $\longrightarrow$ liquid $\longrightarrow$ gas
Dissolving of salt:	crystalline solid + liquid $\longrightarrow$ ions in solution
Chemical change:	crystalline solids $\longrightarrow$ gases + ions in solution

(Recall from Chapter 13 that entropy is a factor in the formation of solutions.)

Entropy as a State Function If a change results in a greater number of microstates, there are more ways to disperse the energy of the system and the entropy increases:

$$S_{\text{more microstates}} > S_{\text{fewer microstates}}$$

If a change results in a lower number of microstates, the entropy decreases.

Like internal energy (E) and enthalpy (H), *entropy is a state function,* so it depends only on the present state of the system, not on how it arrived at that state (Section 6.1). Therefore, the change in entropy of the system (ΔS_{sys}) depends only on the *difference* between its final and initial values:

$$\Delta S_{\text{sys}} = S_{\text{final}} - S_{\text{initial}}$$

Like any state function, $\Delta S_{\text{sys}} > 0$ when its value increases during a change. For example, for the phase change when dry ice sublimes, the entropy increases:

$$CO_2(s) \longrightarrow CO_2(g) \qquad \Delta S_{\text{sys}} = S_{\text{final}} - S_{\text{initial}} = S_{\text{gaseous } CO_2} - S_{\text{solid } CO_2} > 0$$

And $\Delta S_{\text{sys}} < 0$ when the entropy decreases, as when water vapor condenses:

$$H_2O(g) \longrightarrow H_2O(l) \qquad \Delta S_{\text{sys}} = S_{\text{liquid } H_2O} - S_{\text{gaseous } H_2O} < 0$$

As an example of a reaction during which entropy increases, consider the decomposition of dinitrogen tetroxide (written as $O_2N{-}NO_2$):

$$O_2N{-}NO_2(g) \longrightarrow 2NO_2(g)$$

When the N—N bond in 1 mol of dinitrogen tetroxide breaks, the 2 mol of NO_2 molecules have many more possible motions; thus, at any instant, the energy of the system is dispersed into any one of a larger number of microstates. Thus, the change in entropy of the system, which is the change in entropy of the reaction (ΔS_{rxn}), goes up:

$$\Delta S_{\text{sys}} = \Delta S_{\text{rxn}} = S_{\text{final}} - S_{\text{initial}} = S_{\text{products}} - S_{\text{reactants}} = 2S_{NO_2} - S_{N_2O_4} > 0$$

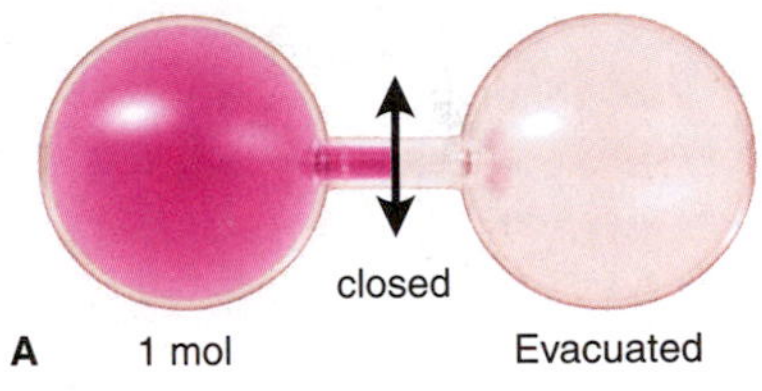

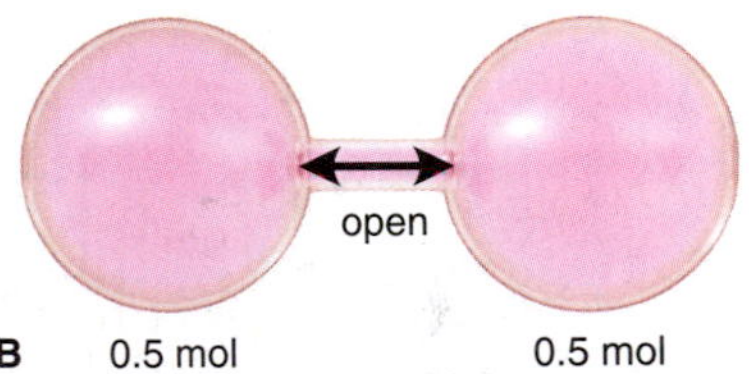

Figure 20.1 Spontaneous expansion of a gas. **A,** With the stopcock closed, the left flask contains 1 mol of Ne. **B,** With the stopcock open, the gas expands and each flask contains 0.5 mol of Ne.

Quantitative Meaning of an Entropy Change There are two approaches for quantifying an entropy change, but they give the same result. The first is based on the number of microstates possible for a system, and the second on the heat absorbed (or released) by the system. We'll examine both with a system of 1 mol of a gas expanding from 1 L to 2 L and behaving ideally, much as neon does at 298 K:

$$1 \text{ mol neon (initial: 1 L and 298 K)} \longrightarrow 1 \text{ mol neon (final: 2 L and 298 K)}$$

1. *Quantifying ΔS_{sys} from the number of microstates.* Figure 20.1 shows two flasks connected by a stopcock—the right flask is evacuated, and the left flask contains 1 mol of neon. When we open the stopcock, the gas expands until each flask contains 0.5 mol—but *why*? Opening the stopcock increases the volume, which increases the number of translational energy levels the particles can occupy as they move to more locations. Thus, the number of microstates—and the entropy—increases.

Figure 20.2 presents this idea with particles on energy levels in a box of changeable volume. When the stopcock opens, there are more energy levels, and they are closer together on average, so more distributions of particles are possible.

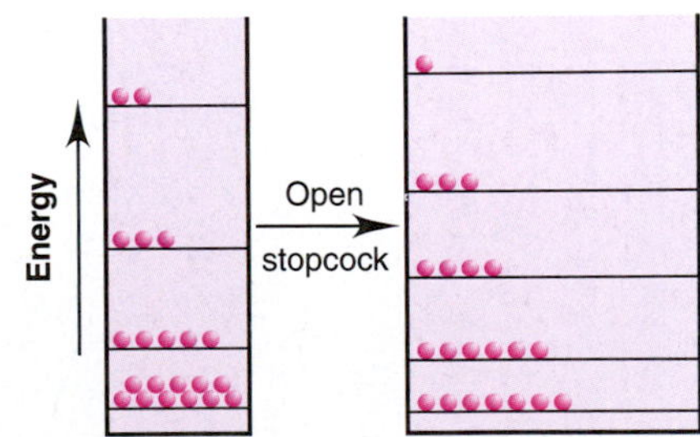

Figure 20.2 The entropy increase due to expansion of a gas. Energy levels are shown as lines in a box of narrow width *(left).* Each distribution of energies for the 21 particles is one microstate. When the stopcock is opened, the box is wider (volume increases, *right*), and the particles have more energy levels available.

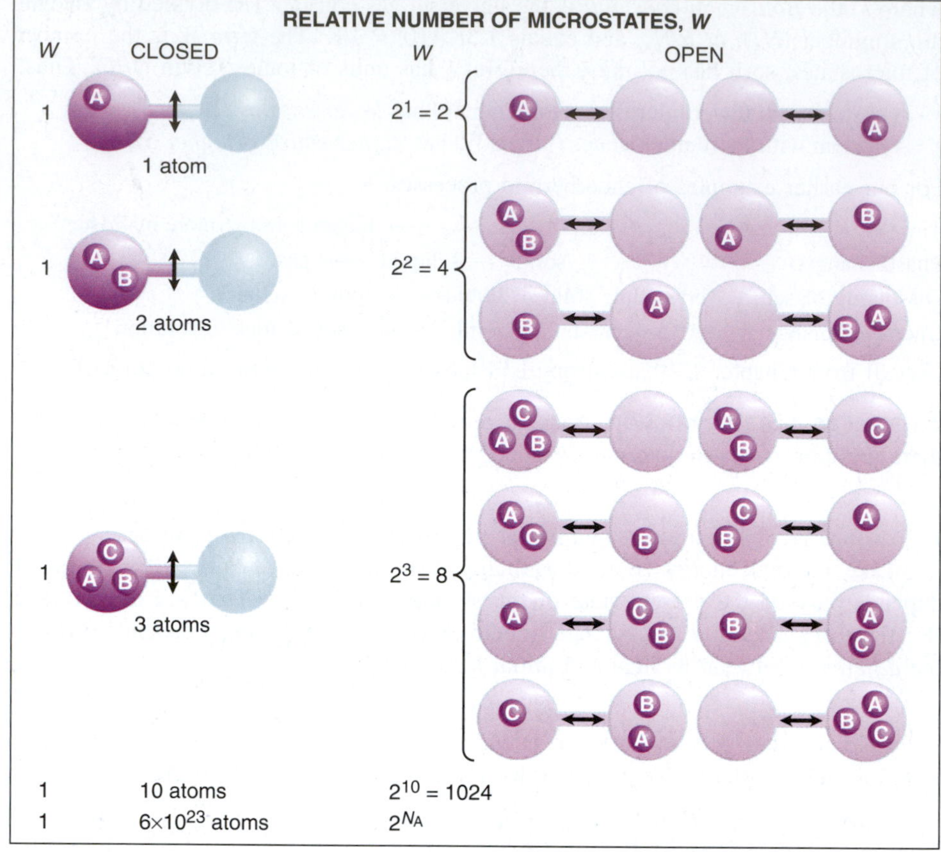

Figure 20.3 **Expansion of a gas and the increase in number of microstates.** Each set of particle locations represents a different microstate. When the volume increases (stopcock opens), the number of microstates is 2^n, where n is the number of particles.

In Figure 20.3 the number of microstates is represented by the placement of particles in the left and/or right flasks:

- *One Ne atom.* At a given instant, an Ne atom in the left flask has its energy in one of some number (W) of microstates. Opening the stopcock increases the volume, which increases the number of possible locations and the number of translational energy levels. Thus, the system has 2^1, or 2, times as many microstates available when the atom moves through both flasks (final state, W_{final}) as when it is confined to the left flask (initial state, $W_{initial}$).
- *Two Ne atoms.* For atoms A and B moving through both flasks, there are 2^2, or 4, times as many microstates as when the atoms were initially in the left flask—some number of microstates with both A and B in the left, that many with A in the left and B in the right or with B in the left and A in the right, and that many with both in the right.
- *Three Ne atoms.* Add another atom, and there are 2^3, or 8, times as many microstates when the stopcock is open.
- *Ten Ne atoms.* With 10 Ne atoms, there are 2^{10}, or 1024, times as many microstates for the atoms in both flasks as there were for the 10 atoms in the left flask.
- *One mole of Ne atoms.* With 1 mol (N_A) of Ne, there are 2^{N_A} times as many microstates for the atoms in the larger volume (W_{final}) than in the smaller ($W_{initial}$):

$$\frac{W_{final}}{W_{initial}} = 2^{N_A}$$

Now let's find ΔS_{sys} through the Boltzmann equation, $S = k \ln W$. From the properties of logarithms (Appendix A), we know that $\ln A - \ln B = \ln A/B$. Thus,

$$\Delta S_{sys} = S_{final} - S_{initial} = k \ln W_{final} - k \ln W_{initial} = k \ln \frac{W_{final}}{W_{initial}} = k \ln 2^{N_A}$$

Also, from Appendix A, $\ln A^y = y \ln A$; thus, with $k = R/N_A$, we have

$$\Delta S_{sys} = \frac{R}{N_A} \ln 2^{N_A} = \left(\frac{R}{N_A}\right) N_A \ln 2 = R \ln 2$$

$$= (8.314 \text{ J/mol}\cdot\text{K})(0.693) = 5.76 \text{ J/mol}\cdot\text{K}$$

2. *Quantifying ΔS_{sys} from the changes in heat.* Now we compare the ΔS_{sys} we just found for a gas expanding into an evacuated flask with the ΔS_{sys} for a gas that is heated and does work on the surroundings. This approach uses the relationship

$$\Delta S_{sys} = \frac{q_{rev}}{T} \tag{20.2}$$

where T is the temperature at which the heat change occurs and q is the heat absorbed. The subscript "rev" refers to a *reversible process,* one that occurs in such tiny increments that the system remains at equilibrium and the direction of the change can be reversed by an *infinitesimal* reversal of conditions.

We can approximate a reversible expansion by placing Ne gas in a piston-cylinder assembly within a heat reservoir (to maintain a constant T of 298 K) and start by confining it to a volume of 1 L by the "pressure" of a beaker of sand on the piston (Figure 20.4). We remove one grain of sand (an "infinitesimal" decrease in pressure) with a pair of tweezers, and the gas expands a tiny amount, raising the piston and doing work, $-w$. Assuming Ne behaves ideally, it absorbs an equivalent tiny increment of heat, q, from the heat reservoir. With each grain of sand removed, the expanding gas absorbs another tiny increment of heat. This process simulates a reversible expansion because we can reverse it by putting back a grain of sand, which causes the surroundings to do a tiny quantity of work compressing the gas, thus releasing a tiny quantity of heat to the reservoir.

Figure 20.4 Simulating a reversible process.

If we continue this nearly reversible expansion to 2 L and use calculus to integrate the tiny increments of heat together, q_{rev} is 1718 J. From Equation 20.2,

$$\Delta S_{sys} = \frac{q_{rev}}{T} = \frac{1718 \text{ J}}{298 \text{ K}} = 5.76 \text{ J/K}$$

This is the *same value* of ΔS_{sys} we obtained based on the number of microstates.

Entropy and the Second Law of Thermodynamics

While it's true that the change in entropy determines the direction of a spontaneous process, we must consider more than the entropy change of the system. After all, some processes, such as ice melting, occur spontaneously and ΔS_{sys} goes up, while others, such as water freezing, occur spontaneously and ΔS_{sys} goes down. But when we consider *both* the system *and* its surroundings, we find that *all real processes occur spontaneously in the direction that increases the entropy of the universe (system plus surroundings).* This is one way to state the **second law of thermodynamics.**

The second law says that *either* the entropy change of the system *or* of the surroundings may be negative. But, for a process to be spontaneous, the *sum* of the two entropy changes must be positive. If the entropy of the system decreases, the entropy of the surroundings must increase even more to offset that decrease, so that the entropy of the universe (system *plus* surroundings) increases. A quantitative statement of the second law is, for any real spontaneous process,

$$\Delta S_{univ} = \Delta S_{sys} + \Delta S_{surr} > 0 \tag{20.3}$$

Standard Molar Entropies and the Third Law

Entropy and enthalpy are state functions, but their values differ in a fundamental way:

- For *enthalpy,* there is no zero point, so we can measure only *changes.*
- For *entropy,* there *is* a zero point, and we can determine *absolute* values by applying the **third law of thermodynamics:** *a perfect crystal has zero entropy at absolute zero:* $S_{sys} = 0$ at 0 K.

A "perfect" crystal means all the particles are aligned flawlessly. At absolute zero, the particles have minimum energy, so there is only one microstate. Thus, in Equation 20.1,

$$W = 1 \quad \text{so} \quad S = k \ln 1 = 0$$

When we warm the crystal to any temperature above 0 K, the total energy increases, so it can be dispersed into more than one microstate. Thus,

$$W > 1 \quad \text{and} \quad \ln W > 0 \quad \text{so} \quad S > 0$$

In principle, to find S of a substance at a given temperature, we cool it as close to 0 K as possible. Then we heat it in small increments, dividing q by T to get the increase in S for each increment, and add up all the entropy increases to the temperature of interest, usually 298 K. Therefore, S of a substance at a given temperature is an *absolute* value. As with other thermodynamic variables,

- We compare entropy values for substances at the temperature of interest in their *standard states: 1 atm for gases, 1 M for solutions, and the pure substance in its most stable form for solids or liquids.*
- Because entropy is an *extensive* property—one that depends on the amount of substance—we specify the **standard molar entropy ($S°$),** in units of J/mol·K (or J mol^{-1} K^{-1}). ($S°$ values at 298 K for many elements, compounds, and ions appear, with other thermodynamic variables, in Appendix B.)

Predicting Relative $S°$ of a System

Let's see how the standard molar entropy of a substance is affected by several parameters: temperature, physical state, dissolution, and atomic size or molecular complexity. (Unless stated otherwise, the $S°$ values refer to the system at 298 K.)

Temperature Changes Temperature has a very important effect on entropy. For any substance, $S°$ *increases as T rises,* as these values for copper metal show:

T (K):	273	295	298
$S°$:	31.0	32.9	33.2

As heat is absorbed ($q > 0$), temperature, which is a measure of the average kinetic energy of the particles, increases. Recall that the kinetic energies of gas particles are distributed over a range, which becomes wider as T rises (Figure 5.13); liquids and solids behave the same. Thus, at any instant, there are more microstates available in which the energy can be dispersed, so the entropy of the substance goes up. Figure 20.5 presents three ways to view the effect of temperature on entropy.

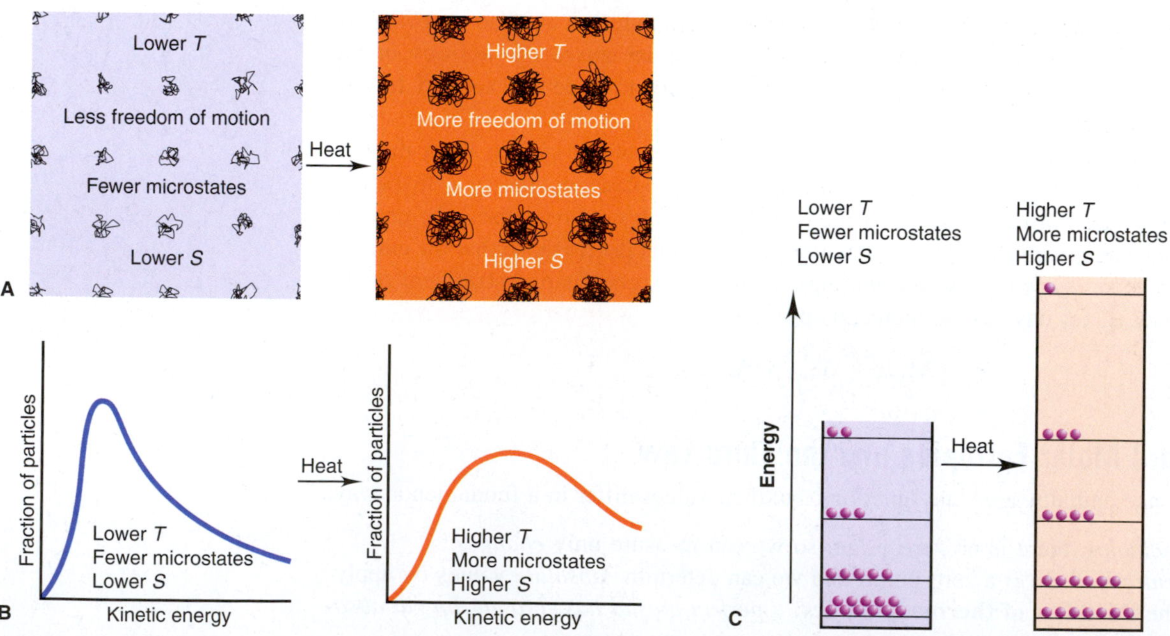

Figure 20.5 Visualizing the effect of temperature on entropy. **A,** Computer simulations show each particle in a crystal moving about its lattice position. Adding heat increases T and the total energy, so the particles have greater freedom of motion, and their energy is more dispersed. Thus, S increases. **B,** At any T, there is a range of occupied energy levels and, thus, a certain number of microstates. Adding heat increases the total energy *(area under curve),* so the range of occupied energy levels becomes greater, as does the number of microstates (higher S). **C,** A system of 21 particles occupy energy levels *(lines)* in a box whose height represents the total energy. When heat is added, the total energy increases *(box is higher) and* becomes more dispersed *(more lines),* so S increases.

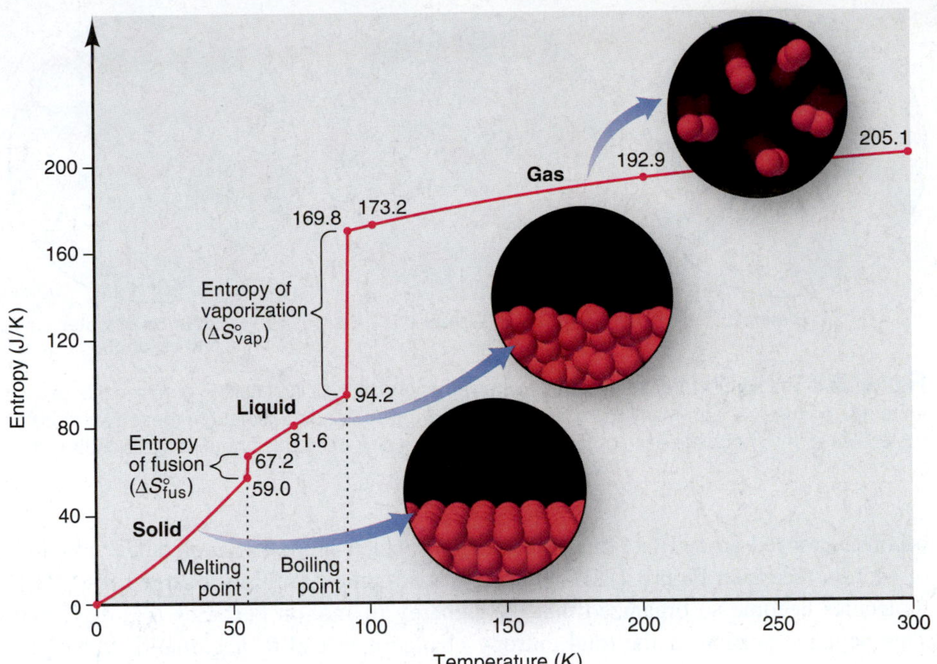

Figure 20.6 The increase in entropy during phase changes from solid to liquid to gas.

Physical States and Phase Changes In melting or vaporizing, heat is absorbed ($q > 0$). The particles have more freedom of motion, and their energy is more dispersed. Thus, S° *increases as the physical state of a substance changes from solid to liquid to gas:*

	Na	H_2O	C(graphite)
$S^\circ(s \text{ or } l)$:	51.4(s)	69.9(l)	5.7(s)
$S^\circ(g)$:	153.6	188.7	158.0

Figure 20.6 plots standard molar entropy versus temperature as solid O_2 is heated and changes to liquid and then to gas, with S° values at various points; this behavior is typical of many substances. At the molecular scale, several stages occur:

- Particles in the solid vibrate about their positions but, on average, remain fixed. The energy of the solid is dispersed least, that is, has the fewest microstates, so the solid has the lowest entropy.
- As T rises, the entropy increases gradually as the particles' kinetic energy increases.
- When the solid melts, the particles move much more freely between and around each other, so there is an abrupt increase in entropy (ΔS°_{fus}).
- Further heating of the liquid increases the speed of the particles, and the entropy increases gradually.
- When the liquid vaporizes and becomes a gas, the particles undergo a much larger, abrupt entropy increase (ΔS°_{vap}); *the increase in entropy from liquid to gas is much larger than from solid to liquid:* $\Delta S^\circ_{vap} >> \Delta S^\circ_{fus}$.
- Finally, with further heating of the gas, the entropy increases gradually.

Dissolving a Solid or a Liquid in Water Recall from Chapter 13 that, in general, entropy increases when a solute dissolves in a solvent: $S_{soln} > (S_{solute} + S_{solvent})$. But, when water is the solvent, the entropy may also depend on the nature of solute *and* solvent interactions and includes two opposing events:

	NaCl	$AlCl_3$	CH_3OH
$S^\circ(s \text{ or } l)$:	72.1(s)	167(s)	127(l)
$S^\circ(aq)$:	115.1	−148	132

1. *For ionic solutes,* when the crystal dissolves in water, the ions have much more freedom of motion and their energy is dispersed into more microstates. That is, the entropy of the ions themselves is greater in the solution. However, some water molecules

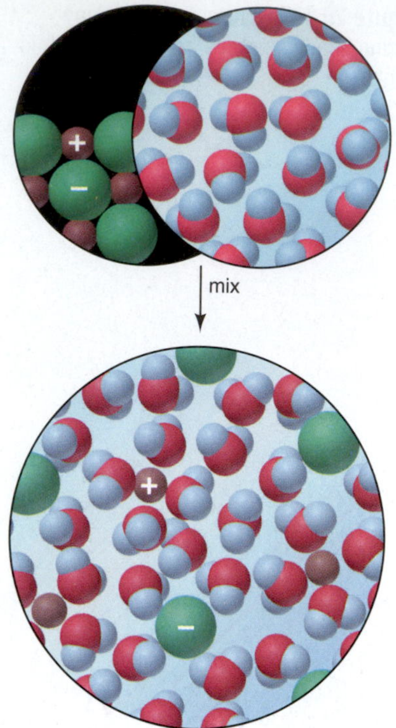

Figure 20.7 **The entropy change accompanying the dissolution of a salt.** The entropy of a salt solution is usually *greater* than that of the solid and of water, but it is affected by water molecules becoming organized around each ion.

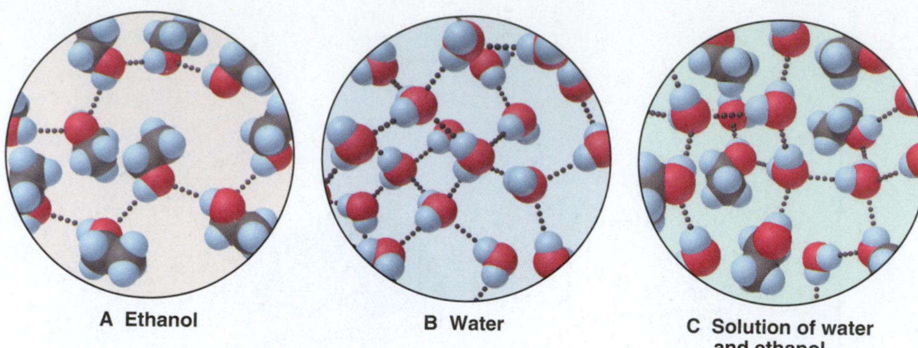

Figure 20.8 **The small increase in entropy when ethanol dissolves in water.** In pure ethanol **(A)** and water **(B)**, molecules form many H bonds to other like molecules. **C,** In solution, these two kinds of molecules form H bonds to each other, so their freedom of motion does not change significantly.

become arranged around the ions (Figure 20.7), which limits the molecules' freedom of motion (see also Figure 13.2). In fact, around small, multiply charged ions, H_2O molecules become so organized that their energy of motion becomes *less* dispersed. This negative portion of the total entropy change can lead to *negative* $S°$ values for ions in solution. In the case of $AlCl_3$, the $Al^{3+}(aq)$ ion has such a negative $S°$ value (−313 J/mol·K) that when $AlCl_3$ dissolves in water, even though $S°$ of $Cl^-(aq)$ is positive, the entropy of aqueous $AlCl_3$ is lower than that of solid $AlCl_3$.*

2. *For molecular solutes,* the increase in entropy upon dissolving is typically much smaller than for ionic solutes. After all, for a solid such as glucose, there is no separation into ions, and for a liquid such as methanol or ethanol (Figure 20.8), there is not even the breakdown of a crystal structure. Furthermore, in these small alcohols and in pure water, the molecules form many H bonds, so there is relatively little change in their freedom of motion either before or after they are mixed.

Dissolving a Gas in a Liquid The particles in a gas already have so much freedom of motion—and, thus, such highly dispersed energy—that they actually lose some when they dissolve in a liquid or solid. Therefore, the entropy of a solution of a gas in a liquid or in a solid is *less* than the entropy of the gas itself. For instance, when gaseous O_2 [$S°(g)$ = 205.0 J/mol·K] dissolves in water, its entropy decreases considerably [$S°(aq)$ = 110.9 J/mol·K] (Figure 20.9). When a gas dissolves in another gas, however, the entropy increases as a result of the separation and mixing of the molecules.

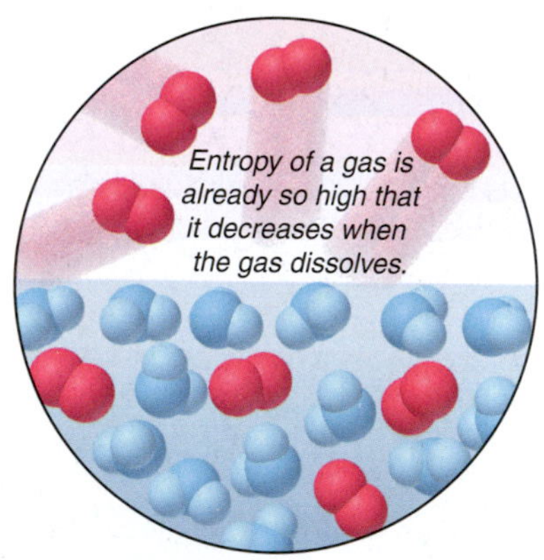

Figure 20.9 The entropy of a gas dissolved in a liquid.

Atomic Size or Molecular Complexity Differences in $S°$ values for substances in the same phase are usually based on atomic size and molecular complexity.

1. *Within a periodic group,* energy levels become closer together for larger, heavier atoms, so the number of microstates, and thus the molar entropy, increases:

	Li	Na	K	Rb	Cs
Atomic radius (pm):	152	186	227	248	265
Molar mass (g/mol):	6.941	22.99	39.10	85.47	132.9
$S°(s)$:	29.1	51.4	64.7	69.5	85.2

The same trend of increasing entropy holds for similar compounds down a group:

	HF	HCl	HBr	HI
Molar mass (g/mol):	20.01	36.46	80.91	127.9
$S°(g)$:	173.7	186.8	198.6	206.3

*An $S°$ value for a hydrated ion can be negative because it is relative to the $S°$ value for the hydrated proton, $H^+(aq)$, which is assigned a value of 0. In other words, $Al^{3+}(aq)$ has a lower entropy than $H^+(aq)$.

2. *For different forms of an element (allotropes),* the entropy is *higher* in the form that allows the atoms more freedom of motion. For example, $S°$ of graphite is 5.69 J/mol·K, whereas $S°$ of diamond is 2.44 J/mol·K. In graphite, covalent bonds extend within a two-dimensional sheet, and the sheets move past each other easily; in diamond, covalent bonds extend in three dimensions, allowing the atoms little movement (Table 12.6).

3. *For compounds,* entropy increases with chemical complexity (that is, with number of atoms in the formula), and this trend holds for ionic and covalent substances:

	NaCl	$AlCl_3$	P_4O_{10}	NO	NO_2	N_2O_4
$S°(s)$:	72.1	167	229			
$S°(g)$:				211	240	304

The trend is also based on the different types of motion. Thus, for example, among the nitrogen oxides listed, the number of different vibrational motions increases with the number of atoms in the molecule (Figure 20.10).

For larger compounds, we also consider the motion of different parts of a molecule. A long hydrocarbon chain can rotate and vibrate in more ways than a short one, so *entropy increases with chain length.* A ring compound, such as cyclopentane (C_5H_{10}), has lower entropy than the corresponding chain compound, pentene (C_5H_{10}), because the ring structure restricts freedom of motion:

	$CH_4(g)$	$C_2H_6(g)$	$C_3H_8(g)$	$C_4H_{10}(g)$	$C_5H_{10}(g)$	C_5H_{10}(cyclo, g)	$C_2H_5OH(l)$
$S°$:	186	230	270	310	348	293	161

Remember, these trends hold only for *substances in the same physical state.* Gaseous methane (CH_4) has higher entropy than liquid ethanol (C_2H_5OH), even though ethanol molecules are more complex. When gases are compared with liquids, *the effect of physical state dominates the effect of molecular complexity.*

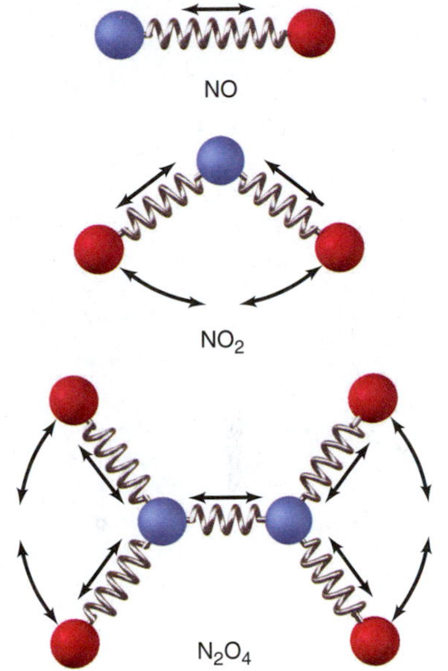

Figure 20.10 Entropy, vibrational motion, and molecular complexity.

Sample Problem 20.1 Predicting Relative Entropy Values

Problem Choose the member with the higher entropy in each of the following pairs, and justify your choice [assume constant temperature, except in part (e)]:

(a) 1 mol of $SO_2(g)$ or 1 mol of $SO_3(g)$
(b) 1 mol of $CO_2(s)$ or 1 mol of $CO_2(g)$
(c) 3 mol of $O_2(g)$ or 2 mol of $O_3(g)$
(d) 1 mol of $KBr(s)$ or 1 mol of $KBr(aq)$
(e) Seawater at 2°C or at 23°C
(f) 1 mol of $CF_4(g)$ or 1 mol of $CCl_4(g)$

Plan In general, particles with more freedom of motion have more microstates in which to disperse their kinetic energy, so they have higher entropy. We know that either raising temperature or having *more* particles increases entropy. We apply the general categories described in the text to choose the member with the higher entropy.

Solution (a) 1 mol of $SO_3(g)$. For equal numbers of moles of substances with the same types of atoms in the same physical state, the more atoms in the molecule, the more types of motion available, and thus the higher the entropy.

(b) 1 mol of $CO_2(g)$. For a given substance, entropy increases in the sequence $s < l < g$.

(c) 3 mol of $O_2(g)$. The two samples contain the same number of oxygen atoms but different numbers of molecules. Despite the greater complexity of O_3, the greater number of molecules dominates because there are many more microstates possible for 3 mol of particles than for 2 mol.

(d) 1 mol of $KBr(aq)$. The two samples have the same number of ions, but their motion is more limited and their energy less dispersed in the solid than in the solution.

(e) Seawater at 23°C. Entropy increases with rising temperature.

(f) 1 mol of $CCl_4(g)$. For similar compounds, entropy increases with molar mass.

FOLLOW-UP PROBLEM 20.1 Select the substance with the higher entropy in each pair, and give the reason for your choice (assume 1 mol of each at the same T):
(a) $PCl_3(g)$ or $PCl_5(g)$ **(b)** $CaF_2(s)$ or $BaCl_2(s)$ **(c)** $Br_2(g)$ or $Br_2(l)$

Summary of Section 20.1

- A change is spontaneous under specified conditions if it occurs without a continuous input of energy.
- Neither the first law of thermodynamics nor the sign of ΔH predicts the direction of a spontaneous change.
- Many spontaneous processes involve an increase in the freedom of motion of the system's particles and, thus, in the dispersal of the system's energy of motion.
- Entropy is a state function that measures the extent of energy dispersed into the number of microstates possible for a system. Each microstate consists of the quantized energy levels of the system at a given instant.
- The second law of thermodynamics states that, in a spontaneous process, the entropy of the universe (system plus surroundings) increases.
- Absolute entropy values can be found because perfect crystals have zero entropy at 0 K (third law of thermodynamics).
- Standard molar entropy, $S°$ (J/mol·K), is affected by temperature, phase changes, dissolution, and atomic size or molecular complexity.

20.2 • CALCULATING THE CHANGE IN ENTROPY OF A REACTION

Chemists are especially interested in learning how to predict the sign *and* calculate the value of the change in entropy that occurs during a reaction.

Entropy Changes in the System: Standard Entropy of Reaction (ΔS°_{rxn})

The **standard entropy of reaction, ΔS°_{rxn}**, is the entropy change that occurs when all reactants and products are in their standard states.

Predicting the Sign of ΔS°_{rxn} A deciding factor in predicting the sign of ΔS°_{rxn} is a change in the amount (mol) of gas, because gases have such great freedom of motion and, thus, high molar entropies: *if the number of moles of gas increases, ΔS°_{rxn} is positive; if the number decreases, ΔS°_{rxn} is negative.* Here are a few examples:

- *Increase in amount of gas.* When gaseous H_2 reacts with solid I_2 to form gaseous HI, the total number of moles of *substance* stays the same. Nevertheless, the sign of ΔS°_{rxn} is positive (entropy increases) because the number of moles of *gas* increases:
$$H_2(g) + I_2(s) \longrightarrow 2HI(g) \qquad \Delta S^\circ_{rxn} = S^\circ_{products} - S^\circ_{reactants} > 0$$
- *Decrease in amount of gas.* When ammonia forms from its elements, 4 mol of gas produce 2 mol of gas, so ΔS°_{rxn} is negative (entropy decreases):
$$N_2(g) + 3H_2(g) \rightleftharpoons 2NH_3(g) \qquad \Delta S^\circ_{rxn} = S^\circ_{products} - S^\circ_{reactants} < 0$$

In general, *we cannot predict the sign of the entropy change unless the reaction involves a change in number of moles of gas.*

Calculating ΔS°_{rxn} from $S°$ Values By applying Hess's law (Section 6.5), we combined ΔH°_f values to find ΔH°_{rxn}. Similarly, we combine $S°$ values to find the standard entropy of reaction, ΔS°_{rxn}:

$$\Delta S^\circ_{rxn} = \Sigma m S^\circ_{products} - \Sigma n S^\circ_{reactants} \qquad (20.4)$$

where m and n are the amounts (mol) of products and reactants, respectively, given by the coefficients in the balanced equation. For the formation of ammonia, we have

$$\Delta S^\circ_{rxn} = [(2 \text{ mol } NH_3)(S^\circ \text{ of } NH_3)] - [(1 \text{ mol } N_2)(S^\circ \text{ of } N_2) + (3 \text{ mol } H_2)(S^\circ \text{ of } H_2)]$$

From Appendix B, we find the $S°$ values:

$$\Delta S^\circ_{rxn} = [(2 \text{ mol})(193 \text{ J/mol·K})] - [(1 \text{ mol})(191.5 \text{ J/mol·K}) + (3 \text{ mol})(130.6 \text{ J/mol·K})]$$
$$= -197 \text{ J/K}$$

As we predicted from the decrease in number of moles of gas, $\Delta S^\circ_{rxn} < 0$.

Sample Problem 20.2 Calculating the Standard Entropy of Reaction, ΔS°_{rxn}

Problem Predict the sign of ΔS°_{rxn}, if possible, and calculate its value for the combustion of 1 mol of propane at 25°C:

$$C_3H_8(g) + 5O_2(g) \longrightarrow 3CO_2(g) + 4H_2O(l)$$

Plan We use the change in the number of moles of gas to predict the sign of ΔS°_{rxn}. The amount (mol) of gas decreases (6 mol yields 3 mol), so the entropy should decrease ($\Delta S^\circ_{rxn} < 0$). To find ΔS°_{rxn}, we apply Equation 20.4.

Solution Calculating ΔS°_{rxn}: Using Appendix B values,

$$\begin{aligned}\Delta S^\circ_{rxn} &= [(3 \text{ mol } CO_2)(S^\circ \text{ of } CO_2) + (4 \text{ mol } H_2O)(S^\circ \text{ of } H_2O)] \\ &\quad - [(1 \text{ mol } C_3H_8)(S^\circ \text{ of } C_3H_8) + (5 \text{ mol } O_2)(S^\circ \text{ of } O_2)] \\ &= [(3 \text{ mol})(213.7 \text{ J/mol·K}) + (4 \text{ mol})(69.9 \text{ J/mol·K})] \\ &\quad - [(1 \text{ mol})(269.9 \text{ J/mol·K}) + (5 \text{ mol})(205.0 \text{ J/mol·K})] \\ &= -374 \text{ J/K}\end{aligned}$$

Check $\Delta S^\circ < 0$, so our prediction is correct. Rounding gives $[3(200) + 4(70)] - [270 + 5(200)] = 880 - 1270 = -390$, close to the calculated value.

FOLLOW-UP PROBLEM 20.2 Balance the following equations, predict the sign of ΔS°_{rxn} if possible, and calculate its value at 25°C:

(a) $NaOH(s) + CO_2(g) \longrightarrow Na_2CO_3(s) + H_2O(l)$

(b) $Fe(s) + H_2O(g) \longrightarrow Fe_2O_3(s) + H_2(g)$

Entropy Changes in the Surroundings: The Other Part of the Total

In the synthesis of ammonia, the combustion of propane, and many other spontaneous reactions, the entropy of the system decreases ($\Delta S^\circ_{rxn} < 0$). Remember that the second law dictates that, for a spontaneous process, a ***decrease*** *in the entropy of the system is outweighed by an* ***increase*** *in the entropy of the surroundings.* In this section, we examine the influence of the surroundings—in particular, the addition (or removal) of heat and the temperature at which this heat flow occurs—on the *total* entropy change.

The Role of the Surroundings In essence, the surroundings *add heat to or remove heat from the system.* That is, the surroundings function as an enormous heat source or heat sink, one so large that its temperature remains constant, even though its entropy changes. The surroundings participate in the two types of enthalpy changes as follows:

1. *In an exothermic change, heat released by the system is absorbed by the surroundings.* More heat increases the freedom of motion of the particles and makes the energy more dispersed, so the entropy of the surroundings increases:

   For an exothermic change: $q_{sys} < 0$, $q_{surr} > 0$, and $\Delta S_{surr} > 0$

2. *In an endothermic change, heat absorbed by the system is released by the surroundings.* Less heat reduces the freedom of motion of the particles and makes the energy less dispersed, so the entropy of the surroundings decreases:

   For an endothermic change: $q_{sys} > 0$, $q_{surr} < 0$, and $\Delta S_{surr} < 0$

Temperature at Which Heat Is Transferred The *temperature* of the surroundings at the time the heat is transferred also affects ΔS_{surr}. Consider the effect of an exothermic reaction at a low or at a high temperature:

- At a low T, such as 20 K, there is little motion in the surroundings because there is little energy. This means there are few energy levels in each microstate and few microstates in which to disperse the energy. Transferring a given quantity of heat to these surroundings causes a relatively large change in how much energy is dispersed.
- At a high T, such as 298 K, the surroundings have a large quantity of energy dispersed. There are more energy levels in each microstate and a greater number of microstates. Transferring the same given quantity of heat to these surroundings causes a relatively small change in how much energy is dispersed.

In other words, ΔS_{surr} is greater when heat is added at a lower T (see analogy, *next page*).

THINK OF IT THIS WAY
Balancing Your Checkbook and Heating the Surroundings

A financial analogy may clarify the relative changes in entropy from heating the surroundings. If you have \$10 in your checking account, a \$10 deposit represents a 100% increase in your net worth. But if you have \$1000 in the account, the same \$10 deposit represents only a 1% increase. Thus, a given addition, whether of heat to the surroundings or money to your bank account, has a greater effect at a lower initial state than at a higher initial state.

Putting these ideas together, ΔS_{surr} *is directly related to an opposite change in the heat of the system* (q_{sys}) *and inversely related to the temperature at which the heat is transferred:*

$$\Delta S_{\text{surr}} = -\frac{q_{\text{sys}}}{T}$$

For a process at *constant pressure,* the heat (q_P) is ΔH (Section 6.2), so

$$\Delta S_{\text{surr}} = -\frac{\Delta H_{\text{sys}}}{T} \qquad \textbf{(20.5)}$$

Thus, we find ΔS_{surr} by measuring ΔH_{sys} and T at which the change takes place.

The main point: If a spontaneous reaction has a negative ΔS_{sys} (fewer microstates into which energy is dispersed), ΔS_{surr} must be positive enough (even more microstates into which energy is dispersed) for ΔS_{univ} to be positive (net increase in number of microstates for dispersing the energy).

Sample Problem 20.3 Determining Reaction Spontaneity

Problem At 298 K, the formation of ammonia has a negative $\Delta S^\circ_{\text{sys}}$:

$$N_2(g) + 3H_2(g) \longrightarrow 2NH_3(g) \qquad \Delta S^\circ_{\text{sys}} = -197 \text{ J/K}$$

Calculate ΔS_{univ}, and state whether the reaction occurs spontaneously at this temperature.

Plan For the reaction to occur spontaneously, $\Delta S_{\text{univ}} > 0$, so ΔS_{surr} must be greater than +197 J/K. To find ΔS_{surr}, we need $\Delta H^\circ_{\text{sys}}$, which is the same as $\Delta H^\circ_{\text{rxn}}$. We use $\Delta H^\circ_{\text{f}}$ values from Appendix B to find $\Delta H^\circ_{\text{rxn}}$. Then, we divide $\Delta H^\circ_{\text{rxn}}$ by the given T (298 K) to find ΔS_{surr}. To find ΔS_{univ}, we add the calculated ΔS_{surr} to the given $\Delta S^\circ_{\text{sys}}$ (−197 J/K).

Solution Calculating $\Delta H^\circ_{\text{sys}}$:

$$\begin{aligned}\Delta H^\circ_{\text{sys}} &= \Delta H^\circ_{\text{rxn}}\\ &= [(2 \text{ mol NH}_3)(-45.9 \text{ kJ/mol})] - [(3 \text{ mol H}_2)(0 \text{ kJ/mol}) + (1 \text{ mol N}_2)(0 \text{ kJ/mol})]\\ &= -91.8 \text{ kJ}\end{aligned}$$

Calculating ΔS_{surr}:

$$\Delta S_{\text{surr}} = -\frac{\Delta H^\circ_{\text{sys}}}{T} = -\frac{-91.8 \text{ kJ} \times \dfrac{1000 \text{ J}}{1 \text{ kJ}}}{298 \text{ K}} = 308 \text{ J/K}$$

Determining ΔS_{univ}:

$$\Delta S_{\text{univ}} = \Delta S^\circ_{\text{sys}} + \Delta S_{\text{surr}} = -197 \text{ J/K} + 308 \text{ J/K} = 111 \text{ J/K}$$

$\Delta S_{\text{univ}} > 0$, so the reaction occurs spontaneously at 298 K *(see margin).*

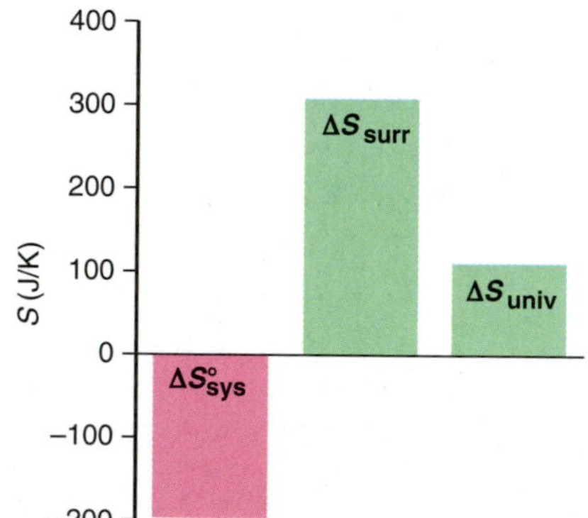

Check Rounding to check the math, we have

$$\begin{aligned}\Delta H^\circ_{\text{rxn}} &\approx 2(-45 \text{ kJ}) = -90 \text{ kJ}\\ \Delta S_{\text{surr}} &\approx -(-90{,}000 \text{ J})/300 \text{ K} = 300 \text{ J/K}\\ \Delta S_{\text{univ}} &\approx -200 \text{ J/K} + 300 \text{ J/K} = 100 \text{ J/K}\end{aligned}$$

Given the negative $\Delta H^\circ_{\text{rxn}}$, Le Châtelier's principle says that low T favors NH_3 formation, so the answer is reasonable (see Section 17.6).

Comment 1. Because ΔH° has units of kJ, and ΔS has units of J/K, don't forget to convert kJ to J, or you'll introduce a large error.

2. This example highlights the distinction between thermodynamics and kinetics. NH_3 forms spontaneously, but so slowly that catalysts are required to achieve a practical rate.

FOLLOW-UP PROBLEM 20.3 Does the oxidation of FeO(s) to Fe_2O_3(s) occur spontaneously at 298 K?

The Entropy Change and the Equilibrium State

For a process approaching equilibrium, $\Delta S_{univ} > 0$. When the process reaches equilibrium, there is no further *net* change, $\Delta S_{univ} = 0$, because any entropy change in the system is balanced by an opposite entropy change in the surroundings:

$$\text{At equilibrium:} \quad \Delta S_{univ} = \Delta S_{sys} + \Delta S_{surr} = 0 \quad \text{so} \quad \Delta S_{sys} = -\Delta S_{surr}$$

As an example, let's calculate ΔS_{univ} for the vaporization-condensation of 1 mol of water at 100°C (373 K),

$$H_2O(l; 373\ K) \rightleftharpoons H_2O(g; 373\ K)$$

First, we find ΔS°_{sys} for the forward change (vaporization) of 1 mol of water:

$$\Delta S^\circ_{sys} = \Sigma m S^\circ_{products} - \Sigma n S^\circ_{reactants} = S^\circ \text{ of } H_2O(g; 373\ K) - S^\circ \text{ of } H_2O(l; 373\ K)$$
$$= 195.9\ J/K - 86.8\ J/K = 109.1\ J/K$$

As we expect, the entropy of the system increases ($\Delta S^\circ_{sys} > 0$) as the liquid absorbs heat and changes to a gas.

For ΔS_{surr} of the vaporization step, we have

$$\Delta S_{surr} = -\frac{\Delta H^\circ_{sys}}{T}$$

where $\Delta H^\circ_{sys} = \Delta H^\circ_{vap}$ at 373 K = 40.7 kJ/mol = 40.7×10^3 J/mol. For 1 mol of water, we have

$$\Delta S_{surr} = -\frac{\Delta H^\circ_{vap}}{T} = -\frac{40.7\times10^3\ J}{373\ K} = -109\ J/K$$

The surroundings lose heat, and the negative sign means that the entropy of the surroundings decreases. The two entropy changes have the same magnitude but opposite signs, so they cancel:

$$\Delta S_{univ} = 109\ J/K + (-109\ J/K) = 0$$

For the reverse change (condensation), ΔS_{univ} also equals zero, but ΔS°_{sys} and ΔS_{surr} have signs opposite those for vaporization.

A similar treatment of a chemical change shows the same result: the entropy change of the forward reaction is *equal in magnitude but opposite in sign* to the entropy change of the reverse reaction. Thus, *when a system reaches equilibrium, neither the forward nor the reverse reaction is spontaneous,* and so there is no net reaction in either direction.

Spontaneous Exothermic and Endothermic Changes

No matter what its *enthalpy* change, a reaction occurs because the total *entropy* of the reacting system *and* its surroundings increases. There are two ways this can happen:

1. *In an exothermic reaction* ($\Delta H_{sys} < 0$), the heat released by the system increases the freedom of motion and dispersal of energy in the surroundings; thus, $\Delta S_{surr} > 0$.

- If the entropy of the products is *more* than that of the reactants ($\Delta S_{sys} > 0$), the total entropy change ($\Delta S_{sys} + \Delta S_{surr}$) will be positive (Figure 20.11A). For example, in the oxidation of glucose, an essential reaction for all higher organisms,

$$C_6H_{12}O_6(s) + 6O_2(g) \longrightarrow 6CO_2(g) + 6H_2O(g) + \textit{heat}$$

6 mol of gas yields 12 mol of gas; thus, $\Delta S_{sys} > 0$, $\Delta S_{surr} > 0$, and $\Delta S_{univ} > 0$.

Figure 20.11 Components of ΔS_{univ} for spontaneous reactions. For a reaction to occur spontaneously, ΔS_{univ} must be positive. **A,** An exothermic reaction in which ΔS_{sys} increases; the size of ΔS_{surr} is not important. **B,** An exothermic reaction in which ΔS_{sys} decreases; ΔS_{surr} must be larger than ΔS_{sys}. **C,** An endothermic reaction in which ΔS_{sys} increases; ΔS_{surr} must be smaller than ΔS_{sys}.

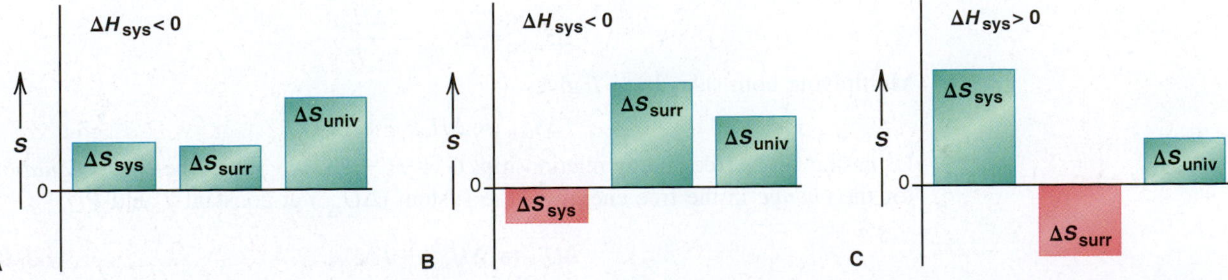

- If the entropy of the products is *less* than that of the reactants ($\Delta S_{sys} < 0$), the entropy of the surroundings must increase even more ($\Delta S_{surr} >> 0$) to make the total ΔS positive (Figure 20.11B). For example, when calcium oxide and carbon dioxide form calcium carbonate, the amount (mol) of gas decreases from 1 to 0:

$$CaO(s) + CO_2(g) \longrightarrow CaCO_3(s) + \textbf{\textit{heat}}$$

However, even though the system's entropy goes down, the heat released increases the entropy of the surroundings even more; thus, $\Delta S_{sys} < 0$, but $\Delta S_{surr} >> 0$, so $\Delta S_{univ} > 0$.

2. *In an endothermic reaction* ($\Delta H_{sys} > 0$), the heat absorbed by the system decreases molecular freedom of motion and dispersal of energy in the surroundings; thus, $\Delta S_{surr} < 0$. Therefore, the only way an endothermic reaction can occur spontaneously is if ΔS_{sys} is positive and large enough to outweigh the negative ΔS_{surr} (Figure 20.11C).

Summary of Section 20.2

- The standard entropy of reaction, ΔS°_{rxn}, is calculated from S° values.
- When the amount (mol) of gas increases in a reaction, usually $\Delta S^\circ_{rxn} > 0$.
- ΔS_{surr} is related directly to ΔH°_{sys} and inversely to the T at which a change occurs.
- In a spontaneous change, the entropy of the system can decrease only if the entropy of the surroundings increases even more, so that $\Delta S_{univ} > 0$.
- For a system at equilibrium, $\Delta S_{univ} = 0$, so $\Delta S^\circ_{sys} = -\Delta S_{surr}$.
- Even if $\Delta S^\circ_{sys} < 0$, an exothermic reaction ($\Delta H^\circ_{rxn} < 0$) is spontaneous ($\Delta S_{univ} > 0$) if $\Delta S_{surr} >> 0$; an endothermic reaction ($\Delta H^\circ_{rxn} > 0$) is spontaneous only if $\Delta S^\circ_{sys} > \Delta S_{surr}$.

20.3 • ENTROPY, FREE ENERGY, AND WORK

By measuring both ΔS_{sys} *and* ΔS_{surr}, we can predict whether a reaction will be spontaneous at a particular temperature. It would be useful, however, to have *one* criterion for spontaneity that we can determine by measuring the system only. The Gibbs free energy, or simply **free energy (*G*),** combines the system's enthalpy and entropy:

$$G = H - TS$$

Named for Josiah Willard Gibbs (1839–1903), the great, but largely unrecognized, American physicist who proposed it and laid much of the foundation for chemical thermodynamics, this function provides the criterion for spontaneity we've been seeking.

Free Energy Change and Reaction Spontaneity

The free energy change (ΔG) is a measure of the spontaneity of a process and of the useful energy available from it.

Deriving the Gibbs Equation Let's derive ΔG from the second law. By definition, the entropy change of the universe is the sum of the entropy changes of the system and the surroundings:

$$\Delta S_{univ} = \Delta S_{sys} + \Delta S_{surr}$$

At constant pressure,

$$\Delta S_{surr} = -\frac{\Delta H_{sys}}{T}$$

Substituting for ΔS_{surr} gives a relationship that relies solely on the system:

$$\Delta S_{univ} = \Delta S_{sys} - \frac{\Delta H_{sys}}{T}$$

Multiplying both sides by $-T$ gives

$$-T\Delta S_{univ} = \Delta H_{sys} - T\Delta S_{sys}$$

Using the Gibbs free energy relationship, $G = H - TS$, we obtain the *Gibbs equation* for the *change* in the free energy of the system (ΔG_{sys}) at constant T and P:

$$\Delta G_{sys} = \Delta H_{sys} - T\Delta S_{sys} \quad (20.6)$$

Combining Equation 20.6 with the one preceding it shows that

$$-T\Delta S_{univ} = \Delta H_{sys} - T\Delta S_{sys} = \Delta G_{sys}$$

Significance of the Sign of ΔG Let's see how the *sign* of ΔG tells if a reaction is spontaneous. According to the second law,

- $\Delta S_{univ} > 0$ for a spontaneous process
- $\Delta S_{univ} < 0$ for a nonspontaneous process
- $\Delta S_{univ} = 0$ for a process at equilibrium

Since the absolute temperature is always positive, for a spontaneous process,

$$T\Delta S_{univ} > 0 \quad \text{so} \quad -T\Delta S_{univ} < 0$$

From our derivation above, $\Delta G = -T\Delta S_{univ}$, so we have

- $\Delta G < 0$ for a spontaneous process
- $\Delta G > 0$ for a nonspontaneous process
- $\Delta G = 0$ for a process at equilibrium

Thus, we have not incorporated any new ideas, but we can predict reaction spontaneity from one variable (ΔG_{sys}) rather that two (ΔS_{sys} and ΔS_{surr}).

Calculating Standard Free Energy Changes

The *sign* of ΔG reveals *whether* a reaction is spontaneous, but the *magnitude* of ΔG tells *how* spontaneous it is. Because free energy (G) combines three state functions, *H, S,* and *T,* it is also a state function. As we do with enthalpy, we focus on the free energy *change* (ΔG). And, as we do with other thermodynamic variables, to compare the free energy changes of different reactions, we calculate the ***standard* free energy change ($\Delta G°$),** which occurs when all components of the system are in their standard states.

Using the Gibbs Equation to Find $\Delta G°$ One way to calculate $\Delta G°$ is by writing the Gibbs equation (20.6) at standard state conditions and using Appendix B to find ΔH°_{sys} and ΔS°_{sys}. Adapting the Gibbs equation, we have

$$\Delta G^\circ_{sys} = \Delta H^\circ_{sys} - T\Delta S^\circ_{sys} \quad (20.7)$$

This important relationship is used to find any one of these three variables, given the other two, as in Sample Problem 20.4.

Sample Problem 20.4 Calculating ΔG°_{rxn} from Enthalpy and Entropy Values

Problem Potassium chlorate, a common oxidizing agent in fireworks and matchheads, undergoes a solid-state disproportionation reaction when heated:

$$4\overset{+5}{KClO_3}(s) \xrightarrow{\Delta} 3\overset{+7}{KClO_4}(s) + \overset{-1}{KCl}(s)$$

Use ΔH°_f and $S°$ values to calculate ΔG°_{sys} (which is ΔG°_{rxn}) at 25°C for this reaction.

Plan To solve for $\Delta G°$, we need values from Appendix B. We use ΔH°_f values to calculate ΔH°_{rxn} (ΔH°_{sys}), use $S°$ values to calculate ΔS°_{rxn} (ΔS°_{sys}), and then apply Equation 20.7.

Solution Calculating ΔH°_{sys} from ΔH°_f values (with Equation 6.9):

$$\begin{aligned}\Delta H^\circ_{sys} &= \Delta H^\circ_{rxn} = \Sigma m\Delta H^\circ_{f(products)} - \Sigma n\Delta H^\circ_{f(reactants)}\\ &= [(3\text{ mol }KClO_4)(\Delta H^\circ_f\text{ of }KClO_4) + (1\text{ mol }KCl)(\Delta H^\circ_f\text{ of }KCl)]\\ &\quad - [(4\text{ mol }KClO_3)(\Delta H^\circ_f\text{ of }KClO_3)]\\ &= [(3\text{ mol})(-432.8\text{ kJ/mol}) + (1\text{ mol})(-436.7\text{ kJ/mol})]\\ &\quad - [(4\text{ mol})(-397.7\text{ kJ/mol})]\\ &= -144\text{ kJ}\end{aligned}$$

Calculating ΔS°_{sys} from S° values (with Equation 20.4):

$$\begin{aligned}\Delta S^\circ_{sys} = \Delta S^\circ_{rxn} &= \Sigma m S^\circ_{products} - \Sigma n S^\circ_{reactants}\\ &= [(3 \text{ mol } KClO_4)(S^\circ \text{ of } KClO_4) + (1 \text{ mol } KCl)(S^\circ \text{ of } KCl)]\\ &\quad - [(4 \text{ mol } KClO_3)(S^\circ \text{ of } KClO_3)]\\ &= [(3 \text{ mol})(151.0 \text{ J/mol·K}) + (1 \text{ mol})(82.6 \text{ J/mol·K})]\\ &\quad - [(4 \text{ mol})(143.1 \text{ J/mol·K})]\\ &= -36.8 \text{ J/K}\end{aligned}$$

Calculating ΔG°_{sys} at 298 K:

$$\Delta G^\circ_{sys} = \Delta H^\circ_{sys} - T\Delta S^\circ_{sys} = -144 \text{ kJ} - \left[(298 \text{ K})(-36.8 \text{ J/K})\left(\frac{1 \text{ kJ}}{1000 \text{ J}}\right)\right] = -133 \text{ kJ}$$

Check Rounding to check the math:

$$\Delta H^\circ \approx [3(-433 \text{ kJ}) + (-440 \text{ kJ})] - [4(-400 \text{ kJ})] = -1740 \text{ kJ} + 1600 \text{ kJ} = -140 \text{ kJ}$$

$$\Delta S^\circ \approx [3(150 \text{ J/K}) + 85 \text{ J/K}] - [4(145 \text{ J/K})] = 535 \text{ J/K} - 580 \text{ J/K} = -45 \text{ J/K}$$

$$\Delta G^\circ \approx -140 \text{ kJ} - 300 \text{ K}(-0.04 \text{ kJ/K}) = -140 \text{ kJ} + 12 \text{ kJ} = -128 \text{ kJ}$$

Comment **1.** Recall from Section 20.1 that reaction spontaneity indicates nothing about rate. Even though this reaction is spontaneous, the rate is very low in the solid. When $KClO_3$ is heated slightly above its melting point, the ions can move and the reaction occurs readily. **2.** For a spontaneous reaction under *any* conditions, the free energy change, ΔG, is negative. Under standard-state conditions, a spontaneous reaction has a negative *standard* free energy change: $\Delta G^\circ < 0$.

FOLLOW-UP PROBLEM 20.4 Determine the standard free energy change at 298 K for the reaction $2NO(g) + O_2(g) \longrightarrow 2NO_2(g)$.

Using Standard Free Energies of Formation to Find ΔG°_{rxn} Another way to calculate ΔG°_{rxn} is with values for the **standard free energy of formation (ΔG°_f)** of the components; ΔG°_f is the free energy change that occurs when 1 mol of compound is made *from its elements,* with all components in their standard states. Because free energy is a state function, we can combine ΔG°_f values of reactants and products to calculate ΔG°_{rxn}, no matter how the reaction takes place:

$$\Delta G^\circ_{rxn} = \Sigma m \Delta G^\circ_{f(products)} - \Sigma n \Delta G^\circ_{f(reactants)} \qquad (20.8)$$

ΔG°_f values have properties similar to ΔH°_f values:

- ΔG°_f of an element in its standard state is zero.
- An equation coefficient (*m* or *n* above) multiplies ΔG°_f by that number.
- Reversing a reaction changes the sign of ΔG°_f.

Many ΔG°_f values appear along with those for ΔH°_f and S° in Appendix B.

Sample Problem 20.5 Calculating ΔG°_{rxn} from ΔG°_f Values

Problem Use ΔG°_f values to calculate ΔG°_{rxn} for the reaction in Sample Problem 20.4:

$$4KClO_3(s) \longrightarrow 3KClO_4(s) + KCl(s)$$

Plan We apply Equation 20.8 to calculate ΔG°_{rxn}.

Solution Applying Equation 20.8 with values from Appendix B:

$$\begin{aligned}\Delta G^\circ_{rxn} &= \Sigma m \Delta G^\circ_{f(products)} - \Sigma n \Delta G^\circ_{f(reactants)}\\ &= [(3 \text{ mol } KClO_4)(\Delta G^\circ_f \text{ of } KClO_4) + (1 \text{ mol } KCl)(\Delta G^\circ_f \text{ of } KCl)]\\ &\quad - [(4 \text{ mol } KClO_3)(\Delta G^\circ_f \text{ of } KClO_3)]\\ &= [(3 \text{ mol})(-303.2 \text{ kJ/mol}) + (1 \text{ mol})(-409.2 \text{ kJ/mol})]\\ &\quad - [(4 \text{ mol})(-296.3 \text{ kJ/mol})]\\ &= -134 \text{ kJ}\end{aligned}$$

Check Rounding to check the math:

$$\begin{aligned}\Delta G^\circ_{rxn} &\approx [3(-300 \text{ kJ}) + 1(-400 \text{ kJ})] - 4(-300 \text{ kJ})\\ &= -1300 \text{ kJ} + 1200 \text{ kJ} = -100 \text{ kJ}\end{aligned}$$

Comment The discrepancy between this answer and the one obtained in Sample Problem 20.4 is due to rounding. As you can see, when ΔG_f° values are available, this method is simpler arithmetically than the approach in Sample Problem 20.4.

FOLLOW-UP PROBLEM 20.5 Use ΔG_f° values to calculate the free energy change at 25°C for each of the following reactions:
(a) $2NO(g) + O_2(g) \longrightarrow 2NO_2(g)$ (from Follow-up Problem 20.4)
(b) $2C(\text{graphite}) + O_2(g) \longrightarrow 2CO(g)$

The Free Energy Change and the Work a System Can Do

Thermodynamics developed after the invention of the steam engine, a major advance that spawned a new generation of machines. Thus, some of the field's key ideas applied the relationships between the free energy change and the work a system can do:

- ΔG is the *maximum useful work* done **by** a system during a *spontaneous* process at constant T and P:

$$\Delta G = w_{\text{max}} \qquad \textbf{(20.9)}$$

- ΔG is the *minimum work* done ***to*** a system to make a *nonspontaneous* process occur at constant T and P.

The free energy change is the maximum work a system can *possibly* do. But the work the system *actually* does depends on how the free energy is released. The maximum work can be done only in an infinite number of steps; that is, *the maximum work is done by a spontaneous process only if it is carried out reversibly* (recall Figure 20.4). In any *real* process, work is done irreversibly—in a finite number of steps—so *we can never obtain the maximum work, and the free energy not used for work is lost as heat.*

Consider the work done by a *battery,* a packaged spontaneous redox reaction that releases free energy to the surroundings (flashlight, radio, motor, or other device). If we connect the battery terminals to each other through a short piece of wire, ΔG_{sys} is released all at once but does no work—it just heats the wire and battery and outside air, which increases the freedom of motion of the particles in the universe. If we connect the battery terminals to a motor, ΔG_{sys} is released more slowly, and much of it runs the motor; however, some is still lost as heat. Only if a battery could discharge infinitely slowly could we obtain the maximum work. This is the compromise that all engineers must face—*no real process uses all the available free energy to do work because some is always "wasted" as heat.*

Let's summarize the relationship between the free energy change of a reaction and the work it can do:

- A spontaneous reaction ($\Delta G_{\text{sys}} < 0$) will do work on the surroundings ($-w$). In any real machine, the actual work done is *always less than the maximum* because some of the ΔG is released as heat.
- A nonspontaneous reaction ($\Delta G_{\text{sys}} > 0$) will occur only if the surroundings do work on the system ($+w$). In any real machine, the actual work done on the system is *always more than the minimum* because some of the added free energy is wasted as heat.
- A reaction at equilibrium ($\Delta G_{\text{sys}} = 0$) can no longer do any work.

The Effect of Temperature on Reaction Spontaneity

In most cases, the enthalpy contribution (ΔH) to the free energy change (ΔG) is much *larger* than the entropy contribution ($T\Delta S$). In fact, the reason most exothermic reactions are spontaneous is that the large negative ΔH makes ΔG negative. However, the *temperature of a reaction influences the magnitude of the $T\Delta S$ term,* so, for many reactions, the overall spontaneity depends on the temperature.

From the signs of ΔH and ΔS, we can predict how the temperature affects the sign of ΔG. The values we'll use for the thermodynamic variables in the following examples are standard state values from Appendix B, but we show them without the degree sign to emphasize that the relationships among ΔG, ΔH, and ΔS are valid at any conditions. Also, we assume that ΔH and ΔS change little with temperature, which is true as long as no phase change occurs.

Let's examine the four combinations of positive and negative ΔH and ΔS—two that are independent of temperature and two that are dependent on temperature:

- *Temperature-independent cases.* When ΔH and ΔS have *opposite* signs, the reaction occurs spontaneously either at all temperatures or at none (nonspontaneous).
  1. *Reaction is spontaneous at all temperatures:* $\Delta H < 0$, $\Delta S > 0$. Since ΔS is positive, $-T\Delta S$ is negative; thus, both contributions favor a negative ΔG. Most combustion reactions are in this category. The decomposition of hydrogen peroxide, a common disinfectant, is also spontaneous at all temperatures:

$$2H_2O_2(l) \longrightarrow 2H_2O(l) + O_2(g)$$
$$\Delta H = -196 \text{ kJ} \quad \text{and} \quad \Delta S = 125 \text{ J/K}$$

  2. *Reaction is nonspontaneous at all temperatures:* $\Delta H > 0$, $\Delta S < 0$. Both contributions oppose spontaneity: ΔH is positive and ΔS is negative, so $-T\Delta S$ is positive; thus, ΔG is always positive. The formation of ozone from oxygen requires a continual energy input, so it is not spontaneous at any temperature:

$$3O_2(g) \longrightarrow 2O_3(g)$$
$$\Delta H = 286 \text{ kJ} \quad \text{and} \quad \Delta S = -137 \text{ J/K}$$

- *Temperature-dependent cases.* When ΔH and ΔS have the *same* sign, the relative magnitudes of $-T\Delta S$ and ΔH determine the sign of ΔG. In these cases, the *direction* of the change in T is crucial.
  3. *Reaction becomes spontaneous as temperature increases:* $\Delta H > 0$ *and* $\Delta S > 0$. With a positive ΔH, the reaction will occur spontaneously only when $-T\Delta S$ becomes large enough to make ΔG negative, which will happen as temperatures rise. For example,

$$2N_2O(g) + O_2(g) \longrightarrow 4NO(g) \qquad \Delta H = 197.1 \text{ kJ} \quad \text{and} \quad \Delta S = 198.2 \text{ J/K}$$

  The oxidation of N_2O occurs spontaneously at any $T > 994$ K.
  4. *Reaction becomes spontaneous as temperature decreases:* $\Delta H < 0$ *and* $\Delta S < 0$. Here, ΔH favors spontaneity, but ΔS does not ($-T\Delta S > 0$). The reaction will occur spontaneously only when $-T\Delta S$ becomes smaller than ΔH, and this happens as temperatures drop. For example,

$$4Fe(s) + 3O_2(g) \longrightarrow 2Fe_2O_3(s) \qquad \Delta H = -1651 \text{ kJ} \quad \text{and} \quad \Delta S = -549.4 \text{ J/K}$$

  The production of iron(III) oxide occurs spontaneously at any $T < 3005$ K.

Table 20.1 summarizes these four possible combinations of ΔH and ΔS, and Sample Problem 20.6 applies them.

Table 20.1 Reaction Spontaneity and the Signs of ΔH, ΔS, and ΔG

ΔH	ΔS	$-T\Delta S$	ΔG	Description
−	+	−	−	Spontaneous at all T
+	−	+	+	Nonspontaneous at all T
+	+	−	+ or −	Spontaneous at higher T; nonspontaneous at lower T
−	−	+	+ or −	Spontaneous at lower T; nonspontaneous at higher T

Sample Problem 20.6 Using Molecular Scenes to Determine the Signs of ΔH, ΔS, and ΔG

Problem The following scenes represent a familiar phase change for water *(blue spheres)*:

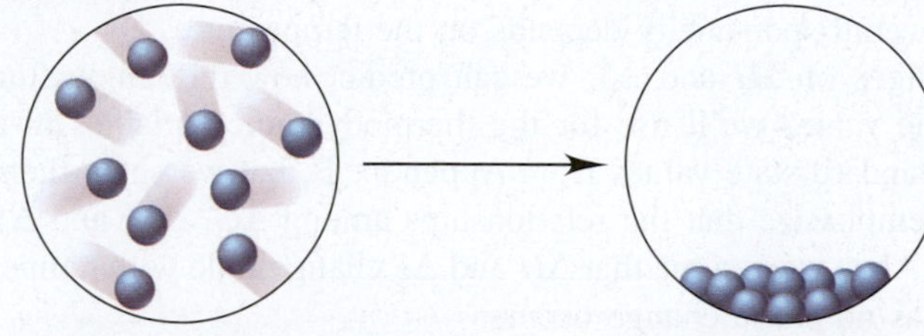

(a) What are the signs of ΔH and ΔS for this process? Explain.
(b) Is the process spontaneous at all T, no T, low T, or high T? Explain.

Plan (a) From the scenes, we determine any change in amount of gas, which indicates the sign of ΔS, and any change in the freedom of motion of the particles, which indicates whether heat is absorbed or released, and thus the sign of ΔH. **(b)** The question refers to the sign of ΔG (+ or −) at the different temperature possibilities, so we apply Equation 20.6 and refer to the previous text discussion and Table 20.1.

Solution (a) The scene represents the condensation of water vapor, so the amount of gas decreases dramatically, and the separated molecules give up energy as they come closer together. Therefore, $\Delta S < 0$ and $\Delta H < 0$. **(b)** With ΔS negative, the $-T\Delta S$ term is positive. In order for $\Delta G < 0$, the magnitude of T must be small. Therefore, the process is spontaneous at low T.

Check The answer in part (b) seems reasonable based on our analysis in part (a). The answer makes sense because we know from everyday experience that water condenses spontaneously, and it does so at low temperatures.

FOLLOW-UP PROBLEM 20.6 The following molecular scenes represent the gas-phase decomposition of X_2Y_2 to X_2 *(red)* and Y_2 *(blue)*:

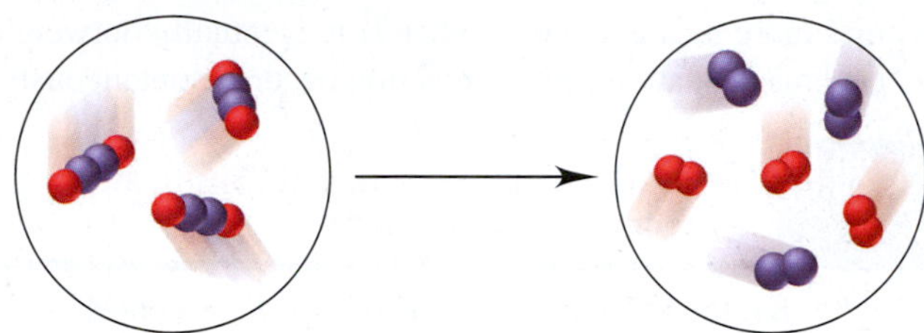

(a) What is the sign of ΔS for the reaction?
(b) If the reaction is spontaneous only above 325°C, what is the sign of ΔH? Explain.

As you saw in Sample Problem 20.4, one way to calculate ΔG is from enthalpy and entropy changes. As long as phase changes don't occur, ΔH and ΔS usually change little with temperature, so we use their values at 298 K in Sample Problem 20.7 to examine the effect of T on ΔG and, thus, on reaction spontaneity.

Sample Problem 20.7 Determining the Effect of Temperature on ΔG

Problem A key step in the production of sulfuric acid is the oxidation of $SO_2(g)$ to $SO_3(g)$:

$$2SO_2(g) + O_2(g) \longrightarrow 2SO_3(g)$$

At 298 K, $\Delta G = -141.6$ kJ; $\Delta H = -198.4$ kJ; and $\Delta S = -187.9$ J/K.
(a) Use the data to decide if this reaction is spontaneous at 25°C, and predict how ΔG will change with increasing T.
(b) Assuming that ΔH and ΔS are constant with T (no phase change occurs), is the reaction spontaneous at 900.°C?

Plan (a) We note the sign of ΔG to see if the reaction is spontaneous and the signs of ΔH and ΔS to see the effect of T. **(b)** We use Equation 20.7 to calculate ΔG from the given ΔH and ΔS at the higher T (in K).

Solution (a) $\Delta G < 0$, so the reaction is spontaneous at 298 K: SO_2 and O_2 will form SO_3 spontaneously. With $\Delta S < 0$, the term $-T\Delta S > 0$, and this term will become more positive at higher T. Therefore, ΔG will become less negative, and the reaction less spontaneous, with increasing T.
(b) Calculating ΔG at 900.°C (T = 273 + 900. = 1173 K):

$$\Delta G = \Delta H - T\Delta S = -198.4 \text{ kJ} - [(1173 \text{ K})(-187.9 \text{ J/K})(1 \text{ kJ}/1000 \text{ J})] = 22.0 \text{ kJ}$$

$\Delta G > 0$, so the reaction is nonspontaneous at the higher T.

Check The answer in part (b) seems reasonable based on our prediction in part (a). The arithmetic seems correct, given considerable rounding:

$$\Delta G \approx -200 \text{ kJ} - [(1200 \text{ K})(-200 \text{ J/K})/1000 \text{ J}] = +40 \text{ kJ}$$

FOLLOW-UP PROBLEM 20.7 A reaction is nonspontaneous at room temperature but *is* spontaneous at −40°C. What can you say about the signs and relative magnitudes of ΔH, ΔS, and $-T\Delta S$?

The Temperature at Which a Reaction Becomes Spontaneous As you've just seen, when the signs are the same for ΔH and ΔS of a reaction, it can be nonspontaneous at one temperature and spontaneous at another. The "crossover" temperature occurs when a positive ΔG switches to a negative ΔG because of the magnitude of the $-T\Delta S$ term. We find this temperature by setting ΔG equal to zero and solving for T:

$$\Delta G = \Delta H - T\Delta S = 0$$

Therefore,

$$\Delta H = T\Delta S \quad \text{and} \quad T = \frac{\Delta H}{\Delta S} \qquad \textbf{(20.10)}$$

Consider the reaction of copper(I) oxide with carbon. It does *not* occur at low temperature but does at high temperature and is used to extract copper from one of its ores:

$$Cu_2O(s) + C(s) \xrightarrow{\Delta} 2Cu(s) + CO(g)$$

We predict that this reaction has a positive ΔS because the number of moles of gas increases; in fact, $\Delta S = 165$ J/K. Furthermore, because the reaction is *non*spontaneous at lower temperatures, it must have a positive ΔH; the actual value is 58.1 kJ. As the $-T\Delta S$ term becomes more negative with higher T, it eventually outweighs the positive ΔH term, so ΔG becomes negative and the reaction occurs spontaneously.

Sample Problem 20.8 Finding the Temperature at Which a Reaction Becomes Spontaneous

Problem At 25°C (298 K), the reduction of copper(I) oxide to copper is nonspontaneous ($\Delta G = 8.9$ kJ). Calculate the temperature at which the reaction becomes spontaneous.

Plan As just discussed, we want the temperature at which ΔG crosses over from a positive to a negative value. We set ΔG equal to zero, and use Equation 20.10 to solve for T, using the ΔH (58.1 kJ) and ΔS (165 J/K) values from the text.

Solution From $\Delta G = \Delta H - T\Delta S = 0$, we have

$$T = \frac{\Delta H}{\Delta S} = \frac{58.1\ \text{kJ} \times \dfrac{1000\ \text{J}}{1\ \text{kJ}}}{165\ \text{J/K}} = 352\ \text{K}$$

Thus, at any temperature above 352 K (79°C), which is a moderate temperature for extracting a metal from its ore, $\Delta G < 0$, so the reaction becomes spontaneous.

Check Rounding to quickly check the math gives

$$T = \frac{60000\ \text{J}}{150\ \text{J/K}} = 400\ \text{K}$$

which is close to the answer.

FOLLOW-UP PROBLEM 20.8 Use Appendix B values to find the temperature at which the following reaction becomes spontaneous (assume ΔH and ΔS are constant with T):

$$CaO(s) + CO_2(g) \longrightarrow CaCO_3(s)$$

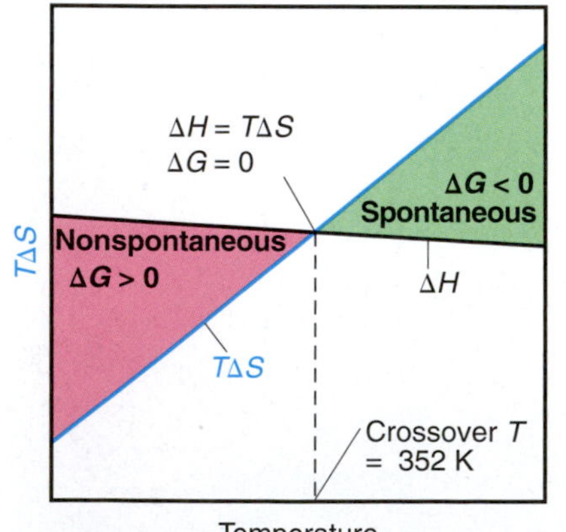

Figure 20.12 The effect of temperature on reaction spontaneity. At low T, $\Delta G > 0$ because ΔH dominates. At some temperature, $\Delta H = T\Delta S$, so $\Delta G = 0$. At any higher T, $\Delta G < 0$ because $-T\Delta S$ dominates.

For reactions with both ΔH and $\Delta S > 0$, Figure 20.12 shows that the line for $T\Delta S$ rises steadily (and thus the $-T\Delta S$ term becomes more negative) with increasing temperature. In Sample Problem 20.8, this line crosses the relatively constant ΔH line at 352 K. At any higher T, the $-T\Delta S$ term is greater than the ΔH term, so ΔG is negative.

Coupling of Reactions to Drive a Nonspontaneous Change

In a complex, multistep reaction, we often see a nonspontaneous step driven by a spontaneous step. In such a **coupling of reactions,** *one step supplies enough free energy for the other to occur,* just as burning gasoline supplies enough free energy to move a car. We'll consider coupling in an inorganic and a biochemical system.

1. Reducing Cu_2O with C. Look again at the reduction of copper(I) oxide by carbon. In Sample Problem 20.8, we found that the *overall* reaction becomes spontaneous at any temperature above 352 K. Dividing the reaction into two steps, however, we find that even at a higher temperature, such as 375 K, copper(I) oxide does not spontaneously decompose to its elements:

$$Cu_2O(s) \longrightarrow 2Cu(s) + \tfrac{1}{2}O_2(g) \qquad \Delta G_{375} = 140.0 \text{ kJ}$$

However, the oxidation of carbon to CO at 375 K is quite spontaneous:

$$C(s) + \tfrac{1}{2}O_2(g) \longrightarrow CO(g) \qquad \Delta G_{375} = -143.8 \text{ kJ}$$

Coupling these reactions allows the reaction with the larger negative ΔG to "drive" the one with the smaller positive ΔG. Adding the reactions together gives

$$Cu_2O(s) + C(s) \longrightarrow 2Cu(s) + CO(g) \qquad \Delta G_{375} = -3.8 \text{ kJ}$$

2. *Phosphorylating glucose with ATP.* Many biochemical reactions—including key steps in the syntheses of proteins, nucleic acids, and fatty acids, the maintenance of ion balance, and the breakdown of nutrients—have nonspontaneous steps. Coupling these steps to spontaneous ones is a strategy common to all organisms.

A key spontaneous biochemical reaction is the hydrolysis of a high-energy molecule called **adenosine triphosphate (ATP)** to adenosine diphosphate (ADP):

$$ATP^{4-} + H_2O \rightleftharpoons ADP^{3-} + HPO_4^{2-} + H^+ \qquad \Delta G^{\circ\prime} = -30.5 \text{ kJ}$$

(For biochemical systems, the standard-state concentration of H^+ is 10^{-7} *M*, not the usual 1 *M*, and the standard free energy change has the symbol $\Delta G^{\circ\prime}$.) In the metabolic breakdown of glucose, for example, the result of the initial step, which is nonspontaneous, is the phosphorylation (addition of a phosphate group) to glucose:

$$\text{Glucose} + HPO_4^{2-} + H^+ \rightleftharpoons [\text{glucose phosphate}]^- + H_2O \qquad \Delta G^{\circ\prime} = 13.8 \text{ kJ}$$

Coupling this reaction to ATP hydrolysis makes the overall process spontaneous. If we add the two reactions, HPO_4^{2-}, H^+, and H_2O cancel, and we obtain

$$\text{Glucose} + ATP^{4-} \rightleftharpoons [\text{glucose phosphate}]^- + ADP^{3-} \qquad \Delta G^{\circ\prime} = -16.7 \text{ kJ}$$

Coupling of the two reactions is accomplished through an enzyme (Section 16.7) that simultaneously binds glucose and ATP and catalyzes the transfer of the phosphate group. The ADP is combined with phosphate to regenerate ATP in reactions catalyzed by other enzymes. Thus, there is a continuous cycling of ATP to ADP and back to ATP again to supply energy to cells (Figure 20.13).

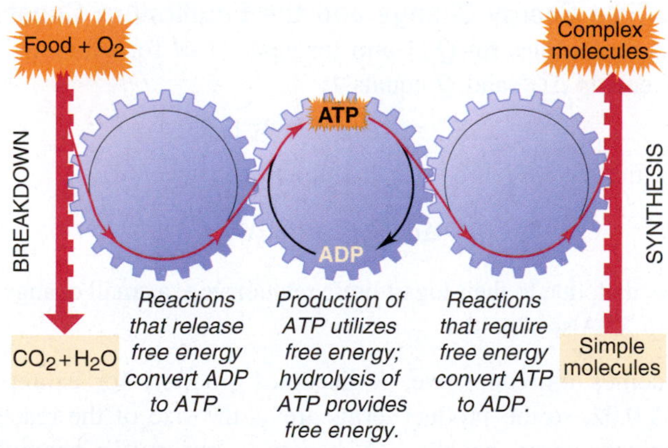

Figure 20.13 **The cycling of metabolic free energy through ATP.** Processes that release free energy are coupled to the formation of ATP from ADP, whereas those that require free energy are coupled to the hydrolysis of ATP to ADP.

Summary of Section 20.3

- The sign of the free energy change, $\Delta G = \Delta H - T\Delta S$, is directly related to reaction spontaneity: a negative ΔG corresponds to a positive ΔS_{univ}.
- We use the standard free energy of formation (ΔG_f°) to calculate ΔG_{rxn}° at 298 K.
- The maximum work a system can do is never obtained from a real (irreversible) process because some free energy is always converted to heat.
- The magnitude of T influences the spontaneity of a temperature-dependent reaction (same signs of ΔH and ΔS) by affecting the size of $T\Delta S$. For such a reaction, the T at which the reaction becomes spontaneous can be found by setting $\Delta G = 0$.
- A nonspontaneous reaction ($\Delta G > 0$) can be coupled to a more spontaneous one ($\Delta G << 0$) to make it occur. For example, in organisms, the hydrolysis of ATP drives many reactions that have a positive ΔG.

20.4 • FREE ENERGY, EQUILIBRIUM, AND REACTION DIRECTION

As you know from discussions in earlier chapters, the sign of ΔG is not the only way to predict reaction direction. In Chapter 17, we did so by comparing the values of the reaction quotient (Q) and the equilibrium constant (K). Recall that

- If $Q < K$ ($Q/K < 1$): reaction proceeds spontaneously to the right.
- If $Q > K$ ($Q/K > 1$): reaction proceeds spontaneously to the left.
- If $Q = K$ ($Q/K = 1$): reaction has attained equilibrium and no longer proceeds spontaneously in either direction.

It is easier to see the relation between these two ways to predict reaction spontaneity—the sign of ΔG and the magnitude of Q/K—when we compare the sign of the natural logarithm of Q/K (ln Q/K) with the sign of ΔG (refer to Appendix A if necessary):

- If $Q/K < 1$, then $\ln Q/K < 0$: reaction proceeds spontaneously to the right ($\Delta G < 0$).
- If $Q/K > 1$, then $\ln Q/K > 0$: reaction proceeds spontaneously to the left ($\Delta G > 0$).
- If $Q/K = 1$, then $\ln Q/K = 0$: reaction is at equilibrium ($\Delta G = 0$).

Note that the signs of ΔG and of ln Q/K are the same for a given direction; in fact, ΔG equals ln Q/K multiplied by the proportionality constant RT:

$$\Delta G = RT \ln \frac{Q}{K} = RT \ln Q - RT \ln K \quad \textbf{(20.11)}$$

Q represents the concentrations (or pressures) of a system's components *at any time* during the reaction, and K represents these quantities *at equilibrium*. Therefore, Equation 20.11 says that ΔG is a measure of how different the ratio of concentrations, Q, is from their ratio at equilibrium, K:

- If Q and K are very different, the reaction releases (or absorbs) a *lot* of free energy.
- If Q and K are nearly the same, the reaction releases (or absorbs) relatively *little*.

The Standard Free Energy Change and the Equilibrium Constant When we choose standard-state values for Q (1 atm for gases, 1 M for solutions, etc.) in Equation 20.11, ΔG equals ΔG° and Q equals 1:

$$\Delta G^\circ = RT \ln 1 - RT \ln K$$

Since ln 1 = 0, the term RT ln Q drops out, and we have

$$\Delta G^\circ = -RT \ln K \quad \textbf{(20.12)}$$

Table 20.2 shows that, due to their logarithmic relationship, a small change in ΔG° causes a large change in K. Also note that

- As ΔG° becomes more positive, K becomes smaller; for example, if $\Delta G^\circ = +10$ kJ, K is 0.02, so the product terms are $\frac{1}{50}$ the size of the reactant terms.
- As ΔG° becomes more negative, K becomes larger; for example, if $\Delta G^\circ = -10$ kJ, the product terms are 50 times larger than the reactant terms.

Finding the Free Energy Change Under Any Conditions In reality, reactions rarely begin with all components in their standard states. By substituting the relationship between ΔG° and K (Equation 20.12) into the expression for ΔG (Equation 20.11), we obtain a relationship that applies to *any starting concentrations:*

$$\Delta G = \Delta G^\circ + RT \ln Q \quad \textbf{(20.13)}$$

Sample Problem 20.9 uses molecular scenes to explore these ideas, and Sample Problem 20.10 applies them to an important industrial reaction.

Table 20.2 The Relationship Between $\Delta G°$ and K at 298 K

$\Delta G°$ (kJ)	K	Significance
200	9×10^{-36}	Essentially no forward reaction; reverse reaction goes to completion.
100	3×10^{-18}	
50	2×10^{-9}	
10	2×10^{-2}	
1	7×10^{-1}	
0	1	Forward and reverse reactions proceed to same extent.
−1	1.5	
−10	5×10^{1}	
−50	6×10^{8}	
−100	3×10^{17}	
−200	1×10^{35}	Forward reaction goes to completion; essentially no reverse reaction.

FORWARD REACTION

REVERSE REACTION

Sample Problem 20.9 Using Molecular Scenes to Find ΔG for a Reaction at Nonstandard Conditions

Problem These molecular scenes represent three mixtures in which A_2 *(black)* and B_2 *(green)* are forming AB. Each molecule represents 0.10 atm. The equation is

$$A_2(g) + B_2(g) \rightleftharpoons 2AB(g) \qquad \Delta G° = -3.4\ \text{kJ/mol}$$

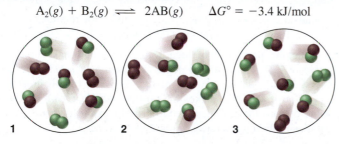

1 2 3

(a) If mixture 1 is at equilibrium, calculate K.
(b) Which mixture has the most negative ΔG, and which has the most positive?
(c) Is the reaction spontaneous at the standard state, that is, $P_{A_2} = P_{B_2} = P_{AB} = 1.0$ atm?

Plan **(a)** Mixture 1 is at equilibrium, so we first write the expression for Q and then find the partial pressure of each substance from the numbers of molecules and calculate K. **(b)** To find ΔG, we apply Equation 20.13. We are given $\Delta G°$ (−3.4 kJ/mol) and know R (8.314 J/mol·K), but we still need to find T. We calculate T from Equation 20.12 using K from part (a), and substitute the partial pressure of each substance (by counting particles) to get Q. **(c)** Since 1.0 atm is the standard state, we substitute it for P_{AB}, P_{A_2}, and P_{B_2} in Q in Equation 20.13 and solve for ΔG. Because these pressures are standard state values, ΔG equals $\Delta G°$.

Solution **(a)** Writing the expression for Q and calculating K:

$$A_2(g) + B_2(g) \rightleftharpoons 2AB(g) \qquad Q = \frac{P_{AB}^{\ 2}}{P_{A_2} \times P_{B_2}} \qquad K = \frac{(0.40)^2}{(0.20)(0.20)} = 4.0$$

(b) Calculating T from Equation 20.12 for use in Equation 20.13:

$$\Delta G° = -RT\ln K = -\frac{3.4\ \text{kJ}}{\text{mol}} = -\left(\frac{8.314\ \text{J}}{\text{mol·K}}\right) T \ln 4.0$$

$$T = \frac{\dfrac{-3.4\ \text{kJ}}{\text{mol}}\left(\dfrac{1000\ \text{J}}{1\ \text{kJ}}\right)}{-\left(\dfrac{8.314\ \text{J}}{\text{mol·K}}\right)\ln 4.0} = 295\ \text{K}$$

Calculating ΔG from Equation 20.13 for each reaction mixture:
Mixture 1:

$$\Delta G = \Delta G^\circ + RT \ln Q = -3.4 \text{ kJ} + RT \ln 4.0$$
$$= -3.4 \text{ kJ}\left(\frac{1000 \text{ J}}{1 \text{ kJ}}\right) + \left(\frac{8.314 \text{ J}}{\text{mol·K}}\right)(295 \text{ K}) \ln 4.0$$
$$= -3400 \text{ J} + 3400 \text{ J} = 0.0 \text{ J}$$

Mixture 2:

$$\Delta G = -3.4 \text{ kJ} + RT \ln \frac{(0.20)^2}{(0.30)(0.30)}$$
$$= -3.4 \text{ kJ}\left(\frac{1000 \text{ J}}{1 \text{ kJ}}\right) + \left(\frac{8.314 \text{ J}}{\text{mol·K}}\right)(295 \text{ K}) \ln 0.44 = -5.4\times10^3 \text{ J}$$

Mixture 3:

$$\Delta G = -3.4 \text{ kJ/mol} + RT \ln \frac{(0.60)^2}{(0.10)(0.10)}$$
$$= -3.4 \text{ kJ/mol}\left(\frac{1000 \text{ J}}{1 \text{ kJ}}\right) + \left(\frac{8.314 \text{ J}}{\text{mol·K}}\right)(295 \text{ K}) \ln 36 = 5.4\times10^3 \text{ J}$$

Mixture 2 has the most negative ΔG, and mixture 3 has the most positive ΔG.

(c) Finding ΔG when $P_{A_2} = P_{B_2} = P_{AB} = 1.0$ atm:

$$\Delta G = \Delta G^\circ + RT \ln Q = -3.4 \text{ kJ/mol} + RT \ln \frac{(1.0)^2}{(1.0)(1.0)}$$
$$= -3.4 \text{ kJ/mol} + RT \ln 1.0 = -3.4 \text{ kJ/mol}$$

Yes, the reaction is spontaneous when the components are in their standard states.

Check In (b), round to check the arithmetic; for example, for mixture 3, $\Delta G \approx -3000 \text{ J} + (8 \text{ J/mol·K})(300 \text{ K})4 \approx 7000 \text{ J}$, which is in the correct ballpark.

Comment 1. By using the properties of logarithms, we didn't have to calculate T and ΔG in (b). For mixture 2, $Q < 1$, so $\ln Q$ is negative, which makes ΔG more negative. Also, note that $Q(0.44) < K(4.0)$, so $\Delta G < 0$. For mixture 3, $Q > 1$ (and is greater than it is for mixture 1), so $\ln Q$ is positive, which makes ΔG positive. Also, $Q(36) > K(4.0)$, so $\Delta G > 0$. **2.** In (b), the value of zero for ΔG of the equilibrium mixture (mixture 1) makes sense, because a system at equilibrium has released all of its free energy.

FOLLOW-UP PROBLEM 20.9 The scenes below depict mixtures in which X_2 *(tan)* and Y_2 *(blue)* are forming XY_2. Each molecule represents 0.10 mol, and the volume is 1 L. The equation is $X_2(g) + 2Y_2(g) \rightleftharpoons 2XY_2(g)$; $\Delta G^\circ = -1.3$ kJ/mol.

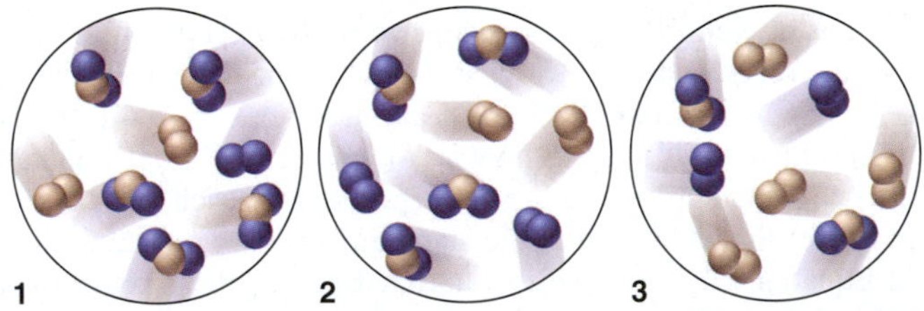

(a) If $K = 2$, which mixture is at equilibrium?
(b) Rank the three mixtures from the lowest (most negative) ΔG to highest (most positive) ΔG.
(c) What is the sign of ΔG for the change that occurs as each nonequilibrium mixture approaches equilibrium?

Sample Problem 20.10 Calculating ΔG at Nonstandard Conditions

Problem The oxidation of $SO_2(g)$, $2SO_2(g) + O_2(g) \longrightarrow 2SO_3(g)$, is too slow at 298 K to be useful in the manufacture of sulfuric acid, so the reaction is run at high T.
(a) Calculate K at 298 K and at 973 K. ($\Delta G^\circ_{298} = -141.6$ kJ/mol of reaction as written; using ΔH° and ΔS° values at 973 K, $\Delta G^\circ_{973} = -12.12$ kJ/mol of reaction as written.)

(b) Two containers are filled with 0.500 atm of SO_2, 0.0100 atm of O_2, and 0.100 atm of SO_3; one is kept at 25°C and the other at 700.°C. In which direction, if any, will the reaction proceed to reach equilibrium at each temperature?
(c) Calculate ΔG for the system in part (b) at each temperature.

Plan **(a)** We know $\Delta G°$, T, and R, so we can calculate the K's from Equation 20.12. **(b)** To determine if a net reaction will occur, we find Q from the given partial pressures and compare it with each K from part (a). **(c)** These are *not* standard-state pressures, so we find ΔG at each T with Equation 20.13 from the values of $\Delta G°$ (given) and Q [from part (b)].

Solution **(a)** Calculating K at the two temperatures:

$$\Delta G° = -RT \ln K \qquad \text{so} \qquad K = e^{-(\Delta G°/RT)}$$

At 298 K, the exponent is

$$-(\Delta G°/RT) = -\left(\frac{-141.6 \text{ kJ/mol} \times \frac{1000 \text{ J}}{1 \text{ kJ}}}{8.314 \text{ J/mol·K} \times 298 \text{ K}}\right) = 57.2$$

so $$K = e^{-(\Delta G°/RT)} = e^{57.2} = 7\times10^{24}$$

At 973 K, the exponent is

$$-(\Delta G°/RT) = -\left(\frac{-12.12 \text{ kJ/mol} \times \frac{1000 \text{ J}}{1 \text{ kJ}}}{8.314 \text{ J/mol·K} \times 973 \text{ K}}\right) = 1.50$$

so $$K = e^{-(\Delta G°/RT)} = e^{1.50} = 4.5$$

(b) Calculating the value of Q:

$$Q = \frac{P^2_{SO_3}}{P^2_{SO_2} \times P_{O_2}} = \frac{0.100^2}{0.500^2 \times 0.0100} = 4.00$$

Because $Q < K$ at both temperatures, the denominator will decrease and the numerator will increase—more SO_3 will form—until Q equals K. To reach equilibrium, the reaction will proceed far to the right at 298 K and slightly to the right at 973 K.
(c) Calculating ΔG, the nonstandard free energy change, at 298 K:

$$\begin{aligned} \Delta G_{298} &= \Delta G° + RT \ln Q \\ &= -141.6 \text{ kJ/mol} + \left(8.314 \text{ J/mol·K} \times \frac{1 \text{ kJ}}{1000 \text{ J}} \times 298 \text{ K} \times \ln 4.00\right) \\ &= -138.2 \text{ kJ/mol} \end{aligned}$$

Calculating ΔG at 973 K:

$$\begin{aligned} \Delta G_{973} &= \Delta G° + RT \ln Q \\ &= -12.12 \text{ kJ/mol} + \left(8.314 \text{ J/mol·K} \times \frac{1 \text{ kJ}}{1000 \text{ J}} \times 973 \text{ K} \times \ln 4.00\right) \\ &= -0.9 \text{ kJ/mol} \end{aligned}$$

Check Note that in parts (a) and (c), we made the free energy units (kJ) consistent with the units in R (J). For significant figures in addition and subtraction, we retain one digit to the right of the decimal place in part (c).

Comment For these starting gas pressures at 973 K, the process is barely spontaneous ($\Delta G = -0.9$ kJ/mol), so why use a higher temperature? As in the synthesis of NH_3 (Section 17.6), where the *yield* is greater at a lower temperature, this process is carried out at higher temperature *with a catalyst* to attain a higher *rate*.

FOLLOW-UP PROBLEM 20.10 At 298 K, hypobromous acid (HBrO) dissociates in water with a K_a of 2.3×10^{-9}.
(a) Calculate $\Delta G°$ for the dissociation of HBrO.
(b) Calculate ΔG if $[H_3O^+] = 6.0\times10^{-4}$ *M*, $[BrO^-] = 0.10$ *M*, and $[HBrO] = 0.20$ *M*.

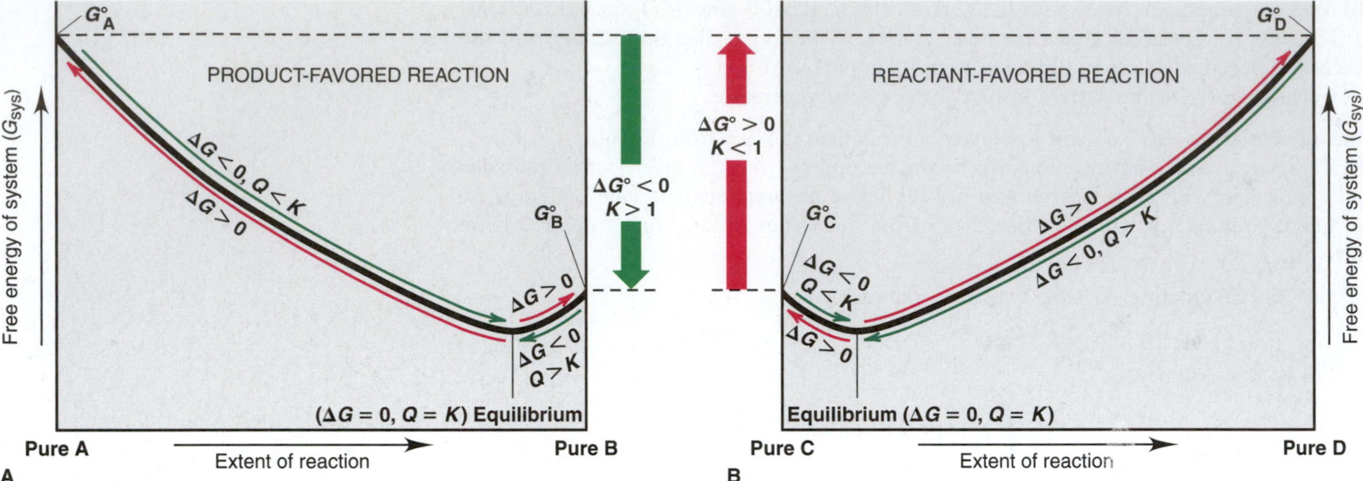

Figure 20.14 Free energy and the extent of reaction. Each reaction proceeds spontaneously *(curved green arrows)* from reactant (A or C) or product (B or D) to the equilibrium mixture, at which point $\Delta G = 0$. After that, the reaction is nonspontaneous *(curved red arrows)*. **A,** For the product-favored reaction $A \rightleftharpoons B$, $G^\circ_A > G^\circ_B$, so $\Delta G^\circ < 0$ and $K > 1$. **B,** For the reactant-favored reaction $C \rightleftharpoons D$, $G^\circ_D > G^\circ_C$, so $\Delta G^\circ > 0$ and $K < 1$.

Another Look at the Meaning of Spontaneity At this point, we introduce two terms related to *spontaneous* and *nonspontaneous:*

1. *Product-favored reaction.* For the general reaction

$$A \rightleftharpoons B \qquad K = [B]/[A] > 1$$

and therefore, the reaction proceeds largely from left to right (Figure 20.14A). From pure A to equilibrium, $Q < K$ and the curved *green* arrow in the figure indicates that the reaction is spontaneous ($\Delta G < 0$). From there on, the curved *red* arrow indicates that the reaction is nonspontaneous ($\Delta G > 0$). Similarly, from pure B to equilibrium, $Q > K$ and the reaction is also spontaneous ($\Delta G < 0$), but not thereafter. In either case, *free energy decreases until the reaction reaches a minimum at the equilibrium mixture:* $Q = K$ *and* $\Delta G = 0$. For the overall reaction $A \rightleftharpoons B$ (starting with all components in their standard states), G°_B is smaller than G°_A, so ΔG° is negative, which corresponds to $K > 1$. We call this a *product-favored* reaction because in its final state the system contains mostly product.

2. *Reactant-favored reaction.* For the opposite reaction,

$$C \rightleftharpoons D \qquad K = [D]/[C] < 1$$

and the reaction proceeds slightly from left to right (Figure 20.14B). Here, too, whether we start with pure C or pure D, the reaction is spontaneous ($\Delta G < 0$) until equilibrium. In this case, however, the equilibrium mixture contains mostly C (the reactant), so we say the reaction is *reactant favored.* Here, G°_D is *larger* than G°_C, so ΔG° is *positive,* which corresponds to $K < 1$.

Thus, "*spontaneous*" refers to that *portion* of a reaction in which the free energy decreases—from the starting mixture to the equilibrium mixture. A product-favored reaction goes predominantly, but *not* completely, toward product, and a reactant-favored reaction goes relatively little toward product (see Table 20.2).

Summary of Section 20.4

- Two ways of predicting reaction spontaneity are from the sign of ΔG and from the value of Q/K. These variables are related to each other by $\Delta G = RT \ln Q/K$. When $Q = K$, $Q/K = 1$ and $\ln Q/K = 0$. Thus, the system is at equilibrium and can release no more free energy.
- Beginning with Q at the standard state, the free energy change is ΔG° and is related to the equilibrium constant: $\Delta G^\circ = -RT \ln K$.
- Any nonequilibrium mixture of reactants and products moves spontaneously ($\Delta G < 0$) toward the equilibrium mixture.
- A product-favored reaction goes predominantly toward product and, thus, has $K > 1$ and $\Delta G^\circ < 0$; a reactant-favored reaction has $K < 1$ and $\Delta G^\circ > 0$.

CHAPTER REVIEW GUIDE

The following sections provide many aids to help you study this chapter. (Numbers in parentheses refer to pages, unless noted otherwise.)

Learning Objectives

These are concepts and skills to review after studying this chapter.

Related section (§), sample problem (SP), and upcoming end-of-chapter problem (EP) numbers are listed in parentheses.

1. Discuss the meaning of a spontaneous change, and explain why the first law or the sign of ΔH° cannot predict its direction (§20.1) (EPs 20.1–20.3, 20.8, 20.9)
2. Understand the meaning of entropy (S) in terms of the number of microstates into which a system's energy is dispersed; describe how the second law provides the criterion for spontaneity, how the third law allows us to find absolute values of standard molar entropies (S°), and how conditions and properties of substances influence S° (§20.1) (SP 20.1) (EPs 20.4–20.7, 20.10–20.23)
3. Calculate ΔS°_{rxn} from S° of reactants and products, understand the influence of ΔS°_{surr} on ΔS°_{rxn}, and describe the relationships between ΔS_{surr} and ΔH_{sys} and between ΔS_{univ} and K (§20.2) (SPs 20.2, 20.3) (EPs 20.24–20.35)
4. Derive the free energy change (ΔG) from the second law, and explain how ΔG is related to work; explain why temperature (T) affects the spontaneity of some reactions but not others; describe how a spontaneous change drives a nonspontaneous one; calculate ΔG°_{rxn} from ΔH°_f and S° values or from ΔG°_f values and quantify the effect of T on ΔG°; find T at which a reaction becomes spontaneous (§20.3) (SPs 20.4–20.8) (EPs 20.36–20.51)
5. Know the relationships of ΔG to Q/K, ΔG° to K, and ΔG to ΔG° and Q, and understand why ΔG decreases as a reaction moves toward equilibrium and the distinction between product-favored and reactant-favored reactions (§20.4) (SPs 20.9, 20.10) (EPs 20.52–20.67)

Key Terms

These important terms appear in boldface in the chapter and are defined again in the Glossary.

Section 20.1
spontaneous change (654)
entropy (S) (656)
second law of thermodynamics (659)
third law of thermodynamics (659)
standard molar entropy (S°) (660)

Section 20.2
standard entropy of reaction (ΔS°_{rxn}) (664)

Section 20.3
free energy (G) (668)
standard free energy change (ΔG°) (669)
standard free energy of formation (ΔG°_f) (670)
coupling of reactions (674)
adenosine triphosphate (ATP) (675)

Key Equations and Relationships

Numbered and screened concepts are listed for you to refer to or memorize.

20.1 Quantifying entropy in terms of the number of microstates (W) over which the energy of a system can be distributed (656):

$$S = k \ln W$$

20.2 Quantifying the entropy change in terms of heat absorbed (or released) in a reversible process (659):

$$\Delta S_{sys} = \frac{q_{rev}}{T}$$

20.3 Stating the second law of thermodynamics, for a spontaneous process (659):

$$\Delta S_{univ} = \Delta S_{sys} + \Delta S_{surr} > 0$$

20.4 Calculating the standard entropy of reaction from the standard molar entropies of reactants and products (664):

$$\Delta S^\circ_{rxn} = \Sigma m S^\circ_{products} - \Sigma n S^\circ_{reactants}$$

20.5 Relating the entropy change in the surroundings to the enthalpy change of the system and the temperature (666):

$$\Delta S_{surr} = -\frac{\Delta H_{sys}}{T}$$

20.6 Expressing the free energy change of the system in terms of its component enthalpy and entropy changes (Gibbs equation) (668):

$$\Delta G_{sys} = \Delta H_{sys} - T\Delta S_{sys}$$

20.7 Calculating the standard free energy change from standard enthalpy and entropy changes (669):

$$\Delta G^\circ_{sys} = \Delta H^\circ_{sys} - T\Delta S^\circ_{sys}$$

20.8 Calculating the standard free energy change from the standard free energies of formation (670):

$$\Delta G^\circ_{rxn} = \Sigma m \Delta G^\circ_{f(products)} - \Sigma n \Delta G^\circ_{f(reactants)}$$

20.9 Relating the free energy change to the maximum work a system can do (671):

$$\Delta G = w_{max}$$

20.10 Finding the temperature at which a reaction becomes spontaneous (674):

$$T = \frac{\Delta H}{\Delta S}$$

20.11 Expressing the free energy change in terms of Q and K (676):

$$\Delta G = RT \ln \frac{Q}{K} = RT \ln Q - RT \ln K$$

20.12 Expressing the free energy change when Q is at standard state conditions (676):

$$\Delta G^\circ = -RT \ln K$$

20.13 Expressing the free energy change for nonstandard initial conditions (676):

$$\Delta G = \Delta G^\circ + RT \ln Q$$

BRIEF SOLUTIONS TO FOLLOW-UP PROBLEMS

Compare your own solutions to these calculation steps and answers.

20.1 (a) $PCl_5(g)$: higher molar mass and more complex molecule; (b) $BaCl_2(s)$: higher molar mass; (c) $Br_2(g)$: gases have more freedom of motion and dispersal of energy than liquids.

20.2 (a) $2NaOH(s) + CO_2(g) \longrightarrow Na_2CO_3(s) + H_2O(l)$

$\Delta n_{gas} = -1$, so ΔS°_{rxn} should be < 0

$$\begin{aligned}\Delta S^\circ_{rxn} &= [(1 \text{ mol } H_2O)(69.9 \text{ J/mol·K}) \\ &\quad + (1 \text{ mol } Na_2CO_3)(139 \text{ J/mol·K})] \\ &\quad - [(1 \text{ mol } CO_2)(213.7 \text{ J/mol·K}) \\ &\quad + (2 \text{ mol NaOH})(64.5 \text{ J/mol·K})] \\ &= -134 \text{ J/K}\end{aligned}$$

(b) $2Fe(s) + 3H_2O(g) \longrightarrow Fe_2O_3(s) + 3H_2(g)$

$\Delta n_{gas} = 0$, so you cannot predict the sign of ΔS°_{rxn}

$$\begin{aligned}\Delta S^\circ_{rxn} &= [(1 \text{ mol } Fe_2O_3)(87.4 \text{ J/mol·K}) \\ &\quad + (3 \text{ mol } H_2)(130.6 \text{ J/mol·K})] \\ &\quad - [(2 \text{ mol Fe})(27.3 \text{ J/mol·K}) \\ &\quad + (3 \text{ mol } H_2O)(188.7 \text{ J/mol·K})] \\ &= -141.5 \text{ J/K}\end{aligned}$$

20.3 $2FeO(s) + \frac{1}{2}O_2(g) \longrightarrow Fe_2O_3(s)$

$$\begin{aligned}\Delta S^\circ_{sys} &= (1 \text{ mol } Fe_2O_3)(87.4 \text{ J/mol·K}) \\ &\quad - [(2 \text{ mol FeO})(60.75 \text{ J/mol·K}) \\ &\quad + (\tfrac{1}{2} \text{ mol } O_2)(205.0 \text{ J/mol·K})] \\ &= -136.6 \text{ J/K}\end{aligned}$$

$$\begin{aligned}\Delta H^\circ_{sys} &= (1 \text{ mol } Fe_2O_3)(-825.5 \text{ kJ/mol}) \\ &\quad - [(2 \text{ mol FeO})(-272.0 \text{ kJ/mol}) + (\tfrac{1}{2} \text{ mol } O_2)(0 \text{ kJ/mol})] \\ &= -281.5 \text{ kJ}\end{aligned}$$

$$\Delta S_{surr} = -\frac{\Delta H^\circ_{sys}}{T} = -\frac{(-281.5 \text{ kJ} \times 1000 \text{ J/kJ})}{298 \text{ K}} = +945 \text{ J/K}$$

$$\begin{aligned}\Delta S_{univ} &= \Delta S^\circ_{sys} + \Delta S_{surr} = -136.6 \text{ J/K} + 945 \text{ J/K} \\ &= 808 \text{ J/K; reaction is spontaneous at 298 K}\end{aligned}$$

20.4 Using ΔH°_f and S° values from Appendix B,

$\Delta H^\circ_{rxn} = -114.2$ kJ and $\Delta S^\circ_{rxn} = -146.5$ J/K

$$\begin{aligned}\Delta G^\circ_{rxn} &= \Delta H^\circ_{rxn} - T\Delta S^\circ_{rxn} \\ &= -114.2 \text{ kJ} - [(298 \text{ K})(-146.5 \text{ J/K})(1 \text{ kJ}/1000 \text{ J})] \\ &= -70.5 \text{ kJ}\end{aligned}$$

20.5 (a)
$$\begin{aligned}\Delta G^\circ_{rxn} &= (2 \text{ mol } NO_2)(51 \text{ kJ/mol}) \\ &\quad - [(2 \text{ mol NO})(86.60 \text{ kJ/mol}) \\ &\quad + (1 \text{ mol } O_2)(0 \text{ kJ/mol})] \\ &= -71 \text{ kJ}\end{aligned}$$

(b)
$$\begin{aligned}\Delta G^\circ_{rxn} &= (2 \text{ mol CO})(-137.2 \text{ kJ/mol}) - [(2 \text{ mol C})(0 \text{ kJ/mol}) \\ &\quad + (1 \text{ mol } O_2)(0 \text{ kJ/mol})] \\ &= -274.4 \text{ kJ}\end{aligned}$$

20.6 (a) More moles of gas are present after the reaction, so $\Delta S > 0$. (b) The problem says the reaction is spontaneous ($\Delta G < 0$) only above 325°C, which implies high T. If $\Delta S > 0$, $-T\Delta S < 0$, so ΔG will become negative at higher T only if $\Delta H > 0$.

20.7 ΔG becomes negative at lower T, so $\Delta H < 0$, $\Delta S < 0$, and $-T\Delta S > 0$. At lower T, the negative ΔH value becomes larger than the positive $-T\Delta S$ value.

20.8

$$\begin{aligned}\Delta H &= \Delta H^\circ_f \text{ of } CaCO_3 - (\Delta H^\circ_f \text{ of CaO} + \Delta H^\circ_f \text{ of } CO_2) \\ &= -1206.9 \text{ kJ} - (-635.1 \text{ kJ} - 393.5 \text{ kJ}) = -178.3 \text{ kJ} \\ \Delta S &= S^\circ \text{ of } CaCO_3 - (S^\circ \text{ of CaO} + S^\circ \text{ of } CO_2) \\ &= 92.9 \text{ J/K} - (38.2 \text{ J/K} + 213.7 \text{ J/K}) = -159.0 \text{ J/K}\end{aligned}$$

$$T = \frac{\Delta H}{\Delta S} = \frac{-178.3 \text{ kJ} \times \dfrac{1000 \text{ J}}{1 \text{ kJ}}}{-159.0 \text{ J/K}} = 1121 \text{ K}$$

Reaction becomes spontaneous ($\Delta G < 0$) at any $T < 1121$ K.

20.9 (a) Mixture 2 is at equilibrium.
(b) 3 (most negative) $< 2 < 1$ (most positive)
(c) Any reaction mixture moves spontaneously toward equilibrium, so both changes have a negative ΔG.

20.10 (a)
$$\begin{aligned}\Delta G^\circ &= -RT \ln K = -8.314 \text{ J/mol·K} \times \frac{1 \text{ kJ}}{1000 \text{ J}} \times 298 \text{ K} \\ &\quad \times \ln (2.3\times10^{-9}) \\ &= 49 \text{ kJ/mol}\end{aligned}$$

(b)
$$Q = \frac{[H_3O^+][BrO^-]}{[HBrO]} = \frac{(6.0\times10^{-4})(0.10)}{0.20} = 3.0\times10^{-4}$$

$$\begin{aligned}\Delta G &= \Delta G^\circ + RT \ln Q \\ &= 49 \text{ kJ/mol} \\ &\quad + \left[8.314 \text{ J/mol·K} \times \frac{1 \text{ kJ}}{1000 \text{ J}} \times 298 \text{ K} \times \ln (3.0\times10^{-4})\right] \\ &= 29 \text{ kJ/mol}\end{aligned}$$

PROBLEMS

Problems with **colored** numbers are answered in Appendix E. Sections match the text and provide the numbers of relevant sample problems. Bracketed problems are grouped in pairs (indicated by a short rule) that cover the same concept. Comprehensive Problems are based on material from any section or previous chapter.

Note: Unless stated otherwise, problems refer to systems at 298 K (25°C). Solving these problems may require values from Appendix B.

The Second Law of Thermodynamics: Predicting Spontaneous Change

(Sample Problem 20.1)

20.1 Distinguish between the terms *spontaneous* and *instantaneous*. Give an example of a process that is spontaneous but very slow, and one that is very fast but not spontaneous.

20.2 Distinguish between the terms *spontaneous* and *nonspontaneous*. Can a nonspontaneous process occur? Explain.

20.3 State the first law of thermodynamics in terms of (a) the energy of the universe; (b) the creation or destruction of energy; (c) the energy change of system and surroundings. Does the first law reveal the direction of spontaneous change? Explain.

20.4 State qualitatively the relationship between entropy and freedom of particle motion. Use this idea to explain why you will probably never (a) be suffocated because all the air near you has moved to the other side of the room; (b) see half the water in your cup of tea freeze while the other half boils.

20.5 Why is ΔS_{vap} of a substance always larger than ΔS_{fus}?

20.6 How does the entropy of the surroundings change during an exothermic reaction? An endothermic reaction? Other than the examples on p. 655, describe a spontaneous endothermic process.

20.7 (a) What is the entropy of a perfect crystal at 0 K?
(b) Does entropy increase or decrease as the temperature rises?
(c) Why is $\Delta H_f^\circ = 0$ but $S^\circ > 0$ for an element?
(d) Why does Appendix B list ΔH_f° values but not ΔS_f° values?

20.8 Which of the following processes are spontaneous?
(a) Water evaporates from a puddle. (b) A lion chases an antelope. (c) An isotope undergoes radioactive disintegration.

20.9 Which of the following processes are spontaneous?
(a) Methane burns in air. (b) A teaspoonful of sugar dissolves in a cup of hot coffee. (c) A soft-boiled egg becomes raw.

20.10 Predict the sign of ΔS_{sys} for each process: (a) A piece of wax melts. (b) Silver chloride precipitates from solution. (c) Dew forms on a lawn in the morning.

20.11 Predict the sign of ΔS_{sys} for each process: (a) Alcohol evaporates. (b) A solid explosive converts to a gas. (c) Perfume vapors diffuse through a room.

20.12 Without using Appendix B, predict the sign of ΔS° for
(a) $2K(s) + F_2(g) \longrightarrow 2KF(s)$
(b) $NH_3(g) + HBr(g) \longrightarrow NH_4Br(s)$
(c) $NaClO_3(s) \longrightarrow Na^+(aq) + ClO_3^-(aq)$

20.13 Without using Appendix B, predict the sign of ΔS° for
(a) $H_2S(g) + \frac{1}{2}O_2(g) \longrightarrow \frac{1}{8}S_8(s) + H_2O(g)$
(b) $HCl(aq) + NaOH(aq) \longrightarrow NaCl(aq) + H_2O(l)$
(c) $2NO_2(g) \longrightarrow N_2O_4(g)$

20.14 Without using Appendix B, predict the sign of ΔS° for
(a) $CaCO_3(s) + 2HCl(aq) \longrightarrow CaCl_2(aq) + H_2O(l) + CO_2(g)$
(b) $2NO(g) + O_2(g) \longrightarrow 2NO_2(g)$
(c) $2KClO_3(s) \longrightarrow 2KCl(s) + 3O_2(g)$

20.15 Without using Appendix B, predict the sign of ΔS° for
(a) $Ag^+(aq) + Cl^-(aq) \longrightarrow AgCl(s)$
(b) $KBr(s) \longrightarrow KBr(aq)$
(c) $CH_3CH{=}CH_2(g) \longrightarrow$ cyclopropane, $H_2C{-}CH_2(g)$ bridged by CH_2

20.16 Predict the sign of ΔS for each process:
(a) $C_2H_5OH(g)$ (350 K and 500 torr) $\longrightarrow$ $C_2H_5OH(g)$ (350 K and 250 torr)
(b) $N_2(g)$ (298 K and 1 atm) $\longrightarrow$ $N_2(aq)$ (298 K and 1 atm)
(c) $O_2(aq)$ (303 K and 1 atm) $\longrightarrow$ $O_2(g)$ (303 K and 1 atm)

20.17 Predict the sign of ΔS for each process:
(a) $O_2(g)$ (1.0 L at 1 atm) $\longrightarrow$ $O_2(g)$ (0.10 L at 10 atm)
(b) $Cu(s)$ (350°C and 2.5 atm) $\longrightarrow$ $Cu(s)$ (450°C and 2.5 atm)
(c) $Cl_2(g)$ (100°C and 1 atm) $\longrightarrow$ $Cl_2(g)$ (10°C and 1 atm)

20.18 Predict which substance has greater molar entropy. Explain.
(a) Butane $CH_3CH_2CH_2CH_3(g)$ or 2-butene $CH_3CH{=}CHCH_3(g)$
(b) $Ne(g)$ or $Xe(g)$ (c) $CH_4(g)$ or $CCl_4(l)$

20.19 Predict which substance has greater molar entropy. Explain.
(a) $CH_3OH(l)$ or $C_2H_5OH(l)$ (b) $KClO_3(s)$ or $KClO_3(aq)$
(c) $Na(s)$ or $K(s)$

20.20 Without consulting Appendix B, arrange each group in order of *increasing* standard molar entropy (S°). Explain.
(a) Graphite, diamond, charcoal
(b) Ice, water vapor, liquid water
(c) O_2, O_3, O atoms

20.21 Without consulting Appendix B, arrange each group in order of *increasing* standard molar entropy (S°). Explain.
(a) Glucose ($C_6H_{12}O_6$), sucrose ($C_{12}H_{22}O_{11}$), ribose ($C_5H_{10}O_5$)
(b) $CaCO_3$, $Ca + C + \frac{3}{2}O_2$, $CaO + CO_2$
(c) $SF_6(g)$, $SF_4(g)$, $S_2F_{10}(g)$

20.22 Without consulting Appendix B, arrange each group in order of *decreasing* standard molar entropy (S°). Explain.
(a) $ClO_4^-(aq)$, $ClO_2^-(aq)$, $ClO_3^-(aq)$
(b) $NO_2(g)$, $NO(g)$, $N_2(g)$
(c) $Fe_2O_3(s)$, $Al_2O_3(s)$, $Fe_3O_4(s)$

20.23 Without consulting Appendix B, arrange each group in order of *decreasing* standard molar entropy (S°). Explain.
(a) Mg metal, Ca metal, Ba metal
(b) Hexane (C_6H_{14}), benzene (C_6H_6), cyclohexane (C_6H_{12})
(c) $PF_2Cl_3(g)$, $PF_5(g)$, $PF_3(g)$

Calculating the Change in Entropy of a Reaction

(Sample Problems 20.2 and 20.3)

20.24 For the reaction depicted in the molecular scenes, X is red and Y is green.

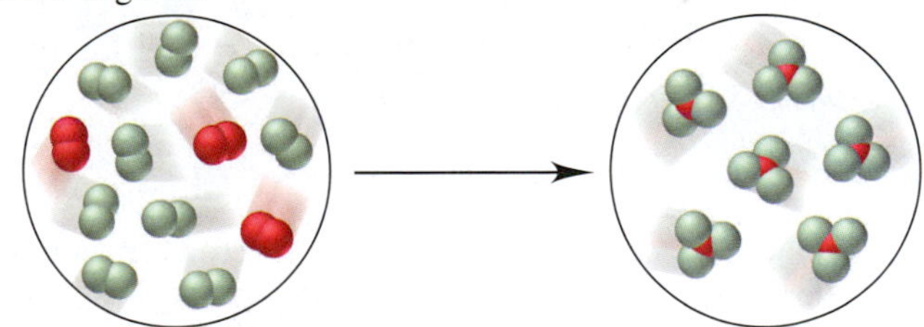

(a) Write a balanced equation.
(b) Determine the sign of ΔS.
(c) Which species has the highest molar entropy?

20.25 What property of entropy allows Hess's law to be used in the calculation of entropy changes?

20.26 Describe the equilibrium condition in terms of the entropy changes of a system and its surroundings. What does this description mean about the entropy change of the universe?

20.27 For the reaction $H_2O(g) + Cl_2O(g) \longrightarrow 2HClO(g)$, you know ΔS°_{rxn} and S° of $HClO(g)$ and of $H_2O(g)$. Write an expression that can be used to determine S° of $Cl_2O(g)$.

20.28 For each reaction, predict the sign and find the value of ΔS°:
(a) $3NO(g) \longrightarrow N_2O(g) + NO_2(g)$
(b) $3H_2(g) + Fe_2O_3(s) \longrightarrow 2Fe(s) + 3H_2O(g)$
(c) $P_4(s) + 5O_2(g) \longrightarrow P_4O_{10}(s)$

20.29 For each reaction, predict the sign and find the value of ΔS°:
(a) $3NO_2(g) + H_2O(l) \longrightarrow 2HNO_3(l) + NO(g)$
(b) $N_2(g) + 3F_2(g) \longrightarrow 2NF_3(g)$
(c) $C_6H_{12}O_6(s) + 6O_2(g) \longrightarrow 6CO_2(g) + 6H_2O(g)$

20.30 Find $\Delta S°$ for the combustion of ethane (C_2H_6) to carbon dioxide and gaseous water. Is the sign of $\Delta S°$ as expected?

20.31 Find $\Delta S°$ for the reaction of nitrogen monoxide with hydrogen to form ammonia and water vapor. Is the sign of $\Delta S°$ as expected?

20.32 Find $\Delta S°$ for the formation of $Cu_2O(s)$ from its elements.

20.33 Find $\Delta S°$ for the formation of $CH_3OH(l)$ from its elements.

20.34 Sulfur dioxide is released in the combustion of coal. Scrubbers use aqueous slurries of calcium hydroxide to remove the SO_2 from flue gases. Write a balanced equation for this reaction and calculate $\Delta S°$ at 298 K [$S°$ of $CaSO_3(s)$ = 101.4 J/mol·K].

20.35 Oxyacetylene welding is used to repair metal structures, including bridges, buildings, and even the Statue of Liberty. Calculate $\Delta S°$ for the combustion of 1 mol of acetylene (C_2H_2).

Entropy, Free Energy, and Work

(Sample Problems 20.4 to 20.8)

20.36 What is the advantage of calculating free energy changes rather than entropy changes to determine reaction spontaneity?

20.37 Given that $\Delta G_{sys} = -T\Delta S_{univ}$, explain how the sign of ΔG_{sys} correlates with reaction spontaneity.

20.38 (a) Is an endothermic reaction more likely to be spontaneous at higher temperatures or lower temperatures? Explain.
(b) The change depicted below occurs at constant pressure. Explain your answers to each of the following: (1) What is the sign of ΔH? (2) What is the sign of ΔS? (3) What is the sign of ΔS_{surr}? (4) How does the sign of ΔG vary with temperature?

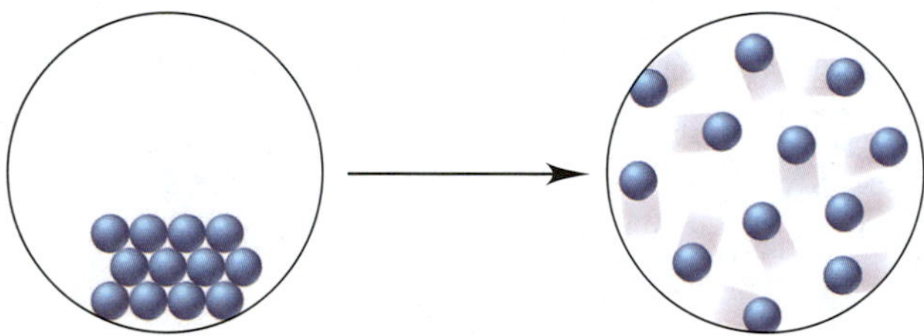

20.39 With its components in their standard states, a certain reaction is spontaneous only at high T. What do you know about the signs of $\Delta H°$ and $\Delta S°$? Describe a process for which this is true.

20.40 Calculate $\Delta G°$ for each reaction using $\Delta G_f°$ values:
(a) $2Mg(s) + O_2(g) \longrightarrow 2MgO(s)$
(b) $2CH_3OH(g) + 3O_2(g) \longrightarrow 2CO_2(g) + 4H_2O(g)$
(c) $BaO(s) + CO_2(g) \longrightarrow BaCO_3(s)$

20.41 Calculate $\Delta G°$ for each reaction using $\Delta G_f°$ values:
(a) $H_2(g) + I_2(s) \longrightarrow 2HI(g)$
(b) $MnO_2(s) + 2CO(g) \longrightarrow Mn(s) + 2CO_2(g)$
(c) $NH_4Cl(s) \longrightarrow NH_3(g) + HCl(g)$

20.42 Find $\Delta G°$ for the reactions in Problem 20.40 using $\Delta H_f°$ and $S°$ values.

20.43 Find $\Delta G°$ for the reactions in Problem 20.41 using $\Delta H_f°$ and $S°$ values.

20.44 Consider the oxidation of carbon monoxide:

$$CO(g) + \tfrac{1}{2}O_2(g) \longrightarrow CO_2(g)$$

(a) Predict the signs of $\Delta S°$ and $\Delta H°$. Explain.
(b) Calculate $\Delta G°$ by two different methods.

20.45 Consider the combustion of butane gas:

$$C_4H_{10}(g) + \tfrac{13}{2}O_2(g) \longrightarrow 4CO_2(g) + 5H_2O(g)$$

(a) Predict the signs of $\Delta S°$ and $\Delta H°$. Explain.
(b) Calculate $\Delta G°$ by two different methods.

20.46 One reaction used to produce small quantities of pure H_2 is

$$CH_3OH(g) \rightleftharpoons CO(g) + 2H_2(g)$$

(a) Determine $\Delta H°$ and $\Delta S°$ for the reaction at 298 K.
(b) Assuming that these values are relatively independent of temperature, calculate $\Delta G°$ at 28°C, 128°C, and 228°C.
(c) What is the significance of the different values of $\Delta G°$?

20.47 A reaction that occurs in the internal combustion engine is

$$N_2(g) + O_2(g) \rightleftharpoons 2NO(g)$$

(a) Determine $\Delta H°$ and $\Delta S°$ for the reaction at 298 K.
(b) Assuming that these values are relatively independent of temperature, calculate $\Delta G°$ at 100.°C, 2560.°C, and 3540.°C.
(c) What is the significance of the different values of $\Delta G°$?

20.48 Use $\Delta H°$ and $\Delta S°$ values for the following process at 1 atm to find the normal boiling point of Br_2:

$$Br_2(l) \rightleftharpoons Br_2(g)$$

20.49 Use $\Delta H°$ and $\Delta S°$ values to find the temperature at which these sulfur allotropes reach equilibrium at 1 atm:

$$S(\text{rhombic}) \rightleftharpoons S(\text{monoclinic})$$

20.50 As a fuel, $H_2(g)$ produces only nonpolluting $H_2O(g)$ when it burns. Moreover, it combines with $O_2(g)$ in a fuel cell (Chapter 21) to provide electrical energy.
(a) Calculate $\Delta H°$, $\Delta S°$, and $\Delta G°$ per mole of H_2 at 298 K.
(b) Is the spontaneity of this reaction dependent on T? Explain.
(c) At what temperature does the reaction become spontaneous?

20.51 The U.S. government requires automobile fuels to contain a renewable component. Fermentation of glucose from corn yields ethanol, which is added to gasoline to fulfill this requirement:

$$C_6H_{12}O_6(s) \longrightarrow 2C_2H_5OH(l) + 2CO_2(g)$$

Calculate $\Delta H°$, $\Delta S°$, and $\Delta G°$ for the reaction at 25°C. Is the spontaneity of this reaction dependent on T? Explain.

Free Energy, Equilibrium, and Reaction Direction

(Sample Problems 20.9 and 20.10)

20.52 (a) If $K << 1$ for a reaction, what do you know about the sign and magnitude of $\Delta G°$? (b) If $\Delta G° << 0$ for a reaction, what do you know about the magnitude of K? Of Q?

20.53 How is the free energy change of a process related to the work that can be obtained from the process? Is this quantity of work obtainable in practice? Explain.

20.54 The scenes and the graph relate to the reaction of $X_2(g)$ *(black)* with $Y_2(g)$ *(orange)* to form $XY(g)$. (a) If reactants and products are in their standard states, what quantity is represented on the graph by x? (b) Which scene represents point 1? Explain. (c) Which scene represents point 2? Explain.

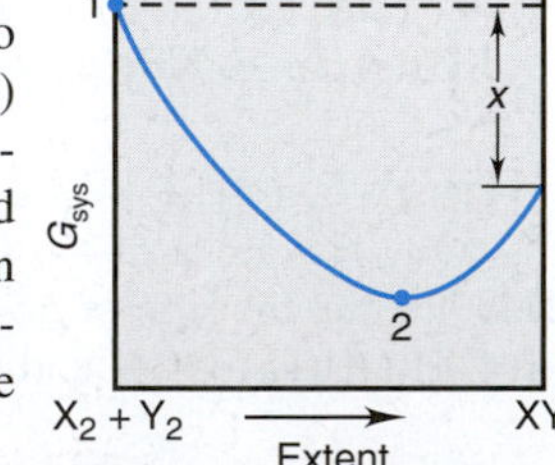

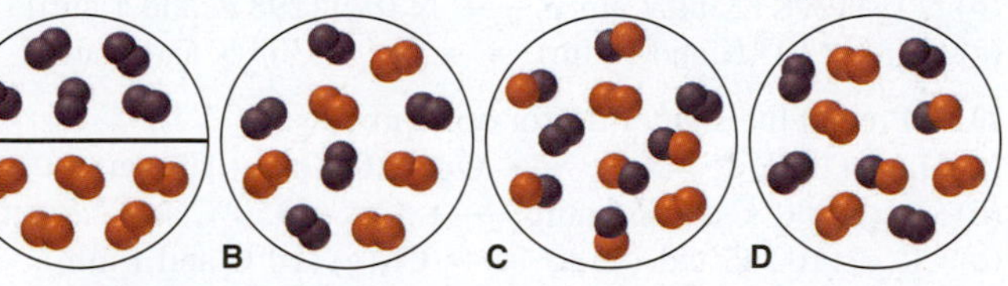

20.55 What is the difference between $\Delta G°$ and ΔG? Under what circumstances does $\Delta G = \Delta G°$?

20.56 Calculate K at 298 K for each reaction:
(a) $NO(g) + \frac{1}{2}O_2(g) \rightleftharpoons NO_2(g)$
(b) $2HCl(g) \rightleftharpoons H_2(g) + Cl_2(g)$
(c) $2C(\text{graphite}) + O_2(g) \rightleftharpoons 2CO(g)$

20.57 Calculate K at 298 K for each reaction:
(a) $2H_2S(g) + 3O_2(g) \rightleftharpoons 2H_2O(g) + 2SO_2(g)$
(b) $H_2SO_4(l) \rightleftharpoons H_2O(l) + SO_3(g)$
(c) $HCN(aq) + NaOH(aq) \rightleftharpoons NaCN(aq) + H_2O(l)$

20.58 Use Appendix B to determine the K_{sp} of Ag_2S.

20.59 Use Appendix B to determine the K_{sp} of CaF_2.

20.60 For the reaction $I_2(g) + Cl_2(g) \rightleftharpoons 2ICl(g)$, calculate K_p at 25°C [$\Delta G_f°$ of $ICl(g) = -6.075$ kJ/mol].

20.61 For the reaction $CaCO_3(s) \rightleftharpoons CaO(s) + CO_2(g)$, calculate the equilibrium P_{CO_2} at 25°C.

20.62 The K_{sp} of $PbCl_2$ is 1.7×10^{-5} at 25°C. What is $\Delta G°$? Is it possible to prepare a solution that contains $Pb^{2+}(aq)$ and $Cl^-(aq)$, at their standard-state concentrations?

20.63 The K_{sp} of ZnF_2 is 3.0×10^{-2} at 25°C. What is $\Delta G°$? Is it possible to prepare a solution that contains $Zn^{2+}(aq)$ and $F^-(aq)$ at their standard-state concentrations?

20.64 The equilibrium constant for the reaction

$$2Fe^{3+}(aq) + Hg_2^{2+}(aq) \rightleftharpoons 2Fe^{2+}(aq) + 2Hg^{2+}(aq)$$

is $K_c = 9.1\times10^{-6}$ at 298 K.
(a) What is $\Delta G°$ at this temperature?
(b) If standard-state concentrations of the reactants and products are mixed, in which direction does the reaction proceed?
(c) Calculate ΔG when $[Fe^{3+}] = 0.20\ M$, $[Hg_2^{2+}] = 0.010\ M$, $[Fe^{2+}] = 0.010\ M$, and $[Hg^{2+}] = 0.025\ M$. In which direction will the reaction proceed to achieve equilibrium?

20.65 The formation constant for the reaction

$$Ni^{2+}(aq) + 6NH_3(aq) \rightleftharpoons Ni(NH_3)_6^{2+}(aq)$$

is $K_f = 5.6\times10^8$ at 25°C.
(a) What is $\Delta G°$ at this temperature?
(b) If standard-state concentrations of the reactants and products are mixed, in which direction does the reaction proceed?
(c) Determine ΔG when $[Ni(NH_3)_6^{2+}] = 0.010\ M$, $[Ni^{2+}] = 0.0010\ M$, and $[NH_3] = 0.0050\ M$. In which direction will the reaction proceed to achieve equilibrium?

20.66 High levels of ozone (O_3) cause rubber to deteriorate, green plants to turn brown, and many people to have difficulty breathing.
(a) Is the formation of O_3 from O_2 favored at all T, no T, high T, or low T?
(b) Calculate $\Delta G°$ for this reaction at 298 K.
(c) Calculate ΔG at 298 K for this reaction in urban smog where $[O_2] = 0.21$ atm and $[O_3] = 5\times10^{-7}$ atm.

20.67 A $BaSO_4$ slurry is ingested before the gastrointestinal tract is x-rayed because it is opaque to x-rays and defines the contours of the tract. Ba^{2+} ion is toxic, but the compound is nearly insoluble. If $\Delta G°$ at 37°C (body temperature) is 59.1 kJ/mol for the process

$$BaSO_4(s) \rightleftharpoons Ba^{2+}(aq) + SO_4^{2-}(aq)$$

what is $[Ba^{2+}]$ in the intestinal tract? (Assume that the only source of SO_4^{2-} is the ingested slurry.)

Comprehensive Problems

20.68 According to advertisements, "a diamond is forever."
(a) Calculate $\Delta H°$, $\Delta S°$, and $\Delta G°$ at 298 K for the phase change

$$\text{Diamond} \longrightarrow \text{graphite}$$

(b) Given the conditions under which diamond jewelry is normally kept, argue for and against the statement in the ad.
(c) Given the answers in part (a), what would need to be done to make synthetic diamonds from graphite?
(d) Assuming $\Delta H°$ and $\Delta S°$ do not change with temperature, can graphite be converted to diamond spontaneously at 1 atm?

20.69 Replace each question mark with the correct information:

	ΔS_{rxn}	ΔH_{rxn}	ΔG_{rxn}	Comment
(a)	+	−	−	?
(b)	?	0	−	Spontaneous
(c)	−	+	?	Not spontaneous
(d)	0	?	−	Spontaneous
(e)	?	0	+	?
(f)	+	+	?	$T\Delta S > \Delta H$

20.70 What is the change in entropy when 0.200 mol of potassium freezes at 63.7°C ($\Delta H_{fus} = 2.39$ kJ/mol)?

20.71 Hemoglobin carries O_2 from the lungs to tissue cells, where the O_2 is released. The protein is represented as Hb in its unoxygenated form and as $Hb{\cdot}O_2$ in its oxygenated form. One reason CO is toxic is that it competes with O_2 in binding to Hb:

$$Hb{\cdot}O_2(aq) + CO(g) \rightleftharpoons Hb{\cdot}CO(aq) + O_2(g)$$

(a) If $\Delta G° \approx -14$ kJ at 37°C (body temperature), what is the ratio of $[Hb{\cdot}CO]$ to $[Hb{\cdot}O_2]$ at 37°C with $[O_2] = [CO]$?
(b) How is Le Châtelier's principle used to treat CO poisoning?

20.72 Magnesia (MgO) is used for fire brick, crucibles, and furnace linings because of its high melting point. It is produced by decomposing magnesite ($MgCO_3$) at around 1200°C.
(a) Write a balanced equation for magnesite decomposition.
(b) Use $\Delta H°$ and $S°$ values to find $\Delta G°$ at 298 K.
(c) Assuming $\Delta H°$ and $S°$ do not change with temperature, find the minimum temperature at which the reaction is spontaneous.
(d) Calculate the equilibrium P_{CO_2} above $MgCO_3$ at 298 K.
(e) Calculate the equilibrium P_{CO_2} above $MgCO_3$ at 1200 K.

20.73 Methanol, a major industrial feedstock, is made by several catalyzed reactions, such as $CO(g) + 2H_2(g) \longrightarrow CH_3OH(l)$.
(a) Show that this reaction is thermodynamically feasible.
(b) Is it favored at low or at high temperatures?
(c) One concern about using CH_3OH as an auto fuel is its oxidation in air to yield formaldehyde, $CH_2O(g)$, which poses a health hazard. Calculate $\Delta G°$ at 100.°C for this oxidation.

20.74 (a) Write a balanced equation for the gaseous reaction between N_2O_5 and F_2 to form NF_3 and O_2. (b) Determine $\Delta G°_{rxn}$. (c) Find ΔG_{rxn} at 298 K if $P_{N_2O_5} = P_{F_2} = 0.20$ atm, $P_{NF_3} = 0.25$ atm, and $P_{O_2} = 0.50$ atm.

20.75 Consider the following reaction:

$$2NOBr(g) \rightleftharpoons 2NO(g) + Br_2(g) \qquad K = 0.42 \text{ at } 373 \text{ K}$$

Given that $S°$ of $NOBr(g) = 272.6$ J/mol·K and that $\Delta S°_{rxn}$ and $\Delta H°_{rxn}$ are constant with temperature, find
(a) $\Delta S°_{rxn}$ at 298 K (b) $\Delta G°_{rxn}$ at 373 K
(c) $\Delta H°_{rxn}$ at 373 K (d) $\Delta H_f°$ of NOBr at 298 K
(e) $\Delta G°_{rxn}$ at 298 K (f) $\Delta G_f°$ of NOBr at 298 K

20.76 Find K for (a) the hydrolysis of ATP, (b) the dehydration-condensation to form glucose phosphate, and (c) the overall coupled reaction between ATP and glucose. (d) How does each K change when T changes from 25°C to 37°C?

20.77 Calculate the equilibrium constants for decomposition of the hydrogen halides at 298 K: $2HX(g) \rightleftharpoons H_2(g) + X_2(g)$. What do these values indicate about the extent of decomposition of HX at 298 K? Suggest a reason for this trend.

20.78 The key process in a blast furnace during the production of iron is the reaction of Fe_2O_3 and carbon to yield Fe and CO_2.
(a) Calculate $\Delta H°$ and $\Delta S°$. [Assume C(graphite).]
(b) Is the reaction spontaneous at low or at high T? Explain.
(c) Is the reaction spontaneous at 298 K?
(d) At what temperature does the reaction become spontaneous?

20.79 Bromine monochloride is formed from the elements:

$$Cl_2(g) + Br_2(g) \rightleftharpoons 2BrCl(g)$$

$$\Delta H°_{rxn} = -1.35 \text{ kJ/mol} \qquad \Delta G°_f = -0.88 \text{ kJ/mol}$$

Calculate (a) $\Delta H°_f$ and (b) $S°$ of BrCl(g).

20.80 Solid N_2O_5 reacts with water to form liquid HNO_3. Consider the reaction with all substances in their standard states.
(a) Is the reaction spontaneous at 25°C?
(b) The solid decomposes to NO_2 and O_2 at 25°C. Is the decomposition spontaneous at 25°C? At what T is it spontaneous?
(c) At what T does *gaseous* N_2O_5 decompose spontaneously? Explain the difference between this T and that in part (b).

20.81 Energy from ATP hydrolysis drives many nonspontaneous cell reactions:

$$ATP^{4-}(aq) + H_2O(l) \rightleftharpoons ADP^{3-}(aq) + HPO_4^{2-}(aq) + H^+(aq) \qquad \Delta G°' = -30.5 \text{ kJ}$$

Energy for the reverse process comes ultimately from glucose metabolism:

$$C_6H_{12}O_6(s) + 6O_2(g) \longrightarrow 6CO_2(g) + 6H_2O(l)$$

(a) Find K for the hydrolysis of ATP at 37°C.
(b) Find $\Delta G°'_{rxn}$ for metabolism of 1 mol of glucose.
(c) How many moles of ATP can be produced by metabolism of 1 mol of glucose?
(d) If 36 mol of ATP is formed, what is the actual yield?

20.82 The molecular scene depicts a gaseous equilibrium mixture at 460°C for the reaction of H_2 *(blue)* and I_2 *(purple)* to form HI. Each molecule represents 0.010 mol and the container volume is 1.0 L. (a) Is K_c >, =, or < 1? (b) Is K_p >, =, or < K_c? (c) Calculate $\Delta G°_{rxn}$. (d) How would the value of $\Delta G°_{rxn}$ change if the purple molecules represented H_2 and the blue I_2? Explain.

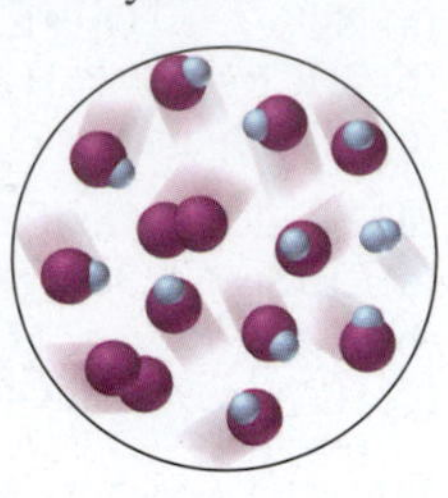

20.83 A chemical reaction, such as HI forming from its elements, can reach equilibrium at many temperatures. In contrast, a phase change, such as ice melting, is in equilibrium at a given pressure only at the melting point. Each of the graphs below depicts G_{sys} vs. extent of change. (a) Which graph depicts how G_{sys} changes for the formation of HI? Explain. (b) Which graph depicts how G_{sys} changes as ice melts at 1°C and 1 atm? Explain.

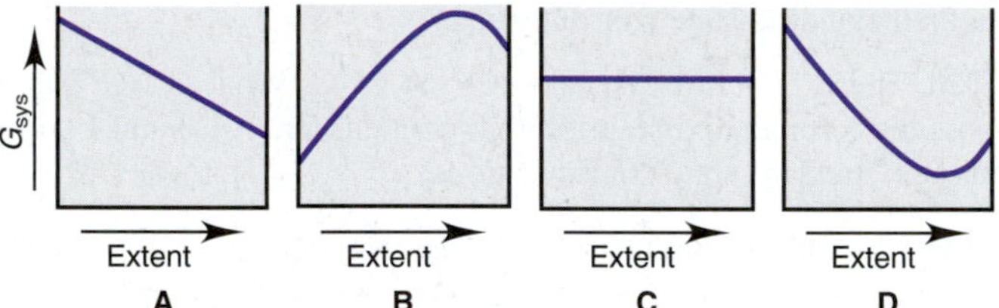

20.84 Consider the formation of ammonia:

$$N_2(g) + 3H_2(g) \rightleftharpoons 2NH_3(g)$$

(a) Assuming that $\Delta H°$ and $\Delta S°$ are constant with temperature, find the temperature at which $K_p = 1.00$.
(b) Find K_p at 400.°C, a typical temperature for NH_3 production.
(c) Given the lower K_p at the higher temperature, why are these conditions used industrially?

Electrochemistry: Chemical Change and Electrical Work

Everyday Electrochemistry As you'll learn in this chapter, batteries, like the one that powers this electronic device and the many others of modern life, apply the same principles that we use to extract aluminum from its ore and chlorine from the sea.

Outline

Key Principles
to focus on while studying this chapter

- *Oxidation-reduction (redox) reactions* involve the movement of electrons from one reactant to another. The *half-reaction method* of balancing redox reactions separates the overall reaction into two *half-reactions,* which mimics the actual separation of an *electrochemical cell* into two *half-cells.* Two types of electrochemical cells are distinguished by whether they generate electrical energy *(voltaic)* or use it *(electrolytic).* In both types of cell, electrodes dip into an electrolyte solution, the oxidation half-reaction occurs at the *anode,* and the reduction half-reaction occurs at the *cathode.* ***(Section 21.1)***
- A *voltaic cell* houses a spontaneous redox reaction ($\Delta G < 0$) and releases electrical energy. Electrons flow from anode to cathode through an external circuit, and ions flow through a *salt bridge* to complete the circuit and balance the charge within the cell. ***(Section 21.2)***
- The anode has a greater ability to give up electrons than the cathode, and the *cell potential,* or *voltage* (E_{cell}), is related to this difference. A negative ΔG (spontaneous reaction) correlates with a positive E_{cell}. Under standard-state conditions, each half-reaction is associated with a *standard electrode potential* ($E^\circ_{half\text{-}cell}$). Pairs of half-reactions can be combined to determine unknown electrode potentials and to write spontaneous redox reactions. ***(Section 21.3)***
- The standard free energy change (ΔG°), the standard cell potential (E°_{cell}), and the equilibrium constant (K) are interrelated. ***(Section 21.4)***
- E_{cell} changes during operation of a cell. The *Nernst equation* shows that E_{cell} depends on E°_{cell} and a term that corrects for nonstandard concentrations. During operation, the reactant concentration starts out *higher* than the product concentration and gradually *decreases* until $Q = K$ and the cell can do no more work. ***(Section 21.4)***
- In a *concentration cell,* both half-cells house the same half-reaction, but the solution concentrations are different. During operation, the solution in the anode half-cell spontaneously becomes *more* concentrated and that in the cathode half-cell becomes *less* concentrated until the concentrations are equal. ***(Section 21.4)***
- A *battery* is a group of voltaic cells arranged in series. In a *primary battery,* reactants become products until equilibrium is reached, at which point the battery is "dead." A *secondary battery* can be recharged by using an external source of electricity to convert the products back into reactants. In a *fuel cell,* reactants enter and products leave continually. ***(Section 21.5)***
- *Corrosion* is a natural, spontaneous, electrochemical process with similarities to the operation of a voltaic cell. It is a major economic problem because the anode is typically a metal tool or structure. ***(Section 21.6)***
- In an *electrolytic cell,* an external source of electricity drives a *non*spontaneous redox reaction ($\Delta G > 0$). In the electrolysis of a molten binary salt, the cation is reduced to the metal and the anion is oxidized to the nonmetal. For an aqueous salt solution, the products depend on whether water or one of the ions requires less energy to be reduced or oxidized. ***(Section 21.7)***
- Electrolysis is employed industrially to isolate elements from their ores. In some cases, a product (usually a gas) may form so slowly (high activation energy) that a higher-than-expected voltage *(overvoltage)* must be applied. ***(Section 21.7)***
- The amount of product formed is proportional to the quantity of charge flowing through the cell *(Faraday's law).* ***(Section 21.7)***

CONCEPTS & SKILLS TO REVIEW
before studying this chapter

- redox terminology (Section 4.5)
- activity series of the metals (Section 4.6)
- trends in ionization energy (Section 8.3) and electronegativity (Section 9.5)
- free energy, work, and equilibrium (Sections 20.3 and 20.4)
- Q vs. K (Section 17.4) and ΔG vs. ΔG° (Section 20.4)

Thermodynamics has countless applications other than the expansion of gases to run steam engines. In fact, probably within your reach right now are a laptop computer, MP3 player, cell phone, and/or wristwatch, which represent one type of application. Metal-plated jewelry and silverware represent the other. These are a few of the objects you use every day that rely on a major field in applied thermodynamics—**electrochemistry,** the study of the relationship between chemical change and electrical work. We typically study this relationship with **electrochemical cells,** systems that incorporate a redox reaction to produce or utilize electrical energy. In this chapter, we examine the essential features of the two types of electrochemical cells as well as the quantitative relationship between free energy and electrical work.

21.1 • REDOX REACTIONS AND ELECTROCHEMICAL CELLS

To begin, we review the oxidation-reduction process and describe the *half-reaction method* of balancing redox reactions. Then we see how redox reactions are used in the two types of electrochemical cells.

A Quick Review of Oxidation-Reduction Concepts

All electrochemical processes involve the *movement of electrons from one chemical species to another* in an oxidation-reduction (redox) reaction. In any redox process,

- *Oxidation* is the loss of electrons, and *reduction* is the gain of electrons. Oxidation and reduction occur *simultaneously*.
- The *oxidizing agent* does the oxidizing by *taking electrons* from the substance being oxidized. The *reducing agent* does the reducing by *giving electrons* to the substance being reduced.
- Therefore, *the oxidizing agent is reduced, and the reducing agent is oxidized.*
- The oxidized substance ends up with a *higher* (more positive or less negative) oxidation number (O.N.), and the reduced substance ends up with a *lower* (less positive or more negative) O.N.
- The total number of electrons gained by the atoms/ions of the oxidizing agent equals the total number lost by the atoms/ions of the reducing agent.

Figure 21.1 presents these ideas for the aqueous reaction between zinc metal and a strong acid. Be sure you can identify the oxidation and reduction parts of a redox process. You may want to review the full discussion in Section 4.5.

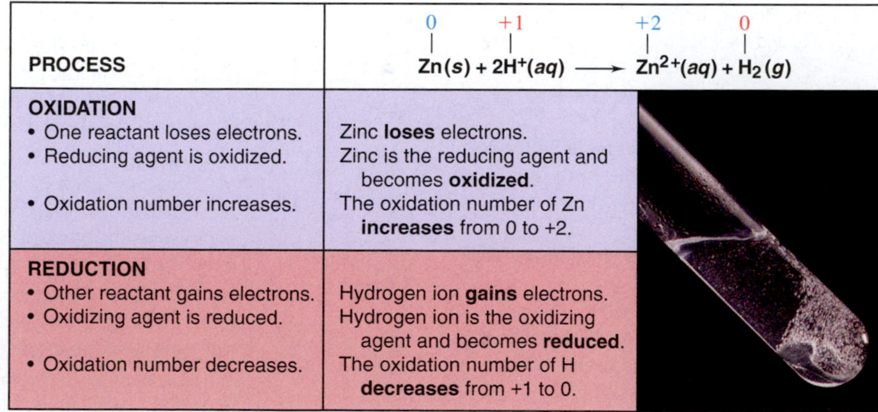

PROCESS	$\overset{0}{Zn(s)} + \overset{+1}{2H^+(aq)} \longrightarrow \overset{+2}{Zn^{2+}(aq)} + \overset{0}{H_2(g)}$
OXIDATION • One reactant loses electrons. • Reducing agent is oxidized. • Oxidation number increases.	Zinc **loses** electrons. Zinc is the reducing agent and becomes **oxidized**. The oxidation number of Zn **increases** from 0 to +2.
REDUCTION • Other reactant gains electrons. • Oxidizing agent is reduced. • Oxidation number decreases.	Hydrogen ion **gains** electrons. Hydrogen ion is the oxidizing agent and becomes **reduced**. The oxidation number of H **decreases** from +1 to 0.

Figure 21.1 A summary of redox terminology, as applied to the reaction of zinc with hydrogen ion.

Half-Reaction Method for Balancing Redox Reactions

The **half-reaction method** for balancing redox reactions is commonly used for studying electrochemistry for several reasons:

- It *separates the overall redox reaction* into oxidation and reduction *half-reactions,* which reflect their actual physical separation in electrochemical cells.
- It is easier to apply to reactions in acidic or basic solution, which is common in cells.
- It (usually) does *not* require assigning O.N.s. (In cases where the half-reactions are not obvious, we assign O.N.s to determine which atoms undergo a change and write half-reactions with the species that contain those atoms.)

Steps in the Half-Reaction Method The balancing process begins with a "skeleton" ionic reaction that consists of only species that are oxidized and reduced. Here are the steps in the half-reaction method:

Step 1. Divide the skeleton reaction into two half-reactions, each of which contains the oxidized and reduced forms of one of the species: *if the oxidized form of a species is on the left side, the reduced form must be on the right, and vice versa.*

Step 2. Balance the atoms and charges in each half-reaction.

- Atoms are balanced *in this order:* atoms other than O and H, then O, then H.
- Charge is balanced by *adding* electrons (e^-) *to the left side in the reduction half-reaction* because the reactant gains them and *to the right side in the oxidation half-reaction* because the reactant loses them.

Step 3. If necessary, multiply one or both half-reactions by an integer so that

$$\text{number of } e^- \text{ gained in reduction} = \text{number of } e^- \text{ lost in oxidation}$$

Step 4. Add the balanced half-reactions, and include states of matter.

Step 5. Check that the atoms and charges are balanced.

Let's balance a redox reaction that occurs in acidic solution and then see how to balance one in basic solution in Sample Problem 21.1.

Balancing Redox Reactions in Acidic Solution For a reaction in acidic solution, H_2O molecules and H^+ ions are present for the balancing. Even though we've usually used H_3O^+ to indicate the proton in water, we use H^+ in this chapter because it makes the balancing simpler.

The reaction between dichromate ion and iodide ion to form chromium(III) ion and solid iodine occurs in acidic solution *(see photo).* The skeleton ionic reaction is

$$Cr_2O_7^{2-}(aq) + I^-(aq) \longrightarrow Cr^{3+}(aq) + I_2(s) \qquad [\text{acidic solution}]$$

Dichromate ion *(left)* and iodide ion *(center)* form chromium(III) ion and solid iodine.

Step 1. Divide the reaction into half-reactions. Each half-reaction contains the oxidized and reduced forms of one species:

$$Cr_2O_7^{2-} \longrightarrow Cr^{3+}$$
$$I^- \longrightarrow I_2$$

Step 2. Balance atoms and charges in each half-reaction. We use H_2O to balance O, H^+ to balance H, and e^- to balance charges.

- For the $Cr_2O_7^{2-}/Cr^{3+}$ half-reaction:
  a. *Balance atoms other than O and H.* We balance the two Cr on the left with a coefficient of 2 on the right:
  $$Cr_2O_7^{2-} \longrightarrow 2Cr^{3+}$$
  b. *Balance O atoms by adding H_2O molecules.* Each H_2O has one O atom, so we add seven H_2O on the right to balance the seven O in $Cr_2O_7^{2-}$:
  $$Cr_2O_7^{2-} \longrightarrow 2Cr^{3+} + 7H_2O$$
  c. *Balance H atoms by adding H^+ ions.* Each H_2O contains two H, and we added seven H_2O, so we add $14H^+$ ions on the left:
  $$14H^+ + Cr_2O_7^{2-} \longrightarrow 2Cr^{3+} + 7H_2O$$

d. *Balance charge by adding electrons.* Each H^+ ion has a 1+ charge, and $14H^+$ plus $Cr_2O_7^{2-}$ gives 12+ on the left. Two Cr^{3+} give 6+ on the right. There is an excess of 6+ on the left, so we add six e^- on the left:

$$6e^- + 14H^+ + Cr_2O_7^{2-} \longrightarrow 2Cr^{3+} + 7H_2O$$

This half-reaction is balanced. It is the *reduction* because electrons appear on the *left, as reactants:* $Cr_2O_7^{2-}$ gains electrons (is reduced), so $Cr_2O_7^{2-}$ is the *oxidizing agent.* The O.N. of Cr decreases from +6 on the left to +3 on the right.

- For the I^-/I_2 half-reaction:

a. *Balance atoms other than O and H.* Two I atoms on the right require a coefficient of 2 on the left:

$$2I^- \longrightarrow I_2$$

b. *Balance O atoms with H_2O.* Not needed; there are no O atoms.

c. *Balance H atoms with H^+.* Not needed; there are no H atoms.

d. *Balance charge with e^-.* To balance the 2− on the left, we add two e^- on the right:

$$2I^- \longrightarrow I_2 + 2e^-$$

This half-reaction is balanced. It is the *oxidation* because electrons appear on the *right, as products:* the reactant I^- loses electrons (is oxidized), so I^- is the *reducing agent.* The O.N. of I increases from −1 to 0.

Step 3. Multiply each half-reaction, if necessary, by an integer so that the number of e^- lost in the oxidation equals the number of e^- gained in the reduction. Two e^- are lost in the oxidation and six e^- are gained in the reduction, so we multiply the oxidation by 3:

$$3(2I^- \longrightarrow I_2 + 2e^-)$$
$$6I^- \longrightarrow 3I_2 + 6e^-$$

Step 4. Add the half-reactions, canceling substances that appear on both sides, and include states of matter. In this example, only the electrons cancel:

$$\cancel{6e^-} + 14H^+ + Cr_2O_7^{2-} \longrightarrow 2Cr^{3+} + 7H_2O$$
$$6I^- \longrightarrow 3I_2 + \cancel{6e^-}$$
$$6I^-(aq) + 14H^+(aq) + Cr_2O_7^{2-}(aq) \longrightarrow 3I_2(s) + 7H_2O(l) + 2Cr^{3+}(aq)$$

Step 5. Check that atoms and charges balance:

Reactants (6I, 14H, 2Cr, 7O; 6+) ⟶ products (6I, 14H, 2Cr, 7O; 6+)

Balancing Redox Reactions in Basic Solution In acidic solution, H_2O molecules and H^+ ions are available to balance a redox reaction, but in basic solution, H_2O molecules and OH^- ions are available.

We need only one additional step to balance a redox reaction that takes place in basic solution. It appears after we balance the half-reactions *as if* they occurred in acidic solution and were combined (step 4). At this point, *we add one OH^-* ***to both sides of the equation*** *for every H^+ present.* (This step is labeled "4 Basic" in Sample Problem 21.1.) The OH^- ions added on the side with H^+ ions combine with them to form H_2O molecules, while the OH^- on the other side remain in the equation, and then excess H_2O is canceled.

Sample Problem 21.1 Balancing a Redox Reaction in Basic Solution

Problem Permanganate ion reacts in basic solution with oxalate ion to form carbonate ion and solid manganese dioxide. Balance the skeleton ionic equation for the reaction between $NaMnO_4$ and $Na_2C_2O_4$ in basic solution:

$$MnO_4^-(aq) + C_2O_4^{2-}(aq) \longrightarrow MnO_2(s) + CO_3^{2-}(aq) \quad [\text{basic solution}]$$

Plan We follow the numbered steps as described in text, and proceed through step 4 as if this reaction occurs in acidic solution. Then, we add the appropriate number of OH^- ions and cancel excess H_2O molecules (step 4 Basic).

Solution

1. Divide into half-reactions.

$MnO_4^- \longrightarrow MnO_2$	$C_2O_4^{2-} \longrightarrow CO_3^{2-}$

2. Balance.

a. Atoms other than O and H,	a. Atoms other than O and H,
Not needed	$C_2O_4^{2-} \longrightarrow 2CO_3^{2-}$
b. O atoms with H_2O,	b. O atoms with H_2O,
$MnO_4^- \longrightarrow MnO_2 + 2H_2O$	$2H_2O + C_2O_4^{2-} \longrightarrow 2CO_3^{2-}$
c. H atoms with H^+,	c. H atoms with H^+,
$4H^+ + MnO_4^- \longrightarrow MnO_2 + 2H_2O$	$2H_2O + C_2O_4^{2-} \longrightarrow 2CO_3^{2-} + 4H^+$
d. Charge with e^-,	d. Charge with e^-,
$3e^- + 4H^+ + MnO_4^- \longrightarrow MnO_2 + 2H_2O$	$2H_2O + C_2O_4^{2-} \longrightarrow 2CO_3^{2-} + 4H^+ + 2e^-$
[reduction]	[oxidation]

3. Multiply each half-reaction, if necessary, by some integer to make e^- lost equal e^- gained.

$2(3e^- + 4H^+ + MnO_4^- \longrightarrow MnO_2 + 2H_2O)$	$3(2H_2O + C_2O_4^{2-} \longrightarrow 2CO_3^{2-} + 4H^+ + 2e^-)$
$6e^- + 8H^+ + 2MnO_4^- \longrightarrow 2MnO_2 + 4H_2O$	$6H_2O + 3C_2O_4^{2-} \longrightarrow 6CO_3^{2-} + 12H^+ + 6e^-$

4. Add half-reactions, and cancel substances appearing on both sides. The six e^- cancel, eight H^+ cancel to leave four H^+ on the right, and four H_2O cancel to leave two H_2O on the left:

$$\cancel{6e^-} + \cancel{8H^+} + 2MnO_4^- \longrightarrow 2MnO_2 + \cancel{4H_2O}$$
$$2\,\cancel{6}H_2O + 3C_2O_4^{2-} \longrightarrow 6CO_3^{2-} + 4\,\cancel{12H^+} + \cancel{6e^-}$$
$$2MnO_4^- + 2H_2O + 3C_2O_4^{2-} \longrightarrow 2MnO_2 + 6CO_3^{2-} + 4H^+$$

4 Basic. Add OH^- to both sides to neutralize H^+, and cancel H_2O.
Adding four OH^- to both sides forms four H_2O on the right. Two of those cancel the two H_2O on the left and leave two H_2O on the right:

$$2MnO_4^- + 2H_2O + 3C_2O_4^{2-} + 4OH^- \longrightarrow 2MnO_2 + 6CO_3^{2-} + [4H^+ + 4OH^-]$$
$$2MnO_4^- + \cancel{2H_2O} + 3C_2O_4^{2-} + 4OH^- \longrightarrow 2MnO_2 + 6CO_3^{2-} + 2\,\cancel{4}H_2O$$

Including states of matter gives the final balanced equation:

$$2MnO_4^-(aq) + 3C_2O_4^{2-}(aq) + 4OH^-(aq) \longrightarrow 2MnO_2(s) + 6CO_3^{2-}(aq) + 2H_2O(l)$$

5. Check that atoms and charges balance.

$$(2Mn, 24O, 6C, 4H; 12-) \longrightarrow (2Mn, 24O, 6C, 4H; 12-)$$

Comment As a final step, let's see how to obtain the balanced *molecular* equation for this reaction. We note the amount (mol) of each anion in the balanced ionic equation and add the correct amount (mol) of spectator ions (in this case, Na^+, as given in the problem statement) to obtain neutral compounds. The balanced molecular equation is

$$2NaMnO_4(aq) + 3Na_2C_2O_4(aq) + 4NaOH(aq) \longrightarrow 2MnO_2(s) + 6Na_2CO_3(aq) + 2H_2O(l)$$

FOLLOW-UP PROBLEM 21.1 Write a balanced molecular equation for the reaction between $KMnO_4$ and KI in basic solution. The skeleton ionic reaction is

$$MnO_4^-(aq) + I^-(aq) \longrightarrow MnO_4^{2-}(aq) + IO_3^-(aq) \quad \text{[basic solution]}$$

An Overview of Electrochemical Cells

The fundamental difference between the two types of electrochemical cells is based on whether the overall redox reaction in the cell is spontaneous (free energy is released) or nonspontaneous (free energy is absorbed):

1. A **voltaic cell** (also called a **galvanic cell**) uses a *spontaneous* reaction ($\Delta G < 0$) to generate electrical energy. In the cell reaction, some of the difference in free energy between higher energy reactants and lower energy products is converted into electrical energy, which operates the load (surroundings)—flashlight, MP3 player, car starter motor, and so forth. Thus, *the system does work on the surroundings.* All batteries contain voltaic cells.
2. An **electrolytic cell** uses electrical energy to drive a *nonspontaneous* reaction ($\Delta G > 0$). In the cell reaction, an *external* source supplies free energy to convert lower energy reactants into higher energy products. Thus, *the surroundings do work on the system.* Electroplating and the recovery of metals from ores utilize electrolytic cells.

The two types of cell have several similarities. Two **electrodes,** which conduct the electricity between cell and surroundings, dip into an **electrolyte,** a mixture of ions (usually in aqueous solution) that is involved in the reaction or that carries the charge (Figure 21.2). An electrode is identified as either **anode** or **cathode** depending on the half-reaction that takes place there:

- *The oxidation half-reaction occurs at the anode.* Electrons lost by the substance being oxidized (reducing agent) *leave the oxidation half-cell* at the anode.
- *The reduction half-reaction occurs at the cathode.* Electrons gained by the substance being reduced (oxidizing agent) *enter the reduction half-cell* at the cathode.

Note that, for reasons we discuss shortly, *the relative charges of the electrodes are* ***opposite in sign*** *in the two types of cell.*

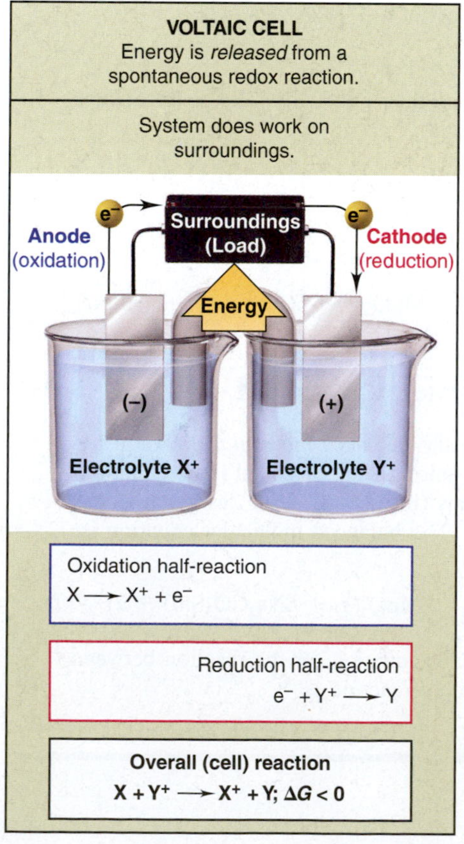

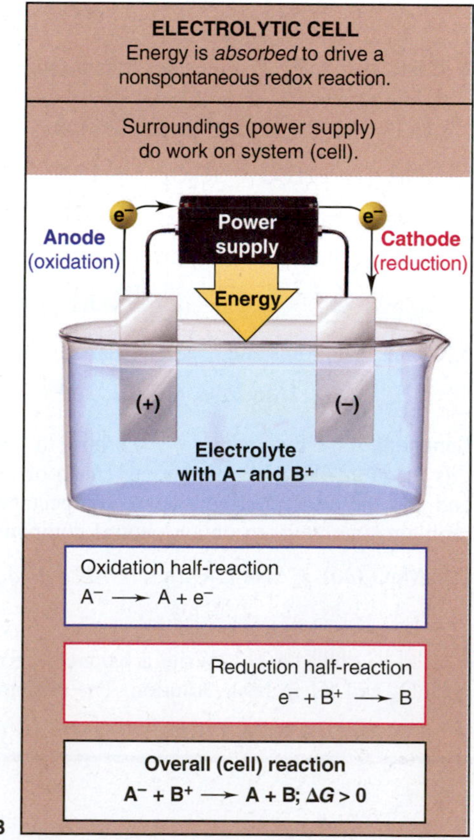

Figure 21.2 General characteristics of **(A)** voltaic cells and **(B)** electrolytic cells.

Here are some memory aids to help you connect the half-reaction with its electrode:

1. The words *anode* and *oxidation* start with vowels; the words *cathode* and *reduction* start with consonants.
2. Alphabetically, the *A* in anode comes before the *C* in cathode, and the *O* in oxidation comes before the *R* in reduction.
3. Look at the first syllables and use your imagination:

ANode, OXidation; REDuction, CAThode $\Longrightarrow$ AN OX and a RED CAT

THINK OF IT THIS WAY

Which Half-Reaction Occurs at Which Electrode?

Summary of Section 21.1

- An oxidation-reduction (redox) reaction involves the transfer of electrons from a reducing agent to an oxidizing agent.
- The half-reaction method of balancing divides the overall reaction into half-reactions that are balanced separately and then recombined.
- There are two types of electrochemical cells. In a voltaic cell, a spontaneous reaction generates electricity and does work on the surroundings. In an electrolytic cell, the surroundings supply electricity that does work to drive a nonspontaneous reaction.
- In both types of cell, two electrodes dip into electrolyte solutions; oxidation occurs at the anode, and reduction occurs at the cathode.

21.2 • VOLTAIC CELLS: USING SPONTANEOUS REACTIONS TO GENERATE ELECTRICAL ENERGY

When you put a strip of zinc metal in a solution of Cu^{2+} ion, the blue color of the solution fades and a brown-black crust of Cu metal forms on the Zn strip (Figure 21.3). During this spontaneous reaction, Cu^{2+} ion is reduced to Cu metal, while Zn metal is oxidized to Zn^{2+} ion. The overall reaction consists of two half-reactions:

$$Cu^{2+}(aq) + 2e^- \longrightarrow Cu(s) \quad \text{[reduction]}$$
$$Zn(s) \longrightarrow Zn^{2+}(aq) + 2e^- \quad \text{[oxidation]}$$
$$Zn(s) + Cu^{2+}(aq) \longrightarrow Zn^{2+}(aq) + Cu(s) \quad \text{[overall reaction]}$$

Let's examine this spontaneous reaction as the basis of a voltaic cell.

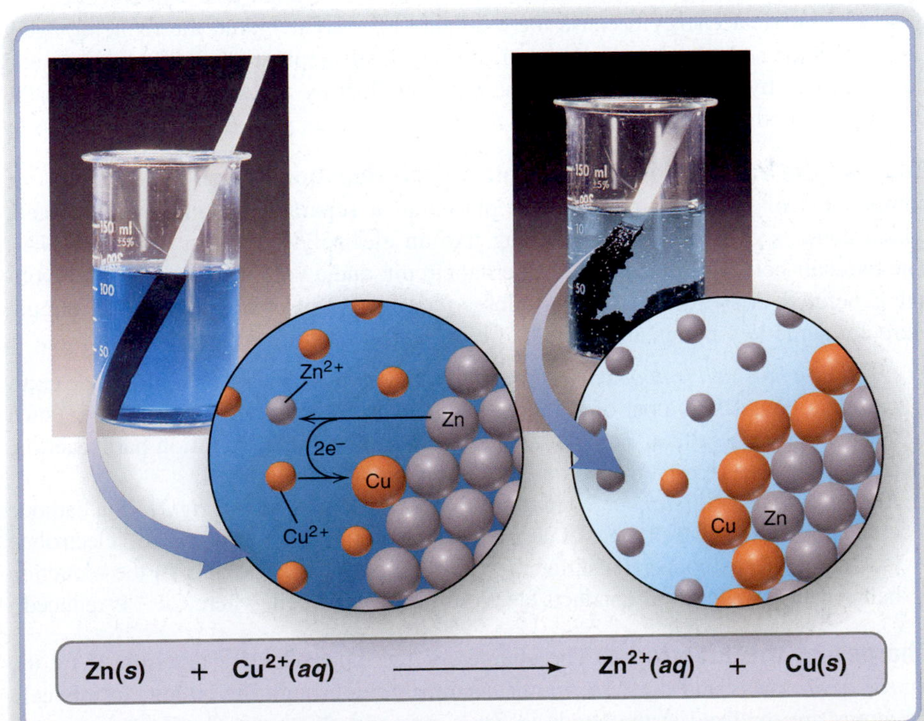

Figure 21.3 The spontaneous reaction between zinc and copper(II) ion. When zinc metal is placed in a solution of Cu^{2+} ion *(left)*, zinc is oxidized to Zn^{2+}, and Cu^{2+} is reduced to copper metal *(right)*. (The very finely divided Cu appears black.)

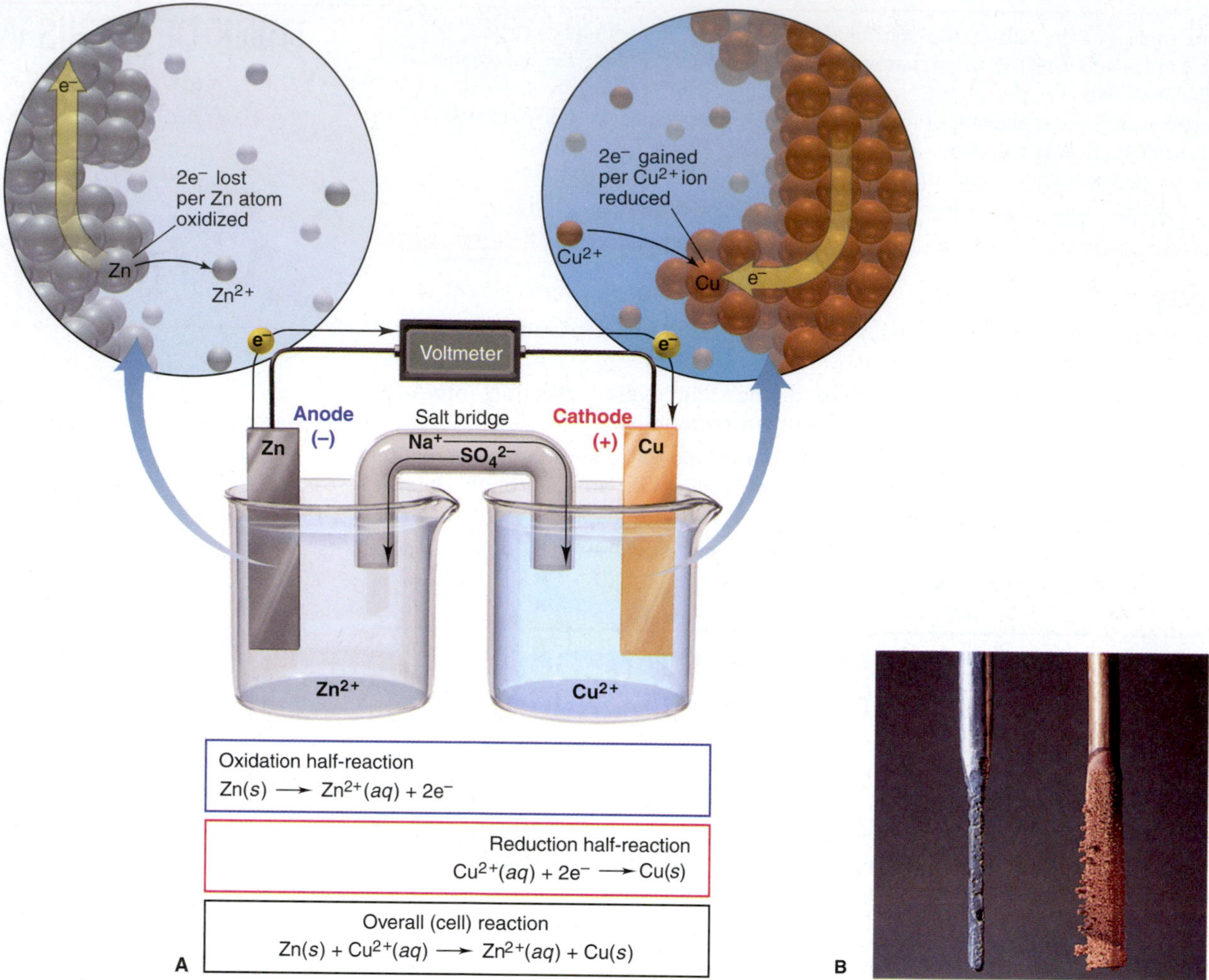

Figure 21.4 A voltaic cell based on the zinc-copper reaction. **A,** In the anode half-cell (oxidation; *left*), two electrons from a Zn atom move through the Zn bar as the Zn^{2+} ion enters the solution. In the cathode half-cell (reduction; *right*), the electrons reduce Cu^{2+} ions to Cu atoms. **B,** After several hours, the Zn anode weighs less and the Cu cathode weighs more.

Construction and Operation of a Voltaic Cell

Electrons are being transferred in the Zn/Cu^{2+} reaction, but the system does not generate electrical energy because the oxidizing agent (Cu^{2+}) and the reducing agent (Zn) are in the *same* beaker. If, however, we physically separate the half-reactions and connect them by an external circuit, the electrons lost by the zinc travel through the circuit and produce an electric current as the copper ions gain them.

Basis of the Voltaic Cell: Separation of Half-Reactions In any voltaic cell, the components of each half-reaction are placed in a separate container, or **half-cell,** which consists of one electrode dipping into an electrolyte solution (Figure 21.4A). The two half-cells are joined by the external circuit, and a voltmeter measures the voltage generated. A switch (not shown) closes (completes) or opens (breaks) the circuit. Here are some key points about the half-cells of the Zn/Cu^{2+} voltaic cell:

1. *Oxidation half-cell (anode compartment, always shown on the left).* The anode compartment consists of a bar of zinc (the anode) immersed in a Zn^{2+} electrolyte (such as aqueous zinc sulfate, $ZnSO_4$). Zinc is the reactant in the oxidation half-reaction, and the bar loses electrons *and* conducts them *out* of this half-cell.
2. *Reduction half-cell (cathode compartment, always shown on the right).* The cathode compartment consists of a bar of copper (the cathode) immersed in a Cu^{2+} electrolyte [such as aqueous copper(II) sulfate, $CuSO_4$]. Copper is the product in the reduction half-reaction, and the bar conducts electrons *into* its half-cell, where Cu^{2+} is reduced.

Charges of the Electrodes The charges of the electrodes are determined by the *source of electrons* and the *direction of electron flow* through the circuit. In this cell, Zn atoms are oxidized at the anode to Zn^{2+} ions and electrons. The Zn^{2+} ions enter the half-cell electrolyte, while the electrons move through the bar and into the wire.

The electrons flow left to right through the wire to the cathode, where Cu^{2+} ions in this electrolyte are reduced to Cu atoms. As the cell operates, electrons are continuously generated at the anode and consumed at the cathode. Therefore, the anode has an excess of electrons and the cathode has a deficit of electrons: *in any* ***voltaic*** *cell, the anode has a negative charge and the cathode has a positive charge.*

Completing the Circuit with a Salt Bridge A cell cannot operate unless the circuit is complete. The oxidation half-cell originally contains a neutral solution of Zn^{2+} and SO_4^{2-} ions, but as Zn atoms in the bar lose electrons, the Zn^{2+} ions that form enter the solution and would give it a net positive charge. Similarly, in the reduction half-cell, the neutral solution of Cu^{2+} and SO_4^{2-} ions would develop a net negative charge as Cu^{2+} ions leave the solution and form Cu atoms.

Such a charge imbalance between the half-cells would stop cell operation, and this situation is avoided by use of a **salt bridge.** It joins the half-cells and acts like a "liquid wire," allowing ions to flow through both compartments and complete the circuit. The salt bridge is an inverted U tube containing nonreacting ions, such as Na^+ and SO_4^{2-}, dissolved in a gel, which does not flow out of the tube but allows ions to diffuse into or out of the half-cells:

- *Maintaining a neutral reduction half-cell (right; cathode compartment).* As Cu^{2+} ions change to Cu atoms, Na^+ ions move from the salt bridge into the electrolyte solution (and some SO_4^{2-} ions move from the solution into the salt bridge).
- *Maintaining a neutral oxidation half-cell (left; anode compartment).* As Zn atoms change to Zn^{2+} ions, SO_4^{2-} ions move from the salt bridge into the electrolyte solution (and some Zn^{2+} ions move from the solution into the salt bridge).

Thus, the wire and the salt bridge complete the circuit:

- *Electrons move left to right* through the wire.
- *Anions move right to left* through the salt bridge.
- *Cations move left to right* through the salt bridge.

Active vs. Inactive Electrodes The electrodes in the Zn/Cu^{2+} cell are *active* because the metals themselves are components of the half-reactions. As the cell operates, the mass of the zinc bar gradually decreases, as the $[Zn^{2+}]$ in the anode half-cell increases. At the same time, the mass of the copper bar increases as the $[Cu^{2+}]$ in the cathode half-cell decreases and the ions form atoms that "plate out" on the electrode (Figure 21.4B).

In many cases, however, there are no reaction components that can be physically used as an electrode, so *inactive* electrodes are used. Most commonly, inactive electrodes are rods of *graphite* or *platinum*. In a voltaic cell based on the following half-reactions, for instance, the reacting species cannot be made into electrodes:

$$2I^-(aq) \longrightarrow I_2(s) + 2e^- \quad \text{[anode; oxidation]}$$

$$MnO_4^-(aq) + 8H^+(aq) + 5e^- \longrightarrow Mn^{2+}(aq) + 4H_2O(l) \quad \text{[cathode; reduction]}$$

Each half-cell consists of an inactive electrode immersed in an electrolyte that contains *all the reactant species involved in that half-reaction* (Figure 21.5):

- In the *anode half-cell*, I^- ions are oxidized to solid I_2, and the released electrons flow into the graphite electrode (C) and through the wire.
- From the wire, the electrons enter the graphite electrode in the *cathode half-cell* and reduce MnO_4^- ions to Mn^{2+} ions. (A KNO_3 salt bridge is used.)

Diagram of a Voltaic Cell As Figures 21.4A and 21.5 show, there are certain consistent features in the *diagram* of any voltaic cell:

- Components of the half-cells include electrode materials, electrolyte ions, and other species involved in the reaction.
- Electrode name (anode or cathode) and charge are shown. By convention, the anode compartment appears *on the left.*
- Each half-reaction and the overall cell reaction are given.
- Direction of electron flow in the external circuit is from left to right.
- Nature of ions and direction of ion flow in the salt bridge are shown, with cations moving right and anions moving left.

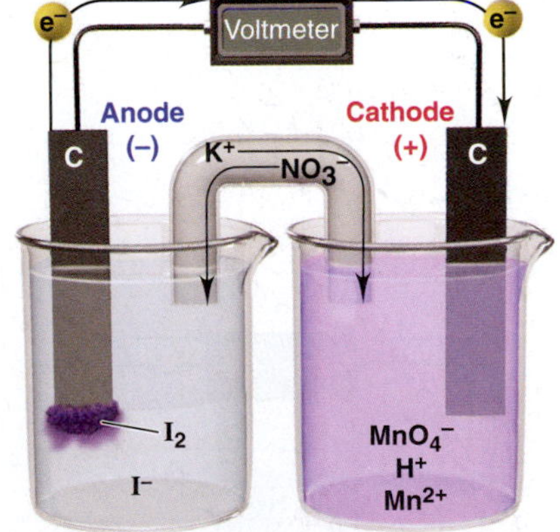

Oxidation half-reaction

$2I^-(aq) \longrightarrow I_2(s) + 2e^-$

Reduction half-reaction

$MnO_4^-(aq) + 8H^+(aq) + 5e^- \longrightarrow Mn^{2+}(aq) + 4H_2O(l)$

Overall (cell) reaction

$2MnO_4^-(aq) + 16H^+(aq) + 10I^-(aq) \longrightarrow 2Mn^{2+}(aq) + 5I_2(s) + 8H_2O(l)$

Figure 21.5 A voltaic cell using inactive electrodes.

Notation for a Voltaic Cell

A useful shorthand notation describes the components of a voltaic cell. The notation for the Zn/Cu^{2+} cell is

$$Zn(s) \mid Zn^{2+}(aq) \parallel Cu^{2+}(aq) \mid Cu(s)$$

Key parts of the notation are as follows:

- The components of the anode compartment (oxidation half-cell) are written *to the left* of the components of the cathode compartment (reduction half-cell).
- A double vertical line indicates the half-cells are physically separated.
- A single vertical line represents a phase boundary. For example, $Zn(s) \mid Zn^{2+}(aq)$ indicates that *solid* Zn is a *different* phase from *aqueous* Zn^{2+}.
- A comma separates the half-cell components that are in the *same* phase. For example, the notation for the voltaic cell shown in Figure 21.5 is

  $$\text{graphite} \mid I^-(aq) \mid I_2(s) \parallel MnO_4^-(aq), H^+(aq), Mn^{2+}(aq) \mid \text{graphite}$$

  That is, in the cathode compartment, MnO_4^-, H^+, and Mn^{2+} ions are in an aqueous solution with solid graphite immersed in it.
- If needed, the concentrations of dissolved components are given in parentheses; for example, if the concentrations of Zn^{2+} and Cu^{2+} are 1 *M*, we write

  $$Zn(s) \mid Zn^{2+}(1\ M) \parallel Cu^{2+}(1\ M) \mid Cu(s)$$
- Half-cell components usually appear in the same order as in the half-reaction, and electrodes appear at the far left (anode) and far right (cathode) of the notation.
- Ions in the salt bridge are not part of the reaction so they are not in the notation.

Sample Problem 21.2 Describing a Voltaic Cell with Diagram and Notation

Problem Draw a diagram, show balanced equations, and write the notation for a voltaic cell that consists of one half-cell with a Cr bar in a $Cr(NO_3)_3$ solution, another half-cell with an Ag bar in an $AgNO_3$ solution, and a KNO_3 salt bridge. Measurement indicates that the Cr electrode is negative relative to the Ag electrode.

Plan From the given contents of the half-cells, we write the half-reactions. To determine which is the anode compartment (oxidation) and which is the cathode (reduction), we note the relative electrode charges (which are based on the direction of the spontaneous redox reaction). Electrons are released into the anode during oxidation, so it has a negative charge. We are told that Cr is negative, so it is the anode and Ag is the cathode.

Solution Writing the balanced half-reactions. The Ag half-reaction consumes e^-:

$$Ag^+(aq) + e^- \longrightarrow Ag(s) \qquad \text{[reduction; cathode]}$$

The Cr half-reaction releases e^-:

$$Cr(s) \longrightarrow Cr^{3+}(aq) + 3e^- \qquad \text{[oxidation; anode]}$$

Writing the balanced overall cell equation. We triple the reduction half-reaction to balance e^- and combine the half-reactions:

$$Cr(s) + 3Ag^+(aq) \longrightarrow Cr^{3+}(aq) + 3Ag(s)$$

Determining direction of electron and ion flow. The released e^- in the Cr electrode (negative) flow through the external circuit to the Ag electrode (positive). As Cr^{3+} ions enter the anode electrolyte, NO_3^- ions enter from the salt bridge to maintain neutrality. As Ag^+ ions leave the cathode electrolyte and plate out on the Ag electrode, K^+ ions enter from the salt bridge to maintain neutrality. The cell diagram is shown in the margin.

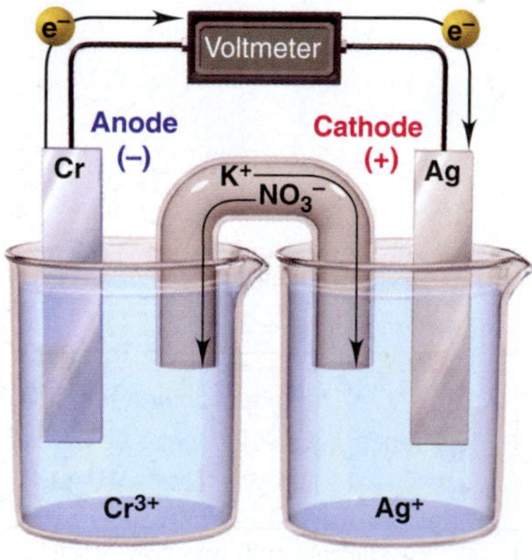

Writing the cell notation:

$$Cr(s) \mid Cr^{3+}(aq) \parallel Ag^+(aq) \mid Ag(s)$$

Check Always be sure that the half-reactions and the cell reaction are balanced, the half-cells contain *all* components of the half-reactions, and the electron and ion flow are shown. You should be able to write the half-reactions from the cell notation as a check.

Comment The diagram of a voltaic cell relies on the *direction of the spontaneous reaction* to give the oxidation (anode; negative) and reduction (cathode; positive) half-reactions.

FOLLOW-UP PROBLEM 21.2 In one compartment of a voltaic cell, a graphite rod dips into an acidic solution of $K_2Cr_2O_7$ and $Cr(NO_3)_3$; in the other compartment, a tin bar dips into a $Sn(NO_3)_2$ solution. A KNO_3 salt bridge joins them. The tin electrode is negative relative to the graphite. Diagram the cell and write the balanced equations and the cell notation.

Summary of Section 21.2

- A voltaic cell consists of oxidation (anode) and reduction (cathode) half-cells, connected by a wire to conduct electrons and a salt bridge to maintain charge neutrality.
- Electrons move from the anode (left) to the cathode (right), while cations move from the salt bridge into the cathode half-cell and anions move from the salt bridge into the anode half-cell.
- The cell notation shows the species and their phases in each half-cell, as well as the direction of current flow.

21.3 • CELL POTENTIAL: OUTPUT OF A VOLTAIC CELL

A voltaic cell converts the free energy change of a spontaneous reaction into the kinetic energy of electrons moving through an external circuit (electrical energy). The electrode with the higher electrical potential to give up its electrons is designated the anode, and the other electrode is the cathode. The electrical energy the cell produces is proportional to the *difference in electrical potential between the two electrodes,* which is called the **cell potential (E_{cell})**; it is also called the **voltage** of the cell or the **electromotive force (emf).**

Electrons flow spontaneously from the negative to the positive electrode, that is, toward the electrode with the more positive electrical potential (anode to cathode). Thus, when the cell operates *spontaneously,* there is a *positive* cell potential:

$$E_{cell} > 0 \text{ for a spontaneous process} \quad (21.1)$$

- A *positive* E_{cell} arises from a spontaneous reaction. The more positive it is, the more work the cell can do, and the farther the reaction proceeds to the right as written.
- A *negative* E_{cell} is associated with a *nonspontaneous* cell reaction.
- If $E_{cell} = 0$, the reaction has reached equilibrium and the cell can do no more work.

Standard Cell Potentials

Cell potential refers to energy available to do the work of moving a charge between electrodes. The SI unit of electrical potential is the **volt (V),** and the SI unit of electrical charge is the **coulomb (C).** By definition, if two electrodes differ by 1 volt of electrical potential, 1 joule of energy is released (that is, 1 joule of work can be done) for each coulomb of charge that moves between the electrodes. That is,

$$1 \text{ V} = 1 \text{ J/C} \quad (21.2)$$

Table 21.1 lists the voltages of some commercial and natural voltaic cells.

The *measured* cell potential is affected by changes in concentration as the reaction proceeds and by energy losses from heating the cell and external circuit. Therefore, as with other thermodynamic quantities, to compare potentials of different cells, we obtain a **standard cell potential (E°_{cell}),** which is measured at a specified temperature (usually 298 K) with no current flowing* and *all components in their standard states:* 1 atm for gases, 1 *M* for solutions, the pure solid for electrodes. For

Table 21.1 Voltages of Some Voltaic Cells

Voltaic Cell	Voltage (V)
Common alkaline flashlight battery	1.5
Lead-acid car battery (6 cells ≈ 12 V)	2.1
Calculator battery (mercury)	1.3
Lithium-ion laptop battery	3.7
Electric eel (~5000 cells in 6-ft eel = 750 V)	0.15
Nerve of giant squid (across cell membrane)	0.070

*The tiny current required to operate modern digital voltmeters makes a negligible difference in the value of E°_{cell}.

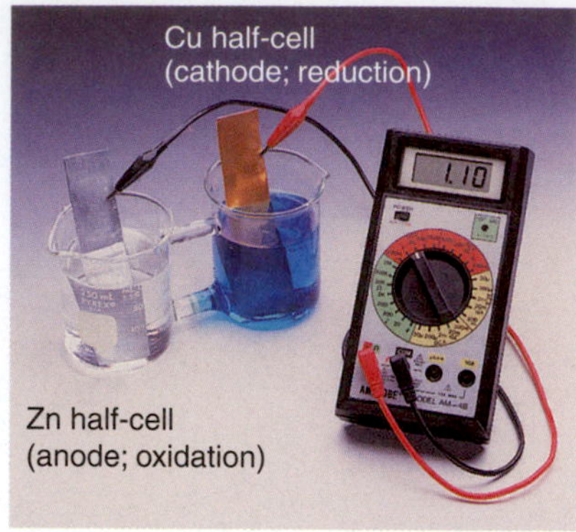

Figure 21.6 Measuring the standard cell potential of a zinc-copper cell.

example, the zinc-copper cell produces 1.10 V when it operates at 298 K with $[Zn^{2+}] = [Cu^{2+}] = 1\ M$ (Figure 21.6):

$$Zn(s) + Cu^{2+}(aq; 1\ M) \longrightarrow Zn^{2+}(aq; 1\ M) + Cu(s) \qquad E^\circ_{cell} = 1.10\ V$$

Standard Electrode (Half-Cell) Potentials Just as each half-reaction makes up part of the overall reaction, the potential of each half-cell makes up a part of the overall cell potential. The **standard electrode potential** ($E^\circ_{half\text{-}cell}$) is the potential of a given half-reaction (electrode compartment) with all components in their standard states.

By convention (and as they appear in reference tables), *a standard electrode potential refers to the half-reaction written as a* ***reduction.*** For the zinc-copper reaction, both the zinc half-reaction (E°_{zinc}, anode compartment) and the copper half-reaction (E°_{copper}, cathode compartment) are written as reductions:

$$Zn^{2+}(aq) + 2e^- \longrightarrow Zn(s) \qquad E^\circ_{zinc}\ (E^\circ_{anode}) \qquad [\text{reduction}]$$
$$Cu^{2+}(aq) + 2e^- \longrightarrow Cu(s) \qquad E^\circ_{copper}\ (E^\circ_{cathode}) \qquad [\text{reduction}]$$

But, since the overall cell reaction involves the *oxidation* of zinc, not the *reduction* of Zn^{2+}, we reverse the zinc half-reaction:

$$Zn(s) \longrightarrow Zn^{2+}(aq) + 2e^- \qquad [\text{oxidation}]$$
$$Cu^{2+}(aq) + 2e^- \longrightarrow Cu(s) \qquad [\text{reduction}]$$

Calculating E°_{cell} from $E^\circ_{half\text{-}cell}$ The overall redox reaction for the zinc-copper cell is the sum of its half-reactions:

$$Zn(s) + Cu^{2+}(aq) \longrightarrow Zn^{2+}(aq) + Cu(s)$$

Because electrons flow spontaneously from the negative to the positive electrode, the copper electrode (cathode) has a more positive $E^\circ_{half\text{-}cell}$ than the zinc electrode (anode). Arithmetically, to obtain a positive E°_{cell} for this spontaneous redox reaction, we must subtract E°_{zinc} from E°_{copper}:

$$E^\circ_{cell} = E^\circ_{copper} - E^\circ_{zinc}$$

We can generalize this result for any voltaic cell: *the standard cell potential is the difference between the standard electrode potential of the cathode (reduction) half-cell and the standard electrode potential of the anode (oxidation) half-cell:*

$$E^\circ_{cell} = E^\circ_{cathode\ (reduction)} - E^\circ_{anode\ (oxidation)} \qquad (21.3)$$

For a *spontaneous* reaction at standard conditions, this calculation gives $E^\circ_{cell} > 0$.

Determining $E^\circ_{half\text{-}cell}$ with the Standard Hydrogen Electrode To compare half-cell potentials, we need to know the portion of E°_{cell} contributed by each half-cell. But, how can we find potentials of the individual half-cells if we can measure only the potential of the overall cell?

Half-cell potentials, such as E°_{zinc} and E°_{copper}, are *relative* to a standard reference half-cell, *which has a standard electrode potential defined as zero* ($E^\circ_{reference} \equiv 0.00$ V). The **standard reference half-cell** is a **standard hydrogen electrode,** which consists of a platinum electrode that has H_2 gas at 1 atm bubbling through it and is immersed in 1 *M* strong acid, $H^+(aq)$ [or $H_3O^+(aq)$]. Thus, the reference half-reaction is

$$2H^+(aq; 1\ M) + 2e^- \rightleftharpoons H_2(g; 1\ atm) \qquad E^\circ_{reference} = 0.00\ V$$

To find an unknown standard electrode potential ($E^\circ_{unknown}$), we construct a voltaic cell consisting of this reference half-cell and the unknown half-cell. Since $E^\circ_{reference}$ is zero, the overall E°_{cell} gives $E^\circ_{unknown}$.

Depending on the unknown half-cell, the reference half-cell can be the anode or the cathode:

- When H_2 is oxidized, the reference half-cell is the anode, and so *reduction* occurs at the unknown half-cell:

$$E^\circ_{cell} = E^\circ_{cathode} - E^\circ_{anode} = E^\circ_{unknown} - E^\circ_{reference} = E^\circ_{unknown} - 0.00\ V = E^\circ_{unknown}$$

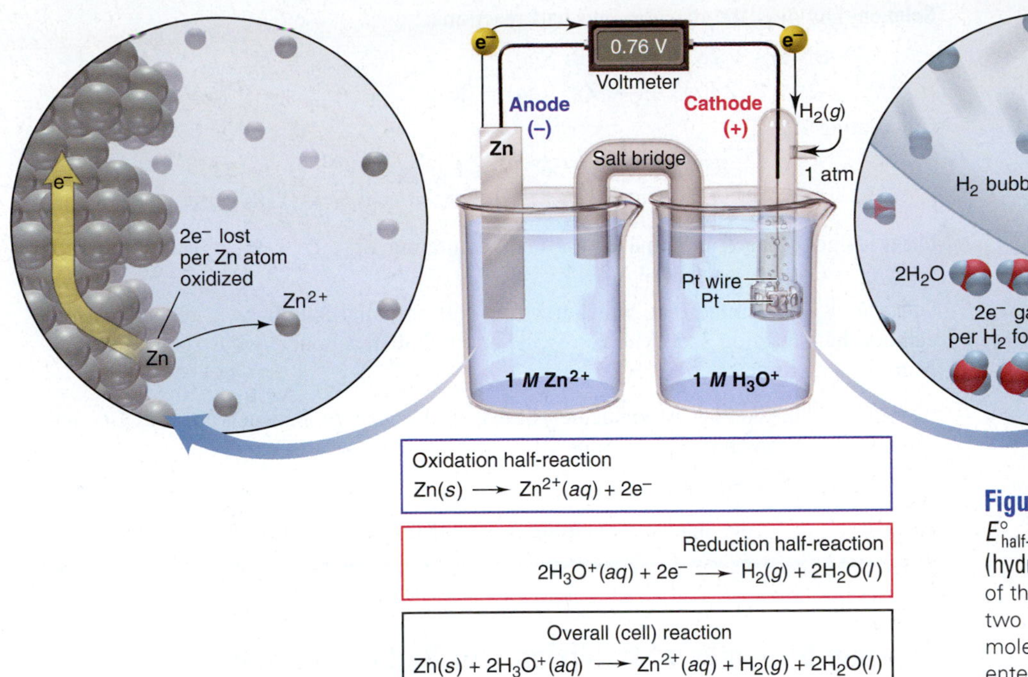

Figure 21.7 Determining an unknown $E^\circ_{\text{half-cell}}$ with the standard reference (hydrogen) electrode. The magnified view of the hydrogen half-reaction *(right)* shows two H_3O^+ ions being reduced to two H_2O molecules and an H_2 molecule, which enters the H_2 bubble.

- When H^+ is reduced, the reference half-cell is the cathode, and so *oxidation* occurs at the unknown half-cell:

$$E^\circ_{\text{cell}} = E^\circ_{\text{cathode}} - E^\circ_{\text{anode}} = E^\circ_{\text{reference}} - E^\circ_{\text{unknown}} = 0.00\text{ V} - E^\circ_{\text{unknown}} = -E^\circ_{\text{unknown}}$$

Figure 21.7 shows a voltaic cell that has the Zn/Zn^{2+} half-reaction in one compartment and the H^+/H_2 (or H_3O^+/H_2) half-reaction in the other. The zinc electrode is negative relative to the hydrogen electrode, so we know that the zinc is being oxidized and is the anode. The measured E°_{cell} is +0.76 V (reaction is spontaneous), and we use this value to find the unknown standard electrode potential, E°_{zinc}:

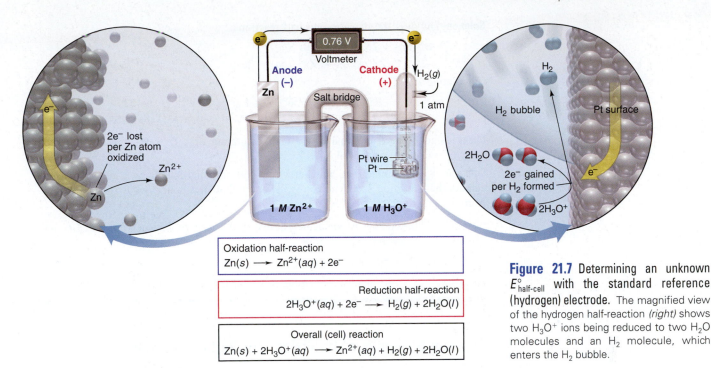

$$\begin{array}{lll} 2H^+(aq) + 2e^- \longrightarrow H_2(g) & E^\circ_{\text{reference}} = 0.00\text{ V} & [\text{cathode; reduction}] \\ Zn(s) \longrightarrow Zn^{2+}(aq) + 2e^- & E^\circ_{\text{zinc}} = ?\text{ V} & [\text{anode; oxidation}] \\ \hline Zn(s) + 2H^+(aq) \longrightarrow Zn^{2+}(aq) + H_2(g) & E^\circ_{\text{cell}} = 0.76\text{ V} & \end{array}$$

$$E^\circ_{\text{cell}} = E^\circ_{\text{cathode}} - E^\circ_{\text{anode}} = E^\circ_{\text{reference}} - E^\circ_{\text{zinc}}$$

$$E^\circ_{\text{zinc}} = E^\circ_{\text{reference}} - E^\circ_{\text{cell}} = 0.00\text{ V} - 0.76\text{ V} = -0.76\text{ V}$$

Now we can return to the zinc-copper cell and use the measured E°_{cell} (1.10 V) and the value we just found for E°_{zinc} to calculate E°_{copper}:

$$E^\circ_{\text{cell}} = E^\circ_{\text{cathode}} - E^\circ_{\text{anode}} = E^\circ_{\text{copper}} - E^\circ_{\text{zinc}}$$

$$E^\circ_{\text{copper}} = E^\circ_{\text{cell}} + E^\circ_{\text{zinc}} = 1.10\text{ V} + (-0.76\text{ V}) = 0.34\text{ V}$$

By continuing this process of constructing cells with one known and one unknown electrode potential, we find other standard electrode potentials.

Sample Problem 21.3 Calculating an Unknown $E^\circ_{\text{half-cell}}$ from E°_{cell}

Problem A voltaic cell houses the reaction between aqueous bromine and zinc metal:

$$Br_2(aq) + Zn(s) \longrightarrow Zn^{2+}(aq) + 2Br^-(aq) \qquad E^\circ_{\text{cell}} = 1.83\text{ V}$$

Calculate E°_{bromine}, given $E^\circ_{\text{zinc}} = -0.76$ V.

Plan E°_{cell} is positive, so the reaction is spontaneous as written. By examining the overall reaction and dividing it into half-reactions, we see that Br_2 is reduced and Zn is oxidized; thus, the zinc half-cell contains the anode. We use Equation 21.3 and the known E°_{zinc} to find E°_{unknown} (E°_{bromine}).

Solution Dividing the reaction into half-reactions:

$$Br_2(aq) + 2e^- \longrightarrow 2Br^-(aq) \qquad E^\circ_{unknown} = E^\circ_{bromine} = ?\ V$$
$$Zn(s) \longrightarrow Zn^{2+}(aq) + 2e^- \qquad E^\circ_{zinc} = -0.76\ V$$

Calculating $E^\circ_{bromine}$:

$$E^\circ_{cell} = E^\circ_{cathode} - E^\circ_{anode} = E^\circ_{bromine} - E^\circ_{zinc}$$
$$E^\circ_{bromine} = E^\circ_{cell} + E^\circ_{zinc} = 1.83\ V + (-0.76\ V) = 1.07\ V$$

Check A good check is to make sure that calculating $E^\circ_{bromine} - E^\circ_{zinc}$ gives E°_{cell}: $1.07\ V - (-0.76\ V) = 1.83\ V$.

Comment Keep in mind that, whichever half-cell potential is unknown, reduction is the cathode half-reaction and oxidation is the anode half-reaction. *Always* subtract E°_{anode} from $E^\circ_{cathode}$ to get E°_{cell}.

FOLLOW-UP PROBLEM 21.3 A voltaic cell based on the reaction between aqueous Br_2 and vanadium(III) ions has $E^\circ_{cell} = 1.39$ V:

$$Br_2(aq) + 2V^{3+}(aq) + 2H_2O(l) \longrightarrow 2VO^{2+}(aq) + 4H^+(aq) + 2Br^-(aq)$$

What is $E^\circ_{vanadium}$, the standard electrode potential for the reduction of VO^{2+} to V^{3+}?

Relative Strengths of Oxidizing and Reducing Agents

We learn the relative strengths of oxidizing and reducing agents from measuring cell potentials. Three oxidizing agents just discussed are Cu^{2+}, H^+, and Zn^{2+}. Let's rank their relative oxidizing strengths from high *(top)* to low *(bottom)* by writing each half-reaction as a reduction (gain of electrons), with its corresponding standard electrode potential:

$$Cu^{2+}(aq) + 2e^- \longrightarrow Cu(s) \qquad E^\circ = 0.34\ V$$
$$2H^+(aq) + 2e^- \longrightarrow H_2(g) \qquad E^\circ = 0.00\ V$$
$$Zn^{2+}(aq) + 2e^- \longrightarrow Zn(s) \qquad E^\circ = -0.76\ V$$

The more positive the E° value, the more readily the reaction (as written) occurs; thus, Cu^{2+} gains two e^- more readily than H^+, which gains them more readily than Zn^{2+}:

- Strength as an oxidizing agent *decreases* top to bottom: $Cu^{2+} > H^+ > Zn^{2+}$.
- Strength as a reducing agent *increases* top to bottom: $Zn > H_2 > Cu$.

By continuing this process with other half-cells, we create a list of reduction half-reactions in *decreasing* order of standard electrode potential (from most positive to most negative). Such a list, called an *emf series* or a *table of standard electrode potentials,* appears in Appendix D; a few examples are presented in Table 21.2. There are several key points to keep in mind:

- All values are relative to the standard hydrogen (reference) electrode.
- Since the half-reactions are written as *reductions, reactants are oxidizing agents* and *products are reducing agents.*
- The more positive the $E^\circ_{half\text{-}cell}$, the more readily the half-reaction occurs, as written.
- Half-reactions are shown with an equilibrium arrow because each can occur as a reduction (at the cathode) or an oxidation (at the anode), depending on the $E^\circ_{half\text{-}cell}$ of the other half-reaction.
- As Appendix D (or Table 21.2) is arranged, the strength of the oxidizing agent (reactant) *increases going up (bottom to top),* and the strength of the reducing agent (product) *increases going down (top to bottom).*

Thus, $F_2(g)$ is the strongest oxidizing agent (has the largest positive E°), which means $F^-(aq)$ is the weakest reducing agent. Similarly, $Li^+(aq)$ is the weakest oxidizing agent (has the most negative E°), which means $Li(s)$ is the strongest reducing agent. This makes sense because F_2 is very electronegative (gains electrons easily; Section 9.5), and Li has a low ionization energy (loses electrons easily; Section 8.3). The key point is that a *strong oxidizing agent forms a weak reducing agent,* and vice versa.

Table 21.2 Selected Standard Electrode Potentials (298 K)

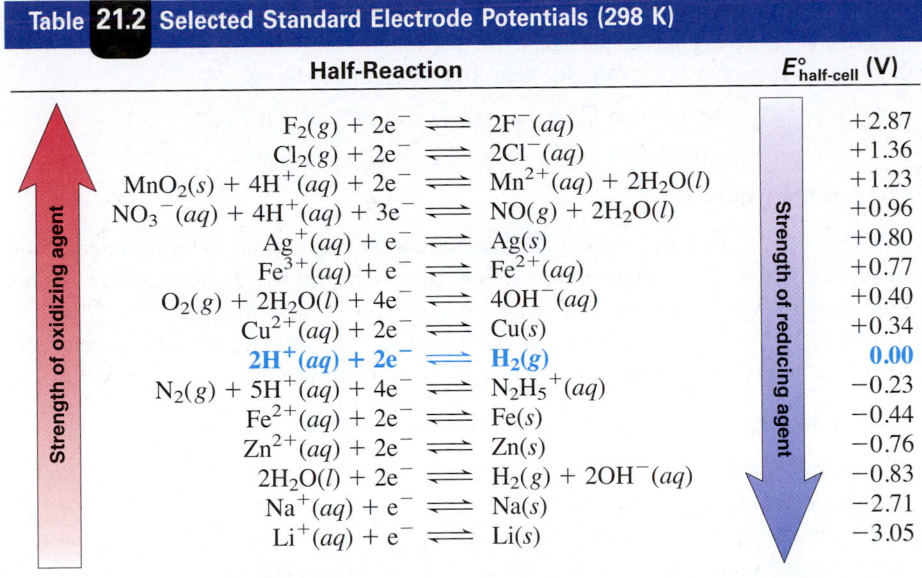

Half-Reaction	$E^\circ_{half\text{-}cell}$ (V)
$F_2(g) + 2e^- \rightleftharpoons 2F^-(aq)$	+2.87
$Cl_2(g) + 2e^- \rightleftharpoons 2Cl^-(aq)$	+1.36
$MnO_2(s) + 4H^+(aq) + 2e^- \rightleftharpoons Mn^{2+}(aq) + 2H_2O(l)$	+1.23
$NO_3^-(aq) + 4H^+(aq) + 3e^- \rightleftharpoons NO(g) + 2H_2O(l)$	+0.96
$Ag^+(aq) + e^- \rightleftharpoons Ag(s)$	+0.80
$Fe^{3+}(aq) + e^- \rightleftharpoons Fe^{2+}(aq)$	+0.77
$O_2(g) + 2H_2O(l) + 4e^- \rightleftharpoons 4OH^-(aq)$	+0.40
$Cu^{2+}(aq) + 2e^- \rightleftharpoons Cu(s)$	+0.34
$2H^+(aq) + 2e^- \rightleftharpoons H_2(g)$	**0.00**
$N_2(g) + 5H^+(aq) + 4e^- \rightleftharpoons N_2H_5^+(aq)$	−0.23
$Fe^{2+}(aq) + 2e^- \rightleftharpoons Fe(s)$	−0.44
$Zn^{2+}(aq) + 2e^- \rightleftharpoons Zn(s)$	−0.76
$2H_2O(l) + 2e^- \rightleftharpoons H_2(g) + 2OH^-(aq)$	−0.83
$Na^+(aq) + e^- \rightleftharpoons Na(s)$	−2.71
$Li^+(aq) + e^- \rightleftharpoons Li(s)$	−3.05

Writing Spontaneous Redox Reactions

Every redox reaction is the sum of two half-reactions, so there is a reducing agent and an oxidizing agent on each side. In the zinc-copper reaction, for instance, Zn and Cu are the reducing agents, and Cu^{2+} and Zn^{2+} are the oxidizing agents. The stronger oxidizing and reducing agents react spontaneously to form the weaker oxidizing and reducing agents:

$$\underset{\text{stronger reducing agent}}{Zn(s)} + \underset{\text{stronger oxidizing agent}}{Cu^{2+}(aq)} \longrightarrow \underset{\text{weaker oxidizing agent}}{Zn^{2+}(aq)} + \underset{\text{weaker reducing agent}}{Cu(s)}$$

Based on the order of the E° values in Appendix D, and as we just saw for the Cu^{2+}/Cu, H^+/H_2, and Zn^{2+}/Zn redox pairs (or redox *couples*), *the stronger oxidizing agent (species on the left) has a half-reaction with a larger (more positive or less negative) E° value, and the stronger reducing agent (species on the right) has a half-reaction with a smaller (less positive or more negative) E° value.* Therefore, we can use Appendix D to choose a redox reaction for constructing a voltaic cell.

Writing a Spontaneous Reaction With Appendix D A spontaneous reaction ($E^\circ_{cell} > 0$) will occur between an oxidizing agent and any reducing agent that lies *below* it in the emf series, listed in Appendix D. In other words, *for a spontaneous reaction to occur, the half-reaction higher in the list proceeds at the cathode as written, and the half-reaction lower in the list proceeds at the anode in reverse.* This pairing ensures that the stronger oxidizing agent (higher on the left) and stronger reducing agent (lower on the right) will be the reactants. For example, two half-reactions in the order they appear in Appendix D are

$$Cl_2(g) + 2e^- \longrightarrow 2Cl^-(aq) \qquad E^\circ_{chlorine} = 1.36\ V$$
$$Ni^{2+}(aq) + 2e^- \longrightarrow Ni(s) \qquad E^\circ_{nickel} = -0.25\ V$$

We reverse the nickel half-reaction (lower in the list); note, however, that we do *not* reverse the sign of $E^\circ_{half\text{-}cell}$ because *the minus sign in Equation 21.3 does that.* Next, we have to be sure that e^- lost equal e^- gained. If they do not, we multiply one or both half-reactions by coefficients to accomplish this. In this case, they do, so we can skip this step. Adding the half-reactions and applying Equation 21.3 gives the balanced equation and E°_{cell}:

$$Cl_2(g) + Ni(s) \longrightarrow Ni^{2+}(aq) + 2Cl^-(aq)$$
$$E^\circ_{cell} = E^\circ_{cathode} - E^\circ_{anode} = E^\circ_{chlorine} - E^\circ_{nickel} = 1.36\ V - (-0.25\ V) = 1.61\ V$$

Writing a Spontaneous Reaction Without Appendix D Even when a list like Appendix D is not available, we can write a spontaneous redox reaction from a given pair of half-reactions. For example, here are two half-reactions:

$$Sn^{2+}(aq) + 2e^- \longrightarrow Sn(s) \qquad E^\circ_{tin} = -0.14\text{ V}$$
$$Ag^+(aq) + e^- \longrightarrow Ag(s) \qquad E^\circ_{silver} = 0.80\text{ V}$$

Two steps are required:

1. Reverse one of the half-reactions into an oxidation step so that the difference of the electrode potentials (cathode *minus* anode) gives a *positive* E°_{cell}. (Remember that, as above, when we reverse the half-reaction, we do *not* reverse the sign of $E^\circ_{half\text{-}cell}$.)
2. Multiply by coefficients to make e^- lost equal e^- gained, add the rearranged half-reactions to get a balanced overall equation, and cancel species common to both sides.

We want the reactants to be the stronger oxidizing and reducing agents:

- The larger (more positive) E° value for the silver half-reaction means that Ag^+ is a stronger oxidizing agent (gains electrons more readily) than Sn^{2+}.
- The smaller (more negative) E° value for the tin half-reaction means that Sn is a stronger reducing agent (loses electrons more readily) than Ag.

For step 1, we reverse the tin half-reaction (but *not* the sign of E°_{tin}):

$$Sn(s) \longrightarrow Sn^{2+}(aq) + 2e^- \qquad E^\circ_{tin} = -0.14\text{ V}$$

because when we subtract $E^\circ_{half\text{-}cell}$ of the tin half-reaction (anode, oxidation) from $E^\circ_{half\text{-}cell}$ of the silver half-reaction (cathode, reduction), we get a positive E°_{cell}:

$$0.80\text{ V} - (-0.14\text{ V}) = 0.94\text{ V}$$

For step 2, the number of electrons lost in the oxidation must equal the number gained in the reduction, so we double the silver (reduction) half-reaction. Adding the half-reactions and applying Equation 21.3 gives

$$2Ag^+(aq) + 2e^- \longrightarrow 2Ag(s) \qquad E^\circ_{silver} = 0.80\text{ V} \qquad \text{[reduction]}$$
$$Sn(s) \longrightarrow Sn^{2+}(aq) + 2e^- \qquad E^\circ_{tin} = -0.14\text{ V} \qquad \text{[oxidation]}$$
$$Sn(s) + 2Ag^+(aq) \longrightarrow Sn^{2+}(aq) + 2Ag(s) \qquad E^\circ_{cell} = E^\circ_{silver} - E^\circ_{tin} = 0.94\text{ V}$$

$E^\circ_{half\text{-}cell}$ as an Intensive Property A *very important point* to note is that, when we double the coefficients of the silver half-reaction, we do *not* double its $E^\circ_{half\text{-}cell}$. *Changing the coefficients of a half-reaction does* **not** *change its* $E^\circ_{half\text{-}cell}$ because a standard electrode potential is an *intensive* property, one that does *not* depend on the amount of substance. Let's see why. The potential is the *ratio* of energy to charge. When we change the coefficients to change the amounts, the energy *and* the charge change proportionately, so their ratio stays the same. (Similarly, density does not change with the amount of substance because the mass *and* the volume change proportionately.)

Sample Problem 21.4 Writing Spontaneous Redox Reactions

Problem Combine the following three half-reactions into three balanced equations for spontaneous reactions (A, B, and C), and calculate E°_{cell} for each.

(1) $NO_3^-(aq) + 4H^+(aq) + 3e^- \longrightarrow NO(g) + 2H_2O(l) \qquad E^\circ = 0.96\text{ V}$

(2) $N_2(g) + 5H^+(aq) + 4e^- \longrightarrow N_2H_5^+(aq) \qquad E^\circ = -0.23\text{ V}$

(3) $MnO_2(s) + 4H^+(aq) + 2e^- \longrightarrow Mn^{2+}(aq) + 2H_2O(l) \qquad E^\circ = 1.23\text{ V}$

Plan To write the redox equations, we combine the possible pairs of half-reactions: (1) and (2) give reaction A, (1) and (3) give B, and (2) and (3) give C. They are all written as reductions, so the oxidizing agents appear as reactants and the reducing agents appear as products. In each pair, we reverse the half-reaction that has the smaller (less positive or more negative) E° value to an oxidation to obtain a positive E°_{cell}. We make e^- lost equal e^- gained, without changing the $E°$ value, add the half-reactions together, and then apply Equation 21.3 to find $E°_{cell}$.

Solution Combining half-reactions (1) and (2) gives equation (A). The $E°$ value for half-reaction (1) is larger (more positive) than that for (2), so we reverse (2) to obtain a positive $E°_{cell}$:

$$\text{(1) } NO_3^-(aq) + 4H^+(aq) + 3e^- \longrightarrow NO(g) + 2H_2O(l) \qquad E° = 0.96\ V$$
$$\text{(rev 2) } N_2H_5^+(aq) \longrightarrow N_2(g) + 5H^+(aq) + 4e^- \qquad E° = -0.23\ V$$

To make e^- lost equal e^- gained, we multiply (1) by four and the reversed (2) by three; then add half-reactions and cancel appropriate numbers of common species (H^+ and e^-):

$$4NO_3^-(aq) + 16H^+(aq) + 12e^- \longrightarrow 4NO(g) + 8H_2O(l) \qquad E° = 0.96\ V$$
$$3N_2H_5^+(aq) \longrightarrow 3N_2(g) + 15H^+(aq) + 12e^- \qquad E° = -0.23\ V$$

$$\text{(A) } 3N_2H_5^+(aq) + 4NO_3^-(aq) + H^+(aq) \longrightarrow 3N_2(g) + 4NO(g) + 8H_2O(l)$$
$$E°_{cell} = 0.96\ V - (-0.23\ V) = 1.19\ V$$

Combining half-reactions (1) and (3) gives equation (B). Half-reaction (1) has a smaller $E°$, so it is reversed:

$$\text{(rev 1) } NO(g) + 2H_2O(l) \longrightarrow NO_3^-(aq) + 4H^+(aq) + 3e^- \qquad E° = 0.96\ V$$
$$\text{(3) } MnO_2(s) + 4H^+(aq) + 2e^- \longrightarrow Mn^{2+}(aq) + 2H_2O(l) \qquad E° = 1.23\ V$$

We multiply reversed (1) by two and (3) by three, then add and cancel:

$$2NO(g) + 4H_2O(l) \longrightarrow 2NO_3^-(aq) + 8H^+(aq) + 6e^- \qquad E° = 0.96\ V$$
$$3MnO_2(s) + 12H^+(aq) + 6e^- \longrightarrow 3Mn^{2+}(aq) + 6H_2O(l) \qquad E° = 1.23\ V$$

$$\text{(B) } 3MnO_2(s) + 4H^+(aq) + 2NO(g) \longrightarrow 3Mn^{2+}(aq) + 2H_2O(l) + 2NO_3^-(aq)$$
$$E°_{cell} = 1.23\ V - 0.96\ V = 0.27\ V$$

Combining half-reactions (2) and (3) gives equation (C). Half-reaction (2) has a smaller $E°$, so it is reversed:

$$\text{(rev 2) } N_2H_5^+(aq) \longrightarrow N_2(g) + 5H^+(aq) + 4e^- \qquad E° = -0.23\ V$$
$$\text{(3) } MnO_2(s) + 4H^+(aq) + 2e^- \longrightarrow Mn^{2+}(aq) + 2H_2O(l) \qquad E° = 1.23\ V$$

We multiply reaction (3) by two, add the half-reactions, and cancel:

$$N_2H_5^+(aq) \longrightarrow N_2(g) + 5H^+(aq) + 4e^- \qquad E° = -0.23\ V$$
$$2MnO_2(s) + 8H^+(aq) + 4e^- \longrightarrow 2Mn^{2+}(aq) + 4H_2O(l) \qquad E° = 1.23\ V$$

$$\text{(C) } N_2H_5^+(aq) + 2MnO_2(s) + 3H^+(aq) \longrightarrow N_2(g) + 2Mn^{2+}(aq) + 4H_2O(l)$$
$$E°_{cell} = 1.23\ V - (-0.23\ V) = 1.46\ V$$

Check As always, check that atoms and charges balance on both sides of the equation. A good way to check the ranking and equations is to list the given half-reactions in order of decreasing $E°$ value:

$$MnO_2(s) + 4H^+(aq) + 2e^- \longrightarrow Mn^{2+}(aq) + 2H_2O(l) \qquad E° = 1.23\ V$$
$$NO_3^-(aq) + 4H^+(aq) + 3e^- \longrightarrow NO(g) + 2H_2O(l) \qquad E° = 0.96\ V$$
$$N_2(g) + 5H^+(aq) + 4e^- \longrightarrow N_2H_5^+(aq) \qquad E° = -0.23\ V$$

Then the oxidizing agents (reactants) decrease in strength going down the list, so the reducing agents (products) decrease in strength going up. Moreover, each of the three spontaneous reactions (A, B, and C) should combine a reactant with a product that is lower down on this list.

FOLLOW-UP PROBLEM 21.4 Is the following reaction spontaneous as written?

$$3Fe^{2+}(aq) \longrightarrow Fe(s) + 2Fe^{3+}(aq)$$

If not, write the equation for the spontaneous reaction, calculate $E°_{cell}$, and rank the three species of iron in order of decreasing strength as a reducing agent.

Explaining the Activity Series of the Metals

In Chapter 4, we discussed the activity series of the metals (see Figure 4.17), which ranks metals by their ability to "displace" one another from aqueous solution. Now you'll see *why* this displacement occurs, as well as why many, but not all, metals react with acid to form H_2, and why a few metals form H_2 even in water.

1. *Metals that can displace H_2 from acid.* The standard hydrogen half-reaction represents the reduction of H^+ ions from an acid to H_2:

$$2H^+(aq) + 2e^- \longrightarrow H_2(g) \qquad E^\circ = 0.00\ \text{V}$$

To see which metals reduce H^+ (referred to as "displacing H_2") from acids, choose a metal, write its half-reaction as an oxidation, combine this half-reaction with the hydrogen half-reaction, and see if E°_{cell} is positive. We find that the metals Li through Pb, those that lie *below* the standard hydrogen (reference) half-reaction in Appendix D, give a positive E°_{cell} when reducing H^+. Iron, for example, reduces H^+ from an acid to H_2:

$$\begin{array}{lll} Fe(s) \longrightarrow Fe^{2+}(aq) + 2e^- & E^\circ = -0.44\ \text{V} & \text{[anode; oxidation]} \\ 2H^+(aq) + 2e^- \longrightarrow H_2(g) & E^\circ = 0.00\ \text{V} & \text{[cathode; reduction]} \\ \hline Fe(s) + 2H^+(aq) \longrightarrow H_2(g) + Fe^{2+}(aq) & E^\circ_{\text{cell}} = 0.00\ \text{V} - (-0.44\ \text{V}) = 0.44\ \text{V} & \end{array}$$

The lower the metal in the list, the stronger it is as a reducing agent; therefore, *if E°_{cell} for the reduction of H^+ is more positive with metal A than with metal B, metal A is a stronger reducing agent than metal B and a more **active** metal.*

2. *Metals that cannot displace H_2 from acid.* For metals that are *above* the standard hydrogen (reference) half-reaction, the E°_{cell} is negative when we reverse the metal half-reaction, so the reaction does not occur. For example, the coinage metals—copper, silver, and gold, which are in Group 1B(11)—are not strong enough reducing agents to reduce H^+ from acids:

$$\begin{array}{lll} Ag(s) \longrightarrow Ag^+(aq) + e^- & E^\circ = 0.80\ \text{V} & \text{[anode; oxidation]} \\ 2H^+(aq) + 2e^- \longrightarrow H_2(g) & E^\circ = 0.00\ \text{V} & \text{[cathode; reduction]} \\ \hline 2Ag(s) + 2H^+(aq) \longrightarrow 2Ag^+(aq) + H_2(g) & E^\circ_{\text{cell}} = 0.00\ \text{V} - 0.80\ \text{V} = -0.80\ \text{V} & \end{array}$$

The *higher* the metal in the list, the *more negative* is its E°_{cell} for the reduction of H^+ to H_2, the *lower* is its reducing strength, and the *less active* it is. Thus, gold is less active than silver, which is less active than copper.

3. *Metals that can displace H_2 from water.* Metals active enough to reduce H_2O *lie below that half-reaction:*

$$2H_2O(l) + 2e^- \longrightarrow H_2(g) + 2OH^-(aq) \qquad E = -0.42\ \text{V}$$

(The value shown here is the *nonstandard* electrode potential because, in pure water, $[OH^-]$ is 1.0×10^{-7} *M*, not the standard-state value of 1 *M*.) For example, consider the reaction of sodium in water (with the Na^+/Na half-reaction reversed and doubled):

$$\begin{array}{lll} 2Na(s) \longrightarrow 2Na^+(aq) + 2e^- & E^\circ = -2.71\ \text{V} & \text{[anode; oxidation]} \\ 2H_2O(l) + 2e^- \longrightarrow H_2(g) + 2OH^-(aq) & E = -0.42\ \text{V} & \text{[cathode; reduction]} \\ \hline 2Na(s) + 2H_2O(l) \longrightarrow 2Na^+(aq) + H_2(g) + 2OH^-(aq) & & \\ & E_{\text{cell}} = -0.42\ \text{V} - (-2.71\ \text{V}) = 2.29\ \text{V} & \end{array}$$

The alkali metals [Group 1A(1)] and the larger alkaline earth metals [Group 2A(2)] can reduce water (displace H_2 from H_2O). Calcium is shown doing this in Figure 21.8.

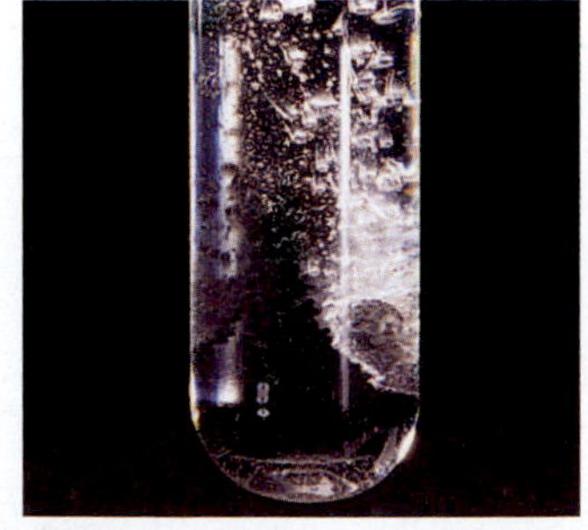

Oxidation half-reaction
$Ca(s) \longrightarrow Ca^{2+}(aq) + 2e^-$

Reduction half-reaction
$2H_2O(l) + 2e^- \longrightarrow H_2(g) + 2OH^-(aq)$

Overall (cell) reaction
$Ca(s) + 2H_2O(l) \longrightarrow Ca(OH)_2(aq) + H_2(g)$

Figure 21.8 The reaction of calcium in water.

4. *Metals that displace other metals from solution.* We can also predict whether one metal can reduce the aqueous ion of another metal. Any metal that is lower in the list in Appendix D can reduce the ion of a metal that is higher up, and thus displace that metal from solution. For example, zinc can displace iron from solution:

$$\begin{array}{lll} Zn(s) \longrightarrow Zn^{2+}(aq) + 2e^- & E^\circ = -0.76\ \text{V} & \text{[anode; oxidation]} \\ Fe^{2+}(aq) + 2e^- \longrightarrow Fe(s) & E^\circ = -0.44\ \text{V} & \text{[cathode; reduction]} \\ \hline Zn(s) + Fe^{2+}(aq) \longrightarrow Zn^{2+}(aq) + Fe(s) & E^\circ_{\text{cell}} = -0.44\ \text{V} - (-0.76\ \text{V}) = 0.32\ \text{V} & \end{array}$$

This particular reaction has tremendous economic importance in protecting iron from rusting, as we'll discuss in Section 21.6.

A common incident involving the reducing power of metals occurs when you bite down with a filled tooth on a scrap of aluminum foil left on a piece of food. The foil acts as an active anode ($E^\circ_{\text{aluminum}} = -1.66$ V), saliva as the electrolyte, and the filling (usually a silver/tin/mercury alloy) as an inactive cathode at which O_2 is reduced to water. The circuit between the foil and the filling creates a current that is sensed as pain by the nerve of the tooth (Figure 21.9).

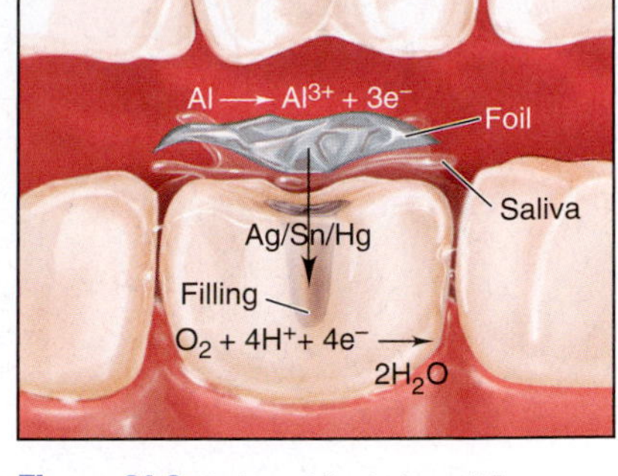

Figure 21.9 A dental "voltaic cell."

Summary of Section 21.3

- The output of a cell is the cell potential (E_{cell}), measured in volts (1 V = 1 J/C).
- With all substances in their standard states, the output is the standard cell potential (E°_{cell}).
- $E^\circ_{\text{cell}} > 0$ for a spontaneous reaction at standard-state conditions.
- By convention, a standard electrode potential ($E^\circ_{\text{half-cell}}$) refers to the *reduction* half-reaction.
- E°_{cell} equals $E^\circ_{\text{half-cell}}$ of the cathode *minus* $E^\circ_{\text{half-cell}}$ of the anode.
- Using a standard hydrogen (reference) electrode ($E^\circ_{\text{reference}} = 0$ V), $E^\circ_{\text{half-cell}}$ values can be measured and used to rank oxidizing (or reducing) agents.
- Spontaneous redox reactions combine stronger oxidizing and reducing agents to form weaker reducing and oxidizing agents, respectively.
- A metal can reduce another species (H^+, H_2O, or an ion of another metal) if E°_{cell} for the overall reaction is positive.

21.4 • FREE ENERGY AND ELECTRICAL WORK

Following up on our discussion in Chapter 20, in this section, we examine the relationship of useful work, free energy, and the equilibrium constant in the context of electrochemical cells and see the effect of concentration on cell potential.

Standard Cell Potential and the Equilibrium Constant

The signs of ΔG and E_{cell} are *opposite* for a spontaneous reaction: a *negative* free energy change ($\Delta G < 0$; Section 20.3) and a *positive* cell potential ($E_{\text{cell}} > 0$). These two indicators of spontaneity are proportional to each other:

$$\Delta G \propto -E_{\text{cell}}$$

Let's determine this proportionality constant. The electrical work done (w, in joules) is the product of the potential (E_{cell}, in volts) and the charge that flows (in coulombs). Since E_{cell} is measured with no current flowing and, thus, no energy lost as heat, it is the maximum voltage possible, and, thus, the maximum work possible (w_{max}).* Work done *by* the cell *on* the surroundings has a negative sign:

$$w_{\text{max}} = -E_{\text{cell}} \times \text{charge}$$

The maximum work done *on* the surroundings is equal to ΔG (Equation 20.9):

$$w_{\text{max}} = \Delta G = -E_{\text{cell}} \times \text{charge}$$

The charge that flows through the cell equals the amount (mol) of electrons transferred (n) times the charge of 1 mol of electrons (which has the symbol F):

$$\text{Charge} = \text{amount (mol) of e}^- \times \frac{\text{charge}}{\text{mol e}^-} \qquad \text{or} \qquad \text{charge} = nF$$

The charge of 1 mol of electrons is the **Faraday constant (*F*),** named for Michael Faraday, the 19th-century British scientist who pioneered the study of electrochemistry:

$$F = \frac{96{,}485 \text{ C}}{\text{mol e}^-}$$

*Recall from Chapter 20 that only a reversible process can do maximum work. For no current to flow and the process to be reversible, E_{cell} must be opposed by an equal potential in the measuring circuit: if the opposing potential is infinitesimally smaller, the cell reaction goes forward; if it is infinitesimally larger, the reaction goes backward.

Because 1 V = 1 J/C, we have 1 C = 1 J/V, and

$$F = 9.65\times10^4 \frac{\text{J}}{\text{V·mol e}^-} \quad \text{(3 sf)} \tag{21.4}$$

Substituting for charge, the proportionality constant is nF:

$$\Delta G = -nFE_{\text{cell}} \tag{21.5}$$

And, when all components are in their standard states, we have

$$\Delta G^\circ = -nFE^\circ_{\text{cell}} \tag{21.6}$$

Using this relationship, we can relate the standard cell potential to the equilibrium constant of the redox reaction. Recall from Equation 20.12 that

$$\Delta G^\circ = -RT \ln K$$

Substituting for ΔG° from Equation 21.6 gives

$$-nFE^\circ_{\text{cell}} = -RT \ln K$$

Solving for E°_{cell} gives

$$E^\circ_{\text{cell}} = \frac{RT}{nF} \ln K \tag{21.7}$$

Figure 21.10 summarizes the interconnections among the standard free energy change, the equilibrium constant, and the standard cell potential. In Chapter 20, we determined K from ΔG°, which we found either from ΔH° and ΔS° values or from ΔG°_f values. Now, for redox reactions, we have a direct *experimental* method for determining K *and* ΔG°: measure E°_{cell}.

In calculations, we adjust Equation 21.7 as follows:

- Substitute 8.314 J/(mol rxn·K) for the constant R.
- Substitute 9.65×10^4 J/(V·mol e$^-$) for the constant F.
- Substitute 298.15 K for T, keeping in mind that the cell can operate at other temperatures.
- Multiply by 2.303 to convert natural to common (base-10) logarithms. This conversion shows that a *10-fold change in K makes* E°_{cell} *change by 1*.

Thus, when n moles of e$^-$ are transferred per mole of reaction based on the balanced equation, we have

$$E^\circ_{\text{cell}} = \frac{RT}{nF} \ln K = 2.303 \times \frac{8.314 \dfrac{\text{J}}{\text{mol rxn·K}} \times 298.15\ \text{K}}{\dfrac{n\ \text{mol e}^-}{\text{mol rxn}}\left(9.65\times10^4 \dfrac{\text{J}}{\text{V·mol e}^-}\right)} \log K$$

And this becomes

$$E^\circ_{\text{cell}} = \frac{0.0592\ \text{V}}{n} \log K \quad \text{or} \quad \log K = \frac{nE^\circ_{\text{cell}}}{0.0592\ \text{V}} \quad \text{(at 298.15 K)} \tag{21.8}$$

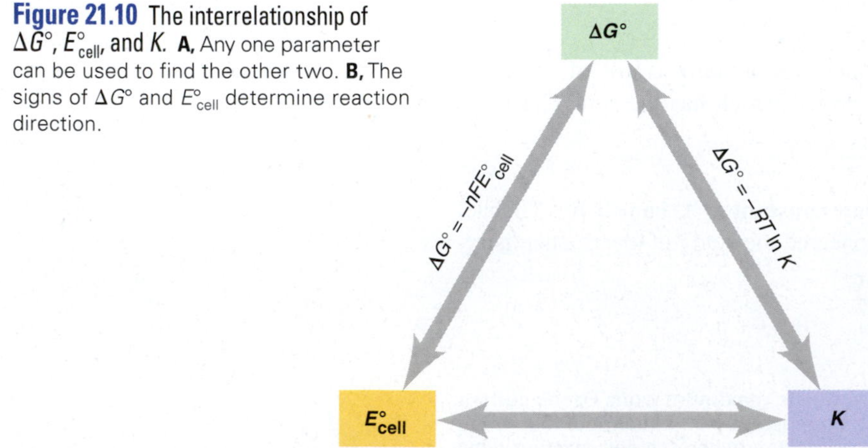

Reaction Parameters at the Standard State			
ΔG°	K	E°_{cell}	Reaction at standard-state conditions
<0	>1	>0	Spontaneous
0	1	0	At equilibrium
>0	<1	<0	Nonspontaneous

B

Figure 21.10 The interrelationship of ΔG°, E°_{cell}, and K. **A,** Any one parameter can be used to find the other two. **B,** The signs of ΔG° and E°_{cell} determine reaction direction.

Sample Problem 21.5 Calculating K and $\Delta G°$ from $E°_{cell}$

Problem Lead can displace silver from solution:

$$Pb(s) + 2Ag^+(aq) \longrightarrow Pb^{2+}(aq) + 2Ag(s)$$

Silver occurs in trace amounts in some ores of lead. As a consequence, silver is a valuable byproduct in the industrial extraction of lead from its ores. Calculate K and $\Delta G°$ at 298.15 K for this reaction.

Plan We divide the spontaneous redox equation into the half-reactions and use values from Appendix D to calculate $E°_{cell}$. Then, we substitute this result into Equation 21.8 to find K and into Equation 21.6 to find $\Delta G°$.

Solution Writing the half-reactions with their $E°$ values:

$$(1)\quad Ag^+(aq) + e^- \longrightarrow Ag(s) \qquad E° = 0.80\ V$$
$$(2)\quad Pb^{2+}(aq) + 2e^- \longrightarrow Pb(s) \qquad E° = -0.13\ V$$

Calculating $E°_{cell}$: We double (1), reverse (2), add the half-reactions, and subtract $E°_{lead}$ from $E°_{silver}$:

$$2Ag^+(aq) + 2e^- \longrightarrow 2Ag(s) \qquad E° = 0.80\ V$$
$$Pb(s) \longrightarrow Pb^{2+}(aq) + 2e^- \qquad E° = -0.13\ V$$
$$Pb(s) + 2Ag^+(aq) \longrightarrow Pb^{2+}(aq) + 2Ag(s) \qquad E°_{cell} = 0.80\ V - (-0.13\ V) = 0.93\ V$$

Calculating K with Equations 21.7 and 21.8: The adjusted half-reactions show that 2 mol of e^- are transferred per mole of reaction as written, so $n = 2$. Then, performing the substitutions for R and F that we just discussed, changing to common logarithms, and running the cell at 25°C (298.15 K), we have

$$E°_{cell} = \frac{RT}{nF} \ln K = \frac{0.0592\ V}{2} \log K = 0.93\ V$$

So,
$$\log K = \frac{0.93\ V \times 2}{0.0592\ V} = 31.42 \qquad \text{and} \qquad K = 2.6\times10^{31}$$

Calculating $\Delta G°$ (Equation 21.6):

$$\Delta G° = -nFE°_{cell} = -\frac{2\ mol\ e^-}{mol\ rxn} \times \frac{96.5\ kJ}{V\cdot mol\ e^-} \times 0.93\ V = -1.8\times10^2\ kJ/mol\ rxn$$

Check The three variables are consistent with the reaction being spontaneous at standard-state conditions: $E°_{cell} > 0$, $\Delta G° < 0$, and $K > 1$. Be sure to round and check the order of magnitude: to find $\Delta G°$, for instance, $\Delta G° \approx -2 \times 100 \times 1 = -200$, so the overall math seems right. Another check would be to obtain $\Delta G°$ directly from its relation with K:

$$\Delta G° = -RT \ln K = -8.314\ J/mol\ rxn\cdot K \times 298.15\ K \times \ln(2.6\times10^{31})$$
$$= -1.8\times10^5\ J/mol\ rxn = -1.8\times10^2\ kJ/mol\ rxn$$

FOLLOW-UP PROBLEM 21.5 When cadmium metal reduces Cu^{2+} in solution, Cd^{2+} forms in addition to copper metal. Given that $\Delta G° = -143$ kJ/mol rxn, calculate K at 25°C. What is $E°_{cell}$ of a voltaic cell that uses this reaction?

The Effect of Concentration on Cell Potential

So far, we've considered cells at standard-state conditions, but most cells don't start with those concentrations, and even if they did, concentrations change as the cell operates. Moreover, in all batteries, reactant concentrations are far from the standard state.

To determine E_{cell}, the cell potential under *nonstandard* conditions, we'll derive an expression for the relation between E_{cell} and concentration based on the relation between ΔG and concentration. Recall from Chapter 20 (Equation 20.13) that

$$\Delta G = \Delta G° + RT \ln Q$$

ΔG is related to E_{cell} and $\Delta G°$ to $E°_{cell}$ (Equations 21.5 and 21.6), so we substitute for them and get

$$-nFE_{cell} = -nFE°_{cell} + RT \ln Q$$

Dividing both sides by $-nF$, we obtain the equation developed by the great German chemist Walther Hermann Nernst in 1889, when he was only 25 years old. (In his career, which culminated in the 1920 Nobel Prize, he also formulated the third law of thermodynamics and established the concept of the solubility product.) The **Nernst equation** says that E_{cell} depends on E°_{cell} *and* a term for the potential at any ratio of concentrations:

$$E_{\text{cell}} = E^\circ_{\text{cell}} - \frac{RT}{nF}\ln Q \quad \textbf{(21.9)}$$

How do changes in Q affect cell potential? From Equation 21.9, we see that

- When $Q < 1$ and thus [reactant] > [product], $\ln Q < 0$, so $E_{\text{cell}} > E^\circ_{\text{cell}}$.
- When $Q = 1$ and thus [reactant] = [product], $\ln Q = 0$, so $E_{\text{cell}} = E^\circ_{\text{cell}}$.
- When $Q > 1$ and thus [reactant] < [product], $\ln Q > 0$, so $E_{\text{cell}} < E^\circ_{\text{cell}}$.

As before, to obtain a form for calculations, we substitute known values of R and F, operate the cell at 298.15 K, and convert to common (base-10) logarithms:

$$E_{\text{cell}} = E^\circ_{\text{cell}} - \frac{RT}{nF}\ln Q$$

$$= E^\circ_{\text{cell}} - 2.303 \times \frac{8.314\,\frac{\cancel{\text{J}}}{\cancel{\text{mol rxn}}\cdot\cancel{\text{K}}} \times 298.15\,\cancel{\text{K}}}{n\,\frac{\cancel{\text{mol e}^-}}{\cancel{\text{mol rxn}}}\left(9.65\times10^4\,\frac{\cancel{\text{J}}}{\text{V}\cdot\cancel{\text{mol e}^-}}\right)}\log Q$$

We obtain:

$$E_{\text{cell}} = E^\circ_{\text{cell}} - \frac{0.0592\text{ V}}{n}\log Q \quad (\text{at } 298.15\text{ K}) \quad \textbf{(21.10)}$$

Remember that the expression for Q *contains only those species with concentrations (and/or pressures) that can vary;* thus, solids do not appear, even when they are the electrodes. For example, in the reaction between cadmium and silver ion, the Cd and Ag electrodes do not appear in the expression for Q:

$$\text{Cd}(s) + 2\text{Ag}^+(aq) \longrightarrow \text{Cd}^{2+}(aq) + 2\text{Ag}(s) \qquad Q = \frac{[\text{Cd}^{2+}]}{[\text{Ag}^+]^2}$$

Sample Problem 21.6 Using the Nernst Equation to Calculate E_{cell}

Problem In a test of a new reference electrode, a chemist constructs a voltaic cell consisting of a Zn/Zn^{2+} half-cell and an H_2/H^+ half-cell under the following conditions:

$$[\text{Zn}^{2+}] = 0.010\,M \qquad [\text{H}^+] = 2.5\,M \qquad P_{\text{H}_2} = 0.30\text{ atm}$$

Calculate E_{cell} at 298.15 K.

Plan To apply the Nernst equation and determine E_{cell}, we must know E°_{cell} and Q. We write the spontaneous reaction and calculate E°_{cell} from standard electrode potentials (Appendix D). To have consistent units, we convert the given pressure to molarity with the ideal gas law and find Q. Then we substitute into Equation 21.10.

Solution Determining the cell reaction and E°_{cell}:

$$2\text{H}^+(aq) + 2\text{e}^- \longrightarrow \text{H}_2(g) \qquad E^\circ = 0.00\text{ V}$$
$$\text{Zn}(s) \longrightarrow \text{Zn}^{2+}(aq) + 2\text{e}^- \qquad E^\circ = -0.76\text{ V}$$
$$2\text{H}^+(aq) + \text{Zn}(s) \longrightarrow \text{H}_2(g) + \text{Zn}^{2+}(aq) \qquad E^\circ_{\text{cell}} = 0.00\text{ V} - (-0.76\text{ V}) = 0.76\text{ V}$$

Converting pressure to molarity: $PV = nRT$, or

$$\frac{n}{V} = \frac{P}{RT} = \frac{0.30\text{ atm}}{0.0821\,\frac{\text{atm}\cdot\text{L}}{\text{mol}\cdot\text{K}} \times 298.15\text{ K}} = 1.2\times10^{-2}\text{ mol/L} = 0.012\,M$$

Calculating Q:

$$Q = \frac{[\text{H}_2] \times [\text{Zn}^{2+}]}{[\text{H}^+]^2} = \frac{0.012 \times 0.010}{2.5^2} = 1.9\times10^{-5}$$

Solving for E_{cell} at 25°C (298.15 K), with $n = 2$:

$$E_{\text{cell}} = E^\circ_{\text{cell}} - \frac{0.0592\text{ V}}{n}\log Q$$

$$= 0.76\text{ V} - \left[\frac{0.0592\text{ V}}{2}\log(1.9\times10^{-5})\right] = 0.76\text{ V} - (-0.14\text{ V}) = 0.90\text{ V}$$

Check After you check the arithmetic, reason through the answer: $E_{cell} > E^\circ_{cell}$ ($0.90 > 0.76$) because the log Q term was negative, which is consistent with $Q < 1$.

FOLLOW-UP PROBLEM 21.6 Consider a voltaic cell based on the following reaction: $Fe(s) + Cu^{2+}(aq) \longrightarrow Fe^{2+}(aq) + Cu(s)$. If $[Cu^{2+}] = 0.30\ M$, what must $[Fe^{2+}]$ be to increase E_{cell} by 0.25 V above E°_{cell} at 25°C?

Changes in Potential During Cell Operation

As with any voltaic cell, the potential of the zinc-copper cell changes during cell operation as the concentrations of the components change. With two of the four components solids, the only variables are $[Cu^{2+}]$ and $[Zn^{2+}]$:

$$Zn(s) + Cu^{2+}(aq) \longrightarrow Zn^{2+}(aq) + Cu(s) \qquad Q = \frac{[Zn^{2+}]}{[Cu^{2+}]}$$

In this section, we follow the potential as the zinc-copper cell operates.

1. *Starting point of cell operation.* The positive E°_{cell} (1.10 V) means that this reaction proceeds *spontaneously* to the right at standard-state conditions, $[Zn^{2+}] = [Cu^{2+}] = 1\ M$ ($Q = 1$). However, if we start the cell when $[Zn^{2+}] < [Cu^{2+}]$ ($Q < 1$), for example, $[Zn^{2+}] = 1.0\times10^{-4}\ M$ and $[Cu^{2+}] = 2.0\ M$, the cell potential starts out *higher* than the standard cell potential:

$$E_{cell} = E^\circ_{cell} - \frac{0.0592\text{ V}}{2}\log\frac{[Zn^{2+}]}{[Cu^{2+}]} = 1.10\text{ V} - \left(\frac{0.0592\text{ V}}{2}\log\frac{1.0\times10^{-4}}{2.0}\right)$$

$$= 1.10\text{ V} - \left[\frac{0.0592\text{ V}}{2}(-4.30)\right] = 1.10\text{ V} + 0.127\text{ V} = 1.23\text{ V}$$

2. *Key stages during cell operation.* Using Equation 21.10, we identify four key stages of operation. Figure 21.11A shows the first three. The main point is that *as the cell operates, its potential decreases.* As $[Zn^{2+}]$ increases and $[Cu^{2+}]$ decreases, Q becomes larger, the $[(0.0592\text{ V}/n)\log Q]$ term becomes less negative (more positive), and E_{cell} decreases:

Stage 1. $E_{cell} > E^\circ_{cell}$ when $Q < 1$: When the cell begins operation, $[Cu^{2+}] > [Zn^{2+}]$, so $(0.0592\text{ V}/n)\log Q < 0$ and $E_{cell} > E^\circ_{cell}$.

Stage 2. $E_{cell} = E^\circ_{cell}$ when $Q = 1$: At the point when $[Cu^{2+}] = [Zn^{2+}]$, $Q = 1$, so $(0.0592\text{ V}/n)\log Q = 0$ and $E_{cell} = E^\circ_{cell}$.

Stage 3. $E_{cell} < E^\circ_{cell}$ when $Q > 1$: As the $[Zn^{2+}]/[Cu^{2+}]$ ratio continues to increase, $(0.0592\text{ V}/n)\log Q > 0$, so $E_{cell} < E^\circ_{cell}$.

Stage 4. $E_{cell} = 0$ when $Q = K$: Eventually, $(0.0592\text{ V}/n)\log Q$ becomes so large that it equals E°_{cell}, which means that E_{cell} is zero. This occurs *when the system reaches* ***equilibrium:*** *no more free energy is released, so the cell can do no more work.* At this point, a battery is "dead."

Figure 21.11B summarizes these four key stages during operation of a voltaic cell.

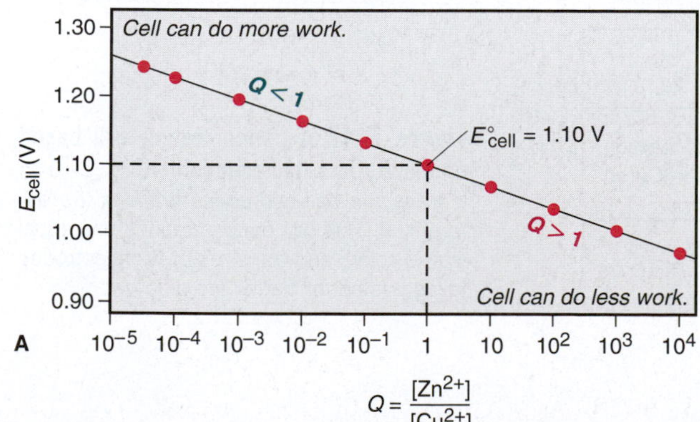

B

Changes in E_{cell} and Concentration

Stage in cell operation	Q	Relative [P] and [R]	$\frac{0.0592\text{ V}}{n}\log Q$
1. $E > E^\circ$	<1	[P] < [R]	<0
2. $E = E^\circ$	=1	[P] = [R]	=0
3. $E < E^\circ$	>1	[P] > [R]	>0
4. $E = 0$	=K	[P] ≫ [R]	=E°

Figure 21.11 The relation between E_{cell} and log Q for the zinc-copper cell. **A,** A plot of E_{cell} vs. Q (on a logarithmic scale) decreases linearly. When $Q < 1$ *(left),* the cell does relatively more work. When $Q = 1$, $E_{cell} = E^\circ_{cell}$. When $Q > 1$ *(right),* the cell does relatively less work. **B,** A summary of the changes in E_{cell} as this or any voltaic cell operates. ($[Zn^{2+}]$ is [P] for [product], and $[Cu^{2+}]$ is [R] for [reactant].)

3. *Q/K and the work the cell can do.* At equilibrium, Equation 21.10 becomes

$$0 = E^\circ_{\text{cell}} - \left(\frac{0.0592\text{ V}}{n}\right)\log K, \qquad \text{which rearranges to} \qquad E^\circ_{\text{cell}} = \frac{0.0592\text{ V}}{n}\log K$$

Note that this result is identical to Equation 21.8, which we obtained from ΔG°. Solving for K of the zinc-copper cell ($E^\circ_{\text{cell}} = 1.10$ V),

$$\log K = \frac{2 \times E^\circ_{\text{cell}}}{0.0592\text{ V}} \qquad \text{so} \qquad K = 10^{(2\times1.10\text{ V})/0.0592\text{ V}} = 10^{37.16} = 1.4\times10^{37}$$

As you can see, this cell does work until $[Zn^{2+}]/[Cu^{2+}]$ is *very* high.

Concentration Cells

If you mix concentrated and dilute solutions of a substance, the final solution has an intermediate concentration. A **concentration cell** employs this simple, spontaneous change to generate electrical energy. The two solutions are in separate half-cells, so they do not mix, but their concentrations become equal as the cell operates.

Finding E_{cell} for a Concentration Cell Suppose a voltaic cell has the Cu/Cu^{2+} half-reaction in both compartments. The cell reaction is the sum of identical half-reactions, written in opposite directions. The *standard* cell potential, E°_{cell}, is zero because the *standard* electrode potentials are both based on 1 *M* Cu^{2+}, so they cancel. *In a concentration cell, however, the concentrations are **different**.* Thus, even though E°_{cell} is still zero, the *nonstandard* cell potential, E_{cell}, depends on the *ratio of concentrations,* so it is *not* zero.

For the final concentration to be equal, a concentration cell must have the dilute solution in the anode compartment and the concentrated solution in the cathode compartment. For example, let's use 0.10 *M* Cu^{2+} in the anode half-cell and 1.0 *M* Cu^{2+}, a 10-fold higher concentration, in the cathode half-cell (Figure 21.12A):

$$Cu(s) \longrightarrow Cu^{2+}(aq;\ 0.10\ M) + 2e^- \qquad \text{[anode; oxidation]}$$

$$Cu^{2+}(aq;\ 1.0\ M) + 2e^- \longrightarrow Cu(s) \qquad \text{[cathode; reduction]}$$

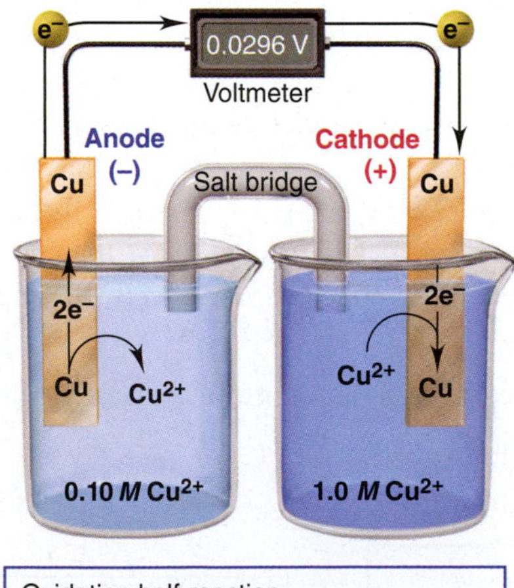

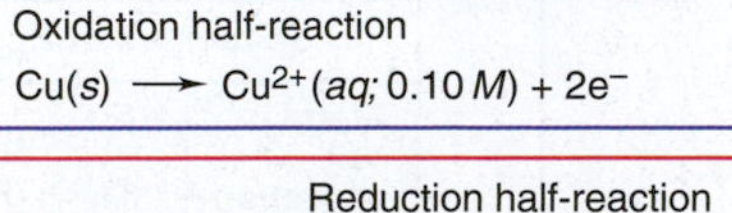

Reduction half-reaction
$Cu^{2+}(aq;\ 1.0\ M) + 2e^- \longrightarrow Cu(s)$

Overall (cell) reaction
$Cu^{2+}(aq;\ 1.0\ M) \longrightarrow Cu^{2+}(aq;\ 0.10\ M)$

A

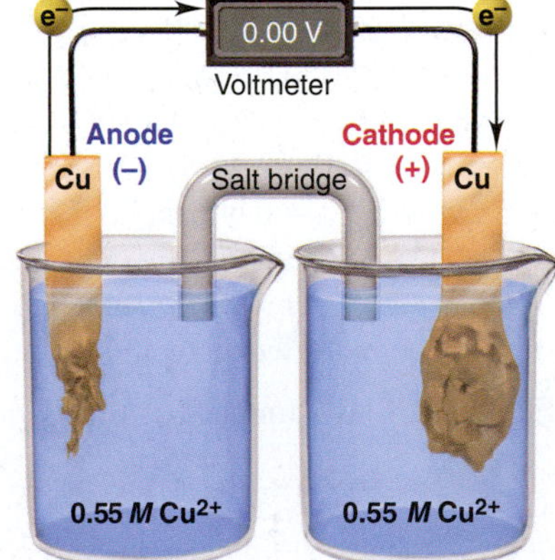

Figure 21.12 A concentration cell based on the Cu/Cu^{2+} half-reaction. **A,** $E_{\text{cell}} > 0$ as long as the half-cell *concentrations* are different. **B,** Half-cell concentrations are equal (same color), and the sizes of the electrodes (exaggerated for clarity) are different.

The overall cell reaction is the sum of the half-reactions:

$$Cu^{2+}(aq;\, 1.0\, M) \longrightarrow Cu^{2+}(aq;\, 0.10\, M) \qquad E_{cell} = ?$$

The cell potential at the initial concentrations of 0.10 *M* (dilute) and 1.0 *M* (concentrated), with $n = 2$, is obtained from the Nernst equation:

$$E_{cell} = E^\circ_{cell} - \frac{0.0592\text{ V}}{2}\log\frac{[Cu^{2+}]_{dil}}{[Cu^{2+}]_{conc}} = 0\text{ V} - \left(\frac{0.0592\text{ V}}{2}\log\frac{0.10\, M}{1.0\, M}\right)$$

$$= 0\text{ V} - \left[\frac{0.0592\text{ V}}{2}(-1.00)\right] = 0.0296\text{ V}$$

Since E°_{cell} is zero, E_{cell} depends entirely on the $[(0.0592\text{ V}/n)\log Q]$ term.

How a Concentration Cell Works Let's see what is happening as this cell operates:

- *In the anode (dilute) half-cell,* Cu atoms in the electrode give up electrons and the resulting Cu^{2+} ions enter the solution and make it *more* concentrated.
- *In the cathode (concentrated) half-cell,* Cu^{2+} ions gain the electrons and the resulting Cu atoms plate out on the electrode, which makes that solution *less* concentrated.

As in any voltaic cell, E_{cell} decreases until equilibrium is attained, which happens when $[Cu^{2+}]$ is the same in both half-cells (Figure 21.12B). The same final concentration would result if we mixed the two solutions, but no electrical work would be done.

Sample Problem 21.7 Calculating the Potential of a Concentration Cell

Problem A concentration cell consists of two Ag/Ag^+ half-cells. In half-cell A, the electrolyte is 0.010 *M* $AgNO_3$; in half-cell B, it is 4.0×10^{-4} *M* $AgNO_3$. What is the cell potential at 298.15 K?

Plan The standard half-cell reactions are identical, so E°_{cell} is zero, and we find E_{cell} from the Nernst equation. Half-cell A has a higher $[Ag^+]$, so Ag^+ ions are reduced and plate out on electrode A. In half-cell B, Ag atoms of the electrode are oxidized and Ag^+ ions enter the solution. As in all voltaic cells, reduction occurs at the cathode, so it is positive.

Solution Writing the spontaneous reaction: The $[Ag^+]$ decreases in half-cell A and increases in half-cell B, so the spontaneous reaction is

$$Ag^+(aq;\, 0.010\, M)\,[\text{half-cell A}] \longrightarrow Ag^+(aq;\, 4.0\times10^{-4}\, M)\,[\text{half-cell B}]$$

Calculating E_{cell}, with $n = 1$:

$$E_{cell} = E^\circ_{cell} - \frac{0.0592\text{ V}}{1}\log\frac{[Ag^+]_{dil}}{[Ag^+]_{conc}} = 0\text{ V} - \left(0.0592\text{ V}\log\frac{4.0\times10^{-4}}{0.010}\right)$$

$$= 0.0828\text{ V}$$

FOLLOW-UP PROBLEM 21.7 A concentration cell is built using two Au/Au^{3+} half-cells. In half-cell A, $[Au^{3+}] = 7.0\times10^{-4}$ *M*, and in half-cell B, $[Au^{3+}] = 2.5\times10^{-2}$ *M*. What is E_{cell}, and which electrode is negative?

Applications of Concentration Cells The principle of a concentration cell has many applications. Here, we discuss one used in the lab and one used in the field:

1. *Measuring pH.* The most important laboratory application of this principle is in measuring $[H^+]$. If we construct a concentration cell in which the cathode compartment is the standard hydrogen electrode and the anode compartment has the same

apparatus dipping into a solution of unknown $[H^+]$, the half-reactions and overall reaction are

$$\begin{array}{rll} H_2(g;\ 1\ \text{atm}) & \longrightarrow 2H^+(aq;\ \text{unknown}) + 2e^- & [\text{anode; oxidation}] \\ 2H^+(aq;\ 1\ M) + 2e^- & \longrightarrow H_2(g;\ 1\ \text{atm}) & [\text{cathode; reduction}] \\ \hline 2H^+(aq;\ 1\ M) & \longrightarrow 2H^+(aq;\ \text{unknown}) & E_{\text{cell}} = ? \end{array}$$

E°_{cell} is zero, but E_{cell} is *not* because the half-cells differ in $[H^+]$. From the Nernst equation, with $n = 2$, we have

$$E_{\text{cell}} = E^\circ_{\text{cell}} - \frac{0.0592\ \text{V}}{2} \log \frac{[H^+]^2_{\text{unknown}}}{[H^+]^2_{\text{standard}}}$$

Substituting 1 M for $[H^+]_{\text{standard}}$ and 0 V for E°_{cell} gives

$$E_{\text{cell}} = 0\ \text{V} - \frac{0.0592\ \text{V}}{2} \log \frac{[H^+]^2_{\text{unknown}}}{1^2} = -\frac{0.0592\ \text{V}}{2} \log\ [H^+]^2_{\text{unknown}}$$

Because $\log x^2 = 2 \log x$ (see Appendix A), we obtain

$$E_{\text{cell}} = -\left[\frac{0.0592\ \text{V}}{2}(2 \log\ [H^+]_{\text{unknown}})\right] = -0.0592\ \text{V} \times \log\ [H^+]_{\text{unknown}}$$

Substituting $-\log\ [H^+] = \text{pH}$, we have

$$E_{\text{cell}} = 0.0592\ \text{V} \times \text{pH}$$

Thus, by measuring E_{cell}, we can find the pH.

For routine lab measurement of pH, a concentration cell made of two hydrogen electrodes is too bulky and difficult to maintain. Instead, we use a pH meter (Figure 21.13A). In a common, but older, design, two separate electrodes dip into the solution being tested:

- The *glass electrode* consists of an Ag/AgCl half-reaction immersed in HCl solution (usually 1.000 M) and enclosed by a thin (~0.05 mm) membrane made of a glass that is very sensitive to H^+ ions.
- The *reference electrode,* usually a *saturated calomel electrode,* consists of a platinum wire immersed in calomel (Hg_2Cl_2) paste, liquid Hg, and saturated KCl solution.

The glass electrode monitors the solution's $[H^+]$ relative to its own fixed internal $[H^+]$, and the instrument converts the potential difference between the glass and reference electrodes into a measure of pH. In modern instruments, a *combination* electrode houses both electrodes in one tube (Figure 21.13B) and can be miniaturized for fieldwork.

2. *Measuring other ions selectively*. The pH electrode is one type of *ion-selective (or ion-specific) electrode;* such electrodes selectively measure certain ion concentrations in a mixture of many, as in natural waters and soils. Biologists implant a tiny ion-selective electrode in a single cell to study ion channels and receptors. Recent advances allow measurement in the femtomolar (10^{-15} M) range. Table 21.3 shows a few of the ions studied.

Table 21.3 Some Ions Measured with Ion-Specific Electrodes

Species Detected	Typical Sample
NH_3/NH_4^+	Industrial wastewater, seawater
CO_2/HCO_3^-	Blood, groundwater
F^-	Drinking water, urine, soil, industrial stack gases
Br^-	Grain, plant tissue
I^-	Milk, pharmaceuticals
NO_3^-	Soil, fertilizer, drinking water
K^+	Blood serum, soil, wine
H^+	Laboratory solutions, soil, natural waters

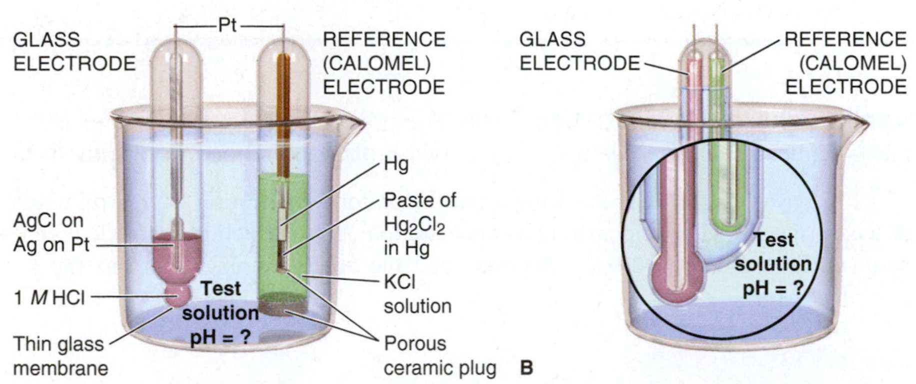

Figure 21.13 Laboratory measurement of pH. **A,** An older style pH meter includes a glass electrode *(left)* and a reference calomel electrode *(right)*. **B,** Modern pH meters use a combination electrode.

Summary of Section 21.4

- A spontaneous process has a negative ΔG and a positive E_{cell}, $\Delta G = -nFE_{cell}$. The ΔG of the cell reaction represents the maximum electrical work the cell can do.
- The standard free energy change, $\Delta G°$, is related to $E°_{cell}$ and to K.
- For nonstandard conditions, the Nernst equation shows that E_{cell} depends on $E°_{cell}$ and a correction term based on Q. E_{cell} is high when Q is small (high [reactant]), and it decreases as the cell operates. At equilibrium, ΔG and E_{cell} are zero, which means that $Q = K$.
- Concentration cells have identical half-reactions, but solutions of differing concentration. They generate electrical energy as the concentrations become equal.
- Ion-specific electrodes, such as the pH electrode, measure the concentration of one species.

21.5 • ELECTROCHEMICAL PROCESSES IN BATTERIES

Because of their compactness and mobility, batteries play a major role in everyday life, and in our increasingly wireless world, that role is growing. In general, a **battery** consists of self-contained voltaic cells arranged in series (plus-to-minus-to-plus, and so on), so that their individual voltages are added. In this section, we examine the three categories of batteries—primary, secondary, and fuel cells (flow batteries).

Primary (Nonrechargeable) Batteries

A *primary battery* cannot be recharged, so it is discarded when the cell reaction has reached equilibrium, that is, when the battery is "dead." We'll discuss the alkaline battery and mercury and silver "button" batteries.

Alkaline Battery The ubiquitous alkaline battery has a zinc anode case that houses a mixture of MnO_2 (the oxidizing agent) and an alkaline paste of KOH and water. The cathode is an inactive graphite rod (Figure 21.14). The half-reactions are

Anode (oxidation): $$Zn(s) + 2OH^-(aq) \longrightarrow ZnO(s) + H_2O(l) + 2e^-$$

Cathode (reduction): $$MnO_2(s) + 2H_2O(l) + 2e^- \longrightarrow Mn(OH)_2(s) + 2OH^-(aq)$$

Overall (cell) reaction:

$$Zn(s) + MnO_2(s) + H_2O(l) \longrightarrow ZnO(s) + Mn(OH)_2(s) \quad E_{cell} = 1.5\ V$$

The alkaline battery powers portable radios, toys, flashlights, and so on, is safe, has a long shelf life, and comes in many sizes.

Figure 21.14 Alkaline battery.

Mercury and Silver (Button) Batteries Both mercury and silver batteries use a zinc container as the anode (reducing agent) in a basic medium. The mercury battery employs HgO as the oxidizing agent, the silver uses Ag_2O, and both use a steel can around the cathode. The solid reactants are compacted with KOH and separated with moist paper. The half-reactions are

Anode (oxidation): $Zn(s) + 2OH^-(aq) \longrightarrow ZnO(s) + H_2O(l) + 2e^-$

Cathode (reduction) (mercury): $HgO(s) + H_2O(l) + 2e^- \longrightarrow Hg(l) + 2OH^-(aq)$

Cathode (reduction) (silver): $Ag_2O(s) + H_2O(l) + 2e^- \longrightarrow 2Ag(s) + 2OH^-(aq)$

Overall (cell) reaction (mercury):

$$Zn(s) + HgO(s) \longrightarrow ZnO(s) + Hg(l) \qquad E_{cell} = 1.3\text{ V}$$

Overall (cell) reaction (silver):

$$Zn(s) + Ag_2O(s) \longrightarrow ZnO(s) + 2Ag(s) \qquad E_{cell} = 1.6\text{ V}$$

Both cells are manufactured as button-sized batteries. The mercury cell is used in calculators and the silver cell (Figure 21.15) in watches, cameras, heart pacemakers, and hearing aids because of its very steady output. Disadvantages are the toxicity of discarded mercury and the high cost of silver.

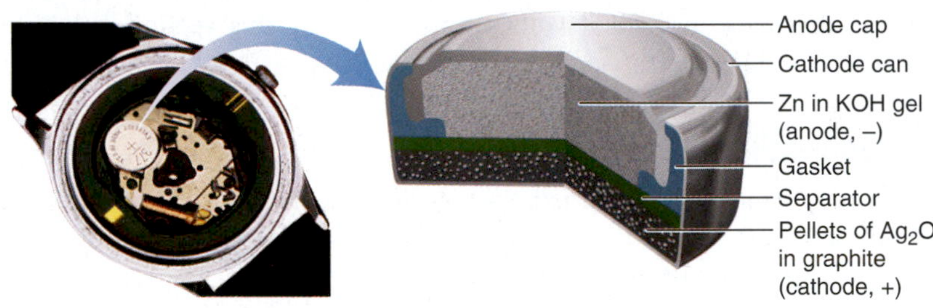

Figure 21.15 Silver button battery.

Secondary (Rechargeable) Batteries

In contrast to a primary battery, a *secondary battery* is *rechargeable;* when it runs down, *electrical energy is supplied to reverse the cell reaction* and form more reactant. In other words, in a secondary battery, the voltaic cells are periodically converted to electrolytic cells to restore *nonequilibrium* concentrations of the cell components. We'll discuss the common car battery, the nickel–metal hydride battery, and the lithium-ion battery.

Lead-Acid Battery A typical lead-acid car battery has six cells connected in series, each of which delivers about 2.1 V for a total of about 12 V. Each cell contains two lead grids packed with high-surface-area (spongy) Pb in the anode and high-surface-area PbO_2 in the cathode. The grids are immersed in a solution of ~4.5 *M* H_2SO_4. Fiberglass sheets between the grids prevent shorting due to physical contact (Figure 21.16).

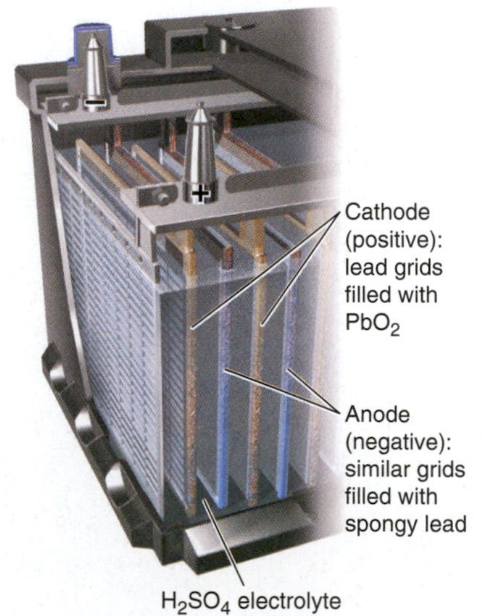

Figure 21.16 Lead-acid battery.

1. *Discharging.* When the cell discharges as a voltaic cell, it generates electrical energy:

Anode (oxidation): $Pb(s) + HSO_4^-(aq) \longrightarrow PbSO_4(s) + H^+ + 2e^-$

Cathode (reduction):

$$PbO_2(s) + 3H^+(aq) + HSO_4^-(aq) + 2e^- \longrightarrow PbSO_4(s) + 2H_2O(l)$$

Both half-reactions form Pb^{2+} ions, one through oxidation of Pb, the other through reduction of PbO_2. The Pb^{2+} forms $PbSO_4(s)$ at both electrodes by reacting with HSO_4^-.

Overall (cell) reaction (discharge):

$$PbO_2(s) + Pb(s) + 2H_2SO_4(aq) \longrightarrow 2PbSO_4(s) + 2H_2O(l) \qquad E_{cell} = 2.1\text{ V}$$

2. *Recharging.* When the cell recharges as an electrolytic cell, it uses electrical energy (supplied by the vehicle's charging system) and the half-cell and overall reactions are reversed.

Overall (cell) reaction (recharge):

$$2PbSO_4(s) + 2H_2O(l) \longrightarrow PbO_2(s) + Pb(s) + 2H_2SO_4(aq)$$

Car and truck owners have relied on the lead-acid battery for over a century to provide the large burst of current needed to start the engine—and to do it in hot and cold weather for years. The main problems with the lead-acid battery are loss of capacity due to corrosion of the positive (Pb) grid, detachment of the active material due to normal mechanical bumping, and formation of large $PbSO_4$ crystals that hinder recharging.

Nickel–Metal Hydride (Ni-MH) Battery Concerns about the toxicity of cadmium in the nickel-cadmium (nicad) battery have led to its replacement by the nickel–metal hydride battery. The anode half-reaction oxidizes the hydrogen absorbed within a metal alloy (such as $LaNi_5$; designated M) in a basic (KOH) electrolyte, while nickel(III) in the form of NiO(OH) is reduced at the cathode (Figure 21.17):

Anode (oxidation): $MH(s) + OH^-(aq) \longrightarrow M(s) + H_2O(l) + e^-$

Cathode (reduction): $NiO(OH)(s) + H_2O(l) + e^- \longrightarrow Ni(OH)_2(s) + OH^-(aq)$

Overall (cell) reaction: $MH(s) + NiO(OH)(s) \longrightarrow M(s) + Ni(OH)_2(s)$

$$E_{cell} = 1.4\ V$$

The cell reaction is reversed during recharging. The Ni-MH battery is common in cordless razors, camera flash units, and power tools. It is lightweight, has high power, and is nontoxic, but it discharges significantly during storage.

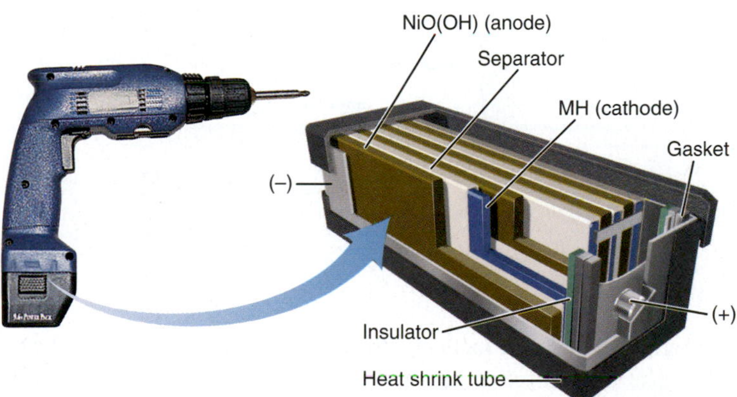

Figure 21.17 Nickel–metal hydride battery.

Lithium-Ion Battery The lithium-ion battery has an anode of Li atoms that lie between sheets of graphite (designated Li_xC_6). The cathode is a lithium metal oxide, such as $LiMn_2O_4$ or $LiCoO_2$, and a typical electrolyte is 1 *M* $LiPF_6$ in an organic solvent, such as dimethyl carbonate mixed with methylethyl carbonate. Electrons flow through the circuit, while solvated Li^+ ions flow from anode to cathode within the cell (Figure 21.18). The cell reactions are

Anode (oxidation): $Li_xC_6 \longrightarrow xLi^+ + xe^- + C_6(s)$

Cathode (reduction): $Li_{1-x}Mn_2O_4(s) + xLi^+ + xe^- \longrightarrow LiMn_2O_4(s)$

Overall (cell) reaction: $Li_xC_6 + Li_{1-x}Mn_2O_4(s) \longrightarrow LiMn_2O_4(s) + C_6(s)$

$$E_{cell} = 3.7\ V$$

The cell reaction is reversed during recharging. The lithium-ion battery powers countless laptop computers, cell phones, and camcorders. Its key drawbacks are cost and flammability of the organic solvent.

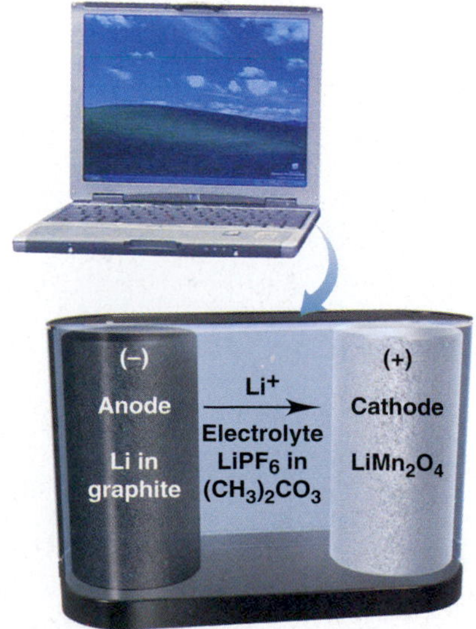

Figure 21.18 Lithium-ion battery.

Fuel Cells

In contrast to primary and secondary batteries, a **fuel cell,** sometimes called a *flow battery,* is not self-contained. The reactants (usually a fuel and oxygen) enter the cell, and the products leave, *generating electricity through controlled combustion* of the fuel. The fuel does not burn because, as in other voltaic cells, the half-reactions are separated, and the electrons move through an external circuit.

A type of fuel cell being developed for use in cars is the *proton exchange membrane (PEM) cell,* which uses H_2 as the fuel and has an operating temperature of around 80°C (Figure 21.19). The cell reactions are

Anode (oxidation):	$2H_2(g)$	$\longrightarrow$	$4H^+(aq) + 4e^-$	
Cathode (reduction):	$O_2(g) + 4H^+(aq) + 4e^-$	$\longrightarrow$	$2H_2O(g)$	
Overall (cell) reaction:	$2H_2(g) + O_2(g)$	$\longrightarrow$	$2H_2O(g)$	$E_{cell} = 1.2$ V

Figure 21.19 Hydrogen fuel cell.

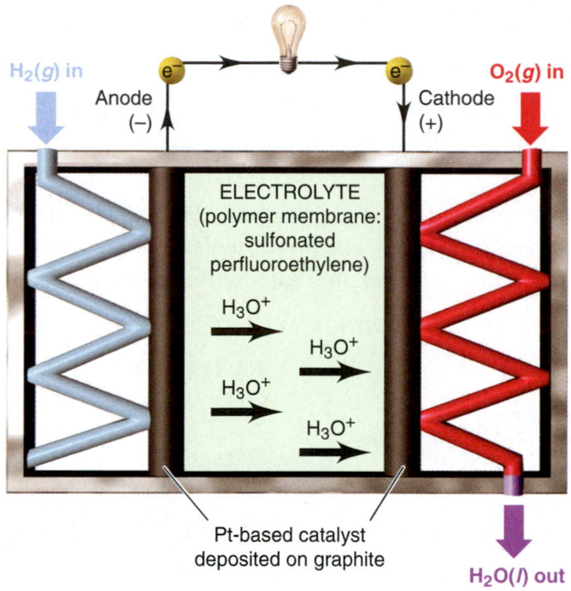

Reaction rates are lower in fuel cells than in other batteries, so an *electrocatalyst* is used to decrease the activation energy (Section 16.7). The PEM cell's electrodes are made of a nanocomposite consisting of a Pt-based catalyst deposited on graphite. These are embedded in a polymer electrolyte membrane having a perfluoroethylene backbone ($-[F_2C-CF_2]_n-$) with attached sulfonic acid groups (RSO_3^-) that play a key role in ferrying protons from anode to cathode.

Hydrogen fuel cells have been used for years to provide electricity *and* water during space flights. Similar ones have begun to supply electric power for residential needs, and every major car manufacturer has a fuel-cell prototype. Fuel cells produce no pollutants and convert about 75% of a fuel's bond energy into power, compared to 40% for a coal-fired power plant and 25% for a gasoline engine.

Summary of Section 21.5

- Batteries are voltaic cells arranged in series and are classified as primary (e.g., alkaline, mercury, and silver), secondary (e.g., lead-acid, nickel–metal hydride, and lithium-ion), or fuel cells.
- Supplying electricity to a rechargeable (secondary) battery reverses the redox reaction, re-forming reactant.
- Fuel cells are not self-contained and generate a current through the controlled oxidation of a fuel such as H_2.

21.6 • CORROSION: AN ENVIRONMENTAL VOLTAIC CELL

If you think all spontaneous electrochemical processes are useful, consider the problem of **corrosion,** which causes tens of billions of dollars of damage to cars, ships, buildings, and bridges each year. This natural process, which oxidizes metals to their oxides and sulfides, shares similarities with the operation of a voltaic cell. We focus on the corrosion of iron, but many other metals, such as copper and silver, also corrode.

The Corrosion of Iron

The most common and economically destructive form of corrosion is the rusting of iron. About 25% of the steel produced in the United States each year is for replacing steel in which the iron has corroded. Rust is *not* a direct product of the reaction between iron and oxygen but arises through a complex electrochemical process. Let's look at the facts of iron corrosion and then use the features of a voltaic cell to explain them:

Fact 1. Iron does not rust in dry air; moisture must be present.
Fact 2. Iron does not rust in air-free water; oxygen must be present.
Fact 3. Iron loss and rust formation occur at *different* places on the *same* object.
Fact 4. Iron rusts more quickly at low pH (high $[H^+]$).
Fact 5. Iron rusts more quickly in ionic solutions.
Fact 6. Iron rusts more quickly in contact with a less active metal (such as Cu) and more slowly in contact with a more active metal (such as Zn).

Two separate redox processes occur during corrosion:

1. *The loss of iron.* Picture the surface of a piece of iron (Figure 21.20A). A strain or dent in contact with water is usually the site of iron loss (fact 1). This site is called an *anodic region* because of the following half-reaction (Figure 21.20B):

$$Fe(s) \longrightarrow Fe^{2+}(aq) + 2e^- \qquad \text{[anodic region; oxidation]}$$

Once the iron atoms lose electrons, the damage to the object has been done, and a pit forms where the iron is lost.

The freed electrons move through the external circuit—the piece of iron itself—until they reach a region of relatively high $[O_2]$ (fact 2), usually the air near the edge of a water droplet that surrounds the newly formed pit. At this *cathodic region,* the electrons released from the iron atoms reduce O_2:

$$O_2(g) + 4H^+(aq) + 4e^- \longrightarrow 2H_2O(l) \qquad \text{[cathodic region; reduction]}$$

This portion of the corrosion process (the sum of these two half-reactions) occurs without any rust forming:

$$2Fe(s) + O_2(g) + 4H^+(aq) \longrightarrow 2Fe^{2+}(aq) + 2H_2O(l) \quad \text{[overall]}$$

2. *The rusting process.* Rust forms in another redox reaction. The Fe^{2+} ions formed at the anodic region disperse through the water and react with O_2, often away from the pit, to form the Fe^{3+} in rust (fact 3). The overall reaction for this step is

$$2Fe^{2+}(aq) + \tfrac{1}{2}O_2(g) + (2 + n)H_2O(l) \longrightarrow Fe_2O_3 \cdot nH_2O(s) + 4H^+(aq)$$

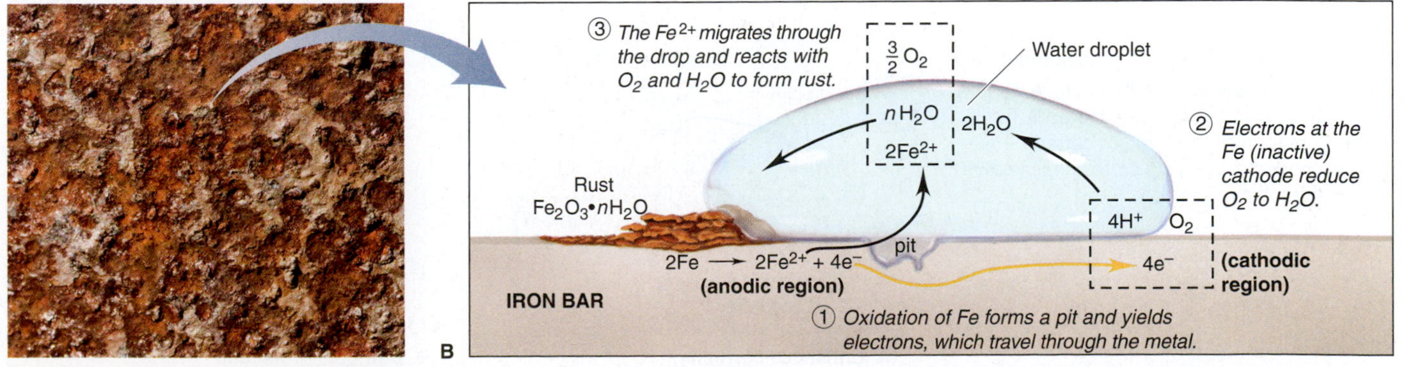

Figure 21.20 The corrosion of iron. **A,** Close-up view of an iron surface. Corrosion usually occurs at a surface irregularity. **B,** A small area of the surface, showing the steps in the corrosion process.

Figure 21.21 Enhanced corrosion at sea. The high ion concentration of seawater enhances the corrosion of iron in hulls and anchors.

[The coefficient n for H_2O appears because rust, $Fe_2O_3 \cdot nH_2O$, has a variable number of waters of hydration.] The rust deposit is incidental to the real damage, which is the loss of iron that weakens the strength of the object. Adding the two previous equations gives the overall equation for the loss and rusting of iron:

$$2Fe(s) + \tfrac{3}{2}O_2(g) + nH_2O(l) + \cancel{4H^+(aq)} \longrightarrow Fe_2O_3 \cdot nH_2O(s) + \cancel{4H^+(aq)}$$

Other species ($2Fe^{2+}$ and $2H_2O$) also cancel, but we showed the canceled H^+ ions to emphasize that they act as a catalyst: they speed the process as they are used up in one step and created in another. For this reason, rusting is faster at low pH (high $[H^+]$) (fact 4). Ionic solutions speed rusting by improving the conductivity of the aqueous medium near the anodic and cathodic regions (fact 5). The effect of ions is especially evident on oceangoing vessels (Figure 21.21) and on the underbodies and around the wheel wells of cars driven in cold climates, where salts are used to melt ice on slippery roads. (We discuss fact 6 in the next subsection.)

Thus, in some key ways, corrosion resembles the operation of a voltaic cell:

- Anodic and cathodic regions are physically separated.
- The regions are connected via an external circuit through which the electrons travel.
- In the anodic region, iron behaves like an active electrode, whereas in the cathodic region, it is inactive.
- The moisture surrounding the pit functions somewhat like an electrolyte and salt bridge, a solution of ions and a means for them to move and keep the solution neutral.

Protecting Against the Corrosion of Iron

Corrosion is prevented by eliminating corrosive factors. Washing road salt off auto bodies removes ions. Painting an object keeps out O_2 and moisture. Plating chromium on plumbing fixtures is a more permanent method, as is "blueing" of gun barrels and other steel objects, in which a coating of Fe_3O_4 (magnetite) is bonded to the surface.

The final point regarding corrosion (fact 6) concerns the relative activity of other metals in contact with iron. The essential idea is that *iron functions as both anode and cathode in the rusting process, but it is lost only at the anode.* Thus,

1. *Corrosion increases when iron behaves more like the anode.* When iron is in contact with a *less* active metal (weaker reducing agent), such as copper, it loses electrons more readily (its anodic function is enhanced; Figure 21.22A). For example, when iron plumbing is connected directly to copper plumbing, the iron pipe corrodes rapidly. Nonconducting rubber or plastic spacers are placed between the metals to avoid this problem.

2. *Corrosion decreases when iron behaves more like the cathode.* In *cathodic protection,* the most effective way to prevent corrosion, iron makes contact with a *more* active metal (stronger reducing agent), such as zinc. The iron becomes the cathode and remains intact, while the zinc acts as the anode and loses electrons (Figure 21.22B). Coating steel with a "sacrificial" layer of zinc is called *galvanizing*. In addition to blocking physical contact with H_2O and O_2, the zinc (or other active metal) is "sacrificed" (oxidized) instead of the iron. Sacrificial anodes are used underwater and underground to protect iron and steel pipes, tanks, oil rigs, and so on. Magnesium and aluminum are often

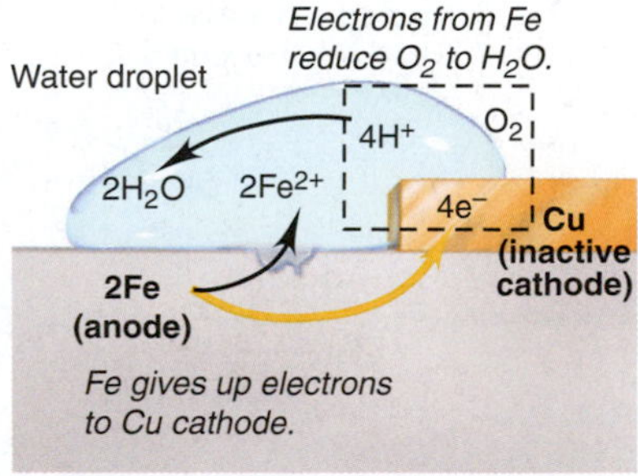

A Enhanced corrosion

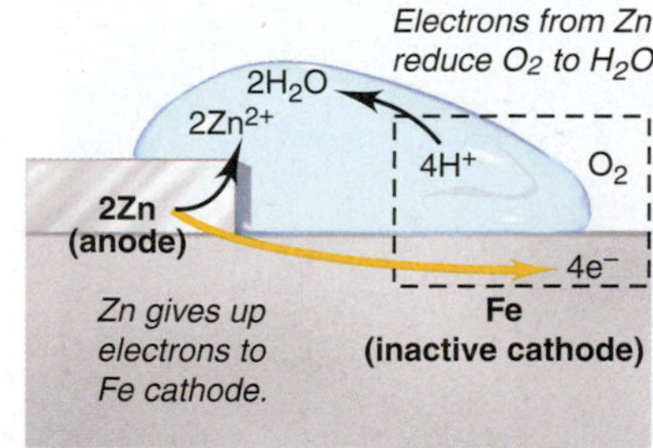

B Cathodic protection

Figure 21.22 The effect of metal-metal contact on the corrosion of iron. **A,** Fe in contact with Cu corrodes faster. **B,** Fe in contact with Zn does not corrode. This method of preventing corrosion is known as *cathodic protection.*

used because they are much more active than iron and, thus, act as the anode (Figure 21.23). Moreover, they form adherent oxide coatings, which slow their own corrosion.

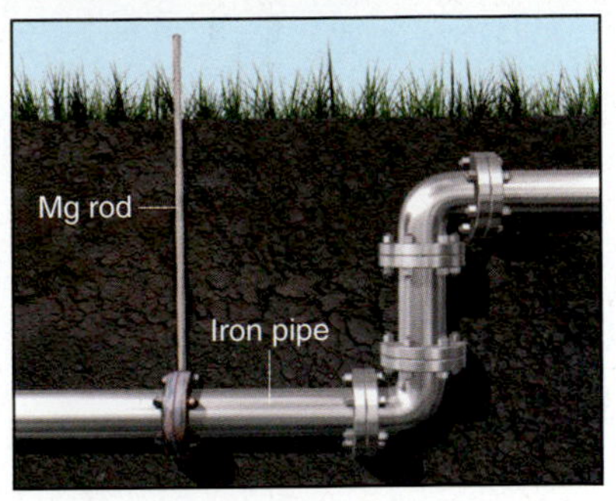

Figure 21.23 **The use of a sacrificial anode to prevent iron corrosion.** In cathodic protection, an active metal, such as zinc, magnesium, or aluminum, acts as the anode and is sacrificed instead of the iron.

Summary of Section 21.6

- Corrosion damages metal structures through a natural electrochemical process.
- Iron corrosion occurs in the presence of oxygen and moisture and is increased by high $[H^+]$, high [ion], or contact with a less active metal, such as Cu.
- Fe is oxidized and O_2 is reduced in one redox reaction, while Fe^{2+} is oxidized and O_2 is reduced to form rust (hydrated form of Fe_2O_3) in another redox reaction that often takes place at a different location.
- Because Fe functions as both anode and cathode in the corrosion process, an iron object can be protected by physically covering it or by joining it to a more active metal (such as Zn, Mg, or Al), which acts as the anode in place of the Fe.

21.7 • ELECTROLYTIC CELLS: USING ELECTRICAL ENERGY TO DRIVE NONSPONTANEOUS REACTIONS

In contrast to a voltaic cell, which generates electrical energy from a spontaneous reaction, *an electrolytic cell requires electrical energy from an external source to drive a nonspontaneous redox reaction.*

Construction and Operation of an Electrolytic Cell

Let's see how an electrolytic cell operates by constructing one from a voltaic cell. Consider a tin-copper voltaic cell (Figure 21.24A). The Sn anode will gradually be oxidized to Sn^{2+} ions, which enter the electrolyte, and the Cu^{2+} ions will gradually be reduced and plate out on the Cu cathode because the cell reaction is spontaneous in that direction:

For the voltaic cell,

$$Sn(s) \longrightarrow Sn^{2+}(aq) + 2e^- \qquad \text{[anode; oxidation]}$$
$$Cu^{2+}(aq) + 2e^- \longrightarrow Cu(s) \qquad \text{[cathode; reduction]}$$
$$Sn(s) + Cu^{2+}(aq) \longrightarrow Sn^{2+}(aq) + Cu(s) \qquad E^\circ_{cell} = 0.48\text{ V and } \Delta G^\circ = -93\text{ kJ}$$

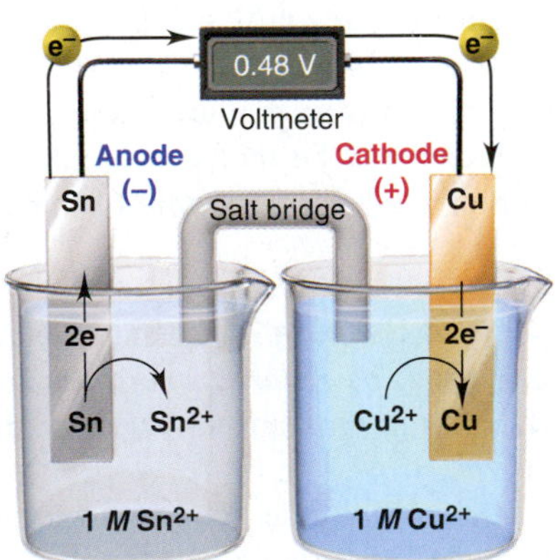

Oxidation half-reaction
$Sn(s) \longrightarrow Sn^{2+}(aq) + 2e^-$

Reduction half-reaction
$Cu^{2+}(aq) + 2e^- \longrightarrow Cu(s)$

Overall (cell) reaction
$Sn(s) + Cu^{2+}(aq) \longrightarrow Sn^{2+}(aq) + Cu(s)$

A Voltaic cell

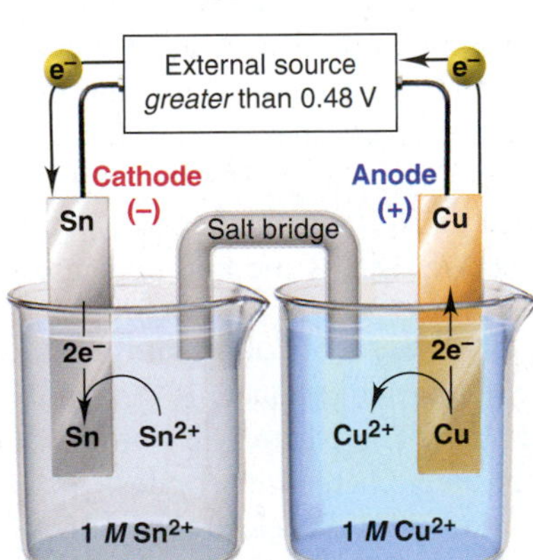

Oxidation half-reaction
$Cu(s) \longrightarrow Cu^{2+}(aq) + 2e^-$

Reduction half-reaction
$Sn^{2+}(aq) + 2e^- \longrightarrow Sn(s)$

Overall (cell) reaction
$Cu(s) + Sn^{2+}(aq) \longrightarrow Cu^{2+}(aq) + Sn(s)$

B Electrolytic cell

Figure 21.24 **The tin-copper reaction as the basis of a voltaic and an electrolytic cell.** **A,** The spontaneous reaction between Sn and Cu^{2+} generates 0.48 V in a voltaic cell. **B,** If more than 0.48 V is supplied, the nonspontaneous (reverse) reaction between Cu and Sn^{2+} occurs. Note the changes in electrode charge and direction of electron flow.

Therefore, the *reverse* cell reaction is *non*spontaneous and never happens on its own. But, we can make it happen by supplying an electric potential *greater than* E°_{cell} from an external source. In effect, we convert the voltaic cell into an electrolytic cell—anode becomes cathode, and cathode becomes anode (Figure 21.24B):

For the electrolytic cell,

$$Cu(s) \longrightarrow Cu^{2+}(aq) + 2e^- \quad \text{[anode; oxidation]}$$
$$Sn^{2+}(aq) + 2e^- \longrightarrow Sn(s) \quad \text{[cathode; reduction]}$$
$$Cu(s) + Sn^{2+}(aq) \longrightarrow Cu^{2+}(aq) + Sn(s) \quad E^\circ_{cell} = -0.48 \text{ V and } \Delta G^\circ = 93 \text{ kJ}$$

In an electrolytic cell, as in a voltaic cell, *oxidation takes place at the anode and reduction takes place at the cathode.* Note, however, that *the direction of electron flow and the signs of the electrodes are reversed.*

To understand these differences, we focus on the *cause* of the electron flow:

- In a voltaic cell, electrons are generated *in* the anode, so it is *negative,* and *removed from* the cathode, so it is *positive.*
- In an electrolytic cell, an external power source supplies electrons *to* the cathode, so it is *negative,* and removes them *from* the anode, so it is *positive.*

Table 21.4 summarizes the processes and signs in the two types of cells.

Table 21.4 Comparison of Voltaic and Electrolytic Cells

			Electrode		
Cell Type	**ΔG**	**E_{cell}**	**Name**	**Process**	**Sign**
Voltaic	<0	>0	Anode	Oxidation	−
Voltaic	<0	>0	Cathode	Reduction	+
Electrolytic	>0	<0	Anode	Oxidation	+
Electrolytic	>0	<0	Cathode	Reduction	−

Predicting the Products of Electrolysis

Electrolysis is the splitting (lysing) of a substance by the input of electrical energy and is often used to convert a compound to its elements, as you'll see for sodium, chlorine, copper, and aluminum in this and the next subsection. Water was first electrolyzed to H_2 and O_2 in 1800, and the process is still used to produce these gases in ultrahigh purity. The electrolyte in an electrolytic cell can be the pure compound (such as H_2O or a molten salt), an aqueous solution of a salt, or a mixture of molten salts. We discuss the first two cases next.

Electrolysis of Molten Salts and the Production of Sodium If the salt is pure, predicting the products is straightforward: *the cation will be reduced and the anion oxidized.* The electrolyte is the molten salt itself, and the ions are attracted by the oppositely charged electrodes.

Consider the electrolysis of molten (fused) calcium chloride. The two species present are Ca^{2+} and Cl^-, so Ca^{2+} ion is reduced and Cl^- ion is oxidized:

$$2Cl^-(l) \longrightarrow Cl_2(g) + 2e^- \quad \text{[anode; oxidation]}$$
$$Ca^{2+}(l) + 2e^- \longrightarrow Ca(s) \quad \text{[cathode; reduction]}$$
$$Ca^{2+}(l) + 2Cl^-(l) \longrightarrow Ca(s) + Cl_2(g) \quad \text{[overall]}$$

Calcium is prepared industrially this way, as are several other active metals, such as Na and Mg, and the halogens Cl_2 and Br_2.

The industrial production of sodium involves electrolysis of molten NaCl. The sodium ore is *halite* (largely NaCl), which is obtained either by evaporation of concen-

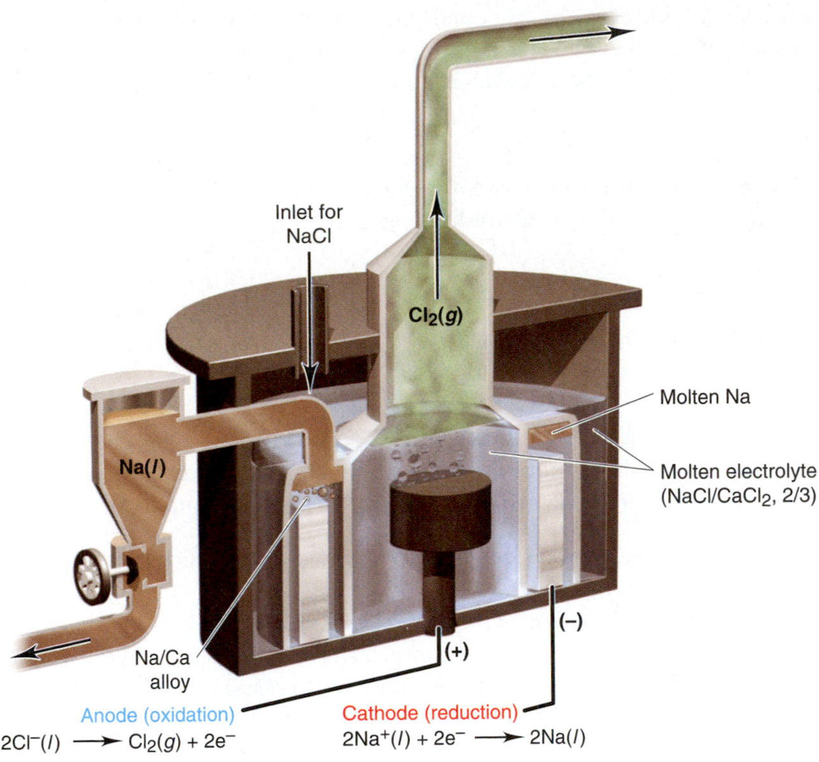

Figure 21.25 The Downs cell for production of sodium.

trated salt solutions (brines) or by mining vast salt deposits formed from the evaporation of prehistoric seas.

Following an evaporation step (if brines are used), the dry solid is crushed and fused (melted) in an electrolytic apparatus called the **Downs cell** (Figure 21.25). To reduce heating costs, the NaCl (mp = 801°C) is mixed in a 2/3 ratio with $CaCl_2$ to form a mixture that melts at only 580°C. Reduction of the metal ions to Na and Ca takes place at a cylindrical steel cathode, with the molten metals floating on the denser molten salt mixture. As the metals rise through a short collecting pipe, the liquid Na is siphoned off, while a higher melting Na/Ca alloy solidifies and falls back into the molten electrolyte. Chloride ions are oxidized to Cl_2 gas at a large anode within an inverted cone-shaped chamber. The cell design separates the metals from the Cl_2 to prevent their explosive recombination. The Cl_2 gas is collected, purified, and sold as a valuable byproduct.

Electrolysis of Water and Nonstandard Half-Cell Potentials Before we analyze the electrolysis products of aqueous salt solutions, let's examine the electrolysis of water itself. Very pure water is difficult to electrolyze because few ions are present to conduct a current. However, if we add a small amount of a salt that cannot be electrolyzed (such as Na_2SO_4), electrolysis proceeds rapidly. An electrolytic cell with separate compartments for H_2 and O_2 is used (Figure 21.26). At the anode, water is oxidized; note the oxidation number (O.N.) of O changes from −2 to 0:

$$2H_2O(l) \longrightarrow O_2(g) + 4H^+(aq) + 4e^- \qquad E = 0.82\text{ V} \qquad \text{[anode; oxidation]}$$

At the cathode, water is reduced; note the O.N. of H changes from +1 to 0:

$$2H_2O(l) + 2e^- \longrightarrow H_2(g) + 2OH^-(aq) \qquad E = -0.42\text{ V} \qquad \text{[cathode; reduction]}$$

Doubling the cathode half-reaction to make e^- loss equal e^- gain, adding the half-reactions (which involves combining the H^+ and OH^- into H_2O and canceling e^- and excess H_2O), and calculating E_{cell} gives the overall reaction:

$$2H_2O(l) \longrightarrow 2H_2(g) + O_2(g) \qquad E_{cell} = -0.42\text{ V} - 0.82\text{ V} = -1.24\text{ V} \quad \text{[overall]}$$

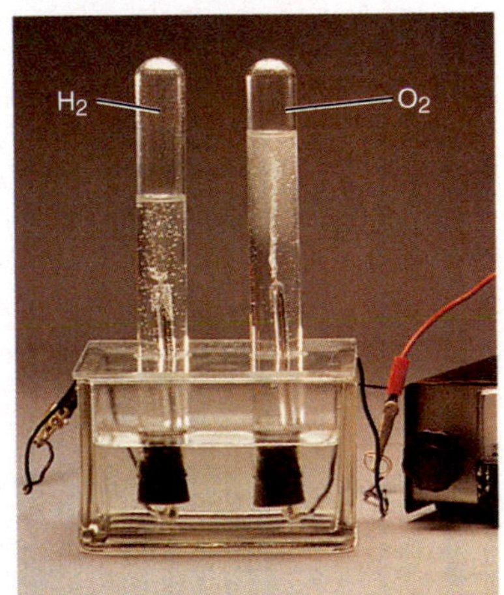

Oxidation half-reaction
$2H_2O(l) \longrightarrow O_2(g) + 4H^+(aq) + 4e^-$

Reduction half-reaction
$2H_2O(l) + 2e^- \longrightarrow H_2(g) + 2OH^-(aq)$

Overall (cell) reaction
$2H_2O(l) \longrightarrow 2H_2(g) + O_2(g)$

Figure 21.26 The electrolysis of water. Twice as much H_2 forms as O_2.

Notice that these electrode potentials are *not* standard electrode potentials (and not designated with $E°$). The $[H^+]$ and $[OH^-]$ in water are 1.0×10^{-7} *M* rather than the standard-state value of 1 *M*. These *E* values are obtained by applying the Nernst equation. For example, the calculation for the anode potential (with $n = 4$) is

$$E_{cell} = E^\circ_{cell} - \frac{0.0592\ V}{4}\log\,(P_{O_2} \times [H^+]^4)$$

The standard potential for the *oxidation* of water is −1.23 V (from Appendix D) and $P_{O_2} \approx 1$ atm in the half-cell, so we have

$$E_{cell} = -1.23\ V - \left\{\frac{0.0592\ V}{4} \times [\log 1 + 4\log(1.0\times10^{-7})]\right\} = -0.82\ V$$

In aqueous ionic solutions, $[H^+]$ and $[OH^-]$ are also approximately 10^{-7} *M*, so we use these nonstandard E_{cell} values to predict electrode products.

Electrolysis of Aqueous Salt Solutions; Overvoltage and the Chlor-Alkali Process Aqueous salt solutions are mixtures of ions *and* water, so we have to compare the various electrode potentials to predict the electrode products.

1. *Predicting the electrode products.* When two half-reactions are possible at an electrode,

- *The reduction with the less negative (more positive) electrode potential occurs.*
- *The oxidation with the less positive (more negative) electrode potential occurs.*

What happens, for instance, when a solution of potassium iodide is electrolyzed?

- The possible *oxidizing agents* are K^+ and H_2O; their reduction half-reactions are

$$K^+(aq) + e^- \longrightarrow K(s) \qquad E^\circ = -2.93\ V$$

$$2H_2O(l) + 2e^- \longrightarrow H_2(g) + 2OH^-(aq) \qquad E = -0.42\ V \qquad \text{[reduction]}$$

The less *negative* electrode potential for water means that it is much easier to reduce than K^+, so H_2 forms at the cathode.

- The possible *reducing agents* are I^- and H_2O; their oxidation half-reactions are

$$2I^-(aq) \longrightarrow I_2(s) + 2e^- \qquad E^\circ = 0.53\ V \qquad \text{[oxidation]}$$

$$2H_2O(l) \longrightarrow O_2(g) + 4H^+(aq) + 4e^- \qquad E = 0.82\ V$$

The less *positive* electrode potential for I^- means that it is easier to oxidize than H_2O, so I_2 forms at the anode.

2. *The requirement for an overvoltage in the production of chlorine.* The products predicted from a comparison of electrode potentials are not always the actual products. For gases to be produced at metal electrodes, additional voltage is required. This increment above the expected voltage is the **overvoltage.** It is 0.4 to 0.6 V for $H_2(g)$ or $O_2(g)$ and is due to the large activation energy (Section 16.5) needed to form gases at the electrode.

Overvoltage has major practical significance. A multibillion-dollar example is the industrial production of chlorine and several other substances from concentrated NaCl solution by the **chlor-alkali process.** Water is easier to reduce than Na^+, so H_2 forms at the cathode, even *with* an overvoltage of 0.6 V:

$$Na^+(aq) + e^- \longrightarrow Na(s) \qquad E^\circ = -2.71\ V$$

$$2H_2O(l) + 2e^- \longrightarrow H_2(g) + 2OH^-(aq) \qquad E = -0.42\ V\ (\approx -1\ V \text{ with overvoltage}) \qquad \text{[reduction]}$$

But Cl_2 *does* form at the anode, even though a comparison of the electrode potentials would lead us to predict that O_2 should form:

$$2H_2O(l) \longrightarrow O_2(g) + 4H^+(aq) + 4e^- \qquad E = 0.82\ V\ (\sim 1.4\ V \text{ with overvoltage})$$

$$2Cl^-(aq) \longrightarrow Cl_2(g) + 2e^- \qquad E^\circ = 1.36\ V \qquad \text{[oxidation]}$$

An overvoltage of ~0.6 V for O_2 makes Cl_2 the product that is easier to form. Keeping $[Cl^-]$ high also favors this step. Thus, Cl_2, one of the 10 most heavily produced chemicals, is formed from plentiful natural sources of aqueous sodium chloride.

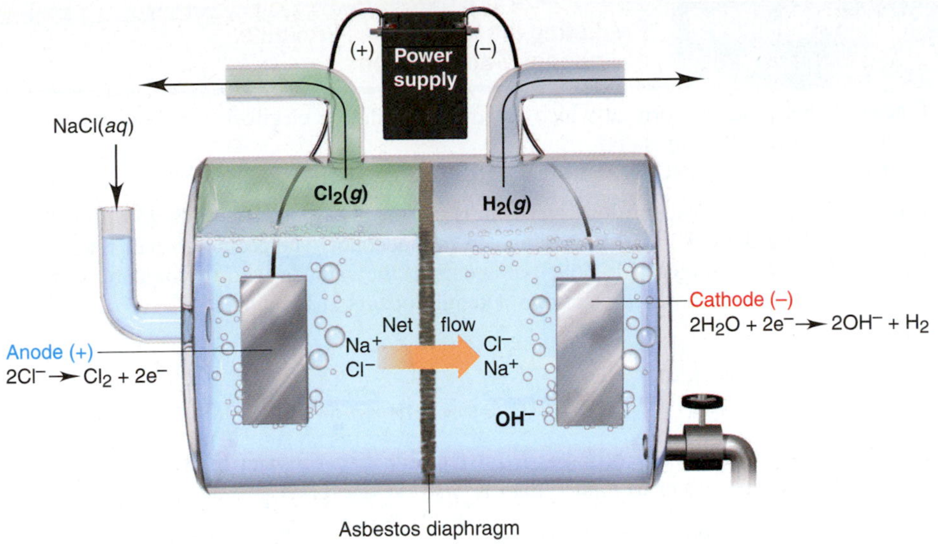

Figure 21.27 A diaphragm cell for the chlor-alkali process.

The half-reactions for electrolysis of aqueous NaCl are

$$\begin{array}{lll} 2Cl^-(aq) \longrightarrow Cl_2(g) + 2e^- & E^\circ = 1.36\text{ V} & \text{[anode; oxidation]} \\ 2H_2O(l) + 2e^- \longrightarrow 2OH^-(aq) + H_2(g) & E = -1.0\text{ V} & \text{[cathode; reduction]} \\ \hline 2Cl^-(aq) + 2H_2O(l) \longrightarrow 2OH^-(aq) + H_2(g) + Cl_2(g) & E_{cell} = -1.0\text{ V} - 1.36\text{ V} = -2.4\text{ V} & \end{array}$$

To obtain commercial amounts of Cl_2, a voltage almost twice this value and a current in excess of 3×10^4 A are used.

When we include the spectator ion Na^+, the total ionic equation shows another important product:

$$2Na^+(aq) + 2Cl^-(aq) + 2H_2O(l) \longrightarrow 2Na^+(aq) + 2OH^-(aq) + H_2(g) + Cl_2(g)$$

As Figure 21.27 shows, the sodium salts in the cathode compartment exist as an aqueous mixture of NaCl and NaOH; the NaCl is removed by fractional crystallization, which separates the compounds by differences in solubility. Thus, in this version of the chlor-alkali process, which uses an *asbestos diaphragm* to separate the anode and cathode compartments, the products are Cl_2, H_2, and industrial-grade NaOH, an important base.

Like other reactive products, H_2 and Cl_2 are kept apart to prevent explosive recombination. Note the higher liquid level in the anode compartment. This slight hydrostatic pressure difference minimizes backflow of NaOH, which prevents the disproportionation (self–oxidation-reduction) reactions of Cl_2 that occur in the presence of OH^-, such as

$$Cl_2(g) + 2OH^-(aq) \longrightarrow Cl^-(aq) + ClO^-(aq) + H_2O(l)$$

A newer chlor-alkali *membrane-cell process,* in which the diaphragm is replaced by a polymeric membrane to separate the cell compartments, has been adopted in much of the industrialized world. The membrane allows only cations to move through it and only from anode to cathode compartments. Thus, as Cl^- ions are removed at the anode through oxidation to Cl_2, Na^+ ions in the anode compartment move through the membrane to the cathode compartment and form an NaOH solution. In addition to forming purer NaOH than the older diaphragm-cell method, the membrane-cell process uses less electricity.

3. *A summary: which product at which electrode*. Experiments have shown the elements that can be prepared electrolytically from aqueous solutions of their salts:

- Cations of less active metals, including gold, silver, copper, chromium, platinum, and cadmium, *are* reduced to the metal.
- Cations of more active metals, including those in Groups 1A(1) and 2A(2) and Al from 3A(13), *are not* reduced. Water is reduced to H_2 and OH^- instead.
- Anions that *are* oxidized, because of overvoltage from O_2 formation, include the halides ($[Cl^-]$ must be high), except for F^-.
- Anions that *are not* oxidized include F^- and common oxoanions, such as SO_4^{2-}, CO_3^{2-}, NO_3^-, and PO_4^{3-}, because the central nonmetal in these oxoanions is already in its highest oxidation state. Water is oxidized to O_2 and H^+ instead.

Sample Problem 21.8 Predicting the Electrolysis Products of Aqueous Salt Solutions

Problem What products form at which electrode during the electrolysis of aqueous solutions of the following salts?
(a) KBr **(b)** $AgNO_3$ **(c)** $MgSO_4$

Plan We identify the reacting ions and compare their electrode potentials with those of water, taking the 0.4 to 0.6 V overvoltage into account. The reduction half-reaction with the less negative electrode potential occurs at the cathode, and the oxidation half-reaction with the less positive electrode potential occurs at the anode.

Solution

(a)

$$K^+(aq) + e^- \longrightarrow K(s) \qquad E° = -2.93 \text{ V}$$
$$2H_2O(l) + 2e^- \longrightarrow H_2(g) + 2OH^-(aq) \qquad E = -0.42 \text{ V}$$

Despite the overvoltage, which makes E for the reduction of water between -0.8 and -1.0 V, H_2O is still easier to reduce than K^+, so $H_2(g)$ forms at the cathode.

$$2Br^-(aq) \longrightarrow Br_2(l) + 2e^- \qquad E° = 1.07 \text{ V}$$
$$2H_2O(l) \longrightarrow O_2(g) + 4H^+(aq) + 4e^- \qquad E = 0.82 \text{ V}$$

Because of the overvoltage, which makes E for the oxidation of water between 1.2 and 1.4 V, Br^- is easier to oxidize than water, so $Br_2(l)$ forms at the anode *(see photo).*

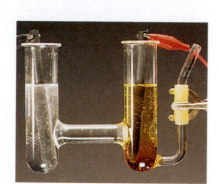

Electrolysis of aqueous KBr.

(b)

$$Ag^+(aq) + e^- \longrightarrow Ag(s) \qquad E° = 0.80 \text{ V}$$
$$2H_2O(l) + 2e^- \longrightarrow H_2(g) + 2OH^-(aq) \qquad E = -0.42 \text{ V}$$

As the cation of an inactive metal, Ag^+ is a better oxidizing agent than H_2O, so $Ag(s)$ forms at the cathode. NO_3^- cannot be oxidized, because N is already in its highest (+5) oxidation state. Thus, $O_2(g)$ forms at the anode:

$$2H_2O(l) \longrightarrow O_2(g) + 4H^+(aq) + 4e^-$$

(c)

$$Mg^{2+}(aq) + 2e^- \longrightarrow Mg(s) \qquad E° = -2.37 \text{ V}$$

Like K^+ in part (a), Mg^{2+} cannot be reduced in the presence of water, so $H_2(g)$ forms at the cathode. The SO_4^{2-} ion cannot be oxidized because S is in its highest (+6) oxidation state. Thus, H_2O is oxidized, and $O_2(g)$ forms at the anode:

$$2H_2O(l) \longrightarrow O_2(g) + 4H^+(aq) + 4e^-$$

FOLLOW-UP PROBLEM 21.8 Write half-reactions for the products you predict will form in the electrolysis of aqueous $AuBr_3$.

Purifying Copper and Isolating Aluminum

In addition to the Downs cell for sodium production and the diaphragm (or membrane) cell for chlorine manufacture, other electrolytic cells are used industrially to obtain metals and nonmetals from their ores or to purify them for later use. Here we focus on two key metals: copper and aluminum.

Electrorefining of Copper The most common copper ore is chalcopyrite, $CuFeS_2$, a mixed sulfide of FeS and CuS. Most remaining deposits contain less than 0.5% Cu by mass. To "win" this small amount of copper from the ore requires several steps. After removing the iron(II) sulfide and reducing the copper(II) sulfide, the copper obtained is usable for plumbing, but it must be purified further for electrical wiring, its most important application.

Purification of copper is accomplished by *electrorefining,* which involves the oxidation of Cu to form Cu^{2+} ions in solution, followed by their reduction and the plating out of Cu metal (Figure 21.28). To do this, impure copper is cast into plates to be used as anodes, and cathodes are made from already purified copper. The electrodes are immersed in acidified $CuSO_4$ solution, and a controlled voltage is applied that accomplishes two tasks simultaneously:

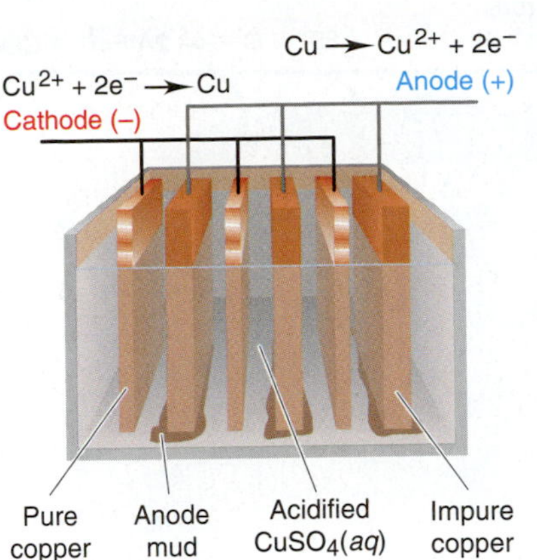

Figure 21.28 The electrorefining of copper.

1. Copper and the more active, but undesirable, impurities (Fe, Ni) are oxidized to their cations, while the less active, but valuable, impurities (Ag, Au, Pt) are not. As the anode slabs react, these unoxidized metals fall off as "anode mud" and are purified separately. Sale of these precious metals in the anode mud nearly offsets the cost of electricity to operate the cell, making Cu wire inexpensive.
2. Because Cu is much less active than the Fe and Ni impurities, Cu^{2+} ions are reduced at the cathode, but Fe^{2+} and Ni^{2+} ions remain in solution:

$$\begin{aligned} Cu^{2+}(aq) + 2e^- &\longrightarrow Cu(s) \qquad E^\circ = 0.34\ V \\ Ni^{2+}(aq) + 2e^- &\longrightarrow Ni(s) \qquad E^\circ = -0.25\ V \\ Fe^{2+}(aq) + 2e^- &\longrightarrow Fe(s) \qquad E^\circ = -0.44\ V \end{aligned}$$

The copper obtained by electrorefining is over 99.99% pure.

Isolation of Aluminum: The Hall-Heroult Process Aluminum, the most abundant metal in Earth's crust by mass, is found in numerous aluminosilicate minerals. Through eons of weathering, certain of these became *bauxite,* a mixed oxide-hydroxide that is the major ore of aluminum. In general terms, the isolation of aluminum is a two-stage, multistep process that combines several physical and chemical separations. In the first stage, the mineral oxide, Al_2O_3, is separated from bauxite, which also contains as impurities SiO_2, TiO_2, and Fe_2O_3. In the second stage, which we focus on here, the oxide is converted to the metal.

Aluminum is an active metal and much too strong a reducing agent to be formed at the cathode from aqueous solution, so the oxide itself is electrolyzed. However, the melting point of Al_2O_3 is very high (2030°C), so major energy (and cost) savings are realized by dissolving the oxide in molten *cryolite* (Na_3AlF_6) to give a mixture that is electrolyzed at ~1000°C. The electrolytic step, called the *Hall-Heroult process,* takes place in a graphite-lined furnace, with the lining itself acting as the cathode. Anodes of graphite dip into the molten Al_2O_3-Na_3AlF_6 mixture (Figure 21.29, *next page*). The cell typically operates at a moderate voltage of 4.5 V, but with an enormous current flow of 1.0×10^5 to 2.5×10^5 A.

Because the process is complex and still not entirely known, the following reactions are chosen from among several other possibilities. Molten cryolite contains several ions (including AlF_6^{3-}, AlF_4^-, and F^-), which react with Al_2O_3 to form fluoro-oxy ions (including $AlOF_3^{2-}$, $Al_2OF_6^{2-}$, and $Al_2O_2F_4^{2-}$) that dissolve in the mixture. For example,

$$2Al_2O_3(s) + 2AlF_6^{3-}(l) \longrightarrow 3Al_2O_2F_4^{2-}(l)$$

Al forms at the cathode (reduction), shown here with AlF_6^{3-} as reactant:

$$AlF_6^{3-}(l) + 3e^- \longrightarrow Al(l) + 6F^-(l) \qquad \text{[cathode; reduction]}$$

The graphite anodes are oxidized and form carbon dioxide gas. Using one of the fluoro-oxy species as an example, the anode reaction is

$$Al_2O_2F_4^{2-}(l) + 8F^-(l) + C(graphite) \longrightarrow 2AlF_6^{3-}(l) + CO_2(g) + 4e^- \qquad \text{[anode; oxidation]}$$

The *anodes are consumed in this half-reaction* and must be replaced frequently.

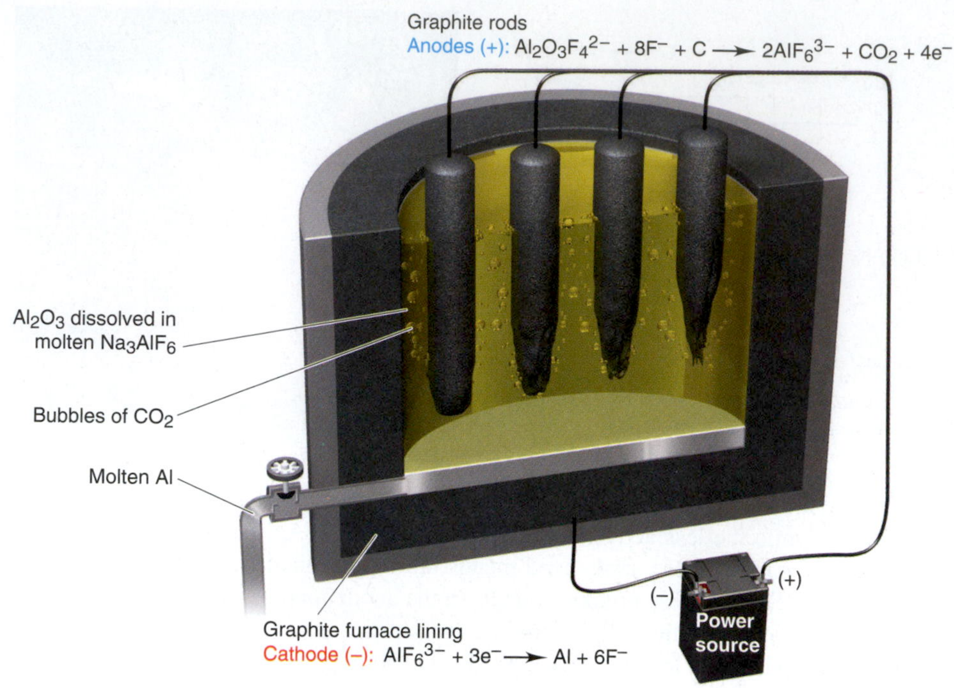

Figure 21.29 The electrolytic cell in the manufacture of aluminum.

Combining the three previous equations and making sure that e^- gained at the cathode equals e^- lost at the anode gives the overall reaction:

$$2Al_2O_3(\text{in } Na_3AlF_6) + 3C(\text{graphite}) \longrightarrow 4Al(l) + 3CO_2(g) \qquad \text{[overall (cell) reaction]}$$

Aluminum production accounts for more than 5% of total U.S. electrical usage! Estimates for the entire manufacturing process (including mining, maintaining operating conditions, and so forth) show that aluminum recycling requires less than 1% as much energy as manufacturing it from the ore, which explains why recycling has become so widespread.

Stoichiometry of Electrolysis: The Relation Between Amounts of Charge and Products

Since charge flowing through an electrolytic cell yields products at the electrodes, it makes sense that the more charge that flows, the more product that forms. In fact, this relationship was first determined experimentally in the 1830s by Michael Faraday and is referred to as *Faraday's law of electrolysis: the amount of substance produced at each electrode is directly proportional to the quantity of charge flowing through the cell.*

Each balanced half-reaction shows the amounts (mol) of reactant, electrons, and product involved in the change, so it contains the information we need to answer such questions as "How much material will form from a given quantity of charge?" or, conversely, "How much charge is needed to produce a given amount of material?" To apply Faraday's law,

1. Balance the half-reaction to find the amount (mol) of electrons needed per mole of product.
2. Use the Faraday constant ($F = 9.65\times10^4$ C/mol e^-) to find the quantity of charge.
3. Use the molar mass to find the charge needed for a given mass of product.

Measuring Current to Find Charge We cannot measure charge directly, but we *can* measure *current,* the charge flowing per unit time. The SI unit of current is the **ampere (A),** which is defined as a charge of 1 coulomb flowing through a conductor in 1 second:

$$1 \text{ ampere} = 1 \text{ coulomb/second} \quad \text{or} \quad 1 \text{ A} = 1 \text{ C/s} \qquad \textbf{(21.11)}$$

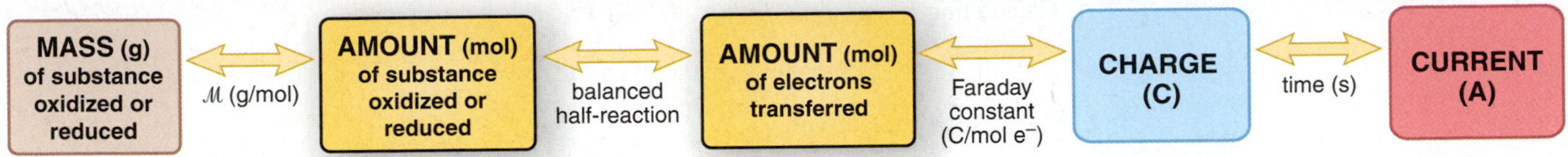

Figure 21.30 A summary diagram for the stoichiometry of electrolysis.

Thus, the current multiplied by the time gives the charge:

$$\text{Current} \times \text{time} = \text{charge} \qquad \text{or} \qquad \text{A} \times \text{s} = \frac{\text{C}}{\text{s}} \times \text{s} = \text{C}$$

Therefore, by measuring the current *and* the time during which the current flows, we find the charge, which relates to the amount of product (Figure 21.30).

Problems Involving Stoichiometry of Electrolysis Problems based on Faraday's law often ask you to calculate current, mass of material, or time. As we said, the electrode half-reaction provides the key to solving these problems because it is related to the mass for a certain quantity of charge.

Here is a typical problem in practical electrolysis: how long does it take to produce 3.0 g of $Cl_2(g)$ by electrolysis of aqueous NaCl using a power supply with a current of 12 A? The problem asks for the time needed to produce a certain mass, so let's first relate mass to amount (mol) of electrons to find the charge needed. Then, we'll relate the charge to the current to find the time.

The half-reaction tells us that 2 mol of electrons are lost to form 1 mol of Cl_2 and we'll use this relationship as a conversion factor:

$$2Cl^-(aq) \longrightarrow Cl_2(g) + 2e^-$$

We convert the given mass of Cl_2 to amount of Cl_2, use the conversion factor from the half-reaction, and multiply by the Faraday constant to find the total charge:

$$\text{Charge (C)} = 3.0 \text{ g } Cl_2 \times \frac{1 \text{ mol } Cl_2}{70.90 \text{ g } Cl_2} \times \frac{2 \text{ mol } e^-}{1 \text{ mol } Cl_2} \times \frac{9.65{\times}10^4 \text{ C}}{1 \text{ mol } e^-} = 8.2{\times}10^3 \text{ C}$$

Now we use the relationship between charge and current to find the time needed:

$$\text{Time(s)} = \frac{\text{charge (C)}}{\text{current (A, or C/s)}} = 8.2{\times}10^3 \text{ C} \times \frac{1 \text{ s}}{12 \text{ C}} = 6.8{\times}10^2 \text{ s} \ ({\sim}11 \text{ min})$$

Note that the entire calculation follows Figure 21.29 until the last step, which asks for the time that the given current must flow:

grams of Cl_2 ⟹ moles of Cl_2 ⟹ moles of e^- ⟹ coulombs ⟹ seconds

Sample Problem 21.9 applies these ideas in an important industrial setting.

Sample Problem 21.9 Applying the Relationship Among Current, Time, and Amount of Substance

Problem A technician plates a faucet with 0.86 g of Cr metal by electrolysis of aqueous $Cr_2(SO_4)_3$. If 12.5 min is allowed for the plating, what current is needed?

Plan To find the current, we divide charge by time, so we need to find the charge. We write the half-reaction for Cr^{3+} reduction to get the amount (mol) of e^- transferred per mole of Cr. To find the charge, we convert the mass of Cr needed (0.86 g) to amount (mol) of Cr. Then, we use the Faraday constant ($9.65{\times}10^4$ C/mol e^-) to find the charge and divide by the time (12.5 min, converted to seconds) to obtain the current (see the road map).

Road Map

Mass (g) of Cr needed
↓ divide by ℳ (g/mol)
Amount (mol) of Cr needed
↓ 3 mol e^- = 1 mol Cr
Amount (mol) of e^- transferred
↓ 1 mol e^- = $9.65{\times}10^4$ C
Charge (C)
↓ divide by time (convert min to s)
Current (A)

Solution Writing the balanced half-reaction:

$$Cr^{3+}(aq) + 3e^- \longrightarrow Cr(s)$$

Combining steps to find amount (mol) of e^- transferred for mass of Cr needed:

$$\text{Amount (mol) of } e^- \text{ transferred} = 0.86 \text{ g Cr} \times \frac{1 \text{ mol Cr}}{52.00 \text{ g Cr}} \times \frac{3 \text{ mol } e^-}{1 \text{ mol Cr}} = 0.050 \text{ mol } e^-$$

Calculating the charge:

$$\text{Charge (C)} = 0.050 \text{ mol e}^- \times \frac{9.65\times10^4 \text{ C}}{1 \text{ mol e}^-} = 4.8\times10^3 \text{ C}$$

Calculating the current:

$$\text{Current (A)} = \frac{\text{charge (C)}}{\text{time (s)}} = \frac{4.8\times10^3 \text{ C}}{12.5 \text{ min}} \times \frac{1 \text{ min}}{60 \text{ s}} = 6.4 \text{ C/s} = 6.4 \text{ A}$$

Check Rounding gives

$$(\sim 0.9 \text{ g})(1 \text{ mol Cr}/50 \text{ g})(3 \text{ mol e}^-/1 \text{ mol Cr}) = 5\times10^{-2} \text{ mol e}^-$$

then

$$(5\times10^{-2} \text{ mol e}^-)(\sim 1\times10^5 \text{ C/mol e}^-) = 5\times10^3 \text{ C}$$

and

$$(5\times10^3 \text{ C}/12 \text{ min})(1 \text{ min}/60 \text{ s}) = 7 \text{ A}$$

Comment For the sake of introducing Faraday's law, we neglected some details about actual electroplating. In practice, electroplating chromium is only 30% to 40% efficient and must be run at a specific temperature range for the plate to appear bright. Nearly 10,000 metric tons (2×10^8 mol) of chromium are used annually for electroplating.

FOLLOW-UP PROBLEM 21.9 Using a current of 4.75 A, how many minutes does it take to plate a sculpture with 1.50 g of Cu from a $CuSO_4$ solution?

Summary of Section 21.7

- An electrolytic cell uses electrical energy to drive a nonspontaneous reaction.
- Oxidation occurs at the anode and reduction at the cathode, but the direction of electron flow and the charges of the electrodes are opposite those in voltaic cells.
- In electrolysis of a pure molten salt, the metal cation is reduced at the cathode, and the nonmetal anion is oxidized at the anode.
- The reduction or oxidation of water takes place at nonstandard conditions.
- Overvoltage causes the actual voltage required to be unexpectedly high (especially for gases, such as H_2 and O_2) and can affect the product that forms at each electrode.
- The amount of product that forms depends on the quantity of charge flowing through the cell, which is related to the time the charge flows and the current.

CHAPTER REVIEW GUIDE

The following sections provide many aids to help you study this chapter. (Numbers in parentheses refer to pages, unless noted otherwise.)

Learning Objectives

These are concepts and skills to review after studying this chapter.

Related section (§), sample problem (SP), and upcoming end-of-chapter problem (EP) numbers are listed in parentheses.

1. Understand the meanings of oxidation, reduction, oxidizing agent, and reducing agent; balance redox reactions by the half-reaction method; distinguish between voltaic and electrolytic cells in terms of the sign of ΔG (§21.1) (SP 21.1) (EPs 21.1–21.12)
2. Describe the physical makeup of a voltaic cell, and explain the direction of electron flow; draw a diagram and write the notation for a voltaic cell (§21.2) (SP 21.2) (EPs 21.13–21.23)
3. Describe how standard electrode potentials ($E^\circ_{\text{half-cell}}$ values) are combined to give E°_{cell} and how the standard reference electrode is used to find an unknown $E^\circ_{\text{half-cell}}$; explain how the reactivity of a metal is related to its $E^\circ_{\text{half-cell}}$; write spontaneous redox reactions using an emf series like that in Appendix D (§21.3) (SPs 21.3, 21.4) (EPs 21.24–21.40)
4. Understand how E_{cell} is related to ΔG and the charge flowing through the cell; use the interrelationship of ΔG°, E°_{cell}, and K to calculate any one of these variables; explain how E_{cell} changes as the cell operates (Q changes), and use the Nernst equation to find E_{cell}; describe how a concentration cell works and calculate its E_{cell} (§21.4) (SPs 21.5–21.7) (EPs 21.41–21.56)
5. Understand how a battery operates, and describe the components of primary and secondary batteries and fuel cells (§21.5) (EPs 21.57–21.59)
6. Explain how corrosion occurs and is prevented (§21.6) (EPs 21.60–21.62)
7. Understand the basis of an electrolytic cell; describe the Downs cell for the production of Na, the chlor-alkali process and the importance of overvoltage for the production of Cl_2, the electrorefining of Cu, and the use of cryolite in the production of Al; know how water influences the products at the electrodes during electrolysis of aqueous salt solutions (§21.7) (SP 21.8) (EPs 21.63–21.75, 21.82)
8. Understand the relationship between charge and amount of product, and calculate the current (or time) needed to produce a given amount of product or vice versa (§21.7) (SP 21.9) (EPs 21.76–21.81, 21.83, 21.84)

Key Terms

These important terms appear in boldface in the chapter and are defined again in the Glossary.

electrochemistry (688)
electrochemical cell (688)

Section 21.1
half-reaction method (689)
voltaic (galvanic) cell (692)
electrolytic cell (692)
electrode (692)
electrolyte (692)
anode (692)
cathode (692)

Section 21.2
half-cell (694)
salt bridge (695)

Section 21.3
cell potential (E_{cell}) (697)
voltage (697)
electromotive force (emf) (697)
volt (V) (697)
coulomb (C) (697)
standard cell potential (E°_{cell}) (697)
standard electrode (half-cell) potential ($E^\circ_{half\text{-}cell}$) (698)
standard reference half-cell (standard hydrogen electrode) (698)

Section 21.4
Faraday constant (F) (705)
Nernst equation (708)
concentration cell (710)

Section 21.5
battery (713)
fuel cell (716)

Section 21.6
corrosion (717)

Section 21.7
electrolysis (720)
Downs cell (721)
overvoltage (722)
chlor-alkali process (722)
ampere (A) (726)

Key Equations and Relationships

Numbered and screened concepts are listed for you to refer to or memorize.

21.1 Relating spontaneity of a process to the sign of the cell potential (697):

$$E_{cell} > 0 \text{ for a spontaneous process}$$

21.2 Relating electric potential to energy and charge in SI units (697):

$$\text{Potential} = \text{energy/charge} \quad \text{or} \quad 1\text{ V} = 1\text{ J/C}$$

21.3 Relating standard cell potential to standard electrode potentials in a voltaic cell (698):

$$E^\circ_{cell} = E^\circ_{cathode\ (reduction)} - E^\circ_{anode\ (oxidation)}$$

21.4 Defining the Faraday constant (706):

$$F = 9.65\times10^4 \frac{\text{J}}{\text{V·mol e}^-} \quad (3\text{ sf})$$

21.5 Relating the free energy change to the cell potential (706):

$$\Delta G = -nFE_{cell}$$

21.6 Finding the standard free energy change from the standard cell potential (706):

$$\Delta G^\circ = -nFE^\circ_{cell}$$

21.7 Finding the equilibrium constant from the standard cell potential (706):

$$E^\circ_{cell} = \frac{RT}{nF}\ln K$$

21.8 Substituting known values of R, F, and T into Equation 21.7 and converting to common logarithms (706):

$$E^\circ_{cell} = \frac{0.0592\text{ V}}{n}\log K \quad \text{or} \quad \log K = \frac{nE^\circ_{cell}}{0.0592\text{ V}} \quad (\text{at } 298.15\text{ K})$$

21.9 Calculating the nonstandard cell potential (Nernst equation) (708):

$$E_{cell} = E^\circ_{cell} - \frac{RT}{nF}\ln Q$$

21.10 Substituting known values of R, F, and T into the Nernst equation and converting to common logarithms (708):

$$E_{cell} = E^\circ_{cell} - \frac{0.0592\text{ V}}{n}\log Q \quad (\text{at } 298.15\text{ K})$$

21.11 Relating current to charge and time (726):

$$\text{Current (A)} = \text{charge (C)/time (s)} \quad \text{and} \quad 1\text{ A} = 1\text{ C/s}$$

BRIEF SOLUTIONS TO FOLLOW-UP PROBLEMS

Compare your own solutions to these calculation steps and answers.

21.1 $6KMnO_4(aq) + 6KOH(aq) + KI(aq) \longrightarrow 6K_2MnO_4(aq) + KIO_3(aq) + 3H_2O(l)$

21.2

$$Sn(s) \longrightarrow Sn^{2+}(aq) + 2e^- \quad \text{[anode; oxidation]}$$

$$6e^- + 14H^+(aq) + Cr_2O_7^{2-}(aq) \longrightarrow 2Cr^{3+}(aq) + 7H_2O(l) \quad \text{[cathode; reduction]}$$

$$3Sn(s) + Cr_2O_7^{2-}(aq) + 14H^+(aq) \longrightarrow 3Sn^{2+}(aq) + 2Cr^{3+}(aq) + 7H_2O(l) \quad \text{[overall]}$$

Cell notation:

$Sn(s) \mid Sn^{2+}(aq) \parallel H^+(aq), Cr_2O_7^{2-}(aq), Cr^{3+}(aq) \mid$ graphite

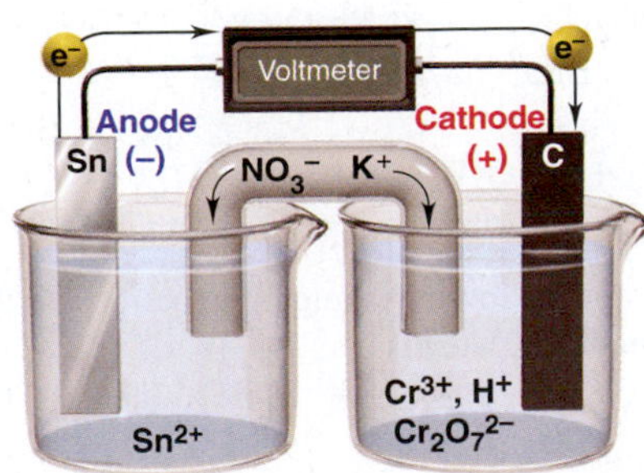

21.3 $Br_2(aq) + 2e^- \longrightarrow 2Br^-(aq) \quad E^\circ_{bromine} = 1.07\text{ V}$ [cathode]

$2V^{3+}(aq) + 2H_2O(l) \longrightarrow 2VO^{2+}(aq) + 4H^+(aq) + 2e^- \quad E^\circ_{vanadium} = ?$ [anode]

$$E^\circ_{vanadium} = E^\circ_{bromine} - E^\circ_{cell} = 1.07\text{ V} - 1.39\text{ V} = -0.32\text{ V}$$

21.4

$$Fe^{2+}(aq) + 2e^- \longrightarrow Fe(s) \quad E^\circ = -0.44\text{ V}$$

$$2[Fe^{2+}(aq) \longrightarrow Fe^{3+}(aq) + e^-] \quad E^\circ = 0.77\text{ V}$$

$$3Fe^{2+}(aq) \longrightarrow 2Fe^{3+}(aq) + Fe(s)$$

$$E^\circ_{cell} = -0.44\text{ V} - 0.77\text{ V} = -1.21\text{ V}$$

The reaction is nonspontaneous. The spontaneous reaction is

$$2Fe^{3+}(aq) + Fe(s) \longrightarrow 3Fe^{2+}(aq) \quad E^\circ_{cell} = 1.21\text{ V}$$

$Fe > Fe^{2+} > Fe^{3+}$

21.5 $Cd(s) + Cu^{2+}(aq) \longrightarrow Cd^{2+}(aq) + Cu(s)$

$$\Delta G^\circ = -RT\ln K = -8.314\text{ J/mol rxn·K} \times 298\text{ K} \times \ln K = -143\text{ kJ/mol rxn};\ K = 1.2\times10^{25}$$

$$E^\circ_{cell} = \frac{0.0592\text{ V}}{2}\log(1.2\times10^{25}) = 0.742\text{ V}$$

BRIEF SOLUTIONS TO FOLLOW-UP PROBLEMS (CONTINUED)

21.6
$$\begin{array}{lll} Fe(s) \longrightarrow Fe^{2+}(aq) + 2e^- & E° = -0.44\ V \\ Cu^{2+}(aq) + 2e^- \longrightarrow Cu(s) & E° = 0.34\ V \\ \hline Fe(s) + Cu^{2+}(aq) \longrightarrow Fe^{2+}(aq) + Cu(s) & E°_{cell} = 0.78\ V \end{array}$$

So $E_{cell} = 0.78\ V + 0.25\ V = 1.03\ V$

$$1.03\ V = 0.78\ V - \frac{0.0592\ V}{2} \log \frac{[Fe^{2+}]}{[Cu^{2+}]}$$

$$\frac{[Fe^{2+}]}{[Cu^{2+}]} = 3.6\times10^{-9}$$

$[Fe^{2+}] = 3.6\times10^{-9} \times 0.30\ M = 1.1\times10^{-9}\ M$

21.7 $Au^{3+}(aq;\ 2.5\times10^{-2}\ M)\ [B] \longrightarrow Au^{3+}(aq;\ 7.0\times10^{-4}\ M)\ [A]$

$$E_{cell} = 0\ V - \left(\frac{0.0592\ V}{3} \times \log \frac{7.0\times10^{-4}}{2.5\times10^{-2}}\right) = 0.0306\ V$$

The electrode in A is negative, so it is the anode.

21.8 The reduction half-reaction with the more positive electrode potential is

$Au^{3+}(aq) + 3e^- \longrightarrow Au(s);\ E° = 1.50\ V$ [cathode; reduction]

Because of overvoltage, O_2 will not form at the anode, so Br_2 will form:

$2Br^-(aq) \longrightarrow Br_2(l) + 2e^-;\ E° = 1.07\ V$ [cathode; oxidation]

21.9 $Cu^{2+}(aq) + 2e^- \longrightarrow Cu(s)$; therefore,

$2\ mol\ e^-/1\ mol\ Cu = 2\ mol\ e^-/63.55\ g\ Cu$

$$\text{Time (min)} = 1.50\ g\ Cu \times \frac{2\ mol\ e^-}{63.55\ g\ Cu} \times \frac{9.65\times10^4\ C}{1\ mol\ e^-} \times \frac{1\ s}{4.75\ C} \times \frac{1\ min}{60\ s} = 16.0\ min$$

PROBLEMS

Problems with **colored** numbers are answered in Appendix E. Sections match the text and provide the numbers of relevant sample problems. Bracketed problems are grouped in pairs (indicated by a short rule) that cover the same concept. Comprehensive Problems are based on material from any section or previous chapter.

Note: Unless stated otherwise, all problems refer to systems at 298.15 K (25°C).

Redox Reactions and Electrochemical Cells

(Sample Problem 21.1)

21.1 Define *oxidation* and *reduction* in terms of electron transfer and change in oxidation number.

21.2 Can one half-reaction in a redox process take place independently of the other? Explain.

21.3 Which type of electrochemical cell has $\Delta G_{sys} < 0$? Which type shows an increase in free energy?

21.4 Which statements are true? Correct any that are false.
(a) In a voltaic cell, the anode is negative relative to the cathode.
(b) Oxidation occurs at the anode of a voltaic or electrolytic cell.
(c) Electrons flow into the cathode of an electrolytic cell.
(d) In a voltaic cell, the surroundings do work on the system.
(e) A metal that plates out of an electrolytic cell appears on the cathode.
(f) The cell electrolyte provides a solution of mobile electrons.

21.5 Consider the following balanced redox reaction:

$$16H^+(aq) + 2MnO_4^-(aq) + 10Cl^-(aq) \longrightarrow 2Mn^{2+}(aq) + 5Cl_2(g) + 8H_2O(l)$$

(a) Which species is being oxidized?
(b) Which species is being reduced?
(c) Which species is the oxidizing agent?
(d) Which species is the reducing agent?
(e) From which species to which does electron transfer occur?
(f) Write the balanced molecular equation, with K^+ and SO_4^{2-} as the spectator ions.

21.6 Consider the following balanced redox reaction:

$$2CrO_2^-(aq) + 2H_2O(l) + 6ClO^-(aq) \longrightarrow 2CrO_4^{2-}(aq) + 3Cl_2(g) + 4OH^-(aq)$$

(a) Which species is being oxidized?
(b) Which species is being reduced?
(c) Which species is the oxidizing agent?
(d) Which species is the reducing agent?
(e) From which species to which does electron transfer occur?
(f) Write the balanced molecular equation, with Na^+ as the spectator ion.

21.7 Balance the following skeleton reactions and identify the oxidizing and reducing agents:
(a) $ClO_3^-(aq) + I^-(aq) \longrightarrow I_2(s) + Cl^-(aq)$ [acidic]
(b) $MnO_4^-(aq) + SO_3^{2-}(aq) \longrightarrow MnO_2(s) + SO_4^{2-}(aq)$ [basic]
(c) $MnO_4^-(aq) + H_2O_2(aq) \longrightarrow Mn^{2+}(aq) + O_2(g)$ [acidic]

21.8 Balance the following skeleton reactions and identify the oxidizing and reducing agents:
(a) $O_2(g) + NO(g) \longrightarrow NO_3^-(aq)$ [acidic]
(b) $CrO_4^{2-}(aq) + Cu(s) \longrightarrow Cr(OH)_3(s) + Cu(OH)_2(s)$ [basic]
(c) $AsO_4^{3-}(aq) + NO_2^-(aq) \longrightarrow AsO_2^-(aq) + NO_3^-(aq)$ [basic]

21.9 Balance the following skeleton reactions and identify the oxidizing and reducing agents:
(a) $Sb(s) + NO_3^-(aq) \longrightarrow Sb_4O_6(s) + NO(g)$ [acidic]
(b) $Mn^{2+}(aq) + BiO_3^-(aq) \longrightarrow MnO_4^-(aq) + Bi^{3+}(aq)$ [acidic]
(c) $Fe(OH)_2(s) + Pb(OH)_3^-(aq) \longrightarrow Fe(OH)_3(s) + Pb(s)$ [basic]

21.10 Balance the following skeleton reactions and identify the oxidizing and reducing agents:
(a) $BH_4^-(aq) + ClO_3^-(aq) \longrightarrow H_2BO_3^-(aq) + Cl^-(aq)$ [basic]

(b) $CrO_4^{2-}(aq) + N_2O(g) \longrightarrow Cr^{3+}(aq) + NO(g)$ [acidic]
(c) $Br_2(l) \longrightarrow BrO_3^-(aq) + Br^-(aq)$ [basic]

21.11 In many residential water systems, the aqueous Fe^{3+}concentration is high enough to stain sinks and turn drinking water light brown. The iron content is analyzed by first reducing the Fe^{3+} to Fe^{2+} and then titrating with MnO_4^- in acidic solution. Balance the skeleton reaction of the titration step:

$$Fe^{2+}(aq) + MnO_4^-(aq) \longrightarrow Mn^{2+}(aq) + Fe^{3+}(aq)$$

21.12 *Aqua regia,* a mixture of concentrated HNO_3 and HCl, was used centuries ago as a means to "dissolve" gold. The process is a redox reaction with this simplified skeleton reaction:

$$Au(s) + NO_3^-(aq) + Cl^-(aq) \longrightarrow AuCl_4^-(aq) + NO_2(g)$$

(a) Balance the reaction by the half-reaction method.
(b) What are the oxidizing and reducing agents?
(c) What is the function of HCl in aqua regia?

Voltaic Cells: Using Spontaneous Reactions to Generate Electrical Energy

(Sample Problem 21.2)

21.13 Consider the following general voltaic cell:

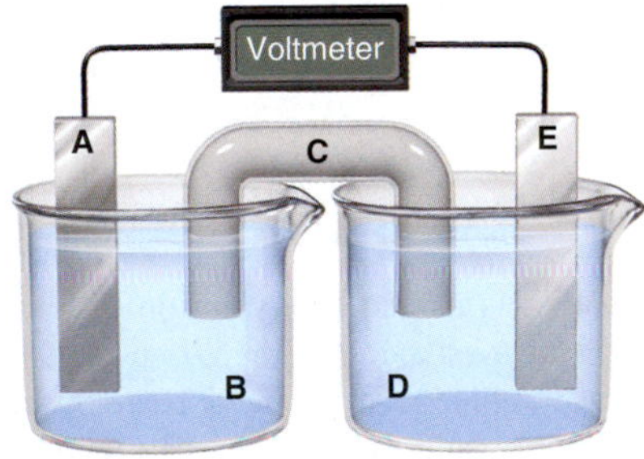

Identify the (a) anode, (b) cathode, (c) salt bridge, (d) electrode at which e^- leave the cell, (e) electrode with a positive charge, and (f) electrode that gains mass as the cell operates (assuming that a metal plates out).

21.14 Why does a voltaic cell not operate unless the two compartments are connected through an external circuit?

21.15 What purpose does the salt bridge serve in a voltaic cell, and how does it accomplish this purpose?

21.16 What is the difference between an active and an inactive electrode? Why are inactive electrodes used? Name two substances commonly used for inactive electrodes.

21.17 When a piece of metal A is placed in a solution containing ions of metal B, metal B plates out on the piece of A.
(a) Which metal is being oxidized?
(b) Which metal is being displaced?
(c) Which metal would you use as the anode in a voltaic cell incorporating these two metals?
(d) If bubbles of H_2 form when B is placed in acid, will they form if A is placed in acid? Explain.

21.18 A voltaic cell is constructed with an Sn/Sn^{2+} half-cell and a Zn/Zn^{2+} half-cell. The zinc electrode is negative.
(a) Write balanced half-reactions and the overall reaction.
(b) Diagram the cell, labeling electrodes with their charges and showing the directions of electron flow in the circuit and of cation and anion flow in the salt bridge.

21.19 A voltaic cell is constructed with an Ag/Ag^+ half-cell and a Pb/Pb^{2+} half-cell. The silver electrode is positive.
(a) Write balanced half-reactions and the overall reaction.
(b) Diagram the cell, labeling electrodes with their charges and showing the directions of electron flow in the circuit and of cation and anion flow in the salt bridge.

21.20 Consider the following voltaic cell:

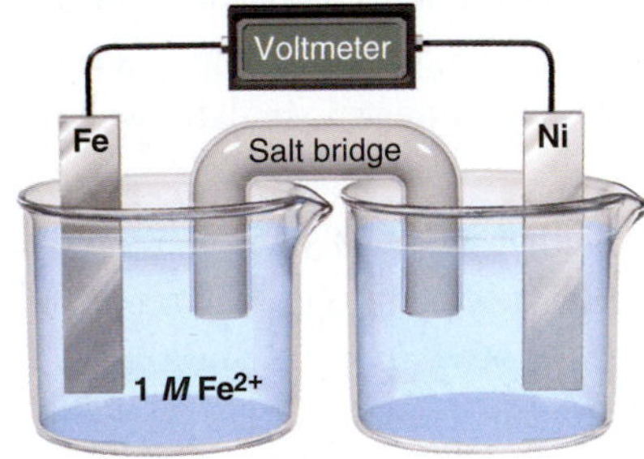

(a) In which direction do electrons flow in the external circuit?
(b) In which half-cell does oxidation occur?
(c) In which half-cell do electrons enter the cell?
(d) At which electrode are electrons consumed?
(e) Which electrode is negatively charged?
(f) Which electrode decreases in mass during cell operation?
(g) Suggest a solution for the cathode electrolyte.
(h) Suggest a pair of ions for the salt bridge.
(i) For which electrode could you use an inactive material?
(j) In which direction do anions within the salt bridge move to maintain charge neutrality?
(k) Write balanced half-reactions and the overall cell reaction.

21.21 Consider the following voltaic cell:

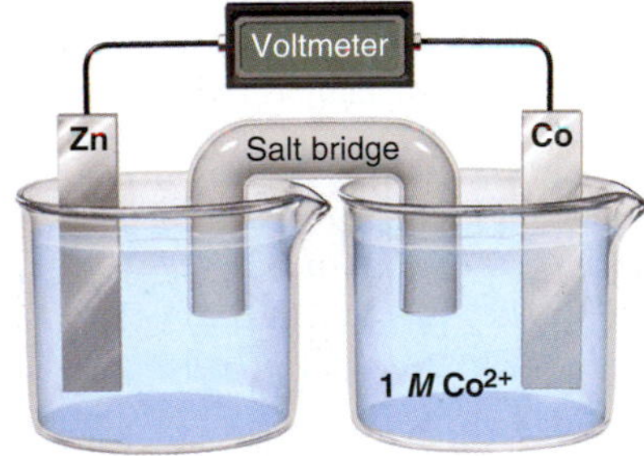

(a) In which direction do electrons flow in the external circuit?
(b) In which half-cell does reduction occur?
(c) In which half-cell do electrons leave the cell?
(d) At which electrode are electrons generated?
(e) Which electrode is positively charged?
(f) Which electrode increases in mass during cell operation?
(g) Suggest a solution for the anode electrolyte.
(h) Suggest a pair of ions for the salt bridge.
(i) For which electrode could you use an inactive material?
(j) In which direction do cations within the salt bridge move to maintain charge neutrality?
(k) Write balanced half-reactions and the overall cell reaction.

21.22 Write the cell notation for the voltaic cell that incorporates each of the following redox reactions:
(a) $Al(s) + Cr^{3+}(aq) \longrightarrow Al^{3+}(aq) + Cr(s)$
(b) $Cu^{2+}(aq) + SO_2(g) + 2H_2O(l) \longrightarrow Cu(s) + SO_4^{2-}(aq) + 4H^+(aq)$

21.23 Write a balanced equation from each cell notation:
(a) $Mn(s) \mid Mn^{2+}(aq) \parallel Cd^{2+}(aq) \mid Cd(s)$
(b) $Fe(s) \mid Fe^{2+}(aq) \parallel NO_3^-(aq) \mid NO(g) \mid Pt(s)$

Cell Potential: Output of a Voltaic Cell

(Sample Problems 21.3 and 21.4)

21.24 How is a standard reference electrode used to determine unknown $E^\circ_{\text{half-cell}}$ values?

21.25 What does a negative E°_{cell} indicate about a redox reaction? What does it indicate about the reverse reaction?

21.26 The standard cell potential is a thermodynamic state function. How are E° values treated similarly to ΔH°, ΔG°, and S° values? How are they treated differently?

21.27 In basic solution, Se^{2-} and SO_3^{2-} ions react spontaneously:

$$2Se^{2-}(aq) + 2SO_3^{2-}(aq) + 3H_2O(l) \longrightarrow 2Se(s) + 6OH^-(aq) + S_2O_3^{2-}(aq) \quad E^\circ_{\text{cell}} = 0.35\text{ V}$$

(a) Write balanced half-reactions for the process.
(b) If E°_{sulfite} is -0.57 V, calculate $E^\circ_{\text{selenium}}$.

21.28 In acidic solution, O_3 and Mn^{2+} ion react spontaneously:

$$O_3(g) + Mn^{2+}(aq) + H_2O(l) \longrightarrow O_2(g) + MnO_2(s) + 2H^+(aq) \quad E^\circ_{\text{cell}} = 0.84\text{ V}$$

(a) Write the balanced half-reactions.
(b) Using Appendix D to find E°_{ozone}, calculate $E^\circ_{\text{manganese}}$.

21.29 Use the emf series (Appendix D) to arrange the species.
(a) In order of *decreasing* strength as *oxidizing* agents: Fe^{3+}, Br_2, Cu^{2+}
(b) In order of *increasing* strength as *oxidizing* agents: Ca^{2+}, $Cr_2O_7^{2-}$, Ag^+

21.30 Use the emf series (Appendix D) to arrange the species.
(a) In order of *decreasing* strength as *reducing* agents: SO_2, $PbSO_4$, MnO_2
(b) In order of *increasing* strength as *reducing* agents: Hg, Fe, Sn

21.31 Balance each skeleton reaction, calculate E°_{cell}, and state whether the reaction is spontaneous:
(a) $Co(s) + H^+(aq) \longrightarrow Co^{2+}(aq) + H_2(g)$
(b) $Mn^{2+}(aq) + Br_2(l) \longrightarrow MnO_4^-(aq) + Br^-(aq)$ [acidic]
(c) $Hg_2^{2+}(aq) \longrightarrow Hg^{2+}(aq) + Hg(l)$

21.32 Balance each skeleton reaction, calculate E°_{cell}, and state whether the reaction is spontaneous:
(a) $Cl_2(g) + Fe^{2+}(aq) \longrightarrow Cl^-(aq) + Fe^{3+}(aq)$
(b) $Mn^{2+}(aq) + Co^{3+}(aq) \longrightarrow MnO_2(s) + Co^{2+}(aq)$ [acidic]
(c) $AgCl(s) + NO(g) \longrightarrow Ag(s) + Cl^-(aq) + NO_3^-(aq)$ [acidic]

21.33 Balance each skeleton reaction, calculate E°_{cell}, and state whether the reaction is spontaneous:
(a) $Ag(s) + Cu^{2+}(aq) \longrightarrow Ag^+(aq) + Cu(s)$
(b) $Cd(s) + Cr_2O_7^{2-}(aq) \longrightarrow Cd^{2+}(aq) + Cr^{3+}(aq)$
(c) $Ni^{2+}(aq) + Pb(s) \longrightarrow Ni(s) + Pb^{2+}(aq)$

21.34 Balance each skeleton reaction, calculate E°_{cell}, and state whether the reaction is spontaneous:
(a) $Cu^+(aq) + PbO_2(s) + SO_4^{2-}(aq) \longrightarrow PbSO_4(s) + Cu^{2+}(aq)$ [acidic]
(b) $H_2O_2(aq) + Ni^{2+}(aq) \longrightarrow O_2(g) + Ni(s)$ [acidic]
(c) $MnO_2(s) + Ag^+(aq) \longrightarrow MnO_4^-(aq) + Ag(s)$ [basic]

21.35 Use the following half-reactions to write three spontaneous reactions, calculate E°_{cell} for each reaction, and rank the strengths of the oxidizing and reducing agents:
(1) $Al^{3+}(aq) + 3e^- \longrightarrow Al(s) \quad E^\circ = -1.66\text{ V}$
(2) $N_2O_4(g) + 2e^- \longrightarrow 2NO_2^-(aq) \quad E^\circ = 0.867\text{ V}$
(3) $SO_4^{2-}(aq) + H_2O(l) + 2e^- \longrightarrow SO_3^{2-}(aq) + 2OH^-(aq) \quad E^\circ = 0.93\text{ V}$

21.36 Use the following half-reactions to write three spontaneous reactions, calculate E°_{cell} for each reaction, and rank the strengths of the oxidizing and reducing agents:
(1) $Au^+(aq) + e^- \longrightarrow Au(s) \quad E^\circ = 1.69\text{ V}$
(2) $N_2O(g) + 2H^+(aq) + 2e^- \longrightarrow N_2(g) + H_2O(l) \quad E^\circ = 1.77\text{ V}$
(3) $Cr^{3+}(aq) + 3e^- \longrightarrow Cr(s) \quad E^\circ = -0.74\text{ V}$

21.37 Use the following half-reactions to write three spontaneous reactions, calculate E°_{cell} for each reaction, and rank the strengths of the oxidizing and reducing agents:
(1) $2HClO(aq) + 2H^+(aq) + 2e^- \longrightarrow Cl_2(g) + 2H_2O(l) \quad E^\circ = 1.63\text{ V}$
(2) $Pt^{2+}(aq) + 2e^- \longrightarrow Pt(s) \quad E^\circ = 1.20\text{ V}$
(3) $PbSO_4(s) + 2e^- \longrightarrow Pb(s) + SO_4^{2-}(aq) \quad E^\circ = -0.31\text{ V}$

21.38 Use the following half-reactions to write three spontaneous reactions, calculate E°_{cell} for each reaction, and rank the strengths of the oxidizing and reducing agents:
(1) $I_2(s) + 2e^- \longrightarrow 2I^-(aq) \quad E^\circ = 0.53\text{ V}$
(2) $S_2O_8^{2-}(aq) + 2e^- \longrightarrow 2SO_4^{2-}(aq) \quad E^\circ = 2.01\text{ V}$
(3) $Cr_2O_7^{2-}(aq) + 14H^+(aq) + 6e^- \longrightarrow 2Cr^{3+}(aq) + 7H_2O(l) \quad E^\circ = 1.33\text{ V}$

21.39 When metal A is placed in a solution of a salt of metal B, the surface of metal A changes color. When metal B is placed in acid solution, gas bubbles form on the surface of the metal. When metal A is placed in a solution of a salt of metal C, no change is observed in the solution or on the surface of metal A. Will metal C cause formation of H_2 when placed in acid solution? Rank metals A, B, and C in order of *decreasing* reducing strength.

21.40 When a clean iron nail is placed in an aqueous solution of copper(II) sulfate, the nail becomes coated with a brownish black material.
(a) What is the material coating the iron?
(b) What are the oxidizing and reducing agents?
(c) Can this reaction be made into a voltaic cell?
(d) Write the balanced equation for the reaction.
(e) Calculate E°_{cell} for the process.

Free Energy and Electrical Work

(Sample Problems 21.5 to 21.7)

21.41 (a) How do the relative magnitudes of Q and K relate to the signs of ΔG and E_{cell}? Explain.
(b) Can a cell do work when $Q/K > 1$ or $Q/K < 1$? Explain.

21.42 A voltaic cell consists of A/A^+ and B/B^+ half-cells, where A and B are metals and the A electrode is negative. The initial $[A^+]/[B^+]$ is such that $E_{\text{cell}} > E^\circ_{\text{cell}}$.
(a) How do $[A^+]$ and $[B^+]$ change as the cell operates?
(b) How does E_{cell} change as the cell operates?
(c) What is $[A^+]/[B^+]$ when $E_{\text{cell}} = E^\circ_{\text{cell}}$? Explain.
(d) Is it possible for E_{cell} to be less than E°_{cell}? Explain.

21.43 Explain whether E_{cell} of a voltaic cell will increase or decrease with each of the following changes:
(a) Decrease in cell temperature
(b) Increase in [active ion] in the anode compartment
(c) Increase in [active ion] in the cathode compartment

(d) Increase in pressure of a gaseous reactant in the cathode compartment

21.44 In a concentration cell, is the more concentrated electrolyte in the cathode or the anode compartment? Explain.

21.45 What is the value of the equilibrium constant for the reaction between each pair at 25°C?
(a) $Ni(s)$ and $Ag^+(aq)$ (b) $Fe(s)$ and $Cr^{3+}(aq)$

21.46 What is the value of the equilibrium constant for the reaction between each pair at 25°C?
(a) $Al(s)$ and $Cd^{2+}(aq)$ (b) $I_2(s)$ and $Br^-(aq)$

21.47 Calculate $\Delta G°$ for each of the reactions in Problem 21.45.

21.48 Calculate $\Delta G°$ for each of the reactions in Problem 21.46.

21.49 What are $E°_{cell}$ and $\Delta G°$ of a redox reaction at 25°C for which $n = 1$ and $K = 5.0\times10^4$?

21.50 What are $E°_{cell}$ and $\Delta G°$ of a redox reaction at 25°C for which $n = 2$ and $K = 0.065$?

21.51 A voltaic cell consists of a standard reference half-cell and a Cu/Cu^{2+} half-cell. Calculate $[Cu^{2+}]$ when E_{cell} is 0.22 V.

21.52 A voltaic cell consists of an Mn/Mn^{2+} half-cell and a Pb/Pb^{2+} half-cell. Calculate $[Pb^{2+}]$ when $[Mn^{2+}]$ is 1.4 M and E_{cell} is 0.44 V.

21.53 A voltaic cell with Ni/Ni^{2+} and Co/Co^{2+} half-cells has the following initial concentrations: $[Ni^{2+}] = 0.80\ M$; $[Co^{2+}] = 0.20\ M$.
(a) What is the initial E_{cell}?
(b) What is $[Ni^{2+}]$ when E_{cell} reaches 0.03 V?
(c) What are the equilibrium concentrations of the ions?

21.54 A voltaic cell with Mn/Mn^{2+} and Cd/Cd^{2+} half-cells has the following initial concentrations: $[Mn^{2+}] = 0.090\ M$; $[Cd^{2+}] = 0.060\ M$.
(a) What is the initial E_{cell}?
(b) What is E_{cell} when $[Cd^{2+}]$ reaches 0.050 M?
(c) What is $[Mn^{2+}]$ when E_{cell} reaches 0.055 V?
(d) What are the equilibrium concentrations of the ions?

21.55 A concentration cell consists of two H_2/H^+ half-cells. Half-cell A has H_2 at 0.95 atm bubbling into 0.10 M HCl. Half-cell B has H_2 at 0.60 atm bubbling into 2.0 M HCl. Which half-cell houses the anode? What is the voltage of the cell?

21.56 A concentration cell consists of two Sn/Sn^{2+} half-cells, A and B. The electrolyte in A is 0.13 M $Sn(NO_3)_2$. The electrolyte in B is 0.87 M $Sn(NO_3)_2$. Which half-cell houses the cathode? What is the voltage of the cell?

Electrochemical Processes in Batteries

21.57 What is the direction of electron flow with respect to the anode and the cathode in a battery? Explain.

21.58 Both a D-sized and an AAA-sized alkaline battery have an output of 1.5 V. What property of the cell potential allows this to occur? What is different about these two batteries?

21.59 Many common electrical devices require the use of more than one battery.
(a) How many alkaline batteries must be placed in series to light a flashlight with a 6.0-V bulb?
(b) What is the voltage requirement of a camera that uses six silver batteries?
(c) How many volts can a car battery deliver if two of its anode/cathode cells are shorted?

Corrosion: An Environmental Voltaic Cell

21.60 During reconstruction of the Statue of Liberty, Teflon spacers were placed between the iron skeleton and the copper plates that cover the statue. What purpose do these spacers serve?

21.61 Why do steel bridge-supports rust at the waterline but not above or below it?

21.62 Which of the following metals are suitable for use as sacrificial anodes to protect against corrosion of underground iron pipes? If any are not suitable, explain why:
(a) Aluminum (b) Magnesium (c) Sodium (d) Lead
(e) Nickel (f) Zinc (g) Chromium

Electrolytic Cells: Using Electrical Energy to Drive Nonspontaneous Reactions

(Sample Problems 21.8 and 21.9)

Note: Unless stated otherwise, assume that the electrolytic cells in the following problems operate at 100% efficiency.

21.63 Consider the following general electrolytic cell:

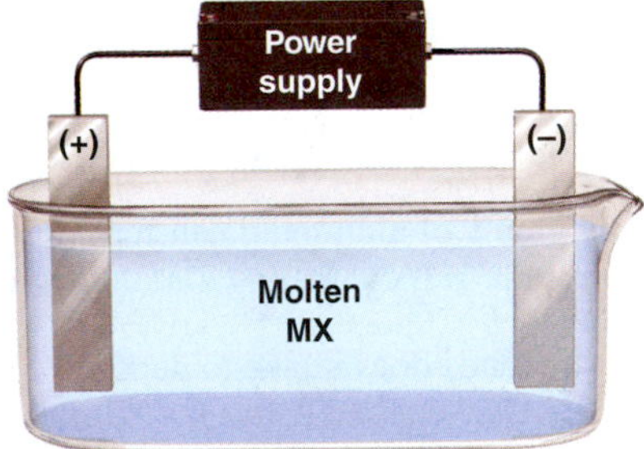

(a) At which electrode does oxidation occur?
(b) At which electrode does elemental M form?
(c) At which electrode are electrons being released by ions?
(d) At which electrode are electrons entering the cell?

21.64 A voltaic cell consists of Cr/Cr^{3+} and Cd/Cd^{2+} half-cells with all components in their standard states. After 10 minutes of operation, a thin coating of cadmium metal has plated out on the cathode. Describe what will happen if you attach the negative terminal of a dry cell (1.5 V) to the cell cathode and the positive terminal to the cell anode.

21.65 Why are $E_{half\text{-}cell}$ values for the oxidation and reduction of water different from $E°_{half\text{-}cell}$ values for the same processes?

21.66 In an aqueous electrolytic cell, nitrate ions never react at the anode, but nitrite ions do. Explain.

21.67 How does overvoltage influence the products in the electrolysis of aqueous salts?

21.68 What property allows copper to be purified in the presence of iron and nickel impurities? Explain.

21.69 What is the practical reason for using cryolite in the electrolysis of aluminum oxide?

21.70 In the electrolysis of molten NaBr,
(a) What product forms at the anode?
(b) What product forms at the cathode?

21.71 In the electrolysis of molten BaI_2,
(a) What product forms at the negative electrode?
(b) What product forms at the positive electrode?

21.72 Identify the elements that can be prepared by electrolysis of their aqueous salts: copper, barium, aluminum, bromine.

21.73 Identify the elements that can be prepared by electrolysis of their aqueous salts: strontium, gold, tin, chlorine.

21.74 What product forms at each electrode in the aqueous electrolysis of the following salts: (a) LiF; (b) $SnSO_4$?

21.75 What product forms at each electrode in the aqueous electrolysis of the following salts: (a) $Cr(NO_3)_3$; (b) $MnCl_2$?

21.76 Electrolysis of molten $MgCl_2$ is the final production step in the isolation of magnesium from seawater by the Dow process. Assuming that 45.6 g of Mg metal forms,
(a) How many moles of electrons are required?
(b) How many coulombs are required?
(c) How many amps will produce this amount in 3.50 h?

21.77 Electrolysis of molten NaCl in a Downs cell is the major isolation step in the production of sodium metal. Assuming that 215 g of Na metal forms,
(a) How many moles of electrons are required?
(b) How many coulombs are required?
(c) How many amps will produce this amount in 9.50 h?

21.78 How many grams of radium can form by passing 235 C through an electrolytic cell containing a molten radium salt?

21.79 How many grams of aluminum can form by passing 305 C through an electrolytic cell containing a molten aluminum salt?

21.80 How many seconds does it take to deposit 65.5 g of Zn on a steel gate when 21.0 A is passed through a $ZnSO_4$ solution?

21.81 How many seconds does it take to deposit 1.63 g of Ni on a decorative drawer handle when 13.7 A is passed through a $Ni(NO_3)_2$ solution?

21.82 A professor adds Na_2SO_4 to water to facilitate its electrolysis in a lecture demonstration. (a) What is the purpose of the Na_2SO_4? (b) Why is the water electrolyzed instead of the salt?

21.83 A Downs cell operating at 75.0 A produces 30.0 kg of Na.
(a) What volume of $Cl_2(g)$ is produced at 1.0 atm and 580.°C?
(b) How many coulombs were passed through the cell?
(c) How long did the cell operate?

21.84 Zinc plating (galvanizing) is an important means of corrosion protection. Although the process is done customarily by dipping the object into molten zinc, the metal can also be electroplated from aqueous solutions. How many grams of zinc can be deposited on a steel tank from a $ZnSO_4$ solution when a 0.855-A current flows for 2.50 days?

Comprehensive Problems

21.85 The MnO_2 used in alkaline batteries can be produced by an electrochemical process of which one half-reaction is

$$Mn^{2+}(aq) + 2H_2O(l) \longrightarrow MnO_2(s) + 4H^+(aq) + 2e^-$$

If a current of 25.0 A is used, how many hours are needed to produce 1.00 kg of MnO_2? At which electrode is the MnO_2 formed?

21.86 The overall cell reaction occurring in an alkaline battery is

$$Zn(s) + MnO_2(s) + H_2O(l) \longrightarrow ZnO(s) + Mn(OH)_2(s)$$

(a) How many moles of electrons flow per mole of reaction?
(b) If 4.50 g of zinc is oxidized, how many grams of manganese dioxide and of water are consumed?
(c) What is the total mass of reactants consumed in part (b)?
(d) How many coulombs are produced in part (b)?
(e) In practice, voltaic cells of a given capacity (coulombs) are heavier than the calculation in part (c) indicates. Explain.

21.87 Brass, an alloy of copper and zinc, can be produced by simultaneously electroplating the two metals from a solution containing their 2+ ions. If 65.0% of the total current is used to plate copper, while 35.0% goes to plating zinc, what is the mass percent of copper in the brass?

21.88 Compare and contrast a voltaic cell and an electrolytic cell with respect to each of the following:
(a) Sign of the free energy change
(b) Nature of the half-reaction at the anode
(c) Nature of the half-reaction at the cathode
(d) Charge on the electrode labeled "anode"
(e) Electrode from which electrons leave the cell

21.89 A thin circular-disk earring 4.00 cm in diameter is plated with a coating of gold 0.25 mm thick from an Au^{3+} bath.
(a) How many days does it take to deposit the gold on one side of one earring if the current is 0.013 A (d of gold = 19.3 g/cm^3)?
(b) How many days does it take to deposit the gold on both sides of the pair of earrings?
(c) If the price of gold in 2010 is \$1210 per troy ounce (31.10 g), what is the total cost of the gold plating?

21.90 (a) How many minutes does it take to form 10.0 L of O_2 measured at 99.8 kPa and 28°C from water if a current of 1.3 A passes through the electrolytic cell? (b) What mass of H_2 forms?

21.91 A silver button battery used in a watch contains 0.75 g of zinc and can run until 80% of the zinc is consumed.
(a) How many days can the battery run at a current of 0.85 microamps (10^{-6} amps)?
(b) When the battery dies, 95% of the Ag_2O has been consumed. How many grams of Ag were used to make the battery?
(c) If Ag costs \$23.00 per troy ounce (31.10 g) in 2010, what is the cost of the Ag consumed each day the watch runs?

21.92 If a chlor-alkali cell used a current of 3×10^4 A, how many pounds of Cl_2 would be produced in a typical 8-h operating day?

21.93 To improve conductivity in the electroplating of automobile bumpers, a thin coating of copper separates the steel from a heavy coating of chromium.
(a) What mass of Cu is deposited on an automobile trim piece if plating continues for 1.25 h at a current of 5.0 A?
(b) If the area of the trim piece is 50.0 cm^2, what is the thickness of the Cu coating (d of Cu = 8.95 g/cm^3)?

21.94 Commercial electrolytic cells for producing aluminum operate at 5.0 V and 100,000 A.
(a) How long does it take to produce exactly 1 metric ton (1000 kg) of aluminum?
(b) How much electrical power (in kilowatt-hours, kW·h) is used [1 W = 1 J/s; 1 kW·h = 3.6×10^3 kJ]?
(c) If electricity costs \$0.123 per kW·h and cell efficiency is 90.%, what is the cost of electricity to produce exactly 1 lb of aluminum?

21.95 Magnesium bars are connected electrically to underground iron pipes to serve as sacrificial anodes.
(a) Do electrons flow from the bar to the pipe or the reverse?
(b) A 12-kg Mg bar is attached to an iron pipe, and it takes 8.5 yr for the Mg to be consumed. What is the average current flowing between the Mg and the Fe during this period?

21.96 Bubbles of H_2 form when metal D is placed in hot H_2O. No reaction occurs when D is placed in a solution of a salt of metal E, but D is discolored and coated immediately when placed in a solution of a salt of metal F. What happens if E is placed in a solution of a salt of metal F? Rank metals D, E, and F in order of *increasing* reducing strength.

21.97 The following reactions are used in batteries:

I $\quad 2H_2(g) + O_2(g) \longrightarrow 2H_2O(l) \qquad E_{cell} = 1.23\text{ V}$

II $\quad Pb(s) + PbO_2(s) + 2H_2SO_4(aq) \longrightarrow 2PbSO_4(s) + 2H_2O(l) \qquad E_{cell} = 2.04\text{ V}$

III $\quad 2Na(l) + FeCl_2(s) \longrightarrow 2NaCl(s) + Fe(s) \qquad E_{cell} = 2.35\text{ V}$

Reaction I is used in fuel cells, II in the automobile lead-acid battery, and III in an experimental high-temperature battery for powering electric vehicles. The aim is to obtain as much work as possible from a cell, while keeping its weight to a minimum.
(a) In each cell, find the moles of electrons transferred and ΔG.
(b) Calculate the ratio, in kJ/g, of w_{max} to mass of reactants for each of the cells. Which has the highest ratio, which the lowest, and why? (*Note:* For simplicity, ignore the masses of cell components that do not appear in the cell as reactants, including electrode materials, electrolytes, separators, cell casing, wiring, etc.)

21.98 From the skeleton equations below, create a list of balanced half-reactions in which the strongest oxidizing agent is on top and the weakest is on the bottom:

$$U^{3+}(aq) + Cr^{3+}(aq) \longrightarrow Cr^{2+}(aq) + U^{4+}(aq)$$
$$Fe(s) + Sn^{2+}(aq) \longrightarrow Sn(s) + Fe^{2+}(aq)$$
$$Fe(s) + U^{4+}(aq) \longrightarrow \text{no reaction}$$
$$Cr^{3+}(aq) + Fe(s) \longrightarrow Cr^{2+}(aq) + Fe^{2+}(aq)$$
$$Cr^{2+}(aq) + Sn^{2+}(aq) \longrightarrow Sn(s) + Cr^{3+}(aq)$$

21.99 Use Appendix D to calculate the K_{sp} of AgCl.

21.100 Calculate the K_f of $Ag(NH_3)_2^+$ from

$$Ag^+(aq) + e^- \rightleftharpoons Ag(s) \qquad E° = 0.80\text{ V}$$
$$Ag(NH_3)_2^+(aq) + e^- \rightleftharpoons Ag(s) + 2NH_3(aq) \qquad E° = 0.37\text{ V}$$

21.101 Use Appendix D to create an activity series of Mn, Fe, Ag, Sn, Cr, Cu, Ba, Al, Na, Hg, Ni, Li, Au, Zn, and Pb. Rank these metals in order of decreasing reducing strength, and divide them into three groups: those that displace H_2 from water, those that displace H_2 from acid, and those that cannot displace H_2.

21.102 The overall cell reaction for aluminum production is

$$2Al_2O_3(\text{in } Na_3AlF_6) + 3C(\text{graphite}) \longrightarrow 4Al(l) + 3CO_2(g)$$

(a) Assuming 100% efficiency, how many metric tons (t) of Al_2O_3 are consumed per metric ton of Al produced?
(b) Assuming 100% efficiency, how many metric tons of the graphite anode are consumed per metric ton of Al produced?
(c) Actual conditions in an aluminum plant require 1.89 t of Al_2O_3 and 0.45 t of graphite per metric ton of Al. What is the percent yield of Al with respect to Al_2O_3?
(d) What is the percent yield of Al with respect to graphite?
(e) What volume of CO_2 (in m^3) is produced per metric ton of Al at operating conditions of 960.°C and exactly 1 atm?

21.103 Two concentration cells are prepared, both with 90.0 mL of 0.0100 *M* $Cu(NO_3)_2$ and a Cu bar in each half-cell.
(a) In the first concentration cell, 10.0 mL of 0.500 *M* NH_3 is added to one half-cell; the complex ion $Cu(NH_3)_4^{2+}$ forms, and E_{cell} is 0.129 V. Calculate K_f for the formation of the complex ion.
(b) Calculate E_{cell} when an additional 10.0 mL of 0.500 *M* NH_3 is added.
(c) In the second concentration cell, 10.0 mL of 0.500 *M* NaOH is added to one half-cell; the precipitate $Cu(OH)_2$ forms ($K_{sp} = 2.2\times10^{-20}$). Calculate $E°_{cell}$.
(d) What would the molarity of NaOH have to be for the addition of 10.0 mL to result in an $E°_{cell}$ of 0.340 V?

21.104 A voltaic cell has one half-cell with a Cu bar in a 1.00 *M* Cu^{2+}salt, and the other half-cell with a Cd bar in the same volume of a 1.00 *M* Cd^{2+}salt. (a) Find $E°_{cell}$, $\Delta G°$, and *K*. (b) As the cell operates, $[Cd^{2+}]$ increases; find E_{cell} and ΔG when $[Cd^{2+}]$ is 1.95 *M*. (c) Find E_{cell}, ΔG, and $[Cu^{2+}]$ at equilibrium.

21.105 If the E_{cell} of the following cell is 0.915 V, what is the pH in the anode compartment?

$$Pt(s) \mid H_2(1.00\text{ atm}) \mid H^+(aq) \parallel Ag^+(0.100\ M) \mid Ag(s)$$

22 Transition Elements and Their Coordination Compounds

Key Principles
to focus on while studying this chapter

- In the *transition elements* (*d* block) and *inner transition elements* (*f* block), inner orbitals are being filled, which results in horizontal and vertical trends in atomic properties that differ markedly from those of the main-group elements. ***(Section 22.1)***
- Outer *ns* electrons are close in energy to the inner $(n - 1)d$ electrons, which allows transition elements to use several different numbers of electrons in bonding. For this reason, transition elements have *multiple oxidation states,* in which the lower states display more *metallic* behavior (ionic bonding and basic oxides). The compounds of ions with a partially filled *d* sublevel are *colored* and *paramagnetic.* ***(Section 22.1)***
- Many transition elements form *coordination compounds,* which consist of a *complex ion* and *counter ions.* A complex ion has a *central metal ion* and surrounding molecular or anionic *ligands.* The number of ligands bonded to the metal ion determines the *shape* of the complex ion. Different positions and bonding arrangements of ligands lead to various types of *isomerism.* ***(Section 22.2)***
- According to *valence bond theory,* the shapes of complex ions arise from *hybridization* of different combinations of $(n - 1)d$, *ns,* and *np* orbitals. ***(Section 22.3)***
- According to *crystal field theory,* ligands approaching a metal ion *split the energies of its d orbitals,* creating two sets of orbitals. Each type of ligand causes a characteristic difference (*crystal field splitting energy,* Δ) between the energies of the two sets, which allows us to rank ligands in a *spectrochemical series.* The energy difference between the two sets of *d* orbitals is related to the color of the compound. It also determines the *electron occupancy* of the two sets, which influences the magnetic behavior of the compound. ***(Section 22.3)***

Useful Metals with Fascinating Compounds Iron (shown being made into steel), chromium, and copper are a few of the essential metals of modern life. As we highlight in this chapter, these elements occupy most of the periodic table, and their compounds provide a deeper understanding of structure and bonding.

Outline

Though almost at the end of the text, we still haven't discussed the majority of the elements *and* some of the most familiar. The **transition elements** *(transition metals)* make up the *d* block (B groups) and *f* block *(inner transition elements)* (Figure 22.1) and have crucial uses in industry and biology.

Some indispensable transition elements are iron (steel), copper (wiring), chromium (plumbing fixtures), gold and silver (jewelry and electronics), platinum (catalytic converters), titanium (bicycle and aircraft parts), nickel (coins and catalysts), and zinc (batteries), to mention a few of the better known ones. There are also the lesser known zirconium (nuclear-reactor liners), vanadium (axles and crankshafts), molybdenum (boiler plates), tantalum (organ-replacement parts), palladium (telephone-relay contacts)—the list goes on and on. As ions, quite a few of these elements also play vital roles in organisms.

In this chapter, we focus almost entirely on the *d*-block elements. We first discuss some properties of the elements and then focus on the most distinctive feature of their chemistry, the formation of coordination compounds—substances that contain complex ions and offer new insights into chemical bonding.

CONCEPTS & SKILLS TO REVIEW before studying this chapter

- properties of light (Section 7.1)
- electron shielding of nuclear charge (Section 8.1)
- electron configuration, ionic size, and magnetic behavior (Sections 8.2 to 8.5)
- valence bond theory (Section 11.1)
- constitutional, geometric, and optical isomerism (Section 15.2)
- Lewis acid-base concepts (Section 18.8)
- complex-ion formation (Section 19.4)
- redox behavior and standard electrode potentials (Section 21.3)

22.1 • PROPERTIES OF THE TRANSITION ELEMENTS

The transition elements differ considerably in physical and chemical behavior from the main-group elements:

- *All transition elements are metals,* whereas main-group elements in each period change from metal to nonmetal.
- *Many transition metal compounds are colored and paramagnetic,* whereas most main-group ionic compounds are colorless and diamagnetic.

We first discuss electron configurations of the atoms and ions, and then examine key properties of transition elements to see how they contrast with the same properties among the main-group elements.

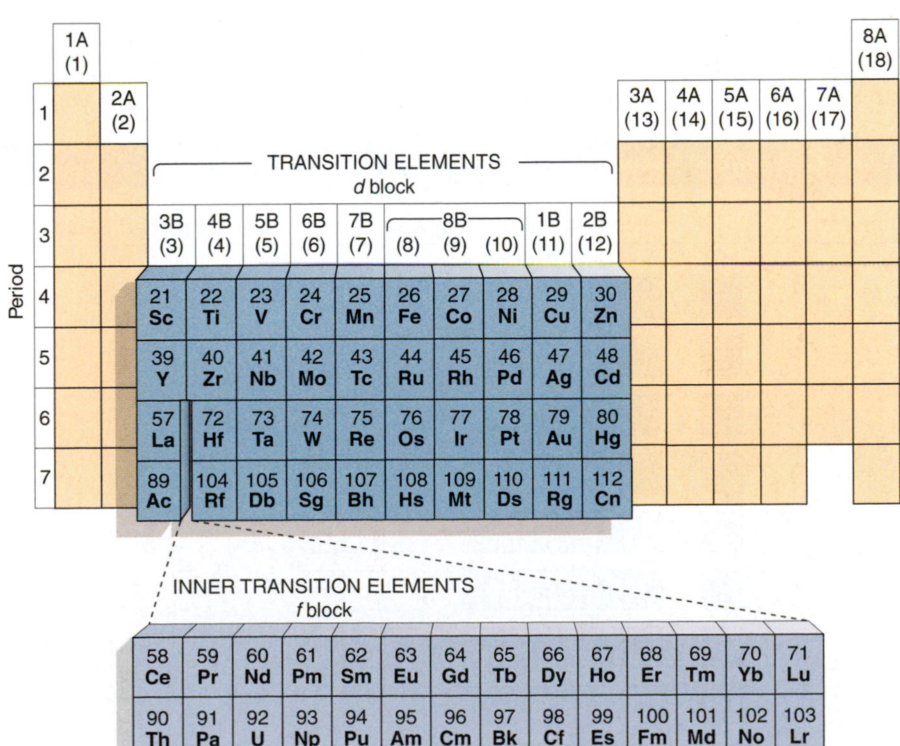

Figure 22.1 The transition elements (*d* block) and inner transition elements (*f* block) in the periodic table.

Scandium, Sc; 3B(3)

Titanium, Ti; 4B(4)

Vanadium, V; 5B(5)

Chromium, Cr; 6B(6)

Manganese, Mn; 7B(7)

Figure 22.2 The Period 4 transition metals. The ten elements appear in periodic-table order.

Electron Configurations of the Transition Metals and Their Ions

As Figure 22.1 shows, *d*-block (B-group) elements occur in four series that lie within Periods 4 through 7. Each series represents the filling of five *d* orbitals and, thus, contains ten elements, for a total of 40 transition elements. The first of these series occurs in Period 4 and consists of scandium (Sc) through zinc (Zn) (Figure 22.2). Lying between the first and second members of the series in Periods 6 and 7 are the inner transition elements, in which *f* orbitals are filled. Two points are important to review:

1. *Electron configurations of the atoms*. Despite several exceptions, in general, the *condensed* ground-state electron configuration for the atoms in each *d*-block series is

$$[\text{noble gas}]\, ns^2(n-1)d^x, \text{ with } n = 4 \text{ to } 7 \text{ and } x = 1 \text{ to } 10$$

In Periods 6 and 7, the condensed configuration includes the *f* sublevel:

$$[\text{noble gas}]\, ns^2(n-2)f^{14}(n-1)d^x, \text{ with } n = 6 \text{ or } 7$$

The *partial* (valence-level) electron configuration for the *d*-block elements excludes the noble gas core and the filled inner *f* sublevel:

$$ns^2(n-1)d^x$$

2. *Electron configurations of the ions*. Transition metal ions form through *the loss of ns electrons before (n − 1)d electrons*. Thus, the electron configuration of Ti^{2+} is [Ar] $3d^2$, *not* [Ar] $4s^2$, and Ti^{2+} is referred to as a d^2 ion. Ions of different metals with the same configuration often have similar properties: for example, both Mn^{2+} and Fe^{3+} (d^5 ions) have pale colors in solution and form complex ions with similar magnetic properties.

Table 22.1 shows partial orbital box diagrams for the Period 4 elements. In general, for all periods, the number of *unpaired* electrons (or half-filled orbitals) *increases*

Table 22.1 Orbital Occupancy of the Period 4 Transition Metals

Element	4s	3d					4p			Unpaired Electrons
Sc	↑↓	↑								1
Ti	↑↓	↑	↑							2
V	↑↓	↑	↑	↑						3
Cr	↑	↑	↑	↑	↑	↑				6
Mn	↑↓	↑	↑	↑	↑	↑				5
Fe	↑↓	↑↓	↑	↑	↑	↑				4
Co	↑↓	↑↓	↑↓	↑	↑	↑				3
Ni	↑↓	↑↓	↑↓	↑↓	↑	↑				2
Cu	↑	↑↓	↑↓	↑↓	↑↓	↑↓				1
Zn	↑↓	↑↓	↑↓	↑↓	↑↓	↑↓				0

Iron, Fe; 8B(8)

Cobalt, Co; 8B(9)

Nickel, Ni; 8B(10)

Copper, Cu; 1B(11)

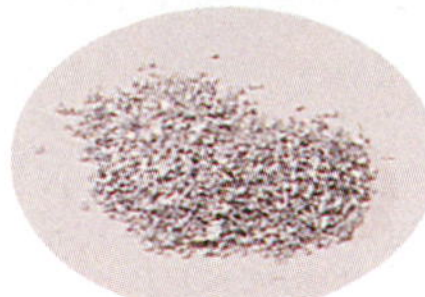
Zinc, Zn; 2B(12)

in the first half of the series and, when pairing begins, *decreases in the second half.* As we would expect, it is the electron configuration of the transition metal *atom* that correlates with the properties of the *element,* such as density and magnetic behavior, whereas it is the electron configuration of the *ion* that correlates with the color and magnetic properties of the *compounds.*

Sample Problem 22.1 Writing Electron Configurations of Transition Metal Atoms and Ions

Problem Write *condensed* electron configurations for the following: **(a)** Zr; **(b)** V^{3+}; **(c)** Mo^{3+}. (Assume that elements in higher periods behave like those in Period 4.)

Plan We locate the element in the periodic table and count its position in the respective transition series. These elements are in Periods 4 and 5, so the general configuration is [noble gas] $ns^2(n-1)d^x$. For the ions, we recall that *ns* electrons are lost first.

Solution **(a)** Zr is the second element in the 4*d* series: [Kr] $5s^24d^2$.
(b) V is the third element in the 3*d* series: [Ar] $4s^23d^3$. In forming V^{3+}, three electrons are lost (two 4*s* and one 3*d*), so V^{3+} is a d^2 ion: [Ar] $3d^2$.
(c) Mo lies below Cr in Group 6B(6). Recall from Section 8.2 that Cr has a configuration with half-filled 4*s* and 3*d* orbitals, so we expect the same pattern for Mo. Thus, Mo is [Kr] $5s^14d^5$. To form the ion, Mo loses the one 5*s* and two of the 4*d* electrons, so Mo^{3+} is a d^3 ion: [Kr] $4d^3$.

Check Figure 8.5 shows we're correct for the atoms. Be sure that charge plus number of *d* electrons in the ion equals the sum of outer *s* and *d* electrons in the atom.

FOLLOW-UP PROBLEM 22.1 Write *partial* electron configurations (no noble-gas core or filled inner sublevels) for the following: **(a)** Ag^+; **(b)** Cd^{2+}; **(c)** Ir^{3+}.

Atomic and Physical Properties of the Transition Elements

Properties of the transition elements contrast in several ways with properties of the main-group elements.

Trends Across a Period Consider the variations in atomic size, electronegativity, and ionization energy across Period 4 (Figure 22.3):

1. *Atomic size.* Atomic size decreases overall across the period (Figure 22.3A). There is a smooth, steady decrease across the main groups because the electrons are added to

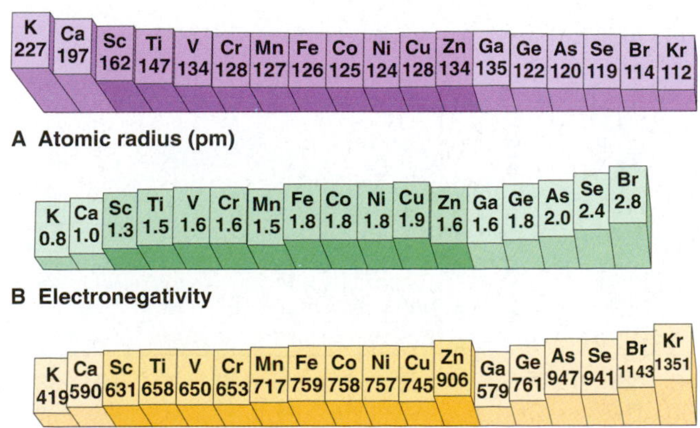

Figure 22.3 Trends in key atomic properties of Period 4 elements. Atomic radius **(A)**, electronegativity **(B)**, and first ionization energy **(C)** of all Period 4 elements are shown as heights of posts, with darker shading for the transition series.

outer orbitals, which shield the increasing nuclear charge poorly. This decrease is not smooth in the transition series, where *atomic size decreases at first but then remains relatively constant.* The reason is that the *d* electrons fill *inner* orbitals, so they shield outer electrons very efficiently and, thus, the outer 4*s* electrons are *not* pulled closer.

2. *Electronegativity (EN).* Electronegativity generally increases across the period, but, once again, the main groups show a steady, steep increase between the metal potassium (0.8) and the nonmetal bromine (2.8), whereas the transition elements have *relatively constant EN* (Figure 22.3B), consistent with their relatively constant size. The transition elements all have intermediate EN values, close to the large, metallic members of Groups 3A(13) to 5A(15).

3. *Ionization energy (IE_1).* The ionization energies of the Period 4 main-group elements rise steeply from left to right, more than tripling from potassium (419 kJ/mol) to krypton (1351 kJ/mol), as electrons become more difficult to remove from the poorly shielded, increasing nuclear charge. In the transition metals, *IE_1 values increase relatively little* because the inner 3*d* electrons shield more effectively (Figure 22.3C). [Recall from Section 8.3 that the drop at Group 3A(13) occurs because it is relatively easy to remove the first electron from the outer *np* orbital.]

Trends Within a Group Vertical trends for transition elements are also different from main-group trends.

1. *Atomic size.* As expected, atomic size increases from Period 4 to 5, as it does for the main-group elements, but there is virtually *no size increase from Period 5 to 6* (Figure 22.4A). Since the lanthanides, with their buried 4*f* sublevel, appear between the 4*d* (Period 5) and 5*d* (Period 6) series, an element in Period 6 is separated from the one above it by 32 elements (ten 4*d*, six 5*p*, two 6*s*, and fourteen 4*f* orbitals) instead of just 18. The extra shrinkage from the increase in nuclear charge due to 14 additional protons is called the **lanthanide contraction.** By coincidence, this size *decrease* is about equal to the normal *increase* between periods, so Periods 5 and 6 transition elements have about the same atomic sizes.

Figure 22.4 Vertical trends in key properties within the transition elements. The trends are unlike those for the main-group elements in several ways: **A,** The second and third members of a transition metal group are nearly the same size. **B,** Electronegativity increases down a transition group. **C,** First ionization energies are highest at the bottom of a transition group. **D,** Densities increase down a transition group because mass increases faster than volume.

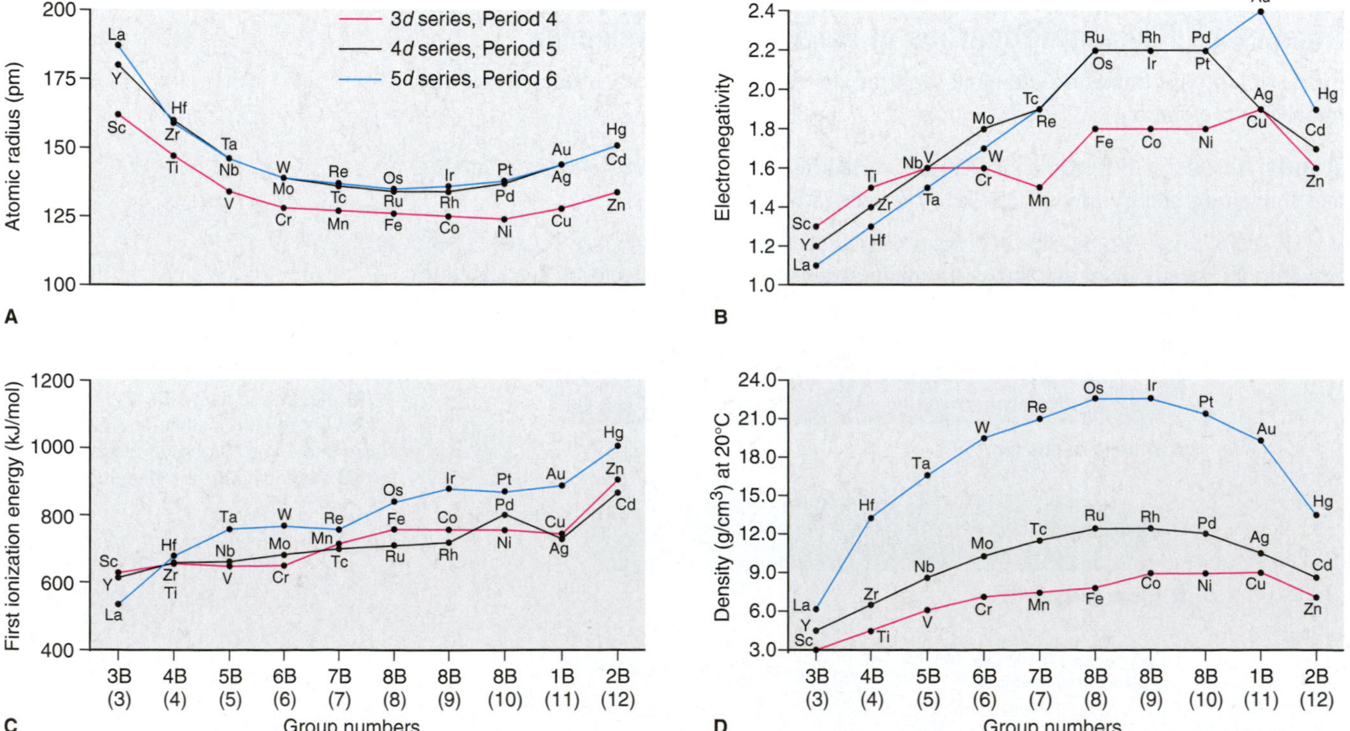

2. *Electronegativity.* The vertical trend in electronegativity seen in most transition groups is opposite the decreasing trend in the main groups. Electronegativity *increases* from Period 4 to Period 5, but there is no further increase in Period 6 (Figure 22.4B). The heavier elements, especially gold (EN = 2.4), become quite electronegative, with values higher than most metalloids and even some nonmetals (e.g., EN of Te and of P = 2.1).

3. *Ionization energy.* The small increase in size combined with the large increase in nuclear charge also explains why *IE_1 values generally increase* down a transition group (Figure 22.4C). This trend also runs counter to the main group trend, where heavier members are so much larger that the outer electron is easier to remove.

4. *Density.* Atomic size, and therefore volume, is inversely related to density. Across a period, densities increase, then level off, and finally dip a bit at the end of a series (Figure 22.4D). Down a transition group, densities increase dramatically because atomic volumes change little from Period 5 to 6, but atomic masses increase significantly. As a result, the Period 6 series contains some of the densest elements known: tungsten, rhenium, osmium, iridium, platinum, and gold have densities about 20 times that of water and twice that of lead.

Chemical Properties of the Transition Elements

Like their atomic and physical properties, the chemical properties of transition elements are very different from those of main-group elements. We examine key properties in the Period 4 series.

Multiple Oxidation States One of the most characteristic chemical properties of the transition metals is the occurrence of *multiple oxidation states;* main-group metals, display one or, at most, two states. For example, vanadium has two common oxidation states, chromium and manganese have three (Figure 22.5A), and many others are seen less often (Table 22.2). Since the *ns* and $(n-1)d$ electrons are so close in energy, transition elements can use all or most of these electrons in bonding.

The highest oxidation state of elements in Groups 3B(3) through 7B(7) equals the group number. These states are seen when the elements combine with highly electronegative oxygen or fluorine. For instance, Figure 22.5B shows vanadium as vanadate ion (VO_4^{3-}; O.N. of V = +5), chromium as dichromate ($Cr_2O_7^{2-}$; O.N. of Cr = +6), and manganese as permanganate (MnO_4^-; O.N. of Mn = +7).

Elements in Groups 8B(8), 8B(9), and 8B(10) exhibit fewer oxidation states, and the highest state is less common and never equal to the group number. For example, we never encounter iron in the +8 state and only rarely in the +6 state. The +2 and +3 states are the most common ones for iron and cobalt, and the +2 state is most common for nickel, copper, and zinc. *The +2 oxidation state is common because ns^2 electrons are readily lost.*

A

B

Figure 22.5 Aqueous oxoanions of transition elements. A, Mn has common +2 (Mn^{2+}, *left*), +6 (MnO_4^{2-}, *middle*), and +7 oxidation states (MnO_4^-, *right*). **B,** The +5 state in VO_4^{3-} *(left)*, the +6 in $Cr_2O_7^{2-}$ *(middle)*, and the +7 in MnO_4^- *(right)*.

Table 22.2 Oxidation States and *d*-Orbital Occupancy of the Period 4 Transition Metals*

Oxidation State	3B (3) Sc	4B (4) Ti	5B (5) V	6B (6) Cr	7B (7) Mn	8B (8) Fe	8B (9) Co	8B (10) Ni	1B (11) Cu	2B (12) Zn
0	d^1	d^2	d^3	d^5	d^5	d^6	d^7	d^8	d^{10}	d^{10}
+1			d^3	d^5	d^5	d^6	d^7	d^8	d^{10}	
+2		d^2	d^3	d^4	d^5	d^6	d^7	d^8	d^9	d^{10}
+3	d^0	d^1	d^2	d^3	d^4	d^5	d^6	d^7	d^8	
+4		d^0	d^1	d^2	d^3	d^4	d^5	d^6		
+5			d^0	d^1	d^2		d^4			
+6				d^0	d^1	d^4				
+7					d^0					

*The most important orbital occupancies are in color.

Metallic Behavior and Oxide Acidity Atomic size and oxidation state have a major effect on the nature of bonding in transition metal compounds. Like the metals in Groups 3A(13), 4A(14), and 5A(15), the transition elements in their *lower* oxidation states behave chemically more like metals. That is, *ionic bonding is more prevalent for the lower oxidation states, and covalent bonding is more prevalent for the higher states.* For example, at room temperature, $TiCl_2$ (O.N. of Ti = +2) is an ionic solid, whereas $TiCl_4$ (O.N. of Ti = +4) is a molecular liquid. In the higher oxidation states, the atoms have higher charge densities, so they polarize the electron clouds of the nonmetal ions more strongly and the bonding becomes more covalent. For the same reason, the oxides become less basic as the oxidation state increases: TiO is weakly basic in water, whereas TiO_2 is amphoteric (reacts with both acid and base).

Table 22.3 Standard Electrode Potentials of Period 4 M^{2+} Ions

Half-Reaction	$E°$ (V)
$Ti^{2+}(aq) + 2e^- \rightleftharpoons Ti(s)$	−1.63
$V^{2+}(aq) + 2e^- \rightleftharpoons V(s)$	−1.19
$Cr^{2+}(aq) + 2e^- \rightleftharpoons Cr(s)$	−0.91
$Mn^{2+}(aq) + 2e^- \rightleftharpoons Mn(s)$	−1.18
$Fe^{2+}(aq) + 2e^- \rightleftharpoons Fe(s)$	−0.44
$Co^{2+}(aq) + 2e^- \rightleftharpoons Co(s)$	−0.28
$Ni^{2+}(aq) + 2e^- \rightleftharpoons Ni(s)$	−0.25
$Cu^{2+}(aq) + 2e^- \rightleftharpoons Cu(s)$	0.34
$Zn^{2+}(aq) + 2e^- \rightleftharpoons Zn(s)$	−0.76

Reducing Strength Table 22.3 shows the standard electrode potentials of the Period 4 transition metals in their +2 oxidation state in acid solution. Note that, in general, reducing strength decreases across the series. All the Period 4 transition metals, except copper, are active enough to reduce H^+ from aqueous acid to form hydrogen gas. In contrast to the rapid reaction at room temperature of the Group 1A(1) and 2A(2) metals with water, however, most transition metals have an oxide coating that allows rapid reaction only with hot water or steam.

Color and Magnetism of Compounds *Most main-group ionic compounds are colorless* because the metal ion has a filled outer level (ns^2 or ns^2np^6: noble gas electron configuration). With only *much* higher energy orbitals available to receive an excited electron, the ion does not absorb visible light. In contrast, electrons in a partially filled *d* sublevel can absorb visible wavelengths and move to *slightly* higher energy orbitals. As a result, *many transition metal compounds have striking colors.* Exceptions are the compounds of scandium, titanium(IV), and zinc, which are colorless because their metal ions have either an empty *d* sublevel (Sc^{3+} or Ti^{4+}: [Ar] $3d^0$) or a filled one (Zn^{2+}: [Ar] $3d^{10}$) (Figure 22.6).

Magnetic properties are also related to sublevel occupancy (Section 8.5). *Most main-group metal ions are diamagnetic* for the same reason they are colorless: all their electrons are paired. In contrast, *many transition metal compounds are paramagnetic because of their unpaired d electrons.* Of course, transition metal ions with a d^0 or d^{10} configuration are colorless and diamagnetic because all the electrons are paired.

Table 22.4 Some Properties of Group 6B(6) Elements

	IE_1 (kJ/mol)	$E°$ (V) for $M^{3+}(aq) \mid M(s)$
Cr	653	−0.74
Mo	685	−0.20
W	770	−0.11

Chemical Behavior Within a Group The *increase* in reactivity going down a group of main-group metals does *not* occur with the transition metals. The chromium (Cr) group [6B(6)] shows a typical pattern (Table 22.4). Because IE_1 *increases* down the group, the two heavier members are *less* reactive than the lightest one. Chromium is also a much stronger reducing agent than molybdenum (Mo) or tungsten (W).

Figure 22.6 Colors of representative compounds of the Period 4 transition metals. Staggered from left to right, the compounds are scandium oxide *(white)*, titanium(IV) oxide *(white)*, vanadyl sulfate dihydrate *(light blue)*, sodium chromate *(yellow)*, manganese(II) chloride tetrahydrate *(light pink)*, potassium ferricyanide *(red-orange)*, cobalt(II) chloride hexahydrate *(violet)*, nickel(II) nitrate hexahydrate *(green)*, copper(II) sulfate pentahydrate *(blue)*, and zinc sulfate heptahydrate *(white)*.

Summary of Section 22.1

- All transition elements are metals.
- Atoms of the *d*-block elements have $(n - 1)d$ orbitals being filled, and their ions have an empty *ns* orbital.
- Compared to the trends for the main-group elements, atomic size, electronegativity, and first ionization energy change relatively little across a transition series. Because of the lanthanide contraction, atomic size changes little from Period 5 to 6 in a transition metal group; thus, electronegativity, first ionization energy, and density *increase* down such a group.
- Transition metals typically have several oxidation states, with the +2 state most common. The elements exhibit less metallic behavior in their higher states.
- Most Period 4 transition metals are active enough to reduce H^+ ion from acid solution.
- Many transition metal compounds are colored because orbitals are available to accept excited electrons; they are paramagnetic because the metal ion has unpaired *d* electrons.
- In contrast to main-group metals, transition metals show decreasing reactivity down a group.

22.2 • COORDINATION COMPOUNDS

The most distinctive feature of transition metal chemistry is the common occurrence of **coordination compounds** (or *complexes*). These species contain at least one **complex ion,** which consists of *a central metal ion bonded to molecules and/or anions called* ***ligands*** *(see margin).* To maintain charge neutrality, the complex ion is associated with **counter ions.** In the coordination compound $[Co(NH_3)_6]Cl_3$ (Figure 22.7A), the complex ion (in square brackets) is $[Co(NH_3)_6]^{3+}$, the six NH_3 molecules bonded to the Co^{3+} are neutral ligands, and the three Cl^- ions are counter ions.

A coordination compound is an electrolyte in water: the complex ion and counter ions separate, but the complex ion behaves like a polyatomic ion because *the ligands and central metal ion remain attached.* Thus, as Figure 22.7A shows, 1 mol of $[Co(NH_3)_6]Cl_3$ yields 1 mol of $[Co(NH_3)_6]^{3+}$ ions and 3 mol of Cl^- ions. This section covers the structure, naming, and properties of complex ions.

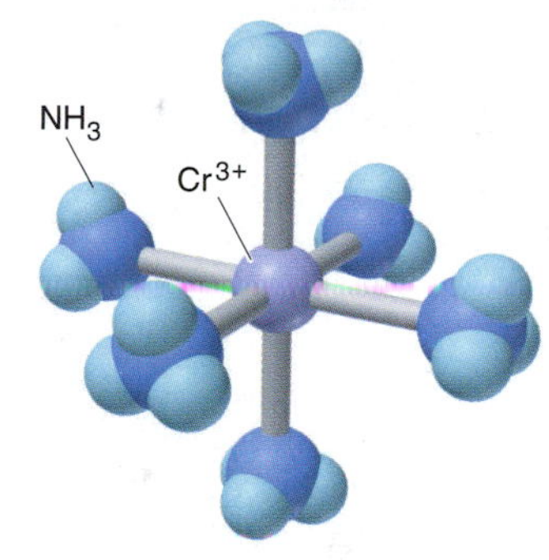

The complex ion $Cr(NH_3)_6^{3+}$ has a central Cr^{3+} ion bonded to six NH_3 ligands.

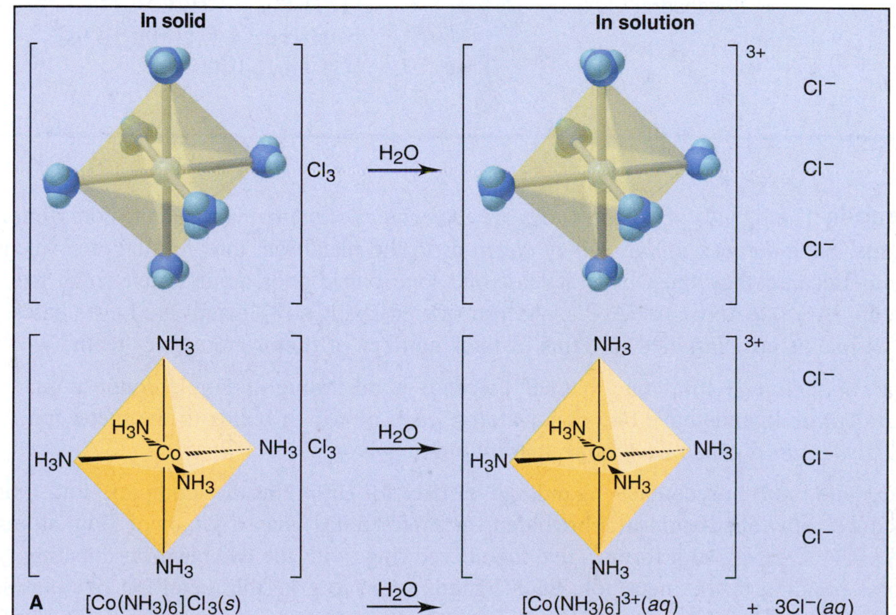

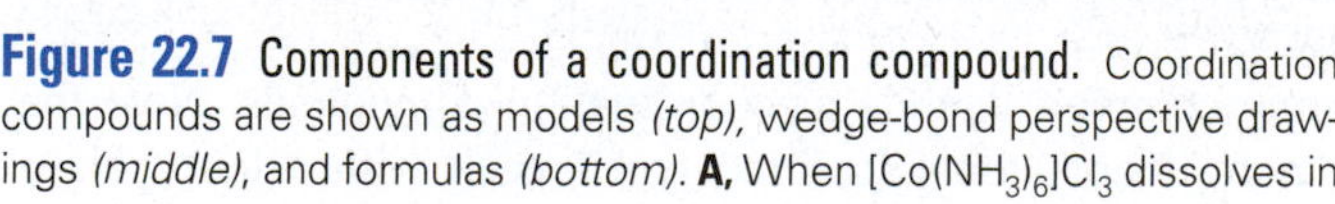

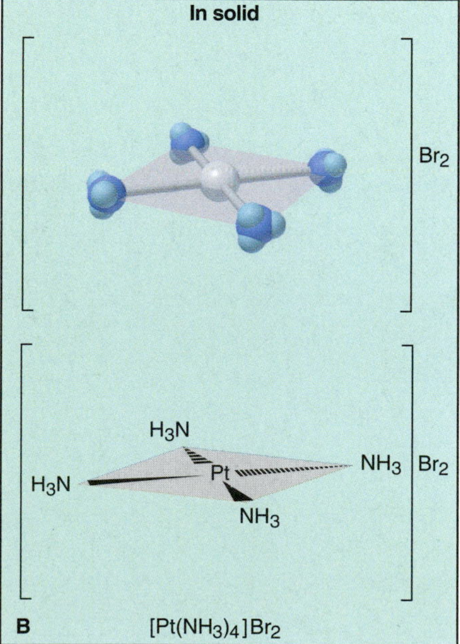

Figure 22.7 Components of a coordination compound. Coordination compounds are shown as models *(top)*, wedge-bond perspective drawings *(middle)*, and formulas *(bottom)*. **A,** When $[Co(NH_3)_6]Cl_3$ dissolves in water, the six ligands remain bound in the complex ion. **B,** $[Pt(NH_3)_4]Br_2$ has four NH_3 ligands and two Br^- counter ions.

Complex Ions: Coordination Numbers, Geometries, and Ligands

A complex ion is described by the metal ion and the number and types of ligands attached to it. The ion's geometry is related to its **coordination number,** the *number of ligand atoms* bonded directly to the central metal ion, which is *specific* for a given metal ion in a particular oxidation state and compound. In general, *the most common coordination number in complex ions is 6,* but 2 and 4 are also seen often, and some higher ones are known.

Geometries The geometry (shape) depends on *the coordination number and the metal ion.* Table 22.5 shows the geometries for coordination numbers 2, 4, and 6. A complex ion whose metal ion has a coordination number of 2, such as $[Ag(NH_3)_2]^+$, is *linear.* The coordination number 4 gives rise to either of two geometries. Most d^8 metal ions form *square planar* complex ions (Figure 22.7B). The d^{10} ions form *tetrahedral* complex ions. A coordination number of 6 results in an *octahedral* geometry, as shown by $[Co(NH_3)_6]^{3+}$ (Figure 22.7A). Note the similarity with some of the molecular shapes in VSEPR theory (Section 10.2).

Table 22.5 Coordination Numbers and Shapes of Some Complex Ions

Coordination Number	Shape	Examples
2	Linear	$[CuCl_2]^-$, $[Ag(NH_3)_2]^+$, $[AuCl_2]^-$
4	Square planar	$[Ni(CN)_4]^{2-}$, $[PdCl_4]^{2-}$, $[Pt(NH_3)_4]^{2+}$, $[Cu(NH_3)_4]^{2+}$
4	Tetrahedral	$[Cu(CN)_4]^{3-}$, $[Zn(NH_3)_4]^{2+}$, $[CdCl_4]^{2-}$, $[MnCl_4]^{2-}$
6	Octahedral	$[Ti(H_2O)_6]^{3+}$, $[V(CN)_6]^{4-}$, $[Cr(NH_3)_4Cl_2]^+$, $[Mn(H_2O)_6]^{2+}$, $[FeCl_6]^{3-}$, $[Co(en)_3]^{3+}$

Ligands The ligands of complex ions are *molecules or anions* with one or more **donor atoms.** Each *donates a lone pair of electrons* to the metal ion, thus forming a covalent bond. Because they must have at least one lone pair, donor atoms often come from Group 5A(15), 6A(16), or 7A(17). (As you saw in Chapter 18, ligands are Lewis bases.)

Ligands are classified in terms of their number of donor atoms, or "teeth":

- *Monodentate* (Latin, "one-toothed") ligands bond through a single donor atom.
- *Bidentate* ligands have two donor atoms, each of which bonds to the metal ion.
- *Polydentate* ligands have more than two donor atoms.

Bidentate and polydentate ligands give rise to *rings* in the complex ion. For instance, ethylenediamine (abbreviated *en* in formulas) has a chain of four atoms (:N—C—C—N:), so it forms a five-membered ring, with the two electron-donating N atoms bonding to the metal ion. Such ligands seem to grab the metal ion like claws, so a complex ion that contains them is also called a **chelate** (pronounced "KEY-late"; Greek *chela,* "crab's claw").

Table 22.6 shows some common ligands in coordination compounds. Note that each ligand has one or more donor atoms (colored type), each with a lone pair of electrons to donate and form a covalent bond to the metal ion.

Table 22.6 Some Common Ligands in Coordination Compounds

Ligand Type	Examples
Monodentate	$H_2\ddot{O}$: water; :NH_3 ammonia; $:\ddot{F}:^-$ fluoride ion; $:\ddot{Cl}:^-$ chloride ion; $[:C\equiv N:]^-$ cyanide ion; $[:\ddot{S}=C=\ddot{N}:]^-$ (or) thiocyanate ion; $[:\ddot{O}-H]^-$ hydroxide ion; $[:\ddot{O}-\ddot{N}=\ddot{O}:]^-$ (or) nitrite ion
Bidentate	$H_2N-CH_2-CH_2-NH_2$ ethylenediamine (en); $[^-O(O=)C-C(=O)O^-]^{2-}$ oxalate ion
Polydentate	$H_2N-CH_2-CH_2-NH-CH_2-CH_2-NH_2$ diethylenetriamine; $[O_3P-O-PO_2-O-PO_3]^{5-}$ triphosphate ion; $[(^-OOC-CH_2)_2N-CH_2-CH_2-N(CH_2-COO^-)_2]^{4-}$ ethylenediaminetetraacetate ion ($EDTA^{4-}$)

Formulas and Names of Coordination Compounds

The specific combination of ions in a coordination compound is the key to writing its formula and name. A coordination compound can consist of a complex cation with simple anionic counter ions, a complex anion with simple cationic counter ions, or even a complex cation with complex anion as counter ion.

Determining the Charge of the Metal Ion To know the number of cations and anions, we must know the ion charges, and we use basic arithmetic to find them:

- For a compound with a *complex anion,* say, $K_2[Co(NH_3)_2Cl_4]$, we know that the K^+ ions balance the charge of the complex anion, which contains two NH_3 molecules and four Cl^- ions as ligands. The two NH_3 are neutral and the four Cl^- have a total charge of 4−, so the entire complex ion must have a charge of 2− to balance the two K^+. The charge of the central metal ion is

$$\text{Charge of complex ion} = \text{Charge of metal ion} + \text{total charge of ligands}$$
$$2- = \text{Charge of metal ion} + [(2 \times 0) + (4 \times 1-)]$$

So, $\quad$ Charge of metal ion = (2−) − (4−) = 2+; that is, Co^{2+}

- For a compound with a *complex cation,* say $[Co(NH_3)_4Cl_2]Cl$, the complex ion is $[Co(NH_3)_4Cl_2]^+$ and one Cl^- is the counter ion. The four NH_3 ligands are neutral, the two Cl^- ligands have a total charge of 2−, and the complex cation has a charge of 1+, so the central metal ion must be Co^{3+}:

$$\text{Charge of metal ion} = \text{charge of complex ion} - \text{total charge of ligands}$$
$$= 1+ - [(4 \times 0) + (2 \times 1-)] = 3+$$

Rules for Writing Formulas There are three rules for writing formulas of coordination compounds (the first two are the same as those for the formula of any ionic compound):

1. *The cation is written before the anion.*
2. *The charge of the cation(s) is balanced by the charge of the anion(s).*
3. *For the complex ion, neutral ligands are written before anionic ligands, and the formula of the whole ion is placed in brackets.*

Rules for Naming Coordination Compounds Originally named after their discoverer or color, coordination compounds are now named systematically with a set of rules. Let's see how to name $[Co(NH_3)_4Cl_2]Cl$. As we go through the naming steps, refer to Tables 22.7 and 22.8 on the next page.

Table 22.7 Names of Some Neutral and Anionic Ligands

Neutral		Anionic	
Name	**Formula**	**Name**	**Formula**
Aqua	H_2O	Fluoro	F^-
Ammine	NH_3	Chloro	Cl^-
Carbonyl	CO	Bromo	Br^-
Nitrosyl	NO	Iodo	I^-
		Hydroxo	OH^-
		Cyano	CN^-

Table 22.8 Names of Some Metal Ions in Complex Anions

Metal	Name in Anion
Iron	Ferrate
Copper	Cuprate
Lead	Plumbate
Silver	Argentate
Gold	Aurate
Tin	Stannate

1. *The cation is named before the anion.* Thus, we name the $[Co(NH_3)_4Cl_2]^+$ ion before the Cl^- ion.
2. *Within the complex ion, the ligands are named in alphabetical order* ***before*** *the metal ion.* In the $[Co(NH_3)_4Cl_2]^+$ ion that we are discussing, four NH_3 and two Cl^- are named before Co^{3+}.
3. *Neutral ligands generally have the molecule name,* but there are a few exceptions (Table 22.7). *Anionic ligands drop the -ide and add -o after the root name;* thus, the anion name *fluoride* for an F^- ion becomes the ligand name *fluoro.* The two ligands in $[Co(NH_3)_4Cl_2]^+$ are *ammine* (NH_3) and *chloro* (Cl^-) with *ammine* coming before *chloro* alphabetically.
4. *A numerical prefix indicates the number of ligands of a particular type.* For example, *tetra*ammine denotes *four* NH_3, and *di*chloro denotes *two* Cl^-. Other prefixes are *tri-, penta-,* and *hexa-*. But prefixes do *not* affect the alphabetical order: *tetraammine* comes before *dichloro.* For ligand names that include a numerical prefix (such as ethylene*di*amine), we use *bis* (2), *tris* (3), or *tetrakis* (4) to indicate the number of such ligands, followed by the ligand name in parentheses. For example, a complex ion that has two ethylenediamine ligands has *bis(ethylenediamine)* in its name.
5. *The oxidation state of the metal ion has a Roman numeral (in parentheses) only if the metal ion can have more than one state.* Since cobalt can have +2 and +3 states, we add a III to name the complex ion. Thus, the compound is

   tetraamminedichlorocobalt(III) chloride

   The only space in the name comes between cation and anion.
6. *If the complex ion is an anion, we drop the ending of the metal name and add -ate.* Thus, the name for $K[Pt(NH_3)Cl_5]$ is potassium amminepentachloroplatinate(IV). There is one K^+ counter ion, so the complex anion has a charge of 1−. The five Cl^- ligands have a total charge of 5−, so Pt must be in the +4 oxidation state. For some metals, we use the Latin root with the *-ate* ending (Table 22.8). For example, the name for $Na_4[FeBr_6]$ is sodium hexabromoferrate(II).

Sample Problem 22.2 Writing Names and Formulas of Coordination Compounds

Problem **(a)** What is the systematic name of $Na_3[AlF_6]$?
(b) What is the systematic name of $[Co(en)_2Cl_2]NO_3$?
(c) What is the formula of tetraamminebromochloroplatinum(IV) chloride?
(d) What is the formula of hexaamminecobalt(III) tetrachloroferrate(III)?

Plan We use the rules that were presented above and Tables 22.7 and 22.8.

Solution **(a)** The complex ion is $[AlF_6]^{3-}$. There are six *(hexa-)* F^- ions *(fluoro)* as ligands, so we have *hexafluoro.* The complex ion is an anion, so the ending of the metal ion (aluminum) must be changed to *-ate:* hexafluoroaluminate. Aluminum has only the +3 oxidation state, so we do *not* use a Roman numeral. The positive counter ion is named first and separated from the anion by a space: sodium hexafluoroaluminate.
(b) Listed alphabetically, there are two Cl^- *(dichloro)* and two en [*bis(ethylenediamine)*] as ligands. The complex ion is a cation, so the metal name is unchanged, but we specify its oxidation state because cobalt can have several. One NO_3^- balances the 1+ cation charge.

With 2− for two Cl^- and 0 for two en, the metal must be *cobalt(III)*. The word *nitrate* follows a space: dichlorobis(ethylenediamine)cobalt(III) nitrate.
(c) The central metal ion is written first, followed by the neutral ligands and then (in alphabetical order) by the negative ligands. *Tetraammine* is four NH_3, *bromo* is one Br^-, *chloro* is one Cl^-, and *platinate(IV)* is Pt^{4+}, so the complex ion is $[Pt(NH_3)_4BrCl]^{2+}$. Its 2+ charge is the sum of 4+ for Pt^{4+}, 0 for four NH_3, 1− for one Br^-, and 1− for one Cl^-. To balance the 2+ charge, we need two Cl^- counter ions: $[Pt(NH_3)_4BrCl]Cl_2$.
(d) This compound consists of two different complex ions. In the cation, *hexaammine* is six NH_3 and *cobalt(III)* is Co^{3+}, so the cation is $[Co(NH_3)_6]^{3+}$. The 3+ charge is the sum of 3+ for Co^{3+} and 0 for six NH_3. In the anion, *tetrachloro* is four Cl^-, and *ferrate(III)* is Fe^{3+}, so the anion is $[FeCl_4]^-$. The 1− charge is the sum of 3+ for Fe^{3+} and 4− for four Cl^-. In the neutral compound, one 3+ cation must be balanced by three 1− anions: $[Co(NH_3)_6][FeCl_4]_3$.

Check Reverse the process to be sure you obtain the name or formula in the problem.

FOLLOW-UP PROBLEM 22.2 **(a)** What is the name of $[Cr(H_2O)_5Br]Cl_2$?
(b) What is the formula of barium hexacyanocobaltate(III)?

Isomerism in Coordination Compounds

Isomers are compounds with the same chemical formula but different properties (Section 3.2). Recall the discussion of isomerism in organic compounds (Section 15.2); coordination compounds exhibit the same two broad categories—constitutional isomers and stereoisomers.

Constitutional Isomers: Atoms Connected Differently Compounds with the same formula, but with the atoms connected differently, are **constitutional (structural) isomers.** Coordination compounds exhibit two types:

1. **Coordination isomers** occur when the composition of the complex ion, but not of the compound, is different. This type of isomerism occurs in two ways:

- *Exchange of ligand and counter ion*. For example, in $[Pt(NH_3)_4Cl_2](NO_2)_2$, the Cl^- ions are the ligands, and the NO_2^- ions are counter ions; in $[Pt(NH_3)_4(NO_2)_2]Cl_2$, the roles are reversed. A common test to see whether Cl^- is a ligand (bound) or a counter ion (free) is to treat a solution of the compound with $AgNO_3$; the Cl^- counter ion will form a white precipitate of AgCl, but the Cl^- ligand won't. Thus, a solution of $[Pt(NH_3)_4Cl_2](NO_2)_2$ does not form AgCl because Cl^- is bound to the metal ion, but a $[Pt(NH_3)_4(NO_2)_2]Cl_2$ solution forms 2 mol of AgCl per mole of compound.
- *Exchange of ligands*. This type of isomerism occurs in compounds consisting of two complex ions. For example, in $[Cr(NH_3)_6][Co(CN)_6]$ and $[Co(NH_3)_6][Cr(CN)_6]$, NH_3 is a ligand of Cr^{3+} in one compound and of Co^{3+} in the other.

2. **Linkage isomers** occur when the composition of the complex ion is the same but the ligand donor atom is different. Some ligands can bind to the metal ion through *either of two donor atoms*. For example, the nitrite ion can bind through the N atom (*nitro,* O_2N:) or either of the O atoms (*nitrito,* ONO:) to give linkage isomers, such as pentaammine*nitro*cobalt(III) chloride $[Co(NH_3)_5(NO_2)]Cl_2$ and pentaammine*nitrito*cobalt(III) chloride $[Co(NH_3)_5(ONO)]Cl_2$ (Figure 22.8).

Figure 22.8 A pair of linkage (constitutional) isomers.

Other examples of ligands that have more than one donor atom are the cyanate ion, which can attach via the O atom (*cyanato,* NCO:) or the N atom (*isocyanato,* OCN:), and the thiocyanate ion, which can attach via the S atom or the N atom:

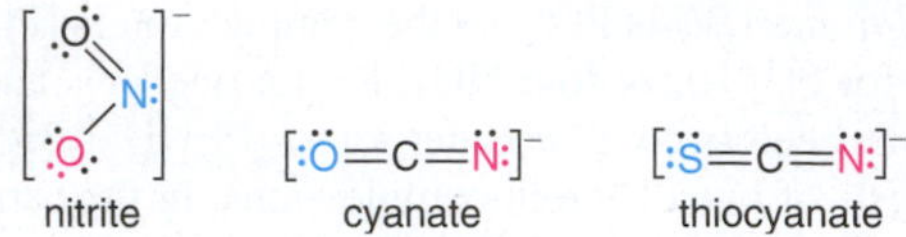

Stereoisomers: Atoms Arranged Differently in Space **Stereoisomers** are compounds that have the same atomic connections but different spatial arrangements of the atoms. *Geometric* and *optical* isomers, which we discussed for organic compounds, occur with coordination compounds as well:

1. **Geometric isomers** (also called ***cis-trans*** **isomers** and *diastereomers*) occur when atoms or groups of atoms are arranged differently in space relative to the central metal ion. For example, the square planar $[Pt(NH_3)_2Cl_2]$ has two arrangements, which give rise to two compounds with *very* different properties (Figure 22.9A). The isomer with identical ligands *next* to each other is *cis*-diamminedichloroplatinum(II), and the one with identical ligands *across* from each other is *trans*-diamminedichloroplatinum(II); the *cis* isomer has significant antitumor activity, but the *trans* isomer has none! Octahedral complexes also exhibit *cis-trans* isomerism (Figure 22.9B). The *cis* isomer of the $[Co(NH_3)_4Cl_2]^+$ ion has the two Cl^- ligands at any two adjacent positions of the ion's octahedral shape, whereas the *trans* isomer has these ligands across from each other.

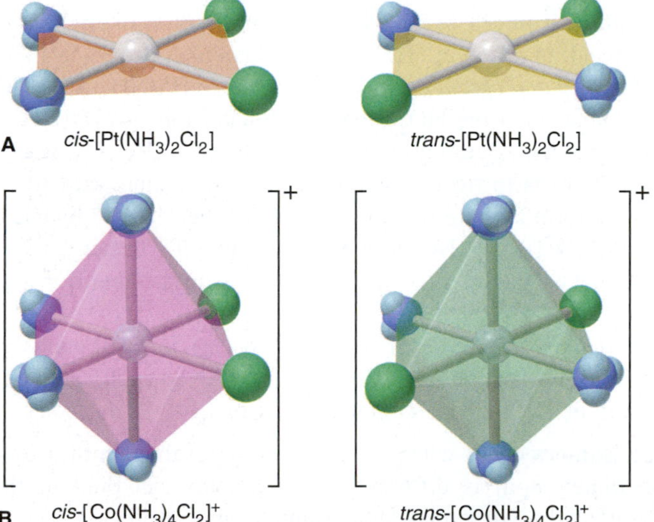

Figure 22.9 Geometric (*cis-trans*) isomerism. **A,** The *cis* and *trans* isomers of $[Pt(NH_3)_2Cl_2]$. **B,** The *cis* and *trans* isomers of $[Co(NH_3)_4Cl_2]^+$. The colored shapes indicate the actual colors of the species.

2. **Optical isomers** (also called *enantiomers*) occur when a molecule and its mirror image cannot be superimposed (see Figures 15.8 and 15.9). Unlike other types of isomers, which have distinct physical properties, optical isomers are physically identical except for *the direction in which they rotate the plane of polarized light.* Many octahedral complex ions show optical isomerism, which we can determine by rotating one isomer and seeing if it is superimposable on the other isomer (its mirror image). For example, in Figure 22.10A, the two structures (I and II) of *cis*-$[Co(en)_2Cl_2]^+$, the *cis*-dichlorobis(ethylenediamine)cobalt(III) ion, are mirror images of each other. Rotate structure I 180° around a vertical axis, and you obtain III. The Cl^- ligands of III match those of II, but the en ligands do not: II and III (which is I rotated) are not superimposable: they are optical isomers. One isomer

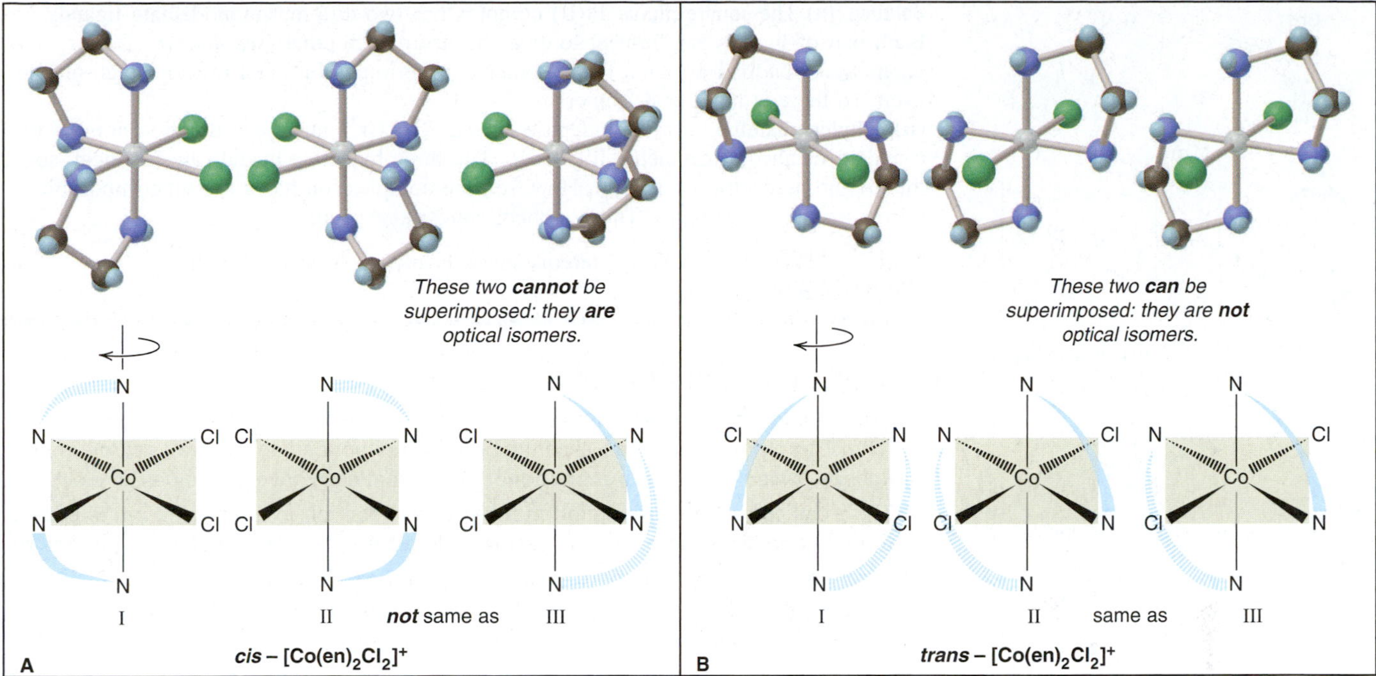

Figure 22.10 Optical isomerism in an octahedral complex ion. **A,** Structure I and its mirror image, structure II, are optical isomers of *cis*-$[Co(en)_2Cl_2]^+$. (The curved wedges represent the bidentate ligand ethylenediamine, $H_2N—CH_2—CH_2—NH_2$.) **B,** The *trans* isomer does *not* have optical isomers.

is designated *d*-$[Co(en)_2Cl_2]^+$ and the other is *l*-$[Co(en)_2Cl_2]^+$, depending on whether it rotates polarized light to the right (*d*- for "dextro-") or to the left (*l*- for "levo-"). (The *d*- or *l*- designation can only be determined by experiment.)

In contrast, as shown in Figure 22.10B, the two structures of the *trans*-dichlorobis(ethylenediamine)cobalt(III) ion are *not* optical isomers: rotate structure I 90° around a vertical axis and you obtain III, which *is* superimposable on II.

Figure 22.11 is an overview of the most common types of isomerism in coordination compounds.

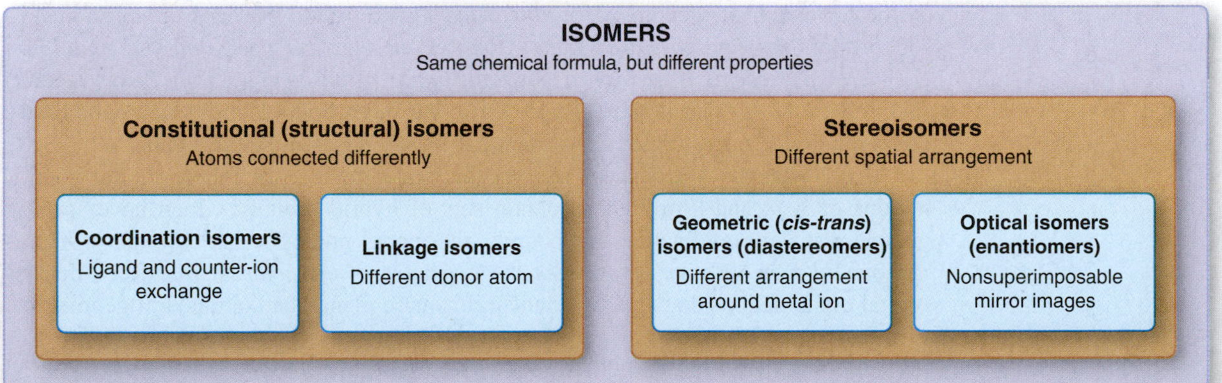

Figure 22.11 Important types of isomerism in coordination compounds.

Sample Problem 22.3 Determining the Type of Stereoisomerism

Problem Draw stereoisomers for each of the following and state the type of isomerism:
(a) $[Pt(NH_3)_2Br_2]$ (square planar) **(b)** $[Cr(en)_3]^{3+}$ (en = $H_2\ddot{N}CH_2CH_2\ddot{N}H_2$)

Plan We determine the geometry around each metal ion and the nature of the ligands. If there are different ligands that can be placed in different positions relative to each other, geometric *(cis-trans)* isomerism occurs. Then, we see whether the mirror image of an isomer is superimposable on the original. If it is *not,* optical isomerism also occurs.

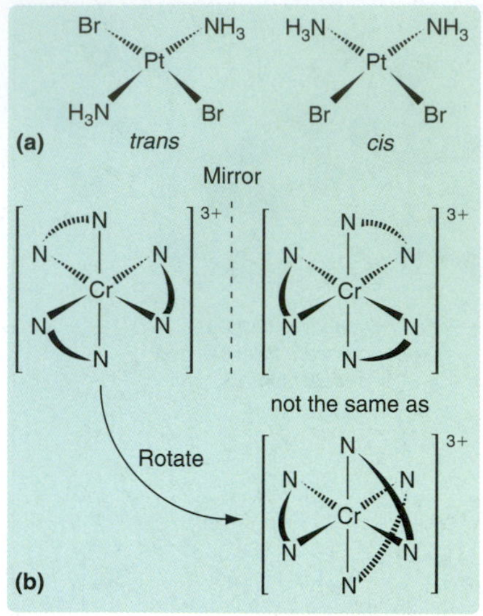

Solution (a) The square planar Pt(II) complex has two different monodentate ligands. Each pair of ligands can lie next to or across from each other *(see margin)*. Thus, geometric isomerism occurs. Each isomer *is* superimposable on a mirror image of itself, so there is no optical isomerism.

(b) Ethylenediamine (en) is a bidentate ligand. The Cr^{3+} has a coordination number of 6 and an octahedral geometry, like Co^{3+}. The three bidentate ligands are identical, so there is no geometric isomerism. However, the complex ion has a nonsuperimposable mirror image *(see margin)*. Thus, optical isomerism occurs.

FOLLOW-UP PROBLEM 22.3 What stereoisomers, if any, are possible for the $[Co(NH_3)_2(en)Cl_2]^+$ ion?

Summary of Section 22.2

- Coordination compounds consist of a complex ion and charge-balancing counter ions. The complex ion has a central metal ion bonded to neutral and/or anionic ligands, which have one or more donor atoms that each have a lone pair of electrons.
- The most common coordination number of a metal ion in a complex ion is 6; thus, the most common complex-ion geometry is octahedral (six ligand atoms bonding).
- Formulas and names of coordination compounds follow systematic rules.
- Coordination compounds can exhibit constitutional isomerism (coordination and linkage) and stereoisomerism (geometric and optical).

22.3 • THEORETICAL BASIS FOR THE BONDING AND PROPERTIES OF COMPLEXES

In this section, we see that valence bond theory addresses how metal-ligand bonds form and why certain geometries are preferred, while a new model—crystal field theory—addresses why complexes are brightly colored and often paramagnetic.

Applying Valence Bond Theory to Complex Ions

Valence bond (VB) theory, which helped explain bonding and structure in main-group compounds (Section 11.1), is also used to describe bonding in complex ions. In the formation of a complex ion, the *filled* ligand orbital overlaps an *empty* metal-ion orbital: *the ligand (Lewis base) donates an electron pair, and the metal ion (Lewis acid) accepts it to form a covalent bond in the complex ion (Lewis adduct)* (Section 18.8). A bond in which one atom contributes both electrons is a **coordinate covalent bond;** once formed, it is identical to any covalent single bond.

Recall that the VB concept of hybridization proposes mixing particular combinations of *s*, *p*, and *d* orbitals to obtain sets of hybrid orbitals, which have specific geometries. For coordination compounds, the model proposes that *the type of metal-ion orbital hybridization determines the geometry of the complex ion.* Let's discuss orbital combinations that lead to octahedral, square planar, and tetrahedral geometries.

Octahedral Complexes The hexaamminechromium(III) ion, $[Cr(NH_3)_6]^{3+}$, is an *octahedral complex* (Figure 22.12). The six lowest-energy, *empty* orbitals of the Cr^{3+}

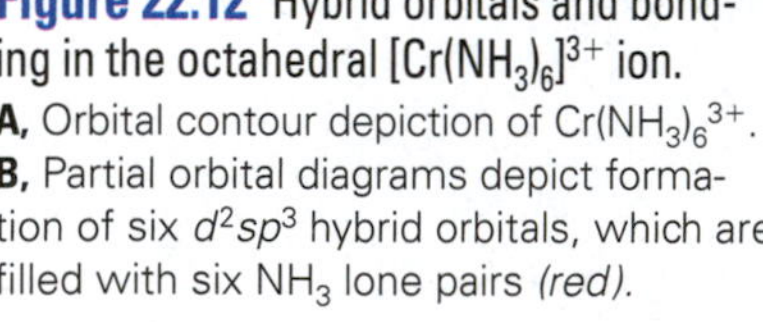
Figure 22.12 Hybrid orbitals and bonding in the octahedral $[Cr(NH_3)_6]^{3+}$ ion.
A, Orbital contour depiction of $Cr(NH_3)_6{}^{3+}$.
B, Partial orbital diagrams depict formation of six d^2sp^3 hybrid orbitals, which are filled with six NH_3 lone pairs *(red)*.

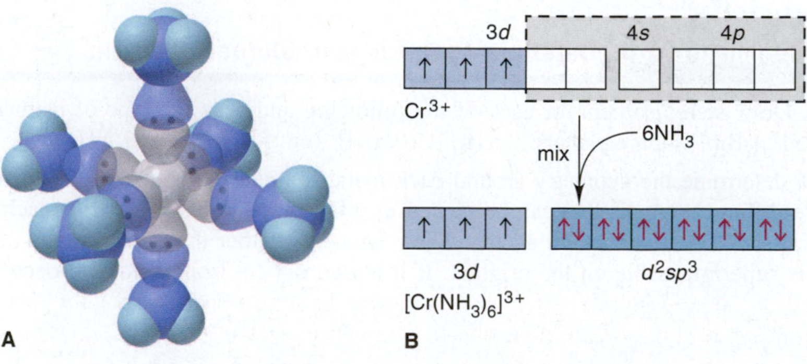

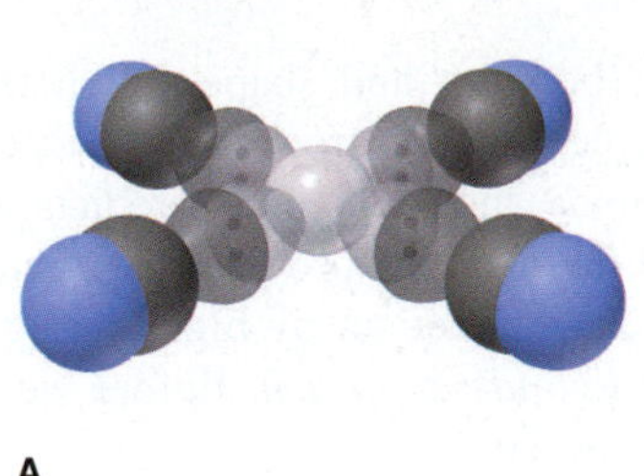

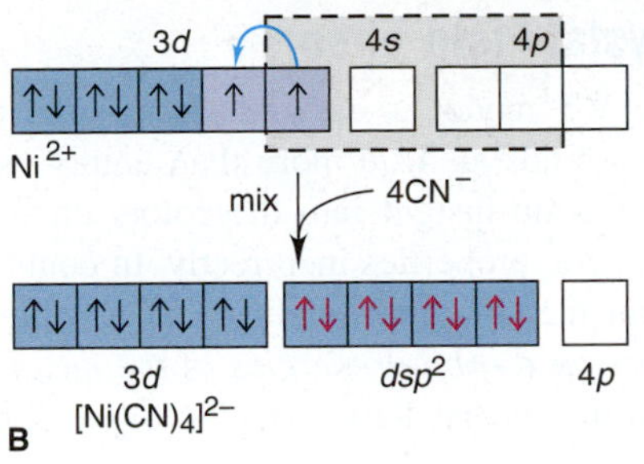

A

Figure 22.13 Hybrid orbitals and bonding in the square planar $[Ni(CN)_4]^{2-}$ ion. **A,** Contour depiction of $[Ni(CN)_4]^{2-}$. **B,** Two lone 3*d* electrons pair up, which frees one 3*d* orbital for hybridization. Four dsp^2 orbitals are filled with lone pairs *(red)* from four CN^- ligands.

ion—two 3*d*, one 4*s*, and three 4*p*—mix and become six equivalent d^2sp^3 hybrid orbitals that point toward the corners of an octahedron.* Six NH_3 molecules donate lone pairs from their N atoms to form six metal-ligand bonds. Three unpaired 3*d* electrons of the central Cr^{3+} ion ([Ar] $3d^3$) remain in unhybridized orbitals and make the complex ion paramagnetic.

Square Planar Complexes Metal ions with a d^8 configuration usually form *square planar complexes* (Figure 22.13). In the $[Ni(CN)_4]^{2-}$ ion, for example, the model proposes that one 3*d*, one 4*s*, and two 4*p* orbitals of Ni^{2+} mix and form four dsp^2 hybrid orbitals, which point to the corners of a square and accept one electron pair from each of four CN^- ligands.

A look at the ground-state electron configuration of the Ni^{2+} ion, however, raises a key question: how can the Ni^{2+} ion ([Ar] $3d^8$) offer an empty 3*d* orbital for accepting a lone pair, if its eight 3*d* electrons lie in three filled and two half-filled orbitals? Apparently, in the d^8 configuration of Ni^{2+}, electrons in the half-filled orbitals *pair up* and leave one 3*d* orbital empty. This explanation is consistent with the fact that the complex is diamagnetic (no unpaired electrons). Moreover, it means that the energy *gained* by using a 3*d* orbital for bonding in the hybrid orbital is greater than the energy *required* to overcome repulsions from pairing the 3*d* electrons.

Tetrahedral Complexes Metal ions that have a filled *d* sublevel, such as Zn^{2+} ([Ar] $3d^{10}$), often form diamagnetic *tetrahedral complexes* (Figure 22.14). In the $[Zn(OH)_4]^{2-}$ ion, for example, VB theory proposes that the lowest *empty* Zn^{2+} orbitals—one 4*s* and three 4*p*—mix to become four sp^3 hybrid orbitals that point to the corners of a tetrahedron and are occupied by lone pairs, one from each of four OH^- ligands.

*Note the distinction between hybrid-orbital designations here and in Chapter 11. Both designations give the orbitals in energy order. In $[Cr(NH_3)_6]^{3+}$, the 3*d* orbitals have a *lower n* value than the 4*s* and 4*p* orbitals, so the hybrid orbitals are d^2sp^3. But, for SF_6, the 3*d* orbitals have the *same n* value as the 3*s* and 3*p* orbitals of S, so the designation is sp^3d^2.

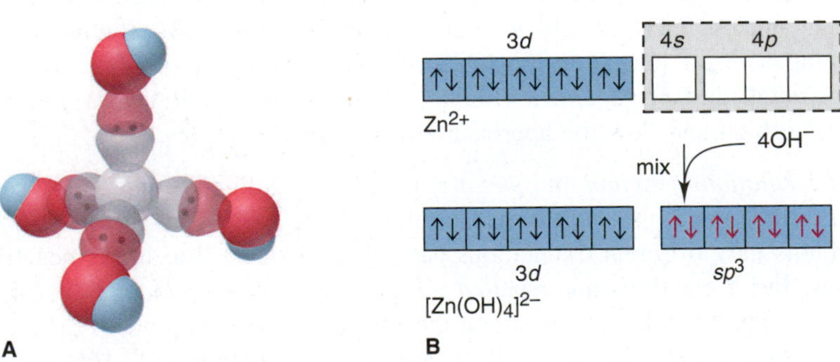

Figure 22.14 Hybrid orbitals and bonding in the tetrahedral $[Zn(OH)_4]^{2-}$ ion. **A,** Contour depiction of $[Zn(OH)_4]^{2-}$. **B,** Formation and filling of four sp^3 hybrid orbitals.

Crystal Field Theory

The VB model is easy to picture and rationalizes bonding and shape, but it treats the orbitals as little more than empty "slots" for accepting electron pairs. Moreover, it gives no insight into the colors of these compounds and sometimes predicts their magnetic properties incorrectly. In contrast, **crystal field theory** provides little insight about metal-ligand bonding, but it explains color and magnetism by highlighting the *effect on d-orbital energies of the metal ion as the ligands approach.* Before we discuss the theory, let's consider why a substance is colored.

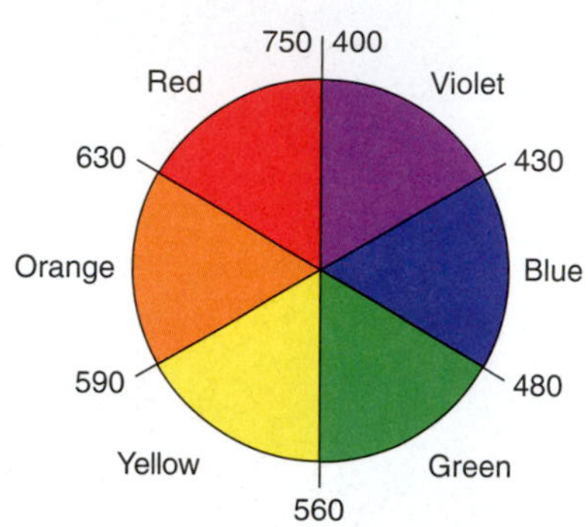

Figure 22.15 An artist's wheel. Colors, with approximate wavelength ranges (in nm), are shown as wedges.

What Is Color? White light consists of all wavelengths (λ) in the visible range (Section 7.1) and can be dispersed into colors of a narrower wavelength range. Objects appear colored in white light because they absorb only certain wavelengths: an opaque object *reflects* the other wavelengths, and a clear one *transmits* them. If an object absorbs all visible wavelengths, it appears black; if it reflects all, it appears white.

Each color has a *complementary* color; for example, green and red are complementary colors. Figure 22.15 shows these relationships on an artist's color wheel in which complementary colors are wedges opposite each other. A mixture of complementary colors absorbs all visible wavelengths and appears black.

An object has a particular color for one of two reasons:

- It reflects (or transmits) light of *that* color. Thus, if an object absorbs all wavelengths *except* green, the reflected (or transmitted) light is seen as green.
- It absorbs light of the *complementary* color. Thus, if the object absorbs only red, the *complement* of green, the remaining mixture of reflected (or transmitted) wavelengths is also seen as green.

Table 22.9 lists the color absorbed and the resulting color observed.

Table 22.9 Relation Between Absorbed and Observed Colors

Absorbed Color	λ (nm)	Observed Color	λ (nm)
Violet	400	Green-yellow	560
Blue	450	Yellow	600
Blue-green	490	Red	620
Yellow-green	570	Violet	410
Yellow	580	Dark blue	430
Orange	600	Blue	450
Red	650	Green	520

Splitting of *d* Orbitals in an Octahedral Field of Ligands The crystal field model explains that the properties of complexes result from the splitting of *d*-orbital energies, which arises from *electrostatic attractions between the metal cation and the negative charge of the ligands.* This negative charge is either partial, as in a polar covalent ligand like NH_3, or full, as in an anionic ligand like Cl^-. Picture six ligands approaching a metal ion along the mutually perpendicular x, y, and z axes, which forms an octahedral arrangement (Figure 22.16A). Let's follow the orientation of ligand and orbital and how the approach affects orbital energies.

1. *Orientation of ligand and metal ion orbitals.* As ligands approach, their electron pairs repel electrons in the five *d* orbitals of the metal ion. In the isolated ion, the *d* orbitals have different orientations but *equal* energies. But, in the negative field of ligands, the *d* electrons are *repelled unequally because of their different orbital orientations.* The ligands moving *along* the x, y, and z axes approach

- *Directly toward* the lobes of the $d_{x^2-y^2}$ and d_{z^2} orbitals (Figure 22.16B and C).
- *Between* the lobes of the d_{xy}, d_{xz}, and d_{yz} orbitals (Figure 22.16D to F).

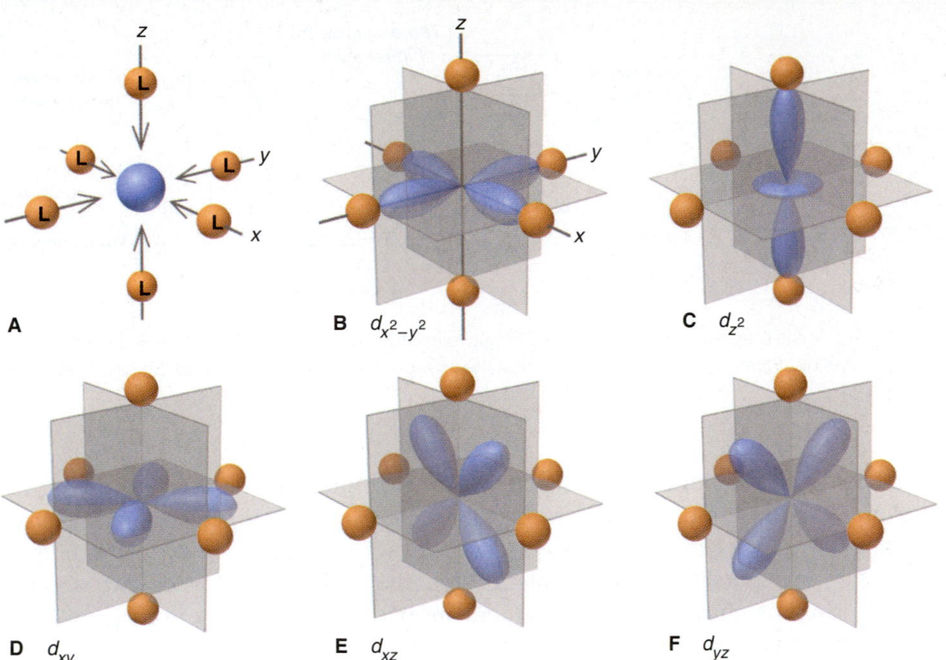

Figure 22.16 The five *d* orbitals in an octahedral field of ligands. **A,** Ligands approach along the *x, y,* and *z* axes. **B** and **C,** The $d_{x^2-y^2}$ and d_{z^2} orbitals point *directly* at some of the ligands. **D** to **F,** The d_{xy}, d_{xz}, and d_{yz} orbitals point *between* the ligands.

2. *Effect on d-orbital energies*. As a result of these different orientations, electrons in the $d_{x^2-y^2}$ and d_{z^2} orbitals experience *stronger* repulsions than electrons in the d_{xy}, d_{xz}, and d_{yz} orbitals do. An energy diagram shows that the five *d* orbitals are most stable in the free ion and their *average* energy is higher in the ligand field. But *the orbital energies split, with two d orbitals higher in energy and three lower* (Figure 22.17):

- The two higher energy orbitals are e_g **orbitals,** and they arise from the $d_{x^2-y^2}$ and d_{z^2}.
- The three lower energy orbitals are t_{2g} **orbitals,** and they arise from the d_{xy}, d_{xz}, and d_{yz}.

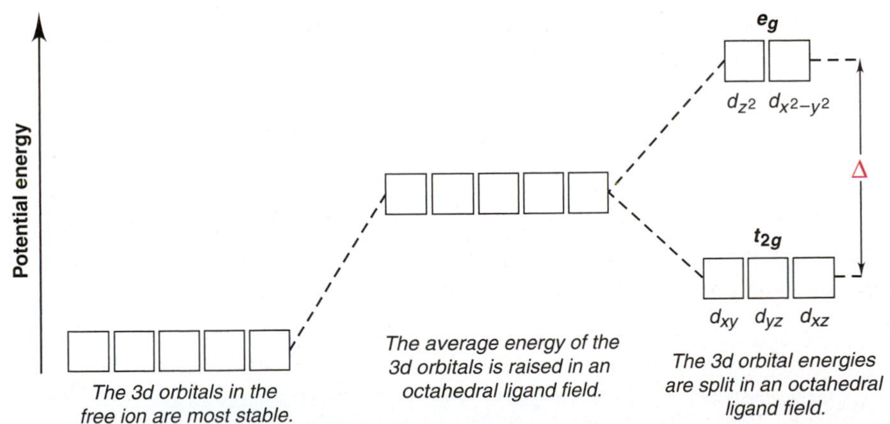

Figure 22.17 Splitting of *d*-orbital energies in an octahedral field of ligands.

3. *The crystal field effect*. This splitting of orbital energies is called the *crystal field effect,* and the energy difference between e_g and t_{2g} orbitals is the **crystal field splitting energy (Δ).** Different ligands create crystal fields of different strength:

- **Strong-field ligands** lead to a *larger* splitting energy (larger Δ).
- **Weak-field ligands** lead to a *smaller* splitting energy (smaller Δ).

For instance, H_2O is a weak-field ligand, and CN^- is a strong-field ligand (Figure 22.18). Note the different orbital occupancies; we discuss the reason for these differences shortly.

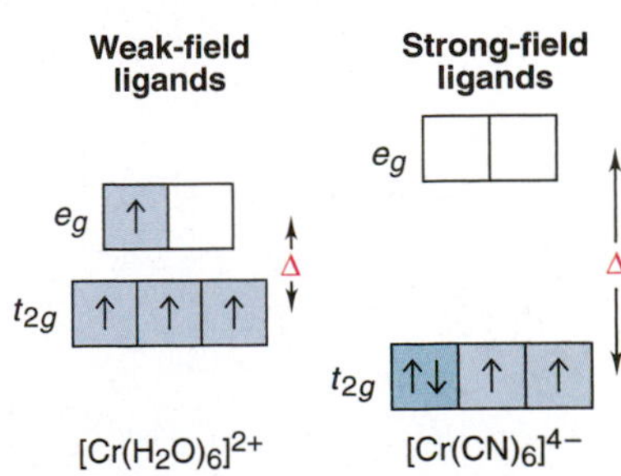

Figure 22.18 The effect of ligands and splitting energy orbital occupancy.

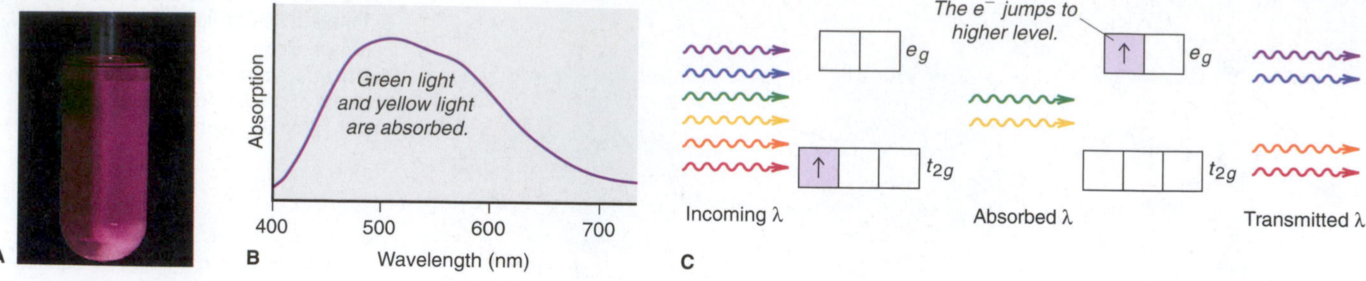

Figure 22.19 The color of $[Ti(H_2O)_6]^{3+}$. **A,** The hydrated Ti^{3+} ion is purple. **B,** An absorption spectrum shows that green light and yellow light are absorbed and other wavelengths are transmitted. **C,** An orbital diagram depicts the colors absorbed when the *d* electron jumps to the higher level.

Explaining the Colors of Transition Metal Complexes The color of a coordination compound is determined by Δ of its complex ion. When the ion absorbs radiant energy, electrons can move from the lower energy t_{2g} level to the higher energy e_g level. Recall from Chapter 7 that the *difference* between two atomic energy levels is equal to the energy (and inversely related to the wavelength) of the absorbed photon:

$$\Delta E_{\text{electron}} = E_{\text{photon}} = h\nu = hc/\lambda$$

Consider the $[Ti(H_2O)_6]^{3+}$ ion, which appears purple in aqueous solution (Figure 22.19). Hydrated Ti^{3+} has its one *d* electron in one of the three lower energy t_{2g} orbitals. The energy difference (Δ) between the t_{2g} and e_g orbitals in this ion corresponds to photons between the green and yellow range. When white light shines on the solution, these colors of light are absorbed, and the electron jumps to one of the e_g orbitals. Red, blue, and violet light are transmitted, so the solution appears purple.

Absorption spectra can show the wavelengths absorbed by (1) a metal ion with different ligands and (2) different metal ions with the same ligand. Such data allow us to relate the energy of the absorbed light to Δ and make two key observations:

- *For a given ligand,* color depends on the *oxidation state of the metal ion.* In Figure 22.20A, aqueous $[V(H_2O)_6]^{2+}$ *(left)* is violet and $[V(H_2O)_6]^{3+}$ *(right)* is yellow.
- *For a given metal ion,* color depends on the *ligand.* A single ligand substitution can affect the wavelengths absorbed and, thus, the color (Figure 22.20B).

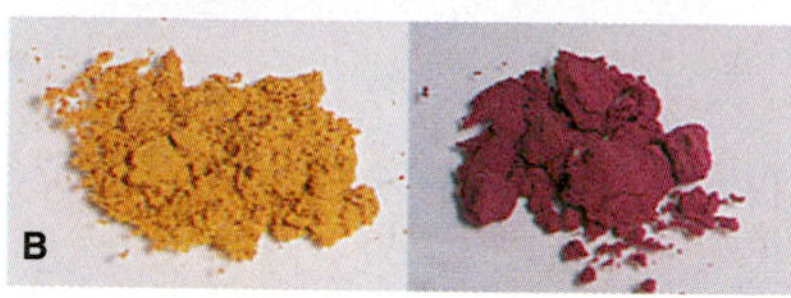

Figure 22.20 **Effects of oxidation state and ligand on color.** **A,** Solutions of two hydrated vanadium ions: O.N. of V is +2 *(left)*; O.N. of V is +3 *(right)*. **B,** A change in one ligand can influence the color. $[Cr(NH_3)_6]^{3+}$ is yellow *(left)*, and $[Cr(NH_3)_5Cl]^{2+}$ is purple *(right)*.

The Spectrochemical Series The fact that color depends on the ligand allows us to create a **spectrochemical series,** which ranks the ability of a ligand to split *d*-orbital energies (Figure 22.21). Using this series, we can predict the *relative* magnitude of Δ for a series of octahedral complexes of a *given* metal ion. Although we cannot predict the actual color of a given complex, we can determine whether a complex will absorb longer or shorter wavelengths than other complexes in the series.

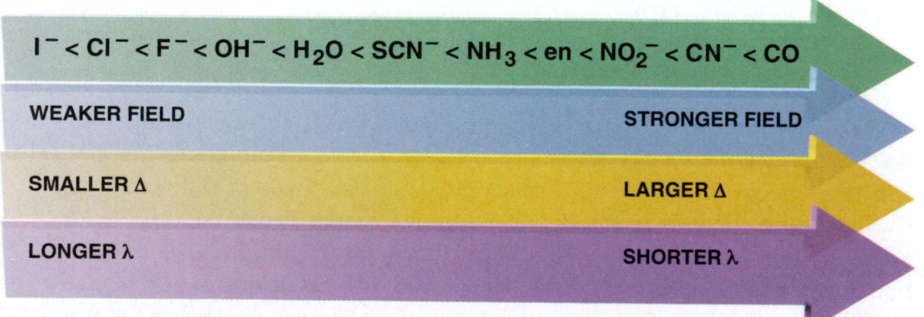

Figure 22.21 **The spectrochemical series.** As Δ increases, shorter wavelengths (higher energies) of light must be absorbed to excite electrons. For reference, water is a weak-field ligand.

Sample Problem 22.4 Ranking Crystal Field Splitting Energies (Δ) for Complex Ions of a Metal

Problem Rank $[Ti(H_2O)_6]^{3+}$, $[Ti(CN)_6]^{3-}$, and $[Ti(NH_3)_6]^{3+}$ in terms of Δ and of the energy of visible light absorbed.

Plan The formulas show that Ti has an oxidation state of +3 in the three ions. From Figure 22.21, we rank the ligands by crystal field strength: the stronger the ligand, the greater the splitting, and the higher the energy of light absorbed.

Solution The ligand field strength is in the order $CN^- > NH_3 > H_2O$, so the relative size of Δ and energy of light absorbed is

$$Ti(CN)_6^{3-} > Ti(NH_3)_6^{3+} > Ti(H_2O)_6^{3+}$$

FOLLOW-UP PROBLEM 22.4 Which complex ion absorbs visible light of higher energy, $[V(H_2O)_6]^{3+}$ or $[V(NH_3)_6]^{3+}$?

Explaining the Magnetic Properties of Transition Metal Complexes Splitting of energy levels gives rise to magnetic properties based on the number of *unpaired* electrons in the metal ion's *d* orbitals. Based on Hund's rule, electrons occupy orbitals of equal energy one at a time. When all lower energy orbitals are half-filled,

- The next electron can enter a half-filled orbital and pair up by overcoming a repulsive *pairing energy* ($E_{pairing}$).
- The next electron can enter an empty, higher energy orbital by overcoming Δ.

Thus, *the relative sizes of $E_{pairing}$ and Δ determine the occupancy of d orbitals,* which determines the number of unpaired electrons and, thus, magnetic behavior of the ion.

As an example, the isolated Mn^{2+} ion ([Ar] $3d^5$) has five unpaired electrons of equal energy (Figure 22.22A). In an octahedral field of ligands, orbital occupancy is affected by the ligand in one of two ways:

- *Weak-field ligands and high-spin complexes.* Weak-field ligands, such as H_2O in $[Mn(H_2O)_6]^{2+}$, cause a *small* splitting energy, so it takes *less* energy for *d* electrons to jump to the e_g set and stay unpaired than to pair up in the t_{2g} set (Figure 22.22B). Thus, for weak-field ligands, $E_{pairing} > \Delta$. Therefore, *the number of unpaired electrons in the complex ion is the* ***same*** *as in the free ion:* weak-field ligands create **high-spin complexes,** those with the *maximum* number of unpaired electrons.
- *Strong-field ligands and low-spin complexes.* In contrast, strong-field ligands, such as CN^- in $[Mn(CN)_6]^{4-}$, cause a *large* splitting energy, so it takes *more* energy for electrons to jump to the e_g set than it takes for them to pair up in the t_{2g} set (Figure 22.22C). Thus, for strong-field ligands, $E_{pairing} < \Delta$. Therefore, *the number of unpaired electrons in the complex ion is* ***less*** *than in the free ion.* Strong-field ligands create **low-spin complexes,** those with *fewer* unpaired electrons.

Orbital diagrams for d^1 through d^9 ions in octahedral complexes show that high-spin *and* low-spin options are possible only for d^4, d^5, d^6, and d^7 ions (Figure 22.23). With three t_{2g} orbitals available, d^1, d^2, and d^3 ions always form

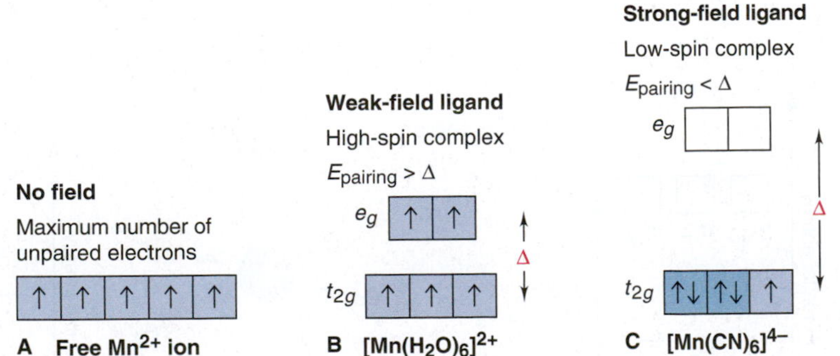

Figure 22.22 High-spin and low-spin octahedral complex ions of Mn^{2+}. **A,** Free Mn^{2+} has five unpaired electrons. **B,** Bonded to a weak-field ligand, Mn^{2+} still has five unpaired electrons (high-spin complex). **C,** Bonded to a strong-field ligand, Mn^{2+} has one unpaired electron (low-spin complex).

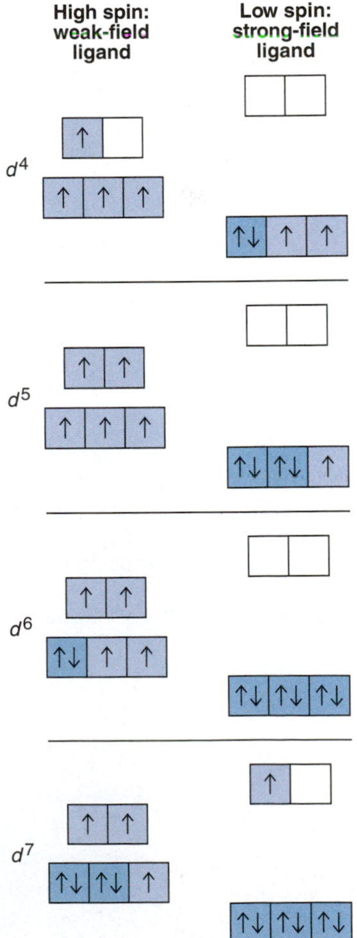

Figure 22.23 Orbital occupancy for high-spin and low-spin octahedral complexes of d^4 through d^7 metal ions.

high-spin complexes because there is no need to pair up. Similarly, d^8 and d^9 ions always form high-spin complexes because the t_{2g} set is filled with six electrons, so the e_g orbitals *must* have either two (d^8) or one (d^9) unpaired electron(s).

Sample Problem 22.5 Identifying High-Spin and Low-Spin Complex Ions

Problem Iron(II) forms a complex in hemoglobin. For each of the two octahedral complex ions $[Fe(H_2O)_6]^{2+}$ and $[Fe(CN)_6]^{4-}$, draw an energy diagram showing orbital splitting, predict the number of unpaired electrons, and identify the ion as low spin or high spin.

Plan The Fe^{2+} electron configuration shows the number of *d* electrons, and the spectrochemical series (Figure 22.21) shows the relative ligand strengths. We draw energy diagrams and separate the t_{2g} and e_g orbital sets more for the strong-field ligand. Then we add electrons, noting that a weak-field ligand gives the *maximum* number of unpaired electrons and a high-spin complex, whereas a strong-field ligand gives the *minimum* number of unpaired electrons and a low-spin complex.

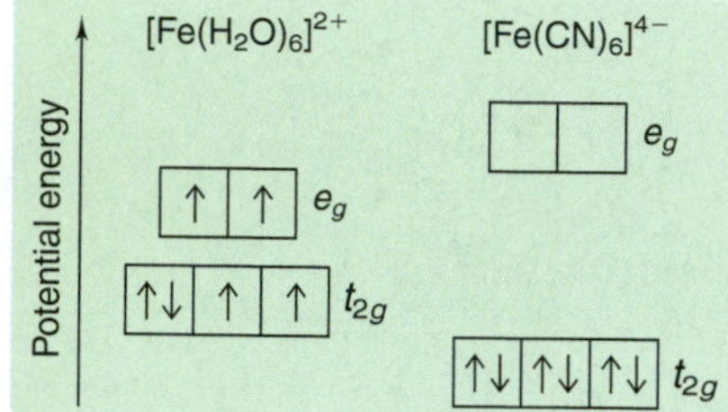

Solution Fe^{2+} has the $[Ar]3d^6$ configuration. H_2O produces smaller splitting than CN^-. The energy diagrams are shown at left. The $[Fe(H_2O)_6]^{2+}$ ion has four unpaired electrons (high spin), and the $[Fe(CN)_6]^{4-}$ ion has no unpaired electrons (low spin).

Comment **1.** H_2O is a weak-field ligand, so it forms high-spin complexes. **2.** We cannot confidently predict the spin of a complex without having actual values for Δ and $E_{pairing}$. **3.** Cyanide ions and carbon monoxide are toxic because they bind to the iron complexes in proteins involved in cellular energy.

FOLLOW-UP PROBLEM 22.5 How many unpaired electrons do you expect for $[Mn(CN)_6]^{3-}$? Is this a high-spin or low-spin complex ion?

Crystal Field Splitting in Tetrahedral and Square Planar Complexes Four ligands around a metal ion also cause *d*-orbital splitting, but the magnitude and pattern of the splitting depend on whether the ligands approach from a tetrahedral or a square planar orientation.

1. *Tetrahedral complexes.* When the ligands approach from the corners of a tetrahedron, none of the five *d* orbitals is directly in their paths (Figure 22.24A). Thus, the overall attraction of ligand and metal ion is weaker, so the splitting of *d*-orbital energies is *less* in a tetrahedral than an octahedral complex with the same ligands:

$$\Delta_{\text{tetrahedral}} < \Delta_{\text{octahedral}}$$

Repulsions are minimized when the ligands approach the d_{xy}, d_{xz}, and d_{yz} orbitals closer than they do the $d_{x^2-y^2}$ and d_{z^2} orbitals. This situation is the *opposite of the octahedral case,* so the relative *d*-orbital energies are reversed: the d_{xy}, d_{xz}, and d_{yz} orbitals become *higher* in energy than the $d_{x^2-y^2}$ and d_{z^2} orbitals. *Only high-spin tetrahedral complexes are known* because the magnitude of Δ is so small.

2. *Square planar complexes.* The effects of the ligand field in the square planar case are easier to picture if we imagine starting with an octahedral geometry and then removing the two ligands along the *z*-axis (Figure 22.24B). With no *z*-axis interactions present,

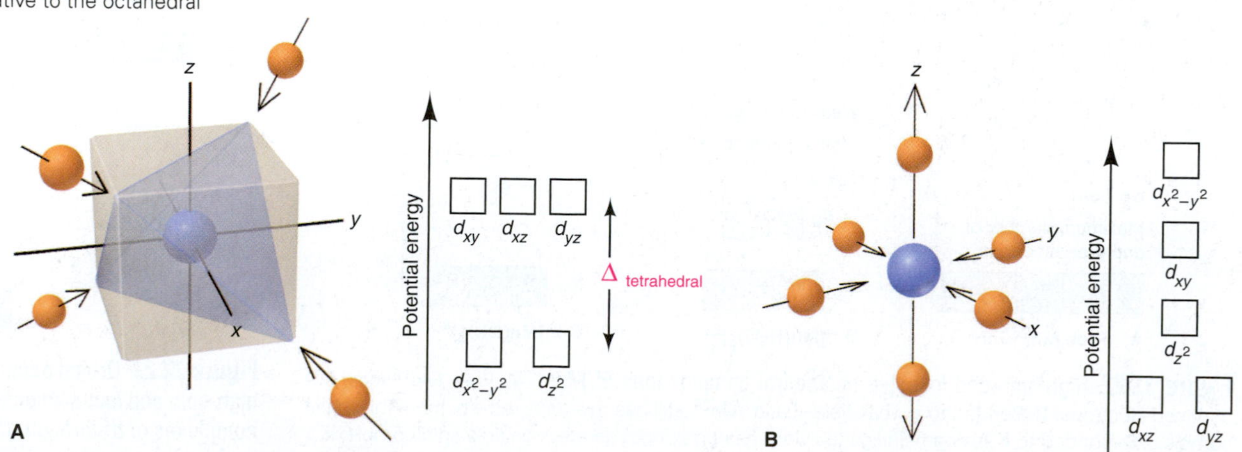

Figure 22.24 Splitting of *d*-orbital energies. **A,** The pattern of splitting by a tetrahedral field of ligands is the opposite of the octahedral pattern. **B,** Splitting by a square planar field of ligands decreases the energies of d_{xz}, d_{yz}, and especially d_{z^2} orbitals relative to the octahedral pattern.

any d orbital with a z-axis component has lower energy, with the d_{z^2} orbital decreasing most, and the d_{xz} and d_{yz}, also decreasing. In contrast, the two d orbitals in the xy-plane interact strongly with the ligands, and because the $d_{x^2-y^2}$ orbital has its lobes *on* the axes, its energy is highest. The d^8 metal ions, such as $[PdCl_4]^{2-}$, form square planar complexes. They are *low spin* and usually *diamagnetic* because the four pairs of d electrons fill the four lowest-energy orbitals.

Transition Metal Complexes in Biological Systems

In addition to four *building-block elements* (C, O, H, and N) and seven elements known as *macro*nutrients (Na, Mg, P, S, Cl, K, and Ca), organisms contain a large number of *trace elements,* most of which are transition metals. With the exception of Sc, Ti, and Ni (in most species), the Period 4 transition elements are essential to many organisms (Table 22.10), and plants require Mo (from Period 5) as well. The principles of bonding and d-orbital splitting are the same in complex biomolecules containing transition metals as in simple inorganic systems. We focus here on an iron-containing complex.

Iron plays a crucial role in oxygen transport in all vertebrates. The O_2-transporting protein hemoglobin (Figure 22.25A) consists of four folded chains, each cradling the Fe-containing complex *heme.* Heme consists of iron(II) bonded to four N lone pairs of a tetradentate ring ligand known as a *porphin* to give a square planar complex. (Porphins are common biological ligands that are also found in chlorophyll, with Mg^{2+} at the center, and in vitamin B_{12}, with Co^{3+} at the center.) In hemoglobin (Figure 22.25B), the complex is *octahedral,* with the fifth ligand of iron(II) being an N atom from a nearby amino acid (histidine), and the sixth an O atom from either an O_2 (shown) or an H_2O molecule.

In the arteries and lungs, the Fe^{2+} ion in heme binds to O_2; in the veins and tissues, O_2 is replaced by H_2O. Because H_2O is a weak-field ligand, the d^6 Fe^{2+} ion is part of a high-spin complex, and the relatively small d-orbital splitting makes venous blood absorb light at the red (low-energy) end of the spectrum and look purplish blue. On the other hand, O_2 is a strong-field ligand, so it increases the splitting energy and gives a low-spin complex. Thus, arterial blood absorbs at the blue (high-energy) end of the spectrum, which accounts for its bright red color.

Carbon monoxide is toxic because it binds to Fe^{2+} ion in heme about 200 times more strongly than O_2, which prevents the heme group from functioning:

$$\text{heme—O}_2 + \text{CO} \rightleftharpoons \text{heme—CO} + \text{O}_2$$

Table 22.10 Essential Transition Metals in Humans

Element	Function
Vanadium	Fat metabolism
Chromium	Glucose utilization
Manganese	Cell respiration
Iron	Oxygen transport; ATP formation
Cobalt	Component of vitamin B_{12}; development of red blood cells
Copper	Hemoglobin synthesis; ATP formation
Zinc	Elimination of CO_2; protein digestion

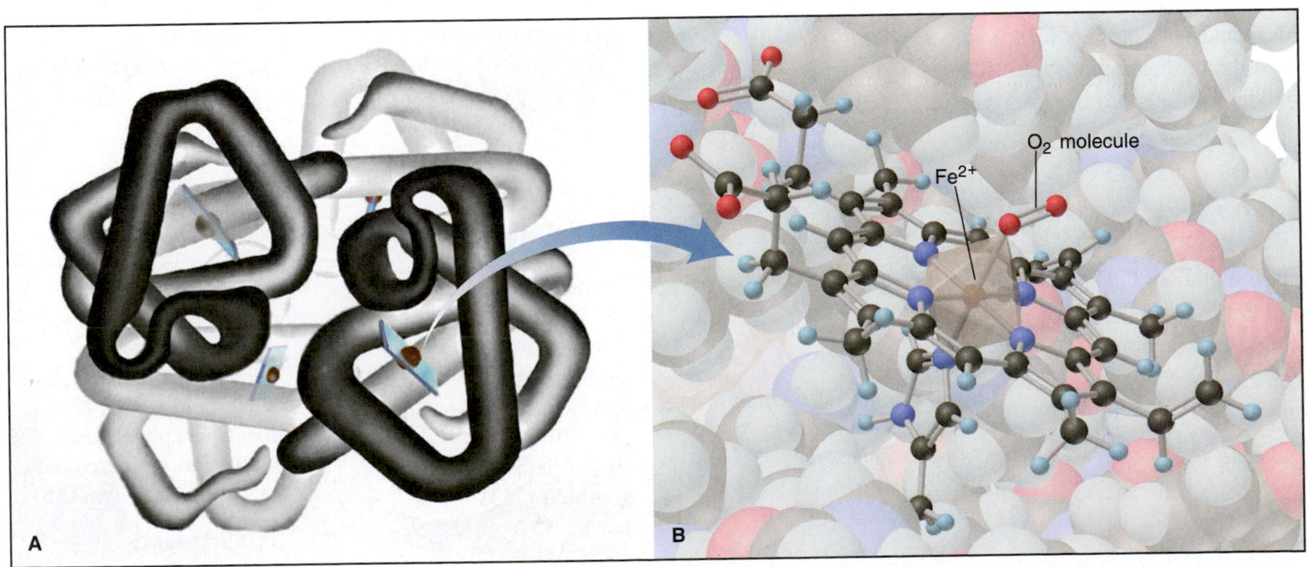

Figure 22.25 Hemoglobin and the octahedral complex in heme. **A,** Hemoglobin consists of four protein chains, each with a bound heme. (Illustration by Irving Geis. Rights owned by Howard Hughes Medical Institute. Not to be used without permission.) **B,** In oxyhemoglobin, the octahedral complex in heme has an O_2 molecule as the sixth ligand for iron(II).

Like O_2, CO is a strong-field ligand, which results in a bright red color of the blood. Because the binding is an equilibrium process, breathing extremely high concentrations of O_2 displaces CO from the heme and reverses CO poisoning:

$$\text{heme—CO} + \mathbf{O_2} \rightleftharpoons \text{heme—O}_2 + \text{CO}$$

Summary of Section 22.3

- According to valence bond theory, complex ions have coordinate covalent bonds between ligands (Lewis bases) and metal ions (Lewis acids).
- Ligand lone pairs occupy hybridized metal-ion orbitals, leading to the characteristic shapes of complex ions.
- In crystal field theory, the surrounding field of ligands splits the metal ion's *d*-orbital energies. The crystal field splitting energy (Δ) depends on the charge of the metal ion and the crystal field strength of the ligands.
- The size of Δ influences the color (energy of the photons absorbed) and paramagnetism (number of unpaired *d* electrons). Strong-field ligands create a large Δ and produce low-spin complexes that absorb light of higher energy (shorter wavelength); the reverse is true of weak-field ligands.
- Several transition metals, including iron, are essential to organisms in trace amounts.

CHAPTER REVIEW GUIDE

The following sections provide many aids to help you study this chapter. (Numbers in parentheses refer to pages, unless noted otherwise.)

Learning Objectives

These are concepts and skills to review after studying this chapter.

Related section (§), sample problem (SP), and upcoming end-of-chapter problem (EP) numbers are listed in parentheses.

1. Write electron configurations of transition metal atoms and ions; compare periodic trends in atomic properties of transition elements with those of main-group elements; explain why transition elements have multiple oxidation states, how their metallic behavior (type of bonding and oxide acidity) changes with oxidation state, and why many of their compounds are colored and paramagnetic (§22.1) (SP 22.1) (EPs 22.1–22.17)
2. Be familiar with the coordination numbers, geometries, and ligands of complex ions; name and write formulas for coordination compounds; describe the types of constitutional and stereoisomerism they exhibit (§22.2) (SPs 22.2, 22.3) (EPs 22.18–22.39)
3. Correlate the shape of a complex ion with the number and type of hybrid orbitals of the central metal ion (§22.3) (EPs 22.40, 22.41, 22.47, 22.48)
4. Describe how approaching ligands cause *d*-orbital energies to split and give rise to octahedral, tetrahedral, and square-planar complexes; explain crystal field splitting energy (Δ) and how it accounts for the colors of complexes; explain how the relative magnitudes of pairing energy and Δ determine the magnetic properties of complexes; use a spectrochemical series to rank complex ions in terms of Δ, and determine if a complex is high spin or low spin (§22.3) (SPs 22.4, 22.5) (EPs 22.42–22.46, 22.49–22.57)

Key Terms

These important terms appear in boldface in the chapter and are defined again in the Glossary.

transition elements (737)

Section 22.1

lanthanide contraction (740)

Section 22.2

coordination compound (743)
complex ion (743)
ligand (743)
counter ion (743)
coordination number (744)
donor atom (744)
chelate (744)
isomer (747)
constitutional (structural) isomers (747)
coordination isomers (747)
linkage isomers (747)
stereoisomers (748)
geometric (*cis-trans*) isomers (748)
optical isomers (748)

Section 22.3

coordinate covalent bond (750)
crystal field theory (752)
e_g orbital (753)
t_{2g} orbital (753)
crystal field splitting energy (Δ) (753)
strong-field ligand (753)
weak-field ligand (753)
spectrochemical series (754)
high-spin complex (755)
low-spin complex (755)

BRIEF SOLUTIONS TO FOLLOW-UP PROBLEMS

Compare your own solutions to these calculation steps and answers.

22.1 (a) Ag^+: $4d^{10}$ (b) Cd^{2+}: $4d^{10}$ (c) Ir^{3+}: $5d^6$

22.2 (a) Pentaaquabromochromium(III) chloride
(b) $Ba_3[Co(CN)_6]_2$

22.3 Two sets of *cis-trans* isomers, and the two *cis* isomers are optical isomers.

trans / cis (geometric); cis / cis (optical); trans / cis (geometric)

22.4 Both metal ions are V^{3+}; in terms of ligand field energy, $NH_3 > H_2O$, so $[V(NH_3)_6]^{3+}$ absorbs light of higher energy.

22.5 The metal ion is Mn^{3+}: [Ar] $3d^4$.

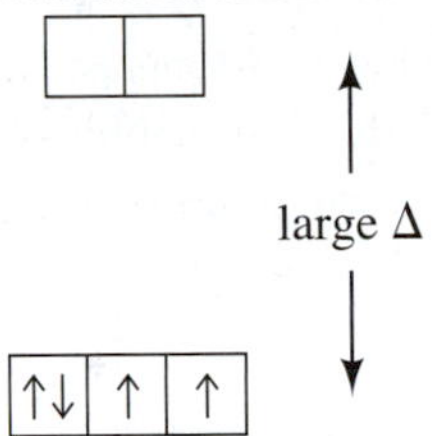

Two unpaired *d* electrons; low-spin complex

PROBLEMS

Problems with **colored** numbers are answered in Appendix E. Sections match the text and provide the numbers of relevant sample problems. Bracketed problems are grouped in pairs (indicated by a short rule) that cover the same concept. Comprehensive Problems are based on material from any section or previous chapter.

Note: In these problems, the term *electron configuration* refers to the condensed, ground-state electron configuration.

Properties of the Transition Elements

(Sample Problem 22.1)

22.1 Write the general electron configuration of a transition element (a) in Period 5; (b) in Period 6.

22.2 What is the general rule concerning the order in which electrons are removed from a transition metal atom to form an ion? Give an example from Group 5B(5). Name two types of measurements used to study electron configurations of ions.

22.3 How does the variation in atomic size across a transition series contrast with the change across the main-group elements of the same period? Why?

22.4 (a) What is the lanthanide contraction? (b) How does it affect atomic size down a group of transition elements? (c) How does it influence the densities of the Period 6 transition elements?

22.5 (a) What is the range in electronegativity across the first (3*d*) transition series? (b) What is the range across Period 4 of main-group elements? (c) Explain the difference.

22.6 (a) Explain the major difference between the number of oxidation states of most transition elements and that of most main-group elements. (b) Why is the +2 oxidation state so common among transition elements?

22.7 (a) What difference in behavior distinguishes a paramagnetic substance from a diamagnetic one?
(b) Why are paramagnetic ions common among the transition elements but not the main-group elements?
(c) Why are colored solutions of metal ions common among the transition elements but not the main-group elements?

22.8 Using the periodic table to locate each element, write the electron configuration of (a) V; (b) Y; (c) Hg.

22.9 Using the periodic table to locate each element, write the electron configuration of (a) Ru; (b) Cu; (c) Ni.

22.10 Give the electron configuration and the number of unpaired electrons for (a) Sc^{3+}; (b) Cu^{2+}; (c) Fe^{3+}; (d) Nb^{3+}.

22.11 Give the electron configuration and the number of unpaired electrons for: (a) Cr^{3+}; (b) Ti^{4+}; (c) Co^{3+}; (d) Ta^{2+}.

22.12 Which transition metals have a maximum O.N. of +6?

22.13 Which transition metals have a maximum O.N. of +4?

22.14 In which compound does Cr exhibit greater metallic behavior, CrF_2 or CrF_6? Explain.

22.15 VF_5 is a liquid that boils at 48°C, whereas VF_3 is a solid that melts above 800°C. Explain this difference in properties.

22.16 Which oxide, CrO_3 or CrO, is more acidic in water? Why?

22.17 Which oxide, Mn_2O_3 or Mn_2O_7, is more basic in water? Why?

Coordination Compounds

(Sample Problems 22.2 and 22.3)

22.18 Describe the makeup of a complex ion, including the nature of the ligands and their interaction with the central metal ion. Explain how a complex ion can be positive or negative and how it occurs as part of a neutral coordination compound.

22.19 What is the coordination number of a metal ion in a complex ion? How does it differ from oxidation number?

22.20 What structural feature is characteristic of a chelate?

22.21 What geometries are associated with the coordination numbers 2, 4, and 6?

22.22 In what sense is a complex ion the adduct of a Lewis acid-base reaction?

22.23 Is a linkage isomer a type of constitutional isomer or stereoisomer? Explain.

22.24 Give systematic names for the following formulas:
(a) $[Ni(H_2O)_6]Cl_2$ (b) $[Cr(en)_3](ClO_4)_3$ (c) $K_4[Mn(CN)_6]$

22.25 Give systematic names for the following formulas:
(a) $[Co(NH_3)_4(NO_2)_2]Cl$ (b) $[Cr(NH_3)_6][Cr(CN)_6]$
(c) $K_2[CuCl_4]$

22.26 What are the charge and coordination number of the central metal ion(s) in each compound of Problem 22.24?

22.27 What are the charge and coordination number of the central metal ion(s) in each compound of Problem 22.25?

22.28 Give systematic names for the following formulas:
(a) $K[Ag(CN)_2]$ (b) $Na_2[CdCl_4]$ (c) $[Co(NH_3)_4(H_2O)Br]Br_2$

22.29 Give systematic names for the following formulas:
(a) $K[Pt(NH_3)Cl_5]$ (b) $[Cu(en)(NH_3)_2][Co(en)Cl_4]$
(c) $[Pt(en)_2Br_2](ClO_4)_2$

22.30 Give formulas corresponding to the following names:
(a) Tetraamminezinc sulfate
(b) Pentaamminechlorochromium(III) chloride
(c) Sodium bis(thiosulfato)argentate(I)

22.31 Give formulas corresponding to the following names:
(a) Dibromobis(ethylenediamine)cobalt(III) sulfate
(b) Hexaamminechromium(III) tetrachlorocuprate(II)
(c) Potassium hexacyanoferrate(II)

22.32 What is the coordination number of the metal ion and the number of individual ions per formula unit in each of the compounds in Problem 22.30?

22.33 What is the coordination number of the metal ion and the number of individual ions per formula unit in each of the compounds in Problem 22.31?

22.34 Which of these ligands can participate in linkage isomerism: (a) NO_2^-; (b) SO_2; (c) NO_3^-? Explain with Lewis structures.

22.35 Which of these ligands can participate in linkage isomerism: (a) SCN^-; (b) $S_2O_3^{2-}$ (thiosulfate); (c) HS^-? Explain with Lewis structures.

22.36 For any of the following that can exist as isomers, state the type of isomerism and draw the structures:
(a) $[Pt(CH_3NH_2)_2Br_2]$ (b) $[Pt(NH_3)_2FCl]$ (c) $[Pt(H_2O)(NH_3)FCl]$

22.37 For any of the following that can exist as isomers, state the type of isomerism and draw the structures:
(a) $[Zn(en)F_2]$ (b) $[Zn(H_2O)(NH_3)FCl]$ (c) $[Pd(CN)_2(OH)_2]^{2-}$

22.38 For any of the following that can exist as isomers, state the type of isomerism and draw the structures:
(a) $[PtCl_2Br_2]^{2-}$ (b) $[Cr(NH_3)_5(NO_2)]^{2+}$ (c) $[Pt(NH_3)_4I_2]^{2+}$

22.39 For any of the following that can exist as isomers, state the type of isomerism and draw the structures:
(a) $[Co(NH_3)_5Cl]Br_2$ (b) $[Pt(CH_3NH_2)_3Cl]Br$
(c) $[Fe(H_2O)_4(NH_3)_2]^{2+}$

Theoretical Basis for the Bonding and Properties of Complexes

(Sample Problems 22.4 and 22.5)

22.40 According to valence bond theory, what set of orbitals is used by a Period 4 metal ion in forming (a) a square planar complex; (b) a tetrahedral complex?

22.41 A metal ion uses d^2sp^3 orbitals when forming a complex. What is its coordination number and the shape of the complex?

22.42 A complex in solution absorbs green light. What is the color of the solution?

22.43 (a) What is the crystal field splitting energy (Δ)?
(b) How does it arise for an octahedral field of ligands?
(c) How is it different for a tetrahedral field of ligands?

22.44 What is the distinction between a weak-field ligand and a strong-field ligand? Give an example of each.

22.45 How do the relative magnitudes of $E_{pairing}$ and Δ affect the paramagnetism of a complex?

22.46 Why are there both high-spin and low-spin octahedral complexes but only high-spin tetrahedral complexes?

22.47 Give the number of d electrons (n of d^n) for the central metal ion in (a) $[TiCl_6]^{2-}$; (b) $K[AuCl_4]$; (c) $[RhCl_6]^{3-}$.

22.48 Give the number of d electrons (n of d^n) for the central metal ion in (a) $[Cr(H_2O)_6](ClO_3)_2$; (b) $[Mn(CN)_6]^{2-}$; (c) $[Ru(NO)(en)_2Cl]Br$.

22.49 Which of these ions *cannot* form both high- and low-spin octahedral complexes: (a) Ti^{3+}; (b) Co^{2+}; (c) Fe^{2+}; (d) Cu^{2+}?

22.50 Which of these ions *cannot* form both high- and low-spin octahedral complexes: (a) Mn^{3+}; (b) Nb^{3+}; (c) Ru^{3+}; (d) Ni^{2+}?

22.51 Draw orbital-energy splitting diagrams and use the spectrochemical series to show the orbital occupancy for each of the following (assuming that H_2O is a weak-field ligand):
(a) $[Cr(H_2O)_6]^{3+}$ (b) $[Cu(H_2O)_4]^{2+}$ (c) $[FeF_6]^{3-}$

22.52 Draw orbital-energy splitting diagrams and use the spectrochemical series to show the orbital occupancy for each of the following (assuming that H_2O is a weak-field ligand):
(a) $[Cr(CN)_6]^{3-}$ (b) $[Rh(CO)_6]^{3+}$ (c) $[Co(OH)_6]^{4-}$

22.53 Rank the following in order of *increasing* Δ and energy of light absorbed: $[Cr(NH_3)_6]^{3+}$, $[Cr(H_2O)_6]^{3+}$, $[Cr(NO_2)_6]^{3-}$.

22.54 Rank the following in order of *decreasing* Δ and energy of light absorbed: $[Cr(en)_3]^{3+}$, $[Cr(CN)_6]^{3-}$, $[CrCl_6]^{3-}$.

22.55 A complex, ML_6^{2+}, is violet. The same metal forms a complex with another ligand, Q, that creates a weaker field. What color might MQ_6^{2+} be expected to show? Explain.

22.56 $[Cr(H_2O)_6]^{2+}$ is violet. Another CrL_6 complex is green. Can ligand L be CN^-? Can it be Cl^-? Explain.

22.57 Three of the complex ions that are formed by Co^{3+} are $[Co(H_2O)_6]^{3+}$, $[Co(NH_3)_6]^{3+}$, and $[CoF_6]^{3-}$. These ions have the observed colors (listed in arbitrary order) yellow-orange, green, and blue. Match each complex with its color. Explain.

Comprehensive Problems

22.58 How many different formulas are there for octahedral complexes with a metal M and four ligands A, B, C, and D? Give the number of isomers for each formula and describe the isomers.

22.59 Correct each name that has an error:
(a) $Na[FeBr_4]$, sodium tetrabromoferrate(II)
(b) $[Ni(NH_3)_6]^{2+}$, nickel hexaammine ion
(c) $[Co(NH_3)_3I_3]$, triamminetriiodocobalt(III)
(d) $[V(CN)_6]^{3-}$, hexacyanovanadium(III) ion
(e) $K[FeCl_4]$, potassium tetrachloroiron(III)

22.60 For the compound $[Co(en)_2Cl_2]Cl$, give:
(a) The coordination number of the metal ion
(b) The oxidation number of the central metal ion
(c) The number of individual ions per formula unit
(d) The moles of AgCl that precipitate when 1 mol of compound is dissolved in water and treated with $AgNO_3$

22.61 Consider the square planar complex shown at right. Which of the structures A–F are geometric isomers of it?

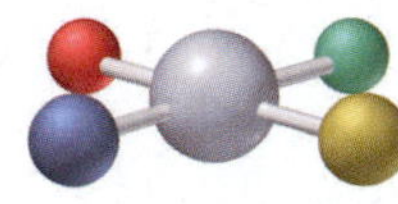

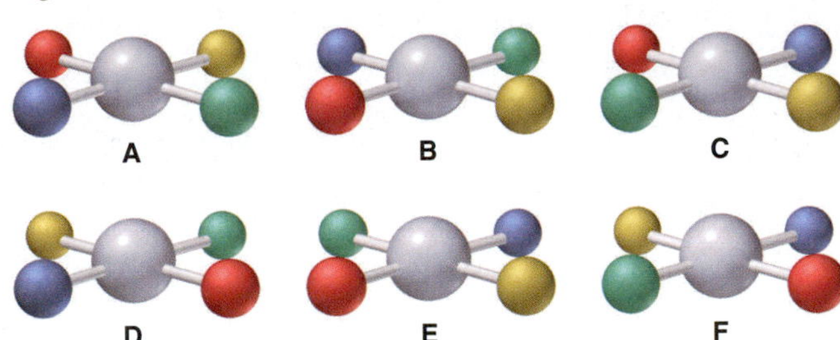

22.62 Hexafluorocobaltate(III) ion is a high-spin complex. Draw the orbital-energy splitting diagram for its *d* orbitals.

22.63 A salt of each of the ions in Table 22.3 (p. 742) is dissolved in water. A Pt electrode is immersed in each solution and connected to a 0.38-V battery. All of the electrolytic cells are run for the same amount of time with the same current.
(a) In which cell(s) will a metal plate out? Explain.
(b) Which cell will plate out the least mass of metal? Explain.

22.64 In many species, a transition metal has an unusually high or low oxidation state. Write balanced equations for the following and find the oxidation state of the transition metal in the product:
(a) Iron(III) ion reacts with hypochlorite ion in basic solution to form ferrate ion (FeO_4^{2-}), Cl^-, and water.
(b) Potassium hexacyanomanganate(II) reacts with K metal to form $K_6[Mn(CN)_6]$.
(c) Heating sodium superoxide (NaO_2) with Co_3O_4 produces Na_4CoO_4 and O_2 gas.
(d) Vanadium(III) chloride reacts with Na metal under a CO atmosphere to produce $Na[V(CO)_6]$ and NaCl.
(e) Barium peroxide reacts with nickel(II) ions in basic solution to produce $BaNiO_3$.
(f) Bubbling CO through a basic solution of cobalt(II) ion produces $[Co(CO)_4]^-$, CO_3^{2-}, and water.
(g) Heating cesium tetrafluorocuprate(II) with F_2 gas under pressure gives Cs_2CuF_6.
(h) Heating tantalum(V) chloride with Na metal produces NaCl and Ta_6Cl_{15}, in which half of the Ta is in the +2 state.
(i) Potassium tetracyanonickelate(II) reacts with hydrazine (N_2H_4) in basic solution to form $K_4[Ni_2(CN)_6]$ and N_2 gas.

22.65 An octahedral complex with three different ligands (A, B, and C) can have formulas with three different ratios of the ligands:
$[MA_4BC]^{n+}$, such as $[Co(NH_3)_4(H_2O)Cl]^{2+}$
$[MA_3B_2C]^{n+}$, such as $[Cr(H_2O)_3Br_2Cl]$
$[MA_2B_2C_2]^{n+}$, such as $[Cr(NH_3)_2(H_2O)_2Br_2]^+$
For each example, give the name, state the type(s) of isomerism present, and draw all isomers.

22.66 In $[Cr(NH_3)_6]Cl_3$, the $[Cr(NH_3)_6]^{3+}$ ion absorbs visible light in the blue-violet range, and the compound is yellow-orange. In $[Cr(H_2O)_6]Br_3$, the $[Cr(H_2O)_6]^{3+}$ ion absorbs visible light in the red range, and the compound is blue-gray. Explain these differences in light absorbed and color of the compound.

22.67 The orbital occupancies for the *d* orbitals of several complex ions are diagrammed below.

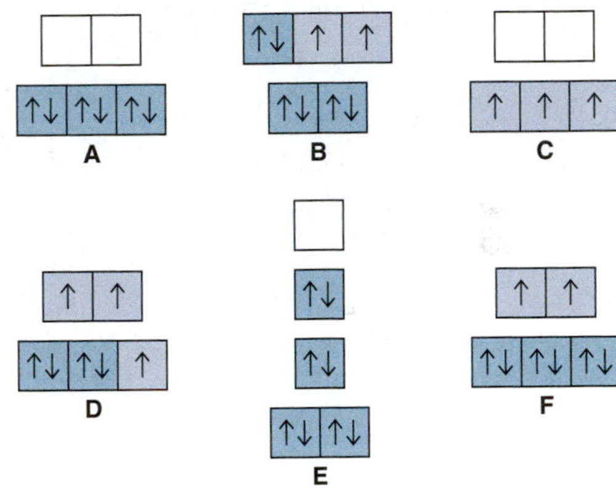

(a) Which diagram corresponds to the orbital occupancy of the cobalt ion in $[Co(CN)_6]^{3-}$?
(b) If diagram D depicts the orbital occupancy of the cobalt ion in $[CoF_6]^n$, what is the value of *n*?
(c) $[NiCl_4]^{2-}$ is paramagnetic and $[Ni(CN)_4]^{2-}$ is diamagnetic. Which diagrams correspond to the orbital occupancies of the nickel ions in these species?
(d) Diagram C shows the orbital occupancy of V^{2+} in the octahedral complex VL_6. Can you determine whether L is a strong- or weak-field ligand? Explain.

22.68 Ionic liquids have many applications in engineering and materials science. The dissolution of the metavanadate ion in chloroaluminate ionic liquids has been studied:

$$VO_3^- + AlCl_4^- \longrightarrow VO_2Cl_2^- + AlOCl_2^-$$

(a) What is the oxidation number of V and Al in each ion?
(b) In reactions of V_2O_5 with HCl, acid concentration affects the product. At low acid concentration, $VO_2Cl_2^-$ and VO_3^- form:

$$V_2O_5 + HCl \longrightarrow VO_2Cl_2^- + VO_3^- + H^+$$

At high acid concentration, $VOCl_3$ forms:

$$V_2O_5 + HCl \longrightarrow VOCl_3 + H_2O$$

Balance each equation, and state which, if either, is a redox process.
(c) What mass of $VO_2Cl_2^-$ or $VOCl_3$ can form from 12.5 g of V_2O_5 and the appropriate concentration of acid?

22.69 Several coordination isomers, with both Co and Cr as 3+ ions, have the molecular formula $CoCrC_6H_{18}N_{12}$. (a) Give the name and formula of the isomer in which the Co complex ion has six NH_3 groups. (b) Give the name and formula of the isomer in which the Co complex ion has one CN and five NH_3 groups.

22.70 The enzyme carbonic anhydrase has zinc in a tetrahedral complex at its active site. Suggest a structural reason why carbonic anhydrase synthesized with Ni^{2+}, Fe^{2+}, or Mn^{2+} in place of Zn^{2+} gives an enzyme with less catalytic efficiency.

22.71 The effect of entropy on reactions appears in the stabilities of certain complexes. (a) In terms of number of product particles, predict which of the following will be favored in terms of ΔS°_{rxn}:

$$[Cu(NH_3)_4]^{2+}(aq) + 4H_2O(l) \longrightarrow [Cu(H_2O)_4]^{2+}(aq) + 4NH_3(aq)$$

$$[Cu(H_2NCH_2CH_2NH_2)_2]^{2+}(aq) + 4H_2O(l) \longrightarrow [Cu(H_2O)_4]^{2+}(aq) + 2en(aq)$$

(b) Given that the Cu—N bond strength is approximately the same in both complexes, which complex will be more stable with respect to ligand exchange in water? Explain.

22.72 The extent of crystal field splitting is often determined from spectra. (a) Given the wavelength (λ) of maximum absorption, find the crystal field splitting energy (Δ), in kJ/mol, for each of the following complex ions:

Ion	λ (nm)	Ion	λ (nm)
$[Cr(H_2O)_6]^{3+}$	562	$[Fe(H_2O)_6]^{2+}$	966
$[Cr(CN)_6]^{3-}$	381	$[Fe(H_2O)_6]^{3+}$	730
$[CrCl_6]^{3-}$	735	$[Co(NH_3)_6]^{3+}$	405
$[Cr(NH_3)_6]^{3+}$	462	$[Rh(NH_3)_6]^{3+}$	295
$[Ir(NH_3)_6]^{3+}$	244		

(b) Write a spectrochemical series for the ligands in Cr complexes. (c) Use the Fe data to state how oxidation state affects Δ. (d) Use the Co, Rh, and Ir data to state how period number affects Δ.

Nuclear Reactions and Their Applications

At the Core This nuclear reactor is used to study the effects of radiation on materials and to produce medical and industrial isotopes. (The blue glow occurs when particles travel faster than light through the cooling water.) In this chapter, we explore the energy and radiation arising from nuclear reactions, as well as their potential benefit and harm.

Outline

Key Principles
to focus on while studying this chapter

- *Nuclear reactions* differ markedly from chemical reactions in several ways: (1) Elements typically *do* change in a nuclear reaction. (2) Nuclear particles participate, but electrons do so much less often. (3) Nuclear reactions release so much energy that the mass *does* change. (4) Rates of nuclear reactions are *not* affected by temperature or catalysts. ***(Introduction)***
- In a balanced nuclear reaction, the total mass number (A) and total charge (Z) of the reactants must equal those of the products. ***(Section 23.1)***
- Protons and neutrons are called *nucleons*. A *nuclide* is any nucleus with specific numbers of each type of nucleon. A plot of number of neutrons (N) versus number of protons (Z) for all nuclei shows a narrow *band of stability*. Unstable nuclei undergo various modes of radioactive decay. The mode can often be predicted from a nuclide's mass relative to the atomic mass and N/Z ratio. Nuclear stability is associated with filled *nucleon energy levels*. Certain heavy nuclei undergo a *decay series* to reach stability. ***(Section 23.1)***
- Radioactive decay is a *first-order process*, which means the decay rate *(activity)* depends *only* on the number of nuclei (i.e., the number raised to the first power). Thus, the *half-life*, or time required for half the nuclei present to decay, does *not* depend on the number of nuclei. ***(Section 23.2)***
- In *radiocarbon dating*, the age of an object is determined by comparing its ^{14}C activity with that of living things. ***(Section 23.2)***
- *Particle accelerators* change one element into another *(nuclear transmutation)* by bombarding nuclei with high-energy particles. ***(Section 23.3)***
- *Ionizing radiation* causes chemical changes in matter. The harm caused in living matter depends on the ionizing ability and penetrating power of the radiation. Cosmic rays and decay of radioactive minerals give rise to a natural *background radiation*. ***(Section 23.4)***
- *Isotopes* of an element have nearly identical chemical properties. A small amount of a radioactive isotope of an element mixed with a large amount of the stable isotope acts as a *tracer* for studying reaction pathways, the physical movement of substances, and medical problems. ***(Section 23.5)***
- The mass of a nucleus is *less* than the sum of its nucleon masses, and Einstein's equation gives the energy equivalent to this *mass difference*, which is the *nuclear binding energy*. The *binding energy per nucleon* is a measure of nuclide stability. Heavy nuclides split *(fission)* and light nuclides join *(fusion)* to release energy, thus increasing the binding energy per nucleon. ***(Section 23.6)***
- Nuclear power plants employ a fission *chain reaction* to create steam that generates electricity. Safety concerns center on leaks and long-term disposal of waste. Commercial energy from fusion is still not practical. ***(Section 23.7)***

CONCEPTS & SKILLS TO REVIEW before studying this chapter

- discovery of the atomic nucleus (Section 2.4)
- protons, neutrons, mass number, and the $^A_Z X$ notation (Section 2.5)
- half-life and first-order reaction rate (Section 16.4)

Far below the outer fringes of its cloud of electrons lies the atom's tiny, dense core. Up to this point, we have focused on the electrons, treating the nucleus as their electrostatic anchor and examining the effect of its positive charge on atomic properties and, ultimately, chemical behavior. Scientists studying the structure and behavior of the nucleus see enormous potential applications as well as great mystery and wonder.

But society is ambivalent about some applications of nuclear research. The promise of abundant energy and treatments for disease is offset by the threat of nuclear waste contamination, reactor accidents, and unimaginable destruction from nuclear war or terrorism. Can the power of the nucleus be harnessed for our benefit, or are the risks too great? By studying this chapter, you'll learn the principles that can help you consider this vital question.

The changes that occur in atomic nuclei differ strikingly from chemical changes. In chemical reactions, electrons are shared or transferred to form *compounds,* while nuclei remain unchanged. In nuclear reactions, the roles are reversed: electrons take part much less often, while nuclei undergo changes that, in nearly every case, form different *elements.* Nuclear reactions are often accompanied by energy changes a million times greater than those for chemical reactions, energy changes so large that changes in mass *are* detectable. Moreover, nuclear reaction yields and rates are *not* subject to the effects of temperature and catalysis that chemical reactions are. Table 23.1 summarizes these general differences.

23.1 • RADIOACTIVE DECAY AND NUCLEAR STABILITY

A stable nucleus remains intact indefinitely, but *the great majority of nuclei are unstable.* An unstable nucleus exhibits **radioactivity,** emission of radiation due to its spontaneous disintegration. In Section 23.2, you'll see that each type of unstable nucleus has its own characteristic *rate* of radioactive decay. In this section, we cover important terms and notation for nuclei, define the types of emission, and describe various modes of radioactive decay and how to predict which occurs for a given nucleus; in the process, you'll learn how to write nuclear equations.

The Components of the Nucleus: Terms and Notation

Recall from Chapter 2 that the nucleus contains essentially all the atom's mass but is only about 10^{-5} times its radius (or 10^{-15} times its volume), making the nucleus incredibly dense: about 10^{14} g/mL. *Protons* and *neutrons,* the elementary particles that make up the nucleus, are called **nucleons.** A **nuclide** is a nucleus with a particular composition, that is, with specific numbers of the two types of nucleons. Most elements occur in nature as a mixture of **isotopes,** atoms with the characteristic number of protons of the element but different numbers of neutrons. Each isotope of an ele-

Table 23.1 Comparison of Chemical and Nuclear Reactions

Chemical Reactions	Nuclear Reactions
1. One substance is converted into another, but atoms never change identity.	1. Atoms of one element typically are converted into atoms of another element.
2. Electrons in orbitals are involved as bonds break and form; nuclear particles do not take part.	2. Protons, neutrons, and other nuclear particles are involved; electrons in orbitals take part much less often.
3. Reactions are accompanied by relatively small changes in energy and no measurable changes in mass.	3. Reactions are accompanied by relatively large changes in energy and measurable changes in mass.
4. Reaction rates are influenced by temperature, concentration, catalysts, and the compound in which an element occurs.	4. Reaction rates depend on number of nuclei, but are not affected by temperature, catalysts, or, except on rare occasions, the compound in which an element occurs.

ment thus has a different nuclide. For example, oxygen has three naturally occurring isotopes—the most abundant contains eight protons and eight neutrons, whereas the least abundant contains eight protons and nine neutrons.

The relative mass and charge of a particle—nucleon, elementary particle, or nuclide—is described by the notation ${}^{A}_{Z}X$, where X is the *symbol* for the particle, A is the *mass number,* or the total number of nucleons, and Z is the *charge* of the particle; for nuclei, A is the *sum of protons and neutrons* and Z is the *number of protons* (atomic number). In this notation, the three subatomic elementary particles are

$${}^{0}_{-1}e \text{ (electron)}, {}^{1}_{1}p \text{ (proton), and } {}^{1}_{0}n \text{ (neutron)}$$

(A proton is also sometimes represented as ${}^{1}_{1}H^{+}$.) The number of neutrons (N) in a nucleus is the mass number (A) minus the atomic number (Z): $N = A - Z$. The two naturally occurring stable isotopes of chlorine, for example, have 17 protons ($Z = 17$), but one has 18 neutrons (${}^{35}_{17}Cl$, also written ${}^{35}Cl$) and the other has 20 (${}^{37}_{17}Cl$, or ${}^{37}Cl$). Nuclides can also be designated with the element name followed by the mass number, for example, chlorine-35 and chlorine-37. In naturally occurring samples of an element or its compounds, *the isotopes of the element are present in specific proportions* that vary only very slightly. Thus, in a sample of sodium chloride (or any Cl-containing substance), 75.77% of the Cl atoms are chlorine-35 and the remaining 24.23% are chlorine-37.

To understand this chapter, you need to be comfortable with nuclear notations, so please take a moment to review Sample Problem 2.4 and Problems 2.26 to 2.35.

Modes of Radioactive Decay; Balancing Nuclear Equations

When a nuclide of one element decays, it emits radiation and usually changes into a nuclide of a different element. The three natural types of radioactive emission are

- **Alpha particles** (symbolized α, ${}^{4}_{2}\alpha$, or ${}^{4}_{2}He^{2+}$) are identical to helium-4 nuclei.
- **Beta particles** (symbolized β, β^{-}, or sometimes ${}^{0}_{-1}\beta$) are high-speed electrons. (The emission of electrons from the nucleus may seem strange, but as you'll see shortly, they result from a nuclear reaction.)
- **Gamma rays** (symbolized γ, or sometimes ${}^{0}_{0}\gamma$) are very high-energy photons.

Figure 23.1 illustrates the behavior of these emissions in an electric field: the positively charged α particles curve to a small extent toward the negative plate, the negatively charged β particles curve to a greater extent toward the positive plate (because they have lower mass), and the uncharged γ rays are not affected by the electric field.

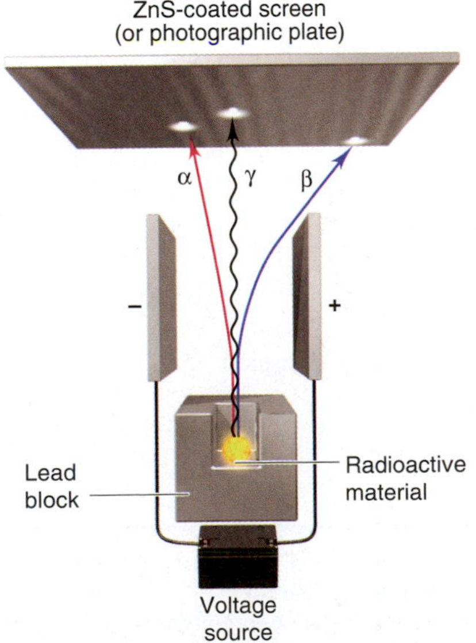

Figure 23.1 How the three types of radioactive emissions behave in an electric field.

When a nuclide decays, it forms a nuclide of lower energy, and the excess energy is carried off by the emitted radiation and the recoiling nucleus. The decaying, or reactant, nuclide is called the *parent;* the product nuclide is called the *daughter.* Nuclides can decay in several ways. As each of the major modes of decay (summarized in Table 23.2 on the next page) is introduced, we'll show examples of that mode and apply the key principle used to balance nuclear reactions: *the total Z (charge, number of protons) and the total A (sum of protons and neutrons) of the reactants equal those of the products:*

$$\frac{\text{Total } A}{\text{Total } Z}\text{ Reactants} = \frac{\text{Total } A}{\text{Total } Z}\text{ Products} \tag{23.1}$$

1. **Alpha (α) decay** involves the loss of an α particle from a nucleus. For each α particle emitted by the parent, *A decreases by 4 and Z decreases by 2* in the daughter. Every element beyond bismuth (Bi; Z = 83) is radioactive and exhibits α decay, which is *the most common means for a heavy, unstable nucleus to become more stable.* For example, radium undergoes α decay to yield radon (Rn; Z = 86):

$${}^{226}_{88}Ra \longrightarrow {}^{222}_{86}Rn + {}^{4}_{2}\alpha$$

Note that the A value for Ra equals the sum of the A values for Rn and α ($226 = 222 + 4$), and that the Z value for Ra equals the sum of the Z values for Rn and α ($88 = 86 + 2$).

Table 23.2 Modes of Radioactive Decay*

Mode	Emission	Decay Process	Change in A	Change in Z	Change in N
α Decay	$\alpha(^4_2He^{2+})$	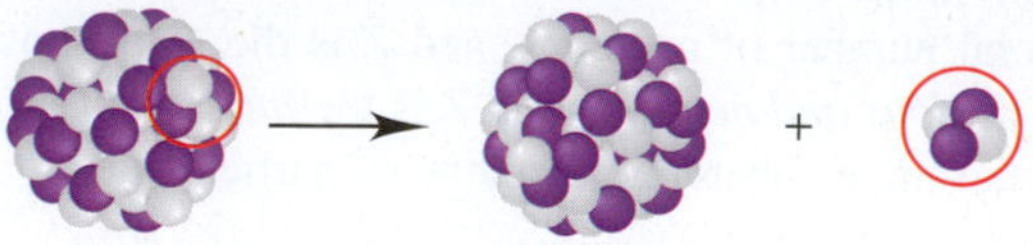 nucleus → nucleus + α expelled	-4	-2	-2
β^- Decay†	$\beta^-(^{\;0}_{-1}\beta)$	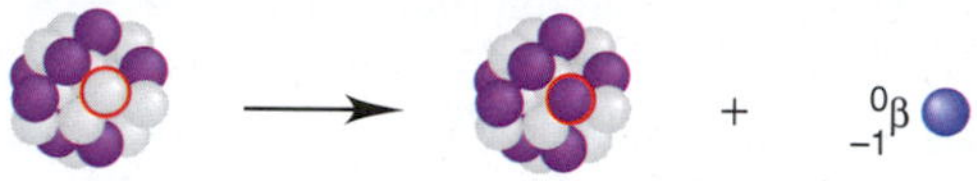 nucleus with xp^+ and yn^0 → nucleus with $(x+1)p^+$ and $(y-1)n^0$ + $^{\;0}_{-1}\beta$ β^- expelled	0	$+1$	-1
	Net:	^1_0n in nucleus → ^1_1p in nucleus + $^{\;0}_{-1}\beta$ β^- expelled			
Positron (β^+) emission†	$\beta^+(^0_1\beta)$	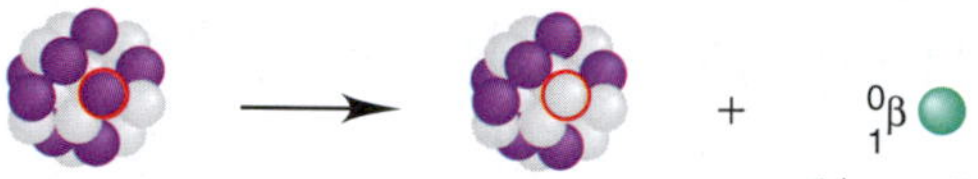 nucleus with xp^+ and yn^0 → nucleus with $(x-1)p^+$ and $(y+1)n^0$ + $^0_1\beta$ β^+ expelled	0	-1	$+1$
	Net:	^1_1p in nucleus → ^1_0n in nucleus + $^0_1\beta$ β^+ expelled			
Electron (e^-) capture (EC)†	x-ray	low-energy orbital; 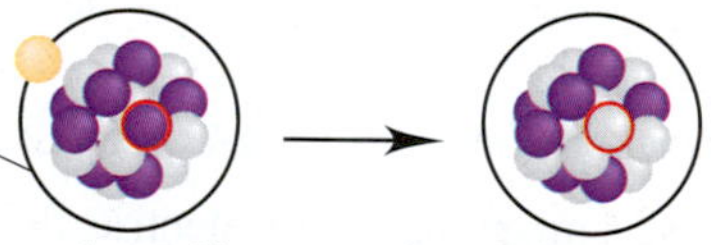 nucleus with xp^+ and yn^0 → nucleus with $(x-1)p^+$ and $(y+1)n^0$	0	-1	$+1$
	Net:	$^{\;0}_{-1}e$ absorbed from low-energy orbital + ^1_1p in nucleus → ^1_0n in nucleus			
Gamma (γ) emission	γ	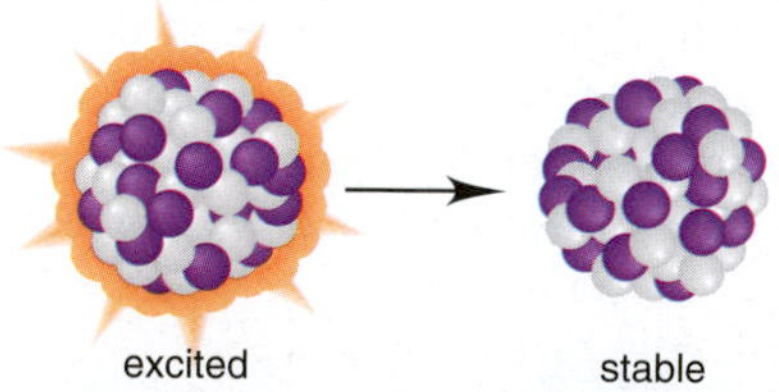 excited nucleus → stable nucleus + γ γ photon radiated	0	0	0

*Nuclear chemists consider β^- decay, positron emission, and electron capture to be three decay modes of the more general process known as beta decay (see text).

†Neutrinos (ν) or antineutrinos ($\bar{\nu}$) are also formed during the three modes of beta decay. Although we will not include them in other equations in the chapter, antineutrinos are always expelled during β^- decay, and neutrinos are expelled during β^+ emission and e^- capture. The masses of these particles are estimated to be less than 10^{-4} the mass of an electron.

2. **Beta (β) decay** is a more general class of radioactive decay that includes three modes: β^- decay, β^+ emission, and electron capture.

- **β^- decay** (or *negatron emission*) occurs through the ejection of a β^- particle from the nucleus. This change does not involve expulsion of a β^- particle that was in the nucleus; rather, *a neutron is converted into a proton, which remains in the nucleus, and a β^- particle, which is expelled immediately:*

$$ {}^{1}_{0}\text{n} \longrightarrow {}^{1}_{1}\text{p} + {}^{\ 0}_{-1}\beta $$

As always, the totals of the *A* and the *Z* values for reactant and products are equal. Radioactive nickel-63 becomes stable copper-63 through β^- decay:

$$ {}^{63}_{28}\text{Ni} \longrightarrow {}^{63}_{29}\text{Cu} + {}^{\ 0}_{-1}\beta $$

Another example is the β^- decay of carbon-14, used in radiocarbon dating:

$$ {}^{14}_{6}\text{C} \longrightarrow {}^{14}_{7}\text{N} + {}^{\ 0}_{-1}\beta $$

Note that *β^- decay results in a product nuclide with the same A but with Z one higher (one more proton) than in the reactant nuclide.* In other words, an atom of the element with the next *higher* atomic number is formed.

- **Positron (β^+) emission** is the emission of a β^+ particle from the nucleus. A key idea of modern physics is that most fundamental particles have corresponding *antiparticles* with the same mass but opposite charge. The **positron** is the antiparticle of the electron. Positron emission occurs through a process in which *a proton in the nucleus is converted into a neutron, and a positron is expelled:*

$$ {}^{1}_{1}\text{p} \longrightarrow {}^{1}_{0}\text{n} + {}^{0}_{1}\beta $$

In terms of the effect on *A* and *Z, positron emission has the opposite effect of β^- decay: the daughter has the same A but Z is one lower (one fewer proton) than the parent.* Thus, an atom of the element with the next *lower* atomic number forms. Carbon-11, a synthetic radioisotope, decays to a stable boron isotope through β^+ emission:

$$ {}^{11}_{6}\text{C} \longrightarrow {}^{11}_{5}\text{B} + {}^{0}_{1}\beta $$

- **Electron (e^-) capture (EC)** occurs when the nucleus interacts with an electron in a low atomic energy level. The net effect is that *a proton is transformed into a neutron:*

$$ {}^{1}_{1}\text{p} + {}^{\ 0}_{-1}\text{e} \longrightarrow {}^{1}_{0}\text{n} $$

(We use the symbol "e" to distinguish an orbital electron from a beta particle, β.) The orbital vacancy is quickly filled by an electron that moves down from a higher energy level, and that process continues through still higher energy levels, with x-ray photons and neutrinos carrying off the energy difference in each step. Radioactive iron forms stable manganese through electron capture:

$$ {}^{55}_{26}\text{Fe} + {}^{\ 0}_{-1}\text{e} \longrightarrow {}^{55}_{25}\text{Mn} + h\nu \text{ (x-rays and neutrinos)} $$

Even though the processes are different, *electron capture has the same net effect as positron emission: Z* lower by 1, *A* unchanged.

3. **Gamma (γ) emission** involves the radiation of high-energy γ photons (also called γ rays) from an excited nucleus. Just as an atom in an excited *electronic* state reduces its energy by emitting photons, usually in the UV and visible ranges (see Section 7.2), a nucleus in an excited state lowers its energy by emitting γ photons, which are of much higher energy (much shorter wavelength) than UV photons. Many nuclear processes leave the nucleus in an excited state, so γ *emission accompanies many other (mostly β) types of decay.* Several γ photons of different energies can be emitted from an excited nucleus as it returns to the ground state:

$$ {}^{215}_{84}\text{Po} \longrightarrow {}^{211}_{82}\text{Pb} + {}^{4}_{2}\alpha \text{ (several } \gamma \text{ emitted)} $$

Gamma emission often accompanies β^- decay:

$$ {}^{99}_{43}\text{Tc} \longrightarrow {}^{99}_{44}\text{Ru} + {}^{\ 0}_{-1}\beta \text{ (several } \gamma \text{ emitted)} $$

Because γ rays have no mass or charge, *γ emission does not change A or Z.* Two gamma rays are emitted when a particle and an antiparticle annihilate each other. In the medical technique known as *positron-emission tomography* (which we discuss in Section 23.5), a positron and an electron annihilate each other (with all A and Z values shown):

$$^{0}_{1}\beta + {}^{0}_{-1}e \longrightarrow 2\,^{0}_{0}\gamma$$

Sample Problem 23.1 Writing Equations for Nuclear Reactions

Problem Write balanced equations for the following nuclear reactions:
(a) Naturally occurring thorium-232 undergoes α decay.
(b) Zirconium-86 undergoes electron capture.

Plan We first write a skeleton equation that includes the mass numbers, atomic numbers, and symbols of all the particles on the correct sides of the equation, showing the unknown product particle as $^{A}_{Z}X$. Then, because the total of mass numbers and the total of charges on the left side and the right side must be equal, we solve for A and Z, and use Z to determine X from the periodic table.

Solution (a) Writing the skeleton equation, with the α particle as a product:

$$^{232}_{90}Th \longrightarrow {}^{A}_{Z}X + {}^{4}_{2}\alpha$$

Solving for A and Z and balancing the equation: For A, $232 = A + 4$, so $A = 228$. For Z, $90 = Z + 2$, so $Z = 88$. From the periodic table, we see that the element with $Z = 88$ is radium (Ra). Thus, the balanced equation is

$$^{232}_{90}Th \longrightarrow {}^{228}_{88}Ra + {}^{4}_{2}\alpha$$

(b) Writing the skeleton equation, with the captured electron as a reactant:

$$^{86}_{40}Zr + {}^{0}_{-1}e \longrightarrow {}^{A}_{Z}X$$

Solving for A and Z and balancing the equation: For A, $86 + 0 = A$, so $A = 86$. For Z, $40 + (-1) = Z$, so $Z = 39$. The element with $Z = 39$ is yttrium (Y), so we have

$$^{86}_{40}Zr + {}^{0}_{-1}e \longrightarrow {}^{86}_{39}Y$$

Check Always read across superscripts and then across subscripts, with the yield arrow as an equal sign, to check your arithmetic. In part (a), for example, $232 = 228 + 4$, and $90 = 88 + 2$.

FOLLOW-UP PROBLEM 23.1 Write a balanced equation for the reaction in which a nuclide undergoes β^- decay and changes to cesium-133.

Nuclear Stability and the Mode of Decay

Can we predict how, and whether, an unstable nuclide will decay? Our knowledge of the nucleus is much less complete than that of the whole atom, but some patterns emerge by observing naturally occurring nuclides.

The Band of Stability Two key factors determine the stability of a nuclide:

1. The number of neutrons (N), the number of protons (Z), and their ratio (N/Z), which we calculate from $(A - Z)/Z$. This factor relates primarily to nuclides that undergo one of the three modes of β decay.
2. The total mass of the nuclide, which mostly relates to nuclides that undergo α decay.

Figure 23.2A is a plot of number of neutrons vs. number of protons for all *stable* nuclides. Note the following:

- The points form a narrow **band of stability** that gradually curves above the line for $N = Z$ ($N/Z = 1$).
- The only stable nuclides with $N/Z < 1$ are $^{1}_{1}H$ and $^{3}_{2}He$.
- Many lighter nuclides with $N = Z$ are stable, such as $^{4}_{2}He$, $^{12}_{6}C$, $^{16}_{8}O$, and $^{20}_{10}Ne$; the heaviest of these is $^{40}_{20}Ca$. Thus, for lighter nuclides, one neutron for each proton ($N = Z$) is enough to provide stability.

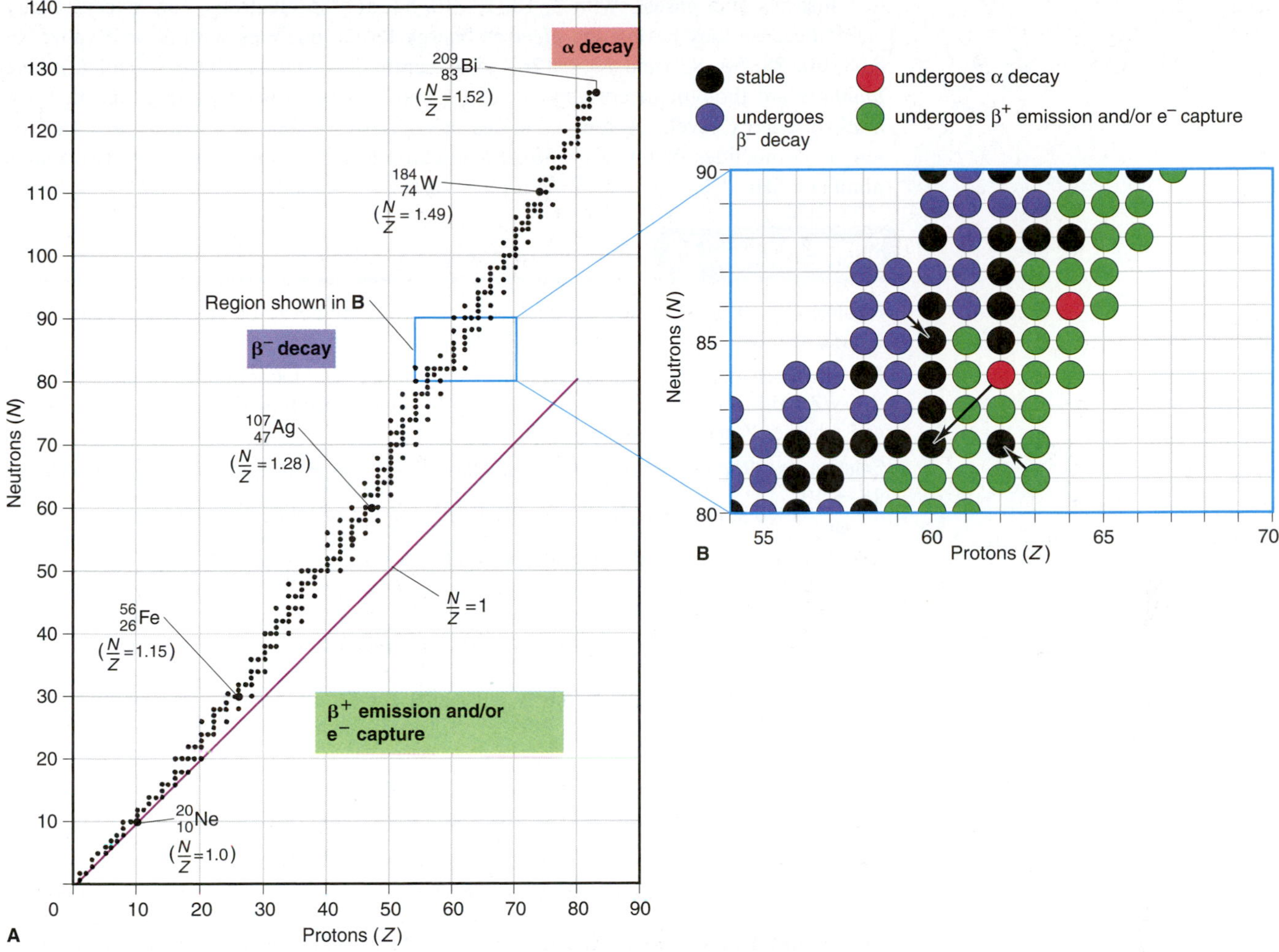

Figure 23.2 A plot of number of neutrons vs. number of protons for the stable nuclides.

- The N/Z ratio of stable nuclides gradually increases as Z increases. For example, for $^{56}_{26}$Fe, $N/Z = 1.15$; for $^{107}_{47}$Ag, $N/Z = 1.28$; for $^{184}_{74}$W, $N/Z = 1.49$, and, finally, for $^{209}_{83}$Bi, $N/Z = 1.52$. Thus, for heavier stable nuclides, $N > Z$ ($N/Z > 1$), and N increases faster than Z. As we discuss below, and show in Figure 23.2B, if N/Z of a nuclide is either too high (above the band) or not high enough (below the band), the nuclide is unstable and undergoes one of the three modes of beta decay.
- All nuclides with $Z > 83$ are unstable. Thus, the largest members of main groups 1A(1) through 8A(18), actinium and the actinides (Z = 89–103), and the other elements of the fourth (6*d*) transition series (Z = 104–112), are radioactive and (as discussed below) undergo α decay.

Stability and Nuclear Structure The oddness or evenness of N and Z values is related to some important patterns of nuclear stability. Two interesting points become apparent when we classify the known stable nuclides:

1. Elements with an even Z (number of protons) usually have a larger number of stable nuclides than elements with an odd Z. Table 23.3 demonstrates this point for cadmium ($Z = 48$) through xenon ($Z = 54$).
2. Well over half the stable nuclides have *both* even N and even Z. Only four nuclides with odd N and odd Z are stable: $^{2}_{1}$H, $^{6}_{3}$Li, $^{10}_{5}$B, and $^{14}_{7}$N.

One model of nuclear structure that attempts to explain the stability of even values of N and Z postulates that protons and neutrons lie in *nucleon energy levels,* and that greater stability results from the *pairing of spins* of like nucleons. (Note the analogy to electron energy levels and the stability from pairing of electron spins.)

Table 23.3 Number of Stable Nuclides for Elements 48 to 54*

Element	Atomic No. (Z)	No. of Nuclides
Cd	**48**	**8**
In	49	2
Sn	**50**	**10**
Sb	51	2
Te	**52**	**8**
I	53	1
Xe	**54**	**9**

*Even Z shown in boldface.

Just as noble gases—with 2, 10, 18, 36, 54, and 86 electrons—are exceptionally stable because they have filled *electron* energy levels, nuclides with N or Z values of 2, 8, 20, 28, 50, 82 (and $N = 126$) are exceptionally stable. These so-called *magic numbers* are thought to correspond to the numbers of protons or neutrons in filled *nucleon* energy levels. A few examples are $^{50}_{22}\text{Ti}$ ($N = 28$), $^{88}_{38}\text{Sr}$ ($N = 50$), and the ten stable nuclides of tin ($Z = 50$). Some extremely stable nuclides have two magic numbers: $^{4}_{2}\text{He}$, $^{16}_{8}\text{O}$, $^{40}_{20}\text{Ca}$, and $^{208}_{82}\text{Pb}$ ($N = 126$).

Sample Problem 23.2 Predicting Nuclear Stability

Problem Which of the following nuclides would you predict to be stable and which radioactive: **(a)** $^{18}_{10}\text{Ne}$; **(b)** $^{32}_{16}\text{S}$; **(c)** $^{236}_{90}\text{Th}$; **(d)** $^{123}_{56}\text{Ba}$? Explain.

Plan In order to evaluate the stability of each nuclide, we determine the N and Z values, the N/Z ratio from $(A - Z)/Z$, the value of Z, stable N/Z ratios (from Figure 23.2A), and whether Z and N are even or odd.

Solution **(a)** Radioactive. This nuclide has $N = 8\ (18 - 10)$ and $Z = 10$, so $N/Z = \dfrac{18 - 10}{10} = 0.8$. Except for hydrogen-1 and helium-3, no nuclides with $N < Z$ are stable; despite its even N and Z, this nuclide has too few neutrons to be stable.

(b) Stable. This nuclide has $N = Z = 16$, so $N/Z = 1.0$. With $Z < 20$ and even N and Z, this nuclide is most likely stable.

(c) Radioactive. This nuclide has $Z = 90$, and every nuclide with $Z > 83$ is radioactive.

(d) Radioactive. This nuclide has $N = 67$ and $Z = 56$, so $N/Z = 1.20$. For Z values of 55 to 60, Figure 23.2A shows $N/Z \geq 1.3$, so this nuclide has too few neutrons to be stable.

Check By consulting a table of isotopes, such as the one in the *CRC Handbook of Chemistry and Physics,* we find that our predictions are correct.

FOLLOW-UP PROBLEM 23.2 Why is $^{31}_{15}\text{P}$ stable but $^{30}_{15}\text{P}$ unstable?

Predicting the Mode of Decay An unstable nuclide generally decays in a mode that shifts its N/Z ratio toward the band of stability. This fact is illustrated in Figure 23.2B, which expands the small region of $Z = 54$ to 70 in Figure 23.2A to include *both* the stable and many of the unstable nuclides in that region, as well as their modes of decay. Note the following points:

1. *Neutron-rich nuclides.* Nuclides with too many neutrons for stability (a high N/Z) lie above the band of stability. They undergo *β^- decay,* which converts a neutron into a proton, thus reducing the value of N/Z.
2. *Proton-rich nuclides.* Nuclides with too many protons for stability (a low N/Z) lie below the band. They undergo *β^+ emission* and/or *e^- capture,* both of which convert a proton into a neutron, thus increasing the value of N/Z. (The rate of e^- capture increases with Z, so β^+ emission is more common among lighter elements and e^- capture more common among heavier elements.)
3. *Heavy nuclides.* Nuclides with $Z > 83$ (shown in Figure 23.2A) are too heavy to be stable and undergo *α decay,* which reduces their Z and N values by two units per emission.

With the information in Figure 23.2, predicting the mode of decay of an unstable nuclide means just comparing its N/Z ratio with those in the nearby region of the band of stability. But, even without Figure 23.2, we can often make an educated guess about mode of decay. The atomic mass of an element is the weighted average of its naturally occurring isotopes. Therefore,

- The mass number A of a *stable* nuclide will be relatively close to the atomic mass.
- If an *unstable* nuclide of an element (given Z) has an A value much higher than the atomic mass, it is neutron rich and will probably decay by β^- emission.
- If, on the other hand, the unstable nuclide has an A value much lower than the atomic mass, it is proton rich and will probably decay by β^+ emission and/or e^- capture.

In the next sample problem, we predict the mode of decay of some unstable nuclides.

Sample Problem 23.3 Predicting the Mode of Nuclear Decay

Problem Use the atomic mass of the element to predict the mode(s) of decay of the following radioactive nuclides: **(a)** $^{12}_{5}B$; **(b)** $^{234}_{92}U$; **(c)** $^{81}_{33}As$; **(d)** $^{127}_{57}La$.

Plan If the nuclide is too heavy to be stable ($Z > 83$), it undergoes α decay. For other cases, we use the *Z* value to obtain its atomic mass from the periodic table. If the mass number of the nuclide is higher than the atomic mass, the nuclide has too many neutrons: *N* too high ⇒ β^- decay. If the mass number is lower than the atomic mass, the nuclide has too many protons: *Z* too high ⇒ β^+ emission and/or e^- capture.

Solution **(a)** This nuclide has $Z = 5$, which is boron (B), and the atomic mass is 10.81. The nuclide's *A* value of 12 is significantly higher than its atomic mass, so this nuclide is neutron rich. It will probably undergo β^- decay.
(b) This nuclide has $Z = 92$, so it will undergo α decay and decrease its total mass.
(c) This nuclide has $Z = 33$, which is arsenic (As), and the atomic mass is 74.92. The *A* value of 81 is much higher, so this nuclide is neutron rich and will probably undergo β^- decay.
(d) This nuclide has $Z = 57$, which is lanthanum (La), and the atomic mass is 138.9. The *A* value of 127 is much lower, so this nuclide is proton rich and will probably undergo β^+ emission and/or e^- capture.

Check To confirm our predictions in (a), (c), and (d), let's compare each nuclide's *N/Z* ratio to those in the band of stability. **(a)** This nuclide has $N = 7$ and $Z = 5$, so $N/Z = 1.40$, which is too high for this region of the band, so it will undergo β^- decay. **(c)** This nuclide has $N = 48$ and $Z = 33$, so $N/Z = 1.45$, which is too high for this region of the band, so it undergoes β^- decay. **(d)** This nuclide has $N = 70$ and $Z = 57$, so $N/Z = 1.23$, which is too low for this region of the band, so it undergoes β^+ emission and/or e^- capture. Our predictions based on *N/Z values* are the same as those based on atomic mass.

Comment Both possible modes of decay are observed for the nuclide in part (d).

FOLLOW-UP PROBLEM 23.3 Use the *A* value for the nuclide and the atomic mass in the periodic table to predict the mode of decay of **(a)** $^{61}_{26}Fe$; **(b)** $^{241}_{95}Am$.

Decay Series A parent nuclide may undergo a series of decay steps before a stable daughter nuclide forms. The succession of steps is called a **decay series,** or **disintegration series,** and is typically depicted on a gridlike display. Figure 23.3 shows the decay series from uranium-238 to lead-206. Numbers of neutrons (*N*) are plotted against numbers of protons (*Z*) to form the grid, which displays a series of α and β^- decays. The typical zigzag pattern arises because $N > Z$, which means that α decay, which reduces both *N* and *Z* by two units, decreases *Z* slightly more than it does *N*. Therefore, α decays result in neutron-rich daughters, which undergo β^- decay to gain more stability. Note that a given nuclide can undergo both modes of decay. (Gamma emission accompanies many of these steps but does not affect the type of nuclide.)

The series in Figure 23.3 is one of three that occur in nature. All end with isotopes of lead, whose nuclides all have one ($Z = 82$) or two ($N = 126$, $Z = 82$) magic numbers. A second series begins with uranium-235 and ends with lead-207, and a third begins with thorium-232 and ends with lead-208. (Neptunium-237 began a fourth series, but its half-life is so much less than the age of Earth that only traces of it remain today.)

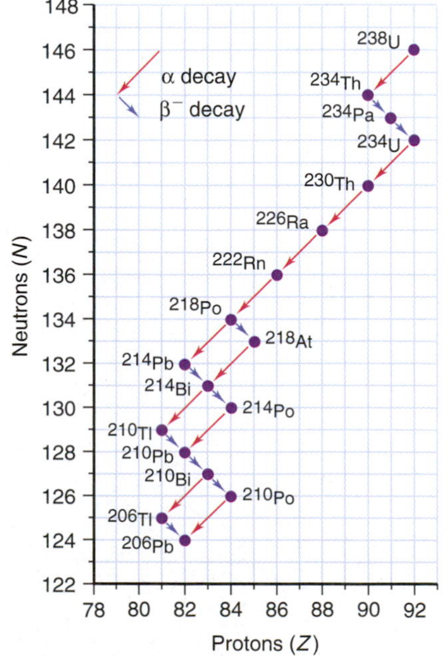

Figure 23.3 The ^{238}U decay series.

Summary of Section 23.1

- In general, nuclear reactions are not affected by reaction conditions or chemical composition and release much more energy than chemical reactions.
- To become more stable, a radioactive nuclide may emit α particles ($^4_2He^{2+}$; helium-4 nuclei), β particles (β^- or $^{\ 0}_{-1}\beta$; high-speed electrons), positrons (β^+ or $^0_1\beta$), or γ rays (high-energy photons) or may capture an orbital electron ($^{\ 0}_{-1}e$).

- A narrow band of neutron-to-proton ratios (N/Z) includes those of all the stable nuclides.
- Even values of N and Z are associated with stable nuclides, as are certain "magic numbers" of neutrons and protons.
- By comparing a nuclide's mass number with the atomic mass and its N/Z ratio with those in the band of stability, we can predict that, in general, neutron-rich nuclides undergo β^- decay and proton-rich nuclides undergo β^+ emission and/or e^- capture. Heavy nuclides ($Z > 83$) undergo α decay.
- Three naturally occurring decay series all end in isotopes of lead.

23.2 • THE KINETICS OF RADIOACTIVE DECAY

Both chemical and nuclear systems change to maximize stability. Just as the concentrations in a chemical system change in a predictable direction to give a stable equilibrium ratio, the type and number of nucleons in an unstable nucleus change in a predictable direction to give a stable N/Z ratio. As you know, however, the tendency of a chemical system to become more stable tells nothing about how long that process will take, and the same holds true for nuclear systems. In this section, we examine the kinetics of nuclear change; later, we'll examine the energetics.

The Rate of Radioactive Decay

Radioactive nuclei decay at a characteristic rate, regardless of the chemical substance in which they occur. The *decay rate,* or **activity ($\mathcal{A}$),** of a radioactive sample is the change in number of nuclei ($\mathcal{N}$) divided by the change in time (t). As with chemical reaction rates (Section 16.2), because the number of nuclei is *decreasing,* a minus sign precedes the expression to obtain a positive decay rate:

$$\text{Decay rate } (\mathcal{A}) = -\frac{\Delta\mathcal{N}}{\Delta t}$$

The SI unit of radioactivity is the **becquerel (Bq),** defined as one disintegration per second (d/s): 1 Bq = 1 d/s. A much larger and more common unit of radioactivity is the **curie (Ci),** which was originally defined as the number of disintegrations per second in 1 g of radium-226, but is now a fixed quantity:

$$1 \text{ Ci} = 3.70\times10^{10} \text{ d/s} \quad \textbf{(23.2)}$$

Because the curie is so large, the millicurie (mCi) and microcurie (μCi) are more commonly used. The radioactivity of a sample is often given as a *specific activity,* the decay rate per gram.

An activity is meaningful only when we consider the large number of nuclei in a macroscopic sample. Suppose there are 1×10^{15} radioactive nuclei of a particular type in a sample and they decay at a rate of 10% per hour. Although any particular nucleus in the sample might decay in a microsecond or in a million hours, the *average* of all decays results in 10% of the entire collection of nuclei disintegrating each hour. During the first hour, 10% of the *original* number, or 1×10^{14} nuclei, will decay. During the next hour, 10% of the remaining 9×10^{14} nuclei, or 9×10^{13} nuclei, will decay. During the next hour, 10% of those remaining will decay, and so forth. Thus, for a large collection of radioactive nuclei, *the number decaying per unit time is proportional to the number present:*

$$\text{Decay rate } (\mathcal{A}) \propto \mathcal{N} \quad \text{or} \quad \mathcal{A} = k\mathcal{N}$$

where k is called the **decay constant** and is characteristic of each type of nuclide. The larger the value of k, the higher is the decay rate: larger $k \Rightarrow$ higher $\mathcal{A}$.

Combining the two expressions for decay rate just given, we obtain

$$\mathcal{A} = -\frac{\Delta\mathcal{N}}{\Delta t} = k\mathcal{N} \quad \textbf{(23.3)}$$

Note that the activity depends only on $\mathcal{N}$ raised to the first power (and on the constant value of k). Therefore, *radioactive decay is a first-order process* (Section 16.4), but with respect to the *number* of nuclei rather than their concentration.

Half-Life of Radioactive Decay Decay rates are also commonly expressed in terms of the fraction of nuclei that decay over a given time interval. The **half-life ($t_{1/2}$)** of a nuclide has the same meaning as for a chemical change (Section 16.4) and can be expressed in terms of number of nuclei, mass of radioactive substance, and activity:

- *Number of nuclei.* Half-life is the time it takes for half the nuclei in a sample to decay—*the number of nuclei remaining is halved after each half-life.* Figure 23.4 shows the decay of carbon-14, which has a half-life of 5730 years, in terms of number of ^{14}C nuclei remaining:

$$^{14}_{6}C \longrightarrow\ ^{14}_{7}N +\ ^{0}_{-1}\beta$$

- *Mass.* As ^{14}C decays, the mass of ^{14}C decreases while the mass of ^{14}N increases. If we start with 1.0 g of ^{14}C, half that mass of ^{14}C (0.50 g) will be left after 5730 years, half of that mass (0.25 g) after another 5730 years, and so on.
- *Activity.* The activity depends on the number of nuclei, so the activity is halved after each succeeding half-life.

We determine the half-life of a nuclear reaction from its rate constant. Rearranging Equation 23.3 and integrating over time gives an expression for finding the number of nuclei remaining, $\mathcal{N}_t$, after a given time t:

$$\ln \frac{\mathcal{N}_t}{\mathcal{N}_0} = -kt \quad \text{or} \quad \mathcal{N}_t = \mathcal{N}_0 e^{-kt} \quad \text{and} \quad \ln \frac{\mathcal{N}_0}{\mathcal{N}_t} = kt \tag{23.4}$$

where $\mathcal{N}_0$ is the number of nuclei at $t = 0$. (Note the similarity to Equation 16.4.) To calculate the half-life ($t_{1/2}$), we set $\mathcal{N}_t$ equal to $\frac{1}{2}\mathcal{N}_0$ and solve for $t_{1/2}$:

$$\ln \frac{\mathcal{N}_0}{\frac{1}{2}\mathcal{N}_0} = kt_{1/2} \quad \text{so} \quad t_{1/2} = \frac{\ln 2}{k} \tag{23.5}$$

Exactly analogous to the half-life of a first-order chemical change, *this half-life is **not** dependent on the number of nuclei and is inversely related to the decay constant:*

$$\text{large } k \Rightarrow \text{short } t_{1/2} \quad \text{and} \quad \text{small } k \Rightarrow \text{long } t_{1/2}$$

The decay constants and half-lives of many radioactive nuclides vary over a very wide range, even for the nuclides of a given element (Table 23.4).

Table 23.4 Decay Constants (k) and Half-Lives ($t_{1/2}$) of Beryllium Isotopes

Nuclide	k	$t_{1/2}$
$^{7}_{4}Be$	1.30×10^{-2}/day	53.3 days
$^{8}_{4}Be$	1.0×10^{16}/s	6.7×10^{-17} s
$^{9}_{4}Be$	Stable	
$^{10}_{4}Be$	4.3×10^{-7}/yr	1.6×10^{6} yr
$^{11}_{4}Be$	5.02×10^{-2}/s	13.8 s

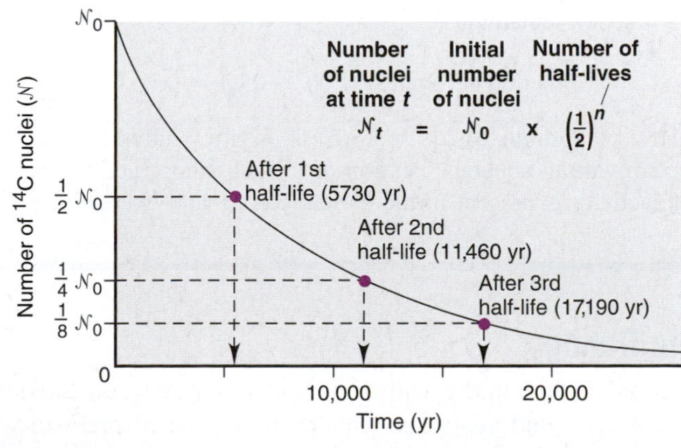

Figure 23.4 Decrease in number of ^{14}C nuclei over time.

Sample Problem 23.4 Finding the Number of Radioactive Nuclei

Problem Strontium-90 is a radioactive byproduct of nuclear reactors that behaves biologically like calcium, the element above it in Group 2A(2). When ^{90}Sr is ingested by mammals, it is found in their milk and eventually in the bones of those drinking the milk. If a sample of ^{90}Sr has an activity of 1.2×10^{12} d/s, what are the activity and the fraction of nuclei that have decayed after 59 yr ($t_{1/2}$ of ^{90}Sr = 29 yr)?

Plan The fraction of nuclei that have decayed is the change in number of nuclei, expressed as a fraction of the starting number. The activity of the sample ($\mathcal{A}$) is proportional to the number of nuclei ($\mathcal{N}$), so we know that

$$\text{Fraction decayed} = \frac{\mathcal{N}_0 - \mathcal{N}_t}{\mathcal{N}_0} = \frac{\mathcal{A}_0 - \mathcal{A}_t}{\mathcal{A}_0}$$

We are given $\mathcal{A}_0$ (1.2×10^{12} d/s), so we find $\mathcal{A}_t$ from the integrated form of the first-order rate equation (Equation 23.4), in which t is 59 yr. To solve that equation, we first need k, which we can calculate from the given $t_{1/2}$ (29 yr).

Solution Calculating the decay constant k:

$$t_{1/2} = \frac{\ln 2}{k} \quad \text{so} \quad k = \frac{\ln 2}{t_{1/2}} = \frac{0.693}{29 \text{ yr}} = 0.024 \text{ yr}^{-1}$$

Applying Equation 23.4 to calculate $\mathcal{A}_t$, the activity remaining at time t:

$$\ln \frac{\mathcal{N}_0}{\mathcal{N}_t} = \ln \frac{\mathcal{A}_0}{\mathcal{A}_t} = kt \quad \text{or} \quad \ln \mathcal{A}_0 - \ln \mathcal{A}_t = kt$$

So,
$$\ln \mathcal{A}_t = -kt + \ln \mathcal{A}_0 = -(0.024 \text{ yr}^{-1} \times 59 \text{ yr}) + \ln (1.2\times10^{12} \text{ d/s})$$
$$\ln \mathcal{A}_t = -1.4 + 27.81 = 26.4$$
$$\mathcal{A}_t = 2.9\times10^{11} \text{ d/s}$$

(All the data contain two significant figures, so we retained two in the answer.)
Calculating the fraction decayed:

$$\text{Fraction decayed} = \frac{\mathcal{A}_0 - \mathcal{A}_t}{\mathcal{A}_0} = \frac{1.2\times10^{12} \text{ d/s} - 2.9\times10^{11} \text{ d/s}}{1.2\times10^{12} \text{ d/s}} = 0.76$$

Check The answer is reasonable: t is about 2 half-lives, so $\mathcal{A}_t$ should be about $\frac{1}{4}\mathcal{A}_0$, or about 0.3×10^{12}; therefore, the activity should have decreased by about $\frac{3}{4}$.

Comment 1. A useful substitution of Equation 23.4 for finding $\mathcal{A}_t$, the activity at time t, is $\mathcal{A}_t = \mathcal{A}_0 e^{-kt}$.

2. Another way to find the fraction of activity (or nuclei) remaining incorporates the number of half-lives ($t/t_{1/2}$). By combining Equations 23.4 and 23.5 and substituting ($\ln 2)/t_{1/2}$ for k, we obtain

$$\ln \frac{\mathcal{N}_0}{\mathcal{N}_t} = \left(\frac{\ln 2}{t_{1/2}}\right) t = \frac{t}{t_{1/2}} \ln 2 = \ln 2^{t/t_{1/2}}$$

Inverting the ratio gives
$$\ln \frac{\mathcal{N}_t}{\mathcal{N}_0} = \ln \left(\frac{1}{2}\right)^{t/t_{1/2}}$$

Taking the antilog gives
$$\text{Fraction remaining} = \frac{\mathcal{N}_t}{\mathcal{N}_0} = \left(\frac{1}{2}\right)^{t/t_{1/2}} = \left(\frac{1}{2}\right)^{59/29} = 0.24$$

So,
$$\text{Fraction decayed} = 1.00 - 0.24 = 0.76$$

FOLLOW-UP PROBLEM 23.4 Sodium-24 ($t_{1/2} = 15$ h) is used to study blood circulation. If a patient is injected with an aqueous solution of ^{24}NaCl whose activity is 2.5×10^9 d/s, how much of the activity is present in the patient's body and excreted fluids after 4.0 days?

Radioisotopic Dating

The historical record fades rapidly with time and is virtually nonexistent for events of more than a few thousand years ago. Much knowledge of prehistory comes from **radioisotopic dating,** which uses **radioisotopes** to determine the ages of objects. This technique supplies data in fields such as art history, archeology, geology, and paleontology.

The technique of *radiocarbon dating,* for which the American chemist Willard F. Libby won the Nobel Prize in chemistry in 1960, is based on measuring the amounts of ^{14}C and ^{12}C in materials of biological origin. The accuracy of the method falls off after about six half-lives of ^{14}C ($t_{1/2}$ = 5730 yr), so it is used to date objects up to about 36,000 years old.

Here is how radiocarbon dating works:

1. High-energy cosmic rays, consisting mainly of protons, enter the atmosphere from outer space and initiate a cascade of nuclear reactions, some of which produce neutrons that bombard ordinary ^{14}N atoms to form ^{14}C:

$$^{14}_{7}\text{N} + ^{1}_{0}\text{n} \longrightarrow ^{14}_{6}\text{C} + ^{1}_{1}\text{p}$$

Through the competing processes of this formation and radioactive decay, the amount of ^{14}C in the atmosphere has remained nearly constant.

2. The ^{14}C atoms combine with O_2, diffuse throughout the lower atmosphere, and enter the total carbon pool as gaseous $^{14}CO_2$ and aqueous $H^{14}CO_3^-$. They mix with ordinary $^{12}CO_2$ and $H^{12}CO_3^-$, reaching a constant $^{12}C/^{14}C$ ratio of about $10^{12}/1$.
3. CO_2 is taken up by plants during photosynthesis, and then taken up and excreted by animals that eat the plants. Thus, the $^{12}C/^{14}C$ ratio of a living organism is the same as the ratio in the environment.
4. When an organism dies, it no longer absorbs or releases CO_2, so the $^{12}C/^{14}C$ ratio steadily increases because *the amount of ^{14}C decreases as it decays:*

$$^{14}_{6}\text{C} \longrightarrow ^{14}_{7}\text{N} + ^{0}_{-1}\beta$$

The difference between the $^{12}C/^{14}C$ ratio in a dead organism and the ratio in living organisms reflects the time elapsed since the organism died.

As you saw in Sample Problem 23.4, the first-order rate equation can be expressed in terms of a ratio of activities:

$$\ln \frac{\mathcal{N}_0}{\mathcal{N}_t} = \ln \frac{\mathcal{A}_0}{\mathcal{A}_t} = kt$$

We use this expression in radiocarbon dating, where $\mathcal{A}_0$ is the activity in a living organism and $\mathcal{A}_t$ is the activity in the object whose age is unknown. Solving for t gives the age of the object:

$$t = \frac{1}{k} \ln \frac{\mathcal{A}_0}{\mathcal{A}_t} \qquad \textbf{(23.6)}$$

To determine the ages of more ancient objects or of objects that do not contain carbon, different radioisotopes are measured. For example, by comparing the ratio of ^{238}U ($t_{1/2}$ = 4.5×10^9 yr) to its final decay product, ^{206}Pb, geochemists found that the oldest known surface rocks on Earth—granite in western Greenland—are about 3.7 billion years old. The $^{238}U/^{206}Pb$ ratio in samples from meteorites gives 4.65 billion years for the age of the Solar System, and thus, of Earth.

Sample Problem 23.5 Applying Radiocarbon Dating

Problem The charred bones of a sloth in a cave in Chile represent the earliest evidence of human presence at the southern tip of South America. A sample of the bone has a specific activity of 5.22 disintegrations per minute per gram of carbon (d/min·g). If the $^{12}C/^{14}C$ ratio for living organisms results in a specific activity of 15.3 d/min·g, how old are the bones ($t_{1/2}$ of ^{14}C = 5730 yr)?

Plan We calculate k from the given $t_{1/2}$ (5730 yr). Then we apply Equation 23.6 to find the age (t) of the bones, using the given activities of the bones ($\mathcal{A}_t$ = 5.22 d/min·g) and of a living organism ($\mathcal{A}_0$ = 15.3 d/min·g).

Solution Calculating k for ^{14}C decay:

$$k = \frac{\ln 2}{t_{1/2}} = \frac{0.693}{5730 \text{ yr}} = 1.21\times10^{-4} \text{ yr}^{-1}$$

Calculating the age (t) of the bones:

$$t = \frac{1}{k} \ln \frac{\mathcal{A}_0}{\mathcal{A}_t} = \frac{1}{1.21\times10^{-4}\ \text{yr}^{-1}} \ln \left(\frac{15.3\ \text{d/min}\cdot\text{g}}{5.22\ \text{d/min}\cdot\text{g}}\right) = 8.89\times10^3\ \text{yr}$$

The bones are about 8900 years old.

Check The activity of the bones is between $\frac{1}{2}$ and $\frac{1}{4}$ the activity of a living organism, so the age should be between one and two half-lives (5730 to 11,460 yr).

FOLLOW-UP PROBLEM 23.5 A sample of wood from an Egyptian mummy case has a specific activity of 9.41 d/min·g. How old is the case?

Summary of Section 23.2

- The decay rate (activity) of a sample is proportional to the number of radioactive nuclei. Nuclear decay is a first-order process, so half-life is constant.
- Radioisotopic methods, such as ^{14}C dating, determine the age of an object by measuring the ratio of specific isotopes in it.

23.3 • NUCLEAR TRANSMUTATION: INDUCED CHANGES IN NUCLEI

The alchemists' dream of changing base metals into gold was never realized, but in the early 20th century, nuclear physicists *did* change one element into another. Research into **nuclear transmutation,** the *induced* conversion of the nucleus of one element into the nucleus of another, was closely linked to research into atomic structure and led to the discovery of the neutron and the production of artificial radioisotopes. Later, high-energy bombardment of nuclei in particle accelerators began the ongoing effort to create new nuclides and new elements, and, most recently, to understand fundamental questions of matter and energy.

During the 1930s and 1940s, researchers probing the nucleus bombarded elements with neutrons, α particles, protons, and *deuterons* (nuclei of the stable hydrogen isotope deuterium, ^2H). Neutrons are especially useful as projectiles because they have no charge and thus are not repelled as they approach a target nucleus. The other particles are all positive, so early researchers found it difficult to give them enough energy to overcome their repulsion by the target nuclei. Beginning in the 1930s, however, **particle accelerators** were invented to impart high kinetic energies to particles by placing them in an electric field, usually in combination with a magnetic field.

Some Types of Accelerators A major advance was the *linear accelerator,* a series of separated tubes of increasing length that, through a source of alternating voltage, change their charge from positive to negative in synchrony with the movement of the particle through them (Figure 23.5). A proton, for example, exits the first tube just when that tube becomes positive and the next tube negative. Repelled by the first tube and attracted by the second, the proton accelerates across the gap between them. The process occurs in stages to achieve high particle energies without having to apply a single high voltage. A 40-ft linear accelerator with 46 tubes, built in California after World War II, accelerated protons to speeds several million times faster than earlier accelerators.

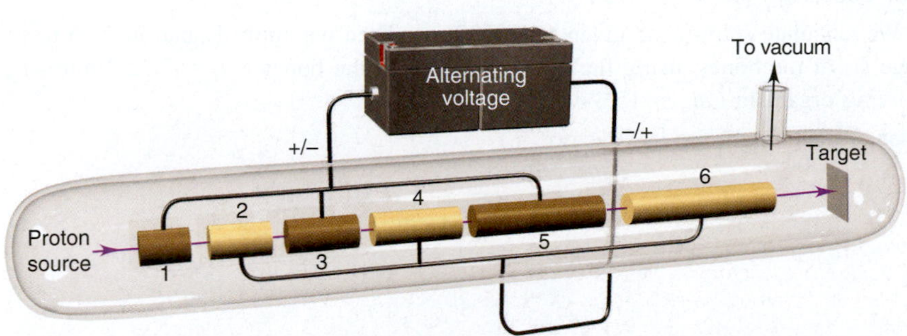

Figure 23.5 Schematic diagram of a linear accelerator.

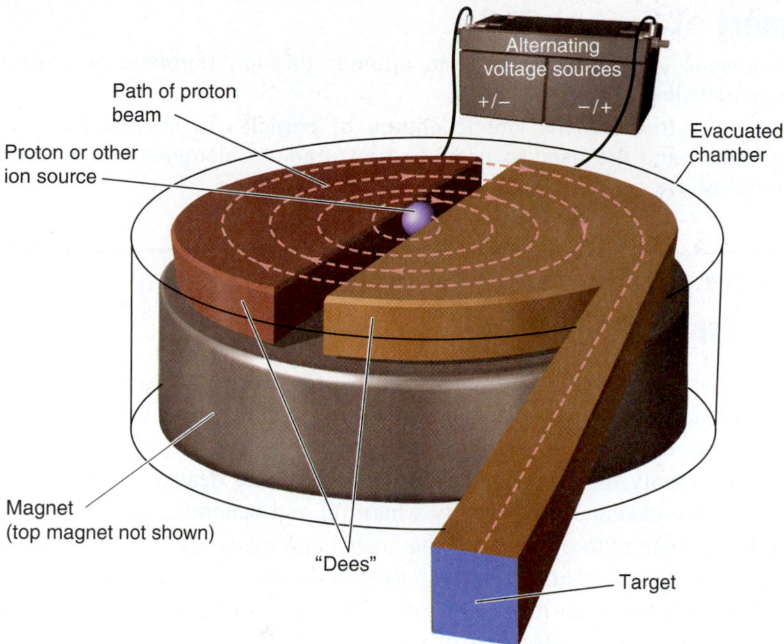

Figure 23.6 Schematic diagram of a cyclotron.

The *cyclotron* (Figure 23.6), invented by E. O. Lawrence in 1930, applies the principle of the linear accelerator but uses electromagnets to give the particle a spiral path to save space. The *synchrotron* uses a synchronously increasing magnetic field to make the particle's path circular rather than spiral.

Some very powerful accelerators include both a linear section and a synchrotron section. The *tevatron* at the Fermi Lab near Chicago, recently shut down, could accelerate particles to an energy slightly less than 1×10^{12} electron volts (1 TeV) before colliding them together. The world's most powerful accelerator is the Large Hadron Collider (LHC) near Geneva, Switzerland, which began operations in early 2010. The LHC can accelerate protons to an energy of 7 TeV, giving them a final speed about 99.99% of the speed of light before they collide with other protons. Physicists are studying the results of these high-energy subatomic collisions for information on the nature of matter and the early universe.

Synthesizing the Transuranium Elements Scientists use accelerators for many applications, from producing radioisotopes used in medical applications to studying the fundamental nature of matter. Perhaps the most specific application for chemists is the synthesis of **transuranium elements,** those with atomic numbers higher than uranium, the heaviest naturally occurring element. Some reactions that were used to form several of these elements appear in Table 23.5.

Table 23.5 Formation of Some Transuranium Nuclides*

Reaction	Half-Life of Product
$^{239}_{94}\text{Pu} + 2^{1}_{0}\text{n} \longrightarrow {}^{241}_{95}\text{Am} + {}^{0}_{-1}\beta$	432 yr
$^{239}_{94}\text{Pu} + {}^{4}_{2}\alpha \longrightarrow {}^{242}_{96}\text{Cm} + {}^{1}_{0}\text{n}$	163 days
$^{241}_{95}\text{Am} + {}^{4}_{2}\alpha \longrightarrow {}^{243}_{97}\text{Bk} + 2^{1}_{0}\text{n}$	4.5 h
$^{242}_{96}\text{Cm} + {}^{4}_{2}\alpha \longrightarrow {}^{245}_{98}\text{Cf} + {}^{1}_{0}\text{n}$	45 min
$^{253}_{99}\text{Es} + {}^{4}_{2}\alpha \longrightarrow {}^{256}_{101}\text{Md} + {}^{1}_{0}\text{n}$	76 min
$^{243}_{95}\text{Am} + {}^{18}_{8}\text{O} \longrightarrow {}^{256}_{103}\text{Lr} + 5^{1}_{0}\text{n}$	28 s

*Like chemical reactions, nuclear reactions may occur in several steps. For example, the first reaction here is actually an overall process that occurs in three steps:
(1) $^{239}_{94}\text{Pu} + {}^{1}_{0}\text{n} \longrightarrow {}^{240}_{94}\text{Pu}$ (2) $^{240}_{94}\text{Pu} + {}^{1}_{0}\text{n} \longrightarrow {}^{241}_{94}\text{Pu}$ (3) $^{241}_{94}\text{Pu} \longrightarrow {}^{241}_{95}\text{Am} + {}^{0}_{-1}\beta$

Summary of Section 23.3

- One nucleus can be transmuted to another through bombardment with high-energy particles.
- Accelerators increase the kinetic energy of particles in nuclear bombardment experiments and are used to produce transuranium elements and radioisotopes for medical use.

23.4 • EFFECTS OF NUCLEAR RADIATION ON MATTER

In 1986, an accident at the Chernobyl nuclear facility in the former Soviet Union released radioactivity that, according to the World Health Organization, will eventually cause thousands of cancer deaths. In the same year, isotopes used in medical treatment emitted radioactivity that prevented thousands of cancer deaths. In this section and Section 23.5, we examine radioactivity's harmful and beneficial effects.

The key to both of these effects is that *nuclear changes cause chemical changes in surrounding matter.* In other words, even though the nucleus of an atom may undergo a reaction with little or no involvement of the atom's electrons, the emissions from that reaction *do* affect the electrons of nearby atoms.

Virtually all radioactivity causes **ionization** in surrounding matter, as the emissions collide with atoms and dislodge electrons:

$$\text{Atom} \xrightarrow{\text{ionizing radiation}} \text{ion}^+ + e^-$$

From each ionization event, a cation and a free electron result, and the number of such *cation-electron pairs* produced is directly related to the energy of the incoming **ionizing radiation.**

Effects of Ionizing Radiation on Living Tissue

Ionizing radiation has a destructive effect on living tissue, and if the ionized atom is part of a key biological macromolecule or cell membrane, the results can be devastating to the cell and perhaps the organism.

Units of Radiation Dose and Its Effects To measure the effects of ionizing radiation, we need a unit for radiation dose. Units of radioactive decay, such as the becquerel and curie, measure the number of decay events in a given time but not their energy or absorption by matter. However, the number of cation-electron pairs produced in a given amount of living tissue *is* a measure of the energy absorbed by the tissue. The SI unit is the **gray (Gy),** equal to 1 joule of energy absorbed per kilogram of body tissue: 1 Gy = 1 J/kg. A more widely used unit is the **rad (*r*adiation-*a*bsorbed *d*ose),** which is one-hundredth as much:

$$1 \text{ rad} = 0.01 \text{ J/kg} = 0.01 \text{ Gy}$$

To measure actual tissue damage, we must account for differences in the strength of the radiation, the exposure time, and the type of tissue. To do this, we multiply the number of rads by a *r*elative *b*iological *e*ffectiveness (RBE) factor, which depends on the effect a given type of radiation has on a given tissue or body part. The product is the **rem (*r*oentgen *e*quivalent for *m*an),** the unit of radiation dosage equivalent to a given amount of tissue damage in a human:

$$\text{no. of rems} = \text{no. of rads} \times \text{RBE}$$

Doses are often expressed in millirems (10^{-3} rem). The SI unit for dosage equivalent is the **sievert (Sv).** It is defined in the same way as the rem but with absorbed dose in grays; thus, 1 rem = 0.01 Sv.

Penetrating Power of Emissions The effect on living tissue of a radiation dose depends on the penetrating power *and* ionizing ability of the radiation. Since water is the main component of living tissue, penetrating power is often measured in terms of the depth of water that stops 50% of incoming radiation. In

Figure 23.7, the average values of the penetrating distances are shown in actual size for the three common types of emissions. Note, in general, that *penetrating power is inversely related to the mass, charge, and energy of the emission.* In other words, if a particle interacts strongly with matter, it penetrates only slightly, and vice versa:

1. *Alpha particles.* Alpha particles are massive and highly charged, which means that they interact with matter most strongly of the three common emissions. As a result, they penetrate so little that a piece of paper, light clothing, or the outer layer of skin can stop α radiation from an external source. However, if ingested, an α emitter can cause grave localized internal damage through extensive ionization.

2. *Beta particles and positrons.* Beta particles (β^-) and positrons (β^+) have less charge and much less mass than α particles, so they interact less strongly with matter. Even though a given particle has less chance of causing ionization, a β^- (or β^+) emitter is a more destructive *external* source because the particles penetrate deeper. Specialized heavy clothing or a thick (0.5 cm) piece of metal is required to stop these particles.

3. *Gamma rays.* Neutral, massless γ rays interact least with matter and, thus, penetrate most. A block of lead several inches thick is needed to stop them. Therefore, an external γ ray source is the most dangerous because the energy can ionize many layers of living tissue.

Figure 23.7 Penetrating power of radioactive emissions.

Sources of Ionizing Radiation

We are continuously exposed to ionizing radiation from natural and artificial sources (Table 23.6). Indeed, life evolved in the presence of natural ionizing radiation, called **background radiation.** Ionizing radiation can alter bonds in DNA and cause harmful mutations but also causes beneficial ones that allow species to evolve.

Table 23.6 Typical Radiation Doses from Natural and Artificial Sources

Source of Radiation	Average Adult Exposure
Natural	
Cosmic radiation	30–50 mrem/yr
Radiation from the ground	
From clay soil and rocks	~25–170 mrem/yr
In wooden houses	10–20 mrem/yr
In brick houses	60–70 mrem/yr
In concrete (cinder block) houses	60–160 mrem/yr
Radiation from the air (mainly radon)	
Outdoors, average value	20 mrem/yr
In wooden houses	70 mrem/yr
In brick houses	130 mrem/yr
In concrete (cinder block) houses	260 mrem/yr
Internal radiation from minerals in tap water and daily intake of food (^{40}K, ^{14}C, Ra)	~40 mrem/yr
Artificial	
Diagnostic x-ray methods	0.04–0.2 rad/film
Lung (local)	1.5–3 rad/film
Kidney (local)	≤1 rad/film
Dental (dose to the skin)	Locally ≤10,000 rad
Therapeutic radiation treatment	
Other sources	
Jet flight (4 h)	~1 mrem
Nuclear testing	<4 mrem/yr
Nuclear power industry	<1 mrem/yr
Total average value	100–200 mrem/yr

Background radiation has several natural sources:

- *Cosmic radiation* increases with altitude because of decreased absorption by the atmosphere. Thus, people in Denver absorb twice as much cosmic radiation as people in Los Angeles; even a jet flight involves measurable absorption.
- Thorium and uranium minerals are present in rocks and soil. Radon, the heaviest noble gas [Group 8A(18)], is a radioactive product of uranium and thorium decay. Its concentration in the air we breathe is related to the presence of trace minerals in building materials and to the uranium content of local soil and rocks. Radon's radioactive decay products cause most of the damage.
- About 150 g of K^+ ions is dissolved in the water in the tissues of an average adult, and 0.0118% of these ions are radioactive ^{40}K. The presence of these substances and of atmospheric $^{14}CO_2$ makes *all* food, water, clothing, and building materials slightly radioactive.

The largest artificial source of radiation, and the one that is easiest to control, is from medical diagnostic techniques, especially x-rays. The radiation dosage from nuclear testing and radioactive waste disposal is miniscule for most people, but exposures for those living near test sites or disposal areas may be much higher.

Summary of Section 23.4

- All radioactive emissions cause ionization.
- The effect of ionizing radiation on living matter depends on the quantity of energy absorbed and the extent of ionization in a given type of tissue. Radiation dose for the human body is measured in rem.
- All organisms are exposed to varying quantities of natural ionizing radiation.

23.5 • APPLICATIONS OF RADIOISOTOPES

Radioisotopes are powerful tools for studying processes in biochemistry, medicine, materials science, environmental studies, and many other scientific and industrial fields. Such uses depend on the fact that *isotopes of an element exhibit* ***very*** *similar chemical and physical behavior.* In other words, except for having a less stable nucleus, a radioisotope has nearly the same properties as a nonradioactive isotope of the same element.* For example, the fact that $^{14}CO_2$ is utilized by a plant in the same way as $^{12}CO_2$ forms the basis of radiocarbon dating.

Radioactive Tracers

A tiny amount of a radioisotope mixed with a large amount of a stable isotope of the same element can act as a **tracer,** a chemical "beacon" emitting radiation that signals the presence of the substance.

Reaction Pathways Tracers are used to understand both simple and complex reaction pathways.

1. *Inorganic systems: the periodate-iodide reaction.* One well-studied inorganic example is the reaction between periodate and iodide ions:

$$IO_4^-(aq) + 2I^-(aq) + H_2O(l) \longrightarrow I_2(s) + IO_3^-(aq) + 2OH^-(aq)$$

Is IO_3^- the result of IO_4^- reduction or I^- oxidation? When we add "cold" (nonradioactive) IO_4^- to a solution of I^- that contains some "hot" (radioactive, indicated in red) $^{131}I^-$, we find that the I_2 is radioactive, not the IO_3^-:

$$IO_4^-(aq) + 2\,^{131}I^-(aq) + H_2O(l) \longrightarrow {}^{131}I_2(s) + IO_3^-(aq) + 2OH^-(aq)$$

*Although this statement is correct for nearly every element, in some cases, differences in isotopic mass *can* influence bond strengths and therefore reaction rates. Such behavior is particularly important for isotopes of hydrogen—^1H, ^2H, and ^3H—because their masses differ by such large proportions.

These results show that IO_3^- forms through the reduction of IO_4^-, and that I_2 forms through the oxidation of I^-. To confirm this pathway, we add IO_4^- containing some $^{131}IO_4^-$ to a solution of I^-. As we expected, the IO_3^- is radioactive, not the I_2:

$$^{131}IO_4^-(aq) + 2I^-(aq) + H_2O(l) \longrightarrow I_2(s) + {}^{131}IO_3^-(aq) + 2OH^-(aq)$$

Thus, tracers act like "handles" we can "hold" to follow the changing reactants.

2. *Biochemical pathways: photosynthesis.* Far more complex pathways can be followed with tracers as well. The photosynthetic pathway, the most essential and widespread metabolic process on Earth, in which energy from sunlight is used to form the chemical bonds of glucose, has an overall reaction that looks quite simple:

$$6CO_2(g) + 6H_2O(l) \xrightarrow[\text{chlorophyll}]{\text{light}} C_6H_{12}O_6(s) + 6O_2(g)$$

However, the actual process is extremely complex: 13 enzyme-catalyzed steps are required to incorporate each C atom from CO_2, so the six CO_2 molecules incorporated to form a molecule of $C_6H_{12}O_6$ require six repetitions of the pathway. Melvin Calvin and his coworkers took seven years to determine the pathway, using ^{14}C in CO_2 as the tracer. Calvin won the Nobel Prize in chemistry in 1961 for this remarkable achievement.

Material Flow Tracers are used in studies of solid surfaces and the flow of materials. Metal atoms hundreds of layers deep within a solid have been shown to exchange with metal ions from the surrounding solution within a matter of minutes. Chemists and engineers use tracers to study material movement in semiconductor chips, paint, and metal plating, in detergent action, and in the process of corrosion, to mention just a few of many applications.

Hydrologic engineers use tracers to study the volume and flow of large bodies of water. By following radionuclides that formed during atmospheric nuclear bomb tests (^3H in H_2O, $^{90}Sr^{2+}$, and $^{137}Cs^+$), scientists have mapped the flow of water from land to lakes and streams to oceans. They also use tracers to study the surface and deep ocean currents that circulate around the globe, the mechanisms of hurricane formation, and the mixing of the troposphere and stratosphere. Industries employ tracers to study material flow during manufacturing processes, such as the flow of ore pellets in smelting kilns, the paths of wood chips and bleach in paper mills, the diffusion of fungicide into lumber, and in a particularly important application, the porosity and leakage of oil and gas wells in geological formations.

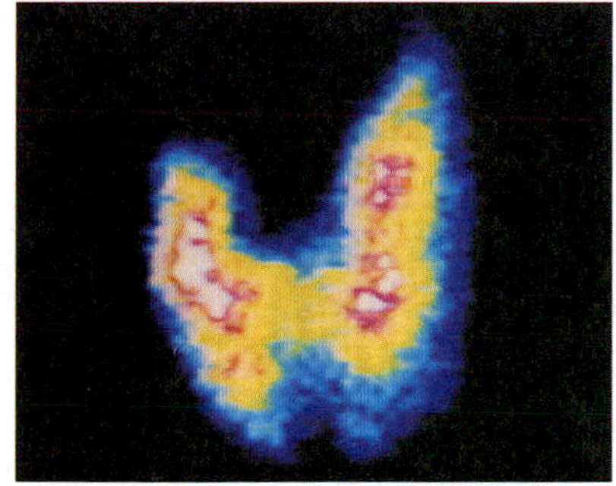

Figure 23.8 **The use of radioisotopes to image the thyroid gland.** This ^{131}I scan shows an asymmetric image that is indicative of disease.

Activation Analysis Another use of tracers is in *neutron activation analysis* (NAA). In this method, neutrons bombard a nonradioactive sample, converting a small fraction of its atoms to radioisotopes, which exhibit characteristic decay patterns, such as γ-ray spectra, that reveal the elements present. Unlike chemical analysis, NAA leaves the sample virtually intact, so the method can be used to determine the composition of a valuable object or a very small sample. For example, a painting thought to be a 16th-century Dutch masterpiece was shown through NAA to be a 20th-century forgery, because a microgram-sized sample of its pigment contained much less silver and antimony than the pigments used by the Dutch masters. Forensic chemists use NAA to detect traces of ammunition on a suspect's hand or traces of arsenic in the hair of a victim of poisoning.

Medical Diagnosis The largest use of radioisotopes is in medical science. In fact, over 25% of U.S. hospital admissions are for diagnoses based on data from radioisotopes. Tracers with half-lives of a few minutes to a few days are employed to observe specific organs and body parts. For example, a healthy thyroid gland incorporates dietary I^- into iodine-containing hormones at a known rate. To assess thyroid function, the patient drinks a solution containing a trace amount of $Na^{131}I$, and a scanning monitor follows the uptake of $^{131}I^-$ by the thyroid gland (Figure 23.8).

Tracers are also used to measure physiological processes, such as blood flow. The rate at which the heart pumps blood, for example, can be observed by injecting ^{59}Fe, which becomes incorporated into the hemoglobin of blood cells. Several radioisotopes used in medical diagnosis are listed in Table 23.7.

Table 23.7 Some Radioisotopes Used as Medical Tracers

Isotope	Body Part or Process
^{11}C, ^{18}F, ^{13}N, ^{15}O	PET studies of brain, heart
^{60}Co, ^{192}Ir	Cancer therapy
^{64}Cu	Metabolism of copper
^{59}Fe	Blood flow, spleen
^{67}Ga	Tumor imaging
^{123}I, ^{131}I	Thyroid
^{111}In	Brain, colon
^{42}K	Blood flow
^{81m}Kr	Lung
^{99m}Tc	Heart, thyroid, liver, lung, bone
^{201}Tl	Heart muscle
^{90}Y	Cancer, arthritis

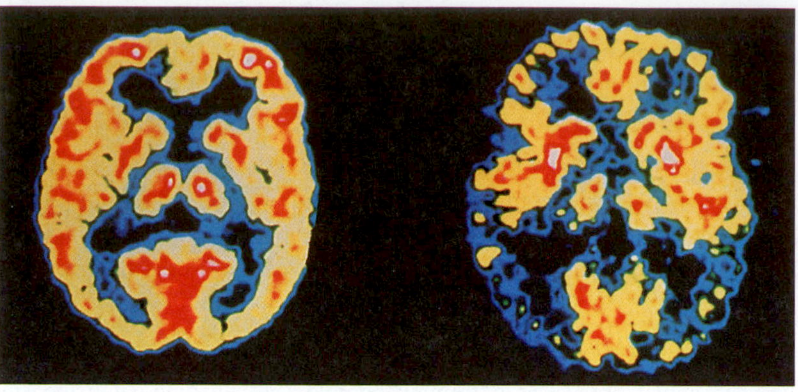

Figure 23.9 PET and brain activity. These PET scans show brain activity in a normal person *(left)* and in a patient with Alzheimer's disease *(right)*. Red and yellow indicate relatively high activity within a region.

Positron-emission tomography (PET) is a powerful imaging method for observing brain structure and function. A biological substance is synthesized with one of its atoms replaced by an isotope that emits positrons. The substance is injected into a patient's bloodstream, from which it is taken up into the brain. The isotope emits positrons, each of which annihilates a nearby electron. As the annihilation occurs, two γ photons are emitted simultaneously 180° from each other:

$$^{0}_{1}\beta + ^{0}_{-1}e \longrightarrow 2^{0}_{0}\gamma$$

An array of detectors around the patient's head pinpoints the sites of γ emission, and the image is analyzed by computer. Two of the isotopes used are ^{15}O, injected as $H_2{}^{15}O$ to measure blood flow, and ^{18}F bonded to a glucose analog to measure glucose uptake, which is an indicator of energy metabolism. Among many fascinating PET findings are those that show how changes in blood flow and glucose uptake accompany normal or abnormal brain activity (Figure 23.9). Also, substances incorporating ^{11}C and ^{15}O are being investigated using PET to learn how molecules interact with and move along the surface of a catalyst.

Additional Applications of Ionizing Radiation

Many other uses of radioisotopes involve higher-energy ionizing radiation.

- *Radiation therapy.* Cancer cells divide more rapidly than normal cells, so radioisotopes that interfere with cell division kill more cancer cells than normal ones. Implants of ^{198}Au or of ^{90}Sr, which decays to the γ-emitting ^{90}Y, have been used to destroy pituitary and breast tumor cells, and γ rays from ^{60}Co have been used to destroy tumors of the brain and other body parts.
- *Destruction of microbes.* Irradiation of food increases shelf life by killing microorganisms that cause rotting or spoilage (Figure 23.10), but the practice is controversial. Advocates point to the benefits of preserving fresh foods, grains, and seeds for long periods, whereas opponents suggest that irradiation might lower the food's nutritional content or produce harmful byproducts. Irradiation also provides a way to destroy newer, more resistant bacterial strains that survive the increasing use of the more common antibiotics in animal feed. The United Nations has approved irradiation for potatoes, wheat, chicken, and strawberries, and, in 2003, the U.S. Food and Drug Administration approved it as well.
- *Insect control.* Ionizing radiation has been used to control harmful insects. Captured males are sterilized by radiation and released to mate, thereby reducing the number of offspring. This method has been used to control the Mediterranean fruit fly in California and disease-causing insects, such as the tsetse fly and malarial mosquito, in other parts of the world.

Figure 23.10 The increased shelf life of irradiated food.

Summary of Section 23.5

- Radioisotopic tracers have been used to study reaction mechanisms, material flow, elemental composition, and medical conditions.
- Ionizing radiation has been used in devices that destroy cancer tissue, kill organisms that spoil food, and control insect populations.

23.6 • THE INTERCONVERSION OF MASS AND ENERGY

Most of the nuclear processes we've considered so far have involved radioactive decay, in which a nucleus emits one or a few small particles or photons to become a more stable, slightly lighter nucleus. Two other nuclear processes cause much greater mass changes. In nuclear **fission,** a heavy nucleus splits into two much lighter nuclei. In the opposite process, nuclear **fusion,** two lighter nuclei combine to form a heavier one. Both fission and fusion release enormous quantities of energy. Let's take a look at the origins of this energy by first examining the change in mass that accompanies the breakup of a nucleus into its nucleons and then considering the energy that is equivalent to this mass change.

The Mass Difference Between a Nucleus and Its Nucleons

For most of the 20th century, we knew that mass and energy are interconvertible. The separate mass and energy conservation laws are combined to state that *the total quantity of mass-energy in the universe is constant.* Therefore, when *any* reacting system releases or absorbs energy, it also loses or gains mass.

Mass Difference in Chemical Reactions The interconversion of mass and energy is not important for chemical reactions because the energy changes involved in breaking or forming chemical bonds are so small that the mass changes are negligible. For example, when 1 mol of water breaks up into its atoms, heat is absorbed and we have:

$$H_2O(g) \longrightarrow 2H(g) + O(g) \qquad \Delta H^\circ_{rxn} = 2 \times \text{BE of O—H} = 934 \text{ kJ}$$

We find the mass that is equivalent to this energy from *Einstein's equation:*

$$E = mc^2 \quad \text{or} \quad \Delta E = \Delta mc^2 \quad \text{so} \quad \Delta m = \frac{\Delta E}{c^2} \qquad \textbf{(23.7)}$$

where Δm is the mass difference between reactants and products:

$$\Delta m = m_{\text{products}} - m_{\text{reactants}}$$

Substituting the enthalpy of reaction (in J/mol) for ΔE and the numerical value for c (2.9979×10^8 m/s), we obtain

$$\Delta m = \frac{9.34\times10^5 \text{ J/mol}}{(2.9979\times10^8 \text{ m/s})^2} = 1.04\times10^{-11} \text{ kg/mol} = 1.04\times10^{-8} \text{ g/mol}$$

(Units of kg/mol are obtained because the joule includes the kilogram: 1 J = 1 kg·m^2/s^2.) The mass of 1 mol of H_2O molecules (reactant) is about 10 ng *less* than the combined masses of 2 mol of H atoms and 1 mol of O atoms (products), a change difficult to measure with even the most sophisticated balance. Such minute mass changes when bonds break or form allow us to assume that, for all practical purposes, mass is conserved in *chemical* reactions.

Mass Difference in Nuclear Reactions The much larger mass change that accompanies a *nuclear* process is related to the enormous energy required to bind the nucleons together in a nucleus. In an analogy with the calculation above for the water molecule, consider the change in mass that occurs when one ^{12}C nucleus breaks apart into its nucleons—six protons and six neutrons:

$$^{12}C \longrightarrow 6\,^1_1p + 6\,^1_0n$$

We calculate this mass difference in a special way. By combining the mass of six H *atoms* and six neutrons and then subtracting the mass of one ^{12}C *atom,* the masses of the electrons cancel: six e^- (in six ^1H atoms) cancel six e^- (in one ^{12}C atom). The mass of one ^1H atom is 1.007825 amu, and the mass of one neutron is 1.008665 amu, so

$$\begin{aligned} \text{Mass of six } ^1\text{H atoms} &= 6 \times 1.007825 \text{ amu} = 6.046950 \text{ amu} \\ \text{Mass of six neutrons} &= 6 \times 1.008665 \text{ amu} = 6.051990 \text{ amu} \\ \hline \text{Total mass} &= 12.098940 \text{ amu} \end{aligned}$$

The mass of the reactant, one ^{12}C atom, is 12 amu (exactly). The mass difference (Δm) we obtain is the total mass of the nucleons minus the mass of the nucleus:

$$\Delta m = 12.098940 \text{ amu} - 12.000000 \text{ amu}$$
$$= 0.098940 \text{ amu}/^{12}\text{C} = 0.098940 \text{ g/mol } ^{12}\text{C}$$

Two key points emerge from these calculations:

- *The mass of the nucleus is* ***less*** *than the combined masses of its nucleons:* there is *always* a mass decrease when nucleons form a nucleus.
- The mass change of this nuclear process (9.89×10^{-2} g/mol) is nearly 10 million times that of the chemical process (10.4×10^{-9} g/mol) we saw earlier and *can* be observed on any laboratory balance.

Nuclear Binding Energy and the Binding Energy per Nucleon

Einstein's equation for the relation between mass and energy allows us to find the energy that is equivalent to any mass change. For 1 mol of ^{12}C, using the value of Δm above and converting grams to kilograms, we have

$$\Delta E = \Delta mc^2 = (9.8940\times10^{-5} \text{ kg/mol})(2.9979\times10^{8} \text{ m/s})^2$$
$$= 8.8921\times10^{12} \text{ J/mol} = 8.8921\times10^{9} \text{ kJ/mol}$$

This quantity of energy is called the **nuclear binding energy** for carbon-12. The nuclear binding energy is the energy required to *break 1 mol of nuclei into individual nucleons:*

$$\text{Nucleus} + \text{nuclear binding energy} \longrightarrow \text{nucleons}$$

Thus, the nuclear binding energy is qualitatively analogous to the sum of bond energies of a covalent compound or the lattice energy of an ionic compound. But, quantitatively, nuclear binding energies are typically several million times greater.

The Electron Volt as a Unit of Energy We use joules to express the binding energy per mole of nuclei, but the joule is much too large a unit to express the binding energy of a single nucleus. Instead, nuclear scientists use the **electron volt (eV),** the energy an electron acquires when it moves through a potential difference of 1 volt:

$$1 \text{ eV} = 1.602\times10^{-19} \text{ J}$$

Binding energies are commonly expressed in millions of electron volts, that is, in *mega–electron volts* (MeV):

$$1 \text{ MeV} = 10^{6} \text{ eV} = 1.602\times10^{-13} \text{ J}$$

A particularly useful factor converts the atomic mass unit to its energy equivalent in electron volts:

$$1 \text{ amu} = 931.5\times10^{6} \text{ eV} = 931.5 \text{ MeV} \qquad \textbf{(23.8)}$$

Nuclear Stability and Binding Energy per Nucleon Earlier we found the mass change when ^{12}C breaks apart into its nucleons to be 0.098940 amu. The binding energy per ^{12}C nucleus, expressed in MeV, is

$$\frac{\text{Binding energy}}{^{12}\text{C nucleus}} = 0.098940 \text{ amu} \times \frac{931.5 \text{ MeV}}{1 \text{ amu}} = 92.16 \text{ MeV}$$

We can compare the stability of nuclides of different elements by determining the *binding energy per nucleon.* For ^{12}C, we have

$$\text{Binding energy per nucleon} = \frac{\text{binding energy}}{\text{no. of nucleons}} = \frac{92.16 \text{ MeV}}{12 \text{ nucleons}} = 7.680 \text{ MeV/nucleon}$$

The point is *the greater the binding energy per nucleon, the more stable the nuclide is.*

Sample Problem 23.6 Calculating the Binding Energy per Nucleon

Problem Iron-56 is an extremely stable nuclide. Compute the binding energy per nucleon for ^{56}Fe and compare it with that for ^{12}C (mass of ^{56}Fe atom = 55.934939 amu; mass of ^{1}H atom = 1.007825 amu; mass of neutron = 1.008665 amu).

Plan Iron-56 has 26 protons and 30 neutrons. We calculate the mass difference, Δm, when the nucleus forms by subtracting the given mass of one ^{56}Fe atom from the sum of the masses of 26 ^{1}H atoms and 30 neutrons. To find the binding energy per nucleon, we multiply Δm by the equivalent in MeV (931.5 MeV/amu) and divide by the number of nucleons (56).

Solution Calculating the mass difference, Δm:

$$\text{Mass difference} = [(26 \times \text{mass } ^{1}\text{H atom}) + (30 \times \text{mass neutron})] - \text{mass } ^{56}\text{Fe atom}$$
$$= [(26)(1.007825 \text{ amu}) + (30)(1.008665 \text{ amu})] - 55.934939 \text{ amu}$$
$$= 0.52846 \text{ amu}$$

Calculating the binding energy per nucleon:

$$\text{Binding energy per nucleon} = \frac{0.52846 \text{ amu} \times 931.5 \text{ MeV/amu}}{56 \text{ nucleons}} = 8.790 \text{ MeV/nucleon}$$

An ^{56}Fe nucleus would require more energy per nucleon to break up into its nucleons than would ^{12}C (7.680 MeV/nucleon), so ^{56}Fe is more stable than ^{12}C.

Check The answer is consistent with the great stability of ^{56}Fe. Given the number of decimal places in the values, rounding to check the math is useful only to find a *major* error. The number of nucleons (56) is an exact number, so we retain four significant figures.

FOLLOW-UP PROBLEM 23.6 Uranium-235 is an essential component of the fuel in nuclear power plants. Calculate the binding energy per nucleon for ^{235}U. Is this nuclide more or less stable than ^{12}C (mass of ^{235}U atom = 235.043924 amu)?

Fission or Fusion: Increasing the Binding Energy per Nucleon Calculations similar to those in Sample Problem 23.6 for other nuclides show that the binding energy per nucleon varies considerably.

Figure 23.11 shows a plot of the binding energy per nucleon vs. mass number. It provides information about nuclide stability and the two possible processes nuclides can undergo to form more stable nuclides. Most nuclides with fewer than 10 nucleons have a relatively small binding energy per nucleon. The ^{4}He nucleus is an exception—it is stable enough to be emitted intact as an α particle. Above A = 12, the binding energy per nucleon varies from about 7.6 to 8.8 MeV.

The most important observation is that *the binding energy per nucleon peaks at elements with $A \approx 60$.* In other words, nuclides become more stable with increasing mass number up to around 60 nucleons and then become less stable with higher

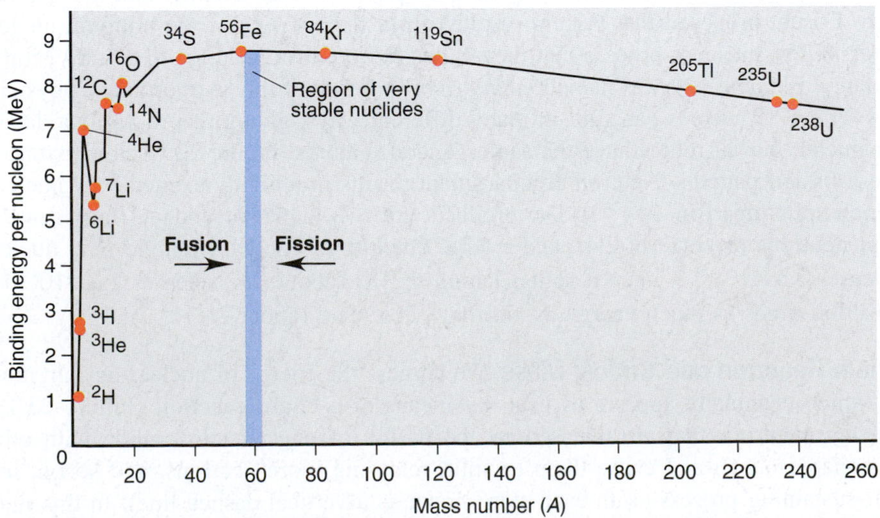

Figure 23.11 The variation in binding energy per nucleon.

numbers of nucleons. The existence of a peak of stability suggests that there are two ways nuclides can increase their binding energy per nucleon:

- *Fission.* A heavier nucleus can *split into lighter ones (closer to* $A \approx 60$*)* by undergoing fission. The product nuclei have greater binding energy per nucleon (are more stable) than the reactant nucleus, and the difference in *energy is released.* Nuclear power plants generate energy through fission (Section 23.7), as do atomic bombs.
- *Fusion.* Lighter nuclei, on the other hand, can *combine to form a heavier one (closer to* $A \approx 60$*)* by undergoing fusion. Once again, the product is more stable than the reactants, and *energy is released.* The Sun and other stars generate energy through fusion, as do thermonuclear (hydrogen) bombs. In these examples and in all current research efforts for developing fusion as a useful energy source, hydrogen nuclei fuse to form the very stable helium-4 nucleus.

In Section 23.7, we examine fission and fusion and their applications.

Summary of Section 23.6

- The mass of a nucleus is less than the sum of the masses of its nucleons. The energy equivalent to this mass difference is the nuclear binding energy, often expressed in units of MeV.
- The binding energy per nucleon is a measure of nuclide stability and varies with the number of nucleons. Nuclides with $A \approx 60$ are most stable.
- Lighter nuclides join (fusion) and heavier nuclides split (fission) to create more stable products.

23.7 • APPLICATIONS OF FISSION AND FUSION

Of the many beneficial applications of nuclear reactions, the greatest is the potential for abundant quantities of energy, which is based on the multimillion-fold increase in energy yield of nuclear reactions over chemical reactions. Our experience with nuclear energy from power plants, however, has shown that we must improve ways to tap this energy source safely and economically and deal with the waste generated. In this section, we discuss how fission and fusion occur and how we are applying them.

The Process of Nuclear Fission

During the mid-1930s, scientists bombarded uranium ($Z = 92$) with neutrons in an attempt to synthesize transuranium elements. Many of the unstable nuclides produced were tentatively identified as having $Z > 92$, and eventually, one was shown to be an isotope of barium ($Z = 56$). The Austrian physicist Lise Meitner and her nephew, Otto Frisch, proposed that barium resulted from the *splitting* of the uranium nucleus into *smaller* nuclei, a process that they called *fission* as an analogy to cell division in biology. Element 109 was named *meitnerium* in honor of this extraordinary physicist.

The ^{235}U nucleus can split in many different ways, giving rise to various daughter nuclei, but all routes have the same general features. Figure 23.12 depicts one of these fission patterns. Neutron bombardment results in a highly excited ^{236}U nucleus, which splits apart in 10^{-14} s. The products are two nuclei of unequal mass, two to four neutrons (average of 2.4), and a large quantity of energy. A single ^{235}U nucleus releases 3.5×10^{-11} J when it splits; 1 mol of ^{235}U (about $\frac{1}{2}$ lb) releases 2.1×10^{13} J—a billion times as much energy as burning $\frac{1}{2}$ lb of coal (about 2×10^{4} J)!

Chain Reaction and Critical Mass We harness the energy of nuclear fission, much of which eventually appears as heat, by means of a **chain reaction** (Figure 23.13): the few neutrons that are released by the fission of one nucleus collide with other fissionable nuclei and cause them to split, releasing more neutrons, and so on, in a self-sustaining process (with each step shown as a vertical dashed line). In this man-

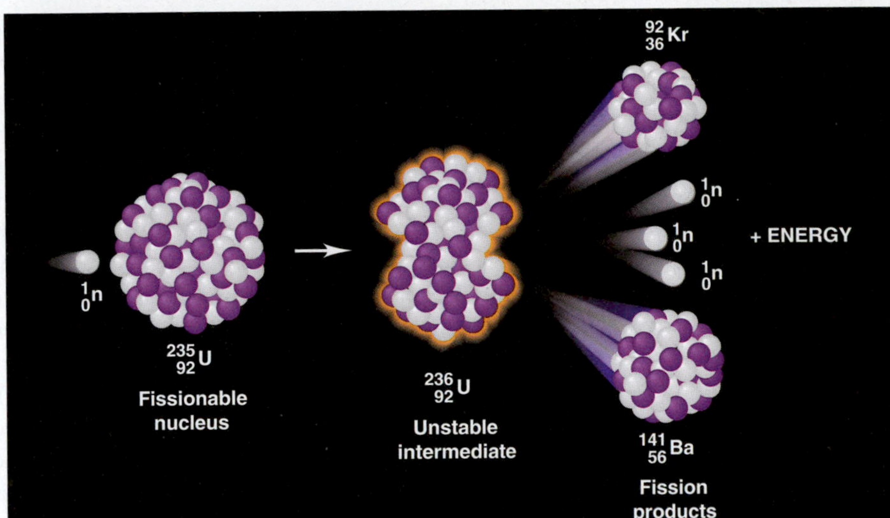

Figure 23.12 Fission of ^{235}U caused by neutron bombardment.

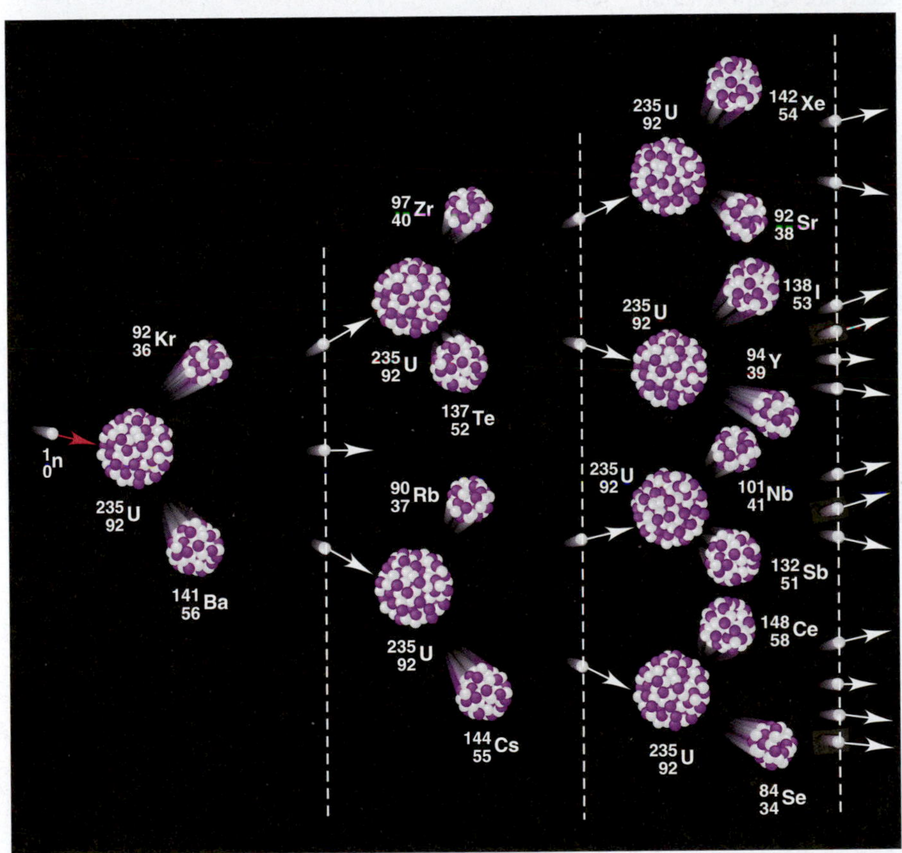

Figure 23.13 A chain reaction involving fission of ^{235}U.

ner, the energy released increases rapidly because each fission event in a chain reaction releases about two-and-a-half times as much energy as the preceding one.

Whether a chain reaction occurs depends on the mass (and thus the volume) of the fissionable sample. If the piece of uranium is large enough, the product neutrons strike another fissionable nucleus *before* flying out of the sample, and a chain reaction takes place. The mass required to achieve a chain reaction is called the **critical mass.** If the sample has less than the critical mass (a *subcritical mass*), too many product neutrons leave the sample before they collide with and cause the fission of another ^{235}U nucleus, and thus a chain reaction does not occur.

Nuclear Energy Reactors Like a coal-fired power plant, *a nuclear power plant generates heat to produce steam, which turns a turbine attached to an electric generator.* But a nuclear plant has the potential to produce electric power much more cleanly than by the combustion of coal.

1. *Operation of a nuclear power plant.* Heat generation takes place in the **reactor core** of a nuclear plant (Figure 23.14). The core contains the *fuel rods,* which consist of fuel enclosed in tubes of a corrosion-resistant zirconium alloy. The fuel is uranium(IV) oxide (UO_2) that has been *enriched* from 0.7% ^{235}U, the natural abundance of this fissionable isotope, to the 3% to 4% ^{235}U required to sustain a chain reaction in a practical volume. (Enrichment of nuclear fuel is the most important application of Graham's law, see Section 5.5.) Sandwiched between the fuel rods are movable *control rods* made of cadmium or boron (or, in nuclear submarines, hafnium), substances that absorb neutrons very efficiently. When the control rods are lowered between the fuel rods, the chain reaction slows because fewer neutrons are available to bombard uranium atoms; when they are raised, the chain reaction speeds up. Neutrons that leave the fuel-rod assembly collide with a *reflector,* usually made of a beryllium alloy, which absorbs very few neutrons. Reflecting the neutrons back to the fuel rods speeds the chain reaction.

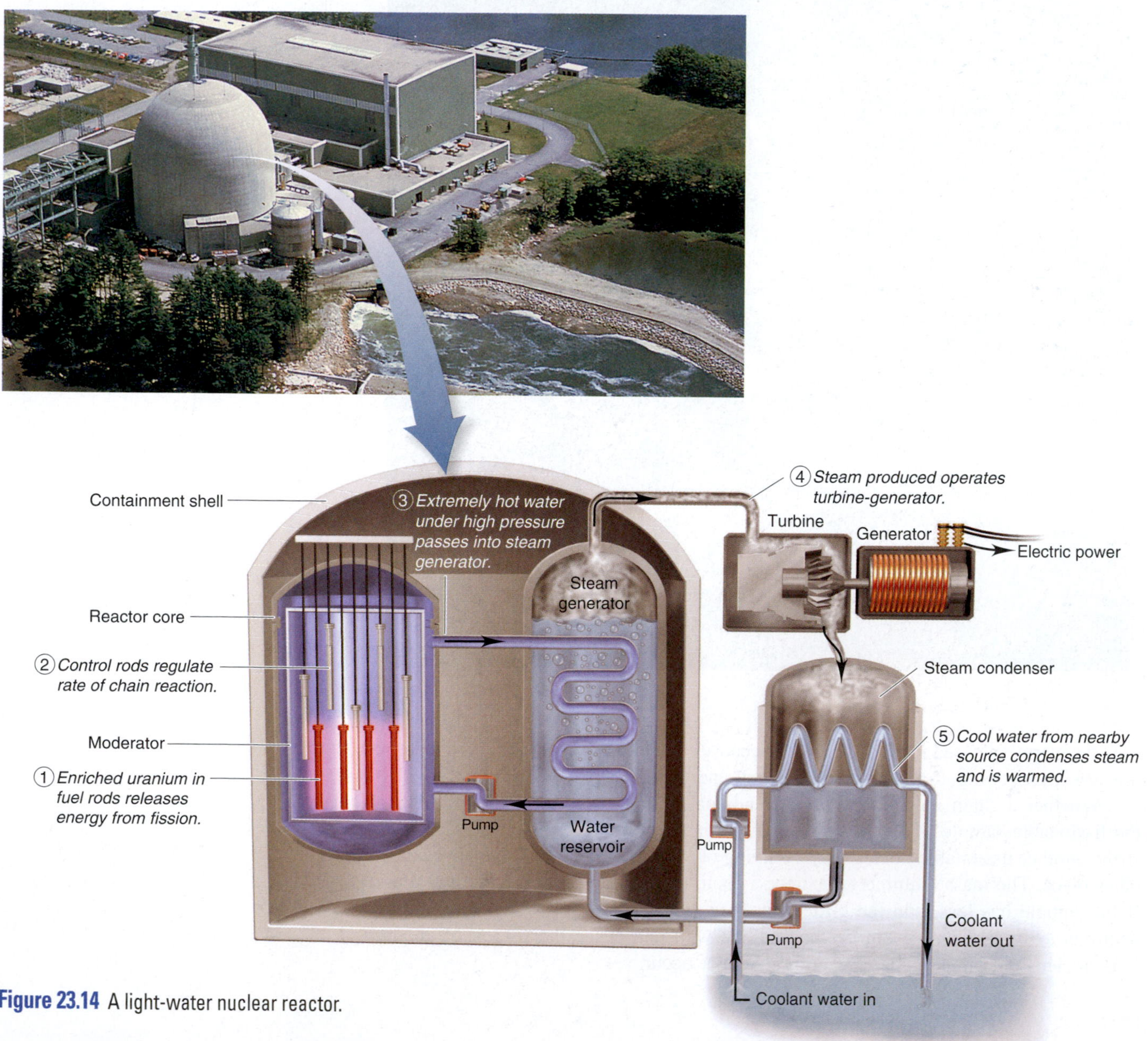

Figure 23.14 A light-water nuclear reactor.

Flowing around the fuel and control rods in the reactor core is the *moderator,* a substance that slows the neutrons, making them much better at causing fission than the fast ones emerging directly from the fission event. In most modern reactors, the moderator also acts as the *coolant,* the fluid that transfers the released heat to the steam-producing region. *Light-water reactors* use H_2O as the moderator because ^1H absorbs neutrons to some extent; in heavy-water reactors, D_2O is used. The advantage of D_2O is that it absorbs very few neutrons, leaving more available for fission, so heavy-water reactors can use uranium that has been *less enriched.* As the coolant flows around the encased fuel, pumps circulate it through coils that transfer its heat to the water reservoir. Steam formed in the reservoir turns the turbine that runs the generator. The steam then enters a condenser, where water from a lake or river is used to cool it back to liquid that is returned to the water reservoir.

2. *Power plant accidents and health concerns.* Some major accidents at nuclear plants have caused decidedly negative public reactions. In 1979, malfunctions of coolant pumps and valves at the Three-Mile Island facility in Pennsylvania led to melting of some of the fuel and damage to the reactor core, but the release of only a very small amount (about 1 Ci) of radioactive gases into the atmosphere. In 1986, a million times as much radioactivity (1 MCi) was released when a cooling system failure at the Chernobyl plant in Ukraine caused a much greater melting of fuel and an uncontrolled reaction. High-pressure steam and ignited graphite moderator rods caused the reactor building to explode and expel radioactive debris. Carried by prevailing winds, the radioactive particles contaminated vegetables and milk in much of Europe. Severe health problems have resulted. And, in 2011, an earthquake and subsequent tsunami caused vast destruction in northern Japan, affecting the Fukushima Dai-ichi nuclear facility and causing the breakdown of cooling-water pumps and the melting of three reactor cores. Large amounts of radioactive steam and water were released into the air and sea, and serious, long-term health effects are expected.

3. *Current use of nuclear power.* Despite potential safety problems, nuclear power remains an important source of electricity. In the late 1990s, nearly every European country employed nuclear power, and it is the major power source in some countries—Sweden creates 50% of its electricity this way and France almost 80%. Currently, the United States obtains about 20% of its electricity from nuclear power, and Canada slightly less. As our need for energy grows and climate change from fossil-fuel consumption worsens, safer reactors will be designed and built. Yet, as a result of the disaster in Japan, many countries are reevaluating their commitment to nuclear energy.

4. *Thermal pollution and waste disposal.* Even a smoothly operating plant has certain inherent problems. The problem of *thermal pollution* is common to all power plants. Water used to condense the steam is several degrees warmer when returned to its source, which can harm aquatic organisms (Section 13.3). A more serious problem is *nuclear waste disposal.* Many of the fission products formed in nuclear reactors have long half-lives, and no satisfactory plan for their permanent disposal has yet been devised. Proposals to place the waste in containers and bury them in deep bedrock cannot possibly be field-tested for the thousands of years the material will remain harmful. Leakage of radioactive material into groundwater is a danger, and earthquakes can occur even in geologically stable regions. It remains to be seen whether we can operate fission reactors *and* dispose of the waste safely and economically.

The Promise of Nuclear Fusion

Nuclear fusion in the Sun is the ultimate source of nearly all the energy—and chemical elements—on Earth. In fact, *all the elements heavier than hydrogen were formed in fusion and decay processes within stars.*

Much research is being devoted to making nuclear fusion a practical, direct source of energy on Earth. To understand the advantages of fusion, let's consider one of the most discussed fusion reactions, in which deuterium and tritium react:

$$^{2}_{1}\text{H} + ^{3}_{1}\text{H} \longrightarrow ^{4}_{2}\text{He} + ^{1}_{0}\text{n}$$

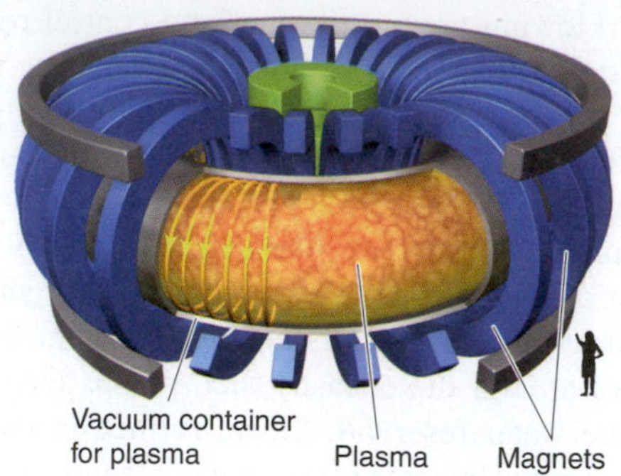

Figure 23.15 The tokamak design for magnetic containment of a fusion plasma.

This reaction produces 1.7×10^9 kJ/mol, an enormous quantity of energy with no radioactive byproducts. Moreover, the reactant nuclei are relatively easy to come by. Thus, in principle, fusion seems promising and may represent an ideal source of power. However, some extremely difficult technical problems remain. Fusion requires enormous energy in the form of heat to give the positively charged nuclei enough kinetic energy to force themselves together. The fusion of deuterium and tritium, for example, occurs at practical rates at about 10^8 K, a temperature hotter than the Sun's core! How can such conditions be achieved?

Two current research approaches have promise. In one, atoms are stripped of their electrons at high temperatures, which results in a gaseous *plasma,* a neutral mixture of positive nuclei and electrons. Because of the extreme temperatures needed for fusion, no *material* can contain the plasma. The most successful approach to date has been to enclose the plasma within a magnetic field. The *tokamak* design has a donut-shaped container in which a helical magnetic field confines the plasma and prevents it from contacting the walls (Figure 23.15). Scientists at the Princeton University Plasma Physics facility have achieved some success in generating energy from fusion this way. In another approach, the high temperature is reached by using many focused lasers to compress and heat the fusion reactants. In any event, one or more major breakthroughs are needed before fusion will be a practical, everyday source of energy.

Summary of Section 23.7

- In nuclear fission, neutron bombardment causes a nucleus to split into two smaller nuclei and release neutrons that split other nuclei, giving rise to a chain reaction.
- A nuclear power plant controls the rate of the chain reaction to produce heat that creates steam, which is used to generate electricity.
- Potential hazards, such as radiation leaks, thermal pollution, and disposal of nuclear waste, remain current concerns.
- Nuclear fusion holds great promise as a source of clean, abundant energy, but it requires extremely high temperatures and is not yet practical.

CHAPTER REVIEW GUIDE

The following sections provide many aids to help you study this chapter. (Numbers in parentheses refer to pages, unless noted otherwise.)

Learning Objectives

These are concepts and skills to review after studying this chapter.

Related section (§), sample problem (SP), and upcoming end-of-chapter problem (EP) numbers are listed in parentheses.

1. Describe the differences between nuclear and chemical changes (Introduction)
2. Identify the three types of radioactive emissions and the types of radioactive decay, and know how each changes *A* and *Z*; explain how a decay series leads to a stable nuclide; write and balance nuclear equations; use the *N/Z* ratio to predict nuclear stability and the type of decay a nuclide undergoes (§23.1) (SPs 23.1–23.3) (EPs 23.1–23.16)
3. Understand why radioactive decay is a first-order process and the meaning of half-life; convert among units of radioactivity, and calculate specific activity, decay constant, half-life, and number of nuclei; estimate the age of an object from its specific activity (§23.2) (SPs 23.4, 23.5) (EPs 23.17–23.30)

4. Describe how particle accelerators are used to synthesize new nuclides (§23.3) (EPs 23.31–23.35)
5. Describe the effect of ionizing radiation on matter; convert among units of radiation dose, and understand the penetrating power of emissions and how ionizing radiation is used beneficially (§23.4) (EPs 23.36–23.42)
6. Describe how radioisotopes are used in research, activation analysis, and diagnosis (§23.5) (EPs 23.43–23.45)
7. Explain the mass difference and how it is related to nuclear binding energy; understand how nuclear stability is related to binding energy per nucleon and why unstable nuclides undergo either fission or fusion; use Einstein's equation to find mass-energy equivalence in J and eV; compare nuclide stability from binding energy per nucleon (§23.6) (SP 23.6) (EPs 23.46–23.52)
8. Discuss the methodology and the pros and cons of power generation by nuclear fission, and evaluate the potential of nuclear fusion (§23.7) (EPs 23.53–23.59)

Key Terms

These important terms appear in boldface in the chapter and are defined again in the Glossary.

Section 23.1
radioactivity (764)
nucleon (764)
nuclide (764)
isotope (764)
alpha (α) particle (765)
beta (β) particle (765)
gamma (γ) ray (765)
alpha (α) decay (765)
beta (β) decay (767)
β^- decay (767)
positron (β^+) emission (767)
positron (767)
electron (e^-) capture (EC) (767)
gamma (γ) emission (767)
band of stability (768)
decay (disintegration) series (771)

Section 23.2
activity ($\mathscr{A}$) (772)
becquerel (Bq) (772)
curie (Ci) (772)
decay constant (772)
half-life ($t_{1/2}$) (773)
radioisotopic dating (774)
radioisotope (774)

Section 23.3
nuclear transmutation (776)
particle accelerator (776)
transuranium element (777)

Section 23.4
ionization (778)
ionizing radiation (778)
gray (Gy) (778)
rad (*r*adiation-*a*bsorbed *d*ose) (778)
rem (*r*oentgen *e*quivalent for *m*an) (778)
sievert (Sv) (778)
background radiation (779)

Section 23.5
tracer (780)

Section 23.6
fission (783)
fusion (783)
nuclear binding energy (784)
electron volt (eV) (784)

Section 23.7
chain reaction (786)
critical mass (787)
reactor core (788)

Key Equations and Relationships

Numbered and screened concepts are listed for you to refer to or memorize.

23.1 Balancing a nuclear equation (765):

$$^{\text{Total } A}_{\text{Total } Z}\text{Reactants} = {}^{\text{Total } A}_{\text{Total } Z}\text{Products}$$

23.2 Defining the unit of radioactivity (curie, Ci) (772):

$$1 \text{ Ci} = 3.70\times10^{10} \text{ disintegrations per second (d/s)}$$

23.3 Expressing the decay rate (activity) for radioactive nuclei (772):

$$\text{Decay rate } (\mathscr{A}) = -\frac{\Delta\mathscr{N}}{\Delta t} = k\mathscr{N}$$

23.4 Finding the number of nuclei remaining after a given time, $\mathscr{N}_t$ (773):

$$\ln\frac{\mathscr{N}_t}{\mathscr{N}_0} = -kt \quad \text{or} \quad \mathscr{N}_t = \mathscr{N}_0 e^{-kt}$$

23.5 Finding the half-life of a radioactive nuclide (773):

$$t_{1/2} = \frac{\ln 2}{k}$$

23.6 Calculating the time to reach a given specific activity (age of an object in radioisotopic dating) (775):

$$t = \frac{1}{k}\ln\frac{\mathscr{A}_0}{\mathscr{A}_t}$$

23.7 Adapting Einstein's equation to calculate mass difference and/or nuclear binding energy (783):

$$\Delta m = \frac{\Delta E}{c^2} \quad \text{or} \quad \Delta E = \Delta mc^2$$

23.8 Relating the atomic mass unit to its energy equivalent in MeV (784):

$$1 \text{ amu} = 931.5\times10^6 \text{ eV} = 931.5 \text{ MeV}$$

BRIEF SOLUTIONS TO FOLLOW-UP PROBLEMS

Compare your own solutions to these calculation steps and answers.

23.1 ${}^{133}_{54}Xe \longrightarrow {}^{133}_{55}Cs + {}^{0}_{-1}\beta$

23.2 ${}^{31}P$ has an even N (16), but ${}^{30}P$ has both N and Z odd. ${}^{31}P$ also has a slightly higher N/Z ratio that is closer to the band of stability.

23.3 (a) $Z = 26$ is iron (Fe), and the atomic mass is 55.85. The A value of 61 is higher: β^- decay; (b) $Z > 83$, which is too high for stability: α decay

23.4 $\ln \mathcal{A}_t = -kt + \ln \mathcal{A}_0$

$$= -\left(\frac{\ln 2}{15\ h} \times 4.0\ \text{days} \times \frac{24\ h}{1\ \text{day}}\right) + \ln(2.5\times10^9)$$

$$= 17.20$$

$$\mathcal{A} = 3.0\times10^7\ \text{d/s}$$

23.5 $t = \frac{1}{k} \ln \frac{\mathcal{A}_0}{\mathcal{A}_t} = \frac{5730\ \text{yr}}{\ln 2} \ln\left(\frac{15.3\ \text{d/min·g}}{9.41\ \text{d/min·g}}\right)$

$$= 4.02\times10^3\ \text{yr}$$

The mummy case is about 4000 years old.

23.6 ${}^{235}U$ has 92 ${}^{1}_{1}p$ and 143 ${}^{1}_{0}n$.

$$\Delta m = [(92 \times 1.007825\ \text{amu}) + (143 \times 1.008665\ \text{amu})] - 235.043924\ \text{amu} = 1.9151\ \text{amu}$$

$$\frac{\text{Binding energy}}{\text{nucleon}} = \frac{1.9151\ \text{amu} \times \frac{931.5\ \text{MeV}}{1\ \text{amu}}}{235\ \text{nucleons}} = 7.591\ \text{MeV/nucleon}$$

Therefore, ${}^{235}U$ is less stable than ${}^{12}C$.

PROBLEMS

Problems with **colored** numbers are answered in Appendix E. Sections match the text and provide the numbers of relevant sample problems. Bracketed problems are grouped in pairs (indicated by a short rule) that cover the same concept. Comprehensive Problems are based on material from any section or previous chapter.

Radioactive Decay and Nuclear Stability

(Sample Problems 23.1 to 23.3)

23.1 How do chemical and nuclear reactions differ in
(a) Magnitude of the energy change?
(b) Effect on rate of increasing temperature?
(c) Effect on rate of higher reactant concentration?
(d) Effect on yield of higher reactant concentration?

23.2 Which of the following produce an atom of a *different* element: (a) α decay; (b) β^- decay; (c) γ emission; (d) β^+ emission; (e) e^- capture? Show how Z and N change, if at all, with each process.

23.3 Why is ${}^{3}_{2}He$ stable, but ${}^{2}_{2}He$ has never been detected?

23.4 How do the modes of decay differ for a neutron-rich nuclide and a proton-rich nuclide?

23.5 Why can't you use the position of a nuclide's N/Z ratio relative to the band of stability to predict whether it is more likely to decay by positron emission or by electron capture?

23.6 Write balanced nuclear equations for the following:
(a) Alpha decay of ${}^{234}_{92}U$
(b) Electron capture by neptunium-232
(c) Positron emission by ${}^{12}_{7}N$

23.7 Write balanced nuclear equations for the following:
(a) β^- decay of sodium-26
(b) β^- decay of francium-223
(c) Alpha decay of ${}^{212}_{83}Bi$

23.8 Write balanced nuclear equations for the following:
(a) Formation of ${}^{48}_{22}Ti$ through positron emission
(b) Formation of silver-107 through electron capture
(c) Formation of polonium-206 through α decay

23.9 Write balanced nuclear equations for the following:
(a) Formation of ${}^{241}_{95}Am$ through β^- decay
(b) Formation of ${}^{228}_{89}Ac$ through β^- decay
(c) Formation of ${}^{203}_{83}Bi$ through α decay

23.10 Which nuclide(s) would you predict to be stable? Why?
(a) ${}^{20}_{8}O$ (b) ${}^{59}_{27}Co$ (c) ${}^{9}_{3}Li$

23.11 Which nuclide(s) would you predict to be stable? Why?
(a) ${}^{146}_{60}Nd$ (b) ${}^{114}_{48}Cd$ (c) ${}^{88}_{42}Mo$

23.12 What is the most likely mode of decay for each?
(a) ${}^{238}_{92}U$ (b) ${}^{48}_{24}Cr$ (c) ${}^{50}_{25}Mn$

23.13 What is the most likely mode of decay for each?
(a) ${}^{111}_{47}Ag$ (b) ${}^{41}_{17}Cl$ (c) ${}^{110}_{44}Ru$

23.14 Why is ${}^{52}_{24}Cr$ the most stable isotope of chromium?

23.15 Why is ${}^{40}_{20}Ca$ the most stable isotope of calcium?

23.16 Neptunium-237 is the parent nuclide of a decay series that starts with α emission, followed by β^- emission, and then two more α emissions. Write a balanced nuclear equation for each step.

The Kinetics of Radioactive Decay

(Sample Problems 23.4 and 23.5)

23.17 What is the reaction order of radioactive decay? Explain.

23.18 After 1 minute, three radioactive nuclei remain from an original sample of six. Is it valid to conclude that $t_{1/2}$ equals 1 minute? Is this conclusion valid if the original sample contained 6×10^{12} nuclei and 3×10^{12} remain after 1 minute? Explain.

23.19 Radioisotopic dating depends on the constant rate of decay and formation of various nuclides in a sample. How is the proportion of ${}^{14}C$ kept relatively constant in living organisms?

23.20 What is the specific activity (in Ci/g) if 1.65 mg of an isotope emits 1.56×10^6 α particles per second?

23.21 What is the specific activity (in Bq/g) if 8.58 μg of an isotope emits 7.4×10^4 α particles per minute?

23.22 If 1.00×10^{-12} mol of ^{135}Cs emits 1.39×10^5 β^- particles in 1.00 yr, what is the decay constant?

23.23 If 6.40×10^{-9} mol of ^{176}W emits 1.07×10^{15} β^+ particles in 1.00 h, what is the decay constant?

23.24 The isotope $^{212}_{83}Bi$ has a half-life of 1.01 yr. What mass (in mg) of a 2.00-mg sample will remain after 3.75×10^3 h?

23.25 The half-life of radium-226 is 1.60×10^3 yr. How many hours will it take for a 2.50-g sample to decay to the point where 0.185 g of the isotope remains?

23.26 A rock contains 270 μmol of ^{238}U ($t_{1/2} = 4.5 \times 10^9$ yr) and 110 μmol of ^{206}Pb. Assuming that all the ^{206}Pb comes from decay of the ^{238}U, estimate the rock's age.

23.27 A fabric remnant from a burial site has a $^{14}C/^{12}C$ ratio of 0.735 of the original value. How old is the fabric?

23.28 Due to decay of ^{40}K, cow's milk has a specific activity of about 6×10^{-11} mCi per milliliter. How many disintegrations of ^{40}K nuclei are there per minute in an 8.0-oz glass of milk?

23.29 Plutonium-239 ($t_{1/2} = 2.41 \times 10^4$ yr) represents a serious nuclear waste hazard. If seven half-lives are required to reach a tolerable level of radioactivity, how long must ^{239}Pu be stored?

23.30 A volcanic eruption melts a large chunk of rock, and all gases are expelled. After cooling, $^{40}_{18}Ar$ accumulates from the ongoing decay of $^{40}_{19}K$ in the rock ($t_{1/2} = 1.25 \times 10^9$ yr). When a piece of rock is analyzed, it is found to contain 1.38 mmol of ^{40}K and 1.14 mmol of ^{40}Ar. How long ago did the rock cool?

Nuclear Transmutation: Induced Changes in Nuclei

23.31 Why must the electrical polarity of the tubes in a linear accelerator be reversed at very short time intervals?

23.32 Why does bombardment with protons usually require higher energies than bombardment with neutrons?

23.33 Name the unidentified species in each transmutation, and write a balanced nuclear equation:
(a) Bombardment of ^{10}B with an α particle yields a neutron and a nuclide.
(b) Bombardment of ^{28}Si with ^2H yields ^{29}P and another particle.
(c) Bombardment of a nuclide with an α particle yields two neutrons and ^{244}Cf.

23.34 Name the unidentified species, and write a balanced nuclear equation for each transmutation: (a) gamma irradiation of a nuclide yields a proton, a neutron, and ^{29}Si; (b) bombardment of ^{252}Cf with ^{10}B yields five neutrons and a nuclide; (c) bombardment of ^{238}U with a particle yields three neutrons and ^{239}Pu.

23.35 Elements 104, 105, and 106 have been named rutherfordium (Rf), dubnium (Db), and seaborgium (Sg), respectively. These elements are synthesized from californium-249 by bombarding with carbon-12, nitrogen-15, and oxygen-18 nuclei, respectively. Four neutrons are formed in each reaction as well. Write balanced nuclear equations for the formation of these elements.

Effects of Nuclear Radiation on Matter

23.36 The effects on matter of γ rays and α particles differ. Explain.

23.37 Suggest a reason why ionizing radiation may be more harmful to children than adults.

23.38 A 135-lb person absorbs 3.3×10^{-7} J of energy from radioactive emissions. (a) How many rads does she receive? (b) How many grays (Gy) does she receive?

23.39 A 3.6-kg laboratory animal receives a single dose of 8.92×10^{-4} Gy. (a) How many rads did the animal receive? (b) How many joules did the animal absorb?

23.40 A 70.-kg person exposed to ^{90}Sr absorbs 6.0×10^5 β^- particles, each with an energy of 8.74×10^{-14} J. (a) How many grays does the person receive? (b) If the RBE is 1.0, how many millirems is this? (c) What is the equivalent dose in sieverts (Sv)?

23.41 A laboratory rat weighs 265 g and absorbs 1.77×10^{10} β^- particles, each with an energy of 2.20×10^{-13} J. (a) How many rads does the animal receive? (b) What is this dose in Gy? (c) If the RBE is 0.75, what is the equivalent dose in Sv?

23.42 A small region of a cancer patient's brain is exposed for 24.0 min to 475 Bq of radioactivity from ^{60}Co for treatment of a tumor. If the brain mass exposed is 1.858 g and each β^- particle emitted has an energy of 5.05×10^{-14} J, what is the dose in rads?

Applications of Radioisotopes

23.43 What two ways are radioactive tracers used in organisms?

23.44 Why is neutron activation analysis (NAA) useful to art historians and criminologists?

23.45 The oxidation of methanol to formaldehyde can be accomplished by reaction with chromic acid:

$$6H^+(aq) + 3CH_3OH(aq) + 2H_2CrO_4(aq) \longrightarrow 3CH_2O(aq) + 2Cr^{3+}(aq) + 8H_2O(l)$$

The reaction can be studied with the stable isotope tracer ^{18}O and mass spectrometry. When a small amount of $CH_3{}^{18}OH$ is present in the alcohol reactant, $CH_2{}^{18}O$ forms. When a small amount of $H_2Cr^{18}O_4$ is present, $H_2{}^{18}O$ forms. Does chromic acid or methanol supply the O atom to the aldehyde? Explain.

Interconversion of Mass and Energy

(Sample Problem 23.6)

Note: Data for problems in this section: mass of ^1H atom = 1.007825 amu; mass of neutron = 1.008665 amu.

23.46 What is a mass difference, and how does it arise?

23.47 What is the binding energy per nucleon? Why is the binding energy per nucleon, rather than per nuclide, used to compare nuclide stability?

23.48 A ^3H nucleus decays with an energy of 0.01861 MeV. Convert this energy into (a) electron volts; (b) joules.

23.49 Arsenic-84 decays with an energy of 1.57×10^{-15} kJ per nucleus. Convert this energy into (a) eV; (b) MeV.

23.50 Cobalt-59 is the only stable isotope of this transition metal. One ^{59}Co atom has a mass of 58.933198 amu. Calculate the binding energy (a) per nucleon in MeV; (b) per atom in MeV; (c) per mole in kJ.

23.51 Iodine-131 is one of the most important isotopes used in the diagnosis of thyroid cancer. One atom has a mass of 130.906114 amu. Calculate the binding energy (a) per nucleon in MeV; (b) per atom in MeV; (c) per mole in kJ.

23.52 The ^{80}Br nuclide decays either by β^- decay or by electron capture. (a) What is the product of each process? (b) Which process releases more energy? (Masses of atoms: ^{80}Br = 79.918528 amu; ^{80}Kr = 79.916380 amu; ^{80}Se = 79.916520 amu; neglect the mass of electrons involved because these are atomic, not nuclear, masses.)

Applications of Fission and Fusion

23.53 In what main way is fission different from radioactive decay? Are all fission events in a chain reaction identical? Explain.

23.54 What is the purpose of enrichment in the preparation of fuel rods? How is it accomplished?

23.55 Describe the nature and purpose of these components of a nuclear reactor: (a) control rods; (b) moderator; (c) reflector.

23.56 State an advantage and a disadvantage of heavy-water reactors compared to light-water reactors.

23.57 What are the expected advantages of fusion reactors over fission reactors?

23.58 The reaction that will probably power the first commercial fusion reactor is

$$^3_1H + ^2_1H \longrightarrow ^4_2He + ^1_0n$$

How much energy would be produced per mole of reaction? (Masses of atoms: ^3_1H = 3.01605 amu; ^2_1H = 2.0140 amu; ^4_2He = 4.00260 amu; mass of ^1_0n = 1.008665 amu.)

Comprehensive Problems

23.59 Seaborgium-263 (Sg), the first isotope of element 106 synthesized, was made, together with four neutrons, by bombarding californium-249 with oxygen-18. It then underwent a series of decays starting with three α emissions. Write balanced equations for the synthesis and the three α emissions of ^{263}Sg.

23.60 Some $^{243}_{95}Am$ was present when Earth formed, but it all decayed in the next billion years. The first three steps in this decay series are emissions of an α particle, a β^- particle, and another α particle. What other isotopes were present on the young Earth in a rock that contained some $^{243}_{95}Am$?

23.61 The scene below depicts a neutron bombarding ^{235}U:

$$^1_0 + ^{235}_{92} \longrightarrow ^{144}_{55} + ? + 2^1_0$$

(a) Is this an example of fission or of fusion? (b) Identify the other nuclide formed. (c) What is the most likely mode of decay of the nuclide with $Z = 55$?

23.62 Curium-243 undergoes α decay to plutonium-239:

$$^{243}Cm \longrightarrow ^{239}Pu + \alpha$$

(a) Find the change in mass, Δm (in kg). (Masses: ^{243}Cm = 243.0614 amu; ^{239}Pu = 239.0522 amu; ^4He = 4.0026 amu; 1 amu = 1.661×10^{-24} g.)
(b) Find the energy released in joules.
(c) Find the energy released in kJ/mol of reaction, and comment on the difference between this value and a typical heat of reaction for a chemical change, which is a few hundred kJ/mol.

23.63 Plutonium "triggers" for nuclear weapons were manufactured at the Rocky Flats plant in Colorado. An 85-kg worker inhaled a dust particle containing 1.00 μg of $^{239}_{94}Pu$, which resided in his body for 16 h ($t_{1/2}$ of $^{239}Pu = 2.41\times10^4$ yr; each disintegration released 5.15 MeV).
(a) How many rads did he receive? (b) How many grays?

23.64 Archeologists removed some charcoal from a Native American campfire, burned it in O_2, and bubbled the CO_2 formed into $Ca(OH)_2$ solution (limewater). The $CaCO_3$ that precipitated was filtered and dried. If 4.58 g of the $CaCO_3$ had a radioactivity of 3.2 d/min, how long ago was the campfire?

23.65 ^{238}U ($t_{1/2} = 4.5\times10^9$ yr) begins a decay series that ultimately forms ^{206}Pb. The scene below depicts the relative number of ^{238}U atoms *(red)* and ^{206}Pb atoms *(green)* in a mineral. If all the Pb comes from ^{238}U, calculate the age of the sample.

23.66 A 5.4-μg sample of $^{226}RaCl_2$ has a radioactivity of 1.5×10^5 Bq. Calculate $t_{1/2}$ of ^{226}Ra.

23.67 The major reaction taking place during hydrogen fusion in a young star is $4^1_1H \longrightarrow ^4_2He + 2^{\ 0}_{1}\beta + 2^0_0\gamma$ + energy. How much energy (in MeV) is released per He nucleus formed? Per mole of He? (Masses: ^1_1H atom = 1.007825 amu; ^4_2He atom = 4.00260 amu; positron = 5.48580×10^{-4} amu.)

23.68 A sample of AgCl emits 175 nCi/g. A saturated solution prepared from the solid emits 1.25×10^{-2} Bq/mL due to radioactive Ag^+ ions. What is the molar solubility of AgCl?

23.69 In the 1950s, radioactive material was spread over the land from aboveground nuclear tests. A woman drinks some contaminated milk and ingests 0.0500 g of ^{90}Sr, which is taken up by bones and teeth and not eliminated. (a) How much ^{90}Sr ($t_{1/2}$ = 29 yr) is present in her body after 10 yr? (b) How long will it take for 99.9% of the ^{90}Sr to decay?

23.70 Technetium-99m is a metastable nuclide used in numerous cancer diagnostic and treatment programs. It is prepared just before use because it decays rapidly through γ emission:

$$^{99m}Tc \longrightarrow ^{99}Tc + \gamma$$

Use the data below to determine (a) the half-life of ^{99m}Tc; (b) the percentage of the isotope that is lost if it takes 2.0 h to prepare and administer the dose.

Time (h)	γ Emission (photons/s)
0	5000.
4	3150.
8	2000.
12	1250.
16	788
20	495

23.71 What volume of radon will be produced per hour at STP from 1.000 g of ^{226}Ra ($t_{1/2}$ = 1599 yr; 1 yr = 8766 h; mass of one ^{226}Ra atom = 226.025402 amu)?

23.72 Which isotope in each pair is more stable? Why?
(a) $^{140}_{55}Cs$ or $^{133}_{55}Cs$ (b) $^{79}_{35}Br$ or $^{78}_{35}Br$
(c) $^{28}_{12}Mg$ or $^{24}_{12}Mg$ (d) $^{14}_7N$ or $^{18}_7N$

23.73 The scene below represents a reaction (with neutrons gray and protons purple) that occurs during the lifetime of a star. (a) Write a balanced nuclear equation for the reaction. (b) If the mass difference is 7.7×10^{-2} amu, find the energy (kJ) released.

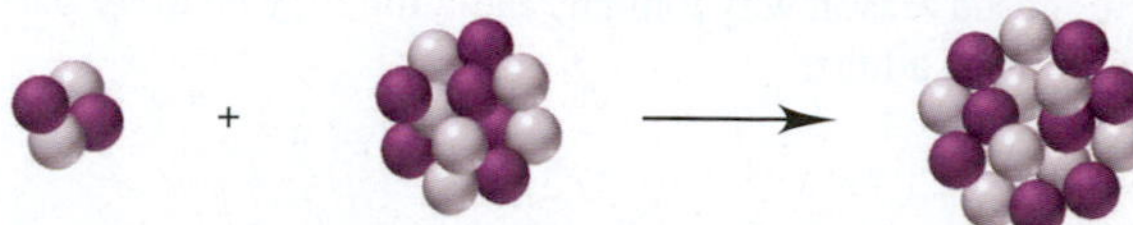

23.74 The 23rd-century starship *Enterprise* uses a substance called "dilithium crystals" as its fuel.
(a) Assuming this material is the result of fusion, what is the product of the fusion of two ^{6}Li nuclei?
(b) How much energy is released per kilogram of dilithium formed? (Mass of one ^{6}Li atom is 6.015121 amu.)
(c) When four ^{1}H atoms fuse to form ^{4}He, how many positrons are released?
(d) To determine the energy potential of the fusion processes in parts (b) and (c), compare the changes in mass per kilogram of dilithium and of ^{4}He.
(e) Compare the change in mass per kilogram in part (b) to that for the formation of ^{4}He by the method used in fusion reactors. (For masses, see Problem 23.58.)
(f) Using early 21st-century fusion technology, how much tritium can be produced per kilogram of ^{6}Li in the following reaction: $^{6}_{3}\text{Li} + ^{1}_{0}\text{n} \longrightarrow ^{4}_{2}\text{He} + ^{3}_{1}\text{H}$? When this amount of tritium is fused with deuterium, what is the change in mass? How does this quantity compare with the use of dilithium in part (b)?

23.75 Nuclear disarmament could be accomplished if weapons were not "replenished." The tritium in warheads decays to helium with a half-life of 12.26 yr and must be replaced or the weapon is useless. What fraction of the tritium is lost in 5.50 yr?

23.76 Gadolinium-146 undergoes electron capture. Identify the product, and use Figure 23.2 to find the modes of decay and the other two nuclides in the series below:

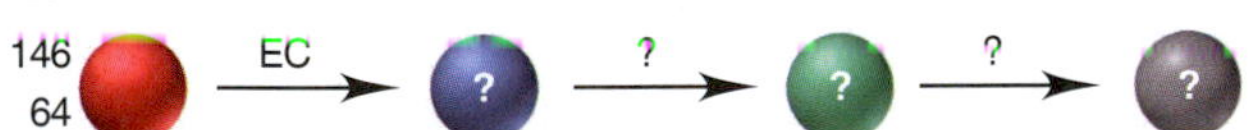

23.77 A decay series starts with the synthetic isotope $^{239}_{92}$U. The first four steps are emissions of a β^- particle, another β^-, an α particle, and another α. Write a balanced nuclear equation for each step. Which natural series could start by this sequence?

23.78 An earthquake in the area of present-day San Francisco is to be dated by measuring the ^{14}C activity ($t_{1/2}$ = 5730 yr) of parts of a tree uprooted during the event. The tree parts have an activity of 12.9 d/min·g C, and a living tree has an activity of 15.3 d/min·g C. How long ago did the earthquake occur?

23.79 Carbon from the remains of an extinct Australian marsupial, called *Diprotodon,* has a specific activity of 0.61 pCi/g. Modern carbon has a specific activity of 6.89 pCi/g. How long ago did the *Diprotodon* apparently become extinct?

23.80 Using 21st-century technology, hydrogen fusion requires temperatures around 10^8 K. But, lower initial temperatures are used if the hydrogen is compressed. In the late 24th century, the starship *Leinad* uses such methods to fuse hydrogen at 10^6 K.
(a) What is the kinetic energy of an H atom at 1.00×10^6 K?
(b) How many H atoms are heated to 1.00×10^6 K from the energy of one H and one anti-H atom annihilating each other?
(c) If these H atoms fuse into ^{4}He atoms (with the loss of two positrons per ^{4}He formed), how much energy (in J) is generated?
(d) How much more energy is generated by the fusion in (c) than by the hydrogen-antihydrogen collision in (b)?
(e) Should the captain of the *Leinad* change the technology and produce ^{3}He (mass = 3.01603 amu) instead of ^{4}He?

23.81 Representations of three nuclei (with neutrons gray and protons purple) are shown below. Nucleus 1 is stable, but 2 and 3 are not. (a) Write the symbol for each isotope. (b) What is (are) the most likely mode(s) of decay for 2 and 3?

1

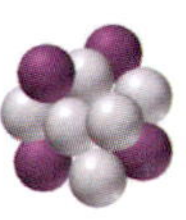
2

3

Common Mathematical Operations in Chemistry

In addition to basic arithmetic and algebra, four mathematical operations are used frequently in general chemistry: manipulating logarithms, using exponential notation, solving quadratic equations, and graphing data.

MANIPULATING LOGARITHMS

Meaning and Properties of Logarithms

A *logarithm* is an exponent. Specifically, if $x^n = A$, we can say that the logarithm to the base x of the number A is n, and we can denote it as

$$\log_x A = n$$

Because logarithms are exponents, they have the following properties:

$$\log_x 1 = 0$$
$$\log_x (A \times B) = \log_x A + \log_x B$$
$$\log_x \frac{A}{B} = \log_x A - \log_x B$$
$$\log_x A^y = y \log_x A$$

Types of Logarithms

Common and natural logarithms are used in chemistry and the other sciences.

1. For *common* logarithms, the base (x in the examples above) is 10, but they are written without specifying the base. That is, $\log_{10} A$ is written more simply as $\log A$; thus, the notation *log* means base 10. The common logarithm of 1000 is 3; in other words, you must raise 10 to the 3rd power to obtain 1000:

$$\log 1000 = 3 \qquad \text{or} \qquad 10^3 = 1000$$

Similarly, we have

$$\log 10 = 1 \qquad \text{or} \qquad 10^1 = 10$$
$$\log 1{,}000{,}000 = 6 \qquad \text{or} \qquad 10^6 = 1{,}000{,}000$$
$$\log 0.001 = -3 \qquad \text{or} \qquad 10^{-3} = 0.001$$
$$\log 853 = 2.931 \qquad \text{or} \qquad 10^{2.931} = 853$$

The last example illustrates an important point about significant figures with all logarithms: *the number of significant figures in the number equals the number of digits to the right of the decimal point in the logarithm.* That is, the number 853 has three significant figures, and the logarithm 2.931 has three digits to the right of the decimal point.

To find a common logarithm with an electronic calculator, enter the number and press the LOG button.

2. For *natural* logarithms, the base is the number e, which is 2.71828 . . . , and $\log_e A$ is written $\ln A$; thus, the notation *ln* means base e. The relationship between the common and natural logarithms is easily obtained:

$$\log 10 = 1 \qquad \text{and} \qquad \ln 10 = 2.303$$

Therefore, we have

$$\ln A = 2.303 \log A$$

To find a natural logarithm with an electronic calculator, enter the number and press the LN button. If your calculator does not have an LN button, enter the number, press the LOG button, and multiply by 2.303.

Antilogarithms

The *antilogarithm* is the base raised to the logarithm:

$$\text{antilogarithm (antilog) of } n \text{ is } 10^n$$

Using two of the earlier examples, the antilog of 3 is 1000, and the antilog of 2.931 is 853. To obtain the antilog with a calculator, enter the number and press the 10^x button. Similarly, to obtain the natural antilogarithm, enter the number and press the e^x button. [On some calculators, you need to enter the number and first press INV and then the LOG (or LN) button.]

USING EXPONENTIAL (SCIENTIFIC) NOTATION

Many quantities in chemistry are very large or very small. For example, in the conventional way of writing numbers, the number of gold atoms in 1 gram of gold is

59,060,000,000,000,000,000,000 atoms (to four significant figures)

As another example, the mass in grams of one gold atom is

0.0000000000000000000003272 g (to four significant figures)

Exponential (scientific) notation provides a much more practical way of writing such numbers. In exponential notation, we express numbers in the form

$$A \times 10^n$$

where A (the coefficient) is greater than or equal to 1 and less than 10 (that is, $1 \leq A < 10$), and n (the exponent) is an integer.

If the number we want to express in exponential notation is larger than 1, the exponent is positive ($n > 0$); if the number is smaller than 1, the exponent is negative ($n < 0$). The size of n tells the number of places the decimal point (in conventional notation) must be moved to obtain a coefficient A greater than or equal to 1 and less than 10 (in exponential notation). In exponential notation, 1 gram of gold contains 5.906×10^{22} atoms, and each gold atom has a mass of 3.272×10^{-22} g.

Changing Between Conventional and Exponential Notation

In order to use exponential notation, you must be able to convert to it from conventional notation, and vice versa.

1. To change a number from conventional to exponential notation, move the decimal point to the left for numbers equal to or greater than 10 and to the right for numbers between 0 and 1:

   75,000,000 changes to 7.5×10^7 (decimal point 7 places to the left)
   0.006042 changes to 6.042×10^{-3} (decimal point 3 places to the right)

2. To change a number from exponential to conventional notation, move the decimal point the number of places indicated by the exponent to the right for numbers with positive exponents and to the left for numbers with negative exponents:

   1.38×10^5 changes to 138,000 (decimal point 5 places to the right)
   8.41×10^{-6} changes to 0.00000841 (decimal point 6 places to the left)

3. An exponential number with a coefficient greater than 10 or less than 1 can be changed to the standard exponential form by converting the coefficient to the standard form and adding the exponents:

$$582.3\times10^6 \text{ changes to } 5.823\times10^2\times10^6 = 5.823\times10^{(2+6)} = 5.823\times10^8$$
$$0.0043\times10^{-4} \text{ changes to } 4.3\times10^{-3}\times10^{-4} = 4.3\times10^{[(-3)+(-4)]} = 4.3\times10^{-7}$$

Using Exponential Notation in Calculations

In calculations, you can treat the coefficient and exponents separately and apply the properties of exponents (see earlier section on logarithms).

1. To multiply exponential numbers, multiply the coefficients, add the exponents, and reconstruct the number in standard exponential notation:

$$(5.5\times10^3)(3.1\times10^5) = (5.5\times3.1)\times10^{(3+5)} = 17\times10^8 = 1.7\times10^9$$
$$(9.7\times10^{14})(4.3\times10^{-20}) = (9.7\times4.3)\times10^{[14+(-20)]} = 42\times10^{-6} = 4.2\times10^{-5}$$

2. To divide exponential numbers, divide the coefficients, subtract the exponents, and reconstruct the number in standard exponential notation:

$$\frac{2.6\times10^6}{5.8\times10^2} = \frac{2.6}{5.8}\times10^{(6-2)} = 0.45\times10^4 = 4.5\times10^3$$
$$\frac{1.7\times10^{-5}}{8.2\times10^{-8}} = \frac{1.7}{8.2}\times10^{[(-5)-(-8)]} = 0.21\times10^3 = 2.1\times10^2$$

3. To add or subtract exponential numbers, change all numbers so that they have the same exponent, then add or subtract the coefficients:

$$(1.45\times10^4) + (3.2\times10^3) = (1.45\times10^4) + (0.32\times10^4) = 1.77\times10^4$$
$$(3.22\times10^5) - (9.02\times10^4) = (3.22\times10^5) - (0.902\times10^5) = 2.32\times10^5$$

SOLVING QUADRATIC EQUATIONS

A *quadratic equation* is one in which the highest power of x is 2. The general form of a quadratic equation is

$$ax^2 + bx + c = 0$$

where a, b, and c are numbers. For given values of a, b, and c, the values of x that satisfy the equation are called *solutions* of the equation. We calculate x with the quadratic formula:

$$x = \frac{-b \pm \sqrt{b^2 - 4ac}}{2a}$$

We commonly require the quadratic formula when solving for some concentration in an equilibrium problem. For example, we might have an expression that is rearranged into the quadratic equation

$$\underset{a}{4.3}x^2 + \underset{b}{0.65}x - \underset{c}{8.7} = 0$$

Applying the quadratic formula, with $a = 4.3$, $b = 0.65$, and $c = -8.7$, gives

$$x = \frac{-0.65 \pm \sqrt{(0.65)^2 - 4(4.3)(-8.7)}}{2(4.3)}$$

The "plus or minus" sign ($\pm$) indicates that there are always two possible values for x. In this case, they are

$$x = 1.3 \quad \text{and} \quad x = -1.5$$

In any real physical system, however, only one of the values will have any meaning. For example, if x were $[H_3O^+]$, the negative value would give a negative concentration, which has no physical meaning.

GRAPHING DATA IN THE FORM OF A STRAIGHT LINE

Visualizing changes in variables by means of a graph is used throughout science. In many cases, it is most useful if the data can be graphed in the form of a straight line. Any equation will appear as a straight line if it has, or can be rearranged to have, the following general form:

$$y = mx + b$$

where y is the dependent variable (typically plotted along the vertical axis), x is the independent variable (typically plotted along the horizontal axis), m is the slope of the line, and b is the intercept of the line on the y axis. The intercept is the value of y when $x = 0$:

$$y = m(0) + b = b$$

The slope of the line is the change in y for a given change in x:

$$\text{Slope } (m) = \frac{y_2 - y_1}{x_2 - x_1} = \frac{\Delta y}{\Delta x}$$

The *sign* of the slope tells the *direction* of the line. If y increases as x increases, m is positive, and the line slopes upward with higher values of x; if y decreases as x increases, m is negative, and the line slopes downward with higher values of x. The *magnitude* of the slope indicates the *steepness* of the line. A line with $m = 3$ is three times as steep (y changes three times as much for a given change in x) as a line with $m = 1$.

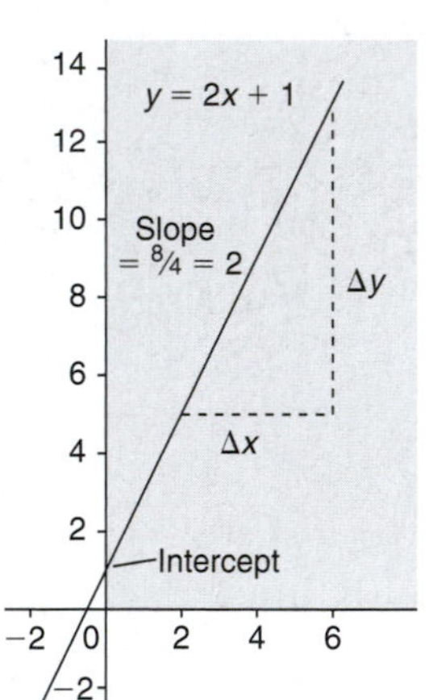

Figure A.1

Consider the linear equation $y = 2x + 1$. A graph of this equation is shown in Figure A.1. In practice, you can find the slope by drawing a right triangle to the line, using the line as the hypotenuse. Then, one leg gives Δy, and the other gives Δx. In the figure, $\Delta y = 8$ and $\Delta x = 4$.

At several places in the text, an equation is rearranged into the form of a straight line in order to determine information from the slope and/or the intercept. For example, in Chapter 16, we obtained the following expression:

$$\ln \frac{[A]_0}{[A]_t} = kt$$

Based on the properties of logarithms, we have

$$\ln [A]_0 - \ln [A]_t = kt$$

Rearranging into the form of an equation for a straight line gives

$$\begin{aligned} \ln [A]_t &= -kt + \ln [A]_0 \\ y &= mx + b \end{aligned}$$

Thus, a plot of $\ln [A]_t$ vs. t is a straight line, from which you can see that the slope is $-k$ (the negative of the rate constant) and the intercept is $\ln [A]_0$ (the natural logarithm of the initial concentration of A).

At many other places in the text, linear relationships occur that were not shown in graphical terms. For example, the conversion of temperature scales in Chapter 1 can also be expressed in the form of a straight line:

$$\begin{aligned} {}^\circ\text{F} &= \tfrac{9}{5}{}^\circ\text{C} + 32 \\ y &= m\,x + b \end{aligned}$$

Standard Thermodynamic Values for Selected Substances*

Substance or Ion	ΔH_f° (kJ/mol)	ΔG_f° (kJ/mol)	S° (J/mol·K)
$e^-(g)$	0	0	20.87
Aluminum			
$Al(s)$	0	0	28.3
$Al^{3+}(aq)$	−524.7	−481.2	−313
$AlCl_3(s)$	−704.2	−628.9	110.7
$Al_2O_3(s)$	−1676	−1582	50.94
Barium			
$Ba(s)$	0	0	62.5
$Ba(g)$	175.6	144.8	170.28
$Ba^{2+}(g)$	1649.9	—	—
$Ba^{2+}(aq)$	−538.36	−560.7	13
$BaCl_2(s)$	−806.06	−810.9	126
$BaCO_3(s)$	−1219	−1139	112
$BaO(s)$	−548.1	−520.4	72.07
$BaSO_4(s)$	−1465	−1353	132
Boron			
B(β-rhombohedral)	0	0	5.87
$BF_3(g)$	−1137.0	−1120.3	254.0
$BCl_3(g)$	−403.8	−388.7	290.0
$B_2H_6(g)$	35	86.6	232.0
$B_2O_3(s)$	−1272	−1193	53.8
$H_3BO_3(s)$	−1094.3	−969.01	88.83
Bromine			
$Br_2(l)$	0	0	152.23
$Br_2(g)$	30.91	3.13	245.38
$Br(g)$	111.9	82.40	174.90
$Br^-(g)$	−218.9	—	—
$Br^-(aq)$	−120.9	−102.82	80.71
$HBr(g)$	−36.3	−53.5	198.59
Cadmium			
$Cd(s)$	0	0	51.5
$Cd(g)$	112.8	78.20	167.64
$Cd^{2+}(aq)$	−72.38	−77.74	−61.1
$CdS(s)$	−144	−141	71
Calcium			
$Ca(s)$	0	0	41.6
$Ca(g)$	192.6	158.9	154.78
$Ca^{2+}(g)$	1934.1	—	—
$Ca^{2+}(aq)$	−542.96	−553.04	−55.2
$CaF_2(s)$	−1215	−1162	68.87
$CaCl_2(s)$	−795.0	−750.2	114
$CaCO_3(s)$	−1206.9	−1128.8	92.9
$CaO(s)$	−635.1	−603.5	38.2
$Ca(OH)_2(s)$	−986.09	−898.56	83.39
$Ca_3(PO_4)_2(s)$	−4138	−3899	263
$CaSO_4(s)$	−1432.7	−1320.3	107
Carbon			
C(graphite)	0	0	5.686
C(diamond)	1.896	2.866	2.439
$C(g)$	715.0	669.6	158.0
$CO(g)$	−110.5	−137.2	197.5
$CO_2(g)$	−393.5	−394.4	213.7
$CO_2(aq)$	−412.9	−386.2	121
$CO_3^{2-}(aq)$	−676.26	−528.10	−53.1
$HCO_3^-(aq)$	−691.11	587.06	95.0
$H_2CO_3(aq)$	−698.7	−623.42	191
$CH_4(g)$	−74.87	−50.81	186.1
$C_2H_2(g)$	227	209	200.85
$C_2H_4(g)$	52.47	68.36	219.22
$C_2H_6(g)$	−84.667	−32.89	229.5
$C_3H_8(g)$	−105	−24.5	269.9
$C_4H_{10}(g)$	−126	−16.7	310
$C_6H_6(l)$	49.0	124.5	172.8
$CH_3OH(g)$	−201.2	−161.9	238
$CH_3OH(l)$	−238.6	−166.2	127
$HCHO(g)$	−116	−110	219
$HCOO^-(aq)$	−410	−335	91.6
$HCOOH(l)$	−409	−346	129.0
$HCOOH(aq)$	−410	−356	164
$C_2H_5OH(g)$	−235.1	−168.6	282.6
$C_2H_5OH(l)$	−277.63	−174.8	161
$CH_3CHO(g)$	−166	−133.7	266
$CH_3COOH(l)$	−487.0	−392	160
$C_6H_{12}O_6(s)$	−1273.3	−910.56	212.1
$C_{12}H_{22}O_{11}(s)$	−2221.7	−1544.3	360.24
$CN^-(aq)$	151	166	118
$HCN(g)$	135	125	201.7
$HCN(l)$	105	121	112.8
$HCN(aq)$	105	112	129
$CS_2(g)$	117	66.9	237.79
$CS_2(l)$	87.9	63.6	151.0
$CH_3Cl(g)$	−83.7	−60.2	234
$CH_2Cl_2(l)$	−117	−63.2	179

*All values at 298 K.

(continued)

Substance or Ion	ΔH°_f (kJ/mol)	ΔG°_f (kJ/mol)	S° (J/mol·K)
$CHCl_3(l)$	−132	−71.5	203
$CCl_4(g)$	−96.0	−53.7	309.7
$CCl_4(l)$	−139	−68.6	214.4
$COCl_2(g)$	−220	−206	283.74
Cesium			
$Cs(s)$	0	0	85.15
$Cs(g)$	76.7	49.7	175.5
$Cs^+(g)$	458.5	427.1	169.72
$Cs^+(aq)$	−248	−282.0	133
$CsF(s)$	−554.7	−525.4	88
$CsCl(s)$	−442.8	−414	101.18
$CsBr(s)$	−395	−383	121
$CsI(s)$	−337	−333	130
Chlorine			
$Cl_2(g)$	0	0	223.0
$Cl(g)$	121.0	105.0	165.1
$Cl^-(g)$	−234	−240	153.25
$Cl^-(aq)$	−167.46	−131.17	55.10
$HCl(g)$	−92.31	−95.30	186.79
$HCl(aq)$	−167.46	−131.17	55.06
$ClO_2(g)$	102	120	256.7
$Cl_2O(g)$	80.3	97.9	266.1
Chromium			
$Cr(s)$	0	0	23.8
$Cr^{3+}(aq)$	−1971	—	—
$CrO_4^{2-}(aq)$	−863.2	−706.3	38
$Cr_2O_7^{2-}(aq)$	−1461	−1257	214
Copper			
$Cu(s)$	0	0	33.1
$Cu(g)$	341.1	301.4	166.29
$Cu^+(aq)$	51.9	50.2	−26
$Cu^{2+}(aq)$	64.39	64.98	−98.7
$Cu_2O(s)$	−168.6	−146.0	93.1
$CuO(s)$	−157.3	−130	42.63
$Cu_2S(s)$	−79.5	−86.2	120.9
$CuS(s)$	−53.1	−53.6	66.5
Fluorine			
$F_2(g)$	0	0	202.7
$F(g)$	78.9	61.8	158.64
$F^-(g)$	−255.6	−262.5	145.47
$F^-(aq)$	−329.1	−276.5	−9.6
$HF(g)$	−273	−275	173.67
Hydrogen			
$H_2(g)$	0	0	130.6
$H(g)$	218.0	203.30	114.60
$H^+(aq)$	0	0	0
$H^+(g)$	1536.3	1517.1	108.83
Iodine			
$I_2(s)$	0	0	116.14
$I_2(g)$	62.442	19.38	260.58
$I(g)$	106.8	70.21	180.67
$I^-(g)$	−194.7	—	—
$I^-(aq)$	−55.94	−51.67	109.4
$HI(g)$	25.9	1.3	206.33
Iron			
$Fe(s)$	0	0	27.3
$Fe^{3+}(aq)$	−47.7	−10.5	−293
$Fe^{2+}(aq)$	−87.9	−84.94	113
$FeCl_2(s)$	−341.8	−302.3	117.9
$FeCl_3(s)$	−399.5	−334.1	142
$FeO(s)$	−272.0	−251.4	60.75
$Fe_2O_3(s)$	−825.5	−743.6	87.400
$Fe_3O_4(s)$	−1121	−1018	145.3
Lead			
$Pb(s)$	0	0	64.785
$Pb^{2+}(aq)$	1.6	−24.3	21
$PbCl_2(s)$	−359	−314	136
$PbO(s)$	−218	−198	68.70
$PbO_2(s)$	−276.6	−219.0	76.6
$PbS(s)$	−98.3	−96.7	91.3
$PbSO_4(s)$	−918.39	−811.24	147
Lithium			
$Li(s)$	0	0	29.10
$Li(g)$	161	128	138.67
$Li^+(g)$	687.163	649.989	132.91
$Li^+(aq)$	−278.46	−293.8	14
$LiF(s)$	−616.9	−588.7	35.66
$LiCl(s)$	−408	−384	59.30
$LiBr(s)$	−351	−342	74.1
$LiI(s)$	−270	−270	85.8
Magnesium			
$Mg(s)$	0	0	32.69
$Mg(g)$	150	115	148.55
$Mg^{2+}(g)$	2351	—	—
$Mg^{2+}(aq)$	−461.96	−456.01	118
$MgCl_2(s)$	−641.6	−592.1	89.630
$MgCO_3(s)$	−1112	−1028	65.86
$MgO(s)$	−601.2	−569.0	26.9
$Mg_3N_2(s)$	−461	−401	88
Manganese			
$Mn(s, \alpha)$	0	0	31.8
$Mn^{2+}(aq)$	−219	−223	−84
$MnO_2(s)$	−520.9	−466.1	53.1
$MnO_4^-(aq)$	−518.4	−425.1	190
Mercury			
$Hg(l)$	0	0	76.027
$Hg(g)$	61.30	31.8	174.87
$Hg^{2+}(aq)$	171	164.4	−32
$Hg_2^{2+}(aq)$	172	153.6	84.5
$HgCl_2(s)$	−230	−184	144
$Hg_2Cl_2(s)$	−264.9	−210.66	196
$HgO(s)$	−90.79	−58.50	70.27
Nitrogen			
$N_2(g)$	0	0	191.5
$N(g)$	473	456	153.2
$N_2O(g)$	82.05	104.2	219.7
$NO(g)$	90.29	86.60	210.65
$NO_2(g)$	33.2	51	239.9
$N_2O_4(g)$	9.16	97.7	304.3
$N_2O_5(g)$	11	118	346
$N_2O_5(s)$	−43.1	114	178
$NH_3(g)$	−45.9	−16	193
$NH_3(aq)$	−80.83	26.7	110
$N_2H_4(l)$	50.63	149.2	121.2

(continued)

Substance or Ion	ΔH_f° (kJ/mol)	ΔG_f° (kJ/mol)	S° (J/mol·K)
$NO_3^-(aq)$	−206.57	−110.5	146
$HNO_3(l)$	−173.23	−79.914	155.6
$HNO_3(aq)$	−206.57	−110.5	146
$NF_3(g)$	−125	−83.3	260.6
$NOCl(g)$	51.71	66.07	261.6
$NH_4Cl(s)$	−314.4	−203.0	94.6
Oxygen			
$O_2(g)$	0	0	205.0
$O(g)$	249.2	231.7	160.95
$O_3(g)$	143	163	238.82
$OH^-(aq)$	−229.94	−157.30	−10.54
$H_2O(g)$	−241.826	−228.60	188.72
$H_2O(l)$	−285.840	−237.192	69.940
$H_2O_2(l)$	−187.8	−120.4	110
$H_2O_2(aq)$	−191.2	−134.1	144
Phosphorus			
$P_4(s$, white)	0	0	41.1
$P(g)$	314.6	278.3	163.1
$P(s$, red)	−17.6	−12.1	22.8
$P_2(g)$	144	104	218
$P_4(g)$	58.9	24.5	280
$PCl_3(g)$	−287	−268	312
$PCl_3(l)$	−320	−272	217
$PCl_5(g)$	−402	−323	353
$PCl_5(s)$	−443.5	—	—
$P_4O_{10}(s)$	−2984	−2698	229
$PO_4^{3-}(aq)$	−1266	−1013	−218
$HPO_4^{2-}(aq)$	−1281	−1082	−36
$H_2PO_4^-(aq)$	−1285	−1135	89.1
$H_3PO_4(aq)$	−1277	−1019	228
Potassium			
$K(s)$	0	0	64.672
$K(g)$	89.2	60.7	160.23
$K^+(g)$	514.197	481.202	154.47
$K^+(aq)$	−251.2	−282.28	103
$KF(s)$	−568.6	−538.9	66.55
$KCl(s)$	−436.7	−409.2	82.59
$KBr(s)$	−394	−380	95.94
$KI(s)$	−328	−323	106.39
$KOH(s)$	−424.8	−379.1	78.87
$KClO_3(s)$	−397.7	−296.3	143.1
$KClO_4(s)$	−432.75	−303.2	151.0
Rubidium			
$Rb(s)$	0	0	69.5
$Rb(g)$	85.81	55.86	169.99
$Rb^+(g)$	495.04	—	—
$Rb^+(aq)$	−246	−282.2	124
$RbF(s)$	−549.28	—	—
$RbCl(s)$	−435.35	−407.8	95.90
$RbBr(s)$	−389.2	−378.1	108.3
$RbI(s)$	−328	−326	118.0
Silicon			
$Si(s)$	0	0	18.0
$SiF_4(g)$	−1614.9	−1572.7	282.4
$SiO_2(s)$	−910.9	−856.5	41.5
Silver			
$Ag(s)$	0	0	42.702
$Ag(g)$	289.2	250.4	172.892
$Ag^+(aq)$	105.9	77.111	73.93
$AgF(s)$	−203	−185	84
$AgCl(s)$	−127.03	−109.72	96.11
$AgBr(s)$	−99.51	−95.939	107.1
$AgI(s)$	−62.38	−66.32	114
$AgNO_3(s)$	−45.06	19.1	128.2
$Ag_2S(s)$	−31.8	−40.3	146
Sodium			
$Na(s)$	0	0	51.446
$Na(g)$	107.76	77.299	153.61
$Na^+(g)$	609.839	574.877	147.85
$Na^+(aq)$	−239.66	−261.87	60.2
$NaF(s)$	−575.4	−545.1	51.21
$NaCl(s)$	−411.1	−384.0	72.12
$NaBr(s)$	−361	−349	86.82
$NaOH(s)$	−425.609	−379.53	64.454
$Na_2CO_3(s)$	−1130.8	−1048.1	139
$NaHCO_3(s)$	−947.7	−851.9	102
$NaI(s)$	−288	−285	98.5
Strontium			
$Sr(s)$	0	0	54.4
$Sr(g)$	164	110	164.54
$Sr^{2+}(g)$	1784	—	—
$Sr^{2+}(aq)$	−545.51	−557.3	−39
$SrCl_2(s)$	−828.4	−781.2	117
$SrCO_3(s)$	−1218	−1138	97.1
$SrO(s)$	−592.0	−562.4	55.5
$SrSO_4(s)$	−1445	−1334	122
Sulfur			
S(rhombic)	0	0	31.9
S(monoclinic)	0.3	0.096	32.6
$S(g)$	279	239	168
$S_2(g)$	129	80.1	228.1
$S_8(g)$	101	49.1	430.211
$S^{2-}(aq)$	41.8	83.7	22
$HS^-(aq)$	−17.7	12.6	61.1
$H_2S(g)$	−20.2	−33	205.6
$H_2S(aq)$	−39	−27.4	122
$SO_2(g)$	−296.8	−300.2	248.1
$SO_3(g)$	−396	−371	256.66
$SO_4^{2-}(aq)$	−907.51	−741.99	17
$HSO_4^-(aq)$	−885.75	−752.87	126.9
$H_2SO_4(l)$	−813.989	−690.059	156.90
$H_2SO_4(aq)$	−907.51	−741.99	17
Tin			
Sn(white)	0	0	51.5
Sn(gray)	3	4.6	44.8
$SnCl_4(l)$	−545.2	−474.0	259
$SnO_2(s)$	−580.7	−519.7	52.3
Titanium			
$Ti(s)$	0	0	30.7
$TiCl_4(l)$	−804.2	−737.2	252.3
$TiO_2(s)$	−944.0	−888.8	50.6
Zinc			
$Zn(s)$	0	0	41.6
$Zn(g)$	130.5	94.93	160.9
$Zn^{2+}(aq)$	−152.4	−147.21	−106.5
$ZnO(s)$	−348.0	−318.2	43.9
$ZnS(s$, zinc blende)	−203	−198	57.7

Appendix C

Equilibrium Constants for Selected Substances*

Dissociation (Ionization) Constants (K_a) of Selected Acids

Name and Formula	Lewis Structure†	K_{a1}	K_{a2}	K_{a3}
Acetic acid CH_3COOH		1.8×10^{-5}		
Acetylsalicylic acid $CH_3COOC_6H_4COOH$		3.6×10^{-4}		
Adipic acid $HOOC(CH_2)_4COOH$		3.8×10^{-5}	3.8×10^{-6}	
Arsenic acid H_3AsO_4		6×10^{-3}	1.1×10^{-7}	3×10^{-12}
Ascorbic acid $H_2C_6H_6O_6$		1.0×10^{-5}	5×10^{-12}	
Benzoic acid C_6H_5COOH		6.3×10^{-5}		
Carbonic acid H_2CO_3		4.5×10^{-7}	4.7×10^{-11}	
Chloroacetic acid $ClCH_2COOH$		1.4×10^{-3}		
Chlorous acid $HClO_2$		1.1×10^{-2}		

*All values at 298 K, except for acetylsalicylic acid, which is at 37°C (310 K) in 0.15 *M* NaCl.

†Acidic (ionizable) proton(s) shown in red. Structures have lowest formal charges. Benzene rings show one resonance form.

(*continued*)

Dissociation (Ionization) Constants (K_a) of Selected Acids

Name and Formula	Lewis Structure†	K_{a1}	K_{a2}	K_{a3}
Citric acid $HOOCCH_2C(OH)(COOH)CH_2COOH$	H—O—C—C—C—C—C—O—H	7.4×10^{-4}	1.7×10^{-5}	4.0×10^{-7}
Formic acid $HCOOH$	H—C(=O)—O—H	1.8×10^{-4}		
Glyceric acid $HOCH_2CH(OH)COOH$	H—O—C—C—C—O—H	2.9×10^{-4}		
Glycolic acid $HOCH_2COOH$	H—O—C—C—O—H	1.5×10^{-4}		
Glyoxylic acid $HC(O)COOH$	O=C—C—O—H	3.5×10^{-4}		
Hydrocyanic acid HCN	H—C≡N:	6.2×10^{-10}		
Hydrofluoric acid HF	H—F:	6.8×10^{-4}		
Hydrosulfuric acid H_2S	H—S—H	9×10^{-8}	1×10^{-17}	
Hypobromous acid $HBrO$	H—O—Br:	2.3×10^{-9}		
Hypochlorous acid $HClO$	H—O—Cl:	2.9×10^{-8}		
Hypoiodous acid HIO	H—O—I:	2.3×10^{-11}		
Iodic acid HIO_3	H—O—I=O	1.6×10^{-1}		
Lactic acid $CH_3CH(OH)COOH$	H—C—C—C—O—H	1.4×10^{-4}		
Maleic acid $HOOCCH=CHCOOH$	O=C—C=C—C=O	1.2×10^{-2}	4.7×10^{-7}	

(continued)

Dissociation (Ionization) Constants (K_a) of Selected Acids (*continued*)

Name and Formula	Lewis Structure†	K_{a1}	K_{a2}	K_{a3}
Malonic acid $HOOCCH_2COOH$	H—Ö—C(=O)—CH₂—C(=O)—Ö—H	1.4×10^{-3}	2.0×10^{-6}	
Nitrous acid HNO_2	H—Ö—N̈=Ö	7.1×10^{-4}		
Oxalic acid $HOOCCOOH$	H—Ö—C(=O)—C(=O)—Ö—H	5.6×10^{-2}	5.4×10^{-5}	
Phenol C_6H_5OH	C_6H_5—Ö—H	1.0×10^{-10}		
Phenylacetic acid $C_6H_5CH_2COOH$	C_6H_5—CH₂—C(=O)—Ö—H	4.9×10^{-5}		
Phosphoric acid H_3PO_4 [or $PO(OH)_3$]	H—Ö—P(=O)(—Ö—H)—Ö—H	7.2×10^{-3}	6.3×10^{-8}	4.2×10^{-13}
Phosphorous acid H_3PO_3 [or $HPO(OH)_2$]	H—P(=O)(—Ö—H)—Ö—H	3×10^{-2}	1.7×10^{-7}	
Propanoic acid CH_3CH_2COOH	H—CH₂—CH₂—C(=O)—Ö—H	1.3×10^{-5}		
Pyruvic acid $CH_3C(O)COOH$	H—CH₂—C(=O)—C(=O)—Ö—H	2.8×10^{-3}		
Succinic acid $HOOCCH_2CH_2COOH$	H—Ö—C(=O)—CH₂—CH₂—C(=O)—Ö—H	6.2×10^{-5}	2.3×10^{-6}	
Sulfuric acid H_2SO_4	H—Ö—S(=O)(=O)—Ö—H	Very large	1.0×10^{-2}	
Sulfurous acid H_2SO_3	H—Ö—S(=O)—Ö—H	1.4×10^{-2}	6.5×10^{-8}	

(*continued*)

Dissociation (Ionization) Constants (K_b) of Selected Amine Bases

Name and Formula	Lewis Structure†	K_{b1}	K_{b2}
Ammonia NH_3		1.76×10^{-5}	
Aniline $C_6H_5NH_2$		4.0×10^{-10}	
Diethylamine $(CH_3CH_2)_2NH$		8.6×10^{-4}	
Dimethylamine $(CH_3)_2NH$		5.9×10^{-4}	
Ethanolamine $HOCH_2CH_2NH_2$		3.2×10^{-5}	
Ethylamine $CH_3CH_2NH_2$		4.3×10^{-4}	
Ethylenediamine $H_2NCH_2CH_2NH_2$		8.5×10^{-5}	7.1×10^{-8}
Methylamine CH_3NH_2		4.4×10^{-4}	
tert-Butylamine $(CH_3)_3CNH_2$		4.8×10^{-4}	
Piperidine $C_5H_{10}NH$		1.3×10^{-3}	
n-Propylamine $CH_3CH_2CH_2NH_2$		3.5×10^{-4}	

†Blue type indicates the basic nitrogen and its lone pair.

(*continued*)

Dissociation (Ionization) Constants (K_b) of Selected Amine Bases (*continued*)

Name and Formula	Lewis Structure†	K_{b1}	K_{b2}
Isopropylamine $(CH_3)_2CHNH_2$		4.7×10^{-4}	
1,3-Propylenediamine $H_2NCH_2CH_2CH_2NH_2$		3.1×10^{-4}	3.0×10^{-6}
Pyridine C_5H_5N		1.7×10^{-9}	
Triethylamine $(CH_3CH_2)_3N$		5.2×10^{-4}	
Trimethylamine $(CH_3)_3N$		6.3×10^{-5}	

Dissociation (Ionization) Constants (K_a) of Some Hydrated Metal Ions

Free Ion	Hydrated Ion	K_a
Fe^{3+}	$Fe(H_2O)_6^{3+}(aq)$	6×10^{-3}
Sn^{2+}	$Sn(H_2O)_6^{2+}(aq)$	4×10^{-4}
Cr^{3+}	$Cr(H_2O)_6^{3+}(aq)$	1×10^{-4}
Al^{3+}	$Al(H_2O)_6^{3+}(aq)$	1×10^{-5}
Cu^{2+}	$Cu(H_2O)_6^{2+}(aq)$	3×10^{-8}
Pb^{2+}	$Pb(H_2O)_6^{2+}(aq)$	3×10^{-8}
Zn^{2+}	$Zn(H_2O)_6^{2+}(aq)$	1×10^{-9}
Co^{2+}	$Co(H_2O)_6^{2+}(aq)$	2×10^{-10}
Ni^{2+}	$Ni(H_2O)_6^{2+}(aq)$	1×10^{-10}

Formation Constants (K_f) of Some Complex Ions

Complex Ion	K_f
$Ag(CN)_2^-$	3.0×10^{20}
$Ag(NH_3)_2^+$	1.7×10^{7}
$Ag(S_2O_3)_2^{3-}$	4.7×10^{13}
AlF_6^{3-}	4×10^{19}
$Al(OH)_4^-$	3×10^{33}
$Be(OH)_4^{2-}$	4×10^{18}
CdI_4^{2-}	1×10^{6}
$Co(OH)_4^{2-}$	5×10^{9}
$Cr(OH)_4^-$	8.0×10^{29}
$Cu(NH_3)_4^{2+}$	5.6×10^{11}
$Fe(CN)_6^{4-}$	3×10^{35}
$Fe(CN)_6^{3-}$	4.0×10^{43}
$Hg(CN)_4^{2-}$	9.3×10^{38}
$Ni(NH_3)_6^{2+}$	2.0×10^{8}
$Pb(OH)_3^-$	8×10^{13}
$Sn(OH)_3^-$	3×10^{25}
$Zn(CN)_4^{2-}$	4.2×10^{19}
$Zn(NH_3)_4^{2+}$	7.8×10^{8}
$Zn(OH)_4^{2-}$	3×10^{15}

Solubility-Product Constants (K_{sp}) of Slightly Soluble Ionic Compounds

Name, Formula	K_{sp}
Carbonates	
Barium carbonate, $BaCO_3$	2.0×10^{-9}
Cadmium carbonate, $CdCO_3$	1.8×10^{-14}
Calcium carbonate, $CaCO_3$	3.3×10^{-9}
Cobalt(II) carbonate, $CoCO_3$	1.0×10^{-10}
Copper(II) carbonate, $CuCO_3$	$3\ \ \times10^{-12}$
Lead(II) carbonate, $PbCO_3$	7.4×10^{-14}
Magnesium carbonate, $MgCO_3$	3.5×10^{-8}
Mercury(I) carbonate, Hg_2CO_3	8.9×10^{-17}
Nickel(II) carbonate, $NiCO_3$	1.3×10^{-7}
Strontium carbonate, $SrCO_3$	5.4×10^{-10}
Zinc carbonate, $ZnCO_3$	1.0×10^{-10}
Chromates	
Barium chromate, $BaCrO_4$	2.1×10^{-10}
Calcium chromate, $CaCrO_4$	$1\ \ \times10^{-8}$
Lead(II) chromate, $PbCrO_4$	2.3×10^{-13}
Silver chromate, Ag_2CrO_4	2.6×10^{-12}
Cyanides	
Mercury(I) cyanide, $Hg_2(CN)_2$	$5\ \ \times10^{-40}$
Silver cyanide, AgCN	2.2×10^{-16}
Halides	
Fluorides	
Barium fluoride, BaF_2	1.5×10^{-6}
Calcium fluoride, CaF_2	3.2×10^{-11}
Lead(II) fluoride, PbF_2	3.6×10^{-8}
Magnesium fluoride, MgF_2	7.4×10^{-9}
Strontium fluoride, SrF_2	2.6×10^{-9}
Chlorides	
Copper(I) chloride, CuCl	1.9×10^{-7}
Lead(II) chloride, $PbCl_2$	1.7×10^{-5}
Silver chloride, AgCl	1.8×10^{-10}
Bromides	
Copper(I) bromide, CuBr	$5\ \ \times10^{-9}$
Silver bromide, AgBr	5.0×10^{-13}
Iodides	
Copper(I) iodide, CuI	$1\ \ \times10^{-12}$
Lead(II) iodide, PbI_2	7.9×10^{-9}
Mercury(I) iodide, Hg_2I_2	4.7×10^{-29}
Silver iodide, AgI	8.3×10^{-17}
Hydroxides	
Aluminum hydroxide, $Al(OH)_3$	$3\ \ \times10^{-34}$
Cadmium hydroxide, $Cd(OH)_2$	7.2×10^{-15}
Calcium hydroxide, $Ca(OH)_2$	6.5×10^{-6}
Cobalt(II) hydroxide, $Co(OH)_2$	1.3×10^{-15}
Copper(II) hydroxide, $Cu(OH)_2$	2.2×10^{-20}
Iron(II) hydroxide, $Fe(OH)_2$	4.1×10^{-15}
Iron(III) hydroxide, $Fe(OH)_3$	1.6×10^{-39}
Magnesium hydroxide, $Mg(OH)_2$	6.3×10^{-10}
Manganese(II) hydroxide, $Mn(OH)_2$	1.6×10^{-13}
Nickel(II) hydroxide, $Ni(OH)_2$	$6\ \ \times10^{-16}$
Zinc hydroxide, $Zn(OH)_2$	$3\ \ \times10^{-16}$
Iodates	
Barium iodate, $Ba(IO_3)_2$	1.5×10^{-9}
Calcium iodate, $Ca(IO_3)_2$	7.1×10^{-7}
Lead(II) iodate, $Pb(IO_3)_2$	2.5×10^{-13}
Silver iodate, $AgIO_3$	3.1×10^{-8}
Strontium iodate, $Sr(IO_3)_2$	3.3×10^{-7}
Zinc iodate, $Zn(IO_3)_2$	3.9×10^{-6}
Oxalates	
Barium oxalate dihydrate, $BaC_2O_4\cdot 2H_2O$	1.1×10^{-7}
Calcium oxalate monohydrate, $CaC_2O_4\cdot H_2O$	2.3×10^{-9}
Strontium oxalate monohydrate, $SrC_2O_4\cdot H_2O$	5.6×10^{-8}
Phosphates	
Calcium phosphate, $Ca_3(PO_4)_2$	1.2×10^{-29}
Magnesium phosphate, $Mg_3(PO_4)_2$	5.2×10^{-24}
Silver phosphate, Ag_3PO_4	2.6×10^{-18}
Sulfates	
Barium sulfate, $BaSO_4$	1.1×10^{-10}
Calcium sulfate, $CaSO_4$	2.4×10^{-5}
Lead(II) sulfate, $PbSO_4$	1.6×10^{-8}
Radium sulfate, $RaSO_4$	$2\ \ \times10^{-11}$
Silver sulfate, Ag_2SO_4	1.5×10^{-5}
Strontium sulfate, $SrSO_4$	3.2×10^{-7}
Sulfides	
Cadmium sulfide, CdS	1.0×10^{-24}
Copper(II) sulfide, CuS	$8\ \ \times10^{-34}$
Iron(II) sulfide, FeS	$8\ \ \times10^{-16}$
Lead(II) sulfide, PbS	$3\ \ \times10^{-25}$
Manganese(II) sulfide, MnS	$3\ \ \times10^{-11}$
Mercury(II) sulfide, HgS	$2\ \ \times10^{-50}$
Nickel(II) sulfide, NiS	$3\ \ \times10^{-16}$
Silver sulfide, Ag_2S	$8\ \ \times10^{-48}$
Tin(II) sulfide, SnS	1.3×10^{-23}
Zinc sulfide, ZnS	2.0×10^{-22}

Appendix D

Standard Electrode (Half-Cell) Potentials*

Half-Reaction	$E°$ (V)
$F_2(g) + 2e^- \rightleftharpoons 2F^-(aq)$	+2.87
$O_3(g) + 2H^+(aq) + 2e^- \rightleftharpoons O_2(g) + H_2O(l)$	+2.07
$Co^{3+}(aq) + e^- \rightleftharpoons Co^{2+}(aq)$	+1.82
$H_2O_2(aq) + 2H^+(aq) + 2e^- \rightleftharpoons 2H_2O(l)$	+1.77
$PbO_2(s) + 3H^+(aq) + HSO_4^-(aq) + 2e^- \rightleftharpoons PbSO_4(s) + 2H_2O(l)$	+1.70
$Ce^{4+}(aq) + e^- \rightleftharpoons Ce^{3+}(aq)$	+1.61
$MnO_4^-(aq) + 8H^+(aq) + 5e^- \rightleftharpoons Mn^{2+}(aq) + 4H_2O(l)$	+1.51
$Au^{3+}(aq) + 3e^- \rightleftharpoons Au(s)$	+1.50
$Cl_2(g) + 2e^- \rightleftharpoons 2Cl^-(aq)$	+1.36
$Cr_2O_7^{2-}(aq) + 14H^+(aq) + 6e^- \rightleftharpoons 2Cr^{3+}(aq) + 7H_2O(l)$	+1.33
$MnO_2(s) + 4H^+(aq) + 2e^- \rightleftharpoons Mn^{2+}(aq) + 2H_2O(l)$	+1.23
$O_2(g) + 4H^+(aq) + 4e^- \rightleftharpoons 2H_2O(l)$	+1.23
$Br_2(l) + 2e^- \rightleftharpoons 2Br^-(aq)$	+1.07
$NO_3^-(aq) + 4H^+(aq) + 3e^- \rightleftharpoons NO(g) + 2H_2O(l)$	+0.96
$2Hg^{2+}(aq) + 2e^- \rightleftharpoons Hg_2^{2+}(aq)$	+0.92
$Hg_2^{2+}(aq) + 2e^- \rightleftharpoons 2Hg(l)$	+0.85
$Ag^+(aq) + e^- \rightleftharpoons Ag(s)$	+0.80
$Fe^{3+}(aq) + e^- \rightleftharpoons Fe^{2+}(aq)$	+0.77
$O_2(g) + 2H^+(aq) + 2e^- \rightleftharpoons H_2O_2(aq)$	+0.68
$MnO_4^-(aq) + 2H_2O(l) + 3e^- \rightleftharpoons MnO_2(s) + 4OH^-(aq)$	+0.59
$I_2(s) + 2e^- \rightleftharpoons 2I^-(aq)$	+0.53
$O_2(g) + 2H_2O(l) + 4e^- \rightleftharpoons 4OH^-(aq)$	+0.40
$Cu^{2+}(aq) + 2e^- \rightleftharpoons Cu(s)$	+0.34
$AgCl(s) + e^- \rightleftharpoons Ag(s) + Cl^-(aq)$	+0.22
$SO_4^{2-}(aq) + 4H^+(aq) + 2e^- \rightleftharpoons SO_2(g) + 2H_2O(l)$	+0.20
$Cu^{2+}(aq) + e^- \rightleftharpoons Cu^+(aq)$	+0.15
$Sn^{4+}(aq) + 2e^- \rightleftharpoons Sn^{2+}(aq)$	+0.13
$2H^+(aq) + 2e^- \rightleftharpoons H_2(g)$	0.00
$Pb^{2+}(aq) + 2e^- \rightleftharpoons Pb(s)$	−0.13
$Sn^{2+}(aq) + 2e^- \rightleftharpoons Sn(s)$	−0.14
$N_2(g) + 5H^+(aq) + 4e^- \rightleftharpoons N_2H_5^+(aq)$	−0.23
$Ni^{2+}(aq) + 2e^- \rightleftharpoons Ni(s)$	−0.25
$Co^{2+}(aq) + 2e^- \rightleftharpoons Co(s)$	−0.28
$PbSO_4(s) + H^+(aq) + 2e^- \rightleftharpoons Pb(s) + HSO_4^-(aq)$	−0.31
$Cd^{2+}(aq) + 2e^- \rightleftharpoons Cd(s)$	−0.40
$Fe^{2+}(aq) + 2e^- \rightleftharpoons Fe(s)$	−0.44
$Cr^{3+}(aq) + 3e^- \rightleftharpoons Cr(s)$	−0.74
$Zn^{2+}(aq) + 2e^- \rightleftharpoons Zn(s)$	−0.76
$2H_2O(l) + 2e^- \rightleftharpoons H_2(g) + 2OH^-(aq)$	−0.83
$Mn^{2+}(aq) + 2e^- \rightleftharpoons Mn(s)$	−1.18
$Al^{3+}(aq) + 3e^- \rightleftharpoons Al(s)$	−1.66
$Mg^{2+}(aq) + 2e^- \rightleftharpoons Mg(s)$	−2.37
$Na^+(aq) + e^- \rightleftharpoons Na(s)$	−2.71
$Ca^{2+}(aq) + 2e^- \rightleftharpoons Ca(s)$	−2.87
$Sr^{2+}(aq) + 2e^- \rightleftharpoons Sr(s)$	−2.89
$Ba^{2+}(aq) + 2e^- \rightleftharpoons Ba(s)$	−2.90
$K^+(aq) + e^- \rightleftharpoons K(s)$	−2.93
$Li^+(aq) + e^- \rightleftharpoons Li(s)$	−3.05

*All values at 298 K. Written as reductions; $E°$ value refers to all components in their standard states: 1 *M* for dissolved species; 1 atm pressure for the gas behaving ideally; the pure substance for solids and liquids.

Appendix E

Answers to Selected Problems

Chapter 1

1.2 Gas molecules fill the entire container; the volume of a gas is the volume of the container. Solids and liquids have a definite volume. The volume of the container does not affect the volume of a solid or liquid. (a) gas (b) liquid (c) liquid **1.3** Physical property: a characteristic shown by a substance itself, without any interaction with or change into other substances. Chemical property: a characteristic of a substance that appears as it interacts with, or transforms into, other substances. (a) Color (yellow-green and silvery to white) and physical state (gas and metal to crystals) are physical properties. The interaction between chlorine gas and sodium metal is a chemical property. (b) Color and magnetism are physical properties. No chemical changes. **1.5**(a) Physical change; there is only a temperature change. (b) Chemical change; the changes in color and texture after heating indicate an irreversible chemical change. (c) Physical change; there is only a change in size, not composition. (d) Chemical change; the wood (and air) become different substances with different compositions. **1.7**(a) fuel (b) wood **1.11** A well-designed experiment must have the following essential features: (1) There must be at least two variables that are to be related; (2) there must be a way to control the variables, so that one at a time may be changed; (3) the results must be reproducible. **1.14**(a) $(2.54\ cm)^2/(1\ in)^2$ and $(1\ m)^2/(100\ cm)^2$ (b) $(1000\ m)^2/(1\ km)^2$ and $(100\ cm)^2/(1\ m)^2$ (c) $(1.609\times10^3\ m/mi)$ and $(1\ h/3600\ s)$ (d) $(1000\ g/2.205\ lb)$ and $(3.531\times10^{-5}\ ft^3/cm^3)$ **1.16** An extensive property depends on the amount of material present. An intensive property is the same regardless of how much material is present. (b) and (d) are intensive properties **1.18**(a) increases (b) remains the same (c) decreases (d) increases (e) remains the same **1.21** 1.43 nm **1.23**(a) $2.07\times10^{-9}\ km^2$ (b) \$10.43 **1.25**(a) $5.52\times10^3\ kg/m^3$ (b) $345\ lb/ft^3$ **1.27**(a) $2.56\times10^{-9}\ mm^3$ (b) 10^{-10} L **1.29**(a) 9.626 cm^3 (b) 64.92 g **1.31** 2.70 g/cm^3 **1.33**(a) 20.°C; 293 K (b) 109 K; −263°F (c) −273°C; −460.°F **1.37**(a) 2.47×10^{-7} m (b) 67.6 Å **1.43** Leading zeros are never significant; internal zeros are always significant; terminal zeros to the right of a decimal point are significant; terminal zeros to the left of a decimal point are significant only if they were measured. **1.44**(a) none (b) none (c) $0.041\underline{0}$ (d) $4.\underline{0}1\underline{00}\times10^4$ **1.46**(a) 1.34 m (b) 21,621 mm^3 (c) 443 cm **1.48**(a) 1.310000×10^5 (b) 4.7×10^{-4} (c) 2.10006×10^5 (d) 2.1605×10^3 **1.50**(a) 5550 (b) 10,070. (c) 0.000000885 (d) 0.003004 **1.52**(a) 4.06×10^{-19} J (b) 1.61×10^{24} molecules (c) 1.82×10^5 J/mol **1.54**(a) Height measured, not exact. (b) Planets counted, exact. (c) Number of grams in a pound is not a unit definition, not exact. (d) Definition of "millimeter," exact. **1.56** 7.50 ± 0.05 cm **1.58**(a) I_{avg} = 8.72 g; II_{avg} = 8.72 g; III_{avg} = 8.50 g; IV_{avg} = 8.56 g; sets I and II are most accurate. (b) Set III is the most precise, but is the least accurate. (c) Set I has the best combination of high accuracy and high precision. (d) Set IV has both low accuracy and low precision. **1.61**(a)

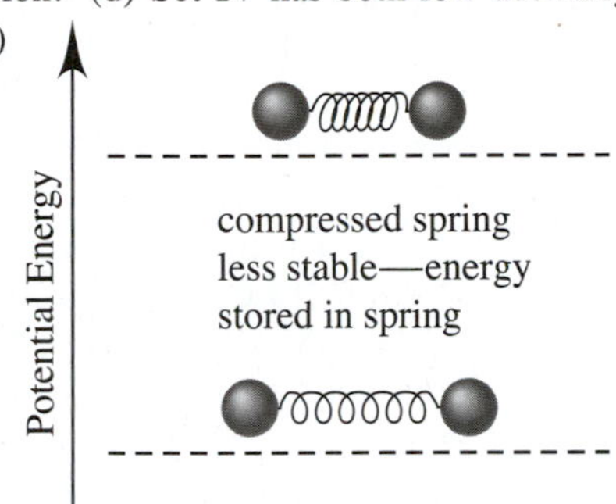

(b)

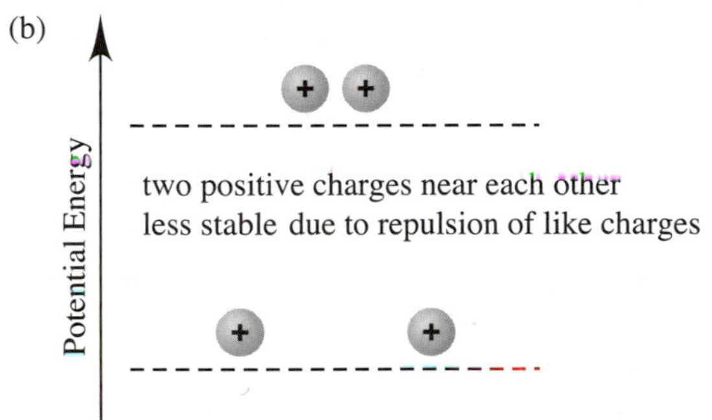

1.65(a) \$19.4 before price increase; \$33.9 after price increase (b) 51.7 coins (c) 21.0 coins **1.66** 7.7/1 **1.68**(a) density = 0.21 g/L, will float (b) CO_2 is denser than air, will sink (c) density = 0.30 g/L, will float (d) O_2 is denser than air, will sink (e) density = 1.38 g/L, will sink (f) 0.55 g for empty ball; 0.50 g for ball filled with hydrogen **1.71**(a) 8.0×10^{12} g (b) $4.1\times10^5\ m^3$ (c) 9.5×10^{13} dollars **1.73**(a) −195.79°C (b) −320.42°F (c) 5.05 L **1.76** 7.1×10^7 microparticles in the room; 1.2×10^3 microparticles in a breath **1.78** 2.3×10^{25} g oxygen; 1.4×10^{25} g silicon; 5×10^{15} g ruthenium and rhodium **1.80** freezing point = −3.7°X; boiling point = 63.3°X

Chapter 2

• **2.1** Compounds contain different types of atoms; there is only one type of atom in an element. **2.4**(a) The presence of more than one element means calcium chloride is a compound. (b) There is only one kind of atom, so sulfur is an element. (c) The presence of more than one compound means baking powder is a mixture. (d) The presence of more than one type

of atom means cytosine cannot be an element. The specific, not variable, arrangement means it is a compound. **2.6**(a) elements, compounds, and mixtures (b) compounds (c) compounds **2.7**(a) Law of definite composition: the composition is the same regardless of its source. (b) Law of mass conservation: the total quantity of matter does not change. (c) Law of multiple proportions: two elements can combine to form two different compounds that have different proportions of those elements. **2.9**(a) No, the percent by mass of each element in a compound is fixed. (b) Yes, the *mass* of each element in a compound depends on the amount of compound. **2.11** The two experiments demonstrate the law of definite composition. The unknown compound decomposes the same way both times. The experiments also demonstrate the law of conservation of mass since the total mass before reaction equals the total mass after reaction. **2.13**(a) 1.34 g F (b) 0.514 Ca; 0.486 F (c) 51.4 mass % Ca; 48.6 mass % F **2.15** 3.498×10^6 g Cu; 1.766×10^6 g S **2.17** compound 1: 0.905 S/Cl; compound 2: 0.451 S/Cl; ratio: 2.00/1.00 **2.20** Coal A **2.21** Dalton postulated that atoms of an element are identical and that compounds result from the chemical combination of specific ratios of different elements. **2.22** If you know the ratio of any two quantities and the value of one of them, the other can always be calculated; in this case, the charge and the mass/charge ratio were known. **2.26** All three isotopes have 18 protons and 18 electrons. Their respective mass numbers are 36, 38, and 40, with the respective numbers of neutrons being 18, 20, and 22. **2.28**(a) These have the same number of protons and electrons, but different numbers of neutrons; same *Z*. (b) These have the same number of neutrons, but different numbers of protons and electrons; same *N*. (c) These have different numbers of protons, neutrons, and electrons; same *A*.
2.30(a) $^{38}_{18}Ar$ (b) $^{55}_{25}Mn$ (c) $^{109}_{47}Ag$
2.32(a) $^{48}_{22}Ti$ (b) $^{79}_{34}Se$ (c) $^{11}_{5}B$

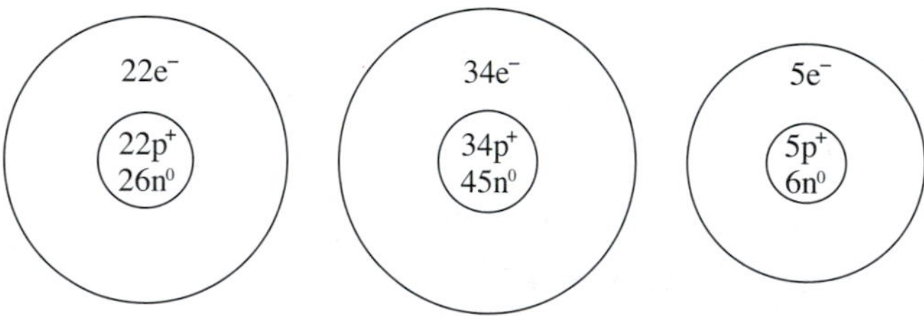

2.34 69.72 amu **2.36** ^{35}Cl = 75.774%, ^{37}Cl = 24.226%
2.38(a) In the modern periodic table, the elements are arranged in order of increasing atomic number. (b) Elements in a group (or family) have similar chemical properties. (c) Elements can be classified as metals, metalloids, or nonmetals. **2.41**(a) germanium; Ge; 4A(14); metalloid (b) phosphorus; P; 5A(15); nonmetal (c) helium; He; 8A(18); nonmetal (d) lithium; Li; 1A(1); metal (e) molybdenum; Mo; 6B(6); metal **2.43**(a) Ra; 88 (b) Si; 14 (c) Cu; 63.55 amu (d) Br; 79.90 amu
2.45 Atoms of these two kinds of substances will form ionic bonds, in which one or more electrons are transferred from the metal atom to the nonmetal atom to form a cation and an anion, respectively. **2.47** Coulomb's law states the energy of attraction in an ionic bond is directly proportional to the product of charges and inversely proportional to the distance between charges. The product of charges in MgO [(2+) (2−)] is greater than that in LiF [(1+) (1−)]. Thus, MgO has stronger ionic bonding. **2.50** K^+; I^- **2.52**(a) oxygen; 17; 6A(16); 2 (b) fluorine; 19; 7A(17); 2 (c) calcium; 40; 2A(2); 4
2.54 Lithium forms the Li^+ ion; oxygen forms the O^{2-} ion. Number of O^{2-} ions = 4.2×10^{21} O^{2-} ions. **2.56** NaCl
2.59 The two samples are similar in that both contain 20 billion oxygen atoms and 20 billion hydrogen atoms. They differ in that they contain different types of molecules: H_2O_2 molecules in the hydrogen peroxide sample, and H_2 and O_2 molecules in the mixture. In addition, the mixture contains 20 billion molecules (10 billion H_2 and 10 billion O_2), while the hydrogen peroxide sample contains 10 billion molecules. **2.62**(a) Na_3N, sodium nitride (b) SrO, strontium oxide (c) $AlCl_3$, aluminum chloride **2.64**(a) MgF_2, magnesium fluoride (b) ZnS, zinc sulfide (c) $SrCl_2$, strontium chloride **2.66**(a) $SnCl_4$ (b) iron(III) bromide (c) CuBr (d) manganese(III) oxide **2.68**(a) BaO (b) $Fe(NO_3)_2$ (c) MgS **2.70**(a) H_2SO_4; sulfuric acid (b) HIO_3; iodic acid (c) HCN; hydrocyanic acid (d) H_2S; hydrosulfuric acid **2.72** Disulfur tetrafluoride, S_2F_4 **2.74**(a) 12 oxygen atoms; 342.2 amu (b) 9 hydrogen atoms; 132.06 amu (c) 8 oxygen atoms; 344.6 amu **2.76**(a) $(NH_4)_2SO_4$; 132.15 amu (b) NaH_2PO_4; 119.98 amu (c) $KHCO_3$; 100.12 amu **2.78**(a) SO_3; sulfur trioxide; 80.07 amu (b) C_3H_8; propane; 44.09 amu **2.82** Separating the components of a mixture requires physical methods only; that is, no chemical changes (no changes in composition) take place, and the components maintain their chemical identities and properties throughout. Separating the components of a compound requires a chemical change (change in composition). **2.85**(a) compound (b) homogeneous mixture (c) heterogeneous mixture (d) homogeneous mixture (e) homogeneous mixture **2.87**(a) fraction of volume = 5.2×10^{-13} (b) mass of nucleus = 6.64466×10^{-24} g; fraction of mass = 0.999726
2.89(a) I = NO; II = N_2O_3; III = N_2O_5 (b) I has 1.14 g O per 1.00 g N; II, 1.71 g O; III, 2.86 g O **2.91**(a) Cl^-, 1.898 mass %; Na^+, 1.056 mass %; SO_4^{2-}, 0.265 mass %; Mg^{2+}, 0.127 mass %; Ca^{2+}, 0.04 mass %; K^+, 0.038 mass %; HCO_3^-, 0.014 mass % (b) 30.72% (c) Alkaline earth metal ions, total mass % = 0.17%; alkali metal ions, total mass % = 1.094%
(d) Anions (2.177 mass %) make up a larger mass fraction than cations (1.26 mass %). **2.94** Molecular formula, $C_4H_6O_4$; molecular mass, 118.09 amu; 40.68% by mass C; 5.122% by mass H; 54.20% by mass O **2.97**(a) Formulas and masses in amu: $^{15}N_2{}^{18}O$, 48; $^{15}N_2{}^{16}O$, 46; $^{14}N_2{}^{18}O$, 46; $^{14}N_2{}^{16}O$, 44; $^{15}N^{14}N^{18}O$, 47; $^{15}N^{14}N^{16}O$, 45 (b) $^{15}N_2{}^{18}O$, least common; $^{14}N_2{}^{16}O$, most common **2.102** 58.091 amu **2.103** nitroglycerin, 39.64 mass % NO; isoamyl nitrate, 22.54 mass % NO
2.104 0.370 lb C; 0.0222 lb H; 0.423 lb O; 0.185 lb N
2.106(1) chemical change (2) physical change (3) chemical change (4) chemical change (5) physical change

Chapter 3

3.2(a) 12 mol C atoms (b) 1.445×10^{25} C atoms **3.6**(a) left for balances A, C, D and right for balance B (b) right for balances A, C, D and left for balance B (c) left for balances A, C, D and right for balance B (d) neither for all balances
3.7(a) 121.64 g/mol (b) 76.02 g/mol (c) 106.44 g/mol (d) 152.00 g/mol **3.9**(a) 134.7 g/mol (b) 175.3 g/mol

(c) 342.17 g/mol (d) 125.84 g/mol **3.11**(a) 1.1×10^2 g $KMnO_4$ (b) 0.188 mol O atoms (c) 1.5×10^{20} O atoms **3.13**(a) 9.73 g $MnSO_4$ (b) 44.6 mol $Fe(ClO_4)_3$ (c) 1.74×10^{21} N atoms **3.15**(a) 1.56×10^3 g Cu_2CO_3 (b) 0.0725 g N_2O_5 (c) 0.644 mol $NaClO_4$; 3.88×10^{23} formula units $NaClO_4$ (d) 3.88×10^{23} Na^+ ions; 3.88×10^{23} ClO_4^- ions; 3.88×10^{23} Cl atoms; 1.55×10^{24} O atoms **3.17**(a) 6.375 mass % H (b) 71.52 mass % O **3.19**(a) 0.9507 mol cisplatin (b) 3.5×10^{24} H atoms **3.21** $CO(NH_2)_2 > NH_4NO_3 > (NH_4)_2SO_4 > KNO_3$ **3.22**(a) 883 mol PbS (b) 1.88×10^{25} Pb atoms **3.24**(b) From the mass percent, determine the empirical formula. Add up the total number of atoms in the empirical formula, and divide that number into the total number of atoms in the molecule. The result is the multiplier that converts the empirical formula into the molecular formula.

Road Map

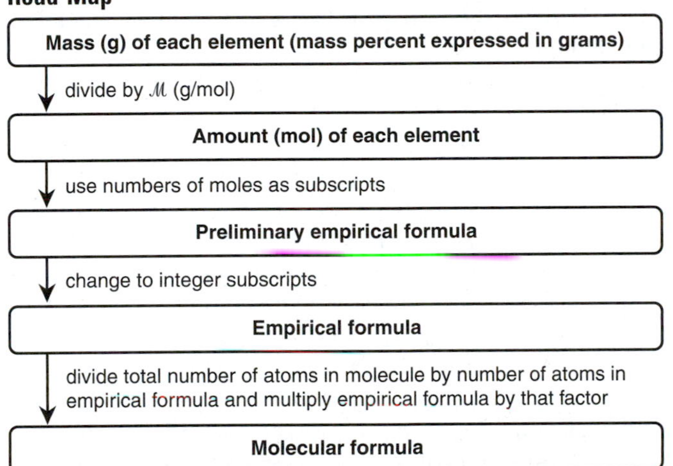

(c) Find the empirical formula from the mass percents. Compare the number of atoms given for the one element to the number in the empirical formula. Multiply the empirical formula by the factor that is needed to obtain the given number of atoms for that element.

Road Map

First three steps are the same as in road map for part (b).

Empirical formula

↓ divide the number of atoms given for the one element by the number of atoms of that element in the empirical formula and multiply the empirical formula by that factor

Molecular formula

(e) Count the numbers of the various types of atoms in the structural formula and put these into a molecular formula.

Road Map

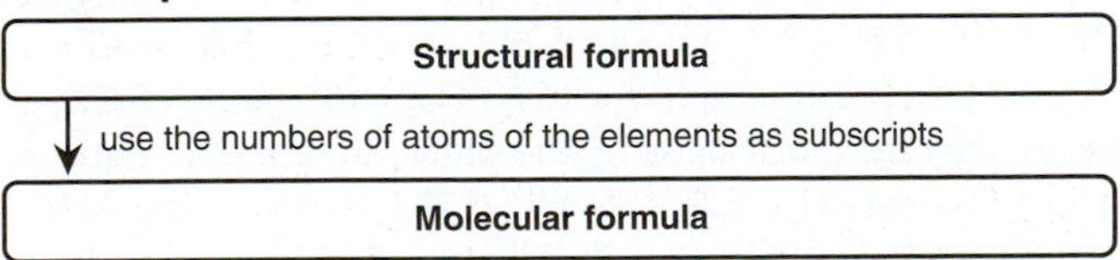

3.25(a) CH_2; 14.03 g/mol (b) CH_3O; 31.03 g/mol (c) N_2O_5; 108.02 g/mol (d) $Ba_3(PO_4)_2$; 601.8 g/mol (e) TeI_4; 635.2 g/mol **3.27**(a) C_3H_6 (b) N_2H_4 (c) N_2O_4 (d) $C_5H_5N_5$ **3.29**(a) Cl_2O_7 (b) $SiCl_4$ (c) CO_2 **3.31**(a) 1.20 mol F (b) 24.0 g M (c) calcium **3.33** $C_{21}H_{30}O_5$ **3.34** $C_{10}H_{20}O$ **3.36** b **3.37**(a) $16Cu(s) + S_8(s) \longrightarrow 8Cu_2S(s)$
(b) $P_4O_{10}(s) + 6H_2O(l) \longrightarrow 4H_3PO_4(l)$
(c) $B_2O_3(s) + 6NaOH(aq) \longrightarrow 2Na_3BO_3(aq) + 3H_2O(l)$
(d) $4CH_3NH_2(g) + 9O_2(g) \longrightarrow 4CO_2(g) + 10H_2O(g) + 2N_2(g)$
3.39(a) $4Ga(s) + 3O_2(g) \longrightarrow 2Ga_2O_3(s)$
(b) $2C_6H_{14}(l) + 19O_2(g) \longrightarrow 12CO_2(g) + 14H_2O(g)$
(c) $3CaCl_2(aq) + 2Na_3PO_4(aq) \longrightarrow Ca_3(PO_4)_2(s) + 6NaCl(aq)$
3.42(a) 1.42×10^3 mol KNO_3 (b) 1.43×10^5 g KNO_3 **3.44** 195.8 g H_3BO_3; 19.16 g H_2 **3.46** 2.60×10^3 g Cl_2 **3.48**(a) 0.105 mol CaO (b) 0.175 mol CaO (c) calcium (d) 5.88 g CaO **3.50** 1.36 mol HIO_3, 239 g HIO_3; 44.9 g H_2O in excess **3.52** 4.40 g CO_2; 4.80 g O_2 in excess **3.54** 12.2 g $Al(NO_2)_3$, no NH_4Cl, 48.7 g $AlCl_3$, 30.7 g N_2, 39.5 g H_2O **3.56** 50.% **3.58** 90.5% **3.60** 24.0 g CH_3Cl **3.62**(a) 39.7 g CF_4 **3.63**(a) Scene A (b) $2Cl_2O(g) \longrightarrow 2Cl_2(g) + O_2(g)$ (c) 1.8×10^{23} Cl_2O molecules **3.67**(a) C (b) B (c) C (d) B **3.68**(a) 7.85 g $Ca(C_2H_3O_2)_2$ (b) 0.254 *M* KI (c) 124 mol NaCN **3.70**(a) 0.0617 *M* KCl (b) 0.00363 *M* $(NH_4)_2SO_4$ (c) 0.138 *M* Na^+ **3.72**(a) 987 g HNO_3/L (b) 15.7 *M* HNO_3 **3.74** 845 mL **3.76** 0.88 g $BaSO_4$ **3.78**(a) Instructions: Be sure to wear goggles to protect your eyes! Pour approximately 2.0 gal of water into the container. Add to the water, slowly and with mixing, 0.90 gal of concentrated HCl. Dilute to 3.0 gal with more water. (b) 22.6 mL **3.79** $x = 3$ **3.80** ethane > propane > cetyl palmitate > ethanol > benzene **3.84** 89.8% **3.85**(a) $2AB_2 + B_2 \longrightarrow 2AB_3$ (b) AB_2 (c) 5.0 mol AB_3 (d) 0.5 mol B_2 **3.88**(a) C (b) B (c) D **3.92** 0.071 *M* KBr **3.95**(a) 586 g CO_2 (b) 10.5% CH_4 by mass **3.96** 10/0.66/1.0 **3.102**(a) 192.12 g/mol; $C_6H_8O_7$ (b) 0.549 mol **3.104**(a) 0.039 g heme (b) 6.3×10^{-5} mol heme (c) 3.5×10^{-3} g Fe (d) 4.1×10^{-2} g hemin **3.108**(a) 46.65 mass % N in urea; 31.98 mass % N in arginine; 21.04 mass % N in ornithine (b) 28.45 g N **3.109** A

Chapter 4

4.2 Ions must be present, and they come from ionic compounds or from acids and bases. **4.5** B **4.8**(a) Benzene is likely to be insoluble in water because it is nonpolar and water is polar. (b) Sodium hydroxide, an ionic compound, is likely to be very soluble in water. (c) Ethanol (CH_3CH_2OH) is likely to be soluble in water because the alcohol group (—OH) is very polar, like the water molecule. (d) Potassium acetate, an ionic compound, is likely to be very soluble in water. **4.10**(a) Yes, CsBr is an ionic compound. (b) Yes, HI forms an acid. **4.12**(a) 3.0 mol (b) 7.57×10^{-5} mol (c) 0.148 mol **4.14**(a) 0.058 mol Al^{3+}; 3.5×10^{22} Al^{3+} ions; 0.18 mol Cl^-; 1.1×10^{23} Cl^- ions (b) 4.62×10^{-4} mol Li^+; 2.78×10^{20} Li^+ ions; 2.31×10^{-4} mol SO_4^{2-}; 1.39×10^{20} SO_4^{2-} ions (c) 1.50×10^{-2} mol K^+; 9.02×10^{21} K^+ ions; 1.50×10^{-2} mol Br^-; 9.02×10^{21} Br^- ions **4.16**(a) 0.35 mol H^+ (b) 6.3×10^{-3} mol H^+ (c) 0.22 mol H^+ **4.23** Assuming that the left beaker contains $AgNO_3$ (because it has gray Ag^+ ions), the right must contain NaCl. Then, NO_3^- is blue, Na^+ is brown, and Cl^- is green.

Molecular equation: $AgNO_3(aq) + NaCl(aq) \longrightarrow AgCl(s) + NaNO_3(aq)$
Total ionic equation: $Ag^+(aq) + NO_3^-(aq) + Na^+(aq) + Cl^-(aq) \longrightarrow AgCl(s) + Na^+(aq) + NO_3^-(aq)$
Net ionic equation: $Ag^+(aq) + Cl^-(aq) \longrightarrow AgCl(s)$

4.24(a) Molecular: $Hg_2(NO_3)_2(aq) + 2KI(aq) \longrightarrow Hg_2I_2(s) + 2KNO_3(aq)$
Total ionic: $Hg_2^{2+}(aq) + 2NO_3^-(aq) + 2K^+(aq) + 2I^-(aq) \longrightarrow Hg_2I_2(s) + 2K^+(aq) + 2NO_3^-(aq)$
Net ionic: $Hg_2^{2+}(aq) + 2I^-(aq) \longrightarrow Hg_2I_2(s)$
Spectator ions are K^+ and NO_3^-.
(b) Molecular: $FeSO_4(aq) + Sr(OH)_2(aq) \longrightarrow Fe(OH)_2(s) + SrSO_4(s)$
Total ionic: $Fe^{2+}(aq) + SO_4^{2-}(aq) + Sr^{2+}(aq) + 2OH^-(aq) \longrightarrow Fe(OH)_2(s) + SrSO_4(s)$
Net ionic: This is the same as the total ionic equation, because there are no spectator ions.

4.26(a) No precipitate will form. (b) A precipitate will form because silver ions, Ag^+, and bromide ions, Br^-, will combine to form a solid salt, silver bromide, AgBr. The ammonium and nitrate ions do not form a precipitate.
Molecular: $NH_4Br(aq) + AgNO_3(aq) \longrightarrow AgBr(s) + NH_4NO_3(aq)$
Total ionic: $NH_4^+(aq) + Br^-(aq) + Ag^+(aq) + NO_3^-(aq) \longrightarrow AgBr(s) + NH_4^+(aq) + NO_3^-(aq)$
Net ionic: $Ag^+(aq) + Br^-(aq) \longrightarrow AgBr(s)$

4.28 0.0354 M Pb^{2+} **4.30**(a) $PbSO_4$ (b) $Pb^{2+}(aq) + SO_4^{2-}(aq) \longrightarrow PbSO_4(s)$ (c) 1.5 g $PbSO_4$
4.32 2.206 mass % Cl^-

4.38(a) Molecular equation: $KOH(aq) + HBr(aq) \longrightarrow KBr(aq) + H_2O(l)$
Total ionic equation: $K^+(aq) + OH^-(aq) + H^+(aq) + Br^-(aq) \longrightarrow K^+(aq) + Br^-(aq) + H_2O(l)$
Net ionic equation: $OH^-(aq) + H^+(aq) \longrightarrow H_2O(l)$
The spectator ions are $K^+(aq)$ and $Br^-(aq)$.
(b) Molecular equation: $NH_3(aq) + HCl(aq) \longrightarrow NH_4Cl(aq)$
Total ionic equation: $NH_3(aq) + H^+(aq) + Cl^-(aq) \longrightarrow NH_4^+(aq) + Cl^-(aq)$
NH_3, a weak base, is written in the molecular (undissociated) form. HCl, a strong acid, is written as dissociated ions. NH_4Cl is a soluble compound because all ammonium compounds are soluble.
Net ionic equation: $NH_3(aq) + H^+(aq) \longrightarrow NH_4^+(aq)$
Cl^- is the only spectator ion.

4.40 Total ionic equation: $CaCO_3(s) + 2H^+(aq) + 2Cl^-(aq) \longrightarrow Ca^{2+}(aq) + 2Cl^-(aq) + H_2O(l) + CO_2(g)$
Net ionic equation: $CaCO_3(s) + 2H^+(aq) \longrightarrow Ca^{2+}(aq) + H_2O(l) + CO_2(g)$

4.42 0.05839 M CH_3COOH **4.49**(a) S has O.N. = +6 in SO_4^{2-} (i.e., H_2SO_4) and O.N. = +4 in SO_2, so S has been reduced (and I^- oxidized); H_2SO_4 acts as an oxidizing agent. (b) The oxidation numbers remain constant throughout; because H_2SO_4 transfers an H^+ to F^- to produce HF, it acts as an acid. **4.50**(a) −1 (b) +2 (c) −3 (d) +3 **4.52**(a) −3 (b) +5 (c) +3 **4.54**(a) +6 (b) +3 (c) +7 **4.56**(a) MnO_4^- is the oxidizing agent; $H_2C_2O_4$ is the reducing agent. (b) Cu is the reducing agent; NO_3^- is the oxidizing agent. **4.58**(a) Oxidizing agent is NO_3^-; reducing agent is Sn. (b) Oxidizing agent is MnO_4^-; reducing agent is Cl^-.

4.64(a) $2Sb(s) + 3Cl_2(g) \longrightarrow 2SbCl_3(s)$; combination
(b) $2AsH_3(g) \longrightarrow 2As(s) + 3H_2(g)$; decomposition
(c) $Zn(s) + Fe(NO_3)_2(aq) \longrightarrow Zn(NO_3)_2(aq) + Fe(s)$; displacement

4.66(a) $N_2(g) + 3H_2(g) \longrightarrow 2NH_3(g)$
(b) $2NaClO_3(s) \xrightarrow{\Delta} 2NaCl(s) + 3O_2(g)$
(c) $Ba(s) + 2H_2O(l) \longrightarrow Ba(OH)_2(aq) + H_2(g)$

4.68(a) $2Cs(s) + I_2(s) \longrightarrow 2CsI(s)$
(b) $2Al(s) + 3MnSO_4(aq) \longrightarrow Al_2(SO_4)_3(aq) + 3Mn(s)$
(c) $2SO_2(g) + O_2(g) \longrightarrow 2SO_3(g)$
(d) $2C_4H_{10}(g) + 13O_2(g) \longrightarrow 8CO_2(g) + 10H_2O(g)$
(e) $2Al(s) + 3Mn^{2+}(aq) \longrightarrow 2Al^{3+}(aq) + 3Mn(s)$

4.70 315 g O_2; 3.95 kg Hg **4.72**(a) O_2 is in excess. (b) 0.117 mol Li_2O (c) 0 g Li, 3.49 g Li_2O, and 4.63 g O_2
4.75 2.79 kg Fe

4.76(a) $Fe(s) + 2H^+(aq) \longrightarrow Fe^{2+}(aq) + H_2(g)$
O.N.: 0 +1 +2 0
(b) 3.1×10^{21} Fe^{2+} ions

4.77 5.11 g C_2H_5OH; 24.9 L CO_2
4.82(a) Step 1: oxidizing agent is O_2; reducing agent is NH_3. Step 2: oxidizing agent is O_2; reducing agent is NO. Step 3: oxidizing agent is NO_2; reducing agent is NO_2. (b) 1.2×10^4 kg NH_3 **4.85** 627 L air **4.89** 3.0×10^{-3} mol CO_2 (b) 0.11 L CO_2 **4.91**(a) $C_7H_5O_4Bi$ (b) $C_{21}H_{15}O_{12}Bi_3$ (c) $Bi(OH)_3(s) + 3HC_7H_5O_3(aq) \longrightarrow Bi(C_7H_5O_3)_3(s) + 3H_2O(l)$ (d) 0.490 mg $Bi(OH)_3$ **4.93**(a) Ethanol: $C_2H_5OH(l) + 3O_2(g) \longrightarrow 2CO_2(g) + 3H_2O(l)$ Gasoline: $2C_8H_{18}(l) + 25O_2(g) \longrightarrow 16CO_2(g) + 18H_2O(g)$ (b) 2.50×10^3 g O_2 (c) 1.75×10^3 L O_2 (d) 8.38×10^3 L air **4.95** yes **4.96**(a) Reaction (2) is a redox process (b) 2.00×10^5 g Fe_2O_3; 4.06×10^5 g $FeCl_3$ (c) 2.09×10^5 g Fe; 4.75×10^5 g $FeCl_2$ (d) 0.313

Chapter 5

• **5.1**(a) The volume of the liquid remains constant, but the volume of the gas increases to the volume of the larger container. (b) The volume of the container holding the gas sample increases when heated, but the volume of the container holding the liquid sample remains essentially constant when heated. (c) The volume of the liquid remains essentially constant, but the volume of the gas is reduced. **5.4** 990 cmH_2O **5.6**(a) 566 mmHg (b) 1.32 bar (c) 3.60 atm (d) 107 kPa **5.12** At constant temperature and volume, the pressure of a gas is directly proportional to number of moles of the gas. **5.14**(a) Volume decreases to one-third of the original volume. (b) Volume increases by a factor of 3.0. (c) Volume increases by a factor of 4. **5.16** −144°C **5.18** 35.3 L **5.20** 0.085 mol Cl_2 **5.22** 1.16 g ClF_3 **5.26** Beaker is inverted for H_2 and upright for CO_2. The molar mass of CO_2 is greater than the molar mass of air, which, in turn, has a greater molar mass than H_2. **5.30** 5.86 g/L **5.32** 1.78×10^{-3} mol AsH_3; 3.48 g/L **5.34** 51.1 g/mol **5.36** 1.33 atm **5.38** 39.2 g P_4 **5.40** 41.2 g PH_3 **5.42** 0.0249 g Al **5.45** C_5H_{12} **5.47**(a) 0.90 mol (b) 6.76 torr **5.48** 286 mL SO_2 **5.49** 0.0997 atm SiF_4 **5.53** At STP, the volume occupied by a mole of any gas is the same.

At the same temperature, all gases have the same average kinetic energy, resulting in the same pressure. **5.54**(a) $P_A > P_B > P_C$ (b) $E_A = E_B = E_C$ (c) $rate_A > rate_B > rate_C$ (d) total $E_A >$ total $E_B >$ total E_C (e) $d_A = d_B = d_C$ (f) collision frequency in A > collision frequency in B > collision frequency in C **5.55** 13.21 **5.57**(a) curve 1 (b) curve 1 (c) curve 1; fluorine (F_2) and argon have about the same molar mass **5.59** 14.9 min **5.61** 4 atoms per molecule **5.64** negative deviations; $N_2 < Kr < CO_2$ **5.66** At 1 atm; at lower pressures, the gas molecules are farther apart and intermolecular forces are less important. **5.68** 6.81×10^4 g/mol **5.70**(a) 22.1 atm (b) 20.9 atm **5.73**(a) 597 torr N_2; 159 torr O_2; 0.3 torr CO_2; 3.5 torr H_2O (b) 74.9 mol % N_2; 13.7 mol % O_2; 5.3 mol % CO_2; 6.2 mol % H_2O (c) 1.6×10^{21} molecules O_2 **5.75**(a) 4×10^2 mL (b) 0.013 mol N_2 **5.76** 35.7 L NO_2 **5.79** Al_2Cl_6 **5.80** 1.52×10^{-2} mol SO_3 **5.85**(a) 9 volumes of $O_2(g)$ (b) CH_5N **5.88** The lungs would expand by a factor of 4.86; the diver can safely ascend 52.5 ft to a depth of 73 ft. **5.95** 17.2 g CO_2; 17.8 g Kr

5.98(a) $\frac{1}{2}m\overline{u^2} = \frac{3}{2}\left(\frac{R}{N_A}\right)T$

$$m\overline{u^2} = 3\left(\frac{R}{N_A}\right)T$$

$$\overline{u^2} = \frac{3RT}{mN_A}$$

$$u_{rms} = \sqrt{\frac{3RT}{\mathcal{M}}} \quad \text{where } \mathcal{M} = mN_A$$

(b) $\overline{E}_k = \frac{1}{2}m_1\overline{u_1^2} = \frac{1}{2}m_2\overline{u_2^2}$

$$m_1\overline{u_1^2} = m_2\overline{u_2^2}$$

$$\frac{m_1}{m_2} = \frac{\overline{u_2^2}}{\overline{u_1^2}}; \text{ so } \frac{\sqrt{m_1}}{\sqrt{m_2}} = \frac{\overline{u_2}}{\overline{u_1}}$$

Substitute molar mass, $\mathcal{M}$, for m:

$$\frac{\sqrt{\mathcal{M}_1}}{\sqrt{\mathcal{M}_2}} = \frac{rate_2}{rate_1}$$

5.102(a) 16.5 L CO_2 (b) $P_{H_2O} = 48.8$ torr; $P_{O_2} = P_{CO_2} = 3.7\times10^2$ torr **5.106** 332 steps **5.107** 1.4

Chapter 6

• **6.5** 0 J **6.7**(a) 6.6×10^7 kJ (b) 1.6×10^7 kcal (c) 6.3×10^7 Btu **6.10**(a) exothermic (b) endothermic (c) exothermic (d) exothermic (e) endothermic (f) endothermic (g) exothermic

6.12 [Diagram: H axis; Reactants (top) → Products (bottom); $\Delta H = (-)$, (exothermic)]

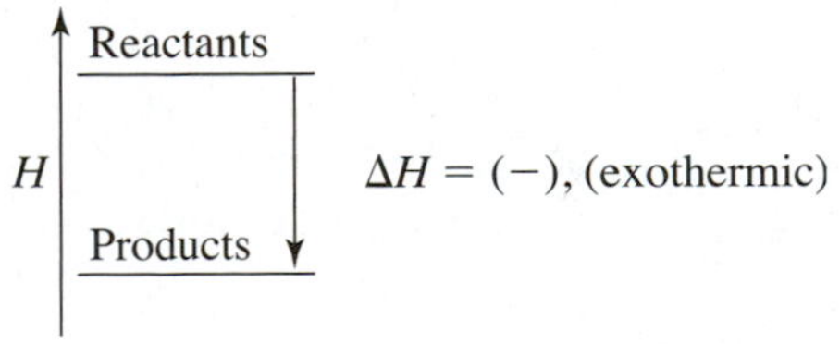

6.14(a) Combustion of ethane: $2C_2H_6(g) + 7O_2(g) \longrightarrow 4CO_2(g) + 6H_2O(g) + \text{heat}$

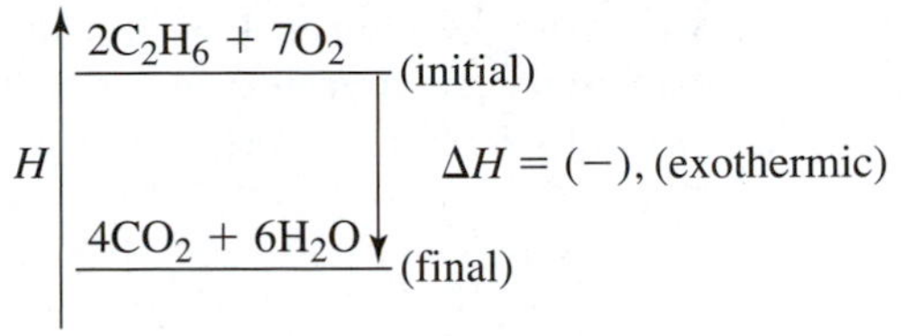

[Diagram: H axis; $2C_2H_6 + 7O_2$ (initial) → $4CO_2 + 6H_2O$ (final); $\Delta H = (-)$, (exothermic)]

(b) Freezing of water: $H_2O(l) \longrightarrow H_2O(s) + \text{heat}$

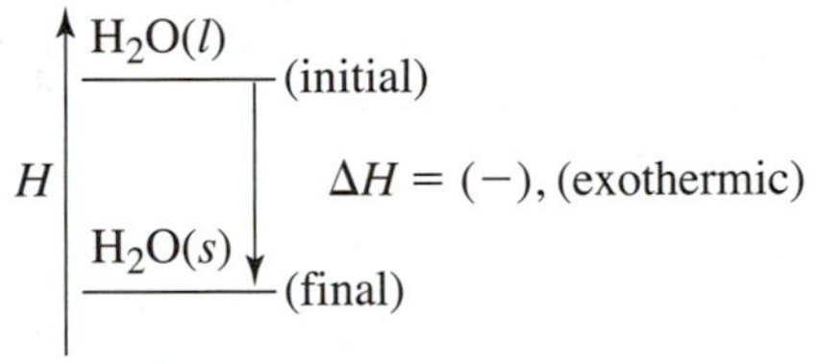

[Diagram: H axis; $H_2O(l)$ (initial) → $H_2O(s)$ (final); $\Delta H = (-)$, (exothermic)]

6.16(a) $2CH_3OH(l) + 3O_2(g) \longrightarrow 2CO_2(g) + 4H_2O(g) + \text{heat}$

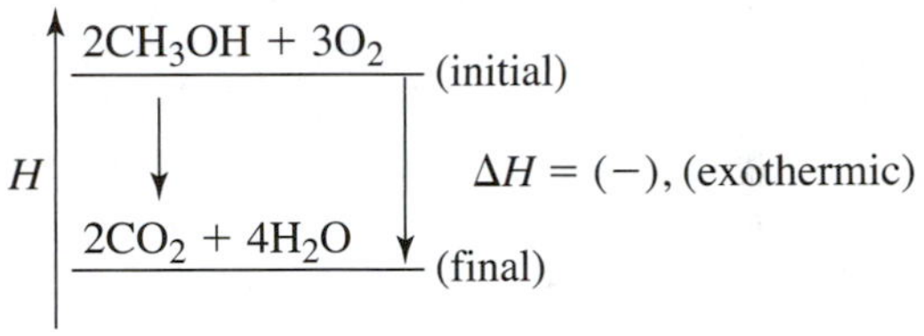

[Diagram: H axis; $2CH_3OH + 3O_2$ (initial) → $2CO_2 + 4H_2O$ (final); $\Delta H = (-)$, (exothermic)]

(b) $\frac{1}{2}N_2(g) + O_2(g) + \text{heat} \longrightarrow NO_2(g)$

[Diagram: H axis; $\frac{1}{2}N_2 + O_2$ (initial) → NO_2 (final); $\Delta H = (+)$, (endothermic)]

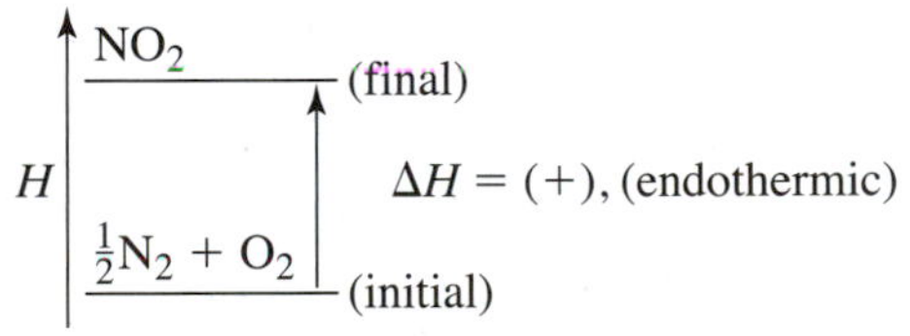

6.18(a) This is a phase change from the solid phase to the gas phase. Heat is absorbed by the system so q_{sys} is positive. (b) The volume of the system is expanding as more moles of gas are present after the phase change than were present before the phase change. So the system has done work of expansion, and w is negative. Since $\Delta E_{sys} = q + w$, the sign of ΔE_{sys} cannot be predicted: it will be positive if $q > w$ and negative if $q < w$. (c) $\Delta E_{univ} = 0$. If the system loses energy, the surroundings gain an equal amount of energy. The sum of the energy of the system and the energy of the surroundings remains constant. **6.20** To determine the specific heat capacity of a substance, you need its mass, the heat added (or lost), and the change in temperature. **6.22** 6.9×10^3 J **6.24** 295°C **6.26** 77.5°C **6.28** 45°C **6.33** The reaction has a positive ΔH, because it requires the input of energy to break the oxygen-oxygen bond. **6.34** ΔH is negative; it is opposite in sign and half of the value for the vaporization of 2 mol of H_2O. **6.35**(a) exothermic (b) 20.2 kJ (c) -4.2×10^2 kJ (d) -15.7 kJ

6.37(a) $\frac{1}{2}N_2(g) + \frac{1}{2}O_2(g) \longrightarrow NO(g)$; $\Delta H = 90.29$ kJ (b) -10.5 kJ

6.39 -1.88×10^6 kJ

6.41(a) $C_2H_4(g) + 3O_2(g) \longrightarrow 2CO_2(g) + 2H_2O(g)$; $\Delta H_{rxn} = -1411$ kJ (b) 1.39 g C_2H_4

6.44 -813.4 kJ **6.46** $N_2(g) + 2O_2(g) \longrightarrow 2NO_2(g)$; $\Delta H_{overall} = 66.4$ kJ; equation 1 is A, equation 2 is B, and equation 3 is C.

6.49 The standard enthalpy of reaction, ΔH°_{rxn}, is the enthalpy change for a reaction in which all substances are in their standard states. The standard enthalpy of formation, ΔH°_f, is the enthalpy change that accompanies the formation of one mole of a compound in its standard state from elements in their standard states.

6.50(a) $\frac{1}{2}Cl_2(g) + Na(s) \longrightarrow NaCl(s)$
(b) $H_2(g) + \frac{1}{2}O_2(g) \longrightarrow H_2O(g)$
(c) no changes

6.51(a) $Ca(s) + Cl_2(g) \longrightarrow CaCl_2(s)$
(b) $Na(s) + \frac{1}{2}H_2(g) + C(\text{graphite}) + \frac{3}{2}O_2(g) \longrightarrow NaHCO_3(s)$
(c) $C(\text{graphite}) + 2Cl_2(g) \longrightarrow CCl_4(l)$
(d) $\frac{1}{2}H_2(g) + \frac{1}{2}N_2(g) + \frac{3}{2}O_2(g) \longrightarrow HNO_3(l)$

6.53(a) -1036.9 kJ (b) -433 kJ **6.55** -157.3 kJ/mol **6.58**(a) 503.9 kJ (b) $-\Delta H_1 + 2\Delta H_2 = 504$ kJ **6.59**(a) $C_{18}H_{36}O_2(s) + 26O_2(g) \longrightarrow 18CO_2(g) + 18H_2O(g)$ (b) $-10{,}488$ kJ (c) -36.9 kJ; -8.81 kcal (d) 8.81 kcal/g $\times$ 11.0 g = 96.9 kcal; this is consistent **6.60**(a) initial = 23.6 L/mol; final = 24.9 L/mol (b) 187 J (c) -1.2×10^2 J (d) 3.1×10^2 J (e) 310 J (f) $\Delta H = \Delta E + P\Delta V = \Delta E - w = (q + w) - w = q_P$ **6.70** 721 kJ **6.78**(a) $\Delta H^\circ_{rxn1} = -657.0$ kJ; $\Delta H^\circ_{rxn2} = 32.9$ kJ (b) -106.6 kJ **6.81**(a) -6.81×10^3 J (b) $+243$°C **6.82** -22.2 kJ **6.84**(a) -1.25×10^3 kJ (b) 2.24×10^3 °C

Chapter 7

• **7.2**(a) x-ray < ultraviolet < visible < infrared < microwave < radio waves (b) radio < microwave < infrared < visible < ultraviolet < x-ray (c) radio < microwave < infrared < visible < ultraviolet < x-ray **7.4** The energy of an atom is not continuous, but quantized. It exists only in certain fixed amounts called *quanta.* **7.7** 316 m; 3.16×10^{11} nm; 3.16×10^{12} Å **7.9** 2.5×10^{-23} J **7.11** red < yellow < blue **7.14**(a) 1.24×10^{15} s^{-1}; 8.21×10^{-19} J (b) 1.4×10^{15} s^{-1}; 9.0×10^{-19} J **7.17**(a) absorption (b) emission (c) emission (d) absorption **7.19** 434.17 nm **7.21** -2.76×10^5 J/mol **7.23** d < a < c < b **7.25** $n = 4$ **7.29** Macroscopic objects do exhibit a wavelike motion, but the wavelength is too small for humans to perceive. **7.31**(a) 7.10×10^{-37} m (b) 1×10^{-35} m **7.33** 2.2×10^{-26} m/s **7.35** 3.75×10^{-36} kg **7.39**(a) principal determinant of the electron's energy or distance from the nucleus (b) determines the shape of the orbital (c) determines the orientation of the orbital in three-dimensional space **7.40**(a) one (b) five (c) three (d) nine **7.42**(a) m_l: $-2, -1, 0, +1, +2$ (b) m_l: 0 (if $n = 1$, then $l = 0$) (c) m_l: $-3, -2, -1, 0, +1, +2, +3$

7.44

Sublevel	Allowable m_l values	No. of orbitals
(a) d ($l = 2$)	$-2, -1, 0, +1, +2$	5
(b) p ($l = 1$)	$-1, 0, +1$	3
(c) f ($l = 3$)	$-3, -2, -1, 0, +1, +2, +3$	7

7.46(a) $n = 5$ and $l = 0$; one orbital (b) $n = 3$ and $l = 1$; three orbitals (c) $n = 4$ and $l = 3$; seven orbitals **7.48**(a) no; $n = 2, l = 1, m_l = -1$; $n = 2, l = 0, m_l = 0$ (b) allowed (c) allowed (d) no; $n = 5, l = 3, m_l = +3$; $n = 5, l = 2, m_l = 0$ **7.50**(a) The attraction of the nucleus for the electrons must be overcome. (b) The electrons in silver are more tightly held by the nucleus. (c) silver (d) Once the electron is freed from the atom, its energy increases in proportion to the frequency of the light. **7.53** Li^{2+} **7.56**(a) Ba; 462 nm (b) 278 to 292 nm **7.58**(a) 2.7×10^2 s (b) 3.6×10^8 m **7.62** 6.4×10^{27} photons **7.64**(a) 7.56×10^{-18} J; 2.63×10^{-8} m (b) 5.122×10^{-17} J; 3.881×10^{-9} m (c) 1.2×10^{-18} J; 1.66×10^{-7} m **7.66**(a) red; green (b) 5.89 kJ (Sr); 5.83 kJ (Ba) **7.68**(a) This is the wavelength of maximum absorbance, so it gives the highest sensitivity. (b) ultraviolet region (c) 1.93×10^{-2} g vitamin A/g oil **7.72** 1.0×10^{18} photons/s

Chapter 8

8.1 Elements are listed in the periodic table in an ordered, systematic way that correlates with a periodicity of their chemical and physical properties. The theoretical basis for the table in terms of atomic number and electron configuration does not allow for a "new element" between Sn and Sb. **8.3**(a) predicted atomic mass = 54.23 amu (b) predicted melting point = 6.3°C **8.5** The quantum number m_s relates to just the electron; all the others describe the orbital. **8.8** Shielding occurs when electrons protect, or shield, other electrons from the full nuclear attraction. The effective nuclear charge is the nuclear charge an electron actually experiences. As the number of electrons, especially inner electrons, increases, the effective nuclear charge decreases. **8.10**(a) 6 (b) 10 (c) 2 **8.12**(a) 6 (b) 2 (c) 14 **8.15** Hund's rule states that electrons will occupy empty orbitals in a given sublevel, with parallel spins, before filling half-filled orbitals. The lowest energy arrangement has the maximum number of unpaired electrons with parallel spins.

N: $1s^22s^22p^3$ [↑↓] [↑↓] [↑|↑|↑]
1s 2s 2p

8.17 Main-group elements from the same group have similar outer electron configurations, and the (old) group number equals the number of outer electrons. Outer electron configurations vary in a periodic manner within a period, with each succeeding element having an additional electron. **8.18**(a) $n = 5$, $l = 0$, $m_l = 0$, and $m_s = +\frac{1}{2}$ (b) $n = 3$, $l = 1$, $m_l = +1$, and $m_s = -\frac{1}{2}$ (c) $n = 5$, $l = 0$, $m_l = 0$, and $m_s = +\frac{1}{2}$ (d) $n = 2$, $l = 1$, $m_l = +1$, and $m_s = -\frac{1}{2}$

8.20(a) Rb: $1s^22s^22p^63s^23p^64s^23d^{10}4p^65s^1$
(b) Ge: $1s^22s^22p^63s^23p^64s^23d^{10}4p^2$
(c) Ar: $1s^22s^22p^63s^23p^6$

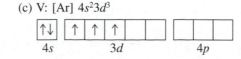

8.22(a) Ti: [Ar] $4s^23d^2$

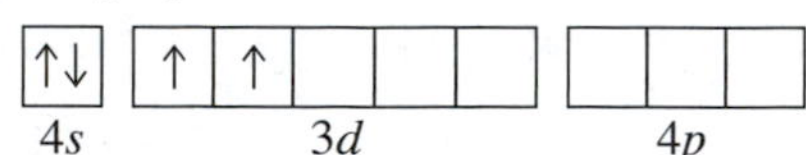

(b) Cl: [Ne] $3s^23p^5$

[↑↓] [↑↓|↑↓|↑]
3s 3p

(c) V: [Ar] $4s^23d^3$

[↑↓] [↑|↑|↑| |] [| |]
4s 3d 4p

8.24(a) O; Group 6A(16); Period 2

↑↓		↑↓	↑	↑
2*s*			2*p*	

(b) P; Group 5A(15); Period 3

↑↓		↑	↑	↑
3*s*			3*p*	

8.26(a) [Ar] $4s^23d^{10}4p^1$; Group 3A(13) (b) [He] $2s^22p^6$; Group 8A(18)

8.28

	Inner Electrons	Outer Electrons	Valence Electrons
(a) O	2	6	6
(b) Sn	46	4	4
(c) Ca	18	2	2
(d) Fe	18	2	8
(e) Se	28	6	6

8.30(a) B; Al, Ga, In, and Tl (b) S; O, Se, Te, and Po (c) La; Sc, Y, and Ac **8.33** Atomic size increases down a group. Ionization energy decreases down a group. Outer electrons are more easily removed as the atom gets larger. **8.38** A high IE_1 and a very negative EA_1 suggest that the elements are halogens, in Group 7A(17), which form 1− ions. **8.40**(a) K < Rb < Cs (b) O < C < Be (c) Cl < S < K (d) Mg < Ca < K **8.42**(a) Ba < Sr < Ca (b) B < N < Ne (c) Rb < Se < Br (d) Sn < Sb < As **8.44** $1s^22s^22p^1$ (boron, B) **8.46**(a) Na (b) Na (c) Be **8.48**(1) Metals conduct electricity; nonmetals do not. (2) Metal ions have a positive charge; nonmetal ions have a negative charge. (3) Metal oxides are mostly ionic and act as bases in water; nonmetal oxides are mostly covalent and act as acids in water. **8.49** Metallic character increases down a group and decreases to the right across a period. These trends are the same as those for atomic size and opposite those for ionization energy. **8.53**(a) Rb (b) Ra (c) I **8.55**(a) Cl^-: $1s^22s^22p^63s^23p^6$ (b) Na^+: $1s^22s^22p^6$ (c) Ca^{2+}: $1s^22s^22p^63s^23p^6$ **8.57**(a) 0 (b) 3 (c) 0 (d) 1 **8.59**(a) V^{3+}: [Ar] $3d^2$, paramagnetic (b) Cd^{2+}: [Kr] $4d^{10}$, diamagnetic (c) Co^{3+}: [Ar] $3d^6$, paramagnetic (d) Ag^+: [Kr] $4d^{10}$, diamagnetic **8.61** For palladium to be diamagnetic, all of its electrons must be paired. (a) You might first write the condensed electron configuration for Pd as [Kr] $5s^24d^8$. However, the partial orbital diagram is not consistent with diamagnetism.

↑↓		↑↓	↑↓	↑↓	↑	↑				
5*s*				4*d*					5*p*	

(b) This is the only configuration that supports diamagnetism, [Kr] $4d^{10}$.

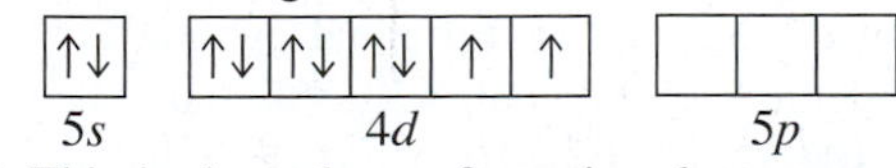

(c) Promoting an *s* electron into the *d* sublevel still leaves two electrons unpaired.

↑		↑↓	↑↓	↑↓	↑↓	↑				
5*s*				4*d*					5*p*	

8.63(a) $Li^+ < Na^+ < K^+$ (b) $Rb^+ < Br^- < Se^{2-}$ (c) $F^- < O^{2-} < N^{3-}$ **8.66**(a) Cl_2O, dichlorine monoxide (b) Cl_2O_3, dichlorine trioxide (c) Cl_2O_5, dichlorine pentoxide (d) Cl_2O_7, dichlorine heptoxide (e) SO_3, sulfur trioxide (f) SO_2, sulfur dioxide (g) N_2O_5, dinitrogen pentoxide (h) N_2O_3, dinitrogen trioxide (i) CO_2, carbon dioxide (j) P_4O_{10}, tetraphosphorus decoxide **8.68**(a) $SrBr_2$, strontium bromide (b) CaS, calcium sulfide (c) ZnF_2, zinc fluoride (d) LiF, lithium fluoride **8.69** All ions except Fe^{8+} and Fe^{14+} are paramagnetic; Fe^+ and Fe^{3+} would be most attracted.

Chapter 9

• **9.1**(a) Greater ionization energy decreases metallic character. (b) Larger atomic radius increases metallic character. (c) Higher number of outer electrons decreases metallic character. (d) Larger effective nuclear charge decreases metallic character. **9.4**(a) Cs (b) Rb (c) As **9.6**(a) ionic (b) covalent (c) metallic **9.8**(a) Rb· (b) ·Ṣi· (c) :Ï· **9.10**(a) 6A(16); [noble gas] ns^2np^4 (b) 3(A)13; [noble gas] ns^2np^1 **9.16**(a) Ba^{2+}, [Xe]; Cl^-, [Ne] $3s^23p^6$, $:\ddot{Cl}:^-$; $BaCl_2$ (b) Sr^{2+}, [Kr]; O^{2-}, [He] $2s^22p^6$, $:\ddot{O}:^{2-}$; SrO (c) Al^{3+}, [Ne]; F^-, [He] $2s^22p^6$, $:\ddot{F}:^-$; AlF_3 (d) Rb^+, [Kr]; O^{2-}, [He] $2s^22p^6$, $:\ddot{O}:^{2-}$; Rb_2O. **9.18**(a) 3A(13) (b) 2A(2) (c) 6A(16) **9.20**(a) BaS; Ba^{2+} is larger than Ca^{2+}. (b) NaF; the charge on each ion is less than the charge on Mg and O. **9.23** When two chlorine atoms are far apart, there is no interaction between them. As the atoms move closer together, the nucleus of each atom attracts the electrons of the other atom. The closer the atoms, the greater this attraction; however, the repulsions of the two nuclei also increase at the same time. The final internuclear distance is the distance at which maximum attraction is achieved in spite of the repulsion. **9.24** The bond energy is the energy required to break the bond between H atoms and Cl atoms in one mole of HCl molecules in the gaseous state. Energy is needed to break bonds, so bond breaking is always endothermic and $\Delta H^\circ_{\text{bond breaking}}$ is positive. The quantity of energy needed to break the bond is released upon its formation, so $\Delta H^\circ_{\text{bond forming}}$ has the same magnitude as $\Delta H^\circ_{\text{bond breaking}}$ but is opposite in sign (always exothermic and negative). **9.28**(a) I—I < Br—Br < Cl—Cl (b) S—Br < S—Cl < S—H (c) C—N < C═N < C≡N **9.30**(a) C—O < C═O; the C═O bond (bond order = 2) is stronger than the C—O bond (bond order = 1). (b) C—H < O—H; O is smaller than C so the O—H bond is shorter and stronger than the C—H bond. **9.33** Less energy is required to break weak bonds. **9.35** Both are one-carbon molecules. Since methane contains fewer carbon-oxygen bonds, it will have the greater enthalpy of reaction per mole for combustion. **9.36** −168 kJ **9.38** −22 kJ **9.40** Electronegativity increases from left to right and increases from bottom to top within a group. Fluorine and oxygen are the two most electronegative elements. Cesium and francium are the two least electronegative elements. **9.42** The H—O bond in water is polar covalent. A nonpolar covalent bond occurs between two atoms with identical electronegativities. A polar covalent bond occurs when the atoms have differing electronegativities. Ionic bonds result from electron transfer between atoms.

9.45(a) Si < S < O (b) Mg < As < P

9.47(a) N—B (arrow ←+ over bond) (b) N—O (+→) (c) C—S (none) (d) S—O (+→) (e) N—H (←+) (f) Cl—O (+→)

9.49 a, d, and e **9.51**(a) nonpolar covalent (b) ionic (c) polar covalent (d) polar covalent (e) nonpolar covalent (f) polar covalent; $SCl_2 < SF_2 < PF_3$

9.53(a) H—I < H—Br < H—Cl

(b) H—C < H—O < H—F

(c) S—Cl < P—Cl < Si—Cl

9.57(a) 800. kJ/mol, which is lower than the value in Table 9.2 (b) -2.417×10^4 kJ (c) 1690. g CO_2 (d) 65.2 L O_2 **9.58**(a) −125 kJ (b) yes, since ΔH_f° is negative (c) −392 kJ (d) No, ΔH_f° for $MgCl_2$ is much more negative than that for MgCl. **9.59**(a) 406 nm (b) 2.93×10^{-19} J (c) 1.87×10^4 m/s **9.62** C—Cl: 3.53×10^{-7} m; bond in O_2: 2.40×10^{-7} m **9.63** XeF_2: 132 kJ/mol; XeF_4: 150. kJ/mol; XeF_6: 146 kJ/mol **9.65**(a) The presence of the very electronegative fluorine atoms bonded to one of the carbons makes the C—C bond polar. This polar bond will tend to undergo heterolytic rather than homolytic cleavage. More energy is required to achieve heterolytic cleavage. (b) 1420 kJ **9.68** 8.70×10^{14} s^{-1}; 3.45×10^{-7} m; the ultraviolet region of the electromagnetic spectrum **9.70**(a) $CH_3OCH_3(g)$: −326 kJ; $CH_3CH_2OH(g)$: −369 kJ (b) The formation of gaseous ethanol is more exothermic. (c) 43 kJ

Chapter 10

• **10.1** He cannot serve as a central atom because it does not bond. H cannot because it forms only one bond. F cannot because it needs only one electron to complete its valence level, and it does not have *d* orbitals available to expand its valence level. Thus, it can bond to only one other atom. **10.3** All the structures obey the octet rule except c and g.

10.5(a) :F̤: / :F̤—Si—F̤: / :F̤: (b) :C̤l̤—S̤e—C̤l̤: (c) :F̤—C(=O:)—F̤:

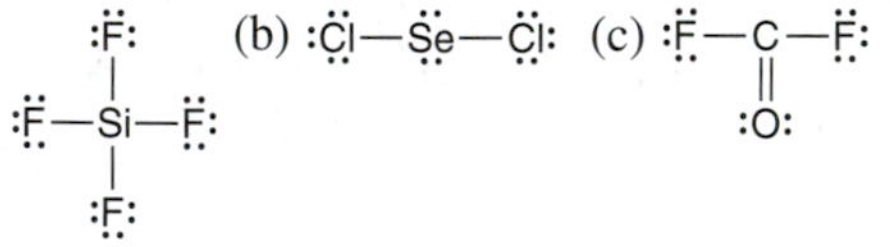

10.7(a) :F̤—P̤(—F̤:)—F̤: (b) H—Ö—C(=O:)—Ö—H (c) :S̤=C=S̤:

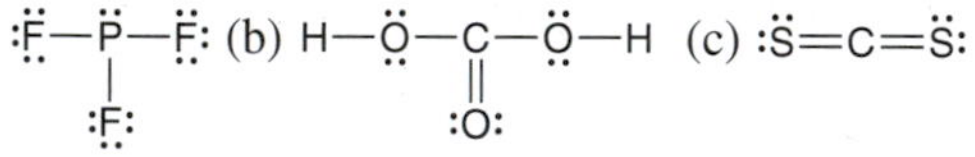

10.9(a) $[\ddot{O}{=}N{=}\ddot{O}]^+$

(b) :F̤—N(—Ö:)=Ö ⟷ :F̤—N(=Ö)—Ö:

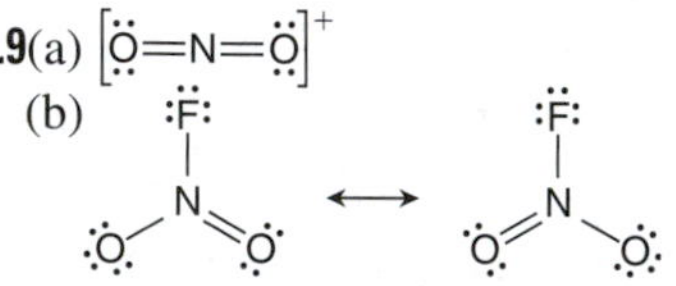

10.11(a) $[:\ddot{N}{=}N{=}\ddot{N}:]^-$ ⟷ $[:\ddot{N}{-}N{\equiv}N:]^-$ ⟷ $[:N{\equiv}N{-}\ddot{N}:]^-$

(b) $[:\ddot{O}{=}N{-}\ddot{O}:]^-$ ⟷ $[:\ddot{O}{-}N{=}\ddot{O}:]^-$

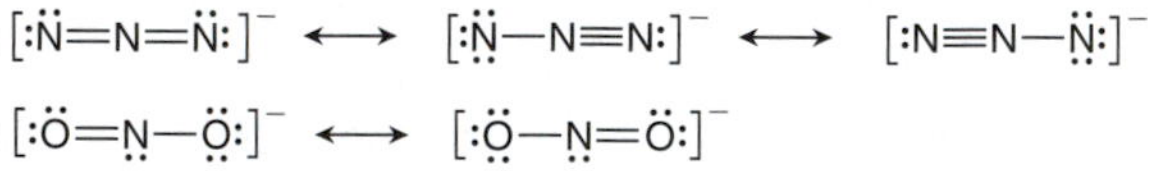

10.13(a) IF_5 (square pyramidal Lewis structure) formal charges: I = 0, F = 0

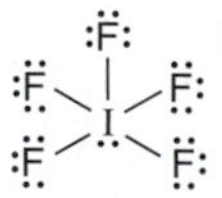

(b) $[AlH_4]^-$ formal charges: H = 0, Al = −1

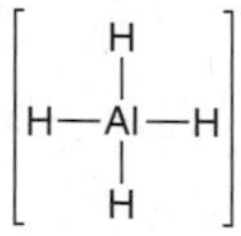

10.15(a) $[\ddot{O}{=}Br(-\ddot{O}:){=}\ddot{O}]^-$ formal charges: Br = 0, doubly bonded O = 0, singly bonded O = −1 O.N.: Br = +5; O = −2

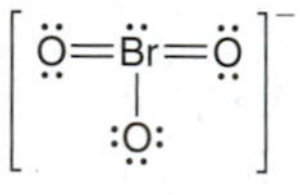

(b) $[:\ddot{O}{-}\ddot{S}(=O:){-}\ddot{O}:]^{2-}$ formal charges: S = 0, singly bonded O = −1, doubly bonded O = 0 O.N.: S = +4; O = −2

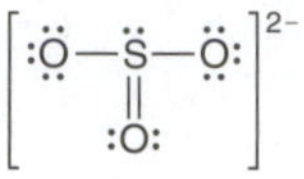

10.17(a) B has 6 valence electrons in BH_3, so the molecule is electron deficient. (b) As has an expanded valence level with 10 electrons. (c) Se has an expanded valence level with 10 electrons.

(a) BH_3 (b) $[AsF_4]^-$ (c) $SeCl_4$

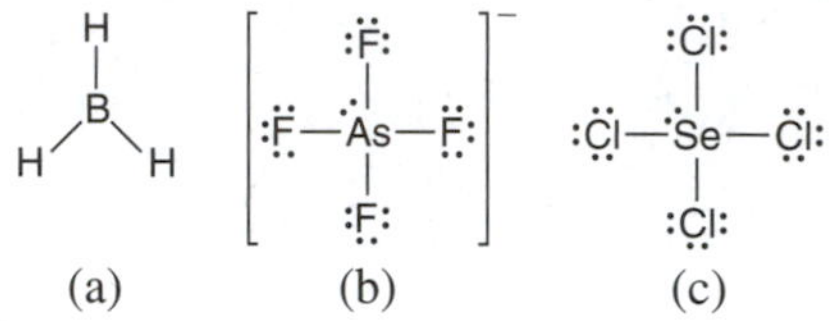

10.19(a) Br expands its valence level to 10 electrons. (b) I has an expanded valence level of 10 electrons. (c) Be has only 4 valence electrons in BeF_2, so the molecule is electron deficient.

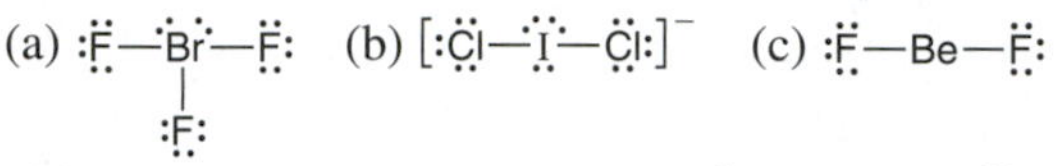

(a) :F̤—B̤r—F̤: with :F̤: below (b) $[:\ddot{Cl}{-}\ddot{I}{-}\ddot{Cl}:]^-$ (c) :F̤—Be—F̤:

10.21

$:\ddot{Cl}{-}Be{-}\ddot{Cl}: + 2\,[:\ddot{Cl}:]^- \longrightarrow [BeCl_4]^{2-}$

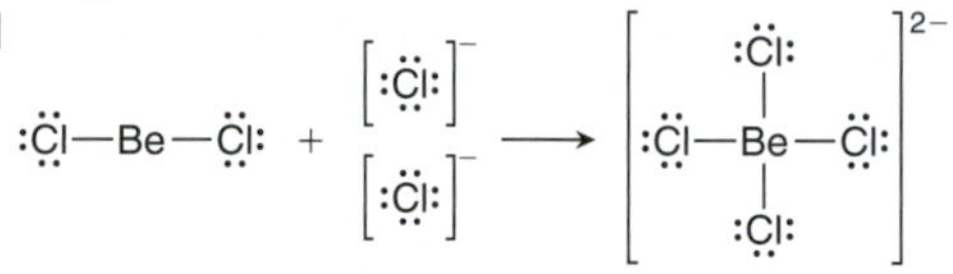

10.24 structure A **10.26** The molecular shape and the electron-group arrangement are the same when no lone pairs are present on the central atom. **10.28** tetrahedral, AX_4; trigonal pyramidal, AX_3E; bent or V shaped, AX_2E_2

10.31(a)–(f) (structures of AX_n shapes)

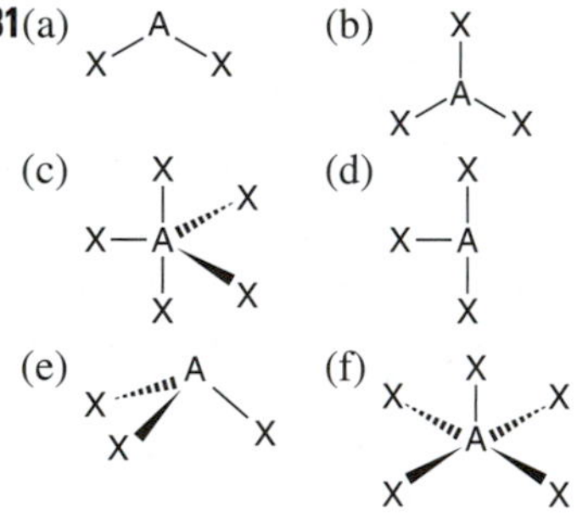

10.33(a) trigonal planar, bent, 120° (b) tetrahedral, trigonal pyramidal, 109.5° (c) tetrahedral, trigonal pyramidal, 109.5° **10.35**(a) trigonal planar, trigonal planar, 120° (b) trigonal planar, bent, 120° (c) tetrahedral, tetrahedral, 109.5° **10.37**(a) trigonal planar, AX_3, 120° (b) trigonal pyramidal, AX_3E, 109.5° (c) trigonal bipyramidal, AX_5, 90° and 120° **10.39**(a) bent, 109.5°, less than 109.5° (b) trigonal bipyramidal, 90° and 120°, angles are ideal (c) seesaw, 90° and 120°, less than ideal (d) linear, 180°, angle is ideal **10.41**(a) C: tetrahedral, 109.5°; O: bent, < 109.5° (b) N: trigonal planar, 120° **10.43**(a) C in CH_3: tetrahedral, 109.5°; C with C=O: trigonal planar, 120°; O with H: bent, < 109.5° (b) O: bent, < 109.5° **10.45** $OF_2 < NF_3 < CF_4 < BF_3 < BeF_2$ **10.47**(a) The C and N each have three groups, so the ideal angles are 120°; the O has four groups, so the ideal angle is 109.5°. The N and O have lone pairs, so the angles are less than ideal. (b) All central atoms have four pairs, so the ideal angles are 109.5°. The lone pairs on the O reduce this value. (c) The B has three groups and an ideal bond angle of 120°. All the O's have four groups (ideal bond angles of

109.5°), two of which are lone pairs that reduce the angle.

10.50

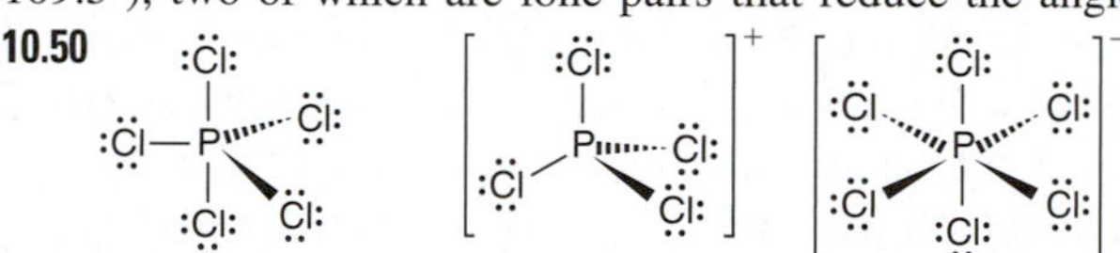

In the gas phase, PCl_5 is AX_5, so the shape is trigonal bipyramidal, and the bond angles are 120° and 90°. The PCl_4^+ ion is AX_4, so the shape is tetrahedral, and the bond angles are 109.5°. The PCl_6^- ion is AX_6, so the shape is octahedral, and the bond angles are 90°. **10.52**(a) CF_4 (b) BrCl and SCl_2 **10.54**(a) SO_2, because it is polar and SO_3 is not. (b) IF has a greater electronegativity difference between its atoms. (c) SF_4, because it is polar and SiF_4 is not. (d) H_2O has a greater electronegativity difference between its atoms.

10.56

X: (H)(:Cl:)C=C(:Cl:)(H)
Y: (H)(H)C=C(:Cl:)(:Cl:)
Z: (:Cl:)(H)C=C(:Cl:)(H)

Yes, compound Y has a dipole moment.

10.57(a) H—N(H)—N(H)—H H—N=N—H :N≡N:

Hydrazine Diazene Nitrogen

The single N—N bond (bond order = 1) is weaker and longer than the others. The triple bond (bond order = 3) is stronger and shorter than the others. The double bond (bond order = 2) has an intermediate strength and length.

(b) $\Delta H^\circ_{rxn} = -367$ kJ

H—N(H)—N=N—N(H)—H ⟶ H—N(H)—N(H)—H + :N≡N:

10.60(a) formal charges: Al = −1, end Cl = 0, bridging Cl = +1; I = −1, end Cl = 0, bridging Cl = +1 (b) The iodine atoms are each AX_4E_2, and the shape around each is square planar. These square planar portions are adjacent, giving a planar molecule. **10.68**(a) −1267 kJ/mol (b) −1226 kJ/mol (c) −1234.8 kJ/mol. The two answers differ by less than 10 kJ/mol. This is very good agreement since average bond energies were used to calculate answers (a) and (b). (d) −37 kJ

10.70 H—Ö—C(=:O:)—C(=:O:)—Ö—H

10.71(a) The O in the OH species has only 7 valence electrons, which is less than an octet, and 1 electron is unpaired. (b) 426 kJ (c) 508 kJ **10.73**(a) The F atoms will substitute at the axial positions first. (b) PF_5 and PCl_3F_2 **10.77** 22 kJ **10.79** Trigonal planar molecules are nonpolar, so AY_3 cannot be that shape. Trigonal pyramidal molecules and T-shaped molecules are polar, so AY_3 could have either of these shapes.

Chapter 11

11.1(a) sp^2 (b) sp^3d^2 (c) sp (d) sp^3 (e) sp^3d **11.3** C has only 2*s* and 2*p* atomic orbitals, allowing for a maximum of four hybrid orbitals. Si has 3*s*, 3*p*, and 3*d* atomic orbitals, allowing it to form more than four hybrid orbitals. **11.5**(a) six, sp^3d^2 (b) four, sp^3 **11.7**(a) sp^2 (b) sp^2 (c) sp^2 **11.9**(a) sp^3 (b) sp^3 (c) sp^3 **11.11**(a) Si: one *s* and three *p* atomic orbitals form four sp^3 hybrid orbitals. (b) C: one *s* and one *p* atomic orbital forms two *sp* hybrid orbitals. (c) S: one *s*, three *p*, and one *d* atomic orbital mix to form five sp^3d hybrid orbitals. (d) N: one *s* and three *p* atomic orbitals mix to form four sp^3 hybrid orbitals. **11.13**(a) B ($sp^3 \longrightarrow sp^3$) (b) A ($sp^2 \longrightarrow sp^3$)

11.15 (a) 4*s* [↑↓] 4*p* [↑][↑][ ] ⟶ sp^3 [↑][↑][↑][↑]

(b) 2*s* [↑↓] 2*p* [↑][ ][ ] ⟶ sp^2 [↑][↑][↑] 2*p* [ ]

(c) 2*s* [↑↓] 2*p* [↑][↑][ ] ⟶ sp^2 [↑][↑][↑] 2*p* [ ] + e⁻

11.17 (a) 4*s* [↑↓] 4*p* [↑↓][↑][↑] ⟶ sp^3 [↑↓][↑↓][↑][↑]

(b) 2*s* [↑↓] 2*p* [↑↓][↑][↑] ⟶ sp^3 [↑↓][↑][↑][↑] + e⁻

(c) 5*s* [↑↓] 5*p* [↑↓][↑↓][↑] 5*d* [ ][ ][ ][ ][ ] + e⁻ ⟶ sp^3d^2 [↑↓][↑↓][↑][↑][↑][↑] 5*d* [ ][ ][ ]

11.20(a) False. A double bond is one σ and one π bond. (b) False. A triple bond consists of one σ and two π bonds. (c) True (d) True (e) False. A π bond consists of a second pair of electrons after a σ bond has been previously formed. (f) False. End-to-end overlap results in a bond with electron density along the bond axis. **11.21**(a) Nitrogen uses sp^2 to form three σ bonds and one π bond. (b) Carbon uses *sp* to form two σ bonds and two π bonds. (c) Carbon uses sp^2 to form three σ bonds and one π bond.

11.23(a) N: sp^2, forming 2 σ bonds and 1 π bond
:F̤—N̈=Ö:

(b) C: sp^2, forming 3 σ bonds and 1 π bond
(:F:)(:F:)C=C(:F:)(:F:)

(c) C: *sp*, forming 2 σ bonds and 2 π bonds
:N≡C—C≡N:

11.25 Four MOs form from the four *p* atomic orbitals. The total number of MOs must equal the number of atomic orbitals. **11.27**(a) Bonding MOs have lower energy than antibonding MOs. Lower energy means they are more stable. (b) Bonding MOs do not have a nodal plane perpendicular to the bond. (c) Bonding MOs have higher electron density between nuclei than antibonding MOs. **11.29**(a) two (b) two (c) four **11.31**(a) A is π^*_{2p}, B is σ_{2p}, C is π_{2p}, and D is σ^*_{2p}. (b) π^*_{2p} (A), σ_{2p} (B), and π_{2p} (C) have at least one electron. (c) π^*_{2p} (A) has only one electron.

11.33

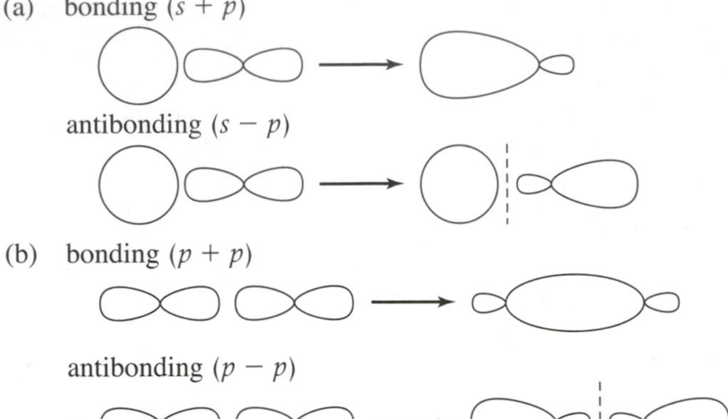

11.35(a) stable (b) paramagnetic (c) $(\sigma_{2s})^2(\sigma^*_{2s})^1$ **11.37**(a) $C_2^+ < C_2 < C_2^-$ (b) $C_2^- < C_2 < C_2^+$ **11.41**(a) C (ring): sp^2; C (all others): sp^3; O (all): sp^3; N: sp^3 (b) 26 (c) 6 **11.43**(a) 17 (b) All carbons are sp^2; the ring N is sp^2, the other N's are sp^3. **11.45**(a) B changes from sp^2 to sp^3. (b) P changes from sp^3 to sp^3d. (c) C changes from sp to sp^2. Two electron groups surround C in C_2H_2, and three electron groups surround C in C_2H_4. (d) Si changes from sp^3 to sp^3d^2. (e) no change for S **11.46** P: tetrahedral, sp^3; N: trigonal pyramid, sp^3; C_1 and C_2: tetrahedral, sp^3; C_3: trigonal planar, sp^2 **11.51**(a) B and D are present. (b) Yes, sp hybrid orbitals. (c) Two sets of sp orbitals, four sets of sp^2 orbitals, and three sets of sp^3 orbitals. **11.52** The central C is sp hybridized, and the other two C atoms are sp^2 hybridized

H, H, C=C=C, H, H

11.52 Through resonance, the C—N bond gains some double-bond character, which hinders rotation about that bond.

:O: ‖ —C—N̈(H)— ⟷ :Ö:⁻ | —C=N⁺(H)—

11.56(a) C in $—CH_3$: sp^3; all other C atoms: sp^2; O in two C—O bonds: sp^3; O in two C=O bonds: sp^2 (b) two (c) eight; one

Chapter 12

• **12.1** In a solid, the energy of attraction of the particles is greater than their energy of motion; in a gas, it is less. Gases have high compressibility and the ability to flow, while solids have neither. **12.4**(a) Because the intermolecular forces are only partially overcome when fusion occurs but need to be totally overcome in vaporization. (b) Because solids have greater intermolecular forces than liquids do. (c) $\Delta H_{vap} = -\Delta H_{cond}$ **12.5**(a) condensation (b) fusion (c) vaporization **12.7** The gas molecules slow down as the gas is compressed. Therefore, much of the kinetic energy lost by the propane molecules is released to the surroundings. **12.11** At first, the vaporization of liquid molecules from the surface predominates, which increases the number of gas molecules and hence the vapor pressure. As more molecules enter the gas phase, more gas molecules hit the surface of the liquid and "stick" more frequently, so the condensation rate increases. When the vaporization and condensation rates become equal, the vapor pressure becomes constant. **12.15** 7.67×10^3 J **12.17** 0.777 atm **12.19**

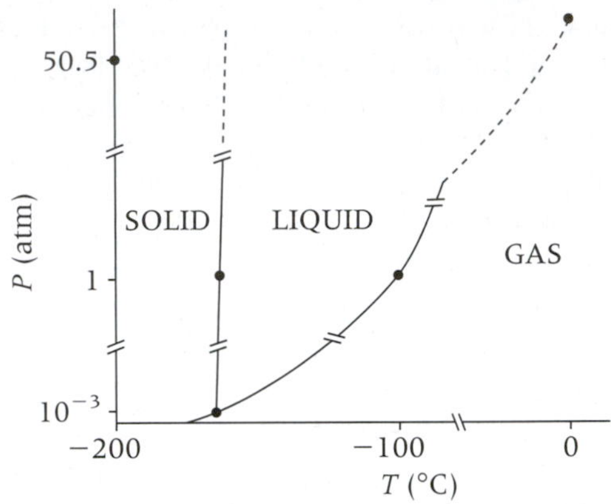

Solid ethylene is more dense than liquid ethylene. **12.21** 32 atm **12.26** O is smaller and more electronegative than Se; so the electron density on O is greater, which attracts H more strongly. **12.27** All particles (atoms and molecules) exhibit dispersion forces, but the total force is weak for small molecules. Dipole-dipole forces between small polar molecules dominate the dispersion forces. **12.30**(a) hydrogen bonding (b) dispersion forces (c) dispersion forces **12.32**(a) dipole-dipole forces (b) dispersion forces (c) hydrogen bonding **12.34**(a)

CH_3—CH(CH_3)—Ö:····H—Ö—CH(CH_3)—CH_3 (with H on the second O)

(b) H—F:····H—F:····H—F:

12.36(a) I^- (b) $CH_2{=}CH_2$ (c) H_2Se. In (a) and (c) the larger particle has the higher polarizability. In (b), the less tightly held π electron clouds are more easily distorted. **12.38**(a) C_2H_6; it is a smaller molecule exhibiting weaker dispersion forces than C_4H_{10}. (b) CH_3CH_2F; it has no H—F bonds, so it exhibits only dipole-dipole forces, which are weaker than the hydrogen bonds of CH_3CH_2OH. (c) PH_3; it has weaker intermolecular forces (dipole-dipole) than NH_3 (hydrogen bonding). **12.40**(a) HCl; it has dipole-dipole forces, and there are stronger ionic bonds in LiCl. (b) PH_3; it has dipole-dipole forces, and there is stronger hydrogen bonding in NH_3. (c) Xe; it exhibits weaker dispersion forces since its smaller size results in lower polarizability than the larger I_2 molecules. **12.42**(a) C_4H_8 (cyclobutane), because it is more compact than C_4H_{10}. (b) PBr_3; the dipole-dipole forces in PBr_3 are weaker than the ionic bonds in NaBr. (c) HBr; the dipole-dipole forces in HBr are weaker than the hydrogen bonds in water. **12.47** The cohesive forces in water and mercury are stronger than the adhesive forces to the nonpolar wax on the floor. Weak adhesive forces result in spherical drops. The adhesive forces overcome the even weaker cohesive forces in the oil, and so the oil drop spreads out. **12.49** $CH_3CH_2CH_2OH < HOCH_2CH_2OH < HOCH_2CH(OH)CH_2OH$. More hydrogen bonding means more attraction between molecules, so more energy is needed to increase surface area. **12.53** Water is a good solvent for polar and ionic substances and a poor solvent for nonpolar substances. Water is a polar molecule and dissolves polar substances whose intermolecular forces are of similar strength. **12.54** A single water molecule can form four H bonds. The two hydrogen atoms each form one H bond to oxygen atoms on neighboring water molecules. The two lone pairs on the oxygen atom form H bonds with hydrogen atoms on two neighboring molecules. **12.56** Water exhibits strong capillary action, which allows it to be easily absorbed by the plant's roots and transported upward to the leaves. **12.62** A solid metal is shiny, conducts heat, is malleable, and melts at high temperatures. (Other answers include a relatively high boiling point and good conduction of electricity.) **12.65** The energy gap is the energy difference between the highest filled energy level (valence band) and the lowest unfilled energy level (conduction band). In conductors and superconductors, the energy gap is zero because the valence band overlaps the conduction band. In semiconductors, the energy gap is small. In insulators, the gap is large. **12.66**(a) face-centered cubic (b) body-centered cubic (c) face-centered cubic **12.68**(a) The change in unit cell is from a sodium chloride structure in CdO to a zinc blende structure in CdSe. (b) Yes, the coordination number of Cd changes from 6 in CdO to 4 in CdSe. **12.70**(a) Nickel forms a metallic solid since its atoms are held together by metallic bonds. (b) Fluorine forms a molecular solid since the F_2 molecules are held together by dispersion forces. (c) Methanol forms a molecular solid since the CH_3OH molecules are held together by dispersion forces and hydrogen bonds. (d) Tin forms a metallic solid since its atoms are held together by metallic bonds. (e) Silicon is directly under carbon in Group 4A(14), so it exhibits similar bonding properties. Since diamond and graphite are both network covalent solids, it makes sense that Si forms a network covalent solid as well. (f) Xe is an atomic solid since its individual atoms are held together by dispersion forces. **12.72** four **12.74**(a) four Se^{2-} ions, four Zn^{2+} ions (b) 577.48 amu (c) 1.77×10^{-22} cm^3 (d) 5.61×10^{-8} cm **12.76**(a) insulator (b) conductor (c) semiconductor **12.77**(a) Conductivity increases. (b) Conductivity increases. (c) Conductivity decreases. **12.82** 259 K **12.84**(a) simple (b) 3.99 g/cm^3 **12.89**(a) furfuryl alcohol

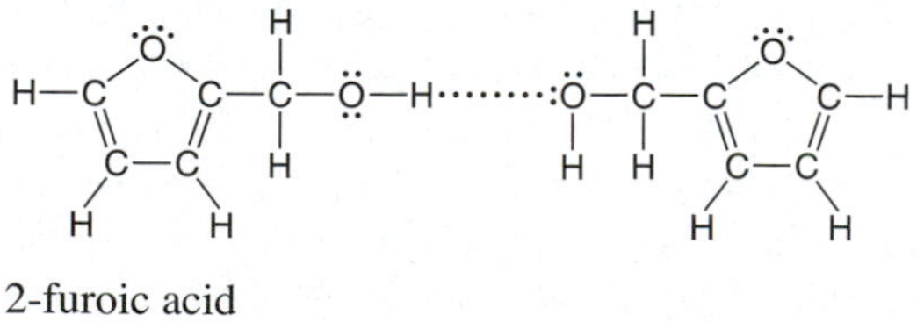

2-furoic acid

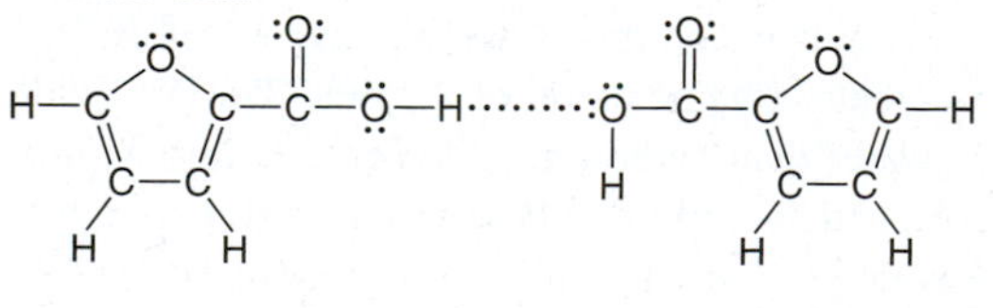

(b) furfuryl alcohol 2-furoic acid

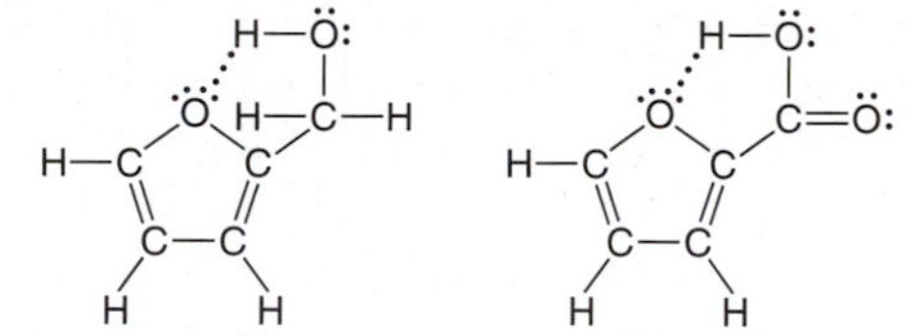

12.93 (a) $4r$ (b) $\sqrt{2}a$ (c) $a = 4r/\sqrt{3}$ (d) two (e) 0.68017

Chapter 13

• **13.2** When a salt such as NaCl dissolves, ion-dipole forces cause the ions to separate, and many water molecules cluster around each ion in hydration shells. Ion-dipole forces bind the first shell to an ion. The water molecules in that shell form H bonds to others to create the next shell, and so on. **13.4** KNO_3 is an ionic compound and is therefore more soluble in water. **13.6**(a) ion-dipole forces (b) hydrogen bonding (c) dipole–induced dipole forces **13.8**(a) hydrogen bonding (b) dipole–induced dipole forces (c) dispersion forces **13.10**(a) HCl(g), because the molecular interactions (dipole-dipole forces) in ether are like those in HCl but not like the ionic bonding in NaCl. (b) $CH_3CHO(l)$, because the molecular interactions with ether (dipole-dipole) can replace those between CH_3CHO, but not the H bonds in water. (c) $CH_3CH_2MgBr(s)$, because the molecular interactions (dipole-dipole and dispersion forces) are greater than between ether and the ions in $MgBr_2$. **13.12** Gluconic acid is soluble in water due to extensive hydrogen bonding by the —OH groups attached to five of its carbons. The dispersion forces in the nonpolar tail of caproic acid are more similar to the dispersion forces in hexane; thus, caproic acid is soluble in hexane. **13.17** Very soluble, because a decrease in enthalpy and an increase in entropy both favor the formation of a solution **13.18**

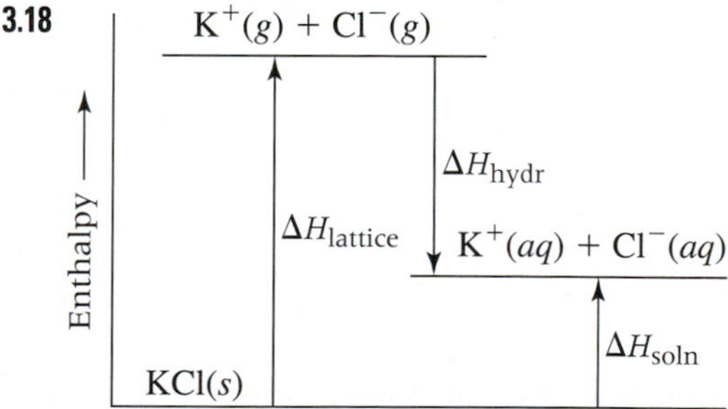

13.20(a) The volume of Na^+ is smaller. (b) Sr^{2+} has a larger ionic charge and a smaller volume. (c) Na^+ is smaller than Cl^-. (d) O^{2-} has a larger ionic charge with a similar ion volume. (e) OH^- has a smaller volume than SH^-. (f) Mg^{2+} has a smaller volume. (g) Mg^{2+} has both a smaller volume and a larger ionic charge. (h) CO_3^{2-} has both a smaller volume and a larger ionic charge. **13.22**(a) Na^+ (b) Sr^{2+} (c) Na^+ (d) O^{2-} (e) OH^- (f) Mg^{2+} (g) Mg^{2+} (h) CO_3^{2-} **13.24**(a) −704 kJ/mol (b) The K^+ ion contributes more because it is smaller and, therefore, has a greater charge density. **13.26**(a) increases (b) decreases (c) increases **13.29** Add a pinch of X to each solution. Addition of a "seed" crystal of solute to a supersaturated solution causes the excess solute to crystallize immediately, leaving behind a saturated solution. The solution in which the added X dissolves is the unsaturated solution. The solution in which the added X remains undissolved is the saturated solution. **13.31**(a) increase (b) decrease **13.33**(a) 0.102 g O_2 (b) 0.0214 g O_2 **13.36** 0.20 mol/L **13.39** With just this information, you can convert between molality and molarity, but you need to know the molar mass of the solvent to convert to mole fraction. **13.41**(a) 0.944 M $C_{12}H_{22}O_{11}$ (b) 0.167 M $LiNO_3$ **13.43**(a) 0.0749 M NaOH (b) 0.36 M HNO_3 **13.45**(a) Add 4.25 g KH_2PO_4 to enough water to make 365 mL of aqueous solution. (b) Add 125 mL of 1.25 M NaOH to enough water to make 465 mL of solution. **13.47**(a) 0.896 m glycine (b) 1.21 m glycerol **13.49** 4.48 m C_6H_6 **13.51**(a) Add 2.39 g $C_2H_6O_2$ to 308 g H_2O. (b) Add 0.0508 kg of 52.0% HNO_3 by mass to 1.15 kg H_2O to make 1.20 kg of 2.20% HNO_3 by mass. **13.53**(a) 0.29 (b) 58 mass % (c) 23 m C_3H_7OH **13.55** 5.11 m NH_3; 4.53 M NH_3; mole fraction = 0.0843 **13.57** 2.5 ppm Ca^{2+}; 0.56 ppm Mg^{2+} **13.60** The boiling point is higher and the freezing point is lower for the solution than for the solvent. **13.64**(a) strong electrolyte (b) strong electrolyte (c) nonelectrolyte (d) weak electrolyte **13.66**(a) 0.6 mol of solute particles (b) 0.13 mol (c) 2×10^{-4} mol (d) 0.06 mol **13.68**(a) CH_3OH in H_2O (b) H_2O in CH_3OH solution **13.70**(a) $\Pi_{II} < \Pi_{I} < \Pi_{III}$ (b) $bp_{II} < bp_{I} < bp_{III}$ (c) $fp_{III} < fp_{I} < fp_{II}$ (d) $vp_{III} < vp_{I} < vp_{II}$ **13.72** 23.4 torr **13.74** -0.467°C **13.76** 79.5°C **13.78** 1.18×10^4 g $C_2H_6O_2$ **13.80**(a) NaCl: 0.173 m and $i = 1.84$ (b) CH_3COOH: 0.0837 m and $i = 1.02$ **13.83** 209 torr for CH_2Cl_2; 48.1 torr for CCl_4 **13.85** 3.4×10^9 L **13.90** 0.0°C: 4.53×10^{-4} M O_2; 20.0°C: 2.83×10^{-4} M O_2; 40.0°C: 2.00×10^{-4} M O_2 **13.91**(a) 89.9 g/mol (b) C_2H_5O; $C_4H_{10}O_2$

(c) Forms H bonds:

```
     H   H   H   H
     |   |   |   |
HO—C—C—C—C—OH
     |   |   |   |
     H   H   H   H
```

Does not form H bonds:

```
   H       H       H   H
   |       |       |   |
H—C—O—C—O—C—C—H
   |       |       |   |
   H       H       H   H
```

13.96(a) 68 g/mol (b) 2.1×10^2 g/mol (c) The molar mass of CaN_2O_6 is 164.10 g/mol. This value is less than the 2.1×10^2 g/mol calculated when the compound is assumed to be a strong electrolyte and is greater than the 68 g/mol calculated when the compound is assumed to be a nonelectrolyte. Thus, the compound forms a nonideal solution because the ions interact but do not dissociate completely in solution. (d) $i = 2.4$ **13.99**(a) 1.82×10^4 g/mol (b) 3.41×10^{-5} °C **13.105**(a) CH_4N_2O (b) 60. g/mol; CH_4N_2O **13.108**(a) 4.3×10^2 g NaCl (b) 5.5×10^2 g $CaCl_2$

Chapter 14

14.1 The outermost electron is attracted by a smaller effective nuclear charge in Li because of shielding by the inner electrons, and it is farther from the nucleus in Li. Both of these factors lead to a lower ionization energy for Li.

14.2(a) $2Al(s) + 6HCl(aq) \longrightarrow 2AlCl_3(aq) + 3H_2(g)$
(b) $LiH(s) + H_2O(l) \longrightarrow LiOH(aq) + H_2(g)$
14.4(a) $NaBH_4$: +1 for Na, +3 for B, −1 for H
$Al(BH_4)_3$: +3 for Al, +3 for B, −1 for H
$LiAlH_4$: +1 for Li, +3 for Al, −1 for H
(b) tetrahedral

```
[    H    ]−
[    |    ]
[ H—B—H ]
[    |    ]
[    H    ]
```

14.7(a) reducing agent (b) Alkali metals have relatively low ionization energies, which means they easily lose the outermost electron.
(c) $2Na(s) + 2H_2O(l) \longrightarrow 2Na^+(aq) + 2OH^-(aq) + H_2(g)$
$2Na(s) + Cl_2(g) \longrightarrow 2NaCl(s)$
14.10 For Groups 1A(1) to 4A(14), the number of covalent bonds equals the (old) group number. For Groups 5A(15) to 7A(17), it equals 8 minus the (old) group number. There are exceptions in Period 3 to Period 6 because it is possible for the 3A(13) to 7A(17) elements to use d orbitals and form more bonds. **14.11** Density and ionic size increase down a group; the other three properties decrease down a group. **14.13** $2Na(s) + O_2(g) \longrightarrow Na_2O_2(s)$ **14.15** $K_2CO_3(s) + 2HI(aq) \longrightarrow 2KI(aq) + H_2O(l) + CO_2(g)$ **14.19** Group 2A(2) metals have an additional valence electron to increase the strength of metallic bonding, which leads to higher melting points, higher boiling points, greater hardness, and greater density.
14.20(a) $CaO(s) + H_2O(l) \longrightarrow Ca(OH)_2(s)$
(b) $2Ca(s) + O_2(g) \longrightarrow 2CaO(s)$
14.22(a) $BeO(s) + H_2O(l) \longrightarrow$ no reaction
(b) $BeCl_2(l) + 2Cl^-(\text{solvated}) \longrightarrow BeCl_4^{2-}(\text{solvated})$
Be behaves like other Group 2A(2) elements in reaction (b). **14.25** The electron removed from Group 2A(2) atoms occupies the outer s orbital, whereas in Group 3A(13) atoms, the electron occupies the outer p orbital. For example, the electron configuration for Be is $1s^22s^2$ and for B it is $1s^22s^22p^1$. It is easier to remove the p electron of B than an s electron of Be, because the energy of a p orbital is higher. Thus, even though atomic size decreases because of increasing Z_{eff}, IE decreases from 2A(2) to 3A(13). **14.26** $In_2O_3 < Ga_2O_3 < Al_2O_3$ **14.28** Apparent O.N., +3; actual O.N., +1. $[:\ddot{I}-\ddot{I}-\ddot{I}:]^-$ The anion I_3^- has the general formula AX_2E_3 and bond angles of 180°. $(Tl^{3+})(I^-)_3$ does not exist because of the low strength of the Tl—I bond. **14.33** In general, network solids have very high melting and boiling points and are very hard, while molecular solids have low melting and boiling points and are soft. The properties of network solids reflect the necessity of breaking covalent bonds throughout the substances, whereas the properties of molecular solids reflect the weaker intermolecular forces between individual molecules. **14.34** Basicity in water is greater for the oxide of a metal. Tin(IV) oxide is more basic in water than carbon dioxide because tin has more metallic character than carbon. **14.36**(a) Ionization energy generally decreases down a group. (b) The deviations (increases) from the expected trend are due to the presence of the first transition series between Si and Ge and of the lanthanides between Sn

and Pb. (c) Group 3A(13) **14.39** Atomic size increases down a group. As atomic size increases, ionization energy decreases and so it is easier to form a positive ion. An atom that is easier to ionize exhibits greater metallic character.

14.41(a)

```
[      :Ö:        :Ö:      ]8−
        |          |
  :Ö—Si—Ö—Si—Ö:
        |          |
       :Ö:        :Ö:
        |          |
  :Ö—Si—Ö—Si—Ö:
        |          |
       :Ö:        :Ö:      ]
```

(b)

```
     H   H
     |   |
  H—C—C—H
     |   |
  H—C—C—H
     |   |
     H   H
```

14.44(a) diamond, C (b) calcium carbonate, $CaCO_3$ (c) carbon dioxide, CO_2 (d) carbon monoxide, CO (e) silicon, Si **14.48**(a) −3 to +5 (b) For a group of nonmetals, the oxidation states range from the lowest, group number − 8, or 5 − 8 = −3 for Group 5A, to the highest, equal to the group number, or +5 for Group 5A. **14.49**(a) The greater the electronegativity of the element, the more covalent the bonding is in its oxide. (b) The more electronegative the element, the more acidic the oxide is. **14.52**(a) $4As(s) + 5O_2(g) \longrightarrow 2As_2O_5(s)$

(b) $2Bi(s) + 3F_2(g) \longrightarrow 2BiF_3(s)$

(c) $Ca_3As_2(s) + 6H_2O(l) \longrightarrow 3Ca(OH)_2(s) + 2AsH_3(g)$

14.58(a) Boiling point and conductivity vary in similar ways down both groups. (b) Degree of metallic character and types of bonding vary in similar ways down both groups. (c) Both P and S have allotropes, and both bond covalently with almost every other nonmetal. (d) Both N and O are diatomic gases at normal temperatures and pressures. (e) O_2 is a very reactive gas, whereas N_2 is not. Nitrogen has six oxidation states, whereas oxygen has two.

14.60(a) $NaHSO_4(aq) + NaOH(aq) \longrightarrow Na_2SO_4(aq) + H_2O(l)$

(b) $S_8(s) + 24F_2(g) \longrightarrow 8SF_6(g)$

(c) $FeS(s) + 2HCl(aq) \longrightarrow H_2S(g) + FeCl_2(aq)$

(d) $Te(s) + 2I_2(s) \longrightarrow TeI_4(s)$

14.62(a) acidic (b) acidic (c) basic (d) amphoteric (e) basic **14.64**(a) O_3, ozone (b) SO_3, sulfur trioxide (c) SO_2, sulfur dioxide (d) $Na_2S_2O_3{\cdot}5H_2O$, sodium thiosulfate pentahydrate **14.65** $S_2F_{10}(g) \longrightarrow SF_4(g) + SF_6(g)$; O.N. of S in S_2F_{10} is +5; O.N. of S in SF_4 is +4; O.N. of S in SF_6 is +6. **14.66**(a) Polarity is the molecular property that is responsible for the difference in boiling points between iodine monochloride (polar) and bromine (nonpolar). It arises from different EN values of the bonded atoms. (b) The boiling point of polar ICl is higher than the boiling point of Br_2. **14.68**(a) −1, +1, +3, +5, +7 (b) The electron configuration for Cl is [Ne] $3s^23p^5$. By gaining one electron, Cl achieves an octet. By forming covalent bonds, Cl completes or expands its valence level by maintaining electron pairs in bonds or as lone pairs. (c) Fluorine has only the −1 oxidation state because its small size and absence of *d* orbitals prevent it from forming more than one covalent bond. **14.69**(a) Cl—Cl bond is stronger than Br—Br bond. (b) Br—Br bond is stronger than I—I bond. (c) Cl—Cl bond is stronger than F—F bond. The fluorine atoms are so small that electron-electron repulsion of the lone pairs decreases the strength of the bond.

14.70 $3Br_2(l) + 6OH^-(aq) \longrightarrow 5Br^-(aq) + BrO_3^-(aq) + 3H_2O(l)$

14.71(a) $I_2(s) + H_2O(l) \longrightarrow HI(aq) + HIO(aq)$

(b) $Br_2(l) + 2I^-(aq) \longrightarrow I_2(s) + 2Br^-(aq)$

(c) $CaF_2(s) + H_2SO_4(l) \longrightarrow CaSO_4(s) + 2HF(g)$

14.74 helium; argon **14.75** Only dispersion forces hold atoms of noble gases together. **14.79**(a) Second ionization energies for alkali metals are so high because the electron being removed is from the next lower energy level and these are very tightly held by the nucleus. Also, the alkali metal would lose its noble gas electron configuration. (b) $2CsF_2(s) \longrightarrow 2CsF(s) + F_2(g)$; −405 kJ/mol

14.81(a) hyponitrous acid, $H_2N_2O_2$; nitroxyl, HNO

(b) H—Ö—N̈=N̈—Ö—H H—N̈=Ö:

(c) In both species the shape is bent about the N atoms.

(d)

```
[    N̈=N̈     ]2−      [        :Ö: ]2−
[   /     \   ]        [        /   ]
[ :Ö:     :Ö: ]        [   ˙N=N:    ]
                       [   /        ]
                       [ :Ö:        ]
      cis                   trans
```

14.84 In a disproportionation reaction, a substance acts as both a reducing agent and an oxidizing agent in that atoms of an element within the substance attain both higher and lower oxidation states in the products. The disproportionation reactions are b, c, d, e, and f. **14.85**(a) Group 5A(15) (b) Group 7A(17) (c) Group 6A(16) (d) Group 1A(1) (e) Group 3A(13) (f) Group 8A(18)

14.86 117.2 kJ

14.87

```
[       :C̈l:       :C̈l:       ]−
[        |          |         ]
[ :C̈l—Al—C̈l—Al—C̈l:  ]
[        |          |         ]
[       :C̈l:       :C̈l:       ]
```

(a) sp^3 orbitals (b) tetrahedral (c) Since the ion is linear, the central Cl atom must be *sp* hybridized. (d) The *sp* hybridization means there are no lone pairs on the central Cl atom. Instead, the extra four electrons interact with the empty *d* orbitals on the Al atoms to form double bonds between the chlorine and each aluminum atom.

14.89 [Ö=N̈—Ö:]⁻ ⟷ [:Ö—N̈=Ö]⁻

Ö=N—Ö: ⟷ :Ö—N=Ö

[Ö=N=Ö]⁺

The nitronium ion (NO_2^+) has a linear shape because the central N atom has two surrounding electron groups, which achieve maximum repulsion at 180°. The nitrite ion (NO_2^-) bond angle is more compressed than the nitrogen dioxide (NO_2) bond angle because the lone pair of electrons takes up more space than the lone electron. **14.91**(a) 39.96 mass % in $CuHAsO_3$; As, 62.42 mass % in $(CH_3)_3As$ (b) 0.35 g $CuHAsO_3$

Chapter 15

• **15.1**(a) Carbon's electronegativity is midway between the most metallic and most nonmetallic elements of Period 2. To attain a filled outer shell, carbon forms covalent bonds to other atoms in molecules, network covalent solids, and polyatomic ions. (b) Since carbon has four valence shell electrons, it forms four covalent bonds to attain an octet. (c) To reach the He electron configuration, a carbon atom must lose four electrons, requiring too much energy to form the C^{4+} cation. To reach the Ne electron configuration, the carbon atom must gain four electrons, also requiring too much energy to form the C^{4-} anion. (d) Carbon is able to bond to itself extensively because its small size allows for close approach and great orbital overlap. The extensive orbital overlap results in a strong, stable bond. (e) The C—C σ bond is short enough to allow sideways overlap of unhybridized *p* orbitals of neighboring C atoms. The sideways overlap of *p* orbitals results in the π bonds that are part of double and triple bonds. **15.2**(a) H, O, N, P, S, and halogens (b) Heteroatoms are atoms of any element other than carbon and hydrogen. (c) More electronegative than C: N, O, F, Cl, and Br; less electronegative than C: H and P. Sulfur and iodine have the same electronegativity as carbon. (d) Since carbon can bond to a wide variety of heteroatoms and to carbon atoms, it can form many different compounds. **15.4** The C—H and C—C bonds are unreactive because electronegativities are close and the bonds are short. The C—I bond is somewhat reactive because it is long and weak. The C=O bond is reactive because oxygen is more electronegative than carbon and the electron-rich π bond makes it attract electron-poor atoms. The C—Li bond is also reactive because the bond polarity results in an electron-rich region around carbon and an electron-poor region around lithium. **15.5**(a) An alkane and a cycloalkane are organic compounds that consist of carbon and hydrogen and have only single bonds. A cycloalkane has a ring of carbon atoms. An alkene is a hydrocarbon with at least one double bond. An alkyne is a hydrocarbon with at least one triple bond. (b) alkane is C_nH_{2n+2}, cycloalkane is C_nH_{2n}, alkene is C_nH_{2n}, alkyne is C_nH_{2n-2} (c) Alkanes and cycloalkanes are saturated hydrocarbons. **15.8** a, c, and f

15.9(a)

```
C=C—C—C—C—C        C=C—C—C—C—C
  |                     |
  C                     C

C=C—C—C—C—C        C=C—C—C—C—C
      |                     |
      C                     C

C—C=C—C—C—C        C—C=C—C—C—C
  |                     |
  C                     C

C—C=C—C—C—C        C—C=C—C—C—C
      |                     |
      C                     C

C—C—C=C—C—C        C—C—C=C—C—C
  |                     |
  C                     C

C—C—C—C—C—C
    ||
    C
```

(b)

```
C=C—C—C—C          C=C—C—C—C
  | |                  |   |
  C C                  C   C

    C
    |
C=C—C—C—C          C=C—C—C—C
    |                  | |
    C                  C C

      C
      |
C=C—C—C—C          C=C—C—C—C
      |                |
      C                C—C

C—C=C—C—C          C—C=C—C—C
  | |                  |   |
  C C                  C   C

                           C
                           |
C—C=C—C—C          C—C=C—C—C
    | |                    |
    C C                    C

C—C=C—C—C
    |
    C—C
```

(c)

```
 C   C           C               C
  \ /            |               |
   C             C               C            C
  / \           / \            /   \        /   \
 C   C         C   C—C        C     C      C     C—C—C
  \ /           \ /            \   /        \   /
  C—C           C—C             C—C          C—C
                                   \
                                    C
```

15.11(a)

$$CH_2{=}\underset{\substack{|\\CH_3}}{C}{-}CH_2{-}CH_2{-}CH_2{-}CH_3 \qquad CH_2{=}CH{-}\underset{\substack{|\\CH_3}}{CH}{-}CH_2{-}CH_2{-}CH_3$$

$$CH_2{=}CH{-}CH_2{-}\underset{\substack{|\\CH_3}}{CH}{-}CH_2{-}CH_3 \qquad CH_2{=}CH{-}CH_2{-}CH_2{-}\underset{\substack{|\\CH_3}}{CH}{-}CH_3$$

$$CH_3{-}\underset{\substack{|\\CH_3}}{C}{=}CH{-}CH_2{-}CH_2{-}CH_3 \qquad CH_3{-}CH{=}\underset{\substack{|\\CH_3}}{C}{-}CH_2{-}CH_2{-}CH_3$$

$$CH_3{-}CH{=}CH{-}\underset{\substack{|\\CH_3}}{CH}{-}CH_2{-}CH_3 \qquad CH_3{-}CH{=}CH{-}CH_2{-}\underset{\substack{|\\CH_3}}{CH}{-}CH_3$$

$$CH_3{-}\underset{\substack{|\\CH_3}}{CH}{-}CH{=}CH{-}CH_2{-}CH_3 \qquad CH_3{-}CH_2{-}\underset{\substack{|\\CH_3}}{C}{=}CH{-}CH_2{-}CH_3$$

$$CH_3{-}CH_2{-}\underset{\substack{\|\\CH_2}}{C}{-}CH_2{-}CH_2{-}CH_3$$

(b) $CH_2{=}C(CH_3){-}CH(CH_3){-}CH_2{-}CH_3$ $\quad$ $CH_2{=}C(CH_3){-}CH_2{-}CH(CH_3){-}CH_3$

$CH_2{=}CH{-}C(CH_3)_2{-}CH_2{-}CH_3$ $\quad$ $CH_2{=}CH{-}CH(CH_3){-}CH(CH_3){-}CH_3$

$CH_2{=}CH{-}CH_2{-}C(CH_3)_2{-}CH_3$ $\quad$ $CH_2{=}CH{-}CH(CH_2{-}CH_3){-}CH_2{-}CH_3$

$CH_3{-}C(CH_3){=}C(CH_3){-}CH_2{-}CH_3$ $\quad$ $CH_3{-}C(CH_3){=}CH{-}CH(CH_3){-}CH_3$

$CH_3{-}CH{=}C(CH_3){-}CH(CH_3){-}CH_3$ $\quad$ $CH_3{-}CH{=}CH{-}C(CH_3)_2{-}CH_3$

$CH_3{-}CH{=}C(CH_2{-}CH_3){-}CH_2{-}CH_3$

(c) cyclopentane ring (CH_2, CH_2, H_2C, CH_2, CH) with $CH{-}CH_2{-}CH_3$ $\quad$ ring (CH_2, CH_2, H_2C) with $CH{-}CH_3$ and $CH{-}CH_3$

ring (CH_2, H_2C, CH_2) with $CH{-}CH_3$ and $H_3C{-}HC$ $\quad$ ring (CH_2, H_2C, $H_2C{-}CH_2$) with $C(CH_3)_2$

15.13(a) $CH_3{-}C(CH_3)_2{-}CH_2{-}CH_3$ $\quad$ (b) $H_2C{=}CH{-}CH_2{-}CH_3$

(c) $HC{\equiv}C{-}CH(CH_2{-}CH_3){-}CH_3$ $\quad$ (d) Structure is correct.

15.15(a) $CH_3{-}CH(CH_3){-}CH(CH_3){-}CH_2{-}CH_2{-}CH_2{-}CH_2{-}CH_3$

(b) cyclohexane ring numbered 1–6, with $CH_2{-}CH_3$ at C-1 and CH_3 at C-3

(c) 3,4-dimethylheptane (d) 2,2-dimethylbutane

15.17(a) $CH_3{-}CH_2{-}CH_2{-}CH(CH_3){-}CH_2{-}CH_3$

Correct name is 3-methylhexane.

(b) $CH_3{-}CH_2{-}CH_2{-}CH(CH_2{-}CH_3){-}CH_3$

Correct name is 3-methylhexane.

(c) cyclohexane ring with CH_3

Correct name is methylcyclohexane.

(d) $CH_3{-}CH_2{-}C(CH_3)_2{-}CH(CH_2{-}CH_3){-}CH_2{-}CH_2{-}CH_2{-}CH_3$

Correct name is 4-ethyl-3,3-dimethyloctane.

15.19(a) cyclopropane ring: C(H,H); Ⓒ(H, Cl); Ⓒ(CH_3, H)

(b) $CH_3{-}CH_2{-}$Ⓒ$H(CH_3){-}CH(CH_3){-}CH_3$

15.21(a) 3-Bromohexane is optically active.

$CH_3{-}CH_2{-}$Ⓒ$H(Br){-}CH_2{-}CH_2{-}CH_3$

(b) 3-Chloro-3-methylpentane is not optically active.

$CH_3{-}CH_2{-}C(CH_3)(Cl){-}CH_2{-}CH_3$

(c) 1,2-Dibromo-2-methylbutane is optically active.

$CH_2(Br){-}$Ⓒ$(CH_3)(Br){-}CH_2{-}CH_3$

15.23(a)

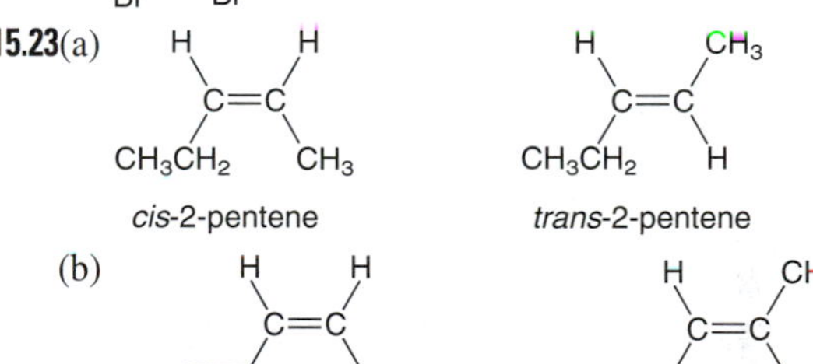

cis-2-pentene $\quad$ trans-2-pentene

(b)

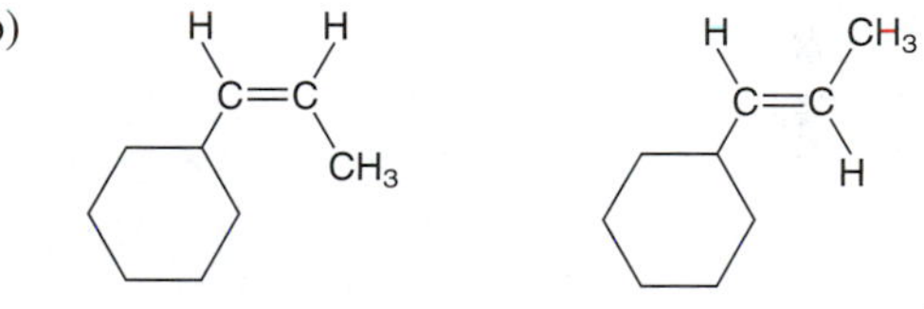

cis-1-cyclohexylpropene $\quad$ trans-1-cyclohexylpropene

(c) no geometric isomers

15.25(a) no geometric isomers

(b)

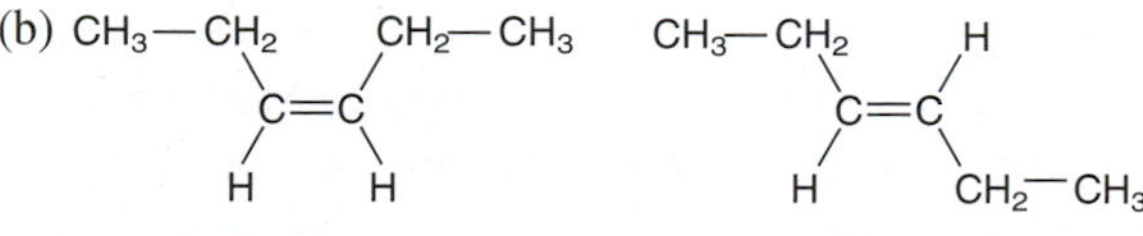

cis-3-hexene $\quad$ trans-3-hexene

(c) no geometric isomers

(d) Cl, Cl / C=C / H, H $\quad$ Cl, H / C=C / H, Cl

cis-1,2-dichloroethene $\quad$ trans-1,2-dichloroethene

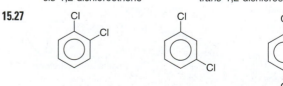

15.27 benzene rings with two Cl substituents (ortho, meta, para)

1,2-dichlorobenzene (o-dichlorobenzene) $\quad$ 1,3-dichlorobenzene (m-dichlorobenzene) $\quad$ 1,4-dichlorobenzene (p-dichlorobenzene)

15.29 2,6-di-*tert*-butyl-4-methylphenol: $(CH_3)_3C$—, OH, —$C(CH_3)_3$ on ring, with CH_3 para to OH

15.30

$CH_3—CH(CH_3)—CH{=}CH—CH_2—CH_3$ (both H on the same side)

cis-2-methyl-3-hexene

$CH_3—CH(CH_3)—CH{=}CH—CH_2—CH_3$ (H atoms on opposite sides)

trans-2-methyl-3-hexene

The compound 2-methyl-2-hexene does not have *cis-trans* isomers. **15.32**(a) elimination (b) addition

15.34(a) $CH_3CH_2CH{=}CHCH_2CH_3 + H_2O \xrightarrow{H^+} CH_3CH_2CH_2CH(OH)CH_2CH_3$

(b) $CH_3CHBrCH_3 + CH_3CH_2OK \longrightarrow CH_3CH{=}CH_2 + CH_3CH_2OH + KBr$

(c) $CH_3CH_3 + 2Cl_2 \xrightarrow{h\nu} CHCl_2CH_3 + 2HCl$

15.37(a) Methylethylamine is more soluble because it has the ability to form H bonds with water molecules. (b) 1-Butanol has a higher melting point because it can form intermolecular H bonds. (c) Propylamine has a higher boiling point because it contains N—H bonds that allow H bonding, and trimethylamine cannot form H bonds. **15.39** Both groups react by addition to the π bond. The very polar C=O bond attracts the electron-rich O of water to the partially positive C. There is no such polarity in the alkene, so either C atom can be attacked, and two products result.

$CH_3—CH_2—C({=}O)—CH_2—CH_3 + H_2O \xrightarrow{H^+} CH_3—CH_2—C(OH)_2—CH_2—CH_3$

$CH_3—CH_2—CH{=}CH—CH_3 + H_2O \xrightarrow{H^+} CH_3—CH_2—CH(OH)—CH_2—CH_3 + CH_3—CH_2—CH_2—CH(OH)—CH_3$

15.41 Esters and acid anhydrides form through dehydration-condensation reactions, and water is the other product.

15.43(a) alkyl halide (b) nitrile (c) carboxylic acid (d) aldehyde

15.45(a) $CH_3—CH{=}CH—CH_2—OH$ (CH=CH: alkene; CH₂—OH: alcohol)

alkene alcohol

(b) $Cl—CH_2—C_6H_4—C({=}O)—OH$

haloalkane carboxylic acid

(c) cyclohexene ring—C(=O)—NH—CH_3

alkene amide

(d) $N{\equiv}C—CH_2—C({=}O)—CH_3$

nitrile ketone

(e) cyclobutyl—C(=O)—O—CH_2—CH_3

ester

15.47 $H_3C—CH_2—CH_2—CH_2—CH_2—OH$

$H_3C—CH_2—CH_2—CH(OH)—CH_3$

$H_3C—CH(CH_3)—CH(OH)—CH_3$ $H_3C—C(OH)(CH_3)—CH_2—CH_3$

$H_3C—CH_2—CH(OH)—CH_2—CH_3$ $H_3C—CH(CH_3)—CH_2—CH_2—OH$

$H_3C—CH_2—CH(CH_3)—CH_2—OH$ $H_3C—C(CH_3)_2—CH_2—OH$

15.49 $H_3C—CH_2—CH_2—CH_2—NH_2$ $H_3C—CH_2—CH(NH_2)—CH_3$

$H_3C—CH(CH_3)—CH_2—NH_2$ $H_3C—C(CH_3)_2—NH_2$

$H_3C—CH_2—NH—CH_2—CH_3$ $H_3C—CH_2—CH_2—NH—CH_3$

$H_3C—CH_2—N(CH_3)—CH_3$ $H_3C—CH(CH_3)—NH—CH_3$

15.51(a) $CH_3—C({=}O)—NH—CH_3$

(b) $CH_3—CH_2—CH_2—C({=}O)—O—CH(CH_3)—CH_3$

(c) $H—C({=}O)—O—CH_2—CH(CH_3)—CH_3$

15.53(a) $CH_3—(CH_2)_4—C({=}O)—OH$ and $HO—CH_2—CH_3$

(b) $C_6H_5—C({=}O)—OH$ and $HO—CH_2—CH_2—CH_3$

(c) $CH_3—CH_2—OH$ and $HO—C({=}O)—CH_2—CH_2—C_6H_5$

15.55(a) $CH_3—CH_2—OH$

$CH_3—CH_2—C({=}O)—O—CH_2—CH_3$

(b) $CH_3—CH_2—CH(C{\equiv}N)—CH_3$

$CH_3—CH_2—CH(C(=O)OH)—CH_3$

15.59 addition reactions and condensation reactions **15.61** Dispersion forces strongly attract the long, unbranched chains of high-density polyethylene (HDPE). Low-density polyethylene (LDPE) has branching in the chains that prevents packing and weakens the attractions. **15.63** An amine and a carboxylic acid react to form a nylon; a carboxylic acid and an alcohol form a polyester.

15.64(a) $\left[\begin{array}{cc} \mathrm{H} & \mathrm{H} \\ | & | \\ -\mathrm{C} - & \mathrm{C}- \\ | & | \\ \mathrm{H} & \mathrm{Cl} \end{array}\right]_n$ (b) $\left[\begin{array}{cc} \mathrm{H} & \mathrm{H} \\ | & | \\ -\mathrm{C} - & \mathrm{C}- \\ | & | \\ \mathrm{H} & \mathrm{CH_3} \end{array}\right]_n$

15.66 $n\mathrm{HO{-}\overset{O}{\overset{\|}{C}}{-}C_6H_4{-}\overset{O}{\overset{\|}{C}}{-}OH} + n\mathrm{HO{-}CH_2{-}CH_2{-}OH} \longrightarrow$

$\mathrm{HO}\left[\mathrm{\overset{O}{\overset{\|}{C}}{-}C_6H_4{-}\overset{O}{\overset{\|}{C}}{-}O{-}CH_2{-}CH_2{-}O}\right]_n\mathrm{H} + 2n - 1\mathrm{H_2O}$

15.68(a) condensation (b) addition (c) condensation (d) condensation **15.70** The amino acid sequence in a protein determines its shape and structure, which determine its function.

15.72(a) $\mathrm{CH_3}$– (b) imidazolium ring (HN, $\overset{+}{\mathrm{N}}\mathrm{H}$) –$\mathrm{CH_2}$– (c) $\mathrm{CH_3{-}S{-}CH_2{-}CH_2}$–

15.74(a) $\mathrm{H_3\overset{+}{N}{-}CH(CH_2{-}C(=O){-}O^-){-}C(=O){-}NH{-}CH(CH_2{-}imidazolium){-}C(=O){-}NH{-}CH(CH_2{-}indolyl){-}C(=O){-}O^-}$

(b) $\mathrm{H_3\overset{+}{N}{-}CH_2{-}C(=O){-}NH{-}CH(CH_2{-}SH){-}C(=O){-}NH{-}CH(CH_2{-}C_6H_4{-}OH){-}C(=O){-}O^-}$

15.76(a) AATCGG (b) TCTGTA **15.78**(a) Both R groups are from cysteine, which can form a disulfide bond (covalent bond). (b) Lysine and aspartic acid give a salt link. (c) Asparagine and serine will hydrogen bond. (d) Valine and phenylalanine interact through dispersion forces.

15.80 $\mathrm{CH_3{-}CH{=}CH{-}CH_3}$

15.81(a) Structure with labeled groups: alkene ($\mathrm{CH_2{=}CH}$, carbon 1), carbon 2 (CH), carbon 3 (CH) bonded to $\mathrm{CH_3{-}C{-}CH_3}$, ester ($\mathrm{C(=O){-}O}$, carbon 4), carbon 5 (CH), alkene ($\mathrm{C{=}C}$, carbon 6, with $\mathrm{CH_3}$), ester ($\mathrm{C(=O){-}O{-}CH_3}$), ketone ($\mathrm{CH_2{-}C{=}O}$, carbon 7).

(b) Carbon 1 is sp^2 hybridized. Carbon 2 is sp^3 hybridized. Carbon 3 is sp^3 hybridized. Carbon 4 is sp^2 hybridized. Carbon 5 is sp^3 hybridized. Carbons 6 and 7 are sp^2 hybridized.

(c) Carbons 2, 3, and 5 are chiral centers, as they are each bonded to four different groups.

Chapter 16

• **16.2** Reaction rate is proportional to concentration. An increase in pressure will increase the concentration, resulting in an increased rate. **16.3** The addition of water will lower the concentrations of all dissolved solutes, and the rate will decrease. **16.5** An increase in temperature increases the rate by increasing the number of collisions between particles, but more importantly, it increases the energy of collisions. **16.8**(a) The instantaneous rate is the rate at one point in time during the reaction. The average rate is the average over a period of time. On a graph of reactant concentration vs. time, the instantaneous rate is the slope of the tangent to the curve at any point. The average rate is the slope of the line connecting two points on the curve. The closer together the two points (the shorter the time interval), the closer the average rate is to the instantaneous rate. (b) The initial rate is the instantaneous rate at $t = 0$, that is, when reactants are mixed.

16.10

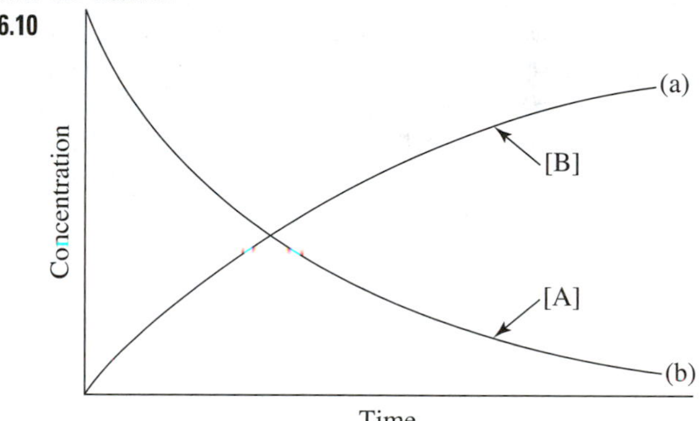

16.12(a) $\text{rate} = -\left(\frac{1}{2}\right)\frac{\Delta[\mathrm{AX_2}]}{\Delta t}$

$= -\left(\frac{1}{2}\right)\frac{(0.0088\ M - 0.0500\ M)}{(20.0\ \mathrm{s} - 0\ \mathrm{s})}$

$= 0.0010\ M/\mathrm{s}$

(b)

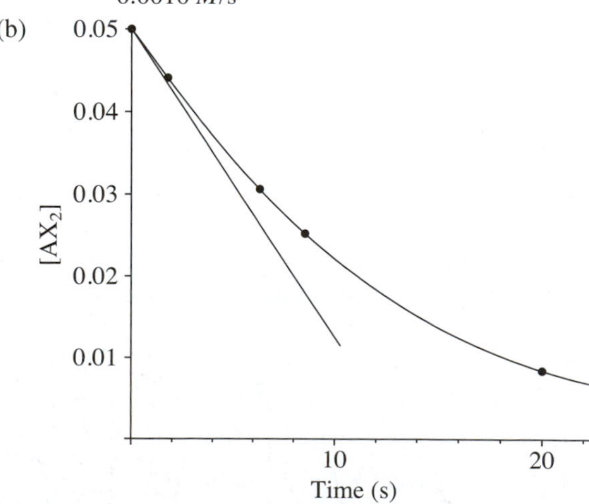

The initial rate is higher than the average rate because the rate decreases as reactant concentration decreases.

16.14 $\text{rate} = -\frac{\Delta[\mathrm{A}]}{\Delta t} = -\frac{1}{2}\frac{\Delta[\mathrm{B}]}{\Delta t} = \frac{\Delta[\mathrm{C}]}{\Delta t}$; 0.2 mol/L·s

16.16 $2\mathrm{N_2O_5}(g) \longrightarrow 4\mathrm{NO_2}(g) + \mathrm{O_2}(g)$

16.19(a) $\text{rate} = -\frac{1}{3}\frac{\Delta[\mathrm{O_2}]}{\Delta t} = \frac{1}{2}\frac{[\Delta\mathrm{O_3}]}{\Delta t}$

(b) 1.45×10^{-5} mol/L·s

16.20(a) k is the rate constant, the proportionality constant in the rate law; it is reaction and temperature specific. (b) m represents the order of the reaction with respect to [A], and n represents the order of the reaction with respect to [B]. The order of a reactant does not necessarily equal its stoichiometric coefficient in the balanced equation. (c) $L^2/mol^2{\cdot}min$ **16.21**(a) Rate doubles. (b) Rate decreases by a factor of 4. (c) Rate increases by a factor of 9. **16.22** first order in BrO_3^-; first order in Br^-; second order in H^+; fourth order overall **16.24**(a) Rate doubles. (b) Rate is halved. (c) The rate increases by a factor of 16. **16.26**(a) second order in A; first order in B (b) rate $= k[A]^2[B]$ (c) $5.00{\times}10^3\ L^2/mol^2{\cdot}min$ **16.29**(a) first order (b) second order (c) zero order **16.31** 7 s **16.33**(a) $k = 0.0660\ min^{-1}$ (b) 21.0 min **16.36** No, other factors that affect the rate are the energy and orientation of the collisions. **16.39** Measure the rate constant at a series of temperatures and plot ln k versus $1/T$. The slope of the line equals $-E_a/R$. **16.42** At the same temperature, both reaction mixtures have the same average kinetic energy, but the reactant molecules do not have the same average velocity. The trimethylamine molecule has greater mass than the ammonia molecule, so trimethylamine molecules will collide less often with HCl. Moreover, the bulky groups bonded to nitrogen in trimethylamine mean that collisions with HCl having the correct orientation occur less frequently. Therefore, the rate of the reaction between NH_3 and HCl is higher. **16.43** 12 unique collisions **16.45** $2.96{\times}10^{-18}$ **16.47** $0.033\ s^{-1}$

16.49(a)

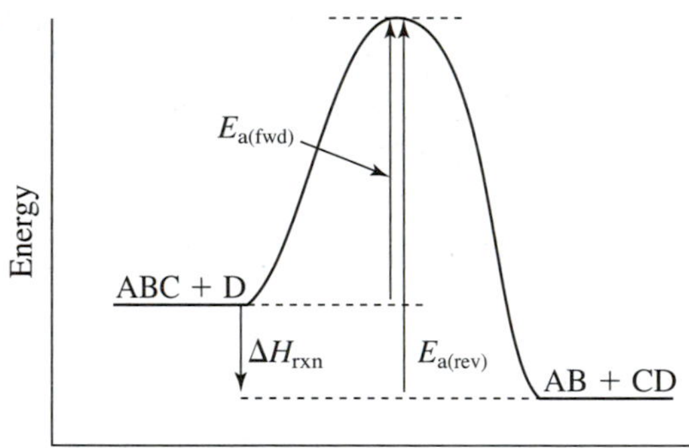

(b) $2.70{\times}10^2$ kJ/mol

(c) A—B••••C••••D (bond forming; bond breaking)

16.52(a) Because the enthalpy change is positive, the reaction is endothermic.

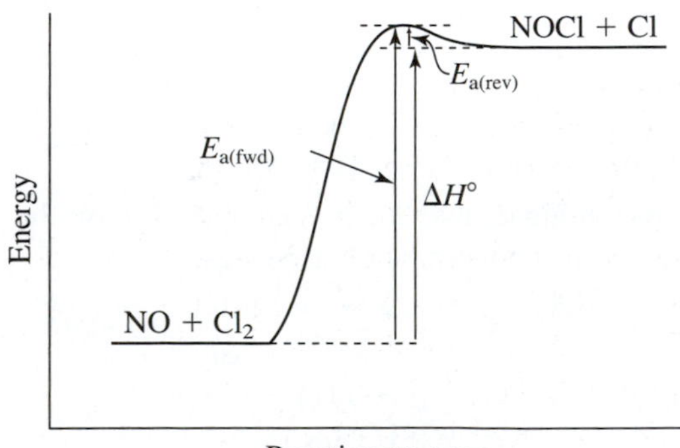

(b) 3 kJ (c) :Cl̤····Cl̤····N̈=O:

16.53 The rate of an overall reaction depends on the rate of the slowest step. The rate of the overall reaction will be lower than the average of the individual rates because the average includes higher rates as well. **16.57** The probability of three particles colliding with one another with the proper energy and orientation is much less than the probability for two particles.
16.58 No, the overall rate law must contain only reactants (no intermediates), and the overall rate is determined by the slow step.

16.59(a) $CO_2(aq) + 2OH^-(aq) \longrightarrow CO_3^{2-}(aq) + H_2O(l)$
(b) $HCO_3^-(aq)$
(c) (1) molecularity = 2; rate$_1 = k_1[CO_2][OH^-]$
(2) molecularity = 2; rate$_2 = k_2[OH_3^-][OH^-]$
(d) Yes

16.61(a) $A(g) + B(g) + C(g) \longrightarrow D(g)$
(b) X and Y are intermediates.
(c)

Step	Molecularity	Rate Law
$A(g) + B(g) \rightleftharpoons X(g)$	bimolecular	rate$_1 = k_1[A][B]$
$X(g) + C(g) \longrightarrow Y(g)$	bimolecular	rate$_2 = k_2[X][C]$
$Y(g) \longrightarrow D(g)$	unimolecular	rate$_3 = k_3[Y]$

(d) yes (e) yes **16.63** The proposed mechanism is valid because the individual steps are chemically reasonable, they add to give the overall equation, and the rate law for the mechanism matches the observed rate law. **16.66** No. A catalyst changes the mechanism of a reaction to one with lower activation energy. Lower activation energy means a faster reaction. An increase in temperature does not influence the activation energy, but increases the fraction of collisions with sufficient energy to equal or exceed the activation energy. **16.69** $4.61{\times}10^4$ J/mol **16.72**(a) Rate increases 2.5 times. (b) Rate is halved. (c) Rate decreases to 0.01 of the original rate. (d) Rate does not change. **16.75** 57 yr **16.76**(a) $0.21\ h^{-1}$; 3.3 h (b) 6.6 h (c) If the concentration of sucrose is relatively low, the concentration of water remains nearly constant even with small changes in the amount of water. This gives an apparent zero-order reaction with respect to water. Thus, the reaction is first order overall because the rate does not change with changes in the amount of water. **16.82**(a) $2.4{\times}10^{-15}\ M$ (b) $2.4{\times}10^{-11}$ mol/L·s **16.83** $7.3{\times}10^3$ J/mol

Chapter 17

• **17.1** If the change is one of concentrations, it results temporarily in more products and less reactants. After equilibrium is re-established, K_c remains unchanged because the ratio of products and reactants remains the same. **17.7** The equilibrium constant expression is $K = [O_2]$. If the temperature remains constant, K remains constant. If the initial amount of Li_2O_2 present is sufficient to reach equilibrium, the amount of O_2 obtained will be constant.

17.8(a) $Q = \frac{[HI]^2}{[H_2][I_2]}$

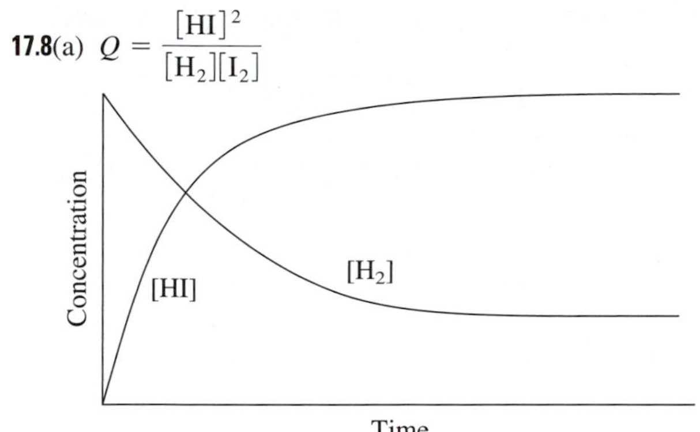

The value of Q increases as a function of time until it reaches the value of K. (b) no **17.11** Yes. If Q_1 is for the formation of 1 mol NH_3 from H_2 and N_2, and Q_2 is for the formation of NH_3 from H_2 and 1 mol of N_2, then $Q_2 = Q_1{}^2$.
17.12(a) $4NO(g) + O_2(g) \rightleftharpoons 2N_2O_3(g)$;

$$Q_c = \frac{[N_2O_3]^2}{[NO]^4[O_2]}$$

(b) $SF_6(g) + 2SO_3(g) \rightleftharpoons 3SO_2F_2(g)$;

$$Q_c = \frac{[SO_2F_2]^3}{[SF_6][SO_3]^2}$$

(c) $2SClF_5(g) + H_2(g) \rightleftharpoons S_2F_{10}(g) + 2HCl(g)$;

$$Q_c = \frac{[S_2F_{10}][HCl]^2}{[SClF_5]^2[H_2]}$$

17.14(a) 7.9 (b) 3.2×10^{-5}
17.16(a) $2Na_2O_2(s) + 2CO_2(g) \rightleftharpoons 2Na_2CO_3(s) + O_2(g)$;

$$Q_c = \frac{[O_2]}{[CO_2]^2}$$

(b) $H_2O(l) \rightleftharpoons H_2O(g)$; $Q_c = [H_2O(g)]$
(c) $NH_4Cl(s) \rightleftharpoons NH_3(g) + HCl(g)$; $Q_c = [NH_3][HCl]$
17.19(a)

(1)	$Cl_2(g) + F_2(g) \rightleftharpoons 2ClF(g)$
(2)	$2ClF(g) + 2F_2(g) \rightleftharpoons 2ClF_3(g)$
overall:	$Cl_2(g) + 3F_2(g) \rightleftharpoons 2ClF_3(g)$

(b) $$Q_{overall} = Q_1Q_2 = \frac{\cancel{[ClF]^2}}{[Cl_2][F_2]} \times \frac{[ClF_3]^2}{\cancel{[ClF]^2}[F_2]^2} = \frac{[ClF_3]^2}{[Cl_2][F_2]^3}$$

17.21 K_c and K_p are equal when $\Delta n_{gas} = 0$. **17.22**(a) smaller (b) Assuming that $RT > 1$ ($T > 12.2$ K), $K_p > K_c$ if there are more moles of products than reactants at equilibrium, and $K_p < K_c$ if there are more moles of reactants than products. **17.23**(a) 3 (b) −1 (c) 3 **17.25**(a) 3.2 (b) 28.5 **17.27** The reaction quotient (Q) and equilibrium constant (K) are determined by the ratio [products]/[reactants]. When $Q < K$, the reaction proceeds to the right to form more products. **17.29** no; to the left **17.33**(a) The approximation applies when the change in concentration from initial concentration to equilibrium concentration is so small that it is insignificant; this occurs when K is small and initial concentration is large. (b) This approximation should not be used when the change in concentration is greater than 5%. This can occur when $[\text{reactant}]_{\text{initial}}$ is very small or when change in [reactant] is relatively large due to a large K. **17.34** 50.8

17.36

Concentration (M)	$PCl_5(g)$	$\rightleftharpoons$	$PCl_3(g)$ +	$Cl_2(g)$
Initial	0.075		0	0
Change	$-x$		$+x$	$+x$
Equilibrium	$0.075 - x$		x	x

17.38 28 atm **17.40** 0.33 atm **17.42** 3.5×10^{-3} M **17.44** $[I_2]_{eq} = [Cl_2]_{eq} = 0.0199$ M; $[ICl]_{eq} = 0.0601$ M **17.46** 6.01×10^{-6} **17.48** Equilibrium position refers to the specific concentrations or pressures of reactants and products that exist at equilibrium, whereas equilibrium constant is the overall ratio of equilibrium concentrations or pressures. Equilibrium position changes as a result of a change in reactant and product concentrations. **17.49**(a) B, because the amount of product increases with temperature (b) A, because the lowest temperature will give the least product **17.52** (a) An exothermic reaction can be written as: reactants $\rightleftharpoons$ products + heat. A rise in temperature (increase in heat) shifts the equilibrium position to the left to absorb the added heat. Since K = (products)/(reactants), the addition of heat increases the denominator and decreases the numerator, making K_2 smaller than K_1. Since $K = k_f/k_r$ and an increase in temperature increases k_r to a greater extent, the value of K is lower at the increased temperature. (b) An endothermic reaction can be written as: reactants + heat $\rightleftharpoons$ products. A rise in temperature (increase in heat) favors the forward direction of the reaction, i.e., the formation of products and consumption of reactants. Since K = [products]/[reactants], the addition of heat increases the numerator and decreases the denominator, making K_2 larger than K_1. For an endothermic reaction, k_f increases more than k_r with an increase in temperature and $K_2 = k_f/k_r$ is larger than K_1. **17.54**(a) shifts toward products (b) shifts toward products (c) does not shift (d) shifts toward reactants **17.56**(a) more F and less F_2 (b) more C_2H_2 and H_2 and less CH_4 **17.58**(a) no change (b) increase volume **17.60**(a) amount decreases (b) amounts increase (c) amounts increase (d) amount decreases **17.63**(a) lower temperature, higher pressure (b) Q decreases; no change in K (c) Reaction rates are lower at lower temperatures, so a catalyst is used to speed up the reaction. **17.64**(a) $P_{N_2} = 31$ atm; $P_{H_2} = 93$ atm; $P_{total} = 174$ atm (b) $P_{N_2} = 18$ atm; $P_{H_2} = 111$ atm; $P_{total} = 179$ atm; not a valid argument **17.65**(a) 3×10^{-3} atm (b) high pressure; low temperature (c) No, because water condenses at a higher temperature. **17.69**(a) 0.016 atm (b) $K_c = 5.6\times10^2$; $P_{O_2} = 0.16$ atm **17.70** 12.5 g $CaCO_3$ **17.73**(a) 3.0×10^{-14} atm (b) 0.013 pg CO/L **17.78**(a) 98.0% (b) 99.0% (c) 2.60×10^5 J/mol **17.79**(a) $2CH_4(g) + O_2(g) + 2H_2O(g) \rightleftharpoons 2CO_2(g) + 6H_2(g)$ (b) 1.76×10^{29} (c) 3.19×10^{23} (d) 48 atm **17.80**(a) 4.0×10^{-21} atm (b) 5.5×10^{-8} atm (c) 29 N atoms/L; 4.0×10^{14} H atoms/L (d) The more reasonable step is $N_2(g) + H(g) \longrightarrow NH(g) + N(g)$. With only 29 N atoms in 1.0 L, the first reaction would produce virtually no $NH(g)$ molecules. There are many orders of magnitude more N_2 molecules than N atoms, so the second reaction is the more reasonable step. **17.83**(a) $P_{N_2} = 0.780$ atm; $P_{O_2} = 0.210$ atm; $P_{NO} = 2.67\times10^{-16}$ atm (b) 0.990 atm (c) $K_c = K_p = 4.35\times10^{-31}$ **17.87**(a) 1.52 (b) 0.9626 atm (c) 0.2000 mol CO (d) 0.01093 M

Chapter 18

18.2 All Arrhenius acids have H in their formula and produce hydronium ion (H_3O^+) in aqueous solution. All Arrhenius bases have OH in their formula and produce hydroxide ion (OH^-) in aqueous solution. Neutralization occurs when each H_3O^+ ion combines with an OH^- ion to form two molecules of H_2O. Chemists found the reaction of any strong base with any strong acid always had a ΔH of -56 kJ/mol H_2O produced.
18.4 Strong acids and bases dissociate completely into ions when dissolved in water. Weak acids and bases dissociate only partially. The characteristic property of all weak acids is that a great majority of the acid molecules are undissociated.
18.5 a, c, and d

18.7(a) $K_a = \dfrac{[NO_2^-][H_3O^+]}{[HNO_2]}$

(b) $K_a = \dfrac{[CH_3COO^-][H_3O^+]}{[CH_3COOH]}$

(c) $K_a = \dfrac{[BrO_2^-][H_3O^+]}{[HBrO_2]}$

18.9 $CH_3COOH < HF < HIO_3 < HI$ **18.11**(a) weak acid (b) strong base (c) weak acid (d) strong acid **18.15**(a) The acid with the smaller K_a (4×10^{-5}) has the higher pH, because less dissociation yields fewer hydronium ions. (b) The acid with the larger pK_a (3.5) has the higher pH, because it has a smaller K_a and, thus, lower $[H_3O^+]$. (c) Lower concentration (0.01 *M*) contains fewer hydronium ions. (d) A 0.1 *M* weak acid solution contains fewer hydronium ions. (e) The 0.01 *M* base solution contains more hydroxide ions, so fewer hydronium ions. (f) The solution that has pOH = 6.0 has the higher pH: pH = 14.0 − 6.0 = 8.0.
18.16(a) 12.05; basic (b) 11.13; acidic
18.18(a) $[H_3O^+] = 1.4\times10^{-10}\ M, [OH^-] = 7.1\times10^{-5}\ M,$ pOH = 4.15

(b) $[H_3O^+] = 2.7\times10^{-5}\ M, [OH^-] = 3.7\times10^{-10}\ M,$ pH = 4.57

18.20 1.4×10^{-4} mol OH^-
18.24(a) Rising temperature increases the value of K_w. (b) $K_w = 2.5\times10^{-14}$; pOH = 6.80; $[OH^-] = 1.6\times10^{-7}\ M$
18.25 The Brønsted-Lowry theory defines acids as proton donors and bases as proton acceptors, while the Arrhenius definition looks at acids as containing ionizable hydrogen atoms and at bases as containing hydroxide ions. In both definitions, an acid produces H_3O^+ ions and a base produces OH^- ions when added to water. Ammonia and carbonate ion are two Brønsted-Lowry bases that are not Arrhenius bases because they do not contain OH^- ions. Brønsted-Lowry acids must contain an ionizable hydrogen atom in order to be proton donors, so a Brønsted-Lowry acid is also an Arrhenius acid. **18.28** An amphiprotic species is one that can lose a proton to act as an acid or gain a proton to act as a base. The dihydrogen phosphate ion, $H_2PO_4^-$, is an example.

$H_2PO_4^-(aq) + OH^-(aq) \longrightarrow H_2O(l) + HPO_4^{2-}(aq)$
$H_2PO_4^-(aq) + HCl(aq) \longrightarrow H_3PO_4(aq) + Cl^-(aq)$

18.29(a) Cl^- (b) HCO_3^- (c) OH^- **18.31**(a) NH_4^+ (b) NH_3 (c) $C_{10}H_{14}N_2H^+$
18.33(a) $\underset{\text{base}}{NH_3} + \underset{\text{acid}}{H_3PO_4} \rightleftharpoons \underset{\text{acid}}{NH_4^+} + \underset{\text{base}}{H_2PO_4^-}$

Conjugate acid-base pairs: $H_3PO_4/H_2PO_4^-$ and NH_4^+/NH_3

(b) $\underset{\text{base}}{CH_3O^-} + \underset{\text{acid}}{NH_3} \rightleftharpoons \underset{\text{acid}}{CH_3OH} + \underset{\text{base}}{NH_2^-}$

Conjugate acid-base pairs: NH_3/NH_2^- and CH_3OH/CH_3O^-

(c) $\underset{\text{base}}{HPO_4^{2-}} + \underset{\text{acid}}{HSO_4^-} \rightleftharpoons \underset{\text{acid}}{H_2PO_4^-} + \underset{\text{base}}{SO_4^{2-}}$

Conjugate acid-base pairs: HSO_4^-/SO_4^{2-} and $H_2PO_4^-/HPO_4^{2-}$

18.35(a) $OH^-(aq) + H_2PO_4^-(aq) \rightleftharpoons H_2O(l) + HPO_4^{2-}(aq)$

Conjugate acid-base pairs: $H_2PO_4^-/HPO_4^{2-}$ and H_2O/OH^-

(b) $HSO_4^-(aq) + CO_3^{2-}(aq) \rightleftharpoons SO_4^{2-}(aq) + HCO_3^-(aq)$

Conjugate acid-base pairs: HSO_4^-/SO_4^{2-} and HCO_3^-/CO_3^{2-}

18.37 $K_c > 1$: $HS^- + HCl \rightleftharpoons H_2S + Cl^-$
$K_c < 1$: $H_2S + Cl^- \rightleftharpoons HS^- + HCl$
18.39 $K_c > 1$ for both a and b
18.41(a) A strong acid is 100% dissociated, so the acid concentration will be very different after dissociation. (b) A weak acid dissociates to a very small extent, so the acid concentration before and after dissociation is nearly the same. (c) same as (b), but with the extent of dissociation greater. (d) same as (a).
18.44 1.5×10^{-5} **18.46** $[H_3O^+] = [NO_2^-] = 2.1\times10^{-2}\ M$; $[OH^-] = 4.8\times10^{-13}\ M$ **18.48** $[H_3O^+] = [ClCH_2COO^-] = 0.041\ M$; $[ClCH_2COOH] = 1.21\ M$; pH = 1.39
18.50(a) $[H_3O^+] = 6.0\times10^{-3}\ M$; pH = 2.22; $[OH^-] = 1.7\times10^{-12}\ M$; pOH = 11.78 (b) 1.9×10^{-4} **18.52**(a) 2.37 (b) 11.51 **18.55** 1.5% **18.56** All Brønsted-Lowry bases contain at least one lone pair of electrons, which binds an H^+ and allows the base to act as a proton acceptor.
18.59(a) $C_5H_5N(aq) + H_2O(l) \rightleftharpoons OH^-(aq) + C_5H_5NH^+(aq)$

$K_b = \dfrac{[C_5H_5NH^+][OH^-]}{[C_5H_5N]}$

(b) $CO_3^{2-}(aq) + H_2O(l) \rightleftharpoons OH^-(aq) + HCO_3^-(aq)$

$K_b = \dfrac{[HCO_3^-][OH^-]}{[CO_3^{2-}]}$

18.61 11.79 **18.63**(a) 12.04 (b) 10.77 **18.65**(a) 11.19 (b) 5.56 **18.67** $[OH^-] = 5.5\times10^{-4}\ M$; pH = 10.74 **18.69** As a nonmetal becomes more electronegative, the acidity of its binary hydride increases. The electronegative nonmetal attracts the electrons more strongly in the polar bond, shifting the electron density away from H, thus making H^+ more easily transferred to a water molecule to form H_3O^+. **18.72** Chlorine is more electronegative than iodine, and $HClO_4$ has more oxygen atoms than HIO. **18.73**(a) H_2Se (b) $B(OH)_3$ (c) $HBrO_2$ **18.75**(a) 0.5 *M* $AlBr_3$ (b) 0.3 *M* $SnCl_2$ **18.78** NaF contains the anion of the weak acid HF, so F^- acts as a base. NaCl contains the anion of the strong acid HCl.
18.80(a) $KBr(s) \xrightarrow{H_2O} K^+(aq) + Br^-(aq)$; neutral

(b) $NH_4I(s) \xrightarrow{H_2O} NH_4^+(aq) + I^-(aq)$
$NH_4^+(aq) + H_2O(l) \rightleftharpoons NH_3(aq) + H_3O^+(aq)$; acidic
(c) $KCN(s) \xrightarrow{H_2O} K^+(aq) + CN^-(aq)$
$CN^-(aq) + H_2O(l) \rightleftharpoons HCN(aq) + OH^-(aq)$; basic
18.82(a) $Fe(NO_3)_2 < KNO_3 < K_2SO_3 < K_2S$
(b) $NaHSO_4 < NH_4NO_3 < NaHCO_3 < Na_2CO_3$
18.85 A Lewis acid is an electron-pair acceptor while a Brønsted-Lowry acid is a proton donor. The proton of a Brønsted-Lowry acid is a Lewis acid because it accepts an electron pair when it bonds with a base. All Lewis acids are not Brønsted-Lowry acids. A Lewis base is an electron-pair donor and a Brønsted-Lowry base is a proton acceptor. All Brønsted-Lowry bases are Lewis bases, and vice versa. **18.86**(a) No, $Ni(H_2O)_6^{2+}(aq) + 6NH_3(aq) \rightleftharpoons Ni(NH_3)_6^{2+} + 6H_2O(l)$; NH_3 is a weak Brønsted-Lowry base, but a strong Lewis base. (b) cyanide ion and water (c) cyanide ion **18.89**(a) Lewis acid (b) Lewis base (c) Lewis acid (d) Lewis base **18.91**(a) Lewis acid: Na^+; Lewis base: H_2O (b) Lewis acid: CO_2; Lewis base: H_2O (c) Lewis acid: BF_3; Lewis base: F^- **18.93**(a) Lewis (b) Brønsted-Lowry and Lewis (c) none (d) Lewis **18.95** 3.5×10^{-8} to 4.5×10^{-8} M H_3O^+; 5.2×10^{-7} to 6.6×10^{-7} M OH^- **18.97**(a) $SnCl_4$ is the Lewis acid; $(CH_3)_3N$ is the Lewis base (b) 5*d* **18.99** pH = 5.00, 6.00, 6.79, 6.98, 7.00 **18.101** H_3PO_4 **18.105** 1.47×10^{-3} **18.107**(a) Ca^{2+} does not react with water; $CH_3CH_2COO^-(aq) + H_2O(l) \rightleftharpoons CH_3CH_2COOH(aq) + OH^-(aq)$; basic (b) 9.08 **18.115**(a) The concentration of oxygen is higher in the lungs, so the equilibrium shifts to the right. (b) In an oxygen-deficient environment, the equilibrium shifts to the left to release oxygen. (c) A decrease in $[H_3O^+]$ shifts the equilibrium to the right. More oxygen is absorbed, but it will be more difficult to remove the O_2. (d) An increase in $[H_3O^+]$ shifts the equilibrium to the left. Less oxygen is bound to Hb, but it will be easier to remove it. **18.117**(a) 10.0 (b) The pK_b for the 3° amine group is much smaller than that for the aromatic ring, thus the K_b is significantly larger (yielding a much greater amount of OH^-). (c) 4.7 (d) 5.1

Chapter 19

• **19.2** The acid component neutralizes added base, and the base component neutralizes added acid, so the pH of the buffer solution remains relatively constant. The components of a buffer do not neutralize one another because they are a conjugate acid-base pair. **19.5** The pH of a buffer decreases only slightly with added H_3O^+. **19.8** The buffer range, the pH over which the buffer acts effectively, is greatest when the buffer-component concentration ratio is 1; the range decreases as this ratio deviates from 1. **19.10** $[H_3O^+] = 5.6\times10^{-6}$ M; pH = 5.25 **19.12** 10.03 **19.14** 9.47 **19.16** 3.6 **19.18** 3.37 **19.20**(a) 4.81 (b) 0.66 g KOH **19.22**(a) $HOOC(CH_2)_4COOH/HOOC(CH_2)_4COO^-$ or $C_6H_5NH_3^+/C_6H_5NH_2$ (b) $H_2PO_4^-/HPO_4^{2-}$ or $H_2AsO_4^-/HAsO_4^{2-}$ **19.25** 1.6 **19.27** To see a distinct color in a mixture of two colors, you need one to have about 10 times the intensity of the other. For this to be the case, the concentration ratio $[HIn]/[In^-]$ has to be greater than 10/1 or less than 1/10. This occurs when pH = $pK_a - 1$ or pH = $pK_a + 1$, respectively, giving a pH range of about 2 units. **19.29** The equivalence point in a titration is the point at which the number of moles of base is stoichiometrically equivalent to the number of moles of acid. The endpoint is the point at which the added indicator changes color. If an appropriate indicator is selected, the endpoint is close to the equivalence point, but they are usually not the same. The pH at the endpoint, or color change, may precede or follow the pH at the equivalence point, depending on the indicator chosen.
19.32(a) initial pH: *strong acid*–strong base < *weak acid*–strong base < *weak base*–strong acid (b) pH at equivalence point: *weak base*–strong acid < *strong acid*–strong base < *weak acid*–strong base **19.34** At the center of the buffer region, the concentrations of weak acid and conjugate base are equal, so the pH = pK_a of the acid. **19.35** pH range from 7.5 to 9.5 **19.37**(a) bromthymol blue (b) thymol blue or phenolphthalein **19.39**(a) 1.00 (b) 1.64 (c) 2.90 (d) 3.90 (e) 7.00 (f) 10.10 (g) 12.05 **19.41**(a) 2.91 (b) 4.81 (c) 5.29 (d) 6.09 (e) 7.41 (f) 8.76 (g) 10.10 (h) 12.05 **19.43**(a) 8.54 and 59.0 mL (b) 7.13 and 66.0 mL, 9.69 and total 132.1 mL **19.46** Fluoride ion is the conjugate base of a weak acid and reacts with H_2O: $F^-(aq) + H_2O(l) \rightleftharpoons HF(aq) + OH^-(aq)$. As the pH increases, the equilibrium shifts to the left and $[F^-]$ increases. As the pH decreases, the equilibrium shifts to the right and $[F^-]$ decreases. The changes in $[F^-]$ influence the solubility of BaF_2. Chloride ion is the conjugate base of a strong acid, so it does not react with water and its concentration is not influenced by pH. **19.47** The compound precipitates. **19.48**(a) $K_{sp} = [Ag^+]^2[CO_3^{2-}]$ (b) $K_{sp} = [Ba^{2+}][F^-]^2$ (c) $K_{sp} = [Cu^{2+}][HS^-][OH^-]$ **19.50** 1.3×10^{-4} **19.52** 2.8×10^{-11} **19.54**(a) 2.3×10^{-5} M (b) 4.2×10^{-9} M **19.56**(a) 1.7×10^{-3} M (b) 2.0×10^{-4} M **19.58**(a) $Mg(OH)_2$ (b) PbS (c) Ag_2SO_4
19.60(a) $AgCl(s) \rightleftharpoons Ag^+(aq) + Cl^-(aq)$. The chloride ion is the anion of a strong acid, so it does not react with H_3O^+. No change with pH. (b) $SrCO_3(s) \rightleftharpoons Sr^{2+}(aq) + CO_3^{2-}(aq)$. The strontium ion is the cation of a strong base, so pH will not affect its solubility. The carbonate ion acts as a base: $CO_3^{2-}(aq) + H_2O(l) \rightleftharpoons HCO_3^-(aq) + OH^-(aq)$; also $CO_2(g)$ forms and escapes: $CO_3^{2-}(aq) + 2H_3O^+(aq) \longrightarrow CO_2(g) + 3H_2O(l)$. Therefore, the solubility of $SrCO_3$ will increase with addition of H_3O^+ (decreasing pH).
19.62 yes **19.67** No, because it indicates that a complex ion forms between the lead ion and hydroxide ions:
$Pb^{2+}(aq) + nOH^-(aq) \rightleftharpoons Pb(OH)_n^{2-n}(aq)$
19.68 $Hg(H_2O)_4^{2+}(aq) + 4CN^-(aq) \rightleftharpoons Hg(CN)_4^{2-}(aq) + 4H_2O(l)$
19.70 $Ag(H_2O)_2^+(aq) + 2S_2O_3^{2-}(aq) \rightleftharpoons Ag(S_2O_3)_2^{3-}(aq) + 2H_2O(l)$
19.72 0.05 M **19.77** 1.3×10^{-4} M **19.79**(a) 14 (b) 1 g from the second beaker **19.81**(a) 0.088 (b) 0.14

19.85(a)

V (mL)	pH	$\Delta pH/\Delta V$	V_{avg} (mL)
0.00	1.00		
10.00	1.22	0.022	5.00
20.00	1.48	0.026	15.00
30.00	1.85	0.037	25.00
35.00	2.18	0.066	32.50
39.00	2.89	0.18	37.00
39.50	3.20	0.62	39.25
39.75	3.50	1.2	39.63
39.90	3.90	2.7	39.83
39.95	4.20	6	39.93
39.99	4.90	18	39.97
V (mL)	**pH**	**$\Delta pH/\Delta V$**	**V_{avg} (mL)**
40.00	7.00	200	40.00
40.01	9.40	200	40.01
40.05	9.80	10	40.03
40.10	10.40	10	40.08
40.25	10.50	0.67	40.18
40.50	10.79	1.2	40.38
41.00	11.09	0.60	40.75
45.00	11.76	0.17	43.00
50.00	12.05	0.058	47.50
60.00	12.30	0.025	55.00
70.00	12.43	0.013	65.00
80.00	12.52	0.009	75.00

(b)

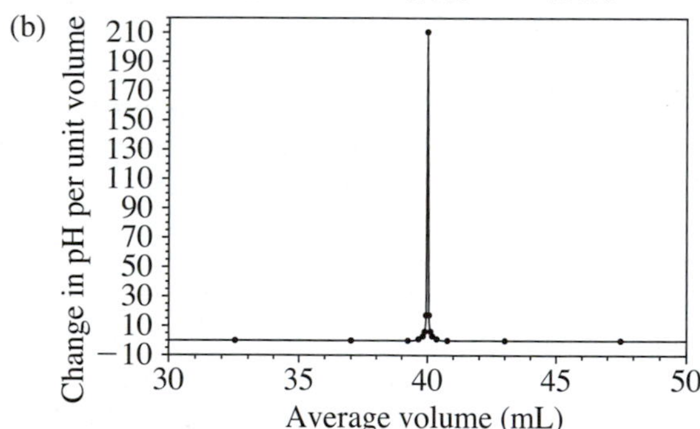

Maximum slope (equivalence point) is at V_{avg} = 40.00 mL.

19.88 $K_b = \dfrac{[BH^+][OH^-]}{[B]}$

Rearranging to isolate $[OH^-]$: $[OH^-] = K_b \dfrac{[B]}{[BH^+]}$

Taking the negative log:

$$-\log [OH^-] = -\log K_b - \log \frac{[B]}{[BH^+]}$$

Therefore, $pOH = pK_b + \log \dfrac{[BH^+]}{[B]}$

19.96(a) 65 mol (b) 6.28 (c) 4.0×10^3 g

19.99(a) A and D (b) pH_A = 4.35; pH_B = 8.67; pH_C = 2.67; pH_D = 4.57 (c) C, A, D, B (d) B

Chapter 20

• **20.2** A spontaneous process occurs by itself, whereas a nonspontaneous process requires a continuous input of energy to make it happen. It is possible to cause a nonspontaneous process to occur, but the process stops once the energy source is removed. A reaction that is nonspontaneous under one set of conditions may be spontaneous under a different set of conditions. **20.5** The transition from liquid to gas involves a greater increase in dispersal of energy and freedom of motion than does the transition from solid to liquid. **20.6** In an exothermic reaction, $\Delta S_{surr} > 0$. In an endothermic reaction, $\Delta S_{surr} < 0$. A chemical cold pack for injuries is an example of an application using a spontaneous endothermic process. **20.8** a, b, and c **20.10**(a) positive (b) negative (c) negative **20.12**(a) negative (b) negative (c) positive **20.14**(a) positive (b) negative (c) positive **20.16**(a) positive (b) negative (c) positive **20.18**(a) Butane. The double bond in 2-butene restricts freedom of rotation. (b) Xe(g). It has the greater molar mass. (c) $CH_4(g)$. Gases have greater entropy than liquids. **20.20**(a) Diamond < graphite < charcoal. Freedom of motion is least in the network solid; more freedom between graphite sheets; most freedom in amorphous solid. (b) Ice < liquid water < water vapor. Entropy increases as a substance changes from solid to liquid to gas. (c) O atoms < O_2 < O_3. Entropy increases with molecular complexity. **20.22**(a) $ClO_4^-(aq) > ClO_3^-(aq) > ClO_2^-(aq)$; decreasing molecular complexity (b) $NO_2(g) > NO(g) > N_2(g)$. N_2 has lower standard molar entropy because it consists of two of the same atoms; the other species have two different types of atoms. NO_2 is more complex than NO. (c) $Fe_3O_4(s) > Fe_2O_3(s) > Al_2O_3(s)$. Fe_3O_4 is more complex and more massive. Fe_2O_3 is more massive than Al_2O_3. **20.26** For a system at equilibrium, $\Delta S_{univ} = \Delta S_{sys} + \Delta S_{surr} = 0$. For a system moving to equilibrium, $\Delta S_{univ} > 0$.
20.27 $S^\circ_{Cl_2O(g)} = 2S^\circ_{HClO(g)} - S^\circ_{H_2O(g)} - \Delta S^\circ_{rxn}$
20.28(a) negative; $\Delta S^\circ = -172.4$ J/K (b) positive; ΔS° = 141.6 J/K (c) negative; $\Delta S^\circ = -837$ J/K **20.30** ΔS° = 93.1 J/K; yes, the positive sign of ΔS is expected because there is a net increase in the number of gas molecules. **20.32** −75.6 J/K **20.35** −97.2 J/K **20.37** A spontaneous process has $\Delta S_{univ} > 0$. Since the absolute temperature is always positive, ΔG_{sys} must be negative ($\Delta G_{sys} < 0$) for a spontaneous process. **20.39** ΔH° is positive and ΔS° is positive. Melting is an example. **20.40**(a) −1138.0 kJ (b) −1379.4 kJ (c) −224 kJ **20.42**(a) −1138 kJ (b) −1379 kJ (c) −226 kJ **20.44**(a) Entropy decreases (ΔS° is negative) because the number of moles of gas decreases. The combustion of CO releases energy (ΔH° is negative). (b) −257.2 kJ or −257.3 kJ, depending on the method **20.46**(a) ΔH°_{rxn} = 90.7 kJ; ΔS°_{rxn} = 221 J/K (b) At 28°C, ΔG° = 24.3 kJ; at 128°C, ΔG° = 2.2 kJ; at 228°C, $\Delta G^\circ = -19.9$ kJ (c) For the substances in their standard states, the reaction is nonspontaneous at 28°C, near equilibrium at 128°C, and spontaneous at 228°C. **20.48** ΔH° = 30910 J, ΔS° = 93.15 J/K, T = 331.8 K **20.50**(a) $\Delta H^\circ_{rxn} = -241.826$ kJ, $\Delta S^\circ_{rxn} = -44.4$ J/K, $\Delta G^\circ_{rxn} = -228.60$ kJ (b) Yes. The reaction will become nonspontaneous at higher temperatures. (c) The reaction is spontaneous below 5.45×10^3 K. **20.52**(a) ΔG° is a relatively large positive value. (b) $K >> 1$. Q depends on initial conditions, not equilibrium conditions. **20.55** The standard free energy change, ΔG°, applies when all components of the system are in their standard states; $\Delta G^\circ = \Delta G$ when all concentrations equal 1 M and all partial pressures equal 1 atm.

20.56(a) 1.7×10^6 (b) 3.89×10^{-34} (c) 1.26×10^{48} **20.58** 4.89×10^{-51} **20.60** 3.36×10^5 **20.62** 2.7×10^4 J/mol; no **20.64**(a) 2.9×10^4 J/mol (b) The reverse direction, formation of reactants, is spontaneous, so the reaction proceeds to the left. (c) 7.0×10^3 J/mol; the reaction proceeds to the left to reach equilibrium. **20.66**(a) no T (b) 163 kJ (c) 1×10^2 kJ/mol **20.69**(a) spontaneous (b) + (c) + (d) − (e) −, not spontaneous (f) − **20.71**(a) 2.3×10^2 (b) The treatment is to administer oxygen-rich air; increasing the concentration of oxygen shifts the equilibrium to the left, in the direction of Hb·O_2. **20.74**(a) $2N_2O_5(g) + 6F_2(g) \longrightarrow 4NF_3(g) + 5O_2(g)$ (b) $\Delta G^\circ_{rxn} = -569$ kJ (c) $\Delta G_{rxn} = -5.60\times10^2$ kJmol **20.78**(a) $\Delta H^\circ_{rxn} = 470.5$ kJ; $S^\circ_{rxn} = 558.4$ J/K (b) The reaction will be spontaneous at high T, because the $-T\Delta S$ term will be larger in magnitude than ΔH. (c) no (d) 842.5 K **20.80**(a) yes, negative Gibbs free energy (b) Yes. It becomes spontaneous at 270.8 K. (c) 234 K. The temperature is different because the ΔH and ΔS values for N_2O_5 vary with physical state. **20.84**(a) 465 K (b) 6.59×10^{-4} (c) The reaction rate is higher at the higher temperature. The shorter time required (kinetics) overshadows the lower yield (thermodynamics).

Chapter 21

• **21.1** Oxidation is the loss of electrons and results in a higher oxidation number; reduction is the gain of electrons and results in a lower oxidation number. **21.2** No, one half-reaction cannot take place independently because there is a transfer of electrons from one substance to another. If one substance loses electrons, another substance must gain them. **21.3** Spontaneous reactions, for which $\Delta G_{sys} < 0$, take place in voltaic cells (also called galvanic cells). Nonspontaneous reactions, for which $\Delta G_{sys} > 0$, take place in electrolytic cells. **21.5**(a) Cl^- (b) MnO_4^- (c) MnO_4^- (d) Cl^- (e) from Cl^- to MnO_4^- (f) $8H_2SO_4(aq) + 2KMnO_4(aq) + 10KCl(aq) \longrightarrow 2MnSO_4(aq) + 5Cl_2(g) + 8H_2O(l) + 6K_2SO_4(aq)$

21.7(a) $ClO_3^-(aq) + 6H^+(aq) + 6I^-(aq) \longrightarrow Cl^-(aq) + 3H_2O(l) + 3I_2(s)$
Oxidizing agent is ClO_3^- and reducing agent is I^-.
(b) $2MnO_4^-(aq) + H_2O(l) + 3SO_3^{2-}(aq) \longrightarrow 2MnO_2(s) + 3SO_4^{2-}(aq) + 2OH^-(aq)$
Oxidizing agent is MnO_4^- and reducing agent is SO_3^{2-}.
(c) $2MnO_4^-(aq) + 6H^+(aq) + 5H_2O_2(aq) \longrightarrow 2Mn^{2+}(aq) + 8H_2O(l) + 5O_2(g)$
Oxidizing agent is MnO_4^- and reducing agent is H_2O_2.

21.9(a) $4NO_3^-(aq) + 4H^+(aq) + 4Sb(s) \longrightarrow 4NO(g) + 2H_2O(l) + Sb_4O_6(s)$
Oxidizing agent is NO_3^- and reducing agent is Sb.
(b) $5BiO_3^-(aq) + 14H^+(aq) + 2Mn^{2+}(aq) \longrightarrow 5Bi^{3+}(aq) + 7H_2O(l) + 2MnO_4^-(aq)$
Oxidizing agent is BiO_3^- and reducing agent is Mn^{2+}.
(c) $Pb(OH)_3^-(aq) + 2Fe(OH)_2(s) \longrightarrow Pb(s) + 2Fe(OH)_3(s) + OH^-(aq)$
Oxidizing agent is $Pb(OH)_3^-$ and reducing agent is $Fe(OH)_2$.

21.12(a) $Au(s) + 3NO_3^-(aq) + 4Cl^-(aq) + 6H^+(aq) \longrightarrow AuCl_4^-(aq) + 3NO_2(g) + 3H_2O(l)$ (b) Oxidizing agent is NO_3^- and reducing agent is Au. (c) HCl provides chloride ions that combine with the gold(III) ion to form the stable $AuCl_4^-$ ion. **21.13**(a) A (b) E (c) C (d) A (e) E (f) E **21.16** An active electrode is a reactant or product in the cell reaction. An inactive electrode does not take part in the reaction and is present only to conduct a current. Platinum and graphite are commonly used as inactive electrodes. **21.17**(a) A (b) B (c) A (d) Hydrogen bubbles will form when metal A is placed in acid. Metal A is a better reducing agent than metal B, so if metal B reduces H^+ in acid, then metal A will also.

21.18(a) Oxidation: $Zn(s) \longrightarrow Zn^{2+}(aq) + 2e^-$
Reduction: $Sn^{2+}(aq) + 2e^- \longrightarrow Sn(s)$
Overall: $Zn(s) + Sn^{2+}(aq) \longrightarrow Zn^{2+}(aq) + Sn(s)$
(b)

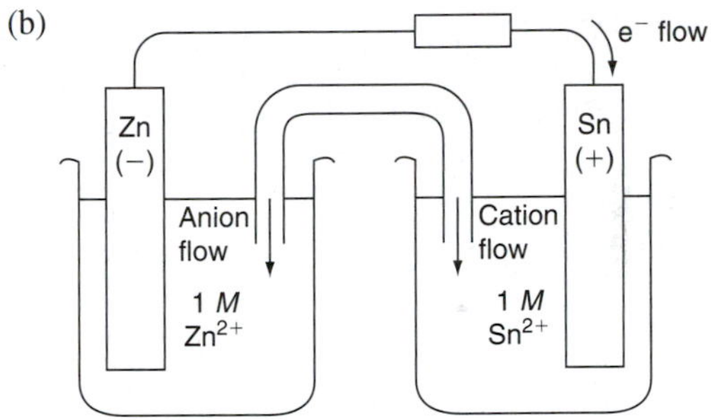

21.20(a) left to right (b) left (c) right (d) Ni (e) Fe (f) Fe (g) 1 *M* $NiSO_4$ (h) K^+ and NO_3^- (i) neither (j) from right to left
(k) Oxidation: $Fe(s) \longrightarrow Fe^{2+}(aq) + 2e^-$
Reduction: $Ni^{2+}(aq) + 2e^- \longrightarrow Ni(s)$
Overall: $Fe(s) + Ni^{2+}(aq) \longrightarrow Fe^{2+}(aq) + Ni(s)$

21.22(a) $Al(s) \mid Al^{3+}(aq) \parallel Cr^{3+}(aq) \mid Cr(s)$
(b) $Pt(s) \mid SO_2(g) \mid SO_4^{2-}(aq), H^+(aq) \parallel Cu^{2+}(aq) \mid Cu(s)$
21.25 A negative E°_{cell} indicates that the redox reaction is not spontaneous at the standard state, that is, $\Delta G^\circ > 0$. The reverse reaction is spontaneous with $E^\circ_{cell} > 0$. **21.26** Similar to other state functions, E° changes sign when a reaction is reversed. Unlike ΔG°, ΔH°, and S°, E° (the ratio of energy to charge) is an intensive property. When the coefficients in a reaction are multiplied by a factor, the values of ΔG°, ΔH°, and S° are multiplied by that factor. However, E° does not change because both the energy and charge are multiplied by the factor and thus their ratio remains unchanged.

21.27(a) Oxidation: $Se^{2-}(aq) \longrightarrow Se(s) + 2e^-$
Reduction: $2SO_3^{2-}(aq) + 3H_2O(l) + 4e^- \longrightarrow S_2O_3^{2-}(aq) + 6OH^-(aq)$
(b) $E^\circ_{anode} = E^\circ_{cathode} - E^\circ_{cell} = -0.57\text{ V} - 0.35\text{ V} = -0.92\text{ V}$

21.29(a) $Br_2 > Fe^{3+} > Cu^{2+}$ (b) $Ca^{2+} < Ag^+ < Cr_2O_7^{2-}$

21.31(a) $Co(s) + 2H^+(aq) \longrightarrow Co^{2+}(aq) + H_2(g)$
$E^\circ_{cell} = 0.28$ V; spontaneous
(b) $2Mn^{2+}(aq) + 5Br_2(l) + 8H_2O(l) \longrightarrow 2MnO_4^-(aq) + 10Br^-(aq) + 16H^+(aq)$
$E^\circ_{cell} = -0.44$ V; not spontaneous
(c) $Hg_2^{2+}(aq) \longrightarrow Hg^{2+}(aq) + Hg(l)$
$E^\circ_{cell} = -0.07$ V; not spontaneous

21.33(a) $2Ag(s) + Cu^{2+}(aq) \longrightarrow 2Ag^{+}(aq) + Cu(s)$
$E^\circ_{cell} = -0.46$ V; not spontaneous
(b) $Cr_2O_7^{2-}(aq) + 3Cd(s) + 14H^{+}(aq) \longrightarrow 2Cr^{3+}(aq) + 3Cd^{2+}(aq) + 7H_2O(l)$
$E^\circ_{cell} = 1.73$ V; spontaneous
(c) $Pb(s) + Ni^{2+}(aq) \longrightarrow Pb^{2+}(aq) + Ni(s)$
$E^\circ_{cell} = -0.12$ V; not spontaneous
21.35 $3N_2O_4(g) + 2Al(s) \longrightarrow 6NO_2^{-}(aq) + 2Al^{3+}(aq)$
$E^\circ_{cell} = 0.867\text{ V} - (-1.66\text{ V}) = 2.53$ V
$2Al(s) + 3SO_4^{2-}(aq) + 3H_2O(l) \longrightarrow 2Al^{3+}(aq) + 3SO_3^{2-}(aq) + 6OH^{-}(aq)$
$E^\circ_{cell} = 2.59$ V
$SO_4^{2-}(aq) + 2NO_2^{-}(aq) + H_2O(l) \longrightarrow SO_3^{2-}(aq) + N_2O_4(g) + 2OH^{-}(aq)$
$E^\circ_{cell} = 0.06$ V
Oxidizing agents: $Al^{3+} < N_2O_4 < SO_4^{2-}$
Reducing agents: $SO_3^{2-} < NO_2^{-} < Al$
21.37 $2HClO(aq) + Pt(s) + 2H^{+}(aq) \longrightarrow Cl_2(g) + Pt^{2+}(aq) + 2H_2O(l)$
$E^\circ_{cell} = 0.43$ V
$2HClO(aq) + Pb(s) + SO_4^{2-}(aq) + 2H^{+}(aq) \longrightarrow Cl_2(g) + PbSO_4(s) + 2H_2O(l)$
$E^\circ_{cell} = 1.94$ V
$Pt^{2+}(aq) + Pb(s) + SO_4^{2-}(aq) \longrightarrow Pt(s) + PbSO_4(s)$
$E^\circ_{cell} = 1.51$ V
Oxidizing agents: $PbSO_4 < Pt^{2+} < HClO$
Reducing agents: $Cl_2 < Pt < Pb$
21.39 yes; C > A > B **21.42** $A(s) + B^{+}(aq) \longrightarrow A^{+}(aq) + B(s)$ with $Q = [A^{+}]/[B^{+}]$. (a) $[A^{+}]$ increases and $[B^{+}]$ decreases. (b) E_{cell} decreases. (c) $E_{cell} = E^\circ_{cell} - (RT/nF) \ln([A^{+}]/[B^{+}])$; $E_{cell} = E^\circ_{cell}$ when $(RT/nF)\ln([A^{+}]/[B^{+}]) = 0$. This occurs when $\ln([A^{+}]/[B^{+}]) = 0$, that is, $[A^{+}]$ equals $[B^{+}]$. (d) Yes, when $[A^{+}] > [B^{+}]$. **21.44** In a concentration cell, the overall reaction decreases the concentration of the more concentrated electrolyte because it is reduced in the cathode compartment. **21.45**(a) 3×10^{35} (b) 4×10^{-31} **21.47**(a) -2.03×10^{5} J (b) 1.7×10^{5} J **21.49** $E^\circ = 0.28$ V; $\Delta G^\circ = -2.7\times10^{4}$ J **21.51** 8.8×10^{-5} *M* **21.53**(a) 0.05 V (b) 0.50 *M* (c) $[Co^{2+}] = 0.91$ *M*; $[Ni^{2+}] = 0.09$ *M* **21.55** A; 0.083 V **21.57** Electrons flow from the anode, where oxidation occurs, to the cathode, where reduction occurs. The electrons always flow from the anode to the cathode no matter what type of battery. **21.58** A D-sized alkaline battery is larger than an AAA-sized one, so it contains greater amounts of the cell components. The cell potential is an intensive property and does not depend on the amounts of the cell components. The total charge, however, does depend on the amount of cell components, so the D-sized battery produces more charge.
21.60 The Teflon spacers keep the two metals separated so that the copper cannot conduct electrons that would promote the corrosion (rusting) of the iron skeleton. **21.62** Sacrificial anodes are made of metals with E° more negative than that of iron, -0.44 V, so they are more easily oxidized than iron. Only (b), (f), and (g) will work for iron: (a) will form an oxide coating that prevents further oxidation; (c) will react with groundwater quickly; (d) and (e) are less easily oxidized than iron. **21.64** To reverse the reaction requires 0.34 V with the cell in its standard state. A 1.5 V cell supplies more than enough potential, so the Cd metal is oxidized to Cd^{2+} and Cr metal plates out. **21.66** The oxidation number of N in NO_3^{-} is +5, the maximum O.N. for N. In the nitrite ion, NO_2^{-}, the O.N. of N is +3, so nitrogen can be further oxidized. **21.68** Iron and nickel are more easily oxidized and less easily reduced than copper. During the purification process, all three metals are in solution, but only Cu^{2+} ions are reduced at the cathode to form Cu(*s*). **21.70**(a) Br_2 (b) Na **21.72** copper and bromine
21.74(a) Anode: $2H_2O(l) \longrightarrow O_2(g) + 4H^{+}(aq) + 4e^{-}$
Cathode: $2H_2O(l) + 2e^{-} \longrightarrow H_2(g) + 2OH^{-}(aq)$
(b) Anode: $2H_2O(l) \longrightarrow O_2(g) + 4H^{+}(aq) + 4e^{-}$
Cathode: $Sn^{2+}(aq) + 2e^{-} \longrightarrow Sn(s)$
21.76(a) 3.75 mol e^{-} (b) 3.62×10^{5} C (c) 28.7 A
21.78 0.275 g Ra **21.80** 9.20×10^{3} s **21.82**(a) The sodium and sulfate ions conduct a current, facilitating electrolysis. Pure water, which contains very low (10^{-7} *M*) concentrations of H^{+} and OH^{-}, conducts electricity very poorly. (b) The reduction of H_2O has a more positive half-potential than does the reduction of Na^{+}; the oxidation of H_2O is the only reaction possible because SO_4^{2-} cannot be oxidized. Thus, it is easier to reduce H_2O than Na^{+} and easier to oxidize H_2O than SO_4^{2-}.
21.83(a) 4.6×10^{4} L (b) 1.26×10^{8} C (c) 1.68×10^{6} s
21.84 62.6 g Zn **21.87** 64.3 mass % Cu **21.89**(a) 8 days (b) 32 days (c) \$940 **21.91** (a) 2.4×10^{4} days (b) 2.1 g (c) 6.1×10^{-5} dollars **21.92** 7×10^{2} lb Cl_2
21.94(a) 1.073×10^{5} s (b) 1.5×10^{4} kW·h (c) 0.92¢
21.96 If metal E and a salt of metal F are mixed, the salt is reduced, producing metal F because E has the greatest reducing strength of the three metals; F < D < E.
21.97(a) Cell I: 4 mol electrons; $\Delta G = -4.75\times10^{5}$ J
Cell II: 2 mol electrons; $\Delta G = -3.94\times10^{5}$ J
Cell III: 2 mol electrons; $\Delta G = -4.53\times10^{5}$ J
(b) Cell I: -13.2 kJ/g
Cell II: -0.613 kJ/g
Cell III: -2.63 kJ/g
Cell I has the highest ratio (most energy released per gram) because the reactants have very low mass, while Cell II has the lowest ratio because the reactants have large masses.
21.98 $Sn^{2+}(aq) + 2e^{-} \longrightarrow Sn(s)$
$Cr^{3+}(aq) + e^{-} \longrightarrow Cr^{2+}(aq)$
$Fe^{2+}(aq) + 2e^{-} \longrightarrow Fe(s)$
$U^{4+}(aq) + e^{-} \longrightarrow U^{3+}(aq)$

21.101 Li > Ba > Na > Al > Mn > Zn > Cr > Fe > Ni > Sn > Pb > Cu > Ag > Hg > Au. Metals with potentials lower than that of water (-0.83 V) can displace H_2 from water: Li, Ba, Na, Al, and Mn. Metals with potentials lower than that of hydrogen (0.00 V) can displace H_2 from acid: Li, Ba, Na, Al, Mn, Zn, Cr, Fe, Ni, Sn, and Pb. Metals with potentials greater than that of hydrogen (0.00 V) cannot displace H_2: Cu, Ag, Hg, and Au. **21.102**(a) 1.890 t Al_2O_3 (b) 0.3339 t C (c) 100% (d) 74% (e) 2.813×10^{3} m^3
21.103(a) 5.3×10^{-11} (b) 0.20 V (c) 0.43 V (d) 8.2×10^{-4} *M* NaOH **21.105** 2.94

Chapter 22

• **22.1**(a) $1s^2 2s^2 2p^6 3s^2 3p^6 4s^2 3d^{10} 4p^6 5s^2 4d^x$
(b) $1s^2 2s^2 2p^6 3s^2 3p^6 4s^2 3d^{10} 4p^6 5s^2 4d^{10} 5p^6 6s^2 4f^{14} 5d^x$ **22.4**(a) The elements should increase in size as they increase in mass from Period 5 to Period 6. Because there are 14 inner transition elements in Period 6, the effective nuclear charge increases significantly; so the atomic size decreases, or "contracts." This effect is significant enough that Zr^{4+} and Hf^{4+} are almost the same size but differ greatly in atomic mass. (b) The atomic size increases from Period 4 to Period 5, but stays fairly constant from Period 5 to Period 6. (c) Atomic mass increases significantly from Period 5 to Period 6, but atomic radius (and thus volume) increases slightly, so Period 6 elements are very dense. **22.7**(a) A paramagnetic substance is attracted to a magnetic field, while a diamagnetic substance is unaffected (or slightly repelled) by one. (b) Ions of transition elements often have half-filled *d* orbitals whose unpaired electrons make the ions paramagnetic. Ions of main-group elements usually have a noble-gas configuration with no partially filled levels.
(c) Some *d* orbitals in the transition element ions are empty, which allows an electron from one *d* orbital to move to one with a slightly higher energy. The energy required for this transition is small and falls in the visible wavelength range. All orbitals are filled in ions of main-group elements, so enough energy would have to be added to move an electron to the next principal energy level, not just another orbital within the same energy level. This amount of energy is very large and much greater than the visible range of wavelengths.
22.8(a) $1s^2 2s^2 2p^6 3s^2 3p^6 4s^2 3d^3$
(b) $1s^2 2s^2 2p^6 3s^2 3p^6 4s^2 3d^{10} 4p^6 5s^2 4d^1$ (c) [Xe] $6s^2 4f^{14} 5d^{10}$
22.10(a) [Ar], no unpaired electrons (b) [Ar] $3d^9$, one unpaired electron (c) [Ar] $3d^5$, five unpaired electrons (d) [Kr] $4d^2$, two unpaired electrons **22.12** Cr, Mo, and W **22.14** in CrF_2, because the chromium is in a lower oxidation state **22.16** CrO_3, with Cr in a higher oxidation state, yields a more acidic aqueous solution. **22.19** The coordination number indicates the number of ligand atoms bonded to the metal ion. The oxidation number represents the number of electrons lost to form the ion. The coordination number is unrelated to the oxidation number. **22.21** 2, linear; 4, tetrahedral or square planar; 6, octahedral **22.24**(a) hexaaquanickel(II) chloride
(b) tris(ethylenediamine)chromium(III) perchlorate (c) potassium hexacyanomanganate(II) **22.26**(a) 2+, 6 (b) 3+, 6
(c) 2+, 6 **22.28**(a) potassium dicyanoargentate(I) (b) sodium tetrachlorocadmate(II) (c) tetraammineaquabromocobalt(III) bromide **22.30**(a) $[Zn(NH_3)_4]SO_4$ (b) $[Cr(NH_3)_5Cl]Cl_2$
(c) $Na_3[Ag(S_2O_3)_2]$ **22.32**(a) 4, two ions (b) 6, three ions
(c) 2, four ions **22.34**(a) The nitrite ion forms linkage isomers because it can bind to the metal ion through the lone pair on the N atom or any lone pair on either O atom.

[:Ö—N̈=Ö]⁻

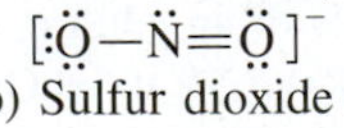

(b) Sulfur dioxide molecules form linkage isomers because the lone pair on the S atom or any lone pair on either O atom can bind the central metal ion.

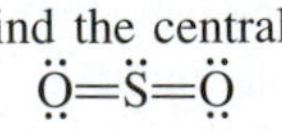

Ö=S̈=Ö

(c) Nitrate ions have an N atom with no lone pair and three O atoms, all with lone pairs that can bind to the metal ion. But all of the O atoms are equivalent, so these ions do not form linkage isomers.

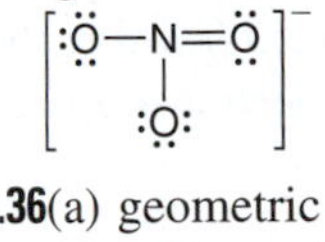

[:Ö—N=Ö (with :Ö: on N)]⁻

23.36(a) geometric isomerism

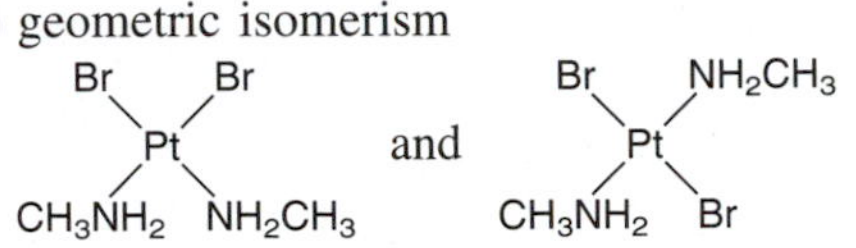

Pt with Br, Br, CH_3NH_2, NH_2CH_3 (*cis*) and Pt with Br, NH_2CH_3, CH_3NH_2, Br (*trans*)

(b) geometric isomerism

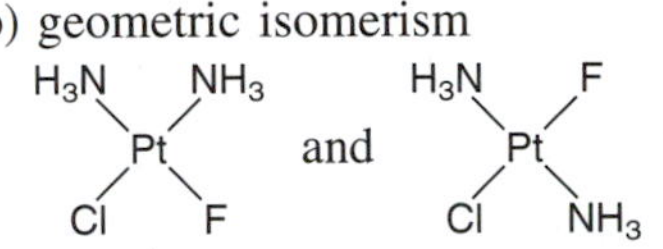

Pt with H_3N, NH_3, Cl, F (*cis*) and Pt with H_3N, F, Cl, NH_3 (*trans*)

(c) geometric isomerism

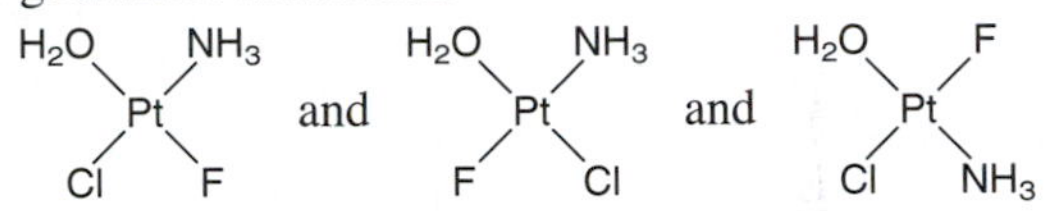

Pt with H_2O, NH_3, Cl, F and Pt with H_2O, NH_3, F, Cl and Pt with H_2O, F, Cl, NH_3

23.38(a) geometric isomerism

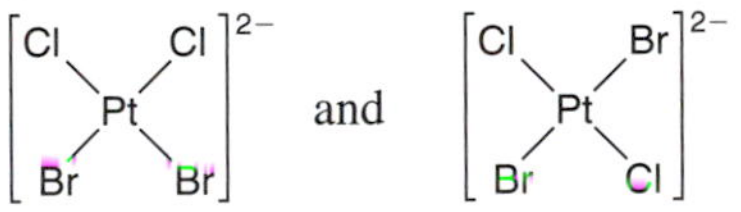

$[PtCl_2Br_2]^{2-}$: Cl, Cl, Br, Br (*cis*) and Cl, Br, Br, Cl (*trans*)

(b) linkage isomerism

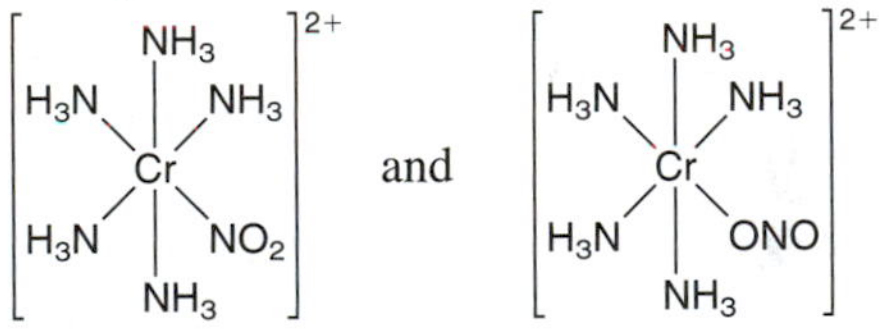

$[Cr(NH_3)_5(NO_2)]^{2+}$ and $[Cr(NH_3)_5(ONO)]^{2+}$

(c) geometric isomerism

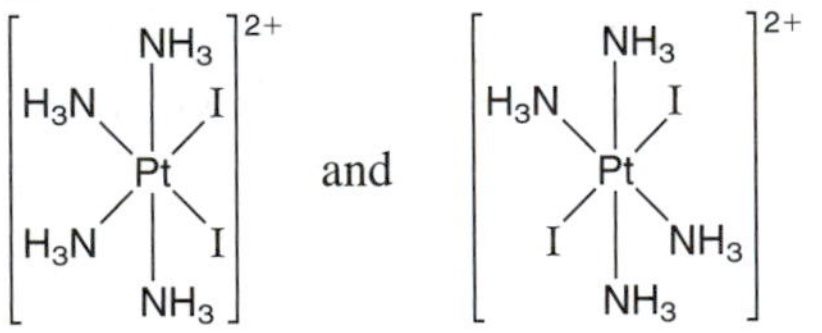

$[Pt(NH_3)_4I_2]^{2+}$ (*cis*) and $[Pt(NH_3)_4I_2]^{2+}$ (*trans*)

22.40(a) dsp^2 (b) sp^3 **22.43**(a) The crystal field splitting energy (Δ) is the energy difference between the two sets of *d* orbitals that result from electrostatic effects of ligands on a central transition metal atom. (b) In an octahedral field of ligands, the ligands approach along the *x*, *y*, and *z* axes. The $d_{x^2-y^2}$ and d_{z^2} orbitals are located *along* the *x*, *y*, and *z* axes, so ligand interaction is higher in energy. The other orbital-ligand interactions are lower in energy because the d_{xy}, d_{yz}, and d_{xz} orbitals are located *between* the *x*, *y*, and *z* axes. (c) In a tetrahedral field of ligands, the ligands do not approach along the *x*, *y*, and *z* axes. The ligand interaction is greater for the d_{xy}, d_{yz}, and d_{xz} orbitals and lesser for the $d_{x^2-y^2}$ and d_{z^2} orbitals. Therefore, the crystal field splitting is reversed, and the d_{xy}, d_{yz}, and d_{xz} orbitals are higher in energy than the $d_{x^2-y^2}$ and d_{z^2} orbitals. **22.45** If Δ is greater than $E_{pairing}$, electrons will pair their spins in the lower energy set of *d* orbitals before entering the higher energy set of d orbitals as unpaired electrons. If Δ is less than $E_{pairing}$, electrons will occupy the higher energy set of d orbitals as unpaired electrons before pairing in the lower energy set of d orbitals. The first case gives a complex that is low-spin and less paramagnetic than the high-spin complex formed in the latter case. **22.47**(a) no d electrons (b) eight d electrons (c) six d electrons

22.49 a and d

22.51

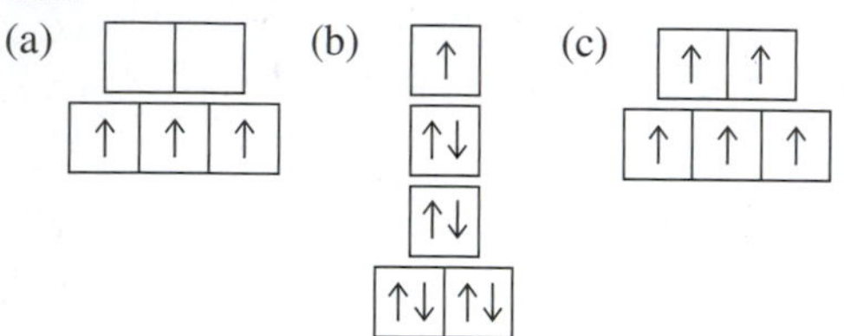

22.53 $[Cr(H_2O)_6]^{3+} < [Cr(NH_3)_6]^{3+} < [Cr(NO_2)_6]^{3-}$

22.55 A violet complex absorbs yellow-green light. The light absorbed by a complex with a weaker field ligand would be at a lower energy and higher wavelength. Light of lower energy than yellow-green light is yellow, orange, or red. The color observed would be blue or green. **22.60**(a) 6 (b) +3 (c) two (d) 1 mol

22.65 $[Co(NH_3)_4(H_2O)Cl]^{2+}$

tetraammineaquachlorocobalt(III) ion

2 geometric isomers

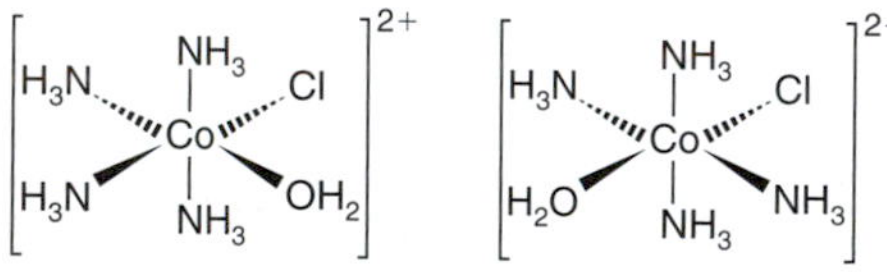

cis *trans*

$[Cr(H_2O)_3Br_2Cl]$

triaquadibromochlorochromium(III)

3 geometric isomers

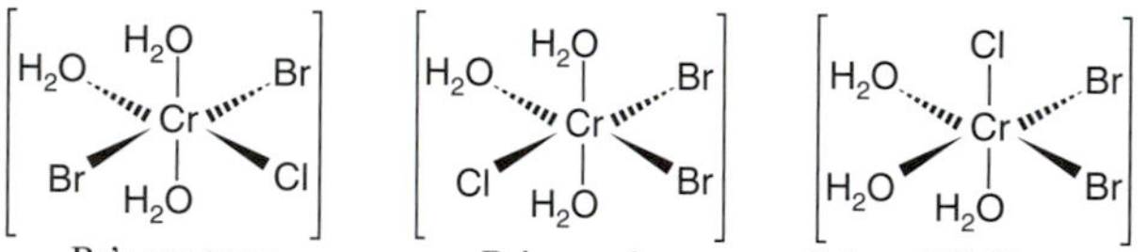

Br's are *trans* Br's are *cis* H₂O's are *cis* and *trans* Br's and H₂O's are *cis*

$[Cr(NH_3)_2(H_2O)_2Br_2]^+$

diamminediaquadibromochromium(III) ion

6 isomers (5 geometric)

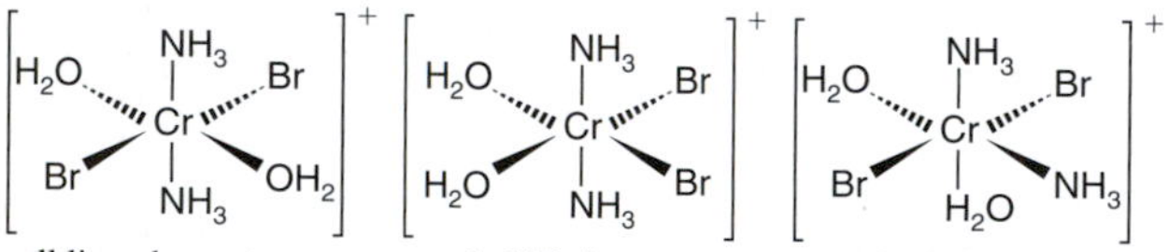

all ligands are *trans* only NH_3 is *trans* only Br is *trans*

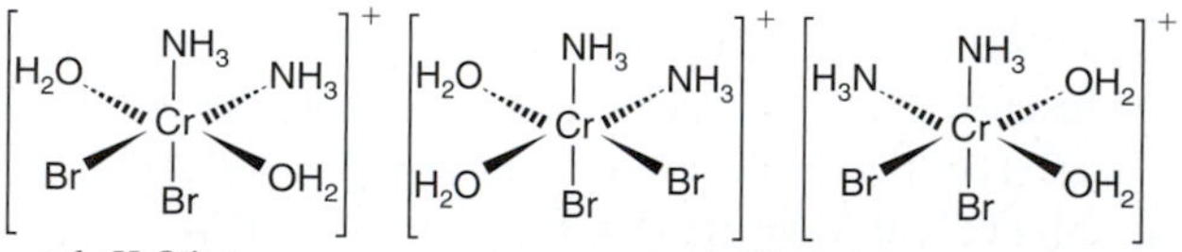

only H_2O is *trans* optical isomers all three ligands are *cis*

22.71(a) The first reaction shows no change in the number of particles. In the second reaction, the number of reactant particles is greater than the number of product particles. A decrease in the number of particles means a decrease in entropy. Based on entropy change only, the first reaction is favored. (b) The ethylenediamine complex will be more stable with respect to ligand exchange in water because the entropy change for that exchange is unfavorable (negative).

Chapter 23

• **23.1**(a) Chemical reactions are accompanied by relatively small changes in energy; nuclear reactions are accompanied by relatively large changes in energy. (b) Increasing temperature increases the rate of a chemical reaction but has no effect on a nuclear reaction. (c) Both chemical and nuclear reaction rates increase with higher reactant concentrations. (d) If the reactant is limiting in a chemical reaction, then more reactant produces more product and the yield increases. The presence of more radioactive reactant results in more decay product, so a higher reactant concentration increases the yield. **23.2**(a) Z down by 2, N down by 2 (b) Z up by 1, N down by 1 (c) no change in Z or N (d) Z down by 1, N up by 1 (e) Z down by 1, N up by 1 A different element is produced in all cases except (c). **23.4** A neutron-rich nuclide decays by beta (minus) decay. A neutron-poor nuclide undergoes positron decay or electron capture. **23.6**(a) ${}^{234}_{92}U \longrightarrow {}^{4}_{2}\alpha + {}^{230}_{90}Th$ (b) ${}^{232}_{93}Np + {}^{0}_{-1}e \longrightarrow {}^{232}_{92}U$ (c) ${}^{12}_{7}N \longrightarrow {}^{0}_{1}\beta + {}^{12}_{6}C$ **23.8**(a) ${}^{48}_{23}V \longrightarrow {}^{48}_{22}Ti + {}^{0}_{1}\beta$ (b) ${}^{107}_{48}Cd + {}^{0}_{-1}e \longrightarrow {}^{107}_{47}Ag$ (c) ${}^{210}_{86}Rn \longrightarrow {}^{206}_{84}Po + {}^{4}_{2}\alpha$ **23.10**(a) Appears stable because its N and Z values are both magic numbers, but its N/Z ratio (1.50) is too high; it is unstable. (b) Appears unstable because its Z value is an odd number, but its N/Z ratio (1.19) is in the band of stability, so it is stable. (c) Unstable because its N/Z ratio is too high. **23.12**(a) alpha decay (b) positron decay or electron capture (c) positron decay or electron capture **23.14** Stability results from a favorable N/Z ratio, even numbered N and/or Z, and the occurrence of magic numbers. The N/Z ratio of ${}^{52}Cr$ is 1.17, which is within the band of stability. The fact that Z is even does not account for the variation in stability because all isotopes of chromium have the same Z. However, ${}^{52}Cr$ has 28 neutrons, so N is both an even number and a magic number for this isotope only.

23.18 No, it is not valid to conclude that $t_{1/2}$ equals 1 min because the number of nuclei is so small. Decay rate is an average rate and is only meaningful when the sample is macroscopic and contains a large number of nuclei. For the sample containing 6×10^{12} nuclei, the conclusion is valid. **23.20** 2.56×10^{-2} Ci/g **23.22** 2.31×10^{-7} yr^{-1} **23.24** 1.49 mg **23.26** 2.2×10^{9} yr **23.28** 30 dpm **23.32** Protons are repelled from the target nuclei due to interaction with like (positive) charges. Higher energy is required to overcome the repulsion. **23.33**(a) ${}^{10}_{5}B + {}^{4}_{2}\alpha \longrightarrow {}^{1}_{0}n + {}^{13}_{7}N$ (b) ${}^{28}_{14}Si + {}^{2}_{1}H \longrightarrow {}^{1}_{0}n + {}^{29}_{15}P$ (c) ${}^{242}_{96}Cm + {}^{4}_{2}\alpha \longrightarrow 2{}^{1}_{0}n + {}^{244}_{98}Cf$ **23.37** Ionizing radiation is more dangerous to children because their cells are rapidly dividing and increasing in number, unlike cells of an adult. **23.38**(a) 5.4×10^{-7} rad (b) 5.4×10^{-9} Gy **23.40**(a) 7.5×10^{-10} Gy (b) 7.5×10^{-5} mrem (c) 7.5×10^{-10} Sv **23.42** 1.86×10^{-3} rad **23.44** NAA does not destroy the sample, while chemical analyses do. Neutrons bombard a nonradioactive sample, inducing some atoms within the sample to be radioactive. The radioisotopes decay by emitting radiation characteristic of each isotope. **23.48**(a) 1.861×10^4 eV (b) 2.981×10^{-15} J **23.50**(a) 8.768 MeV/nucleon (b) 517.3 MeV/atom (c) 4.99×10^{10} kJ/mol **23.53** Radioactive decay is a spontaneous process in which unstable nuclei emit radioactive particles and energy. Fission occurs as the result of high-energy bombardment of nuclei with small particles that cause the nuclei to break into smaller nuclides, radioactive particles, and energy. All fission events are not the same. The nuclei split in a number of ways to produce several different products. **23.56** The water serves to slow the neutrons so that they are better able to cause a fission reaction. Heavy water is a better moderator because it does not absorb neutrons as well as light water does; thus, more neutrons are available to initiate the fission process. However, D_2O does not occur naturally in great abundance, so its production adds to the cost of a heavy-water reactor. **23.62**(a) 1.1×10^{-29} kg (b) 9.9×10^{-13} J (c) 5.9×10^8 kJ/mol; this is approximately 1 million times larger than a typical enthalpy of reaction. **23.64** 8.0×10^3 yr **23.68** 1.35×10^{-5} *M* **23.70** (a) 5.99 h (b) 21% **23.71** 4.904×10^{-9} L/h **23.75** 0.267 **23.80** (a) 2.07×10^{-17} J (b) 1.45×10^7 H atoms (c) 1.4960×10^{-5} J (d) 1.4959×10^{-5} J (e) No, the captain should continue using the current technology.

Numbers in parentheses refer to the page(s) on which a term is introduced and/or discussed.

A

absolute scale (also *Kelvin scale*) The preferred temperature scale in scientific work, which has absolute zero (0 K, or −273.15°C) as the lowest temperature. (18) [See also *kelvin* (*K*).]
absorption spectrum The spectrum produced when atoms absorb specific wavelengths of incoming light and become excited from lower to higher energy levels. (228)
accuracy The closeness of a measurement to the actual value. (24)
acid In common laboratory terms, any species that produces H^+ ions when dissolved in water. (126) [See also *Arrhenius, Brønsted-Lowry,* and *Lewis acid-base definitions.*]
acid anhydride A compound, usually formed by a dehydration-condensation reaction of an oxoacid, that yields two molecules of the acid when it reacts with water. (480)
acid-base buffer (also *buffer*) A solution that resists changes in pH when a small amount of either strong acid or strong base is added. (618)
acid-base indicator An organic molecule whose color is different in acid and in base and is used to monitor the equivalence point of a titration or the pH of a solution. (588)
acid-base reaction Any reaction between an acid and a base. (126) (See also *neutralization reaction.*)
acid-base titration curve A plot of the pH of a solution of acid (or base) versus the volume of base (or acid) added to the solution. (626)
acid-dissociation (acid-ionization) constant (K_a) An equilibrium constant for the dissociation of an acid (HA) in H_2O to yield the conjugate base (A^-) and H_3O^+:

$$K_a = \frac{[H_3O^+][A^-]}{[HA]} \qquad (582)$$

actinides The Period 7 elements that constitute the second inner transition series (5*f* block), which includes thorium (Th; $Z = 90$) through lawrencium (Lr; $Z = 103$). (256)
activated complex (See *transition state.*)
activation energy (E_a) The minimum energy with which molecules must collide to react. (519)
active site The region of an enzyme formed by specific amino acid side chains at which catalysis occurs. (532)
activity ($\mathcal{A}$) (also *decay rate*) The change in number of nuclei ($\mathcal{N}$) of a radioactive sample divided by the change in time (t). (772)
activity series of the metals A listing of metals arranged in order of decreasing strength of the metal as a reducing agent in aqueous reactions. (138)
actual yield The amount of product actually obtained in a chemical reaction. (97)
addition polymer (also *chain-reaction,* or *chain-growth, polymer*) A polymer formed when monomers (usually containing C=C) combine through an addition reaction. (484)
addition reaction A type of organic reaction in which atoms linked by a multiple bond become bonded to more atoms. (473)
adduct The product of a Lewis acid-base reaction characterized by the formation of a new covalent bond. (607)
adenosine triphosphate (ATP) A high-energy molecule that serves most commonly as a store and source of energy in organisms. (675)
alcohol An organic compound (ending, *-ol*) that contains a —C—Ö—H functional group. (476)
aldehyde An organic compound (ending, *-al*) that contains the carbonyl functional group (C=Ö) in which the carbonyl C is also bonded to H. (479)
alkane A hydrocarbon that contains only single bonds and no rings (general formula, C_nH_{2n+2}). (464)
alkene A hydrocarbon that contains at least one C=C bond (general formula, C_nH_{2n}). (469)
alkyl group A saturated hydrocarbon chain with one bond available. (472)
alkyl halide (See *haloalkane.*)
alkyne A hydrocarbon that contains at least one C≡C bond (general formula, C_nH_{2n-2}). (470)
allotrope One of two or more crystalline or molecular forms of an element. In general, one allotrope is more stable than another at a particular pressure and temperature. (434)
alloy A mixture of elements with metallic properties that is typically a solid-solid solution. (396)
alpha (α) decay A radioactive process in which an alpha particle is emitted from a nucleus. (765)
alpha particle (α or $^4_2He^{2+}$) A positively charged particle, identical to a helium-4 nucleus, that is one of the common types of radioactive emissions. (765)
amide An organic compound that contains the —C(=:O:)—N̈— functional group. (480)
amine An organic compound (general formula, —C—N̈—) derived structurally by replacing one or more H atoms of ammonia with organic groups; a weak organic base. (477)
amino acid An organic compound [general formula, H_2N—CH(R)—COOH] with at least one carboxyl and one amine group on the same molecule; the monomer unit of a protein. (487)
amorphous solid A solid that has a poorly defined shape because it lacks extensive molecular-level ordering of its particles. (373)
ampere (A) The SI unit of electric current; 1 ampere of current results when 1 coulomb flows through a conductor in 1 second. (726)
amphoteric Able to act as either an acid or a base. (265)
amplitude The height of the crest (or depth of the trough) of a wave; related to the intensity of the energy (brightness of the light). (217)
angular momentum quantum number (l) An integer from 0 to $n - 1$ that is related to the shape of an atomic orbital. (235)
anion A negatively charged ion. (49)

anode The electrode at which oxidation occurs in an electrochemical cell. Electrons are given up by the reducing agent and leave the cell at the anode. (692)

antibonding MO A molecular orbital formed when wave functions are subtracted from each other, which decreases electron density between the nuclei and leaves a node. Electrons occupying such an orbital destabilize the molecule. (339)

aqueous solution A solution in which water is the solvent. (61)

aromatic hydrocarbon A compound of C and H with one or more rings of C atoms (often drawn with alternating C—C and C=C bonds), in which there is extensive delocalization of π electrons. (471)

Arrhenius acid-base definition A model of acid-base behavior in which an acid is a substance that has H in its formula and dissociates in water to yield H_3O^+, and a base is a substance that has OH in its formula and produces OH^- in water. (580)

Arrhenius equation An equation that expresses the exponential relationship between temperature and the rate constant: $k = Ae^{-E_a/RT}$. (519)

atmosphere (See *standard atmosphere.*)

atom The smallest particle of an element that retains the chemical nature of the element. A neutral, spherical entity composed of a positively charged central nucleus surrounded by one or more negatively charged electrons. (38)

atomic mass (also *atomic weight*) The average of the masses of the naturally occurring isotopes of an element weighted according to their abundances. (45)

atomic mass unit (amu) (also *dalton, Da*) A mass exactly equal to $\frac{1}{12}$ the mass of a carbon-12 atom. (44)

atomic number (*Z*) The unique number of protons in the nucleus of each atom of an element (equal to the number of electrons in the neutral atom). An integer that expresses the positive charge of a nucleus or subatomic particle in multiples of the electronic charge. (43)

atomic orbital (also *wave function*) A mathematical expression that describes the motion of the electron's matter-wave in terms of time and position in the region of the nucleus. The term is used qualitatively to mean the region of space in which there is a high probability of finding the electron. (232)

atomic size One-half the distance between nuclei of two adjacent atoms in a sample of the element. (258) Also refers to the extent of the *spherical contour* in which an atom's electrons spend 90% of their time. (234)

atomic solid A solid consisting of individual atoms held together by dispersion forces; the frozen noble gases are the only examples. (379)

atomic symbol (also *element symbol*) A one- or two-letter abbreviation for the English, Latin, or Greek name of an element. (43)

atomic weight (See *atomic mass.*)

aufbau principle The conceptual approach for building up atoms by adding one proton at a time to the nucleus and one electron around it to obtain the ground-state electron configurations of the elements. (249)

autoionization (also *self-ionization*) A reaction in which two molecules of a substance react to give ions. The most important example is for water:

$$2H_2O(l) \rightleftharpoons H_3O^+(aq) + OH^-(aq) \qquad (584)$$

average rate The change in concentration of reactants (or products) divided by a finite time period. (501)

Avogadro's law The gas law stating that, at fixed temperature and pressure, equal volumes of any ideal gas contain equal numbers of particles, and, therefore, the volume of a gas is directly proportional to its amount (mol): $V \propto n$. (156)

Avogadro's number A number (6.022×10^{23} to four significant figures) equal to the number of atoms in exactly 12 g of carbon-12; the number of atoms, molecules, or formula units in one mole of an element or compound. (72)

axial group An atom (or group) that lies above or below the trigonal plane of a trigonal bipyramidal molecule, or a similar structural feature in a molecule. (316)

B

background radiation Natural ionizing radiation, the most important form of which is cosmic radiation. (779)

balancing coefficient (also *stoichiometric coefficient*) A numerical multiplier of all the atoms in the formula immediately following it in a balanced chemical equation. (86)

band of stability The band of stable nuclides that appears on a plot of number of neutrons versus number of protons for all nuclides. (768)

band theory An extension of molecular orbital (MO) theory that explains many properties of metals and other solids, in particular, the differences in electrical conductivity of conductors, semiconductors, and insulators. (382)

barometer A device used to measure atmospheric pressure. Most commonly, a tube open at one end, which is filled with mercury and inverted into a dish of mercury. (150)

base In common laboratory terms, any species that produces OH^- ions when dissolved in water. (126) [See also *Arrhenius, Brønsted-Lowry,* and *Lewis acid-base definitions.*]

base-dissociation (base-ionization) constant (K_b) An equilibrium constant for the reaction of a base (B) with H_2O to yield the conjugate acid (BH^+) and OH^-:

$$K_b = \frac{[BH^+][OH^-]}{[B]} \qquad (598)$$

base pair Two complementary mononucleotide bases that are H bonded to each other; guanine (G) always pairs with cytosine (C), and adenine (A) always pairs with thymine (T) (or uracil, U). (490)

base unit (also *fundamental unit*) A unit that defines the standard for one of the seven physical quantities in the International System of Units (SI). (14)

battery A group of voltaic cells arranged in series; primary and secondary types are self-contained, but flow batteries are not. (713)

becquerel (Bq) The SI unit of radioactivity; 1 Bq = 1 d/s (disintegration per second). (772)

bent shape (also *V shape*) A molecular shape that arises when a central atom is bonded to two other atoms and has one or two lone pairs; occurs as the AX_2E shape class (bond angle $< 120°$) in the trigonal planar arrangement and as the AX_2E_2 shape class (bond angle $< 109.5°$) in the tetrahedral arrangement. (314)

beta decay A radioactive change that encompasses any of three specific processes— β^- decay, β^+ emission, and e^- capture. (767)

β^- decay A radioactive process in which a beta particle is emitted from a nucleus. (767)

beta particle (β, β^-, or $^{0}_{-1}\beta$) A negatively charged particle, identified as a high-speed electron, that is one of the common types of radioactive emissions. (765)
bimolecular reaction An elementary reaction involving the collision of two reactant species. (525)
binary covalent compound A compound that consists of atoms of two elements in which bonding occurs primarily through electron sharing. (58)
binary ionic compound A compound that consists of the oppositely charged ions of two elements. (49)
body-centered cubic unit cell A unit cell in which a particle lies at each corner and in the center of a cube. (374)
boiling point (bp or T_b) The temperature at which the vapor pressure of a gas equals the external (atmospheric) pressure. (359)
boiling point elevation (ΔT_b) The increase in the boiling point of a solvent caused by the presence of dissolved solute. (410)
bond angle The angle formed by the nuclei of two surrounding atoms with the nucleus of the central atom at the vertex. (313)
bond energy (BE) (also *bond enthalpy* or *bond strength*) The enthalpy change (always >0) accompanying the breakage of a given bond in 1 mol of gaseous molecules. (286)
bond length The distance between the nuclei of two bonded atoms. (286)
bond order The number of electron pairs shared by two bonded atoms. (285)
bonding MO A molecular orbital formed when wave functions are added to each other, which increases electron density between the nuclei. Electrons occupying such an orbital stabilize the molecule. (338)
bonding pair (also *shared pair*) An electron pair shared by two nuclei; the mutual attraction between the nuclei and the electron pair forms a covalent bond. (285)
Boyle's law The gas law stating that, at constant temperature and amount of gas, the volume occupied by a gas is inversely proportional to the applied (external) pressure: $V \propto 1/P$. (154)
Brønsted-Lowry acid-base definition A model of acid-base behavior based on proton transfer, in which an acid and a base are defined, respectively, as species that donate and accept a proton. (588)
buffer (See *acid-base buffer.*)
buffer capacity A measure of the ability of a buffer to resist a change in pH; related to the total concentrations and relative proportions of buffer components. (623)
buffer range The pH range over which a buffer acts effectively. (624)

C

calibration The process of correcting for systematic error of a measuring device by comparing it to a known standard. (25)
calorie (cal) A unit of energy defined as exactly 4.184 joules; originally defined as the heat needed to raise the temperature of 1 g of water 1°C (from 14.5°C to 15.5°C). (193)
calorimeter A device used to measure the heat released or absorbed by a physical or chemical process taking place within it. (198)
capillarity (or *capillary action*) A property that results in a liquid rising through a narrow space against the pull of gravity. (370)
carbonyl group The C=O grouping of atoms. (479)
carboxylic acid An organic compound (ending, *-oic acid*) that contains the —C(=O)—OH group. (480)
catalyst A substance or mixture that increases the rate of a reaction without being used up in the process. (530)
catenation The process by which atoms of an element bond to each other in chains; most common with carbon in organic compounds but also occurs with boron, silicon, sulfur, and several other elements. (460)
cathode The electrode at which reduction occurs in an electrochemical cell. Electrons enter the cell and are acquired by the oxidizing agent at the cathode. (692)
cathode ray The ray of light emitted by the cathode (negative electrode) in a gas discharge tube; travels in straight lines, unless deflected by magnetic or electric fields. (40)
cation A positively charged ion. (49)
cell potential (E_{cell}) (also *electromotive force,* or *emf; cell voltage*) The potential difference between the electrodes of an electrochemical cell when no current flows. (697)
Celsius scale (formerly *centigrade scale*) A temperature scale in which the freezing and boiling points of water are defined as 0°C and 100°C, respectively. (18)
chain reaction In nuclear fission, a self-sustaining process in which neutrons released by splitting of one nucleus cause other nuclei to split, which releases more neutrons, and so on. (786)
change in enthalpy (ΔH) The change in internal energy plus the product of the constant pressure and the change in volume: $\Delta H = \Delta E + P\Delta V$; the heat lost or gained at constant pressure: $\Delta H = q_P$. (195)
charge density The ratio of the charge of an ion to its volume. (398)
Charles's law The gas law stating that at constant pressure, the volume occupied by a fixed amount of gas is directly proportional to its absolute temperature: $V \propto T$. (154)
chelate A complex ion in which the metal ion is bonded to a bidentate or polydentate ligand. (744)
chemical bond The force that holds two atoms together in a molecule (or formula unit). (49)
chemical change (also *chemical reaction*) A change in which one or more substances are converted into one or more substances with different composition and properties. (4)
chemical equation A statement that uses chemical formulas to express the identities and quantities of the substances involved in a chemical or physical change. (85)
chemical formula A notation of atomic symbols and numerical subscripts that shows the type and number of each atom in a molecule or formula unit of a substance. (53)
chemical kinetics The study of the rates and mechanisms of reactions. (499)
chemical property A characteristic of a substance that appears as it interacts with, or transforms into, other substances. (4)
chemical reaction (See *chemical change.*)
chemistry The scientific study of matter and its properties, the changes it undergoes, and the energy associated with those changes. (3)
chiral molecule One that is not superimposable on its mirror image; an optically active molecule. In organic compounds, a chiral molecule typically contains a C atom bonded to four different groups (asymmetric C). (468)
chlor-alkali process An industrial method that electrolyzes concentrated aqueous NaCl and produces Cl_2, H_2, and NaOH. (722)

cis-trans isomers (See *geometric isomers.*)
Clausius-Clapeyron equation An equation that expresses the linear relationship between vapor pressure P of a liquid and temperature T; in two-point form, it is

$$\ln \frac{P_2}{P_1} = \frac{-\Delta H_{vap}}{R}\left(\frac{1}{T_2} - \frac{1}{T_1}\right) \quad (358)$$

colligative property A property of a solution that depends on the number, not the identity, of solute particles. (408) (See also *boiling point elevation, freezing point depression, osmotic pressure,* and *vapor pressure lowering.*)
collision theory A model that explains reaction rate as based on the number, energy, and orientation of colliding particles. (518)
combustion The process of burning in air, often with release of heat and light. (10)
combustion analysis A method for determining the formula of a compound from the amounts of its combustion products; used commonly for organic compounds. (83)
common-ion effect The shift in the position of an ionic equilibrium away from an ion involved in the process that is caused by the addition or presence of that ion. (618)
complex (See *coordination compound.*)
complex ion An ion consisting of a central metal ion bonded covalently to molecules and/or anions called ligands (643, 743)
composition The types and amounts of simpler substances that make up a sample of matter. (3)
compound A substance composed of two or more elements that are chemically combined in fixed proportions. (34)
concentration A measure of the quantity of solute dissolved in a given quantity of solution (or solvent). (99)
concentration cell A voltaic cell in which both compartments contain the same components but at different concentrations. (710)
condensation The process of a gas changing into a liquid. (352)
condensation polymer A polymer formed by monomers with two functional groups that are linked together in a dehydration-condensation reaction. (484)
conduction band In band theory, the empty, higher energy portion of the band of molecular orbitals into which electrons move when conducting heat and electricity. (383)
conductor A substance (usually a metal) that conducts an electric current well. (384)
conjugate acid-base pair Two species related to each other through the gain or loss of a proton; the acid has one more proton than its conjugate base. (589)
constitutional isomers (also *structural isomers*) Compounds with the same molecular formula but different arrangements of atoms. (467, 747)
controlled experiment An experiment that measures the effect of one variable at a time by keeping other variables constant. (9)
conversion factor A ratio of equivalent quantities that is equal to 1 and used to convert the units of a quantity. (10)
coordinate covalent bond A covalent bond formed when one atom donates both electrons to provide the shared pair and, once formed, is identical to any covalent single bond. (750)
coordination compound (also *complex*) A substance containing at least one complex ion and counter ion(s). (743)
coordination isomers Two or more coordination compounds with the same composition in which the complex ions have different ligand arrangements. (747)
coordination number In a crystal, the number of nearest neighbors surrounding a particle. (374) In a complex, the number of ligand atoms bonded to the central metal ion. (744)
core electrons (See *inner electrons.*)
corrosion The natural redox process that results in unwanted oxidation of a metal. (717)
coulomb (C) The SI unit of electric charge. One coulomb is the charge of 6.242×10^{18} electrons; one electron possesses a charge of 1.602×10^{-19} C. (697)
Coulomb's law A law stating that the electrostatic energy between particles A and B is directly proportional to the product of their charges and inversely proportional to the distance between them:

$$\text{electrostatic energy} \propto \frac{\text{charge A} \times \text{charge B}}{\text{distance}} \quad (282)$$

counter ion A simple ion associated with a complex ion in a coordination compound. (743)
coupling of reactions The pairing of reactions of which one releases enough free energy for the other to occur. (674)
covalent bond A type of bond in which atoms are bonded through the sharing of electrons; the mutual attraction of the nuclei and an electron pair that holds atoms together in a molecule. (51, 285)
covalent bonding The idealized bonding type that is based on localized electron-pair sharing between two atoms with little difference in their tendencies to lose or gain electrons (most commonly nonmetals). (278)
covalent compound A compound that consists of atoms bonded together by shared electron pairs. (49)
covalent radius One-half the shortest distance between nuclei of identical covalently bonded atoms. (258)
critical mass The minimum mass needed to achieve a chain reaction. (787)
critical point The point on a phase diagram above which the vapor cannot be condensed to a liquid; the end of the liquid-gas curve. (361)
crystal field splitting energy (Δ) The difference in energy between two sets of metal-ion d orbitals that results from electrostatic interactions with the surrounding ligands. (753)
crystal field theory A model that explains the color and magnetism of coordination compounds based on the effects of ligands on metal-ion d-orbital energies. (752)
crystalline solid Solid with a well-defined shape because of the orderly arrangement of the atoms, molecules, or ions. (373)
cubic closest packing A crystal structure based on the face-centered cubic unit cell in which the layers have an *abcabc . . .* pattern. (376)
cubic meter (m^3) The SI-derived unit of volume. (15)
curie (Ci) The most common unit of radioactivity, defined as the number of nuclei disintegrating each second in 1 g of radium-226; 1 Ci = 3.70×10^{10} d/s (disintegrations per second). (772)
cyclic hydrocarbon A hydrocarbon with one or more rings in its structure. (466)

D

***d* orbital** An atomic orbital with $l = 2$. (238)
dalton (Da) A unit of mass identical to *atomic mass unit.* (44)
Dalton's law of partial pressures A gas law stating that, in a mixture of unreacting gases, the total pressure is the sum of the partial pressures of the individual gases: $P_{total} = P_1 + P_2 + \cdots$. (165)

data Pieces of quantitative information obtained by observation. (9)
de Broglie wavelength The wavelength of a moving particle obtained from the de Broglie equation: $\lambda = h/mu$. (229)
decay constant The rate constant k for radioactive decay. (772)
decay series (also *disintegration series*) The succession of steps a parent nucleus undergoes as it decays into a stable daughter nucleus. (771)
dehydration-condensation reaction A reaction in which an H_2O molecule is lost for every pair of OH groups that join. (444)
delocalization (See *electron-pair delocalization.*)
density (*d*) An intensive physical property of a substance at a given temperature and pressure, defined as the ratio of the mass to the volume: $d = m/V$. (17)
deposition The process of changing directly from gas to solid. (352)
derived unit Any of various combinations of the seven SI base units. (14)
diagonal relationship Physical and chemical similarities between a Period 2 element and one located diagonally down and to the right in Period 3. (432)
diamagnetism The tendency of a species not to be attracted (or to be slightly repelled) by a magnetic field as a result of its electrons being paired. (269)
diastereomers (See *geometric isomers.*)
diffraction The phenomenon in which a wave striking the edge of an object bends around it. A wave passing through a slit as wide as its wavelength forms a circular wave. (220)
diffusion The movement of one fluid through another. (176)
dipole-dipole force The intermolecular attraction between oppositely charged poles of nearby polar molecules. (363)
dipole–induced dipole force The intermolecular attraction between a polar molecule and the oppositely charged pole it induces in a nearby molecule. (393)
dipole moment (μ) A measure of molecular polarity; the magnitude of the partial charges on the ends of a molecule (in coulombs) times the distance between them (in meters). (321)
disaccharide An organic compound formed by a dehydration-condensation reaction between two simple sugars (monosaccharides). (486)
disintegration series (See *decay series.*)
dispersion force (also *London force*) The intermolecular attraction between all particles as a result of instantaneous polarizations of their electron clouds; the intermolecular force primarily responsible for the condensed states of nonpolar substances. (366)
disproportionation A reaction in which a given substance is both oxidized and reduced. (442)
donor atom An atom that donates a lone pair of electrons to form a covalent bond, usually from ligand to metal ion in a complex. (744)
double bond A covalent bond that consists of two bonding pairs; two atoms sharing four electrons in the form of one σ and one π bond. (285)
double helix The two intertwined polynucleotide strands held together by H bonds that form the structure of DNA (deoxyribonucleic acid). (491)
Downs cell An industrial apparatus that electrolyzes molten NaCl to produce sodium and chlorine. (721)
dynamic equilibrium The condition at which the forward and reverse reactions are taking place at the same rate, so there is no net change in the amounts of reactants or products. (357)

E

e_g orbitals The set of orbitals (composed of $d_{x^2-y^2}$ and d_{z^2}) that results when the energies of the metal-ion d orbitals are split by a ligand field. This set is higher in energy than the other (t_{2g}) set in an octahedral field of ligands and lower in energy in a tetrahedral field. (753)
effective collision A collision in which the particles meet with sufficient energy and an orientation that allows them to react. (521)
effective nuclear charge (Z_{eff}) The nuclear charge an electron actually experiences as a result of shielding effects due to the presence of other electrons. (248)
effusion The process by which a gas escapes from its container through a tiny hole into an evacuated space. (174)
electrochemical cell A system that incorporates a redox reaction to produce or use electrical energy. (688)
electrochemistry The study of the relationship between chemical change and electrical work. (688)
electrode The part of an electrochemical cell that conducts the electricity between the cell and the surroundings. (692)
electrolysis The nonspontaneous lysing (splitting) of a substance, often to its component elements, by supplying electrical energy. (720)
electrolyte A substance that conducts a current when it dissolves in water. (117, 408) A mixture of ions, in which the electrodes of an electrochemical cell are immersed, that conducts a current. (692)
electrolytic cell An electrochemical system that uses electrical energy to drive a nonspontaneous chemical reaction ($\Delta G > 0$). (692)
electromagnetic radiation (also *electromagnetic energy* or *radiant energy*) Oscillating, perpendicular electric and magnetic fields moving simultaneously through space as waves and manifested as visible light, x-rays, microwaves, radio waves, and so on. (217)
electromagnetic spectrum The continuum of wavelengths of radiant energy. (218)
electromotive force (emf) (See *cell potential.*)
electron (e^-) A subatomic particle that possesses a unit negative charge (-1.60218×10^{-19} C) and occupies the space around the atomic nucleus. (43)
electron affinity (EA) The energy change (in kJ) accompanying the addition of one mole of electrons to one mole of gaseous atoms or ions. (263)
electron capture (EC) A type of radioactive decay in which a nucleus draws in an orbital electron, usually one from the lowest energy level, and releases energy. (767)
electron cloud depiction An imaginary representation of an electron's rapidly changing position around the nucleus over time. (233)
electron configuration The distribution of electrons within the orbitals of the atoms of an element; also the notation for such a distribution. (246)
electron deficient Referring to a bonded atom, such as Be or B, that has fewer than eight valence electrons. (309)
electron density diagram (also *electron probability density diagram*) The pictorial representation for a given energy sublevel of the quantity ψ^2 (the probability density of the electron lying within a particular tiny volume) as a function of r (distance from the nucleus). (233)

electron-pair delocalization (also *delocalization*) The process by which electron density is spread over several atoms rather than remaining between two. (307)

electron-sea model A qualitative description of metallic bonding proposing that metal atoms pool their valence electrons into a delocalized "sea" of electrons in which the metal cores (metal ions) are submerged in an orderly array. (382)

electron volt (eV) The energy (in joules, J) that an electron acquires when it moves through a potential difference of 1 volt; $1\ \text{eV} = 1.602\times10^{-19}$ J. (784)

electronegativity (EN) The relative ability of a bonded atom to attract shared electrons. (293)

electronegativity difference (ΔEN) The difference in electronegativities between the atoms in a bond. (295)

element The simplest type of substance with unique physical and chemical properties. An element consists of only one kind of atom, so it cannot be broken down into simpler substances. (33)

element symbol (See *atomic symbol.*)

elementary reaction (also *elementary step*) A simple reaction that describes a single molecular event in a proposed reaction mechanism. (525)

elimination reaction A type of organic reaction in which C atoms are bonded to fewer atoms in the product than in the reactant, which leads to multiple bonding. (473)

emission spectrum The line spectrum produced when excited atoms return to lower energy levels and emit photons characteristic of the element. (228)

empirical formula A chemical formula that shows the lowest relative number of atoms of each element in a compound. (80)

enantiomers (See *optical isomers.*)

end point The point in a titration at which the indicator changes color. (130, 628)

endothermic process A process that occurs with an absorption of heat from the surroundings and therefore an increase in the enthalpy of the system ($\Delta H > 0$). (196)

energy The capacity to do work, that is, to move matter. (7) [See also *kinetic energy* (E_k) and *potential energy* (E_p).]

enthalpy (*H*) A thermodynamic quantity that is the sum of the internal energy plus the product of the pressure and volume. (195)

enthalpy diagram A graphic depiction of the enthalpy change of a system. (196)

enthalpy of formation (ΔH_f) (also *heat of formation*) The enthalpy change occurring when 1 mol of a compound forms from its elements. When all components are in their standard states, this is called the *standard enthalpy of formation* (ΔH°_f). (205)

enthalpy of fusion (See *heat of fusion.*)

enthalpy of hydration (See *heat of hydration.*)

enthalpy of reaction (ΔH_{rxn}) (also *heat of reaction*) The enthalpy change that occurs during a reaction. When all components are in their standard states, this is called the *standard enthalpy of reaction* (ΔH°_{rxn}). (205)

enthalpy of solution (See *heat of solution.*)

enthalpy of sublimation (See *heat of sublimation.*)

enthalpy of vaporization (See *heat of vaporization.*)

entropy (*S*) A thermodynamic quantity related to the number of ways the energy of a system can be dispersed through the motions of its particles. (399, 656)

enzyme A biological macromolecule (usually a protein) that acts as a catalyst. (532)

equatorial group An atom (or group) that lies in the trigonal plane of a trigonal bipyramidal molecule, or a similar structural feature in a molecule. (316)

equilibrium constant (*K*) The value obtained when equilibrium concentrations are substituted into the reaction quotient. (544)

equilibrium vapor pressure (See *vapor pressure.*)

equivalence point The point in a titration when the number of moles of the added species is stoichiometrically equivalent to the original number of moles of the other species. (130, 628)

ester An organic compound that contains the $-\overset{\overset{\displaystyle :\ddot{O}:}{\|}}{C}-\ddot{\underset{\cdot\cdot}{O}}-\underset{|}{\overset{|}{C}}-$ group. (480)

exact number A quantity, usually obtained by counting or based on a unit definition, that has no uncertainty associated with it and, therefore, contains as many significant figures as a calculation requires. (23)

exchange reaction (See *metathesis reaction.*)

excited state Any electron configuration of an atom or molecule other than the lowest energy (ground) state. (224)

exclusion principle A principle developed by Wolfgang Pauli stating that no two electrons in an atom can have the same set of four quantum numbers. The principle arises from the fact that an orbital has a maximum occupancy of two electrons and their spins are paired. (247)

exothermic process A process that occurs with a release of heat to the surroundings and therefore a decrease in the enthalpy of the system ($\Delta H < 0$). (196)

expanded valence shell A valence level that can accommodate more than eight electrons by using available *d* orbitals; occurs only with central nonmetal atoms from Period 3 or higher. (310)

experiment A set of procedural steps that tests a hypothesis. (9)

extensive property A property, such as mass, that depends on the quantity of substance present. (20)

F

face-centered cubic unit cell A unit cell in which a particle occurs at each corner and in the center of each face of a cube. (374)

Faraday constant (*F*) The physical constant representing the charge of 1 mol of electrons: $F = 96{,}485$ C/mol e^-. (705)

fatty acid A carboxylic acid that has a long hydrocarbon chain and is derived from a natural source. (480)

first law of thermodynamics (See *law of conservation of energy.*)

fission The process by which a heavier nucleus splits into lighter nuclei with the release of energy. (783)

formal charge The hypothetical charge on an atom in a molecule or ion, equal to the number of valence electrons minus the sum of all the unshared and half the shared valence electrons. (308)

formation constant (K_f) An equilibrium constant for the formation of a complex ion from the hydrated metal ion and ligands. (644)

formation equation An equation in which 1 mole of a compound forms from its elements. (205)

formula mass The sum (in amu) of the atomic masses of a formula unit of a (usually ionic) compound. (59)

formula unit The chemical unit of a compound that contains the relative numbers of the types of atoms or ions expressed in the chemical formula. (53)

fossil fuel Any fuel, including coal, petroleum, and natural gas, derived from the products of the decay of dead organisms. (207)

fraction by mass (also *mass fraction*) The portion of a compound's mass contributed by an element; the mass of an element in a compound divided by the mass of the compound. (36)
free energy (G) A thermodynamic quantity that is the difference between the enthalpy and the product of the absolute temperature and the entropy: $G = H - TS$. (668)
free energy change (ΔG) The change in free energy that occurs during a reaction. (827)
free radical A molecular or atomic species with one or more unpaired electrons, which typically make it very reactive. (310)
freezing The process of cooling a liquid until it solidifies. (352)
freezing point depression (ΔT_f) The lowering of the freezing point of a solvent caused by the presence of dissolved solute particles. (411)
frequency (ν) The number of cycles a wave undergoes per second, expressed in units of 1/second, or s^{-1} [also called *hertz* (Hz)]; related inversely to wavelength. (217)
frequency factor (A) The product of the collision frequency Z and an orientation probability factor p that is specific for a reaction. (521)
fuel cell (also *flow battery*) A battery that is not self-contained and in which electricity is generated by the controlled oxidation of a fuel. (716)
functional group A specific combination of atoms, typically containing a carbon-carbon multiple bond and/or carbon-heteroatom bond, that reacts in a characteristic way no matter what molecule it occurs in. (461)
fundamental unit (See *base unit.*)
fusion (See *melting.*)
fusion (nuclear) The process by which light nuclei combine to form a heavier nucleus with the release of energy. (783)

G

galvanic cell (See *voltaic cell.*)
gamma emission The type of radioactive decay in which gamma rays are emitted from an excited nucleus. (767)
gamma (γ) ray A very high-energy photon. (767)
gas One of the three states of matter. A gas fills its container regardless of the shape because its particles are far apart. (6)
genetic code The set of three-base sequences that is translated into specific amino acids during the process of protein synthesis. (492)
geometric isomers (also *cis-trans isomers* or *diastereomers*) Stereoisomers in which the molecules have the same connections between atoms but differ in the spatial arrangements of the atoms. The *cis* isomer has similar groups on the same side of a double bond; the *trans* isomer has them on opposite sides. (469, 748)
Graham's law of effusion A gas law stating that the rate of effusion of a gas is inversely proportional to the square root of its density (or molar mass):

$$\text{rate} \propto \frac{1}{\sqrt{\mathcal{M}}} \qquad (174)$$

gray (Gy) The SI unit of absorbed radiation dose; 1 Gy = 1 J/kg tissue. (778)
ground state The electron configuration of an atom or ion that is lowest in energy. (224)
group A vertical column in the periodic table; elements in a group usually have the same outer electron configuration and, thus, similar chemical behavior. (47)

H

H bond (See *hydrogen bond.*)
Haber process An industrial process used to form ammonia from its elements. (570)
half-cell A portion of an electrochemical cell in which a half-reaction takes place. (694)
half-life ($t_{1/2}$) In chemical processes, the time required for half the initial reactant concentration to be consumed. (515) In nuclear processes, the time required for half the initial number of nuclei in a sample to decay. (773)
half-reaction method A method of balancing redox reactions by treating the oxidation and reduction half-reactions separately. (688)
haloalkane (also *alkyl halide*) A hydrocarbon with one or more halogen atoms (X) in place of H; contains a $-\overset{|}{\underset{|}{C}}-\ddot{\underset{..}{X}}:$ group. (476)
heat (q) (also *thermal energy*) The energy transferred between objects because of a difference in their temperatures only. (18, 190)
heat capacity The quantity of heat required to change the temperature of an object by 1 K. (197)
heat of formation (See *enthalpy of formation.*)
heat of fusion (ΔH_{fus}) (also *enthalpy of fusion*) The enthalpy change occurring when 1 mol of a solid substance melts; designated ΔH°_{fus} at the standard state. (352)
heat of hydration (ΔH_{hydr}) (also *enthalpy of hydration*) The enthalpy change occurring when 1 mol of a gaseous species (often an ion) is hydrated. The sum of the enthalpies from separating water molecules and mixing the gaseous species with them; designated ΔH°_{hydr} at the standard state. (398)
heat of reaction (ΔH_{rxn}) (See *enthalpy of reaction.*)
heat of solution (ΔH_{soln}) (also *enthalpy of solution*) The enthalpy change occurring when a solution forms from solute and solvent. The sum of the enthalpies from separating solute and solvent substances and mixing them; designated ΔH°_{soln} at the standard state. (397)
heat of sublimation (ΔH_{subl}) (also *enthalpy of sublimation*) The enthalpy change occurring when 1 mol of a solid substance changes directly to a gas. The sum of the heats of fusion and vaporization; designated ΔH°_{subl} at the standard state. (353)
heat of vaporization (ΔH_{vap}) (also *enthalpy of vaporization*) The enthalpy change occurring when 1 mol of a liquid substance vaporizes; designated ΔH°_{vap} at the standard state. (352)
heating-cooling curve A plot of temperature vs. time for a substance when heat is absorbed or released by the system at a constant rate. (354)
Henderson-Hasselbalch equation An equation for calculating the pH of a buffer system:

$$\text{pH} = \text{p}K_a + \log\left(\frac{[\text{base}]}{[\text{acid}]}\right) \qquad (623)$$

Henry's law A law stating that the solubility of a gas in a liquid is directly proportional to the partial pressure of the gas above the liquid: $S_{gas} = k_H \times P_{gas}$. (403)
Hess's law A law stating that the enthalpy change of an overall process is the sum of the enthalpy changes of the individual steps. (203)
heteroatom Any atom in an organic compound other than C or H. (461)

heterogeneous catalyst A catalyst that occurs in a different phase from the reactants, usually a solid interacting with gaseous or liquid reactants. (531)

heterogeneous mixture A mixture that has one or more visible boundaries among its components. (61)

hexagonal closest packing A crystal structure based on the hexagonal unit cell in which the layers have an *abab* . . . pattern. (376)

high-spin complex Complex ion that has the same number of unpaired electrons as in the isolated metal ion; contains weak-field ligands. (755)

homogeneous catalyst A catalyst (gas, liquid, or soluble solid) that exists in the same phase as the reactants. (530)

homogeneous mixture (also *solution*) A mixture that has no visible boundaries among its components. (61)

homologous series A series of organic compounds in which each member differs from the next by a $-CH_2-$ (methylene) group. (464)

homonuclear diatomic molecule A molecule composed of two identical atoms. (341)

Hund's rule A principle stating that when orbitals of equal energy are available, the electron configuration of lowest energy has the maximum number of unpaired electrons with parallel spins. (250)

hybrid orbital An atomic orbital postulated to form during bonding by the mathematical mixing of specific combinations of nonequivalent orbitals in a given atom. (329)

hybridization A postulated process of orbital mixing to form hybrid orbitals. (329)

hydrate A compound in which a specific number of water molecules are associated with each formula unit. (56)

hydration Solvation in water. (398)

hydration shell The oriented cluster of water molecules that surrounds an ion in aqueous solution. (393)

hydrocarbon An organic compound that contains only H and C atoms. (462)

hydrogen bond (H bond) A type of dipole-dipole force that arises between molecules that have an H atom bonded to a small, highly electronegative atom with lone pairs, usually N, O, or F. (364)

hydrogenation The addition of hydrogen to a carbon-carbon multiple bond to form a carbon-carbon single bond. (531)

hydrolysis Cleaving a molecule by reaction with water, in which one part of the molecule bonds to the water —OH and the other to the water H. (481)

hydronium ion (H_3O^+) A proton covalently bonded to a water molecule. (580)

hypothesis A testable proposal made to explain an observation. If inconsistent with experimental results, a hypothesis is revised or discarded. (9)

I

ideal gas A hypothetical gas that exhibits linear relationships among volume, pressure, temperature, and amount (mol) at all conditions; approximated by simple gases at ordinary conditions. (153)

ideal gas law (also *ideal gas equation*) An equation that expresses the relationships among volume, pressure, temperature, and amount (mol) of an ideal gas: $PV = nRT$. (157)

ideal solution A solution whose vapor pressure equals the mole fraction of the solvent times the vapor pressure of the pure solvent; approximated only by very dilute solutions. (409) (See also *Raoult's law.*)

indicator (See *acid-base indicator.*)

infrared (IR) The region of the electromagnetic spectrum between the microwave and visible regions. (218)

infrared (IR) spectroscopy An instrumental technique for determining the types of bonds in a covalent molecule by measuring the absorption of IR radiation. (289)

initial rate The instantaneous rate occurring as soon as the reactants are mixed, that is, at $t = 0$. (502)

inner electrons (also *core electrons*) Electrons that fill all the energy levels of an atom except the valence level; electrons also present in atoms of the previous noble gas and any completed transition series. (256)

inner transition elements The elements of the periodic table in which *f* orbitals are being filled; the lanthanides and actinides. (256)

instantaneous rate The reaction rate at a particular time, given by the slope of a tangent to a plot of reactant concentration vs. time. (502)

insulator A substance (usually a nonmetal) that does not conduct an electric current. (384)

integrated rate law A mathematical expression for reactant concentration as a function of time. (512)

intensive property A property, such as density, that does not depend on the quantity of substance present. (20)

intermolecular forces (also *interparticle forces*) The attractive and repulsive forces among the particles—molecules, atoms, or ions—in a sample of matter. (351)

internal energy (E) The sum of the kinetic and potential energies of all the particles in a system. (190)

ion A charged particle that forms from an atom (or covalently bonded group of atoms) when it gains or loses one or more electrons. (49)

ion-dipole force The intermolecular attractive force between an ion and a polar molecule (dipole). (362)

ion–induced dipole force The attractive force between an ion and the dipole it induces in the electron cloud of a nearby nonpolar molecule. (393)

ion pair A pair of ions that form a gaseous ionic molecule; sometimes formed when a salt boils. (283)

ion-product constant for water (K_w) The equilibrium constant for the autoionization of water; equal to 1.0×10^{-14} at 298 K.

$$K_w = [H_3O^+][OH^-] \qquad (584)$$

ionic atmosphere A cluster of ions of net opposite charge surrounding a given ion in solution. (415)

ionic bonding The idealized type of bonding based on the attraction of oppositely charged ions that arise through electron transfer between atoms with large differences in their tendencies to lose or gain electrons (typically metals and nonmetals). (277)

ionic compound A compound that consists of oppositely charged ions. (49)

ionic radius The size of an ion as measured by the distance between the nuclei of adjacent ions in a crystalline ionic compound. (269)

ionic solid A solid whose unit cell contains cations and anions. (380)

ionization In nuclear chemistry, the process by which a substance absorbs energy from high-energy radioactive particles and loses an electron to become ionized. (778)
ionization energy (IE) The energy (in kJ) required to remove completely one mole of electrons from one mole of gaseous atoms or ions. (260)
ionizing radiation The high-energy radiation from natural and artificial sources that forms ions in a substance by causing electron loss. (778)
isoelectronic Having the same number and configuration of electrons as another species. (266)
isomer One of two or more compounds with the same molecular formula but different properties, often as a result of different arrangements of atoms. (84, 467, 747)
isotopes Atoms of a given atomic number (that is, of a specific element) that have different numbers of neutrons and therefore different mass numbers. (44, 764)
isotopic mass The mass (in amu) of an isotope relative to the mass of carbon-12. (45)

J

joule (J) The SI unit of energy; $1\ J = 1\ kg{\cdot}m^2/s^2$. (193)

K

kelvin (K) The SI base unit of temperature. The kelvin is the same size as the Celsius degree. (18)
Kelvin scale (See *absolute scale.*)
ketone An organic compound (ending, *-one*) that contains a carbonyl group bonded to two other C atoms, $-\overset{|}{\underset{|}{C}}-\overset{:\ddot{O}:}{\overset{\|}{C}}-\overset{|}{\underset{|}{C}}-$. (479)
kilogram (kg) The SI base unit of mass. (16)
kinetic energy (E_k) The energy an object has because of its motion. (7)
kinetic-molecular theory The model that explains gas behavior in terms of particles in random motion whose volumes and interactions are negligible. (170)

L

lanthanide contraction The additional decrease in atomic and ionic size, beyond the expected trend, caused by the poor shielding of the increasing nuclear charge by *f* electrons in the elements following the lanthanides. (740)
lanthanides (also *rare earths*) The Period 6 (4*f*) series of inner transition elements, which includes cerium (Ce; $Z = 58$) through lutetium (Lu; $Z = 71$). (256)
lattice The three-dimensional arrangement of points created by choosing each point to be at the same location within each particle of a crystal; thus, the lattice consists of all points with identical surroundings. (374)
lattice energy ($\Delta H^\circ_{lattice}$) The enthalpy change (always positive) that accompanies the separation of 1 mol of a solid ionic compound into gaseous ions. (281)
law (See *natural law.*)
law of chemical equilibrium (also *law of mass action*) The law stating that when a system reaches equilibrium at a given temperature, the ratio of quantities that make up the reaction quotient has a constant numerical value. (545)
law of conservation of energy (also *first law of thermodynamics*) A basic observation that the total energy of the universe is constant; thus, $\Delta E_{universe} = \Delta E_{system} + \Delta E_{surroundings} = 0$. (193)
law of definite (or constant) composition A mass law stating that, no matter what its source, a particular compound is composed of the same elements in the same parts (fractions) by mass. (36)
law of mass action (See *law of chemical equilibrium.*)
law of mass conservation A mass law stating that the total mass of substances does not change during a chemical reaction. (35)
law of multiple proportions A mass law stating that if elements A and B react to form two or more compounds, the different masses of B that combine with a fixed mass of A can be expressed as a ratio of small whole numbers. (37)
Le Châtelier's principle A principle stating that if a system in a state of equilibrium is disturbed, it will undergo a change that shifts its equilibrium position in a direction that reduces the effect of the disturbance. (563)
level (also *shell*) A specific energy state of an atom given by the principal quantum number *n*. (235)
Lewis acid-base definition A model of acid-base behavior in which acids and bases are defined, respectively, as species that accept and donate an electron pair. (607)
Lewis electron-dot symbol A notation in which the element symbol represents the nucleus and inner electrons, and surrounding dots represent the valence electrons. (278)
Lewis structure (also *Lewis formula*) A structural formula consisting of electron-dot symbols, with lines as bonding pairs and dots as lone pairs. (303)
ligand A molecule or anion bonded to a central metal ion in a complex ion. (643, 743)
like-dissolves-like rule An empirical observation stating that substances having similar kinds of intermolecular forces dissolve in each other. (393)
limiting reactant (also *limiting reagent*) The reactant that is consumed when a reaction occurs and, therefore, the one that determines the maximum amount of product that can form. (93)
line spectrum A series of separated lines of different colors representing photons whose wavelengths are characteristic of an element. (223) (See also *emission spectrum.*)
linear arrangement The geometric arrangement obtained when two electron groups maximize their separation around a central atom. (313)
linear shape A molecular shape formed by three atoms lying in a straight line, with a bond angle of 180° (shape class AX_2 or AX_2E_3). (313)
linkage isomers Coordination compounds with the same composition but with different ligand donor atoms linked to the central metal ion. (747)
lipid Any of a class of biomolecules, including fats, that are soluble in nonpolar solvents and not soluble in water. (481)
liquid One of the three states of matter. A liquid fills a container to the extent of its own volume and thus forms a surface. (6)
liter (L) A non-SI unit of volume equivalent to 1 cubic decimeter ($0.001\ m^3$). (15)
London force (See *dispersion force.*)
lone pair (also *unshared pair*) An electron pair that is part of an atom's valence level but not involved in covalent bonding. (285)
low-spin complex Complex ion that has fewer unpaired electrons than in the free metal ion because of the presence of strong-field ligands. (755)

M

macromolecule (See *polymer*.)
magnetic quantum number (m_l) An integer from $-l$ through 0 to $+l$ that specifies the orientation of an atomic orbital in the three-dimensional space about the nucleus. (235)
mass The quantity of matter an object contains. Balances are designed to measure mass. (16)
mass-action expression (See *reaction quotient*.)
mass fraction (See *fraction by mass*.)
mass number (A) The total number of protons and neutrons in the nucleus of an atom. (43)
mass percent (also *mass %* or *percent by mass*) The fraction by mass expressed as a percentage. (36) A concentration term [% (w/w)] expressed as the mass of solute dissolved in 100. parts by mass of solution. (405)
mass spectrometry An instrumental method for measuring the relative masses of particles in a sample by creating charged particles and separating them according to their mass-charge ratio. (45)
matter Anything that possesses mass and occupies volume. (3)
melting (also *fusion*) The change of a substance from a solid to a liquid. (352)
melting point (mp or T_f) The temperature at which the solid and liquid forms of a substance are at equilibrium. (360)
metal A substance or mixture that is relatively shiny and malleable and is a good conductor of heat and electricity. In reactions, metals tend to transfer electrons to nonmetals and form ionic compounds. (47)
metallic bonding An idealized type of bonding based on the attraction between metal ions and their delocalized valence electrons. (278) (See also *electron-sea model* and *bond theory.*)
metallic radius One-half the shortest distance between the nuclei of adjacent individual atoms in a crystal of an element. (258)
metallic solid A solid whose individual atoms are held together by metallic bonding. (381)
metalloid (also *semimetal*) An element with properties between those of metals and nonmetals. (48)
metathesis reaction (also *exchange reaction*) A reaction in which atoms or ions of two compounds exchange bonding partners. (123)
meter (m) The SI base unit of length. The distance light travels in a vacuum in 1/299,792,458 second. (14)
milliliter (mL) A volume (0.001 L) equivalent to 1 cm^3. (15)
millimeter of mercury (mmHg) A unit of pressure based on the difference in the heights of mercury in a barometer or manometer. Renamed the *torr* in honor of Torricelli. (151)
miscible Soluble in any proportion. (392)
mixture A group of two or more elements and/or compounds that are physically intermingled. (34)
MO bond order One-half the difference between the number of electrons in bonding and antibonding MOs. (340)
model (also *theory*) A simplified conceptual picture based on experiment that explains how a natural phenomenon occurs. (9)
molality (m) A concentration term expressed as number of moles of solute dissolved in 1000 g (1 kg) of solvent. (404)
molar heat capacity (C) The quantity of heat required to change the temperature of 1 mol of a substance by 1 K. (198)
molar mass ($\mathcal{M}$) The mass of 1 mol of entities (atoms, molecules, or formula units) of a substance, in units of g/mol. (73)
molarity (M) A concentration term expressed as the moles of solute dissolved in 1 L of solution. (99)
mole (mol) The SI base unit for amount of a substance. The amount that contains a number of objects equal to the number of atoms in exactly 12 g of carbon-12 (which is 6.022×10^{23}). (72)
mole fraction (X) A concentration term expressed as the ratio of moles of one component of a mixture to the total moles present. (165, 405)
molecular equation A chemical equation showing a reaction in solution in which reactants and products appear as intact, undissociated compounds. (120)
molecular formula A formula that shows the actual number of atoms of each element in a molecule of a compound. (60, 80)
molecular mass (also *molecular weight*) The sum (in amu) of the atomic masses of the elements in a molecule (or formula unit) of a compound. (59)
molecular orbital (MO) An orbital of given energy and shape that extends over a molecule and can be occupied by no more than two paired electrons. (338)
molecular orbital (MO) diagram A depiction of the relative energy and number of electrons in each MO, as well as the atomic orbitals from which the MOs form. (339)
molecular orbital (MO) theory A model that describes a molecule as a collection of nuclei and electrons in which the electrons occupy orbitals that extend over the entire molecule. (338)
molecular polarity The overall distribution of electronic charge in a molecule, determined by its shape and bond polarities. (320)
molecular shape The three-dimensional structure defined by the relative positions of the atomic nuclei in a molecule. (312)
molecular solid A solid held together by intermolecular forces between individual molecules. (379)
molecular weight (See *molecular mass.*)
molecularity The number of reactant particles involved in an elementary step. (525)
molecule A structure consisting of two or more atoms that are bound chemically and behave as an independent unit. (33)
monatomic ion An ion derived from a single atom. (49)
monomer A small molecule, linked covalently to others of the same or similar type to form a polymer; the repeat unit of the polymer. (483)
mononucleotide A monomer unit of a nucleic acid, consisting of an N-containing base, a sugar, and a phosphate group. (490)
monosaccharide A simple sugar; a polyhydroxy ketone or aldehyde with three to nine C atoms. (486)

N

natural law (also *law*) A summary, often in mathematical form, of a universal observation. (9)
Nernst equation An equation stating that the voltage of an electrochemical cell under any conditions depends on the standard cell voltage and the concentrations of the cell components:

$$E_{\text{cell}} = E^\circ_{\text{cell}} - \frac{RT}{nF}\ln Q \qquad (708)$$

net ionic equation A chemical equation of a reaction in solution in which spectator ions have been eliminated to show the actual chemical change. (121)
network covalent solid A solid in which all the atoms are bonded covalently so that individual molecules are not present. (381)
neutralization Process that occurs when an H^+ ion from an acid combines with an OH^- ion from a base to form H_2O. (580)

neutralization reaction An acid-base reaction that yields water and a solution of a salt; when the H^+ ions of a strong acid react with an equivalent amount of the OH^- ions of a strong base, the solution is neutral. (145)
neutron (n^0) An uncharged subatomic particle found in the nucleus, with a mass slightly greater than that of a proton. (43)
nitrile An organic compound containing the —C≡N: group. (482)
node A region of an orbital where the probability of finding the electron is zero. (238)
nonelectrolyte A substance whose aqueous solution does not conduct an electric current. (120, 408)
nonmetal An element that lacks metallic properties. In reactions, nonmetals tend to share electrons with each other to form covalent compounds or accept electrons from metals to form ionic compounds. (48)
nonpolar covalent bond A covalent bond between identical atoms such that the bonding pair is shared equally. (295)
nuclear binding energy The energy required to break 1 mole of nuclei of an element into individual nucleons. (784)
nuclear transmutation The induced conversion of one nucleus into another by bombardment with a particle. (776)
nucleic acid An unbranched polymer consisting of mononucleotides that occurs as two types, DNA and RNA (deoxyribonucleic and ribonucleic acids), which differ chemically in the nature of the sugar portion of the mononucleotides. (490)
nucleon A subatomic particle found in the nucleus; a proton or neutron. (764)
nucleus The tiny central region of the atom that contains all the positive charge and essentially all the mass. (42)
nuclide A nuclear species with specified numbers of protons and neutrons. (764)

O

observation A fact obtained with the senses, often with the aid of instruments. Quantitative observations provide data that can be compared. (9)
octahedral arrangement The geometric arrangement obtained when six electron groups maximize their space around a central atom; when all six groups are bonding groups, the molecular shape is octahedral (AX_6; ideal bond angle = 90°). (317)
octet rule The observation that when atoms bond, they often lose, gain, or share electrons to attain a filled outer level of eight electrons (or two for H and Li). (279)
optical isomers (also *enantiomers*) A pair of stereoisomers consisting of a molecule and its mirror image that cannot be superimposed on each other. (468, 748)
optically active Able to rotate the plane of polarized light. (469)
orbital diagram A depiction of orbital occupancy in terms of electron number and spin shown by means of arrows in a series of small boxes, lines, or circles. (249)
organic compound A compound in which carbon is nearly always bonded to itself and to hydrogen, and often to other elements. (460)
osmosis The process by which solvent flows through a semipermeable membrane from a dilute to a concentrated solution. (412)
osmotic pressure (Π) The pressure that results from the ability of solvent, but not solute, particles to cross a semipermeable membrane. The pressure required to prevent the net movement of solvent across the membrane. (413)
outer electrons Electrons that occupy the highest energy level (highest n value) and are, on average, farthest from the nucleus. (256)
overvoltage The additional voltage, usually associated with gaseous products forming at an electrode, that is required above the standard cell voltage to accomplish electrolysis. (722)
oxidation The loss of electrons by a species, accompanied by an increase in oxidation number. (133)
oxidation number (O.N.) (also *oxidation state*) A number equal to the magnitude of the charge an atom would have if its shared electrons were held completely by the atom that attracts them more strongly. (133)
oxidation-reduction reaction (also *redox reaction*) A process in which there is a net movement of electrons from one reactant (reducing agent) to another (oxidizing agent). (132)
oxidation state (See *oxidation number.*)
oxidizing agent The substance that accepts electrons in a reaction and undergoes a decrease in oxidation number. (133)
oxoanion An anion in which an element is bonded to one or more oxygen atoms. (56)

P

***p* orbital** An atomic orbital with $l = 1$. (238)
packing efficiency The percentage of the available volume occupied by atoms, ions, or molecules in a unit cell. (376)
paramagnetism The tendency of a species with unpaired electrons to be attracted by an external magnetic field. (269)
partial ionic character An estimate of the actual charge separation in a bond (caused by the electronegativity difference of the bonded atoms) relative to complete separation. (295)
partial pressure The portion of the total pressure contributed by a gas in a mixture of gases. (165)
particle accelerator A device used to impart high kinetic energies to nuclear particles. (776)
pascal (Pa) The SI unit of pressure; 1 Pa = 1 N/m^2. (151)
penetration The process by which an outer electron moves through the region occupied by the core electrons to spend part of its time closer to the nucleus; penetration increases the average effective nuclear charge for that electron. (248)
percent by mass (mass %) (See *mass percent.*)
percent yield (% yield) The actual yield of a reaction expressed as a percentage of the theoretical yield. (98)
period A horizontal row of the periodic table. (47)
periodic law A law stating that when the elements are arranged by atomic number, they exhibit a periodic recurrence of properties. (246)
periodic table of the elements A table in which the elements are arranged by atomic number into columns (groups) and rows (periods). (47)
pH The negative common logarithm of [H_3O^+]. (585)
phase A physically distinct and homogeneous part of a system. (351)
phase change A physical change from one phase to another, usually referring to a change in physical state. (351)
phase diagram A diagram used to describe the stable phases and phase changes of a substance as a function of temperature and pressure. (360)
photoelectric effect The observation that when monochromatic light of sufficient frequency shines on a metal, an electric current is produced. (221)

photon A quantum of electromagnetic radiation. (221)
physical change A change in which the physical form (or state) of a substance, but not its composition, is altered. (3)
physical property A characteristic shown by a substance itself, without interacting with or changing into other substances. (3)
pi (π) bond A covalent bond formed by sideways overlap of two atomic orbitals that has two regions of electron density, one above and one below the internuclear axis. (336)
pi (π) MO A molecular orbital formed by combination of two atomic (usually *p*) orbitals whose orientations are perpendicular to the internuclear axis. (341)
Planck's constant (*h*) A proportionality constant relating the energy and frequency of a photon, equal to 6.626×10^{-34} J·s. (221)
polar covalent bond A covalent bond in which the electron pair is shared unequally, so the bond has partially negative and partially positive poles. (294)
polar molecule A molecule with an unequal distribution of charge as a result of its polar bonds and shape. (116)
polarizability The ease with which a particle's electron cloud can be distorted. (366)
polyatomic ion An ion in which two or more atoms are bonded covalently. (52)
polymer (also *macromolecule*) An extremely large molecule that results from the covalent linking of many simpler molecular units (monomers). (483)
polyprotic acid An acid with more than one ionizable proton. (597)
polysaccharide A macromolecule composed of many simple sugars linked covalently. (486)
positron (β^+) The antiparticle of an electron. (767)
positron emission A type of radioactive decay in which a positron is emitted from a nucleus. (767)
potential energy (E_p) The energy an object has as a result of its position relative to other objects or because of its composition. (7)
precipitate The insoluble product of a precipitation reaction. (122)
precipitation reaction A reaction in which two soluble ionic compounds form an insoluble product, a precipitate. (122)
precision (also *reproducibility*) The closeness of a measurement to other measurements of the same phenomenon in a series of experiments. (24)
pressure (*P*) The force exerted per unit of surface area. (150)
pressure-volume work (*PV* work) A type of work in which a volume change occurs against an external pressure. (195)
principal quantum number (*n*) A positive integer that specifies the energy and relative size of an atomic orbital; a number that specifies an energy level in an atom. (234)
probability contour A shape that defines the volume around an atomic nucleus within which an electron spends a given percentage of its time. (234)
product A substance formed in a chemical reaction. (86)
property A characteristic that gives a substance its unique identity. (3)
protein A natural, linear polymer composed of any of about 20 types of amino acid monomers linked together by peptide bonds. (487)
proton (p^+) A subatomic particle found in the nucleus that has a unit positive charge (1.60218×10^{-19} C). (43)
proton acceptor A substance that accepts an H^+ ion; a Brønsted-Lowry base. (588)
proton donor A substance that donates an H^+ ion; a Brønsted-Lowry acid. (588)
pseudo–noble gas configuration The $(n-1)d^{10}$ configuration of a *p*-block metal ion that has an empty outer energy level. (267)

Q

quantum A packet of energy equal to $h\nu$. The smallest quantity of energy that can be emitted or absorbed. (221)
quantum mechanics The branch of physics that examines the wave motion of objects on the atomic scale. (232)
quantum number A number that specifies a property of an orbital or an electron. (221)

R

rad (*r*adiation-*a*bsorbed *d*ose) The quantity of radiation that results in 0.01 J of energy being absorbed per kilogram of tissue; 1 rad = 0.01 J/kg tissue = 10^{-2} Gy. (778)
radial probability distribution plot The graphic depiction of the total probability distribution (sum of ψ^2) of an electron in the region near the nucleus. (234)
radioactivity The emissions resulting from the spontaneous disintegration of an unstable nucleus. (764)
radioisotope An isotope with an unstable nucleus that decays through radioactive emissions. (774)
radioisotopic dating A method for determining the age of an object based on the rate of decay of a particular radioactive nuclide relative to a stable nuclide. (774)
random error Error that occurs in all measurements (with its size depending on the measurer's skill and the instrument's precision) and results in values *both* higher and lower than the actual value. (24)
Raoult's law A law stating that the vapor pressure of solvent above a solution equals the mole fraction of solvent times the vapor pressure of pure solvent: $P_{\text{solvent}} = X_{\text{solvent}} \times P^\circ_{\text{solvent}}$. (409)
rare earths (See *lanthanides.*)
rate constant (*k*) The proportionality constant that relates reaction rate to reactant (and product) concentrations. (504)
rate-determining step (also *rate-limiting step*) The slowest step in a reaction mechanism and therefore the step that limits the overall rate. (526)
rate law (also *rate equation*) An equation that expresses the rate of a reaction as a function of reactant (and product) concentrations. (504)
reactant A starting substance in a chemical reaction. (86)
reaction energy diagram A graph that shows the potential energy of a reacting system as it progresses from reactants to products. (522)
reaction intermediate A substance that is formed and used up during the overall reaction and therefore does not appear in the overall equation. (527)
reaction mechanism A series of elementary steps that sum to the overall reaction and is consistent with the rate law. (525)
reaction order The exponent of a reactant concentration in a rate law that shows how the rate is affected by changes in that concentration. (505)

reaction quotient (Q) (also *mass-action expression*) A ratio of terms for a given reaction consisting of product concentrations multiplied together and divided by reactant concentrations multiplied together, each raised to the power of their balancing coefficient. The value of Q changes until the system reaches equilibrium, at which point it equals K. (545)
reaction rate The change in the concentrations of reactants (or products) with time. (499)
reactor core The part of a nuclear reactor that contains the fuel rods and generates heat from fission. (788)
redox reaction (See *oxidation-reduction reaction.*)
reducing agent The substance that donates electrons in a redox reaction and undergoes an increase in oxidation number. (133)
reduction The gain of electrons by a species, accompanied by a decrease in oxidation number. (133)
refraction A phenomenon in which a wave changes its speed and therefore its direction as it passes through a phase boundary into a different medium. (219)
rem (*r*oentgen *e*quivalent for *m*an) The unit of radiation dosage for a human based on the product of the number of rads and a factor related to the biological tissue; 1 rem $= 10^{-2}$ Sv. (778)
reproducibility (See *precision.*)
resonance hybrid The weighted average of the resonance structures of a molecule. (306)
resonance structure (also *resonance form*) One of two or more Lewis structures for a molecule that cannot be adequately depicted by a single structure. Resonance structures differ only in the position of bonding and lone electron pairs. (306)
rms (root-mean-square) speed (u_{rms}) The speed of a molecule having the average kinetic energy; very close to the most probable speed. (174)
round off The process of removing digits based on a series of rules to obtain an answer with the proper number of significant figures (or decimal places). (22)

S

***s* orbital** An atomic orbital with $l = 0$. (237)
salt An ionic compound that results from an Arrhenius acid-base reaction after solvent is removed. (128)
salt bridge An inverted U tube containing a solution of a nonreacting electrolyte that connects the compartments of a voltaic cell and maintains neutrality by allowing ions to flow between compartments. (695)
saturated hydrocarbon A hydrocarbon in which each C is bonded to four other atoms. (464)
saturated solution A solution that contains the maximum amount of dissolved solute at a given temperature (prepared with undissolved solute present). (401)
Schrödinger equation An equation that describes how the electron matter-wave changes in space around the nucleus. Solutions of the equation provide allowable energy levels of the atom. (232)
scientific method A process of creative thinking and testing aimed at objective, verifiable discoveries of the causes of natural events. (9)
screening (See *shielding.*)
second (s) The SI base unit of time. (20)
second law of thermodynamics A law stating that a process occurs spontaneously in the direction that increases the entropy of the universe. (659)
seesaw shape A molecular shape caused by the presence of one equatorial lone pair in a trigonal bipyramidal arrangement (AX_4E). (316)
self-ionization (See *autoionization.*)
semiconductor A substance whose electrical conductivity is poor at room temperature but increases significantly with rising temperature. (384)
semimetal (See *metalloid.*)
semipermeable membrane A membrane that allows solvent, but not solute, to pass through. (412)
shared pair (See *bonding pair.*)
shell (See *level.*)
shielding (also *screening*) The ability of other electrons, especially those occupying inner orbitals, to lessen the nuclear attraction for an outer electron. (248)
SI unit A unit composed of one or more of the base units of the Système International d'Unités, a revised metric system. (14)
side reaction An undesired chemical reaction that consumes some of the reactant and reduces the overall yield of the desired product. (97)
sievert (Sv) The SI unit of human radiation dosage; 1 Sv = 100 rem. (778)
sigma (σ) bond A type of covalent bond that arises through end-to-end orbital overlap and has most of its electron density along an imaginary line joining the nuclei. (335)
sigma (σ) MO A molecular orbital that is cylindrically symmetrical about an imaginary line that runs through the nuclei of the component atoms. (339)
significant figures The digits obtained in a measurement. The greater the number of significant figures, the greater the certainty of the measurement. (21)
silicate A type of compound found throughout rocks and soil and consisting of repeating —Si—O groupings and, in most cases, metal cations. (438)
silicone A type of synthetic polymer containing —Si—O repeat units, with organic groups and crosslinks. (438)
simple cubic unit cell A unit cell in which a particle occupies each corner of a cube. (374)
single bond A bond that consists of one electron pair. (285)
solid One of the three states of matter. A solid has a fixed shape that does not conform to the container shape. (5)
solubility (S) The maximum amount of solute that dissolves in a fixed quantity of a particular solvent at a specified temperature. (392)
solubility-product constant (K_{sp}) An equilibrium constant for a slightly soluble ionic compound dissolving in water. (634)
solute The substance that dissolves in the solvent. (99, 392)
solution (See *homogeneous mixture.*)
solvated Surrounded closely by solvent molecules. (117)
solvation The process of surrounding a solute particle with solvent particles. (397)
solvent The substance in which the solute(s) dissolve. (99, 392)
***sp* hybrid orbital** An orbital formed by the mixing of one s and one p orbital of a central atom. (330)
sp^2 hybrid orbital An orbital formed by the mixing of one s and two p orbitals of a central atom. (330)
sp^3 hybrid orbital An orbital formed by the mixing of one s and three p orbitals of a central atom. (332)
sp^3d hybrid orbital An orbital formed by the mixing of one s, three p, and one d orbital of a central atom. (333)

sp^3d^2 hybrid orbital An orbital formed by the mixing of one *s*, three *p*, and two *d* orbitals of a central atom. (333)
specific heat capacity (*c*) The quantity of heat required to change the temperature of 1 gram of a material by 1 K. (197)
spectator ion An ion that is present as part of a reactant but is not involved in the chemical change. (120)
spectrochemical series A ranking of ligands in terms of their ability to split *d*-orbital energies. (754)
spectrometry Any instrumental technique that uses a portion of the electromagnetic spectrum to measure the atomic and molecular energy levels of a substance. (228)
speed of light (*c*) A fundamental constant giving the speed at which electromagnetic radiation travels in a vacuum: $c = 2.99792458\times10^8$ m/s. (217)
spin quantum number (m_s) A number, either $+\frac{1}{2}$ or $-\frac{1}{2}$, that indicates the direction of electron spin. (247)
spontaneous change A change that occurs by itself, that is, without an ongoing input of external energy. (654)
square planar shape A molecular shape (AX_4E_2) caused by the presence of two lone pairs at opposite vertices in an octahedral arrangement. (317)
square pyramidal shape A molecular shape (AX_5E) caused by the presence of one lone pair in an octahedral arrangement. (317)
standard atmosphere (atm) The average atmospheric pressure measured at sea level and 0°C, defined as 1.01325×10^5 Pa. (151)
standard cell potential (E°_{cell}) The potential of a cell measured with all components in their standard states and no current flowing. (697)
standard electrode potential ($E^\circ_{half\text{-}cell}$) (also *standard half-cell potential*) The standard potential of a half-cell, with the half-reaction written as a reduction. (698)
standard enthalpy of formation (ΔH°_f) (also *standard heat of formation*) The enthalpy change occurring when 1 mol of a compound forms from its elements with all components in their standard states. (205)
standard enthalpy of reaction (ΔH°_{rxn}) (also *standard heat of reaction*) The enthalpy change that occurs during a reaction when all components are in their standard states. (205)
standard entropy of reaction (ΔS°_{rxn}) The entropy change that occurs when all components are in their standard states. (664)
standard free energy change (ΔG°) The free energy change that occurs when all components are in their standard states. (669)
standard free energy of formation (ΔG°_f) The standard free energy change that occurs when 1 mol of a compound is made from its elements with all components in their standard states. (670)
standard half-cell potential (See *standard electrode potential.*)
standard heat of formation (See *standard enthalpy of formation.*)
standard heat of reaction (See *standard enthalpy of reaction.*)
standard hydrogen electrode (See *standard reference half-cell.*)
standard molar entropy (S°) The entropy of 1 mol of a substance in its standard state. (660)
standard molar volume The volume of 1 mol of an ideal gas at standard temperature and pressure: 22.4141 L. (157)
standard reference half-cell (also *standard hydrogen electrode*) A specially prepared platinum electrode immersed in 1 *M* $H^+(aq)$ through which H_2 gas at 1 atm is bubbled. $E^\circ_{half\text{-}cell}$ is defined as 0 V. (698)
standard state A set of specifications used to compare thermodynamic data: 1 atm for gases behaving ideally, 1 *M* for dissolved species, or the pure substance for liquids and solids. (205)
standard temperature and pressure (STP) The reference conditions for a gas:

$$0°\text{C } (273.15 \text{ K}) \text{ and } 1 \text{ atm } (760 \text{ torr}) \qquad (156)$$

state function A property of the system determined by its current state, regardless of how it arrived at that state. (194)
state of matter One of the three physical forms of matter: solid, liquid, or gas. (5)
stationary state In the Bohr model, one of the allowable energy levels of the atom in which it does not release or absorb energy. (224)
stereoisomers Molecules with the same connections of atoms but different orientations of groups in space. (468, 748) (See also *geometric isomers* and *optical isomers.*)
stoichiometric coefficient (See *balancing coefficient.*)
stoichiometry The study of the mass-mole-number relationships of chemical formulas and reactions. (72)
strong-field ligand A ligand that causes larger crystal field splitting energy and therefore is part of a low-spin complex. (753)
structural formula A formula that shows the actual number of atoms, their relative placement, and the bonds between them. (60, 80)
structural isomers (See *constitutional isomers.*)
sublevel (also *subshell*) An energy substate of an atom within a level. Given by the *n* and *l* values, the sublevel designates the size and shape of the atomic orbitals. (235)
sublimation The process by which a solid changes directly into a gas. (352)
substance A type of matter, either an element or a compound, that has a fixed composition. (33)
substitution reaction An organic reaction that occurs when an atom (or group) from one reactant substitutes for one in another reactant. (473)
superconductivity The ability to conduct a current with no loss of energy to resistive heating. (384)
supersaturated solution An unstable solution in which more solute is dissolved than in a saturated solution. (401)
surface tension The energy required to increase the surface area of a liquid by a given amount. (369)
surroundings All parts of the universe other than the system being considered. (189)
system The defined part of the universe under study. (189)
systematic error A type of error producing values that are all either higher or lower than the actual value, often caused by faulty equipment or a consistent flaw in technique. (24)

T

t_{2g} orbitals The set of orbitals (composed of d_{xy}, d_{yz}, and d_{xz}) that results when the energies of the metal-ion *d* orbitals are split by a ligand field. This set is lower in energy than the other (e_g) set in an octahedral field and higher in energy in a tetrahedral field. (753)
T shape A molecular shape caused by the presence of two equatorial lone pairs in a trigonal bipyramidal arrangement (AX_3E_2). (317)
temperature (*T*) A measure of how hot or cold a substance is relative to another substance. (18) A measure of the average kinetic energy of the particles in a sample. (208)
tetrahedral arrangement The geometric arrangement formed when four electron groups maximize their separation around a central atom; when all four groups are bonding groups, the molecular shape is tetrahedral (AX_4; ideal bond angle 109.5°). (315)

theoretical yield The amount of product predicted by the stoichiometrically equivalent molar ratio in the balanced equation. (97)
theory (See *model.*)
thermochemical equation A chemical equation that shows the heat involved for the amounts of substances specified. (201)
thermochemistry The branch of thermodynamics that focuses on the heat involved in chemical and physical change. (189)
thermodynamics The study of heat (thermal energy) and its interconversions. (189)
thermometer A device for measuring temperature that contains a fluid that expands or contracts within a graduated tube. (18)
third law of thermodynamics A law stating that the entropy of a perfect crystal is zero at 0 K. (659)
titration A method of determining the concentration of a solution by monitoring relative amounts during its reaction with a solution of known concentration. (130)
torr A unit of pressure identical to 1 mmHg. (151)
total ionic equation An equation for an aqueous reaction that shows all the soluble ionic substances dissociated into ions. (120)
tracer A radioisotope that signals the presence of the species of interest by radioactive emissions. (780)
transition element An element that occupies the *d* block of the periodic table; one whose *d* orbitals are being filled. (253, 737)
transition state (also *activated complex*) An unstable species formed in an effective collision of reactants that exists momentarily when the system is highest in energy and that can either form products or re-form reactants. (522)
transition state theory A model that explains how the energy of reactant collisions is used to form a high-energy transitional species that can change to reactant or product. (522)
transuranium element An element with atomic number higher than that of uranium ($Z = 92$). (777)
trigonal bipyramidal arrangement The geometric arrangement formed when five electron groups maximize their separation around a central atom. When all five groups are bonding groups, the molecular shape is trigonal bipyramidal (AX_5; ideal bond angles, axial-center-equatorial 90° and equatorial-center-equatorial 120°). (316)
trigonal planar arrangement The geometric arrangement formed when three electron groups maximize their separation around a central atom. When all three groups are bonding groups, the molecular shape is trigonal planar (AX_3; ideal bond angle 120°). (314)
trigonal pyramidal shape A molecular shape (AX_3E) caused by the presence of one lone pair in a tetrahedral arrangement. (315)
triple bond A covalent bond that consists of three bonding pairs, two atoms sharing six electrons; one σ and two π bonds. (286)
triple point The pressure and temperature at which three phases of a substance are in equilibrium. In a phase diagram, the point at which three phase-transition curves meet. (360)

U

ultraviolet (UV) Radiation in the region of the electromagnetic spectrum between the visible and the x-ray regions. (218)
uncertainty A characteristic of every measurement that results from the inexactness of the measuring device and the need to estimate when taking a reading. (21)
uncertainty principle The principle stated by Werner Heisenberg that it is impossible to know simultaneously the exact position and velocity of a particle; the principle becomes important only for particles of very small mass. (231)
unimolecular reaction An elementary reaction that involves the decomposition or rearrangement of a single particle. (525)
unit cell The smallest portion of a crystal that, if repeated in all three directions, gives the crystal. (374)
universal gas constant (*R*) A proportionality constant that relates the energy, amount of substance, and temperature of a system; $R = 0.0820578$ atm·L/mol·K = 8.31447 J/mol·K. (157)
unsaturated hydrocarbon A hydrocarbon with at least one carbon-carbon multiple bond; one in which at least two C atoms are bonded to fewer than four atoms. (469)
unsaturated solution A solution in which more solute can be dissolved at a given temperature. (401)
unshared pair (See *lone pair.*)

V

V shape (See *bent shape.*)
valence band In band theory, the lower energy portion of the band of molecular orbitals, which is filled with valence electrons. (383)
valence bond (VB) theory A model that attempts to reconcile the shapes of molecules with those of atomic orbitals through the concepts of orbital overlap and hybridization. (329)
valence electrons The electrons involved in compound formation; in main-group elements, the electrons in the valence (outer) level. (256)
valence-shell electron-pair repulsion (VSEPR) theory A model explaining that the shapes of molecules and ions result from minimizing electron-pair repulsions around a central atom. (312)
van der Waals constants Experimentally determined positive numbers used in the van der Waals equation to account for the intermolecular attractions and molecular volume of real gases. (178)
van der Waals equation An equation that accounts for the behavior of real gases. (178)
van der Waals radius One-half of the shortest distance between the nuclei of identical nonbonded atoms. (362)
vapor pressure (also *equilibrium vapor pressure*) The pressure exerted by a vapor at equilibrium with its liquid in a closed system. (357)
vapor pressure lowering (ΔP) The decrease in the vapor pressure of a solvent caused by the presence of dissolved solute particles. (408)
vaporization The process of changing from a liquid to a gas. (352)
variable A quantity that can have more than a single value. (9) (See also *controlled experiment.*)
viscosity A measure of the resistance of a liquid to flow. (370)
volt (V) The SI unit of electric potential: 1 V = 1 J/C. (697)
voltage (See *cell potential.*)
voltaic cell (also *galvanic cell*) An electrochemical cell that uses a spontaneous redox reaction to generate electric energy. (691)
volume (*V*) The space occupied by a sample of matter. (15)
volume percent [% (v/v)] A concentration term defined as the volume of solute in 100. volumes of solution. (405)

W

wave function (See *atomic orbital.*)
wave-particle duality The principle stating that both matter and energy have wavelike and particle-like properties. (231)
wavelength (λ) The distance between any point on a wave and the corresponding point on the next wave, that is, the distance a wave travels during one cycle. (217)

weak-field ligand A ligand that causes smaller crystal field splitting energy and therefore is part of a high-spin complex. (753)

weight The force exerted by a gravitational field on an object that is directly proportional to its mass. (16)

work (*w*) The energy transferred when an object is moved by a force. (190)

x-ray diffraction analysis An instrumental technique used to determine dimensions of a crystal structure by measuring the diffraction patterns caused by x-rays impinging on the crystal. (376)

Credits

Chapter 1

Figure 1.1a: © Paul Morrell/Stone/Getty Images; 1.1b, 1.5, 1.8a-b: © The McGraw-Hill Companies, Inc. Stephen Frisch, photographer.

Chapter 2

Opener: © jokerpro/Shutterstock.com; Table 2.1:© The McGraw-Hill Companies, Inc. Stephen Frisch, photographer; 2.2(statue): © Punchstock RF; 2.2(coral): © Alexander Cherednichenko/Shutterstock.com; 2.10a, 2.10e: © The McGraw-Hill Companies, Inc. Stephen Frisch, photographer; 2.14: © Mark Schneider/Visuals Unlimited; 2.17a,b: © The McGraw-Hill Companies, Inc. Stephen Frisch, photographer.

Chapter 3

Opener: © Ray Pfortner/Peter Arnold/Photolibrary.com; 3.1, 3.6, 3.10: © The McGraw-Hill Companies, Inc. Stephen Frisch, Photographer.

Chapter 4

Opener: © Richard Megna/Fundamental Photographs, NYC; 4.3a-c, 4.4: © The McGraw-Hill Companies, Inc. Stephen Frisch, photographer; 4.5: © The McGraw-Hill Companies, Inc. Richard Megna, photographer; 4.6, 4.7, 4.9a-c, 4.13(all), 4.14, 4.15(both): © The McGraw-Hill Companies, Inc. Stephen Frisch, photographer.

Chapter 5

Opener: © Lena Johansson/Nordic Photos/Getty Images; 5.1a-c, 5.2a-b: © The McGraw-Hill Companies, Inc. Stephen Frisch, photographer; 5.9: © The McGraw-Hill Companies, Inc. Charles Winters photographer; Page 168: © The McGraw-Hill Companies, Inc. Stephen Frisch photographer.

Chapter 6

Opener: © Mauricio Inostroza Fotografo (www.photographersdirect.com).

Chapter 7

Opener: © Joe Drivas/Ionica/Getty Images; 7.5A: © Richard Megna/ Fundamental Photographs, NYC; 7.13a-b: PSSC Physics © 1965, Education Development Center, Inc.; D.C. Heath & Company/Education Development Center, Inc.

Chapter 8

Opener: © Geanina Bechea/Shutterstock.com.

Chapter 9

Opener: © Olivier Polet/Corbis; 9.6a-b, 9.8a, 9.9(all), 9.13, 9.23: © The McGraw-Hill Companies, Inc. Stephen Frisch, photographer.

Chapter 10

Opener: © amlet/Shutterstock.com; 10.2(all), 10.6(both): © The McGraw-Hill Companies, Inc. Stephen Frisch, photographer.

Chapter 11

Figure 11.21:© The McGraw-Hill Companies, Inc. Charles Winters photographer.

Chapter 12

Opener: © Alan Majchrowicz/age fotostock; Page 352: © Christopher Meder Photography/Shutterstock.com; 12.18a,b: © The McGraw-Hill Companies, Inc. Stephen Frisch, photographer; 12.20b: © Scott Camazine/Photo Researchers, Inc.; 12.21a: © Jeffrey A. Scovil Photography (Wayne Thompson Minerals); 12.21b: © Mark Schneider/Visuals Unlimited; 12.21c: © Jeffrey A. Scovil Photography; 12.21d: © Jeffrey A. Scovil Photography (Francis Benjamin Collection); Page 376(fruit): © Bryan Busovicki/Shutterstock.com; 12.31: © The McGraw-Hill Companies, Inc./Stephen Frisch, Photographer.

Chapter 13

Opener: © Jill Braaten; Table 13.1: © A. B. Dowsett/SPL/Photo Researchers, Inc.; 13.8a-c: © The McGraw-Hill Companies, Inc. Stephen Frisch, photographer.

Chapter 14

Opener: © Digital Vision/Getty RF; Page 428 (all), Page 431 (all), Page 433 (all), Page 435 (all): © The McGraw-Hill Companies, Inc. Stephen Frisch, photographer; 14.5a: © Gordon Vrdoljak, UC Berkeley Electron Microscope Lab; Page 440 (all), Page 445 (all), Page 449 (all), 14.13b(both), Page 453 (all): © The McGraw-Hill Companies, Inc. Stephen Frisch, photographer.

Chapter 15

Opener: © Jill Braaten; 15.8: © Martin Bough/Fundamental Photographs, NYC

Chapter 16

Opener: © Getty RF; 16.20a-b: © 2009 Richard Megna, Fundamental Photographs, NYC.

Chapter 17

Opener: © Richard Megna, Fundamental Photographs, NYC; 17.1a-d: © The McGraw-Hill Companies, Inc. Stephen Frisch, photographer.

Chapter 18

Opener: © Jill Braaten; 18.6a,b: © The McGraw-Hill Companies, Inc. Stephen Frisch, photographer.

Chapter 19

Opener: © Getty RF; 19.1a-b, 19.2a-b, 19.6: © The McGraw-Hill Companies, Inc. Stephen Frisch, photographer; Page 636: © Jeffrey A. Scovil Photography; 19.10a,b, 19.11, Page 640: © The McGraw-Hill Companies, Inc. Stephen Frisch, photographer.

Chapter 20

Opener: © Brand X Pictures/PunchStock RF.

Chapter 21

Opener: © Don Emmert/AFP/Getty Images; 21.1, Page 689, 21.3(both), 21.4b:© The McGraw-Hill Companies, Inc. Stephen Frisch, photographer; 21.6: © Richard Megna/Fundamental Photographs, NYC; 21.8: © The McGraw-Hill Companies, Inc. Stephen Frisch, photographer; 21.14: © Jill Braaten; 21.15: © The McGraw-Hill Companies, Inc. Pat Watson, photographer; 21.17: © The McGraw-Hill Companies, Inc. Stephen Frisch, photographer; 21.20a: © commons/Shutterstock.com; 21.21: © David Weintraub/ Photo Researchers, Inc.; 21.26, Page 724: © The McGraw-Hill Companies, Inc. Stephen Frisch, photographer; 21.28: © Tom Hollyman/Photo Researchers, Inc.

Chapter 22

Opener: © Louis Veiga/The Image Bank/Getty Images; 22.2(all), 22.5a,b, 22.6: © The McGraw-Hill Companies, Inc. Stephen Frisch, photographer; 22.8a,b: © Richard Megna/Fundamental Photographs, NYC; 22.19: © The McGraw-Hill Companies, Inc. Pat Watson, photographer; 22.20a-b: © The McGraw-Hill Companies, Inc. Stephen Frisch, photographer.

Chapter 23

Opener: © Science Source/Photo Researchers, Inc.; 23.8: © Scott Camazine/ Photo Researchers, Inc.; 23.9: © Dr. Robert Friedland/SPL/Photo Researchers, Inc.; 23.10: © Dr. Dennis Olson/Meat Lab, Iowa State University, Ames, IA; 23.14: © Albert Copley/Visuals Unlimited; 23.15: © Dietmar Krause/ Princeton Plasma Physics Lab.

Index

Page numbers followed by *f* indicate figures; *n*, footnotes; and *t*, tables

A

B

C

D

E

F

H

I

J

K

L

M

N

O

P

Q

R

S

Fundamental Physical Constants (six significant figures)

Avogadro's number	N_A	$= 6.02214 \times 10^{23}$/mol
atomic mass unit	amu	$= 1.66054 \times 10^{-27}$ kg
charge of the electron (or proton)	e	$= 1.60218 \times 10^{-19}$ C
Faraday constant	F	$= 9.64853 \times 10^{4}$ C/mol
mass of the electron	m_e	$= 9.10939 \times 10^{-31}$ kg
mass of the neutron	m_n	$= 1.67493 \times 10^{-27}$ kg
mass of the proton	m_p	$= 1.67262 \times 10^{-27}$ kg
Planck's constant	h	$= 6.62607 \times 10^{-34}$ J·s
speed of light in a vacuum	c	$= 2.99792 \times 10^{8}$ m/s
standard acceleration of gravity	g	$= 9.80665$ m/s^2
universal gas constant	R	$= 8.31447$ J/(mol·K)
		$= 8.20578 \times 10^{-2}$ (atm·L)/(mol·K)

SI Unit Prefixes

p	n	μ	m	c	d	k	M	G
pico-	nano-	micro-	milli-	centi-	deci-	kilo-	mega-	giga-
10^{-12}	10^{-9}	10^{-6}	10^{-3}	10^{-2}	10^{-1}	10^{3}	10^{6}	10^{9}

Conversions and Relationships

Length

SI unit: meter, m

1 km = 1000 m
= 0.62 mile (mi)
1 inch (in) = 2.54 cm
1 m = 1.094 yards (yd)
1 pm = 10^{-12} m = 0.01 Å

Volume

SI unit: cubic meter, m^3

1 dm^3 = 10^{-3} m^3
= 1 liter (L)
= 1.057 quarts (qt)
1 cm^3 = 1 mL
1 m^3 = 35.3 ft^3

Pressure

SI unit: pascal, Pa

1 Pa = 1 N/m^2
= 1 kg/m·s^2
1 atm = 1.01325×10^5 Pa
= 760 torr
1 bar = 1×10^5 Pa

Mass

SI unit: kilogram, kg

1 kg = 10^3 g
= 2.205 lb
1 metric ton (t) = 10^3 kg

Energy

SI unit: joule, J

1 J = 1 kg·m^2/s^2
= 1 coulomb·volt (1 C·V)
1 cal = 4.184 J
1 eV = 1.602×10^{-19} J

Math relationships

$\pi = 3.1416$
volume of sphere = $\frac{4}{3}\pi r^3$
volume of cylinder = $\pi r^2 h$

Temperature

SI unit: kelvin, K

0 K = −273.15°C
mp of H_2O = 0°C (273.15 K)
bp of H_2O = 100°C (373.15 K)
T (K) = T (°C) + 273.15
T (°C) = [T (°F) − 32]$\frac{5}{9}$
T (°F) = $\frac{9}{5}T$ (°C) + 32